College Algebra

SECOND EDITION

CYNTHIA
YOUNG

WHARTON COUNTY JUNIOR COLLEGE

WILEY Custom
LEARNING SOLUTIONS

To order books or for customer service, please call 1(800)-CALL-WILEY (225-5945).

Printed in the United States of America.

ISBN 978-1-118-10333-3

Printed and bound by Victor Graphics, Inc.

10 9 8 7 6 5 4 3 2 1

DEFINITIONS, RULES, FORMULAS, AND GRAPHS

ARITHMETIC OPERATIONS

$$ab + ac = a(b + c) \qquad \frac{a}{b} + \frac{c}{d} = \frac{ad + bc}{bd} \qquad \frac{a + b}{c} = \frac{a}{c} + \frac{b}{c} \qquad \frac{\left(\frac{a}{b}\right)}{\left(\frac{c}{d}\right)} = \frac{ad}{bc}$$

$$a\left(\frac{b}{c}\right) = \frac{ab}{c} \qquad \frac{a - b}{c - d} = \frac{b - a}{d - c} \qquad \frac{ab + ac}{a} = b + c,\ a \neq 0$$

$$\frac{\left(\frac{a}{b}\right)}{c} = \frac{a}{bc} \qquad \frac{a}{\left(\frac{b}{c}\right)} = \frac{ac}{b}$$

EXPONENTS AND RADICALS

$$a^0 = 1,\ a \neq 0 \qquad \frac{a^x}{a^y} = a^{x-y} \qquad \left(\frac{a}{b}\right)^x = \frac{a^x}{b^x} \qquad \sqrt[n]{a^m} = a^{m/n} = \left(\sqrt[n]{a}\right)^m$$

$$a^{-x} = \frac{1}{a^x} \qquad (a^x)^y = a^{xy} \qquad \sqrt{a} = a^{1/2} \qquad \sqrt[n]{ab} = \sqrt[n]{a}\sqrt[n]{b}$$

$$a^x a^y = a^{x+y} \qquad (ab)^x = a^x b^x \qquad \sqrt[n]{a} = a^{1/n} \qquad \sqrt[n]{\left(\frac{a}{b}\right)} = \frac{\sqrt[n]{a}}{\sqrt[n]{b}}$$

ABSOLUTE VALUE

1. $|x| = \begin{cases} x & \text{if } x \geq 0 \\ -x & \text{if } x > 0 \end{cases}$
2. If $|x| = c$, then $x = c$ or $x = -c$. $(c > 0)$
3. If $|x| < c$, then $-c < x < c$. $(c > 0)$
4. If $|x| > c$, then $x < -c$ or $x > c$. $(c > 0)$

PROPERTIES OF LOGARITHMS

1. $\log_b(MN) = \log_b M + \log_b N$
2. $\log_b\left(\frac{M}{N}\right) = \log_b M - \log_b N$
3. $\log_b M^p = p\log_b M$
4. $\log_b M = \frac{\log_a M}{\log_a b} = \frac{\ln M}{\ln b} = \frac{\log M}{\log b}$
5. $\log_b b^x = x$; $\ln e^x = x$
6. $b^{\log_b x} = x$; $e^{\ln x} = x \quad x > 0$

SPECIAL FACTORIZATIONS

1. *Difference of two squares:*
   $A^2 - B^2 = (A + B)(A - B)$
2. *Perfect square trinomials:*
   $A^2 + 2AB + B^2 = (A + B)^2$
   $A^2 - 2AB + B^2 = (A - B)^2$
3. *Sum of two cubes:*
   $A^3 + B^3 = (A + B)(A^2 - AB + B^2)$
4. *Difference of two cubes:*
   $A^3 - B^3 = (A - B)(A^2 + AB + B^2)$

SYMMETRY

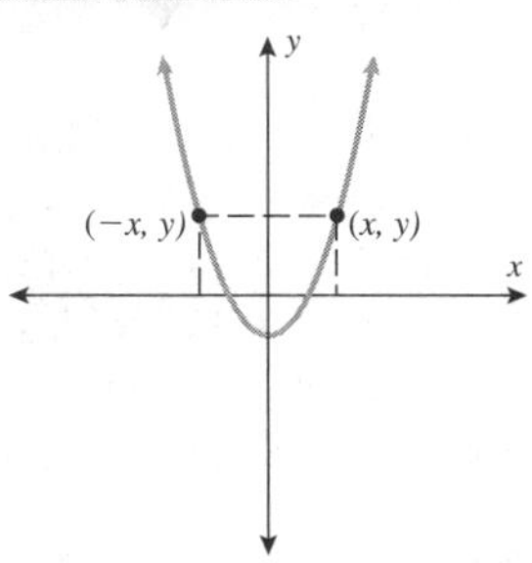

y-Axis Symmetry

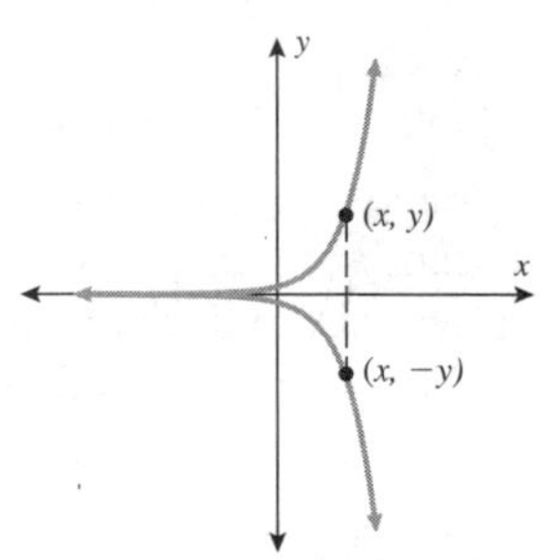

x-Axis Symmetry

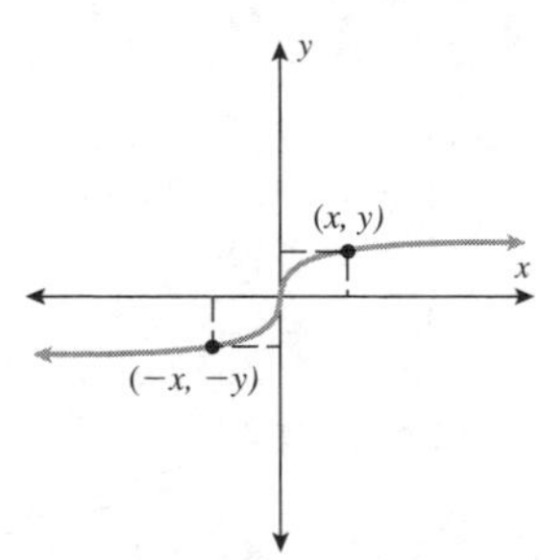

Origin Symmetry

Formulas/Equations

Distance Formula The distance from (x_1, y_1) to (x_2, y_2) is $\sqrt{(x_2 - x_1)^2 + (y_2 - y_1)^2}$.

Midpoint Formula The midpoint of the line segment with endpoints (x_1, y_1) and (x_2, y_2) is $\left(\frac{x_1 + x_2}{2}, \frac{y_1 + y_2}{2}\right)$.

Standard Equation of a Circle The standard equation of a circle of radius r with center at (h, k) is

$$(x - h)^2 + (y - k)^2 = r^2$$

Slope Formula The slope m of the line containing the points (x_1, y_1) and (x_2, y_2) is

$$\text{slope } (m) = \frac{\text{change in } y}{\text{change in } x} = \frac{y_2 - y_1}{x_2 - x_1} \quad (x_1 \neq x_2)$$

m is undefined if $x_1 = x_2$

Slope-Intercept Equation of a Line The equation of a line with slope m and y-intercept $(0, b)$ is $y = mx + b$

Point-Slope Equation of a Line The equation of a line with slope m containing the point (x_1, y_1) is $y - y_1 = m(x - x_1)$

Quadratic Formula The solutions of the equation $ax^2 + bx + c = 0$, $a \neq 0$, are $x = \frac{-b \pm \sqrt{b^2 - 4ac}}{2a}$

If $b^2 - 4ac > 0$, there are two distinct real solutions.

If $b^2 - 4ac = 0$, there is a repeated real solution.

If $b^2 - 4ac < 0$, there are two complex solutions (complex conjugates).

Geometry Formulas

Circle

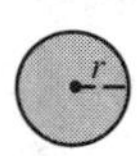

r = Radius, A = Area, C = Circumference

$A = \pi r^2 \quad C = 2\pi r$

Triangle

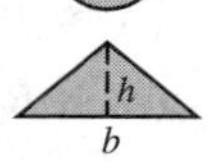

b = Base, h = Height (Altitude), A = area

$A = \frac{1}{2}bh$

Rectangle

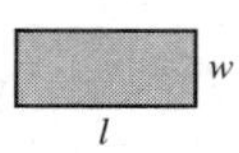

l = Length, w = Width, A = area, P = perimeter

$A = lw \quad P = 2l + 2w$

Rectangular Box

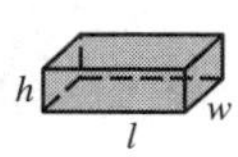

l = Length, w = Width, h = Height, V = Volume, S = Surface area

$V = lwh \quad S = 2lw + 2lh + 2wh$

Sphere

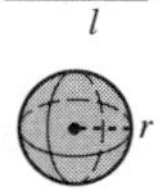

r = Radius, V = Volume, S = Surface area

$V = \frac{4}{3}\pi r^3 \quad S = 4\pi r^2$

Right Circular Cylinder

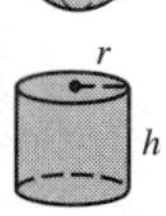

r = Radius, h = Height, V = Volume, S = Surface area

$V = \pi r^2 h \quad S = 2\pi r^2 + 2\pi rh$

Conversion Table

1 centimeter ≈ 0.394 inch
1 meter ≈ 39.370 inches
≈ 3.281 feet
1 kilometer ≈ 0.621 mile
1 liter ≈ 0.264 gallon
1 newton ≈ 0.225 pound

1 joule ≈ 0.738 foot-pound
1 gram ≈ 0.035 ounce
1 kilogram ≈ 2.205 pounds
1 inch ≈ 2.540 centimeters
1 foot ≈ 30.480 centimeters
≈ 0.305 meter

1 mile ≈ 1.609 kilometers
1 gallon ≈ 3.785 liters
1 pound ≈ 4.448 newtons
1 foot-lb ≈ 1.356 Joules
1 ounce ≈ 28.350 grams
1 pound ≈ 0.454 kilogram

FUNCTIONS

Constant Function	$f(x) = b$
Linear Function	$f(x) = mx + b$, where m is the slope and b is the y-intercept
Quadratic Function	$f(x) = ax^2 + bx + c,\ a \neq 0$ or $f(x) = a(x - h)^2 + k$ parabola vertex (h, k)
Polynomial Function	$f(x) = a_n x^n + a_{n-1}x^{n-1} + \cdots + a_1 x + a_0$
Rational Function	$R(x) = \dfrac{n(x)}{d(x)} = \dfrac{a_n x^n + a_{n-1}x^{n-1} + \cdots + a_1 x + a_0}{b_m x^m + a_{m-1}x^{m-1} + \cdots + b_1 x + b_0}$
Exponential Function	$f(x) = b^x, b > 0, b \neq 1$
Logarithmic Function	$f(x) = \log_b x, b > 0, b \neq 1$

GRAPHS OF COMMON FUNCTIONS

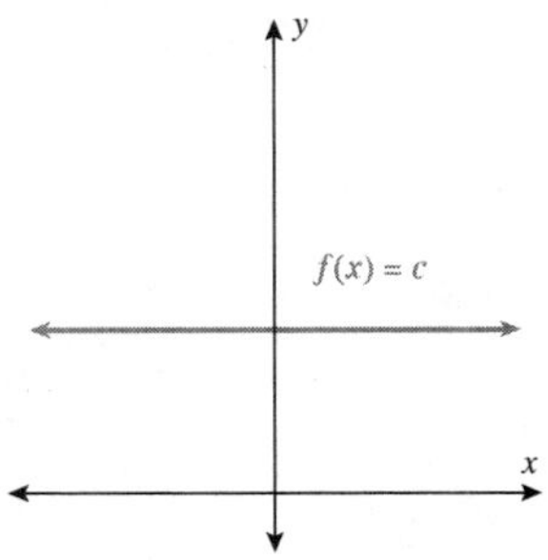

Constant Function

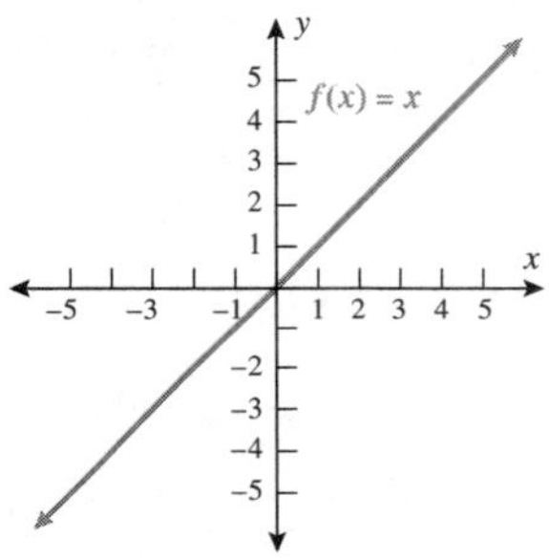

Identity Function

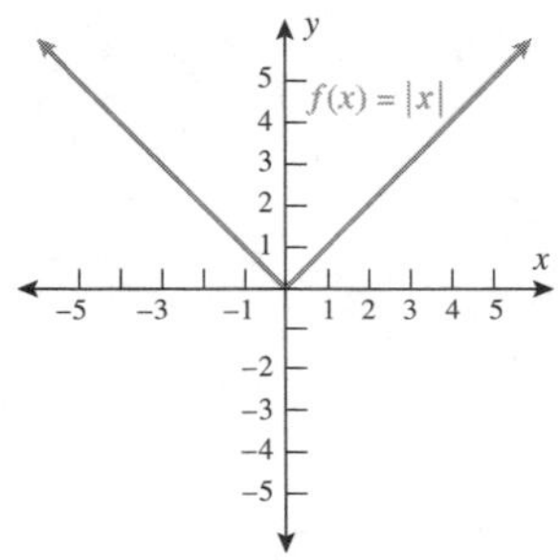

Absolute Value Function

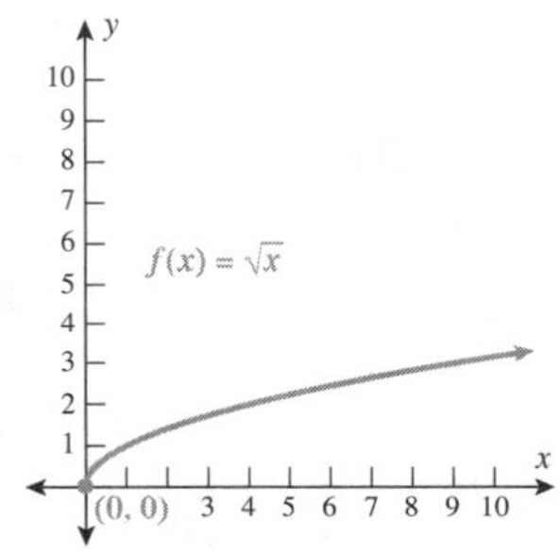

Square Root Function

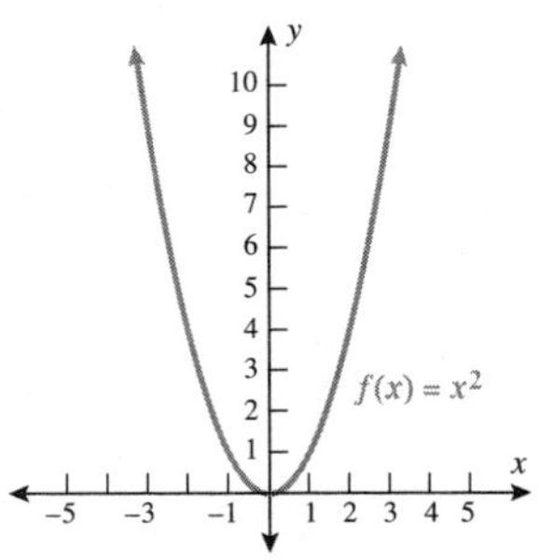

Quadratic Function

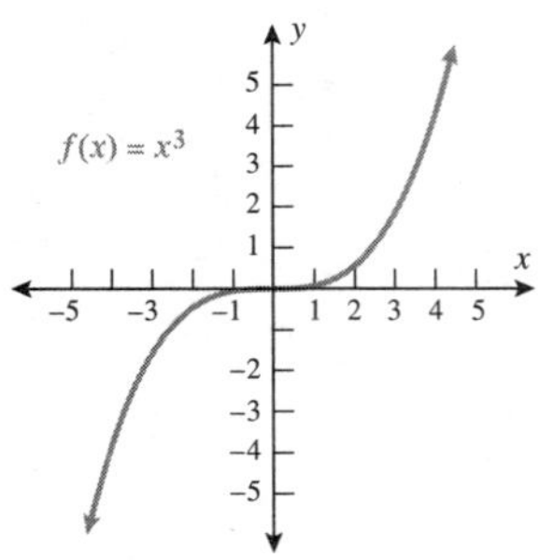

Cubic Function

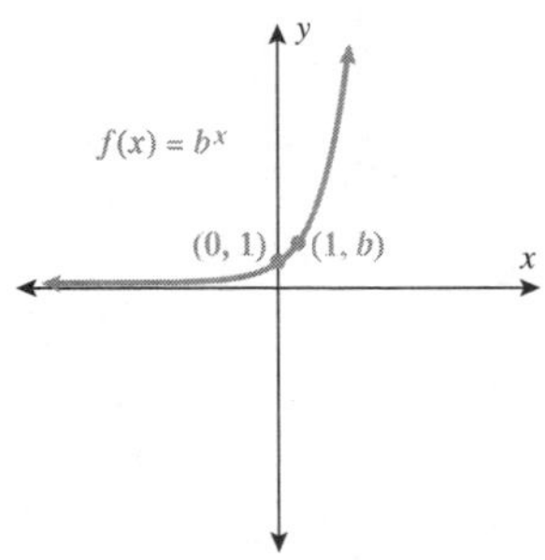

Exponential Function

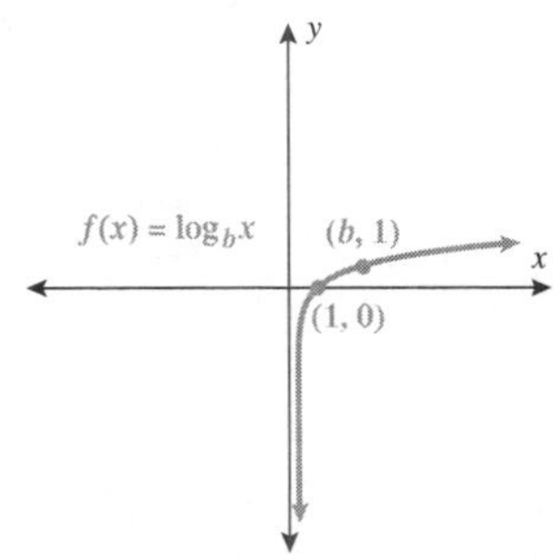

Logarithmic Function

TRANSFORMATIONS

In each case, c represents a positive real number.

Function		**Draw the graph of f and:**
Vertical translations	$y = f(x) + c$	Shift f upward c units.
	$y = f(x) - c$	Shift f downward c units.
Horizontal translations	$y = f(x - c)$	Shift f to the right c units.
	$y = f(x + c)$	Shift f to the left c units.
Reflections	$y = -f(x)$	Reflect f about the x-axis.
	$y = f(-x)$	Reflect f about the y-axis.

HERON'S FORMULA FOR AREA

If the semiperimeter, s, of a triangle is

$$s = \frac{a + b + c}{2}$$

then the area of that triangle is

$$A = \sqrt{s(s - a)(s - b)(s - c)}$$

Why WileyPLUS for Math and Statistics?

WileyPLUS is designed to help Math students bridge the gap between class work and homework; so they spend quality time learning – and you spend quality time teaching in a dynamic digital learning environment. Personalized technology that is there for your students when you are not, *WileyPLUS* provides the instant feedback and guidance they need – when they need it, empowering students to build the skills that will allow them to perform with confidence in your course.

"WileyPLUS really did help because it was always there."
– Student Jamie Farrington, CCAC South Campus

- ***Algorithmic problems in WileyPLUS allow students to practice the same problem with differing variables to build prerequisite basic skills, as well as the confidence they need to work toward conceptual understanding.***

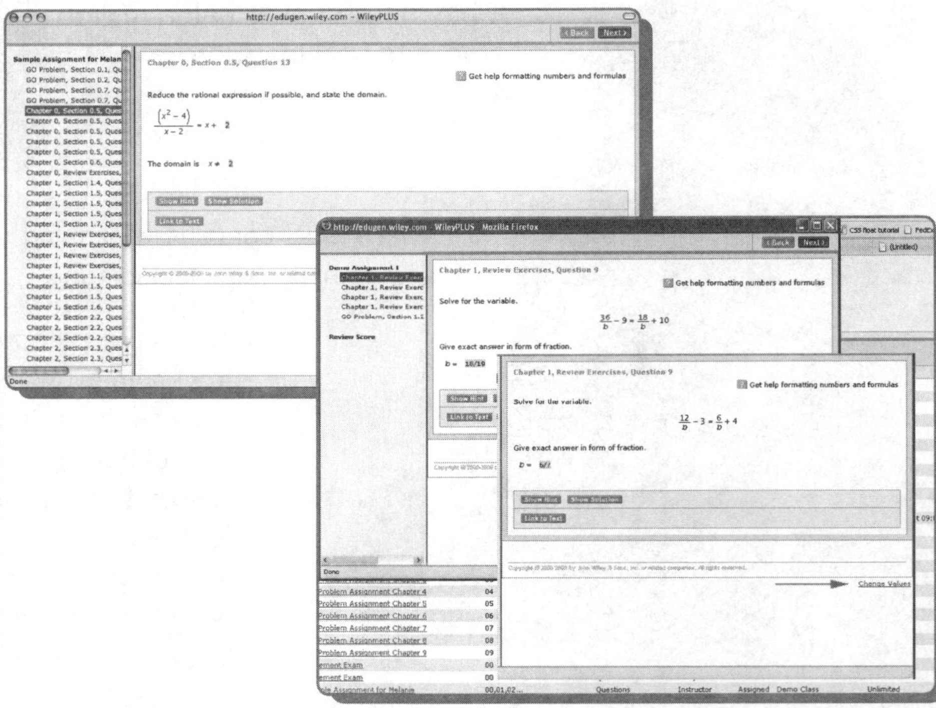

- ***The diagnostic tests and exercises in WileyPLUS help students assess their understanding and interpretation of key topics, helping them to focus their study on the areas in which they need it most.***

- ***Your WileyPLUS gradebook provides results to instantly assess your students' comprehension and customize lectures and homework assignments.***

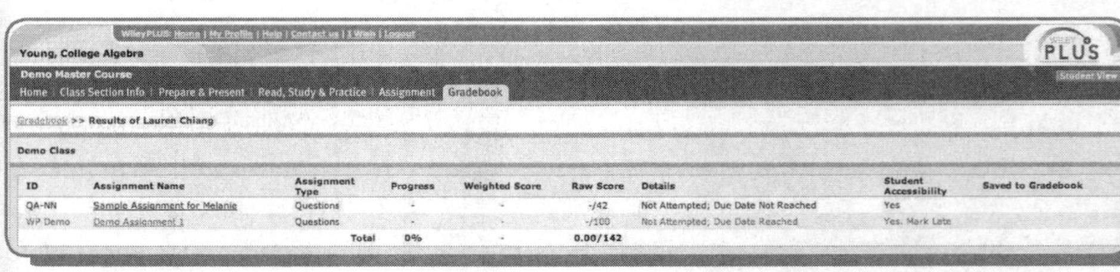

"I understood more clearly what had to be done... I was able to learn, do the work and succeed faster than ever before. WileyPLUS helped me learn better than ever before."
– Student Tyler Everett, Blinn College-Bryan

WILEY PLUS
www.wileyplus.com

See and Try WileyPLUS in action!
Details and Demo: www.wileyplus.com

WileyPLUS combines robust course management tools with the complete online text and all of the interactive teaching and learning resources you and your students need in one easy to use system.

When you assign homework in *WileyPLUS*, students can easily and immediately identify the concepts they have not yet mastered—and then choose the instructor-controlled help they need to remedy their weakness. This structured environment empowers students to focus their study time with far less frustration and far greater efficiency.

"WileyPLUS made studying more convenient...I was actually able to spend less time on the material and retain more."
– Student Brandon Combs, Bluegrass Community College

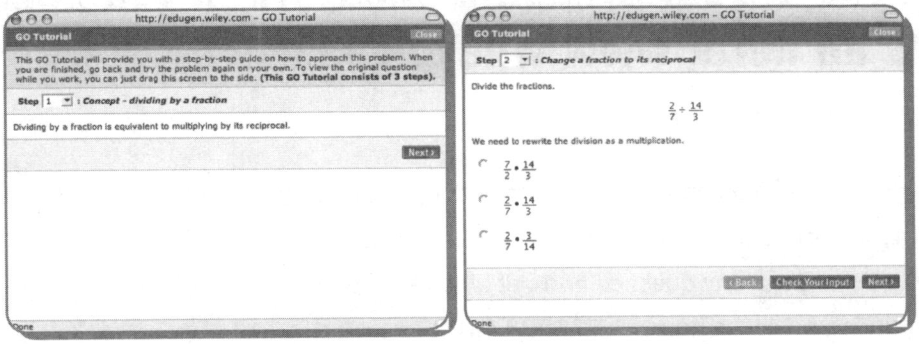

⊕ *Assignable GO™ Tutorials show students how to break-down complex problems into simplified, logical steps; with immediate feedback provided with each and every step.*

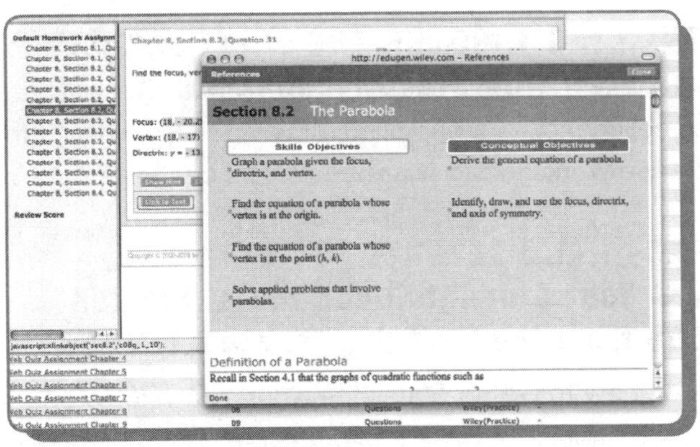

⊕ *Links to specific portions of the entire online text contained in WileyPLUS provide students with the contextual help they need—when they need it.*

⊕ *In section by section mini-lecture videos, Cynthia Young walks students through worked examples of the concepts with which they most often struggle, giving them twenty-four hour access to a professor.*

WILEY PLUS
www.wileyplus.com

College Algebra

Second Edition

College Algebra

Second Edition

CYNTHIA Y. YOUNG | *Professor of Mathematics*
UNIVERSITY OF CENTRAL FLORIDA

WILEY
John Wiley & Sons, Inc.

VICE PRESIDENT AND PUBLISHER	Laurie Rosatone
ACQUISITIONS EDITOR	Jessica Jacobs
SENIOR DEVELOPMENT EDITOR	Ellen Ford
ASSOCIATE EDITORS	Michael Shroff, Will Art
SENIOR EDITORIAL ASSISTANT	Jeffrey Benson
SENIOR PRODUCTION EDITOR	Sujin Hong
V.P DIRECTOR OF MARKETING	Susan Elbe
MARKETING MANAGER	Jaclyn Elkins
COVER PHOTO	© Brian Bailey/Stone/Getty Images
SENIOR ILLUSTRATION EDITOR	Sigmund Malinowski
PHOTO EDITOR	Sarah Wilkin
MEDIA EDITOR	Melissa Edwards
DEVELOPMENT ASSISTANT	Marcus Van Harpen
DIRECTOR, CREATIVE SERVICES	Harry Nolan
COVER DESIGNER	Jeof Vita
INTERIOR DESIGNER	Brian Salisbury

This book was set in 10/12 Times by Aptara and printed and bound by Quebecor. The cover was printed by Quebecor.

This book is printed on acid free paper. ∞

ISBN-978-0-470-22272-0

Printed in the United States of America

10 9 8 7 6 5 4 3 2 1

The Wiley Faculty Network

Where Faculty Connect

The Wiley Faculty Network is a faculty-to-faculty network promoting the effective use of technology to enrich the teaching experience. The Wiley Faculty Network facilitates the exchange of best practices, connects teachers with technology, and helps to enhance instructional efficiency and effectiveness. The network provides technology training and tutorials, including *WileyPLUS* training, online seminars, peer-to-peer exchanges of experiences and ideas, personalized consulting, and sharing of resources.

Connect with a Colleague

Wiley Faculty Network mentors are faculty like you, from educational institutions around the country, who are passionate about enhancing instructional efficiency and effectiveness through best practices. You can engage a faculty mentor in an online conversation at www.wherefacultyconnect.com

Participate in a Faculty-Led Online Seminar

The Wiley Faculty Network provides you with virtual seminars led by faculty using the latest teaching technologies. In these seminars, faculty share their knowledge and experiences on discipline-specific teaching and learning issues. All you need to participate in a virtual seminar is high-speed internet access and a phone line. To register for a seminar, go to www.wherefacultyconnect.com

Connect with the Wiley Faculty Network

Web: www.wherefacultyconnect.com

Phone: 1-866-4FACULTY

WILEY FACULTY NETWORK

This book is dedicated to teachers and students.

About the Author

Cynthia Y. Young is a native of Tampa, Florida. She is currently a Professor of Mathematics at the University of Central Florida (UCF) and author of *College Algebra, Trigonometry,* and *Algebra and Trigonometry*. She holds a B.A. degree in Secondary Mathematics Education from the University of North Carolina, an M.S. degree in Mathematical Sciences from UCF, and was awarded a graduate fellowship sponsored by Kennedy Space Center (KSC) to pursue doctoral studies at the University of Washington (UW). While earning her M.S. in Electrical Engineering and Ph.D. in Applied Mathematics from UW, she worked in the failure analysis and optics labs at KSC as a graduate intern and taught developmental mathematics at Shoreline Community College in Seattle, Washington. The summer after completing graduate studies at UW, she worked in the optics branch at Boeing Space and Defense and simultaneously earned her scuba diving certification and private pilot's license. Dr. Young turned down several industrial offers as a senior scientist in laser propagation to accept a faculty position at the University of Central Florida in order to pursue her passion for teaching. In 2001, the Office of Naval Research selected her for the Young Investigator Award, and in 2006 she was named a Fellow of the International Society for Optical Engineers. Although she excels in her research field, she considers teaching her true calling. She is a recipient of the UCF Excellence In Undergraduate Teaching Award, the UCF Scholarship of Teaching and Learning Award, and a two-time recipient of the UCF Teaching Incentive Program. Dr. Young is committed to improving student learning and has shared her techniques and experiences with colleagues around the country through talks at colleges, universities, and conferences.

Dr. Young and her husband, Dr. Christopher Parkinson, enjoy spending time outdoors and competing in Field Trials with their Labrador retrievers Ellie and Wiley in both the United States and Canada.

Dr. Young is pictured with Wiley (left) and Ellie (right).

Preface

To the Instructor

"I understand you in class but when I get home I am lost."
"I can't read the book."

How many times have we heard these statements from our students? Students come together in a College Algebra course with varied backgrounds; many have some level of math anxiety, some may have poor study habits, insufficient prerequisites, or different learning styles. As a result, there is a gap that exists between when we have our students in our classrooms and when they are on their own. My goal in writing this College Algebra book was to bridge that gap and to help them succeed. I recognize that it is difficult to get students to love mathematics, but you can get students to love *succeeding* at mathematics. In order to accomplish this goal, I wrote with two main objectives: that the voice of the book should be your voice (*seamless with how you teach*) so that your students hear you when they read this book, and to enable the students to become independent, successful learners by giving them a book *they can read* (without sacrificing the rigor needed for concept mastery).

New to the Second Edition

When instructors using the first edition were asked what they wanted in the second edition, there were four consistent answers:

- More Exercises
- More Modeling
- More Technology
- The Addition of Cumulative Tests

We have added hundreds more exercises to the second edition spanning all categories: Skills, Applications, Catch the Mistake, Conceptual, Challenge, and Technology. Several of the new Applications exercises require the student to model applications.

An end of chapter, themed modeling project called *Modeling Your World,* has been created for Chapters 1–9. At UCF, our Undergraduate Freshman/Sophomore level common theme is Global Climate Change. As instructors, we are asked to incorporate Global Climate Change into our curriculum. I started by assigning a project where students use the functions we learn in this class such as polynomial and exponential functions to model global mean temperatures and CO_2 emissions. What I found the first few times I assigned these projects surprised me: the "C" level students shined in this activity. Although they are not as confident in their mathematics as some of the "A" students, these students enjoyed using the mathematics they had learned to model environmental phenomena and become informed participants in debates. This feature is at the end of every chapter and its purpose is to empower students to apply what they learn in this class to something they care about—the world outside of the classroom.

To facilitate the use of calculators, additional Technology Tips and Technology Exercises were added throughout all sections, as well as in Review Exercises, Practice Tests, and Cumulative Tests. Many more Technology Tips are posted on the book companion website.

Lastly, a Cumulative Test was added at the end of every chapter that assesses students' understanding of material covered to date in the course. In my classroom, I have demonstrated a doubling of mastery rates when I used cumulative assessment in the course; so I encourage you to try and assign these to your students whenever possible.

Distinguishing Features

The following features which are carried over from the first edition and have been well received continue to support my goal and distinguish this book from others.

Parallel Words and Math

In most mathematics books, the mathematics is performed in the center of the page and if a student doesn't understand a step, they look to the right for some brief description or hint. But, that's not how we teach. In class we don't write the mathematics on the board and then say "Class, guess what I just did." Instead, we lead them, telling them what step is coming, and then performing that mathematical operation together. I have tried to parallel our classroom experience in this book. In examples, the student is told the step that is coming, then they see the step—reading naturally from left to right.

Skills and Conceptual Objectives and Exercises

In order to give students a true understanding of any college mathematics course, it is important to emphasize both *skills* and *concepts*. In my experience, skills help students at the micro level while concepts help students at the macro level of understanding. Therefore, I have separated section objectives into *skills objectives* and *conceptual objectives* and have labeled exercises as *skill* or *conceptual* questions.

Catch the Mistake

When a student comes to your office hours and brings you a problem they have unsuccessfully tried to solve, the first question they generally ask is "Where did I go wrong?" Instead of painstakingly going over their work to see where their solution went awry, we often find it far easier to simply start over with a fresh piece of paper. Catch the Mistake Exercises are found in every section, facilitating a higher level of mastery by putting students into the role of teacher.

Features and Benefits at a Glance

Each of the features in this book has been designed to help instructors bridge the gap that exisits between class work and homework.

FEATURE	BENEFIT TO STUDENT
Chapter Opening Vignette	Peaks the student's interest with a real-world application of material presented in that chapter. Later in the chapter, the same concept from the vignette is reinforced.
Chapter Overview/Flowchart	Students see the big picture of how topics relate to each other.
Skills and Conceptual Objectives	Emphasize conceptual understanding and push students to think at a more global level on the topics presented.
Clear, Concise, and Inviting Writing Style, Tone, and Layout	Students are able to **read** this book, which reduces math anxiety and encourages student success.
Parallel Words and Math	Increases students' ability to read and understand examples.
Common Mistake/ Correct vs. Incorrect Boxes	Addresses different learning styles (by counter example). Demonstrates common mistakes so students understand why a step is incorrect and reinforce the correct mathematics.
Color for Pedagogical Reasons	A student sees that a function is written in red in the text, then its graph is also red, whereas another function is written in blue and its graph is blue. Particularly helpful for visual learners in piecewise defined functions, graphing by transformations, and graphing inverse functions.
Study Tips	Serves to reinforce notes you would write on the board in class.
Technology Tips	Demonstrates use of technology to confirm analytic solutions. Kept in the margins so that faculty can choose to use at their discretion.
Your Turn	Engages students during class, build student confidence, and assist instructor in assessing class understanding.
Catch the Mistake Exercises	Encourages students to take the role of teacher–demonstrates higher mastery level.
Conceptual Exercises	Teaches students to think globally. For example, students have to answer questions about all types of functions, not a specific function.
Modeling Your World	Engages students through timely issues: global climate change. When students use mathematical models they develop as a result of this course, they value the course more.
Chapter Review (Key Ideas)	Key formulas and ideas are presented section by section in a chart format–improves study skills.
Chapter Review Exercises	Improves study skills.
Chapter Practice Test	Offers self-assessment and improves study skills.
Cumulative Test	Improves retention.

Acknowledgments

I first want to thank all of the instructors and students who used the first edition and sent me valuable feedback. The first edition was *my* book and, with your help, this second edition is *our* book. I want to especially thank Mark McKibben who has written the solutions manuals for this series and always stayed in sync with me on a day to day basis throughout the process. It's so refreshing to have a colleague to whom I can send an email at 6 AM and get an immediate response! Special thanks to Pauline Chow for her Herculean effort on the Technology Tips, Technology Exercises, and accuracy checking. I want to thank Jean Shutters for writing the Cumulative Tests and for teaching me the value of cumulative assessment—Jean, my students thank you; Mike Kirby for his thoughtful suggestions; and William "Rudy" Radulovich on whom I could always count on for his feedback, support, and good sports banter. I appreciate all of you for your responsiveness, honesty, humor, judgment, and commitment to this book and its ability to improve student learning.

I'd like to thank the entire Wiley *team*. Every year that I am with Wiley, I realize that I have the best partner in this market. To my editor, Jessica Jacobs, thank you for adopting this series as your own. We may be opposites, English goddess, but we complement each others' styles well and you have helped this book grow. Thanks to my developmental editor, Ellen Ford, who coordinated this book from manuscript through production and who somehow kept us all on schedule and could always be counted on for a laugh. To Michael Shroff and Will Art, associate editors, who coordinated the AIE, reviewers, accuracy checkers, and all of the supplements. Thanks to Jaclyn Elkins for your novel marketing efforts. A special thanks to Laurie Rosatone, publisher, who has shared my vision for this series since its inception. To my production editor, Sujin Hong, and the illustration and photo editors, Sigmund Malinowski and Sarah Wilkin, thank you for your hard work, dedication, and patience with schedules and revisions. And, finally, I want to thank the entire Wiley sales force for your unwavering commitment and constant enthusiasm for my series of books.

To the greatest man I know, my husband Christopher, thanks for all your support.

My sincere appreciation goes out to all the reviewers, class testers, and accuracy checkers on the second edition. Their input, comments, and detailed feedback have been invaluable in improving and fine tuning this edition.

Gerardo Aladro, *Victoria College*
Kara Allara, *Central Piedmont Community College*
Dionne Bailey, *Angelo State University*
Abigail Baumgardner, *Blinn College*
Shari Beck, *Navarro College*
Becky Berthiaume, *Edison College*
David Bregenzer, *Utah State University*
Denise Brown, *Collin County Community College*
Todd Cadwallader Olsker, *California State University—Fullerton*
Elsie M. Campbell, *Angelo State University*
Scott Carter, *Florida Community College of Jacksonville*
Stanley Cauthers, *Orange County Community College*
Martha Chalhoub, *Collin County Community College*
O. Pauline Chow, *Harrisburg Area Community College*
Madeleine Chowdhury, *Maricopa Community College*
Kathy Cousins-Cooper, *North Carolina A&T State University*
Charles Cunningham, *James Madison University*
Kathy Davis, *Kirkwood Community College*
Michelle DeDeo, *University of North Florida*
Ruth Dellinger, *Florida Community College of Jacksonville*
Judy Fethe, *Pellissippi State Technical Community College*
Jeff Feuer, *Dowling College*
Michael Frantz, *University of La Verne*
John Garlow, *Tarrant County Junior College*

Lee Gibson, *University of Louisville*
Megan Granich, *Loyola Marymount University*
De Galdra Gurley, *Tarrant County Junior College*
Tom Hayes, *Montana State University*
Celeste Hernandez, *Richland College*
Tammy Higson, *Hillsborough Community College*
Brian Hons, *San Antonio College*
Christy Jackson, *University of Arkansas—Little Rock*
Nancy Johnson, *Manatee Community College*
Martha Jones, *South Plains College*
Mike Kirby, *Tidewater Community College*
Rebecca Lanier, *Bluegrass Community and Technical College*
Anahita Lorestani, *San Antonio College*
Babette Lowe, *Victoria College*
Mark McKibben, *Goucher College*
Owen Mertens, *Missouri State University*
Rita Marie O'Brien, *Navarro College*
Joni Pirnot, *Manatee Community College*
Leela Rakesh, *Central Michigan University*
Sharon Sanders, *Georgia Perimeter College*
Brunilda Santiago, *Indian River Community College*
Alicia Serfaty, *Miami Dade College*
Donna Sherrill, *Missouri State University*
Jean Shutters, *Harrisburg Area Community College*
Jacqui Stone, *University of Maryland*
Jo Temple, *Texas Tech University*
Yan Tian, *Palomar College*
Robert Vaughn, *Graceland University*
Philip Veer, *Johnson County Community College*
Jim Voss, *Front Range Community College*
Christine Walker, *Utah Valley University*
Elmira Yakutova-Lorentz, *Brevard Community College*
Naveed Zaman, *West Virginia State University*

My continued appreciation is extended to those reviewers, contributors, focus group participants, accuracy checkers, and class testers of the first edition, as it was only with their assistance that I was able to get this work from idea to completion. Their input is the foundation on which this edition is built.

Ray Abbasi, *Montgomery College*
Marwan Abu-Sawwa, *Florida Community College at Jacksonville*
Scott Adamson, *Chandler Gilbert Community College*
Ebrahim Ahmadizadeh, *Northampton Community College*
Margo Alexander, *Georgia State University*
David Arreazola, *Laredo Community College*
Kathy Autrey, *Northwestern State University*
Abigail Baungardner, *Blinn College*
Ignacio Bello, *University of South Florida–Tampa*
Nicholas Belloit, *Florida Community College at Jacksonville*
Rebecca Berthiaume, *Edison College*
Jean Bevis, *Georgia State University*
Ben Brink, *Wharton County Junior College*
Lee Ann Brown, *University of Central Oklahoma*
Connie Buller, *Metropolitan Community College*
Warren Burch, *Brevard Community College*
Elsie Campbell, *Angelo State University*
Sandra Campbell, *Armstrong Atlantic State University*
Nancy Carpenter, *Johnson County Community College*
James Chesla, *Grand Rapids Community College*
O. Pauline Chow, *Harrisburg Area Community College*
Lynn Cleaveland, *University of Arkansas*
Scott Collins, *Pima Community College*
Suzanne Doviak, *Old Dominion University*
Lori Dunlop-Pyle, *University of Central Florida*
Gay Ellis, *Southwest Missouri State University*
Judith Fethe, *Pellissippi State Technical Community College*
Lee Gibson, *University of Louisville*
Perry Gillespie, *Fayetteville State University*
Tammy Higson, *Hillsborough Community College*
Debra Hill, *University of North Carolina–Charlotte*
Laura Hillerbrand, *Broward Community College*
Joe Howe, *St. Charles Community College*
Nancy Johnson, *Manatee Community College*
David Keller, *Kirkwood Community College*
Susan Kellicut, *Seminole Community College*
Esmarie Kennedy, *University of Texas—San Antonio*
John Khoury, *Brevard Community College*
Julie Killingbeck, *Ball State University*
Michael Lanstrum, *Cuyahoga Community College*
William Livingston, *Missouri Southern State University*
Stephanie Lochbaum, *Austin Community College*
Wanda Long, *St. Charles Community College*
Catherine Louchart, *Northern Arizona University*
Rudy Maglio, *Northeastern Illinois University*
James McGlothin, *Lower Columbia College*
M. Dee Medley, *Augusta State University*
Margaret Michener, *University of Nebraska–Kearney*
Debbie Millard, *Florida Community College at Jacksonville*
Michael Montano, *Riverside Community College*
Linda Myers, *Harrisburg Area Community College*
Shai Neumann, *Brevard Community College*

Ron Palcic, *Johnson County Community College*
Vic Perera, *Kent State University*
Mihaela Poplicher, *University of Cincinnati*
Jonathan Prewett, *University of Wyoming*
William Radulovich, *Florida Community College at Jacksonville*
Virginia Starkenburg, *San Diego City College*
Stan Stascinsky, *Tarrant County College*
James Stein, *California State University–Long Beach*
Mary Jane Sterling, *Bradley University*
Jacqueline Stone, *University of Maryland*
Matt Strom, *Minnesota State University–Mankato*
Todd Timmons, *University of Arkansas–Fort Smith*
Cari Van Tuinen, *Purdue University*
Mark Turner, *Cuesta College*
Draga Vidakovic, *Georgia State University*
Kendall Walden, *Arkansas State University*
Natalie Weaver, *Daytona Beach Community College*
Jan Wehr, *University of Arizona*
Cheryl Whitelaw, *Southern Utah University*
Sandra Wray-McAfee, *University of Michigan—Dearborn*
Elmira Yakutova-Lorentz, *Brevard Community College*

Commitment to Accuracy

From the beginning, the editorial team and I have been committed to providing an accurate and error-free text. In this goal, we have benefited from the help of many people. In addition to the reviewer feedback on this issue and normal proofreading, we enlisted the help of several extra sets of eyes during the production stages and during video filming. I wish to thank them for their dedicated reading of various rounds of pages to help me insure the highest level of accuracy.

Marc Campbell, *Daytona Beach Community College*
O. Pauline Chow, *Harrisburg Area Community College*
De Cook, *Okaloosa-Walton College Collegiate High School*
Jean Davis, *Texas State University—San Marcos*
Victoria Green, *University of Vermont*
Rebecca Hubiak, *Tidewater Community College*
Esmarie Kennedy, *University of Texas—San Antonio*
Frank Purcell, *Twin Prime Editorial*
William Radulovich, *Florida Community College at Jacksonville*
Lauri Semarne
Marie Vanisko, *Carroll College*
Philip Veer, *Johnson County Community College*
Randall Wills, *Southeastern Louisiana University*

Supplements

Instructor Supplements

INSTRUCTOR'S SOLUTIONS MANUAL (ISBN: 9780470417089)

- Contains worked out solutions to all exercises in the text.

INSTRUCTOR'S MANUAL (ISBN: 9780470417980)

Authored by Cynthia Young, the manual provides practical advice on teaching with the text, including:

- sample lesson plans and homework assignments
- suggestions for the effective utilization of additional resources and supplements

- sample syllabi
- Cynthia Young's Top 10 Teaching Tips & Tricks
- online component featuring the author presenting these Tips & Tricks

Annotated Instructor's Edition (ISBN: 9780471791362)

- Displays answers to all exercise questions, which can be found in the back of the book.
- Provides additional classroom examples within the standard difficulty range of the in-text exercises, as well as challenge problems to assess your students mastery of the material.

PowerPoint Slides

- For each section of the book, a corresponding set of lecture notes and worked out examples are presented as PowerPoint slides, available on the Book Companion Site (www.wiley.com/college/young) and WileyPLUS.

Test Bank (ISBN: 9780470417089)

Contains approximately 900 questions and answers from every section of the text.

Computerized Test Bank

Electonically enhanced version of the Test Bank that

- contains approximately 900 algorithmically-generated questions.
- allows instructors to freely edit, randomize, and create questions.
- allows instructors to create and print different versions of a quiz or exam.
- recognizes symbolic notation.
- allows for partial credit if used within WileyPLUS.

Book companion website (www.wiley.com/college/young)

- Contains all instructor supplements listed plus a selection of personal response system questions.

WileyPLUS

- Features a full-service, digital learning environment, including additional resources for instructors, such as assignable homework exercises, tutorials, gradebook, and integrated links between the online version of the text and supplements.

Student Supplements

Student Solutions Manual (ISBN: 9780470417034)

- Includes worked out solutions for all odd problems in the text.

Book companion website (www.wiley.com/college/young)

- Provides additional resources for students, including web quizzes, video clips, and audio clips.

WileyPLUS

- Features a full-service, digital learning environment, including additional resources for students, such as additional self-practice exercises, tutorials, and integrated links between the online version of the text and supplements.

WileyPLUS

Helping Teachers Teach and Students Learn www.wiley.com/college/young

This title is available with *WileyPLUS,* a powerful online tool that provides instructors and students with an integrated suite of teaching and learning resources in one easy-to-use website. *WileyPLUS* is organized around the essential activities you and your students perform in class:

For Instructors:

- **Prepare & Present:** Create class presentations using a wealth of Wiley-provided resources—such as an online version of the textbook, PowerPoint slides, and interactive simulations—making your preparation time more efficient. You may easily adapt, customize, and add to this content to meet the needs of your course.
- **Create Assignments:** Automate the assigning and grading of homework or quizzes by using Wiley-provided question banks, or by writing your own. Student results will be automatically graded and recorded in your gradebook. WileyPLUS can link homework problems to the relevant section of the online text, providing students with context-sensitive help.
- **Track Student Progress:** Keep track of your students' progress via an instructor's gradebook, which allows you to analyze individual and overall class results to determine their progress and level of understanding
- **Administer Your Course:** *WileyPLUS* can easily be integrated with another course management system, gradebook, or other resources you are using in your class, providing you with the flexibility to build your course, your way.

For Students:

WileyPLUS provides immediate feedback on student assignments and a wealth of support materials. This powerful study tool will help your students develop their conceptual understanding of the class material and increase their ability to solve problems.

- **A "Study and Practice"** area links directly to text content, allowing students to review the text while they study and complete homework assignments. Additional resources can include interactive simulations, study guide and solutions manual material, and other problem-solving resources.
- **An "Assignment"** area keeps all the work you want your students to complete in one location, making it easy for them to stay "on task." Students will have access to a variety of interactive problem-solving tools, as well as other resources for building their confidence and understanding. In addition, many homework problems contain a link to the relevant section of the multimedia book, providing students with context-sensitive help that allows them to conquer problem-solving obstacles as they arise.
- **A Personal Gradebook** for each student will allow students to view their results from past assignments at any time.

Please view our online demo at www.wiley.com/college/wileyplus or contact your local Wiley representative for more details. Here you will find additional information about the features and benefits of *WileyPLUS*, how to request a "test drive" of *WileyPLUS* for this title, and how to adopt it for class use.

The Wiley Faculty Network—Where Faculty Connect

The Wiley Faculty Network is a faculty-to-faculty network promoting the effective use of technology to enrich the teaching experience. The Wiley Faculty Network facilitates the exchange of best practices, connects teachers with technology, and helps to enhance instructional efficiency and effectiveness. The network provides technology training and tutorials, including *WileyPLUS* training, online seminars, peer-to-peer exchanges of experiences and ideas, personalized consulting, and sharing of resources.

Connect with a Colleague

Wiley Faculty Network mentors are faculty like you, from educational institutions around the country, who are passionate about enhancing instructional efficiency and effectiveness through best practices. You can engage a faculty mentor in an online conversation at www.wherefacultyconnect.com.

Participate in a Faculty-Led Online Seminar

The Wiley Faculty Network provides you with virtual seminars led by faculty using the latest teaching technologies. In these seminars, faculty share their knowledge and experiences on discipline-specific teaching and learning issues. All you need to participate in a virtual seminar is high-speed Internet access and a phone line. To register for a seminar, go to www.wherefacultyconnect.com.

Connect with the Wiley Faculty Network

Web: www.wherefacultyconnect.com
Phone: 1-866-4FACULTY

Table of Contents

4 Polynomial and Rational Functions 324

5 Exponential and Logarithmic Functions 420

6 Systems of Linear Equations and Inequalities 490

7 Matrices 560

8 Conics and Systems of Nonlinear Equations and Inequalities 636

9 Sequences, Series, and Probability 704

Answers to Odd Numbered Exercises 781

Applications Index 841

Subject Index 845

A Note from the Author to the Student

I wrote this text with careful attention to ways in which to make your learning experience more successful. If you take full advantage of the unique features and elements of this textbook, I believe your experience will be fulfilling and enjoyable. Let's walk through some of the special book features that will help you in your study of College Algebra.

Prerequisites and Review (Chapter 0)

A comprehensive review of prerequisite knowledge (intermediate algebra topics) in Chapter 0 provides a brush up on knowledge and skills necessary for success in the course. In addition, "just in time reviews" of prerequisite concepts are provided throughout chapters on an as-needed basis.

Clear, Concise, and Inviting Writing

Special attention has been made to present an engaging, clear, precise narrative in a layout that is easy to use and designed to reduce any math anxiety you may have.

Chapter Introduction, Flow Chart, Section Headings, and Objectives

An opening vignette, flow chart, list of chapter sections, and targeted chapter objectives give you an overview of the chapter to help you see the big picture, the relationships between topics, and clear objectives for learning in the chapter.

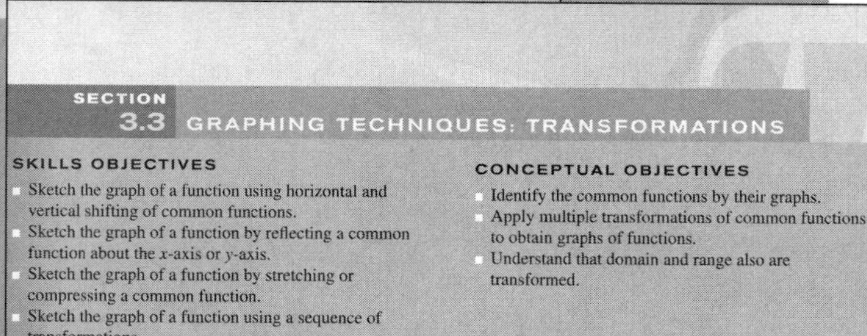

Skills and Conceptual Objectives

For every section, objectives are further divided by skills *and* concepts so you can see the difference between solving problems and truly understanding concepts.

Examples

Examples pose a specific problem using concepts already presented and then work through the solution. These serve to enhance your understanding of the subject matter.

EXAMPLE 1 Operations on Functions: Determining Domains of New Functions

For the functions $f(x) = \sqrt{x - 1}$ and $g(x) = \sqrt{4 - x}$, determine the sum function, difference function, product function, and quotient function. State the domain of these four new functions.

Solution:

Sum function: $f(x) + g(x) = \sqrt{x - 1} + \sqrt{4 - x}$

Difference function: $f(x) - g(x) = \sqrt{x - 1} - \sqrt{4 - x}$

Product function: $f(x) \cdot g(x) = \sqrt{x - 1} \cdot \sqrt{4 - x}$
$= \sqrt{(x - 1)(4 - x)} = \sqrt{-x^2 + 5x - 4}$

Quotient function: $\dfrac{f(x)}{g(x)} = \dfrac{\sqrt{x - 1}}{\sqrt{4 - x}} = \sqrt{\dfrac{x - 1}{4 - x}}$

The domain of the square root function is determined by setting the argument under the radical greater than or equal to zero.

Domain of $f(x)$: $[1, \infty)$

Domain of $g(x)$: $(-\infty, 4]$

The domain of the sum, difference, and product functions is

$$[1, \infty) \cap (-\infty, 4] = [1, 4]$$

The quotient function has the additional constraint that the denominator cannot be zero. This implies that $x \neq 4$, so the domain of the quotient function is $[1, 4)$.

■ **YOUR TURN** Given the function $f(x) = \sqrt{x + 3}$ and $g(x) = \sqrt{1 - x}$, find $(f + g)(x)$ and state its domain.

Your Turn

Immediately following many examples, you are given a similar problem to reinforce and check your understanding. This helps build confidence as you progress in the chapter. These are ideal for in-class activity or for preparing for homework later. Answers are provided in the margin for a quick check of your work.

COMMON MISTAKE

The most common mistake in calculating slope is writing the coordinates in the wrong order, which results in the slope being opposite in sign.

Find the slope of the line containing the two points (1, 2) and (3, 4).

CORRECT

Label the points.

$(x_1, y_1) = (1, 2) \qquad (x_2, y_2) = (3, 4)$

Write the slope formula.

$$m = \frac{y_2 - y_1}{x_2 - x_1}$$

Substitute the coordinates.

$$m = \frac{4 - 2}{3 - 1}$$

Simplify. $m = \dfrac{2}{2} = \boxed{1}$

INCORRECT

The **ERROR** is interchanging the coordinates of the first and second points.

$$m = \frac{4 - 2}{1 - 3}$$

The calculated slope is **INCORRECT** by a negative sign.

$$m = \frac{2}{-2} = -1$$

Common Mistake/ Correct vs. Incorrect

In addition to standard examples, some problems are worked out both correctly and incorrectly to highlight common errors students make. Counter examples like these are often an effective learning approach for many students.

Parallel Words and Math

This text reverses the common textbook presentation of examples by placing the explanation in words *on the left* and the mathematics in parallel *on the right*. This makes it easier for students to read through examples as the material flows more naturally from left to right and as commonly presented in class.

WORDS

Step 1: In a table, list several pairs of coordinates that make the equation true.

MATH

x	$y = x^2$	(x, y)
0	0	(0, 0)
−1	1	(−1, 1)
1	1	(1, 1)
−2	4	(−2, 4)
2	4	(2, 4)

Step 2: Plot these points on a graph and connect the points with a smooth curve. Use arrows to indicate that the graph continues.

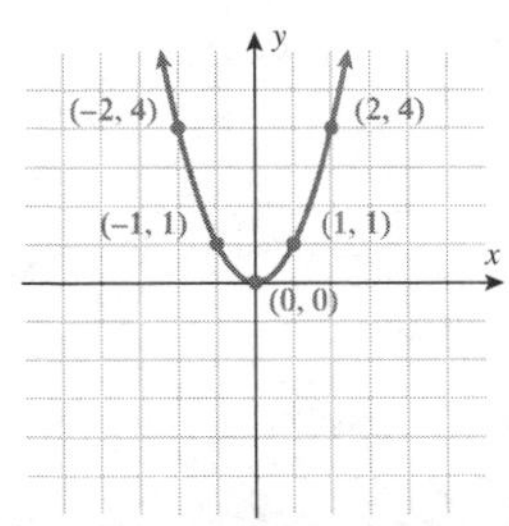

Study Tips and Caution Notes

These marginal reminders call out important hints or warnings to be aware of related to the topic or problem.

Study Tip

Shifts outside the function are vertical shifts with the sign.

Up (+)
Down (−)

CAUTION

Interchanging the coordinates can result in a sign error in the slope.

Technology Tip

Use a graphing utility to display graphs of $y_1 = H(x + 1) = (x + 1)^2 + 2(x + 1)$ and $y_2 = H(x) + H(1) = x^2 + 2x + 3$.

Technology Tips

These marginal notes provide problem solving instructions and visual examples using graphing calculators.

IN THIS CHAPTER you will find that functions are part of our everyday thinking: converting from degrees Celsius to degrees Fahrenheit, DNA testing in forensic science, determining stock values and the sale price of a shirt. We will develop a more complete, thorough understanding of functions. First, we will establish what a relation is, and then we will determine whether a relation is a function. We will discuss common functions, domain and range of functions, and graphs of functions. We will determine whether a function is increasing or decreasing on an interval and calculate the average rate of change of a function. We will perform operations on functions and composition of functions. We ... functions. Finally, we will model applications with functions using variation.

EXAMPLE 10 Determining the Domain of a Function

State the domain of the given functions.

a. $F(x) = \dfrac{3}{x^2 - 25}$ **b.** $H(x) = \sqrt[4]{9 - 2x}$ **c.** $G(x) = \sqrt[3]{x - 1}$

Solution (a):

Write the original equation. $F(x) = \dfrac{3}{x^2 - 25}$

Video icons

Video icons appear on all chapter and section introductions, chapter and section reviews, as well as selected examples throughout the chapter to indicate that the author has created a video segment for that element. These video clips help you work through the selected examples with the author as your "private tutor."

SECTION 3.5 SUMMARY

One-to-One Functions
Each input in the domain corresponds to exactly one output in the range, and no two inputs map to the same output. There are three ways to test a function to determine whether it is a one-to-one function.

1. Discrete points: For the set of all points (a, b) verify that no y-values are repeated.
2. Algebraic equations: Let $f(x_1) = f(x_2)$; if it can be shown

Properties of Inverse Functions

1. If f is a one-to-one function, then f^{-1} exists.
2. Domain and range
   - Domain of f = range of f^{-1}
   - Domain of f^{-1} = range of f
3. Composition of inverse functions
   - $f^{-1}(f(x)) = x$ for all x in the domain of f.

CHAPTER 3 REVIEW

Section	Concept	Pages	Review Exercises	Key Ideas/Point
3.1	**Functions**	226–238	1–30	
	Functions defined by equations	228–230	5–8	A vertical line can intersect a function in at most one point.
	Function notation	231–235	11–22	Placeholder notation j Difference quotient:

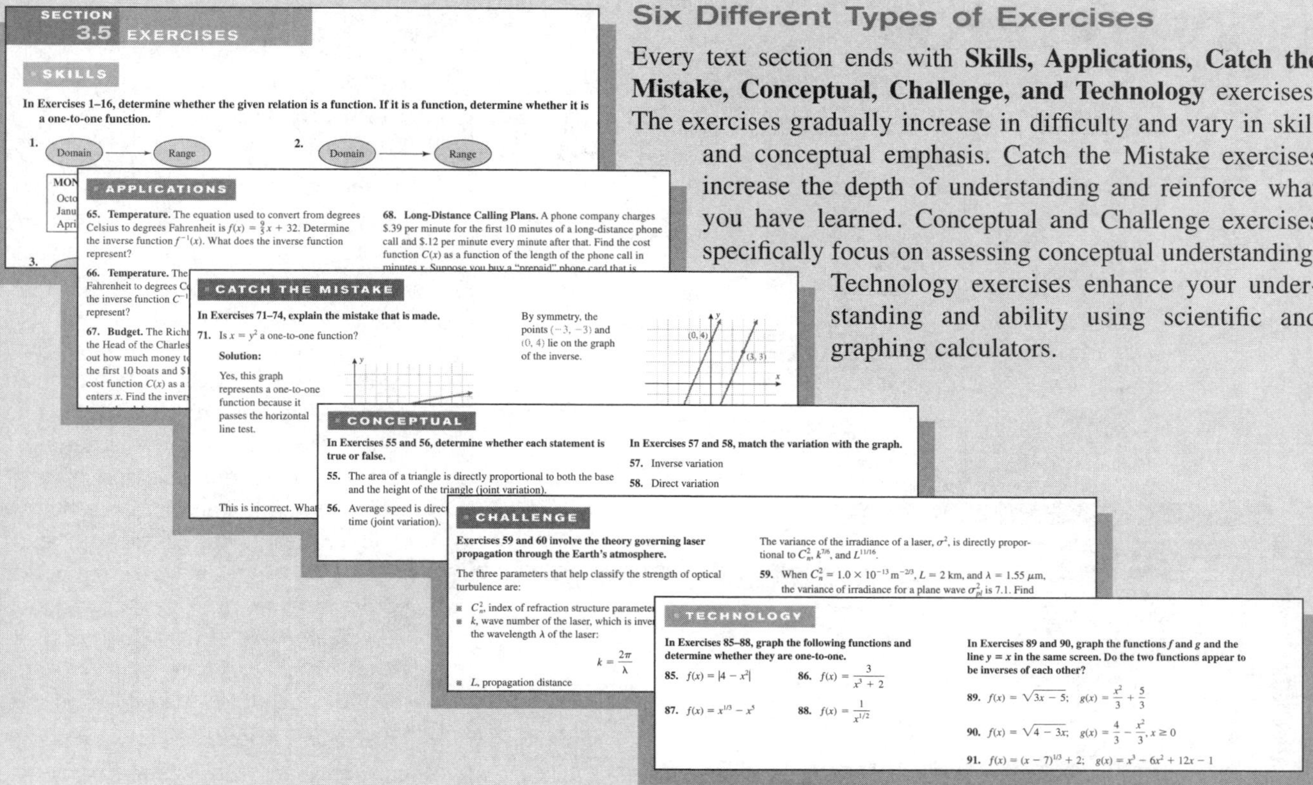
SECTION 3.5 EXERCISES

SKILLS

In Exercises 1–16, determine whether the given relation is a function. If it is a function, determine whether it is a one-to-one function.

APPLICATIONS

65. **Temperature.** The equation used to convert from degrees Celsius to degrees Fahrenheit is $f(x) = \frac{9}{5}x + 32$. Determine the inverse function $f^{-1}(x)$. What does the inverse function represent?

68. **Long-Distance Calling Plans.** A phone company charges \$.39 per minute for the first 10 minutes of a long-distance phone call and \$.12 per minute every minute after that. Find the cost function $C(x)$ as a function of the length of the phone call in minutes x.

CATCH THE MISTAKE

In Exercises 71–74, explain the mistake that is made.

71. Is $x = y^2$ a one-to-one function?

Solution:

Yes, this graph represents a one-to-one function because it passes the horizontal line test.

This is incorrect. What

By symmetry, the points $(-3, -3)$ and $(0, 4)$ lie on the graph of the inverse.

CONCEPTUAL

In Exercises 55 and 56, determine whether each statement is true or false.

55. The area of a triangle is directly proportional to both the base and the height of the triangle (joint variation).

56. Average speed is directly ... time (joint variation).

In Exercises 57 and 58, match the variation with the graph.

57. Inverse variation

58. Direct variation

CHALLENGE

Exercises 59 and 60 involve the theory governing laser propagation through the Earth's atmosphere.

The three parameters that help classify the strength of optical turbulence are:

- C_n^2, index of refraction structure parameter
- k, wave number of the laser, which is inversely proportional to the wavelength λ of the laser: $k = \frac{2\pi}{\lambda}$
- L, propagation distance

The variance of the irradiance of a laser, σ^2, is directly proportional to C_n^2, $k^{7/6}$, and $L^{11/6}$.

59. When $C_n^2 = 1.0 \times 10^{-13}\,\text{m}^{-2/3}$, $L = 2$ km, and $\lambda = 1.55\,\mu\text{m}$, the variance of irradiance for a plane wave σ_{pl}^2 is 7.1. Find

TECHNOLOGY

In Exercises 85–88, graph the following functions and determine whether they are one-to-one.

85. $f(x) = |4 - x^2|$

86. $f(x) = \frac{3}{x^3 + 2}$

87. $f(x) = x^{1/3} - x^5$

88. $f(x) = \frac{1}{x^{1/2}}$

In Exercises 89 and 90, graph the functions f and g and the line $y = x$ in the same screen. Do the two functions appear to be inverses of each other?

89. $f(x) = \sqrt{3x - 5}$; $g(x) = \frac{x^2}{3} + \frac{5}{3}$

90. $f(x) = \sqrt{4 - 3x}$; $g(x) = \frac{4}{3} - \frac{x^2}{3}, x \geq 0$

91. $f(x) = (x - 7)^{1/3} + 2$; $g(x) = x^3 - 6x^2 + 12x - 1$

Six Different Types of Exercises

Every text section ends with **Skills, Applications, Catch the Mistake, Conceptual, Challenge, and Technology** exercises. The exercises gradually increase in difficulty and vary in skill and conceptual emphasis. Catch the Mistake exercises increase the depth of understanding and reinforce what you have learned. Conceptual and Challenge exercises specifically focus on assessing conceptual understanding. Technology exercises enhance your understanding and ability using scientific and graphing calculators.

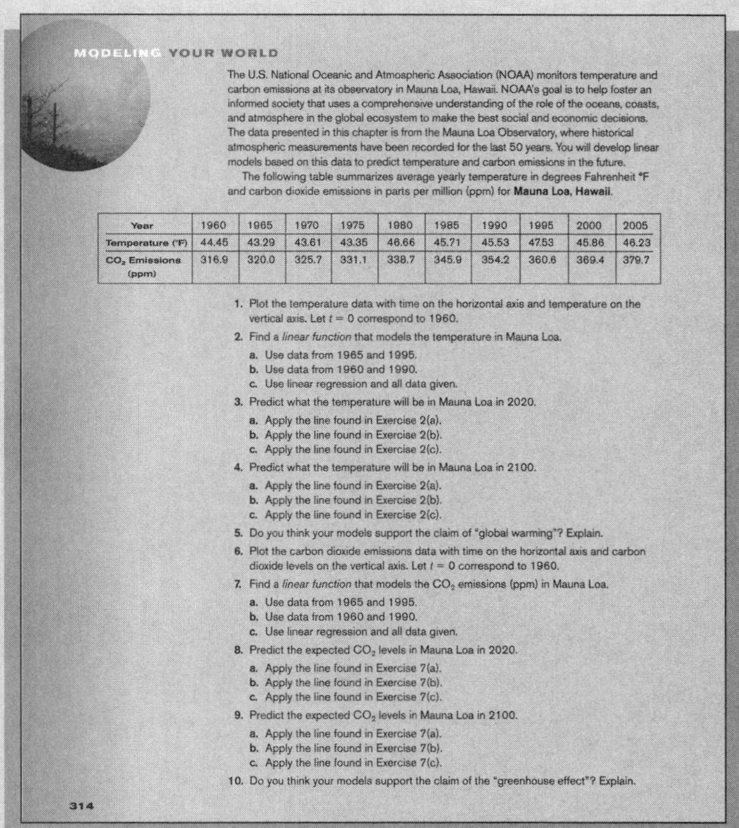
MODELING YOUR WORLD

The U.S. National Oceanic and Atmospheric Association (NOAA) monitors temperature and carbon emissions at its observatory in Mauna Loa, Hawaii. NOAA's goal is to help foster an informed society that uses a comprehensive understanding of the role of the oceans, coasts, and atmosphere in the global ecosystem to make the best social and economic decisions. The data presented in this chapter is from the Mauna Loa Observatory, where historical atmospheric measurements have been recorded for the last 50 years. You will develop linear models based on this data to predict temperature and carbon emissions in the future.

The following table summarizes average yearly temperature in degrees Fahrenheit °F and carbon dioxide emissions in parts per million (ppm) for **Mauna Loa, Hawaii**.

Year	1960	1965	1970	1975	1980	1985	1990	1995	2000	2005
Temperature (°F)	44.45	43.29	43.61	43.35	46.66	45.71	45.53	47.53	45.86	46.23
CO_2 Emissions (ppm)	316.9	320.0	325.7	331.1	338.7	345.9	354.2	360.6	369.4	379.7

1. Plot the temperature data with time on the horizontal axis and temperature on the vertical axis. Let $t = 0$ correspond to 1960.
2. Find a *linear function* that models the temperature in Mauna Loa.
   a. Use data from 1965 and 1995.
   b. Use data from 1960 and 1990.
   c. Use linear regression and all data given.
3. Predict what the temperature will be in Mauna Loa in 2020.
   a. Apply the line found in Exercise 2(a).
   b. Apply the line found in Exercise 2(b).
   c. Apply the line found in Exercise 2(c).
4. Predict what the temperature will be in Mauna Loa in 2100.
   a. Apply the line found in Exercise 2(a).
   b. Apply the line found in Exercise 2(b).
   c. Apply the line found in Exercise 2(c).
5. Do you think your models support the claim of "global warming"? Explain.
6. Plot the carbon dioxide emissions data with time on the horizontal axis and carbon dioxide levels on the vertical axis. Let $t = 0$ correspond to 1960.
7. Find a *linear function* that models the CO_2 emissions (ppm) in Mauna Loa.
   a. Use data from 1965 and 1995.
   b. Use data from 1960 and 1990.
   c. Use linear regression and all data given.
8. Predict the expected CO_2 levels in Mauna Loa in 2020.
   a. Apply the line found in Exercise 7(a).
   b. Apply the line found in Exercise 7(b).
   c. Apply the line found in Exercise 7(c).
9. Predict the expected CO_2 levels in Mauna Loa in 2100.
   a. Apply the line found in Exercise 7(a).
   b. Apply the line found in Exercise 7(b).
   c. Apply the line found in Exercise 7(c).
10. Do you think your models support the claim of the "greenhouse effect"? Explain.

314

Modeling Your World

These unique end of chapter exercises provide a fun and interesting way to take what you have learned and model a real world problem. By using global warming as the continuous theme, you can develop more advanced modeling skills with each chapter while seeing how modeling can help you better understand the world around you.

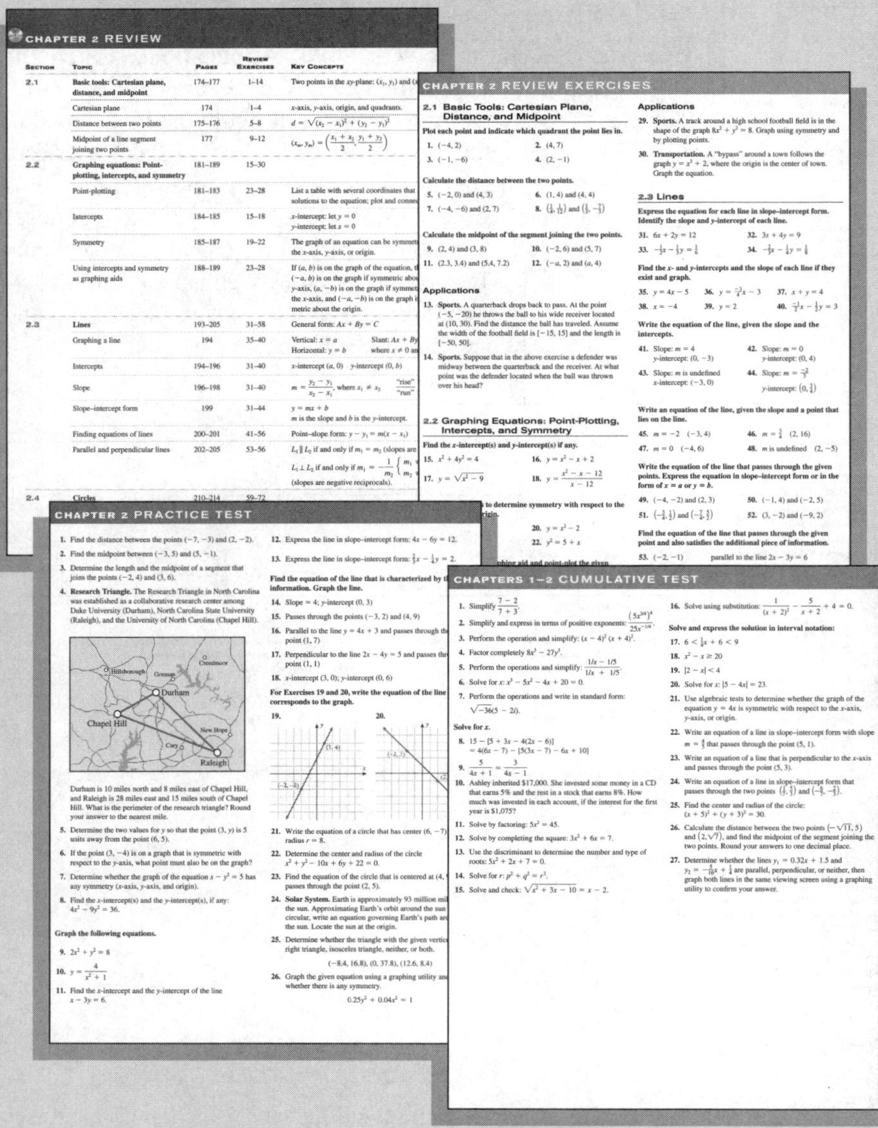

Chapter Review, Review Exercises, Practice Test, Cumulative Test

At the end of every chapter, a summary review chart organizes the key learning concepts in an easy to use one or two-page layout. This feature includes key ideas and formulas, as well as indicating relevant pages and review exercises so that you can quickly summarize a chapter and study smarter. Review Exercises, arranged by section heading, are provided for extra study and practice. A Practice Test, without section headings, offers even more self practice before moving on. A new Cumulative Test feature offers study questions based on all previous chapters' content, thus helping you build upon previously learned concepts.

College Algebra

Second Edition

0

Prerequisites and Review

Would you be able to walk successfully along a tightrope? Most people probably would say no because the foundation is "shaky." Would you be able to walk successfully along a beam (4 inches wide)? Most people would probably say yes—even though for some of us it is still challenging. Think of this chapter as the foundation for your walk. The more solid your foundation is now, the more successful your walk through *College Algebra* will be.

The purpose of this chapter is to review concepts and skills that you already have learned in a previous course. Some students will have taken intermediate algebra at a college or university prior to this class, while others coming from high school will have taken Algebra I and II. Mathematics is a cumulative subject in that it requires a solid foundation to proceed to the next level. Use this chapter to reaffirm your current knowledge base before jumping into the course.

IN THIS CHAPTER you will first review real numbers. Integer exponents and scientific notation will be discussed, followed by rational exponents and radicals. Simplification of radicals and rationalization of denominators will be reviewed. Basic operations such as addition, subtraction, and multiplication of polynomials will be discussed followed by a review of how to factor polynomials. Rational expressions will be discussed and a brief overview of solving simple algebraic equations will be given. After reviewing all these aspects of real numbers, we will conclude this chapter with a review of complex numbers.

REVIEW

NUMBERS
- Real numbers
  - Rational
  - Irrational
- Complex numbers
  - Real
  - Imaginary
  - Complex conjugate

EXPONENTS
- Integer
- Rational
- Radical
  - Simplify
  - Rationalize
- Positive/Negative

POLYNOMIALS
- Basic operations
  - Adding
  - Subtracting
  - Multiplying
- Factoring
- Solving equations

RATIONAL EXPRESSIONS
- Ratios of polynomials
- Domain restrictions
- Basic operations

SECTION

0.1 REAL NUMBERS

0.2 INTEGER EXPONENTS AND SCIENTIFIC NOTATION

0.3 POLYNOMIALS: BASIC OPERATIONS

0.4 FACTORING POLYNOMIALS

0.5 RATIONAL EXPRESSIONS

0.6 RATIONAL EXPONENTS AND RADICALS

0.7 COMPLEX NUMBERS

CHAPTER OBJECTIVES

- Understand that rational and irrational numbers together constitute the real numbers.
- Round and truncate real numbers.
- Apply properties of exponents.
- Use scientific notation to represent a real number.
- Simplify expressions that contain rational exponents.
- Simplify radicals.
- Perform operations on polynomials.
- Multiply two binomials.
- Factor polynomials.
- Perform operations on rational expressions.
- Write complex numbers in standard form.

SECTION

0.1 REAL NUMBERS

SKILLS OBJECTIVES

- Classify real numbers as rational or irrational.
- Round or truncate real numbers.
- Evaluate algebraic expressions.
- Apply properties of real numbers.

CONCEPTUAL OBJECTIVES

- Understand that rational and irrational numbers are mutually exclusive and complementary subsets of real numbers.
- Learn the order of operations for real numbers.

The Set of Real Numbers

A **set** is a group or collection of objects that are called **members** or **elements** of the set. If *every* member of set B is also a member of set A, then we say B is a **subset** of A and denote it as $B \subset A$.

For example, the starting lineup on a baseball team is a subset of the entire team. The set of **natural numbers**, $\{1, 2, 3, 4, \ldots\}$, is a subset of the set of **whole numbers**, $\{0, 1, 2, 3, 4, \ldots\}$, which is a subset of the set of **integers**, $\{\ldots, -4, -3, -2, -1, 0, 1, 2, 3, \ldots\}$, which is a subset of the set of *rational numbers*, which is a subset of the set of *real numbers*. The three dots, called an **ellipsis**, indicate that the pattern continues indefinitely.

If a set has no elements, it is called the **empty set**, or **null set**, and is denoted by the symbol $\varnothing$. The **set of real numbers** consists of two main subsets: *rational* and *irrational* numbers.

DEFINITION **Rational Number**

A **rational number** is a number that can be expressed as a quotient (ratio) of two integers, $\frac{a}{b}$, where the integer a is called the **numerator** and the integer b is called the **denominator** and where $b \neq 0$.

Rational numbers include all integers or all fractions that are ratios of integers. Note that any integer can be written as a ratio whose denominator is equal to 1. In decimal form, the rational numbers are those that terminate or are nonterminating with a repeated decimal pattern, which we represent with an overbar. Those decimals that do not repeat and do not terminate are **irrational numbers**. The numbers

$$5, \quad -17, \quad \frac{1}{3}, \quad \sqrt{2}, \quad \pi, \quad 1.37, \quad 0, \quad -\frac{19}{17}, \quad 3.66\overline{6}, \quad 3.2179\ldots$$

are examples of **real** numbers, where 5, -17, $\frac{1}{3}$, 1.37, 0, $-\frac{19}{17}$, and $3.66\overline{6}$ are rational numbers, and $\sqrt{2}$, π, and $3.2179\ldots$ are irrational numbers. It is important to note that the ellipsis following the last decimal digit denotes continuing in an irregular fashion, whereas the absence of such dots to the right of the last decimal digit implies the decimal expansion terminates.

RATIONAL NUMBER (FRACTION)	CALCULATOR DISPLAY	DECIMAL REPRESENTATION	DESCRIPTION
$\frac{7}{2}$	3.5	3.5	*Terminates*
$\frac{15}{12}$	1.25	1.25	*Terminates*
$\frac{2}{3}$	0.666666666	$0.\overline{6}$	*Repeats*
$\frac{1}{11}$	0.09090909	$0.\overline{09}$	*Repeats*

Notice that the overbar covers the entire repeating pattern. The following figure and table illustrate the subset relationship and examples of different types of real numbers.

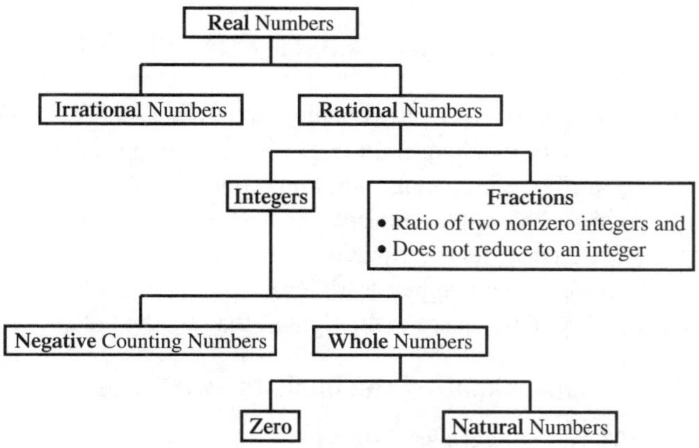

SYMBOL	NAME	DESCRIPTION	EXAMPLES
$\mathbb{N}$	Natural numbers	Counting numbers	$1, 2, 3, 4, 5, \ldots$
$\mathbb{W}$	Whole numbers	Natural numbers and zero	$0, 1, 2, 3, 4, 5, \ldots$
$\mathbb{Z}$	Integers	Whole numbers and negative natural numbers	$\ldots, -5, -4, -3, -2, -1, 0, 1, 2, 3, 4, 5, \ldots$
$\mathbb{Q}$	Rational numbers	Ratios of integers: $\frac{a}{b}$ $(b \neq 0)$ • Decimal representation terminates, or • Decimal representation repeats	$-17, -\frac{19}{7}, 0, \frac{1}{3}, 1.37, 3.6\overline{66}, 5$
$\mathbb{I}$	Irrational numbers	Numbers whose decimal representation does *not* terminate or repeat	$\sqrt{2}, 1.2179\ldots, \pi$
$\mathbb{R}$	Real numbers	Rational and irrational numbers	$\pi, 5, -\frac{2}{3}, 17.25, \sqrt{7}$

Study Tip

Every real number is either a rational number *or* an irrational number.

Since the set of real numbers can be formed by combining the set of rational numbers and the set of irrational numbers, then every real number is either rational or irrational. We say that the set of rational numbers and the set of irrational numbers are mutually exclusive (no shared elements) and complementary sets. The **real number line** is a graph used to represent the set of all real numbers.

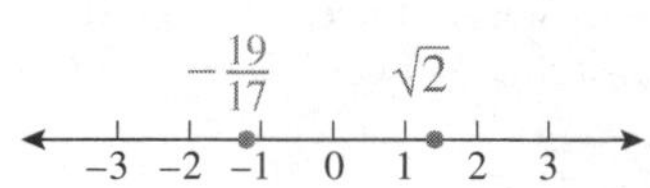

EXAMPLE 1 Classifying Real Numbers

Classify the following real numbers as rational or irrational:

$$-3, \quad 0, \quad \tfrac{1}{4}, \quad \sqrt{3}, \quad \pi, \quad 7.51, \quad \tfrac{1}{3}, \quad -\tfrac{8}{5}, \quad 6.666\overline{6}$$

Solution:

Rational: $-3, \quad 0, \quad \frac{1}{4}, \quad 7.51, \quad \frac{1}{3}, \quad -\frac{8}{5}, \quad -6.666\overline{6}$ Irrational: $\sqrt{3}, \pi$

■ **YOUR TURN** Classify the following real numbers as rational or irrational:

$$-\tfrac{7}{3}, 5.999\overline{9}, 12, 0, -5.27, \sqrt{5}, 2.010010001\ldots$$

■ **Answer:**
Rational: $-\frac{7}{3}, 5.999\overline{9}, 12, 0, -5.27$
Irrational: $\sqrt{5}, 2.010010001\ldots$

Approximations: Rounding and Truncation

Every real number can be represented by a decimal. When a real number is in decimal form we often approximate the number by either *rounding off* or *truncating* to a given decimal place. **Truncation** is "cutting off" or eliminating everything to the right of a certain decimal place. **Rounding** means looking to the right of the specified decimal place and making a judgment. If the digit to the right is greater than or equal to 5, then the specified digit is rounded up, or increased by 1 unit. If the digit to the right is less than 5, then the specified digit stays the same. In both of these cases, all decimal places to the right of the specified place are removed.

Study Tip

When rounding, look to the right of the specified decimal place and use that digit (do not round that digit first). 5.23491 rounded to two decimal places is 5.23 (do not round the 4 to a 5 first).

EXAMPLE 2 Approximating Decimals to Two Places

Approximate 17.368204 to two decimal places by

a. truncation **b.** rounding

Solution:

a. To truncate, eliminate all digits to the right of the 6. $\boxed{17.36}$

b. To round, look to the right of the 6.
Because "8" is greater than 5, we round up (add 1 to the 6). $\boxed{17.37}$

■ **YOUR TURN** Approximate 23.02492 to two decimal places by

a. truncation **b.** rounding

■ **Answer:** **a.** Truncation: 23.02
b. Rounding: 23.02

EXAMPLE 3 Approximating Decimals to Four Places

Approximate 7.293516 to four decimal places by

a. truncation **b.** rounding

Solution:

The "5" is in the fourth decimal place.

a. To truncate, eliminate all digits to the right of 5. $\boxed{7.2935}$

b. To round, look to the right of the 5.
Because "1" is less than 5, the 5 remains the same. $\boxed{7.2935}$

■ **YOUR TURN** Approximate -2.381865 to four decimal places by

a. truncation **b.** rounding

■ **Answer:** **a.** Truncation: -2.3818
b. Rounding: -2.3819

In Examples 2 and 3, we see that *rounding and truncation sometimes yield the same approximation, but not always.*

Order of Operations

Addition, subtraction, multiplication, and division are called arithmetic operations. The results of these operations are called the sum, difference, product, and quotient, respectively. These four operations are summarized in the following table.

OPERATION	NOTATION	RESULT
Addition	$a + b$	Sum
Subtraction	$a - b$	Difference
Multiplication	$a \cdot b$ or ab or $(a)(b)$	Product
Division	$\frac{a}{b}$ or a/b $(b \neq 0)$	Quotient (Ratio)

Since algebra involves *variables* such as x, in order to avoid confusion we avoid the traditional multiplication sign $\times$ when working with symbols. Three alternatives are shown in the preceding table. Similarly, the arithmetic sign for division, $\div$, is often represented by vertical or slanted fractions.

The symbol $=$ is called the **equal sign**, and is pronounced "equals" or "is," and it implies that the expression on one side of the equal sign is equivalent to (has the same value as) the expression on the other side of the equal sign.

WORDS	MATH
The sum of seven and eleven equals eighteen:	$7 + 11 = 18$
Three times five is fifteen:	$3 \cdot 5 = 15$
Four times six equals twenty-four:	$4(6) = 24$
Eight divided by two is four:	$\frac{8}{2} = 4$
Three subtracted from five is two:	$5 - 3 = 2$

When evaluating expressions involving real numbers, it is important to remember the correct *order of operations*. For example, how do we simplify the expression $3 + 2 \cdot 5$? Do we multiply first and then add, or add first and then multiply? In mathematics, conventional order implies multiplication first, and then addition: $3 + 2 \cdot 5 = 3 + 10 = 13$. Parentheses imply grouping of terms, and the necessary operations should always be performed inside them first. If there are nested parentheses, always start with the innermost parentheses and work your way out. Within parentheses follow the conventional order of operations. Exponents are an important part of order of operations and will be discussed in Section 0.2.

ORDER OF OPERATIONS

1. Start with the innermost parentheses (grouping symbols) and work outward.
2. Perform all indicated multiplications and divisions, working from left to right.
3. Perform all additions and subtractions, working from left to right.

EXAMPLE 4 **Simplifying Expressions Using the Correct Order of Operations**

Simplify the expressions.

a. $4 + 3 \cdot 2 - 7 \cdot 5 + 6$ **b.** $\dfrac{7 - 6}{2 \cdot 3 + 8}$

Solution (a):

Perform multiplication first. $4 + \underbrace{3 \cdot 2}_{6} - \underbrace{7 \cdot 5}_{35} + 6$

Then perform the indicated additions and subtractions. $= 4 + 6 - 35 + 6 = \boxed{-19}$

Solution (b):

The numerator and the denominator are similar to expressions in parentheses. Simplify these separately first, following the correct order of operations.

Perform multiplication in the denominator first. $\dfrac{7 - 6}{\underbrace{2 \cdot 3}_{6} + 8}$

Then perform subtraction in the numerator and addition in the denominator. $= \dfrac{7 - 6}{6 + 8} = \boxed{\dfrac{1}{14}}$

■ **YOUR TURN** Simplify the expressions.

a. $-7 + 4 \cdot 5 - 2 \cdot 6 + 9$ **b.** $\dfrac{9 - 6}{2 \cdot 5 + 6}$

■ **Answer: a.** 10 **b.** $\frac{3}{16}$

Parentheses ( ) and brackets [ ] are the typical notations for grouping and are often used interchangeably. When nested (groups within groups), we use parentheses on the innermost and then brackets on the outermost.

EXAMPLE 5 **Simplifying Expressions That Involve Grouping Signs Using the Correct Order of Operations**

Simplify the expression $3[5 \cdot (4 - 2) - 2 \cdot 7]$.

Solution:

Simplify the inner parentheses: $(4 - 2) = 2$. $= 3[5 \cdot 2 - 2 \cdot 7]$

Inside the brackets, perform the multiplication $5 \cdot 2 = 10$ and $2 \cdot 7 = 14$. $= 3[10 - 14]$

Inside the brackets, perform the subtraction. $= 3[-4]$

Multiply. $= \boxed{-12}$

■ **YOUR TURN** Simplify the expression $2[-3 \cdot (13 - 5) + 4 \cdot 3]$.

■ **Answer:** -24

Algebraic Expressions

Everything we have discussed until now has involved real numbers (explicitly). In algebra, however, numbers are often represented by letters (such as x and y), which we call **variables.** A **constant** is a fixed (known) number such as 5. A **coefficient** is the constant that is multiplied by a variable. Quantities within the *algebraic expression* that are separated by addition or subtraction are referred to as **terms**.

DEFINITION **Algebraic Expression**

An **algebraic expression** is the combination of variables and constants using basic operations such as addition, subtraction, multiplication, and division. Each term is separated by addition or subtraction.

Algebraic Expression	Variable Term	Constant Term	Coefficient
$5x + 3$	$5x$	3	5

When we know the value of the variables, we can **evaluate an algebraic expression** using the **substitution principle**:

Algebraic expression: $5x + 3$
Value of the variable: $x = 2$
Substitute $x = 2$: $5(2) + 3 = 10 + 3 = 13$

EXAMPLE 6 **Evaluating Algebraic Expressions**

Evaluate the algebraic expression $7x + 2$ for $x = 3$.

Solution:

Start with the algebraic expression.	$7x + 2$
Substitute $x = 3$.	$7(3) + 2$
Perform multiplication.	$= 21 + 2$
Perform addition.	$= \boxed{23}$

■ **YOUR TURN** Evaluate the algebraic expression $6y + 4$ for $y = 2$.

■ **Answer:** 16

In Example 6, the value for the variable was specified in order for us to evaluate the algebraic expression. What if the value of the variable is not specified; can we simplify an expression like $3(2x - 5y)$? In this case, we cannot subtract $5y$ from $2x$. Instead, we rely on the basic *properties of real numbers*, or the *basic rules of algebra*.

Properties of Real Numbers

You probably already know many properties of real numbers. For example, if you add up four numbers, it does not matter in which order you add them. If you multiply five numbers, it does not matter what order you multiply them. If you add 0 to a real number or multiply a real number by 1, the result yields the original real number. **Basic properties of real numbers** are summarized in the following table. Because these properties are true for variables and algebraic expressions, these properties are often called the **basic rules of algebra**.

PROPERTIES OF REAL NUMBERS (BASIC RULES OF ALGEBRA)

NAME	DESCRIPTION	MATH (LET *a*, *b*, AND *c* EACH BE ANY REAL NUMBER)	EXAMPLE
Commutative property of addition	Two real numbers can be added in any order.	$a + b = b + a$	$3x + 5 = 5 + 3x$
Commutative property of multiplication	Two real numbers can be multiplied in any order.	$ab = ba$	$y \cdot 3 = 3y$
Associative property of addition	When three real numbers are added, it does not matter which two numbers are added first.	$(a + b) + c = a + (b + c)$	$(x + 5) + 7 = x + (5 + 7)$
Associative property of multiplication	When three real numbers are multiplied, it does not matter which two numbers are multiplied first.	$(ab)c = a(bc)$	$(-3x)y = -3(xy)$
Distributive property	Multiplication is distributed over *all* the terms of the sums or differences within the parentheses.	$a(b + c) = ab + ac$	$5(x + 2) = 5x + 5 \cdot 2$
Additive identity property	Adding zero to any real number yields the same real number.	$a + 0 = a$ $0 + a = a$	$7y + 0 = 7y$
Multiplicative identity property	Multiplying any real number by 1 yields the same real number.	$a \cdot 1 = a$ $1 \cdot a = a$	$(8x)(1) = 8x$
Additive inverse property	The sum of a real number and its additive inverse (opposite) is zero.	$a + (-a) = 0$	$4x + (-4x) = 0$
Multiplicative inverse property	The product of a nonzero real number and its multiplicative inverse (reciprocal) is 1.	$a \cdot \frac{1}{a} = 1 \quad a \neq 0$	$(x + 2) \cdot \left(\frac{1}{x + 2}\right) = 1$ $x \neq -2$

The properties in the previous table govern addition and multiplication. Subtraction can be defined in terms of addition of the *additive inverse*, and division can be defined in terms of multiplication by *the multiplicative inverse (reciprocal).*

SUBTRACTION AND DIVISION

Let *a* and *b* be real numbers.

	MATH	TYPE OF INVERSE	WORDS
Subtraction	$a - b = a + (-b)$	$-b$ is the **additive inverse** or **opposite** of *b*	Subtracting a real number is equal to adding its opposite
Division	$a \div b = a \cdot \frac{1}{b}$, $b \neq 0$	$\frac{1}{b}$ is the **multiplicative inverse** or **reciprocal** of *b*	Dividing by a real number is equal to multiplying by its reciprocal

In the fractional form, $\frac{a}{b}$, *a* is called the **numerator** and *b* is called the **denominator**.

It is important to note that the subtraction form of the distributive property is

$$a(b - c) = ab - ac$$

EXAMPLE 7 Using the Distributive Property

Use the distributive property to eliminate the parentheses.

a. $3(x + 5)$ **b.** $2(y - 6)$

Solution (a):

Use the distributive property. $3(x + 5) = 3(x) + 3(5)$

Perform the multiplication. $= \boxed{3x + 15}$

Solution (b):

Use the distributive property. $2(y - 6) = 2(y) - 2(6)$

Perform the multiplication. $= \boxed{2y - 12}$

■ **YOUR TURN** Use the distributive property to eliminate the parentheses.

a. $2(x + 3)$ **b.** $5(y - 3)$

■ **Answer: a.** $2x + 6$ **b.** $5y - 15$

You also probably know the rules that apply when multiplying a negative real number. For example, "a negative times a negative is a positive."

PROPERTIES OF NEGATIVES

DESCRIPTION	MATH (LET *a* AND *b* BE POSITIVE REAL NUMBERS)	EXAMPLE
A *negative* quantity *times* a *positive* quantity is a *negative* quantity.	$(-a)(b) = -ab$	$(-8)(3) = -24$
A *negative* quantity *divided* by a *positive* quantity is a *negative* quantity. or A *positive* quantity *divided* by a *negative* quantity is a *negative* quantity.	$\frac{-a}{b} = -\frac{a}{b}$ or $\frac{a}{-b} = -\frac{a}{b}$	$\frac{-16}{4} = -4$ or $\frac{15}{-3} = -5$
A *negative* quantity times a *negative* quantity is a *positive* quantity.	$(-a)(-b) = ab$	$(-2x)(-5) = 10x$
A *negative* quantity *divided* by a *negative* quantity is a *positive* quantity.	$\frac{-a}{-b} = \frac{a}{b}$	$\frac{-12}{-3} = 4$
The opposite of a negative quantity is a positive quantity (Subtracting a negative quantity is equivalent to adding a positive quantity).	$-(-a) = a$	$-(-9) = 9$
A negative sign preceding an expression is distributed throughout the expression.	$-(a + b) = -a - b$ $-(a - b) = -a + b$	$-3(x + 5) = -3x - 15$ $-3(x - 5) = -3x + 15$

EXAMPLE 8 Using Properties of Negatives

Eliminate the parentheses and perform the operations.

a. $-5 + 7 - (-2)$ **b.** $-(-3)(-4)(-6)$

Solution:

a. Distribute the negative. $-5 + 7 - \underbrace{(-2)}_{2}$

$= -5 + 7 + 2$

Combine the three quantities. $= \boxed{4}$

b. Group the terms. $[-(-3)][(-4)(-6)]$

Perform the multiplication inside the [ ]. $= [3][24]$

Multiply. $= \boxed{72}$

Technology Tip

a.

```
-5+7--2
                4
```

b. Here are the calculator keystrokes for $-(-3)(-4)(-6)$.

```
--3*-4*-6
               72
```

We use properties of negatives to define the *absolute value* of any real number. The **absolute value** of a real number a, denoted $|a|$, is its magnitude. On a number line this is the distance from the origin, 0, to the point. For example, algebraically, the absolute value of 5 is 5, that is, $|5| = 5$; and the absolute value of -5 is 5, or $|-5| = 5$. Graphically, the distance on the real number line from 0 to either -5 or 5 is 5.

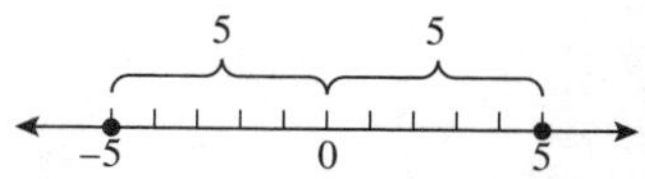

Notice that the absolute value does not change a positive real number, but changes a negative real number to a positive number. A negative number becomes a positive number if it is multiplied by -1.

IF a IS A	$\|a\|$	EXAMPLE
Positive real number	$\|a\| = a$	$\|5\| = 5$
Negative real number	$\|a\| = -a$	$\|-5\| = -(-5) = 5$
Zero	$\|a\| = a$	$\|0\| = 0$

EXAMPLE 9 Finding the Absolute Value of a Real Number

Evaluate the expressions.

a. $|-3 + 7|$ **b.** $|2 - 8|$

Solution:

a. $|-3 + 7| = |4| = \boxed{4}$

b. $|2 - 8| = |-6| = -(-6) = \boxed{6}$

Properties of the absolute value will be discussed in Section 1.7.

EXAMPLE 10 Using Properties of Negatives and the Distributive Property

Eliminate the parentheses $-(2x - 3y)$.

COMMON MISTAKE

A common mistake is applying a negative only to the first term.

CORRECT

$-(2x - 3y)$

$= -(2x) - (-3y)$

$= \boxed{-2x + 3y}$

INCORRECT

Error: $-2x - 3y$

The negative $(-)$ was not distributed through the second term.

YOUR TURN Eliminate the parentheses.

a. $-2(x + 5y)$ **b.** $-(3 - 2b)$

Answer: a. $-2x - 10y$
b. $-3 + 2b$

What is the product of any real number and zero? The answer is zero. This property also leads to the zero product property, which is the basis for factoring (one of the methods used to solve quadratic equations, which will be discussed in Section 1.3).

PROPERTIES OF ZERO

DESCRIPTION	MATH (LET *a* BE A REAL NUMBER)	EXAMPLE
A real number multiplied by zero is zero.	$a \cdot 0 = 0$	$0 \cdot x = 0$
Zero divided by a nonzero real number is zero.	$\frac{0}{a} = 0 \quad a \neq 0$	$\frac{0}{3 - x} = 0 \quad x \neq 3$
A real number divided by zero is undefined.	$\frac{a}{0}$ is undefined	$\frac{x + 2}{0}$ is undefined

ZERO PRODUCT PROPERTY

DESCRIPTION	MATH	EXAMPLE
If the product of two real numbers is zero, then one of those numbers has to be zero.	If $ab = 0$, then $a = 0$ or $b = 0$	If $x(x + 2) = 0$, then $x = 0$ or $x + 2 = 0$

Note: If *a* and *b* are *both* equal to zero, then the product is still zero.

Fractions always seem to intimidate students. In fact, many instructors teach students to eliminate fractions in algebraic equations. It is important to realize that you can never divide by zero. Therefore, in the following table of fractional properties it is assumed that no denominators are zero.

FRACTIONAL PROPERTIES

DESCRIPTION	MATH	ZERO CONDITION	EXAMPLE
Equivalent fractions	$\frac{a}{b} = \frac{c}{d}$ if and only if $ad = bc$	$b \neq 0$ and $d \neq 0$	$\frac{y}{2} = \frac{6y}{12}$ since $12y = 12y$
Multiplying two fractions	$\frac{a}{b} \cdot \frac{c}{d} = \frac{ac}{bd}$	$b \neq 0$ and $d \neq 0$	$\frac{3}{5} \cdot \frac{x}{7} = \frac{3x}{35}$
Adding fractions that have the same denominator	$\frac{a}{b} + \frac{c}{b} = \frac{a + c}{b}$	$b \neq 0$	$\frac{x}{3} + \frac{2}{3} = \frac{x + 2}{3}$
Subtracting fractions that have the same denominator	$\frac{a}{b} - \frac{c}{b} = \frac{a - c}{b}$	$b \neq 0$	$\frac{7}{3} - \frac{5}{3} = \frac{2}{3}$
Adding fractions with different denominators using a least common denominator	$\frac{a}{b} + \frac{c}{d} = \frac{ad}{bd} + \frac{cb}{bd} = \frac{ad + bc}{bd}$	$b \neq 0$ and $d \neq 0$	$\frac{1}{2} + \frac{5}{3} = \frac{(1)(3) + (5)(2)}{6} = \frac{13}{6}$
Subtracting fractions with different denominators using a least common denominator	$\frac{a}{b} - \frac{c}{d} = \frac{ad}{bd} - \frac{cb}{bd} = \frac{ad - bc}{bd}$	$b \neq 0$ and $d \neq 0$	$\frac{1}{3} - \frac{1}{4} = \frac{(1)(4) - (1)(3)}{12} = \frac{1}{12}$
Dividing by a fraction is equivalent to multiplying by its reciprocal	$\frac{a}{b} \div \frac{c}{d} = \frac{a}{b} \cdot \frac{d}{c}$	$b \neq 0$, $c \neq 0$, and $d \neq 0$	$\frac{x}{3} \div \frac{2}{7} = \frac{x}{3} \cdot \frac{7}{2} = \frac{7x}{6}$

The **least common multiple** of two or more integers is the smallest integer that is evenly divisible by each of the integers. For example, the least common multiple (LCM) of 3 and 4 is 12. The LCM of 8 and 6 is 24. The reason the LCM of 8 and 6 is not 48 is that 8 and 6 have a common factor of 2. When adding and subtracting fractions, a common denominator can be found by multiplying the denominators. When there are common factors in the denominators, the ***least*** **common denominator** is the LCM of the original denominators.

Technology Tip

To change the decimal number to a fraction, press: MATH 1: ▶ Frac ENTER ENTER

a.

b.

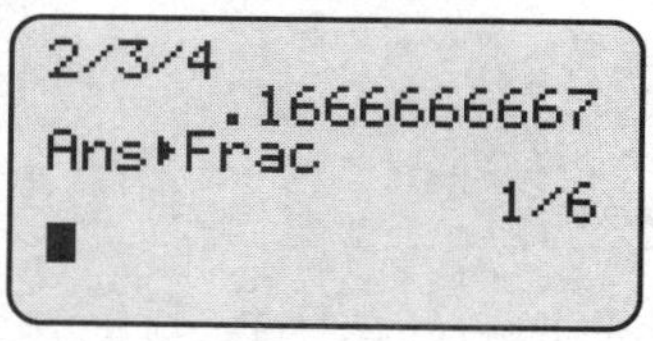

EXAMPLE 11 Performing Operations with Fractions

Perform the indicated operations involving fractions and simplify.

a. $\frac{2}{3} - \frac{1}{4}$ **b.** $\frac{2}{3} \div 4$ **c.** $\frac{x}{2} + \frac{3}{5}$

Solution (a):

Determine the least common denominator. $3 \cdot 4 = 12$

Rewrite fractions applying the least common denominator. $\frac{2}{3} - \frac{1}{4} = \frac{2 \cdot 4}{3 \cdot 4} - \frac{1 \cdot 3}{4 \cdot 3}$

$= \frac{2(4) - 1(3)}{3(4)}$

Eliminate the parentheses. $= \frac{8 - 3}{12}$

Combine terms in the numerator. $= \boxed{\frac{5}{12}}$

Solution (b):

Rewrite 4 with an understood 1 in the denominator. $= \frac{2}{3} \div \frac{4}{1}$

Dividing by a fraction is equivalent to multiplying by its reciprocal. $= \frac{2}{3} \cdot \frac{1}{4}$

Multiply numerators and denominators, respectively. $= \frac{2}{12}$

Reduce the fraction to simplest form. $= \boxed{\frac{1}{6}}$

Solution (c):

Determine the least common denominator. $2 \cdot 5 = 10$

Rewrite fractions in terms of the LCD. $\frac{x}{2} + \frac{3}{5} = \frac{5x + 3(2)}{10}$

Simplify the numerator. $= \boxed{\frac{5x + 6}{10}}$

YOUR TURN Perform the indicated operations involving fractions.

a. $\frac{3}{5} + \frac{1}{2}$ **b.** $\frac{1}{5} \div \frac{3}{10}$ **c.** $\frac{2}{3} - \frac{x}{5}$

Answer:

a. $\frac{11}{10}$ **b.** $\frac{2}{3}$ **c.** $\frac{10 - 3x}{15}$

SECTION 0.1 SUMMARY

In this section, real numbers were defined as the set of all rational and irrational numbers. Decimals are approximated by either truncating or rounding.

- *Truncating*: Eliminate all values after a particular digit.
- *Rounding*: Look to the right of a particular digit. If the number is 5 or greater, increase the digit by 1; otherwise, leave it as is and eliminate all digits to the right.

The *order* in which we perform operations is

1. parentheses (grouping); work from inside outward.
2. multiplication/division; work from left to right.
3. addition/subtraction; work from left to right.

The *properties of real numbers* are employed as the basic rules of algebra when dealing with algebraic expressions.

- Commutative property of addition: $a + b = b + a$
- Commutative property of multiplication: $ab = ba$
- Associative property of addition: $(a + b) + c = a + (b + c)$
- Associative property of multiplication: $(ab)c = a(bc)$
- Distributive property: $a(b + c) = ab + ac$ or $a(b - c) = ab - ac$
- Additive identity: $a + 0 = a$
- Multiplicative identity: $a \cdot 1 = a$
- Additive inverse (opposite): $a + (-a) = 0$
- Multiplicative inverse (reciprocal): $a \cdot \frac{1}{a} = 1 \qquad a \neq 0$

Subtraction and division can be defined in terms of addition and multiplication.

- *Subtraction:* $a - b = a + (-b)$ (add the opposite)
- *Division:* $a \div b = a \cdot \frac{1}{b}$, where $b \neq 0$ (multiply by the reciprocal)

Properties of negatives were reviewed. If a and b are positive real numbers, then:

- $(-a)(b) = -ab$
- $(-a)(-b) = ab$
- $-(-a) = a$
- $-(a + b) = -a - b$ and $-(a - b) = -a + b$
- $\frac{-a}{b} = -\frac{a}{b}$
- $\frac{-a}{-b} = \frac{a}{b}$

Absolute value of real numbers: $|a| = a$ if a is nonnegative, and $|a| = -a$ if a is negative. Properties of zero were reviewed.

- $a \cdot 0 = 0$ and $\frac{0}{a} = 0 \qquad a \neq 0$
- $\frac{a}{0}$ is undefined
- Zero product property: If $ab = 0$, then $a = 0$ or $b = 0$

Properties of fractions were also reviewed.

- $\frac{a}{b} \pm \frac{c}{d} = \frac{ad \pm bc}{bd} \qquad b \neq 0$ and $d \neq 0$
- $\frac{a}{b} \div \frac{c}{d} = \frac{a}{b} \cdot \frac{d}{c} \qquad b \neq 0, c \neq 0$, and $d \neq 0$

SECTION 0.1 EXERCISES

SKILLS

In Exercises 1–8, classify the following real numbers as rational or irrational.

1. $\frac{11}{3}$ **2.** $\frac{22}{3}$ **3.** 2.07172737. . . **4.** π

5. $2.77667\overline{76677}$ **6.** $5.22222\overline{2}$ **7.** $\sqrt{5}$ **8.** $\sqrt{17}$

In Exercises 9–16, approximate the real number to three decimal places by (a) rounding and (b) truncation.

9. 7.3471 **10.** 9.2549 **11.** 2.9949 **12.** 6.9951

13. 0.234492 **14.** 1.327491 **15.** 5.238473 **16.** 2.118465

In Exercises 17–40, perform the indicated operations in the correct order.

17. $5 + 2 \cdot 3 - 7$

18. $2 + 5 \cdot 4 + 3 \cdot 6$

19. $2 \cdot (5 + 7 \cdot 4 - 20)$

20. $-3 \cdot (2 + 7) + 8 \cdot (7 - 2 \cdot 1)$

21. $2 - 3[4(2 \cdot 3 + 5)]$

22. $4 \cdot 6(5 - 9)$

23. $8 - (-2) + 7$

24. $-10 - (-9)$

25. $-3 - (-6)$

26. $-5 + 2 - (-3)$

27. $x - (-y) - z$

28. $-a + b - (-c)$

29. $-(3x + y)$

30. $-(4a - 2b)$

31. $\dfrac{-3}{(5)(-1)}$

32. $-\dfrac{12}{(-3)(-4)}$

33. $-4 - 6[(5 - 8)(4)]$

34. $\dfrac{-14}{5 - (-2)}$

35. $-(6x - 4y) - (3x + 5y)$

36. $\dfrac{-4x}{6 - (-2)}$

37. $-(3 - 4x) - (4x + 7)$

38. $2 - 3[(4x - 5) - 3x - 7]$

39. $\dfrac{-4(5) - 5}{-5}$

40. $-6(2x + 3y) - [3x - (2 - 5y)]$

In Exercises 41–56, write as a single fraction and simplify.

41. $\dfrac{1}{3} + \dfrac{5}{4}$

42. $\dfrac{1}{2} - \dfrac{1}{5}$

43. $\dfrac{5}{6} - \dfrac{1}{3}$

44. $\dfrac{7}{3} - \dfrac{1}{6}$

45. $\dfrac{3}{2} + \dfrac{5}{12}$

46. $\dfrac{1}{3} + \dfrac{5}{9}$

47. $\dfrac{1}{9} - \dfrac{2}{27}$

48. $\dfrac{3}{7} - \dfrac{(-4)}{3} - \dfrac{5}{6}$

49. $\dfrac{x}{5} + \dfrac{2x}{15}$

50. $\dfrac{y}{3} - \dfrac{y}{6}$

51. $\dfrac{x}{3} - \dfrac{2x}{7}$

52. $\dfrac{y}{10} - \dfrac{y}{15}$

53. $\dfrac{4y}{15} - \dfrac{(-3y)}{4}$

54. $\dfrac{6x}{12} - \dfrac{7x}{20}$

55. $\dfrac{3}{40} + \dfrac{7}{24}$

56. $\dfrac{-3}{10} - \left(\dfrac{-7}{12}\right)$

In Exercises 57–68, perform the indicated operation and simplify, if possible.

57. $\dfrac{2}{7} \cdot \dfrac{14}{3}$

58. $\dfrac{2}{3} \cdot \dfrac{9}{10}$

59. $\dfrac{2}{7} \div \dfrac{10}{3}$

60. $\dfrac{4}{5} \div \dfrac{7}{10}$

61. $\dfrac{4b}{9} \div \dfrac{a}{27} \quad a \neq 0$

62. $\dfrac{3a}{7} \div \dfrac{b}{21} \quad b \neq 0$

63. $\dfrac{3x}{10} \div \dfrac{6x}{15} \quad x \neq 0$

64. $4\dfrac{1}{5} \div 7\dfrac{1}{20}$

65. $\dfrac{3x}{4} \div \dfrac{9}{16y} \quad y \neq 0$

66. $\dfrac{14m}{2} \cdot \dfrac{4}{7}$

67. $\dfrac{6x}{7} \div \dfrac{3y}{28} \quad y \neq 0$

68. $2\dfrac{1}{3} \cdot 7\dfrac{5}{6}$

In Exercises 69–72, evaluate the algebraic expression for the specified values.

69. $\dfrac{-c}{2d}$ for $c = -4, d = 3$

70. $2l + 2w$ for $l = 5, w = 10$

71. $\dfrac{m_1 \cdot m_2}{r^2}$ for $m_1 = 3, m_2 = 4, r = 10$

72. $\dfrac{x - \mu}{\sigma}$ for $x = 100, \mu = 70, \sigma = 15$

APPLICATIONS

On December 16, 2007, the United States debt was estimated at $9,176,366,494,947, and at that time the estimated population was 303,818,361 citizens.

73. U.S. National Debt. Round the debt to the nearest million.

74. U.S. Population. Round the number of citizens to the nearest thousand.

75. U.S. Debt. If the debt is distributed evenly to all citizens, what is the national debt per citizen? Round your answer to the nearest dollar.

76. U.S. Debt. If the debt is distributed evenly to all citizens, what is the national debt per citizen? Round your answer to the nearest cent.

CATCH THE MISTAKE

In Exercises 77–80, explain the mistake that is made.

77. Round 13.2749 to two decimal places.

Solution:

The 9, to the right of the 4, causes the 4 to round to 5.	13.275
The 5, to the right of the 7, causes the 7 to be rounded to 8.	13.28

This is incorrect. What mistake was made?

78. Simplify the expression $\frac{2}{3} + \frac{1}{9}$.

Solution:

Add the numerators and denominators. $\frac{2 + 1}{3 + 9} = \frac{3}{12}$

Reduce. $= \frac{1}{4}$

This is incorrect. What mistake was made?

79. Simplify the expression $3(x + 5) - 2(4 + y)$.

Solution:

Eliminate parentheses. $3x + 15 - 8 + y$

Simplify. $3x + 7 + y$

This is incorrect. What mistake was made?

80. Simplify the expression $-3(x + 2) - (1 - y)$.

Solution:

Eliminate parentheses. $-3x - 6 - 1 - y$

Simplify. $-3x - 7 - y$

This is incorrect. What mistake was made?

CONCEPTUAL

In Exercises 81–84, determine whether each of the following statements is true or false.

81. Student athletes are a subset of the students in the honors program.

82. The students who are members of fraternities or sororities are a subset of the entire student population.

83. Every integer is a rational number.

84. A real number can be both rational and irrational.

85. What restrictions are there on x for the following to be true:

$$\frac{3}{x} \div \frac{5}{x} = \frac{3}{5}$$

86. What restrictions are there on x for the following to be true:

$$\frac{x}{2} \div \frac{x}{6} = 3$$

CHALLENGE

In Exercises 87 and 88, simplify the expressions.

87. $-2[3(x - 2y) + 7] + [3(2 - 5x) + 10] - 7[-2(x - 3) + 5]$

88. $-2\{-5(y - x) - 2[3(2x - 5) + 7(2) - 4] + 3\} + 7$

TECHNOLOGY

89. Use your calculator to evaluate $\sqrt{1260}$. Does the answer appear to be a rational or an irrational number? Why?

90. Use your calculator to evaluate $\sqrt{\frac{144}{25}}$. Does the answer appear to be a rational or an irrational number? Why?

91. Use your calculator to evaluate $\sqrt{4489}$. Does the answer appear to be a rational or an irrational number? Why?

92. Use your calculator to evaluate $\sqrt{\frac{882}{49}}$. Does the answer appear to be a rational or an irrational number? Why?

SECTION 0.2 INTEGER EXPONENTS AND SCIENTIFIC NOTATION

SKILLS OBJECTIVES

- Evaluate expressions involving integer exponents.
- Apply properties of exponents.
- Use scientific notation.

CONCEPTUAL OBJECTIVES

- Visualize negative exponents as reciprocals.
- Understand that scientific notation is an effective way to represent very large or very small real numbers.

Integer Exponents

Exponents represent repeated multiplication. For example, $2 \cdot 2 \cdot 2 \cdot 2 \cdot 2 = 2^5$. The 2 that is repeatedly multiplied is called the *base*, and the small number 5 above and to the right of the 2 is called the *exponent*.

Study Tip

a^n: "*a* raised to the *n*th power"
a^2: "*a* squared"
a^3: "*a* cubed"

DEFINITION Natural-Number Exponent

Let a be a real number and n be a natural number (positive integer); then a^n is defined as

$$a^n = \underbrace{a \cdot a \cdot a \cdots a}_{n \text{ factors}} \qquad (a \text{ appears as a factor } n \text{ times})$$

where n is the **exponent**, or **power**, and a is the **base**.

Technology Tip

```
4^3
              64
8^1
               8
```

```
5^4
             625
(1/2)^5
          .03125
Ans▸Frac
            1/32
```

EXAMPLE 1 Evaluating Expressions Involving Natural-Number Exponents

Evaluate the expressions.

a. 4^3 **b.** 8^1 **c.** 5^4 **d.** $\left(\frac{1}{2}\right)^5$

Solution:

a. $4^3 = 4 \cdot 4 \cdot 4 = \boxed{64}$

b. $8^1 = \boxed{8}$

c. $5^4 = 5 \cdot 5 \cdot 5 \cdot 5 = \boxed{625}$

d. $\left(\frac{1}{2}\right)^5 = \frac{1}{2} \cdot \frac{1}{2} \cdot \frac{1}{2} \cdot \frac{1}{2} \cdot \frac{1}{2} = \boxed{\frac{1}{32}}$

■ YOUR TURN Evaluate the expressions.

a. 6^3 **b.** $\left(\frac{1}{3}\right)^4$

■ **Answer: a.** 216 **b.** $\frac{1}{81}$

We now include exponents in our order of operations:

1. Parentheses
2. Exponents
3. Multiplication/Division
4. Addition/Subtraction

EXAMPLE 2 **Evaluating Expressions Involving Natural-Number Exponents**

Evaluate the expressions.

a. $(-3)^4$ **b.** -3^4 **c.** $(-2)^3 \cdot 5^2$

Solution:

a. $(-3)^4 = (-3)(-3)(-3)(-3) = \boxed{81}$

b. $-3^4 = -\underbrace{3 \cdot 3 \cdot 3 \cdot 3}_{81} = \boxed{-81}$

c. $(-2)^3 \cdot 5^2 = \underbrace{(-2)(-2)(-2)}_{-8} \cdot \underbrace{5 \cdot 5}_{25} = \boxed{-200}$

■ **YOUR TURN** Evaluate the expression $-4^3 \cdot 2^3$.

Technology Tip

```
(-3)^4
                81
-3^4
               -81
(-2)^3*5²
              -200
```

■ **Answer:** -512

So far, we have discussed only exponents that are natural numbers (positive integers). When the exponent is a negative integer we use the following property.

Study Tip

A negative exponent implies a reciprocal.

NEGATIVE-INTEGER EXPONENT PROPERTY

Let a be any nonzero real number and n be a natural number (positive integer); then

$$a^{-n} = \frac{1}{a^n} \qquad a \neq 0$$

In other words, a base raised to a negative-integer exponent is equivalent to the reciprocal of the base raised to the opposite (positive) integer exponent.

EXAMPLE 3 **Evaluating Expressions Involving Negative-Integer Exponents**

Evaluate the expressions.

a. 2^{-4} **b.** $\frac{1}{3^{-3}}$ **c.** $4^{-3} \cdot \frac{1}{2^{-4}}$ **d.** $-2^3 \cdot \frac{1}{(-6)^{-2}}$

Solution:

a. $2^{-4} = \frac{1}{2^4} = \boxed{\frac{1}{16}}$

b. $\frac{1}{3^{-3}} = \frac{1}{\left(\frac{1}{3^3}\right)} = 1 \div \left(\frac{1}{3^3}\right) = 1 \cdot \frac{3^3}{1} = 3^3 = \boxed{27}$

c. $4^{-3} \cdot \frac{1}{2^{-4}} = \frac{1}{4^3} \cdot 2^4 = \frac{16}{64} = \boxed{\frac{1}{4}}$

d. $-2^3 \cdot \frac{1}{(-6)^{-2}} = \underbrace{-2^3}_{-8} \cdot \underbrace{(-6)^2}_{36} = (-8)(36) = \boxed{-288}$

■ **YOUR TURN** Evaluate the expressions.

a. $-\frac{1}{5^{-2}}$ **b.** $\frac{1}{3^{-2}} \cdot 6^{-2}$

Technology Tip

```
2^(-4)
             .0625
Ans►Frac
              1/16
1/3^(-3)
                27
```

```
4^(-3)*1/2^(-4)
               .25
Ans►Frac
               1/4
-2^3*1/(-6)^(-2)
              -288
```

■ **Answer:** **a.** -25 **b.** $\frac{1}{4}$

Now we can evaluate expressions involving positive and negative exponents. How do we evaluate an expression with a zero exponent? We define any nonzero real number raised to the zero power as 1.

ZERO-EXPONENT PROPERTY

Let a be any nonzero real number; then

$$a^0 = 1 \qquad a \neq 0$$

EXAMPLE 4 Evaluating Expressions Involving Zero Exponents

Evaluate the expressions.

a. 5^0 **b.** $\frac{1}{2^0}$ **c.** $(-3)^0$ **d.** -4^0

Solution:

a. $5^0 = \boxed{1}$ **b.** $\frac{1}{2^0} = \frac{1}{1} = \boxed{1}$ **c.** $(-3)^0 = \boxed{1}$ **d.** $-\underbrace{4^0}_{1} = \boxed{-1}$

We now can evaluate expressions involving integer (positive, negative, or zero) exponents. What about when expressions involving integer exponents are multiplied, divided, or raised to a power?

WORDS	MATH
When expressions with the *same base* are *multiplied*, the exponents are *added.*	$\underbrace{2^3}_{2\cdot2\cdot2} \cdot \underbrace{2^4}_{2\cdot2\cdot2\cdot2} = \overbrace{2\cdot2\cdot2}^{2^3}\cdot\overbrace{2\cdot2\cdot2\cdot2}^{2^4} = 2^{3+4} = 2^7$
When expressions with the *same base* are *divided*, the exponents are *subtracted.*	$\frac{2^5}{2^3} = \frac{2\cdot2\cdot2\cdot2\cdot2}{2\cdot2\cdot2} = \frac{2\cdot2}{1} = 2^2$ or $2^{5-3} = 2^2$
When an expression involving an exponent is *raised to a power*, the exponents are *multiplied.*	$(2^3)^2 = (8)^2 = 64$ or $(2^3)^2 = 2^{3\cdot2} = 2^6 = 64$

The following table summarizes the *properties of integer exponents.*

PROPERTIES OF INTEGER EXPONENTS

NAME	DESCRIPTION	MATH (LET a AND b BE NONZERO REAL NUMBERS AND m AND n BE INTEGERS)	EXAMPLE
Product property	When multiplying exponentials with the same base, **add** exponents.	$a^m \cdot a^n = a^{m+n}$	$x^2 \cdot x^5 = x^{2+5} = x^7$
Quotient property	When dividing exponentials with the same base, **subtract** the exponents, (numerator − denominator).	$\frac{a^m}{a^n} = a^{m-n}$	$\frac{x^5}{x^3} = x^{5-3} = x^2 \quad x \neq 0$
Power property	When raising an exponential to a power, **multiply** exponents.	$(a^m)^n = a^{mn}$	$(x^2)^4 = x^{2\cdot4} = x^8$
Product to a power property	A product raised to a power is equal to the product of each factor raised to the power.	$(ab)^n = a^n b^n$	$(2x)^3 = 2^3 \cdot x^3 = 8x^3$
Quotient to a power property	A quotient raised to a power is equal to the quotient of the factors raised to the power.	$\left(\frac{a}{b}\right)^n = \frac{a^n}{b^n}$	$\left(\frac{x}{y}\right)^4 = \frac{x^4}{y^4} \quad y \neq 0$

Common Errors Made Using Properties of Exponents

INCORRECT	CORRECT	ERROR
$x^4 \cdot x^3 = x^{12}$	$x^4 \cdot x^3 = x^7$	Exponents should be added (not multiplied).
$\dfrac{x^{18}}{x^6} = x^3$	$\dfrac{x^{18}}{x^6} = x^{12};\ x \neq 0$	Exponents should be subtracted (not divided).
$(x^2)^3 = x^8$	$(x^2)^3 = x^6$	Exponents should be multiplied (not raised to power).
$(2x)^3 = 2x^3$	$(2x)^3 = 8x^3$	Both factors (the 2 and the x) should be cubed.
$2^3 \cdot 2^4 = 4^7$	$2^3 \cdot 2^4 = 2^7$	The original common base should be retained.
$2^3 \cdot 3^5 = 6^8$	$2^3 \cdot 3^5$	The properties of integer exponents require the *same* base.

We will now use properties of integer exponents to *simplify exponential expressions*. An exponential expression is **simplified** when:

- All parentheses (groupings) have been eliminated.
- A base appears only once.
- No powers are raised to other powers.
- All exponents are positive.

EXAMPLE 5 Simplifying Exponential Expressions

Simplify the expressions (assume all variables are nonzero).

a. $(-2x^2y^3)(5x^3y)$ **b.** $(2x^2yz^3)^3$ **c.** $\dfrac{25x^3y^6}{-5x^5y^4}$

Solution (a):

Parentheses imply multiplication.

Group the same bases together. $(-2x^2y^3)(5x^3y) = (-2)(5)x^2x^3y^3y$

Apply the product property. $= (-2)(5)\underbrace{x^2x^3}_{x^{2+3}}\underbrace{y^3y}_{y^{3+1}}$

Multiply the constants. $= \boxed{-10x^5y^4}$

Solution (b):

Apply the product to a power property. $(2x^2yz^3)^3 = (2)^3(x^2)^3(y)^3(z^3)^3$

Apply the power property. $= 8x^{2\cdot3}y^{1\cdot3}z^{3\cdot3}$

Simplify. $= \boxed{8x^6y^3z^9}$

Solution (c):

Group the same bases together. $\dfrac{25x^3y^6}{-5x^5y^4} = \left(\dfrac{25}{-5}\right)\left(\dfrac{x^3}{x^5}\right)\left(\dfrac{y^6}{y^4}\right)$

Apply the quotient property. $= (-5)x^{3-5}y^{6-4}$

$= -5x^{-2}y^2$

Apply the negative exponent property. $= \boxed{\dfrac{-5y^2}{x^2}}$

Study Tip

It is customary not to leave negative exponents. Instead we use the negative exponent property to write exponential expressions with only positive exponents.

YOUR TURN Simplify the expressions (assume all variables are nonzero).

a. $(-3x^3y^2)(4xy^3)$ **b.** $(-3xy^3z^2)^3$ **c.** $\dfrac{-16x^4y^3}{4xy^7}$

Answer:
a. $-12x^4y^5$ **b.** $-27x^3y^9z^6$ **c.** $\dfrac{-4x^3}{y^4}$

EXAMPLE 6 Simplifying Exponential Expressions

Write each expression so that all exponents are positive (assume all variables are nonzero).

a. $\left(3x^2z^{-4}\right)^{-3}$ **b.** $\dfrac{\left(x^2y^{-3}\right)^2}{\left(x^{-1}y^4\right)^{-3}}$ **c.** $\dfrac{\left(-2xy^2\right)^3}{-\left(6xz^3\right)^2}$

Solution (a):

Apply the product to a power property. $\left(3x^2z^{-4}\right)^{-3} = (3)^{-3}\left(x^2\right)^{-3}\left(z^{-4}\right)^{-3}$

Apply the power property. $= 3^{-3}x^{-6}z^{12}$

Apply the negative-integer exponent property. $= \dfrac{z^{12}}{3^3x^6}$

Evaluate 3^3. $= \boxed{\dfrac{z^{12}}{27x^6}}$

Solution (b):

Apply the product to a power property. $\dfrac{\left(x^2y^{-3}\right)^2}{\left(x^{-1}y^4\right)^{-3}} = \dfrac{x^4y^{-6}}{x^3y^{-12}}$

Apply the quotient property. $= x^{4-3}y^{-6-(-12)}$

Simplify. $= \boxed{xy^6}$

Solution (c):

Apply the product to a power property on both the numerator and denominator. $\dfrac{\left(-2xy^2\right)^3}{-\left(6xz^3\right)^2} = \dfrac{(-2)^3(x)^3(y^2)^3}{-(6)^2(x)^2(z^3)^2}$

Apply the power property. $= \dfrac{-8x^3y^6}{-36x^2z^6}$

Group constant terms and x terms. $= \left(\dfrac{-8}{-36}\right)\left(\dfrac{x^3}{x^2}\right)\left(\dfrac{y^6}{z^6}\right)$

Apply the quotient property. $= \left(\dfrac{8}{36}\right)\left(x^{3-2}\right)\left(\dfrac{y^6}{z^6}\right)$

Simplify. $= \boxed{\dfrac{2xy^6}{9z^6}}$

■ **Answer:** $\dfrac{2t}{s^3}$

■ **YOUR TURN** Simplify the exponential expression and express it in terms of positive exponents $\dfrac{\left(ts^2\right)^{-3}}{\left(2t^4s^3\right)^{-1}}$.

Scientific Notation

You are already familiar with base 10 raised to positive-integer powers. However, it can be inconvenient to write all the zeros out, so we give certain powers of 10 particular names: thousand, million, billion, trillion, and so on. For example, we say there are 300 million U.S. citizens as opposed to writing out 300,000,000 citizens. Or, we say that the national debt is \$9 trillion as opposed to writing out \$9,000,000,000,000. The following table contains scientific notation for positive exponents and examples of some common prefixes and abbreviations. One of the fundamental applications of scientific notation is *measurement*.

EXPONENTIAL FORM	REAL NUMBER	NUMBER OF ZEROS FOLLOWING THE 1	PREFIX	ABBREVIATION	EXAMPLE
10^1	10	1			
10^2	100	2			
10^3	1,000 (one thousand)	3	kilo-	k	The relay-for-life team ran a total of 80 km (kilometers).
10^4	10,000	4			
10^5	100,000	5			
10^6	1,000,000 (one million)	6	mega-	M	Modern high-powered diesel–electric railroad locomotives typically have a peak power output of 3 to 5 MW (megawatts).
10^7	10,000,000	7			
10^8	100,000,000	8			
10^9	1,000,000,000 (one billion)	9	giga-	G	A flash drive typically has 1 to 4 GB (gigabytes) of storage.
10^{10}	10,000,000,000	10			
10^{11}	100,000,000,000	11			
10^{12}	1,000,000,000,000 (one trillion)	12	tera-	T	Laser systems offer higher frequencies on the order of THz (terahertz).

Notice that 10^8 is a 1 followed by 8 zeros; alternatively, you can start with 1.0 and move the decimal point 8 places to the right (insert zeros). The same type of table can be made for negative-integer powers with base 10. To find the real number associated with exponential form, start with 1.0 and move the decimal a certain number of places to the left (fill in missing decimal places with zeros).

EXPONENTIAL FORM	REAL NUMBER	NUMBER OF PLACES DECIMAL (1.0) MOVES TO THE LEFT	PREFIX	ABBREVIATION	EXAMPLE
10^{-1}	0.1	1			
10^{-2}	0.01	2			
10^{-3}	0.001 (one thousandth)	3	milli-	m	Excedrin Extra Strength tablets each have 250 mg (milligrams) of acetaminophen.
10^{-4}	0.0001	4			
10^{-5}	0.00001	5			
10^{-6}	0.000001 (one millionth)	6	micro-	μ	A typical laser has a wavelength of 1.55 μm (micrometers*).
10^{-7}	0.0000001	7			
10^{-8}	0.00000001	8			
10^{-9}	0.000000001 (one billionth)	9	nano-	n	PSA levels less than 4 ng/ml (nanogram per milliliter of blood) represents low risk for prostate cancer.
10^{-10}	0.0000000001	10			
10^{-11}	0.00000000001	11			
10^{-12}	0.000000000001 (one trillionth)	12	pico-	p	A single yeast cell weighs 44 pg (picograms).

*In optics a micrometer is called a micron.

Study Tip

Scientific notation is a number between 1 and 10 that is multiplied by 10 to a power.

Study Tip

Real numbers greater than 1 correspond to positive exponents in scientific notation, whereas real numbers greater than 0 but less than 1 correspond to negative exponents in scientific notation.

SCIENTIFIC NOTATION

A positive real number can be written in **scientific notation** with the form $c \times 10^n$, where $1 \le c < 10$ and n is an integer.

Note that c is a real number between 1 and 10. Therefore, 22.5×10^3 is not in scientific notation, but we can convert it to scientific notation: 2.25×10^4.

For example, there are approximately 50 trillion cells in the human body. We write 50 trillion as 50 followed by 12 zeros 50,000,000,000,000. An efficient way of writing such a large number is using **scientific notation**. Notice that 50,000,000,000,000 is **5** followed by **13** zeros, or in scientific notation, $\mathbf{5 \times 10^{13}}$. Very small numbers can also be written using scientific notation. For example, in laser communications a pulse width is 2 femtoseconds, or 0.000000000000002 second. Notice that if we start with **2**.0 and move the decimal point **15** places to the left (adding zeros in between), the result is 0.000000000000002, or in scientific notation, $\mathbf{2 \times 10^{-15}}$.

Technology Tip

```
3856000000000000
             3.856E15
.00000275
              2.75E-6
```

EXAMPLE 7 Expressing a Positive Real Number in Scientific Notation

Express the numbers in scientific notation.

a. 3,856,000,000,000,000 **b.** 0.00000275

Solution:

a. Rewrite the number with the implied decimal point. 3,856,000,000,000,000.

Move the decimal point to the left 15 places. $= \boxed{3.856 \times 10^{15}}$

b. Move the decimal point to the right 6 places. $0.00000275 = \boxed{2.75 \times 10^{-6}}$

■ **YOUR TURN** Express the numbers in scientific notation.

a. 4,520,000,000 **b.** 0.00000043

■ **Answer: a.** 4.52×10^9
b. 4.3×10^{-7}

Technology Tip

```
2.869E5
               286900
1.03E-3
               .00103
```

EXAMPLE 8 Converting from Scientific Notation to Decimals

Write each number as a decimal.

a. 2.869×10^5 **b.** 1.03×10^{-3}

Solution:

a. Move the decimal point 5 places to the right (add zeros in between). 286,900. or $\boxed{286,900}$

b. Move the decimal point 3 places to the left (add zeros in between). $\boxed{0.00103}$

■ **YOUR TURN** Write each number as a decimal.

a. 8.1×10^4 **b.** 3.7×10^{-8}

■ **Answer: a.** 81,000
b. 0.000000037

SECTION 0.2 SUMMARY

In this section we discussed properties of exponents.

Integer Exponents

The following table summarizes integer exponents. Let a be any real number and n be a natural number.

NAME	DESCRIPTION	MATH
Natural-number exponent	Multiply n factors of a.	$a^n = \underbrace{a \cdot a \cdot a \cdots a}_{n \text{ factors}}$
Negative-integer exponent property	A negative exponent implies a reciprocal.	$a^{-n} = \frac{1}{a^n} \quad a \neq 0$
Zero-exponent property	Any nonzero real number raised to the zero power is equal to one.	$a^0 = 1 \quad a \neq 0$

Properties of Integer Exponents

The following table summarizes properties of integer exponents. Let a and b be nonzero real numbers and m and n be integers.

NAME	DESCRIPTION	MATH
Product property	When multiplying exponentials with the same base, **add** exponents.	$a^m \cdot a^n = a^{m+n}$
Quotient property	When dividing exponentials with the same base, **subtract** the exponents, (numerator − denominator).	$\frac{a^m}{a^n} = a^{m-n}$
Power property	When raising an exponential to a power, **multiply** exponents.	$(a^m)^n = a^{mn}$
Product to a power property	A product raised to a power is equal to the product of each factor raised to the power.	$(ab)^n = a^n b^n$
Quotient to a power property	A quotient raised to a power is equal to the quotient of the factors raised to the power.	$\left(\frac{a}{b}\right)^n = \frac{a^n}{b^n}$

Scientific Notation

Scientific notation is a convenient way of using exponents to represent either very small or very large numbers. Real numbers greater than 1 correspond to positive exponents in scientific notation, whereas real numbers greater than 0 but less than 1 correspond to negative exponents in scientific notation. Scientific notation offers the convenience of multiplying and dividing real numbers by applying properties of exponents.

REAL NUMBER (DECIMAL FORM)	PROCESS	SCIENTIFIC NOTATION
2,357,000,000	Move the implied decimal point to the *left* 9 places	2.357×10^9
0.00000465	Move the decimal point to the *right* 6 places	4.65×10^{-6}

SECTION 0.2 EXERCISES

SKILLS

In Exercises 1–20, evaluate each expression.

1. 4^4 **2.** 5^3 **3.** $(-3)^5$ **4.** $(-4)^2$ **5.** -5^2

6. -7^2 **7.** $-2^2 \cdot 4$ **8.** $-3^2 \cdot 5$ **9.** 9^0 **10.** $-8x^0$

11. 10^{-1} **12.** a^{-1} **13.** 8^{-2} **14.** 3^{-4} **15.** $-6 \cdot 5^2$

16. $-2 \cdot 4^2$ **17.** $8 \cdot 2^{-3} \cdot 5$ **18.** $5 \cdot 2^{-4} \cdot 32$ **19.** $-6 \cdot 3^{-2} \cdot 81$ **20.** $6 \cdot 4^2 \cdot 4^{-4}$

In Exercises 21–50, simplify and write the resulting expression with only positive exponents.

21. $x^2 \cdot x^3$

22. $y^3 \cdot y^5$

23. x^2x^{-3}

24. $y^3 \cdot y^{-7}$

25. $(x^2)^3$

26. $(y^3)^2$

27. $(4a)^3$

28. $(4x^2)^3$

29. $(-2t)^3$

30. $(-3b)^4$

31. $(5xy^2)^2(3x^3y)$

32. $(4x^2y)(2xy^3)^2$

33. $\dfrac{x^5y^3}{x^7y}$

34. $\dfrac{y^5x^2}{y^{-2}x^{-5}}$

35. $\dfrac{(2xy)^2}{(-2xy)^3}$

36. $\dfrac{(-3x^3y)}{-4(x^2y^3)^3}$

37. $\left(\dfrac{b}{2}\right)^{-4}$

38. $\left(\dfrac{c}{3}\right)^{-2}$

39. $(9a^{-2}b^3)^{-2}$

40. $(-9x^{-3}y^2)^{-4}$

41. $\dfrac{a^{-2}b^3}{a^4b^5}$

42. $\dfrac{x^{-3}y^2}{y^{-4}x^5}$

43. $\dfrac{(x^3y^{-1})^2}{(xy^2)^{-2}}$

44. $\dfrac{(x^3y^{-2})^2}{(x^4y^3)^{-3}}$

45. $\dfrac{3(x^2y)^3}{12(x^{-2}y)^4}$

46. $\dfrac{(-4x^{-2})^2y^3z}{(2x^3)^{-2}(y^{-1}z)^4}$

47. $\dfrac{(x^{-4}y^5)^{-2}}{[-2(x^3)^2y^{-4}]^5}$

48. $-2x^2(-2x^3)^5$

49. $\left[\dfrac{a^2(-xy^4)^3}{x^4(-a^3y^2)^2}\right]^3$

50. $\left[\dfrac{b^{-3}(-x^3y^2)^4}{y^2(-b^2x^5)^3}\right]^5$

51. Write $2^8 \cdot 16^3 \cdot (64)$ as a power of $2{:}2^?$

52. Write $3^9 \cdot 81^5 \cdot (9)$ as a power of $3{:}3^?$

In Exercises 53–60, express the given number in scientific notation.

53. 27,600,000

54. 144,000,000,000

55. 93,000,000

56. 1,234,500,000

57. 0.0000000567

58. 0.00000828

59. 0.000000123

60. 0.000000005

In Exercises 61–66, write the number as a decimal.

61. 4.7×10^7

62. 3.9×10^5

63. 2.3×10^4

64. 7.8×10^{-3}

65. 4.1×10^{-5}

66. 9.2×10^{-8}

APPLICATIONS

67. Astronomy. The distance from Earth to Mars on a particular day can be 200 million miles. Express this distance in scientific notation.

68. Astronomy. The distance from Mars to the sun on a particular day can be 142 million miles. Express this distance in scientific notation.

69. Lasers. The wavelength of a typical laser used for communication systems is 1.55 microns (or 1.55×10^{-6} meters). Express the wavelength in decimal representation in terms of meters.

70. Lasers. A ruby-red laser has a wavelength of 694 nanometers (or 6.93×10^{-7} meters). Express the wavelength in decimal representation in terms of meters.

CATCH THE MISTAKE

In Exercises 71–74, explain the mistake that is made.

71. Simplify $(-2y^3)(3x^2y^2)$.

Group like factors together. $(-2)(3)x^2y^3y^2$

Use the product property. $-6x^2y^6$

This is incorrect. What mistake was made?

72. Simplify $(2xy^2)^3$.

Eliminate the parentheses. $(2xy^2)^3 = 2x^3y^6$

This is incorrect. What mistake was made?

73. Simplify $(-2xy^3)^2(5x^2y)^2$.

Apply the product to a power property. $= (-2)^2x^2(y^3)^2(5)^2(x^2)^2y^2$

Apply the power rule. $= 4x^2y^9 \cdot 25x^4y^2$

Group like factors. $= (4)(25)x^2x^4y^9y^2$

Apply the product property. $= 100x^6y^{11}$

This is incorrect. What mistake was made?

74. Simplify $\dfrac{-4x^{16}y^9}{8x^2y^3}$.

Group like factors. $= \left(\dfrac{-4}{8}\right)\left(\dfrac{x^{16}}{x^2}\right)\left(\dfrac{y^9}{y^3}\right)$

Use the quotient property. $= -\dfrac{1}{2}x^8y^3$

This is incorrect. What mistake was made?

CONCEPTUAL

In Exercises 75–78, determine whether each of the following statements is true or false.

75. $-2^n = (-2)^n$, if n is an integer.

76. Any nonzero real number raised to the zero power is one.

77. $\dfrac{x^{n+1}}{x^n} = x$ for $x =$ any real number.

78. $x^{-1} + x^{-2} = x^{-3}$

79. Simplify $((a^m)^n)^k$.

80. Simplify $((a^{-m})^{-n})^{-k}$.

In Exercises 81–84, evaluate the expression for the given value.

81. $-a^2 + 2ab$ for $a = -2, b = 3$

82. $2a^3 - 7a^2$ for $a = 4$

83. $-16t^2 + 100t$ for $t = 3$

84. $\dfrac{a^3 - 27}{a - 4}$ for $a = -2$

CHALLENGE

85. The Earth's population is approximately 6.6×10^9 people, and there are approximately 1.5×10^8 square kilometers of land on the surface of the Earth. If one square kilometer is approximately 247 acres, how many acres per person are there on Earth? Round to the nearest tenth of an acre.

86. The population of the United States is approximately 3.0×10^8 people, and there are approximately 3.79×10^6 square miles of land in the United States. If one square mile is approximately 640 acres, how many acres per person are there in the United States? Round to the nearest tenth of an acre.

87. Evaluate: $\dfrac{(4 \times 10^{-23})(3 \times 10^{12})}{(6 \times 10^{-10})}$. Express your answer in both scientific and decimal notation.

88. Evaluate: $\dfrac{(2 \times 10^{-17})(5 \times 10^{13})}{(1 \times 10^{-6})}$. Express your answer in both scientific and decimal notation.

TECHNOLOGY

Scientific calculators have an EXP button that is used for scientific notation. For example, 2.5×10^3 can be input into the calculator by pressing 2.5 EXP 3.

89. Repeat Exercise 85 and confirm your answer with a calculator.

90. Repeat Exercise 86 and confirm your answer with a calculator.

In Exercises 91 and 92, use a graphing utility or scientific calculator to evaluate the expression. Express your answer in scientific notation.

91. $\dfrac{(7.35 \times 10^{-26})(2.19 \times 10^{19})}{(3.15 \times 10^{-21})}$

92. $\dfrac{(1.6849 \times 10^{32})}{(8.12 \times 10^{16})(3.32 \times 10^{-9})}$

SECTION 0.3 POLYNOMIALS: BASIC OPERATIONS

SKILLS OBJECTIVES

- Add and subtract polynomials.
- Multiply polynomials.

CONCEPTUAL OBJECTIVES

- Recognize like terms.
- Learn formulas for special products.

The expressions

$$3x^2 - 7x - 1 \qquad 4y^3 - y \qquad 5z$$

are all examples of *polynomials* in one variable. A monomial in one variable, ax^k, is the product of a constant and a variable raised to a nonnegative-integer power. The constant a

is called the **coefficient** of the monomial, and k is called the **degree** of the monomial. A **polynomial** is the sum of monomials. The monomials that are part of a polynomial are called **terms**.

DEFINITION Polynomial

A **polynomial in x** is an algebraic expression of the form

$$a_n x^n + a_{n-1}x^{n-1} + a_{n-2}x^{n-2} + \cdots + a_2 x^2 + a_1 x + a_0$$

where $a_0, a_1, a_2, \ldots, a_n$ are real numbers, with $a_n \neq 0$, and n is a nonnegative integer. The polynomial is of **degree** n, a_n is the **leading coefficient**, and a_0 is the **constant term**.

Polynomials with one, two, and three terms are called **monomials**, **binomials**, and **trinomials**, respectively. Polynomials are typically written in **standard form** in order of decreasing degrees, and the **degree** of the polynomial is determined by the highest degree (exponent) of any single term.

POLYNOMIAL	STANDARD FORM	SPECIAL NAME	DEGREE	DESCRIPTION
$4x^3 - 5x^7 + 2x - 6$	$-5x^7 + 4x^3 + 2x - 6$	Polynomial	7	A *seventh*-degree polynomial in x
$5 + 2y^3 - 4y$	$2y^3 - 4y + 5$	Trinomial	3	A *third*-degree polynomial in y
$7z^2 + 2$	$7z^2 + 2$	Binomial	2	A *second*-degree polynomial in z
$-17x^5$	$-17x^5$	Monomial	5	A *fifth*-degree monomial in x

EXAMPLE 1 Writing Polynomials in Standard Form

Write the polynomials in standard form and state their degree, leading coefficient, and constant term.

a. $4x - 9x^5 + 2$ **b.** $3 - x^2$
c. $3x^2 - 8 + 14x^3 - 20x^8 + x$ **d.** $-7x^3 + 25x$

Solution:

	Standard Form	Degree	Leading Coefficient	Constant Term
a.	$-9x^5 + 4x + 2$	5	-9	2
b.	$-x^2 + 3$	2	-1	3
c.	$-20x^8 + 14x^3 + 3x^2 + x - 8$	8	-20	-8
d.	$-7x^3 + 25x$	3	-7	0

YOUR TURN Write the polynomial in standard form and state its degree, leading coefficient, and constant term.

$$17x^2 - 4x^3 + 5 - x$$

Answer: $-4x^3 + 17x^2 - x + 5$
Degree: 3
Leading coefficient: -4
Constant term: 5

Adding and Subtracting Polynomials

Polynomials are added and subtracted by combining *like terms*. **Like terms** are terms having the same variables and exponents. Like terms can be combined by adding their coefficients.

Words	**Math**
Identify like terms.	$\underline{3x^2} + 2x + \underline{4x^2} + 5$
Add coefficients of like terms.	$7x^2 + 2x + 5$

Note: The $2x$ and 5 could not be combined because they are *not* like terms.

EXAMPLE 2 Adding Polynomials

Find the sum and simplify $(5x^2 - 2x + 3) + (3x^3 - 4x^2 + 7)$.

Solution:

Eliminate parentheses.	$5x^2 - 2x + 3 + 3x^3 - 4x^2 + 7$
Identify like terms.	$\underline{5x^2} - 2x + \underline{\underline{3}} + 3x^3 - \underline{4x^2} + \underline{\underline{7}}$
Combine like terms.	$x^2 - 2x + 10 + 3x^3$
Write in standard form.	$\boxed{3x^3 + x^2 - 2x + 10}$

YOUR TURN Find the sum and simplify:

$$(3x^2 + 5x - 2x^5) + (6x^3 - x^2 + 11)$$

Answer: $-2x^5 + 6x^3 + 2x^2 + 5x + 11$

EXAMPLE 3 Subtracting Polynomials

Find the difference and simplify $(3x^3 - 2x + 1) - (x^2 + 5x - 9)$.

Common Mistake

Distributing the negative to only the first term in the second polynomial.

CORRECT

Eliminate the parentheses.

$$3x^3 - 2x + 1 - x^2 - 5x + 9$$

Identify like terms.

$$3x^3 - \underline{2x} + \underline{\underline{1}} - x^2 - \underline{5x} + \underline{\underline{9}}$$

Combine like terms.

$$\boxed{3x^3 - x^2 - 7x + 10}$$

INCORRECT

ERROR:

$$3x^3 - 2x + 1 - x^2 + 5x - 9$$

Don't forget to distribute the negative through the entire second polynomial.

YOUR TURN Find the difference and simplify:

$$(-7x^2 - x + 5) - (2 - x^3 + 3x)$$

Study Tip

When subtracting polynomials, it is important to distribute the negative through *all* of the terms in the *second* polynomial.

Answer: $x^3 - 7x^2 - 4x + 3$

Multiplying Polynomials

The product of two monomials is found by using the properties of exponents (Section 0.2). For example,

$$(-5x^3)(9x^2) = (-5)(9)x^{3+2} = -45x^5$$

To multiply a monomial and a polynomial we use the distributive property (Section 0.1).

EXAMPLE 4 Multiplying a Monomial and a Polynomial

Find the product and simplify $5x^2(3x^5 - x^3 + 7x - 4)$.

Solution:

Use the distributive property.	$5x^2(3x^5 - x^3 + 7x - 4)$
	$= 5x^2(3x^5) - 5x^2(x^3) + 5x^2(7x) - 5x^2(4)$
Multiply each set of monomials.	$= \boxed{15x^7 - 5x^5 + 35x^3 - 20x^2}$

■ **YOUR TURN** Find the product and simplify $3x^2(4x^2 - 2x + 1)$.

■ **Answer:** $12x^4 - 6x^3 + 3x^2$

How do we multiply two polynomials if neither one is a monomial? For example, how do we find the product of a binomial and a trinomial such as $(2x - 5)(x^2 - 2x + 3)$? Notice that the binomial is a combination of two monomials. Therefore, we treat each monomial, $2x$ and -5, separately and then combine our results. In other words, use the distributive property repeatedly.

WORDS	MATH
Apply the distributive property.	$(2x - 5)(x^2 - 2x + 3) = 2x(x^2 - 2x + 3) - 5(x^2 - 2x + 3)$
Apply the distributive property.	$= (2x)(x^2) + (2x)(-2x) + (2x)(3) - 5(x^2) - 5(-2x) - 5(3)$
Multiply the monomials.	$= 2x^3 - 4x^2 + 6x - 5x^2 + 10x - 15$
Combine like terms.	$= 2x^3 - 9x^2 + 16x - 15$

EXAMPLE 5 Multiplying Two Polynomials

Multiply and simplify $(2x^2 - 3x + 1)(x^2 - 5x + 7)$.

Solution:

Multiply each term of the first trinomial by the entire second trinomial.	$= 2x^2(x^2 - 5x + 7) - 3x(x^2 - 5x + 7) + 1(x^2 - 5x + 7)$
Identify like terms.	$= 2x^4 - \underline{10x^3} + \underline{\underline{14x^2}} - \underline{3x^3} + \underline{\underline{15x^2}} - \underline{\underline{21x}} + \underline{x^2} - \underline{\underline{5x}} + 7$
Combine like terms.	$= \boxed{2x^4 - 13x^3 + 30x^2 - 26x + 7}$

■ **YOUR TURN** Multiply and simplify $(-x^3 + 2x - 4)(3x^2 - x + 5)$.

■ **Answer:** $-3x^5 + x^4 + x^3 - 14x^2 + 14x - 20$

Special Products

The method outlined for multiplying polynomials works for *all* products of polynomials. For the special case when both polynomials are binomials, the **FOIL method** can also be used.

Words	**Math**
Apply the distributive property.	$(5x - 1)(2x + 3) = 5x(2x + 3) - 1(2x + 3)$
Apply the distributive property.	$= 5x(2x) + 5x(3) - 1(2x) - 1(3)$
Multiply each set of monomials.	$= 10x^2 + 15x - 2x - 3$
Combine like terms.	$= 10x^2 + 13x - 3$

The FOIL method finds the products of the **F**irst terms, **O**uter terms, **I**nner terms, and **L**ast terms.

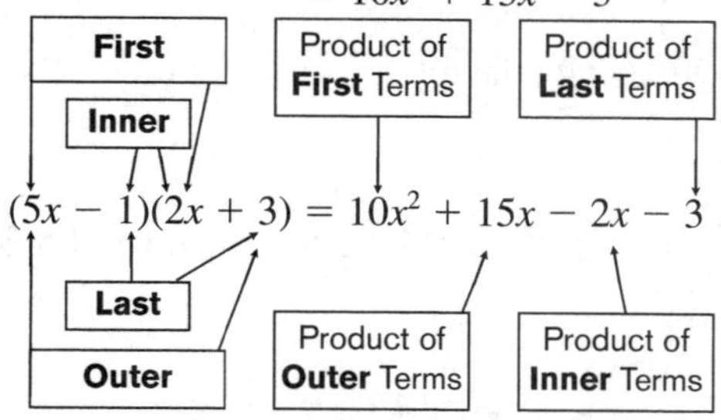

Study Tip

When the binomials are of the form $(ax + b)(cx + d)$, the outer and inner terms will be like terms and can be combined.

EXAMPLE 6 Multiplying Binomials Using the FOIL Method

Multiply $(3x + 1)(2x - 5)$ using the FOIL method.

Solution:

Multiply the **first** terms.	$(3x)(2x) = 6x^2$
Multiply the **outer** terms.	$(3x)(-5) = -15x$
Multiply the **inner** terms.	$(1)(2x) = 2x$
Multiply the **last** terms.	$(1)(-5) = -5$
Add the first, outer, inner, and last terms, and identify the like terms.	$(3x + 1)(2x - 5) = 6x^2 \underline{- 15x} + \underline{2x} - 5$
Combine like terms.	$= \boxed{6x^2 - 13x - 5}$

■ **YOUR TURN** Multiply $(2x - 3)(5x - 2)$.

■ **Answer:** $10x^2 - 19x + 6$

There are some products of binomials that occur frequently in algebra and are given special names. Example 7 illustrates the *difference of two squares* and *perfect squares*.

EXAMPLE 7 Multiplying Binomials Resulting in Special Products

Find the following: **a.** $(x - 5)(x + 5)$ **b.** $(x + 5)^2$ **c.** $(x - 5)^2$

Solution:

a. $(x - 5)(x + 5) = \overbrace{x^2}^{\text{First}} \underbrace{+ 5x}_{\text{Outer}} \overbrace{- 5x}^{\text{Inner}} \underbrace{- 5^2}_{\text{Last}} = \overbrace{x^2 - 5^2}^{\text{Difference of two squares}} = \boxed{x^2 - 25}$

b. $(x + 5)^2 = (x + 5)(x + 5) = \overbrace{x^2}^{\text{First}} \underbrace{+ 5x}_{\text{Outer}} \overbrace{+ 5x}^{\text{Inner}} \underbrace{+ 5^2}_{\text{Last}} = x^2 + 2(5x) + 5^2 = \boxed{x^2 + 10x + 25}$

c. $(x - 5)^2 = (x - 5)(x - 5) = \overbrace{x^2}^{\text{First}} \underbrace{- 5x}_{\text{Outer}} \overbrace{- 5x}^{\text{Inner}} \underbrace{+ 5^2}_{\text{Last}} = x^2 - 2(5x) + 5^2 = \boxed{x^2 - 10x + 25}$

Let a and b be any real number, variable, or algebraic expression in the following special products.

DIFFERENCE OF TWO SQUARES

$$(a + b)(a - b) = a^2 - b^2$$

PERFECT SQUARES

Square of a binomial sum: $(a + b)^2 = (a + b)(a + b) = a^2 + 2ab + b^2$

Square of a binomial difference: $(a - b)^2 = (a - b)(a - b) = a^2 - 2ab + b^2$

EXAMPLE 8 Finding the Square of a Binomial Sum

Find $(x + 3)^2$.

COMMON MISTAKE

Forgetting the middle term, which is *twice* the product of the two terms in the binomial.

CORRECT

$$(x + 3)^2 = (x + 3)(x + 3)$$
$$= x^2 + 3x + 3x + 9$$
$$= x^2 + 6x + 9$$

INCORRECT

ERROR:

$(x + 3)^2 \neq x^2 + 9$

Don't forget the middle term, which is twice the product of the two terms in the binomial.

EXAMPLE 9 Using Special Product Formulas

Find the following:

a. $(2x - 1)^2$ **b.** $(3 + 2y)^2$ **c.** $(4x + 3)(4x - 3)$

Solution (a):

Write the square of a binomial difference formula. $(a - b)^2 = a^2 - 2ab + b^2$

Let $a = 2x$ and $b = 1$. $(2x - 1)^2 = (2x)^2 - 2(2x)(1) + 1^2$

Simplify. $= \boxed{4x^2 - 4x + 1}$

Solution (b):

Write the square of a binomial sum formula. $(a + b)^2 = a^2 + 2ab + b^2$

Let $a = 3$ and $b = 2y$. $(3 + 2y)^2 = (3)^2 + 2(3)(2y) + (2y)^2$

Simplify. $= 9 + 12y + 4y^2$

Write in standard form. $= \boxed{4y^2 + 12y + 9}$

Solution (c):

Write the difference of two squares formula. $(a + b)(a - b) = a^2 - b^2$

Let $a = 4x$ and $b = 3$. $(4x + 3)(4x - 3) = (4x)^2 - 3^2$

Simplify. $= \boxed{16x^2 - 9}$

■ **YOUR TURN** Find the following:

a. $(3x + 1)^2$ **b.** $(1 - 3y)^2$ **c.** $(3x + 2)(3x - 2)$

■ **Answer:** **a.** $9x^2 + 6x + 1$
b. $9y^2 - 6y + 1$
c. $9x^2 - 4$

EXAMPLE 10 Cubing a Binomial

Find the following:

a. $(x + 2)^3$ **b.** $(x - 2)^3$

Solution (a):

Write the cube as a product of three binomials. $(x + 2)^3 = (x + 2)\underbrace{(x + 2)(x + 2)}_{(x + 2)^2}$

Apply the perfect square formula. $= (x + 2)(x^2 + 4x + 4)$

Apply the distributive property. $= x(x^2 + 4x + 4) + 2(x^2 + 4x + 4)$

Apply the distributive property. $= x^3 + 4x^2 + 4x + 2x^2 + 8x + 8$

Combine like terms. $= \boxed{x^3 + 6x^2 + 12x + 8}$

Solution (b):

Write the cube as a product of three binomials. $(x - 2)^3 = (x - 2)\underbrace{(x - 2)(x - 2)}_{(x - 2)^2}$

Apply the perfect square formula. $= (x - 2)(x^2 - 4x - 4)$

Apply the distributive property. $= x(x^2 - 4x + 4) - 2(x^2 - 4x + 4)$

Apply the distributive property. $= x^3 - 4x^2 + 4x - 2x^2 + 8x - 8$

Combine like terms. $= \boxed{x^3 - 6x^2 + 12x - 8}$

Study Tip

$(x + 2)^3 \neq x^3 + 8$

$(x - 2)^3 \neq x^3 - 8$

The formulas are for *cubes of binomials* (*perfect cubes*), as well as a *sum* or *difference of cubes*.

PERFECT CUBES

Cube of a binomial sum:	$(a + b)^3 = a^3 + 3a^2b + 3ab^2 + b^3$
Cube of a binomial difference:	$(a - b)^3 = a^3 - 3a^2b + 3ab^2 - b^3$

EXAMPLE 11 Applying the Special Product Formulas

Find the following:

a. $(2x + 1)^3$ **b.** $(2x - 5)^3$

Solution (a):

Write the cube of a binomial sum formula. $(a + b)^3 = a^3 + 3a^2b + 3ab^2 + b^3$

Let $a = 2x$ and $b = 1$. $(2x + 1)^3 = (2x)^3 + 3(2x)^2(1) + 3(2x)(1)^2 + 1^3$

Simplify. $= \boxed{8x^3 + 12x^2 + 6x + 1}$

Solution (b):

Write the cube of a binomial difference formula. $(a - b)^3 = a^3 - 3a^2b + 3ab^2 - b^3$

Let $a = 2x$ and $b = 5$. $(2x - 5)^3 = (2x)^3 - 3(2x)^2(5) + 3(2x)(5)^2 - 5^3$

Simplify. $= \boxed{8x^3 - 60x^2 + 150x - 125}$

■ YOUR TURN Find $(3x - 4)^3$.

■ **Answer:** $27x^3 - 108x^2 + 144x + 64$

EXAMPLE 12 Applying the Special Product Formulas for Binomials in Two Variables

Find $(2x - 3y)^2$.

Solution:

Write the square of a binomial difference formula. $(a - b)^2 = (a - b)(a - b) = a^2 - 2ab + b^2$

Let $a = 2x$ and $b = 3y$. $(2x - 3y)^2 = (2x)^2 - 2(2x)(3y) + (3y)^2$

Simplify. $= \boxed{4x^2 - 12xy + 9y^2}$

■ YOUR TURN Find $(3x - 2y)^2$.

■ **Answer:** $9x^2 - 12xy + 4y^2$

SECTION 0.3 SUMMARY

In this section, polynomials were defined. Polynomials with one, two, and three terms are called monomials, binomials, and trinomials, respectively. Polynomials are added and subtracted by combining like terms. Polynomials are multiplied by distributing the monomials in the first polynomial throughout the second polynomial. In the special case of the product of two binomials, the FOIL method can also be used. The following are special products of binomials.

Difference of Two Squares

$$(a + b)(a - b) = a^2 - b^2$$

Perfect Squares

Square of a binomial sum.

$$(a + b)^2 = (a + b)(a + b) = a^2 + 2ab + b^2$$

Square of a binomial difference.

$$(a - b)^2 = (a - b)(a - b) = a^2 - 2ab + b^2$$

Perfect Cubes

Cube of a binomial sum.

$$(a + b)^3 = a^3 + 3a^2b + 3ab^2 + b^3$$

Cube of a binomial difference.

$$(a - b)^3 = a^3 - 3a^2b + 3ab^2 - b^3$$

SECTION 0.3 EXERCISES

SKILLS

In Exercises 1–8, write the polynomial in standard form and state the degree of the polynomial.

1. $5x^2 - 2x^3 + 16 - 7x^4$

2. $7x^3 - 9x^2 + 5x - 4$

3. $4x + 3 - 6x^3$

4. $5x^5 - 7x^3 + 8x^4 - x^2 + 10$

5. 15

6. -14

7. $y - 2$

8. $x - 5$

In Exercises 9–24, add or subtract the polynomials, gather like terms, and write the simplified expression in standard form.

9. $(2x^2 - x + 7) + (-3x^2 + 6x - 2)$

10. $(3x^2 + 5x + 2) + (2x^2 - 4x - 9)$

11. $(-7x^2 - 5x - 8) - (-4x - 9x^2 + 10)$

12. $(8x^3 - 7x^2 - 10) - (7x^3 + 8x^2 - 9x)$

13. $(2x^4 - 7x^2 + 8) - (3x^2 - 2x^4 + 9)$

14. $(4x^2 - 9x - 2) - (5 - 3x - 5x^2)$

15. $(7z^2 - 2) - (5z^2 - 2z + 1)$

16. $(25y^3 - 7y^2 + 9y) - (14y^2 - 7y + 2)$

17. $(3y^3 - 7y^2 + 8y - 4) - (14y^3 - 8y + 9y^2)$

18. $(2x^2 + 3xy) - (x^2 + 8xy - 7y^2)$

19. $(6x - 2y) - 2(5x - 7y)$

20. $3a - [2a^2 - (5a - 4a^2 + 3)]$

21. $(2x^2 - 2) - (x + 1) - (x^2 - 5)$

22. $(3x^3 + 1) - (3x^2 - 1) - (5x - 3)$

23. $(4t - t^2 - t^3) - (3t^2 - 2t + 2t^3) + (3t^3 - 1)$

24. $(-z^3 - 2z^2) + (z^2 - 7z + 1) - (4z^3 + 3z^2 - 3z + 2)$

In Exercises 25–64, multiply the polynomials and write the expressions in standard form.

25. $5xy^2(7xy)$

26. $6z(4z^3)$

27. $2x^3(1 - x + x^2)$

28. $-4z^2(2 + z - z^2)$

29. $-2x^2(5 + x - 5x^2)$

30. $-\frac{1}{2}z(2z + 4z^2 - 10)$

31. $(x^2 + x - 2)2x^3$

32. $(x^2 - x + 2)3x^3$

33. $2ab^2(a^2 + 2ab - 3b^2)$

34. $bc^3d^2(b^2c + cd^3 - b^2d^4)$

35. $(2x + 1)(3x - 4)$

36. $(3z - 1)(4z + 7)$

37. $(x + 2)(x - 2)$

38. $(y - 5)(y + 5)$

39. $(2x + 3)(2x - 3)$

40. $(5y + 1)(5y - 1)$

41. $(2x - 1)(1 - 2x)$

42. $(4b - 5y)(4b + 5y)$

43. $(2x^2 - 3)(2x^2 + 3)$

44. $(4xy - 9)(4xy + 9)$

45. $(7y - 2y^2)(y - y^2 + 1)$

46. $(4 - t^2)(6t + 1 - t^2)$

47. $(x + 1)(x^2 - 2x + 3)$

48. $(x + 3)(x^2 - 3x + 9)$

49. $(t - 2)^2$

50. $(t - 3)^2$

51. $(z + 2)^2$

52. $(z + 3)^2$

53. $[(x + y) - 3]^2$

54. $(2x^2 + 3y)^2$

55. $(5x - 2)^2$

56. $(x + 1)(x^2 + x + 1)$

57. $y(3y + 4)(2y - 1)$

58. $p^2(p + 1)(p - 2)$

59. $(x^2 + 1)(x^2 - 1)$

60. $(t - 5)^2(t + 5)^2$

61. $(b - 3a)(a + 2b)(b + 3a)$

62. $(x - 2y)(x^2 + 2xy + 4y^2)$

63. $(x + y - z)(2x - 3y + 5z)$

64. $(5b^2 - 2b + 1)(3b - b^2 + 2)$

APPLICATIONS

In Exercises 65–68, profit is equal to revenue minus cost: $P = R - C$.

65. Profit. Donna decides to sell fabric cord covers on eBay for \$20 a piece. The material for each cord cover costs \$9 and it costs her \$100 a month to advertise on eBay. Let x be the number of cord covers sold. Write a polynomial representing her monthly profit.

66. Profit. Calculators are sold for \$25.00 each. Advertising costs are \$75.00 per month. Let x be the number of calculators sold. Write a polynomial representing the monthly profit earned by selling x calculators.

67. Profit. If the revenue associated with selling x units of a product is $R = -x^2 + 100x$, and the cost associated with producing x units of the product is $C = -100x + 7500$, find the polynomial that represents the profit of making and selling x units.

68. Profit. A business sells a certain quantity x of items. The revenue generated by selling x items is given by the equation $R = -\frac{1}{2}x^2 + 50x$. The costs are given by $C = 8000 - 150x$. Find a polynomial representing the net profit of this business when x items are sold.

69. Volume of a Box. A rectangular sheet of cardboard is to be used in the construction of a box by cutting out squares of side length x from each corner and turning up the sides. Suppose the dimensions of the original rectangle are 15 inches by 8 inches. Determine a polynomial in x that would give the volume of the box.

70. Volume of a Box. Suppose a box is to be constructed from a square piece of material of side length x by cutting out a 2-inch square from each corner and turning up the sides. Express the volume of the box as a polynomial in the variable x.

71. Geometry. Suppose a running track is constructed of a rectangular portion that measures $2x$ feet wide by $2x + 5$ feet long. Each end of the rectangular portion consists of a semicircle whose diameter is $2x$. Write a polynomial that determines the

a. perimeter of the track in terms of the variable x.
b. area of the track in terms of x.

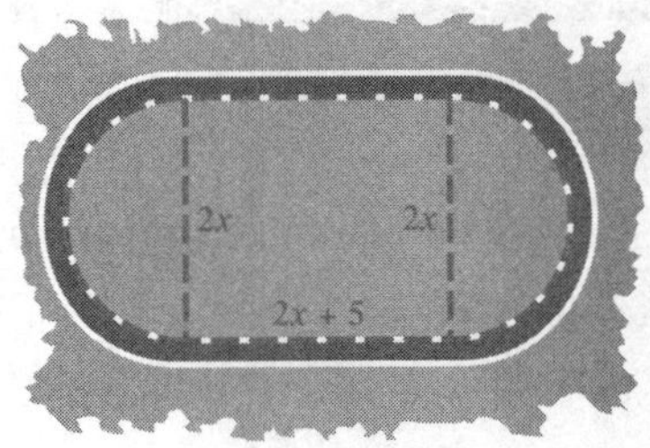

72. Geometry. A right circular cylinder whose radius is r and whose height is $2r$ is surmounted by a hemisphere of radius r.

a. Find a polynomial in the variable r that represents the volume of the "silo" shape.
b. Find a polynomial in r that represents the total surface area of the "silo."

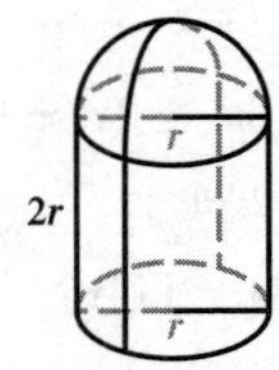

73. Engineering. The force of an electrical field is given by the equation $F = k\frac{q_1 q_2}{r^2}$. Suppose $q_1 = x$, $q_2 = 3x$, and $r = 10x$. Find a polynomial representing the force of the electrical field in terms of the variable x.

74. Engineering. If a football (or other projectile) is thrown upward, its height above the ground is given by the equation $s = 16t^2 + v_0 t + s_0$, where v_0 and s_0 are the initial velocity and initial height of the football, respectively, and t is the time in seconds. Suppose the football is thrown from the top of a building that is 192 feet tall, with an initial speed of 96 feet per second.

a. Write the polynomial that gives the height of the football in terms of the variable t (time).
b. What is the height of the football after 2 seconds have elapsed? Will the football hit the ground after 2 seconds?

CATCH THE MISTAKE

In Exercises 75 and 76, explain the mistake that is made.

75. Subtract and simplify $(2x^2 - 5) - (3x - x^2 + 1)$.

Solution:

Eliminate the parentheses. $2x^2 - 5 - 3x - x^2 + 1$

Collect like terms. $x^2 - 3x - 4$

This is incorrect. What mistake was made?

76. Simplify $(2 + x)^2$.

Solution:

Write the square of the binomial as the sum of the squares. $(2 + x)^2 = 2^2 + x^2$

Simplify. $= x^2 + 4$

This is incorrect. What mistake was made?

CONCEPTUAL

In Exercises 77–80, determine whether each of the following statements is true or false.

77. All binomials are polynomials.

78. The product of two monomials is a binomial.

79. $(x + y)^3 = x^3 - y^3$

80. $(x - y)^2 = x^2 + y^2$

In Exercises 81 and 82, let *m* and *n* be real numbers and *m* > *n*.

81. What degree is the *product* of a polynomial of degree n and a polynomial of degree m?

82. What degree is the *sum* of a polynomial of degree n and a polynomial of degree m?

CHALLENGE

In Exercises 83–86, perform the indicated operations and simplify.

83. $(7x - 4y^2)^2(7x + 4y^2)^2$

84. $(3x - 5y^2)^2(3x + 5y^2)^2$

85. $(x - a)(x^2 + ax + a^2)$

86. $(x + a)(x^2 - ax + a^2)$

TECHNOLOGY

87. Use a graphing utility to plot the graphs of the three expression $(2x + 3)(x - 4)$, $2x^2 + 5x - 12$, and $2x^2 - 5x - 12$. Which two graphs agree with each other?

88. Use a graphing utility to plot the graphs of the three expressions $(x + 5)^2$, $x^2 + 25$, and $x^2 + 10x + 25$. Which two graphs agree with each other?

SECTION 0.4 FACTORING POLYNOMIALS

SKILLS OBJECTIVES

- Factor out the greatest common factor.
- Factor the difference of two squares.
- Factor perfect squares.
- Factor the sum or difference of two cubes.
- Factor a trinomial as a product of binomials.
- Factor by grouping.

CONCEPTUAL OBJECTIVES

- Identify common factors.
- Understand that factoring has its basis in the distributive property.
- Identify prime (irreducible) polynomials.
- Develop a general strategy for factoring polynomials.

In Section 0.3 we discussed multiplying polynomials. In this section we examine the reverse of that process, which is called *factoring*. Consider the following product:

$$(x + 3)(x + 1) = x^2 + 4x + 3$$

To *factor* the resulting polynomial, you reverse the process to undo the multiplication:

$$x^2 + 4x + 3 = (x + 3)(x + 1)$$

The polynomials $(x + 3)$ and $(x + 1)$ are called **factors** of the polynomial $x^2 + 4x + 3$. The process of writing a polynomial as a product is called **factoring**. In Chapter 1 we will solve linear and quadratic equations by factoring.

In this section we will restrict our discussion to factoring polynomials with integer coefficients, which is called **factoring over the integers**. If a polynomial cannot be factored using integer coefficients, then it is **prime** or irreducible over the integers. When a polynomial is written as a product of prime polynomials, then the polynomial is said to be **factored completely**.

Greatest Common Factor

The simplest type of factoring of polynomials occurs when there is a factor common to every term of the polynomial. This **common factor** is a monomial that can be "factored out" by applying the distributive property in reverse:

$$ab + ac = a(b + c)$$

For example, $4x^2 - 6x$ can be written as $2x(x) - 2x(3)$. Notice that $2x$ is a common factor to both terms, so the distributive property tells us we can factor this polynomial to yield $2x(x - 3)$. Although 2 is a common factor and x is a common factor, the monomial $2x$ is called the *greatest common factor*.

GREATEST COMMON FACTOR

The monomial ax^k is called the **greatest common factor (GCF)** of a polynomial in x with integer coefficients if *both* of the following are true:

- a is the *greatest* integer factor common to all of the polynomial coefficients.
- k is the *smallest* exponent on x found in all of the terms of the polynomial.

POLYNOMIAL	GCF	WRITE EACH TERM AS A PRODUCT OF GCF AND REMAINING FACTOR	FACTORED FORM
$7x + 21$	7	$7(x) + 7(3)$	$7(x + 3)$
$3x^2 + 12x$	$3x$	$3x(x) + 3x(4)$	$3x(x + 4)$
$4x^3 + 2x + 6$	2	$2(2x^3) + 2x + 2(3)$	$2(2x^3 + x + 3)$
$6x^4 - 9x^3 + 12x^2$	$3x^2$	$3x^2(2x^2) - 3x^2(3x) + 3x^2(4)$	$3x^2(2x^2 - 3x + 4)$
$-5x^4 + 25x^3 - 20x^2$	$-5x^2$	$-5x^2x^2 - 5x^2(-5x) - 5x^2(4)$	$-5x^2(x^2 - 5x + 4)$

EXAMPLE 1 Factoring Polynomials by Extracting the Greatest Common Factor

Factor:

a. $6x^5 - 18x^4$ **b.** $6x^5 - 10x^4 - 8x^3 + 12x^2$

Solution (a):

Identify the greatest common factor. $6x^4$

Write each term as a product with the GCF as a factor. $6x^5 - 18x^4 = 6x^4(x) - 6x^4(3)$

Factor out the GCF. $= \boxed{6x^4(x - 3)}$

Solution (b):

Identify the greatest common factor. $2x^2$

Write each term as a product with the GCF as a factor.

$$6x^5 - 10x^4 - 8x^3 + 12x^2 = 2x^2(3x^3) - 2x^2(5x^2) - 2x^2(4x) + 2x^2(6)$$

Factor out the GCF. $= \boxed{2x^2(3x^3 - 5x^2 - 4x + 6)}$

■ **YOUR TURN** Factor:

a. $12x^3 - 4x$ **b.** $3x^5 - 9x^4 + 12x^3 - 6x^2$

■ **Answer: a.** $4x(3x^2 - 1)$
b. $3x^2(x^3 - 3x^2 + 4x - 2)$

Factoring Formulas: Special Polynomial Forms

The first step in factoring polynomials is to look for a common factor. If there is no common factor, then we look for special polynomial forms that we learned were special products in Section 0.3 and reverse the process.

Difference of two squares	$a^2 - b^2 = (a + b)(a - b)$
Perfect squares	$a^2 + 2ab + b^2 = (a + b)^2$
	$a^2 - 2ab + b^2 = (a - b)^2$
Sum of two cubes	$a^3 + b^3 = (a + b)(a^2 - ab + b^2)$
Difference of two cubes	$a^3 - b^3 = (a - b)(a^2 + ab + b^2)$

EXAMPLE 2 Factoring the Difference of Two Squares

Factor:

a. $x^2 - 9$ **b.** $4x^2 - 25$ **c.** $x^4 - 16$

Solution (a):

Rewrite as the difference of two squares. $x^2 - 9 = x^2 - 3^2$

Let $a = x$ and $b = 3$ in $a^2 - b^2 = (a + b)(a - b)$. $= \boxed{(x + 3)(x - 3)}$

Solution (b):

Rewrite as the difference of two squares. $4x^2 - 25 = (2x)^2 - 5^2$

Let $a = 2x$ and $b = 5$ in $a^2 - b^2 = (a + b)(a - b)$. $= \boxed{(2x + 5)(2x - 5)}$

Solution (c):

Rewrite as the difference of two squares. $x^4 - 16 = (x^2)^2 - 4^2$

Let $a = x^2$ and $b = 4$ in $a^2 - b^2 = (a + b)(a - b)$. $= (x^2 + 4)(x^2 - 4)$

Note that $x^2 - 4$ is also a difference of two squares (part a). $= \boxed{(x + 2)(x - 2)(x^2 + 4)}$

■ **YOUR TURN** Factor:

a. $x^2 - 4$ **b.** $9x^2 - 16$ **c.** $x^4 - 81$

■ **Answer: a.** $(x + 2)(x - 2)$
b. $(3x + 4)(3x - 4)$
c. $(x^2 + 9)(x - 3)(x + 3)$

A trinomial is a perfect square if it has the form $a^2 \pm 2ab + b^2$. Notice that:

- The first term and third term are perfect squares.
- The middle term is twice the product of the bases of these two perfect squares.
- The sign of the middle term determines the sign of the factored form:

$$a^2 \pm 2ab + b^2 = (a \pm b)^2$$

EXAMPLE 3 Factoring Trinomials That Are Perfect Squares

Factor:

a. $x^2 + 6x + 9$ **b.** $x^2 - 10x + 25$ **c.** $9x^2 - 12x + 4$

Solution (a):

Rewrite the trinomial so that the first and third terms are perfect squares. $x^2 + 6x + 9 = x^2 + 6x + 3^2$

Notice that if we let $a = x$ and $b = 3$ in $a^2 + 2ab + b^2 = (a + b)^2$, then the middle term $6x$ is $2ab$. $x^2 + 6x + 9 = x^2 + 2(3x) + 3^2 = \boxed{(x + 3)^2}$

Solution (b):

Rewrite the trinomial so that the first and third terms are perfect squares. $x^2 - 10x + 25 = x^2 - 10x + 5^2$

Notice that if we let $a = x$ and $b = 5$ in $a^2 - 2ab + b^2 = (a - b)^2$, then the middle term $-10x$ is $-2ab$. $x^2 - 10x + 25 = x^2 - 2(5x) + 5^2 = \boxed{(x - 5)^2}$

Solution (c):

Rewrite the trinomial so that the first and third terms are perfect squares. $9x^2 - 12x + 4 = (3x)^2 - 2(3x)(2) + 2^2$

Notice that if we let $a = 3x$ and $b = 2$ in $a^2 - 2ab + b^2 = (a - b)^2$, then the middle term $-12x$ is $-2ab$. $9x^2 - 12x + 4 = (3x)^2 - 2(3x)(2) + 2^2 = \boxed{(3x - 2)^2}$

YOUR TURN Factor:

a. $x^2 + 8x + 16$ **b.** $x^2 - 4x + 4$ **c.** $25x^2 - 50x + 4$

Answer: a. $(x + 4)^2$ **b.** $(x - 2)^2$ **c.** $(5x - 2)^2$

EXAMPLE 4 Factoring the Sum of Two Cubes

Factor $x^3 + 27$.

Solution:

Rewrite as the sum of two cubes. $x^3 + 27 = x^3 + 3^3$

Write the sum of two cubes formula. $a^3 + b^3 = (a + b)(a^2 - ab + b^2)$

Let $a = x$ and $b = 3$. $x^3 + 27 = x^3 + 3^3 = \boxed{(x + 3)(x^2 - 3x + 9)}$

EXAMPLE 5 Factoring the Difference of Two Cubes

Factor $x^3 - 125$.

Solution:

Rewrite as the difference of two cubes. $x^3 - 125 = x^3 - 5^3$

Write the difference of two cubes formula. $a^3 - b^3 = (a - b)(a^2 + ab + b^2)$

Let $a = x$ and $b = 5$. $x^3 - 125 = x^3 - 5^3 = \boxed{(x - 5)(x^2 + 5x + 25)}$

■ **YOUR TURN** Factor:

a. $x^3 + 8$ **b.** $x^3 - 64$

■ **Answer: a.** $(x + 2)(x^2 - 2x + 4)$
b. $(x - 4)(x^2 + 4x + 16)$

Factoring a Trinomial as a Product of Two Binomials

The first step in factoring is to look for a common factor. If there is no common factor, look to see whether the polynomial is a special form for which we know the factoring formula. If it is not of such a special form, then if it is a trinomial, we proceed with a general factoring strategy.

We know that $(x + 3)(x + 2) = x^2 + 5x + 6$, so we say the **factors** of $x^2 + 5x + 6$ are $\mathbf{(x + 3)}$ and $\mathbf{(x + 2)}$. In factored form we have $x^2 + 5x + 6 = (x + 3)(x + 2)$. Recall the FOIL method from Section 0.3. The product of the last terms (3 and 2) is 6 and the sum of the products of the inner terms ($3x$) and the outer terms ($2x$) is $5x$. Let's pretend for a minute that we didn't know this factored form but had to work with the general form:

$$x^2 + 5x + 6 = (x + a)(x + b)$$

The goal is to find a and b. We start by multiplying the two binomials on the right.

$$x^2 + 5x + 6 = (x + a)(x + b) = x^2 + ax + bx + ab = x^2 + (a + b)x + ab$$

Compare the expression we started with on the left with the expression on the far right $x^2 + 5x + 6 = x^2 + (a + b)x + ab$. We see that $ab = 6$ and $(a + b) = 5$. *Start* with the possible combinations of a and b whose product is 6, and *then* look among those for the combination whose sum is 5.

$ab = 6$	a, b:	1, 6	−1, −6	2, 3	−2, −3
$a + b$		7	−7	**5**	−5

All of the possible a, b combinations in the first row have a product equal to 6, but only one of those has a sum equal to 5. Therefore the factored form is

$$x^2 + 5x + 6 = (x + a)(x + b) = (x + 2)(x + 3)$$

EXAMPLE 6 Factoring a Trinomial

Factor $x^2 + 10x + 9$.

Solution:

Write the trinomial as a product of two binomials in general form. $x^2 + 10x + 9 = (x + \square)(x + \square)$

Write all of the integers whose product is 9.

Integers whose product is 9	1, 9	−1, −9	3, 3	−3, −3

Determine the sum of the integers.

Integers whose product is 9	**1, 9**	−1, −9	3, 3	−3,−3
Sum	**10**	−10	9	−9

Select 1, 9 because the product is 9 (last term of the trinomial) and the sum is 10 (middle term coefficient of the trinomial).

$$\boxed{x^2 + 10x + 9 = (x + 9)(x + 1)}$$

Check: $(x + 9)(x + 1) = x^2 + 9x + 1x + 9 = x^2 + 10x + 9$ ✓

■ **Answer:** $(x + 4)(x + 5)$

■ **YOUR TURN** Factor $x^2 + 9x + 20$.

In Example 6, all terms in the trinomial are positive. When the constant term is negative, then (regardless of whether the middle term is positive or negative) the factors will be opposite in sign, as illustrated in Example 7.

EXAMPLE 7 Factoring a Trinomial

Factor $x^2 - 3x - 28$.

Solution:

Write the trinomial as a product of two binomials in general form.

$$x^2 - 3x + -28 = (x + \square)(x - \square)$$

Write all of the integers whose product is −28.

Integers whose product is −28	1, −28	−1, 28	2, −14	−2, 14	4, −7	−4, 7

Determine the sum of the integers.

Integers whose product is −28	1, −28	−1, 28	2, −14	−2, 14	**4, −7**	−4, 7
Sum	−27	27	−12	12	**−3**	3

Select 4, −7 because the product is −28 (last term of the trinomial) and the sum is −3 (middle-term coefficient of the trinomial).

$$\boxed{x^2 - 3x - 28 = (x + 4)(x - 7)}$$

Check: $(x + 4)(x - 7) = x^2 - 7x + 4x + -28 = x^2 - 3x + -28$ ✓

■ **Answer:** $(x + 6)(x - 3)$

■ **YOUR TURN** Factor $x^2 + 3x - 18$.

When the leading coefficient of the trinomial is not equal to 1, then we consider all possible factors using the following procedure, which is based on the FOIL method in reverse.

Factoring a Trinomial Whose Leading Coefficient Is Not 1

Factors of a

$$ax^2 + bx + c = (\square x + \square)(\square x + \square)$$

Factors of c

Step 1: Find two **F**irst terms whose product is the first term of the trinomial.

Step 2: Find two **L**ast terms whose product is the last term of the trinomial.

Step 3: Consider all possible combinations found in Steps 1 and 2 until the sum of the **O**uter and **I**nner products are equal to the middle term of the trinomial.

EXAMPLE 8 Factoring a Trinomial Whose Leading Coefficient Is Not 1

Factor $5x^2 + 9x - 2$.

Solution:

STEP 1 Start with the first term. Note that $5x \cdot x = 5x^2$. $\quad (5x \pm \square)(x \pm \square)$

STEP 2 The product of the last terms should yield -2. $\quad -1, 2$ or $1, -2$

STEP 3 Consider all possible factors based on Steps 1 and 2.

$(5x - 1)(x + 2)$
$(5x + 1)(x - 2)$
$(5x + 2)(x - 1)$
$(5x - 2)(x + 1)$

Since the outer and inner products must sum to $9x$, the factored form must be: $\boxed{5x^2 + 9x - 2 = (5x - 1)(x + 2)}$

Check: $(5x - 1)(x + 2) = 5x^2 + 10x - 1x - 2 = 5x^2 + 9x - 2 \quad \checkmark$

■ **YOUR TURN** Factor $2t^2 + t - 3$.

■ **Answer:** $(2t + 3)(t - 1)$

EXAMPLE 9 Factoring a Trinomial Whose Leading Coefficient Is Not 1

Factor $15x^2 - x - 6$.

Solution:

STEP 1 Start with the first term. $\quad (5x \pm \square)(3x \pm \square)$ or $(15x \pm \square)(x \pm \square)$

STEP 2 The product of the last terms should yield -6. $\quad -1, 6$ or $1, -6$ or $2, -3$ or $-2, 3$

STEP 3 Consider all possible factors based on Steps 1 and 2.

$(5x - 1)(3x + 6)$	$(15x - 1)(x + 6)$
$(5x + 6)(3x - 1)$	$(15x + 6)(x - 1)$
$(5x + 1)(3x - 6)$	$(15x + 1)(x - 6)$
$(5x - 6)(3x + 1)$	$(15x - 6)(x + 1)$
$(5x + 2)(3x - 3)$	$(15x + 2)(x - 3)$
$(5x - 3)(3x + 2)$	$(15x - 3)(x + 2)$
$(5x - 2)(3x + 3)$	$(15x - 2)(x + 3)$
$(5x + 3)(3x - 2)$	$(15x + 3)(x - 2)$

Study Tip

In Example 9, Step 3, we can eliminate any factors that have a common factor since there is no common factor to the terms in the trinomial.

Since the outer and inner products must sum to $-x$, the factored form must be: $15x^2 - x - 6 = (5x + 3)(3x - 2)$

Check: $(5x + 3)(3x - 2) = 15x^2 - 10x + 9x - 6 = 15x^2 - x - 6$ ✓

■ **YOUR TURN** Factor $6x^2 + x - 12$.

■ **Answer:** $(3x - 4)(2x + 3)$

EXAMPLE 10 Identifying Prime (Irreducible) Polynomials

Factor $x^2 + x - 8$.

Solution:

Write the trinomial as a product of two binomials in general form. $x^2 + x - 8 = (x + \square)(x - \square)$

Write all of the integers whose product is -8.

Integers whose product is -8	$1, -8$	$-1, 8$	$4, -2$	$-4, 2$

Determine the sum of the integers.

Integers whose product is -8	$1, -8$	$-1, 8$	$4, -2$	$-4, 2$
Sum	-7	7	2	-2

The middle term of the trinomial is $-x$, so we look for the sum of the integers that equals -1. Since no sum exists for the given combinations, we say that this polynomial is *prime* (irreducible) over the integers.

Factoring by Grouping

Much of our attention in this section has been on factoring trinomials. For polynomials with more than three terms we first look for a common factor to all terms. If there is no common factor to all terms of the polynomial, we look for a group of terms that have a common factor. This strategy is called **factoring by grouping**.

EXAMPLE 11 Factoring a Polynomial by Grouping

Factor $x^3 - x^2 + 2x - 2$.

Solution:

Group the terms that have a common factor. $= (x^3 - x^2) + (2x - 2)$

Factor out the common factor in each pair of parentheses. $= x^2(x - 1) + 2(x - 1)$

Use the distributive property. $= (x^2 + 2)(x - 1)$

EXAMPLE 12 Factoring a Polynomial by Grouping

Factor $2x^2 + 2x - x - 1$.

Solution:

Group the terms that have a common factor. $= (2x^2 + 2x) + (-x - 1)$

Factor out the common factor in each pair of parentheses. $= 2x(x + 1) - 1(x + 1)$

Use the distributive property. $= (2x - 1)(x + 1)$

■ **YOUR TURN** Factor $x^3 + x^2 - 3x - 3$.

■ **Answer:** $(x + 1)(x^2 - 3)$

A Strategy for Factoring Polynomials

The first step in factoring a polynomial is to look for the greatest common factor. When specifically factoring trinomials look for special known forms: a perfect square or a difference of two squares. A general approach to factoring a trinomial uses the FOIL method in reverse. Finally, we look for factoring by grouping. The following strategy for factoring polynomials is based on the techniques discussed in this section.

STRATEGY FOR FACTORING POLYNOMIALS

1. Factor out the greatest common factor (monomial).
2. Identify any special polynomial forms and apply factoring formulas.
3. Factor a trinomial into a product of two binomials: $(ax + b)(cx + d)$.
4. Factor by grouping.

Study Tip

When factoring always start by factoring out the GCF.

EXAMPLE 13 Factoring Polynomials

Factor:

a. $3x^2 - 6x + 3$

b. $-4x^3 + 2x^2 + 6x$

c. $15x^2 + 7x - 2$

d. $x^3 - x + 2x^2 - 2$

Solution (a):

Factor out the greatest common factor. $3x^2 - 6x + 3 = 3(x^2 - 2x + 1)$

The trinomial is a perfect square. $= \boxed{3(x-1)^2}$

Solution (b):

Factor out the greatest common factor. $-4x^3 + 2x^2 + 6x = -2x(2x^2 - x - 3)$

Use the FOIL method in reverse to factor the trinomial. $= \boxed{-2x(2x-3)(x+1)}$

Solution (c):

There is no common factor. $15x^2 + 7x - 2$

Use the FOIL method in reverse to factor the trinomial. $= \boxed{(3x+2)(5x-1)}$

Solution (d):

Factor by grouping.

$$\begin{aligned} &= (x^3 - x) + (2x^2 - 2) \\ &= x(x^2 - 1) + 2(x^2 - 1) \\ &= (x + 2)(x^2 - 1) \\ &= \boxed{(x+2)(x-1)(x+1)} \end{aligned}$$

SECTION 0.4 SUMMARY

In this section, we discussed factoring polynomials, which is the reverse process of multiplying polynomials. Four main techniques were discussed.

Greatest Common Factor: ax^k

- a is the greatest common factor for all coefficients of the polynomial.
- k is the smallest exponent found on all of the terms in the polynomial.

Factoring Formulas: Special Polynomial Forms

Difference of two squares:	$a^2 - b^2 = (a + b)(a - b)$
Perfect squares:	$a^2 + 2ab + b^2 = (a + b)^2$
	$a^2 - 2ab + b^2 = (a - b)^2$
Sum of two cubes:	$a^3 + b^3 = (a + b)(a^2 - ab + b^2)$
Difference of two cubes:	$a^3 - b^3 = (a - b)(a^2 + ab + b^2)$

Factoring a Trinomial as a Product of Two Binomials

- $x^2 + bx + c = (x + ?)(x + ?)$
  1. Find all possible combinations of factors whose product is c.
  2. Of the combinations in Step 1, look for the sum of factors that equals b.
- $ax^2 + bx + c = (?x + ?)(?x + ?)$
  1. Find all possible combinations of the first terms whose product is ax^2.
  2. Find all possible combinations of the last terms whose product is c.
  3. Consider asll possible factors based on Steps 1 and 2.

Factoring by Grouping

- Group terms that have a common factor.
- Use the distributive property.

SECTION 0.4 EXERCISES

SKILLS

In Exercises 1–12, factor each expression. Start by finding the greatest common factor (GCF).

1. $5x + 25$ **2.** $x^2 + 2x$ **3.** $4t^2 - 2$ **4.** $16z^2 - 20z$

5. $2x^3 - 50x$ **6.** $4x^2y - 8xy^2 + 16x^2y^2$ **7.** $3x^3 - 9x^2 + 12x$ **8.** $14x^4 - 7x^2 + 21x$

9. $x^3 - 3x^2 - 40x$ **10.** $-9y^2 + 45y$ **11.** $4x^2y^3 + 6xy$ **12.** $3z^3 - 6z^2 + 18z$

In Exercises 13–20, factor the difference of two squares.

13. $x^2 - 9$ **14.** $x^2 - 25$ **15.** $4x^2 - 9$ **16.** $1 - x^4$

17. $2x^2 - 98$ **18.** $144 - 81y^2$ **19.** $225x^2 - 169y^2$ **20.** $121y^2 - 49x^2$

In Exercises 21–32, factor the perfect squares.

21. $x^2 + 8x + 16$ **22.** $y^2 - 10y + 25$ **23.** $x^4 - 4x^2 + 4$ **24.** $1 - 6y + 9y^2$

25. $4x^2 + 12xy + 9y^2$ **26.** $x^2 - 6xy + 9y^2$ **27.** $9 - 6x + x^2$ **28.** $25x^2 - 20xy + 4y^2$

29. $x^4 + 2x^2 + 1$ **30.** $x^6 - 6x^3 + 9$ **31.** $p^2 + 2pq + q^2$ **32.** $p^2 - 2pq + q^2$

In Exercises 33–42, factor the sum or difference of two cubes.

33. $t^3 + 27$ **34.** $z^3 + 64$ **35.** $y^3 - 64$ **36.** $x^3 - 1$

37. $8 - x^3$ **38.** $27 - y^3$ **39.** $y^3 + 125$ **40.** $64x - x^4$

41. $27 + x^3$ **42.** $216x^3 - y^3$

In Exercises 43–52, factor each trinomial into a product of two binomials.

43. $x^2 - 6x + 5$
44. $t^2 - 5t - 6$
45. $y^2 - 2y - 3$
46. $y^2 - 3y - 10$
47. $2y^2 - 5y - 3$
48. $2z^2 - 4z - 6$
49. $3t^2 + 7t + 2$
50. $4x^2 - 2x - 12$
51. $-6t^2 + t + 2$
52. $-6x^2 - 17x + 10$

In Exercises 53–60, factor by grouping.

53. $x^3 - 3x^2 + 2x - 6$
54. $x^5 + 5x^3 - 3x^2 - 15$
55. $a^4 + 2a^3 - 8a - 16$
56. $x^4 - 3x^3 - x + 3$
57. $3xy - 5rx - 10rs + 6sy$
58. $6x^2 - 10x + 3x - 5$
59. $20x^2 + 8xy - 5xy - 2y^2$
60. $9x^5 - a^2x^3 - 9x^2 + a^2$

In Exercises 61–92, factor each of the polynomials completely, if possible. If the polynomial cannot be factored, state that it is prime.

61. $x^2 - 4y^2$
62. $a^2 + 5a + 6$
63. $3a^2 + a - 14$
64. $ax + b + bx + a$
65. $x^2 + 16$
66. $x^2 + 49$
67. $4z^2 + 25$
68. $\frac{1}{16} - b^4$
69. $6x^2 + 10x + 4$
70. $x^2 + 7x + 5$
71. $6x^2 + 13xy - 5y^2$
72. $15x + 15xy$
73. $36s^2 - 9t^2$
74. $3x^3 - 108x$
75. $a^2b^2 - 25c^2$
76. $2x^3 + 54$
77. $4x^2 - 3x - 10$
78. $10x - 25 - x^2$
79. $3x^3 - 5x^2 - 2x$
80. $2y^3 + 3y^2 - 2y$
81. $x^3 - 9x$
82. $w^3 - 25w$
83. $xy - x - y + 1$
84. $a + b + ab + b^2$
85. $x^4 + 5x^2 + 6$
86. $x^6 - 7x^3 - 8$
87. $x^2 - 2x - 24$
88. $25x^2 + 30x + 9$
89. $x^4 + 125x$
90. $x^4 - 1$
91. $x^4 - 81$
92. $10x^2 - 31x + 15$

APPLICATIONS

93. Geometry. A rectangle has a length of $2x + 4$ and a width of x. Express the perimeter of the rectangle as a factored polynomial in x.

94. Geometry. The volume of a box is given by the expression $x^3 + 7x^2 + 12x$. Express the volume as a factored polynomial in the variable x.

95. Business. The profit of a business is given by the expression $P = 2x^2 - 15x + 4x - 30$. Express the profit as a factored polynomial in the variable x.

96. Business. The break-even point for a company is given by solving the equation $3x^2 + 9x - 4x - 12 = 0$. Factor the polynomial on the left side of the equation.

97. Engineering. The height of a projectile is given by the equation $s = -16t^2 - 78t + 10$. Factor the expression on the right side of the equal sign.

98. Engineering. The electrical field at a point P between two charges is given by $k = \dfrac{10x - x^2}{100}$. Factor the numerator of this expression.

CATCH THE MISTAKE

In Exercises 99 and 100, explain the mistake that is made.

99. Factor $x^3 - x^2 - 9x + 9$.

Solution:

Group terms with common factors.	$(x^3 - x^2) + (-9x + 9)$
Factor out common factors.	$x^2(x - 1) - 9(x - 1)$
Distributive property.	$(x - 1)(x^2 - 9)$
Factor $x^2 - 9$.	$(x - 1)(x - 3)^2$

This is incorrect. What mistake was made?

100. Factor $4x^2 + 12x - 40$.

Solution:

Factor the trinomial into a product of binomials.	$(2x - 4)(2x + 10)$
Factor out a 2.	$= 2(x - 2)(x + 5)$

This is incorrect. What mistake was made?

CONCEPTUAL

In Exercises 101–104, determine whether each of the following statements is true or false.

101. All trinomials can be factored into a product of two binomials.

102. All polynomials can be factored into prime factors with respect to the integers.

103. $x^2 - y^2 = (x - y)(x + y)$

104. $x^2 + y^2 = (x + y)^2$

CHALLENGE

105. Factor $a^{2n} - b^{2n}$ completely, assuming a, b, and n are positive integers.

106. Find all the values of c such that the trinomial $x^2 + cx - 14$ can be factored.

TECHNOLOGY

107. Use a graphing utility to plot the graphs of the three expressions $8x^3 + 1$, $(2x + 1)(4x^2 - 2x + 1)$, and $(2x - 1)(4x^2 + 2x + 1)$. Which two graphs agree with each other?

108. Use a graphing utility to plot the graphs of the three expressions $27x^3 - 1$, $(3x - 1)^3$, and $(3x - 1)(9x^2 + 3x + 1)$. Which two graphs agree with each other?

SECTION 0.5 RATIONAL EXPRESSIONS

SKILLS OBJECTIVES

- Find the domain of an algebraic expression.
- Reduce a rational expression to lowest terms.
- Multiply and divide rational expressions.
- Add and subtract rational expressions.
- Simplify complex rational expressions.

CONCEPTUAL OBJECTIVES

- Understand why rational expressions have domain restrictions.
- Understand the least common denominator method for rational expressions.

Rational Expressions and Domain Restrictions

Recall that a rational number is the ratio of two integers with the denominator not equal to zero. Similarly, the ratio, or quotient, of two polynomials is a **rational expression**.

Rational Numbers: $\dfrac{3}{7} \quad \dfrac{5}{9} \quad \dfrac{9}{11}$

Rational Expressions: $\dfrac{3x + 2}{x - 5} \quad \dfrac{5x^2}{x^2 + 1} \quad \dfrac{9}{3x - 2}$

As with rational numbers, the denominators of rational expressions are never equal to zero. In the above first and third rational expressions, there are values of the variable that would correspond to a denominator equal to zero; these values are not permitted:

$$\frac{3x + 2}{x - 5} \quad x \neq 5 \qquad\qquad \frac{9}{3x - 2} \quad x \neq \frac{2}{3}$$

In the second rational expression, $\frac{5x^2}{x^2 + 1}$, there are no real numbers that will correspond to a zero denominator.

The set of real numbers for which an algebraic expression is *defined* is called the **domain**. Since a rational expression is not defined if its denominator is zero, we must eliminate from the domain those values of the variable that would result in a zero denominator.

To find the domain of an algebraic expression we ask the question, "What can x (the variable) be?" For rational expressions the answer in general is "any values except those that make the denominator equal to zero."

EXAMPLE 1 Finding the Domain of an Algebraic Expression

Find the domain of the expressions.

a. $2x^2 - 5x + 3$ **b.** $\frac{2x + 1}{x - 4}$ **c.** $\frac{x}{x^2 + 1}$ **d.** $\frac{3x + 1}{x}$

Solution:

ALGEBRAIC EXPRESSION	TYPE	DOMAIN	NOTE
a. $2x^2 - 5x + 3$	Polynomial	All real numbers	The domain of all polynomials is the set of all real numbers.
b. $\frac{2x + 1}{x - 4}$	Rational expression	All real numbers except $x = 4$	When $x = 4$, the rational expression is undefined.
c. $\frac{x}{x^2 + 1}$	Rational expression	All real numbers	There are no real numbers that will result in the denominator being equal to zero.
d. $\frac{3x + 1}{x}$	Rational expression	All real numbers except $x = 0$	When $x = 0$, the rational expression is undefined.

YOUR TURN Find the domain of the expressions.

a. $\frac{3x - 1}{x + 1}$ **b.** $\frac{5x - 1}{x}$ **c.** $3x^2 + 2x - 7$ **d.** $\frac{2x + 5}{x^2 + 4}$

Answer: **a.** $x \neq -1$
b. $x \neq 0$
c. all real numbers
d. all real numbers

In this text, it will be assumed that the domain is the set of all real numbers except the real numbers shown to be excluded.

EXAMPLE 2 Excluding Values from the Domain of Rational Expressions

Determine what real numbers must be excluded from the domain of the following rational expressions.

a. $\frac{7x + 5}{x^2 - 4}$ **b.** $\frac{3x + 2}{x^2 - 5x}$

Solution (a):

Factor the denominator.

$$\frac{7x + 5}{x^2 - 4} = \frac{7x + 5}{(x + 2)(x - 2)}$$

Determine the values of x that will make the denominator equal to zero.

$\boxed{x = -2}$ and $\boxed{x = 2}$ must be excluded from the domain.

Solution (b):

Factor the denominator. $\dfrac{3x + 2}{x^2 - 5x} = \dfrac{3x + 2}{x(x - 5)}$

Determine the values of x that will make the denominator equal to zero. $\boxed{x = 0}$ and $\boxed{x = 5}$ must be excluded from the domain.

In this section, we will simplify rational expressions and perform operations on rational expressions such as multiplication, division, addition, and subtraction. The resulting expressions may not have *explicit* domain restrictions, but it is important to note that there are *implicit* domain restrictions, because the domain restrictions on the original rational expression still apply.

Simplifying Rational Expressions

Recall that a fraction is *reduced* when it is written with no common factors.

$$\frac{16}{12} = \frac{4 \cdot 4}{4 \cdot 3} = \left(\frac{4}{4}\right) \cdot \left(\frac{4}{3}\right) = (1) \cdot \left(\frac{4}{3}\right) = \frac{4}{3} \quad \text{or} \quad \frac{16}{12} = \frac{\cancel{4} \cdot 4}{\cancel{4} \cdot 3} = \frac{4}{3}$$

Similarly, rational expressions are **reduced to lowest terms**, or **simplified**, if the numerator and denominator have no common factors other than ± 1. As with real numbers, the ability to write fractions in reduced form is dependent upon your ability to factor.

Technology Tip

Compare the graphs of

$Y_1 = \dfrac{x^2 - x - 2}{2x + 2}$ and

$Y_2 = \dfrac{x - 2}{2}, x \neq -1.$

To show the graph and table of Y_1, unhighlight [=] for Y_2 and press [GRAPH].

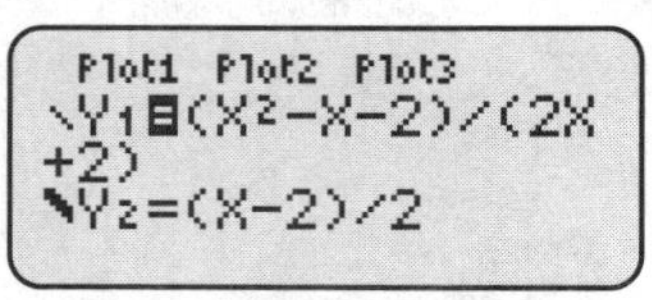

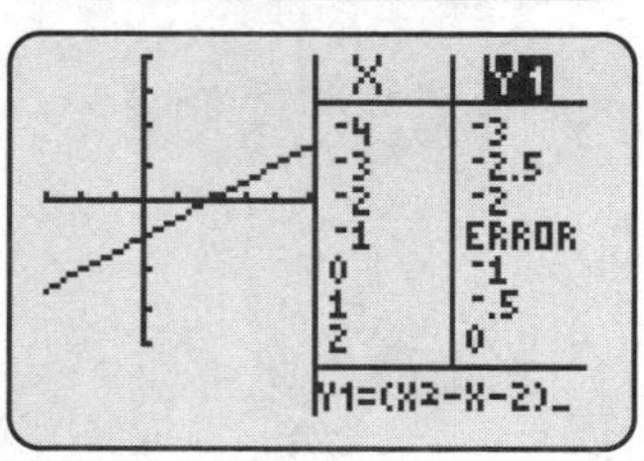

Note that $x = -1$ is not defined.

Reducing a Rational Expression to Lowest Terms (Simplifying)

1. Factor the numerator and denominator completely.
2. State any domain restrictions.
3. Cancel (divide out) the common factors in the numerator and denominator.

EXAMPLE 3 Reducing a Rational Expression to Lowest Terms

Simplify $\dfrac{x^2 - x - 2}{2x + 2}$ and state any domain restrictions.

Solution:

Factor the numerator and denominator. $\dfrac{x^2 - x - 2}{2x + 2} = \dfrac{(x - 2)(x + 1)}{2(x + 1)}$

State any domain restrictions. $x \neq -1$

Cancel (divide out) the common factor, $x + 1$. $= \dfrac{(x - 2)\cancel{(x + 1)}}{2\cancel{(x + 1)}}$

The rational expression is now in lowest terms (simplified). $= \boxed{\dfrac{x - 2}{2} \quad x \neq -1}$

■ **YOUR TURN** Simplify $\dfrac{x^2 + x - 2}{2x - 2}$ and state any domain restrictions.

■ **Answer:** $\dfrac{x + 2}{2} \quad x \neq 1$

The following table summarizes two **common mistakes** made with rational expressions.

CORRECT	INCORRECT	COMMENT
$\frac{x+5}{y+5}$ is already simplified.	**Error:** $\frac{x+\cancel{5}}{y+\cancel{5}} = \frac{x}{y}$	*Factors* can be divided out (canceled). Terms or parts of terms cannot be divided out. Remember to factor the numerator and denominator first, and then divide out common factors.
$\frac{x}{x^2+x}$ $= \frac{\cancel{x}}{\cancel{x}(x+1)} \quad x \neq 0, x \neq -1$ $= \boxed{\frac{1}{x+1} \quad x \neq 0, x \neq -1}$	**Error:** $\frac{x}{x^2+x} = \frac{1}{x+1} \quad x \neq -1$ *Note:* Missing $x \neq 0$.	Determine the domain restrictions *before* dividing out common factors.

Study Tip

Factors can be divided out (canceled). Factor the numerator and denominator first, and then divide out common factors.

Study Tip

Determine domain restrictions of a rational expression *before* dividing out (canceling) common factors.

EXAMPLE 4 Simplifying Rational Expressions

Reduce $\frac{x^2 - x - 6}{x^2 + x - 2}$ to lowest terms and state any domain restrictions.

Solution:

Factor the numerator and denominator. $\quad \frac{x^2 - x - 6}{x^2 + x - 2} = \frac{(x-3)(x+2)}{(x-1)(x+2)}$

State domain restrictions. $\quad x \neq -2, x \neq 1$

Divide out the common factor, $x + 2$. $\quad = \frac{(x-3)\cancel{(x+2)}}{(x-1)\cancel{(x+2)}}$

Simplify. $\quad = \boxed{\frac{x-3}{x-1} \quad x \neq -2, x \neq 1}$

■ **YOUR TURN** Reduce $\frac{x^2 + x - 6}{x^2 + 2x - 3}$ to lowest terms and state any domain restrictions.

■ **Answer:** $\frac{x-2}{x-1} \quad x \neq -3, x \neq 1$

EXAMPLE 5 Simplifying Rational Expressions

Reduce $\frac{x^2 - 4}{2 - x}$ to lowest terms and state any domain restrictions.

Solution:

Factor the numerator and denominator. $\quad \frac{x^2 - 4}{2 - x} = \frac{(x-2)(x+2)}{(2-x)}$

State domain restrictions. $\quad x \neq 2$

Factor out a negative in the denominator. $= \dfrac{(x-2)(x+2)}{-(x-2)}$

Cancel (divide out) the common factor, $x - 2$. $= \dfrac{\cancel{(x-2)}(x+2)}{-\cancel{(x-2)}}$

Simplify. $= \boxed{-(x+2) \qquad x \neq 2}$

■ **Answer:** $-(x+5) \qquad x \neq 5$

■ **YOUR TURN** Reduce $\dfrac{x^2-25}{5-x}$ to lowest terms and state any domain restrictions.

Multiplying and Dividing Rational Expressions

The same rules that apply to multiplying and dividing rational numbers also apply to rational expressions.

PROPERTY	RESTRICTION	DESCRIPTION
$\dfrac{a}{b}\cdot\dfrac{c}{d}=\dfrac{ac}{bd}$	$b \neq 0, d \neq 0$	Multiply numerators and denominators, respectively.
$\dfrac{a}{b}\div\dfrac{c}{d}=\dfrac{a}{b}\cdot\dfrac{d}{c}$	$b \neq 0, d \neq 0, c \neq 0$	Dividing is equivalent to multiplying by a reciprocal.

Multiplying Rational Expressions

1. Factor all numerators and denominators completely.
2. State any domain restrictions.
3. Divide the numerators and denominators by any common factors.
4. Multiply the remaining numerators and denominators, respectively.

Dividing Rational Expressions

1. Factor all numerators and denominators completely.
2. State any domain restrictions.
3. Rewrite division as multiplication by a reciprocal.
4. State any additional domain restrictions.
5. Divide the numerators and denominators by any common factors.
6. Multiply the remaining numerators and denominators, respectively.

EXAMPLE 6 Multiplying Rational Expressions

Multiply and simplify $\dfrac{3x+1}{4x^2+4x}\cdot\dfrac{x^3+3x^2+2x}{9x+3}$.

Solution:

Factor the numerators and denominators. $= \dfrac{(3x+1)}{4x(x+1)}\cdot\dfrac{x(x+1)(x+2)}{3(3x+1)}$

State any domain restrictions. $x \neq 0, x \neq -1, x \neq -\dfrac{1}{3}$

Divide the numerators and denominators by common factors.	$= \dfrac{\cancel{(3x+1)}}{4\cancel{x}\cancel{(x+1)}} \cdot \dfrac{\cancel{x}\cancel{(x+1)}(x+2)}{3\cancel{(3x+1)}}$
Simplify.	$= \boxed{\dfrac{x+2}{12} \quad x \neq 0, \quad x \neq -1, \quad x \neq -\dfrac{1}{3}}$

■ YOUR TURN Multiply and simplify $\dfrac{2x+1}{3x^2-3x} \cdot \dfrac{x^3+2x^2-3x}{8x+4}$.

■ Answer: $\dfrac{x+3}{12} \quad x \neq 0, x \neq 1, x \neq -\frac{1}{2}$

EXAMPLE 7 Dividing Rational Expressions

Divide and simplify $\dfrac{x^2-4}{x} \div \dfrac{3x^3-12x}{5x^3}$.

Solution:

Factor numerators and denominators.	$\dfrac{(x-2)(x+2)}{x} \div \dfrac{3x(x-2)(x+2)}{5x^3}$
State any domain restrictions.	$x \neq 0$
Write the quotient as a product.	$= \dfrac{(x-2)(x+2)}{x} \cdot \dfrac{5x^3}{3x(x-2)(x+2)}$
State any additional domain restrictions.	$x \neq -2, x \neq 2$
Divide out the common factors.	$= \dfrac{\cancel{(x-2)}\cancel{(x+2)}}{\cancel{x}} \cdot \dfrac{5x^{\cancel{3}}}{3\cancel{x}\,\cancel{(x-2)}\cancel{(x+2)}}$
Simplify.	$= \boxed{\dfrac{5x}{3} \quad x \neq -2, x \neq 0, x \neq 2}$

■ YOUR TURN Divide and simplify $\dfrac{x^2-9}{x} \div \dfrac{2x^3-18x}{7x^4}$.

■ Answer: $\dfrac{7x^2}{2} \quad x \neq -3, x \neq 0, x \neq 3$

Adding and Subtracting Rational Expressions

The same rules that apply to adding and subtracting rational numbers also apply to rational expressions.

PROPERTY	RESTRICTION	DESCRIPTION
$\dfrac{a}{b} \pm \dfrac{c}{b} = \dfrac{a \pm c}{b}$	$b \neq 0$	Adding or subtracting rational expressions when the denominators are the same
$\dfrac{a}{b} \pm \dfrac{c}{d} = \dfrac{ad \pm bc}{bd}$	$b \neq 0$ and $d \neq 0$	Adding or subtracting rational expressions when the denominators are different

EXAMPLE 8 Adding and Subtracting Rational Expressions: Equal Denominators

Perform the indicated operation and simplify.

a. $\dfrac{x+7}{(x+2)^2} + \dfrac{3x+1}{(x+2)^2}$ **b.** $\dfrac{6x+7}{2x-1} - \dfrac{2x+9}{2x-1}$

Solution (a):

Write as a single expression. $= \dfrac{x + 7 + 3x + 1}{(x + 2)^2}$

State any domain restrictions. $x \neq -2$

Combine like terms in the numerator. $= \dfrac{4x + 8}{(x + 2)^2}$

Factor out the common factor in the numerator. $= \dfrac{4(x + 2)}{(x + 2)^2}$

Cancel (divide out) the common factor, $x + 2$. $= \boxed{\dfrac{4}{x + 2} \quad x \neq -2}$

Study Tip

When subtracting a rational expression, distribute the negative of the quantity to be subtracted over all terms in the numerator.

Solution (b):

Write as a single expression. Use parentheses around the second numerator to ensure that the negative will be distributed throughout all terms. $= \dfrac{6x + 7 - (2x + 9)}{2x - 1}$

State any domain restrictions. $x \neq \dfrac{1}{2}$

Eliminate parentheses. Distribute the negative. $= \dfrac{6x + 7 - 2x - 9}{2x - 1}$

Combine like terms in the numerator. $= \dfrac{4x - 2}{2x - 1}$

Factor out the common factor in the numerator. $= \dfrac{2(2x - 1)}{2x - 1}$

Divide out (cancel) the common factor, $2x - 1$. $= \boxed{2 \quad x \neq \dfrac{1}{2}}$

EXAMPLE 9 Adding and Subtracting Rational Expressions: No Common Factors in Denominators

Perform the indicated operation and simplify.

a. $\dfrac{3 - x}{2x + 1} + \dfrac{x}{x - 1}$ **b.** $\dfrac{1}{x^2} - \dfrac{2}{x + 1}$

Solution (a):

The common denominator is the product of the denominators.

$$\frac{3 - x}{2x + 1} + \frac{x}{x - 1} = \frac{3(x - 1)}{(2x + 1)(x - 1)} + \frac{x(2x + 1)}{(x - 1)(2x + 1)}$$

$$= \frac{(3 - x)(x - 1) + x(2x + 1)}{(2x + 1)(x - 1)}$$

Eliminate parentheses in the numerator. $= \dfrac{3x - 3 - x^2 + x + 2x^2 + x}{(2x + 1)(x - 1)}$

Combine like terms in the numerator. $= \boxed{\dfrac{x^2 + 5x - 3}{(2x + 1)(x - 1)} \quad x \neq -\dfrac{1}{2}, x \neq 1}$

Solution (b):

The common denominator is the product of the denominators.

$$\frac{1}{x^2} - \frac{2}{x+1} = \frac{(1)(x+1) - 2\left(x^2\right)}{x^2(x+1)}$$

Eliminate parentheses in the numerator.

$$= \frac{x+1-2x^2}{x^2(x+1)} = \frac{-2x^2+x+1}{x^2(x+1)} = \frac{-\left(2x^2-x-1\right)}{x^2(x+1)}$$

Write the numerator in factored form to ensure no further simplification is possible.

$$= \boxed{\frac{-(2x+1)(x-1)}{x^2(x+1)} \qquad x \neq -1, x \neq 0}$$

YOUR TURN Perform the indicated operation and simplify.

a. $\frac{2x-1}{x+3} + \frac{x}{2x+1}$ **b.** $\frac{2}{x^2+1} - \frac{1}{x}$

Study Tip

When adding or subtracting rational expressions whose denominators have no common factors, the least common denominator is the product of the two denominators.

Answer:

a. $\frac{5x^2+3x-1}{(x+3)(2x+1)}$ $x \neq -3, x \neq -\frac{1}{2}$

b. $\frac{-(x-1)^2}{x\left(x^2+1\right)}$ $x \neq 0$

When combining two or more fractions through addition or subtraction recall that the **least common multiple**, or **least common denominator** (**LCD**), is the smallest real number that all of the denominators divide into evenly (that is, the smallest of which all are factors). For example,

$$\frac{2}{3} + \frac{1}{6} - \frac{4}{9}$$

To find the LCD of these three fractions, factor the denominators into prime factors:

$$\begin{aligned} 3 &= \mathbf{3} \\ 6 &= \mathbf{3 \cdot 2} \\ 9 &= \mathbf{3} \quad \mathbf{\cdot 3} \\ \hline &\mathbf{3 \cdot 2 \cdot 3 = 18} \end{aligned}$$

$$\frac{2}{3} + \frac{1}{6} - \frac{4}{9} = \frac{12+3-8}{18} = \frac{7}{18}$$

Rational expressions follow this same procedure, only now variables are also considered:

$$\frac{1}{2x} + \frac{1}{x^3} = \frac{1}{2x} + \frac{1}{x \cdot x \cdot x} \qquad \text{LCD} = 2x^3$$

$$\frac{2}{x+1} - \frac{x}{2x+1} + \frac{3-x}{(x+1)^2} = \frac{2}{(x+1)} - \frac{x}{(2x+1)} + \frac{3-x}{(x+1)(x+1)} \qquad \text{LCD} = (x+1)^2(2x+1)$$

The following box summarizes the LCD method for adding and subtracting rational expressions whose denominators have common factors.

THE LCD METHOD FOR ADDING AND SUBTRACTING RATIONAL EXPRESSIONS

1. Factor each of the denominators completely.
2. The LCD is the product of each of these distinct factors raised to the highest power to which that factor appears in any of the denominators.
3. Write each rational expression using the LCD for each denominator.
4. Add or subtract the resulting numerators.
5. Factor the resulting numerator to check for common factors.

EXAMPLE 10 Subtracting Rational Expressions: Common Factors in Denominators (LCD)

Perform the indicated operation and write in simplified form.

$$\frac{5x}{2x-6} - \frac{7x-2}{x^2-x-6}$$

Solution:

Factor the denominators. $= \frac{5x}{2(x-3)} - \frac{7x-2}{(x-3)(x+2)}$

Identify the least common denominator. $\text{LCD} = 2(x-3)(x+2)$

Write each expression using the LCD as the denominator. $= \frac{5x(x+2)}{2(x-3)(x+2)} - \frac{2(7x-2)}{2(x-3)(x+2)}$

Combine into one expression. Distribute the negative through the entire second numerator. $= \frac{5x^2+10x-14x+4}{2(x-3)(x+2)}$

Simplify. $= \boxed{\frac{5x^2-4x+4}{2(x-3)(x+2)} \quad x \neq -2, x \neq 3}$

■ **YOUR TURN** Perform the indicated operation and write in simplified form.

$$\frac{2x}{3x-6} - \frac{5x+1}{x^2+2x-8}$$

■ **Answer:** $\frac{2x^2-7x-3}{3(x-2)(x+4)} \quad x \neq -4,\ x \neq 2$

Complex Rational Expressions

A rational expression that contains another rational expression in either its numerator or denominator is called a **complex rational expression**. The following are examples of complex rational expressions.

$$\frac{\frac{1}{x}-5}{2+x} \qquad \frac{2-x}{4+\frac{3}{x-1}} \qquad \frac{\frac{3}{x}-7}{\frac{6}{2x-5}-1}$$

TWO METHODS FOR SIMPLIFYING COMPLEX RATIONAL EXPRESSIONS

Write a sum or difference of rational expressions that appear in either the numerator or denominator as a single rational expression. Once the complex rational expression contains a single rational expression in the numerator and one in the denominator, then rewrite the division as multiplication by the reciprocal.

OR

Find the LCD of all rational expressions contained in both the numerator and denominator. Multiply the numerator and denominator by this LCD and simplify.

EXAMPLE 11 Simplifying a Complex Rational Expression

Write the rational expression in simplified form.

$$\frac{\frac{2}{x}+1}{1+\frac{1}{x+1}}$$

Solution:

State the domain restrictions. $\quad \frac{\boxed{\frac{2}{x}}+1}{\boxed{1+\boxed{\frac{1}{x+1}}}} \qquad \boxed{x \neq 0}, \boxed{x \neq -1}, \text{ and } \boxed{x \neq -2}$

Procedure 1:

Add the expressions in both the numerator and denominator. $\quad = \frac{\frac{2}{x}+\frac{x}{x}}{\frac{x+1}{x+1}+\frac{1}{x+1}} = \frac{\frac{2+x}{x}}{\frac{(x+1)+1}{x+1}}$

Simplify. $\quad = \frac{\frac{2+x}{x}}{\frac{x+2}{x+1}}$

Express the quotient as a product. $\quad = \frac{2+x}{x} \cdot \frac{x+1}{x+2}$

Divide out the common factors. $\quad = \frac{\cancel{2+x}}{x} \cdot \frac{x+1}{\cancel{x+2}}$

Write in simplified form. $\quad = \boxed{\frac{x+1}{x} \qquad x \neq -2, x \neq -1, x \neq 0}$

Procedure 2:

Find the LCD of the numerator and denominator. $\quad \frac{\frac{2}{x}+1}{1+\frac{1}{x+1}}$

Identify the LCD's.

Numerator LCD: x
Denominator LCD: $x+1$
Combined LCD: $x(x+1)$

Multiply both numerator and denominator by their combined LCD. $\quad = \frac{\frac{2}{x}+1}{1+\frac{1}{x+1}} \cdot \frac{x(x+1)}{x(x+1)}$

Multiply the numerators and denominators, respectively, applying the distributive property. $\quad = \frac{\frac{2}{x} \cdot x(x+1) + 1x(x+1)}{1 \cdot x(x+1) + \frac{1}{x+1} \cdot x(x+1)}$

Divide out common factors. $\quad = \frac{\frac{2}{\cancel{x}} \cdot \cancel{x}(x+1) + 1x(x+1)}{x(x+1) + \frac{1}{\cancel{x+1}} \cdot x(\cancel{x+1})}$

Simplify. $\quad = \frac{2(x+1)+x(x+1)}{x(x+1)+x}$

Technology Tip

Compare the graphs and tables of values of $Y_1 = \frac{2/x+1}{1+1/(x+1)}$, $x \neq -2, x \neq -1, x \neq 0$ and $Y_2 = \frac{x+1}{x}$, $x \neq -2, x \neq -1, x \neq 0$.

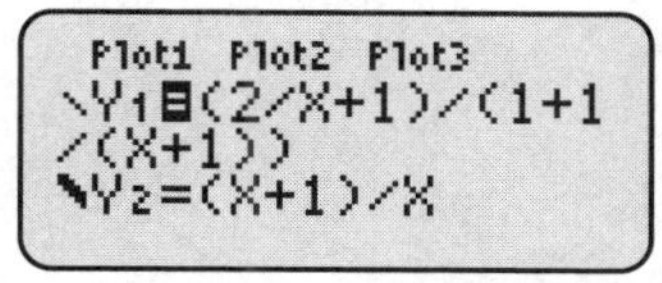

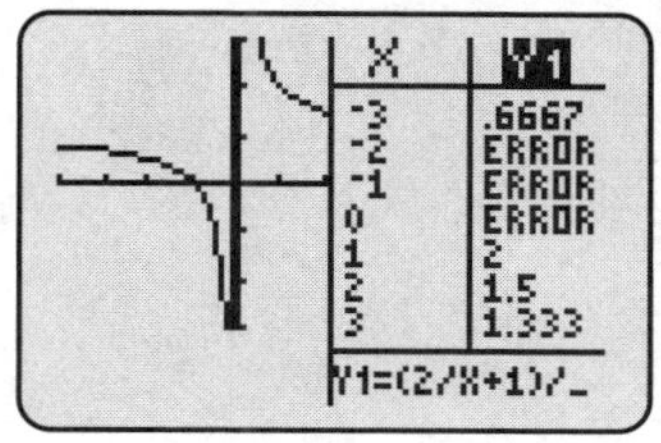

Note that $x = -1$, $x = 0$, and $x = -2$ are not defined.

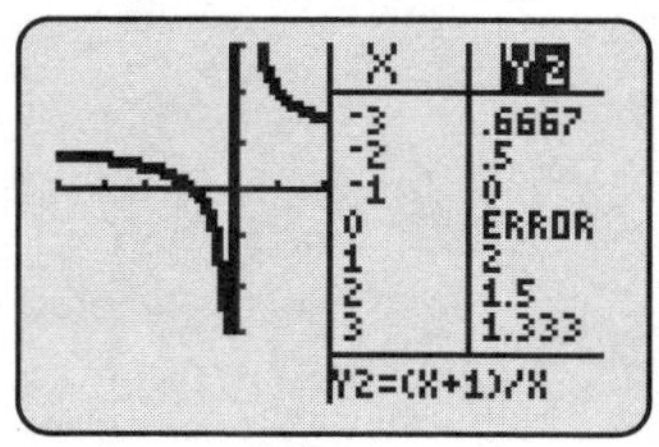

Note that $x = 0$ is not defined. Be careful not to forget that the original domain restrictions ($x \neq -2, -1, 0$) still hold.

Apply the distributive property. $= \dfrac{2x + 2 + x^2 + x}{x^2 + x + x}$

Combine like terms. $= \dfrac{x^2 + 3x + 2}{x^2 + 2x}$

Factor the numerator and denominator. $= \dfrac{(x + 2)(x + 1)}{x(x + 2)}$

Divide out the common factor. $= \dfrac{\cancel{(x + 2)}(x + 1)}{x\cancel{(x + 2)}}$

Write in simplified form. $= \boxed{\dfrac{x + 1}{x} \quad x \neq -2, x \neq -1, x \neq 0}$

EXAMPLE 12 Simplifying a Complex Rational Expression

Write the rational expression in simplified form.

$$\frac{\dfrac{1}{x^2 - 9} + 3}{1 - \dfrac{x}{2x + 6}} \qquad x \neq -6, -3, 3$$

Solution:

Factor the respective denominators. $\dfrac{\dfrac{1}{(x - 3)(x + 3)} + 3}{1 - \dfrac{x}{2(x + 3)}}$

Identify the LCDs.

Numerator LCD: $(x - 3)(x + 3)$
Denominator LCD: $2(x + 3)$
Combined LCD: $2(x - 3)(x + 3)$

Multiply both the numerator and the denominator by the combined LCD. $= \dfrac{\dfrac{1}{(x - 3)(x + 3)} + 3}{1 - \dfrac{x}{2(x + 3)}} \cdot \dfrac{2(x + 3)(x - 3)}{2(x + 3)(x - 3)}$

Multiply the numerators and denominators, respectively, applying the distributive property. $= \dfrac{\dfrac{2(x + 3)(x - 3)}{(x - 3)(x + 3)} + 3 \cdot 2(x + 3)(x - 3)}{1 \cdot 2(x + 3)(x - 3) - \dfrac{x \cdot 2(x + 3)(x - 3)}{2(x + 3)}}$

Simplify. $= \dfrac{2 + 6(x + 3)(x - 3)}{1 \cdot 2(x + 3)(x - 3) - x(x - 3)}$

Eliminate the parentheses. $= \dfrac{2 + 6x^2 - 54}{2x^2 - 18 - x^2 + 3x}$

Combine like terms. $= \dfrac{6x^2 - 52}{x^2 + 3x - 18}$

Factor the numerator and denominator to make sure there are no common factors. $= \boxed{\dfrac{2(3x^2 - 26)}{(x + 6)(x - 3)} \quad x \neq -6, x \neq -3, x \neq 3}$

SECTION 0.5 SUMMARY

In this section, rational expressions were defined as quotients of polynomials. The domain of any polynomial is the set of all real numbers. Since rational expressions are ratios of polynomials, the domain of rational expressions is the set of all real numbers *except those values that make the denominator equal to zero*. In this section, rational expressions were simplified (written with no common factors), multiplied, divided, added, and subtracted.

OPERATION	EXAMPLE	NOTE
Multiplying rational expressions	$\frac{2}{x+1}\cdot\frac{3x}{x-1}=\frac{6x}{x^2-1} \quad x \neq \pm 1$	State domain restrictions.
Dividing rational expressions	$\frac{2}{x+1}\div\frac{3x}{x-1} \quad x \neq \pm 1$ $=\frac{2}{x+1}\cdot\frac{x-1}{3x} \quad x \neq 0$ $=\frac{2(x-1)}{3x(x+1)} \quad x \neq 0, x \neq \pm 1$	When dividing rational expressions, remember to check for additional domain restrictions once the division is rewritten as multiplication by a reciprocal.
Adding/subtracting rational expressions with no common factors	$\frac{2}{x+1}+\frac{3x}{x-1} \quad x \neq \pm 1$ $\text{LCD} = (x+1)(x-1)$ $=\frac{2(x-1)+3x(x+1)}{(x+1)(x-1)}$ $=\frac{2x-2+3x^2+3x}{(x+1)(x-1)}$ $=\frac{3x^2+5x-2}{(x+1)(x-1)}$ $=\frac{(3x-1)(x+2)}{(x+1)(x-1)} \quad x \neq \pm 1$	The least common denominator (LCD) is the product of the two denominators.
Adding/subtracting rational expressions with common factors	$\frac{3}{x(x+1)}-\frac{2}{x(x+2)} \quad x \neq -2, -1, 0$ $\text{LCD} = x(x+1)(x+2)$ $=\frac{3(x+2)-2(x+1)}{x(x+1)(x+2)}=\frac{3x+6-2x-2}{x(x+1)(x+2)}$ $=\frac{x+4}{x(x+1)(x+2)} \quad x \neq -2, -1, 0$	The LCD is the product of each of these distinct factors raised to the highest power that appears in any of the denominators.

Complex rational expressions are simplified one of two ways:

1. Combine the sum or difference of rational expressions in a numerator or denominator as a single rational expression. The result is a rational expression in the numerator and a rational expression in the denominator. Then write the division as multiplication by the reciprocal.
2. Multiply the numerator and denominator by the overall LCD (LCD for all rational expressions that appear). The result is a single rational expression. Then simplify, if possible.

SECTION 0.5 EXERCISES

SKILLS

In Exercises 1–10, state any real numbers that must be excluded from the domain of each rational expression.

1. $\frac{3}{x}$
2. $\frac{5}{x}$
3. $\frac{3}{x-1}$
4. $\frac{6}{y-1}$
5. $\frac{5x-1}{x+1}$
6. $\frac{2x}{3-x}$
7. $\frac{2p^2}{p^2-1}$
8. $\frac{3t}{t^2-9}$
9. $\frac{3p-1}{p^2+1}$
10. $\frac{2t-2}{t^2+4}$

In Exercises 11–30, reduce the rational expression to lowest terms and state any real numbers that must be excluded from the domain.

11. $\frac{(x+3)(x-9)}{2(x+3)(x+9)}$
12. $\frac{4y(y-8)(y+7)}{8y(y+7)(y+8)}$
13. $\frac{(x-3)(x+1)}{2(x+1)}$
14. $\frac{(2x+1)(x-3)}{3(x-3)}$
15. $\frac{2(3y+1)(2y-1)}{3(2y-1)(3y)}$
16. $\frac{7(2y+1)(3y-1)}{5(3y-1)(2y)}$
17. $\frac{(5y-1)(y+1)}{25y-5}$
18. $\frac{(2t-1)(t+2)}{4t+8}$
19. $\frac{(3x+7)(x-4)}{4x-16}$
20. $\frac{(t-7)(2t+5)}{3t-21}$
21. $\frac{x^2-4}{x-2}$
22. $\frac{t^3-t}{t-1}$
23. $\frac{x+7}{x+7}$
24. $\frac{2y+9}{2y+9}$
25. $\frac{x^2+9}{2x+9}$
26. $\frac{x^2+4}{2x+4}$
27. $\frac{x^2+5x+6}{x^2-3x-10}$
28. $\frac{x^2+19x+60}{x^2+8x+16}$
29. $\frac{6x^2-x-1}{2x^2+9x-5}$
30. $\frac{15x^2-x-2}{5x^2+13x-6}$

In Exercises 31–48, multiply the rational expressions and simplify. State any real numbers that must be excluded from the domain.

31. $\frac{x-2}{x+1} \cdot \frac{3x+5}{x-2}$
32. $\frac{4x+5}{x-2} \cdot \frac{3x+4}{4x+5}$
33. $\frac{5x+6}{x} \cdot \frac{2x}{5x-6}$
34. $\frac{4(x-2)(x+5)}{8x} \cdot \frac{16x}{(x-5)(x+5)}$
35. $\frac{2x-2}{3x} \cdot \frac{x^2+x}{x^2-1}$
36. $\frac{5x-5}{10x} \cdot \frac{x^2+x}{x^2-1}$
37. $\frac{3x^2-12}{x} \cdot \frac{x^2+5x}{x^2+3x-10}$
38. $\frac{4x^2-32x}{x} \cdot \frac{x^2+3x}{x^2-5x-24}$
39. $\frac{t+2}{3t-9} \cdot \frac{t^2-6t+9}{t^2+4t+4}$
40. $\frac{y+3}{3y+9} \cdot \frac{y^2-10y+25}{y^2+3y-40}$
41. $\frac{t^2+4}{t-3} \cdot \frac{3t}{t+2}$
42. $\frac{7a^2+21a}{14(a^2-9)} \cdot \frac{a+3}{7}$
43. $\frac{y^2-4}{y-3} \cdot \frac{3y}{y+2}$
44. $\frac{t^2+t-6}{t^2-4} \cdot \frac{8t}{2t^2}$
45. $\frac{3x^2-15x}{2x^3-50x} \cdot \frac{2x^2-7x-15}{3x^2+15x}$
46. $\frac{5t-1}{4t} \cdot \frac{4t^2+3t}{16t^2-9}$
47. $\frac{6x^2-11x-35}{8x^2-22x-21} \cdot \frac{4x^2-49}{9x^2-25}$
48. $\frac{3x^2-2x}{12x^3-8x^2} \cdot \frac{x^2-7x-18}{2x^2-162}$

In Exercises 49–66, divide the rational expressions and simplify. State any real numbers that must be excluded from the domain.

49. $\frac{3}{x} \div \frac{12}{x^2}$
50. $\frac{5}{x^2} \div \frac{10}{x^3}$
51. $\frac{6}{x-2} \div \frac{12}{(x-2)(x+2)}$
52. $\frac{5(x+6)}{10(x-6)} \div \frac{20(x+6)}{8}$
53. $\frac{1}{x-1} \div \frac{5}{x^2-1}$
54. $\frac{5}{3x-4} \div \frac{10}{9x^2-16}$

55. $\dfrac{2-p}{p^2-1} \div \dfrac{2p-4}{p+1}$

56. $\dfrac{4-x}{x^2-16} \div \dfrac{12-3x}{x-4}$

57. $\dfrac{36-n^2}{n^2-9} \div \dfrac{n+6}{n+3}$

58. $\dfrac{49-y^2}{y^2-25} \div \dfrac{7+y}{2y+10}$

59. $\dfrac{3t^3-6t^2-9t}{5t-10} \div \dfrac{6+6t}{4t-8}$

60. $\dfrac{x^3+8x^2+12x}{5x^2-10x} \div \dfrac{4x+8}{x^2-4}$

61. $\dfrac{w^2-w}{w} \div \dfrac{w^3-w}{5w^3}$

62. $\dfrac{y^2-3y}{2y} \div \dfrac{y^3-3y^2}{8y}$

63. $\dfrac{x^2+4x-21}{x^2+3x-10} \div \dfrac{x^2-2x-63}{x^2+x-20}$

64. $\dfrac{2y^2-5y-3}{2y^2-9y-5} \div \dfrac{3y-9}{y^2-5y}$

65. $\dfrac{20x^2-3x-2}{25x^2-4} \div \dfrac{12x^2+23x+5}{3x^2+5x}$

66. $\dfrac{x^2-6x-27}{2x^2+13x-7} \div \dfrac{2x^2-15x-27}{2x^2+9x-5}$

In Exercises 67–82, add or subtract the rational expression and simplify. State any real numbers that must be excluded from the domain.

67. $\dfrac{3}{x} - \dfrac{2}{5x}$

68. $\dfrac{5}{7x} - \dfrac{3}{x}$

69. $\dfrac{3}{p-2} + \dfrac{5p}{p+1}$

70. $\dfrac{4}{9+x} - \dfrac{5x}{x-2}$

71. $\dfrac{2x+1}{5x-1} - \dfrac{3-2x}{1-5x}$

72. $\dfrac{7}{2x-1} - \dfrac{5}{1-2x}$

73. $\dfrac{3y^2}{y+1} + \dfrac{1-2y}{y-1}$

74. $\dfrac{3}{1-x} + \dfrac{4}{x-1}$

75. $\dfrac{3x}{x^2-4} + \dfrac{3+x}{x+2}$

76. $\dfrac{x-1}{4-x^2} - \dfrac{x+1}{2+x}$

77. $\dfrac{x-1}{x-2} + \dfrac{x-6}{x^2-4}$

78. $\dfrac{2}{y-3} + \dfrac{7}{y+2}$

79. $\dfrac{5a}{a^2-b^2} - \dfrac{7}{b-a}$

80. $\dfrac{1}{y} + \dfrac{4}{y^2-4} - \dfrac{2}{y^2-2y}$

81. $7 + \dfrac{1}{x-3}$

82. $\dfrac{3}{5y+6} - \dfrac{4}{y-2} + \dfrac{y^2-y}{5y^2-4y-12}$

In Exercises 83–90, simplify the complex rational expressions. State any real numbers that must be excluded from the domain.

83. $\dfrac{\frac{1}{x}-1}{1-\frac{2}{x}}$

84. $\dfrac{\frac{3}{y}-5}{4-\frac{2}{y}}$

85. $\dfrac{3+\frac{1}{x}}{9-\frac{1}{x^2}}$

86. $\dfrac{\frac{1}{x}+\frac{2}{x^2}}{\frac{9}{x}-\frac{5}{x^2}}$

87. $\dfrac{\frac{1}{x-1}+1}{1-\frac{1}{x+1}}$

88. $\dfrac{\frac{7}{y+7}}{\frac{1}{y+7}-\frac{1}{y}}$

89. $\dfrac{\frac{1}{x-1}+1}{\frac{1}{x+1}+1}$

90. $\dfrac{\frac{3}{x+1}-\frac{3}{x-1}}{\frac{5}{x^2-1}}$

APPLICATIONS

91. **Finance.** The amount of payment made on a loan is given by the formula $A = \dfrac{pi}{1 - 1/(1+i)^n}$, where p is the principal (amount borrowed), and $i = \dfrac{r}{n}$, where r is the interest rate expressed as a decimal and n is the number of payments per year. Suppose $n = 5$. Simplify the formula as much as possible.

92. **Finance.** Use the formula $A = \dfrac{pi}{1 - \dfrac{1}{(1+i)^{nt}}}$ to calculate the amount your monthly payment will be on a loan of \$150,000 at an interest rate of 6.5% for 30 years ($nt = 360$).

93. **Circuits.** If two resistances are connected in parallel, the combined resistance is given by the formula $= \dfrac{1}{1/R_1 + 1/R_2}$, where R_1 and R_2 are the individual resistances. Simplify the formula.

94. **Optics.** The focal length of a lens can be calculated by applying the formula $f = \dfrac{1}{1/p + 1/q}$, where p is the distance that the object is from the lens and q is the distance that the image is from the lens. Simplify the formula.

CATCH THE MISTAKE

In Exercises 95 and 96, explain the mistake that is made.

95. Simplify $\dfrac{x^2 + 2x + 1}{x + 1}$.

Solution:

Factor the numerator. $\dfrac{x^2 + 2x + 1}{x + 1} = \dfrac{(x + 1)(x + 1)}{(x + 1)}$

Cancel the common factor, $x + 1$. $= \dfrac{(x + 1)\cancel{(x + 1)}}{\cancel{(x + 1)}}$

Write in simplified form. $= x + 1$

This is incorrect. What mistake was made?

96. Simplify $\dfrac{x + 1}{x^2 + 2x + 1}$.

Solution:

Cancel the common 1s. $\dfrac{x \cancel{+ 1}}{x^2 + 2x \cancel{+ 1}}$

Factor the denominator. $= \dfrac{x}{x(x + 2)}$

Cancel the common x. $= \dfrac{\cancel{x}}{\cancel{x}(x + 2)}$

Write in simplified form. $= \dfrac{1}{x + 2} \quad x \neq -2$

This is incorrect. What mistake was made?

CONCEPTUAL

In Exercises 97–100, determine whether each of the statements is true or false.

97. $\dfrac{x^2 - 81}{x - 9} = x + 9$

98. $\dfrac{x - 9}{x^2 - 81} = \dfrac{1}{x + 9} \quad x \neq -9, 9$

99. When adding or subtracting rational expressions, the LCD is always the product of all the denominators.

100. $\dfrac{x - c}{c - x} = -1$ for all values of x.

CHALLENGE

101. Perform the operation and simplify (remember to state domain restrictions).

$$\frac{x + a}{x + b} \div \frac{x + c}{x + d}$$

102. Write the numerator as the product of two binomials. Divide out any common factors of the numerator and denominator.

$$\frac{a^{2n} - b^{2n}}{a^n - b^n}$$

TECHNOLOGY

103. Utilizing a graphing technology, plot the expression $y = \dfrac{x + 7}{x + 7}$. Zoom in near $x = -7$. Does this agree with what you found in Exercise 23?

104. Utilizing a graphing technology, plot the expression $y = \dfrac{x^2 - 4}{x - 2}$. Zoom in near $x = 2$. Does this agree with what you found in Exercise 21?

In Exercises 105 and 106, for each given expression: (a) simplify the expression, (b) use a graphing utility to plot the expression and the answer in (a) in the same viewing window, and (c) determine the domain restriction(s) where the graphs will agree with each other.

105. $\dfrac{1 + \dfrac{1}{x - 2}}{1 - \dfrac{1}{x + 2}}$

106. $\dfrac{1 - \dfrac{2}{x + 3}}{1 + \dfrac{1}{x + 4}}$

55. $\frac{2-p}{p^2-1} \div \frac{2p-4}{p+1}$

56. $\frac{4-x}{x^2-16} \div \frac{12-3x}{x-4}$

57. $\frac{36-n^2}{n^2-9} \div \frac{n+6}{n+3}$

58. $\frac{49-y^2}{y^2-25} \div \frac{7+y}{2y+10}$

59. $\frac{3t^3-6t^2-9t}{5t-10} \div \frac{6+6t}{4t-8}$

60. $\frac{x^3+8x^2+12x}{5x^2-10x} \div \frac{4x+8}{x^2-4}$

61. $\frac{w^2-w}{w} \div \frac{w^3-w}{5w^3}$

62. $\frac{y^2-3y}{2y} \div \frac{y^3-3y^2}{8y}$

63. $\frac{x^2+4x-21}{x^2+3x-10} \div \frac{x^2-2x-63}{x^2+x-20}$

64. $\frac{2y^2-5y-3}{2y^2-9y-5} \div \frac{3y-9}{y^2-5y}$

65. $\frac{20x^2-3x-2}{25x^2-4} \div \frac{12x^2+23x+5}{3x^2+5x}$

66. $\frac{x^2-6x-27}{2x^2+13x-7} \div \frac{2x^2-15x-27}{2x^2+9x-5}$

In Exercises 67–82, add or subtract the rational expression and simplify. State any real numbers that must be excluded from the domain.

67. $\frac{3}{x} - \frac{2}{5x}$

68. $\frac{5}{7x} - \frac{3}{x}$

69. $\frac{3}{p-2} + \frac{5p}{p+1}$

70. $\frac{4}{9+x} - \frac{5x}{x-2}$

71. $\frac{2x+1}{5x-1} - \frac{3-2x}{1-5x}$

72. $\frac{7}{2x-1} - \frac{5}{1-2x}$

73. $\frac{3y^2}{y+1} + \frac{1-2y}{y-1}$

74. $\frac{3}{1-x} + \frac{4}{x-1}$

75. $\frac{3x}{x^2-4} + \frac{3+x}{x+2}$

76. $\frac{x-1}{4-x^2} - \frac{x+1}{2+x}$

77. $\frac{x-1}{x-2} + \frac{x-6}{x^2-4}$

78. $\frac{2}{y-3} + \frac{7}{y+2}$

79. $\frac{5a}{a^2-b^2} - \frac{7}{b-a}$

80. $\frac{1}{y} + \frac{4}{y^2-4} - \frac{2}{y^2-2y}$

81. $7 + \frac{1}{x-3}$

82. $\frac{3}{5y+6} - \frac{4}{y-2} + \frac{y^2-y}{5y^2-4y-12}$

In Exercises 83–90, simplify the complex rational expressions. State any real numbers that must be excluded from the domain.

83. $\frac{\frac{1}{x}-1}{1-\frac{2}{x}}$

84. $\frac{\frac{3}{y}-5}{4-\frac{2}{y}}$

85. $\frac{3+\frac{1}{x}}{9-\frac{1}{x^2}}$

86. $\frac{\frac{1}{x}+\frac{2}{x^2}}{\frac{9}{x}-\frac{5}{x^2}}$

87. $\frac{\frac{1}{x-1}+1}{1-\frac{1}{x+1}}$

88. $\frac{\frac{7}{y+7}}{\frac{1}{y+7}-\frac{1}{y}}$

89. $\frac{\frac{1}{x-1}+1}{\frac{1}{x+1}+1}$

90. $\frac{\frac{3}{x+1}-\frac{3}{x-1}}{\frac{5}{x^2-1}}$

APPLICATIONS

91. Finance. The amount of payment made on a loan is given by the formula $A = \frac{pi}{1-1/(1+i)^n}$, where p is the principal (amount borrowed), and $i = \frac{r}{n}$, where r is the interest rate expressed as a decimal and n is the number of payments per year. Suppose $n = 5$. Simplify the formula as much as possible.

92. Finance. Use the formula $A = \frac{pi}{1-\frac{1}{(1+i)^{nt}}}$ to calculate the amount your monthly payment will be on a loan of \$150,000 at an interest rate of 6.5% for 30 years ($nt = 360$).

93. Circuits. If two resistances are connected in parallel, the combined resistance is given by the formula $= \frac{1}{1/R_1 + 1/R_2}$, where R_1 and R_2 are the individual resistances. Simplify the formula.

94. Optics. The focal length of a lens can be calculated by applying the formula $f = \frac{1}{1/p + 1/q}$, where p is the distance that the object is from the lens and q is the distance that the image is from the lens. Simplify the formula.

CATCH THE MISTAKE

In Exercises 95 and 96, explain the mistake that is made.

95. Simplify $\frac{x^2 + 2x + 1}{x + 1}$.

Solution:

Factor the numerator. $\frac{x^2 + 2x + 1}{x + 1} = \frac{(x + 1)(x + 1)}{(x + 1)}$

Cancel the common factor, $x + 1$. $= \frac{(x + 1)\cancel{(x + 1)}}{\cancel{(x + 1)}}$

Write in simplified form. $= x + 1$

This is incorrect. What mistake was made?

96. Simplify $\frac{x + 1}{x^2 + 2x + 1}$.

Solution:

Cancel the common 1s. $\frac{x \cancel{+ 1}}{x^2 + 2x \cancel{+ 1}}$

Factor the denominator. $= \frac{x}{x(x + 2)}$

Cancel the common x. $= \frac{\cancel{x}}{\cancel{x}(x + 2)}$

Write in simplified form. $= \frac{1}{x + 2} \quad x \neq -2$

This is incorrect. What mistake was made?

CONCEPTUAL

In Exercises 97–100, determine whether each of the statements is true or false.

97. $\frac{x^2 - 81}{x - 9} = x + 9$

98. $\frac{x - 9}{x^2 - 81} = \frac{1}{x + 9} \quad x \neq -9, 9$

99. When adding or subtracting rational expressions, the LCD is always the product of all the denominators.

100. $\frac{x - c}{c - x} = -1$ for all values of x.

CHALLENGE

101. Perform the operation and simplify (remember to state domain restrictions).

$$\frac{x + a}{x + b} \div \frac{x + c}{x + d}$$

102. Write the numerator as the product of two binomials. Divide out any common factors of the numerator and denominator.

$$\frac{a^{2n} - b^{2n}}{a^n - b^n}$$

TECHNOLOGY

103. Utilizing a graphing technology, plot the expression $y = \frac{x + 7}{x + 7}$. Zoom in near $x = -7$. Does this agree with what you found in Exercise 23?

104. Utilizing a graphing technology, plot the expression $y = \frac{x^2 - 4}{x - 2}$. Zoom in near $x = 2$. Does this agree with what you found in Exercise 21?

In Exercises 105 and 106, for each given expression: (a) simplify the expression, (b) use a graphing utility to plot the expression and the answer in (a) in the same viewing window, and (c) determine the domain restriction(s) where the graphs will agree with each other.

105. $\dfrac{1 + \dfrac{1}{x - 2}}{1 - \dfrac{1}{x + 2}}$ **106.** $\dfrac{1 - \dfrac{2}{x + 3}}{1 + \dfrac{1}{x + 4}}$

SECTION 0.6 RATIONAL EXPONENTS AND RADICALS

SKILLS OBJECTIVES

- Simplify square roots.
- Simplify radicals.
- Add and subtract radicals.
- Rationalize denominators containing radicals.
- Apply properties of rational exponents.

CONCEPTUAL OBJECTIVES

- Understand that radicals are equivalent to rational exponents.
- Understand that a radical implies one number (the principal root), not two ($\pm$ the principal root).

In Section 0.2, we discussed integer exponents and their properties. For example, $4^2 = 16$ and $x^2 \cdot x^3 = x^5$. In this section we expand our discussion of exponents to include any rational numbers. For example, $16^{1/2} = ?$ and $\left(x^{1/2}\right)^{3/4} = ?$. We will first start with a more familiar notation (*roots*) and discuss operations on *radicals*, and then *rational exponents* will be discussed.

Square Roots

DEFINITION **Principal Square Root**

Let a be any nonnegative real number; then the nonnegative real number b is called the **principal square root of a**, denoted $b = \sqrt{a}$, if $b^2 = a$. The symbol $\sqrt{\ }$ is called a **radical sign**, and a is called the **radicand**.

It is important to note that the principal square root b is nonnegative. "The principal square root of 16 is 4" implies $4^2 = 16$. Although it is also true that $(-4)^2 = 16$, the principal square root is defined to be nonnegative.

It is also important to note that negative real numbers do not have real square roots. For example $\sqrt{-9}$ is not a real number because there are no real numbers that when squared yield -9. Since principal square roots are defined to be nonnegative, this means they must be zero or positive. The square root of zero is equal to zero: $\sqrt{0} = 0$. All other nonnegative principal square roots are positive.

Study Tip

Incorrect: $\sqrt{25} = \pm 5$

Correct: $\sqrt{25} = 5$

The principal square root, $\sqrt{\ }$, is defined as a nonnegative real number.

EXAMPLE 1 **Evaluating Square Roots**

Evaluate the square roots, if possible.

a. $\sqrt{169}$ **b.** $\sqrt{\frac{4}{9}}$ **c.** $\sqrt{-36}$

Solution:

a. What positive real number squared results in 169? $\sqrt{169} = \boxed{13}$

Check: $13^2 = 169$

b. What positive real number squared results in $\frac{4}{9}$? $\sqrt{\frac{4}{9}} = \boxed{\frac{2}{3}}$

Check: $\left(\frac{2}{3}\right)^2 = \frac{4}{9}$

c. What positive real number squared results in -36? $\boxed{\text{No real number}}$

SQUARE ROOTS OF PERFECT SQUARES

Let a be any real number; then:

$$\sqrt{a^2} = |a|$$

EXAMPLE 2 Finding Square Roots of Perfect Squares

Evaluate the following:

a. $\sqrt{6^2}$ **b.** $\sqrt{(-7)^2}$ **c.** $\sqrt{x^2}$

Solution:

a. $\sqrt{6^2} = \sqrt{36} = \boxed{6}$ **b.** $\sqrt{(-7)^2} = \sqrt{49} = \boxed{7}$ **c.** $\sqrt{x^2} = \boxed{|x|}$

Simplifying Square Roots

So far only square roots of perfect squares have been discussed. Now we consider how to simplify square roots such as $\sqrt{12}$. We rely on the following properties.

PROPERTIES OF SQUARE ROOTS

Let a and b be nonnegative real numbers, then:

Property	Description	Example
$\sqrt{a \cdot b} = \sqrt{a} \cdot \sqrt{b}$	The square root of a product is the product of the square roots.	$\sqrt{20} = \sqrt{4} \cdot \sqrt{5} = 2\sqrt{5}$
$\sqrt{\frac{a}{b}} = \frac{\sqrt{a}}{\sqrt{b}}$ $b \neq 0$	The square root of a quotient is the quotient of the square roots.	$\sqrt{\frac{40}{49}} = \frac{\sqrt{40}}{\sqrt{49}} = \frac{\sqrt{4} \cdot \sqrt{10}}{7} = \frac{2\sqrt{10}}{7}$

EXAMPLE 3 Simplifying Square Roots

Simplify:

a. $\sqrt{48x^2}$ **b.** $\sqrt{28x^3}$ **c.** $\sqrt{12x} \cdot \sqrt{6x}$ **d.** $\dfrac{\sqrt{45x^3}}{\sqrt{5x}}$

Solution:

a. $\sqrt{48x^2} = \sqrt{48} \cdot \sqrt{x^2} = \sqrt{16 \cdot 3} \cdot \sqrt{x^2} = \underbrace{\sqrt{16}}_{4} \cdot \sqrt{3} \cdot \underbrace{\sqrt{x^2}}_{|x|} = \boxed{4\,|x|\,\sqrt{3}}$

b. $\sqrt{28x^3} = \sqrt{28} \cdot \sqrt{x^3} = \sqrt{4 \cdot 7} \cdot \sqrt{x^2 \cdot x} = \underbrace{\sqrt{4}}_{2} \cdot \sqrt{7} \cdot \underbrace{\sqrt{x^2}}_{|x|} \cdot \sqrt{x}$

$= 2|x|\sqrt{7}\,\sqrt{x} = 2|x|\sqrt{7x} = \boxed{2x\sqrt{7x}}$ since $x \geq 0$

c. $\sqrt{12x} \cdot \sqrt{6x} = \sqrt{72x^2} = \sqrt{36 \cdot 2 \cdot x^2} = \underbrace{\sqrt{36}}_{6} \cdot \sqrt{2} \cdot \underbrace{\sqrt{x^2}}_{|x|} = 6|x|\sqrt{2}$

$= \boxed{6x\sqrt{2}}$ since $x \geq 0$

d. $\dfrac{\sqrt{45x^3}}{\sqrt{5x}} = \sqrt{\dfrac{45x^3}{5x}} = \sqrt{9x^2} = \underbrace{\sqrt{9}}_{3} \cdot \underbrace{\sqrt{x^2}}_{|x|} = 3\,|x| = \boxed{3x}$ since $x > 0$

Note: $x \neq 0$, $x > 0$

Study Tip

If you have a single odd power under the square root, like $\sqrt{x^3}$, the variable is forced to be nonnegative so the absolute value is not necessary.

■ **YOUR TURN** Simplify: **a.** $\sqrt{60x^3}$ **b.** $\dfrac{\sqrt{125x^5}}{\sqrt{25x^3}}$

■ **Answer:**
a. $2x\sqrt{15x}$ **b.** $x\sqrt{5}$

Other (*n*th) Roots

We now expand our discussion from square roots to other *n*th roots.

DEFINITION **Principal *n*th Root**

Let a be a real number and n be a positive integer. Then the real number b is called the **principal *n*th root of *a***, denoted $b = \sqrt[n]{a}$, if $b^n = a$. If n is even, then a and b are nonnegative real numbers. The positive integer n is called the **index**. The square root corresponds to $n = 2$, and the **cube root** corresponds to $n = 3$.

n	*a*	*b*	EXAMPLE
Even	Positive	Positive	$\sqrt[4]{16} = 2$ because $2^4 = 16$
Even	Negative	Not a real number	$\sqrt[4]{-16}$ is not a real number
Odd	Positive	Positive	$\sqrt[3]{27} = 3$ because $3^3 = 27$
Odd	Negative	Negative	$\sqrt[3]{-125} = -5$ because $(-5)^3 = -125$

A radical sign, $\sqrt{\ }$, combined with a radicand is called a **radical**.

PROPERTIES OF RADICALS

Let a and b be real numbers, then

PROPERTY	DESCRIPTION	EXAMPLE
$\sqrt[n]{ab} = \sqrt[n]{a} \cdot \sqrt[n]{b}$ if $\sqrt[n]{a}$ and $\sqrt[n]{b}$ both exist	The *n*th root of a product is the product of the *n*th roots.	$\sqrt[3]{16} = \sqrt[3]{8} \cdot \sqrt[3]{2} = 2\sqrt[3]{2}$
$\sqrt[n]{\dfrac{a}{b}} = \dfrac{\sqrt[n]{a}}{\sqrt[n]{b}}$ $\quad b \neq 0$ if $\sqrt[n]{a}$ and $\sqrt[n]{b}$ both exist	The *n*th root of a quotient is the quotient of the *n*th roots.	$\sqrt[4]{\dfrac{81}{16}} = \dfrac{\sqrt[4]{81}}{\sqrt[4]{16}} = \dfrac{3}{2}$
$\sqrt[n]{a^m} = (\sqrt[n]{a})^m$	The *n*th root of a power is the power of the *n*th root.	$\sqrt[3]{8^2} = (\sqrt[3]{8})^2 = (2)^2 = 4$
$\sqrt[n]{a^n} = a$ $\quad n$ is odd	When n is odd, the *n*th root of a raised to the *n*th power is a.	$\sqrt[3]{x^5} = \underbrace{\sqrt[3]{x^3}}_{x} \cdot \sqrt[3]{x^2} = x\sqrt[3]{x^2}$
$\sqrt[n]{a^n} = \lvert a \rvert$ $\quad n$ is even	When n is even, the *n*th root of a raised to the *n*th power is the absolute value of a.	$\sqrt[4]{x^5} = \underbrace{\sqrt[4]{x^4}}_{\lvert x \rvert} \cdot \sqrt[4]{x} = \lvert x \rvert \sqrt[4]{x}$

EXAMPLE 4 Simplifying Radicals

Simplify:

a. $\sqrt[3]{-24x^5}$ **b.** $\sqrt[4]{32x^5}$

Solution:

a. $\sqrt[3]{-24x^5} = \sqrt[3]{(-8)(3)x^3x^2} = \underbrace{\sqrt[3]{-8}}_{-2} \cdot \sqrt[3]{3} \cdot \underbrace{\sqrt[3]{x^3}}_{x} \cdot \sqrt[3]{x^2} = \boxed{-2x\sqrt[3]{3x^2}}$

b. $\sqrt[4]{32x^5} = \sqrt[4]{16 \cdot 2 \cdot x^4 \cdot x} = \underbrace{\sqrt[4]{16}}_{2} \cdot \sqrt[4]{2} \cdot \underbrace{\sqrt[4]{x^4}}_{|x|} \cdot \sqrt[4]{x} = 2|x|\sqrt[4]{2x} = \boxed{2x\sqrt[4]{2x}}$ since $x > 0$

Combining Like Radicals

We have already discussed properties for multiplying and dividing radicals. Now we focus on combining (adding or subtracting) radicals. Radicals with the same index and radicand are called **like radicals**. Only like radicals can be added or subtracted.

EXAMPLE 5 Combining Like Radicals

Combine the radicals if possible.

a. $4\sqrt{3} - 6\sqrt{3} + 7\sqrt{3}$ **b.** $2\sqrt{5} - 3\sqrt{7} + 6\sqrt{3}$

c. $3\sqrt{5} + \sqrt{20} - 2\sqrt{45}$ **d.** $\sqrt[4]{10} - 2\sqrt[3]{10} + 3\sqrt{10}$

Solution (a):

Use the distributive property. $4\sqrt{3} - 6\sqrt{3} + 7\sqrt{3} = (4 - 6 + 7)\sqrt{3}$

Eliminate the parentheses. $= \boxed{5\sqrt{3}}$

Solution (b):

None of these radicals are alike.
The expression is in simplified form. $\boxed{2\sqrt{5} - 3\sqrt{7} + 6\sqrt{3}}$

Solution (c):

Write the radicands as products with a factor of 5. $3\sqrt{5} + \sqrt{20} - 2\sqrt{45} = 3\sqrt{5} + \sqrt{4 \cdot 5} - 2\sqrt{9 \cdot 5}$

The square root of a product is the product of square roots. $3\sqrt{5} + \sqrt{4} \cdot \sqrt{5} - 2\sqrt{9} \cdot \sqrt{5}$

Simplify the square roots of perfect squares. $3\sqrt{5} + 2\sqrt{5} - \underbrace{2(3)}_{6}\sqrt{5}$

All three radicals are now like radicals. $3\sqrt{5} + 2\sqrt{5} - 6\sqrt{5}$

Use the distributive property. $(3 + 2 - 6)\sqrt{5}$

Simplify. $\boxed{-\sqrt{5}}$

Solution (d):

None of these radicals are alike because they have different indices. The expression is in simplified form. $\boxed{\sqrt[4]{10} - 2\sqrt[3]{10} + 3\sqrt{10}}$

YOUR TURN Combine the radicals.

a. $4\sqrt[3]{7} - 6\sqrt[3]{7} + 9\sqrt[3]{7}$ **b.** $5\sqrt{24} - 2\sqrt{54}$

Answer:

a. $7\sqrt[3]{7}$ **b.** $4\sqrt{6}$

Rationalizing Denominators

When radicals appear in a quotient, it is customary to write the quotient with no radicals in the denominator. This process is called **rationalizing the denominator** and involves multiplying by an expression that will eliminate the radical in the denominator.

For example, the expression $\frac{1}{\sqrt{3}}$ contains a single radical in the denominator. In a case like this, multiply the numerator and denominator by an appropriate radical expression, so that the resulting denominator will be radical free:

$$\frac{1}{\sqrt{3}} \cdot \underbrace{\frac{\sqrt{3}}{\sqrt{3}}}_{1} = \frac{\sqrt{3}}{\sqrt{3} \cdot \sqrt{3}} = \frac{\sqrt{3}}{3}$$

If the denominator contains a sum of the form $a + \sqrt{b}$, multiply both the numerator and the denominator by the **conjugate** of the denominator, $a - \sqrt{b}$, which uses the difference of two squares to eliminate the radical term. Similarly, if the denominator contains a difference of the form $a - \sqrt{b}$, multiply both the numerator and the denominator by the conjugate of the denominator, $a + \sqrt{b}$. For example, to rationalize $\frac{1}{3 - \sqrt{5}}$, take the conjugate of the denominator, which is $3 + \sqrt{5}$:

$$\frac{1}{(3 - \sqrt{5})} \cdot \frac{(3 + \sqrt{5})}{(3 + \sqrt{5})} = \frac{3 + \sqrt{5}}{3^2 \underbrace{+ 3\sqrt{5} - 3\sqrt{5}}_{\text{like terms}} - (\sqrt{5})^2} = \frac{3 + \sqrt{5}}{9 - 5} = \frac{3 + \sqrt{5}}{4}$$

In general we apply the difference of two squares:

$$(\sqrt{a} + \sqrt{b})(\sqrt{a} - \sqrt{b}) = (\sqrt{a})^2 - (\sqrt{b})^2 = a - b$$

Notice that the product does not contain a radical. Therefore, to simplify the expression

$$\frac{1}{(\sqrt{a} + \sqrt{b})}$$

multiply the numerator and denominator by $(\sqrt{a} - \sqrt{b})$:

$$\frac{1}{(\sqrt{a} + \sqrt{b})} \cdot \frac{(\sqrt{a} - \sqrt{b})}{(\sqrt{a} - \sqrt{b})}$$

The denominator now contains no radicals:

$$\frac{(\sqrt{a} - \sqrt{b})}{(a - b)}$$

EXAMPLE 6 Rationalizing Denominators

Rationalize the denominators and simplify.

a. $\frac{2}{3\sqrt{10}}$ **b.** $\frac{5}{3 - \sqrt{2}}$ **c.** $\frac{\sqrt{5}}{\sqrt{2} - \sqrt{7}}$

Solution (a):

Multiply the numerator and denominator by $\sqrt{10}$. $= \frac{2}{3\sqrt{10}} \cdot \frac{\sqrt{10}}{\sqrt{10}}$

Simplify. $= \frac{2\sqrt{10}}{3(\sqrt{10})^2} = \frac{2\sqrt{10}}{3(10)} = \frac{2\sqrt{10}}{30}$

Divide out the common 2 in the numerator and denominator. $= \boxed{\frac{\sqrt{10}}{15}}$

Solution (b):

Multiply the numerator and denominator by the conjugate, $3 + \sqrt{2}$.

$$= \frac{5}{(3-\sqrt{2})} \cdot \frac{(3+\sqrt{2})}{(3+\sqrt{2})}$$

$$= \frac{5(3+\sqrt{2})}{(3-\sqrt{2})(3+\sqrt{2})}$$

The denominator now contains no radicals.

$$= \frac{15 + 5\sqrt{2}}{9 - 2}$$

Simplify.

$$= \boxed{\frac{15 + 5\sqrt{2}}{7}}$$

Solution (c):

Multiply the numerator and denominator by the conjugate, $\sqrt{2} + \sqrt{7}$.

$$= \frac{\sqrt{5}}{(\sqrt{2}-\sqrt{7})} \cdot \frac{(\sqrt{2}+\sqrt{7})}{(\sqrt{2}+\sqrt{7})}$$

Multiply the numerators and denominators, respectively.

$$= \frac{\sqrt{5}(\sqrt{2}+\sqrt{7})}{(\sqrt{2}-\sqrt{7})(\sqrt{2}+\sqrt{7})}$$

The denominator now contains no radicals.

$$= \frac{\sqrt{10} + \sqrt{35}}{2 - 7}$$

Simplify.

$$= \boxed{-\frac{\sqrt{10} + \sqrt{35}}{5}}$$

■ **Answer:** $-\frac{7(1+\sqrt{3})}{2}$

■ **YOUR TURN** Write the expression $\frac{7}{1 - \sqrt{3}}$ in simplified form.

SIMPLIFIED FORM OF A RADICAL EXPRESSION

A radical expression is in **simplified form** if

- No factor in the radicand is raised to a power greater than or equal to the index.
- The power of the radicand does not share a common factor with the index.
- The denominator does not contain a radical.
- The radical does not contain a fraction.

EXAMPLE 7 Expressing a Radical Expression in Simplified Form

Express the radical expression in simplified form: $\sqrt[3]{\frac{16x^5}{81y^7}}$ $\quad x \geq 0, y > 0$

Solution:

Rewrite the expression so that the radical does not contain a fraction.

$$\sqrt[3]{\frac{16x^5}{81y^7}} = \frac{\sqrt[3]{16x^5}}{\sqrt[3]{81y^7}}$$

Let $16 = 2^4$ and $81 = 3^4$.

$$= \frac{\sqrt[3]{2^4 \cdot x^5}}{\sqrt[3]{3^4 \cdot y^7}}$$

Factors in both radicands are raised to powers greater than the index (3). Rewrite the expression so that each power in the radicand is less than the index.

$$= \frac{2x\sqrt[3]{2x^2}}{3y^2\sqrt[3]{3y}}$$

The denominator contains a radical. In order to eliminate the radical in the denominator, we multiply the numerator and denominator by $\sqrt[3]{9y^2}$.

$$= \frac{2x\sqrt[3]{2x^2}}{3y^2\sqrt[3]{3y}} \cdot \frac{\sqrt[3]{9y^2}}{\sqrt[3]{9y^2}}$$

$$= \frac{2x\sqrt[3]{18x^2y^2}}{3y^2\sqrt[3]{27y^3}}$$

$$= \frac{2x\sqrt[3]{18x^2y^2}}{9y^3}$$

The radical expression now satisfies the conditions for simplified form.

$$= \boxed{\frac{2x\sqrt[3]{18x^2y^2}}{9y^3}}$$

Rational Exponents

We now use radicals to define rational exponents.

RATIONAL EXPONENTS: $\frac{1}{n}$

Let a be any real number and n be a positive integer, then

$$a^{1/n} = \sqrt[n]{a}$$

where $\frac{1}{n}$ is the **rational exponent** of a.

- When n is even and a is negative, then $a^{1/n}$ and $\sqrt[n]{a}$ are not real numbers.
- Furthermore, if m is a positive integer with m and n having no common factors, then

$$a^{m/n} = \left(a^{1/n}\right)^m = \left(a^m\right)^{1/n} = \sqrt[n]{a^m}$$

Note: Any of the four notations can be used.

EXAMPLE 8 Simplifying Expressions with Rational Exponents

Simplify:

a. $16^{3/2}$ **b.** $(-8)^{2/3}$

Solution:

a. $16^{3/2} = \left(16^{1/2}\right)^3 = \left(\sqrt{16}\right)^3 = 4^3 = 64$

b. $(-8)^{2/3} = \left[(-8)^{1/3}\right]^2 = (-2)^2 = 4$

■ **YOUR TURN** Simplify $27^{2/3}$.

Technology Tip

```
16^(3/2)
                64
(-8)^(2/3)
                 4
```

■ **Answer:** 9

The properties of exponents that hold for integers also hold for rational numbers:

$$a^{-1/n} = \frac{1}{a^{1/n}} \quad \text{and} \quad a^{-m/n} = \frac{1}{a^{m/n}} \qquad a \neq 0$$

EXAMPLE 9 **Simplifying Expressions with Negative Rational Exponents**

Simplify $\dfrac{(9x)^{-1/2}}{4x^{-3/2}}$ $\quad x > 0$.

Solution:

Negative exponents correspond to positive exponents in the reciprocal. $\quad \dfrac{(9x)^{-1/2}}{4x^{-3/2}} = \dfrac{x^{3/2}}{4 \cdot (9x)^{1/2}}$

Eliminate the parentheses. $\quad = \dfrac{x^{3/2}}{4 \cdot 9^{1/2}x^{1/2}}$

Apply the quotient property on x. $\quad = \dfrac{x^{3/2-1/2}}{4 \cdot \underbrace{9^{1/2}}_{3}}$

Simplify. $\quad = \boxed{\dfrac{x}{12}}$

■ **Answer:** $18x^2$

■ **YOUR TURN** Simplify $\dfrac{9x^{3/2}}{(4x)^{-1/2}}$ $\quad x > 0$.

EXAMPLE 10 **Simplifying Algebraic Expressions with Rational Exponents**

Simplify $\dfrac{(-8x^2y)^{1/3}}{(9xy^4)^{1/2}}$ $\quad x > 0, y > 0$.

Solution:

$$\frac{(-8x^2y)^{1/3}}{(9xy^4)^{1/2}} = \frac{(-8)^{1/3}(x^2)^{1/3}y^{1/3}}{9^{1/2}x^{1/2}(y^4)^{1/2}} = \frac{(-8)^{1/3}x^{2/3}y^{1/3}}{9^{1/2}x^{1/2}y^2} = \left(\frac{-2}{3}\right)x^{2/3-1/2}y^{1/3-2} = -\frac{2}{3}x^{1/6}y^{-5/3}$$

Write in terms of positive exponents. $\quad = \boxed{-\dfrac{2x^{1/6}}{3y^{5/3}}}$

■ **Answer:** $\dfrac{4x^{5/6}}{3y^{1/2}}$

■ **YOUR TURN** Simplify $\dfrac{(16x^3y)^{1/2}}{(27x^2y^3)^{1/3}}$ and write your answer with only positive exponents.

EXAMPLE 11 **Factoring Expressions with Negative Rational Exponents**

Factor completely $x^{8/3} - 5x^{5/3} - 6x^{2/3}$.

Solution:

Factor out the greatest common factor $x^{2/3}$. $\quad \underbrace{x^{8/3}}_{x^{2/3}x^{6/3}} - \underbrace{5x^{5/3}}_{x^{2/3}x^{3/3}} - 6x^{2/3} = x^{2/3}(x^2 - 5x - 6)$

Factor the trinomial. $\quad = \boxed{x^{2/3}(x - 6)(x + 1)}$

■ **Answer:** $x^{1/3}(x - 2)(x + 1)$

■ **YOUR TURN** Factor completely $x^{7/3} - x^{4/3} - 2x^{1/3}$.

SECTION 0.6 SUMMARY

In this section, we defined radicals as "$b = \sqrt[n]{a}$ means $a = b^n$" for a and b positive real numbers when n is a positive even integer, and a and b any real numbers when n is a positive odd integer.

Properties of Radicals

PROPERTY		EXAMPLE
$\sqrt[n]{ab} = \sqrt[n]{a} \cdot \sqrt[n]{b}$		$\sqrt[3]{16} = \sqrt[3]{8} \cdot \sqrt[3]{2} = 2\sqrt[3]{2}$
$\sqrt[n]{\dfrac{a}{b}} = \dfrac{\sqrt[n]{a}}{\sqrt[n]{b}}$	$b \neq 0$	$\sqrt[4]{\dfrac{81}{16}} = \dfrac{\sqrt[4]{81}}{\sqrt[4]{16}} = \dfrac{3}{2}$
$\sqrt[n]{a^m} = \left(\sqrt[n]{a}\right)^m$		$\sqrt[3]{8^2} = \left(\sqrt[3]{8}\right)^2 = (2)^2 = 4$
$\sqrt[n]{a^n} = a$	n is odd	$\sqrt[3]{x^5} = \underbrace{\sqrt[3]{x^3}}_{x} \cdot \sqrt[3]{x^2} = x\sqrt[3]{x^2}$
$\sqrt[n]{a^n} = \lvert a \rvert$	n is even	$\sqrt[4]{x^6} = \sqrt[4]{x^4} \cdot \sqrt[4]{x^2} = \lvert x \rvert \sqrt[4]{x^2}$

Radicals can be combined only if they are like radicals (same radicand and index). Quotients with radicals in the denominator are usually rewritten with no radicals in the denominator.

Rational exponents were defined in terms of radicals: $a^{1/n} = \sqrt[n]{a}$. The properties for integer exponents we learned in Section 0.2 also hold true for rational exponents:

$$a^{m/n} = \left(a^{1/n}\right)^m = \left(\sqrt[n]{a}\right)^m \quad \text{and} \quad a^{m/n} = \left(a^m\right)^{1/n} = \sqrt[n]{a^m}$$

Negative rational exponents: $a^{-m/n} = \dfrac{1}{a^{m/n}}$, for m and n positive integers with no common factors.

SECTION 0.6 EXERCISES

SKILLS

In Exercises 1–24, evaluate each expression or state that it is not a real number.

1. $\sqrt{100}$
2. $\sqrt{121}$
3. $-\sqrt{144}$
4. $\sqrt{-169}$
5. $\sqrt[3]{-216}$
6. $\sqrt[3]{-125}$
7. $\sqrt[3]{343}$
8. $-\sqrt[3]{-27}$
9. $\sqrt[8]{1}$
10. $\sqrt[7]{-1}$
11. $\sqrt[3]{0}$
12. $\sqrt[5]{0}$
13. $\sqrt{-16}$
14. $\sqrt[5]{-1}$
15. $(-27)^{1/3}$
16. $(-64)^{1/3}$
17. $8^{2/3}$
18. $(-64)^{2/3}$
19. $(-32)^{1/5}$
20. $(-243)^{1/3}$
21. $(-1)^{1/3}$
22. $1^{5/2}$
23. $9^{3/2}$
24. $(27)^{2/3}$

In Exercises 25–40, simplify (if possible) the radical expressions.

25. $\sqrt{2} - 5\sqrt{2}$
26. $3\sqrt{5} - 7\sqrt{5}$
27. $3\sqrt{5} - 2\sqrt{5} + 7\sqrt{5}$
28. $6\sqrt{7} + 7\sqrt{7} - 10\sqrt{7}$
29. $\sqrt{12} \cdot \sqrt{2}$
30. $2\sqrt{5} \cdot 3\sqrt{40}$
31. $\sqrt[3]{12} \cdot \sqrt[3]{4}$
32. $\sqrt[4]{8} \cdot \sqrt[4]{4}$
33. $\sqrt{3}\sqrt{7}$
34. $\sqrt{5}\sqrt{2}$
35. $8\sqrt{25x^2}$
36. $16\sqrt{36y^4}$
37. $\sqrt{4x^2y}$
38. $\sqrt{16x^3y}$
39. $\sqrt[3]{-81x^6y^8}$
40. $\sqrt[5]{-32x^{10}y^8}$

In Exercises 41–56, rationalize the denominators.

41. $\sqrt{\dfrac{1}{3}}$
42. $\sqrt{\dfrac{2}{5}}$
43. $\dfrac{2}{3\sqrt{11}}$
44. $\dfrac{5}{3\sqrt{2}}$
45. $\dfrac{3}{1 - \sqrt{5}}$
46. $\dfrac{2}{1 + \sqrt{3}}$
47. $\dfrac{1 + \sqrt{2}}{1 - \sqrt{2}}$
48. $\dfrac{3 - \sqrt{5}}{3 + \sqrt{5}}$

49. $\dfrac{3}{\sqrt{2}-\sqrt{3}}$

50. $\dfrac{5}{\sqrt{2}+\sqrt{5}}$

51. $\dfrac{4}{3\sqrt{2}+2\sqrt{3}}$

52. $\dfrac{7}{2\sqrt{3}+3\sqrt{2}}$

53. $\dfrac{4+\sqrt{5}}{3+2\sqrt{5}}$

54. $\dfrac{6}{3\sqrt{2}+4}$

55. $\dfrac{\sqrt{7}+3}{\sqrt{2}-\sqrt{5}}$

56. $\dfrac{\sqrt{y}}{\sqrt{x}-\sqrt{y}}$

In Exercises 57–64, simplify by applying the properties of rational exponents. Express your answers in terms of positive exponents.

57. $\left(x^{1/2}y^{2/3}\right)^6$

58. $\left(y^{2/3}\cdot y^{1/4}\right)^{12}$

59. $\dfrac{\left(x^{1/3}y^{1/2}\right)^{-3}}{\left(x^{-1/2}y^{1/4}\right)^2}$

60. $\dfrac{\left(x^{-2/3}\cdot y^{-3/4}\right)^{-2}}{\left(x^{1/3}\cdot y^{1/4}\right)^4}$

61. $\dfrac{x^{1/2}y^{1/5}}{x^{-2/3}y^{-9/5}}$

62. $\dfrac{\left(y^{-3/4}\cdot x^{-2/3}\right)^{12}}{\left(y^{1/4}\cdot x^{7/3}\right)^{24}}$

63. $\dfrac{\left(2x^{2/3}\right)^3}{\left(4x^{-1/3}\right)^2}$

64. $\dfrac{\left(2x^{-2/3}\right)^3}{\left(4x^{-4/3}\right)^2}$

In Exercises 65–68, factor each expression completely.

65. $x^{7/3}-x^{4/3}-2x^{1/3}$

66. $8x^{1/4}+4x^{5/4}$

67. $7x^{3/7}-14x^{6/7}+21x^{10/7}$

68. $7x^{-1/3}+70x$

APPLICATIONS

69. **Gravity.** If a penny is dropped off a building, the time it takes (seconds) to fall d feet is given by $\sqrt{\dfrac{d}{16}}$. If a penny is dropped off a 1280-foot-tall building, how long will it take until it hits the ground? Round to the nearest second.

70. **Gravity.** If a ball is dropped off a building, the time it takes (seconds) to fall d meters is approximately given by $\sqrt{\dfrac{d}{5}}$. If a ball is dropped off a 600-meter-tall building, how long will it take until it hits the ground? Round to the nearest second.

71. **Kepler's Law.** The square of the period p (in years) of a planet's orbit around the sun is equal to the cube of the planet's maximum distance from the sun, d, (in astronomical units or AU). This relationship can be expressed mathematically as $p^2 = d^3$. If this formula is solved for d, the resulting equation is $d = p^{2/3}$. If Saturn has an orbital period of 29.46 Earth years, calculate Saturn's maximum distance from the sun to the nearest hundredth of an AU.

72. **Period of a Pendulum.** The period (in seconds) of a pendulum of length L (in meters) is given by $P = 2\cdot\pi\cdot\left(\frac{L}{9.8}\right)^{1/2}$. If a certain pendulum has a length of 19.6 meters, determine the period P of this pendulum to the nearest tenth of a second.

CATCH THE MISTAKE

In Exercises 73 and 74, explain the mistake that is made.

73. Simplify $\left(4x^{1/2}y^{1/4}\right)^2$

Solution:

Use properties of exponents. $4\left(x^{1/2}\right)^2\left(y^{1/4}\right)^2$

Simplify. $4xy^{1/2}$

This is incorrect. What mistake was made?

74. Simplify $\dfrac{2}{5-\sqrt{11}}$.

Solution:

Multiply numerator and denominator by $5-\sqrt{11}$. $\dfrac{2}{5-\sqrt{11}}\cdot\dfrac{(5-\sqrt{11})}{(5-\sqrt{11})}$

Multiply numerators and denominators. $\dfrac{2(5-\sqrt{11})}{25-11}$

Simplify. $\dfrac{2(5-\sqrt{11})}{14}=\dfrac{5-\sqrt{11}}{7}$

This is incorrect. What mistake was made?

CONCEPTUAL

In Exercises 75–78, determine whether each statement is true or false.

75. $\sqrt{121} = \pm 11$

76. $\sqrt{x^2} = x$, where x is any real number.

77. $\sqrt{a^2 + b^2} = \sqrt{a} + \sqrt{b}$

78. $\sqrt{-4} = -2$

In Exercises 79 and 80, *a*, *m*, *n*, and *k* are any positive real numbers.

79. Simplify $\left(\left(a^m\right)^n\right)^k$.

80. Simplify $\left(a^{-k}\right)^{-1/k}$.

In Exercises 81 and 82, evaluate each algebraic expression for the specified values.

81. $\dfrac{\sqrt{b^2 - 4ac}}{2a}$ for $a = 1, b = 7, c = 12$

82. $\sqrt{b^2 - 4ac}$ for $a = 1, b = 7, c = 12$

CHALLENGE

83. Rationalize the denominator and simplify: $\dfrac{1}{\left(\sqrt{a} + \sqrt{b}\right)^2}$.

84. Rationalize the denominator and simplify: $\dfrac{\sqrt{a+b} - \sqrt{a}}{\sqrt{a+b} + \sqrt{a}}$.

TECHNOLOGY

85. Use a calculator to approximate $\sqrt{11}$ to three decimal places.

86. Use a calculator to approximate $\sqrt[3]{7}$ to three decimal places.

87. Given $\dfrac{4}{5\sqrt{2} + 4\sqrt{3}}$

a. Rationalize the denominator.
b. Use a graphing utility to evaluate the expression and the answer.
c. Do they agree?

88. Given $\dfrac{2}{4\sqrt{5} - 3\sqrt{6}}$

a. Rationalize the denominator.
b. Use a graphing utility to evaluate the expression and the answer.
c. Do they agree?

SECTION 0.7 COMPLEX NUMBERS

SKILLS OBJECTIVES

- Write radicals with negative radicands as imaginary numbers.
- Add and subtract complex numbers.
- Multiply complex numbers.
- Divide complex numbers.
- Raise complex numbers to powers.

CONCEPTUAL OBJECTIVES

- Understand that real numbers and imaginary numbers are subsets of complex numbers.
- Understand how to eliminate imaginary numbers in denominators.

In Section 1.3, we will be studying equations whose solutions sometimes involve the square roots of negative numbers. In Section 0.6, when asked to evaluate the square root of a negative number, like $\sqrt{-16}$, we said "it is not a real number," because there is no real number such that $x^2 = -16$. To include such roots in the number system, mathematicians created a new expanded set of numbers, called the *complex numbers*. The foundation of this new set of numbers is the imaginary unit i.

DEFINITION **The Imaginary Unit i**

The **imaginary unit** is denoted by the letter i and is defined as

$$i = \sqrt{-1}$$

where $i^2 = -1$.

Recall that for positive real numbers a and b we defined the principal square root as

$$b = \sqrt{a} \quad \text{which means} \quad b^2 = a$$

Similarly, we define the *principal square root of a negative number* as $\sqrt{-a} = i\sqrt{a}$, since $\left(i\sqrt{a}\right)^2 = i^2 a = -a$.

If $-a$ is a negative real number, then the **principal square root** of $-a$ is

$$\sqrt{-a} = i\sqrt{a}$$

where i is the imaginary unit and $i^2 = -1$.

Technology Tip

Be sure to put the graphing calculator in $\boxed{a + bi}$ mode.

a. $\sqrt{-9}$ **b.** $\sqrt{-8}$

■ **Answer:** $12i$

We write $i\sqrt{a}$ instead of $\sqrt{a}\,i$ to avoid any confusion.

EXAMPLE 1 **Using Imaginary Numbers to Simplify Radicals**

Simplify using imaginary numbers.

a. $\sqrt{-9}$ **b.** $\sqrt{-8}$

Solution:

a. $\sqrt{-9} = i\sqrt{9} = \boxed{3i}$ **b.** $\sqrt{-8} = i\sqrt{8} = i \cdot 2\sqrt{2} = \boxed{2i\sqrt{2}}$

■ **YOUR TURN** Simplify $\sqrt{-144}$.

DEFINITION **Complex Number**

A **complex number** in standard form is defined as

$$a + bi$$

where a and b are real numbers and i is the imaginary unit. We denote a as the **real part** of the complex number and b as the **imaginary part** of the complex number.

A complex number written as $a + bi$ is said to be in **standard form**. If $a = 0$ and $b \neq 0$, then the resulting complex number bi is called a **pure imaginary number**. If $b = 0$, then $a + bi$ is a real number. The set of all real numbers and the set of all imaginary numbers are both subsets of the set of complex numbers.

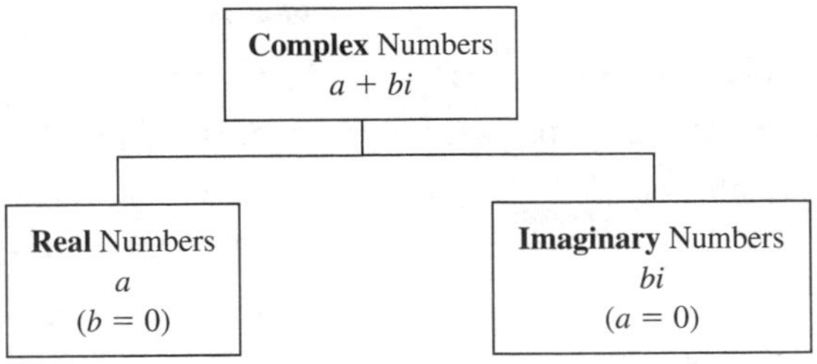

The following are examples of complex numbers.

$$17 \qquad 2 - 3i \qquad -5 + i \qquad 3 - i\sqrt{11} \qquad -9i$$

DEFINITION **Equality of Complex Numbers**

The complex numbers $a + bi$ and $c + di$ are **equal** if and only if $a = c$ and $b = d$. In other words, two complex numbers are equal if and only if both real parts are equal *and* both imaginary parts are equal.

Adding and Subtracting Complex Numbers

Complex numbers in the standard form $a + bi$ are treated in much the same way as binomials of the form $a + bx$. We can add, subtract, and multiply complex numbers the same way we performed these operations on binomials. When adding or subtracting complex numbers, combine real parts with real parts and combine imaginary parts with imaginary parts.

EXAMPLE 2 **Adding and Subtracting Complex Numbers**

Perform the indicated operation and simplify.

a. $(3 - 2i) + (-1 + i)$ **b.** $(2 - i) - (3 - 4i)$

Solution (a):

Eliminate the parentheses. $= 3 - 2i - 1 + i$

Group real and imaginary numbers, respectively. $= (3 - 1) + (-2i + i)$

Simplify. $= \boxed{2 - i}$

Solution (b):

Eliminate the parentheses (distribute the negative). $= 2 - i - 3 + 4i$

Group real and imaginary numbers, respectively. $= (2 - 3) + (-i + 4i)$

Simplify. $= \boxed{-1 + 3i}$

YOUR TURN Perform the indicated operation and simplify: $(4 + i) - (3 - 5i)$.

Technology Tip

Be sure to put the graphing calculator in $\boxed{a + bi}$ mode.

a. $(3 - 2i) + (-1 + i)$
b. $(2 - i) - (3 - 4i)$

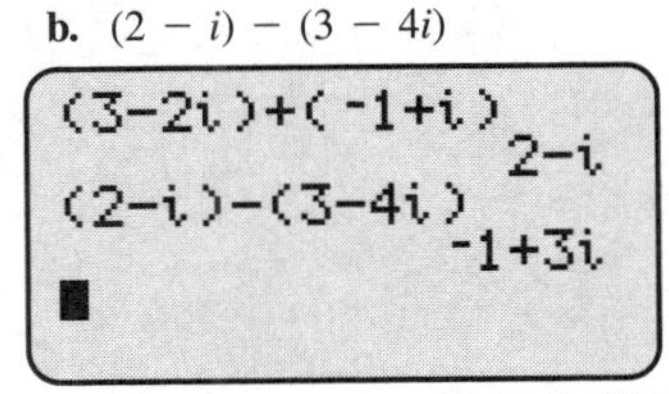

Answer: $1 + 6i$

Multiplying Complex Numbers

When multiplying complex numbers, you apply all the same methods as you did when multiplying binomials. It is important to remember that $i^2 = -1$.

Study Tip

When multiplying complex numbers, remember that $i^2 = -1$.

WORDS	MATH
Multiply the complex numbers.	$(5 - i)(3 - 4i)$
Multiply using the distributive property.	$= 5(3) + 5(-4i) - i(3) - (i)(-4i)$
Eliminate the parentheses.	$= 15 - 20i - 3i + 4i^2$
Let $i^2 = -1$.	$= 15 - 20i - 3i + 4(-1)$
Simplify.	$= 15 - 20i - 3i - 4$
Combine real parts and imaginary parts, respectively.	$= 11 - 23i$

Technology Tip

Be sure to put the graphing calculator in $\boxed{a + bi}$ mode.

a. $(3 - i)(2 + i)$
b. $i(-3 + i)$

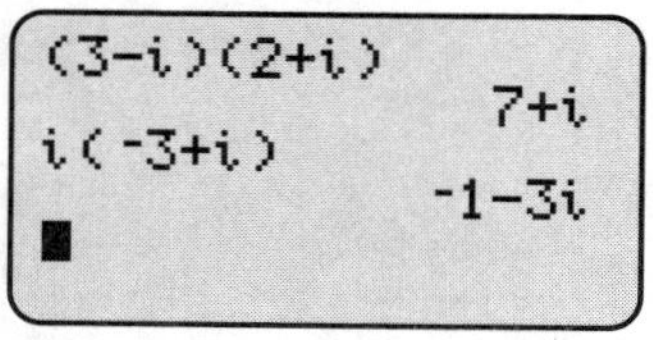

Answer: $2 + 11i$

EXAMPLE 3 Multiplying Complex Numbers

Multiply the complex numbers and express the result in standard form, $a \pm bi$.

a. $(3 - i)(2 + i)$ **b.** $i(-3 + i)$

Solution (a):

Use the distributive property.	$(3 - i)(2 + i) = 3(2) + 3(i) - i(2) - i(i)$
Eliminate the parentheses.	$= 6 + 3i - 2i - i^2$
Substitute $i^2 = -1$.	$= 6 + 3i - 2i - (-1)$
Group like terms.	$= (6 + 1) + (3i - 2i)$
Simplify.	$= \boxed{7 + i}$

Solution (b):

Use the distributive property.	$i(-3 + i) = -3i + i^2$
Substitute $i^2 = -1$.	$= -3i - 1$
Write in standard form.	$= \boxed{-1 - 3i}$

YOUR TURN Multiply the complex numbers and express the result in standard form, $a \pm bi$: $(4 - 3i)(-1 + 2i)$.

Dividing Complex Numbers

Recall the special product that produces a difference of two squares, $(a + b)(a - b) = a^2 - b^2$. This special product has only first and last terms because the outer and inner terms cancel each other out. Similarly, if we multiply complex numbers in the same manner, the result is a real number because the imaginary terms cancel each other out.

COMPLEX CONJUGATE

The product of a complex number, $z = a + bi$, and its **complex conjugate**, $\bar{z} = a - bi$, is a real number.

$$z\bar{z} = (a + bi)(a - bi) = a^2 - b^2i^2 = a^2 - b^2(-1) = a^2 + b^2$$

In order to write a quotient of complex numbers in standard form, $a + bi$, multiply the numerator and the denominator by the complex conjugate of the denominator. It is important to note that if i is present in the denominator, then the complex number is *not* in standard form.

EXAMPLE 4 Dividing Complex Numbers

Write the quotient in standard form: $\dfrac{2-i}{1+3i}$.

Solution:

Multiply numerator and denominator by the complex conjugate of the denominator, $1 - 3i$.

$$\left(\frac{2-i}{1+3i}\right)\left(\frac{1-3i}{1-3i}\right)$$

Multiply the numerators and denominators, respectively.

$$= \frac{(2-i)(1-3i)}{(1+3i)(1-3i)}$$

Use the FOIL method (or distributive property).

$$= \frac{2-6i-i+3i^2}{1-3i+3i-9i^2}$$

Combine imaginary parts.

$$= \frac{2-7i+3i^2}{1-9i^2}$$

Substitute $i^2 = -1$.

$$= \frac{2-7i-3}{1-9(-1)}$$

Simplify the numerator and denominator.

$$= \frac{-1-7i}{10}$$

Write in standard form. Recall that $\dfrac{a+b}{c} = \dfrac{a}{c} + \dfrac{b}{c}$.

$$= \boxed{-\frac{1}{10} - \frac{7}{10}i}$$

■ **YOUR TURN** Write the quotient in standard form: $\dfrac{3+2i}{4-i}$.

Technology Tip

Be sure to put the graphing calculator in [a + bi] mode.

$$\frac{2-i}{1+3i}$$

To change the answer to the fraction form, press [MATH], highlight [1: ▶Frac], press [ENTER] and [ENTER].

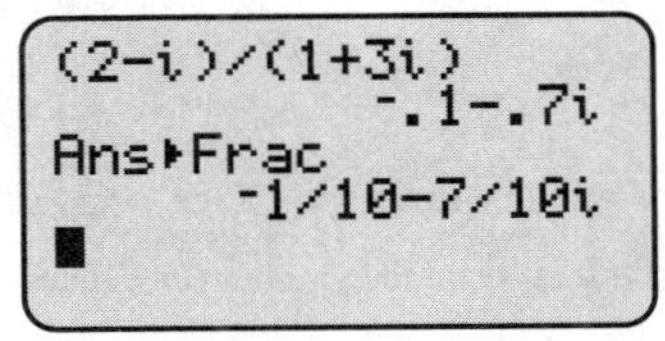

■ **Answer:** $\frac{10}{17} + \frac{11}{17}i$

Raising Complex Numbers to Integer Powers

Note that i raised to the fourth power is 1. In simplifying imaginary numbers, we factor out i raised to the largest multiple of 4.

$$\begin{aligned}
i &= \sqrt{-1} \\
i^2 &= -1 \\
i^3 &= i^2 \cdot i = (-1)i = -i \\
i^4 &= i^2 \cdot i^2 = (-1)(-1) = 1 \\
\hline
i^5 &= i^4 \cdot i = (1)(i) = i \\
i^6 &= i^4 \cdot i^2 = (1)(-1) = -1 \\
i^7 &= i^4 \cdot i^3 = (1)(-i) = -i \\
i^8 &= \left(i^4\right)^2 = 1
\end{aligned}$$

Technology Tip

```
i^7
             -3E-13-i
```

This is due to rounding off error in the programming. Since -3×10^{-13} can be approximated as 0, $i^7 = -i$.

```
i^13
             -3E-13+i
i^100
                    1
```

■ **Answer:** $-i$

EXAMPLE 5 Raising the Imaginary Unit to Integer Powers

Simplify:

a. i^7 **b.** i^{13} **c.** i^{100}

Solution:

a. $i^7 = i^4 \cdot i^3 = (1)(-i) = \boxed{-i}$

b. $i^{13} = i^{12} \cdot i = (i^4)^3 \cdot i = 1^3 \cdot i = \boxed{i}$

c. $i^{100} = (i^4)^{25} = 1^{25} = \boxed{1}$

■ **YOUR TURN** Simplify i^{27}.

Technology Tip

Be sure to put the graphing calculator in $\boxed{a + bi}$ mode.

$(2 - i)^3$

```
(2-i)^3
               2-11i
```

■ **Answer:** $2 + 11i$

EXAMPLE 6 Raising a Complex Number to an Integer Power

Write $(2 - i)^3$ in standard form.

Solution:

Recall the formula for cubing a binomial. $\quad (a - b)^3 = a^3 - 3a^2b + 3ab^2 - b^3$

Let $a = 2$ and $b = i$. $\quad (2 - i)^3 = 2^3 - 3(2)^2(i) + 3(2)(i)^2 - i^3$

Let $i^2 = -1$ and $i^3 = -i$. $\quad = 2^3 - 3(2)^2(i) + (3)(2)(-1) - (-i)$

Eliminate parentheses. $\quad = 8 - 6 - 12i + i$

Combine the real parts and imaginary parts, respectively. $\quad = \boxed{2 - 11i}$

■ **YOUR TURN** Write $(2 + i)^3$ in standard form.

SECTION 0.7 SUMMARY

The Imaginary Unit *i*

- $i = \sqrt{-1}$
- $i^2 = -1$

Complex Numbers

- *Standard Form*: $a + bi$, where a is the real part and b is the imaginary part.
- The set of real numbers and the set of pure imaginary numbers are subsets of the set of complex numbers.

Adding and Subtracting Complex Numbers

- $(a + bi) + (c + di) = (a + c) + (b + d)i$
- $(a + bi) - (c + di) = (a - c) + (b - d)i$
- To add or subtract complex numbers, add or subtract the real parts and imaginary parts, respectively.

Multiplying Complex Numbers

- $(a + bi)(c + di) = (ac - bd) + (ad + bc)i$
- Apply the same methods for multiplying binomials. It is important to remember that $i^2 = -1$.

Dividing Complex Numbers

- Complex conjugate of $a + bi$ is $a - bi$.
- In order to write a quotient of complex numbers in standard form, multiply the numerator and the denominator by the complex conjugate of the denominator:

$$\frac{a + bi}{c + di} \cdot \frac{(c - di)}{(c - di)}$$

SECTION 0.7 EXERCISES

SKILLS

In Exercises 1–12, write each expression as a complex number in standard form. Some expressions simplify to either a real number or a pure imaginary number.

1. $\sqrt{-16}$ **2.** $\sqrt{-100}$ **3.** $\sqrt{-20}$ **4.** $\sqrt{-24}$

5. $\sqrt[3]{-64}$ **6.** $\sqrt[3]{-27}$ **7.** $\sqrt{-64}$ **8.** $\sqrt{-27}$

9. $3 - \sqrt{-100}$ **10.** $4 - \sqrt{-121}$ **11.** $-10 - \sqrt{-144}$ **12.** $7 - \sqrt[3]{-125}$

In Exercises 13–40, perform the indicated operation, simplify, and express in standard form.

13. $(3 - 7i) + (-1 - 2i)$ **14.** $(1 + i) + (9 - 3i)$ **15.** $(3 - 4i) + (7 - 10i)$ **16.** $(5 + 7i) + (-10 - 2i)$

17. $(4 - 5i) - (2 - 3i)$ **18.** $(-2 + i) - (1 - i)$ **19.** $(-3 + i) - (-2 - i)$ **20.** $(4 + 7i) - (5 + 3i)$

21. $3(4 - 2i)$ **22.** $4(7 - 6i)$ **23.** $12(8 - 5i)$ **24.** $-3(16 + 4i)$

25. $-3(16 - 9i)$ **26.** $5(-6i + 3)$ **27.** $-6(17 - 5i)$ **28.** $-12(8 + 3i)$

29. $(1 - i)(3 + 2i)$ **30.** $(-3 + 2i)(1 - 3i)$ **31.** $(5 - 7i)(-3 + 4i)$ **32.** $(16 - 5i)(-2 - i)$

33. $(7 - 5i)(6 + 9i)$ **34.** $(-3 - 2i)(7 - 4i)$ **35.** $(12 - 18i)(-2 + i)$ **36.** $(-4 + 3i)(-4 - 3i)$

37. $\left(\frac{1}{2} + 2i\right)\left(\frac{4}{9} - 3i\right)$ **38.** $\left(-\frac{3}{4} + \frac{9}{16}i\right)\left(\frac{2}{3} + \frac{4}{9}i\right)$ **39.** $(-i + 17)(2 + 3i)$ **40.** $(-3i - 2)(-2 - 3i)$

For Exercises 41–48, for each complex number z, write the complex conjugate $\bar{z}$ and find $z\bar{z}$.

41. $z = 4 + 7i$ **42.** $z = 2 + 5i$ **43.** $z = 2 - 3i$ **44.** $z = 5 - 3i$

45. $z = 6 + 4i$ **46.** $z = -2 + 7i$ **47.** $z = -2 - 6i$ **48.** $z = -3 - 9i$

For Exercises 49–64, write each quotient in standard form.

49. $\frac{2}{i}$ **50.** $\frac{3}{i}$ **51.** $\frac{1}{3 - i}$ **52.** $\frac{2}{7 - i}$ **53.** $\frac{1}{3 + 2i}$ **54.** $\frac{1}{4 - 3i}$ **55.** $\frac{2}{7 + 2i}$ **56.** $\frac{8}{1 + 6i}$

57. $\frac{1 - i}{1 + i}$ **58.** $\frac{3 - i}{3 + i}$ **59.** $\frac{2 + 3i}{3 - 5i}$ **60.** $\frac{2 + i}{3 - i}$ **61.** $\frac{4 - 5i}{7 + 2i}$ **62.** $\frac{7 + 4i}{9 - 3i}$ **63.** $\frac{8 + 3i}{9 - 2i}$ **64.** $\frac{10 - i}{12 + 5i}$

For Exercises 65–76, simplify.

65. i^{15} **66.** i^{99} **67.** i^{40} **68.** i^{18} **69.** $(5 - 2i)^2$ **70.** $(3 - 5i)^2$

71. $(2 + 3i)^2$ **72.** $(4 - 9i)^2$ **73.** $(3 + i)^3$ **74.** $(2 + i)^3$ **75.** $(1 - i)^3$ **76.** $(4 - 3i)^3$

APPLICATIONS

Electrical impedance is the ratio of voltage to current in ac circuits. Let Z represent the total impedance of an electrical circuit. If there are two resistors in a circuit, let $Z_1 = 3 - 6i$ ohms and $Z_2 = 5 + 4i$.

77. Electrical Circuits in Series. When the resistors in the circuit are placed in series, the total impedance is the sum of the two impedances $Z = Z_1 + Z_2$. Find the total impedance of the electrical circuit in series.

78. Electrical Circuits in Parallel. When the resistors in the circuit are placed in parallel, the total impedance is given by $\frac{1}{Z} = \frac{1}{Z_1} + \frac{1}{Z_2}$. Find the total impedance of the electrical circuit in parallel.

CATCH THE MISTAKE

In Exercises 79 and 80, explain the mistake that is made.

79. Write the quotient in standard form: $\frac{2}{4 - i}$.

Solution:

Multiply the numerator and the denominator by $4 - i$. $\quad \frac{2}{4 - i} \cdot \frac{4 - i}{4 - i}$

Multiply the numerator using the distributive property and the denominator using the FOIL method. $\quad \frac{8 - 2i}{16 - 1}$

Simplify. $\quad \frac{8 - 2i}{15}$

Write in standard form. $\quad \frac{8}{15} - \frac{2}{15}i$

This is incorrect. What mistake was made?

80. Write the product in standard form: $(2 - 3i)(5 + 4i)$.

Solution:

Use the FOIL method to multiply the complex numbers. $\quad 10 - 7i - 12i^2$

Simplify. $\quad -2 - 7i$

This is incorrect. What mistake was made?

CONCEPTUAL

In Exercises 81–84, determine whether each statement is true or false.

81. The product is a real number: $(a + bi)(a - bi)$.

82. Imaginary numbers are a subset of the complex numbers.

83. Real numbers are a subset of the complex numbers.

84. There is no complex number that equals its conjugate.

CHALLENGE

85. Factor completely over the complex numbers: $x^4 + 2x^2 + 1$.

86. Factor completely over the complex numbers: $x^4 + 18x^2 + 81$.

TECHNOLOGY

In Exercises 87–90, apply a graphing utility to simplify the expression. Write your answer in standard form.

87. $(1 + 2i)^5$ **88.** $(3 - i)^6$

89. $\frac{1}{(2 - i)^3}$ **90.** $\frac{1}{(4 + 3i)^2}$

CHAPTER 0 REVIEW

Section	Concept	Pages	Review Exercises	Key Ideas/Points
0.1	**Real numbers**	4–14	1–10	
	The set of real numbers	4–6	1–10	*Rational:* $\frac{a}{b}$, where a and b are integers or a decimal that terminates or repeats. *Irrational:* nonrepeating/nonterminating decimal.
	Approximations: Rounding and truncation	6–7	1–2	*Rounding:* Examine the digit to the right of the last desired digit. Digit < 5: keep last desired digit as is. Digit ≥ 5: round the last desired digit up 1. *Truncating:* Eliminate all digits to the right of the desired digit.
	Order of operations	7–9	3–6	1. Parentheses 2. Multiplication and Division 3. Addition and Subtraction
	Properties of real numbers	9–14	3–10	■ $a(b + c) = ab + ac$ ■ If $xy = 0$, then $x = 0$ or $y = 0$ ■ $\frac{a}{b} \pm \frac{c}{d} = \frac{ad \pm bc}{bd}$ $\quad b \neq 0$ and $d \neq 0$ ■ $\frac{a}{b} \div \frac{c}{d} = \frac{a}{b} \cdot \frac{d}{c}$ $\quad b \neq 0, c \neq 0,$ and $d \neq 0$
0.2	**Integer exponents and scientific notation**	18–24	11–16	$a^n = \underbrace{a \cdot a \cdot a \cdots a}_{n \text{ factors}}$
	Integer exponents	18–22	11–14	$a^m \cdot a^n = a^{m+n}$ $\quad \frac{a^m}{a^n} = a^{m-n}$ $\quad a^0 = 1$ $(a^m)^n = a^{mn}$ $\quad a^{-n} = \frac{1}{a^n} = \left(\frac{1}{a}\right)^n$ $\quad a \neq 0$
	Scientific notation	22–24	15–16	$c \times 10^n$ where c is a positive real number $1 \leq c < 10$ and n is an integer.
0.3	**Polynomials: Basic operations**	27–34	17–28	
	Adding and subtracting polynomials	29	17–20	Combine like terms.
	Multiplying polynomials	30	21–28	Distributive Property
	Special products	31–34	25–28	$(x + a)(x + b) = x^2 + (ax + bx) + ab$ *Perfect Squares* $(a + b)^2 = (a + b)(a + b) = a^2 + 2ab + b^2$ $(a - b)^2 = (a - b)(a - b) = a^2 - 2ab + b^2$ *Difference of Two Squares* $(a + b)(a - b) = a^2 - b^2$ *Perfect Cubes* $(a + b)^3 = a^3 + 3a^2b + 3ab^2 + b^3$ $(a - b)^3 = a^3 - 3a^2b + 3ab^2 - b^3$

Section	Concept	Pages	Review Exercises	Key Ideas/Points
0.4	**Factoring polynomials**	37–45	29–40	
	Greatest common factor	38–39	29–30	Factor out using Distributive Property: ax^k
	Factoring formulas: Special polynomial forms	39–41	35–36	*Difference of Two Squares* $a^2 - b^2 = (a + b)(a - b)$ *Perfect Squares* $a^2 + 2ab + b^2 = (a + b)^2$ $a^2 - 2ab + b^2 = (a - b)^2$ *Sum of Two Cubes* $a^3 + b^3 = (a + b)(a^2 - ab + b^2)$ *Difference of Two Cubes* $a^3 - b^3 = (a - b)(a^2 + ab + b^2)$
	Factoring a trinomial as a product of two binomials	41–44	31–34 37–38	■ $x^2 + bx + c = (x + ?)(x + ?)$ ■ $ax^2 + bx + c = (?x + ?)(?x + ?)$
	Factoring by grouping	44–45	39–40	Group terms with common factors.
0.5	**Rational expressions**	48–58	41–52	
	Rational expressions and domain restrictions	48–50	41–42	Note domain restrictions when denominator is equal to zero.
	Simplifying rational expressions	50–52	43–46	
	Multiplying and dividing rational expressions	52–53	47–48	■ Use properties of rational numbers. ■ State additional domain restrictions once division is rewritten as multiplication of a reciprocal.
	Adding and subtracting rational expressions	53–56	49–50	Least common denominator (LCD)
	Complex rational expressions	56–58	51–52	Two strategies: 1. Write sum/difference in numerator/ denominator as a rational expression. 2. Multiply by the LCD of the numerator and denominator.
0.6	**Rational exponents and radicals**	63–70	53–56	
	Square roots	63–64	53–60	$\sqrt{25} = 5$
	Other (nth) roots	65–69	63–66	$b = \sqrt[n]{a}$ means $a = b^n$ for a and b positive real numbers and n a positive even integer, or for a and b any real numbers and n a positive odd integer. $\sqrt[n]{ab} = \sqrt[n]{a} \cdot \sqrt[n]{b}$ $\quad \sqrt[n]{\frac{a}{b}} = \frac{\sqrt[n]{a}}{\sqrt[n]{b}} \quad b \neq 0$ $\sqrt[n]{a^m} = \left(\sqrt[n]{a}\right)^m$ $\sqrt[n]{a^n} = a \quad n$ is odd $\sqrt[n]{a^n} = \lvert a \rvert \quad n$ is even

CHAPTER REVIEW

SECTION	CONCEPT	PAGES	REVIEW EXERCISES	KEY IDEAS/POINTS
	Rational exponents	69–70	63–66	$a^{1/n} = \sqrt[n]{a}$ $a^{m/n} = \left(a^{1/n}\right)^m = \left(\sqrt[n]{a}\right)^m$ $a^{-m/n} = \dfrac{1}{a^{m/n}}$ for m and n positive integers with no common factors $a \neq 0$.
0.7	**Complex numbers**	73–78	67–80	$a + bi$; where a and b are real numbers.
	Adding and subtracting complex numbers	75	71–74	Combine real parts with real parts and imaginary parts with imaginary parts.
	Multiplying complex numbers	75–76	71–74	Use the FOIL method and $i^2 = -1$ to simplify.
	Dividing complex numbers	76–77	75–80	If $a + bi$ is in the denominator, then multiply the numerator and the denominator by $a - bi$. The result is a real number in the denominator.
	Raising complex numbers to integer powers	77–78	69–70	$i = \sqrt{-1}$ $\quad i^2 = -1$ $\quad i^3 = -i$ $\quad i^4 = 1$

0.1 Real Numbers

Approximate to two decimal places by (a) rounding and (b) truncating.

1. 5.21597
2. 7.3623

Simplify.

3. $7 - 2 \cdot 5 + 4 \cdot 3 - 5$
4. $-2(5 + 3) + 7(3 - 2 \cdot 5)$
5. $-\dfrac{16}{(-2)(-4)}$
6. $-3(x - y) + 4(3x - 2y)$

Perform the indicated operation and simplify.

7. $\dfrac{x}{4} - \dfrac{x}{3}$
8. $\dfrac{y}{3} + \dfrac{y}{5} - \dfrac{y}{6}$
9. $\dfrac{12}{7} \cdot \dfrac{21}{4}$
10. $\dfrac{a^2}{b^3} \div \dfrac{2a}{b^2}$

0.2 Integer Exponents and Scientific Notation

Simplify using properties of exponents.

11. $(-2z)^3$
12. $(-4z^2)^3$
13. $\dfrac{(3x^3y^2)^2}{2(x^2y)^4}$
14. $\dfrac{(2x^2y^3)^2}{(4xy)^3}$
15. Express 0.00000215 in scientific notation.
16. Express 7.2×10^9 as a real number.

0.3 Polynomials: Basic Operations

Perform the indicated operation and write the results in standard form.

17. $(14z^2 + 2) + (3z - 4)$
18. $(27y^2 - 6y + 2) - (y^2 + 3y - 7)$
19. $(36x^2 - 4x - 5) - (6x - 9x^2 + 10)$
20. $[2x - (4x^2 - 7x)] - [3x - (2x^2 + 5x - 4)]$
21. $5xy^2(3x - 4y)$
22. $-2st^2(-t + s - 2st)$
23. $(x - 7)(x + 9)$
24. $(2x + 1)(3x - 2)$
25. $(2x - 3)^2$
26. $(5x - 7)(5x + 7)$
27. $(x^2 + 1)^2$
28. $(1 - x^2)^2$

0.4 Factoring Polynomials

Factor out the common factor.

29. $14x^2y^2 - 10xy^3$
30. $30x^4 - 20x^3 + 10x^2$

Factor the trinomial into a product of two binomials.

31. $2x^2 + 9x - 5$
32. $6x^2 - 19x - 7$
33. $16x^2 - 25$
34. $9x^2 - 30x + 25$

Factor the sum or difference of two cubes.

35. $x^3 + 125$
36. $1 - 8x^3$

Factor into a product of three polynomials.

37. $2x^3 + 4x^2 - 30x$
38. $6x^3 - 5x^2 + x$

Factor into a product of two binomials by grouping.

39. $x^3 + x^2 - 2x - 2$
40. $2x^3 - x^2 + 6x - 3$

0.5 Rational Expressions

State the domain restrictions on each of the rational expressions.

41. $\dfrac{4x^2 - 3}{x^2 - 9}$
42. $\dfrac{1}{x^2 + 1}$

Simplify.

43. $\dfrac{x^2 - 4}{x - 2}$
44. $\dfrac{x - 5}{x - 5}$
45. $\dfrac{t^2 + t - 6}{t^2 - t - 2}$
46. $\dfrac{z^3 - z}{z^2 + z}$

Perform the indicated operation and simplify.

47. $\dfrac{x^2 + 3x - 10}{x^2 + 2x - 3} \cdot \dfrac{x^2 + x - 2}{x^2 + x - 6}$
48. $\dfrac{x^2 - x - 2}{x^3 + 3x^2} \div \dfrac{x + 1}{x^2 + 2x}$
49. $\dfrac{1}{x + 1} - \dfrac{1}{x + 3}$
50. $\dfrac{1}{x} - \dfrac{1}{x + 1} + \dfrac{1}{x + 2}$

Simplify.

51. $\dfrac{2 + \dfrac{1}{x - 3}}{\dfrac{1}{5x - 15} + 4}$
52. $\dfrac{\dfrac{1}{x} + \dfrac{2}{x^2}}{3 - \dfrac{1}{x^2}}$

0.6 Rational Exponents and Radicals

Simplify.

53. $\sqrt{20}$
54. $\sqrt{80}$
55. $\sqrt[3]{-125x^5y^4}$
56. $\sqrt[4]{32x^4y^5}$
57. $3\sqrt{20} + 5\sqrt{80}$
58. $4\sqrt{27x} - 8\sqrt{12x}$
59. $(2 + \sqrt{5})(1 - \sqrt{5})$
60. $(3 + \sqrt{x})(4 - \sqrt{x})$
61. $\dfrac{1}{2 - \sqrt{3}}$
62. $\dfrac{1}{3 - \sqrt{x}}$

REVIEW EXERCISES

63. $\dfrac{(3x^{2/3})^2}{(4x^{1/3})^2}$

64. $\dfrac{(4x^{3/4})^2}{(2x^{-1/3})^2}$

65. $\dfrac{5^{1/2}}{5^{1/3}}$

66. $(x^{-2/3}y^{1/4})^{12}$

0.7 Complex Numbers

Simplify.

67. $\sqrt{-169}$

68. $\sqrt{-32}$

69. i^{19}

70. i^9

Express the product in standard form.

71. $(2 + 2i)(3 - 3i)$

72. $(1 + 6i)(1 + 5i)$

73. $(4 + 7i)^2$

74. $(7 - i)^2$

Express the quotient in standard form.

75. $\dfrac{1}{2 - i}$

76. $\dfrac{1}{3 + i}$

77. $\dfrac{7 + 2i}{4 + 5i}$

78. $\dfrac{6 - 5i}{3 - 2i}$

79. $\dfrac{10}{3i}$

80. $\dfrac{7}{2i}$

Technology

Section 0.1

81. Use your calculator to evaluate $\sqrt{272.25}$. Does the answer appear to be a rational or an irrational number? Why?

82. Use your calculator to evaluate $\sqrt{\dfrac{1053}{81}}$. Does the answer appear to be a rational or an irrational number? Why?

Section 0.2

Use a graphing utility to evaluate the expression. Express your answer in scientific notation.

83. $\dfrac{(8.2 \times 10^{11})(1.167 \times 10^{-35})}{(4.92 \times 10^{-18})}$

84. $\dfrac{(1.4805 \times 10^{21})}{(5.64 \times 10^{26})(1.68 \times 10^{-9})}$

Section 0.3

85. Use a graphing utility to plot the graphs of the three expressions $(2x + 3)^3$, $8x^3 + 27$, and $8x^3 + 36x^2 + 54x + 27$. Which two graphs agree with each other?

86. Use a graphing utility to plot the graphs of the three expressions $(x - 3)^2$, $8x^2 + 9$, and $x^2 - 6x + 9$. Which two graphs agree with each other?

Section 0.4

87. Use a graphing utility to plot the graphs of the three expressions $x^2 - 3x + 18$, $(x + 6)(x - 3)$, and $(x - 6)(x + 3)$. Which two graphs agree with each other?

88. Use a graphing utility to plot the graphs of the three expressions $x^2 - 8x + 16$, $(x + 4)^2$, and $(x - 4)^2$. Which two graphs agree with each other?

Section 0.5

For each given expression: (a) simplify the expression, (b) use a graphing utility to plot the expression and the answer in (a) in the same viewing window, and (c) determine the domain restriction(s) where the graphs will agree with each other.

89. $\dfrac{1 - 4/x}{1 - 4/x^2}$

90. $\dfrac{1 - 3/x}{1 + 9/x^2}$

Section 0.6

91. Given $\dfrac{6}{\sqrt{5} - \sqrt{2}}$

a. Rationalize the denominator.
b. Use a graphing utility to evaluate the expression and the answer.
c. Do they agree?

92. Given $\dfrac{11}{2\sqrt{6} + \sqrt{13}}$

a. Rationalize the denominator.
b. Use a graphing utility to evaluate the expression and the answer.
c. Do they agree?

Section 0.7

In Exercises 93 and 94, apply a graphing utility to simplify the expression. Write your answer in standard form.

93. $(3 + 5i)^5$

94. $\dfrac{1}{(1 + 3i)^4}$

95. Use a graphing utility to evaluate the expression. Express your answer in scientific notation.

$$\frac{(5.6 \times 10^{-12})(1.2 \times 10^{21})}{(4.2 \times 10^{-5})}$$

96. Apply a graphing utility to simplify the expression and write your answer in standard form.

$$\frac{1}{(6 + 2i)^4}$$

Simplify.

1. $\sqrt{16}$
2. $\sqrt[3]{54x^6}$
3. $-3(2 + 5^2) + 2(3 - 7) - (3^2 - 1)$
4. $\sqrt[5]{-32}$
5. $\sqrt{-12x^2}$
6. i^{17}
7. $\dfrac{(x^2y^{-3}z^{-1})^{-2}}{(x^{-1}y^2z^3)^{1/2}}$
8. $3\sqrt{x} - 4\sqrt{x} + 5\sqrt{x}$
9. $3\sqrt{18} - 4\sqrt{32}$
10. $(5\sqrt{6} - 2\sqrt{2})(\sqrt{6} + 3\sqrt{2})$

Perform the indicated operation and simplify.

11. $(3y^2 - 5y + 7) - (y^2 + 7y - 13)$
12. $(2x - 3)(5x + 7)$

Factor.

13. $x^2 - 16$
14. $3x^2 + 15x + 18$
15. $4x^2 + 12xy + 9y^2$
16. $x^4 - 2x^2 + 1$
17. $2x^2 - x - 1$
18. $6y^2 - y - 1$
19. $2t^3 - t^2 - 3t$
20. $2x^3 - 5x^2 - 3x$
21. $x^2 - 3yx + 4yx - 12y^2$
22. $x^4 + 5x^2 - 3x^2 - 15$
23. $81 + 3x^3$
24. $27x - x^4$

Perform the indicated operations and simplify.

25. $\dfrac{2}{x} + \dfrac{3}{x - 1}$
26. $\dfrac{5x}{x^2 - 7x + 10} - \dfrac{4}{x^2 - 25}$
27. $\dfrac{x - 1}{x^2 - 1} \cdot \dfrac{x^2 + x + 1}{x^3 - 1}$
28. $\dfrac{4x^2 - 9}{x^2 - 11x - 60} \cdot \dfrac{x^2 - 16}{2x + 3}$
29. $\dfrac{x - 3}{2x - 5} \div \dfrac{x^2 - 9}{5 - 2x}$
30. $\dfrac{1 - t}{3t + 1} \div \dfrac{t^2 - 2t + 1}{7t + 21t^2}$

Write the resulting expression in standard form.

31. $(1 - 3i)(7 - 5i)$
32. $\dfrac{2 - 11i}{4 + i}$
33. Rationalize the denominator: $\dfrac{7 - 2\sqrt{3}}{4 - 5\sqrt{3}}$.
34. Represent 0.0000155 in scientific notation.
35. Simplify $\dfrac{\frac{1}{x} - \frac{2}{x + 1}}{x - 1}$ and state any domain restrictions.
36. For the given expression:

$$\frac{1 + \frac{5}{x}}{1 - \frac{25}{x^2}}$$

   a. Simplify the expression.
   b. Use a graphing utility to plot the expression and the answer in (a) in the same viewing window.
   c. Determine the domain restriction(s) where the graphs will agree with each other.

37. Apply a graphing utility to evaluate the expression. Round your answer to three decimal places.

$$\frac{\sqrt{5}}{\sqrt{13} - \sqrt{7}}$$

1

Equations and Inequalities

RPM Pictures/Getty Images

Golf courses usually charge both greens fees (cost of playing the course) and cart fees (cost of renting a golf cart). Two friends who enjoy playing golf decide to investigate becoming members at a golf course.

The course they enjoy playing the most charges \$40 for greens fees and \$15 for cart rental (per person), so it currently costs each of them \$55 every time they play. The membership offered at that course costs \$160 per month with no greens fees, but there is still the per person cart rental fee.

How many times a month would they have to play golf in order for the membership option to be the better deal?* This is just one example of how the real world can be modeled with equations and inequalities.

---

*See Section 1.5, Exercise 101.

IN THIS CHAPTER you will solve linear and quadratic equations. You will then solve more complicated equations (polynomial, rational, radical, and absolute value) by first transforming them into linear or quadratic equations. Then you will solve linear, quadratic, polynomial, rational, and absolute value inequalities. Throughout this chapter you will solve applications of equations and inequalities.

EQUATIONS AND INEQUALITIES

EQUATIONS

- Linear
- Quadratic
- Radical
- Quadratic in form
- Polynomial
- Rational
- Absolute value

APPLICATIONS

- **Equations**
  - Linear
  - Quadratic
  - Absolute value
- **Inequalities**
  - Linear
  - Polynomial
  - Rational
  - Absolute value

INEQUALITIES

- Linear
- Polynomial
- Rational
- Absolute value

SECTION

CHAPTER OBJECTIVES

- Classify equations as linear, quadratic, polynomial, rational, radical, quadratic in form, or absolute value.
- Solve linear equations.
- Solve quadratic equations.
- Solve rational equations.
- Solve polynomial equations.
- Solve radical equations.
- Solve equations involving rational exponents.
- Solve equations that are quadratic in form.
- Solve absolute value equations.
- Eliminate extraneous solutions.
- Eliminate solutions that are not possible values of the independent variable.
- Solve linear, quadratic, polynomial, rational, and absolute value inequalities.
- Solve application problems involving equations and inequalities.

SECTION
1.1 LINEAR EQUATIONS

SKILLS OBJECTIVES

- Solve linear equations in one variable.
- Solve rational equations that are reducible to linear equations.

CONCEPTUAL OBJECTIVE

- Eliminate values that result in a denominator being equal to zero.

An **algebraic expression** (see Chapter 0) consists of one or more terms that are combined through basic operations such as addition, subtraction, multiplication, or division; for example:

$$3x + 2 \qquad 5 - 2y \qquad x + y$$

An **equation** is a statement that says two expressions are equal. For example, the following are all equations in one variable, x:

$$x + 7 = 11 \qquad x^2 = 9 \qquad 7 - 3x = 2 - 3x \qquad 4x + 7 = x + 2 + 3x + 5$$

To **solve** an equation means to find all the values of x that make the equation true. These values are called **solutions**, or **roots**, of the equation. The first of these statements shown above, $x + 7 = 11$, is true when $x = 4$ and false for any other values of x. We say that $x = 4$ is the solution to the equation. Sometimes an equation can have more than one solution, as in $x^2 = 9$. In this case, there are actually two values of x that make this equation true, $x = -3$ and $x = 3$. We say the **solution set** of this equation is $\{-3, 3\}$. In the third equation, $7 - 3x = 2 - 3x$, no values of x make the statement true. Therefore, we say this equation has **no solution**. And the fourth equation, $4x + 7 = x + 2 + 3x + 5$, is true for any values of x. An equation that is true for any value of the variable x is called an **identity**. In this case, we say the solution set is the **set of all real numbers**.

Two equations that have the same solution set are called **equivalent equations**. For example,

$$3x + 7 = 13 \qquad 3x = 6 \qquad x = 2$$

are all equivalent equations because each of them has the solution set $\{2\}$. Note that $x^2 = 4$ is not equivalent to these three equations because it has the solution set $\{-2, 2\}$.

When solving equations it helps to find a simpler equivalent equation in which the variable is isolated (alone). The following table summarizes the procedures for generating equivalent equations.

Generating Equivalent Equations

ORIGINAL EQUATION	DESCRIPTION	EQUIVALENT EQUATION
$3(x - 6) = 6x - x$	■ Eliminate parentheses. ■ Combine like terms on one or both sides of an equation.	$3x - 18 = 5x$
$7x + 8 = 29$	Add (or subtract) the same quantity to (from) *both* sides of an equation. $7x + 8 - \mathbf{8} = 29 - \mathbf{8}$	$7x = 21$
$5x = 15$	Multiply (or divide) both sides of an equation by the same nonzero quantity: $\frac{5x}{5} = \frac{15}{5}$.	$x = 3$
$-7 = x$	Interchange the two sides of the equation.	$x = -7$

Solving Linear Equations in One Variable

You probably already know how to solve simple linear equations. Solving a linear equation in one variable is done by finding an equivalent equation. In generating an equivalent equation, remember that whatever operation is performed on one side of an equation must also be performed on the other side of the equation.

EXAMPLE 1 Solving a Linear Equation

Solve the equation $3x + 4 = 16$.

Solution:

Subtract 4 from both sides of the equation.

$$\begin{array}{rcl} 3x + 4 &=& 16 \\ -4 && -4 \\ \hline 3x &=& 12 \end{array}$$

Divide both sides by 3.

$$\frac{3x}{3} = \frac{12}{3}$$

The solution is $x = 4$.

$$\boxed{x = 4}$$

The solution set is $\{4\}$.

■ **YOUR TURN** Solve the equation $2x + 3 = 9$.

Technology Tip

Use a graphing utility to display graphs of $y_1 = 3x + 4$ and $y_2 = 16$.

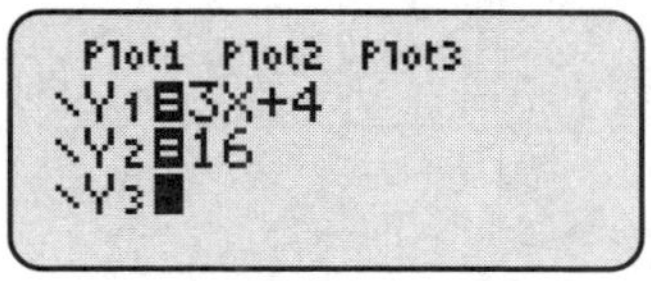

The x-coordinate of the point of intersection is the solution to the equation $3x + 4 = 16$.

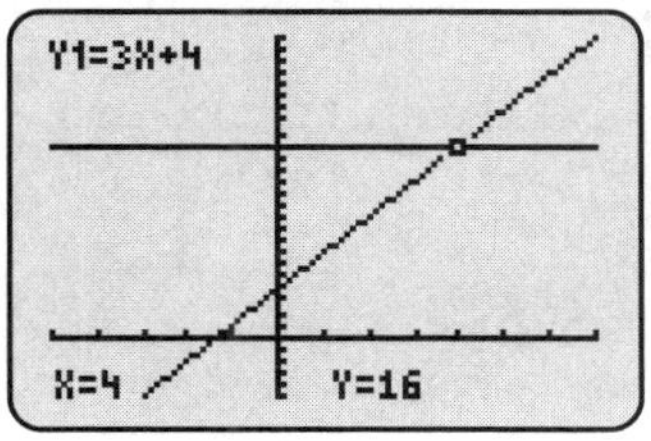

■ **Answer:** The solution is $x = 3$. The solution set is $\{3\}$.

Example 1 illustrates solving linear equations in one variable. What is a linear equation in one variable?

DEFINITION **Linear Equation**

A **linear equation in one variable**, x, can be written in the form

$$ax + b = 0$$

where a and b are real numbers and $a \neq 0$.

What makes this equation linear is that x is raised to the first power. We can also classify a linear equation as a **first-degree** equation.

Equation	Degree	General Name
$x - 7 = 0$	First	Linear
$x^2 - 6x - 9 = 0$	Second	Quadratic
$x^3 + 3x^2 - 8 = 0$	Third	Cubic

Technology Tip

Use a graphing utility to display graphs of $y_1 = 5x - (7x - 4) - 2$ and $y_2 = 5 - (3x + 2)$.

```
Plot1 Plot2 Plot3
\Y1■5X-(7X-4)-2
\Y2■5-(3X+2)
```

The x-coordinate of the point of intersection is the solution to the equation $5x - (7x - 4) - 2 = 5 - (3x + 2)$.

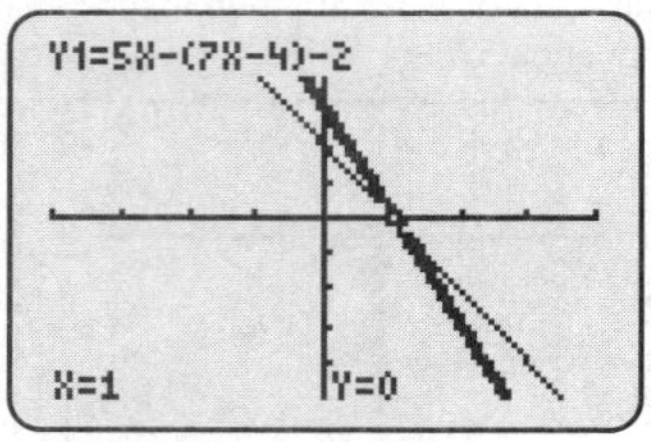

■ **Answer:** The solution is $x = 2$. The solution set is $\{2\}$.

Study Tip

Prime Factors

$$\begin{aligned} 2 &= 2 \\ 6 &= 2 \cdot 3 \\ 5 &= \quad\;\; \cdot 5 \\ \hline \text{LCD} &= 2 \cdot 3 \cdot 5 = 30 \end{aligned}$$

Technology Tip

Use a graphing utility to display graphs of $y_1 = \frac{1}{2}p - 5$ and $y_2 = \frac{3}{4}p$.

```
Plot1 Plot2 Plot3
\Y1■1/2X-5
\Y2■3/4X
```

The x-coordinate of the point of intersection is the solution.

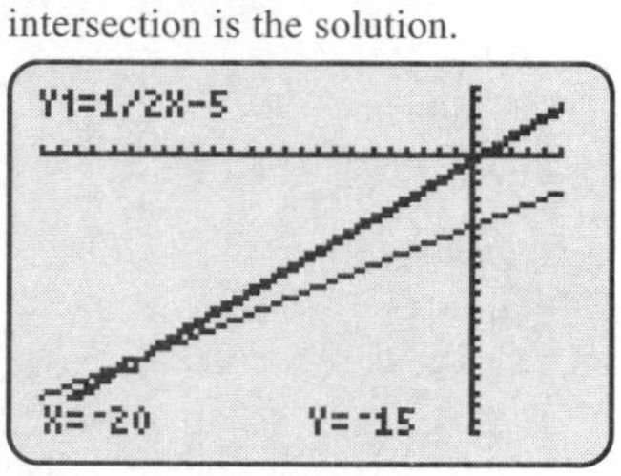

■ **Answer:** The solution is $m = -18$. The solution set is $\{-18\}$.

EXAMPLE 2 Solving a Linear Equation

Solve the equation $5x - (7x - 4) - 2 = 5 - (3x + 2)$.

Solution:

Eliminate the parentheses. $\qquad 5x - (7x - 4) - 2 = 5 - (3x + 2)$

Don't forget to distribute the negative sign through *both* terms inside the parentheses. $\qquad 5x - 7x + 4 - 2 = 5 - 3x - 2$

Combine x terms on the left, constants on the right. $\qquad -2x + 2 = 3 - 3x$

Add $3x$ to both sides.

$$\begin{array}{r} -2x + 2 = 3 - 3x \\ +3x \qquad\quad + 3x \\ \hline x + 2 = 3 \end{array}$$

Subtract 2 from both sides.

$$\begin{array}{r} -2 \;\; -2 \\ \hline x = 1 \end{array}$$

Check to verify that $x = 1$ is a solution to the original equation.

$$\begin{aligned} 5 \cdot 1 - (7 \cdot 1 - 4) - 2 &= 5 - (3 \cdot 1 + 2) \\ 5 - (7 - 4) - 2 &= 5 - (3 + 2) \\ 5 - (3) - 2 &= 5 - (5) \\ 0 &= 0 \end{aligned}$$

Since the solution $x = 1$ makes the equation true, the solution set is $\{1\}$.

■ **YOUR TURN** Solve the equation $4(x - 1) - 2 = x - 3(x - 2)$.

To solve a linear equation involving fractions, find the least common denominator (LCD) of all terms and multiply both sides of the equation by the LCD. We will first review how to find the LCD.

To add the fractions $\frac{1}{2} + \frac{1}{6} + \frac{2}{5}$, we must first find a common denominator. Some people are taught to find the lowest number that 2, 6, and 5 all divide evenly into. Others prefer a more systematic approach in terms of prime factors.

EXAMPLE 3 Solving a Linear Equation Involving Fractions

Solve the equation $\frac{1}{2}p - 5 = \frac{3}{4}p$.

Solution:

Write the equation. $\qquad \dfrac{1}{2}p - 5 = \dfrac{3}{4}p$

Multiply each term in the equation by the LCD, 4. $\qquad (4)\dfrac{1}{2}p - (4)5 = (4)\dfrac{3}{4}p$

The result is a linear equation with no fractions. $\qquad 2p - 20 = 3p$

Subtract $2p$ from both sides.

$$\begin{array}{r} -2p \qquad -2p \\ \hline -20 = p \end{array}$$

$$\boxed{p = -20}$$

Since $p = -20$ satisfies the original equation, the solution set is $\{-20\}$.

■ **YOUR TURN** Solve the equation $\frac{1}{4}m = \frac{1}{12}m - 3$.

Solving a Linear Equation in One Variable

Step	Description	Example
1	Simplify the algebraic expressions on both sides of the equation.	$-3(x-2)+5=7(x-4)-1$ $-3x+6+5=7x-28-1$ $-3x+11=7x-29$
2	Gather all variable terms on one side of the equation and all constant terms on the other side.	$-3x+11=7x-29$ $+3x \qquad +3x$ $11=10x-29$ $+29 \qquad +29$ $40=10x$
3	Isolate the variable.	$10x=40$ $x=\frac{40}{10}$ $\boxed{x=4}$

Solving Rational Equations That Are Reducible to Linear Equations

A **rational equation** is an equation that contains one or more rational expressions (Chapter 0). Some rational equations can be transformed into linear equations that you can then solve, but as you will see momentarily, you must be certain that the solution to the linear equation also satisfies the original rational equation.

EXAMPLE 4 Solving a Rational Equation That Can Be Reduced to a Linear Equation

Solve the equation $\frac{2}{3x}+\frac{1}{2}=\frac{4}{x}+\frac{4}{3}$.

Solution:

State the excluded values (those which make any denominator equal 0). $\qquad \frac{2}{3x}+\frac{1}{2}=\frac{4}{x}+\frac{4}{3} \quad x \neq 0$

Multiply *each term* by the LCD, $6x$. $\qquad 6x\left(\frac{2}{3x}\right)+6x\left(\frac{1}{2}\right)=6x\left(\frac{4}{x}\right)+6x\left(\frac{4}{3}\right)$

Simplify both sides. $\qquad 4+3x=24+8x$

Subtract 4. $\qquad -4 \qquad -4$

$$3x=20+8x$$

Subtract $8x$. $\qquad -8x \qquad -8x$

$$-5x=20$$

Divide by -5. $\qquad \boxed{x=-4}$

Since $x=-4$ satisfies the original equation, the solution set is $\{-4\}$.

■ **YOUR TURN** Solve the equation $\frac{3}{y}+2=\frac{7}{2y}$.

Technology Tip

Use a graphing utility to display graphs of $y_1=\frac{2}{3x}+\frac{1}{2}$ and $y_2=\frac{4}{x}+\frac{4}{3}$.

The x-coordinate of the point of intersection is the solution to the equation $\frac{2}{3x}+\frac{1}{2}=\frac{4}{x}+\frac{4}{3}$.

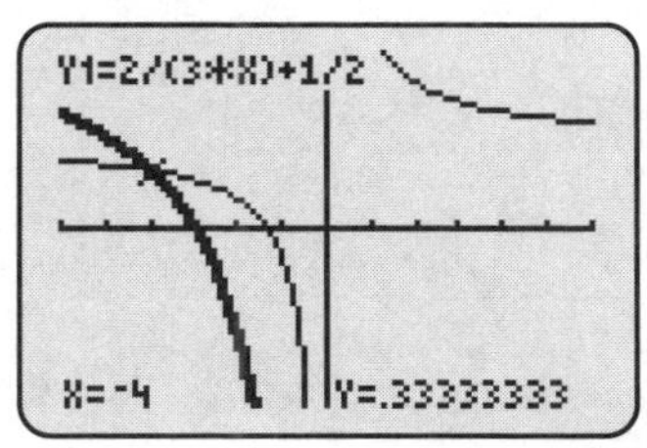

Study Tip

Since dividing by 0 is not defined, we exclude values of the variable that correspond to a denominator equaling 0.

■ **Answer:** The solution is $y=\frac{1}{4}$. The solution set is $\left\{\frac{1}{4}\right\}$.

Technology Tip

Use a graphing utility to display graphs of $y_1 = \frac{3x}{x-1} + 2$ and $y_2 = \frac{3}{x-1}$.

The x-coordinate of the point of intersection is the solution to the equation $\frac{3x}{x-1} + 2 = \frac{3}{x-1}$.

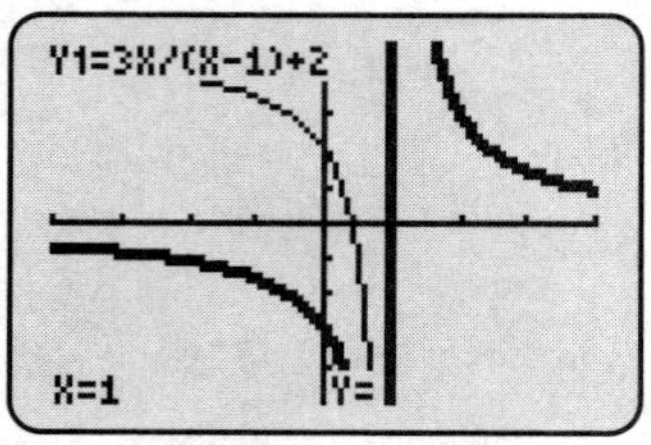

No intersection implies no solution.

Study Tip

When a variable is in the denominator of a fraction, the LCD will contain the variable. This sometimes results in an extraneous solution.

■ **Answer:** no solution

Extraneous solutions are solutions that satisfy a transformed equation but do not satisfy the original equation. It is important to first state any values of the variable that must be eliminated based on the original rational equation. Once the rational equation is transformed to a linear equation and solved, remove any excluded values of the variable.

EXAMPLE 5 Solving Rational Equations That Can Be Reduced to Linear Equations

Solve the equation $\frac{3x}{x-1} + 2 = \frac{3}{x-1}$.

Solution:

State the excluded values (those which make any denominator equal 0). $\quad \frac{3x}{x-1} + 2 = \frac{3}{x-1} \quad x \neq 1$

Eliminate the fractions by multiplying each term by the LCD, $x - 1$. $\quad \frac{3x}{x-1} \cdot (x-1) + 2 \cdot (x-1) = \frac{3}{x-1} \cdot (x-1)$

Simplify. $\quad \frac{3x}{\cancel{x-1}} \cdot \cancel{(x-1)} + 2 \cdot (x-1) = \frac{3}{\cancel{x-1}} \cdot \cancel{(x-1)}$

$$3x + 2(x-1) = 3$$

Distribute the 2. $\quad 3x + 2x - 2 = 3$

Combine x terms on the left. $\quad 5x - 2 = 3$

Add 2 to both sides. $\quad 5x = 5$

Divide both sides by 5. $\quad x = 1$

It may seem that $x = 1$ is the solution. However, the original equation had the restriction $x \neq 1$. Therefore, $x = 1$ is an extraneous solution and must be eliminated as a possible solution.

Thus, the equation $\frac{3x}{x-1} + 2 = \frac{3}{x-1}$ has **no solution**.

■ **YOUR TURN** Solve the equation $\frac{2x}{x-2} - 3 = \frac{4}{x-2}$.

We have reviewed finding the least common denominator (LCD) for real numbers. Now let us consider finding the LCD for rational equations that have different denominators. We multiply the denominators in order to get a common denominator.

Rational expression: $\frac{1}{x} + \frac{2}{x-1} \qquad$ LCD: $x(x-1)$

In order to find a *least* common denominator, it is useful to first factor the denominators to identify common multiples.

Rational equation: $\quad \frac{1}{3x-3} + \frac{1}{2x-2} = \frac{1}{x^2-x}$

Factor the denominators: $\quad \frac{1}{3(x-1)} + \frac{1}{2(x-1)} = \frac{1}{x(x-1)}$

LCD: $\quad 6x(x-1)$

EXAMPLE 6 Solving Rational Equations

Solve the equation $\frac{1}{3x+18} - \frac{1}{2x+12} = \frac{1}{x^2+6x}$.

Solution:

Factor the denominators. $\frac{1}{3(x+6)} - \frac{1}{2(x+6)} = \frac{1}{x(x+6)}$

State the excluded values. $x \neq 0, -6$

Multiply the equation by the LCD, $6x(x+6)$.

$$6x(x+6) \cdot \frac{1}{3(x+6)} - 6x(x+6) \cdot \frac{1}{2(x+6)} = 6x(x+6) \cdot \frac{1}{x(x+6)}$$

Divide out the common factors.

$$6x\cancel{(x+6)} \cdot \frac{1}{3\cancel{(x+6)}} - 6x\cancel{(x+6)} \cdot \frac{1}{2\cancel{(x+6)}} = 6\cancel{x(x+6)} \cdot \frac{1}{\cancel{x(x+6)}}$$

Simplify. $2x - 3x = 6$

Solve the linear equation. $x = -6$

Since one of the excluded values is $x \neq -6$, we say that $x = -6$ is an extraneous solution. Therefore, this rational equation has $\boxed{\text{no solution}}$.

YOUR TURN Solve the equation $\frac{2}{x} + \frac{1}{x+1} = -\frac{1}{x(x+1)}$.

Answer: no solution

EXAMPLE 7 Solving Rational Equations

Solve the equation $\frac{2}{x-3} = \frac{-3}{2-x}$.

Solution:

What values make *either* denominator equal to zero? The values $x = 2$ and $x = 3$ must be excluded from possible solutions to the equation. $\frac{2}{x-3} = \frac{-3}{2-x} \quad x \neq 2, x \neq 3$

Multiply the equation by the LCD, $(x-3)(2-x)$. $\frac{2}{x-3}(x-3)(2-x) = \frac{-3}{2-x}(x-3)(2-x)$

Divide out the common factors. $2(2-x) = -3(x-3)$

Eliminate the parentheses. $4 - 2x = -3x + 9$

Collect x terms on the left, constants on the right. $\boxed{x = 5}$

Since $x = 5$ satisfies the original equation, the solution set is $\{5\}$.

YOUR TURN Solve the equation $\frac{-4}{x+8} = \frac{3}{x-6}$.

Technology Tip

Use a graphing utility to display graphs of $y_1 = \frac{2}{x-3}$ and $y_2 = \frac{-3}{2-x}$.

The x-coordinate of the point of intersection is the solution to the equation $\frac{2}{x-3} = \frac{-3}{2-x}$.

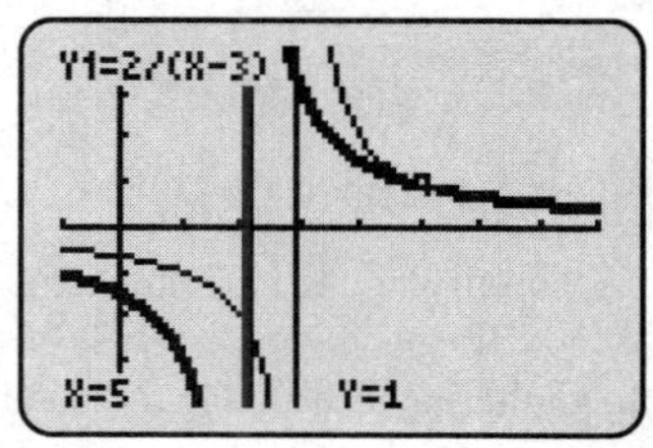

Answer: The solution is $x = 0$. The solution set is $\{0\}$.

EXAMPLE 8 Automotive Service

A car dealership charges for parts and an hourly rate for labor. If parts cost \$273, labor is \$53 per hour, and the total bill is \$458.50, how many hours did the dealership spend working on your car?

Solution:

Let x equal the number of hours the dealership worked on your car.

Write the cost equation. $\overbrace{53x}^{\text{labor}} + \overbrace{273}^{\text{parts}} = \overbrace{458.50}^{\text{total cost}}$

Subtract 273 from both sides of the equation. $53x = 185.50$

Divide both sides of the equation by 53. $\boxed{x = 3.5}$

$\boxed{\text{The dealership charged for 3.5 hours of labor.}}$

EXAMPLE 9 Grades

Dante currently has the following 3 test scores: 82, 79, and 90. If the score on the final exam is worth 2 test scores and his goal is to earn an 85 for his class average, what score on the final exam does Dante need to achieve his course goal?

Solution:

Let x equal final exam grade.

Write the equation that determines the course grade.

$$\frac{\overbrace{82 + 79 + 90}^{\text{scores 1, 2, and 3}} + \overbrace{2x}^{\text{final is worth 2 test scores}}}{\underbrace{5}_{\text{total of 5 test scores}}} = \underbrace{85}_{\text{average}}$$

Simplify the numerator. $\frac{251 + 2x}{5} = 85$

Multiply the equation by 5 (or cross multiply). $251 + 2x = 425$

Solve the linear equation. $\boxed{x = 87}$

$\boxed{\text{Dante needs to score at least an 87 on the final exam.}}$

SECTION 1.1 SUMMARY

Linear equations, $Ax + B = C$, are solved by:

1. Simplifying the algebraic expressions on both sides of the equation.
2. Gathering all variable terms on one side of the equation and all constant terms on the other side.
3. Isolating the variable.

Rational equations are solved by:

1. Determining any excluded values (denominator equals 0).
2. Multiplying the equation by the LCD.
3. Solving the resulting linear equation.
4. Eliminating any extraneous solutions.

SECTION 1.1 EXERCISES

SKILLS

In Exercises 1–36, solve for the indicated variable.

1. $5x = 35$
2. $4t = 32$
3. $-3 + n = 12$
4. $4 = -5 + y$
5. $24 = -3x$
6. $-50 = -5t$
7. $\frac{1}{5}n = 3$
8. $6 = \frac{1}{3}p$
9. $3x - 5 = 7$
10. $4p + 5 = 9$
11. $9m - 7 = 11$
12. $2x + 4 = 5$
13. $5t + 11 = 18$
14. $7x + 4 = 21 + 24x$
15. $3x - 5 = 25 + 6x$
16. $5x + 10 = 25 + 2x$
17. $20n - 30 = 20 - 5n$
18. $14c + 15 = 43 + 7c$
19. $4(x - 3) = 2(x + 6)$
20. $5(2y - 1) = 2(4y - 3)$
21. $-3(4t - 5) = 5(6 - 2t)$
22. $2(3n + 4) = -(n + 2)$
23. $2(x - 1) + 3 = x - 3(x + 1)$
24. $4(y + 6) - 8 = 2y - 4(y + 2)$
25. $5p + 6(p + 7) = 3(p + 2)$
26. $3(z + 5) - 5 = 4z + 7(z - 2)$
27. $7x - (2x + 3) = x - 2$
28. $3x - (4x + 2) = x - 5$
29. $2 - (4x + 1) = 3 - (2x - 1)$
30. $5 - (2x - 3) = 7 - (3x + 5)$
31. $2a - 9(a + 6) = 6(a + 3) - 4a$
32. $25 - [2 + 5y - 3(y + 2)] = -3(2y - 5) - [5(y - 1) - 3y + 3]$
33. $32 - [4 + 6x - 5(x + 4)] = 4(3x + 4) - [6(3x - 4) + 7 - 4x]$
34. $12 - [3 + 4m - 6(3m - 2)] = -7(2m - 8) - 3[(m - 2) + 3m - 5]$
35. $20 - 4[c - 3 - 6(2c + 3)] = 5(3c - 2) - [2(7c - 8) - 4c + 7]$
36. $46 - [7 - 8y + 9(6y - 2)] = -7(4y - 7) - 2[6(2y - 3) - 4 + 6y]$

Exercises 37–48 involve fractions. Clear the fractions by first multiplying by the least common denominator, and then solve the resulting linear equation.

37. $\frac{1}{5}m = \frac{1}{60}m + 1$
38. $\frac{1}{12}z = \frac{1}{24}z + 3$
39. $\frac{x}{7} = \frac{2x}{63} + 4$
40. $\frac{a}{11} = \frac{a}{22} + 9$
41. $\frac{1}{3}p = 3 - \frac{1}{24}p$
42. $\frac{3x}{5} - x = \frac{x}{10} - \frac{5}{2}$
43. $\frac{5y}{3} - 2y = \frac{2y}{84} + \frac{5}{7}$
44. $2m - \frac{5m}{8} = \frac{3m}{72} + \frac{4}{3}$
45. $p + \frac{p}{4} = \frac{5}{2}$
46. $\frac{c}{4} - 2c = \frac{5}{4} - \frac{c}{2}$
47. $\frac{x - 3}{3} - \frac{x - 4}{2} = 1 - \frac{x - 6}{6}$
48. $1 - \frac{x - 5}{3} = \frac{x + 2}{5} - \frac{6x - 1}{15}$

In Exercises 49–70, specify any values that must be excluded from the solution set and then solve the equation.

49. $\frac{4}{y} - 5 = \frac{5}{2y}$
50. $\frac{4}{x} + 10 = \frac{2}{3x}$
51. $7 - \frac{1}{6x} = \frac{10}{3x}$
52. $\frac{7}{6t} = 2 + \frac{5}{3t}$
53. $\frac{2}{a} - 4 = \frac{4}{3a}$
54. $\frac{4}{x} - 2 = \frac{5}{2x}$
55. $\frac{x}{x - 2} + 5 = \frac{2}{x - 2}$
56. $\frac{n}{n - 5} + 2 = \frac{n}{n - 5}$
57. $\frac{2p}{p - 1} = 3 + \frac{2}{p - 1}$
58. $\frac{4t}{t + 2} = 3 - \frac{8}{t + 2}$
59. $\frac{3x}{x + 2} - 4 = \frac{2}{x + 2}$
60. $\frac{5y}{2y - 1} - 3 = \frac{12}{2y - 1}$

61. $\frac{1}{n} + \frac{1}{n+1} = \frac{-1}{n(n+1)}$

62. $\frac{1}{x} + \frac{1}{x-1} = \frac{1}{x(x-1)}$

63. $\frac{3}{a} - \frac{2}{a+3} = \frac{9}{a(a+3)}$

64. $\frac{1}{c-2} + \frac{1}{c} = \frac{2}{c(c-2)}$

65. $\frac{n-5}{6n-6} = \frac{1}{9} - \frac{n-3}{4n-4}$

66. $\frac{5}{m} + \frac{3}{m-2} = \frac{6}{m(m-2)}$

67. $\frac{2}{5x+1} = \frac{1}{2x-1}$

68. $\frac{3}{4n-1} = \frac{2}{2n-5}$

69. $\frac{t-1}{1-t} = \frac{3}{2}$

70. $\frac{2-x}{x-2} = \frac{3}{4}$

APPLICATIONS

71. Temperature. To calculate temperature in degrees Fahrenheit we use the formula $F = \frac{9}{5}C + 32$, where F is degrees Fahrenheit and C is degrees Celsius. Find the formula to convert from Fahrenheit to Celsius.

72. Geometry. The perimeter P of a rectangle is related to the length L and width W of the rectangle through the equation $P = 2L + 2W$. Determine the width in terms of the perimeter and length.

73. Costs: Cellular Phone. Your cell phone plan charges \$15 a month plus 12 cents per minute. If your monthly bill is \$25.08, how many minutes did you use?

74. Costs: Rental Car. Becky rented a car on her Ft. Lauderdale vacation. The car was \$25 a day plus 10 cents per mile. She kept the car for 5 days and her bill was \$185. How many miles did she drive the car?

75. Costs: Internet. When traveling in London, Charlotte decided to check her e-mail at an Internet café. There was a flat charge of \$2 plus a charge of 10 cents a minute. How many minutes was she logged on if her bill was \$3.70?

76. Sales: Income. For a summer job, Dwayne decides to sell magazine subscriptions. He will be paid \$20 a day plus \$1 for each subscription he sells. If he works for 25 days and makes \$645, how many subscriptions did he sell?

77. Speed of Light. The frequency f of an optical signal in hertz (Hz) is related to the wavelength λ in meters (m) of a laser through the equation $f = \frac{c}{\lambda}$, c where c is the speed of light in a vacuum and is typically taken to be $c = 3.0 \times 10^8$ meters per second (m/s). What values must be eliminated from the wavelengths?

78. Optics. For an object placed near a lens, an image forms on the other side of the lens at a distinct position determined by the distance from the lens to the object. The position of the image is found using the thin lens equation:

$$\frac{1}{f} = \frac{1}{d_o} + \frac{1}{d_i}$$

where d_o is the distance from the object to the lens, d_i is the distance from the lens to the image, and f is the *focal length* of the lens. Solve for the object distance d_o in terms of the focal length and image distance.

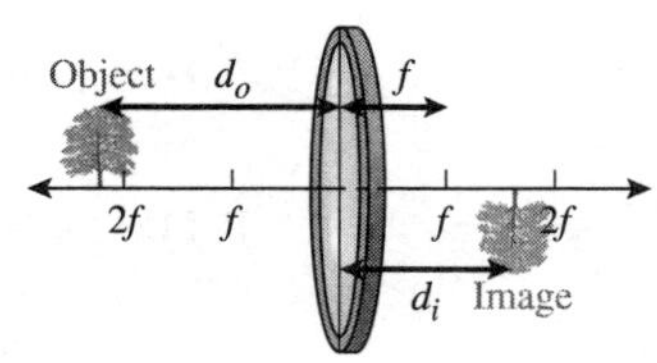

CATCH THE MISTAKE

In Exercises 79–82, explain the mistake that is made.

79. Solve the equation $4x + 3 = 6x - 7$.

Solution:

Subtract $4x$ and add 7 to the equation. $3 = 6x$

Divide by 3. $x = 2$

This is incorrect. What mistake was made?

80. Solve the equation $3(x + 1) + 2 = x - 3(x - 1)$.

Solution:

$$3x + 3 + 2 = x - 3x - 3$$
$$3x + 5 = -2x - 3$$
$$5x = -8$$
$$x = -\frac{8}{5}$$

This is incorrect. What mistake was made?

81. Solve the equation $\frac{4}{p} - 3 = \frac{2}{5p}$.

Solution:

Cross multiply.

$$(p - 3)2 = 4(5p)$$
$$2p - 6 = 20p$$
$$-6 = 18p$$
$$p = -\frac{6}{18}$$
$$p = -\frac{1}{3}$$

This is incorrect. What mistake was made?

82. Solve the equation $\frac{1}{x} + \frac{1}{x - 1} = \frac{1}{x(x - 1)}$.

Solution:

Multiply by the LCD, $x(x - 1)$.

$$\frac{x(x-1)}{x} + \frac{x(x-1)}{x-1} = \frac{x(x-1)}{x(x-1)}$$

Simplify.

$$(x - 1) + x = 1$$
$$x - 1 + x = 1$$
$$2x = 2$$
$$x = 1$$

This is incorrect. What mistake was made?

CONCEPTUAL

In Exercises 83–86, determine whether each of the statements is true or false.

83. The solution to the equation $x = \frac{1}{1/x}$ is the set of all real numbers.

84. The solution to the equation

$$\frac{1}{(x - 1)(x + 2)} = \frac{1}{x^2 + x - 2}$$

is the set of all real numbers.

85. $x = -1$ is a solution to the equation $\frac{x^2 - 1}{x - 1} = x + 1$.

86. $x = 1$ is a solution to the equation $\frac{x^2 - 1}{x - 1} = x + 1$.

87. Solve for x, given that a, b, and c are real numbers and $a \neq 0$:

$$ax + b = c$$

88. Solve for x, given that a, b, and c are real numbers and $c \neq 0$:

$$\frac{a}{x} - \frac{b}{x} = c$$

CHALLENGE

89. Solve the equation for x: $\frac{b + c}{x + a} = \frac{b - c}{x - a}$. Are there any restrictions given that $a \neq 0, x \neq 0$?

90. Solve the equation for y: $\frac{1}{y - a} + \frac{1}{y + a} = \frac{2}{y - 1}$. Does y have any restrictions?

91. Solve for x: $\frac{1 - 1/x}{1 + 1/x} = 1$.

92. Solve for t: $\frac{t + 1/t}{1/t - 1} = 1$.

93. Solve the equation for x in terms of y:

$$y = \frac{a}{1 + b/x + c}$$

94. Find the number a for which $y = 2$ is a solution of the equation $y - a = y + 5 - 3ay$.

TECHNOLOGY

In Exercises 95–102, graph the function represented by each side of the equation in the same viewing rectangle and solve for x.

95. $3(x + 2) - 5x = 3x - 4$

96. $-5(x - 1) - 7 = 10 - 9x$

97. $2x + 6 = 4x - 2x + 8 - 2$

98. $10 - 20x = 10x - 30x + 20 - 10$

99. $\frac{x(x - 1)}{x^2} = 1$

100. $\frac{2x(x + 3)}{x^2} = 2$

101. $0.035x + 0.029(8706 - x) = 285.03$

102. $\frac{1}{0.75x} - \frac{0.45}{x} = \frac{1}{9}$

SECTION 1.2 APPLICATIONS INVOLVING LINEAR EQUATIONS

SKILLS OBJECTIVES

- Develop mathematical models for real-life problems.
- Solve application problems involving common formulas.
- Solve geometry problems.
- Solve interest problems.
- Solve mixture problems.
- Solve distance–rate–time problems.

CONCEPTUAL OBJECTIVE

- Understand the mathematical modeling process.

Solving Application Problems Using Mathematical Models

In this section, we will use algebra to solve problems that occur in our day-to-day lives. You typically will read the problem in words, develop a mathematical model (equation) for the problem, solve the equation, and write the answer in words.

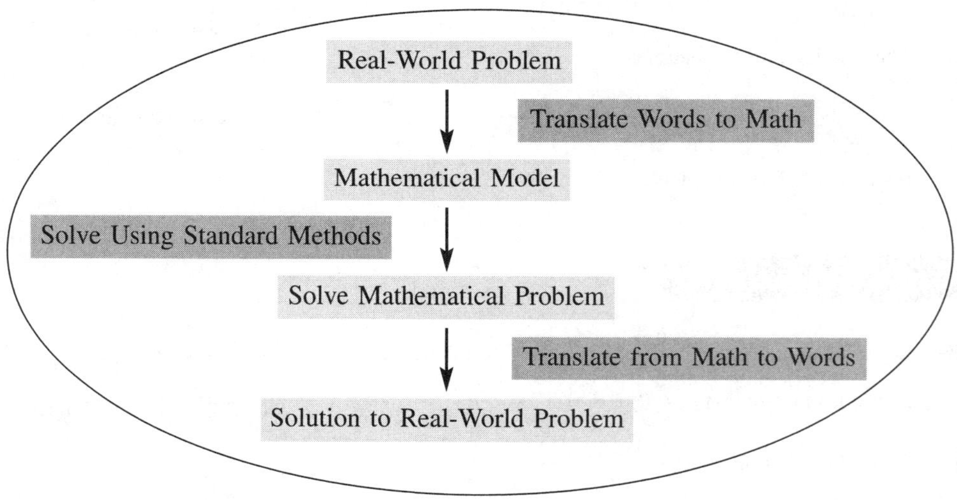

You will have to come up with a unique formula to solve each kind of word problem, but there is a universal *procedure* for approaching all word problems.

PROCEDURE FOR SOLVING WORD PROBLEMS

Step 1: **Identify the question.** Read the problem *one* time and note what you are asked to find.
Step 2: **Make notes.** Read until you can note something (an amount, a picture, anything). Continue reading and making notes until you have read the problem a second* time.
Step 3: **Set up an equation.** Assign a variable to represent what you are asked to find.
Step 4: **Solve the equation.**
Step 5: **Check the solution.** Run the solution past the "common sense department" using estimation.

*Step 2 often requires multiple readings of the problem.

EXAMPLE 1 How Long Was the Trip?

During a camping trip in North Bay, Ontario, a couple went one-third of the way by boat, 10 miles by foot, and one-sixth of the way by horse. How long was the trip?

Solution:

STEP 1 Identify the question.
How many miles was the trip?

STEP 2 Make notes.

Read	Write
... one-third of the way by boat	BOAT: $\frac{1}{3}$ of the trip
... 10 miles by foot	FOOT: 10 miles
... one-sixth of the way by horse	HORSE: $\frac{1}{6}$ of the trip

STEP 3 Set up an equation.
The total distance of the trip is the sum of all the distances by boat, foot, and horse.

Distance by boat + Distance by foot + Distance by horse = Total distance of trip

Distance of total trip in miles $= x$

Distance by boat $= \frac{1}{3}x$

Distance by foot $=$ 10 miles

Distance by horse $= \frac{1}{6}x$

$$\underbrace{\frac{1}{3}x}_{\text{boat}} + \underbrace{10}_{\text{foot}} + \underbrace{\frac{1}{6}x}_{\text{horse}} = \underbrace{x}_{\text{total}}$$

STEP 4 Solve the equation. $\frac{1}{3}x + 10 + \frac{1}{6}x = x$

Multiply by the LCD, 6. $2x + 60 + x = 6x$

Collect x terms on the right. $60 = 3x$

Divide by 3. $20 = x$

The trip was 20 miles. $x = 20$

STEP 5 Check the solution.
Estimate: The boating distance, $\frac{1}{3}$ of 20 miles, is approximately 7 miles; the riding distance on horse, $\frac{1}{6}$ of 20 miles, is approximately 3 miles. Adding these two distances to the 10 miles by foot gives a trip distance of 20 miles.

■ **YOUR TURN** A family arrives at the Walt Disney World parking lot. To get from their car in the parking lot to the gate at the Magic Kingdom they walk $\frac{1}{4}$ mile, take a tram for $\frac{1}{3}$ of their total distance, and take a monorail for $\frac{1}{2}$ of their total distance. How far is it from their car to the gate of Magic Kingdom?

■ **Answer:** The distance from their car to the gate is 1.5 miles.

EXAMPLE 2 Find the Numbers

Find three consecutive even integers so that the sum of the three numbers is 2 more than twice the third.

Solution:

STEP 1 Identify the question.
What are the three consecutive even integers?

STEP 2 Make notes.
Examples of three consecutive even integers are 14, 16, 18 or 20, 22, 24 or 2, 4, 6. Let n represent the first even integer. The next consecutive even integer is $n + 2$ and the next consecutive even integer after that is $n + 4$.

$$\begin{aligned} n &= \text{1st integer} \\ n + 2 &= \text{2nd consecutive even integer} \\ n + 4 &= \text{3rd consecutive even integer} \end{aligned}$$

Read	Write
... sum of the first three	$n + (n + 2) + (n + 4)$
... is	$=$
... two more than	$+2$
... twice the third	$2(n + 4)$

STEP 3 Set up an equation.

$$\underbrace{n + (n + 2) + (n + 4)}_{\text{sum of the first three}} \underbrace{=}_{\text{is}} \underbrace{2\ +}_{\text{2 more than}} \underbrace{2(n + 4)}_{\text{twice the third}}$$

STEP 4 Solve the equation. $n + (n + 2) + (n + 4) = 2 + 2(n + 4)$

Eliminate the parentheses. $n + n + 2 + n + 4 = 2 + 2n + 8$

Simplify both sides. $3n + 6 = 2n + 10$

Collect n terms on the left and constants on the right. $n = 4$

The three consecutive even integers are 4, 6, and 8.

STEP 5 Check the solution.
Substitute the solution into the problem to see whether it makes sense. The sum of the three integers $(4 + 6 + 8)$ is 18. Twice the third is 16. Since 2 more than twice the third is 18, the solution checks.

YOUR TURN Find three consecutive odd integers so that the sum of the three integers is 5 less than 4 times the first.

Answer: The three consecutive odd integers are 11, 13 and 15.

Geometry Problems

Some problems require geometric formulas in order to be solved. The following geometric formulas may be useful.

Geometric Formulas

Rectangle	Perimeter	Area
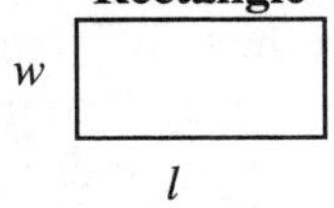	$P = 2l + 2w$	$A = l \cdot w$
Circle	**Circumference**	**Area**
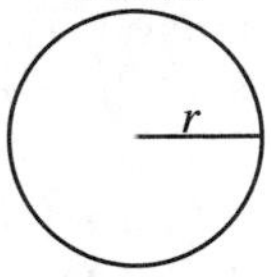	$C = 2\pi r$	$A = \pi r^2$
Triangle	**Perimeter**	**Area**
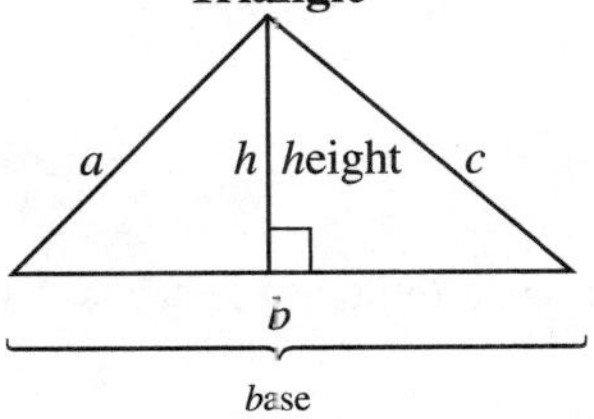	$P = a + b + c$	$A = \frac{1}{2}bh$

EXAMPLE 3 Geometry

A rectangle 24 meters long has the same area as a square with 12 meter sides. What are the dimensions of the rectangle?

Solution:

STEP 1 Identify the question.
What are the dimensions (length and width) of the rectangle?

STEP 2 Make notes.

Read	Write/Draw
A rectangle 24 meters long	 $l = 24$
... has the same area	rectangle area $= 1 \cdot w = 24w$
... as a square that is 12 meters long	area of square $= 12 \cdot 12 = 144$

STEP 3 Set up an equation.

The area of the rectangle is equal to the area of the square. rectangle area = square area

Substitute in known quantities. $24w = 144$

STEP 4 Solve the equation.

Divide by 24. $w = \frac{144}{24} = 6$

The rectangle is 24 meters long and 6 meters wide.

STEP 5 Check the solution.
A 24 m by 6 m rectangle has an area of 144 m^2.

■ YOUR TURN A rectangle 3 inches wide has the same area as a square with 9 inch sides. What are the dimensions of the rectangle?

■ **Answer:** The rectangle is 27 inches long and 3 inches wide.

Interest Problems

In our personal or business financial planning, a particular concern we have is interest. **Interest** is money paid for the use of money; it is the cost of borrowing money. The total amount borrowed is called the **principal**. The principal can be the price of our new car; we pay the bank interest for loaning us money. The principal can also be the amount we keep in a CD or money market account; the bank uses this money and pays us interest. Typically interest rate, expressed as a percentage, is the amount charged for the use of the principal for a given time, usually in years.

Simple interest is interest that is paid only on principal during a period of time. Later we will discuss *compound interest*, which is interest paid on both principal and the interest accrued over a period of time.

DEFINITION **Simple Interest**

If a principal of P dollars is borrowed for a period of t years at an annual interest rate r (expressed in decimal form), the interest I charged is

$$I = Prt$$

This is the formula for **simple interest**.

EXAMPLE 4 Simple Interest

Through a summer job Morgan is able to save \$2,500. If she puts that money into a 6-month certificate of deposit (CD) that pays a simple interest rate of 3% a year, how much money will she have in her CD at the end of the 6 months?

Solution:

STEP 1 Identify the question.
How much money does Morgan have after 6 months?

STEP 2 Make notes.
The principal is \$2,500.

The annual interest rate is 3%, which in decimal form is 0.03.

The time the money spends accruing interest is 6 months, or $\frac{1}{2}$ of a year.

STEP 3 Set up an equation.
Write the simple interest formula. $I = Prt$

Label the known quantities. $P = 2500$, $r = 0.03$, and $t = 0.5$

STEP 4 Solve the equation. $I = Prt$

$$I = (2500)(0.03)(0.5) = 37.5$$

The interest paid on the CD is \$37.50. Adding this to the principal gives a total of

$$\$2,500 + \$37.50 = \boxed{\$2,537.50}$$

STEP 5 Check the solution.
This answer agrees with our intuition. Had we made a mistake, say, of moving one decimal place to the right, then the interest would have been \$375, which is much larger than we would expect on a principal of only \$2,500.

EXAMPLE 5 Multiple Investments

Theresa earns a full athletic scholarship for college, and her parents have given her the \$20,000 they had saved to pay for her college tuition. She decides to invest that money with an overall goal of earning 11% interest. She wants to put some of the money in a low-risk investment that has been earning 8% a year and the rest of the money in a medium-risk investment that typically earns 12% a year. How much money should she put in each investment to reach her goal?

Solution:

STEP 1 Identify the question.
How much money is invested in each (the 8% and the 12%) account?

STEP 2 Make notes.

Read

Theresa has \$20,000 to invest.

If part is invested at 8% and the rest at 12%, how much should be invested at each rate to yield 11% on the total amount invested?

Write/Draw

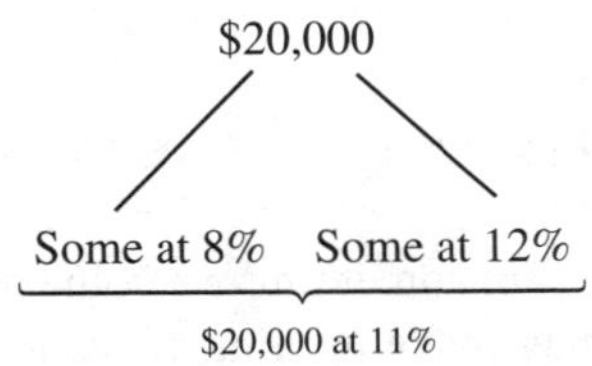

STEP 3 Set up an equation.
If we let x represent the amount Theresa puts into the 8% investment, how much of the \$20,000 is left for her to put in the 12% investment?

Amount in the 8% investment: x

Amount in the 12% investment: $20{,}000 - x$

Simple interest formula: $I = Prt$

INVESTMENT	PRINCIPAL	RATE	TIME (YR)	INTEREST
8% Account	x	0.08	1	$0.08x$
12% Account	$20{,}000 - x$	0.12	1	$0.12(20{,}000 - x)$
Total	20,000	0.11	1	$0.11(20{,}000)$

Adding the interest earned in the 8% investment to the interest earned in the 12% investment should earn an average of 11% on the total investment.

$$0.08x + 0.12(20{,}000 - x) = 0.11(20{,}000)$$

STEP 4 Solve the equation.

Eliminate the parentheses. $\quad 0.08x + 2400 - 0.12x = 2200$

Collect x terms on the left, constants on the right. $\quad -0.04x = -200$

Divide by -0.04. $\quad x = 5000$

Calculate the amount at 12%. $\quad 20{,}000 - 5000 = 15{,}000$

Theresa should invest \$5,000 at 8% and \$15,000 at 12% to reach her goal.

STEP 5 Check the solution.
If money is invested at 8% and 12% with a goal of averaging 11%, our intuition tells us that more should be invested at 12% than 8%, which is what we found. The exact check is as follows:

$$0.08(5000) + 0.12(15{,}000) = 0.11(20{,}000)$$
$$400 + 1800 = 2200$$
$$2200 = 2200$$

■ **YOUR TURN** You win $24,000 and you decide to invest the money in two different investments: one paying 18% and the other paying 12%. A year later you have $27,480 total. How much did you originally invest in each account?

■ **Answer:** $10,000 is invested at 18% and $14,000 is invested at 12%.

Mixture Problems

Mixtures are something we come across every day. Different candies that sell for different prices may make up a movie snack. New blends of coffees are developed by coffee connoisseurs. Chemists mix different concentrations of acids in their labs. Whenever two or more distinct ingredients are combined the result is a **mixture**.

Our choice at a gas station is typically 87, 89, and 93 octane. The octane number is the number that represents the percentage of iso-octane in fuel. 89 octane is significantly overpriced. Therefore, if your car requires 89 octane, it would be more cost effective to mix 87 and 93 octane.

EXAMPLE 6 Mixture Problem

The manual for your new car suggests using gasoline that is 89 octane. In order to save money, you decide to use some 87 octane and some 93 octane in combination with the 89 octane currently in your tank in order to have an approximate 89 octane mixture. Assuming you have 1 gallon of 89 octane remaining in your tank (your tank capacity is 16 gallons), how many gallons of 87 and 93 octane should be used to fill up your tank to achieve a mixture of 89 octane?

Solution:

STEP 1 Identify the question.
How many gallons of 87 octane and how many gallons of 93 octane should be used?

STEP 2 Make notes.

Read

Assuming you have one gallon of 89 octane remaining in your tank (your tank capacity is 16 gallons), how many gallons of 87 and 93 octane should you add?

Write/Draw

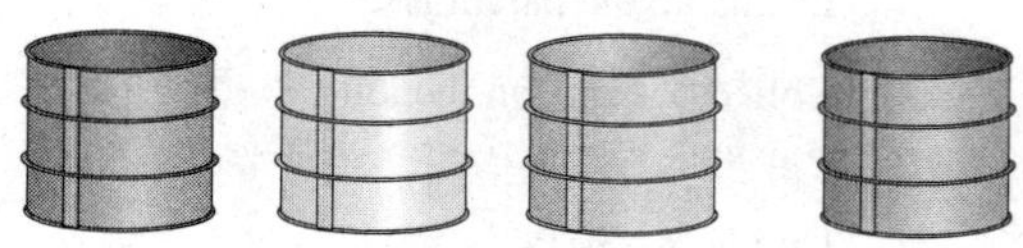

89 octane [1 gallon] + 87 octane [? gallons] + 93 octane [? gallons] = 89 octane [16 gallons]

STEP 3 Set up an equation.

x = gallons of 87 octane gasoline added at the pump

$15 - x$ = gallons of 93 octane gasoline added at the pump

1 = gallons of 89 octane gasoline already in the tank

$$0.89(1) + 0.87x + 0.93(15 - x) = 0.89(16)$$

STEP 4 Solve the equation. $0.89(1) + 0.87x + 0.93(15 - x) = 0.89(16)$

Eliminate the parentheses. $0.89 + 0.87x + 13.95 - 0.93x = 14.24$

Collect x terms on the left side. $-0.06x + 14.84 = 14.24$

Subtract 14.84 from both sides of the equation. $-0.06x = -0.6$

Divide both sides by -0.06. $x = 10$

Calculate the amount of 93 octane. $15 - 10 = 5$

Add 10 gallons of 87 octane and 5 gallons of 93 octane.

STEP 5 Check the solution.

Estimate: Our intuition tells us that if the desired mixture is 89 octane, then we should add approximately 1 part 93 octane and 2 parts 87 octane. The solution we found, 10 gallons of 87 octane and 5 gallons of 93 octane, agrees with this.

YOUR TURN For a certain experiment, a student requires 100 ml of a solution that is 11% HCl (hydrochloric acid). The storeroom has only solutions that are 5% HCl and 15% HCl. How many milliliters of each available solution should be mixed to get 100 ml of 11% HCl?

Answer: 40 ml of 5% HCl and 60 ml of 15% HCl

Distance–Rate–Time Problems

The next example deals with distance, rate, and time. On a road trip, you see a sign that says your destination is 90 miles away, and your speedometer reads 60 miles per hour. Dividing 90 miles by 60 miles per hour tells you that your arrival will be in 1.5 hours. Here is how you know.

If the rate, or speed, is assumed to be constant, then the equation that relates distance (d), rate (r), and time (t) is given by $d = r \cdot t$. In the above driving example,

$$d = 90 \text{ miles} \qquad r = 60 \frac{\text{miles}}{\text{hour}}$$

Substituting these into $d = r \cdot t$, we arrive at

$$90 \text{ miles} = \left[60 \frac{\text{miles}}{\text{hour}}\right] \cdot t$$

Solving for t, we get

$$t = \frac{90 \text{ miles}}{60 \frac{\text{miles}}{\text{hour}}} = 1.5 \text{ hours}$$

EXAMPLE 7 Distance–Rate–Time

It takes 8 hours to fly from Orlando to London and 9.5 hours to return. If an airplane averages 550 mph in still air, what is the average rate of the wind blowing in the direction from Orlando to London? Assume the speed of the wind (jet stream) is constant and the same for both legs of the trip.

Solution:

STEP 1 Identify the question.
At what rate in mph is the wind blowing?

STEP 2 Make notes.

Read	Write/Draw
It takes 8 hours to fly from Orlando to London and 9.5 hours to return.	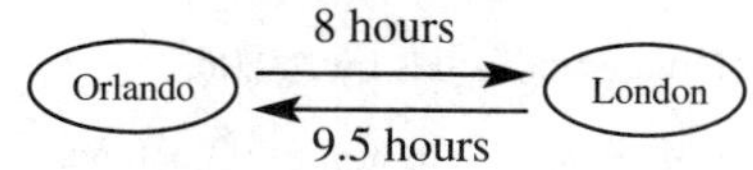
If the airplane averages 550 mph in still air...	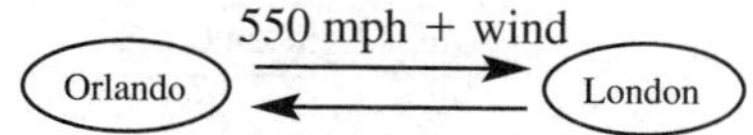

STEP 3 Set up an equation.
The formula relating distance, rate, and time is $d = r \cdot t$. The distance d of each flight is the same. On the Orlando to London flight the time is 8 hours due to an increased speed from a tailwind. On the London to Orlando flight the time is 9.5 hours and the speed is decreased due to the headwind. Let w represent the wind speed.

Orlando to London: $d = (550 + w)8$

London to Orlando: $d = (550 - w)9.5$

These distances are the same, so set them equal to each other:

$$(550 + w)8 = (550 - w)9.5$$

STEP 4 Solve the equation.

Eliminate the parentheses. $4400 + 8w = 5225 - 9.5w$

Collect w terms on the left, constants on the right. $17.5w = 825$

Divide by 17.5. $w = 47.1429 \approx 47$

The wind is blowing approximately $\boxed{47 \text{ mph}}$ in the direction from Orlando to London.

STEP 5 Check the solution.
Estimate: Going from Orlando to London, the tailwind is approximately 50 mph, which added to the plane's 550 mph speed yields a ground speed of 600 mph. The Orlando to London route took 8 hours. The distance of that flight is (600 mph)(8 hr), which is 4800 miles. The return trip experienced a headwind of approximately 50 mph, so subtracting the 50 from 550 gives an average speed of 500 mph. That route took 9.5 hours, so the distance of the London to Orlando flight was (500 mph)(9.5 hr), which is 4750 miles. Note that the estimates of 4800 and 4750 miles are close.

■ **YOUR TURN** A Cessna 150 averages 150 mph in still air. With a tailwind it is able to make a trip in $2\frac{1}{3}$ hours. Because of the headwind, it is only able to make the return trip in $3\frac{1}{2}$ hours. What is the average wind speed?

■ **Answer:** The wind is blowing 30 mph.

EXAMPLE 8 Work

Connie can clean her house in 2 hours. If Alvaro helps her, they can clean the house in 1 hour and 15 minutes together. How long would it take Alvaro to clean the house by himself?

Solution:

- Connie can clean her house in 2 hours, so Connie can clean $\frac{1}{2}$ of the house per hour.
- Together Connie and Alvaro can clean the house in 1 hour and 15 minutes, or $\frac{5}{4}$ of an hour. Therefore together they can clean $\frac{1}{5/4} = \frac{4}{5}$ of the house per hour.
- Let x = number of hours it takes Alvaro to clean the house by himself. So Alvaro can clean $\frac{1}{x}$ of the house per hour.

	Amount of Time to Do One Job	Amount of Job Done per Unit of Time
Connie	2	$\frac{1}{2}$
Alvaro	x	$\frac{1}{x}$
Together	$\frac{5}{4}$	$\frac{4}{5}$

Set up the equation.

$$\underbrace{\frac{1}{2}}_{\text{Amount of house Connie can clean per hour}} + \underbrace{\frac{1}{x}}_{\text{Amount of house Alvaro can clean per hour}} = \underbrace{\frac{4}{5}}_{\text{Amount of house they can clean per hour if they work together}}$$

Multiply by the LCD, $10x$. $\quad 5x + 10 = 8x$

Solve for x. $\quad x = \frac{10}{3} = 3\frac{1}{3}$

It takes Alvaro 3 hours and 20 minutes to clean the house by himself.

SECTION 1.2 SUMMARY

In the real world many kinds of application problems can be solved through modeling with linear equations. The following five-step procedure will help you develop the model. Some problems require development of a mathematical model, while others rely on common formulas.

1. Identify the quantity you are to determine.
2. Make notes on any clues that will help you set up an equation.
3. Set up the equation.
4. Solve the equation.
5. Check the solution against your intuition.

SECTION 1.2 EXERCISES

APPLICATIONS

1. **Discount Price.** Donna redeems a 10% off coupon at her local nursery. After buying azaleas, bougainvillea, and bags of potting soil, her checkout price before tax is \$217.95. How much would she have paid without the coupon?

2. **Discount Price.** The original price of a pair of binoculars is \$74. The sale price is \$51.80. How much was the markdown?

3. **Cost: Fair Share.** Jeff, Tom, and Chelsea order a large pizza. They decide to split the cost according to how much they will eat. Tom pays \$5.16, Chelsea eats $\frac{1}{8}$ of the pizza, and Jeff eats $\frac{1}{2}$ of the pizza. How much did the pizza cost?

4. **Event Planning.** A couple decide to analyze their monthly spending habits. The monthly bills are 50% of their take-home pay, and they invest 20% of their take-home pay. They spend \$560 on groceries, and 23% goes to miscellaneous. How much is their take-home pay per month?

5. **Discounts.** A builder of tract homes reduced the price of a model by 15%. If the new price is \$125,000, what was its original price? How much can be saved by purchasing the model?

6. **Markups.** A college bookstore marks up the price it pays the publisher for a book by 25%. If the selling price of a book is \$79, how much did the bookstore pay for the book?

7. **Puzzle.** Angela is on her way from home in Jersey City into New York City for dinner. She walks 1 mile to the train station, takes the train $\frac{3}{4}$ of the way, and takes a taxi $\frac{1}{6}$ of the way to the restaurant. How far does Angela live from the restaurant?

8. **Puzzle.** An employee at Kennedy Space Center (KSC) lives in Daytona Beach and works in the vehicle assembly building (VAB). She carpools to work with a colleague. On the days that her colleague drives the car pool, she drives 7 miles to the park-and-ride with her colleague to the KSC headquarters building, and then takes the KSC shuttle from the headquarters building to the VAB. The drive from the park-and-ride to the headquarters building is $\frac{5}{6}$ of her total trip and the shuttle ride is $\frac{1}{20}$ of her total trip. How many miles does she travel from her house to the VAB on days when her colleague drives?

9. **Puzzle.** A typical college student spends $\frac{1}{3}$ of her waking time in class, $\frac{1}{5}$ of her waking time eating, $\frac{1}{10}$ of her waking time working out, 3 hours studying, and $2\frac{1}{2}$ hours doing other things. How many hours of sleep does the typical college student get?

10. **Diet.** A particular 1550-calories-per-day diet suggests eating breakfast, lunch, dinner, and two snacks. Dinner is twice the calories of breakfast. Lunch is 100 calories more than breakfast. The two snacks are 100 and 150 calories. How many calories are each meal?

11. **Budget.** A company has a total of \$20,000 allocated for monthly costs. Fixed costs are \$15,000 per month and variable costs are \$18.50 per unit. How many units can be manufactured in a month?

12. **Budget.** A woman decides to start a small business making monogrammed cocktail napkins. She can set aside \$1,870 for monthly costs. Fixed costs are \$1,329.50 per month and variable costs are \$3.70 per set of napkins. How many sets of napkins can she afford to make per month?

13. **Numbers.** Find a number such that 10 less than $\frac{2}{3}$ the number is $\frac{1}{4}$ the number.

14. **Numbers.** Find a positive number such that 10 times the number is 16 more than twice the number.

15. **Numbers.** Find two consecutive even integers such that 4 times the smaller number is two more than 3 times the larger number.

16. **Numbers.** Find three consecutive integers such that the sum of the three is equal to 2 times the sum of the first two integers.

17. **Geometry.** Find the perimeter of a triangle if one side is 11 inches, another side is $\frac{1}{5}$ the perimeter, and the third side is $\frac{1}{4}$ the perimeter.

18. **Geometry.** Find the dimensions of a rectangle whose length is one foot longer than twice its width and whose perimeter is 20 feet.

19. **Geometry.** An NFL playing field is a rectangle. The length of the field (excluding the end zones) is 40 more yards than twice the width. The perimeter of the playing field is 260 yards. What are the dimensions of the field in yards?

20. **Geometry.** The length of a rectangle is 2 more than 3 times the width, and the perimeter is 28 inches. What are the dimensions of the rectangle?

21. **Geometry.** Consider two circles, a smaller one and a larger one. If the larger one has a radius that is 3 feet larger than that of the smaller circle and the ratio of the circumferences is 2:1, what are the radii of the two circles?

22. **Geometry.** The perimeter of a semicircle is doubled when the radius is increased by 1. Find the radius of the semicircle.

23. **Home Improvement.** A man wants to remove a tall pine tree from his yard. Before he goes to Home Depot, he needs to know how tall an extension ladder he needs to purchase. He measures the shadow of the tree to be 225 feet long. At the same time he measures the shadow of a 4 foot stick to be 3 feet. Approximately how tall is the pine tree?

24. **Home Improvement.** The same man in Exercise 23 realizes he also wants to remove a dead oak tree. Later in the day he measures the shadow of the oak tree to be 880 feet long and the 4 foot stick now has a shadow of 10 feet. Approximately how tall is the oak tree?

25. **Biology: Alligators.** It is common to see alligators in ponds, lakes, and rivers in Florida. The ratio of head size (back of the head to the end of the snout) to the full body length of an alligator is typically constant. If a $3\frac{1}{2}$ foot alligator has a head length of 6 inches, how long would you expect an alligator to be whose head length is 9 inches?

26. **Biology: Snakes.** In the African rainforest there is a snake called a Gaboon viper. The fang size of this snake is proportional to the length of the snake. A 3 foot snake typically has 2 inch fangs. If a herpetologist finds Gaboon viper fangs that are 2.6 inches long, how big a snake would she expect to find?

27. **Investing.** Ashley has $120,000 to invest and decides to put some in a CD that earns 4% interest per year and the rest in a low-risk stock that earns 7%. How much did she invest in each to earn $7,800 interest in the first year?

28. **Investing.** You inherit $13,000 and you decide to invest the money in two different investments: one paying 10% and the other paying 14%. A year later your investments are worth $14,580. How much did you originally invest in each account?

29. **Investing.** Wendy was awarded a volleyball scholarship to the University of Michigan, so on graduation her parents gave her the $14,000 they had saved for her college tuition. She opted to invest some money in a privately held company that pays 10% per year and evenly split the remaining money between a money market account yielding 2% and a high-risk stock that yielded 40%. At the end of the first year she had $16,610 total. How much did she invest in each of the three?

30. **Interest.** A high school student was able to save $5,000 by working a part-time job every summer. He invested half the money in a money market account and half the money in a stock that paid three times as much interest as the money market account. After a year he earned $150 in interest. What were the interest rates of the money market account and the stock?

31. **Budget: Home Improvement.** When landscaping their yard, a couple budgeted $4,200. The irrigation system costs $2,400 and the sod costs $1,500. The rest they will spend on trees and shrubs. Trees each cost $32 and shrubs each cost $4. They plant a total of 33 trees and shrubs. How many of each did they plant in their yard?

32. **Budget: Shopping.** At the deli Jennifer bought spicy turkey and provolone cheese. The turkey costs $6.32 per pound and the cheese costs $4.27 per pound. In total, she bought 3.2 pounds and the price was $17.56. How many pounds of each did she buy?

33. **Chemistry.** For a certain experiment, a student requires 100 ml of a solution that is 8% HCl (hydrochloric acid). The storeroom has only solutions that are 5% HCl and 15% HCl. How many milliliters of each available solution should be mixed to get 100 ml of 8% HCl?

34. **Chemistry.** How many gallons of pure alcohol must be mixed with 5 gallons of a solution that is 20% alcohol to make a solution that is 50% alcohol?

35. **Automobiles.** A mechanic has tested the amount of antifreeze in your radiator. He says it is only 40% antifreeze and the remainder is water. How many gallons must be drained from your 5 gallon radiator and replaced with pure antifreeze to make the mixture in your radiator 80% antifreeze?

36. **Costs: Overhead.** A professor is awarded two research grants, each having different overhead rates. The research project conducted on campus has a rate of 42.5% overhead, and the project conducted in the field, off campus, has a rate of 26% overhead. If she was awarded $1,170,000 total for the two projects with an average overhead rate of 39%, how much was the research project on campus and how much was the research project off campus?

37. **Theater.** On the way to the movies a family picks up a custom-made bag of candies. The parents like caramels ($1.50/lb) and the children like gummy bears ($2.00/lb). They bought a 1.25 lb bag of combined candies that cost $2.50. How much of each candy did they buy?

38. **Coffee.** Joy is an instructional assistant in one of the college labs. She is on a very tight budget. She loves Jamaican Blue Mountain coffee, but it costs $12 a pound. She decides to blend this with regular coffee beans that costs $4.20 a pound. If she spends $14.25 on 2 pounds of coffee, how many pounds of each did she purchase?

39. **Communications.** The speed of light is approximately 3.0×10^8 meters per second (670,616,629 mph). The distance from Earth to Mars varies because of the orbits of the planets around the Sun. On average Mars is 100 million miles from Earth. If we use laser communication systems, what will be the delay between Houston and NASA astronauts on Mars?

40. **Speed of Sound.** The speed of sound is approximately 760 mph in air. If a gun is fired $\frac{1}{2}$ mile away, how long will it take the sound to reach you?

41. **Boating.** A motorboat can maintain a constant speed of 16 miles per hour relative to the water. The boat makes a trip upstream to a marina in 20 minutes. The return trip takes 15 minutes. What is the speed of the current?

42. **Aviation.** A Cessna 175 can average 130 mph. If a trip takes 2 hours one way and the return takes 1 hour and 15 minutes, find the wind speed, assuming it is constant.

43. **Exercise.** A jogger and a walker cover the same distance. The jogger finishes in 40 minutes. The walker takes an hour. How fast is each exerciser moving if the jogger runs 2 mph faster than the walker?

44. **Travel.** A high school student in Seattle, Washington, attended the University of Central Florida. On the way to UCF he took a southern route. After graduation he returned to Seattle via a northern trip. On both trips he had the same average speed. If the southern trek took 45 hours and the northern trek took 50 hours, and the northern trek was 300 miles longer, how long was each trip?

45. Distance–Rate–Time. College roommates leave for their first class in the same building. One walks at 2 mph and the other rides his bike at a slow 6 mph pace. How long will it take each to get to class if the walker takes 12 minutes longer to get to class and they travel on the same path?

46. Distance–Rate–Time. A long-distance delivery service sends out a truck with a package at 7 A.M. At 7:30 the manager realizes there was another package going to the same location. He sends out a car to catch the truck. If the truck travels at an average speed of 50 mph and the car travels at 70 mph, how long will it take the car to catch the truck?

47. Work. Christopher can paint the interior of his house in 15 hours. If he hires Cynthia to help him, they can do the same job together in 9 hours. If he lets Cynthia work alone, how long will it take her to paint the interior of his house?

48. Work. Jay and Morgan work in the summer for a landscaper. It takes Jay 3 hours to complete the company's largest yard alone. If Morgan helps him, it takes only 1 hour. How much time would it take Morgan alone?

49. Work. Tracey and Robin deliver Coke products to local convenience stores. Tracey can complete the deliveries in 4 hours alone. Robin can do it in 6 hours alone. If they decide to work together on a Saturday, how long will it take?

50. Work. Joshua can deliver his newspapers in 30 minutes. It takes Amber 20 minutes to do the same route. How long would it take them to deliver the newspapers if they worked together?

51. Music. A major chord in music is composed of notes whose frequencies are in the ratio 4:5:6. If the first note of a chord has a frequency of 264 hertz (middle C on the piano), find the frequencies of the other two notes. (*Hint:* Set up two proportions using 4:5 and 4:6.)

52. Music. A minor chord in music is composed of notes whose frequencies are in the ratio 10:12:15. If the first note of a minor chord is A, with a frequency of 220 hertz, what are the frequencies of the other two notes?

53. Grades. Danielle's test scores are 86, 80, 84, and 90. The final exam will count as $\frac{2}{3}$ of the final grade. What score does Danielle need on the final in order to earn a B, which requires an average score of 80? What score does she need to earn an A, which requires an average of 90?

54. Grades. Sam's final exam will count as two tests. His test scores are 80, 83, 71, 61, and 95. What score does Sam need on the final in order to have an average score of 80?

55. Sports. In Super Bowl XXXVII, the Tampa Bay Buccaneers scored a total of 48 points. All of their points came from field goals and touchdowns. Field goals are worth 3 points and each touchdown was worth 7 points (Martin Gramatica was successful in every extra point attempt). They scored a total of 8 times. How many field goals and touchdowns were scored?

56. Sports. A tight end can run the 100-yard dash in 12 seconds. A defensive back can do it in 10 seconds. The tight end catches a pass at his own 20 yard line with the defensive back at the 15 yard line. If no other players are nearby, at what yard line will the defensive back catch up to the tight end?

57. Recreation. How do two children of different weights balance on a seesaw? The heavier child sits closer to the center and the lighter child sits further away. When the product of the weight of the child and the distance from the center is equal on both sides, the seesaw should be horizontal to the ground. Suppose Max weighs 42 lbs and Maria weighs 60 lbs. If Max sits 5 feet from the center, how far should Maria sit from the center in order to balance the seesaw horizontal to the ground?

58. Recreation. Refer to Exercise 57. Suppose Martin, who weighs 33 lbs, sits on the side of the seesaw with Max. If their average distance to the center is 4 feet, how far should Maria sit from the center in order to balance the seesaw horizontal to the ground?

59. Recreation. If a seesaw has an adjustable bench, then the board can slide along the fulcrum. Maria and Max in Exercise 57 decide to sit on the very edge of the board on each side. Where should the fulcrum be placed along the board in order to balance the seesaw horizontally to the ground? Give the answer in terms of the distance from each child's end.

60. Recreation. Add Martin (Exercise 58) to Max's side of the seesaw and recalculate Exercise 59.

For Exercises 61–64, refer to this lens law. (See Exercise 74 in Section 1.1.)

The position of the image is found using the thin lens equation:

$$\frac{1}{f} = \frac{1}{d_o} + \frac{1}{d_i},$$

where d_o is the distance from the object to the lens, d_i is the distance from the lens to the image, and f is the focal length of the lens.

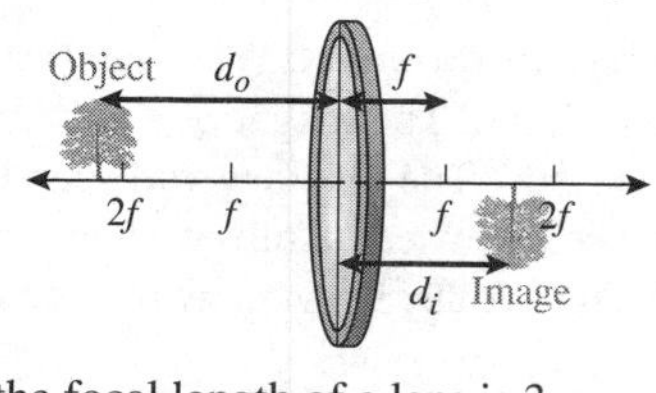

61. Optics. If the focal length of a lens is 3 cm and the image distance is 5 cm from the lens, what is the distance from the object to the lens?

62. Optics. If the focal length of the lens is 8 cm and the image distance is 2 cm from the lens, what is the distance from the object to the lens?

63. Optics. The focal length of a lens is 2 cm. If the image distance from the lens is half the distance from the object to the lens, find the object distance.

64. Optics. The focal length of a lens is 8 cm. If the image distance from the lens is half the distance from the object to the lens, find the object distance.

CONCEPTUAL

In Exercises 65–72, solve each formula for the specified variable.

65. $P = 2l + 2w$ for w

66. $P = 2l + 2w$ for l

67. $A = \frac{1}{2}bh$ for h

68. $C = 2\pi r$ for r

69. $A = lw$ for w

70. $d = rt$ for t

71. $V = lwh$ for h

72. $V = \pi r^2 h$ for h

CHALLENGE

73. Tricia and Janine are roommates and leave Houston on Interstate 10 at the same time to visit their families for a long weekend. Tricia travels west and Janine travels east. If Tricia's average speed is 12 mph faster than Janine's, find the speed of each if they are 320 miles apart in 2 hours and 30 minutes.

74. Rick and Mike are roommates and leave Gainesville on Interstate 75 at the same time to visit their girlfriends for a long weekend. Rick travels north and Mike travels south. If Mike's average speed is 8 mph faster than Rick's, find the speed of each if they are 210 miles apart in 1 hour and 30 minutes.

TECHNOLOGY

75. Suppose you bought a house for \$132,500 and sold it 3 years later for \$168,190. Plot these points using a graphing utility. Assuming a linear relationship, how much could you have sold the house for had you waited 2 additional years?

76. Suppose you bought a house for \$132,500 and sold it 3 years later for \$168,190. Plot these points using a graphing utility. Assuming a linear relationship, how much could you have sold the house for had you sold it 1 year after buying it?

77. A golf club membership has two options. Option A is a \$300 monthly fee plus \$15 cart fee every time you play. Option B has a \$150 monthly fee and a \$42 fee every time you play. Write a mathematical model for monthly costs for each plan and graph both in the same viewing rectangle using a graphing utility. Explain when Option A is the better deal and when Option B is the better deal.

78. A phone provider offers two calling plans. Plan A has a \$30 monthly charge and a \$.10 per minute charge on every call. Plan B has a \$50 monthly charge and a \$.03 per minute charge on every call. Explain when Plan A is the better deal and when Plan B is the better deal.

SECTION 1.3 QUADRATIC EQUATIONS

SKILLS OBJECTIVES

- Solve quadratic equations by factoring.
- Solve quadratic equations by means of the square root method.
- Complete the square to solve quadratic equations.
- Solve quadratic equations by means of the quadratic formula.
- Solve real-world problems that involve quadratic equations.

CONCEPTUAL OBJECTIVES

- Choose appropriate methods for solving quadratic equations.
- Interpret different types of solution sets (real, imaginary, complex conjugates, repeated roots).
- Derive the quadratic formula.

In a linear equation, the variable is raised only to the first power in any term where it occurs. In a *quadratic equation*, the variable is raised to the second power in at least one term. Examples of *quadratic equations*, also called second-degree equations, are:

$$x^2 + 3 = 7 \qquad 5x^2 + 4x - 7 = 0 \qquad x^2 - 3 = 0$$

DEFINITION **Quadratic Equation**

A **quadratic equation** in x is an equation that can be written in the **standard form**

$$ax^2 + bx + c = 0$$

where a, b, and c are real numbers and $a \neq 0$.

There are several methods for solving quadratic equations: *factoring*, the *square root method*, *completing the square*, and the *quadratic formula.*

Factoring

FACTORING METHOD

The **factoring method** applies the **zero product property:**

WORDS	MATH
If a product is zero, then at least one of its factors has to be zero.	If $B \cdot C = 0$, then $B = 0$ or $C = 0$ or both.

Consider $(x - 3)(x + 2) = 0$. The zero product property says that $x - 3 = 0$ or $x + 2 = 0$, which leads to $x = -2$ or $x = 3$. The solution set is $\{-2, 3\}$.

When a quadratic equation is written in the standard form $ax^2 + bx + c = 0$ it may be possible to factor the left side of the equation as a product of two first-degree polynomials. We use the zero product property and set each linear factor equal to zero. We solve the resulting two linear equations to obtain the solutions of the quadratic equation.

EXAMPLE 1 Solving a Quadratic Equation by Factoring

Solve the equation $x^2 - 6x - 16 = 0$.

Solution:

The quadratic equation is already in standard form.	$x^2 - 6x - 16 = 0$
Factor the left side into a product of two linear factors.	$(x - 8)(x + 2) = 0$
If a product equals zero, one of its factors has to be equal to zero.	$x - 8 = 0$ or $x + 2 = 0$
Solve both linear equations.	$\boxed{x = 8 \text{ or } x = -2}$

The solution set is $\boxed{\{-2, 8\}}$.

■ **YOUR TURN** Solve the quadratic equation $x^2 + x - 20 = 0$ by factoring.

■ **Answer:** The solution is $x = -5, 4$. The solution set is $\{-5, 4\}$.

EXAMPLE 2 Solving a Quadratic Equation by Factoring

Solve the equation $x^2 - 6x + 5 = -4$.

COMMON MISTAKE

A common mistake is to forget to put the equation in standard form first and then use the zero product property incorrectly.

CORRECT

Write the original equation.

$$x^2 - 6x + 5 = -4$$

Write the equation in standard form by adding 4 to both sides.

$$x^2 - 6x + 9 = 0$$

Factor the left side.

$$(x - 3)(x - 3) = 0$$

Use the zero product property and set each factor equal to zero.

$$x - 3 = 0 \text{ or } x - 3 = 0$$

Solve each linear equation.

$$\boxed{x = 3}$$

INCORRECT

Factor the left side.

$$(x - 5)(x - 1) = -4$$

The **error** occurs here.

$$x - 5 = -4 \text{ or } x - 1 = -4$$

▸**Don't forget to put the quadratic equation in standard form first.**

Note: The equation has one solution, or root, which is 3. The solution set is $\{3\}$. Since the linear factors were the same, or repeated, we say that 3 is a **double root**, or **repeated root**.

■ **YOUR TURN** Solve the quadratic equation $9p^2 = 24p - 16$ by factoring.

▼ **CAUTION**

Don't forget to put the quadratic equation in standard form first.

Technology Tip

Use a graphing utility to display graphs of $y_1 = x^2 - 6x + 5$ and $y_2 = -4$.

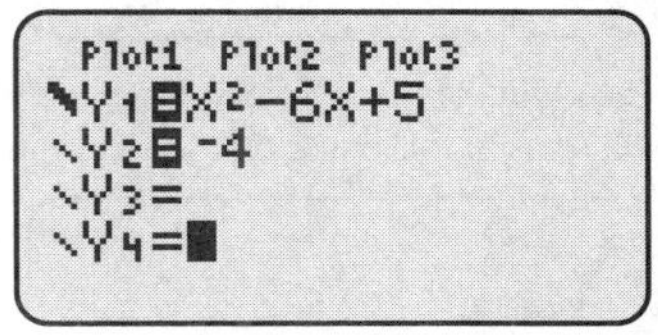

The point of intersection is the solution to this equation.

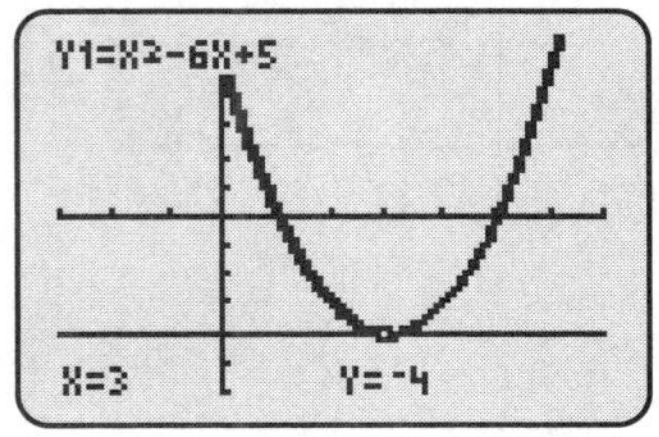

■ **Answer:** The solution is $p = \frac{4}{3}$, which is a double root. The solution set is $\left\{\frac{4}{3}\right\}$.

EXAMPLE 3 Solving a Quadratic Equation by Factoring

Solve the equation $2x^2 = 3x$.

COMMON MISTAKE

The common mistake here is dividing both sides by x, which is not allowed because x might be 0.

CORRECT

Write the equation in standard form by subtracting $3x$.

$$2x^2 - 3x = 0$$

Factor the left side.

$$x(2x - 3) = 0$$

INCORRECT

Write the original equation.

$$2x^2 = 3x$$

The **error** occurs here when both sides are divided by x.

$$2x = 3$$

▼ **CAUTION**

Do not divide by a variable (because the value of that variable may be zero). Bring all terms to one side first and then factor.

Technology Tip

Use a graphing utility to display graphs of $y_1 = 2x^2$ and $y_2 = 3x$.

The points of intersection are the solutions to this equation.

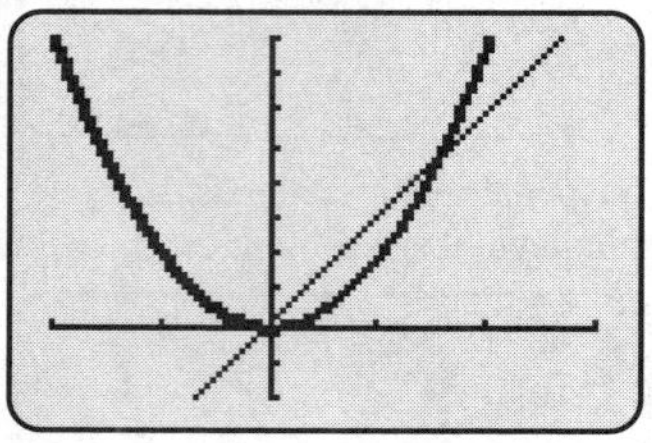

Use the zero product property and set each factor equal to zero.

$$x = 0 \text{ or } 2x - 3 = 0$$

Solve each linear equation.

$$\boxed{x = 0 \text{ or } x = \frac{3}{2}}$$

The solution set is $\boxed{\{0, \frac{3}{2}\}}$.

In Example 3, the root $x = 0$ is lost when the original quadratic equation is divided by x. Remember to put the equation in standard form first and then factor.

Square Root Method

The square root of 16, $\sqrt{16}$, is 4, *not* ± 4. In the review (Chapter 0) the **principal square root** was discussed. The solutions to $x^2 = 16$, however, are $x = -4$ and $x = 4$. Let us now investigate quadratic equations that do not have a first-degree term. They have the form

$$ax^2 + c = 0 \quad a \neq 0$$

The method we use to solve such equations uses the square root property.

SQUARE ROOT PROPERTY

WORDS	**MATH**
If an expression squared is equal to a constant, then that expression is equal to the positive or negative square root of the constant.	If $x^2 = P$, then $x = \pm\sqrt{P}$.

Note: The variable squared must be isolated first (coefficient equal to 1).

EXAMPLE 4 Using the Square Root Property

Solve the equation $3x^2 - 27 = 0$.

Solution:

Add 27 to both sides. $\quad 3x^2 = 27$

Divide both sides by 3. $\quad x^2 = 9$

Apply the square root property. $\quad x = \pm\sqrt{9} = \pm 3$

The solution set is $\boxed{\{-3, 3\}}$.

If we alter Example 4 by changing subtraction to addition, we see in Example 5 that we get imaginary roots (as opposed to real roots), which we discussed in Chapter 0.

EXAMPLE 5 Using the Square Root Property

Solve the equation $3x^2 + 27 = 0$.

Solution:

Subtract 27 from both sides. $3x^2 = -27$

Divide by 3. $x^2 = -9$

Apply the square root property. $x = \pm\sqrt{-9}$

Simplify. $x = \pm i\sqrt{9} = \pm 3i$

The solution set is $\boxed{\{-3i, 3i\}}$.

YOUR TURN Solve the equations $y^2 - 147 = 0$ and $v^2 + 64 = 0$.

Answer: The solution is $y = \pm 7\sqrt{3}$. The solution set is $\{-7\sqrt{3}, 7\sqrt{3}\}$. The solution is $v = \pm 8i$. The solution set is $\{-8i, 8i\}$.

EXAMPLE 6 Using the Square Root Property

Solve the equation $(x - 2)^2 = 16$.

Solution:

If an expression squared is 16, then the expression equals $\pm\sqrt{16}$. $(x - 2) = \pm\sqrt{16}$

Separate into two equations. $x - 2 = \sqrt{16}$ or $x - 2 = -\sqrt{16}$

$x - 2 = 4 \qquad x - 2 = -4$

$x = 6 \qquad x = -2$

The solution set is $\boxed{\{-2, 6\}}$.

It is acceptable notation to keep the equations together.

$$(x - 2) = \pm\sqrt{16}$$
$$x - 2 = \pm 4$$
$$x = 2 \pm 4$$
$$\boxed{x = -2, 6}$$

Completing the Square

Factoring and the square root method are two efficient, quick procedures for solving many quadratic equations. However, some equations, such as $x^2 - 10x - 3 = 0$, cannot be solved directly by these methods. A more general procedure to solve this kind of equation is called **completing the square**. The idea behind completing the square is to transform any standard quadratic equation $ax^2 + bx + c = 0$ into the form $(x + A)^2 = B$, where A and B are constants and the left side, $(x + A)^2$, has the form of a **perfect square**. This last equation can then be solved by the square root method. How do we transform the first equation into the second equation?

Note that the above-mentioned example, $x^2 - 10x - 3 = 0$, cannot be factored into expressions in which all numbers are integers (or even rational numbers). We can, however, transform this quadratic equation into a form that contains a perfect square.

WORDS	MATH
Write the original equation.	$x^2 - 10x - 3 = 0$
Add 3 to both sides.	$x^2 - 10x = 3$
Add 25 to both sides.*	$x^2 - 10x + 25 = 3 + 25$
The left side can be written as a perfect square.	$(x - 5)^2 = 28$
Apply the square root method.	$x - 5 = \pm\sqrt{28}$
Add 5 to both sides.	$x = 5 \pm 2\sqrt{7}$

*Why did we add 25 to both sides? Recall that $(x - c)^2 = x^2 - 2xc + c^2$. In this case $c = 5$ in order for $-2xc = -10x$. Therefore the desired perfect square $(x - 5)^2$ results in $x^2 - 10x + 25$. Applying this product we see that $+25$ is needed. A systematic approach is to take the coefficient of the first degree term $x^2 - 10x - 3 = 0$, which is -10. Take half of (-10) which is (-5) and then square it $(-5)^2 = 25$.

SOLVING A QUADRATIC EQUATION BY COMPLETING THE SQUARE

WORDS	MATH
Express the quadratic equation in the following form.	$x^2 + bx = c$
Divide b by 2 and square the result, then add the square to both sides.	$x^2 + bx + \left(\frac{b}{2}\right)^2 = c + \left(\frac{b}{2}\right)^2$
Write the left side of the equation as a perfect square.	$\left(x + \frac{b}{2}\right)^2 = c + \left(\frac{b}{2}\right)^2$
Solve using the square root method.	

Technology Tip

Graph $y_1 = x^2 + 8x - 3$.

```
Plot1 Plot2 Plot3
\Y1=X²+8X-3
```

The x-intercepts are the solutions to this equation.

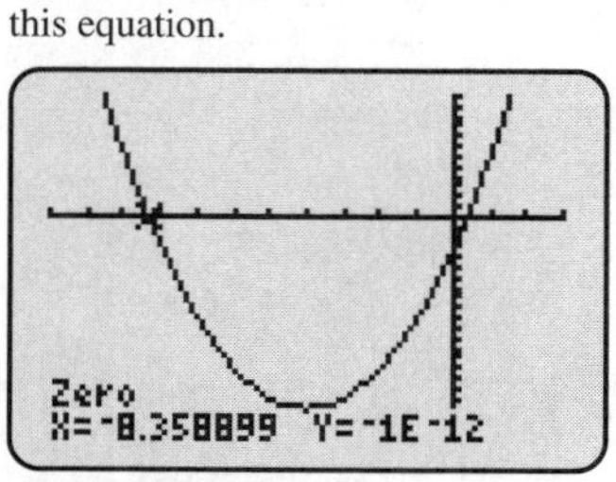


```
-4+√(19)
          .3588989435
-4-√(19)
         -8.358898944
```

EXAMPLE 7 Completing the Square

Solve the quadratic equation $x^2 + 8x - 3 = 0$ by completing the square.

Solution:

Add 3 to both sides.	$x^2 + 8x = 3$
Add $\left(\frac{1}{2}\cdot 8\right)^2 = 4^2$ to both sides.	$x^2 + 8x + 4^2 = 3 + 4^2$
Write the left side as a perfect square and simplify the right side.	$(x + 4)^2 = 19$
Apply the square root method to solve.	$x + 4 = \pm\sqrt{19}$
Subtract 4 from both sides.	$\boxed{x = -4 \pm \sqrt{19}}$

The solution set is $\boxed{\{-4 - \sqrt{19}, -4 + \sqrt{19}\}}$.

In Example 7, the leading coefficient (the coefficient of the x^2 term) is 1. When the leading coefficient is not 1, start by first dividing the equation by that leading coefficient.

EXAMPLE 8 Completing the Square When the Leading Coefficient Is Not Equal to 1

Solve the equation $3x^2 - 12x + 13 = 0$ by completing the square.

Solution:

Divide by the leading coefficient, 3. $$x^2 - 4x + \frac{13}{3} = 0$$

Collect variables to one side of the equation and constants to the other side. $$x^2 - 4x = -\frac{13}{3}$$

Add $\left(-\frac{4}{2}\right)^2 = 4$ to both sides. $$x^2 - 4x + 4 = -\frac{13}{3} + 4$$

Write the left side of the equation as a perfect square and simplify the right side. $$(x - 2)^2 = -\frac{1}{3}$$

Solve using the square root method. $$x - 2 = \pm\sqrt{-\frac{1}{3}}$$

Simplify. $$x = 2 \pm i\sqrt{\frac{1}{3}}$$

Rationalize the denominator (Chapter 0). $$\boxed{x = 2 - \frac{i\sqrt{3}}{3},\ x = 2 + \frac{i\sqrt{3}}{3}}$$

The solution set is $\boxed{\left\{2 - \frac{i\sqrt{3}}{3}, 2 + \frac{i\sqrt{3}}{3}\right\}}$.

■ **YOUR TURN** Solve the equation $2x^2 - 4x + 3 = 0$ by completing the square.

Technology Tip

Graph $y_1 = 3x^2 - 12x + 13$.

The graph does not cross the x-axis, so there is no real solution to this equation.

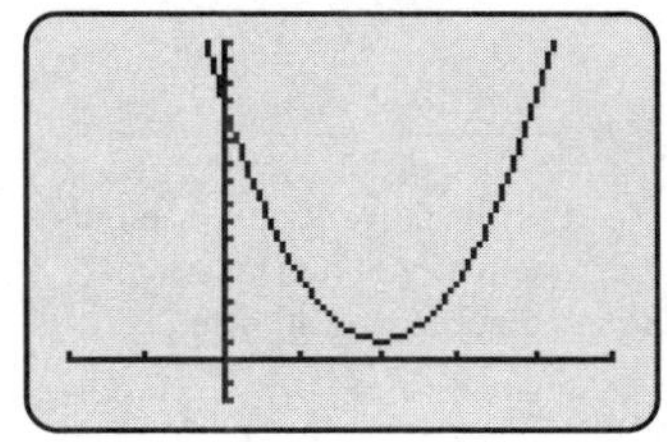

Study Tip

When the leading coefficient is not 1, start by first dividing the equation by that leading coefficient.

■ **Answer:** The solution is $x = 1 \pm \frac{i\sqrt{2}}{2}$. The solution set is $\left\{1 - \frac{i\sqrt{2}}{2}, 1 + \frac{i\sqrt{2}}{2}\right\}$.

Quadratic Formula

Let us now consider the most general quadratic equation:

$$ax^2 + bx + c = 0 \quad a \neq 0$$

We can solve this equation by completing the square.

WORDS	MATH
Divide the equation by the leading coefficient a.	$x^2 + \frac{b}{a}x + \frac{c}{a} = 0$
Subtract $\frac{c}{a}$ from both sides.	$x^2 + \frac{b}{a}x = -\frac{c}{a}$
Square half of $\frac{b}{a}$ and add the result $\left(\frac{b}{2a}\right)^2$ to both sides.	$x^2 + \frac{b}{a}x + \left(\frac{b}{2a}\right)^2 = \left(\frac{b}{2a}\right)^2 - \frac{c}{a}$
Write the left side of the equation as a perfect square and the right side as a single fraction.	$\left(x + \frac{b}{2a}\right)^2 = \frac{b^2 - 4ac}{4a^2}$

Solve using the square root method. $x + \frac{b}{2a} = \pm\sqrt{\frac{b^2 - 4ac}{4a^2}}$

Subtract $\frac{b}{2a}$ from both sides and simplify the radical. $x = -\frac{b}{2a} \pm \frac{\sqrt{b^2 - 4ac}}{2a}$

Write as a single fraction. $x = \frac{-b \pm \sqrt{b^2 - 4ac}}{2a}$

We have derived the **quadratic formula**.

Study Tip

$x = \frac{-b \pm \sqrt{b^2 - 4ac}}{2a}$

Read as "negative b plus or minus the square root of the quantity b squared minus $4ac$ all over $2a$."

QUADRATIC FORMULA

If $ax^2 + bx + c = 0$, $a \neq 0$, then the solution is

$$x = \frac{-b \pm \sqrt{b^2 - 4ac}}{2a}$$

Note: The quadratic equation must be in standard form ($ax^2 + bx + c = 0$) in order to identify the parameters:

a—coefficient of x^2 b—coefficient of x c—constant

Study Tip

The quadratic formula works for *any* quadratic equation.

We read this formula as *negative b plus or minus the square root of the quantity b squared minus 4ac all over 2a.* It is important to note that negative b could be positive (if b is negative). For this reason, an alternate form is "opposite b. . . ." The quadratic formula should be memorized and used when simpler methods (factoring and the square root method) cannot be used. The quadratic formula works for *any* quadratic equation.

Study Tip

Using parentheses as place holders helps avoid ± errors.

$x = \frac{-b \pm \sqrt{b^2 - 4ac}}{2a}$

$x = \frac{-(\square) \pm \sqrt{(\square)^2 - 4(\square)(\square)}}{2(\square)}$

EXAMPLE 9 Using the Quadratic Formula and Finding Two Distinct Real Roots

Use the quadratic formula to solve the quadratic equation $x^2 - 4x - 1 = 0$.

Solution:

For this problem $a = 1$, $b = -4$, and $c = -1$.

Write the quadratic formula. $x = \frac{-b \pm \sqrt{b^2 - 4ac}}{2a}$

Use parentheses to avoid losing a minus sign. $x = \frac{-(\square) \pm \sqrt{(\square)^2 - 4(\square)(\square)}}{2(\square)}$

Substitute values for a, b, and c into the parentheses. $x = \frac{-(-4) \pm \sqrt{(-4)^2 - 4(1)(-1)}}{2(1)}$

Simplify. $x = \frac{4 \pm \sqrt{16 + 4}}{2} = \frac{4 \pm \sqrt{20}}{2} = \frac{4 \pm 2\sqrt{5}}{2} = \frac{4}{2} \pm \frac{2\sqrt{5}}{2} = \boxed{2 \pm \sqrt{5}}$

The solution set $\boxed{\{2 - \sqrt{5}, 2 + \sqrt{5}\}}$ contains two distinct real numbers.

YOUR TURN Use the quadratic formula to solve the quadratic equation $x^2 + 6x - 2 = 0$.

Answer: The solution is $x = -3 \pm \sqrt{11}$. The solution set is $\{-3 - \sqrt{11}, -3 + \sqrt{11}\}$.

EXAMPLE 10 Using the Quadratic Formula and Finding Two Complex Roots

Use the quadratic formula to solve the quadratic equation $x^2 + 8 = 4x$.

Solution:

Write this equation in standard form $x^2 - 4x + 8 = 0$ in order to identify $a = 1$, $b = -4$, and $c = 8$.

Write the quadratic formula. $$x = \frac{-b \pm \sqrt{b^2 - 4ac}}{2a}$$

Use parentheses to avoid overlooking a minus sign. $$x = \frac{-(\square) \pm \sqrt{(\square)^2 - 4(\square)(\square)}}{2(\square)}$$

Substitute the values for a, b, and c into the parentheses. $$x = \frac{-(-4) \pm \sqrt{(-4)^2 - 4(1)(8)}}{2(1)}$$

Simplify. $$x = \frac{4 \pm \sqrt{16 - 32}}{2} = \frac{4 \pm \sqrt{-16}}{2} = \frac{4 \pm 4i}{2} = \frac{4}{2} \pm \frac{4i}{2} = \boxed{2 \pm 2i}$$

The solution set $\boxed{\{2 - 2i, 2 + 2i\}}$ contains two complex numbers. Note that they are complex conjugates of each other.

YOUR TURN Use the quadratic formula to solve the quadratic equation $x^2 + 2 = 2x$.

Technology Tip

Set the graphing utility in complex number mode. Keystrokes:

MODE NORMAL $a + bi$

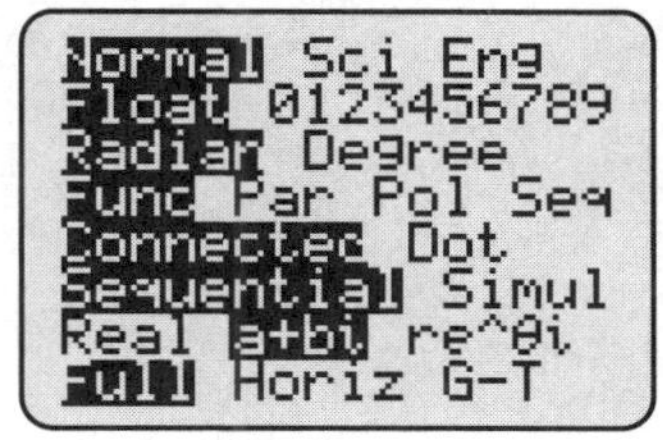

Answer: The solution set is $\{1 - i, 1 + i\}$.

EXAMPLE 11 Using the Quadratic Formula and Finding One Repeated Real Root

Use the quadratic formula to solve the quadratic equation $4x^2 - 4x + 1 = 0$.

Solution:

Identify a, b, and c. $a = 4, b = -4, c = 1$

Write the quadratic formula. $$x = \frac{-b \pm \sqrt{b^2 - 4ac}}{2a}$$

Use parentheses to avoid losing a minus sign. $$x = \frac{-(\square) \pm \sqrt{(\square)^2 - 4(\square)(\square)}}{2(\square)}$$

Substitute values $a = 4$, $b = -4$, $c = 1$. $$x = \frac{-(-4) \pm \sqrt{(-4)^2 - 4(4)(1)}}{2(4)}$$

Simplify. $$x = \frac{4 \pm \sqrt{16 - 16}}{8} = \frac{4 \pm 0}{8} = \frac{1}{2}$$

The solution set is a repeated real root $\boxed{\left\{\frac{1}{2}\right\}}$.

Note: This quadratic also could have been solved by factoring: $(2x - 1)^2 = 0$.

YOUR TURN Use the quadratic formula to solve the quadratic equation $9x^2 - 6x + 1 = 0$.

Answer: $\left\{\frac{1}{3}\right\}$

TYPES OF SOLUTIONS

The term inside the radical, $b^2 - 4ac$, is called the **discriminant**. The discriminant gives important information about the corresponding solutions or roots of $ax^2 + bx + c = 0$, where a, b, and c are real numbers.

$b^2 - 4ac$	SOLUTIONS (ROOTS)
Positive	Two distinct real roots
0	One real root (a double or repeated root)
Negative	Two complex roots (complex conjugates)

In Example 9, the discriminant is positive and the solution has two distinct real roots. In Example 10, the discriminant is negative and the solution has two complex (conjugate) roots. In Example 11, the discriminant is zero and the solution has one repeated real root.

Applications Involving Quadratic Equations

In Section 1.2, we developed a procedure for solving word problems involving linear equations. The procedure is the same for applications involving quadratic equations. The only difference is that the mathematical equations will be quadratic, as opposed to linear.

EXAMPLE 12 Stock Value

From 1999 to 2001 the price of Abercrombie & Fitch's (ANF) stock was approximately given by $P = 0.2t^2 - 5.6t + 50.2$, where P is the price of stock in dollars, t is in months, and $t = 1$ corresponds to January 1999. When was the value of the stock worth \$30?

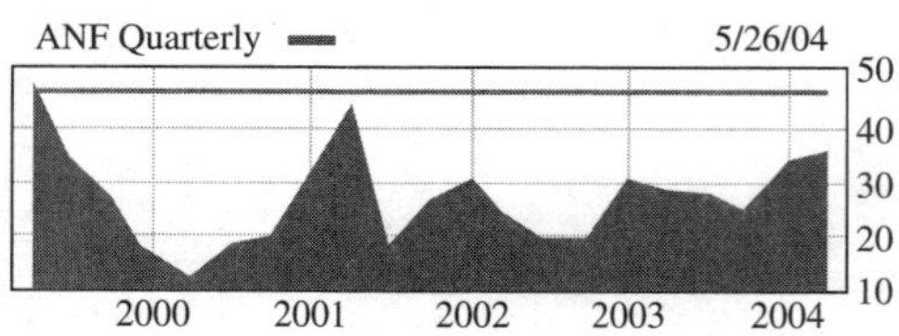

Technology Tip

The graphing utility screen for $\dfrac{-(-5.6) \pm \sqrt{(-5.6)^2 - 4(0.2)(20.2)}}{2(0.2)}$

```
(-(-5.6)-√((-5.6
)²-4*.2*20.2))/(
2*.2)
         4.253205655
```

```
(-(-5.6)+√((-5.6
)²-4*.2*20.2))/(
2*.2)
         23.74679434
```

Solution:

STEP 1 Identify the question.
When is the price of the stock equal to \$30?

STEP 2 Make notes.
Stock price: $P = 0.2t^2 - 5.6t + 50.2$

$P = 30$

STEP 3 Set up an equation. $0.2t^2 - 5.6t + 50.2 = 30$

STEP 4 Solve the equation.
Subtract 30 from both sides. $0.2t^2 - 5.6t + 20.2 = 0$

Solve for t using the quadratic formula.

$$t = \frac{-(-5.6) \pm \sqrt{(-5.6)^2 - 4(0.2)(20.2)}}{2(0.2)}$$

Simplify. $t \approx \dfrac{5.6 \pm 3.9}{0.4} \approx 4.25, 23.75$

Rounding these two numbers, we find that $t \approx 4$ and $t \approx 24$. Since $t = 1$ corresponds to January 1999, these two solutions correspond to $\boxed{\text{April 1999 and December 2000}}$.

STEP 5 Check the solution.
Look at the figure. The horizontal axis represents the year (2000 corresponds to January 2000), and the vertical axis represents the stock price. Estimating when the stock price is approximately \$30, we find April 1999 and December 2000.

EXAMPLE 13 Pythagorean Theorem

Hitachi makes a 60 inch HDTV that has a 60 inch diagonal. If the width of the screen is approximately 52 inches, what is the approximate height of the screen?

Solution:

STEP 1 Identify the question.
What is the approximate height of the HDTV screen?

STEP 2 Make notes.

STEP 3 Set up an equation.

Recall the Pythagorean theorem. $a^2 + b^2 = c^2$

Substitute in the known values. $h^2 + 52^2 = 60^2$

STEP 4 Solve the equation.

Simplify the constants. $h^2 + 2704 = 3600$

Subtract 2704 from both sides. $h^2 = 896$

Solve using the square root method. $h = \pm\sqrt{896} \approx \pm 30$

Distance is positive, so the negative value is eliminated.
The height is approximately $\boxed{30 \text{ inches}}$.

STEP 5 Check the solution.

$$30^2 + 52^2 \stackrel{?}{=} 60^2$$

$$900 + 2704 \stackrel{?}{=} 3600$$

$$3604 \approx 3600$$

Study Tip

Dimensions such as length and width are distances, which are defined as positive quantities. Although the mathematics may yield both positive and negative values, the negative values are excluded.

SECTION 1.3 SUMMARY

The four methods for solving quadratic equations are factoring, the square root method, completing the square, and the quadratic formula. Factoring and the square root method are the quickest and easiest but cannot always be used. The quadratic formula and completing the square work for all quadratic equations and can yield three types of solutions: two distinct real roots, one real root (repeated), or two complex roots (conjugates of each other).

SECTION 1.3 EXERCISES

SKILLS

In Exercises 1–22, solve by factoring.

1. $x^2 - 5x + 6 = 0$
2. $v^2 + 7v + 6 = 0$
3. $p^2 - 8p + 15 = 0$
4. $u^2 - 2u - 24 = 0$
5. $x^2 = 12 - x$
6. $11x = 2x^2 + 12$
7. $16x^2 + 8x = -1$
8. $3x^2 + 10x - 8 = 0$
9. $9y^2 + 1 = 6y$
10. $4x = 4x^2 + 1$
11. $8y^2 = 16y$
12. $3A^2 = -12A$
13. $9p^2 = 12p - 4$
14. $4u^2 = 20u - 25$
15. $x^2 - 9 = 0$
16. $16v^2 - 25 = 0$
17. $x(x + 4) = 12$
18. $3t^2 - 48 = 0$
19. $2p^2 - 50 = 0$
20. $5y^2 - 45 = 0$
21. $3x^2 = 12$
22. $7v^2 = 28$

In Exercises 23–34, solve using the square root method.

23. $p^2 - 8 = 0$
24. $y^2 - 72 = 0$
25. $x^2 + 9 = 0$
26. $v^2 + 16 = 0$
27. $(x - 3)^2 = 36$
28. $(x - 1)^2 = 25$
29. $(2x + 3)^2 = -4$
30. $(4x - 1)^2 = -16$
31. $(5x - 2)^2 = 27$
32. $(3x + 8)^2 = 12$
33. $(1 - x)^2 = 9$
34. $(1 - x)^2 = -9$

In Exercises 35–44, what number should be added to complete the square of each expression?

35. $x^2 + 6x$
36. $x^2 - 8x$
37. $x^2 - 12x$
38. $x^2 + 20x$
39. $x^2 - \frac{1}{2}x$
40. $x^2 - \frac{1}{3}x$
41. $x^2 + \frac{2}{5}x$
42. $x^2 + \frac{4}{5}x$
43. $x^2 - 2.4x$
44. $x^2 + 1.6x$

In Exercises 45–56, solve by completing the square.

45. $x^2 + 2x = 3$
46. $y^2 + 8y - 2 = 0$
47. $t^2 - 6t = -5$
48. $x^2 + 10x = -21$
49. $y^2 - 4y + 3 = 0$
50. $x^2 - 7x + 12 = 0$
51. $2p^2 + 8p = -3$
52. $2x^2 - 4x + 3 = 0$
53. $2x^2 - 7x + 3 = 0$
54. $3x^2 - 5x - 10 = 0$
55. $\frac{x^2}{2} - 2x = \frac{1}{4}$
56. $\frac{t^2}{3} + \frac{2t}{3} + \frac{5}{6} = 0$

In Exercises 57–68, solve using the quadratic formula.

57. $t^2 + 3t - 1 = 0$
58. $t^2 + 2t = 1$
59. $s^2 + s + 1 = 0$
60. $2s^2 + 5s = -2$
61. $3x^2 - 3x - 4 = 0$
62. $4x^2 - 2x = 7$
63. $x^2 - 2x + 17 = 0$
64. $4m^2 + 7m + 8 = 0$
65. $5x^2 + 7x = 3$
66. $3x^2 + 5x = -11$
67. $\frac{1}{4}x^2 + \frac{2}{3}x - \frac{1}{2} = 0$
68. $\frac{1}{4}x^2 - \frac{2}{3}x - \frac{1}{3} = 0$

In Exercises 69–74, determine whether the discriminant is positive, negative, or zero, and indicate the number and type of root to expect. Do not solve.

69. $x^2 - 22x + 121 = 0$
70. $x^2 - 28x + 196 = 0$
71. $2y^2 - 30y + 68 = 0$
72. $-3y^2 + 27y + 66 = 0$
73. $9x^2 - 7x + 8 = 0$
74. $-3x^2 + 5x - 7 = 0$

In Exercises 75–94, solve using any method.

75. $v^2 - 8v = 20$
76. $v^2 - 8v = -20$
77. $t^2 + 5t - 6 = 0$
78. $t^2 + 5t + 6 = 0$
79. $(x + 3)^2 = 16$
80. $(x + 3)^2 = -16$
81. $(p - 2)^2 = 4p$
82. $(u + 5)^2 = 16u$
83. $8w^2 + 2w + 21 = 0$
84. $8w^2 + 2w - 21 = 0$
85. $3p^2 - 9p + 1 = 0$
86. $3p^2 - 9p - 1 = 0$

87. $\frac{2}{3}t^2 - \frac{4}{3}t = \frac{1}{5}$

88. $\frac{1}{2}x^2 + \frac{2}{3}x = \frac{2}{5}$

89. $x + \frac{12}{x} = 7$

90. $x - \frac{10}{x} = -3$

91. $\frac{4(x-2)}{x-3} + \frac{3}{x} = \frac{-3}{x(x-3)}$

92. $\frac{5}{y+4} = 4 + \frac{3}{y-2}$

93. $x^2 - 0.1x = 0.12$

94. $y^2 - 0.5y = -0.06$

APPLICATIONS

95. Stock Value. From June 2003 until April 2004 JetBlue airlines stock (JBLU) was approximately worth $P = -4t^2 + 80t - 360$, where P denotes the price of the stock in dollars and t corresponds to months, with $t = 1$ corresponding to January 2003. During what months was the stock equal to \$24?

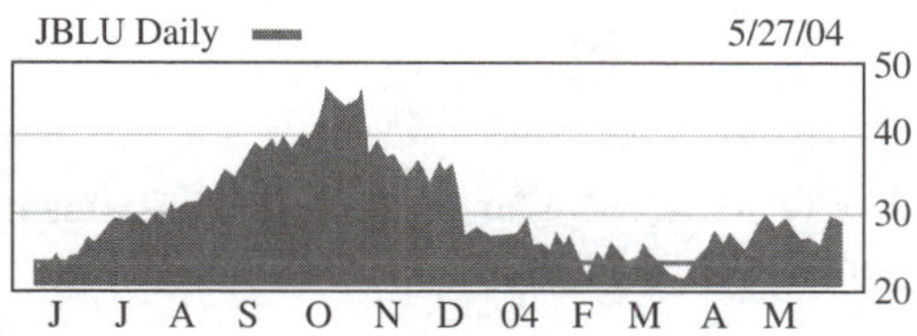

96. Stock Value. From November 2003 until March 2004 Wal-Mart Stock (WMT) was approximately worth $P = 2t^2 - 12t + 70$, where P denotes the price of the stock in dollars and t corresponds to months, with $t = 1$ corresponding to November 2003. During what months was the stock equal to \$60?

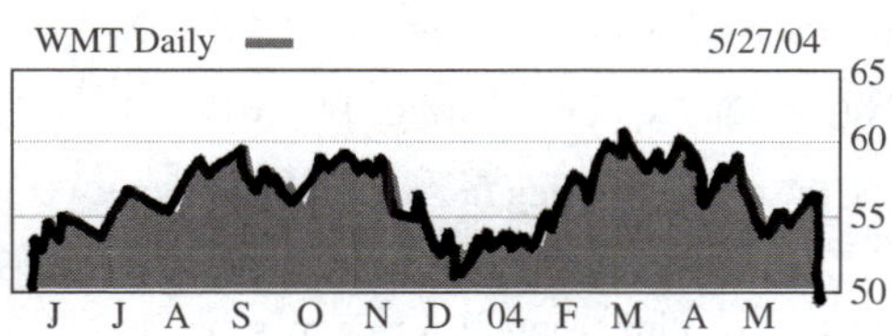

97. Television. A standard 32-in. television has a 32 in. diagonal and a 25 in. width. What is the height of the 32-in. television?

98. Television. A 42-in. LCD television has a 42 in. diagonal and a 20 in. height. What is the width of the 42-in. LCD television?

99. Numbers. Find two consecutive numbers such that their sum is 35 and their product is 306.

100. Numbers. Find two consecutive odd integers such that their sum is 24 and their product is 143.

101. Geometry. The area of a rectangle is 135 square feet. The width is 6 feet less than the length. Find the dimensions of the rectangle.

102. Geometry. A rectangle has an area of 31.5 square meters. If the length is 2 more than twice the width, find the dimensions of the rectangle.

103. Geometry. A triangle has a height that is 2 more than 3 times the base and an area of 60 square units. Find the base and height.

104. Geometry. A square's side is increased by 3 yards, which corresponds to an increase in the area by 69 square yards. How many yards is the side of the initial square?

105. Falling Objects. If a person drops a water balloon off the rooftop of a 100 foot building, the height of the water balloon is given by the equation $h = -16t^2 + 100$, where t is in seconds. When will the water balloon hit the ground?

106. Falling Objects. If the person in Exercise 105 throws the water balloon downward with a speed of 5 feet per second, the height of the water balloon is given by the equation$h = -16t^2 - 5t + 100$, where t is in seconds. When will the water balloon hit the ground?

107. Gardening. A square garden has an area of 900 square feet. If a sprinkler (with a circular pattern) is placed in the center of the garden, what is the minimum radius of spray the sprinkler would need in order to water all of the garden?

108. Sports. A baseball diamond is a square. The distance from base to base is 90 feet. What is the distance from home plate to second base?

109. Volume. A flat square piece of cardboard is used to construct an open box. Cutting a 1 foot by 1 foot square off of each corner and folding up the edges will yield an open box (assuming these edges are taped together). If the desired volume of the box is 9 cubic feet, what are the dimensions of the original square piece of cardboard?

110. Volume. A rectangular piece of cardboard whose length is twice its width is used to construct an open box. Cutting a 1 foot by 1 foot square off of each corner and folding up the edges will yield an open box. If the desired volume is 12 cubic feet, what are the dimensions of the original rectangular piece of cardboard?

111. Gardening. A landscaper has planted a rectangular garden that measures 8 feet by 5 feet. He has ordered 1 cubic yard (27 cubic feet) of stones for a border along the outside of the garden. If the border needs to be 4 inches deep and he wants to use all of the stones, how wide should the border be?

112. Gardening. A gardener has planted a semicircular rose garden with a radius of 6 feet, and 2 cubic yards of mulch (1 cubic yard = 27 cubic feet) is being delivered. Assuming she uses all of the mulch, how deep will the layer of mulch be?

113. Work. Lindsay and Kimmie, working together, can balance the financials for the Kappa Kappa Gamma sorority in 6 days. Lindsay by herself can complete the job in 5 days less than Kimmie. How long will it take Lindsay to complete the job by herself?

114. Work. When Jack cleans the house, it takes him 4 hours. When Ryan cleans the house, it takes him 6 hours. How long would it take both of them if they worked together?

CATCH THE MISTAKE

In Exercises 115–118, explain the mistake that is made.

115.
$$t^2 - 5t - 6 = 0$$
$$(t-3)(t-2) = 0$$
$$t = 2, 3$$

116.
$$(2y-3)^2 = 25$$
$$2y - 3 = 5$$
$$2y = 8$$
$$y = \frac{5}{4}$$
$$y = 4$$

117.
$$16a^2 + 9 = 0$$
$$16a^2 = -9$$
$$a^2 = -\frac{9}{16}$$
$$a = \pm\sqrt{\frac{9}{16}}$$
$$a = \pm\frac{3}{4}$$

118.
$$2x^2 - 4x = 3$$
$$2(x^2 - 2x) = 3$$
$$2(x^2 - 2x + 1) = 3 + 1$$
$$2(x-1)^2 = 4$$
$$(x-1)^2 = 2$$
$$x - 1 = \pm\sqrt{2}$$
$$x = 1 \pm \sqrt{2}$$

CONCEPTUAL

In Exercises 119–122, determine whether the following statements are true or false.

119. The equation $(3x + 1)^2 = 16$ has the same solution set as the equation $3x + 1 = 4$.

120. The quadratic equation $ax^2 + bx + c = 0$ can be solved by the square root method only if $b = 0$.

121. All quadratic equations can be solved exactly.

122. The quadratic formula can be used to solve any quadratic equation.

123. Write a quadratic equation in general form that has $x = a$ as a repeated real root.

124. Write a quadratic equation in general form that has $x = bi$ as a root.

125. Write a quadratic equation in general form that has the solution set $\{2, 5\}$.

126. Write a quadratic equation in general form that has the solution set $\{-3, 0\}$.

In Exercises 127–130, solve for the indicated variable in terms of other variables.

127. Solve $s = \frac{1}{2}gt^2$ for t.

128. Solve $A = P(1 + r)^2$ for r.

129. Solve $a^2 + b^2 = c^2$ for c.

130. Solve $P = EI - RI^2$ for I.

131. Solve the equation by factoring: $x^4 - 4x^2 = 0$.

132. Solve the equation by factoring: $3x - 6x^2 = 0$.

133. Solve the equation using factoring by grouping: $x^3 + x^2 - 4x - 4 = 0$.

134. Solve the equation using factoring by grouping: $x^3 + 2x^2 - x - 2 = 0$.

CHALLENGE

135. Show that the sum of the roots of a quadratic equation is equal to $-\frac{b}{a}$.

136. Show that the product of the roots of a quadratic equation is equal to $\frac{c}{a}$.

137. Write a quadratic equation in general form whose solution set is $\{3 + \sqrt{5}, 3 - \sqrt{5}\}$.

138. Write a quadratic equation in general form whose solution set is $\{2 - i, 2 + i\}$.

139. Aviation. An airplane takes 1 hour longer to go a distance of 600 miles flying against a headwind than on the return trip with a tailwind. If the speed of the wind is 50 mph, find the speed of the plane in still air.

140. Boating. A speedboat takes 1 hour longer to go 24 miles up a river than to return. If the boat cruises at 10 mph in still water, what is the rate of the current?

141. Find a quadratic equation whose two distinct real roots are the negatives of the two distinct real roots of the equation $ax^2 + bx + c = 0$.

142. Find a quadratic equation whose two distinct real roots are the reciprocals of the two distinct real roots of the equation $ax^2 + bx + c = 0$.

143. A small jet and a 757 leave Atlanta at 1 P.M. The small jet is traveling due west. The 757 is traveling due south. The speed of the 757 is 100 mph faster than the small jet. At 3 P.M. the planes are 1000 miles apart. Find the average speed of each plane.

144. Two boats leave Key West at noon. The smaller boat is traveling due west. The larger boat is traveling due south. The speed of the larger boat is 10 mph faster than the speed of the smaller boat. At 3 P.M. the boats are 150 miles apart. Find the average speed of each boat.

TECHNOLOGY

145. Solve the equation $x^2 - x = 2$ by first writing in standard form and then factoring. Now plot both sides of the equation in the same viewing screen ($y_1 = x^2 - x$ and $y_2 = 2$). At what x-values do these two graphs intersect? Do those points agree with the solution set you found?

146. Solve the equation $x^2 - 2x = -2$ by first writing in standard form and then using the quadratic formula. Now plot both sides of the equation in the same viewing screen ($y_1 = x^2 - 2x$ and $y_2 = -2$). Do these graphs intersect? Does this agree with the solution set you found?

147. a. Solve the equation $x^2 - 2x = b$, $b = 8$ by first writing in standard form. Now plot both sides of the equation in the same viewing screen ($y_1 = x^2 - 2x$ and $y_2 = b$). At what x values do these two graphs intersect? Do those points agree with the solution set you found?

b. Repeat (a) for $b = -3, -1, 0$, and 5.

148. a. Solve the equation $x^2 + 2x = b$, $b = 8$ by first writing in standard form. Now plot both sides of the equation in the same viewing screen ($y_1 = x^2 + 2x$ and $y_2 = b$). At what x values do these two graphs intersect? Do those points agree with the solution set you found?

b. Repeat (a) for $b = -3, -1, 0$, and 5.

SECTION 1.4 OTHER TYPES OF EQUATIONS

SKILLS OBJECTIVES

- Solve radical equations.
- Solve equations that are quadratic in form.
- Solve equations that are factorable.

CONCEPTUAL OBJECTIVES

- Transform a difficult equation into a simpler linear or quadratic equation.
- Recognize the need to check solutions when the transformation process may produce extraneous solutions.
- Realize that not all polynomial equations are factorable.

Radical Equations

Radical equations are equations in which the variable is inside a radical (that is, under a square root, cube root, or higher root). Examples of radical equations follow.

$$\sqrt{x - 3} = 2 \qquad \sqrt{2x + 3} = x \qquad \sqrt{x + 2} + \sqrt{7x + 2} = 6$$

Until now your experience has been with linear and quadratic equations. Often you can transform a radical equation into a simple linear or quadratic equation. Sometimes the transformation process yields **extraneous solutions**, or apparent solutions that may solve the transformed problem but are not solutions of the original radical equation. Therefore, it is very important to check your answers.

EXAMPLE 1 Solving an Equation Involving a Radical

Solve the equation $\sqrt{x-3} = 2$.

Solution:

Square both sides of the equation. $\left[\sqrt{x-3}\right]^2 = 2^2$

Simplify. $x - 3 = 4$

Solve the resulting linear equation. $\boxed{x = 7}$

The solution set is $\{7\}$.

Check: $\sqrt{7-3} = \sqrt{4} = 2$

■ **YOUR TURN** Solve the equation $\sqrt{3p+4} = 5$.

■ **Answer:** $p = 7$ or $\{7\}$

Study Tip

Extraneous solutions are common when we deal with radical equations, so remember to check your answers.

When both sides of an equation are squared, extraneous solutions can arise. For example, take the equation

$$x = 2$$

If we square both sides of this equation, then the resulting equation, $x^2 = 4$, has two solutions: $x = -2$ and $x = 2$. Notice that the value $x = -2$ is not in the solution set of the original equation $x = 2$. Therefore, we say that $x = -2$ is an extraneous solution.

In solving a radical equation we square both sides of the equation and then solve the resulting equation. The solutions to the resulting equation can sometimes be extraneous in that they do not satisfy the *original* radical equation.

Technology Tip

Use a graphing utility to display graphs of $y_1 = \sqrt{2x+3}$ and $y_2 = x$.

The x-coordinate of the point of intersection is the solution to the equation $\sqrt{2x+3} = x$.

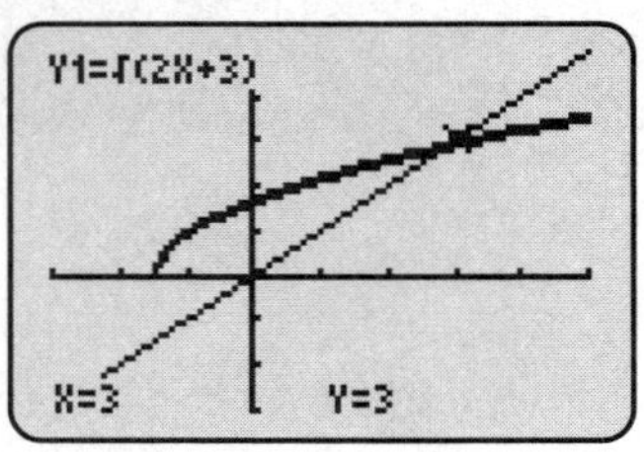

EXAMPLE 2 Solving an Equation Involving a Radical

Solve the equation $\sqrt{2x+3} = x$.

Solution:

Square both sides of the equation. $\left(\sqrt{2x+3}\right)^2 = x^2$

Simplify. $2x + 3 = x^2$

Write the quadratic equation in standard form. $x^2 - 2x - 3 = 0$

Factor. $(x-3)(x+1) = 0$

Use the zero product property. $x = 3$ or $x = -1$

Check these values to see whether they *both* make the original equation statement true.

$x = 3$: $\sqrt{2(3)+3} = 3 \Rightarrow \sqrt{6+3} = 3 \Rightarrow \sqrt{9} = 3 \Rightarrow 3 = 3$ (True)

$x = -1$: $\sqrt{2(-1)+3} = -1 \Rightarrow \sqrt{-2+3} = -1 \Rightarrow \sqrt{1} = -1 \Rightarrow 1 \neq -1$ (False)

The solution is $\boxed{x = 3}$. The solution set is $\{3\}$.

■ **YOUR TURN** Solve the equation $\sqrt{12+t} = t$.

■ **Answer:** $t = 4$ or $\{4\}$

■ **YOUR TURN** Solve the equation $\sqrt{2x+6} = x + 3$.

■ **Answer:** $x = -1$ and $x = -3$ or $\{-3, -1\}$

What happened in Example 2? When we transformed the radical equation into a quadratic equation, we created an **extraneous solution**, $x = -1$, a solution that appears to solve the original equation but does not. When solving radical equations, you must take the time to check answers to avoid including extraneous solutions in your solution set.

Examples 1 and 2 contained only one radical each. We transformed the radical equation into a linear (Example 1) or quadratic (Example 2) equation with one step. The next example contains two radicals. Our technique will be to isolate one radical on one side of the equation with the other radical on the other side of the equation.

CORRECT

Square the expression.

$$(3 + \sqrt{x + 2})^2$$

Write the square as a product of two factors.

$$(3 + \sqrt{x + 2})(3 + \sqrt{x + 2})$$

Use the FOIL method.

$$9 + 6\sqrt{x + 2} + (x + 2)$$

INCORRECT

Square the expression.

$$(3 + \sqrt{x + 2})^2$$

The **error** occurs here when only individual terms are squared.

$$\neq 9 + (x + 2)$$

EXAMPLE 3 Solving an Equation with More Than One Radical

Solve the equation $\sqrt{x + 2} + \sqrt{7x + 2} = 6$.

Solution:

Subtract $\sqrt{x + 2}$ from both sides. $\qquad \sqrt{7x + 2} = 6 - \sqrt{x + 2}$

Square both sides. $\qquad (\sqrt{7x + 2})^2 = (6 - \sqrt{x + 2})^2$

Simplify. $\qquad 7x + 2 = (6 - \sqrt{x + 2})(6 - \sqrt{x + 2})$

Multiply the expressions on the right side of the equation. $\qquad 7x + 2 = 36 - 12\sqrt{x + 2} + (x + 2)$

Isolate the term with the radical on the left side. $\qquad 12\sqrt{x + 2} = 36 + x + 2 - 7x - 2$

Combine like terms on the right side. $\qquad 12\sqrt{x + 2} = 36 - 6x$

Divide by 6. $\qquad 2\sqrt{x + 2} = 6 - x$

Square both sides. $\qquad 4(x + 2) = (6 - x)^2$

Simplify. $\qquad 4x + 8 = 36 - 12x + x^2$

Rewrite the quadratic equation in standard form. $\qquad x^2 - 16x + 28 = 0$

Factor. $\qquad (x - 14)(x - 2) = 0$

Solve. $\qquad x = 14$ and $x = 2$

The apparent solutions are 2 and 14. Note that $x = 14$ does not satisfy the original equation; therefore it is extraneous. The solution is $\boxed{x = 2}$. The solution set is $\{2\}$.

YOUR TURN Solve the equation $\sqrt{x - 4} = 5 - \sqrt{x + 1}$.

Technology Tip

Use a graphing utility to display graphs of

$y_1 = \sqrt{x + 2} + \sqrt{7x + 2}$

and $y_2 = 6$.

The x-coordinate of the point of intersection is the solution to the equation

$\sqrt{x + 2} + \sqrt{7x + 2} = 6$.

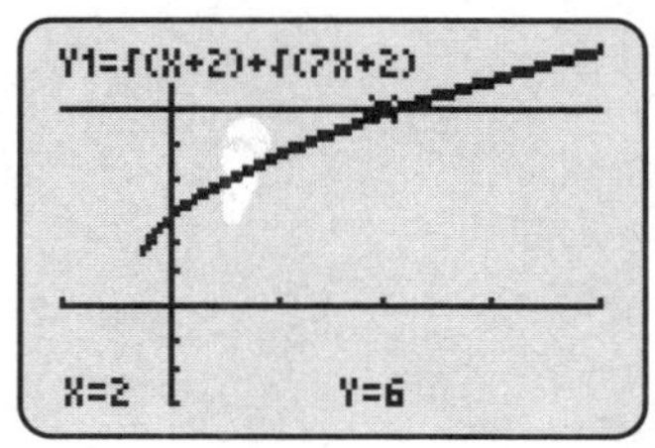

Answer: $x = 8$ or $\{8\}$

PROCEDURE FOR SOLVING RADICAL EQUATIONS

Step 1: Isolate the term with a radical on one side.
Step 2: Raise both (*entire*) sides of the equation to the power that will eliminate this radical, and simplify the equation.
Step 3: If a radical remains, repeat Steps 1 and 2.
Step 4: Solve the resulting linear or quadratic equation.
Step 5: Check the solutions and eliminate any extraneous solutions.

Note: If there is more than one radical in the equation, it does not matter which radical is isolated first.

Equations Quadratic in Form: *u*-Substitution

Equations that are higher order or that have fractional powers often can be transformed into a quadratic equation by introducing a u-substitution. When this is the case we say that equations are **quadratic in form**. In the table below, the two original equations are quadratic in form because they can be transformed into a quadratic equation given the correct substitution.

ORIGINAL EQUATION	SUBSTITUTION	NEW EQUATION
$x^4 - 3x^2 - 4 = 0$	$u = x^2$	$u^2 - 3u - 4 = 0$
$t^{2/3} + 2t^{1/3} + 1 = 0$	$u = t^{1/3}$	$u^2 + 2u + 1 = 0$

For example, the equation $x^4 - 3x^2 - 4 = 0$ is a fourth-degree equation in x. How did we know that $u = x^2$ would transform the original equation into a quadratic equation? If we rewrite the original equation as $(x^2)^2 - 3(x^2) - 4 = 0$, the expression in parentheses is the u-substitution.

Let us introduce the substitution $u = x^2$. Note that squaring both sides implies $u^2 = x^4$. We then replace x^2 in the original equation with u, and x^4 in the original equation with u^2, which leads to a quadratic equation in u: $u^2 - 3u - 4 = 0$.

WORDS	MATH
Solve for x.	$x^4 - 3x^2 - 4 = 0$
Introduce u-substitution.	$u = x^2$ [Note that $u^2 = x^4$.]
Write the quadratic equation in u.	$u^2 - 3u - 4 = 0$
Factor.	$(u - 4)(u + 1) = 0$
Solve for u.	$u = 4$ or $u = -1$
Transform back to x, $u = x^2$.	$x^2 = 4$ or $x^2 = -1$
Solve for x.	$\boxed{x = \pm 2 \text{ or } x = \pm i}$

The solution set is $\{\pm 2, \pm i\}$.

It is important to correctly determine the appropriate substitution in order to arrive at an equation quadratic in form. For example, $t^{2/3} + 2t^{1/3} + 1 = 0$ is an original equation given in the above table. If we rewrite this equation as $(t^{1/3})^2 + 2(t^{1/3}) + 1 = 0$, then it becomes apparent that the correct substitution is $u = t^{1/3}$, which transforms the equation in t into a quadratic equation in u: $u^2 + 2u + 1 = 0$.

Procedure for Solving Equations Quadratic in Form

Step 1: Identify the substitution.
Step 2: Transform the equation into a quadratic equation.
Step 3: Solve the quadratic equation.
Step 4: Apply the substitution to rewrite the solution in terms of the original variable.
Step 5: Solve the resulting equation.
Step 6: Check the solutions in the original equation.

EXAMPLE 4 Solving an Equation Quadratic in Form with Negative Exponents

Find the solutions to the equation $x^{-2} - x^{-1} - 12 = 0$.

Solution:

Rewrite the original equation. $(x^{-1})^2 - (x^{-1}) - 12 = 0$

Determine the u-substitution. $u = x^{-1}$ [Note that $u^2 = x^{-2}$.]

The original equation in x corresponds to a quadratic equation in u. $u^2 - u - 12 = 0$

Factor. $(u - 4)(u + 3) = 0$

Solve for u. $u = 4$ or $u = -3$

The most common mistake is forgetting to transform back to x.

Transform back to x. Let $u = x^{-1}$. $x^{-1} = 4$ or $x^{-1} = -3$

Write x^{-1} as $\frac{1}{x}$. $\frac{1}{x} = 4$ or $\frac{1}{x} = -3$

Solve for x. $\boxed{x = \frac{1}{4}}$ or $\boxed{x = -\frac{1}{3}}$

The solution set is $\left\{-\frac{1}{3}, \frac{1}{4}\right\}$.

YOUR TURN Find the solutions to the equation $x^{-2} - x^{-1} - 6 = 0$.

Technology Tip

Use a graphing utility to graph $y_1 = x^{-2} - x^{-1} - 12$.

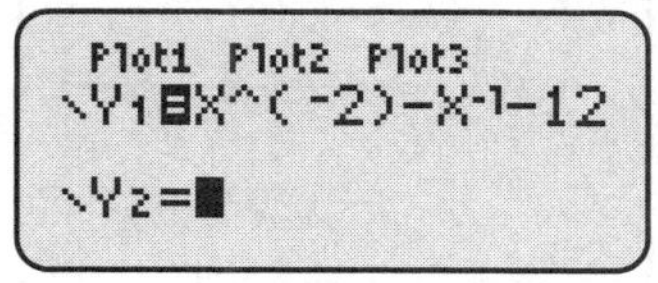

The x-intercepts are the solutions to this equation.

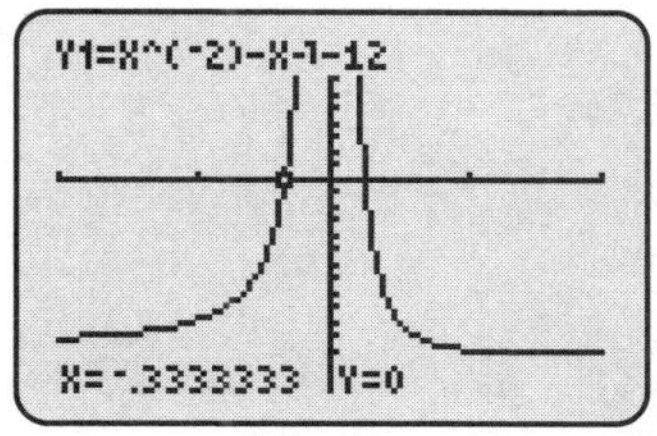

Study Tip

Remember to transform back to the original variable.

Answer: The solution is $x = -\frac{1}{2}$ or $x = \frac{1}{3}$. The solution set is $\left\{-\frac{1}{2}, \frac{1}{3}\right\}$.

EXAMPLE 5 Solving an Equation Quadratic in Form with Fractional Exponents

Find the solutions to the equation $x^{2/3} - 3x^{1/3} - 10 = 0$.

Solution:

Rewrite the original equation. $(x^{1/3})^2 - 3x^{1/3} - 10 = 0$

Identify the substitution as $u = x^{1/3}$. $u^2 - 3u - 10 = 0$

Factor. $(u - 5)(u + 2) = 0$

Solve for u. $u = 5$ or $u = -2$

Let $u = x^{1/3}$ again. $x^{1/3} = 5$ $\quad x^{1/3} = -2$

Cube both sides of the equations. $(x^{1/3})^3 = (5)^3$ $\quad (x^{1/3})^3 = (-2)^3$

Simplify. $\boxed{x = 125}$ $\quad \boxed{x = -8}$

The solution set is $\boxed{\{-8, 125\}}$, which a check will confirm.

YOUR TURN Find the solution to the equation $2t - 5t^{1/2} - 3 = 0$.

Answer: $t = 9$ or $\{9\}$.

Factorable Equations

Some equations (both polynomial and with rational exponents) that are factorable can be solved using the zero product property.

EXAMPLE 6 Solving an Equation with Rational Exponents by Factoring

Solve the equation $x^{7/3} - 3x^{4/3} - 4x^{1/3} = 0$.

Solution:

Factor the left side of the equation. $x^{1/3}\left(x^2 - 3x - 4\right) = 0$

Factor the quadratic expression. $x^{1/3}(x - 4)(x + 1) = 0$

Apply the zero product property. $x^{1/3} = 0$ or $x - 4 = 0$ or $x + 1 = 0$

Solve for x. $\boxed{x = 0}$ or $\boxed{x = 4}$ or $\boxed{x = -1}$

The solution set is $\{-1, 0, 4\}$.

Technology Tip

Use a graphing utility to graph $y_1 = x^{7/3} - 3x^{4/3} - 4x^{1/3}$.

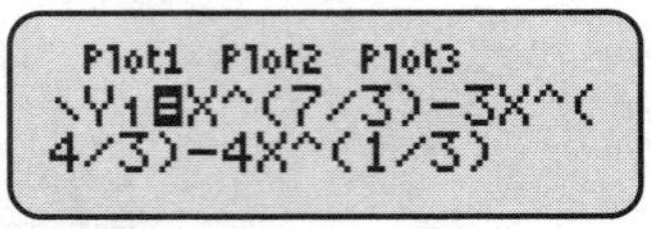

The x-intercepts are the solutions to this equation.

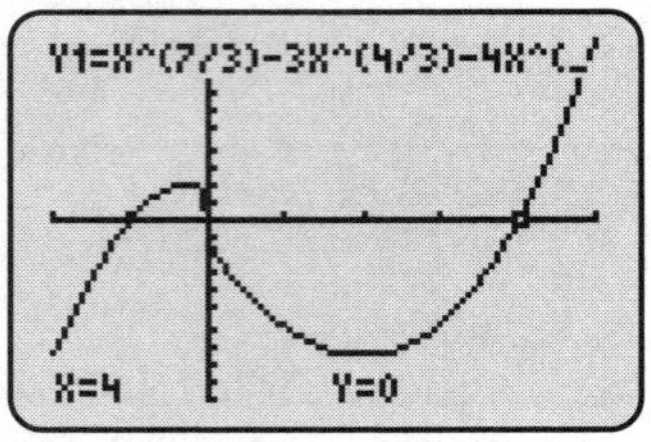

EXAMPLE 7 Solving a Polynomial Equation Using Factoring by Grouping

Solve the equation $x^3 + 2x^2 - x - 2 = 0$.

Solution:

Factor by grouping (Chapter 0). $\left(x^3 - x\right) + \left(2x^2 - 2\right) = 0$

Identify the common factors. $x\left(x^2 - 1\right) + 2\left(x^2 - 1\right) = 0$

Factor. $(x + 2)\left(x^2 - 1\right) = 0$

Factor the quadratic expression. $(x + 2)(x - 1)(x + 1) = 0$

Apply the zero product property. $x + 2 = 0$ or $x - 1 = 0$ or $x + 1 = 0$

Solve for x. $\boxed{x = -2}$ or $\boxed{x = 1}$ or $\boxed{x = -1}$

The solution set is $\{-2, -1, 1\}$.

■ **YOUR TURN** Solve the equation $x^3 + x^2 - 4x - 4 = 0$.

■ **Answer:** $x = -1$ or $x = \pm 2$ or $\{-2, -1, 2\}$

SECTION 1.4 SUMMARY

Radical equations, equations quadratic in form, and factorable equations can often be solved by transforming them into simpler linear or quadratic equations.

- **Radical Equations:** Isolate the term containing a radical and raise it to the appropriate power that will eliminate the radical. If there is more than one radical, it does not matter which radical is isolated first. Raising radical equations to powers may cause extraneous solutions, so check each solution.
- **Equations Quadratic in Form:** Identify the u-substitution that transforms the equation into a quadratic equation. Solve the quadratic equation and then remember to transform back to the original variable.
- **Factorable Equations:** Look for a factor common to all terms or factor by grouping.

SECTION 1.4 EXERCISES

SKILLS

In Exercises 1–30, solve the radical equation for the given variable.

1. $\sqrt{t-5}=2$
2. $\sqrt{2t-7}=3$
3. $(4p-7)^{1/2}=5$
4. $11=(21-p)^{1/2}$
5. $\sqrt{u+1}=-4$
6. $-\sqrt{3-2u}=9$
7. $\sqrt[3]{5x+2}=3$
8. $\sqrt[3]{1-x}=-2$
9. $(4y+1)^{1/3}=-1$
10. $(5x-1)^{1/3}=4$
11. $\sqrt{12+x}=x$
12. $x=\sqrt{56-x}$
13. $y=5\sqrt{y}$
14. $\sqrt{y}=\frac{y}{4}$
15. $s=3\sqrt{s-2}$
16. $-2s=\sqrt{3-s}$
17. $\sqrt{2x+6}=x+3$
18. $\sqrt{8-2x}=2x-2$
19. $\sqrt{1-3x}=x+1$
20. $\sqrt{2-x}=x-2$
21. $\sqrt{x^2-4}=x-1$
22. $\sqrt{25-x^2}=x+1$
23. $\sqrt{x^2-2x-5}=x+1$
24. $\sqrt{2x^2-8x+1}=x-3$
25. $\sqrt{2x-1}-\sqrt{x-1}=1$
26. $\sqrt{8-x}=2+\sqrt{2x+3}$
27. $\sqrt{3x-5}=7-\sqrt{x+2}$
28. $\sqrt{x+5}=1+\sqrt{x-2}$
29. $\sqrt{2+\sqrt{x}}=\sqrt{x}$
30. $\sqrt{2-\sqrt{x}}=\sqrt{x}$

In Exercises 31–56, solve the equations by introducing a substitution that transforms these equations to quadratic form.

31. $x^{2/3}+2x^{1/3}=0$
32. $x^{1/2}-2x^{1/4}=0$
33. $x^4-3x^2+2=0$
34. $x^4-8x^2+16=0$
35. $2x^4+7x^2+6=0$
36. $x^8-17x^4+16=0$
37. $(2x+1)^2+5(2x+1)+4=0$
38. $(x-3)^2+6(x-3)+8=0$
39. $4(t-1)^2-9(t-1)=-2$
40. $2(1-y)^2+5(1-y)-12=0$
41. $x^{-8}-17x^{-4}+16=0$
42. $2u^{-2}+5u^{-1}-12=0$
43. $3y^{-2}+y^{-1}-4=0$
44. $5a^{-2}+11a^{-1}+2=0$
45. $z^{2/5}-2z^{1/5}+1=0$
46. $2x^{1/2}+x^{1/4}-1=0$
47. $6t^{-2/3}-t^{-1/3}-1=0$
48. $t^{-2/3}-t^{-1/3}-6=0$
49. $3=\frac{1}{(x+1)^2}+\frac{2}{(x+1)}$
50. $\frac{1}{(x+1)^2}+\frac{4}{(x+1)}+4=0$
51. $\left(\frac{1}{2x-1}\right)^2+\left(\frac{1}{2x-1}\right)-12=0$
52. $\frac{5}{(2x+1)^2}-\frac{3}{(2x+1)}=2$
53. $u^{4/3}-5u^{2/3}=-4$
54. $u^{4/3}+5u^{2/3}=-4$
55. $t=\sqrt[4]{t^2+6}$
56. $u=\sqrt[4]{-2u^2-1}$

In Exercises 57–72, solve by factoring.

57. $x^3-x^2-12x=0$
58. $2y^3-11y^2+12y=0$
59. $4p^3-9p=0$
60. $25x^3=4x$
61. $u^5-16u=0$
62. $t^5-81t=0$
63. $x^3-5x^2-9x+45=0$
64. $2p^3-3p^2-8p+12=0$
65. $y(y-5)^3-14(y-5)^2=0$
66. $v(v+3)^3-40(v+3)^2=0$
67. $x^{9/4}-2x^{5/4}-3x^{1/4}=0$
68. $u^{7/3}+u^{4/3}-20u^{1/3}=0$
69. $t^{5/3}-25t^{-1/3}=0$
70. $4x^{9/5}-9x^{-1/5}=0$
71. $y^{3/2}-5y^{1/2}+6y^{-1/2}=0$
72. $4p^{5/3}-5p^{2/3}-6p^{-1/3}=0$

APPLICATIONS

73. Insurance: Health. Cost for health insurance with a private policy is given by $C = \sqrt{10 + a}$, where C is the cost per day and a is the insured's age in years. Health insurance for a 6 year old, $a = 6$, is \$4 a day (or \$1,460 per year). At what age would someone be paying \$9 a day (or \$3,285 per year)?

74. Insurance: Life. Cost for life insurance is given by $C = \sqrt{5a + 1}$, where C is the cost per day and a is the insured's age in years. Life insurance for a newborn, $a = 0$, is \$1 a day (or \$365 per year). At what age would someone be paying \$20 a day (or \$7,300 per year)?

75. Stock Value. The stock price of MGI Pharmaceutical (MOGN) from March 2004 to June 2004 can be approximately modeled by the equation $P = 5\sqrt{t^2 + 1} + 50$, where P is the price of the stock in dollars and t is the month with $t = 0$ corresponding to March 2004. Assuming this trend continues, when would the stock be worth \$85?

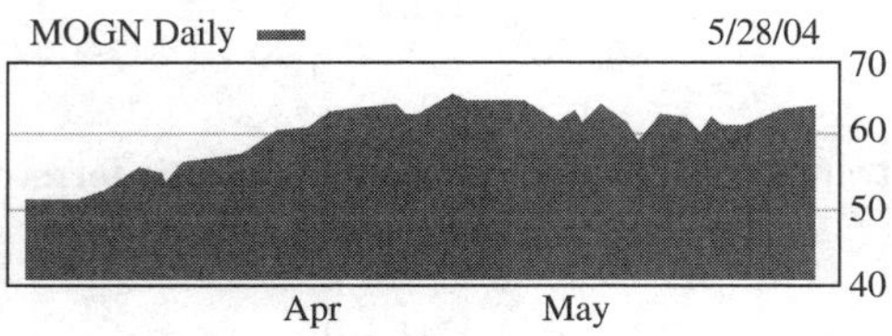

76. Grades. The average combined math and verbal SAT score of incoming freshmen at a university is given by the equation $S = 1000 + 10\sqrt{2t}$, where t is in years and $t = 0$ corresponds to 1990. What year will the incoming class have an average SAT score of 1230?

77. Speed of Sound. A man buys a house with an old well but does not know how deep the well is. To get an estimate he decides to drop a rock at the opening of the well and count how long it takes until he hears the splash. The total elapsed time T given by $T = t_1 + t_2$, is the sum of the time it takes for the rock to reach the water, t_1, and the time it takes for the sound of the splash to travel to the top of the well, t_2. The time (seconds) that it takes for the rock to reach the water is given by $t_1 = \frac{\sqrt{d}}{4}$, where d is the depth of the well in feet. Since the speed of sound is 1100 ft/s, the time (seconds) it takes for the sound to reach the top of the well is $t_2 = \frac{d}{1100}$. If the splash is heard after 3 seconds, how deep is the well?

78. Speed of Sound. If the owner of the house in Exercise 77 forgot to account for the speed of sound, what would he have calculated the depth of the well to be?

79. Physics: Pendulum. The period (T) of a pendulum is related to the length (L) of the pendulum and acceleration due to gravity (g) by the formula $T = 2\pi\sqrt{\frac{L}{g}}$. If gravity is 9.8 m/s^2 and the period is 1 second, find the approximate length of the pendulum. Round to the nearest centimeter. (*Note:* 100 cm = 1 m)

80. Physics: Pendulum. The period (T) of a pendulum is related to the length (L) of the pendulum and acceleration due to gravity (g) by the formula $T = 2\pi\sqrt{\frac{L}{g}}$. If gravity is 32 ft/s^2 and the period is 1 second, find the approximate length of the pendulum. Round to the nearest inch. (*Note:* 12 in. = 1 ft)

For Exercises 81 and 82, refer to the following:

Einstein's special theory of relativity states that time is relative: Time speeds up or slows down, depending on how fast one object is moving with respect to another. For example, a space probe traveling at a velocity v near the speed of light c will have "clocked" a time t hours, but for a stationary observer on Earth that corresponds to a time t_0. The formula governing this relativity is given by

$$t = t_0\sqrt{1 - \frac{v^2}{c^2}}$$

81. Physics: Special Theory of Relativity. If the time elapsed on a space probe mission is 18 years but the time elapsed on Earth during that mission is 30 years, how fast is the space probe traveling? Give your answer relative to the speed of light.

82. Physics: Special Theory of Relativity. If the time elapsed on a space probe mission is 5 years but the time elapsed on Earth during that mission is 30 years, how fast is the space probe traveling? Give your answer relative to the speed of light.

CATCH THE MISTAKE

In Exercises 83–86, explain the mistake that is made.

83. Solve the equation $\sqrt{3t + 1} = -4$.

Solution:

$$3t + 1 = 16$$
$$3t = 15$$
$$t = 5$$

This is incorrect. What mistake was made?

84. Solve the equation $x = \sqrt{x + 2}$.

Solution:

$$x^2 = x + 2$$
$$x^2 - x - 2 = 0$$
$$(x - 2)(x + 1) = 0$$
$$x = -1, x = 2$$

This is incorrect. What mistake was made?

85. Solve the equation $x^{2/3} - x^{1/3} - 20 = 0$.

Solution:

$$u = x^{1/3}$$
$$u^2 - u - 20 = 0$$
$$(u - 5)(u + 4) = 0$$
$$x = 5, x = -4$$

This is incorrect. What mistake was made?

86. Solve the equation $x^4 - 2x^2 = 3$.

Solution:

$$x^4 - 2x^2 - 3 = 0$$
$$u = x^2$$
$$u^2 - 2u - 3 = 0$$
$$(u - 3)(u + 1) = 0$$
$$u = -1, u = 3$$
$$u = x^2$$
$$x^2 = -1, x^2 = 3$$
$$x = \pm 1, x = \pm 3$$

This is incorrect. What mistake was made?

CONCEPTUAL

In Exercises 87–90, determine whether each statement is true or false.

87. The equation $(2x - 1)^6 + 4(2x - 1)^3 + 3 = 0$ is quadratic in form.

88. The equation $t^{25} + 2t^5 + 1 = 0$ is quadratic in form.

89. If two solutions are found and one does not check, then the other does not check.

90. Squaring both sides of $\sqrt{x + 2} + \sqrt{x} = \sqrt{x + 5}$ leads to $x + 2 + x = x + 5$.

CHALLENGE

91. Solve $\sqrt{x^2} = x$.

92. Solve $\sqrt{x^2} = -x$.

93. Solve the equation $3x^2 + 2x = \sqrt{3x^2 + 2x}$ *without squaring* both sides.

94. Solve the equation $3x^{7/12} - x^{5/6} - 2x^{1/3} = 0$.

95. Solve the equation $\sqrt{x + 6} + \sqrt{11 + x} = 5\sqrt{3 + x}$.

96. Solve the equation $\sqrt[4]{2x\sqrt[3]{x\sqrt{x}}} = 2$.

TECHNOLOGY

97. Solve the equation $\sqrt{x - 3} = 4 - \sqrt{x + 2}$. Plot both sides of the equation in the same viewing screen, $y_1 = \sqrt{x - 3}$ and $y_2 = 4 - \sqrt{x + 2}$, and zoom in on the x-coordinate of the point of intersection. Does the graph agree with your solution?

98. Solve the equation $2\sqrt{x + 1} = 1 + \sqrt{3 - x}$. Plot both sides of the equation in the same viewing screen, $y_1 = 2\sqrt{x + 1}$ and $y_2 = 1 + \sqrt{3 - x}$, and zoom in on the x-coordinate of the points of intersection. Does the graph agree with your solution?

99. Solve the equation $-4 = \sqrt{x + 3}$. Plot both sides of the equation in the same viewing screen, $y_1 = -4$ and $y_2 = \sqrt{x + 3}$. Does the graph agree or disagree with your solution?

100. Solve the equation $x^{1/4} = -4x^{1/2} + 21$. Plot both sides of the equation in the same viewing screen, $y_1 = x^{1/4}$ and $y_2 = -4x^{1/2} + 21$. Does the point(s) of intersection agree with your solution?

101. Solve the equation $x^{1/2} = -4x^{1/4} + 21$. Plot both sides of the equation in the same viewing screen, $y_1 = x^{1/2}$ and $y_2 = -4x^{1/4} + 21$. Does the point(s) of intersection agree with your solution?

102. Solve the equation $x^{-1} = 3x^{-2} - 10$. Plot both sides of the equation in the same viewing screen, $y_1 = x^{-1}$ and $y_2 = 3x^{-2} - 10$. Does the point(s) of intersection agree with your solution?

103. Solve the equation $x^{-2} = 3x^{-1} - 10$. Plot both sides of the equation in the same viewing screen, $y_1 = x^{-2}$ and $y_2 = 3x^{-1} - 10$. Does the point(s) of intersection agree with your solution?

SECTION 1.5 LINEAR INEQUALITIES

SKILLS OBJECTIVES

- Use interval notation.
- Solve linear inequalities.
- Solve application problems involving linear inequalities.

CONCEPTUAL OBJECTIVES

- Apply intersection and union concepts.
- Compare and contrast equations and inequalities.
- Understand that linear inequalities may have one solution, no solution, or an interval solution.

An example of a linear equation is $3x - 2 = 7$. On the other hand, $3x - 2 \leq 7$ is an example of a **linear inequality**. One difference between a linear equation and a linear inequality is that the equation has at most only one solution, or value of x, that makes the statement true, whereas the inequality can have a range or continuum of numbers that make the statement true. For example, the inequality $x \leq 4$ denotes all real numbers x that are less than or equal to 4. Four inequality symbols are used.

Symbol	In Words
$<$	Less than
$>$	Greater than
$\leq$	Less than or equal to
$\geq$	Greater than or equal to

We call $<$ and $>$ **strict inequalities**. For any two real numbers a and b, one of three things *must* be true:

$$a < b \qquad \text{or} \qquad a = b \qquad \text{or} \qquad a > b$$

This property is called the **trichotomy property** of real numbers.

If x is less than 5 ($x < 5$) and x is greater than or equal to -2 ($x \geq -2$), then we can represent this as a **double** (or **combined**) **inequality**, $-2 \leq x < 5$, which means that x is greater than or equal to -2 and less than 5.

Graphing Inequalities and Interval Notation

We will express solutions to inequalities four ways: an inequality, a solution set, an interval, and a graph. The following are ways of expressing all real numbers greater than or equal to a and less than b.

Inequality Notation	Solution Set	Interval Notation	Graph/Number Line
$a \leq x < b$	$\{x \mid a \leq x < b\}$	$[a, b)$	 or 

In this example, a is referred to as the **left endpoint** and b is referred to as the **right endpoint**. If an inequality is a strict inequality ($<$ or $>$), then the graph and interval notation use *parentheses*. If it includes an endpoint ($\geq$ or $\leq$), then the graph and interval notation use *brackets*. Number lines are drawn with either closed/open circles or brackets/parentheses. In this text the brackets/parentheses notation will be used. Intervals are classified as follows:

Open (,)   Closed [,]   Half open (,] or [,)

Let x be a real number. x is...	**Inequality**	**Set Notation**	**Interval**	**Graph**
greater than a and less than b	$a < x < b$	$\{x \mid a < x < b\}$	(a, b)	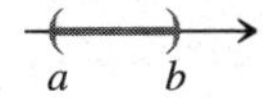
greater than or equal to a and less than b	$a \le x < b$	$\{x \mid a \le x < b\}$	$[a, b)$	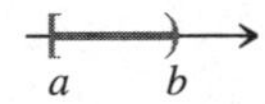
greater than a and less than or equal to b	$a < x \le b$	$\{x \mid a < x \le b\}$	$(a, b]$	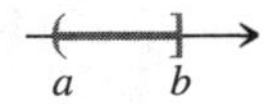
greater than or equal to a and less than or equal to b	$a \le x \le b$	$\{x \mid a \le x \le b\}$	$[a, b]$	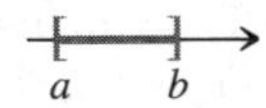
less than a	$x < a$	$\{x \mid x < a\}$	$(-\infty, a)$	
less than or equal to a	$x \le a$	$\{x \mid x \le a\}$	$(-\infty, a]$	
greater than b	$x > b$	$\{x \mid x > b\}$	(b, ∞)	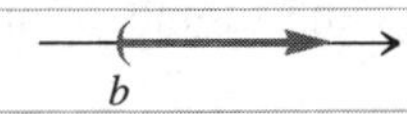
greater than or equal to b	$x \ge b$	$\{x \mid x \ge b\}$	$[b, \infty)$	
all real numbers	$\mathbb{R}$	$\mathbb{R}$	$(-\infty, \infty)$	

1. *Infinity* (∞) is not a number. It is a symbol that means continuing indefinitely to the right on the number line. Similarly, *negative infinity* ($-\infty$) means continuing indefinitely to the left on the number line. Since both are unbounded, we use a parenthesis, never a bracket.
2. In interval notation, the lower number is always written to the left.

Write the inequality in interval notation: $-1 \le x < 3$.

CORRECT $[-1, 3)$ **INCORRECT** $(3, -1]$

EXAMPLE 1 **Expressing Inequalities Using Interval Notation and a Graph**

Express the following as an inequality, an interval, and a graph.

a. x is greater than -3.
b. x is less than or equal to 5.
c. x is greater than or equal to -1 and less than 4.
d. x is greater than or equal to 0 and less than or equal to 4.

Solution:

Inequality	**Interval**	**Graph**
a. $x > -3$	$(-3, \infty)$	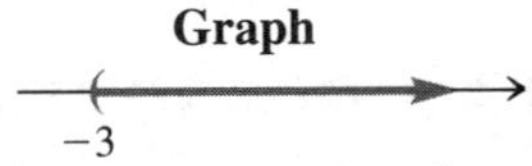
b. $x \le 5$	$(-\infty, 5]$	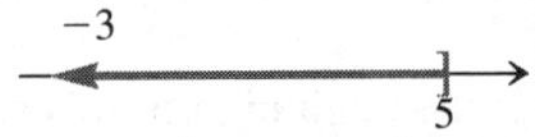
c. $-1 \le x < 4$	$[-1, 4)$	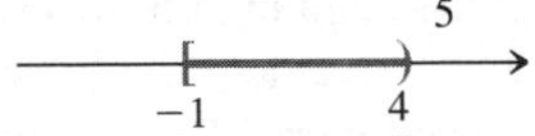
d. $0 \le x \le 4$	$[0, 4]$	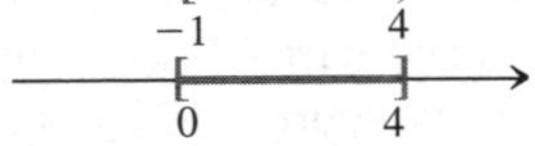

Since the solutions to inequalities are sets of real numbers, it is useful to discuss two operations on sets called **intersection** and **union**.

DEFINITION **Union and Intersection**

The **union** of sets A and B, denoted $A \cup B$, is the set formed by combining all the elements in A with all the elements in B.

$$A \cup B = \{x | x \text{ is in } A \textbf{ or } B \textbf{ or } \text{both}\}$$

The **intersection** of sets A and B, denoted $A \cap B$, is the set formed by the elements that are in both A and B.

$$A \cap B = \{x | x \text{ is in } A \textbf{ and } B\}$$

The notation "$x \mid x$ is in" is read "all x such that x is in." The vertical line represents "such that."

As an example of intersection and union, consider the following sets of people:

$A = \{$Austin, Brittany, Jonathan$\}$ $\quad B = \{$Anthony, Brittany, Elise$\}$

Intersection: $A \cap B = \{$Brittany$\}$

Union: $A \cup B = \{$Anthony, Austin, Brittany, Elise, Jonathan$\}$

EXAMPLE 2 Determining Unions and Intersections: Intervals and Graphs

If $A = [-3, 2]$ and $B = (1, 7)$, determine $A \cup B$ and $A \cap B$. Write these sets in interval notation, and graph.

Solution:

Set	Interval notation	Graph
A	$[-3, 2]$	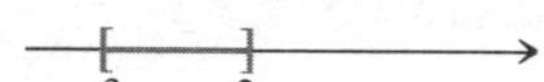
B	$(1, 7)$	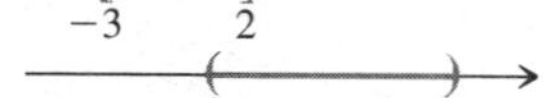
$A \cup B$	$[-3, 7)$	
$A \cap B$	$(1, 2]$	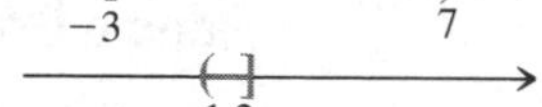

YOUR TURN If $C = [-3, 3)$ and $D = (0, 5]$, find $C \cup D$ and $C \cap D$. Express the intersection and union in interval notation, and graph.

Answer:

$C \cup D = [-3, 5]$

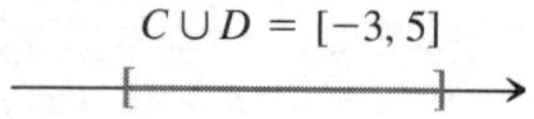

$C \cap D = (0, 3)$

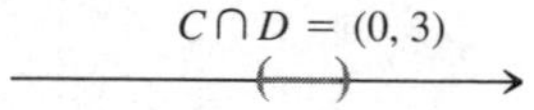

Solving Linear Inequalities

As mentioned at the beginning of this section, if we were to solve the equation $3x - 2 = 7$, we would add 2 to both sides, divide by 3, and find that $x = 3$ is the solution, the *only* value that makes the equation true. If we were to solve the linear inequality $3x - 2 \leq 7$, we would follow the same procedure: add 2 to both sides, divide by 3, and find that $x \leq 3$, which is an *interval* or *range* of numbers that make the inequality true.

In solving linear inequalities we follow the same procedures that we used in solving linear equations with one general exception: *if you multiply or divide an inequality by a*

Study Tip

If you multiply or divide an inequality by a negative number, remember to change the direction of the inequality sign.

negative number, then you must change the direction of the inequality sign. For example, if $-2x < -10$, then the solution set includes real numbers such as $x = 6$ and $x = 7$. Note that real numbers such as $x = -6$ and $x = -7$ are not included in the solution set. Therefore, when this inequality is divided by -2, the inequality sign must also be reversed: $x > 5$. If $a < b$, then $ac < bc$ if $c > 0$ and $ac > bc$ if $c < 0$.

The most common mistake that occurs when solving an inequality is forgetting to change the direction of, or reverse, the inequality symbol when the inequality is multiplied or divided by a negative number.

INEQUALITY PROPERTIES

Procedures That Do Not Change the Inequality Sign

1. Simplifying by eliminating parentheses and collecting like terms. $3(x - 6) < 6x - x$; $3x - 18 < 5x$
2. Adding or subtracting the same quantity on both sides. $7x + 8 \geq 29$; $7x \geq 21$
3. Multiplying or dividing by the same *positive* real number. $5x \leq 15$; $x \leq 3$

Procedures That Change (Reverse) the Inequality Sign

1. Interchanging the two sides of the inequality. $x \leq 4$ is equivalent to $4 \geq x$
2. Multiplying or dividing by the same *negative* real number. $-5x \leq 15$ is equivalent to $x \geq -3$

EXAMPLE 3 Solving a Linear Inequality

Solve and graph the inequality $5 - 3x < 23$.

Solution:

Write the original inequality.	$5 - 3x < 23$
Subtract 5 from both sides.	$-3x < 18$
Divide both sides by -3 and reverse the inequality sign.	$\frac{-3x}{-3} > \frac{18}{-3}$
Simplify.	$x > -6$

Solution set: $\{x \mid x > -6\}$ Interval notation: $(-6, \infty)$ Graph:

■ **YOUR TURN** Solve the inequality $5 \leq 3 - 2x$. Express the solution in set and interval notation, and graph.

Technology Tip

Use a graphing utility to display graphs of $y_1 = 5 - 3x$ and $y_2 = 23$.

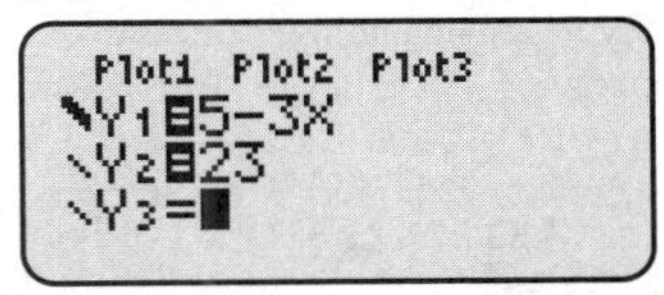

The solutions are the x-values such that the graph of $y_1 = 5 - 3x$ is below that of $y_2 = 23$.

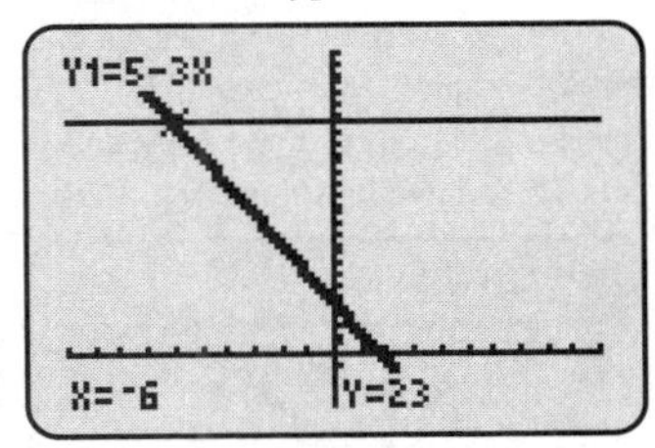

■ **Answer:** Solution set: $\{x \mid x \leq -1\}$
Interval notation: $(-\infty, -1]$
Graph:

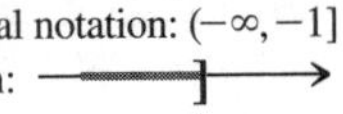

EXAMPLE 4 Solving Linear Inequalities with Fractions

Solve the inequality $\dfrac{5x}{3} \le \dfrac{4 + 3x}{2}$.

COMMON MISTAKE

A common mistake is using cross multiplication to solve inequalities. The reason cross multiplication should not be used is because it is not clear which inequality sign to use.

CORRECT

Eliminate the fractions by multiplying by the LCD, 6.

$$6\left(\frac{5x}{3}\right) \le 6\left(\frac{4 + 3x}{2}\right)$$

Simplify.

$$10x \le 3(4 + 3x)$$

Eliminate the parentheses.

$$10x \le 12 + 9x$$

Subtract $9x$ from both sides.

$$\boxed{x \le 12}$$

INCORRECT

Cross multiply. $3(4 + 3x) \le 2(5x)$

The error is in cross multiplying.

CAUTION

Cross multiplication cannot be used in solving inequalities.

Although it is not possible to "check" inequalities since the solutions are often intervals, it is possible to confirm that some points that lie in your solution do satisfy the inequality. It is important to remember that cross multiplication cannot be used in solving inequalities.

Technology Tip

Use a graphing utility to display graphs of $y_1 = -2$, $y_2 = 3x + 4$, and $y_3 = 16$.

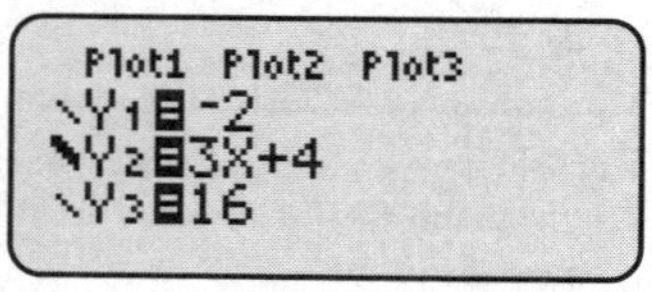

The solutions are the x-values such that the graph of $y_2 = 3x + 4$ is between the graphs of $y_1 = -2$ and $y_3 = 16$ and overlaps that of $y_3 = 16$.

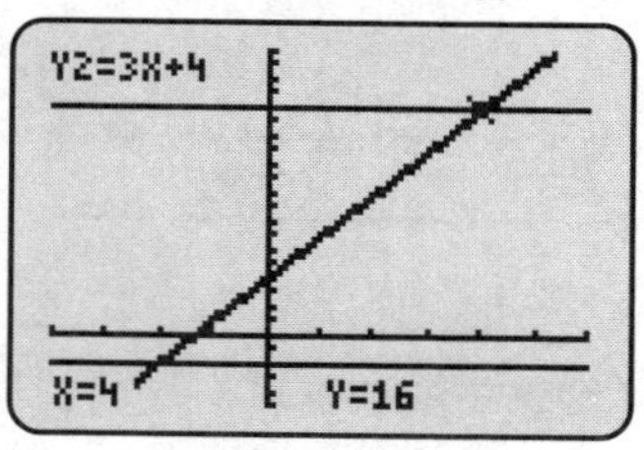

EXAMPLE 5 Solving a Double Linear Inequality

Solve the inequality $-2 < 3x + 4 \le 16$.

Solution:

This double inequality can be written as two inequalities. $\qquad -2 < 3x + 4 \le 16$

Both inequalities must be satisfied. $\qquad -2 < 3x + 4$ and $3x + 4 \le 16$

Subtract 4 from both sides of each inequality. $\qquad -6 < 3x$ and $3x \le 12$

Divide each inequality by 3. $\qquad -2 < x$ and $x \le 4$

Combining these two inequalities gives us $-2 < x \le 4$ in inequality notation; in interval notation we have $(-2, \infty) \cap (-\infty, 4]$ or $(-2, 4]$.

Notice that the steps we took in solving these inequalities individually were identical. This leads us to a **shortcut method** in which we solve them together:

Write the combined inequality. $\qquad -2 < 3x + 4 \le 16$

Subtract 4 from each part. $\qquad -6 < 3x \le 12$

Divide each part by 3. $\qquad -2 < x \le 4$

Interval notation: $\boxed{(-2, 4]}$

For the remainder of this section we will use the shortcut method for solving inequalities.

EXAMPLE 6 Solving a Double Linear Inequality

Solve the inequality $1 \le \frac{-2 - 3x}{7} < 4$. Express the solution set in interval notation, and graph.

Solution:

Write the original double inequality.	$1 \le \frac{-2 - 3x}{7} < 4$
Multiply each part by 7.	$7 \le -2 - 3x < 28$
Add 2 to each part.	$9 \le -3x < 30$
Divide each part by -3 and reverse the inequality signs.	$-3 \ge x > -10$
Write in standard form.	$-10 < x \le -3$

Interval notation: $(-10, -3]$ Graph:

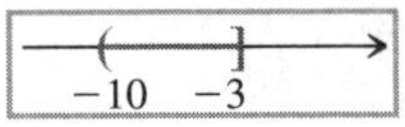

Applications Involving Linear Inequalities

EXAMPLE 7 Temperature Ranges

New York City on average has a yearly temperature range of 23 degrees Fahrenheit to 86 degrees Fahrenheit. What is the range in degrees Celsius given that the conversion relation is $F = 32 + \frac{9}{5}C$?

Solution:

The temperature ranges from 23 °F to 86 °F.	$23 \le F \le 86$
Replace F using the Celsius conversion.	$23 \le 32 + \frac{9}{5}C \le 86$
Subtract 32 from all three parts.	$-9 \le \frac{9}{5}C \le 54$
Multiply all three parts by $\frac{5}{9}$.	$-5 \le C \le 30$

New York City has an average yearly temperature range of **−5 °C to 30 °C**.

Technology Tip

Use a graphing utility to display graphs of $y_1 = 23$, $y_2 = 32 + \frac{9}{5}x$, and $y_3 = 86$.

The solutions are the x-values such that the graph of $y_2 = 32 + \frac{9}{5}x$ is between the graphs of $y_1 = 23$ and $y_3 = 86$ and overlaps both graphs.

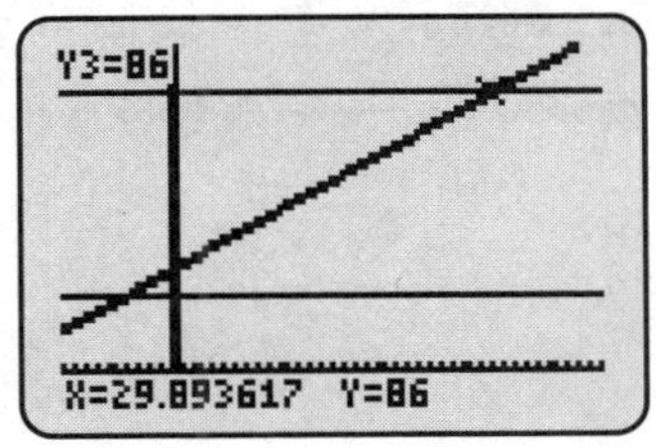

EXAMPLE 8 Comparative Shopping

Two car rental companies have advertised weekly specials on full-size cars. Hertz is advertising an $80 rental fee plus an additional $.10 per mile. Thrifty is advertising $60 and $.20 per mile. How many miles must you drive for the rental car from Hertz to be the better deal?

Solution:

Let x = number of miles driven during the week.

Write the cost for the Hertz rental.	$80 + 0.1x$
Write the cost for the Thrifty rental.	$60 + 0.2x$
Write the inequality if Hertz is less than Thrifty.	$80 + 0.1x < 60 + 0.2x$
Subtract $0.1x$ from both sides.	$80 < 60 + 0.1x$
Subtract 60 from both sides.	$20 < 0.1x$
Divide both sides by 0.1.	$200 < x$

You must drive more than 200 miles for Hertz to be the better deal.

SECTION 1.5 SUMMARY

The solutions of linear inequalities are solution sets that can be expressed four ways:

1. Inequality notation $a < x \leq b$
2. Set notation $\{x \mid a < x \leq b\}$
3. Interval notation $(a, b]$
4. Graph (number line)

Linear inequalities are solved using the same procedures as linear equations with one exception: **when you multiply or divide by a negative number, you must reverse the inequality sign.**

Note: Cross multiplication cannot be used with inequalities.

SECTION 1.5 EXERCISES

SKILLS

In Exercises 1–16, rewrite in interval notation and graph.

1. $x \geq 3$ **2.** $x < -2$ **3.** $x \leq -5$ **4.** $x > -7$

5. $-2 \leq x < 3$ **6.** $-4 \leq x \leq -1$ **7.** $-3 < x \leq 5$ **8.** $0 < x < 6$

9. $0 \leq x \leq 0$ **10.** $-7 \leq x \leq -7$ **11.** $x \leq 6$ and $x \geq 4$ **12.** $x > -3$ and $x \leq 2$

13. $x \leq -6$ and $x \geq -8$ **14.** $x < 8$ and $x < 2$ **15.** $x > 4$ and $x \leq -2$ **16.** $x \geq -5$ and $x < -6$

In Exercises 17–24, rewrite in set notation.

17. $[0, 2)$ **18.** $(0, 3]$ **19.** $(-7, -2)$ **20.** $[-3, 2]$

21. $(-\infty, 6]$ **22.** $(5, \infty)$ **23.** $(-\infty, \infty)$ **24.** $[4, 4]$

In Exercises 25–32, write in inequality and interval notation.

25.

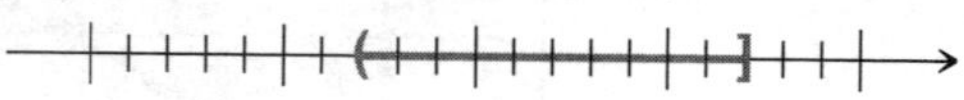

26.

27.

28.

29.

30.

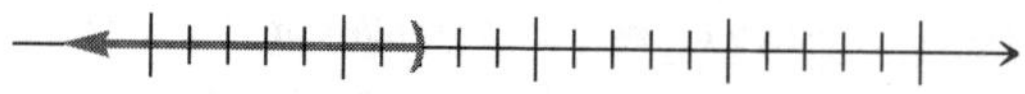

31.

32.

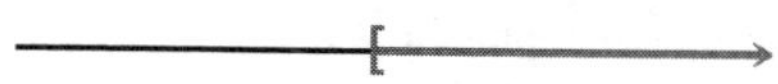

In Exercises 33–50, graph the indicated set and write as a single interval, if possible.

33. $(-5, 2] \cup (-1, 3)$ **34.** $(2, 7) \cup [-5, 3)$ **35.** $[-6, 4) \cup [-2, 5)$ **36.** $[-3, 1) \cup [-6, 0)$

37. $(-\infty, 1] \cap [-1, \infty)$ **38.** $(-\infty, -5) \cap (-\infty, 7]$ **39.** $(-\infty, 4) \cap [1, \infty)$ **40.** $(-3, \infty) \cap [-5, \infty)$

41. $[-5, 2) \cap [-1, 3]$ **42.** $[-4, 5) \cap [-2, 7)$ **43.** $(-\infty, 4) \cup (4, \infty)$ **44.** $(-\infty, -3] \cup [-3, \infty)$

45. $(-\infty, -3] \cup [3, \infty)$ **46.** $(-2, 2) \cap [-3, 1]$ **47.** $(-\infty, \infty) \cap (-3, 2]$ **48.** $(-\infty, \infty) \cup (-4, 7)$

49. $(-6, -2) \cap [1, 4)$ **50.** $(-\infty, -2) \cap (-1, \infty)$

In Exercises 51–58, write in interval notation.

51. 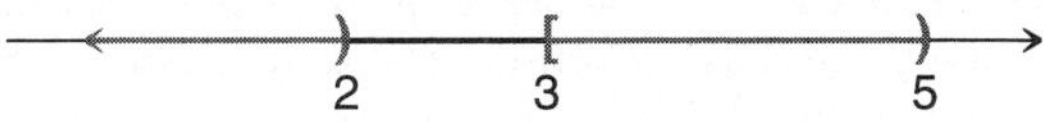

52. 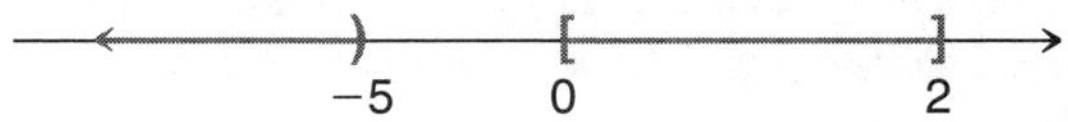

53. 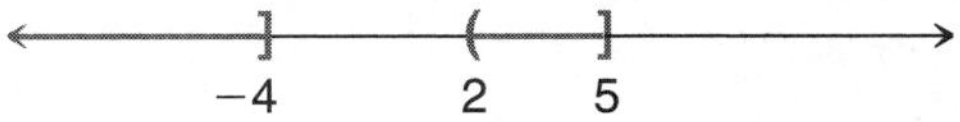

54. 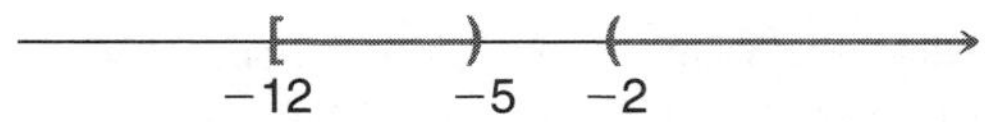

55. 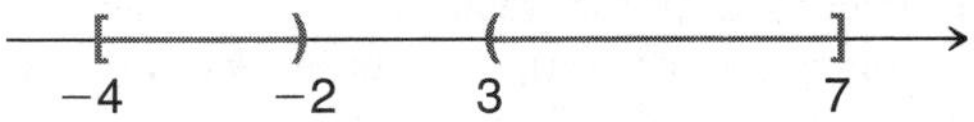

56. 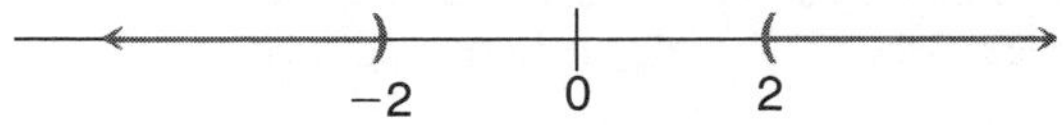

57. 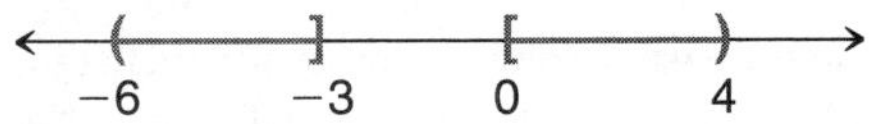

58. 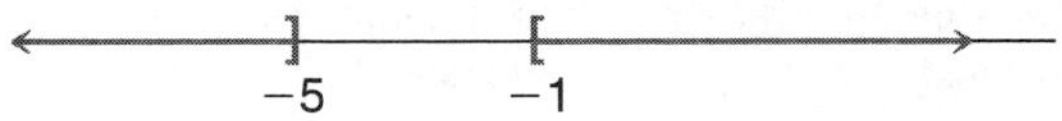

In Exercises 59–86, solve and express the solution in interval notation.

59. $x - 3 < 7$

60. $x + 4 > 9$

61. $3x - 2 \le 4$

62. $3x + 7 \ge -8$

63. $-5p \ge 10$

64. $-4u < 12$

65. $3 - 2x \le 7$

66. $4 - 3x > -17$

67. $3(t + 1) > 2t$

68. $2(y + 5) \le 3(y - 4)$

69. $7 - 2(1 - x) > 5 + 3(x - 2)$

70. $4 - 3(2 + x) < 5$

71. $\frac{x+2}{3} - 2 \ge \frac{x}{2}$

72. $\frac{y-3}{5} - 2 \le \frac{y}{4}$

73. $\frac{t-5}{3} \le -4$

74. $\frac{2p+1}{5} > -3$

75. $\frac{2}{3}y - \frac{1}{2}(5 - y) < \frac{5y}{3} - (2 + y)$

76. $\frac{s}{2} - \frac{(s-3)}{3} > \frac{s}{4} - \frac{1}{12}$

77. $-2 < x + 3 < 5$

78. $1 < x + 6 < 12$

79. $-8 \le 4 + 2x < 8$

80. $0 < 2 + x \le 5$

81. $-3 < 1 - x \le 9$

82. $3 \le -2 - 5x \le 13$

83. $0 < 2 - \frac{1}{3}y < 4$

84. $3 < \frac{1}{2}A - 3 < 7$

85. $\frac{1}{2} \le \frac{1+y}{3} \le \frac{3}{4}$

86. $-1 < \frac{2-z}{4} \le \frac{1}{5}$

APPLICATIONS

87. Weight. A healthy weight range for a woman is given by the following formula:

- 110 lb for the first 5 feet (tall)
- 2–6 lb per inch for every inch above 5 feet

Write an inequality representing a healthy weight, w, for a 5 foot 9 inch woman.

88. Weight. NASA has more stringent weight allowances for its astronauts. Write an inequality representing allowable weight for a female 5 foot 9 inch mission specialist given 105 lb for the first 5 feet, and 1–5 lb/in. for every additional inch.

89. Profit. A seamstress decides to open a dress shop. Her fixed costs are \$4,000 per month, and it costs her \$20 to make each dress. If the price of each dress is \$100, how many dresses does she have to sell per month to make a profit?

90. Profit. Labrador retrievers that compete in field trials typically cost \$2,000 at birth. Professional trainers charge \$400 to \$1,000 per month to train the dogs. If the dog is a champion by age 2, it sells for \$30,000. What is the range of profit for a champion at age 2?

91. Cost: Cell Phones. A cell phone company charges \$50 for an 800-minute monthly plan, plus an additional \$0.22/minute for every minute over 800. If a customer's bill ranged from a low of \$67.16 to a high of \$96.86 over a 6-month period, what were the most minutes used in a single month? What were the least?

92. Cost: Internet. An Internet provider charges \$30 per month for 1000 minutes of DSL service plus \$0.08 for each additional minute. In a one-year period the customer's bill ranged from \$36.40 to \$47.20. What were the most and least minutes used?

93. Grades. In your general biology class, your first three test scores are 67, 77, and 84. What is the lowest score you can get on the fourth test to earn at least a B for the course? Assume that each test is of equal weight and the minimum score required to earn a B is an 80.

94. Grades. In your Economics I class there are four tests and a final exam, all of which count equally. Your four test grades are 96, 87, 79, and 89. What grade on your exam is needed to earn between 80 and 90 for the course?

95. Markups. Typical markup on new cars is 15%–30%. If the sticker price is \$27,999, write an inequality that gives the range of the invoice price (what the dealer paid the manufacturer for the car).

96. Markups. Repeat Exercise 95 with a sticker price of \$42,599.

97. Lasers. A circular laser beam with a radius r_T is transmitted from one tower to another tower. If the received beam radius r_R fluctuates 10% from the transmitted beam radius due to atmospheric turbulence, write an inequality representing the received beam radius.

98. Electronics: Communications. Communication systems are often evaluated based on their signal-to-noise ratio (SNR), which is the ratio of the average power of received signal, *S*, to average power of noise, *N*, in the system. If the SNR is required to be at least 2 at all times, write an inequality representing the received signal power if the noise can fluctuate 10%.

99. Real Estate. The Aguileras are listing their house with a real estate agent. They are trying to determine a listing price, *L*, for the house. Their realtor advises them that most buyers traditionally offer a buying price, *B*, that is 85%–95% of the listing price. Write an inequality that relates the buying price to the listing price.

100. Humidity. The National Oceanic and Atmospheric Administration (NOAA) has stations on buoys in the oceans to measure atmosphere and ocean characteristics such as temperature, humidity, and wind. The humidity sensors have an error of 5%. Write an inequality relating the measured humidity h_m, and the true humidity h_t.

101. Recreation: Golf. Two friends enjoy playing golf. Their favorite course charges \$40 for greens fees (to play the course) and a \$15 cart rental (per person), so it currently costs each of them \$55 every time they play. The membership offered at that course is \$160 per month. The membership allows them to play as much as they want (no greens fees), but does still charge a cart rental fee of \$10 every time they play. What is the least number of times they should play a month in order for the membership to be the better deal?

102. Recreation: Golf. The same friends in Exercise 101 have a second favorite course. That golf course charges \$30 for greens fees (to play the course) and a \$10 cart rental (per person), so it currently costs each of them \$40 every time they play. The membership offered at that course is \$125 per month. The membership allows them to play as much as they want (no greens fees), but does still charge a cart rental fee of \$10. What is the least number of times they should play a month in order for the membership to be the better deal?

The following table is the 2007 Federal Tax Rate Schedule for people filing as *single*:

TAX BRACKET #	IF TAXABLE INCOME IS OVER–	BUT NOT OVER–	THE TAX IS:
I	\$0	\$7,825	10% of the amount over \$0
II	\$7,825	\$31,850	\$782.50 plus 15% of the amount over 7,825
III	\$31,850	\$77,100	\$4,386.25 plus 25% of the amount over 31,850
IV	\$77,100	\$160,850	\$15,698.75 plus 28% of the amount over 77,100
V	\$160,850	\$349,700	\$39,148.75 plus 33% of the amount over 160,850
VI	\$349,700	No limit	\$101,469.25 plus 35% of the amount over 349,700

103. Federal Income Tax. What is the range of federal income taxes a person in tax bracket III will pay the IRS?

104. Federal Income Tax. What is the range of federal income taxes a person in tax bracket IV will pay the IRS?

CATCH THE MISTAKE

In Exercises 105–108, explain the mistake that is made.

105. Rewrite in interval notation.

$-1 \le x < 4$

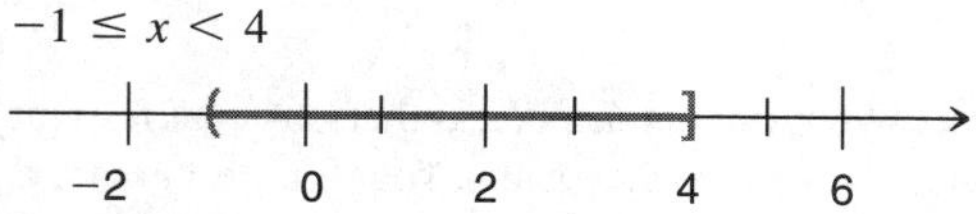

$(-1, 4]$

This is incorrect. What mistake was made?

106. Graph the indicated set and write as a single interval if possible.

$[-2, 4) \cap (3, 6]$

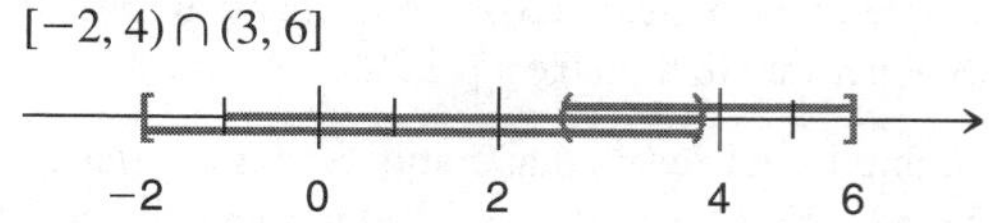

$[-2, 6]$

This is incorrect. What mistake was made?

107. Solve the inequality $2 - 3p \leq -4$ and express the solution in interval notation.

Solution:

$$\begin{aligned} 2 - 3p &\leq -4 \\ -3p &\leq -6 \\ p &\leq 2 \\ &(-\infty, 2] \end{aligned}$$

This is incorrect. What mistake was made?

108. Solve the inequality $3 - 2x \leq 7$ and express the solution in interval notation.

Solution:

$$\begin{aligned} 3 - 2x &\leq 7 \\ -2x &\leq 4 \\ x &\geq -2 \\ &(-\infty, -2] \end{aligned}$$

This is incorrect. What mistake was made?

CONCEPTUAL

In Exercises 109 and 110, determine whether each statement is true or false.

109. If $x < a$, then $a > x$.

110. If $-x \geq a$, then $x \geq -a$.

In Exercises 111–114, select any of the statements $a - d$ that could be true.

a. $m > 0$ and $n > 0$

b. $m < 0$ and $n < 0$

c. $m > 0$ and $n < 0$

d. $m < 0$ and $n > 0$

111. $mn > 0$

112. $mn < 0$

113. $\frac{m}{n} > 0$

114. $\frac{m}{n} < 0$

In Exercises 115 and 116, select any of the statements $a - c$ that could be true.

a. $n = 0$

b. $n > 0$

c. $n < 0$

115. $m + n < m - n$

116. $m + n \geq m - n$

CHALLENGE

117. Solve the inequality $x \leq -x$ mentally (without doing any algebraic manipulation).

118. Solve the inequality $x > -x$ mentally (without doing any algebraic manipulation).

119. Solve the inequality $ax + b < ax - c$, where $0 < b < c$.

120. Solve the inequality $-ax + b < -ax + c$, where $0 < b < c$.

TECHNOLOGY

121. a. Solve the inequality $2.7x + 3.1 < 9.4x - 2.5$.

b. Graph each side of the inequality in the same viewing screen. Find the range of x-values when the graph of the left side lies *below* the graph of the right side.

c. Do (a) and (b) agree?

122. a. Solve the inequality $-0.5x + 2.7 > 4.1x - 3.6$.

b. Graph each side of the inequality in the same viewing screen. Find the range of x-values when the graph of the left side lies *above* the graph of the right side.

c. Do (a) and (b) agree?

123. a. Solve the inequality $x - 3 < 2x - 1 < x + 4$.

b. Graph all three expressions of the inequality in the same viewing screen. Find the range of x-values when the graph of the middle expression lies above the graph of the left side and below the graph of the right side.

c. Do (a) and (b) agree?

124. a. Solve the inequality $x - 2 < 3x + 4 \leq 2x + 6$.

b. Graph all three expressions of the inequality in the same viewing screen. Find the range of x-values when the graph of the middle expression lies above the graph of the left side and on top of and below the graph of the right side.

c. Do (a) and (b) agree?

125. a. Solve the inequality $x + 3 < x + 5$.

b. Graph each side of the inequality in the same viewing screen. Find the range of x-values when the graph of the left side lies *below* the graph of the right side.

c. Do (a) and (b) agree?

126. a. Solve the inequality $\frac{1}{2}x - 3 > -\frac{2}{3}x + 1$.

b. Graph each side of the inequality in the same viewing screen. Find the range of x-values when the graph of the left side lies above the graph of the right side.

c. Do (a) and (b) agree?

SECTION 1.6 POLYNOMIAL AND RATIONAL INEQUALITIES

SKILLS OBJECTIVES

- Solve polynomial inequalities.
- Solve rational inequalities.

CONCEPTUAL OBJECTIVES

- Understand zeros and test intervals.
- Realize that a rational inequality has an implied domain restriction on the variable.

Polynomial Inequalities

In this section we will focus primarily on quadratic inequalities, but the procedures outlined are also valid for higher degree polynomial inequalities. An example of a quadratic inequality is $x^2 + x - 2 < 0$. This statement is true when the value of the polynomial on the left side is negative. For any value of x, a polynomial has either a positive, negative, or zero value. A polynomial must pass through zero before its value changes from positive to negative or from negative to positive. **Zeros** of a polynomial are the values of x that make the polynomial equal to zero. These zeros divide the real number line into **test intervals** where the value of the polynomial is either positive or negative. For example, if we set the above polynomial equal to zero and solve:

$$x^2 + x - 2 = 0$$
$$(x + 2)(x - 1) = 0$$
$$x = -2 \text{ or } x = 1$$

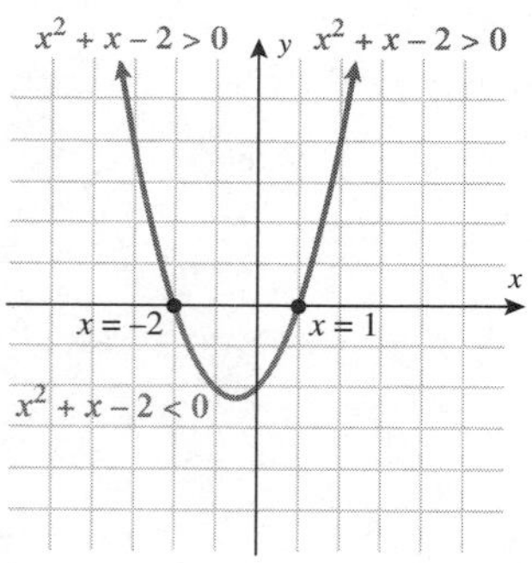

we find that $x = -2$ and $x = 1$ are the zeros. These zeros divide the real number line into three test intervals: $(-\infty, -2)$, $(-2, 1)$, and $(1, \infty)$.

Since the polynomial is equal to zero at $x = -2$ and $x = 1$, the value of the polynomial in each of these three intervals is either positive or negative. We select one real number that lies in each of the three intervals and test to see whether the value of the polynomial at each point is either positive or negative. In this example, we select the real numbers: $x = -3$, $x = 0$, and $x = 2$. At this point, there are two ways we can determine whether the value of the polynomial is positive or negative on the interval. One approach is to substitute each of the test points into the polynomial $x^2 + x - 2$.

$x = -3$	$(-3)^2 + (-3) - 2 = 9 - 3 - 2 = 4$	Positive
$x = 0$	$(0)^2 + (0) - 2 = 0 - 0 - 2 = -2$	Negative
$x = 2$	$(2)^2 + (2) - 2 = 4 + 2 - 2 = 4$	Positive

The second approach is to simply determine the sign of the result as opposed to actually calculating the exact number. This alternate approach is often used when the expressions or test points get more complicated to evaluate. The polynomial is written as the product $(x + 2)(x - 1)$; therefore, we simply look for the sign in each set of parentheses.

$$(x + 2)(x - 1)$$

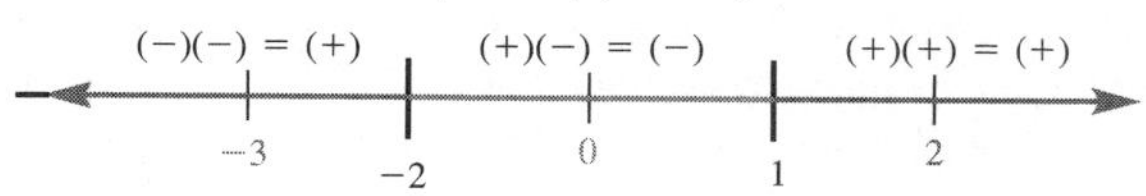

In this second approach we find the same result: $(-\infty, -2)$ and $(1, \infty)$ correspond to a positive value of the polynomial, and $(-2, 1)$ corresponds to a negative value of the polynomial.

In this example, the statement $x^2 + x - 2 < 0$ is true when the value of the polynomial (in factored form), $(x + 2)(x - 1)$, is negative. In the interval $(-2, 1)$, the value of the polynomial is negative. Thus, the solution to the inequality $x^2 + x - 2 < 0$ is $(-2, 1)$. To check the solution, select any number in the interval and substitute it into the original inequality to make sure it makes the statement true. The value $x = -1$ lies in the interval $(-2, 1)$. Upon substituting into the original inequality, we find that $x = -1$ satisfies the inequality $(-1)^2 + (-1) - 2 = -2 < 0$.

PROCEDURE FOR SOLVING POLYNOMIAL INEQUALITIES

Step 1: Write the inequality in *standard form.*
Step 2: Identify zeros.
Step 3: Draw the number line with zeros labeled.
Step 4: Determine the sign of the polynomial in each interval.
Step 5: Identify which interval(s) make the inequality true.
Step 6: Write the solution in interval notation.

Study Tip

If the original polynomial is < 0, then the interval(s) that yield(s) *negative* products should be selected. If the original polynomial is > 0, then the interval(s) that yield(s) *positive* products should be selected.

Note: Be careful in Step 5. If the original polynomial is < 0, then the interval(s) that correspond(s) to the value of the polynomial being negative should be selected. If the original polynomial is > 0, then the interval(s) that correspond(s) to the value of the polynomial being positive should be selected.

EXAMPLE 1 Solving a Quadratic Inequality

Solve the inequality $x^2 - x > 12$.

Solution:

Write the inequality in standard form. $x^2 - x - 12 > 0$

Factor the left side. $(x + 3)(x - 4) > 0$

Identify the zeros. $(x + 3)(x - 4) = 0$

$x = -3$ or $x = 4$

Draw the number line with the zeros labeled.

Test each interval. $(x + 3)(x - 4)$

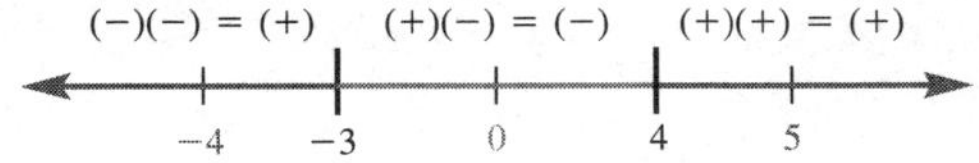

Intervals in which the value of the polynomial is *positive* make this inequality true. $(-\infty, -3)$ or $(4, \infty)$

Write the solution in interval notation. $\boxed{(-\infty, -3) \cup (4, \infty)}$

■ **YOUR TURN** Solve the inequality $x^2 - 5x \leq 6$ and express the solution in interval notation.

Technology Tip

Use a graphing utility to display graphs of $y_1 = x^2 - x$ and $y_2 = 12$.

The solutions are the x-values such that the graph of y_1 lies above the graph of y_2.

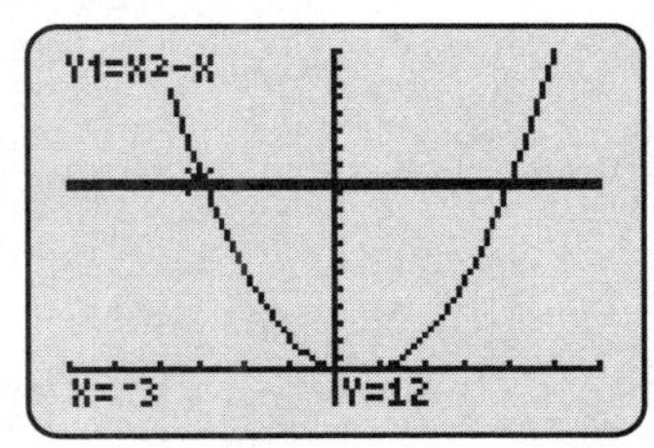

■ **Answer:** $[-1, 6]$

The inequality in Example 1, $x^2 - x > 12$, is a strict inequality, so we use parentheses when we express the solution in interval notation $(-\infty, -3) \cup (4, \infty)$. It is important to note that if we change the inequality sign from $>$ to $\geq$, then the zeros $x = -3$ and $x = 4$ also make the inequality true. Therefore the solution to $x^2 - x \geq 12$ is $(-\infty, -3] \cup [4, \infty)$.

CAUTION

The square root method cannot be used for quadratic inequalities.

EXAMPLE 2 Solving a Quadratic Inequality

Solve the inequality $x^2 \le 4$.

COMMON MISTAKE

Do not take the square root of both sides. You must write the inequality in standard form and factor.

CORRECT

Write the inequality in standard form.

$$x^2 - 4 \le 0$$

Factor.

$$(x - 2)(x + 2) \le 0$$

Identify the zeros.

$$(x - 2)(x + 2) = 0$$

$$x = 2 \text{ and } x = -2$$

Draw the number line and test the intervals.

$$(x - 2)(x + 2)$$

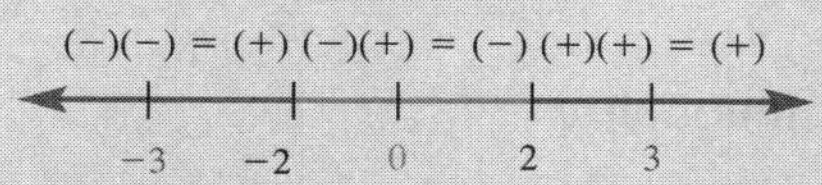

Intervals in which the value of the polynomial is *negative* make the inequality true.

$$(-2, 2)$$

The endpoints, $x = -2$ and $x = 2$, satisfy the inequality, so they are included in the solution and expressed with brackets.

$$\boxed{[-2, 2]}$$

When solving quadratic inequalities, you must first write the inequality in standard form and then factor to identify zeros.

INCORRECT

ERROR:

Take the square root of both sides.

$$x \le \pm 2$$

Not all inequalities have a solution. For example, $x^2 < 0$ has no real solution. Any real number squared is always nonnegative, so there are no real values that when squared will yield a negative number. The zero is $x = 0$, which divides the real number line into two intervals: $(-\infty, 0)$ and $(0, \infty)$. Both of these intervals, however, correspond to the value of x^2 being positive, so there are no intervals that satisfy the inequality. We say that this inequality has no real solution.

EXAMPLE 3 Solving a Quadratic Inequality

Solve the inequality $x^2 + 2x \geq -3$.

Solution:

Write the inequality in standard form. $x^2 + 2x + 3 \geq 0$

Identify the zeros. $x^2 + 2x + 3 = 0$

Apply the quadratic formula. $x = \dfrac{-2 \pm \sqrt{2^2 - 4(1)(3)}}{2(1)}$

Simplify. $x = \dfrac{-2 \pm \sqrt{-8}}{2} = \dfrac{-2 \pm 2i\sqrt{2}}{2} = -1 \pm i\sqrt{2}$

Since there are no real zeros, the quadratic expression $x^2 + 2x + 3$ never equals zero, and hence its value is either always positive or always negative. If we select any value for x, say, $x = 0$, we find that $(0)^2 + 2(0) + 3 \geq 0$. Therefore the quadratic expression is always positive, so the solution is the set of all real numbers, $\boxed{(-\infty, \infty)}$.

Study Tip

When solving quadratic inequalities you must first write the inequality in standard form and then factor to identify zeros.

EXAMPLE 4 Solving a Quadratic Inequality

Solve the inequality $x^2 > -5x$.

COMMON MISTAKE

A common mistake is to divide by x. Never divide by a variable, because the value of the variable might be zero. Always start by writing the inequality in standard form and then factor to determine the zeros.

✪ CORRECT

Write the inequality in standard form.

$$x^2 + 5x > 0$$

Factor.

$$x(x + 5) > 0$$

Identify the zeros.

$$x = 0, x = -5$$

Draw the number line and test the intervals.

$x(x + 5)$

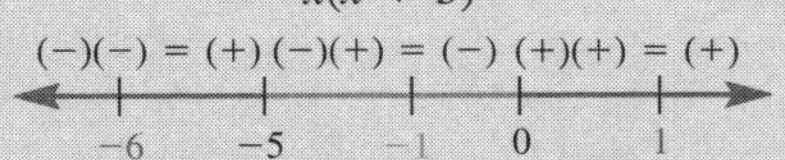

Intervals in which the value of the polynomial is *positive* satisfy the inequality.

$(-\infty, -5)$ and $(0, \infty)$

Express the solution in interval notation.

$$\boxed{(-\infty, -5) \cup (0, \infty)}$$

☒ INCORRECT

Write the original inequality.

$$x^2 > -5x$$

ERROR:
Divide both sides by x.

$$x > -5$$

Dividing by x is the mistake. If x is negative, the inequality sign must be reversed. What if x is zero?

▼ CAUTION

Do not divide inequalities by a variable.

Technology Tip

Using a graphing utility, graph $y_1 = x^2 + 2x$ and $y_2 = 1$.

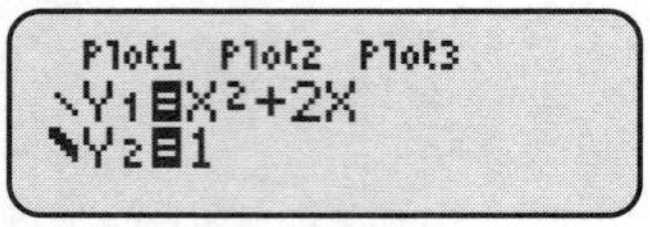

The solutions are the x-values such that the graph of y_1 lies below the graph of y_2.

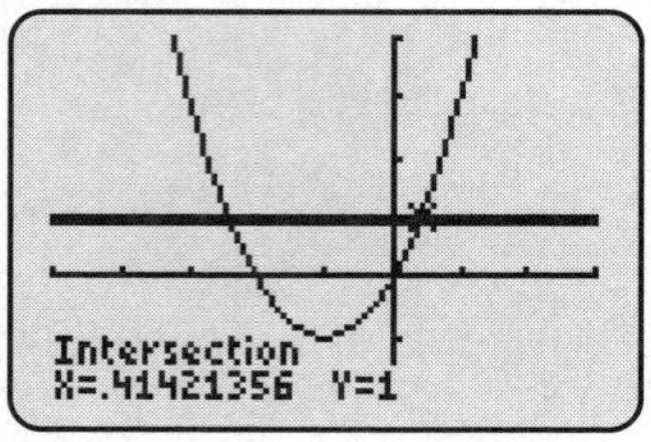

Note that:

EXAMPLE 5 Solving a Quadratic Inequality

Solve the inequality $x^2 + 2x < 1$.

Solution:

Write the inequality in standard form. $x^2 + 2x - 1 < 0$

Identify the zeros. $x^2 + 2x - 1 = 0$

Apply the quadratic formula. $x = \dfrac{-2 \pm \sqrt{2^2 - 4(1)(-1)}}{2(1)}$

Simplify. $x = \dfrac{-2 \pm \sqrt{8}}{2} = \dfrac{-2 \pm 2\sqrt{2}}{2} = -1 \pm \sqrt{2}$

Draw the number line with the intervals labeled.

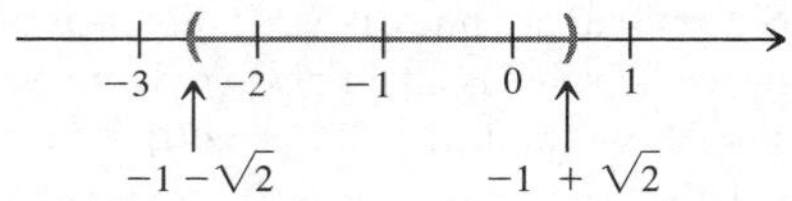

Note: $-1 - \sqrt{2} \approx -2.41$
$-1 + \sqrt{2} \approx 0.41$

Test each interval.

$(-\infty, -1 - \sqrt{2})$ $\quad x = -3$: $\quad (-3)^2 + 2(-3) - 1 = 2 > 0$

$(-1 - \sqrt{2}, -1 + \sqrt{2})$ $\quad x = 0$: $\quad (0)^2 + 2(0) - 1 = -1 < 0$

$(-1 + \sqrt{2}, \infty)$ $\quad x = 1$: $\quad (1)^2 + 2(1) - 1 = 2 > 0$

Intervals in which the value of the polynomial is *negative* make this inequality true. $\boxed{(-1 - \sqrt{2}, -1 + \sqrt{2})}$

YOUR TURN Solve the inequality $x^2 - 2x \geq 1$.

■ **Answer:** $(-\infty, 1 - \sqrt{2}\,] \cup [\,1 + \sqrt{2}, \infty)$

EXAMPLE 6 Solving a Polynomial Inequality

Solve the inequality $x^3 - 3x^2 \geq 10x$.

Solution:

Write the inequality in standard form. $x^3 - 3x^2 - 10x \geq 0$

Factor. $x(x - 5)(x + 2) \geq 0$

Identify the zeros. $x = 0, x = 5, x = -2$

Draw the number line with the zeros (intervals) labeled.

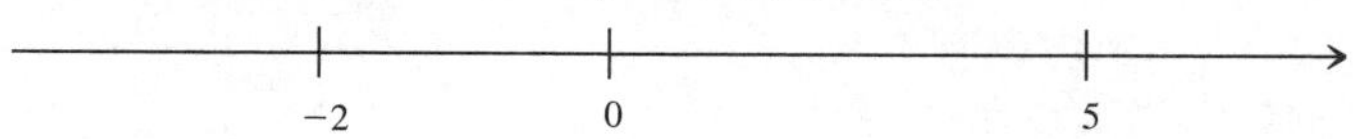

Test each interval.

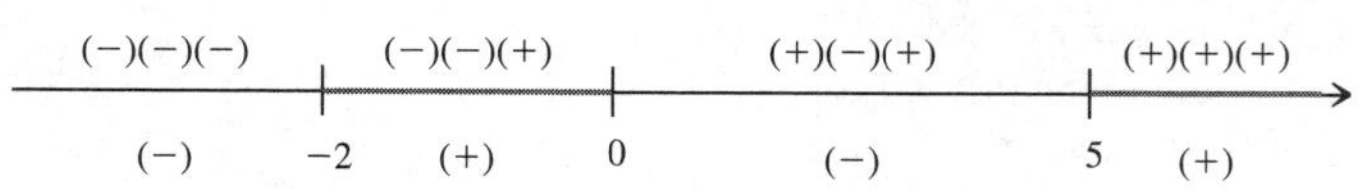

Intervals in which the value of the polynomial is *positive* make this inequality true. $\boxed{[-2, 0] \cup [5, \infty)}$

YOUR TURN Solve the inequality $x^3 - x^2 - 6x < 0$.

■ **Answer:** $(-\infty, -2) \cup (0, 3)$

Rational Inequalities

Rational expressions have numerators and denominators. Recalling the properties of negative real numbers (Chapter 0), we see that the following possible combinations correspond to either positive or negative rational expressions.

$$\frac{(+)}{(+)} = (+) \qquad \frac{(-)}{(+)} = (-) \qquad \frac{(-)}{(-)} = (+) \qquad \frac{(+)}{(-)} = (-)$$

A rational expression can change signs if either the numerator or denominator changes signs. In order to go from positive to negative or vice versa, you must pass through zero. Therefore, to solve rational inequalities such as $\frac{x-3}{x^2-4} \geq 0$ we use a similar procedure to the one used for solving polynomial inequalities, with one exception. You must eliminate values for x that make the denominator equal to zero. In this example, we must eliminate $x = -2$ and $x = 2$ because these values make the denominator equal to zero. Rational inequalities have implied domains. In this example, $x \neq \pm 2$ is a domain restriction and these values ($x = -2$ and $x = 2$) must be eliminated from a possible solution.

We will proceed with a similar procedure involving zeros and test intervals that was outlined for polynomial inequalities. However, in rational inequalities once expressions are combined into a single fraction, any values that make *either* the numerator *or* the denominator equal to zero divide the number line into intervals.

Study Tip

Values that make the denominator equal to zero are always excluded.

EXAMPLE 7 Solving a Rational Inequality

Solve the inequality $\frac{x-3}{x^2-4} \geq 0$.

Solution:

Factor the denominator. $\frac{(x-3)}{(x-2)(x+2)} \geq 0$

State the domain restrictions on the variable. $x \neq 2, x \neq -2$

Identify the zeros of numerator and denominator. $x = -2, x = 2, x = 3$

Draw the number line and divide into intervals.

Test the intervals.

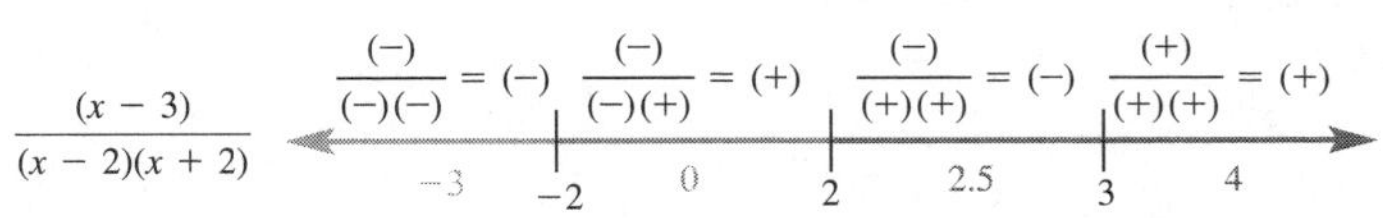

Intervals in which the value of the polynomial is *positive* satisfy this inequality. $(-2, 2)$ and $(3, \infty)$

Since this inequality is greater than or equal to, we include $x = 3$ in our solution because it satisfies the inequality. However, $x = -2$ and $x = 2$ are not included in the solution because they make the denominator equal to zero.

The solution is $\boxed{(-2, 2) \cup [3, \infty)}$.

■ **YOUR TURN** Solve the inequality $\frac{x+2}{x-1} \leq 0$.

Technology Tip

Use a graphing utility to display the graph of $y_1 = \frac{x-3}{x^2-4}$.

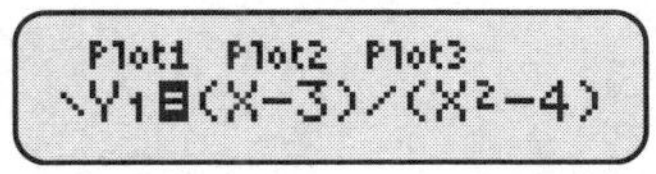

The solutions are the x-values such that the graph of y_1 lies on top and above the x-axis, excluding $x = \pm 2$.

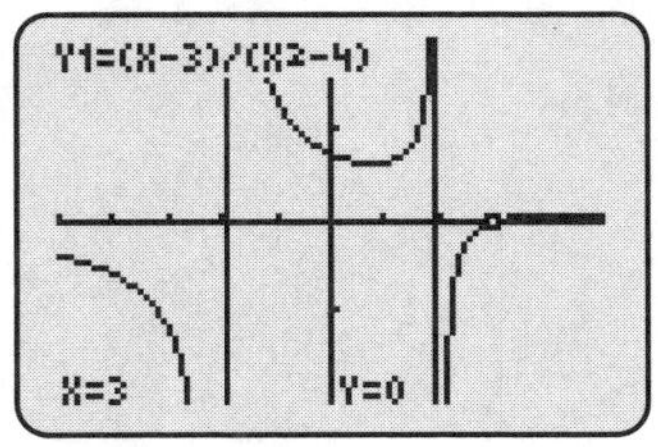

■ **Answer:** $[-2, 1)$

EXAMPLE 8 Solving a Rational Inequality

Solve the inequality $\dfrac{x}{x+2} \le 3$.

▼ CAUTION

Rational inequalities cannot be solved by cross multiplication.

COMMON MISTAKE

Do not cross multiply.

CORRECT

Subtract 3 from both sides.

$$\frac{x}{x+2} - 3 \le 0$$

Write as a single rational expression.

$$\frac{x - 3(x+2)}{x+2} \le 0$$

Eliminate the parentheses.

$$\frac{x - 3x - 6}{x+2} \le 0$$

Simplify the numerator.

$$\frac{-2x - 6}{x+2} \le 0$$

Factor the numerator.

$$\frac{-2(x+3)}{x+2} \le 0$$

Identify the zeros of the numerator and the denominator.

$$x = -3 \text{ and } x = -2$$

Draw the number line and test the intervals.

$$\frac{-2(x+3)}{x+2} \le 0$$

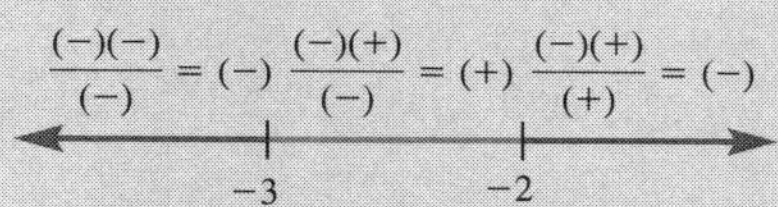

Intervals in which the value of the polynomial is *negative* satisfy the inequality. $(-\infty, -3]$ and $(-2, \infty)$. Note that $x = -2$ is not included in the solution because it makes the denominator zero, and $x = -3$ is included because it satisfies the inequality.

The solution is:

$$\boxed{(-\infty, -3] \cup (-2, \infty)}$$

INCORRECT

ERROR:
Do not cross multiply.

$$x \le 3(x+2)$$

Applications

EXAMPLE 9 Stock Prices

From November 2003 until March 2004 Wal-Mart stock (WMT) was worth approximately $P = 2t^2 - 12t + 70$, where P denotes the price of the stock in dollars and t corresponds to months. November 2003 is represented by $t = 1$, December 2003 by $t = 2$, January 2004 by $t = 3$, and so on. During what months was the stock value at least \$54?

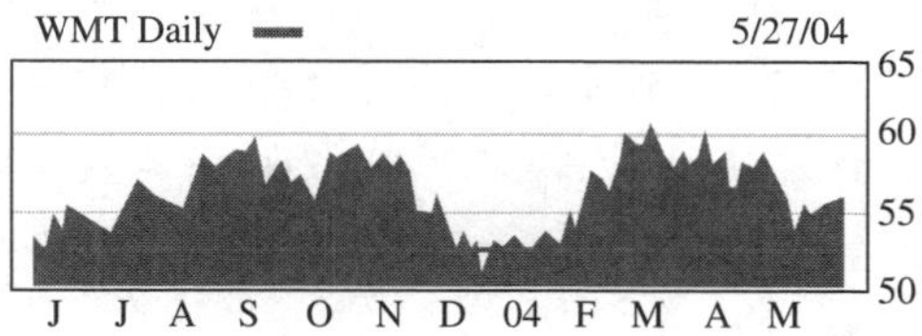

Solution:

Set the price greater than or equal to 54. $2t^2 - 12t + 70 \geq 54 \quad 1 \leq t \leq 5$

Write in standard form. $2t^2 - 12t + 16 \geq 0$

Divide by 2. $(t^2 - 6t + 8) \geq 0$

Factor. $(t - 4)(t - 2) \geq 0$

Identify the zeros. $t = 4$ and $t = 2$

Test the intervals. $(t - 4)(t - 2) \geq 0$

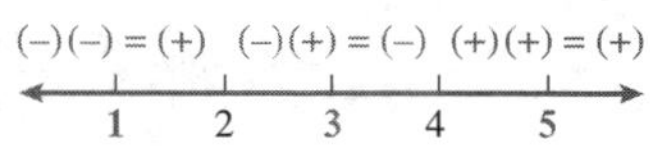

Positive intervals satisfy the inequality. $[1, 2]$ and $[4, 5]$

The Wal-Mart stock price was at least \$54 during November 2003, December 2003, February 2004, and March 2004.

SECTION 1.6 SUMMARY

The following procedure can be used for solving polynomial and rational inequalities.

1. Write in standard form—zero on one side.
2. Determine the zeros; if it is a rational function, note the domain restrictions.
   - Polynomial Inequality
     - Factor if possible.
     - Otherwise, use the quadratic formula.
   - Rational Inequality
     - Write as a single fraction.
     - Determine values that make the numerator or denominator equal to zero. Always exclude values that make the denominator = 0.
3. Draw the number line labeling the intervals.
4. Test the intervals to determine whether they are positive or negative.
5. Select the intervals according to the sign of the inequality.
6. Write the solution in interval notation.

SECTION 1.6 EXERCISES

SKILLS

In Exercises 1–28, solve the polynomial inequality and express the solution set in interval notation.

1. $x^2 - 3x - 10 \geq 0$
2. $x^2 + 2x - 3 < 0$
3. $u^2 - 5u - 6 \leq 0$
4. $u^2 - 6u - 40 > 0$
5. $p^2 + 4p < -3$
6. $p^2 - 2p \geq 15$
7. $2t^2 - 3 \leq t$
8. $3t^2 \geq -5t + 2$
9. $5v - 1 > 6v^2$
10. $12t^2 < 37t + 10$
11. $2s^2 - 5s \geq 3$
12. $8s + 12 \leq -s^2$
13. $y^2 + 2y \geq 4$
14. $y^2 + 3y \leq 1$
15. $x^2 - 4x < 6$
16. $x^2 - 2x > 5$
17. $u^2 \geq 3u$
18. $u^2 \leq -4u$
19. $-2x \leq -x^2$
20. $-3x \leq x^2$
21. $x^2 > 9$
22. $x^2 \geq 16$
23. $t^2 < 81$
24. $t^2 \leq 49$
25. $z^2 > -16$
26. $z^2 \geq -2$
27. $y^2 < -4$
28. $y^2 \leq -25$

In Exercises 29–58, solve the rational inequality and graph the solution on the real number line.

29. $-\frac{3}{x} \leq 0$
30. $\frac{3}{x} \leq 0$
31. $\frac{y}{y+3} > 0$
32. $\frac{y}{2-y} \leq 0$
33. $\frac{t+3}{t-4} \geq 0$
34. $\frac{2t-5}{t-6} < 0$
35. $\frac{s+1}{4-s^2} \geq 0$
36. $\frac{s+5}{4-s^2} \leq 0$
37. $\frac{x-3}{x^2-25} \geq 0$
38. $\frac{1-x}{x^2-9} \leq 0$
39. $\frac{2u^2+u}{3} < 1$
40. $\frac{u^2-3u}{3} \geq 6$
41. $\frac{3t^2}{t+2} \geq 5t$
42. $\frac{-2t-t^2}{4-t} \geq t$
43. $\frac{3p-2p^2}{4-p^2} < \frac{3+p}{2-p}$
44. $-\frac{7p}{p^2-100} \leq \frac{p+2}{p+10}$
45. $\frac{x^2}{5+x^2} < 0$
46. $\frac{x^2}{5+x^2} \leq 0$
47. $\frac{x^2+10}{x^2+16} > 0$
48. $-\frac{x^2+2}{x^2+4} < 0$
49. $\frac{v^2-9}{v-3} \geq 0$
50. $\frac{v^2-1}{v+1} \leq 0$
51. $\frac{2}{t-3} + \frac{1}{t+3} \geq 0$
52. $\frac{1}{t-2} + \frac{1}{t+2} \leq 0$
53. $\frac{3}{x+4} - \frac{1}{x-2} \leq 0$
54. $\frac{2}{x-5} - \frac{1}{x-1} \geq 0$
55. $\frac{1}{p+4} + \frac{1}{p-4} > \frac{p^2-48}{p^2-16}$
56. $\frac{1}{p-3} - \frac{1}{p+3} \leq 2$
57. $\frac{1}{p-2} - \frac{1}{p+2} \geq \frac{3}{p^2-4}$
58. $\frac{2}{2p-3} - \frac{1}{p+1} \leq \frac{1}{2p^2-p-3}$

APPLICATIONS

59. Profit. A Web-based embroidery company makes monogrammed napkins. The profit associated with producing x orders of napkins is governed by the equation

$$P(x) = -x^2 + 130x - 3000$$

Determine the range of orders the company should accept in order to make a profit.

60. Profit. Repeat Exercise 59 using $P(x) = x^2 - 130x + 3600$.

61. Car Value. The term "upside down" on car payments refers to owing more than a car is worth. Assume you buy a new car and finance 100% over 5 years. The difference between the value of the car and what is owed on the car is governed by the expression $\frac{t}{t-3}$ where t is age (in years) of the car. Determine the time period when the car is worth more than you owe $\left(\frac{t}{t-3} > 0\right)$. When do you owe more than it's worth $\left(\frac{t}{t-3} < 0\right)$?

62. Car Value. Repeat Exercise 61 using the expression $-\frac{2-t}{4-t}$.

63. Bullet Speed. A .22 caliber gun fires a bullet at a speed of 1200 feet per second. If a .22 caliber gun is fired straight upward into the sky, the height of the bullet in feet is given by the equation $h = -16t^2 + 1200t$, where t is the time in seconds with $t = 0$ corresponding to the instant the gun is fired. How long is the bullet in the air?

64. Bullet Speed. A .38 caliber gun fires a bullet at a speed of 600 feet per second. If a .38 caliber gun is fired straight upward into the sky, the height of the bullet in feet is given by the equation $h = -16t^2 + 600t$. How many seconds is the bullet in the air?

65. Geometry. A rectangular area is fenced in with 100 feet of fence. If the minimum area enclosed is to be 600 square feet, what is the range of feet allowed for the length of the rectangle?

66. Stock Value. From June 2003 until April 2004 JetBlue airlines stock (JBLU) was approximately worth $P = -4t^2 + 80t - 360$ where P denotes the price of the stock in dollars and t corresponds to months, with $t = 1$ corresponding to January 2003. During what months was the stock value at least $36?

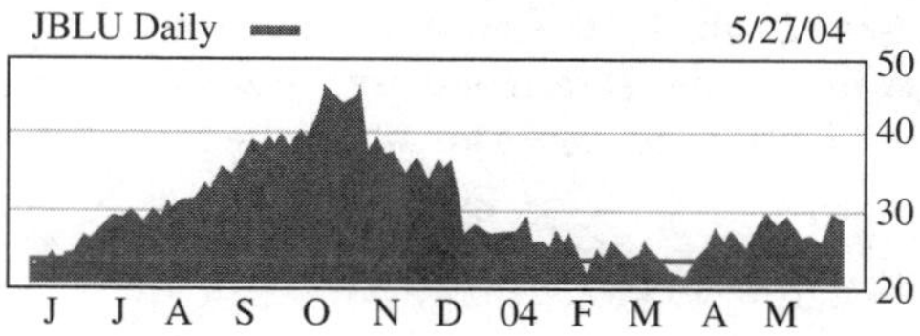

67. Real Estate. A woman is selling a piece of land that she advertises as 400 acres (±7 acres) for $1.36 million. If you pay that price, what is the range of dollars per acre you have paid? Round to the nearest dollar.

68. Real Estate. A woman is selling a piece of land that she advertises as 1000 acres (±10 acres) for $1 million. If you pay that price, what is the range of dollars per acre you have paid? Round to the nearest dollar.

CATCH THE MISTAKE

In Exercises 69–72, explain the mistake that is made.

69. Solve the inequality $3x < x^2$.

Solution:

Divide by x.	$3 < x$
Write the solution in interval notation.	$(3, \infty)$

This is incorrect. What mistake was made?

70. Solve the inequality $u^2 < 25$.

Solution:

Take the square root of both sides.	$u < -5$
Write the solution in interval notation.	$(-\infty, -5)$

This is incorrect. What mistake was made?

71. Solve the inequality $\dfrac{x^2 - 4}{x + 2} > 0$.

Solution:

Factor the numerator and denominator.	$\dfrac{(x - 2)(x + 2)}{(x + 2)} < 0$
Cancel the $(x + 2)$ common factor.	$x - 2 > 0$
Solve.	$x > 2$

This is incorrect. What mistake was made?

72. Solve the inequality $\dfrac{x + 4}{x} < -\dfrac{1}{3}$.

Solution:

Cross multiply.	$3(x + 4) < -1(x)$
Eliminate the parentheses.	$3x + 12 < -x$
Combine like terms.	$4x < -12$
Divide both sides by 4.	$x < -3$

This is incorrect. What mistake was made?

CONCEPTUAL

In Exercises 73 and 74, determine whether each statement is true or false. Assume that a is a positive real number.

73. If $x < a^2$, then the solution is $(-\infty, a)$.

74. If $x \geq a^2$, then the solution is $[a, \infty)$.

75. Assume the quadratic inequality $ax^2 + bx + c < 0$. If $b^2 - 4ac < 0$, then describe the solution.

76. Assume the quadratic inequality $ax^2 + bx + c > 0$. If $b^2 - 4ac < 0$, then describe the solution.

CHALLENGE

In Exercises 77–80, solve for x given that a and b are both positive real numbers.

77. $-x^2 \leq a^2$

78. $\dfrac{x^2 - b^2}{x + b} < 0$

79. $\dfrac{x^2 + a^2}{x^2 + b^2} \geq 0$

80. $\dfrac{a}{x^2} < -b$

TECHNOLOGY

In Exercises 81–84, plot the left side and the right side of each inequality in the same screen and use the zoom feature to determine the range of values for which the inequality is true.

81. $1.4x^2 - 7.2x + 5.3 > -8.6x + 3.7$

82. $17x^2 + 50x - 19 < 9x^2 + 2$

83. $11x^2 < 8x + 16$

84. $0.1x + 7.3 > 0.3x^2 - 4.1$

87. $\dfrac{2p}{5 - p} > 1$

85. $x < x^2 - 3x < 6 - 2x$

86. $x^2 + 3x - 5 \geq -x^2 + 2x + 10$

88. $\dfrac{3p}{4 - p} < 1$

SECTION 1.7 ABSOLUTE VALUE EQUATIONS AND INEQUALITIES

SKILLS OBJECTIVES

- Solve absolute value equations.
- Solve absolute value inequalities.

CONCEPTUAL OBJECTIVE

- Understand absolute value in terms of distance on the number line.

The **absolute value** of a real number can be interpreted algebraically and graphically. Algebraically, the absolute value of 5 is 5, or in mathematical notation, $|5| = 5$; and the absolute value of -5 is 5 or $|-5| = 5$. Graphically, the absolute value of a real number is the distance on the real number line between the real number and the origin; thus the distance from 0 to either -5 or 5 is 5.

DEFINITION **Absolute Value**

The **absolute value** of a real number a, denoted by the symbol $|a|$, is defined by

$$|a| = \begin{cases} a, & \text{if } a \geq 0 \\ -a, & \text{if } a < 0 \end{cases}$$

The absolute value of a real number is never negative. When $a = -5$, this definition says $|-5| = -(-5) = 5$.

PROPERTIES OF ABSOLUTE VALUE

For all real numbers a and b,

1. $|a| \geq 0$ **2.** $|-a| = |a|$ **3.** $|ab| = |a|\,|b|$ **4.** $\left|\dfrac{a}{b}\right| = \dfrac{|a|}{|b|} \quad b \neq 0$

Absolute value can be used to define the distance between two points on the real number line.

DISTANCE BETWEEN TWO POINTS ON THE REAL NUMBER LINE

If a and b are real numbers, the **distance between a and b** is the absolute value of their difference given by $|a - b|$ or $|b - a|$.

EXAMPLE 1 Finding the Distance Between Two Points on the Number Line

Find the distance between -4 and 3 on the real number line.

Solution:

The distance between -4 and 3 is given by the absolute value of the difference. $|-4 - 3| = |-7| = 7$

Note that if we reverse the numbers the result is the same. $|3 - (-4)| = |7| = 7$

We check this by counting the units between -4 and 3 on the number line.

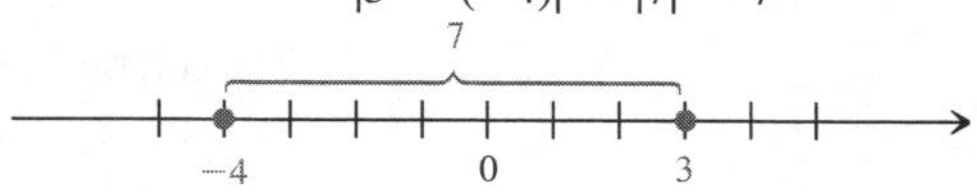

Equations Involving Absolute Value

When absolute value is involved in algebraic equations, we interpret the definition of absolute value as follows.

DEFINITION Absolute Value Equation

If $|x| = a$, then $x = -a$ or $x = a$, where $a \geq 0$.

In words, "if the absolute value of a number is a, then that number equals $-a$ or a." For example, the equation $|x| = 7$ is true if $x = -7$ or $x = 7$. We say the equation $|x| = 7$ has the solution set $\{-7, 7\}$. *Note:* $|x| = -3$ does not have a solution because there is no value of x such that its absolute value is -3.

EXAMPLE 2 Solving an Absolute Value Equation

Solve the equation $|x - 3| = 8$ algebraically and graphically.

Solution:

Using the absolute value equation definition, we see that if the absolute value of an expression is 8, then that expression is either -8 or 8. Rewrite as two equations:

$$x - 3 = -8 \quad \text{or} \quad x - 3 = 8$$
$$x = -5 \qquad\qquad x = 11$$

The solution set is $\boxed{\{-5, 11\}}$.

Graph: The absolute value equation $|x - 3| = 8$ is interpreted as "what numbers are 8 units away from 3 on the number line?" We find that 8 units to the right of 3 is 11 and 8 units to the left of 3 is -5.

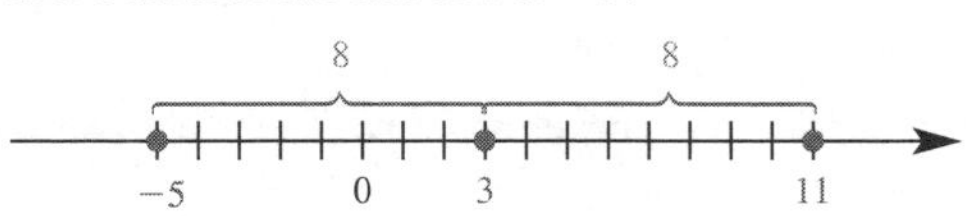

■ YOUR TURN Solve the equation $|x + 5| = 7$.

Technology Tip

Use a graphing utility to display graphs of $y_1 = |x - 3|$ and $y_2 = 8$.

The x-coordinate of the point of intersection is the solution to $|x - 3| = 8$.

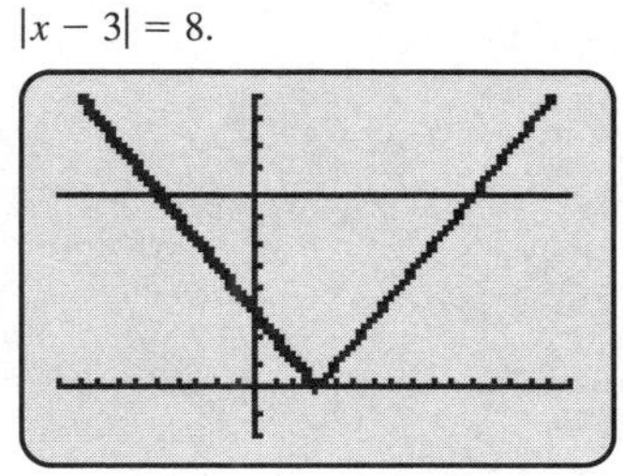

■ **Answer:** $x = 2$ or $x = -12$. The solution set is $\{-12, 2\}$.

Study Tip

Rewrite an absolute value equation as two equations.

EXAMPLE 3 Solving an Absolute Value Equation

Solve the equation $|1 - 3x| = 7$.

Solution:

If the absolute value of an expression is 7, then that expression is -7 or 7.

$$1 - 3x = -7 \text{ or } 1 - 3x = 7$$
$$-3x = -8 \qquad -3x = 6$$
$$x = \frac{8}{3} \qquad x = -2$$

The solution set is $\boxed{\{-2, \frac{8}{3}\}}$.

■ **YOUR TURN** Solve the equation $|1 + 2x| = 5$.

■ **Answer:** $x = -3$ or $x = 2$. The solution set is $\{-3, 2\}$.

EXAMPLE 4 Solving an Absolute Value Equation

Solve the equation $2 - 3|x - 1| = -4|x - 1| + 7$.

Solution:

Isolate the absolute value expressions to one side.

Add $4|x - 1|$ to both sides. $\qquad 2 + |x - 1| = 7$

Subtract 2 from both sides. $\qquad |x-1| = 5$

If the absolute value of an expression is equal to 5, then the expression is equal to either -5 or 5.

$$x - 1 = -5 \text{ or } x - 1 = 5$$
$$x = -4 \qquad x = 6$$

The solution set is $\boxed{\{-4, 6\}}$.

■ **YOUR TURN** Solve the equation $3 - 2|x - 4| = -3|x - 4| + 11$.

■ **Answer:** $x = -4$ or $x = 12$. The solution set is $\{-4, 12\}$.

EXAMPLE 5 Finding That an Absolute Value Equation Has No Solution

Solve the equation $|1 - 3x| = -7$.

Solution:

The absolute value of an expression is never negative. Therefore no values of x make this equation true.

$\boxed{\text{No solution}}$

EXAMPLE 6 Solving a Quadratic Absolute Value Equation

Solve the equation $|5 - x^2| = 1$.

Solution:

If the absolute value of an expression is 1, that expression is either -1 or 1, which leads to two equations.

$$5 - x^2 = -1 \text{ or } 5 - x^2 = 1$$
$$-x^2 = -6 \qquad -x^2 = -4$$
$$x^2 = 6 \qquad x^2 = 4$$
$$x = \pm\sqrt{6} \qquad x = \pm\sqrt{4} = \pm 2$$

The solution set is $\boxed{\{\pm 2, \pm\sqrt{6}\}}$.

YOUR TURN Solve the equation $|7 - x^2| = 2$.

Answer: $x = \pm\sqrt{5}$ or $x = \pm 3$. The solution set is $\{\pm\sqrt{5}, \pm 3\}$.

Inequalities Involving Absolute Value

To solve the inequality $|x| < 3$, look for all real numbers that make this statement true. Some numbers that make it true are -2, -1, 0, 1, and 2. Some numbers that make it false are -7, -5, -3, 3, and 4. If we interpret this inequality as distance, we ask *what numbers are less than 3 units from the origin?* We can represent the solution in the following ways.

Inequality notation: $-3 < x < 3$

Interval notation: $(-3, 3)$

Graph:

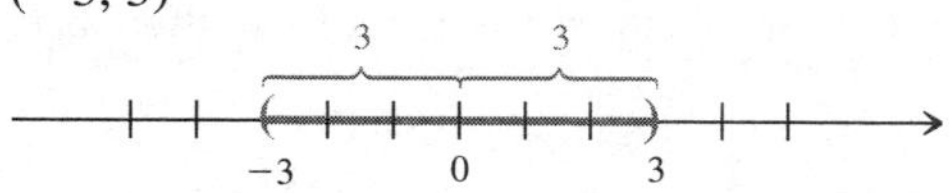

Similarly, to solve the inequality $|x| \geq 3$, look for all real numbers that make the statement true. If we interpret this inequality as a distance, we ask *what numbers are at least 3 units from the origin?* We can represent the solution in the following three ways.

Inequality notation: $x \leq -3$ or $x \geq 3$

Interval notation: $(-\infty, -3] \cup [3, \infty)$

Graph:

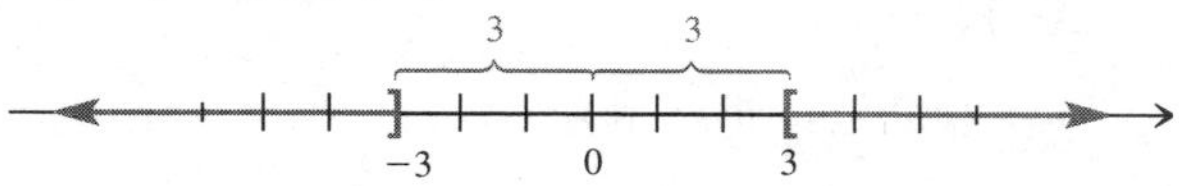

This discussion leads us to the following equivalence relations.

PROPERTIES OF ABSOLUTE VALUE INEQUALITIES

1. $|x| < a$ is equivalent to $-a < x < a$

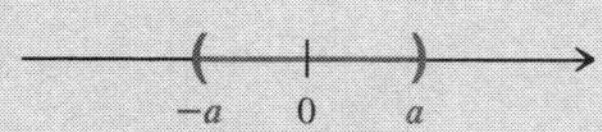

2. $|x| \leq a$ is equivalent to $-a \leq x \leq a$

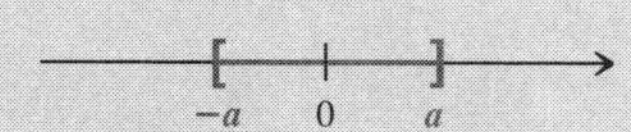

3. $|x| > a$ is equivalent to $x < -a$ or $x > a$

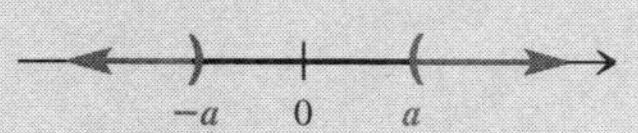

4. $|x| \geq a$ is equivalent to $x \leq -a$ or $x \geq a$

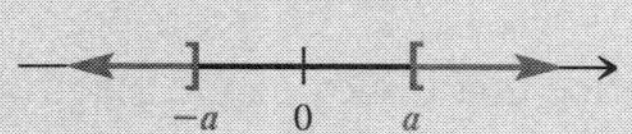

It is important to realize that in the above four properties the variable x can be any algebraic expression.

Technology Tip

Use a graphing utility to display graphs of $y_1 = |3x - 2|$ and $y_2 = 7$.

The values of x where the graph of y_1 lies on top and below the graph of y_2 are the solutions to this inequality.

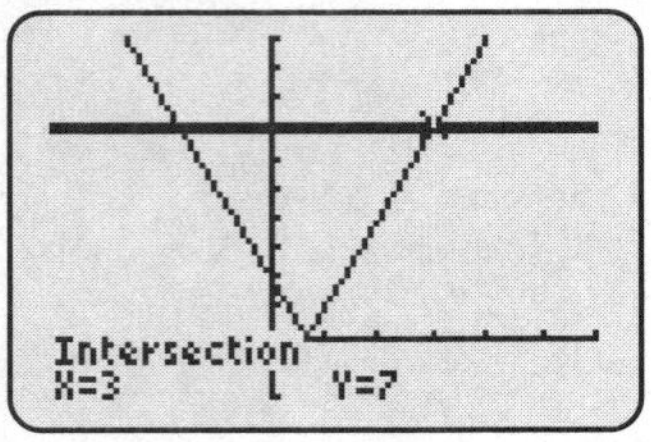

EXAMPLE 7 Solving an Inequality Involving an Absolute Value

Solve the inequality $|3x - 2| \leq 7$.

Solution:

We apply property (2) and squeeze the absolute value expression between -7 and 7. $\quad -7 \leq 3x - 2 \leq 7$

Add 2 to all three parts. $\quad -5 \leq 3x \leq 9$

Divide all three parts by 3. $\quad -\frac{5}{3} \leq x \leq 3$

The solution in interval notation is $\boxed{\left[-\frac{5}{3}, 3\right]}$.

Graph:

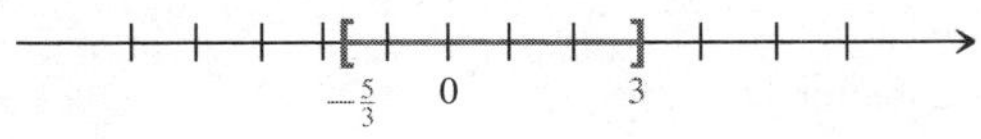

■ **YOUR TURN** Solve the inequality $|2x + 1| < 11$.

■ **Answer:** Inequality notation: $-6 < x < 5$. Interval notation: $(-6, 5)$.

It is often helpful to note that for absolute value inequalities,

- *less than* inequalities can be written as a single statement (see Example 7).
- *greater than* inequalities must be written as two statements (see Example 8).

Study Tip

Less than inequalities can be written as a single statement.

Greater than inequalities must be written as two statements.

EXAMPLE 8 Solving an Inequality Involving an Absolute Value

Solve the inequality $|1 - 2x| > 5$.

Solution:

Apply property (3). $\quad 1 - 2x < -5 \text{ or } 1 - 2x > 5$

Subtract 1 from all expressions. $\quad -2x < -6 \qquad -2x > 4$

Divide by -2 and reverse the inequality sign. $\quad x > 3 \qquad x < -2$

Express the solution in interval notation. $\quad \boxed{(-\infty, -2) \cup (3, \infty)}$

Graph:

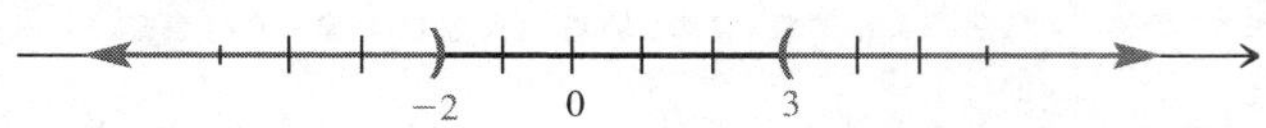

■ **YOUR TURN** Solve the inequality $|5 - 2x| \geq 1$.

■ **Answer:** Inequality notation: $x \leq 2$ or $x \geq 3$. Interval notation: $(-\infty, 2] \cup [3, \infty)$.

Notice that if we change the problem in Example 8 to $|1 - 2x| > -5$, the answer is all real numbers because the absolute value of any expression is greater than or equal to zero. Similarly, $|1 - 2x| < -5$ would have no solution because the absolute value of an expression can never be negative.

EXAMPLE 9 Solving an Inequality Involving an Absolute Value

Solve the inequality $2 - |3x| < 1$.

Solution:

Subtract 2 from both sides. $-|3x| < -1$

Multiply by (-1) and reverse the inequality sign. $|3x| > 1$

Apply property (3). $3x < -1$ or $3x > 1$

Divide both inequalities by 3. $x < -\frac{1}{3}$ $x > \frac{1}{3}$

Express in interval notation. $\boxed{\left(-\infty, -\frac{1}{3}\right) \cup \left(\frac{1}{3}, \infty\right)}$

Graph.

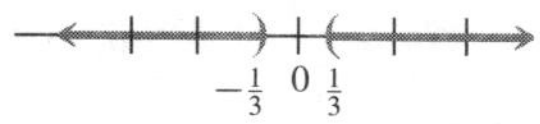

SECTION 1.7 SUMMARY

Absolute value equations and absolute value inequalities are solved by writing the equations or inequalities in terms of two equations or inequalities.

Equations

$|x| = A$ is equivalent to $x = -A$ or $x = A$

Inequalities

$|x| < A$ is equivalent to $-A < x < A$

$|x| > A$ is equivalent to $x < -A$ or $x > A$

SECTION 1.7 EXERCISES

SKILLS

In Exercises 1–34, solve the equation.

1. $|x| = 3$
2. $|x| = 2$
3. $|x| = -4$
4. $|x| = -2$
5. $|t + 3| = 2$
6. $|t - 3| = 2$
7. $|p - 7| = 3$
8. $|p + 7| = 3$
9. $|4 - y| = 1$
10. $|2 - y| = 11$
11. $|3x| = 9$
12. $|5x| = 50$
13. $|2x + 7| = 9$
14. $|2x - 5| = 7$
15. $|3t - 9| = 3$
16. $|4t + 2| = 2$
17. $|7 - 2x| = 9$
18. $|6 - 3y| = 12$
19. $|1 - 3y| = 1$
20. $|5 - x| = 2$
21. $|4.7 - 2.1x| = 3.3$
22. $|5.2x + 3.7| = 2.4$
23. $\left|\frac{2}{3}x - \frac{4}{7}\right| = \frac{5}{3}$
24. $\left|\frac{1}{2}x + \frac{3}{4}\right| = \frac{1}{16}$
25. $|x - 5| + 4 = 12$
26. $|x + 3| - 9 = 2$
27. $2\,|p + 3| - 15 = 5$
28. $8 - 3\,|p - 4| = 2$
29. $5|y - 2| - 10 = 4|y - 2| - 3$
30. $3 - |y + 9| = 11 - 3\,|y + 9|$
31. $|4 - x^2| = 1$
32. $|7 - x^2| = 3$
33. $|x^2 + 1| = 5$
34. $|x^2 - 1| = 5$

In Exercises 35–62, solve the inequality and express the solution in interval notation.

35. $|x| < 7$
36. $|y| < 9$
37. $|y| \geq 5$
38. $|x| \geq 2$
39. $|x + 3| < 7$
40. $|x + 2| \leq 4$
41. $|x - 4| > 2$
42. $|x - 1| < 3$
43. $|4 - x| \leq 1$
44. $|1 - y| < 3$
45. $|2x| > -3$
46. $|2x| < -3$
47. $|2t + 3| < 5$
48. $|3t - 5| > 1$
49. $|7 - 2y| \geq 3$
50. $|6 - 5y| \leq 1$
51. $|4 - 3x| \geq 0$
52. $|4 - 3x| \geq 1$
53. $2|4x| - 9 \geq 3$
54. $5|x - 1| + 2 \leq 7$
55. $9 - |2x| < 3$
56. $4 - |x + 1| > 1$
57. $|1 - 2x| < \frac{1}{2}$
58. $\left|\frac{2 - 3x}{5}\right| \geq \frac{2}{5}$
59. $|2.6x + 5.4| < 1.8$
60. $|3.7 - 5.5x| > 4.3$
61. $|x^2 - 1| \leq 8$
62. $|x^2 + 4| \geq 29$

In Exercises 63–68, write an inequality that fits the description.

63. Any real numbers less than 7 units from 2.
64. Any real numbers more than 3 units from -2.
65. Any real numbers at least $\frac{1}{2}$ unit from $\frac{3}{2}$.
66. Any real number no more than $\frac{5}{3}$ units from $\frac{11}{3}$.
67. Any real numbers no more than 2 units from a.
68. Any real number at least a units from -3.

APPLICATIONS

69. **Temperature.** If the average temperature in Hawaii is 83 degrees Fahrenheit plus or minus 15 degrees, write an absolute value inequality representing the temperature in Hawaii.

70. **Temperature.** If the average temperature of a human is 97.8 (±1.2) degrees Fahrenheit, write an absolute value inequality describing normal human body temperature.

71. **Sports.** Two women tee off the green of a par-3 hole on a golf course. They are playing "closest to the pin." If the first woman tees off and lands exactly 4 feet from the hole, write an inequality that describes where the second woman must land in order to win the hole. What equation would suggest a tie? Let d = the distance from where the second woman lands to the tee.

72. **Electronics.** A band-pass filter in electronics allows certain frequencies within a range (or band) to pass through to the receiver and eliminates all other frequencies. Write an absolute value inequality that allows any frequency f within 15 Hz of the carrier frequency f_c to pass.

CATCH THE MISTAKE

In Exercises 73–76, explain the mistake that is made.

73. Solve the absolute value equation $|x - 3| = 7$.

Solution:

Eliminate the absolute value symbols.	$x - 3 = 7$
Add 3 to both sides.	$x = 10$
Check.	$\lvert 10 - 3\rvert = 7$

This is incorrect. What mistake was made?

74. Solve the inequality $|x - 3| < 7$.

Solution:

Eliminate the absolute value symbols.	$x - 3 < -7$ or	$x - 3 > 7$
Add 3 to both sides.	$x < -4$	$x > 10$

The solution is $(-\infty, -4) \cup (10, \infty)$.

This is incorrect. What mistake was made?

75. Solve the inequality $|5 - 2x| \leq 1$.

Solution:

Eliminate the absolute value symbols.	$-1 \leq 5 - 2x \leq 1$
Subtract 5.	$-6 \leq -2x \leq -4$
Divide by -2.	$3 \leq x \leq 2$
Write the solution in interval notation.	$(-\infty, 2] \cup [3, \infty)$

This is incorrect. What mistake was made?

76. Solve the equation $|5 - 2x| = -1$.

Solution: $5 - 2x = -1$ or $5 - 2x = 1$

$-2x = -6 \qquad -2x = -4$

$x = 3 \qquad x = 2$

The solution is $\{2, 3\}$.

This is incorrect. What mistake was made?

CONCEPTUAL

In Exercises 77–80, determine whether each statement is true or false.

77. $-|m| \le m \le |m|$

78. $|n^2| = n^2$

79. $|m + n| = |m| + |n|$ is true only when m and n are both nonnegative.

80. For what values of x does the absolute value equation $|x - 7| = x - 7$ hold?

In Exercises 81–86, assuming a and b are real positive numbers, solve the equation or inequality and express the solution in interval notation.

81. $|x - a| < b$

82. $|a - x| > b$

83. $|x| \ge -a$

84. $|x| \le -b$

85. $|x - a| = b$, where a and b are both positive, real numbers.

86. $|x - a| = -b$, where a and b are both positive, real numbers.

CHALLENGE

87. For what values of x does the absolute value equation $|x + 1| = 4 + |x - 2|$ hold?

88. Solve the inequality $|3x^2 - 7x + 2| > 8$.

TECHNOLOGY

89. Graph $y_1 = |x - 7|$ and $y_2 = x - 7$ in the same screen. Do the x-values where these two graphs coincide agree with your result in Exercise 80?

90. Graph $y_1 = |x + 1|$ and $y_2 = |x - 2| + 4$ in the same screen. Do the x-values where these two graphs coincide agree with your result in Exercise 87?

91. Graph $y_1 = |3x^2 - 7x + 2|$ and $y_2 = 8$ in the same screen. Do the x-values where y_1 lies above y_2 agree with your result in Exercise 88?

92. Solve the inequality $|2.7x^2 - 7.9x + 5| \le |5.3x^2 - 9.2|$ by graphing both sides of the inequality and identify which x-values make this statement true.

93. Solve the inequality $\left|\dfrac{x}{x+1}\right| < 1$ by graphing both sides of the inequality, and identify which x-values make this statement true.

94. Solve the inequality $\left|\dfrac{x}{x+1}\right| < 2$ by graphing both sides of the inequality, and identify which x-values make this statement true.

MODELING YOUR WORLD

Used in fields of study that range from engineering to economics to sociology, a mathematical model is a tool that uses mathematical language to describe a system. There are many types of models, which help us not only better understand the world as it is, but by projecting different scenarios based on available data, allow us glimpses into possible futures.

Current changes in the environment, which will affect all of our futures, have brought about a fierce debate. Some scientists believe that human activities have played a large part in bringing about global warming, which impacts not only day-to-day temperatures, but species extinction, loss of glacial ice, and the quality of the air we breathe. Others feel that current changes in the climate are simply part of a natural cycle and are not a cause for concern. The "Modeling Your World" feature at the end of every chapter allows you to use modeling to explore the topic of global climate change and become an informed participant in this debate.

The Earth's near surface global air temperature increased approximately 0.74 ± 0.18 °C over 1905–2005.*

**Climate Change 2007: The Physical Science Basis. Contribution of Working Group 1 to the Fourth Assessment Report of the Intergovernmental Panel on Climate Change.* Intergovernmental Panel on Climate Change (2007-02-05).

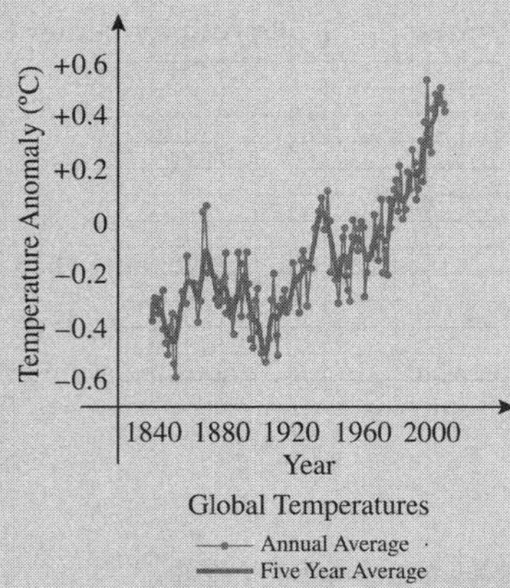

1. Write an absolute value equation that models the increase in the Earth's near surface air temperature from 1905 to 2005 in degrees Celsius °C. Let t represent the increase in temperature.
2. Use the temperature scale conversion $F = \frac{9}{5}C + 32$ equation to write an absolute value equation that models the increase in the Earth's near surface air temperature from 1905 to 2005 in degrees Fahrenheit °F. Let t represent the increase in temperature. The following chart illustrates different global warming projections for the next 100 years.

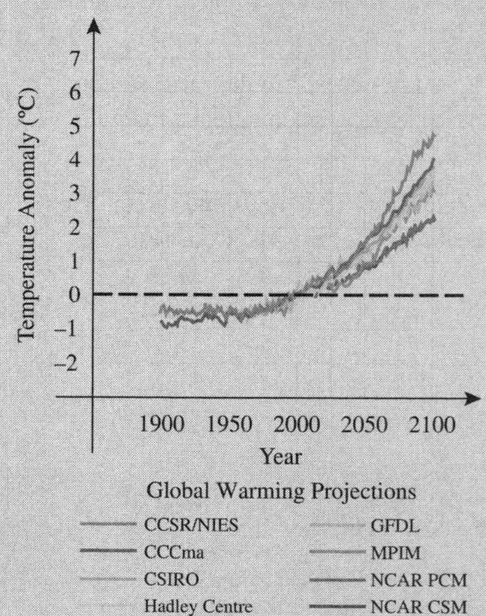

3. Let t represent the increase in temperature in °C (let $t = 0$ correspond to the year 2000) and write an approximate absolute value inequality such that the NCAR PCM projection is the lowest possible temperature anomaly and the CCSR/NIES projection is the highest possible temperature anomaly.
4. Repeat problem 3 for temperature in °F.

CHAPTER 1 REVIEW

SECTION	CONCEPT	PAGES	REVIEW EXERCISES	KEY IDEAS/POINTS
1.1	**Linear equations**	90–96	1–22	$ax + b = 0$
	Solving linear equations in one variable	91–93	1–12	Isolate variable on one side and constants on the other side.
	Solving rational equations that are reducible to linear equations	93–96	13–22	Any values that make the denominator equal to 0 must be eliminated as possible solutions.
1.2	**Applications involving linear equations**	100–109	23–34	
	Solving application problems using mathematical models	100–102		Five-step procedure: Step 1: Identify the question. Step 2: Make notes. Step 3: Set up an equation. Step 4: Solve the equation. Step 5: Check the solution.
	Geometry problems	102–103	27–28	Formulas for rectangles, triangles, and circles
	Interest problems	104–106	29–30	Simple interest: $I = Prt$
	Mixture problems	106–107	31–32	Whenever two *distinct* quantities are mixed, the result is a mixture.
	Distance–rate–time problems	107–109		$d = r \cdot t$
1.3	**Quadratic equations**	113–123	35–62	$ax^2 + bx + c = 0$
	Factoring	114–116	35–38	If $(x - h)(x - k) = 0$, then $x = h$ or $x = k$.
	Square root method	116–117	39–42	If $x^2 = P$, then $x = \pm\sqrt{P}$.
	Completing the square	117–119	43–46	Find half of b; square that quantity; add the result to both sides.
	Quadratic formula	119–123	47–50	$x = \dfrac{-b \pm \sqrt{b^2 - 4ac}}{2a}$
1.4	**Other types of equations**	127–132	63–94	
	Radical equations	127–130	63–74	Check solutions to avoid extraneous solutions.
	Equations quadratic in form: u-substitution	130–131	75–86	Use a u-substitution to write the equation in quadratic form.
	Factorable equations	132	87–94	Extract common factor or factor by grouping.
1.5	**Linear inequalities**	136–141	95–118	Solutions are a range of real numbers.
	Graphing inequalities and interval notation	136–138	95–108	■ $a < x < b$ is equivalent to (a, b). ■ $x \leq a$ is equivalent to $(-\infty, a]$. ■ $x > a$ is equivalent to (a, ∞).
	Solving linear inequalities	138–141	109–118	If an inequality is multiplied or divided by a *negative* number, the inequality sign must be reversed.
1.6	**Polynomial and rational inequalities**	146–153	119–134	
	Polynomial inequalities	146–150	119–126	Zeros are values that make the polynomial equal to 0.
	Rational inequalities	151–153	127–134	The number line is divided into intervals. The end points of these intervals are values that make either the numerator or denominator equal to 0. Always exclude values that make the denominator = 0.

Section	Concept	Pages	Review Exercises	Key Ideas/Points
1.7	**Absolute value equations and inequalities**	156–161	135–148	$\lvert b - a\rvert$ is the distance between the points a and b on the number line.
	Equations involving absolute value	157–159	135–138	If $\lvert x\rvert = a$, then $x = -a$ or $x = a$.
	Inequalities involving absolute value	159–161	139–148	■ $\lvert x\rvert \le a$ is equivalent to $-a \le x \le a$. ■ $\lvert x\rvert > a$ is equivalent to $x < -a$ or $x > a$.

CHAPTER 1 REVIEW EXERCISES

1.1 Linear Equations

Solve for the variable.

1. $7x - 4 = 12$
2. $13d + 12 = 7d + 6$
3. $20p + 14 = 6 - 5p$
4. $4(x - 7) - 4 = 4$
5. $3(x + 7) - 2 = 4(x - 2)$
6. $7c + 3(c - 5) = 2(c + 3) - 14$
7. $14 - [-3(y - 4) + 9] = [4(2y + 3) - 6] + 4$
8. $[6 - 4x + 2(x - 7)] - 52 = 3(2x - 4) + 6[3(2x - 3) + 6]$
9. $\dfrac{12}{b} - 3 = \dfrac{6}{b} + 4$
10. $\dfrac{g}{3} + g = \dfrac{7}{9}$
11. $\dfrac{13x}{7} - x = \dfrac{x}{4} - \dfrac{3}{14}$
12. $5b + \dfrac{b}{6} = \dfrac{b}{3} - \dfrac{29}{6}$

Specify any values that must be excluded from the solution set and then solve.

13. $\dfrac{1}{x} - 4 = 3(x - 7) + 5$
14. $\dfrac{4}{x + 1} - \dfrac{8}{x - 1} = 3$
15. $\dfrac{2}{t + 4} - \dfrac{7}{t} = \dfrac{6}{t(t + 4)}$
16. $\dfrac{3}{2x - 7} = \dfrac{-2}{3x + 1}$
17. $\dfrac{3}{2x} - \dfrac{6}{x} = 9$
18. $\dfrac{3 - 5/m}{2 + 5/m} = 1$
19. $7x - (2 - 4x) = 3[-6 + (4 - 2x + 7)] + 12$
20. $\dfrac{x}{5} - \dfrac{x - 3}{15} = -6$

Solve for the specified variable.

21. Solve for x in terms of y:
$3x - 2[(y + 4)3 - 7] = y - 2x + 6(x - 3)$.
22. If $y = \dfrac{x + 3}{1 + 2x}$, find $\dfrac{y + 2}{1 - 2y}$ in terms of x.

1.2 Applications Involving Linear Equations

23. **Transportation.** Maria is on her way from her home near Orlando to the Sundome in Tampa for a rock concert. She drives 16 miles to the Orlando partk-n-ride, takes a bus $\frac{3}{4}$ of the way to a bus station in Tampa, and then take a cab $\frac{1}{12}$ of the way to the Sundome. How far does Maria live from the Sundome?
24. **Diet.** A particular 2000 calorie per day diet suggests eating breakfast, lunch, dinner, and four snacks. Each snack is $\frac{1}{4}$ the calories of lunch. Lunch has 100 calories less than dinner. Dinner has 1.5 times as many calories as breakfast. How many calories are in each meal and snack?
25. **Numbers.** Find a number such that 12 more than $\frac{1}{4}$ the number is $\frac{1}{3}$ the number.
26. **Numbers.** Find four consecutive odd integers such that the sum of the four is equal to 3 more than 3 times the fourth integer.
27. **Geometry.** The length of a rectangle is 1 more than 2 times the width, and the perimeter is 20 inches. What are the dimensions of the rectangle?
28. **Geometry.** Find the perimeter of a triangle if one side is 10 inches, another side is $\frac{1}{3}$ of the perimeter, and the third side is $\frac{1}{6}$ of the perimeter.
29. **Investments.** You win \$25,000 and you decide to invest the money in two different investments: one paying 20% and the other paying 8%. A year later you have \$27,600 total. How much did you originally invest in each account?
30. **Investments.** A college student on summer vacation was able to make \$5,000 by working a full-time job every summer. He invested half the money in a mutual fund and half the money in a stock that yielded four times as much interest as the mutual fund. After a year he earned \$250 in interest. What were the interest rates of the mutual fund and the stock?

31. **Chemistry.** For an experiment, a student requires 150 ml of a solution that is 8% NaCl (sodium chloride). The storeroom has only solutions that are 10% NaCl and 5% NaCl. How many milliliters of each available solution should be mixed to get 150 ml of 8% NaCl?

32. **Chemistry.** A mixture containing 8% salt is to be mixed with 4 ounces of a mixture that is 20% salt, in order to obtain a solution that is 12% salt. How much of the first solution must be used?

33. **Grades.** Going into the College Algebra final exam, which will count as two tests, Danny has test scores of 95, 82, 90, and 77. If his final exam is higher than his lowest test score, then it will count for the final exam and replace the lowest test score. What score does Danny need on the final in order to have an average score of at least 90?

34. **Car Value.** A car salesperson reduced the price of a model car by 20%. If the new price is \$25,000, what was its original price? How much can be saved by purchasing the model?

1.3 Quadratic Equations

Solve by factoring.

35. $b^2 = 4b + 21$
36. $x(x - 3) = 54$
37. $x^2 = 8x$
38. $6y^2 - 7y - 5 = 0$

Solve by the square root method.

39. $q^2 - 169 = 0$
40. $c^2 + 36 = 0$
41. $(2x - 4)^2 = -64$
42. $(d + 7)^2 - 4 = 0$

Solve by completing the square.

43. $x^2 - 4x - 12 = 0$
44. $2x^2 - 5x - 7 = 0$
45. $\frac{x^2}{2} = 4 + \frac{x}{2}$
46. $8m = m^2 + 15$

Solve by the quadratic formula.

47. $3t^2 - 4t = 7$
48. $4x^2 + 5x + 7 = 0$
49. $8f^2 - \frac{1}{3}f = \frac{7}{6}$
50. $x^2 = -6x + 6$

Solve by any method.

51. $5q^2 - 3q - 3 = 0$
52. $(x - 7)^2 = -12$
53. $2x^2 - 3x - 5 = 0$
54. $(g - 2)(g + 5) = -7$
55. $7x^2 = -19x + 6$
56. $7 = (2b^2 + 1)$

Solve for the indicated variable.

57. $S = \pi r^2 h$ for r
58. $V = \frac{\pi r^3 h}{3}$ for r
59. $h = vt - 16t^2$ for v
60. $A = 2\pi r^2 + 2\pi rh$ for h

61. **Geometry.** Find the base and height of a triangle with an area of 2 square feet if its base is 3 feet longer than its height.

62. **Falling Objects.** A man is standing on top of a building 500 feet tall. If he drops a penny off the roof, the height of the penny is given by $h = -16t^2 + 500$, where t is in seconds. Determine how many seconds it takes until the penny hits the ground.

1.4 Other Types of Equations

Solve the radical equation for the given variable.

63. $\sqrt[3]{2x - 4} = 2$
64. $\sqrt{x - 2} = -4$
65. $(2x - 7)^{1/5} = 3$
66. $x = \sqrt{7x - 10}$
67. $x - 4 = \sqrt{x^2 + 5x + 6}$
68. $\sqrt{2x - 7} = \sqrt{x + 3}$
69. $\sqrt{x + 3} = 2 - \sqrt{3x + 2}$
70. $4 + \sqrt{x - 3} = \sqrt{x - 5}$
71. $x - 2 = \sqrt{49 - x^2}$
72. $\sqrt{2x - 5} - \sqrt{x + 2} = 3$
73. $-x = \sqrt{3 - x}$
74. $\sqrt{15 + 2\sqrt{x - 4} + \sqrt{x}} = 5$

Solve the equation by introducing a substitution that transforms the equation to quadratic form.

75. $-28 = (3x - 2)^2 - 11(3x - 2)$
76. $x^4 - 6x^2 + 9 = 0$
77. $\left(\frac{x}{1 - x}\right)^2 = 15 - 2\left(\frac{x}{1 - x}\right)$
78. $3(x - 4)^4 - 11(x - 4)^2 - 20 = 0$
79. $y^{-2} - 5y^{-1} + 4 = 0$
80. $p^{-2} + 4p^{-1} = 12$
81. $3x^{1/3} + 2x^{2/3} = 5$
82. $2x^{2/3} - 3x^{1/3} - 5 = 0$
83. $x^{-2/3} + 3x^{-1/3} + 2 = 0$
84. $y^{-1/2} - 2y^{-1/4} + 1 = 0$
85. $x^4 + 5x^2 = 36$
86. $3 - 4x^{-1/2} + x^{-1} = 0$

Solve the equation by factoring.

87. $x^3 + 4x^2 - 32x = 0$
88. $9t^3 - 25t = 0$
89. $p^3 - 3p^2 - 4p + 12 = 0$
90. $4x^3 - 9x^2 + 4x - 9 = 0$
91. $p(2p - 5)^2 - 3(2p - 5) = 0$
92. $2(t^2 - 9)^3 - 20(t^2 - 9)^2 = 0$
93. $y - 81y^{-1} = 0$
94. $9x^{3/2} - 37x^{1/2} + 4x^{-1/2} = 0$

1.5 Linear Inequalities

Rewrite using interval notation.

95. $x \leq -4$ **96.** $-1 < x \leq 7$

97. $2 \leq x \leq 6$ **98.** $x > -1$

Rewrite using inequality notation.

99. $(-6, \infty)$ **100.** $(-\infty, 0]$

101. $[-3, 7]$ **102.** $(-5, 2]$

Express each interval using inequality and interval notation.

103. 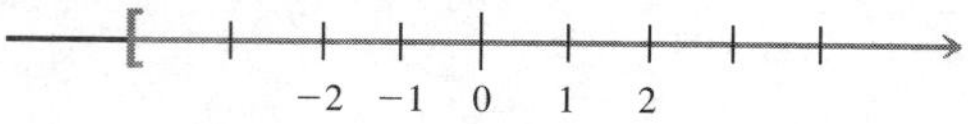

104. 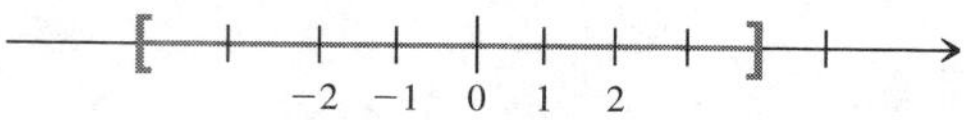

Graph the indicated set and write as a single interval, if possible.

105. $(4, 6] \cup [5, \infty)$ **106.** $(-\infty, -3) \cup [-7, 2]$

107. $(3, 12] \cap [8, \infty)$ **108.** $(-\infty, -2) \cap [-2, 9)$

Solve and graph.

109. $2x < 5 - x$ **110.** $6x + 4 \leq 2$

111. $4(x - 1) > 2x - 7$ **112.** $\frac{x + 3}{3} \geq 6$

113. $6 < 2 + x \leq 11$ **114.** $-6 \leq 1 - 4(x + 2) \leq 16$

115. $\frac{2}{3} \leq \frac{1 + x}{6} \leq \frac{3}{4}$ **116.** $\frac{x}{3} + \frac{x + 4}{9} > \frac{x}{6} - \frac{1}{3}$

Applications

117. Grades. In your algebra class your first four exam grades are 72, 65, 69, and 70. What is the lowest score you can get on the fifth exam to earn a C for the course? Assume that each exam is equal in weight and a C is any score greater than or equal to 70.

118. Profit. A tailor decided to open a men's custom suit business. His fixed costs are $8,500 per month and it costs him $50 for the materials to make each suit. If the price he charges per suit is $300, how many suits does he have to tailor per month to make a profit?

1.6 Polynomial and Rational Inequalities

Solve the polynomial inequality and express the solution set using interval notation.

119. $x^2 \leq 36$ **120.** $6x^2 - 7x < 20$

121. $4x \leq x^2$ **122.** $-x^2 \geq 9x + 14$

123. $-x^2 < -7x$ **124.** $x^2 < -4$

125. $4x^2 - 12 > 13x$ **126.** $3x \leq x^2 + 2$

Solve the rational inequality and express the solution set using interval notation.

127. $\frac{x}{x - 3} < 0$ **128.** $\frac{x - 1}{x - 4} > 0$

129. $\frac{x^2 - 3x}{3} \geq 18$ **130.** $\frac{x^2 - 49}{x - 7} \geq 0$

131. $\frac{3}{x - 2} - \frac{1}{x - 4} \leq 0$ **132.** $\frac{4}{x - 1} \leq \frac{2}{x + 3}$

133. $\frac{x^2 + 9}{x - 3} \geq 0$ **134.** $x < \frac{5x + 6}{x}$

1.7 Absolute Value Equations and Inequalities

Solve the equation.

135. $|x - 3| = -4$ **136.** $|2 + x| = 5$

137. $|3x - 4| = 1.1$ **138.** $|x^2 - 6| = 3$

Solve the inequality and express the solution using interval notation.

139. $|x| < 4$ **140.** $|x - 3| < 6$

141. $|x + 4| > 7$ **142.** $|-7 + y| \leq 4$

143. $|2x| > 6$ **144.** $\left|\frac{4 + 2x}{3}\right| \geq \frac{1}{7}$

145. $|2 + 5x| \geq 0$ **146.** $|1 - 2x| \leq 4$

Applications

147. Temperature. If the average temperature in Phoenix is 85 degrees Fahrenheit plus or minus 10 degrees, write an inequality representing the average temperature T in Phoenix.

148. Blood Alcohol Level. If a person registers a 0.08 blood alcohol level, he will be issued a DUI ticket in the state of Florida. If the test is accurate within 0.007,write a linear inequality representing an actual blood alcohol level that will not be issued a ticket.

Technology Exercises

Section 1.1

Graph the function represented by each side of the question in the same viewing rectangle, and solve for x.

149. $0.031x + 0.017(4000 - x) = 103.14$

150. $\frac{1}{0.16x} - \frac{0.2}{x} = \frac{1}{4}$

Section 1.3

151. a. Solve the equation $x^2 + 4x = b$, $b = 5$ by first writing in standard form and then factoring. Now plot both sides of the equation in the same viewing screen ($y_1 = x^2 + 4x$ and $y_2 = b$). At what x-values do these two graphs intersect? Do those points agree with the solution set you found?

b. Repeat (a) for $b = -5, 0, 7$, and 12.

152. a. Solve the equation $x^2 - 4x = b$, $b = 5$ by first writing in standard form and then factoring. Now plot both sides of the equation in the same viewing screen ($y_1 = x^2 - 4x$ and $y_2 = b$). At what x-values do these two graphs intersect? Do those points agree with the solution set you found?

b. Repeat (a) for $b = -5, 0, 7$, and 12.

Section 1.4

153. Solve the equation $2x^{1/4} = -x^{1/2} + 6$. Round your answer to two decimal places. Plot both sides of the equation in the same viewing screen, $y_1 = 2x^{1/4}$ and $y_2 = -x^{1/2} + 6$. Does the point(s) of intersection agree with your solution?

154. Solve the equation $2x^{-1/2} = x^{-1/4} + 6$. Plot both sides of the equation in the same viewing screen, $y_1 = 2x^{-1/2}$ and $y_2 = x^{-1/4} + 6$. Does the point(s) of intersection agree with your solution?

Section 1.5

155. a. Solve the inequality $-0.61x + 7.62 > 0.24x - 5.47$. Express the solution set using interval notation.

b. Graph each side of the inequality in the same viewing screen. Find the range of x-values when the graph of the left side lies above the graph of the right side.

c. Do (a) and (b) agree?

156. a. Solve the inequality $-\frac{1}{2}x + 7 < \frac{3}{4}x - 5$. Express the solution set using interval notation.

b. Graph each side of the inequality in the same viewing screen. Find the range of x-values when the graph of the left side lies below the graph of the right side.

c. Do (a) and (b) agree?

Section 1.6

Plot the left side and the right side of each inequality in the same screen, and use the zoom feature to determine the range of values for which the inequality is true.

157. $0.2x^2 - 2 > 0.05x + 3.25$

158. $12x^2 - 7x - 10 < 2x^2 + 2x - 1$

159. $\dfrac{3p}{7 - 2p} > 1$

160. $\dfrac{7p}{15 - 2p} < 1$

Section 1.7

161. Solve the inequality $|1.6x^2 - 4.5| < 3.2$ by graphing both sides of the inequality, and identify which x-values make this statement true. Express the solution using interval notation and round to 2 decimal places.

162. Solve the inequality $|0.8x^2 - 5.4x| > 4.5$ by graphing both sides of the inequality, and identify which x-values make this statement true. Express the solution using interval notation and round to 2 decimal places.

Solve the equation.

1. $4p - 7 = 6p - 1$
2. $-2(z - 1) + 3 = -3z + 3(z - 1)$
3. $3t = t^2 - 28$
4. $8x^2 - 13x = 6$
5. $6x^2 - 13x = 8$
6. $\dfrac{3}{x - 1} = \dfrac{5}{x + 2}$
7. $\dfrac{5}{y - 3} + 1 = \dfrac{30}{y^2 - 9}$
8. $x^4 - 5x^2 - 36 = 0$
9. $\sqrt{2x + 1} + x = 7$
10. $2x^{2/3} + 3x^{1/3} - 2 = 0$
11. $\sqrt{3y - 2} = 3 - \sqrt{3y + 1}$
12. $x(3x - 5)^3 - 2(3x - 5)^2 = 0$
13. $x^{7/3} - 8x^{4/3} + 12x^{1/3} = 0$

Solve for the specified variable.

14. $F = \frac{9}{5}C + 32$ for C
15. $P = 2L + 2W$ for L

Solve the inequality and express the solution in interval notation.

16. $7 - 5x > -18$
17. $3x + 19 \geq 5(x - 3)$
18. $-1 \leq 3x + 5 < 26$
19. $\dfrac{2}{5} < \dfrac{x + 8}{4} \leq \dfrac{1}{2}$
20. $3x \geq 2x^2$
21. $3p^2 \geq p + 4$
22. $|5 - 2x| > 1$
23. $\dfrac{x - 3}{2x + 1} \leq 0$
24. $\dfrac{x + 4}{x^2 - 9} \geq 0$

25. **Puzzle.** A piling supporting a bridge sits so that $\frac{1}{4}$ of the piling is in the sand, 150 feet is in the water, and $\frac{3}{5}$ of the piling is in the air. What is the total height of the piling?

26. **Real Estate.** As a realtor you earn 7% of the sale price. The owners of a house you have listed at \$150,000 will entertain offers within 10% of the list price. Write an inequality that models the commission you could make on this sale.

27. **Costs: Cell Phones.** A cell phone company charges \$49 for a 600-minute monthly plan, plus an additional \$0.17/minute for every minute over 600. If a customer's bill ranged from a low of \$53.59 to a high of \$69.74 over a 6-month period, write an inequality expressing the number of monthly minutes used over the 6-month period.

28. **Television.** Television and film formats are classified as ratios of width to height. Traditional televisions have a 4:3 ratio (1.33:1), and movies are typically made in widescreen format with a 21:9 ratio (2.35:1). If you own a traditional 25 inch television (20 inch × 15 inch screen) and you play a widescreen DVD on it, there will be black bars above and below the image. What are the dimensions of the movie and of the black bars?

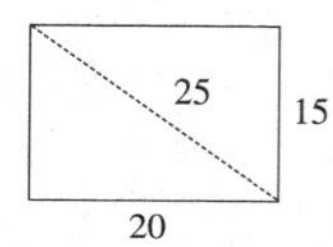

29. Solve the equation $\dfrac{1}{0.75x} - \dfrac{0.45}{x} = \dfrac{1}{9}$. Graph the function represented by each side in the same viewing rectangle and solve for x.

Andy Washnik

30. Solve the inequality $0.3 + |2.4x^2 - 1.5| \leq 6.3$ by graphing both sides of the inequality, and identify which x-values make this statement true. Express the solution using interval notation.

CHAPTER 1 CUMULATIVE TEST

Simplify.

1. $5 \cdot (7 - 3 \cdot 4 + 2)$

Simplify and express in terms of positive exponents.

2. $(4x^{-3}b^{4})^{-3}$

3. $\dfrac{(x^{2}y^{-2})^{3}}{(x^{2}y)^{-3}}$

Perform the operations and simplify.

4. $(-x^4 + 2x^3) + (x^3 - 5x - 6) - (5x^4 + 4x^3 - 6x + 8)$

5. $x^2(x + 5)(x - 3)$

Factor completely.

6. $3x^3 - 3x^2 - 60x$

7. $2a^3 + 2000$

Perform the operations and simplify.

8. $\dfrac{3 - x}{x^2 - 1} \div \dfrac{5x - 15}{x + 1}$

9. $\dfrac{6x}{x - 2} - \dfrac{5x}{x + 2}$

Solve for x.

10. $x^3 - x^2 - 30x = 0$

11. $\frac{2}{7}x = \frac{1}{8}x + 9$

12. Perform the operation and express in standard form: $\dfrac{45}{6 - 3i}$.

Solve for x.

13. $\dfrac{6x}{5} - \dfrac{8x}{3} = 4 - \dfrac{7x}{15}$

14. $\dfrac{x - 6}{6 - x} = \dfrac{3}{2}$

15. Tim can paint the interior of a condo in 9 hours. If Chelsea is hired to help him, they can do a similar condo in 5 hours. Working alone, how long will it take Chelsea to paint a similar condo?

16. Solve using the square root method: $y^2 + 36 = 0$.

17. Solve by completing the square: $x^2 + 12x + 40 = 0$.

18. Solve using the quadratic formula: $x^2 + x + 9 = 0$.

19. Solve and check: $\sqrt{4 - x} = x - 4$.

20. Solve using substitution: $3x^{-2} + 8x^{-1} + 4 = 0$.

Solve and express the solution in interval notation.

21. $0 < 4 - x \leq 7$

22. $4x^2 < 9x - 11$

23. $\dfrac{x + 2}{9 - x^2} \geq 0$

24. $\left|\dfrac{4 - 5x}{7}\right| \geq \dfrac{3}{14}$

25. Solve for x: $\left|\frac{1}{5}x + \frac{2}{3}\right| = \frac{7}{15}$.

26. Solve the equation $x^6 + \frac{37}{8}x^3 = 27$. Plot both sides of the equation in the same viewing screen, $y_1 = x^6 + \frac{37}{8}x^3$ and $y_2 = 27$. Does the point(s) of intersection agree with your solution?

27. Solve the inequality $\left|\dfrac{3x}{x - 2}\right| < 1$ by graphing both sides of the inequality, and identify which x-values make this statement true.

2

Graphs

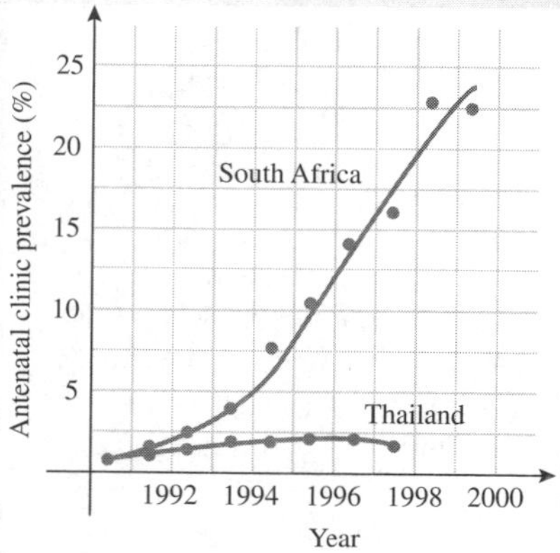

HIV infection rates

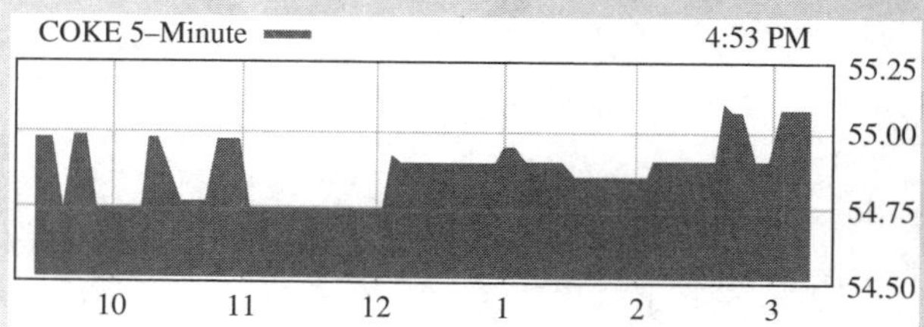

Stock prices fluctuating throughout the day

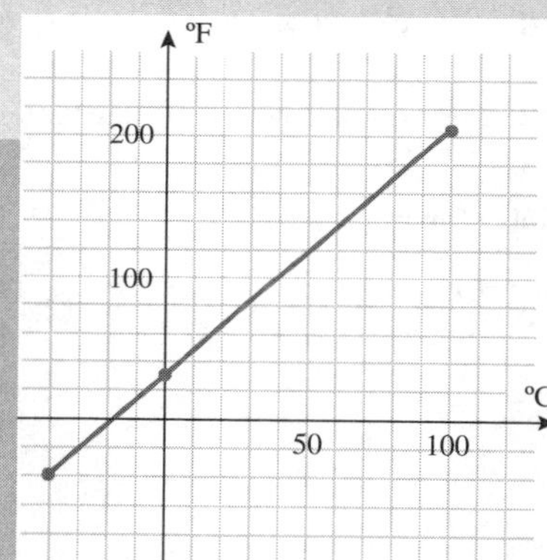

The conversion between degrees Fahrenheit and degrees Celsius is a linear relationship. Notice that 0 °C corresponds to 32 °F.

Johnny Johnson/Getty Images

Emperor penguins walking in a line, Weddell Sea, Antarctica

Graphs are used in many ways. There is only one temperature that yields the same number in degrees Celsius and degrees Fahrenheit. Do you know what it is?* The penguins are a clue.

*See Section 2.3, Exercise 101.

IN THIS CHAPTER you will review the Cartesian plane. You will calculate the distance between two points and find the midpoint of a line segment joining two points. You will then apply point-plotting techniques to sketch graphs of equations. Special attention is given to two types of equations: lines and circles.

GRAPHS

CARTESIAN PLANE

- Axes
- Coordinates
- Distance
- Midpoint

GRAPHING EQUATIONS

- Point-plotting
- Intercepts
  - *x*-intercepts
  - *y*-intercepts
- Symmetry
  - *x*-axis
  - *y*-axis
  - Origin

COMMON GRAPHS

- Lines
  - Slope
  - *y*-intercept
- Circles
  - Center
  - Radius

SECTION

CHAPTER OBJECTIVES

- Plot points in the Cartesian plane.
- Calculate the distance between two points.
- Find the midpoint of a line segment joining two points.
- Sketch the graph of an equation by point-plotting.
- Find intercepts of a graph.
- Determine whether a graph has symmetry.
- Use intercepts and symmetry as graphing aids.
- Find the slope of a line.
- Graph lines and circles.
- Determine equations of lines and circles.

SECTION 2.1 BASIC TOOLS: CARTESIAN PLANE, DISTANCE, AND MIDPOINT

SKILLS OBJECTIVES

- Plot points in the Cartesian plane.
- Calculate the distance between two points.
- Find the midpoint of a line segment joining two points.

CONCEPTUAL OBJECTIVE

- Expand the concept of a one-dimensional number line to a two-dimensional plane.

Cartesian Plane

HIV infection rates, stock prices, and temperature conversions are all examples of relationships between two quantities that can be expressed in a two-dimensional graph. Because it is two dimensional, such a graph lies in a **plane**.

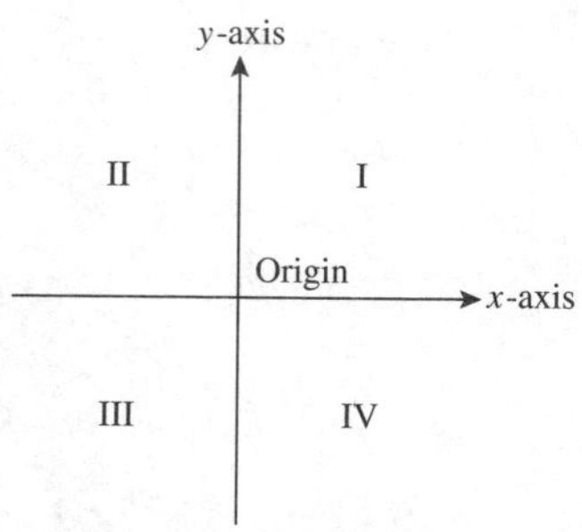

Two perpendicular real number lines, known as the **axes** in the plane, intersect at a point we call the **origin**. Typically, the horizontal axis is called the ***x*-axis** and the vertical axis is denoted as the ***y*-axis**. The axes divide the plane into four **quadrants**, numbered by Roman numerals and ordered counterclockwise.

Points in the plane are represented by **ordered pairs**, denoted (x, y). The first number of the ordered pair indicates the position in the horizontal direction and is often called the x-coordinate or **abscissa**. The second number indicates the position in the vertical direction and is often called the y-coordinate or **ordinate**. The origin is denoted $(0, 0)$.

Examples of other coordinates are given on the graph to the right.

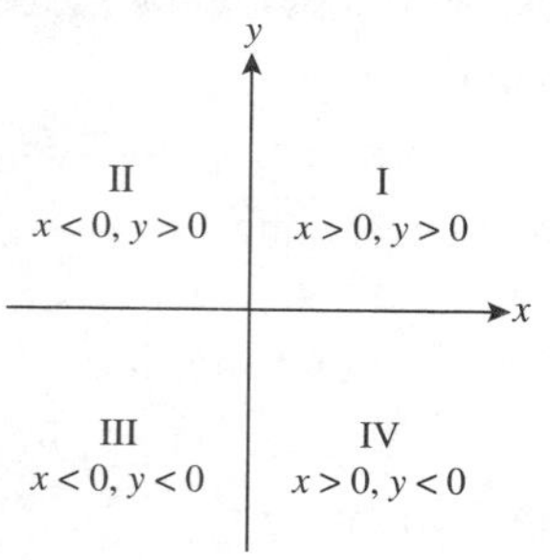

The point $(2, 4)$ lies in quadrant I. To **plot** this point, start at the origin $(0, 0)$ and move to the right two units and up four units.

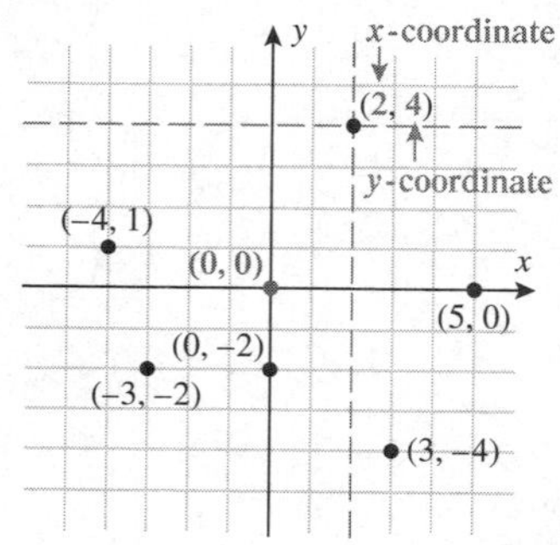

All points in quadrant I have positive coordinates, and all points in quadrant III have negative coordinates. Quadrant II has negative x-coordinates and positive y-coordinates; quadrant IV has positive x-coordinates and negative y-coordinates.

This representation is called the **rectangular coordinate system** or **Cartesian coordinate system**, named after the French mathematician René Descartes.

EXAMPLE 1 Plotting Points in a Cartesian Plane

a. Plot and label the points $(-1, -4)$, $(2, 2)$, $(-2, 3)$, $(2, -3)$, $(0, 5)$, and $(-3, 0)$ in the Cartesian plane.

b. List the points and corresponding quadrant or axis in a table.

Solution:

a.

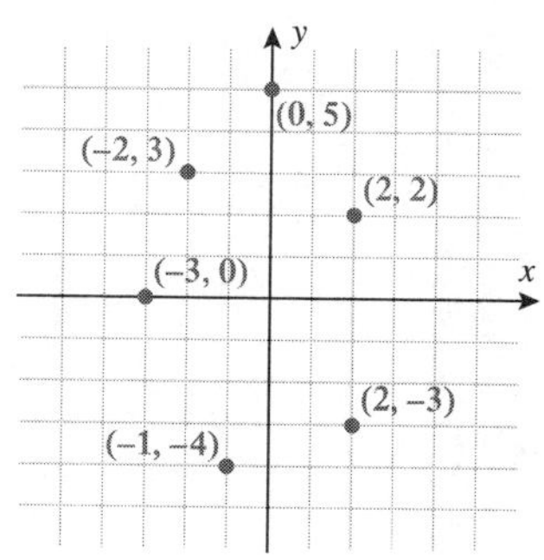

b.

POINT	QUADRANT
$(2, 2)$	I
$(-2, 3)$	II
$(-1, -4)$	III
$(2, -3)$	IV
$(0, 5)$	y-axis
$(-3, 0)$	x-axis

Distance Between Two Points

Suppose you want to find the distance between any two points in the plane. In the previous graph, to find the distance between the points (2, −3) and (2, 2), count the units between the two points. The distance is 5. What if the two points do not lie along a horizontal or vertical line? Example 2 uses the Pythagorean theorem to help find the distance between any two points.

EXAMPLE 2 Finding the Distance Between Two Points

Find the distance between the points (−2, −1) and (1, 3).

Solution:

STEP 1 Plot and label the two points in the Cartesian plane and draw a segment indicating the distance d between the two points.

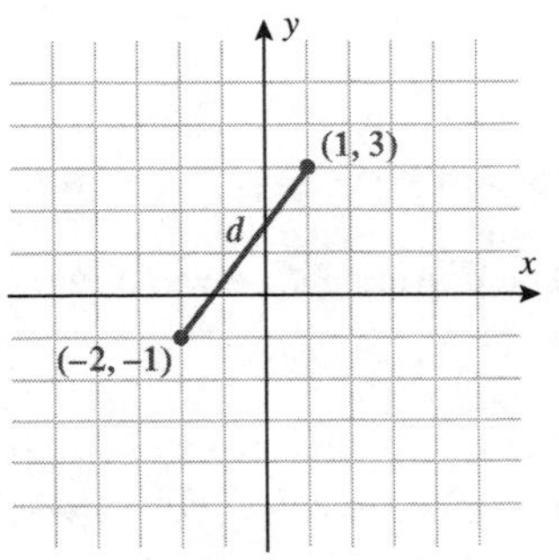

STEP 2 Form a right triangle by connecting the points to a third point, (1, −1).

STEP 3 Calculate the length of the horizontal segment. $3 = |1 - (-2)|$

Calculate the length of the vertical segment. $4 = |3 - (-1)|$

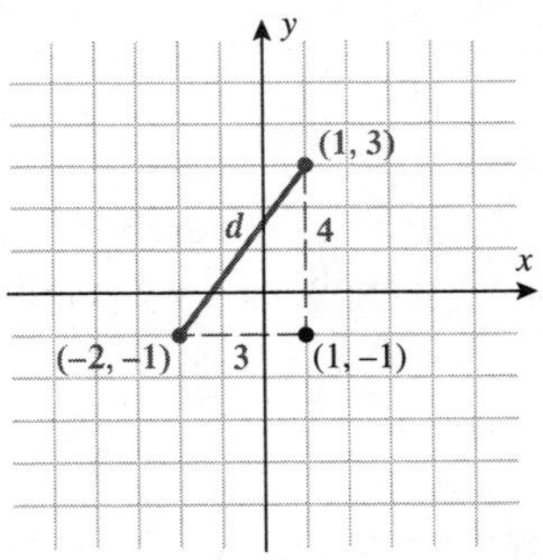

STEP 4 Use the Pythagorean theorem to calculate the distance d.

$$d^2 = 3^2 + 4^2$$
$$d^2 = 9 + 16 = 25$$
$$\boxed{d = 5}$$

WORDS

For any two points, (x_1, y_1) and (x_2, y_2):

MATH

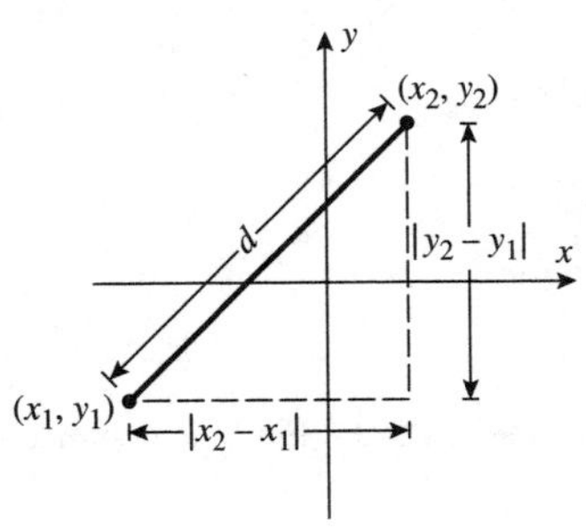

The distance along the horizontal segment is the absolute value of the difference between the x-values.	$\lvert x_2 - x_1 \rvert$
The distance along the vertical segment is the absolute value of the difference between the y-values.	$\lvert y_2 - y_1 \rvert$
Use the Pythagorean theorem to calculate the distance d.	$d^2 = \lvert x_2 - x_1 \rvert^2 + \lvert y_2 - y_1 \rvert^2$
$\lvert a \rvert^2 = a^2$ for all real numbers a.	$d^2 = (x_2 - x_1)^2 + (y_2 - y_1)^2$
Use the square root property.	$d = \pm\sqrt{(x_2 - x_1)^2 + (y_2 - y_1)^2}$
Distance can be only positive.	$d = \sqrt{(x_2 - x_1)^2 + (y_2 - y_1)^2}$

Study Tip

It does not matter which point is taken to be the first point or the second point.

DEFINITION **Distance Formula**

The **distance *d*** between two points $P_1 = (x_1, y_1)$ and $P_2 = (x_2, y_2)$ is given by

$$d = \sqrt{(x_2 - x_1)^2 + (y_2 - y_1)^2}$$

The distance between two points is the square root of the sum of the square of the distance between the x-coordinates and the square of the distance between the y-coordinates.

You will prove in the exercises that it does not matter which point you take to be the first point when applying the distance formula.

EXAMPLE 3 Using the Distance Formula to Find the Distance Between Two Points

Find the distance between $(-3, 7)$ and $(5, -2)$.

Solution:

Write the distance formula.	$d = \sqrt{[x_2 - x_1]^2 + [y_2 - y_1]^2}$
Substitute $(x_1, y_1) = (-3, 7)$ and $(x_2, y_2) = (5, -2)$.	$d = \sqrt{[5 - (-3)]^2 + [-2 - 7]^2}$
Simplify.	$d = \sqrt{[5 + 3]^2 + [-2 - 7]^2}$
	$d = \sqrt{8^2 + (-9)^2} = \sqrt{64 + 81} = \sqrt{145}$
Solve for d.	$\boxed{d = \sqrt{145}}$

YOUR TURN Find the distance between $(4, -5)$ and $(-3, -2)$.

Answer: $d = \sqrt{58}$

Midpoint of a Line Segment Joining Two Points

The **midpoint**, (x_m, y_m), of a line segment joining two points (x_1, y_1) and (x_2, y_2) is defined as the point that lies on the segment which has the same distance d from both points. In other words, the midpoint of a segment lies halfway between the given endpoints. The coordinates of the midpoint are found by averaging the x-coordinates and averaging the y-coordinates.

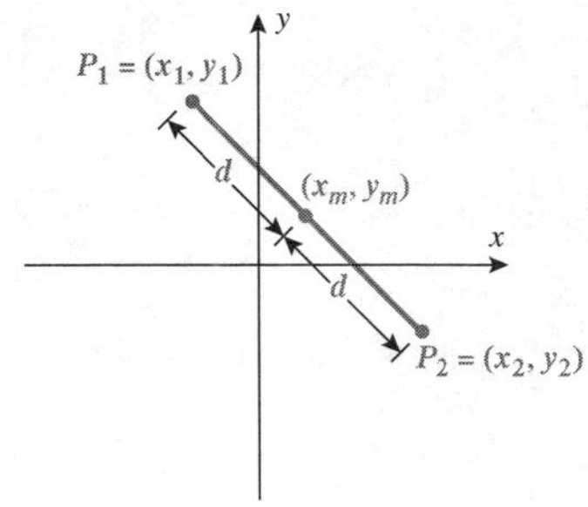

DEFINITION **Midpoint Formula**

The **midpoint**, (x_m, y_m), of the line segment with endpoints (x_1, y_1) and (x_2, y_2) is given by

$$(x_m, y_m) = \left(\frac{x_1 + x_2}{2}, \frac{y_1 + y_2}{2}\right)$$

The midpoint can be found by averaging the x-coordinates and averaging the y-coordinates.

EXAMPLE 4 Finding the Midpoint of a Line Segment

Find the midpoint of the line segment joining the points (2, 6) and (−4, −2).

Solution:

Write the midpoint formula. $\quad (x_m, y_m) = \left(\frac{x_1 + x_2}{2}, \frac{y_1 + y_2}{2}\right)$

Substitute $(x_1, y_1) = (2, 6)$ and $(x_2, y_2) = (-4, -2)$. $\quad (x_m, y_m) = \left(\frac{2 + (-4)}{2}, \frac{6 + (-2)}{2}\right)$

Simplify. $\quad \boxed{(x_m, y_m) = (-1, 2)}$

One way to verify your answer is to plot the given points and the midpoint to make sure your answer looks reasonable.

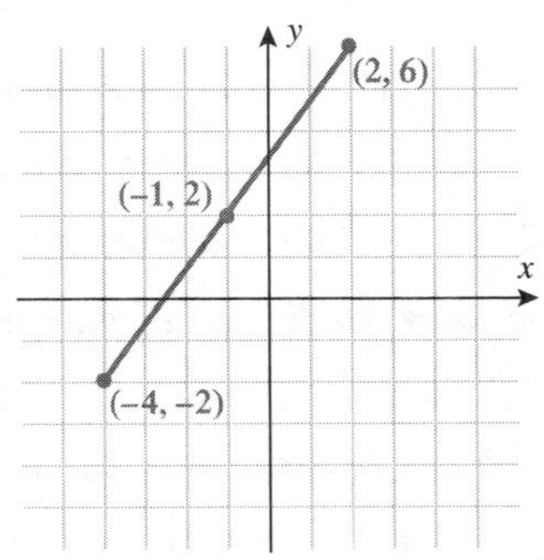

■ **YOUR TURN** Find the midpoint of the line segment joining the points (3, −4) and (5, 8).

Technology Tip

Show a screen display of how to enter $\frac{2 + (-4)}{2}$ and $\frac{6 + (-2)}{2}$.

Scientific calculators:

[(] 2 [+] [(-)] 4 [)] [÷] 2 [=]

[(] 6 [+] [(-)] 2 [)] [÷] 2 [=]

Or,

[(] 2 [+] 4 [+/−] [)] [÷] 2 [=]

[(] 6 [+] 2 [+/−] [)] [÷] 2 [=]

```
(2+-4)/2
                -1
(6+-2)/2
                 2
```

■ **Answer:** Midpoint = (4, 2)

SECTION 2.1 SUMMARY

Cartesian Plane

- Plotting coordinates: (x, y)
- Quadrants: I, II, III, and IV
- Origin: $(0, 0)$

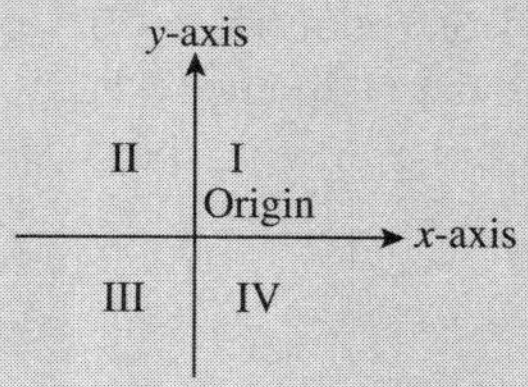

Distance Between Two Points

$$d = \sqrt{(x_2 - x_1)^2 + (y_2 - y_1)^2}$$

Midpoint of Segment Joining Two Points

$$\text{Midpoint} = (x_m, y_m) = \left(\frac{x_1 + x_2}{2}, \frac{y_1 + y_2}{2}\right)$$

SECTION 2.1 EXERCISES

SKILLS

In Exercises 1–6, give the coordinates for each point labeled.

1. Point A
2. Point B
3. Point C
4. Point D
5. Point E
6. Point F

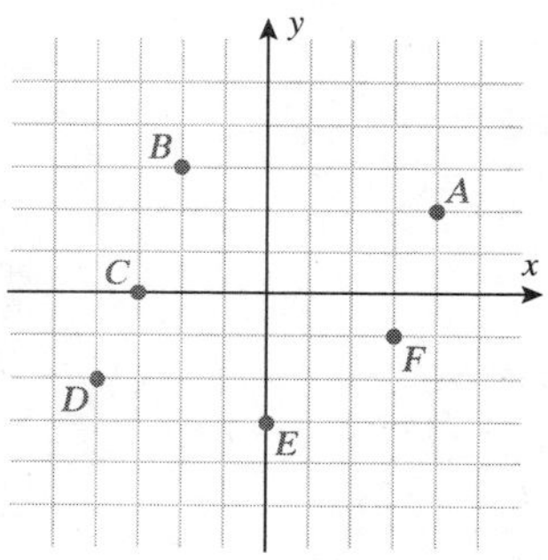

In Exercises 7 and 8, plot each point in the Cartesian plane and indicate in which quadrant or on which axis the point lies.

7. A: $(-2, 3)$ B: $(1, 4)$ C: $(-3, -3)$ D: $(5, -1)$ E: $(0, -2)$ F: $(4, 0)$
8. A: $(-1, 2)$ B: $(1, 3)$ C: $(-4, -1)$ D: $(3, -2)$ E: $(0, 5)$ F: $(-3, 0)$
9. Plot the points $(-3, 1)$, $(-3, 4)$, $(-3, -2)$, $(-3, 0)$, $(-3, -4)$. Describe the line containing points of the form $(-3, y)$.
10. Plot the points $(-1, 2)$, $(-3, 2)$, $(0, 2)$, $(3, 2)$, $(5, 2)$. Describe the line containing points of the form $(x, 2)$.

In Exercises 11–32, calculate the distance between the given points, and find the midpoint of the segment joining them.

11. $(1, 3)$ and $(5, 3)$
12. $(-2, 4)$ and $(-2, -4)$
13. $(-1, 4)$ and $(3, 0)$
14. $(-3, -1)$ and $(1, 3)$
15. $(-10, 8)$ and $(-7, -1)$
16. $(-2, 12)$ and $(7, 15)$
17. $(-3, -1)$ and $(-7, 2)$
18. $(-4, 5)$ and $(-9, -7)$
19. $(-6, -4)$ and $(-2, -8)$
20. $(0, -7)$ and $(-4, -5)$
21. $\left(-\frac{1}{2}, \frac{1}{3}\right)$ and $\left(\frac{7}{2}, \frac{10}{3}\right)$
22. $\left(\frac{1}{5}, \frac{7}{3}\right)$ and $\left(\frac{9}{5}, -\frac{2}{3}\right)$
23. $\left(-\frac{2}{3}, -\frac{1}{5}\right)$ and $\left(\frac{1}{4}, \frac{1}{3}\right)$
24. $\left(\frac{7}{5}, \frac{1}{9}\right)$ and $\left(\frac{1}{2}, -\frac{7}{3}\right)$
25. $(-1.5, 3.2)$ and $(2.1, 4.7)$
26. $(-1.2, -2.5)$ and $(3.7, 4.6)$
27. $(-14.2, 15.1)$ and $(16.3, -17.5)$
28. $(1.1, 2.2)$ and $(3.3, 4.4)$
29. $\left(\sqrt{3}, 5\sqrt{2}\right)$ and $\left(\sqrt{3}, \sqrt{2}\right)$
30. $\left(3\sqrt{5}, -3\sqrt{3}\right)$ and $\left(-\sqrt{5}, -\sqrt{3}\right)$
31. $\left(1, \sqrt{3}\right)$ and $\left(-\sqrt{2}, -2\right)$
32. $\left(2\sqrt{5}, 4\right)$ and $\left(1, 2\sqrt{3}\right)$

In Exercises 33 and 34, calculate (to two decimal places) the perimeter of the triangle with the following vertices:

33. Points A, B, and C

34. Points C, D, and E

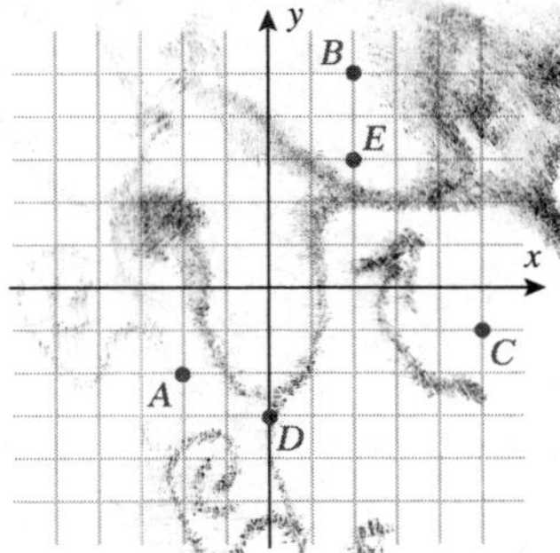

In Exercises 35–38, determine whether the triangle with the given vertices is a right triangle, an isosceles triangle, neither, or both. (Recall that a right triangle satisfies the Pythagorean theorem and an isosceles triangle has at least two sides of equal length.)

35. $(0, -3)$, $(3, -3)$, and $(3, 5)$

36. $(0, 2)$, $(-2, -2)$, and $(2, -2)$

37. $(1, 1)$, $(3, -1)$, and $(-2, -4)$

38. $(-3, 3)$, $(3, 3)$, and $(-3, -3)$

APPLICATIONS

39. Cell Phones. A cellular phone company currently has three towers: one in Tampa, one in Orlando, and one in Gainesville to serve the central Florida region. If Orlando is 80 miles east of Tampa and Gainesville is 100 miles north of Tampa, what is the distance from Orlando to Gainesville?

40. Cell Phones. The same cellular phone company in Exercise 39 has decided to add additional towers at each "halfway" between cities. How many miles from Tampa is each "halfway" tower?

41. Travel. A retired couple who live in Columbia, South Carolina, decide to take their motor home and visit two children who live in Atlanta and in Savannah, Georgia. Savannah is 160 miles south of Columbia and Atlanta is 215 miles west of Columbia. How far apart do the children live from each other?

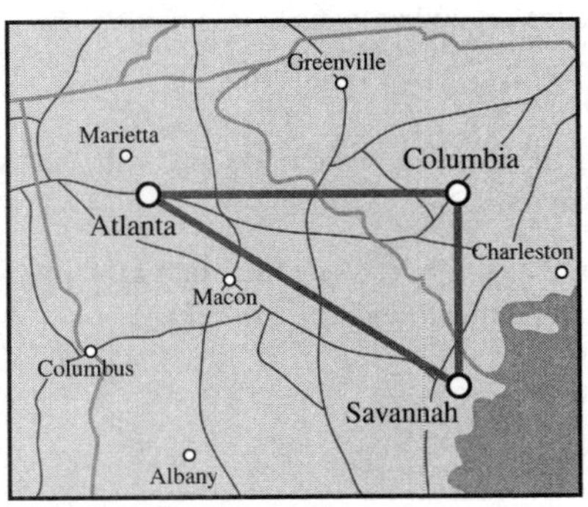

42. Sports. In the 1984 Orange Bowl, Doug Flutie, the 5 foot 9 inch quarterback for Boston College, shocked the world as he threw a "hail Mary" pass that was caught in the end zone with no time left on the clock, defeating the Miami Hurricanes 47–45. Although the record books have it listed as a 48 yard pass, what was the actual distance the ball was thrown? The following illustration depicts the path of the ball.

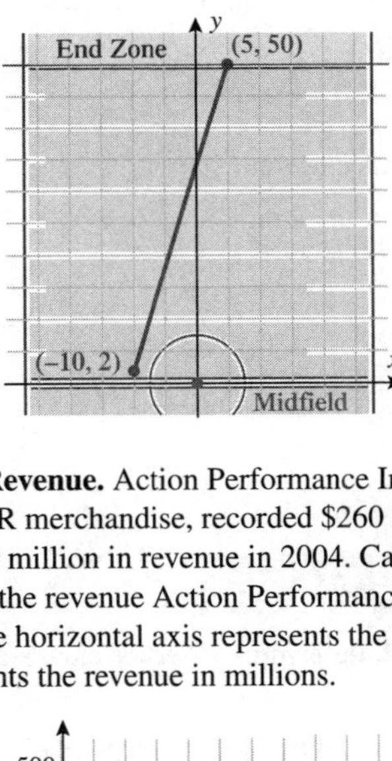

43. NASCAR Revenue. Action Performance Inc., the leading seller of NASCAR merchandise, recorded \$260 million in revenue in 2002 and \$400 million in revenue in 2004. Calculate the midpoint to estimate the revenue Action Performance Inc. recorded in 2003. Assume the horizontal axis represents the year and the vertical axis represents the revenue in millions.

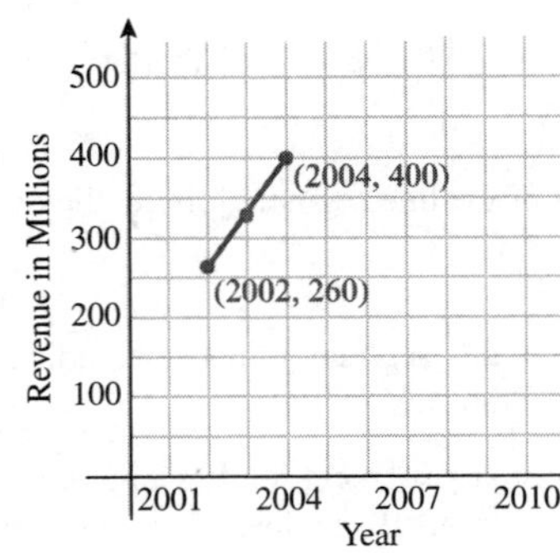

44. **Ticket Price.** In 1993, the average Miami Dolphins ticket price was \$28 and in 2001 the average price was \$56. Find the midpoint of the segment joining these two points to estimate the ticket price in 1997.

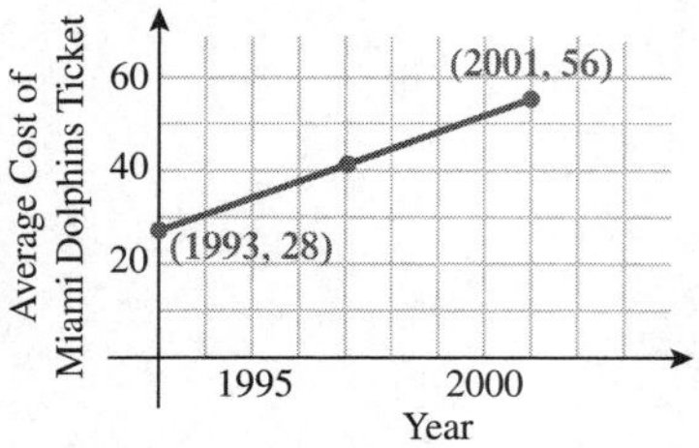

CATCH THE MISTAKE

In Exercises 45–48, explain the mistake that is made.

45. Calculate the distance between (2, 7) and (9, 10).

Solution:

Write the distance formula. $d = \sqrt{(x_2 - x_1)^2 + (y_2 - y_1)^2}$

Substitute (2, 7) and (9, 10). $d = \sqrt{(7-2)^2 + (10-9)^2}$

Simplify. $d = \sqrt{(5)^2 + (1)^2} = \sqrt{26}$

This is incorrect. What mistake was made?

46. Calculate the distance between (−2, 1) and (3, −7).

Solution:

Write the distance formula. $d = \sqrt{(x_2 - x_1)^2 + (y_2 - y_1)^2}$

Substitute (−2, 1) and (3, −7). $d = \sqrt{(3-2)^2 + (-7-1)^2}$

Simplify. $d = \sqrt{(1)^2 + (-8)^2} = \sqrt{65}$

This is incorrect. What mistake was made?

47. Compute the midpoint of the segment with endpoints (−3, 4) and (7, 9).

Solution:

Write the midpoint formula. $(x_m, y_m) = \left(\frac{x_1 + x_2}{2}, \frac{y_1 + y_2}{2}\right)$

Substitute (−3, 4) and (7, 9). $(x_m, y_m) = \left(\frac{-3+4}{2}, \frac{7+9}{2}\right)$

Simplify. $(x_m, y_m) = \left(\frac{1}{2}, \frac{16}{2}\right) = \left(\frac{1}{2}, 4\right)$

This is incorrect. What mistake was made?

48. Compute the midpoint of the segment with endpoints (−1, −2) and (−3, −4).

Solution:

Write the midpoint formula. $(x_m, y_m) = \left(\frac{x_1 - x_2}{2}, \frac{y_1 - y_2}{2}\right)$

Substitute (−1, −2) and (−3, −4). $(x_m, y_m) = \left(\frac{-1-(-3)}{2}, \frac{-2-(-4)}{2}\right)$

Simplify. $(x_m, y_m) = (1, 1)$

This is incorrect. What mistake was made?

CONCEPTUAL

In Exercises 49–52, determine whether each statement is true or false.

49. The distance from the origin to the point (a, b) is $d = \sqrt{a^2 + b^2}$.

50. The midpoint of the line segment joining the origin and the point (a, a) is $\left(\frac{a}{2}, \frac{a}{2}\right)$.

51. The midpoint of any segment joining two points in quadrant I also lies in quadrant I.

52. The midpoint of any segment joining a point in quadrant I to a point in quadrant III also lies in either quadrant I or III.

53. Calculate the length and the midpoint of the line segment joining the points (a, b) and (b, a).

54. Calculate the length and the midpoint of the line segment joining the points (a, b) and $(-a, -b)$.

CHALLENGE

55. Assume that two points (x_1, y_1) and (x_2, y_2) are connected by a segment. Prove that the distance from the midpoint of the segment to either of the two points is the same.

56. Prove that the diagonals of a parallelogram in the figure intersect at their midpoints.

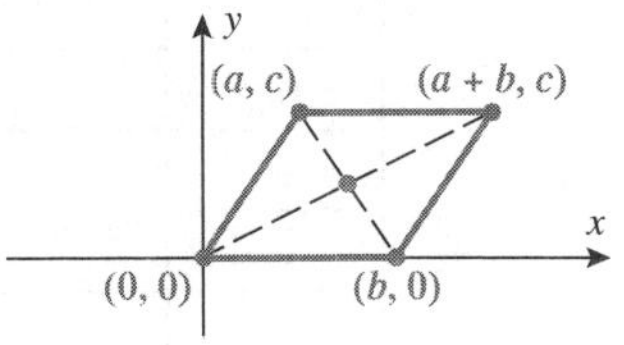

57. Assume that two points (a, b) and (c, d) are the endpoints of a line segment. Calculate the distance between the two points. Prove that it does not matter which point is labeled as the "first" point in the distance formula.

58. Show that the points $(-1, -1)$, $(0, 0)$, and $(2, 2)$ are collinear (lie on the same line) by showing that the sum of the distance from $(-1, -1)$ to $(0, 0)$ and the distance from $(0, 0)$ to $(2, 2)$ is equal to the distance from $(-1, -1)$ to $(2, 2)$.

TECHNOLOGY

In Exercises 59–62, calculate the distance between the two points. Use a graphing utility to graph the segment joining the two points and find the midpoint of the segment.

59. $(-2.3, 4.1)$ and $(3.7, 6.2)$

60. $(-4.9, -3.2)$ and $(5.2, 3.4)$

61. $(1.1, 2.2)$ and $(3.3, 4.4)$

62. $(-1.3, 7.2)$ and $(2.3, -4.5)$

SECTION 2.2 GRAPHING EQUATIONS: POINT-PLOTTING, INTERCEPTS, AND SYMMETRY

SKILLS OBJECTIVES

- Sketch graphs of equations by plotting points.
- Find intercepts for graphs of equations.
- Use intercepts and symmetry as graphing aids.
- Conduct a test for symmetry about the x-axis, y-axis, and origin.

CONCEPTUAL OBJECTIVE

- Relate symmetry graphically and algebraically.

In this section, you will learn how to graph equations by plotting points. However, when we discuss graphing principles in Chapter 3, you will see that other techniques can be more efficient.

Point-Plotting

Most equations in two variables, such as $y = x^2$, have an infinite number of ordered pairs as solutions. For example, $(0, 0)$ is a solution to $y = x^2$ because when $x = 0$ and $y = 0$, the equation is true. Two other solutions are $(-1, 1)$ and $(1, 1)$.

The **graph of an equation** in two variables, x and y, consists of all the points in the xy-plane whose coordinates (x, y) satisfy the equation. A procedure for plotting the graphs of equations is outlined below and is illustrated with the example $y = x^2$.

WORDS

Step 1: In a table, list several pairs of coordinates that make the equation true.

MATH

x	$y = x^2$	(x, y)
0	0	(0, 0)
−1	1	(−1, 1)
1	1	(1, 1)
−2	4	(−2, 4)
2	4	(2, 4)

Step 2: Plot these points on a graph and connect the points with a smooth curve. Use arrows to indicate that the graph continues.

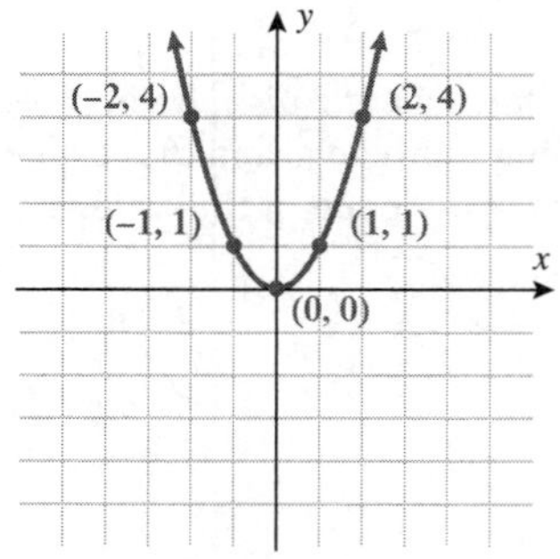

In graphing an equation, first select arbitrary values for x and then use the equation to find the corresponding value of y, or vice versa.

EXAMPLE 1 Graphing an Equation of a Line by Plotting Points

Graph the equation $y = 2x - 1$.

Solution:

STEP 1 In a table, list several pairs of coordinates that make the equation true.

x	$y = 2x - 1$	(x, y)
0	−1	(0, −1)
−1	−3	(−1, −3)
1	1	(1, 1)
−2	−5	(−2, −5)
2	3	(2, 3)

STEP 2 Plot these points on a graph and connect the points, resulting in a line.

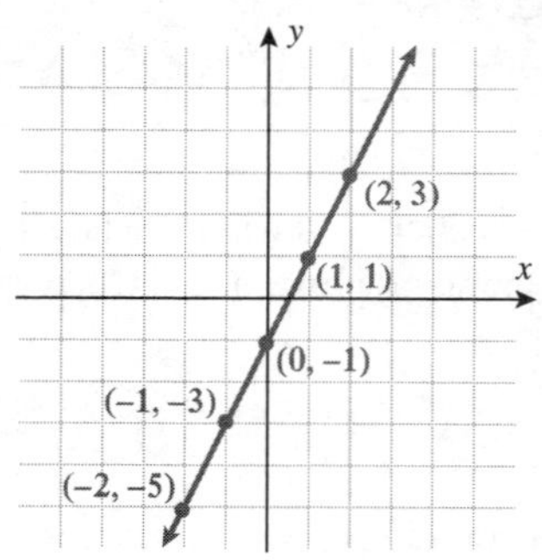

■ **YOUR TURN** The graph of the equation $y = -x + 1$ is a line. Graph the line.

■ **Answer:**

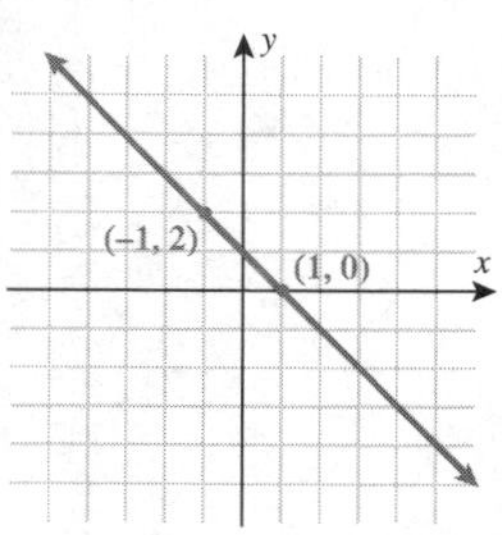

EXAMPLE 2 Graphing an Equation by Plotting Points

Graph the equation $y = x^2 - 5$.

Solution:

STEP 1 In a table, list several pairs of coordinates that make the equation true.

x	$y = x^2 - 5$	(x, y)
0	−5	(0, −5)
−1	−4	(−1, −4)
1	−4	(1, −4)
−2	−1	(−2, −1)
2	−1	(2, −1)
−3	4	(−3, 4)
3	4	(3, 4)

STEP 2 Plot these points on a graph and connect the points with a smooth curve, indicating with arrows that the curve continues.

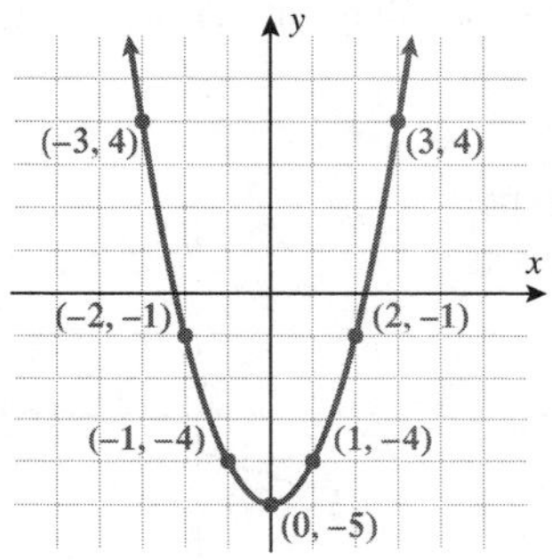

This graph is called a *parabola* and will be discussed in further detail in Chapter 8.

■ **YOUR TURN** Graph the equation $y = x^2 - 1$.

Technology Tip

Graph of $y_1 = x^2 - 5$ is shown in a $[-5, 5]$ by $[-2, 8]$ viewing rectangle.

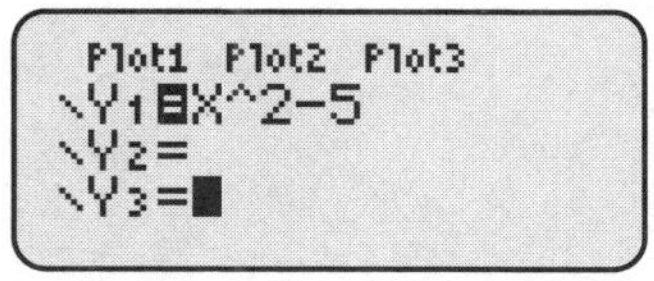

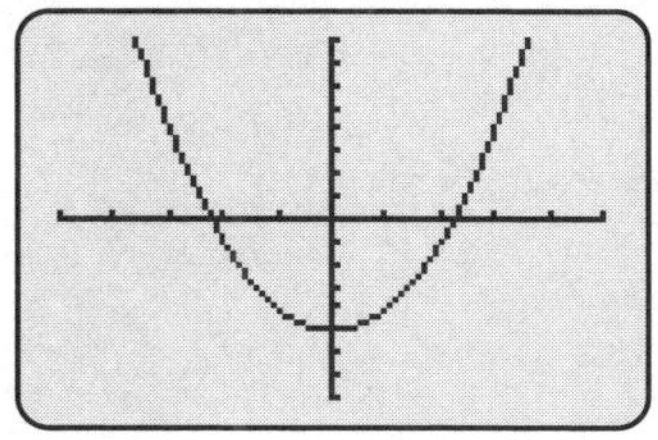

■ **Answer:**

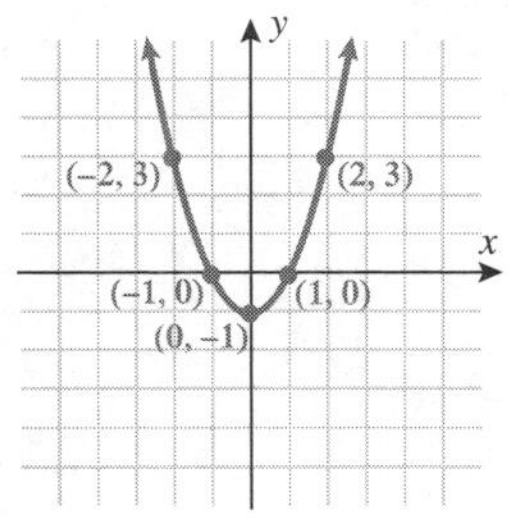

EXAMPLE 3 Graphing an Equation by Plotting Points

Graph the equation $y = x^3$.

Solution:

STEP 1 In a table, list several pairs of coordinates that satisfy the equation.

x	$y = x^3$	(x, y)
0	0	(0, 0)
−1	−1	(−1, −1)
1	1	(1, 1)
−2	−8	(−2, −8)
2	8	(2, 8)

STEP 2 Plot these points on a graph and connect the points with a smooth curve, indicating with arrows that the curve continues in both the positive and negative directions.

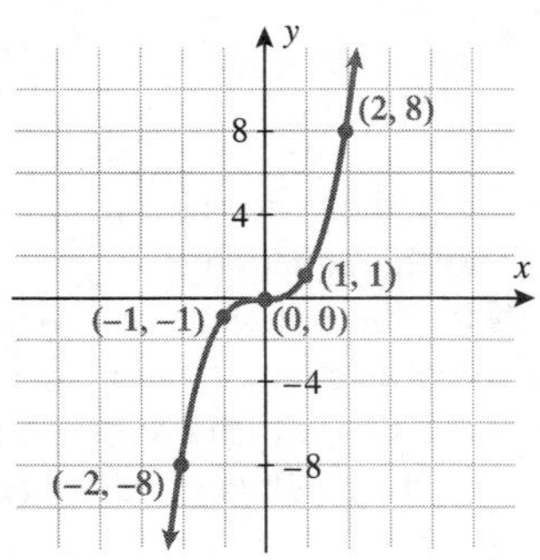

Intercepts

Study Tip

Identifying the points help define the graph unmistakenly.

When point-plotting graphs of equations, which points should be selected? Points where a graph crosses (or touches) either the x-axis or y-axis are called *intercepts* and identifying these points helps define the graph unmistakably.

An ***x*-intercept** of a graph is a point where the graph intersects the x-axis. Specifically, an x-intercept is the x-coordinate of such a point. For example, if a graph intersects the x-axis at the point (3, 0), then we say that 3 is the x-intercept. Since the value for y along the x-axis is zero, all points corresponding to x-intercepts have the form $(a, 0)$.

A ***y*-intercept** of a graph is a point where the graph intersects the y-axis. Specifically, a y-intercept is the y-coordinate of such a point. For example, if a graph intersects the y-axis at the point (0, 2), then we say that 2 is the y-intercept. Since the value for x along the y-axis is zero, all points corresponding to y-intercepts have the form $(0, b)$.

It is important to note that graphs of equations do not have to have intercepts, and if they do have intercepts, they can have one or more of each type.

One x-intercept
Two y-intercepts

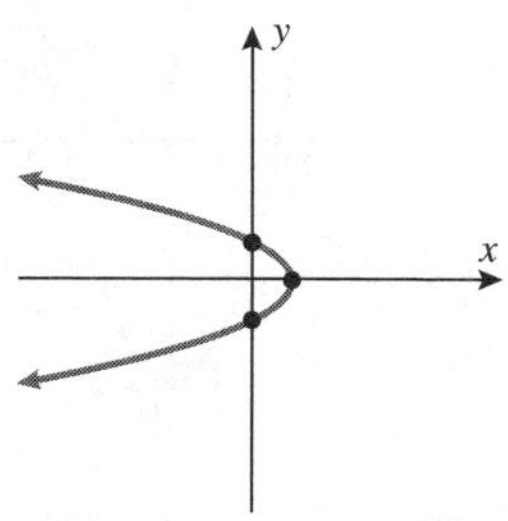

No x-intercepts
One y-intercept

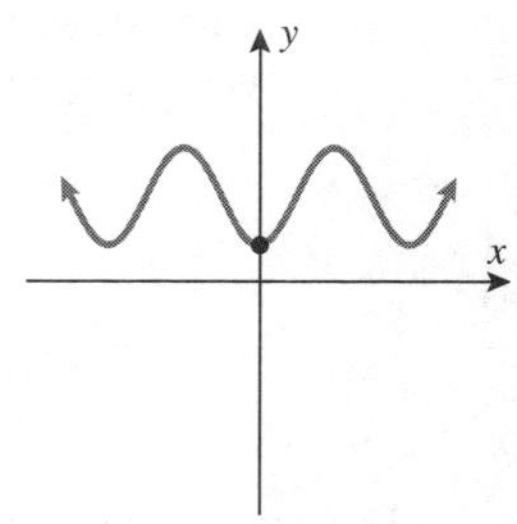

No x-intercepts
No y-intercepts

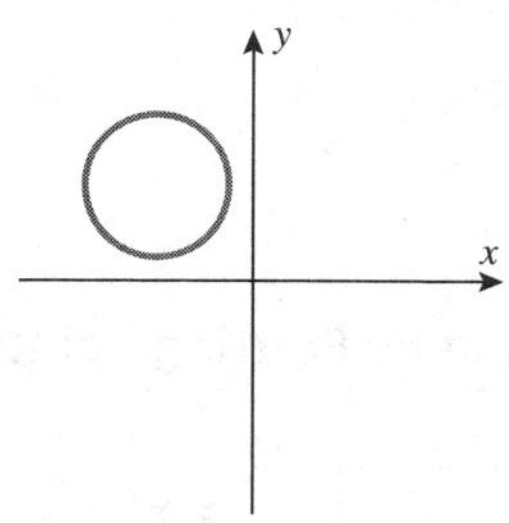

Three x-intercepts
One y-intercept

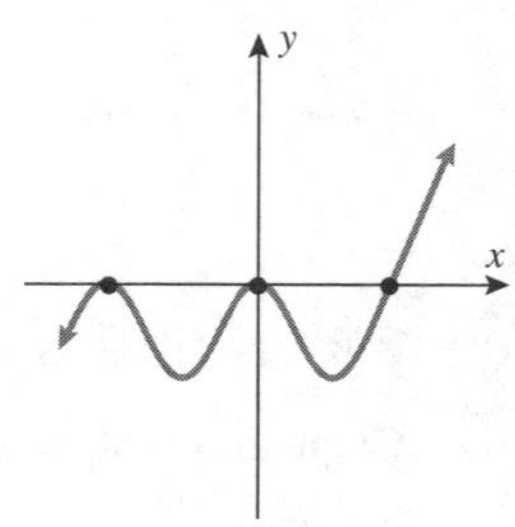

Note: The origin (0, 0) corresponds to both an x-intercept and a y-intercept.

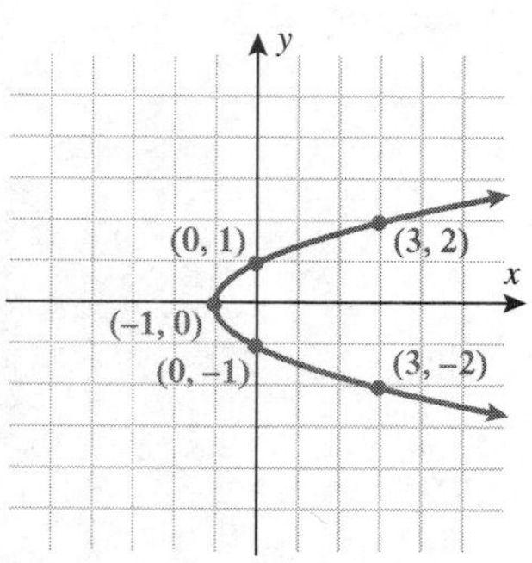

The graph given to the left has two y-intercepts and one x-intercept.

- The x-intercept is -1, which corresponds to the point $(-1, 0)$.
- The y-intercepts are -1 and 1, which correspond to the points $(0, -1)$ and $(0, 1)$, respectively.

Algebraically, how do we find intercepts from an equation? The graph in the margin corresponds to the equation $x = y^2 - 1$. The x-intercepts are located on the x-axis, which corresponds to $y = 0$. If we let $y = 0$ in the equation $x = y^2 - 1$ and solve for x, the result is $x = -1$. This corresponds to the x-intercept we identified above. Similarly, the y-intercepts are located on the y-axis, which corresponds to $x = 0$. If we let $x = 0$ in the equation $x = y^2 - 1$ and solve for y, the result is $y = \pm 1$. These correspond to the y-intercepts we identified above.

EXAMPLE 4 Finding Intercepts from an Equation

Given the equation $y = x^2 + 4$, find the indicated intercepts of its graph, if any.

a. x-intercept(s) **b.** y-intercept(s)

Solution (a):

Let $y = 0$. $0 = x^2 + 1$

Solve for x. $x^2 = -1$ no real solution

There are no x-intercepts.

Solution (b):

Let $x = 0$. $y = 0^2 + 1$

Solve for y. $y = 1$

The y-intercept is located at the point $(0, 1)$.

YOUR TURN For the equation $y = x^2 - 4$,

a. find the x-intercept(s), if any. **b.** find the y-intercept(s), if any.

Answer:
a. x-intercepts: -2 and 2
b. y-intercept: -4

Symmetry

The word **symmetry** conveys balance. Suppose you have two pictures to hang on a wall. If you space them equally apart on the wall, then you prefer a symmetric décor. This is an example of symmetry about a line. The word (water) written below is identical if you rotate the word 180 degrees (or turn the page upside down). This is an example of symmetry about a point. Symmetric graphs have the characteristic that their mirror image can be obtained about a reference, typically a line or a point.

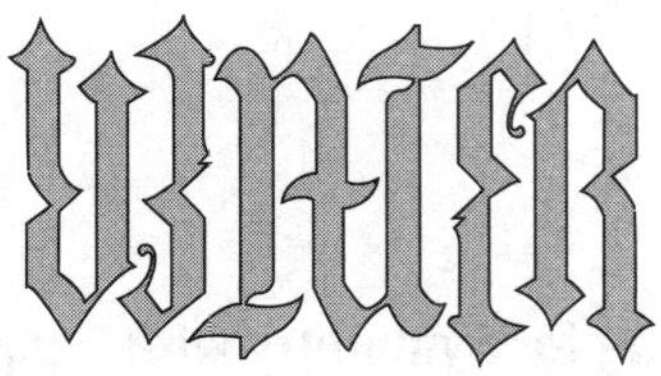

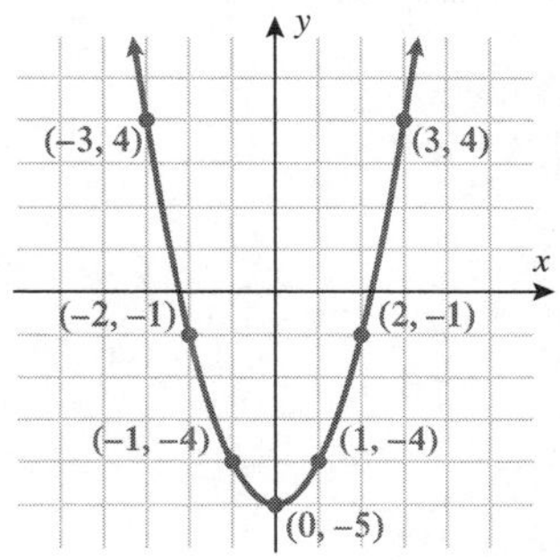

In Example 2, the points $(-2, -1)$ and $(2, -1)$ both lie on the graph, as do the points $(-1, -4)$ and $(1, -4)$. Notice that the graph on the right side of the y-axis is a mirror image of the part of the graph to the left of the y-axis. This graph illustrates *symmetry* with respect to the *y-axis* (the line $x = 0$).

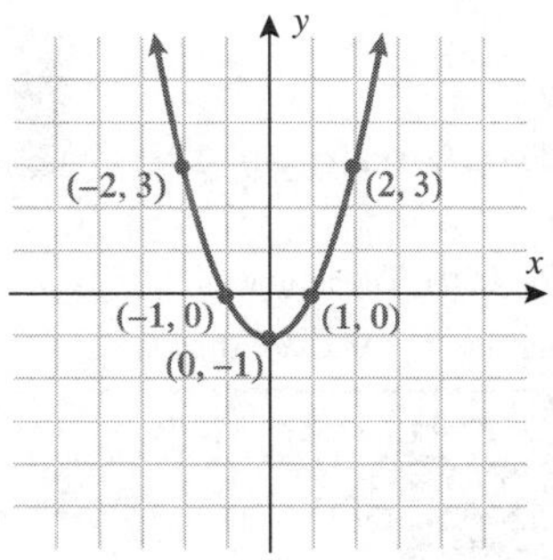

In the Your Turn following Example 2, the points $(0, 1)$ and $(0, -1)$ both lie on the graph, as do the points $(3, 2)$ and $(3, -2)$. Notice that the part of the graph above the x-axis is a mirror image of the part of the graph below the x-axis. This graph illustrates *symmetry* with respect to the *x-axis* (the line $y = 0$).

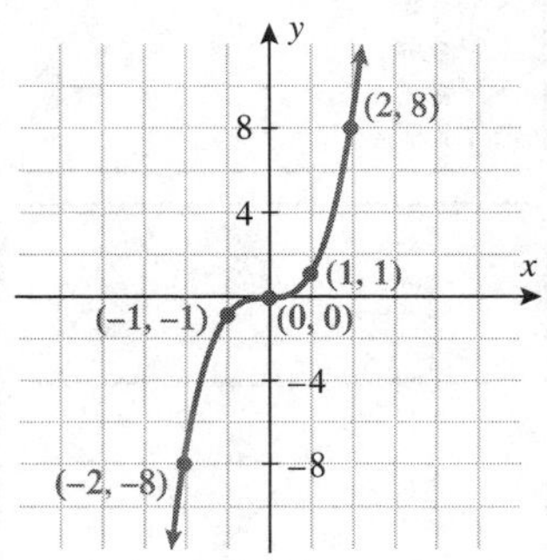

In Example 3, the points $(-1, -1)$ and $(1, 1)$ both lie on the graph. Notice that rotating this graph 180 degrees (or turning your page upside down) results in an identical graph. This is an example of *symmetry* with respect to the *origin* $(0, 0)$.

Symmetry aids in graphing by giving information "for free." For example, if a graph is symmetric about the y-axis, then once the graph to the right of the y-axis is found, the left side of the graph is the mirror image of that. If a graph is symmetric about the origin, then once the graph is known in quadrant I, the graph in quadrant III is found by rotating the known graph 180 degrees.

It would be beneficial to know whether a graph of an equation is symmetric about a line or point before the graph of the equation is sketched. Although a graph can be symmetric about any line or point, we will discuss only symmetry about the x-axis, y-axis, and origin. These types of symmetry and the algebraic procedures for testing for symmetry are outlined below.

Types and Tests for Symmetry

Type of Symmetry	Graph	If the Point (a, b) Is on the Graph, Then the Point . . .	Algebraic Test for Symmetry
Symmetric with respect to the **x-axis**	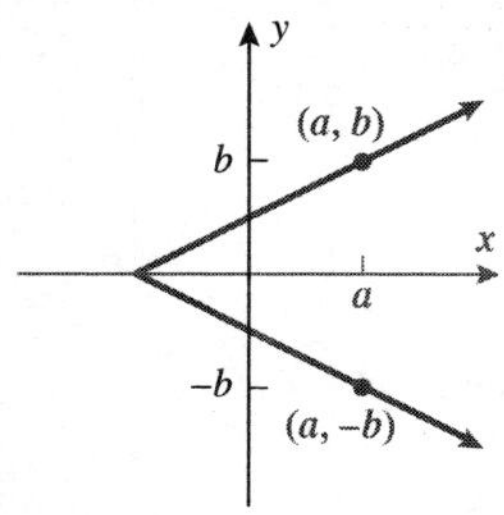	$(a, -b)$ is on the graph.	Replacing y with $-y$ leaves the equation unchanged.
Symmetric with respect to the **y-axis**	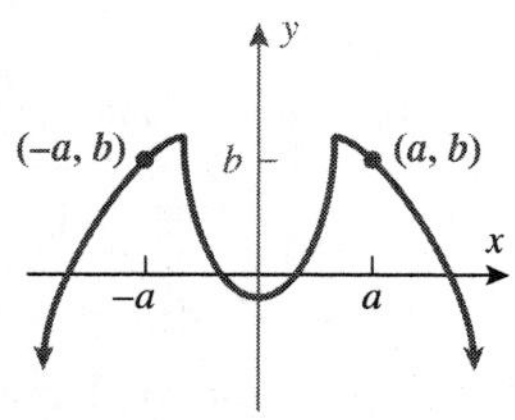	$(-a, b)$ is on the graph.	Replacing x with $-x$ leaves the equation unchanged.
Symmetric with respect to the **origin**	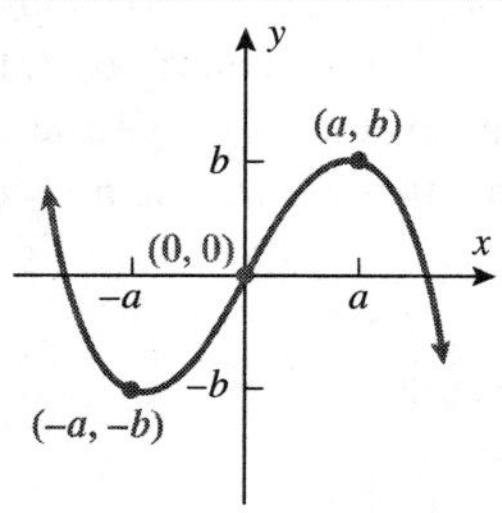	$(-a, -b)$ is on the graph.	Replacing x with $-x$ and y with $-y$ leaves the equation unchanged.

Study Tip

Symmetry gives us information about the graph "for free."

Technology Tip

To enter the graph of $y^2 = x^3$, solve for y first. The graphs of $y_1 = \sqrt{x^3}$ and $y_2 = -\sqrt{x^3}$ are shown.

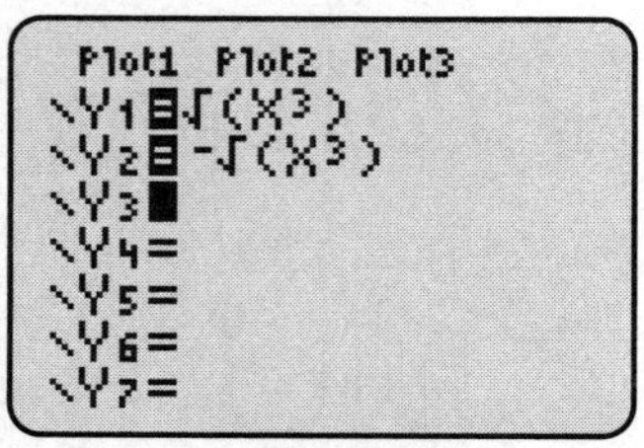

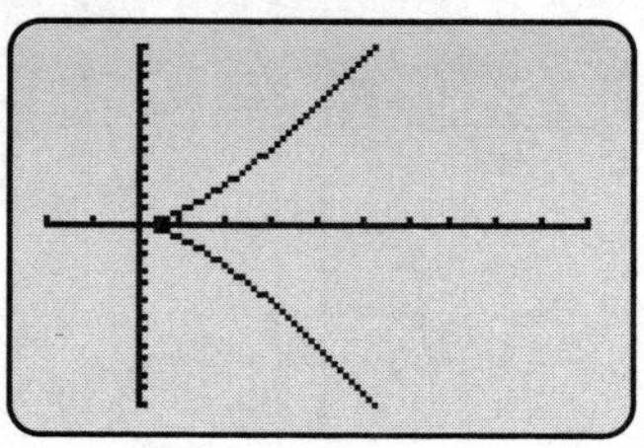

EXAMPLE 5 Testing for Symmetry with Respect to the Axes

Test the equation $y^2 = x^3$ for symmetry with respect to the axes.

Solution:

Test for symmetry with respect to the x-axis.

Replace y with $-y$. $\qquad (-y)^2 = x^3$

Simplify. $\qquad y^2 = x^3$

The resulting equation is the same as the original equation, $y^2 = x^3$.

Therefore $y^2 = x^3$ is **symmetric with respect to the x-axis.**

Test for symmetry with respect to the y-axis.

Replace x with $-x$. $\qquad y^2 = (-x)^3$

Simplify. $\qquad y^2 = -x^3$

The resulting equation, $y^2 = -x^3$, is not the same as the original equation, $y^2 = x^3$.

Therefore $y^2 = x^3$ is **not** symmetric with respect to the y-axis.

When testing for symmetry about the x-axis, y-axis, and origin, there are *five* possibilities:

- No symmetry
- Symmetry with respect to the x-axis
- Symmetry with respect to the y-axis
- Symmetry with respect to the origin
- Symmetry with respect to the x-axis, y-axis, and origin

EXAMPLE 6 Testing for Symmetry

Determine what type of symmetry (if any) the graphs of the equations exhibit.

a. $y = x^2 + 1$ **b.** $y = x^3 + 1$

Solution (a):

Replace x with $-x$. $y = (-x)^2 + 1$

Simplify. $y = x^2 + 1$

The resulting equation is equivalent to the original equation, so the graph of the equation $y = x^2 + 1$ is symmetric with respect to the y-axis.

Replace y with $-y$. $(-y) = x^2 + 1$

Simplify. $y = -x^2 - 1$

The resulting equation $y = -x^2 - 1$ is not equivalent to the original equation $y = x^2 + 1$, so the graph of the equation $y = x^2 + 1$ is not symmetric with respect to the x-axis.

Replace x with $-x$ and y with $-y$. $(-y) = (-x)^2 + 1$

Simplify. $-y = x^2 + 1$

$y = -x^2 - 1$

The resulting equation $y = -x^2 - 1$ is not equivalent to the original equation $y = x^2 + 1$, so the graph of the equation $y = x^2 + 1$ is not symmetric with respect to the origin.

The graph of the equation $y = x^2 + 1$ is **symmetric with respect to the y-axis.**

Solution (b):

Replace x with $-x$. $y = (-x)^3 + 1$

Simplify. $y = -x^3 + 1$

The resulting equation $y = -x^3 + 1$ is not equivalent to the original equation $y = x^3 + 1$. Therefore, the graph of the equation $y = x^3 + 1$ is not symmetric with respect to the y-axis.

Replace y with $-y$. $(-y) = x^3 + 1$

Simplify. $y = -x^3 - 1$

The resulting equation $y = -x^3 - 1$ is not equivalent to the original equation $y = x^3 + 1$. Therefore, the graph of the equation $y = x^3 + 1$ is not symmetric with respect to the x-axis.

Replace x with $-x$ and y with $-y$. $(-y) = (-x)^3 + 1$

Simplify. $-y = -x^3 + 1$

$y = x^3 - 1$

The resulting equation $y = x^3 - 1$ is not equivalent to the original equation $y = x^3 + 1$. Therefore, the graph of the equation $y = x^3 + 1$ is not symmetric with respect to the origin.

The graph of the equation $y = x^3 + 1$ exhibits **no symmetry**.

YOUR TURN Determine the symmetry (if any) for $x = y^2 - 1$.

Technology Tip

Graph of $y_1 = x^2 + 1$ is shown.

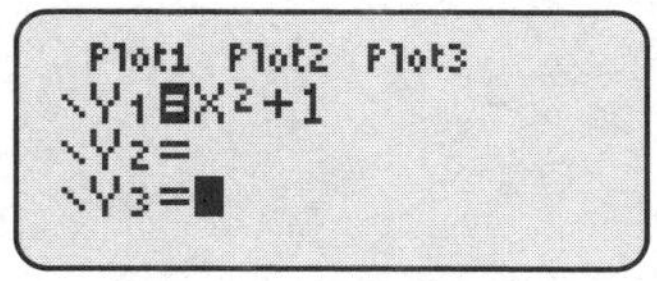

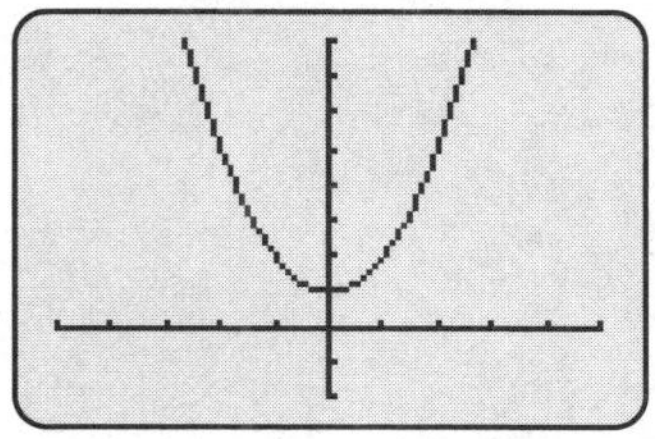

Graph of $y_1 = x^3 + 1$ is shown.

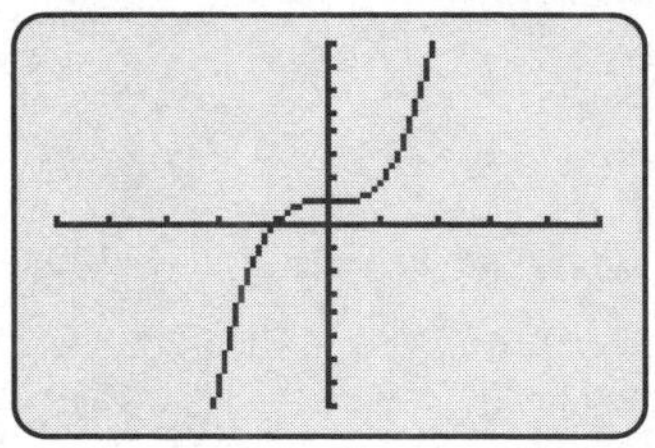

- **Answer:** The graph of the equation is symmetric with respect to the x-axis.

Using Intercepts and Symmetry as Graphing Aids

How can we use intercepts and symmetry to assist us in graphing? Intercepts are a good starting point—though not the only one. For symmetry, look back at Example 2, $y = x^2 - 5$. We selected seven x-coordinates and solved the equation to find the corresponding y-coordinates. If we had known that this graph was symmetric with respect to the y-axis, then we would have had to find the solutions to only the positive x-coordinates, since we get the negative x-coordinates for free. For example, we found the point $(1, -4)$ to be a solution to the equation. The rules of symmetry tell us that $(-1, -4)$ is also on the graph.

EXAMPLE 7 Using Intercepts and Symmetry as Graphing Aids

For the equation $x^2 + y^2 = 25$, use intercepts and symmetry to help you graph the equation using the point-plotting technique.

Solution:

STEP 1 Find the intercepts.

For the x-intercepts, let $y = 0$. $\quad x^2 + 0^2 = 25$

Solve for x. $\quad x = \pm 5$

The two x-intercepts correspond to the points $(-5, 0)$ and $(5, 0)$.

For the y-intercepts, let $x = 0$. $\quad 0^2 + y^2 = 25$

Solve for y. $\quad y = \pm 5$

The two y-intercepts correspond to the points $(0, -5)$ and $(0, 5)$.

STEP 2 Identify the points on the graph corresponding to the intercepts.

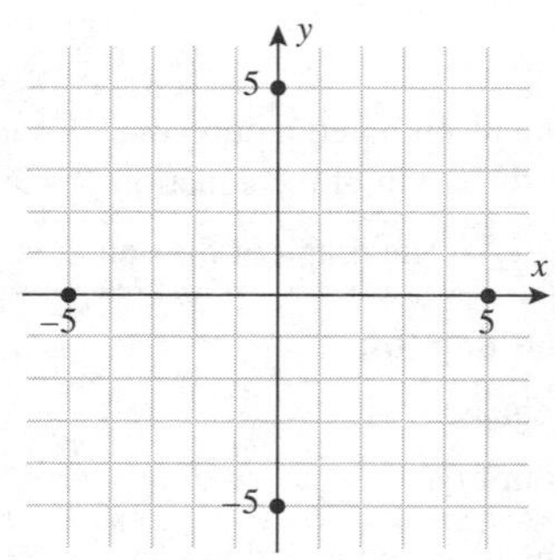

STEP 3 Test for symmetry with respect to the y-axis, x-axis, and origin.

Test for symmetry with respect to the y-axis.

Replace x with $-x$. $\quad (-x)^2 + y^2 = 25$

Simplify. $\quad x^2 + y^2 = 25$

The resulting equation is equivalent to the original, so the graph of $x^2 + y^2 = 25$ is symmetric with respect to the y-axis.

Test for symmetry with respect to the x-axis.

Replace y with $-y$. $\quad x^2 + (-y)^2 = 25$

Simplify. $\quad x^2 + y^2 = 25$

The resulting equation is equivalent to the original, so the graph of $x^2 + y^2 = 25$ is symmetric with respect to the x-axis.

Test for symmetry with respect to the origin.

Replace x with $-x$ and y with $-y$. $\qquad (-x)^2 + (-y)^2 = 25$

Simplify. $\qquad x^2 + y^2 = 25$

The resulting equation is equivalent to the original, so the graph of $x^2 + y^2 = 25$ is symmetric with respect to the origin.

Since the graph is symmetric with respect to the y-axis, x-axis, and origin, we need to determine solutions to the equation on only the positive x- and y-axes and in quadrant I because of the following symmetries:

- Symmetry with respect to the y-axis gives the solutions in quadrant II.
- Symmetry with respect to the origin gives the solutions in quadrant III.
- Symmetry with respect to the x-axis yields solutions in quadrant IV.

Solutions to $x^2 + y^2 = 25$.

Quadrant I: (3, 4), (4, 3)

Additional points due to symmetry:

Quadrant II: (−3, 4), (−4, 3)

Quadrant III: (−3, −4), (−4, −3)

Quadrant IV: (3, −4), (4, −3)

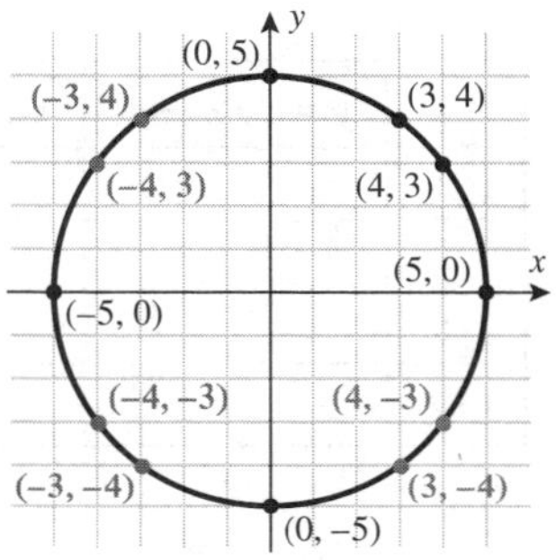

Connecting the points with a smooth curve yields a **circle**. We will discuss circles in more detail in Section 2.4.

Technology Tip

To enter the graph of $x^2 + y^2 = 25$, solve for y first. The graphs of $y_1 = \sqrt{25 - x^2}$ and $y_2 = -\sqrt{25 - x^2}$ are shown.

```
Plot1 Plot2 Plot3
\Y1=√(25-X²)
\Y2=-√(25-X²)
```

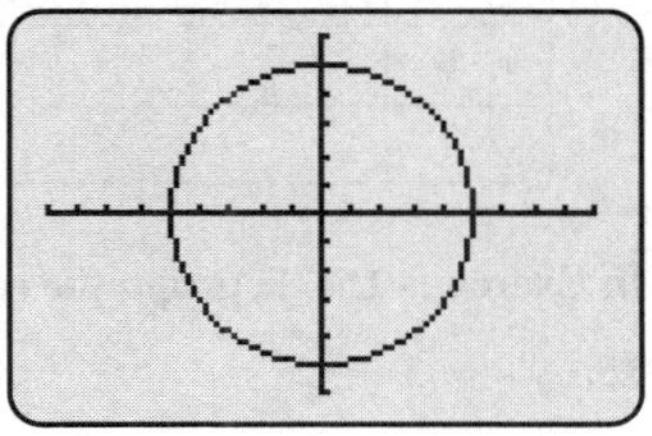

SECTION 2.2 SUMMARY

Sketching graphs of equations can be accomplished using a point-plotting technique. Intercepts are defined as points where a graph intersects an axis or the origin.

Symmetry about the x-axis, y-axis, and origin is defined both algebraically and graphically. Intercepts and symmetry provide much of the information useful for sketching graphs of equations.

Intercepts	Point where the graph intersects the...	How to find intercepts	Point on graph
x-intercept	x-axis	Let $y = 0$ and solve for x.	$(a, 0)$
y-intercept	y-axis	Let $x = 0$ and solve for y.	$(0, b)$

SECTION 2.2 EXERCISES

SKILLS

In Exercises 1–8, determine whether each point lies on the graph of the equation.

1. $y = 3x - 5$ **a.** (1, 2) **b.** (−2, −11)

2. $y = -2x + 7$ **a.** (−1, 9) **b.** (2, −4)

3. $y = \frac{2}{5}x - 4$ **a.** (5, −2) **b.** (−5, 6)

4. $y = -\frac{3}{4}x + 1$ **a.** (8, 5) **b.** (−4, 4)

5. $y = x^2 - 2x + 1$ a. $(-1, 4)$ b. $(0, -1)$

6. $y = x^3 - 1$ a. $(-1, 0)$ b. $(-2, -9)$

7. $y = \sqrt{x + 2}$ a. $(7, 3)$ b. $(-6, 4)$

8. $y = 2 + |3 - x|$ a. $(9, -4)$ b. $(-2, 7)$

In Exercises 9–14, complete the table and use the table to sketch a graph of the equation.

9.

x	$y = 2 + x$	(x, y)
-2		
0		
1		

10.

x	$y = 3x - 1$	(x, y)
-1		
0		
2		

11.

x	$y = x^2 - x$	(x, y)
-1		
0		
$\frac{1}{2}$		
1		
2		

12.

x	$y = 1 - 2x - x^2$	(x, y)
-3		
-2		
-1		
0		
1		

13.

x	$y = \sqrt{x - 1}$	(x, y)
1		
2		
5		
10		

14.

x	$y = -\sqrt{x + 2}$	(x, y)
-2		
-1		
2		
7		

In Exercises 15–22, graph the equation by plotting points.

15. $y = -3x + 2$

16. $y = 4 - x$

17. $y = x^2 - x - 2$

18. $y = x^2 - 2x + 1$

19. $x = y^2 - 1$

20. $x = |y + 1| + 2$

21. $y = \frac{1}{2}x - \frac{3}{2}$

22. $y = 0.5\,|x - 1|$

In Exercises 23–32, find the *x*-intercept(s) and *y*-intercepts(s) (if any) of the graphs of the given equations.

23. $2x - y = 6$

24. $4x + 2y = 10$

25. $y = x^2 - 9$

26. $y = 4x^2 - 1$

27. $y = \sqrt{x - 4}$

28. $y = \sqrt[3]{x - 8}$

29. $y = \dfrac{1}{x^2 + 4}$

30. $y = \dfrac{x^2 - x - 12}{x}$

31. $4x^2 + y^2 = 16$

32. $x^2 - y^2 = 9$

In Exercises 33–38, match the graph with the corresponding symmetry.

a. No symmetry

b. Symmetry with respect to the *x*-axis

c. Symmetry with respect to the *y*-axis

d. Symmetry with respect to the origin

e. Symmetry with respect to the *x*-axis, *y*-axis, and origin

33.

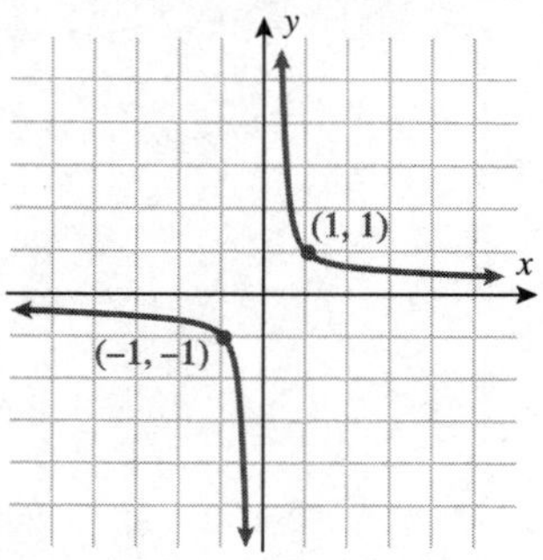

34.

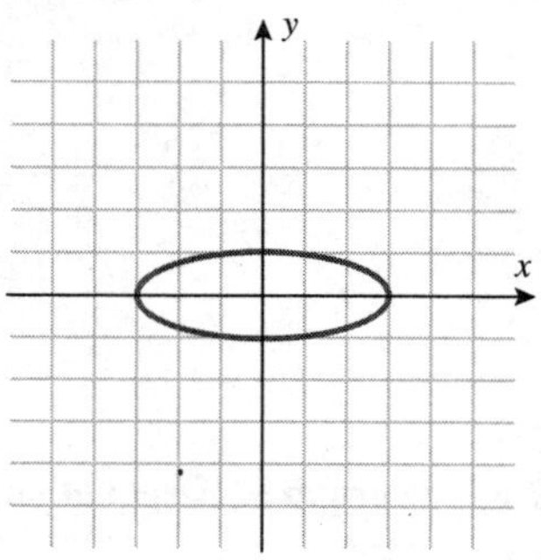

35.

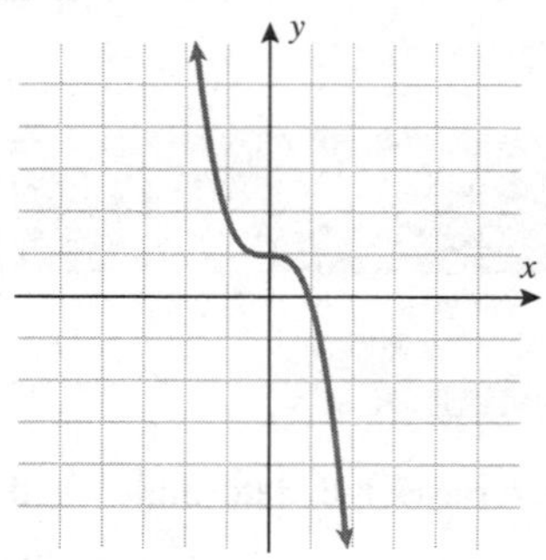

36.

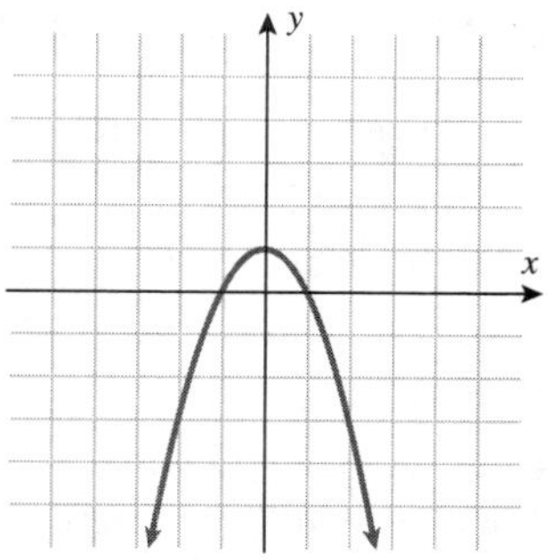

37.

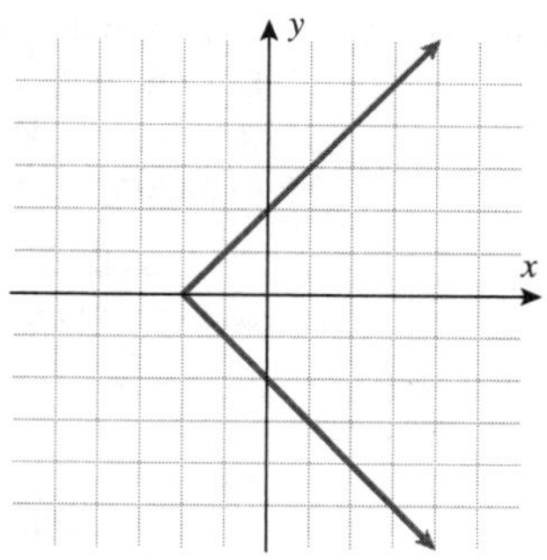

38.

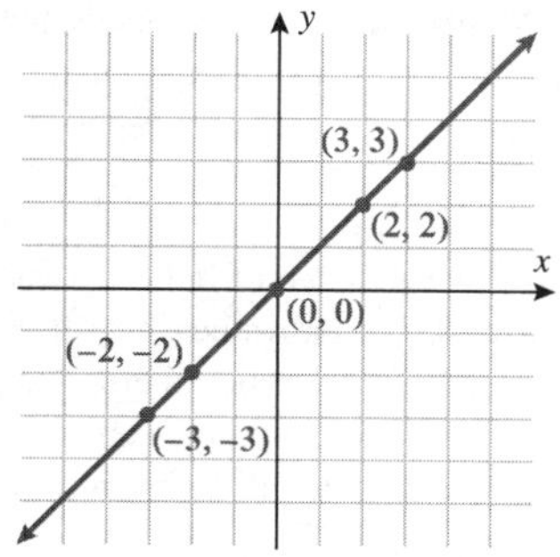

In Exercises 39–44, a point that lies on a graph is given along with that graph's symmetry. State the other known points that must also lie on the graph.

	Point on a Graph	The Graph Is Symmetric About the		Point on a Graph	The Graph Is Symmetric About the
39.	$(-1, 3)$	x-axis	**40.**	$(-2, 4)$	y-axis
41.	$(7, -10)$	origin	**42.**	$(-1, -1)$	origin
43.	$(3, -2)$	x-axis, y-axis, and origin	**44.**	$(-1, 7)$	x-axis, y-axis, and origin

In Exercises 45–58, test algebraically to determine whether the equation's graph is symmetric with respect to the x-axis, y-axis, or origin.

45. $x = y^2 + 4$ **46.** $x = 2y^2 + 3$ **47.** $y = x^3 + x$ **48.** $y = x^5 + 1$ **49.** $x = |y|$

50. $x = |y| - 2$ **51.** $x^2 - y^2 = 100$ **52.** $x^2 + 2y^2 = 30$ **53.** $y = x^{2/3}$ **54.** $x = y^{2/3}$

55. $x^2 + y^3 = 1$ **56.** $y = \sqrt{1 + x^2}$ **57.** $y = \frac{2}{x}$ **58.** $xy = 1$

In Exercises 59–72, plot the graph of the given equation.

59. $y = x$ **60.** $y = -\frac{1}{2}x + 3$ **61.** $y = x^2 - 1$ **62.** $y = 9 - 4x^2$ **63.** $y = \frac{x^3}{2}$

64. $x = y^2 + 1$ **65.** $y = \frac{1}{x}$ **66.** $xy = -1$ **67.** $y = |x|$ **68.** $|x| = |y|$

69. $x^2 + y^2 = 16$ **70.** $\frac{x^2}{4} + \frac{y^2}{9} = 1$ **71.** $x^2 - y^2 = 16$ **72.** $x^2 - \frac{y^2}{25} = 1$

APPLICATIONS

73. Sprinkler. A sprinkler will water a grassy area in the shape of $x^2 + y^2 = 9$. Apply symmetry to draw the watered area, assuming the sprinkler is located at the origin.

74. Sprinkler. A sprinkler will water a grassy area in the shape of $x^2 + \frac{y^2}{9} = 1$. Apply symmetry to draw the watered area, assuming the sprinkler is located at the origin.

75. Electronic Signals: Radio Waves. The received power of an electromagnetic signal is a fraction of the power transmitted. The relationship is given by

$$P_{\text{received}} = P_{\text{transmitted}} \cdot \frac{1}{R^2}$$

where R is the distance that the signal has traveled in meters. Plot the percentage of transmitted power that is received for $R = 100$ m, 1 km, and 10,000 km.

76. Electronic Signals: Laser Beams. The wavelength λ and the frequency f of a signal are related by the equation

$$f = \frac{c}{\lambda}$$

where c is the speed of light in a vacuum, $c = 3.0 \times 10^8$ meters per second. For the values, $\lambda = 0.001$, $\lambda = 1$, and $\lambda = 100$ mm, plot the points corresponding to frequency, f. What do you notice about the relationship between frequency and wavelength? Note that the frequency will have units Hz = 1/seconds.

77. Profit. The profit associated with making a particular product is given by the equation

$$y = -x^2 + 6x - 8$$

where y represents the profit in millions of dollars and x represents the number of thousands of units sold. ($x = 1$ corresponds to 1000 units and $y = 1$ corresponds to \$1M.) Graph this equation and determine how many units must be sold to break even (profit = 0). Determine the range of units sold that correspond to making a profit.

78. Profit. The profit associated with making a particular product is given by the equation

$$y = -x^2 + 4x - 3$$

where y represents the profit in millions of dollars and x represents the number of thousands of units sold. ($x = 1$ corresponds to 1000 units and $y = 1$ corresponds to \$1M.) Graph this equation and determine how many units must be sold to break even (profit = 0). Determine the range of units sold that correspond to making a profit.

CATCH THE MISTAKE

In Exercises 79–82, explain the mistake that is made.

79. Graph the equation $y = x^2 + 1$.

Solution:

x	$y = x^2 + 1$	(x, y)
0	1	(0, 1)
1	2	(1, 2)

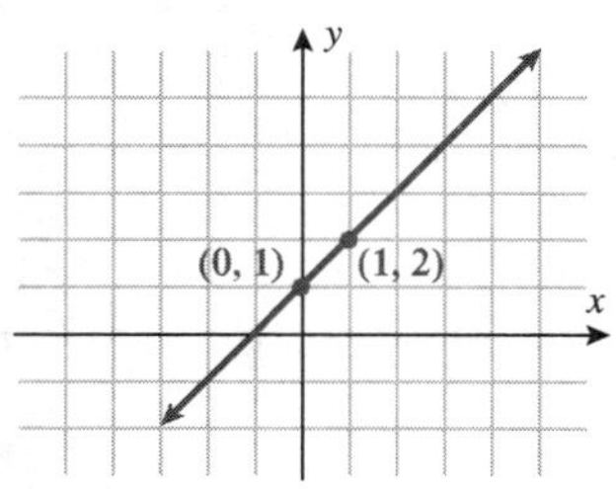

This is incorrect. What mistake was made?

80. Test $y = -x^2$ for symmetry with respect to the y-axis.

Solution:

Replace x with $-x$. $\quad y = -(-x)^2$

Simplify. $\quad y = x^2$

The resulting equation is not equivalent to the original equation; $y = -x^2$ is not symmetric with respect to the y-axis.

This is incorrect. What mistake was made?

81. Test $x = |y|$ for symmetry with respect to the y-axis.

Solution:

Replace y with $-y$. $\quad x = |-y|$

Simplify. $\quad x = |y|$

The resulting equation is equivalent to the original equation; $x = |y|$ is symmetric with respect to the y-axis.

This is incorrect. What mistake was made?

82. Use symmetry to help you graph $x^2 = y - 1$.

Solution:

Replace x with $-x$. $\quad (-x)^2 = y - 1$

Simplify. $\quad x^2 = y - 1$

$x^2 = y - 1$ is symmetric with respect to the x-axis.

Determine points that lie on the graph in quadrant I.

y	$x^2 = y - 1$	(x, y)
1	0	(0, 1)
2	1	(1, 2)
5	2	(2, 5)

Symmetry with respect to the x-axis implies that $(0, -1)$, $(1, -2)$, and $(2, -5)$ are also points that lie on the graph.

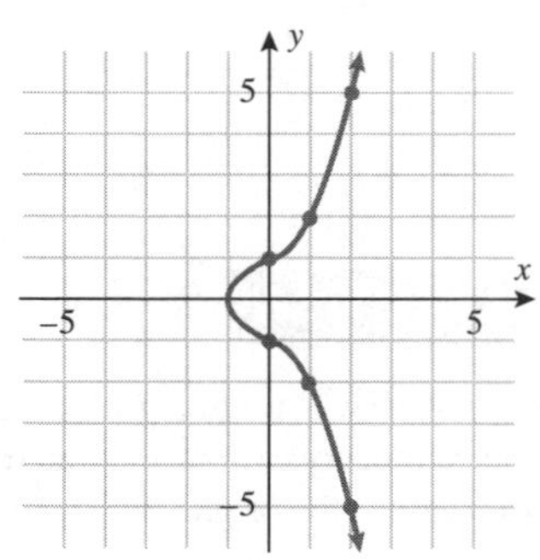

This is incorrect. What mistake was made?

CONCEPTUAL

Determine whether each statement is true or false.

83. If the point (a, b) lies on a graph that is symmetric about the x-axis, then the point $(-a, b)$ also must lie on the graph.

84. If the point (a, b) lies on a graph that is symmetric about the y-axis, then the point $(-a, b)$ also must lie on the graph.

85. If the point $(a, -b)$ lies on a graph that is symmetric about the x-axis, y-axis, and origin, then the points (a, b), $(-a, -b)$, and $(-a, b)$ must also lie on the graph.

86. Two points are all that is needed to plot the graph of an equation.

CHALLENGE

87. Determine whether the graph of $y = \dfrac{ax^2 + b}{cx^3}$ has any symmetry, where a, b, and c are real numbers.

88. Find the intercepts of $y = (x - a)^2 - b^2$, where a and b are real numbers.

TECHNOLOGY

In Exercises 89–94, graph the equation using a graphing utility and state whether there is any symmetry.

89. $y = 16.7x^4 - 3.3x^2 + 7.1$

90. $y = 0.4x^5 + 8.2x^3 - 1.3x$

91. $2.3x^2 = 5.5\,|y|$

92. $3.2x^2 - 5.1y^2 = 1.3$

93. $1.2x^2 + 4.7y^2 = 19.4$

94. $2.1y^2 = 0.8\,|x + 1|$

SECTION 2.3 LINES

SKILLS OBJECTIVES

- Graph a line.
- Determine x- and y-intercepts.
- Calculate the slope of a line.
- Find an equation of a line.
- Graph lines that are parallel or perpendicular to given lines.
- Solve application problems involving linear equations.

CONCEPTUAL OBJECTIVES

- Classify lines as rising, falling, horizontal, and vertical.
- Understand slope as a rate of change.
- Associate two lines having the same slope with the graph of parallel lines.
- Associate two lines having negative reciprocal slopes with the graph of perpendicular lines.

What is the shortest path between two points? The answer is a *straight line.* In this section, we will discuss characteristics of lines such as slope and intercepts. We will also discuss types of lines such as horizontal, vertical, falling, and rising, and recognize relationships between lines such as perpendicular and parallel. At the end of this section, you should be able to find the equation of a line, given two specific pieces of information about the line.

Graphing a Line

First-degree equations, such as

$$y = -2x + 4, \qquad 3x + y = 6, \qquad y = 2, \qquad \text{and} \qquad x = -3,$$

have graphs that are straight lines. The first two equations given represent inclined or "slant" lines, whereas $y = 2$ represents a horizontal line and $x = -3$ represents a vertical line. One way of writing an equation of a straight line is called *general form.*

EQUATION OF A STRAIGHT LINE: GENERAL* FORM

If A, B, and C are constants and x and y are variables, then the equation

$$Ax + By = C$$

is in **general form** and its graph is a straight line.

Note: A or B (but not both) can be zero.

The equation $2x - y = -2$ is a first-degree equation, so its graph is a straight line. To graph this line, list two solutions in a table, plot those points, and use a straight edge to draw the line.

x	y	(x, y)
-2	-2	$(-2, -2)$
1	4	$(1, 4)$

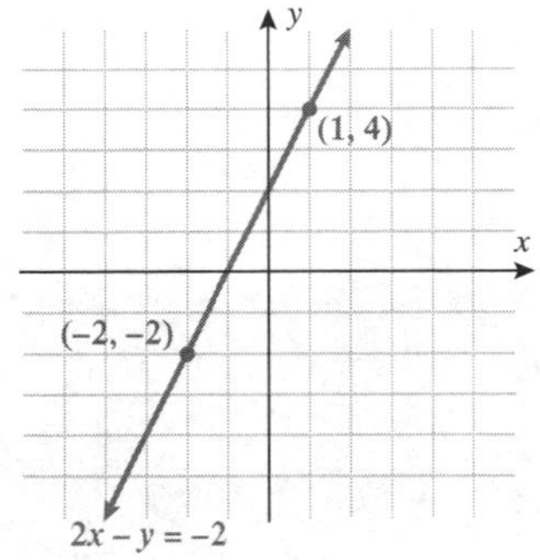

Intercepts

Study Tip

x-intercept: Set $y = 0$.
y-intercept: Set $x = 0$.

The point where a line crosses, or intersects, the x-axis is called the ***x*-intercept**. The point where a line crosses, or intersects, the y-axis is called the ***y*-intercept**. By inspecting the graph of the previous line, we see that the ***x*-intercept** is $(-1, 0)$ and the ***y*-intercept** is $(0, 2)$.

Intercepts were discussed in Section 2.2. There, we found that the graphs of some equations could have non-intercepts or one or multiple intercepts. Slant lines, however, have exactly one x-intercept and exactly one y-intercept.

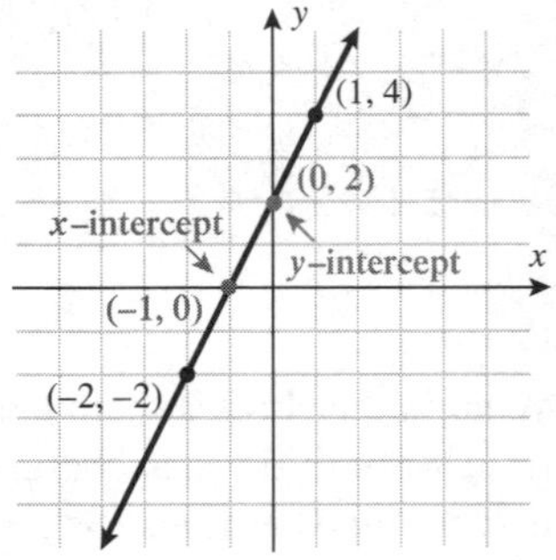

*Some books refer to this as standard form.

DETERMINING INTERCEPTS

	Coordinates	Axis Crossed	Algebraic Method
x-intercept:	$(a, 0)$	x-axis	Set $y = 0$ and solve for x.
y-intercept:	$(0, b)$	y-axis	Set $x = 0$ and solve for y.

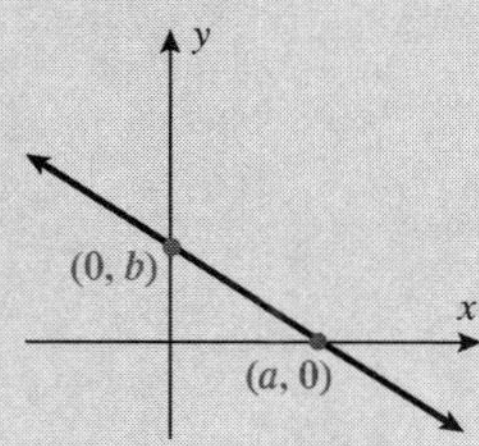

Horizontal lines and vertical lines, however, each have only one intercept.

TYPE OF LINE	EQUATION	x-INTERCEPT	y-INTERCEPT	GRAPH
Horizontal	$y = b$	None	b	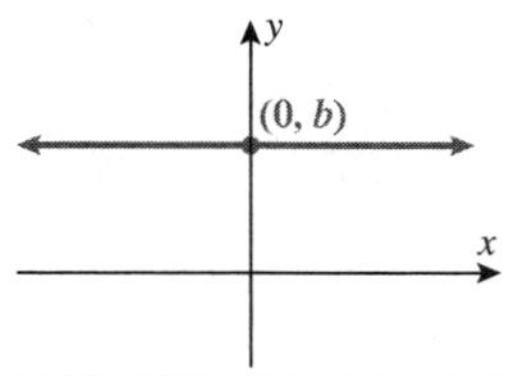
Vertical	$x = a$	a	None	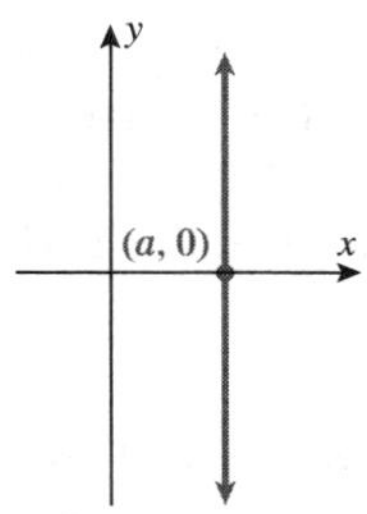

Note: The special cases of $x = 0$ (y-axis) and $y = 0$ (x-axis) have infinitely many y-intercepts and x-intercepts, respectively.

EXAMPLE 1 Determining x- and y-Intercepts

Determine the x- and y-intercepts (if they exist) for the lines given by the following equations.

a. $2x + 4y = 10$ **b.** $x = -2$

Solution (a): $2x + 4y = 10$

To find the x-intercept, set $y = 0$. $2x + 4(0) = 10$

Solve for x. $2x = 10$

$x = 5$

The x-intercept corresponds to the point $\boxed{(5, 0)}$.

To find the y-intercept, set $x = 0$. $2(0) + 4y = 10$

Solve for y. $4y = 10$

$y = \frac{5}{2}$

The y-intercept corresponds to the point $\boxed{\left(0, \frac{5}{2}\right)}$.

Solution (b): $x = -2$

This vertical line consists of all points $(-2, y)$.

The graph shows that the x-intercept is -2.

We also find that the line never crosses the y-axis, so the y-intercept does not exist.

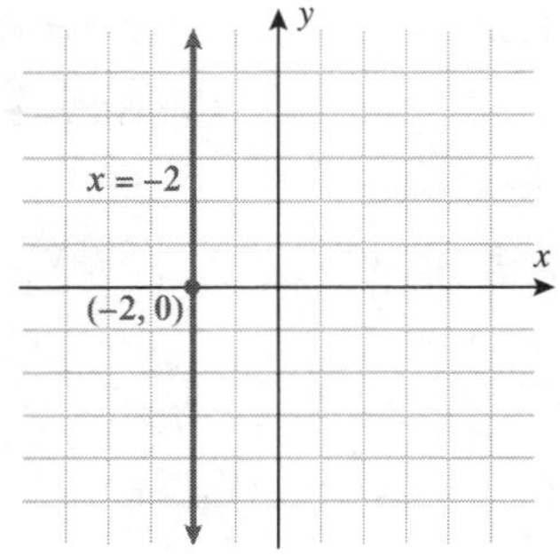

■ **Answer:** **a.** x-intercept: $\left(\frac{2}{3}, 0\right)$; y-intercept: $(0, -2)$
b. x-intercept does not exist; y-intercept: $(0, 5)$

■ **YOUR TURN** Determine the x- and y-intercepts (if they exist) for the lines given by the following equations.

a. $3x - y = 2$ **b.** $y = 5$

Slope

If the graph of $2x - y = -2$ represented an incline that you were about to walk on, would you classify that incline as steep? In the language of mathematics, we use the word **slope** as a measure of steepness. Slope is the ratio of the change in y over the change in x. An easy way to remember this is *rise over run.*

Study Tip

To get the correct sign ($\pm$) for the slope, remember to start with the same point for both x and y.

SLOPE OF A LINE

A nonvertical line passing through two points (x_1, y_1) and (x_2, y_2) has slope, m, given by the formula

$$m = \frac{y_2 - y_1}{x_2 - x_1}, \text{ where } x_1 \neq x_2 \text{ or}$$

$$m = \frac{\text{rise}}{\text{run}} = \frac{\text{vertical change}}{\text{horizontal change}}$$

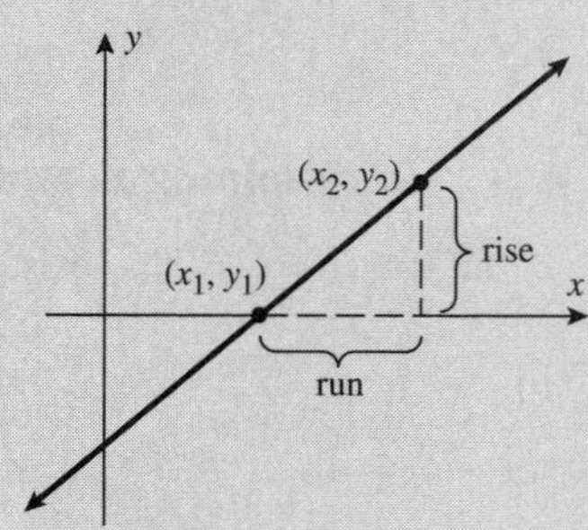

Note: Always start with the same point for both the x-coordinates and the y-coordinates.

Let's find the slope of our graph $2x - y = -2$. We'll let $(x_1, y_1) = (-2, -2)$ and $(x_2, y_2) = (1, 4)$ in the slope formula:

$$m = \frac{y_2 - y_1}{x_2 - x_1} = \frac{[4 - (-2)]}{[1 - (-2)]} = \frac{6}{3} = 2$$

Notice that if we had chosen the two intercepts $(x_1, y_1) = (0, 2)$ and $(x_2, y_2) = (-1, 0)$ instead, we still would have found the slope to be $m = 2$.

COMMON MISTAKE

The most common mistake in calculating slope is writing the coordinates in the wrong order, which results in the slope being opposite in sign.

Find the slope of the line containing the two points (1, 2) and (3, 4).

CORRECT

Label the points.

$(x_1, y_1) = (1, 2) \qquad (x_2, y_2) = (3, 4)$

Write the slope formula.

$$m = \frac{y_2 - y_1}{x_2 - x_1}$$

Substitute the coordinates.

$$m = \frac{4 - 2}{3 - 1}$$

Simplify. $m = \frac{2}{2} = \boxed{1}$

INCORRECT

The **ERROR** is interchanging the coordinates of the first and second points.

$$m = \frac{4 - 2}{1 - 3}$$

The calculated slope is **INCORRECT** by a negative sign.

$$m = \frac{2}{-2} = -1$$

CAUTION

Interchanging the coordinates can result in a sign error in the slope.

When interpreting slope, always read the graph from *left to right.* Since we have determined the slope to be 2, or $\frac{2}{1}$, we can interpret this as rising 2 units and running (to the right) 1 unit. If we start at the point $(-2, -2)$ and move 2 units up and 1 unit to the right, we end up at the x-intercept, $(-1, 0)$. Again, moving 2 units up and 1 unit to the right puts us at the y-intercept, $(0, 2)$. Another rise of 2 and run of 1 takes us to the point $(1, 4)$. See the figure on the left.

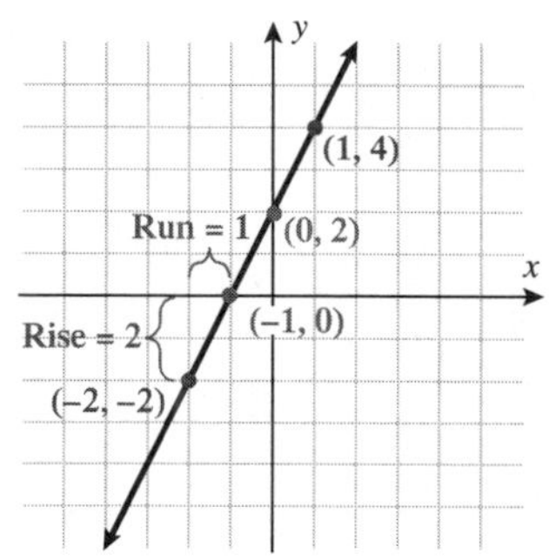

Lines fall into one of 4 categories: rising, falling, horizontal, or vertical.

Line	Slope
Rising	Positive ($m > 0$)
Falling	Negative ($m < 0$)
Horizontal	Zero ($m = 0$), hence $y = b$
Vertical	Undefined, hence $x = a$

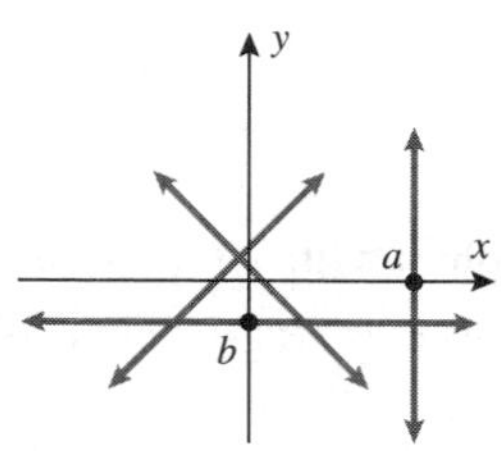

The slope of a horizontal line is 0 because the y-coordinates of any two points are the same. The change in y in the slope formula's numerator is 0, hence $m = 0$. The slope of a vertical line is undefined because the x-coordinates of any two points are the same. The change in x in the slope formula's denominator is zero; hence m is undefined.

EXAMPLE 2 Graph, Classify the Line, and Determine the Slope

Sketch a line through each pair of points, classify the line as rising, falling, vertical, or horizontal, and determine its slope.

a. $(-1, -3)$ and $(1, 1)$ **b.** $(-3, 3)$ and $(3, 1)$
c. $(-1, -2)$ and $(3, -2)$ **d.** $(1, -4)$ and $(1, 3)$

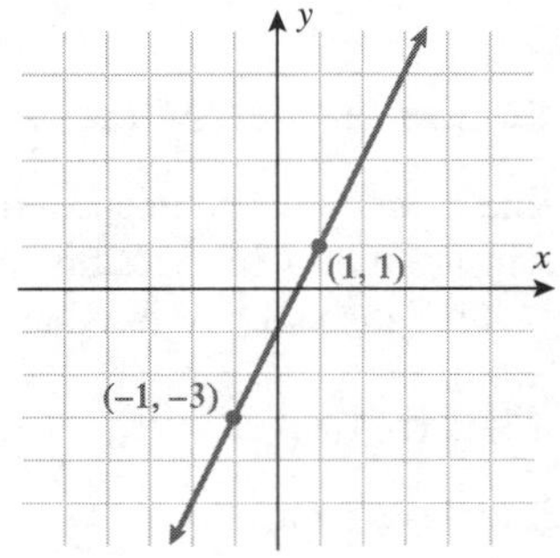

Solution (a): $(-1, -3)$ 1and $(1, 1)$

This line is rising, so its slope is positive.

$$m = \frac{1 - (-3)}{1 - (-1)} = \frac{4}{2} = \frac{2}{1} = 2$$

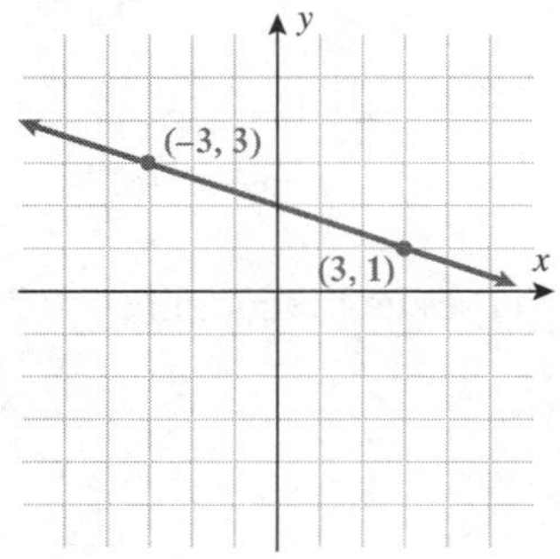

Solution (b): $(-3, 3)$ and $(3, 1)$

This line is falling, so its slope is negative.

$$m = \frac{3 - 1}{-3 - 3} = -\frac{2}{6} = -\frac{1}{3}$$

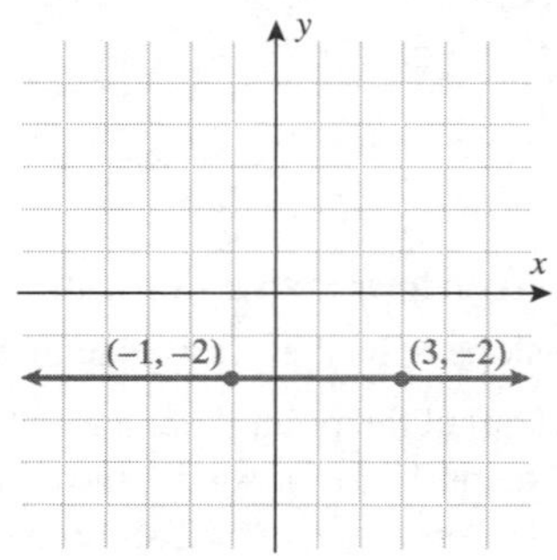

Solution (c): $(-1, -2)$ and $(3, -2)$

This is a horizontal line, so its slope is zero.

$$m = \frac{-2 - (-2)}{3 - (-1)} = \frac{0}{4} = 0$$

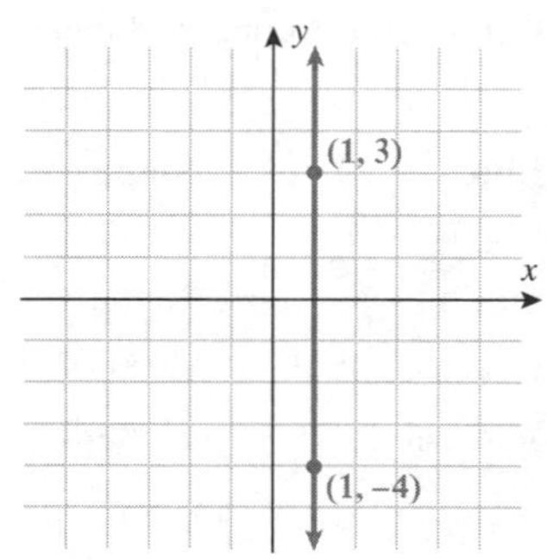

Solution (d): $(1, -4)$ and $(1, 3)$

This is a vertical line, so its slope is undefined.

$$m = \frac{3 - (-4)}{1 - 1} = \frac{7}{0}, \text{ which is undefined.}$$

YOUR TURN For each pair of points classify the line that passes through them as rising, falling, vertical, or horizontal, and determine its slope. Do not graph.

a. $(2, 0)$ and $(1, 5)$ **b.** $(-2, -3)$ and $(2, 5)$
c. $(-3, -1)$ and $(-3, 4)$ **d.** $(-1, 2)$ and $(3, 2)$

Answer:
a. $m = -5$, falling
b. $m = 2$, rising
c. slope is undefined, vertical
d. $m = 0$, horizontal

Slope–Intercept Form

As mentioned earlier, the general form for an equation of a line is $Ax + By = C$. A more standard way to write an equation of a line is in slope–intercept form, because it identifies the slope and the y-intercept.

EQUATION OF A STRAIGHT LINE: SLOPE–INTERCEPT FORM

The **slope–intercept form** for the equation of a nonvertical line is

$$y = mx + b$$

Its graph has slope m and y-intercept b.

For example, $2x - y = -3$ is in general form. To write this equation in **slope–intercept form**, we isolate the y variable:

$$y = 2x + 3$$

The **slope** of this line is **2** and the **y-intercept** is **3**.

EXAMPLE 3 Using Slope–Intercept Form to Graph an Equation of a Line

Write $2x - 3y = 15$ in slope–intercept form and graph it.

Solution:

STEP 1 *Write in slope–intercept form.*

Subtract $2x$ from both sides. $\quad -3y = -2x + 15$

Divide both sides by -3. $\quad \boxed{y = \frac{2}{3}x - 5}$

STEP 2 *Graph.*

Identify the slope and y-intercept. $\quad$ Slope: $m = \frac{2}{3}$ $\quad$ y-intercept: $b = -5$

Plot the point corresponding to the y-intercept $(0, -5)$.

From the point $(0, -5)$, rise 2 units and run (to the right) 3 units, which corresponds to the point $(3, -3)$.

Draw the line passing through the two points.

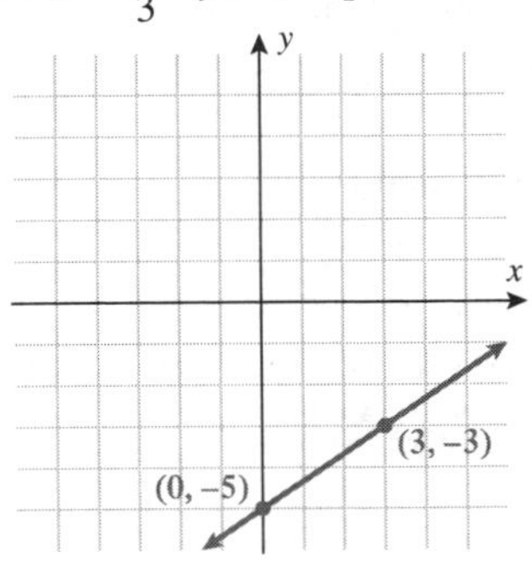

■ **YOUR TURN** Write $3x - 2y = 12$ in slope–intercept form and graph it.

Technology Tip

To graph the equation $2x - 3y = 15$, solve for y first. The graph of $y_1 = \frac{2}{3}x - 5$ is shown.

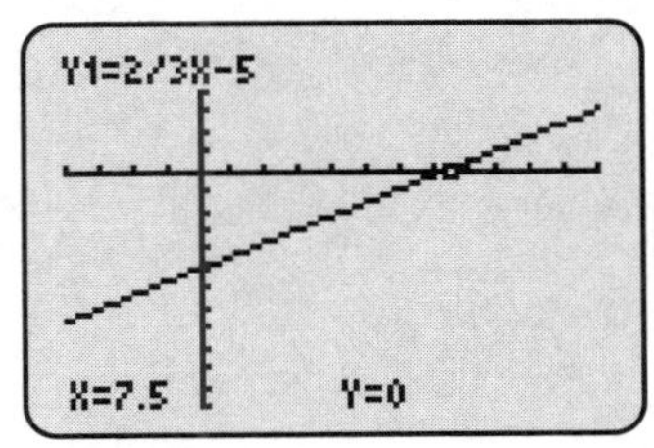

■ **Answer:** $y = \frac{3}{2}x - 6$

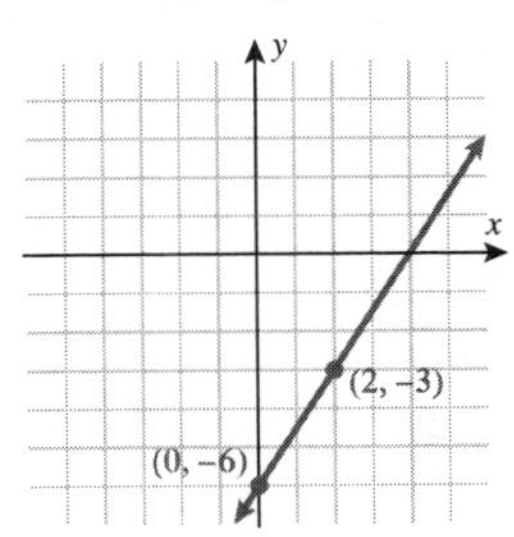

Finding Equations of Lines

Instead of starting with equations of lines and characterizing them, let us now start with particular features of a line and derive its governing equation. Suppose that you are given the y-intercept and the slope of a line. Using the slope–intercept form of an equation of a line, $y = mx + b$, you could find its equation.

EXAMPLE 4 Using Slope–Intercept Form to Find the Equation of a Line

Find the equation of a line that has slope $\frac{2}{3}$ and y-intercept (0, 1).

Solution:

Write the slope–intercept form of an equation of a line. $y = mx + b$

Label the slope. $m = \frac{2}{3}$

Label the y-intercept. $b = 1$

The equation of the line in slope–intercept form is $\boxed{y = \frac{2}{3}x + 1}$.

■ **YOUR TURN** Find the equation of the line that has slope $-\frac{3}{2}$ and y-intercept (0, 2).

■ **Answer:** $y = -\frac{3}{2}x + 2$

Now, suppose that the two pieces of information you are given about an equation are its slope and one point that lies on its graph. You still have enough information to write an equation of the line. Recall the formula for slope:

$$m = \frac{y_2 - y_1}{x_2 - x_1}, \qquad \text{where } x_2 \neq x_1$$

We are given the slope m, and we know a particular point that lies on the line (x_1, y_1). We refer to all other points that lie on the line as (x, y). Substituting these values into the slope formula gives us

$$m = \frac{y - y_1}{x - x_1}$$

Cross multiplying yields

$$y - y_1 = m(x - x_1)$$

This is called the *point–slope form* of an equation of a line.

EQUATION OF A STRAIGHT LINE: POINT–SLOPE FORM

The **point–slope form** for the equation of a line is

$$y - y_1 = m(x - x_1)$$

Its graph passes through the point (x_1, y_1), and its slope is m.

Note: This formula does not hold for vertical lines since their slope is undefined.

EXAMPLE 5 **Using Point–Slope Form to Find the Equation of a Line**

Find the equation of the line that has slope $-\frac{1}{2}$ and passes through the point $(-1, 2)$.

Solution:

Write the point–slope form of an equation of a line. $y - y_1 = m(x - x_1)$

Substitute the values $m = -\frac{1}{2}$ and $(x_1, y_1) = (-1, 2)$. $y - 2 = -\frac{1}{2}(x - (-1))$

Distribute. $y - 2 = -\frac{1}{2}x - \frac{1}{2}$

Isolate y. $y = -\frac{1}{2}x + \frac{3}{2}$

We can also express the equation in general form $\boxed{x + 2y = 3}$.

■ **YOUR TURN** Derive the equation of the line that has slope $\frac{1}{4}$ and passes through the point $\left(1, -\frac{1}{2}\right)$. Give the answer in slope–intercept form.

■ **Answer:** $y = \frac{1}{4}x - \frac{3}{4}$ or $-x + 4y = -3$

Suppose the slope of a line is not given at all. Instead, two points that lie on the line are given. If we know two points that lie on the line, then we can calculate the slope. Then, using the slope and *either* of the two points, the equation of the line can be derived.

EXAMPLE 6 **Finding the Equation of a Line Given Two Points**

Find the equation of the line that passes through the points $(-2, -1)$ and $(3, 2)$.

Solution:

Write the equation of a line. $y = mx + b$

Calculate the slope. $m = \frac{y_2 - y_1}{x_2 - x_1}$

Substitute $(x_1, y_1) = (-2, -1)$ and $(x_2, y_2) = (3, 2)$. $m = \frac{2 - (-1)}{3 - (-2)} = \frac{3}{5}$

Substitute $\frac{3}{5}$ for the slope. $y = \frac{3}{5}x + b$

Let $(x, y) = (3, 2)$. (Either point satisfies the equation.) $2 = \frac{3}{5}(3) + b$

Solve for b. $b = \frac{1}{5}$

Write the equation in slope–intercept form. $y = \frac{3}{5}x + \frac{1}{5}$

Write the equation in general form. $\boxed{-3x + 5y = 1}$

■ **YOUR TURN** Find the equation of the line that passes through the points $(-1, 3)$ and $(2, -4)$.

Study Tip

When two points that lie on a line are given, first calculate the slope of the line, then use either point and the slope–intercept form (shown in Example 6) or the point–slope form:

$$m = \tfrac{3}{5}, (3, 2)$$
$$y - y_1 = m(x - x_1)$$
$$y - 2 = \tfrac{3}{5}(x - 3)$$
$$5y - 10 = 3(x - 3)$$
$$5y - 10 = 3x - 9$$
$$\boxed{-3x + 5y = 1}$$

■ **Answer:** $y = -\frac{7}{3}x + \frac{2}{3}$ or $7x + 3y = 2$

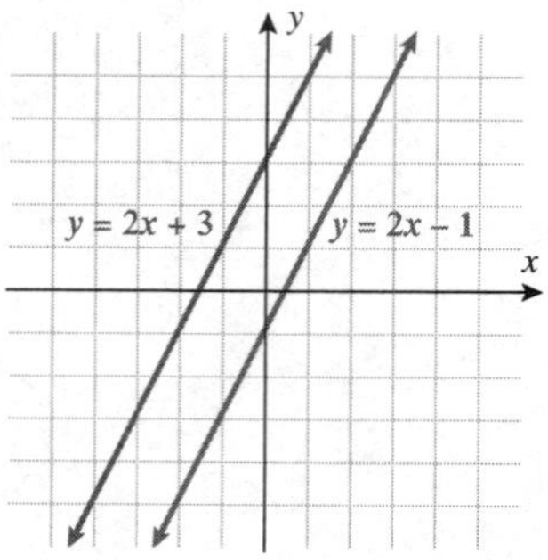

Parallel and Perpendicular Lines

Two distinct nonintersecting lines in a plane are *parallel.* How can we tell whether the two lines in the graph on the left are parallel? Parallel lines must have the same steepness. In other words, parallel lines must have the same slope. The two lines shown on the left are parallel because they have the same slope, 2.

DEFINITION **Parallel Lines**

Two distinct lines in a plane are **parallel** if and only if their slopes are equal.

In other words, if two lines in a plane are parallel, then their slopes are equal, and if the slopes of two lines in a plane are equal, then the lines are parallel.

WORDS	**MATH**
Lines L_1 and L_2 are parallel.	$L_1 \| L_2$
Two parallel lines have the same slope.	$m_1 = m_2$

EXAMPLE 7 Determining Whether Two Lines Are Parallel

Determine whether the lines $-x + 3y = -3$ and $y = \frac{1}{3}x - 6$ are parallel.

Solution:

Write the first line in slope–intercept form. $\quad -x + 3y = -3$

Add x to both sides. $\quad 3y = x - 3$

Divide by 3. $\quad y = \frac{1}{3}x - 1$

Compare the two lines. $\quad y = \frac{1}{3}x - 1$ and $y = \frac{1}{3}x - 6$

Both lines have the same slope, $\frac{1}{3}$. These are distinct lines because they have different y-intercepts. Thus, the **two lines are parallel**.

EXAMPLE 8 Finding an Equation of a Parallel Line

Find the equation of the line that passes through the point (1, 1) and is parallel to the line $y = 3x + 1$.

Solution:

Write the slope–intercept equation of a line. $\quad y = mx + b$

Parallel lines have equal slope. $\quad m = 3$

Substitute the slope into the equation of the line. $\quad y = 3x + b$

Since the line passes through (1, 1), this point must satisfy the equation. $\quad 1 = 3(1) + b$

Solve for b. $\quad b = -2$

The equation of the line is $\boxed{y = 3x - 2}$.

■ **YOUR TURN** Find the equation of the line parallel to $y = 2x - 1$ that passes through the point $(-1, 3)$.

■ **Answer:** $y = 2x + 5$

Two *perpendicular* lines form a right angle at their point of intersection. Notice the slopes of the two perpendicular lines in the figure to the right. They are $-\frac{1}{2}$ and 2, negative reciprocals of each other. It turns out that almost all perpendicular lines share this property. Horizontal ($m = 0$) and vertical (m undefined) lines do not share this property.

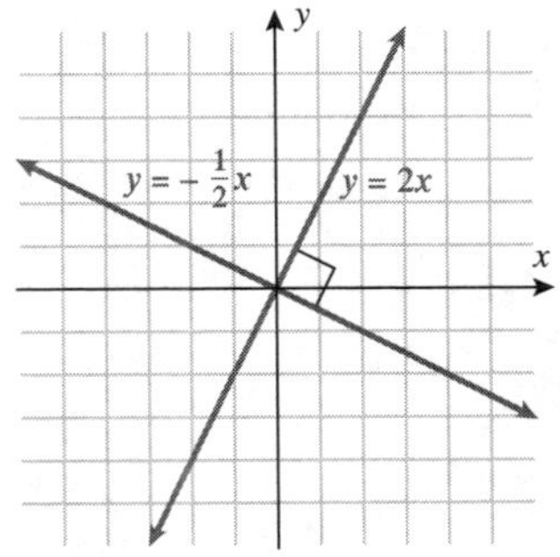

DEFINITION **Perpendicular Lines**

Except for the special case of a vertical and a horizontal line, two lines in a plane are **perpendicular** if and only if their slopes are negative reciprocals of each other.

In other words, if two lines in a plane are perpendicular, their slopes are negative reciprocals, provided their slopes are defined. Similarly, if the slopes of two lines in a plane are negative reciprocals, then the lines are perpendicular.

WORDS	MATH
Lines L_1 and L_2 are perpendicular.	$L_1 \perp L_2$
Two perpendicular lines have negative reciprocal slopes.	$m_1 = -\frac{1}{m_2} \quad m_1 \neq 0, m_2 \neq 0$

Study Tip

If a line has slope equal to 3, then a line perpendicular to it has slope $-\frac{1}{3}$.

EXAMPLE 9 **Finding an Equation of a Line That Is Perpendicular to Another Line**

Find the equation of the line that passes through the point (3, 0) and is perpendicular to the line $y = 3x + 1$.

Solution:

Identify the slope of the given line $y = 3x + 1$. $\quad m_1 = 3$

The slope of a line perpendicular to the given line is the negative reciprocal of the slope of the given line. $\quad m_2 = -\frac{1}{m_1} = -\frac{1}{3}$

Write the equation of the line we are looking for in slope–intercept form. $\quad y = m_2 x + b$

Substitute $m_2 = -\frac{1}{3}$ into $y = m_2 x + b$. $\quad y = -\frac{1}{3}x + b$

Since the desired line passes through (3, 0), this point must satisfy the equation.

$$0 = -\frac{1}{3}(3) + b$$

$$0 = -1 + b$$

Solve for b. $\quad b = 1$

The equation of the line is $\boxed{y = -\frac{1}{3}x + 1}$.

■ **YOUR TURN** Find the equation of the line that passes through the point $(1, -5)$ and is perpendicular to the line $y = -\frac{1}{2}x + 4$.

■ **Answer:** $y = 2x - 7$

Applications Involving Linear Equations

Slope is the ratio of the change in y over the change in x. In applications, slope can often be interpreted as the **rate of change**, as illustrated in the next example.

EXAMPLE 10 Slope as a Rate of Change

The average age that a person first marries has been increasing over the last several decades. In 1970 the average age was 20 and in 1990 it was 25 years old, and in 2010 it is expected that the average age will be 30 years old at the time of a person's first marriage. Find the slope of the line passing through these points. Describe what that slope represents.

Solution:

If we let x represent the year and y represent the age, then two points* that lie on the line are (1970, 20) and (2010, 30).

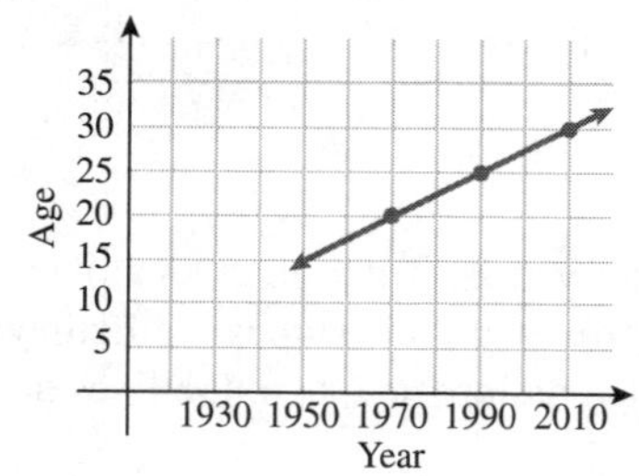

Write the slope formula. $m = \dfrac{\text{change in } y}{\text{change in } x}$

Substitute the points into the slope formula. $m = \dfrac{30 - 20}{2010 - 1970} = \dfrac{10}{40} = \dfrac{1}{4}$

The slope is $\frac{1}{4}$ and can be interpreted as the rate of change of the average age when a person is first married. Every 4 years the average age at the first marriage is 1 year older.

EXAMPLE 11 Service Charges

Suppose that your two neighbors both use the same electrician. One neighbor had a 2-hour job, which cost her \$100, and another neighbor had a 3-hour job that cost him \$130. Assuming that a linear equation governs the service charge of this electrician, what will your cost be for a 5-hour job?

Solution:

STEP 1 Identify the question.
Determine the linear equation for this electrician's service charge and calculate the charge for a 5-hour job.

STEP 2 Make notes.
A 2-hour job costs \$100 and a 3-hour job costs \$130.

STEP 3 Set up an equation.
Let x equal the number of hours and y equal the service charge in dollars.

Linear equation: $y = mx + b$

Two points that must satisfy this equation are (2, 100) and (3, 130).

STEP 4 Solve the equation.

Calculate the rate of change (slope). $m = \dfrac{130 - 100}{3 - 2} = \dfrac{30}{1} = 30$

Substitute the slope into the linear equation. $y = 30x + b$

*In Example 10 we chose to use the points (1970, 20) and (2010, 30). We could have also used the point (1990, 25) with either of the other points.

In Exercises 61–80, write the equation of the line that passes through the given points. Express the equation in slope–intercept form or in the form $x = a$ or $y = b$.

61. $(-2, -1)$ and $(3, 2)$
62. $(-4, -3)$ and $(5, 1)$
63. $(-3, -1)$ and $(-2, -6)$
64. $(-5, -8)$ and $(7, -2)$
65. $(20, -37)$ and $(-10, -42)$
66. $(-8, 12)$ and $(-20, -12)$
67. $(-1, 4)$ and $(2, -5)$
68. $(-2, 3)$ and $(2, -3)$
69. $\left(\frac{1}{2}, \frac{3}{4}\right)$ and $\left(\frac{3}{2}, \frac{9}{4}\right)$
70. $\left(-\frac{2}{3}, -\frac{1}{2}\right)$ and $\left(\frac{7}{3}, \frac{1}{2}\right)$
71. $(3, 5)$ and $(3, -7)$
72. $(-5, -2)$ and $(-5, 4)$
73. $(3, 7)$ and $(9, 7)$
74. $(-2, -1)$ and $(3, -1)$
75. $(0, 6)$ and $(-5, 0)$
76. $(0, -3)$ and $(0, 2)$
77. $(-6, 8)$ and $(-6, -2)$
78. $(-9, 0)$ and $(-9, 2)$
79. $\left(\frac{2}{5}, -\frac{3}{4}\right)$ and $\left(\frac{2}{5}, \frac{1}{2}\right)$
80. $\left(\frac{1}{3}, \frac{2}{5}\right)$ and $\left(\frac{1}{3}, \frac{1}{2}\right)$

In Exercises 81–86, write the equation corresponding to each line. Express the equation in slope–intercept form.

81.

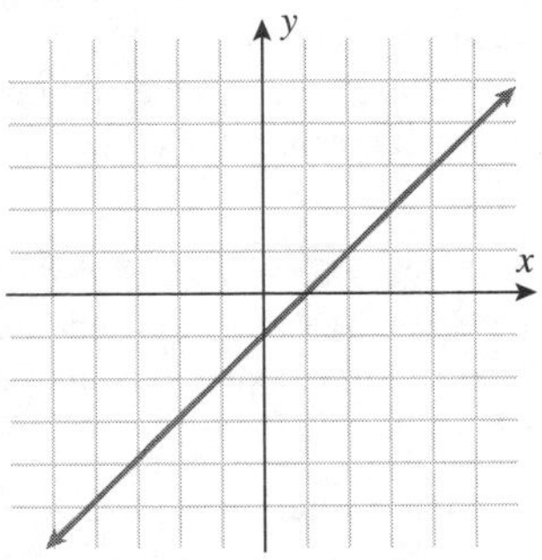

82.

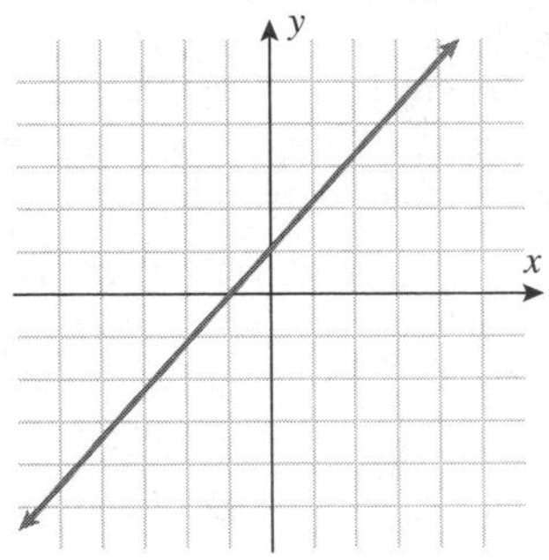

83.

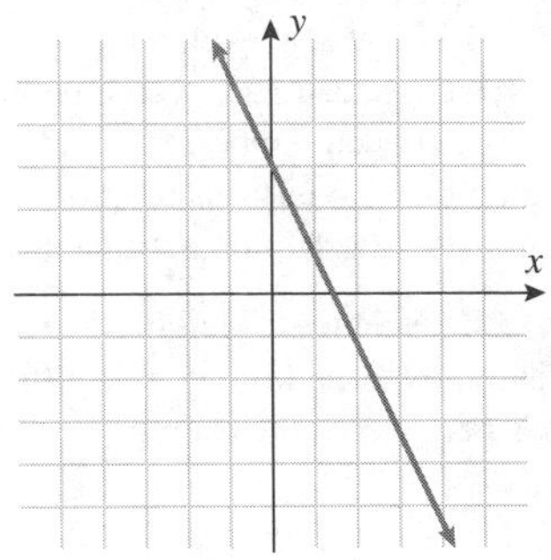

84.

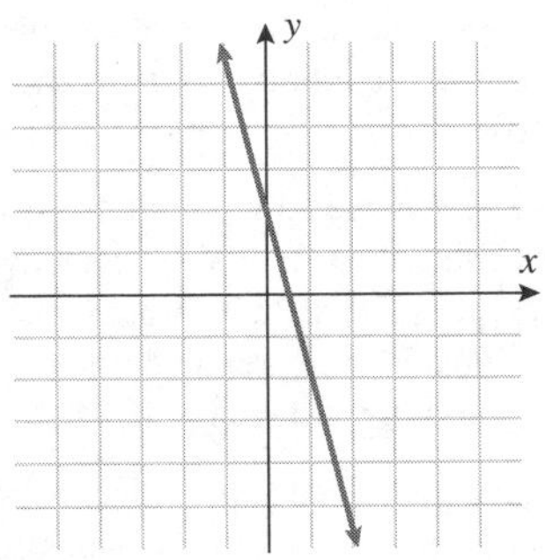

85.

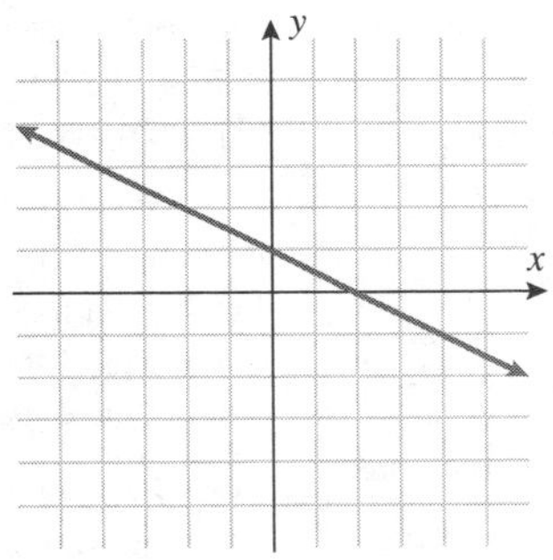

86.

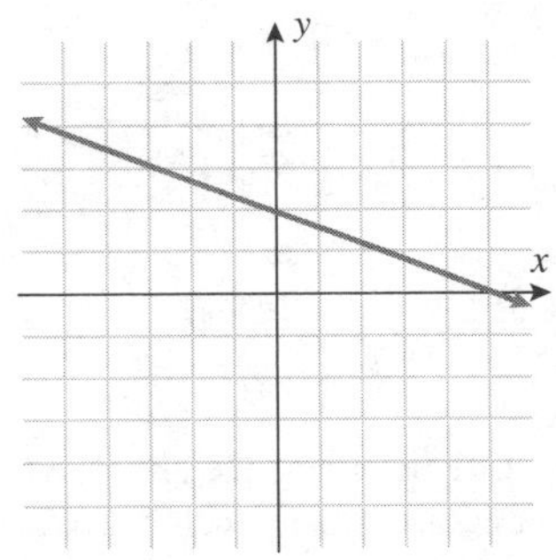

In Exercises 87–96, find the equation of the line that passes through the given point and also satisfies the additional piece of information. Express your answer in slope–intercept form, if possible.

87. $(-3, 1)$; parallel to the line $y = 2x - 1$
88. $(1, 3)$; parallel to the line $y = -x + 2$
89. $(0, 0)$; perpendicular to the line $2x + 3y = 12$
90. $(0, 6)$; perpendicular to the line $x - y = 7$
91. $(3, 5)$; parallel to the x-axis
92. $(3, 5)$; parallel to the y-axis
93. $(-1, 2)$; perpendicular to the y-axis
94. $(-1, 2)$; perpendicular to the x-axis
95. $(-2, -7)$; parallel to the line $\frac{1}{2}x - \frac{1}{3}y = 5$
96. $(1, 4)$; perpendicular to the line $-\frac{2}{3}x + \frac{3}{2}y = -2$

APPLICATIONS

97. Budget: Home Improvement. The cost of having your bathroom remodeled is the combination of material costs and labor costs. The materials (tile, grout, toilet, fixtures, etc.) cost is \$1,200 and the labor cost is \$25 per hour. Write an equation that models the total cost C of having your bathroom remodeled as a function of hours h. How much will the job cost if the worker estimates 32 hours?

98. Budget: Rental Car. The cost of a one-day car rental is the sum of the rental fee, \$50, plus \$.39 per mile. Write an equation that models the total cost associated with the car rental.

99. Monthly Driving Costs. The monthly costs associated with driving a new Honda Accord are the monthly loan payment plus \$25 every time you fill up with gasoline. If you fill up 5 times in a month, your total monthly cost is \$500. How much is your loan payment?

100. Budget: Monthly Driving Costs. The monthly costs associated with driving a Ford Explorer are the monthly loan payment plus the cost of filling up your tank with gasoline. If you fill up 3 times in a month, your total monthly cost is \$520. If you fill up 5 times in a month, your total monthly cost is \$600. How much is your monthly loan, and how much does it cost every time you fill up with gasoline?

101. Weather: Temperature. The National Oceanic and Atmospheric Administration (NOAA) has an online conversion chart that relates degrees Fahrenheit, °F, to degrees Celsius, °C. 77 °F is equivalent to 25 °C, and 68 °F is equivalent to 20 °C. Assuming the relationship is linear, write the equation relating degrees Celsius C to degrees Fahrenheit F. What temperature is the same in both degrees Celsius and degrees Fahrenheit?

102. Weather: Temperature. According to NOAA, a "standard day" is 15 °C at sea level, and every 500 feet elevation above sea level corresponds to a 1 °C temperature drop. Assuming the relationship between temperature and elevation is linear, write an equation that models this relationship. What is the expected temperature at 2500 feet on a "standard day"?

103. Life Sciences: Height. The average height of a man has increased over the last century. What is the rate of change in inches per year of the average height of men?

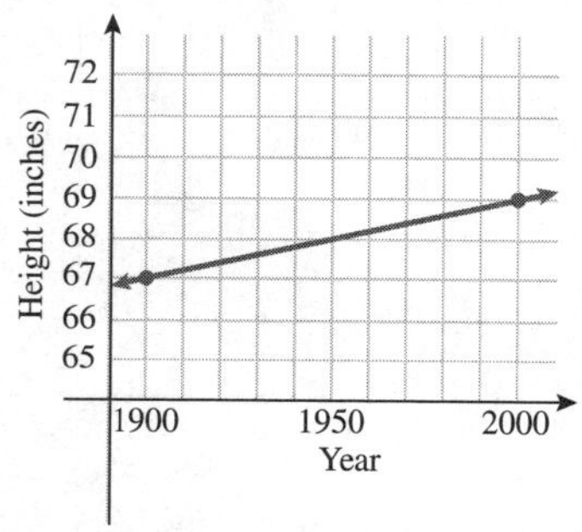

104. Life Sciences: Height. The average height of a woman has increased over the last century. What is the rate of change in inches per year of the average height of women?

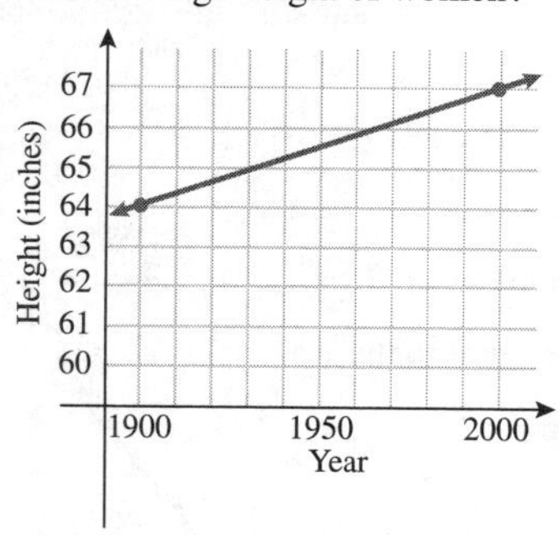

105. Life Sciences: Weight. The average weight of a baby born in 1900 was 6 pounds 4 ounces. In 2000, the average weight of a newborn was 6 pounds 10 ounces. What is the rate of change of birth weight in ounces per year? What do we expect babies to weigh at birth in 2040?

106. Sports. The fastest a man could run a mile in 1906 was 4 minutes and 30 seconds. In 1957, Don Bowden became the first American to break the 4-minute mile. Calculate the rate of change in mile speed per year.

107. Monthly Phone Costs. Mike's home phone plan charges a flat monthly fee plus a charge of \$0.05 per minute for long-distance calls. The total monthly charge is represented by $y = 0.05x + 35$, $x \geq 0$, where y is the total monthly charge and x is the number of long-distance minutes used. Interpret the meaning of the y-intercept.

108. Cost: Automobile. The value of a Daewoo car is given by $y = 11{,}100 - 1850x$, $x \geq 0$, where y is the value of the car and x is the age of the car in years. Find the x-intercept and y-intercept and interpret the meaning of each.

109. Weather: Rainfall. The average rainfall in Norfolk, Virginia, for July was 5.2 inches in 2003. The average July rainfall for Norfolk was 3.8 inches in 2007. What is the rate of change of rainfall in inches per year? If this trend continues, what is the expected average rainfall in 2010?

110. Weather: Temperature. The average temperature for Boston in January 2005 was 43 degrees Fahrenheit. In 2007 the average January temperature was 44.5 degrees Fahrenheit. What is the rate of change of the temperature per year? If this trend continues, what is the expected average temperature in January 2010?

111. Environment. In 2000, Americans used approximately 380 billion plastic bags. In 2005, approximately 392 billion were used. What is the rate of change of plastic bags used per year? How many plastic bags will be expected to be used in 2010?

112. Finance: Debt. According to the Federal Reserve, Americans individually owed \$744 in revolving credit in 2004. In 2006, they owed approximately \$788. What is the rate of change of the amount of revolving credit owed per year? How much should Americans be expected to owe in 2008?

CATCH THE MISTAKE

In Exercises 113–116, explain the mistake that is made.

113. Find the x- and y-intercepts of the line with equation $2x - 3y = 6$.

Solution:

x-intercept: set $x = 0$ and solve for y. $\quad -3y = 6$

$y = -2$

The x-intercept is $(0, -2)$.

y-intercept: set $y = 0$ and solve for x. $\quad 2x = 6$

$x = 3$

The y-intercept is $(3, 0)$.

This is incorrect. What mistake was made?

114. Find the slope of the line that passes through the points $(-2, 3)$ and $(4, 1)$.

Solution:

Write the slope formula. $m = \dfrac{y_2 - y_1}{x_2 - x_1}$

Substitute $(-2, 3)$ and $(4, 1)$. $m = \dfrac{1 - 3}{-2 - 4} = \dfrac{-2}{-6} = \dfrac{1}{3}$

This is incorrect. What mistake was made?

115. Find the slope of the line that passes through the points $(-3, 4)$ and $(-3, 7)$.

Solution:

Write the slope formula. $m = \dfrac{y_2 - y_1}{x_2 - x_1}$

Substitute $(-3, 4)$ and $(-3, 7)$. $m = \dfrac{-3 - (-3)}{4 - 7} = 0$

This is incorrect. What mistake was made?

116. Given the slope, classify the line as rising, falling, horizontal, or vertical.

a. $m = 0$

b. m undefined

c. $m = 2$

d. $m = -1$

Solution:

a. vertical line

b. horizontal line

c. rising

d. falling

These are incorrect. What mistakes were made?

CONCEPTUAL

In Exercises 117–122, determine whether each statement is true or false.

117. A line can have at most one x-intercept.

118. A line must have at least one y-intercept.

119. If the slopes of two lines are $-\frac{1}{5}$ and 5, then the lines are parallel.

120. If the slopes of two lines are -1 and 1, then the lines are perpendicular.

121. If a line has slope equal to zero, describe a line that is perpendicular to it.

122. If a line has no slope, describe a line that is parallel to it.

CHALLENGE

123. Find an equation of a line that passes through the point $(-B, A + 1)$ and is parallel to the line $Ax + By = C$.

124. Find an equation of a line that passes through the point $(B, A - 1)$ and is parallel to the line $Ax + By = C$. Assume that B is not equal to zero.

125. Find an equation of a line that passes through the point $(-A, B - 1)$ and is perpendicular to the line $Ax + By = C$. Assume that A and B are both nonzero.

126. Find an equation of a line that passes through the point $(A, B + 1)$ and is perpendicular to the line $Ax + By = C$.

127. Show that two lines with equal slopes and different y-intercepts have no point in common. *Hint:* Let $y_1 = mx + b_1$ and $y_2 = mx + b_2$ with $b_1 \neq b_2$. What equation must be true for there to be a point of intersection? Show that this leads to a contradiction.

128. Let $y_1 = m_1x + b_1$ and $y_2 = m_2x + b_2$ be two nonparallel lines $(m_1 \neq m_2)$. What is the x-coordinate of the point where they intersect?

TECHNOLOGY

For Exercises 129–134, determine whether the lines are parallel, perpendicular, or neither, and then graph both lines in the same viewing screen using a graphing utility to confirm your answer.

129. $y_1 = 17x + 22$
$y_2 = -\frac{1}{17}x - 13$

130. $y_1 = 0.35x + 2.7$
$y_2 = 0.35x - 1.2$

131. $y_1 = 0.25x + 3.3$
$y_2 = -4x + 2$

132. $y_1 = \frac{1}{2}x + 5$
$y_2 = 2x - 3$

133. $y_1 = 0.16x + 2.7$
$y_2 = 6.25x - 1.4$

134. $y_1 = -3.75x + 8.2$
$y_2 = \frac{4}{15}x + \frac{5}{6}$

SECTION 2.4 CIRCLES

SKILLS OBJECTIVES

- Identify the center and radius of a circle from the standard equation.
- Graph a circle.
- Transform equations of circles to the standard form by completing the square.

CONCEPTUAL OBJECTIVE

- Understand algebraic and graphical representations of circles.

Standard Equation of a Circle

Most people understand the shape of a circle. The goal in this section is to develop the equation of a circle.

DEFINITION **Circle**

A **circle** is the set of all points in a plane that are a fixed distance from a point, the **center**. The center, *C*, is typically denoted by (h, k), and the fixed distance, or **radius**, is denoted by r.

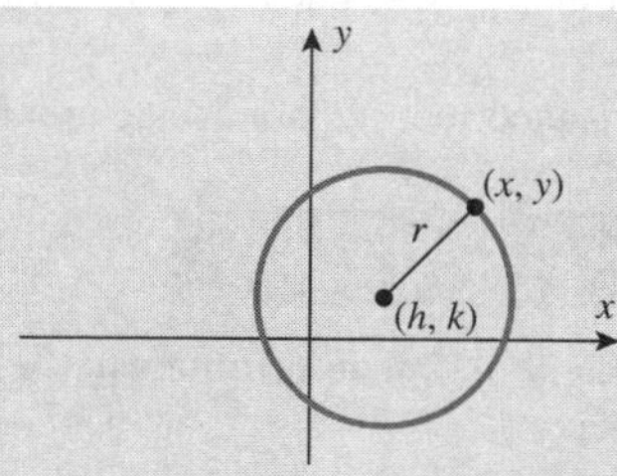

What is the equation of a circle? We'll use the distance formula from Section 2.1.

$$\text{Distance formula: } d = \sqrt{(x_2 - x_1)^2 + (y_2 - y_1)^2}$$

The distance between the center (h, k) and any point (x, y) on the circle is the radius r. Substitute these values $d = r$, $(x_1, y_1) = (h, k)$, and $(x_2, y_2) = (x, y)$ into the distance formula $r = \sqrt{(x - h)^2 + (y - k)^2}$ and square both sides: $(x - h)^2 + (y - k)^2 = r^2$. All circles can be written in *standard form*, which makes it easy to identify the center and radius.

EQUATION OF A CIRCLE

The standard form of the equation of a **circle** with **radius** r and **center** (h, k) is

$$(x - h)^2 + (y - k)^2 = r^2$$

For the special case of a circle with center at the origin $(0, 0)$, the equation simplifies to $x^2 + y^2 = r^2$.

UNIT CIRCLE

A circle with radius 1 and center $(0, 0)$ is called the **unit circle**:

$$x^2 + y^2 = 1$$

The unit circle plays an important role in the study of trigonometry. Note that if $x^2 + y^2 = 0$, the radius is 0, so the "circle" is just a point.

EXAMPLE 1 Finding the Center and Radius of a Circle

Identify the center and radius of the given circle and graph.

$$(x - 2)^2 + (y + 1)^2 = 4$$

Solution:

Rewrite this equation in standard form. $[x - 2]^2 + [y - (-1)]^2 = 2^2$

Identify h, k, and r by comparing this equation with the standard form of a circle: $(x - h)^2 + (y - k)^2 = r^2$. $h = 2, k = -1$, and $r = 2$

Center $(2, -1)$ and $r = 2$

To draw the circle, label the center $(2, -1)$. Label four additional points 2 units (the radius) away from the center: $(4, -1)$, $(0, -1)$, $(2, 1)$, and $(2, -3)$.

Note that the easiest four points to get are those obtained by going out from the center both horizontally and vertically. Connect those four points with a smooth curve.

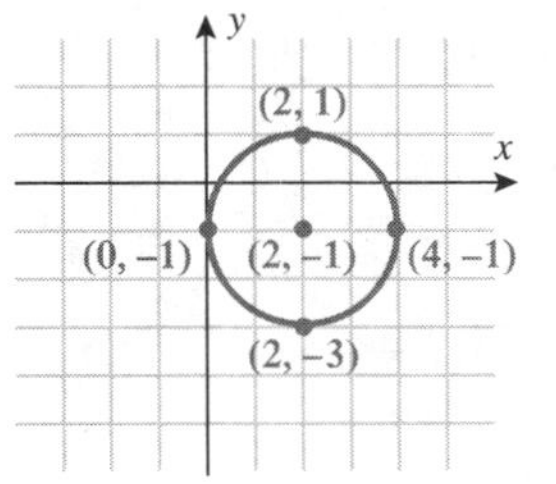

■ **YOUR TURN** Identify the center and radius of the given circle and graph.

$$(x + 1)^2 + (y + 2)^2 = 9$$

Technology Tip

To enter the graph of $(x - 2)^2 + (y + 1)^2 = 4$, solve for y first. The graphs of $y_1 = \sqrt{4 - (x - 2)^2} - 1$ and $y_2 = -\sqrt{4 - (x - 2)^2} - 1$ are shown.

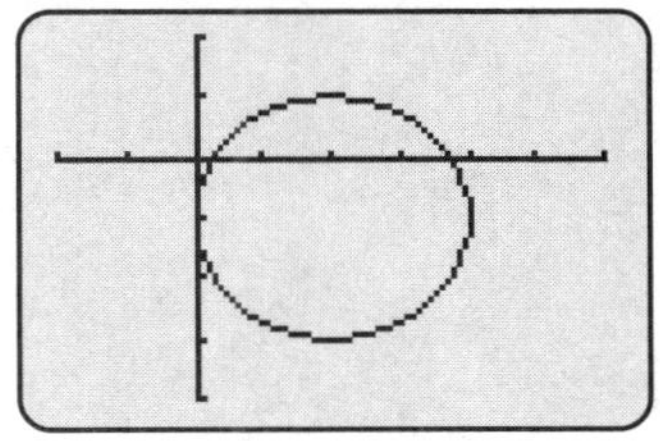

■ **Answer:** Center: $(-1, -2)$
Radius: 3

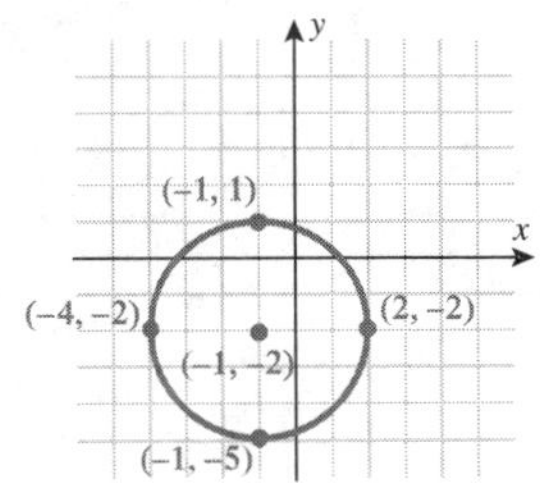

EXAMPLE 2 Graphing a Circle: Fractions and Radicals

Identify the center and radius of the given circle and sketch its graph.

$$\left(x - \frac{1}{2}\right)^2 + \left(y + \frac{1}{3}\right)^2 = 20$$

Solution:

If $r^2 = 20$, then $r = \sqrt{20} = 2\sqrt{5}$. Write the equation in standard form. $\left(x - \frac{1}{2}\right)^2 + \left[y - \left(-\frac{1}{3}\right)\right]^2 = (2\sqrt{5})^2$

Identify the center and radius. **Center $\left(\frac{1}{2}, -\frac{1}{3}\right)$ and $r = 2\sqrt{5}$**

To graph the circle, we'll use decimal approximations of the fractions and radicals: $(0.5, -0.3)$ for the center and 4.5 for the radius. Four points on the circle that are 4.5 units from the center are $(-4, -0.3)$, $(5, -0.3)$, $(0.5, 4.2)$, and $(0.5, -4.8)$. Connect them with a smooth curve.

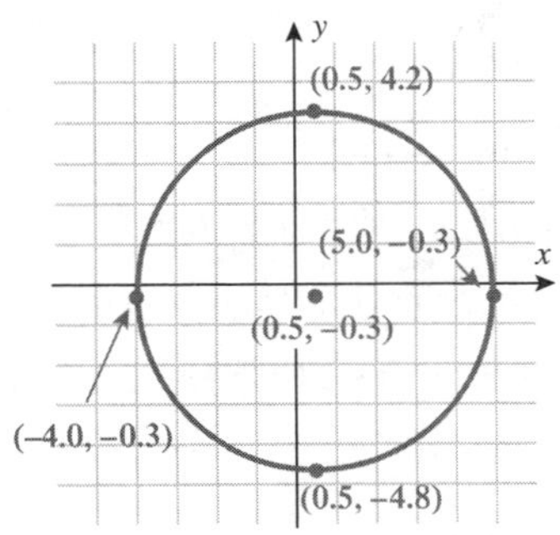

Technology Tip

To enter the graph of $\left(x - \frac{1}{2}\right)^2 + \left(y + \frac{1}{3}\right)^2 = 20$, solve for y first. The graphs of $y_1 = \sqrt{20 - \left(x - \frac{1}{2}\right)^2} - \frac{1}{3}$ and $y_2 = -\sqrt{20 - \left(x - \frac{1}{2}\right)^2} - \frac{1}{3}$ are shown.

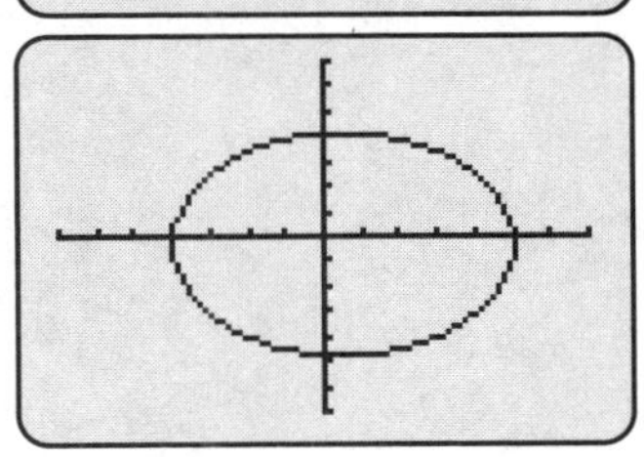

EXAMPLE 3 **Determining the Equation of a Circle Given the Center and Radius**

Find the equation of a circle with radius 5 and center $(-2, 3)$. Graph the circle.

Solution:

Substitute $(h, k) = (-2, 3)$ and $r = 5$ into the standard equation of a circle. $[x - (-2)]^2 + (y - 3)^2 = 5^2$

Simplify. $\boxed{(x + 2)^2 + (y - 3)^2 = 25}$

To graph the circle, plot the center $(-2, 3)$ and four points 5 units away from the center: $(-7, 3)$, $(3, 3)$, $(-2, -2)$, and $(-2, 8)$. Connect them with a smooth curve.

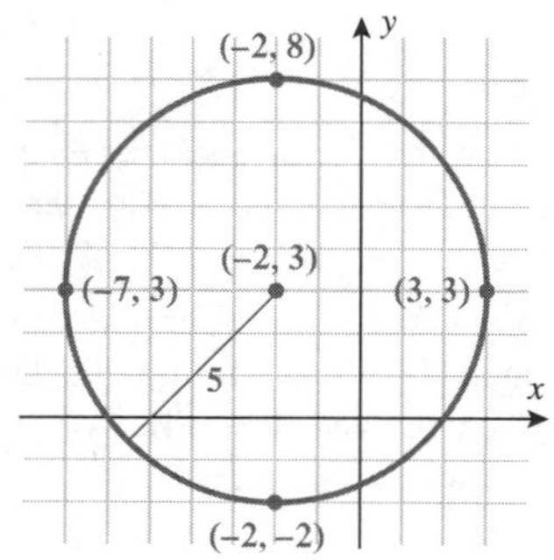

■ **YOUR TURN** Find the equation of a circle with radius 3 and center $(0, 1)$ and graph.

■ **Answer:** $x^2 + (y - 1)^2 = 9$

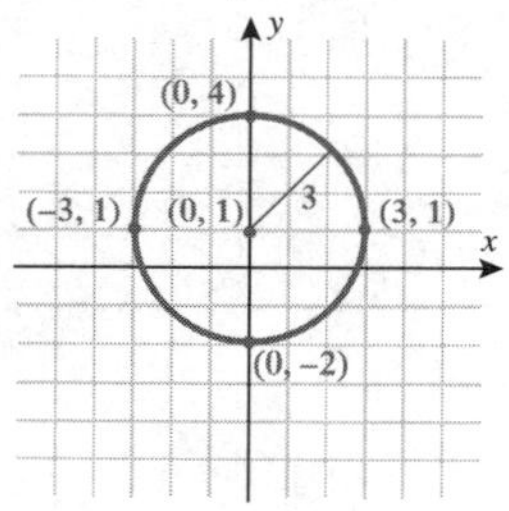

Let's change the look of the equation given in Example 1.

In Example 1 the equation of the circle was given as: $(x - 2)^2 + (y + 1)^2 = 4$

Eliminate the parentheses. $x^2 - 4x + 4 + y^2 + 2y + 1 = 4$

Group like terms and subtract 4 from both sides. $x^2 + y^2 - 4x + 2y + 1 = 0$

We have written the *general form* of the equation of the circle in Example 1.

> The **general form** of the **equation of a circle** is
>
> $$x^2 + y^2 + ax + by + c = 0$$

Suppose you are given a point that lies on a circle and the center of the circle. Can you find the equation of the circle?

EXAMPLE 4 **Finding the Equation of a Circle Given Its Center and One Point**

The point $(10, -4)$ lies on a circle centered at $(7, -8)$. Find the equation of the circle in general form.

Solution:

This circle is centered at $(7, -8)$, so its standard equation is $(x - 7)^2 + (y + 8)^2 = r^2$.

Since the point $(10, -4)$ lies on the circle, it must satisfy the equation of the circle.

Substitute $(x, y) = (10, -4)$. $(10 - 7)^2 + (-4 + 8)^2 = r^2$

Simplify. $3^2 + 4^2 = r^2$

The distance from $(10, -4)$ to $(7, -8)$ is 5 units. $r = 5$

Substitute $r = 5$ into the standard equation. $(x - 7)^2 + (y + 8)^2 = 5^2$

Eliminate the parentheses and simplify. $x^2 - 14x + 49 + y^2 + 16y + 64 = 25$

Write in **general form**. $\boxed{x^2 + y^2 - 14x + 16y + 88 = 0}$

■ **YOUR TURN** The point (1, 11) lies on a circle centered at $(-5, 3)$. Find the equation of the circle in general form.

■ **Answer:** $x^2 + y^2 + 10x - 6y - 66 = 0$

Transforming Equations of Circles to the Standard Form by Completing the Square

If the equation of a circle is given in general form, it must be rewritten in standard form in order to identify its center and radius. To transform equations of circles from general to standard form, complete the square (Section 1.3) on both the x- and y-variables.

Technology Tip

To graph $x^2 - 8x + y^2 + 20y + 107 = 0$ without transforming it into a standard form, solve for y using the quadratic formula:

$$y = \frac{-20 \pm \sqrt{20^2 - 4(1)(x^2 - 8x + 107)}}{2}$$

Next, set the window to $[-5, 30]$ by $[-30, 5]$ and use ZSquare under ZOOM to adjust the window variable to make the circle look circular. The graphs of

$$y_1 = \frac{-20 + \sqrt{20^2 - 4(x^2 - 8x + 107)}}{2}$$

and

$$y_2 = \frac{-20 - \sqrt{20^2 - 4(x^2 - 8x + 107)}}{2}$$

are shown.

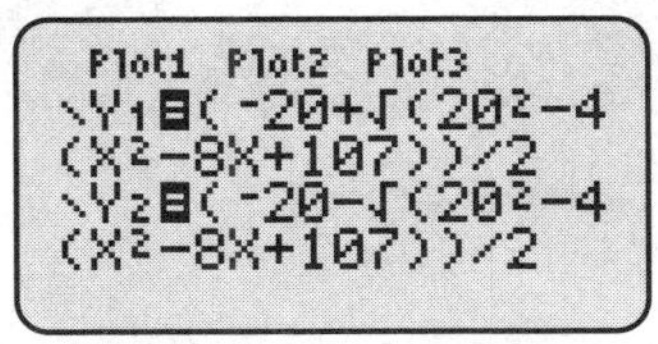

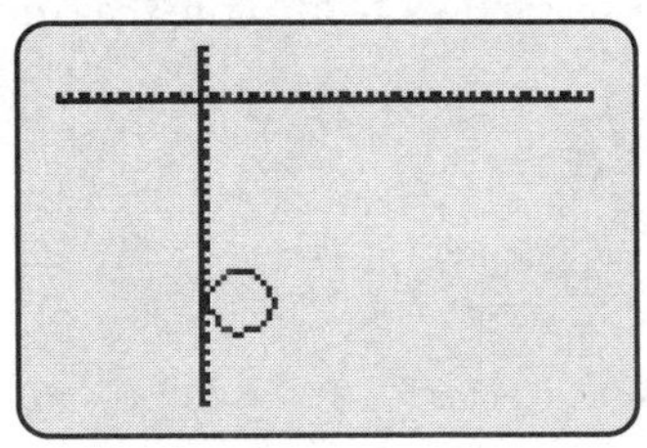

EXAMPLE 5 Finding the Center and Radius of a Circle by Completing the Square

Find the center and radius of the circle with the equation:

$$x^2 - 8x + y^2 + 20y + 107 = 0$$

Solution:

Our goal is to transform this equation into standard form $(x - h)^2 + (y - k)^2 = r^2$

Group x and y terms, respectively, on the left side of the equation; move constants to the right side. $(x^2 - 8x) + (y^2 + 20y) = -107$

Complete the square on both the x and y expressions. $(x^2 - 8x + \square) + (y^2 + 20y + \square) = -107$

Add $\left(-\frac{8}{2}\right)^2 = 16$ and $\left(\frac{20}{2}\right)^2 = 100$ to both sides.

$$(x^2 - 8x + 16) + (y^2 + 20y + 100) = -107 + 16 + 100$$

Factor the perfect squares on the left side and simplify the right side. $(x - 4)^2 + (y + 10)^2 = 9$

Write in standard form. $(x - 4)^2 + [y - (-10)]^2 = 3^2$

$\boxed{\text{The center is } (4, -10) \text{ and the radius is } 3.}$

■ **YOUR TURN** Find the center and radius of the circle with the equation:

$$x^2 + y^2 + 4x - 6y - 12 = 0$$

■ **Answer:** Center: $(-2, 3)$
Radius: 5

CAUTION

Don't forget to add *both* constants to each side of the equation when completing the square for x and y.

COMMON MISTAKE

A common mistake is forgetting to add *both* constants to the right side of the equation. Identify the center and radius of the circle with the equation:

$$x^2 + y^2 + 16x + 8y + 44 = 0$$

CORRECT

$$x^2 + y^2 + 16x + 8y + 44 = 0$$

$$(x^2 + 16x) + (y^2 + 8y) = -44$$

$$(x^2 + 16x + \square) + (y^2 + 8y + \square) = -44$$

$$(x^2 + 16x + 64) + (y^2 + 8y + 16) = -44 + 64 + 16$$

$$(x + 8)^2 + (y + 4)^2 = 36$$

Center: $(-8, -4)$ Radius: 6

INCORRECT

$$(x^2 + 16x + 64) + (y^2 + 8y + 16) = -44$$

ERROR: Don't forget to add 16 and 64 to the right.

SECTION 2.4 SUMMARY

The equation of a circle is given by

- Standard form: $(x - h)^2 + (y - k)^2 = r^2$.
  - Center: (h, k)
  - Radius: r
- General form: $x^2 + y^2 + ax + by + c = 0$.
  - Complete the square to transform the equation to standard form.

SECTION 2.4 EXERCISES

SKILLS

In Exercises 1–18, write the equation of the circle in standard form.

1. Center $(1, 2)$ $r = 3$
2. Center $(3, 4)$ $r = 5$
3. Center $(-3, -4)$ $r = 10$
4. Center $(-1, -2)$ $r = 4$
5. Center $(5, 7)$ $r = 9$
6. Center $(2, 8)$ $r = 6$
7. Center $(-11, 12)$ $r = 13$
8. Center $(6, -7)$ $r = 8$
9. Center $(0, 0)$ $r = 2$
10. Center $(0, 0)$ $r = 3$
11. Center $(0, 0)$ $r = \sqrt{2}$
12. Center $(-1, 2)$ $r = \sqrt{7}$
13. Center $(5, -3)$ $r = 2\sqrt{3}$
14. Center $(-4, -1)$ $r = 3\sqrt{5}$
15. Center $\left(\frac{2}{3}, -\frac{3}{5}\right)$ $r = \frac{1}{4}$
16. Center $\left(-\frac{1}{3}, -\frac{2}{7}\right)$ $r = \frac{2}{5}$
17. Center $(1.3, 2.7)$ $r = 3.2$
18. Center $(-3.1, 4.2)$ $r = 5.5$

In Exercises 19–30, find the center and radius of the circle with the given equations.

19. $(x - 1)^2 + (y - 3)^2 = 25$

20. $(x + 1)^2 + (y + 3)^2 = 11$

21. $(x - 2)^2 + (y + 5)^2 = 49$

22. $(x + 3)^2 + (y - 7)^2 = 81$

23. $(x - 4)^2 + (y - 9)^2 = 20$

24. $(x + 1)^2 + (y + 2)^2 = 8$

25. $\left(x - \frac{2}{5}\right)^2 + \left(y - \frac{1}{7}\right)^2 = \frac{4}{9}$

26. $\left(x - \frac{1}{2}\right)^2 + \left(y - \frac{1}{3}\right)^2 = \frac{9}{25}$

27. $(x - 1.5)^2 + (y + 2.7)^2 = 1.69$

28. $(x + 3.1)^2 + (y - 7.4)^2 = 56.25$

29. $x^2 + y^2 - 50 = 0$

30. $x^2 + y^2 - 8 = 0$

In Exercises 31–46, state the center and radius of each circle.

31. $x^2 + y^2 + 4x + 6y - 3 = 0$

32. $x^2 + y^2 + 2x + 10y + 17 = 0$

33. $x^2 + y^2 + 6x + 8y - 75 = 0$

34. $x^2 + y^2 + 2x + 4y - 9 = 0$

35. $x^2 + y^2 - 10x - 14y - 7 = 0$

36. $x^2 + y^2 - 4x - 16y + 32 = 0$

37. $x^2 + y^2 - 2x - 6y + 1 = 0$

38. $x^2 + y^2 - 8x - 6y + 21 = 0$

39. $x^2 + y^2 - 10x + 6y + 22 = 0$

40. $x^2 + y^2 + 8x + 2y - 28 = 0$

41. $x^2 + y^2 - 6x - 4y + 1 = 0$

42. $x^2 + y^2 - 2x - 10y + 2 = 0$

43. $x^2 + y^2 - x + y + \frac{1}{4} = 0$

44. $x^2 + y^2 - \frac{x}{2} - \frac{3y}{2} + \frac{3}{8} = 0$

45. $x^2 + y^2 - 2.6x - 5.4y - 1.26 = 0$

46. $x^2 + y^2 - 6.2x - 8.4y - 3 = 0$

In Exercises 47–52, find the equation of each circle.

47. Centered at $(-1, -2)$ and passing through the point $(1, 0)$.

48. Centered at $(4, 9)$ and passing through the point $(2, 5)$.

49. Centered at $(-2, 3)$ and passing through the point $(3, 7)$.

50. Centered at $(1, 1)$ and passing through the point $(-8, -5)$.

51. Centered at $(-2, -5)$ and passing through the point $(1, -9)$.

52. Centered at $(-3, -4)$ and passing through the point $(-1, -8)$.

APPLICATIONS

53. Cell Phones. If a cellular phone tower has a reception radius of 100 miles and you live 95 miles north and 33 miles east of the tower, can you use your cell phone while at home?

54. Cell Phones. Repeat Exercise 53, assuming you live 45 miles south and 87 miles west of the tower.

55. Construction/Home Improvement. A couple and their dog moved into a new house that does not have a fenced-in backyard. The backyard is square with dimensions 100 ft × 100 ft. If they put a stake in the center of the backyard with a long leash, write the equation of the circle that will map out the dog's outer perimeter.

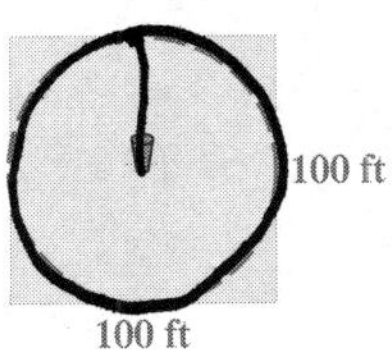

56. Construction/Home Improvement. Repeat Exercise 55 except that the couple put in a pool and a garden and want to restrict the dog to quadrant I. What coordinates represent the center of the circle? What is the radius?

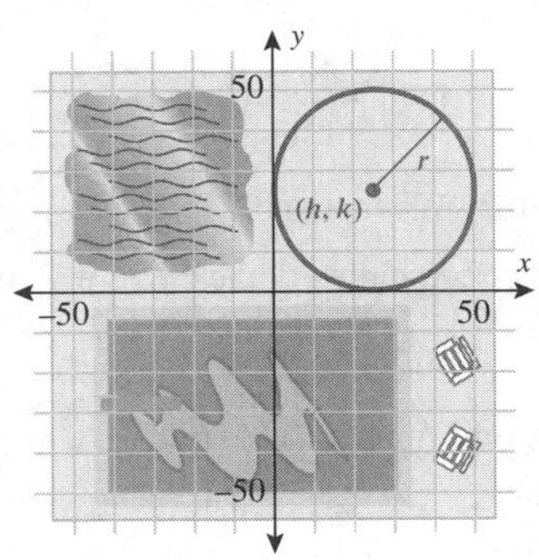

57. Design. A university designs its campus with a master plan of two concentric circles. All of the academic buildings are within the inner circle (so that students can get between classes in less than 10 minutes), and the outer circle contains all the dormitories, the Greek park, cafeterias, the gymnasium, and intramural fields. Assuming the center of campus is the origin, write an equation for the inner circle if the diameter is 3000 feet.

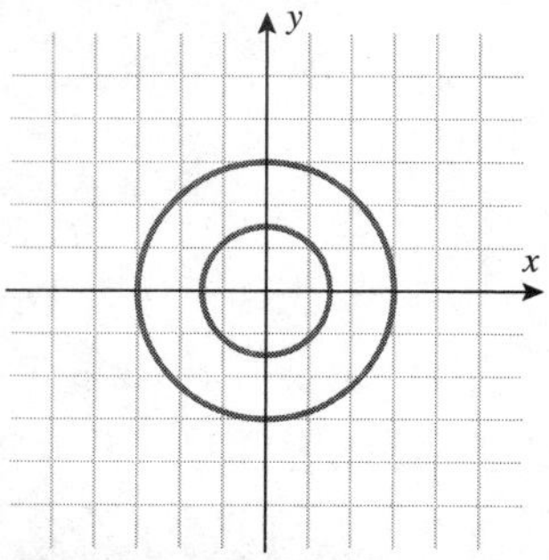

58. Design. Repeat Exercise 57 for the outer circle with a diameter of 6000 feet.

59. Cell Phones. A cellular phone tower has a reception radius of 200 miles. Assuming the tower is located at the origin, write the equation of the circle that represents the reception area.

60. Environment. In a state park, a fire has spread in the form of a circle. If the radius is 2 miles, write an equation for the circle.

CATCH THE MISTAKE

In Exercises 61–64, explain the mistake that is made.

61. Identify the center and radius of the circle with equation $(x - 4)^2 + (y + 3)^2 = 25$.

Solution: The center is (4, 3) and the radius is 5.

This is incorrect. What mistake was made?

62. Identify the center and radius of the circle with equation $(x - 2)^2 + (y + 3)^2 = 2$.

Solution: The center is $(2, -3)$ and the radius is 2.

This is incorrect. What mistake was made?

63. Graph the solution to the equation $(x - 1)^2 + (y + 2)^2 = -16$.

Solution: The center is $(1, -2)$ and the radius is 4.

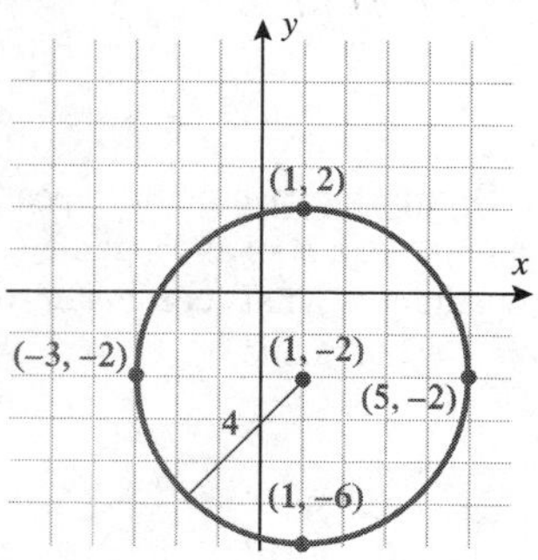

This is incorrect. What mistake was made?

64. Find the center and radius of the circle with the equation $x^2 + y^2 - 6x + 4y - 3 = 0$.

Solution:

Group like terms. $(x^2 - 6x) + (y^2 + 4y) = 3$

Complete the square. $(x^2 - 6x + 9) + (y^2 + 4y + 4) = 12$

$(x - 3)^2 + (y + 2)^2 = (2\sqrt{3})^2$

The center is $(3, -2)$ and the radius is $2\sqrt{3}$.

This is incorrect. What mistake was made?

CONCEPTUAL

In Exercises 65–68, determine whether each statement is true or false.

65. The equation whose graph is depicted has infinitely many solutions.

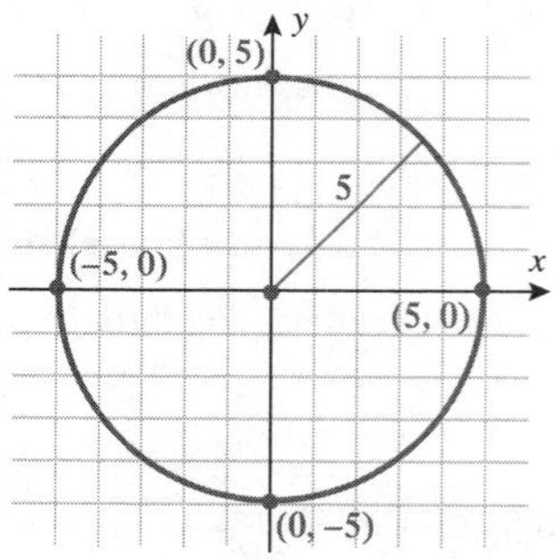

66. The equation $(x - 7)^2 + (y + 15)^2 = -64$ has no solution.

67. The equation $(x - 2)^2 + (y + 5)^2 = -20$ has no solution.

68. The equation $(x - 1)^2 + (y + 3)^2 = 0$ has only one solution.

69. Describe the graph (if it exists) of:

$$x^2 + y^2 + 10x - 6y + 34 = 0$$

70. Describe the graph (if it exists) of:

$$x^2 + y^2 - 4x + 6y + 49 = 0$$

CHALLENGE

71. Find the equation of a circle that has a diameter with endpoints (5, 2) and (1, −6).

72. Find the equation of a circle that has a diameter with endpoints (3, 0) and (−1, −4).

73. For the equation $x^2 + y^2 + ax + by + c = 0$, specify conditions on a, b, and c so that the graph is a single point.

74. For the equation $x^2 + y^2 + ax + by + c = 0$, specify conditions on a, b, and c so that there is no corresponding graph.

TECHNOLOGY

In Exercises 75–78, use a graphing utility to graph each equation. Does this agree with the answer you gave in the Conceptual section?

75. $(x - 2)^2 - (y + 5)^2 = -20$ (See Exercise 67 for comparison.)

76. $(x - 1)^2 + (y + 3)^2 = 0$ (See Exercise 68 for comparison.)

77. $x^2 + y^2 + 10x - 6y + 34 = 0$ (See Exercise 69 for comparison.)

78. $x^2 + y^2 - 4x + 6y + 49 = 0$ (See Exercise 70 for comparison.)

In Exercises 79 and 80, (a) with the equation of the circle in standard form, state the center and radius, and graph; (b) use the quadratic formula to solve for y; and (c) use a graphing utility to graph each equation found in (b). Does the graph in (a) agree with the graphs in (c)?

79. $x^2 + y^2 - 11x + 3y - 7.19 = 0$

80. $x^2 + y^2 + 1.2x - 3.2y + 2.11 = 0$

MODELING YOUR WORLD

The Intergovernmental Panel on Climate Change (IPCC) claims that Carbon Dioxide (CO_2) production from industrial activity (such as fossil fuel burning and other human activities) has increased the CO_2 concentrations in the atmosphere. Because it is a greenhouse gas, elevated CO_2 levels will increase global mean (average) temperature. In this section, we will examine the increasing rate of carbon emissions on Earth.

In 1955, there were (globally) 2 billion tons of carbon emitted per year. In 2005, the carbon emissions has more than tripled to reach approximately 7 billion tons of carbon emitted per year. Currently, we are on the path to doubling our current carbon emissions in the next 50 years.

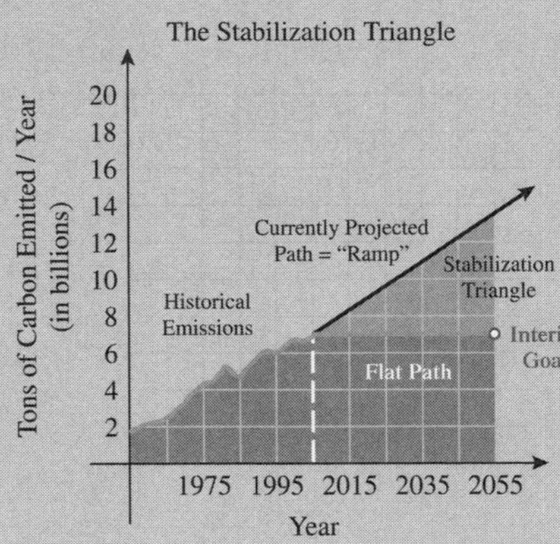

Two Princeton professors* (Stephen Pacala and Rob Socolow) introduced the Climate Carbon Wedge concept. A "wedge" is a strategy to reduce carbon emissions over a 50-year time period from zero to 1.0 GtC/yr (gigatons of carbon per year).

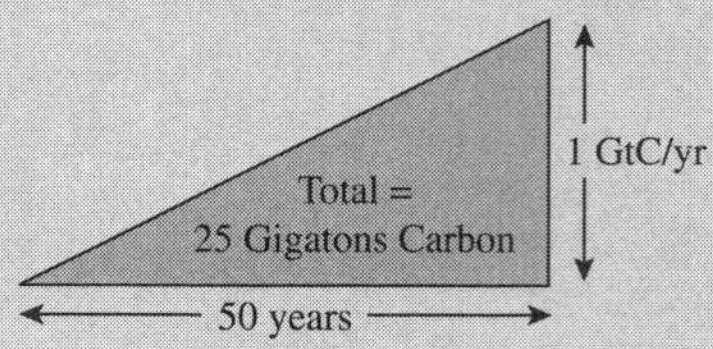

1. Draw the Cartesian plane. Label the vertical axis C, where C represents the number of gigatons (billions of tons) of carbon emitted, and label the horizontal axis t, where t is the number of years. Let $t = 0$ correspond to 2005.

2. Find the equations of the flat path and the seven lines corresponding to the seven wedges.

   a. Flat path (no increase) over 50 years (2005 to 2055)
   b. Increase of 1 GtC over 50 years (2005 to 2055)
   c. Increase of 2 GtC over 50 years (2005 to 2055)
   d. Increase of 3 GtC over 50 years (2005 to 2055)
   e. Increase of 4 GtC over 50 years (2005 to 2055)
   f. Increase of 5 GtC over 50 years (2005 to 2055)
   g. Increase of 6 GtC over 50 years (2005 to 2055)
   h. Increase of 7 GtC over 50 years (2005 to 2055) [projected path]

3. For each of the seven wedges and the flat path, determine how many **total** gigatons of carbon will be reduced over a 50-year period. In other words, how many gigatons of carbon would the world have to reduce in each of the eight cases?

   a. Flat path
   b. Increase of 1GtC over 50 years
   c. Increase of 2 GtC over 50 years
   d. Increase of 3 GtC over 50 years
   e. Increase of 4 GtC over 50 years
   f. Increase of 5 GtC over 50 years
   g. Increase of 6 GtC over 50 years
   h. Increase of 7 GtC over 50 years (projected path)

4. Research the "climate carbon wedge" concept and discuss the types of changes (transportation efficiency, transportation conservation, building efficiency, efficiency in electricity production, alternate energies, etc.) the world would have to make that would correspond to each of the seven wedges and the flat path.

   a. Flat path
   b. Wedge 1
   c. Wedge 2
   d. Wedge 3
   e. Wedge 4
   f. Wedge 5
   g. Wedge 6

*S. Pacala and R. Socolow, "Stabilization Wedges: Solving the Climate Problem for the Next 50 Years with Current Technologies," *Science*, Vol. 305 (2004).

Section	Topic	Pages	Review Exercises	Key Concepts
2.1	**Basic tools: Cartesian plane, distance, and midpoint**	174–177	1–14	Two points in the xy-plane: (x_1, y_1) and (x_2, y_2).
	Cartesian plane	174	1–4	x-axis, y-axis, origin, and quadrants.
	Distance between two points	175–176	5–8	$d = \sqrt{(x_2 - x_1)^2 + (y_2 - y_1)^2}$
	Midpoint of a line segment joining two points	177	9–12	$(x_m, y_m) = \left(\frac{x_1 + x_2}{2}, \frac{y_1 + y_2}{2}\right)$
2.2	**Graphing equations: Point-plotting, intercepts, and symmetry**	181–189	15–30	
	Point-plotting	181–183	23–28	List a table with several coordinates that are solutions to the equation; plot and connect.
	Intercepts	184–185	15–18	x-intercept: let $y = 0$ y-intercept: let $x = 0$
	Symmetry	185–187	19–22	The graph of an equation can be symmetric about the x-axis, y-axis, or origin.
	Using intercepts and symmetry as graphing aids	188–189	23–28	If (a, b) is on the graph of the equation, then $(-a, b)$ is on the graph if symmetric about the y-axis, $(a, -b)$ is on the graph if symmetric about the x-axis, and $(-a, -b)$ is on the graph if symmetric about the origin.
2.3	**Lines**	193–205	31–58	General form: $Ax + By = C$
	Graphing a line	194	35–40	Vertical: $x = a$ Slant: $Ax + By = C$ Horizontal: $y = b$ where $x \neq 0$ and $y \neq 0$
	Intercepts	194–196	31–40	x-intercept $(a, 0)$ y-intercept $(0, b)$
	Slope	196–198	31–40	$m = \frac{y_2 - y_1}{x_2 - x_1}$, where $x_1 \neq x_2$ $\frac{\text{"rise"}}{\text{"run"}}$
	Slope–intercept form	199	31–44	$y = mx + b$ m is the slope and b is the y-intercept.
	Finding equations of lines	200–201	41–56	Point–slope form: $y - y_1 = m(x - x_1)$
	Parallel and perpendicular lines	202–205	53–56	$L_1 \parallel L_2$ if and only if $m_1 = m_2$ (slopes are equal). $L_1 \perp L_2$ if and only if $m_1 = -\frac{1}{m_2}$ $\begin{cases} m_1 \neq 0 \\ m_2 \neq 0 \end{cases}$ (slopes are negative reciprocals).
2.4	**Circles**	210–214	59–72	
	Standard equation of a circle	210–213	59–65	$(x - h)^2 + (y - k)^2 = r^2$ $C: (h, k)$
	Transforming equations of circles to the standard form by completing the square	213–214	66–70	General form: $x^2 + y^2 + ax + by + c = 0$

2.1 Basic Tools: Cartesian Plane, Distance, and Midpoint

Plot each point and indicate which quadrant the point lies in.

1. $(-4, 2)$ **2.** $(4, 7)$

3. $(-1, -6)$ **4.** $(2, -1)$

Calculate the distance between the two points.

5. $(-2, 0)$ and $(4, 3)$ **6.** $(1, 4)$ and $(4, 4)$

7. $(-4, -6)$ and $(2, 7)$ **8.** $\left(\frac{1}{4}, \frac{1}{12}\right)$ and $\left(\frac{1}{3}, -\frac{7}{3}\right)$

Calculate the midpoint of the segment joining the two points.

9. $(2, 4)$ and $(3, 8)$ **10.** $(-2, 6)$ and $(5, 7)$

11. $(2.3, 3.4)$ and $(5.4, 7.2)$ **12.** $(-a, 2)$ and $(a, 4)$

Applications

13. Sports. A quarterback drops back to pass. At the point $(-5, -20)$ he throws the ball to his wide receiver located at $(10, 30)$. Find the distance the ball has traveled. Assume the width of the football field is $[-15, 15]$ and the length is $[-50, 50]$.

14. Sports. Suppose that in the above exercise a defender was midway between the quarterback and the receiver. At what point was the defender located when the ball was thrown over his head?

2.2 Graphing Equations: Point-Plotting, Intercepts, and Symmetry

Find the *x*-intercept(s) and *y*-intercept(s) if any.

15. $x^2 + 4y^2 = 4$ **16.** $y = x^2 - x + 2$

17. $y = \sqrt{x^2 - 9}$ **18.** $y = \dfrac{x^2 - x - 12}{x - 12}$

Use algebraic tests to determine symmetry with respect to the *x*-axis, *y*-axis, or origin.

19. $x^2 + y^3 = 4$ **20.** $y = x^2 - 2$

21. $xy = 4$ **22.** $y^2 = 5 + x$

Use symmetry as a graphing aid and point-plot the given equations.

23. $y = x^2 - 3$ **24.** $y = |x| - 4$ **25.** $y = \sqrt[3]{x}$

26. $x = y^2 - 2$ **27.** $y = x\sqrt{9 - x^2}$ **28.** $x^2 + y^2 = 36$

Applications

29. Sports. A track around a high school football field is in the shape of the graph $8x^2 + y^2 = 8$. Graph using symmetry and by plotting points.

30. Transportation. A "bypass" around a town follows the graph $y = x^3 + 2$, where the origin is the center of town. Graph the equation.

2.3 Lines

Express the equation for each line in slope–intercept form. Identify the slope and *y*-intercept of each line.

31. $6x + 2y = 12$ **32.** $3x + 4y = 9$

33. $-\frac{1}{2}x - \frac{1}{3}y = \frac{1}{6}$ **34.** $-\frac{2}{3}x - \frac{1}{4}y = \frac{1}{8}$

Find the *x*- and *y*-intercepts and the slope of each line if they exist and graph.

35. $y = 4x - 5$ **36.** $y = \frac{-3}{4}x - 3$ **37.** $x + y = 4$

38. $x = -4$ **39.** $y = 2$ **40.** $\frac{-1}{2}x - \frac{1}{2}y = 3$

Write the equation of the line, given the slope and the intercepts.

41. Slope: $m = 4$; *y*-intercept: $(0, -3)$

42. Slope: $m = 0$; *y*-intercept: $(0, 4)$

43. Slope: m is undefined; *x*-intercept: $(-3, 0)$

44. Slope: $m = \frac{-2}{3}$; *y*-intercept: $\left(0, \frac{3}{4}\right)$

Write an equation of the line, given the slope and a point that lies on the line.

45. $m = -2$ $(-3, 4)$ **46.** $m = \frac{3}{4}$ $(2, 16)$

47. $m = 0$ $(-4, 6)$ **48.** m is undefined $(2, -5)$

Write the equation of the line that passes through the given points. Express the equation in slope–intercept form or in the form of $x = a$ or $y = b$.

49. $(-4, -2)$ and $(2, 3)$ **50.** $(-1, 4)$ and $(-2, 5)$

51. $\left(-\frac{3}{4}, \frac{1}{2}\right)$ and $\left(-\frac{7}{4}, \frac{5}{2}\right)$ **52.** $(3, -2)$ and $(-9, 2)$

Find the equation of the line that passes through the given point and also satisfies the additional piece of information.

53. $(-2, -1)$ parallel to the line $2x - 3y = 6$

54. $(5, 6)$ perpendicular to the line $5x - 3y = 0$

55. $\left(-\frac{3}{4}, \frac{5}{2}\right)$ perpendicular to the line $\frac{2}{3}x - \frac{1}{2}y = 12$

56. $(a + 2, b - 1)$ parallel to the line $Ax + By = C$

Applications

57. Grades. For a GRE prep class, a student must take a pretest and then a posttest after the completion of the course. Two students' results are shown below. Give a linear equation to represent the given data.

PRETEST	POSTTEST
1020	1324
950	1240

58. Budget: Car Repair. The cost of having the air conditioner in your car repaired is the combination of material costs and labor costs. The materials (tubing, coolant, etc.) are \$250 and the labor costs \$38 per hour. Write an equation that models the total cost C of having your air conditioner repaired as a function of hours t. Graph this equation with t as the horizontal axis and C representing the vertical axis. How much will the job cost if the mechanic works 1.5 hours?

2.4 Circles

Write the equation of the circle in standard form.

59. center $(-2, 3)$
$r = 6$

60. center $(-6, -8)$
$r = 3\sqrt{6}$

61. center $\left(\frac{3}{4}, \frac{5}{2}\right)$
$r = \frac{2}{5}$

62. center $(1.2, -2.4)$
$r = 3.6$

Find the center and the radius of the circle given by the equation.

63. $(x + 2)^2 + (y + 3)^2 = 81$

64. $(x - 4)^2 + (y + 2)^2 = 32$

65. $\left(x + \frac{3}{4}\right)^2 + \left(y - \frac{1}{2}\right)^2 = \frac{16}{36}$

66. $x^2 + y^2 + 4x - 2y = 0$

67. $x^2 + y^2 + 2y - 4x + 11 = 0$

68. $3x^2 + 3y^2 - 6x - 7 = 0$

69. $9x^2 + 9y^2 - 6x + 12y - 76 = 0$

70. $x^2 + y^2 + 3.2x - 6.6y - 2.4 = 0$

71. Find the equation of a circle centered at $(2, 7)$ and passing through $(3, 6)$.

72. Find the equation of a circle that has the diameter with endpoints $(-2, -1)$ and $(5, 5)$.

Technology Exercises

Section 2.1

Determine whether the triangle with the given vertices is a right triangle, isosceles triangle, neither, or both.

73. $(-10, -5), (20, -45), (10, 10)$

74. $(4.2, 8.4), (-4.2, 2.1), (6.3, -10.5)$

Section 2.2

Graph the equation using a graphing utility and state whether there is any symmetry.

75. $y^2 = |x^2 - 4|$

76. $0.8x^2 - 1.5y^2 = 4.8$

Section 2.3

Determine whether the lines are parallel, perpendicular, or neither, then graph both lines in the same viewing screen using a graphing utility to confirm your answer.

77. $y_1 = 0.875x + 1.5$
$y_2 = -\frac{8}{7}x - \frac{9}{14}$

78. $y_1 = -0.45x - 2.1$
$y_2 = \frac{5}{6} - \frac{9}{20}x$

Section 2.4

79. Use the quadratic formula to solve for y, and use a graphing utility to graph each equation. Do the graphs agree with the graph in Exercise 69?

$$9x^2 + 9y^2 - 6x + 12y - 76 = 0$$

80. Use the quadratic formula to solve for y, and use a graphing utility to graph each equation. Do the graphs agree with the graph in Exercise 70?

$$x^2 + y^2 + 3.2x - 6.6y - 2.4 = 0$$

1. Find the distance between the points $(-7, -3)$ and $(2, -2)$.
2. Find the midpoint between $(-3, 5)$ and $(5, -1)$.
3. Determine the length and the midpoint of a segment that joins the points $(-2, 4)$ and $(3, 6)$.
4. **Research Triangle.** The Research Triangle in North Carolina was established as a collaborative research center among Duke University (Durham), North Carolina State University (Raleigh), and the University of North Carolina (Chapel Hill).

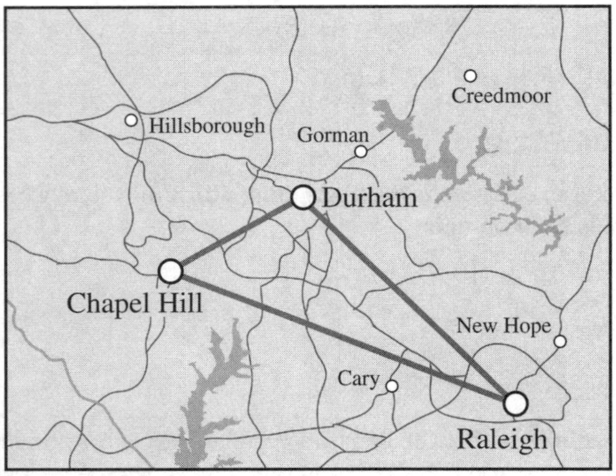

Durham is 10 miles north and 8 miles east of Chapel Hill, and Raleigh is 28 miles east and 15 miles south of Chapel Hill. What is the perimeter of the research triangle? Round your answer to the nearest mile.

5. Determine the two values for y so that the point $(3, y)$ is 5 units away from the point $(6, 5)$.
6. If the point $(3, -4)$ is on a graph that is symmetric with respect to the y-axis, what point must also be on the graph?
7. Determine whether the graph of the equation $x - y^2 = 5$ has any symmetry (x-axis, y-axis, and origin).
8. Find the x-intercept(s) and the y-intercept(s), if any: $4x^2 - 9y^2 = 36$.

Graph the following equations.

9. $2x^2 + y^2 = 8$
10. $y = \dfrac{4}{x^2 + 1}$
11. Find the x-intercept and the y-intercept of the line $x - 3y = 6$.
12. Express the line in slope–intercept form: $4x - 6y = 12$.
13. Express the line in slope–intercept form: $\frac{2}{3}x - \frac{1}{4}y = 2$.

Find the equation of the line that is characterized by the given information. Graph the line.

14. Slope $= 4$; y-intercept $(0, 3)$
15. Passes through the points $(-3, 2)$ and $(4, 9)$
16. Parallel to the line $y = 4x + 3$ and passes through the point $(1, 7)$
17. Perpendicular to the line $2x - 4y = 5$ and passes through the point $(1, 1)$
18. x-intercept $(3, 0)$; y-intercept $(0, 6)$

For Exercises 19 and 20, write the equation of the line that corresponds to the graph.

19.

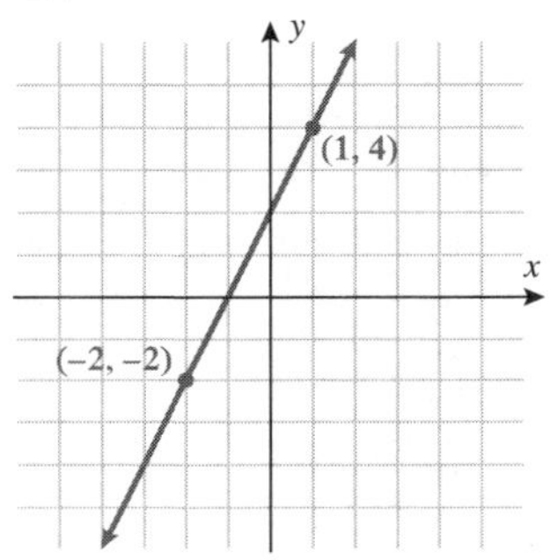

20.

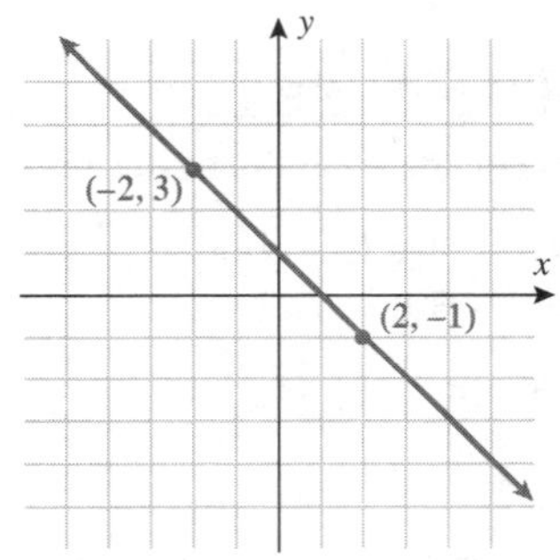

21. Write the equation of a circle that has center $(6, -7)$ and radius $r = 8$.
22. Determine the center and radius of the circle $x^2 + y^2 - 10x + 6y + 22 = 0$.
23. Find the equation of the circle that is centered at $(4, 9)$ and passes through the point $(2, 5)$.
24. **Solar System.** Earth is approximately 93 million miles from the sun. Approximating Earth's orbit around the sun as circular, write an equation governing Earth's path around the sun. Locate the sun at the origin.
25. Determine whether the triangle with the given vertices is a right triangle, isosceles triangle, neither, or both.

$$(-8.4, 16.8), (0, 37.8), (12.6, 8.4)$$

26. Graph the given equation using a graphing utility and state whether there is any symmetry.

$$0.25y^2 + 0.04x^2 = 1$$

1. Simplify $\frac{7-2}{7+3}$.

2. Simplify and express in terms of positive exponents: $\frac{(5x^{3/4})^4}{25x^{-1/4}}$.

3. Perform the operation and simplify: $(x-4)^2(x+4)^2$.

4. Factor completely $8x^3 - 27y^3$.

5. Perform the operations and simplify: $\frac{1/x - 1/5}{1/x + 1/5}$.

6. Solve for x: $x^3 - 5x^2 - 4x + 20 = 0$.

7. Perform the operations and write in standard form: $\sqrt{-36}(5-2i)$.

Solve for *x*.

8. $15 - [5 + 3x - 4(2x - 6)] = 4(6x - 7) - [5(3x - 7) - 6x + 10]$

9. $\frac{5}{4x+1} = \frac{3}{4x-1}$

10. Ashley inherited \$17,000. She invested some money in a CD that earns 5% and the rest in a stock that earns 8%. How much was invested in each account, if the interest for the first year is \$1,075?

11. Solve by factoring: $5x^2 = 45$.

12. Solve by completing the square: $3x^2 + 6x = 7$.

13. Use the discriminant to determine the number and type of roots: $5x^2 + 2x + 7 = 0$.

14. Solve for r: $p^2 + q^2 = r^2$.

15. Solve and check: $\sqrt{x^2 + 3x - 10} = x - 2$.

16. Solve using substitution: $\frac{1}{(x+2)^2} - \frac{5}{x+2} + 4 = 0$.

Solve and express the solution in interval notation:

17. $6 < \frac{1}{4}x + 6 < 9$

18. $x^2 - x \geq 20$

19. $|2 - x| < 4$

20. Solve for x: $|5 - 4x| = 23$.

21. Use algebraic tests to determine whether the graph of the equation $y = 4x$ is symmetric with respect to the x-axis, y-axis, or origin.

22. Write an equation of a line in slope–intercept form with slope $m = \frac{4}{5}$ that passes through the point (5, 1).

23. Write an equation of a line that is perpendicular to the x-axis and passes through the point (5, 3).

24. Write an equation of a line in slope–intercept form that passes through the two points $\left(\frac{1}{7}, \frac{5}{3}\right)$ and $\left(-\frac{6}{7}, -\frac{2}{3}\right)$.

25. Find the center and radius of the circle: $(x+5)^2 + (y+3)^2 = 30$.

26. Calculate the distance between the two points $(-\sqrt{11}, 5)$ and $(2, \sqrt{7})$, and find the midpoint of the segment joining the two points. Round your answers to one decimal place.

27. Determine whether the lines $y_1 = 0.32x + 1.5$ and $y_2 = -\frac{5}{16}x + \frac{1}{4}$ are parallel, perpendicular, or neither, then graph both lines in the same viewing screen using a graphing utility to confirm your answer.

3

Functions and Their Graphs

On a sales rack of clothes at a department store, you see a shirt you like. The original price of the shirt was \$100, but it has been discounted 30%. As a preferred shopper, you get an automatic additional 20% off the sale price at the register. How much will you pay for the shirt?

Naïve shoppers might be lured into thinking this shirt will cost \$50 because they add the 20% and 30% to get 50% off, but they will end up paying more than that. Experienced shoppers know that they first take 30% off of \$100, which results in a price of \$70, and then they take an additional 20% off of the sale price, \$70, which results in a final discounted price of \$56. Experienced shoppers have already learned *composition of functions.*

A *composition of functions* can be thought of as a function of a function. One function takes an input (original price, \$100) and maps it to an output (sale price, \$70), and then another function takes that output as its input (sale price, \$70) and maps that to an output (checkout price, \$56).

IN THIS CHAPTER you will find that functions are part of our everyday thinking: converting from degrees Celsius to degrees Fahrenheit, DNA testing in forensic science, determining stock values and the sale price of a shirt. We will develop a more complete, thorough understanding of functions. First, we will establish what a relation is, and then we will determine whether a relation is a function. We will discuss common functions, domain and range of functions, and graphs of functions. We will determine whether a function is increasing or decreasing on an interval and calculate the average rate of change of a function. We will perform operations on functions and composition of functions. We will discuss one-to-one functions and inverse functions. Finally, we will model applications with functions using variation.

FUNCTIONS AND THEIR GRAPHS

DEFINITION OF A FUNCTION
- Determine whether a relation is a function
- Determine the domain and range of a function

GRAPHS OF FUNCTIONS
- Common functions
- Increasing/decreasing/constant
- Graphing aids

TYPES OF FUNCTIONS
- Common functions and their properties
- Piecewise-defined
- Even/odd
- One-to-one
- Inverse

OPERATIONS ON FUNCTIONS
- Addition/subtraction
- Multiplication/division
- Composition

MODELING WITH FUNCTIONS
- Piecewise-defined
- Composite
- Variation

SECTION

3.1 FUNCTIONS

3.2 GRAPHS OF FUNCTIONS; PIECEWISE-DEFINED FUNCTIONS; INCREASING AND DECREASING FUNCTIONS; AVERAGE RATE OF CHANGE

3.3 GRAPHING TECHNIQUES: TRANSFORMATIONS

3.4 OPERATIONS ON FUNCTIONS AND COMPOSITION OF FUNCTIONS

3.5 ONE-TO-ONE FUNCTIONS AND INVERSE FUNCTIONS

3.6 MODELING FUNCTIONS USING VARIATION

CHAPTER OBJECTIVES

- Determine whether a relation is a function.
- Use function notation.
- Find the domain and range of a function.
- Perform operations on functions.
- Sketch the graphs of common functions.
- Determine whether a function is increasing, decreasing, or constant on an interval.
- Find the average rate of change of a function.
- Classify a function as even, odd, or neither.
- Sketch graphs of general functions employing translations of common functions.
- Perform composition of functions.
- Determine whether functions are one-to-one.
- Find the inverse of a function.
- Model applications with functions using variation.

SECTION 3.1 FUNCTIONS

SKILLS OBJECTIVES

- Determine whether a relation is a function.
- Determine whether an equation represents a function.
- Use function notation.
- Find the value of a function.
- Determine the domain and range of a function.

CONCEPTUAL OBJECTIVES

- Think of function notation as a placeholder or mapping.
- Understand that all functions are relations but not all relations are functions.

What do the following pairs have in common?

- Every person has a blood type.
- Temperature at a particular time of day.
- Every working household phone in the United States has a 10-digit phone number.
- First-class postage rates correspond to the weight of a letter.
- Certain times of the day are start times of sporting events at a university.

They all describe a particular correspondence between two groups. A **relation** is a correspondence between two sets. The first set is called the **domain** and the corresponding second set is called the **range**. Members of these sets are called **elements**.

DEFINITION **Relation**

A **relation** is a correspondence between two sets where each element in the first set, called the **domain**, corresponds to *at least* one element in the second set, called the **range**.

A relation is a set of ordered pairs. The domain is the set of all the first components of the ordered pairs, and the range is the set of all the second components of the ordered pairs.

PERSON	BLOOD TYPE	ORDERED PAIR
Michael	A	(Michael, A)
Tania	A	(Tania, A)
Dylan	AB	(Dylan, AB)
Trevor	O	(Trevor, O)
Megan	O	(Megan, O)

WORDS	MATH
The domain is the set of all the first components.	{Michael, Tania, Dylan, Trevor, Megan}
The range is the set of all the second components.	{A, AB, O}

A relation in which each element in the domain corresponds to exactly one element in the range is a **function**.

DEFINITION Function

A **function** is a correspondence between two sets where each element in the first set, called the **domain**, corresponds to *exactly* one element in the second set, called the **range**.

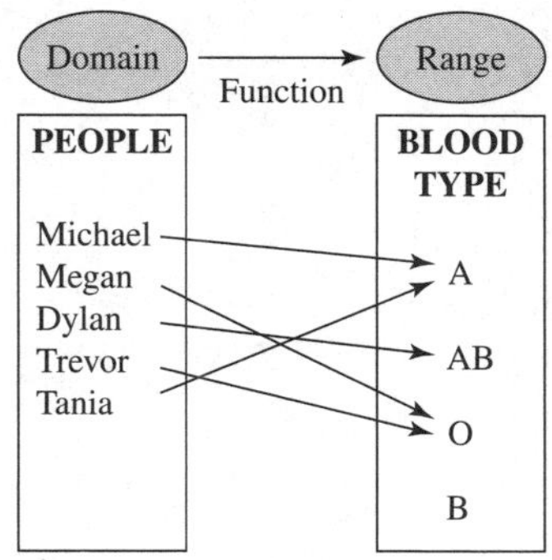

Note that the definition of a function is more restrictive than the definition of a relation. For a relation, each input corresponds to *at least* one output, whereas, for a function, each input corresponds to *exactly* one output. The blood-type example given is both a relation and a function.

Also note that the range (set of values to which the elements of the domain correspond) is a subset of the set of all blood types. However, although all functions are relations, not all relations are functions.

For example, at a university, four primary sports typically overlap in the late fall: football, volleyball, soccer, and basketball. On a given Saturday, the following table indicates the start times for the competitions.

TIME OF DAY	COMPETITION
1:00 P.M.	Football
2:00 P.M.	Volleyball
7:00 P.M.	Soccer
7:00 P.M.	Basketball

WORDS	MATH
The 1:00 start time corresponds to exactly one event, Football.	(1:00 P.M., Football)
The 2:00 start time corresponds to exactly one event, Volleyball.	(2:00 P.M., Volleyball)
The 7:00 start time corresponds to two events, Soccer and Basketball.	(7:00 P.M., Soccer) (7:00 P.M., Basketball)

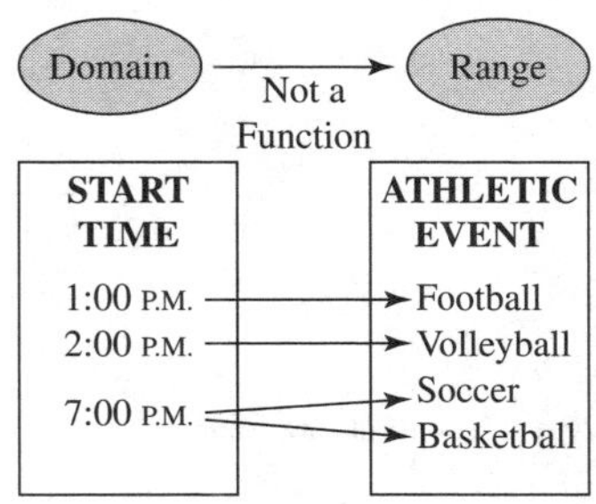

Because an element in the domain, 7:00 P.M., corresponds to more than one element in the range, Soccer and Basketball, this is not a function. It is, however, a relation.

Study Tip

All functions are relations but not all relations are functions.

EXAMPLE 1 Determining Whether a Relation Is a Function

Determine whether the following relations are functions.

a. $\{(-3, 4), (2, 4), (3, 5), (6, 4)\}$
b. $\{(-3, 4), (2, 4), (3, 5), (2, 2)\}$
c. Domain = Set of all items for sale in a grocery store; Range = Price

Solution:

a. No x-value is repeated. Therefore, each x-value corresponds to exactly one y-value.

This relation is a function.

b. The value $x = 2$ corresponds to *both* $y = 2$ and $y = 4$. This relation is not a function.

c. Each item in the grocery store corresponds to exactly one price. This relation is a function.

■ **YOUR TURN** Determine whether the following relations are functions.

a. $\{(1, 2), (3, 2), (5, 6), (7, 6)\}$
b. $\{(1, 2), (1, 3), (5, 6), (7, 8)\}$
c. $\{$(11:00 A.M., 83°F), (2:00 P.M., 89°F), (6:00 P.M., 85°F)$\}$

■ **Answer: a.** function
b. not a function
c. function

All of the examples we have discussed thus far are **discrete** sets in that they represent a countable set of distinct pairs of (x, y). A function can also be defined algebraically by an equation.

Functions Defined by Equations

Let's start with the equation $y = x^2 - 3x$, where x can be any real number. This equation assigns to each x-value exactly one corresponding y-value.

x	$y = x^2 - 3x$	y
1	$y = (1)^2 - 3(1)$	-2
5	$y = (5)^2 - 3(5)$	10
$-\frac{2}{3}$	$y = \left(-\frac{2}{3}\right)^2 - 3\left(-\frac{2}{3}\right)$	$\frac{22}{9}$
1.2	$y = (1.2)^2 - 3(1.2)$	-2.16

Since the variable y *depends* on what value of x is selected, we denote y as the **dependent variable**. The variable x can be any number in the domain; therefore, we denote x as the **independent variable**.

Although functions are defined by equations, it is important to recognize that *not all equations are functions.* The requirement for an equation to define a function is that each element in the domain corresponds to exactly one element in the range. Throughout the ensuing discussion, we assume x to be the independent variable and y to be the dependent variable.

CAUTION

Not all equations are functions.

Equations that represent functions of x:	$y = x^2$	$y = \lvert x\rvert$	$y = x^3$
Equations that do not represent functions of x:	$x = y^2$	$x^2 + y^2 = 1$	$x = \lvert y\rvert$

Study Tip

We say that $x = y^2$ is not a function of x. However, if we reverse the independent and dependent variables, then $x = y^2$ is a function of y.

In the "equations that represent functions of x," every x-value corresponds to exactly one y-value. Some ordered pairs that correspond to these functions are

$y = x^2$: $(-1, 1)\ (0, 0)\ (1, 1)$

$y = |x|$: $(-1, 1)\ (0, 0)\ (1, 1)$

$y = x^3$: $(-1, -1)\ (0, 0)\ (1, 1)$

The fact that $x = -1$ and $x = 1$ both correspond to $y = 1$ in the first two examples does not violate the definition of a function.

In the "equations that do not represent functions of x," some x-values correspond to *more than one* y-value. Some ordered pairs that correspond to these equations are

$x = y^2$:	$(\mathbf{1}, \mathbf{-1})\ (0, 0)\ (\mathbf{1}, \mathbf{1})$	$x = \mathbf{1}$ maps to **both** $y = \mathbf{-1}$ and $y = \mathbf{1}$
$x^2 + y^2 = 1$:	$(\mathbf{0}, \mathbf{-1})\ (\mathbf{0}, \mathbf{1})\ (-1, 0)\ (1, 0)$	$x = \mathbf{0}$ maps to **both** $y = \mathbf{-1}$ and $y = \mathbf{1}$
$x = \lvert y\rvert$:	$(\mathbf{1}, \mathbf{-1})\ (0, 0)\ (\mathbf{1}, \mathbf{1})$	$x = \mathbf{1}$ maps to **both** $y = \mathbf{-1}$ and $y = \mathbf{1}$

Let's look at the graphs of the three **functions of *x***:

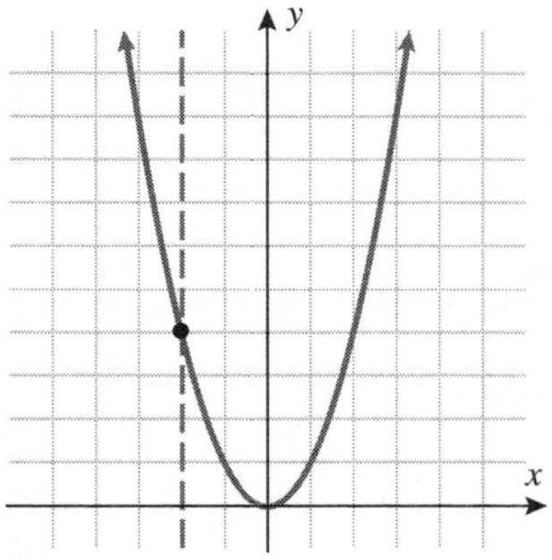

$y = x^2$

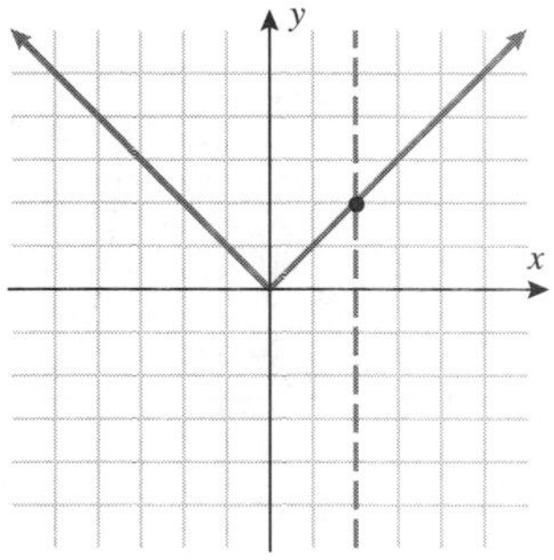

$y = |x|$

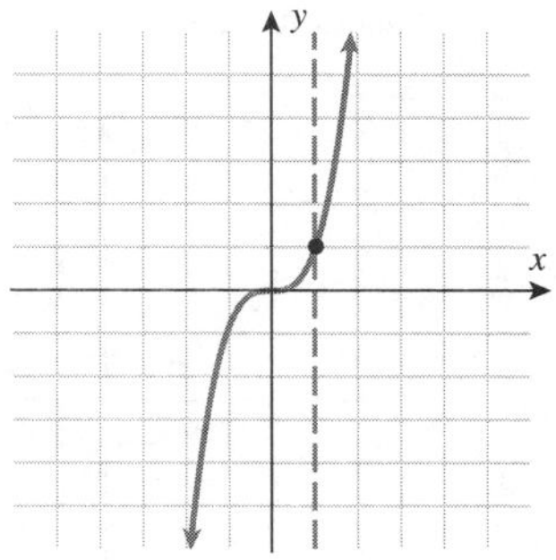

$y = x^3$

Let's take any value for x, say $x = a$. The graph of $x = a$ corresponds to a vertical line. A function of x maps each x-value to exactly one y-value; therefore, there should be at most one point of intersection with any vertical line. We see in the three graphs of the functions above that if a vertical line is drawn at any value of x on any of the three graphs, the vertical line only intersects the graph in one place. Look at the graphs of the three equations that do **not** represent **functions of *x***.

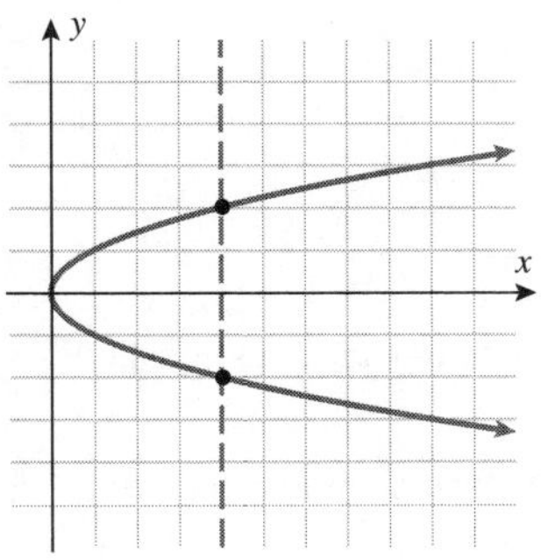

$x = y^2$

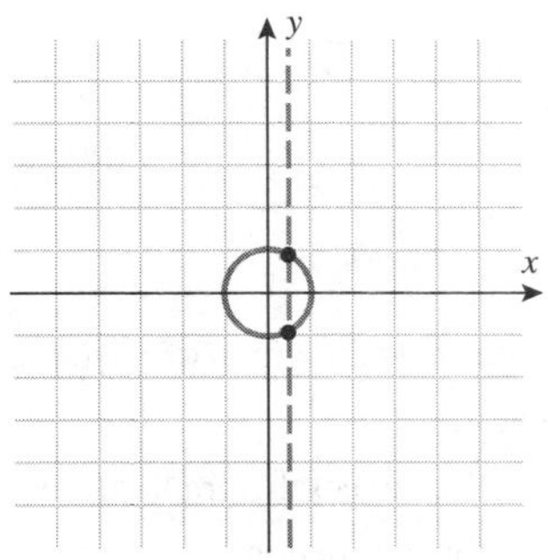

$x^2 + y^2 = 1$

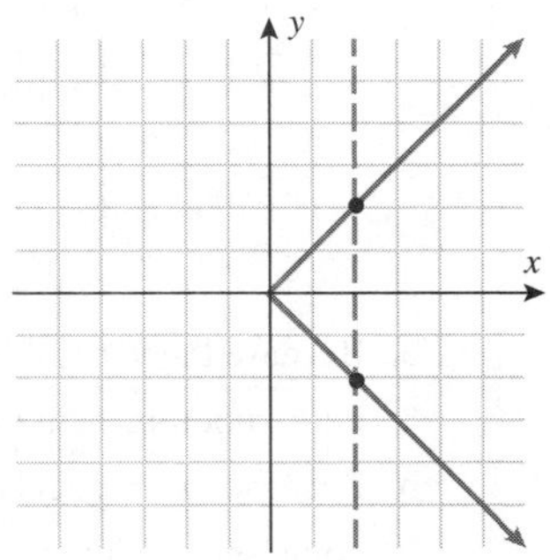

$x = |y|$

A vertical line can be drawn on any of the three graphs such that the vertical line will intersect each of these graphs at two points. Thus, there are two y-values that correspond to some x-value in the domain, which is why these equations do not define functions of x.

DEFINITION **Vertical Line Test**

Given the graph of an equation, if any vertical line that can be drawn intersects the graph at no more than one point, the equation defines a function of x. This test is called the **vertical line test**.

Study Tip

If any x-value corresponds to more than one y-value, then y is **not** a function of x.

EXAMPLE 2 Using the Vertical Line Test

Use the vertical line test to determine whether the graphs of equations define functions of x.

a.

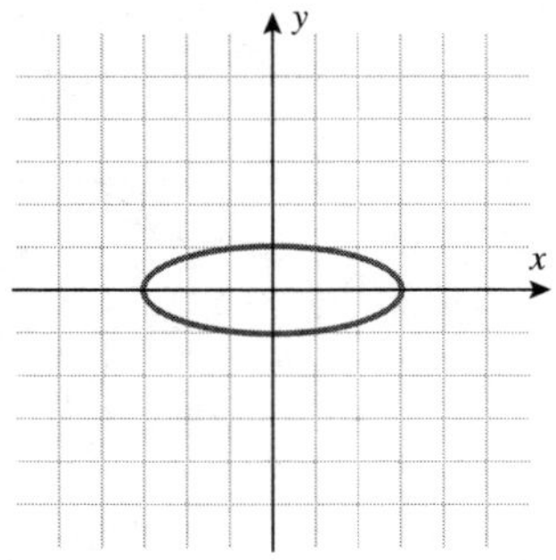

b.

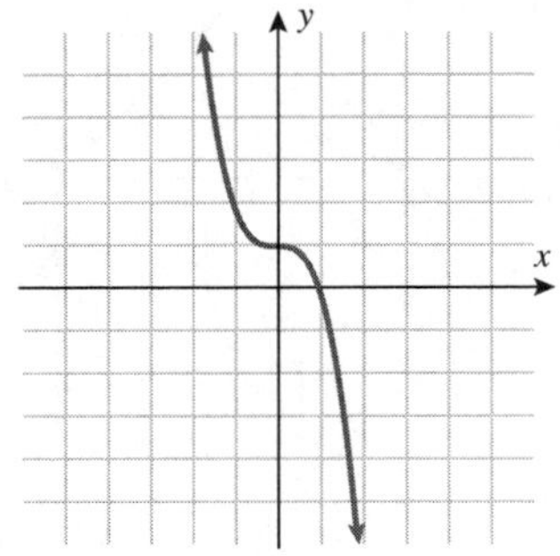

Solution:

Apply the vertical line test.

a.

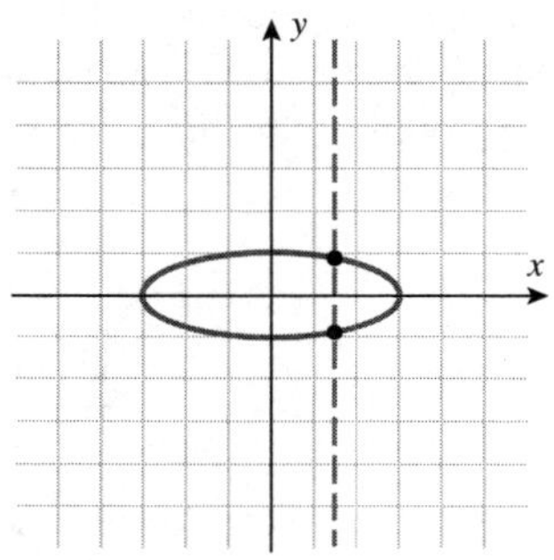

b.

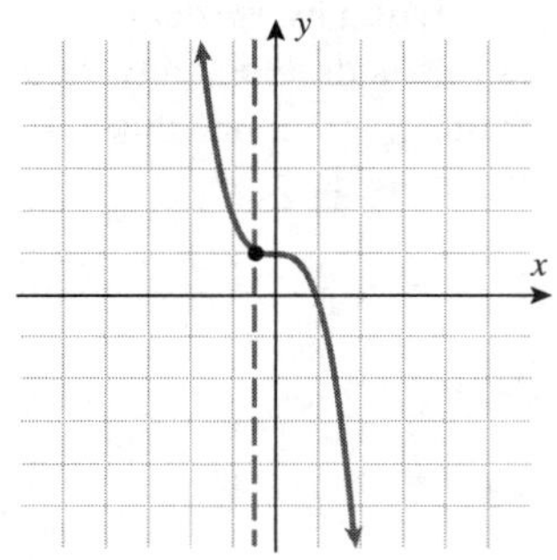

a. Because the vertical line intersects the graph of the equation at two points, this equation does not represent a function.

b. Because any vertical line will intersect the graph of this equation at no more than one point, this equation represents a function.

■ **YOUR TURN** Determine whether the equation $(x - 3)^2 + (y + 2)^2 = 16$ is a function of x.

■ **Answer:** The graph of the equation is a circle, which does not pass the vertical line test. Therefore, the equation does not define a function.

To recap, a function can be expressed one of four ways: verbally, numerically, algebraically, and graphically. This is sometimes called the Rule of 4.

Expressing a Function

VERBALLY	NUMERICALLY	ALGEBRAICALLY	GRAPHICALLY
Every real number has a corresponding absolute value.	$\{(-3, 3), (-1, 1), (0, 0), (1, 1), (5, 5)\}$	$y = \lvert x \rvert$	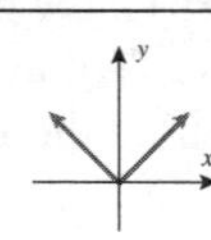

Function Notation

We know that the equation $y = 2x + 5$ is a function because its graph is a nonvertical line and thus passes the vertical line test. We can select x-values (input) and determine unique corresponding y-values (output). The output is found by taking 2 times the input and then adding 5. If we give the function a name, say, "f", then we can use **function notation**:

$$f(x) = 2x + 5$$

The symbol $f(x)$ is read "f evaluated at x" or "f of x" and represents the y-value that corresponds to a particular x-value. In other words, $y = f(x)$.

INPUT	FUNCTION	OUTPUT	EQUATION
x	f	$f(x)$	$f(x) = 2x + 5$
Independent variable	Mapping	Dependent variable	Mathematical rule

It is important to note that f is the function name, whereas $f(x)$ is the value of the function. In other words, the function f maps some value x in the domain to some value $f(x)$ in the range.

x	$f(x) = 2x + 5$	$f(x)$
0	$2(0) + 5$	$f(0) = 5$
1	$2(1) + 5$	$f(1) = 7$
2	$2(2) + 5$	$f(2) = 9$

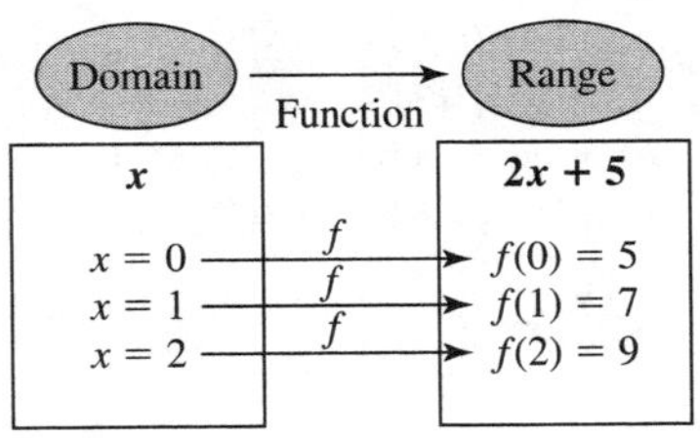

The independent variable is also referred to as the **argument** of a function. To evaluate functions, it is often useful to think of the independent variable or argument as a placeholder. For example, $f(x) = x^2 - 3x$ can be thought of as

$$f(\square) = (\square)^2 - 3(\square)$$

In other words, "f of the argument is equal to the argument squared minus 3 times the argument." Any expression can be substituted for the argument:

$$f(1) = (1)^2 - 3(1)$$
$$f(x + 1) = (x + 1)^2 - 3(x + 1)$$
$$f(-x) = (-x)^2 - 3(-x)$$

It is important to note:

- $f(x)$ does *not* mean f times x.
- The most common function names are f and F since the word function begins with an "f". Other common function names are g and G, but any letter can be used.
- The letter most commonly used for the independent variable is x. The letter t is also common because in real-world applications it represents time, but any letter can be used.
- Although we can think of y and $f(x)$ as interchangeable, the function notation is useful when we want to consider two or more functions of the same independent variable.

Study Tip

It is important to note that $f(x)$ does not mean f times x.

EXAMPLE 3 Evaluating Functions by Substitution

Given the function $f(x) = 2x^3 - 3x^2 + 6$, find $f(-1)$.

Solution:

Consider the independent variable x to be a placeholder. $f(\square) = 2(\square)^3 - 3(\square)^2 + 6$

To find $f(-1)$, substitute $x = -1$ into the function. $f(-1) = 2(-1)^3 - 3(-1)^2 + 6$

Evaluate the right side. $f(-1) = -2 - 3 + 6$

Simplify. $\boxed{f(-1) = 1}$

EXAMPLE 4 Finding Function Values from the Graph of a Function

The graph of f is given on the right.

a. Find $f(0)$.
b. Find $f(1)$.
c. Find $f(2)$.
d. Find $4f(3)$.
e. Find x such that $f(x) = 10$.
f. Find x such that $f(x) = 2$.

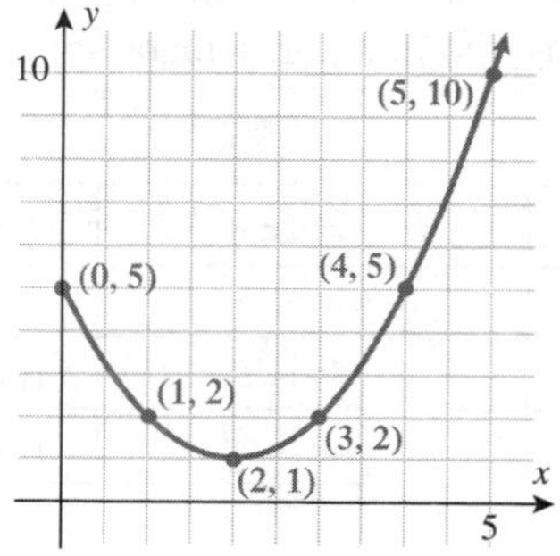

Solution (a): The value $x = 0$ corresponds to the value $y = 5$. $\boxed{f(0) = 5}$

Solution (b): The value $x = 1$ corresponds to the value $y = 2$. $\boxed{f(1) = 2}$

Solution (c): The value $x = 2$ corresponds to the value $y = 1$. $\boxed{f(2) = 1}$

Solution (d): The value $x = 3$ corresponds to the value $y = 2$. $4f(3) = 4 \cdot 2 = \boxed{8}$

Solution (e): The value $y = 10$ corresponds to the value $\boxed{x = 5}$.

Solution (f): The value $y = 2$ corresponds to the values $\boxed{x = 1}$ and $\boxed{x = 3}$.

■ **YOUR TURN** For the following graph of a function, find:

a. $f(-1)$ **b.** $f(0)$ **c.** $3f(2)$
d. the value of x that corresponds to $f(x) = 0$

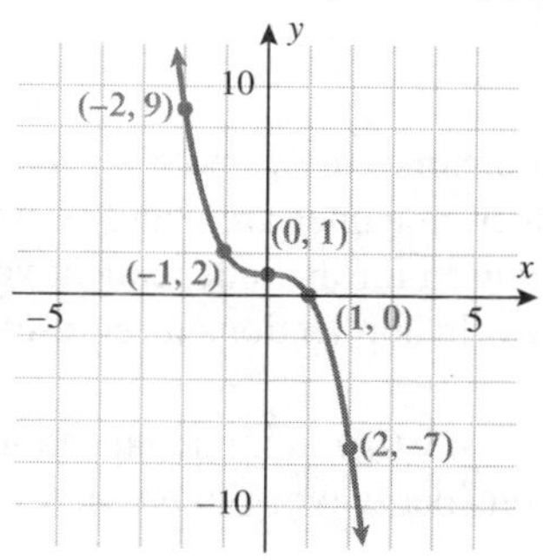

■ **Answer:** **a.** $f(-1) = 2$
b. $f(0) = 1$
c. $3f(2) = -21$
d. $x = 1$

EXAMPLE 5 Evaluating Functions with Variable Arguments (Inputs)

For the given function $f(x) = x^2 - 3x$, evaluate $f(x + 1)$.

COMMON MISTAKE

A common misunderstanding is to interpret the notation $f(x + 1)$ as a sum: $f(x + 1) \neq f(x) + f(1)$.

CORRECT

Write the original function.

$$f(x) = x^2 - 3x$$

Replace the argument x with a placeholder.

$$f(\square) = (\square)^2 - 3(\square)$$

Substitute $x + 1$ for the argument.

$$f(x + 1) = (x + 1)^2 - 3(x + 1)$$

Eliminate the parentheses.

$$f(x + 1) = x^2 + 2x + 1 - 3x - 3$$

Combine like terms.

$$\boxed{f(x + 1) = x^2 - x - 2}$$

INCORRECT

The **ERROR** is in interpreting the notation as a sum.

$$f(x + 1) \neq f(x) + f(1)$$
$$\neq x^2 - 3x - 2$$

CAUTION

$f(x + 1) \neq f(x) + f(1)$

YOUR TURN For the given function $g(x) = x^2 - 2x + 3$, evaluate $g(x - 1)$.

Answer: $g(x - 1) = x^2 - 4x + 6$

EXAMPLE 6 Evaluating Functions: Sums

For the given function $H(x) = x^2 + 2x$, evaluate:

a. $H(x + 1)$ **b.** $H(x) + H(1)$

Solution (a):

Write the function H in placeholder notation. $H(\square) = (\square)^2 + 2(\square)$

Substitute $x + 1$ for the argument of H. $H(x + 1) = (x + 1)^2 + 2(x + 1)$

Eliminate the parentheses on the right side. $H(x + 1) = x^2 + 2x + 1 + 2x + 2$

Combine like terms on the right side. $\boxed{H(x + 1) = x^2 + 4x + 3}$

Solution (b):

Write $H(x)$. $H(x) = x^2 + 2x$

Evaluate H at $x = 1$. $H(1) = (1)^2 + 2(1) = 3$

Evaluate the sum $H(x) + H(1)$. $H(x) + H(1) = x^2 + 2x + 3$

$$\boxed{H(x) + H(1) = x^2 + 2x + 3}$$

Note: Comparing the results of part (a) and part (b), we see that $H(x + 1) \neq H(x) + H(1)$.

Technology Tip

Use a graphing utility to display graphs of $y_1 = H(x + 1) = (x + 1)^2 + 2(x + 1)$ and $y_2 = H(x) + H(1) = x^2 + 2x + 3$.

The graphs are not the same.

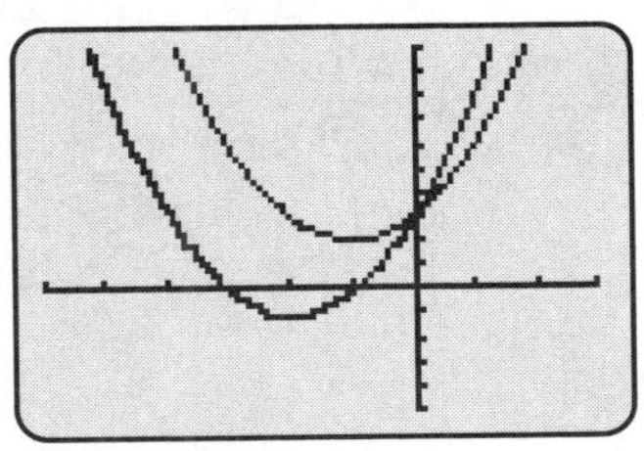

Technology Tip

Use a graphing utility to display graphs of $y_1 = G(-x) = (-x)^2 - (-x)$ and $y_2 = -G(x) = -(x^2 - x)$.

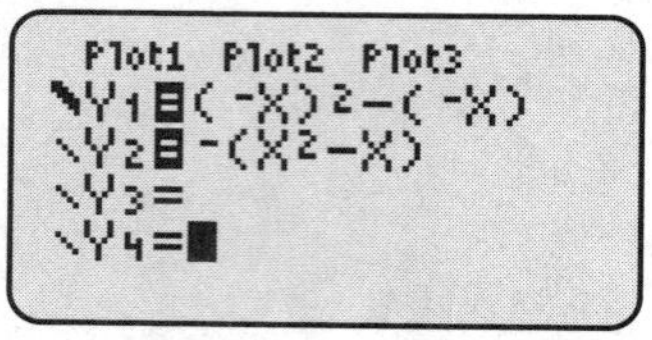

The graphs are not the same.

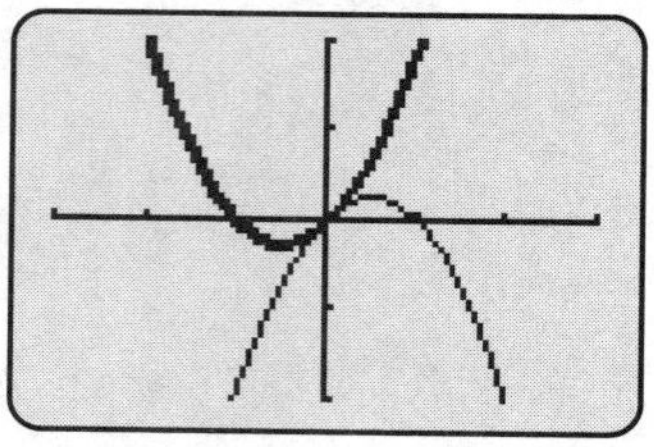

EXAMPLE 7 Evaluating Functions: Negatives

For the given function $G(t) = t^2 - t$, evaluate:

a. $G(-t)$ **b.** $-G(t)$

Solution (a):

Write the function G in placeholder notation. $G(\square) = (\square)^2 - (\square)$

Substitute $-t$ for the argument of G. $G(-t) = (-t)^2 - (-t)$

Eliminate the parentheses on the right side. $\boxed{G(-t) = t^2 + t}$

Solution (b):

Write $G(t)$. $G(t) = t^2 - t$

Multiply by -1. $-G(t) = -(t^2 - t)$

Eliminate the parentheses on the right side. $\boxed{-G(t) = -t^2 + t}$

Note: Comparing the results of part (a) and part (b), we see that $G(-t) \neq -G(t)$.

EXAMPLE 8 Evaluating Functions: Quotients

For the given function $F(x) = 3x + 5$, evaluate:

a. $F\left(\frac{1}{2}\right)$ **b.** $\frac{F(1)}{F(2)}$

Solution (a):

Write F in placeholder notation. $F(\square) = 3(\square) + 5$

Replace the argument with $\frac{1}{2}$. $F\left(\frac{1}{2}\right) = 3\left(\frac{1}{2}\right) + 5$

Simplify the right side. $\boxed{F\left(\frac{1}{2}\right) = \frac{13}{2}}$

Solution (b):

Evaluate $F(1)$. $F(1) = 3(1) + 5 = 8$

Evaluate $F(2)$. $F(2) = 3(2) + 5 = 11$

Divide $F(1)$ by $F(2)$. $\boxed{\frac{F(1)}{F(2)} = \frac{8}{11}}$

Note: Comparing the results of part (a) and part (b), we see that $F\left(\frac{1}{2}\right) \neq \frac{F(1)}{F(2)}$.

CAUTION

$$f\left(\frac{a}{b}\right) \neq \frac{f(a)}{f(b)}$$

YOUR TURN Given the function $G(t) = 3t - 4$, evaluate:

a. $G(t - 2)$ **b.** $G(t) - G(2)$ **c.** $\frac{G(1)}{G(3)}$ **d.** $G\left(\frac{1}{3}\right)$

Answer: **a.** $G(t - 2) = 3t - 10$
b. $G(t) - G(2) = 3t - 6$
c. $\frac{G(1)}{G(3)} = -\frac{1}{5}$
d. $G\left(\frac{1}{3}\right) = -3$

Examples 6, 7, and 8 illustrate the following:

$$f(a + b) \neq f(a) + f(b) \qquad f(-t) \neq -f(t) \qquad f\left(\frac{a}{b}\right) \neq \frac{f(a)}{f(b)}$$

Now that we have shown that $f(x + h) \neq f(x) + f(h)$, we turn our attention to one of the fundamental expressions in calculus: the **difference quotient**.

$$\frac{f(x+h) - f(x)}{h} \qquad h \neq 0$$

Example 9 illustrates the difference quotient, which will be discussed in detail in Section 3.2. For now, we will concentrate on the algebra involved when finding the difference quotient. In Section 3.2, the application of the difference quotient will be the emphasis.

EXAMPLE 9 Evaluating the Difference Quotient

For the function $f(x) = x^2 - x$, find $\dfrac{f(x+h) - f(x)}{h}$.

Solution:

Use placeholder notation for the function $f(x) = x^2 - x$. $\quad f(\square) = (\square)^2 - (\square)$

Calculate $f(x + h)$. $\quad f(x+h) = (x+h)^2 - (x+h)$

Write the difference quotient. $\quad \dfrac{f(x+h) - f(x)}{h}$

Let $f(x + h) = (x + h)^2 - (x + h)$ and $f(x) = x^2 - x$.

$$\frac{f(x+h) - f(x)}{h} = \frac{\overbrace{\left[(x+h)^2 - (x+h)\right]}^{f(x+h)} - \overbrace{\left[x^2 - x\right]}^{f(x)}}{h} \qquad h \neq 0$$

Eliminate the parentheses inside the first set of brackets.

$$= \frac{[x^2 + 2xh + h^2 - x - h] - [x^2 - x]}{h}$$

Eliminate the brackets in the numerator.

$$= \frac{x^2 + 2xh + h^2 - x - h - x^2 + x}{h}$$

Combine like terms.

$$= \frac{2xh + h^2 - h}{h}$$

Factor the numerator.

$$= \frac{h(2x + h - 1)}{h}$$

Divide out the common factor, h.

$$= \boxed{2x + h - 1} \qquad h \neq 0$$

YOUR TURN Evaluate the difference quotient for $f(x) = x^2 - 1$.

Answer: $2x + h$

Domain of a Function

Sometimes the domain of a function is stated *explicitly*. For example,

$$f(x) = |x| \qquad \underbrace{x < 0}_{\text{domain}}$$

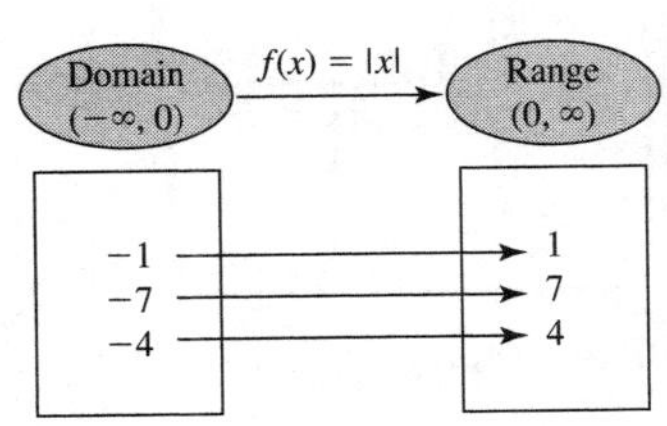

Here, the **explicit domain** is the set of all negative real numbers, $(-\infty, 0)$. Every negative real number in the domain is mapped to a positive real number in the range through the absolute value function.

If the expression that defines the function is given but the domain is not stated explicitly, then the domain is implied. The **implicit domain** is the largest set of real numbers for which the function is defined and the output value $f(x)$ is a real number. For example,

$$f(x) = \sqrt{x}$$

does not have the domain explicitly stated. There is, however, an implicit domain. Note that if the argument is negative, that is, if $x < 0$, then the result is an imaginary number. In order for the output of the function, $f(x)$, to be a real number, we must restrict the domain to nonnegative numbers, that is, if $x \geq 0$.

FUNCTION	IMPLICIT DOMAIN
$f(x) = \sqrt{x}$	$[0, \infty)$

In general, we ask the question, "what can x be?" The implicit domain of a function excludes values that cause a function to be undefined or have outputs that are not real numbers.

EXPRESSION THAT DEFINES THE FUNCTION	EXCLUDED x-VALUES	EXAMPLE	IMPLICIT DOMAIN
Polynomial	None	$f(x) = x^3 - 4x^2$	All real numbers
Rational	x-values that make the denominator equal to 0	$g(x) = \dfrac{2}{x^2 - 9}$	$x \neq \pm 3$ or $(-\infty, -3) \cup (-3, 3) \cup (3, \infty)$
Radical	x-values that result in a square (even) root of a negative number	$h(x) = \sqrt{x - 5}$	$x \geq 5$ or $[5, \infty)$

Technology Tip

To visualize the domain of each function, ask the question: What are the excluded x-values in the graph?

a. Graph of $F(x) = \dfrac{3}{x^2 - 25}$ is shown.

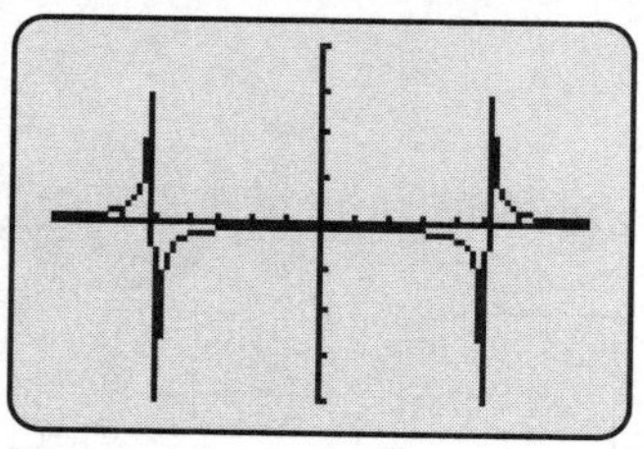

The excluded x-values are -5 and 5.

EXAMPLE 10 Determining the Domain of a Function

State the domain of the given functions.

a. $F(x) = \dfrac{3}{x^2 - 25}$ **b.** $H(x) = \sqrt[4]{9 - 2x}$ **c.** $G(x) = \sqrt[3]{x - 1}$

Solution (a):

Write the original equation. $F(x) = \dfrac{3}{x^2 - 25}$

Determine any restrictions on the values of x. $x^2 - 25 \neq 0$

Solve the restriction equation. $x^2 \neq 25$ or $x \neq \pm\sqrt{25} = \pm 5$

State the domain restrictions. $x \neq \pm 5$

Write the domain in interval notation. $\boxed{(-\infty, -5) \cup (-5, 5) \cup (5, \infty)}$

Solution (b):

Write the original equation. $H(x) = \sqrt[4]{9 - 2x}$

Determine any restrictions on the values of x. $9 - 2x \geq 0$

Solve the restriction equation. $9 \geq 2x$

State the domain restrictions. $x \leq \frac{9}{2}$

Write the domain in interval notation. $\boxed{\left(-\infty, \frac{9}{2}\right]}$

Solution (c):

Write the original equation. $G(x) = \sqrt[3]{x - 1}$

Determine any restrictions on the values of x. no restrictions

State the domain. $\mathbb{R}$

Write the domain in interval notation. $\boxed{(-\infty, \infty)}$

YOUR TURN State the domain of the given functions.

a. $f(x) = \sqrt{x - 3}$ **b.** $g(x) = \frac{1}{x^2 - 4}$

Answer: a. $x \geq 3$ or $[3, \infty)$
b. $x \neq \pm 2$ or $(-\infty, -2) \cup (-2, 2) \cup (2, \infty)$

Applications

Functions that are used in applications often have restrictions on the domains due to physical constraints. For example, the volume of a cube is given by the function $V(x) = x^3$, where x is the length of a side. The function $f(x) = x^3$ has no restrictions on x, and therefore the domain is the set of all real numbers. However, the volume of any cube has the restriction that the length of a side can never be negative or zero.

EXAMPLE 11 Price of Gasoline

Following the capture of Saddam Hussein in Iraq in 2003, gas prices in the United States escalated and then finally returned to their precapture prices. Over a 6-month period, the average price of a gallon of 87 octane gasoline was given by the function $C(x) = -0.05x^2 + 0.3x + 1.7$, where C is the cost function and x represents the number of months after the capture.

a. Determine the domain of the cost function.
b. What was the average price of gas per gallon 3 months after the capture?

Solution (a):

Since the cost function $C(x) = -0.05x^2 + 0.3x + 1.7$ modeled the price of gas only for 6 months after the capture, the domain is $0 \leq x \leq 6$ or $\boxed{[0, 6]}$.

Solution (b):

Write the cost function. $C(x) = -0.05x^2 + 0.3x + 1.7 \quad 0 \leq x \leq 6$

Find the value of the function when $x = 3$. $C(3) = -0.05(3)^2 + 0.3(3) + 1.7$

Simplify. $C(3) = 2.15$

The average price per gallon 3 months after the capture was $\boxed{\$2.15}$.

EXAMPLE 12 The Dimensions of a Pool

Express the volume of a 30 ft × 10 ft rectangular swimming pool as a function of its depth.

Solution:

The volume of any rectangular box is $V = lwh$, where V is the volume, l is the length, w is the width, and h is the height. In this example, the length is 30 ft, the width is 10 ft, and the height represents the depth d of the pool.

Write the volume as a function of depth d. $V(d) = (30)(10)d$

Simplify. $\boxed{V(d) = 300d}$

Determine any restrictions on the domain. $d > 0$

SECTION 3.1 SUMMARY

Relations and Functions (Let x represent the independent variable and y the dependent variable.)

TYPE	MAPPING/CORRESPONDENCE	EQUATION	GRAPH
Relation	Every x-value in the domain maps to **at least one** y-value in the range.	$x = y^2$	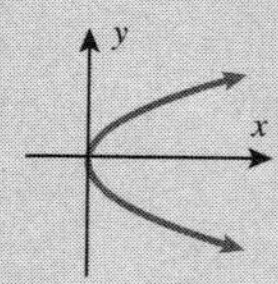
Function	Every x-value in the domain maps to **exactly one** y-value in the range.	$y = x^2$	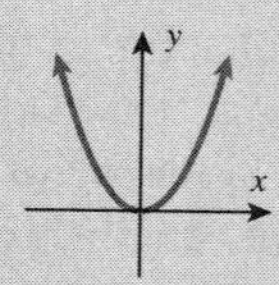 Passes vertical line test

All functions are relations, but not all relations are functions. Functions can be represented by equations. In the following table, each column illustrates an alternative notation.

INPUT	CORRESPONDENCE	OUTPUT	EQUATION
x	Function	y	$y = 2x + 5$
Independent variable	Mapping	Dependent variable	Mathematical rule
Argument	f	$f(x)$	$f(x) = 2x + 5$

The **domain** is the set of all inputs (x-values) and the **range** is the set of all corresponding outputs (y-values). Placeholder notation is useful when evaluating functions.

$$f(x) = 3x^2 + 2x$$

$$f(\square) = 3(\square)^2 + 2(\square)$$

Explicit domain is stated, whereas **implicit domain** is found by *excluding* x-values that:

- make the function undefined (denominator = 0).
- result in a nonreal output (even roots of negative real numbers).

SECTION 3.1 EXERCISES

SKILLS

In Exercises 1–24, determine whether each relation is a function. Assume that the coordinate pair (x, y) represents the independent variable x and the dependent variable y.

1.

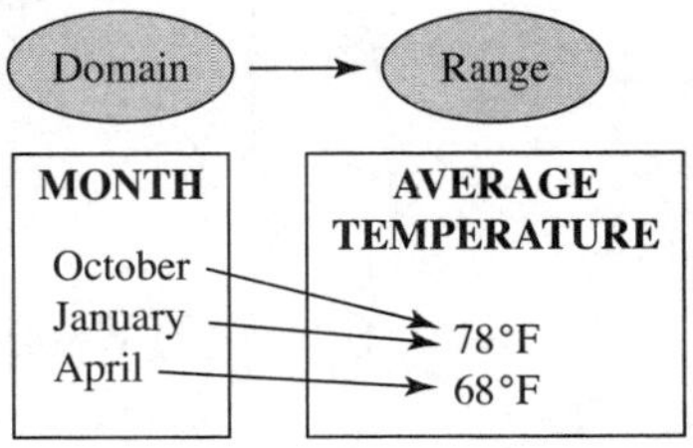

2.

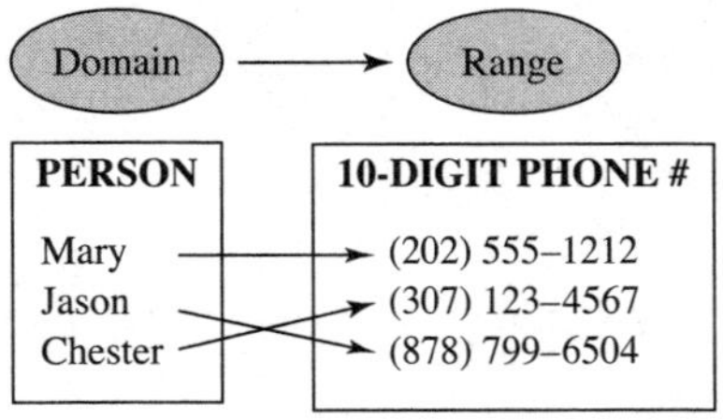

3.

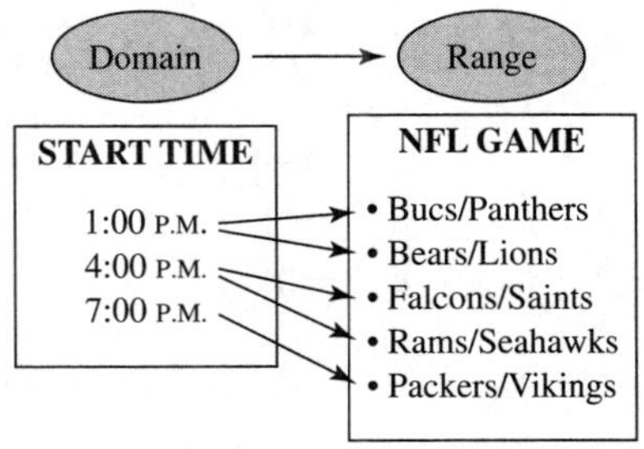

4.

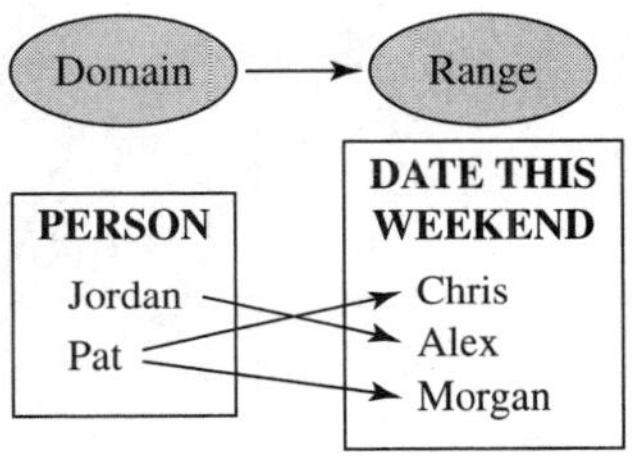

5.

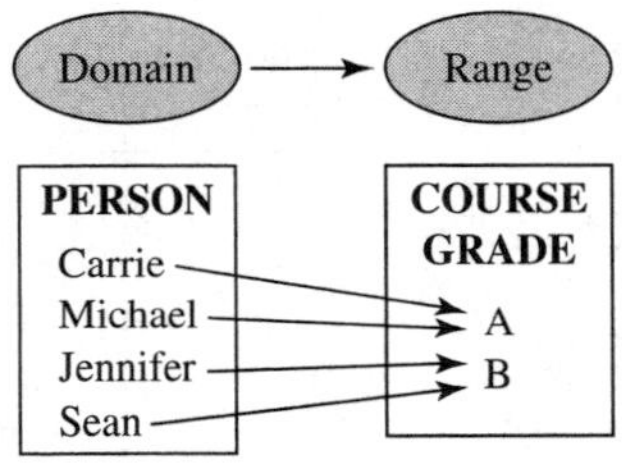

6.

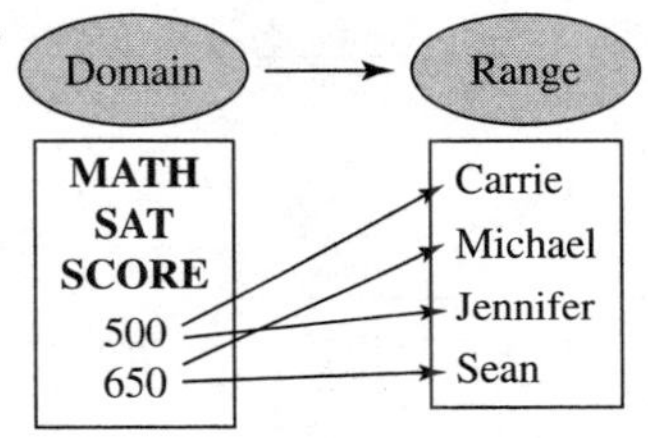

7. $\{(0, -3), (0, 3), (-3, 0), (3, 0)\}$

8. $\{(2, -2), (2, 2), (5, -5), (5, 5)\}$

9. $\{(0, 0), (9, -3), (4, -2), (4, 2), (9, 3)\}$

10. $\{(0, 0), (-1, -1), (-2, -8), (1, 1), (2, 8)\}$

11. $\{(0, 1), (1, 0), (2, 1), (-2, 1), (5, 4), (-3, 4)\}$

12. $\{(0, 1), (1, 1), (2, 1), (3, 1)\}$

13. $x^2 + y^2 = 9$

14. $x = |y|$

15. $x = y^2$

16. $y = x^3$

17. $y = |x - 1|$

18. $y = 3$

19.

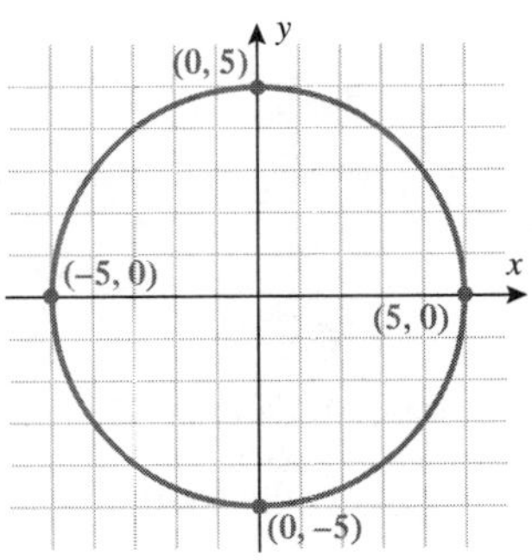

20.

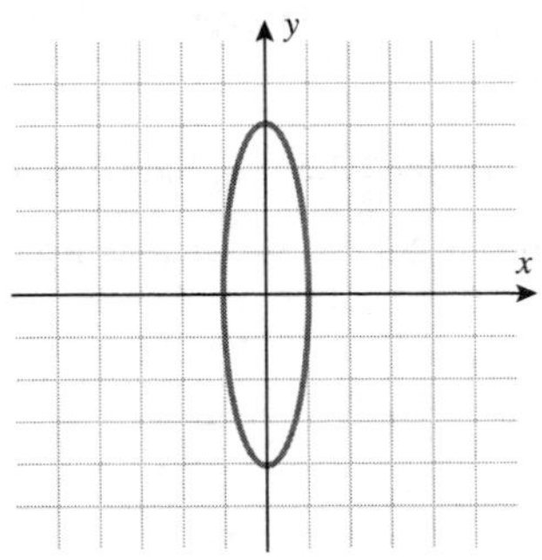

21.

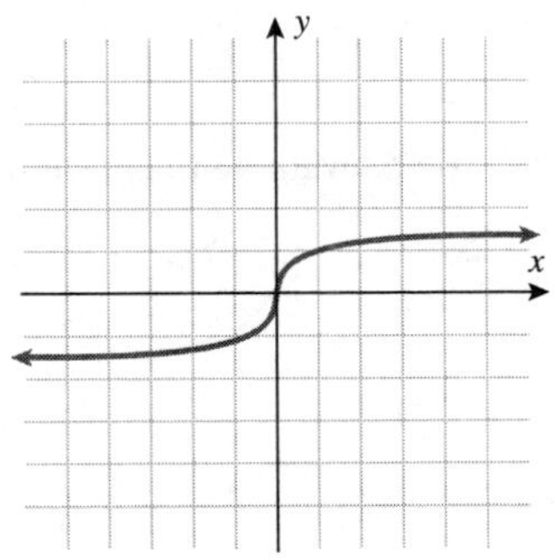

22.

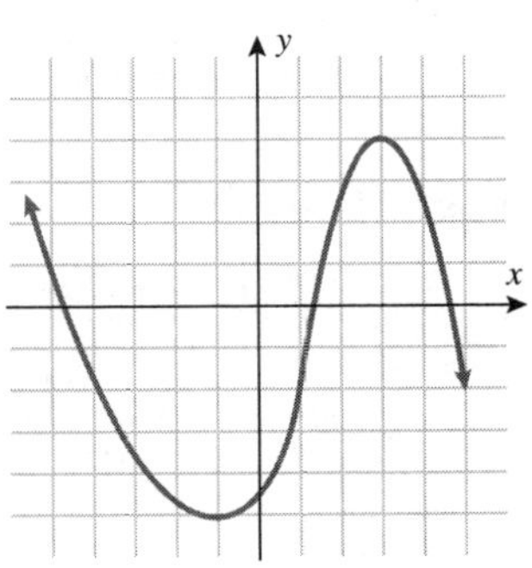

23.

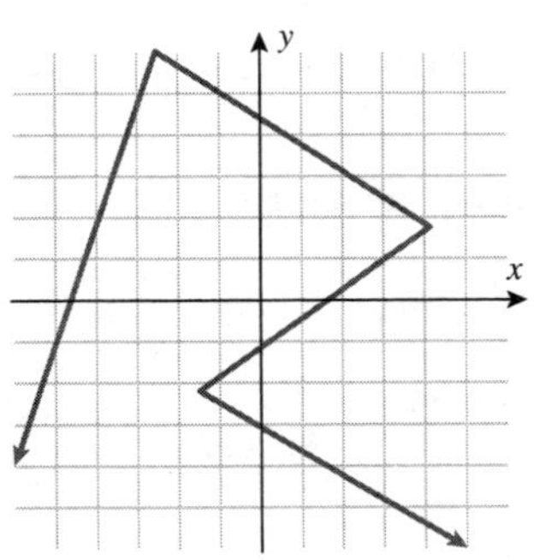

24.

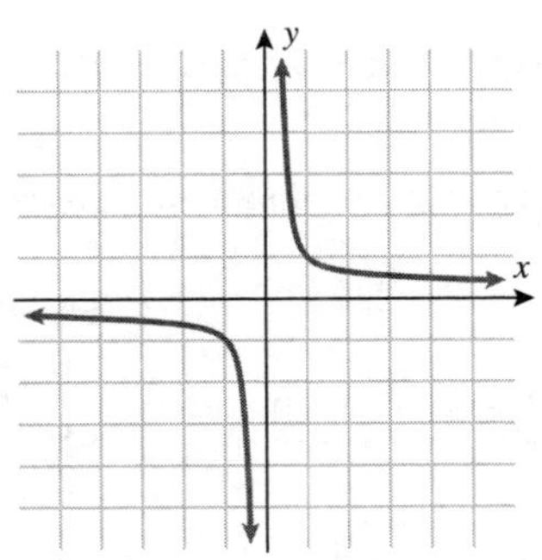

In Exercises 25–32, use the given graphs to evaluate the functions.

25. $y = f(x)$

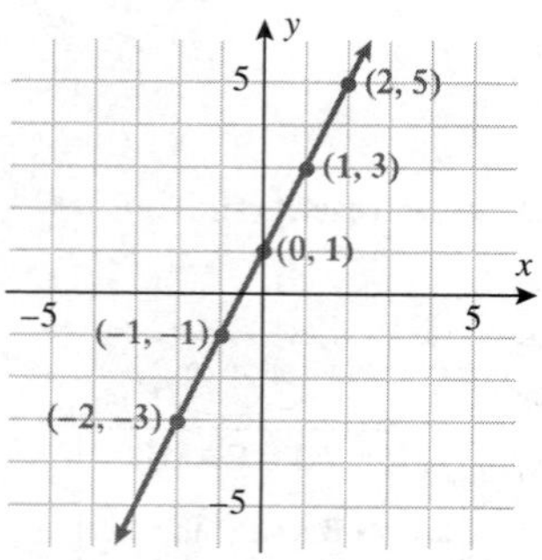

a. $f(2)$ **b.** $f(0)$ **c.** $f(-2)$

26. $y = g(x)$

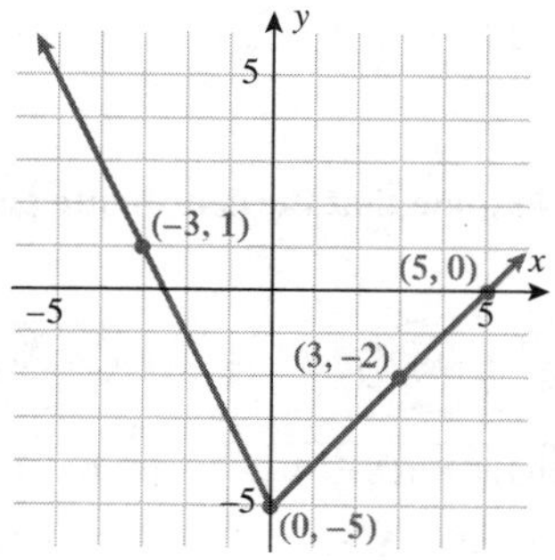

a. $g(-3)$ **b.** $g(0)$ **c.** $g(5)$

27. $y = p(x)$

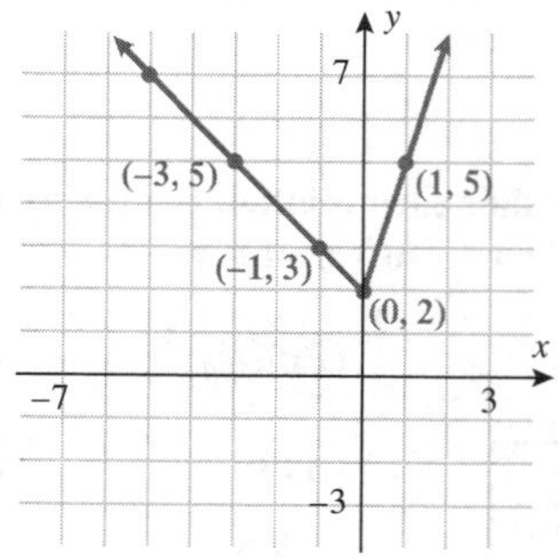

a. $p(-1)$ **b.** $p(0)$ **c.** $p(1)$

28. $y = r(x)$

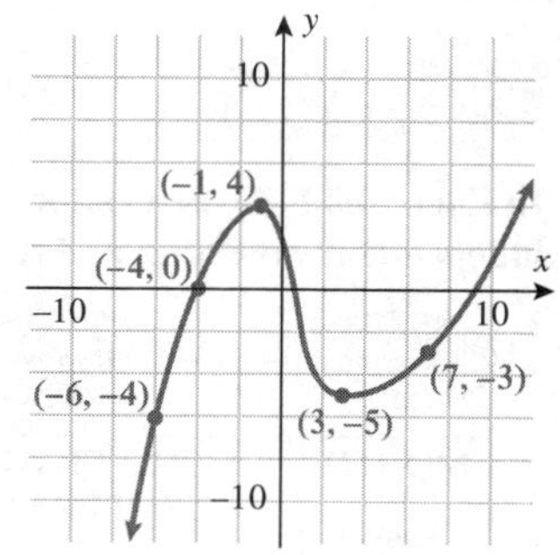

a. $r(-4)$ **b.** $r(-1)$ **c.** $r(3)$

29. $y = C(x)$

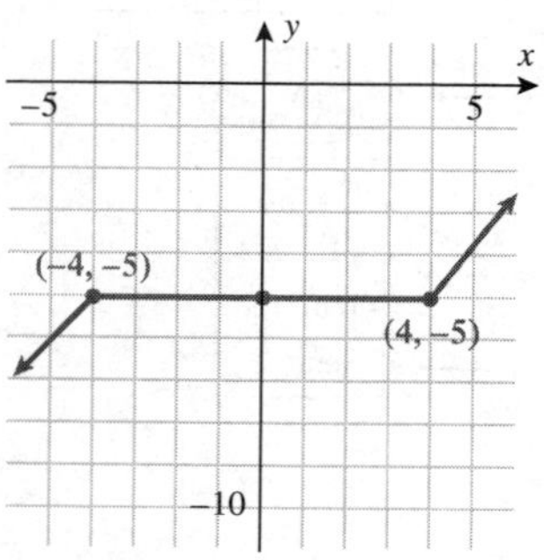

a. $C(2)$ **b.** $C(0)$ **c.** $C(-2)$

30. $y = q(x)$

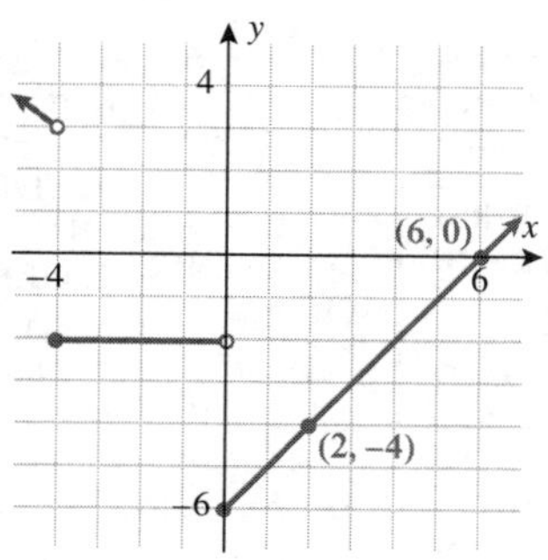

a. $q(-4)$ **b.** $q(0)$ **c.** $q(2)$

31. $y = S(x)$

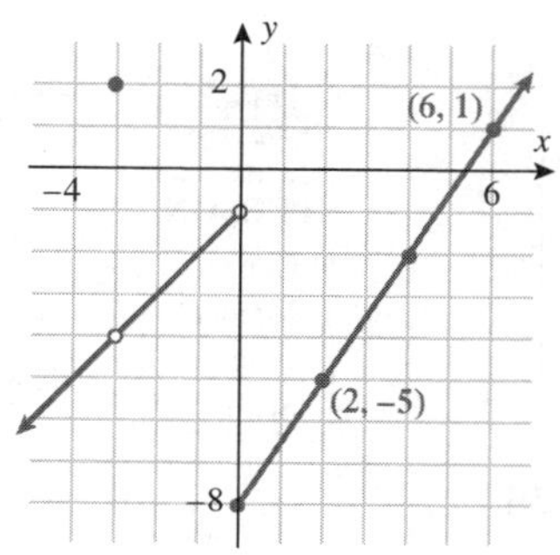

a. $S(-3)$ **b.** $S(0)$ **c.** $S(2)$

32. $y = T(x)$

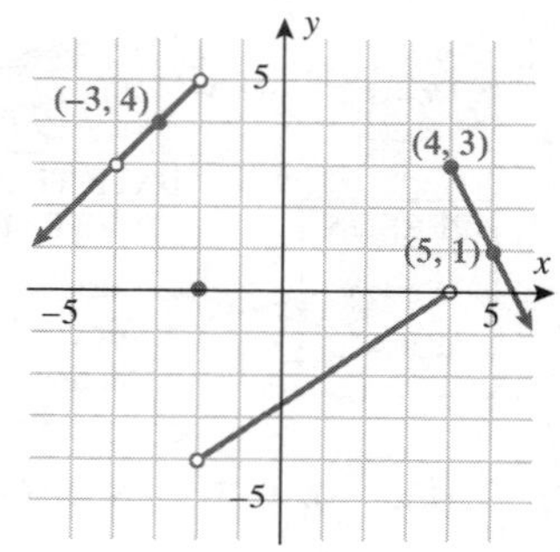

a. $T(-5)$ **b.** $T(-2)$ **c.** $T(4)$

33. Find x if $f(x) = 3$ in Exercise 25.

34. Find x if $g(x) = -2$ in Exercise 26.

35. Find x if $p(x) = 5$ in Exercise 27.

36. Find x if $C(x) = -7$ in Exercise 29.

37. Find x if $C(x) = -5$ in Exercise 29.

38. Find x if $q(x) = -2$ in Exercise 30.

39. Find x if $S(x) = 1$ in Exercise 31.

40. Find x if $T(x) = 4$ in Exercise 32.

In Exercises 41–56, evaluate the given quantities applying the following four functions.

$f(x) = 2x - 3$ $\quad F(t) = 4 - t^2$ $\quad g(t) = 5 + t$ $\quad G(x) = x^2 + 2x - 7$

41. $f(-2)$

42. $G(-3)$

43. $g(1)$

44. $F(-1)$

45. $f(-2) + g(1)$

46. $G(-3) - F(-1)$

47. $3f(-2) - 2g(1)$

48. $2F(-1) - 2G(-3)$

49. $\dfrac{f(-2)}{g(1)}$

50. $\dfrac{G(-3)}{F(-1)}$

51. $\dfrac{f(0) - f(-2)}{g(1)}$

52. $\dfrac{G(0) - G(-3)}{F(-1)}$

53. $f(x + 1) - f(x - 1)$

54. $F(t + 1) - F(t - 1)$

55. $g(x + a) - f(x + a)$

56. $G(x + b) + F(b)$

In Exercises 57–64, evaluate the difference quotients using the same *f*, *F*, *G*, and *g* given for Exercises 41–56.

57. $\dfrac{f(x + h) - f(x)}{h}$

58. $\dfrac{F(t + h) - F(t)}{h}$

59. $\dfrac{g(t + h) - g(t)}{h}$

60. $\dfrac{G(x + h) - G(x)}{h}$

61. $\dfrac{f(-2 + h) - f(-2)}{h}$

62. $\dfrac{F(-1 + h) - F(-1)}{h}$

63. $\dfrac{g(1 + h) - g(1)}{h}$

64. $\dfrac{G(-3 + h) - G(-3)}{h}$

In Exercises 65–96, find the domain of the given function. Express the domain in interval notation.

65. $f(x) = 2x - 5$

66. $f(x) = -2x - 5$

67. $g(t) = t^2 + 3t$

68. $h(x) = 3x^4 - 1$

69. $P(x) = \dfrac{x+5}{x-5}$

70. $Q(t) = \dfrac{2-t^2}{t+3}$

71. $T(x) = \dfrac{2}{x^2-4}$

72. $R(x) = \dfrac{1}{x^2-1}$

73. $F(x) = \dfrac{1}{x^2+1}$

74. $G(t) = \dfrac{2}{t^2+4}$

75. $q(x) = \sqrt{7-x}$

76. $k(t) = \sqrt{t-7}$

77. $f(x) = \sqrt{2x+5}$

78. $g(x) = \sqrt{5-2x}$

79. $G(t) = \sqrt{t^2-4}$

80. $F(x) = \sqrt{x^2-25}$

81. $F(x) = \dfrac{1}{\sqrt{x-3}}$

82. $G(x) = \dfrac{2}{\sqrt{5-x}}$

83. $f(x) = \sqrt[3]{1-2x}$

84. $g(x) = \sqrt[5]{7-5x}$

85. $P(x) = \dfrac{1}{\sqrt[5]{x+4}}$

86. $Q(x) = \dfrac{x}{\sqrt[3]{x^2-9}}$

87. $R(x) = \dfrac{x+1}{\sqrt[4]{3-2x}}$

88. $p(x) = \dfrac{x^2}{\sqrt{25-x^2}}$

89. $H(t) = \dfrac{t}{\sqrt{t^2-t-6}}$

90. $f(t) = \dfrac{t-3}{\sqrt[4]{t^2+9}}$

91. $f(x) = (x^2-16)^{1/2}$

92. $g(x) = (2x-5)^{1/3}$

93. $r(x) = x^2(3-2x)^{-1/2}$

94. $p(x) = (x-1)^2(x^2-9)^{-3/5}$

95. $f(x) = \frac{2}{5}x - \frac{2}{4}$

96. $g(x) = \frac{2}{3}x^2 - \frac{1}{6}x - \frac{3}{4}$

97. Let $g(x) = x^2 - 2x - 5$ and find the values of x that correspond to $g(x) = 3$.

98. Let $g(x) = \frac{5}{6}x - \frac{3}{4}$ and find the value of x that corresponds to $g(x) = \frac{2}{3}$.

99. Let $f(x) = 2x(x-5)^3 - 12(x-5)^2$ and find the values of x that correspond to $f(x) = 0$.

100. Let $f(x) = 3x(x+3)^2 - 6(x+3)^3$ and find the values of x that correspond to $f(x) = 0$.

APPLICATIONS

101. Budget: Event Planning. The cost associated with a catered wedding reception is \$45 per person for a reception for more than 75 people. Write the cost of the reception in terms of the number of guests and state any domain restrictions.

102. Budget: Long-Distance Calling. The cost of a local home phone plan is \$35 for basic service and \$.10 per minute for any domestic long-distance calls. Write the cost of monthly phone service in terms of the number of monthly long-distance minutes and state any domain restrictions.

103. Temperature. The average temperature in Tampa, Florida, in the springtime is given by the function $T(x) = -0.7x^2 + 16.8x - 10.8$, where T is the temperature in degrees Fahrenheit and x is the time of day in military time and is restricted to $6 \le x \le 18$ (sunrise to sunset). What is the temperature at 6 A.M.? What is the temperature at noon?

104. Falling Objects: Firecrackers. A firecracker is launched straight up, and its height is a function of time, $h(t) = -16t^2 + 128t$, where h is the height in feet and t is the time in seconds with $t = 0$ corresponding to the instant it launches. What is the height 4 seconds after launch? What is the domain of this function?

105. Collectibles. The price of a signed Alex Rodriguez baseball card is a function of how many are for sale. When Rodriguez was traded from the Texas Rangers to the New York Yankees in 2004, the going rate for a signed baseball card on eBay was $P(x) = 10 + \sqrt{400{,}000 - 100x}$, where x represents the number of signed cards for sale. What was the value of the card when there were 10 signed cards for sale? What was the value of the card when there were 100 signed cards for sale?

106. Collectibles. In Exercise 105, what was the lowest price on eBay, and how many cards were available then? What was the highest price on eBay, and how many cards were available then?

107. Volume. An open box is constructed from a square 10-inch piece of cardboard by cutting squares of length x inches out of each corner and folding the sides up. Express the volume of the box as a function of x, and state the domain.

108. Volume. A cylindrical water basin will be built to harvest rainwater. The basin is limited in that the largest radius it can have is 10 feet. Write a function representing the volume of water V as a function of height h. How many additional gallons of water will be collected if you increase the height by 2 feet? (*Hint:* 1 cubic foot = 7.48 gallons.)

Refer to the table below for Exercises 109 and 110. It illustrates the average federal funds rate for the month of January (2000 to 2008).

YEAR	FED. RATE
2000	5.45
2001	5.98
2002	1.73
2003	1.24
2004	1.00
2005	2.25
2006	4.50
2007	5.25
2008	3.50

109. Finance. Is the relation whose domain is the year and whose range is the average federal funds rate for the month of January a function? Explain.

110. Finance. Write five ordered pairs whose domain is the set of even years from 2000 to 2008 and whose range is the set of corresponding average federal funds rate for the month of January.

For Exercises 111 and 112, use the following figure:

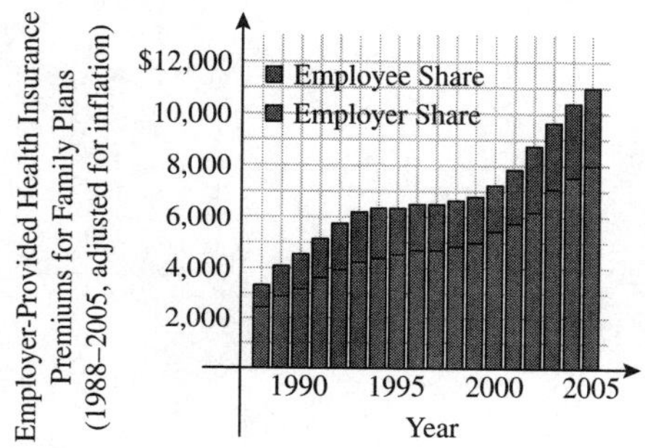

Source: Kaiser Family Foundation Health Research and Education Trust. *Note:* The following years were interpolated: 1989–1992; 1994–1995; 1997–1998.

111. Health Care Costs: Fill in the following table. Round dollars to the nearest $1,000.

YEAR	TOTAL HEALTH CARE COST FOR FAMILY PLANS
1989	
1993	
1997	
2001	
2005	

Write the five ordered pairs resulting from the table.

112. Health Care Costs. Using the table found in Exercise 111, let the years correspond to the domain and the total costs correspond to the range. Is this relation a function? Explain.

For Exercises 113 and 114, use the following information:

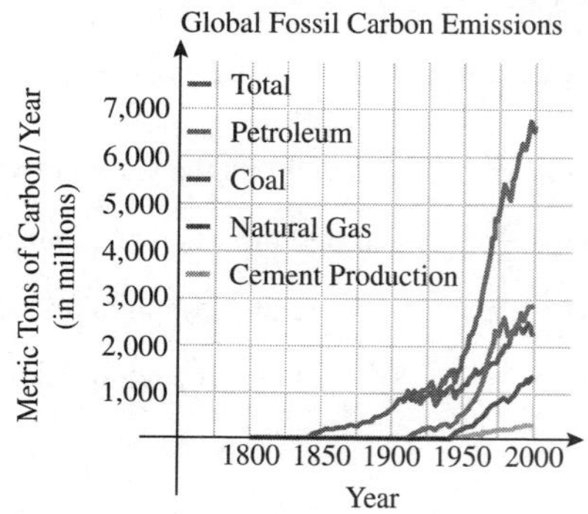

Source: http:/www.naftc.wvu.edu

Let the functions f, F, g, G, and H represent the number of tons of carbon emitted per year as a function of year corresponding to cement production, natural gas, coal, petroleum, and the total amount, respectively. Let t represent the year, with $t = 0$ corresponding to 1900.

113. Environment: Global Climate Change. Estimate (to the nearest thousand) the value of

a. $F(50)$ **b.** $g(50)$ **c.** $H(50)$

114. Environment: Global Climate Change. Explain what the sum $F(100) + g(100) + G(100)$ represents.

CATCH THE MISTAKE

In Exercises 115–120, explain the mistake that is made.

115. Determine whether the relationship is a function.

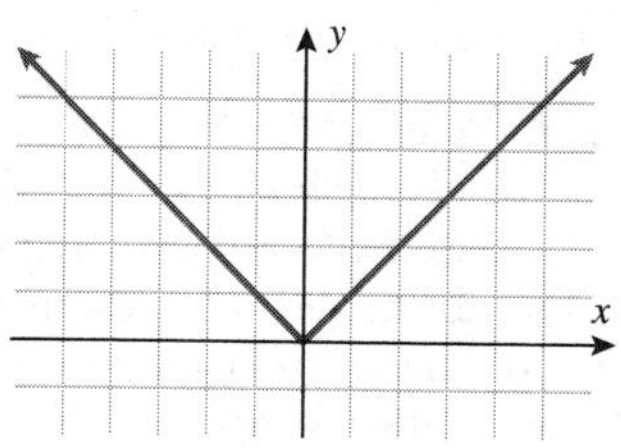

Solution:

Apply the horizontal line test.

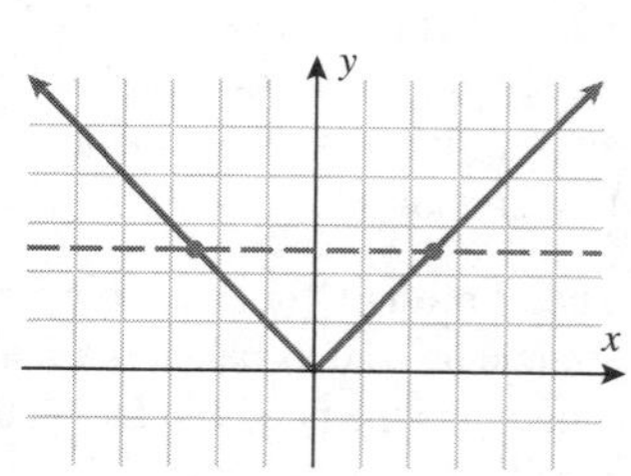

Because the horizontal line intersects the graph in two places, this is not a function.

This is incorrect. What mistake was made?

116. Given the function $H(x) = 3x - 2$, evaluate the quantity $H(3) - H(-1)$.

Solution: $H(3) - H(-1) = H(3) + H(1) = 7 + 1 = 8$

This is incorrect. What mistake was made?

117. Given the function $f(x) = x^2 - x$, evaluate the quantity $f(x + 1)$.

Solution: $f(x + 1) = f(x) + f(1) = x^2 - x + 0$
$f(x + 1) = x^2 - x$

This is incorrect. What mistake was made?

118. Determine the domain of the function $g(t) = \sqrt{3 - t}$ and express it in interval notation.

Solution:

What can t be? Any nonnegative real number.
$3 - t > 0$
$3 > t$ or $t < 3$
Domain: $(-\infty, 3)$

This is incorrect. What mistake was made?

119. Given the function $G(x) = x^2$, evaluate $\dfrac{G(-1 + h) - G(-1)}{h}$.

Solution:

$$\frac{G(-1 + h) - G(-1)}{h} = \frac{G(-1) + G(h) - G(-1)}{h}$$
$$= \frac{G(h)}{h} = \frac{h^2}{h} = h$$

This is incorrect. What mistake was made?

120. Given the functions $f(x) = |x - A| - 1$ and $f(1) = -1$, find A.

Solution:

Since $f(1) = -1$, the point $(-1, 1)$ must satisfy the function. $-1 = |-1 - A| - 1$

Add 1 to both sides of the equation. $|-1 - A| = 0$

The absolute value of zero is zero, so there is no need for the absolute value signs: $-1 - A = 0 \Rightarrow A = -1$.

This is incorrect. What mistake was made?

CONCEPTUAL

In Exercises 121–124, determine whether each statement is true or false.

121. If a vertical line does not intersect the graph of an equation, then that equation does not represent a function.

122. If a horizontal line intersects a graph of an equation more than once, the equation does not represent a function.

123. If $f(-a) = f(a)$, then f does not represent a function.

124. If $f(-a) = f(a)$, then f may or may not represent a function.

125. If $f(x) = Ax^2 - 3x$ and $f(1) = -1$, find A.

126. If $g(x) = \dfrac{1}{b - x}$ and $g(3)$ is undefined, find b.

CHALLENGE

127. If $F(x) = \dfrac{C - x}{D - x}$, $F(-2)$ is undefined, and $F(-1) = 4$, find C and D.

128. Construct a function that is undefined at $x = 5$ and whose graph passes through the point $(1, -1)$.

In Exercises 129 and 130, find the domain of each function, where a is any positive real number.

129. $f(x) = \dfrac{-100}{x^2 - a^2}$

130. $f(x) = -5\sqrt{x^2 - a^2}$

TECHNOLOGY

131. Using a graphing utility, graph the temperature function in Exercise 103. What time of day is it the warmest? What is the temperature? Looking at this function, explain why this model for Tampa, Florida, is valid only from sunrise to sunset (6 to 18).

132. Using a graphing utility, graph the height of the firecracker in Exercise 104. How long after liftoff is the firecracker airborne? What is the maximum height that the firecracker attains? Explain why this height model is valid only for the first 8 seconds.

133. Using a graphing utility, graph the price function in Exercise 105. What are the lowest and highest prices of the cards? Does this agree with what you found in Exercise 106?

134. The makers of malted milk balls are considering increasing the size of the spherical treats. The thin chocolate coating on a malted milk ball can be approximated by the surface area, $S(r) = 4\pi r^2$. If the radius is increased 3 mm, what is the resulting increase in required chocolate for the thin outer coating?

135. Let $f(x) = x^2 + 1$. Graph $y_1 = f(x)$ and $y_2 = f(x - 2)$ in the same viewing window. Describe how the graph of y_2 can be obtained from the graph of y_1.

136. Let $f(x) = 4 - x^2$. Graph $y_1 = f(x)$ and $y_2 = f(x + 2)$ in the same viewing window. Describe how the graph of y_2 can be obtained from the graph of y_1.

SECTION 3.2

GRAPHS OF FUNCTIONS; PIECEWISE-DEFINED FUNCTIONS; INCREASING AND DECREASING FUNCTIONS; AVERAGE RATE OF CHANGE

SKILLS OBJECTIVES

- Recognize and graph common functions.
- Classify functions as even, odd, or neither.
- Determine whether functions are increasing, decreasing, or constant.
- Calculate the average rate of change of a function.
- Evaluate the difference quotient for a function.
- Graph piecewise-defined functions.

CONCEPTUAL OBJECTIVES

- Identify common functions.
- Develop and graph piecewise-defined functions.
  - Identify and graph points of discontinuity.
  - State the domain and range.
- Understand that even functions have graphs that are symmetric about the y-axis.
- Understand that odd functions have graphs that are symmetric about the origin.

Common Functions

Point-plotting techniques were introduced in Section 2.2, and we noted there that we would explore some more efficient ways of graphing functions in Chapter 3. The nine main functions you will read about in this section will constitute a "library" of functions that you should commit to memory. We will draw on this library of functions in the next section when graphing transformations are discussed. Several of these functions have been shown previously in this chapter, but now we will classify them specifically by name and identify properties that each function exhibits.

In Section 2.3, we discussed equations and graphs of lines. All lines (with the exception of vertical lines) pass the vertical line test, and hence are classified as functions. Instead of the traditional notation of a line, $y = mx + b$, we use function notation and classify a function whose graph is a *line* as a *linear* function.

LINEAR FUNCTION

$f(x) = mx + b$ $\quad$ m and b are real numbers.

The domain of a linear function $f(x) = mx + b$ is the set of all real numbers $\mathbb{R}$. The graph of this function has slope m and y-intercept b.

LINEAR FUNCTION: $f(x) = mx + b$	SLOPE: m	y-INTERCEPT: b
$f(x) = 2x - 7$	$m = 2$	$b = -7$
$f(x) = -x + 3$	$m = -1$	$b = 3$
$f(x) = x$	$m = 1$	$b = 0$
$f(x) = 5$	$m = 0$	$b = 5$

One special case of the linear function is the *constant function* ($m = 0$).

CONSTANT FUNCTION

$f(x) = b$ b is any real number.

The graph of a constant function $f(x) = b$ is a horizontal line. The y-intercept corresponds to the point $(0, b)$. The domain of a constant function is the set of all real numbers $\mathbb{R}$. The range, however, is a single value b. In other words, all x-values correspond to a single y-value.

Points that lie on the graph of a constant function $f(x) = b$ are

$(-5, b)$
$(-1, b)$
$(0, b)$
$(2, b)$
$(4, b)$
...
(x, b)

Domain: $(-\infty, \infty)$ Range: $[b, b]$ or $\{b\}$

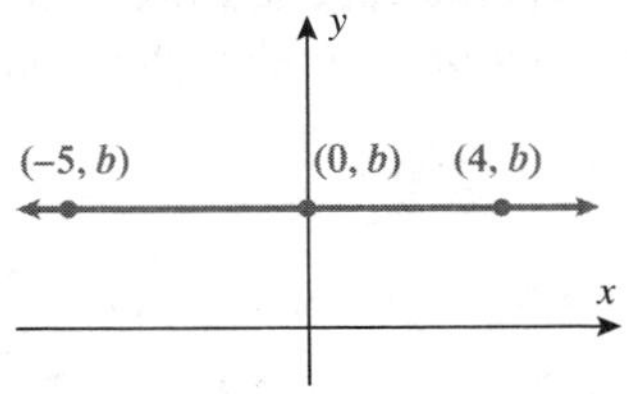

Another specific example of a linear function is the function having a slope of one ($m = 1$) and a y-intercept of zero ($b = 0$). This special case is called the *identity function.*

IDENTITY FUNCTION

$$f(x) = x$$

Identity Function
Domain: $(-\infty, \infty)$ Range: $(-\infty, \infty)$

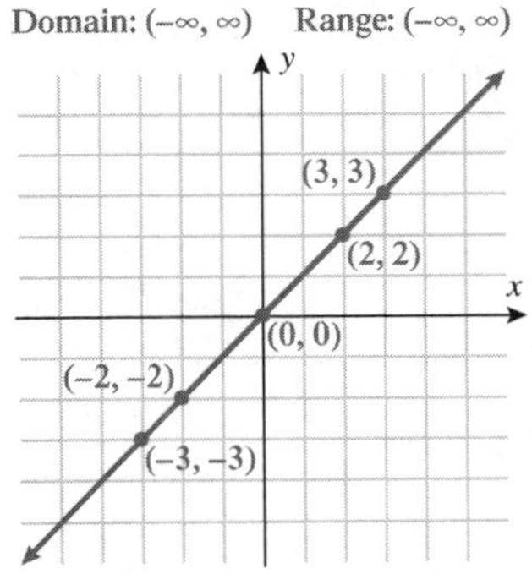

The graph of the identity function has the following properties: It passes through the origin, and every point that lies on the line has equal x- and y-coordinates. Both the domain and the range of the identity function are the set of all real numbers $\mathbb{R}$.

A function that squares the input is called the *square function.*

SQUARE FUNCTION

$$f(x) = x^2$$

Square Function
Domain: $(-\infty, \infty)$ Range: $[0, \infty)$

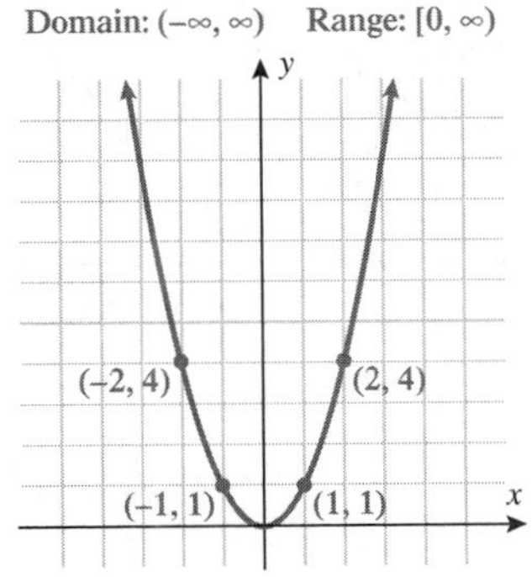

The graph of the square function is called a parabola and will be discussed in further detail in Chapters 4 and 8. The domain of the square function is the set of all real numbers $\mathbb{R}$. Because squaring a real number always yields a positive number or zero, the range of the square function is the set of all nonnegative numbers. Note that the intercept is the origin and the square function is symmetric about the y-axis. This graph is contained in quadrants I and II.

Cube Function
Domain: $(-\infty, \infty)$ Range: $(-\infty, \infty)$

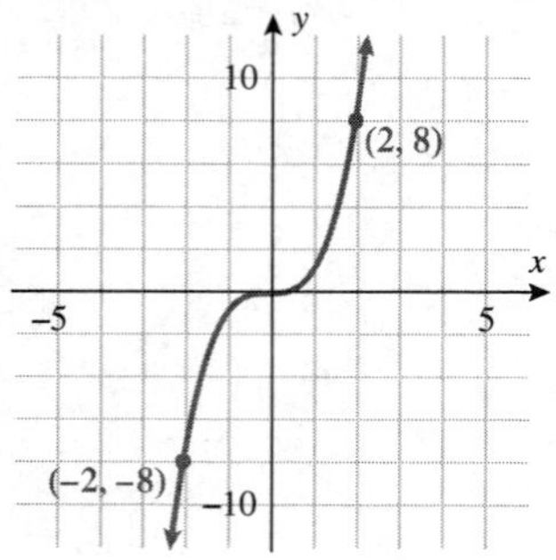

A function that cubes the input is called the *cube function.*

CUBE FUNCTION

$$f(x) = x^3$$

The domain of the cube function is the set of all real numbers $\mathbb{R}$. Because cubing a negative number yields a negative number, cubing a positive number yields a positive number, and cubing 0 yields 0, the range of the cube function is also the set of all real numbers $\mathbb{R}$. Note that the only intercept is the origin and the cube function is symmetric about the origin. This graph extends only into quadrants I and III.

The next two functions are counterparts of the previous two functions: square root and cube root. When a function takes the square root of the input or the cube root of the input, the function is called the *square root function* or the *cube root function*, respectively.

Square Root Function
Domain: $[0, \infty)$ Range: $[0, \infty)$

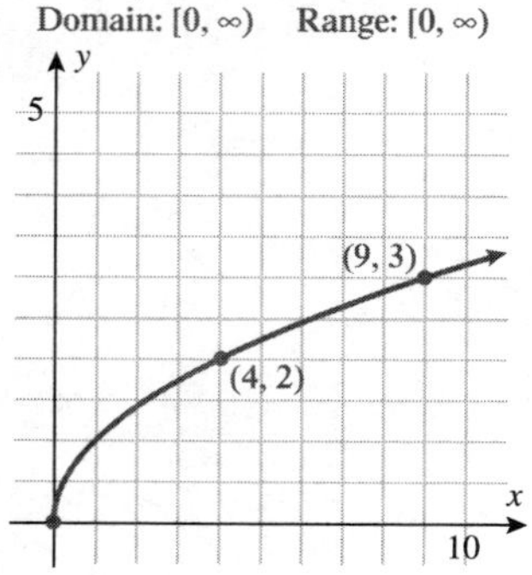

SQUARE ROOT FUNCTION

$$f(x) = \sqrt{x} \quad \text{or} \quad f(x) = x^{1/2}$$

In Section 3.1, we found the domain to be $[0, \infty)$. The output of the function will be all real numbers greater than or equal to zero. Therefore, the range of the square root function is $[0, \infty)$. The graph of this function will be contained in quadrant I.

Cube Root Function
Domain: $(-\infty, \infty)$ Range: $(-\infty, \infty)$

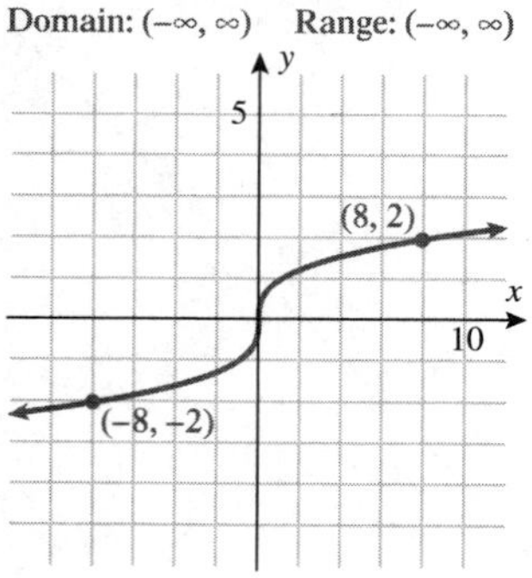

CUBE ROOT FUNCTION

$$f(x) = \sqrt[3]{x} \quad \text{or} \quad f(x) = x^{1/3}$$

In Section 3.1, we stated the domain of the cube root function to be $(-\infty, \infty)$. We see by the graph that the range is also $(-\infty, \infty)$. This graph is contained in quadrants I and III and passes through the origin. This function is symmetric about the origin.

In Section 1.7, you read about absolute value equations and inequalities. Now we shift our focus to the graph of the *absolute value function.*

Absolute Value Function
Domain: $(-\infty, \infty)$ Range: $[0, \infty)$

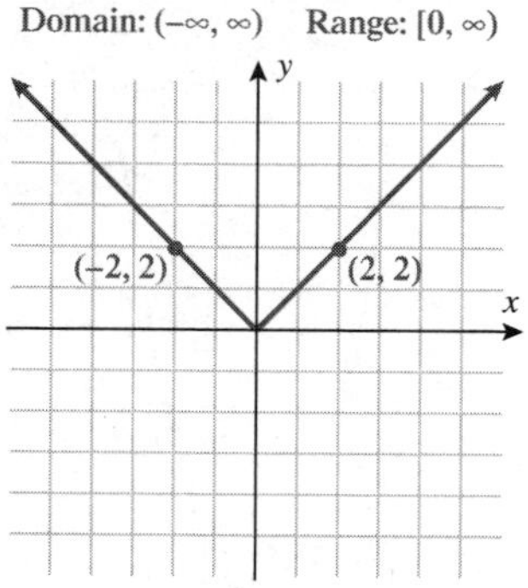

ABSOLUTE VALUE FUNCTION

$$f(x) = |x|$$

Some points that are on the graph of the absolute value function are $(-1, 1)$, $(0, 0)$, and $(1, 1)$. The domain of the absolute value function is the set of all real numbers $\mathbb{R}$, yet the range is the set of nonnegative real numbers. The graph of this function is symmetric with respect to the y-axis and is contained in quadrants I and II.

A function whose output is the reciprocal of its input is called the *reciprocal function.*

Reciprocal Function
Domain: $(-\infty, 0) \cup (0, \infty)$
Range: $(-\infty, 0) \cup (0, \infty)$

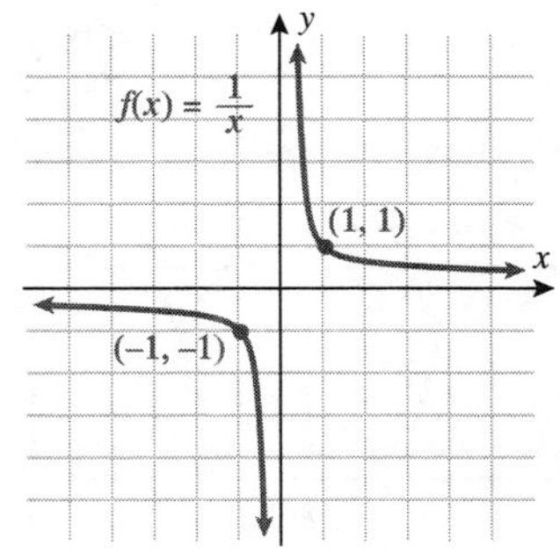

RECIPROCAL FUNCTION

$$f(x) = \frac{1}{x} \qquad x \neq 0$$

The only restriction on the domain of the reciprocal function is that $x \neq 0$. Therefore, we say the domain is the set of all real numbers excluding zero. The graph of the reciprocal function illustrates that its range is also the set of all real numbers except zero. Note that the reciprocal function is symmetric with respect to the origin and is contained in quadrants I and III.

Even and Odd Functions

Of the nine functions discussed above, several have similar properties of symmetry. The constant function, square function, and absolute value function are all symmetric with respect to the y-axis. The identity function, cube function, cube root function, and reciprocal function are all symmetric with respect to the origin. The term **even** is used to describe functions that are symmetric with respect to the y-axis, or vertical axis, and the term **odd** is used to describe functions that are symmetric with respect to the origin. Recall from Section 2.2 that symmetry can be determined both graphically and algebraically. The box below summarizes the graphic and algebraic characteristics of even and odd functions.

EVEN AND ODD FUNCTIONS

Function	Symmetric with Respect to	On Replacing x with $-x$
Even	y-axis or vertical axis	$f(-x) = f(x)$
Odd	origin	$f(-x) = -f(x)$

The algebraic method for determining symmetry with respect to the y-axis, or vertical axis, is to substitute in $-x$ for x. If the result is an equivalent equation, the function is symmetric with respect to the y-axis. Some examples of even functions are $f(x) = b$, $f(x) = x^2$, $f(x) = x^4$, and $f(x) = |x|$. In any of these equations, if $-x$ is substituted for x, the result is the same, that is, $f(-x) = f(x)$. Also note that, with the exception of the absolute value function, these examples are all even-degree polynomial equations. All constant functions are degree zero and are even functions.

The algebraic method for determining symmetry with respect to the origin is to substitute $-x$ for x. If the result is the negative of the original function, that is, if $f(-x) = -f(x)$, then the function is symmetric with respect to the origin and, hence, classified as an odd function. Examples of odd functions are $f(x) = x$, $f(x) = x^3$, $f(x) = x^5$, and $f(x) = x^{1/3}$. In any of these functions, if $-x$ is substituted in for x, the result is the negative of the original function. Note that with the exception of the cube root function, these equations are odd-degree polynomials.

Be careful, though, because functions that are combinations of even- and odd-degree polynomials can turn out to be neither even nor odd, as we will see in Example 1.

EXAMPLE 1 Determining Whether a Function Is Even, Odd, or Neither

Determine whether the functions are even, odd, or neither.

a. $f(x) = x^2 - 3$ **b.** $g(x) = x^5 + x^3$ **c.** $h(x) = x^2 - x$

Solution (a):

Original function. $f(x) = x^2 - 3$

Replace x with $-x$. $f(-x) = (-x)^2 - 3$

Simplify. $f(-x) = x^2 - 3 = f(x)$

Because $f(-x) = f(x)$, we say that $f(x)$ is an *even* function.

Solution (b):

Original function. $g(x) = x^5 + x^3$

Replace x with $-x$. $g(-x) = (-x)^5 + (-x)^3$

Simplify. $g(-x) = -x^5 - x^3 = -(x^5 + x^3) = -g(x)$

Because $g(-x) = -g(x)$, we say that $g(x)$ is an *odd* function.

Solution (c):

Original function. $h(x) = x^2 - x$

Replace x with $-x$. $h(-x) = (-x)^2 - (-x)$

Simplify. $h(-x) = x^2 + x$

$h(-x)$ is neither $-h(x)$ nor $h(x)$; therefore the function $h(x)$ is neither even nor odd.

In parts (*a*), (*b*), and (*c*), we classified these functions as either even, odd, or neither, using the algebraic test. Look back at them now and reflect on whether these classifications agree with your intuition. In part (a), we combined two functions: the square function and the constant function. Both of these functions are even, and adding even functions yields another even function. In part (b), we combined two odd functions: the fifth-power function and the cube function. Both of these functions are odd, and adding two odd functions yields another odd function. In part (c), we combined two functions: the square function and the identity function. The square function is even, and the identity function is odd. In this part, combining an even function with an odd function yields a function that is neither even nor odd and, hence, has no symmetry with respect to the vertical axis or the origin.

YOUR TURN Classify the functions as even, odd, or neither.

a. $f(x) = |x| + 4$ **b.** $f(x) = x^3 - 1$

Technology Tip

a. Graph $y_1 = f(x) = x^2 - 3$.

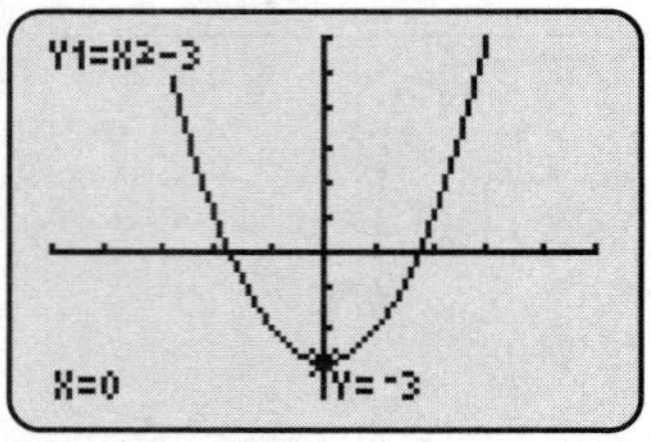

Even; symmetric with respect to origin.

b. Graph $y_1 = g(x) = x^5 + x^3$.

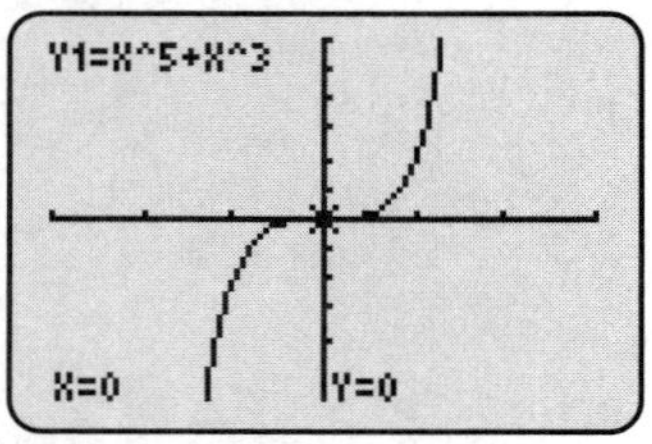

Odd; symmetric with respect to origin.

c. Graph $y_1 = h(x) = x^2 - x$.

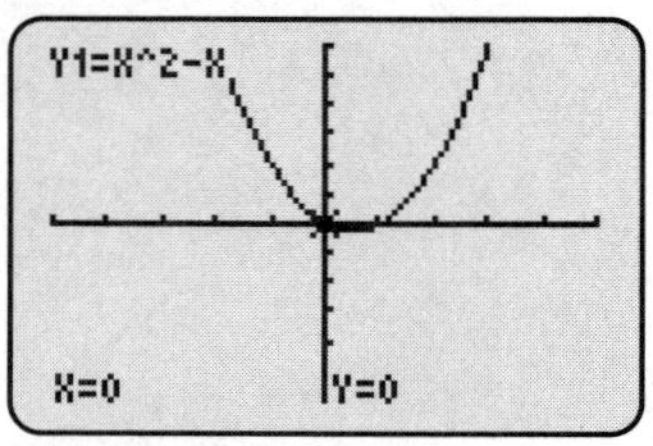

No symmetry with respect to y-axis or origin.

Answer: a. even **b.** neither

Increasing and Decreasing Functions

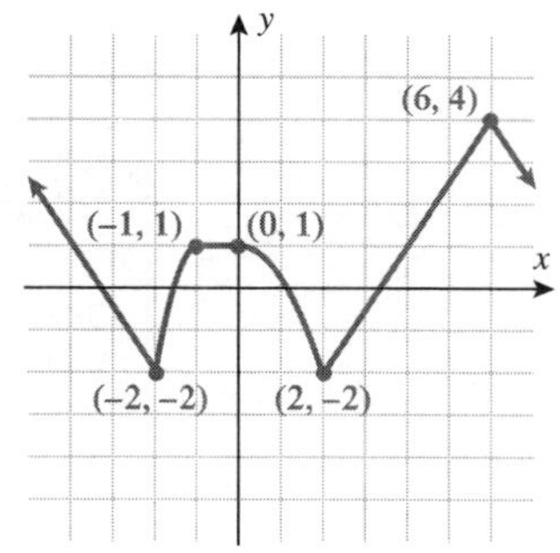

Look at the figure on the right. Graphs are read from **left to right**. If we start at the left side of the graph and trace the red curve with our pen, we see that the function values (values in the vertical direction) are decreasing until arriving at the point $(-2, -2)$. Then, the function values increase until arriving at the point $(-1, 1)$. The values then remain constant ($y = 1$) between the points $(-1, 1)$ and $(0, 1)$. Proceeding beyond the point $(0, 1)$, the function values decrease again until the point $(2, -2)$. Beyond the point $(2, -2)$, the function values increase again until the point $(6, 4)$. Finally, the function values decrease and continue to do so.

When specifying a function as increasing, decreasing, or constant, the **intervals are classified according to the *x*-coordinate**. For instance, in this graph, we say the function is increasing when x is between $x = -2$ and $x = -1$ and again when x is between $x = 2$ and $x = 6$. The graph is classified as decreasing when x is less than -2 and again when x is between 0 and 2 and again when x is greater than 6. The graph is classified as constant when x is between -1 and 0. In interval notation, this is summarized as

Decreasing	**Increasing**	**Constant**
$(-\infty, -2) \cup (0, 2) \cup (6, \infty)$	$(-2, -1) \cup (2, 6)$	$(-1, 0)$

Study Tip

- Graphs are read from left to right.
- Intervals correspond to the *x*-coordinates.

An algebraic test for determining whether a function is increasing, decreasing, or constant is to compare the value $f(x)$ of the function for particular points in the intervals.

INCREASING, DECREASING, AND CONSTANT FUNCTIONS

1. A function f is **increasing** on an open interval I if for any x_1 and x_2 in I, where $x_1 < x_2$, then $f(x_1) < f(x_2)$.
2. A function f is **decreasing** on an open interval I if for any x_1 and x_2 in I, where $x_1 < x_2$, then $f(x_1) > f(x_2)$.
3. A function f is **constant** on an open interval I if for any x_1 and x_2 in I, then $f(x_1) = f(x_2)$.

In addition to classifying a function as increasing, decreasing, or constant, we can also determine the domain and range of a function by inspecting its graph from left to right:

- The domain is the set of all x-values (from left to right) where the function is defined.
- The range is the set of all y-values (from bottom to top) that the graph of the function corresponds to.
- A solid dot on the left or right end of a graph indicates that the graph terminates there and the point is included in the graph.
- An open dot indicates that the graph terminates there and the point is not included in the graph.
- Unless a dot is present, it is assumed that a graph continues indefinitely in the same direction. (An arrow is used in some books to indicate direction.)

EXAMPLE 2 Finding Intervals When a Function Is Increasing or Decreasing

Given the graph of a function:

a. State the domain and range of the function.

b. Find the intervals when the function is increasing, decreasing, or constant.

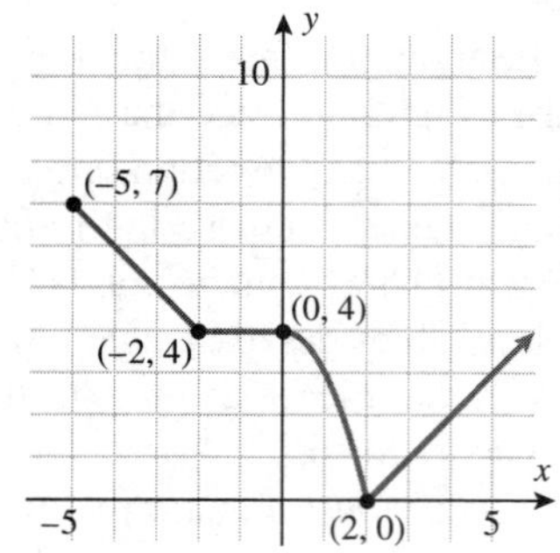

Solution:

Domain: $[-5, \infty)$

Range: $[0, \infty)$

Reading the graph from **left to right**, we see that the graph

- decreases from the point $(-5, 7)$ to the point $(-2, 4)$.
- is constant from the point $(-2, 4)$ to the point $(0, 4)$.
- decreases from the point $(0, 4)$ to the point $(2, 0)$.
- increases from the point $(2, 0)$ on.

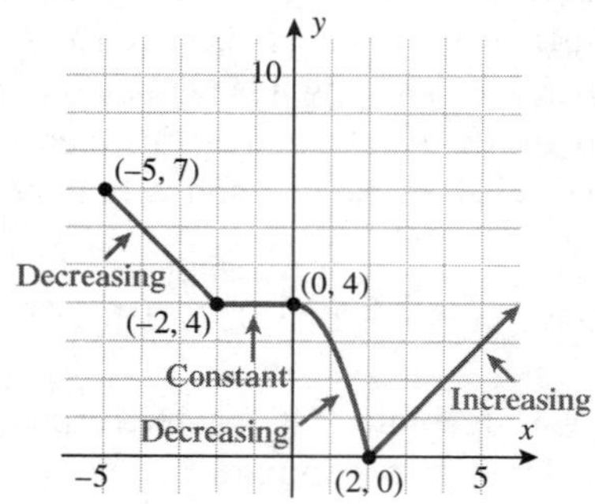

The intervals of increasing and decreasing correspond to the ***x*-coordinates**.

We say that this function is

- increasing on the interval $(2, \infty)$.
- decreasing on the interval $(-5, -2) \cup (0, 2)$.
- constant on the interval $(-2, 0)$.

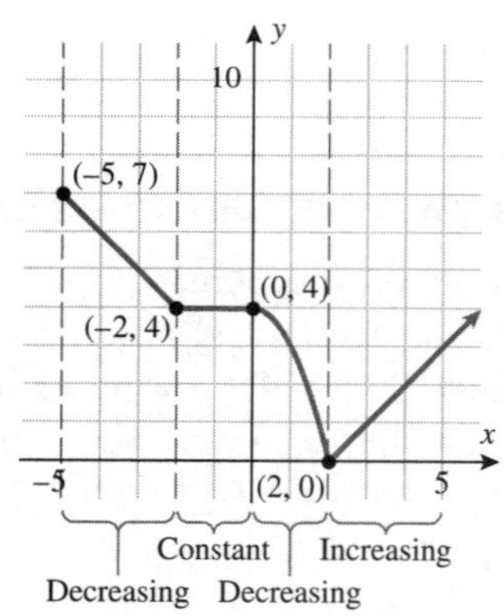

Note: The intervals of increasing or decreasing are defined on *open* intervals. This should not be confused with the domain. For example, the point $x = -5$ is included in the domain of the function but not in the interval where the function is classified as decreasing.

Average Rate of Change

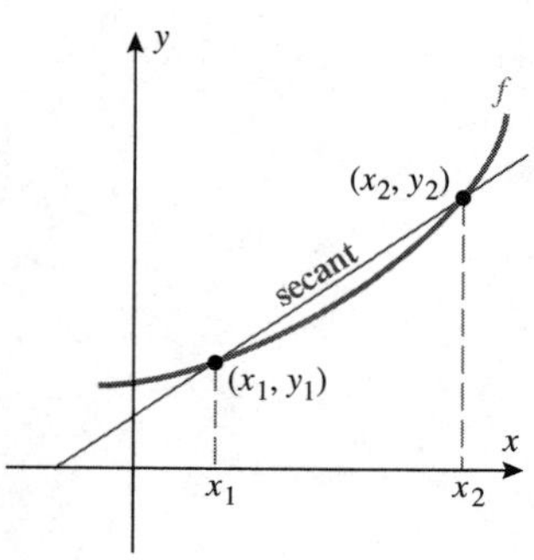

How do we know *how much* a function is increasing or decreasing? For example, is the price of a stock slightly increasing or is it doubling every week? One way we determine how much a function is increasing or decreasing is by calculating its *average rate of change.*

Let (x_1, y_1) and (x_2, y_2) be two points that lie on the graph of a function f. Draw the line that passes through these two points (x_1, y_1) and (x_2, y_2). This line is called a **secant line.**

Note that the slope of the secant line is given by $m = \dfrac{y_2 - y_1}{x_2 - x_1}$, and recall that the slope of a line is the rate of change of that line. The **slope of the secant line** is used to represent the *average rate of change* of the function.

AVERAGE RATE OF CHANGE

Let $(x_1, f(x_1))$ and $(x_2, f(x_2))$ be two distinct points, $(x_1 \neq x_2)$, on the graph of the function f. The **average rate of change** of f between x_1 and x_2 is given by

$$\textbf{Average rate of change} = \frac{f(x_2) - f(x_1)}{x_2 - x_1}$$

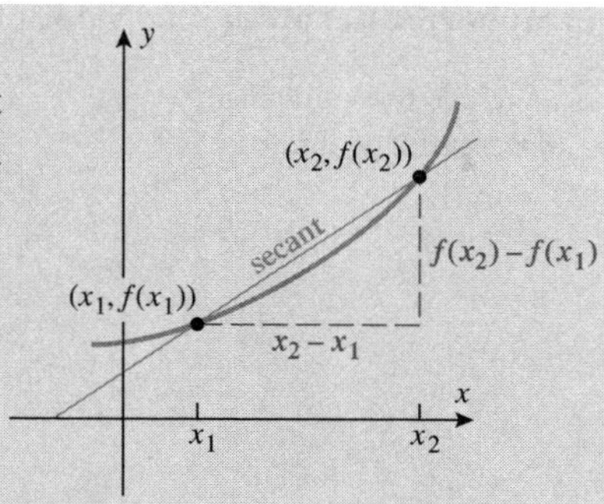

EXAMPLE 3 Average Rate of Change

Find the average rate of change of $f(x) = x^4$ from:

a. $x = -1$ to $x = 0$ **b.** $x = 0$ to $x = 1$ **c.** $x = 1$ to $x = 2$

Solution (a):

Write the average rate of change formula. $\frac{f(x_2) - f(x_1)}{x_2 - x_1}$

Let $x_1 = -1$ and $x_2 = 0$. $= \frac{f(0) - f(-1)}{0 - (-1)}$

Substitute $f(-1) = (-1)^4 = 1$ and $f(0) = 0^4 = 0$. $= \frac{0 - 1}{0 - (-1)}$

Simplify. $= \boxed{-1}$

Solution (b):

Write the average rate of change formula. $\frac{f(x_2) - f(x_1)}{x_2 - x_1}$

Let $x_1 = 0$ and $x_2 = 1$. $= \frac{f(1) - f(0)}{1 - 0}$

Substitute $f(0) = 0^4 = 0$ and $f(1) = (1)^4 = 1$. $= \frac{1 - 0}{1 - 0}$

Simplify. $= \boxed{1}$

Solution (c):

Write the average rate of change formula. $\frac{f(x_2) - f(x_1)}{x_2 - x_1}$

Let $x_1 = 1$ and $x_2 = 2$. $= \frac{f(2) - f(1)}{2 - 1}$

Substitute $f(1) = 1^4 = 1$ and $f(2) = (2)^4 = 16$. $= \frac{16 - 1}{2 - 1}$

Simplify. $= \boxed{15}$

Graphical Interpretation: Slope of the Secant Line

a. Average rate of change:
Decreasing at a rate of 1

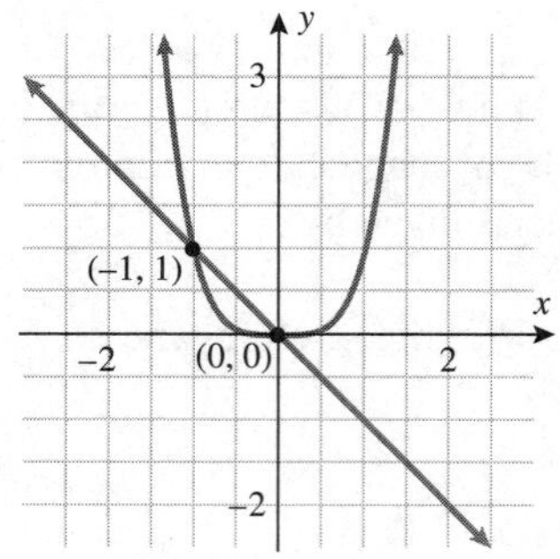

b. Average rate of change:
Increasing at a rate of 1

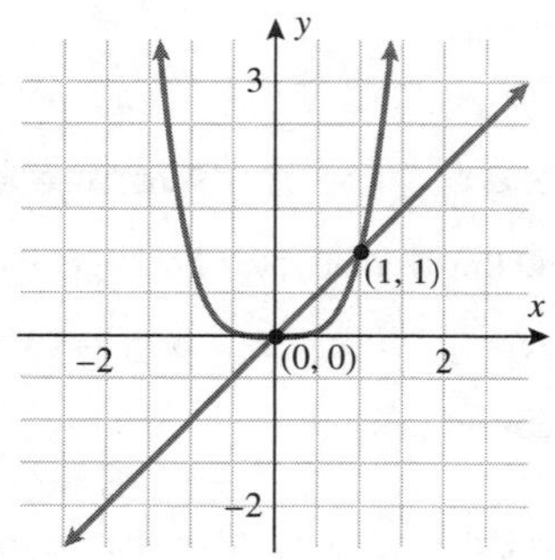

c. Average rate of change:
Increasing at a rate of 15

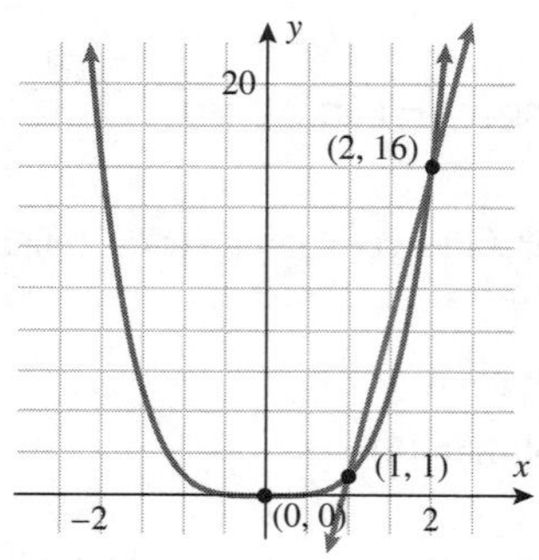

Answer: a. -2 **b.** 2

YOUR TURN Find the average rate of change of $f(x) = x^2$ from:

a. $x = -2$ to $x = 0$ **b.** $x = 0$ to $x = 2$

The average rate of change can also be written in terms of the difference quotient.

WORDS	MATH
Let the distance between x_1 and x_2 be h.	$x_2 - x_1 = h$
Solve for x_2.	$x_2 = x_1 + h$
Substitute $x_2 - x_1 = h$ into the denominator and $x_2 = x_1 + h$ into the numerator of the average rate of change.	Average rate of change $= \dfrac{f(x_2) - f(x_1)}{x_2 - x_1}$ $= \dfrac{f(x_1 + h) - f(x_1)}{h}$
Let $x_1 = x$.	$= \boxed{\dfrac{f(x + h) - f(x)}{h}}$

When written in this form, the average rate of change is called the **difference quotient.**

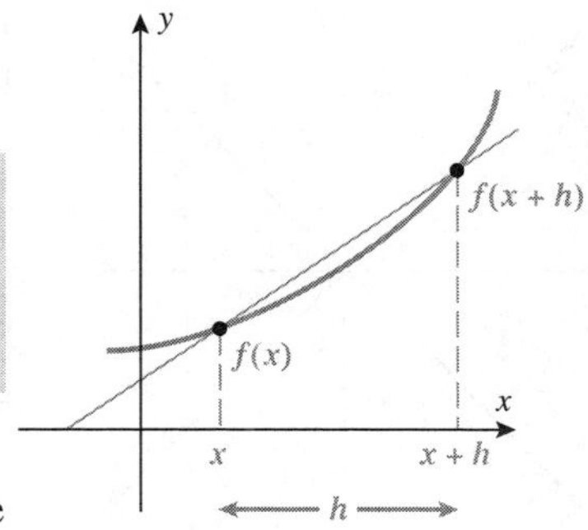

DEFINITION **Difference Quotient**

The expression $\frac{f(x + h) - f(x)}{h}$, where $h \neq 0$, is called the **difference quotient**.

The difference quotient is more meaningful when h is small. In calculus the difference quotient is used to define a derivative.

EXAMPLE 4 **Calculating the Difference Quotient**

Calculate the difference quotient for the function $f(x) = 2x^2 + 1$.

Solution:

Find $f(x + h)$.

$$f(x + h) = 2(x + h)^2 + 1$$
$$= 2(x^2 + 2xh + h^2) + 1$$
$$= 2x^2 + 4xh + 2h^2 + 1$$

Find the difference quotient.

$$\frac{f(x + h) - f(x)}{h} = \frac{\overbrace{2x^2 + 4xh + 2h^2 + 1}^{f(x+h)} - \overbrace{(2x^2 + 1)}^{f(x)}}{h}$$

Simplify.

$$\frac{f(x + h) - f(x)}{h} = \frac{\cancel{2x^2} + 4xh + 2h^2 \cancel{+1} - \cancel{2x^2} \cancel{-1}}{h}$$

$$\frac{f(x + h) - f(x)}{h} = \frac{4xh + 2h^2}{h}$$

Factor the numerator.

$$\frac{f(x + h) - f(x)}{h} = \frac{h(4x + 2h)}{h}$$

Cancel (divide out) the common h.

$$\frac{f(x + h) - f(x)}{h} = \boxed{4x + 2h} \qquad h \neq 0$$

■ YOUR TURN Calculate the difference quotient for the function $f(x) = -x^2 + 2$.

Study Tip

Use brackets or parentheses around $f(x)$ to avoid forgetting to distribute the negative sign:

$$\frac{f(x + h) - [f(x)]}{h}$$

■ Answer:
$$\frac{f(x + h) - f(x)}{h} = -2x - h$$

Piecewise-Defined Functions

Most of the functions that we have seen in this text are functions defined by polynomials. Sometimes the need arises to define functions in terms of *pieces*. For example, most plumbers charge a flat fee for a house call and then an additional hourly rate for the job. For instance, if a particular plumber charges \$100 to drive out to your house and work for 1 hour and then an additional \$25 an hour for every additional hour he or she works on your job, we would define this function in pieces. If we let h be the number of hours worked, then the charge is defined as

$$\text{Plumbing charge} = \begin{cases} 100 & h \leq 1 \\ 100 + 25(h - 1) & h > 1 \end{cases}$$

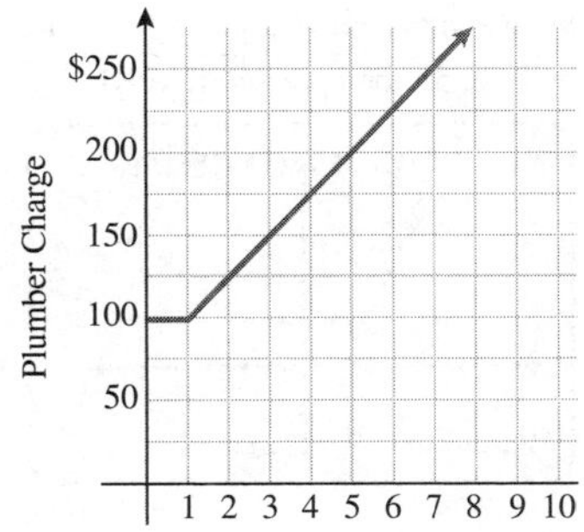

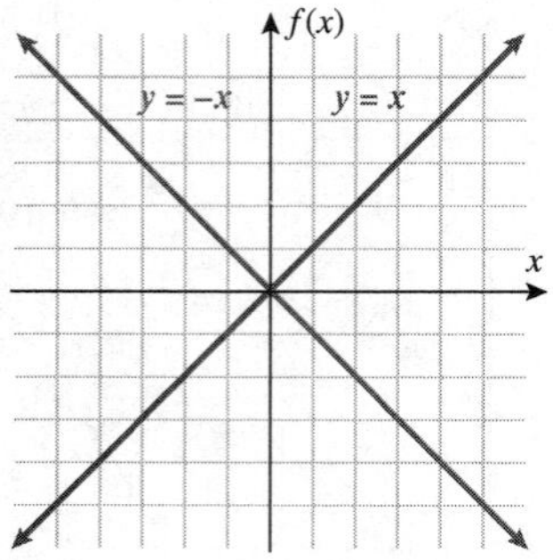

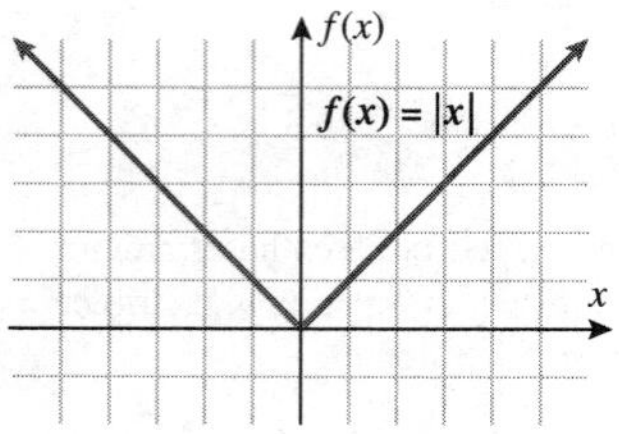

If we were to graph this function, we would see that there is 1 hour that is constant and after that the function continually increases.

Another piecewise-defined function is the absolute value function. The absolute value function can be thought of as two pieces: the line $y = -x$ (when x is negative) and the line $y = x$ (when x is nonnegative). We start by graphing these two lines on the same graph.

The absolute value function behaves like the line $y = -x$ when x is negative (erase the blue graph in quadrant IV) and like the line $y = x$ when x is positive (erase the red graph in quadrant III).

Absolute value piecewise-defined function

$$f(x) = |x| = \begin{cases} -x & x < 0 \\ x & x \geq 0 \end{cases}$$

The next example is a piecewise-defined function given in terms of functions in our "library of functions." Because the function is defined in terms of pieces of other functions, we draw the graph of each individual function and, then, for each function darken the piece corresponding to its part of the domain. This is like the procedure above for the absolute value function.

Technology Tip

Plot a piecewise-defined function using the [TEST] menu operations to define the inequalities in the function. Press:

[Y=] [(] [x, T, θ, n] [x^2] [)] [(] [x, T, θ, n] [2nd] [TEST] [5:<] [ENTER] [(−)] 1 [)] [+] [(] 1 [)] [(] [x, T, θ, n] [2nd] [TEST] [4:≥] [ENTER] [(−)] [1] [)] [(] [x, T, θ, n] [2nd] [TEST] [6:≤] [ENTER] 1 [)] [+] [(] [x, T, θ, n] [)] [(] [x, T, θ, n] [2nd] [TEST] [3:>] [ENTER] 1 [)]

Set the viewing rectangle as [−4, 4] by [−2, 5]; then press [GRAPH].

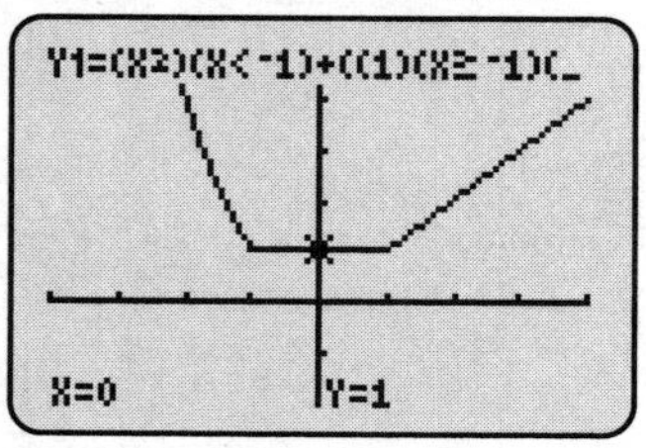

EXAMPLE 5 Graphing Piecewise-Defined Functions

Graph the piecewise-defined function, and state the domain, range, and intervals when the function is increasing, decreasing, or constant.

$$G(x) = \begin{cases} x^2 & x < -1 \\ 1 & -1 \leq x \leq 1 \\ x & x > 1 \end{cases}$$

Solution:

Graph each of the functions on the same plane.

Square function:
$f(x) = x^2$

Constant function:
$f(x) = 1$

Identity function:
$f(x) = x$

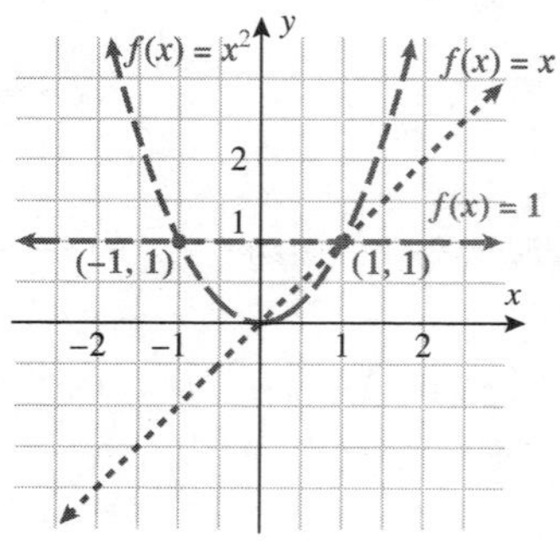

The points to focus on in particular are the x-values where the pieces change over—that is, $x = -1$ and $x = 1$.

Let's now investigate each piece. When $x < -1$, this function is defined by the square function, $f(x) = x^2$, so darken that particular function to the left of $x = -1$. When $-1 \leq x \leq 1$, the function is defined by the constant function, $f(x) = 1$, so darken that particular function between the x values of -1 and 1. When $x > 1$, the function is defined by the identity function, $f(x) = x$, so darken that function to the right of $x = 1$. Erase everything that is not darkened, and the resulting graph of the piecewise-defined function is given on the right.

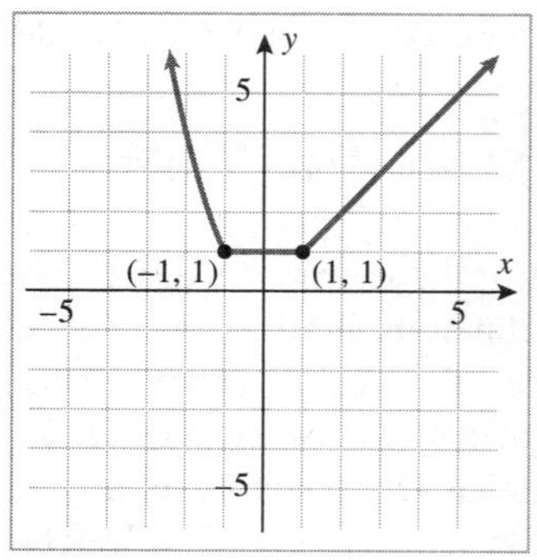

This function is defined for all real values of x, so the domain of this function is the set of all real numbers. The values that this function yields in the vertical direction are all real numbers greater than or equal to 1. Hence, the range of this function is $[1, \infty)$. The intervals of increasing, decreasing, and constant are as follows:

$$\text{Decreasing: } (-\infty, -1)$$

$$\text{Constant: } (-1, 1)$$

$$\text{Increasing: } (1, \infty)$$

The term **continuous** implies that there are no holes or jumps and that the graph can be drawn without picking up your pencil. A function that does have holes or jumps and cannot be drawn in one motion without picking up your pencil is classified as **discontinuous**, and the points where the holes or jumps occur are called *points of discontinuity.*

The previous example illustrates a *continuous* piecewise-defined function. At the $x = -1$ junction, the square function and constant function both pass through the point $(-1, 1)$. At the $x = 1$ junction, the constant function and the identity function both pass through the point $(1, 1)$. Since the graph of this piecewise-defined function has no holes or jumps, we classify it as a continuous function.

The next example illustrates a *discontinuous* piecewise-defined function.

EXAMPLE 6 Graphing a Discontinuous Piecewise-Defined Function

Graph the piecewise-defined function, and state the intervals where the function is increasing, decreasing, or constant, along with the domain and range.

$$f(x) = \begin{cases} 1 - x & x < 0 \\ x & 0 \le x < 2 \\ -1 & x > 2 \end{cases}$$

Solution:

Graph these functions on the same plane.

Linear function:
$f(x) = 1 - x$

Identity function:
$f(x) = x$

Constant function:
$f(x) = -1$

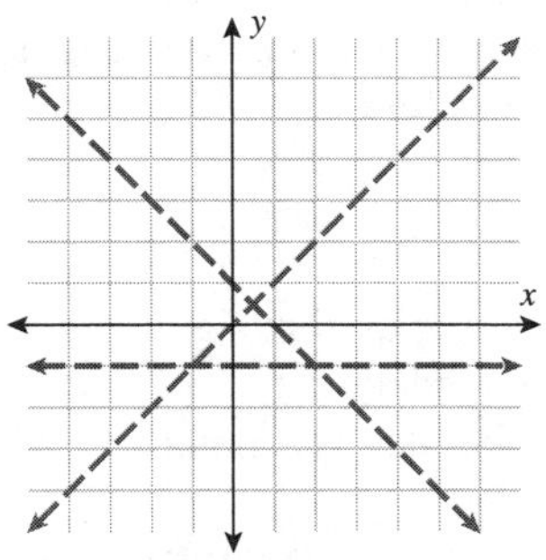

Darken the piecewise-defined function on the graph. For all values less than zero $(x < 0)$ the function is defined by the linear function. Note the use of an open circle, indicating up to but not including $x = 0$. For values $0 \le x < 2$, the function is defined by the identity function.

The circle is filled in at the left endpoint, $x = 0$. An open circle is used at $x = 2$. For all values greater than 2, $x > 2$, the function is defined by the constant function. Because this interval does not include the point $x = 2$, an open circle is used.

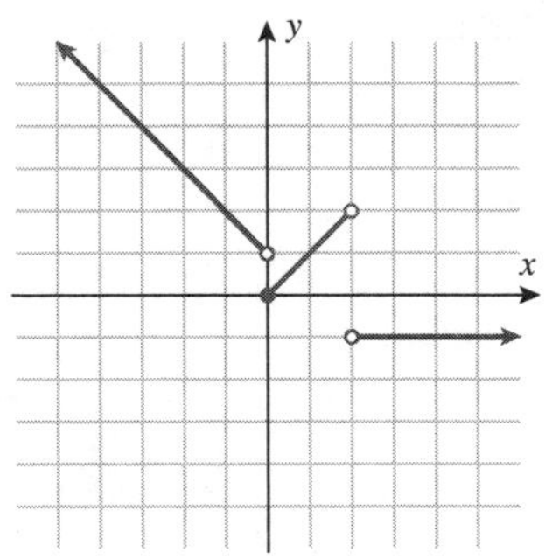

Technology Tip

To plot a piecewise-defined function using a graphing utility, use the [TEST] menu operations to define the inequalities in the function. Press:

[Y=] [(] 1 [−] [x, T, θ, n] [)] [(] [x, T, θ, n] [2nd] [TEST] [5:<] [ENTER] 0 [)] [+] [(] [x, T, θ, n] [)] [(] [x, T, θ, n] [2nd] [TEST] [4:≥] [ENTER] 0 [)] [(] [x, T, θ, n] [2nd] [TEST] [5:<] [ENTER] 2 [)] [+] [(] [(−)] 1 [)] [(] [x, T, θ, n] [2nd] [TEST] [3:>] [ENTER] 2 [)]

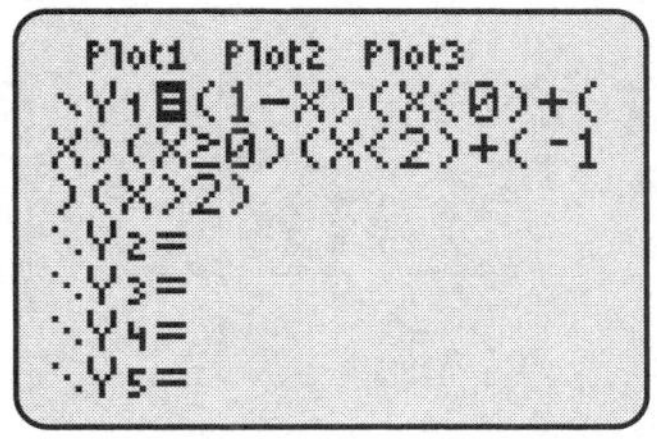

To avoid connecting graphs of the pieces, press [MODE] and [Dot]. Set the viewing rectangle as $[-3, 4]$ by $[-2, 5]$; then press [GRAPH].

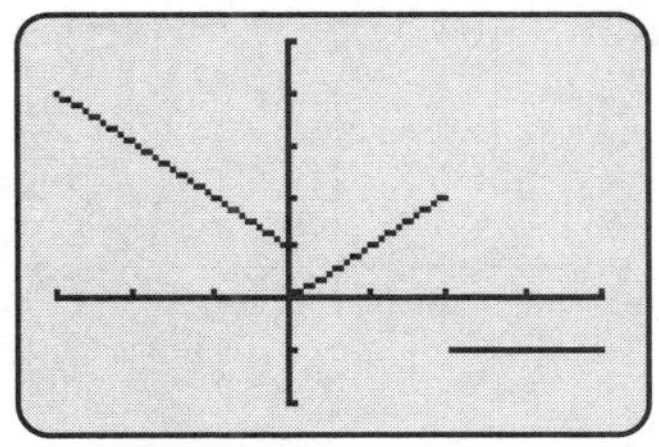

Be sure to include the open circle and closed circle at the appropriate endpoints of each piece in the function.

The table of values supports the graph, except at $x = 2$. The function is not defined at $x = 2$.

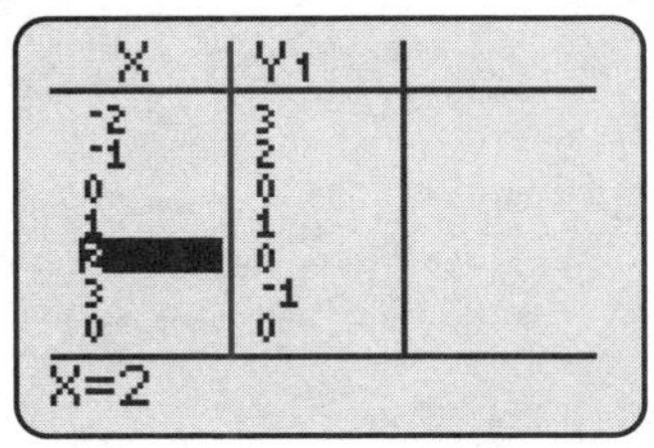

At what intervals is the function increasing, decreasing, or constant? Remember that the intervals correspond to the x-values.

Decreasing: $(-\infty, 0)$ Increasing: $(0, 2)$ Constant: $(2, \infty)$

The function is defined for all values of x except $x = 2$.

Domain: $(-\infty, 2) \cup (2, \infty)$

The output of this function (vertical direction) takes on the y-values $y \geq 0$ and the additional single value $y = -1$.

Range: $[-1, -1] \cup [0, \infty)$ or $\{-1\} \cup [0, \infty)$

We mentioned earlier that a discontinuous function has a graph that exhibits holes or jumps. In Example 6, the point $x = 0$ corresponds to a jump, because you would have to pick up your pencil to continue drawing the graph. The point $x = 2$ corresponds to both a hole and a jump. The hole indicates that the function is not defined at that point, and there is still a jump because the identity function and the constant function do not meet at the same y-value at $x = 2$.

■ YOUR TURN Graph the piecewise-defined function, and state the intervals where the function is increasing, decreasing, or constant, along with the domain and range.

$$f(x) = \begin{cases} -x & x \leq -1 \\ 2 & -1 < x < 1 \\ x & x > 1 \end{cases}$$

■ **Answer:** Increasing: $(1, \infty)$
Decreasing: $(-\infty, -1)$
Constant: $(-1, 1)$
Domain: $(-\infty, 1) \cup (1, \infty)$
Range: $[1, \infty)$

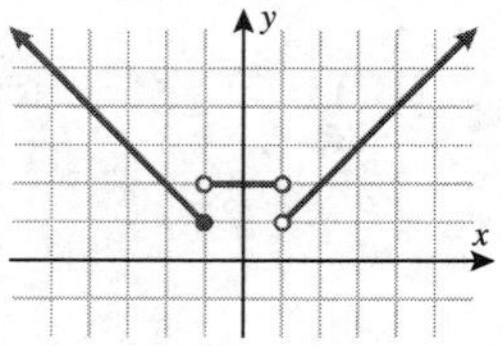

Piecewise-defined functions whose "pieces" are constants are called **step functions**. The reason for this name is that the graph of a step function looks like steps of a staircase. A common step function used in engineering is the **Heaviside step function** (also called the **unit step function**):

$$H(t) = \begin{cases} 0 & t < 0 \\ 1 & t \geq 0 \end{cases}$$

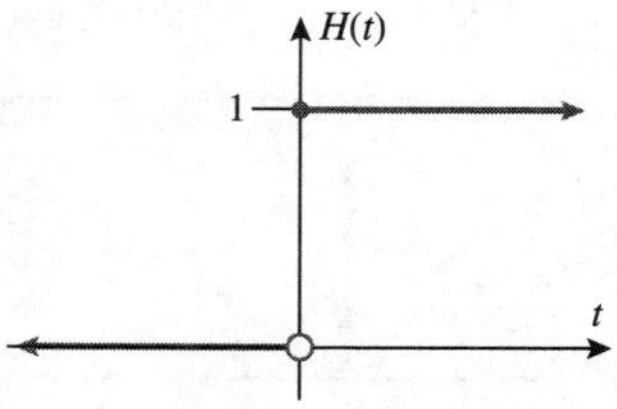

This function is used in signal processing to represent a signal that turns on at some time and stays on indefinitely.

A common step function used in business applications is the *greatest integer function.*

GREATEST INTEGER FUNCTION

$$f(x) = [[x]] = \text{greatest integer less than or equal to } x.$$

x	1.0	1.3	1.5	1.7	1.9	2.0
$f(x) = [[x]]$	1	1	1	1	1	2

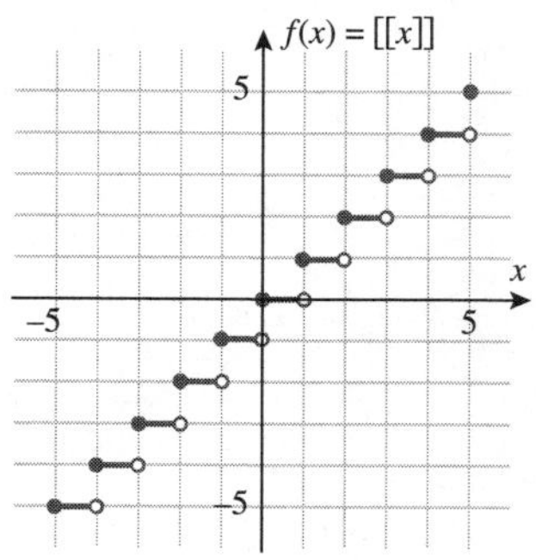

SECTION 3.2 SUMMARY

NAME	FUNCTION	DOMAIN	RANGE	GRAPH	EVEN/ODD
Linear	$f(x) = mx + b, m \neq 0$	$(-\infty, \infty)$	$(-\infty, \infty)$	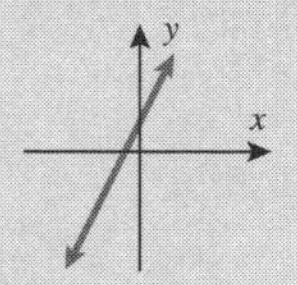	Neither (unless $y = x$)
Constant	$f(x) = c$	$(-\infty, \infty)$	$[c, c]$ or $\{c\}$	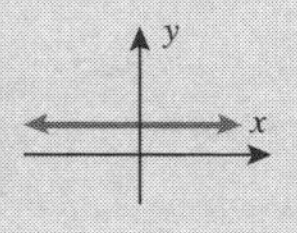	Even
Identity	$f(x) = x$	$(-\infty, \infty)$	$(-\infty, \infty)$	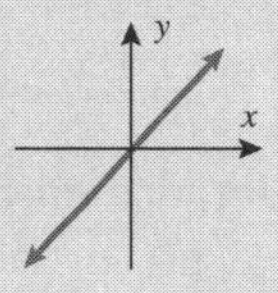	Odd
Square	$f(x) = x^2$	$(-\infty, \infty)$	$[0, \infty)$	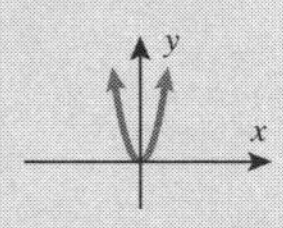	Even
Cube	$f(x) = x^3$	$(-\infty, \infty)$	$(-\infty, \infty)$	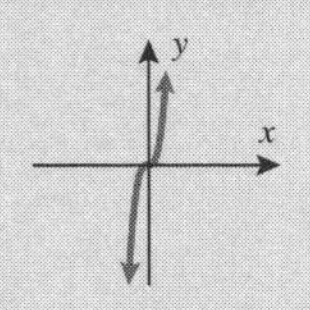	Odd
Square Root	$f(x) = \sqrt{x}$	$[0, \infty)$	$[0, \infty)$	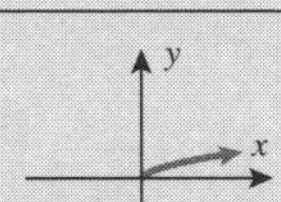	Neither
Cube Root	$f(x) = \sqrt[3]{x}$	$(-\infty, \infty)$	$(-\infty, \infty)$	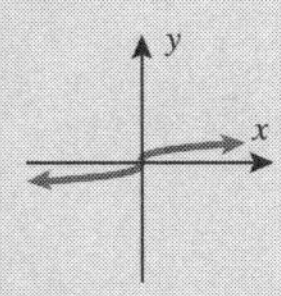	Odd
Absolute Value	$f(x) = \lvert x \rvert$	$(-\infty, \infty)$	$[0, \infty)$	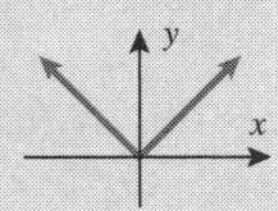	Even
Reciprocal	$f(x) = \dfrac{1}{x}$	$(-\infty, 0) \cup (0, \infty)$	$(-\infty, 0) \cup (0, \infty)$	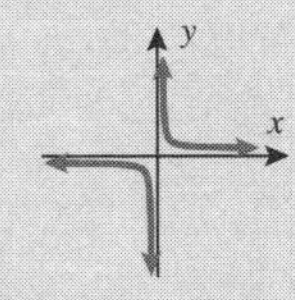	Odd

Domain and Range of a Function

- Implied domain: Exclude any values that lead to the function being undefined (dividing by zero) or imaginary outputs (square root of a negative real number).
- Inspect the graph to determine the set of all inputs (domain) and the set of all outputs (range).

Finding Intervals Where a Function Is Increasing, Decreasing, or Constant

- Increasing: Graph of function rises from left to right.
- Decreasing: Graph of function falls from left to right.
- Constant: Graph of function does not change height from left to right.

Average Rate of Change $\dfrac{f(x_2) - f(x_1)}{x_2 - x_1}$ $\quad x_1 \neq x_2$

Difference Quotient $\dfrac{f(x + h) - f(x)}{h}$ $\quad h \neq 0$

Piecewise-Defined Functions

- Continuous: You can draw the graph of a function without picking up the pencil.
- Discontinuous: Graph has holes and/or jumps.

SECTION 3.2 EXERCISES

SKILLS

In Exercises 1–24, determine whether the function is even, odd, or neither.

1. $G(x) = x + 4$
2. $h(x) = 3 - x$
3. $f(x) = 3x^2 + 1$
4. $F(x) = x^4 + 2x^2$
5. $g(t) = 5t^3 - 3t$
6. $f(x) = 3x^5 + 4x^3$
7. $h(x) = x^2 + 2x$
8. $G(x) = 2x^4 + 3x^3$
9. $h(x) = x^{1/3} - x$
10. $g(x) = x^{-1} + x$
11. $f(x) = |x| + 5$
12. $f(x) = |x| + x^2$
13. $f(x) = |x|$
14. $f(x) = |x^3|$
15. $G(t) = |t - 3|$
16. $g(t) = |t + 2|$
17. $G(t) = \sqrt{t - 3}$
18. $f(x) = \sqrt{2 - x}$
19. $g(x) = \sqrt{x^2 + x}$
20. $f(x) = \sqrt{x^2 + 2}$
21. $h(x) = \dfrac{1}{x} + 3$
22. $h(x) = \dfrac{1}{x} - 2x$
23.

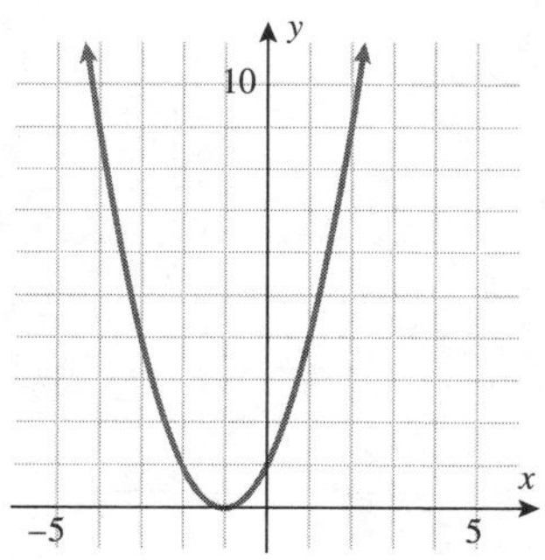

24.

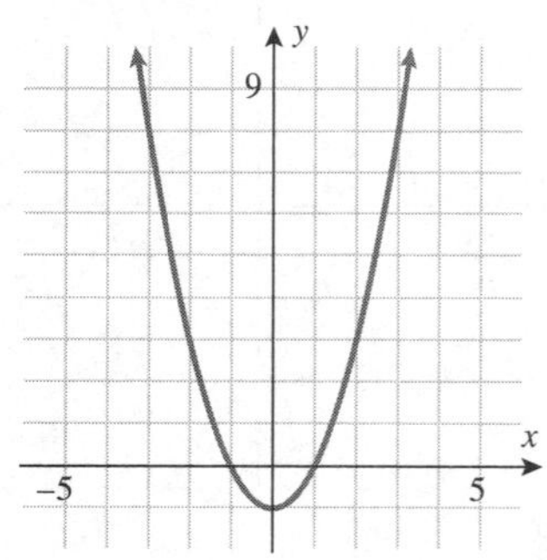

In Exercises 25–36, state the (a) domain, (b) range, and (c) *x*-interval(s) where the function is increasing, decreasing, or constant. Find the values of (d) $f(0)$, (e) $f(-2)$, and (f) $f(2)$.

25.

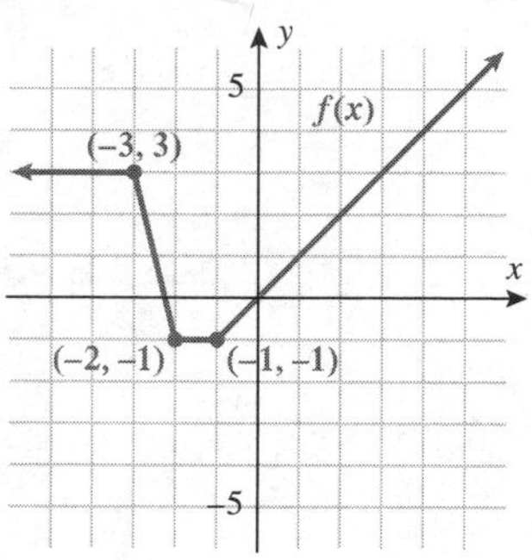

26.

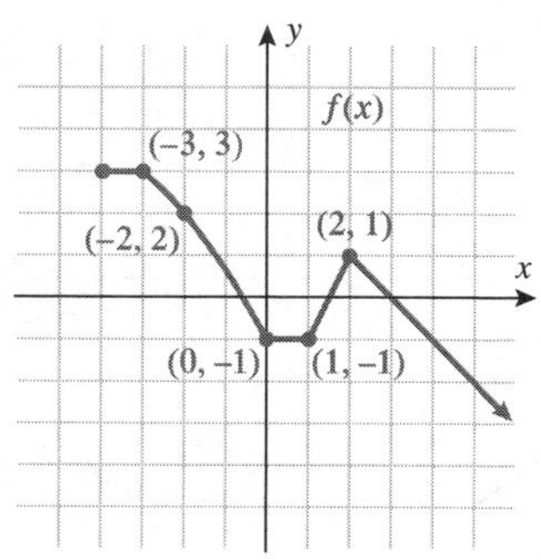

27.

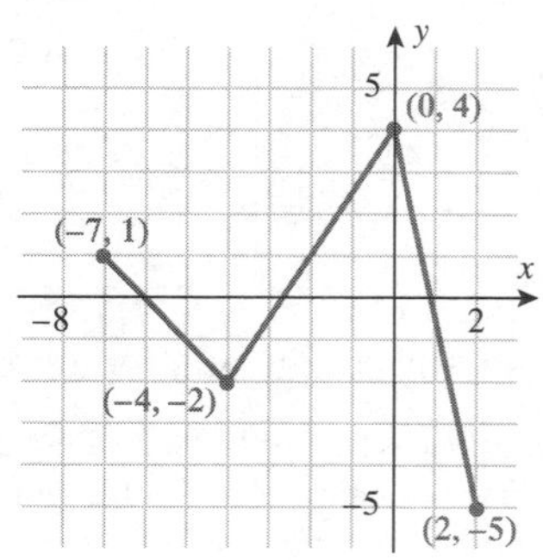

28. 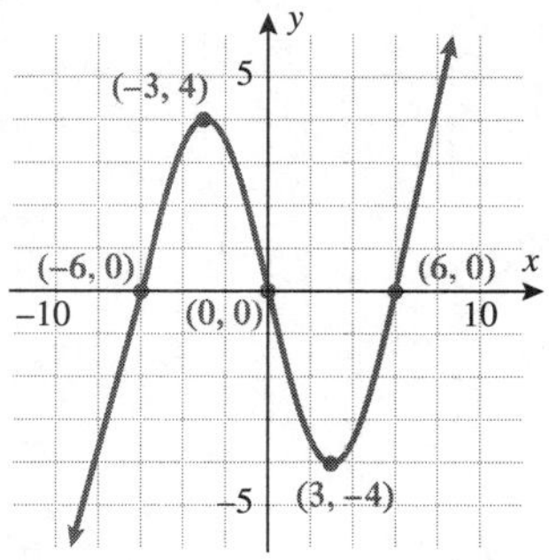

29. 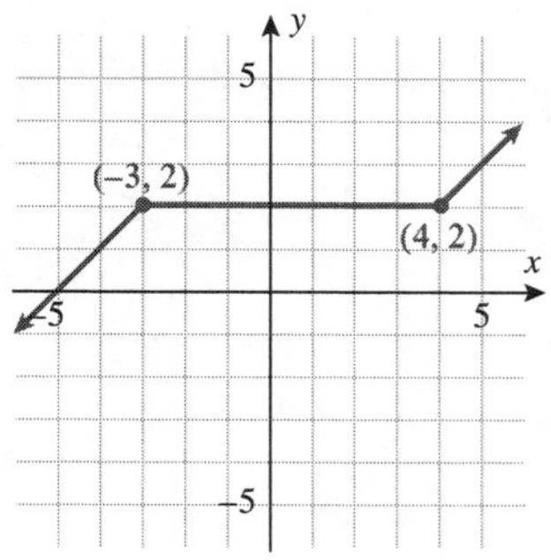

30. 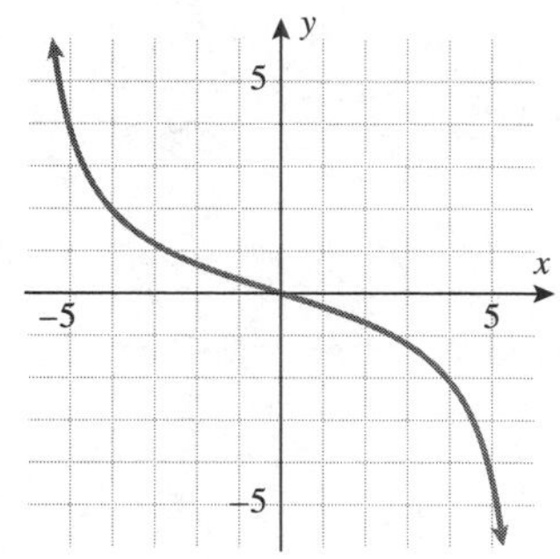

31. 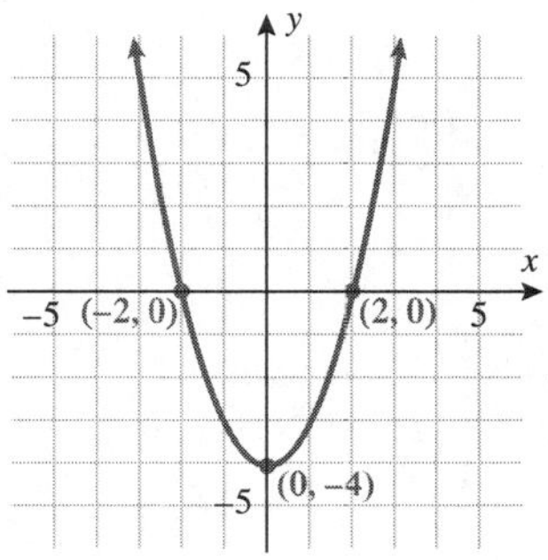

32. 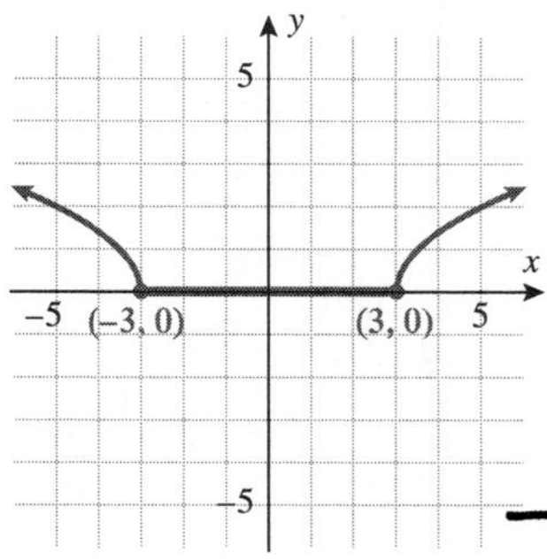

33. 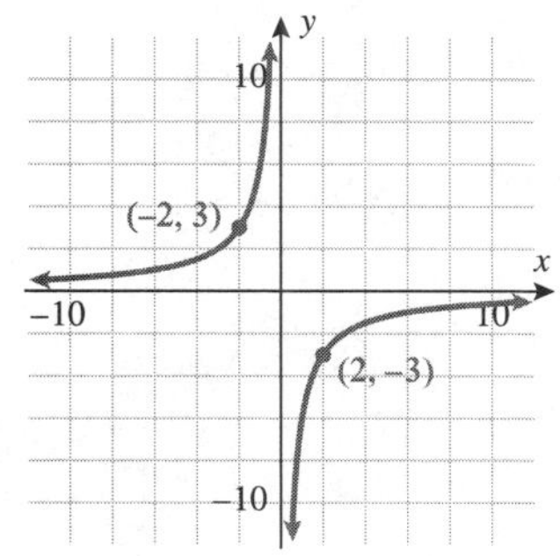

34. 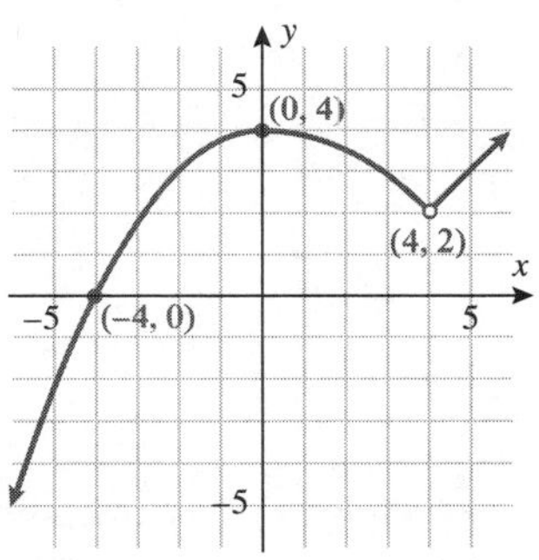

35. 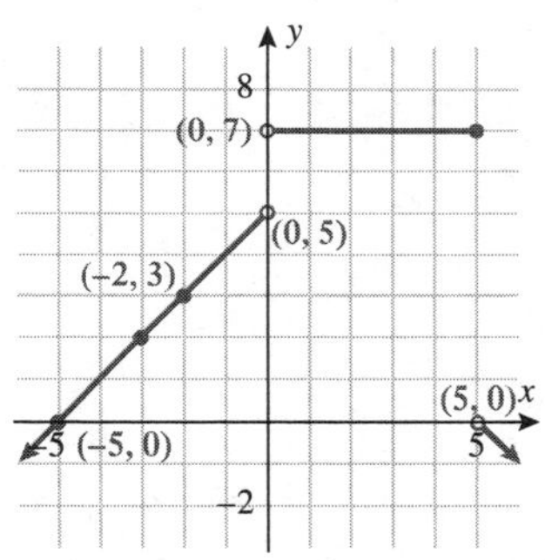

36. 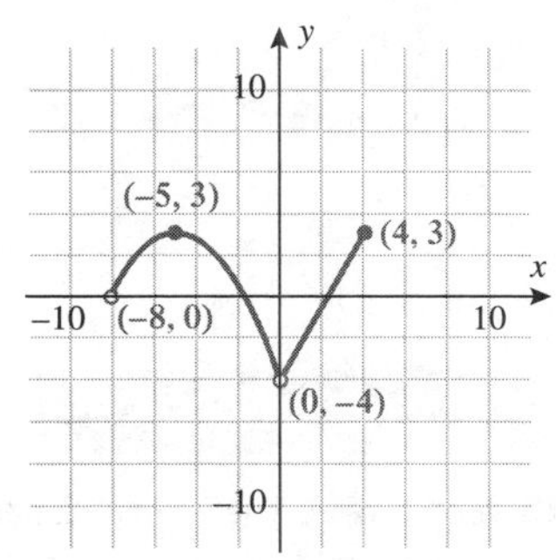

In Exercises 37–44, find the difference quotient $\dfrac{f(x+h)-f(x)}{h}$ for each function.

37. $f(x) = x^2 - x$
38. $f(x) = x^2 + 2x$
39. $f(x) = 3x + x^2$
40. $f(x) = 5x - x^2$
41. $f(x) = x^2 - 3x + 2$
42. $f(x) = x^2 - 2x + 5$
43. $f(x) = -3x^2 + 5x - 4$
44. $f(x) = -4x^2 + 2x - 3$

In Exercises 45–52, find the average rate of change of the function from $x = 1$ to $x = 3$.

45. $f(x) = x^3$
46. $f(x) = \dfrac{1}{x}$
47. $f(x) = |x|$
48. $f(x) = 2x$
49. $f(x) = 1 - 2x$
50. $f(x) = 9 - x^2$
51. $f(x) = |5 - 2x|$
52. $f(x) = \sqrt{x^2 - 1}$

In Exercises 53–76, graph the piecewise-defined functions. State the domain and range in interval notation. Determine the intervals where the function is increasing, decreasing, or constant.

53. $f(x) = \begin{cases} x & x < 2 \\ 2 & x \geq 2 \end{cases}$
54. $f(x) = \begin{cases} -x & x < -1 \\ -1 & x \geq -1 \end{cases}$
55. $f(x) = \begin{cases} 1 & x < -1 \\ x^2 & x \geq -1 \end{cases}$
56. $f(x) = \begin{cases} x^2 & x < 2 \\ 4 & x \geq 2 \end{cases}$
57. $f(x) = \begin{cases} x & x < 0 \\ x^2 & x \geq 0 \end{cases}$
58. $f(x) = \begin{cases} -x & x \leq 0 \\ x^2 & x > 0 \end{cases}$

59. $f(x) = \begin{cases} -x+2 & x < 1 \\ x^2 & x \geq 1 \end{cases}$

60. $f(x) = \begin{cases} 2+x & x \leq -1 \\ x^2 & x > -1 \end{cases}$

61. $G(x) = \begin{cases} -1 & x < -1 \\ x & -1 \leq x \leq 3 \\ 3 & x > 3 \end{cases}$

62. $G(x) = \begin{cases} -1 & x < -1 \\ x & -1 < x < 3 \\ 3 & x > 3 \end{cases}$

63. $G(t) = \begin{cases} 1 & t < 1 \\ t^2 & 1 \leq t \leq 2 \\ 4 & t > 2 \end{cases}$

64. $G(t) = \begin{cases} 1 & t < 1 \\ t^2 & 1 < t < 2 \\ 4 & t > 2 \end{cases}$

65. $f(x) = \begin{cases} -x-1 & x < -2 \\ x+1 & -2 < x < 1 \\ -x+1 & x \geq 1 \end{cases}$

66. $f(x) = \begin{cases} -x-1 & x \leq -2 \\ x+1 & -2 < x < 1 \\ -x+1 & x > 1 \end{cases}$

67. $G(x) = \begin{cases} 0 & x < 0 \\ \sqrt{x} & x \geq 0 \end{cases}$

68. $G(x) = \begin{cases} 1 & x < 1 \\ \sqrt[3]{x} & x > 1 \end{cases}$

69. $G(x) = \begin{cases} 0 & x = 0 \\ \dfrac{1}{x} & x \neq 0 \end{cases}$

70. $G(x) = \begin{cases} 0 & x = 0 \\ -\dfrac{1}{x} & x \neq 0 \end{cases}$

71. $G(x) = \begin{cases} -\sqrt[3]{x} & x \leq -1 \\ x & -1 < x < 1 \\ -\sqrt{x} & x > 1 \end{cases}$

72. $G(x) = \begin{cases} -\sqrt[3]{x} & x < -1 \\ x & -1 \leq x < 1 \\ \sqrt{x} & x > 1 \end{cases}$

73. $f(x) = \begin{cases} x+3 & x \leq -2 \\ |x| & -2 < x < 2 \\ x^2 & x \geq 2 \end{cases}$

74. $f(x) = \begin{cases} |x| & x < -1 \\ 1 & -1 < x < 1 \\ |x| & x > 1 \end{cases}$

75. $f(x) = \begin{cases} x & x \leq -1 \\ x^3 & -1 < x < 1 \\ x^2 & x > 1 \end{cases}$

76. $f(x) = \begin{cases} x^2 & x \leq -1 \\ x^3 & -1 < x < 1 \\ x & x \geq 1 \end{cases}$

APPLICATIONS

77. Budget: Costs. The Kappa Kappa Gamma sorority decides to order custom-made T-shirts for its *Kappa Krush* mixer with the Sigma Alpha Epsilon fraternity. If the sorority orders 50 or fewer T-shirts, the cost is \$10 per shirt. If it orders more than 50 but fewer than 100, the cost is \$9 per shirt. If it orders more than 100, the cost is \$8 per shirt. Find the cost function $C(x)$ as a function of the number of T-shirts x ordered.

78. Budget: Costs. The marching band at a university is ordering some additional uniforms to replace existing uniforms that are worn out. If the band orders 50 or fewer, the cost is \$176.12 per uniform. If it orders more than 50 but fewer than 100, the cost is \$159.73 per uniform. Find the cost function $C(x)$ as a function of the number of new uniforms x ordered.

79. Budget: Costs. The Richmond rowing club is planning to enter the *Head of the Charles* race in Boston and is trying to figure out how much money to raise. The entry fee is \$250 per boat for the first 10 boats and \$175 for each additional boat. Find the cost function $C(x)$ as a function of the number of boats x the club enters.

80. Phone Cost: Long-Distance Calling. A phone company charges \$.39 per minute for the first 10 minutes of an international long-distance phone call and \$.12 per minute every minute after that. Find the cost function $C(x)$ as a function of the length of the phone call x in minutes.

81. Event Planning. A young couple are planning their wedding reception at a yacht club. The yacht club charges a flat rate of \$1,000 to reserve the dining room for a private party. The cost of food is \$35 per person for the first 100 people and \$25 per person for every additional person beyond the first 100. Write the cost function $C(x)$ as a function of the number of people x attending the reception.

82. Home Improvement. An irrigation company gives you an estimate for an eight-zone sprinkler system. The parts are \$1,400, and the labor is \$25 per hour. Write a function $C(x)$ that determines the cost of a new sprinkler system if you choose this irrigation company.

83. Sales. A famous author negotiates with her publisher the monies she will receive for her next suspense novel. She will receive \$50,000 up front and a 15% royalty rate on the first 100,000 books sold, and 20% on any books sold beyond that. If the book sells for \$20 and royalties are based on the selling price, write a royalties function $R(x)$ as a function of total number x of books sold.

84. Sales. Rework Exercise 83 if the author receives \$35,000 up front, 15% for the first 100,000 books sold, and 25% on any books sold beyond that.

85. Profit. A group of artists are trying to decide whether they will make a profit if they set up a Web-based business to market and sell stained glass that they make. The costs associated with this business are \$100 per month for the website and \$700 per month for the studio they rent. The materials cost \$35 for each work in stained glass, and the artists charge \$100 for each unit they sell. Write the monthly profit as a function of the number of stained-glass units they sell.

86. Profit. Philip decides to host a shrimp boil at his house as a fundraiser for his daughter's AAU basketball team. He orders gulf shrimp to be flown in from New Orleans. The shrimp costs \$5 per pound. The shipping costs \$30. If he charges \$10 per person, write a function $F(x)$ that represents either his loss or profit as a function of the number of people x that attend. Assume that each person will eat 1 pound of shrimp.

87. Postage Rates. The following table corresponds to first-class postage rates for the U.S. Postal Service. Write a piecewise-defined function in terms of the greatest integer function that models this cost of mailing flat envelopes first class.

WEIGHT LESS THAN (OUNCES)	FIRST-CLASS RATE (FLAT ENVELOPES)
1	\$0.80
2	\$0.97
3	\$1.14
4	\$1.31
5	\$1.48
6	\$1.65
7	\$1.82
8	\$1.99
9	\$2.16
10	\$2.33
11	\$2.50
12	\$2.67
13	\$2.84

88. Postage Rates. The following table corresponds to first-class postage rates for the U.S. Postal Service. Write a piecewise-defined function in terms of the greatest integer function that models this cost of mailing parcels first class.

WEIGHT LESS THAN (OUNCES)	FIRST-CLASS RATE (PARCELS)
1	\$1.13
2	\$1.30
3	\$1.47
4	\$1.64
5	\$1.81
6	\$1.98
7	\$2.15
8	\$2.32
9	\$2.49
10	\$2.66
11	\$2.83
12	\$3.00
13	\$3.17

A square wave is a waveform used in electronic circuit testing and signal processing. A square wave alternates regularly and instantaneously between two levels.

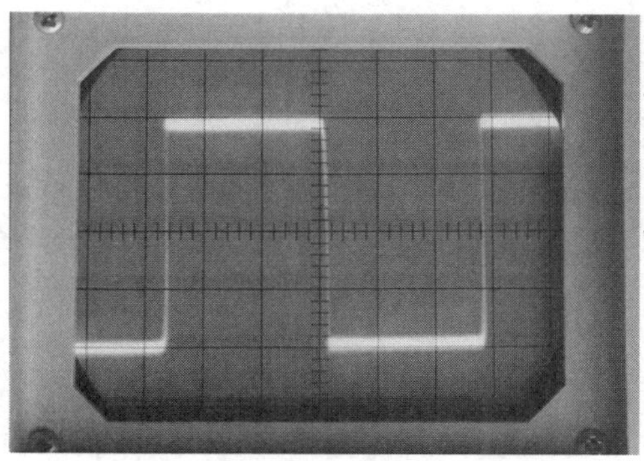

sciencephotos/Alamy

89. Electronics: Signals. Write a step function $f(t)$ that represents the following square wave.

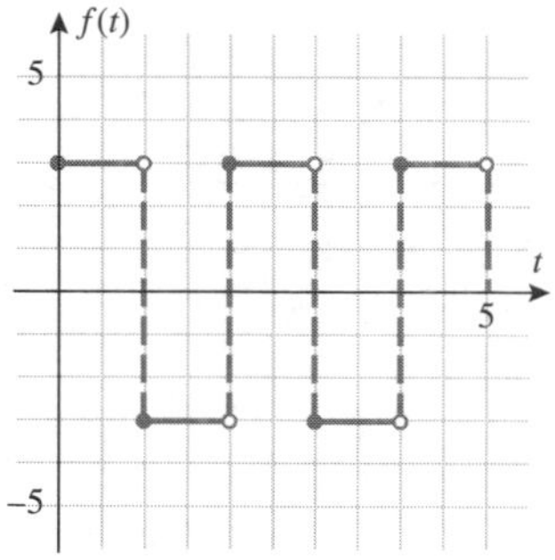

90. Electronics: Signals. Write a step function $f(x)$ that represents the following square wave, where x represents frequency in Hz.

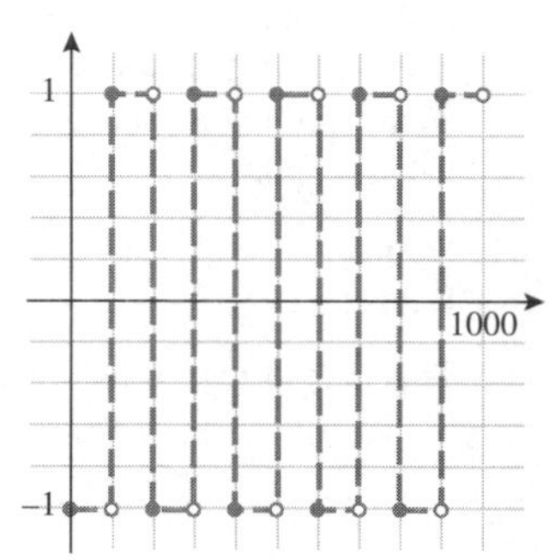

For Exercises 91 and 92, refer to the following table:

Global Carbon Emissions from Fossil Fuel Burning

YEAR	MILLIONS OF TONS OF CARBON
1900	500
1925	1000
1950	1500
1975	5000
2000	7000

91. Climate Change: Global Warming. What is the average rate of change in global carbon emissions from fossil fuel burning from

a. 1900 to 1950
b. 1950 to 2000?

92. Climate Change: Global Warming. What is the average rate of change in global carbon emissions from fossil fuel burning from

a. 1950 to 1975
b. 1975 to 2000?

For Exercises 93 and 94, use the following information:

The height (in feet) of a falling object with an initial velocity of 48 feet per second launched straight upward from the ground is given by $h(t) = -16t^2 + 48t$, where t is time (in seconds).

93. Falling Objects. What is the average rate of change of the height as a function of time from $t = 1$ to $t = 2$?

94. Falling Objects. What is the average rate of change of the height as a function of time from $t = 1$ to $t = 3$?

CATCH THE MISTAKE

In Exercises 95–98, explain the mistake that is made.

95. Graph the piecewise-defined function. State the domain and range.

$$f(x) = \begin{cases} -x & x < 0 \\ x & x > 0 \end{cases}$$

Solution:

Draw the graphs of $f(x) = -x$ and $f(x) = x$.

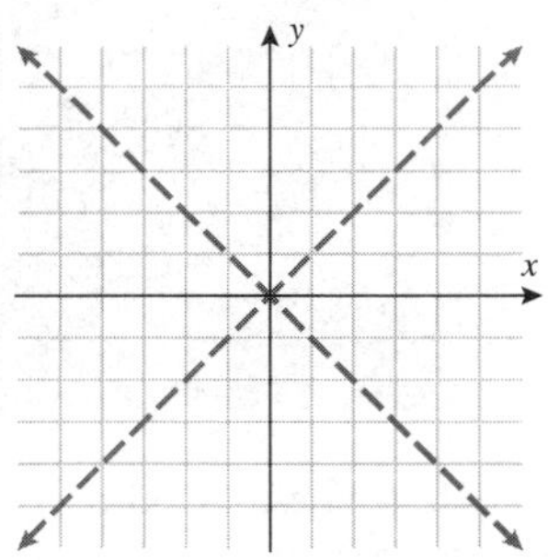

Darken the function $f(x) = -x$ when $x < 0$ and the function $f(x) = x$ when $x > 0$. This gives us the familiar absolute value graph.

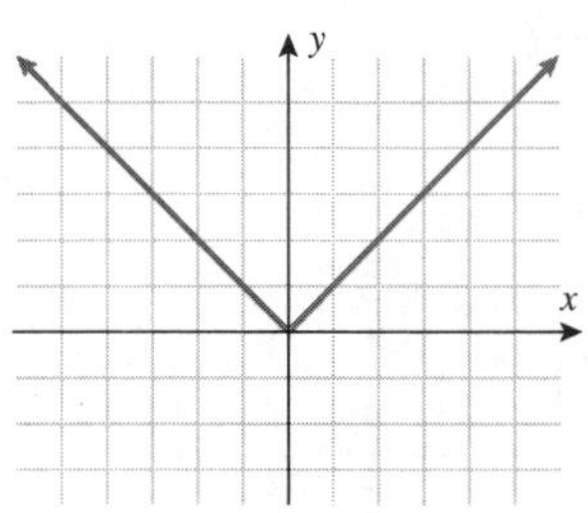

Domain: $(-\infty, \infty)$ or $\mathbb{R}$
Range: $[0, \infty)$

This is incorrect. What mistake was made?

96. Graph the piecewise-defined function. State the domain and range.

$$f(x) = \begin{cases} -x & x \leq 1 \\ x & x > 1 \end{cases}$$

Solution:

Draw the graphs of $f(x) = -x$ and $f(x) = x$.

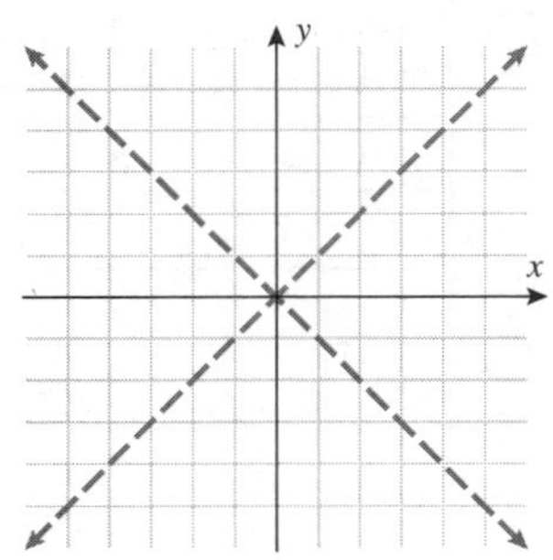

Darken the function $f(x) = -x$ when $x < 1$ and the function $f(x) = x$ when $x > 1$.

The resulting graph is as shown.

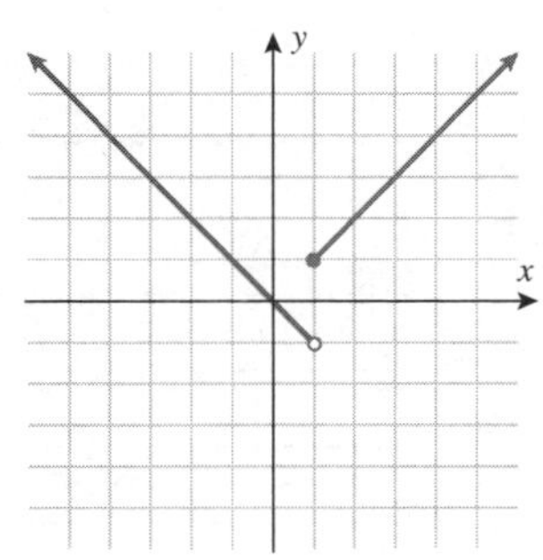

Domain: $(-\infty, \infty)$ or $\mathbb{R}$
Range: $(-1, \infty)$

This is incorrect. What mistake was made?

97. The cost of airport internet access is \$15 for the first 30 minutes and \$1 per minute for each additional minute. Write a function describing the cost of the service as a function of minutes used online.

Solution: $C(x) = \begin{cases} 15 & x \leq 30 \\ 15 + x & x > 30 \end{cases}$

This is incorrect. What mistake was made?

98. Most money market accounts pay a higher interest with a higher principal. If the credit union is offering 2% on accounts with less than or equal to \$10,000 and 4% on the additional money over \$10,000, write the interest function $I(x)$ that represents the interest earned on an account as a function of dollars in the account.

Solution: $I(x) = \begin{cases} 0.02x & x \leq 10{,}000 \\ 0.02(10{,}000) + 0.04x & x > 10{,}000 \end{cases}$

This is incorrect. What mistake was made?

CONCEPTUAL

In Exercises 99–102, determine whether each statement is true or false.

99. The identity function is a special case of the linear function.

100. The constant function is a special case of the linear function.

101. If an odd function has an interval where the function is increasing, then it also has to have an interval where the function is decreasing.

102. If an even function has an interval where the function is increasing, then it also has to have an interval where the function is decreasing.

CHALLENGE

In Exercises 103 and 104, for *a* and *b* real numbers, can the function given ever be a continuous function? If so, specify the value for *a* and *b* that would make it so.

103. $f(x) = \begin{cases} ax & x \leq 2 \\ bx^2 & x > 2 \end{cases}$

104. $f(x) = \begin{cases} -\dfrac{1}{x} & x < a \\ \dfrac{1}{x} & x \geq a \end{cases}$

TECHNOLOGY

105. In trigonometry you will learn about the sine function, sin x. Plot the function $f(x) = \sin x$, using a graphing utility. It should look like the graph on the right. Is the sine function even, odd, or neither?

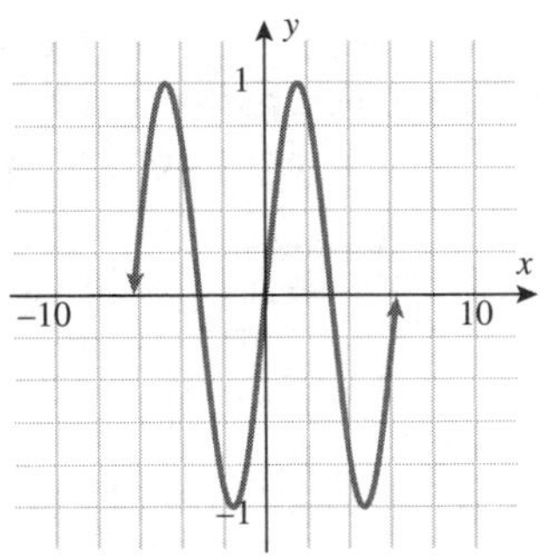

106. In trigonometry you will learn about the cosine function, cos x. Plot the function $f(x) = \cos x$, using a graphing utility. It should look like the graph on the right. Is the cosine function even, odd, or neither?

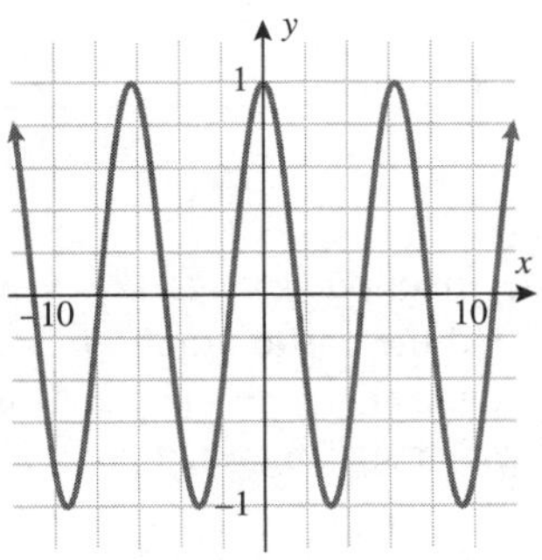

107. In trigonometry you will learn about the tangent function, tan x. Plot the function $f(x) = \tan x$, using a graphing utility. If you restrict the values of x so that $-\frac{\pi}{2} < x < \frac{\pi}{2}$, the graph should resemble the graph below. Is the tangent function even, odd, or neither?

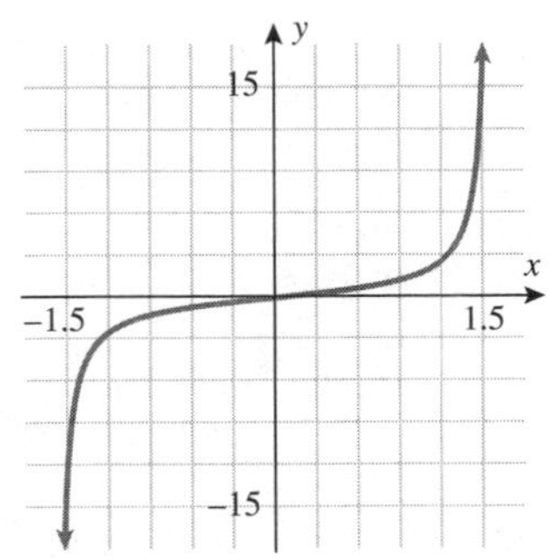

108. Plot the function $f(x) = \dfrac{\sin x}{\cos x}$. What function is this?

109. Graph the function $f(x) = [[3x]]$ using a graphing utility. State the domain and range.

110. Graph the function $f(x) = \left[\left[\frac{1}{3}x\right]\right]$ using a graphing utility. State the domain and range.

SECTION

3.3 GRAPHING TECHNIQUES: TRANSFORMATIONS

SKILLS OBJECTIVES

- Sketch the graph of a function using horizontal and vertical shifting of common functions.
- Sketch the graph of a function by reflecting a common function about the x-axis or y-axis.
- Sketch the graph of a function by stretching or compressing a common function.
- Sketch the graph of a function using a sequence of transformations.

CONCEPTUAL OBJECTIVES

- Identify the common functions by their graphs.
- Apply multiple transformations of common functions to obtain graphs of functions.
- Understand that domain and range also are transformed.

The focus of the previous section was to learn the graphs that correspond to particular functions such as identity, square, cube, square root, cube root, absolute value, and reciprocal. Therefore, at this point, you should be able to recognize and generate the graphs of $y = x$, $y = x^2$, $y = x^3$, $y = \sqrt{x}$, $y = \sqrt[3]{x}$, $y = |x|$, and $y = \frac{1}{x}$. In this section, we will discuss how to sketch the graphs of functions that are very simple modifications of these functions. For instance, a common function may be shifted (horizontally or vertically), reflected, or stretched (or compressed). Collectively, these techniques are called **transformations**.

Horizontal and Vertical Shifts

Let's take the absolute value function as an example. The graph of $f(x) = |x|$ was given in the last section. Now look at two examples that are much like this function: $g(x) = |x| + 2$ and $h(x) = |x - 1|$. Graphing these functions by point plotting yields

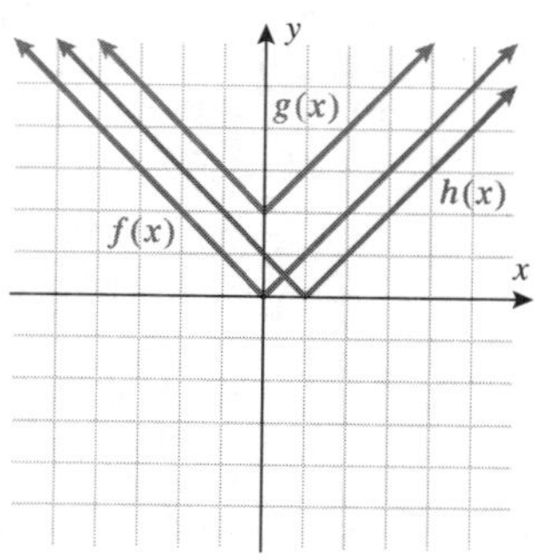

x	$f(x)$
−2	2
−1	1
0	0
1	1
2	2

x	$g(x)$
−2	4
−1	3
0	2
1	3
2	4

x	$h(x)$
−2	3
−1	2
0	1
1	0
2	1

Instead of point-plotting the function $g(x) = |x| + 2$, we could have started with the function $f(x) = |x|$ and shifted the entire graph *up* 2 units. Similarly, we could have generated the graph of the function $h(x) = |x - 1|$ by shifting the function $f(x) = |x|$ to the *right* 1 unit. In both cases, the base or starting function is $f(x) = |x|$. Why did we go up for $g(x)$ and to the right for $h(x)$?

Note that we could rewrite the functions $g(x)$ and $h(x)$ in terms of $f(x)$:

$$g(x) = |x| + 2 = f(x) + 2$$

$$h(x) = |x - 1| = f(x - 1)$$

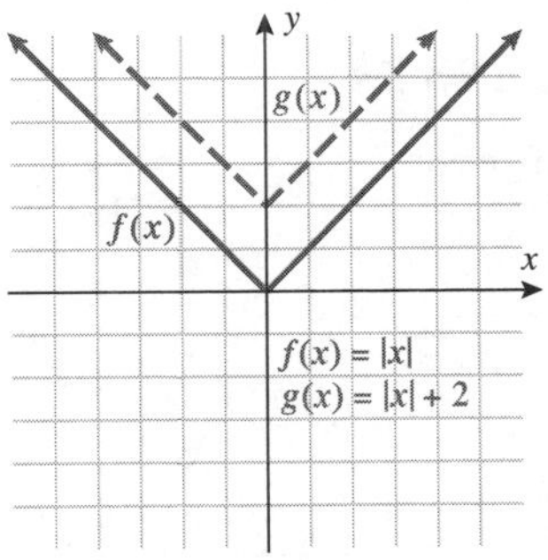

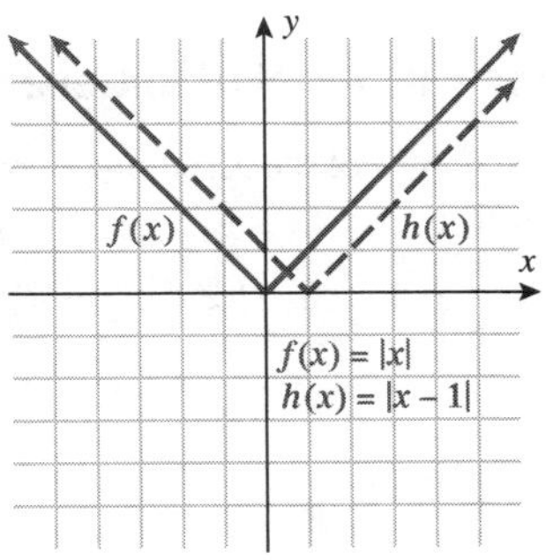

In the case of $g(x)$, the shift $(+2)$ occurs "outside" the function—that is, outside the parentheses showing the argument. Therefore, the output for $g(x)$ is 2 more than the typical output for $f(x)$. Because the output corresponds to the vertical axis, this results in a shift *upward* of 2 units. In general, shifts that occur *outside* the function correspond to a *vertical* shift corresponding to the sign of the shift. For instance, had the function been $G(x) = |x| - 2$, this graph would have started with the graph of the function $f(x)$ and shifted down 2 units.

In the case of $h(x)$, the shift occurs "inside" the function—that is, inside the parentheses showing the argument. Note that the point $(0, 0)$ that lies on the graph of $f(x)$ was shifted to the point $(1, 0)$ on the graph of the function $h(x)$. The y-value remained the same, but the x-value shifted to the right 1 unit. Similarly, the points $(-1, 1)$ and $(1, 1)$ were shifted to the points $(0, 1)$ and $(2, 1)$, respectively. In general, shifts that occur *inside* the function correspond to a *horizontal* shift opposite the sign. In this case, the graph of the function $h(x) = |x - 1|$ shifted the graph of the function $f(x)$ to the right 1 unit. If, instead, we had the function $H(x) = |x + 1|$, this graph would have started with the graph of the function $f(x)$ and shifted to the left 1 unit.

VERTICAL SHIFTS

Assuming that c is a positive constant,

To Graph	Shift the Graph of $f(x)$
$f(x) + c$	c units upward
$f(x) - c$	c units downward

Adding or subtracting a constant **outside** the function corresponds to a **vertical** shift that goes **with the sign.**

Study Tip

Shifts outside the function are vertical shifts with the sign.

Up (+)
Down (−)

HORIZONTAL SHIFTS

Assuming that c is a positive constant,

To Graph	Shift the Graph of $f(x)$
$f(x + c)$	c units to the left
$f(x - c)$	c units to the right

Adding or subtracting a constant **inside** the function corresponds to a **horizontal** shift that goes **opposite the sign.**

Study Tip

Shifts inside the function are horizontal shifts opposite the sign.

Left (+)
Right (−)

EXAMPLE 1 Horizontal and Vertical Shifts

Sketch the graphs of the given functions using horizontal and vertical shifts:

a. $g(x) = x^2 - 1$ **b.** $H(x) = (x + 1)^2$

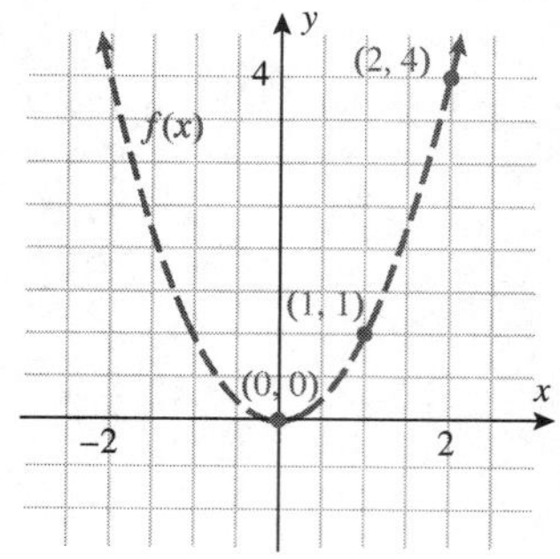

Solution:

In both cases, the function to start with is $f(x) = x^2$.

a. $g(x) = x^2 - 1$ can be rewritten as
$g(x) = f(x) - 1$.

1. The shift (1 unit) occurs *outside* of the function. Therefore, we expect a vertical shift that goes with the sign.
2. Since the sign is *negative*, this corresponds to a *downward* shift.
3. Shifting the graph of the function $f(x) = x^2$ down 1 unit yields the graph of $g(x) = x^2 - 1$.

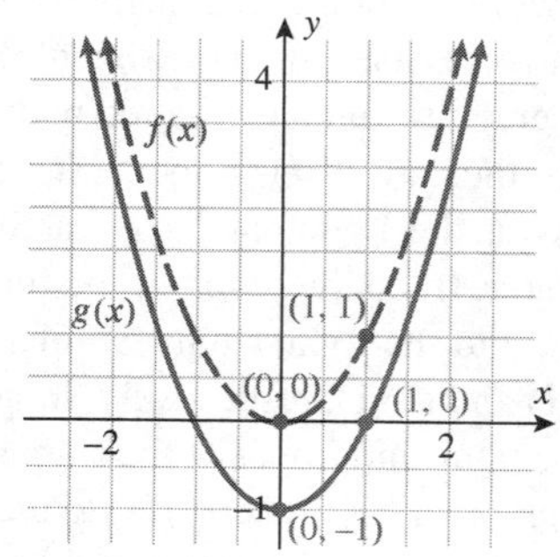

b. $H(x) = (x + 1)^2$ can be rewritten as
$H(x) = f(x + 1)$.

1. The shift (1 unit) occurs *inside* of the function. Therefore, we expect a horizontal shift that goes *opposite* the sign.
2. Since the sign is *positive*, this corresponds to a shift to the *left*.
3. Shifting the graph of the function $f(x) = x^2$ to the left 1 unit yields the graph of $H(x) = (x + 1)^2$.

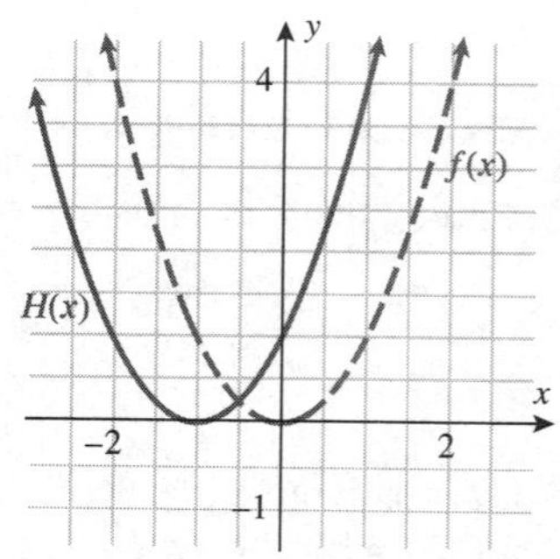

YOUR TURN Sketch the graphs of the given functions using horizontal and vertical shifts.

a. $g(x) = x^2 + 1$ **b.** $H(x) = (x - 1)^2$

Technology Tip

a. Graphs of $y_1 = x^2$ and $y_2 = g(x) = x^2 - 1$ are shown.

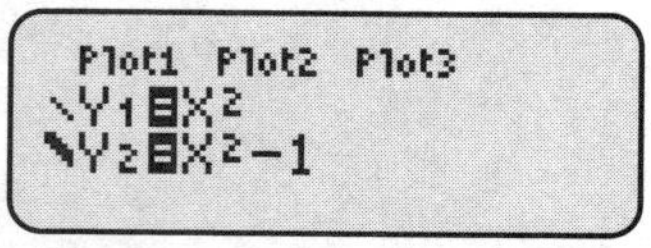

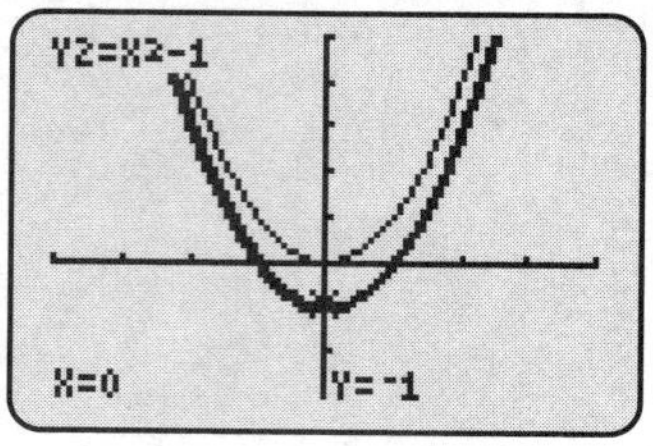

b. Graphs of $y_1 = x^2$ and $y_2 = H(x) = (x + 1)^2$ are shown.

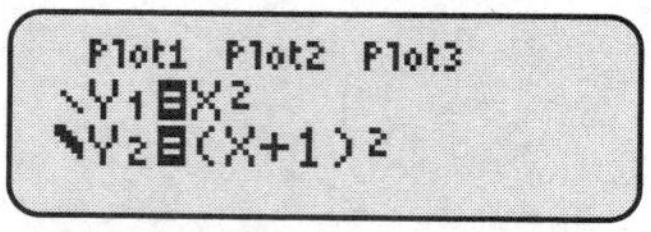

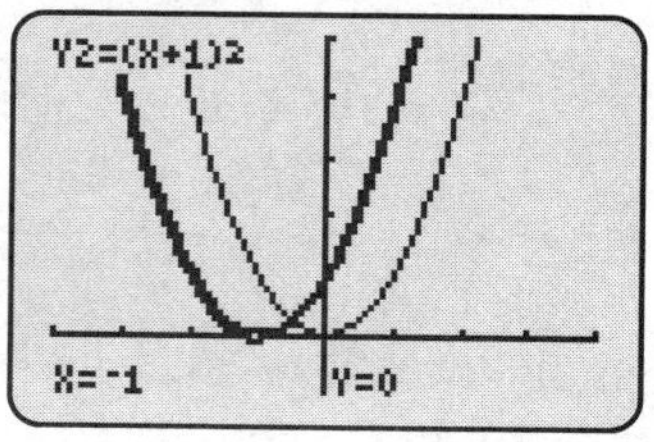

Answer:

a.

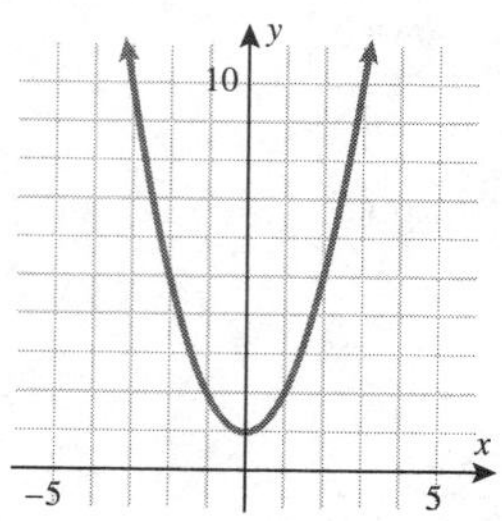

b.

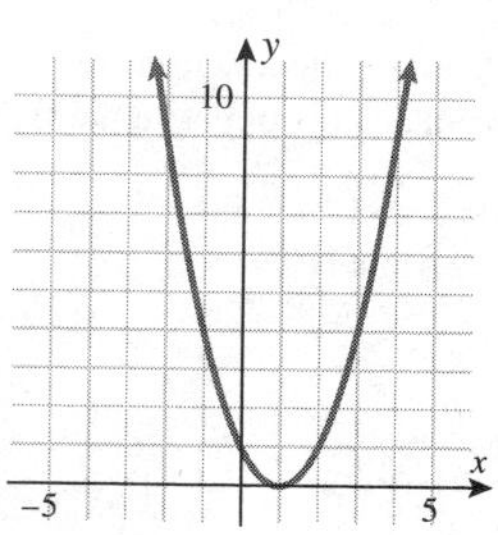

It is important to note that the domain and range of the resulting function can be thought of as also being shifted. Shifts in the domain correspond to horizontal shifts, and shifts in the range correspond to vertical shifts.

EXAMPLE 2 Horizontal and Vertical Shifts and Changes in the Domain and Range

Graph the functions using translations and state the domain and range of each function.

a. $g(x) = \sqrt{x+1}$ **b.** $G(x) = \sqrt{x} - 2$

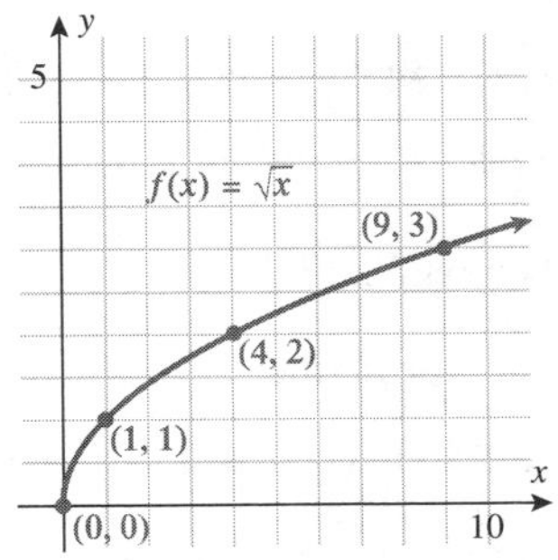

Solution:

In both cases the function to start with is $f(x) = \sqrt{x}$.

Domain: $[0, \infty)$

Range: $[0, \infty)$

a. $g(x) = \sqrt{x+1}$ can be rewritten as $g(x) = f(x+1)$.

1. The shift (1 unit) is *inside* the function, which corresponds to a *horizontal* shift *opposite the sign.*

2. Shifting the graph of $f(x) = \sqrt{x}$ to the *left* 1 unit yields the graph of $g(x) = \sqrt{x+1}$. Notice that the point $(0, 0)$, which lies on the graph of $f(x)$, gets shifted to the point $(-1, 0)$ on the graph of $g(x)$.

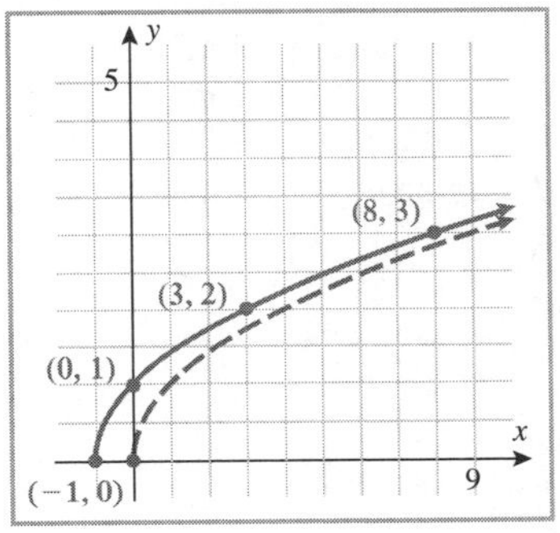

Although the original function $f(x) = \sqrt{x}$ had an implicit restriction on the domain $[0, \infty)$, the function $g(x) = \sqrt{x+1}$ has the implicit restriction that $x \geq -1$. We see that the output or range of $g(x)$ is the same as the output of the original function $f(x)$.

Domain: $[-1, \infty)$ Range: $[0, \infty)$

b. $G(x) = \sqrt{x} - 2$ can be rewritten as $G(x) = f(x) - 2$.

1. The shift (2 units) is *outside* the function, which corresponds to a *vertical* shift *with the sign.*

2. The graph of $G(x) = \sqrt{x} - 2$ is found by shifting $f(x) = \sqrt{x}$ down 2 units. Note that the point $(0, 0)$, which lies on the graph of $f(x)$, gets shifted to the point $(0, -2)$ on the graph of $G(x)$.

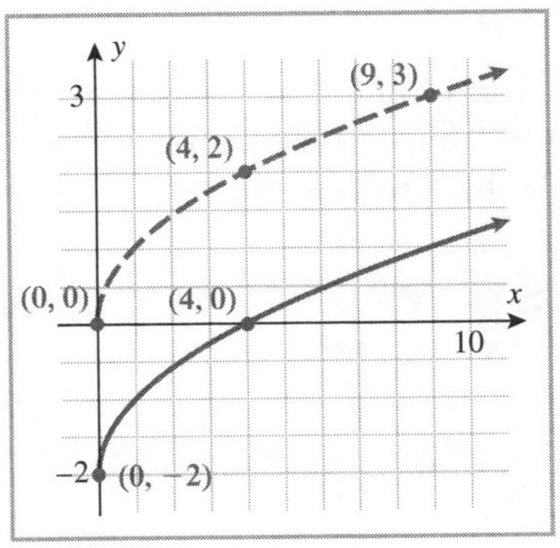

The original function $f(x) = \sqrt{x}$ has an implicit restriction on the domain: $[0, \infty)$. The function $G(x) = \sqrt{x} - 2$ also has the implicit restriction that $x \geq 0$. The output or range of $G(x)$ is always 2 units less than the output of the original function $f(x)$.

Domain: $[0, \infty)$ Range: $[-2, \infty)$

YOUR TURN Sketch the graph of the functions using shifts and state the domain and range.

a. $G(x) = \sqrt{x-2}$ **b.** $h(x) = |x| + 1$

Answer:

a. $G(x) = \sqrt{x-2}$

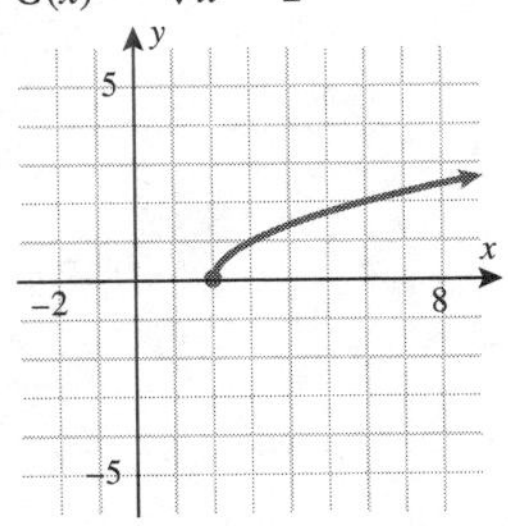

Domain: $[2, \infty)$ Range: $[0, \infty)$

b. $h(x) = |x| + 1$

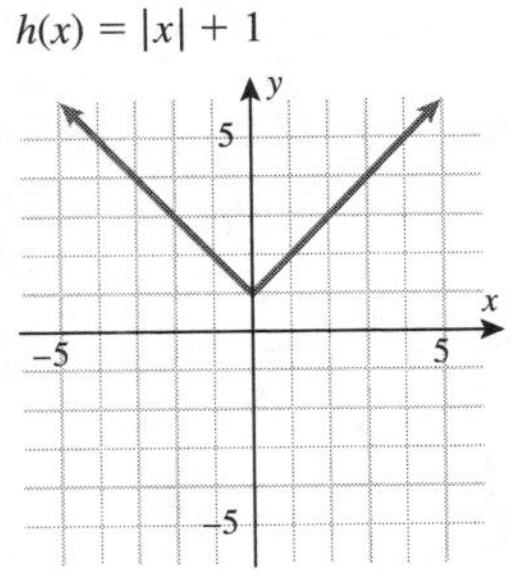

Domain: $(-\infty, \infty)$ Range: $[1, \infty)$

The previous examples have involved graphing functions by shifting a known function either in the horizontal or vertical direction. Let us now look at combinations of horizontal and vertical shifts.

Technology Tip

a. Graphs of $y_1 = x^2$, $y_2 = (x + 1)^2$, and $y_3 = F(x) = (x + 1)^2 - 2$ are shown.

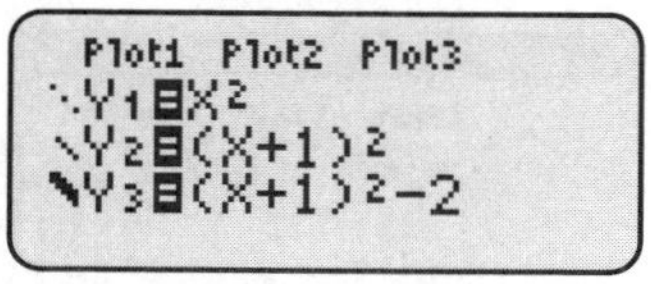

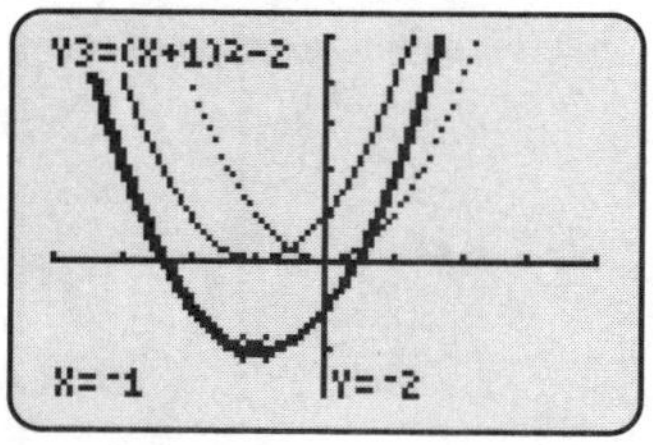

EXAMPLE 3 Combining Horizontal and Vertical Shifts

Sketch the graph of the function $F(x) = (x + 1)^2 - 2$. State the domain and range of F.

Solution:

The base function is $y = x^2$.

1. The shift (1 unit) is *inside* the function, so it represents a *horizontal* shift *opposite the sign.*
2. The -2 shift is *outside* the function, which represents a *vertical* shift *with the sign.*
3. Therefore, we shift the graph of $y = x^2$ to the left 1 unit and down 2 units. For instance, the point $(0, 0)$ on the graph of $y = x^2$ shifts to the point $(-1, -2)$ on the graph of $F(x) = (x + 1)^2 - 2$.

Domain: $(-\infty, \infty)$ Range: $[-2, \infty)$

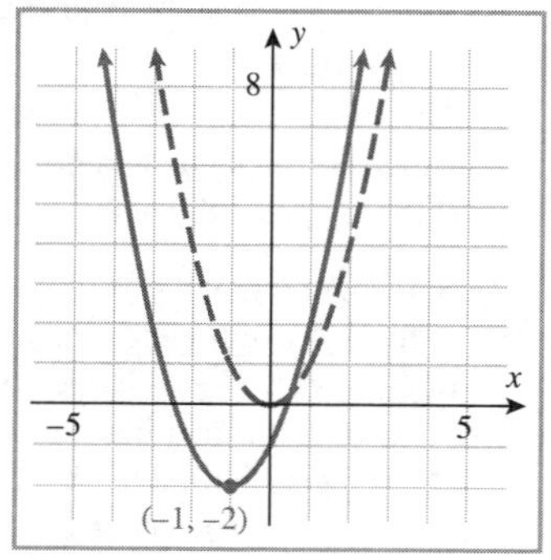

■ **YOUR TURN** Sketch the graph of the function $f(x) = |x - 2| + 1$. State the domain and range of f.

■ **Answer:**
$f(x) = |x - 2| + 1$
$f(x) = |x|$
Domain: $(-\infty, \infty)$
Range: $[1, \infty)$

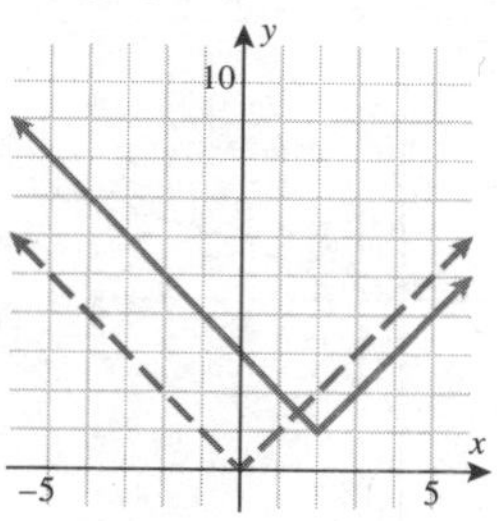

All of the previous transformation examples involve starting with a common function and shifting the function in either the horizontal or the vertical direction (or a combination of both). Now, let's investigate *reflections* of functions about the x-axis or y-axis.

Reflection About the Axes

To sketch the graphs of $f(x) = x^2$ and $g(x) = -x^2$ start by first listing points that are on each of the graphs and then connecting the points with smooth curves.

x	$f(x)$
-2	4
-1	1
0	0
1	1
2	4

x	$g(x)$
-2	-4
-1	-1
0	0
1	-1
2	-4

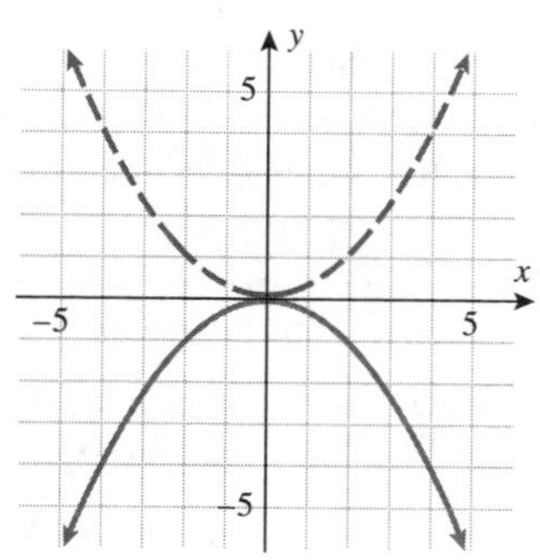

Note that if the graph of $f(x) = x^2$ is reflected about the x-axis, the result is the graph of $g(x) = -x^2$. Also note that the function $g(x)$ can be written as the negative of the function $f(x)$; that is $g(x) = -f(x)$. In general, **reflection about the x-axis** is produced by multiplying a function by -1.

Let's now investigate reflection about the y-axis. To sketch the graphs of $f(x) = \sqrt{x}$ and $g(x) = \sqrt{-x}$ start by listing points that are on each of the graphs and then connecting the points with smooth curves.

x	$f(x)$
0	0
1	1
4	2
9	3

x	$g(x)$
−9	3
−4	2
−1	1
0	0

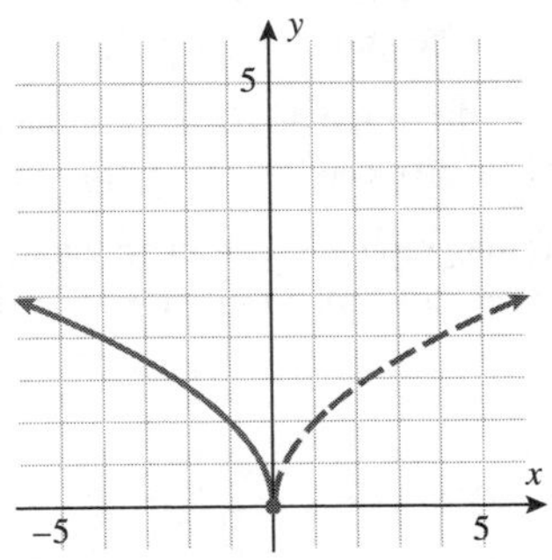

Note that if the graph of $f(x) = \sqrt{x}$ is reflected about the y-axis, the result is the graph of $g(x) = \sqrt{-x}$. Also note that the function $g(x)$ can be written as $g(x) = f(-x)$. In general, **reflection about the y-axis** is produced by replacing x with $-x$ in the function. Notice that the domain of f is $[0, \infty)$, whereas the domain of g is $(-\infty, 0]$.

Reflection About the Axes

The graph of $-f(x)$ is obtained by reflecting the graph of $f(x)$ about the x-axis.
The graph of $f(-x)$ is obtained by reflecting the graph of $f(x)$ about the y-axis.

EXAMPLE 4 Sketching the Graph of a Function Using Both Shifts and Reflections

Sketch the graph of the function $G(x) = -\sqrt{x+1}$.

Solution:

Start with the square root function. $f(x) = \sqrt{x}$

Shift the graph of $f(x)$ to the left 1 unit to arrive at the graph of $f(x+1)$. $f(x+1) = \sqrt{x+1}$

Reflect the graph of $f(x+1)$ about the x-axis to arrive at the graph of $-f(x+1)$. $-f(x+1) = -\sqrt{x+1}$

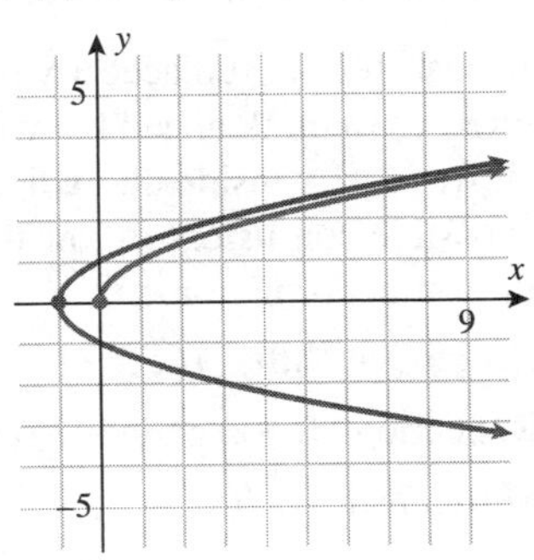

Technology Tip

a. Graphs of $y_1 = \sqrt{x}$, $y_2 = \sqrt{x+2}$, $y_3 = \sqrt{-x+2}$, and $y_4 = f(x) = \sqrt{2-x} + 1$ are shown.

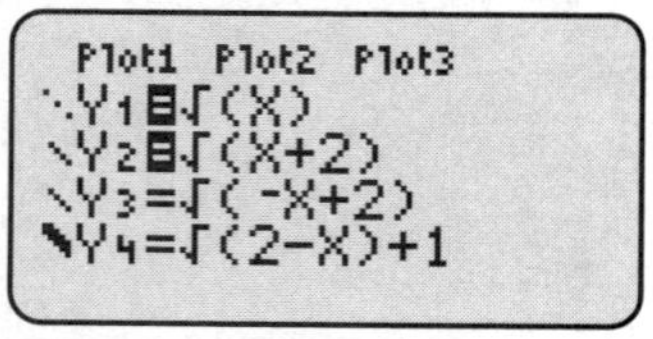

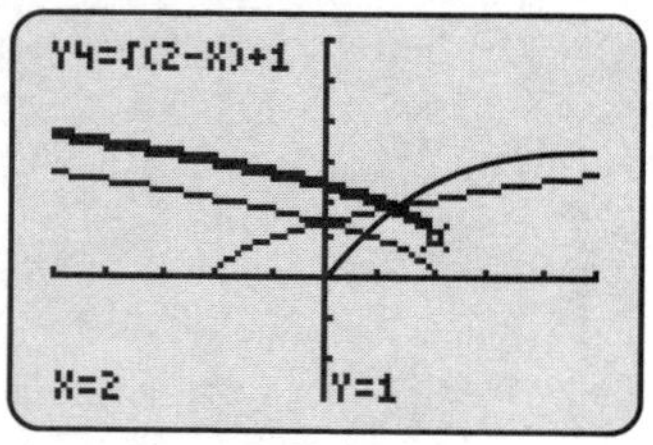

EXAMPLE 5 Sketching the Graph of a Function Using Both Shifts and Reflections

Sketch the graph of the function $f(x) = \sqrt{2-x} + 1$.

Solution:

Start with the square root function.	$g(x) = \sqrt{x}$
Shift the graph of $g(x)$ to the left 2 units to arrive at the graph of $g(x+2)$.	$g(x+2) = \sqrt{x+2}$
Reflect the graph of $g(x+2)$ about the y-axis to arrive at the graph of $g(-x+2)$.	$g(-x+2) = \sqrt{-x+2}$
Shift the graph $g(-x+2)$ up 1 unit to arrive at the graph of $g(-x+2)+1$.	$g(-x+2) + 1 = \sqrt{2-x} + 1$

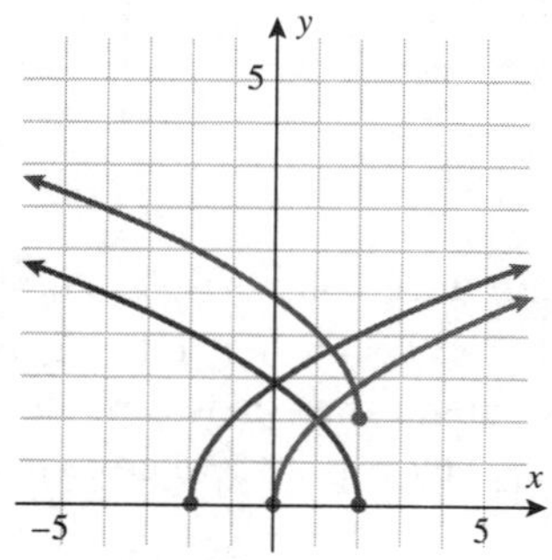

YOUR TURN Use shifts and reflections to sketch the graph of the function $f(x) = -\sqrt{x-1} + 2$. State the domain and range of $f(x)$.

Answer:
Domain: $[1, \infty)$
Range: $(-\infty, 2]$

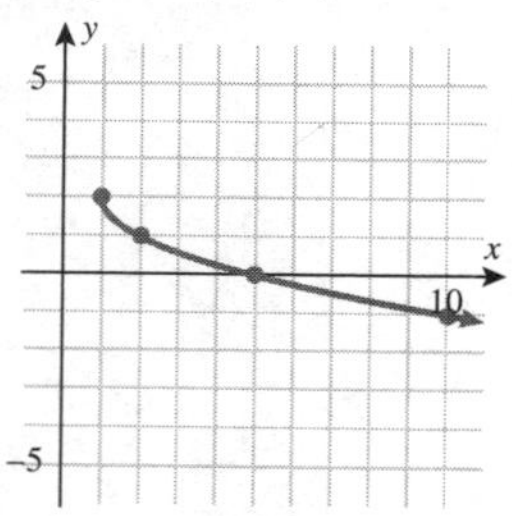

Look back at the order in which transformations were performed in Example 5: horizontal shift, reflection, and then vertical shift. Let us consider an alternate order of transformations.

Words	Math
Start with the square root function.	$g(x) = \sqrt{x}$
Shift the graph of $g(x)$ up 1 unit to arrive at the graph of $g(x) + 1$.	$g(x) + 1 = \sqrt{x} + 1$
Reflect the graph of $g(x) + 1$ about the y-axis to arrive at the graph of $g(-x) + 1$.	$g(-x) + 1 = \sqrt{-x} + 1$
Replace x with $x - 2$, which corresponds to a shift of the graph of $g(-x) + 1$ to the right 2 units to arrive at the graph of $g[-(x-2)] + 1$.	$g(-x+2) + 1 = \sqrt{2-x} + 1$

In the last step we replaced x with $x - 2$, which required us to think ahead knowing the desired result was $2 - x$ inside the radical. To avoid any possible confusion, follow this order of transformations:

1. Horizontal shifts: $f(x \pm c)$
2. Reflection: $f(-x)$ and/or $-f(x)$
3. Vertical shifts: $f(x) \pm c$

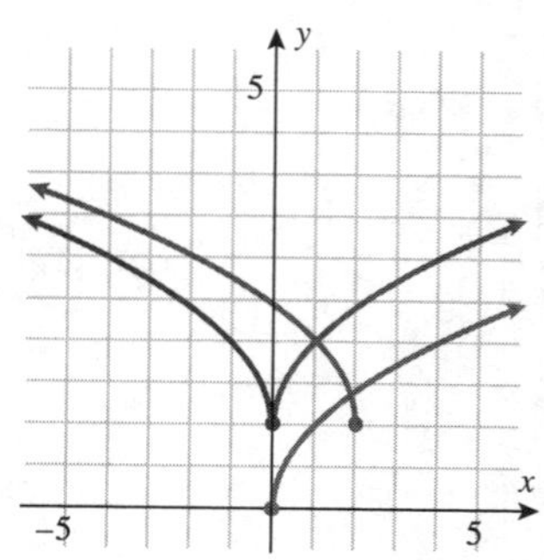

Stretching and Compressing

Horizontal shifts, vertical shifts, and reflections change only the position of the graph in the Cartesian plane, leaving the basic shape of the graph unchanged. These transformations (shifts and reflections) are called **rigid transformations** because they alter only the *position.* **Nonrigid transformations**, on the other hand, distort the *shape* of the original graph. We now consider *stretching* and *compressing* of graphs in both the vertical and the horizontal direction.

A vertical stretch or compression of a graph occurs when the function is multiplied by a positive constant. For example, the graphs of the functions $f(x) = x^2$, $g(x) = 2f(x) = 2x^2$, and $h(x) = \frac{1}{2}f(x) = \frac{1}{2}x^2$ are illustrated below. Depending on if the constant is larger than 1 or smaller than 1 will determine whether it corresponds to a stretch (expansion) or compression (contraction) in the vertical direction.

x	$f(x)$
-2	4
-1	1
0	0
1	1
2	4

x	$g(x)$
-2	8
-1	2
0	0
1	2
2	8

x	$h(x)$
-2	2
-1	$\frac{1}{2}$
0	0
1	$\frac{1}{2}$
2	2

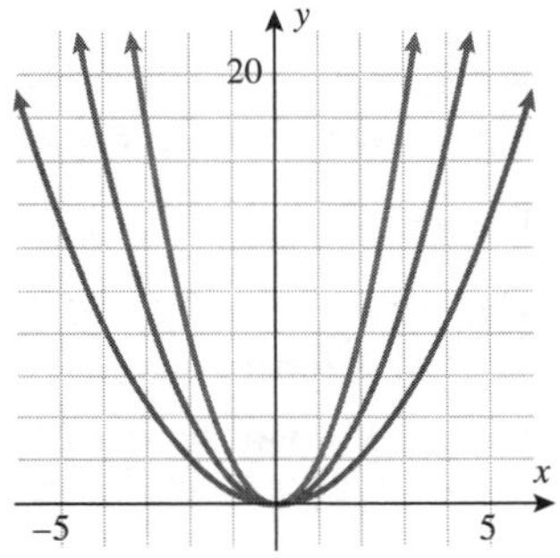

Note that when the function $f(x) = x^2$ is multiplied by 2, so that $g(x) = 2f(x) = 2x^2$, the result is a graph stretched in the vertical direction. When the function $f(x) = x^2$ is multiplied by $\frac{1}{2}$, so that $h(x) = \frac{1}{2}f(x) = \frac{1}{2}x^2$, the result is a graph that is compressed in the vertical direction.

VERTICAL STRETCHING AND VERTICAL COMPRESSING OF GRAPHS

The graph of $cf(x)$ is found by:

- **Vertically stretching** the graph of $f(x)$ if $c > 1$
- **Vertically compressing** the graph of $f(x)$ if $0 < c < 1$

Note: c is any positive real number.

EXAMPLE 6 Vertically Stretching and Compressing Graphs

Graph the function $h(x) = \frac{1}{4}x^3$.

Solution:

1. Start with the cube function. $f(x) = x^3$
2. Vertical compression is expected because $\frac{1}{4}$ is less than 1. $h(x) = \dfrac{1}{4}x^3$

3. Determine a few points that lie on the graph of h.

$(0, 0) \quad (2, 2) \quad (-2, -2)$

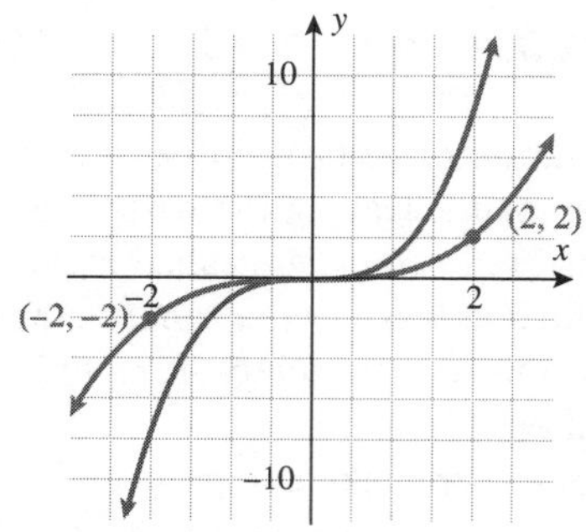

Conversely, if the argument x of a function f is multiplied by a positive real number c, then the result is a *horizontal* stretch of the graph of f if $0 < c < 1$. If $c > 1$, then the result is a *horizontal* compression of the graph of f.

HORIZONTAL STRETCHING AND HORIZONTAL COMPRESSING OF GRAPHS

The graph of $f(cx)$ is found by:

- **Horizontally stretching** the graph of $f(x)$ if $0 < c < 1$
- **Horizontally compressing** the graph of $f(x)$ if $c > 1$

Note: c is any positive real number.

EXAMPLE 7 Vertically Stretching and Horizontally Compressing Graphs

Given the graph of $f(x)$, graph:

a. $2f(x)$ **b.** $f(2x)$

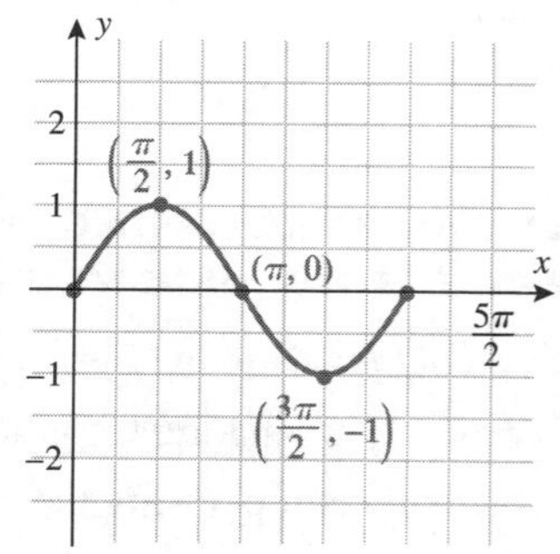

Solution (a):

Since the function is multiplied (on the outside) by 2, the result is that each **y-value** of $f(x)$ is ***multiplied* by 2**, which corresponds to vertical stretching.

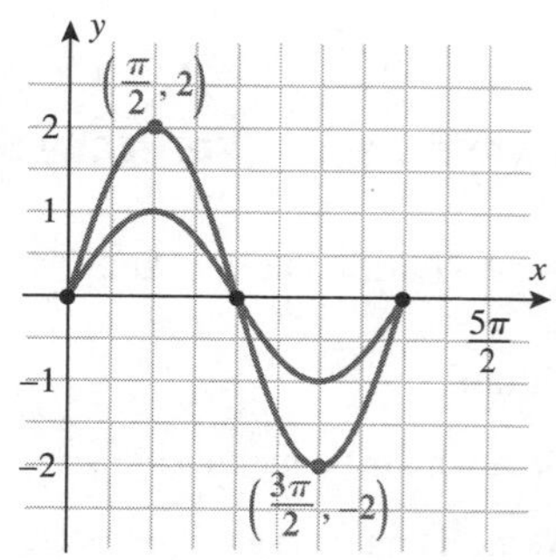

Solution (b):

Since the argument of the function is multiplied (on the inside) by 2, the result is that each ***x*-value** of $f(x)$ is **divided by 2**, which corresponds to horizontal compression.

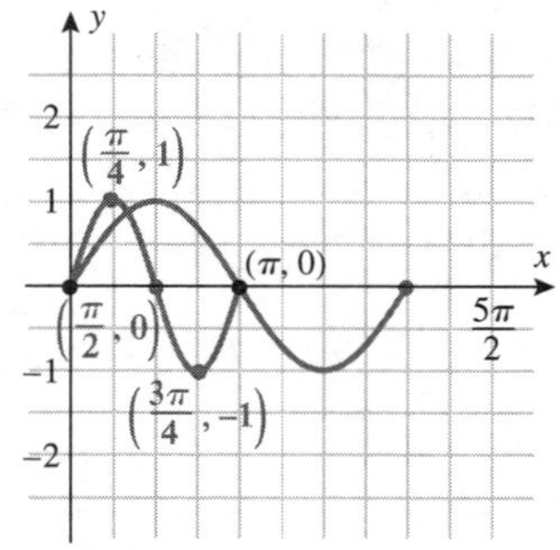

■ **YOUR TURN** Graph the function $g(x) = 4x^3$.

■ **Answer:** *Expansion* of the graph $f(x) = x^3$.

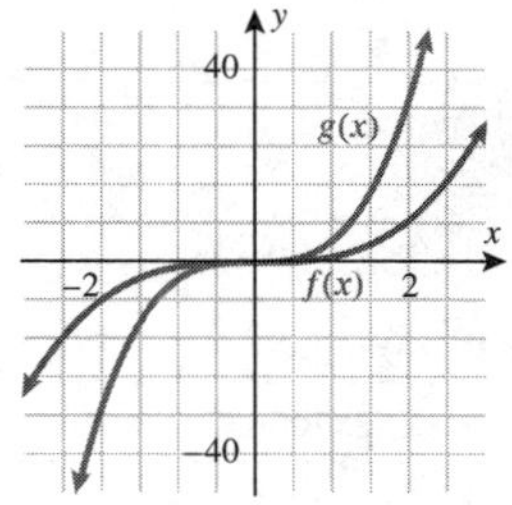

EXAMPLE 8 Sketching the Graph of a Function Using Multiple Transformations

Sketch the graph of the function $H(x) = -2(x - 3)^2$.

Solution:

Start with the square function. $f(x) = x^2$

Shift the graph of $f(x)$ to the right 3 units to arrive at the graph of $f(x - 3)$. $f(x - 3) = (x - 3)^2$

Vertically stretch the graph of $f(x - 3)$ by a factor of 2 to arrive at the graph of $2f(x - 3)$. $2f(x - 3) = 2(x - 3)^2$

Reflect the graph $2f(x - 3)$ about the x-axis to arrive at the graph of $-2f(x - 3)$. $-2f(x - 3) = -2(x - 3)^2$

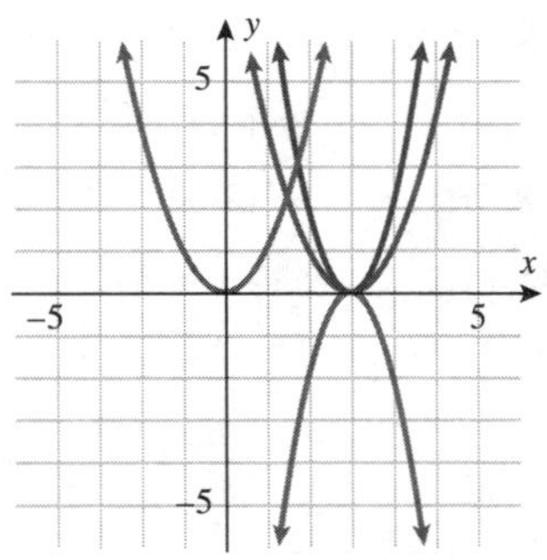

Technology Tip

Graphs of $y_1 = x^2$, $y_2 = (x - 3)^2$, $y_3 = 2(x - 3)^2$, and $y_4 = H(x) = -2(x - 3)^2$ are shown.

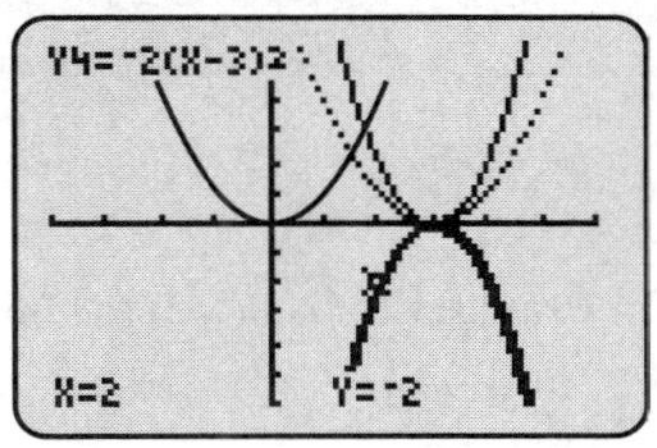

In Example 8 we followed the same "inside out" approach with the functions to determine the order for the transformations: horizontal shift, vertical stretch, and reflection.

SECTION 3.3 SUMMARY

TRANSFORMATION	TO GRAPH THE FUNCTION...	DRAW THE GRAPH OF *f* AND THEN...	DESCRIPTION
Horizontal shifts	$f(x + c)$ $f(x - c)$	Shift the graph of f to the left c units. Shift the graph of f to the right c units.	Replace x by $x + c$. Replace x by $x - c$.
Vertical shifts	$f(x) + c$ $f(x) - c$	Shift the graph of f up c units. Shift the graph of f down c units.	Add c to $f(x)$. Subtract c from $f(x)$.
Reflection about the x-axis	$-f(x)$	Reflect the graph of f about the x-axis.	Multiply $f(x)$ by -1.
Reflection about the y-axis	$f(-x)$	Reflect the graph of f about the y-axis.	Replace x by $-x$.
Vertical stretch	$cf(x)$, where $c > 1$	Vertically stretch the graph of f.	Multiply $f(x)$ by c.
Vertical compression	$cf(x)$, where $0 < c < 1$	Vertically compress the graph of f.	Multiply $f(x)$ by c.
Horizontal stretch	$f(cx)$, where $0 < c < 1$	Horizontally stretch the graph of f.	Replace x by cx.
Horizontal compression	$f(cx)$, where $c > 1$	Horizontally compress the graph of f.	Replace x by cx.

SECTION 3.3 EXERCISES

SKILLS

In Exercises 1–12, match the function to the graph.

1. $f(x) = x^2 + 1$
2. $f(x) = (x - 1)^2$
3. $f(x) = -(1 - x)^2$
4. $f(x) = -x^2 - 1$
5. $f(x) = -(x + 1)^2$
6. $f(x) = -(1 - x)^2 + 1$
7. $f(x) = \sqrt{x - 1} + 1$
8. $f(x) = -\sqrt{x} - 1$
9. $f(x) = \sqrt{1 - x} - 1$
10. $f(x) = \sqrt{-x} + 1$
11. $f(x) = -\sqrt{-x} + 1$
12. $f(x) = -\sqrt{1 - x} - 1$

a.

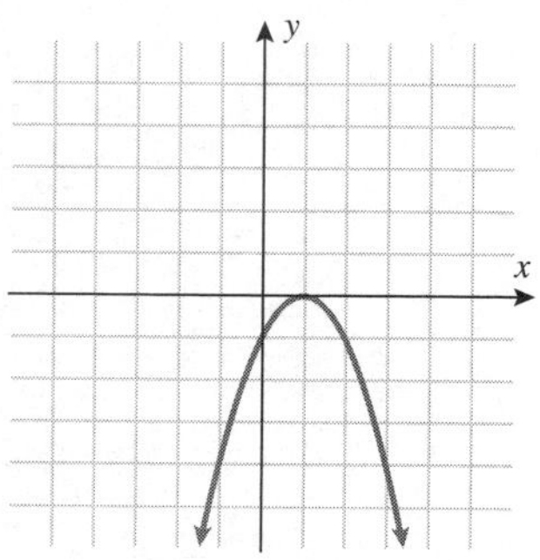

b.

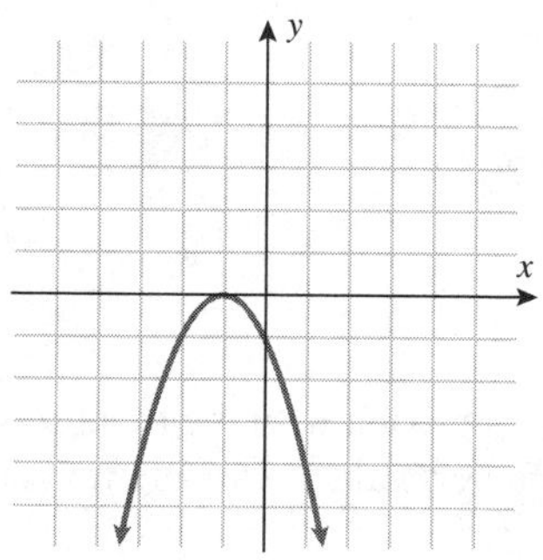

c.

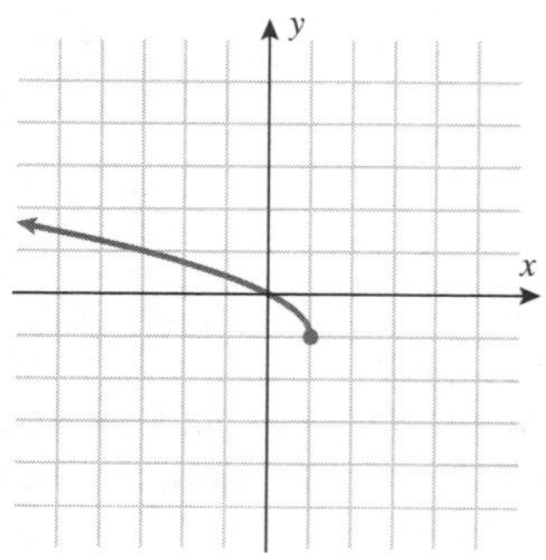

d.

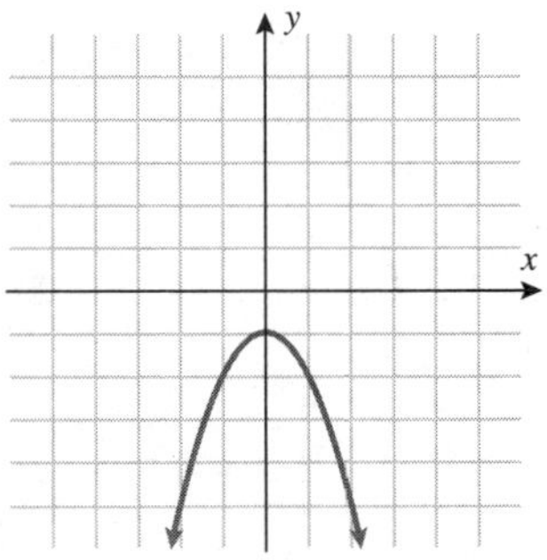

e.

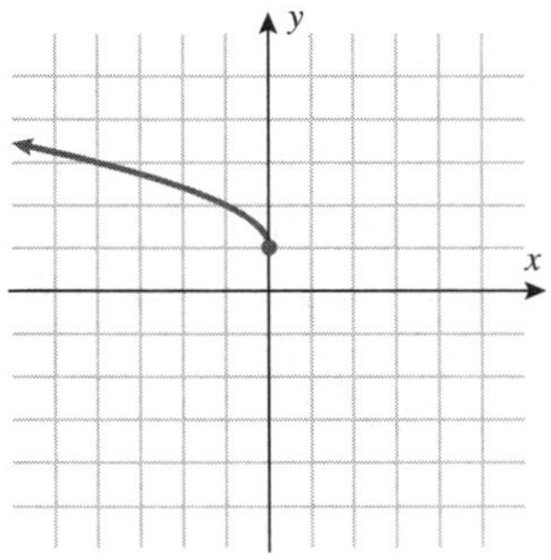

f.

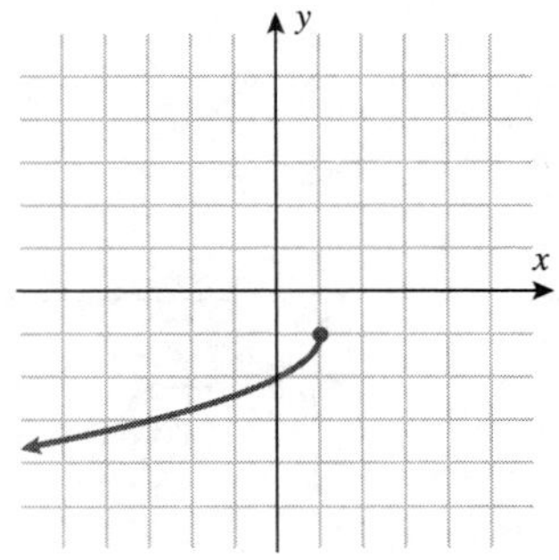

g.

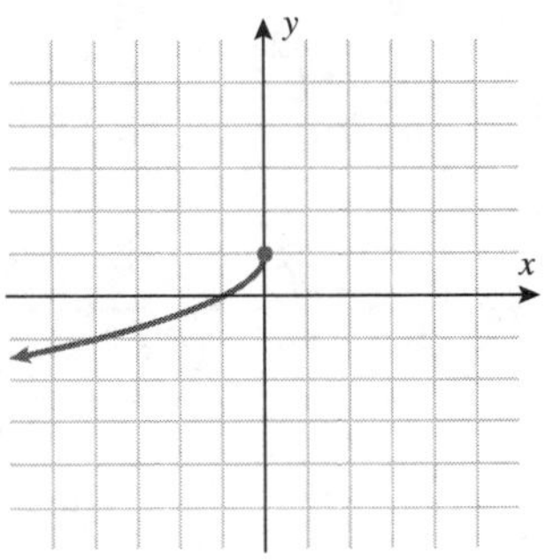

h.

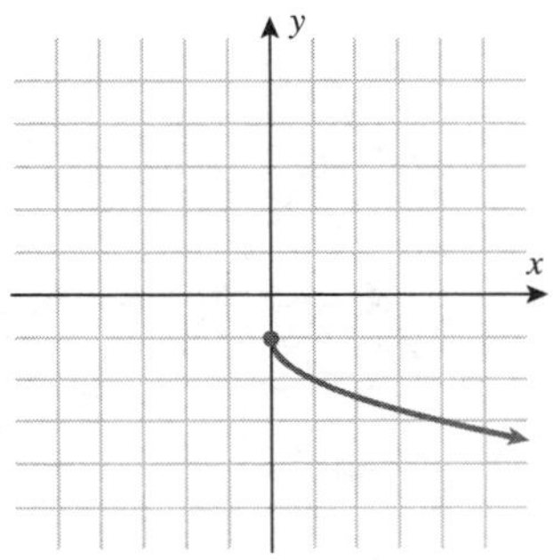

i.

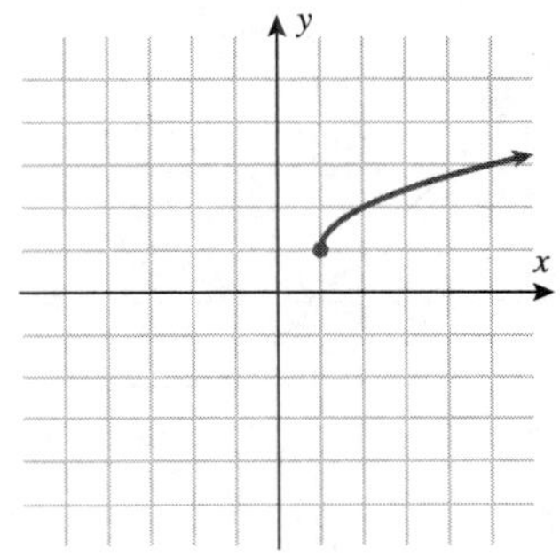

j.

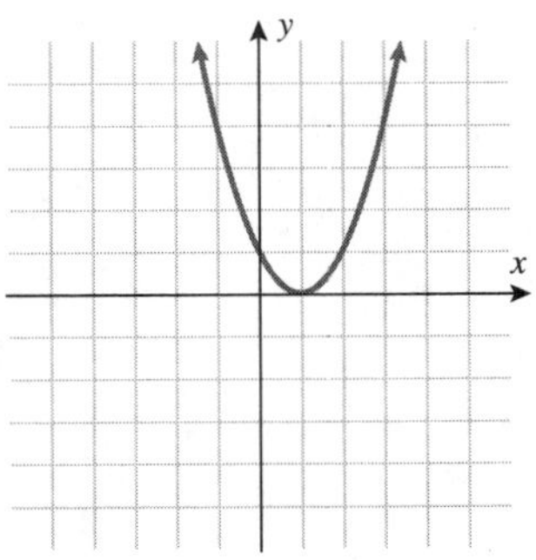

k.

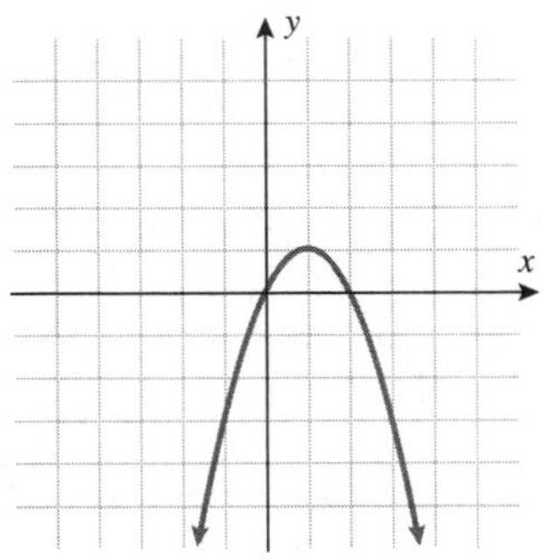

l.

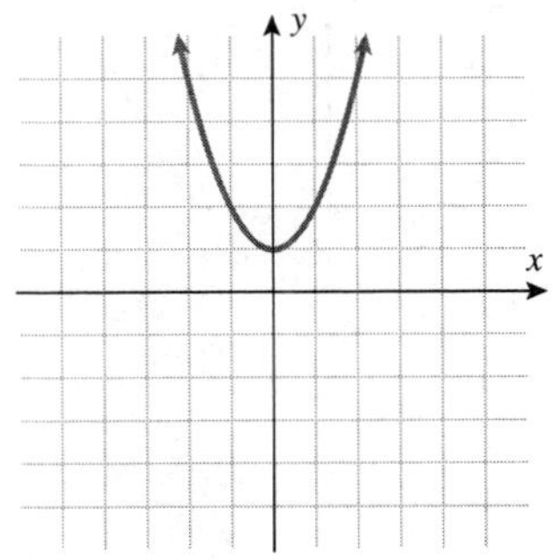

In Exercises 13–18, write the function whose graph is the graph of $y = |x|$, but is transformed accordingly.

13. Shifted up 3 units

14. Shifted to the left 4 units

15. Reflected about the y-axis

16. Reflected about the x-axis

17. Vertically stretched by a factor of 3

18. Vertically compressed by a factor of 3

In Exercises 19–24, write the function whose graph is the graph of $y = x^3$, but is transformed accordingly.

19. Shifted down 4 units

20. Shifted to the right 3 units

21. Shifted up 3 units and to the left 1 unit

22. Reflected about the x-axis

23. Reflected about the y-axis

24. Reflected about both the x-axis and the y-axis

In Exercises 25–48, use the given graph to sketch the graph of the indicated functions.

25.

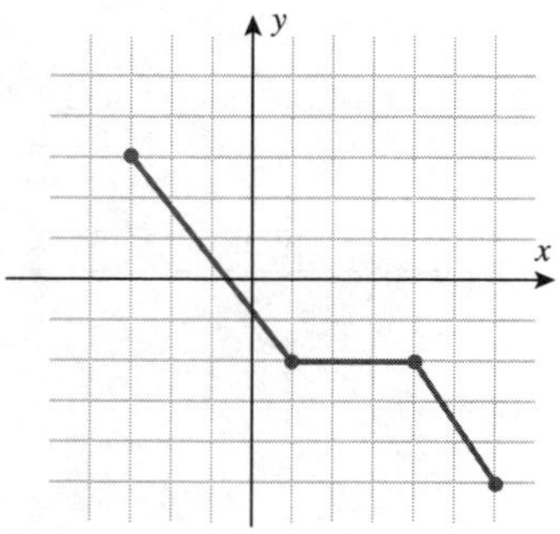

a. $y = f(x - 2)$
b. $y = f(x) - 2$

26.

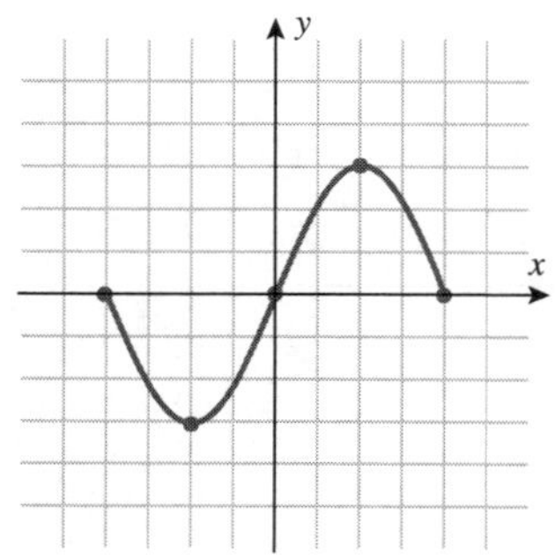

a. $y = f(x + 2)$
b. $y = f(x) + 2$

27.

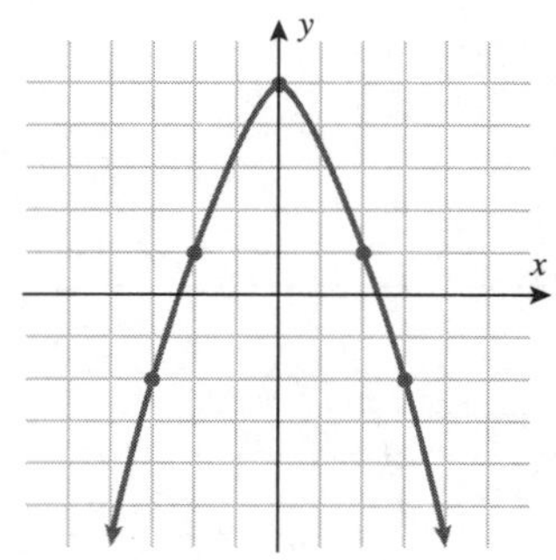

a. $y = f(x) - 3$
b. $y = f(x - 3)$

28.

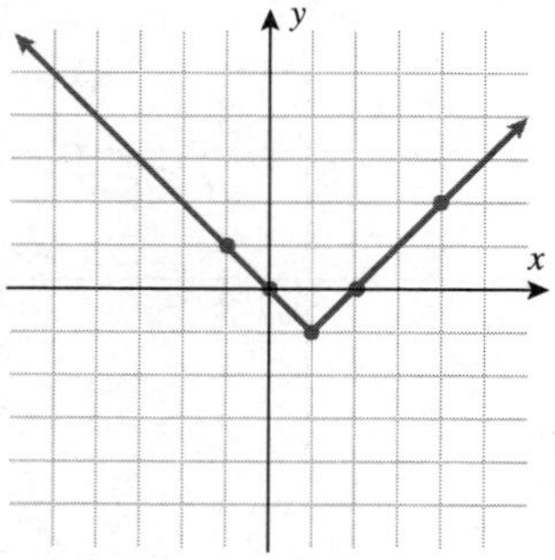

a. $y = f(x) + 3$
b. $y = f(x + 3)$

29.

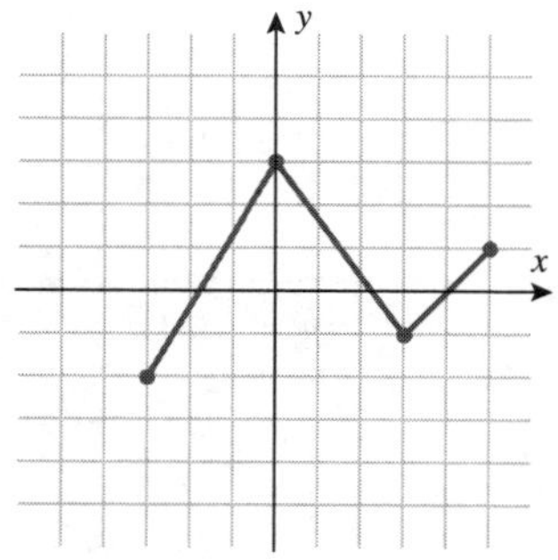

a. $y = -f(x)$
b. $y = f(-x)$

30.

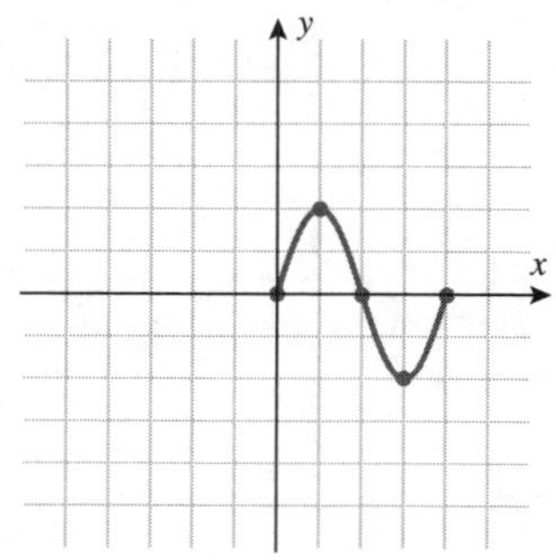

a. $y = -f(x)$
b. $y = f(-x)$

31.

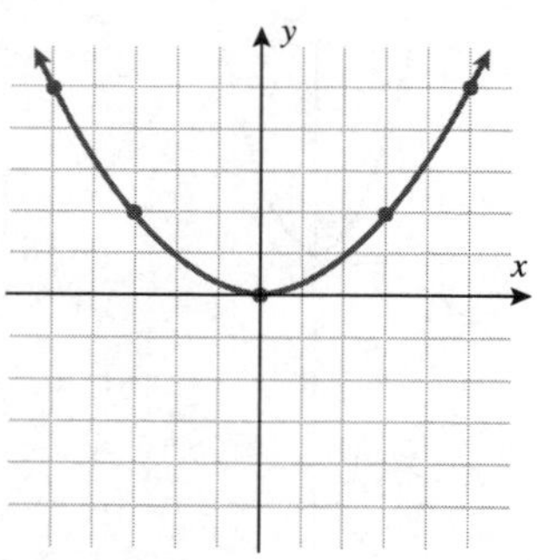

a. $y = 2f(x)$
b. $y = f(2x)$

32.

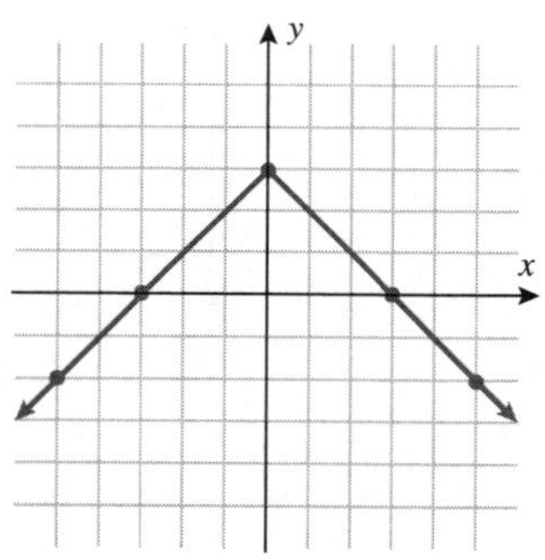

a. $y = 2f(x)$
b. $y = f(2x)$

33. $y = f(x - 2) - 3$

34. $y = f(x + 1) - 2$

35. $y = -f(x - 1) + 2$

36. $y = -2f(x) + 1$

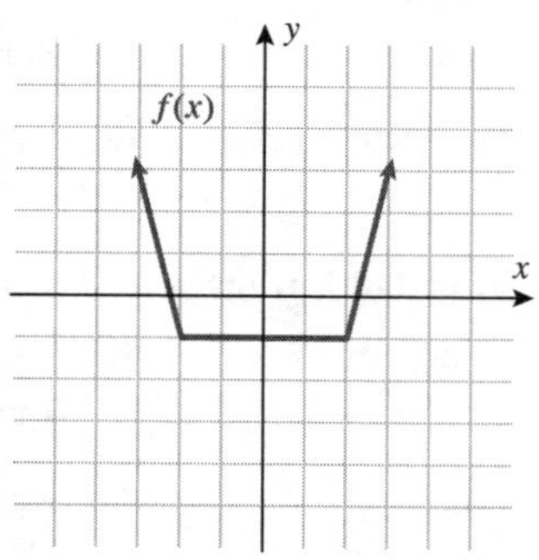

37. $y = -\frac{1}{2}g(x)$

38. $y = \frac{1}{4}g(-x)$

39. $y = -g(2x)$

40. $y = g\left(\frac{1}{2}x\right)$

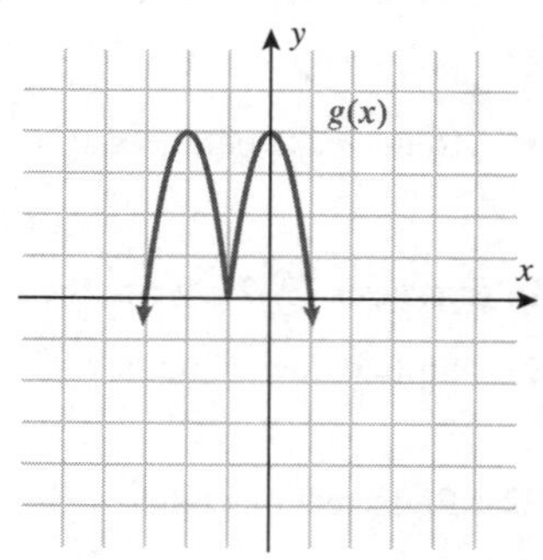

41. $y = \frac{1}{2}F(x - 1) + 2$

42. $y = \frac{1}{2}F(-x)$

43. $y = -F(1 - x)$

44. $y = -F(x - 2) - 1$

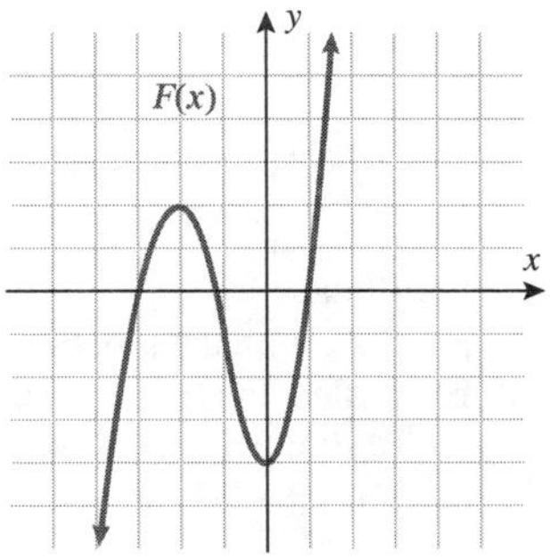

45. $y = 2G(x + 1) - 4$

46. $y = 2G(-x) + 1$

47. $y = -2G(x - 1) + 3$

48. $y = -G(x - 2) - 1$

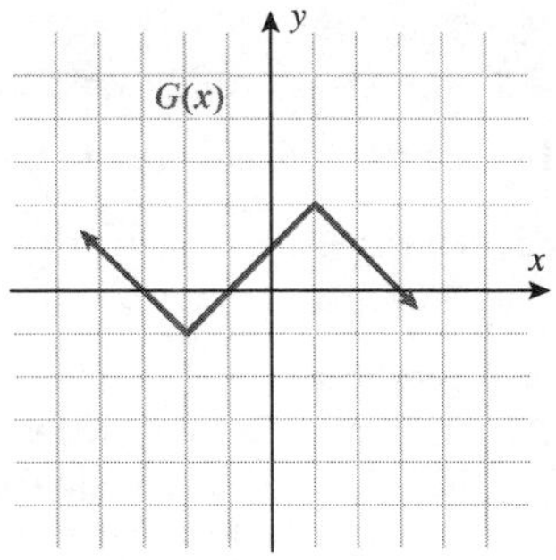

In Exercises 49–74, graph the function using transformations.

49. $y = x^2 - 2$

50. $y = x^2 + 3$

51. $y = (x + 1)^2$

52. $y = (x - 2)^2$

53. $y = (x - 3)^2 + 2$

54. $y = (x + 2)^2 + 1$

55. $y = -(1 - x)^2$

56. $y = -(x + 2)^2$

57. $y = |-x|$

58. $y = -|x|$

59. $y = -|x + 2| - 1$

60. $y = |1 - x| + 2$

61. $y = 2x^2 + 1$

62. $y = 2|x| + 1$

63. $y = -\sqrt{x - 2}$

64. $y = \sqrt{2 - x}$

65. $y = -\sqrt{2 + x} - 1$

66. $y = \sqrt{2 - x} + 3$

67. $y = \sqrt[3]{x - 1} + 2$

68. $y = \sqrt[3]{x + 2} - 1$

69. $y = \dfrac{1}{x + 3} + 2$

70. $y = \dfrac{1}{3 - x}$

71. $y = 2 - \dfrac{1}{x + 2}$

72. $y = 2 - \dfrac{1}{1 - x}$

73. $y = 5\sqrt{-x}$

74. $y = -\frac{1}{5}\sqrt{x}$

In Exercises 75–80, transform the function into the form $f(x) = c(x - h)^2 + k$, where c, k, and h are constants, by completing the square. Use graph-shifting techniques to graph the function.

75. $y = x^2 - 6x + 11$

76. $f(x) = x^2 + 2x - 2$

77. $f(x) = -x^2 - 2x$

78. $f(x) = -x^2 + 6x - 7$

79. $f(x) = 2x^2 - 8x + 3$

80. $f(x) = 3x^2 - 6x + 5$

APPLICATIONS

81. Salary. A manager hires an employee at a rate of \$10 per hour. Write the function that describes the current salary of the employee as a function of the number of hours worked per week, x. After a year, the manager decides to award the employee a raise equivalent to paying him for an additional 5 hours per week. Write a function that describes the salary of the employee after the raise.

82. Profit. The profit associated with St. Augustine sod in Florida is typically $P(x) = -x^2 + 14{,}000x - 48{,}700{,}000$, where x is the number of pallets sold per year in a normal year. In rainy years Sod King gives away 10 free pallets per year. Write the function that describes the profit of x pallets of sod in rainy years.

83. Taxes. Every year in the United States each working American typically pays in taxes a percentage of his or her earnings (minus the standard deduction). Karen's 2005 taxes were calculated based on the formula $T(x) = 0.22(x - 6500)$. That year the standard deduction was \$6,500 and her tax bracket paid 22% in taxes. Write the function that will determine her 2006 taxes, assuming she receives the raise that places her in the 33% bracket.

84. Medication. The amount of medication that an infant requires is typically a function of the baby's weight. The number of milliliters of an antiseizure medication A is given by $A(x) = \sqrt{x} + 2$, where x is the weight of the infant in ounces. In emergencies there is often not enough time to weigh the infant, so nurses have to estimate the baby's weight. What is the function that represents the actual amount of medication the infant is given if his weight is overestimated by 3 ounces?

CATCH THE MISTAKE

In Exercises 85–88, explain the mistake that is made.

85. Describe a procedure for graphing the function $f(x) = \sqrt{x - 3} + 2$.

Solution:

a. Start with the function $f(x) = \sqrt{x}$.
b. Shift the function to the left 3 units.
c. Shift the function up 2 units.

This is incorrect. What mistake was made?

86. Describe a procedure for graphing the function $f(x) = -\sqrt{x + 2} - 3$.

Solution:

a. Start with the function $f(x) = \sqrt{x}$.
b. Shift the function to the left 2 units.
c. Reflect the function about the y-axis.
d. Shift the function down 3 units.

This is incorrect. What mistake was made?

87. Describe a procedure for graphing the function $f(x) = |3 - x| + 1$.

Solution:

a. Start with the function $f(x) = |x|$.
b. Reflect the function about the y-axis.
c. Shift the function to the left 3 units.
d. Shift the function up 1 unit.

This is incorrect. What mistake was made?

88. Describe a procedure for graphing the function $f(x) = -2x^2 + 1$.

Solution:

a. Start with the function $f(x) = x^2$.
b. Reflect the function about the y-axis.
c. Shift the function up 1 unit.
d. Expand in the vertical direction by a factor of 2.

This is incorrect. What mistake was made?

CONCEPTUAL

In Exercises 89–92, determine whether each statement is true or false.

89. The graph of $y = |-x|$ is the same as the graph of $y = |x|$.

90. The graph of $y = \sqrt{-x}$ is the same as the graph of $y = \sqrt{x}$.

91. If the graph of an odd function is reflected around the x-axis and then the y-axis, the result is the graph of the original odd function.

92. If the graph of $y = \dfrac{1}{x}$ is reflected around the x-axis, it produces the same graph as if it had been reflected about the y-axis.

CHALLENGE

93. The point (a, b) lies on the graph of the function $y = f(x)$. What point is guaranteed to lie on the graph of $f(x - 3) + 2$?

94. The point (a, b) lies on the graph of the function $y = f(x)$. What point is guaranteed to lie on the graph of $-f(-x) + 1$?

TECHNOLOGY

95. Use a graphing utility to graph:

a. $y = x^2 - 2$ and $y = |x^2 - 2|$

b. $y = x^3 - 1$ and $y = |x^3 + 1|$

What is the relationship between $f(x)$ and $|f(x)|$?

96. Use a graphing utility to graph:

a. $y = x^2 - 2$ and $y = |x|^2 - 2$

b. $y = x^3 + 1$ and $y = |x|^3 + 1$

What is the relationship between $f(x)$ and $f(|x|)$?

97. Use a graphing utility to graph:

a. $y = \sqrt{x}$ and $y = \sqrt{0.1x}$

b. $y = \sqrt{x}$ and $y = \sqrt{10x}$

What is the relationship between $f(x)$ and $f(ax)$, assuming that a is positive?

98. Use a graphing utility to graph:

a. $y = \sqrt{x}$ and $y = 0.1\sqrt{x}$

b. $y = \sqrt{x}$ and $y = 10\sqrt{x}$

What is the relationship between $f(x)$ and $af(x)$, assuming that a is positive?

99. Use a graphing utility to graph $y = f(x) = [[0.5x]] + 1$. Use transforms to describe the relationship between $f(x)$ and $y = [[x]]$.

100. Use a graphing utility to graph $y = g(x) = 0.5\,[[x]] + 1$. Use transforms to describe the relationship between $g(x)$ and $y = [[x]]$.

SECTION 3.4 OPERATIONS ON FUNCTIONS AND COMPOSITION OF FUNCTIONS

SKILLS OBJECTIVES

- Add, subtract, multiply, and divide functions.
- Evaluate composite functions.
- Determine domain of functions resulting from operations on and composition of functions.

CONCEPTUAL OBJECTIVES

- Understand domain restrictions when dividing functions.
- Realize that the domain of a composition of functions excludes values that are not in the domain of the inside function.

Two different functions can be combined using mathematical operations such as addition, subtraction, multiplication, and division. Also, there is an operation on functions called composition, which can be thought of as a function of a function. When we combine functions, we do so algebraically. Special attention must be paid to the domain and range of the combined functions.

Adding, Subtracting, Multiplying, and Dividing Functions

Consider the two functions $f(x) = x^2 + 2x - 3$ and $g(x) = x + 1$. The domain of both of these functions is the set of all real numbers. Therefore, we can add, subtract, or multiply these functions for any real number x.

$$\text{Addition: } f(x) + g(x) = x^2 + 2x - 3 + x + 1 = x^2 + 3x - 2$$

The result is in fact a new function, which we denote:

$$(f + g)(x) = x^2 + 3x - 2 \qquad \text{This is the } \textbf{sum function}.$$

$$\text{Subtraction: } f(x) - g(x) = x^2 + 2x - 3 - (x + 1) = x^2 + x - 4$$

The result is in fact a new function, which we denote:

$$(f - g)(x) = x^2 + x - 4 \qquad \text{This is the } \textbf{difference function}.$$

$$\text{Multiplication: } f(x) \cdot g(x) = (x^2 + 2x - 3)(x + 1) = x^3 + 3x^2 - x - 3$$

The result is in fact a new function, which we denote:

$$(f \cdot g)(x) = x^3 + 3x^2 - x - 3 \qquad \text{This is the } \textbf{product function}.$$

Although both f and g are defined for all real numbers x, we must restrict x so that $x \neq -1$ to form the quotient $\frac{f}{g}$.

$$\text{Division: } \frac{f(x)}{g(x)} = \frac{x^2 + 2x - 3}{x + 1}, \quad x \neq -1$$

The result is in fact a new function, which we denote:

$$\left(\frac{f}{g}\right)(x) = \frac{x^2 + 2x - 3}{x + 1}, \quad x \neq -1 \qquad \text{This is called the } \textbf{quotient function}.$$

Two functions can be added, subtracted, and multiplied. The resulting function domain is therefore the intersection of the domains of the two functions. However, for division, any value of x (input) that makes the denominator equal to zero must be eliminated from the domain.

The previous examples involved polynomials. The domain of any polynomial is the set of all real numbers. Adding, subtracting, and multiplying polynomials result in other polynomials, which have domains of all real numbers. Let's now investigate operations applied to functions that have a restricted domain.

The domain of the sum function, difference function, or product function is the *intersection* of the individual domains of the two functions. The quotient function has a similar domain in that it is the intersection of the two domains. However, any values that make the denominator zero must also be eliminated.

Function	Notation	Domain
Sum	$(f+g)(x) = f(x) + g(x)$	$\{\text{domain of } f\} \cap \{\text{domain of } g\}$
Difference	$(f-g)(x) = f(x) - g(x)$	$\{\text{domain of } f\} \cap \{\text{domain of } g\}$
Product	$(f \cdot g)(x) = f(x) \cdot g(x)$	$\{\text{domain of } f\} \cap \{\text{domain of } g\}$
Quotient	$\left(\frac{f}{g}\right)(x) = \frac{f(x)}{g(x)}$	$\{\text{domain of } f\} \cap \{\text{domain of } g\} \cap \{g(x) \neq 0\}$

We can think of this in the following way: Any number that is in the domain of *both* the functions is in the domain of the combined function. The exception to this is the quotient function, which also eliminates values that make the denominator equal to zero.

EXAMPLE 1 Operations on Functions: Determining Domains of New Functions

For the functions $f(x) = \sqrt{x-1}$ and $g(x) = \sqrt{4-x}$, determine the sum function, difference function, product function, and quotient function. State the domain of these four new functions.

Solution:

Sum function: $f(x) + g(x) = \sqrt{x-1} + \sqrt{4-x}$

Difference function: $f(x) - g(x) = \sqrt{x-1} - \sqrt{4-x}$

Product function: $f(x) \cdot g(x) = \sqrt{x-1} \cdot \sqrt{4-x}$

$$= \sqrt{(x-1)(4-x)} = \sqrt{-x^2+5x-4}$$

Quotient function: $\frac{f(x)}{g(x)} = \frac{\sqrt{x-1}}{\sqrt{4-x}} = \sqrt{\frac{x-1}{4-x}}$

The domain of the square root function is determined by setting the argument under the radical greater than or equal to zero.

Domain of $f(x)$: $[1, \infty)$

Domain of $g(x)$: $(-\infty, 4]$

The domain of the sum, difference, and product functions is

$$[1, \infty) \cap (-\infty, 4] = [1, 4]$$

The quotient function has the additional constraint that the denominator cannot be zero. This implies that $x \neq 4$, so the domain of the quotient function is $[1, 4)$.

YOUR TURN Given the function $f(x) = \sqrt{x+3}$ and $g(x) = \sqrt{1-x}$, find $(f+g)(x)$ and state its domain.

Answer:
$(f+g)(x) = \sqrt{x+3} + \sqrt{1-x}$
Domain: $[-3, 1]$

EXAMPLE 2 Quotient Function and Domain Restrictions

Given the functions $F(x) = \sqrt{x}$ and $G(x) = |x - 3|$, find the quotient function, $\left(\frac{F}{G}\right)(x)$, and state its domain.

Solution:

The quotient function is written as

$$\left(\frac{F}{G}\right)(x) = \frac{F(x)}{G(x)} = \frac{\sqrt{x}}{|x - 3|}$$

Domain of $F(x)$: $[0, \infty)$ Domain of $G(x)$: $(-\infty, \infty)$

The real numbers that are in both the domain for $F(x)$ and the domain for $G(x)$ are represented by the intersection $[0, \infty) \cap (-\infty, \infty) = [0, \infty)$. Also, the denominator of the quotient function is equal to zero when $x = 3$, so we must eliminate this value from the domain.

$$\text{Domain of } \left(\frac{F}{G}\right)(x): [0, 3) \cup (3, \infty)$$

■ **YOUR TURN** For the functions given in Example 2, determine the quotient function $\left(\frac{G}{F}\right)(x)$, and state its domain.

Technology Tip

The graphs of $y_1 = F(x) = \sqrt{x}$, $y_2 = G(x) = |x - 3|$, and $y_3 = \frac{F(x)}{G(x)} = \frac{\sqrt{x}}{|x - 3|}$ are shown.

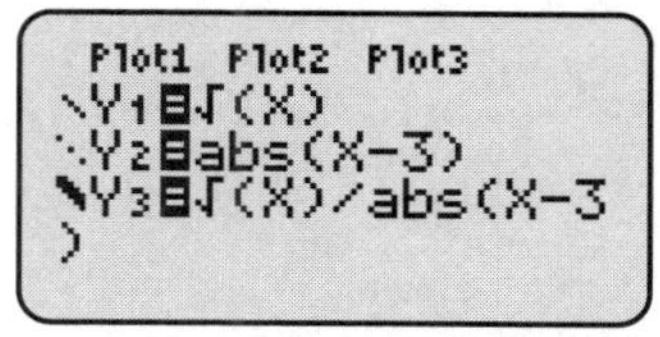

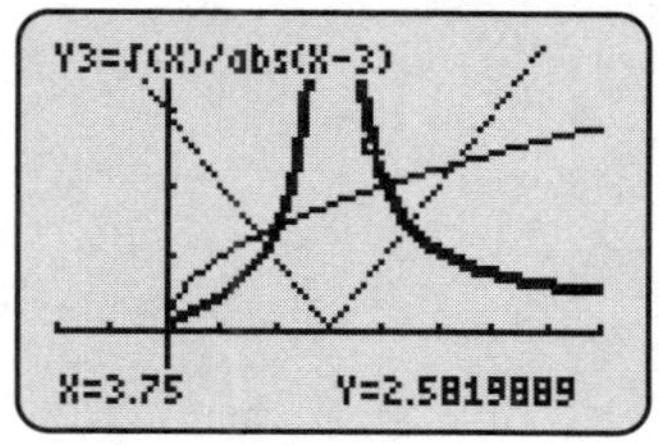

■ **Answer:**

$\left(\frac{G}{F}\right)(x) = \frac{G(x)}{F(x)} = \frac{|x - 3|}{\sqrt{x}}$

Domain: $(0, \infty)$

Composition of Functions

Recall that a function maps every element in the domain to exactly one corresponding element in the range as shown in the figure on the right.

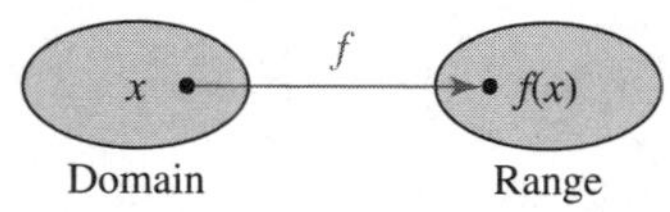

Suppose there is a sales rack of clothes in a department store. Let x correspond to the original price of each item on the rack. These clothes have recently been marked down 20%. Therefore, the function $g(x) = 0.80x$ represents the current sale price of each item. You have been invited to a special sale that lets you take 10% off the current sale price and an additional \$5 off every item at checkout. The function $f(g(x)) = 0.90f(x) - 5$ determines the checkout price. Note that the output of the function g is the input of the function f as shown in the figure below.

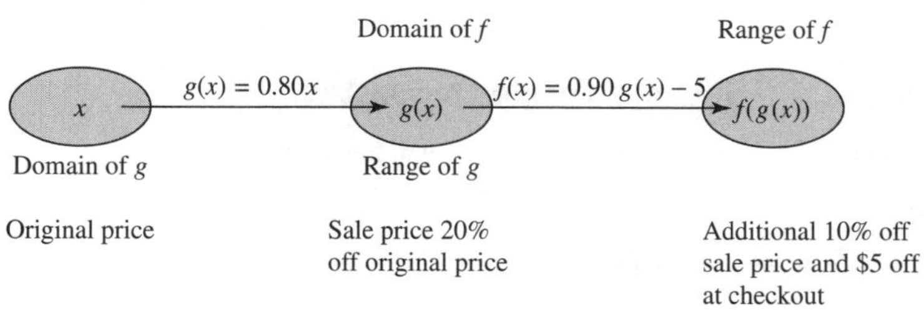

This is an example of a **composition of functions**, when the output of one function is the input of another function. It is commonly referred to as a function of a function.

An algebraic example of this is the function $y = \sqrt{x^2 - 2}$. Suppose we let $g(x) = x^2 - 2$ and $f(x) = \sqrt{x}$. Recall that the independent variable in function notation is a placeholder. Since $f(\square) = \sqrt{(\square)}$, then $f(g(x)) = \sqrt{(g(x))}$. Substituting the expression for $g(x)$, we find $f(g(x)) = \sqrt{x^2 - 2}$. The function $y = \sqrt{x^2 - 2}$ is said to be a composite function, $y = f(g(x))$.

Note that the domain of $g(x)$ is the set of all real numbers, and the domain of $f(x)$ is the set of all nonnegative numbers. The domain of a composite function is the set of all x such that $g(x)$ is in the domain of f. For instance, in the composite function $y = f(g(x))$, we know that the allowable inputs into f are all numbers greater than or equal to zero. Therefore, we restrict the outputs of $g(x) \geq 0$ and find the corresponding x-values. Those x-values are the only allowable inputs and constitute the domain of the composite function $y = f(g(x))$.

The symbol that represents composition of functions is a small open circle; thus $(f \circ g)(x) = f(g(x))$ and is read aloud as "f of g." It is important not to confuse this with the multiplication sign: $(f \cdot g)(x) = f(x)g(x)$.

CAUTION

$f \circ g \neq f \cdot g$

Composition of Functions

Given two functions f and g, there are two **composite functions** that can be formed.

Notation	Words	Definition	Domain
$f \circ g$	f composed with g	$f(g(x))$	The set of all real numbers x in the domain of g such that $g(x)$ is also in the domain of f.
$g \circ f$	g composed with f	$g(f(x))$	The set of all real numbers x in the domain of f such that $f(x)$ is also in the domain of g.

Study Tip

Order is important:

$(f \circ g)(x) = f(g(x))$
$(g \circ f)(x) = g(f(x))$

It is important to realize that there are two "filters" that allow certain values of x into the domain. The first filter is $g(x)$. If x is not in the domain of $g(x)$, it cannot be in the domain of $(f \circ g)(x) = f(g(x))$. Of those values for x that are in the domain of $g(x)$, only some pass through, because we restrict the output of $g(x)$ to values that are allowable as input into f. This adds an additional filter.

The domain of $f \circ g$ is always a subset of the domain of g, and the range of $f \circ g$ is always a subset of the range of f.

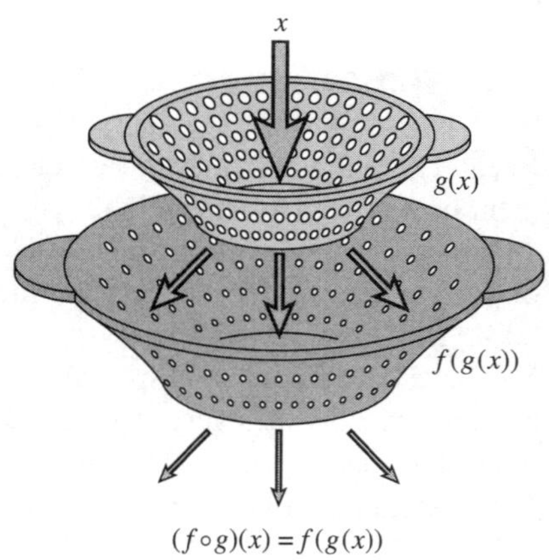

Study Tip

The domain of $f \circ g$ is always a subset of the domain of g, and the range of $f \circ g$ is always a subset of the range of f.

EXAMPLE 3 Finding a Composite Function

Given the functions $f(x) = x^2 + 1$ and $g(x) = x - 3$, find $(f \circ g)(x)$.

Solution:

Write $f(x)$ using placeholder notation. $\quad f(\square) = (\square)^2 + 1$

Express the composite function $f \circ g$. $\quad f(g(x)) = (g(x))^2 + 1$

Substitute $g(x) = x - 3$ into f. $\quad f(g(x)) = (x - 3)^2 + 1$

Eliminate the parentheses on the right side. $\quad f(g(x)) = x^2 - 6x + 10$

$$\boxed{(f \circ g)(x) = f(g(x)) = x^2 - 6x + 10}$$

YOUR TURN Given the functions in Example 3, find $(g \circ f)(x)$.

Answer: $g \circ f = g(f(x)) = x^2 - 2$

EXAMPLE 4 Determining the Domain of a Composite Function

Given the functions $f(x) = \dfrac{1}{x-1}$ and $g(x) = \dfrac{1}{x}$, determine $f \circ g$, and state its domain.

Solution:

Write $f(x)$ using placeholder notation. $f(\square) = \dfrac{1}{(\square) - 1}$

Express the composite function $f \circ g$. $f(g(x)) = \dfrac{1}{g(x) - 1}$

Substitute $g(x) = \dfrac{1}{x}$ into f. $f(g(x)) = \dfrac{1}{\frac{1}{x} - 1}$

Multiply the right side by $\dfrac{x}{x}$. $f(g(x)) = \dfrac{1}{\frac{1}{x} - 1} \cdot \dfrac{x}{x} = \dfrac{x}{1-x}$

$$\boxed{(f \circ g) = f(g(x)) = \dfrac{x}{1-x}}$$

What is the domain of $(f \circ g)(x) = f(g(x))$? By inspecting the final result of $f(g(x))$, we see that the denominator is zero when $x = 1$. Therefore, $x \neq 1$. Are there any other values for x that are not allowed? The function $g(x)$ has the domain $x \neq 0$; therefore we must also exclude zero.

The domain of $(f \circ g)(x) = f(g(x))$ excludes $x = 0$ and $x = 1$ or, in interval notation,

$$\boxed{(-\infty, 0) \cup (0, 1) \cup (1, \infty)}$$

YOUR TURN For the functions f and g given in Example 4, determine the composite function $g \circ f$ and state its domain.

Technology Tip

The graphs of $y_1 = f(x) = \dfrac{1}{x-1}$, $y_2 = g(x) = \dfrac{1}{x}$, and $y_3 = (f \circ g)(x) = \dfrac{1}{1/x - 1} = \dfrac{x}{1-x}$ are shown.

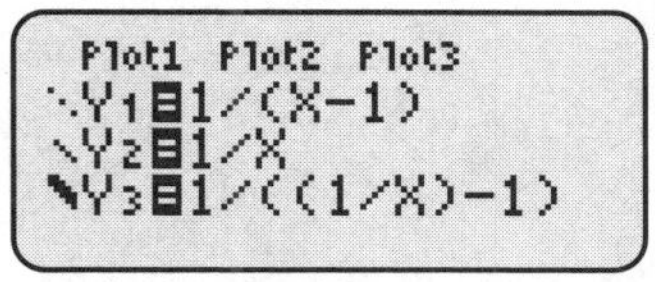

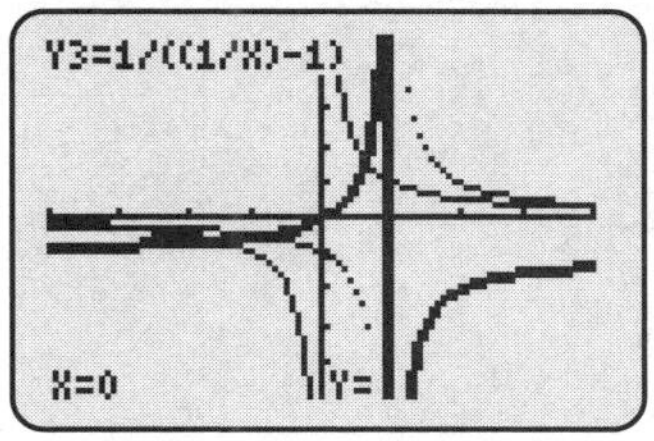

■ **Answer:** $g(f(x)) = x - 1$. Domain of $g \circ f$ is $x \neq 1$, or in interval notation, $(-\infty, 1) \cup (1, \infty)$.

The domain of the composite function cannot always be determined by examining the final form of $f \circ g$.

CAUTION

The domain of the composite function cannot always be determined by examining the final form of $f \circ g$.

EXAMPLE 5 Determining the Domain of a Composite Function (Without Finding the Composite Function)

Let $f(x) = \dfrac{1}{x-2}$ and $g(x) = \sqrt{x+3}$. Find the domain of $f(g(x))$. Do not find the composite function.

Solution:

Find the domain of g. $[-3, \infty)$

Find the range of g. $[0, \infty)$

In $f(g(x))$, the output of g becomes the input for f. Since the domain of f is the set of all real numbers except 2, we eliminate any values of x in the domain of g that correspond to $g(x) = 2$.

Let $g(x) = 2$. $\sqrt{x+3} = 2$

Square both sides. $x + 3 = 4$

Solve for x. $x = 1$

Eliminate $x = 1$ from the domain of g, $[-3, \infty)$.

State the domain of $f(g(x))$. $\boxed{[-3, 1) \cup (1, \infty)}$

EXAMPLE 6 Evaluating a Composite Function

Given the functions $f(x) = x^2 - 7$ and $g(x) = 5 - x^2$, evaluate:

a. $f(g(1))$ **b.** $f(g(-2))$ **c.** $g(f(3))$ **d.** $g(f(-4))$

Solution:

One way of evaluating these composite functions is to calculate the two individual composites in terms of x: $f(g(x))$ and $g(f(x))$. Once those functions are known, the values can be substituted for x and evaluated.

Another way of proceeding is as follows:

a. Write the desired quantity. $f(g(1))$

Find the value of the inner function g. $g(1) = 5 - 1^2 = 4$

Substitute $g(1) = 4$ into f. $f(g(1)) = f(4)$

Evaluate $f(4)$. $f(4) = 4^2 - 7 = 9$

$$\boxed{f(g(1)) = 9}$$

b. Write the desired quantity. $f(g(-2))$

Find the value of the inner function g. $g(-2) = 5 - (-2)^2 = 1$

Substitute $g(-2) = 1$ into f. $f(g(-2)) = f(1)$

Evaluate $f(1)$. $f(1) = 1^2 - 7 = -6$

$$\boxed{f(g(-2)) = -6}$$

c. Write the desired quantity. $g(f(3))$

Find the value of the inner function f. $f(3) = 3^2 - 7 = 2$

Substitute $f(3) = 2$ into g. $g(f(3)) = g(2)$

Evaluate $g(2)$. $g(2) = 5 - 2^2 = 1$

$$\boxed{g(f(3)) = 1}$$

d. Write the desired quantity. $g(f(-4))$

Find the value of the inner function f. $f(-4) = (-4)^2 - 7 = 9$

Substitute $f(-4) = 9$ into g. $g(f(-4)) = g(9)$

Evaluate $g(9)$. $g(9) = 5 - 9^2 = -76$

$$\boxed{g(f(-4)) = -76}$$

■ **Answer:** $f(g(1)) = 5$ and $g(f(1)) = -7$

■ **YOUR TURN** Given the functions $f(x) = x^3 - 3$ and $g(x) = 1 + x^3$, evaluate $f(g(1))$ and $g(f(1))$.

Application Problems

Recall the example at the beginning of this chapter regarding the clothes that are on sale. Often, real-world applications are modeled with composite functions. In the clothes example, x is the original price of each item. The first function maps its input (original price) to an output (sale price). The second function maps its input (sale price) to an output (checkout price). Example 7 is another real-world application of composite functions.

Three temperature scales are commonly used:

- The degree Celsius (°C) scale
  - This scale was devised by dividing the range between the freezing (0 °C) and boiling (100 °C) points of pure water at sea level into 100 equal parts. This scale is used in science and is one of the standards of the "metric" (SI) system of measurements.

- The Kelvin (K) temperature scale
  - This scale shifts the Celsius scale down so that the zero point is equal to absolute zero (about -273.15 °C), a hypothetical temperature at which there is a complete absence of heat energy.
  - Temperatures on this scale are called **kelvins**, *not* degrees kelvin, and kelvin is not capitalized. The symbol for the kelvin is K.
- The degree Fahrenheit (°F) scale
  - This scale evolved over time and is still widely used mainly in the United States, although Celsius is the preferred "metric" scale.
  - With respect to pure water at sea level, the **degrees Fahrenheit** are gauged by the spread from 32 °F (freezing) to 212 °F (boiling).

The equations that relate these temperature scales are

$$F = \frac{9}{5}C + 32 \qquad C = K - 273.15$$

EXAMPLE 7 Applications Involving Composite Functions

Determine degrees Fahrenheit as a function of kelvins.

Solution:

Degrees Fahrenheit is a function of degrees Celsius.

$$F = \frac{9}{5}C + 32$$

Now substitute $C = K - 273.15$ into the equation for F.

$$F = \frac{9}{5}(K - 273.15) + 32$$

Simplify.

$$F = \frac{9}{5}K - 491.67 + 32$$

$$\boxed{F = \frac{9}{5}K - 459.67}$$

SECTION 3.4 SUMMARY

Operations on Functions

Function	Notation
Sum	$(f + g)(x) = f(x) + g(x)$
Difference	$(f - g)(x) = f(x) - g(x)$
Product	$(f \cdot g)(x) = f(x) \cdot g(x)$
Quotient	$\left(\frac{f}{g}\right)(x) = \frac{f(x)}{g(x)} \qquad g(x) \neq 0$

The domain of the sum, difference, and product functions is the intersection of the domains, or common domain shared by both f and g. The domain of the quotient function is also the intersection of the domain shared by both f and g with an additional restriction that $g(x) \neq 0$.

Composition of Functions

$$(f \circ g)(x) = f(g(x))$$

The domain restrictions cannot always be determined simply by inspecting the final form of $f(g(x))$. Rather, the domain of the composite function is a subset of the domain of $g(x)$. Values of x must be eliminated if their corresponding values of $g(x)$ are not in the domain of f.

SECTION 3.4 EXERCISES

SKILLS

In Exercises 1–10, given the functions f and g, find $f + g$, $f - g$, $f \cdot g$, and $\frac{f}{g}$, and state the domain of each.

1. $f(x) = 2x + 1$, $g(x) = 1 - x$
2. $f(x) = 3x + 2$, $g(x) = 2x - 4$
3. $f(x) = 2x^2 - x$, $g(x) = x^2 - 4$
4. $f(x) = 3x + 2$, $g(x) = x^2 - 25$
5. $f(x) = \frac{1}{x}$, $g(x) = x$
6. $f(x) = \frac{2x + 3}{x - 4}$, $g(x) = \frac{x - 4}{3x + 2}$
7. $f(x) = \sqrt{x}$, $g(x) = 2\sqrt{x}$
8. $f(x) = \sqrt{x - 1}$, $g(x) = 2x^2$
9. $f(x) = \sqrt{4 - x}$, $g(x) = \sqrt{x + 3}$
10. $f(x) = \sqrt{1 - 2x}$, $g(x) = \frac{1}{x}$

In Exercises 11–20, for the given functions f and g, find the composite functions $f \circ g$ and $g \circ f$, and state their domains.

11. $f(x) = 2x + 1$, $g(x) = x^2 - 3$
12. $f(x) = x^2 - 1$, $g(x) = 2 - x$
13. $f(x) = \frac{1}{x - 1}$, $g(x) = x + 2$
14. $f(x) = \frac{2}{x - 3}$, $g(x) = 2 + x$
15. $f(x) = |x|$, $g(x) = \frac{1}{x - 1}$
16. $f(x) = |x - 1|$, $g(x) = \frac{1}{x}$
17. $f(x) = \sqrt{x - 1}$, $g(x) = x + 5$
18. $f(x) = \sqrt{2 - x}$, $g(x) = x^2 + 2$
19. $f(x) = x^3 + 4$, $g(x) = (x - 4)^{1/3}$
20. $f(x) = \sqrt[3]{x^2 - 1}$, $g(x) = x^{2/3} + 1$

In Exercises 21–38, evaluate the functions for the specified values, if possible.

$$f(x) = x^2 + 10 \qquad g(x) = \sqrt{x - 1}$$

21. $(f + g)(2)$
22. $(f + g)(10)$
23. $(f - g)(2)$
24. $(f - g)(5)$
25. $(f \cdot g)(4)$
26. $(f \cdot g)(5)$
27. $\left(\frac{f}{g}\right)(10)$
28. $\left(\frac{f}{g}\right)(2)$
29. $f(g(2))$
30. $f(g(1))$
31. $g(f(-3))$
32. $g(f(4))$
33. $f(g(0))$
34. $g(f(0))$
35. $f(g(-3))$
36. $g\left(f\left(\sqrt{7}\right)\right)$
37. $(f \circ g)(4)$
38. $(g \circ f)(-3)$

In Exercises 39–50, evaluate $f(g(1))$ and $g(f(2))$, if possible.

39. $f(x) = \frac{1}{x}$, $g(x) = 2x + 1$
40. $f(x) = x^2 + 1$, $g(x) = \frac{1}{2 - x}$
41. $f(x) = \sqrt{1 - x}$, $g(x) = x^2 + 2$
42. $f(x) = \sqrt{3 - x}$, $g(x) = x^2 + 1$
43. $f(x) = \frac{1}{|x - 1|}$, $g(x) = x + 3$
44. $f(x) = \frac{1}{x}$, $g(x) = |2x - 3|$
45. $f(x) = \sqrt{x - 1}$, $g(x) = x^2 + 5$
46. $f(x) = \sqrt[3]{x - 3}$, $g(x) = \frac{1}{x - 3}$
47. $f(x) = \frac{1}{x^2 - 3}$, $g(x) = \sqrt{x - 3}$
48. $f(x) = \frac{x}{2 - x}$, $g(x) = 4 - x^2$
49. $f(x) = (x - 1)^{1/3}$, $g(x) = x^2 + 2x + 1$
50. $f(x) = (1 - x^2)^{1/2}$, $g(x) = (x - 3)^{1/3}$

In Exercises 51–60, show that $f(g(x)) = x$ and $g(f(x)) = x$.

51. $f(x) = 2x + 1$, $g(x) = \frac{x - 1}{2}$
52. $f(x) = \frac{x - 2}{3}$, $g(x) = 3x + 2$
53. $f(x) = \sqrt{x - 1}$, $g(x) = x^2 + 1$ for $x \geq 1$
54. $f(x) = 2 - x^2$, $g(x) = \sqrt{2 - x}$ for $x \leq 2$

55. $f(x) = \frac{1}{x}, \quad g(x) = \frac{1}{x}$ for $x \neq 0$

56. $f(x) = (5 - x)^{1/3}, \quad g(x) = 5 - x^3$

57. $f(x) = 4x^2 - 9, \quad g(x) = \frac{\sqrt{x+9}}{2}$ for $x \geq 0$

58. $f(x) = \sqrt[3]{8x - 1}, \quad g(x) = \frac{x^3 + 1}{8}$

59. $f(x) = \frac{1}{x-1}, \quad g(x) = \frac{x+1}{x}$ for $x \neq 0,\ x \neq 1$

60. $f(x) = \sqrt{25 - x^2}, \quad g(x) = \sqrt{25 - x^2}$ for $0 \leq x \leq 5$

In Exercises 61–66, write the function as a composite of two functions *f* and *g*. (More than one answer is correct.)

61. $f(g(x)) = 2(3x - 1)^2 + 5(3x - 1)$

62. $f(g(x)) = \frac{1}{1 + x^2}$

63. $f(g(x)) = \frac{2}{|x - 3|}$

64. $f(g(x)) = \sqrt{1 - x^2}$

65. $f(g(x)) = \frac{3}{\sqrt{x+1} - 2}$

66. $f(g(x)) = \frac{\sqrt{x}}{3\sqrt{x} + 2}$

APPLICATIONS

Exercises 67 and 68 depend on the relationship between degrees Fahrenheit, degrees Celsius, and kelvins:

$$F = \frac{9}{5}C + 32 \qquad C = K - 273.15$$

67. Temperature. Write a composite function that converts kelvins into degrees Fahrenheit.

68. Temperature. Convert the following degrees Fahrenheit to kelvins: 32 °F and 212 °F.

69. Dog Run. Suppose that you want to build a *square* fenced-in area for your dog. Fencing is purchased in linear feet.

a. Write a composite function that determines the area of your dog pen as a function of how many linear feet are purchased.
b. If you purchase 100 linear feet, what is the area of your dog pen?
c. If you purchase 200 linear feet, what is the area of your dog pen?

70. Dog Run. Suppose that you want to build a *circular* fenced-in area for your dog. Fencing is purchased in linear feet.

a. Write a composite function that determines the area of your dog pen as a function of how many linear feet are purchased.
b. If you purchase 100 linear feet, what is the area of your dog pen?
c. If you purchase 200 linear feet, what is the area of your dog pen?

71. Market Price. Typical supply and demand relationships state that as the number of units for sale increases, the market price decreases. Assume that the market price p and the number of units for sale x are related by the demand equation:

$$p = 3000 - \frac{1}{2}x$$

Assume that the cost $C(x)$ of producing x items is governed by the equation

$$C(x) = 2000 + 10x$$

and the revenue $R(x)$ generated by selling x units is governed by

$$R(x) = 100x$$

a. Write the cost as a function of price p.
b. Write the revenue as a function of price p.
c. Write the profit as a function of price p.

72. Market Price. Typical supply and demand relationships state that as the number of units for sale increases, the market price decreases. Assume that the market price p and the number of units for sale x are related by the demand equation:

$$p = 10{,}000 - \frac{1}{4}x$$

Assume that the cost $C(x)$ of producing x items is governed by the equation

$$C(x) = 30{,}000 + 5x$$

and the revenue $R(x)$ generated by selling x units is governed by

$$R(x) = 1000x$$

a. Write the cost as a function of price p.
b. Write the revenue as a function of price p.
c. Write the profit as a function of price p.

73. Environment: Oil Spill. An oil spill makes a circular pattern around a ship such that the radius in feet grows as a function of time in hours $r(t) = 150\sqrt{t}$. Find the area of the spill as a function of time.

74. Pool Volume. A 20 ft × 10 ft rectangular pool has been built. If 50 cubic feet of water is pumped into the pool per hour, write the water-level height (feet) as a function of time (hours).

75. Fireworks. A family is watching a fireworks display. If the family is 2 miles from where the fireworks are being launched and the fireworks travel vertically, what is the distance between the family and the fireworks as a function of height above ground?

76. Real Estate. A couple are about to put their house up for sale. They bought the house for \$172,000 a few years ago, and when they list it with a realtor they will pay a 6% commission. Write a function that represents the amount of money they will make on their home as a function of the asking price p.

CATCH THE MISTAKE

In Exercises 77–82, for the functions $f(x) = x + 2$ and $g(x) = x^2 - 4$, find the indicated function and state its domain. Explain the mistake that is made in each problem.

77. $\dfrac{g}{f}$

Solution:

$$\frac{g(x)}{f(x)} = \frac{x^2 - 4}{x + 2} = \frac{(x-2)(x+2)}{x+2} = x - 2$$

Domain: $(-\infty, \infty)$

This is incorrect. What mistake was made?

78. $\dfrac{f}{g}$

Solution:

$$\frac{f(x)}{g(x)} = \frac{x + 2}{x^2 - 4} = \frac{x+2}{(x-2)(x+2)} = \frac{1}{x-2} = \frac{1}{x-2}$$

Domain: $(-\infty, 2) \cup (2, \infty)$

This is incorrect. What mistake was made?

79. $f \circ g$

Solution:

$$f \circ g = f(x)g(x) = (x+2)(x^2-4) = x^3 + 2x^2 - 4x - 8$$

Domain: $(-\infty, \infty)$

This is incorrect. What mistake was made?

80. Given the function $f(x) = x^2 + 7$ and $g(x) = \sqrt{x - 3}$, find $f \circ g$, and state the domain.

Solution: $f \circ g = f(g(x)) = \left(\sqrt{x-3}\right)^2 + 7$

$$= f(g(x)) = x - 3 + 7 = x - 4$$

Domain: $(-\infty, \infty)$

This is incorrect. What mistake was made?

81. $(f + g)(2) = (x + 2 + x^2 - 4)(2)$

$$= (x^2 + x - 2)(2) = 2x^2 + 2x - 4$$

Domain: $(-\infty, \infty)$

This is incorrect. What mistake was made?

82. $f(x) - g(x) = x + 2 - x^2 - 4$

$$= -x^2 + x - 2$$

Domain: $(-\infty, \infty)$

This is incorrect. What mistake was made?

CONCEPTUAL

In Exercises 83–86, determine whether each statement is true or false.

83. When adding, subtracting, multiplying, or dividing two functions, the domain of the resulting function is the union of the domains of the individual functions.

84. For any functions f and g, $f(g(x)) = g(f(x))$ for all values of x that are in the domain of both f and g.

85. For any functions f and g, $(f \circ g)(x)$ exists for all values of x that are in the domain of $g(x)$, provided the range of g is a subset of the domain of f.

86. The domain of a composite function can be found by inspection, without knowledge of the domain of the individual functions.

CHALLENGE

87. For the functions $f(x) = x + a$ and $g(x) = \dfrac{1}{x - a}$, find $g \circ f$ and state its domain.

88. For the functions $f(x) = ax^2 + bx + c$ and $g(x) = \dfrac{1}{x - c}$, find $g \circ f$ and state its domain.

89. For the functions $f(x) = \sqrt{x + a}$ and $g(x) = x^2 - a$ find $g \circ f$ and state its domain.

90. For the functions $f(x) = \dfrac{1}{x^a}$ and $g(x) = \dfrac{1}{x^b}$, find $g \circ f$ and state its domain. Assume $a > 1$ and $b > 1$.

TECHNOLOGY

91. Using a graphing utility, plot $y_1 = \sqrt{x+7}$ and $y_2 = \sqrt{9-x}$. Plot $y_3 = y_1 + y_2$. What is the domain of y_3?

92. Using a graphing utility, plot $y_1 = \sqrt[3]{x+5}$, $y_2 = \dfrac{1}{\sqrt{3-x}}$, and $y_3 = \dfrac{y_1}{y_2}$. What is the domain of y_3?

93. Using a graphing utility, plot $y_1 = \sqrt{x^2 - 3x - 4}$, $y_2 = \dfrac{1}{x^2 - 14}$, and $y_3 = \dfrac{1}{y_1^2 - 14}$. If y_1 represents a function f and y_2 represents a function g, then y_3 represents the composite function $g \circ f$. The graph of y_3 is only defined for the domain of $g \circ f$. State the domain of $g \circ f$.

94. Using a graphing utility, plot $y_1 = \sqrt{1-x}$, $y_2 = x^2 + 2$, and $y_3 = y_1^2 + 2$. If y_1 represents a function f and y_2 represents a function g, then y_3 represents the composite function $g \circ f$. The graph of y_3 is only defined for the domain of $g \circ f$. State the domain of $g \circ f$.

SECTION 3.5 ONE-TO-ONE FUNCTIONS AND INVERSE FUNCTIONS

SKILLS OBJECTIVES

- Determine whether a function is a one-to-one function
  - algebraically,
  - graphically.
- Find the inverse of a function.
- Graph the inverse function given the graph of the function.

CONCEPTUAL OBJECTIVES

- Visualize the relationships between domain and range of a function and the domain and range of its inverse.
- Understand why functions and their inverses are symmetric about $y = x$.

Every human being has a blood type, and every human being has a DNA sequence. These are examples of functions, where a person is the input and the output is blood type or DNA sequence. These relationships are classified as functions because each person can have one and only one blood type or DNA strand. The difference between these functions is that many people have the same blood type, but DNA is unique to each individual. Can we map backwards? For instance, if you know the blood type, do you know specifically which person it came from? No, but, if you know the DNA sequence, you know exactly to which person it corresponds. When a function has a one-to-one correspondence, like the DNA example, then mapping backwards is possible. The map back is called the *inverse function.*

Determine Whether a Function Is One-to-One

In Section 3.1, we defined a function as a relationship that maps an input (contained in the domain) to exactly one output (found in the range). Algebraically, each value for x can correspond to only a single value for y. Recall the square, identity, absolute value, and reciprocal functions from our library of functions in Section 3.3.

All of the graphs of these functions satisfy the vertical line test. Although the square function and the absolute value function map each value of x to exactly one value for y, these two functions map two values of x to the same value for y. For example, $(-1, 1)$ and $(1, 1)$ lie on both graphs. The identity and reciprocal functions, on the other hand, map each x to a single value for y, and no two x-values map to the same y-value. These two functions are examples of *one-to-one functions.*

DEFINITION **One-to-One Function**

A function $f(x)$ is **one-to-one** if no two elements in the domain correspond to the same element in the range; that is,

$$\text{if } x_1 \neq x_2, \text{ then } f(x_1) \neq f(x_2).$$

In other words, it is one-to-one if no two inputs map to the same output.

EXAMPLE 1 Determining Whether a Function Defined as a Set of Points Is a One-to-One Function

For each of the three relations, determine whether the relation is a function. If it is a function, determine whether it is a one-to-one function.

$$f = \{(0, 0), (1, 1), (1, -1)\}$$
$$g = \{(-1, 1), (0, 0), (1, 1)\}$$
$$h = \{(-1, -1), (0, 0), (1, 1)\}$$

Solution:

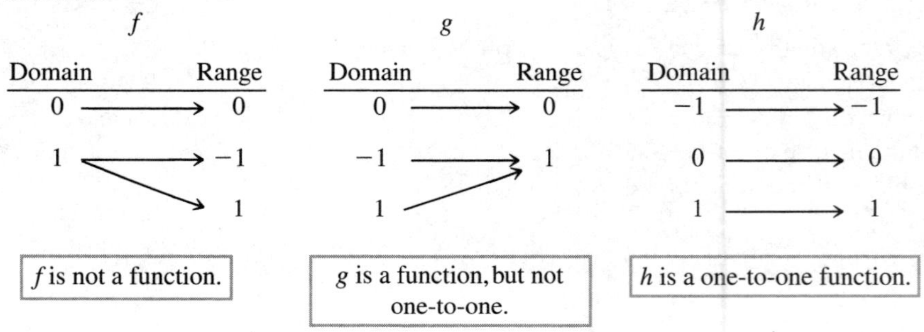

Just as there is a graphical test for functions, the vertical line test, there is a graphical test for one-to-one functions, the *horizontal line test.* Note that a horizontal line can be drawn on the square and absolute value functions so that it intersects the graph of each function at two points. The identity and reciprocal functions, however, will intersect a horizontal line in at most only one point. This leads us to the horizontal line test for one-to-one functions.

DEFINITION **Horizontal Line Test**

If every horizontal line intersects the graph of a function in at most one point, then the function is classified as a one-to-one function.

EXAMPLE 2 Using the Horizontal Line Test to Determine Whether a Function Is One-to-One

For each of the three relations, determine whether the relation is a function. If it is a function, determine whether it is a one-to-one function. Assume that x is the independent variable and y is the dependent variable.

$$x = y^2 \qquad y = x^2 \qquad y = x^3$$

Solution:

$x = y^2$

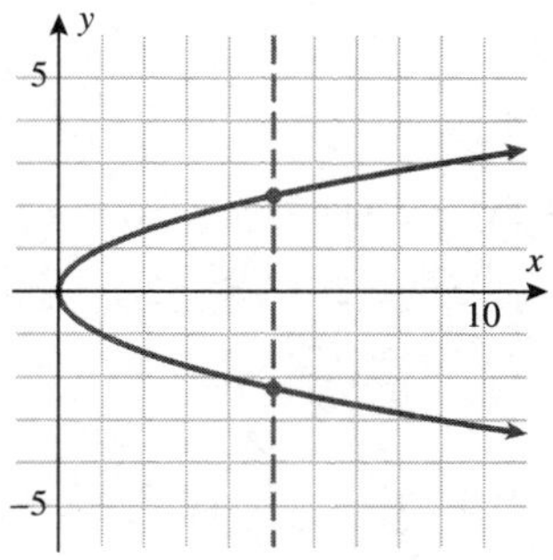

Not a function

(fails vertical line test)

$y = x^2$

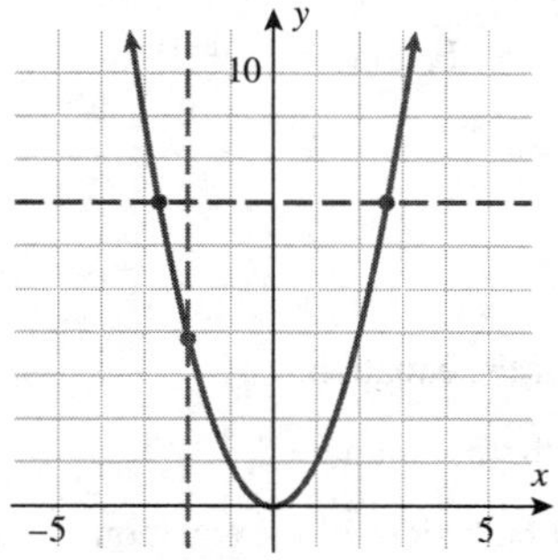

Function, but not one-to-one

(passes vertical line test, but fails horizontal line test)

$y = x^3$

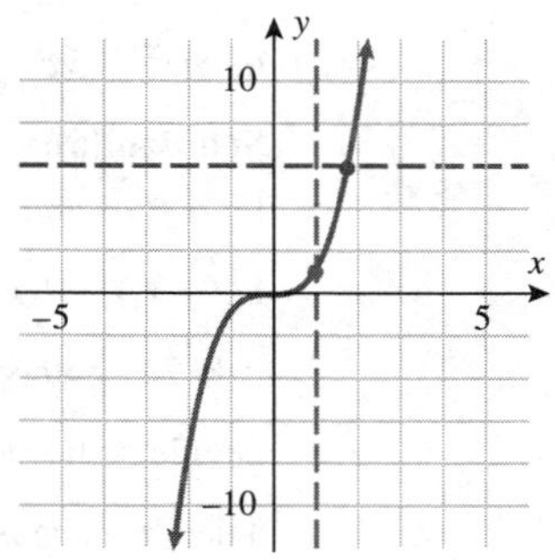

One-to-one function

(passes both horizontal and vertical line tests)

■ **YOUR TURN** Determine whether each of the functions is a one-to-one function.

a. $f(x) = x + 2$ **b.** $f(x) = x^2 + 1$

■ **Answer:**
a. yes **b.** no

Another way of writing the definition of a one-to-one function is:

$$\text{If } f(x_1) = f(x_2), \text{ then } x_1 = x_2.$$

In the Your Turn following Example 2, we found (using the horizontal line test) that $f(x) = x + 2$ is a one-to-one function, but that $f(x) = x^2 + 1$ is not a one-to-one function. We can also use this alternative definition to determine algebraically whether a function is one-to-one.

Words	**Math**
State the function.	$f(x) = x + 2$
Let there be two real numbers, x_1 and x_2, such that $f(x_1) = f(x_2)$.	$x_1 + 2 = x_2 + 2$
Subtract 2 from both sides of the equation.	$x_1 = x_2$

$f(x) = x + 2$ is a one-to-one function.

Words	**Math**
State the function.	$f(x) = x^2 + 1$
Let there be two real numbers, x_1 and x_2, such that $f(x_1) = f(x_2)$.	$x_1^2 + 1 = x_2^2 + 1$
Subtract 1 from both sides of the equation.	$x_1^2 = x_2^2$
Solve for x_1.	$x_1 = \pm x_2$

$f(x) = x^2 + 2$ is *not* a one-to-one function.

EXAMPLE 3 Determining Algebraically Whether a Function Is One-to-One

Determine algebraically whether the following functions are one-to-one:

a. $f(x) = 5x^3 - 2$ **b.** $f(x) = |x + 1|$

Solution (a):

Find $f(x_1)$ and $f(x_2)$. $f(x_1) = 5x_1^3 - 2$ and $f(x_2) = 5x_2^3 - 2$

Let $f(x_1) = f(x_2)$. $5x_1^3 - 2 = 5x_2^3 - 2$

Add 2 to both sides of the equation. $5x_1^3 = 5x_2^3$

Divide both sides of the equation by 5. $x_1^3 = x_2^3$

Take the cube root of both sides of the equation. $(x_1^3)^{1/3} = (x_2^3)^{1/3}$

Simplify. $x_1 = x_2$

$f(x) = 5x^3 - 2$ is a one-to-one function.

Solution (b):

Find $f(x_1)$ and $f(x_2)$. $f(x_1) = |x_1 + 1|$ and $f(x_2) = |x_2 + 1|$

Let $f(x_1) = f(x_2)$. $|x_1 + 1| = |x_2 + 1|$

Solve the absolute value equation. $(x_1 + 1) = (x_2 + 1)$ or $(x_1 + 1) = -(x_2 + 1)$

$x_1 = x_2$ or $x_1 = -x_2 - 2$

$f(x) = |x + 1|$ is **not** a one-to-one function.

Inverse Functions

If a function is one-to-one, then the function maps each x to exactly one y, and no two x-values map to the same y-value. This implies that there is a one-to-one correspondence between the inputs (domain) and outputs (range) of a one-to-one function $f(x)$. In the special case of a one-to-one function, it would be possible to map from the output (range of f) back to the input (domain of f), and this mapping would also be a function. The function that maps the output back to the input of a function f is called the **inverse function** and is denoted $f^{-1}(x)$.

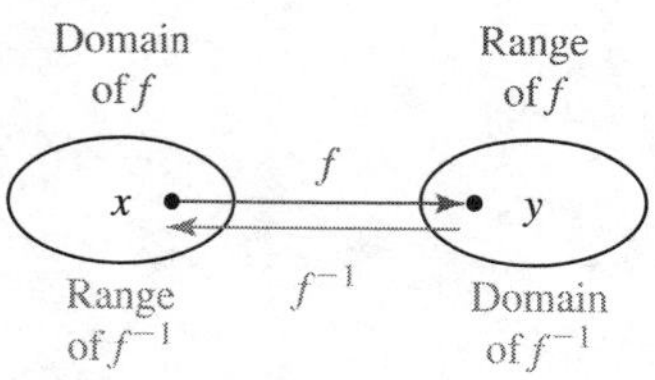

A one-to-one function f maps every x in the domain to a unique and distinct corresponding y in the range. Therefore, the inverse function f^{-1} maps every y back to a unique and distinct x.

The function notations $f(x) = y$ and $f^{-1}(y) = x$ indicate that if the point (x, y) satisfies the function, then the point (y, x) satisfies the inverse function.

For example, let the function $h(x) = \{(-1, 0), (1, 2), (3, 4)\}$.

$$h = \{(-1, 0), (1, 2), (3, 4)\}$$

Domain		Range
−1	⇄	**0**
1	⇄	**2**
3	⇄	**4**
Range		Domain

h is a one-to-one function

$$h^{-1} = \{(0, -1), (2, 1), (4, 3)\}$$

The inverse function undoes whatever the function does. For example, if $f(x) = 5x$, then the function f maps any value x in the domain to a value $5x$ in the range. If we want to map backwards or undo the $5x$, we develop a function called the inverse function that takes $5x$ as input and maps back to x as output. The inverse function is $f^{-1}(x) = \frac{1}{5}x$. Note that if we input $5x$ into the inverse function, the output is x: $f^{-1}(5x) = \frac{1}{5}(5x) = x$.

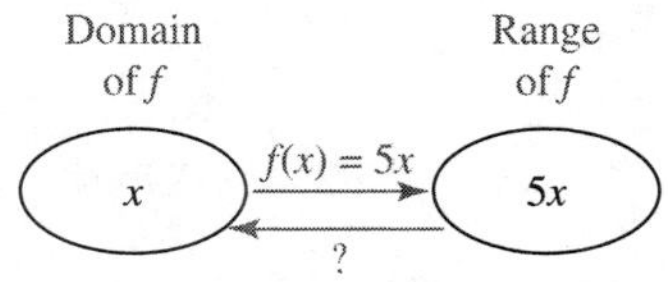

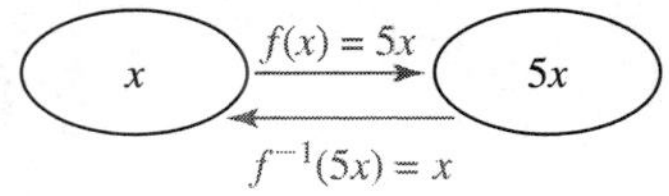

DEFINITION **Inverse Function**

If f and g denote two one-to-one functions such that

$$f(g(x)) = x \text{ for every } x \text{ in the domain of } g$$
and
$$g(f(x)) = x \text{ for every } x \text{ in the domain of } f,$$

then g is the **inverse** of the function f. The function g is denoted by f^{-1} (read "f-inverse").

CAUTION

$$f^{-1} \neq \frac{1}{f}$$

Note: f^{-1} is used to denote the inverse of f. The -1 is not used as an exponent and, therefore, does not represent the reciprocal of f: $\frac{1}{f}$.

Two properties hold true relating one-to-one functions to their inverses: (1) the range of the function is the domain of the inverse, and the range of the inverse is the domain of the function, and (2) the composite function that results with a function and its inverse (and vice versa) is the identity function x.

$$\text{Domain of } f = \text{range of } f^{-1} \text{ and range of } f = \text{domain of } f^{-1}$$

$$f^{-1}(f(x)) = x \quad \text{and} \quad f(f^{-1}(x)) = x$$

EXAMPLE 4 **Verifying Inverse Functions**

Verify that $f^{-1}(x) = \frac{1}{2}x - 2$ is the inverse of $f(x) = 2x + 4$.

Solution:

Show that $f^{-1}(f(x)) = x$ and $f(f^{-1}(x)) = x$.

Write f^{-1} using placeholder notation. $\quad f^{-1}(\square) = \frac{1}{2}(\square) - 2$

Substitute $f(x) = 2x + 4$ into f^{-1}. $\quad f^{-1}(f(x)) = \frac{1}{2}(2x + 4) - 2$

Simplify. $\quad f^{-1}(f(x)) = x + 2 - 2 = x$

$f^{-1}(f(x)) = x$

Write f using placeholder notation. $\quad f(\square) = 2(\square) + 4$

Substitute $f^{-1}(x) = \frac{1}{2}x - 2$ into f. $\quad f(f^{-1}(x)) = 2\left(\frac{1}{2}x - 2\right) + 4$

Simplify. $\quad f(f^{-1}(x)) = x - 4 + 4 = x$

$f(f^{-1}(x)) = x$

Note the relationship between the domain and range of f and f^{-1}.

	DOMAIN	RANGE
$f(x) = 2x + 4$	$(-\infty, \infty)$	$(-\infty, \infty)$
$f^{-1}(x) = \frac{1}{2}x - 2$	$(-\infty, \infty)$	$(-\infty, \infty)$

EXAMPLE 5 Verifying Inverse Functions with Domain Restrictions

Verify that $f^{-1}(x) = x^2$, for $x \geq 0$, is the inverse of $f(x) = \sqrt{x}$.

Solution:

Show that $f^{-1}(f(x)) = x$ and $f(f^{-1}(x)) = x$.

Write f^{-1} using placeholder notation. $\quad f^{-1}(\square) = (\square)^2$

Substitute $f(x) = \sqrt{x}$ into f^{-1}. $\quad f^{-1}(f(x)) = (\sqrt{x})^2 = x$

$$f^{-1}(f(x)) = x \text{ for } x \geq 0$$

Write f using placeholder notation. $\quad f(\square) = \sqrt{(\square)}$

Substitute $f^{-1}(x) = x^2, x \geq 0$ into f. $\quad f(f^{-1}(x)) = \sqrt{x^2} = x, x \geq 0$

$$f(f^{-1}(x)) = x \text{ for } x \geq 0$$

	DOMAIN	RANGE
$f(x) = \sqrt{x}$	$[0, \infty)$	$[0, \infty)$
$f^{-1}(x) = x^2, x \geq 0$	$[0, \infty)$	$[0, \infty)$

Graphical Interpretation of Inverse Functions

In Example 4, we showed that $f^{-1}(x) = \frac{1}{2}x - 2$ is the inverse of $f(x) = 2x + 4$. Let's now investigate the graphs that correspond to the function f and its inverse f^{-1}.

$f(x)$

x	y
−3	−2
−2	0
−1	2
0	4

$f^{-1}(x)$

x	y
−2	−3
0	−2
2	−1
4	0

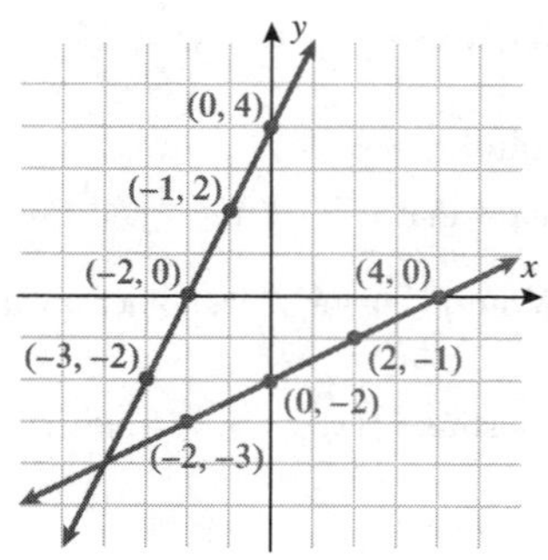

Study Tip

If the point (a, b) is on the function, then the point (b, a) is on the inverse. Notice the interchanging of the x- and y-coordinates.

Note that the point $(-3, -2)$ lies on the function and the point $(-2, -3)$ lies on the inverse. In fact, every point (a, b) that lies on the function corresponds to a point (b, a) that lies on the inverse.

Draw the line $y = x$ on the graph. In general, the point (b, a) on the inverse $f^{-1}(x)$ is the reflection (about $y = x$) of the point (a, b) on the function $f(x)$.

In general, if the point (a, b) is on the graph of a function, then the point (b, a) is on the graph of its inverse.

EXAMPLE 6 Graphing the Inverse Function

Given the graph of the function $f(x)$, plot the graph of its inverse $f^{-1}(x)$.

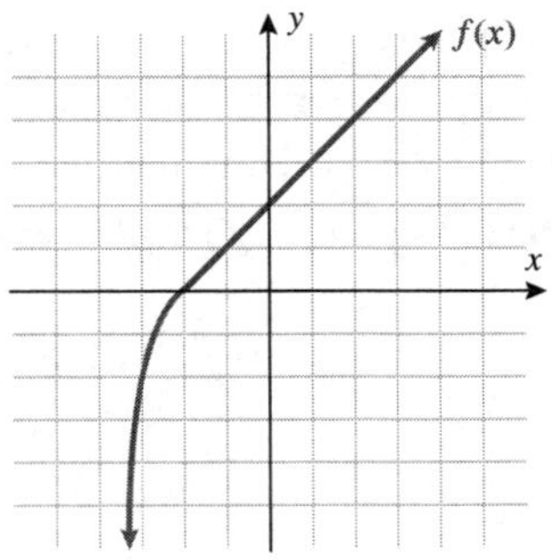

Solution:

Because the points $(-3, -2)$, $(-2, 0)$, $(0, 2)$, and $(2, 4)$ lie on the graph of f, then the points $(-2, -3)$, $(0, -2)$, $(2, 0)$, and $(4, 2)$ lie on the graph of f^{-1}.

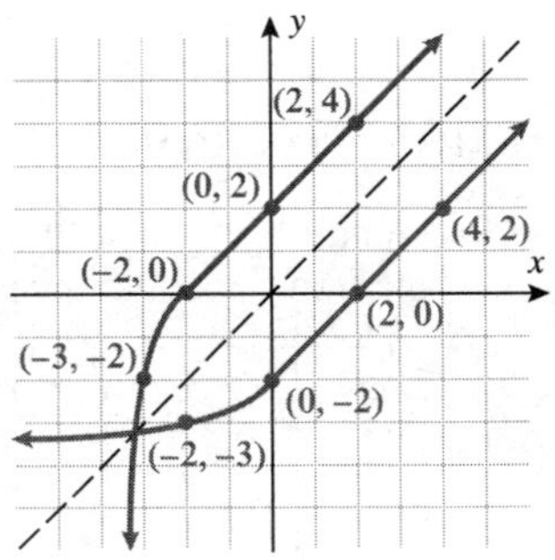

■ **YOUR TURN** Given the graph of a function f, plot the inverse function.

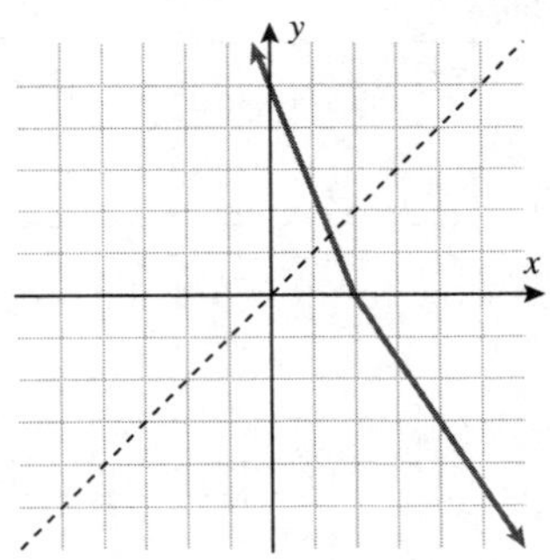

■ **Answer:**

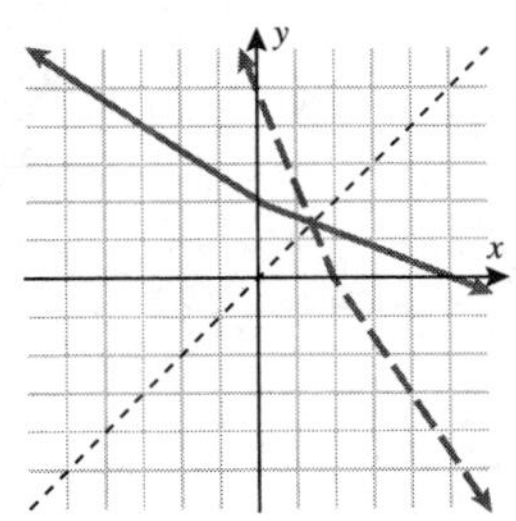

We have developed the definition of an inverse function, and properties of inverses. At this point, you should be able to determine whether two functions are inverses of one another. Let's turn our attention to another problem: How do you find the inverse of a function?

Finding the Inverse Function

If the point (a, b) lies on the graph of a function, then the point (b, a) lies on the graph of the inverse function. The symmetry about the line $y = x$ tells us that the roles of x and y interchange. Therefore, if we start with every point (x, y) that lies on the graph of a function, then every point (y, x) lies on the graph of its inverse. Algebraically, this corresponds to interchanging x and y. Finding the inverse of a finite set of ordered pairs is easy: simply interchange the x- and y-coordinates. Earlier, we found that if $h(x) = \{(-1, 0), (1, 2), (3, 4)\}$, then $h^{-1}(x) = \{(0, -1), (2, 1), (4, 3)\}$. But how do we find the inverse of a function defined by an equation?

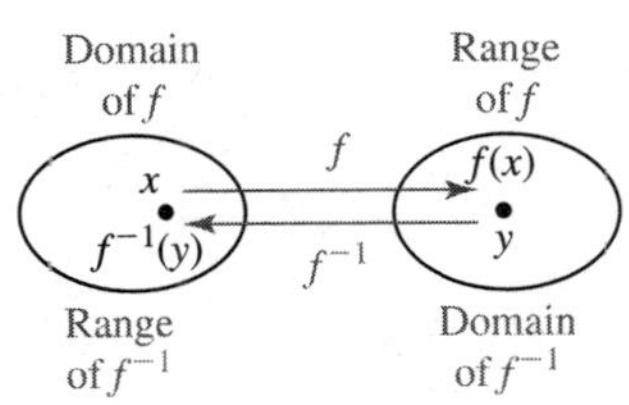

Recall the mapping relationship if f is a one-to-one function. This relationship implies that $f(x) = y$ and $f^{-1}(y) = x$. Let's use these two identities to find the inverse. Now consider the function defined by $f(x) = 3x - 1$. To find f^{-1}, we let $f(x) = y$, which yields $y = 3x - 1$. Solve for the variable x: $x = \frac{1}{3}y + \frac{1}{3}$.

Recall that $f^{-1}(y) = x$, so we have found the inverse to be $f^{-1}(y) = \frac{1}{3}y + \frac{1}{3}$. It is customary to write the independent variable as x, so we write the inverse as $f^{-1}(x) = \frac{1}{3}x + \frac{1}{3}$. Now that we have found the inverse, let's confirm that the properties $f^{-1}(f(x)) = x$ and $f(f^{-1}(x)) = x$ hold.

$$f(f^{-1}(x)) = 3\left(\frac{1}{3}x + \frac{1}{3}\right) - 1 = x + 1 - 1 = x$$

$$f^{-1}(f(x)) = \frac{1}{3}(3x - 1) + \frac{1}{3} = x - \frac{1}{3} + \frac{1}{3} = x$$

Finding the Inverse of a Function

Let f be a one-to-one function, then the following procedure can be used to find the inverse function f^{-1} if the inverse exists.

Step	Procedure	Example
1	Let $y = f(x)$.	$f(x) = -3x + 5$ $y = -3x + 5$
2	Solve the resulting equation for x in terms of y (if possible).	$3x = -y + 5$ $x = -\frac{1}{3}y + \frac{5}{3}$
3	Let $x = f^{-1}(y)$.	$f^{-1}(y) = -\frac{1}{3}y + \frac{5}{3}$
4	Let $y = x$ (interchange x and y).	$f^{-1}(x) = -\frac{1}{3}x + \frac{5}{3}$

The same result is found if we first interchange x and y and then solve for y in terms of x.

Step	Procedure	Example
1	Let $y = f(x)$.	$f(x) = -3x + 5$ $y = -3x + 5$
2	Interchange x and y.	$x = -3y + 5$
3	Solve for y in terms of x.	$3y = -x + 5$ $y = -\frac{1}{3}x + \frac{5}{3}$
4	Let $y = f^{-1}(x)$	$f^{-1}(x) = -\frac{1}{3}x + \frac{5}{3}$

Note the following:

- Verify first that a function is one-to-one prior to finding an inverse (if it is not one-to-one, then the inverse does not exist).
- State the domain restrictions on the inverse function. The domain of f is the range of f^{-1} and vice versa.
- To verify that you have found the inverse, show that $f(f^{-1}(x)) = x$ for all x in the domain of f^{-1} and $f^{-1}(f(x)) = x$ for all x in the domain of f.

EXAMPLE 7 The Inverse of a Square Root Function

Find the inverse of the function $f(x) = \sqrt{x + 2}$. State the domain and range of both f and f^{-1}.

Solution:

$f(x)$ is a one-to-one function because it passes the horizontal line test.

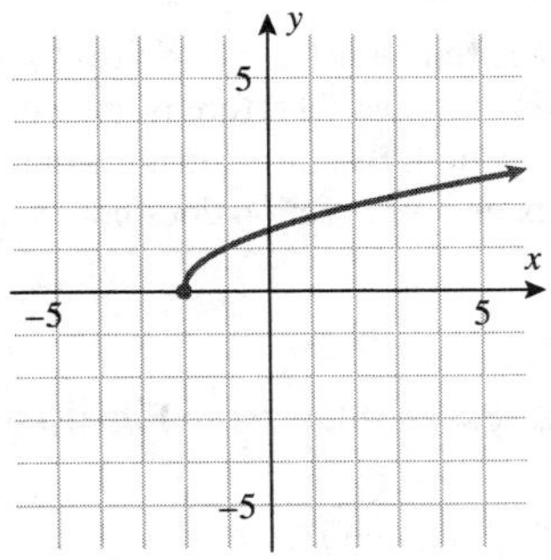

STEP 1 Let $y = f(x)$. $\quad y = \sqrt{x + 2}$

STEP 2 Interchange x and y. $\quad x = \sqrt{y + 2}$

STEP 3 Solve for y.

Square both sides of the equation. $\quad x^2 = y + 2$

Subtract 2 from both sides. $\quad x^2 - 2 = y$ or $y = x^2 - 2$

STEP 4 Let $y = f^{-1}(x)$. $\quad f^{-1}(x) = x^2 - 2$

Note any domain restrictions. (State the domain and range of both f and f^{-1}.)

f:	Domain: $[-2, \infty)$	Range: $[0, \infty)$
f^{-1}:	Domain: $[0, \infty)$	Range: $[-2, \infty)$

The inverse of $f(x) = \sqrt{x + 2}$ is $\boxed{f^{-1}(x) = x^2 - 2 \text{ for } x \geq 0}$.

Check.

$f^{-1}(f(x)) = x$ for all x in the domain of f.

$$f^{-1}(f(x)) = \left(\sqrt{x + 2}\right)^2 - 2$$
$$= x + 2 - 2 \text{ for } x \geq -2$$
$$= x$$

$f(f^{-1}(x)) = x$ for all x in the domain of f^{-1}.

$$f(f^{-1}(x)) = \sqrt{(x^2 - 2) + 2}$$
$$= \sqrt{x^2} \text{ for } x \geq 0$$
$$= x$$

Note that the function $f(x) = \sqrt{x + 2}$ and its inverse $f^{-1}(x) = x^2 - 2$ for $x \geq 0$ are symmetric about the line $y = x$.

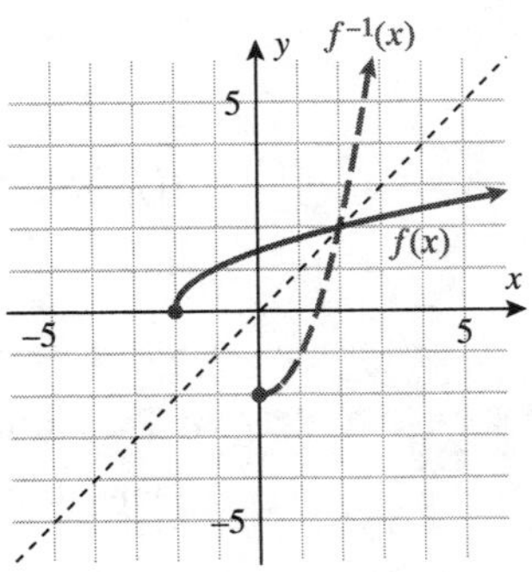

YOUR TURN Find the inverse of the given function. State the domain and range of the inverse function.

a. $f(x) = 7x - 3$ $\qquad$ **b.** $g(x) = \sqrt{x - 1}$

Technology Tip

Using a graphing utility, plot $y_1 = f(x) = \sqrt{x + 2}$, $y_2 = f^{-1}(x) = x^2 - 2$, and $y_3 = x$.

Note that the function $f(x)$ and its inverse $f^{-1}(x)$ are symmetric about the line $y = x$.

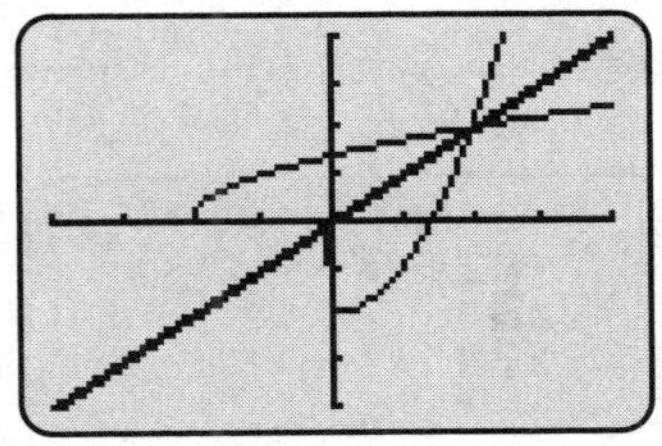

Study Tip

Had we ignored the domain and range in Example 7, we would have found the inverse function to be the square function $f(x) = x^2 - 2$, which is not a one-to-one function. It is only when we restrict the domain of the square function that we get a one-to-one function.

Answers:

a. $f^{-1}(x) = \dfrac{x + 3}{7}$, Domain: $(-\infty, \infty)$, Range: $(-\infty, \infty)$

b. $g^{-1}(x) = x^2 + 1$, Domain: $[0, \infty)$, Range: $[1, \infty)$

EXAMPLE 8 A Function That Does Not Have an Inverse Function

Find the inverse of the function $f(x) = |x|$ if it exists.

Solution:

The function $f(x) = |x|$ fails the horizontal line test and therefore is not a one-to-one function. Because f is not a one-to-one function, its inverse function does not exist.

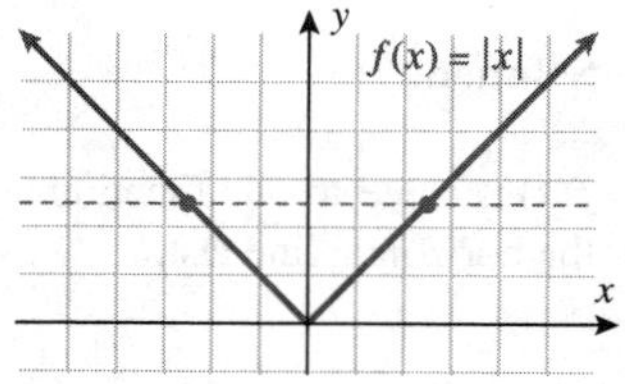

Technology Tip

The graphs of $y_1 = f(x) = \frac{2}{x+3}$, $x \neq -3$, and $y_2 = f^{-1}(x) = -3 + \frac{2}{x}$, $x \neq 0$, are shown.

Note that the function $f(x)$ and its inverse $f^{-1}(x)$ are symmetric about the line $y = x$.

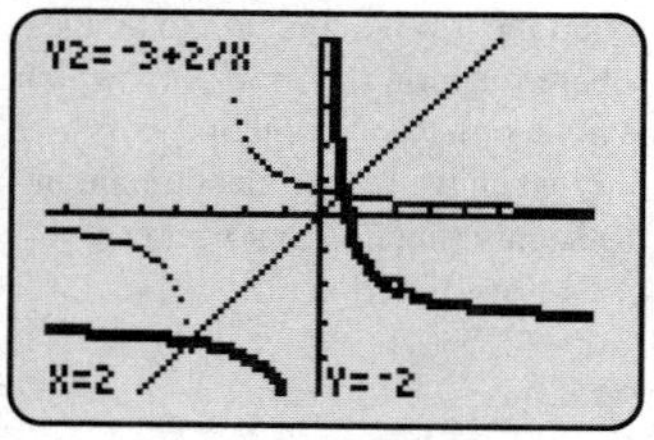

EXAMPLE 9 Finding the Inverse Function

The function $f(x) = \frac{2}{x+3}$, $x \neq -3$, is a one-to-one function. Find its inverse.

Solution:

STEP 1 Let $y = f(x)$. $\qquad y = \frac{2}{x+3}$

STEP 2 Interchange x and y. $\qquad x = \frac{2}{y+3}$

STEP 3 Solve for y.

Multiply the equation by $(y + 3)$. $\qquad x(y + 3) = 2$

Eliminate the parentheses. $\qquad xy + 3x = 2$

Subtract $3x$ from both sides. $\qquad xy = -3x + 2$

Divide the equation by x. $\qquad y = \frac{-3x + 2}{x} = -3 + \frac{2}{x}$

STEP 4 Let $y = f^{-1}(x)$. $\qquad f^{-1}(x) = -3 + \frac{2}{x}$

Note any domain restrictions on $f^{-1}(x)$. $\qquad x \neq 0$

The inverse of the function $f(x) = \frac{2}{x+3}$, $x \neq -3$, is $\boxed{f^{-1}(x) = -3 + \frac{2}{x},\ x \neq 0}$.

Check.

$$f^{-1}(f(x)) = -3 + \frac{2}{\left(\frac{2}{x+3}\right)} = -3 + (x + 3) = x,\ x \neq -3$$

$$f(f^{-1}(x)) = \frac{2}{\left(-3 + \frac{2}{x}\right) + 3} = \frac{2}{\left(\frac{2}{x}\right)} = x,\ x \neq 0$$

Study Tip

The range of the function is equal to the domain of its inverse function.

YOUR TURN The function $f(x) = \frac{4}{x-1}$, $x \neq 1$, is a one-to-one function. Find its inverse.

■ **Answer:** $f^{-1}(x) = 1 + \frac{4}{x}$, $x \neq 0$

Note in Example 9 that the domain of f is $(-\infty, -3) \cup (-3, \infty)$ and the domain of f^{-1} is $(-\infty, 0) \cup (0, \infty)$. Therefore, we know that the range of f is $(-\infty, 0) \cup (0, \infty)$, and the range of f^{-1} is $(-\infty, -3) \cup (-3, \infty)$.

SECTION 3.5 SUMMARY

One-to-One Functions

Each input in the domain corresponds to exactly one output in the range, and no two inputs map to the same output. There are three ways to test a function to determine whether it is a one-to-one function.

1. Discrete points: For the set of all points (a, b) verify that no y-values are repeated.
2. Algebraic equations: Let $f(x_1) = f(x_2)$; if it can be shown that $x_1 = x_2$, then the function is one-to-one.
3. Graphs: Use the horizontal line test; if any horizontal line intersects the graph of the function in more than one point, then the function is not one-to-one.

Properties of Inverse Functions

1. If f is a one-to-one function, then f^{-1} exists.
2. Domain and range
   - Domain of f = range of f^{-1}
   - Domain of f^{-1} = range of f
3. Composition of inverse functions
   - $f^{-1}(f(x)) = x$ for all x in the domain of f.
   - $f(f^{-1}(x)) = x$ for all x in the domain of f^{-1}.
4. The graphs of f and f^{-1} are symmetric with respect to the line $y = x$.

Procedure for Finding the Inverse of a Function

1. Let $y = f(x)$.
2. Interchange x and y.
3. Solve for y.
4. Let $y = f^{-1}(x)$.

SECTION 3.5 EXERCISES

SKILLS

In Exercises 1–16, determine whether the given relation is a function. If it is a function, determine whether it is a one-to-one function.

1.

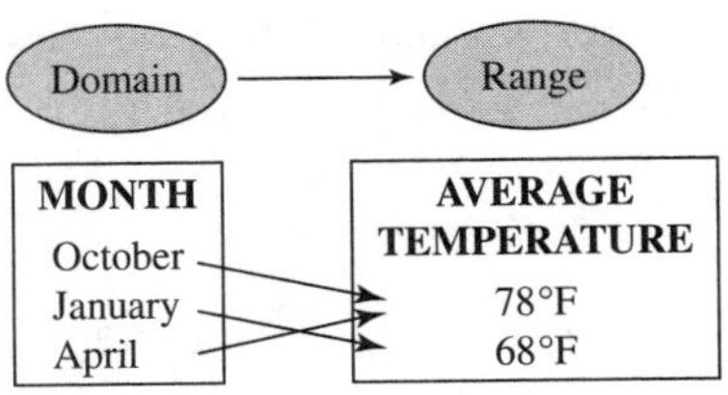

2.

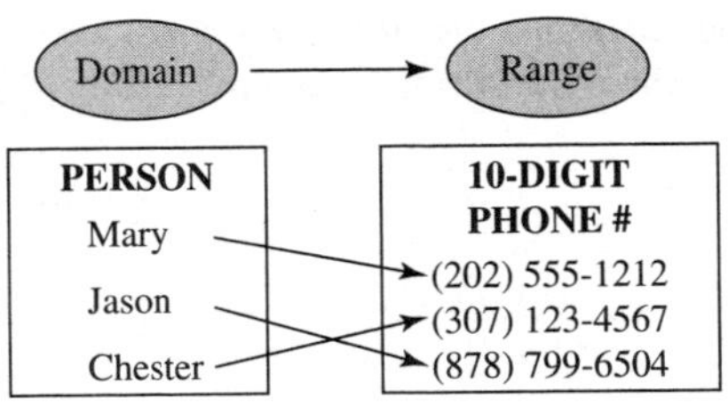

3.

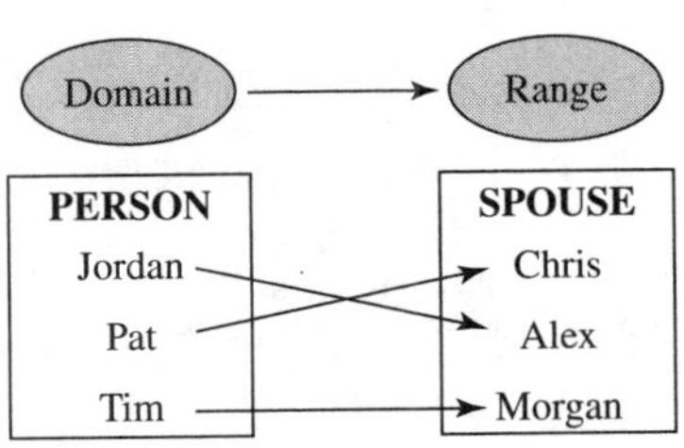

4.

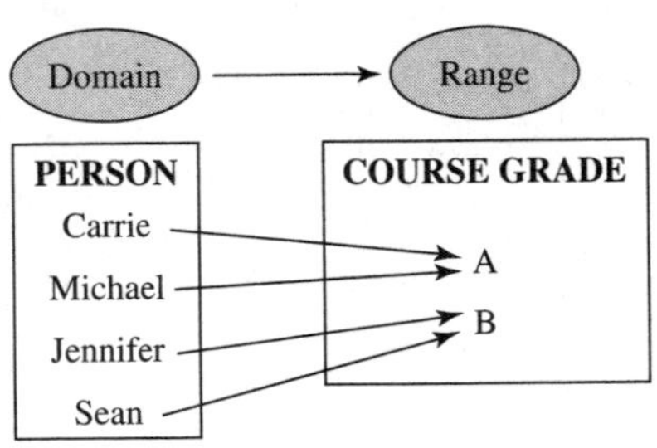

5. {(0, 1), (1, 2), (2, 3), (3, 4)}

6. {(0, −2), (2, 0), (5, 3), (−5, −7)}

7. {(0, 0), (9, −3), (4, −2), (4, 2), (9, 3)}

8. {(0, 1), (1, 1), (2, 1), (3, 1)}

9. {(0, 1), (1, 0), (2, 1), (−2, 1), (5, 4), (−3, 4)}

10. {(0, 0), (−1, −1), (−2, −8), (1, 1), (2, 8)}

11.
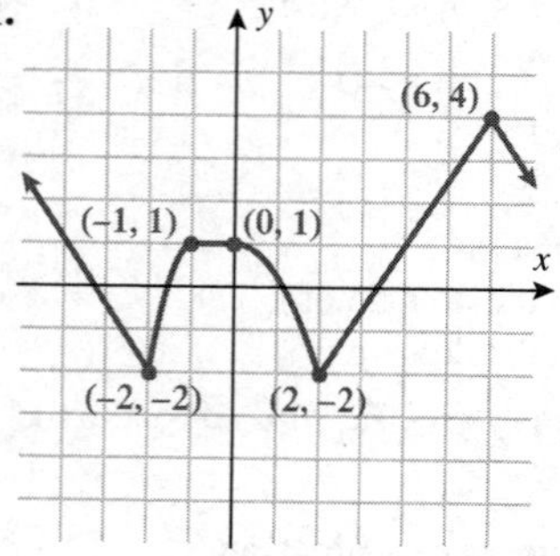

12.
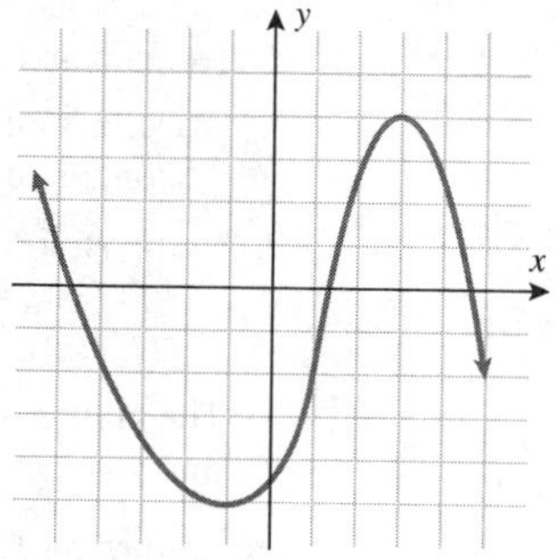

13.
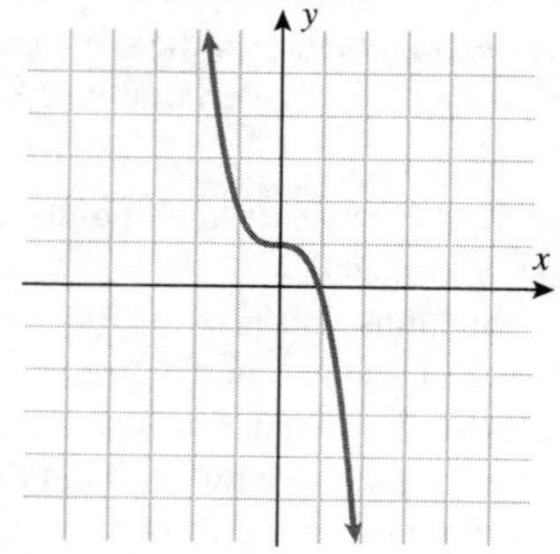

14.
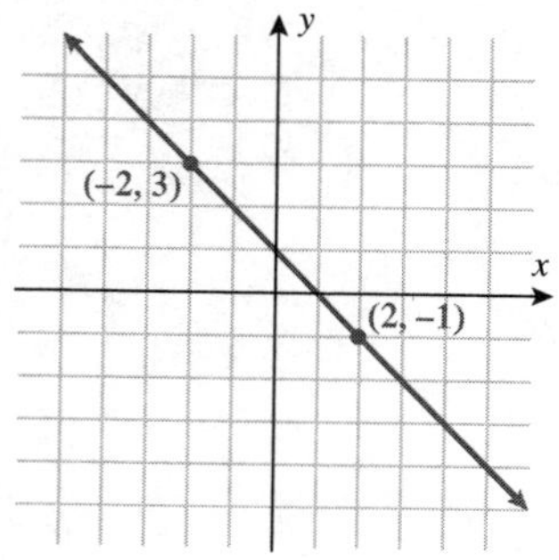

15.
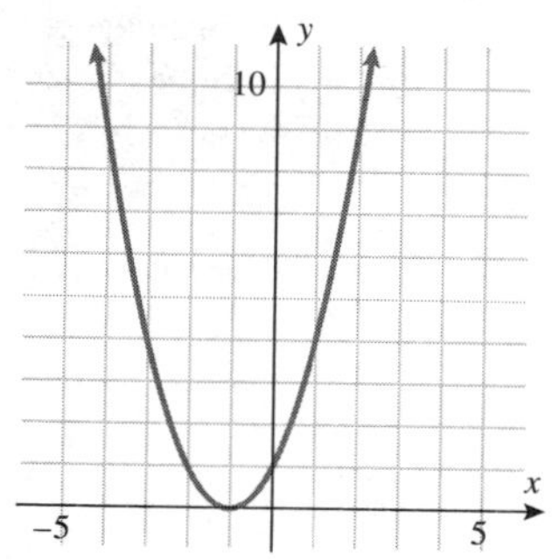

16.
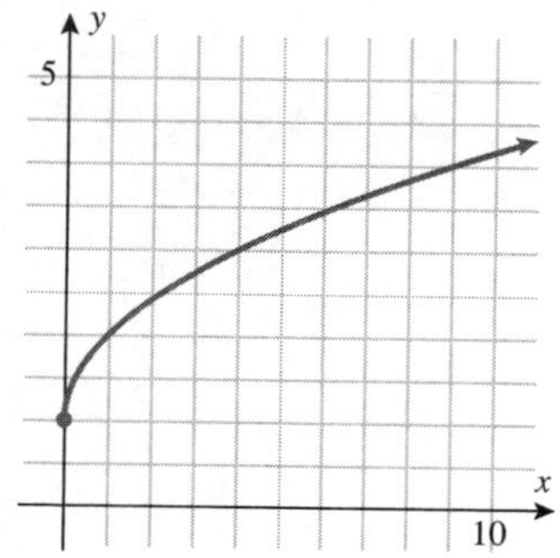

In 17–24, determine algebraically and graphically whether the function is one-to-one.

17. $f(x) = |x - 3|$

18. $f(x) = (x - 2)^2 + 1$

19. $f(x) = \dfrac{1}{x - 1}$

20. $f(x) = \sqrt[3]{x}$

21. $f(x) = x^2 - 4$

22. $f(x) = \sqrt{x + 1}$

23. $f(x) = x^3 - 1$

24. $f(x) = \dfrac{1}{x + 2}$

In Exercises 25–34, verify that the function $f^{-1}(x)$ is the inverse of $f(x)$ by showing that $f(f^{-1}(x)) = x$ and $f^{-1}(f(x)) = x$. Graph $f(x)$ and $f^{-1}(x)$ on the same axes to show the symmetry about the line $y = x$.

25. $f(x) = 2x + 1;\ f^{-1}(x) = \dfrac{x - 1}{2}$

26. $f(x) = \dfrac{x - 2}{3};\ f^{-1}(x) = 3x + 2$

27. $f(x) = \sqrt{x - 1}, x \geq 1;\ f^{-1}(x) = x^2 + 1, x \geq 0$

28. $f(x) = 2 - x^2, x \geq 0;\ f^{-1}(x) = \sqrt{2 - x}, x \leq 2$

29. $f(x) = \dfrac{1}{x};\ f^{-1}(x) = \dfrac{1}{x}, x \neq 0$

30. $f(x) = (5 - x)^{1/3};\ f^{-1}(x) = 5 - x^3$

31. $f(x) = \dfrac{1}{2x + 6}, x \neq -3;\ f^{-1}(x) = \dfrac{1}{2x} - 3, x \neq 0$

32. $f(x) = \dfrac{3}{4 - x}, x \neq 4;\ f^{-1}(x) = 4 - \dfrac{3}{x}, x \neq 0$

33. $f(x) = \dfrac{x + 3}{x + 4}, x \neq -4;\ f^{-1}(x) = \dfrac{3 - 4x}{x - 1}, x \neq 1$

34. $f(x) = \dfrac{x - 5}{3 - x}, x \neq 3;\ f^{-1}(x) = \dfrac{3x + 5}{x + 1}, x \neq -1$

In Exercises 35–42, graph the inverse of the one-to-one function that is given.

35.

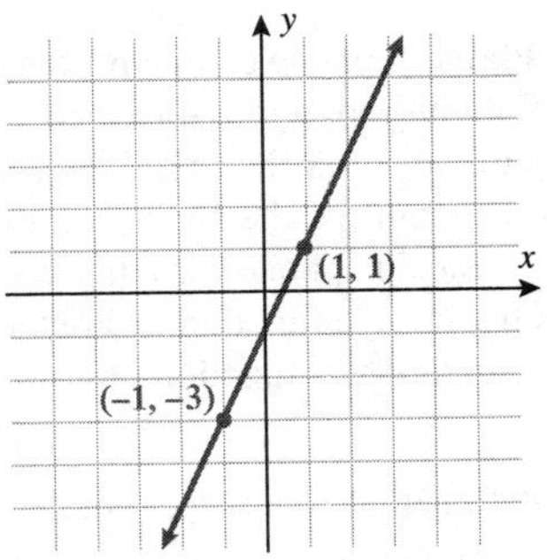

36.

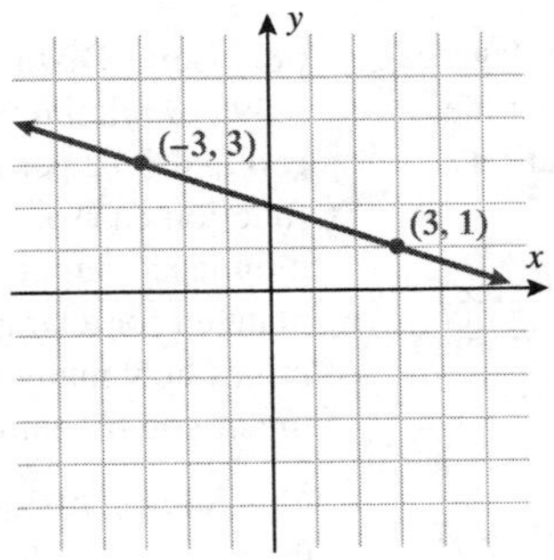

37.

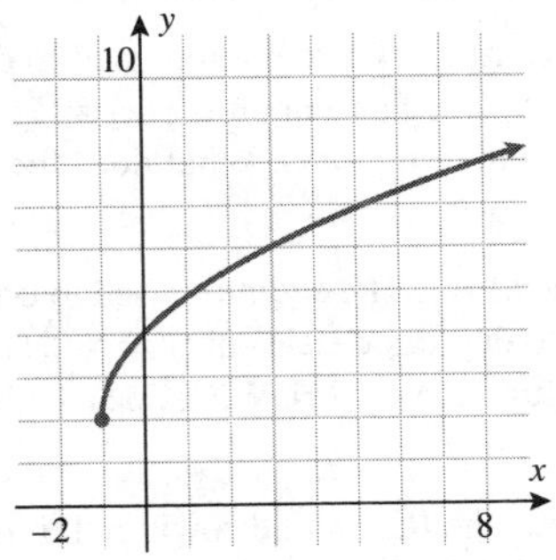

38.

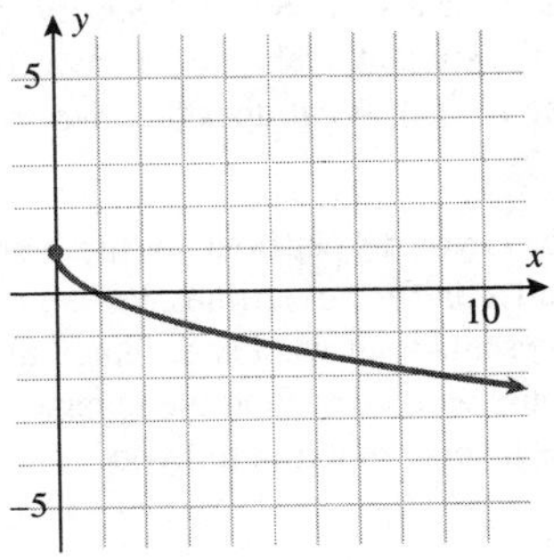

39.

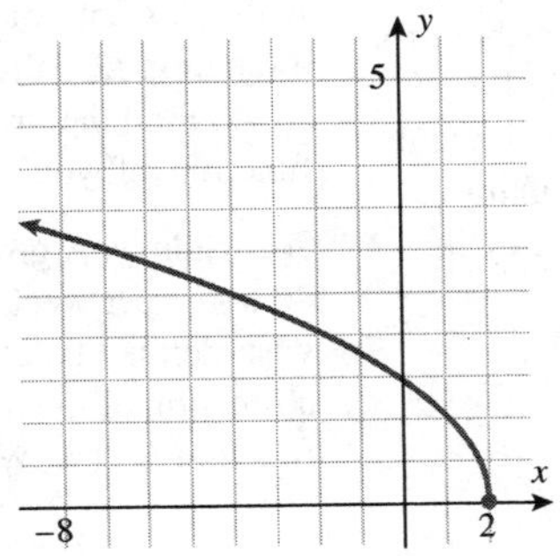

40.

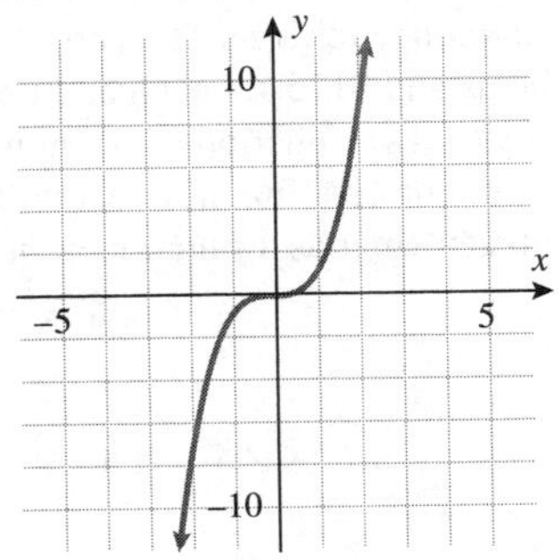

41.

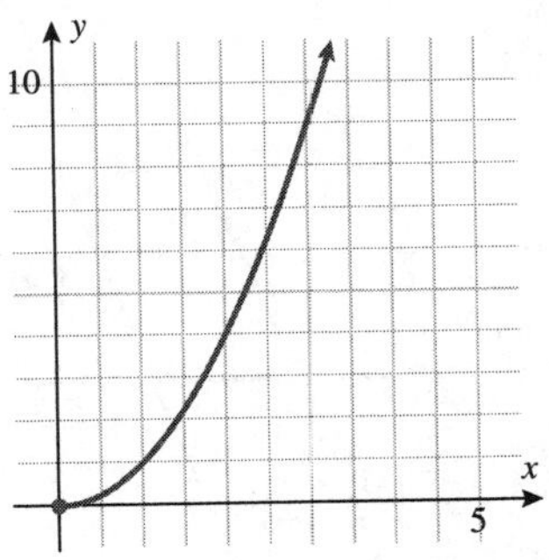

42.

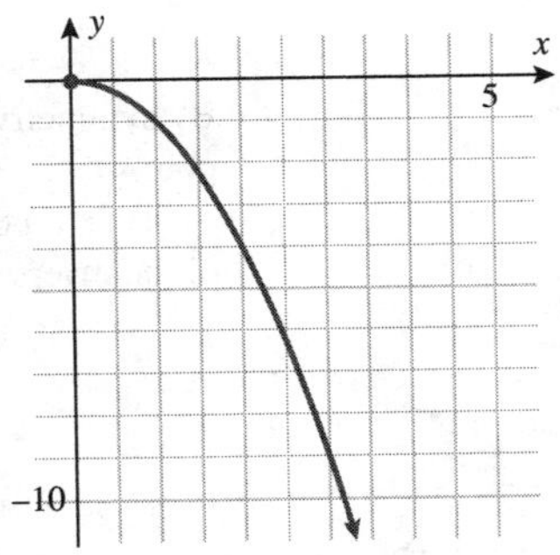

In Exercises 43–60, the function f is one-to-one. Find its inverse, and check your answer. State the domain and range of both f and f^{-1}.

43. $f(x) = x - 1$

44. $f(x) = 7x$

45. $f(x) = -3x + 2$

46. $f(x) = 2x + 3$

47. $f(x) = x^3 + 1$

48. $f(x) = x^3 - 1$

49. $f(x) = \sqrt{x - 3}$

50. $f(x) = \sqrt{3 - x}$

51. $f(x) = x^2 - 1, x \geq 0$

52. $f(x) = 2x^2 + 1, x \geq 0$

53. $f(x) = (x + 2)^2 - 3, x \geq -2$

54. $f(x) = (x - 3)^2 - 2, x \geq 3$

55. $f(x) = \dfrac{2}{x}$

56. $f(x) = -\dfrac{3}{x}$

57. $f(x) = \dfrac{2}{3 - x}$

58. $f(x) = \dfrac{7}{x + 2}$

59. $f(x) = \dfrac{7x + 1}{5 - x}$

60. $f(x) = \dfrac{2x + 5}{7 + x}$

In Exercises 61–64, graph the piecewise-defined function to determine whether it is a one-to-one function. If it is a one-to-one function, find its inverse.

61. $G(x) = \begin{cases} 0 & x < 0 \\ \sqrt{x} & x \geq 0 \end{cases}$

62. $G(x) = \begin{cases} \dfrac{1}{x} & x < 0 \\ \sqrt{x} & x \geq 0 \end{cases}$

63. $f(x) = \begin{cases} x & x \leq -1 \\ x^3 & -1 < x < 1 \\ x & x \geq 1 \end{cases}$

64. $f(x) = \begin{cases} x + 3 & x \leq -2 \\ |x| & -2 < x < 2 \\ x^2 & x \geq 2 \end{cases}$

APPLICATIONS

65. Temperature. The equation used to convert from degrees Celsius to degrees Fahrenheit is $f(x) = \frac{9}{5}x + 32$. Determine the inverse function $f^{-1}(x)$. What does the inverse function represent?

66. Temperature. The equation used to convert from degrees Fahrenheit to degrees Celsius is $C(x) = \frac{5}{9}(x - 32)$. Determine the inverse function $C^{-1}(x)$. What does the inverse function represent?

67. Budget. The Richmond rowing club is planning to enter the Head of the Charles race in Boston and is trying to figure out how much money to raise. The entry fee is \$250 per boat for the first 10 boats and \$175 for each additional boat. Find the cost function $C(x)$ as a function of the number of boats the club enters x. Find the inverse function that will yield how many boats the club can enter as a function of how much money it will raise.

68. Long-Distance Calling Plans. A phone company charges \$.39 per minute for the first 10 minutes of a long-distance phone call and \$.12 per minute every minute after that. Find the cost function $C(x)$ as a function of the length of the phone call in minutes x. Suppose you buy a "prepaid" phone card that is planned for a single call. Find the inverse function that determines how many minutes you can talk as a function of how much you prepaid.

69. Salary. A student works at Target making \$7 per hour and the weekly number of hours worked per week x varies. If Target withholds 25% of his earnings for taxes and Social Security, write a function $E(x)$ that expresses the student's take-home pay each week. Find the inverse function $E^{-1}(x)$. What does the inverse function tell you?

70. Salary. A grocery store pays you \$8 per hour for the first 40 hours per week and time and a half for overtime. Write a piecewise-defined function that represents your weekly earnings $E(x)$ as a function of the number of hours worked x. Find the inverse function $E^{-1}(x)$. What does the inverse function tell you?

CATCH THE MISTAKE

In Exercises 71–74, explain the mistake that is made.

71. Is $x = y^2$ a one-to-one function?

Solution:

Yes, this graph represents a one-to-one function because it passes the horizontal line test.

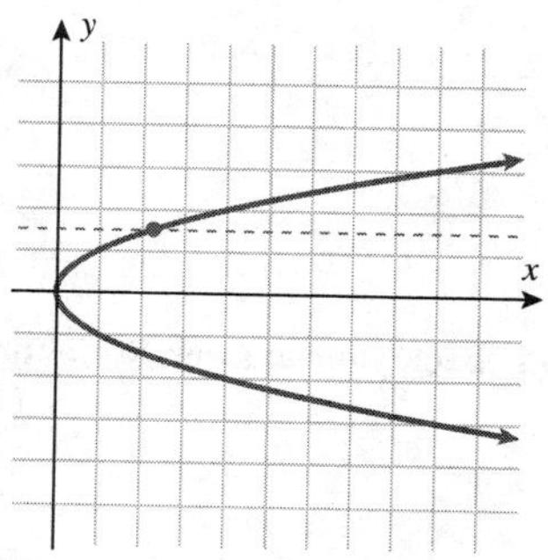

This is incorrect. What mistake was made?

72. A linear one-to-one function is graphed below. Draw its inverse.

Solution:

Note that the points (3, 3) and (0, −4) lie on the graph of the function.

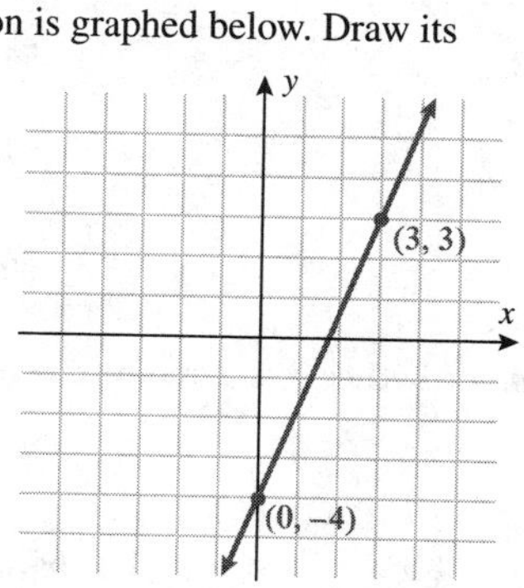

By symmetry, the points (−3, −3) and (0, 4) lie on the graph of the inverse.

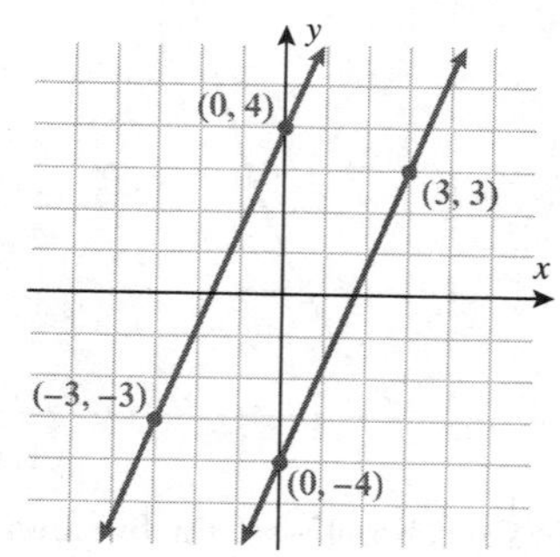

This is incorrect. What mistake was made?

73. Given the function $f(x) = x^2$, find the inverse function $f^{-1}(x)$.

Solution:

Step 1: Let $y = f(x)$. $\quad y = x^2$

Step 2: Solve for x. $\quad x = \sqrt{y}$

Step 3: Interchange x and y. $\quad y = \sqrt{x}$

Step 4: Let $y = f^{-1}(x)$. $\quad f^{-1}(x) = \sqrt{x}$

Check: $f(f^{-1}(x)) = (\sqrt{x})^2 = x$ and $f^{-1}(f(x)) = \sqrt{x^2} = x$.

The inverse of $f(x) = x^2$ is $f^{-1}(x) = \sqrt{x}$.

This is incorrect. What mistake was made?

74. Given the function $f(x) = \sqrt{x - 2}$, find the inverse function $f^{-1}(x)$, and state the domain restrictions on $f^{-1}(x)$.

Solution:

Step 1: Let $y = f(x)$. $\quad y = \sqrt{x - 2}$

Step 2: Interchange x and y. $\quad x = \sqrt{y - 2}$

Step 3: Solve for y. $\quad y = x^2 + 2$

Step 4: Let $f^{-1}(x) = y$. $\quad f^{-1}(x) = x^2 + 2$

Step 5: Domain restrictions: $f(x) = \sqrt{x - 2}$ has the domain restriction that $x \geq 2$.

The inverse of $f(x) = \sqrt{x - 2}$ is $f^{-1}(x) = x^2 + 2$.

The domain of $f^{-1}(x)$ is $x \geq 2$.

This is incorrect. What mistake was made?

CONCEPTUAL

In Exercises 75–78, determine whether each statement is true or false.

75. Every even function is a one-to-one function.

76. Every odd function is a one-to-one function.

77. It is not possible that $f = f^{-1}$.

78. A function f has an inverse. If the function lies in quadrant II, then its inverse lies in quadrant IV.

79. If $(0, b)$ is the y-intercept of a one-to-one function f, what is the x-intercept of the inverse f^{-1}?

80. If $(a, 0)$ is the x-intercept of a one-to-one function f, what is the y-intercept of the inverse f^{-1}?

CHALLENGE

81. The unit circle is not a function. If we restrict ourselves to the semicircle that lies in quadrants I and II, the graph represents a function, but it is not a one-to-one function. If we further restrict ourselves to the quarter circle lying in quadrant I, the graph does represent a one-to-one function. Determine the equations of both the one-to-one function and its inverse. State the domain and range of both.

82. Find the inverse of $f(x) = \frac{c}{x}$, $c \neq 0$.

83. Under what conditions is the linear function $f(x) = mx + b$ a one-to-one function?

84. Assuming that the conditions found in Exercise 83 are met, determine the inverse of the linear function.

TECHNOLOGY

In Exercises 85–88, graph the following functions and determine whether they are one-to-one.

85. $f(x) = |4 - x^2|$

86. $f(x) = \dfrac{3}{x^3 + 2}$

87. $f(x) = x^{1/3} - x^5$

88. $f(x) = \dfrac{1}{x^{1/2}}$

In Exercises 89 and 90, graph the functions f and g and the line $y = x$ in the same screen. Do the two functions appear to be inverses of each other?

89. $f(x) = \sqrt{3x - 5}$; $\quad g(x) = \dfrac{x^2}{3} + \dfrac{5}{3}$

90. $f(x) = \sqrt{4 - 3x}$; $\quad g(x) = \dfrac{4}{3} - \dfrac{x^2}{3}, x \geq 0$

91. $f(x) = (x - 7)^{1/3} + 2$; $\quad g(x) = x^3 - 6x^2 + 12x - 1$

92. $f(x) = \sqrt[3]{x + 3} - 2$; $\quad g(x) = x^3 + 6x^2 + 12x + 6$

SECTION
3.6 MODELING FUNCTIONS USING VARIATION

SKILLS OBJECTIVES

- Develop mathematical models using direct variation.
- Develop mathematical models using inverse variation.
- Develop mathematical models using combined variation.
- Develop mathematical models using joint variation.

CONCEPTUAL OBJECTIVES

- Understand the difference between direct variation and inverse variation.
- Understand the difference between combined variation and joint variation.

In this section we discuss mathematical models for different applications. Two quantities in the real world often *vary* with respect to one another. Sometimes, they vary *directly*. For example, the more money we make, the more total dollars of federal income tax we expect to pay. Sometimes, quantities vary *inversely*. For example, when interest rates on mortgages decrease, we expect the number of homes purchased to increase, because a buyer can afford "more house" with the same mortgage payment when rates are lower. In this section we discuss quantities varying *directly*, *inversely*, or *jointly*.

Direct Variation

When one quantity is a constant multiple of another quantity, we say that the quantities are *directly proportional* to one another.

DIRECT VARIATION

Let x and y represent two quantities. The following are equivalent statements:

- $y = kx$, where k is a nonzero constant.
- y **varies directly** with x.
- y is **directly proportional** to x.

The constant k is called the **constant of variation** or the **constant of proportionality**.

In 2005, the national average cost of residential electricity was 9.53 ¢/kWh (cents per kilowatt-hour). For example, if a residence used 3400 kWh, then the bill would be \$324, and if a residence used 2500 kWh, then the bill would be \$238.25.

EXAMPLE 1 Finding the Constant of Variation

In the United States, the cost of electricity is directly proportional to the number of kilowatt · hours (kWh) used. If a household in Tennessee on average used 3098 kWh per month and had an average monthly electric bill of \$179.99, find a mathematical model that gives the cost of electricity in Tennessee in terms of the number of kilowatt · hours used.

Solution:

Write the direct variation model. $y = kx$

Label the variables and constant.

$x =$ number of kWh
$y =$ cost (dollars)
$k =$ cost per kWh

Substitute the given data $x = 3098$ kWh and $y = \$179.99$ into $y = kx$. $179.99 = 3098k$

Solve for k. $k = \frac{179.99}{3098} = 0.05810$

$y = 0.0581x$

In Tennessee the cost of electricity is $\boxed{5.81\ ¢/\text{kWh}}$.

■ **YOUR TURN** Find a mathematical model that describes the cost of electricity in California if the cost is directly proportional to the number of kWh used and a residence that consumes 4000 kWh is billed $480.

■ **Answer:** $y = 0.12x$; the cost of electricity in California is 12 ¢/kWh.

Not all variation we see in nature is direct variation. Isometric growth, where the various parts of an organism grow in direct proportion to each other, is rare in living organisms. If organisms grew isometrically, young children would look just like adults, only smaller. In contrast, most organisms grow nonisometrically; the various parts of organisms do not increase in size in a one-to-one ratio. The relative proportions of a human body change dramatically as the human grows. Children have proportionately larger heads and shorter legs than adults. *Allometric growth* is the pattern of growth whereby different parts of the body grow at different rates with respect to each other. Some human body characteristics vary directly, and others can be mathematically modeled by *direct variation with powers*.

DIRECT VARIATION WITH POWERS

Let x and y represent two quantities. The following are equivalent statements:

- $y = kx^n$, where k is a nonzero constant.
- y **varies directly with the** n**th power** of x.
- y **is directly proportional to the** n**th power** of x.

One example of direct variation with powers is height and weight of humans. Weight (in pounds) is directly proportional to the cube of height (feet).

$$W = kH^3$$

EXAMPLE 2 Direct Variation with Powers

The following is a personal ad:

Single professional male (6 ft/194 lbs) seeks single professional female for long-term relationship. Must be athletic, smart, like the movies and dogs, and have height and weight similarly proportioned to mine.

Find a mathematical equation that describes the height and weight of the male who wrote the ad. How much would a 5′6″ woman weigh who has the same proportionality as the male?

Solution:

Write the direct variation (cube) model for height versus weight. $W = kH^3$

Substitute the given data $W = 194$ and $H = 6$ into $W = kH^3$. $\quad 194 = k(6)^3$

Solve for k. $\quad k = \frac{194}{216} = 0.898148 \approx 0.90$

$W = 0.9H^3$

Let $H = 5.5$ ft. $\quad W = 0.9(5.5)^3 = 149.73$

A woman 5′6″ tall with the same height and weight proportionality as the male would weigh $\boxed{150 \text{ lb}}$.

■ **Answer:** 200 pounds

■ **YOUR TURN** A brother and sister both have weight (pounds) that varies as the cube of height (feet) and they share the same proportionality constant. The sister is 6′ tall and weighs 170 pounds. Her brother is 6′4″. How much does he weigh?

Inverse Variation

Two fundamental topics covered in economics are supply and demand. Supply is the quantity that producers are willing to sell at a given price. For example, an artist may be willing to paint and sell 5 portraits if each sells for \$50, but that same artist may be willing to sell 100 portraits if each sells for \$10,000. Demand is the quantity of a good that consumers are not only willing to purchase but also have the capacity to buy at a given price. For example, consumers may purchase 1 billion Big Macs from McDonald's every year, but perhaps only 1 million filets mignons are sold at Outback. There may be 1 billion people who want to buy the filet mignon but don't have the financial means to do so. Economists study the equilibrium between supply and demand.

Demand can be modeled with an *inverse variation* of price: when the price increases, demand decreases, and vice versa.

INVERSE VARIATION

Let x and y represent two quantities. The following are equivalent statements:

- $y = \frac{k}{x}$, where k is a nonzero constant.
- y **varies inversely** with x.
- y is **inversely proportional** to x.

The constant k is called the **constant of variation** or the **constant of proportionality**.

EXAMPLE 3 Inverse Variation

The number of potential buyers of a house decreases as the price of the house increases (see graph on the right). If the number of potential buyers of a house in a particular city is inversely proportional to the price of the house, find a mathematical equation that describes the demand for houses as it relates to price. How many potential buyers will there be for a \$2 million house?

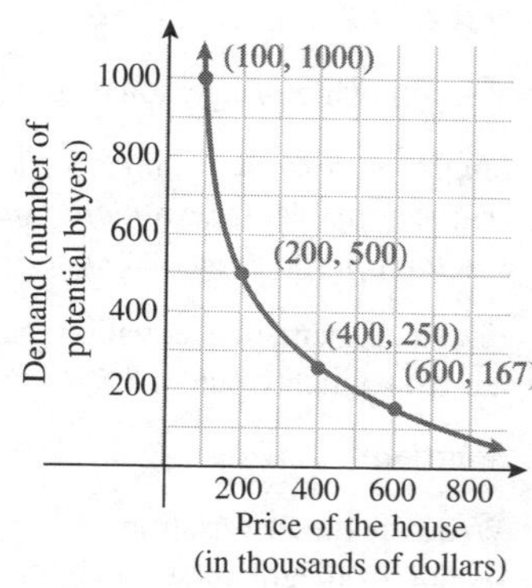

Solution:

Write the inverse variation model. $y = \frac{k}{x}$

Label the variables and constant. x = price of house in thousands of dollars
y = number of buyers

Select *any* point that lies on the curve. (200, 500)

Substitute the given data $x = 200$ and $y = 500$ into $y = \frac{k}{x}$. $500 = \frac{k}{200}$

Solve for k. $k = 200 \cdot 500 = 100{,}000$

$$y = \frac{100{,}000}{x}$$

Let $x = 2000$. $y = \frac{100{,}000}{2000} = 50$

There are only 50 potential buyers for a \$2 million house in this city.

YOUR TURN In New York City, the number of potential buyers in the housing market is inversely proportional to the price of a house. If there are 12,500 potential buyers for a \$2 million condominium, how many potential buyers are there for a \$5 million condominium?

Answer: 5000

Two quantities can vary inversely with the nth power of x.

If x and y are related by the equation $y = \frac{k}{x^n}$, then we say that y varies **inversely with the nth power of x**, or y is inversely **proportional to the nth power of x**.

Joint Variation and Combined Variation

We now discuss combinations of variations. When one quantity is proportional to the product of two or more other quantities, the variation is called **joint variation**. When direct variation and inverse variation occur at the same time, the variation is called **combined variation**.

An example of a **joint variation** is simple interest (Section 1.2), which is defined as

$$I = Prt$$

where

- I is the interest in dollars
- P is the principal (initial) dollars
- r is the interest rate (expressed in decimal form)
- t is time in years

The interest earned is proportional to the product of three quantities (principal, interest rate, and time). Note that if the interest rate increases, then the interest earned also increases.

Similarly, if either the initial investment (principal) or the time the money is invested increases, then the interest earned also increases.

An example of **combined variation** is the combined gas law in chemistry,

$$P = k\frac{T}{V}$$

where

- P is pressure
- T is temperature (kelvins)
- V is volume
- k is a gas constant

This relation states that the pressure of a gas is directly proportional to the temperature and inversely proportional to the volume containing the gas. For example, as the temperature increases, the pressure increases, but when the volume decreases, pressure increases.

As an example, the gas in the headspace of a soda bottle has a fixed volume. Therefore, as temperature increases, the pressure increases. Compare the different pressures of opening a twist-off cap on a bottle of soda that is cold versus one that is hot. The hot one feels as though it "releases more pressure."

EXAMPLE 4 Combined Variation

The gas in the headspace of a soda bottle has a volume of 9.0 ml, pressure of 2 atm (atmospheres), and a temperature of 298 K (standard room temperature of 77 °F). If the soda bottle is stored in a refrigerator, the temperature drops to approximately 279 K (42 °F). What is the pressure of the gas in the headspace once the bottle is chilled?

Solution:

Write the combined gas law. $P = k\frac{T}{V}$

Let $P = 2$ atm, $T = 298$ K, and $V = 9.0$ ml. $2 = k\frac{298}{9}$

Solve for k. $k = \frac{18}{298}$

Let $k = \frac{18}{298}$, $T = 279$, and $V = 9.0$ in $P = k\frac{T}{V}$. $P = \frac{18}{298} \cdot \frac{279}{9} \approx 1.87$

Since we used the same physical units for both the chilled and room-temperature soda bottles, the pressure is in atmospheres. $\boxed{P = 1.87 \text{ atm}}$

SECTION 3.6 SUMMARY

Direct, inverse, joint, and combined variation can be used to model the relationship between two quantities. For two quantities x and y, we say that

- y is directly proportional to x if $y = kx$.
- y is inversely proportional to x if $y = \frac{k}{x}$.

Joint variation occurs when one quantity is directly proportional to two or more quantities. Combined variation occurs when one quantity is directly proportional to one or more quantities and inversely proportional to one or more other quantities.

SECTION 3.6 EXERCISES

SKILLS

In Exercises 1–16, write an equation that describes each variation. Use *k* as the constant of variation.

1. y varies directly with x.

2. s varies directly with t.

3. V varies directly with x^3.

4. A varies directly with x^2.

5. z varies directly with m.

6. h varies directly with $\sqrt{t}$.

7. f varies inversely with λ.

8. P varies inversely with r^2.

9. F varies directly with w and inversely with L.

10. V varies directly with T and inversely with P.

11. v varies directly with both g and t.

12. S varies directly with both t and d.

13. R varies inversely with both P and T.

14. y varies inversely with both x and z.

15. y is directly proportional to the square root of x.

16. y is inversely proportional to the cube of t.

In Exercises 17–36, write an equation that describes each variation.

17. d is directly proportional to t. $d = r$ when $t = 1$.

18. F is directly proportional to m. $F = a$ when $m = 1$.

19. V is directly proportional to both l and w. $V = 6h$ when $w = 3$ and $l = 2$.

20. A is directly proportional to both b and h. $A = 10$ when $b = 5$ and $h = 4$.

21. A varies directly with the square of r. $A = 9\pi$ when $r = 3$.

22. V varies directly with the cube of r. $V = 36\pi$ when $r = 3$.

23. V varies directly with both h and r^2. $V = 1$ when $r = 2$ and $h = \frac{4}{\pi}$.

24. W is directly proportional to both R and the square of I. $W = 4$ when $R = 100$ and $I = 0.25$.

25. V varies inversely with P. $V = 1000$ when $P = 400$.

26. I varies inversely with the square of d. $I = 42$ when $d = 16$.

27. F varies inversely with both λ and L. $F = 20\pi$ when $\lambda = 1\ \mu\text{m}$ and $L = 100$ km.

28. y varies inversely with both x and z. $y = 32$ when $x = 4$ and $z = 0.05$.

29. t varies inversely with s. $t = 2.4$ when $s = 8$.

30. W varies inversely with the square of d. $W = 180$ when $d = 0.2$.

31. R varies inversely with the square of I. $R = 0.4$ when $I = 3.5$.

32. y varies inversely with both x and the square root of z. $y = 12$ when $x = 0.2$ and $z = 4$.

33. R varies directly with L and inversely with A. $r = 0.5$ when $L = 20$ and $A = 0.4$.

34. F varies directly with m and inversely with d. $F = 32$ when $m = 20$ and $d = 8$.

35. F varies directly with both m_1 and m_2 and inversely with the square of d. $F = 20$ when $m_1 = 8$, $m_2 = 16$, and $d = 0.4$.

36. w varies directly with the square root of g and inversely with the square of t. $w = 20$ when $g = 16$ and $t = 0.5$.

APPLICATIONS

37. Wages. Jason and Valerie both work at Panera Bread and have the following paycheck information for a certain week. Find an equation that shows their wages W varying directly with the number of hours worked H.

EMPLOYEE	HOURS WORKED	WAGES
Jason	23	\$172.50
Valerie	32	\$240.00

38. Sales Tax. The sales tax in Orange and Seminole Counties in Florida differs by only 0.5%. A new resident knows this but doesn't know which of the counties has the higher tax. The resident lives near the border of the counties and is in the market for a new plasma television and wants to purchase it in the county with the lower tax. If the tax on a pair of \$40 sneakers is \$2.60 in Orange County and the tax on a \$12 T-shirt is \$0.84 in Seminole County, write two equations: one for each county that describes the tax T, which is directly proportional to the purchase price P.

For Exercises 39 and 40, refer to the following:

The ratio of the speed of an object to the speed of sound determines the Mach number. Aircraft traveling at a subsonic speed (less than the speed of sound) have a Mach number less than 1. In other words, the speed of an aircraft is directly proportional to its Mach number. Aircraft traveling at a supersonic speed (greater than the speed of sound) have a Mach number greater than 1. The speed of sound at sea level is approximately 760 mph.

39. Military. The U.S. Navy Blue Angels fly F-18 Hornets that are capable of Mach 1.7. How fast can F-18 Hornets fly at sea level?

40. Military. The U.S. Air Force's newest fighter aircraft is the F-22A Raptor, which is capable of Mach 1.5. How fast can an F-22A Raptor fly at sea level?

Exercises 41 and 42 are examples of the golden ratio, or phi, a proportionality constant that appears in nature. The numerical approximate value of phi is 1.618. From www.goldenratio.net.

41. Human Anatomy. The length of your forearm F (wrist to elbow) is directly proportional to the length of your hand H(length from wrist to tip of middle finger). Write the equation that describes this relationship if the length of your forearm is 11 inches and the length of your hand is 6.8 inches.

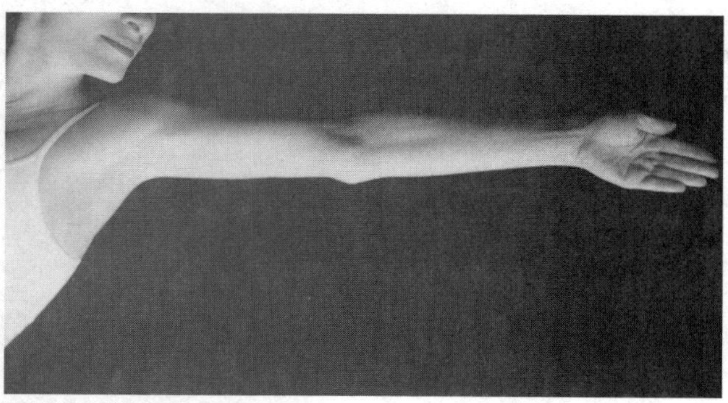

Kim Steele/Getty Images, Inc.

42. Human Anatomy. Each section of your index finger, from the tip to the base of the wrist, is larger than the preceding one by about the golden (Fibonacci) ratio. Find an equation that represents the ratio of each section of your finger related to the previous one if one section is 8 units long and the next section is 5 units long.

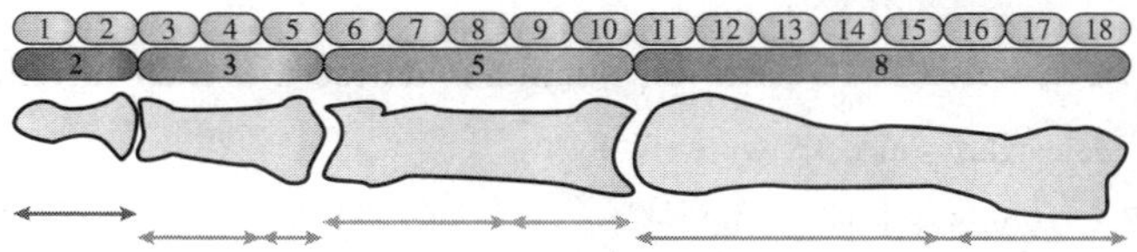

For Exercises 43 and 44, refer to the following:

Hooke's law in physics states that if a spring at rest (equilibrium position) has a weight attached to it, then the distance the spring stretches is directly proportional to the force (weight), according to the formula:

$$F = kx$$

where F is the force in Newtons (N), x is the distance stretched in meters (m), and k is the spring constant (N/m).

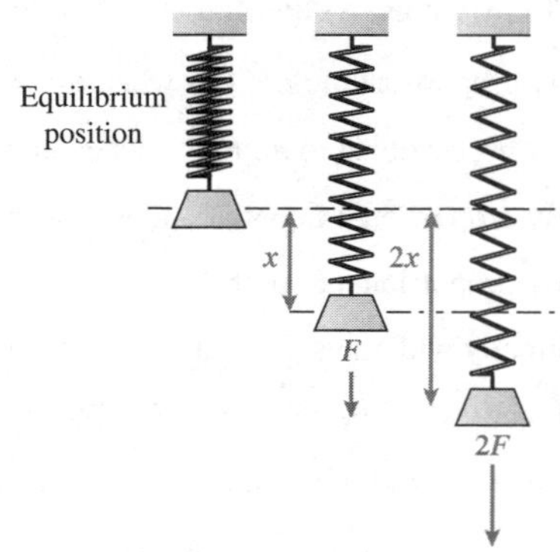

43. Physics. A force of 30 N will stretch the spring 10 cm. How far will a force of 72 N stretch the spring?

44. Physics. A force of 30 N will stretch the spring 10 cm. How much force is required to stretch the spring 18 cm?

45. Sales. Levi's makes jeans in a variety of price ranges for juniors. The Flare 519 jeans sell for about \$20, whereas the 646 Vintage Flare jeans sell for \$300. The demand for Levi's jeans is inversely proportional to the price. If 300,000 pairs of the 519 jeans were bought, approximately how many of the Vintage Flare jeans were bought?

46. Sales. Levi's makes jeans in a variety of price ranges for men. The Silver Tab Baggy jeans sell for about \$30, whereas the Offender jeans sell for about \$160. The demand for Levi's jeans is inversely proportional to the price. If 400,000 pairs of the Silver Tab Baggy jeans were bought, approximately how many of the Offender jeans were bought?

For Exercises 47 and 48, refer to the following:

In physics, the inverse square law states that any physical quantity or strength is inversely proportional to the square of the distance from the source of that physical quantity. In particular, the intensity of light radiating from a point source is inversely proportional to the square of the distance from the source. Below is a table of average distances from the Sun:

PLANET	DISTANCE TO THE SUN
Mercury	58,000 km
Earth	150,000 km
Mars	228,000 km

47. Solar Radiation. The solar radiation on the Earth is approximately 1400 watts per square meter (W/m^2). How much solar radiation is there on Mars? Round to the nearest hundred watts per square meter.

48. Solar Radiation. The solar radiation on the Earth is approximately 1400 watts per square meter. How much solar radiation is there on Mercury? Round to the nearest hundred watts per square meter.

49. Investments. Marilyn receives a $25,000 bonus from her company and decides to put the money toward a new car that she will need in 2 years. Simple interest is directly proportional to the principal and the time invested. She compares two different banks' rates on money market accounts. If she goes with Bank of America, she will earn $750 in interest, but if she goes with the Navy Federal Credit Union, she will earn $1,500. What is the interest rate on money market accounts at both banks?

50. Investments. Connie and Alvaro sell their house and buy a fixer-upper house. They made $130,000 on the sale of their previous home. They know it will take 6 months before the general contractor will start their renovation, and they want to take advantage of a 6-month CD that pays simple interest. What is the rate of the 6-month CD if they will make $3,250 in interest?

51. Chemistry. A gas contained in a 4 ml container at a temperature of 300 K has a pressure of 1 atm. If the temperature decreases to 275 K, what is the resulting pressure?

52. Chemistry. A gas contained in a 4 ml container at a temperature of 300 K has a pressure of 1 atm. If the container changes to a volume of 3 ml, what is the resulting pressure?

CATCH THE MISTAKE

In Exercises 53 and 54, explain the mistake that is made.

53. y varies directly with t and indirectly with x. When $x = 4$ and $t = 2$, then $y = 1$. Find an equation that describes this variation.

Solution:

Write the variation equation. $y = ktx$

Let $x = 4$, $t = 2$, and $y = 1$. $1 = k(2)(4)$

Solve for k. $k = \frac{1}{8}$

Substitute $k = \frac{1}{8}$ into $y = ktx$. $y = \frac{1}{8}tx$

This is incorrect. What mistake was made?

54. y varies directly with t and the square of x. When $x = 4$ and $t = 1$, then $y = 8$. Find an equation that describes this variation.

Solution:

Write the variation equation. $y = kt\sqrt{x}$

Let $x = 4$, $t = 1$, and $y = 8$. $8 = k(1)\sqrt{4}$

Solve for k. $k = 4$

Substitute $k = 4$ into $y = kt\sqrt{x}$. $y = 4t\sqrt{x}$

This is incorrect. What mistake was made?

CONCEPTUAL

In Exercises 55 and 56, determine whether each statement is true or false.

55. The area of a triangle is directly proportional to both the base and the height of the triangle (joint variation).

56. Average speed is directly proportional to both distance and time (joint variation).

In Exercises 57 and 58, match the variation with the graph.

57. Inverse variation

58. Direct variation

a.

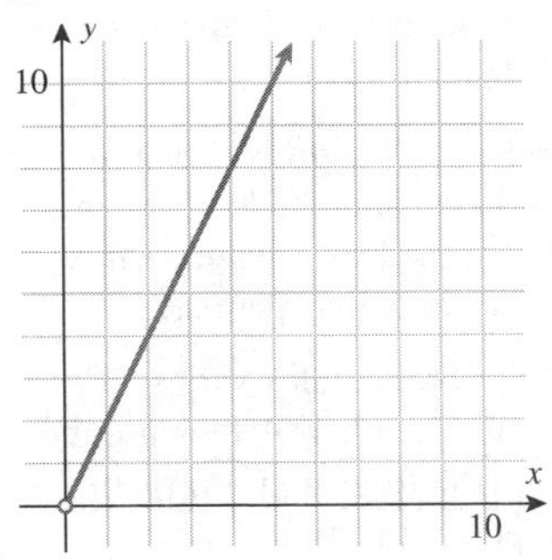

b.

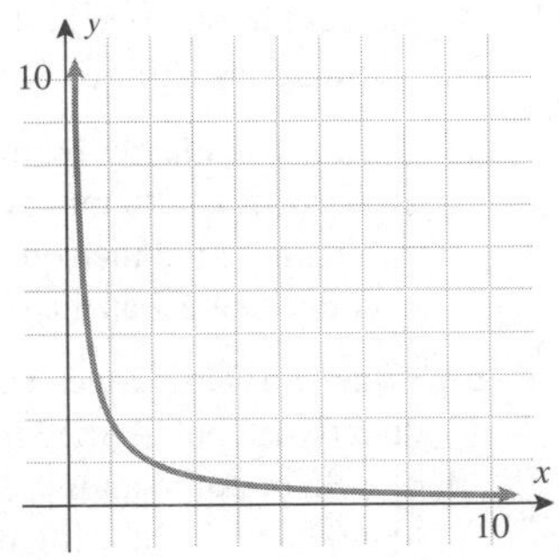

CHALLENGE

Exercises 59 and 60 involve the theory governing laser propagation through the Earth's atmosphere.

The three parameters that help classify the strength of optical turbulence are:

- C_n^2, index of refraction structure parameter
- k, wave number of the laser, which is inversely proportional to the wavelength λ of the laser:

$$k = \frac{2\pi}{\lambda}$$

- L, propagation distance

The variance of the irradiance of a laser, σ^2, is directly proportional to C_n^2, $k^{7/6}$, and $L^{11/16}$.

59. When $C_n^2 = 1.0 \times 10^{-13}\,\text{m}^{-2/3}$, $L = 2$ km, and $\lambda = 1.55\ \mu\text{m}$, the variance of irradiance for a plane wave σ_{pl}^2 is 7.1. Find the equation that describes this variation.

60. When $C_n^2 = 1.0 \times 10^{-13}\,\text{m}^{-2/3}$, $L = 2$ km, and $\lambda = 1.55\ \mu\text{m}$, the variance of irradiance for a spherical wave σ_{sp}^2 is 2.3. Find the equation that describes this variation.

TECHNOLOGY

For Exercises 61–64, refer to the following:

Data from 1995 to 2006 for oil prices in dollars per barrel, the U.S. Dow Jones Utilities Stock Index, New Privately Owned Housing, and 5-year Treasury Constant Maturity Rate are given in the table. (Data are from Forecast Center's Historical Economic and Market Home Page at www.neatideas.com/djutil.htm.)

Use the calculator STAT EDIT commands to enter the table with L_1 as the oil price, L_2 as the utilities stock index, L_3 as number of housing units, and L_4 as the 5-year maturity rate.

January of Each Year	Oil Price, \$ per Barrel	U.S. Dow Jones Utilities Stock Index	New, Privately Owned Housing Units	5-Year Treasury Constant Maturity Rate
1995	17.99	193.12	1407	7.76
1996	18.88	230.85	1467	5.36
1997	25.17	232.53	1355	6.33
1998	16.71	263.29	1525	5.42
1999	12.47	302.80	1748	4.60
2000	27.18	315.14	1636	6.58
2001	29.58	372.32	1600	4.86
2002	19.67	285.71	1698	4.34
2003	32.94	207.75	1853	3.05
2004	32.27	271.94	1911	3.12
2005	46.84	343.46	2137	3.71
2006	65.51	413.84	2265	4.35

61. An increase in oil price in dollars per barrel will drive the U.S. Dow Jones Utilities Stock Index to soar.

a. Use the calculator commands STAT, linReg $(ax + b)$, and STATPLOT to model the data using the least squares regression. Find the equation of the least squares regression line using x as the oil price in dollars per barrel.

b. If the U.S. Dow Jones Utilities Stock Index varies directly as the oil price in dollars per barrel, then use the calculator commands STAT, PwrReg, and STATPLOT to model the data using the power function. Find the variation constant and equation of variation using x as the oil price in dollars per barrel.

c. Use the equations you found in (a) and (b) to predict the stock index when the oil price hits \$72.70 per barrel in September 2006. Which answer is closer to the actual stock index of 417? Round all answers to the nearest whole number.

62. An increase in oil price in dollars per barrel will affect the interest rates across the board—in particular, the 5-year Treasury constant maturity rate.

a. Use the calculator commands STAT, linReg $(ax + b)$, and STATPLOT to model the data using the least squares regression. Find the equation of the least squares regression line using x as the oil price in dollars per barrel.

b. If the 5-year Treasury constant maturity rate varies inversely as the oil price in dollars per barrel, then use the calculator commands STAT, PwrReg, and STATPLOT to model the data using the power function. Find the variation constant and equation of variation using x as the oil price in dollars per barrel.

c. Use the equations you found in (a) and (b) to predict the maturity rate when the oil price hits \$72.70 per barrel in September 2006. Which answer is closer to the actual maturity rate at 5.02%? Round all answers to two decimal places.

63. An increase in interest rates—in particular, the 5-year Treasury constant maturity rate—will affect the number of new, privately owned housing units.

a. Use the calculator commands STAT, linReg $(ax + b)$, and STATPLOT to model the data using the least squares regression. Find the equation of the least squares regression line using x as the 5-year rate.

b. If the number of new privately owned housing units varies inversely as the 5-year Treasury constant maturityrate, then use the calculator commands STAT, PwrReg, and STATPLOT to model the data using the power function. Find the variation constant and equation of variation using x as the 5-year rate.

c. Use the equations you found in (a) and (b) to predict the number of housing units when the maturity rate is 5.02% in September 2006. Which answer is closer to the actual number of new, privately owned housing units of 1861? Round all answers to the nearest unit.

64. An increase in the number of new, privately owned housing units will affect the U.S. Dow Jones Utilities Stock Index.

a. Use the calculator commands STAT, linReg $(ax + b)$, and STATPLOT to model the data using the least squares regression. Find the equation of the least squares regression line using x as the number of housing units.

b. If the U.S. Dow Jones Utilities Stock Index varies directly as the number of new, privately owned housing units, then use the calculator commands STAT, PwrReg, and STATPLOT to model the data using the power function. Find the variation constant and equation of variation using x as the number of housing units.

c. Use the equations you found in (a) and (b) to predict the utilities stock index if there are 1861 new, privately owned housing units in September 2006. Which answer is closer to the actual stock index of 417? Round all answers to the nearest whole number.

For Exercises 65 and 66, refer to the following:

From March 2000 to March 2008, data for retail gasoline price in dollars per gallon are given in the table below. (Data are from Energy Information Administration, Official Energy Statistics from the U.S. Government at http://tonto.eia.doe.gov/oog/info/gdu/gaspump.html.) Use the calculator STAT EDIT command to enter the table below with L_1 as the year ($x - 1$ for year 2000) and L_2 as the gasoline price in dollars per gallon.

MARCH OF EACH YEAR	2000	2001	2002	2003	2004	2005	2006	2007	2008
RETAIL GASOLINE PRICE \$ PER GALLON	1.517	1.409	1.249	1.693	1.736	2.079	2.425	2.563	3.244

65. a. Use the calculator commands STAT LinReg to model the data using the least squares regression. Find the equation of the least squares regression line using x as the year ($x = 1$ for year 2000) and y as the gasoline price in dollars per gallon. Round all answers to three decimal places.

b. Use the equation to predict the gasoline price in March 2006. Round all answers to three decimal places. Is the answer close to the actual price?

c. Use the equation to predict the gasoline price in March 2009. Round all answers to three decimal places.

66. a. Use the calculator commands STAT PwrReg to model the data using the power function. Find the variation constant and equation of variation using x as the year ($x = 1$ for year 2000) and y as the gasoline price in dollars per gallon. Round all answers to three decimal places.

b. Use the equation to predict the gasoline price in March 2006. Round all answers to three decimal places. Is the answer close to the actual price?

c. Use the equation to predict the gasoline price in March 2009. Round all answers to three decimal places.

MODELING YOUR WORLD

The U.S. National Oceanic and Atmospheric Association (NOAA) monitors temperature and carbon emissions at its observatory in Mauna Loa, Hawaii. NOAA's goal is to help foster an informed society that uses a comprehensive understanding of the role of the oceans, coasts, and atmosphere in the global ecosystem to make the best social and economic decisions. The data presented in this chapter is from the Mauna Loa Observatory, where historical atmospheric measurements have been recorded for the last 50 years. You will develop linear models based on this data to predict temperature and carbon emissions in the future.

The following table summarizes average yearly temperature in degrees Fahrenheit °F and carbon dioxide emissions in parts per million (ppm) for **Mauna Loa, Hawaii.**

Year	1960	1965	1970	1975	1980	1985	1990	1995	2000	2005
Temperature (°F)	44.45	43.29	43.61	43.35	46.66	45.71	45.53	47.53	45.86	46.23
CO_2 Emissions (ppm)	316.9	320.0	325.7	331.1	338.7	345.9	354.2	360.6	369.4	379.7

1. Plot the temperature data with time on the horizontal axis and temperature on the vertical axis. Let $t = 0$ correspond to 1960.
2. Find a *linear function* that models the temperature in Mauna Loa.
   a. Use data from 1965 and 1995.
   b. Use data from 1960 and 1990.
   c. Use linear regression and all data given.
3. Predict what the temperature will be in Mauna Loa in 2020.
   a. Apply the line found in Exercise 2(a).
   b. Apply the line found in Exercise 2(b).
   c. Apply the line found in Exercise 2(c).
4. Predict what the temperature will be in Mauna Loa in 2100.
   a. Apply the line found in Exercise 2(a).
   b. Apply the line found in Exercise 2(b).
   c. Apply the line found in Exercise 2(c).
5. Do you think your models support the claim of "global warming"? Explain.
6. Plot the carbon dioxide emissions data with time on the horizontal axis and carbon dioxide levels on the vertical axis. Let $t = 0$ correspond to 1960.
7. Find a *linear function* that models the CO_2 emissions (ppm) in Mauna Loa.
   a. Use data from 1965 and 1995.
   b. Use data from 1960 and 1990.
   c. Use linear regression and all data given.
8. Predict the expected CO_2 levels in Mauna Loa in 2020.
   a. Apply the line found in Exercise 7(a).
   b. Apply the line found in Exercise 7(b).
   c. Apply the line found in Exercise 7(c).
9. Predict the expected CO_2 levels in Mauna Loa in 2100.
   a. Apply the line found in Exercise 7(a).
   b. Apply the line found in Exercise 7(b).
   c. Apply the line found in Exercise 7(c).
10. Do you think your models support the claim of the "greenhouse effect"? Explain.

SECTION	CONCEPT	PAGES	REVIEW EXERCISES	KEY IDEAS/POINT
3.1	**Functions**	226–238	1–30	
	Functions defined by equations	228–230	5–8	A vertical line can intersect a function in at most one point.
	Function notation	231–235	11–22	Placeholder notation j Difference quotient: $\frac{f(x+h)-f(x)}{h}$; $h \neq 0$
	Domain of a function	235–238	23–28	Are there any restrictions on x?
3.2	**Graphs of functions; piecewise-defined functions; increasing and decreasing functions; average rate of change**	244–256	31–48	
	Common functions	244–247		$f(x) = mx + b$, $f(x) = x$, $f(x) = x^2$, $f(x) = x^3$, $f(x) = \sqrt{x}$, $f(x) = \sqrt[3]{x}$, $f(x) = \lvert x \rvert$, $f(x) = \frac{1}{x}$
	Even and odd functions	247–248	31–38	$f(-x) = f(x)$ Symmetry about y-axis $f(-x) = -f(x)$ Symmetry about origin
	Increasing and decreasing functions	249–250	39–40	• Increasing: rises (left to right) • Decreasing: falls (left to right)
	Average rate of change	250–253	41–42	$\frac{f(x_2) - f(x_1)}{x_2 - x_1}$ $x_1 \neq x_2$
	Piecewise-defined functions	253–256	43–46	Points of discontinuity
3.3	**Graphing techniques: Transformations**	264–273	49–66	Shift the graph of $f(x)$.
	Horizontal and vertical shifts	264–268	49–66	$f(x + c)$ c units to the left $f(x - c)$ c units to the right $f(x) + c$ c units upward $f(x) - c$ c units downward
	Reflection about the axes	268–270	49–66	$-f(x)$ Reflection about the x-axis $f(-x)$ Reflection about the y-axis
	Stretching and compressing	271–273	49–66	$cf(x)$ if $c > 1$; stretch vertically $cf(x)$ if $0 < c < 1$; compress vertically $f(cx)$ if $c > 1$; compress horizontally $f(cx)$ if $0 < c < 1$; stretch horizontally

Section	Concept	Pages	Review Exercises	Key Ideas/Point
3.4	**Operations on functions and composition of functions**	279–285	67–90	
	Adding, subtracting, multiplying, and dividing functions	279–281	67–72	$(f + g)(x) = f(x) + g(x)$ $(f - g)(x) = f(x) - g(x)$ $(f \cdot g)(x) = f(x) \cdot g(x)$ Domain of the resulting function is the intersection of the individual domains. $\left(\frac{f}{g}\right)(x) = \frac{f(x)}{g(x)},\ g(x) \neq 0$ Domain of the quotient is the intersection of the domains of f and g, and any points when $g(x) = 0$ must be eliminated.
	Composition of functions	281–285	73–88	$(f \circ g)(x) = f(g(x))$ The domain of the composite function is a subset of the domain of $g(x)$. Values for x must be eliminated if their corresponding values $g(x)$ are not in the domain of f.
3.5	**One-to-one functions and inverse functions**	289–298	91–112	
	Determine whether a function is one-to-one	289–292	91–100	• No two x-values map to the same y-value. If $f(x_1) = f(x_2)$, then $x_1 = x_2$. • A horizontal line may intersect a one-to-one function in at most one point.
	Inverse functions	292–294	101–110	• Only one-to-one functions have inverses. • $f^{-1}(f(x)) = x$ and $f(f^{-1}(x)) = x$. • Domain of f = range of f^{-1}. Range of f = domain of f^{-1}.
	Graphical interpretation of inverse functions	294–295	101–104	• The graph of a function and its inverse are symmetric about the line $y = x$. • If the point (a, b) lies on the graph of a function, then the point (b, a) lies on the graph of its inverse.
	Finding the inverse function	295–298	105–112	1. Let $y = f(x)$. 2. Interchange x and y. 3. Solve for y. 4. Let $y = f^{-1}(x)$.
3.6	**Modeling functions using variation**	304–308	113–118	
	Direct variation	304–306	113–115	$y = kx$
	Inverse variation	306–307	116	$y = \frac{k}{x}$
	Joint variation and combined variation	307–308	117–118	Joint: One quantity is directly proportional to the product of two or more other quantities. Combined: Direct variation and inverse variation occur at the same time.

3.1 Functions

Determine whether each relation is a function.

1.

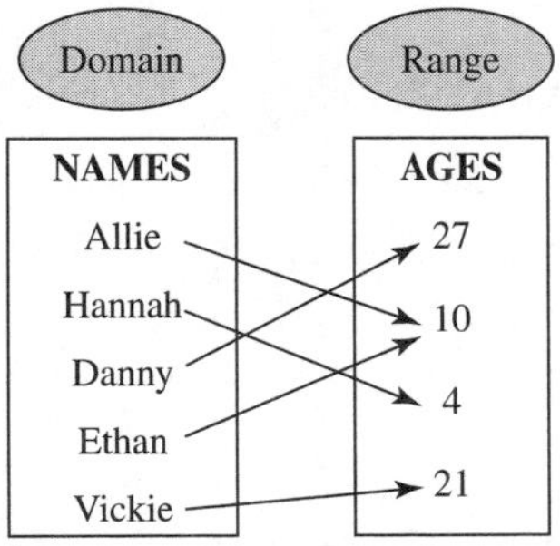

2. $\{(1, 2), (3, 4), (2, 4), (3, 7)\}$

3. $\{(-2, 3), (1, -3), (0, 4), (2, 6)\}$

4. $\{(4, 7), (2, 6), (3, 8), (1, 7)\}$

5. $x^2 + y^2 = 36$

6. $x = 4$

7. $y = |x + 2|$

8. $y = \sqrt{x}$

9.

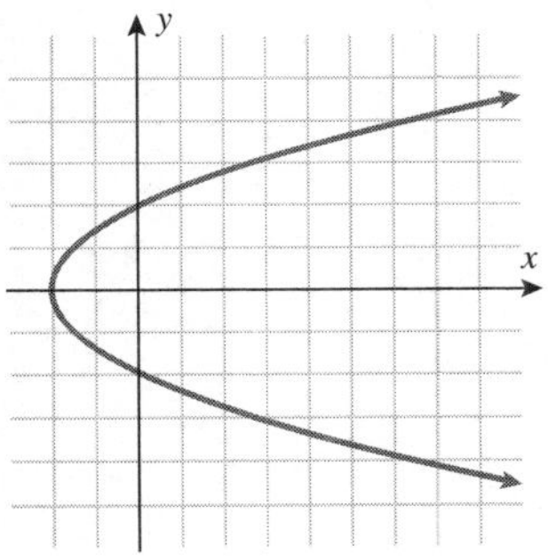

10.

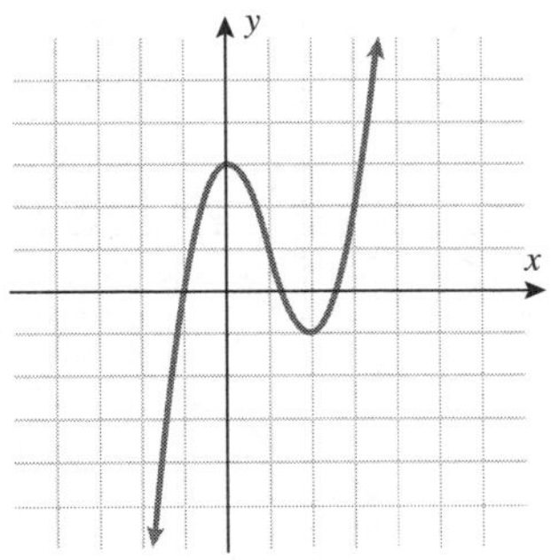

Use the graphs of the functions to find:

11.

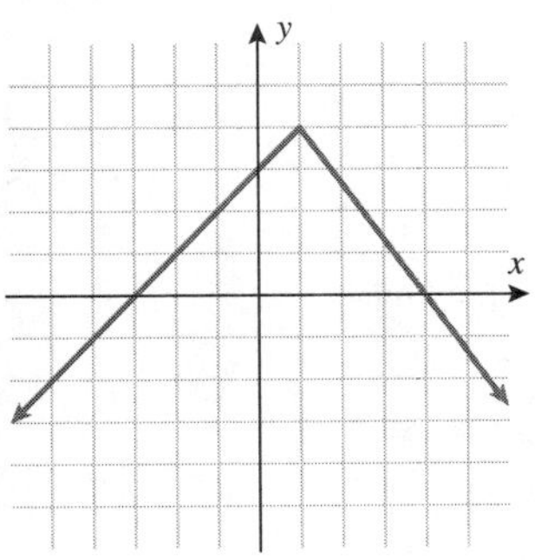

a. $f(-1)$ **b.** $f(1)$

c. x, where $f(x) = 0$

12.

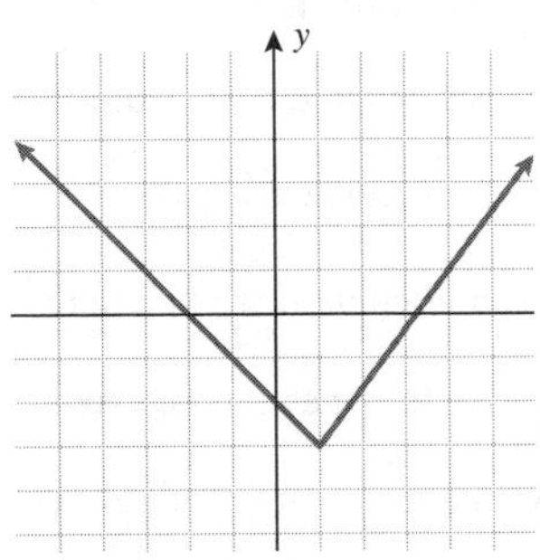

a. $f(-4)$ **b.** $f(0)$

c. x, where $f(x) = 0$

13.

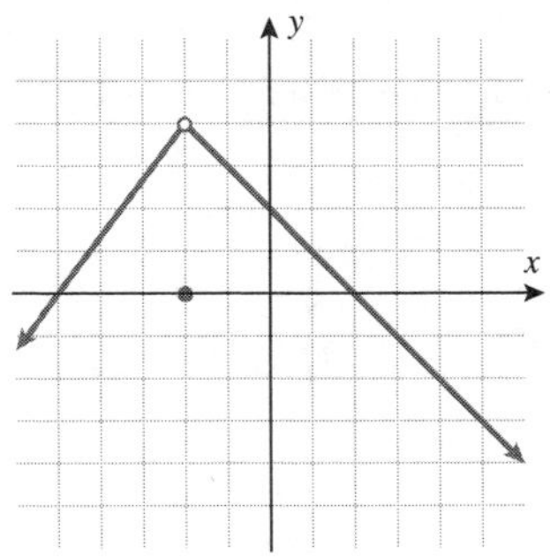

a. $f(-2)$ **b.** $f(4)$

c. x, where $f(x) = 0$

14.

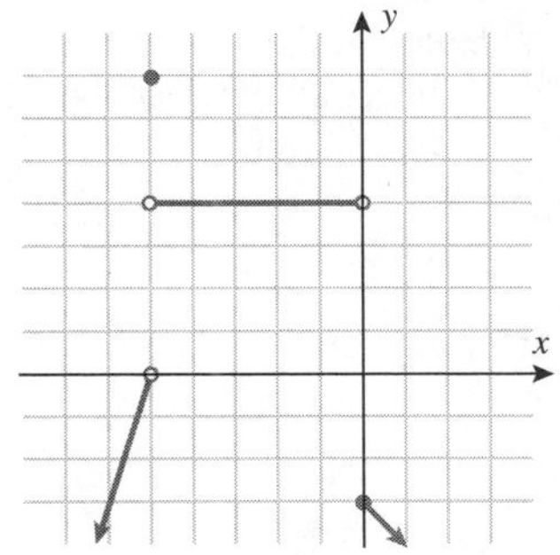

a. $f(-5)$ **b.** $f(0)$

c. x, where $f(x) = 0$

Evaluate the given quantities using the following three functions.

$f(x) = 4x - 7$ $\quad F(t) = t^2 + 4t - 3$ $\quad g(x) = |x^2 + 2x + 4|$

15. $f(3)$

16. $F(4)$

17. $f(-7) \cdot g(3)$

18. $\dfrac{F(0)}{g(0)}$

19. $\dfrac{f(2) - F(2)}{g(0)}$

20. $f(3 + h)$

21. $\dfrac{f(3 + h) - f(3)}{h}$

22. $\dfrac{F(t + h) - F(t)}{h}$

Find the domain of the given function. Express the domain in interval notation.

23. $f(x) = -3x - 4$

24. $g(x) = x^2 - 2x + 6$

25. $h(x) = \dfrac{1}{x + 4}$

26. $F(x) = \dfrac{7}{x^2 + 3}$

27. $G(x) = \sqrt{x - 4}$

28. $H(x) = \dfrac{1}{\sqrt{2x - 6}}$

Challenge

29. If $f(x) = \dfrac{D}{x^2 - 16}$, $f(4)$ and $f(-4)$ are undefined, and $f(5) = 2$, find D.

30. Construct a function that is undefined at $x = -3$ and $x = 2$ such that the point $(0, -4)$ lies on the graph of the function.

3.2 Graphs of Functions

Determine whether the function is even, odd, or neither.

31. $f(x) = 2x - 7$

32. $g(x) = 7x^5 + 4x^3 - 2x$

33. $h(x) = x^3 - 7x$

34. $f(x) = x^4 + 3x^2$

35. $f(x) = x^{1/4} + x$

36. $f(x) = \sqrt{x + 4}$

37. $f(x) = \dfrac{1}{x^3} + 3x$

38. $f(x) = \dfrac{1}{x^2} + 3x^4 + |x|$

Use the graph of the functions to find:

a. Domain
b. Range
c. Intervals on which the function is increasing, decreasing, or constant.

39.

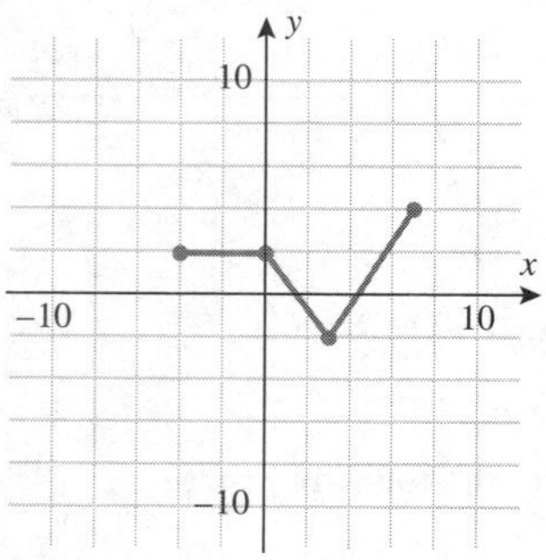

40.

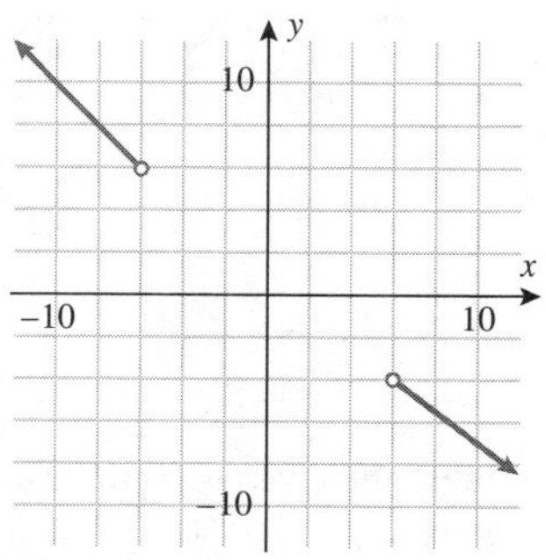

41. Find the average rate of change of $f(x) = 4 - x^2$ from $x = 0$ to $x = 2$.

42. Find the average rate of change of $f(x) = |2x - 1|$ from $x = 1$ to $x = 5$.

Graph the piecewise-defined function. State the domain and range in interval notation.

43. $F(x) = \begin{cases} x^2 & x < 0 \\ 2 & x \geq 0 \end{cases}$

44. $f(x) = \begin{cases} -2x - 3 & x \leq 0 \\ 4 & 0 < x \leq 1 \\ x^2 + 4 & x > 1 \end{cases}$

45. $f(x) = \begin{cases} x^2 & x \leq 0 \\ -\sqrt{x} & 0 < x \leq 1 \\ |x + 2| & x > 1 \end{cases}$

46. $F(x) = \begin{cases} x^2 & x < 0 \\ x^3 & 0 < x < 1 \\ -|x| - 1 & x \geq 1 \end{cases}$

Applications

47. Tutoring Costs. A tutoring company charges \$25 for the first hour of tutoring and \$10.50 for every 30-minute period after that. Find the cost function $C(x)$ as a function of the length of the tutoring session. Let x = number of 30-minute periods.

48. Salary. An employee who makes \$30 per hour also earns time and a half for overtime (any hours worked above the normal 40-hour work week). Write a function $E(x)$ that describes her weekly earnings as a function of the number of hours worked x.

3.3 Graphing Techniques: Transformations

Graph the following functions using graphing aids.

49. $y = -(x - 2)^2 + 4$

50. $y = |-x + 5| - 7$

51. $y = \sqrt[3]{x - 3} + 2$

52. $y = \dfrac{1}{x - 2} - 4$

53. $y = -\frac{1}{2}x^3$

54. $y = 2x^2 + 3$

Use the given graph to graph the following:

55.

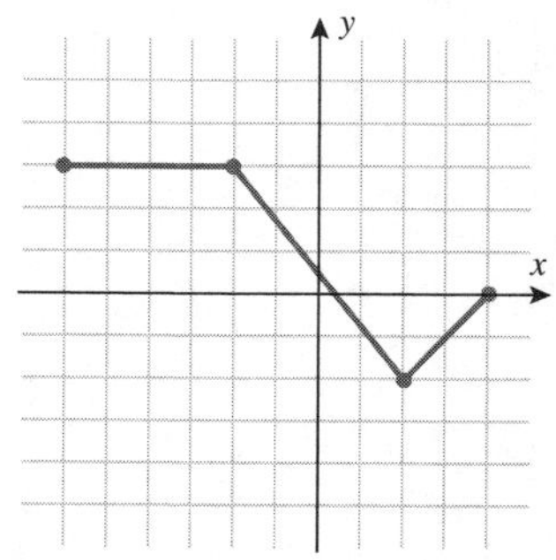

$y = f(x - 2)$

56.

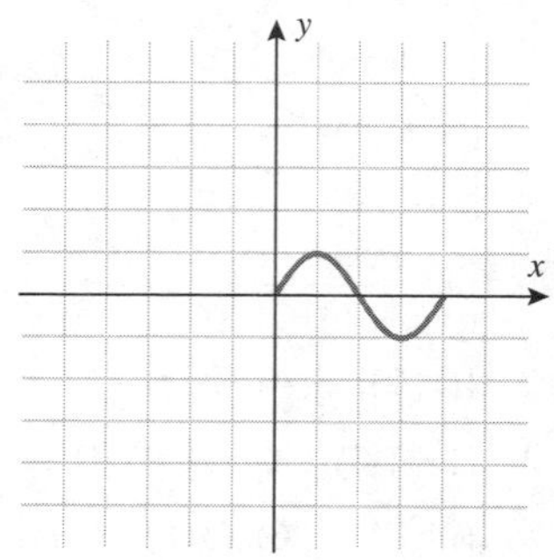

$y = 3f(x)$

57.

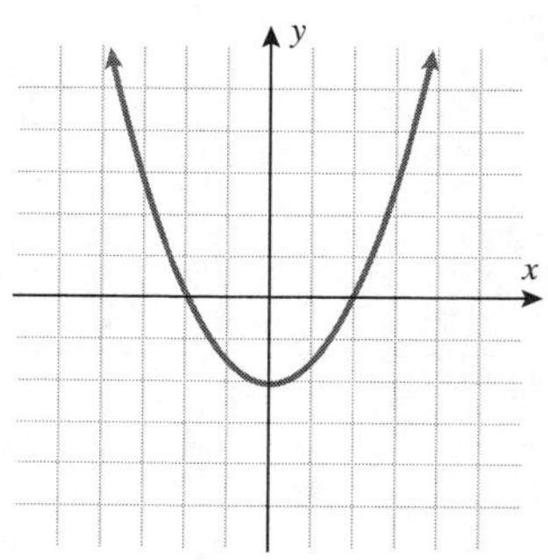

$y = -2f(x)$

58.

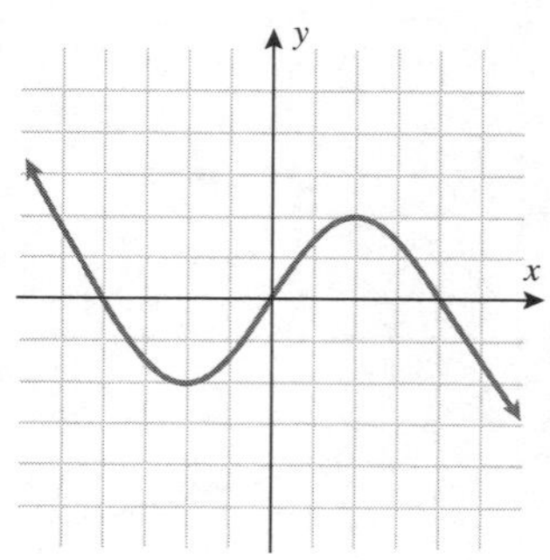

$y = f(x) + 3$

Write the function whose graph is the graph of $y = \sqrt{x}$, but is transformed accordingly, and state the domain of the resulting function.

59. Shifted to the left 3 units

60. Shifted down 4 units

61. Shifted to the right 2 units and up 3 units

62. Reflected about the y-axis

63. Stretched by a factor of 5 and shifted down 6 units

64. Compressed by a factor of 2 and shifted up 3 units

Transform the function into the form $f(x) = c(x - h)^2 + k$ by completing the square and graph the resulting function using transformations.

65. $y = x^2 + 4x - 8$

66. $y = 2x^2 + 6x - 5$

3.4 Operations on Functions and Composition of Functions

Given the functions g and h, find $g + h, g - h, g \cdot h$, and $\frac{g}{h}$, and state the domain.

67. $g(x) = -3x - 4$
$h(x) = x - 3$

68. $g(x) = 2x + 3$
$h(x) = x^2 + 6$

69. $g(x) = \dfrac{1}{x^2}$
$h(x) = \sqrt{x}$

70. $g(x) = \dfrac{x + 3}{2x - 4}$
$h(x) = \dfrac{3x - 1}{x - 2}$

71. $g(x) = \sqrt{x - 4}$
$h(x) = \sqrt{2x + 1}$

72. $g(x) = x^2 - 4$
$h(x) = x + 2$

For the given functions f and g, find the composite functions $f \circ g$ and $g \circ f$, and state the domains.

73. $f(x) = 3x - 4$
$g(x) = 2x + 1$

74. $f(x) = x^3 + 2x - 1$
$g(x) = x + 3$

75. $f(x) = \dfrac{2}{x + 3}$
$g(x) = \dfrac{1}{4 - x}$

76. $f(x) = \sqrt{2x^2 - 5}$
$g(x) = \sqrt{x + 6}$

77. $f(x) = \sqrt{x - 5}$
$g(x) = x^2 - 4$

78. $f(x) = \dfrac{1}{\sqrt{x}}$
$g(x) = \dfrac{1}{x^2 - 4}$

Evaluate $f(g(3))$ and $g(f(-1))$, if possible.

79. $f(x) = 4x^2 - 3x + 2$
$g(x) = 6x - 3$

80. $f(x) = \sqrt{4 - x}$
$g(x) = x^2 + 5$

81. $f(x) = \dfrac{x}{|2x - 3|}$
$g(x) = |5x + 2|$

82. $f(x) = \dfrac{1}{x - 1}$
$g(x) = x^2 - 1$

83. $f(x) = x^2 - x + 10$
$g(x) = \sqrt[3]{x - 4}$

84. $f(x) = \dfrac{4}{x^2 - 2}$
$g(x) = \dfrac{1}{x^2 - 9}$

Write the function as a composite $f(g(x))$ of two functions f and g.

85. $h(x) = 3(x - 2)^2 + 4(x - 2) + 7$

86. $h(x) = \dfrac{\sqrt[3]{x}}{1 - \sqrt[3]{x}}$

87. $h(x) = \dfrac{1}{\sqrt{x^2 + 7}}$

88. $h(x) = \sqrt{|3x + 4|}$

Applications

89. Rain. A rain drop hitting a lake makes a circular ripple. If the radius, in inches, grows as a function of time, in minutes, $r(t) = 25\sqrt{t + 2}$, find the area of the ripple as a function of time.

90. Geometry. Let the area of a rectangle be given by $42 = l \cdot w$, and let the perimeter be $36 = 2 \cdot l + 2 \cdot w$. Express the perimeter in terms of w.

3.5 One-to-One Functions and Inverse Functions

Determine whether the given function is a one-to-one function.

91.

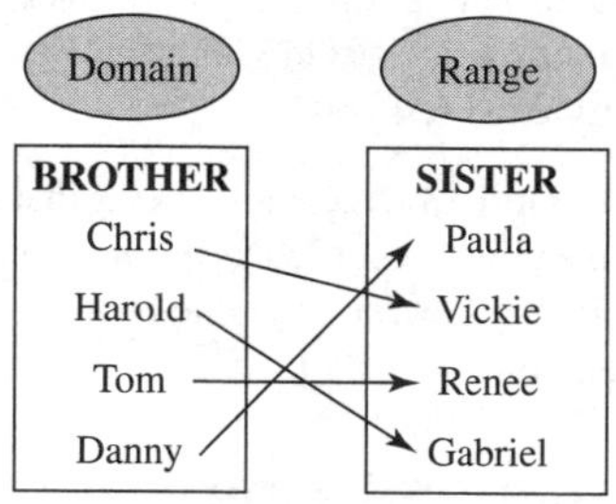

92.

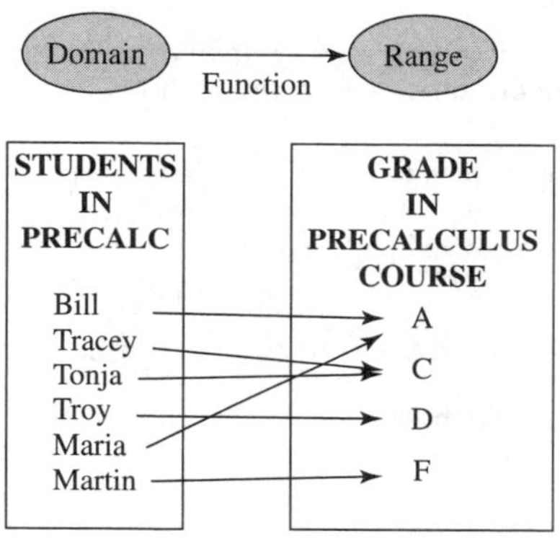

93. $\{(2, 3), (-1, 2), (3, 3), (-3, -4), (-2, 1)\}$

94. $\{(-3, 9), (5, 25), (2, 4), (3, 9)\}$

95. $\{(-2, 0), (4, 5), (3, 7)\}$

96. $\{(-8, -6), (-4, 2), (0, 3), (2, -8), (7, 4)\}$

97. $y = \sqrt{x}$ **98.** $y = x^2$ **99.** $f(x) = x^3$ **100.** $f(x) = \dfrac{1}{x^2}$

Verify that the function $f^{-1}(x)$ is the inverse of $f(x)$ by showing that $f(f^{-1}(x)) = x$. Graph $f(x)$ and $f^{-1}(x)$ on the same graph and show the symmetry about the line $y = x$.

101. $f(x) = 3x + 4; f^{-1}(x) = \dfrac{x - 4}{3}$

102. $f(x) = \dfrac{1}{4x - 7}; f^{-1}(x) = \dfrac{1 + 7x}{4x}$

103. $f(x) = \sqrt{x + 4}; f^{-1}(x) = x^2 - 4 \quad x \geq 0$

104. $f(x) = \dfrac{x + 2}{x - 7}; f^{-1}(x) = \dfrac{7x + 2}{x - 1}$

The function f is one-to-one. Find its inverse and check your answer. State the domain and range of both f and f^{-1}.

105. $f(x) = 2x + 1$

106. $f(x) = x^5 + 2$

107. $f(x) = \sqrt{x + 4}$

108. $f(x) = (x + 4)^2 + 3 \quad x \geq -4$

109. $f(x) = \dfrac{x + 6}{x + 3}$

110. $f(x) = 2\sqrt[3]{x - 5} - 8$

Applications

111. Salary. A pharmaceutical salesperson makes \$22,000 base salary a year plus 8% of the total products sold. Write a function $S(x)$ that represents her yearly salary as a function of the total dollars worth of products sold x. Find $S^{-1}(x)$. What does this inverse function tell you?

112. Volume. Express the volume V of a rectangular box that has a square base of length s and is 3 feet high as a function of the square length. Find V^{-1}. If a certain volume is desired, what does the inverse tell you?

3.6 Modeling Functions Using Variation

Write an equation that describes each variation.

113. C is directly proportional to r. $C = 2\pi$ when $r = 1$.

114. V is directly proportional to both l and w. $V = 12h$ when $w = 6$ and $l = 2$.

115. A varies directly with the square of r. $A = 25\pi$ when $r = 5$.

116. F varies inversely with both λ and L. $F = 20\pi$ when $\lambda = 10\ \mu$m and $L = 10$ km.

Applications

117. Wages. Cole and Dickson both work at the same museum and have the following paycheck information for a certain week. Find an equation that shows their wages (W) varying directly with the number of hours (H) worked.

Employee	Hours Worked	Wage
Cole	27	\$229.50
Dickson	30	\$255.00

118. Sales Tax. The sales tax in two neighboring counties differs by 1%. A new resident knows the difference but doesn't know which county has the higher tax rate. The resident lives near the border of the two counties and wants to buy a new car. If the tax on a \$50 jacket is \$3.50 in County A and the tax on a \$20 calculator is \$1.60 in County B, write two equations (one for each county) that describe the tax (T), which is directly proportional to the purchase price (P).

Technology Exercises

Section 3.1

119. Use a graphing utility to graph the function and find the domain. Express the domain in interval notation.

$$f(x) = \frac{1}{\sqrt{x^2 - 2x - 3}}$$

120. Use a graphing utility to graph the function and find the domain. Express the domain in interval notation.

$$f(x) = \frac{x^2 - 4x - 5}{x^2 - 9}$$

Section 3.2

121. Use a graphing utility to graph the function. State the (a) domain, (b) range, and (c) x intervals where the function is increasing, decreasing, and constant.

$$f(x) = \begin{cases} 1 - x & x < -1 \\ [[x]] & -1 \leq x < 2 \\ x + 1 & x > 2 \end{cases}$$

122. Use a graphing utility to graph the function. State the (a) domain, (b) range, and (c) x intervals where the function is increasing, decreasing, and constant.

$$f(x) = \begin{cases} |x^2 - 1| & -2 < x < 2 \\ \sqrt{x - 2} + 4 & x > 2 \end{cases}$$

Section 3.3

123. Use a graphing utility to graph $f(x) = x^2 - x - 6$ and $g(x) = x^2 - 5x$. Use transforms to describe the relationship between $f(x)$ and $g(x)$?

124. Use a graphing utility to graph $f(x) = 2x^2 - 3x - 5$ and $g(x) = -2x^2 - x + 6$. Use transforms to describe the relationship between $f(x)$ and $g(x)$?

Section 3.4

125. Using a graphing utility, plot $y_1 = \sqrt{2x + 3}$, $y_2 = \sqrt{4 - x}$, and $y_3 = \dfrac{y_1}{y_2}$. What is the domain of y_3?

126. Using a graphing utility, plot $y_1 = \sqrt{x^2 - 4}$, $y_2 = x^2 - 5$, and $y_3 = y_1^2 - 5$. If y_1 represents a function f and y_2 represents a function g, then y_3 represents the composite function $g \circ f$. The graph of y_3 is only defined for the domain of $g \circ f$. State the domain of $g \circ f$.

Section 3.5

127. Use a graphing utility to graph the function and determine if it is one-to-one.

$$f(x) = \frac{6}{\sqrt[5]{x^3} - 1}$$

128. Use a graphing utility to graph the functions f and g and the line $y = x$ in the same screen. Are the two functions inverses of each other?

$$f(x) = \sqrt[4]{x - 3} + 1, \; g(x) = x^4 - 4x^3 + 6x^2 - 4x + 3$$

Section 3.6

From December 1999 to December 2007, data for gold price in dollars per ounce are given in the table below. (Data are from Finfacts Ireland Business & Finance Portal, Ireland's Top Business website at www.finfacts.ie/Private/curency/goldmarketprice.htm.) Use the calculator STAT EDIT commands to enter the table below with L_1 as the year ($x = 1$ for year 1999) and L_2 as the gold price in dollars per ounce.

December of Each Year	Gold Price in $ per Ounce
1999	287.90
2000	272.15
2001	278.70
2002	346.70
2003	414.80
2004	438.10
2005	517.20
2006	636.30
2007	833.20

129. **a.** Use the calculator commands STAT LinReg to model the data using the least squares regression. Find the equation of the least squares regression line using x as the year ($x = 1$ for year 1999) and y as the gold price in dollars per ounce. Round all answers to two decimal places.

b. Use the equation to predict the gold price in December 2005. Round all answers to two decimal places. Is the answer close to the actual price?

c. Use the equation to predict the gold price in December 2008. Round all answers to two decimal places.

130. **a.** Use the calculator commands STAT PwrReg to model the data using the power function. Find the variation constant and equation of variation using x as the year ($x = 1$ for year 1991) and y as the gold price in dollars per ounce. Round all answers to two decimal places.

b. Use the equation to predict the gold price in December 2005. Round all answers to two decimal places. Is the answer close to the actual price?

c. Use the equation to predict the gold price in December 2008. Round all answers to two decimal places.

CHAPTER 3 PRACTICE TEST

Assuming that x represents the independent variable and y represents the dependent variable, classify the relationships as:

a. not a function
b. a function, but not one-to-one
c. a one-to-one function

1. $f(x) = |2x + 3|$ **2.** $x = y^2 + 2$ **3.** $y = \sqrt[3]{x + 1}$

Use $f(x) = \sqrt{x - 2}$ and $g(x) = x^2 + 11$, and determine the desired quantity or expression. In the case of an expression, state the domain.

4. $f(11) - 2g(-1)$

5. $\left(\frac{f}{g}\right)(x)$

6. $\left(\frac{g}{f}\right)(x)$

7. $g(f(x))$

8. $(f + g)(6)$

9. $f(g(\sqrt{7}))$

Determine whether the function is odd, even, or neither.

10. $f(x) = |x| - x^2$ **11.** $f(x) = 9x^3 + 5x - 3$ **12.** $f(x) = \frac{2}{x}$

Graph the functions. State the domain and range of each function.

13. $f(x) = -\sqrt{x - 3} + 2$

14. $f(x) = -2(x - 1)^2$

15. $f(x) = \begin{cases} -x & x < -1 \\ 1 & -1 < x < 2 \\ x^2 & x \geq 2 \end{cases}$

Use the graphs of the function to find:

16.

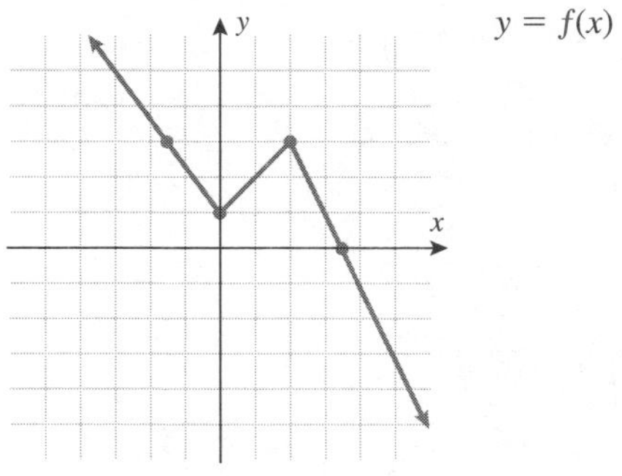

a. $f(3)$ **b.** $f(0)$ **c.** $f(-4)$
d. x, where $f(x) = 3$ **e.** x, where $f(x) = 0$

PRACTICE TEST

17. $y = g(x)$

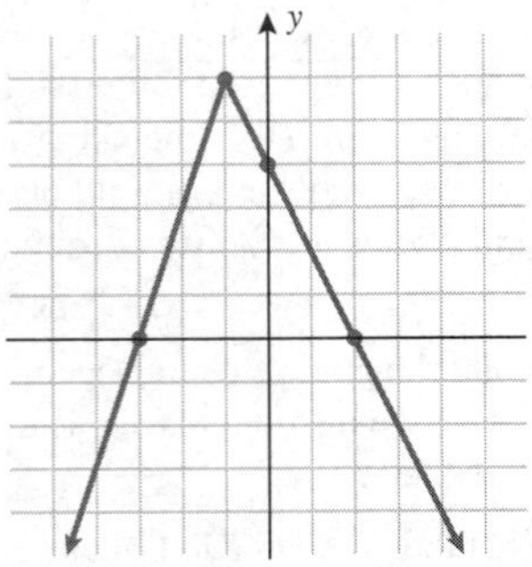

a. $g(3)$ **b.** $g(0)$ **c.** $g(-4)$
d. x, where $g(x) = 0$

18. $y = f(x)$

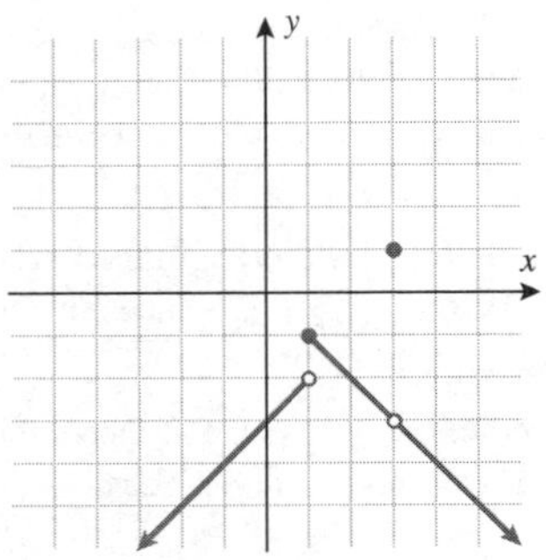

a. $p(0)$ **b.** x, where $p(x) = 0$
c. $p(1)$ **d.** $p(3)$

Find $\dfrac{f(x+h) - f(x)}{h}$ **for:**

19. $f(x) = 3x^2 - 4x + 1$ **20.** $f(x) = 5 - 7x$

Find the average rate of change of the given functions.

21. $f(x) = 64 - 16x^2$ for $x = 0$ to $x = 2$

22. $f(x) = \sqrt{x - 1}$ for $x = 2$ to $x = 10$

Given the function f, find the inverse if it exists. State the domain and range of both f and f^{-1}.

23. $f(x) = \sqrt{x - 5}$ **24.** $f(x) = x^2 + 5$

25. $f(x) = \dfrac{2x + 1}{5 - x}$ **26.** $f(x) = \begin{cases} -x & x \le 0 \\ -x^2 & x > 0 \end{cases}$

27. What domain restriction can be made so that $f(x) = x^2$ has an inverse?

28. If the point $(-2, 5)$ lies on the graph of a function, what point lies on the graph of its inverse function?

29. Discount. Suppose a suit has been marked down 40% off the original price. An advertisement in the newspaper has an "additional 30% off the sale price" coupon. Write a function that determines the "checkout" price of the suit.

30. Temperature. Degrees Fahrenheit (°F), degrees Celsius (°C), and kelvins (K) are related by the two equations: $F = \frac{9}{5}C + 32$ and $K = C + 273.15$. Write a function whose input is kelvins and output is degrees Fahrenheit.

31. Circles. If a quarter circle is drawn by tracing the unit circle in quadrant III, what does the inverse of that function look like? Where is it located?

32. Sprinkler. A sprinkler head malfunctions at midfield in an NFL football field. The puddle of water forms a circular pattern around the sprinkler head with a radius in yards that grows as a function of time, in hours: $r(t) = 10\sqrt{t}$. When will the puddle reach the sidelines? (A football field is 30 yards from sideline to sideline.)

33. Internet. The cost of airport Internet access is $15 for the first 30 minutes and $1 per minute for each minute after that. Write a function describing the cost of the service as a function of minutes used.

Use variation to find a model for the given problem.

34. y varies directly with the square of x. $y = 8$ when $x = 5$

35. F varies directly with m and inversely with p. $F = 20$ when $m = 2$ and $p = 3$

36. Use a graphing utility to graph the function. State the (a) domain, (b) range, and (c) x intervals where the function is increasing, decreasing, and constant.

$$f(x) = \begin{cases} 5 & -4 \le x < -2 \\ |3 + 2x - x^2| & -2 < x \le 4 \end{cases}$$

37. Use a graphing utility to graph the function and determine if it is one-to-one.

$$y = x^3 - 12x^2 + 48x - 65$$

CHAPTERS 1–3 CUMULATIVE TEST

1. Simplify $\dfrac{2}{3 - \sqrt{5}}$.

2. Factor completely: $10x^2 - 29x - 21$.

3. Simplify and state the domain: $\dfrac{x^3 - 4x}{x + 2}$.

4. Solve for x: $\frac{1}{6}x = -\frac{1}{5}x + 11$.

5. Perform the operation, simplify, and express in standard form: $(8 - 9i)(8 + 9i)$.

6. Solve for x, and give any excluded values: $\dfrac{5}{x} - 10 = \dfrac{10}{3x}$.

7. The original price of a hiking stick is \$59.50. The sale price is \$35.70. Find the percent of the markdown.

8. Solve by factoring: $x(6x + 1) = 12$.

9. Solve by completing the square: $\dfrac{x^2}{2} - x = \dfrac{1}{5}$.

10. Solve and check: $\sqrt[3]{x + 2} = -3$.

11. Solve using substitution: $x^4 - x^2 - 12 = 0$.

Solve and express the solution in interval notation.

12. $-7 < 3 - 2x \leq 5$

13. $\dfrac{x}{x - 5} < 0$

14. $|2.7 - 3.2x| \leq 1.3$

15. Calculate the distance and midpoint between the segment joining the points $(-2.7, -1.4)$ and $(5.2, 6.3)$.

16. Find the slope of the line passing through the points $(0.3, -1.4)$ and $(2.7, 4.3)$.

17. Write an equation of a line that passes through the points $(1.2, -3)$ and $(-0.2, -3)$.

18. Transform the equation into standard form by completing the square, and state the center and radius of the circle: $x^2 + y^2 + 12x - 18y - 4 = 0$.

19. Find the equation of a circle with center $(-2, -1)$ and passing through the point $(-4, 3)$.

20. If a cellular phone tower has a reception radius of 100 miles and you live 85 miles north and 23 miles east of the tower, can you use your cell phone at home? Explain.

21. Use interval notation to express the domain of the function $g(x) = \dfrac{1}{x - 1}$.

22. Find the average rate of change for $f(x) = 5x^2$, from $x = 2$ to $x = 4$.

23. Evaluate $g(f(-1))$ for $f(x) = |6 - x|$ and $g(x) = x^2 - 3$.

24. Find the inverse of the function $f(x) = x^2 + 3$ for $x \geq 0$.

25. Write an equation that describes the variation: r is inversely proportional to t. $r = 45$ when $t = 3$.

26. Use a graphing utility to graph the function. State the (a) domain, (b) range, and (c) x intervals where the function is increasing, decreasing, and constant.

$$f(x) = \begin{cases} 1 - |x| & -1 \leq x < 1 \\ 1 - |x - 2| & 1 < x \leq 3 \end{cases}$$

27. Use a graphing utility to graph the function $f(x) = x^2 - 3x$ and $g(x) = x^2 + x - 2$ in the same screen. Find the function h such that $g \circ h = f$.

4

Polynomial and Rational Functions

Ray Guy

Indoor football stadiums are designed so that football punters will not hit the roof with the football. One of the greatest NFL punters of all time was Ray Guy, who played 14 seasons from 1973 to 1986. "In the 1976 Pro Bowl, one of his punts hit the giant TV screen hanging from the rafters in the Louisiana Superdome. Not only did Guy punt high and far—the term "hang time" came into the NFL lexicon during his tenure—once he even had an opponent take a ball he punted and test it for helium!" (www.prokicker.com; Ray Guy Fact Sheet)

The path that punts typically follow is called a *parabola* and is classified as a *quadratic function*. The distance of a punt is measured in the horizontal direction. The yard line where the punt is kicked from and the yard line where the punt either hits the field or is caught are the zeros of the quadratic function. Zeros are the points where the function value is equal to zero.

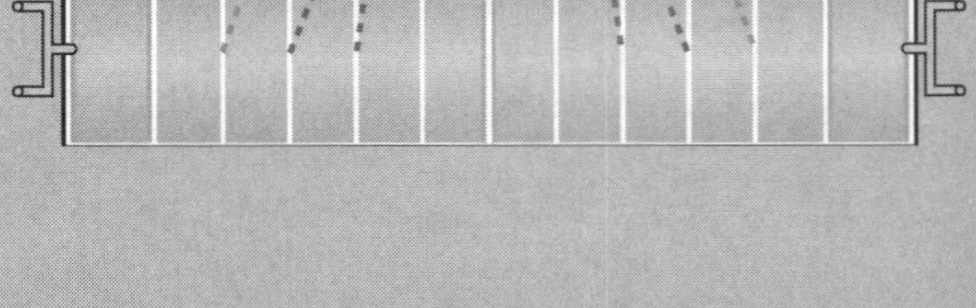

IN THIS CHAPTER we will start by discussing quadratic functions (polynomial functions of degree 2) whose graphs are parabolas. We will find the vertex, which is the maximum or minimum point on the graph. Then we will expand our discussion to higher degree polynomial functions. We will discuss techniques to find zeros of polynomial functions and strategies for graphing polynomial functions. Last we will discuss rational functions, which are ratios of polynomial functions.

POLYNOMIAL AND RATIONAL FUNCTIONS

DIVIDING POLYNOMIALS

- Long Division
- Synthetic Division

ZEROS OF POLYNOMIALS

- How to Find Zeros
  - Rational zero theorem
  - Intermediate value theorem (irrational zeros)
  - Descartes's rule of signs
  - Complex conjugates
- Factoring Polynomials
  - Remainder theorem
  - Fundamental theorem of algebra

GRAPHS OF POLYNOMIALS

- Real Zeros
  - x-intercepts
  - Multiplicity (touch/cross)
- Quadratic Functions (Parabolas)
- Higher Order Polynomial Functions
  - Find intercepts
  - Test points in intervals
  - End behavior

RATIONAL FUNCTIONS

- Ratios of Polynomial Functions
- Graphs
  - Intercepts
  - Asymptotes

SECTION

CHAPTER OBJECTIVES

- Find the vertex (maximum or minimum) of the graph of a quadratic function.
- Graph quadratic functions.
- Divide polynomials using long division.
- Divide a polynomial by a linear factor with synthetic division.
- Understand that real zeros of a polynomial function correspond to x-intercepts on its graph.
- Develop strategies for searching for zeros (both real and complex) of a polynomial function.
- Understand that complex zeros come in conjugate pairs.
- Graph polynomial functions.
- Find the asymptotes on the graph of a rational function.
- Graph rational functions.
- Solve application problems involving polynomial and rational functions.

SECTION 4.1 QUADRATIC FUNCTIONS

SKILLS OBJECTIVES

- Graph a quadratic function in standard form.
- Write a quadratic function in standard form.
- Find the equation of a parabola.
- Solve application problems that involve quadratic functions.

CONCEPTUAL OBJECTIVES

- Recognize characteristics of graphs of quadratic functions (parabolas):
  - whether the parabola opens up or down
  - whether the vertex is a maximum or minimum
  - the axis of symmetry

In Chapter 3, we studied functions in general. In this chapter we will learn about a special group of functions called *polynomial functions.* Polynomial functions are simple functions; often, more complicated functions are approximated by polynomial functions. Polynomial functions model many real-world applications such as the stock market, football punts, business costs, revenues and profits, and the flight path of NASA's "vomit comet." Let's start by defining a polynomial function.

DEFINITION **Polynomial Function**

Let n be a nonnegative integer, and let $a_n, a_{n-1}, \ldots, a_2, a_1, a_0$ be real numbers with $a_n \neq 0$. The function

$$f(x) = a_nx^n + a_{n-1}x^{n-1} + \cdots + a_2x^2 + a_1x + a_0$$

is called a **polynomial function of *x* with degree *n***. The coefficient a_n is called the **leading coefficient**, and a_0 is the constant.

Polynomials of particular degrees have special names. In Chapter 3, the library of functions included the constant function $f(x) = b$, which is a horizontal line; the linear function $f(x) = mx + b$, which is a line with slope m and y-intercept $(0, b)$; the square function $f(x) = x^2$; and the cube function $f(x) = x^3$. These are all special cases of a polynomial function.

Here are more examples of polynomial functions of particular degree together with their names:

Polynomial	Degree	Special Name
$f(x) = 3$	0	Constant function
$f(x) = -2x + 1$	1	Linear function
$f(x) = 7x^2 - 5x + 19$	2	Quadratic function
$f(x) = 4x^3 + 2x - 7$	3	Cubic function

The leading coefficients of these functions are 3, -2, 7, and 4, respectively. The constants of these functions are 3, 1, 19, and -7, respectively. In Section 2.3, we discussed graphs of linear functions, which are first-degree polynomial functions. In this section we will discuss graphs of quadratic functions, which are second-degree polynomial functions.

Graph of a Quadratic Function

In Section 3.2, the library of functions that we compiled included the square function $f(x) = x^2$, whose graph is a parabola. See the graph on the right.

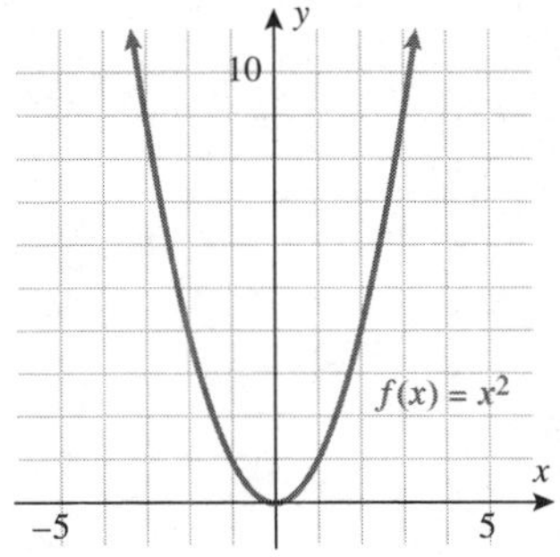

In Section 3.3, we graphed functions using transformation techniques such as $F(x) = (x + 1)^2 - 2$, which can be graphed by starting with the square function $y = x^2$ and shifting 1 unit to the left and down 2 units. See the graph on the right.

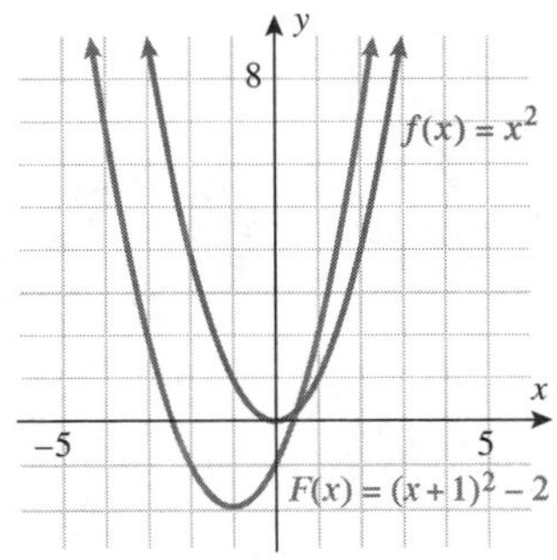

Note that if we eliminate the parentheses in $F(x) = (x + 1)^2 - 2$ to get

$$\begin{aligned} F(x) &= x^2 + 2x + 1 - 2 \\ &= x^2 + 2x - 1 \end{aligned}$$

the result is a function defined by a second-degree polynomial (a polynomial with x^2 as the highest degree term), which is also called a *quadratic function.*

DEFINITION **Quadratic Function**

Let a, b, and c be real numbers with $a \neq 0$. The function

$$f(x) = ax^2 + bx + c$$

is called a **quadratic function.**

The graph of any quadratic function is a **parabola**. If the leading coefficient a is *positive,* then the parabola opens *up.* If the leading coefficient a is *negative,* then the parabola opens *down.* The **vertex** (or turning point) is the minimum point, or low point, on the graph if the parabola opens up, whereas it is the maximum point, or high point, on the graph if the parabola opens down. The vertical line that intersects the parabola at the vertex is called the **axis of symmetry**.

The axis of symmetry is the line $x = h$, and the vertex is located at the point (h, k), as shown in the following two figures.

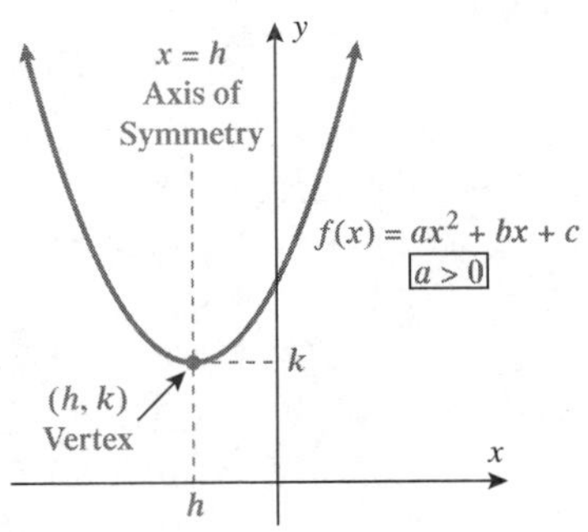

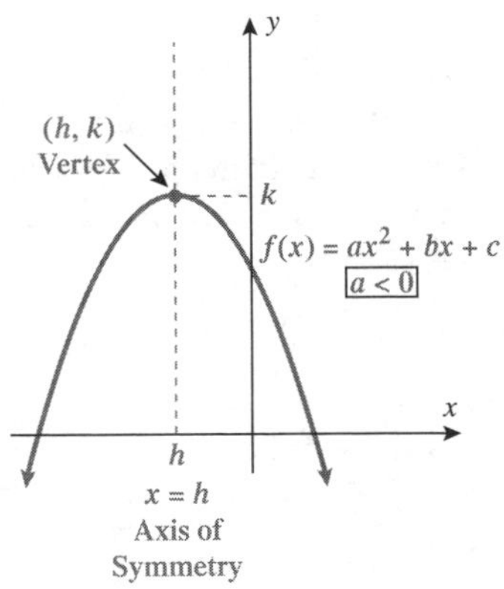

At the start of this section, the function $F(x) = x^2 + 2x - 1$, which can also be written as $F(x) = (x + 1)^2 - 2$, was shown through graph-shifting techniques to have a graph of a parabola. Looking at the graph on top of the margin, we see that the parabola opens up ($a = 1 > 0$), has a vertex at the point $(-1, -2)$, and has an axis of symmetry of $x = -1$.

Graphing Quadratic Functions in Standard Form

In general, writing a quadratic function in the form

$$f(x) = a(x - h)^2 + k$$

allows the vertex (h, k) and the axis of symmetry $x = h$ to be determined by inspection. This form is a convenient way to express a quadratic function in order to quickly determine its corresponding graph. Hence, this form is called *standard form.*

QUADRATIC FUNCTION: STANDARD FORM

The quadratic function

$$f(x) = a(x - h)^2 + k$$

is in **standard form**. The graph of f is a parabola whose vertex is the point (h, k). The parabola is symmetric with respect to the line $x = h$. If $a > 0$, the parabola opens up. If $a < 0$, the parabola opens down.

Recall that graphing linear functions requires finding two points on the line, or a point and the slope of the line. However, for quadratic functions simply knowing two points that lie on its graph is no longer sufficient. Below is a general step-by-step procedure for graphing quadratic functions given in standard form.

GRAPHING QUADRATIC FUNCTIONS

To graph $f(x) = a(x - h)^2 + k$

Step 1: Determine whether the parabola opens up or down.

$$a > 0 \quad \text{up}$$
$$a < 0 \quad \text{down}$$

Step 2: Determine the vertex (h, k).
Step 3: Find the y-intercept (by setting $x = 0$).
Step 4: Find any x-intercepts (by setting $f(x) = 0$ and solving for x).
Step 5: Plot the vertex and intercepts and connect them with a smooth curve.

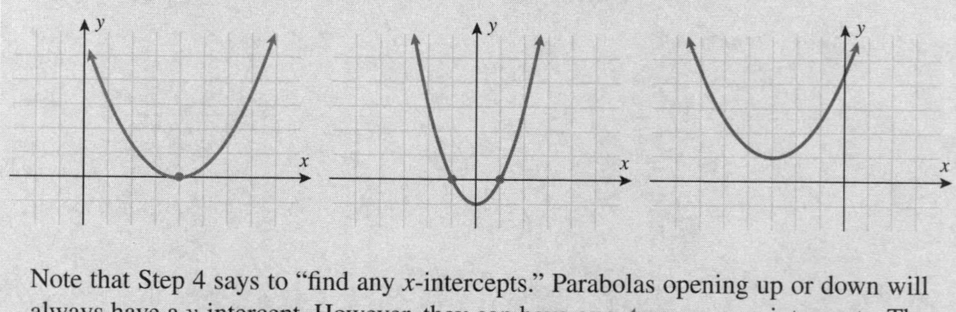

Note that Step 4 says to "find any x-intercepts." Parabolas opening up or down will always have a y-intercept. However, they can have **one**, **two**, or **no** x-intercepts. The figures above illustrate this for parabolas opening up, and the same can be said about parabolas opening down.

EXAMPLE 1 Graphing a Quadratic Function Given in Standard Form

Graph the quadratic function $f(x) = (x - 3)^2 - 1$.

Solution:

STEP 1 The parabola opens up. $a = 1$, so $a > 0$

STEP 2 Determine the vertex. $(h, k) = (3, -1)$

STEP 3 Find the y-intercept. $f(0) = (-3)^2 - 1 = 8$
$(0, 8)$ corresponds to the y-intercept

STEP 4 Find any x-intercepts. $f(x) = (x - 3)^2 - 1 = 0$
$(x - 3)^2 = 1$

Use square-root method. $x - 3 = \pm 1$

Solve. $x = 2$ or $x = 4$
$(2, 0)$ and $(4, 0)$ correspond to the x-intercepts

STEP 5 Plot the vertex and intercepts $(3, -1)$, $(0, 8)$, $(2, 0)$, $(4, 0)$

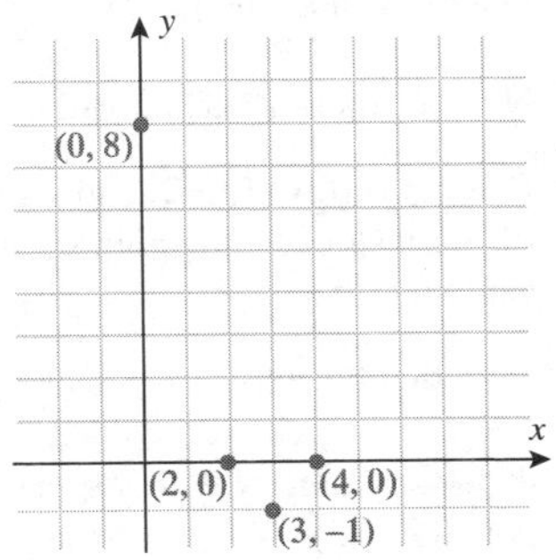

Connect the points with a smooth curve opening up.

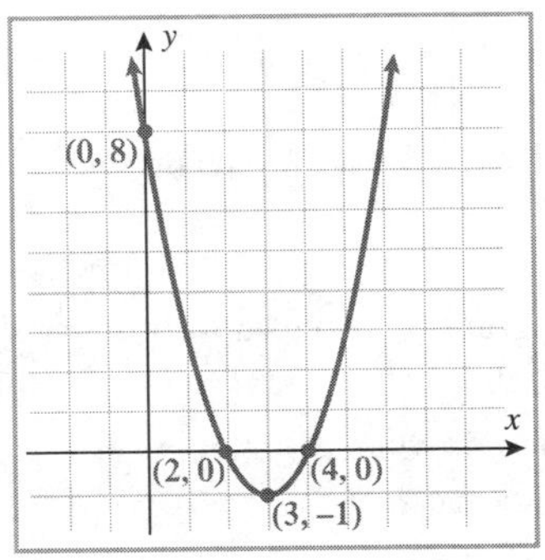

The graph in Example 1 could also have been found by shifting the square function to the right 3 units and down 1 unit.

YOUR TURN Graph the quadratic function $f(x) = (x - 1)^2 - 4$.

Study Tip

A quadratic function given in standard form can be graphed using the transformation techniques shown in Section 3.3 for the square function.

Answer:

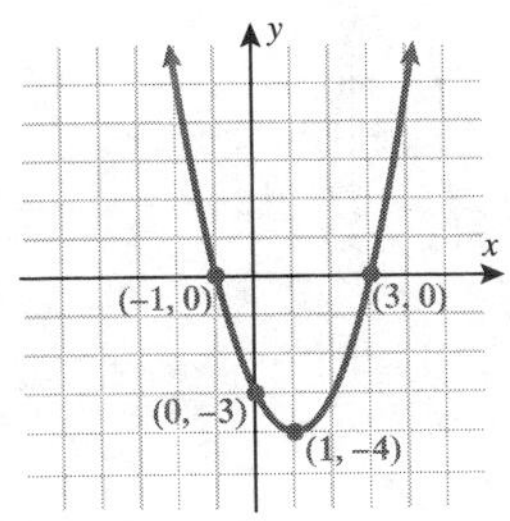

EXAMPLE 2 Graphing a Quadratic Function Given in Standard Form with a Negative Leading Coefficient

Graph the quadratic function $f(x) = -2(x - 1)^2 - 3$.

Solution:

STEP 1 The parabola opens down. $a = -2$, so $a < 0$

STEP 2 Determine the vertex. $(h, k) = (1, -3)$

STEP 3 Find the y-intercept. $f(0) = -2(-1)^2 - 3 = -2 - 3 = -5$

$(0, -5)$ corresponds to the y-intercept

STEP 4 Find any x-intercepts.

$$f(x) = -2(x - 1)^2 - 3 = 0$$
$$-2(x - 1)^2 = 3$$
$$(x - 1)^2 = -\frac{3}{2}$$

The square root of a negative number is not a real number. $x - 1 = \pm\sqrt{-\frac{3}{2}}$

No real solutions. There are no x-intercepts.

STEP 5 Plot the vertex $(1, -3)$ and y-intercept $(0, -5)$. Connect the points with a smooth curve.

Note that the axis of symmetry is $x = 1$. Because the point $(0, -5)$ lies on the parabola, then by symmetry with respect to $x = 1$, the point $(2, -5)$ also lies on the graph.

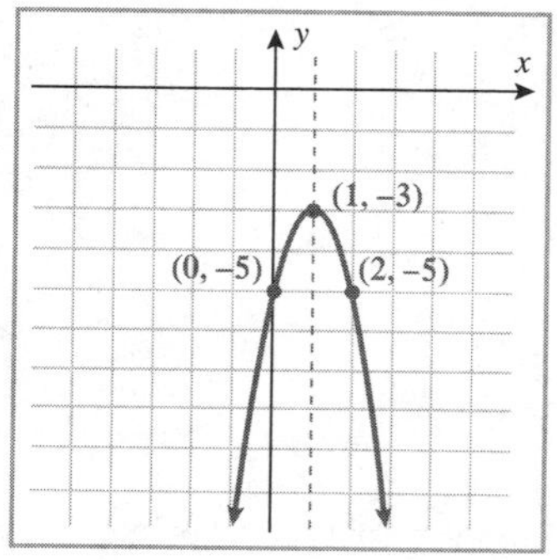

■ **YOUR TURN** Graph the quadratic function $f(x) = -3(x + 1)^2 - 2$.

■ **Answer:**

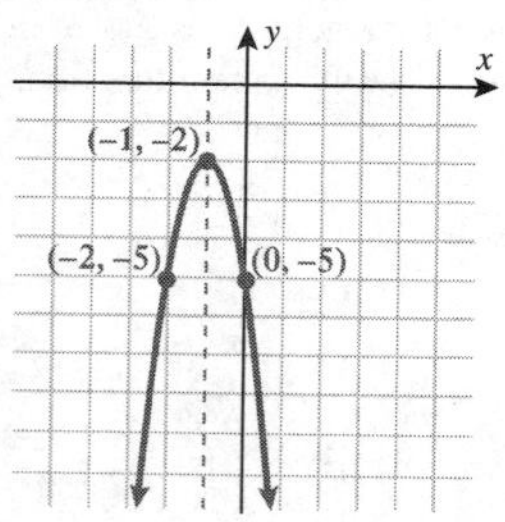

When graphing quadratic functions (parabolas), have *at least 3 points* labeled on the graph.

- When there are x-intercepts (Example 1), label the vertex, y-intercept, and x-intercepts.
- When there are no x-intercepts (Example 2), label the vertex, y-intercept, and another point.

Graphing Quadratic Functions in General Form

A quadratic function is often written in one of two forms:

Standard form: $f(x) = a(x - h)^2 + k$

General form: $f(x) = ax^2 + bx + c$

When the quadratic function is expressed in standard form, the graph is easily obtained by identifying the vertex (h, k) and the intercepts and drawing a smooth curve that opens either up or down, depending on the sign of a.

Typically, quadratic functions are expressed in general form and a graph is the ultimate goal, so we must first express the quadratic function in standard form. One technique for transforming a quadratic function from general form to standard form was introduced in Section 1.3 and is called *completing the square.*

EXAMPLE 3 Graphing a Quadratic Function Given in General Form

Graph the quadratic function $f(x) = x^2 - 6x + 4$.

Solution:

Express the quadratic function in standard form by completing the square.

Write the original function. $f(x) = x^2 - 6x + 4$

Group the variable terms together. $= (x^2 - 6x) + 4$

Complete the square.

Half of -6 is -3; -3 squared is 9.

Add and subtract 9 within the parentheses. $= (x^2 - 6x + 9 - 9) + 4$

Write the -9 outside the parentheses. $= (x^2 - 6x + 9) - 9 + 4$

Write the expression inside the parentheses as a perfect square and simplify. $= (x - 3)^2 - 5$

Now that the quadratic function is written in standard form, $f(x) = (x - 3)^2 - 5$, we follow our step-by-step procedure for graphing a quadratic function in standard form.

STEP 1 The parabola opens up. $a = 1$, so $a > 0$

STEP 2 Determine the vertex. $(h, k) = (3, -5)$

STEP 3 Find the y-intercept. $f(0) = (0)^2 - 6(0) + 4 = 4$
(0, 4) corresponds to the y-intercept

STEP 4 Find any x-intercepts.

$$f(x) = 0$$
$$f(x) = (x - 3)^2 - 5 = 0$$
$$(x - 3)^2 = 5$$
$$x - 3 = \pm\sqrt{5}$$
$$x = 3 \pm \sqrt{5}$$

$(3 + \sqrt{5}, 0)$ and $(3 - \sqrt{5}, 0)$ correspond to the x-intercepts.

STEP 5 Plot the vertex and intercepts $(3, -5)$, $(0, 4)$, $(3 + \sqrt{5}, 0)$, and $(3 - \sqrt{5}, 0)$.

Connect the points with a smooth parabolic curve.

Note: $3 + \sqrt{5} \approx 5.24$ and $3 - \sqrt{5} \approx 0.76$.

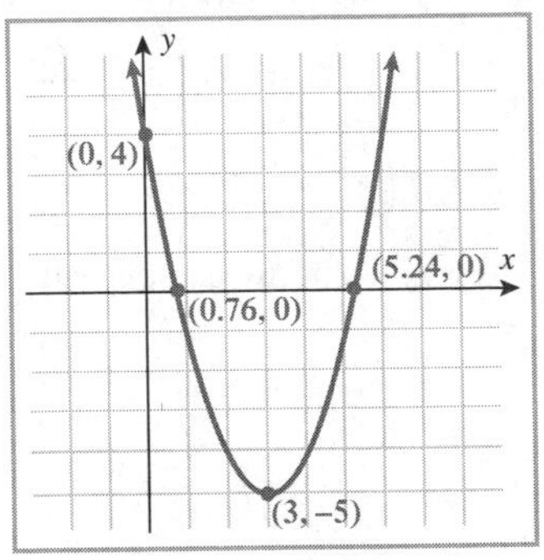

■ **YOUR TURN** Graph the quadratic function $f(x) = x^2 - 8x + 14$.

Technology Tip

Use a graphing utility to graph the function $f(x) = x^2 - 6x + 4$ as y_1.

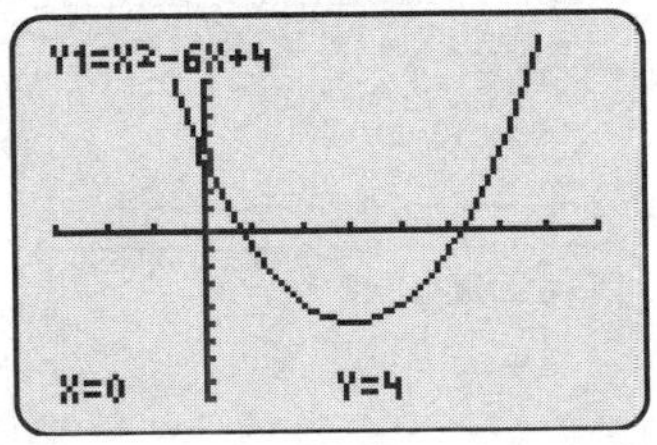

Study Tip

Although either form (standard or general) can be used to find the intercepts, it is often more convenient to use the general form when finding the y-intercept and the standard form when finding the x-intercept.

■ **Answer:**

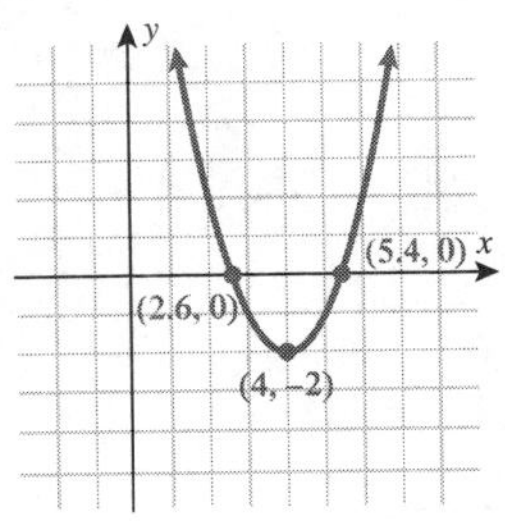

Technology Tip

Use a graphing utility to graph the function $f(x) = -3x^2 + 6x + 2$ as y_1.

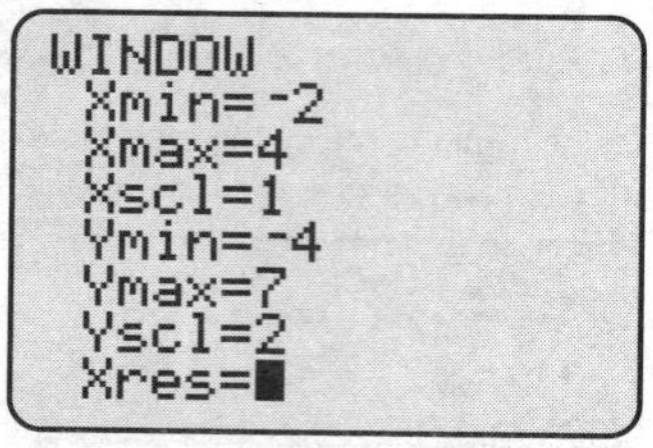

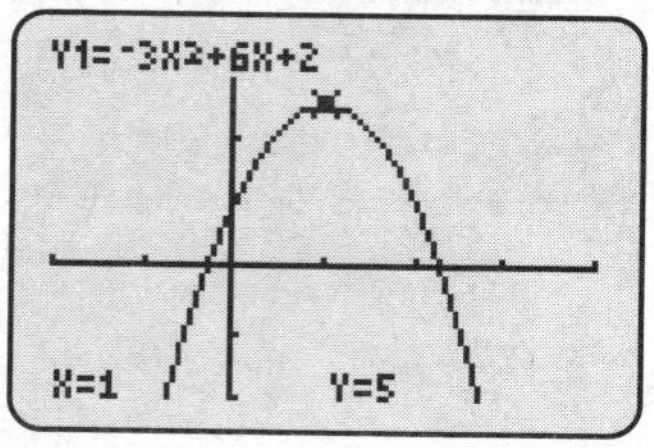

EXAMPLE 4 Graphing a Quadratic Function Given in General Form with a Negative Leading Coefficient

Graph the quadratic function $f(x) = -3x^2 + 6x + 2$.

Solution:

Express the function in standard form by completing the square.

Write the original function.	$f(x) = -3x^2 + 6x + 2$
Group the variable terms together.	$= (-3x^2 + 6x) + 2$
-3 is a common factor.	$= -3(x^2 - 2x) + 2$
Add and subtract 1 inside the parentheses to create a perfect square.	$= -3(x^2 - 2x + 1 - 1) + 2$
Regroup the terms.	$= -3(x^2 - 2x + 1) - 3(-1) + 2$
Write the expression inside the parentheses as a perfect square and simplify.	$= -3(x - 1)^2 + 5$

Now that the quadratic function is written in standard form, $f(x) = -3(x - 1)^2 + 5$, we follow our step-by-step procedure for graphing a quadratic function in standard form.

STEP 1 The parabola opens down. $\quad a = -3$, therefore $a < 0$

STEP 2 Determine the vertex. $\quad (h, k) = (1, 5)$

STEP 3 Find the y-intercept using the general form. $\quad f(0) = -3(0)^2 + 6(0) + 2 = 2$

(0, 2) corresponds to the y-intercept

STEP 4 Find any x-intercepts using the standard form.

$$f(x) = -3(x - 1)^2 + 5 = 0$$
$$-3(x - 1)^2 = -5$$
$$(x - 1)^2 = \frac{5}{3}$$
$$x - 1 = \pm\sqrt{\frac{5}{3}}$$
$$x = 1 \pm \sqrt{\frac{5}{3}}$$
$$= 1 \pm \frac{\sqrt{15}}{3}$$

The x-intercepts are $\left(1 + \frac{\sqrt{15}}{3}, 0\right)$ and $\left(1 - \frac{\sqrt{15}}{3}, 0\right)$.

STEP 5 Plot the vertex and intercepts

$(1, 5)$, $(0, 2)$, $\left(1 + \frac{\sqrt{15}}{3}, 0\right)$, and $\left(1 - \frac{\sqrt{15}}{3}, 0\right)$.

Connect the points with a smooth curve.

Note: $1 + \frac{\sqrt{15}}{3} \approx 2.3$ and

$1 - \frac{\sqrt{15}}{3} \approx -0.3$.

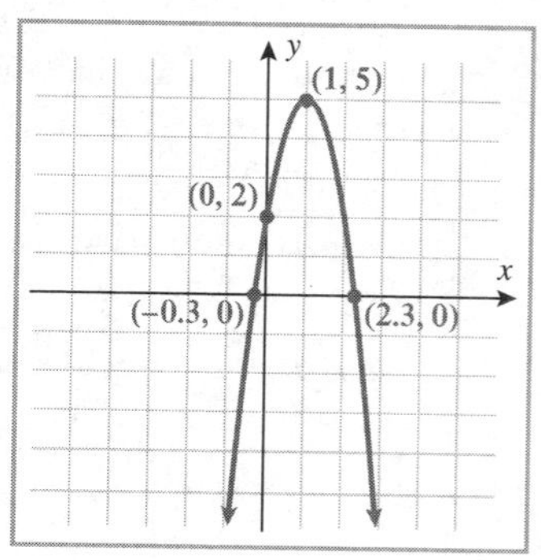

YOUR TURN Graph the quadratic function $f(x) = -2x^2 - 4x + 1$.

Answer:

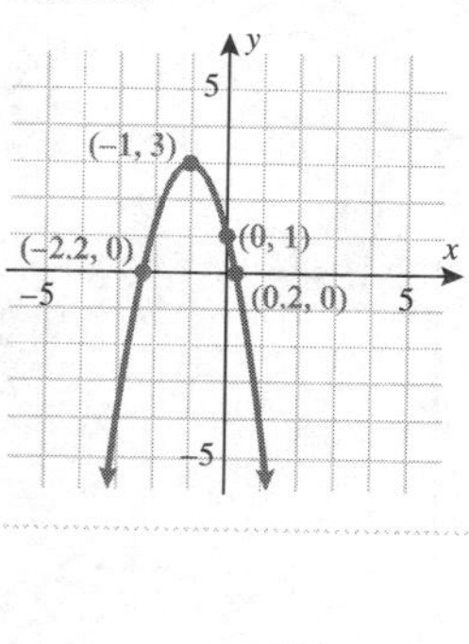

In Examples 3 and 4, the quadratic functions were given in general form and they were transformed into standard form by completing the square. It can be shown (by completing the square) that the vertex of a quadratic function in general form, $f(x) = ax^2 + bx + c$, is located at $x = -\frac{b}{2a}$.

VERTEX OF A PARABOLA

The graph of a quadratic function $f(x) = ax^2 + bx + c$ is a parabola with the **vertex** located at the point

$$\left(-\frac{b}{2a}, f\left(-\frac{b}{2a}\right)\right) = \left(-\frac{b}{2a}, \frac{4ac - b^2}{4a}\right)$$

GRAPHING A QUADRATIC FUNCTION IN GENERAL FORM

Step 1: Find the vertex.
Step 2: Determine whether the parabola opens up or down.
- If $a > 0$, the parabola opens up.
- If $a < 0$, the parabola opens down.

Step 3: Find additional points near the vertex.
Step 4: Sketch the graph with a parabolic curve.

EXAMPLE 5 Graphing a Quadratic Function Given in General Form

Sketch the graph of $f(x) = -2x^2 + 4x + 5$.

Solution: Let $a = -2$, $b = 4$, and $c = 5$.

STEP 1 Find the vertex.

$$x = -\frac{b}{2a} = -\frac{4}{2(-2)} = 1$$
$$f(1) = -2(1)^2 + 4(1) + 5 = 7$$
Vertex $= (1, 7)$

STEP 2 The parabola opens down. $a = -2$

STEP 3 Find additional points near the vertex.

x	−1	0	1	2	3
$f(x)$	$f(-1) = -1$	$f(0) = 5$	$f(1) = 7$	$f(2) = 5$	$f(3) = -1$

STEP 4 Label the vertex and additional points then sketch the graph.

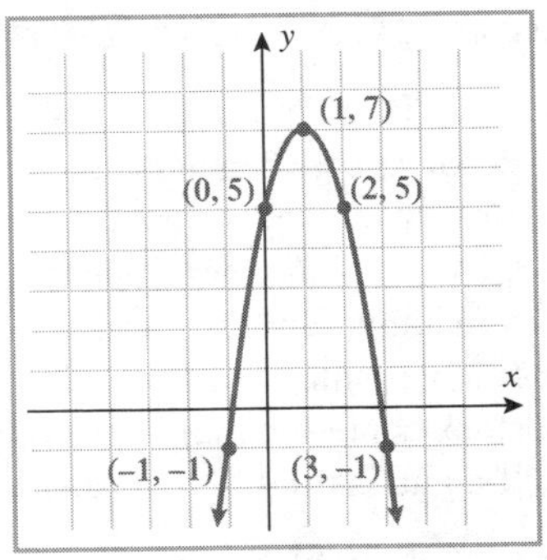

■ **YOUR TURN** Sketch the graph of $f(x) = 3x^2 - 6x + 4$.

■ **Answer:**

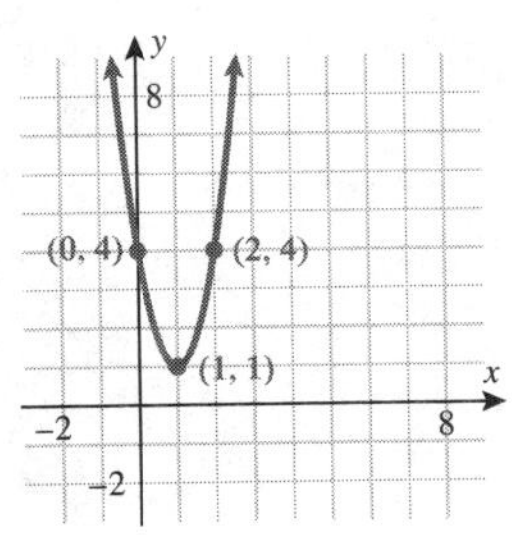

Finding the Equation of a Parabola

It is important to understand that the equation $y = x^2$ is equivalent to the quadratic function $f(x) = x^2$. Both have the same parabolic graph. Thus far, we have been given the equation and then asked to find characteristics (vertex and intercepts) in order to graph. We now turn our attention to the problem of determining the equation, or function, given certain characteristics.

EXAMPLE 6 Finding the Equation of a Parabola Given the Vertex and Another Point

Find the equation of a parabola whose graph has a vertex at (3, 4) and which passes through the point (2, 3). Express the quadratic function in both standard and general forms.

Solution:

Write the standard form of the equation of a parabola. $f(x) = a(x - h)^2 + k$

Substitute the coordinates of the vertex $(h, k) = (3, 4)$. $f(x) = a(x - 3)^2 + 4$

Use the point (2, 3) to find a.

The point (2, 3) implies $f(2) = 3$. $f(2) = a(2 - 3)^2 + 4 = 3$

Solve for a.

$$a(2 - 3)^2 + 4 = 3$$
$$a(-1)^2 + 4 = 3$$
$$a + 4 = 3$$
$$a = -1$$

Write both forms of the equation of this parabola.

Standard form: $f(x) = -(x - 3)^2 + 4$ General form: $f(x) = -x^2 + 6x - 5$

■ **YOUR TURN** Find the standard form of the equation of a parabola whose graph has a vertex at $(-3, -5)$ and which passes through the point $(-2, -4)$.

■ **Answer:** Standard form: $f(x) = (x + 3)^2 - 5$

As we have seen in Example 6, once the vertex is known, the leading coefficient a can be found from any point that lies on the parabola.

Application Problems That Involve Quadratic Functions

- What is the minimum distance a driver has to maintain in order to be at a safe distance between her car and the car in front as a function of speed?
- How many units produced will yield a maximum profit for a company?
- Given the number of linear feet of fence, what rectangular dimensions will yield a maximum fenced-in area?
- If a particular stock price has been shown to follow a quadratic trend, when will the stock achieve a maximum value?
- If a gun is fired into the air, where will the bullet land?

These are all problems that can be solved using quadratic functions. Because the vertex of a parabola represents either the minimum or maximum value of the quadratic function, in application problems it often suffices simply to find the vertex.

In Example 6, the function $F(x) = -x^2 + 6x - 5$, which can also be written as $F(x) = -(x - 3)^2 + 4$, was shown to "open down" and has a vertex at the point (3, 4). Suppose this function represents profit, where x is the number (millions) of units made. The vertex would then represent the *maximum* profit. Instead of rewriting the function in standard form through completing the square, we use the vertex formula $x = -\frac{b}{2a}$.

WORDS	**MATH**
Quadratic function.	$F(x) = -x^2 + 6x - 5$
Coefficients.	$a = -1, b = 6, c = -5$
Find the x-coordinate of the vertex.	$x = -\frac{b}{2a} = -\frac{6}{2(-1)} = 3$
Find the value of the function at $x = 3$.	$f(3) = -(3)^2 + 6(3) - 5 = 4$

Therefore, the vertex is located at the point $\boxed{(3, 4)}$.

EXAMPLE 7 Finding the Minimum Cost of Manufacturing a Motorcycle

A company that produces motorcycles has a daily production cost of

$$C(x) = 2000 - 15x + 0.05x^2$$

where C is the cost in dollars to manufacture a motorcycle and x is the number of motorcycles produced. How many motorcycles can be produced each day in order to minimize the cost of each motorcycle? What is the corresponding minimum cost?

Solution:

The graph of the quadratic function is a parabola.

Rewrite the quadratic function in general form.	$C(x) = 0.05x^2 - 15x + 2000$
The parabola opens up, because a is positive.	$a = 0.05 > 0$
Because the parabola opens up, the vertex of the parabola is a *minimum*.	
Find the x-coordinate of the vertex.	$x = -\frac{b}{2a} = -\frac{(-15)}{2(0.05)} = 150$

The company keeps costs to a minimum when 150 motorcycles are produced each day.

The minimum cost is \$875 per motorcycle.	$C(150) = 875$

YOUR TURN The revenue associated with selling vitamins is

$$R(x) = 500x - 0.001x^2$$

where R is the revenue in dollars and x is the number of bottles of vitamins sold. Determine how many bottles of vitamins should be sold to maximize the revenue.

Technology Tip

Use a graphing utility to graph the cost function $C(x) = 2000 - 15x + 0.05x^2$ as y_1.

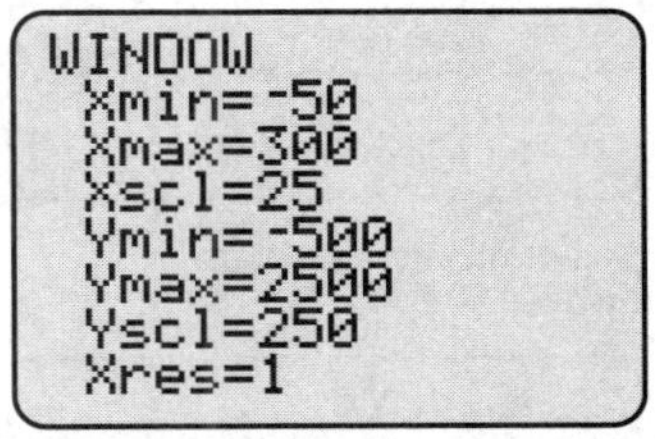

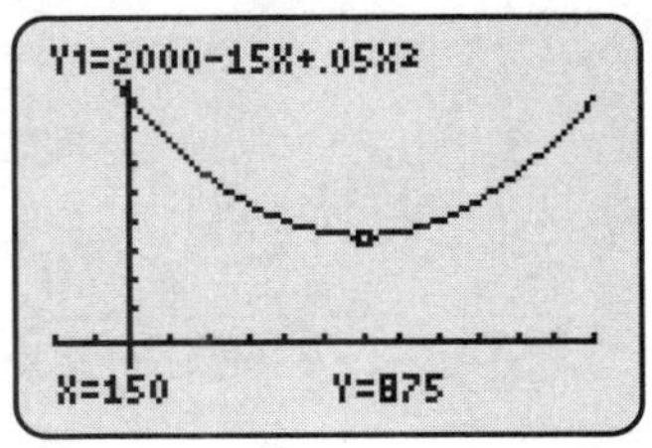

■ **Answer:** 250,000 bottles

Technology Tip

Use a graphing utility to graph the area function $A(x) = -2x^2 + 100x$.

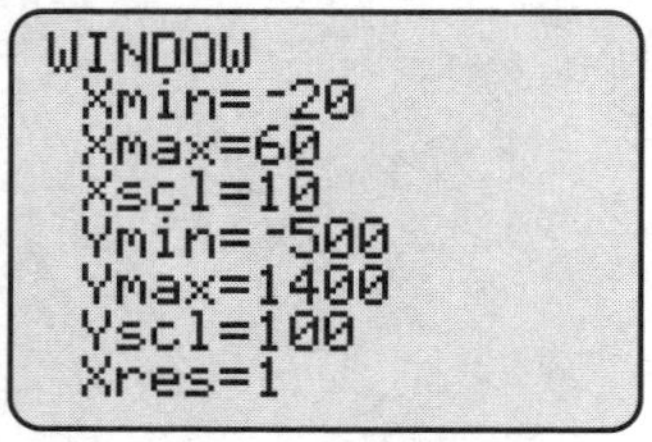

The maximum occurs when $x = 25$. The maximum area is $y = 1250$ square feet.

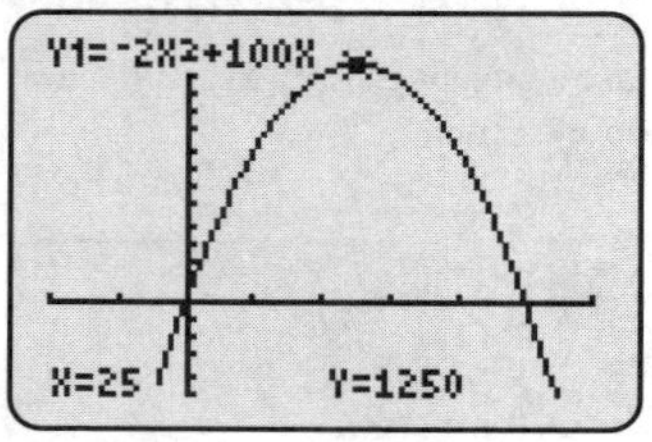

A table of values supports the solution.

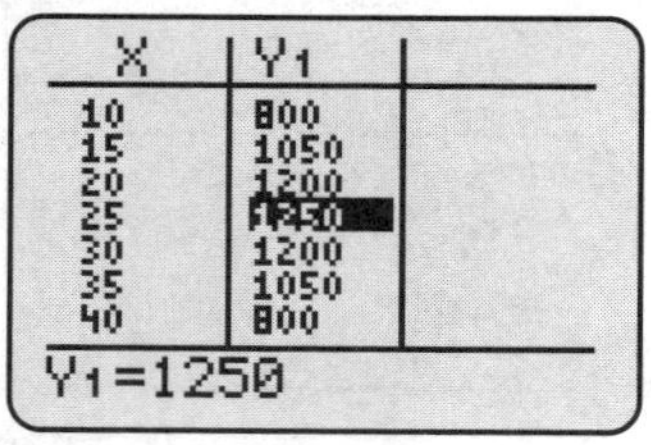

■ **Answer:** 50 feet by 50 feet

EXAMPLE 8 Finding the Dimensions That Yield a Maximum Area

You have just bought a puppy and want to fence in an area in the backyard for her. You buy 100 linear feet of fence from Home Depot and have decided to make a rectangular fenced-in area using the back of your house as one side. Determine the dimensions of the rectangular pen that will maximize the area in which your puppy may roam. What is the maximum area of the rectangular pen?

Solution:

STEP 1 Identify the question.
Find the dimensions of the rectangular pen.

STEP 2 Draw a picture.

STEP 3 Set up a function.
If we let x represent the length of one side of the rectangle, then the opposite side is also of length x. Because there are 100 feet of fence, the remaining fence left for the side opposite the house is $100 - 2x$.

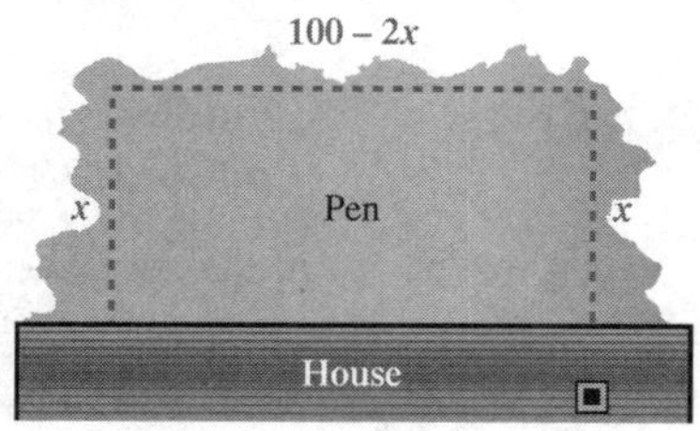

The area of a rectangle is equal to length times width:

$$A(x) = x(100 - 2x)$$

STEP 4 Find the maximum value of the function.

$$A(x) = x(100 - 2x) = -2x^2 + 100x$$

Find the maximum of the parabola that corresponds to the quadratic function for area $A(x) = -2x^2 + 100x$.

$a = -2$ and $b = 100$; therefore, the maximum occurs when

$$x = -\frac{b}{2a} = -\frac{100}{2(-2)} = 25$$

Replacing x with 25 in our original diagram:

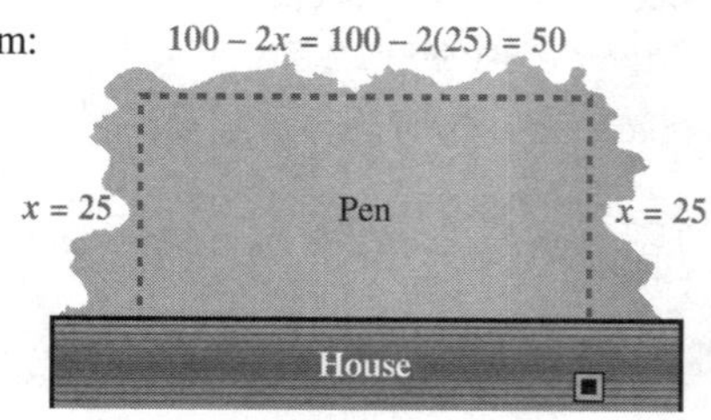

The dimensions of the rectangle are 25 feet by 50 feet.

The maximum area $A(25) = 1250$ is 1250 square feet.

STEP 5 Check the solution.
Two sides are 25 feet and one side is 50 feet, and together they account for all 100 feet of fence.

■ **YOUR TURN** Suppose you have 200 linear feet of fence to enclose a rectangular garden. Determine the dimensions of the rectangle that will yield the greatest area.

EXAMPLE 9 Path of a Punted Football

The path of a particular punt follows the quadratic function: $h(x) = -\frac{1}{8}(x - 5)^2 + 50$, where $h(x)$ is the height of the ball in yards and x corresponds to the horizontal distance in yards. Assume $x = 0$ corresponds to midfield (the 50 yard line). For example, $x = -20$ corresponds to the punter's own 30 yard line, whereas $x = 20$ corresponds to the other team's 30 yard line.

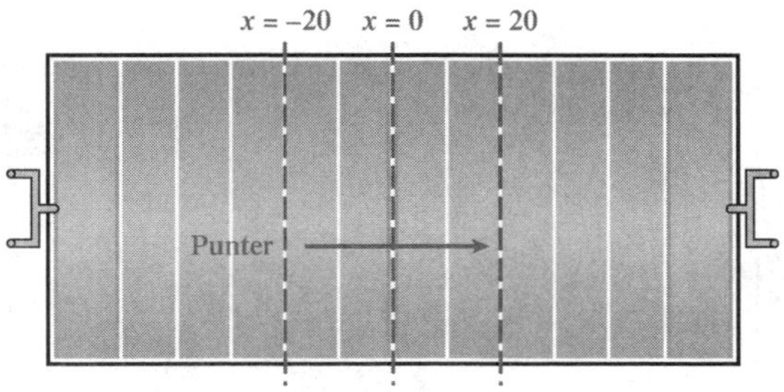

a. Find the maximum height the ball achieves.

b. Find the horizontal distance the ball covers. Assume the height is zero when the ball is kicked and when the ball is caught.

Solution (a):

Identify the vertex since it is given in standard form. $(h, k) = (5, 50)$

The maximum height of the punt occurs at the other team's 45 yard line, and the height the ball achieves is $\boxed{\text{50 yards (150 feet)}}$.

Solution (b):

The height when the ball is kicked or caught is zero. $h(x) = -\frac{1}{8}(x - 5)^2 + 50 = 0$

Solve for x.

$$\frac{1}{8}(x - 5)^2 = 50$$
$$(x - 5)^2 = 400$$
$$(x - 5) = \pm\sqrt{400}$$
$$x = 5 \pm 20$$
$$x = -15 \text{ and } x = 25$$

The horizontal distance is the distance between these two points: $|25 - (-15)| = \boxed{\text{40 yards}}$.

SECTION 4.1 SUMMARY

All quadratic functions $f(x) = ax^2 + bx + c$ or $f(x) = a(x - h)^2 + k$ have graphs that are parabolas:

- If $a > 0$, the parabola opens up.
- If $a < 0$, the parabola opens down.
- The vertex is at the point

$$(h, k) = \left(-\frac{b}{2a}, f\left(-\frac{b}{2a}\right)\right) = \left(-\frac{b}{2a}, \frac{4ac - b^2}{4a}\right)$$

- When the quadratic function is given in general form, completing the square can be used to rewrite the function in standard form.
- At least three points are needed to graph a quadratic function.
  - vertex
  - y-intercept
  - x-intercept(s) or other point(s)

SECTION 4.1 EXERCISES

SKILLS

In Exercises 1–4, match the quadratic function with its graph.

1. $f(x) = 3(x+2)^2 - 5$ **2.** $f(x) = 2(x-1)^2 + 3$ **3.** $f(x) = -\frac{1}{2}(x+3)^2 + 2$ **4.** $f(x) = -\frac{1}{3}(x-2)^2 + 3$

a.

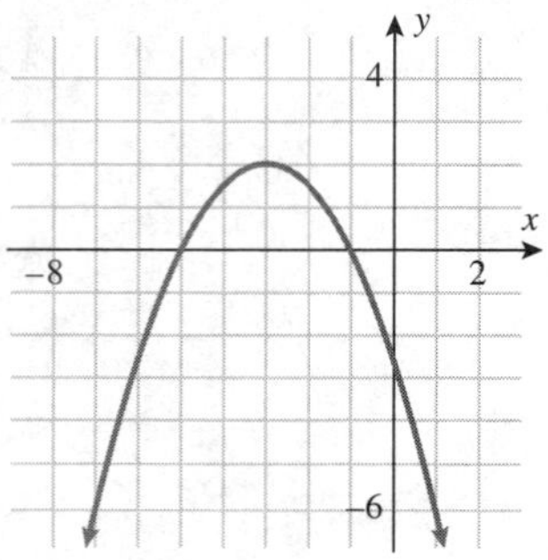

b.

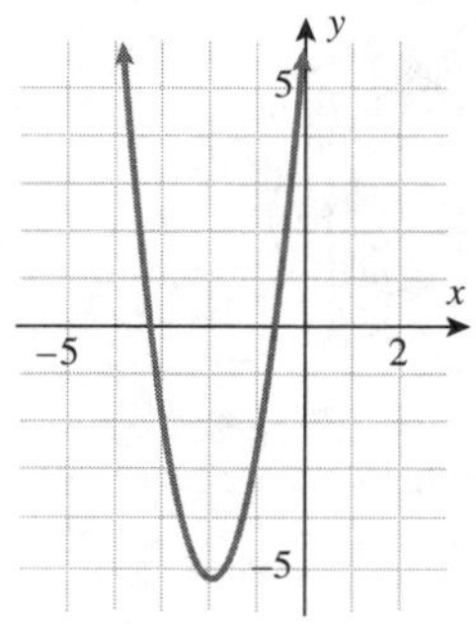

c.

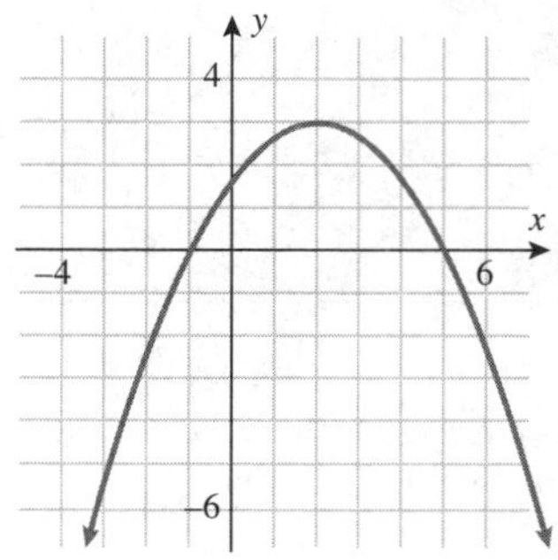

d.

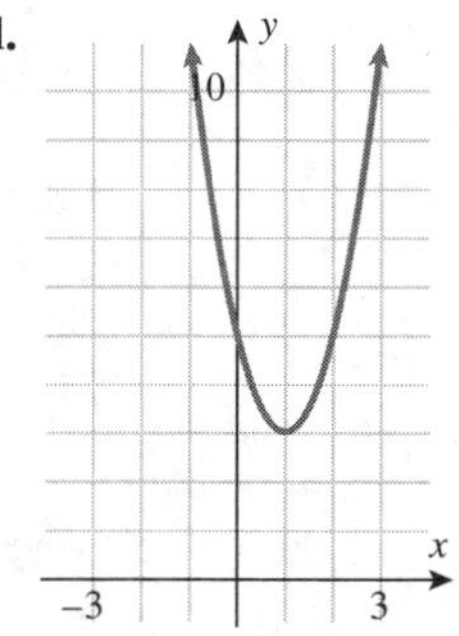

In Exercises 5–8, match the quadratic function with its graph.

5. $f(x) = 3x^2 + 5x - 2$ **6.** $f(x) = 3x^2 - x - 2$ **7.** $f(x) = -x^2 + 2x - 1$ **8.** $f(x) = -2x^2 - x + 3$

a.

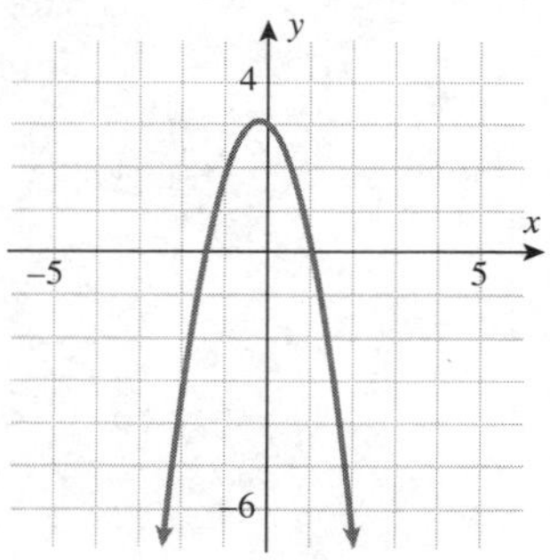

b.

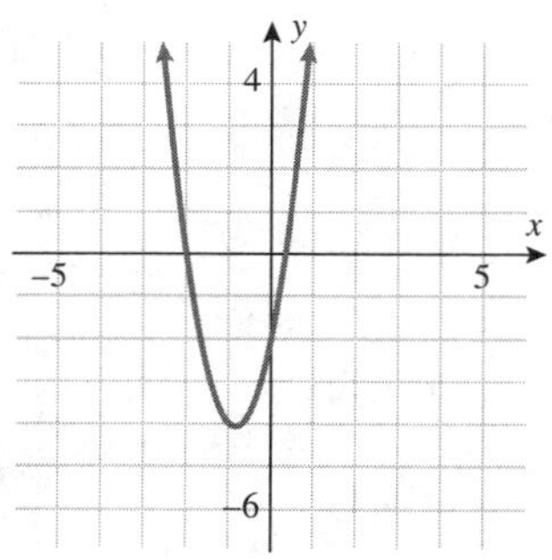

c.

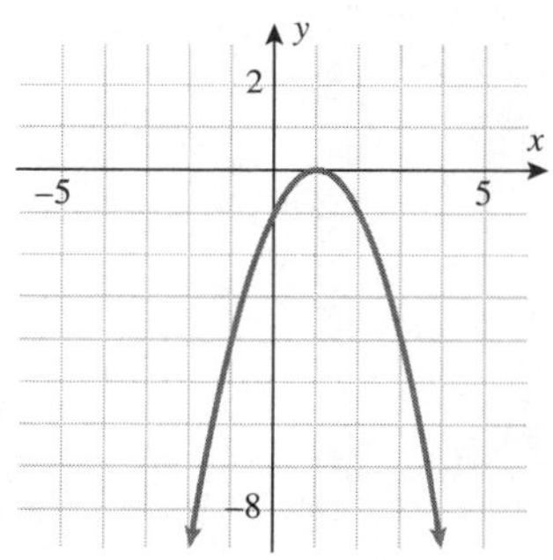

d.

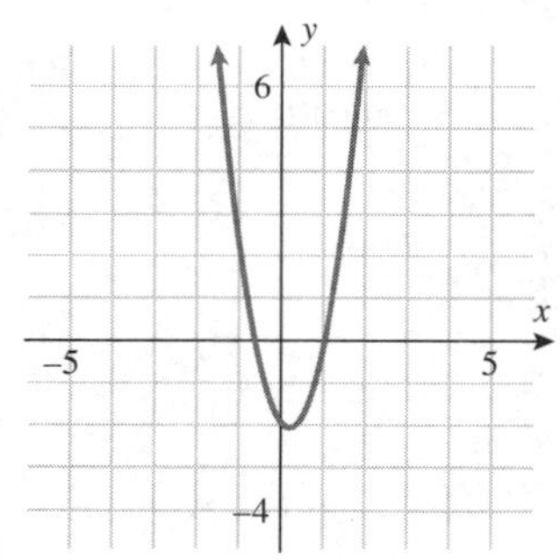

In Exercises 9–22, graph the quadratic function, which is given in standard form.

9. $f(x) = (x+1)^2 - 2$

10. $f(x) = (x+2)^2 - 1$

11. $f(x) = (x-2)^2 - 3$

12. $f(x) = (x-4)^2 + 2$

13. $f(x) = -(x-3)^2 + 9$

14. $f(x) = -(x-5)^2 - 4$

15. $f(x) = -(x+1)^2 - 3$

16. $f(x) = -(x-2)^2 + 6$

17. $f(x) = 2(x-2)^2 + 2$

18. $f(x) = -3(x+2)^2 - 15$

19. $f(x) = \left(x - \frac{1}{3}\right)^2 + \frac{1}{9}$

20. $f(x) = \left(x + \frac{1}{4}\right)^2 - \frac{1}{2}$

21. $f(x) = -0.5(x - 0.25)^2 + 0.75$

22. $f(x) = -0.2(x + 0.6)^2 + 0.8$

In Exercises 23–32, rewrite the quadratic function in standard form by completing the square.

23. $f(x) = x^2 + 6x - 3$

24. $f(x) = x^2 + 8x + 2$

25. $f(x) = -x^2 - 10x + 3$

26. $f(x) = -x^2 - 12x + 6$

27. $f(x) = 2x^2 + 8x - 2$

28. $f(x) = 3x^2 - 9x + 11$

29. $f(x) = -4x^2 + 16x - 7$

30. $f(x) = -5x^2 + 100x - 36$

31. $f(x) = \frac{1}{2}x^2 - 4x + 3$

32. $f(x) = -\frac{1}{3}x^2 + 6x + 4$

In Exercises 33–40, graph the quadratic function.

33. $f(x) = x^2 + 6x - 7$

34. $f(x) = x^2 - 3x + 10$

35. $f(x) = x^2 + 2x - 15$

36. $f(x) = -x^2 + 3x + 4$

37. $f(x) = -2x^2 - 12x - 16$

38. $f(x) = -3x^2 + 12x - 12$

39. $f(x) = \frac{1}{2}x^2 - \frac{1}{2}$

40. $f(x) = -\frac{1}{3}x^2 + \frac{4}{3}$

In Exercises 41–48, find the vertex of the parabola associated with each quadratic function.

41. $f(x) = 33x^2 - 2x + 15$

42. $f(x) = 17x^2 + 4x - 3$

43. $f(x) = \frac{1}{2}x^2 - 7x + 5$

44. $f(x) = -\frac{1}{3}x^2 + \frac{2}{5}x + 4$

45. $f(x) = -0.002x^2 - 0.3x + 1.7$

46. $f(x) = 0.05x^2 + 2.5x - 1.5$

47. $f(x) = -\frac{2}{5}x^2 + \frac{3}{7}x + 2$

48. $f(x) = -\frac{1}{7}x^2 - \frac{2}{3}x + \frac{1}{9}$

In Exercises 49–58, find the quadratic function that has the given vertex and goes through the given point.

49. vertex: $(-1, 4)$ point: $(0, 2)$

50. vertex: $(2, -3)$ point: $(0, 1)$

51. vertex: $(2, 5)$ point: $(3, 0)$

52. vertex: $(1, 3)$ point: $(-2, 0)$

53. vertex: $(-1, -3)$ point: $(-4, 2)$

54. vertex: $(0, -2)$ point: $(3, 10)$

55. vertex: $\left(\frac{1}{2}, -\frac{3}{4}\right)$ point: $\left(\frac{3}{4}, 0\right)$

56. vertex: $\left(-\frac{5}{6}, \frac{2}{3}\right)$ point: $(0, 0)$

57. vertex: $(2.5, -3.5)$ point: $(4.5, 1.5)$

58. vertex: $(1.8, 2.7)$ point: $(-2.2, -2.1)$

APPLICATIONS

Exercises 59 and 60 concern the path of a punted football. Refer to the diagram in Example 9.

59. Sports. The path of a particular punt follows the quadratic function

$$h(x) = -\frac{8}{125}(x + 5)^2 + 40$$

where $h(x)$ is the height of the ball in yards and x corresponds to the horizontal distance in yards. Assume $x = 0$ corresponds to midfield (the 50 yard line). For example, $x = -20$ corresponds to the punter's own 30 yard line, whereas $x = 20$ corresponds to the other team's 30 yard line.

a. Find the maximum height the ball achieves.

b. Find the horizontal distance the ball covers. Assume the height is zero when the ball is kicked and when the ball is caught.

60. Sports. The path of a particular punt follows the quadratic function

$$h(x) = -\frac{5}{40}(x - 30)^2 + 50$$

where $h(x)$ is the height of the ball in yards and x corresponds to the horizontal distance in yards. Assume $x = 0$ corresponds to midfield (the 50 yard line). For example, $x = -20$ corresponds to the punter's own 30 yard line, whereas $x = 20$ corresponds to the other team's 30 yard line.

a. Find the maximum height the ball achieves.

b. Find the horizontal distance the ball covers. Assume the height is zero when the ball is kicked and when the ball is caught.

61. Ranching. A rancher has 10,000 linear feet of fencing and wants to enclose a rectangular field and then divide it into two equal pastures with an internal fence parallel to one of the rectangular sides. What is the maximum area of each pasture? Round to the nearest square foot.

62. Ranching. A rancher has 30,000 linear feet of fencing and wants to enclose a rectangular field and then divide it into four equal pastures with three internal fences parallel to one of the rectangular sides. What is the maximum area of each pasture?

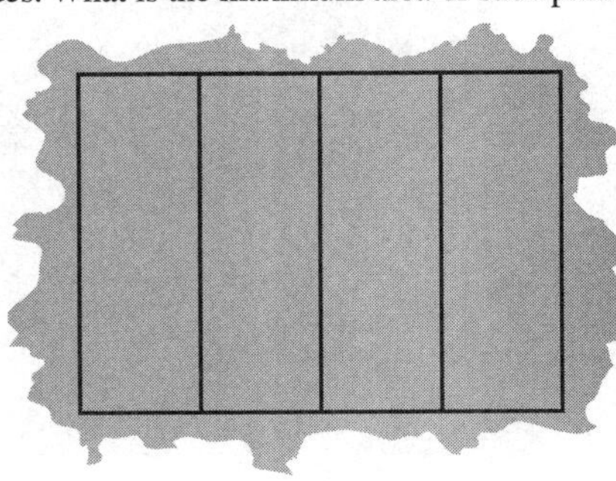

63. Gravity. A person standing near the edge of a cliff 100 feet above a lake throws a rock upward with an initial speed of 32 feet per second. The height of the rock above the lake at the bottom of the cliff is a function of time and is described by

$$h(t) = -16t^2 + 32t + 100$$

a. How many seconds will it take until the rock reaches its maximum height? What is that height?

b. At what time will the rock hit the water?

64. Gravity. A person holds a pistol straight upward and fires. The initial velocity of most bullets is around 1200 feet/second. The height of the bullet is a function of time and is described by

$$h(t) = -16t^2 + 1200t$$

How long, after the gun is fired, does the person have to get out of the way of the bullet falling from the sky?

65. Zero Gravity. As part of their training, astronauts ride the "vomit comet," NASA's reduced gravity KC 135A aircraft that performs parabolic flights to simulate weightlessness. The plane starts at an altitude of 20,000 feet and makes a steep climb at 52° with the horizon for 20–25 seconds and then dives at that same angle back down, repeatedly. The equation governing the altitude of the flight is

$$A(x) = -0.0003x^2 + 9.3x - 46{,}075$$

where $A(x)$ is altitude and x is horizontal distance in feet.

a. What is the maximum altitude the plane attains?

b. Over what horizontal distance is the entire maneuver performed? (Assume the starting and ending altitude is 20,000 feet.)

Courtesy NASA

NASA's "Vomit Comet"

66. Sports. A soccer ball is kicked from the ground at a 45° angle with an initial velocity of 40 feet per second. The height of the soccer ball above the ground is given by $H(x) = -0.0128x^2 + x$, where x is the horizontal distance the ball travels.

a. What is the maximum height the ball reaches?

b. What is the horizontal distance the ball travels?

67. Profit. A small company in Virginia Beach manufactures handcrafted surfboards. The profit of selling x boards is given by

$$P(x) = 20{,}000 + 80x - 0.4x^2$$

a. How many boards should be made to maximize the profit?

b. What is the maximum profit?

68. Environment: Fuel Economy. Gas mileage (miles per gallon, mpg) can be approximated by a quadratic function of speed. For a particular automobile, assume the vertex occurs when the speed is 50 mph (the mpg will be 30).

a. Write a quadratic function that models this relationship, assuming 70 mph corresponds to 25 mpg.

b. What gas mileage would you expect for this car driving 90 mph?

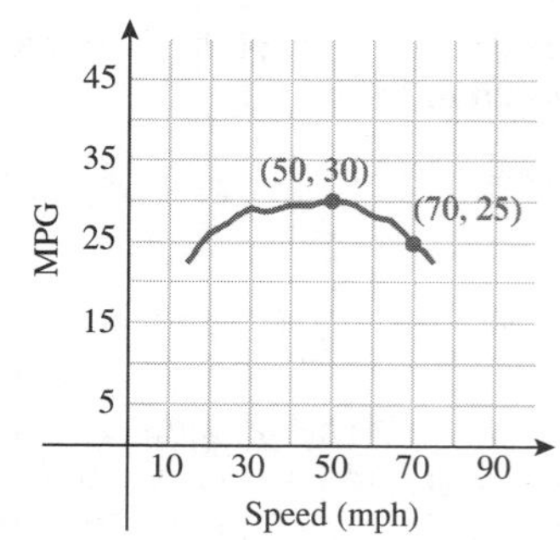

For Exercises 69 and 70, use the following information:

One function of particular interest in economics is the **profit function**. We denote this function by $P(x)$. It is defined to be the difference between revenue $R(x)$ and cost $C(x)$ so that

$$P(x) = R(x) - C(x)$$

The total revenue received from the sale of x goods at price p is given by

$$R(x) = px$$

The total cost function relates the cost of production to the level of output x. This includes both fixed costs C_f and variable costs C_v (costs per unit produced). The total cost in producing x goods is given by

$$C(x) = C_f + C_v x$$

Thus, the profit function is

$$P(x) = px - C_f - C_v x$$

Assume fixed costs are \$1,000, variable costs per unit are \$20, and the demand function is

$$p = 100 - x$$

69. Profit. How many units should the company produce to break even?

70. Profit. What is the maximum profit?

71. Cell Phones. The number of cell phones in the United States can be approximated by a quadratic function. In 1996 there were approximately 16 million cell phones, and in 2005 there were approximately 100 million. Let t be the number of years since 1996. The number of cell phones in 1996 is represented by (0, 16), and the number in 2005 is (9, 100). Let (0, 16) be the vertex.

a. Find a quadratic function that represents the number of cell phones.

b. Based on this model, how many cell phones will be in use in 2010?

72. Underage Smoking. The number of underage cigarette smokers (ages 10–17) has declined in the United States. The peak percent was in 1998 at 49%. In 2006 this had dropped to 36%. Let t be time in years after 1998 ($t = 0$ corresponds to 1998).

a. Find a quadratic function that models the percent of underage smokers as a function of time. Let (0, 49) be the vertex.

b. Now that you have the model, predict the percent of underage smokers in 2010.

73. Drug Concentration. The concentration of a drug in the bloodstream, measured in parts per million, can be modeled with a quadratic function. In 50 minutes the concentration is 93.75 parts per million. The maximum concentration of the drug in the bloodstream occurs in 225 minutes and is 400 parts per million.

a. Find a quadratic function that models the concentration of the drug as a function of time in minutes.

b. In how many minutes will the drug be eliminated from the bloodstream?

74. Revenue. Jeff operates a mobile car washing business. When he charged \$20 a car, he washed 70 cars a month. He raised the price to \$25 a car and his business dropped to 50 cars a month.

a. Find a linear function that represents the demand equation (the price per car as a function of the number of cars washed).

b. Find the revenue function $R(x) = xp$.

c. How many cars should he wash to maximize the revenue?

d. What price should he charge to maximize revenue?

CATCH THE MISTAKE

In Exercises 75–78, explain the mistake that is made. There may be a single mistake or there may be more than one mistake.

75. Plot the quadratic function $f(x) = (x + 3)^2 - 1$.

Solution:

Step 1: The parabola opens up because $a = 1 > 0$.

Step 2: The vertex is $(3, -1)$.

Step 3: The y-intercept is (0, 8).

Step 4: The x-intercepts are (2, 0) and (4, 0).

Step 5: Plot the vertex and intercepts, and connect the points with a smooth curve.

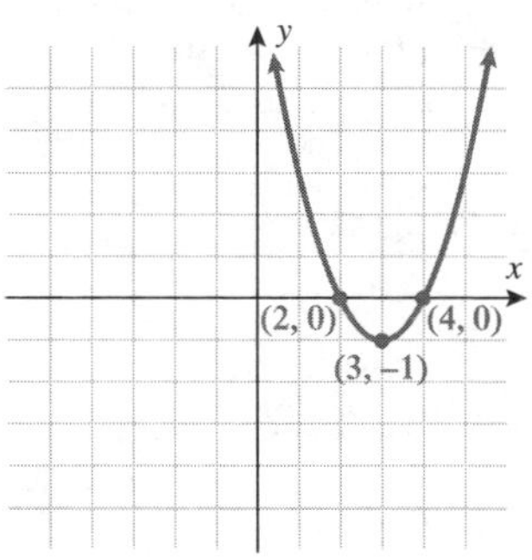

This is incorrect. What mistake(s) was made?

76. Determine the vertex of the quadratic function $f(x) = 2x^2 - 6x - 18$.

Solution:

Step 1: The vertex is given by $(h, k) = \left(-\frac{b}{2a}, f\left(-\frac{b}{2a}\right)\right)$.

In this case, $a = 2$ and $b = 6$.

Step 2: The x-coordinate of the vertex is

$$x = -\frac{6}{2(2)} = -\frac{6}{4} = -\frac{3}{2}$$

Step 3: The y-coordinate of the vertex is

$$\begin{aligned} f\left(-\frac{3}{2}\right) &= -2\left(-\frac{3}{2}\right)^2 + 6\left(-\frac{3}{2}\right) - 18 \\ &= -2\left(\frac{9}{4}\right) - \frac{18}{2} - 18 \\ &= -\frac{9}{2} - 9 - 18 \\ &= -\frac{63}{2} \end{aligned}$$

This is incorrect. What mistake(s) was made?

77. Rewrite the following quadratic function in standard form:

$$f(x) = -x^2 + 2x + 3$$

Solution:

Step 1: Group the variables together. $(-x^2 + 2x) + 3$

Step 2: Factor out a negative. $-(x^2 + 2x) + 3$

Step 3: Add and subtract 1 inside the parentheses. $-(x^2 + 2x + 1 - 1) + 3$

Step 4: Factor out the -1. $-(x^2 + 2x + 1) + 1 + 3$

Step 5: Simplify. $-(x + 1)^2 + 4$

This is incorrect. What mistake(s) was made?

78. Find the quadratic function whose vertex is $(2, -3)$ and whose graph passes through the point $(9, 0)$.

Solution:

Step 1: Write the quadratic function in standard form. $f(x) = a(x - h)^2 + k$

Step 2: Substitute $(h, k) = (2, -3)$. $f(x) = a(x - 2)^2 - 3$

Step 3: Substitute the point $(9, 0)$ and solve for a. $f(0) = a(0 - 2)^2 - 3 = 9$

$$4a - 3 = 9$$
$$4a = 12$$
$$a = 3$$

The quadratic function sought is: $f(x) = 3(x - 2)^2 - 3$.

This is incorrect. What mistake(s) was made?

CONCEPTUAL

In Exercises 79–82, determine whether each statement is true or false.

79. A quadratic function must have a y-intercept.

80. A quadratic function must have an x-intercept.

81. A quadratic function may have more than one y-intercept.

82. A quadratic function may have more than one x-intercept.

CHALLENGE

83. For the general quadratic equation, $f(x) = ax^2 + bx + c$, show that the vertex is $(h, k) = \left(-\frac{b}{2a}, f\left(-\frac{b}{2a}\right)\right)$.

84. Given the quadratic function $f(x) = a(x - h)^2 + k$, determine the x- and y-intercepts in terms of a, h, and k.

85. A rancher has 1000 feet of fence to enclose a pasture.

a. Determine the maximum area if a rectangular fence is used.

b. Determine the maximum area if a circular fence is used.

86. A 600-room hotel in Orlando is filled to capacity every night when the rate is \$90 per night. For every \$5 increase in the rate, 10 fewer rooms are filled. How much should the hotel charge to produce the maximum income? What is the maximum income?

TECHNOLOGY

87. On a graphing calculator, plot the quadratic function $f(x) = -0.002x^2 + 5.7x - 23$.

a. Identify the vertex of this parabola.

b. Identify the y-intercept.

c. Identify the x-intercepts (if any).

d. What is the axis of symmetry?

88. Determine the quadratic function whose vertex is $(-0.5, 1.7)$ and whose graph passes through the point $(0, 4)$.

a. Write the quadratic function in general form.

b. Plot this quadratic function with a graphing calculator.

c. Zoom in on the vertex and y-intercept. Do they agree with the given values?

In Exercises 89 and 90: (a) use the calculator commands STAT QuadReg to model the data using a quadratic function; (b) write the quadratic function in standard form and identify the vertex; (c) plot this quadratic function with a graphing calculator and use the TRACE key to highlight the given points. Do they agree with the given values?

89.

x	-2	2	5
y	-29.28	21.92	18.32

90.

x	-9	-2	4
y	-2.72	-16.18	6.62

SECTION 4.2 POLYNOMIAL FUNCTIONS OF HIGHER DEGREE

SKILLS OBJECTIVES

- Graph polynomial functions using transformations.
- Identify real zeros of a polynomial function and their multiplicities.
- Determine the end behavior of a polynomial function.
- Graph polynomial functions.
  - x-intercepts
  - multiplicity (touch/cross) of each zero
  - end behavior

CONCEPTUAL OBJECTIVES

- Understand that real zeros of polynomial functions correspond to x-intercepts.
- Understand the intermediate value theorem and how it assists in graphing polynomial functions.
- Realize that end behavior is a result of the leading term dominating.
- Understand that zeros correspond to factors of the polynomial.

Polynomial functions model many real-world applications. Often, the input is time and the output of the function can be many things. For example, the number of active duty military personnel in the United States, the number of new cases in the spread of an epidemic, and the stock price as a function of time t can all be modeled with polynomial functions.

DEFINITION **Polynomial Function**

Let n be a nonnegative integer and let $a_n, a_{n-1}, \ldots, a_2, a_1, a_0$ be real numbers with $a_n \neq 0$. The function

$$f(x) = a_n x^n + a_{n-1}x^{n-1} + \cdots + a_2 x^2 + a_1 x + a_0$$

is called a **polynomial function of x with degree n**. The coefficient a_n is called the leading coefficient.

Note: If n is a nonnegative integer, then $n - 1, n - 2, \ldots, 2, 1, 0$ are also nonnegative integers.

EXAMPLE 1 Identifying Polynomials and Their Degree

For each of the functions given, determine whether the function is a polynomial function. If it is a polynomial function, then state the degree of the polynomial. If it is not a polynomial function, justify your answer.

a. $f(x) = 3 - 2x^5$ **b.** $F(x) = \sqrt{x} + 1$ **c.** $g(x) = 2$

d. $h(x) = 3x^2 - 2x + 5$ **e.** $H(x) = 4x^5(2x - 3)^2$ **f.** $G(x) = 2x^4 - 5x^3 - 4x^{-2}$

Solution:

a. $f(x)$ is a polynomial function of degree 5.

b. $F(x)$ is not a polynomial function. The variable x is raised to the power of $\frac{1}{2}$, which is not an integer.

c. $g(x)$ is a polynomial function of degree zero, also known as a constant function. Note that $g(x) = 2$ can also be written as $g(x) = 2x^0$ (assuming $x \neq 0$).

d. $h(x)$ is a polynomial function of degree 2. A polynomial function of degree 2 is called a quadratic function.

e. $H(x)$ is a polynomial function of degree 7. *Note:* $4x^5(4x^2 - 12x + 9) = 16x^7 - 48x^6 + 36x^5$.

f. $G(x)$ is not a polynomial function. $-4x^{-2}$ has an exponent that is negative.

■ **Answer:**
a. $f(x)$ is not a polynomial because x is raised to the power of -1, which is a negative integer.
b. $g(x)$ is a polynomial of degree 13.

■ **YOUR TURN** For each of the functions given, determine whether the function is a polynomial function. If it is a polynomial function, then state the degree of the polynomial. If it is not a polynomial function, justify your answer.

a. $f(x) = \frac{1}{x} + 2$ **b.** $g(x) = 3x^8(x - 2)^2(x + 1)^3$

Graphs of Polynomial Functions

Whenever we have discussed a particular polynomial, we have graphed it too. The graph of a constant function (degree 0) is a horizontal line. The graph of a general linear function (degree 1) is a slant line. The graph of a quadratic function (degree 2) is a parabola. These functions are summarized in the table below.

POLYNOMIAL	DEGREE	SPECIAL NAME	GRAPH
$f(x) = c$	0	Constant function	Horizontal line
$f(x) = mx + b$	1	Linear function	Line • Slope $= m$ • y-intercept: $(0, b)$
$f(x) = ax^2 + bx + c$	2	Quadratic function	Parabola • Opens up if $a > 0$. • Opens down if $a < 0$.

How do we graph polynomial functions that are of degree 3 or higher, and why do we care? Polynomial functions model real-world applications, as mentioned earlier. One example is the percentage of fat in our bodies as we age. We can model the weight of a baby after it comes home from the hospital as a function of time. When a baby comes home from the hospital, it usually experiences weight loss. Then typically there is an increase in the percent of body fat when the baby is nursing. When infants start to walk, the increase in exercise is associated with a drop in the percentage of fat. Growth spurts in children are examples of the percent of body fat increasing and decreasing. Later in life, our metabolism slows down, and typically the percent of body fat increases. We will model this with a polynomial function. Other examples are stock prices, the federal funds rate, and yo-yo dieting as functions of time.

Polynomial functions are considered simple functions. Graphs of all polynomial functions are both *continuous* and *smooth.* A **continuous** graph is one you can draw completely without picking up your pencil (the graph has no jumps or holes). A **smooth** graph has no sharp corners. The following graphs illustrate what it means to be smooth (no sharp corners or cusps) and continuous (no holes or jumps).

The graph is *not continuous.*

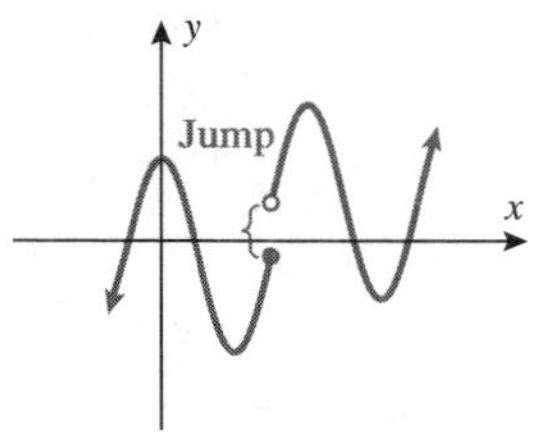

The graph is *not continuous.*

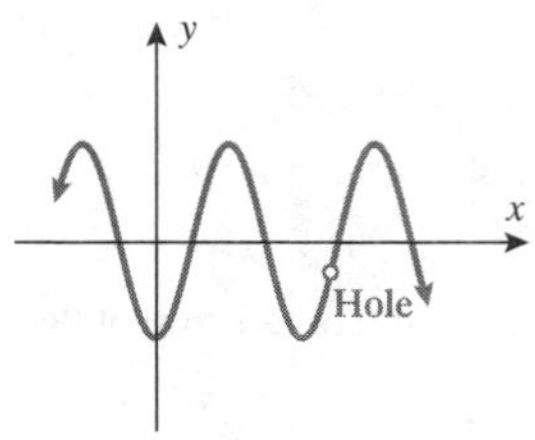

The graph is *continuous* but *not smooth*.

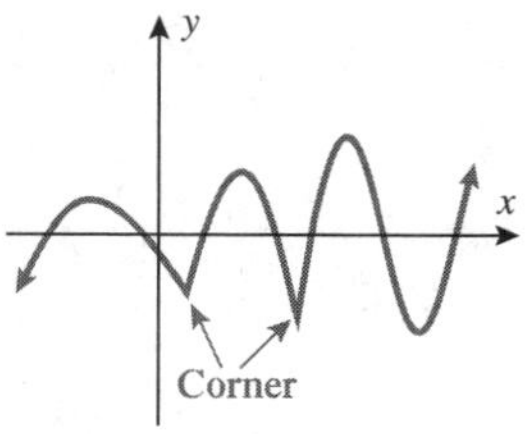

The graph is *continuous* and *smooth*.

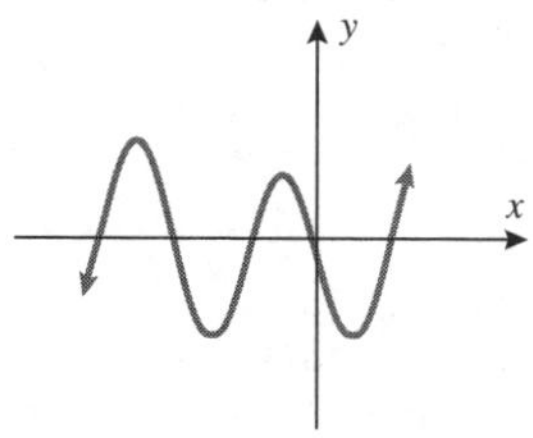

Study Tip

All polynomial functions have graphs that are both continuous and smooth.

All polynomial functions have graphs that are both continuous and smooth. Recall from Chapter 3 that graphs of functions can be drawn by hand using graphing aids such as intercepts and symmetry. The graphs of polynomial functions can be graphed using these same aids. Let's start with the simplest types of polynomial functions called **power functions**. Power functions are monomial functions of the form $f(x) = x^n$, where n is an integer greater than zero.

DEFINITION **Power Function**

Let n be a positive integer and the coefficient $a \neq 0$ be a real number. The function

$$f(x) = ax^n$$

is called a **power function of degree n.**

Power functions with *even* powers look similar to the square function.

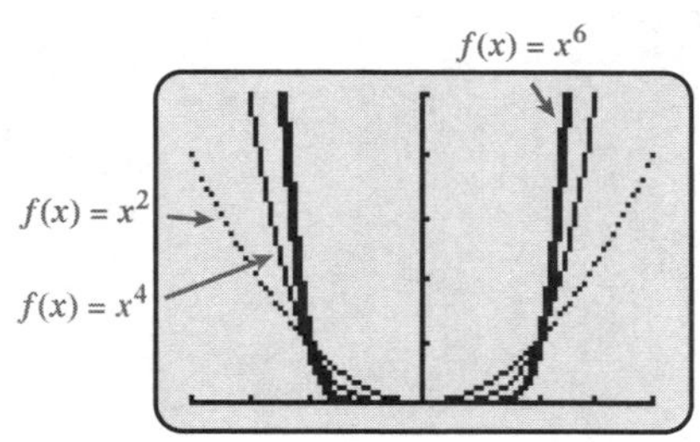

Power functions with *odd* powers (other than $n = 1$) look similar to the cube function.

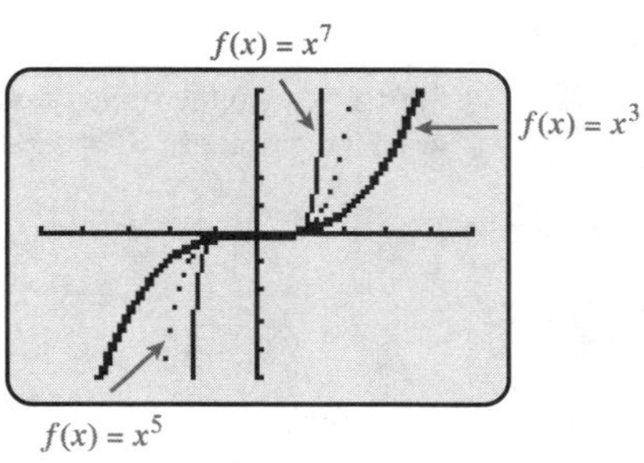

All even power functions have similar characteristics to a quadratic function (parabola), and all odd ($n > 1$) power functions have similar characteristics to a cubic function. For example, all even functions are symmetric with respect to the y-axis, whereas all odd functions are symmetric with respect to the origin. This table summarizes their characteristics.

CHARACTERISTICS OF POWER FUNCTIONS: $f(x) = x^n$

	n EVEN	n ODD
Symmetry	y-axis	Origin
Domain	$(-\infty, \infty)$	$(-\infty, \infty)$
Range	$[0, \infty)$	$(-\infty, \infty)$
Some key points that lie on the graph	$(-1, 1)$, $(0, 0)$, and $(1, 1)$	$(-1, -1)$, $(0, 0)$, and $(1, 1)$
Increasing	$(0, \infty)$	$(-\infty, \infty)$
Decreasing	$(-\infty, 0)$	N/A

Graphing Polynomial Functions Using Transformations of Power Functions

We now have the tools to graph polynomial functions that are transformations of power functions. We will use the power functions summarized above, combined with our graphing techniques such as horizontal and vertical shifting and reflection (Section 3.3).

EXAMPLE 2 Graphing Transformations of Power Functions

Graph the function $f(x) = (x - 1)^3$.

Solution:

STEP 1 Start with the graph of $y = x^3$.

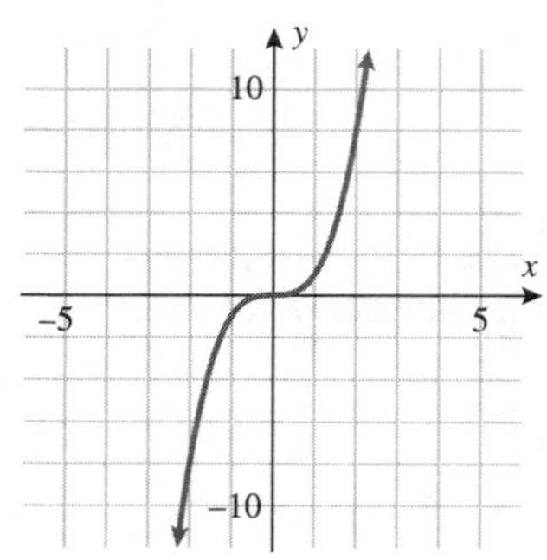

STEP 2 Shift $y = x^3$ to the right 1 unit to yield the graph of $f(x) = (x - 1)^3$.

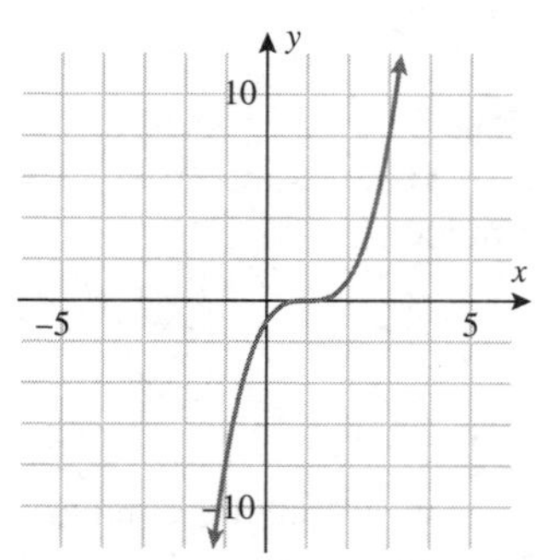

■ **YOUR TURN** Graph the function $f(x) = 1 - x^4$.

■ **Answer:** $f(x) = 1 - x^4$

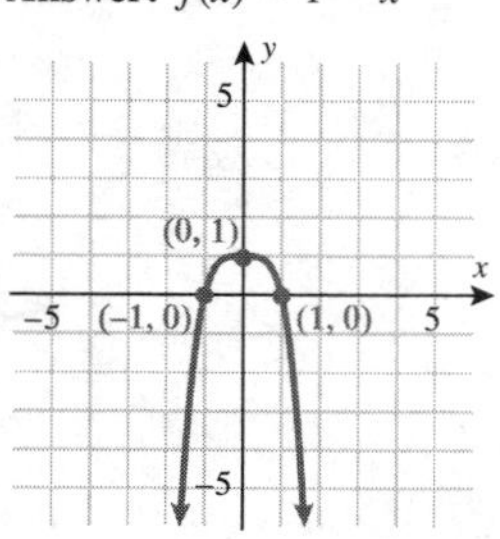

Real Zeros of a Polynomial Function

How do we graph general polynomial functions of degree greater than or equal to 3 if they cannot be written as transformations of power functions? We start by identifying the x-intercepts of the polynomial function. Recall that we determine the x-intercepts by setting the function equal to *zero* and solving for x. Therefore, an alternative name for an x-intercept of a function is a *zero* of the function. In our experience, to set a quadratic function equal to zero, the first step is to factor the quadratic expression into linear factors and then set each factor equal to zero. Therefore, there are four equivalent relationships that are summarized in the following box.

Real zeros of polynomial functions

If $f(x)$ is a polynomial function and a is a *real* number, then the following statements are equivalent.

Relationship 1: $x = a$ is a **solution**, or **root**, of the equation $f(x) = 0$.
Relationship 2: $(a, 0)$ is an **x-intercept** of the graph of $f(x)$.
Relationship 3: $x = a$ is a **zero** of the function $f(x)$.
Relationship 4: $(x - a)$ is a **factor** of $f(x)$.

Study Tip

Real zeros correspond to x-intercepts.

Let's use a simple polynomial function to illustrate these four relationships. We'll focus on the quadratic function $f(x) = x^2 - 1$. The graph of this function is a parabola that opens up and has as its vertex the point $(0, -1)$.

Illustration of Relationship 1

Set the function equal to zero, $f(x) = 0$.	$x^2 - 1 = 0$
Factor.	$(x - 1)(x + 1) = 0$
Solve.	$x = 1$ or $x = -1$

$x = -1$ and $x = 1$ are solutions, or roots, of the equation $x^2 - 1 = 0$.

Illustration of Relationship 2

The **x-intercepts are $(-1, 0)$ and $(1, 0)$**.

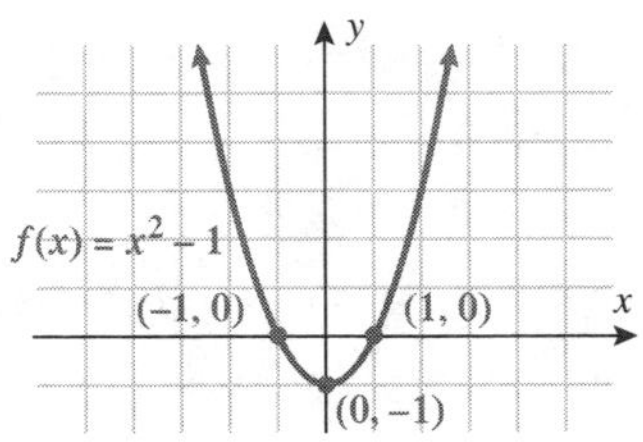

Illustration of Relationship 3

The value of the function at $x = -1$ and $x = 1$ is 0.

$$\mathbf{f(-1)} = (-1)^2 - 1 = \mathbf{0}$$
$$\mathbf{f(1)} = (1)^2 - 1 = \mathbf{0}$$

Illustration of Relationship 4

$$f(x) = x^2 - 1$$
$$= (x - 1)(x + 1)$$

$x = 1$ is a zero of the polynomial, so $(x - 1)$ is a factor.
$x = -1$ is a zero of the polynomial, so $(x + 1)$ is a factor.

We have a good reason for wanting to know the x-intercepts, or zeros. When the value of a continuous function transitions from negative to positive and vice versa, it must pass through zero.

DEFINITION Intermediate Value Theorem

Let a and b be real numbers such that $a < b$ and let f be a polynomial function. If $f(a)$ and $f(b)$ have opposite signs, then there is at least one zero between a and b.

Technology Tip

Graph $y_1 = x^3 + x^2 - 2x$.

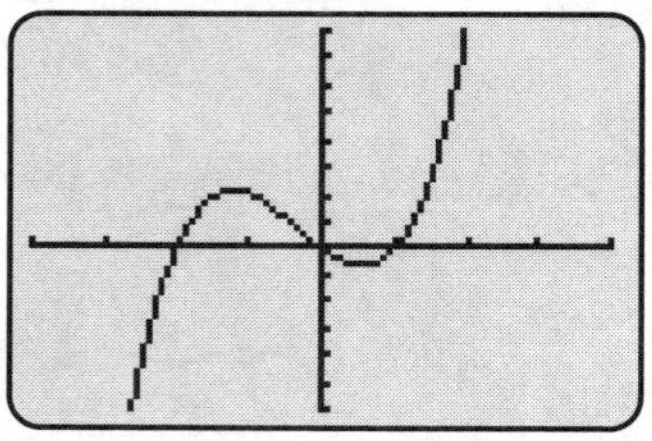

The zeros of the function -2, 0, and 1 correspond to the x-intercepts $(-2, 0)$, $(0, 0)$, and $(1, 0)$.

The table supports the real zeros shown by the graph.

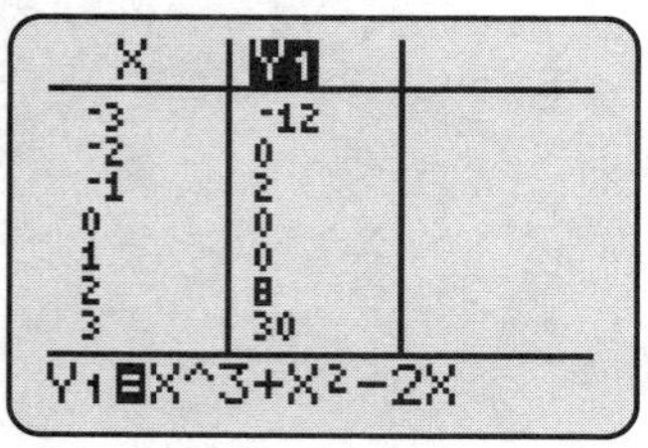

X	Y1
-3	-12
-2	0
-1	2
0	0
1	0
2	8
3	30

Y1■X^3+X²-2X

■ **Answer:** The zeros are 0, 3, and 4.

The intermediate value theorem will be used later in this chapter to assist us in finding real zeros of a polynomial function. For now, it tells us that in order to change signs, the polynomial function must pass through the x-axis. In other words, once we know the zeros, then we know that between two consecutive zeros the polynomial is either entirely above the x-axis or entirely below the x-axis. This enables us to break the x-axis down into intervals that we can test, which will assist us in graphing polynomial functions. Keep in mind, though, that the existence of a zero does not imply that the function will change signs—as you will see in the subsection on graphing general polynomial functions.

EXAMPLE 3 Identifying the Real Zeros of a Polynomial Function

Find the zeros of the polynomial function $f(x) = x^3 + x^2 - 2x$.

Solution:

Set the function equal to zero. $\quad x^3 + x^2 - 2x = 0$

Factor out an x common to all three terms. $\quad x(x^2 + x - 2) = 0$

Factor the quadratic expression inside the parentheses. $\quad x(x + 2)(x - 1) = 0$

Apply the zero product property. $\quad x = 0$ or $(x + 2) = 0$ or $(x - 1) = 0$

Solve. $\quad x = -2, x = 0$, and $x = 1$

The zeros are $\boxed{-2, 0, \text{and } 1}$.

■ **YOUR TURN** Find the zeros of the polynomial function $f(x) = x^3 - 7x^2 + 12x$.

Recall that when we were factoring a quadratic equation, if the factor was raised to a power greater than 1, the corresponding root, or zero, was repeated. For example, the quadratic equation $x^2 - 2x + 1 = 0$ when factored is written as $(x - 1)^2 = 0$. The solution, or root, in this case is $x = 1$, and we say that it is a **repeated** root. Similarly, when determining zeros of higher order polynomial functions, if a factor is repeated, we say that the zero is a repeated, or **multiple**, zero of the function. The number of times that a zero repeats is called its *multiplicity*.

DEFINITION **Multiplicity of a Zero**

If $(x - a)^n$ is a factor of a polynomial f, then a is called a **zero of multiplicity n** of f.

EXAMPLE 4 **Finding the Multiplicities of Zeros of a Polynomial Function**

Find the zeros, and state their multiplicities, of the polynomial function $g(x) = (x - 1)^2\left(x + \frac{3}{5}\right)^7(x + 5)$.

Solution:

1 is a zero of multiplicity 2.
$-\frac{3}{5}$ is a zero of multiplicity 7.
-5 is a zero of multiplicity 1.

Note: Adding the multiplicities yields the degree of the polynomial. The polynomial $g(x)$ is of degree 10, since $2 + 7 + 1 = 10$.

YOUR TURN For the polynomial $h(x)$, determine the zeros and state their multiplicities.

$$h(x) = x^2(x - 2)^3\left(x + \frac{1}{2}\right)^5$$

■ **Answer:** 0 is a zero of multiplicity 2. 2 is a zero of multiplicity 3. $-\frac{1}{2}$ is a zero of multiplicity 5.

EXAMPLE 5 **Finding a Polynomial from Its Zeros**

Find a polynomial of degree 7 whose zeros are:

-2 (multiplicity 2) 0 (multiplicity 4) 1 (multiplicity 1)

Solution:

If $x = a$ is a zero, then $(x - a)$ is a factor. $f(x) = (x + 2)^2(x - 0)^4(x - 1)^1$

Simplify. $= x^4(x + 2)^2(x - 1)$

Square the binomial. $= x^4(x^2 + 4x + 4)(x - 1)$

Multiply the two polynomials. $= x^4(x^3 + 3x^2 - 4)$

Distribute x^4. $= \boxed{x^7 + 3x^6 - 4x^4}$

Graphing General Polynomial Functions

Let's develop a strategy for sketching an approximate graph of any polynomial function. First, we determine the x- and y-intercepts. Then we use the x-intercepts, or zeros, to divide the domain into intervals where the polynomial is positive or negative so that we can find points in those intervals to assist in sketching a smooth and continuous graph. *Note:* It is not always possible to find x-intercepts. Sometimes there are no x-intercepts.

Study Tip

It is not always possible to find x-intercepts. Sometimes there are no x-intercepts.

Technology Tip

The graph of $f(x) = (x + 2)(x - 1)^2$ is shown.

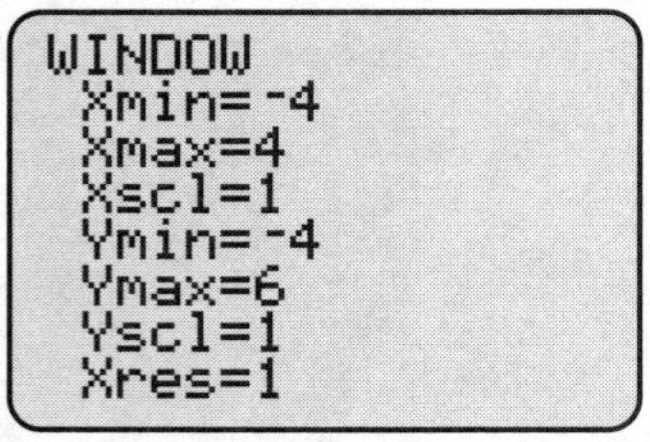

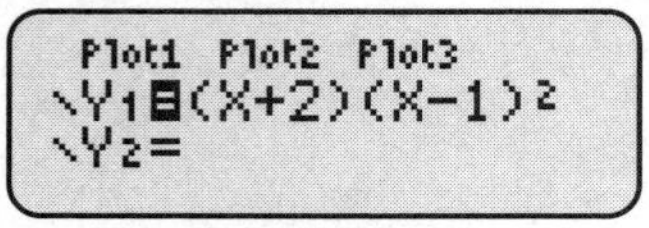

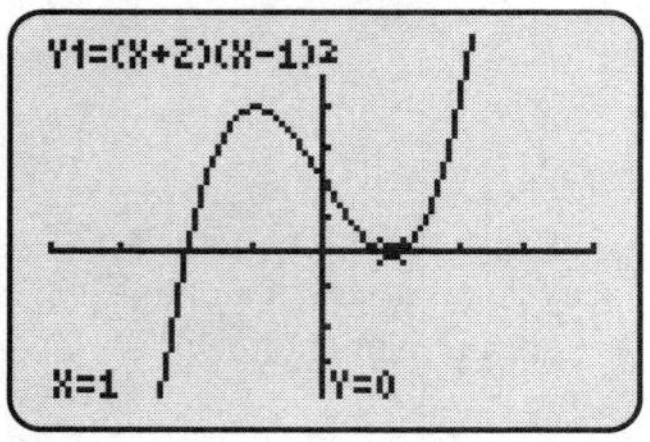

Note: The graph crosses the x-axis at the point $x = -2$ and touches the x-axis at the point $x = 1$.

A table of values supports the graph.

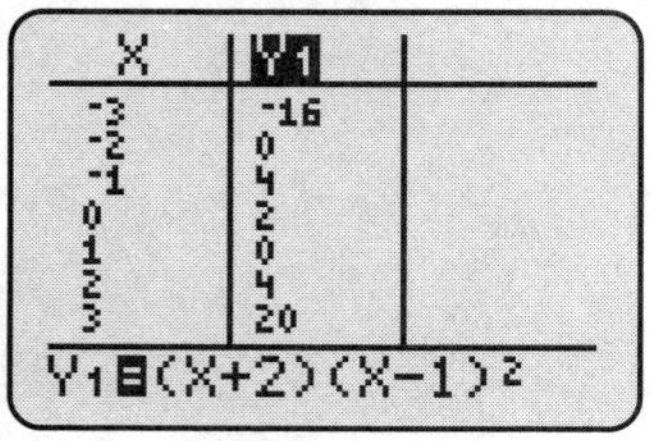

EXAMPLE 6 Using a Strategy for Sketching the Graph of a Polynomial Function

Sketch the graph of $f(x) = (x + 2)(x - 1)^2$.

Solution:

STEP 1 Find the y-intercept. (Let $x = 0$.)
$f(0) = (2)(-1)^2 = 2$
$(0, 2)$ is the y-intercept

STEP 2 Find any x-intercepts. (Set $f(x) = 0$.)
$f(x) = (x + 2)(x - 1)^2 = 0$
$x = -2$ or $x = 1$
$(-2, 0)$ and $(1, 0)$ are the x-intercepts

STEP 3 Plot the intercepts.

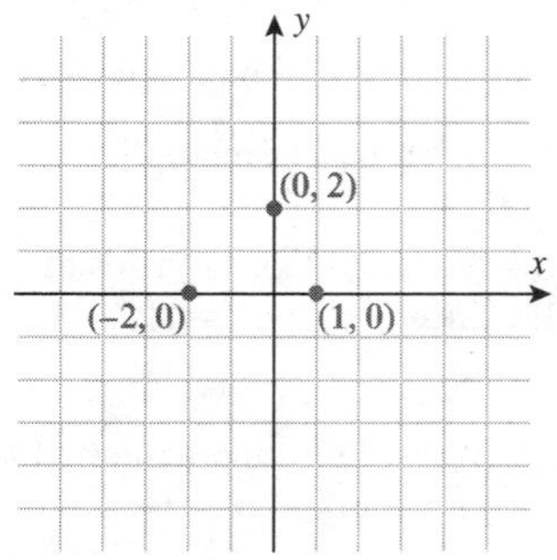

STEP 4 Divide the x-axis into intervals:* $(-\infty, -2)$, $(-2, 1)$, and $(1, \infty)$

*The x-intercepts $(-2, 0)$ and $(1, 0)$ divide the x-axis into three intervals similar to those created by zeros when we studied inequalities in Section 1.5.

STEP 5 Select a number in each interval and test each interval. The function $f(x)$ either *crosses* the x-axis at an x-intercept or *touches* the x-axis at an x-intercept. Therefore, we need to check each of these intervals to determine whether the function is positive (above the x-axis) or negative (below the x-axis). We do so by selecting numbers in the intervals and determining the value of the function at the corresponding points.

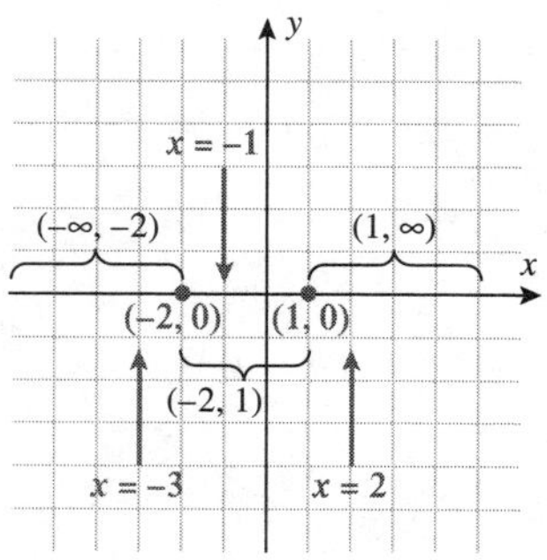

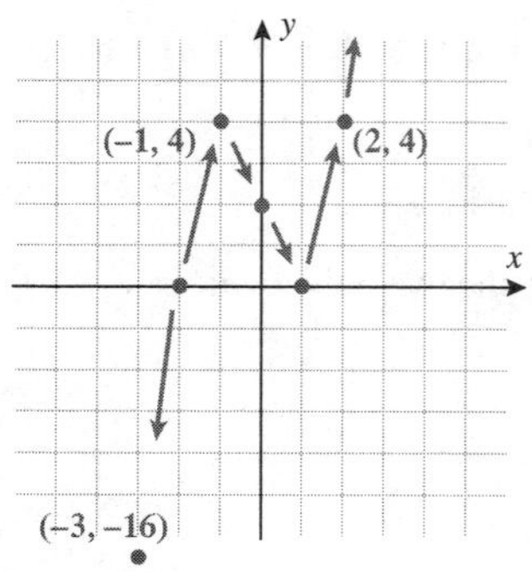

Interval	$(-\infty, -2)$	$(-2, 1)$	$(1, \infty)$
Number Selected in Interval	-3	-1	2
Value of Function	$f(-3) = -16$	$f(-1) = 4$	$f(2) = 4$
Point on Graph	$(-3, -16)$	$(-1, 4)$	$(2, 4)$
Interval Relation to x-Axis	Below x-axis	Above x-axis	Above x-axis

From the table, we find three additional points on the graph: $(-3, -16)$, $(-1, 4)$, and $(2, 4)$. The point $(-2, 0)$ is an intercept where the function *crosses* the x-axis, because it is below the x-axis to the left of -2 and above the x-axis to the right of -2. The point $(1, 0)$ is an intercept where the function *touches* the x-axis, because it is above the x-axis on both sides of $x = 1$. Connecting these points with a smooth curve yields the graph.

STEP 6 Sketch a plot of the function.

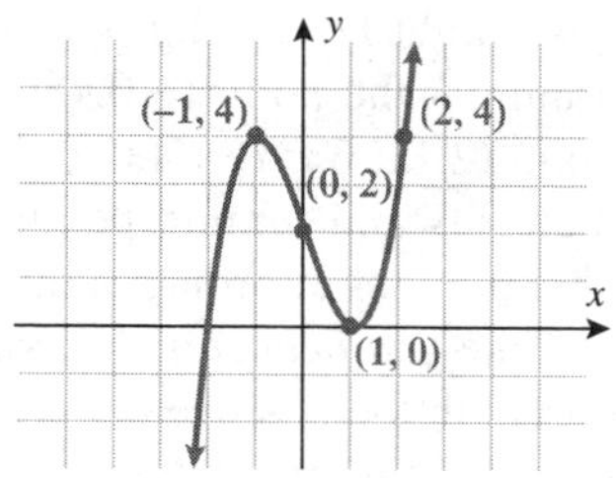

■ **YOUR TURN** Sketch the graph of $f(x) = x^2(x + 3)^2$.

Study Tip

We do not know for sure that the points $(-1, 4)$ and $(1, 0)$ are turning points. A graphing utility can confirm this, and later in calculus you will learn how to find relative maximum points and relative minimum points.

■ **Answer:**

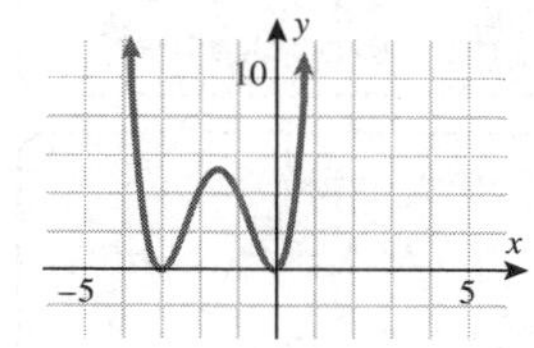

In Example 6, we found that the function crosses the x-axis at the point $(-2, 0)$. Note that -2 is a zero of multiplicity 1. We also found that the function touches the x-axis at the point $(1, 0)$. Note that 1 is a zero of multiplicity 2. In general, zeros with even multiplicity correspond to intercepts where the function touches the x-axis and zeros with odd multiplicity correspond to intercepts where the function crosses the x-axis.

MULTIPLICITY OF A ZERO AND RELATION TO THE GRAPH OF A POLYNOMIAL

If a is a zero of $f(x)$, then:

MULTIPLICITY OF a	$f(x)$ ON EITHER SIDE OF $x = a$	GRAPH OF FUNCTION AT THE INTERCEPT
Even	Does not change sign	Touches the x-axis (turns around) at point $(a, 0)$
Odd	Changes sign	Crosses the x-axis at point $(a, 0)$

Study Tip

In general, zeros with *even* multiplicity correspond to intercepts where the function *touches* the x-axis and zeros with *odd* multiplicity correspond to intercepts where the function *crosses* the x-axis.

Also in Example 6, we know that somewhere in the interval $(-2, 1)$ the function must reach a maximum and then turn back toward the x-axis, because both points $(-2, 0)$ and $(1, 0)$ correspond to x-intercepts. When we sketch the graph, it "appears" that the point $(-1, 4)$ is a *turning point*. The point $(1, 0)$ also corresponds to a turning point. In general, if f is a polynomial of degree n, then the graph of f has at most $n - 1$ turning points.

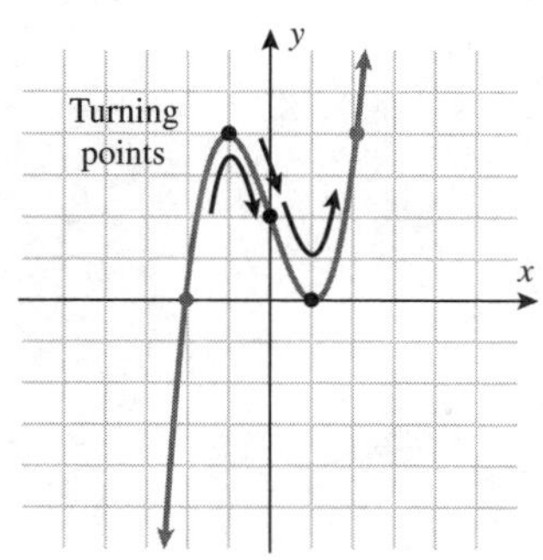

The point $(-1, 4)$, which we call a turning point, is also a "high point" on the graph in the vicinity of the point $(-1, 4)$. Also note that the point $(1, 0)$, which we call a turning point, is a "low point" on the graph in the vicinity of the point $(1, 0)$. We call a "high point" on a graph a **local (relative) maximum** and a "low point" on a graph a **local (relative) minimum**. For quadratic functions we can find the maximum or minimum point by finding the vertex. However, for higher degree polynomial functions, we rely on graphing utilities to locate such points. Later in calculus, techniques will be developed for finding such points exactly. For now, we use the zoom and trace features to locate such points

Study Tip

If f is a polynomial of degree n, then the graph of f has at most $n - 1$ turning points.

Technology Tip

Use TI to graph the function $f(x) = x^3 - 2x^2 - 5x + 6$ as Y_1.

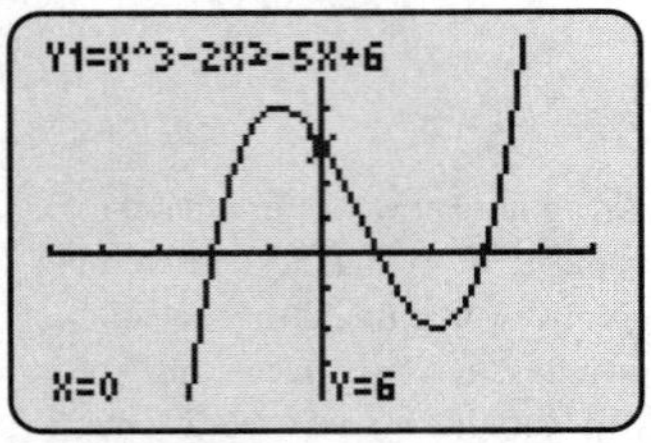

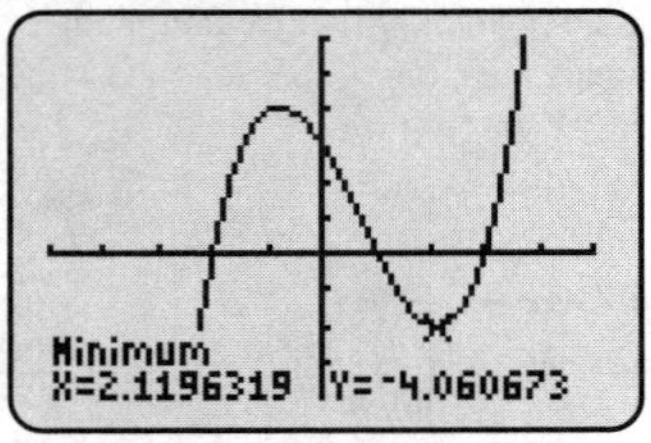

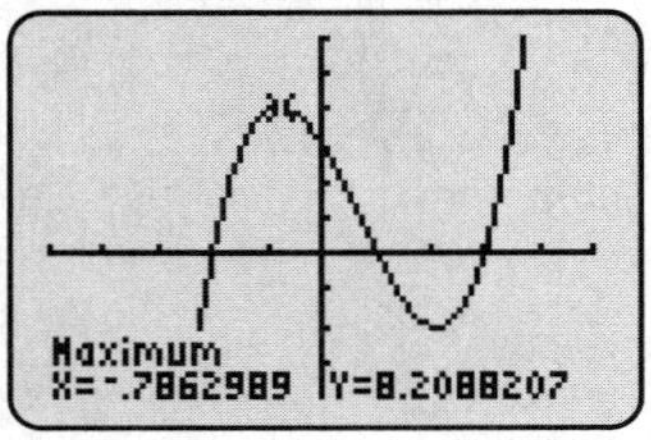

on a graph, and we can use the table feature of a graphing utility to approximate relative minima or maxima.

Let us take the polynomial $f(x) = x^3 - 2x^2 - 5x + 6$. Using methods discussed thus-far we can find that the x-intercepts of its graph are $(-2, 0)$, $(1, 0)$, and $(3, 0)$ and the y-intercept is the point $(0, 6)$. We can also find additional points that lie on the graph such as $(-1, 8)$ and $(2, -4)$. Plotting these points we might "think" that the points $(-1, 8)$ and $(2, -4)$ might be turning points, but a graphing utility reveals an approximate relative maximum at the point $(-0.7863, 8.2088207)$ and an approximate relative minimum at the point $(2.1196331, -4.060673)$.

Intercepts and turning points assist us in sketching graphs of polynomial functions. Another piece of information that will assist us in graphing polynomial functions is knowledge of the *end behavior.* All polynomials eventually rise or fall without bound as x gets large in both the positive $(x \to \infty)$ and negative $(x \to -\infty)$ directions. The highest degree monomial within the polynomial dominates the *end behavior.* In other words, the highest power term is eventually going to overwhelm the other terms as x grows without bound.

END BEHAVIOR

As x gets large in the positive $(x \to \infty)$ and negative $(x \to -\infty)$ directions, the graph of the polynomial

$$f(x) = a_n x^n + a_{n-1} x^{n-1} + \cdots + a_2 x^2 + a_1 x + a_0$$

has the same behavior as the power function

$$y = a_n x^n$$

Power functions behave much like a quadratic function (parabola) for even-degree polynomial functions and much like a cubic function for odd-degree polynomial functions. There are four possibilities because the leading coefficient can be positive or negative with either an odd or even power.

Let $y = a_n x^n$; then

n	Even	Even	Odd	Odd
a_n	Positive	Negative	Negative	Positive
$x \to -\infty$ **(Left)**	The graph of the function *rises*	The graph of the function *falls*	The graph of the function *rises*	The graph of the function *falls*
$x \to \infty$ **(Right)**	The graph of the function *rises*	The graph of the function *falls*	The graph of the function *falls*	The graph of the function *rises*
Graph	$a_n > 0$ 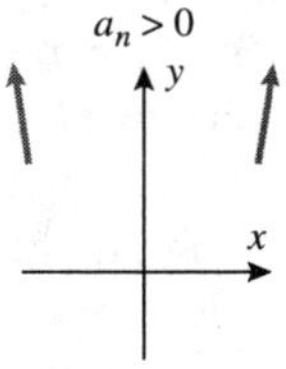	$a_n < 0$ 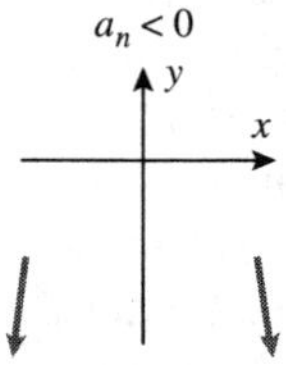	$a_n < 0$ 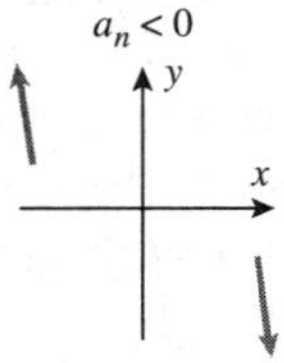	$a_n > 0$ 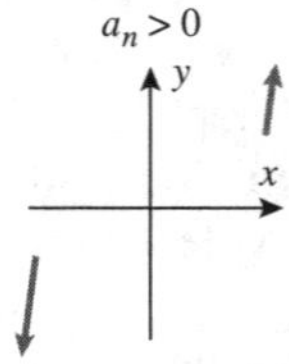

EXAMPLE 7 Graphing a Polynomial Function

Sketch a graph of the polynomial function $f(x) = 2x^4 - 8x^2$.

Solution:

STEP 1 Determine the y-intercept: $(x = 0)$. $f(0) = 0$

The y-intercept corresponds to the point $(0, 0)$.

STEP 2 Find the zeros of the polynomial. $f(x) = 2x^4 - 8x^2$

Factor out the common $2x^2$. $= 2x^2(x^2 - 4)$

Factor the quadratic binomial. $= 2x^2(x - 2)(x + 2)$

Set $f(x) = 0$. $= 2x^2(x - 2)(x + 2) = 0$

0 is a zero of multiplicity 2. The graph will *touch* the x-axis.

2 is a zero of multiplicity 1. The graph will *cross* the x-axis.

-2 is a zero of multiplicity 1. The graph will *cross* the x-axis.

STEP 3 Determine the end behavior. $f(x) = 2x^4 - 8x^2$ behaves like $y = 2x^4$.

$y = 2x^4$ is of even degree, and the leading coefficient is positive, so the graph rises without bound as x gets large in both the positive and negative directions.

STEP 4 Sketch the intercepts and end behavior.

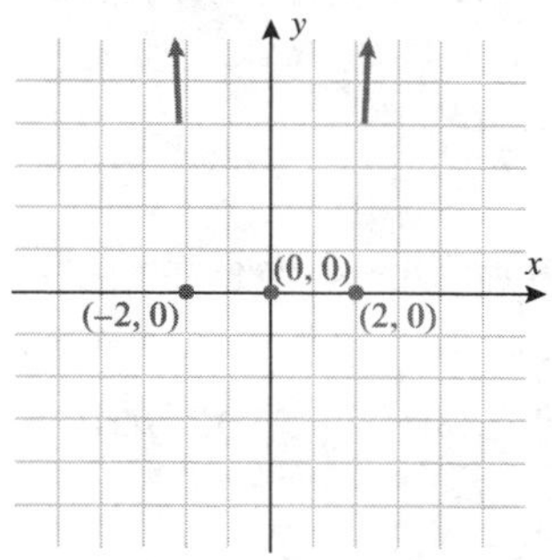

STEP 5 Find additional points.

x	-1	$-\frac{1}{2}$	$\frac{1}{2}$	1
f(x)	-6	$-\frac{15}{8}$	$-\frac{15}{8}$	-6

STEP 6 Sketch the graph.

- estimate additional points
- connect with a smooth curve

Note the symmetry about the y-axis. This function is an even function: $f(-x) = f(x)$.

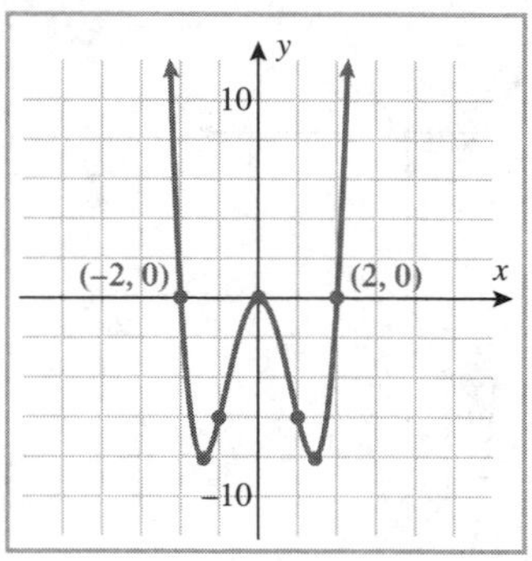

It is important to note that the local minimums occur at $x = \pm\sqrt{2} \approx \pm 1.14$, but at this time can only be illustrated using a graphing utility.

■ **YOUR TURN** Sketch a graph of the polynomial function $f(x) = x^5 - 4x^3$.

Technology Tip

The graph of $f(x) = (x + 2)(x - 1)^2$ is shown.

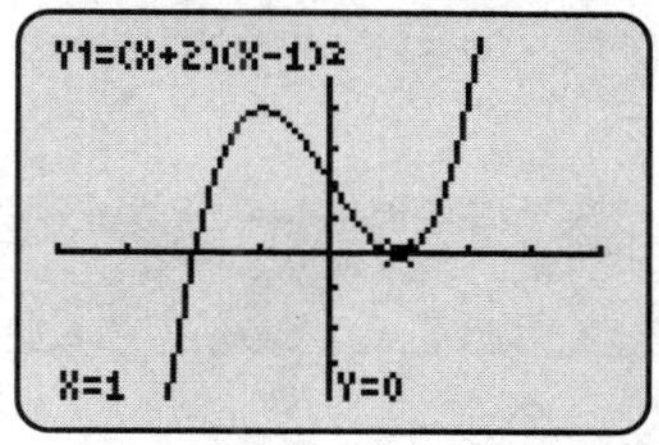

Note: The graph crosses the x-axis at the point $x = -2$ and touches the x-axis at the point $x = 1$. A table of values supports the graph.

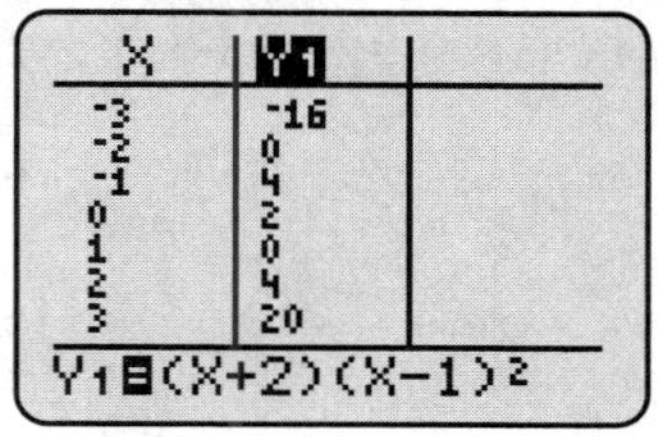

■ **Answer:**

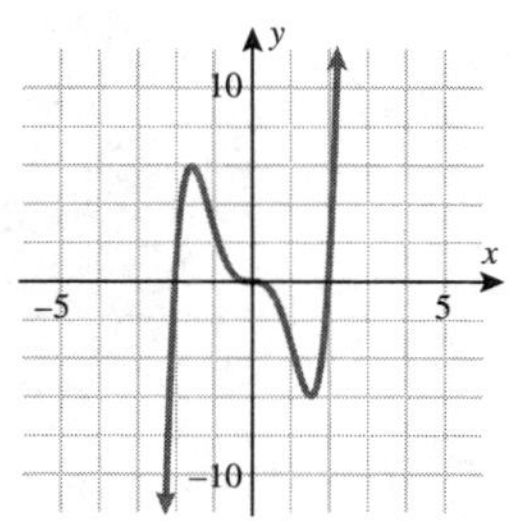

SECTION 4.2 SUMMARY

In general, polynomials can be graphed in one of two ways:

- Use graph-shifting techniques with power functions.
- General polynomial function.
  1. Identify intercepts.
  2. Determine each real zero and its multiplicity, and ascertain whether the graph crosses or touches the x-axis there.
  3. x-intercepts (real zeros) divide the x-axis into intervals. Test points in the intervals to determine whether the graph is above or below the x-axis.
  4. Determine the end behavior by investigating the end behavior of the highest degree monomial.
  5. Sketch the graph with a smooth curve.

SECTION 4.2 EXERCISES

SKILLS

In Exercises 1–10, determine which functions are polynomials, and for those that are, state their degree.

1. $f(x) = -3x^2 + 15x - 7$ **2.** $f(x) = 2x^5 - x^2 + 13$ **3.** $g(x) = (x + 2)^3\left(x - \frac{3}{5}\right)^2$ **4.** $g(x) = x^4(x - 1)^2(x + 2.5)^3$

5. $h(x) = \sqrt{x} + 1$ **6.** $h(x) = (x - 1)^{1/2} + 5x$ **7.** $F(x) = x^{1/3} + 7x^2 - 2$ **8.** $F(x) = 3x^2 + 7x - \dfrac{2}{3x}$

9. $G(x) = \dfrac{x + 1}{x^2}$ **10.** $H(x) = \dfrac{x^2 + 1}{2}$

In Exercises 11–18, match the polynomial function with its graph.

11. $f(x) = -3x + 1$ **12.** $f(x) = -3x^2 - x$ **13.** $f(x) = x^2 + x$ **14.** $f(x) = -2x^3 + 4x^2 - 6x$

15. $f(x) = x^3 - x^2$ **16.** $f(x) = 2x^4 - 18x^2$ **17.** $f(x) = -x^4 + 5x^3$ **18.** $f(x) = x^5 - 5x^3 + 4x$

a.

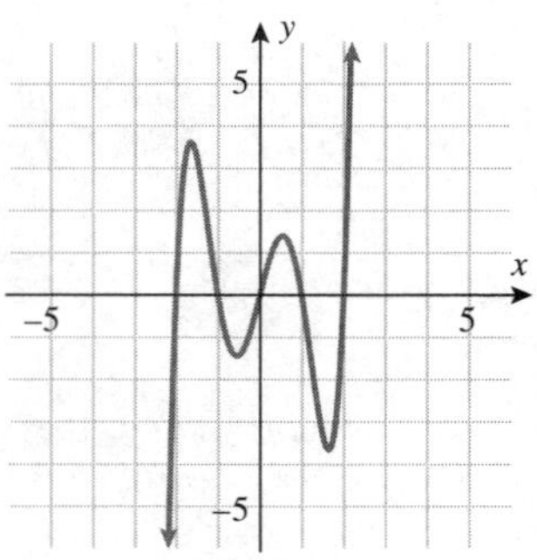

b.

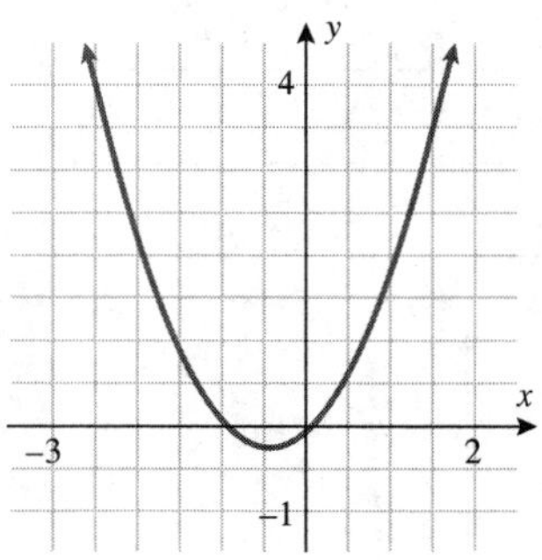

c.

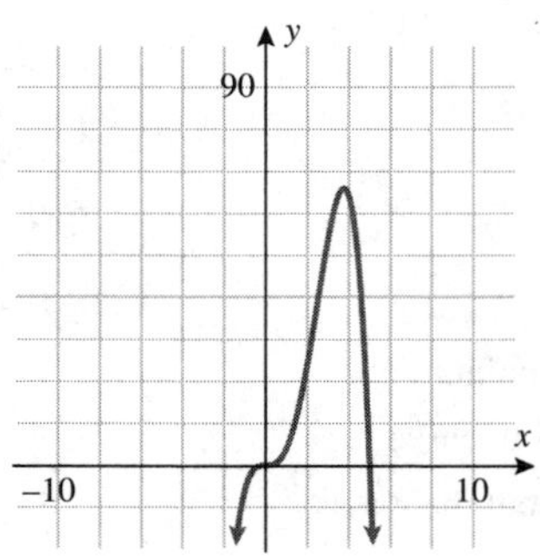

d.

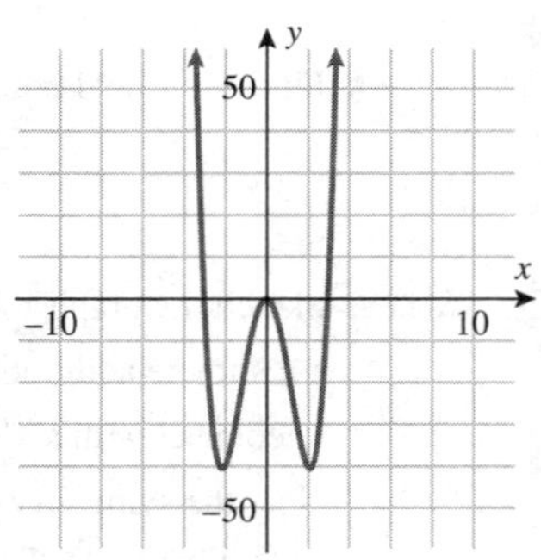

e.

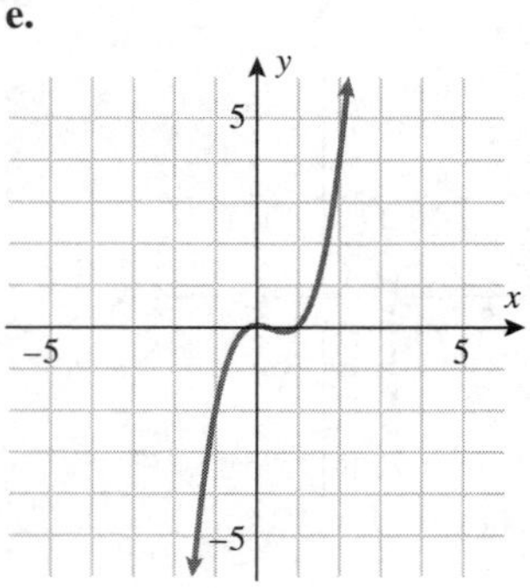

f.

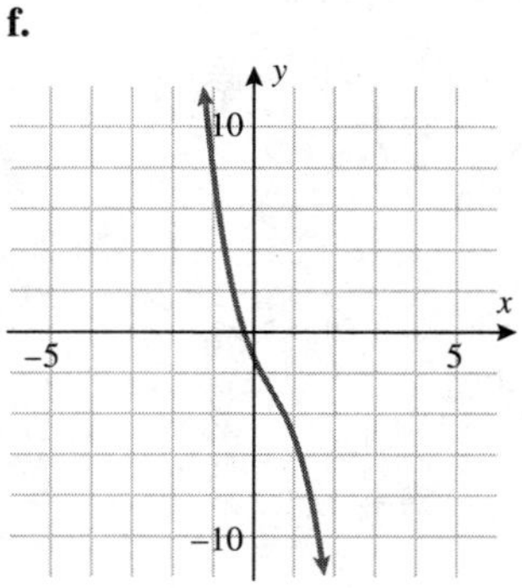

g.

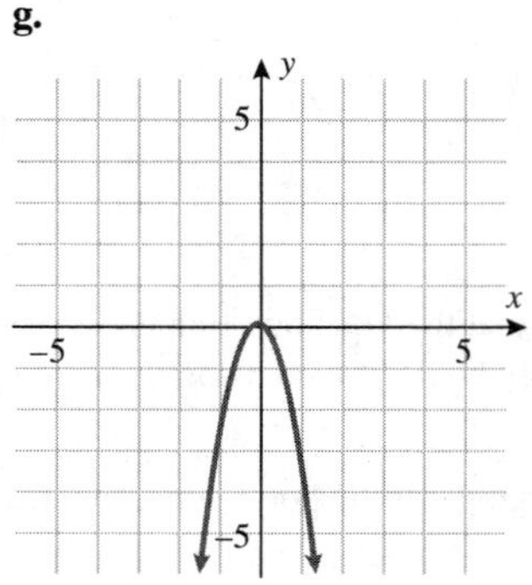

h.

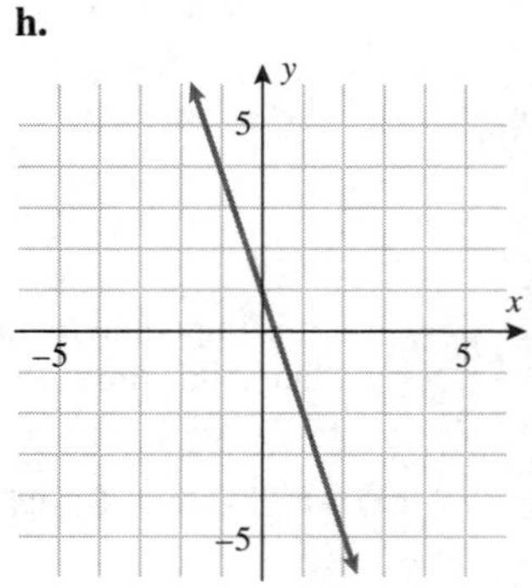

In Exercises 19–26, graph each function by transforming a power function $y = x^n$.

19. $f(x) = -x^5$
20. $f(x) = -x^4$
21. $f(x) = (x - 2)^4$
22. $f(x) = (x + 2)^5$
23. $f(x) = x^5 + 3$
24. $f(x) = -x^4 - 3$
25. $f(x) = 3 - (x + 1)^4$
26. $f(x) = (x - 3)^5 - 2$

In Exercises 27–38, find all the real zeros (and state their multiplicities) of each polynomial function.

27. $f(x) = 2(x - 3)(x + 4)^3$
28. $f(x) = -3(x + 2)^3(x - 1)^2$
29. $f(x) = 4x^2(x - 7)^2(x + 4)$
30. $f(x) = 5x^3(x + 1)^4(x - 6)$
31. $f(x) = 4x^2(x - 1)^2(x^2 + 4)$
32. $f(x) = 4x^2(x^2 - 1)(x^2 + 9)$
33. $f(x) = 8x^3 + 6x^2 - 27x$
34. $f(x) = 2x^4 + 5x^3 - 3x^2$
35. $f(x) = -2.7x^3 - 8.1x^2$
36. $f(x) = 1.2x^6 - 4.6x^4$
37. $f(x) = \frac{1}{3}x^6 + \frac{2}{5}x^4$
38. $f(x) = \frac{2}{7}x^5 - \frac{3}{4}x^4 + \frac{1}{2}x^3$

In Exercises 39–52, find a polynomial (there are many) of minimum degree that has the given zeros.

39. $-3, 0, 1, 2$
40. $-2, 0, 2$
41. $-5, -3, 0, 2, 6$
42. $0, 1, 3, 5, 10$
43. $-\frac{1}{2}, \frac{2}{3}, \frac{3}{4}$
44. $-\frac{3}{4}, -\frac{1}{3}, 0, \frac{1}{2}$
45. $1 - \sqrt{2}, 1 + \sqrt{2}$
46. $1 - \sqrt{3}, 1 + \sqrt{3}$
47. -2 (multiplicity 3), 0 (multiplicity 2)
48. -4 (multiplicity 2), 5 (multiplicity 3)
49. -3 (multiplicity 2), 7 (multiplicity 5)
50. 0 (multiplicity 1), 10 (multiplicity 3)
51. $-\sqrt{3}$ (multiplicity 2), -1 (multiplicity 1), 0 (multiplicity 2), $\sqrt{3}$ (multiplicity 2)
52. $-\sqrt{5}$ (multiplicity 2), 0 (multiplicity 1), 1 (multiplicity 2), $\sqrt{5}$ (multiplicity 2)

In Exercises 53–72, for each polynomial function given: (a) list each real zero and its multiplicity; (b) determine whether the graph touches or crosses at each x-intercept; (c) find the y-intercept and a few points on the graph; (d) determine the end behavior; and (e) sketch the graph.

53. $f(x) = -x^2 - 6x - 9$
54. $f(x) = x^2 + 4x + 4$
55. $f(x) = (x - 2)^3$
56. $f(x) = -(x + 3)^3$
57. $f(x) = x^3 - 9x$
58. $f(x) = -x^3 + 4x^2$
59. $f(x) = -x^3 + x^2 + 2x$
60. $f(x) = x^3 - 6x^2 + 9x$
61. $f(x) = -x^4 - 3x^3$
62. $f(x) = x^5 - x^3$
63. $f(x) = 12x^6 - 36x^5 - 48x^4$
64. $f(x) = 7x^5 - 14x^4 - 21x^3$
65. $f(x) = 2x^5 - 6x^4 - 8x^3$
66. $f(x) = -5x^4 + 10x^3 - 5x^2$
67. $f(x) = x^3 - x^2 - 4x + 4$
68. $f(x) = x^3 - x^2 - x + 1$
69. $f(x) = -(x + 2)^2(x - 1)^2$
70. $f(x) = (x - 2)^3(x + 1)^3$
71. $f(x) = x^2(x - 2)^3(x + 3)^2$
72. $f(x) = -x^3(x - 4)^2(x + 2)^2$

In Exercises 73–76, for each graph given: (a) list each real zero and its smallest possible multiplicity; (b) determine whether the degree of the polynomial is even or odd; (c) determine whether the leading coefficient of the polynomial is positive or negative; (d) find the y-intercept; and (e) write an equation for the polynomial function (assume the least degree possible).

73.

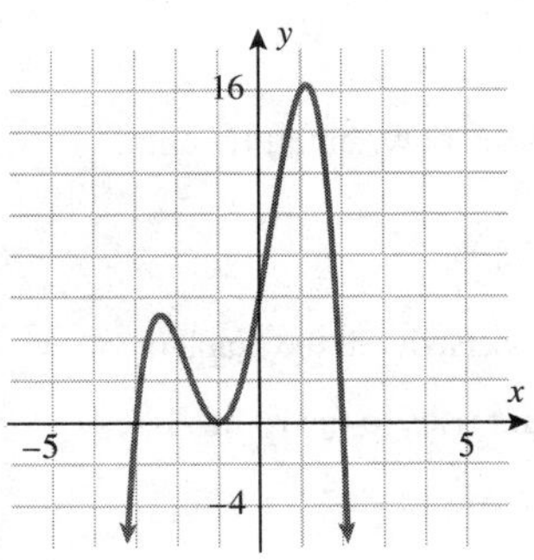

74.

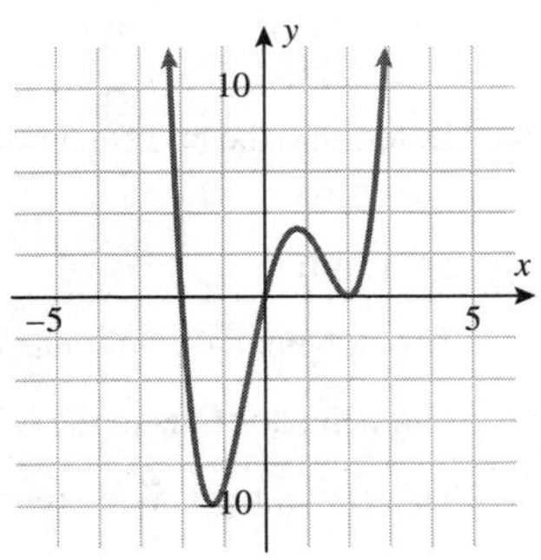

75.

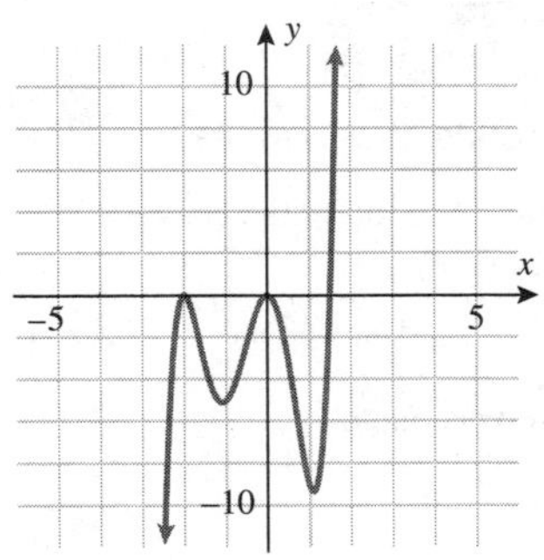

76.

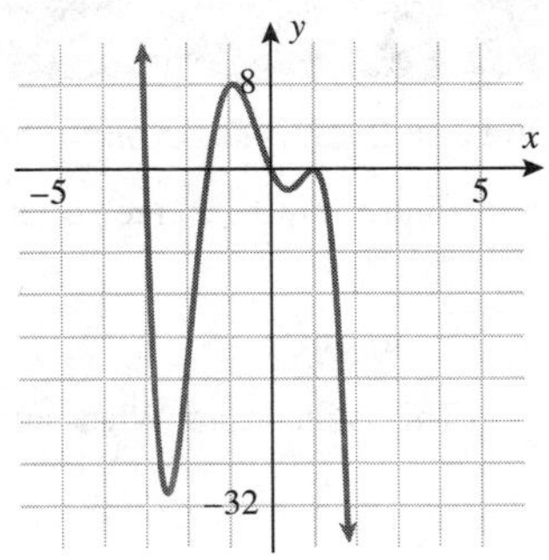

APPLICATIONS

77. Weight. Jennifer has joined a gym to lose weight and feel better. She still likes to cheat a little and will enjoy the occasional bad meal with an ice cream dream dessert and then miss the gym for a couple of days. Given in the table is Jennifer's weight for a period of 8 months. Her weight can be modeled as a polynomial. Plot these data. How many turning points are there? Assuming these are the minimum number of turning points, what is the lowest degree polynomial that can represent Jennifer's weight?

MONTH	WEIGHT
1	169
2	158
3	150
4	161
5	154
6	159
7	148
8	153

78. Stock Value. A day trader checks the stock price of Coca-Cola during a 4-hour period (given below). The price of Coca-Cola stock during this 4-hour period can be modeled as a polynomial function. Plot these data. How many turning points are there? Assuming these are the minimum number of turning points, what is the lowest degree polynomial that can represent the Coca-Cola stock price?

PERIOD WATCHING STOCK MARKET	PRICE
1	$53.00
2	$56.00
3	$52.70
4	$51.50

79. Stock Value. The price of Tommy Hilfiger stock during a 4-hour period is given below. If a third-degree polynomial models this stock, do you expect the stock to go up or down in the fifth period?

PERIOD WATCHING STOCK MARKET	PRICE
1	$15.10
2	$14.76
3	$15.50
4	$14.85

80. Stock Value. The stock prices for Coca-Cola during a 4-hour period on another day yield the following results. If a third-degree polynomial models this stock, do you expect the stock to go up or down in the fifth period?

PERIOD WATCHING STOCK MARKET	PRICE
1	$52.80
2	$53.00
3	$56.00
4	$52.70

For Exercises 81 and 82, the following table graph illustrates the average federal funds rate in the month of January (2000 to 2008).

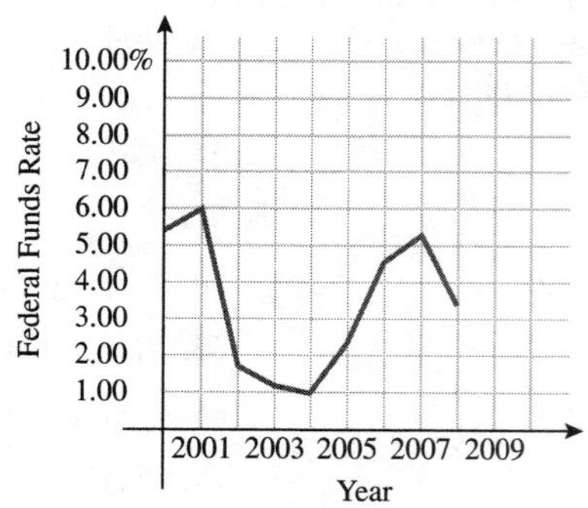

81. Finance. If a polynomial function is used to model the federal funds rate data shown in the graph, determine the degree of the lowest degree polynomial that can be used to model those data.

82. Finance. Should the leading coefficient in the polynomial found in Exercise 81 be positive or negative? Explain.

CATCH THE MISTAKE

In Exercises 83–86, explain the mistake that is made.

83. Find a fourth-degree polynomial function with zeros -2, -1, 3, 4.

Solution: $f(x) = (x - 2)(x - 1)(x + 3)(x + 4)$

This is incorrect. What mistake was made?

84. Determine the end behavior of the polynomial function $f(x) = x(x - 2)^3$.

Solution:

This polynomial has similar end behavior to the graph of $y = x^3$.

End behavior falls to the left and rises to the right.

This is incorrect. What mistake was made?

85. Graph the polynomial function $f(x) = (x - 1)^2(x + 2)^3$.

Solution:

The zeros are -2 and 1, and therefore, the x-intercepts are $(-2, 0)$ and $(1, 0)$.

The y-intercept is $(0, 8)$.

Plotting these points and connecting with a smooth curve yields the graph on the right.

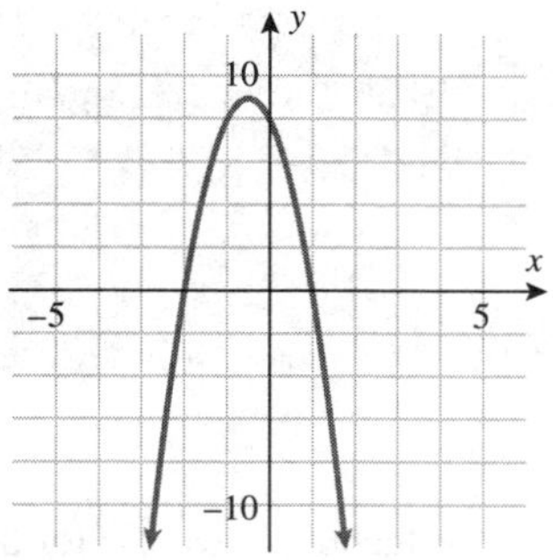

This graph is incorrect. What did we forget to do?

86. Graph the polynomial function $f(x) = (x + 1)^2(x - 1)^2$.

Solution:

The zeros are -1 and 1, so the x-intercepts are $(-1, 0)$ and $(1, 0)$.

The y-intercept is $(0, 1)$.

Plotting these points and connecting with a smooth curve yields the graph on the right.

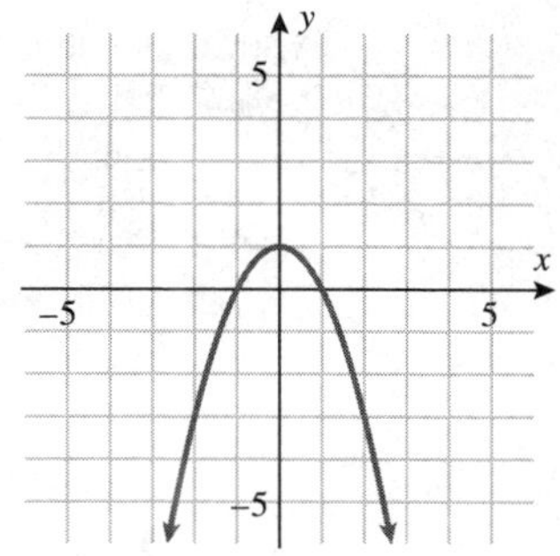

This graph is incorrect. What did we forget to do?

CONCEPTUAL

In Exercises 87–90, determine if each statement is true or false.

87. The graph of a polynomial function might not have any y-intercepts.

88. The graph of a polynomial function might not have any x-intercepts.

89. The domain of all polynomial functions is $(-\infty, \infty)$.

90. The range of all polynomial functions is $(-\infty, \infty)$.

91. What is the maximum number of zeros that a polynomial of degree n can have?

92. What is the maximum number of turning points a graph of an nth-degree polynomial can have?

CHALLENGE

93. Find a seventh-degree polynomial that has the following graph characteristics: The graph touches the x-axis at $x = -1$, and the graph crosses the x-axis at $x = 3$. Plot this polynomial function.

94. Find a fifth-degree polynomial that has the following graph characteristics: The graph touches the x-axis at $x = 0$ and crosses the x-axis at $x = 4$. Plot the polynomial function.

95. Determine the zeros of the polynomial $f(x) = x^3 + (b - a)x^2 - abx$ for the positive real numbers a and b.

96. Graph the function $f(x) = x^2(x - a)^2(x - b)^2$ for the positive real numbers a, b, where $b > a$.

TECHNOLOGY

In Exercises 97 and 98, use a graphing calculator or computer to graph each polynomial. From that graph, estimate the x-intercepts (if any). Set the function equal to zero, and solve for the zeros of the polynomial. Compare the zeros with the x-intercepts.

97. $f(x) = x^4 + 2x^2 + 1$

98. $f(x) = 1.1x^3 - 2.4x^2 + 5.2x$

For each polynomial in Exercises 99 and 100, determine the power function that has similar end behavior. Plot this power function and the polynomial. Do they have similar end behavior?

99. $f(x) = -2x^5 - 5x^4 - 3x^3$

100. $f(x) = x^4 - 6x^2 + 9$

In Exercises 101 and 102, use a graphing calculator or a computer to graph each polynomial. From the graph, estimate the x-intercepts and state the zeros of the function and their multiplicities.

101. $f(x) = x^4 - 15.9x^3 + 1.31x^2 + 292.905x + 445.7025$

102. $f(x) = -x^5 + 2.2x^4 + 18.49x^3 - 29.878x^2 - 76.5x + 100.8$

In Exercises 103 and 104, use a graphing calculator or a computer to graph each polynomial. From the graph, estimate the coordinates of the relative maximum and minimum points. Round your answers to 2 decimal places.

103. $f(x) = 2x^4 + 5x^3 - 10x^2 - 15x + 8$

104. $f(x) = 2x^5 - 4x^4 - 12x^3 + 18x^2 + 16x - 7$

SECTION 4.3 DIVIDING POLYNOMIALS: LONG DIVISION AND SYNTHETIC DIVISION

SKILLS OBJECTIVES

- Divide polynomials with long division.
- Divide polynomials with synthetic division.
- Check the results now that you know the division algorithm.

CONCEPTUAL OBJECTIVES

- Extend long division of real numbers to polynomials.
- Understand *when* synthetic division can be used.

$$\begin{array}{r} 311 \\ 21\overline{)6542} \\ \underline{-63} \\ 24 \\ \underline{-21} \\ 32 \\ \underline{-21} \\ 11 \end{array}$$

To divide polynomials, we rely on the technique we use for dividing real numbers. For example, if you were asked to divide 6542 by 21, the long division method used is illustrated in the margin. This solution can be written two ways: 311 *R*11 or $311 + \frac{11}{21}$.

In this example, the **dividend** is 6542, the **divisor** is 21, the **quotient** is 311, and the **remainder** is 11. We employ a similar technique (dividing the leading terms) when dividing polynomials.

Long Division of Polynomials

Let's start with an example whose answer we already know. We know that a quadratic expression can be factored into the product of two linear factors: $x^2 + 4x - 5 = (x + 5)(x - 1)$. Therefore, if we divide both sides of the equation by $(x - 1)$, we get

$$\frac{x^2 + 4x - 5}{x - 1} = x + 5$$

We can state this by saying $x^2 + 4x - 5$ divided by $x - 1$ is equal to $x + 5$. Confirm this statement by long division:

$$x - 1\overline{)x^2 + 4x - 5}$$

Note that although this is standard division notation, the dividend and the divisor are both polynomials that consist of multiple terms. The *leading* terms of each algebraic expression will guide us.

WORDS	**MATH**
Q: x times what quantity gives x^2?	$x - 1\overline{)x^2 + 4x - 5}$
A: x	$\begin{array}{r} x \\ x - 1\overline{)x^2 + 4x - 5} \end{array}$
Multiply $x(x - 1) = x^2 - x$.	$\begin{array}{r} x \\ x - 1\overline{)x^2 + 4x - 5} \\ x^2 - x \end{array}$
Subtract $(x^2 - x)$ from $x^2 + 4x - 5$. *Note:* $-(x^2 - x) = -x^2 + x$. Bring down the -5.	$\begin{array}{r} x \\ x - 1\overline{)x^2 + 4x - 5} \\ \underline{-x^2 + x } \\ 5x - 5 \end{array}$

WORDS	**MATH**
Q: x times what quantity is $5x$? A: 5 Multiply $5(x - 1) = 5x - 5$.	$$\begin{array}{r} x + 5 \\ x - 1\overline{)x^2 + 4x - 5} \\ \underline{-x^2 + x} \\ 5x - 5 \\ 5x - 5 \end{array}$$
Subtract $(5x - 5)$. *Note:* $-(5x - 5) = -5x + 5$.	$$\begin{array}{r} x + 5 \\ x - 1\overline{)x^2 + 4x - 5} \\ \underline{-x^2 + x} \\ 5x - 5 \\ \underline{-5x + 5} \\ 0 \end{array}$$

As expected, the remainder is 0. By long division we have shown that

$$\boxed{\frac{x^2 + 4x - 5}{x - 1} = x + 5}$$

Check: Multiplying the equation by $x - 1$ yields $x^2 + 4x - 5 = (x + 5)(x - 1)$, which we knew to be true.

EXAMPLE 1 Dividing Polynomials Using Long Division; Zero Remainder

Divide $2x^3 - 9x^2 + 7x + 6$ by $2x + 1$.

Solution:

$$\begin{array}{r} x^2 - 5x + 6 \\ 2x + 1\overline{)2x^3 - 9x^2 + 7x + 6} \\ \underline{-(2x^3 + x^2)} \\ -10x^2 + 7x \\ \underline{-(-10x^2 - 5x)} \\ 12x + 6 \\ \underline{-(12x + 6)} \\ 0 \end{array}$$

Multiply: $x^2(2x + 1)$.

Subtract: Bring down the $7x$.

Multiply: $-5x(2x + 1)$.

Subtract: Bring down the 6.

Multiply: $6(2x + 1)$.

Subtract.

Quotient: $\boxed{x^2 - 5x + 6}$

Check: $(2x + 1)(x^2 - 5x + 6) = 2x^3 - 9x^2 + 7x + 6$.

Note: Since the divisor cannot be equal to zero, $2x + 1 \neq 0$, then we say $x \neq -\frac{1}{2}$.

■ **YOUR TURN** Divide $4x^3 + 13x^2 - 2x - 15$ by $4x + 5$.

■ **Answer:** $x^2 + 2x - 3$, remainder 0.

Technology Tip

A graphing utility can be used to check $(x^2 - 5x + 6)(2x + 1) = 2x^3 - 9x^2 + 7x + 6$ using their graphs.

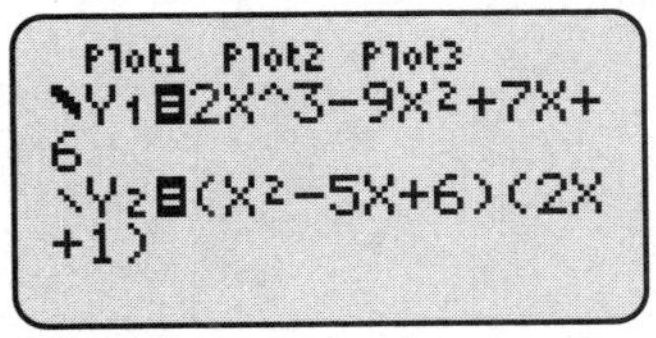

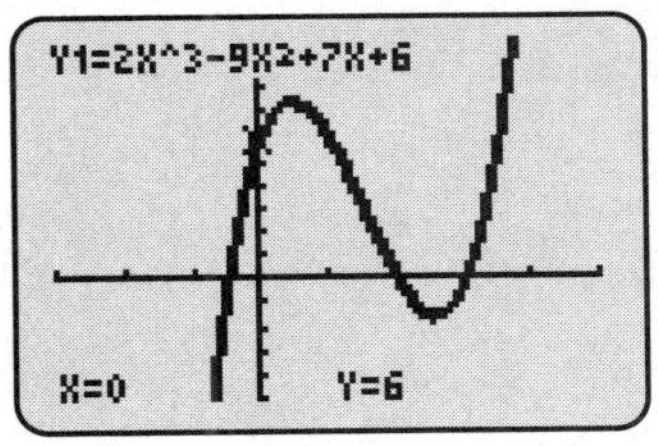

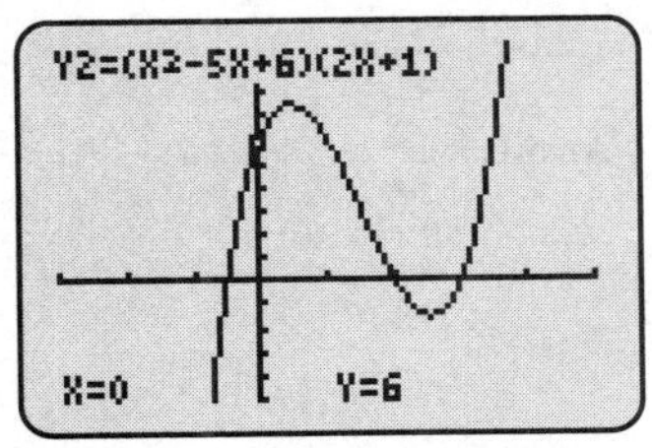

Notice that the graphs are the same.

Why are we interested in dividing polynomials? Because it helps us find zeros of polynomials. In Example 1, using long division, we found that

$$2x^3 - 9x^2 + 7x + 6 = (2x + 1)(x^2 - 5x + 6)$$

Factoring the quadratic expression enables us to write the cubic polynomial as a product of three linear factors:

$$2x^3 - 9x^2 + 7x + 6 = (2x + 1)(x^2 - 5x + 6) = (2x + 1)(x - 3)(x - 2)$$

Set the value of the polynomial equal to zero, $(2x + 1)(x - 3)(x - 2) = 0$, and solve for x. The zeros of the polynomial are $-\frac{1}{2}$, 2, and 3. In Example 1 and in the Your Turn, the remainder was 0. Sometimes there is a nonzero remainder (Example 2).

EXAMPLE 2 Dividing Polynomials Using Long Division; Nonzero Remainder

Divide $6x^2 - x - 2$ by $x + 1$.

Solution:

$$\begin{array}{r} 6x - 7 \\ x + 1\overline{)6x^2 - x - 2} \end{array}$$

Multiply $6x(x + 1)$. $\quad -(6x^2 + 6x)$

Subtract and bring down -2. $\quad -7x - 2$

Multiply $-7(x + 1)$. $\quad -(-7x - 7)$

Subtract and identify the remainder. $\quad +5$

$$\underbrace{\frac{6x^2 - x - 2}{x + 1}}_{\text{Dividend / Divisor}} = \underbrace{6x - 7}_{\text{Quotient}} + \underbrace{\frac{5}{x + 1}}_{\text{Remainder / Divisor}} \qquad \boxed{x \neq -1}$$

Check: Multiply equation by $x + 1$.

$$\begin{aligned} 6x^2 - x - 2 &= (6x - 7)(x + 1) + 5 \\ &= 6x^2 - x - 7 + 5 \\ &= 6x^2 - x - 2 \quad \checkmark \end{aligned}$$

■ **YOUR TURN** Divide $2x^3 + x^2 - 4x - 3$ by $x - 1$.

■ **Answer:** $2x^2 + 3x - 1$ $R{:}\ -4$ or $2x^2 + 3x - 1 - \dfrac{4}{x - 1}$

In general, when a polynomial is divided by another polynomial, we express the result in the following form:

$$\frac{P(x)}{d(x)} = Q(x) + \frac{r(x)}{d(x)}$$

where $P(x)$ is the **dividend**, $d(x) \neq 0$ is the divisor, $Q(x)$ is the quotient, and $r(x)$ is the remainder. Multiplying this equation by the divisor, $d(x)$, leads us to the division algorithm.

THE DIVISION ALGORITHM

If $P(x)$ and $d(x)$ are polynomials with $d(x) \neq 0$, and if the degree of $P(x)$ is greater than or equal to the degree of $d(x)$, then unique polynomials $Q(x)$ and $r(x)$ exist such that

$$P(x) = d(x) \cdot Q(x) + r(x)$$

If the remainder $r(x) = 0$, then we say that $d(x)$ divides $P(x)$ and that $d(x)$ and $Q(x)$ are factors of $P(x)$.

EXAMPLE 3 **Long Division of Polynomials with "Missing" Terms**

Divide $x^3 - 8$ by $x - 2$.

Solution:

	$x^2 + 2x + 4$
Insert $0x^2 + 0x$ for placeholders.	$x - 2\overline{)x^3 + 0x^2 + 0x - 8}$
Multiply $x^2(x - 2) = x^3 - 2x^2$.	$-(x^3 - 2x^2)$
Subtract and bring down $0x$.	$2x^2 + 0x$
Multiply $2x(x - 2) = 2x^2 - 4x$.	$-(2x^2 - 4x)$
Subtract and bring down -8.	$4x - 8$
Multiply $4(x - 2) = 4x - 8$.	$-(4x - 8)$
Subtract and get remainder 0.	0

Since the remainder is 0, $x - 2$ is a factor of $x^3 - 8$.

$$\boxed{\frac{x^3 - 8}{x - 2} = x^2 + 2x + 4, x \neq 2}$$

Check: $x^3 - 8 = (x^2 + 2x + 4)(x - 2) = x^3 + 2x^2 + 4x - 2x^2 - 4x - 8 = x^3 - 8$ ✓

■ YOUR TURN Divide $x^3 - 1$ by $x - 1$.

■ **Answer:** $x^2 + x + 1$

EXAMPLE 4 **Long Division of Polynomials**

Divide $3x^4 + 2x^3 + x^2 + 4$ by $x^2 + 1$.

Solution:

	$3x^2 + 2x - 2$
Insert $0x$ as a placeholder in both the divisor and the dividend.	$x^2 + 0x + 1\overline{)3x^4 + 2x^3 + x^2 + 0x + 4}$
Multiply $3x^2(x^2 + 0x + 1)$.	$-(3x^4 + 0x^3 + 3x^2)$
Subtract and bring down $0x$.	$2x^3 - 2x^2 + 0x$
Multiply $2x(x^2 + 0x + 1)$.	$-(2x^3 + 0x^2 + 2x)$
Subtract and bring down 4.	$-2x^2 - 2x + 4$
Multiply $-2(x^2 - 2x + 1)$.	$-(-2x^2 + 0x - 2)$
Subtract and get remainder $-2x + 6$.	$-2x + 6$

$$\boxed{\frac{3x^4 + 2x^3 + x^2 + 4}{x^2 + 1} = 3x^2 + 2x - 2 + \frac{-2x + 6}{x^2 + 1}}$$

■ YOUR TURN Divide $2x^5 + 3x^2 + 12$ by $x^3 - 3x - 4$.

■ **Answer:**

$2x^2 + 6 + \dfrac{11x^2 + 18x + 36}{x^3 - 3x - 4}$

Synthetic Division of Polynomials

In the special case when the *divisor is a linear factor* of the form $x - a$ or $x + a$, there is another, more efficient way to divide polynomials. This method is called **synthetic division**. It is called synthetic because it is a contrived shorthand way of dividing a polynomial by a linear factor. A detailed step-by-step procedure is given below for synthetic division. Let's divide $x^4 - x^3 - 2x + 2$ by $x + 1$ using synthetic division.

Study Tip

If $(x - a)$ is divisor, then a is the number used in synthetic division.

STEP 1 Write the division in synthetic form.

- List the coefficients of the dividend. **Remember to use 0 for a placeholder.**
- The divisor is $x + 1$. Since $x + 1 = 0$, $x = -1$ is used.

Coefficients of Dividend

$$\begin{array}{r|rrrrr} -1 & 1 & -1 & 0 & -2 & 2 \\ & & & & & \\ \hline \end{array}$$

STEP 2 *Bring down* the first term (1) in the dividend.

$$\begin{array}{r|rrrrr} -1 & 1 & -1 & 0 & -2 & 2 \\ & \downarrow & \text{Bring down the 1} & & & \\ \hline & \mathbf{1} & & & & \end{array}$$

STEP 3 *Multiply* the -1 by this leading coefficient (1), and place the product up and to the right in the second column.

$$\begin{array}{r|rrrrr} -1 & 1 & -1 & 0 & -2 & 2 \\ & & -1 & & & \\ \hline & \mathbf{1} & & & & \end{array}$$

STEP 4 *Add* the values in the second column.

$$\begin{array}{r|rrrrr} -1 & 1 & -1 & 0 & -2 & 2 \\ & & \downarrow -1 \ \text{ADD} & & & \\ \hline & \mathbf{1} & \mathbf{-2} & & & \end{array}$$

STEP 5 Repeat Steps 3 and 4 until all columns are filled.

$$\begin{array}{r|rrrrr} -1 & 1 & -1 & 0 & -2 & 2 \\ & \downarrow & -1\downarrow & 2\downarrow & -2\downarrow & 4\downarrow \\ \hline & \mathbf{1} & \mathbf{-2} & \mathbf{2} & \mathbf{-4} & 6 \end{array}$$

Study Tip

Synthetic division can only be used when the divisor is of the form $x - a$ or $x + a$.

STEP 6 Identify the **quotient** by assigning powers of x in descending order, beginning with $x^{n-1} = x^{4-1} = x^3$. The last term is the remainder.

$$\begin{array}{r|rrrrr} -1 & 1 & -1 & 0 & -2 & 2 \\ & & -1 & 2 & -2 & 4 \\ \hline & \mathbf{1} & \mathbf{-2} & \mathbf{2} & \mathbf{-4} & 6 \end{array}$$

Quotient Coefficients: $x^3 - 2x^2 + 2x - 4$ Remainder: 6

We know that the degree of the first term of the quotient is 3, because a fourth-degree polynomial was divided by a first-degree polynomial. Let's compare dividing $x^4 - x^3 - 2x + 2$ by $x + 1$ using both long division and synthetic division.

Long Division

$$\begin{array}{r} x^3 - 2x^2 + 2x - 4 \\ x + 1 \overline{)\, x^4 - x^3 + 0x^2 - 2x + 2} \\ \underline{x^4 + x^3} \qquad\qquad\qquad\quad \\ -2x^3 + 0x^2 \qquad\qquad \\ \underline{-(-2x^3 - 2x^2)} \qquad\qquad \\ 2x^2 - 2x \qquad \\ \underline{-(2x^2 + 2x)} \qquad \\ -4x + 2 \\ \underline{-(-4x - 4)} \\ +6 \end{array}$$

Synthetic Division

$$\begin{array}{r|rrrrr} -1 & 1 & -1 & 0 & -2 & 2 \\ & & -1 & 2 & -2 & 4 \\ \hline & 1 & -2 & 2 & -4 & 6 \end{array}$$

$x^3 - 2x^2 + 2x - 4$

Both long division and synthetic division yield the same answer.

$$\frac{x^4 - x^3 - 2x + 2}{x + 1} = x^3 - 2x^2 + 2x - 4 + \frac{6}{x + 1}$$

EXAMPLE 5 Synthetic Division

Use synthetic division to divide $3x^5 - 2x^3 + x^2 - 7$ by $x + 2$.

Solution:

STEP 1 Write the division in synthetic form.

- List the coefficients of the dividend. Remember to use 0 for a placeholder.
- The divisor of the original problem is $x + 2$. If we set $x + 2 = 0$ we find that $x = -2$, so -2 is the divisor for synthetic division.

−2	3	0	−2	1	0	−7

STEP 2 Perform the synthetic division steps.

−2	3	0	−2	1	0	−7
		−6	12	−20	38	−76
	3	−6	10	−19	38	−83

STEP 3 Identify the quotient and remainder.

−2	3	0	−2	1	0	−7
		−6	12	−20	38	−76
	3	−6	10	−19	38	[−83]

$3x^4 - 6x^3 + 10x^2 - 19x + 38$

$$\frac{3x^5 - 2x^3 + x^2 - 7}{x + 2} = 3x^4 - 6x^3 + 10x^2 - 19x + 38 - \frac{83}{x + 2}$$

YOUR TURN Use synthetic division to divide $2x^3 - x + 3$ by $x - 1$.

Answer: $2x^2 + 2x + 1 + \frac{4}{x - 1}$

SECTION 4.3 SUMMARY

Division of Polynomials

- Long division can always be used.
- Synthetic division is restricted to when the divisor is of the form $x - a$ or $x + a$.

Expressing Results

- $\frac{\text{Dividend}}{\text{Divisor}} = \text{quotient} + \frac{\text{remainder}}{\text{divisor}}$
- Dividend = (quotient)(divisor) + remainder

When Remainder Is Zero

- Dividend = (quotient)(divisor)
- Quotient and divisor are factors of the dividend.

SECTION 4.3 EXERCISES

SKILLS

In Exercises 1–30, divide the polynomials using long division. Use exact values. Express the answer in the form $Q(x) = ?$, $r(x) = ?$.

1. $(2x^2 + 5x - 3) \div (x + 3)$
2. $(2x^2 + 5x - 3) \div (x - 3)$
3. $(x^2 - 5x + 6) \div (x - 2)$
4. $(2x^2 + 3x + 1) \div (x + 1)$
5. $(3x^2 - 9x - 5) \div (x - 2)$
6. $(x^2 + 4x - 3) \div (x - 1)$
7. $(3x^2 - 13x - 10) \div (x + 5)$
8. $(3x^2 - 13x - 10) \div (x - 5)$
9. $(x^2 - 4) \div (x + 4)$
10. $(x^2 - 9) \div (x - 2)$
11. $(9x^2 - 25) \div (3x - 5)$
12. $(5x^2 - 3) \div (x + 1)$
13. $(4x^2 - 9) \div (2x + 3)$
14. $(8x^3 + 27) \div (2x + 3)$
15. $(11x + 20x^2 + 12x^3 + 2) \div (3x + 2)$
16. $(12x^3 + 2 + 11x + 20x^2) \div (2x + 1)$
17. $(4x^3 - 2x + 7) \div (2x + 1)$
18. $(6x^4 - 2x^2 + 5) \div (-3x + 2)$
19. $(4x^3 - 12x^2 - x + 3) \div \left(x - \frac{1}{2}\right)$
20. $(12x^3 + 1 + 7x + 16x^2) \div \left(x + \frac{1}{3}\right)$
21. $(-2x^5 + 3x^4 - 2x^2) \div (x^3 - 3x^2 + 1)$
22. $(-9x^6 + 7x^4 - 2x^3 + 5) \div (3x^4 - 2x + 1)$
23. $\dfrac{x^4 - 1}{x^2 - 1}$
24. $\dfrac{x^4 - 9}{x^2 + 3}$
25. $\dfrac{40 - 22x + 7x^3 + 6x^4}{6x^2 + x - 2}$
26. $\dfrac{-13x^2 + 4x^4 + 9}{4x^2 - 9}$
27. $\dfrac{-3x^4 + 7x^3 - 2x + 1}{x - 0.6}$
28. $\dfrac{2x^5 - 4x^3 + 3x^2 + 5}{x - 0.9}$
29. $(x^4 + 0.8x^3 - 0.26x^2 - 0.168x + 0.0441) \div (x^2 + 1.4x + 0.49)$
30. $(x^5 + 2.8x^4 + 1.34x^3 - 0.688x^2 - 0.2919x + 0.0882) \div (x^2 - 0.6x + 0.09)$

In Exercises 31–50, divide the polynomial by the linear factor with synthetic division. Indicate the quotient $Q(x)$ and the remainder $r(x)$.

31. $(3x^2 + 7x + 2) \div (x + 2)$
32. $(2x^2 + 7x - 15) \div (x + 5)$
33. $(7x^2 - 3x + 5) \div (x + 1)$
34. $(4x^2 + x + 1) \div (x - 2)$
35. $(3x^2 + 4x - x^4 - 2x^3 - 4) \div (x + 2)$
36. $(3x^2 - 4 + x^3) \div (x - 1)$
37. $(x^4 + 1) \div (x + 1)$
38. $(x^4 + 9) \div (x + 3)$
39. $(x^4 - 16) \div (x + 2)$
40. $(x^4 - 81) \div (x - 3)$
41. $(2x^3 - 5x^2 - x + 1) \div \left(x + \frac{1}{2}\right)$
42. $(3x^3 - 8x^2 + 1) \div \left(x + \frac{1}{3}\right)$
43. $(2x^4 - 3x^3 + 7x^2 - 4) \div \left(x - \frac{2}{3}\right)$
44. $(3x^4 + x^3 + 2x - 3) \div \left(x - \frac{3}{4}\right)$
45. $(2x^4 + 9x^3 - 9x^2 - 81x - 81) \div (x + 1.5)$
46. $(5x^3 - x^2 + 6x + 8) \div (x + 0.8)$
47. $\dfrac{x^7 - 8x^4 + 3x^2 + 1}{x - 1}$
48. $\dfrac{x^6 + 4x^5 - 2x^3 + 7}{x + 1}$
49. $(x^6 - 49x^4 - 25x^2 + 1225) \div (x - \sqrt{5})$
50. $(x^6 - 4x^4 - 9x^2 + 36) \div (x - \sqrt{3})$

In Exercises 51–60, divide the polynomials by either long division or synthetic division.

51. $(6x^2 - 23x + 7) \div (3x - 1)$
52. $(6x^2 + x - 2) \div (2x - 1)$
53. $(x^3 - x^2 - 9x + 9) \div (x - 1)$
54. $(x^3 + 2x^2 - 6x - 12) \div (x + 2)$
55. $(x^5 + 4x^3 + 2x^2 - 1) \div (x - 2)$
56. $(x^4 - x^2 + 3x - 10) \div (x + 5)$
57. $(x^4 - 25) \div (x^2 - 1)$
58. $(x^3 - 8) \div (x^2 - 2)$
59. $(x^7 - 1) \div (x - 1)$
60. $(x^6 - 27) \div (x - 3)$

APPLICATIONS

61. Geometry. The area of a rectangle is $6x^4 + 4x^3 - x^2 - 2x - 1$ square feet. If the length of the rectangle is $2x^2 - 1$ feet, what is the width of the rectangle?

62. Geometry. If the rectangle in Exercise 61 is the base of a rectangular box with volume $18x^5 + 18x^4 + x^3 - 7x^2 - 5x - 1$ cubic feet, what is the height of the box?

63. Travel. If a car travels a distance of $x^3 + 60x^2 + x + 60$ miles at an average speed of $x + 60$ miles per hour, how long does the trip take?

64. Sports. If a quarterback throws a ball $-x^2 - 5x + 50$ yards in $5 - x$ seconds, how fast is the football traveling?

CATCH THE MISTAKE

In Exercises 65–68, explain the mistake that is made.

65. Divide $x^3 - 4x^2 + x + 6$ by $x^2 + x + 1$.

Solution:

$$\begin{array}{r} x - 3 \\ x^2 + x + 1\overline{)x^3 - 4x^2 + x + 6} \\ \underline{x^3 + x^2 + x} \\ -3x^2 + 2x + 6 \\ \underline{-3x^2 - 3x - 3} \\ -x + 3 \end{array}$$

This is incorrect. What mistake was made?

66. Divide $x^4 - 3x^2 + 5x + 2$ by $x - 2$.

Solution:

$$\begin{array}{r|rrrr} -2 & 1 & -3 & 5 & 2 \\ & & -2 & 10 & -30 \\ \hline & 1 & -5 & 15 & \boxed{-28} \end{array}$$

$x^2 - 5x + 15$

This is incorrect. What mistake was made?

67. Divide $x^3 + 4x - 12$ by $x - 3$.

Solution:

$$\begin{array}{r|rrr} 3 & 1 & 4 & -12 \\ & & 3 & 21 \\ \hline & 1 & 7 & \boxed{9} \end{array}$$

$x + 7$

This is incorrect. What mistake was made?

68. Divide $x^3 + 3x^2 - 2x + 1$ by $x^2 + 1$.

Solution:

$$\begin{array}{r|rrrr} -1 & 1 & 3 & -2 & 1 \\ & & -1 & -2 & 4 \\ \hline & 1 & 2 & -4 & \boxed{5} \end{array}$$

$x^2 - 2x - 4$

This is incorrect. What mistake was made?

CONCEPTUAL

In Exercises 69–72, determine whether each statement is true or false.

69. A fifth-degree polynomial divided by a third-degree polynomial will yield a quadratic quotient.

70. A third-degree polynomial divided by a linear polynomial will yield a linear quotient.

71. Synthetic division can be used whenever the degree of the dividend is exactly one more than the degree of the divisor.

72. When the remainder is zero, the divisor is a factor of the dividend.

CHALLENGE

73. Is $x + b$ a factor of $x^3 + (2b - a)x^2 + (b^2 - 2ab)x - ab^2$?

74. Is $x + b$ a factor of $x^4 + (b^2 - a^2)x^2 - a^2b^2$?

75. Divide $x^{3n} + x^{2n} - x^n - 1$ by $x^n - 1$.

76. Divide $x^{3n} + 5x^{2n} + 8x^n + 4$ by $x^n + 1$.

TECHNOLOGY

77. Plot $\dfrac{2x^3 - x^2 + 10x - 5}{x^2 + 5}$. What type of function is it? Perform this division using long division, and confirm that the graph corresponds to the quotient.

78. Plot $\dfrac{x^3 - 3x^2 + 4x - 12}{x - 3}$. What type of function is it? Perform this division using synthetic division, and confirm that the graph corresponds to the quotient.

79. Plot $\dfrac{x^4 + 2x^3 - x - 2}{x + 2}$. What type of function is it? Perform this division using synthetic division, and confirm that the graph corresponds to the quotient.

80. Plot $\dfrac{x^5 - 9x^4 + 18x^3 + 2x^2 - 5x - 3}{x^4 - 6x^3 + 2x + 1}$. What type of function is it? Perform this division using long division, and confirm that the graph corresponds to the quotient.

81. Plot $\dfrac{-6x^3 + 7x^2 + 14x - 15}{2x + 3}$. What type of function is it? Perform this division using long division, and confirm that the graph corresponds to the quotient.

82. Plot $\dfrac{-3x^5 - 4x^4 + 29x^3 + 36x^2 - 18x}{3x^2 + 4x - 2}$. What type of function is it? Perform this division using long division, and confirm that the graph corresponds to the quotient.

SECTION 4.4 THE REAL ZEROS OF A POLYNOMIAL FUNCTION

SKILLS OBJECTIVES

- Apply the remainder theorem to evaluate a polynomial function.
- Apply the factor theorem.
- Use the Rational zero (root) theorem to list possible rational zeros.
- Apply Descartes's rule of signs to determine the possible combination of positive and negative real zeros.
- Utilize the upper and lower bound theorems to narrow the search for real zeros.
- Find the real zeros of a polynomial function.
- Factor a polynomial function.
- Employ the intermediate value theorem to approximate a real zero.

CONCEPTUAL OBJECTIVES

- Understand that a polynomial of degree n has at most n real zeros.
- Understand that a real zero can be either rational or irrational and that irrational zeros will not be listed as possible zeros through the rational zero test.
- Realize that rational zeros can be found exactly, whereas irrational zeros must be approximated.

The zeros of a polynomial function assist us in finding the x-intercepts of the graph of a polynomial function. How do we find the zeros of a polynomial function? For polynomial functions of degree 2, we have the quadratic formula, which allows us to find the two zeros. For polynomial functions whose degree is greater than 2, much more work is required.* In this section we focus our attention on finding the *real* zeros of a polynomial function. Later, in Section 4.5, we expand our discussion to *complex* zeros of polynomial functions.

In this section we start by listing possible rational zeros. As you will see there are sometimes many possibilities. We can then narrow the search using Descartes's rule of signs, which tells us possible combinations of positive and negative real zeros. We can narrow the search even further with the upper and lower bound rules. Once we have tested possible values and determined a zero, we will employ synthetic division to divide the polynomial by the linear factor associated with the zero. We will continue the process until we have factored the polynomial function into a product of either linear factors or irreducible quadratic factors. Last, we will discuss how to find irrational real zeros using the intermediate value theorem.

The Remainder Theorem and the Factor Theorem

If we divide the polynomial function $f(x) = x^3 - 2x^2 + x - 3$ by $x - 2$ using synthetic division, we find the remainder is -1.

$$\begin{array}{r|rrrr} 2 & 1 & -2 & 1 & -3 \\ & & 2 & 0 & 2 \\ \hline & 1 & 0 & 1 & -1 \end{array}$$

Notice that if we evaluate the function at $x = 2$, the result is -1. $\quad f(2) = -1$

This leads us to the *remainder theorem.*

*There are complicated formulas for finding the zeros of polynomial functions of degree 3 and 4, but there are no such formulas for degree 5 and higher polynomials (according to the Abel–Ruffini theorem).

REMAINDER THEOREM

If a polynomial $P(x)$ is divided by $x - a$, then the remainder is $r = P(a)$.

The remainder theorem tells you that polynomial division can be used to evaluate a polynomial function at a particular point.

EXAMPLE 1 Two Methods for Evaluating Polynomials

Let $P(x) = 4x^5 - 3x^4 + 2x^3 - 7x^2 + 9x - 5$ and evaluate $P(2)$ by

a. Evaluating $P(2)$ directly
b. The remainder theorem and synthetic division

Solution:

a.
$$\begin{aligned} P(2) &= 4(2)^5 - 3(2)^4 + 2(2)^3 - 7(2)^2 + 9(2) - 5 \\ &= 4(32) - 3(16) + 2(8) - 7(4) + 9(2) - 5 \\ &= 128 - 48 + 16 - 28 + 18 - 5 \\ &= \boxed{81} \end{aligned}$$

b.
$$\begin{array}{r|rrrrrr} 2 & 4 & -3 & 2 & -7 & 9 & -5 \\ & & 8 & 10 & 24 & 34 & 86 \\ \hline & 4 & 5 & 12 & 17 & 43 & \boxed{81} \end{array}$$

YOUR TURN Let $P(x) = -x^3 + 2x^2 - 5x + 2$ and evaluate $P(-2)$ using the remainder theorem and synthetic division.

Technology Tip

A graphing utility can be used to evaluate $P(2)$. Enter $P(x) = 4x^5 - 3x^4 + 2x^3 - 7x^2 + 9x - 5$ as Y_1.

To evaluate $P(2)$, press VARS Y-VARS 1:Function.. ENTER 1:Y_1 ENTER (2) ENTER

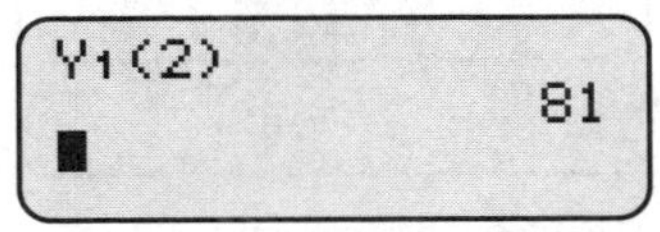

■ **Answer:** $P(-2) = 28$

Recall that when a polynomial is divided by $x - a$, if the remainder is zero, we say that $x - a$ is a factor of the polynomial. Through the remainder theorem, we now know that the remainder is related to evaluation of the polynomial at the point $x = a$. We are then led to the *factor theorem.*

FACTOR THEOREM

If $P(a) = 0$, then $x - a$ is a factor of $P(x)$. Conversely, if $x - a$ is a factor of $P(x)$, then $P(a) = 0$.

EXAMPLE 2 Using the Factor Theorem to Factor a Polynomial

Determine whether $x + 2$ is a factor of $P(x) = x^3 - 2x^2 - 5x + 6$. If so, factor $P(x)$ completely.

Solution:

By the factor theorem, $x + 2$ is a factor of $P(x) = x^3 - 2x^2 - 5x + 6$ if $P(-2) = 0$. By the remainder theorem, if we divide $P(x) = x^3 - 2x^2 - 5x + 6$ by $x + 2$, then the remainder is equal to $P(-2)$.

Technology Tip

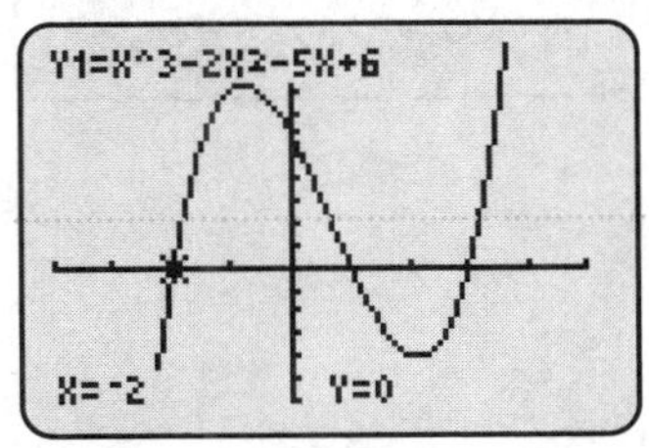

The three zeros of the function give the three factors $x + 2$, $x - 1$, and $x - 3$. A table of values supports the zeros of the graph.

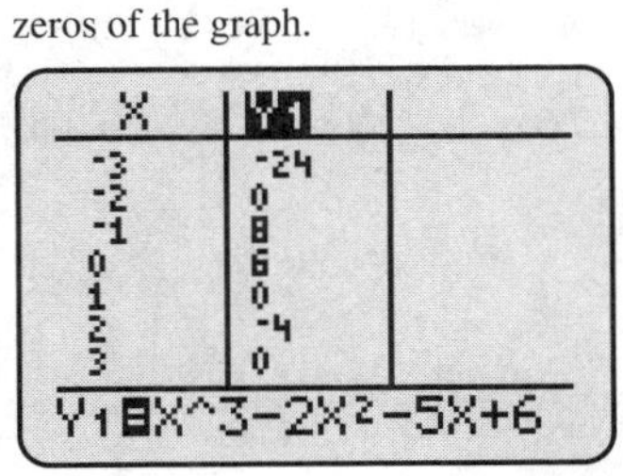

STEP 1 Divide $P(x) = x^3 - 2x^2 - 5x + 6$ by $x + 2$ using synthetic division.

$$\begin{array}{r|rrrr} -2 & 1 & -2 & -5 & 6 \\ & & -2 & 8 & -6 \\ \hline & 1 & -4 & 3 & \boxed{0} \end{array}$$

$$x^2 - 4x + 3$$

Since the remainder is zero, $P(-2) = 0$, $\boxed{x + 2 \text{ is a factor}}$ of $P(x) = x^3 - 2x^2 - 5x + 6$.

STEP 2 Write $P(x)$ as a product. $P(x) = (x + 2)(x^2 - 4x + 3)$

STEP 3 Factor the quadratic polynomial. $\boxed{P(x) = (x + 2)(x - 3)(x - 1)}$

■ **YOUR TURN** Determine whether $x - 1$ is a factor of $P(x) = x^3 - 4x^2 - 7x + 10$. If so, factor $P(x)$ completely.

■ **Answer:** $(x - 1)$ is a factor; $P(x) = (x - 5)(x - 1)(x + 2)$

Technology Tip

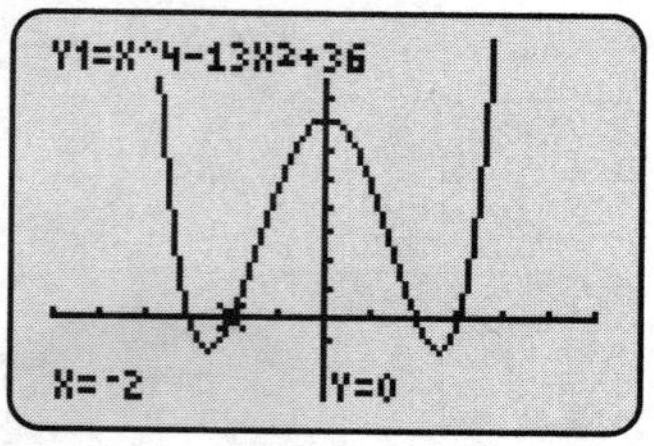

The zeros of the function at $x = -2$ and 3 give the factors $x + 2$ and $x - 3$. A table of values supports the zeros of the graph.

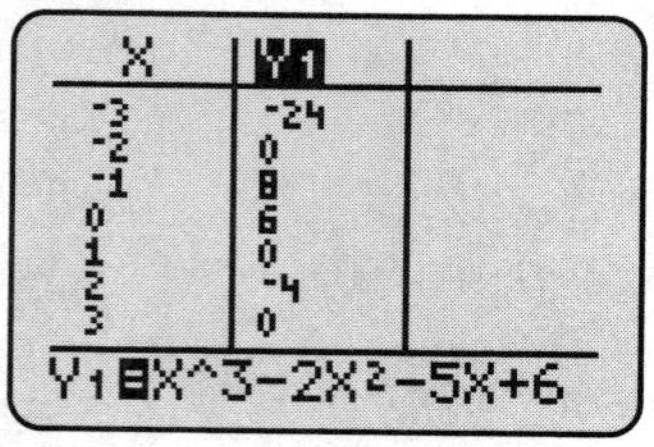

EXAMPLE 3 Using the Factor Theorem to Factor a Polynomial

Determine whether $x - 3$ and $x + 2$ are factors of $P(x) = x^4 - 13x^2 + 36$. If so, factor $P(x)$ completely.

Solution:

STEP 1 With synthetic division divide $P(x) = x^4 - 13x^2 + 36$ by $x - 3$.

$$\begin{array}{r|rrrrr} 3 & 1 & 0 & -13 & 0 & 36 \\ & & 3 & 9 & -12 & -36 \\ \hline & 1 & 3 & -4 & -12 & \boxed{0} \end{array}$$

$$x^3 + 3x^2 - 4x - 12$$

Because the remainder is 0, $\boxed{x - 3 \text{ is a factor}}$, and we can write the polynomial as

$$P(x) = (x - 3)(x^3 + 3x^2 - 4x - 12)$$

STEP 2 With synthetic division divide the remaining cubic polynomial $(x^3 + 3x^2 - 4x - 12)$ by $x + 2$.

$$\begin{array}{r|rrrr} -2 & 1 & 3 & -4 & -12 \\ & & -2 & -2 & 12 \\ \hline & 1 & 1 & -6 & \boxed{0} \end{array}$$

$$x^2 + x - 6$$

Because the remainder is 0, $\boxed{x + 2 \text{ is a factor}}$, and we can now write the polynomial as

$$P(x) = (x - 3)(x + 2)(x^2 + x - 6)$$

STEP 3 Factor the quadratic polynomial: $x^2 + x - 6 = (x + 3)(x - 2)$.

STEP 4 Write $P(x)$ as a product of linear factors:

$$\boxed{P(x) = (x - 3)(x - 2)(x + 2)(x + 3)}$$

■ **YOUR TURN** Determine whether $x - 3$ and $x + 2$ are factors of $P(x) = x^4 - x^3 - 7x^2 + x + 6$. If so, factor $P(x)$ completely.

■ **Answer:** $(x - 3)$ and $(x + 2)$ are factors; $P(x) = (x - 3)(x + 2)(x - 1)(x + 1)$

The Search for Real Zeros

In all of the examples thus far, the polynomial function and one or more real zeros (or linear factors) were given. Now, we will not be given any real zeros to start with. Instead, we will develop methods to search for them.

Each real zero corresponds to a linear factor and each linear factor is of degree 1. Therefore, the largest number of real zeros a polynomial function can have is equal to the degree of the polynomial.

THE NUMBER OF REAL ZEROS

A polynomial function cannot have more real zeros than its degree.

Study Tip

The largest number of zeros a polynomial can have is equal to the degree of the polynomial.

The following functions illustrate that a polynomial function of degree n can have at most n real zeros.

POLYNOMIAL FUNCTION	DEGREE	REAL ZEROS	COMMENTS
$f(x) = x^2 - 9$	2	$x = \pm 3$	**Two** real zeros
$f(x) = x^2 + 4$	2	None	**No** real zeros
$f(x) = x^3 - 1$	3	$x = 1$	**One** real zero
$f(x) = x^3 - x^2 - 6x$	3	$x = -2, 0, 3$	**Three** real zeros

Now that we know the *maximum* number of real zeros a polynomial function can have, let us discuss how to find these zeros.

The Rational Zero Theorem and Descartes's Rule of Signs

When the coefficients of a polynomial are integers, then the *rational zero theorem* (*rational root test*) gives us a list of possible rational zeros. We can then test these possible values to determine whether they really do correspond to actual zeros. *Descartes's rule of signs* tells us the possible combinations of *positive* real zeros and *negative* real zeros. Using Descartes's rule of signs will assist us in narrowing down the large list of possible zeros generated through the rational zero theorem to a (hopefully) shorter list of possible zeros. First, let's look at the rational zero theorem; then we'll turn to Descartes's rule of signs.

THE RATIONAL ZERO THEOREM (RATIONAL ROOT TEST)

If the polynomial function $P(x) = a_n x^n + a_{n-1}x^{n-1} + \cdots + a_2 x^2 + a_1 x + a_0$ has *integer* coefficients, then every rational zero of $P(x)$ has the form:

$$\text{Rational zero} = \frac{\text{factors of } a_0}{\text{factors of } a_n} = \frac{\text{factors of constant term}}{\text{factors of leading coefficient}}$$

$$= \pm\frac{\text{positive factors of constant term}}{\text{positive factors of leading coefficient}}$$

To use this theorem, simply list all combinations of factors of both the constant term a_0 and the leading coefficient term a_n and take all appropriate combinations of ratios. This procedure is illustrated in Example 4. Notice that when the leading coefficient is 1, then the possible rational zeros will simply be the possible factors of the constant term.

EXAMPLE 4 Using the Rational Zero Theorem

Determine possible rational zeros for the polynomial $P(x) = x^4 - x^3 - 5x^2 - x - 6$ by the rational zero theorem. Test each one to find all rational zeros.

Solution:

STEP 1 List factors of the constant and leading coefficient terms.
$a_0 = -6$ $\pm1, \pm2, \pm3, \pm6$
$a_n = 1$ ±1

STEP 2 List possible rational zeros $\frac{a_0}{a_n}$. $\frac{\pm1}{\pm1}, \frac{\pm2}{\pm1}, \frac{\pm3}{\pm1}, \frac{\pm6}{\pm1} = \pm1, \pm2, \pm3, \pm6$

There are three ways to test whether any of these are zeros: Substitute these values into the polynomial to see which ones yield zero, or use either polynomial division or synthetic division to divide the polynomial by these possible zeros, and look for a zero remainder.

STEP 3 Test possible zeros by looking for zero remainders.

1 is not a zero: $P(1) = (1)^4 - (1)^3 - 5(1)^2 - (1) - 6 = -12$

-1 is not a zero: $P(-1) = (-1)^4 - (-1)^3 - 5(-1)^2 - (-1) - 6 = -8$

We could continue testing with direct substitution, but let us now use synthetic division as an alternative.

2 is not a zero:

$$\begin{array}{r|rrrrr} 2 & 1 & -1 & -5 & -1 & -6 \\ & & 2 & 2 & -6 & -14 \\ \hline & 1 & 1 & -3 & -7 & \boxed{-20} \end{array}$$

-2 is a zero:

$$\begin{array}{r|rrrrr} -2 & 1 & -1 & -5 & -1 & -6 \\ & & -2 & 6 & -2 & 6 \\ \hline & 1 & -3 & 1 & -3 & \boxed{0} \end{array}$$

Since -2 is a zero, then $x + 2$ is a factor of $P(x)$, and the remaining quotient is $x^3 - 3x^2 + x - 3$. Therefore, if there are any other real roots remaining, we can now use the simpler $x^3 - 3x^2 + x - 3$ for the dividend. Also note that the rational zero theorem can be applied to the new dividend and possibly shorten the list of possible rational zeros. In this case, the possible rational zeros of $F(x) = x^3 - 3x^2 + x - 3$ are ±1 and ±3.

3 is a zero:

$$\begin{array}{r|rrrr} 3 & 1 & -3 & 1 & -3 \\ & & 3 & 0 & 3 \\ \hline & 1 & 0 & 1 & \boxed{0} \end{array}$$

We now know that $\boxed{-2}$ and $\boxed{3}$ are confirmed zeros. If we continue testing, we will find that the other possible zeros fail. This is a fourth-degree polynomial, and we have found two rational real zeros. We see in the graph on the right that these two real zeros correspond to the x-intercepts.

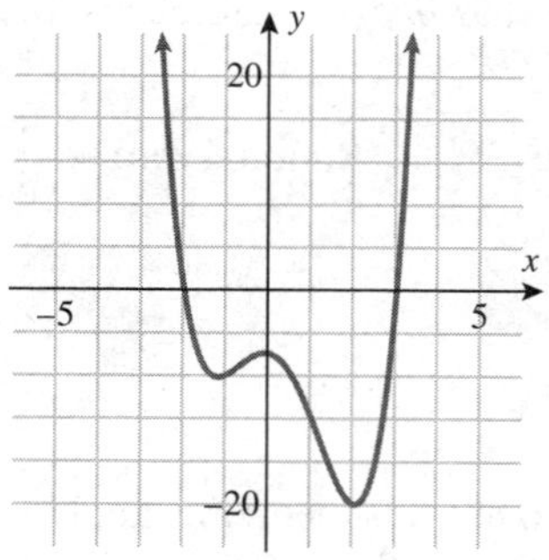

Study Tip

The remainder can be found by evaluating the function or synthetic division. For simple values like $x = \pm1$, it is easier to evaluate the polynomial function. For other values, it is often easier to use synthetic division.

■ **YOUR TURN** List the possible rational zeros of the polynomial $P(x) = x^4 + 2x^3 - 2x^2 + 2x - 3$, and determine rational real zeros.

■ **Answer:** Possible rational zeros: ±1 and ±3. Rational real zeros: 1 and -3.

Notice in Example 4 that the polynomial function $P(x) = x^4 - x^3 - 5x^2 - x - 6$ had two rational real zeros, $x = -2$ and $x = 3$. This implies that $x + 2$ and $x - 3$ are factors of $P(x)$. Also note in the last step when we divided by the zero $x = 3$, the quotient was $x^2 + 1$. Therefore, we can write the polynomial in factored form as

$$P(x) = \underbrace{(x+2)}_{\substack{\text{linear}\\\text{factor}}}\ \underbrace{(x-3)}_{\substack{\text{linear}\\\text{factor}}}\ \underbrace{(x^2+1)}_{\substack{\text{irreducible}\\\text{quadratic}\\\text{factor}}}$$

Notice that the first two factors are of degree 1, so we call them **linear factors**. The third expression, $x^2 + 1$, is of degree 2 and cannot be factored in terms of real numbers. We will discuss complex zeros in the next section. For now, we say that a quadratic expression, $ax^2 + bx + c$, is called **irreducible** if it cannot be factored over the real numbers.

EXAMPLE 5 Factoring a Polynomial Function

Write the following polynomial function as a product of linear and/or irreducible quadratic factors:

$$P(x) = x^4 - 4x^3 + 4x^2 - 36x - 45.$$

Solution:

Use the rational zero theorem to list possible rational roots. $\quad x = \pm1, \pm3, \pm5, \pm9, \pm15, \pm45$

Test possible zeros by evaluating the function or by utilizing synthetic division.

$x = 1$ is not a zero. $\quad P(1) = -80$

$x = -1$ is a zero. $\quad P(-1) = 0$

Divide $P(x)$ by $x + 1$.

$$\begin{array}{r|rrrrr} -1 & 1 & -4 & 4 & -36 & -45 \\ & & -1 & 5 & -9 & 45 \\ \hline & 1 & -5 & 9 & -45 & \boxed{0} \end{array}$$

$x = 5$ is a zero.

$$\begin{array}{r|rrrr} 5 & 1 & -5 & 9 & -45 \\ & & 5 & 0 & 45 \\ \hline & \underbrace{1 \quad 0 \quad 9}_{x^2+9} & & & \boxed{0} \end{array}$$

The factor $x^2 + 9$ is irreducible.

Write the polynomial as a product of linear and/or irreducible quadratic factors. $\quad \boxed{P(x) = (x - 5)(x + 1)(x^2 + 9)}$

Notice that the graph of this polynomial function has x-intercepts at $x = -1$ and $x = 5$.

YOUR TURN Write the following polynomial function as a product of linear and/or irreducible quadratic factors.

$$\boxed{P(x) = x^4 - 2x^3 - x^2 - 4x - 6}$$

Technology Tip

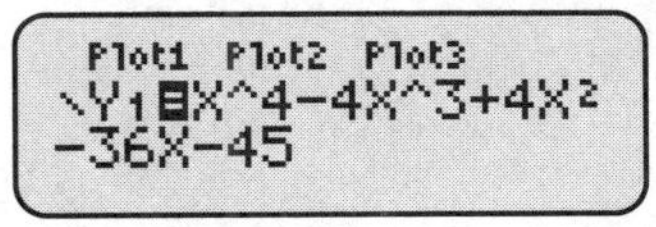

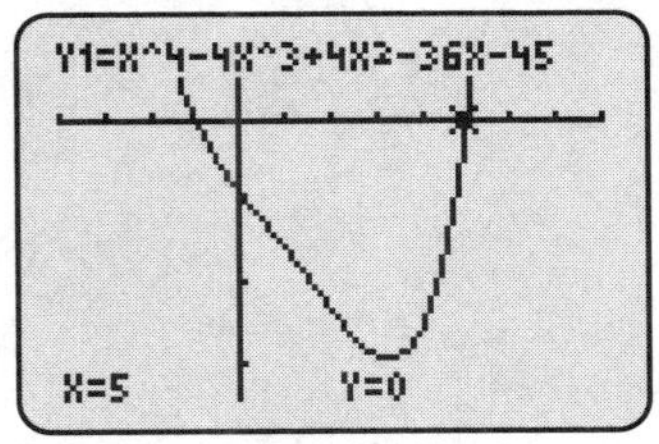

The real zeros of the function at $x = -1$ and $x = 5$ give the linear factors $x + 1$ and $x - 5$. Use the synthetic division to find the irreducible quadratic factors.

A table of values supports the real zeros of the graph.

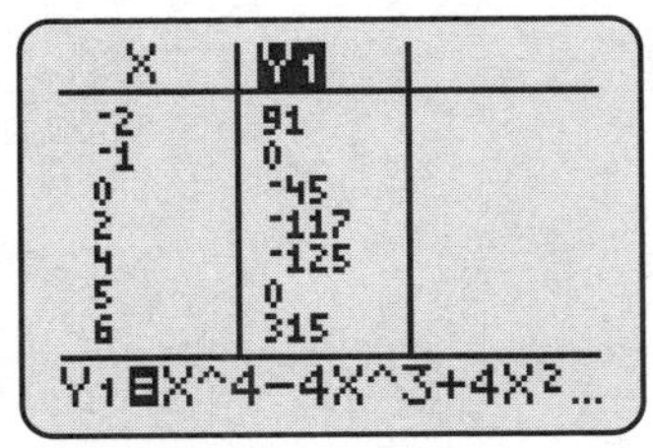

Answer: $P(x) = (x + 1)(x - 3)(x^2 + 2)$

The rational zero theorem lists possible zeros. It would be helpful if we could narrow that list. Descartes's rule of signs determines the possible combinations of positive real zeros and negative real zeros through variations of sign. A *variation in sign* is a sign difference seen between consecutive coefficients.

Sign Change
− to +

$$P(x) = 2x^6 - 5x^5 - 3x^4 + 2x^3 - x^2 - x - 1$$

Sign Change **+ to −** | **Sign Change** **+ to −**

This polynomial experiences three sign changes or variations in sign.

DESCARTES'S RULE OF SIGNS

If the polynomial function $P(x) = a_nx^n + a_{n-1}x^{n-1} + \cdots + a_2x^2 + a_1x + a_0$ has real coefficients and $a_0 \neq 0$, then:

- The number of **positive** real zeros of the polynomial is either equal to the number of variations of sign of $P(x)$ or less than that number by an even integer.
- The number of **negative** real zeros of the polynomial is either equal to the number of variations of sign of $P(-x)$ or less than that number by an even integer.

Descartes's rule of signs narrows our search for real zeros, because we don't have to test all of the possible rational zeros. For example, if we know there is one positive real zero, then if we find a positive rational zero we no longer need to continue to test possible positive zeros.

EXAMPLE 6 Using Descartes's Rule of Signs

Determine the possible combinations of zeros for $P(x) = x^3 - 2x^2 - 5x + 6$.

Solution:

Determine the number of variations of sign in $P(x)$.

Sign Change

$$P(x) = x^3 - 2x^2 - 5x + 6$$

Sign Change

$P(x)$ has 2 variations in sign.

Apply Descartes's rule of signs. $P(x)$ has *either* 2 or 0 **positive** real zeros.

Determine the number of variations of sign in $P(-x)$.

$$(-x)^3 - 2(-x)^2 - 5(-x) + 6$$
$$-x^3 - 2x^2 + 5x + 6$$

Sign Change

$P(-x)$ has **1** variation in sign. $P(x) = -x^3 - 2x^2 + 5x + 6$

Apply Descartes's rule of signs. $P(x)$ must have 1 **negative** real zero.

Since $P(x) = x^3 - 2x^2 - 5x + 6$ is a *third*-degree polynomial, there are at most 3 real zeros. One zero is a negative real number, and there can be either 2 positive real zeros or 0 positive real zeros. Now look back at Example 2 and see that in fact there were 1 negative real zero and 2 positive real zeros.

EXAMPLE 7 Using Descartes's Rule of Signs to Find Possible Combinations of Real Zeros

Determine the possible combinations of real zeros for $P(x) = x^4 - 2x^3 + x^2 + 2x - 2$.

Solution:

$P(x)$ has 3 variations in sign.

$$P(x) = x^4 - 2x^3 + x^2 + 2x - 2$$

Sign Change (between x^4 and $-2x^3$); Sign Change (between $-2x^3$ and $+x^2$); Sign Change (between $+2x$ and -2)

Apply Descartes's rule of signs. $P(x)$ has *either* 3 or 1 **positive** real zero.

Find $P(-x)$. $P(-x) = (-x)^4 - 2(-x)^3 + (-x)^2 + 2(-x) - 2$

$= x^4 + 2x^3 + x^2 - 2x - 2$

$P(-x)$ has 1 variation in sign. $P(-x) = x^4 + 2x^3 + x^2 - 2x - 2$

Sign Change (between $+x^2$ and $-2x$)

Apply Descartes's rule of signs. $P(x)$ has 1 **negative** real zero.

Since $P(x) = x^4 - 2x^3 + x^2 + 2x - 2$ is a *fourth*-degree polynomial, there are at most 4 real zeros. One zero is a negative real number.

$P(x)$ could have 3 positive real zeros or 1 positive real zero.

Look at the graph in the Technology Tip to confirm 1 negative real zero and 1 positive real zero.

YOUR TURN Determine the possible combinations of zeros for:

$$P(x) = x^4 + 2x^3 + x^2 + 8x - 12$$

Technology Tip

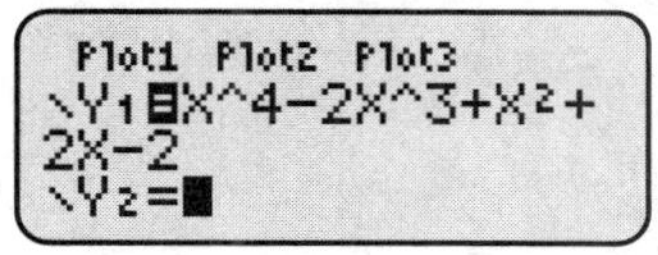

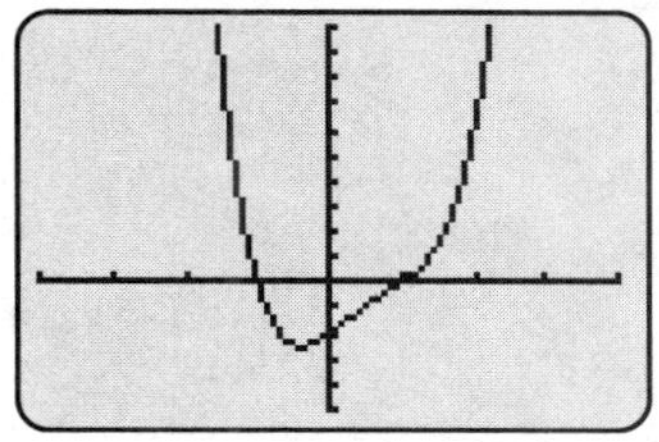

There are 1 negative real zero and 1 posi ive real zero.

The number of negative real zero is 1 and the number of positive real zero is 1. So, it has 2 complex conjugate zeros.

A table of values supports the zeros of the function.

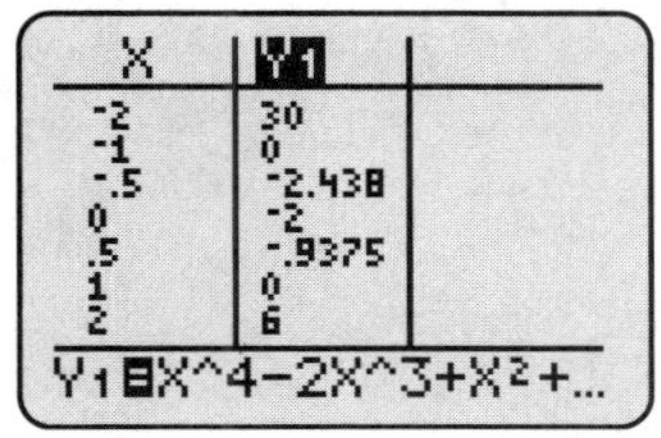

X	Y1
-2	30
-1	0
-.5	-2.438
0	-2
.5	-.9375
1	0
2	6

Y1=X^4-2X^3+X²+...

■ **Answer:** Positive real zeros: 1
Negative real zeros: 3 or 1

Factoring Polynomials

Now let's draw on the tests discussed in this section thus far to help us in finding all real zeros of a polynomial function. Doing so will enable us to factor polynomials.

EXAMPLE 8 Factoring a Polynomial

Write the polynomial $P(x) = x^5 + 2x^4 - x - 2$ as a product of linear and/or irreducible quadratic factors.

Solution:

STEP 1 Determine variations in sign.

$P(x)$ has 1 sign change. $\quad P(x) = x^5 + 2x^4 - x - 2$

$P(-x)$ has 2 sign changes. $\quad P(-x) = -x^5 + 2x^4 + x - 2$

STEP 2 Apply Descartes's rule of signs.

Positive Real Zeros: 1

Negative Real Zeros: 2 or 0

STEP 3 Use the rational zero theorem to determine the possible rational zeros. $\quad \pm1, \pm2$

We know (Step 2) that there is one positive real zero, so test the possible positive rational zeros first.

STEP 4 Test possible rational zeros.

1	1	2	0	0	−1	−2
		1	3	3	3	2
	1	3	3	3	2	[0]

1 is a zero:

Now that we have found *the* positive zero, we can test the other two possible negative zeros—because either they both are zeros or neither is a zero.

−1	1	3	3	3	2
		−1	−2	−1	−2
	1	2	1	2	[0]

−1 is a zero:

At this point, from Descartes's rule of signs we know that −2 must also be a zero, since there are either 2 or 0 negative zeros. Let's confirm this:

−2	1	2	1	2
		−2	0	−2
	1	0	1	[0]

−2 is a zero: $\quad x^2 + 1$

STEP 5 Three of the five have been found to be zero: −1, −2, and 1.

STEP 6 Write the fifth-degree polynomial as a product of 3 linear factors and an irreducible quadratic factor.

$$\boxed{P(x) = (x - 1)(x + 1)(x + 2)(x^2 + 1)}$$

YOUR TURN Write the polynomial $P(x) = x^5 - 2x^4 + x^3 - 2x^2 - 2x + 4$ as a product of linear and/or irreducible quadratic factors.

Answer: $P(x) = (x - 2)(x + 1)(x - 1)(x^2 + 2)$

Study Tip

If $f(x)$ has a common monomial factor, it should be canceled first before applying the bound rules.

The rational zero theorem gives us possible rational zeros of a polynomial, and Descartes's rule of signs gives us possible combinations of positive and negative real zeros. Additional aids that help eliminate possible zeros are the *upper* and *lower bound rules.* These rules can give you an upper and lower bound on the real zeros of a polynomial function. If $f(x)$ has a common monomial factor, you should factor it out first, and then follow the upper and lower bound rules.

UPPER AND LOWER BOUND RULES

Let $f(x)$ be a polynomial with real coefficients and a positive leading coefficient. Suppose $f(x)$ is divided by $x - c$ using synthetic division.

1. If $c > 0$ and each number in the bottom row is either positive or zero, c is an **upper bound** for the real zeros of f.
2. If $c < 0$ and the numbers in the bottom row are alternately positive and negative (zero entries count as either positive or negative), c is a **lower bound** for the real zeros of f.

EXAMPLE 9 Using Upper and Lower Bounds to Eliminate Possible Zeros

Find the real zeros of $f(x) = 4x^3 - x^2 + 36x - 9$.

Solution:

STEP 1 The rational zero theorem gives possible rational zeros.

$$\frac{\text{Factors of 9}}{\text{Factors of 4}} = \frac{\pm 1, \pm 3, \pm 9}{\pm 1, \pm 2, \pm 4}$$

$$= \pm 1, \pm\frac{1}{2}, \pm\frac{1}{4}, \pm\frac{3}{4}, \pm\frac{3}{2}, \pm\frac{9}{4}, \pm 3, \pm\frac{9}{2}, \pm 9$$

STEP 2 Apply Descartes's rule of signs:

$f(x)$ has 3 sign variations. 3 or 1 positive real zeros

$f(-x)$ has no sign variations. no negative real zeros

STEP 3 Try $x = 1$.

$$\begin{array}{r|rrrr} 1 & 4 & -1 & 36 & -9 \\ & & 4 & 3 & 39 \\ \hline & 4 & 3 & 39 & 30 \end{array}$$

$x = 1$ is not a zero, but because the last row contains all positive entries, $x = 1$ is an *upper* bound. Since we know there are no negative real zeros, we restrict our search to between 0 and 1.

STEP 4 Try $x = \frac{1}{4}$.

$$\begin{array}{r|rrrr} \frac{1}{4} & 4 & -1 & 36 & -9 \\ & & 1 & 0 & 9 \\ \hline & 4 & 0 & 36 & 0 \end{array}$$

We are going to take one of our "possible" zeros that didn't prove to be a zero and regard it as the "c" in the preceeding bound rules and test it to see if it is an upper bound or a lower bound.

$\frac{1}{4}$ is a zero and the quotient $4x^2 + 36$ has all positive coefficients; therefore,

$\boxed{\frac{1}{4} \text{ is the only real zero}}$.

Note: If $f(x)$ has a common monomial factor it should be factored out first before applying the bound rules.

The Intermediate Value Theorem

In our search for zeros, we sometimes encounter irrational zeros, as in, for example, the polynomial

$$f(x) = x^5 - x^4 - 1$$

Descartes's rule of signs tells us there is exactly one real positive zero. However, the rational zero test yields only $x = \pm 1$, neither of which are zeros. So if we know there is a real positive zero and we know it's not rational, it must be irrational. Notice that $f(1) = -1$ and $f(2) = 15$. Since polynomial functions are continuous and the function goes from negative to positive between $x = 1$ and $x = 2$, we expect a zero somewhere in that interval. Generating a graph with a graphing utility, we find that there is a zero around $x = 1.3$.

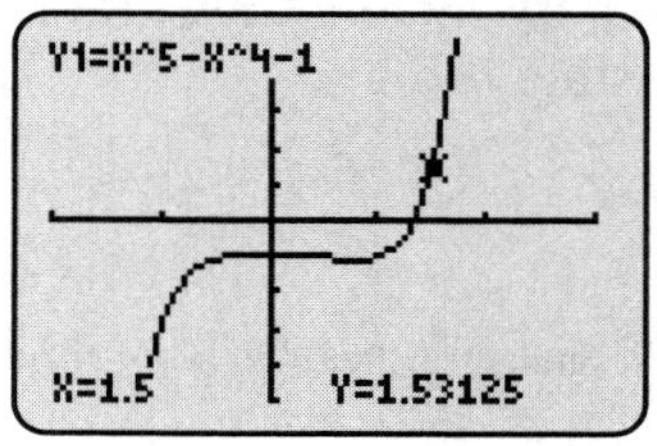

The *intermediate value theorem* is based on the fact that polynomial functions are continuous.

INTERMEDIATE VALUE THEOREM

Let a and b be real numbers such that $a < b$ and $f(x)$ be a polynomial function. If $f(a)$ and $f(b)$ have opposite signs, then there is at least one real zero between a and b.

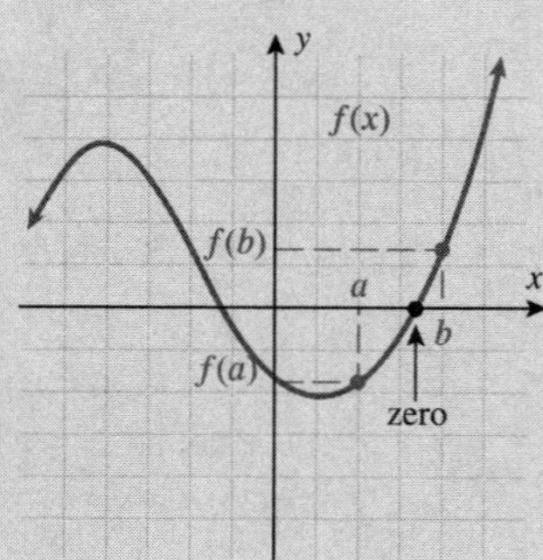

If the intermediate value theorem tells us that there is a real zero in the interval (a, b), how do we approximate that zero? The **bisection method*** is a root-finding algorithm that approximates the solution to the equation $f(x) = 0$. In the bisection method the interval is divided in half and then the subinterval that contains the zero is selected. This is repeated until the bisection method converges to an approximate root of f.

*In calculus you will learn Newton's method, which is a more efficient approximation technique for finding zeros.

EXAMPLE 10 Approximating Real Zeros of a Polynomial Function

Approximate the real zero of $f(x) = x^5 - x^4 - 1$.

Note: Descartes's rule of signs tells us that there are no real negative zeros and there is exactly one real positive zero.

Solution:

Find two consecutive integer values for x that have corresponding function values opposite in sign.

x	1	−1
f(x)	2	15

Note that a graphing utility would have shown an x-intercept between $x = 1$ and $x = 2$.

Apply the bisection method, with $a = 1$ and $b = 2$. $\quad c = \dfrac{a+b}{2} = \dfrac{1+2}{2} = \dfrac{3}{2}$

Evaluate the function at $x = c$. $\quad f(1.5) \approx 1.53$

Compare the values of f at the endpoints and midpoint. $\quad f(1) = -1, f(1.5) \approx 1.53, f(2) = 15$

Select the subinterval corresponding to the *opposite* signs of f. $\quad (1, 1.5)$

Apply the bisection method again (repeat the algorithm). $\quad \dfrac{1 + 1.5}{2} = 1.25$

Evaluate the function at $x = 1.25$. $\quad f(1.25) \approx -0.38965$

Compare the values of f at the endpoints and midpoint. $\quad f(1) = -1, f(1.25) \approx -0.38965, f(1.5) \approx 1.53$

Select the subinterval corresponding to the *opposite* signs of f. $\quad (1.25, 1.5)$

Apply the bisection method again (repeat the algorithm). $\quad \dfrac{1.25 + 1.5}{2} = 1.375$

Evaluate the function at $x = 1.375$. $\quad f(1.375) \approx 0.3404$

Compare the values of f at the endpoints and midpoint. $\quad f(1.25) \approx -0.38965, f(1.375) \approx 0.3404, f(1.5) \approx 1.53$

Select the subinterval corresponding to the *opposite* signs of f. $\quad (1.25, 1.375)$

We can continue this procedure (*applying the bisection method*) to find that the zero is somewhere between $f(1.32) \approx -0.285$ and $f(1.33) \approx 0.0326$.

We find that to three significant digits, $\boxed{1.32}$ is an approximation to the real zero.

Technology Tip

The graph of $f(x) = x^5 - x^4 - 1$ is shown. To find the zero of the function, press

[2nd] [Calc] [▼] [2:zero] [ENTER] [1] [ENTER] [2] [ENTER] [ENTER]

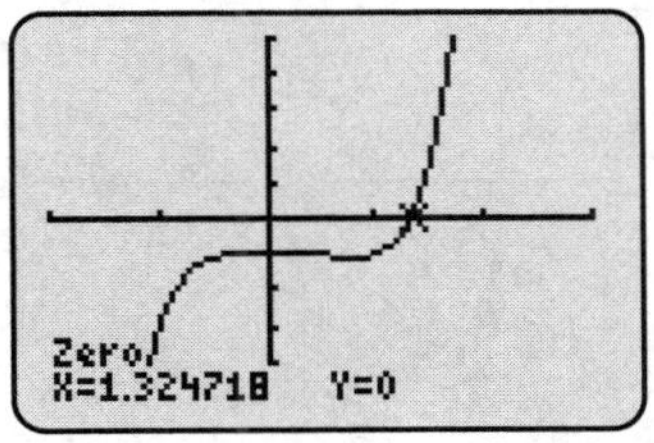

A table of values supports this zero of the function.

Graphing Polynomial Functions

In Section 4.2, we graphed simple polynomial functions that were easily factored. Now that we have procedures for finding real zeros of polynomial functions (rational zero theorem, Descartes's rule of signs, and upper and lower bound rules for rational zeros, and the intermediate value theorem and the bisection method for irrational zeros), let us return to the topic of graphing polynomial functions. Since a real zero of a polynomial function corresponds to an x-intercept of its graph, we now have methods for finding (or estimating) any x-intercepts of the graph of any polynomial function.

Technology Tip

The graph of $f(x) = 2x^4 - 2x^3 + 5x^2 + 17x - 22$ is shown.

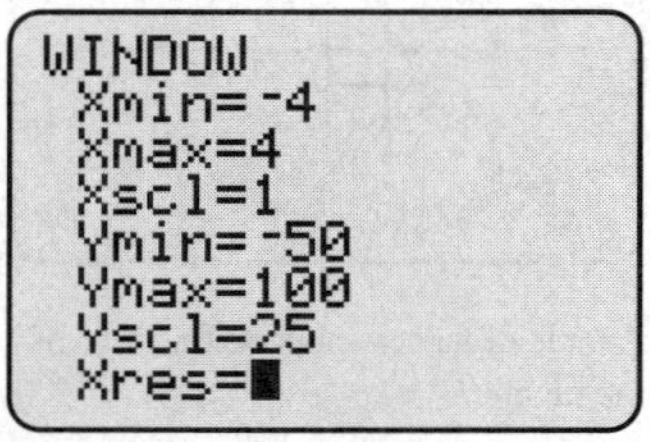

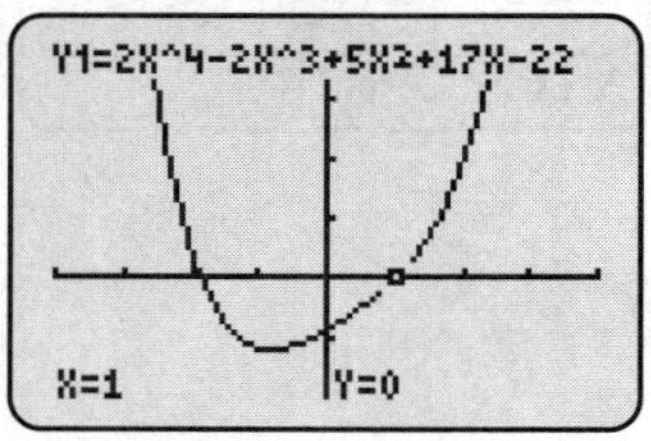

To find the zero of the function, press:

2nd | Calc | ▼ | 2:zero | ENTER | −3 | ENTER | −1 | ENTER | ENTER

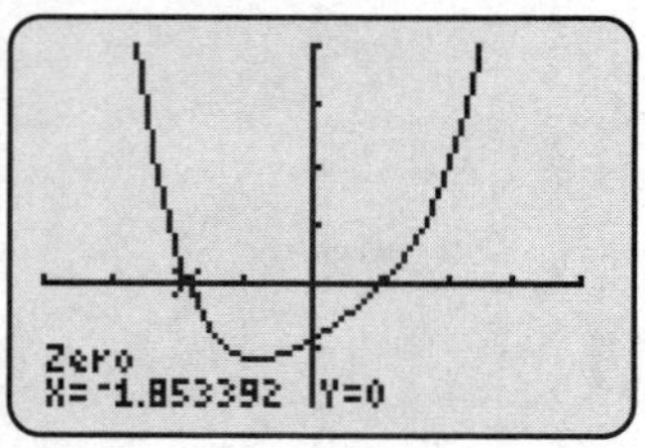

A table of values supports this zero of the function and the graph.

EXAMPLE 11 Graphing a Polynomial Function

Graph the function $f(x) = 2x^4 - 2x^3 + 5x^2 + 17x - 22$.

Solution:

STEP 1 Find the y-intercept. $f(0) = -22$

STEP 2 Find any x-intercepts (real zeros).

Apply Descartes's rule of signs.

3 sign changes correspond to 3 or 1 positive real zeros. $f(x) = 2x^4 - 2x^3 + 5x^2 + 17x - 22$

1 sign change corresponds to 1 negative real zero. $f(-x) = 2x^4 + 2x^3 + 5x^2 - 17x - 22$

Apply the rational zero theorem.

Let $a_0 = -22$ and $a_n = 2$. $\dfrac{\text{Factors of } a_0}{\text{Factors of } a_n} = \pm\dfrac{1}{2}, \pm1, \pm2, \pm\dfrac{11}{2}, \pm11, \pm22$

Test the possible zeros.

$x = 1$ is a zero. $f(1) = 0$

There are no other rational zeros.

Apply the upper bound rule.

1	2	−2	5	17	−22
		2	0	5	22
	2	0	5	22	0

Since $x = 1$ is positive and all of the numbers in the bottom row are positive (or zero), $x = 1$ is an upper bound for the real zeros. We know there is exactly one negative real zero, but none of the possible zeros from the rational zero theorem is a zero. Therefore, the negative real zero is irrational.

Apply the intermediate value theorem and the bisection method.

f is positive at $x = -2$. $f(-2) = 12$

f is negative at $x = -1$. $f(-1) = -30$

Use the bisection method to find the negative real zero between -2 and -1. $x \approx -1.85$

STEP 3 Determine the end behavior. $y = 2x^4$

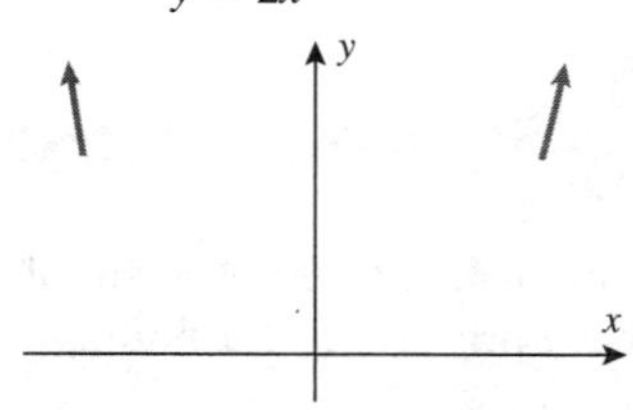

STEP 4 Find additional points.

x	-2	-1.85	-1	0	1	2
$f(x)$	12	0	-30	-22	0	48
Point	$(-2, 12)$	$(-1.85, 0)$	$(-1, -30)$	$(0, -22)$	$(1, 0)$	$(2, 48)$

STEP 5 Sketch the graph.

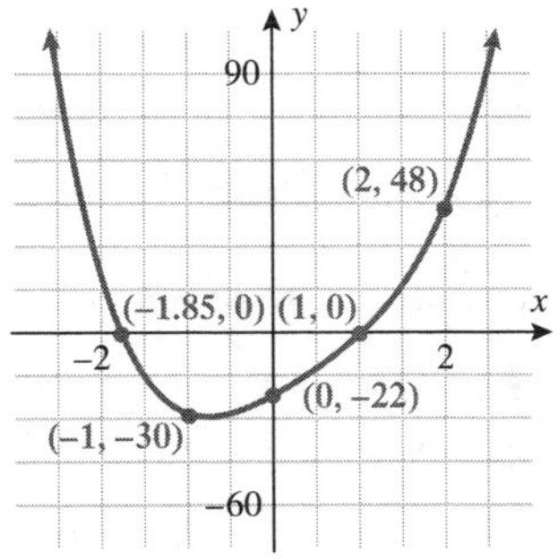

SECTION 4.4 SUMMARY

In this section we discussed how to find the real zeros of a polynomial function. Once real zeros are known, it is possible to write the polynomial function as a product of linear and/or irreducible quadratic factors.

The Number of Zeros

- A polynomial of degree n has *at most n* real zeros.
- *Descartes's rule of signs* determines the possible combinations of positive and negative real zeros.
- *Upper and lower bounds* help narrow the search for zeros.

How to Find Zeros

- *Rational zero theorem*: List possible rational zeros:

$$\frac{\text{Factors of constant, } a_0}{\text{Factors of leading coefficient, } a_n}$$

- *Irrational zeros:* Approximate zeros by determining when the polynomial function changes sign (intermediate value theorem).

Procedure for Factoring a Polynomial Function

- List possible rational zeros (rational zero theorem).
- List possible combinations of positive and negative real zeros (Descartes's rule of signs).
- Test possible values until a zero is found.*
- Once a real zero is found, repeat testing on the quotient until linear and/or irreducible quadratic factors remain.
- If there is a real zero but all possible rational roots have failed, then approximate the zero using the *intermediate value theorem* and the *bisection method.*

*Depending on the form of the quotient, upper and lower bounds may eliminate possible zeros.

SECTION 4.4 EXERCISES

SKILLS

In Exercises 1–6, find the following values by using synthetic division. Check by substituting the value into the function.

$$f(x) = 3x^4 - 2x^3 + 7x^2 - 8 \qquad g(x) = 2x^3 + x^2 + 1$$

1. $f(1)$ **2.** $f(-1)$ **3.** $g(1)$ **4.** $g(-1)$ **5.** $f(-2)$ **6.** $g(2)$

In Exercises 7–10, determine whether the number given is a zero of the polynomial.

7. -7, $P(x) = x^3 + 2x^2 - 29x + 42$

8. 2, $P(x) = x^3 + 2x^2 - 29x + 42$

9. -3, $P(x) = x^3 - x^2 - 8x + 12$

10. 1, $P(x) = x^3 - x^2 - 8x + 12$

In Exercises 11–20, given a real zero of the polynomial, determine all other real zeros, and write the polynomial in terms of a product of linear and/or irreducible quadratic factors.

	Polynomial	Zero		Polynomial	Zero
11.	$P(x) = x^3 - 13x + 12$	1	**12.**	$P(x) = x^3 + 3x^2 - 10x - 24$	3
13.	$P(x) = 2x^3 + x^2 - 13x + 6$	$\frac{1}{2}$	**14.**	$P(x) = 3x^3 - 14x^2 + 7x + 4$	$-\frac{1}{3}$
15.	$P(x) = x^4 - 2x^3 - 11x^2 - 8x - 60$	$-3, 5$	**16.**	$P(x) = x^4 - x^3 + 7x^2 - 9x - 18$	$-1, 2$
17.	$P(x) = x^4 - 5x^2 + 10x - 6$	$1, -3$	**18.**	$P(x) = x^4 - 4x^3 + x^2 + 6x - 40$	$4, -2$
19.	$P(x) = x^4 + 6x^3 + 13x^2 + 12x + 4$	-2 (multiplicity 2)	**20.**	$P(x) = x^4 + 4x^3 - 2x^2 - 12x + 9$	1 (multiplicity 2)

In Exercises 21–28, use the rational zero theorem to list the *possible* rational zeros.

21. $P(x) = x^4 + 3x^2 - 8x + 4$
22. $P(x) = -x^4 + 2x^3 - 5x + 4$
23. $P(x) = x^5 - 14x^3 + x^2 - 15x + 12$
24. $P(x) = x^5 - x^3 - x^2 + 4x + 9$
25. $P(x) = 2x^6 - 7x^4 + x^3 - 2x + 8$
26. $P(x) = 3x^5 + 2x^4 - 5x^3 + x - 10$
27. $P(x) = 5x^5 + 3x^4 + x^3 - x - 20$
28. $P(x) = 4x^6 - 7x^4 + 4x^3 + x - 21$

In Exercises 29–32, list the possible rational zeros, and test to determine all rational zeros.

29. $P(x) = x^4 + 2x^3 - 9x^2 - 2x + 8$
30. $P(x) = x^4 + 2x^3 - 4x^2 - 2x + 3$
31. $P(x) = 2x^3 - 9x^2 + 10x - 3$
32. $P(x) = 3x^3 - 5x^2 - 26x - 8$

In Exercises 33–44, use Descartes's rule of signs to determine the possible number of positive real zeros and negative real zeros.

33. $P(x) = x^4 - 32$
34. $P(x) = x^4 + 32$
35. $P(x) = x^5 - 1$
36. $P(x) = x^5 + 1$
37. $P(x) = x^5 - 3x^3 - x + 2$
38. $P(x) = x^4 + 2x^2 - 9$
39. $P(x) = 9x^7 + 2x^5 - x^3 - x$
40. $P(x) = 16x^7 - 3x^4 + 2x - 1$
41. $P(x) = x^6 - 16x^4 + 2x^2 + 7$
42. $P(x) = -7x^6 - 5x^4 - x^2 + 2x + 1$
43. $P(x) = -3x^4 + 2x^3 - 4x^2 + x - 11$
44. $P(x) = 2x^4 - 3x^3 + 7x^2 + 3x + 2$

For each polynomial in Exercises 45–62: (a) use Descartes's rule of signs to determine the possible combinations of positive real zeros and negative real zeros; (b) use the rational zero test to determine possible rational zeros; (c) test for rational zeros; and (d) factor as a product of linear and/or irreducible quadratic factors.

45. $P(x) = x^3 + 6x^2 + 11x + 6$
46. $P(x) = x^3 - 6x^2 + 11x - 6$
47. $P(x) = x^3 - 7x^2 - x + 7$
48. $P(x) = x^3 - 5x^2 - 4x + 20$
49. $P(x) = x^4 + 6x^3 + 3x^2 - 10x$
50. $P(x) = x^4 - x^3 - 14x^2 + 24x$
51. $P(x) = x^4 - 7x^3 + 27x^2 - 47x + 26$
52. $P(x) = x^4 - 5x^3 + 5x^2 + 25x - 26$
53. $P(x) = 10x^3 - 7x^2 - 4x + 1$
54. $P(x) = 12x^3 - 13x^2 + 2x - 1$
55. $P(x) = 6x^3 + 17x^2 + x - 10$
56. $P(x) = 6x^3 + x^2 - 5x - 2$
57. $P(x) = x^4 - 2x^3 + 5x^2 - 8x + 4$
58. $P(x) = x^4 + 2x^3 + 10x^2 + 18x + 9$
59. $P(x) = x^6 + 12x^4 + 23x^2 - 36$
60. $P(x) = x^4 - x^2 - 16x^2 + 16$
61. $P(x) = 4x^4 - 20x^3 + 37x^2 - 24x + 5$
62. $P(x) = 4x^4 - 8x^3 + 7x^2 + 30x + 50$

In Exercises 63–66, use the information found in Exercises 47, 51, 55, and 61 to assist in sketching a graph of each polynomial function.

63. Exercise 47.
64. Exercise 51.
65. Exercise 55.
66. Exercise 61.

In Exercises 67–72, use the intermediate value theorem to approximate the real zero in the indicated interval. Approximate to two decimal places.

67. $f(x) = x^4 - 3x^3 + 4 \quad [1, 2]$
68. $f(x) = x^5 - 3x^3 + 1 \quad [0, 1]$
69. $f(x) = 7x^5 - 2x^2 + 5x - 1 \quad [0, 1]$
70. $f(x) = -2x^3 + 3x^2 + 6x - 7 \quad [-2, -1]$
71. $f(x) = x^3 - 2x^2 - 8x - 3 \quad [-1, 0]$
72. $f(x) = x^4 + 4x^2 - 7x - 13 \quad [-2, -1]$

APPLICATIONS

73. Profit. A mathematics honorary society wants to sell magazine subscriptions to *Math Weekly*. If there are x hundred subscribers, its monthly revenue and cost are given by:

$$R(x) = 46 - 3x^2 \quad \text{and} \quad C(x) = 20 + 2x$$

a. Determine the profit function. *Hint:* $P = R - C$.

b. Determine the number of subscribers needed in order to *break even*.

74. Profit. Using the profit equation $P(x) = x^3 - 5x^2 + 3x + 6$, when will the company break even if x represents the units sold?

CATCH THE MISTAKE

In Exercises 75 and 76, explain the mistake that is made.

75. Use Descartes's rule of signs to determine the possible combinations of zeros of

$$P(x) = 2x^5 + 7x^4 + 9x^3 + 9x^2 + 7x + 2$$

Solution:

No sign changes, so no positive real zeros.

$$P(x) = 2x^5 + 7x^4 + 9x^3 + 9x^2 + 7x + 2$$

Five sign changes, so five negative real zeros.

$$P(-x) = -2x^5 + 7x^4 - 9x^3 + 9x^2 - 7x + 2$$

This is incorrect. What mistake was made?

76. Determine whether $x - 2$ is a factor of $P(x) = x^3 - 2x^2 - 5x + 6$.

Solution:

$$\begin{array}{r|rrrr} -2 & 1 & -2 & -5 & 6 \\ & & -2 & 8 & -6 \\ \hline & 1 & -4 & 3 & \boxed{0} \end{array}$$

Yes, $x - 2$ is a factor of $P(x)$.

This is incorrect. What mistake was made?

CONCEPTUAL

In Exercises 77–80 determine whether each statement is true or false.

77. All real zeros of a polynomial correspond to x-intercepts.

78. A polynomial of degree n, $n > 0$, must have at least one zero.

79. A polynomial of degree n, $n > 0$, can be written as a product of n linear factors over real numbers.

80. The number of sign changes in a polynomial is equal to the number of positive real zeros of that polynomial.

CHALLENGE

81. Given that $x = a$ is a zero of $P(x) = x^3 - (a + b + c)x^2 + (ab + ac + bc)x - abc$, find the other two zeros, given that a, b, c are real numbers and $a > b > c$.

82. Given that $x = a$ is a zero of $p(x) = x^3 + (-a + b - c)x^2 - (ab + bc - ac)x + abc$, find the other two real zeros, given that a, b, c are real positive numbers.

TECHNOLOGY

In Exercises 83 and 84, determine all possible rational zeros of the polynomial. There are many possibilities. Instead of trying them all, use a graphing calculator or software to graph $P(x)$ to help find a zero to test.

83. $P(x) = x^3 - 2x^2 + 16x - 32$

84. $P(x) = x^3 - 3x^2 + 16x - 48$

In Exercises 85 and 86: (a) determine all possible rational zeros of the polynomial, use a graphing calculator or software to graph $P(x)$ to help find the zeros; and (b) factor as a product of linear and/or irreducible quadratic factors.

85. $P(x) = 12x^4 + 25x^3 + 56x^2 - 7x - 30$

86. $P(x) = -3x^3 - x^2 - 7x - 49$

SECTION 4.5 COMPLEX ZEROS: THE FUNDAMENTAL THEOREM OF ALGEBRA

SKILLS OBJECTIVES

- Find the complex zeros of a polynomial function.
- Use the complex conjugate zeros theorem.
- Factor polynomial functions.

CONCEPTUAL OBJECTIVES

- Extend the domain of polynomial functions to complex numbers.
- Understand how the fundamental theorem of algebra guarantees at least one zero.
- Understand why complex zeros occur in conjugate pairs.

Study Tip

The zeros of a polynomial can be complex numbers. Only when the zeros are real numbers do we interpret zeros as x-intercepts.

In Section 4.4, we found the real zeros of a polynomial function. In this section we find the complex zeros of a polynomial function. In this chapter we assume the coefficients of polynomial functions are real numbers. The domain of polynomial functions thus far has been the set of all real numbers. Now, we consider a more general case. In this section, the coefficients of a polynomial function and the domain of a polynomial function are *complex numbers*. Note that the set of real numbers is a subset of the complex numbers. (Choose the imaginary part to be zero.)

It is important to note, however, that when we are discussing *graphs* of polynomial functions, we restrict the domain to the set of real numbers.

A *zero* of a polynomial $P(x)$ is the *solution* or *root* of the equation $P(x) = 0$. The *zeros of a polynomial can be complex numbers.* However, since the xy-plane represents real numbers, we interpret zeros as x-intercepts only when the zeros are real numbers.

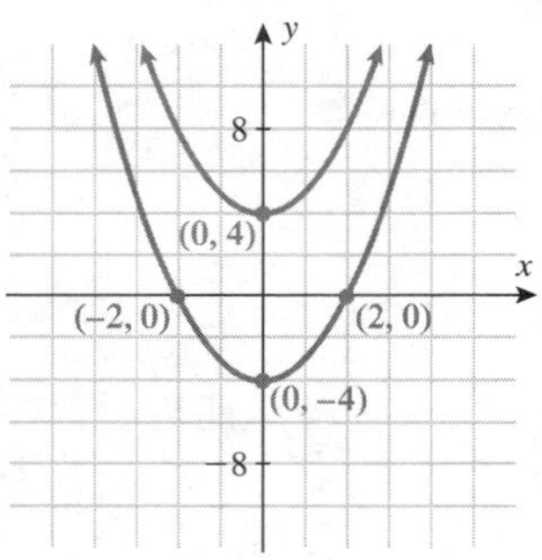

We can illustrate the relationship between real and complex zeros of polynomial functions and their graphs with two similar examples. Let's take the two quadratic functions $f(x) = x^2 - 4$ and $g(x) = x^2 + 4$. The graphs of these two functions are parabolas that open upward with $f(x)$ shifted down 4 units and $g(x)$ shifted up 4 units as shown on the left. Setting each function equal to zero and solving for x, we find that the zeros for $f(x)$ are -2 and 2 and the zeros for $g(x)$ are $-2i$ and $2i$. Notice that the x-intercepts for $f(x)$ are $(-2, 0)$ and $(2, 0)$ and $g(x)$ has no x-intercepts.

The Fundamental Theorem of Algebra

Study Tip

The largest number of zeros a polynomial can have is equal to the degree of the polynomial.

In Section 4.4, we were able to write a polynomial function as a product of linear and/or irreducible quadratic factors. Now, we consider factors over complex numbers. Therefore, what were irreducible quadratic factors over real numbers will now be a product of two linear factors over the complex numbers.

What are the minimum and maximum number of zeros a polynomial can have? Every polynomial has *at least one zero* (provided the degree is greater than zero). The largest number of zeros a polynomial can have is equal to the degree of the polynomial.

THE FUNDAMENTAL THEOREM OF ALGEBRA

Every polynomial $P(x)$ of degree $n > 0$ has *at least one zero* in the complex number system.

The fundamental theorem of algebra and the factor theorem are used to prove the following n zeros theorem.

n ZEROS THEOREM

Every polynomial $P(x)$ of degree $n > 0$ can be expressed as the product of n linear factors. Hence, $P(x)$ has exactly n zeros, not necessarily distinct.

These two theorems are illustrated with five polynomials below.

a. The **first**-degree polynomial $f(x) = x + 3$ has exactly **one** zero: $x = -3$.
b. The **second**-degree polynomial $f(x) = x^2 + 10x + 25 = (x + 5)(x + 5)$ has exactly **two** zeros: $x = -5$ and $x = -5$. It is customary to write this as a single zero of multiplicity 2 or refer to it as a repeated root.
c. The **third**-degree polynomial $f(x) = x^3 + 16x = x(x^2 + 16) = x(x + 4i)(x - 4i)$ has exactly **three** zeros: $x = 0$, $x = -4i$, and $x = 4i$.
d. The **fourth**-degree polynomial $f(x) = x^4 - 1 = (x^2 - 1)(x^2 + 1)$ $= (x - 1)(x + 1)(x - i)(x + i)$ has exactly **four** zeros: $x = 1$, $x = -1$, $x = i$, and $x = -i$.
e. The **fifth**-degree polynomial $f(x) = x^5 = x \cdot x \cdot x \cdot x \cdot x$ has exactly **five** zeros: $x = 0$, which has multiplicity 5.

The fundamental theorem of algebra and the n zeros theorem only tell you that the zeros *exist*—not how to find them. We must rely on techniques discussed in Section 4.4 and additional strategies discussed in this section to determine the zeros.

Complex Zeros

Often, at a grocery store or a drugstore, we see signs for special offers—"buy one, get one free." A similar phenomenon occurs for complex zeros of a polynomial function with real coefficients. If we restrict the coefficients of a polynomial to real numbers, complex zeros always come in conjugate pairs. In other words, if a zero of a polynomial function is a complex number, then another zero will always be its complex conjugate. Look at the third-degree polynomial in the above illustration, part (c), where two of the zeros were $-4i$ and $4i$, and in part (d), where two of the zeros were i and $-i$. In general, if we restrict the coefficients of a polynomial to real numbers, complex zeros always come in conjugate pairs.

Study Tip

If we restrict the coefficients of a polynomial to real numbers, complex zeros always come in conjugate pairs.

COMPLEX CONJUGATE ZEROS THEOREM

If a polynomial $P(x)$ has real coefficients, and if $a + bi$ is a zero of $P(x)$, then its complex conjugate $a - bi$ is also a zero of $P(x)$.

EXAMPLE 1 Zeros That Appear as Complex Conjugates

Find the zeros of the polynomial $P(x) = x^2 - 4x + 13$.

Solution:

Set the polynomial equal to zero. $P(x) = x^2 - 4x + 13 = 0$

Use the quadratic formula to solve for x. $x = \dfrac{-(-4) \pm \sqrt{(-4)^2 - 4(1)(13)}}{2(1)}$

Simplify. $\boxed{x = 2 \pm 3i}$

The zeros are the complex conjugates $2 - 3i$ and $2 + 3i$.

Check: This is a *second-degree* polynomial, so we expect *two* zeros.

Technology Tip

The graph of $P(x) = x^4 - x^3 - 5x^2 - x - 6$ is shown.

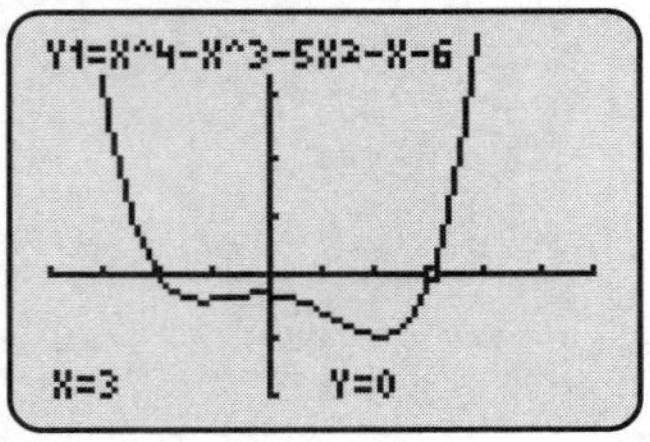

The real zeros of the function at $x = -2$ and $x = 3$ give the factors of $x + 2$ and $x - 3$. Use the synthetic division to find the other factors.

A table of values supports the real zeros of the function and its factors.

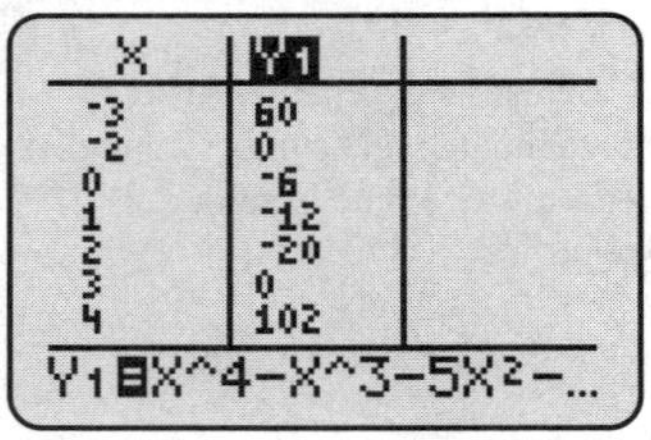

■ **Answer:** $P(x) = (x - 2i)(x + 2i)(x - 1)(x - 2)$
Note: The zeros of $P(x)$ are 1, 2, $2i$, and $-2i$.

In Example 1, we solved a quadratic equation to find the two zeros that are complex conjugates of one another. In the next example, we use the complex zeros theorem to assist us in factoring a higher degree polynomial.

EXAMPLE 2 Factoring a Polynomial with Complex Zeros

Factor the polynomial $P(x) = x^4 - x^3 - 5x^2 - x - 6$ given that i is a zero of $P(x)$.

Since $P(x)$ is a *fourth*-degree polynomial we expect *four* zeros. The goal in this problem is to write $P(x)$ as a product of four linear factors: $P(x) = (x - a)(x - b)(x - c)(x - d)$, where a, b, c, and d are complex numbers and represent the zeros of the polynomial.

Solution:

Write known zeros and linear factors.

Since i is a zero, we know that $-i$ is a zero. $\quad x = i$ and $x = -i$

We now know two linear factors of $P(x)$. $\quad (x - i)$ and $(x + i)$

Write $P(x)$ as a product of four factors. $\quad P(x) = (x - i)(x + i)(x - c)(x - d)$

Multiply the two known factors.

$$\begin{aligned}(x + i)(x - i) &= x^2 - i^2 \\ &= x^2 - (-1) \\ &= x^2 + 1\end{aligned}$$

Rewrite the polynomial. $\quad P(x) = \left(x^2 + 1\right)(x - c)(x - d)$

Divide both sides of the equation by $x^2 + 1$. $\quad \dfrac{P(x)}{x^2 + 1} = (x - c)(x - d)$

Divide $P(x)$ by $x^2 + 1$ using long division.

$$\begin{array}{r} x^2 - x - 6 \\ x^2 + 0x + 1\,\overline{)\,x^4 - x^3 - 5x^2 - x - 6} \\ \underline{-(x^4 + 0x^3 + x^2)} \\ -x^3 - 6x^2 - x \\ \underline{-(-x^3 + 0x^2 - x)} \\ -6x^2 + 0x - 6 \\ \underline{-(-6x^2 + 0x - 6)} \\ 0 \end{array}$$

Since the remainder is 0, $x^2 - x - 6$ is a factor. $\quad P(x) = \left(x^2 + 1\right)\left(x^2 - x - 6\right)$

Factor the quotient $x^2 - x - 6$. $\quad x^2 - x - 6 = (x - 3)(x + 2)$

Write $P(x)$ as a product of four linear factors. $\quad \boxed{P(x) = (x - i)(x + i)(x - 3)(x + 2)}$

Check: $P(x)$ is a *fourth*-degree polynomial and we found *four* zeros, two of which are complex conjugates.

■ **YOUR TURN** Factor the polynomial $P(x) = x^4 - 3x^3 + 6x^2 - 12x + 8$ given that $x - 2i$ is a factor.

EXAMPLE 3 Factoring a Polynomial with Complex Zeros

Factor the polynomial $P(x) = x^4 - 2x^3 + x^2 + 2x - 2$ given that $1 + i$ is a zero of $P(x)$.

Since $P(x)$ is a *fourth*-degree polynomial, we expect *four* zeros. The goal in this problem is to write $P(x)$ as a product of four linear factors: $P(x) = (x - a)(x - b)(x - c)(x - d)$, where a, b, c, and d are complex numbers and represent the zeros of the polynomial.

Solution:

STEP 1 Write known zeros and linear factors.

Since $1 + i$ is a zero, we know that $1 - i$ is a zero. $\quad x = 1 + i$ and $x = 1 - i$

We now know two linear factors of $P(x)$. $\quad [x - (1 + i)]$ and $[x - (1 - i)]$

STEP 2 Write $P(x)$ as a product of four factors. $\quad P(x) = [x - (1 + i)][x - (1 - i)](x - c)(x - d)$

STEP 3 Multiply the first two terms. $\quad [x - (1 + i)][x - (1 - i)]$

First group the real parts together in each bracket. $\quad [(x - 1) - i][(x - 1) + i]$

Use the special product $(a - b)(a + b) = a^2 - b^2$, where a is $(x - 1)$ and b is i.

$$(x - 1)^2 - i^2$$
$$(x^2 - 2x + 1) - (-1)$$
$$x^2 - 2x + 2$$

STEP 4 Rewrite the polynomial. $\quad P(x) = (x^2 - 2x + 2)(x - c)(x - d)$

STEP 5 Divide both sides of the equation by $x^2 - 2x + 2$, and substitute in the original polynomial $P(x) = x^4 - 2x^3 + x^2 + 2x - 2$.

$$\frac{x^4 - 2x^3 + x^2 + 2x - 2}{x^2 - 2x + 2} = (x - c)(x - d)$$

STEP 6 Divide the left side of the equation using long division.

$$\frac{x^4 - 2x^3 + x^2 + 2x - 2}{x^2 - 2x + 2} = x^2 - 1$$

STEP 7 Factor $x^2 - 1$. $\quad (x - 1)(x + 1)$

STEP 8 Write $P(x)$ as a product of four linear factors.

$$\boxed{P(x) = [x - (1 + i)][x - (1 - i)][x - 1][x + 1]}$$

YOUR TURN Factor the polynomial $P(x) = x^4 - 2x^2 + 16x - 15$ given that $1 + 2i$ is a zero.

Answer: $P(x) = [x - (1 + 2i)] \cdot [x - (1 - 2i)](x - 1)(x + 3)$
Note: The zeros of $P(x)$ are $1, -3, 1 + 2i$, and $1 - 2i$.

Because an n-degree polynomial function has exactly n zeros and since complex zeros always come in conjugate pairs, if the degree of the polynomial is **odd**, there is guaranteed to be **at least one zero that is a real number**. If the degree of the polynomial is even, there is no guarantee that a zero will be real—all the zeros could be complex.

Study Tip

Odd-degree polynomials have at least one real zero.

EXAMPLE 4 Finding Possible Combinations of Real and Complex Zeros

List the possible combinations of real and complex zeros for the given polynomials.

a. $17x^5 + 2x^4 - 3x^3 + x^2 - 5$ **b.** $5x^4 + 2x^3 - x + 2$

Solution:

a. Since this is a *fifth*-degree polynomial, there are *five* zeros. Because complex zeros come in conjugate pairs, the table describes the possible five zeros.

REAL ZEROS	COMPLEX ZEROS
1	4
3	2
5	0

Applying Descartes's rule of signs we find that there are 3 or 1 positive real zeros and 2 or 0 negative real zeros.

Positive Real Zeros	Negative Real Zeros	Complex Zeros
1	0	4
3	0	2
1	2	2
3	2	0

b. Because this is a *fourth*-degree polynomial, there are *four* zeros. Since complex zeros come in conjugate pairs, the table describes the possible four zeros.

Real Zeros	Complex Zeros
0	4
2	2
4	0

Applying Descartes's rule of signs we find that there are 2 or 0 positive real zeros and 2 or 0 negative real zeros.

Positive Real Zeros	Negative Real Zeros	Complex Zeros
0	0	4
2	0	2
0	2	2
2	2	0

■ **YOUR TURN** List the possible combinations of real and complex zeros for:

$$P(x) = x^6 - 7x^5 + 8x^3 - 2x + 1$$

■ **Answer:**

Real Zeros	Complex Zeros
0	6
2	4
4	2
6	0

Factoring Polynomials

Now let's draw on the tests discussed in this chapter to help us find all the zeros of a polynomial. Doing so will enable us to write polynomials as a product of linear factors. Before reading Example 5, reread Section 4.4, Example 8.

EXAMPLE 5 Factoring a Polynomial

Factor the polynomial $P(x) = x^5 + 2x^4 - x - 2$.

Solution:

Step 1 Determine variations in sign.

$P(x)$ has 1 sign change. $\qquad P(x) = x^5 + 2x^4 - x - 2$

$P(-x)$ has 2 sign changes. $\qquad P(-x) = -x^5 + 2x^4 + x - 2$

Step 2 Apply Descartes's rule of signs and summarize the results in a table.

Positive Real Zeros	Negative Real Zeros	Complex Zeros
1	2	2
1	0	4

STEP 3 Utilize the rational zero theorem to determine the possible rational zeros. $\pm 1, \pm 2$

STEP 4 Test possible rational zeros.

1 is a zero: $P(1) = (1)^5 + 2(1)^4 - (1) - 2 = 0$

Now that we have found *the* positive zero, we can test the two negative zeros.

-1 is a zero: $P(-1) = (-1)^5 + 2(-1)^4 - (-1) - 2 = 0$

-2 is a zero: $P(-2) = (-2)^5 + 2(-2)^4 - (-2) - 2$
$= -32 + 32 + 2 - 2 = 0$

STEP 5 Three of the five zeros have been found: -1, -2, and 1.

STEP 6 Write the fifth-degree polynomial as a product of five linear factors.

$$P(x) = x^5 + 2x^4 - x - 2 = (x - 1)(x + 1)(x + 2)(x - c)(x - d)$$

We know that the remaining two zeros are complex according to Descartes's rule of signs and that complex roots come in conjugate pairs.

STEP 7 Multiply the three known linear factors.

$$(x - 1)(x + 1)(x + 2) = x^3 + 2x^2 - x - 2$$

STEP 8 Rewrite the polynomial.

$$P(x) = x^5 + 2x^4 - x - 2 = (x^3 + 2x^2 - x - 2)(x - c)(x - d)$$

STEP 9 Divide both sides of the equation by $x^3 + 2x^2 - x - 2$.

$$\frac{x^5 + 2x^4 - x - 2}{x^3 + 2x^2 - x - 2} = (x - c)(x - d)$$

STEP 10 Dividing the left side of the equation with long division yields:

$$\frac{x^5 + 2x^4 - x - 2}{x^3 + 2x^2 - x - 2} = x^2 + 1 = (x - i)(x + i)$$

STEP 11 Write the remaining two zeros. $x = \pm i$

STEP 12 Write $P(x)$ as a product of linear factors.

$$\boxed{P(x) = (x - 1)(x + 1)(x + 2)(x - i)(x + i)}$$

Study Tip

From Step 2 we know there is one positive real zero, so test the positive possible rational zeros first in Step 4.

Study Tip

In Step 4 we could have used synthetic division to show that 1, -1, and -2 are all zeros of $P(x)$.

Technology Tip

The graph of $P(x) = x^5 + 2x^4 - x - 2$ is shown.

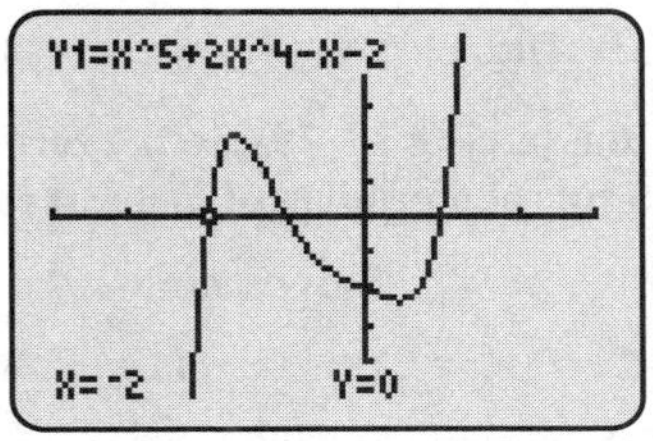

The real zeros of the function at $x = -2$, $x = -1$, and $x = 1$ give the factors of $x + 2$, $x + 1$, and $x - 1$. Use synthetic division to find the other factors.

A table of values supports the real zeros of the function and its factors.

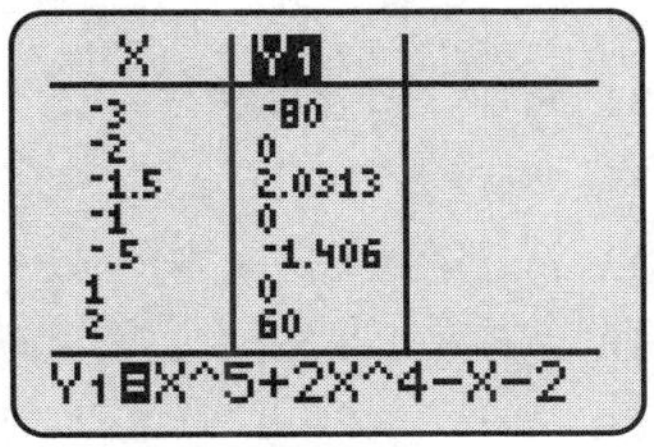

X	Y1
-3	-80
-2	0
-1.5	2.0313
-1	0
-.5	-1.406
1	0
2	60

Y1=X^5+2X^4-X-2

Study Tip

$P(x)$ is a *fifth*-degree polynomial, so we expect *five* zeros.

SECTION 4.5 SUMMARY

In this section we discussed **complex zeros** of polynomial functions. A polynomial function, $P(x)$, of degree n with real coefficients has the following properties:

- $P(x)$ has at least one zero (if $n > 0$) and no more than n zeros.
- If $a + bi$ is a zero, then $a - bi$ is also a zero.
- The polynomial can be written as a product of linear factors, not necessarily distinct.

SECTION 4.5 EXERCISES

SKILLS

In Exercises 1–8, find all zeros (real and complex). Factor the polynomial as a product of linear factors.

1. $P(x) = x^2 + 4$
2. $P(x) = x^2 + 9$
3. $P(x) = x^2 - 2x + 2$
4. $P(x) = x^2 - 4x + 5$
5. $P(x) = x^4 - 16$
6. $P(x) = x^4 - 81$
7. $P(x) = x^4 - 25$
8. $P(x) = x^4 - 9$

In Exercises 9–16, a polynomial function is described. Find all remaining zeros.

9. Degree: 3 Zeros: $-1, i$
10. Degree: 3 Zeros: $1, -i$
11. Degree: 4 Zeros: $2i, 3 - i$
12. Degree: 4 Zeros: $3i, 2 + i$
13. Degree: 6 Zeros: 2 (multiplicity 2), $1 - 3i, 2 + 5i$
14. Degree: 6 Zeros: -2 (multiplicity 2), $1 - 5i, 2 + 3i$
15. Degree: 6 Zeros: $-i, 1 - i$ (multiplicity 2)
16. Degree: 6 Zeros: $2i, 1 + i$ (multiplicity 2)

In Exercises 17–28, given a zero of the polynomial, determine all other zeros (real and complex) and write the polynomial in terms of a product of linear factors.

Polynomial	Zero	Polynomial	Zero
17. $P(x) = x^4 - 2x^3 - 11x^2 - 8x - 60$	$-2i$	18. $P(x) = x^4 - x^3 + 7x^2 - 9x - 18$	$3i$
19. $P(x) = x^4 - 4x^3 + 4x^2 - 4x + 3$	i	20. $P(x) = x^4 - x^3 + 2x^2 - 4x - 8$	$-2i$
21. $P(x) = x^4 - 2x^3 + 10x^2 - 18x + 9$	$-3i$	22. $P(x) = x^4 - 3x^3 + 21x^2 - 75x - 100$	$5i$
23. $P(x) = x^4 - 9x^2 + 18x - 14$	$1 + i$	24. $P(x) = x^4 - 4x^3 + x^2 + 6x - 40$	$1 - 2i$
25. $P(x) = x^4 - 6x^3 + 6x^2 + 24x - 40$	$3 - i$	26. $P(x) = x^4 - 4x^3 + 4x^2 + 4x - 5$	$2 + i$
27. $P(x) = x^4 - 9x^3 + 29x^2 - 41x + 20$	$2 - i$	28. $P(x) = x^4 - 7x^3 + 14x^2 + 2x - 20$	$3 + i$

In Exercises 29–52, factor each polynomial as a product of linear factors.

29. $P(x) = x^3 - x^2 + 9x - 9$
30. $P(x) = x^3 - 2x^2 + 4x - 8$
31. $P(x) = x^3 - 5x^2 + x - 5$
32. $P(x) = x^3 - 7x^2 + x - 7$
33. $P(x) = x^3 + x^2 + 4x + 4$
34. $P(x) = x^3 + x^2 - 2$
35. $P(x) = x^3 - x^2 - 18$
36. $P(x) = x^4 - 2x^3 - 2x^2 - 2x - 3$
37. $P(x) = x^4 - 2x^3 - 11x^2 - 8x - 60$
38. $P(x) = x^4 - x^3 + 7x^2 - 9x - 18$
39. $P(x) = x^4 - 4x^3 - x^2 - 16x - 20$
40. $P(x) = x^4 - 3x^3 + 11x^2 - 27x + 18$
41. $P(x) = x^4 - 7x^3 + 27x^2 - 47x + 26$
42. $P(x) = x^4 - 5x^3 + 5x^2 + 25x - 26$
43. $P(x) = -x^4 - 3x^3 + x^2 + 13x + 10$
44. $P(x) = -x^4 - x^3 + 12x^2 + 26x + 24$
45. $P(x) = x^4 - 2x^3 + 5x^2 - 8x + 4$
46. $P(x) = x^4 + 2x^3 + 10x^2 + 18x + 9$
47. $P(x) = x^6 + 12x^4 + 23x^2 - 36$
48. $P(x) = x^6 - 2x^5 + 9x^4 - 16x^3 + 24x^2 - 32x + 16$
49. $P(x) = 4x^4 - 20x^3 + 37x^2 - 24x + 5$
50. $P(x) = 4x^4 - 44x^3 + 145x^2 - 114x + 26$
51. $P(x) = 3x^5 - 2x^4 + 9x^3 - 6x^2 - 12x + 8$
52. $P(x) = 2x^5 - 5x^4 + 4x^3 - 26x^2 + 50x - 25$

APPLICATIONS

In Exercises 53–56, assume the profit model is given by a polynomial function $P(x)$ where x is the number of units sold by the company per year.

53. **Profit.** If the profit function of a given company has all imaginary zeros and the leading coefficient is positive, would you invest in this company? Explain.

54. **Profit.** If the profit function of a given company has all imaginary zeros and the leading coefficient is negative, would you invest in this company? Explain.

55. **Profit.** If the profit function of a company is modeled by a third-degree polynomial with a negative leading coefficient and this polynomial has two complex conjugates as zeros and one positive real zero, would you invest in this company? Explain.

56. **Profit.** If the profit function of a company is modeled by a third-degree polynomial with a positive leading coefficient and this polynomial has two complex conjugates as zeros and one positive real zero, would you invest in this company? Explain.

CATCH THE MISTAKE

In Exercises 57 and 58, explain the mistake that is made.

57. Given that 1 is a zero of $P(x) = x^3 - 2x^2 - 7x + 6$, find all other zeros.

Solution:

Step 1: $P(x)$ is a third-degree polynomial, so we expect three zeros.

Step 2: Because 1 is a zero, -1 is a zero, so two linear factors are $(x - 1)$ and $(x + 1)$.

Step 3: Write the polynomial as a product of three linear factors.

$$P(x) = (x - 1)(x + 1)(x - c)$$
$$P(x) = (x^2 - 1)(x - c)$$

Step 4: To find the remaining linear factor, we divide $P(x)$ by $x^2 - 1$.

$$\frac{x^3 - 2x^2 - 7x + 6}{x^2 - 1} = x - 2 + \frac{-6x + 4}{x^2 - 1}$$

Which has a nonzero remainder? What went wrong?

58. Factor the polynomial $P(x) = 2x^3 + x^2 + 2x + 1$.

Solution:

Step 1: Since $P(x)$ is an odd-degree polynomial, we are guaranteed one real zero (since complex zeros come in conjugate pairs).

Step 2: Apply the rational zero test to develop a list of potential rational zeros.

Possible zeros: ± 1

Step 3: Test possible zeros.

$$1 \text{ is not a zero: } P(x) = 2(1)^3 + (1)^2 + 2(1) + 1 = 6$$

$$-1 \text{ is not a zero: } P(x) = 2(-1)^3 + (-1)^2 + 2(-1) + 1 = -2$$

Note: $-\frac{1}{2}$ is the real zero. Why did we not find it?

CONCEPTUAL

In Exercises 59–62, determine whether each statement is true or false.

59. If $x = 1$ is a zero of a polynomial function, then $x = -1$ is also a zero of the polynomial function.

60. All zeros of a polynomial function correspond to x-intercepts.

61. A polynomial function of degree n, $n > 0$ must have at least one zero.

62. A polynomial function of degree n, $n > 0$ can be written as a product of n linear factors.

63. Is it possible for an odd-degree polynomial to have all imaginary complex zeros? Explain.

64. Is it possible for an even-degree polynomial to have all imaginary zeros? Explain.

CHALLENGE

In Exercises 65 and 66, assume *a* and *b* are nonzero real numbers.

65. Find a polynomial function that has degree 6, and for which bi is a zero of multiplicity 3.

66. Find a polynomial function that has degree 4, and for which $a + bi$ is a zero of multiplicity 2.

TECHNOLOGY

For Exercises 67 and 68, determine possible combinations of real and complex zeros. Plot $P(x)$ and identify any real zeros with a graphing calculator or software. Does this agree with your list?

67. $P(x) = x^4 + 13x^2 + 36$

68. $P(x) = x^6 + 2x^4 + 7x^2 - 130x - 288$

For Exercises 69 and 70, find all zeros (real and complex). Factor the polynomial as a product of linear factors.

69. $P(x) = -5x^5 + 3x^4 - 25x^3 + 15x^2 - 20x + 12$

70. $P(x) = x^5 + 2.1x^4 - 5x^3 - 5.592x^2 + 9.792x - 3.456$

SECTION

4.6 RATIONAL FUNCTIONS

SKILLS OBJECTIVES

- Find the domain of a rational function.
- Determine vertical, horizontal, and slant asymptotes of rational functions.
- Graph rational functions.

CONCEPTUAL OBJECTIVES

- Understand arrow notation.
- Interpret the behavior of a rational function approaching an asymptote.

So far in this chapter we have discussed polynomial functions. We now turn our attention to *rational functions,* which are *ratios* of polynomial functions. Ratios of integers are called *rational numbers*. Similarly, ratios of polynomial functions are called *rational functions*.

DEFINITION **Rational Function**

A function $f(x)$ is a **rational function** if

$$f(x) = \frac{n(x)}{d(x)} \qquad d(x) \neq 0$$

where the numerator, $n(x)$, and the denominator, $d(x)$, are polynomial functions. The domain of $f(x)$ is the set of all real numbers x such that $d(x) \neq 0$.

Note: If $d(x)$ is a constant, then $f(x)$ is a polynomial function.

Domain of Rational Functions

The domain of any polynomial function is the set of all real numbers. When we divide two polynomial functions, the result is a *rational function,* and we must exclude any values of x that make the denominator equal to zero.

EXAMPLE 1 **Finding the Domain of a Rational Function**

Find the domain of the rational function $f(x) = \dfrac{x+1}{x^2 - x - 6}$. Express the domain in interval notation.

Solution:

Set the denominator equal to zero.	$x^2 - x - 6 = 0$
Factor.	$(x+2)(x-3) = 0$
Solve for x.	$x = -2$ or $x = 3$
Eliminate these values from the domain.	$x \neq -2$ or $x \neq 3$
State the domain in interval notation.	$\boxed{(-\infty, -2) \cup (-2, 3) \cup (3, \infty)}$

■ **YOUR TURN** Find the domain of the rational function $f(x) = \dfrac{x-2}{x^2 - 3x - 4}$. Express the domain in interval notation.

■ **Answer:** The domain is the set of all real numbers such that $x \neq -1$ or $x \neq 4$. Interval notation: $(-\infty, -1) \cup (-1, 4) \cup (4, \infty)$

It is important to note that there are not always restrictions on the domain. For example, if the denominator is never equal to zero, the domain is the set of all real numbers.

EXAMPLE 2 **When the Domain of a Rational Function Is the Set of All Real Numbers**

Find the domain of the rational function $g(x) = \frac{3x}{x^2 + 9}$. Express the domain in interval notation.

Solution:

Set the denominator equal to zero.	$x^2 + 9 = 0$
Subtract 9 from both sides.	$x^2 = -9$
Solve for x.	$x = -3i$ or $x = 3i$
There are no *real* solutions; therefore the domain has no restrictions.	$\mathbb{R}$, the set of all real numbers
State the domain in interval notation.	$\boxed{(-\infty, \infty)}$

YOUR TURN Find the domain of the rational function $g(x) = \frac{5x}{x^2 + 4}$. Express the domain in interval notation.

Answer: The domain is the set of all real numbers. Interval notation: $(-\infty, \infty)$

It is important to note that $f(x) = \frac{x^2 - 4}{x + 2}$, where $x \neq -2$, and $g(x) = x - 2$ are *not* the same function. Although $f(x)$ can be written in the factored form $f(x) = \frac{(x - 2)(x + 2)}{x + 2} = x - 2$, its domain is different. The domain of $g(x)$ is the set of all real numbers, whereas the domain of $f(x)$ is the set of all real numbers such that $x \neq -2$. If we were to plot $f(x)$ and $g(x)$, they would both look like the line $y = x - 2$. However, $f(x)$ would have a hole, or discontinuity, at the point $x = -2$.

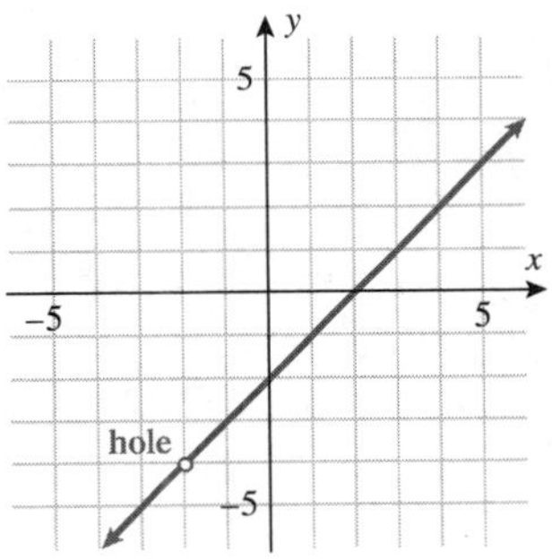

Vertical, Horizontal, and Slant Asymptotes

If a function is not defined at a point, then it is still useful to know how the function behaves near that point. Let's start with a simple rational function, the reciprocal function $f(x) = \frac{1}{x}$. This function is defined everywhere except at $x = 0$.

x	$-\frac{1}{10}$	$-\frac{1}{100}$	$-\frac{1}{1000}$	0	$\frac{1}{1000}$	$\frac{1}{100}$	$\frac{1}{10}$
$f(x) = \frac{1}{x}$	-10	-100	-1000	undefined	1000	100	10

x approaching 0 from the left (first three columns); x approaching 0 from the right (last three columns)

x	$f(x) = \frac{1}{x}$
-10	$-\frac{1}{10}$
-1	-1
1	1
10	$\frac{1}{10}$

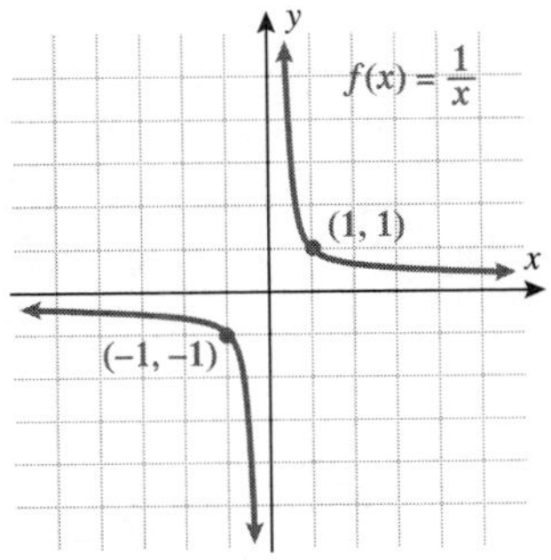

We cannot let $x = 0$, because that point is not in the domain of the function. We should, however, ask the question, "how does $f(x)$ behave as x *approaches* zero?" Let us take values that get closer and closer to $x = 0$, such as $\frac{1}{10}, \frac{1}{100}, \frac{1}{1000}, \ldots$ (See the table above.) We use an *arrow* to represent the word *approach*, a *positive* superscript to represent from the *right*, and a *negative* superscript to represent from the *left*. A plot of this function can be generated using point-plotting techniques. The following are observations of the graph $f(x) = \frac{1}{x}$.

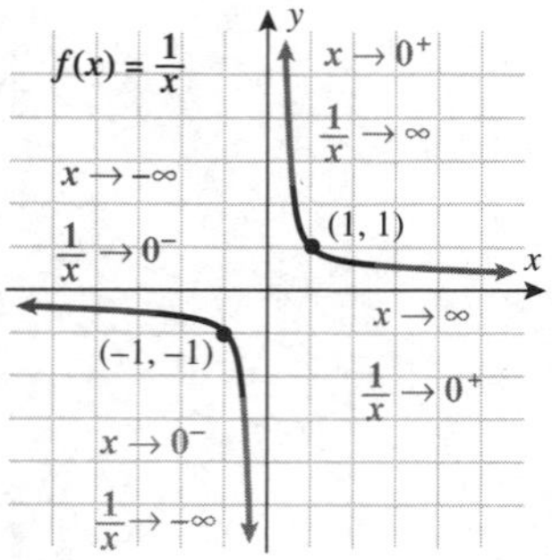

Words	**Math**
As x approaches zero from the *right,* the function $f(x)$ increases without bound.	$x \to 0^+$ $\frac{1}{x} \to \infty$
As x approaches zero from the *left,* the function $f(x)$ decreases without bound.	$x \to 0^-$ $\frac{1}{x} \to -\infty$
As x approaches infinity (increases without bound), the function $f(x)$ approaches zero from *above.*	$x \to \infty$ $\frac{1}{x} \to 0^+$
As x approaches negative infinity (decreases without bound), the function $f(x)$ approaches zero from *below.*	$x \to -\infty$ $\frac{1}{x} \to 0^-$

The symbol ∞ does not represent an actual real number. This symbol represents growing without bound.

1. Notice that the function is not defined at $x = 0$. The y-axis, or the vertical line $x = 0$, represents the *vertical asymptote.*
2. Notice that the value of the function is never equal to zero. The x-axis is never touched by the function. The x-axis, or $y = 0$, is a *horizontal asymptote.*

Asymptotes are lines that the graph of a function approaches. Suppose a football team's defense is its own 8 yard line and the team gets an "offsides" penalty that results in loss of "half the distance to the goal." Then the offense would get the ball on the 4 yard line. Suppose the defense gets another penalty on the next play that results in "half the distance to the goal." The offense would then get the ball on the 2 yard line. If the defense received 10 more penalties all resulting in "half the distance to the goal," would the referees *give* the offense a touchdown? No, because although the offense may appear to be snapping the ball from the goal line, technically it has not actually reached the goal line. Asymptotes utilize the same concept.

We will start with *vertical asymptotes.* Although the function $f(x) = \frac{1}{x}$ had one vertical asymptote, in general, rational functions can have *none, one,* or *several* vertical asymptotes. We will first formally define what a vertical asymptote is and then discuss how to find it.

DEFINITION **Vertical Asymptotes**

The line $x = a$ is a **vertical asymptote** for the graph of a function if $f(x)$ either increases or decreases without bound as x approaches a from either the left or the right.

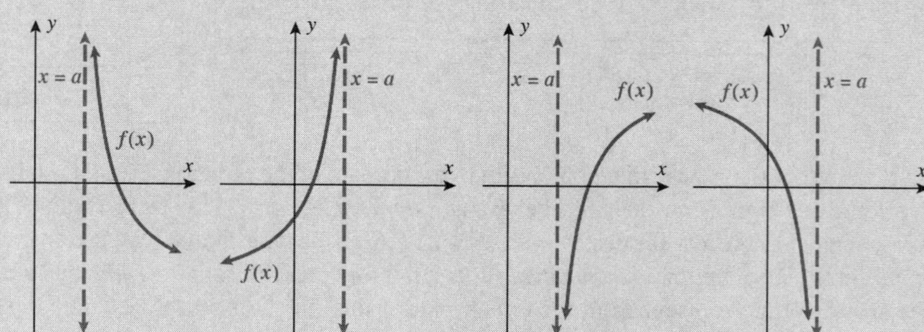

As $x \to a^+, f(x) \to \infty$ As $x \to a^-, f(x) \to \infty$ As $x \to a^+, f(x) \to -\infty$ As $x \to a^-, f(x) \to -\infty$

Vertical asymptotes assist us in graphing rational functions since they essentially "steer" the function in the vertical direction. How do we locate the vertical asymptotes of a rational function? Set the denominator equal to zero. If the numerator and denominator have no common factors, then any numbers that are excluded from the domain of a rational function locate vertical asymptotes.

A rational function $f(x) = \dfrac{n(x)}{d(x)}$ is said to be in **lowest terms** if the numerator $n(x)$ and denominator $d(x)$ have no common factors. Let $f(x) = \dfrac{n(x)}{d(x)}$ be a rational function in lowest terms; then any zeros of the numerator $n(x)$ correspond to x-intercepts of the graph of f, and any zeros of the denominator $d(x)$ correspond to vertical asymptotes of the graph of f. If a rational function does have a common factor (is not in lowest terms), then the common factor(s) should be canceled, resulting in an equivalent rational function $R(x)$ in lowest terms. If $(x - a)^p$ is a factor of the numerator and $(x - a)^q$ is a factor of the denominator, then there is a *hole* in the graph at $x = a$ provided $p \geq q$ and $x = a$ is a vertical asymptote if $p < q$.

Locating Vertical Asymptotes

Let $f(x) = \dfrac{n(x)}{d(x)}$ be a rational function in lowest terms (that is, assume $n(x)$ and $d(x)$ are polynomials with no common factors); then the graph of f has a vertical asymptote at any real zero of the denominator $d(x)$. That is, if $d(a) = 0$, then $x = a$ corresponds to a vertical asymptote on the graph of f.

Note: If f is a rational function that is not in lowest terms, then divide out the common factors, resulting in a rational function R that is in lowest terms. Any common factor $x - a$ of the function f corresponds to a hole in the graph of f at $x = a$ provided the multiplicity of a in the numerator is greater than or equal to the multiplicity of a in the denominator.

Study Tip

The vertical asymptotes of a rational function in *lowest terms* occur at x-values that make the denominator equal to zero.

Example 3 Determining Vertical Asymptotes

Locate any vertical asymptotes of the rational function $f(x) = \dfrac{5x + 2}{6x^2 - x - 2}$.

Solution:

Factor the denominator. $\quad f(x) = \dfrac{5x + 2}{(2x + 1)(3x - 2)}$

The numerator and denominator have no common factors.

Set the denominator equal to zero. $\quad 2x + 1 = 0 \quad$ and $\quad 3x - 2 = 0$

Solve for x. $\quad x = -\dfrac{1}{2} \quad$ and $\quad x = \dfrac{2}{3}$

The vertical asymptotes are $\boxed{x = -\frac{1}{2}}$ and $\boxed{x = \frac{2}{3}}$.

YOUR TURN Locate any vertical asymptotes of the following rational function:

$$f(x) = \frac{3x - 1}{2x^2 - x - 15}$$

Answer: $x = -\frac{5}{2}$ and $x = 3$

EXAMPLE 4 Determining Vertical Asymptotes When the Rational Function Is Not in Lowest Terms

Locate any vertical asymptotes of the rational function $f(x) = \dfrac{x + 2}{x^3 - 3x^2 - 10x}$.

Solution:

Factor the denominator.

$$x^3 - 3x^2 - 10x = x(x^2 - 3x - 10) = x(x - 5)(x + 2)$$

Write the rational function in factored form.

$$f(x) = \frac{(x + 2)}{x(x - 5)(x + 2)}$$

Cancel (divide out) the common factor $(x + 2)$.

$$R(x) = \frac{1}{x(x - 5)} \quad x \neq -2$$

Find the values when the denominator of R is equal to zero.

$$x = 0 \text{ and } x = 5$$

The vertical asymptotes are $\boxed{x = 0}$ and $\boxed{x = 5}$.

Note: $x = -2$ is not in the domain of $f(x)$, even though there is no vertical asymptote there. There is a "hole" in the graph at $x = -2$. Graphing calculators do not always show such "holes."

■ **Answer:** $x = 3$

■ **YOUR TURN** Locate any vertical asymptotes of the following rational function:

$$f(x) = \frac{x^2 - 4x}{x^2 - 7x + 12}$$

We now turn our attention to *horizontal asymptotes.* As we have seen, rational functions can have several vertical asymptotes. However, rational functions can have *at most* one horizontal asymptote. Horizontal asymptotes imply that a function approaches a constant value as x becomes large in the positive or negative direction. Another difference between vertical and horizontal asymptotes is that the graph of a function never touches a vertical asymptote but, as you will see in the next box, the graph of a function may cross a horizontal asymptote, just not at the "ends" $(x \to \pm\infty)$.

DEFINITION Horizontal Asymptote

The line $y = b$ is a **horizontal asymptote** of the graph of a function if $f(x)$ approaches b as x increases or decreases without bound. The following are three examples:

As $x \to \infty, f(x) \to b$

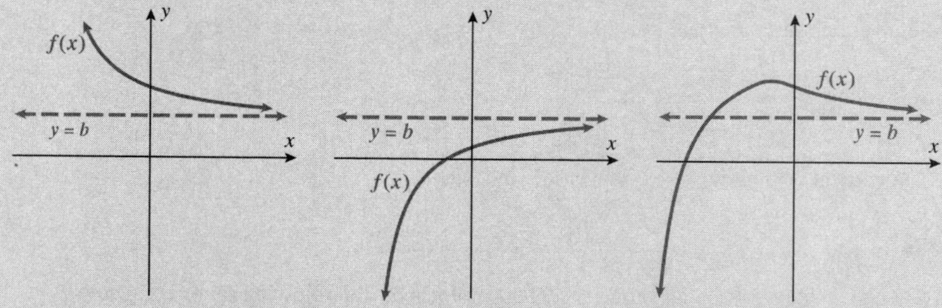

Note: A horizontal asymptote steers a function as x gets large. Therefore, when x is not large, the function may cross the asymptote.

How do we determine whether a horizontal asymptote exists? And, if it does, how do we locate it? We investigate the value of the rational function as $x \to \infty$ or as $x \to -\infty$. One of two things will happen: either the rational function will increase or decrease without bound or the rational function will approach a constant value.

We say that a rational function is **proper** if the degree of the numerator is less than the degree of the denominator. Proper rational functions, like $f(x) = \frac{1}{x}$, approach zero as x gets large. Therefore, all proper rational functions have the specific horizontal asymptote, $y = 0$ (see Example 5a).

We say that a rational function is **improper** if the degree of the numerator is greater than or equal to the degree of the denominator. In this case, we can divide the numerator by the denominator and determine how the quotient behaves as x increases without bound.

- If the quotient is a constant (resulting when the degrees of the numerator and denominator are equal), then as $x \to \infty$ or as $x \to -\infty$, the rational function approaches the constant quotient (see Example 5b).
- If the quotient is a polynomial function of degree 1 or higher, then the quotient depends on x and does not approach a constant value as x increases (see Example 5c). In this case, we say that there is no horizontal asymptote.

We find horizontal asymptotes by comparing the degree of the numerator and the degree of the denominator. There are three cases to consider:

1. The degree of the numerator is less than the degree of the denominator.
2. The degree of the numerator is equal to the degree of the denominator.
3. The degree of the numerator is greater than the degree of the denominator.

LOCATING HORIZONTAL ASYMPTOTES

Let f be a rational function given by

$$f(x) = \frac{n(x)}{d(x)} = \frac{a_n x^n + a_{n-1}x^{n-1} + \cdots + a_1 x + a_0}{b_m x^m + b_{m-1}x^{m-1} + \cdots + b_1 x + b_0}$$

where $n(x)$ and $d(x)$ are polynomials.

1. When $n < m$, the x-axis ($y = 0$) is the horizontal asymptote.
2. When $n = m$, the line $y = \frac{a_n}{b_m}$ (ratio of leading coefficients) is the horizontal asymptote.
3. When $n > m$, there is no horizontal asymptote.

In other words,

1. When the degree of the numerator is less than the degree of the denominator, then $y = 0$ is the horizontal asymptote.
2. When the degree of the numerator is the same as the degree of the denominator, then the horizontal asymptote is the ratio of the leading coefficients.
3. If the degree of the numerator is greater than the degree of the denominator, then there is no horizontal asymptote.

Technology Tip

The following graphs correspond to the rational functions given in Example 5. The horizontal asymptotes are apparent, but are not drawn in the graph.

a. Graph $f(x) = \dfrac{8x + 3}{4x^2 + 1}$.

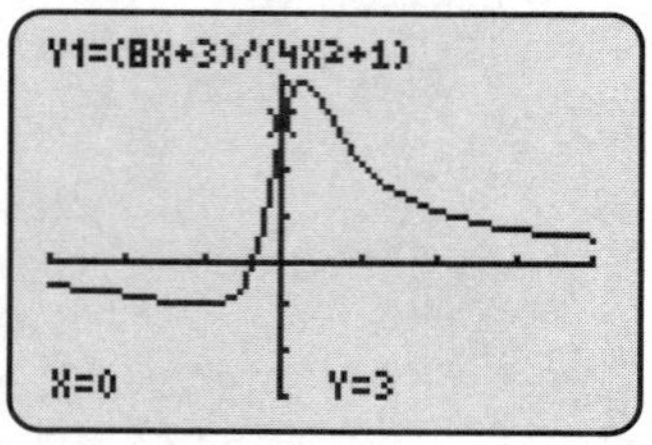

b. Graph $g(x) = \dfrac{8x^2 + 3}{4x^2 + 1}$.

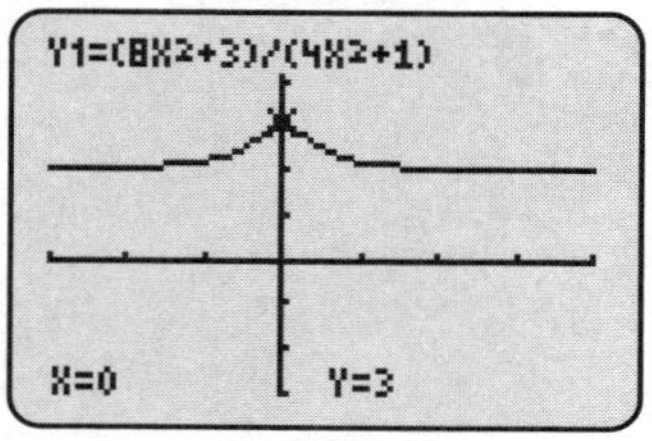

c. Graph $h(x) = \dfrac{8x^3 + 3}{4x^3 + 1}$.

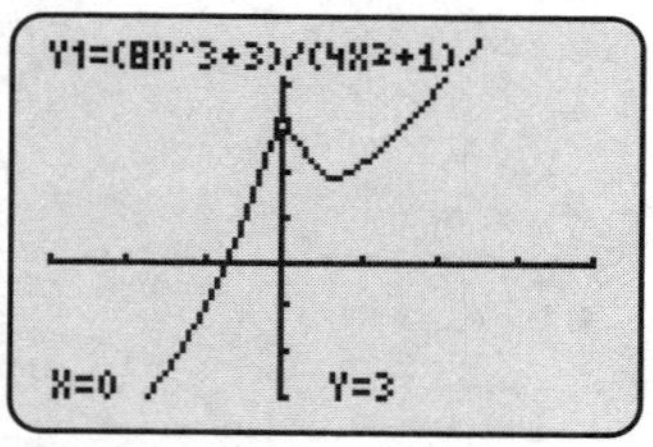

EXAMPLE 5 Finding Horizontal Asymptotes

Determine whether a horizontal asymptote exists for the graph of each of the given rational functions. If it does, locate the horizontal asymptote.

a. $f(x) = \dfrac{8x + 3}{4x^2 + 1}$ **b.** $g(x) = \dfrac{8x^2 + 3}{4x^2 + 1}$ **c.** $h(x) = \dfrac{8x^3 + 3}{4x^2 + 1}$

Solution (a):

The degree of the numerator $8x + 3$ is 1. $n = 1$

The degree of the denominator $4x^2 + 1$ is 2. $m = 2$

The degree of the numerator is less than the degree of the denominator. $n < m$

The x-axis is the horizontal asymptote for the graph of $f(x)$. $y = 0$

The line $\boxed{y = 0}$ is the horizontal asymptote for the graph of $f(x)$.

Solution (b):

The degree of the numerator $8x^2 + 3$ is 2. $n = 2$

The degree of the denominator $4x^2 + 1$ is 2. $m = 2$

The degree of the numerator is equal to the degree of the denominator. $n = m$

The ratio of the leading coefficients is the horizontal asymptote for the graph of $g(x)$. $y = \dfrac{8}{4} = 2$

The line $\boxed{y = 2}$ is the horizontal asymptote for the graph of $g(x)$.

If we divide the numerator by the denominator, the resulting quotient is the constant 2. $g(x) = \dfrac{8x^2 + 3}{4x^2 + 1} = 2 + \dfrac{1}{4x^2 + 1}$

Solution (c):

The degree of the numerator $8x^3 + 3$ is 3. $n = 3$

The degree of the denominator $4x^2 + 1$ is 2. $m = 2$

The degree of the numerator is greater than the degree of the denominator. $n > m$

The graph of the rational function $h(x)$ has $\boxed{\text{no horizontal asymptote}}$.

If we divide the numerator by the denominator, the resulting quotient is a linear function. $h(x) = \dfrac{8x^3 + 3}{4x^2 + 1} = 2x + \dfrac{-2x + 3}{4x^2 + 1}$

YOUR TURN Find the horizontal asymptote (if one exists) for the graph of the rational function $f(x) = \dfrac{7x^3 + x - 2}{-4x^3 + 1}$.

Answer: $y = -\frac{7}{4}$ is the horizontal asymptote.

Study Tip

There are three types of linear asymptotes: horizontal, vertical, and *slant.*

Thus far we have discussed linear asymptotes: vertical and horizontal. There are three types of lines: horizontal (slope is zero), vertical (slope is undefined), and slant (nonzero slope). Similarly, there are three types of linear asymptotes: horizontal, vertical, and *slant.*

Recall that when dividing polynomials the degree of the quotient is always the difference between the degree of the numerator and the degree of the denominator. For example, a cubic (third-degree) polynomial divided by a quadratic (second-degree) polynomial results in a linear (first-degree) polynomial. A fifth-degree polynomial divided by a fourth-degree polynomial results in a first-degree (linear) polynomial. When the degree of the numerator is exactly one more than the degree of the denominator, the quotient is linear and represents a *slant asymptote.*

SLANT ASYMPTOTES

Let f be a rational function given by $f(x) = \frac{n(x)}{d(x)}$, where $n(x)$ and $d(x)$ are polynomials and the degree of $n(x)$ is *one more than* the degree of $d(x)$. On dividing $n(x)$ by $d(x)$, the rational function can be expressed as

$$f(x) = mx + b + \frac{r(x)}{d(x)}$$

where the degree of the remainder $r(x)$ is less than the degree of $d(x)$ and the line $y = mx + b$ is a **slant asymptote** for the graph of f.

Note that as $x \to -\infty$ or $x \to \infty$, $f(x) \to mx + b$.

EXAMPLE 6 Finding Slant Asymptotes

Determine the slant asymptote of the rational function $f(x) = \dfrac{4x^3 + x^2 + 3}{x^2 - x + 1}$.

Solution:

Divide the numerator by the denominator with long division.

$$\begin{array}{r} 4x + 5 \\ x^2 - x + 1 \overline{)\, 4x^3 + x^2 + 0x + 3} \\ -(4x^3 - 4x^2 + 4x) \\ \hline 5x^2 - 4x + 3 \\ -(5x^2 - 5x + 5) \\ \hline x - 2 \end{array}$$

Note that as $x \to \pm\infty$ the rational expression approaches 0.

$$f(x) = 4x + 5 + \underbrace{\frac{x - 2}{x^2 - x + 1}}_{\substack{\to 0 \text{ as} \\ x \to \pm\infty}}$$

The quotient is the slant asymptote.

$$\boxed{y = 4x + 5}$$

■ **YOUR TURN** Find the slant asymptote of the rational function $f(x) = \dfrac{x^2 + 3x + 2}{x - 2}$.

■ **Answer:** $y = x + 5$

Study Tip

Common factors need to be divided out first; then the remaining x-values corresponding to a denominator value of 0 are vertical asymptotes.

Graphing Rational Functions

We can now graph rational functions using asymptotes as graphing aids. The following box summarizes the five-step procedure for graphing rational functions.

GRAPHING RATIONAL FUNCTIONS

Let f be a rational function given by $f(x) = \dfrac{n(x)}{d(x)}$.

Step 1: Find the domain of the rational function f.

Step 2: Find the **intercept(s)**.

- y-intercept: evaluate $f(0)$.
- x-intercept: solve the equation $n(x) = 0$ for x in the domain of f.

Step 3: Find any **holes**.

- Factor the numerator and denominator.
- Divide out common factors.
- A common factor $x - a$ corresponds to a hole in the graph of f at $x = a$ if the multiplicity of a in the numerator is greater than or equal to the multiplicity of a in the denominator.
- The result is an equivalent rational function $R(x) = \dfrac{p(x)}{q(x)}$ in lowest terms.

Step 4: Find any **asymptotes**.

- Vertical asymptotes: solve $q(x) = 0$.
- Compare the degree of the numerator and the degree of the denominator to determine whether either a horizontal or slant asymptote exists. If one exists, find it.

Step 5: Find **additional points** on the graph of f—particularly near asymptotes.

Step 6: **Sketch** the graph; draw the asymptotes, label the intercept(s) and additional points, and complete the graph with a smooth curve between and beyond the vertical asymptotes.

It is important to note that any real number eliminated from the domain of a rational function corresponds to either a vertical asymptote or a hole on its graph.

Study Tip

Any real number excluded from the domain of a rational function corresponds to either a vertical asymptote or a hole on its graph.

EXAMPLE 7 Graphing a Rational Function

Graph the rational function $f(x) = \dfrac{x}{x^2 - 4}$.

Solution:

STEP 1 Find the **domain**.

Set the denominator equal to zero. $x^2 - 4 = 0$

Solve for x. $x = \pm 2$

State the domain. $(-\infty, -2) \cup (-2, 2) \cup (2, \infty)$

STEP 2 Find the **intercepts**.

y-intercept: $f(0) = \dfrac{0}{-4} = 0$ $\quad y = 0$

x-intercepts: $f(x) = \dfrac{x}{x^2 - 4} = 0$ $\quad x = 0$

The only intercept is at the point $\boxed{(0, 0)}$.

STEP 3 Find any holes. $f(x) = \dfrac{x}{(x+2)(x-2)}$

There are no common factors, so f is in lowest terms.
Since there are no common factors, there are no holes on the graph of f.

STEP 4 Find any **asymptotes**.

Vertical asymptotes: $d(x) = (x+2)(x-2) = 0$

$\boxed{x = -2}$ and $\boxed{x = 2}$

Horizontal asymptote: $\dfrac{\text{Degree of numerator}}{\text{Degree of denominator}} = \dfrac{1}{2}$

Degree of numerator < Degree of denominator $\boxed{y = 0}$

STEP 5 Find **additional points** on the graph.

x	-3	-1	1	3
$f(x)$	$-\frac{3}{5}$	$\frac{1}{3}$	$-\frac{1}{3}$	$\frac{3}{5}$

STEP 6 **Sketch** the graph; label the intercepts, asymptotes, and additional points and complete with a smooth curve approaching the asymptotes.

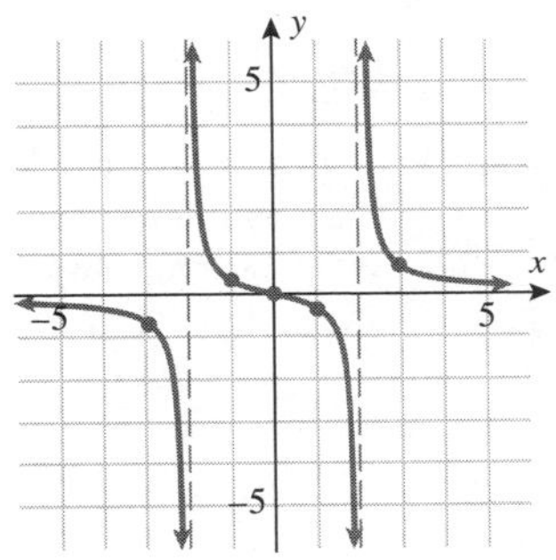

■ **YOUR TURN** Graph the rational function $f(x) = \dfrac{x}{x^2 - 1}$.

■ **Answer:**

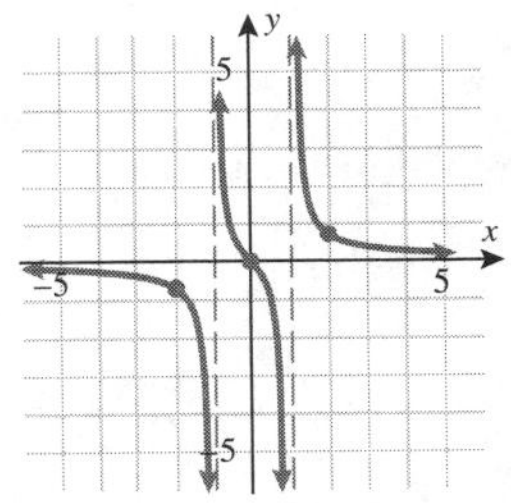

EXAMPLE 8 **Graphing a Rational Function with No Horizontal or Slant Asymptotes**

State the asymptotes (if there are any) and graph the rational function $f(x) = \dfrac{x^4 - x^3 - 6x^2}{x^2 - 1}$.

Solution:

STEP 1 Find the domain.

Set the denominator equal to zero. $x^2 - 1 = 0$

Solve for x. $x = \pm 1$

State the domain. $(-\infty, -1) \cup (-1, 1) \cup (1, \infty)$

STEP 2 Find the **intercepts**.

y-intercept: $f(0) = \dfrac{0}{-1} = 0$

x-intercepts: $n(x) = x^4 - x^3 - 6x^2 = 0$

Factor. $x^2(x-3)(x+2) = 0$

Solve. $x = 0, x = 3$, and $x = -2$

The **intercepts** are the points $(0, 0)$, $(3, 0)$, and $(-2, 0)$.

Technology Tip

The behavior of each function as x approaches ∞ or $-\infty$ can be shown using tables of values.

Graph $f(x) = \dfrac{x^4 - x^3 - 6x^2}{x^2 - 1}$.

```
WINDOW
 Xmin=-4
 Xmax=5
 Xscl=1
 Ymin=-8
 Ymax=8
 Yscl=1
 Xres=1
```

```
Plot1 Plot2 Plot3
\Y1=(X^4-X^3-6X²
)/(X²-1)
\Y2=
```

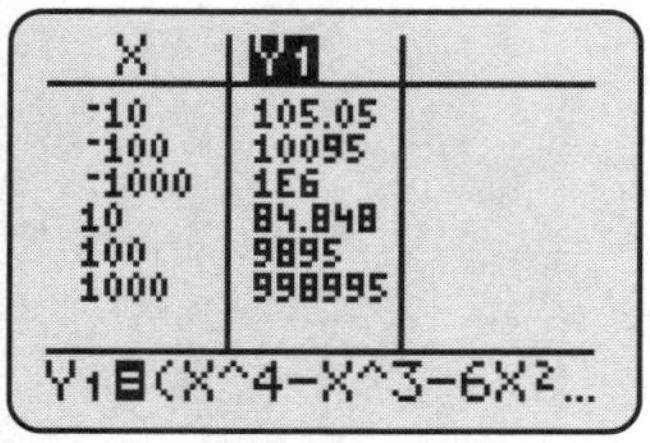

The graph of $f(x)$ shows that the vertical asymptotes are at $x = \pm 1$ and there is no horizontal asymptote or slant asymptote.

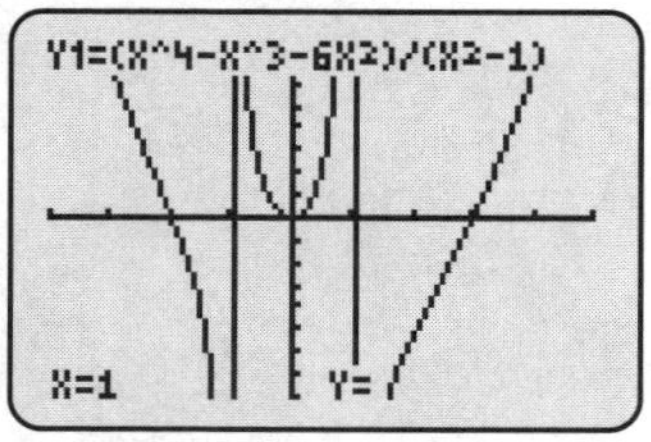

Step 3 Find any **holes**.

$$f(x) = \frac{x^2(x-3)(x+2)}{(x-1)(x+1)}$$

There are no common factors, so f is in lowest terms.
Since there are no common factors, there are no holes on the graph of f.

Step 4 Find the **asymptotes**.

Vertical asymptote: $d(x) = x^2 - 1 = 0$

Factor. $(x+1)(x-1) = 0$

Solve. $x = -1$ and $x = 1$

No horizontal asymptote: degree of $n(x) >$ degree of $d(x)$ $[4 > 2]$

No slant asymptote: degree of $n(x) -$ degree of $d(x) > 1$ $[4 - 2 = 2 > 1]$

The **asymptotes** are $x = -1$ and $x = 1$.

Step 5 Find **additional points** on the graph.

x	-3	-0.5	0.5	2	4
$f(x)$	6.75	1.75	2.08	−5.33	6.4

Step 6 **Sketch** the graph; label the **intercepts** and **asymptotes**, and complete with a smooth curve between and beyond the vertical asymptote.

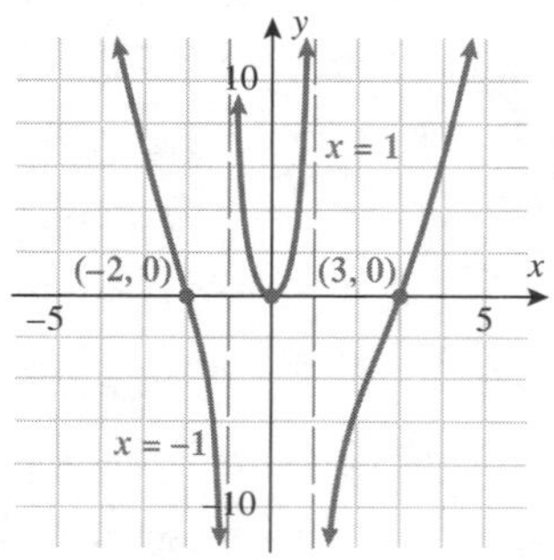

■ **YOUR TURN** State the asymptotes (if there are any) and graph the rational function

$$f(x) = \frac{x^3 - 2x^2 - 3x}{x + 2}.$$

■ **Answer:** Vertical asymptote: $x = -2$. No horizontal or slant asymptotes.

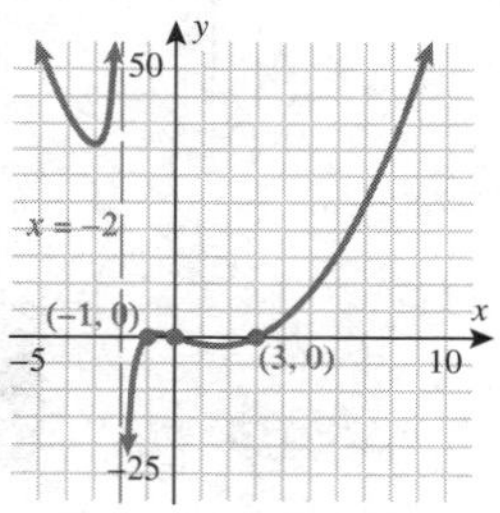

EXAMPLE 9 Graphing a Rational Function with a Horizontal Asymptote

State the asymptotes (if there are any) and graph the rational function

$$f(x) = \frac{4x^3 + 10x^2 - 6x}{8 - x^3}.$$

Solution:

Step 1 Find the **domain**.

Set the denominator equal to zero. $8 - x^3 = 0$

Solve for x. $x = 2$

State the domain. $(-\infty, 2) \cup (2, \infty)$

STEP 2 Find the **intercepts**.

y-intercept: $f(0) = \frac{0}{8} = 0$

x-intercepts: $n(x) = 4x^3 + 10x^2 - 6x = 0$

Factor. $2x(2x - 1)(x + 3) = 0$

Solve. $x = 0, x = \frac{1}{2},$ and $x = -3$

The **intercepts** are the points $(0, 0)$, $\left(\frac{1}{2}, 0\right)$, and $(-3, 0)$.

STEP 3 Find the **holes**. $f(x) = \dfrac{2x(2x - 1)(x + 3)}{(2 - x)(x^2 + 2x + 4)}$

There are no common factors, so f is in lowest terms (no holes).

STEP 4 Find the **asymptotes**.

Vertical asymptote: $d(x) = 8 - x^3 = 0$

Solve. $x = 2$

Horizontal asymptote: degree of $n(x)$ = degree of $d(x)$

Use leading coefficients. $y = \frac{4}{-1} = -4$

The **asymptotes** are $x = 2$ and $y = -4$.

STEP 5 Find **additional points** on the graph.

x	-4	-1	$\frac{1}{4}$	1	3
$f(x)$	-1	1.33	-0.10	1.14	-9.47

STEP 6 **Sketch** the graph; label the intercepts and asymptotes and complete with a smooth curve.

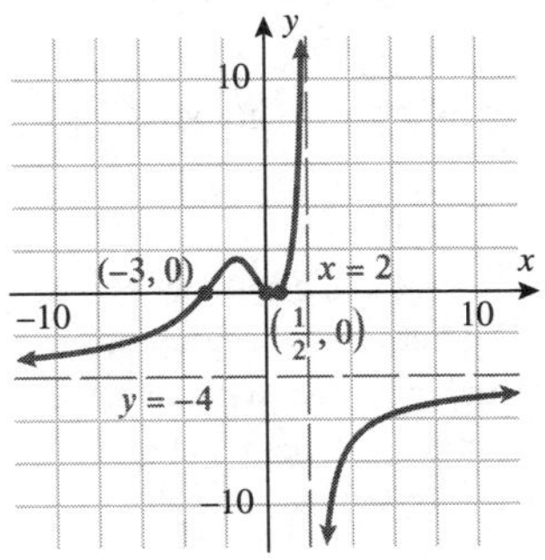

■ YOUR TURN Graph the rational function $f(x) = \dfrac{2x^2 - 7x + 6}{x^2 - 3x - 4}$. Give equations of the vertical and horizontal asymptotes and state the intercepts.

Technology Tip

The behavior of each function as x approaches ∞ or $-\infty$ can be shown using tables of values.

Graph $f(x) = \dfrac{4x^3 - 10x^2 - 6x}{8 - x^3}$.

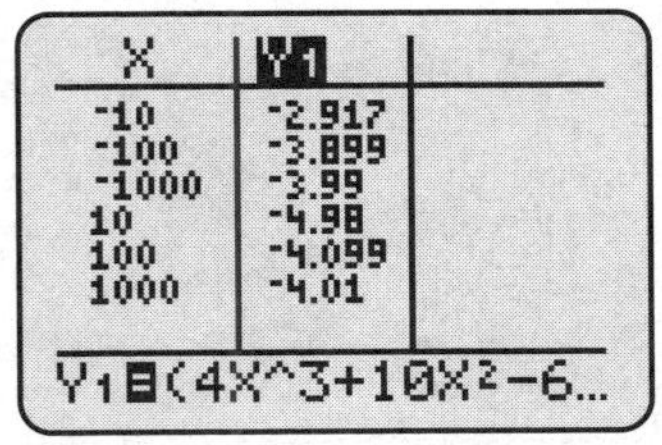

The graph of $f(x)$ shows that the vertical asymptote is at $x = 1$ and the horizontal asymptote is at $y = -4$.

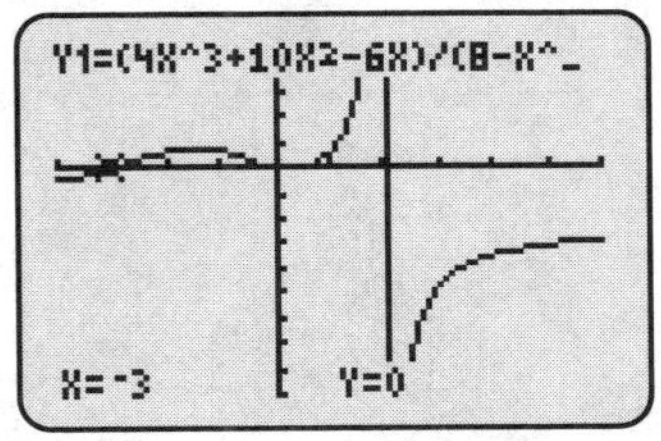

■ **Answer:** Vertical asymptotes: $x = 4, x = -1$
Horizontal asymptote: $y = 2$
Intercepts: $\left(0, -\frac{3}{2}\right)$, $\left(\frac{3}{2}, 0\right)$, $(2, 0)$

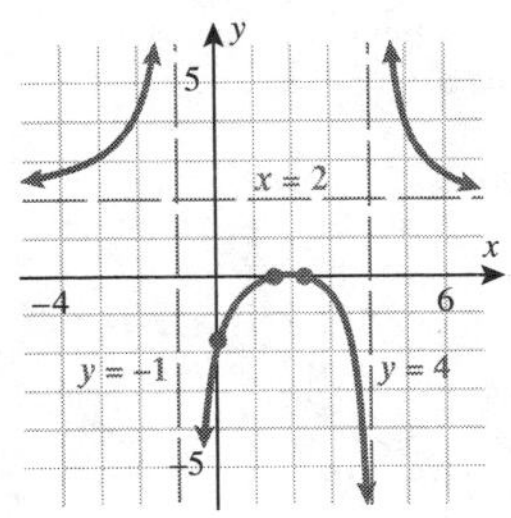

Technology Tip

The behavior of each function as x approaches ∞ or $-\infty$ can be shown using tables of values.

Graph $f(x) = \dfrac{x^2 - 3x - 4}{x + 2}$.

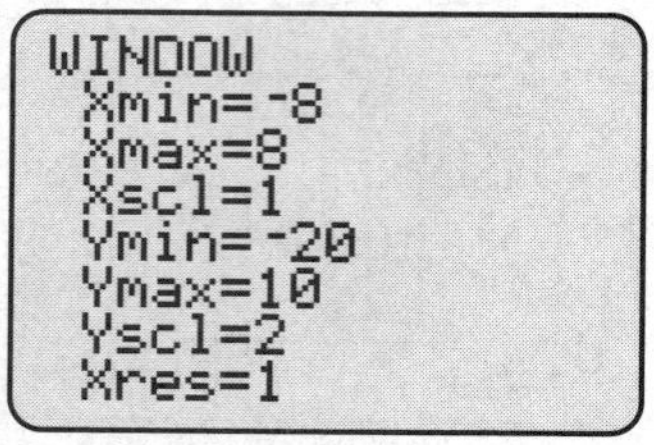

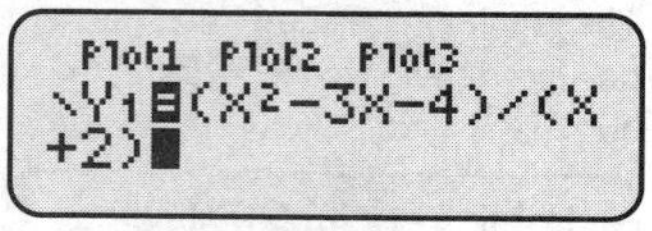

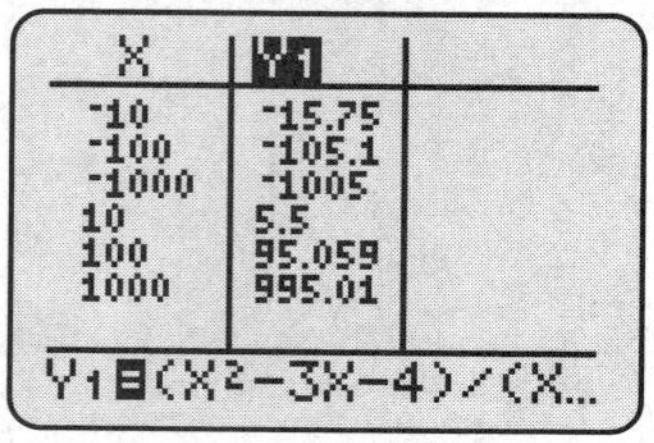

The graph of $f(x)$ shows that the vertical asymptote is at $x = -2$ and the slant asymptote is at $y = x - 5$.

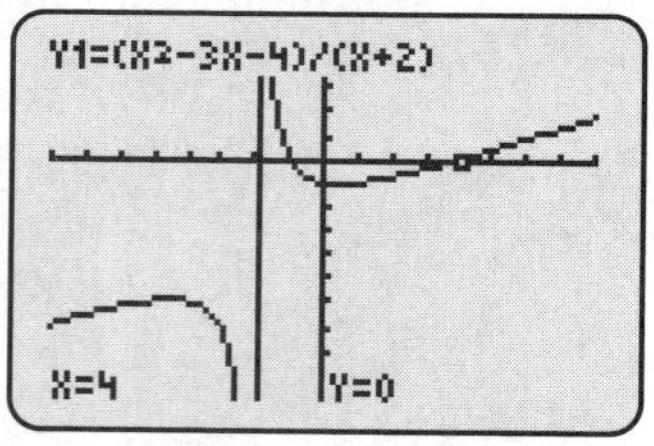

■ **Answer:**
Horizontal asymptote: $x = 3$
Slant asymptote: $y = x + 4$

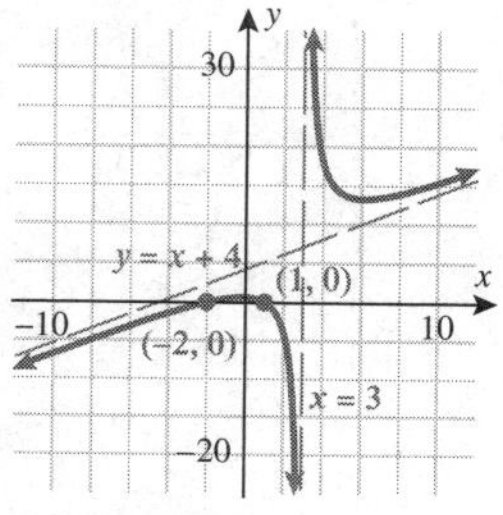

EXAMPLE 10 Graphing a Rational Function with a Slant Asymptote

Graph the rational function $f(x) = \dfrac{x^2 - 3x - 4}{x + 2}$.

Solution:

STEP 1 Find the **domain**.

Set the denominator equal to zero.	$x + 2 = 0$
Solve for x.	$x = -2$
State the domain.	$(-\infty, -2) \cup (-2, \infty)$

STEP 2 Find the **intercepts**.

y-intercept:	$f(0) = -\dfrac{4}{2} = -2$
x-intercepts:	$n(x) = x^2 - 3x - 4 = 0$
Factor.	$(x + 1)(x - 4) = 0$
Solve.	$x = -1$ and $x = 4$

The **intercepts** are the points $(0, -2)$, $(-1, 0)$, and $(4, 0)$.

STEP 3 Find any **holes**. $f(x) = \dfrac{(x - 4)(x + 1)}{(x + 2)}$

There are no common factors, so f is in lowest terms.
Since there are no common factors, there are no holes on the graph of f.

STEP 4 Find the **asymptotes**.

Vertical asymptote:	$d(x) = x + 2 = 0$
Solve.	$x = -2$
Slant asymptote:	degree of $n(x)$ − degree of $d(x) = 1$
Divide $n(x)$ by $d(x)$.	$f(x) = \dfrac{x^2 - 3x - 4}{x + 2} = x - 5 + \dfrac{6}{x + 2}$
Write the equation of the asymptote.	$y = x - 5$

The **asymptotes** are $x = -2$ and $y = x - 5$.

STEP 5 Find **additional points** on the graph.

x	−6	−5	−3	5	6
$f(x)$	−12.5	−12	−14	0.86	1.75

STEP 6 **Sketch** the graph; label the intercepts and asymptotes, and complete with a smooth curve between and beyond the vertical asymptote.

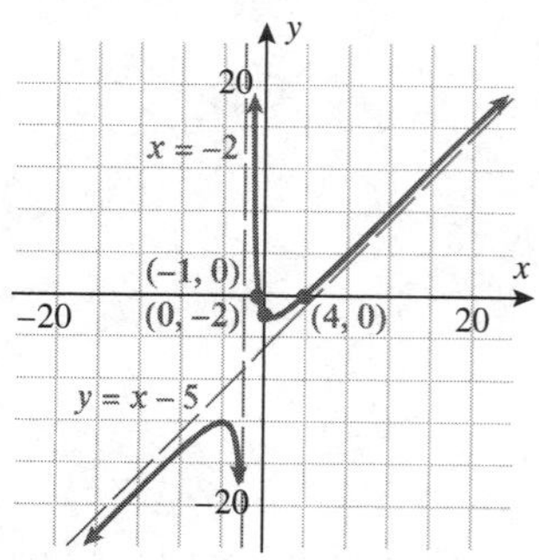

■ **YOUR TURN** For the function $f(x) = \dfrac{x^2 + x - 2}{x - 3}$, state the asymptotes (if any exist) and graph the function.

EXAMPLE 11 Graphing a Rational Function with a Hole in the Graph

Graph the rational function $f(x) = \dfrac{x^2 + x - 6}{x^2 - x - 2}$.

Solution:

STEP 1 Find the **domain**.

Set the denominator equal to zero. $\quad x^2 - x - 2 = 0$

Solve for x. $\quad (x - 2)(x + 1) = 0$

$x = -1$ or $x = 2$

State the domain. $\quad (-\infty, -1) \cup (-1, 2) \cup (2, \infty)$

STEP 2 Find the **intercepts**.

y-intercept: $\quad f(0) = \dfrac{-6}{-2} = 3 \qquad y = 3$

x-intercepts: $\quad n(x) = x^2 + x - 6 = 0$

$(x + 3)(x - 2) = 0$

$x = -3$ or $x = 2$

The intercepts correspond to the points $\boxed{(0, 3)}$ and $\boxed{(-3, 0)}$. The point (2, 0) appears to be an x-intercept; however, $x = 2$ is not in the domain of the function.

STEP 3 Find any **holes**. $\quad f(x) = \dfrac{(x - 2)(x + 3)}{(x - 2)(x + 1)}$

Since $x - 2$ is a common factor, there is a *hole* in the graph of f at $x = 2$.
Dividing out the common factor generates an equivalent rational function in lowest terms. $\quad R(x) = \dfrac{(x + 3)}{(x + 1)}$

STEP 4 Find the **asymptotes**.

Vertical asymptotes: $\quad x + 1 = 0$

$\boxed{x = -1}$

Horizontal asymptote: $\quad \dfrac{\text{Degree of numerator}}{\text{Degree of denominator}} = \overbrace{\dfrac{2}{2}}^{f} = \overbrace{\dfrac{1}{1}}^{R}$

Since the degree of the numerator equals the degree of the denominator, use the leading coefficients. $\quad \boxed{y = \dfrac{1}{1} = 1}$

STEP 5 Find **additional points** on the graph.

x	-4	-2	$-\frac{1}{2}$	1	3
$f(x)$ or $R(x)$	$\frac{1}{3}$	-1	5	2	$\frac{3}{2}$

STEP 6 **Sketch** the graph; label the intercepts, asymptotes, and additional points and complete with a smooth curve approaching asymptotes.

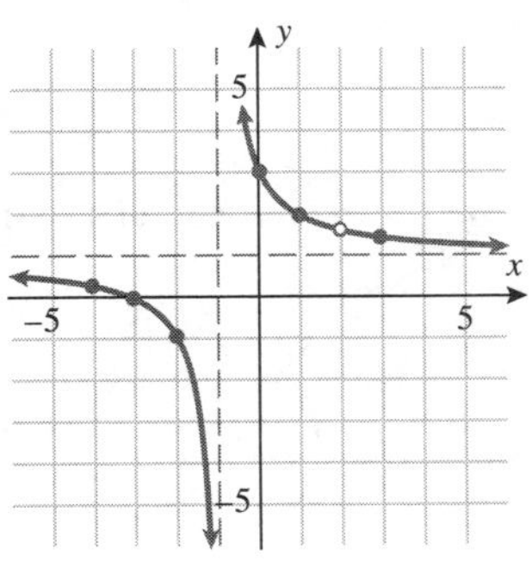

■ **YOUR TURN** Graph the rational function $f(x) = \dfrac{x^2 - x - 2}{x^2 + x - 6}$.

Technology Tip

The behavior of each function as x approaches ∞ or $-\infty$ can be shown using tables of values.

Graph $f(x) = \dfrac{x^2 + x - 6}{x^2 - x - 2}$.

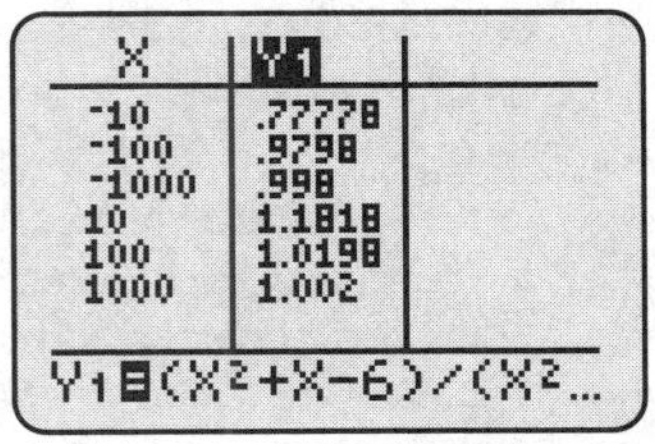

The graph of $f(x)$ shows that the vertical asymptote is at $x = -1$ and the horizontal asymptote is at $y = 1$.

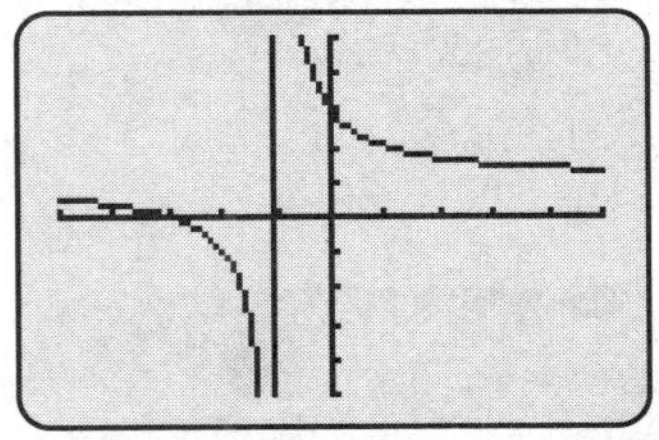

Notice that the hole at $x = 2$ is not apparent in the graph. A table of values supports the graph.

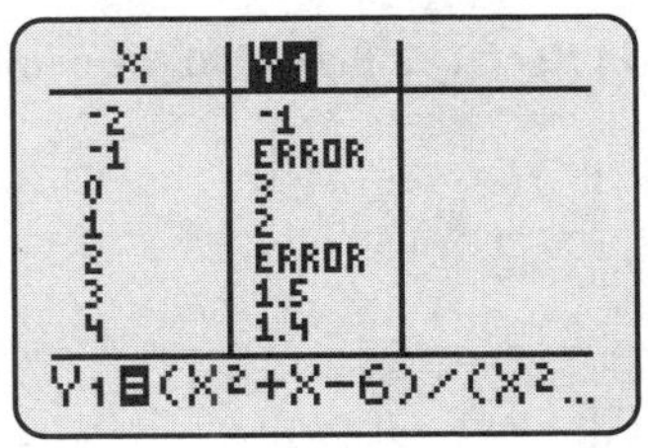

■ **Answer:**

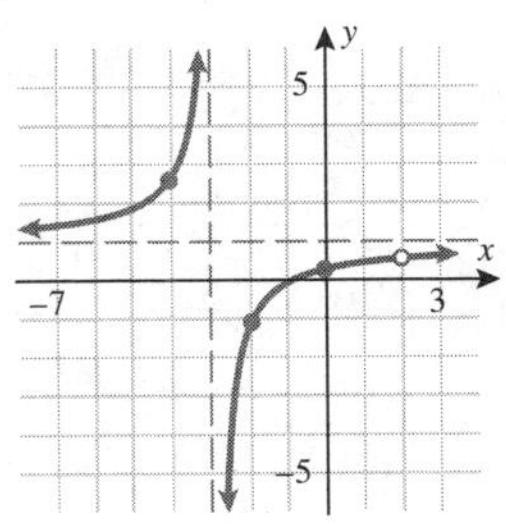

SECTION 4.6 SUMMARY

In this section, rational functions were discussed.

$$f(x) = \frac{n(x)}{d(x)}$$

- **Domain:** All real numbers except the x-values that make the denominator equal to zero, $d(x) = 0$.
- **Vertical Asymptotes:** Vertical lines, $x = a$, where $d(a) = a$, after all common factors have been divided out. Vertical asymptotes steer the graph and are never touched.
- **Horizontal Asymptotes:** Horizontal lines, $y = b$, that steer the graph as $x \to \pm\infty$.
  1. If degree of the numerator < degree of the denominator, then $y = 0$ is a horizontal asymptote.
  2. If degree of the numerator = degree of the denominator, then $y = c$ is a horizontal asymptote where c is the ratio of the leading coefficients of the numerator and denominator, respectively.
  3. If degree of the numerator > degree of the denominator, then there is no horizontal asymptote.
- **Slant Asymptotes:** Slant lines, $y = mx + b$, that steer the graph as $x \to \pm\infty$.
  1. If degree of the numerator − degree of the denominator = 1, then there is a slant asymptote.
  2. Divide the numerator by the denominator. The quotient corresponds to the equation of the line (slant asymptote).

Procedure for Graphing Rational Functions

1. Find the domain of the function.
2. Find the intercept(s).
   - y-intercept
   - x-intercepts (if any)
3. Find any holes.
   - If $x - a$ is a common factor of the numerator and denominator, then $x = a$ corresponds to a hole in the graph of the rational function if the multiplicity of a in the numerator is greater than or equal to the multiplicity of a in the denominator. The result after the common factor is canceled is an equivalent rational function in lowest terms (no common factor).
4. Find any asymptotes.
   - Vertical asymptotes
   - Horizontal/slant asymptotes
5. Find additional points on the graph.
6. Sketch the graph: draw the asymptotes and label the intercepts and points and connect with a smooth curve.

SECTION 4.6 EXERCISES

SKILLS

In Exercises 1–10, find the domain of each rational function.

1. $f(x) = \dfrac{1}{x + 3}$
2. $f(x) = \dfrac{3}{4 - x}$
3. $f(x) = \dfrac{2x + 1}{(3x + 1)(2x - 1)}$
4. $f(x) = \dfrac{5 - 3x}{(2 - 3x)(x - 7)}$
5. $f(x) = \dfrac{x + 4}{x^2 + x - 12}$
6. $f(x) = \dfrac{x - 1}{x^2 + 2x - 3}$
7. $f(x) = \dfrac{7x}{x^2 + 16}$
8. $f(x) = -\dfrac{2x}{x^2 + 9}$
9. $f(x) = -\dfrac{3(x^2 + x - 2)}{2(x^2 - x - 6)}$
10. $f(x) = \dfrac{5(x^2 - 2x - 3)}{(x^2 - x - 6)}$

In Exercises 11–20, find all vertical asymptotes and horizontal asymptotes (if there are any).

11. $f(x) = \dfrac{1}{x + 2}$
12. $f(x) = \dfrac{1}{5 - x}$
13. $f(x) = \dfrac{7x^3 + 1}{x + 5}$
14. $f(x) = \dfrac{2 - x^3}{2x - 7}$
15. $f(x) = \dfrac{6x^5 - 4x^2 + 5}{6x^2 + 5x - 4}$
16. $f(x) = \dfrac{6x^2 + 3x + 1}{3x^2 - 5x - 2}$
17. $f(x) = \dfrac{\frac{1}{3}x^2 + \frac{1}{3}x - \frac{1}{4}}{x^2 + \frac{1}{9}}$
18. $f(x) = \dfrac{\frac{1}{10}(x^2 - 2x + \frac{3}{10})}{2x - 1}$
19. $f(x) = \dfrac{(0.2x - 3.1)(1.2x + 4.5)}{0.7(x - 0.5)(0.2x + 0.3)}$
20. $f(x) = \dfrac{0.8x^4 - 1}{x^2 - 0.25}$

In Exercises 21–26, find the slant asymptote corresponding to the graph of each rational function.

21. $f(x) = \dfrac{x^2 + 10x + 25}{x + 4}$

22. $f(x) = \dfrac{x^2 + 9x + 20}{x - 3}$

23. $f(x) = \dfrac{2x^2 + 14x + 7}{x - 5}$

24. $f(x) = \dfrac{3x^3 + 4x^2 - 6x + 1}{x^2 - x - 30}$

25. $f(x) = \dfrac{8x^4 + 7x^3 + 2x - 5}{2x^3 - x^2 + 3x - 1}$

26. $f(x) = \dfrac{2x^6 + 1}{x^5 - 1}$

In Exercises 27–32, match the function to the graph.

27. $f(x) = \dfrac{3}{x - 4}$

28. $f(x) = \dfrac{3x}{x - 4}$

29. $f(x) = \dfrac{3x^2}{x^2 - 4}$

30. $f(x) = -\dfrac{3x^2}{x^2 + 4}$

31. $f(x) = \dfrac{3x^2}{4 - x^2}$

32. $f(x) = \dfrac{3x^2}{x + 4}$

a.

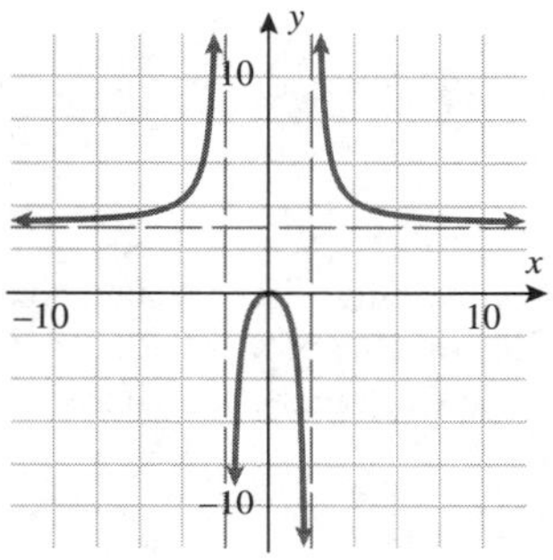

b.

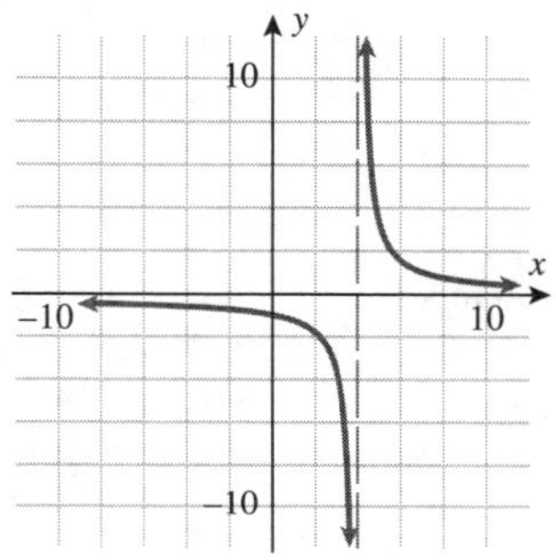

c.

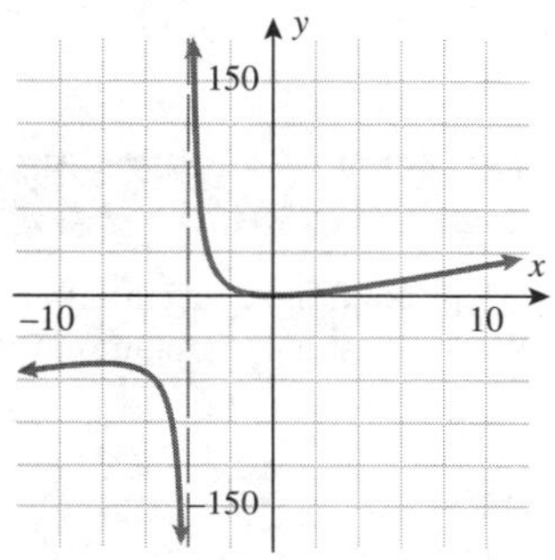

d.

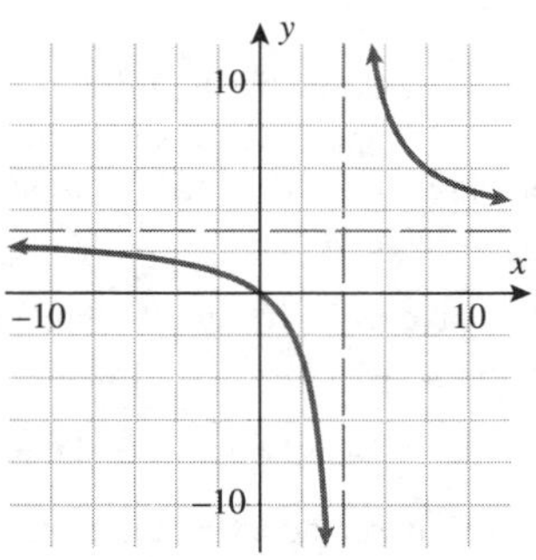

e.

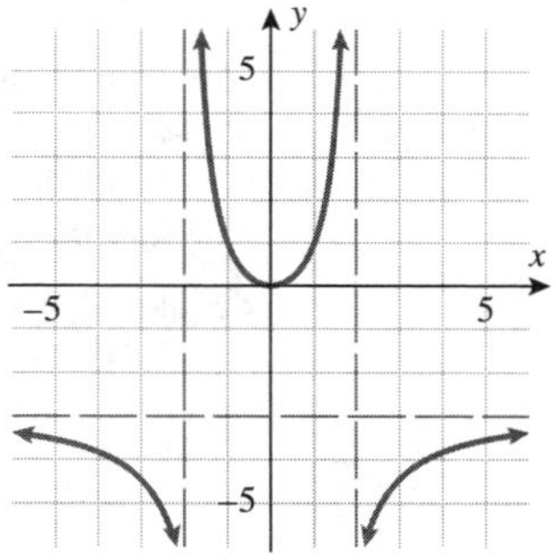

f.

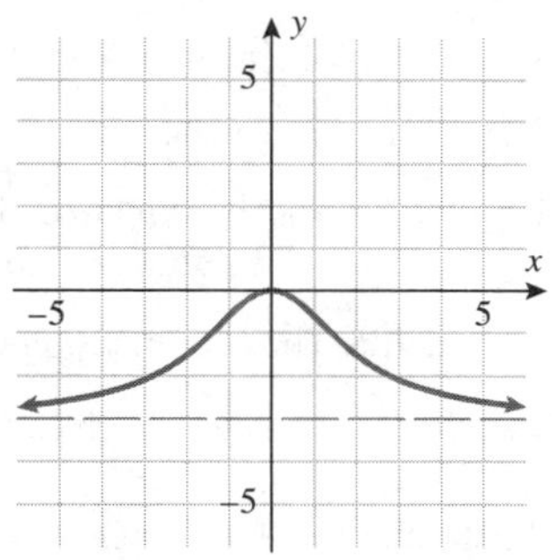

In Exercises 33–58, use the graphing strategy outlined in this section to graph the rational functions.

33. $f(x) = \dfrac{2}{x + 1}$

34. $f(x) = \dfrac{4}{x - 2}$

35. $f(x) = \dfrac{2x}{x - 1}$

36. $f(x) = \dfrac{4x}{x + 2}$

37. $f(x) = \dfrac{x - 1}{x}$

38. $f(x) = \dfrac{2 + x}{x - 1}$

39. $f(x) = \dfrac{2(x^2 - 2x - 3)}{x^2 + 2x}$

40. $f(x) = \dfrac{3(x^2 - 1)}{x^2 - 3x}$

41. $f(x) = \dfrac{x^2}{x + 1}$

42. $f(x) = \dfrac{x^2 - 9}{x + 2}$

43. $f(x) = \dfrac{2x^3 - x^2 - x}{x^2 - 4}$

44. $f(x) = \dfrac{3x^3 + 5x^2 - 2x}{x^2 + 4}$

45. $f(x) = \dfrac{x^2 + 1}{x^2 - 1}$

46. $f(x) = \dfrac{1 - x^2}{x^2 + 1}$

47. $f(x) = \dfrac{7x^2}{(2x + 1)^2}$

48. $f(x) = \dfrac{12x^4}{(3x + 1)^4}$

49. $f(x) = \dfrac{1 - 9x^2}{(1 - 4x^2)^3}$

50. $f(x) = \dfrac{25x^2 - 1}{(16x^2 - 1)^2}$

51. $f(x) = 3x + \dfrac{4}{x}$

52. $f(x) = x - \dfrac{4}{x}$

53. $f(x) = \dfrac{(x - 1)^2}{(x^2 - 1)}$

54. $f(x) = \dfrac{(x + 1)^2}{(x^2 - 1)}$

55. $f(x) = \dfrac{(x - 1)(x^2 - 4)}{(x - 2)(x^2 + 1)}$

56. $f(x) = \dfrac{(x - 1)(x^2 - 9)}{(x - 3)(x^2 + 1)}$

57. $f(x) = \dfrac{3x(x - 1)}{x(x^2 - 4)}$

58. $f(x) = \dfrac{-2x(x - 3)}{x(x^2 + 1)}$

In Exercises 59–62, for each graph of the rational function given determine: (a) all intercepts, (b) all asymptotes, and (c) equation of the rational function.

59.

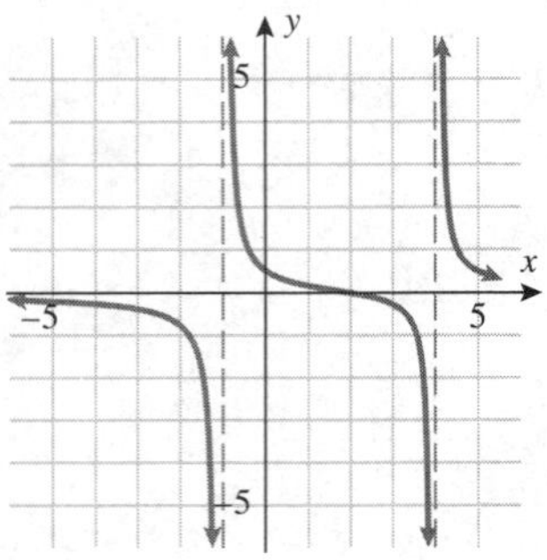

60.

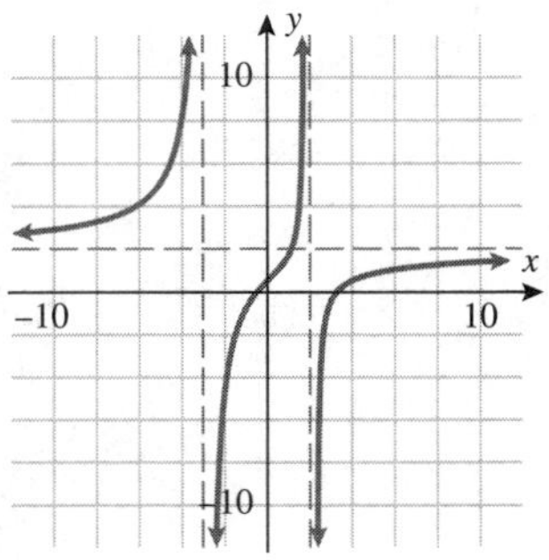

61.

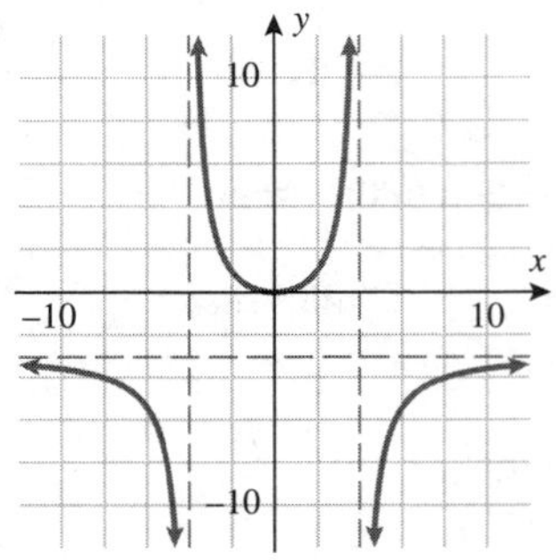

62.

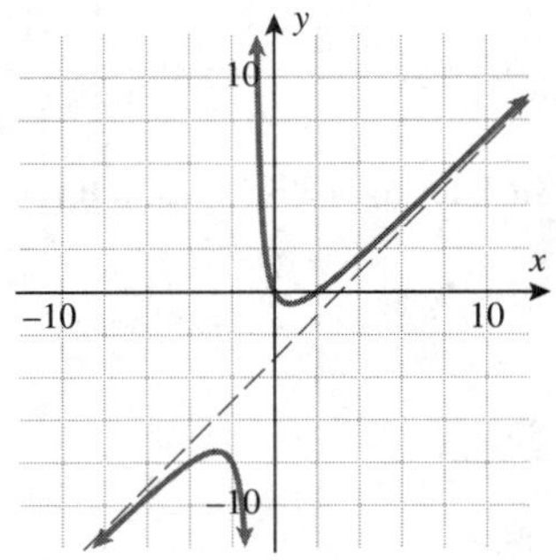

APPLICATIONS

63. Medicine. The concentration C of a particular drug in a person's bloodstream t minutes after injection is given by

$$C(t) = \frac{2t}{t^2 + 100}$$

a. What is the concentration in the bloodstream after 1 minute?

b. What is the concentration in the bloodstream after 1 hour?

c. What is the concentration in the bloodstream after 5 hours?

d. Find the horizontal asymptote of $C(t)$. What do you expect the concentration to be after several days?

64. Medicine. The concentration C of aspirin in the bloodstream t hours after consumption is given by $C(t) = \frac{t}{t^2 + 40}$.

a. What is the concentration in the bloodstream after $\frac{1}{2}$ hour?

b. What is the concentration in the bloodstream after 1 hour?

c. What is the concentration in the bloodstream after 4 hours?

d. Find the horizontal asymptote for $C(t)$. What do you expect the concentration to be after several days?

65. Typing. An administrative assistant is hired after graduating from high school and learns to type on the job. The number of words he can type per minute is given by

$$N(t) = \frac{130t + 260}{t + 5} \qquad t \geq 0$$

where t is the number of months he has been on the job.

a. How many words per minute can he type the day he starts?

b. How many words per minute can he type after 12 months?

c. How many words per minute can he type after 3 years?

d. How many words per minute would you expect him to type if he worked there until he retired?

66. Memorization. A professor teaching a large lecture course tries to learn students' names. The number of names she can remember $N(t)$ increases with each week in the semester t and is given by the rational function:

$$N(t) = \frac{600t}{t + 20}$$

How many students' names does she know by the third week in the semester? How many students' names should she know by the end of the semester (16 weeks)? According to this function, what are the most names she can remember?

67. Food. The amount of food that cats typically eat increases as their weight increases. A rational function that describes this is $F(x) = \frac{10x^2}{x^2 + 4}$, where the amount of food $F(x)$ is given in ounces and the weight of the cat x is given in pounds. Calculate the horizontal asymptote. How many ounces of food will most adult cats eat?

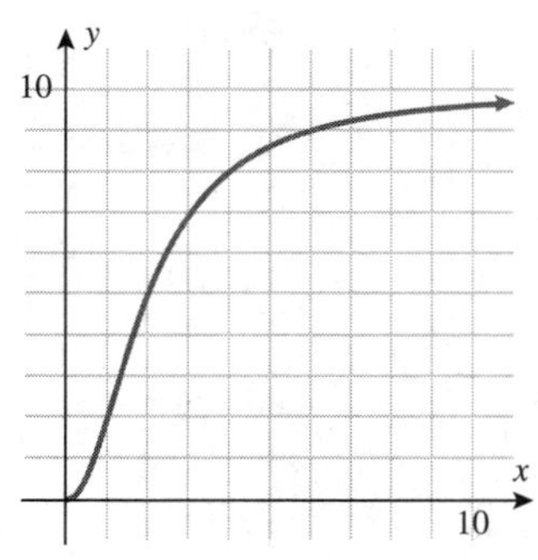

68. Memorization. The *Guinness Book of World Records, 2004* states that Dominic O'Brien (England) memorized on a single sighting a random sequence of 54 separate packs of cards all shuffled together (2808 cards in total) at Simpson's-In-The-Strand, London, England, on May 1, 2002. He memorized the cards in 11 hours 42 minutes, and then recited them in exact sequence in a time of 3 hours 30 minutes. With only a 0.5% margin of error allowed (no more than 14 errors) he broke the record with just 8 errors. If we let x represent the time (hours) it takes to memorize the cards and y represent the number of cards memorized, then a rational function that models this event is given by $y = \frac{2800x^2 + x}{x^2 + 2}$. According to this model how many cards could be memorized in an hour? What is the greatest number of cards that can be memorized?

69. Gardening. A 500-square-foot rectangular garden will be enclosed with fencing. Write a rational function that describes how many linear feet of fence will be needed to enclose the garden as a function of the width of the garden w.

70. Geometry. A rectangular picture has an area of 414 square inches. A border (matting) is used when framing. If the top and bottom borders are each 4 inches and the side borders are 3.5 inches, write a function that represents the area $A(l)$ of the entire frame as a function of the length of the picture l.

CATCH THE MISTAKE

In Exercises 71–74, explain the mistake that is made.

71. Determine the vertical asymptotes of the function $f(x) = \frac{x - 1}{x^2 - 1}$.

Solution:

Set the denominator equal to zero. $x^2 - 1 = 0$

Solve for x. $x = \pm 1$

The vertical asymptotes are $x = -1$ and $x = 1$.

The following is a correct graph of the function. Note that only $x = -1$ is an asymptote. What went wrong?

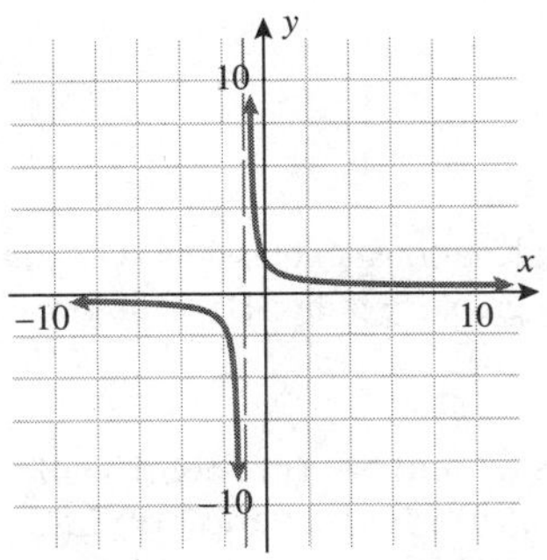

72. Determine the vertical asymptotes of $f(x) = \frac{2x}{x^2 + 1}$.

Solution:

Set the denominator equal to zero. $x^2 + 1 = 0$

Solve for x. $x = \pm 1$

The vertical asymptotes are $x = -1$ and $x = 1$.

The following is a correct graph of the function. Note that there are no vertical asymptotes. What went wrong?

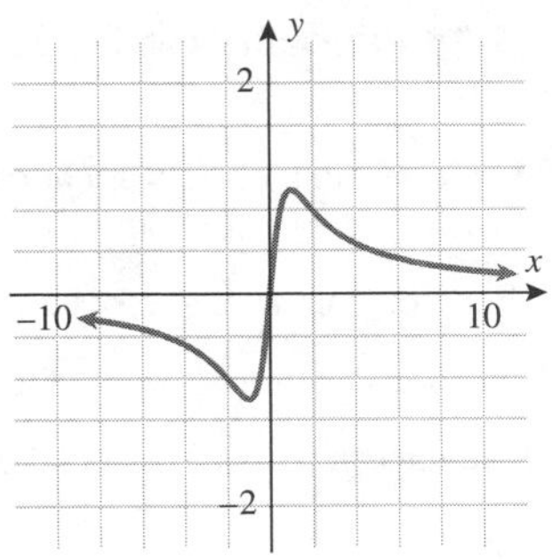

73. Determine whether a horizontal or a slant asymptote exists for the function $f(x) = \frac{9 - x^2}{x^2 - 1}$. If one does, find it.

Solution:

Step 1: The degree of the numerator equals the degree of the denominator, so there is a horizontal asymptote.

Step 2: The horizontal asymptote is the ratio of the lead coefficients: $y = \frac{9}{1} = 9$.

The horizontal asymptote is $y = 9$.

The following is a correct graph of the function. Note that there is no horizontal asymptote at $y = 9$. What went wrong?

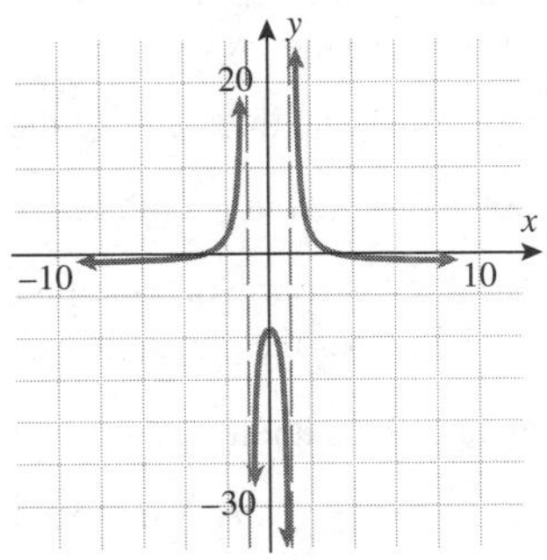

74. Determine whether a horizontal or a slant asymptote exists for the function $f(x) = \dfrac{x^2 + 2x - 1}{3x^3 - 2x^2 - 1}$. If one does, find it.

Solution:

Step 1: The degree of the denominator is exactly one more than the degree of the numerator, so there is a slant asymptote.

Step 2: Divide.

$$\begin{array}{r} 3x - 8 \\ x^2 + 2x - 1 \overline{)3x^3 - 2x^2 + 0x - 1} \\ 3x^3 + 6x^2 - 3x \\ \hline -8x^2 + 3x - 1 \\ 8x^2 - 16x + 8 \\ \hline 19x - 9 \end{array}$$

The slant asymptote is $y = 3x - 8$.

The following is the correct graph of the function. Note that $y = 3x - 8$ is not an asymptote. What went wrong?

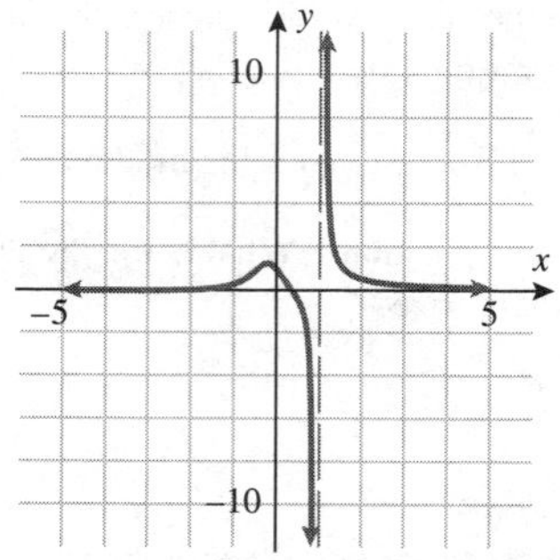

CONCEPTUAL

For Exercises 75–78, determine whether each statement is true or false.

75. A rational function can have either a horizontal asymptote or an oblique asymptote, but not both.

76. A rational function can have at most one vertical asymptote.

77. A rational function can cross a vertical asymptote.

78. A rational function can cross a horizontal or an oblique asymptote.

CHALLENGE

79. Determine the asymptotes of the rational function $f(x) = \dfrac{(x - a)(x + b)}{(x - c)(x + d)}$.

80. Determine the asymptotes of the rational function $f(x) = \dfrac{3x^2 + b^2}{x^2 + a^2}$.

81. Write a rational function that has vertical asymptotes at $x = -3$ and $x = 1$ and a horizontal asymptote at $y = 4$.

82. Write a rational function that has no vertical asymptotes, approaches the x-axis as a horizontal asymptote, and has an x-intercept of (3, 0).

TECHNOLOGY

83. Determine the vertical asymptotes of $f(x) = \dfrac{x - 4}{x^2 - 2x - 8}$. Graph this function utilizing a graphing utility. Does the graph confirm the asymptotes?

84. Determine the vertical asymptotes of $f(x) = \dfrac{2x + 1}{6x^2 + x - 1}$. Graph this function utilizing a graphing utility. Does the graph confirm the asymptotes?

85. Find the asymptotes and intercepts of the rational function $f(x) = \dfrac{1}{3x + 1} - \dfrac{2}{x}$. (*Note:* Combine the two expressions into a single rational expression.) Graph this function utilizing a graphing utility. Does the graph confirm what you found?

86. Find the asymptotes and intercepts of the rational function $f(x) = -\dfrac{1}{x^2 + 1} + \dfrac{1}{x}$. (*Note:* Combine the two expressions into a single rational expression.) Graph this function utilizing a graphing utility. Does the graph confirm what you found?

For Exercises 87 and 88: (a) Identify all asymptotes for each function. (b) Plot $f(x)$ and $g(x)$ in the same window. How does the end behavior of the function f differ from g? (c) Plot $g(x)$ and $h(x)$ in the same window. How does the end behavior of g differ from h? (d) Combine the two expressions into a single rational expression for the functions g and h. Does the strategy of finding horizontal and slant asymptotes agree with your findings in (b) and (c)?

87. $f(x) = \dfrac{1}{x - 3}$, $g(x) = 2 + \dfrac{1}{x - 3}$, $h(x) = -3 + \dfrac{1}{x - 3}$

88. $f(x) = \dfrac{2x}{x^2 - 1}$, $g(x) = x + \dfrac{2x}{x^2 - 1}$, $h(x) = x - 3 + \dfrac{2x}{x^2 - 1}$

MODELING YOUR WORLD

The following table summarizes average yearly temperature in degrees Fahrenheit (°F) and carbon dioxide emissions in parts per million (ppm) for **Mauna Loa, Hawaii.**

YEAR	1960	1965	1970	1975	1980	1985	1990	1995	2000	2005
TEMPERATURE	44.45	43.29	43.61	43.35	46.66	45.71	45.53	47.53	45.86	46.23
CO_2 EMISSIONS (ppm)	316.9	320.0	325.7	331.1	338.7	345.9	354.2	360.6	369.4	379.7

In the Modeling Your World in Chapter 3, the temperature and carbon emissions were modeled with *linear functions*. Now, let us model these same data using *polynomial functions*.

1. Plot the temperature data with time on the horizontal axis and temperature on the vertical axis. Let $t = 0$ correspond to 1960. Adjust the vertical range of the graph to (43, 48). How many turning points (local maxima and minima) do these data exhibit? What is the lowest degree polynomial function whose graph can pass through these data?

2. Find a *polynomial function* that models the temperature in Mauna Loa.
   a. Find a quadratic function: Let the data from 1995 correspond to the vertex of the graph and apply the 2005 data to determine the function.
   b. Find a quadratic function: Let the data from 2000 correspond to the vertex of the graph and apply the 2005 data to determine the function.
   c. Utilize regression and all data given to find a polynomial function whose degree is found in 1.

3. Predict what the temperature will be in Mauna Loa in 2020.
   a. Use the line found in 2a.
   b. Use the line found in 2b.
   c. Use the line found in 2c.

4. Predict what the temperature will be in Mauna Loa in 2100.
   a. Use the line found in 2a.
   b. Use the line found in 2b.
   c. Use the line found in 2c.

5. Do your models support the claim of "global warming"? Explain. Do these models make similar predictions to the linear models found in Chapter 2?

6. Plot the carbon dioxide emissions data with time on the horizontal axis and carbon dioxide emissions on the vertical axis. Let $t = 0$ correspond to 1960. Adjust the vertical range of the graph to (315, 380).

7. Find a *quadratic function* that models the CO_2 emissions (ppm) in Mauna Loa.
   a. Let the data from 1960 correspond to the vertex of the graph and apply the 2005 data to determine the function.
   b. Let the data from 1980 correspond to the vertex of the graph and apply the 2005 data to determine the function.
   c. Utilize regression and all data given.

8. Predict the expected CO_2 levels in Mauna Loa in 2020.
   a. Use the line found in 7a.
   b. Use the line found in 7b.
   c. Use the line found in 7c.

9. Predict the expected CO_2 levels in Mauna Loa in 2100.
   a. Use the line found in 7a.
   b. Use the line found in 7b.
   c. Use the line found in 7c.

10. Do your models support the claim of "global warming"? Explain. Do these models give similar predictions to the linear models found in Chapter 3?

11. Discuss differences in models and predictions found in parts (a), (b), and (c) and also discuss difference in linear and polynomial functions.

CHAPTER 4 REVIEW

SECTION	CONCEPT	PAGES	REVIEW EXERCISES	KEY IDEAS/POINTS
4.1	**Quadratic functions**	326–337	1–28	
	Graph of a quadratic function	327	1–8	Parabolas
	Graphing quadratic functions in standard form	328–330	5–8	Parabola $f(x) = a(x - h)^2 + k$ ■ Vertex: (h, k) ■ Opens Up: $a > 0$ ■ Opens Down: $a < 0$
	Graphing quadratic functions in general form	330–337	9–20	$f(x) = ax^2 + bx + c$, vertex is $(h, k) = \left(-\frac{b}{2a}, f\left(-\frac{b}{2a}\right)\right)$
4.2	**Polynomial functions of higher degree**	343–353	29–56	$P(x) = a_nx^n + a_{n-1}x^{n-1} + \cdots + a_2x^2 + a_1x + a_0$ is a polynomial of degree n.
	Graphs of polynomial functions	344–346	33–36	$y = x^n$ behave similar to: ■ $y = x^2$, when n is even. ■ $y = x^3$, when n is odd.
	Graphing polynomial functions using transformations of power functions	346	37–40	Shift power functions
	Real zeros of a polynomial function	347–349	41–50	$P(x) = (x - a)(x - b)^n = 0$ ■ a is a zero of multiplicity 1. ■ b is a zero of multiplicity n.
	Graphing general polynomial functions	349–353	51–54	Intercepts; zeros and multiplicities; end behavior
4.3	**Dividing polynomials: Long division and synthetic division**	358–363	57–70	Use zero placeholders for missing terms.
	Long division of polynomials	358–361	57–60	Can be used for all polynomial division.
	Synthetic division of polynomials	361–363	61–64	Can only be used when dividing by $(x \pm a)$.
4.4	**The real zeros of a polynomial function**	366–379	71–100	$P(x) = a_nx^n + a_{n-1}x^{n-1} + \cdots + a_2x^2 + a_1x + a_0$ If $P(c) = 0$, then c is a zero of $P(x)$.
	The remainder theorem and the factor theorem	366–369	71–74	If $P(x)$ is divided by $x - a$, then the remainder r is $r = P(a)$.
	The rational zero theorem and Descartes's rule of signs	369–373	79–94	Possible zeros $= \dfrac{\text{Factors of } a_0}{\text{Factors of } a_n}$ Number of positive or negative real zeros is related to the number of sign variations in $P(x)$ or $P(-x)$.

Section	Concept	Pages	Review Exercises	Key Ideas/Points
	Factoring polynomials	373–375	95–100	1. List possible rational zeros (rational zero theorem). 2. List possible combinations of positive and negative real zeros (Descartes's rule of signs). 3. Test possible values until a zero is found. 4. Once a real zero is found, use synthetic division. Then repeat testing on quotient until linear and/or irreducible quadratic factors are reached. 5. If there is a real zero but all possible rational roots have failed, then approximate the real zero using the intermediate value theorem/bisection method.
	The intermediate value theorem	376–377		Intermediate value theorem and the bisection method are used to approximate irrational zeros.
	Graphing polynomial functions	377–379	95–100	1. Find the intercepts. 2. Determine end behavior. 3. Find additional points. 4. Sketch a smooth curve.
4.5	**Complex zeros: The fundamental theorem of algebra**	382–387	101–116	$P(x) = a_n x^n + a_{n-1}x^{n-1} + \cdots + a_2x^2 + a_1x + a_0$ $P(x) = \underbrace{(x - c_1)(x - c_2) \cdots (x - c_n)}_{n \text{ factors}}$ where the c's represent complex (not necessarily distinct) zeros.
	The fundamental theorem of algebra	382–383	101–104	■ $P(x)$ of degree n has at least one zero and at most n zeros.
	Complex zeros	383–386	105–112	■ If $a + bi$ is a zero of $P(x)$, then $a - bi$ is also a zero.
	Factoring polynomials	386–387	109–116	The polynomial can be written as a product of linear factors, not necessarily distinct.
4.6	**Rational functions**	390–403	117–128	$f(x) = \dfrac{n(x)}{d(x)} \quad d(x) \neq 0$
	Domain of rational functions	390–391	117–122	**Domain:** All real numbers except x-values that make the denominator equal to zero; that is, $d(x) = 0$. A rational function $f(x) = \dfrac{n(x)}{d(x)}$ is said to be in *lowest terms* if $n(x)$ and $d(x)$ have no common factors.
	Horizontal, vertical, and slant asymptotes	391–397	117–122	A rational function that has a common factor $x - a$ in both the numerator and denominator has a hole at $x = a$ in its graph if the multiplicity of a in the numerator is greater than or equal to the multiplicity of a in the denominator. *Vertical Asymptotes* A rational function in lowest terms has a vertical asymptote corresponding to any x-values that make the denominator equal to zero.

Section	Concept	Pages	Review Exercises	Key Ideas/Points
				Horizontal Asymptotes ■ $y = 0$ if degree of $n(x) <$ degree of $d(x)$. ■ No horizontal asymptote if degree of $n(x) >$ degree of $d(x)$. ■ $y = \dfrac{\text{Leading coefficient of } n(x)}{\text{Leading coefficient of } d(x)}$ if degree of $n(x) =$ degree of $d(x)$. *Slant Asymptotes* If degree of $n(x) -$ degree of $d(x) = 1$. Divide $n(x)$ by $d(x)$ and the quotient determines the slant asymptote; that is, $y =$ quotient.
	Graphing rational functions	397–403	123–128	1. Find the domain of the function. 2. Find the intercept(s). 3. Find any holes. 4. Find any asymptotes. 5. Find additional points on the graph. 6. *Sketch the graph:* Draw the asymptotes and label the intercepts and points and connect with a smooth curve.

4.1 Quadratic Functions

Match the quadratic function with its graph.

1. $f(x) = -2(x + 6)^2 + 3$ **2.** $f(x) = \frac{1}{4}(x - 4)^2 + 2$

3. $f(x) = x^2 + x - 6$ **4.** $f(x) = -3x^2 - 10x + 8$

a.

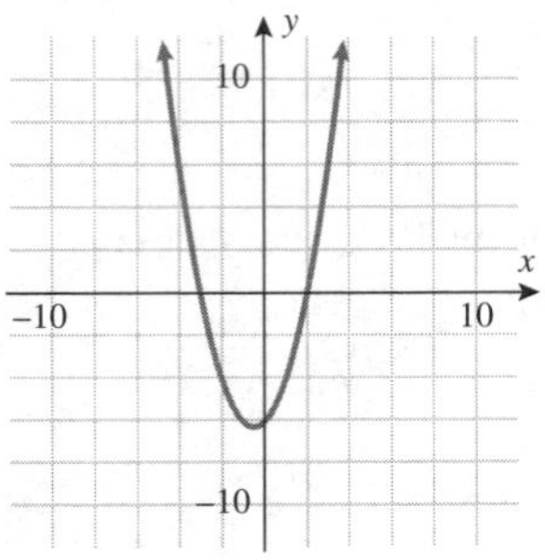

b.

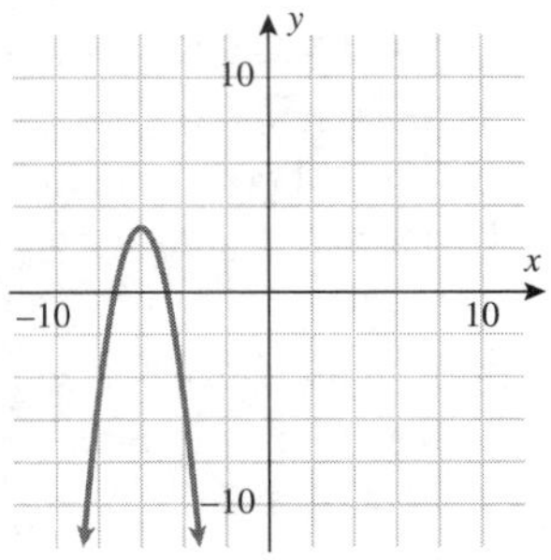

c.

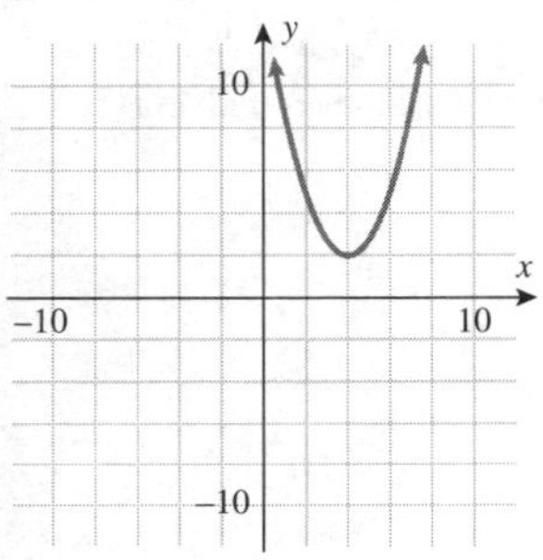

d.

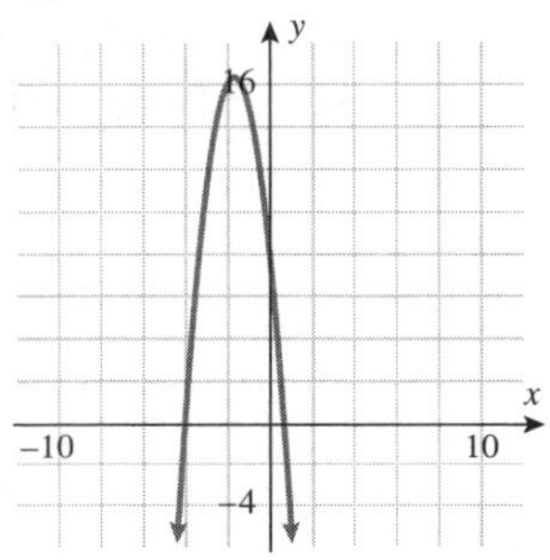

Graph the quadratic function given in standard form.

5. $f(x) = -(x - 7)^2 + 4$ **6.** $f(x) = (x + 3)^2 - 5$

7. $f(x) = -\frac{1}{2}\left(x - \frac{1}{3}\right)^2 + \frac{2}{5}$ **8.** $f(x) = 0.6(x - 0.75)^2 + 0.5$

Rewrite the quadratic function in standard form by completing the square.

9. $f(x) = x^2 - 3x - 10$ **10.** $f(x) = x^2 - 2x - 24$

11. $f(x) = 4x^2 + 8x - 7$ **12.** $f(x) = -\frac{1}{4}x^2 + 2x - 4$

Graph the quadratic function given in general form.

13. $f(x) = x^2 - 3x + 5$

14. $f(x) = -x^2 + 4x + 2$

15. $f(x) = -4x^2 + 2x + 3$

16. $f(x) = -0.75x^2 + 2.5$

Find the vertex of the parabola associated with each quadratic function.

17. $f(x) = 13x^2 - 5x + 12$

18. $f(x) = \frac{2}{5}x^2 - 4x + 3$

19. $f(x) = -0.45x^2 - 0.12x + 3.6$

20. $f(x) = -\frac{3}{4}x^2 + \frac{2}{5}x + 4$

Find the quadratic function that has the given vertex and goes through the given point.

21. vertex: $(-2, 3)$ point: $(1, 4)$

22. vertex: $(4, 7)$ point: $(-3, 1)$

23. vertex: $(2.7, 3.4)$ point: $(3.2, 4.8)$

24. vertex: $\left(-\frac{5}{2}, \frac{7}{4}\right)$ point: $\left(\frac{1}{2}, \frac{3}{5}\right)$

Applications

25. Profit. The revenue and the cost of a local business are given below as functions of the number of units x in thousands produced and sold. Use the cost and the revenue to answer the questions that follow.

$$C(x) = \frac{1}{3}x + 2 \quad \text{and} \quad R(x) = -2x^2 + 12x - 12$$

a. Determine the profit function.

b. State the break-even points.

c. Graph the profit function.

d. What is the range of units to make and sell that will correspond to a profit?

26. Geometry. Given the length of a rectangle is $2x - 4$ and the width is $x + 7$, find the area of the rectangle. What dimensions correspond to the largest area?

27. Geometry. A triangle has a base of $x + 2$ units and a height of $4 - x$ units. Determine the area of the triangle. What dimensions correspond to the largest area?

28. Geometry. A person standing at a ridge in the Grand Canyon throws a penny upward and toward the pit of the canyon. The height of the penny is given by the function:

$$h(t) = -12t^2 + 80t$$

a. What is the maximum height that the penny will reach?

b. How many seconds will it take the penny to hit the ground below?

4.2 Polynomial Functions of Higher Degree

Determine which functions are polynomials, and for those, state their degree.

29. $f(x) = x^6 - 2x^5 + 3x^2 + 9x - 42$

30. $f(x) = (3x - 4)^3(x + 6)^2$

31. $f(x) = 3x^4 - x^3 + x^2 + \sqrt[4]{x} + 5$

32. $f(x) = 5x^3 - 2x^2 + \frac{4x}{7} - 3$

Match the polynomial function with its graph.

33. $f(x) = 2x - 5$

34. $f(x) = -3x^2 + x - 4$

35. $f(x) = x^4 - 2x^3 + x^2 - 6$

36. $f(x) = x^7 - x^5 + 3x^4 + 3x + 7$

a.

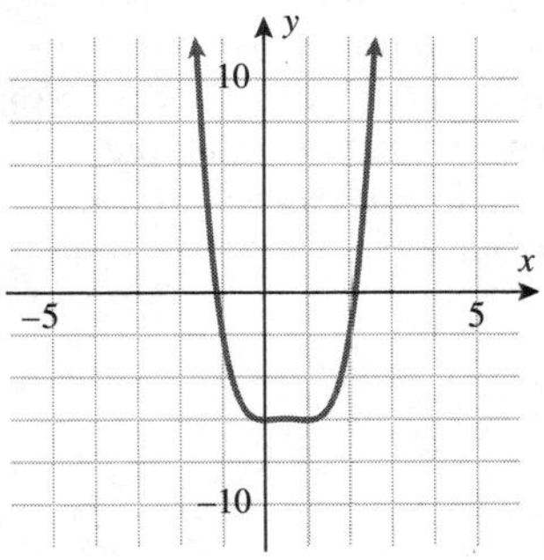

b.

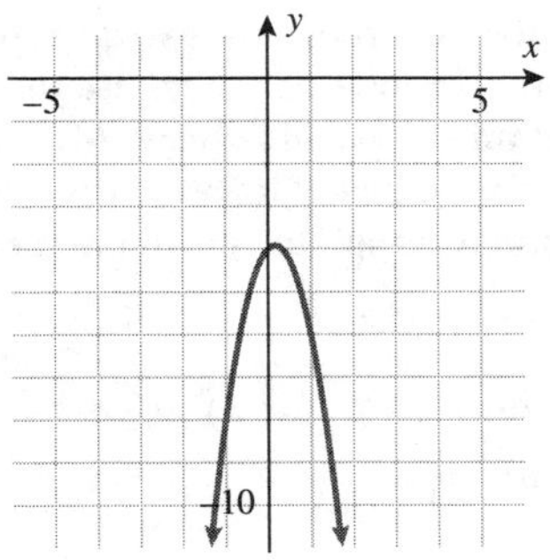

c.

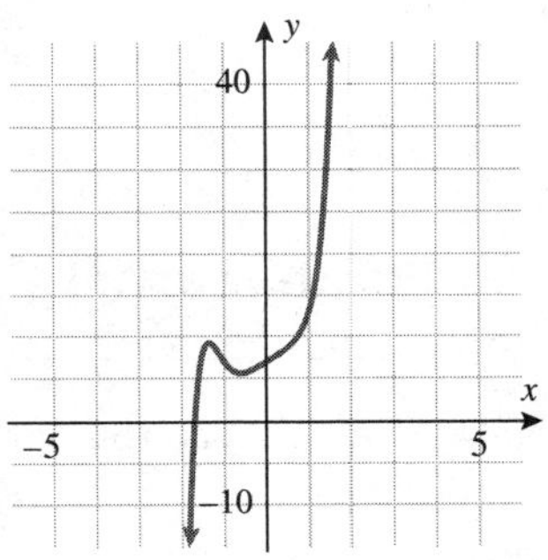

d.

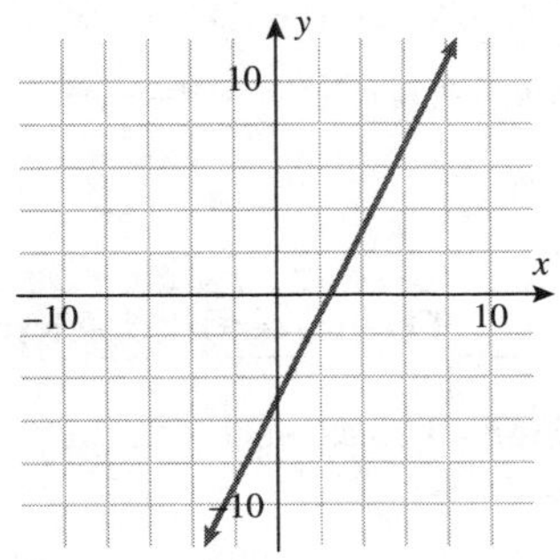

Graph each function by transforming a power function $y = x^n$.

37. $f(x) = -x^7$

38. $f(x) = (x - 3)^3$

39. $f(x) = x^4 - 2$

40. $f(x) = -6 - (x + 7)^5$

Find all the real zeros of each polynomial function, and state their multiplicities.

41. $f(x) = 3(x + 4)^2(x - 6)^5$

42. $f(x) = 7x(2x - 4)^3(x + 5)$

43. $f(x) = x^5 - 13x^3 + 36x$

44. $f(x) = 4.2x^4 - 2.6x^2$

Find a polynomial of minimum degree that has the given zeros.

45. $-3, 0, 4$

46. $2, 4, 6, -8$

47. $-\frac{2}{5}, \frac{3}{4}, 0$

48. $2 - \sqrt{5}, 2 + \sqrt{5}$

49. -2 (multiplicity of 2), 3 (multiplicity of 2)

50. 3 (multiplicity of 2), -1 (multiplicity of 2), 0 (multiplicity of 3)

For each polynomial function given: (a) list each real zero and its multiplicity; (b) determine whether the graph touches or crosses at each *x*-intercept; (c) find the *y*-intercept and a few points on the graph; (d) determine the end behavior; and (e) sketch the graph.

51. $f(x) = x^2 - 5x - 14$

52. $f(x) = -(x - 5)^5$

53. $f(x) = 6x^7 + 3x^5 - x^2 + x - 4$

54. $f(x) = -x^4(3x + 6)^3(x - 7)^3$

Applications

55. Salary. Tiffany has started tutoring students x hours per week. The tutoring job corresponds to the following additional income:

$$f(x) = (x - 1)(x - 3)(x - 7)$$

a. Graph the polynomial function.

b. Give any real zeros that occur.

c. How many hours of tutoring are financially beneficial to Tiffany?

56. Profit. The following function is the profit for Walt Disney World, where $P(x)$ represents profit in millions of dollars and x represents the month ($x = 1$ corresponds to January):

$$P(x) = 3(x - 2)^2(x - 5)^2(x - 10)^2 \qquad 1 \leq x \leq 12$$

Graph the polynomial. When are the peak seasons?

4.3 Dividing Polynomials: Long Division and Synthetic Division

Divide the polynomials with long division. If you choose to use a calculator, do not round off. Keep the exact values instead. Express the answer in the form $Q(x) = ?$, $r(x) = ?$.

57. $(x^2 + 2x - 6) \div (x - 2)$

58. $(2x^2 - 5x - 1) \div (2x - 3)$

59. $(4x^4 - 16x^3 + x - 9 + 12x^2) \div (2x - 4)$

60. $(6x^2 + 2x^3 - 4x^4 + 2 - x) \div (2x^2 + x - 4)$

Use synthetic division to divide the polynomial by the linear factor. Indicate the quotient $Q(x)$ and the remainder $r(x)$.

61. $(x^4 + 4x^3 + 5x^2 - 2x - 8) \div (x + 2)$

62. $(x^3 - 10x + 3) \div (2 + x)$

63. $(x^6 - 64) \div (x + 8)$

64. $(2x^5 + 4x^4 - 2x^3 + 7x + 5) \div (x - \frac{3}{4})$

Divide the polynomials with either long division or synthetic division.

65. $(5x^3 + 8x^2 - 22x + 1) \div (5x^2 - 7x + 3)$

66. $(x^4 + 2x^3 - 5x^2 + 4x + 2) \div (x - 3)$

67. $(x^3 - 4x^2 + 2x - 8) \div (x + 1)$

68. $(x^3 - 5x^2 + 4x - 20) \div (x^2 + 4)$

Applications

69. Geometry. The area of a rectangle is given by the polynomial $6x^4 - 8x^3 - 10x^2 + 12x - 16$. If the width is $2x - 4$, what is the length of the rectangle?

70. **Volume.** A 10 inch by 15 inch rectangular piece of cardboard is used to make a box. Square pieces x inches on a side are cut out from the corners of the cardboard and then the sides are folded up. Find the volume of the box.

4.4 The Real Zeros of a Polynomial Function

Find the following values by applying synthetic division. Check by substituting the value into the function.

$f(x) = 6x^5 + x^4 - 7x^2 + x - 1 \qquad g(x) = x^3 + 2x^2 - 3$

71. $f(-2)$ 72. $f(1)$ 73. $g(1)$ 74. $g(-1)$

Determine whether the number given is a zero of the polynomial.

75. -3, $P(x) = x^3 - 5x^2 + 4x + 2$
76. 2 and -2, $P(x) = x^4 - 16$
77. 1, $P(x) = 2x^4 - 2x$
78. 4, $P(x) = x^4 - 2x^3 - 8x$

Given a zero of the polynomial, determine all other real zeros, and write the polynomial in terms of a product of linear or irreducible factors.

Polynomial	Zero
79. $P(x) = x^4 - 6x^3 + 32x$	-2
80. $P(x) = x^3 - 7x^2 + 36$	3
81. $P(x) = x^5 - x^4 - 8x^3 + 12x^2$	0
82. $P(x) = x^4 - 32x^2 - 144$	6

Use Descartes's rule of signs to determine the possible number of positive real zeros and negative real zeros.

83. $P(x) = x^4 + 3x^3 - 16$
84. $P(x) = x^5 + 6x^3 - 4x - 2$
85. $P(x) = x^9 - 2x^7 + x^4 - 3x^3 + 2x - 1$
86. $P(x) = 2x^5 - 4x^3 + 2x^2 - 7$

Use the rational zero theorem to list the possible rational zeros.

87. $P(x) = x^3 - 2x^2 + 4x + 6$
88. $P(x) = x^5 - 4x^3 + 2x^2 - 4x - 8$
89. $P(x) = 2x^4 + 2x^3 - 36x^2 - 32x + 64$
90. $P(x) = -4x^5 - 5x^3 + 4x + 2$

List the possible rational zeros, and test to determine all rational zeros.

91. $P(x) = 2x^3 - 5x^2 + 1$
92. $P(x) = 12x^3 + 8x^2 - 13x + 3$
93. $P(x) = x^4 - 5x^3 + 20x - 16$
94. $P(x) = 24x^4 - 4x^3 - 10x^2 + 3x - 2$

For each polynomial: (a) use Descartes's rule of signs to determine the possible combinations of positive real zeros and negative real zeros; (b) use the rational zero test to determine possible rational zeros; (c) use the upper and lower bound rules to eliminate possible zeros; (d) test for rational zeros; (e) factor as a product of linear and/or irreducible quadratic factors; and (f) graph the polynomial function.

95. $P(x) = x^3 + 3x - 5$
96. $P(x) = x^3 + 3x^2 - 6x - 8$
97. $P(x) = x^3 - 9x^2 + 20x - 12$
98. $P(x) = x^4 - x^3 - 7x^2 + x + 6$
99. $P(x) = x^4 - 5x^3 - 10x^2 + 20x + 24$
100. $P(x) = x^5 - 3x^3 - 6x^2 + 8x$

4.5 Complex Zeros: The Fundamental Theorem of Algebra

Find all zeros. Factor the polynomial as a product of linear factors.

101. $P(x) = x^2 + 25$
102. $P(x) = x^2 + 16$
103. $P(x) = x^2 - 2x + 5$
104. $P(x) = x^2 + 4x + 5$

A polynomial function is described. Find all remaining zeros.

105. Degree: 4 Zeros: $-2i$, $3 + i$
106. Degree: 4 Zeros: $3i$, $2 - i$
107. Degree: 6 Zeros: i, $2 - i$ (multiplicity 2)
108. Degree: 6 Zeros: $2i$, $1 - i$ (multiplicity 2)

Given a zero of the polynomial, determine all other zeros (real and complex) and write the polynomial in terms of a product of linear factors.

Polynomial	Zero
109. $P(x) = x^4 - 3x^3 - 3x^2 - 3x - 4$	i
110. $P(x) = x^4 - 4x^3 + x^2 + 16x - 20$	$2 - i$
111. $P(x) = x^4 - 2x^3 + 11x^2 - 18x + 18$	$-3i$
112. $P(x) = x^4 - 5x^2 + 10x - 6$	$1 + i$

Factor each polynomial as a product of linear factors.

113. $P(x) = x^4 - 81$
114. $P(x) = x^3 - 6x^2 + 12x$

115. $P(x) = x^3 - x^2 + 4x - 4$

116. $P(x) = x^4 - 5x^3 + 12x^2 - 2x - 20$

4.6 Rational Functions

Determine the vertical, horizontal, or slant asymptotes (if they exist) for the following rational functions.

117. $f(x) = \dfrac{7 - x}{x + 2}$

118. $f(x) = \dfrac{2 - x^2}{(x - 1)^3}$

119. $f(x) = \dfrac{4x^2}{x + 1}$

120. $f(x) = \dfrac{3x^2}{x^2 + 9}$

121. $f(x) = \dfrac{2x^2 - 3x + 1}{x^2 + 4}$

122. $f(x) = \dfrac{-2x^2 + 3x + 5}{x + 5}$

Graph the rational functions.

123. $f(x) = -\dfrac{2}{x - 3}$

124. $f(x) = \dfrac{5}{x + 1}$

125. $f(x) = \dfrac{x^2}{x^2 + 4}$

126. $f(x) = \dfrac{x^2 - 36}{x^2 + 25}$

127. $f(x) = \dfrac{x^2 - 49}{x + 7}$

128. $f(x) = \dfrac{2x^2 - 3x - 2}{2x^2 - 5x - 3}$

Technology

Section 4.1

129. On a graphing calculator, plot the quadratic function:

$f(x) = 0.005x^2 - 4.8x - 59$

a. Identify the vertex of this parabola.

b. Identify the y-intercept.

c. Identify the x-intercepts (if any).

d. What is the axis of symmetry?

130. Determine the quadratic function whose vertex is $(2.4, -3.1)$ and passes through the point $(0, 5.54)$.

a. Write the quadratic function in general form.

b. Plot this quadratic function with a graphing calculator.

c. Zoom in on the vertex and y-intercept. Do they agree with the given values?

Section 4.2

In Exercises 131 and 132, use a graphing calculator or a computer to graph each polynomial. From the graph, estimate the x-intercepts and state the zeros of the function and their multiplicities.

131. $f(x) = 5x^3 - 11x^2 - 10.4x + 5.6$

132. $f(x) = -x^3 - 0.9x^2 + 2.16x - 2.16$

Section 4.3

133. Plot $\dfrac{15x^3 - 47x^2 + 38x - 8}{3x^2 - 7x + 2}$. What type of function is it? Perform this division using long division, and confirm that the graph corresponds to the quotient.

134. Plot $\dfrac{-4x^3 + 14x^2 - x - 15}{x - 3}$. What type of function is it? Perform this division using synthetic division, and confirm that the graph corresponds to the quotient.

Section 4.4

In Exercises 135 and 136: (a) Determine all possible rational zeros of the polynomial. Use a graphing calculator or software to graph $P(x)$ to help find the zeros. (b) Factor as a product of linear and/or irreducible quadratic factors.

135. $P(x) = x^4 - 3x^3 - 12x^2 + 20x + 48$

136. $P(x) = -5x^5 - 18x^4 - 32x^3 - 24x^2 + x + 6$

Section 4.5

Find all zeros (real and complex). Factor the polynomial as a product of linear factors.

137. $P(x) = 2x^3 + x^2 - 2x - 91$

138. $P(x) = -2x^4 + 5x^3 + 37x^2 - 160x + 150$

Section 4.6

In Exercise 139 and 140: (a) graph the function $f(x)$ utilizing a graphing utility to determine if it is a one-to-one function; (b) if it is, find its inverse; and (c) graph both functions in the same viewing window.

139. $f(x) = \dfrac{2x - 3}{x + 1}$

140. $f(x) = \dfrac{4x + 7}{x - 2}$

1. Graph the parabola $y = -(x - 4)^2 + 1$.

2. Write the parabola in standard form $y = -x^2 + 4x - 1$.

3. Find the vertex of the parabola $f(x) = -\frac{1}{2}x^2 + 3x - 4$.

4. Find a quadratic function whose vertex is $(-3, -1)$ and whose graph passes through the point $(-4, 1)$.

5. Find a sixth-degree polynomial function with the given zeros:

   2 of multiplicity 3 1 of multiplicity 2 0 of multiplicity 1

6. For the polynomial function $f(x) = x^4 + 6x^3 - 7x$:

   a. List each real zero and its multiplicity.

   b. Determine whether the graph touches or crosses at each x-intercept.

   c. Find the y-intercept and a few points on the graph.

   d. Determine the end behavior.

   e. Sketch the graph.

7. Divide $-4x^4 + 2x^3 - 7x^2 + 5x - 2$ by $2x^2 - 3x + 1$.

8. Divide $17x^5 - 4x^3 + 2x - 10$ by $x + 2$.

9. Is $x - 3$ a factor of $x^4 + x^3 - 13x^2 - x + 12$?

10. Determine whether -1 is a zero of $P(x) = x^{21} - 2x^{18} + 5x^{12} + 7x^3 + 3x^2 + 2$.

11. Given that $x - 7$ is a factor of $P(x) = x^3 - 6x^2 - 9x + 14$, factor the polynomial in terms of linear factors.

12. Given that $3i$ is a zero of $P(x) = x^4 - 3x^3 + 19x^2 - 27x + 90$, find all other zeros.

13. Can a polynomial have zeros that are not x-intercepts? Explain.

14. Apply Descartes's rule of signs to determine the possible combinations of positive real zeros, negative real zeros, and complex zeros of $P(x) = 3x^5 + 2x^4 - 3x^3 + 2x^2 - x + 1$.

15. From the rational zero test, list all possible rational zeros of $P(x) = 3x^4 - 7x^2 + 3x + 12$.

In Exercises 16–18, determine all zeros of the polynomial function and graph.

16. $P(x) = -x^3 + 4x$

17. $P(x) = 2x^3 - 3x^2 + 8x - 12$

18. $P(x) = x^4 - 6x^3 + 10x^2 - 6x + 9$

19. **Sports.** A football player shows up in August at 300 pounds. After 2 weeks of practice in the hot sun, he is down to 285 pounds. Ten weeks into the season he is up to 315 pounds because of weight training. In the spring he does not work out, and he is back to 300 pounds by the next August. Plot these points on a graph. What degree polynomial could this be?

20. **Profit.** The profit of a company is governed by the polynomial $P(x) = x^3 - 13x^2 + 47x - 35$, where x is the number of units sold in thousands. How many units does the company have to sell to break even?

21. **Interest Rate.** The interest rate for a 30-year fixed mortgage fluctuates with the economy. In 1970, the mortgage interest rate was 8%, and in 1988 it peaked at 13%. In 2002, it dipped down to 4%, and in 2005, it was up to 6%. What is the lowest degree polynomial that can represent this function?

In Exercises 22–25, determine (if any) the:

a. x- and y-intercepts

b. vertical asymptotes

c. horizontal asymptotes

d. slant asymptotes

e. graph

22. $f(x) = \dfrac{2x - 9}{x + 3}$

23. $g(x) = \dfrac{x}{x^2 - 4}$

24. $h(x) = \dfrac{3x^3 - 3}{x^2 - 4}$

25. $F(x) = \dfrac{x - 3}{x^2 - 2x - 8}$

26. **Food.** On eating a sugary snack, the average body almost doubles its glucose level. The percentage increase in glucose level y can be approximated by the rational function $y = \dfrac{25x}{x^2 + 50}$, where x represents the number of minutes after eating the snack. Graph the function.

27. **a.** Use the calculator commands STAT QuadReg to model the data using a quadratic function.

    b. Write the quadratic function in standard form and identify the vertex.

    c. Find the x-intercepts.

    d. Plot this quadratic function with a graphing calculator. Do they agree with the given values?

x	-3	2.2	7.5
y	10.01	-9.75	25.76

28. Find the asymptotes and intercepts of the rational function $f(x) = \dfrac{x(2x - 3)}{x^2 - 3x} + 1$. (*Note:* Combine the two expressions into a single rational expression.) Graph this function utilizing a graphing utility. Does the graph confirm what you found?

1. Simplify $\frac{(5x^{-1}y^{-2})^3}{(-10x^2y^2)^2}$ and express in terms of positive exponents.

2. Factor $2xy - 2x + 3y - 3$.

3. Multiply and simplify $\frac{4x^2 - 36}{x^2} \cdot \frac{x^3 + 6x^2}{x^2 + 9x + 18}$.

4. Solve for x: $|2x - 5| + 3 > 10$.

5. Austin can mow a lawn in 75 minutes. The next week Stacey mows the same lawn in 60 minutes. How long would it take them to mow the lawn working together?

6. Use the discriminant to determine the number and type of roots: $-4x^2 + 3x + 15 = 0$.

7. Solve and check $\sqrt{16 + x^2} = x + 2$.

8. Apply algebraic tests to determine whether the equation's graph is symmetric with respect to the x-axis, y-axis, or origin for $y = |x| - 3$.

9. Write an equation of the line that is parallel to the line $x - 3y = 8$ and has the point $(4, 1)$.

10. Find the x-intercept and y-intercept and sketch the graph for $2y - 6 = 0$.

11. Write the equation of a circle with center $(0, 6)$ and that passes through the point $(1, 5)$.

12. Express the domain of the function $f(x) = \sqrt{6x - 7}$ with interval notation.

13. Determine whether the function $g(x) = \sqrt{x + 10}$ is even, odd, or neither.

14. For the function $y = -(x + 1)^2 + 2$, identify all of the transformations of $y = x^2$.

15. Sketch the graph of $y = \sqrt{x - 1} + 3$ and identify all transformations.

16. Find the composite function $f \circ g$ and state the domain for $f(x) = x^2 - 3$ and $g(x) = \sqrt{x + 2}$.

17. Evaluate $g(f(-1))$ for $f(x) = 7 - 2x^2$ and $g(x) = 2x - 10$.

18. Find the inverse of the function $f(x) = (x - 4)^2 + 2$, where $x \geq 4$.

19. Find a quadratic function that has the vertex $(-2, 3)$ and point $(-1, 4)$.

20. Find all of the real zeros and state the multiplicity of the function $f(x) = -3.7x^4 - 14.8x^3$.

21. Use long division to find the quotient $Q(x)$ and the remainder $r(x)$ of $(-20x^3 - 8x^2 + 7x - 5) \div (-5x + 3)$.

22. Use synthetic division to find the quotient $Q(x)$ and the remainder $r(x)$ of $(2x^3 + 3x^2 - 11x + 6) \div (x - 3)$.

23. List the possible rational zeros, and test to determine all rational zeros for $P(x) = 12x^3 + 29x^2 + 7x - 6$.

24. Given the real zero $x = 5$ of the polynomial $P(x) = 2x^3 - 3x^2 - 32x - 15$, determine all the other zeros and write the polynomial in terms of a product of linear factors.

25. Find all vertical and horizontal asymptotes of $f(x) = \frac{3x - 5}{x^2 - 4}$.

26. Graph the function $f(x) = \frac{2x^3 - x^2 - x}{x^2 - 1}$.

27. Find the asymptotes and intercepts of the rational function $f(x) = \frac{5}{2x - 3} - \frac{1}{x}$. (*Note:* Combine the two expressions into a single rational expression.) Graph this function utilizing a graphing utility. Does the graph confirm what you found?

28. Find the asymptotes and intercepts of the rational function $f(x) = \frac{6x}{3x + 1} - \frac{6x}{4x - 1}$. (*Note:* Combine the two expressions into a single rational expression.) Graph this function utilizing a graphing utility. Does the graph confirm what you found?

5

Exponential and Logarithmic Functions

First human fossil found at Olorgesailie (Kenya) field site; Smithsonian anthropologist makes dramatic discovery.

Working at the Olorgesailie field site in Kenya during the summer of 2003, Rick Potts, director of the Smithsonian's Human Origins Program at the Museum of Natural History, discovered part of a skull. The fossil is about 900,000 years old. It is the first human fossil ever found at the Olorgesailie site and the first well-dated fossil from a 400,000-year gap between 600,000 and 1 million years old in the human fossil record of East Africa.

How do archaeologists and anthropologists date fossils? One method is carbon testing. The ratio of carbon 12 (the more stable kind of carbon) to carbon 14 at the moment of death is the same as the ratio in all living things while they are still alive, but the carbon 14 decays and is not replaced. By looking at the ratio of carbon 12 to carbon 14 in the sample and comparing it with the ratio in a living organism, it is possible to determine the age of a formerly living thing fairly precisely.

The amount of carbon in a fossil is a function of how many years the organism has been dead. The amount is modeled by an *exponential function*, and the inverse of the exponential function, a *logarithmic function*, is used to determine the age of the fossil.

IN THIS CHAPTER, we will discuss exponential functions and their inverses, logarithmic functions. We will graph these functions and use their properties to solve exponential and logarithmic equations. Last, we will discuss particular exponential and logarithmic models that represent phenomena such as compound interest, world populations, conservation biology models, carbon dating, pH values in chemistry, and the bell curve that is fundamental in statistics for describing how quantities vary in the real world.

EXPONENTIAL AND LOGARITHMIC FUNCTIONS

FUNCTIONS

- Exponential Functions
  - Constant base
  - Natural base: *e*
- Logarithmic Functions
  - Common base 10
  - Natural base: *e*
  - Other bases

GRAPHS

- Exponential Functions
  - *y*-intercept (0, 1)
  - Horizontal asymptote: *x*-axis
- Logarithmic Functions
  - *x*-intercept (1, 0)
  - Vertical asymptote: *y*-axis

EQUATIONS

- Solving Exponential and Logarithmic Equations
  - Use properties of one-to-one functions
  - Use inverse properties

MODELS

- Exponential
  - Compound interest
  - Growth (populations)
  - Decay (carbon dating)
  - Gaussian (bell curve)
  - Logistic (conservation biology)
- Logarithmic
  - pH values
  - Richter scale
  - Decibels

SECTION

CHAPTER OBJECTIVES

- Evaluate exponential functions.
- Graph exponential functions.
- Evaluate logarithmic functions.
- Graph logarithmic functions.
- Understand that exponential functions and logarithmic functions are inverses of each other.
- Solve exponential and logarithmic equations.
- Understand domain restrictions on logarithmic functions.
- Check for extraneous solutions when solving logarithmic equations.
- Use exponential growth models and exponential decay models to represent a variety of real-world phenomena.
- Use logarithmic models to represent pH values in chemistry and the Richter scale and decibels in engineering.
- Use logistic growth models to represent conservation biology populations.

SECTION 5.1 EXPONENTIAL FUNCTIONS AND THEIR GRAPHS

SKILLS OBJECTIVES

- Evaluate exponential functions.
- Graph exponential functions.
- Find the domain and range of exponential functions.
- Define the number *e*.
- Solve compound interest problems.

CONCEPTUAL OBJECTIVES

- Understand the difference between algebraic and exponential functions.
- Understand that irrational exponents lead to approximations.

Most of the functions (polynomial, rational, radical, etc.) we have studied thus far have been **algebraic functions**. Algebraic functions involve basic operations, powers, and roots. In this chapter, we discuss *exponential functions* and *logarithmic functions*, which are called **transcendental functions** because they transcend our ability to define them with a finite number of algebraic expressions. The following table illustrates the difference between algebraic functions and *exponential functions*:

FUNCTION	VARIABLE IS IN THE	CONSTANT IS IN THE	EXAMPLE	EXAMPLE
Algebraic	Base	Exponent	$f(x) = x^2$	$g(x) = x^{1/3}$
Exponential	Exponent	Base	$F(x) = 2^x$	$G(x) = \left(\frac{1}{3}\right)^x$

DEFINITION Exponential Function

An **exponential function** with **base *b*** is denoted by

$$f(x) = b^x$$

where b and x are any real numbers such that $b > 0$ and $b \neq 1$.

Note:

- We eliminate $b = 1$ as a value for the base because it merely yields the constant function $f(x) = 1^x = 1$.
- We eliminate negative values for b because they would give non-real-number values such as $(-9)^{1/2} = \sqrt{-9} = 3i$.
- We eliminate $b = 0$ because 0^x corresponds to an undefined value when x is negative.

Sometimes the value of an exponential function for a specific argument can be found by inspection as an *exact* number.

x	-3	-1	0	1	3
$F(x) = 2^x$	$2^{-3} = \frac{1}{2^3} = \frac{1}{8}$	$2^{-1} = \frac{1}{2^1} = \frac{1}{2}$	$2^0 = 1$	$2^1 = 2$	$2^3 = 8$

If an exponential function cannot be evaluated exactly, then we find the decimal *approximation* using a calculator. Most calculators have either a base to a power button $\boxed{x^y}$ or a caret $\boxed{\wedge}$ button for working with exponents.

x	-2.7	$-\frac{4}{5}$	$\frac{5}{7}$	2.7
$F(x) = 2^x$	$2^{-2.7} \approx 0.154$	$2^{-4/5} \approx 0.574$	$2^{5/7} \approx 1.641$	$2^{2.7} \approx 6.498$

The domain of exponential functions, $f(x) = b^x$, is the set of all real numbers. It is important to note that in Chapter 0 we discussed exponents of the form b^x, where x is an integer or a rational number. What happens if x is irrational? We can approximate the irrational number with a decimal approximation such as $b^{\pi} \approx b^{3.14}$ or $b^{\sqrt{2}} \approx b^{1.41}$.

Consider $7^{\sqrt{3}}$, and realize that the irrational number $\sqrt{3}$ is a decimal that never terminates or repeats: $\sqrt{3} = 1.7320508 \ldots$. We can show in advanced mathematics that there is a number $7^{\sqrt{3}}$, and although we cannot write it exactly, we can approximate the number. In fact the closer the exponent is to $\sqrt{3}$, the closer the approximation is to $7^{\sqrt{3}}$.

It is important to note that the properties of exponents (Chapter 0) hold when the exponent is any real number (rational or irrational).

$$7^{1.7} \approx 27.3317$$
$$7^{1.73} \approx 28.9747$$
$$7^{1.732} \approx 29.0877$$
$$\ldots$$
$$7^{\sqrt{3}} \approx 29.0906$$

EXAMPLE 1 Evaluating Exponential Functions

Let $f(x) = 3^x$, $g(x) = \left(\frac{1}{4}\right)^x$, and $h(x) = 10^{x-2}$. Find the following values. If an approximation is required, approximate to four decimal places.

a. $f(2)$ **b.** $f(\pi)$ **c.** $g\left(-\frac{3}{2}\right)$ **d.** $h(2.3)$ **e.** $f(0)$ **f.** $g(0)$

Solution:

a. $f(2) = 3^2 = \boxed{9}$

b. $f(\pi) = 3^{\pi} \approx \boxed{31.5443}$ *

c. $g\left(-\frac{3}{2}\right) = \left(\frac{1}{4}\right)^{-3/2} = 4^{3/2} = \left(\sqrt{4}\right)^3 = 2^3 = \boxed{8}$

d. $h(2.3) = 10^{2.3-2} = 10^{0.3} \approx \boxed{1.9953}$

e. $f(0) = 3^0 = \boxed{1}$

f. $g(0) = \left(\frac{1}{4}\right)^0 = \boxed{1}$

Notice that parts (a) and (c) were evaluated exactly, whereas parts (b) and (d) required approximation using a calculator.

YOUR TURN Let $f(x) = 2^x$ and $g(x) = \left(\frac{1}{9}\right)^x$ and $h(x) = 5^{x-2}$. Find the following values. Evaluate exactly when possible, and round to four decimal places when a calculator is needed.

a. $f(4)$ **b.** $f(\pi)$ **c.** $g\left(-\frac{3}{2}\right)$ **d.** $h(2.9)$

Answer: **a.** 16 **b.** 8.8250 **c.** 27 **d.** 4.2567

*In part (b), the π button on the calculator is selected. If we instead approximate π by 3.14, we get a slightly different approximation for the function value:

$$f(\pi) = 3^{\pi} \approx 3^{3.14} \approx 31.4891$$

Graphs of Exponential Functions

Let's graph two exponential functions, $y = 2^x$ and $y = 2^{-x} = \left(\frac{1}{2}\right)^x$, by plotting points.

x	$y = 2^x$	(x, y)
-2	$2^{-2} = \frac{1}{2^2} = \frac{1}{4}$	$\left(-2, \frac{1}{4}\right)$
-1	$2^{-1} = \frac{1}{2^1} = \frac{1}{2}$	$\left(-1, \frac{1}{2}\right)$
0	$2^0 = 1$	(0, 1)
1	$2^1 = 2$	(1, 2)
2	$2^2 = 4$	(2, 4)
3	$2^3 = 8$	(3, 8)

x	$y = 2^{-x}$	(x, y)
-3	$2^{-(-3)} = 2^3 = 8$	$(-3, 8)$
-2	$2^{-(-2)} = 2^2 = 4$	$(-2, 4)$
-1	$2^{-(-1)} = 2^1 = 2$	$(-1, 2)$
0	$2^0 = 1$	(0, 1)
1	$2^{-1} = \frac{1}{2^1} = \frac{1}{2}$	$\left(1, \frac{1}{2}\right)$
2	$2^{-2} = \frac{1}{2^2} = \frac{1}{4}$	$\left(2, \frac{1}{4}\right)$

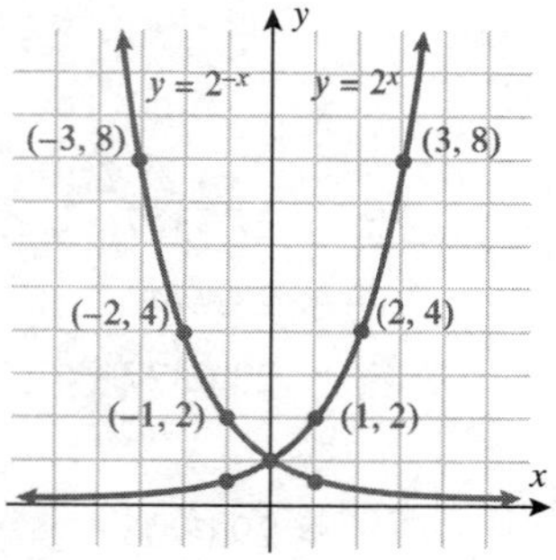

Notice that both graphs' y-intercept is (0, 1) (as shown to the left) and neither graph has an x-intercept. The x-axis is a horizontal asymptote for both graphs. The following box summarizes general characteristics of exponential functions.

CHARACTERISTICS OF GRAPHS OF EXPONENTIAL FUNCTIONS

$$f(x) = b^x, \qquad b > 0, \qquad b \neq 1$$

- Domain: $(-\infty, \infty)$
- Range: $(0, \infty)$
- x-intercepts: none
- y-intercept: (0, 1)
- Horizontal asymptote: x-axis
- The graph passes through $(1, b)$ and $\left(-1, \frac{1}{b}\right)$.
- As x increases, $f(x)$ increases if $b > 1$ and decreases if $0 < b < 1$.
- The function f is one-to-one.

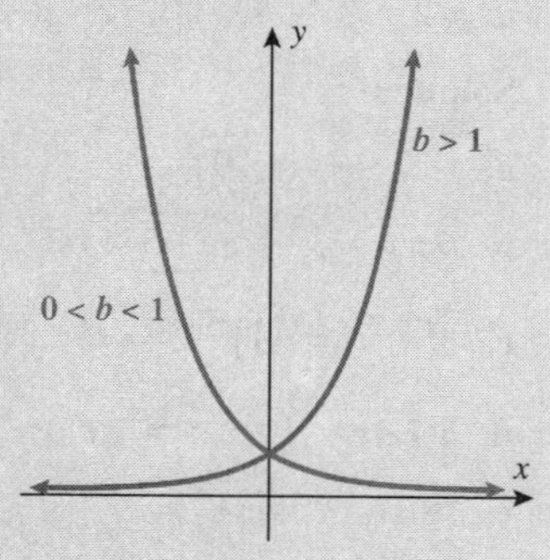

Since exponential functions, $f(x) = b^x$, all go through the point (0, 1) and have the x-axis as a horizontal asymptote, we can find the graph by finding two additional points as outlined in the following procedure.

PROCEDURE FOR GRAPHING $f(x) = b^x$

Step 1: Label the point (0, 1) corresponding to the y-intercept $f(0)$.

Step 2: Find and label two additional points corresponding to $f(-1)$ and $f(1)$.

Step 3: Connect the three points with a *smooth* curve with the x-axis as the horizontal asymptote.

EXAMPLE 2 Graphing Exponential Functions for $b > 1$

Graph the function $f(x) = 5^x$.

Solution:

STEP 1: Label the y-intercept (0, 1). $f(0) = 5^0 = 1$

STEP 2: Label the point (1, 5). $f(1) = 5^1 = 5$

Label the point $(-1, 0.2)$. $f(-1) = 5^{-1} = \frac{1}{5} = 0.2$

STEP 3: Sketch a smooth curve through the three points with the x-axis as a horizontal asymptote.

Domain: $(-\infty, \infty)$

Range: $(0, \infty)$

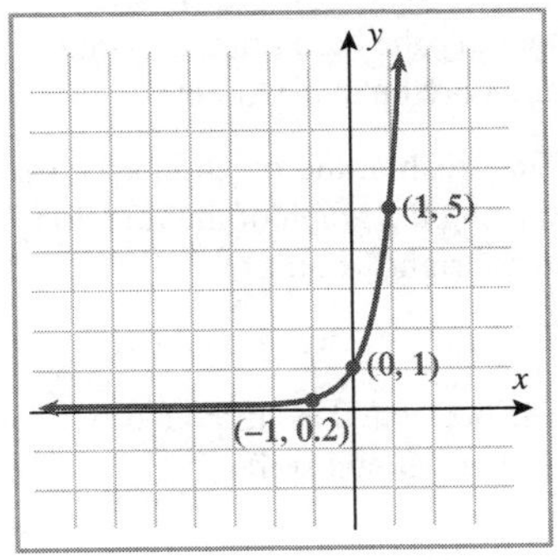

Technology Tip

The graph of $f(x) = 5^x$ is shown.

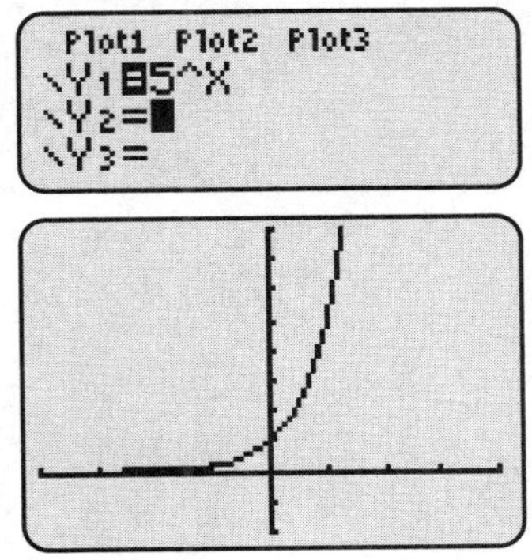

■ **YOUR TURN** Graph the function $f(x) = 5^{-x}$.

■ **Answer:**

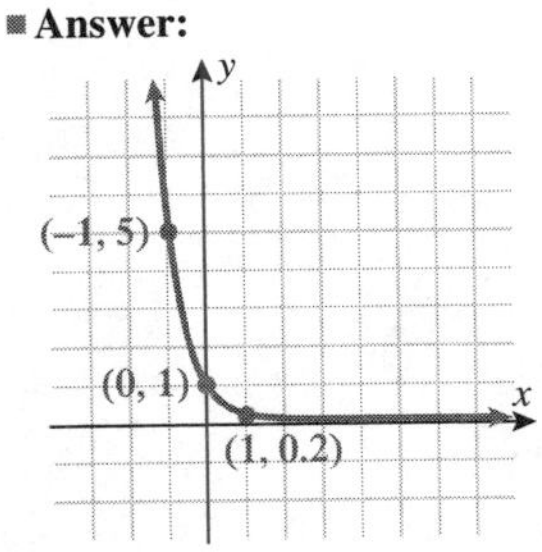

EXAMPLE 3 Graphing Exponential Functions for $b < 1$

Graph the function $f(x) = \left(\frac{2}{5}\right)^x$.

Solution:

STEP 1: Label the y-intercept (0, 1). $f(0) = \left(\frac{2}{5}\right)^0 = 1$

STEP 2: Label the point $(-1, 2.5)$. $f(-1) = \left(\frac{2}{5}\right)^{-1} = \frac{5}{2} = 2.5$

Label the point (1, 0.4). $f(1) = \left(\frac{2}{5}\right)^1 = \frac{2}{5} = 0.4$

STEP 3: Sketch a smooth curve through the three points with the x-axis as a horizontal asymptote.

Domain: $(-\infty, \infty)$

Range: $(0, \infty)$

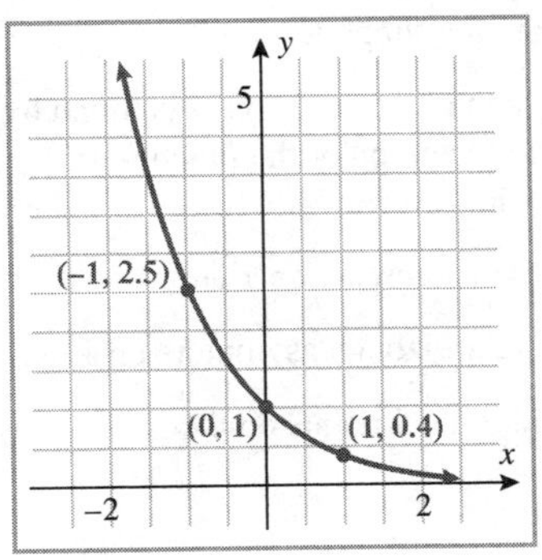

Exponential functions, like all functions, can be graphed by point-plotting. We can also use transformations (horizontal and vertical shifting and reflection; Section 3.3) to graph exponential functions.

EXAMPLE 4 Graphing Exponential Functions Using a Horizontal or Vertical Shift

a. Graph the function $F(x) = 2^{x-1}$. State the domain and range of F.
b. Graph the function $G(x) = 2^x + 1$. State the domain and range of G.

Solution (a):

Identify the base function. $f(x) = 2^x$

Identify the base function y-intercept and horizontal asymptote. $(0, 1)$ and $y = 0$

The graph of the function F is found by shifting the graph of the function f to the right 1 unit. $F(x) = f(x - 1)$

Shift the y-intercept to the right 1 unit. $(0, 1)$ shifts to $(1, 1)$

The horizontal asymptote is not altered by a horizontal shift. $y = 0$

Find additional points on the graph. $F(0) = 2^{0-1} = 2^{-1} = \frac{1}{2}$

$F(2) = 2^{2-1} = 2^1 = 2$

Sketch the graph of $F(x) = 2^{x-1}$ with a *smooth* curve.

Domain: $(-\infty, \infty)$
Range: $(0, \infty)$

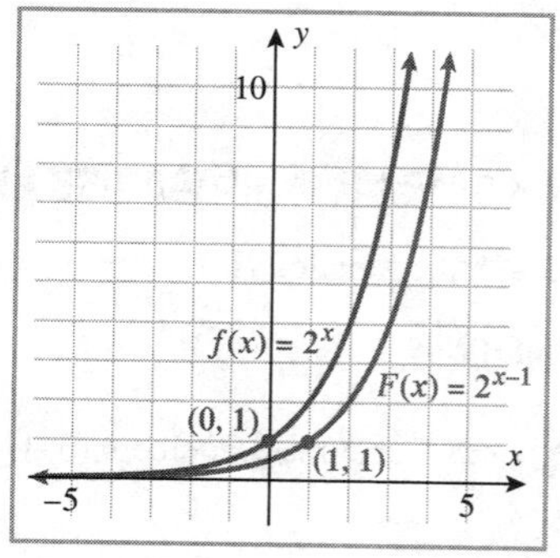

Solution (b):

Identify the base function. $f(x) = 2^x$

Identify the base function y-intercept and horizontal asymptote. $(0, 1)$ and $y = 0$

The graph of the function G is found by shifting the graph of the function f up 1 unit. $G(x) = f(x) + 1$

Shift the y-intercept up 1 unit. $(0, 1)$ shifts to $(0, 2)$

Shift the horizontal asymptote up 1 unit. $y = 0$ shifts to $y = 1$

Find additional points on the graph. $G(1) = 2^1 + 1 = 2 + 1 = 3$

$G(-1) = 2^{-1} + 1 = \frac{1}{2} + 1 = \frac{3}{2}$

Sketch the graph of $G(x) = 2^x + 1$ with a *smooth* curve.

Domain: $(-\infty, \infty)$
Range: $(1, \infty)$

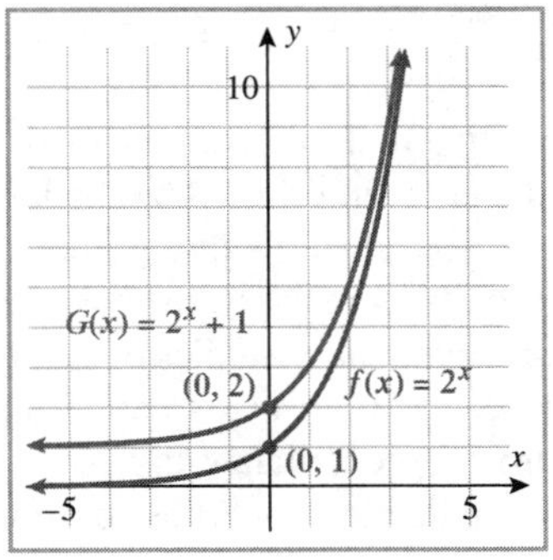

Technology Tip

The graph of $f(x) = 3^{x+1} - 2$ is shown.

```
WINDOW
 Xmin=-5
 Xmax=5
 Xscl=1
 Ymin=-3
 Ymax=8
 Yscl=1
 Xres=1
```

```
Plot1 Plot2 Plot3
\Y1=3^(X)
\Y2=3^(X+1)-2
\Y3=-2
```

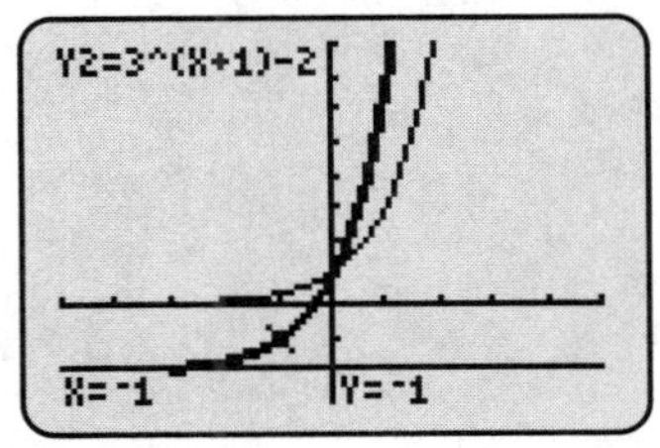

EXAMPLE 5 Graphing Exponential Functions Using Both Horizontal and Vertical Shifts

Graph the function $F(x) = 3^{x+1} - 2$. State the domain and range of F.

Solution:

Identify the base function. $f(x) = 3^x$

Identify the base function y-intercept and horizontal asymptote. $(0, 1)$ and $y = 0$

The graph of the function F is found by shifting the graph of the function f to the left 1 unit and down 2 units. $F(x) = f(x + 1) - 2$

Shift the y-intercept to the left 1 unit and down 2 units. $(0, 1)$ shifts to $(-1, -1)$

Shift the horizontal asymptote down 2 units. $y = 0$ shifts to $y = -2$

Find additional points on the graph. $F(0) = 3^{0+1} - 2 = 3 - 2 = 1$

$F(1) = 3^{1+1} - 2 = 9 - 2 = 7$

Sketch the graph of $F(x) = 3^{x+1} - 2$ with a *smooth* curve.

Domain: $(-\infty, \infty)$
Range: $(-2, \infty)$

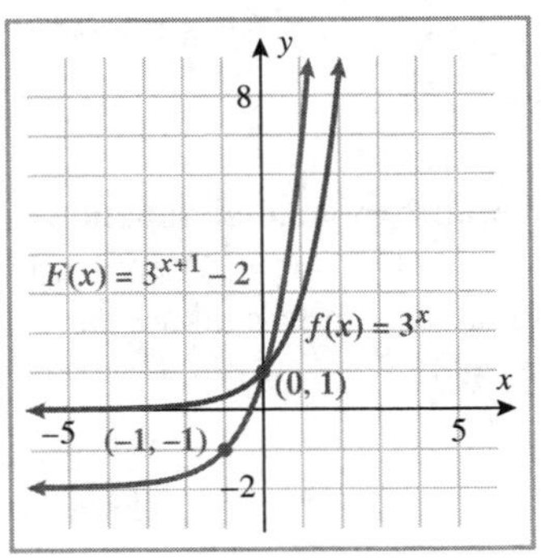

■ **YOUR TURN** Graph $f(x) = 2^{x+3} - 1$. State the domain and range of f.

■ **Answer:** Domain: $(-\infty, \infty)$
Range: $(-1, \infty)$

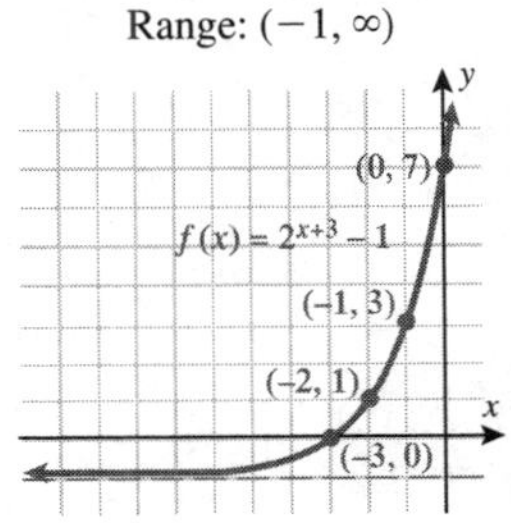

The Natural Base *e*

Technology Tip

Plot the graph of $\left(1 + \frac{1}{x}\right)^x$ as Y_1 as shown.

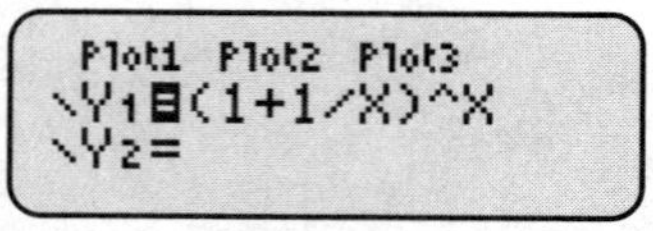

Use a graphing utility to set up the table for $\left(1 + \frac{1}{m}\right)^m$. Enter x for m.

Any positive real number can serve as the base for an exponential function. A particular irrational number, denoted by the letter e, appears as the base in many applications, as you will soon see when we discuss continuous compounded interest. Although you will see 2 and 10 as common bases, the base that appears most often is e, because $\boldsymbol{e}$, as you will come to see in your further studies of mathematics, is the **natural base**. The exponential function with base e, $f(x) = e^x$, is called the **exponential function** or the **natural exponential function**. Mathematicians did not pull this irrational number out of a hat. The number e has many remarkable properties, but most simply, it comes from evaluating the expression $\left(1 + \frac{1}{m}\right)^m$ as m increases without bound.

m	$\left(1 + \frac{1}{m}\right)^m$
1	2
10	2.59374
100	2.70481
1000	2.71692
10,000	2.71815
100,000	2.71827
1,000,000	**2.71828**

$e \approx 2.71828$

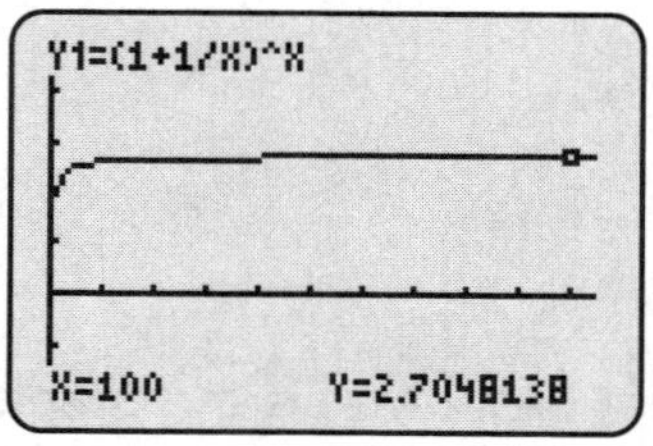

Calculators have an e^x button for approximating the natural exponential function.

EXAMPLE 6 Evaluating the Natural Exponential Function

Evaluate $f(x) = e^x$ for the given x-values. Round your answers to four decimal places.

a. $x = 1$ **b.** $x = -1$ **c.** $x = 1.2$ **d.** $x = -0.47$

Solution:

a. $f(1) = e^1 \approx 2.718281828 \approx \boxed{2.7183}$

b. $f(-1) = e^{-1} \approx 0.367879441 \approx \boxed{0.3679}$

c. $f(1.2) = e^{1.2} \approx 3.320116923 \approx \boxed{3.3201}$

d. $f(-0.47) = e^{-0.47} \approx 0.625002268 \approx \boxed{0.6250}$

Like all exponential functions of the form $f(x) = b^x$, $f(x) = e^x$ and $f(x) = e^{-x}$ have $(0, 1)$ as their y-intercept and the x-axis as a horizontal asymptote as shown in the figure on the right.

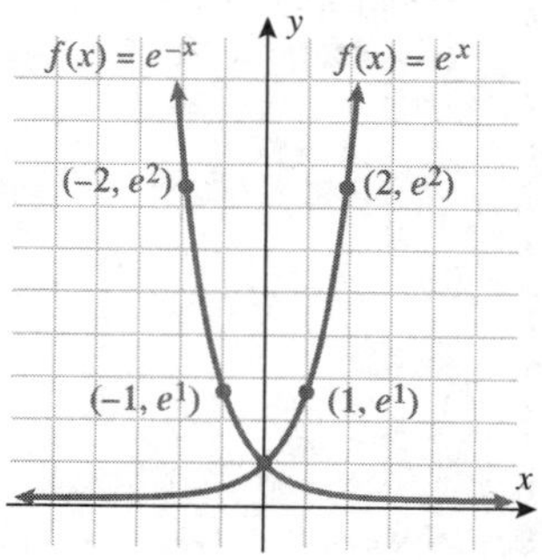

EXAMPLE 7 Graphing Exponential Functions with Base *e*

Graph the function $f(x) = 3 + e^{2x}$.

Solution:

x	$f(x) = 3 + e^{2x}$	(x, y)
-2	3.02	$(-2, 3.02)$
-1	3.14	$(1, 3.14)$
0	4	$(0, 4)$
1	10.39	$(1, 10.39)$
2	57.60	$(2, 57.60)$

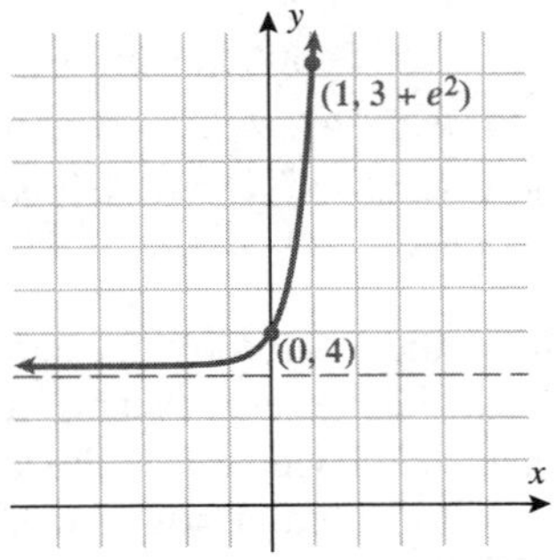

Note: The y-intercept is $(0, 4)$ and the line $y = 3$ is the horizontal asymptote.

■ **YOUR TURN** Graph the function $f(x) = e^{x+1} - 2$.

Technology Tip

The graph of $y_1 = 3 + e^{2x}$ is shown. To setup the split screen, press MODE G-T ENTER GRAPH.

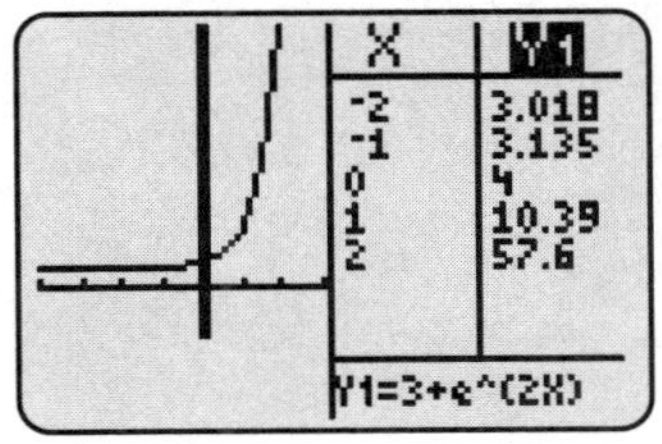

■ **Answer:**

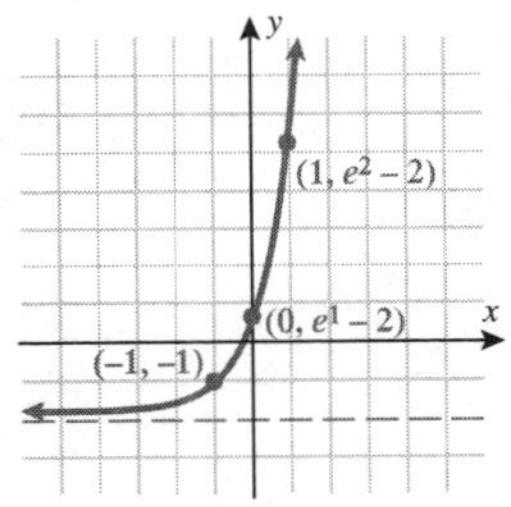

Applications of Exponential Functions

Exponential functions describe either *growth* or *decay.* Populations and investments are often modeled with exponential growth functions, while the declining value of a used car and the radioactive decay of isotopes are often modeled with exponential decay functions. In Section 5.5, various exponential models will be discussed. In this section we discuss doubling time, half-life, and compound interest.

A successful investment program, growing at about 7.2% per year, will double in size every 10 years. Let's assume that you will retire at the age of 65. There is a saying: *It's not the first time your money doubles, it's the last time that makes such a difference.* As you may already know or as you will soon find, it is important to start investing early.

Suppose Maria invests \$5,000 at age 25 and David invests \$5,000 at age 35. Let's calculate how much will accrue from the initial \$5,000 investment by the time they each retire, assuming their money doubles every 10 years.

AGE	MARIA	DAVID
25	\$5,000	
35	\$10,000	\$5,000
45	\$20,000	\$10,000
55	\$40,000	\$20,000
65	**\$80,000**	**\$40,000**

They each made a one-time investment of \$5,000. By investing 10 years sooner, Maria made twice what David made.

A measure of growth rate is the *doubling time*, the time it takes for something to double. Often doubling time is used to describe populations.

DOUBLING TIME GROWTH MODEL

The doubling time growth model is given by

$$P = P_0 2^{t/d}$$

where P = Population at time t
P_0 = Population at time $t = 0$
d = Doubling time

Note that when $t = d$, $P = 2P_0$ (population is equal to twice the original).

The units for P and P_0 are the same and can be any quantity (people, dollars, etc.). The units for t and d must be the same (years, weeks, days, hours, seconds, etc.).

In the investment scenario with Maria and David, $P_0 = \$5{,}000$ and $d = 10$ years, so the model used to predict how much money the original \$5,000 investment yielded is $P = 5000(2)^{t/10}$. Maria retired 40 years after the original investment, $t = 40$, and David retired 30 years after the original investment, $t = 30$.

$$\text{Maria: } P = 5000(2)^{40/10} = 5000(2)^4 = 5000(16) = 80{,}000$$

$$\text{David: } P = 5000(2)^{30/10} = 5000(2)^3 = 5000(8) = 40{,}000$$

EXAMPLE 8 Doubling Time of Populations

In 2004, the population in Kazakhstan, a country in Asia, reached 15 million. It is estimated that the population will double in 30 years. If the population continues to grow at the same rate, what will the population be 20 years from now? Round to the nearest million.

Solution:

Write the doubling model. $\qquad P = P_0 2^{t/d}$

Substitute $P_0 = 15$ million, $d = 30$ years, and $t = 20$ years. $\qquad P = 15(2)^{20/30}$

Simplify. $\qquad P = 15(2)^{2/3} \approx 23.8110$

In 20 years, there will be approximately **24 million people** in Kazakhstan.

■ **YOUR TURN** What will the approximate population in Kazakhstan be in 40 years? Round to the nearest million.

■ **Answer:** 38 million

We now turn our attention from exponential growth to exponential decay, or negative growth. Suppose you buy a brand-new car from a dealership for \$24,000. The value of a car decreases over time according to an exponential decay function. The **half-life** of this particular car, or the time it takes for the car to depreciate 50%, is approximately 3 years. The exponential decay is described by

$$A = A_0\left(\frac{1}{2}\right)^{t/h}$$

where A_0 is the amount the car is worth (in dollars) when new (that is, when $t = 0$), A is the amount the car is worth (in dollars) after t years, and h is the half-life in years. In our car scenario, $A_0 = 24{,}000$ and $h = 3$:

$$A = 24{,}000\left(\frac{1}{2}\right)^{t/3}$$

How much is the car worth after three years? Six years? Nine years? Twenty-four years?

$$t = 3: \quad A = 24{,}000\left(\frac{1}{2}\right)^{3/3} = 24{,}000\left(\frac{1}{2}\right) = 12{,}000$$

$$t = 6: \quad A = 24{,}000\left(\frac{1}{2}\right)^{6/3} = 24{,}000\left(\frac{1}{2}\right)^{2} = 6000$$

$$t = 9: \quad A = 24{,}000\left(\frac{1}{2}\right)^{9/3} = 24{,}000\left(\frac{1}{2}\right)^{3} = 3000$$

$$t = 24: \quad A = 24{,}000\left(\frac{1}{2}\right)^{24/3} = 24{,}000\left(\frac{1}{2}\right)^{8} = 93.75 \approx 100$$

The car that was worth \$24,000 new is worth \$12,000 in 3 years, \$6,000 in 6 years, \$3,000 in 9 years, and about \$100 in the junkyard in 24 years.

EXAMPLE 9 Radioactive Decay

The radioactive isotope of potassium ^{42}K which is used in the diagnosis of brain tumors, has a half-life of 12.36 hours. If 500 milligrams of potassium 42 are taken, how many milligrams will remain after 24 hours? Round to the nearest milligram.

Solution:

Write the half-life formula. $A = A_0\left(\frac{1}{2}\right)^{t/h}$

Substitute $A_0 = 500$ mg, $h = 12.36$ hours, $t = 24$ hours. $A = 500\left(\frac{1}{2}\right)^{24/12.36}$

Simplify. $A \approx 500(0.2603) \approx 130.15$

After 24 hours, there are approximately **130 milligrams** of potassium 42 left.

YOUR TURN How many milligrams of potassium are expected to be left in the body after 1 week?

Answer: 0.04 milligram (less than 1 milligram)

In Section 1.2, *simple interest* was discussed where the interest I is calculated based on the principal P, the annual interest rate r, and the time t in years, using the formula: $I = Prt$.

If the interest earned in a period is then reinvested at the same rate, future interest is earned on both the principal and the reinvested interest during the next period. Interest paid on both the principal and interest is called *compound interest.*

COMPOUND INTEREST

If a **principal** P is invested at an annual **rate** r **compounded** n times a year, then the **amount** A in the account at the end of t years is given by

$$A = P\left(1 + \frac{r}{n}\right)^{nt}$$

The annual interest rate r is expressed as a decimal.

The following list shows the typical number of times interest is compounded.

Annually	$n = 1$	Monthly	$n = 12$
Semiannually	$n = 2$	Weekly	$n = 52$
Quarterly	$n = 4$	Daily	$n = 365$

Technology Tip

Enter $3000\left(1 + \frac{0.03}{4}\right)^{(4)(7)}$ into the graphing calculator.

```
3000(1+0.03/4)^(
4*7)
          3698.135243
```

EXAMPLE 10 Compound Interest

If \$3,000 is deposited in an account paying 3% compounded quarterly, how much will you have in the account in 7 years?

Solution:

Write the compound interest formula. $A = P\left(1 + \frac{r}{n}\right)^{nt}$

Substitute $P = 3000$, $r = 0.03$, $n = 4$, and $t = 7$. $A = 3000\left(1 + \frac{0.03}{4}\right)^{(4)(7)}$

Simplify. $A = 3000(1.0075)^{28} \approx 3698.14$

You will have **\$3,698.14** in the account.

YOUR TURN If \$5,000 is deposited in an account paying 6% compounded annually, how much will you have in the account in 4 years?

Answer: \$6,312.38

Notice in the compound interest formula that as n increases the amount A also increases. In other words, the more times the interest is compounded per year, the more money you make. Ideally, your bank will compound your interest infinitely many times. This is called *compounding continuously*. We will now show the development of the compounding continuously formula, $A = Pe^{rt}$.

WORDS	MATH
Write the compound interest formula.	$A = P\left(1 + \frac{r}{n}\right)^{nt}$
Note that $\frac{r}{n} = \frac{1}{n/r}$ and $nt = \left(\frac{n}{r}\right)rt$.	$A = P\left(1 + \frac{1}{n/r}\right)^{(n/r)rt}$
Let $m = \frac{n}{r}$.	$A = P\left(1 + \frac{1}{m}\right)^{mrt}$
Use the exponential property: $x^{mrt} = (x^m)^{rt}$.	$A = P\left[\left(1 + \frac{1}{m}\right)^m\right]^{rt}$

Earlier in this section, we showed that as m increases, $\left(1 + \frac{1}{m}\right)^m$ approaches e. Therefore, as the number of times the interest is compounded approaches infinity, or as $n \to \infty$, the amount in an account $A = P\left(1 + \frac{r}{n}\right)^{nt}$ approaches $A = Pe^{rt}$.

CONTINUOUS COMPOUND INTEREST

If a **principal** P is invested at an annual **rate** r **compounded continuously**, then the **amount** A in the account at the end of t years is given by

$$A = Pe^{rt}$$

The annual interest rate r is expressed as a decimal.

It is important to note that for a given interest rate, the highest return you can earn is by compounding continuously.

EXAMPLE 11 Continuously Compounded Interest

If \$3,000 is deposited in a savings account paying 3% a year compounded continuously, how much will you have in the account in 7 years?

Solution:

Write the continuous compound interest formula. $A = Pe^{rt}$

Substitute $P = 3000$, $r = 0.03$, and $t = 7$. $A = 3000e^{(0.03)(7)}$

Simplify. $A \approx 3701.034$

There will be **\$3,701.03** in the account in 7 years.

Note: In Example 10, we worked this same problem compounding *quarterly,* and the result was \$3,698.14.

If the number of times per year interest is compounded increases, then the total interest earned that year also increases.

■ **YOUR TURN** If \$5,000 is deposited in an account paying 6% compounded continuously, how much will be in the account in 4 years?

Technology Tip

Enter $3000e^{(0.03)(7)}$ into the graphing calculator.

```
3000e^(0.03*7)
        3701.03418
```

Study Tip

If the number of times per year interest is compounded increases, then the total interest earned that year also increases.

■ **Answer:** \$ 6,356.25

SECTION 5.1 SUMMARY

In this section we discussed exponential functions (constant base, variable exponent).

General Exponential Functions: $f(x) = b^x$, $b \neq 1$, and $b > 0$

1. Evaluating exponential functions
   - Exact (by inspection): $f(x) = 2^x$ $f(3) = 2^3 = 8$.
   - Approximate (with the aid of a calculator): $f(x) = 2^x$ $f(\sqrt{3}) = 2^{\sqrt{3}} \approx 3.322$
2. Graphs of exponential functions
   - Domain: $(-\infty, \infty)$ and range: $(0, \infty)$.
   - The point $(0, 1)$ corresponds to the y-intercept.
   - The graph passes through the points $(1, b)$ and $\left(-1, \frac{1}{b}\right)$.
   - The x-axis is a horizontal asymptote.
   - The function f is one-to-one.

Procedure for Graphing: $f(x) = b^x$

Step 1: Label the point $(0, 1)$ corresponding to the y-intercept $f(0)$.

Step 2: Find and label two additional points corresponding to $f(-1)$ and $f(1)$.

Step 3: Connect the three points with a smooth curve with the x-axis as the horizontal asymptote.

The Natural Exponential Function: $f(x) = e^x$

- The irrational number e is called the natural base.
- $e = \left(1 + \frac{1}{m}\right)^m$ as $m \to \infty$
- $e \approx 2.71828$

Applications of Exponential Functions (all variables expressed in consistent units)

1. Doubling time: $P = P_0 2^{t/d}$
   - d is doubling time.
   - P is population at time t.
   - P_0 is population at time $t = 0$.
2. Half-life: $A = A_0\left(\frac{1}{2}\right)^{t/h}$
   - h is the half-life.
   - A is amount at time t.
   - A_0 is amount at time $t = 0$.
3. Compound interest (P = principal, A = amount after t years, r = interest rate)
   - Compounded n times a year: $A = P\left(1 + \frac{r}{n}\right)^{nt}$
   - Compounded continuously: $A = Pe^{rt}$

SECTION 5.1 EXERCISES

SKILLS

In Exercises 1–10, evaluate *exactly* (without using a calculator). For rational exponents, consider converting to radical form first.

1. 2^4 **2.** 3^4 **3.** 5^{-2} **4.** 4^{-3} **5.** $8^{2/3}$ **6.** $27^{2/3}$ **7.** $\left(\frac{1}{9}\right)^{-3/2}$ **8.** $\left(\frac{1}{16}\right)^{-3/2}$ **9.** -5^0 **10.** -6^0

In Exercises 11–18, approximate with a calculator. Round your answer to four decimal places.

11. $4^{1.2}$ **12.** $4^{-1.2}$ **13.** $5^{\sqrt{2}}$ **14.** $6^{\sqrt{3}}$

15. e^2 **16.** $e^{1/2}$ **17.** $e^{-\pi}$ **18.** $e^{-\sqrt{2}}$

In Exercises 19–26, for the functions $f(x) = 3^x$, $g(x) = \left(\frac{1}{16}\right)^x$, and $h(x) = 10^{x+1}$, find the function value at the indicated points.

19. $f(3)$ **20.** $h(1)$ **21.** $g(-1)$ **22.** $f(-2)$

23. $g\left(-\frac{1}{2}\right)$ **24.** $g\left(-\frac{3}{2}\right)$ **25.** $f(e)$ **26.** $g(\pi)$

In Exercises 27–32, match the graph with the function.

27. $y = 5^{x-1}$ **28.** $y = 5^{1-x}$ **29.** $y = -5^x$

30. $y = -5^{-x}$ **31.** $y = 1 - 5^{-x}$ **32.** $y = 5^x - 1$

a.

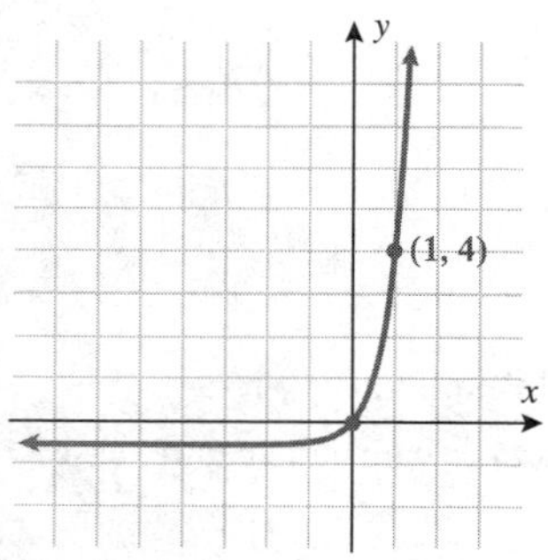

b.

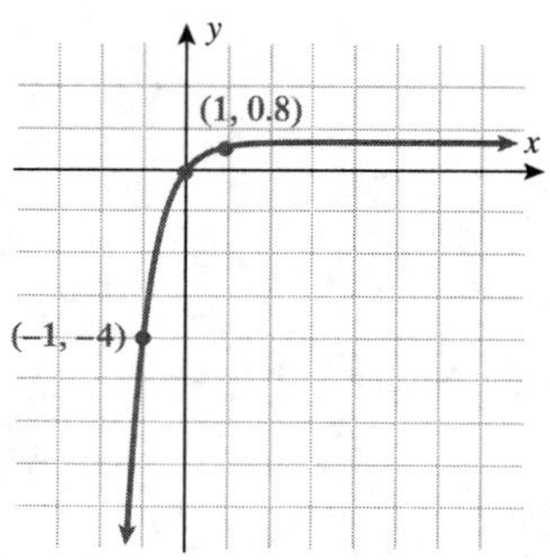

c.

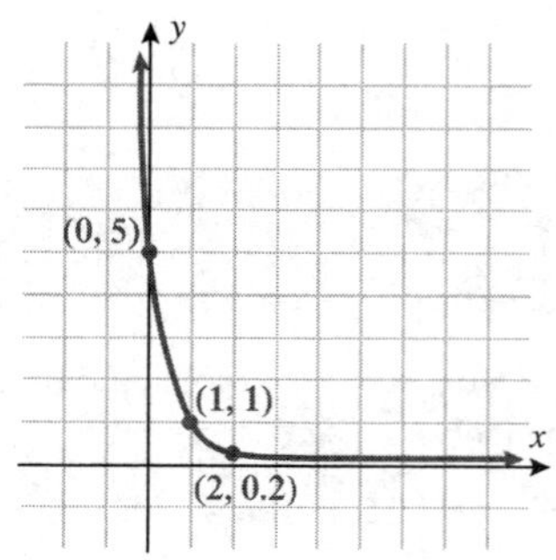

d.

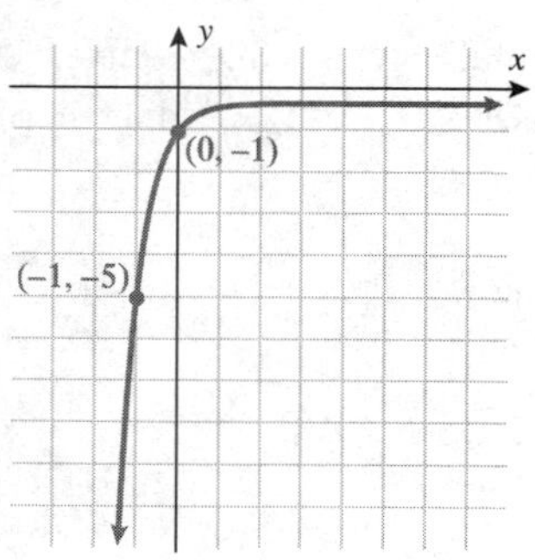

e.

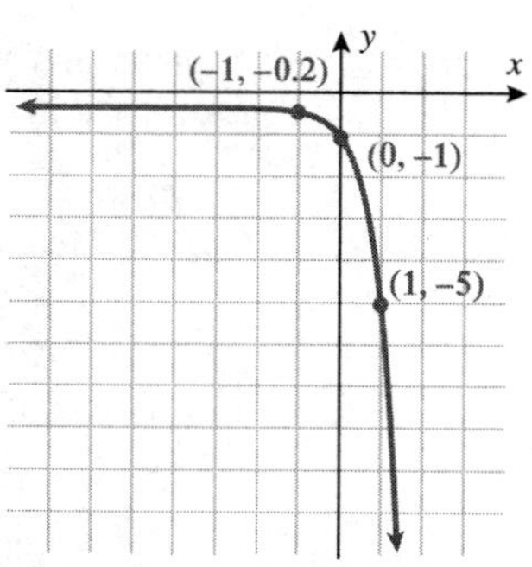

f.

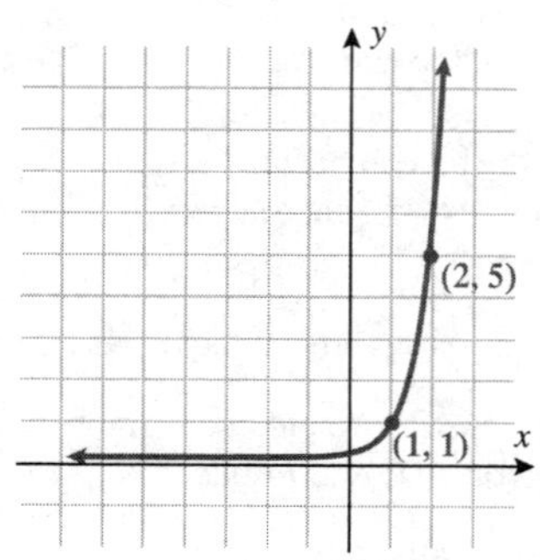

In Exercises 33–48, graph the exponential function using transformations. State the *y*-intercept, two additional points, the domain, the range, and the horizontal asymptote.

33. $f(x) = 6^x$ **34.** $f(x) = 7^x$ **35.** $f(x) = 10^{-x}$ **36.** $f(x) = 4^{-x}$ **37.** $f(x) = e^{-x}$ **38.** $f(x) = -e^x$

39. $f(x) = 2^x - 1$ **40.** $f(x) = 3^x - 1$ **41.** $f(x) = 2 - e^x$ **42.** $f(x) = 1 + e^{-x}$ **43.** $f(x) = e^{x+1} - 4$ **44.** $f(x) = e^{x-1} + 2$

45. $f(x) = 3e^{x/2}$ **46.** $f(x) = 2e^{-x}$ **47.** $f(x) = 1 + \left(\frac{1}{2}\right)^{x-2}$ **48.** $f(x) = 2 - \left(\frac{1}{3}\right)^{x+1}$

APPLICATIONS

49. **Population Doubling Time.** In 2002, there were 7.1 million people living in London, England. If the population is expected to double by 2090, what is the expected population in London in 2050?

50. **Population Doubling Time.** In 2004, the population in Morganton, Georgia, was 43,000. The population in Morganton is expected to double by 2010. If the growth rate remains the same, what is the expected population in Morganton in 2020?

51. **Investments.** Suppose an investor buys land in a rural area for \$1,500 an acre and sells some of it 5 years later at \$3,000 an acre and the rest of it 10 years later at \$6,000. Write a function that models the value of land in that area, assuming the growth rate stays the same. What would the expected cost per acre be 30 years after the initial investment of \$1,500?

52. **Salaries.** Twin brothers, Collin and Cameron, get jobs immediately after graduating from college at the age of 22. Collin opts for the higher starting salary, \$55,000, and stays with the same company until he retires at 65. His salary doubles every 15 years. Cameron opts for a lower starting salary, \$35,000, but moves to a new job every 5 years; he doubles his salary every 10 years until he retires at 65. What is the annual salary of each brother upon retirement?

53. **Radioactive Decay.** A radioactive isotope of selenium ^{75}Se used in the creation of medical images of the pancrea, has a half-life of 119.77 days. If 200 milligrams are given to a patient, how many milligrams are left after 30 days?

54. **Radioactive Decay.** The radioactive isotope indium-111 (^{111}In), used as a diagnostic tool for locating tumors associated with prostate cancer, has a half-life of 2.807 days. If 300 milligrams are given to a patient, how many milligrams will be left after a week?

55. **Depreciation of Furniture.** A couple buy a new bedroom set for \$8,000 and 10 years later sell it for \$4,000. If the depreciation continues at the same rate, how much would the bedroom set be worth in 4 more years?

56. **Depreciation of a Computer.** A student buys a new laptop for \$1,500 when she arrives as a freshman. A year later, the computer is worth approximately \$750. If the depreciation continues at the same rate, how much would she expect to sell her laptop for when she graduates 4 years after she bought it?

57. **Compound Interest.** If you put \$3,200 in a savings account that earns 2.5% interest per year compounded quarterly, how much would you expect to have in that account in 3 years?

58. **Compound Interest.** If you put \$10,000 in a savings account that earns 3.5% interest per year compounded annually, how much would you expect to have in that account in 5 years?

59. **Compound Interest.** How much money should you put in a savings account now that earns 5% a year compounded daily if you want to have \$32,000 in 18 years?

60. **Compound Interest.** How much money should you put in a savings account now that earns 3.0% a year compounded weekly if you want to have \$80,000 in 15 years?

61. **Compound Interest.** If you put \$3,200 in a savings account that pays 2% a year compounded continuously, how much will you have in the account in 15 years?

62. **Compound Interest.** If you put \$7,000 in a money market account that pays 4.3% a year compounded continuously, how much will you have in the account in 10 years?

63. **Compound Interest.** How much money should you deposit into a money market account that pays 5% a year compounded continuously to have \$38,000 in the account in 20 years?

64. **Compound Interest.** How much money should you deposit into a certificate of deposit that pays 6% a year compounded continuously to have \$80,000 in the account in 18 years?

CATCH THE MISTAKE

In Exercises 65–68, explain the mistake that is made.

65. Evaluate the expression $4^{-1/2}$.

Solution: $4^{-1/2} = 4^2 = 16$

The correct value is $\frac{1}{2}$. What mistake was made?

66. Evaluate the function for the given x: $f(x) = 4^x$ for $x = \frac{3}{2}$.

Solution:

$$f\left(\frac{3}{2}\right) = 4^{3/2}$$

$$= \frac{4^3}{4^2} = \frac{64}{16} = 4$$

The correct value is 8. What mistake was made?

67. If \$2,000 is invested in a savings account that earns 2.5% interest compounding continuously, how much will be in the account in one year?

Solution:

Write the compound continuous interest formula. $A = Pe^{rt}$

Substitute $P = 2000$, $r = 2.5$, and $t = 1$. $A = 2000e^{(2.5)(1)}$

Simplify. $A = 24{,}364.99$

This is incorrect. What mistake was made?

68. If \$5,000 is invested in a savings account that earns 3% interest compounding continuously, how much will be in the account in 6 months?

Solution:

Write the compound continuous interest formula. $A = Pe^{rt}$

Substitute $P = 5000$, $r = 0.03$, and $t = 6$. $A = 5000e^{(0.03)(6)}$

Simplify. $A = 5986.09$

This is incorrect. What mistake was made?

CONCEPTUAL

In Exercises 69–72, determine whether each statement is true or false.

69. The function $f(x) = -e^{-x}$ has the y-intercept (0,1).

70. The function $f(x) = -e^{-x}$ has a horizontal asymptote along the x-axis.

71. The functions $y = 3^{-x}$ and $y = \left(\frac{1}{3}\right)^x$ have the same graphs.

72. $e = 2.718$.

73. Plot $f(x) = 3^x$ and its inverse on the same graph.

74. Plot $f(x) = e^x$ and its inverse on the same graph.

CHALLENGE

75. Graph $f(x) = e^{|x|}$.

76. Graph $f(x) = e^{-|x|}$.

77. Find the y-intercept and horizontal asymptote of $f(x) = be^{-x+1} - a$.

78. Find the y-intercept and horizontal asymptote of $f(x) = a + be^{x+1}$.

79. Graph $f(x) = b^{|x|}$, $b > 1$, and state the domain.

80. Graph the function $f(x) = \begin{cases} a^x & x < 0 \\ a^{-x} & x \geq 0 \end{cases}$ where $a > 1$.

TECHNOLOGY

81. Plot the function $y = \left(1 + \frac{1}{x}\right)^x$. What is the horizontal asymptote as x increases?

82. Plot the functions $y = 2^x$, $y = e^x$, and $y = 3^x$ in the same viewing screen. Explain why $y = e^x$ lies between the other two graphs.

83. Plot $y_1 = e^x$ and $y_2 = 1 + x + \frac{x^2}{2} + \frac{x^3}{6} + \frac{x^4}{24}$ in the same viewing screen. What do you notice?

84. Plot $y_1 = e^{-x}$ and $y_2 = 1 - x + \frac{x^2}{2} - \frac{x^3}{6} + \frac{x^4}{24}$ in the same viewing screen. What do you notice?

85. Plot the functions $f(x) = \left(1 + \frac{1}{x}\right)^x$, $g(x) = \left(1 + \frac{2}{x}\right)^x$, and $h(x) = \left(1 + \frac{2}{x}\right)^{2x}$ in the same viewing screen. Compare their horizontal asymptotes as x increases. What can you say about the function values of f, g, and h in terms of the powers of e as x increases?

86. Plot the functions $f(x) = \left(1 + \frac{1}{x}\right)^x$, $g(x) = \left(1 - \frac{1}{x}\right)^x$, and $h(x) = \left(1 - \frac{2}{x}\right)^x$ in the same viewing screen. Compare their horizontal asymptotes as x increases. What can you say about the function values of f, g, and h in terms of the powers of e as x increases?

SECTION 5.2 LOGARITHMIC FUNCTIONS AND THEIR GRAPHS

SKILLS OBJECTIVES

- Convert exponential expressions to logarithmic expressions.
- Convert logarithmic expressions to exponential expressions.
- Evaluate logarithmic expressions exactly by inspection.
- Approximate common and natural logarithms using a calculator.
- Graph logarithmic functions.
- Determine domain restrictions on logarithmic functions.
- Graph functions using a logarithmic scale.

CONCEPTUAL OBJECTIVES

- Interpret logarithmic functions as inverses of exponential functions.
- Understand that logarithmic functions allow very large ranges of numbers in science and engineering applications to be represented on a smaller scale.

Evaluating Logarithms

In Section 5.1, we found that the graph of an exponential function $f(x) = b^x$ passes through the point (0, 1), with the x-axis as a horizontal asymptote. The graph passes both the vertical line test (for a function) and the horizontal line test (for a one-to-one function), and therefore an inverse exists. We will now apply the technique outlined in Section 3.5 to find the inverse of $f(x) = b^x$:

WORDS	**MATH**
Let $y = f(x)$.	$y = b^x$
Interchange x and y.	$x = b^y$
Solve for y.	$y = ?$

We see that y is the exponent that b is raised to in order to obtain x. We call this exponent a **logarithm** (or "log" for short).

WORDS	**MATH**
$x = b^y$ is equivalent to $y = \log_b x$.	$y = \log_b x$
Let $y = f^{-1}(x)$.	$f^{-1}(x) = \log_b x$

DEFINITION **Logarithmic Function**

For $x > 0$, $b > 0$, and $b \neq 1$, the **logarithmic function with base b** is denoted $f(x) = \log_b x$, where

$$y = \log_b x \text{ if and only if } x = b^y$$

We read $\log_b x$ as "log base b of x."

Study Tip

$\log_b x = y$ is equivalent to $b^y = x$

This definition says that $x = b^y$ (**exponential form**) and $y = \log_b x$ (**logarithmic form**) are equivalent. One way to remember this relationship is by adding arrows to the logarithmic form:

$$\log_b x = y \Leftrightarrow b^y = x$$

EXAMPLE 1 Changing from Logarithmic Form to Exponential Form

Write each equation in its equivalent exponential form.

a. $\log_2 8 = 3$ **b.** $\log_9 3 = \frac{1}{2}$ **c.** $\log_5\left(\frac{1}{25}\right) = -2$

Solution:

a. $\log_2 8 = 3$ is equivalent to $\boxed{8 = 2^3}$

b. $\log_9 3 = \frac{1}{2}$ is equivalent to $\boxed{3 = 9^{1/2}} = \sqrt{9}$

c. $\log_5\left(\frac{1}{25}\right) = -2$ is equivalent to $\boxed{\frac{1}{25} = 5^{-2}}$

YOUR TURN Write each equation in its equivalent exponential form.

a. $\log_3 9 = 2$ **b.** $\log_{16} 4 = \frac{1}{2}$ **c.** $\log_2\left(\frac{1}{8}\right) = -3$

Answer: **a.** $9 = 3^2$ **b.** $4 = 16^{1/2}$ **c.** $\frac{1}{8} = 2^{-3}$

EXAMPLE 2 Changing from Exponential Form to Logarithmic Form

Write each equation in its equivalent logarithmic form.

a. $16 = 2^4$ **b.** $9 = \sqrt{81}$ **c.** $\frac{1}{9} = 3^{-2}$ **d.** $x^a = z$

Solution:

a. $16 = 2^4$ is equivalent to $\boxed{\log_2 16 = 4}$

b. $9 = \sqrt{81} = 81^{1/2}$ is equivalent to $\boxed{\log_{81} 9 = \frac{1}{2}}$

c. $\frac{1}{9} = 3^{-2}$ is equivalent to $\boxed{\log_3\left(\frac{1}{9}\right) = -2}$

d. $x^a = z$ is equivalent to $\boxed{\log_x z = a}$

YOUR TURN Write each equation in its equivalent logarithmic form.

a. $81 = 9^2$ **b.** $12 = \sqrt{144}$ **c.** $\frac{1}{49} = 7^{-2}$ **d.** $y^b = w$

Answer: **a.** $\log_9 81 = 2$ **b.** $\log_{144} 12 = \frac{1}{2}$ **c.** $\log_7\left(\frac{1}{49}\right) = -2$ **d.** $\log_y w = b$

Some logarithms can be found exactly, while others must be approximated. Example 3 illustrates how to find the exact value of a logarithm. Example 4 illustrates approximating values of logarithms with a calculator.

EXAMPLE 3 Finding the Exact Value of a Logarithm

Find the exact value of:

a. $\log_3 81$ **b.** $\log_{169} 13$ **c.** $\log_5\left(\frac{1}{5}\right)$

Solution (a):

The logarithm has some value. Let's call it x. $\log_3 81 = x$

Change from logarithmic to exponential form. $3^x = 81$

3 raised to what power is 81? $3^4 = 81$ $x = 4$

Change from exponential to logarithmic form. $\boxed{\log_3 81 = 4}$

Solution (b):

The logarithm has some value. Let's call it x. $\log_{169} 13 = x$

Change from logarithmic to exponential form. $169^x = 13$

169 raised to what power is 13? $169^{1/2} = \sqrt{169} = 13 \quad x = \frac{1}{2}$

Change from exponential to logarithmic form. $\boxed{\log_{169} 13 = \frac{1}{2}}$

Solution (c):

The logarithm has some value. Let's call it x. $\log_5\left(\frac{1}{5}\right) = x$

Change from logarithmic to exponential form. $5^x = \frac{1}{5}$

5 raised to what power is $\frac{1}{5}$? $5^{-1} = \frac{1}{5} \quad x = -1$

Change from exponential to logarithmic form. $\boxed{\log_5\left(\frac{1}{5}\right) = -1}$

■ **YOUR TURN** Evaluate the given logarithms exactly.

a. $\log_2 \frac{1}{2}$ **b.** $\log_{100} 10$ **c.** $\log_{10} 1000$

■ **Answer: a.** $\log_2 \frac{1}{2} = -1$
b. $\log_{100} 10 = \frac{1}{2}$
c. $\log_{10} 1000 = 3$

Common and Natural Logarithms

Two logarithmic bases that arise frequently are base 10 and base e. The logarithmic function of base 10 is called the **common logarithmic function**. Since it is common, $f(x) = \log_{10} x$ is often expressed as $f(x) = \log x$. Thus, if no explicit base is indicated, base 10 is implied. The logarithmic function of base e is called the **natural logarithmic function**. The natural logarithmic function $f(x) = \log_e x$ is often expressed as $f(x) = \ln x$. Both the LOG and LN buttons appear on scientific and graphing calculators. For the ***logarithms*** (not the functions) we say "the log" (for base 10) and "the natural log" (for base e).

Earlier in this section, we evaluated logarithms exactly by converting to exponential form and identifying the exponent. For example, to evaluate $\log_{10} 100$, we ask the question, 10 raised to what power is 100? The answer is 2.

Calculators enable us to approximate logarithms. For example, evaluate $\log_{10} 233$. We are unable to evaluate this exactly by asking the question, 10 raised to what power is 233? Since $10^2 < 10^x < 10^3$, we know the answer x must lie between 2 and 3. Instead, we use a calculator to find an approximate value 2.367.

Study Tip

- $\log_{10} x = \log x$. No explicit base implies base 10.
- $\log_e x = \ln x$

EXAMPLE 4 Using a Calculator to Evaluate Common and Natural Logarithms

Use a calculator to evaluate the common and natural logarithms. Round your answers to four decimal places.

a. $\log 415$ **b.** $\ln 415$ **c.** $\log 1$ **d.** $\ln 1$ **e.** $\log(-2)$ **f.** $\ln(-2)$

Study Tip

Logarithms can only be evaluated for positive arguments.

Solution:

a. $\log(415) \approx 2.618048097 \approx \boxed{2.6180}$ **b.** $\ln(415) \approx 6.02827852 \approx \boxed{6.0283}$

c. $\log(1) = \boxed{0}$ **d.** $\ln(1) = \boxed{0}$

e. $\log(-2)$ $\boxed{\text{undefined}}$ **f.** $\ln(-2)$ $\boxed{\text{undefined}}$

Parts (c) and (d) in Example 4 illustrate that all logarithmic functions pass through the point (1, 0). Parts (e) and (f) in Example 4 illustrate that the domains of logarithmic functions are positive real numbers.

Graphs of Logarithmic Functions

The general logarithmic function $y = \log_b x$ is defined as the inverse of the exponential function $y = b^x$. Therefore, when these two functions are plotted on the same graph, they are symmetric about the line $y = x$. Notice the symmetry about the line $y = x$ when $y = b^x$ and $y = \log_b x$ are plotted on the same graph.

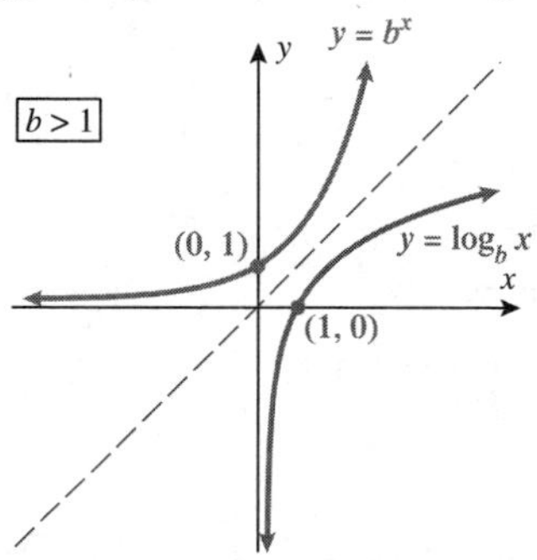

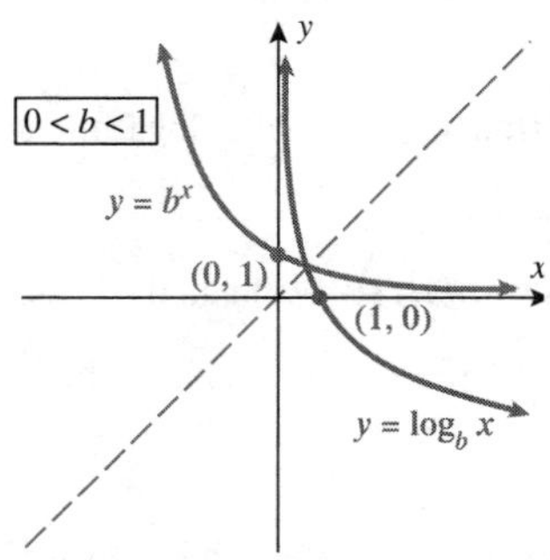

Comparison of Inverse Functions: $f(x) = \log_b x$ and $f^{-1}(x) = b^x$

EXPONENTIAL FUNCTION	LOGARITHMIC FUNCTION
$y = b^x$	$y = \log_b x$
y-intercept (0, 1)	x-intercept (1, 0)
Domain $(-\infty, \infty)$	Domain $(0, \infty)$
Range $(0, \infty)$	Range $(-\infty, \infty)$
Horizontal asymptote: x-axis	Vertical asymptote: y-axis

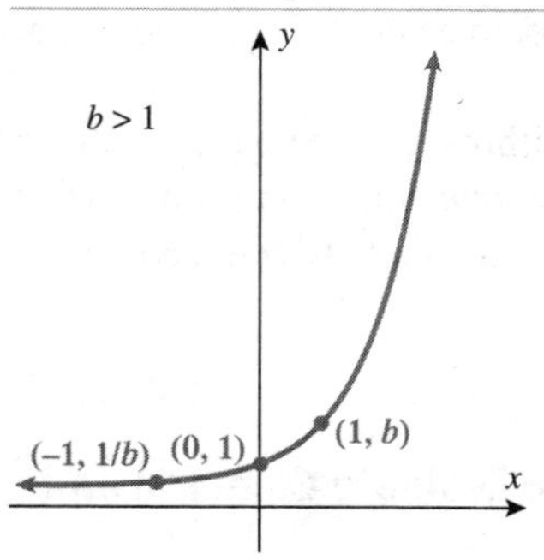

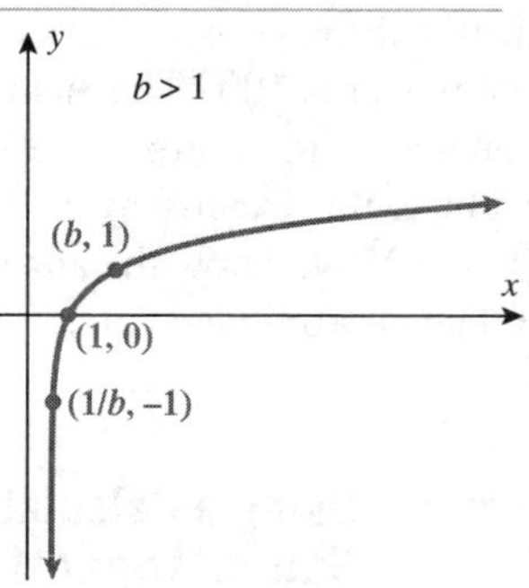

Additionally, the domain of one function is the range of the other, and vice versa. When dealing with logarithmic functions, special attention must be paid to the domain of the function. The domain of $y = \log_b x$ is $(0, \infty)$. In other words, you can only take the log of a positive real number, $x > 0$.

EXAMPLE 5 Finding the Domain of a Shifted Logarithmic Function

Find the domain of each of the given logarithmic functions.

a. $f(x) = \log_b(x - 4)$ **b.** $g(x) = \log_b(5 - 2x)$

Solution (a):

Set the argument greater than zero. $x - 4 > 0$

Solve the inequality. $x > 4$

Write the domain in interval notation. $\boxed{(4, \infty)}$

Solution (b):

Set the argument greater than zero. $5 - 2x > 0$

Solve the inequality. $-2x > -5$

$2x < 5$

$x < \frac{5}{2}$

Write the domain in interval notation. $\boxed{\left(-\infty, \frac{5}{2}\right)}$

YOUR TURN Find the domain of the given logarithmic functions.

a. $f(x) = \log_b(x + 2)$ **b.** $g(x) = \log_b(3 - 5x)$

Answer: a. $(-2, \infty)$ **b.** $\left(-\infty, \frac{5}{3}\right)$

It is important to note that when finding the domain of a logarithmic function, we set the argument strictly greater than zero and solve.

EXAMPLE 6 Finding the Domain of a Logarithmic Function with a Complicated Argument

Find the domain of each of the given logarithmic functions.

a. $\ln\left(x^2 - 9\right)$ **b.** $\log(|x + 1|)$

Solution (a):

Set the argument greater than zero. $x^2 - 9 > 0$

Solve the inequality. $\boxed{(-\infty, -3) \cup (3, \infty)}$

Solution (b):

Set the argument greater than zero. $|x + 1| > 0$

Solve the inequality. $x \neq -1$

Write the domain in interval notation. $\boxed{(-\infty, -1) \cup (-1, \infty)}$

YOUR TURN Find the domain of each of the given logarithmic functions.

a. $\ln\left(x^2 - 4\right)$ **b.** $\log(|x - 3|)$

Answer: a. $(-\infty, -2) \cup (2, \infty)$ **b.** $(-\infty, 3) \cup (3, \infty)$

Recall from Section 3.3 that a technique for graphing general functions is transformations of known functions. For example, to graph $f(x) = (x - 3)^2 + 1$, we start with the known parabola $y = x^2$, whose vertex is at (0, 0), and we shift that graph to the right 3 units and up 1 unit. We use the same techniques for graphing logarithmic functions. To graph $y = \log_b(x + 2) - 1$, we start with the graph of $y = \log_b(x)$ and shift the graph to the left 2 units and down 1 unit.

Technology Tip

a. To graph $\log_2(x - 3)$, use the change-base formula to express it as $\frac{\ln(x - 3)}{\ln(2)}$. Graphs of $Y_1 = \log_2 x$ and $Y_2 = \log_2(x - 3)$ are shown.

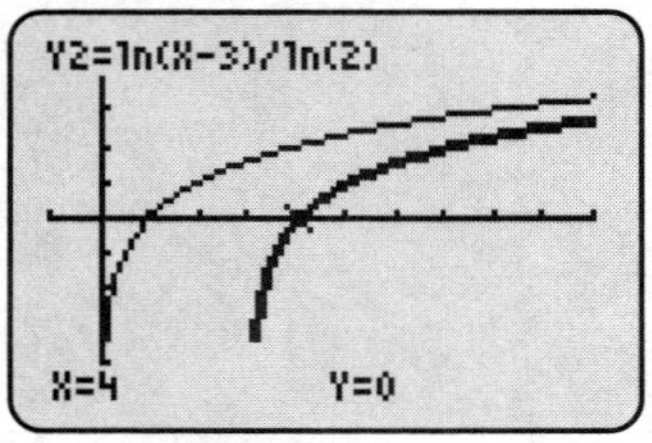

b. Graphs of $Y_1 = \log_2 x$ and $Y_2 = \log_2(x) - 3$ are shown.

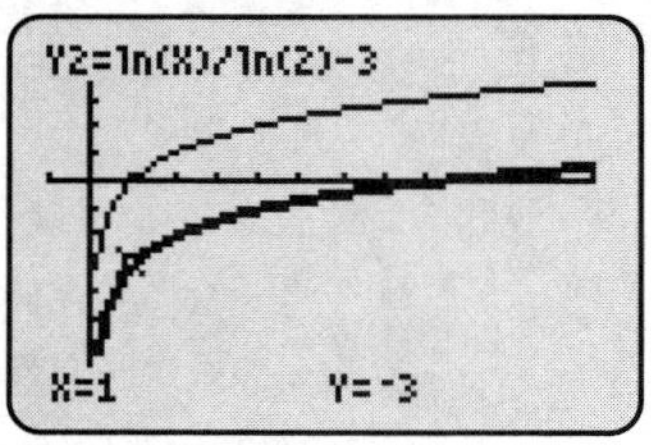

EXAMPLE 7 Graphing Logarithmic Functions Using Horizontal and Vertical Shifts

Graph the functions, and state the domain and range of each.

a. $y = \log_2(x - 3)$ **b.** $\log_2(x) - 3$

Solution:

Identify the base function. $y = \log_2 x$

Label key features of $y = \log_2 x$.

x-intercept: $(1, 0)$

Vertical asymptote: $x = 0$

Additional points: $(2, 1)$, $(4, 2)$

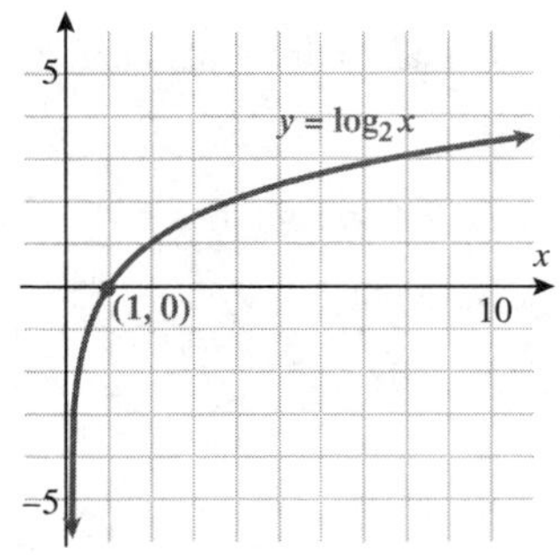

a. Shift base function to the *right* 3 units.

x-intercept: $(4, 0)$

Vertical asymptote: $x = 3$

Additional points: $(5, 1)$, $(7, 2)$

Domain: $(3, \infty)$	Range: $(-\infty, \infty)$

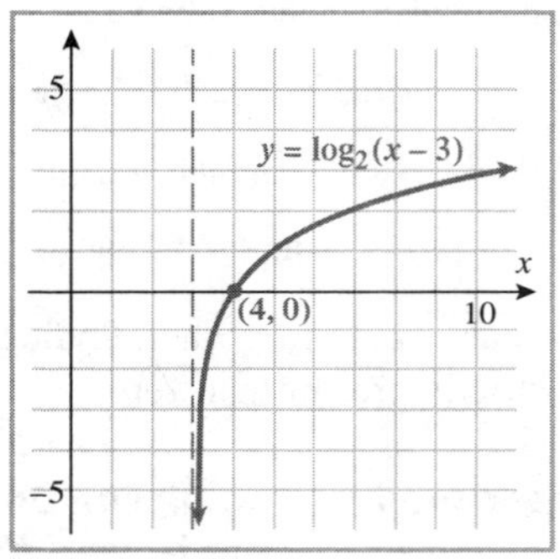

b. Shift base function *down* 3 units.

x-intercept: $(1, -3)$

Vertical asymptote: $x = 0$

Additional points: $(2, -2)$, $(4, -1)$

Domain: $(0, \infty)$	Range: $(-\infty, \infty)$

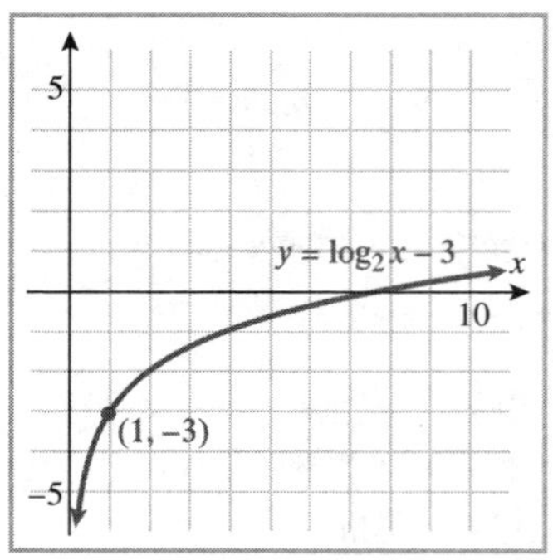

■ **YOUR TURN** Graph the functions and state the domain and range of each.

a. $y = \log_3 x$ **b.** $y = \log_3(x + 3)$ **c.** $\log_3(x) + 1$

■ **Answer:**
a. Domain: $(0, \infty)$ Range: $(-\infty, \infty)$
b. Domain: $(-3, \infty)$ Range: $(-\infty, \infty)$
c. Domain: $(0, \infty)$ Range: $(-\infty, \infty)$

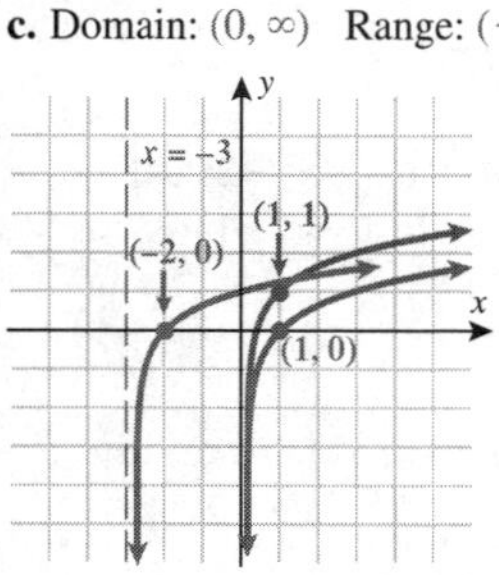

All of the transformation techniques (shifting, reflection, and compression) discussed in Chapter 3 also apply to logarithmic functions. For example, the graphs of $-\log_2 x$ and $\log_2(-x)$ are found by reflecting the graph of $y = \log_2 x$ about the x-axis and y-axis, respectively.

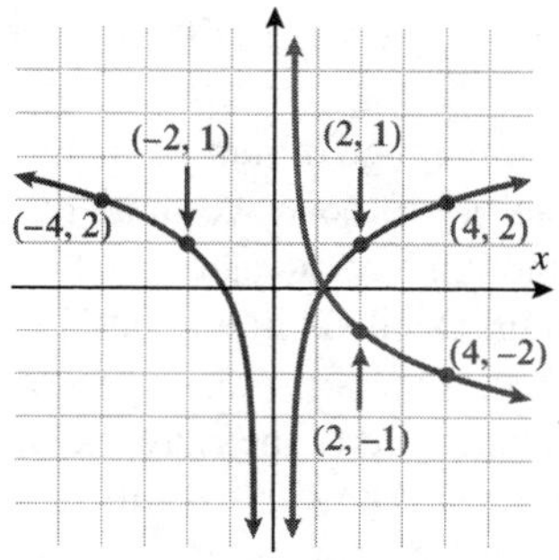

EXAMPLE 8 Graphing Logarithmic Functions Using Transformations

Graph the function $f(x) = -\log_2(x - 3)$ and state its domain and range.

Solution:

Graph $y = \log_2 x$.

x-intercept: $(1, 0)$

Vertical asymptote: $x = 0$

Additional points: $(2, 1)$, $(4, 2)$

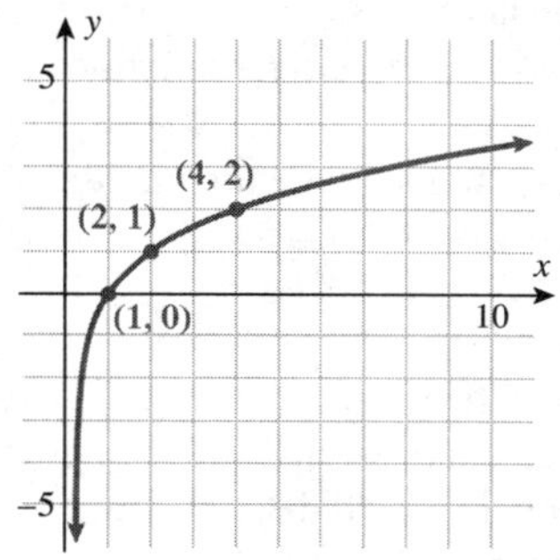

Graph $y = \log_2(x - 3)$ by shifting $y = \log_2 x$ to the *right* 3 units.

x-intercept: $(4, 0)$

Vertical asymptote: $x = 3$

Additional points: $(5, 1)$, $(7, 2)$

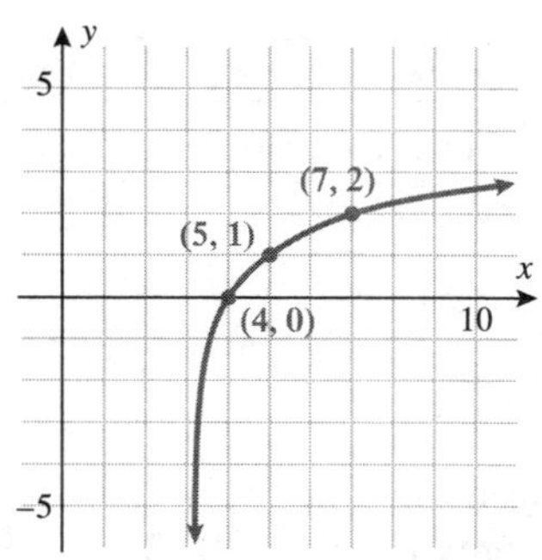

Graph $y = -\log_2(x - 3)$ by reflecting $y = \log_2(x - 3)$ about the x-axis.

x-intercept: $(4, 0)$

Vertical asymptote: $x = 3$

Additional points: $(5, -1)$, $(7, -2)$

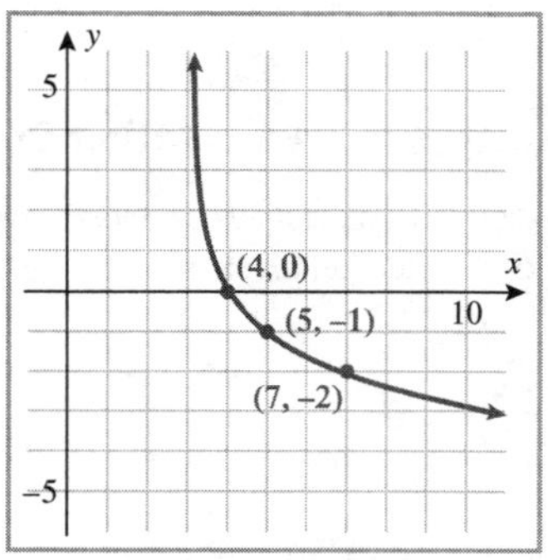

Domain: $(3, \infty)$ Range: $(-\infty, \infty)$

Technology Tip

Graphs $Y_1 = \log_2 x$, $Y_2 = \log_2(x - 3)$, and $Y_3 = -\log_2(x - 3)$ are shown.

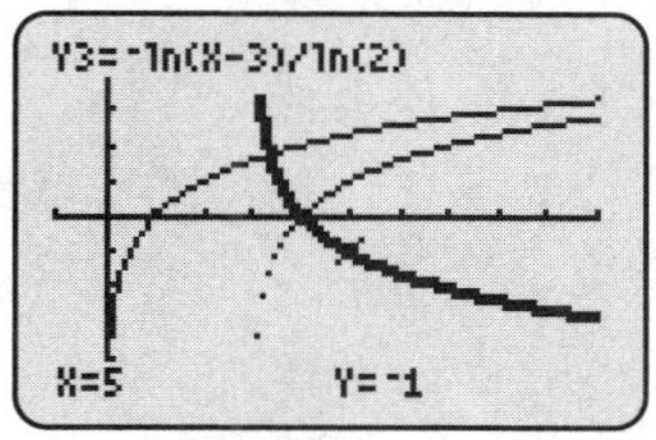

A table of values supports the horizontal shift of the base function $Y_1 = \log_2 x$ right 3 units to obtain the function $Y_2 = \log_2(x) - 3$. The graph of $Y_3 = -\log_2(x - 3)$ is obtained by reflecting $Y_2 = \log_2(x) - 3$ about the x-axis.

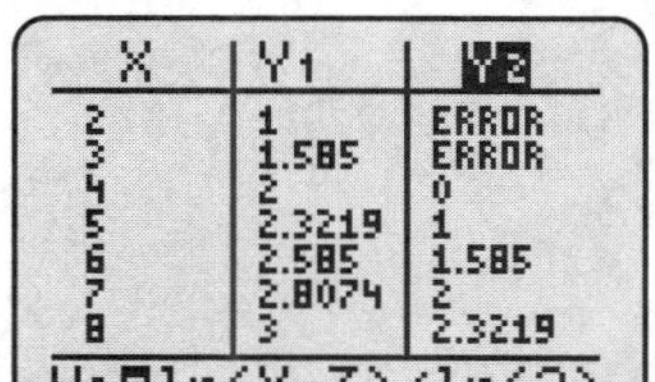

X	Y1	Y2
2	1	ERROR
3	1.585	ERROR
4	2	0
5	2.3219	1
6	2.585	1.585
7	2.8074	2
8	3	2.3219

Y2=ln(X-3)/ln(2)

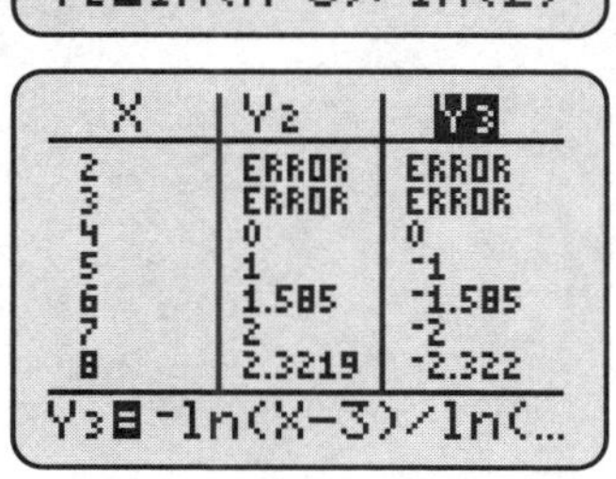

X	Y2	Y3
2	ERROR	ERROR
3	ERROR	ERROR
4	0	0
5	1	-1
6	1.585	-1.585
7	2	-2
8	2.3219	-2.322

Y3=-ln(X-3)/ln(...

Applications of Logarithms

Logarithms are used to make a large range of numbers manageable. For example, to create a scale to measure a human's ability to hear, we must have a way to measure the sound intensity of an explosion, even though that intensity can be more than a trillion (10^{12}) times greater than that of a soft whisper. Decibels in engineering and physics, pH in chemistry, and the Richter scale for earthquakes are all applications of logarithmic functions.

The **decibel** is a logarithmic unit used to measure the magnitude of a physical quantity relative to a specified reference level. The *decibel* (dB) is employed in many engineering and science applications. The most common application is the intensity of sound.

DEFINITION **Decibel (Sound)**

The **decibel** is defined as $$D = 10\log\left(\frac{I}{I_T}\right)$$

where D is the decibel level (dB), I is the intensity of the sound measured in watts per square meter, and I_T is the intensity threshold of the least audible sound a human can hear.

The human average threshold is $I_T = 1 \times 10^{-12}\ \text{W/m}^2$.

Notice that when $I = I_T$, then $D = 10\overbrace{\log 1}^{0} = 0$ dB. People who work professionally with sound, such as acoustics engineers or medical hearing specialists, refer to this threshold level I_T as "0 dB." The following table illustrates typical sounds we hear and their corresponding decibel levels.

SOUND SOURCE	SOUND INTENSITY (W/m²)	DECIBELS (dB)
Threshold of hearing	1.0×10^{-12}	0
Vacuum cleaner	1.0×10^{-4}	80
iPod	1.0×10^{-2}	100
Jet engine	1.0×10^{3}	150

We add logarithms (decibels) as a substitute for multiplying numbers. For example, a whisper (approximately 0 dB) from someone standing next to a jet engine (150 dB) might go unheard because when these are added, we get approximately 150 dB (the jet engine).

EXAMPLE 9 Calculating Decibels of Sounds

Suppose you have seats to a concert given by your favorite musical artist. Calculate the approximate decibel level associated with the typical sound intensity, given $I = 1 \times 10^{-2}\ \text{W/m}^2$.

Solution:

Write the decibel-scale formula. $$D = 10\log\left(\frac{I}{I_T}\right)$$

Substitute $I = 1 \times 10^{-2}\ \text{W/m}^2$ and $I_T = 1 \times 10^{-12}\ \text{W/m}^2$. $$D = 10\log\left(\frac{1 \times 10^{-2}}{1 \times 10^{-12}}\right)$$

Simplify.	$D = 10\log(10^{10})$
Recall that the implied base for log is 10.	$D = 10\log_{10}(10^{10})$
Evaluate the right side. $[\log_{10}(10^{10}) = 10]$	$D = 10 \cdot 10$
	$D = 100$

The typical sound level on the front row of a rock concert is $\boxed{100\text{ dB}}$.

■ YOUR TURN Calculate the approximate decibels associated with a sound so loud it will cause instant perforation of the eardrums, $I = 1 \times 10^4\ \text{W/m}^2$.

■ **Answer:** 160 dB

The Richter scale (earthquakes) is another application of logarithms.

DEFINITION **Richter Scale**

The magnitude M of an earthquake is measured using the **Richter scale**

$$M = \frac{2}{3}\log\left(\frac{E}{E_0}\right)$$

where: M is the magnitude
E is the seismic energy released by the earthquake (in joules)
E_0 is the energy released by a reference earthquake $E_0 = 10^{4.4}$ joules

EXAMPLE 10 Calculating the Magnitude of an Earthquake

On October 17, 1989, just moments before game 3 of the World Series between the Oakland A's and the San Francisco Giants was about to start—with 60,000 fans in Candlestick Park—a devastating earthquake erupted. Parts of interstates and bridges collapsed, and President George H. W. Bush declared the area a disaster zone. The earthquake released approximately 1.12×10^{15} joules. Calculate the magnitude of the earthquake using the Richter scale.

Solution:

Write the Richter scale formula.	$M = \frac{2}{3}\log\left(\frac{E}{E_0}\right)$
Substitute $E = 1.12 \times 10^{15}$ and $E_0 = 10^{4.4}$.	$M = \frac{2}{3}\log\left(\frac{1.12 \times 10^{15}}{10^{4.4}}\right)$
Simplify.	$M = \frac{2}{3}\log\left(1.12 \times 10^{10.6}\right)$
Approximate the logarithm using a calculator.	$M \approx \frac{2}{3}(10.65) \approx 7.1$

The 1989 earthquake in California measured $\boxed{7.1}$ on the Richter scale.

■ YOUR TURN On May 3, 1996, Seattle experienced a moderate earthquake. The energy that the earthquake released was approximately 1.12×10^{12} joules. Calculate the magnitude of the 1996 Seattle earthquake using the Richter scale.

Technology Tip

Enter the number 1.12×10^{15} using the scientific notation key EXP or EE.

■ **Answer:** 5.1

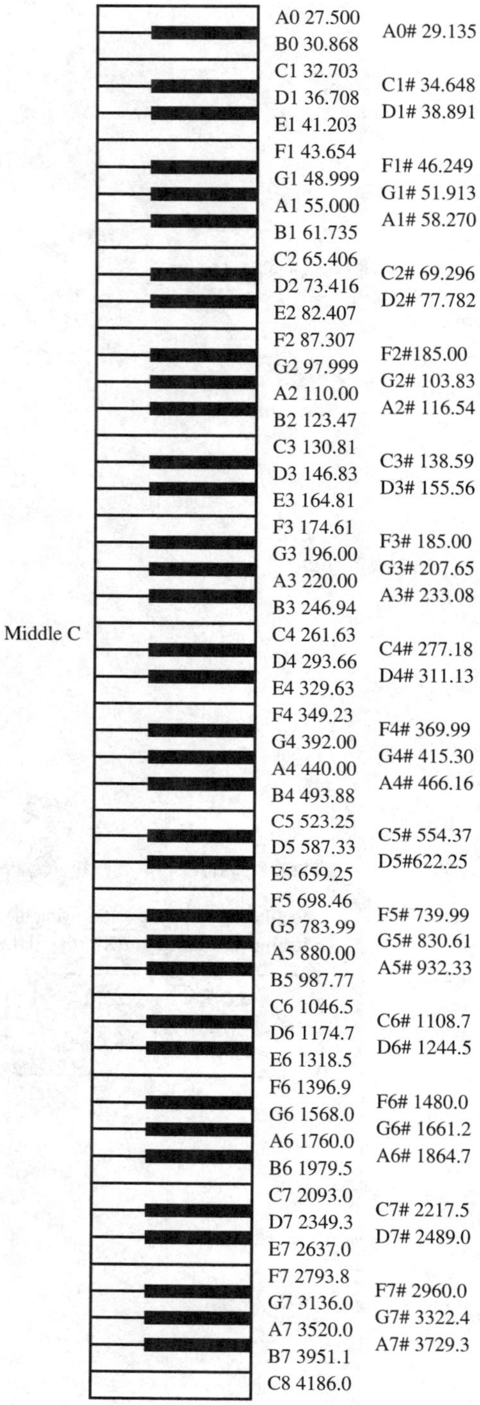

A **logarithmic scale** expresses the logarithm of a physical quantity instead of the quantity itself. In music, the pitch is the perceived fundamental frequency of sound. The note A above middle C on a piano has the pitch associated with a pure tone of 440 hertz (Hz). An octave is the interval between one musical pitch and another with either double or half its frequency. For example, if a note has a frequency of 440 Hz, then the note an octave above it has a frequency of 880 Hz, and the note an octave below it has a frequency of 220 Hz. Therefore, the ratio of two notes an octave apart is 2:1.

The following table lists the frequencies associated with A notes.

NOTE	A_1	A_2	A_3	A_4	A_5	A_6	A_7
Frequency (Hz)	55	110	220	440	880	1760	3520
Octave with respect to A_4	−3	−2	−1	0	+1	+2	+3

We can graph $\frac{\text{Frequency of note}}{440\text{ Hz}}$ on the horizontal axis and the octave (with respect to A_4) on the vertical axis.

If we instead graph the logarithm of this quantity, $\log\frac{\text{Frequency of note}}{440\text{ Hz}}$, we see that using a logarithmic scale expresses octaves linearly (up or down an octave). In other words, an "octave" is a purely logarithmic concept.

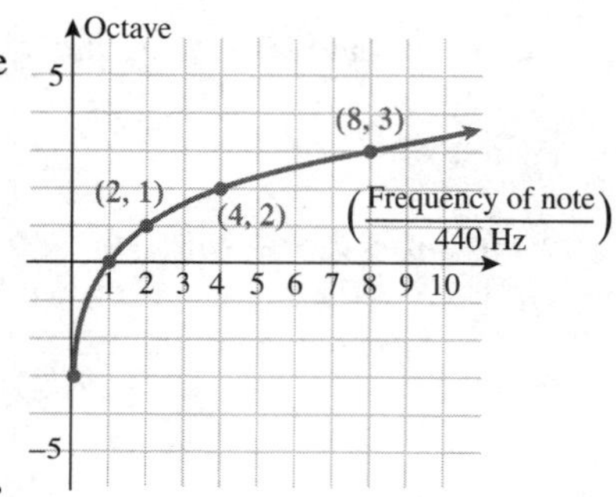

When a logarithmic scale is used we typically classify a graph one of two ways:

- Log-Log plot (both the horizontal and vertical axes use logarithmic scales)
- Semilog plot (one of the axes uses a logarithmic scale)

Semilog Plot

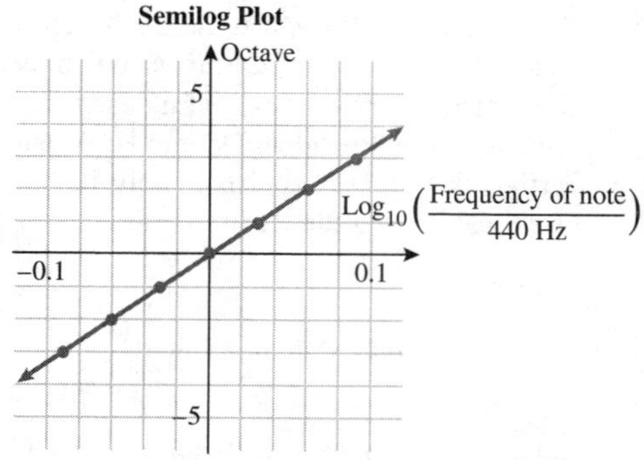

The second graph with octaves on the vertical axis and the log of the ratio of frequencies on the horizontal axis is called a semilog plot. An example of a log-log plot would be frequency versus wavelength in engineering. Frequency is inversely proportional to the wavelength. In a vacuum $f = \frac{c}{\lambda}$, where f is the frequency (in hertz), $c = 3.0 \times 10^8$ m/s^2 is the speed of light in a vacuum, and λ is the wavelength in meters.

Log-Log Plot

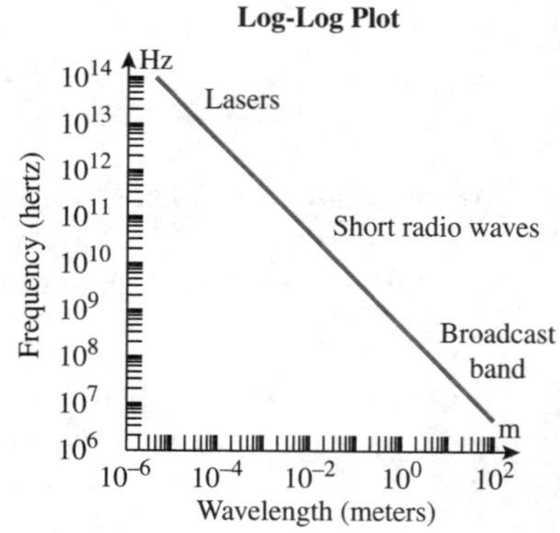

The logarithmic scales allow us to represent a large range of numbers. In this graph, the x-axis ranges from microns, 10^{-6} meters, to hundreds of meters, and the y-axis ranges from megahertz (MHz), 10^6 hertz, to hundreds of terahertz (THz), 10^{12} hertz.

SECTION 5.2 SUMMARY

In this section, logarithmic functions were defined as inverses of exponential functions.

$$y = \log_b x \text{ is equivalent to } x = b^y$$

NAME	EXPLICIT BASE	IMPLICIT BASE
Common logarithm	$f(x) = \log_{10} x$	$f(x) = \log x$
Natural logarithm	$f(x) = \log_e x$	$f(x) = \ln x$

Evaluating Logarithms

- *Exact:* Convert to exponential form first, then evaluate.
- *Approximate:* Natural and common logarithms with calculators.

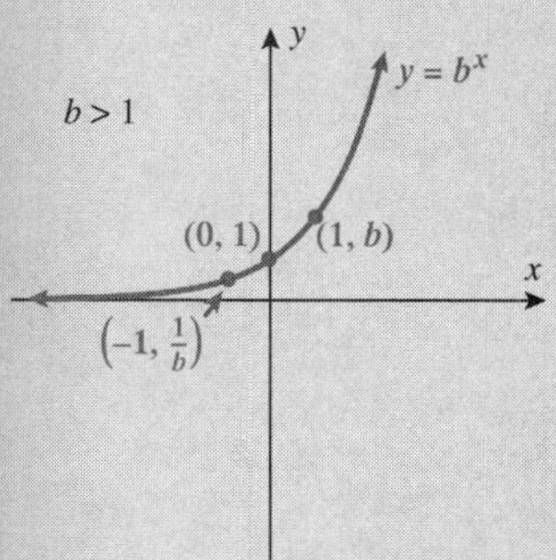

Graphs of Logarithmic Functions

EXPONENTIAL FUNCTION	LOGARITHMIC FUNCTION
$y = b^x$	$y = \log_b x$
y-intercept: (0, 1)	x-intercept: (1, 0)
Domain: $(-\infty, \infty)$	Domain: $(0, \infty)$
Range: $(0, \infty)$	Range: $(-\infty, \infty)$
Horizontal asymptote: x-axis	Vertical asymptote: y-axis

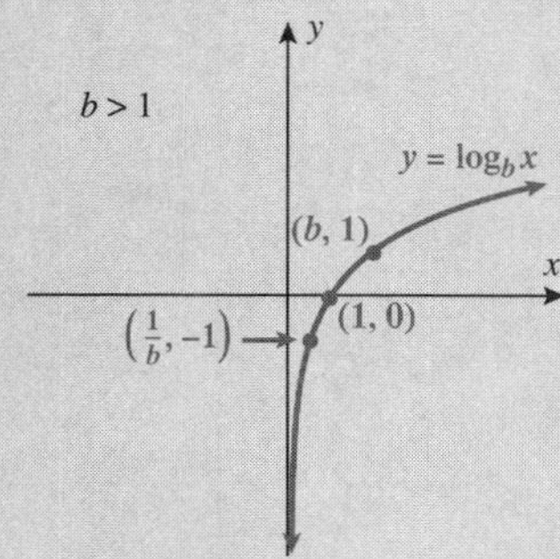

SECTION 5.2 EXERCISES

SKILLS

In Exercises 1–20, write each logarithmic equation in its equivalent exponential form.

1. $\log_5 125 = 3$ **2.** $\log_3 27 = 3$ **3.** $\log_{81} 3 = \frac{1}{4}$ **4.** $\log_{121} 11 = \frac{1}{2}$ **5.** $\log_2\left(\frac{1}{32}\right) = -5$

6. $\log_3\left(\frac{1}{81}\right) = -4$ **7.** $\log 0.01 = -2$ **8.** $\log 0.0001 = -4$ **9.** $\log 10{,}000 = 4$ **10.** $\log 1000 = 3$

11. $\log_{1/4}(64) = -3$ **12.** $\log_{1/6}(36) = -2$ **13.** $-1 = \ln\left(\frac{1}{e}\right)$ **14.** $1 = \ln e$ **15.** $\ln 1 = 0$

16. $\log 1 = 0$ **17.** $\ln 5 = x$ **18.** $\ln 4 = y$ **19.** $z = \log_x y$ **20.** $y = \log_x z$

In Exercises 21–34, write each exponential equation in its equivalent logarithmic form.

21. $8^3 = 512$ **22.** $2^6 = 64$ **23.** $0.00001 = 10^{-5}$ **24.** $100{,}000 = 10^5$ **25.** $15 = \sqrt{225}$

26. $7 = \sqrt[3]{343}$ **27.** $\frac{8}{125} = \left(\frac{2}{5}\right)^3$ **28.** $\frac{8}{27} = \left(\frac{2}{3}\right)^3$ **29.** $3 = \left(\frac{1}{27}\right)^{-1/3}$ **30.** $4 = \left(\frac{1}{1024}\right)^{-1/5}$

31. $e^x = 6$ **32.** $e^{-x} = 4$ **33.** $x = y^z$ **34.** $z = y^x$

In Exercises 35–46, evaluate the logarithms exactly (if possible).

35. $\log_2 1$ **36.** $\log_5 1$ **37.** $\log_5 3125$ **38.** $\log_3 729$

39. $\log 10^7$ **40.** $\log 10^{-2}$ **41.** $\log_{1/4} 4096$ **42.** $\log_{1/7} 2401$

43. $\log 0$ **44.** $\ln 0$ **45.** $\log(-100)$ **46.** $\ln(-1)$

In Exercises 47–54, approximate (if possible) the common and natural logarithms using a calculator. Round to two decimal places.

47. $\log 29$ **48.** $\ln 29$ **49.** $\ln 380$ **50.** $\log 380$

51. $\log 0$ **52.** $\ln 0$ **53.** $\ln 0.0003$ **54.** $\log 0.0003$

In Exercises 55–64, state the domain of the logarithmic function in interval notation.

55. $f(x) = \log_2(x + 5)$ **56.** $f(x) = \log_2(4x - 1)$ **57.** $f(x) = \log_3(5 - 2x)$ **58.** $f(x) = \log_3(5 - x)$

59. $f(x) = \ln(7 - 2x)$ **60.** $f(x) = \ln(3 - x)$ **61.** $f(x) = \log|x|$ **62.** $f(x) = \log|x + 1|$

63. $f(x) = \log\left(x^2 + 1\right)$ **64.** $f(x) = \log\left(1 - x^2\right)$

In Exercises 65–70, match the graph with the function.

65. $y = \log_5 x$ **66.** $y = \log_5(-x)$ **67.** $y = -\log_5(-x)$

68. $y = \log_5(x + 3) - 1$ **69.** $y = \log_5(1 - x) - 2$ **70.** $y = -\log_5(3 - x) + 2$

a.

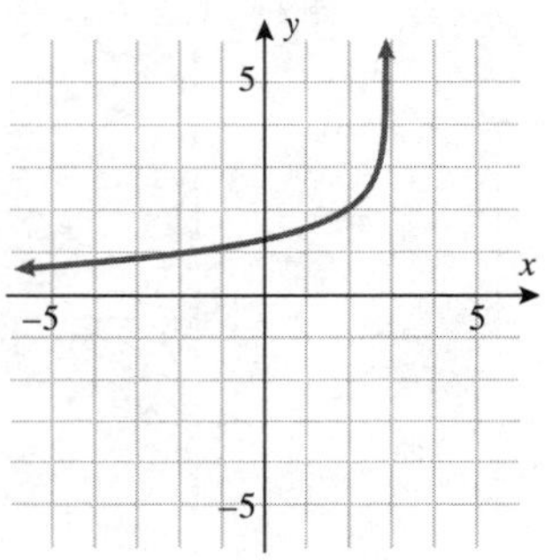

b.

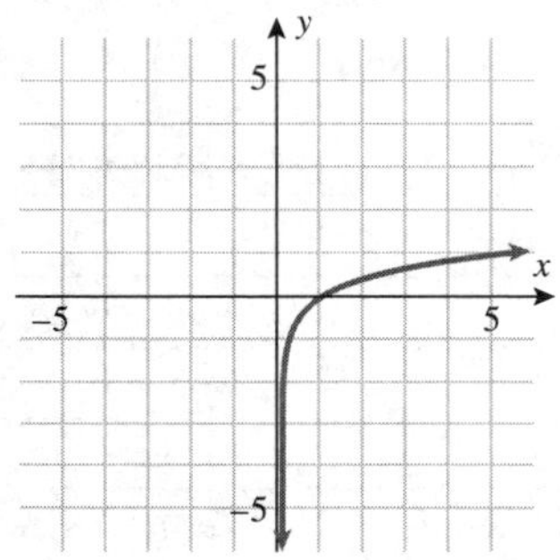

c.

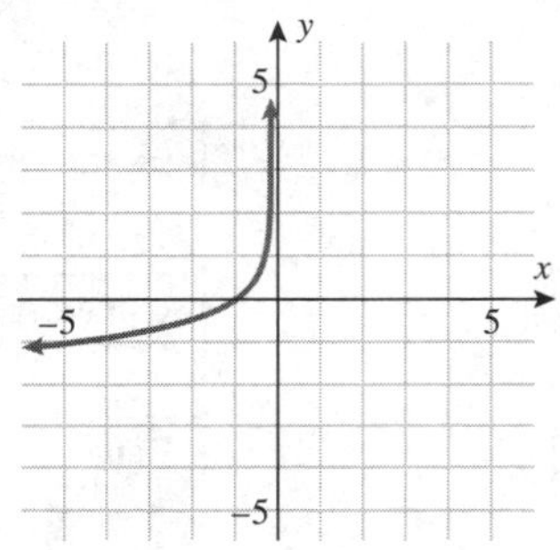

d.

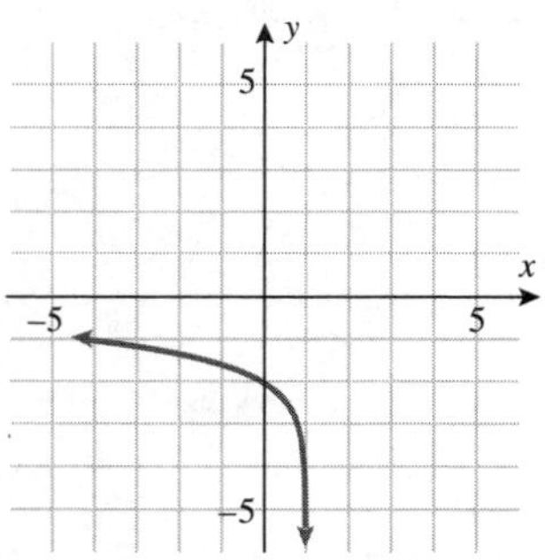

e.

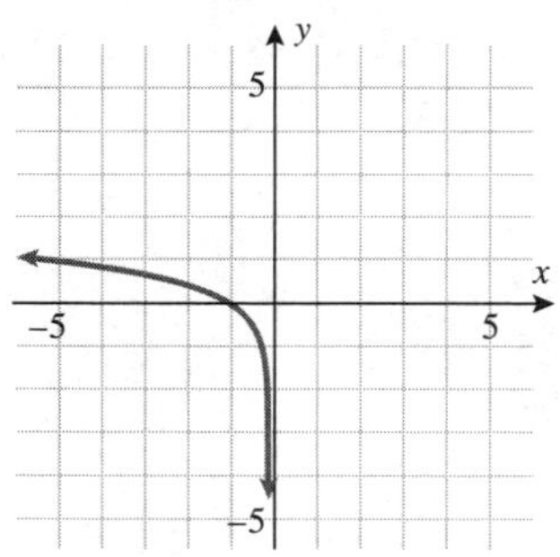

f.

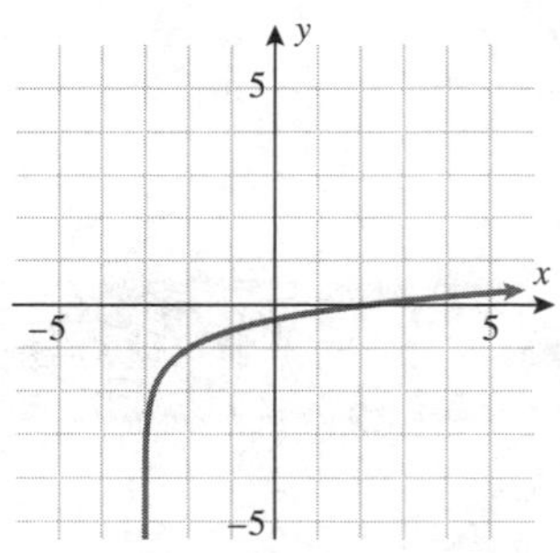

In Exercises 71–82, graph the logarithmic function using transformation techniques. State the domain and range of *f*.

71. $f(x) = \log(x - 1)$ **72.** $f(x) = \log(x + 2)$ **73.** $\ln x + 2$ **74.** $\ln x - 1$

75. $f(x) = \log_3(x + 2) - 1$ **76.** $f(x) = \log_3(x + 1) - 2$ **77.** $f(x) = -\log(x) + 1$ **78.** $f(x) = \log(-x) + 2$

79. $f(x) = \ln(x + 4)$ **80.** $f(x) = \ln(4 - x)$ **81.** $f(x) = \log(2x)$ **82.** $f(x) = 2\ln(-x)$

APPLICATIONS

83. Sound. Calculate the decibels associated with *normal conversation* if the intensity is $I = 1 \times 10^{-6}\ \text{W/m}^2$.

84. Sound. Calculate the decibels associated with the *onset of pain* if the intensity is $I = 1 \times 10^{1}\ \text{W/m}^2$.

85. Sound. Calculate the decibels associated with attending *a football game in a loud college stadium* if the intensity is $I = 1 \times 10^{-0.3}\ \text{W/m}^2$.

86. Sound. Calculate the decibels associated with a *doorbell* if the intensity is $I = 1 \times 10^{-4.5}\ \text{W/m}^2$.

87. Earthquakes. On Good Friday 1964, one of the most severe North American earthquakes ever recorded struck Alaska. The energy released measured 1.41×10^{17} joules. Calculate the magnitude of the 1964 Alaska earthquake using the Richter scale.

88. Earthquakes. On January 22, 2003, Colima, Mexico, experienced a major earthquake. The energy released measured 6.31×10^{15} joules. Calculate the magnitude of the 2003 Mexican earthquake using the Richter scale.

89. Earthquakes. On December 26, 2003, a major earthquake rocked southeastern Iran. In Bam, 30,000 people were killed, and 85% of buildings were damaged or destroyed. The energy released measured 2×10^{14} joules. Calculate the magnitude of the 2003 Iran earthquake with the Richter scale.

90. Earthquakes. On November 1, 1755, Lisbon was destroyed by an earthquake, which killed 90,000 people and destroyed 85% of the city. It was one of the most destructive earthquakes in history. The energy released measured 8×10^{17} joules. Calculate the magnitude of the 1755 Lisbon earthquake with the Richter scale.

For Exercises 91–96, refer to the following:

The pH of a solution is a measure of the molar concentration of hydrogen ions, H^+ in moles per liter, in the solution, which means that it is a measure of the acidity or basicity of the solution. The letters pH stand for "power of hydrogen," and the numerical value is defined as

$$\text{pH} = -\log_{10}[H^+]$$

Very acid corresponds to pH values near 1, neutral corresponds to a pH near 7 (pure water), and very basic corresponds to values near 14. In the next six exercises you will be asked to calculate the pH value of wine, Pepto-Bismol, normal rainwater, bleach, and fruit. List these six liquids and use your intuition to classify them as neutral, acidic, very acidic, basic, or very basic before you calculate their actual pH values.

91. Chemistry. If wine has an approximate hydrogen ion concentration of 5.01×10^{-4}, calculate its pH value.

92. Chemistry. Pepto-Bismol has a hydrogen ion concentration of about 5.01×10^{-11}. Calculate its pH value.

93. Chemistry. Normal rainwater is slightly acidic and has an approximate hydrogen ion concentration of $10^{-5.6}$. Calculate its pH value. Acid rain and tomato juice have similar approximate hydrogen ion concentrations of 10^{-4}. Calculate the pH value of acid rain and tomato juice.

94. Chemistry. Bleach has an approximate hydrogen ion concentration of 5.0×10^{-13}. Calculate its pH value.

95. Chemistry. An apple has an approximate hydrogen ion concentration of $10^{-3.6}$. Calculate its pH value.

96. Chemistry. An orange has an approximate hydrogen ion concentration of $10^{-4.2}$. Calculate its pH value.

97. Archaeology. Carbon dating is a method used to determine the age of a fossil or other organic remains. The age t in years is related to the mass C (in milligrams) of carbon 14 through a logarithmic equation:

$$t = -\frac{\ln\left(\frac{C}{500}\right)}{0.0001216}$$

How old is a fossil that contains 100 milligrams of carbon 14?

98. Archaeology. Repeat Exercise 97, only now the fossil contains 40 milligrams of carbon 14.

99. Broadcasting. Decibels are used to quantify losses associated with atmospheric interference in a communication system. The ratio of the power (watts) received to the power transmitted (watts) is often compared. Often, *watts* are transmitted, but losses due to the atmosphere typically correspond to *milliwatts* being received:

$$\text{dB} = 10 \log\left(\frac{\text{Power received}}{\text{Power transmitted}}\right)$$

If 1 W of power is transmitted and 3 mW is received, calculate the power loss in dB.

100. Broadcasting. Repeat Exercise 99 assuming 3 W of power is transmitted and 0.2 mW is received.

CATCH THE MISTAKE

In Exercises 101–104, explain the mistake that is made.

101. Evaluate the logarithm $\log_2 4$.

Solution:

Set the logarithm equal to x.	$\log_2 4 = x$
Write the logarithm in exponential form.	$x = 2^4$
Simplify.	$x = 16$
Answer:	$\log_2 4 = 16$

This is incorrect. The correct answer is $\log_2 4 = 2$. What went wrong?

102. Evaluate the logarithm $\log_{100} 10$.

Solution:

Set the logarithm equal to x.	$\log_{100} 10 = x$
Express the equation in exponential form.	$10^x = 100$
Solve for x.	$x = 2$
Answer:	$\log_{100} 10 = 2$

This is incorrect. The correct answer is $\log_{100} 10 = \frac{1}{2}$. What went wrong?

103. State the domain of the logarithmic function $f(x) = \log_2(x + 5)$ in interval notation.

Solution:

The domain of all logarithmic functions is $x > 0$.

Interval notation: $(0, \infty)$

This is incorrect. What went wrong?

104. State the domain of the logarithmic function $f(x) = \ln|x|$ in interval notation.

Solution:

Since the absolute value eliminates all negative numbers, the domain is the set of all real numbers.

Interval notation: $(-\infty, \infty)$

This is incorrect. What went wrong?

CONCEPTUAL

In Exercises 105–108, determine whether each statement is true or false.

105. The domain of the standard logarithmic function, $y = \ln x$, is the set of nonnegative real numbers.

106. The horizontal axis is the horizontal asymptote of the graph of $y = \ln x$.

107. The graphs of $y = \log x$ and $y = \ln x$ have the same x-intercept $(1, 0)$.

108. The graphs of $y = \log x$ and $y = \ln x$ have the same vertical asymptote, $x = 0$.

CHALLENGE

109. State the domain, range, and x-intercept of the function $f(x) = -\ln(x - a) + b$ for a and b real positive numbers.

110. State the domain, range, and x-intercept of the function $f(x) = \log(a - x) - b$ for a and b real positive numbers.

111. Graph the function $f(x) = \begin{cases} \ln(-x) & x < 0 \\ \ln(x) & x > 0 \end{cases}$.

112. Graph the function $f(x) = \begin{cases} -\ln(-x) & x < 0 \\ -\ln(x) & x > 0 \end{cases}$.

TECHNOLOGY

113. Apply a graphing utility to graph $y = e^x$ and $y = \ln x$ in the same viewing screen. What line are these two graphs symmetric about?

114. Apply a graphing utility to graph $y = 10^x$ and $y = \log x$ in the same viewing screen. What line are these two graphs symmetric about?

115. Apply a graphing utility to graph $y = \log x$ and $y = \ln x$ in the same viewing screen. What are the two common characteristics?

116. Using a graphing utility, graph $y = \ln|x|$. Is the function defined everywhere?

117. Apply a graphing utility to graph $f(x) = \ln(3x)$, $g(x) = \ln 3 + \ln x$, and $h(x) = (\ln 3)(\ln x)$ in the same viewing screen. Determine the domain where two functions give the same graph.

118. Apply a graphing utility to graph $f(x) = \ln(x^2 - 4)$, $g(x) = \ln(x + 2) + \ln(x - 2)$, and $h(x) = \ln(x + 2)\ln(x - 2)$ in the same viewing screen. Determine the domain where two functions give the same graph.

SECTION

5.3 PROPERTIES OF LOGARITHMS

SKILLS OBJECTIVES

- Write a single logarithm as a sum or difference of logarithms.
- Write a logarithmic expression as a single logarithm.
- Evaluate logarithms of a general base (other than base 10 or e).

CONCEPTUAL OBJECTIVES

- Derive the seven basic logarithmic properties.
- Derive the change-of-base formula.

Since exponential functions and logarithmic functions are inverses of one another, properties of exponents are related to properties of logarithms. We will start by reviewing properties of exponents, and then proceed to properties of logarithms.

In Chapter 0, properties of exponents were discussed.

PROPERTIES OF EXPONENTS

Let a, b, m, and n be any real numbers and $m > 0$ and $n > 0$, then the following are true.

1. $b^m \cdot b^n = b^{m+n}$ **2.** $b^{-m} = \frac{1}{b^m} = \left(\frac{1}{b}\right)^m$ **3.** $\frac{b^m}{b^n} = b^{m-n}$

4. $(b^m)^n = b^{mn}$ **5.** $(ab)^m = a^m \cdot b^m$ **6.** $b^0 = 1$

7. $b^1 = b$

From these properties of exponents we can develop similar properties for logarithms. We list seven basic properties.

PROPERTIES OF LOGARITHMS

If b, M, and N are positive real numbers, where $b \neq 1$ and p and x are real numbers, then the following are true:

1. $\log_b 1 = 0$ **2.** $\log_b b = 1$

3. $\log_b b^x = x$ **4.** $b^{\log_b x} = x \quad x > 0$

5. $\log_b MN = \log_b M + \log_b N$ *Product rule:* Log of a product is the sum of the logs.

6. $\log_b\left(\frac{M}{N}\right) = \log_b M - \log_b N$ *Quotient rule:* Log of a quotient is the difference of the logs.

7. $\log_b M^p = p \log_b M$ *Power rule:* Log of a number raised to an exponent is the exponent times the log of the number.

We will devote this section to proving and illustrating these seven properties.

The first two properties follow directly from the definition of a logarithmic function and properties of exponentials.

Property (1): $\log_b 1 = 0$ since $b^0 = 1$

Property (2): $\log_b b = 1$ since $b^1 = b$

The third and fourth properties follow from the fact that exponential functions and logarithmic functions are inverses of one another. Recall that inverse functions satisfy the relationship that $f^{-1}(f(x)) = x$ for all x in the domain of $f(x)$, and $f\left(f^{-1}(x)\right) = x$ for all x in the domain of f^{-1}. Let $f(x) = b^x$ and $f^{-1}(x) = \log_b x$.

Property (3):

Write the inverse identity. $f^{-1}(f(x)) = x$

Substitute $f^{-1}(x) = \log_b x$. $\log_b(f(x)) = x$

Substitute $f(x) = b^x$. $\log_b b^x = x$

Property (4):

Write the inverse identity. $f\left(f^{-1}(x)\right) = x$

Substitute $f(x) = b^x$. $b^{f^{-1}(x)} = x$

Substitute $f^{-1}(x) = \log_b x$ $x > 0$. $b^{\log_b x} = x$

The first four properties are summarized for common and natural logarithms.

COMMON AND NATURAL LOGARITHM PROPERTIES

Common Logarithm (base 10)	Natural Logarithm (base e)
1. $\log 1 = 0$	**1.** $\ln 1 = 0$
2. $\log 10 = 1$	**2.** $\ln e = 1$
3. $\log 10^x = x$	**3.** $\ln e^x = x$
4. $10^{\log x} = x \quad x > 0$	**4.** $e^{\ln x} = x \quad x > 0$

EXAMPLE 1 Using Logarithmic Properties

Use properties (1)–(4) to simplify the following expressions.

a. $\log_{10} 10$ **b.** $\ln 1$ **c.** $10^{\log(x+8)}$

d. $e^{\ln(2x+5)}$ **e.** $\log 10^{x^2}$ **f.** $\ln e^{x+3}$

Solution:

a. Use property (2). $\log_{10} 10 = \boxed{1}$

b. Use property (1). $\ln 1 = \boxed{0}$

c. Use property (4). $10^{\log(x+8)} = \boxed{x + 8}$ $x > -8$

d. Use property (4). $e^{\ln(2x+5)} = \boxed{2x + 5}$ $x > -\frac{5}{2}$

e. Use property (3). $\log 10^{x^2} = \boxed{x^2}$

f. Use property (3). $\ln e^{x+3} = \boxed{x + 3}$

The fifth through seventh properties follow from the properties of exponents and the definition of logarithms. We will prove the product rule and leave the proofs of the quotient and power rules for the exercises.

$$\text{Property (5):} \quad \log_b MN = \log_b M + \log_b N$$

WORDS	**MATH**
Assume two logs that have the same base.	Let $u = \log_b M$ and $v = \log_b N$ $M > 0, N > 0$
Change to equivalent exponential forms.	$b^u = M$ and $b^v = N$
Write the log of a product.	$\log_b MN$
Substitute $M = b^u$ and $N = b^v$.	$= \log_b(b^u\, b^v)$
Use properties of exponents.	$= \log_b(b^{u+v})$
Apply property 3.	$= u + v$
Substitute $u = \log_b M$, $v = \log_b N$.	$= \log_b M + \log_b N$

$$\boxed{\log_b MN = \log_b M + \log_b N}$$

In other words, the log of a product is the sum of the logs. Let us illustrate this property with a simple example.

$$\overbrace{\log_2 8}^{3} + \overbrace{\log_2 4}^{2} = \overbrace{\log_2 32}^{5}$$

Notice that $\log_2 8 + \log_2 4 \neq \log_2 12$.

EXAMPLE 2 Writing a Logarithmic Expression as a Sum of Logarithms

Use the logarithmic properties to write the expression $\log_b\left(u^2\sqrt{v}\right)$ as a sum of simpler logarithms.

Solution:

Convert the radical to exponential form.	$\log_b\left(u^2\sqrt{v}\right) = \log_b\left(u^2 v^{1/2}\right)$
Use the product property (5).	$= \log_b u^2 + \log_b v^{1/2}$
Use the power property (7).	$= \boxed{2\log_b u + \frac{1}{2}\log_b v}$

YOUR TURN Use the logarithmic properties to write the expression $\log_b\left(x^4\sqrt[3]{y}\right)$ as a sum of simpler logarithms.

Answer: $\log_b\left(x^4\sqrt[3]{y}\right) = 4\log_b x + \frac{1}{3}\log_b y$

EXAMPLE 3 Writing a Sum of Logarithms as a Single Logarithmic Expression: The Right Way and the Wrong Way

Use properties of logarithms to write the expression $2 \log_b 3 + 4 \log_b u$ as a single logarithmic expression.

CAUTION

$\log_b M + \log_b N = \log_b(MN)$

$\log_b M + \log_b N \neq \log_b(M + N)$

COMMON MISTAKE

A common mistake is to write the sum of the logs as a log of the sum.

$$\log_b M + \log_b N \neq \log_b(M + N)$$

CORRECT

Use the power property (7).

$2 \log_b 3 + 4 \log_b u = \log_b 3^2 + \log_b u^4$

Simplify.

$$\log_b 9 + \log_b u^4$$

Use the product property (5).

$$\boxed{= \log_b(9u^4)}$$

INCORRECT

$$\neq \log_b\left(9 + u^4\right) \quad \textbf{ERROR}$$

■ **YOUR TURN** Express $2 \ln x + 3 \ln y$ as a single logarithm.

■ **Answer:** $\ln(x^2y^3)$

EXAMPLE 4 Writing a Logarithmic Expression as a Difference of Logarithms

Write the expression $\ln\left(\frac{x^3}{y^2}\right)$ as a difference of logarithms.

Solution:

Apply the quotient property (6). $\quad \ln\left(\frac{x^3}{y^2}\right) = \ln(x^3) - \ln(y^2)$

Apply the power property (7). $\quad = \boxed{3 \ln x - 2 \ln y}$

■ **YOUR TURN** Write the expression $\log\left(\frac{a^4}{b^5}\right)$ as a difference of logarithms.

■ **Answer:** $4 \log a - 5 \log b$

Another common mistake is misinterpreting the quotient rule.

EXAMPLE 5 Writing the Difference of Logarithms as a Logarithm of a Quotient

Write the expression $\frac{2}{3}\ln x - \frac{1}{2}\ln y$ as a logarithm of a quotient.

COMMON MISTAKE

$$\log_b M - \log_b N \neq \frac{\log_b M}{\log_b N}$$

CORRECT

Use the power property (7).

$\frac{2}{3}\ln x - \frac{1}{2}\ln y = \ln x^{2/3} - \ln y^{1/2}$

Use the quotient property (6).

$$\boxed{\ln\left(\frac{x^{2/3}}{y^{1/2}}\right)}$$

INCORRECT

$$\frac{\ln x^{2/3}}{\ln y^{1/2}} \quad \textbf{ERROR}$$

CAUTION

$$\log_b M - \log_b N = \log_b\left(\frac{M}{N}\right)$$

$$\log_b M - \log_b N \neq \frac{\log_b M}{\log_b N}$$

■ **YOUR TURN** Write the expression $\frac{1}{2}\log a - 3\log b$ as a single logarithm.

■ **Answer:** $\log\left(\frac{a^{1/2}}{b^3}\right)$

EXAMPLE 6 Combining Logarithmic Expressions into a Single Logarithm

Write the expression $3\log_b x + \log_b(2x + 1) - 2\log_b 4$ as a single logarithm.

Solution:

Use the power property (7) on the first and third terms. $\quad = \log_b x^3 + \log_b(2x + 1) - \log_b 4^2$

Use the product property (5) on the first two terms. $\quad = \log_b\left[x^3(2x + 1)\right] - \log_b 16$

Use the quotient property (6). $\quad = \boxed{\log_b\left[\frac{x^3(2x + 1)}{16}\right]}$

■ **YOUR TURN** Write the expression $2\ln x - \ln(3y) + 3\ln z$ as a single logarithm.

■ **Answer:** $\ln\left(\frac{x^2z^3}{3y}\right)$

EXAMPLE 7 Expanding a Logarithmic Expression into a Sum or Difference of Logarithms

Write $\ln\left[\frac{x^2 - x - 6}{x^2 + 7x + 6}\right]$ as a sum or difference of logarithms.

Solution:

Factor the numerator and denominator. $\quad = \ln\left[\frac{(x - 3)(x + 2)}{(x + 6)(x + 1)}\right]$

Use the quotient property (6). $\quad = \ln[(x - 3)(x + 2)] - \ln[(x + 6)(x + 1)]$

Use the product property (5). $\quad = \ln(x - 3) + \ln(x + 2) - [\ln(x + 6) + \ln(x + 1)]$

Eliminate brackets. $\quad = \boxed{\ln(x - 3) + \ln(x + 2) - \ln(x + 6) - \ln(x + 1)}$

Change-of-Base Formula

Recall that in the last section we were able to evaluate logarithms two ways: (1) exactly by writing the logarithm in exponential form and identifying the exponent and (2) using a calculator if the logarithms were base 10 or e. How do we evaluate a logarithm of general base if we cannot identify the exponent? We use the *change-of-base formula.*

EXAMPLE 8 Using Properties of Logarithms to Change the Base to Evaluate a General Logarithm

Evaluate $\log_3 8$. Round the answer to four decimal places.

Solution:

Let $y = \log_3 8$.	$y = \log_3 8$
Write the logarithm in exponential form.	$3^y = 8$
Take the log of both sides.	$\log 3^y = \log 8$
Use the power property (7).	$y \log 3 = \log 8$
Divide both sides by log 3.	$y = \dfrac{\log 8}{\log 3}$
Use a calculator to approximate.	$y \approx 1.892789261$
Substitute $y = \log_3 8$.	$\log_3 8 \approx \boxed{1.8928}$

Example 8 illustrated our ability to use properties of logarithms to change from base 3 to base 10, which our calculators can handle. This leads to the general change-of-base formula.

CHANGE-OF-BASE FORMULA

For any logarithmic bases a and b and any positive number M, the change-of-base formula says that

$$\log_b M = \frac{\log_a M}{\log_a b}$$

In the special case when a is either 10 or e, this relationship becomes

Common Logarithms		Natural Logarithms
$\log_b M = \dfrac{\log M}{\log b}$	or	$\log_b M = \dfrac{\ln M}{\ln b}$

It does not matter what base we select (10, e, or any other base), the ratio will be the same.

Proof of Change-of-Base Formula

WORDS	MATH
Let y be the logarithm we want to evaluate.	$y = \log_b M$
Write $y = \log_b M$ in exponential form.	$b^y = M$
Let a be any positive real number (where $a \neq 1$).	
Take the log of base a of both sides of the equation.	$\log_a b^y = \log_a M$
Use the power rule on the left side of the equation.	$y \log_a b = \log_a M$
Divide both sides of the equation by $\log_a b$.	$y = \dfrac{\log_a M}{\log_a b}$

EXAMPLE 9 **Using the Change-of-Base Formula**

Use the change-of-base formula to evaluate $\log_4 17$. Round to four decimal places.

Solution:

We will illustrate this two ways (choosing common and natural logarithms) using a scientific calculator.

Common Logarithms

Use the change-of-base formula with base 10. $\log_4 17 = \dfrac{\log 17}{\log 4}$

Approximate with a calculator. ≈ 2.043731421

$\approx \boxed{2.0437}$

Natural Logarithms

Use the change-of-base formula with base e. $\log_4 17 = \dfrac{\ln 17}{\ln 4}$

Approximate with a calculator. ≈ 2.043731421

$\approx \boxed{2.0437}$

■ **YOUR TURN** Use the change-of-base formula to approximate $\log_7 34$. Round to four decimal places.

■ **Answer:** $\log_7 34 \approx 1.8122$

SECTION 5.3 SUMMARY

Properties of Logarithms

If b, M, and N are positive real numbers, where $b \neq 1$ and p and x are real numbers, then the following are true:

- Product Property: $\log_b MN = \log_b M + \log_b N$
- Quotient Property: $\log_b\left(\dfrac{M}{N}\right) = \log_b M - \log_b N$
- Power Property: $\log_b M^p = p \log_b M$

General Logarithm	Common Logarithm	Natural Logarithm
$\log_b 1 = 0$	$\log 1 = 0$	$\ln 1 = 0$
$\log_b b = 1$	$\log 10 = 1$	$\ln e = 1$
$\log_b b^x = x$	$\log 10^x = x$	$\ln e^x = x$
$b^{\log_b x} = x \quad x > 0$	$10^{\log x} = x \quad x > 0$	$e^{\ln x} = x \quad x > 0$
Change-of-base formula	$\log_b M = \dfrac{\log M}{\log b}$	$\log_b M = \dfrac{\ln M}{\ln b}$

SECTION 5.3 EXERCISES

SKILLS

In Exercises 1–18, apply the properties of logarithms to simplify each expression. Do not use a calculator.

1. $\log_9 1$ **2.** $\log_{69} 1$ **3.** $\log_{1/2}\left(\frac{1}{2}\right)$ **4.** $\log_{3.3} 3.3$ **5.** $\log_{10} 10^8$

6. $\ln e^3$ **7.** $\log_{10} 0.001$ **8.** $\log_3 3^7$ **9.** $\log_2 \sqrt{8}$ **10.** $\log_5 \sqrt[3]{5}$

11. $8^{\log_8 5}$ **12.** $2^{\log_2 5}$ **13.** $e^{\ln(x+5)}$ **14.** $10^{\log(3x^2+2x+1)}$ **15.** $5^{3\log_5 2}$

16. $7^{2\log_7 5}$ **17.** $7^{-2\log_7 3}$ **18.** $e^{-2\ln 10}$

In Exercises 19–32, write each expression as a sum or difference of logarithms.

Example: $\log(m^2n^5) = 2\log m + 5\log n$

19. $\log_b(x^3y^5)$ **20.** $\log_b(x^{-3}y^{-5})$ **21.** $\log_b(x^{1/2}y^{1/3})$ **22.** $\log_b(\sqrt{r}\sqrt[3]{t})$

23. $\log_b\left(\frac{r^{1/3}}{s^{1/2}}\right)$ **24.** $\log_b\left(\frac{r^4}{s^2}\right)$ **25.** $\log_b\left(\frac{x}{yz}\right)$ **26.** $\log_b\left(\frac{xy}{z}\right)$

27. $\log(x^2\sqrt{x+5})$ **28.** $\log[(x-3)(x+2)]$ **29.** $\ln\left[\frac{x^3(x-2)^2}{\sqrt{x^2+5}}\right]$ **30.** $\ln\left[\frac{\sqrt{x+3}\sqrt[3]{x-4}}{(x+1)^4}\right]$

31. $\log\left[\frac{x^2-2x+1}{x^2-9}\right]$ **32.** $\log\left[\frac{x^2-x-2}{x^2+3x-4}\right]$

In Exercises 33–44, write each expression as a single logarithm.

Example: $2\log m + 5\log n = \log(m^2n^5)$

33. $3\log_b x + 5\log_b y$ **34.** $2\log_b u + 3\log_b v$ **35.** $5\log_b u - 2\log_b v$ **36.** $3\log_b x - \log_b y$

37. $\frac{1}{2}\log_b x + \frac{2}{3}\log_b y$ **38.** $\frac{1}{2}\log_b x - \frac{2}{3}\log_b y$ **39.** $2\log u - 3\log v - 2\log z$ **40.** $3\log u - \log 2v - \log z$

41. $\ln(x+1) + \ln(x-1) - 2\ln(x^2+3)$ **42.** $\ln\sqrt{x-1} + \ln\sqrt{x+1} - 2\ln(x^2-1)$

43. $\frac{1}{2}\ln(x+3) - \frac{1}{3}\ln(x+2) - \ln(x)$ **44.** $\frac{1}{3}\ln(x^2+4) - \frac{1}{2}\ln(x^2-3) - \ln(x-1)$

In Exercises 45–54, evaluate the logarithms using the change-of-base formula. Round to four decimal places.

45. $\log_5 7$ **46.** $\log_4 19$ **47.** $\log_{1/2} 5$ **48.** $\log_5 \frac{1}{2}$ **49.** $\log_{2.7} 5.2$

50. $\log_{7.2} 2.5$ **51.** $\log_\pi 10$ **52.** $\log_\pi 2.7$ **53.** $\log_{\sqrt{3}} 8$ **54.** $\log_{\sqrt{2}} 9$

APPLICATIONS

55. Sound. Sitting in the front row of a rock concert exposes us to a sound pressure (or sound level) of 1×10^{-1} W/m² (or 110 dB), and a normal conversation is typically around 1×10^{-6} W/m² (or 60 dB). How many decibels are you exposed to if a friend is talking in your ear at a rock concert? (*Note:* 160 dB causes perforation of the eardrums.) (*Hint:* Add the sound pressures and convert to dB.)

56. Sound. A whisper corresponds to 1×10^{-10} W/m² (or 20 dB) and a normal conversation is typically around 1×10^{-6} W/m² (or 60 dB). How many decibels are you exposed to if one friend is whispering in your ear while the other one is talking at a normal level? (*Hint:* Add the sound pressures and convert to dB.)

For Exercises 57 and 58, refer to the following:

There are two types of waves associated with an earthquake: *compression* and *shear*. The compression, or longitudinal, waves displace material behind the earthquake's path. Longitudinal waves travel at great speeds and are often called "primary waves" or simply "P" waves. Shear, or transverse, waves displace material at right angles to its path. Transverse waves do not travel as rapidly through the Earth's crust and mantle as do longitudinal waves, and they are called "secondary" or "S" waves.

57. Earthquakes. If a seismologist records the energy of P waves as 4.5×10^{12} joules and the energy of S waves as 7.8×10^{8} joules, what is the total energy? What would the combined effect be on the Richter scale?

58. Earthquakes. Repeat Exercise 57 assuming the energy associated with the P waves is 5.2×10^{11} joules and the energy associated with the S waves is 4.1×10^{9} joules.

CATCH THE MISTAKE

In Exercises 59–62, simplify if possible and explain the mistake that is made.

59. $3 \log 5 - \log 25$

Solution:

Apply the quotient property (6).	$\dfrac{3 \log 5}{\log 25}$
Write $25 = 5^2$.	$\dfrac{3 \log 5}{\log 5^2}$
Apply the power property (7).	$\dfrac{3 \log 5}{2 \log 5}$
Simplify.	$\dfrac{3}{2}$

This is incorrect. The correct answer is log 5. What mistake was made?

60. $\ln 3 + 2 \ln 4 - 3 \ln 2$

Solution:

Apply the power property (7).	$\ln 3 + \ln 4^2 - \ln 2^3$
Simplify.	$\ln 3 + \ln 16 - \ln 8$
Apply property (5).	$\ln(3 + 16 - 8)$
Simplify.	$\ln 11$

This is incorrect. The correct answer is ln 64. What mistake was made?

61. $\log_2 x + \log_3 y - \log_4 z$

Solution:

Apply the product property (5).	$\log_6 xy - \log_4 z$
Apply the quotient property (6).	$\log_{24} xyz$

This is incorrect. What mistake was made?

62. $2(\log 3 - \log 5)$

Solution:

Apply the quotient property (6).	$2\left(\log \dfrac{3}{5}\right)$
Apply the power property (7).	$\left(\log \dfrac{3}{5}\right)^2$
Apply a calculator to approximate.	≈ 0.0492

This is incorrect. What mistake was made?

CONCEPTUAL

In Exercises 63–66, determine whether each statement is true or false.

63. $\log e = \dfrac{1}{\ln 10}$

64. $\ln e = \dfrac{1}{\log 10}$

65. $\ln (xy)^3 = (\ln x + \ln y)^3$

66. $\dfrac{\ln a}{\ln b} = \dfrac{\log a}{\log b}$

CHALLENGE

67. Prove the quotient rule: $\log_b\left(\dfrac{M}{N}\right) = \log_b M - \log_b N.$

Hint: Let $u = \log_b M$ and $v = \log_b N$. Write both in exponential form and find the quotient $\log_b\left(\dfrac{M}{N}\right)$.

68. Prove the power rule: $\log_b M^p = p \log_b M$.

Hint: Let $u = \log_b M$. Write this log in exponential form, and find $\log_b M^p$.

69. Write in terms of simpler logarithmic forms.

$$\log_b\left(\sqrt{\frac{x^2}{y^3 z^{-5}}}\right)^6$$

70. Show that $\log_b\left(\dfrac{1}{x}\right) = -\log_b x$.

TECHNOLOGY

71. Use a graphing calculator to plot $y = \ln(2x)$ and $y = \ln 2 + \ln x$. Are they the same graph?

72. Use a graphing calculator to plot $y = \ln(2 + x)$ and $y = \ln 2 + \ln x$. Are they the same graph?

73. Use a graphing calculator to plot $y = \dfrac{\log x}{\log 2}$ and $y = \log x - \log 2$. Are they the same graph?

74. Use a graphing calculator to plot $y = \log\left(\dfrac{x}{2}\right)$ and $y = \log x - \log 2$. Are they the same graph?

75. Use a graphing calculator to plot $y = \ln(x^2)$ and $y = 2 \ln x$. Are they the same graph?

76. Use a graphing calculator to plot $y = (\ln x)^2$ and $y = 2 \ln x$. Are they the same graph?

77. Use a graphing calculator to plot $y = \ln x$ and $y = \dfrac{\log x}{\log e}$. Are they the same graph?

78. Use a graphing calculator to plot $y = \log x$ and $y = \dfrac{\ln x}{\ln 10}$. Are they the same graph?

SECTION 5.4 EXPONENTIAL AND LOGARITHMIC EQUATIONS

SKILLS OBJECTIVES

- Solve exponential equations.
- Solve logarithmic equations.
- Solve application problems using exponential and logarithmic equations.

CONCEPTUAL OBJECTIVE

- Understand how exponential and logarithmic equations are solved using properties of one-to-one functions and inverses.

In this book you have solved algebraic equations such as $x^2 - 9 = 0$, in which the goal is to solve for x by finding the values of x that make the statement true. Exponential and logarithmic equations have the x buried within an exponent or a logarithm, but the goal is the same. Solve for x.

$$\text{Exponential equation:} \quad e^{2x+1} = 5$$

$$\text{Logarithmic equation:} \quad \log(3x - 1) = 7$$

There are two methods for solving exponential and logarithmic equations that are based on the properties of one-to-one functions and inverses. To solve simple exponential and logarithmic equations, we will use one-to-one properties. To solve more complicated exponential and logarithmic equations, we will use properties of inverses. The following box summarizes the one-to-one and inverse properties that hold true when $b > 0$ and $b \neq 1$.

ONE-TO-ONE PROPERTIES

$b^x = b^y$	if and only if	$x = y$
$\log_b x = \log_b y$	if and only if	$x = y$

INVERSE PROPERTIES

$$b^{\log_b x} = x \qquad x > 0$$

$$\log_b b^x = x$$

We now outline a strategy for solving exponential and logarithmic equations using the one-to-one and inverse properties.

STRATEGIES FOR SOLVING EXPONENTIAL AND LOGARITHMIC EQUATIONS

Strategy for Solving Exponential Equations

TYPE OF EQUATION	STRATEGY	EXAMPLE
Simple	1. Rewrite both sides of the equation in terms of the same base.	$2^{x-3} = 32$ $2^{x-3} = 2^5$
	2. Use the one-to-one property to equate the exponents.	$x - 3 = 5$
	3. Solve for the variable.	$\boxed{x = 8}$
Complicated	1. Isolate the exponential expression.	$3e^{2x} - 2 = 7$ $3e^{2x} = 9$ $e^{2x} = 3$
	2. Take the same logarithm* of both sides.	$\ln e^{2x} = \ln 3$
	3. Simplify using the inverse properties.	$2x = \ln 3$
	4. Solve for the variable.	$\boxed{x = \frac{1}{2}\ln 3}$

*Take the logarithm with base that is equal to the base of the exponent and use the property $\log_b b^x = x$ or take the natural logarithm and use the property $\ln M^p = p \ln M$.

Strategy for Solving Logarithmic Equations

TYPE OF EQUATION	STRATEGY	EXAMPLE
Simple	1. Combine logarithms on each side of the equation using properties.	$\log(x - 3) + \log x = \log 4$ $\log x(x - 3) = \log 4$
	2. Use the one-to-one property to equate the arguments.	$x(x - 3) = 4$
	3. Solve for the variable.	$x^2 - 3x - 4 = 0$ $(x - 4)(x + 1) = 0$ $x = -1, 4$
	4. Eliminate any extraneous solutions.	Eliminate $x = -1$ because $\log -1$ is undefined. $\boxed{x = 4}$
Complicated	1. Combine and isolate the logarithmic expressions.	$\log_5(x + 2) - \log_5 x = 2$ $\log_5\left(\frac{x+2}{x}\right) = 2$
	2. Rewrite the equation in exponential form.	$\frac{x+2}{x} = 5^2$
	3. Solve for the variable.	$x + 2 = 25x$ $24x = 2$ $\boxed{x = \frac{1}{12}}$

Solving Exponential Equations

EXAMPLE 1 Solving a Simple Exponential Equation

Solve the exponential equations using the one-to-one property.

a. $3^x = 81$ **b.** $5^{7-x} = 125$ **c.** $\left(\frac{1}{2}\right)^{4y} = 16$

Solution (a):

Substitute $81 = 3^4$. $3^x = 3^4$

Use the one-to-one property to identify x. $\boxed{x = 4}$

Solution (b):

Substitute $125 = 5^3$. $5^{7-x} = 5^3$

Use the one-to-one property. $7 - x = 3$

Solve for x. $\boxed{x = 4}$

Solution (c):

Substitute $\left(\frac{1}{2}\right)^{4y} = \left(\frac{1}{2^{4y}}\right) = 2^{-4y}$. $2^{-4y} = 16$

Substitute $16 = 2^4$. $2^{-4y} = 2^4$

Use the one-to-one property to identify y. $\boxed{y = -1}$

■ Answer: **a.** $x = 4$ **b.** $y = -3$

■ **YOUR TURN** Solve the following equations.

a. $2^{x-1} = 8$ **b.** $\left(\frac{1}{3}\right)^y = 27$

In Example 1, we were able to rewrite the equation in a form with the same bases so that we could use the one-to-one property. In Example 2, we will not be able to write both sides in a form with the same bases. Instead, we will use properties of inverses.

EXAMPLE 2 Solving a More Complicated Exponential Equation with a Base Other Than 10 or *e*

Solve the exponential equations and round the answers to four decimal places.

a. $5^{3x} = 16$ **b.** $4^{3x+2} = 71$

Solution (a):

Take the natural logarithm of both sides of the equation. $\ln 5^{3x} = \ln 16$

Use the power property on the left side of the equation. $3x \ln 5 = \ln 16$

Divide both sides of the equation by 3 ln 5. $x = \dfrac{\ln 16}{3 \ln 5}$

Use a calculator to approximate x to four decimal places. $\boxed{x \approx 0.5742}$

Solution (b):

Rewrite in logarithmic form. $3x + 2 = \log_4 71$

Subtract 2 from both sides. $3x = \log_4 71 - 2$

Divide both sides by 3. $x = \dfrac{\log_4 71 - 2}{3}$

Use the change-of-base formula, $\log_4 71 = \dfrac{\ln 71}{\ln 4}$. $x = \dfrac{\dfrac{\ln 71}{\ln 4} - 2}{3}$

Use a calculator to approximate x to four decimal places. $x \approx \dfrac{3.0749 - 2}{3} \approx \boxed{0.3583}$

We could have proceeded in an alternative way by taking either the natural log or the common log of both sides and using the power property (instead of using the change-of-base formula) to evaluate the logarithm with base 4.

Take the natural logarithm of both sides. $\ln\left(4^{3x+2}\right) = \ln 71$

Use the power property (7). $(3x + 2)\ln 4 = \ln 71$

Divide by ln 4. $3x + 2 = \dfrac{\ln 71}{\ln 4}$

Subtract 2 and divide by 3. $x = \dfrac{\dfrac{\ln 71}{\ln 4} - 2}{3}$

Use a calculator to approximate x. $x \approx \dfrac{3.0749 - 2}{3} \approx \boxed{0.3583}$

Technology Tip

```
((ln(71)/ln(4)-2
)/3
          .3582911866
```

■ **YOUR TURN** Solve the equation $5^{y^2} = 27$ and round the answer to four decimal places.

■ **Answer:** $y \approx \pm 1.4310$

EXAMPLE 3 Solving a More Complicated Exponential Equation with Base 10 or *e*

Solve the exponential equation $4e^{x^2} = 64$ and round the answer to four decimal places.

Solution:

Divide both sides by 4. $e^{x^2} = 16$

Take the natural logarithm (ln) of both sides. $\ln\left(e^{x^2}\right) = \ln 16$

Simplify the left side with the property of inverses. $x^2 = \ln 16$

Solve for x using the square-root method. $x = \pm\sqrt{\ln 16}$

Use a calculator to approximate x to four decimal places. $\boxed{x \approx \pm 1.6651}$

■ **YOUR TURN** Solve the equation $10^{2x-3} = 7$ and round the answer to four decimal places.

■ **Answer:** $x \approx 1.9225$

Technology Tip

The graph of the function $e^{2x} - 4e^x + 3$ is shown. The x-intercepts are $x \approx 1.10$ and $x = 0$.

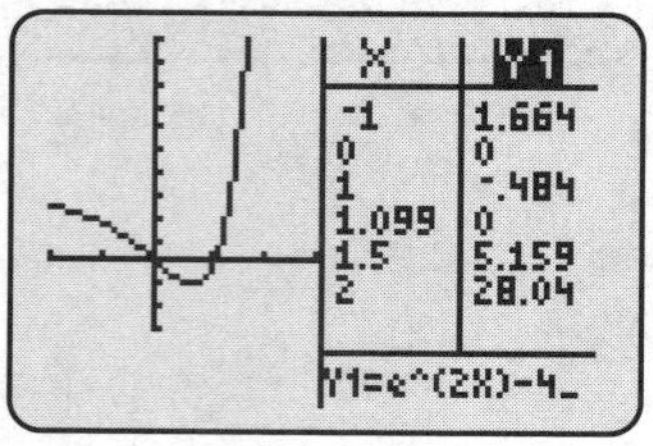

■ **Answer:** $x \approx 0.3010$

EXAMPLE 4 Solving an Exponential Equation Quadratic in Form

Solve the equation $e^{2x} - 4e^x + 3 = 0$ and round the answer to four decimal places.

Solution:

Let $u = e^x$. $\left(\textit{Note: } u^2 = e^x \cdot e^x = e^{2x}.\right)$ $\quad u^2 - 4u + 3 = 0$

Factor. $\quad (u - 3)(u - 1) = 0$

Solve for u. $\quad u = 3 \text{ or } u = 1$

Substitute $u = e^x$. $\quad e^x = 3 \text{ or } e^x = 1$

Take the natural logarithm (ln) of both sides. $\quad \ln(e^x) = \ln 3 \text{ or } \ln(e^x) = \ln 1$

Simplify with the properties of logarithms. $\quad x = \ln 3 \text{ or } x = \ln 1$

Approximate or evaluate exactly the right sides. $\quad \boxed{x \approx 1.0986}$ or $\boxed{x = 0}$

■ **YOUR TURN** Solve the equation $100^x - 10^x - 2 = 0$ and round the answer to four decimal places.

Solving Logarithmic Equations

We can solve simple logarithmic equations using the property of one-to-one functions. For more complicated logarithmic equations we can employ properties of logarithms and properties of inverses. **Solutions must be checked to eliminate extraneous solutions.**

Study Tip

Solutions should be checked in the original equation to eliminate extraneous solutions.

■ **Answer:** $x = 2$

EXAMPLE 5 Solving a Simple Logarithmic Equation

Solve the equation $\log_4(2x - 3) = \log_4(x) + \log_4(x - 2)$.

Solution:

Apply the product property (5) on the right side. $\quad \log_4(2x - 3) = \log_4[x(x - 2)]$

Apply the property of one-to-one functions. $\quad 2x - 3 = x(x - 2)$

Distribute and simplify. $\quad x^2 - 4x + 3 = 0$

Factor. $\quad (x - 3)(x - 1) = 0$

Solve for x. $\quad x = 3 \text{ or } x = 1$

The possible solution $x = 1$ must be eliminated because it is not in the domain of two of the logarithmic functions.

$$x = 1\text{: } \overbrace{\log_4(-1)}^{\text{undefined}} \stackrel{?}{=} \log_4(1) + \overbrace{\log_4(-1)}^{\text{undefined}}$$

$\boxed{x = 3}$

■ **YOUR TURN** Solve the equation $\ln(x + 8) = \ln(x) + \ln(x + 3)$.

EXAMPLE 6 Solving a More Complicated Logarithmic Equation

Solve the equation $\log_3(9x) - \log_3(x - 8) = 4$.

Solution:

Employ the quotient property (6) on the left side. $\quad \log_3\left(\dfrac{9x}{x - 8}\right) = 4$

Write in exponential form. $\quad \log_b x = y \Rightarrow x = b^y \quad \dfrac{9x}{x - 8} = 3^4$

Simplify the right side. $\dfrac{9x}{x - 8} = 81$

Multiply the equation by the LCD, $x - 8$. $9x = 81(x - 8)$

Eliminate parentheses. $9x = 81x - 648$

Solve for x. $-72x = -648$

$\boxed{x = 9}$

Check: $\log_3[9 \cdot 9] - \log_3[9 - 8] = \log_3[81] - \log_3 1 = 4 - 0 = 4$

YOUR TURN Solve the equation $\log_2(4x) - \log_2(2) = 2$.

Answer: $x = 2$

EXAMPLE 7 Solving a Logarithmic Equation with No Solution

Solve the equation $\ln(3 - x^2) = 7$.

Solution:

Write in exponential form. $3 - x^2 = e^7$

Simplify. $x^2 = \underbrace{3 - e^7}$

$3 - e^7$ is negative. $x^2 =$ negative real number

There are no real numbers that when squared yield a negative real number. Therefore, there is $\boxed{\text{no real solution}}$.

Technology Tip

Graphs of $y_1 = \ln(3 - x^2)$ and $y_2 = 7$ are shown.

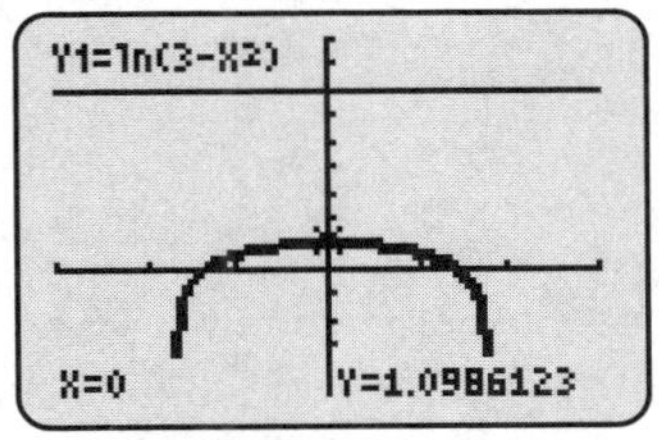

No intersection of the graphs indicates no solution.

Applications

In the chapter opener we saw that archaeologists determine the age of a fossil by how much carbon 14 is present at the time of discovery. The number of grams of carbon 14 based on the radioactive decay of the isotope is given by

$$A = A_0 e^{-0.000124t}$$

where A is the number of grams of carbon 14 at the present time, A_0 is the number of grams of carbon 14 while alive, and t is the number of years since death. Using the inverse properties, we can isolate t.

WORDS	**MATH**
Divide by A_0.	$\dfrac{A}{A_0} = e^{-0.000124t}$
Take the natural logarithm of both sides.	$\ln\left(\dfrac{A}{A_0}\right) = \ln\left(e^{-0.000124t}\right)$
Simplify the right side utilizing properties of inverses.	$\ln\left(\dfrac{A}{A_0}\right) = -0.000124t$
Solve for t.	$t = -\dfrac{1}{0.000124}\ln\left(\dfrac{A}{A_0}\right)$

Let's assume that animals have approximately 1000 mg of carbon 14 in their bodies when they are alive. If a fossil has 200 mg of carbon 14, approximately how old is the fossil? Substituting $A = 200$ and $A_0 = 1000$ into our equation for t, we find

$$t = -\frac{1}{0.000124}\ln\left(\frac{1}{5}\right) \approx 12{,}979.338$$

The fossil is approximately 13,000 years old.

EXAMPLE 8 Calculating How Many Years It Will Take for Money to Double

You save $1,000 from a summer job and put it in a CD earning 5% compounding continuously. How many years will it take for your money to double? Round to the nearest year.

Solution:

Recall the compound continuous interest formula. $A = Pe^{rt}$

Substitute $P = 1000$, $A = 2000$, and $r = 0.05$. $2000 = 1000e^{0.05t}$

Divide by 1000. $2 = e^{0.05t}$

Take the natural logarithm of both sides. $\ln 2 = \ln\left(e^{0.05t}\right)$

Simplify with the property $\ln e^x = x$. $\ln 2 = 0.05t$

Solve for t. $t = \dfrac{\ln 2}{0.05} \approx 13.8629$

It will take almost $\boxed{14 \text{ years}}$ for your money to double.

■ **Answer:** Approximately 11 years

■ **YOUR TURN** How long will it take $1,000 to triple (become $3,000) in a savings account earning 10% a year compounding continuously? Round your answer to the nearest year.

When an investment is compounded continuously, how long will it take for that investment to double?

WORDS	MATH
Write the interest formula for compounding continuously.	$A = Pe^{rt}$
Let $A = 2P$ (investment doubles).	$2P = Pe^{rt}$
Divide both sides of the equation by P.	$2 = e^{rt}$
Take the natural log of both sides of the equation.	$\ln 2 = \ln e^{rt}$
Simplify the right side by applying the property $\ln e^x = x$.	$\ln 2 = rt$
Divide both sides by r.	$t = \dfrac{\ln 2}{r}$
Approximate $\ln 2 \approx 0.7$.	$t \approx \dfrac{0.7}{r}$

Recall that an interest rate such as 4.5% is written in decimal form as $r = 0.045$.

Multiply the numerator and denominator by 100. $t \approx \dfrac{70}{100r}$

This is the "rule of 70."

If we divide 70 by the interest rate (compounding continuously), we get the approximate time for an investment to double. In Example 8, the interest rate (compounding continuously) is 5%. Dividing 70 by 5 yields 14 years.

SECTION 5.4 SUMMARY

Strategy for Solving Exponential Equations

TYPE OF EQUATION	STRATEGY
Simple	1. Rewrite both sides of the equation in terms of the same base. 2. Use the one-to-one property to equate the exponents. 3. Solve for the variable.
Complicated	1. Isolate the exponential expression. 2. Take the same logarithm* of both sides. 3. Simplify using the inverse properties. 4. Solve for the variable.

*Take the logarithm with the base that is equal to the base of the exponent and use the property $\log_b b^x = x$ or take the natural logarithm and use the property $\ln M^p = p \ln M$.

Strategy for Solving Logarithmic Equations

TYPE OF EQUATION	STRATEGY
Simple	1. Combine logarithms on each side of the equation using properties. 2. Use the one-to-one property to equate the arguments. 3. Solve for the variable. 4. Eliminate any extraneous solutions.
Complicated	1. Combine and isolate the logarithmic expressions. 2. Rewrite the equation in exponential form. 3. Solve for the variable.

SECTION 5.4 EXERCISES

SKILLS

In Exercises 1–18, solve the exponential equations exactly for x.

1. $3^x = 81$ **2.** $5^x = 125$ **3.** $7^x = \frac{1}{49}$ **4.** $4^x = \frac{1}{16}$ **5.** $2^{x^2} = 16$

6. $169^x = 13$ **7.** $\left(\frac{2}{3}\right)^{x+1} = \frac{27}{8}$ **8.** $\left(\frac{3}{5}\right)^{x+1} = \frac{25}{9}$ **9.** $e^{2x+3} = 1$ **10.** $10^{x^2-1} = 1$

11. $7^{2x-5} = 7^{3x-4}$ **12.** $125^x = 5^{2x-3}$ **13.** $2^{x^2+12} = 2^{7x}$ **14.** $5^{x^2-3} = 5^{2x}$ **15.** $9^x = 3^{x^2-4x}$

16. $16^{x-1} = 2^{x^2}$ **17.** $e^{5x-1} = e^{x^2+3}$ **18.** $10^{x^2-8} = 100^x$

In Exercises 19–44, solve the exponential equations. Make sure to isolate the base to a power first. Round your answers to three decimal places.

19. $10^{2x-3} = 81$ **20.** $2^{3x+1} = 21$ **21.** $3^{x+1} = 5$ **22.** $5^{2x-1} = 35$

23. $27 = 2^{3x-1}$ **24.** $15 = 7^{3-2x}$ **25.** $3e^x - 8 = 7$ **26.** $5e^x + 12 = 27$

27. $9 - 2e^{0.1x} = 1$ **28.** $21 - 4e^{0.1x} = 5$ **29.** $2(3^x) - 11 = 9$ **30.** $3(2^x) + 8 = 35$

31. $e^{3x+4} = 22$ **32.** $e^{x^2} = 73$ **33.** $3e^{2x} = 18$ **34.** $4(10^{3x}) = 20$

35. $e^{2x} + 7e^x - 3 = 0$ **36.** $e^{2x} - 4e^x - 5 = 0$ **37.** $(3^x - 3^{-x})^2 = 0$ **38.** $(3^x - 3^{-x})(3^x + 3^{-x}) = 0$

39. $\dfrac{2}{e^x - 5} = 1$ **40.** $\dfrac{17}{e^x + 4} = 2$ **41.** $\dfrac{20}{6 - e^{2x}} = 4$ **42.** $\dfrac{4}{3 - e^{3x}} = 8$

43. $\dfrac{4}{10^{2x} - 7} = 2$ **44.** $\dfrac{28}{10^x + 3} = 4$

In Exercises 45–60, solve the logarithmic equations exactly.

45. $\log(2x) = 2$
46. $\log(5x) = 3$
47. $\log_3(2x + 1) = 4$
48. $\log_2(3x - 1) = 3$
49. $\log_2(4x - 1) = -3$
50. $\log_4(5 - 2x) = -2$
51. $\ln x^2 - \ln 9 = 0$
52. $\log x^2 + \log x = 3$
53. $\log_5(x - 4) + \log_5 x = 1$
54. $\log_2(x - 1) + \log_2(x - 3) = 3$
55. $\log(x - 3) + \log(x + 2) = \log(4x)$
56. $\log_2(x + 1) + \log_2(4 - x) = \log_2(6x)$
57. $\log_4(4x) - \log_4\left(\frac{x}{4}\right) = 3$
58. $\log_3(7 - 2x) - \log_3(x + 2) = 2$
59. $\log(2x - 5) - \log(x - 3) = 1$
60. $\log_3(10 - x) - \log_3(x + 2) = 1$

In Exercises 61–76, solve the logarithmic equations. Round your answers to three decimal places.

61. $\ln x^2 = 5$
62. $\log 3x = 2$
63. $\log(2x + 5) = 2$
64. $\ln(4x - 7) = 3$
65. $\ln(x^2 + 1) = 4$
66. $\log(x^2 + 4) = 2$
67. $\ln(2x + 3) = -2$
68. $\log(3x - 5) = -1$
69. $\log(2 - 3x) + \log(3 - 2x) = 1.5$
70. $\log_2(3 - x) + \log_2(1 - 2x) = 5$
71. $\ln(x) + \ln(x - 2) = 4$
72. $\ln(4x) + \ln(2 + x) = 2$
73. $\log_7(1 - x) - \log_7(x + 2) = \log_7 x$
74. $\log_5(x + 1) - \log_5(x - 1) = \log_5(x)$
75. $\ln\sqrt{x + 4} - \ln\sqrt{x - 2} = \ln\sqrt{x + 1}$
76. $\log(\sqrt{1 - x}) - \log(\sqrt{x + 2}) = \log x$

APPLICATIONS

77. Money. If money is invested in a savings account earning 3.5% interest compounded yearly, how many years will pass until the money triples?

78. Money. If money is invested in a savings account earning 3.5% interest compounded monthly, how many years will pass until the money triples?

79. Money. If $7,500 is invested in a savings account earning 5% interest compounded quarterly, how many years will pass until there is $20,000?

80. Money. If $9,000 is invested in a savings account earning 6% interest compounded continuously, how many years will pass until there is $15,000?

81. Earthquakes. On September 25, 2003 an earthquake that measured 7.4 on the Richter scale shook Hokkaido, Japan. How much energy (joules) did the earthquake emit?

82. Earthquakes. Again, on that same day (September 25, 2003), a second earthquake that measured 8.3 on the Richter scale shook Hokkaido, Japan. How much energy (joules) did the earthquake emit?

83. Sound. Matt likes to drive around campus in his classic Mustang with the stereo blaring. If his boom stereo has a sound intensity of 120 dB, how many watts per square meter does the stereo emit?

84. Sound. The New York Philharmonic has a sound intensity of 100 dB. How many watts per square meter does the orchestra emit?

85. Anesthesia. When a person has a cavity filled, the dentist typically administers a local anesthetic. After leaving the dentist's office, one's mouth often remains numb for several more hours. If a shot of anesthesia is injected into the bloodstream at the time of the procedure ($t = 0$), and the amount of anesthesia still in the bloodstream t hours after the initial injection is given by $A = A_0 e^{-0.5t}$, in how many hours will only 10% of the original anesthetic still be in the bloodstream?

86. Investments. Money invested in an account that compounds interest continuously at a rate of 3% a year is modeled by $A = A_0 e^{0.03t}$, where A is the amount in the investment after t years and A_0 is the initial investment. How long will it take the initial investment to double?

87. Biology. The U.S. Fish and Wildlife Service is releasing a population of the endangered Mexican gray wolf in a protected area along the New Mexico and Arizona border. They estimate the population of the Mexican gray wolf to be approximated by

$$P(t) = \frac{200}{1 + 24e^{-0.2t}}$$

How many years will it take for the population to reach 100 wolves?

88. Introducing a New Car Model. If the number of new model Honda Accord hybrids purchased in North America is given by $N = \frac{100{,}000}{1 + 10e^{-2t}}$, where t is the number of weeks after Honda releases the new model, how many weeks will it take after the release until there are 50,000 Honda hybrids from that batch on the road?

89. Earthquakes. A P wave measures 6.2 on the Richter scale, and an S wave measures 3.3 on the Richter scale. What is their combined measure on the Richter scale?

90. Sound. You and a friend get front row seats to a rock concert. The music level is 100 dB, and your normal conversation is 60 dB. If your friend is telling you something during the concert, how many decibels are you subjecting yourself to?

CATCH THE MISTAKE

In Exercises 91–94, explain the mistake that is made.

91. Solve the equation: $4e^x = 9$.

Solution:

Take the natural log of both sides. $\ln(4e^x) = \ln 9$

Apply the property of inverses. $4x = \ln 9$

Solve for x. $x = \dfrac{\ln 9}{4} \approx 0.55$

This is incorrect. What mistake was made?

92. Solve the equation: $\log(x) + \log(3) = 1$.

Solution:

Apply the product property (5). $\log(3x) = 1$

Exponentiate (base 10). $10^{\log(3x)} = 1$

Apply the properties of inverses. $3x = 1$

Solve for x. $x = \dfrac{1}{3}$

This is incorrect. What mistake was made?

93. Solve the equation: $\log(x) + \log(x + 3) = 1$ for x.

Solution:

Apply the product property (5). $\log(x^2 + 3x) = 1$

Exponentiate both sides (base 10). $10^{\log(x^2+3x)} = 10^1$

Apply the property of inverses. $x^2 + 3x = 10$

Factor. $(x + 5)(x - 2) = 0$

Solve for x. $x = -5$ and $x = 2$

This is incorrect. What mistake was made?

94. Solve the equation: $\log x + \log 2 = \log 5$.

Solution:

Combine the logarithms on the left. $\log(x + 2) = \log 5$

Apply the property of one-to-one functions. $x + 2 = 5$

Solve for x. $x = 3$

This is incorrect. What mistake was made?

CONCEPTUAL

In Exercises 95–98, determine whether each statement is true or false.

95. The sum of logarithms with the same base is equal to the logarithm of the product.

96. A logarithm squared is equal to two times the logarithm.

97. $e^{\log x} = x$

98. $e^x = -2$ has no solution.

99. Solve for x in terms of b:

$$\frac{1}{3}\log_b(x^3) + \frac{1}{2}\log_b(x^2 - 2x + 1) = 2$$

100. Solve exactly:

$$2\log_b(x) + 2\log_b(1 - x) = 4.$$

CHALLENGE

101. Solve $y = \dfrac{3000}{1 + 2e^{-0.2t}}$ for t in terms of y.

102. State the range of values of x that the following identity holds: $e^{\ln(x^2-a)} = x^2 - a$.

103. A function called the hyperbolic cosine is defined as the average of exponential growth and exponential decay by $f(x) = \dfrac{e^x + e^{-x}}{2}$. Find its inverse.

104. A function called the hyperbolic sine is defined by $f(x) = \dfrac{e^x - e^{-x}}{2}$. Find its inverse.

TECHNOLOGY

105. Solve the equation $\ln 3x = \ln(x^2 + 1)$. Using a graphing calculator, plot the graphs $y = \ln(3x)$ and $y = \ln(x^2 + 1)$ in the same viewing rectangle. Zoom in on the point where the graphs intersect. Does this agree with your solution?

106. Solve the equation $10^{x^2} = 0.001^x$. Using a graphing calculator, plot the graphs $y = 10^{x^2}$ and $y = 0.001^x$ in the same viewing rectangle. Does this confirm your solution?

107. Use a graphing utility to help solve $3^x = 5x + 2$.

108. Use a graphing utility to help solve $\log x^2 = \ln(x - 3) + 2$.

109. Use a graphing utility to graph $y = \dfrac{e^x + e^{-x}}{2}$. State the domain. Determine if there are any symmetry and asymptote.

110. Use a graphing utility to graph $y = \dfrac{e^x + e^{-x}}{e^x - e^{-x}}$. State the domain. Determine if there are any symmetry and asymptote.

SECTION 5.5 EXPONENTIAL AND LOGARITHMIC MODELS

SKILLS OBJECTIVES

- Apply exponential growth and exponential decay models to biological, demographic, and economic phenomena.
- Represent distributions by means of a Gaussian model.
- Use logistic growth models to represent phenomena involving limits to growth.
- Solve problems such as species populations, credit card payoff, and wearoff of anesthesia through logarithmic models.

CONCEPTUAL OBJECTIVE

- Recognize exponential growth, exponential decay, Gaussian distributions, logistic growth, and logarithmic models.

The following table summarizes the five primary models that involve exponential and logarithmic functions.

NAME	MODEL	GRAPH	APPLICATIONS
Exponential growth	$f(x) = ce^{kx}$ $\quad k > 0$	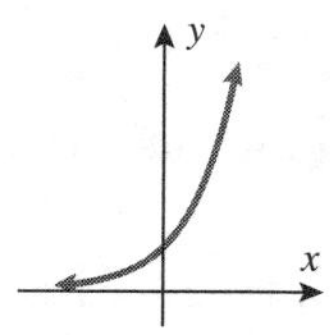	World populations, bacteria growth, appreciation, global spread of the HIV virus
Exponential decay	$f(x) = ce^{-kx}$ $\quad k > 0$	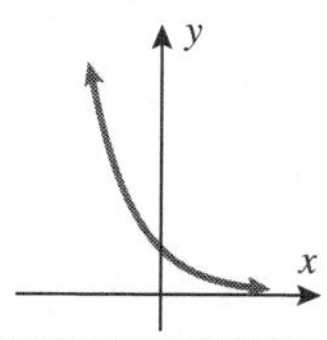	Radioactive decay, carbon dating, depreciation
Gaussian (normal) distribution	$f(x) = ce^{-(x-a)^2/k}$	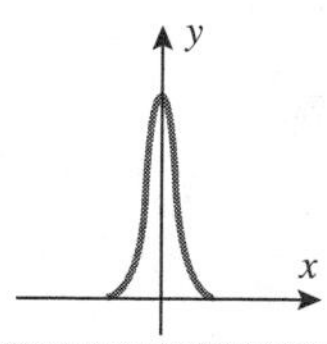	Bell curve (grade distribution), life expectancy, height/weight charts, intensity of a laser beam, IQ tests
Logistic growth	$f(x) = \dfrac{a}{1 + ce^{-kx}}$	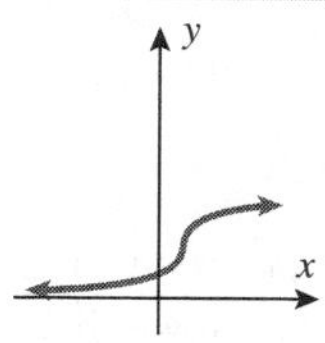	Conservation biology, learning curve, spread of virus on an island, carrying capacity
Logarithmic	$f(x) = a + c \log x$ $f(x) = a + c \ln x$	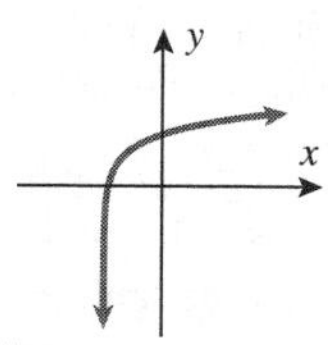	Population of species, anesthesia wearing off, time to pay off credit cards

Exponential Growth Models

Quite often one will hear that something "grows exponentially," meaning that it grows very fast and at increasing speed. In mathematics, the precise meaning of **exponential growth** is a *growth rate of a function that is proportional to its current size*. Let's assume you get a 5% raise every year in a government job. If your starting annual salary out of college is \$40,000, then your first raise will be \$2,000. Fifteen years later your annual salary will be approximately \$83,000 and your next 5% raise will be around \$4,150. The raise is always 5% of the current salary, so the larger the current salary, the larger the raise.

In Section 5.1 we saw that interest that is compounded continuously is modeled by $A = Pe^{rt}$. Here A stands for amount and P stands for principal. There are similar models for populations; these take the form $N(t) = N_0 e^{rt}$, where N_0 represents the number of people at time $t = 0$, r is the growth rate, t is time in years, and N represents the number of people at time t. In general, any model of the form $f(x) = ce^{kx}$, $k > 0$, models exponential growth.

EXAMPLE 1 World Population Projections

The world population is the total number of humans on Earth at a given time. In 2000 the world population was 6.1 billion and in 2005 the world population was 6.5 billion. Find the relative growth rate and determine what year the population will reach 9 billion.

Solution:

Assume an exponential growth model. $N(t) = N_0 e^{rt}$

Let $t = 0$ correspond to 2000. $N(0) = N_0 = 6.1$

In 2005, $t = 5$, the population was 6.5 billion. $6.5 = 6.1e^{5r}$

Solve for r. $\frac{6.5}{6.1} = e^{5r}$

$$\ln\left(\frac{6.5}{6.1}\right) = \ln\left(e^{5r}\right)$$

$$\ln\left(\frac{6.5}{6.1}\right) = 5r$$

$$r \approx 0.012702681$$

The relative growth rate is approximately $\boxed{1.3\%}$ per year.

Assuming the growth rate stays the same, write a population model. $N(t) = 6.1e^{0.013t}$

Let $N(t) = 9$. $9 = 6.1e^{0.013t}$

Solve for t. $e^{0.013t} = \frac{9}{6.1}$

$$\ln(e^{0.013t}) = \ln\left(\frac{9}{6.1}\right)$$

$$0.013t = \ln\left(\frac{9}{6.1}\right)$$

$$t \approx 29.91813894$$

In $\boxed{2030}$ the world population will reach 9 billion if the same growth rate is maintained.

YOUR TURN The population of North America was 300 million in 1995, and in 2005 the North American population was 332 million. Find the relative growth rate and determine what year the population will reach 1 billion.

Answer: 1% per year; 2115

Exponential Decay Models

We mentioned radioactive decay briefly in Section 5.1. Radioactive decay is the process in which radioactive elements (atoms) lose energy by emitting radiation in the form of particles. This results in loss of mass. This process is random, but given a large number of atoms, the decay rate is directly proportional to the mass of the radioactive substance. Since the mass is decreasing, we say this represents *exponential decay*, $m = m_0e^{-rt}$, where m_0 represents the initial mass at time $t = 0$, r is the decay rate, t is time, and m represents the mass at time t. In general, any model of the form $f(x) = ce^{-kx}$, $k > 0$, models **exponential decay**.

Typically, the decay rate r is expressed in terms of the half-life h. Recall (Section 5.1) that half-life is the time it takes for a quantity to decrease by half.

WORDS	MATH
Write the radioactive decay model.	$m = m_0e^{-rt}$
Divide both sides by m_0.	$\frac{m}{m_0} = e^{-rt}$
The remaining mass is half of the initial mass when $t = h$.	$\frac{1}{2} = e^{-rh}$
Solve for r.	
Take the natural logarithm of both sides.	$\ln\left(\frac{1}{2}\right) = \ln\left(e^{-rh}\right)$
Simplify.	$\underbrace{\ln 1}_{0} - \ln 2 = -rh$
	$rh = \ln 2$
	$\boxed{r = \frac{\ln 2}{h}}$

EXAMPLE 2 Radioactive Decay

The radioactive isotope of potassium ^{42}K, which is vital in the diagnosis of brain tumors, has a half-life of 12.36 hours.

a. Determine the exponential decay model that represents the mass of ^{42}K.
b. If 500 milligrams of potassium-42 are taken, how many milligrams will remain after 48 hours?
c. How long will it take for the original 500-milligram sample to decay to a mass of 5 milligrams?

Solution (a):

Write the relationship between rate of decay and half-life. $r = \frac{\ln 2}{h}$

Let $h = 12.36$. $r \approx 0.056$

Write the exponential decay model for the mass of ^{42}K. $\boxed{m = m_0e^{-0.056t}}$

Solution (b):

Let $m_0 = 500$ and $t = 48$. $m = 500e^{-(0.056)(48)} \approx 34.00841855$

There are approximately $\boxed{\text{34 milligrams}}$ of ^{42}K still in the body after 48 hours.

Note: Had we used the full value of $r = 0.056079868$, the resulting mass would have been $m = 33.8782897$, which is approximately 34 mg.

Solution (c):

Write the exponential decay model for the mass of ^{42}K. $\quad m = m_0 e^{-0.056t}$

Let $m = 5$ and $m_0 = 500$. $\quad 5 = 500e^{-0.056t}$

Solve for t.

Divide by 500. $\quad e^{-0.056t} = \dfrac{5}{500} = \dfrac{1}{100}$

Take the natural logarithm of both sides. $\quad \ln\left(e^{-0.056t}\right) = \ln\left(\dfrac{1}{100}\right)$

Simplify. $\quad -0.056t = \ln\left(\dfrac{1}{100}\right)$

Divide by -0.056 and approximate with a calculator. $\quad t \approx 82.2352$

It will take approximately **82 hours** for the original 500-mg substance to decay to a mass of 5 mg.

Technology Tip

Graphs of $Y_1 = 500e^{-0.56x}$ with $x = t$ and $Y_2 = 5$ are shown.

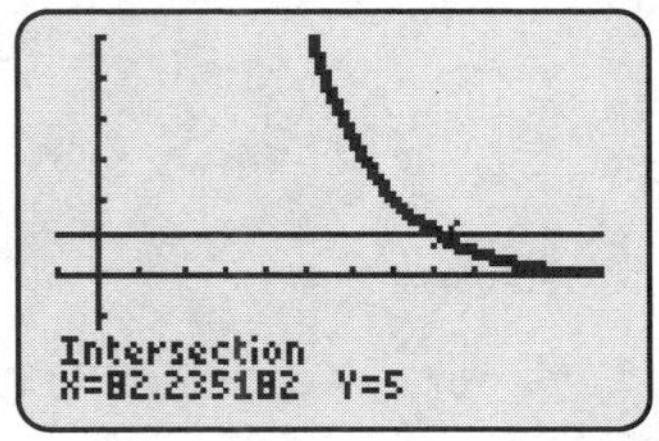

YOUR TURN The radioactive element radon-222 has a half-life of 3.8 days.

a. Determine the exponential decay model that represents the mass of radon-222.

b. How much of a 64-gram sample of radon-222 will remain after 7 days? Round to the nearest gram.

c. How long will it take for the original 64-gram sample to decay to a mass of 4 grams? Round to the nearest day.

Answer: **a.** $m = m_0 e^{-0.1824t}$
b. 18 grams
c. 15 days

Gaussian (Normal) Distribution Models

If your instructor plots the grades from the last test, typically you will see a **Gaussian (normal) distribution** of scores, otherwise known as the *bell-shaped curve*. Other examples of phenomena that tend to follow a Gaussian distribution are SAT scores, height distributions of adults, and standardized tests like IQ assessments.

The graph to the right represents a Gaussian distribution of IQ scores. The average score, which for IQ is 100, is always the maximum point on the curve. The typical probability distribution is

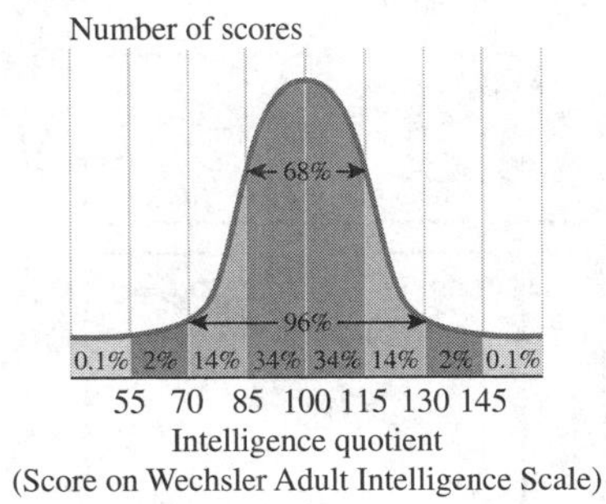

$$F(x) = \frac{1}{\sigma\sqrt{2\pi}} e^{-(x-\mu)^2/2\sigma^2}$$

where μ is the average or mean value and the variance is σ^2.

Any model of the form $f(x) = ce^{-(x-a)^2/k}$ is classified as a **Gaussian model**.

EXAMPLE 3 Weight Distributions

Suppose each member of a Little League football team is weighed and the weight distribution follows the Gaussian model $f(x) = 10e^{-(x-100)^2/25}$.

a. Graph the weight distribution.

b. What is the average weight of a member of this team?

c. Approximately how many boys weigh 95 pounds?

Solution:

a.

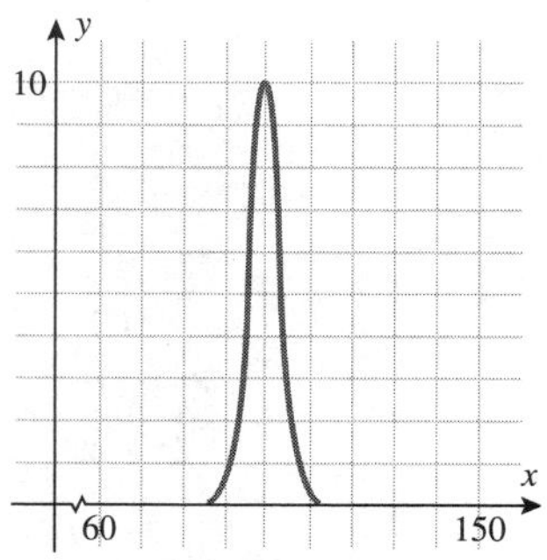

b. 100 pounds

c. $f(95) = 10e^{-(95-100)^2/25}$

$= 10e^{-25/25}$

$= 10e^{-1}$

≈ 3.6788

Approximately 4 boys weigh 95 lb.

Logistic Growth Models

Earlier in this section we discussed exponential growth models for populations that experience uninhibited growth. Now we will turn our attention to *logistic growth*, which models population growth when there are factors that impact the ability to grow, such as food and space. For example, if 10 rabbits are dropped off on an uninhabited island, they will reproduce and the population of rabbits on that island will experience rapid growth. The population will continue to increase rapidly until the rabbits start running out of space or food on the island. In other words, under favorable conditions the growth is not restricted, while under less favorable conditions the growth becomes restricted. This type of growth is represented by **logistic growth models**, $f(x) = \dfrac{a}{1 + ce^{-kx}}$. Ultimately, the population of rabbits reaches the island's *carrying capacity, a*.

Technology Tip

Plot the graph of $Y_1 = \dfrac{500000}{1 + 5e^{-0.12x}}$ with $x = t$. Use the TRACE key to find $f(20)$ and $f(30)$.

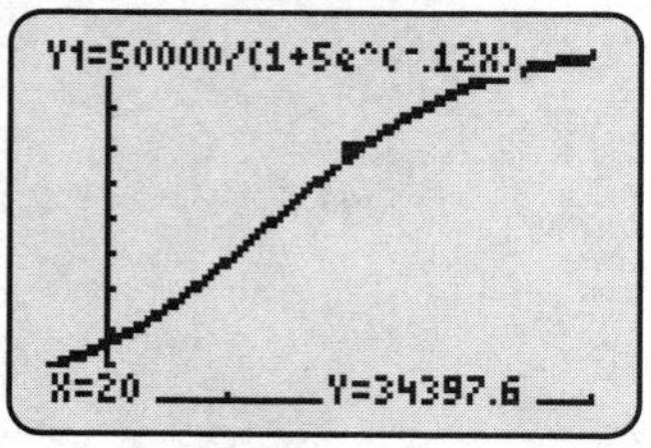

A table of values supports the solution at $t = 20$ and $t = 30$.

EXAMPLE 4 Number of Students on a College Campus

In 2008, the University of Central Florida was the sixth largest university in the country. The number of students can be modeled by the function:

$f(x) = \dfrac{50{,}000}{1 + 5e^{-0.12t}}$, where t is time in years and $t = 0$ corresponds to 1970.

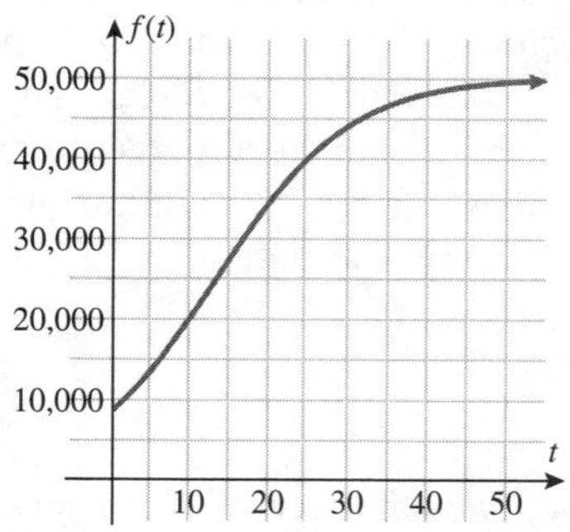

a. How many students attended UCF in 1990? Round to the nearest thousand.
b. How many students attended UCF in 2000?
c. What is the carrying capacity of the UCF main campus?

Round all answers to the nearest thousand.

Solution (a): Let $t = 20$. $\quad f(x) = \dfrac{50{,}000}{1 + 5e^{-0.12(20)}} \approx$ 34,000

Solution (b): Let $t = 30$. $\quad f(x) = \dfrac{50{,}000}{1 + 5e^{-0.12(30)}} \approx$ 44,000

Solution (c): As t increases, the UCF student population approaches 50,000.

Logarithmic Models

Homeowners typically ask the question, "If I increase my payment, how long will it take to pay off my current mortgage?" In general, a loan over t years with an annual interest rate r with n periods per year corresponds to an interest rate per period of $i = \frac{r}{n}$. Typically loans are paid in equal payments consisting of the principal P plus total interest divided by the total number of periods over the life of the loan nt. The periodic payment R is given by

$$R = P\frac{i}{1 - (1 + i)^{-nt}}$$

We can find the time (in years) it will take to pay off the loan as a function of periodic payment by solving for t.

Words	**Math**
Multiply both sides by $1 - (1 + i)^{-nt}$.	$R[1 - (1 + i)^{-nt}] = Pi$
Eliminate the brackets.	$R - R(1 + i)^{-nt} = Pi$
Subtract R.	$-R(1 + i)^{-nt} = Pi - R$
Divide by $-R$.	$(1 + i)^{-nt} = 1 - \frac{Pi}{R}$
Take the natural log of both sides.	$\ln(1 + i)^{-nt} = \ln\left(1 - \frac{Pi}{R}\right)$
Use the power property for logarithms.	$-nt\ln(1 + i) = \ln\left(1 - \frac{Pi}{R}\right)$
Isolate t.	$t = -\frac{\ln(1 - Pi/R)}{n\ln(1 + i)}$
Let $i = \frac{r}{n}$.	$\boxed{t = -\frac{\ln(1 - Pr/nR)}{n\ln(1 + r/n)}}$

Technology Tip

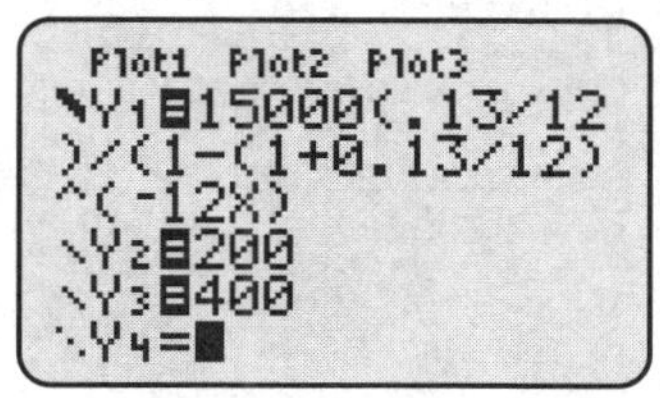

Use the keystrokes [2nd] [Calc] [5:Intersect] to find the points of intersection.

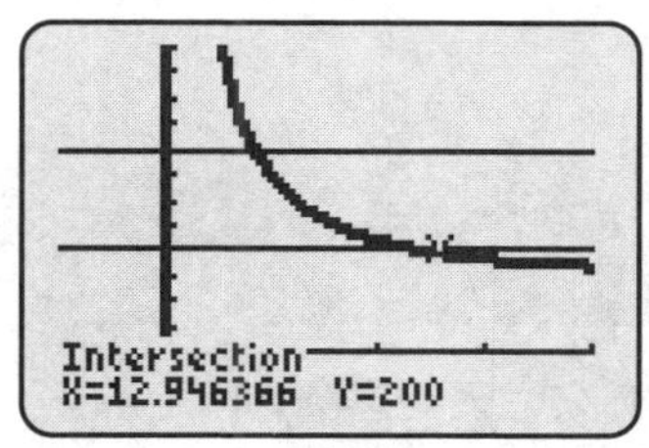

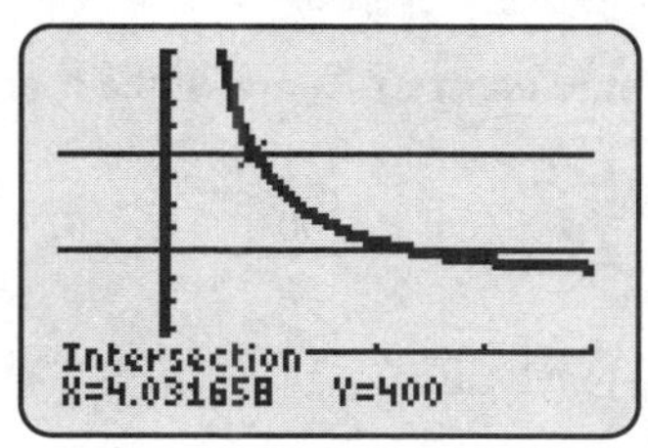

EXAMPLE 5 Paying Off Credit Cards

James owes \$15,000 on his credit card. The annual interest rate is 13% compounded monthly.

a. Find the time it will take to pay off his credit card if he makes payments of \$200 per month.

b. Find the time it will take to pay off his credit card if he makes payments of \$400 per month.

Let $P = 15{,}000$, $r = 0.13$, and $n = 12$.

$$t = -\frac{\ln\left(1 - \frac{15{,}000(0.13)}{12R}\right)}{12\ln\left(1 + \frac{0.13}{12}\right)}$$

Solution (a): Let $R = 200$.

$$t = -\frac{\ln\left(1 - \frac{15{,}000(0.13)}{12(200)}\right)}{12\ln\left(1 + \frac{0.13}{12}\right)} \approx 13$$

\$200 monthly payments will allow James to pay off his credit card in about $\boxed{13 \text{ years}}$.

Solution (b): Let $R = 400$.

$$t = -\frac{\ln\left(1 - \frac{15{,}000(0.13)}{12(400)}\right)}{12\ln\left(1 + \frac{0.13}{12}\right)} \approx 4$$

\$400 monthly payments will allow James to pay off the balance in approximately $\boxed{4 \text{ years}}$. It is important to note that doubling the payment reduced the time to pay off the balance by less than a third.

SECTION 5.5 SUMMARY

In this section we discussed five main types of models that involve exponential and logarithmic functions.

NAME	MODEL	APPLICATIONS
Exponential growth	$f(x) = ce^{kx}, k > 0$	Uninhibited growth (populations/inflation)
Exponential decay	$f(x) = ce^{-kx}, k > 0$	Carbon dating, depreciation
Gaussian (normal) distributions	$f(x) = ce^{-(x-a)^2/k}$	Bell curves (standardized tests, height/weight charts, distribution of power flux of laser beams)
Logistic growth	$f(x) = \dfrac{a}{1 + ce^{-kx}}$	Conservation biology (growth limited by factors like food and space), learning curve
Logarithmic	$f(x) = a + c\log x$ $f(x) = a + c\ln x$	Time to pay off credit cards, annuity planning

SECTION 5.5 EXERCISES

SKILLS

In Exercises 1–6, match the function with the graph (a to f) and the model name (i to v).

1. $f(t) = 5e^{2t}$

2. $N(t) = 28e^{-t/2}$

3. $T(x) = 4e^{-(x-80)^2/10}$

4. $P(t) = \dfrac{200}{1 + 5e^{-0.4t}}$

5. $D(x) = 4 + \log(x - 1)$

6. $h(t) = 2 + \ln(t + 3)$

Model Name

i. Logarithmic **ii.** Logistic **iii.** Gaussian **iv.** Exponential growth **v.** Exponential decay

Graphs

a.

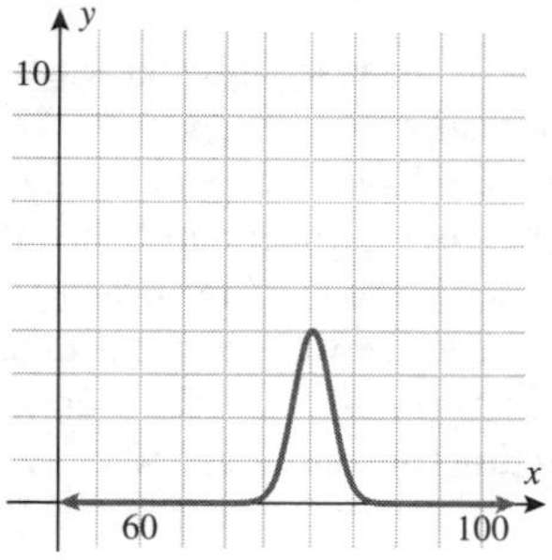

b.

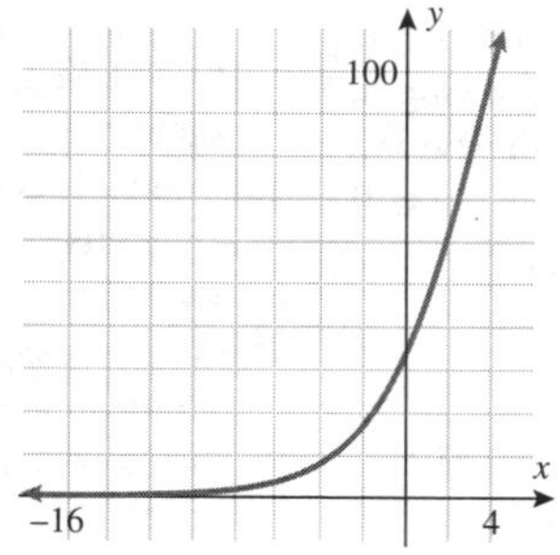

c.

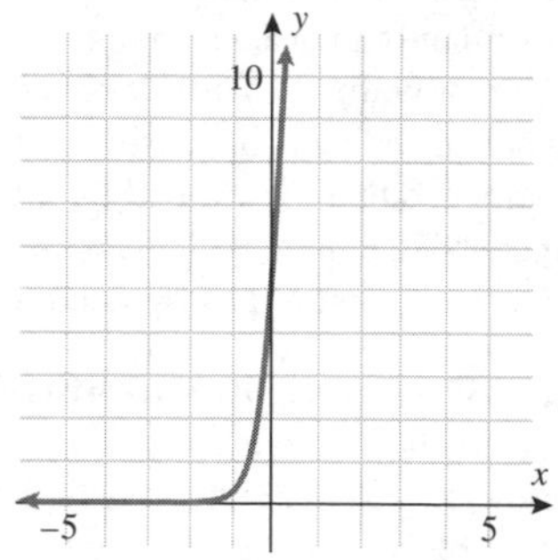

d.

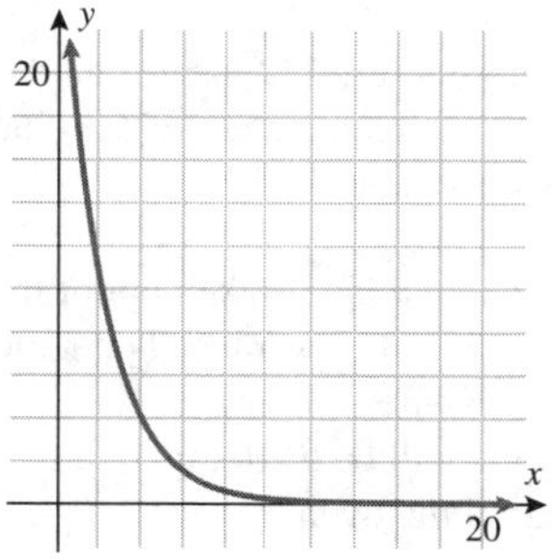

e.

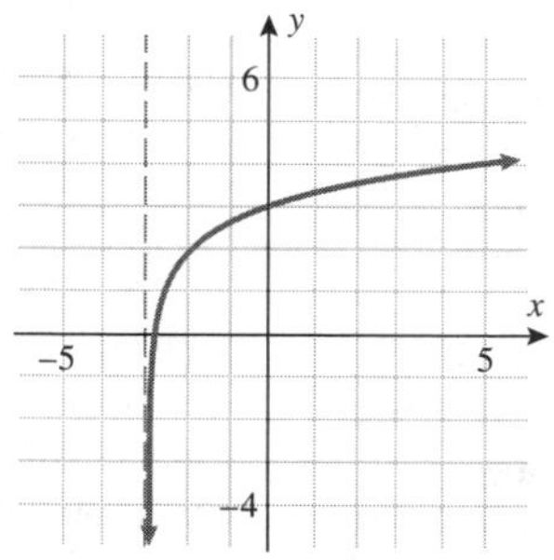

f.

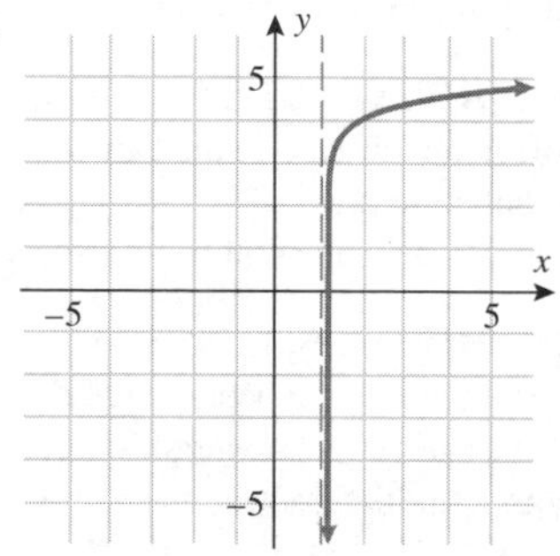

APPLICATIONS

7. Population Growth. The population of the Philippines in 2003 was 80 million. Their population increases 2.36% per year. What is the expected population of the Philippines in 2010? Apply the formula $N = N_0e^{rt}$, where N represents the number of people.

8. Population Growth. China's urban population is growing at 2.5% a year, compounding continuously. If there were 13.7 million people in Shanghai in 1996, approximately how many people will there be in 2016? Apply the formula $N = N_0e^{rt}$, where N represents the number of people.

9. Population Growth. Port St. Lucie, Florida, had the United States' fastest growth rate among cities with a population of 100,000 or more between 2003 and 2004. In 2003, the population was 103,800 and increasing at a rate of 12% per year. In what year should the population reach 200,000? (Let $t = 0$ correspond to 2003.) Apply the formula $N = N_0e^{rt}$, where N represents the number of people.

10. Population Growth. San Francisco's population has been declining since the "dot com" bubble burst. In 2002, the population was 776,000. If the population is declining at a rate of 1.5% per year, in what year will the population be 700,000? (Let $t = 0$ correspond to 2002.) Apply the formula $N = N_0e^{-rt}$, where N represents the number of people.

11. Cellular Phone Plans. The number of cell phones in China is exploding. In 2007, there were 487.4 million cell phone subscribers and the number is increasing at a rate of 16.5% per year. How many cell phone subscribers are expected in 2010? Use the formula $N = N_0e^{rt}$, where N represents the number of cell phone subscribers. Let $t = 0$ correspond to 2007.

12. Bacteria Growth. A colony of bacteria is growing exponentially. Initially 500 bacteria were in the colony. The growth rate is 20% per hour. (a) How many bacteria should be in the colony in 12 hours? (b) How many in one day? Use the formula $N = N_0e^{rt}$, where N represents the number of bacteria.

13. Real Estate Appreciation. In 2004, the average house in Las Vegas cost \$185,000, and real estate prices were increasing at an amazing rate of 30% per year. What is the expected cost of an average house in Las Vegas in 2007? Use the formula $N = N_0e^{rt}$, where N represents the average cost of a home. Round to the nearest thousand.

14. Real Estate Appreciation. The average cost of a single family home in California in 2004 was \$230,000. In 2005, the average cost was \$252,000. If this trend continued, what is the expected cost in 2007? Use the formula $N = N_0e^{rt}$, where N represents the average cost of a home. Round to the nearest thousand.

15. Oceanography (Growth of Phytoplankton). Phytoplankton are microscopic plants that live in the ocean. Phytoplankton grow abundantly in oceans around the world and are the foundation of the marine food chain. One variety of phytoplankton growing in tropical waters is increasing at a rate of 20% per month. If it is estimated that there are 100 million in the water, how many will there be in 6 months? Utilize formula $N = N_0e^{rt}$, where N represents the population of phytoplankton.

16. Oceanography (Growth of Phytoplankton). In Arctic waters there are an estimated 50,000,000 phytoplankton. The growth rate is 12% per month. How many phytoplankton will there be in 3 months? Utilize formula $N = N_0e^{rt}$, where N represents the population of phytoplankton.

17. HIV/AIDS. In 2003, an estimated 1 million people had been infected with HIV in the United States. If the infection rate increases at an annual rate of 2.5% a year compounding continuously, how many Americans will be infected with the HIV virus by 2010?

18. HIV/AIDS. In 2003, there were an estimated 25 million people who have been infected with HIV in sub-Saharan Africa. If the infection rate increases at an annual rate of 9% a year compounding continuously, how many Africans will be infected with the HIV virus by 2010?

19. Anesthesia. When a person has a cavity filled, the dentist typically gives a local anesthetic. After leaving the dentist's office, one's mouth often is numb for several more hours. If 100 ml of anesthesia is injected into the local tissue at the time of the procedure ($t = 0$), and the amount of anesthesia still in the local tissue t hours after the initial injection is given by $A = 100e^{-0.5t}$, how much is in the local tissue 4 hours later?

20. Anesthesia. When a person has a cavity filled, the dentist typically gives a local anesthetic. After leaving the dentist's office, one's mouth often is numb for several more hours. If 100 ml of anesthesia is injected into the local tissue at the time of the procedure ($t = 0$), and the amount of anesthesia still in the local tissue t hours after the initial injection is given by $A = 100e^{-0.5t}$, how much is in the local tissue 12 hours later?

21. Radioactive Decay. Carbon-14 has a half-life of 5730 years. How long will it take 5 grams of carbon-14 to be reduced to 2 grams?

22. Radioactive Decay. Radium-226 has a half-life of 1600 years. How long will it take 5 grams of radium-226 to be reduced to 2 grams?

23. Radioactive Decay. The half-life of uranium-238 is 4.5 billion years. If 98% of uranium-238 remains in a fossil, how old is the fossil?

24. Radioactive Decay. A drug has a half-life of 12 hours. If the initial dosage is 5 milligrams, how many milligrams will be in the patient's body in 16 hours?

In Excercises 25–28, use the following formula for Newton's Law of Cooling.

If you take a hot dinner out of the oven and place it on the kitchen countertop, the dinner cools until it reaches the temperature of the kitchen. Likewise, a glass of ice set on a table in a room eventually melts into a glass of water at that room temperature. The rate at which the hot dinner cools or the ice in the glass melts at any given time is proportional to the difference between its temperature and the temperature of its surroundings (in this case, the room). This is called **Newton's law of cooling** (or warming) and is modeled by

$$T = T_S + (T_0 - T_S)e^{-kt}$$

where T is the temperature of an object at time t, T_s is the temperature of the surrounding medium, T_0 is the temperature of the object at time $t = 0$, t is the time, and k is a constant.

25. Newton's Law of Cooling. An apple pie is taken out of the oven with an internal temperature of 325 °F. It is placed on a rack in a room with a temperature of 72 °F. After 10 minutes, the temperature of the pie is 200 °F. What will be the temperature of the pie 30 minutes after coming out of the oven?

26. Newton's Law of Cooling. A cold drink is taken out of an ice chest with a temperature of 38 °F and placed on a picnic table with a surrounding temperature of 75 °F. After 5 minutes the temperature of the drink is 45 °F. What will the temperature of the drink be 20 minutes after it is taken out of the chest?

27. Forensic Science (Time of Death). A body is discovered in a hotel room. At 7:00 A.M. a police detective found the body's temperature to be 85 °F. At 8:30 A.M. a medical examiner measures the body's temperature to be 82 °F. Assuming the room in which the body was found had a constant temperature of 74 °F, how long has the victim been dead? (Normal body temperature is 98.6 °F.)

28. Forensic Science (Time of Death). At 4 A.M. a body is found in a park. The police measure the body's temperature to be 90 °F. At 5 A.M. the medical examiner arrives and determines the temperature to be 86 °F. Assuming the temperature of the park was constant at 60 °F, how long has the victim been dead?

29. Depreciation of Automobile. A new Lexus IS250 has a book value of \$38,000, and after one year has a book value of \$32,000. What is the car's value in 4 years? Apply the formula $N = N_0e^{-rt}$, where N represents the value of the car. Round to the nearest hundred.

30. Depreciation of Automobile. A new Hyundai Triburon has a book value of \$22,000, and after 2 years a book value of \$14,000. What is the car's value in 4 years? Apply the formula $N = N_0e^{-rt}$, where N represents the value of the car. Round to the nearest hundred.

31. Automotive. A new model BMW convertible coupe is designed and produced in time to appear in North America in the fall. BMW Corporation has a limited number of new models available. The number of new model BMW convertible coupes purchased in North America is given by $N = \dfrac{100,000}{1 + 10e^{-2t}}$, where t is the number of weeks after the BMW is released.

a. How many new model BMW convertible coupes will have been purchased 2 weeks after the new model becomes available?

b. How many after 30 weeks?

c. What is the maximum number of new model BMW convertible coupes that will be sold in North America?

32. iPhone. The number of iPhones purchased is given by $N = \dfrac{2,000,000}{1 + 2e^{-4t}}$, where t is the time in weeks after they are made available for purchase.

a. How many iPhones are purchased within the first 2 weeks?

b. How many iPhones are purchased within the first month?

33. Spread of a Virus. The number of MRSA (methicillin-resistant *Staphylococcus aureus*) cases has been rising sharply in England and Wales since 1997. In 1997, 2422 cases were reported. The number of cases reported in 2003 was 7684. How many cases will be expected in 2010? (Let $t = 0$ correspond to 1997.) Use the formula $N = N_0e^{-rt}$, where N represents the number of cases reported.

34. Spread of a Virus. Dengue fever, an illness carried by mosquitoes, is occurring in one of the worst outbreaks in decades across Latin America and the Caribbean. In 2004, 300,000 cases were reported, and 630,000 cases in 2007. How many cases might be expected in 2010? (Let $t = 0$ be 2004.) Use the formula $N = N_0e^{-rt}$, where N represents the number of cases.

35. Carrying Capacity. The Virginia Department of Fish and Game stock a mountain lake with 500 trout. Officials believe the lake can support no more than 10,000 trout. The number of trout is given by $N = \dfrac{10,000}{1 + 19e^{-1.56t}}$, where t is time in years. How many years will it take for the trout population to reach 5000?

36. Carrying Capacity. The World Wildlife Fund has placed 1000 rare pygmy elephants in a conservation area in Borneo. They believe 1600 pygmy elephants can be supported in this environment. The number of elephants is given by $N = \dfrac{1600}{1 + 0.6e^{-0.14t}}$, where t is time in years. How many years will it take the herd to reach 1200 elephants?

37. Lasers. The intensity of a laser beam is given by the ratio of power to area. A particular laser beam has an intensity function given by $I = e^{-r^2}$ mW/cm^2, where r is the radius off the center axis given in cm. Where is the beam brightest (largest intensity)?

38. Lasers. The intensity of a laser beam is given by the ratio of power to area. A particular laser beam has an intensity function given by $I = e^{-r^2}$ mW/cm^2, where r is the radius off the center axis given in cm. What percentage of the on-axis intensity ($r = 0$) corresponds to $r = 2$ cm?

39. Grade Distribution. Suppose the first test in this class has a normal, or bell-shaped, grade distribution of test scores, with an average score of 75. An approximate function that models your class's grades on test 1 is $N(x) = 10e^{-(x-75)^2/25^2}$, where N represents the number of students who received the score x.

a. Graph this function.

b. What is the average grade?

c. Approximately how many students scored a 50?

d. Approximately how many students scored 100?

40. Grade Distribution. Suppose the final exam in this class has a normal, or bell-shaped, grade distribution of exam scores, with an average score of 80. An approximate function that models your class's grades on the exam is $N(x) = 10e^{-(x-80)^2/16^2}$, where N represents the number of students who received the score x.

a. Graph this function.

b. What is the average grade?

c. Approximately how many students scored a 60?

d. Approximately how many students scored 100?

41. Time to Pay Off Debt. Diana just graduated from medical school owing \$80,000 in student loans. The annual interest rate is 9%.

a. Approximately how many years will it take to pay off her student loan if she makes a monthly payment of \$750?

b. Approximately how many years will it take to pay off her loan if she makes a monthly payment of \$1,000?

42. Time to Pay Off Debt. Victor owes \$20,000 on his credit card. The annual interest rate is 17%.

a. Approximately how many years will it take him to pay off this credit card if he makes a monthly payment of \$300?

b. Approximately how many years will it take him to pay off this credit card if he makes a monthly payment of \$400?

CATCH THE MISTAKE

In Exercises 43 and 44, explain the mistake that is made.

43. The city of Orlando, Florida, has a population that is growing at 7% a year, compounding continuously. If there were 1.1 million people in greater Orlando in 2006, approximately how many people will there be in 2016? Apply the formula $N = N_0e^{rt}$, where N represents the number of people.

Solution:

Use the population growth model. $N = N_0e^{rt}$

Let $N_0 = 1.1$, $r = 7$, and $t = 10$. $N = 1.1e^{(7)(10)}$

Approximate with a calculator. 2.8×10^{30}

This is incorrect. What mistake was made?

44. The city of San Antonio, Texas, has a population that is growing at 5% a year, compounding continuously. If there were 1.3 million people in the greater San Antonio area in 2006, approximately how many people will there be in 2016? Apply the formula $N = N_0e^{rt}$, where N represents the number of people.

Solution:

Use the population growth model. $N = N_0e^{rt}$

Let $N_0 = 1.3$, $r = 5$, and $t = 10$. $N = 1.3e^{(5)(10)}$

Approximate with a calculator. 6.7×10^{21}

This is incorrect. What mistake was made?

CONCEPTUAL

In Exercises 45–48, determine whether each statement is true or false.

45. When a species gets placed on an endangered species list, the species begins to grow rapidly, and then reaches a carrying capacity. This can be modeled by logistic growth.

46. A professor has 400 students one semester. The number of names (of her students) she can memorize can be modeled by a logarithmic function.

47. The spread of lice at an elementary school can be modeled by exponential growth.

48. If you purchase a laptop computer this year ($t = 0$), then the value of the computer can be modeled with exponential decay.

CHALLENGE

In Exercises 49 and 50, refer to the logistic model $f(x) = \dfrac{a}{1 + ce^{-kx}}$, where a is the carrying capacity.

49. As c increases, does the model reach the carrying capacity in less time or more time?

50. As k increases, does the model reach the carrying capacity in less time or more time?

TECHNOLOGY

51. Wing Shan just graduated from dental school owing \$80,000 in student loans. The annual interest is 6%. Her time t to pay off the loan is given by

$$t = -\frac{\ln(1 - 80000(0.06)/nR)}{n\ln(1 + 0.06/n)}$$

where n is the number of payment periods per year and R is the periodic payment.

a. Use a graphing utility to graph

$$t_1 = -\frac{\ln(1 - 80000(0.06)/12x)}{12\ln(1 + 0.06/12)} \text{ as } Y_1 \text{ and}$$

$$t_2 = -\frac{\ln(1 - 80000(0.06)/26x)}{26\ln(1 + 0.06/26)} \text{ as } Y_2.$$

Explain the difference in the two graphs.

b. Use the TRACE key to estimate the number of years that will take Wing Shan to pay off her student loan if she can afford a monthly payment of \$800.

c. If she can make a biweekly payment of \$400, estimate the number of years that will take her to pay off the loan.

d. If she adds \$200 more to her monthly or \$100 more to her biweekly payment, estimate the number of years that will take her to pay off the loan.

52. Hong has a credit card debt in the amount of \$12,000. The annual interest is 18%. His time t to pay off the loan is given by

$$t = -\frac{\ln(1 - 12000(0.18)/nR)}{n\ln(1 + 0.18/n)}$$

where n is the number of payment periods per year and R is the periodic payment.

a. Use a graphing utility to graph

$$t_1 = -\frac{\ln(1 - 12000(0.18)/12x)}{12\ln(1 + 0.18/12)} \text{ as } Y_1 \text{ and}$$

$$t_2 = -\frac{\ln(1 - 12000(0.18)/26x)}{26\ln(1 + 0.18/26)} \text{ as } Y_2.$$

Explain the difference in the two graphs.

b. Use the [TRACE] key to estimate the number of years that will take Hong to pay off his credit card if he can afford a monthly payment of $300.

c. If he can make a biweekly payment of $150, estimate the number of years that will take him to pay off the credit card.

d. If he adds $100 more to his monthly or $50 more to his biweekly payment, estimate the number of years that will take him to pay off the credit card.

MODELING YOUR WORLD

The following table summarizes the average yearly temperature in degrees Fahrenheit (°F) and carbon dioxide emissions in parts per million (ppm) for **Mauna Loa, Hawaii.**

YEAR	1960	1965	1970	1975	1980	1985	1990	1995	2000	2005
TEMPERATURE	44.45	43.29	43.61	43.35	46.66	45.71	45.53	47.53	45.86	46.23
CO_2 EMISSIONS (PPM)	316.9	320.0	325.7	331.1	338.7	345.9	354.2	360.6	369.4	379.7

In the Modeling Your World in Chapters 3 and 4, the temperature and carbon emissions were modeled with *linear functions* and *polynomial functions*, respectively. Now, let us model these same data using *exponential* and *logarithmic functions*.

1. Plot the temperature data, with time on the horizontal axis and temperature on the vertical axis. Let $t = 1$ correspond to 1960.
2. Find a *logarithmic function* with base e, $f(t) = A \ln Bt$, that models the temperature in Mauna Loa.
   - **a.** Apply data from 1965 and 2005.
   - **b.** Apply data from 2000 and 2005.
   - **c.** Apply regression and all data given.
3. Predict what the temperature will be in Mauna Loa in 2020.
   - **a.** Use the line found in 2a.
   - **b.** Use the line found in 2b.
   - **c.** Use the line found in 2c.
4. Predict what the temperature will be in Mauna Loa in 2100.
   - **a.** Use the line found in 2a.
   - **b.** Use the line found in 2b.
   - **c.** Use the line found in 2c.
5. Do your models support the claim of "global warming"? Explain. Do these logarithmic models give similar predictions to the linear models found in Chapter 3 and the polynomial models found in Chapter 4?
6. Plot the carbon dioxide emissions data, with time on the horizontal axis and carbon dioxide emissions on the vertical axis. Let $t = 0$ correspond to 1960.

7. Find an *exponential function* with base e, $f(t) = Ae^{bt}$, that models the CO_2 emissions (ppm) in Mauna Loa.
   a. Apply data from 1960 and 2005.
   b. Apply data from 1960 and 2000.
   c. Apply regression and all data given.

8. Predict the expected CO_2 levels in Mauna Loa in 2020.
   a. Use the line found in 7a.
   b. Use the line found in 7b.
   c. Use the line found in 7c.

9. Predict the expected CO_2 levels in Mauna Loa in 2100.
   a. Use the line found in 7a.
   b. Use the line found in 7b.
   c. Use the line found in 7c.

10. Do your models support the claim of "global warming"? Explain. Do these exponential models give similar predictions to the linear models found in Chapter 3 or the polynomial models found in Chapter 4?

11. Comparing the models developed in Chapters 3, 4, and 5, do you believe that global temperatures are best modeled with a linear, polynomial, or logarithmic function?

12. Comparing the models developed in Chapters 3, 4, and 5, do you believe that CO_2 emissions are best modeled by linear, polynomial, or exponential functions?

CHAPTER 5 REVIEW

SECTION	CONCEPT	PAGES	REVIEW EXERCISES	KEY IDEAS/POINTS
5.1	**Exponential functions and their graphs**	422–433	1–28	
	Graphs of exponential functions	424–427	13–24	y-intercept $(0, 1)$ Horizontal asymptote: $y = 0$; the points $(1, b)$ and $(-1, 1/b)$
	The natural base e	428–429	5–8, 21–24	$f(x) = e^x$
	Applications of exponential functions	429–433	25–28	Doubling time: $P = P_0 2^{t/d}$ Compound interest: $A = P\left(1 + \frac{r}{n}\right)^{nt}$ Compounded continuously: $A = Pe^{rt}$
5.2	**Logarithmic functions and their graphs**	437–447	29–60	$y = \log_b x \quad x > 0$ $b > 0, b \neq 1$
	Evaluating logarithms	437–439	29–40	$y = \log_b x$ and $x = b^y$
	Common and natural logarithms	439–440	41–44	$y = \log x$ Common (base 10) $y = \ln x$ Natural (base e)
	Graphs of logarithmic functions	440–443	49–56	x-intercept $(1, 0)$ Vertical asymptote: $x = 0$; the points $(b, 1)$ and $(1/b, -1)$
	Applications of logarithms	444–447	57–60	Decibel scale: $D = 10 \log\left(\frac{I}{I_T}\right) \quad I_T = 1 \times 10^{-12}\ \text{W/m}^2$ Richter scale: $M = \frac{2}{3}\log\left(\frac{E}{E_0}\right) \quad E_0 = 10^{4.4}$ joules
5.3	**Properties of logarithms**	451–457	61–74	**1.** $\log_b 1 = 0$ **2.** $\log_b b = 1$ **3.** $\log_b b^x = x$ **4.** $b^{\log_b x} = x \quad x > 0$ Product property: **5.** $\log_b MN = \log_b M + \log_b N$ Quotient property: **6.** $\log_b\left(\frac{M}{N}\right) = \log_b M - \log_b N$ Power property: **7.** $\log_b M^p = p \log_b M$
	Change-of-base formula	456–457	71–74	$\log_b M = \frac{\log M}{\log b}$ or $\log_b M = \frac{\ln M}{\ln b}$
5.4	**Exponential and logarithmic equations**	460–466	75–94	
	Solving exponential equations	462–464	75–86	*Simple exponential equations* **1.** Rewrite both sides of the equation in terms of the same base. **2.** Use the one-to-one property to equate the exponents. **3.** Solve for the variable.

CHAPTER REVIEW

Section	Concept	Pages	Review Exercises	Key Ideas/Points
				Complicated exponential equations **1.** Isolate the exponential expression. **2.** Take the same logarithm* of both sides. **3.** Simplify using the inverse properties. **4.** Solve for the variable.
	Solving logarithmic equations	464–466	87–94	*Simple logarithmic equations* **1.** Combine logarithms on each side of the equation using properties. **2.** Use the one-to-one property to equate the exponents. **3.** Solve for the variable. **4.** Eliminate any extraneous solutions *Complicated logarithmic equations* **1.** Combine and isolate the logarithmic expressions. **2.** Rewrite the equation in exponential form. **3.** Solve for the variable.
5.5	**Exponential and logarithmic models**	470–476	95–108	
	Exponential growth models	471	95–102	$f(x) = ce^{kx} \quad k > 0$
	Exponential decay models	472–473	103–106	$f(x) = ce^{-kx} \quad k > 0$
	Gaussian (normal) distribution models	473–474		$f(x) = ce^{-(x-a)^2/k}$
	Logistic growth models	474	107	$f(x) = \dfrac{a}{1 + ce^{-kx}}$
	Logarithmic models	475–476		$f(x) = a + c \log x$ $f(x) = a + c \ln x$

CHAPTER 5 REVIEW EXERCISES

5.1 Exponential Functions and Their Graphs

Approximate each number using a calculator and round your answer to two decimal places.

1. $8^{4.7}$ **2.** $\pi^{2/5}$ **3.** $4 \cdot 5^{0.2}$ **4.** $1.2^{1.2}$

Approximate each number using a calculator and round your answer to two decimal places.

5. $e^{3.2}$ **6.** e^{π} **7.** $e^{\sqrt{\pi}}$ **8.** $e^{-2.5\sqrt{3}}$

Evaluate each exponential function for the given values.

9. $f(x) = 2^{4-x}$ $\quad f(-2.2)$

10. $f(x) = -2^{x+4}$ $\quad f(1.3)$

11. $f(x) = \left(\frac{2}{5}\right)^{1-6x}$ $\quad f\left(\frac{1}{2}\right)$

12. $f(x) = \left(\frac{4}{7}\right)^{5x+1}$ $\quad f\left(\frac{1}{5}\right)$

Match the graph with the function.

13. $y = 2^{x-2}$ **14.** $y = -2^{2-x}$

15. $y = 2 + 3^{x+2}$ **16.** $y = -2 - 3^{2-x}$

a.

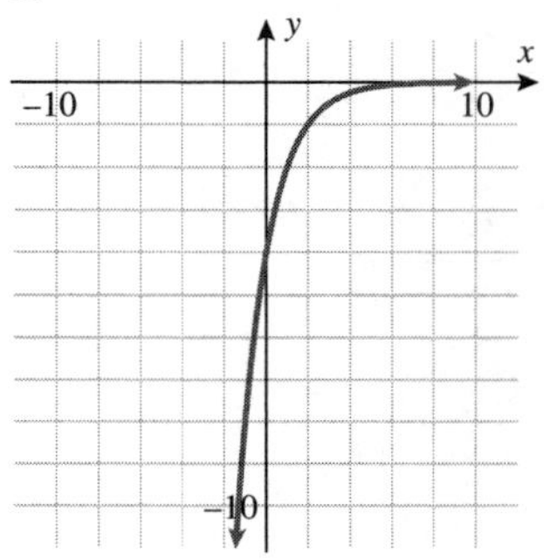

b.

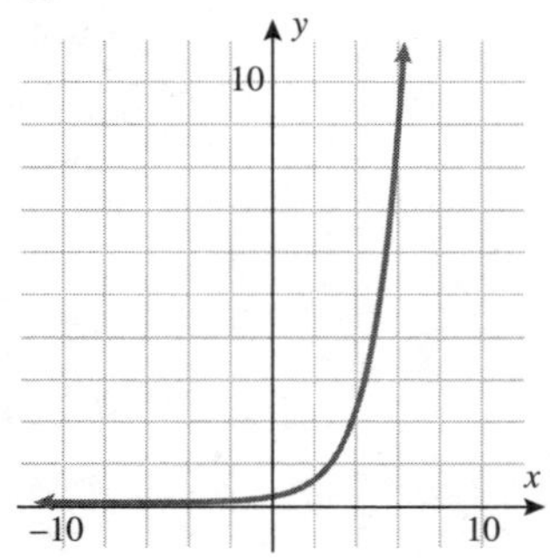

c.

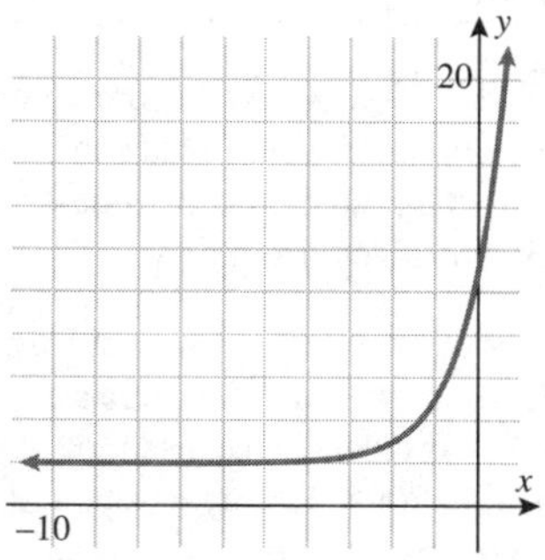

d.

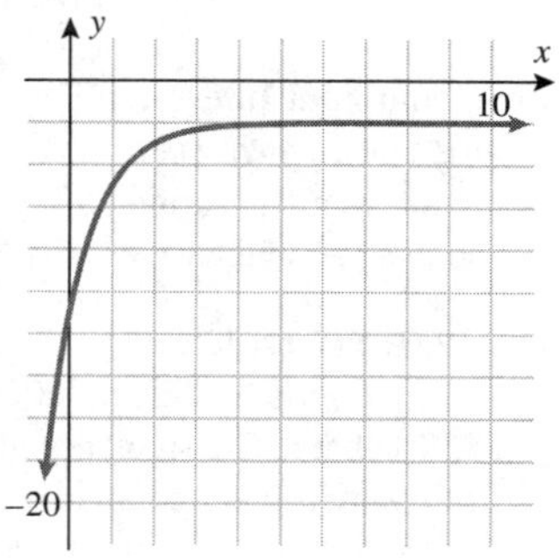

State the *y*-intercept and the horizontal asymptote and graph the exponential function.

17. $y = -6^{-x}$ **18.** $y = 4 - 3^x$

19. $y = 1 + 10^{-2x}$ **20.** $y = 4^x - 4$

State the *y*-intercept and horizontal asymptote, and graph the exponential function.

21. $y = e^{-2x}$ **22.** $y = e^{x-1}$

23. $y = 3.2e^{x/3}$ **24.** $y = 2 - e^{1-x}$

Applications

25. Compound Interest. If \$4,500 is deposited into an account paying 4.5% compounding semiannually, how much will you have in the account in 7 years?

26. Compound Interest. How much money should be put in a savings account now that earns 4.0% a year compounded quarterly if you want \$25,000 in 8 years?

27. Compound Interest. If \$13,450 is put in a money market account that pays 3.6% a year compounded continuously, how much will be in the account in 15 years?

28. Compound Interest. How much money should be invested today in a money market account that pays 2.5% a year compounded continuously if you desire \$15,000 in 10 years?

5.2 Logarithmic Functions and Their Graphs

Write each logarithmic equation in its equivalent exponential form.

29. $\log_4 64 = 3$ **30.** $\log_4 2 = \frac{1}{2}$

31. $\log\left(\frac{1}{100}\right) = -2$ **32.** $\log_{16} 4 = \frac{1}{2}$

Write each exponential equation in its equivalent logarithmic form.

33. $6^3 = 216$ **34.** $10^{-4} = 0.0001$

35. $\frac{4}{169} = \left(\frac{2}{13}\right)^2$ **36.** $\sqrt[3]{512} = 8$

Evaluate the logarithms exactly.

37. $\log_7 1$ **38.** $\log_4 256$

39. $\log_{1/6} 1296$ **40.** $\log 10^{12}$

Approximate the common and natural logarithms utilizing a calculator. Round to two decimal places.

41. $\log 32$ **42.** $\ln 32$

43. $\ln 0.125$ **44.** $\log 0.125$

State the domain of the logarithmic function in interval notation.

45. $f(x) = \log_3(x + 2)$ **46.** $f(x) = \log_2(2 - x)$

47. $f(x) = \log(x^2 + 3)$ **48.** $f(x) = \log(3 - x^2)$

Match the graph with the function.

49. $y = \log_7 x$ **50.** $y = -\log_7(-x)$

51. $y = \log_7(x + 1) - 3$ **52.** $y = -\log_7(1 - x) + 3$

a.

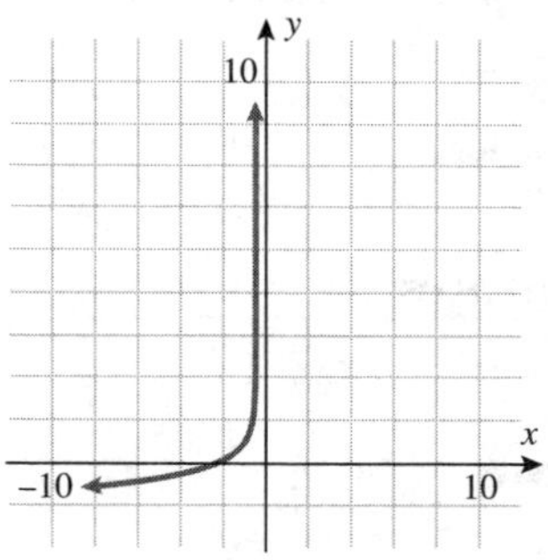

b.

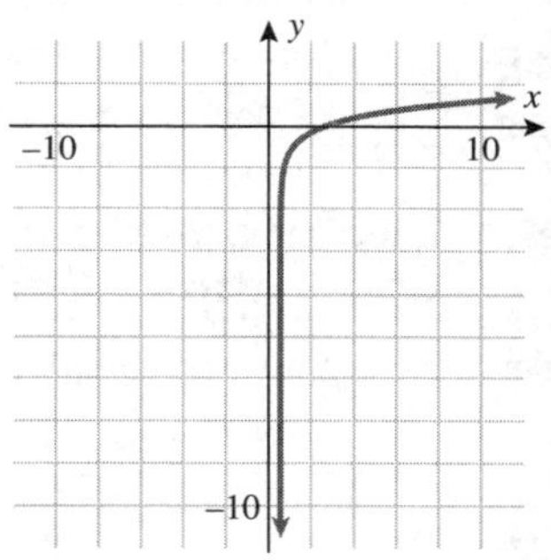

c.

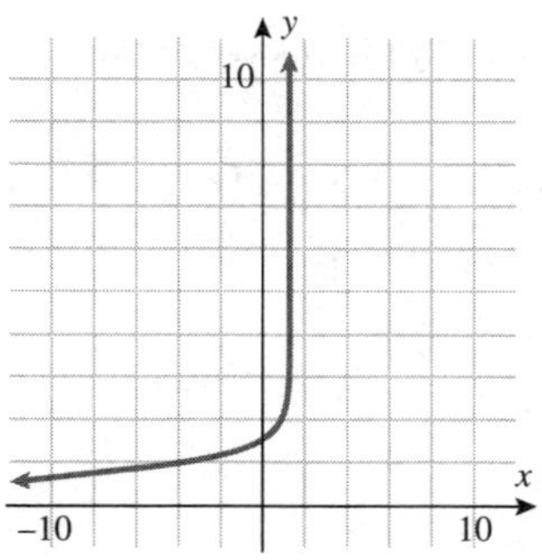

d.

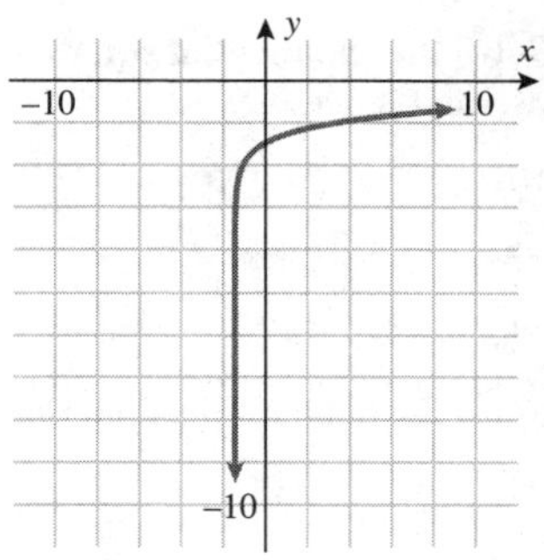

Graph the logarithmic function with transformation techniques.

53. $f(x) = \log_4(x - 4) + 2$ **54.** $f(x) = \log_4(x + 4) - 3$

55. $f(x) = -\log_4(x) - 6$ **56.** $f(x) = -2\log_4(-x) + 4$

Applications

57. Chemistry. Calculate the pH value of milk, assuming it has a concentration of hydrogen ions given by $H^+ = 3.16 \times 10^{-7}$.

58. Chemistry. Calculate the pH value of Coca-Cola, assuming it has a concentration of hydrogen ions given by $H^+ = 2.0 \times 10^{-3}$.

59. Sound. Calculate the decibels associated with a teacher speaking to a medium-sized class if the sound intensity is 1×10^{-7} W/m^2.

60. Sound. Calculate the decibels associated with an alarm clock if the sound intensity is 1×10^{-4} W/m^2.

5.3 Properties of Logarithms

Use the properties of logarithms to simplify each expression.

61. $\log_{2.5} 2.5$

62. $\log_2 \sqrt{16}$

63. $2.5^{\log_{2.5} 6}$

64. $e^{-3 \ln 6}$

Write each expression as a sum or difference of logarithms.

65. $\log_c x^a y^b$

66. $\log_3 x^2 y^{-3}$

67. $\log_j \left(\frac{rs}{t^3}\right)$

68. $\log x^c \sqrt{x+5}$

69. $\log \left[\frac{a^{1/2}}{b^{3/2} c^{2/5}}\right]$

70. $\log_7 \left[\frac{c^3 d^{1/3}}{e^6}\right]^{1/3}$

Evaluate the logarithms using the change-of-base formula.

71. $\log_8 3$

72. $\log_5 \frac{1}{2}$

73. $\log_\pi 1.4$

74. $\log_{\sqrt{3}} 2.5$

5.4 Exponential and Logarithmic Equations

Solve the exponential equations exactly for x.

75. $4^x = \frac{1}{256}$

76. $3^{x^2} = 81$

77. $e^{3x-4} = 1$

78. $e^{\sqrt{x}} = e^{4.8}$

79. $\left(\frac{1}{3}\right)^{x+2} = 81$

80. $100^{x^2-3} = 10$

Solve the exponential equation. Round your answer to three decimal places.

81. $e^{2x+3} - 3 = 10$

82. $2^{2x-1} + 3 = 17$

83. $e^{2x} + 6e^x + 5 = 0$

84. $4e^{0.1x} = 64$

85. $(2^x - 2^{-x})(2^x + 2^{-x}) = 0$

86. $5(2^x) = 25$

Solve the logarithmic equations exactly.

87. $\log(3x) = 2$

88. $\log_3(x+2) = 4$

89. $\log_4 x + \log_4 2x = 8$

90. $\log_6 x + \log_6(2x-1) = \log_6 3$

Solve the logarithmic equations. Round your answers to three decimal places.

91. $\ln x^2 = 2.2$

92. $\ln(3x-4) = 7$

93. $\log_3(2-x) - \log_3(x+3) = \log_3 x$

94. $4\log(x+1) - 2\log(x+1) = 1$

5.5 Exponential and Logarithmic Models

95. Compound Interest. If Tania needs \$30,000 a year from now for a down payment on a new house, how much should she put in a 1-year CD earning 5% a year compounding continuously so that she will have exactly \$30,000 a year from now?

96. Stock Prices. Jeremy is tracking the stock value of Best Buy (BBY on the NYSE). In 2003, he purchased 100 shares at \$28 a share. The stock did not pay dividends because the company reinvested all earnings. In 2005, Jeremy cashed out and sold the stock for \$4,000. What was the annual rate of return on BBY?

97. Compound Interest. Money is invested in a savings account earning 4.2% interest compounded quarterly. How many years will pass until the money doubles?

98. Compound Interest. If \$9,000 is invested in an investment earning 8% interest compounded continuously, how many years will pass until there is \$22,500?

99. Population. Nevada has the fastest growing population according to the U.S. Census Bureau. In 2004, the population of Nevada was 2.62 million and increasing at an annual rate of 3.5%. What is the expected population in 2010? (Let $t = 0$ be 2004.) Apply the formula $N = N_0 e^{rt}$, where N is the population.

100. Population. The Hispanic population in the United States is the fastest growing of any ethnic group. In 1996, there were an estimated 28.3 million Hispanics in the United States, and in 2000 there were an estimated 32.5 million. What is the expected population of Hispanics in the United States in 2010? (Let $t = 0$ be 1996.) Apply the formula $N = N_0 e^{rt}$, where N is the population.

101. Bacteria Growth. Bacteria are growing exponentially. Initially, there were 1000 bacteria; after 3 hours there were 2500. How many bacteria should be expected in 6 hours? Apply the formula $N = N_0 e^{rt}$, where N is the number of cbacteria.

102. Population. In 2003, the population of Phoenix, Arizona, was 1,388,215. In 2004, the population was 1,418,041. What is the expected population in 2010? (Let $t = 0$ be 2003.) Apply the formula $N = N_0 e^{rt}$, where N is the population.

103. Radioactive Decay. Strontium-90 has a half-life of 28 years. How long will it take for 20 grams of this to decay to 5 grams? Apply the formula $N = N_0 e^{-rt}$, where N is the number of grams.

104. Radioactive Decay. Plutonium-239 has a half-life of 25,000 years. How long will it take for 100 grams to decay to 20 grams? Apply the formula $N = N_0 e^{-rt}$, where N is the number of grams.

105. Wild Life Population. The *Boston Globe* reports that the fish population of the Essex River in Massachusetts is declining. In 2003, it was estimated there were 5600 fish in the river, and in 2004, there were only 2420 fish. How many fish should be expected in 2010 if this trend continues? Apply the formula $N = N_0 e^{-rt}$, where N is the number of fish.

106. Car Depreciation. A new Acura TSX costs \$28,200. In 2 years the value will be \$24,500. What is the expected value in 6 years? Apply the formula $N = N_0 e^{-rt}$, where N is the value of the car.

107. Carrying Capacity. The carrying capacity of a species of beach mice in St. Croix is given by $M = 1000\left(1 - e^{-0.035t}\right)$ where M is the number of mice and t is time in years ($t = 0$ corresponds to 1998). How many mice will there be in 2010?

108. Population. The city of Brandon, Florida, had 50,000 residents in 1970, and since the crosstown expressway was built, its population has increased 2.3% per year. If the growth continues at the same rate, how many residents will Brandon have in 2030?

Technology

Section 5.1

109. Use a graphing utility to graph the function $f(x) = \left(1 + \frac{\sqrt{2}}{x}\right)^x$. Determine the horizontal asymptote as x increases.

110. Use a graphing utility to graph the functions $y = e^{-x+2}$ and $y = 3^x + 1$ in the same viewing screen. Estimate the coordinates of the point of intersection. Round your answers to three decimal places.

Section 5.2

111. Use a graphing utility to graph the functions $y = \log_{2.4}(3x - 1)$ and $y = \log_{0.8}(x - 1) + 3.5$ in the same viewing screen. Estimate the coordinates of the point of intersection. Round your answers to three decimal places.

112. Use a graphing utility to graph the functions $y = \log_{2.5}(x - 1) + 2$ and $y = 3.5^{x-2}$ in the same viewing screen. Estimate the coordinates of the point(s) of intersection. Round your answers to three decimal places.

Section 5.3

113. Use a graphing utility to graph $f(x) = \log_2\left(\frac{x^3}{x^2 - 1}\right)$ and $g(x) = 3\log_2 x - \log_2(x + 1) - \log_2(x - 1)$ in the same viewing screen. Determine the domain where the two functions give the same graph.

114. Use a graphing utility to graph $f(x) = \ln\left(\frac{9 - x^2}{x^2 - 1}\right)$ and $g(x) = \ln(3 - x) + \ln(3 + x) - \ln(x + 1) - \ln(x - 1)$ in the same viewing screen. Determine the domain where the two functions give the same graph.

Section 5.4

115. Use a graphing utility to graph $y = \frac{e^x - e^{-x}}{e^x + e^{-x}}$. State the domain. Determine if there are any symmetry and asymptote.

116. Use a graphing utility to graph $y = \frac{1}{e^x - e^{-x}}$. State the domain. Determine if there are any symmetry and asymptote.

Section 5.5

117. A drug with initial dosage of 4 milligrams has a half-life of 18 hours. Let (0, 4) and (18, 2) be two points.

a. Determine the equation of the dosage.

b. Use [STAT] [CALC] [ExpReg] to model the equation of the dosage.

c. Are the equations in (a) and (b) the same?

118. In Exercise 105, let $t = 0$ be 2003 and (0, 5600) and (1, 2420) be the two points.

a. Use [STAT] [CALC] [ExpReg] to model the equation for the fish population.

b. Using the equation found in (a), how many fish should be expected in 2010?

c. Does the answer in (b) agree with the answer in Exercise 105?

1. Simplify $\log 10^{x^3}$.

2. Use a calculator to evaluate $\log_5 326$ (round to two decimal places).

3. Find the exact value of $\log_{1/3} 81$.

4. Rewrite the expression $\ln\left[\dfrac{e^{5x}}{x(x^4+1)}\right]$ in a form with no logarithms of products, quotients, or powers.

In Exercises 5–20, solve for *x*, exactly if possible. If an approximation is required round your answer to three decimal places.

5. $e^{x^2-1} = 42$

6. $e^{2x} - 5e^x + 6 = 0$

7. $27e^{0.2x+1} = 300$

8. $3^{2x-1} = 15$

9. $3\ln(x-4) = 6$

10. $\log(6x+5) - \log 3 = \log 2 - \log x$

11. $\ln(\ln x) = 1$

12. $\log_2(3x-1) - \log_2(x-1) = \log_2(x+1)$

13. $\log_6 x + \log_6(x-5) = 2$

14. $\ln(x+2) - \ln(x-3) = 2$

15. $\ln x + \ln(x+3) = 1$

16. $\log_2\left(\dfrac{2x+3}{x-1}\right) = 3$

17. $\dfrac{12}{1+2e^x} = 6$

18. $\ln x + \ln(x-3) = 2$

19. State the domain of the function $f(x) = \log\left[\dfrac{x}{x^2-1}\right]$.

20. State the range of x values for which the following is true: $10^{\log(4x-a)} = 4x - a$.

In Exercises 21–24, find all intercepts and asymptotes, and graph.

21. $f(x) = 3^{-x} + 1$

22. $f(x) = \left(\frac{1}{2}\right)^x - 3$

23. $f(x) = \ln(2x-3) + 1$

24. $f(x) = \log(1-x) + 2$

25. **Interest.** If \$5,000 is invested at a rate of 6% a year, compounded quarterly, what is the amount in the account after 8 years?

26. **Interest.** If \$10,000 is invested at a rate of 5%, compounded continuously, what is the amount in the account after 10 years?

27. **Sound.** A lawn mower's sound intensity is approximately 1×10^{-3} W/m^2. Assuming your threshold of hearing is 1×10^{-12} W/m^2, calculate the decibels associated with the lawn mower.

28. **Population.** The population in Seattle, Washington, has been increasing at a rate of 5% a year. If the population continues to grow at that rate, and in 2004 there are 800,000 residents, how many residents will there be in 2010?
Hint: $N = N_0e^{rt}$.

29. **Earthquake.** An earthquake is considered moderate if it is between 5 and 6 on the Richter scale. What is the energy range in joules for a moderate earthquake?

30. **Radioactive Decay.** The mass $m(t)$ remaining after t hours from a 50-gram sample of a radioactive substance is given by the equation $m(t) = 50e^{-0.0578t}$. After how long will only 30 grams of the substance remain? Round your answer to the nearest hour.

31. **Bacteria Growth.** The number of bacteria in a culture is increasing exponentially. Initially, there were 200 in the culture. After 2 hours there are 500. How many should be expected in 8 hours? Round your answer to the nearest hundred.

32. **Carbon Decay.** Carbon-14 has a half-life of 5730 years. How long will it take for 100 grams to decay to 40 grams?

33. **Spread of a Virus.** The number of people infected by a virus is given by $N = \dfrac{2000}{1+3e^{-0.4t}}$, where t is time in days. In how many days will 1000 people be infected?

34. **Oil Consumtion.** The world consumption of oil was 76 million barrels per day in 2002. In 2004, the consumption was 83 million barrels per day. How many barrels should be expected to be consumed in 2010?

35. Use a graphing utility to graph $y = \dfrac{e^x - e^{-x}}{2}$. State the domain. Determine if there are any symmetry and asymptotes.

36. Use a graphing utility to help solve the equation $4^{3-x} = 2x - 1$. Round your answer to two decimal places.

1. Simplify $\frac{x^{1/3}y^{2/5}}{x^{-1/2}y^{-8/5}}$ and express in terms of positive exponents.

2. Simplify $\frac{\frac{2}{x-2}+3}{3-\frac{2}{x-2}}$.

3. Solve using the quadratic formula: $5x^2 - 4x = 3$.

4. Solve and check: $\sqrt{2x+13} = 2 + \sqrt{x+3}$.

5. Solve and express the solution in interval notation: $\frac{x+4}{5} \leq -3$.

6. Solve for x: $|x^2 - 4| = 9$.

7. Write an equation of the line that is perpendicular to the line $4x + 3y = 6$ and that passes through the point $(7, 6)$.

8. Using the function $f(x) = 4x - x^2$, evaluate the difference quotient $\frac{f(x+h) - f(x)}{h}$.

9. Given the piecewise-defined function

$$f(x) = \begin{cases} 5 & -2 < x \leq 0 \\ 2 - \sqrt{x} & 0 < x < 4 \\ x - 3 & x \geq 4 \end{cases}$$

find

a. $f(4)$ **b.** $f(0)$ **c.** $f(1)$ **d.** $f(-4)$

e. State the domain and range in interval notation.

f. Determine the intervals where the function is increasing, decreasing, or constant.

10. Sketch the graph of the function $y = \sqrt{1-x}$ and identify all transformations.

11. Determine whether the function $f(x) = \sqrt{x-4}$ is one-to-one.

12. Write an equation that describes the variation: R is inversely proportional to the square of d and $R = 3.8$ when $d = 0.02$.

13. Find the vertex of the parabola associated with the quadratic function $f(x) = -4x^2 + 8x - 5$.

14. Find a polynomial of minimum degree (there are many) that has the zeros $x = -5$ (multiplicity 2) and $x = 9$ (multiplicity 4).

15. Use synthetic division to find the quotient $Q(x)$ and remainder $r(x)$ of $(3x^2 - 4x^3 - x^4 + 7x - 20) \div (x + 4)$.

16. Given the zero $x = 2 + i$ of the polynomial $P(x) = x^4 - 7x^3 + 13x^2 + x - 20$, determine all the other zeros and write the polynomial as the product of linear factors.

17. Find the vertical and slant asymptotes of $f(x) = \frac{x^2+7}{x-3}$.

18. Graph the rational function $f(x) = \frac{3x}{x+1}$. Give all asymptotes.

19. Without employing a calculator, give the exact value of $\left(\frac{1}{25}\right)^{-3/2}$.

20. If \$5,400 is invested at 2.75% compounded monthly, how much is in the account after 4 years?

21. Give the exact value of $\log_3 243$.

22. Write the expression $\frac{1}{2}\ln(x+5) - 2\ln(x+1) - \ln(3x)$ as a single logarithm.

23. Solve the logarithmic equation exactly: $10^{2\log(4x+9)} = 121$.

24. Give an exact solution to the exponential equation $5^{x^2} = 625$.

25. If \$8,500 is invested at 4% compounded continuously, how many years will pass until there is \$12,000?

26. Use a graphing utility to help solve the equation $e^{3-2x} = 2^{x-1}$. Round your answer to two decimal places.

27. Strontium-90 with an initial amount of 6 grams has a half-life of 28 years.

a. Use STAT CALC ExpReg to model the equation of the amount remaining.

b. How many grams will remain after 32 years? Round your answer to two decimal places.

6

Systems of Linear Equations and Inequalities

Hurricane Charley
August 13, 2004

Tim Boyles/Getty Images News and Sport Services

Hurricane Frances
September 4, 2004

©AP/Wide World Photos

Hurricane Ivan
September 16, 2004

Scott Olson/Getty Images News and Sport Services

On Friday, the thirteenth of August, 2004, Hurricane Charley whipped through Florida. Three weeks later, Hurricane Frances pummeled Florida. Twelve days after that, Hurricane Ivan decided to make its appearance. On September 26, 2004, Hurricane Jeanne, the strongest and deadliest of all, made landfall in Florida. There had not been a state hit by four hurricanes in one season since 1886 in Texas.

Hurricane Jeanne, September 26, 2004

Joe Raedle/Getty Images News and Sport Services

The Federal Emergency Management Agency (FEMA) coordinated the relief effort, including trucks that hauled generators, water, and tarps. FEMA was faced with an *optimization* problem in mathematics that involved maximizing the number of people aided by the relief subject to the constraints of the weight of the relief items and how much space they would take on the trucks. For example, how many generators, cases of water, and tarps could each truck haul, and how many of each would bring aid to the greatest number of people?

IN THIS CHAPTER, we will solve *systems* of linear equations in two variables and in three variables. We will then perform partial-fraction decomposition, which will be valuable in calculus and is made possible by techniques of solving systems of linear equations. We will discuss linear inequalities in two variables and systems of linear inequalities. Systems of nonlinear equations and inequalities will be discussed in Chapter 8. Finally, we will discuss the linear programming model, which will enable you to address the FEMA relief problem as a result of Florida's multiple hurricanes.

SYSTEMS OF LINEAR EQUATIONS AND INEQUALITIES

SYSTEMS OF LINEAR EQUATIONS

Solving Systems of Linear Equations in Two Variables
- Substitution
- Elimination
- Graphing (two lines)

Solving Systems of Linear Equations in Three Variables
- Elimination
- Substitution

Types of Solutions
- One solution
- No solution
- Infinitely many solutions

Applications
- Partial fraction decomposition

SYSTEMS OF LINEAR INEQUALITIES

Linear Inequalities in Two Variables
- Line and shaded half-plane

Systems of Linear Inequalities in Two Variables
- Multiple lines and shaded regions
- Overlapping shaded region as the solution of a system

Applications of Systems of Linear Inequalities
- The linear programming model

SECTION

CHAPTER OBJECTIVES

- Solve systems of linear equations in two variables with the substitution method and the elimination method.
- Graph systems of linear equations (lines).
- Understand that systems of linear equations may have one solution, no solution, or infinitely many solutions.
- Solve systems of linear equations in three variables employing a combination of the elimination and substitution methods.
- Understand that graphs of linear equations in three variables correspond to planes.
- Perform partial-fraction decomposition.
- Solve a system of linear inequalities by finding the overlapping shaded regions.
- Use the linear programming model to solve optimization problems subject to constraints.

SECTION 6.1 SYSTEMS OF LINEAR EQUATIONS IN TWO VARIABLES

SKILLS OBJECTIVES

- Solve systems of linear equations in two variables using the substitution method.
- Solve systems of linear equations in two variables using the elimination method.
- Solve systems of linear equations in two variables by graphing.
- Solve applications involving systems of linear equations.

CONCEPTUAL OBJECTIVES

- Understand that a system of linear equations has either one solution, no solution, or infinitely many solutions.
- Visualize two lines that intersect at one point, no points (parallel lines), or infinitely many points (same line).

We learned in Section 2.3 that a linear equation in two variables is given in standard form by

$$Ax + By = C$$

and that the graph of this linear equation is a line, provided that A and B are not both equal to zero. In this section we discuss **systems of linear equations**, which can be thought of as simultaneous equations. To **solve** a system of linear equations means to find the solution that satisfies *both* equations:

$$\begin{aligned} x + 2y &= 6 \\ 3x - y &= 11 \end{aligned}$$

We can interpret the solution to this system of equations both algebraically and graphically.

	ALGEBRAIC	GRAPHICAL
Solution	$x = 4$ and $y = 1$	(4, 1)
Check	**Equation 1**: $x + 2y = 6$; $(4) + 2(1) = 6$ ✓ **Equation 2**: $3x - y = 11$; $3(4) - 1 = 11$ ✓	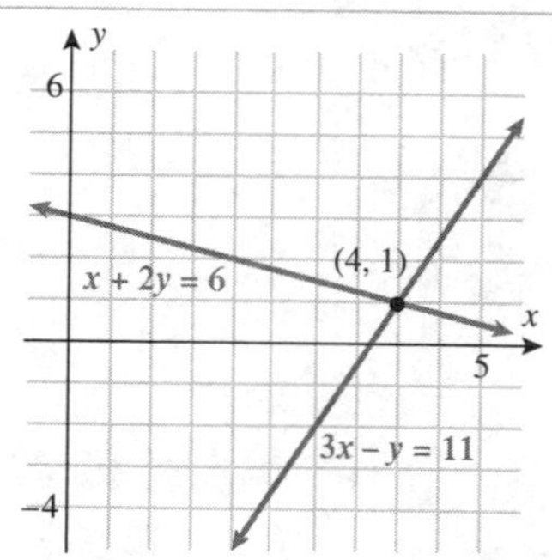
Interpretation	$x = 4$ and $y = 1$ satisfy both equations	The point (4, 1) lies on both lines

This particular example had *one solution*. There are systems of equations that have *no solution* or *infinitely many solutions*. We give these systems special names: **independent**, **inconsistent**, and **dependent**, respectively.

INDEPENDENT SYSTEM	INCONSISTENT SYSTEM	DEPENDENT SYSTEM
One solution	No solution	Infinitely many solutions
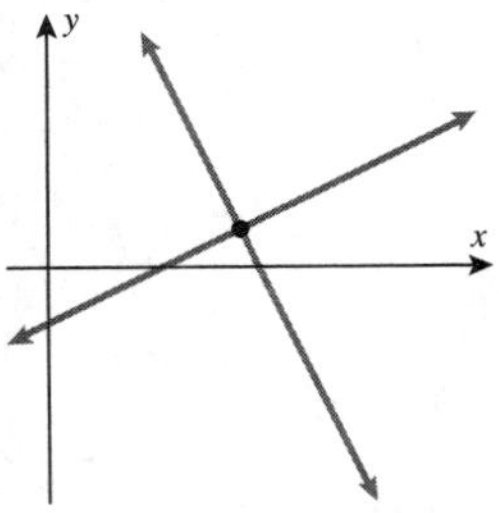	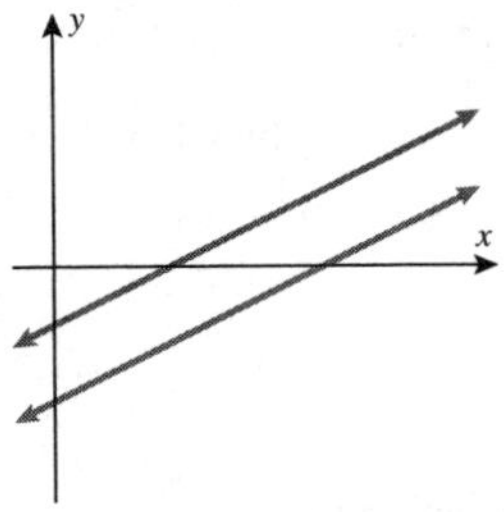	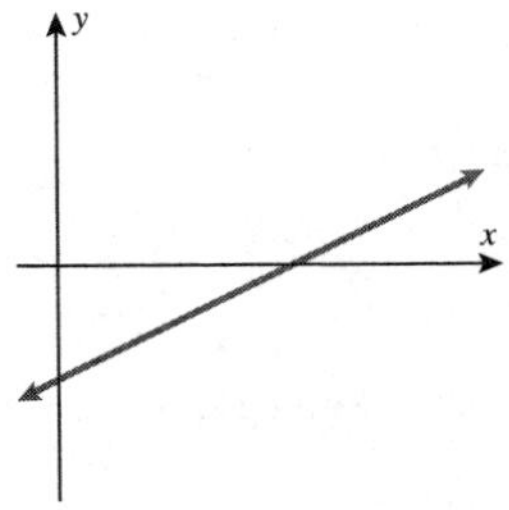
Lines have different slopes	Lines are parallel (same slope and different y-intercepts)	Lines coincide (same slope and same y-intercept)

In this section we discuss three methods for solving systems of linear equations: *substitution*, *elimination*, and *graphing*. We use the algebraic methods—substitution and elimination—to find solutions exactly; we then look at a graphical interpretation of the solution (two lines that intersect at one point, parallel lines, or coinciding lines).

We will illustrate each method with the same example given earlier:

$$x + 2y = 6 \qquad \text{Equation (1)}$$

$$3x - y = 11 \qquad \text{Equation (2)}$$

Substitution Method

The following box summarizes the substitution method for solving systems of linear equations in two variables.

SUBSTITUTION METHOD

Step 1: Solve one of the equations for one variable in terms of the other variable.

Equation (2): $y = 3x - 11$

Step 2: Substitute the expression found in Step 1 into the *other* equation. The result is an equation in one variable.

Equation (1): $x + 2(3x - 11) = 6$

Step 3: Solve the equation obtained in Step 2.

$$x + 6x - 22 = 6$$
$$7x = 28$$
$$\boxed{x = 4}$$

Step 4: Back-substitute the value found in Step 3 into the expression found in Step 1.

$$y = 3(4) - 11$$
$$\boxed{y = 1}$$

Step 5: Check that the solution satisfies *both* equations. Substitute (4, 1) into both equations.

Equation (1): $x + 2y = 6$

$(4) + 2(1) = 6$ ✓

Equation (2): $3x - y = 11$

$3(4) - 1 = 11$ ✓

EXAMPLE 1 **Determining by Substitution That a System Has One Solution**

Use the substitution method to solve the following system of linear equations:

$$x + y = 8 \quad \text{Equation (1)}$$
$$3x - y = 4 \quad \text{Equation (2)}$$

Solution:

STEP 1 Solve Equation (2) for y in terms of x. $\quad y = 3x - 4$

STEP 2 Substitute $y = 3x - 4$ into Equation (1). $\quad x + (3x - 4) = 8$

STEP 3 Solve for x.

$$x + 3x - 4 = 8$$
$$4x = 12$$
$$\boxed{x = 3}$$

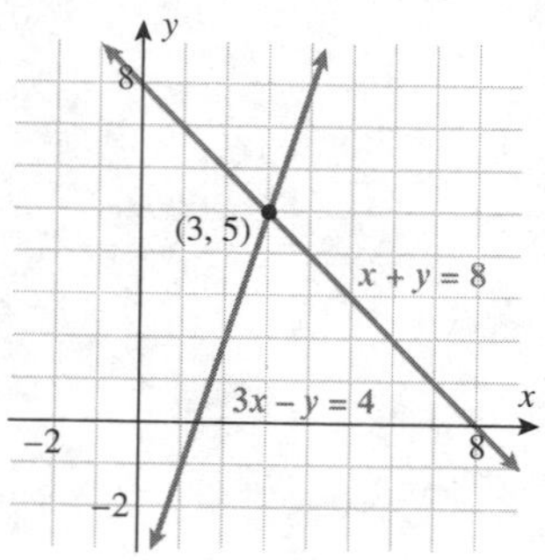

STEP 4 Back-substitute $x = 3$ into Equation (1).

$$3 + y = 8$$
$$\boxed{y = 5}$$

STEP 5 Check that (3, 5) satisfies *both* equations.

Equation (1): $x + y = 8$
$3 + 5 = 8$

Equation (2): $3x - y = 4$
$3(3) - 5 = 4$

Note: The graphs of the two equations are two lines that intersect at the point (3, 5).

EXAMPLE 2 **Determining by Substitution That a System Has No Solution**

Use the substitution method to solve the following system of linear equations:

$$x - y = 2 \quad \text{Equation (1)}$$
$$2x - 2y = 10 \quad \text{Equation (2)}$$

Solution:

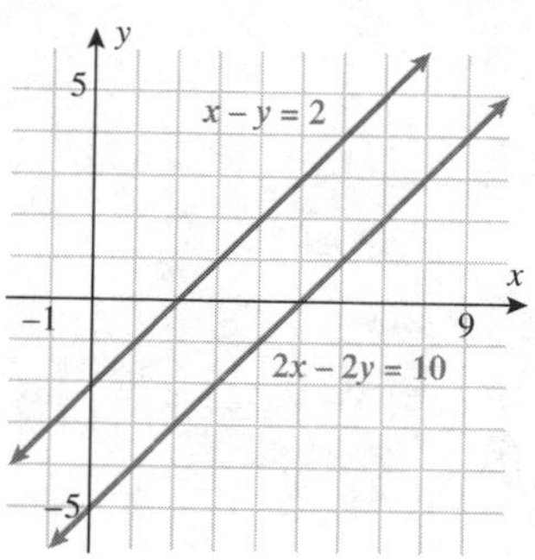

STEP 1 Solve Equation (1) for y in terms of x. $\quad y = x - 2$

STEP 2 Substitute $y = x - 2$ into Equation (2). $\quad 2x - 2(x - 2) = 10$

STEP 3 Solve for x.

$$2x - 2x + 4 = 10$$
$$4 = 10$$

$4 = 10$ is never true, so this is called an inconsistent system. There is $\boxed{\text{no solution}}$ to this system of linear equations.

Note: The graphs of the two equations are parallel lines.

EXAMPLE 3 **Determining by Substitution That a System Has Infinitely Many Solutions**

Use the substitution method to solve the following system of linear equations:

$$x - y = 2 \quad \text{Equation (1)}$$
$$-x + y = -2 \quad \text{Equation (2)}$$

Solution:

STEP 1 Solve Equation (1) for y in terms of x. $y = x - 2$

STEP 2 Substitute $y = x - 2$ into Equation (2). $-x + (x - 2) = -2$

STEP 3 Solve for x. $-x + x - 2 = -2$

$$-2 = -2$$

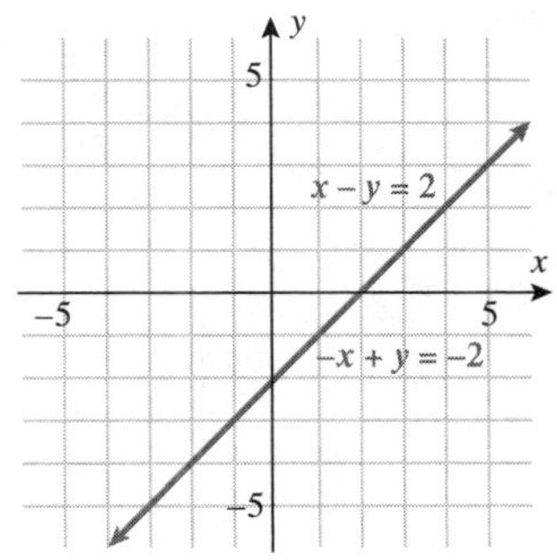

$-2 = -2$ is always true, so this is called a dependent system. Notice for instance, the points (2, 0), (4, 2), and (7, 5) all satisfy both equations. In fact, there are infinitely many solutions to this system of linear equations. All solutions are in the form (x, y), where $y = x - 2$.

Note: The graphs of these two equations are the same line.

YOUR TURN Use the substitution method to solve each system of linear equations.

a. $2x + y = 3$, $4x + 2y = 4$ **b.** $x - y = 2$, $4x - 3y = 10$ **c.** $x + 2y = 1$, $2x + 4y = 2$

Answer: a. no solution **b.** (4, 2) **c.** infinitely many solutions

Elimination Method

We now turn our attention to another method, *elimination*, which is often preferred over substitution and will later be used in higher order systems. In a system of two linear equations in two variables, the equations can be combined, resulting in a third equation in one variable, thus *eliminating* one of the variables. The following is an example of when elimination would be preferred because the y terms sum to zero when the two equations are added together:

$$\begin{array}{rcr} 2x - y &=& 5 \\ -x + y &=& -2 \\ \hline x \quad &=& 3 \end{array}$$

When you cannot eliminate a variable simply by *adding* the two equations, multiply one equation by a constant that will cause the coefficients of some variable in the two equations to match and be opposite in sign.

The following box summarizes the *elimination method*, also called the *addition method*, for solving systems of linear equations in two variables using the same example given earlier:

$$x + 2y = 6 \quad \text{Equation (1)}$$
$$3x - y = 11 \quad \text{Equation (2)}$$

ELIMINATION METHOD

Step 1*: **Multiply** the coefficients of one or both of the equations so that one of the variables will be eliminated when the two equations are added.

Multiply Equation (2) by 2:
$6x - 2y = 22$

Step 2: **Eliminate** one of the variables by adding the expression found in Step 1 to the *other* equation. The result is an equation in one variable.

$$\begin{array}{r} x + 2y = 6 \\ 6x - 2y = 22 \\ \hline 7x \quad\;\; = 28 \end{array}$$

Step 3: **Solve** the equation obtained in Step 2.

$$7x = 28$$
$$\boxed{x = 4}$$

Step 4: **Back-substitute** the value found in Step 3 into either of the two original equations.

$$(4) + 2y = 6$$
$$2y = 2$$
$$\boxed{y = 1}$$

Step 5: **Check** that the solution satisfies *both* equations. Substitute $\boxed{(4, 1)}$ into both equations.

Equation (1):
$$x + 2y = 6$$
$$(4) + 2(1) = 6 \checkmark$$

Equation (2):
$$3x - y = 11$$
$$3(4) - 1 = 11 \checkmark$$

*Step 1 is not necessary in cases where corresponding terms already sum to zero.

EXAMPLE 4 Determining by the Elimination Method That a System Has One Solution

Use the elimination method to solve the following system of linear equations:

$$2x - y = -5 \qquad \text{Equation (1)}$$
$$4x + y = 11 \qquad \text{Equation (2)}$$

Solution:

STEP 1 Not necessary.

STEP 2 Eliminate y by adding Equation (1) to Equation (2).

$$\begin{array}{r} 2x - y = -5 \\ 4x + y = 11 \\ \hline 6x \quad\;\; = 6 \end{array}$$

STEP 3 Solve for x.

$$\boxed{x = 1}$$

STEP 4 Back-substitute $x = 1$ into Equation (2). Solve for y.

$$4(1) + y = 11$$
$$\boxed{y = 7}$$

STEP 5 Check that (1, 7) satisfies both equations.

Equation (1):
$$2x - y = -5$$
$$2(1) - (7) = -5 \checkmark$$

Equation (2):
$$4x + y = 11$$
$$4(1) + (7) = 11 \checkmark$$

Note: The graphs of the two given equations correspond to two lines that intersect at the point (1, 7).

Study Tip

You can eliminate one variable from the system by addition when (1) the coefficients are equal and (2) the signs are opposite.

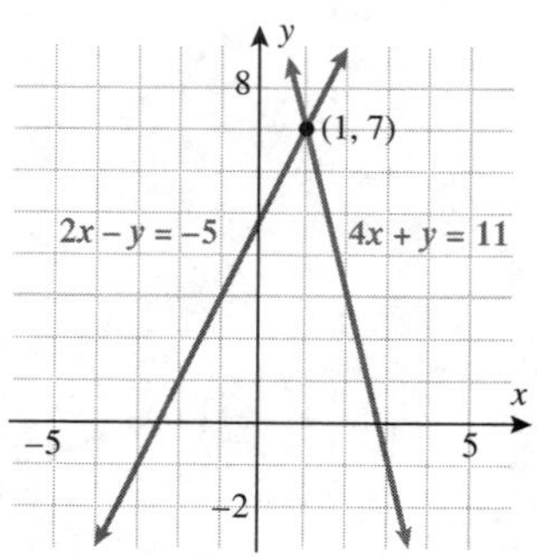

In Example 4, we eliminated the variable y simply by adding the two equations. Sometimes it is necessary to multiply one (Example 5) or both (Example 6) equations by constants prior to adding.

EXAMPLE 5 Applying the Elimination Method When Multiplying One Equation by a Constant Is Necessary

Use the elimination method to solve the following system of linear equations:

$$-4x + 3y = 23 \quad \text{Equation (1)}$$
$$12x + 5y = 1 \quad \text{Equation (2)}$$

Solution:

STEP 1 Multiply Equation (1) by 3.

$$-12x + 9y = 69$$

STEP 2 Eliminate x by adding the modified Equation (1) to Equation (2).

$$\begin{aligned} -12x + 9y &= 69 \\ 12x + 5y &= 1 \\ \hline 14y &= 70 \end{aligned}$$

STEP 3 Solve for y.

$$\boxed{y = 5}$$

STEP 4 Back-substitute $y = 5$ into Equation (2). Solve for x.

$$\begin{aligned} 12x + 5(5) &= 1 \\ 12x + 25 &= 1 \\ 12x &= -24 \end{aligned}$$

$$\boxed{x = -2}$$

STEP 5 Check that $(-2, 5)$ satisfies both equations.

Equation (1): $-4(-2) + 3(5) = 23$
$8 + 15 = 23$ ✓

Equation (2): $12(-2) + 5(5) = 1$
$-24 + 25 = 1$ ✓

Note: The graphs of the two given equations correspond to two lines that intersect at the point $(-2, 5)$.

Study Tip

Be sure to multiply the **entire** equation by the constant.

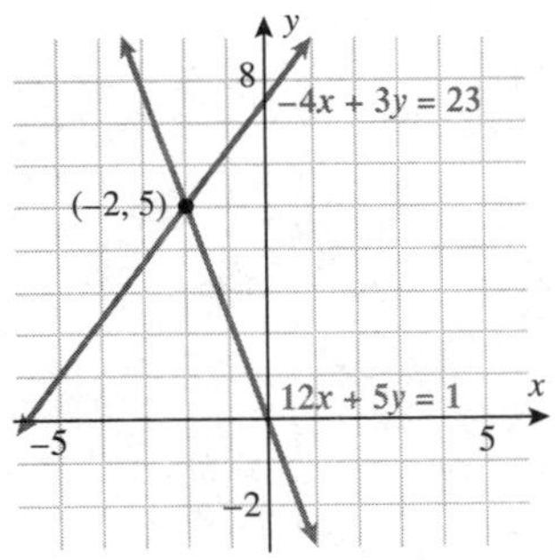

In Example 5, we eliminated x simply by multiplying the first equation by a constant and adding the result to the second equation. In order to eliminate either of the variables in Example 6, we will have to multiply *both* equations by constants prior to adding.

EXAMPLE 6 Applying the Elimination Method When Multiplying Both Equations by Constants Is Necessary

Use the elimination method to solve the following system of linear equations:

$$3x + 2y = 1 \quad \text{Equation (1)}$$
$$5x + 7y = 9 \quad \text{Equation (2)}$$

Solution:

STEP 1 Multiply Equation (1) by 5 and Equation (2) by -3.

$$\begin{aligned} 15x + 10y &= 5 \\ -15x - 21y &= -27 \end{aligned}$$

STEP 2 Eliminate x by adding the modified Equation (1) to the modified Equation (2).

$$\begin{aligned} 15x + 10y &= 5 \\ -15x - 21y &= -27 \\ \hline -11y &= -22 \end{aligned}$$

STEP 3 Solve for y.

$$\boxed{y = 2}$$

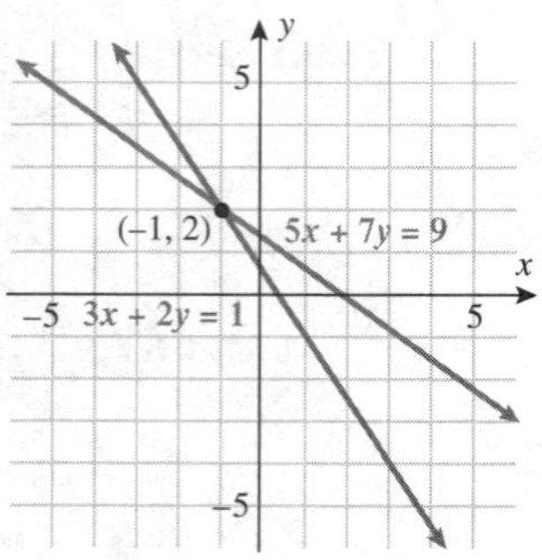

STEP 4 Back-substitute $y = 2$ into Equation (1). Solve for x.

$$\begin{aligned} 3x + 2(2) &= 1 \\ 3x &= -3 \\ \boxed{x = -1} \end{aligned}$$

STEP 5 Check that $(-1, 2)$ satisfies both equations.

Equation (1): $3x + 2y = 1$

$3(-1) + 2(2) = 1$ ✓

Equation (2): $5x + 7y = 9$

$5(-1) + 7(2) = 9$ ✓

Note: The graphs of the two given equations correspond to two lines that intersect at the point $(-1, 2)$.

Notice in Example 6 that we could have also eliminated y by multiplying the first equation by 7 and the second equation by -2. Typically, the choice is dictated by which approach will keep the coefficients as simple as possible. In the event that the coefficients contain fractions or decimals, first rewrite the equations in standard form with integer coefficients and then make the decision.

EXAMPLE 7 Determining by the Elimination Method That a System Has No Solution

Use the elimination method to solve the following system of linear equations:

$$\begin{aligned} -x + y &= 7 \quad \text{Equation (1)} \\ 2x - 2y &= 4 \quad \text{Equation (2)} \end{aligned}$$

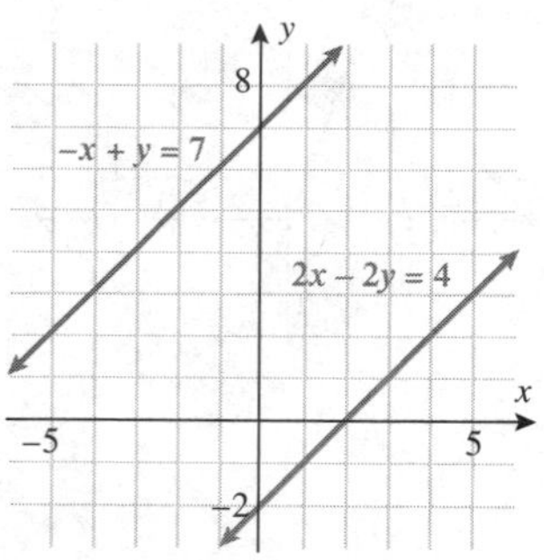

Solution:

STEP 1 Multiply Equation (1) by 2. $\quad -2x + 2y = 14$

STEP 2 Eliminate y by adding the modified Equation (1) found in Step 1 to Equation (2).

$$\begin{aligned} -2x + 2y &= 14 \\ 2x - 2y &= 4 \\ \hline 0 &= 18 \end{aligned}$$

This system is inconsistent since $0 = 18$ is never true. Therefore, there are no values of x and y that satisfy both equations. We say that there is **no solution** to this system of linear equations.

Note: The graphs of the two equations are two parallel lines.

Study Tip

Systems of linear equations in two variables have either one solution, no solution, or infinitely many solutions.

EXAMPLE 8 Determining by the Elimination Method That a System Has Infinitely Many Solutions

Use the elimination method to solve the following system of linear equations:

$$\begin{aligned} 7x + y &= 2 \quad \text{Equation (1)} \\ -14x - 2y &= -4 \quad \text{Equation (2)} \end{aligned}$$

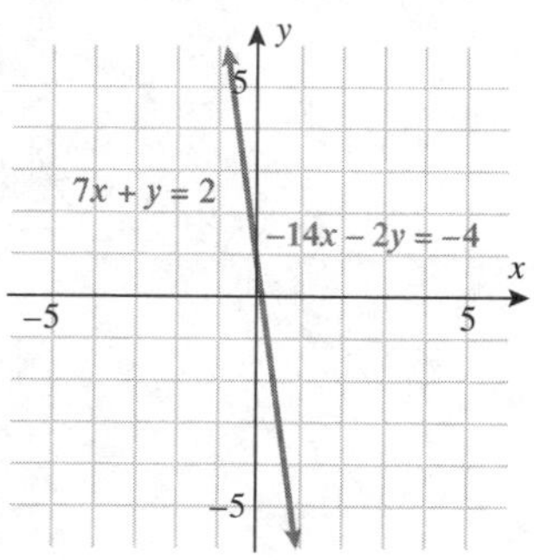

Solution:

STEP 1 Multiply Equation (1) by 2. $\quad 14x + 2y = 4$

STEP 2 Add the modified Equation (1) found in Step 1 to Equation (2).

$$\begin{aligned} 14x + 2y &= 4 \\ -14x - 2y &= -4 \\ \hline 0 &= 0 \end{aligned}$$

This system is dependent since $0 = 0$ is always true. We say that there are infinitely many solutions to this system of linear equations.

Note: The graphs of the two equations are the same line.

YOUR TURN Apply the elimination method to solve each system of linear equations.

a. $\begin{aligned} 2x + 3y &= 1 \\ 4x - 3y &= -7 \end{aligned}$ **b.** $\begin{aligned} x - 5y &= 2 \\ -10x + 50y &= -20 \end{aligned}$ **c.** $\begin{aligned} x - y &= 14 \\ -x + y &= 9 \end{aligned}$

Answer: **a.** $(-1, 1)$ **b.** infinitely many solutions **c.** no solution

Graphing Method

A third way to solve a system of linear equations in two variables is to graph the two lines. If the two lines intersect, then the point of intersection is the solution. Graphing is the most labor-intensive method for solving systems of linear equations in two variables. The graphing method is typically not used to solve systems of linear equations when an exact solution is desired. Instead, it is used to interpret or confirm the solution(s) found by the other two methods (substitution and elimination). If you are using a graphing calculator, however, you will get as accurate an answer using the graphing method as you will when applying the other methods.

The following box summarizes the graphing method for solving systems of linear equations in two variables using the same example given earlier:

$$x + 2y = 6 \qquad \text{Equation (1)}$$
$$3x - y = 11 \qquad \text{Equation (2)}$$

GRAPHING METHOD

Step 1*: Write the equations in slope–intercept form.

Equation (1): $y = -\frac{1}{2}x + 3$ Equation (2): $y = 3x - 11$

Step 2: Graph the two lines.

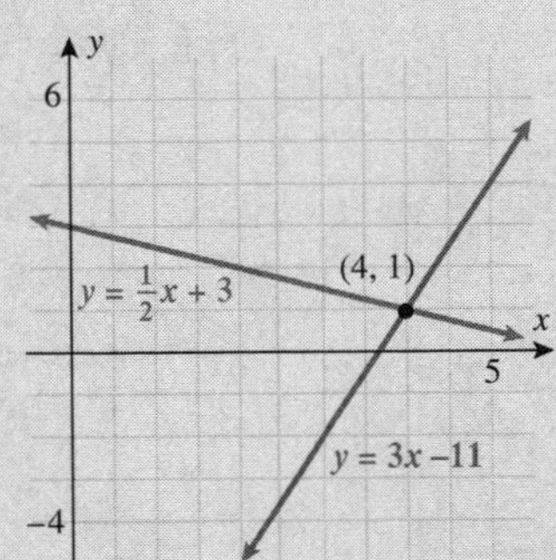

Step 3: Identify the point of intersection.

(4, 1)

Step 4: Check that the solution satisfies *both* equations.

Equation (1): $x + 2y = 6$, $(4) + 2(1) = 6$ ✓

Equation (2): $3x - y = 11$, $3(4) - 1 = 11$ ✓

*Step 1 is not necessary when the lines are already in slope–intercept form.

EXAMPLE 9 Determining by Graphing That a System Has One Solution

Use graphing to solve the following system of linear equations:

$$x + y = 2 \qquad \text{Equation (1)}$$
$$3x - y = 2 \qquad \text{Equation (2)}$$

Solution:

STEP 1 Write each equation in slope–intercept form.

$$y = -x + 2 \qquad \text{Equation (1)}$$
$$y = 3x - 2 \qquad \text{Equation (2)}$$

STEP 2 Plot both lines on the same graph.

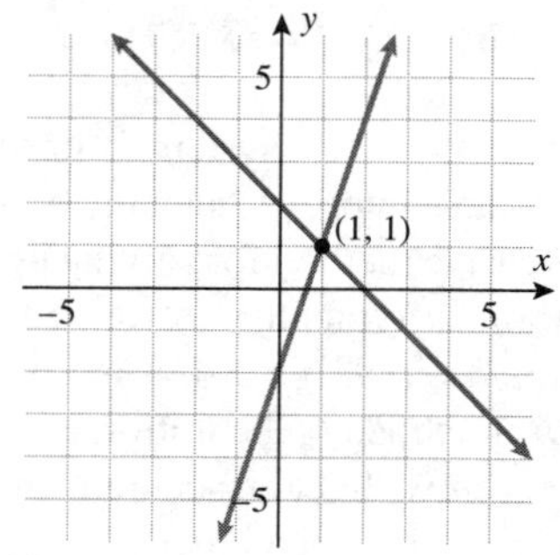

STEP 3 Identify the point of intersection. $\boxed{(1, 1)}$

STEP 4 Check that the point (1, 1) satisfies both equations.

$$x + y = 2$$
$$1 + 1 = 2 \checkmark \qquad \text{Equation (1)}$$

$$3x - y = 2$$
$$3(1) - (1) = 2 \checkmark \qquad \text{Equation (2)}$$

Note: There is one solution, because the two lines intersect at one point.

Technology Tip

First solve for y in each equation. The graphs of $Y_1 = \frac{1}{3}(2x - 9)$ and $Y_2 = \frac{1}{6}(4x + 12)$ are shown.

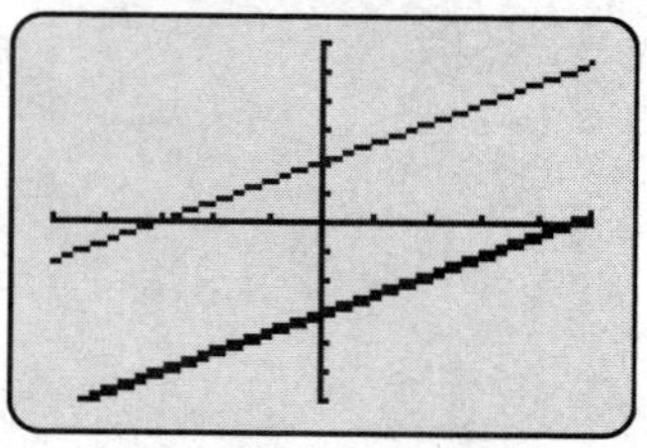

The graphs and table show that there is no solution to the system.

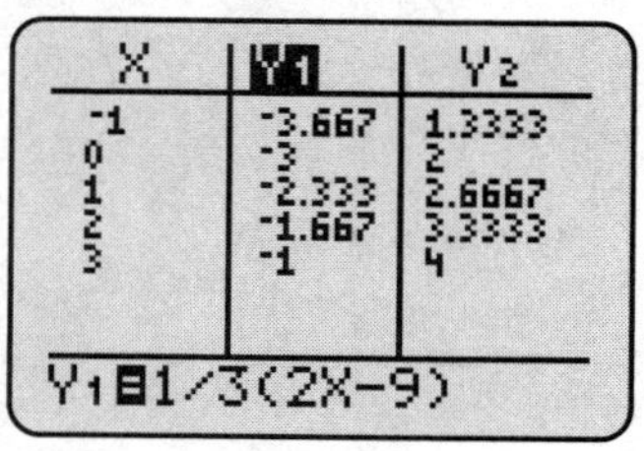

EXAMPLE 10 Determining by Graphing That a System Has No Solution

Use graphing to solve the following system of linear equations:

$$2x - 3y = 9 \qquad \text{Equation (1)}$$
$$-4x + 6y = 12 \qquad \text{Equation (2)}$$

Solution:

STEP 1 Write each equation in slope–intercept form.

$$y = \frac{2}{3}x - 3 \qquad \text{Equation (1)}$$
$$y = \frac{2}{3}x + 2 \qquad \text{Equation (2)}$$

STEP 2 Plot both lines on the same graph.

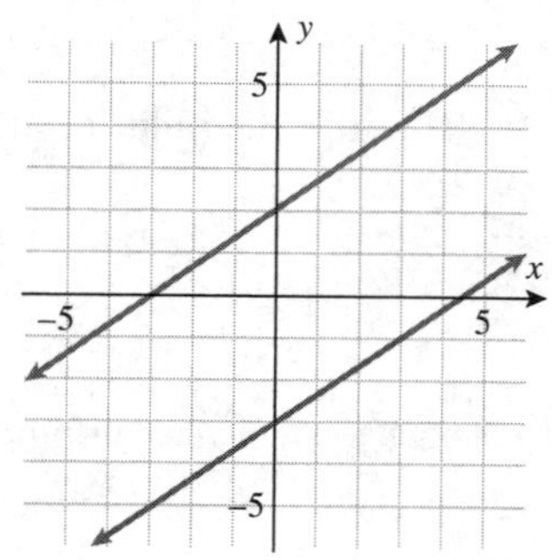

STEP 3 Identify the point of intersection. None

There is no solution, because two parallel lines never intersect.

Note: Both lines have the same slope, but different y-intercepts.

EXAMPLE 11 Determining by Graphing That a System Has Infinitely Many Solutions

Use graphing to solve the following system of linear equations:

$$3x + 4y = 12 \qquad \text{Equation (1)}$$

$$\frac{3}{4}x + y = 3 \qquad \text{Equation (2)}$$

Solution:

STEP 1 Write each equation in slope–intercept form. $y = -\frac{3}{4}x + 3 \quad y = -\frac{3}{4}x + 3$

STEP 2 Plot both lines on the same graph.

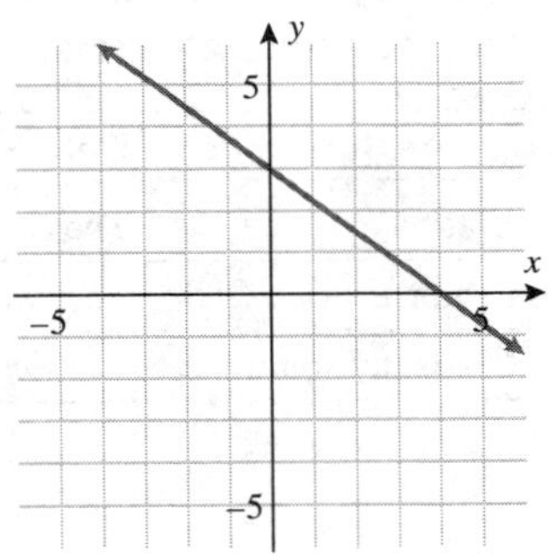

STEP 3 Identify the point of intersection. Infinitely many points

There are infinitely many solutions, since the two lines are identical and coincide.

YOUR TURN Utilize graphing to solve each system of linear equations.

a. $x - 2y = 1$, $2x - 4y = 2$ **b.** $x - 2y = 1$, $2x + y = 7$ **c.** $2x + y = 3$, $2x + y = 7$

Technology Tip

Solve for y in each equation first. The graphs of $Y_1 = \frac{1}{4}(-3x + 12)$ and $Y_2 = -\frac{3}{4}x + 3$ are shown.

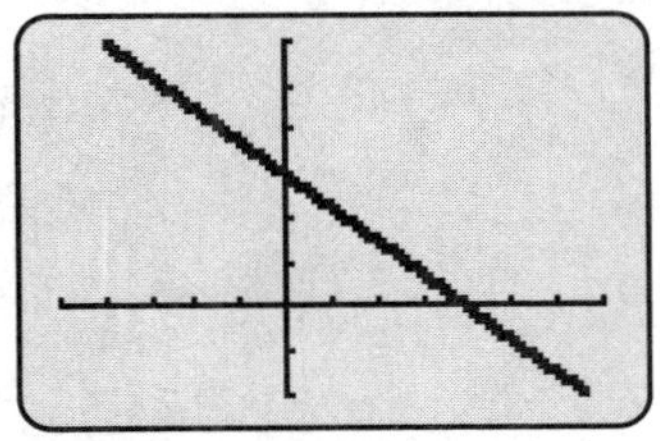

Both lines are on the same graph.

Answer:
a. infinitely many solutions
b. (3, 1)
c. no solution

Three Methods and Three Types of Solutions

Given any system of two linear equations in two variables, any of the three methods (substitution, elimination, or graphing) can be utilized. If you find that it is easy to eliminate a variable by adding multiples of the two equations, then elimination is the preferred choice. If you do not see an obvious elimination, then solve the system by substitution. For exact solutions, choose one of these two algebraic methods. You should typically use graphing to confirm the solution(s) you have found by applying the other two methods or when you are using a graphing utility.

EXAMPLE 12 Identifying Which Method to Use

State which of the two algebraic methods (elimination or substitution) would be the preferred method to solve each system of linear equations.

a. $x - 2y = 1$, $-x + y = 2$ **b.** $x = 2y - 1$, $2x - y = 4$ **c.** $7x - 20y = 1$, $5x + 3y = 18$

Solution:

a. Elimination: Because the x variable is eliminated when the two equations are added.

b. Substitution: Because the first equation is easily substituted into the second equation (for x).

c. Either: There is no preferred method, as both elimination and substitution require substantial work.

Regardless of which method is used to solve systems of linear equations in two variables, in general, we can summarize the three types of solutions both algebraically and graphically.

THREE TYPES OF SOLUTIONS TO SYSTEMS OF LINEAR EQUATIONS

NUMBER OF SOLUTIONS	GRAPHICAL INTERPRETATION
One solution	The two lines intersect at one point.
No solution	The two lines are parallel. (*Same* slope/*different* y-intercepts.)
Infinitely many solutions	The two lines coincide. (*Same* slope/*same* y-intercept.)

Applications

Suppose you have two job offers that require sales. One pays a higher base, while the other pays a higher commission. Which job do you take?

Technology Tip

The graphs of $Y_1 = 20{,}000 + 0.05x$ and $Y_2 = 30{,}000 + 0.03x$ are shown.

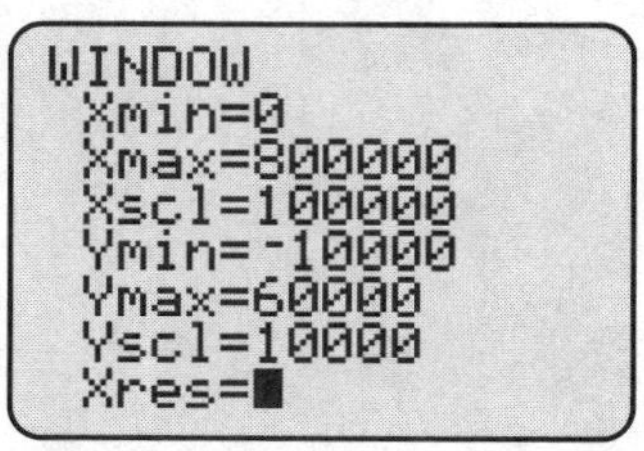

EXAMPLE 13 Deciding Which Job to Take

Suppose that upon graduation you are offered a job selling biomolecular devices to laboratories studying DNA. The Beckman-Coulter Company offers you a job selling its DNA sequencer with an annual base salary of \$20,000 plus 5% commission on total sales. The MJ Research Corporation offers you a job selling its PCR Machine that makes copies of DNA with an annual base salary of \$30,000 plus 3% commission on sales. Determine what the total sales would have to be to make the Beckman-Coulter job the better offer.

Solution:

STEP 1 Identify the question.
When would these two jobs have equal compensations?

STEP 2 Make notes.

Beckman-Coulter	20,000 + 5%
MJ Research	30,000 + 3%

STEP 3 Set up the equations.
Let x = total sales and y = compensation.

Equation (1) Beckman-Coulter: $y = 20{,}000 + 0.05x$

Equation (2) MJ Research: $y = 30{,}000 + 0.03x$

STEP 4 Solve the system of equations.
*Substitution method**

Substitute Equation (1) into Equation (2). $20{,}000 + 0.05x = 30{,}000 + 0.03x$

Solve for x. $0.02x = 10{,}000$

$x = 500{,}000$

If you make \$500,000 worth of sales per year, the jobs will yield equal compensations. If you sell less than \$500,000, the MJ Research job is the better offer, and more than **\$500,000**, the Beckman-Coulter job is the better offer.

*The elimination method could also have been used.

STEP 5 Check the solution.

Equation (1) Beckman-Coulter: $y = 20{,}000 + 0.05(500{,}000) = \$45{,}000$

Equation (2) MJ Research: $y = 30{,}000 + 0.03(500{,}000) = \$45{,}000$

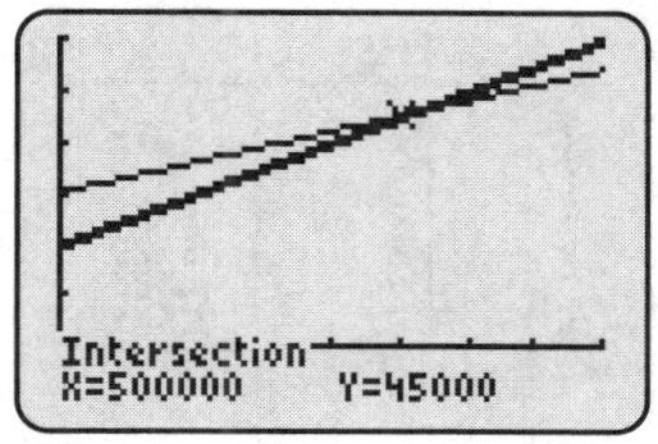

The graphs and table support the solution to the system.

X	Y1	Y2
100000	25000	33000
200000	30000	36000
300000	35000	39000
400000	40000	42000
500000	45000	45000
600000	50000	48000
700000	55000	51000

Y1=20000+0.05X

EXAMPLE 14 Deciding How Many Pounds of Each Meat to Buy at the Deli

The Chi Omega sorority is hosting a party, and the membership committee would like to make sandwiches for the new members. They already have bread and condiments but ask Tara to run to the deli to buy sliced turkey. The membership chair has instructed Tara to buy 5 pounds of the best turkey she can find, and the treasurer has given her \$37 and told her to spend it all. When she arrives at Publix supermarket, she has a choice of two types of turkey: Boar's Head (\$8 per pound) and the store brand (\$7 per pound). How much of each kind should Tara buy to spend the entire \$37 on the best quality of turkey?

Solution:

STEP 1 Identify the question.
How much of each type of sliced turkey should Tara buy?

STEP 2 Make notes.
Store brand turkey costs \$7/pound.
Boar's Head turkey costs \$8/pound.
Tara has \$37 to spend on 5 pounds of turkey.

STEP 3 Set up the equations.
Let x = number of pounds of Boar's Head turkey and y = number of pounds of store brand turkey. The system of equations therefore has to include one equation giving a total of 5 pounds and one giving a total of \$37.

Equation (1): $x + y = 5$

Equation (2): $8x + 7y = 37$

Technology Tip

First solve for y in each equation. The graphs of $Y_1 = -x + 5$ and $Y_2 = \frac{1}{7}(-8x + 37)$ are shown.

WINDOW
Xmin=-2
Xmax=5
Xscl=1
Ymin=-1
Ymax=6
Yscl=1
Xres=1

Plot1 Plot2 Plot3
\Y1=-X+5
\Y2=(-8X+37)/7

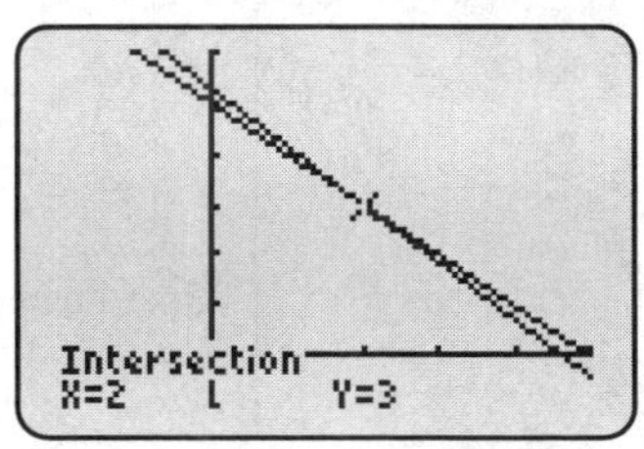

The graphs and table support the solution to the system.

STEP 4 Solve the system of equations.

*Elimination method**

Multiply Equation (1) by −7. $-7x - 7y = -35$

Add this result to Equation (2).

$$\begin{aligned} -7x - 7y &= -35 \\ 8x + 7y &= 37 \\ \hline x &= 2 \end{aligned}$$

Substitute $x = 2$ into the original Equation (1). $2 + y = 5$

Solve for y. $y = 3$

Tara should buy 2 pounds of Boar's Head and 3 pounds of store brand turkey.

*The substitution method could also have been used to solve this system of linear equations.

STEP 5 Check the solution.

Equation (1): $2 + 3 = 5$

Equation (2): $8(2) + 7(3) = 37$

SECTION 6.1 SUMMARY

In this section we discussed two algebraic techniques for solving systems of linear equations in two variables:

- Substitution method
- Elimination method

The algebraic methods are preferred for exact solutions, and the graphing method is typically used to give a visual interpretation and confirmation of the solution. There are three types of solutions to systems of linear equations: one solution, no solution, or infinitely many solutions.

INDEPENDENT SYSTEM	INCONSISTENT SYSTEM	DEPENDENT SYSTEM
One solution	No solution	Infinitely many solutions
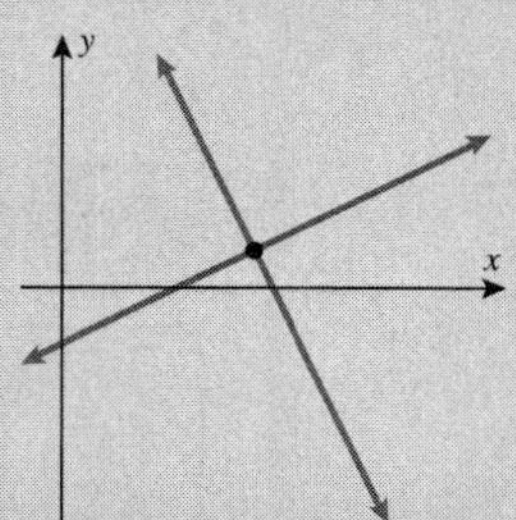	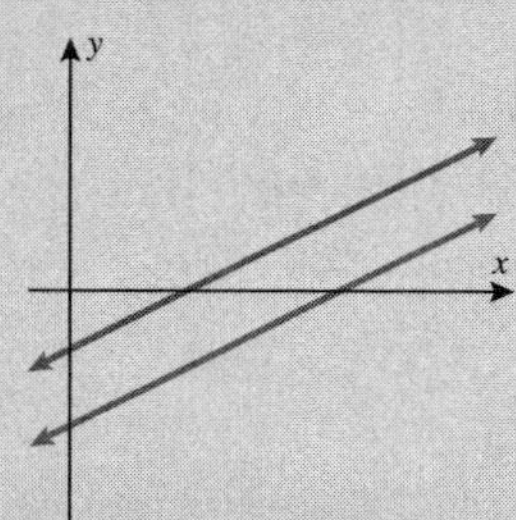	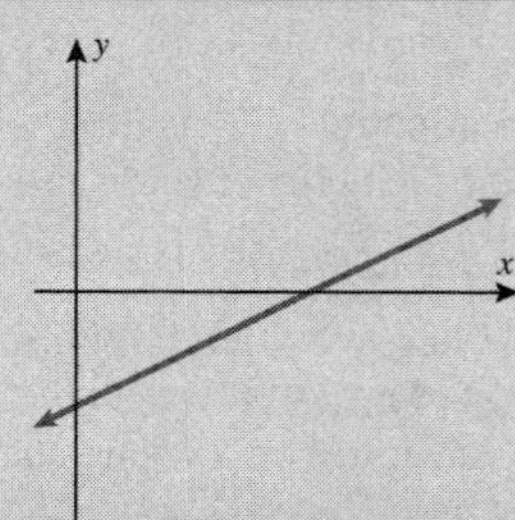
Lines have different slopes	Lines are parallel (same slope and different y-intercepts)	Lines coincide (same slope and same y-intercept)

SECTION 6.1 EXERCISES

SKILLS

In Exercises 1–22, solve each system of linear equations by substitution.

1. $\begin{aligned} x - y &= 1 \\ x + y &= 1 \end{aligned}$

2. $\begin{aligned} x - y &= 2 \\ x + y &= -2 \end{aligned}$

3. $\begin{aligned} x + y &= 7 \\ x - y &= 9 \end{aligned}$

4. $\begin{aligned} x - y &= -10 \\ x + y &= 4 \end{aligned}$

5. $\begin{aligned} 2x - y &= 3 \\ x - 3y &= 4 \end{aligned}$

6. $\begin{aligned} 4x + 3y &= 3 \\ 2x + y &= 1 \end{aligned}$

7. $\begin{aligned} 3x + y &= 5 \\ 2x - 5y &= -8 \end{aligned}$

8. $\begin{aligned} 6x - y &= -15 \\ 2x - 4y &= -16 \end{aligned}$

9. $\begin{aligned} 2u + 5v &= 7 \\ 3u - v &= 5 \end{aligned}$

10. $\begin{aligned} m - 2n &= 4 \\ 3m + 2n &= 1 \end{aligned}$

11. $\begin{aligned} 2x + y &= 7 \\ -2x - y &= 5 \end{aligned}$

12. $\begin{aligned} 3x - y &= 2 \\ 3x - y &= 4 \end{aligned}$

13. $\begin{aligned} 4r - s &= 1 \\ 8r - 2s &= 2 \end{aligned}$

14. $\begin{aligned} -3p + q &= -4 \\ 6p - 2q &= 8 \end{aligned}$

15. $\begin{aligned} 5r - 3s &= 15 \\ -10r + 6s &= -30 \end{aligned}$

16. $\begin{aligned} -5p - 3q &= -1 \\ 10p + 6q &= 2 \end{aligned}$

17. $\begin{aligned} 2x - 3y &= -7 \\ 3x + 7y &= 24 \end{aligned}$

18. $\begin{aligned} 4x - 5y &= -7 \\ 3x + 8y &= 30 \end{aligned}$

19. $\begin{aligned} \tfrac{1}{3}x - \tfrac{1}{4}y &= 0 \\ -\tfrac{2}{3}x + \tfrac{3}{4}y &= 2 \end{aligned}$

20. $\begin{aligned} \tfrac{1}{5}x + \tfrac{2}{3}y &= 10 \\ -\tfrac{1}{2}x - \tfrac{1}{6}y &= -7 \end{aligned}$

21. $\begin{aligned} 7.2x - 4.1y &= 7.0 \\ -3.5x + 16.5y &= 2.4 \end{aligned}$

22. $\begin{aligned} 6.3x + 1.5y &= 10.5 \\ -0.4x + 2.2y &= -8.7 \end{aligned}$

In Exercises 23–42, solve each system of linear equations by elimination.

23. $\begin{aligned} x - y &= 2 \\ x + y &= 4 \end{aligned}$

24. $\begin{aligned} x + y &= 2 \\ x - y &= -2 \end{aligned}$

25. $\begin{aligned} x - y &= -3 \\ x + y &= 7 \end{aligned}$

26. $\begin{aligned} x - y &= -10 \\ x + y &= 8 \end{aligned}$

27. $\begin{aligned} 5x + 3y &= -3 \\ 3x - 3y &= -21 \end{aligned}$

28. $\begin{aligned} -2x + 3y &= 1 \\ 2x - y &= 7 \end{aligned}$

29. $\begin{aligned} 2x - 7y &= 4 \\ 5x + 7y &= 3 \end{aligned}$

30. $\begin{aligned} 3x + 2y &= 6 \\ -3x + 6y &= 18 \end{aligned}$

31. $\begin{aligned} 2x + 5y &= 7 \\ 3x - 10y &= 5 \end{aligned}$

32. $\begin{aligned} 6x - 2y &= 3 \\ -3x + 2y &= -2 \end{aligned}$

33. $\begin{aligned} 2x + 5y &= 5 \\ -4x - 10y &= -10 \end{aligned}$

34. $\begin{aligned} 11x + 3y &= 3 \\ 22x + 6y &= 6 \end{aligned}$

35. $\begin{aligned} 3x - 2y &= 12 \\ 4x + 3y &= 16 \end{aligned}$

36. $\begin{aligned} 5x - 2y &= 7 \\ 3x + 5y &= 29 \end{aligned}$

37. $\begin{aligned} 6x - 3y &= -15 \\ 7x + 2y &= -12 \end{aligned}$

38. $\begin{aligned} 7x - 4y &= -1 \\ 3x - 5y &= 16 \end{aligned}$

39. $\begin{aligned} 0.02x + 0.05y &= 1.25 \\ -0.06x - 0.15y &= -3.75 \end{aligned}$

40. $\begin{aligned} -0.5x + 0.3y &= 0.8 \\ -1.5x + 0.9y &= 2.4 \end{aligned}$

41. $\begin{aligned} \tfrac{1}{3}x + \tfrac{1}{2}y &= 1 \\ \tfrac{1}{5}x + \tfrac{7}{2}y &= 2 \end{aligned}$

42. $\begin{aligned} \tfrac{1}{2}x - \tfrac{1}{3}y &= 0 \\ \tfrac{3}{2}x + \tfrac{1}{2}y &= \tfrac{3}{4} \end{aligned}$

In Exercises 43–46, match the systems of equations with the graphs.

43. $\begin{aligned} 3x - y &= 1 \\ 3x + y &= 5 \end{aligned}$

44. $\begin{aligned} -x + 2y &= -1 \\ 2x + y &= 7 \end{aligned}$

45. $\begin{aligned} 2x + y &= 3 \\ 2x + y &= 7 \end{aligned}$

46. $\begin{aligned} x - 2y &= 1 \\ 2x - 4y &= 2 \end{aligned}$

a.

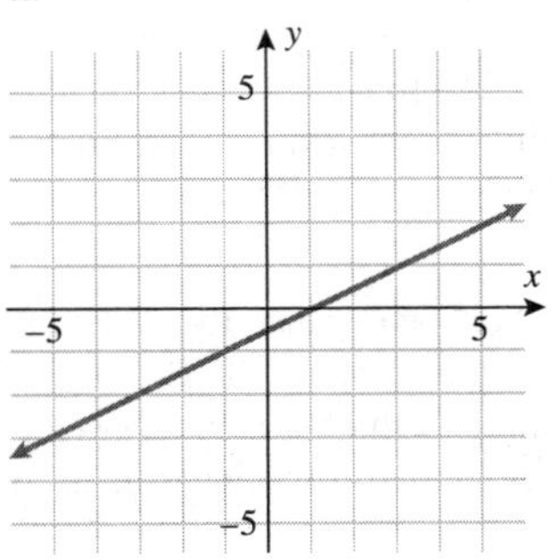

b.

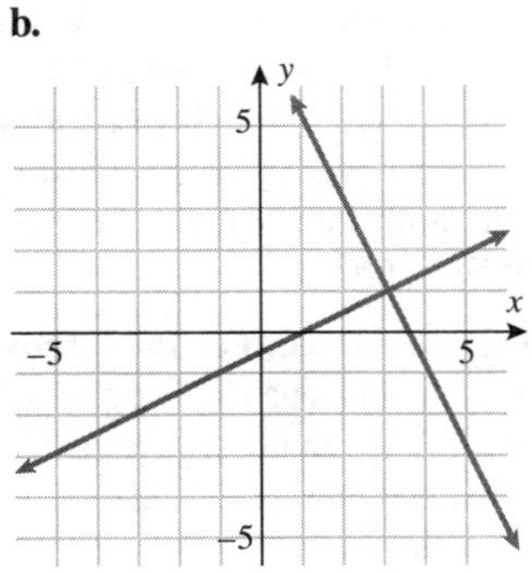

c.

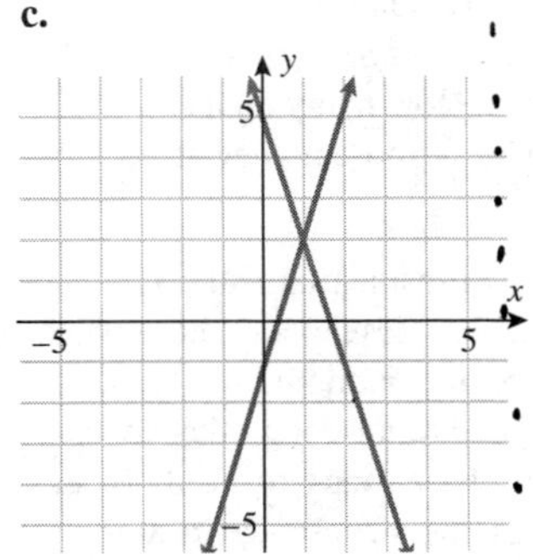

d.

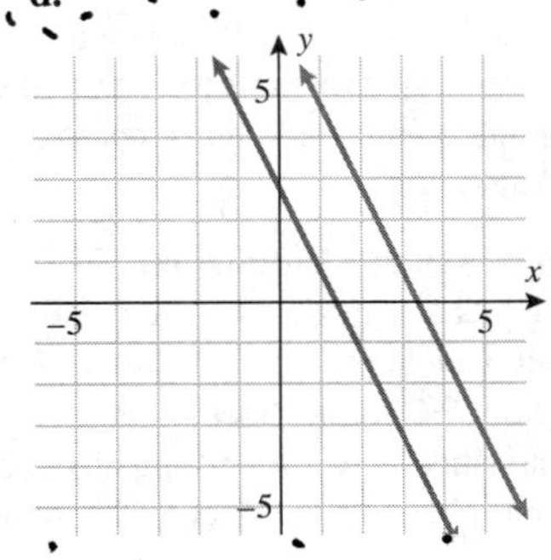

In Exercises 47–54, solve each system of linear equations by graphing.

47. $y = -x$
$y = x$

48. $x - 3y = 0$
$x + 3y = 0$

49. $2x + y = -3$
$x + y = -2$

50. $x - 2y = -1$
$-x - y = -5$

51. $\frac{1}{2}x - \frac{2}{3}y = 4$
$\frac{1}{4}x - y = 6$

52. $\frac{1}{5}x - \frac{5}{2}y = 10$
$\frac{1}{15}x - \frac{5}{6}y = \frac{10}{3}$

53. $1.6x - y = 4.8$
$-0.8x + 0.5y = 1.5$

54. $1.1x - 2.2y = 3.3$
$-3.3x + 6.6y = -6.6$

APPLICATIONS

55. Event Planning. You are in charge of an event planning for a major corporation that is having a reception. Your boss has given you a budget of $1,000.00 for gifts and wants you to order Montblanc pens for VIPs and Cross pens for other guests. A Montblanc pen costs $72.00 and a Cross pen costs $10.00. You must have 69 pens for guests. How many pens of each type should you order in order to stay within the budget?

56. Health Club Management. A fitness club has a budget of $915.00 to purchase two different types of dumbbell sets. One set costs $30.00 each, and the deluxe model dumbbell set costs $45.00 each. The club wants to purchase 24 new dumbbell sets to use in its gym. How many of sets of each type can it purchase?

57. Mixture. In chemistry lab, Stephanie has to make a 37 milliliter solution that is 12% HCl. All that is in the lab is 8% and 15% HCl. How many milliliters of each solution should she use to obtain the desired mix?

58. Mixture. A mechanic has 340 gallons of gasoline and 10 gallons of oil to make gas/oil mixtures. He wants one mixture to be 4% oil and the other mixture to be 2.5% oil. If he wants to use all of the gas and oil, how many gallons of gas and oil are in each of the resulting mixtures?

59. Salary Comparison. Upon graduation with a degree in management of information systems (MIS), you decide to work for a company that buys data from states' departments of motor vehicles and sells to banks and car dealerships customized reports detailing how many cars at each dealership are financed through particular banks. Autocount Corporation offers you a $15,000 base salary and 10% commission on your total annual sales. Polk Corporation offers you a base salary of $30,000 plus a 5% commission on your total annual sales. How many total sales would you have to make per year to make more money at Autocount?

60. Salary Comparison. Two types of residential real estate agents are: those who sell existing houses (resale) and those who sell new homes for developers. Resale of existing homes typically earns 6% commission on every sale, and representing developers in selling new homes typically earns a base salary of $15,000 per year plus an additional 1.5% commission, because agents are required to work 5 days a week on site in a new development. How many total dollars would an agent have to sell per year to make more money in resale than in new homes?

61. Gas Mileage. A Honda Accord gets approximately 26 mpg on the highway and 19 mpg in the city. You drove 349.5 miles on a full tank (16 gallons) of gasoline. Approximately how many miles did you drive in the city and how many on the highway?

62. Wireless Plans. AT&T is offering a 600-minute peak plan with free mobile-to-mobile and weekend minutes at $59 per month plus $0.13 per minute for every minute over 600. The next plan up is the 800-minute plan that costs $79 per month. You think you may go over 600 minutes, but are not sure you need 800 minutes. How many minutes would you have to talk for the 800-minute plan to be the better deal?

63. Distance/Rate/Time. A direct flight on Delta Air Lines from Atlanta to Paris is 4000 miles and takes approximately 8 hours going east (Atlanta to Paris) and 10 hours going west (Paris to Atlanta). Although the plane averages the same airspeed, there is a headwind while traveling west and a tailwind while traveling east, resulting in different air speeds. What is the average air speed of the plane, and what is the average wind speed?

64. Distance/Rate/Time. A private pilot flies a Cessna 172 on a trip that is 500 miles each way. It takes her approximately 3 hours to get there and 4 hours to return. What is the approximate air speed of the Cessna, and what is the approximate wind speed?

65. Investment Portfolio. Leticia has been tracking two volatile stocks. Stock A over the last year has increased 10%, and stock B has increased 14% (using a simple interest model). She has $10,000 to invest and would like to split it between these two stocks. If the stocks continue to perform at the same rate, how much should she invest in each to result in a balance of $11,260?

66. Investment Portfolio. Toby split his savings into two different investments, one earning 5% and the other earning 7%. He put twice as much in the investment earning the higher rate. In 1 year, he earned $665 in interest. How much money did he invest in each account?

67. **Break-Even Analysis.** A company produces CD players for a unit cost of \$15.00 per CD player. The company has fixed costs of \$120.00. If each CD player can be sold for \$30.00, how many CD players must be sold to break even? Determine the cost equation first. Next, determine the revenue equation. Use the two equations you have found to determine the break-even point.

68. **Managing a Lemonade Stand.** An elementary-school-age child wants to have a lemonade stand. She wants to sell each glass of lemonade for \$0.25. She has determined that each glass of lemonade costs about \$0.10 to make (for lemons and sugar). It costs her \$15.00 for materials to make the lemonade stand. How many glasses of lemonade must she sell to break even?

CATCH THE MISTAKE

In Exercises 69–72, explain the mistake that is made.

69. Solve the system of equations by elimination.

$$\begin{aligned} 2x + y &= -3 \\ 3x + y &= 8 \end{aligned}$$

Solution:

Multiply Equation (1) by -1. $\quad 2x - y = -3$

Add the result to Equation (2). $\quad 3x + y = 8$

$$5x = 5$$

Solve for x. $\quad x = 1$

Substitute $x = 1$ into Equation (2). $\quad 3(1) + y = 8$

$$y = 5$$

The answer $(1, 5)$ is incorrect. What mistake was made?

70. Solve the system of equations by elimination.

$$\begin{aligned} 4x - y &= 12 \\ 4x - y &= 24 \end{aligned}$$

Solution:

Multiply Equation (1) by -1. $\quad -4x + y = -12$

Add the result to Equation (2).

$$\begin{aligned} -4x + y &= -12 \\ 4x - y &= 24 \\ \hline 0 &= 12 \end{aligned}$$

Answer: Infinitely many solutions.

This is incorrect. What mistake was made?

71. Solve the system of equations by substitution.

$$\begin{aligned} x + 3y &= -4 \\ -x + 2y &= -6 \end{aligned}$$

Solution:

Solve Equation (1) for x. $\quad x = -3y - 4$

Substitute $x = -3y - 4$ into Equation (2). $\quad -(-3y - 4) + 2y = -6$

Solve for y.

$$\begin{aligned} 3y - 4 + 2y &= -6 \\ 5y &= -2 \\ y &= -\frac{2}{5} \end{aligned}$$

Substitute $y = -\frac{2}{5}$ into Equation (1). $\quad x + 3\left(-\frac{2}{5}\right) = -4$

Solve for x. $\quad x = -\frac{14}{5}$

The answer $\left(-\frac{2}{5}, -\frac{14}{5}\right)$ is incorrect. What mistake was made?

72. Solve the system of equations by graphing.

$$\begin{aligned} 2x + 3y &= 5 \\ 4x + 6y &= 10 \end{aligned}$$

Solution:

Write both equations in slope–intercept form.

$$y = -\frac{2}{3}x + \frac{5}{3}$$

$$y = -\frac{2}{3}x + \frac{5}{3}$$

Since these lines have the same slope, they are parallel lines.

Parallel lines do not intersect, so there is no solution.

This is incorrect. What mistake was made?

CONCEPTUAL

In Exercises 73–76, determine whether each statement is true or false.

73. A system of equations represented by a graph of two lines with the same slope always has no solution.

74. A system of equations represented by a graph of two lines with slopes that are negative reciprocals always has one solution.

75. If two lines do not have exactly one point of intersection, then they must be parallel.

76. The system of equations $Ax - By = 1$ and $-Ax + By = -1$ has no solution.

CHALLENGE

77. The point $(2, -3)$ is a solution to the system of equations

$$\begin{aligned} Ax + By &= -29 \\ Ax - By &= 13 \end{aligned}$$

Find A and B.

78. If you graph the lines

$$\begin{aligned} x - 50y &= 100 \\ x - 48y &= -98 \end{aligned}$$

they appear to be parallel lines. However, there is a unique solution. Explain how this might be possible.

79. Energy Drinks. A nutritionist wishes to market a new vitamin-enriched fruit drink and is preparing two versions of it to distribute at a local health club. She has 100 cups of pineapple juice and 4 cups of super vitamin-enriched pomegranate concentrate. One version of the drink is to contain 2% pomegranate and the other version 4% pomegranate. How much of each drink can she create?

80. Easter Eggs. A family is coloring Easter eggs and wants to make 2 shades of purple, "light purple" and "deep purple." They have 30 tablespoons of deep red solution and 2 tablespoons of blue solution. If "light purple" consists of 2% blue solution and "deep purple" consists of 10% blue solution, how much of each version of purple solution can be created?

TECHNOLOGY

81. Apply a graphing utility to graph the two equations $y = -1.25x + 17.5$ and $y = 2.3x - 14.1$. Approximate the solution to this system of linear equations.

82. Apply a graphing utility to graph the two equations $y = 14.76x + 19.43$ and $y = 2.76x + 5.22$. Approximate the solution to this system of linear equations.

83. Apply a graphing utility to graph the two equations and determine the solution set: $23x + 15y = 7$ and $46x + 30y = 14$.

84. Apply a graphing utility to graph the two equations and determine the solution set: $-3x + 7y = 2$ and $6x - 14y = 3$.

85. Apply a graphing utility to graph the two equations $\frac{1}{3}x - \frac{5}{12}y = \frac{5}{6}$ and $\frac{3}{7}x + \frac{1}{14}y = \frac{29}{28}$. Approximate the solution to this system of linear equations.

86. Apply a graphing utility to graph the two equations $\frac{5}{9}x + \frac{11}{13}y = 2$ and $\frac{3}{4}x + \frac{5}{7}y = \frac{13}{14}$. Approximate the solution to this system of linear equations.

SECTION 6.2 SYSTEMS OF LINEAR EQUATIONS IN THREE VARIABLES

SKILLS OBJECTIVES

- Solve systems of linear equations in three variables using a combination of both the elimination method and the substitution method.
- Solve application problems using systems of linear equations in three variables.

CONCEPTUAL OBJECTIVES

- Understand that a graph of a linear equation in three variables corresponds to a plane.
- Identify three types of solutions: one solution (point), no solution, or infinitely many solutions (a single line in three dimensional space).

In Section 6.1, we solved systems of linear equations in two variables. Graphs of linear equations in two variables correspond to lines. Now we turn our attention to linear equations in *three* variables. A **linear equation in three variables** x, y, and z is given by

$$Ax + By + Cz = D$$

where A, B, C, and D are real numbers that are not all equal to zero. All three variables have degree equal to one, which is why this is called a linear equation in three variables. The graph of any equation in three variables requires a three-dimensional coordinate system.

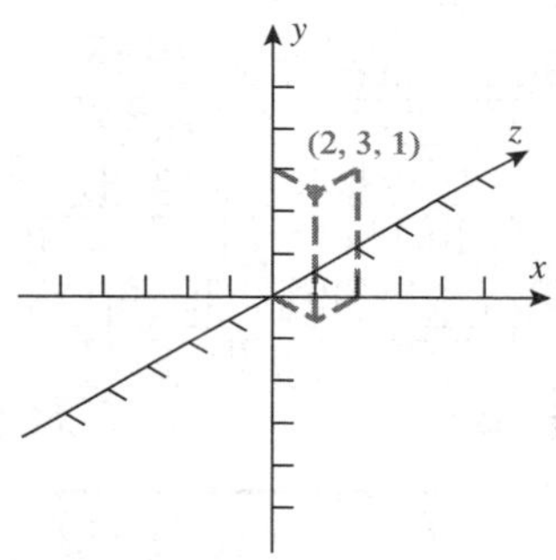

The x-axis, y-axis, and z-axis are each perpendicular to the other two. For the three-dimensional coordinate system on the right, a point $(x, y, z) = (2, 3, 1)$ is found by starting at the origin, moving 2 units to the right, 3 units up, and 1 unit out toward you.

In two variables, the graph of a linear equation is a line. In three variables, however, the graph of a linear equation is a **plane**. A plane can be thought of as an infinite sheet of paper. When solving systems of linear equations in three variables, we find one of three possibilities: one solution, no solution, and infinitely many solutions.

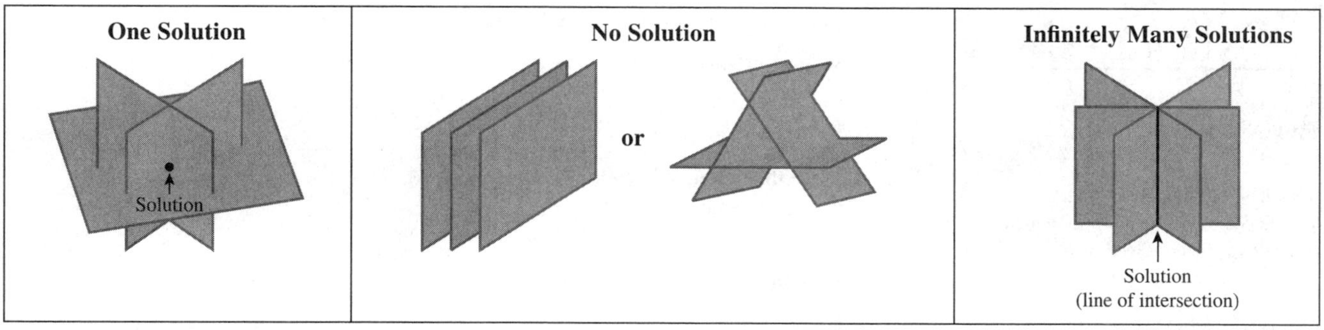

Solving Systems of Linear Equations in Three Variables

There are many ways to solve systems of linear equations in more than two variables. One method is to combine the elimination and substitution methods, which will be discussed in this section. Other methods involve matrices, which will be discussed in the next chapter. We now outline a procedure for solving systems of linear equations in three variables, which can be extended to solve systems of more than three variables. Solutions are usually given as ordered triples of the form (x, y, z).

SOLVING SYSTEMS OF LINEAR EQUATIONS IN THREE VARIABLES USING ELIMINATION AND SUBSTITUTION

Step 1: Reduce the system of three equations in three variables to two equations in two (of the same) variables by applying elimination.
Step 2: Solve the resulting system of two linear equations in two variables by applying elimination or substitution.
Step 3: Substitute the solutions in Step 2 into *any* of the original equations and solve for the third variable.
Step 4: Check that the solution satisfies *all* three original equations.

Technology Tip

The [rref] under the [MATRIX] menu will be used to solve the system of equations by entering the coefficients of x, y, and z.

[2nd] [MATRIX] [▶] [MATH] [▼]
[B:rref(] [ENTER] [2nd] [[] [2nd] [[]
[2] [,] [1] [,] [8] [,] [(−)] [1] [2nd] []]
[2nd] [[] [1] [,] [(−)] [1] [,] [1] [,] [(−)]
[2] [2nd] []] [2nd] [[] [3] [,] [(−)] [2] [,]
[(−)] [2] [,] [2] [2nd] []] [2nd] []] [)]
[ENTER]

```
rref([[2,1,8,-1]
[1,-1,1,-2][3,-2
,-2,2]])
    [[1 0 0 2 ]
     [0 1 0 3 ]
     [0 0 1 -1]]
```

First row gives $x = 2$, second row gives $y = 3$, and third row gives $z = -1$.

Note: The TI function [rref] stands for reduced row echelon form, which will be explained in Section 7.2.

Study Tip

Eliminate the *same* variable.

EXAMPLE 1 **Solving a System of Linear Equations in Three Variables**

Solve the system:

$$2x + y + 8z = -1 \quad \text{Equation (1)}$$
$$x - y + z = -2 \quad \text{Equation (2)}$$
$$3x - 2y - 2z = 2 \quad \text{Equation (3)}$$

Solution:

Inspecting the three equations, we see that y is easily eliminated when Equations (1) and (2) are added, because the coefficients of y, $+1$ and -1, are equal in magnitude and opposite in sign. We can also eliminate y from Equation (3) by adding Equation (3) to *either* 2 times Equation (1) *or* -2 times Equation (2). Therefore, our plan of attack is to eliminate y from the system of equations, so the result will be two equations in two variables x and z.

STEP 1 Eliminate y in Equation (1) and Equation (2).

Equation (1): $2x + y + 8z = -1$
Equation (2): $x - y + z = -2$
Add. $3x + 9z = -3$

Eliminate y in Equation (2) and Equation (3).

Multiply Equation (2) by -2. $-2x + 2y - 2z = 4$
Equation (3): $3x - 2y - 2z = 2$
Add. $x - 4z = 6$

STEP 2 Solve the system of two linear equations in two variables.

$$3x + 9z = -3$$
$$x - 4z = 6$$

Substitution* method: $x = 4z + 6$ $\quad 3(4z + 6) + 9z = -3$

Distribute. $\quad 12z + 18 + 9z = -3$

Combine like terms. $\quad 21z = -21$

Solve for z. $\quad \boxed{z = -1}$

Substitute $z = -1$ into $x = 4z + 6$. $\quad x = 4(-1) + 6 = 2$

$\boxed{x = 2}$ and $\boxed{z = -1}$ are the solutions to the system of two equations.

STEP 3 Substitute $x = 2$ and $z = -1$ into any of the three original equations and solve for y.

Substitute $x = 2$ and $z = -1$ into Equation (2). $\quad 2 - y - 1 = -2$

Solve for y. $\quad \boxed{y = 3}$

STEP 4 Check that $x = 2$, $y = 3$, and $z = -1$ satisfy all three equations.

Equation (1): $2(2) + 3 + 8(-1) = 4 + 3 - 8 = -1$

Equation (2): $2 - 3 - 1 = -2$

Equation (3): $3(2) - 2(3) - 2(-1) = 6 - 6 + 2 = 2$

The solution is $\boxed{x = 2, y = 3, z = -1, \text{ or } (2, 3, -1)}$.

*Elimination method could also be used.

■ YOUR TURN Solve the system:

$$\begin{aligned} 2x - y + 3z &= -1 \\ x + y - z &= 0 \\ 3x + 3y - 2z &= 1 \end{aligned}$$

■ **Answer:**
$x = -1, y = 2, z = 1$

In Example 1 and the Your Turn, the variable y was eliminated by adding the first and second equations. In practice, any of the three variables can be eliminated, but typically we select the most convenient variable to eliminate. If a variable is missing from one of the equations (has a coefficient of 0), then we eliminate that variable from the other two equations.

EXAMPLE 2 Solving a System of Linear Equations in Three Variables When One Variable Is Missing

Solve the system:

$$\begin{aligned} x \qquad + z &= 1 && \text{Equation (1)} \\ 2x + y - z &= -3 && \text{Equation (2)} \\ x + 2y - z &= -1 && \text{Equation (3)} \end{aligned}$$

Solution:

Since y is missing from Equation (1), y is the variable to be eliminated in Equation (2) and Equation (3).

STEP 1 Eliminate y.

Multiply Equation (2) by -2. $\quad -4x - 2y + 2z = 6$

Equation (3): $\quad x + 2y - z = -1$

Add. $\quad -3x \qquad + z = 5$

STEP 2 Solve the system of two equations: Equation (1) and the resulting equation in Step 1.

$$\begin{aligned} x + z &= 1 \\ -3x + z &= 5 \end{aligned}$$

Multiply the second equation by (-1) and add to first equation.

$$\begin{aligned} x + z &= 1 \\ 3x - z &= -5 \\ \hline 4x \qquad &= -4 \end{aligned}$$

Solve for x. $\quad x = -1$

Substitute $x = -1$ into Equation (1). $\quad -1 + z = 1$

Solve for z. $\quad z = 2$

STEP 3 Substitute $x = -1$ and $z = 2$ into one of the original equations (Equation 2 or Equation 3) and solve for y.

Substitute $x = -1$ and $z = 2$ into $x + 2y - z = -1$. $\quad (-1) + 2y - 2 = -1$

Gather like terms. $\quad 2y = 2$

Solve for y. $\quad y = 1$

Technology Tip

The rref under the MATRIX menu will be used to solve the system of equations by entering the coefficients of x, y, and z.

2nd MATRIX ► MATH ▼ B:rref(ENTER 2nd [2nd [1 , 0 , 1 , 1 2nd] 2nd [2 , 1 , (−) 1 , (−) 3 2nd] 2nd [1 , 2 , (−) 1 , (−) 1 2nd] 2nd]) ENTER

```
rref([[1,0,1,1][
2,1,-1,-3][1,2,-
1,-1]])
      [[1 0 0 -1]
       [0 1 0 1 ]
       [0 0 1 2 ]]
```

The first row gives $x = -1$, second row gives $y = 1$, and third row gives $z = 2$.

STEP 4 Check that $x = -1$, $y = 1$, and $z = 2$ satisfy all three equations.

Equation (1): $(-1) + 2 = 1$

Equation (2): $2(-1) + (1) - (2) = -3$

Equation (3): $(-1) + 2(1) - (2) = -1$

The solution is $\boxed{x = -1, y = 1, z = 2}$.

■ **YOUR TURN** Solve the system:

$$\begin{aligned} x + y + z &= 0 \\ 2x \quad + z &= -1 \\ x - y - z &= 2 \end{aligned}$$

■ **Answer:** $x = 1, y = 2, z = -3$

Types of Solutions

There are three types of solutions: independent, dependent, and inconsistent. There are three corresponding outcomes: one solution, infinitely many solutions, or no solution. Examples 1 and 2 each had one solution. Examples 3 and 4 illustrate systems with infinitely many solutions and no solution, respectively.

Technology Tip

The [rref] under the [MATRIX] menu will be used to solve the system of equations by entering the coefficients of x, y, and z.

[2nd] [MATRIX] [▶] [MATH] [▼] [B:rref(] [ENTER] [2nd] [[] [2nd] [[] [2] [,] [1] [,] [(−)] [1] [,] [4] [2nd] []] [2nd] [[] [1] [,] [1] [,] [0] [,] [2] [2nd] []] [2nd] [[] [3] [,] [2] [,] [(−)] [1] [,] [6] [2nd] []] [2nd] []] [)] [ENTER]

```
rref([[2,1,-1,4]
[1,1,0,2][3,2,-1
,6]])
      [[1 0 -1 2]
       [0 1 1  0]
       [0 0 0  0]]
```

The first row gives $x - z = 2$ and second row gives $y + z = 0$. The equations are dependent. Let $z = a$. After solving, $x = a + 2$, $y = -a$, and $z = a$.

EXAMPLE 3 A Dependent System of Linear Equations in Three Variables (Infinitely Many Solutions)

Solve the system:

$$2x + y - z = 4 \qquad \text{Equation (1)}$$
$$x \quad + y = 2 \qquad \text{Equation (2)}$$
$$3x + 2y - z = 6 \qquad \text{Equation (3)}$$

Solution:

Since z is missing from Equation (2), z is the variable to be eliminated from Equation (1) and Equation (3).

STEP 1 Eliminate z.

Multiply Equation (1) by (-1). $\quad -2x - y + z = -4$

Equation (3): $\quad 3x + 2y - z = 6$

Add. $\quad x + y = 2$

STEP 2 Solve the system of two equations:
Equation (2) and the resulting equation in Step 1.

$$x + y = 2$$
$$x + y = 2$$

Multiply the first equation by (-1) and add to second equation.

$$-x - y = -2$$
$$x + y = 2$$
$$0 = 0$$

This statement is always true; therefore, there are $\boxed{\text{infinitely many solutions}}$. The original system has been reduced to a system of two identical linear equations. Therefore the equations are dependent (share infinitely many solutions). Typically, to define those infinitely many solutions, we let $z = a$, where a stands for any real number, and then find x and y in terms of a. The resulting ordered triple showing the three variables in terms of a is called a **parametric representation** of a line in three dimensions.

STEP 3 Let $\boxed{z = a}$ and find x and y in terms of a.

Solve Equation (2) for y. $\qquad y = 2 - x$

Let $y = 2 - x$ and $z = a$ in Equation (1). $\qquad 2x + (2 - x) - a = 4$

Solve for x. $\qquad 2x + 2 - x - a = 4$

$$x - a = 2$$

$$\boxed{x = a + 2}$$

Let $\boxed{x = a + 2}$ in Equation (2). $\qquad (a + 2) + y = 2$

Solve for y. $\qquad \boxed{y = -a}$

The infinitely many solutions are written as $\boxed{(a + 2, -a, a)}$.

STEP 4 Check that $x = a + 2$, $y = -a$, and $z = a$ satisfy all three equations.

Equation (1): $2(a + 2) + (-a) - a = 2a + 4 - a - a = 4 \checkmark$

Equation (2): $(a + 2) + (-a) = a + 2 - a = 2 \checkmark$

Equation (3): $3(a + 2) + 2(-a) - a = 3a + 6 - 2a - a = 6 \checkmark$

■ **YOUR TURN** Solve the system:

$$\begin{aligned} x + y - 2z &= 0 \\ x \qquad - z &= -1 \\ x - 2y + z &= -3 \end{aligned}$$

■ **Answer:** $(a - 1, a + 1, a)$

EXAMPLE 4 **An Inconsistent System of Linear Equations in Three Variables (No Solution)**

Solve the system:

$$\begin{aligned} x + 2y - z &= 3 \qquad \text{Equation (1)} \\ 2x + y + 2z &= -1 \qquad \text{Equation (2)} \\ -2x - 4y + 2z &= 5 \qquad \text{Equation (3)} \end{aligned}$$

Solution:

STEP 1 Eliminate x.

Multiply Equation (1) by -2. $\qquad -2x - 4y + 2z = -6$

Equation (2): $\qquad 2x + y + 2z = -1$

Add. $\qquad -3y + 4z = -7$

Equation (2): $\qquad 2x + y + 2z = -1$

Equation (3): $\qquad -2x - 4y + 2z = 5$

Add. $\qquad -3y + 4z = 4$

STEP 2 Solve the system of two equations:

$$\begin{aligned} -3y + 4z &= -7 \\ -3y + 4z &= 4 \end{aligned}$$

Multiply the top equation by (-1) and add to the second equation.

$$\begin{aligned} 3y - 4z &= 7 \\ -3y + 4z &= 4 \\ \hline 0 &= 11 \end{aligned}$$

This is a contradiction, or inconsistent statement, and therefore there is $\boxed{\text{no solution}}$.

Technology Tip

The [rref] under the [MATRIX] menu will be used to solve the system of equations by entering the coefficients of x, y, and z.

[2nd] [MATRIX] [▶] [MATH] [▼] [B:rref(] [ENTER] [2nd] [[] [2nd] [[] [1] [,] [2] [,] [(−)] [1] [,] [3] [2nd] []] [2nd] [[] [2] [,] [1] [,] [2] [,] [(−)] [1] [2nd] []] [2nd] [[] [(−)] [2] [,] [(−)] [4] [,] [2] [,] [5] [2nd] []] [2nd] []] [)] [ENTER]

```
rref([[1,2,-1,3]
[2,1,2,-1][-2,-4
,2,5]])
[[1 0 1.6666666...
 [0 1 -1.333333...
 [0 0 0        ...
```

To change to fractions, press:

[MATH] [1:Frac] [ENTER] [ENTER]

```
[[1 0 1.6666666...
 [0 1 -1.333333...
 [0 0 0        ...
Ans▸Frac
  [[1 0 5/3  0]
   [0 1 -4/3 0]
   [0 0 0    1]]
```

The third row gives $0 = 1$, which is a contradiction. Therefore, there is no solution.

So far in this section we have discussed only systems of *three* linear equations in *three* variables. What happens if we have a system of *two* linear equations in *three* variables? The two linear equations in three variables will always correspond to two planes in three dimensions. The possibilities are no solution (the two planes are parallel) or infinitely many solutions (the two planes intersect in a line).

NO SOLUTION	INFINITELY MANY SOLUTIONS (LINE)
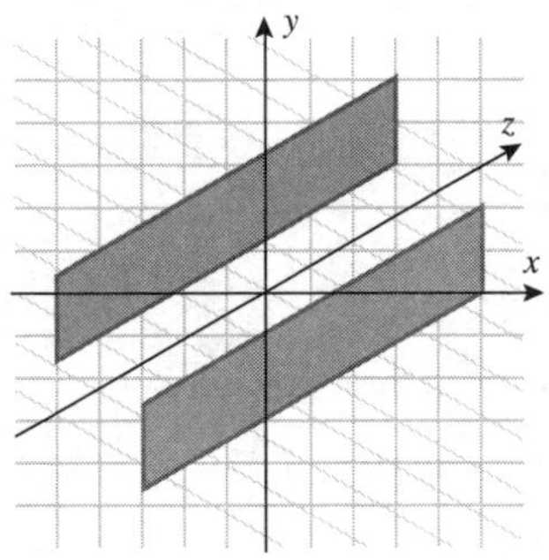	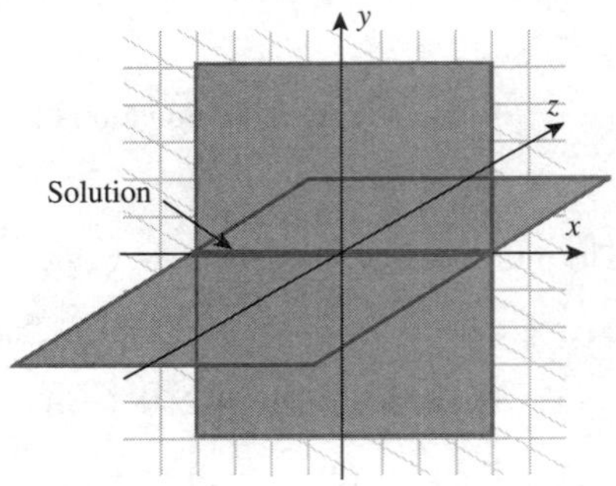

EXAMPLE 5 Solving a System of Two Linear Equations in Three Variables

Solve the system of linear equations:

$$x - y + z = 7 \quad \text{Equation (1)}$$
$$x + y + 2z = 2 \quad \text{Equation (2)}$$

Solution:

Eliminate y by adding the two equations.

$$\begin{aligned} x - y + z &= 7 \\ x + y + 2z &= 2 \\ \hline 2x \quad + 3z &= 9 \end{aligned}$$

Therefore, Equation (1) and Equation (2) are both true if $2x + 3z = 9$. Since we know there is a solution, it must be a line. To define the line of intersection, we again turn to parametric representation.

Let $\boxed{z = a}$, where a is any real number. $2x + 3a = 9$

Solve for x. $\boxed{x = \frac{9}{2} - \frac{3}{2}a}$

Substitute $z = a$ and $x = \frac{9}{2} - \frac{3}{2}a$ into Equation (1). $\left(\frac{9}{2} - \frac{3}{2}a\right) - y + a = 7$

Solve for y. $\boxed{y = -\frac{1}{2}a - \frac{5}{2}}$

The solution is the line in three dimensions given by $\boxed{\left(\frac{9}{2} - \frac{3}{2}a, -\frac{1}{2}a - \frac{5}{2}, a\right)}$, where a is any real number.

Note: Every real number a corresponds to a point on the line of intersection.

a	$\left(\frac{9}{2} - \frac{3}{2}a, -\frac{1}{2}a - \frac{5}{2}, a\right)$
-1	$(6, -2, -1)$
0	$\left(\frac{9}{2}, -\frac{5}{2}, 0\right)$
1	$(3, -3, 1)$

Modeling with a System of Three Linear Equations

Suppose you want to model a stock price as a function of time and based on the data you feel a quadratic model would be the best fit. Therefore the model is given by

$$P(t) = at^2 + bt + c$$

where $P(t)$ is the price of the stock at time t. If we have data corresponding to three distinct points $(t, P(t))$, the result is a system of three linear equations in three variables a, b, and c. We can solve the resulting system of linear equations, which determines the coefficients a, b, and c of the quadratic model for stock price.

EXAMPLE 6 Stock Value

The Oracle Corporation's stock (ORCL) over 3 days (Wednesday, October 13, to Friday, October 15, 2004) can be approximately modeled by a quadratic function: $f(t) = at^2 + bt + c$. If Wednesday corresponds to $t = 1$, where t is in days, then the following data points approximately correspond to the stock value.

t	$f(t)$	DAYS
1	\$12.20	Wednesday
2	\$12.00	Thursday
3	\$12.20	Friday

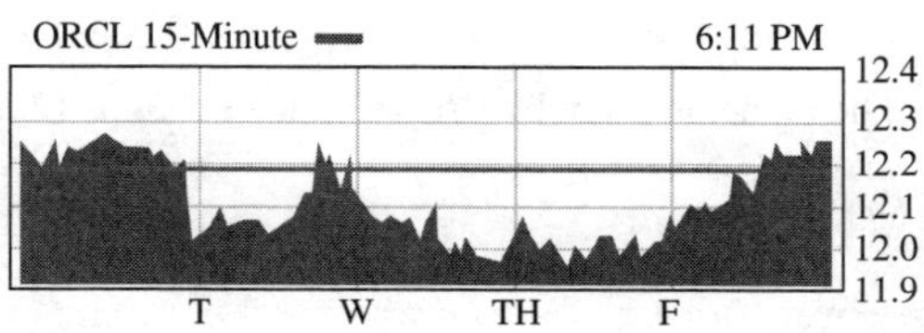

Determine the function that models this behavior.

Solution:

Substitute the points (1, 12.20), (2, 12.00), and (3, 12.20) into $f(t) = at^2 + bt + c$.

$$a(1)^2 + b(1) + c = 12.20$$
$$a(2)^2 + b(2) + c = 12.00$$
$$a(3)^2 + b(3) + c = 12.20$$

Simplify to a system of three equations in three variables (a, b, and c).

$$a + b + c = 12.20 \quad \text{Equation (1)}$$
$$4a + 2b + c = 12.00 \quad \text{Equation (2)}$$
$$9a + 3b + c = 12.20 \quad \text{Equation (3)}$$

Solve for a, b, and c by applying the technique of this section.

STEP 1 Eliminate c.

Multiply Equation (1) by (−1). $\quad -a - b - c = -12.20$

Equation (2): $\quad 4a + 2b + c = 12.20$

Add. $\quad 3a + b = -0.20$

Multiply Equation (1) by −1. $\quad -a - b - c = -12.20$

Equation (3): $\quad 9a + 3b + c = 12.20$

Add. $\quad 8a + 2b = 0$

STEP 2 Solve the system of two equations.

$$3a + b = -0.20$$
$$8a + 2b = 0$$

Multiply the first equation by -2 and add to the second equation.

$$-6a - 2b = 0.40$$
$$8a + 2b = 0$$

Add. $\quad 2a = 0.4$

Solve for a. $\quad a = 0.2$

Substitute $a = 0.2$ into $8a + 2b = 0$. $\quad 8(0.2) + 2b = 0$

Simplify. $\quad 2b = -1.6$

Solve for b. $\quad b = -0.8$

STEP 3 Substitute $a = 0.2$ and $b = -0.8$ into one of the original three equations.

Substitute $a = 0.2$ and $b = -0.8$ into $a + b + c = 12.20$. $\quad 0.2 - 0.8 + c = 12.20$

Gather like terms. $\quad -0.6 + c = 12.20$

Solve for c. $\quad c = 12.80$

STEP 4 Check that $a = 0.2$, $b = -0.8$, and $c = 12.80$ satisfy all three equations.

Equation (1): $\quad a + b + c = 0.2 - 0.8 + 12.8 = 12.2$

Equation (2): $\quad 4a + 2b + c = 4(0.2) + 2(-0.8) + 12.80 = 0.8 - 1.6 + 12.8 = 12.00$

Equation (3): $\quad 9a + 3b + c = 9(0.2) + 3(-0.8) + 12.80 = 1.8 - 2.4 + 12.8 = 12.20$

The model is given by $\boxed{f(t) = 0.2t^2 - 0.8t + 12.80}$.

SECTION 6.2 SUMMARY

Graphs of linear equations in *two* variables are *lines*, whereas graphs of linear equations in *three* variables are *planes*. Systems of linear equations in three variables have one of three outcomes:

- One solution (point)
- No solution (no intersection of all three planes)
- Infinitely many solutions (planes intersect along a line)

When the solution to a system of three linear equations is a line in three dimensions, we use parametric representation to express the solution.

SECTION 6.2 EXERCISES

SKILLS

In Exercises 1–30, solve each system of linear equations.

1. $x - y + z = 6$
$-x + y + z = 3$
$-x - y - z = 0$

2. $-x - y + z = -1$
$-x + y - z = 3$
$x - y - z = 5$

3. $x + y - z = 2$
$-x - y - z = -3$
$-x + y - z = 6$

4. $x + y + z = -1$
$-x + y - z = 3$
$-x - y + z = 8$

5. $\begin{aligned} -x + y - z &= -1 \\ x - y - z &= 3 \\ x + y - z &= 9 \end{aligned}$

6. $\begin{aligned} x - y - z &= 2 \\ -x - y + z &= 4 \\ -x + y - z &= 6 \end{aligned}$

7. $\begin{aligned} 2x - 3y + 4z &= -3 \\ -x + y + 2z &= 1 \\ 5x - 2y - 3z &= 7 \end{aligned}$

8. $\begin{aligned} x - 2y + z &= 0 \\ -2x + y - z &= -5 \\ 13x + 7y + 5z &= 6 \end{aligned}$

9. $\begin{aligned} 3y - 4x + 5z &= 2 \\ 2x - 3y - 2z &= -3 \\ 3z + 4y - 2x &= 1 \end{aligned}$

10. $\begin{aligned} 2y + z - x &= 5 \\ 2x + 3z - 2y &= 0 \\ -2z + y - 4x &= 3 \end{aligned}$

11. $\begin{aligned} x - y + z &= -1 \\ y - z &= -1 \\ -x + y + z &= 1 \end{aligned}$

12. $\begin{aligned} -y + z &= 1 \\ x - y + z &= -1 \\ x - y - z &= -1 \end{aligned}$

13. $\begin{aligned} 3x - 2y - 3z &= -1 \\ x - y + z &= -4 \\ 2x + 3y + 5z &= 14 \end{aligned}$

14. $\begin{aligned} 3x - y + z &= 2 \\ x - 2y + 3z &= 1 \\ 2x + y - 3z &= -1 \end{aligned}$

15. $\begin{aligned} -3x - y - z &= 2 \\ x + 2y - 3z &= 4 \\ 2x - y + 4z &= 6 \end{aligned}$

16. $\begin{aligned} 2x - 3y + z &= 1 \\ x + 4y - 2z &= 2 \\ 3x - y + 4z &= -3 \end{aligned}$

17. $\begin{aligned} 3x + 2y + z &= 4 \\ -4x - 3y - z &= -15 \\ x - 2y + 3z &= 12 \end{aligned}$

18. $\begin{aligned} 3x - y + 4z &= 13 \\ -4x - 3y - z &= -15 \\ x - 2y + 3z &= 12 \end{aligned}$

19. $\begin{aligned} -x + 2y + z &= -2 \\ 3x - 2y + z &= 4 \\ 2x - 4y - 2z &= 4 \end{aligned}$

20. $\begin{aligned} 2x - y &= 1 \\ -x + z &= -2 \\ -2x + y &= -1 \end{aligned}$

21. $\begin{aligned} x - z - y &= 10 \\ 2x - 3y + z &= -11 \\ y - x + z &= -10 \end{aligned}$

22. $\begin{aligned} 2x + z + y &= -3 \\ 2y - z + x &= 0 \\ x + y + 2z &= 5 \end{aligned}$

23. $\begin{aligned} 3x_1 + x_2 - x_3 &= 1 \\ x_1 - x_2 + x_3 &= -3 \\ 2x_1 + x_2 + x_3 &= 0 \end{aligned}$

24. $\begin{aligned} 2x_1 + x_2 + x_3 &= -1 \\ x_1 + x_2 - x_3 &= 5 \\ 3x_1 - x_2 - x_3 &= 1 \end{aligned}$

25. $\begin{aligned} 2x + 5y &= 9 \\ x + 2y - z &= 3 \\ -3x - 4y + 7z &= 1 \end{aligned}$

26. $\begin{aligned} x - 2y + 3z &= 1 \\ -2x + 7y - 9z &= 4 \\ x + z &= 9 \end{aligned}$

27. $\begin{aligned} 2x_1 - x_2 + x_3 &= 3 \\ x_1 - x_2 + x_3 &= 2 \\ -2x_1 + 2x_2 - 2x_3 &= -4 \end{aligned}$

28. $\begin{aligned} x_1 - x_2 - 2x_3 &= 0 \\ -2x_1 + 5x_2 + 10x_3 &= -3 \\ 3x_1 + x_2 &= 0 \end{aligned}$

29. $\begin{aligned} 2x + y - z &= 2 \\ x - y - z &= 6 \end{aligned}$

30. $\begin{aligned} 3x + y - z &= 0 \\ x + y + 7z &= 4 \end{aligned}$

APPLICATIONS

31. Football. On September 1, 2007, the Appalachian State Mountaineers defeated the University of Michigan Wolverines by a score of 34 to 32. The points came from a total of these types of plays: touchdowns (six points), extra points (one point), and field goals (three points). There were a total of 18 scoring plays. There were four more touchdowns than field goals. How many touchdowns, extra points, and field goals were scored in this football game?

32. Basketball. On April 4, 2004, the University of Connecticut Huskies finished the season the same way they started it—as the number one men's basketball team in the NCAA. They defeated the Georgia Tech Yellow Jackets 82–73 in the Final Four championship game. The points came from three types of scoring plays: two-point shots, three-point shots, and one-point free throws. There were seven more two-point shots made than there were one-point free throws completed. The number of successful two-point shots was four more than four times the number of successful three-point shots. How many two-point, three-point, and one-point free throw shots were made in the finals of the 2004 Final Four NCAA tournament?

Exercises 33 and 34 rely on a selection of Subway sandwiches whose nutrition information is given in the table.

Suppose you are going to eat only Subway sandwiches for a week (seven days) for lunch and dinner (total of fourteen meals).

SANDWICH	CALORIES	FAT (GRAMS)
Mediterranean Chicken	350	18
Six Inch Tuna	430	19
Six Inch Roast Beef	290	5

www.subway.com

33. Diet. Your goal is a total of 4840 calories and 190 grams of fat. How many of each sandwich would you eat that week to obtain this goal?

34. Diet. Your goal is a total of 4380 calories and 123 grams of fat. How many of each sandwich would you eat that week to obtain this goal?

Exercises 35 and 36 involve vertical motion and the effect of gravity on an object.

Because of gravity, an object that is projected upward will eventually reach a maximum height and then fall to the ground. The equation that determines the height h of a projectile t seconds after it is shot upward is given by

$$h = \frac{1}{2}at^2 + v_0t + h_0$$

where a is the acceleration due to gravity, h_0 is the initial height of the object at time $t = 0$, and v_0 is the initial velocity of the object at time $t = 0$. Note that a projectile follows the path of a parabola opening down, so $a < 0$.

35. Vertical Motion. An object is thrown upward and the following table depicts the height of the ball t seconds after the projectile is released. Find the initial height, initial velocity, and acceleration due to gravity.

t SECONDS	HEIGHT (FEET)
1	36
2	40
3	12

36. Vertical Motion. An object is thrown upward and the following table depicts the height of the ball t seconds after the projectile is released. Find the initial height, initial velocity, and acceleration due to gravity.

t SECONDS	HEIGHT (FEET)
1	84
2	136
3	156

37. Data Curve-Fitting. The number of minutes that an average person of age x spends driving a car can be modeled by a quadratic function $y = ax^2 + bx + c$, where $a < 0$ and $18 \le x \le 65$. The following table gives the average number of minutes per day that a person spends driving a car. Determine the quadratic function that models this quantity.

AGE	AVERAGE DAILY MINUTES DRIVING
20	30
40	60
60	40

38. Data Curve-Fitting. The average age when a woman gets married began increasing during the last century. In 1930 the average age was 18.6, in 1950 the average age was 20.2, and in 2002 the average age was 25.3. Find a quadratic function $y = ax^2 + bx + c$, where $a > 0$ and $18 < y < 35$, that models the average age y when a woman gets married as a function of the year x ($x = 0$ corresponds to 1930). What will the average age be in 2010?

39. Money. Tara and Lamar decide to place \$20,000 of their savings into investments. They put some in a money market account earning 3% interest, some in a mutual fund that has been averaging 7% a year, and some in a stock that rose 10% last year. If they put \$6,000 more in the money market than in the mutual fund and the mutual fund and stocks have the same growth in the next year as they did in the previous year, they will earn \$1,180 in a year. How much money did they put in each of the three investments?

40. Money. Tara talks Lamar into putting less money in the money market and more money in the stock. They place \$20,000 of their savings into investments. They put some in a money market account earning 3% interest, some in a mutual fund that has been averaging 7% a year, and some in a stock that rose 10% last year. If they put \$6,000 more in the stock than in the mutual fund and the mutual fund and stock have the same growth in the next year as they did in the previous year, they will earn \$1,680 in a year. How much money did they put in each of the three investments?

41. Ski Production. A company produces three types of skis: regular model, trick ski, and slalom ski. They need to fill a customer order of 110 pairs of skis. There are two major production divisions within the company: labor and finishing. Each regular model of skis requires 2 hours of labor and 1 hour of finishing. Each trick ski model requires 3 hours of labor and 2 hours of finishing. Finally, each slalom ski model requires 3 hours of labor and 5 hours of finishing. Suppose the company has only 297 labor hours and 202 finishing hours. How many of each type ski can be made under these restrictions?

42. Automobile Production. An automobile manufacturing company produces three types of automobiles: compact car, intermediate, and luxury model. The company has the capability of producing 500 automobiles. Suppose that each compact-model car requires 200 units of steel and 30 units of rubber, each intermediate model requires 300 units of steel and 20 units of rubber, and each luxury model requires 250 units of steel and 45 units of rubber. The number of units of steel available is 128,750, and the number of units of rubber available is 15,625. How many of each type of automobile can be produced with these restraints?

43. Computer versus Man. *The Seattle Times* reported a story on November 18, 2006 about a game of Scrabble played between a human and a computer. The best Scrabble player in the United States was pitted against a computer program designed to play the game. Remarkably, the human beat the computer in the best of two out of three games competition. The total points scored by both computer and the man for all three games was 2591. The difference between the first game's total and second game's total was 62 points. The difference between the first game's total and the third game's total was only 2 points. Determine the total number of points scored by both computer and the man for all three contests.

44. Brain versus Computer. Can the human brain perform more calculations per second than a supercomputer? The calculating speed of the three top supercomputers, IBM's Blue Gene/L, IBM's BGW, and IBM's ASC Purple, has been determined. The speed of IBM's Blue Gene/L is 245 teraflops more than IBM's BGW. The computing speed of IBM's BGW is 22 teraflops more than IBM's ASC Purple. The combined speed of all three top supercomputers is 568 teraflops. Determine the computing speed (in teraflops) of each supercomputer. A **teraflop** is measure of a computer's speed and can be expressed as one trillion floating-point operations per second. By comparison, it is estimated that the human brain can perform 10 quadrillion calculations per second.

CATCH THE MISTAKE

In Exercises 45 and 46, explain the mistake that is made.

45. Solve the system of equations.

Equation (1): $2x - y + z = 2$

Equation (2): $x - y \quad = 1$

Equation (3): $x \quad + z = 1$

Solution:

Equation (2): $x - y \quad = 1$

Equation (3): $x \quad + z = 1$

Add Equation (2) and Equation (3). $-y + z = 2$

Multiply Equation (1) by (-1). $-2x + y - z = -2$

Add. $-2x \quad = 0$

Solve for x. $x = 0$

Substitute $x = 0$ into Equation (2). $0 - y = 1$

Solve for y. $y = -1$

Substitute $x = 0$ into Equation (3). $0 + z = 1$

Solve for z. $z = 1$

The answer is $x = 0$, $y = -1$, and $z = 1$.

This is incorrect. Although $x = 0$, $y = -1$, and $z = 1$ does satisfy the three original equations, it is only one of infinitely many solutions. What mistake was made?

46. Solve the system of equations.

Equation (1): $x + 3y + 2z = 4$

Equation (2): $3x + 10y + 9z = 17$

Equation (3): $2x + 7y + 7z = 17$

Solution:

Multiply Equation (1) by -3. $-3x - 9y - 6z = -12$

Equation (2): $3x + 10y + 9z = 17$

Add. $y + 3z = 5$

Multiply Equation (1) by -2. $-2x - 6y - 4z = -8$

Equation (3): $2x + 7y + 7z = 17$

Add. $y + 3z = 9$

Solve the system of two equations. $y + 3z = 5$

$y + 3z = 9$

Infinitely many solutions.

Let $z = a$, then $y = 5 - 3a$.

Substitute $z = a$ and $y = 5 - 3a$ into Equation (1). $x + 3y + 2z = 4$

$x + 3(5 - 3a) + 2a = 4$

Eliminate parentheses. $x + 15 - 9a + 2a = 4$

Solve for x. $x = 7a - 11$

The answer is $x = 7a - 11$, $y = 5 - 3a$, and $z = a$.

This is incorrect. There is no solution. What mistake was made?

CONCEPTUAL

In Exercises 47 and 48, determine whether each statement is true or false.

47. A system of linear equations that has more variables than equations cannot have a unique solution.

48. A system of linear equations that has the same number of equations as variables always has a unique solution.

49. Geometry. The circle given by the equation $x^2 + y^2 + ax + by + c = 0$ passes through the points $(-2, 4)$, $(1, 1)$, and $(-2, -2)$. Find a, b, and c.

50. Geometry. The circle given by the equation $x^2 + y^2 + ax + by + c = 0$ passes through the points $(0, 7)$, $(6, 1)$, and $(5, 4)$. Find a, b, and c.

CHALLENGE

51. A fourth-degree polynomial, $f(x) = ax^4 + bx^3 + cx^2 + dx + e$, with $a < 0$, can be used to represent the following data on the number of deaths per year due to lightning strikes. Assume 1999 corresponds to $x = -2$ and 2003 corresponds to $x = 2$. Use the data to determine a, b, c, d, and e.

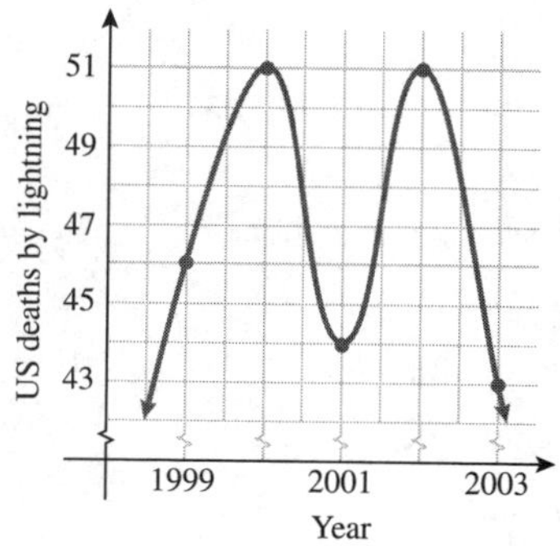

52. A copy machine accepts nickels, dimes, and quarters. After 1 hour, there are 30 coins total and their value is \$4.60. If there are four more quarters than nickels, how many nickels, quarters, and dimes are in the machine?

In Exercises 53–56, solve the system of linear equations.

53. $$\begin{aligned} 2y + z &= 3 \\ 4x - z &= -3 \\ 7x - 3y - 3z &= 2 \\ x - y - z &= -2 \end{aligned}$$

54. $$\begin{aligned} -2x - y + 2z &= 3 \\ 3x - 4z &= 2 \\ 2x + y &= -1 \\ -x + y - z &= -8 \end{aligned}$$

55. $$\begin{aligned} 3x_1 - 2x_2 + x_3 + 2x_4 &= -2 \\ -x_1 + 3x_2 + 4x_3 + 3x_4 &= 4 \\ x_1 + x_2 + x_3 + x_4 &= 0 \\ 5x_1 + 3x_2 + x_3 + 2x_4 &= -1 \end{aligned}$$

56. $$\begin{aligned} 5x_1 + 3x_2 + 8x_3 + x_4 &= 1 \\ x_1 + 2x_2 + 5x_3 + 2x_4 &= 3 \\ 4x_1 + x_3 - 2x_4 &= -3 \\ x_2 + x_3 + x_4 &= 0 \end{aligned}$$

TECHNOLOGY

In Exercises 57 and 58, employ a graphing calculator to solve the system of linear equations (most graphing calculators have the capability of solving linear systems with the user entering the coefficients).

57. $$\begin{aligned} x - z - y &= 10 \\ 2x - 3y + z &= -11 \\ y - x + z &= -10 \end{aligned}$$

58. $$\begin{aligned} 2x + z + y &= -3 \\ 2y - z + x &= 0 \\ x + y + 2z &= 5 \end{aligned}$$

59. Graphing calculators and graphing utilities have the ability to graph in three dimensions (3D) as opposed to the traditional two dimensions (2D). The line must be given in the form $z = ax + by + c$. Rewrite the system of equations in Exercise 57 in this form and graph the three lines in 3D. What is the point of intersection? Compare that with your answer in Exercise 57.

60. Graphing calculators and graphing utilities have the ability to graph in three dimensions (3D) as opposed to the traditional two dimensions (2D). The line must be given in the form $z = ax + by + c$. Rewrite the system of equations in Exercise 58 in this form and graph the three lines in 3D. What is the point of intersection? Compare that with your answer in Exercise 58.

In Exercises 61 and 62, employ a graphing calculator to solve the system of equations.

61. $$\begin{aligned} 0.2x - 0.7y + 0.8z &= 11.2 \\ -1.2x + 0.3y - 1.5z &= 0 \\ 0.8x - 0.1y + 2.1z &= 6.4 \end{aligned}$$

62. $$\begin{aligned} 1.8x - 0.5y + 2.4z &= 1.6 \\ 0.3x - 0.6z &= 0.2 \end{aligned}$$

SECTION

6.3 PARTIAL FRACTIONS

SKILLS OBJECTIVES

- Decompose rational expressions into sums of partial fractions when the denominators contain:
  - Distinct linear factors
  - Repeated linear factors
  - Distinct irreducible quadratic factors
  - Repeated irreducible quadratic factors

CONCEPTUAL OBJECTIVE

- Understand the connection between partial-fraction decomposition and systems of linear equations.

In Chapter 4 we studied polynomial functions, and in Section 4.5 we discussed ratios of polynomial functions, called rational functions. Rational expressions are of the form:

$$\frac{n(x)}{d(x)} \qquad d(x) \neq 0$$

where the numerator $n(x)$ and the denominator $d(x)$ are polynomials. Examples of rational expressions are

$$\frac{4x - 1}{2x + 3} \qquad \frac{2x + 5}{x^2 - 1} \qquad \frac{3x^4 - 2x + 5}{x^2 + 2x + 4}$$

Suppose we are asked to add two rational expressions: $\frac{2}{x + 1} + \frac{5}{x - 3}$.

We already possess the skills to accomplish this. We first identify the least common denominator $(x + 1)(x - 3)$ and combine the fractions into a single expression.

$$\frac{2}{x + 1} + \frac{5}{x - 3} = \frac{2(x - 3) + 5(x + 1)}{(x + 1)(x - 3)} = \frac{2x - 6 + 5x + 5}{(x + 1)(x - 3)} = \frac{7x - 1}{x^2 - 2x - 3}$$

How do we do this in reverse? For example, how do we start with $\frac{7x - 1}{x^2 - 2x - 3}$ and write this expression as a sum of two simpler expressions?

$$\frac{7x - 1}{x^2 - 2x - 3} = \overbrace{\underbrace{\frac{2}{x + 1}}_{\text{Partial Fraction}} + \underbrace{\frac{5}{x - 3}}_{\text{Partial Fraction}}}^{\text{Partial Fraction Decomposition}}$$

Each of the two expressions on the right is called a **partial fraction**. The sum of these fractions is called the **partial-fraction decomposition** of $\frac{7x - 1}{x^2 - 2x - 3}$.

Partial-fraction decomposition is an important tool in calculus. Calculus operations such as differentiation and integration are often made simpler if you apply partial fractions. The reason partial fractions were not discussed until now is because partial-fraction decomposition *requires the ability to solve systems of linear equations*. Since partial-fraction decomposition is made possible by the techniques of solving systems of linear equations, we consider partial fractions an important application of systems of linear equations.

As mentioned earlier, a rational expression is the ratio of two polynomial expressions $n(x)/d(x)$ and we assume that $n(x)$ and $d(x)$ are polynomials with no common factors. If the degree of $n(x)$ is less than the degree of $d(x)$, then the rational expression $n(x)/d(x)$ is said to be **proper**. If the degree of $n(x)$ is greater than or equal to the degree of $d(x)$, the rational expression is said to be **improper**. If the rational expression is improper, it should first be divided using long division.

$$\frac{n(x)}{d(x)} = Q(x) + \frac{r(x)}{d(x)}$$

The result is the sum of a quotient $Q(x)$ and a rational expression, which is the ratio of the remainder $r(x)$ and the divisor $d(x)$. The rational expression $n(x)/d(x)$ is proper, and the techniques outlined in this section can be applied to its partial-fraction decomposition.

Partial-fraction decomposition of proper rational expressions always begins with factoring the denominator $d(x)$. The goal is to write $d(x)$ as a product of distinct linear factors, but that may not always be possible. Sometimes $d(x)$ can be factored into a product of linear factors, where one or more are repeated. And, sometimes the factored form of $d(x)$ contains irreducible quadratic factors, such as $x^2 + 1$. There are times when the irreducible quadratic factors are repeated, such as $(x^2 + 1)^2$. A procedure is now outlined for partial-fraction decomposition.

PARTIAL-FRACTION DECOMPOSITION

To write a rational expression $\frac{n(x)}{d(x)}$ as a sum of partial fractions:

Step 1: Determine whether the rational expression is proper or improper.

- Proper: degree of $n(x) <$ degree of $d(x)$
- Improper: degree of $n(x) \geq$ degree of $d(x)$

Step 2: If proper, proceed to Step 3.

If improper, divide $\frac{n(x)}{d(x)}$ using polynomial (long) division and write the result as $\frac{n(x)}{d(x)} = Q(x) + \frac{r(x)}{d(x)}$ and proceed to Step 3 with $\frac{r(x)}{d(x)}$.

Step 3: Factor $d(x)$. One of four possible cases will arise:

Case 1: Distinct (nonrepeated) *linear* factors: $(ax + b)$

$$d(x) = (3x - 1)(x + 2)$$

Case 2: One or more repeated linear factors: $(ax + b)^m \quad m \geq 2$

$$d(x) = (x + 5)^2(x - 3)$$

Case 3: One or more distinct irreducible ($ax^2 + bx + c = 0$ has no real roots) quadratic factors: $(ax^2 + bx + c)$

$$d(x) = (x^2 + 4)(x + 1)(x - 2)$$

Case 4: One or more repeated irreducible quadratic factors: $(ax^2 + bx + c)^m$

$$d(x) = (x^2 + x + 1)^2(x + 1)(x - 2)$$

Step 4: Decompose the rational expression into a sum of partial fractions according to the procedure outlined in each case in this section.

Step 4 depends on which cases, or types of factors, arise. It is important to note that these four cases are not exclusive and combinations of different types of factors will appear.

Distinct Linear Factors

CASE 1: $d(x)$ HAS ONLY DISTINCT (NONREPEATED) LINEAR FACTORS

If $d(x)$ is a polynomial of degree p, and it can be factored into p linear factors:

$$d(x) = \underbrace{(ax + b)(cx + d)\ldots}_{p \text{ linear factors}}$$

where no two factors are the same, then the partial-fraction decomposition of $\frac{n(x)}{d(x)}$ can be written as

$$\frac{n(x)}{d(x)} = \frac{A}{(ax + b)} + \frac{B}{(cx + d)} + \cdots$$

where the numerators, A, B, and so on are constants to be determined.

The goal is to write a proper rational expression as the sum of proper rational expressions. Therefore, if the denominator is a linear factor (degree 1), then the numerator is a constant (degree 0).

EXAMPLE 1 Partial-Fraction Decomposition with Distinct Linear Factors

Find the partial-fraction decomposition of $\frac{5x + 13}{x^2 + 4x - 5}$.

Solution:

Factor the denominator. $\frac{5x + 13}{(x - 1)(x + 5)}$

Express as a sum of two partial fractions. $\frac{5x + 13}{(x - 1)(x + 5)} = \frac{A}{(x - 1)} + \frac{B}{(x + 5)}$

Multiply the equation by the LCD $(x - 1)(x + 5)$. $5x + 13 = A(x + 5) + B(x - 1)$

Eliminate the parentheses. $5x + 13 = Ax + 5A + Bx - B$

Group the x's and constants on the right. $5x + 13 = (A + B)x + (5A - B)$

Identify like terms. $5x + 13 = (A + B)x + (5A - B)$

Equate the coefficients of x. $5 = A + B$

Equate the **constant** terms. $13 = 5A - B$

Solve the system of two linear equations using any method to solve for A and B. $A = 3, B = 2$

Substitute $A = 3$, $B = 2$ into partial fraction decomposition. $\boxed{\frac{5x + 13}{(x - 1)(x + 5)} = \frac{3}{(x - 1)} + \frac{2}{(x + 5)}}$

Check by adding the partial fractions.

$$\frac{3}{(x - 1)} + \frac{2}{(x + 5)} = \frac{3(x + 5) + 2(x - 1)}{(x - 1)(x + 5)} = \frac{5x + 13}{x^2 + 4x - 5}$$

■ **YOUR TURN** Find the partial-fraction decomposition of $\frac{4x - 13}{x^2 - 3x - 10}$.

Technology Tip

Use TI to check the graph of $Y_1 = \frac{5x + 13}{x^2 + 4x - 5}$ and its partial fraction decomposition $Y_2 = \frac{3}{x - 1} + \frac{2}{x + 5}$. The graphs and tables of values are shown.

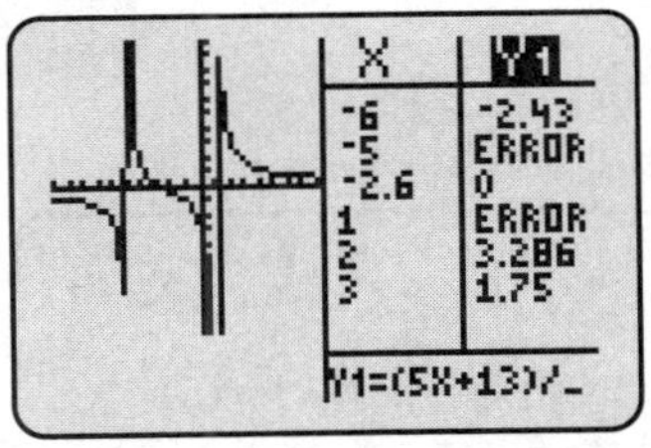

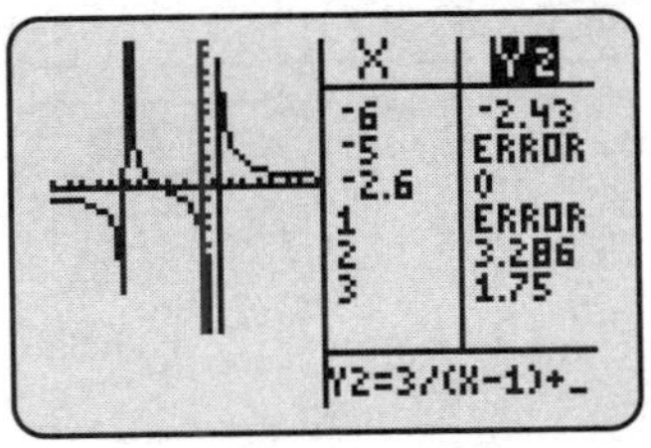

■ **Answer:** $\frac{4x - 13}{x^2 - 3x - 10} = \frac{3}{x + 2} + \frac{1}{x - 5}$

In Example 1, we started with a rational expression that had a numerator of degree 1 and a denominator of degree 2. Partial-fraction decomposition enabled us to write that rational expression as a sum of two rational expressions with degree 0 numerators and degree 1 denominators.

Repeated Linear Factors

CASE 2: $d(x)$ HAS AT LEAST ONE REPEATED LINEAR FACTOR

If $d(x)$ can be factored into a product of linear factors, then the partial-fraction decomposition will proceed as in Case 1, with the exception of a repeated factor $(ax + b)^m$, $m \geq 2$. Any linear factor repeated m times will result in the sum of m partial fractions:

$$\frac{A}{(ax+b)} + \frac{B}{(ax+b)^2} + \frac{C}{(ax+b)^3} + \cdots + \frac{M}{(ax+b)^m}$$

where the numerators, $A, B, C, \ldots, M$ are constants to be determined.

Note that if $d(x)$ is of degree p, the general form of the decomposition will have p partial fractions. If some numerator constants turn out to be zero, then the final decomposition may have fewer than p partial fractions.

Technology Tip

Use TI to check the graph of $Y_1 = \dfrac{-3x^2 + 13x - 12}{x^3 - 4x^2 + 4x}$ and its partial fraction decomposition $Y_2 = \dfrac{-3}{x} + \dfrac{1}{(x-2)^2}$. The graphs and tables of values are shown.

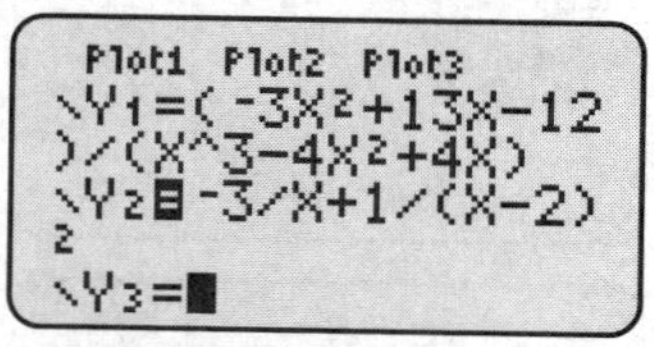

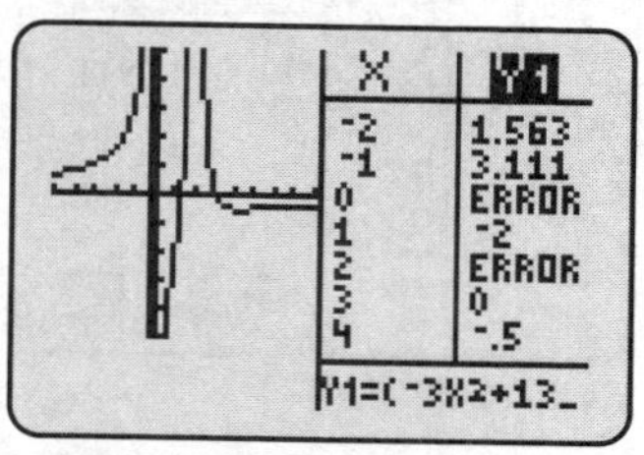

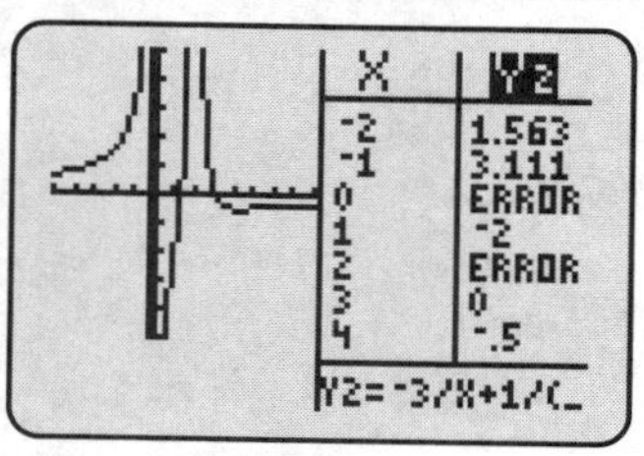

EXAMPLE 2 **Partial-Fraction Decomposition with a Repeated Linear Factor**

Find the partial-fraction decomposition of $\dfrac{-3x^2 + 13x - 12}{x^3 - 4x^2 + 4x}$.

Solution:

Factor the denominator. $\dfrac{-3x^2 + 13x - 12}{x(x-2)^2}$

Express as a sum of three partial fractions. $\dfrac{-3x^2 + 13x - 12}{x(x-2)^2} = \dfrac{A}{x} + \dfrac{B}{(x-2)} + \dfrac{C}{(x-2)^2}$

Multiply the equation by the LCD $x(x-2)^2$. $-3x^2 + 13x - 12 = A(x-2)^2 + Bx(x-2) + Cx$

Eliminate the parentheses. $-3x^2 + 13x - 12 = Ax^2 - 4Ax + 4A + Bx^2 - 2Bx + Cx$

Group like terms on the right. $-3x^2 + 13x - 12 = (A + B)x^2 + (-4A - 2B + C)x + 4A$

Identify like terms on both sides. $-3x^2 + 13x - 12 = (A + B)x^2 + (-4A - 2B + C)x + 4A$

Equate the coefficients of x^2. $-3 = A + B$ (1)

Equate the coefficients of x. $13 = -4A - 2B + C$ (2)

Equate the **constant** terms. $-12 = 4A$ (3)

Solve the system of three equations for A, B, and C.

Solve (3) for A. $A = -3$

Substitute $A = -3$ into (1). $B = 0$

Substitute $A = -3$ and $B = 0$ into (2). $C = 1$

Substitute $A = -3$, $B = 0$, $C = 1$ into the partial-fraction decomposition.

$$\frac{-3x^2 + 13x - 12}{x(x-2)^2} = \frac{-3}{x} + \frac{0}{(x-2)} + \frac{1}{(x-2)^2}$$

$$\boxed{\frac{-3x^2 + 13x - 12}{x^3 - 4x^2 + 4x} = \frac{-3}{x} + \frac{1}{(x-2)^2}}$$

Check by adding the partial fractions.

$$\frac{-3}{x} + \frac{1}{(x-2)^2} = \frac{-3(x-2)^2 + 1(x)}{x(x-2)^2} = \frac{-3x^2 + 13x - 12}{x^3 - 4x^2 + 4x}$$

■ **YOUR TURN** Find the partial-fraction decomposition of $\dfrac{x^2 + 1}{x^3 + 2x^2 + x}$.

■ **Answer:**
$$\frac{x^2 + 1}{x^3 + 2x^2 + x} = \frac{1}{x} - \frac{2}{(x+1)^2}$$

EXAMPLE 3 Partial-Fraction Decomposition with Multiple Repeated Linear Factors

Find the partial-fraction decomposition of $\dfrac{2x^3 + 6x^2 + 6x + 9}{x^4 + 6x^3 + 9x^2}$.

Solution:

Factor the denominator. $\dfrac{2x^3 + 6x^2 + 6x + 9}{x^2(x+3)^2}$

Express as a sum of four partial fractions. $\dfrac{2x^3 + 6x^2 + 6x + 9}{x^2(x+3)^2} = \dfrac{A}{x} + \dfrac{B}{x^2} + \dfrac{C}{(x+3)} + \dfrac{D}{(x+3)^2}$

Multiply the equation by the LCD $x^2(x+3)^2$. $2x^3 + 6x^2 + 6x + 9 = Ax(x+3)^2 + B(x+3)^2 + Cx^2(x+3) + Dx^2$

Eliminate the parentheses. $2x^3 + 6x^2 + 6x + 9 = Ax^3 + 6Ax^2 + 9Ax + Bx^2 + 6Bx + 9B + Cx^3 + 3Cx^2 + Dx^2$

Group like terms on the right. $2x^3 + 6x^2 + 6x + 9 = (A + C)x^3 + (6A + B + 3C + D)x^2 + (9A + 6B)x + 9B$

Identify like terms on both sides. $2x^3 + 6x^2 + 6x + 9 = (A + C)x^3 + (6A + B + 3C + D)x^2 + (9A + 6B)x + 9B$

Equate the coefficients of x^3. $2 = A + C$ (1)

Equate the coefficients of x^2. $6 = 6A + B + 3C + D$ (2)

Equate the coefficients of x. $6 = 9A + 6B$ (3)

Equate the **constant** terms. $9 = 9B$ (4)

Solve the system of four equations for A, B, C, and D.

Solve Equation (4) for B. $B = 1$

Substitute $B = 1$ into Equation (3) and solve for A. $A = 0$

Substitute $A = 0$ into Equation (1) and solve for C. $C = 2$

Substitute $A = 0$, $B = 1$, and $C = 2$ into Equation (2) and solve for D. $D = -1$

Substitute $A = 0, B = 1, C = 2, D = -1$ into the partial-fraction decomposition.

$$\frac{2x^3 + 6x^2 + 6x + 9}{x^2(x + 3)^2} = \frac{0}{x} + \frac{1}{x^2} + \frac{2}{(x + 3)} + \frac{-1}{(x + 3)^2}$$

$$\boxed{\frac{2x^3 + 6x^2 + 6x + 9}{x^2(x + 3)^2} = \frac{1}{x^2} + \frac{2}{(x + 3)} - \frac{1}{(x + 3)^2}}$$

Check by adding the partial fractions.

$$\frac{1}{x^2} + \frac{2}{(x + 3)} - \frac{1}{(x + 3)^2} = \frac{(x + 3)^2 + 2x^2(x + 3) - 1(x^2)}{x^2(x + 3)^2}$$

$$= \frac{2x^3 + 6x^2 + 6x + 9}{x^4 + 6x^3 + 9x^2}$$

■ **YOUR TURN** Find the partial-fraction decomposition of $\frac{2x^3 + 2x + 1}{x^4 + 2x^3 + x^2}$.

■ **Answer:** $\frac{2x^3 + 2x + 1}{x^4 + 2x^3 + x^2} = \frac{1}{x^2} + \frac{2}{(x + 1)} - \frac{3}{(x + 1)^2}$

Distinct Irreducible Quadratic Factors

There will be times when a polynomial cannot be factored into a product of linear factors with real coefficients. For example, $x^2 + 4$, $x^2 + x + 1$, and $9x^2 + 3x + 2$ are all examples of *irreducible quadratic* expressions. The general form of an **irreducible quadratic factor** is given by:

$$ax^2 + bx + c \qquad \text{where } ax^2 + bx + c = 0 \text{ has no real roots}$$

CASE 3: $d(x)$ HAS A DISTINCT IRREDUCIBLE QUADRATIC FACTOR

If the factored form of $d(x)$ contains an irreducible quadratic factor $ax^2 + bx + c$ then the partial-fraction decomposition will contain a term of the form:

$$\frac{Ax + B}{ax^2 + bx + c}$$

where A and B are constants to be determined.

Study Tip

The degree of the numerator is always 1 less than the degree of the denominator.

Recall that for a proper rational expression, the numerator is less than the denominator. For irreducible quadratic (degree 2) denominators we assume a linear (degree 1) numerator. For example,

$$\frac{7x^2 + 2}{(2x + 1)(x^2 + 1)} = \underbrace{\frac{A}{(2x + 1)}}_{\frac{\text{Constant numerator}}{\text{Linear factor}}} + \underbrace{\frac{Bx + C}{(x^2 + 1)}}_{\frac{\text{Linear numerator}}{\text{Quadratic factor}}}$$

A constant is used in the numerator when the denominator consists of a linear expression and a linear expression is used in the numerator when the denominator consists of a quadratic expression.

EXAMPLE 4 Partial-Fraction Decomposition with an Irreducible Quadratic Factor

Find the partial-fraction decomposition of $\dfrac{7x^2 + 2}{(2x + 1)(x^2 + 1)}$.

Solution:

The denominator is already in factored form. $\dfrac{7x^2 + 2}{(2x + 1)(x^2 + 1)}$

Express as a sum of two partial fractions. $\dfrac{7x^2 + 2}{(2x + 1)(x^2 + 1)} = \dfrac{A}{(2x + 1)} + \dfrac{Bx + C}{(x^2 + 1)}$

Multiply the equation by the LCD $(2x + 1)(x^2 + 1)$. $7x^2 + 2 = A(x^2 + 1) + (Bx + C)(2x + 1)$

Eliminate the parentheses. $7x^2 + 2 = Ax^2 + A + 2Bx^2 + Bx + 2Cx + C$

Group like terms on the right. $7x^2 + 2 = (A + 2B)x^2 + (B + 2C)x + (A + C)$

Identify like terms on both sides. $7x^2 + 0x + 2 = (A + 2B)x^2 + (B + 2C)x + (A + C)$

Equate the coefficients of x^2. $7 = A + 2B$

Equate the coefficients of x. $0 = B + 2C$

Equate the **constant** terms. $2 = A + C$

Solve the system of three equations for A, B, and C. $A = 3, B = 2, C = -1$

Substitute $A = 3, B = 2, C = -1$ into the partial-fraction decomposition.

$$\boxed{\frac{7x^2 + 2}{(2x + 1)(x^2 + 1)} = \frac{3}{(2x + 1)} + \frac{2x - 1}{(x^2 + 1)}}$$

Check by adding the partial fractions.

$$\frac{3}{(2x + 1)} + \frac{2x - 1}{(x^2 + 1)} = \frac{3(x^2 + 1) + (2x - 1)(2x + 1)}{(2x + 1)(x^2 + 1)} = \frac{7x^2 + 2}{(2x + 1)(x^2 + 1)}$$

■ **YOUR TURN** Find the partial-fraction decomposition of $\dfrac{-2x^2 + x + 6}{(x - 1)(x^2 + 4)}$.

Technology Tip

Use TI to check the graph of $Y_1 = \dfrac{7x^2 + 2}{(2x + 1)(x^2 + 1)}$ and its partial fraction decomposition $Y_2 = \dfrac{3}{2x + 1} + \dfrac{2x - 1}{x^2 + 1}$. The graphs and tables of values are shown.

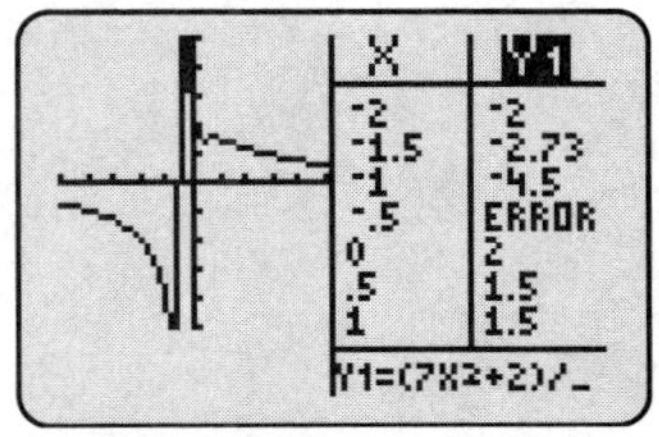

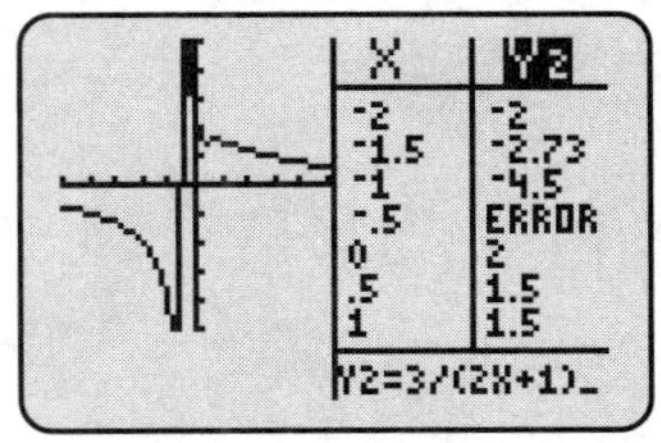

■ **Answer:** $\dfrac{-2x^2 + x + 6}{(x - 1)(x^2 + 4)} = \dfrac{1}{x - 1} - \dfrac{3x + 2}{x^2 + 4}$

Repeated Irreducible Quadratic Factors

CASE 4: $d(x)$ HAS A REPEATED IRREDUCIBLE QUADRATIC FACTOR

If the factored form of $d(x)$ contains an irreducible quadratic factor $(ax^2 + bx + c)^m$, where $b^2 - 4ac < 0$, then the partial-fraction decomposition will contain a series of terms of the form:

$$\frac{A_1x + B_1}{ax^2 + bx + c} + \frac{A_2x + B_2}{(ax^2 + bx + c)^2} + \frac{A_3x + B_3}{(ax^2 + bx + c)^3} + \cdots + \frac{A_mx + B_m}{(ax^2 + bx + c)^m}$$

where A_i and B_i $i = 1, 2, \ldots, m$ are constants to be determined.

EXAMPLE 5 Partial-Fraction Decomposition with a Repeated Irreducible Quadratic Factor

Find the partial-fraction decomposition of $\dfrac{x^3 - x^2 + 3x + 2}{(x^2 + 1)^2}$.

Solution:

The denominator is already in factored form. $\dfrac{x^3 - x^2 + 3x + 2}{(x^2 + 1)^2}$

Express as a sum of two partial fractions. $\dfrac{x^3 - x^2 + 3x + 2}{(x^2 + 1)^2} = \dfrac{Ax + B}{x^2 + 1} + \dfrac{Cx + D}{(x^2 + 1)^2}$

Multiply the equation by the LCD $(x^2 + 1)^2$. $x^3 - x^2 + 3x + 2 = (Ax + B)(x^2 + 1) + Cx + D$

Eliminate the parentheses. $x^3 - x^2 + 3x + 2 = Ax^3 + Bx^2 + Ax + B + Cx + D$

Group like terms on the right. $x^3 - x^2 + 3x + 2 = Ax^3 + Bx^2 + (A + C)x + (B + D)$

Identify like terms on both sides. $x^3 - x^2 + 3x + 2 = Ax^3 + Bx^2 + (A + C)x + (B + D)$

Equate the coefficients of x^3. $1 = A$ (1)

Equate the coefficients of x^2. $-1 = B$ (2)

Equate the coefficients of x. $3 = A + C$ (3)

Equate the **constant** terms. $2 = B + D$ (4)

Substitute $A = 1$ into Equation (3) and solve for C. $C = 2$

Substitute $B = -1$ into Equation (4) and solve for D. $D = 3$

Substitute $A = 1$, $B = -1$, $C = 2$, $D = 3$ into the partial-fraction decomposition.

$$\boxed{\frac{x^3 - x^2 + 3x + 2}{(x^2 + 1)^2} = \frac{x - 1}{x^2 + 1} + \frac{2x + 3}{(x^2 + 1)^2}}$$

Check by adding the partial fractions.

$$\frac{x - 1}{x^2 + 1} + \frac{2x + 3}{(x^2 + 1)^2} = \frac{(x - 1)(x^2 + 1) + (2x + 3)}{(x^2 + 1)^2} = \frac{x^3 - x^2 + 3x + 2}{(x^2 + 1)^2}$$

■ **YOUR TURN** Find the partial-fraction decomposition of $\dfrac{3x^3 + x^2 + 4x - 1}{(x^2 + 4)^2}$.

■ **Answer:** $\dfrac{3x^3 + x^2 + 4x - 1}{(x^2 + 4)^2} = \dfrac{3x + 1}{x^2 + 4} - \dfrac{8x + 5}{(x^2 + 4)^2}$

Combinations of All Four Cases

As you probably can imagine, there are rational expressions that have combinations of all four cases, which can lead to a system of several equations when solving for the numerator constants.

EXAMPLE 6 Partial-Fraction Decomposition

Find the partial-fraction decomposition of $\dfrac{x^5 + x^4 + 4x^3 - 3x^2 + 4x - 8}{x^2(x^2 + 2)^2}$.

Solution:

The denominator is already in factored form.

$$\frac{x^5 + x^4 + 4x^3 - 3x^2 + 4x - 8}{x^2(x^2 + 2)^2}$$

Express as a sum of partial fractions.

There are repeated linear and irreducible quadratic factors.

$$\frac{x^5 + x^4 + 4x^3 - 3x^2 + 4x - 8}{x^2(x^2 + 2)^2} = \frac{A}{x} + \frac{B}{x^2} + \frac{Cx + D}{(x^2 + 2)} + \frac{Ex + F}{(x^2 + 2)^2}$$

Multiply the equation by the LCD $x^2(x^2 + 2)^2$.

$$x^5 + x^4 + 4x^3 - 3x^2 + 4x - 8$$
$$= Ax(x^2 + 2)^2 + B(x^2 + 2)^2 + (Cx + D)x^2(x^2 + 2) + (Ex + F)x^2$$

Eliminate the parentheses.

$$x^5 + x^4 + 4x^3 - 3x^2 + 4x - 8$$
$$= Ax^5 + 4Ax^3 + 4Ax + Bx^4 + 4Bx^2 + 4B + Cx^5 + 2Cx^3 + Dx^4 + 2Dx^2 + Ex^3 + Fx^2$$

Group like terms on the right.

$$x^5 + x^4 + 4x^3 - 3x^2 + 4x - 8$$
$$= (A + C)x^5 + (B + D)x^4 + (4A + 2C + E)x^3 + (4B + 2D + F)x^2 + 4Ax + 4B$$

Equating the coefficients of like terms leads to six equations.

$$A + C = 1$$
$$B + D = 1$$
$$4A + 2C + E = 4$$
$$4B + 2D + F = -3$$
$$4A = 4$$
$$4B = -8$$

Solve this system of equations.

$$A = 1, \quad B = -2, \quad C = 0, \quad D = 3, \quad E = 0, \quad F = -1$$

Substitute $A = 1, B = -2, C = 0, D = 3, E = 0, F = -1$ into the partial-fraction decomposition.

$$\frac{x^5 + x^4 + 4x^3 - 3x^2 + 4x - 8}{x^2(x^2 + 2)^2} = \frac{1}{x} + \frac{-2}{x^2} + \frac{0x + 3}{(x^2 + 2)} + \frac{0x + -1}{(x^2 + 2)^2}$$

$$\boxed{\frac{x^5 + x^4 + 4x^3 - 3x^2 + 4x - 8}{x^2(x^2 + 2)^2} = \frac{1}{x} + \frac{-2}{x^2} + \frac{3}{(x^2 + 2)} - \frac{1}{(x^2 + 2)^2}}$$

Check by adding the partial fractions.

SECTION 6.3 SUMMARY

A rational expression $\frac{n(x)}{d(x)}$ is

- **Proper:** If the degree of the numerator is less than the degree of the denominator.
- **Improper:** If the degree of the numerator is equal to or greater than the degree of the denominator.

Partial-Fraction Decomposition of Proper Rational Expressions

1. Distinct (nonrepeated) linear factors

$$\frac{3x - 10}{(x - 5)(x + 4)} = \frac{A}{x - 5} + \frac{B}{x + 4}$$

2. Repeated linear factors

$$\frac{2x + 5}{(x - 3)^2(x + 1)} = \frac{A}{x - 3} + \frac{B}{(x - 3)^2} + \frac{C}{x + 1}$$

3. Distinct irreducible quadratic factors

$$\frac{1 - x}{(x^2 + 1)} = \frac{Ax + B}{x^2 + 1}$$

4. Repeated irreducible quadratic factors

$$\frac{4x^2 - 3x + 2}{(x^2 + 1)^2} = \frac{Ax + B}{x^2 + 1} + \frac{Cx + D}{(x^2 + 1)^2}$$

SECTION 6.3 EXERCISES

SKILLS

In Exercises 1–6, match the rational expression (1–6) with the form of the partial-fraction decomposition (a–f).

1. $\frac{3x + 2}{x(x^2 - 25)}$
2. $\frac{3x + 2}{x(x^2 + 25)}$
3. $\frac{3x + 2}{x^2(x^2 + 25)}$
4. $\frac{3x + 2}{x^2(x^2 - 25)}$
5. $\frac{3x + 2}{x(x^2 + 25)^2}$
6. $\frac{3x + 2}{x^2(x^2 + 25)^2}$

a. $\frac{A}{x} + \frac{B}{x^2} + \frac{Cx + D}{x^2 + 25}$

b. $\frac{A}{x} + \frac{Bx + C}{x^2 + 25} + \frac{Dx + E}{(x^2 + 25)^2}$

c. $\frac{A}{x} + \frac{Bx + C}{x^2 + 25}$

d. $\frac{A}{x} + \frac{B}{x + 5} + \frac{C}{x - 5}$

e. $\frac{A}{x} + \frac{B}{x^2} + \frac{Cx + D}{x^2 + 25} + \frac{Ex + F}{(x^2 + 25)^2}$

f. $\frac{A}{x} + \frac{B}{x^2} + \frac{C}{x + 5} + \frac{D}{x - 5}$

In Exercises 7–14, write the form of the partial-fraction decomposition. Do not solve for the constants.

7. $\frac{9}{x^2 - x - 20}$
8. $\frac{8}{x^2 - 3x - 10}$
9. $\frac{2x + 5}{x^3 - 4x^2}$
10. $\frac{x^2 + 2x - 1}{x^4 - 9x^2}$
11. $\frac{2x^3 - 4x^2 + 7x + 3}{(x^2 + x + 5)}$
12. $\frac{2x^3 + 5x^2 + 6}{(x^2 - 3x + 7)}$
13. $\frac{3x^3 - x + 9}{(x^2 + 10)^2}$
14. $\frac{5x^3 + 2x^2 + 4}{(x^2 + 13)^2}$

In Exercises 15–40, find the partial-fraction decomposition for each rational function.

15. $\frac{1}{x(x + 1)}$
16. $\frac{1}{x(x - 1)}$
17. $\frac{x}{x(x - 1)}$
18. $\frac{x}{x(x + 1)}$
19. $\frac{9x - 11}{(x - 3)(x + 5)}$
20. $\frac{8x - 13}{(x - 2)(x + 1)}$
21. $\frac{3x + 1}{(x - 1)^2}$
22. $\frac{9y - 2}{(y - 1)^2}$
23. $\frac{4x - 3}{x^2 + 6x + 9}$
24. $\frac{3x + 1}{x^2 + 4x + 4}$
25. $\frac{4x^2 - 32x + 72}{(x + 1)(x - 5)^2}$
26. $\frac{4x^2 - 7x - 3}{(x + 2)(x - 1)^2}$

27. $\dfrac{5x^2 + 28x - 6}{(x + 4)(x^2 + 3)}$

28. $\dfrac{x^2 + 5x + 4}{(x - 2)(x^2 + 2)}$

29. $\dfrac{-2x^2 - 17x + 11}{(x - 7)(3x^2 - 7x + 5)}$

30. $\dfrac{14x^2 + 8x + 40}{(x + 5)(2x^2 - 3x + 5)}$

31. $\dfrac{x^3}{(x^2 + 9)^2}$

32. $\dfrac{x^2}{(x^2 + 9)^2}$

33. $\dfrac{2x^3 - 3x^2 + 7x - 2}{(x^2 + 1)^2}$

34. $\dfrac{-x^3 + 2x^2 - 3x + 15}{(x^2 + 8)^2}$

35. $\dfrac{3x + 1}{x^4 - 1}$

36. $\dfrac{2 - x}{x^4 - 81}$

37. $\dfrac{5x^2 + 9x - 8}{(x - 1)(x^2 + 2x - 1)}$

38. $\dfrac{10x^2 - 5x + 29}{(x - 3)(x^2 + 4x + 5)}$

39. $\dfrac{3x}{x^3 - 1}$

40. $\dfrac{5x + 2}{x^3 - 8}$

APPLICATIONS

41. Optics. The relationship between the distance of an object to a lens d_o, the distance to the image d_i, and the focal length f of the lens is given by

$$\frac{f(d_i + d_o)}{d_i d_o} = 1$$

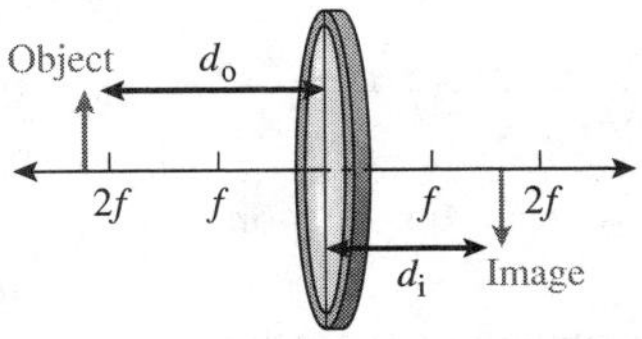

Use partial-fraction decomposition to write the lens law in terms of sums of fractions. What does each term represent?

42. Sums. Find the partial-fraction decomposition of $\dfrac{1}{n(n + 1)}$, and apply it to find the sum of

$$\frac{1}{1\cdot 2} + \frac{1}{2\cdot 3} + \frac{1}{3\cdot 4} + \cdots + \frac{1}{999\cdot 1000}.$$

CATCH THE MISTAKE

In Exercises 43 and 44, explain the mistake that is made.

43. Find the partial-fraction decomposition of $\dfrac{3x^2 + 3x + 1}{x(x^2 + 1)}$.

Solution:

Write the partial-fraction decomposition form. $\dfrac{3x^2 + 3x + 1}{x(x^2 + 1)} = \dfrac{A}{x} + \dfrac{B}{x^2 + 1}$

Multiply by the LCD $x(x^2 + 1)$. $3x^2 + 3x + 1 = A(x^2 + 1) + Bx$

Eliminate the parentheses. $3x^2 + 3x + 1 = Ax^2 + Bx + A$

Matching like terms leads to three equations. $A = 3, B = 3$, and $A = 1$

This is incorrect. What mistake was made?

44. Find the partial-fraction decomposition of $\dfrac{3x^4 - x - 1}{x(x - 1)}$.

Solution:

Write the partial-fraction decomposition form. $\dfrac{3x^4 - x - 1}{x(x - 1)} = \dfrac{A}{x} + \dfrac{B}{x - 1}$

Multiply by the LCD $x(x - 1)$. $3x^4 - x - 1 = A(x - 1) + Bx$

Eliminate the parentheses and group like terms. $3x^4 - x - 1 = (A + B)x - A$

Compare like coefficients. $A = 1, B = -2$

This is incorrect. What mistake was made?

CONCEPTUAL

In Exercises 45 and 46, determine whether each statement is true or false.

45. Partial-fraction decomposition can only be employed when the degree of the numerator is greater than the degree of the denominator.

46. The degree of the denominator of a reducible rational expression is equal to the number of partial fractions in its decomposition.

CHALLENGE

For Exercises 47–52, find the partial-fraction decomposition.

47. $\dfrac{x^2 + 4x - 8}{x^3 - x^2 - 4x + 4}$

48. $\dfrac{ax + b}{x^2 - c^2}$ a, b, c are real numbers.

49. $\dfrac{2x^3 + x^2 - x - 1}{x^4 + x^3}$

50. $\dfrac{-x^3 + 2x - 2}{x^5 - x^4}$

51. $\dfrac{x^5 + 2}{(x^2 + 1)^3}$

52. $\dfrac{x^2 - 4}{(x^2 + 1)^3}$

TECHNOLOGY

53. Apply a graphing utility to graph $y_1 = \dfrac{5x + 4}{x^2 + x - 2}$ and $y_2 = \dfrac{3}{x - 1} + \dfrac{2}{x + 2}$ in the same viewing rectangle. Is y_2 the partial-fraction decomposition of y_1?

54. Apply a graphing utility to graph $y_1 = \dfrac{2x^2 + 2x - 5}{x^3 + 5x}$ and $y_2 = \dfrac{3x + 2}{x^2 + 5} - \dfrac{1}{x}$ in the same viewing rectangle. Is y_2 the partial-fraction decomposition of y_1?

55. Apply a graphing utility to graph $y_1 = \dfrac{x^9 + 8x - 1}{x^5(x^2 + 1)^3}$ and $y_2 = \dfrac{4}{x} - \dfrac{1}{x^5} + \dfrac{2}{x^2 + 1} - \dfrac{3x + 2}{(x^2 + 1)^2}$ in the same viewing rectangle. Is y_2 the partial-fraction decomposition of y_1?

56. Apply a graphing utility to graph $y_1 = \dfrac{x^3 + 2x + 6}{(x + 3)(x^2 - 4)^3}$ and $y_2 = \dfrac{2}{x + 3} + \dfrac{x + 3}{(x^2 - 4)^3}$ in the same viewing rectangle. Is y_2 the partial-fraction decomposition of y_1?

57. Apply a graphing utility to graph $y_1 = \dfrac{2x^3 - 8x + 16}{(x - 2)^2(x^2 + 4)}$ and $y_2 = \dfrac{1}{x - 2} + \dfrac{2}{(x - 2)^2} + \dfrac{x + 4}{x^2 + 4}$ in the same viewing rectangle. Is y_2 the partial-fraction decomposition of y_1?

58. Apply a graphing utility to graph $y_1 = \dfrac{3x^3 + 14x^2 + 6x + 51}{(x^2 + 3x - 4)(x^2 + 2x + 5)}$ and $y_2 = \dfrac{2}{x - 1} - \dfrac{1}{x + 4} + \dfrac{2x - 3}{x^2 + 2x + 5}$ in the same viewing rectangle. Is y_2 the partial-fraction decomposition of y_1?

SECTION 6.4 SYSTEMS OF LINEAR INEQUALITIES IN TWO VARIABLES

SKILLS OBJECTIVES

- Graph a linear inequality in two variables.
- Graph a system of linear inequalities in two variables.

CONCEPTUAL OBJECTIVES

- Interpret the difference between solid and dashed lines.
- Interpret an overlapped shaded region as a solution.

Linear Inequalities in Two Variables

To graph linear inequalities in two variables, we will bridge together two concepts that we have already learned: *linear inequalities* (Section 1.5) and *lines* (Section 2.3). Recall that in Section 1.5 we discussed linear inequalities in one variable. For example, $3x - 1 < 8$ has a solution $x < 3$, which can be represented graphically on a number line where the red colored area to the left of 3 represents the solution.

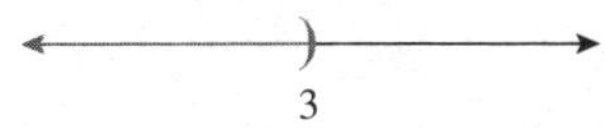

Recall in Section 2.3 that $y = 2x + 1$ is an *equation in two variables* whose graph is a line in the xy-plane. We now turn our attention to **linear inequalities in two variables**. For example, if we change the = in $y = 2x + 1$ to < we get $y < 2x + 1$. The solution to this inequality in two variables is the set of all points (x, y) that make this inequality true. Some solutions to this inequality are $(-2, -5), (0, 0), (3, 4), (5, -1), \ldots$

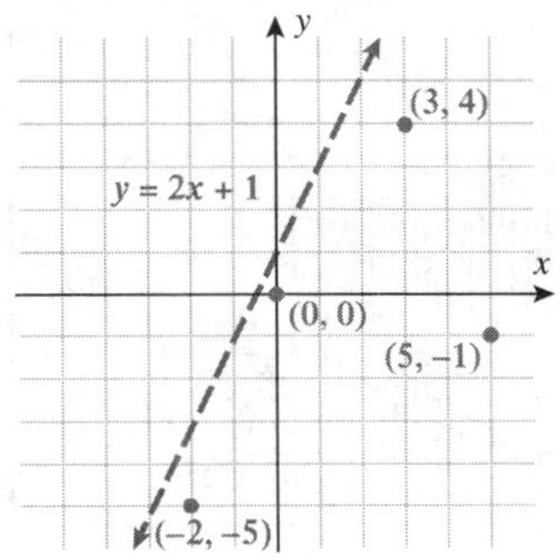

In fact, the entire region *below* the line $y = 2x + 1$ satisfies the inequality $y < 2x + 1$. If we reverse the sign of the inequality to get $y > 2x + 1$, then the entire region *above* the line $y = 2x + 1$ represents the solution to the inequality.

Any line divides the xy-plane into two **half-planes**. For example, the line $y = 2x + 1$ divides the xy-plane into two half-planes represented as $y > 2x + 1$ and $y < 2x + 1$. Recall that with inequalities in one variable we used the notation of parentheses and brackets to denote the type of inequality (strict or nonstrict). We use a similar notation with linear inequalities in two variables. If the inequality is a strict inequality, < or >, then the line is *dashed*, and, if the inequality includes the equal sign, ≤ or ≥, then a *solid* line is used. The following box summarizes the procedure for graphing a linear inequality in two variables.

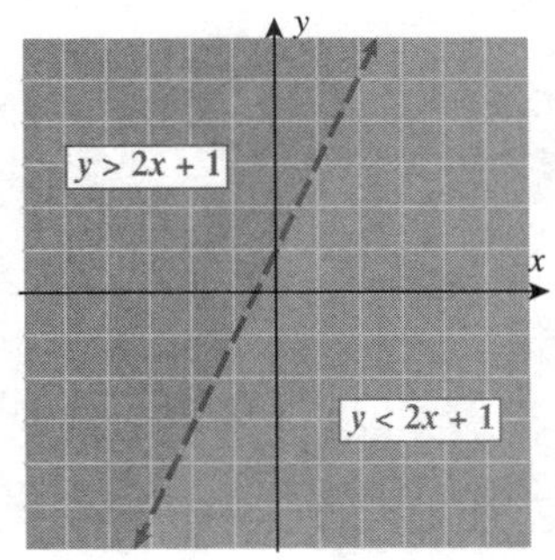

GRAPHING A LINEAR INEQUALITY IN TWO VARIABLES

Step 1: Change the sign. Change the inequality sign, <, ≤, ≥, or >, to an equal sign, =.

Step 2: Draw the line that corresponds to the resulting equation in Step 1.
- If the inequality is strict, < or >, use a **dashed** line.
- If the inequality is not strict, ≤ or ≥, use a **solid** line.

Step 3: Test a point.
- Select a point in one half-plane and test to see whether it satisfies the inequality. If it does, then so do all the points in that region (half-plane). If not, then none of the points in that half-plane satisfy the inequality.
- Repeat this step for the other half-plane.

Step 4: Shade the half-plane that satisfies the inequality.

Study Tip

A dashed line means that the points that lie on the line are not included in the solution of the linear inequality.

Technology Tip

The graphing calculator can be used to help in shading the linear inequality $3x + y < 2$. However, it will not show whether the line is solid or dashed. First solve for y, $y < -3x + 2$. Then, enter $y_1 = -3x + 2$. Since $y_1 < -3x + 2$, the region below the dashed line is shaded.

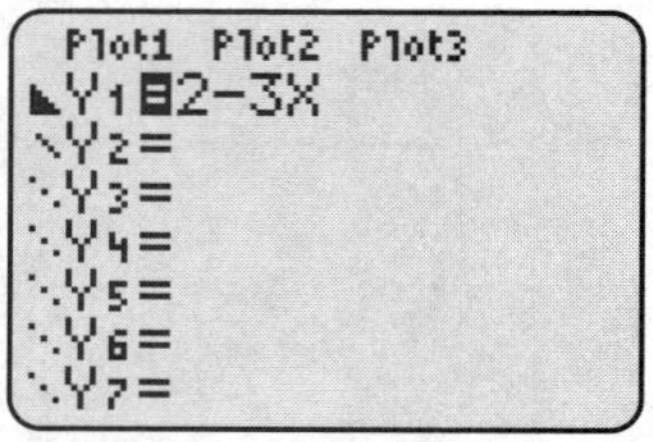

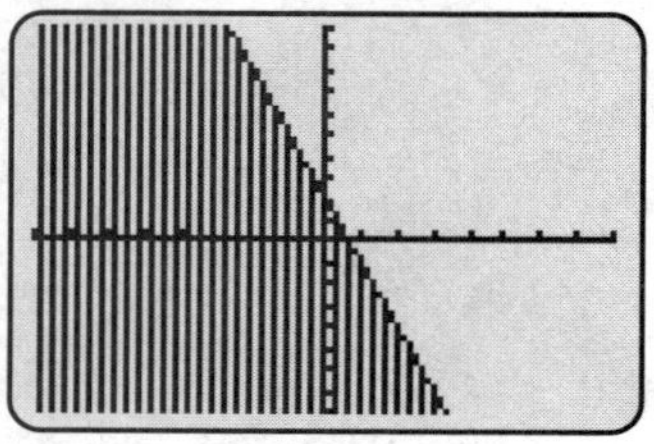

EXAMPLE 1 Graphing a Strict Linear Inequality in Two Variables

Graph the inequality $3x + y < 2$.

Solution:

STEP 1 Change the inequality sign to an equal sign. $\quad 3x + y = 2$

STEP 2 Draw the line.
Convert from standard form to slope–intercept form. $\quad y = -3x + 2$

Since the inequality $<$ is a strict inequality, use a **dashed** line.

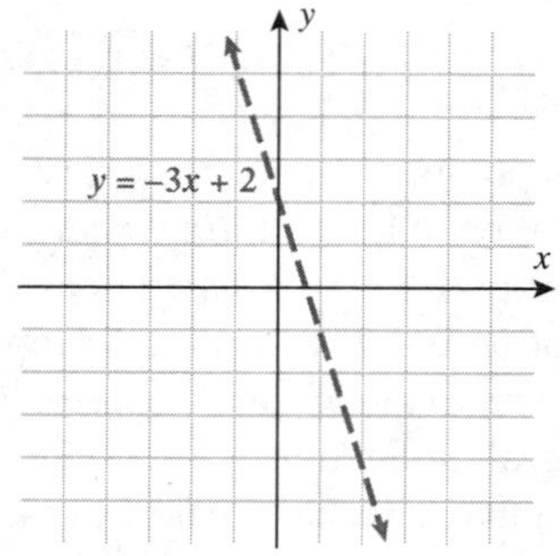

STEP 3 Test points in each half-plane.

Substitute (3, 0) into $3x + y < 2$. $\quad 3(3) + 0 < 2$

The point (3, 0) does not satisfy the inequality. $\quad 9 < 2$

Substitute $(-2, 0)$ into $3x + y < 2$. $\quad 3(-2) + 0 < 2$

The point $(-2, 0)$ does satisfy the inequality. $\quad -6 < 2$

STEP 4 Shade the region containing the point $(-2, 0)$.

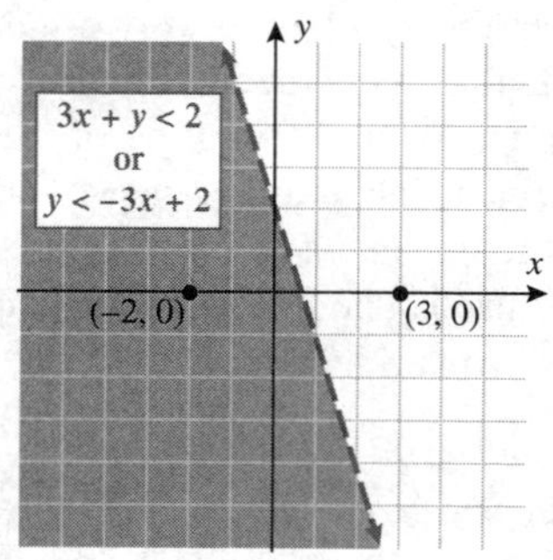

■ **YOUR TURN** Graph the inequality $-x + y > -1$.

■ **Answer:**

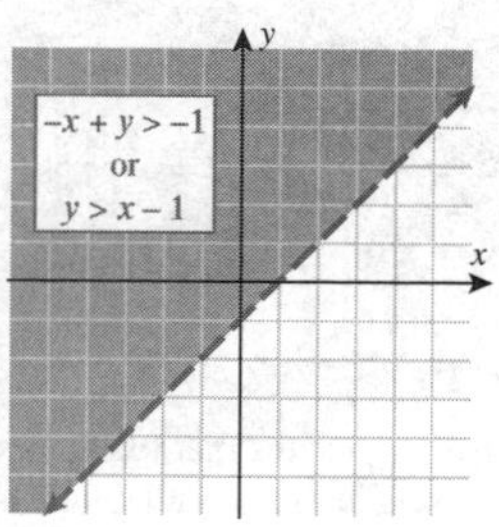

EXAMPLE 2 Graphing a Nonstrict Linear Inequality in Two Variables

Graph the inequality $2x - 3y \geq 6$.

Solution:

Step 1 Change the inequality sign to an equal sign. $2x - 3y = 6$

Step 2 Draw the line.

Convert from standard form to slope–intercept form. $y = \frac{2}{3}x - 2$

Since the inequality $\geq$ is not a strict inequality, use a **solid** line.

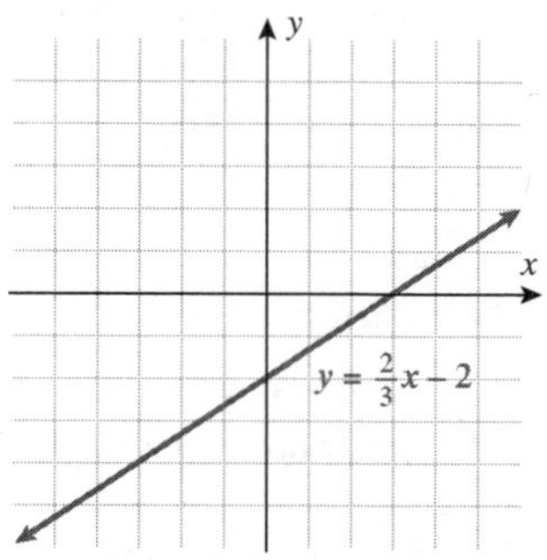

Step 3 Test points in each half-plane.

Substitute (5, 0) into $2x - 3y \geq 6$. $2(5) - 3(0) \geq 6$

The point (5, 0) satisfies the inequality. $10 \geq 6$

Substitute (0, 0) into $2x - 3y \geq 6$. $2(0) - 3(0) \geq 6$

$0 \geq 6$

The point (0, 0) does not satisfy the inequality.

Step 4 Shade the region containing the point (5, 0).

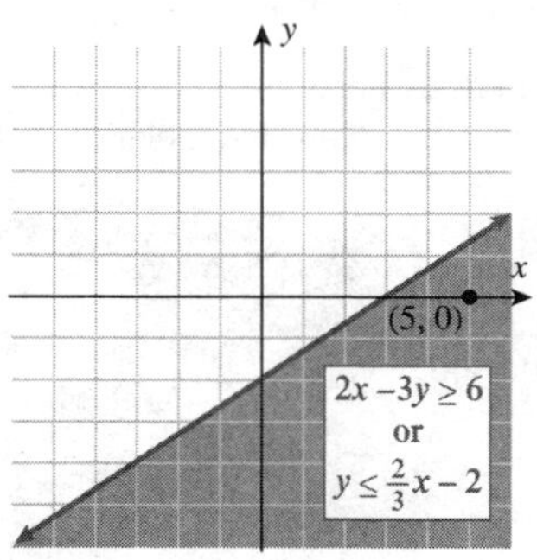

Technology Tip

```
Plot1 Plot2 Plot3
\Y1■1/3(2X-6)
\Y2=
```

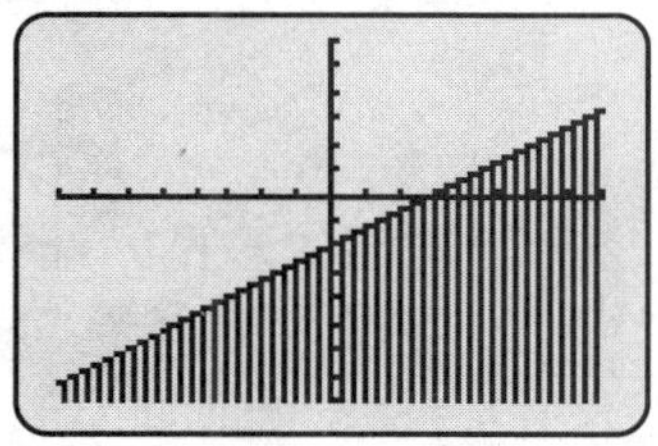

■ **YOUR TURN** Graph the inequality $x - 2y \leq 6$.

■ **Answer:**

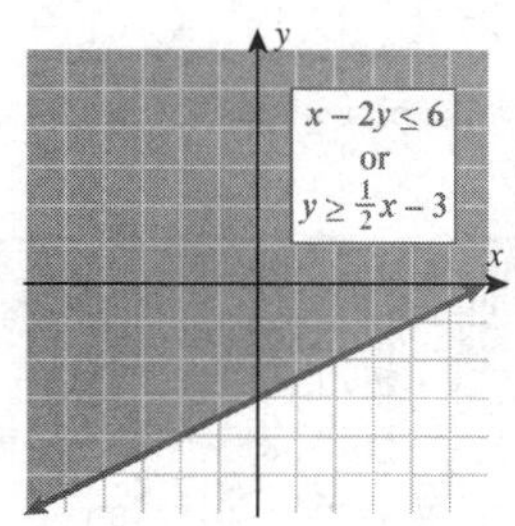

Systems of Linear Inequalities in Two Variables

Systems of linear inequalities are similar to *systems of linear equations.* In systems of linear equations we sought the points that satisfied *all* of the equations. The **solution set of a system of inequalities** contains the points that satisfy *all* of the inequalities. The graph of a system of inequalities can be obtained by simultaneously graphing each individual inequality and finding where the shaded regions intersect (or overlap), if at all.

EXAMPLE 3 Graphing a System of Two Linear Inequalities

Graph the system of inequalities: $x + y \geq -2$
$x + y \leq 2$

Solution:

STEP 1 Change the inequality signs to equal signs.

$x + y = -2$
$x + y = 2$

STEP 2 Draw the two lines.

Because the inequality signs are not strict, use solid lines.

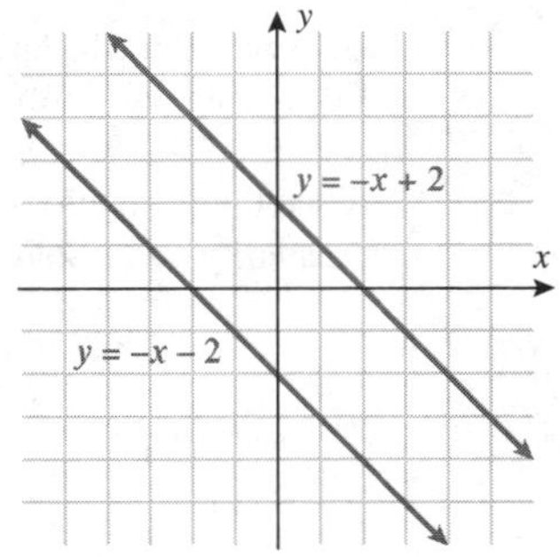

STEP 3 Test points for each inequality.

$x + y \geq -2$

Substitute $(-4, 0)$ into $x + y \geq -2$. $-4 \geq -2$

The point $(-4, 0)$ does not satisfy the inequality.

Substitute $(0, 0)$ into $x + y \geq -2$. $0 \geq -2$

The point $(0, 0)$ does satisfy the inequality.

$x + y \leq 2$

Substitute $(0, 0)$ into $x + y \leq 2$. $0 \leq 2$

The point $(0, 0)$ does satisfy the inequality.

Substitute $(4, 0)$ into $x + y \leq 2$. $4 \leq 2$

The point $(4, 0)$ does not satisfy the inequality.

STEP 4 For $x + y \geq -2$, shade the region *above* that includes $(0, 0)$.

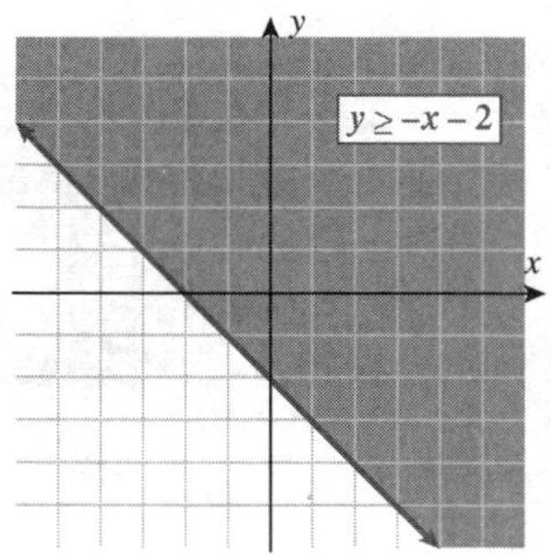

For $x + y \leq 2$, shade the region *below* that includes $(0, 0)$.

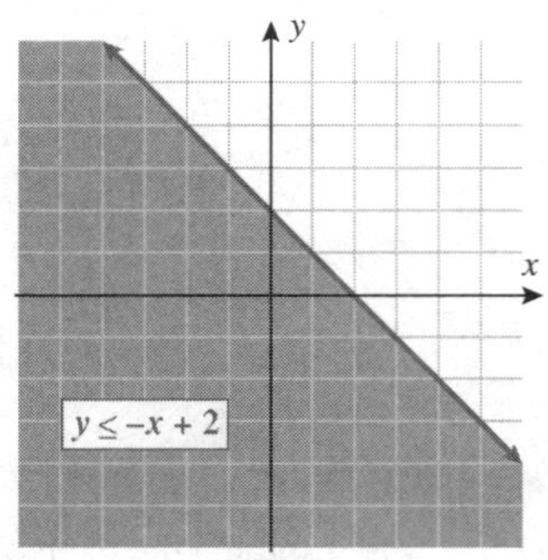

STEP 5 The overlapping region is the solution.

Notice that the points $(0, 0)$, $(-1, 1)$, and $(1, -1)$ all lie in the shaded region and all three satisfy both inequalities.

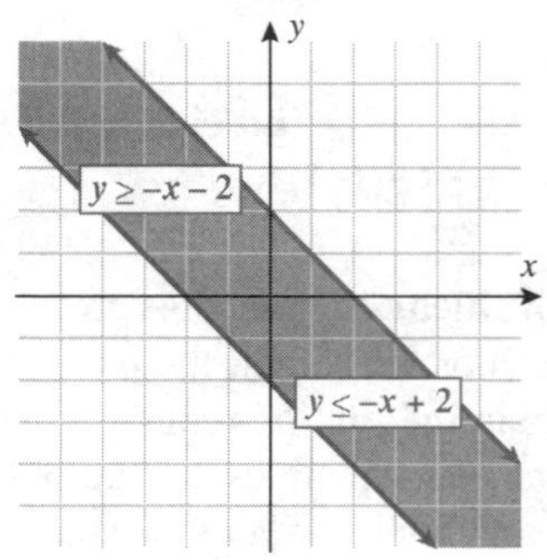

Technology Tip

Solve for y in each inequality first. Enter $y_1 \geq -x - 2$ and $y_2 \leq -x + 2$.

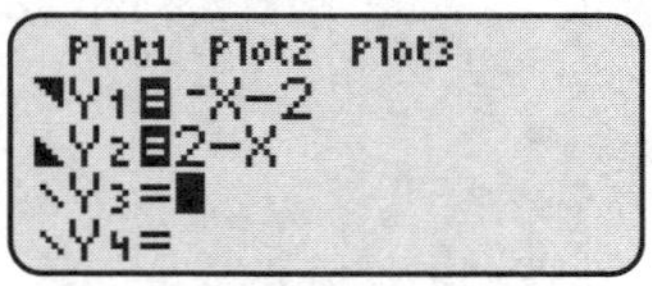

The overlapping region is the solution.

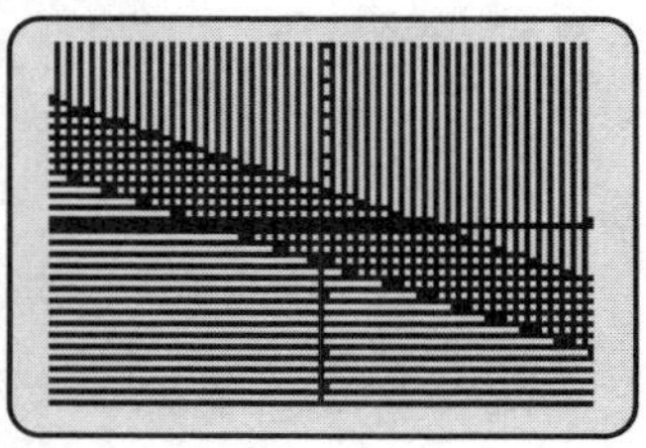

EXAMPLE 4 Graphing a System of Two Linear Inequalities with No Solution

Graph the system of inequalities: $x + y \leq -2$
$x + y \geq 2$

Solution:

STEP 1 Change the inequality signs to equal signs.

$x + y = -2$
$x + y = 2$

STEP 2 Draw the two lines.

Because the inequality signs are not strict, use solid lines.

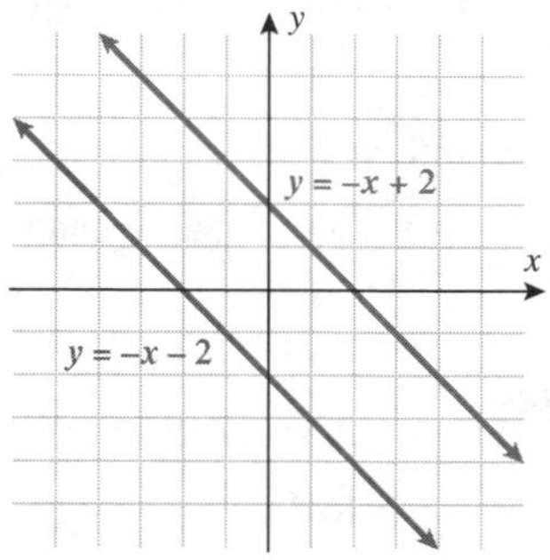

STEP 3 Test points for each inequality.

$x + y \leq -2$

Substitute $(-4, 0)$ into $x + y \leq -2$. $-4 \leq -2$

The point $(-4, 0)$ does satisfy the inequality.

Substitute $(0, 0)$ into $x + y \leq -2$. $0 \leq -2$

The point $(0, 0)$ does not satisfy the inequality.

$x + y \geq 2$

Substitute $(0, 0)$ into $x + y \geq 2$. $0 \geq 2$

The point $(0, 0)$ does not satisfy the inequality.

Substitute $(4, 0)$ into $x + y \geq 2$. $4 \geq 2$

The point $(4, 0)$ does satisfy the inequality.

STEP 4 For $x + y \leq -2$, shade the region *below* that includes $(-4, 0)$. For $x + y \leq 2$, shade the region *above* that includes $(4, 0)$.

STEP 5 There is no overlapping region. Therefore, no points satisfy both inequalities. We say there is **no solution**.

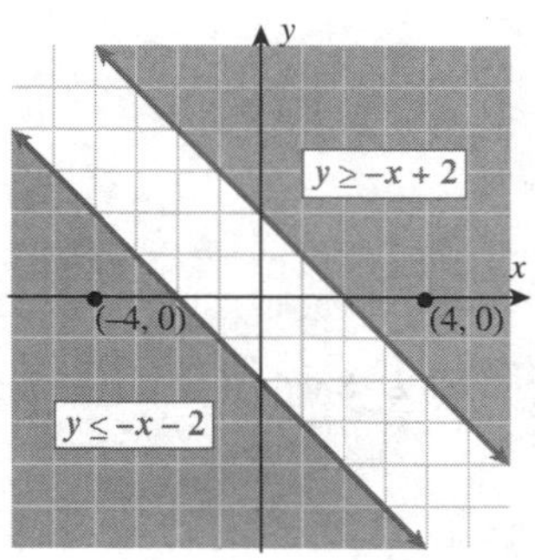

YOUR TURN Graph the solution to the system of inequalities.

a. $y > x + 1$
$y < x - 1$

b. $y < x + 1$
$y > x - 1$

Technology Tip

Solve for y in each inequality first. Enter $y_1 \leq -x - 2$ and $y_2 \geq -x + 2$.

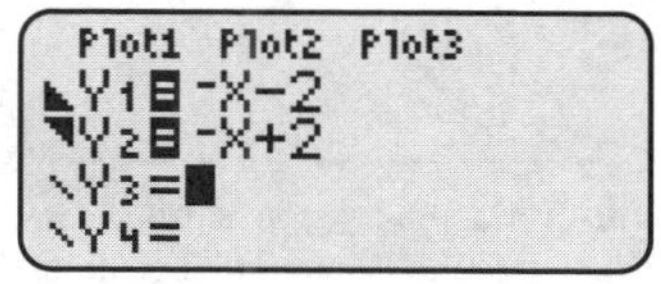

There is no overlapping region. Therefore, there is no solution to the system of inequalities.

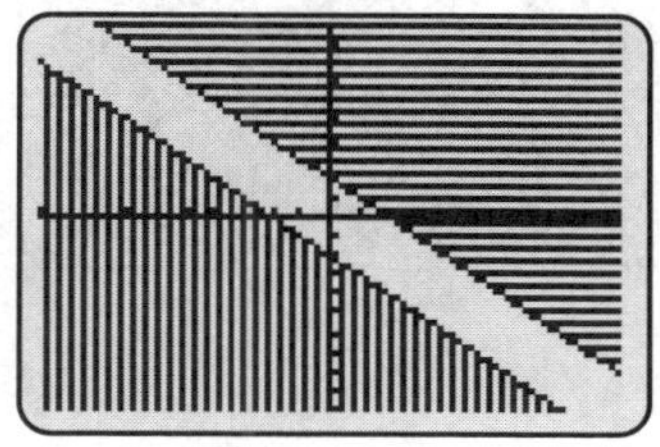

■ **Answer:**

a. No solution.

b.

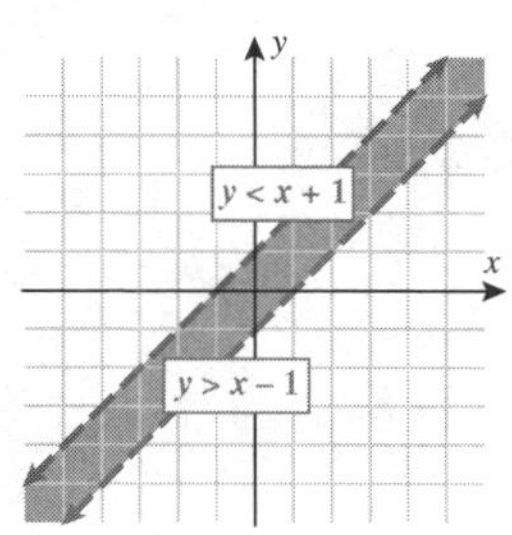

Technology Tip

Solve for y in each inequality first. Enter $y_1 \leq x$, $y_2 \geq -x$, and $y_3 < 3$.

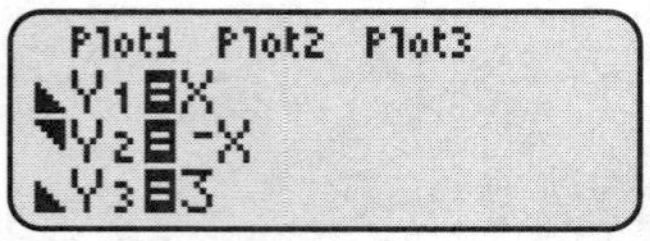

The overlapping region is the solution to the system of inequalities.

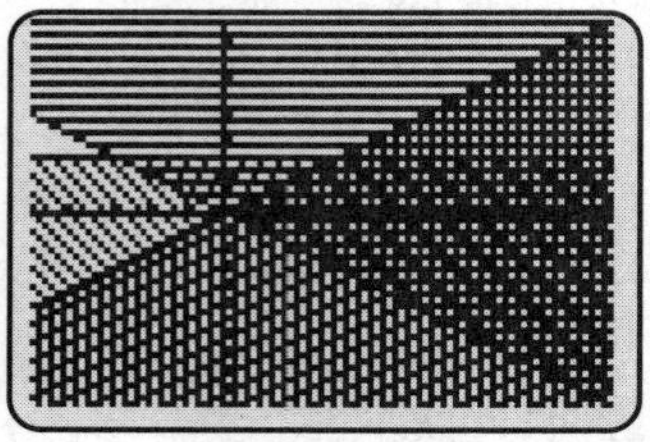

Thus far we have addressed only systems of two linear inequalities. Systems with more than two inequalities are treated in a similar manner. The solution is the set of all points that satisfy *all* of the inequalities. When there are more than two linear inequalities, the solution may be a **bounded** region. We can algebraically determine where the lines intersect by setting the y-values equal to each other.

EXAMPLE 5 Graphing a System of Multiple Linear Inequalities

Solve the system of inequalities:

$$y \leq x \quad y \geq -x \quad y < 3$$

Solution:

STEP 1 Change the inequalities to equal signs.

$$y = x \quad y = -x \quad y = 3$$

STEP 2 Draw the three lines.

To determine the points of intersection, set the y-values equal.

Point where $y = x$ and $y = -x$ intersect: $x = -x$, $x = 0$

Substitute $x = 0$ into $y = x$. $(0, 0)$

Point where $y = -x$ and $y = 3$ intersect: $-x = 3$, $x = -3$, $(-3, 3)$

Point where $y = 3$ and $y = x$ intersect: $x = 3$, $(3, 3)$

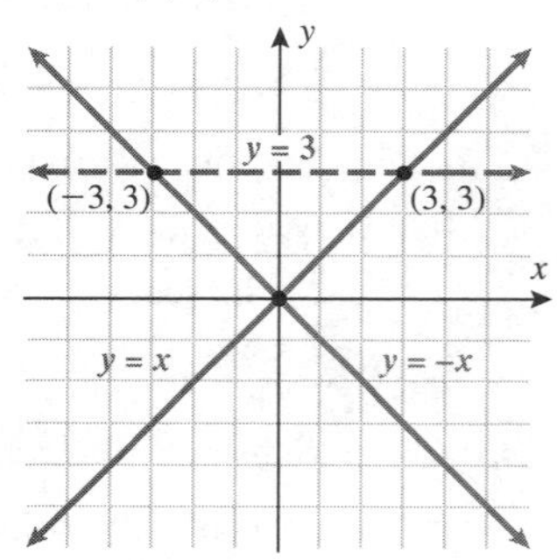

STEP 3 Test points to determine the shaded half-planes corresponding to $y \leq x$, $y \geq -x$, and $y < 3$.

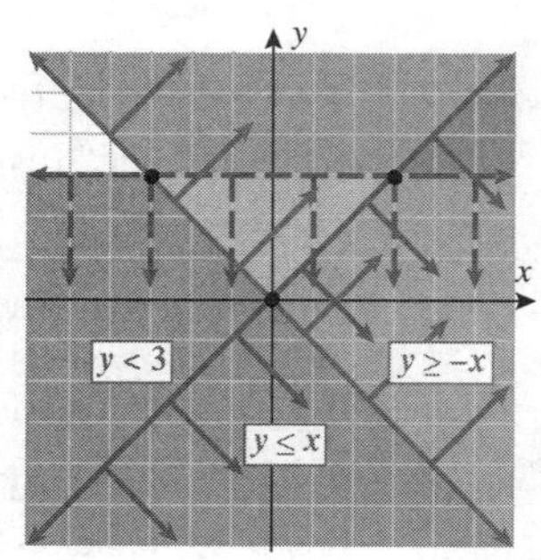

STEP 4 Shade the overlapping region.

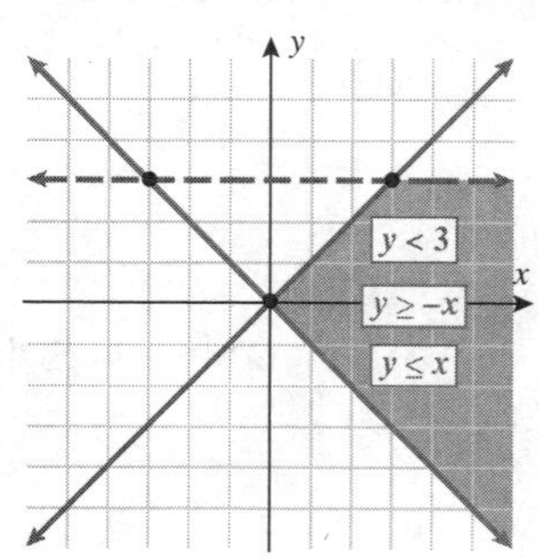

Applications

Systems of linear inequalities arise in many applications, for example, the target zone for heart rate during exercise, normal weight ranges for humans, capacity of a room for an event, and return on investments.

EXAMPLE 6 Cost of a Wedding Reception

A couple has invited 300 guests to their wedding. The fixed costs (such as formal wear, entertainment, flowers, and invitations) are \$7,000, and the variable costs (party favors, chair covers, food, and drinks) range between \$25 and \$50 per person, depending on the menu. Assuming at least 200 and at most 300 people attend, graph the cost of the wedding as a system of inequalities.

Solution:

Let x represent the number of people attending the wedding and y represent the cost of the wedding. There are four linear inequalities:

At least 200 guests attend:	$x \geq 200$
No more than 300 guests attend:	$x \leq 300$
Minimum cost of the wedding:	$y \geq 7000 + 25x$
Maximum cost of the wedding:	$y \leq 7000 + 50x$

Graph the system of inequalities.

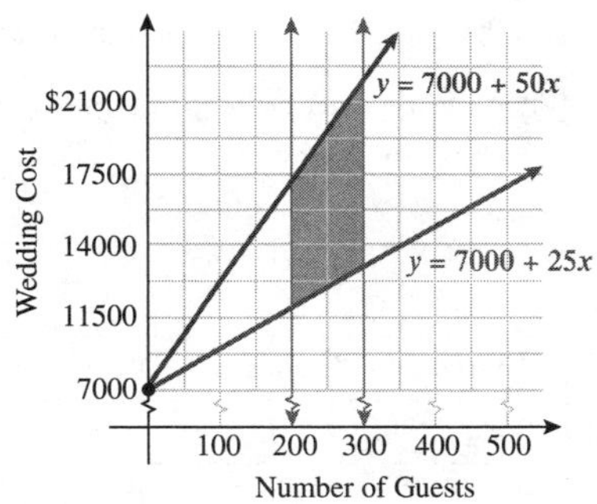

This solution implies that the wedding cost will be approximately between \$12,000 and \$22,000.

In economics, the point where the supply and demand curves intersect is called the **equilibrium point. Consumer surplus** is a measure of the amount that consumers benefit by being able to purchase a product for a price less than the maximum they would be willing to pay. **Producer surplus** is a measure of the amount that producers benefit by selling at a market price that is higher than the least they would be willing to sell for.

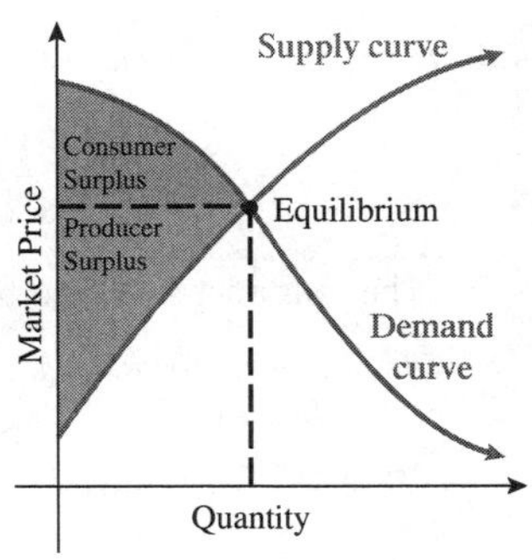

EXAMPLE 7 Consumer Surplus and Producer Surplus

The Tesla Motors Roadster is the first electric car that will be able to travel 220 miles on a single charge. The price of a 2008 model is approximately \$90,000.

Courtesy Tesla Motors

Suppose the supply and demand equations for this electric car are given by

$$P = 90{,}000 - 0.1x \qquad \text{(Demand)}$$
$$P = 10{,}000 + 0.3x \qquad \text{(Supply)}$$

where P is the price in dollars and x is the number of cars produced. Calculate the consumer surplus and the producer surplus for these two equations.

Solution:

Find the equilibrium point.

$$90{,}000 - 0.1x = 10{,}000 + 0.3x$$
$$0.4x = 80{,}000$$
$$x = 200{,}000$$

Let $x = 200{,}000$ in either the supply or demand equation.

$$P = 90{,}000 - 0.1(200{,}000) = 70{,}000$$
$$P = 10{,}000 + 0.3(200{,}000) = 70{,}000$$

According to these models, if the price of a Tesla Motors Roadster is \$70,000, then 200,000 cars will be sold and there will be no surplus.

Write the systems of linear inequalities that correspond to consumer surplus and producer surplus.

CONSUMER SURPLUS	PRODUCER SURPLUS
$P \le 90{,}000 - 0.1x$	$P \ge 10{,}000 + 0.3x$
$P \ge 70{,}000$	$P \le 70{,}000$
$x \ge 0$	$x \ge 0$

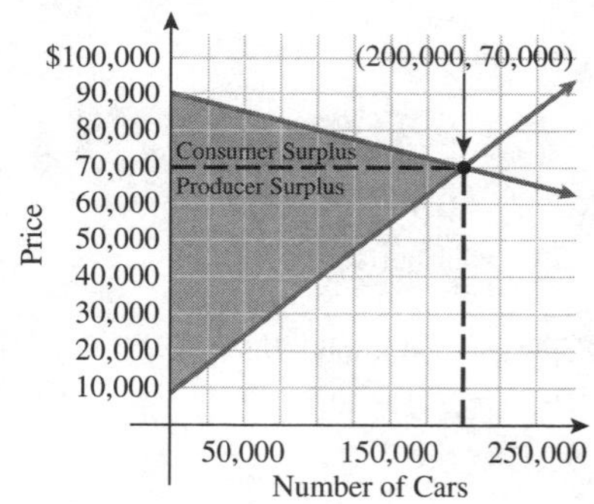

The consumer surplus is the area of the red triangle.

$$A = \frac{1}{2}bh$$
$$= \frac{1}{2}(200{,}000)(20{,}000)$$
$$= 2{,}000{,}000{,}000$$

The consumer surplus is \$2B.

The producer surplus is the area of the blue triangle.

$$A = \frac{1}{2}bh$$
$$= \frac{1}{2}(200{,}000)(60{,}000)$$
$$= 6{,}000{,}000{,}000$$

The producer surplus is \$6B.

The graphs of the systems of linear inequalities in Examples 6 and 7 are said to be **bounded**, whereas the graphs of the systems of linear inequalities in Examples 3, 4, and 5 are said to be **unbounded**. Any points that correspond to boundary lines intersecting are called **corner points** or **vertices**. In Example 7, the vertices corresponding to the consumer surplus are the points (0, 90,000), (0, 70,000), and (200,000, 70,000), and the vertices corresponding to the producer surplus are the points (0, 70,000), (0, 10,000), and (200,000, 70,000).

SECTION 6.4 SUMMARY

Linear Inequality

1. Change the inequality sign to an equal sign.
2. Draw the line. (Dashed for strict inequalities and solid for nonstrict inequalities.)
3. Test a point. (Select a point in one-half plane and test the inequality. Repeat this step for the other half-plane.)
4. Shade the half-plane that satisfies the linear inequality.

System of Linear Inequalities

- Draw the individual linear inequalities.
- The overlapped shaded region is the solution.

SECTION 6.4 EXERCISES

SKILLS

In Exercises 1–4, match the linear inequality with the correct graph.

1. $y > x$
2. $y \geq x$
3. $y < x$
4. $y \leq x$

a.

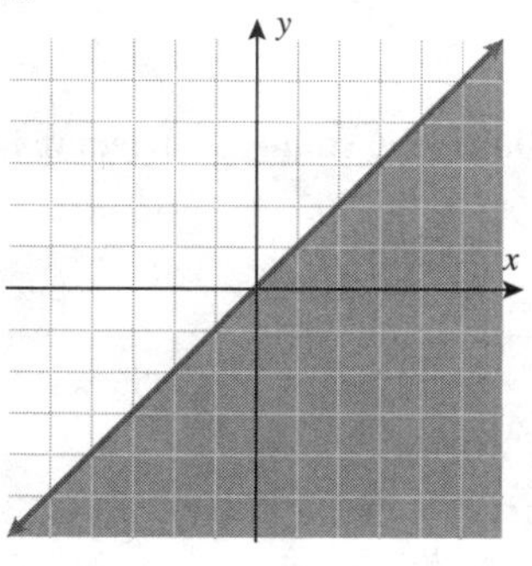

b.

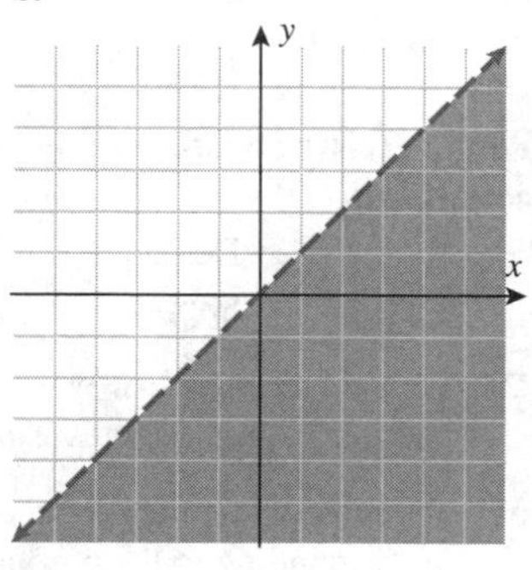

c.

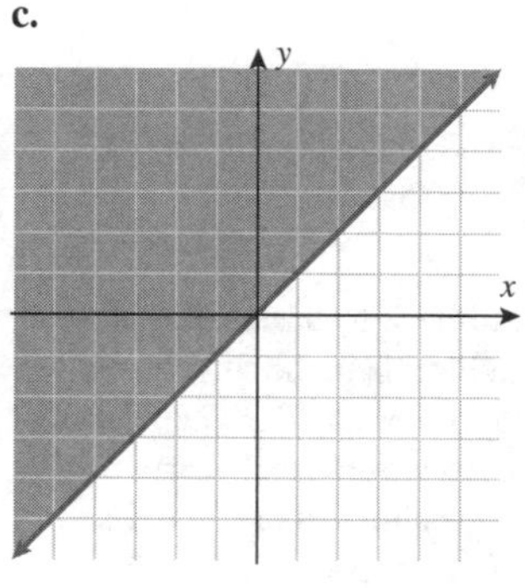

d.

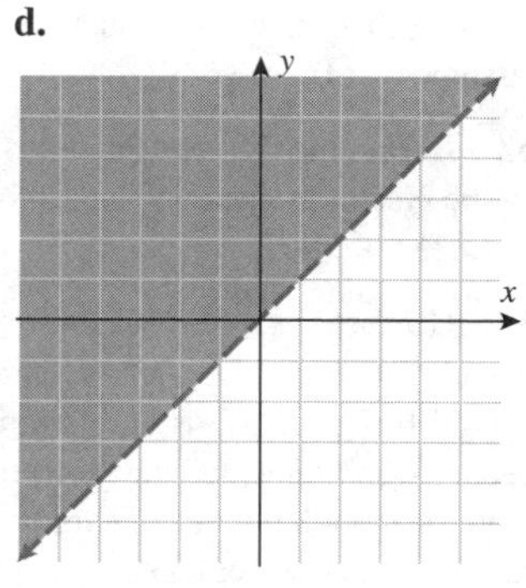

In Exercises 5–20, graph each linear inequality.

5. $y > x - 1$
6. $y \geq -x + 1$
7. $y \leq -x$
8. $y > -x$
9. $y \leq -3x + 2$
10. $y < 2x + 3$
11. $y \leq -2x + 1$
12. $y > 3x - 2$
13. $3x + 4y < 2$
14. $2x + 3y > -6$
15. $5x + 3y < 15$
16. $4x - 5y \leq 20$
17. $4x - 2y \geq 6$
18. $6x - 3y \geq 9$
19. $6x + 4y \leq 12$
20. $5x - 2y \geq 10$

In Exercises 21–48, graph each system of inequalities or indicate that the system has no solution.

21. $y \ge x - 1$
$y \le x + 1$

22. $y > x + 1$
$y < x - 1$

23. $y > 2x + 1$
$y < 2x - 1$

24. $y \le 2x - 1$
$y \ge 2x + 1$

25. $y \ge 2x$
$y \le 2x$

26. $y > 2x$
$y < 2x$

27. $x > -2$
$x < 4$

28. $y < 3$
$y > 0$

29. $x \ge 2$
$y \le x$

30. $y \le 3$
$y \ge x$

31. $y > x$
$x < 0$
$y < 4$

32. $y \le x$
$x \ge 0$
$y \le 1$

33. $x + y > 2$
$y < 1$
$x > 0$

34. $x + y < 4$
$x > 0$
$y \ge 1$

35. $-x + y > 1$
$y < 3$
$x > 0$

36. $x - y > 2$
$y < 4$
$x \ge 0$

37. $x + 3y > 6$
$y < 1$
$x \ge 1$

38. $x + 2y > 4$
$y < 1$
$x \ge 0$

39. $y \ge x - 1$
$y \le -x + 3$
$y < x + 2$

40. $y < 4 - x$
$y > x - 4$
$y > -x - 4$

41. $x + y > -4$
$-x + y < 2$
$y \ge -1$
$y \le 1$

42. $y < x + 2$
$y > x - 2$
$y < -x + 2$
$y > -x - 2$

43. $y < x + 3$
$x + y \ge 1$
$y \ge 1$
$y \le 3$

44. $y \le -x + 2$
$y - x \ge -3$
$y \ge -2$
$y \le 1$

45. $y + x < 2$
$y + x \ge 4$
$y \ge -2$
$y \le 1$

46. $y - x < 3$
$y + x > 3$
$y \le -2$
$y \ge -4$

47. $2x - y < 2$
$2x + y > 2$
$y < 2$

48. $3x - y > 3$
$3x + y < 3$
$y < -2$

APPLICATIONS

49. Area. Find the area enclosed by the system of inequalities.
$y > |x|$
$y < 2$

50. Area. Find the area enclosed by the system of inequalities.
$y < |x|$
$x \ge 0$
$y \ge 0$
$x < 3$

51. Area. Find the area enclosed by the system of linear inequalities (assume $y \ge 0$).
$5x + y \le 10$
$x \ge 0$
$x \le 1$

52. Area. Find the area enclosed by the system of linear inequalities (assume $y \ge 0$).
$-5x + y \le 0$
$x \ge 1$
$x \le 2$

53. Hurricanes. After back-to-back-to-back-to-back hurricanes (Charley, Frances, Ivan, and Jeanne) in Florida in the summer of 2004, FEMA sent disaster relief trucks to Florida. Floridians mainly needed drinking water and generators. Each truck could carry no more than 6000 pounds of cargo or 2400 cubic feet of cargo. Each case of bottled water takes up 1 cubic foot of space and weighs 25 pounds. Each generator takes up 20 cubic feet and weighs 150 pounds. Let x represent the number of cases of water and y represent the number of generators, and write a system of linear inequalities that describes the number of generators and cases of water each truck can haul to Florida.

54. Hurricanes. Repeat Exercise 53 with a smaller truck and different supplies. Suppose the smaller trucks that can haul 2000 pounds and 1500 cubic feet of cargo are used to haul plywood and tarps. A case of plywood is 60 cubic feet and weighs 500 pounds. A case of tarps is 10 cubic feet and weighs 50 pounds. Letting x represent the number of cases of plywood and y represent the number of cases of tarps, write a system of linear inequalities that describes the number of cases of tarps and plywood each truck can haul to Florida. Graph the system of linear inequalities.

For Exercises 55–58, employ the following supply and demand equations:

Demand: $P = 80 - 0.01x$
Supply: $P = 20 + 0.02x$

55. Consumer Surplus. Write a system of linear inequalities corresponding to the consumer surplus.

56. Producer Surplus. Write a system of linear inequalities corresponding to the producer surplus.

57. Consumer Surplus. Calculate the consumer surplus given the supply and demand equations.

58. Producer Surplus. Calculate the producer surplus given the supply and demand equations.

CATCH THE MISTAKE

In Exercises 59 and 60, explain the mistake that is made.

59. Graph the inequality $y \geq 2x + 1$.

Solution:

Graph the line $y = 2x + 1$ with a solid line.

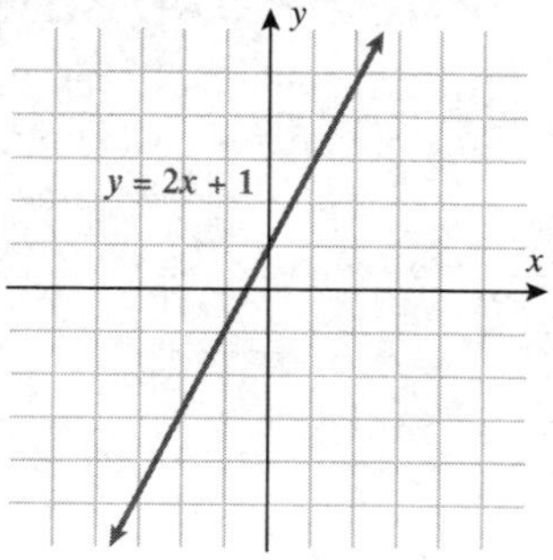

Since the inequality is $\geq$, shade to the *right*.

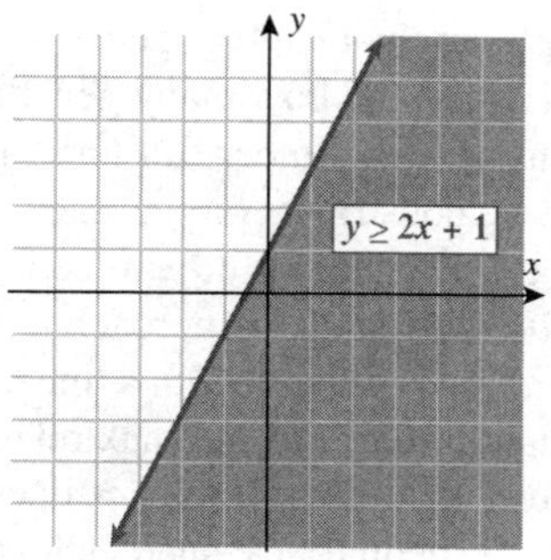

This is incorrect. What mistake was made?

60. Graph the inequality $y < 2x + 1$.

Solution:

Graph the line $y = 2x + 1$ with a solid line.

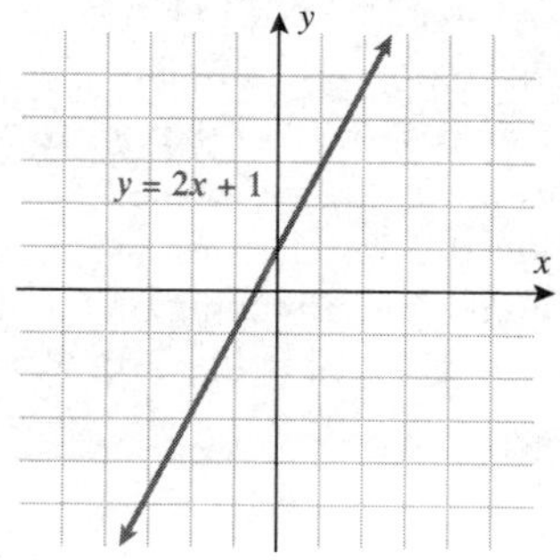

Since the inequality is $<$, shade *below*.

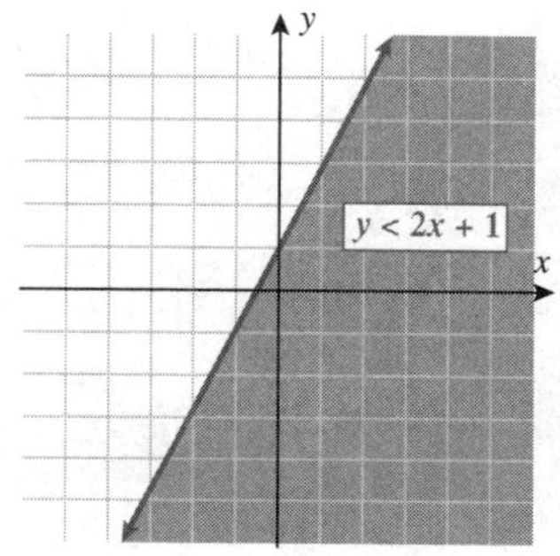

This is incorrect. What mistake was made?

CONCEPTUAL

In Exercises 61–64, determine whether each statement is true or false.

61. A linear inequality always has a solution that is a half-plane.

62. A dashed curve is used for strict inequalities.

63. A solid curve is used for strict inequalities.

64. A system of linear inequalities always has a solution.

CHALLENGE

For the system of linear inequalities (Exercises 65 and 66), assume *a*, *b*, *c*, and *d* are real numbers.

$$\begin{aligned} x &\geq a \\ x &< b \\ y &> c \\ y &\leq d \end{aligned}$$

65. Describe the solution when $a < b$ and $c < d$.

66. What will the solution be if $a > b$ and $c > d$, given the system of linear inequalities?

Use the following system of linear inequalities for Exercises 67 and 68:

$$\begin{aligned} y &\leq ax + b \\ y &\geq -ax + b \end{aligned}$$

67. If a and b are positive real numbers, graph the solution.

68. If a and b are negative real numbers, graph the solution.

TECHNOLOGY

In Exercises 69 and 70, apply a graphing utility to graph the following inequalities.

69. $4x - 2y \geq 6$ (Check with your answer to Exercise 17.)

70. $6x - 3y \geq 9$ (Check with your answer to Exercise 18.)

In Exercises 71 and 72, use a graphing utility to graph each system of inequalities or indicate that the system has no solution.

71. $\begin{aligned} -0.05x + 0.02y &\geq 0.12 \\ 0.01x + 0.08y &\leq 0.08 \end{aligned}$

72. $\begin{aligned} y &\leq 2x + 3 \\ y &> -0.5x + 5 \end{aligned}$

SECTION

6.5 THE LINEAR PROGRAMMING MODEL

SKILLS OBJECTIVES

- Write an objective function that represents a quantity to be minimized or maximized.
- Utilize inequalities to describe constraints.
- Solve the optimization problem, which combines minimizing or maximizing a function subject to constraints, using linear programming.

CONCEPTUAL OBJECTIVES

- Understand that linear programming is a graphical method that solves optimization problems.
- Understand why vertices represent maxima or minima.

Often we seek to maximize or minimize a function subject to constraints. This process is called **optimization**. For example, in the chapter opener about Florida's hurricanes in the summer of 2004, FEMA had to determine how many generators, cases of water, and tarps should be in each truck to maximize the number of Floridians given help, yet at the same time factor in the weight and space constraints on the trucks.

When the function we seek to minimize or maximize is linear and the constraints are given in terms of linear inequalities, a graphing approach to such problems is called **linear programming**. In linear programming, we start with a linear equation, called the **objective function**, that represents the quantity that is to be maximized or minimized, for example, the number of Floridians aided by FEMA. The number of people aided, however, is subject to constraints represented as linear inequalities, such as how much weight each truck can haul and how much space each truck has for cargo.

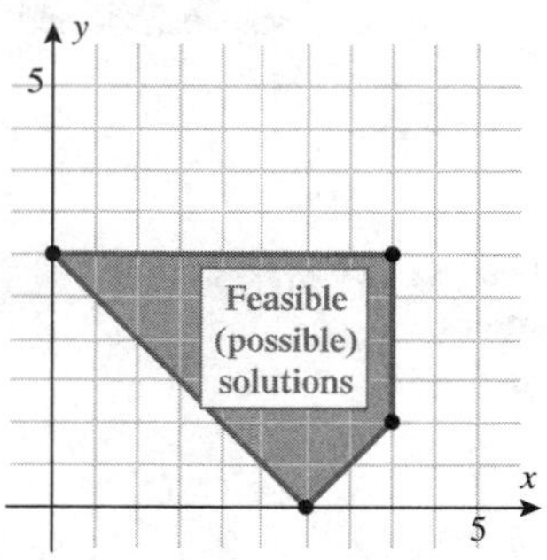

The goal is to minimize or maximize the objective function $z = Ax + By$ subject to *constraints*. In other words, find the points (x, y) that make the value of z the largest (or smallest). The **constraints** are a system of linear inequalities, and the common shaded region represents the **feasible** (possible) **solutions**.

If the constraints form a bounded region, the vertices represent the coordinates (x, y) that correspond to a maximum or minimum value of the objective function $z = Ax + By$. If the region is not bounded, then if an optimal solution exists, it will occur at a vertex. A procedure for solving linear programming problems is outlined below.

SOLVING AN OPTIMIZATION PROBLEM USING LINEAR PROGRAMMING

Step 1: Write the objective function. This expression represents the quantity that is to be minimized or maximized.

Step 2: Write the constraints. This is a system of linear inequalities.

Step 3: Graph the constraints. Graph the system of linear inequalities and shade the common region, which contains the feasible solutions.

Step 4: Identify the vertices. The corner points of the shaded region represent maximum or minimum values of the objective function.

Step 5: Evaluate the objective function for each vertex. For each corner point of the shaded region, substitute the coordinates into the objective function and list the value of the objective function.

Step 6: Identify the optimal solution. The largest (maximum) or smallest (minimum) value of the objective function in Step 5 is the optimal solution.

EXAMPLE 1 Maximizing an Objective Function

Find the maximum value of $z = 2x + y$, subject to the constraints:

$$x \geq 1 \qquad x \leq 4 \qquad x + y \leq 5 \qquad y \geq 0$$

Solution:

STEP 1 Write the objective function. $z = 2x + y$

STEP 2 Write the constraints.

$x \geq 1$
$x \leq 4$
$y \leq -x + 5$
$y \geq 0$

STEP 3 Graph the constraints.

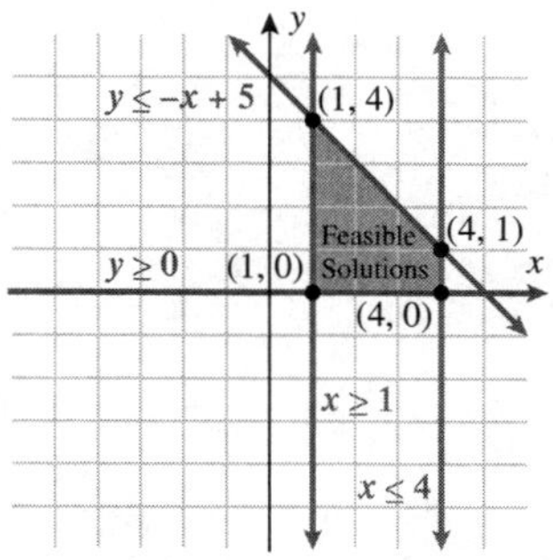

STEP 4 Identify the vertices. (1, 4), (4, 1), (1, 0), (4, 0)

STEP 5 Evaluate the objective function for each vertex.

VERTEX	x	y	OBJECTIVE FUNCTION: $z = 2x + y$
(1, 4)	1	4	$2(1) + 4 = \mathbf{6}$
(4, 1)	4	1	$2(4) + 1 = \mathbf{9}$
(1, 0)	1	0	$2(1) + 0 = \mathbf{2}$
(4, 0)	4	0	$2(4) + 0 = \mathbf{8}$

STEP 6 The maximum value of z is **9**, which occurs when $x = 4$ and $y = 1$.

Study Tip

The bounded region is the region that satisfies *all* of the constraints. Only vertices of the bounded region correspond to possible solutions. Even though $y = -x + 5$ and $y = 0$ intersect at $x = 5$, that point of intersection is outside the shaded region and therefore is *not* one of the vertices.

■ **YOUR TURN** Find the maximum value of $z = x + 3y$ subject to the constraints:

$$x \geq 1 \qquad x \leq 3 \qquad y \leq -x + 3 \qquad y \geq 0$$

■ **Answer:** The maximum value of z is **7**, which occurs when $x = 1$ and $y = 2$.

EXAMPLE 2 Minimizing an Objective Function

Find the minimum value of $z = 4x + 5y$, subject to the constraints:

$$x \geq 0 \qquad 2x + y \leq 6 \qquad x + y \leq 5 \qquad y \geq 0$$

Solution:

STEP 1 Write the objective function. $z = 4x + 5y$

STEP 2 Write the constraints.

$$x \geq 0$$
$$y \leq -2x + 6$$
$$y \leq -x + 5$$
$$y \geq 0$$

STEP 3 Graph the constraints.

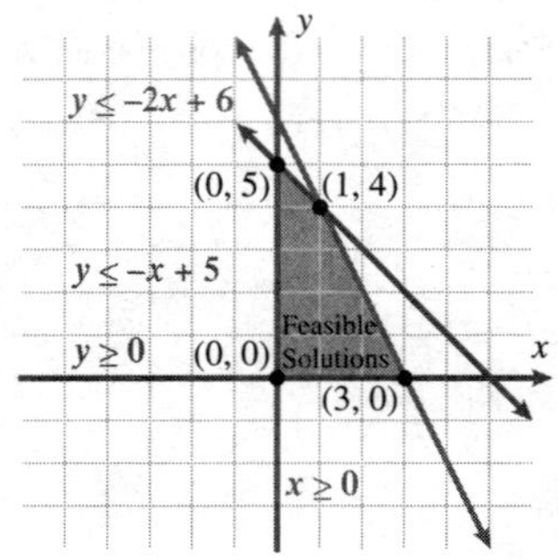

STEP 4 Identify the vertices. (0, 0), (0, 5), (1, 4), (3, 0)

STEP 5 Evaluate the objective function for each vertex.

VERTEX	x	y	OBJECTIVE FUNCTION: $z = 4x + 5y$
(0, 0)	0	0	$4(0) + 5(0) = \mathbf{0}$
(0, 5)	0	5	$4(0) + 5(5) = \mathbf{25}$
(1, 4)	1	4	$4(1) + 5(4) = \mathbf{24}$
(3, 0)	3	0	$4(3) + 5(0) = \mathbf{12}$

STEP 6 The minimum value of z is **0**, which occurs when $x = 0$ and $y = 0$.

Study Tip

Maxima or minima of objective functions occur at the vertices of the shaded region corresponding to the constraints.

■ **YOUR TURN** Find the minimum value of $z = 2x + 3y$ subject to the constraints:

$$x \geq 1 \qquad 2x + y \leq 8 \qquad x + y \geq 4$$

■ **Answer:** The minimum value of z is **8**, which occurs when $x = 4$ and $y = 0$.

EXAMPLE 3 Solving an Optimization Problem Using Linear Programming: Unbounded Region

Find the maximum value and minimum value of $z = 7x + 3y$, subject to the constraints:

$$y \geq 0 \qquad -2x + y \leq 0 \qquad -x + y \geq -4$$

Solution:

STEP 1 Write the objective function. $z = 7x + 3y$

STEP 2 Write the constraints.

$y \geq 0$
$y \leq 2x$
$y \geq x - 4$

STEP 3 Graph the constraints.

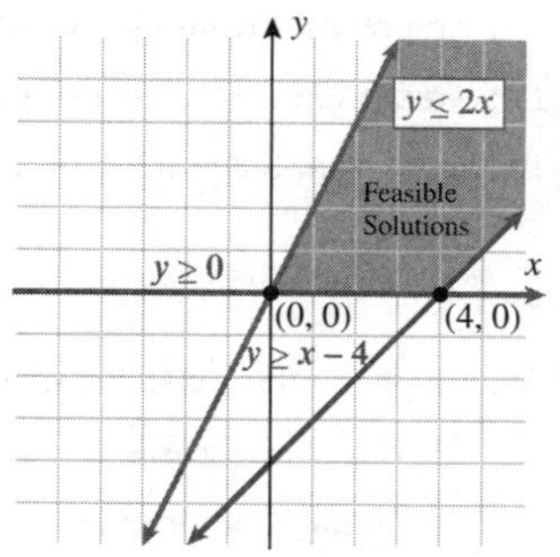

STEP 4 Identify the vertices. (0, 0), (4, 0)

STEP 5 Evaluate the objective function for each vertex.

VERTEX	x	y	OBJECTIVE FUNCTION: $z = 7x + 3y$
(0, 0)	0	0	$7(0) + 3(0) = \mathbf{0}$
(4, 0)	4	0	$7(4) + 3(0) = \mathbf{28}$

STEP 6 The minimum value of z is **0**, which occurs when $x = 0$ and $y = 0$.

There is no maximum value, because if we select a point in the shaded region, say (3, 3), the objective function at (3, 3) is equal to 30, which is greater than 28.

When the feasible solutions are contained in a bounded region, then a maximum and a minimum exist and are each located at one of the vertices. If the feasible solutions are contained in an unbounded region, then if a maximum or minimum exists, it is located at one of the vertices.

EXAMPLE 4 Maximizing the Number of People Aided by a Hurricane Relief Effort

After the hurricanes in Florida in the summer of 2004, FEMA sent disaster relief trucks to the state. Each truck could carry no more than 6000 pounds of cargo. Each case of bottled water weighs 25 pounds and each generator weighs 150 pounds. If each generator helps one household and five cases of water help one household, determine the maximum number of Florida households aided by each truck and how many generators and cases of water should be sent in each truck. Due to the number of trucks and the supply of water and generators nationwide, each truck must contain at least 10 times as many cases of water as generators.

Solution:

Let x represent the number of cases of water.

Let y represent the number of generators.

Let z represent the number of households aided per truck.

STEP 1 Write the objective function. $z = \frac{1}{5}x + y$

STEP 2 Write the constraints.

Number of cases of water is nonnegative. $x \geq 0$

Number of generators is nonnegative. $y \geq 0$

At least 10 times as many cases of water as generators. $x \geq 10y$

Weight capacity of truck: $25x + 150y \leq 6000$

STEP 3 Graph the constraints and determine feasible solutions.

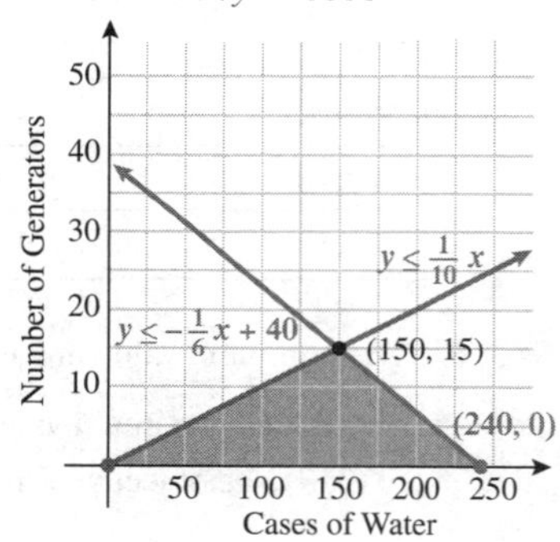

STEP 4 Identify the vertices. (0, 0), (150, 15), (240, 0)

STEP 5 Evaluate the objective function for each vertex.

VERTEX	x	y	OBJECTIVE FUNCTION: $z = \frac{1}{5}x + y$
(0, 0)	0	0	**0**
(150, 15)	150	15	**45**
(240, 0)	240	0	**48**

STEP 6 A maximum of **48 households** are aided by each truck. This maximum occurs when there are **240 cases of water** and **0 generators** on each truck.

SECTION 6.5 SUMMARY

In this section the linear programming model was discussed.

1. Write the objective function.
2. Write the constraints.
3. Graph the constraints.
4. Identify the vertices.
5. Evaluate the objective function for each vertex.
6. Identify the optimal solution.

SECTION 6.5 EXERCISES

SKILLS

In Exercises 1–4, find the value of the objective function at each of the vertices. What is the maximum value of the objective function? What is the minimum value of the objective function?

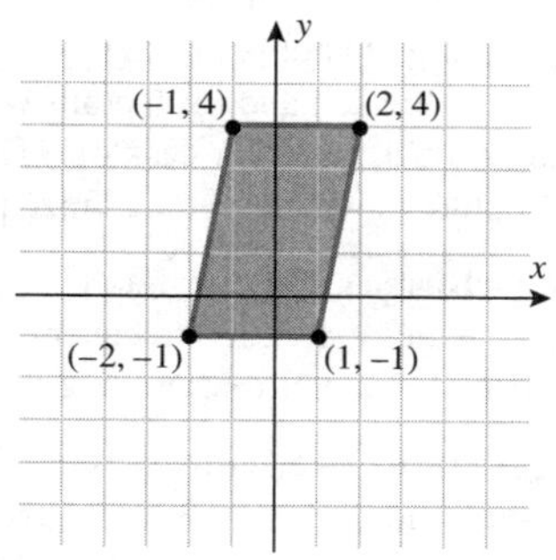

1. Objective function: $z = 2x + 3y$

2. Objective function: $z = 3x + 2y$

3. Objective function: $z = 1.5x + 4.5y$

4. Objective function: $z = \frac{2}{3}x + \frac{3}{5}y$

In Exercises 5–12, minimize or maximize each objective function subject to the constraints.

5. Minimize $z = 7x + 4y$ subject to:

$x \geq 0 \quad y \geq 0 \quad -x + y \leq 4$

6. Maximize $z = 3x + 5y$ subject to:

$x \geq 0 \quad y \geq 0 \quad -x + y \geq 4$

7. Maximize $z = 4x + 3y$ subject to:

$x \geq 0 \quad y \leq -x + 4 \quad y \geq -x$

8. Minimize $z = 4x + 3y$ subject to:

$x \geq 0 \quad y \geq 0$
$x + y \leq 10 \quad x + y \geq 0$

9. Minimize $z = 2.5x + 3.1y$ subject to:

$x \geq 0 \quad y \geq 0 \quad x \leq 4$
$-x + y \leq 2 \quad x + y \leq 6$

10. Maximize $z = 2.5x - 3.1y$ subject to:

$x \geq 1 \quad y \leq 7 \quad x \leq 3$
$-x + y \geq 2 \quad x + y \geq 6$

11. Maximize $z = \frac{1}{4}x + \frac{2}{5}y$ subject to:

$x + y \geq 5 \quad x + y \leq 7$
$-x + y \leq 5 \quad -x + y \geq 3$

12. Minimize $z = \frac{1}{3}x - \frac{2}{5}y$ subject to:

$x + y \geq 6 \quad x + y \leq 8$
$-x + y \leq 6 \quad -x + y \geq 4$

APPLICATIONS

13. Hurricanes. After the 2004 hurricanes in Florida, a student at Valencia Community College decides to create two T-shirts to sell. One T-shirt says, "I survived Charley on Friday the Thirteenth," and the second says, "I survived Charley, Frances, Ivan, and Jeanne." The Charley T-shirt costs him \$7 to make and he sells it for \$13. The other T-shirt costs him \$5 to make and he sells it for \$10. He does not want to invest more than \$1,000. He estimates that the total demand will not exceed 180 T-shirts. Find the number of each type of T-shirt he should make to yield maximum profit.

14. Hurricanes. After Hurricane Charley devastated central Florida unexpectedly, Orlando residents prepared for Hurricane Frances by boarding up windows and filling up their cars with gas. It took 5 hours of standing in line to get plywood, and lines for gas were just as time-consuming. A student at Seminole Community College decides to do a spoof of the "Got Milk" ads and creates two T-shirts: "Got Plywood," with a line of people in a home improvement store, and "Got Gas," with a street lined with cars waiting to pump gasoline. The "Got Plywood" shirts cost \$8 to make, and she sells them for \$13. The "Got Gas" shirts cost \$6 to make, and she sells them for \$10. She decides to limit her costs to \$1,400. She estimates that demand for these T-shirts will not exceed 200 T-shirts. Find the number of each type of T-shirt she should make to yield maximum profit.

15. Computer Business. A computer science major and a business major decide to start a small business that builds and sells a desktop computer and a laptop computer. They buy the parts, assemble them, load the operating system, and sell the computers to other students. The costs for parts, time to assemble the computer, and profit are summarized in the following table:

	DESKTOP	LAPTOP
Cost of Parts	\$700	\$400
Time to Assemble (hours)	5	3
Profit	\$500	\$300

They were able to get a small business loan in the amount of \$10,000 to cover costs. They plan on making these computers over the summer and selling them the first day of class. They can dedicate at most only 90 hours to assembling these computers. They estimate that the demand for laptops will be at least three times as great as the demand for desktops. How many of each type of computer should they make to maximize profit?

16. Computer Business. Repeat Exercise 15 if the two students are able to get a loan for \$30,000 to cover costs and they can dedicate at most 120 hours to assembling the computers.

17. Passenger Ratio. The Eurostar is a high-speed train that travels between London, Brussels, and Paris. There are 30 cars on each departure. Each train car is designated first class or second class. Based on demand for each type of fare, there should always be at least two but no more than four first class train cars. The management wants to claim that the ratio of first class to second class cars never exceeds 1:8. If the profit on each first class train car is twice as much as the profit of each second class train car, find the number of each class of train car that will generate a maximum profit.

18. Passenger Ratio. Repeat Exercise 17. This time, assume that there has to be at least one first class train car and that the profit from each first class train car is 1.2 times as much as the profit from each second class train car. The ratio of first class to second class cannot exceed 1:10.

19. Production. A manufacturer of skis produces two models: a regular ski and a slalom ski. A set of regular skis produces a \$25.00 profit and a set of slalom skis produces a profit of \$50.00. The manufacturer expects a customer demand of at least 200 pairs of regular skis and at least 80 pair of slalom skis. The maximum number of pairs of skis that can be produced by this company is 400. How many of each model of skis should be produced to maximize profits?

20. Donut Inventory. A well-known donut store makes two popular types of donuts: crème-filled and jelly-filled. The manager knows from past statistics that the number of dozens of donuts sold is at least 10, but no more than 30. To prepare the donuts for frying, the baker needs (on the average) 3 minutes for a dozen crème-filled and 2 minutes for jelly-filled. The baker has at most two hours available per day to prepare the donuts. How many dozens of each type should be prepared to maximize the daily profit if there is a \$1.20 profit for each dozen crème-filled and \$1.80 profit for each dozen jelly-filled donuts?

CATCH THE MISTAKE

In Exercises 21 and 22, explain the mistake that is made.

21. Maximize the objective function $z = 2x + y$ subject to the following constraints:

$$x \geq 0 \qquad y \geq 0$$
$$-x + y \geq 0 \qquad x + y \leq 2$$

Solution:

Graph the constraints.

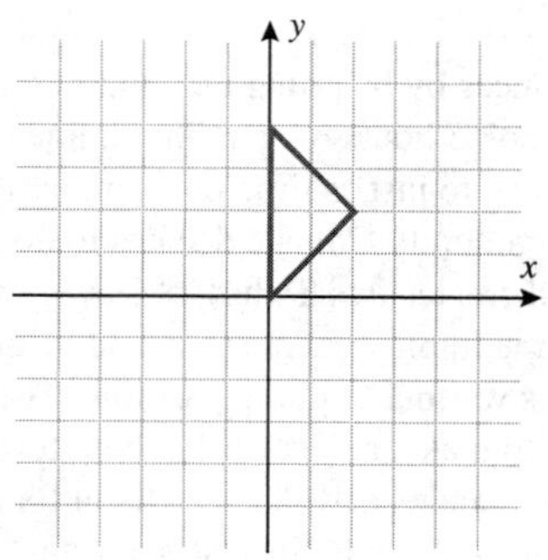

Identify the vertices. (0, 2), (0, 0), (1, 1)

Comparing the y-coordinates of the vertices, the largest y-value is 2.

The maximum value occurs at (0, 2).

The objective function at that point is equal to **2**.

This is incorrect. The maximum value of the objective function is not 2. What mistake was made?

22. Maximize the objective function $z = 2x + y$ subject to the following constraints:

$$x \geq 0 \qquad y \geq 0$$
$$-x + y \leq 0 \qquad x + y \leq 2$$

Solution:

Graph the constraints.

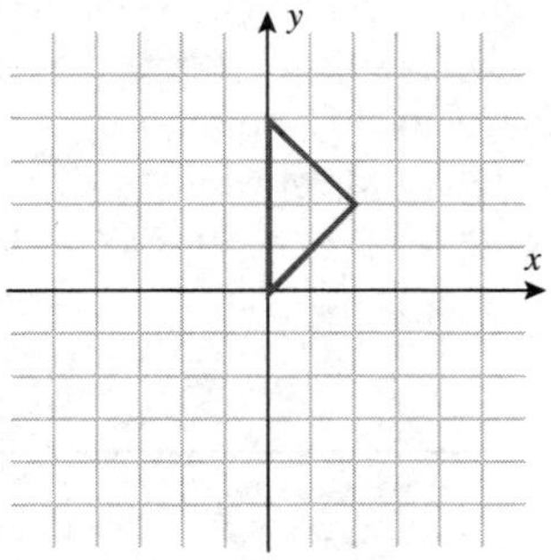

Identify the vertices. (0, 2), (0, 0), (1, 1)

VERTEX	x	y	OBJECTIVE FUNCTION: $z = 2x + y$
(0, 2)	0	2	$2(0) + 2 = \mathbf{2}$
(0, 0)	0	0	$2(0) + 0 = \mathbf{0}$
(1, 1)	1	1	$2(1) + 1 = \mathbf{3}$

The maximum, **3**, is located at the point (1, 1).

This is incorrect. What mistake was made?

CONCEPTUAL

In Exercises 23 and 24, determine whether each statement is true or false.

23. An objective function always has a maximum or minimum.

24. An objective function subject to constraints that correspond to a bounded region always has maximum and a minimum.

CHALLENGE

25. Maximize the objective function $z = 2x + y$ subject to the conditions, where $a > 2$.

$$ax + y \geq -a$$
$$-ax + y \leq a$$
$$ax + y \leq a$$
$$-ax + y \geq -a$$

26. Maximize the objective function $z = x + 2y$ subject to the conditions, where $a > b > 0$.

$$x + y \geq a$$
$$-x + y \leq a$$
$$x + y \leq a + b$$
$$-x + y \geq a - b$$

TECHNOLOGY

In Exercises 27–32, employ a graphing utility to determine the region of feasible solutions, and apply the intersect feature to find the approximate coordinates of the vertices to solve the linear programming problems.

27. Minimize $z = 2.5x + 3.1y$ subject to:

$$x \geq 0 \qquad y \geq 0 \qquad x \leq 4$$
$$-x + y \leq 2 \qquad x + y \leq 6$$

Compare with your answer to Exercise 9.

28. Maximize $z = 2.5x + 3.1y$ subject to:

$$x \geq 0 \qquad y \geq 0 \qquad x \leq 4$$
$$-x + y \leq 2 \qquad x + y \leq 6$$

Compare with your answer to Exercise 10.

29. Maximize $z = 17x + 14y$ subject to:

$$y \geq 4.5 \qquad -x + y \leq 3.7 \qquad x + y \leq 11.2$$

30. Minimize $z = 1.2x + 1.5y$ subject to:

$$2.3x + y \leq 14.7 \qquad -2.3x + y \leq 14.7$$
$$-5.2x + y \leq 3.7 \qquad -2.3x + y \geq 1.5$$

31. Maximize $z = 4.5x + 1.8y$ subject to:

$$-3.1x + y \leq 11.4 \qquad 1.4x + y \geq 1.5 \qquad 5x + y \leq 15$$

32. Minimize $z = 5.4x - 1.6y$ subject to:

$$y \geq 1.8x - 3.6 \qquad y \geq -2.2x + 1.8$$
$$y \leq 1.8x + 4.2 \qquad y \leq -2.2x + 10.8$$

MODELING YOUR WORLD

In 2005, hybrid vehicles were introduced into the U.S. market. The demand for hybrids, which are typically powered by a combination of gasoline and electric batteries, was based on popular recognition of petroleum as an increasingly scarce nonrenewable resource, as well as consumers' need to combat rising prices at the gas pumps. In addition to achieving greater fuel economy than conventional internal combustion engine vehicles (ICEVs), their use also results in reduced emissions.

An online "Gas Mileage Impact Calculator," created by the American Council for an Energy-Efficient Economy (www.aceee.org), was used to generate the following tables comparing a conventional sedan (four-door) and an SUV versus their respective hybrid counterparts.

Gas Mileage Impact Calculator

	Toyota Camry 2.4L 4, Auto $3.75/gallon 15,000 mi/year	Toyota Camry Hybrid 2.4L 4, Auto $3.75/gallon 15,000 mi/year
Gas Consumption	611 gallons	449 gallons
Gas Cost	$2,289.75	$1,681.99
Fuel Economy	25 mpg	33 mpg
Emissions		
Carbon Dioxide (greenhouse gas)	11,601 pounds	8,522 pounds
Carbon Monoxide (poisonous gas)	235 pounds	169 pounds
Nitrogen Oxide (lung irritant and smog)	10 pounds	7 pounds
Particulate Matter (soot)	255 grams	255 grams
Hydrocarbons (smog)	6 pounds	8 pounds

	Toyota Highlander 3.5L 6, Auto Stk $3.75/gallon 15,000 mi/year	Toyota Highlander Hybrid 3.3L 6, Auto AWD $3.75/gallon 15,000 mi/year
Gas Consumption	740 gallons	576 gallons
Gas Cost	$2,773.44	$2,158.33
Fuel Economy	20 mpg	26 mpg
Emissions		
Carbon Dioxide (greenhouse gas)	14,052 pounds	10,936 pounds
Carbon Monoxide (poisonous gas)	229 pounds	187 pounds
Nitrogen Oxide (lung irritant and smog)	11 pounds	8 pounds
Particulate Matter (soot)	320 grams	399 grams
Hydrocarbons (smog)	7 pounds	16 pounds

The MSRP and mileage comparisons for the 2008 models are given below:

	CAMRY	CAMRY HYBRID	HIGHLANDER	HIGHLANDER HYBRID
MSRP	\$19,435	\$26,065	\$28,035	\$34,435
Miles per gallon in city	21	33	18	27
Miles per gallon on highway	31	34	24	25

For the following questions, assume that you drive 15,000 miles per year (all in city) and the price of gasoline is \$3.75 per gallon.

1. Write a linear equation that models the total cost of owning and operating each vehicle *y* as a function of the number of years of ownership *x*.
   a. Camry
   b. Camry Hybrid
   c. Highlander
   d. Highlander Hybrid

2. Write a linear equation that models the total number of pounds of carbon dioxide each vehicle emits *y* as a function of the number of years of ownership *x*.
   a. Camry
   b. Camry Hybrid
   c. Highlander
   d. Highlander Hybrid

3. How many years would you have to own and drive the vehicle for the hybrid to be the better deal?
   a. Camry Hybrid vs. Camry
   b. Highlander Hybrid vs. Highlander

4. How many years would you have to own and drive the vehicle for the hybrid to emit 50% less carbon dioxide than its conventional counterpart?
   a. Camry Hybrid vs. Camry
   b. Highlander Hybrid vs. Highlander

Section	Concept	Pages	Review Exercises	Key Ideas/Points
6.1	**Systems of linear equations in two variables**	492–504	1–22	$A_1x + B_1y = C_1$ $A_2x + B_2y = C_2$
	Substitution method	493–495	1–16	Solve for one variable in terms of the other and substitute that expression into the other equation.
	Elimination method	495–499	1–16	Eliminate a variable by adding multiples of the equations.
	Graphing method	499–501	17–20	Graph the two lines. The solution is the point of intersection. Parallel lines have no solution and identical lines have infinitely many solutions.
	Three methods and three types of solutions	501–504	1–22	One solution, no solution, infinitely many solutions.
6.2	**Systems of linear equations in three variables**	509–516	23–28	Planes in three-dimensional coordinate system.
	Solving systems of linear equations in three variables	509–512	23–28	**Step 1:** Reduce the system to 2 equations and 2 unknowns. **Step 2:** Solve the resulting system from Step 1. **Step 3:** Substitute solutions found in Step 2 into any of the equations to find the third variable. **Step 4:** Check.
	Types of solutions	512–516	23–28	One solution (point), no solution, infinitely many solutions (line).
6.3	**Partial fractions**	521–529	25–32	$\frac{n(x)}{d(x)}$ Factor $d(x)$
	Distinct linear factors	523–524		$\frac{n(x)}{d(x)} = \frac{A}{(ax+b)} + \frac{B}{(cx+d)} + \cdots$
	Repeated linear factors	524–526		$\frac{n(x)}{d(x)} = \frac{A}{(ax+b)} + \frac{B}{(ax+b)^2} + \cdots + \frac{M}{(ax+b)^m}$
	Distinct irreducible quadratic factors	526–527		$\frac{n(x)}{d(x)} = \frac{Ax+B}{ax^2+bx+c}$
	Repeated irreducible quadratic factors	527–529		$\frac{n(x)}{d(x)} = \frac{A_1x+B_1}{ax^2+bx+c} + \frac{A_2x+B_2}{(ax^2+bx+c)^2} + \frac{A_3x+B_3}{(ax^2+bx+c)^3} + \cdots + \frac{A_mx+B_m}{(ax^2+bx+c)^m}$
6.4	**Systems of linear inequalities in two variables**	533–541	45–58	
	Linear inequalities in two variables	533–535		• $\leq$ or $\geq$ use solid lines • $<$ or $>$ use dashed lines
	Systems of linear inequalities in two variables	535–541		Solutions are determined graphically by finding the common shaded regions.
6.5	**The linear programming model**	544–548	59–66	Finding optimal solutions. Minimizing or maximizing a function subject to constraints (linear inequalities).

6.1 Systems of Linear Equations in Two Variables

Solve each system of linear equations.

1. $r - s = 3$
$r + s = 3$

2. $3x + 4y = 2$
$x - y = 6$

3. $-4x + 2y = 3$
$4x - y = 5$

4. $0.25x - 0.5y = 0.6$
$0.5x + 0.25y = 0.8$

5. $x + y = 3$
$x - y = 1$

6. $3x + y = 4$
$2x + y = 1$

7. $4c - 4d = 3$
$c + d = 4$

8. $5r + 2s = 1$
$r - s = -3$

9. $y = -\frac{1}{2}x$
$y = \frac{1}{2}x + 2$

10. $2x + 4y = -2$
$4x - 2y = 3$

11. $1.3x - 2.4y = 1.6$
$0.7x - 1.2y = 1.4$

12. $\frac{1}{4}x - \frac{3}{4}y = 12$
$\frac{1}{2}y + \frac{1}{4}x = \frac{1}{2}$

13. $5x - 3y = 21$
$-2x + 7y = -20$

14. $6x - 2y = -2$
$4x + 3y = 16$

15. $10x - 7y = -24$
$7x + 4y = 1$

16. $\frac{1}{3}x - \frac{2}{9}y = \frac{2}{9}$
$\frac{4}{5}x + \frac{3}{4}y = -\frac{3}{4}$

Match each system of equations with its graph.

17. $2x - 3y = 4$
$x + 4y = 3$

18. $5x - y = 2$
$5x - y = -2$

19. $x + 2y = -6$
$2x + 4y = -12$

20. $5x + 2y = 3$
$4x - 2y = 6$

a.

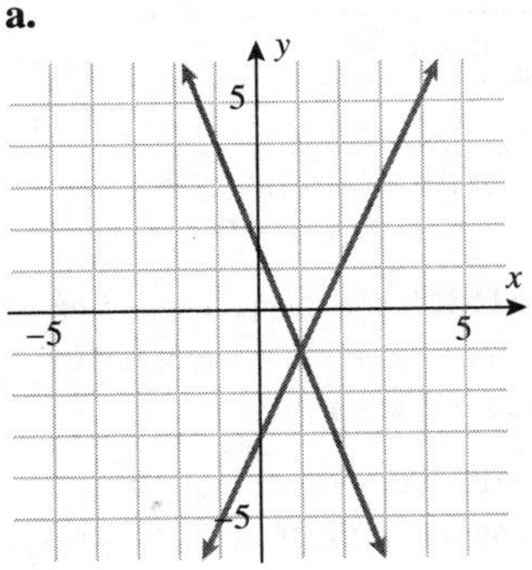

b.

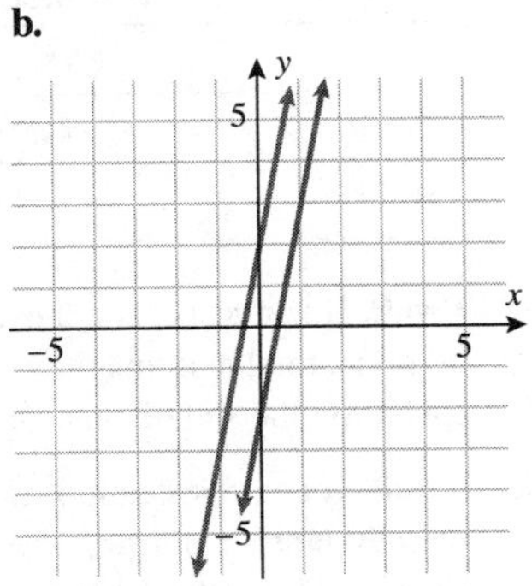

c.

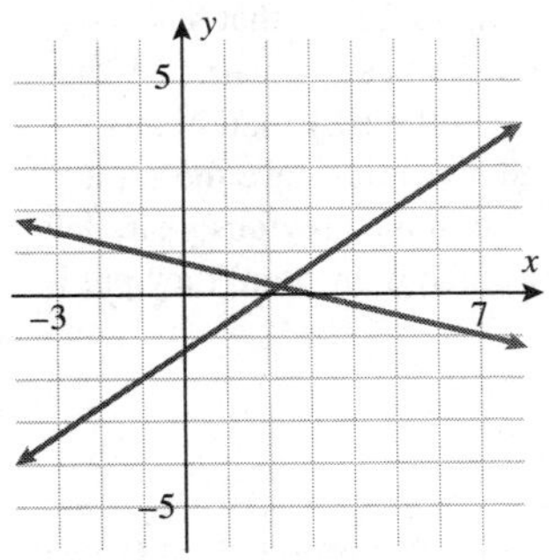

d.

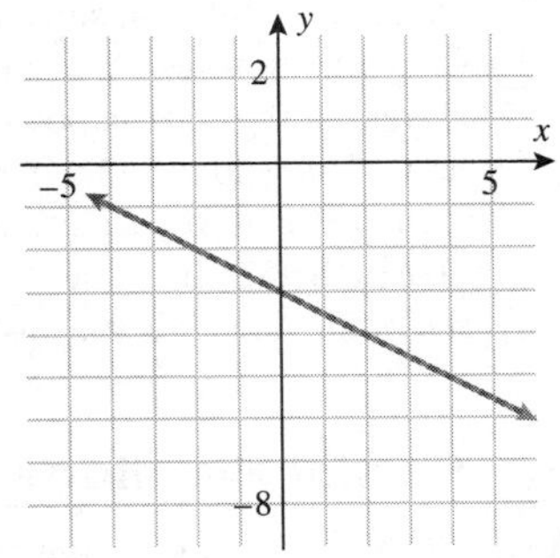

Applications

21. Chemistry. In chemistry lab, Alexandra needs to make a 42 milliliter solution that is 15% NaCl. All that is in the lab is 6% and 18% NaCl. How many milliliters of each solution should she use to obtain the desired mix?

22. Gas Mileage. A Nissan Sentra gets approximately 32 mpg on the highway and 18 mpg in the city. Suppose 265 miles were driven on a full tank (12 gallons) of gasoline. Approximately how many miles were driven in the city and how many on the highway?

6.2 Systems of Linear Equations in Three Variables

Solve each system of linear equations.

23. $x + y + z = 1$
$x - y - z = -3$
$-x + y + z = 3$

24. $x - 2y + z = 3$
$2x - y + z = -4$
$3x - 3y - 5z = 2$

25. $x + y + z = 7$
$x - y - z = 17$
$y + z = 5$

26. $x + z = 3$
$-x + y - z = -1$
$x + y + z = 5$

Applications

27. Fitting a Curve to Data. The average number of flights on a commercial plane that a person takes per year can be modeled by a quadratic function $y = ax^2 + bx + c$, where $a < 0$, and x represents age: $16 \le x \le 65$. The following table gives the average number of flights per year that a person takes on a commercial airline. Determine the quadratic function that models this quantity.

AGE	NUMBER OF FLIGHTS PER YEAR
16	2
40	6
65	4

28. Investment Portfolio. Danny and Paula decide to invest \$20,000 of their savings. They put some in an IRA account earning 4.5% interest, some in a mutual fund that has been averaging 8% a year, and some in a stock that earned 12% last year. If they put \$4,000 more in the IRA than in the mutual fund and the mutual fund and stock have the same growth in the next year as they did in the previous year, they will earn \$1,525 in a year. How much money did they put in each of the three investments?

6.3 Partial Fractions

Write the form of each partial-fraction decomposition. Do not solve for the constants.

29. $\dfrac{4}{(x-1)^2(x+3)(x-5)}$ **30.** $\dfrac{7}{(x-9)(3x+5)^2(x+4)}$

31. $\dfrac{12}{x(4x+5)(2x+1)^2}$ **32.** $\dfrac{2}{(x+1)(x-5)(x-9)^2}$

33. $\dfrac{3}{x^2+x-12}$ **34.** $\dfrac{x^2+3x-2}{x^3+6x^2}$

35. $\dfrac{3x^3+4x^2+56x+62}{(x^2+17)^2}$ **36.** $\dfrac{x^3+7x^2+10}{(x^2+13)^2}$

Find the partial-fraction decomposition for each rational function.

37. $\dfrac{9x+23}{(x-1)(x+7)}$ **38.** $\dfrac{12x+1}{(3x+2)(2x-1)}$

39. $\dfrac{13x^2+90x-25}{2x^3-50x}$ **40.** $\dfrac{5x^2+x+24}{x^3+8x}$

41. $\dfrac{2}{x^2+x}$ **42.** $\dfrac{x}{x(x+3)}$

43. $\dfrac{5x-17}{x^2+4x+4}$ **44.** $\dfrac{x^3}{(x^2+64)^2}$

6.4 Systems of Linear Inequalities in Two Variables

Graph each linear inequality.

45. $y \geq -2x + 3$ **46.** $y < x - 4$

47. $2x + 4y > 5$ **48.** $5x + 2y \leq 4$

49. $y \geq -3x + 2$ **50.** $y < x - 2$

51. $3x + 8y \leq 16$ **52.** $4x - 9y \leq 18$

Graph each system of inequalities or indicate that the system has no solution.

53. $y \geq x + 2$, $y \leq x - 2$

54. $y \geq 3x$, $y \leq 3x$

55. $x \leq -2$, $y > x$

56. $x + 3y \geq 6$, $2x - y \leq 8$

57. $3x - 4y \leq 16$, $5x + 3y > 9$

58. $x + y > -4$, $x - y < 3$, $y \geq -2$, $x \leq 8$

6.5 The Linear Programming Model

Minimize or maximize the objective function subject to the constraints.

59. Minimize $z = 2x + y$ subject to:

$$x \geq 0 \qquad y \geq 0 \qquad x + y \leq 3$$

60. Maximize $z = 2x + 3y$ subject to:

$$x \geq 0 \qquad y \geq 0$$
$$-x + y \leq 0 \qquad x \leq 3$$

61. Maximize $z = 2.5x + 3.2y$ subject to:

$$x \geq 0 \qquad y \geq 0$$
$$x + y \leq 8 \qquad -x + y \geq 0$$

62. Minimize $z = 5x + 11y$ subject to:

$$-x + y \leq 10 \qquad x + y \geq 2 \qquad x \leq 2$$

63. Minimize $z = 3x - 5y$ subject to:

$$2x + y > 6 \qquad 2x - y < 6 \qquad x > 0$$

64. Maximize $z = -2x + 7y$ subject to:

$$3x + y < 7 \qquad x - 2y > 1 \qquad x \geq 0$$

Applications

For Exercises 65 and 66, refer to the following:

An art student decides to hand-paint coasters and sell sets at a flea market. She decides to make two types of coaster sets: an ocean watercolor and black-and-white geometric shapes. The cost, profit, and time it takes her to paint each set are summarized in the table below.

	OCEAN WATERCOLOR	GEOMETRIC SHAPES
Cost	\$4	\$2
Profit	\$15	\$8
Hours	3	2

65. Profit. If her costs cannot exceed \$100 and she can spend only 90 hours total painting the coasters, determine the number of each type she should make to maximize her profit.

66. Profit. If her costs cannot exceed \$300 and she can spend only 90 hours painting, determine the number of each type she should make to maximize her profit.

Technology Exercises

Section 6.1

67. Apply a graphing utility to graph the two equations $0.4x + 0.3y = -0.1$ and $0.5x - 0.2y = 1.6$. Find the solution to this system of linear equations.

68. Apply a graphing utility to graph the two equations $\frac{1}{2}x + \frac{3}{10}y = \frac{1}{5}$ and $-\frac{5}{3}x + \frac{1}{2}y = \frac{4}{3}$. Find the solution to this system of linear equations.

Section 6.2

In Exercises 69 and 70, employ a graphing calculator to solve the system of equations.

69.
$$\begin{aligned} 5x - 3y + 15z &= 21 \\ -2x + 0.8y - 4z &= -8 \\ 2.5x - y + 7.5z &= 12 \end{aligned}$$

70.
$$\begin{aligned} 2x - 1.5y + 3z &= 6.5 \\ 0.5x - 0.375y + 0.75z &= 1.5 \end{aligned}$$

Section 6.3

71. Apply a graphing utility to graph $y_1 = \dfrac{x^2 + 4}{x^4 - x^2}$ and $y_2 = -\dfrac{4}{x^2} + \dfrac{5/2}{x - 1} - \dfrac{5/2}{x + 1}$ in the same viewing rectangle. Is y_2 the partial-fraction decomposition of y_1?

72. Apply a graphing utility to graph $y_1 = \dfrac{x^3 + 6x^2 + 27x + 38}{(x^2 + 8x + 17)(x^2 + 6x + 13)}$ and $y_2 = \dfrac{2x + 1}{x^2 + 8x + 17} - \dfrac{x - 3}{x^2 + 6x + 13}$ in the same viewing rectangle. Is y_2 the partial-fraction decomposition of y_1?

Section 6.4

In Exercises 73 and 74, use a graphing utility to graph each system of inequalities or indicate that the system has no solution.

73.
$$\begin{aligned} 2x + 5y &\geq -15 \\ y &\leq -\tfrac{2}{3}x - 1 \end{aligned}$$

74.
$$\begin{aligned} y &\leq 0.5x \\ y &> -1.5x + 6 \end{aligned}$$

Section 6.5

75. Maximize $z = 6.2x + 1.5y$ subject to:
$$\begin{aligned} 4x - 3y &\leq 5.4 \\ 2x + 4.5y &\leq 6.3 \\ 3x - y &\geq -10.7 \end{aligned}$$

76. Minimize $z = 1.6x - 2.8y$ subject to:
$$\begin{aligned} y &\geq 3.2x - 4.8 \qquad x \geq -2 \\ y &\leq 3.2x + 4.8 \qquad x \leq 4 \end{aligned}$$

Solve each system of linear equations.

1. $\begin{aligned} x - 2y &= 1 \\ -x + 3y &= 2 \end{aligned}$

2. $\begin{aligned} 3x + 5y &= -2 \\ 7x + 11y &= -6 \end{aligned}$

3. $\begin{aligned} x - y &= 2 \\ -2x + 2y &= -4 \end{aligned}$

4. $\begin{aligned} 3x - 2y &= 5 \\ 6x - 4y &= 0 \end{aligned}$

5. $\begin{aligned} x + y + z &= -1 \\ 2x + y + z &= 0 \\ -x + y + 2z &= 0 \end{aligned}$

6. $\begin{aligned} 6x + 9y + z &= 5 \\ 2x - 3y + z &= 3 \\ 10x + 12y + 2z &= 9 \end{aligned}$

7. $\begin{aligned} x - 2y + 3z &= 5 \\ -2x + y + 4z &= 21 \\ 3x - 5y + z &= -14 \end{aligned}$

8. $\begin{aligned} x + z &= 1 \\ x + y &= -1 \\ y + z &= 0 \end{aligned}$

Write each rational expression as a sum of partial fractions.

9. $\dfrac{2x + 5}{x^2 + x}$

10. $\dfrac{3x - 13}{(x - 5)^2}$

11. $\dfrac{7x + 5}{(x + 2)^2}$

12. $\dfrac{1}{2x^2 + 5x - 3}$

13. $\dfrac{5x - 3}{x(x^2 - 9)}$

14. $\dfrac{5x - 3}{x(x^2 + 9)}$

Graph the inequalities.

15. $-2x + y < 6$

16. $4x - y \geq 8$

In Exercises 17–20, graph the system of inequalities.

17. $\begin{aligned} x + y &\leq 4 \\ -x + y &\geq -2 \end{aligned}$

18. $\begin{aligned} x + 3y &\leq 6 \\ 2x - y &\leq 4 \end{aligned}$

19. $\begin{aligned} 5x - 2y &\leq 8 \\ 3x + 4y &< 12 \end{aligned}$

20. $\begin{aligned} -x + y &> 6 \\ x + y &\leq 4 \end{aligned}$

21. Minimize the function $z = 5x + 7y$, subject to the constraints:

$$x \geq 0 \qquad y \geq 0 \qquad x + y \leq 3 \qquad -x + y \geq 1$$

22. Find the maximum value of the objective function $z = 3x + 6y$, given the constraints:

$$x \geq 0 \qquad y \geq 0$$
$$x + y \leq 6 \qquad -x + 2y \leq 4$$

23. **Stock Investment.** On starting a new job, Cameron gets a \$30,000 signing bonus and decides to invest the money. She invests some in a money market account earning 3% and some in two different stocks. The aggressive stock rose 12% last year and the conservative stock rose 6% last year. She invested \$1,000 more in the aggressive stock than in the conservative stock and in 1 year she earned \$1,890 on the investment. How much did she put in the money market, and how much was invested in each stock?

24. **Ranch.** A rancher buys a rectangular parcel of property that is 245,000 square feet. She fences the entire border and then adds three internal fences so there are four equal rectangular pastures. If the fence required 3150 linear feet, what are the dimensions of the entire property?

25. Employ a graphing calculator to solve the system of equations.

$$\begin{aligned} 5x - 3y + 3z &= 1 \\ 2x - y + z &= 4 \\ 4x - 2y + 3z &= 9 \end{aligned}$$

26. Minimize $z = x - 1.2y$ subject to:

$$\begin{aligned} 3.2x + y &\leq 15.9 \\ 0.4x - y &\leq 0.3 \\ x &\geq 1.0 \end{aligned}$$

1. Simplify $\dfrac{6x^2 - 11x + 5}{30x - 25}$.

2. Perform the operation and write in standard form: $(5 - 7i) - (11 - 9i)$.

3. Solve for x and give any excluded values: $\dfrac{5}{x} - \dfrac{5}{x+1} = \dfrac{-5}{x^2 + x}$.

4. Solve for the radius r: $A = \pi r^2$.

5. Solve and express the solution in interval notation: $\dfrac{2t^2}{t-1} \geq 3t$.

6. Calculate the distance and midpoint between the segment joining the points $\left(-\frac{3}{4}, \frac{1}{6}\right)$ and $\left(\frac{1}{5}, -\frac{2}{3}\right)$.

7. Write an equation of a line in slope–intercept form that passes through the two points (5.6, 6.2) and (3.2, 5.0).

8. Use interval notation to express the domain of the function $f(x) = \sqrt{x^2 - 25}$.

9. Find the average rate of change for $f(x) = \dfrac{5}{x}$, from $x = 2$ to $x = 4$.

10. Sketch the graph of the function $y = \dfrac{1}{x-2} + 1$ and identify all transformations.

11. Evaluate $g(f(-1))$ for $f(x) = \sqrt[3]{x - 7}$ and $g(x) = \dfrac{5}{3 - x}$.

12. Find the inverse of the function $f(x) = \dfrac{5x + 2}{x - 3}$.

13. Find the quadratic function that has the vertex (0, 7) and goes through the point (2, −1).

14. Find all of the real zeros and state the multiplicity of the function $f(x) = 4x^5 + x^3$.

15. List the possible rational zeros and test to determine all rational zeros for $P(x) = 2x^4 + 7x^3 - 18x^2 - 13x + 10$.

16. Factor the polynomial $P(x) = 4x^4 - 4x^3 + 13x^2 + 18x + 5$ as a product of linear factors.

17. Graph the exponential function $f(x) = 5^{-x} - 1$; state the y-intercept, the domain, the range, and the horizontal asymptote.

18. How much money should be put in a savings account now that earns 5.2% a year compounded daily if you want to have $50,000 in 16 years?

19. Graph the function $f(x) = \ln(x + 1) - 3$ using transformation techniques.

20. Evaluate $\log_{4.7} 8.9$ using the change-of-base formula. Round the answer to three decimal places.

21. Solve the equation $5(10^{2x}) = 37$ for x. Round the answers to three decimal places.

22. At a coffee shop, one order of 4 tall coffees and 6 donuts is $14.10. A second order of 5 tall coffees and 4 donuts is $13.11. Find the cost of a tall coffee and the cost of a donut.

23. Solve the system of linear equations.

$$\begin{aligned} x + 6y - z &= -3 \\ 2x - 5y + z &= 9 \\ x + 4y + 2z &= 12 \end{aligned}$$

24. Find the partial fraction decomposition for the rational expression $\dfrac{-1}{x^2 - 2x + 1}$.

25. Graph the system of linear inequalities.

$$\begin{aligned} y &> -x \\ y &\geq -3 \\ x &\leq 3 \end{aligned}$$

26. Employ a graphing calculator to solve the system of equations.

$$\begin{aligned} 2x - 9y + 8z &= -21.8 \\ 5x + 6y - 11z &= 41.7 \\ 3x - y + 9z &= -11.7 \end{aligned}$$

27. Apply a graphing utility to graph $y_1 = \dfrac{3x^2 + x + 8}{x^3 + 4x}$ and $y_2 = \dfrac{2}{x} + \dfrac{x+1}{x^2+4}$ in the same viewing rectangle. Is y_2 the partial-fraction decomposition of y_1?

7

Matrices

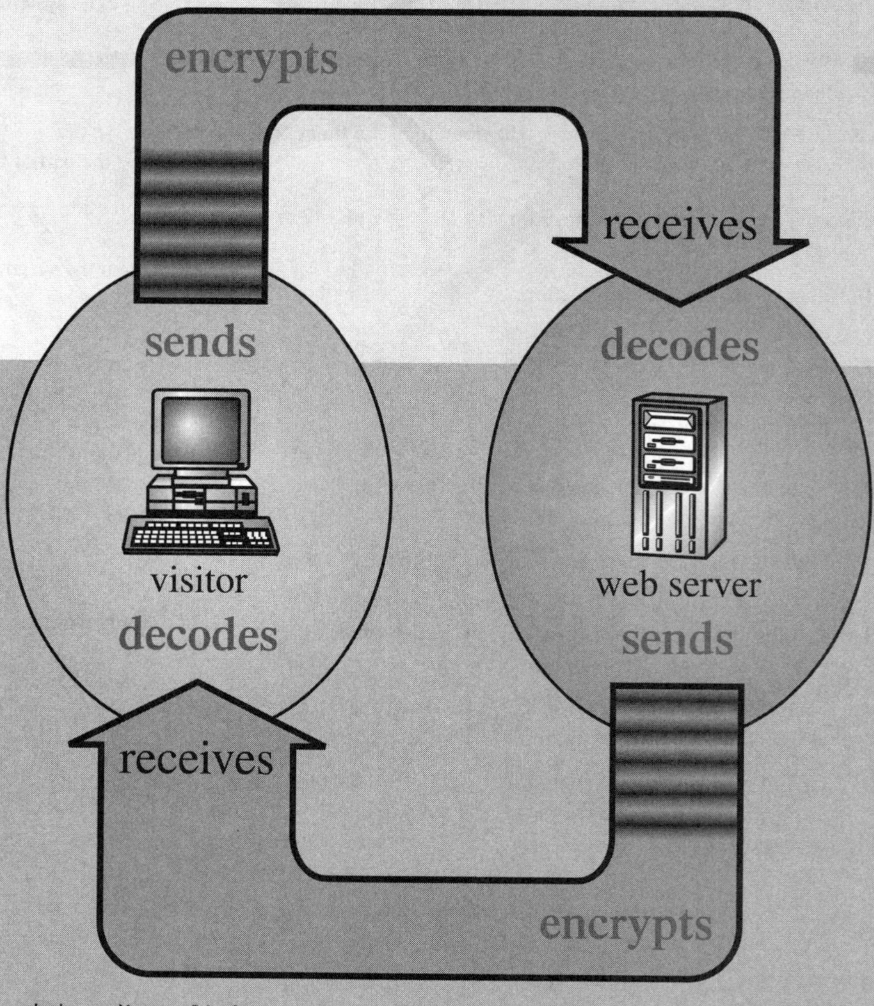

Cryptography is the practice and study of encryption and decryption—encoding data so that it can be decoded only by specific individuals. In other words, it turns a message into gibberish so that only the person who has the deciphering tools can turn that gibberish back into the original message. ATM cards, online shopping sites, and secure military communications all depend on coding and decoding of information. Matrices are used extensively in cryptography. A *matrix* is used as the "key" to encode the data, and then its *inverse matrix* is used as the "key" to decode the data.*

*Section 7.3, Exercises 49–54.

IN THIS CHAPTER we will use matrices to express systems of linear equations. We will discuss three ways that matrices can be used to solve systems of linear equations: augmented matrices (Gaussian elimination and Gauss-Jordan elimination), inverse matrices (matrix algebra), and Cramer's rule (determinants).

MATRICES

AUGMENTED MATRICES

Solve systems of linear equations
- Gaussian elimination with back-substitution
- Gauss–Jordan elimination

MATRIX ALGEBRA

- Equality
- Adding/ subtracting
- Scalar multiplication
- Multiplying matrices

INVERSE OF A SQUARE MATRIX

- Find the inverse of a square matrix
- Solve systems of linear equations: Inverse matrices

DETERMINANT OF A SQUARE MATRIX

- Find the determinant of a square matrix
- Solve systems of linear equations: Cramer's rule

SECTION

CHAPTER OBJECTIVES

- Represent a system of linear equations with an augmented matrix.
- Perform row operations in order to represent augmented matrices in row–echelon form and reduced row–echelon form.
- Apply Gaussian elimination with back-substitution to solve systems of linear equations.
- Apply Gauss–Jordan elimination to solve systems of linear equations.
- Perform matrix operations: addition, subtraction, and multiplication.
- Find the inverse of a square matrix.
- Utilize matrix algebra and inverses of matrices to solve systems of linear equations.
- Find the determinant of a square matrix.
- Use Cramer's rule (determinants) to solve systems of linear equations.

SECTION 7.1 MATRICES AND SYSTEMS OF LINEAR EQUATIONS

SKILLS OBJECTIVES

- Write a system of linear equations as an augmented matrix.
- Perform row operations on an augmented matrix.
- Write a matrix in row–echelon form.
- Solve systems of linear equations using Gaussian elimination with back-substitution.
- Write a matrix in reduced row–echelon form.
- Solve systems of linear equations using Gauss–Jordan elimination.

CONCEPTUAL OBJECTIVES

- Visualize an augmented matrix as a system of linear equations.
- Understand that solving systems with augmented matrices is equivalent to solving by the method of elimination.
- Recognize matrices that correspond to inconsistent and dependent systems.

Some information is best displayed in a table. For example, the number of calories burned per half hour of exercise depends upon the person's weight, as illustrated in the following table. Note that the rows correspond to activities and the columns correspond to weight.

ACTIVITY	127–137 LB	160–170 LB	180–200 LB
Walking/4 mph	156	183	204
Volleyball	267	315	348
Jogging/5 mph	276	345	381

Another example is the driving distance in miles from cities in Arizona (columns) to cities outside the state (rows).

CITY	FLAGSTAFF	PHOENIX	TUCSON	YUMA
Albuquerque, NM	325	465	440	650
Las Vegas, NV	250	300	415	295
Los Angeles, CA	470	375	490	285

If we selected only the numbers in each of the preceding tables and placed brackets around them, the result would be a *matrix.*

$$\text{Calories: } \begin{bmatrix} 156 & 183 & 204 \\ 267 & 315 & 348 \\ 276 & 345 & 381 \end{bmatrix} \qquad \text{Miles: } \begin{bmatrix} 325 & 465 & 440 & 650 \\ 250 & 300 & 415 & 295 \\ 470 & 375 & 490 & 285 \end{bmatrix}$$

A ***matrix*** is a rectangular array of numbers written within brackets.

$$\begin{bmatrix} a_{11} & a_{12} & \cdots & a_{1j} & \cdots & a_{1n} \\ a_{21} & a_{22} & \cdots & a_{2j} & \cdots & a_{2n} \\ \vdots & \vdots & \cdots & \vdots & \cdots & \vdots \\ a_{i1} & a_{i2} & \cdots & a_{ij} & \cdots & a_{in} \\ \vdots & \vdots & \cdots & \vdots & \cdots & \vdots \\ a_{m1} & a_{m2} & \cdots & a_{mj} & \cdots & a_{mn} \end{bmatrix}$$

Each number a_{ij} in the matrix is called an **element** of the matrix. The first subscript i is the **row index**, and the second subscript j is the **column index**. This matrix contains m rows and n columns, and is said to be of **order** $m \times n$.

Study Tip

The order of a matrix is always given as the number of rows by the number of columns.

When the number of rows equals the number of columns (that is, when $m = n$), the matrix is a **square matrix** of order n. In a square matrix, the entries $a_{11}, a_{22}, a_{33}, \ldots, a_{nn}$ are the **main diagonal** entries.

The matrix

$$A_{4\times 3} = \begin{bmatrix} * & * & * \\ * & * & * \\ * & a_{32} & * \\ * & * & * \end{bmatrix}$$

has order (dimensions) 4×3, since there are four rows and three columns. The element a_{32} is in the third row and second column.

EXAMPLE 1 Finding the Order of a Matrix

Determine the order of each matrix given.

a. $\begin{bmatrix} 2 & 1 \\ 3 & 0 \end{bmatrix}$ **b.** $\begin{bmatrix} 1 & -2 & 5 \\ -1 & 3 & 4 \end{bmatrix}$ **c.** $\begin{bmatrix} -2 & 5 & 4 \\ 1 & -\frac{1}{3} & 0 \\ 3 & 8 & 1 \end{bmatrix}$

d. $\begin{bmatrix} 4 & 9 & -\frac{1}{2} & 3 \end{bmatrix}$ **e.** $\begin{bmatrix} 3 & -2 \\ 5 & 1 \\ 0 & -\frac{2}{3} \\ 7 & 6 \end{bmatrix}$

Solution:

a. This matrix has **2** rows and **2** columns, so the order of the matrix is $\boxed{2 \times 2}$.

b. This matrix has **2** rows and **3** columns, so the order of the matrix is $\boxed{2 \times 3}$.

c. This matrix has **3** rows and **3** columns, so the order of the matrix is $\boxed{3 \times 3}$.

d. This matrix has **1** row and **4** columns, so the order of the matrix is $\boxed{1 \times 4}$.

e. This matrix has **4** rows and **2** columns, so the order of the matrix is $\boxed{4 \times 2}$.

A matrix with only one column is called a **column matrix**, and a matrix that has only one row is called a **row matrix**. Notice that in Example 1 the matrices given in parts (a) and (c) are square matrices and the matrix given in part (d) is a row matrix.

You can use matrices as a shorthand way of writing systems of linear equations. There are two ways we can represent systems of linear equations with matrices: as *augmented matrices* or with *matrix equations*. In this section we will discuss *augmented matrices* and solve systems of linear equations using two methods: *Gaussian elimination with back-substitution* and *Gauss–Jordan elimination*.

Augmented Matrices

A particular type of matrix that is used in representing a system of linear equations is an **augmented matrix**. It resembles a matrix with an additional vertical line and column of numbers, hence the name *augmented*. The following table illustrates examples of augmented matrices that represent systems of linear equations.

SYSTEM OF LINEAR EQUATIONS	AUGMENTED MATRIX
$\begin{aligned} 3x + 4y &= 1 \\ x - 2y &= 7 \end{aligned}$	$\left[\begin{array}{cc\|c} 3 & 4 & 1 \\ 1 & -2 & 7 \end{array}\right]$
$\begin{aligned} x - y + z &= 2 \\ 2x + 2y - 3z &= -3 \\ x + y + z &= 6 \end{aligned}$	$\left[\begin{array}{ccc\|c} 1 & -1 & 1 & 2 \\ 2 & 2 & -3 & -3 \\ 1 & 1 & 1 & 6 \end{array}\right]$
$\begin{aligned} x + y + z &= 0 \\ 3x + 2y - z &= 2 \end{aligned}$	$\left[\begin{array}{ccc\|c} 1 & 1 & 1 & 0 \\ 3 & 2 & -1 & 2 \end{array}\right]$

Note the following:

- Each row represents an equation.
- The vertical line represents the equal sign.
- The first column represents the coefficients of the variable x.
- The second column represents the coefficients of the variable y.
- The third column (in the second and third systems) represents the coefficients of the variable z.
- The coefficients of the variables are on the left of the equal sign (vertical line) and the constants are on the right.
- Any variable that does not appear in an equation has an implied coefficient of 0.

EXAMPLE 2 Writing a System of Linear Equations as an Augmented Matrix

Write each system of linear equations as an augmented matrix.

a. $\begin{aligned} 2x - y &= 5 \\ -x + 2y &= 3 \end{aligned}$

b. $\begin{aligned} 3x - 2y + 4z &= 5 \\ y - 3z &= -2 \\ 7x \quad - z &= 1 \end{aligned}$

c. $\begin{aligned} x_1 - x_2 + 2x_3 - 3 &= 0 \\ x_1 + x_2 - 3x_3 + 5 &= 0 \\ x_1 - x_2 + x_3 - 2 &= 0 \end{aligned}$

Solution:

a. $\left[\begin{array}{rr|r} 2 & -1 & 5 \\ -1 & 2 & 3 \end{array}\right]$

b. Note that all missing terms have a 0 coefficient.

$$\begin{aligned} 3x - 2y + 4z &= 5 \\ 0x + y - 3z &= -2 \\ 7x + 0y - z &= 1 \end{aligned}$$

$$\left[\begin{array}{rrr|r} 3 & -2 & 4 & 5 \\ 0 & 1 & -3 & -2 \\ 7 & 0 & -1 & 1 \end{array}\right]$$

c. Write the constants on the right side of the vertical line in the matrix.

$$\begin{aligned} x_1 - x_2 + 2x_3 &= 3 \\ x_1 + x_2 - 3x_3 &= -5 \\ x_1 - x_2 + x_3 &= 2 \end{aligned}$$

$$\left[\begin{array}{rrr|r} 1 & -1 & 2 & 3 \\ 1 & 1 & -3 & -5 \\ 1 & -1 & 1 & 2 \end{array}\right]$$

■ **YOUR TURN** Write each system of linear equations as an augmented matrix.

a. $\begin{aligned} 2x + y - 3 &= 0 \\ x - y &= 5 \end{aligned}$

b. $\begin{aligned} y - x + z &= 7 \\ x - y - z &= 2 \\ z - y &= -1 \end{aligned}$

■ **Answer:**

a. $\left[\begin{array}{rr|r} 2 & 1 & 3 \\ 1 & -1 & 5 \end{array}\right]$

b. $\left[\begin{array}{rrr|r} -1 & 1 & 1 & 7 \\ 1 & -1 & -1 & 2 \\ 0 & -1 & 1 & -1 \end{array}\right]$

Row Operations on a Matrix

Row operations on a matrix are used to solve a system of linear equations when the system is written as an augmented matrix. Recall from the elimination method in Chapter 6 that we could interchange equations, multiply an entire equation by a nonzero constant, and add a multiple of one equation to another equation to produce equivalent systems. Because each row in a matrix represents an equation, the operations that produced equivalent systems of equations that were used in the elimination method will also produce equivalent augmented matrices.

Study Tip

Each missing term in the system of linear equations is represented with a zero in the augmented matrix.

ROW OPERATIONS

The following operations on an augmented matrix will yield an equivalent matrix:

1. Interchange any two rows.
2. Multiply a row by a nonzero constant.
3. Add a multiple of one row to another row.

The following symbols describe these row operations:

1. $R_i \leftrightarrow R_j$ Interchange row i with row j.
2. $cR_i \rightarrow R_i$ Multiply row i by the constant c.
3. $cR_i + R_j \rightarrow R_j$ Multiply row i by the constant c and add to row j, writing the results in row j.

EXAMPLE 3 Applying a Row Operation to an Augmented Matrix

For each matrix, perform the given operation.

a. $\left[\begin{array}{cc|c} 2 & -1 & 3 \\ 0 & 2 & 1 \end{array}\right] \quad R_1 \leftrightarrow R_2$

b. $\left[\begin{array}{ccc|c} -1 & 0 & 1 & -2 \\ 3 & -1 & 2 & 3 \\ 0 & 1 & 3 & 1 \end{array}\right] \quad 2R_3 \rightarrow R_3$

c. $\left[\begin{array}{cccc|c} 1 & 2 & 0 & 2 & 2 \\ 0 & 1 & 2 & 3 & 5 \end{array}\right] \quad R_1 - 2R_2 \rightarrow R_1$

Solution:

a. Interchange the first row with the second row.

$$\left[\begin{array}{cc|c} 2 & -1 & 3 \\ 0 & 2 & 1 \end{array}\right] \quad R_1 \leftrightarrow R_2 \quad \boxed{\left[\begin{array}{cc|c} 0 & 2 & 1 \\ 2 & -1 & 3 \end{array}\right]}$$

b. Multiply the third row by 2.

$$\left[\begin{array}{ccc|c} -1 & 0 & 1 & -2 \\ 3 & -1 & 2 & 3 \\ 0 & 1 & 3 & 1 \end{array}\right] \quad 2R_3 \rightarrow R_3 \quad \boxed{\left[\begin{array}{ccc|c} -1 & 0 & 1 & -2 \\ 3 & -1 & 2 & 3 \\ 0 & 2 & 6 & 2 \end{array}\right]}$$

c. From row 1 subtract 2 times row 2, and write the answer in row 1. Note that finding row 1 minus 2 times row 2 is the same as adding row 1 to the product of -2 with row 2.

$$R_1 - 2R_2 \rightarrow R_1 \quad \left[\begin{array}{cccc|c} 1-2(0) & 2-2(1) & 0-2(2) & 2-2(3) & 2-2(5) \\ 0 & 1 & 2 & 3 & 5 \end{array}\right]$$

$$\boxed{\left[\begin{array}{cccc|c} 1 & 0 & -4 & -4 & -8 \\ 0 & 1 & 2 & 3 & 5 \end{array}\right]}$$

■ **YOUR TURN** Perform the operation $R_1 + 2R_3 \rightarrow R_1$ on the matrix.

$$\left[\begin{array}{ccc|c} 1 & 0 & -2 & -3 \\ 0 & 1 & 2 & 3 \\ 0 & 0 & 1 & 2 \end{array}\right]$$

■ **Answer:**

$$\left[\begin{array}{ccc|c} 1 & 0 & 0 & 1 \\ 0 & 1 & 2 & 3 \\ 0 & 0 & 1 & 2 \end{array}\right]$$

Row–Echelon Form of a Matrix

We can solve systems of linear equations using augmented matrices with two procedures: *Gaussian elimination with back-substitution*, which uses row operations to transform a matrix into *row–echelon form,* and *Gauss–Jordan elimination*, which uses row operations to transform a matrix into *reduced row–echelon form*.

Row–Echelon Form

A matrix is in **row–echelon** form if it has all three of the following properties:

1. Any rows consisting entirely of 0s are at the bottom of the matrix.
2. For each row that does not consist entirely of 0s, the first (leftmost) nonzero entry is 1 (called the leading 1).
3. For two successive nonzero rows, the leading 1 in the higher row is farther to the left than the leading 1 in the lower row.

Reduced Row–Echelon Form

If a matrix in row–echelon form has the following additional property, then the matrix is in **reduced row–echelon form**:

4. Every column containing a leading 1 has zeros in every position above and below the leading 1.

EXAMPLE 4 Determining Whether a Matrix Is in Row–Echelon Form

Determine whether each matrix is in row–echelon form. If it is in row–echelon form, determine whether it is in reduced row–echelon form.

a. $\left[\begin{array}{ccc|c} 1 & 3 & 2 & 3 \\ 0 & 1 & 4 & 2 \\ 0 & 0 & 1 & -1 \end{array}\right]$ **b.** $\left[\begin{array}{ccc|c} 1 & 3 & 2 & 3 \\ 0 & 1 & 1 & 3 \\ 0 & 0 & 0 & 0 \end{array}\right]$ **c.** $\left[\begin{array}{ccc|c} 1 & 0 & 3 & 2 \\ 0 & 1 & -1 & 5 \end{array}\right]$

d. $\left[\begin{array}{cc|c} 1 & 0 & 1 \\ 0 & 3 & 1 \end{array}\right]$ **e.** $\left[\begin{array}{ccc|c} 1 & 0 & 0 & 3 \\ 0 & 1 & 0 & 5 \\ 0 & 0 & 1 & 7 \end{array}\right]$ **f.** $\left[\begin{array}{ccc|c} 1 & 3 & 2 & 3 \\ 0 & 0 & 1 & 2 \\ 0 & 1 & 0 & -3 \end{array}\right]$

Solution:

The matrices in (a), (b), (c), and (e) are in row–echelon form. The matrix in (d) is not in row–echelon form, by condition 2; the leading nonzero entry is not a 1 in each row. If the "3" were a "1," the matrix would be in row–echelon form. The matrix in (f) is not in row–echelon form, because of condition 3; the leading 1 in row 2 is not to the left of the leading 1 in row 3. The matrices in (c) and (e) are in reduced row–echelon form, because in the columns containing the leading 1s there are zeros in every position (above and below the leading 1).

Gaussian Elimination with Back-Substitution

Gaussian elimination with back-substitution is a method that uses row operations to transform an augmented matrix into row–echelon form and then uses back-substitution to find the solution to the system of linear equations.

GAUSSIAN ELIMINATION WITH BACK-SUBSTITUTION

Step 1: Write the system of linear equations as an augmented matrix.
Step 2: Use row operations to rewrite the augmented matrix in row–echelon form.
Step 3: Write the system of linear equations that corresponds to the matrix in row–echelon form found in Step 2.
Step 4: Use the system of linear equations found in Step 3 together with back-substitution to find the solution of the system.

Study Tip

For row–echelon form, get 1s along the main diagonal and 0s below these 1s.

The order in which we perform row operations is important. You should move from left to right. Here is an example of Step 2 in the procedure.

$$\left[\begin{array}{ccc|c} 1 & * & * & * \\ * & * & * & * \\ * & * & * & * \end{array}\right] \rightarrow \left[\begin{array}{ccc|c} 1 & * & * & * \\ 0 & * & * & * \\ 0 & * & * & * \end{array}\right] \rightarrow \left[\begin{array}{ccc|c} 1 & * & * & * \\ 0 & 1 & * & * \\ 0 & * & * & * \end{array}\right] \rightarrow \left[\begin{array}{ccc|c} 1 & * & * & * \\ 0 & 1 & * & * \\ 0 & 0 & * & * \end{array}\right] \rightarrow \left[\begin{array}{ccc|c} 1 & * & * & * \\ 0 & 1 & * & * \\ 0 & 0 & 1 & * \end{array}\right]$$

Matrices are not typically used for systems of linear equations in two variables because the methods from Chapter 6 (substitution and elimination) are more efficient. Example 5 illustrates this procedure with a simple system of linear equations in two variables.

EXAMPLE 5 Using Gaussian Elimination with Back-Substitution to Solve a System of Linear Equations in Two Variables

Apply Gaussian elimination with back-substitution to solve the system of linear equations.

$$\begin{aligned} 2x + y &= -8 \\ x + 3y &= 6 \end{aligned}$$

Solution:

STEP 1 Write the system of linear equations as an augmented matrix. $\left[\begin{array}{cc|c} 2 & 1 & -8 \\ 1 & 3 & 6 \end{array}\right]$

STEP 2 Use row operations to rewrite the matrix in row–echelon form.

Get a 1 in the top left. Interchange rows 1 and 2.

$$\left[\begin{array}{cc|c} 2 & 1 & -8 \\ 1 & 3 & 6 \end{array}\right] \quad R_1 \leftrightarrow R_2 \quad \left[\begin{array}{cc|c} 1 & 3 & 6 \\ 2 & 1 & -8 \end{array}\right]$$

Get a 0 below the leading 1 in row 1.

$$\left[\begin{array}{cc|c} 1 & 3 & 6 \\ 2 & 1 & -8 \end{array}\right] \quad R_2 - 2R_1 \to R_2 \quad \left[\begin{array}{cc|c} 1 & 3 & 6 \\ 0 & -5 & -20 \end{array}\right]$$

Get a leading 1 in row 2. Make the "-5" a "1" by dividing by -5.
Dividing by -5 is the same as multiplying by its reciprocal $-\frac{1}{5}$.

$$\left[\begin{array}{cc|c} 1 & 3 & 6 \\ 0 & -5 & -20 \end{array}\right] \quad -\tfrac{1}{5}R_2 \to R_2 \quad \left[\begin{array}{cc|c} 1 & 3 & 6 \\ 0 & 1 & 4 \end{array}\right]$$

The resulting matrix is in row–echelon form.

STEP 3 Write the system of linear equations corresponding to the row–echelon form of the matrix resulting in Step 2.

$$\left[\begin{array}{cc|c} 1 & 3 & 6 \\ 0 & 1 & 4 \end{array}\right] \quad \to \quad \begin{aligned} x + 3y &= 6 \\ y &= 4 \end{aligned}$$

STEP 4 Use back-substitution to find the solution to the system.

Let $y = 4$ in the first equation $x + 3y = 6$. $\quad x + 3(4) = 6$

Solve for x. $\quad x = -6$

The solution to the system of linear equations is $\boxed{x = -6, y = 4}$.

EXAMPLE 6 Using Gaussian Elimination with Back-Substitution to Solve a System of Linear Equations in Three Variables

Use Gaussian elimination with back-substitution to solve the system of linear equations.

$$\begin{aligned} 2x + y + 8z &= -1 \\ x - y + z &= -2 \\ 3x - 2y - 2z &= 2 \end{aligned}$$

Solution:

STEP 1 Write the system of linear equations as an augmented matrix. $\left[\begin{array}{ccc|c} 2 & 1 & 8 & -1 \\ 1 & -1 & 1 & -2 \\ 3 & -2 & -2 & 2 \end{array}\right]$

STEP 2 Use row operations to rewrite the matrix in row–echelon form.

Get a 1 in the top left. Interchange rows 1 and 2.

$$R_1 \leftrightarrow R_2 \quad \left[\begin{array}{ccc|c} 1 & -1 & 1 & -2 \\ 2 & 1 & 8 & -1 \\ 3 & -2 & -2 & 2 \end{array}\right]$$

Get 0s below the leading 1 in row 1.

$$R_2 - 2R_1 \rightarrow R_2 \quad \left[\begin{array}{ccc|c} 1 & -1 & 1 & -2 \\ 0 & 3 & 6 & 3 \\ 3 & -2 & -2 & 2 \end{array}\right]$$

$$R_3 - 3R_1 \rightarrow R_3 \quad \left[\begin{array}{ccc|c} 1 & -1 & 1 & -2 \\ 0 & 3 & 6 & 3 \\ 0 & 1 & -5 & 8 \end{array}\right]$$

Get a leading 1 in row 2. Make the "3" a "1" by dividing by 3.

$$\tfrac{1}{3}R_2 \rightarrow R_2 \quad \left[\begin{array}{ccc|c} 1 & -1 & 1 & -2 \\ 0 & 1 & 2 & 1 \\ 0 & 1 & -5 & 8 \end{array}\right]$$

Get a zero below the leading 1 in row 2.

$$R_3 - R_2 \rightarrow R_3 \quad \left[\begin{array}{ccc|c} 1 & -1 & 1 & -2 \\ 0 & 1 & 2 & 1 \\ 0 & 0 & -7 & 7 \end{array}\right]$$

Get a leading 1 in row 3. Make the "−7" a "1" by dividing by −7.

$$-\tfrac{1}{7}R_3 \rightarrow R_3 \quad \left[\begin{array}{ccc|c} 1 & -1 & 1 & -2 \\ 0 & 1 & 2 & 1 \\ 0 & 0 & 1 & -1 \end{array}\right]$$

STEP 3 Write the system of linear equations corresponding to the row–echelon form of the matrix resulting in Step 2.

$$\begin{aligned} x - y + z &= -2 \\ y + 2z &= 1 \\ z &= -1 \end{aligned}$$

STEP 4 Use back-substitution to find the solution to the system.

Let $\boxed{z = -1}$ in the second equation $y + 2z = 1$. $\quad y + 2(-1) = 1$

Solve for y. $\quad \boxed{y = 3}$

Let $y = 3$ and $z = -1$ in the first equation $x - y + z = -2$. $\quad x - (3) + (-1) = -2$

Solve for x. $\quad \boxed{x = 2}$

The solution to the system of linear equations is $\boxed{x = 2, y = 3, \text{ and } z = -1}$.

YOUR TURN Use Gaussian elimination with back-substitution to solve the system of linear equations.

$$\begin{aligned} x + y - z &= 0 \\ 2x + y + z &= 1 \\ 2x - y + 3z &= -1 \end{aligned}$$

Answer: $x = -1$, $y = 2$, and $z = 1$

Gauss–Jordan Elimination

In Gaussian elimination with back-substitution, we used row operations to rewrite the matrix in an equivalent row–echelon form. If we continue using row operations until the matrix is in *reduced* row–echelon form, this eliminates the need for back-substitution, and we call this process *Gauss–Jordan elimination*.

GAUSS–JORDAN ELIMINATION

Step 1: Write the system of linear equations as an augmented matrix.

Step 2: Use row operations to rewrite the augmented matrix in *reduced* row–echelon form.

Step 3: Write the system of linear equations that corresponds to the matrix in reduced row–echelon form found in Step 2. The result is the solution to the system.

Study Tip

For reduced row–echelon form, get 1s along the main diagonal and 0s above and below these 1s.

The order in which we perform row operations is important. You should move from left to right. Think of this process as climbing *down* a set of stairs first and then back up the stairs second. On the way *down* the stairs always use operations with rows *above* where you currently are, and on the way back *up* the stairs always use rows *below* where you currently are.

Down the stairs:

$$\left[\begin{array}{ccc|c} 1 & * & * & * \\ * & * & * & * \\ * & * & * & * \end{array}\right] \rightarrow \left[\begin{array}{ccc|c} 1 & * & * & * \\ 0 & * & * & * \\ 0 & * & * & * \end{array}\right] \rightarrow \left[\begin{array}{ccc|c} 1 & * & * & * \\ 0 & 1 & * & * \\ 0 & * & * & * \end{array}\right] \rightarrow \left[\begin{array}{ccc|c} 1 & * & * & * \\ 0 & 1 & * & * \\ 0 & 0 & * & * \end{array}\right] \rightarrow \left[\begin{array}{ccc|c} 1 & * & * & * \\ 0 & 1 & * & * \\ 0 & 0 & 1 & * \end{array}\right]$$

Up the stairs:

$$\left[\begin{array}{ccc|c} 1 & * & * & * \\ 0 & 1 & * & * \\ 0 & 0 & 1 & * \end{array}\right] \rightarrow \left[\begin{array}{ccc|c} 1 & * & * & * \\ 0 & 1 & 0 & * \\ 0 & 0 & 1 & * \end{array}\right] \rightarrow \left[\begin{array}{ccc|c} 1 & * & 0 & * \\ 0 & 1 & 0 & * \\ 0 & 0 & 1 & * \end{array}\right] \rightarrow \left[\begin{array}{ccc|c} 1 & 0 & 0 & * \\ 0 & 1 & 0 & * \\ 0 & 0 & 1 & * \end{array}\right]$$

EXAMPLE 7 Using Gauss–Jordan Elimination to Solve a System of Linear Equations in Three Variables

Apply Gauss–Jordan elimination to solve the system of linear equations.

$$\begin{aligned} x - y + 2z &= -1 \\ 3x + 2y - 6z &= 1 \\ 2x + 3y + 4z &= 8 \end{aligned}$$

Solution:

STEP 1 Write the system as an augmented matrix.

$$\left[\begin{array}{ccc|c} 1 & -1 & 2 & -1 \\ 3 & 2 & -6 & 1 \\ 2 & 3 & 4 & 8 \end{array}\right]$$

STEP 2 Utilize row operations to rewrite the matrix in reduced row–echelon form.

There is already a 1 in the first row/first column.

Get 0s below the leading 1 in row 1.

$$\begin{array}{r} R_2 - 3R_1 \rightarrow R_2 \\ R_3 - 2R_1 \rightarrow R_3 \end{array} \quad \left[\begin{array}{rrr|r} 1 & -1 & 2 & -1 \\ 0 & 5 & -12 & 4 \\ 0 & 5 & 0 & 10 \end{array}\right]$$

Get a 1 in row 2/column 2.

$$R_2 \leftrightarrow R_3 \quad \left[\begin{array}{rrr|r} 1 & -1 & 2 & -1 \\ 0 & 5 & 0 & 10 \\ 0 & 5 & -12 & 4 \end{array}\right]$$

$$\tfrac{1}{5}R_2 \rightarrow R_2 \quad \left[\begin{array}{rrr|r} 1 & -1 & 2 & -1 \\ 0 & 1 & 0 & 2 \\ 0 & 5 & -12 & 4 \end{array}\right]$$

Get a 0 in row 3/column 2.

$$R_3 - 5R_2 \rightarrow R_3 \quad \left[\begin{array}{rrr|r} 1 & -1 & 2 & -1 \\ 0 & 1 & 0 & 2 \\ 0 & 0 & -12 & -6 \end{array}\right]$$

Get a 1 in row 3/column 3.

$$-\tfrac{1}{12}R_3 \rightarrow R_3 \quad \left[\begin{array}{rrr|r} 1 & -1 & 2 & -1 \\ 0 & 1 & 0 & 2 \\ 0 & 0 & 1 & \frac{1}{2} \end{array}\right]$$

Now, go back up the stairs.

Get 0s above the 1 in row 3/column 3.

$$R_1 - 2R_3 \rightarrow R_1 \quad \left[\begin{array}{rrr|r} 1 & -1 & 0 & -2 \\ 0 & 1 & 0 & 2 \\ 0 & 0 & 1 & \frac{1}{2} \end{array}\right]$$

Get a 0 in row 1/column 2.

$$R_1 + R_2 \rightarrow R_1 \quad \left[\begin{array}{rrr|r} 1 & 0 & 0 & 0 \\ 0 & 1 & 0 & 2 \\ 0 & 0 & 1 & \frac{1}{2} \end{array}\right]$$

STEP 3 Identify the solution.

$$\boxed{x = 0,\ y = 2,\ z = \tfrac{1}{2}}$$

YOUR TURN Use an augmented matrix and Gaussian elimination to solve the system of equations.

$$\begin{aligned} x + y - z &= -2 \\ 3x + y - z &= -4 \\ 2x - 2y + 3z &= 3 \end{aligned}$$

Answer: $x = -1$, $y = 2$, and $z = 3$

EXAMPLE 8 Solving a System of Linear Equations in Four Variables

Solve the system of equations with Gauss–Jordan elimination.

$$\begin{aligned} x_1 + x_2 - x_3 + 3x_4 &= 3 \\ 3x_2 - 2x_4 &= 4 \\ 2x_1 - 3x_3 &= -1 \\ 4x_4 + 2x_1 &= -6 \end{aligned}$$

Study Tip

Careful attention should be paid to order of terms, and zeros should be used for missing terms.

Solution:

STEP 1 Write the system as an augmented matrix.

$$\left[\begin{array}{cccc|c} 1 & 1 & -1 & 3 & 3 \\ 0 & 3 & 0 & -2 & 4 \\ 2 & 0 & -3 & 0 & -1 \\ 2 & 0 & 0 & 4 & -6 \end{array}\right]$$

STEP 2 Use row operations to rewrite the matrix in reduced row–echelon form.

There is already a 1 in the first row/first column.

Get 0s below the 1 in row 1/column 1.

$$\begin{array}{r} \\ \\ R_3 - 2R_1 \to R_3 \\ R_4 - 2R_1 \to R_4 \end{array} \left[\begin{array}{cccc|c} 1 & 1 & -1 & 3 & 3 \\ 0 & 3 & 0 & -2 & 4 \\ 0 & -2 & -1 & -6 & -7 \\ 0 & -2 & 2 & -2 & -12 \end{array}\right]$$

Get a 1 in row 2/column 2.

$$R_2 \leftrightarrow R_4 \quad \left[\begin{array}{cccc|c} 1 & 1 & -1 & 3 & 3 \\ 0 & -2 & 2 & -2 & -12 \\ 0 & -2 & -1 & -6 & -7 \\ 0 & 3 & 0 & -2 & 4 \end{array}\right]$$

$$-\tfrac{1}{2}R_2 \leftrightarrow R_2 \quad \left[\begin{array}{cccc|c} 1 & 1 & -1 & 3 & 3 \\ 0 & 1 & -1 & 1 & 6 \\ 0 & -2 & -1 & -6 & -7 \\ 0 & 3 & 0 & -2 & 4 \end{array}\right]$$

Get 0s below the 1 in row 2/column 2.

$$\begin{array}{r} \\ R_3 + 2R_2 \to R_3 \\ R_4 - 3R_2 \to R_4 \\ \\ \end{array} \left[\begin{array}{cccc|c} 1 & 1 & -1 & 3 & 3 \\ 0 & 1 & -1 & 1 & 6 \\ 0 & 0 & -3 & -4 & 5 \\ 0 & 0 & 3 & -5 & -14 \end{array}\right]$$

Get a 1 in row 3/column 3.

$$-\tfrac{1}{3}R_3 \to R_3 \quad \left[\begin{array}{cccc|c} 1 & 1 & -1 & 3 & 3 \\ 0 & 1 & -1 & 1 & 6 \\ 0 & 0 & 1 & \frac{4}{3} & -\frac{5}{3} \\ 0 & 0 & 3 & -5 & -14 \end{array}\right]$$

Get a 0 in row 4/column 3.

$$R_4 - 3R_3 \to R_4 \quad \left[\begin{array}{cccc|c} 1 & 1 & -1 & 3 & 3 \\ 0 & 1 & -1 & 1 & 6 \\ 0 & 0 & 1 & \frac{4}{3} & -\frac{5}{3} \\ 0 & 0 & 0 & -9 & -9 \end{array}\right]$$

Get a 1 in row 4/column 4.

$$-\tfrac{1}{9}R_4 \to R_4 \quad \left[\begin{array}{cccc|c} 1 & 1 & -1 & 3 & 3 \\ 0 & 1 & -1 & 1 & 6 \\ 0 & 0 & 1 & \frac{4}{3} & -\frac{5}{3} \\ 0 & 0 & 0 & 1 & 1 \end{array}\right]$$

Now go back up the stairs.

Get 0s above the 1 in row 4/column 4.

$$\begin{array}{r} R_3 - \tfrac{4}{3}R_4 \to R_3 \\ R_2 - R_4 \to R_2 \\ R_1 - 3R_4 \to R_1 \end{array} \left[\begin{array}{cccc|c} 1 & 1 & -1 & 0 & 0 \\ 0 & 1 & -1 & 0 & 5 \\ 0 & 0 & 1 & 0 & -3 \\ 0 & 0 & 0 & 1 & 1 \end{array}\right]$$

Get 0s above the 1 in row 3/column 3.

$$\begin{array}{l} R_2 + R_3 \rightarrow R_2 \\ R_1 + R_3 \rightarrow R_1 \end{array} \quad \left[\begin{array}{cccc|c} 1 & 1 & 0 & 0 & -3 \\ 0 & 1 & 0 & 0 & 2 \\ 0 & 0 & 1 & 0 & -3 \\ 0 & 0 & 0 & 1 & 1 \end{array}\right]$$

Get a 0 in row 1/column 2.

$$R_1 - R_2 \rightarrow R_1 \quad \left[\begin{array}{cccc|c} 1 & 0 & 0 & 0 & -5 \\ 0 & 1 & 0 & 0 & 2 \\ 0 & 0 & 1 & 0 & -3 \\ 0 & 0 & 0 & 1 & 1 \end{array}\right]$$

STEP 3 Identify the solution.

$$\boxed{x_1 = -5, x_2 = 2, x_3 = -3, x_4 = 1}$$

Inconsistent and Dependent Systems

Recall that systems of linear equations can be independent, inconsistent, or dependent systems and therefore have *one solution*, *no solution*, or *infinitely many solutions*. All of the systems we have solved so far in this section have been independent systems (unique solution). When solving a system of linear equations using Gaussian elimination or Gauss–Jordan elimination, the following will indicate the three possible types of solutions.

SYSTEM	TYPE OF SOLUTION	MATRIX DURING GAUSS–JORDAN ELIMINATION	EXAMPLE
Independent	One (unique) solution	Diagonal entries are all 1s and 0s elsewhere.	$\left[\begin{array}{ccc\|c} 1 & 0 & 0 & 1 \\ 0 & 1 & 0 & -3 \\ 0 & 0 & 1 & 2 \end{array}\right]$ or $\begin{aligned} x &= 1 \\ y &= -3 \\ z &= 2 \end{aligned}$
Inconsistent	No solution	One row will have only zero entries for coefficients and a nonzero entry for the constant.	$\left[\begin{array}{ccc\|c} 1 & 0 & 0 & 1 \\ 0 & 1 & 0 & -3 \\ 0 & 0 & 0 & 2 \end{array}\right]$ or $\begin{aligned} x &= 1 \\ y &= -3 \\ 0 &= 2 \end{aligned}$
Dependent	Infinitely many solutions	One row will be entirely 0s when the number of equations equals the number of variables.	$\left[\begin{array}{ccc\|c} 1 & 0 & -2 & 1 \\ 0 & 1 & 1 & -3 \\ 0 & 0 & 0 & 0 \end{array}\right]$ or $\begin{aligned} x - 2z &= 1 \\ y + z &= -3 \\ 0 &= 0 \end{aligned}$

EXAMPLE 9 Solving an Inconsistent System: No Solution

Solve the system of equations.

$$\begin{aligned} x + 2y - z &= 3 \\ 2x + y + 2z &= -1 \\ -2x - 4y + 2z &= 5 \end{aligned}$$

Solution:

STEP 1 Write the system of equations as an augmented matrix.

$$\left[\begin{array}{ccc|c} 1 & 2 & -1 & 3 \\ 2 & 1 & 2 & -1 \\ -2 & -4 & 2 & 5 \end{array}\right]$$

STEP 2 Apply row operations to rewrite the matrix in row–echelon form.

Get 0s below the 1 in column 1. $\begin{matrix} R_2 - 2R_1 \to R_2 \\ R_3 + 2R_1 \to R_3 \end{matrix} \left[\begin{array}{ccc|c} 1 & 2 & -1 & 3 \\ 0 & -3 & 4 & -7 \\ 0 & 0 & 0 & 11 \end{array}\right]$

There is no need to continue because row 3 is a contradiction. $0x + 0y + 0z = 11$ or $0 = 11$

Since this is inconsistent, there is *no solution* to this system of equations.

EXAMPLE 10 Solving a Dependent System: Infinitely Many Solutions

Solve the system of equations.

$$\begin{aligned} x \quad\;\; + z &= 3 \\ 2x + y + 4z &= 8 \\ 3x + y + 5z &= 11 \end{aligned}$$

Solution:

STEP 1 Write the system of equations as an augmented matrix. $\left[\begin{array}{ccc|c} 1 & 0 & 1 & 3 \\ 2 & 1 & 4 & 8 \\ 3 & 1 & 5 & 11 \end{array}\right]$

STEP 2 Use row operations to rewrite the matrix in reduced row–echelon form.

Get the 0s below the 1 in column 1. $\begin{matrix} R_2 - 2R_1 \to R_2 \\ R_3 - 3R_1 \to R_3 \end{matrix} \left[\begin{array}{ccc|c} 1 & 0 & 1 & 3 \\ 0 & 1 & 2 & 2 \\ 0 & 1 & 2 & 2 \end{array}\right]$

Get a 0 in row 3/column 2. $R_3 - R_2 \to R_3 \left[\begin{array}{ccc|c} 1 & 0 & 1 & 3 \\ 0 & 1 & 2 & 2 \\ 0 & 0 & 0 & 0 \end{array}\right]$

This matrix is in reduced row–echelon form. This matrix corresponds to a dependent system of linear equations and has infinitely many solutions.

STEP 3 Write the augmented matrix as a system of linear equations. $\begin{aligned} x + \;\; z &= 3 \\ y + 2z &= 2 \end{aligned}$

Let $z = a$, where a is any real number, and substitute this into the two equations.

We find that $x = 3 - a$ and $y = 2 - 2a$. The answer is written as $\boxed{x = 3 - a,\ y = 2 - 2a,\ z = a}$ for a any real number.

Study Tip

In a system with three variables, say x, y, and z, we typically let $z = a$ (where a is called a parameter) and then solve for x and y in terms of a.

A common mistake that is made is to identify a unique solution as no solution when one of the variables is equal to zero. For example, what is the difference between the following two matrices?

$$\left[\begin{array}{ccc|c} 1 & 0 & 2 & 1 \\ 0 & 1 & 3 & 2 \\ 0 & 0 & 3 & 0 \end{array}\right] \quad \text{and} \quad \left[\begin{array}{ccc|c} 1 & 0 & 2 & 1 \\ 0 & 1 & 3 & 2 \\ 0 & 0 & 0 & 3 \end{array}\right]$$

The first matrix has a *unique solution,* whereas the second matrix has *no solution*. The third row of the first matrix corresponds to the equation $3z = 0$, which implies that $z = 0$. The third row of the second matrix corresponds to the equation $0x + 0y + 0z = 3$ or $0 = 3$, which is inconsistent, and therefore the system has no solution.

EXAMPLE 11 **Solving a Dependent System: Infinitely Many Solutions**

Solve the system of linear equations.

$$\begin{aligned} 2x + y + z &= 8 \\ x + y - z &= -3 \end{aligned}$$

Solution:

STEP 1 Write the system of equations as an augmented matrix.
$$\left[\begin{array}{ccc|c} 2 & 1 & 1 & 8 \\ 1 & 1 & -1 & -3 \end{array}\right]$$

STEP 2 Use row operations to rewrite the matrix in reduced row–echelon form.

Get a 1 in row 1/column 1. $R_1 \leftrightarrow R_2$
$$\left[\begin{array}{ccc|c} 1 & 1 & -1 & -3 \\ 2 & 1 & 1 & 8 \end{array}\right]$$

Get a 0 in row 2/column 1. $R_2 - 2R_1 \rightarrow R_2$
$$\left[\begin{array}{ccc|c} 1 & 1 & -1 & -3 \\ 0 & -1 & 3 & 14 \end{array}\right]$$

Get a 1 in row 2/column 1. $-R_2 \rightarrow R_2$
$$\left[\begin{array}{ccc|c} 1 & 1 & -1 & -3 \\ 0 & 1 & -3 & -14 \end{array}\right]$$

Get a 0 in row 1/column 2. $R_1 - R_2 \rightarrow R_1$
$$\left[\begin{array}{ccc|c} 1 & 0 & 2 & 11 \\ 0 & 1 & -3 & -14 \end{array}\right]$$

This matrix is in reduced row–echelon form.

STEP 3 Identify the solution.
$$\begin{aligned} x + 2z &= 11 \\ y - 3z &= -14 \end{aligned}$$

Let $z = a$, where a is any real number. Substituting $z = a$ into these two equations gives the infinitely many solutions $\boxed{x = 11 - 2a,\ y = 3a - 14,\ z = a}$.

■ YOUR TURN Solve the system of equations using an augmented matrix.

$$\begin{aligned} x + y + z &= 0 \\ 3x + 2y - z &= 2 \end{aligned}$$

■ **Answer:** $x = 3a + 2$, $y = -4a - 2$, $z = a$, where a is any real number.

Applications

Remember Jared who lost all that weight eating at Subway and is still keeping it off 10 years later? He ate Subway sandwiches for lunch and dinner for one year and lost 235 pounds! The following table gives nutritional information for Subway's 6-inch sandwiches advertised with 6 grams of fat or less.

IVAN CHAVEZ/©AP/ Wide World Photos

SANDWICH	CALORIES	FAT (g)	CARBOHYDRATES (g)	PROTEIN (g)
Veggie Delight	350	18	17	36
Oven-roasted chicken breast	430	19	46	20
Ham (Black Forest without cheese)	290	5	45	19

EXAMPLE 12 Subway Diet

Suppose you are going to eat only Subway 6-inch sandwiches for a week (seven days) for both lunch and dinner (total of 14 meals). If your goal is to eat 388 grams of protein and 4900 calories in those 14 sandwiches, how many of each sandwich should you eat that week?

Solution:

STEP 1 Determine the system of linear equations.

Let three variables represent the number of each type of sandwich you eat in a week.

x = number of Veggie Delight sandwiches
y = number of chicken breast sandwiches
z = number of ham sandwiches

The total number of sandwiches eaten is 14. $x + y + z = 14$

The total number of calories consumed is 4900. $350x + 430y + 290z = 4900$

The total number of grams of protein consumed is 388. $36x + 20y + 19z = 388$

Write an augmented matrix representing this system of linear equations.

$$\left[\begin{array}{ccc|c} 1 & 1 & 1 & 14 \\ 350 & 430 & 290 & 4900 \\ 36 & 20 & 19 & 388 \end{array}\right]$$

STEP 2 Utilize row operations to rewrite the matrix in reduced row–echelon form.

$$\begin{array}{r} R_2 - 350R_1 \to R_2 \\ R_3 - 36R_1 \to R_3 \end{array} \quad \left[\begin{array}{ccc|c} 1 & 1 & 1 & 14 \\ 0 & 80 & -60 & 0 \\ 0 & -16 & -17 & -116 \end{array}\right]$$

$$\tfrac{1}{80}R_2 \to R_2 \quad \left[\begin{array}{ccc|c} 1 & 1 & 1 & 14 \\ 0 & 1 & -\tfrac{3}{4} & 0 \\ 0 & -16 & -17 & -116 \end{array}\right]$$

$$R_3 + 16R_2 \to R_3 \quad \left[\begin{array}{ccc|c} 1 & 1 & 1 & 14 \\ 0 & 1 & -\tfrac{3}{4} & 0 \\ 0 & 0 & -29 & -116 \end{array}\right]$$

$$-\tfrac{1}{29}R_3 \to R_3 \quad \left[\begin{array}{ccc|c} 1 & 1 & 1 & 14 \\ 0 & 1 & -\tfrac{3}{4} & 0 \\ 0 & 0 & 1 & 4 \end{array}\right]$$

$$\begin{array}{r} R_2 + \tfrac{3}{4}R_3 \to R_2 \\ R_1 - R_3 \to R_1 \end{array} \quad \left[\begin{array}{ccc|c} 1 & 1 & 0 & 10 \\ 0 & 1 & 0 & 3 \\ 0 & 0 & 1 & 4 \end{array}\right]$$

$$R_1 - R_2 \to R_1 \quad \left[\begin{array}{ccc|c} 1 & 0 & 0 & 7 \\ 0 & 1 & 0 & 3 \\ 0 & 0 & 1 & 4 \end{array}\right]$$

STEP 3 Identify the solution. $\boxed{x = 7,\ y = 3,\ z = 4}$

You should eat [7 Veggie Delights, 3 oven-roasted chicken breast, and 4 ham sandwiches].

Many times in the real world we see a relationship that looks like a particular function such as a quadratic function and we know particular data points, but we do not know the function. We start with the general function, fit the curve to particular data points, and solve a system of linear equations to determine the specific function parameters.

EXAMPLE 13 Fitting a Curve to Data

The amount of money awarded in medical malpractice suits is rising. This can be modeled with a quadratic function $y = at^2 + bt + c$, where $t > 0$ and $a > 0$. Determine a quadratic function that passes through the three points shown on the graph.

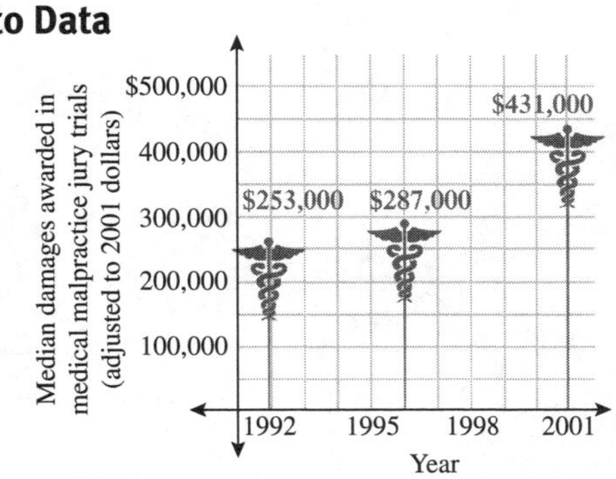

Solution:

Let 1991 correspond to $t = 0$ and y represent the number of dollars awarded for malpractice suits. The following data are reflected in the illustration above:

YEAR	t	y (THOUSANDS OF DOLLARS)	(t, y)
1992	1	253	(1, 253)
1996	5	287	(5, 287)
2001	10	431	(10, 431)

Substitute the three points, (1, 253), (5, 287), and (10, 431) into the general quadratic equation: $y = at^2 + bt + c$.

POINT	$y = at^2 + bt + c$	SYSTEM OF EQUATIONS
(1, 253)	$253 = a(1)^2 + b(1) + c$	$a + b + c = 253$
(5, 287)	$287 = a(5)^2 + b(5) + c$	$25a + 5b + c = 287$
(10, 431)	$431 = a(10)^2 + b(10) + c$	$100a + 10b + c = 431$

STEP 1 Write this system of linear equations as an augmented matrix.

$$\left[\begin{array}{ccc|c} 1 & 1 & 1 & 253 \\ 25 & 5 & 1 & 287 \\ 100 & 10 & 1 & 431 \end{array}\right]$$

STEP 2 Apply row operations to rewrite the matrix in reduced row–echelon form.

$$R_2 - 25R_1 \to R_2 \quad \left[\begin{array}{ccc|c} 1 & 1 & 1 & 253 \\ 0 & -20 & -24 & -6{,}038 \\ 100 & 10 & 1 & 431 \end{array}\right]$$

$$R_3 - 100R_1 \to R_3 \quad \left[\begin{array}{ccc|c} 1 & 1 & 1 & 253 \\ 0 & -20 & -24 & -6{,}038 \\ 0 & -90 & -99 & -24{,}869 \end{array}\right]$$

$$-\tfrac{1}{20}R_2 \to R_2 \quad \left[\begin{array}{ccc|c} 1 & 1 & 1 & 253 \\ 0 & 1 & \frac{6}{5} & \frac{3{,}019}{10} \\ 0 & -90 & -99 & -24{,}869 \end{array}\right]$$

Technology Tip

Enter the matrix into the TI. Press [2nd] [MATRIX]. Use [▶] [EDIT] [ENTER] [ENTER].

Now enter the size by typing the number of rows first and the number of columns second. Enter the elements of the matrix one row at a time by pressing the number and [ENTER] each time.

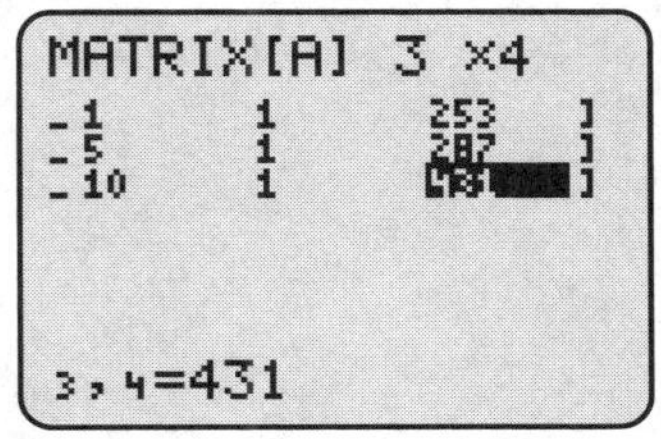

When done, press [2nd] [QUIT].

To show the matrix A, press [2nd] [MATRIX] [ENTER] [ENTER].

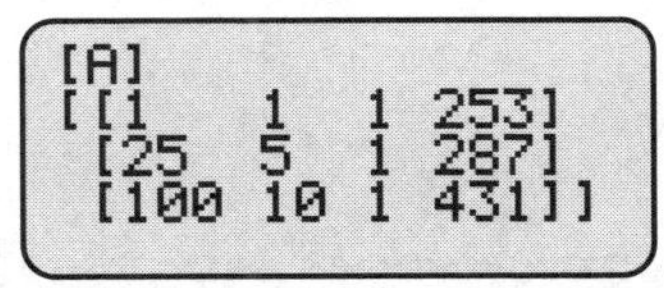

To find the reduced row–echelon form of the original matrix directly, use [rref] (matrix) command. Press [2nd] [MATRIX] [▶] [MATH] [▼] [B:rref(] [ENTER] [2nd] [MATRIX] [ENTER] [)] [▶] [MATH] [1:Frac] [ENTER].

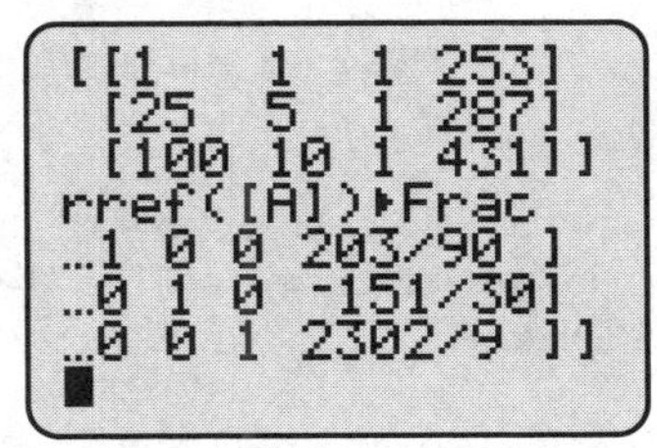

The solution to the system is $a = \frac{203}{90}$, $b = -\frac{151}{30}$, $c = \frac{2302}{9}$.

$$R_3 + 90R_2 \to R_3 \quad \left[\begin{array}{ccc|c} 1 & 1 & 1 & 253 \\ 0 & 1 & \frac{6}{5} & \frac{3{,}019}{10} \\ 0 & 0 & 9 & 2{,}302 \end{array}\right]$$

$$\tfrac{1}{9}R_3 \to R_3 \quad \left[\begin{array}{ccc|c} 1 & 1 & 1 & 253 \\ 0 & 1 & \frac{6}{5} & \frac{3{,}019}{10} \\ 0 & 0 & 1 & \frac{2{,}302}{9} \end{array}\right]$$

$$R_2 - \tfrac{6}{5}R_3 \to R_2 \quad \left[\begin{array}{ccc|c} 1 & 1 & 1 & 253 \\ 0 & 1 & 0 & -\frac{151}{30} \\ 0 & 0 & 1 & \frac{2{,}302}{9} \end{array}\right]$$

$$R_1 - R_3 \to R_1 \quad \left[\begin{array}{ccc|c} 1 & 1 & 0 & -\frac{25}{9} \\ 0 & 1 & 0 & -\frac{151}{30} \\ 0 & 0 & 1 & \frac{2{,}302}{9} \end{array}\right]$$

$$R_1 - R_2 \to R_1 \quad \left[\begin{array}{ccc|c} 1 & 0 & 0 & \frac{203}{90} \\ 0 & 1 & 0 & -\frac{151}{30} \\ 0 & 0 & 1 & \frac{2{,}302}{9} \end{array}\right]$$

STEP 3 Identify the solution. $\quad a = \dfrac{203}{90} \quad b = -\dfrac{151}{30} \quad c = \dfrac{2{,}302}{9}$

Substituting $a = 203/90$, $b = -151/30$, $c = 2302/9$ into $y = at^2 + bt + c$, we find that the thousands of dollars spent on malpractice suits as a function of year is given by

$$\boxed{y = \frac{203}{90}t^2 - \frac{151}{30}t + \frac{2302}{9}} \qquad 1991 \text{ is } t = 0$$

Notice that all three points lie on this curve.

SECTION 7.1 SUMMARY

In this section, we used augmented matrices to represent a system of linear equations.

$$\begin{array}{l} a_1x + b_1y + c_1z = d_1 \\ a_2x + b_2y + c_2z = d_2 \\ a_3x + b_3y + c_3z = d_3 \end{array} \iff \left[\begin{array}{ccc|c} a_1 & b_1 & c_1 & d_1 \\ a_2 & b_2 & c_2 & d_2 \\ a_3 & b_3 & c_3 & d_3 \end{array}\right]$$

Any missing terms correspond to a 0 in the matrix. A matrix is in **row–echelon** form if it has all three of the following properties:

1. Any rows consisting entirely of 0s are at the bottom of the matrix.
2. For each row that does not consist entirely of 0s, the first (leftmost) nonzero entry is 1 (called the leading 1).
3. For two successive nonzero rows, the leading 1 in the higher row is farther to the left than the leading 1 in the lower row.

If a matrix in row–echelon form has the following additional property, then the matrix is in **reduced row–echelon form**:

4. Every column containing a leading 1 has zeros in every position above and below the leading 1.

The two methods used for solving systems of linear equations represented as augmented matrices are Gaussian elimination with back-substitution and Gauss–Jordan elimination. In both cases, we represent the system of linear equations as an augmented matrix and then use row operations to rewrite in row–echelon form. With Gaussian elimination we then stop and perform back-substitution to solve the system, and with Gauss–Jordan elimination we continue with row operations until the matrix is in reduced row–echelon form and then identify the solution to the system.

SECTION 7.1 EXERCISES

SKILLS

In Exercises 1–6, determine the order of each matrix.

1. $\begin{bmatrix} -1 & 3 & 4 \\ 2 & 7 & 9 \end{bmatrix}$ **2.** $\begin{bmatrix} 0 & 1 \\ 3 & 9 \\ 7 & 8 \end{bmatrix}$ **3.** $\begin{bmatrix} 1 & 2 & 3 & 4 \end{bmatrix}$ **4.** $\begin{bmatrix} 3 \\ 7 \\ -1 \\ 10 \end{bmatrix}$ **5.** $[0]$ **6.** $\begin{bmatrix} -1 & 3 & 6 & 8 \\ 2 & 9 & 7 & 3 \\ 5 & 4 & -2 & -10 \\ 6 & 3 & 1 & 5 \end{bmatrix}$

In Exercises 7–14, write the augmented matrix for each system of linear equations.

7. $\begin{aligned} 3x - 2y &= 7 \\ -4x + 6y &= -3 \end{aligned}$

8. $\begin{aligned} -x + y &= 2 \\ x - y &= -4 \end{aligned}$

9. $\begin{aligned} 2x - 3y + 4z &= -3 \\ -x + y + 2z &= 1 \\ 5x - 2y - 3z &= 7 \end{aligned}$

10. $\begin{aligned} x - 2y + z &= 0 \\ -2x + y - z &= -5 \\ 13x + 7y + 5z &= 6 \end{aligned}$

11. $\begin{aligned} x + y &= 3 \\ x - z &= 2 \\ y + z &= 5 \end{aligned}$

12. $\begin{aligned} x - y &= -4 \\ y + z &= 3 \end{aligned}$

13. $\begin{aligned} 3y - 4x + 5z - 2 &= 0 \\ 2x - 3y - 2z &= -3 \\ 3z + 4y - 2x - 1 &= 0 \end{aligned}$

14. $\begin{aligned} 2y + z - x - 3 &= 2 \\ 2x + 3z - 2y &= 0 \\ -2z + y - 4x - 3 &= 0 \end{aligned}$

In Exercises 15–20, write the system of linear equations represented by the augmented matrix. Utilize the variables *x*, *y*, and *z*.

15. $\left[\begin{array}{cc|c} -3 & 7 & 2 \\ 1 & 5 & 8 \end{array}\right]$ **16.** $\left[\begin{array}{ccc|c} -1 & 2 & 4 & 4 \\ 7 & 9 & 3 & -3 \\ 4 & 6 & -5 & 8 \end{array}\right]$ **17.** $\left[\begin{array}{ccc|c} -1 & 0 & 0 & 4 \\ 7 & 9 & 3 & -3 \\ 4 & 6 & -5 & 8 \end{array}\right]$

18. $\left[\begin{array}{ccc|c} 2 & 3 & -4 & 6 \\ 7 & -1 & 5 & 9 \end{array}\right]$ **19.** $\left[\begin{array}{cc|c} 1 & 0 & a \\ 0 & 1 & b \end{array}\right]$ **20.** $\left[\begin{array}{ccc|c} 3 & 0 & 5 & 1 \\ 0 & -4 & 7 & -3 \\ 2 & -1 & 0 & 8 \end{array}\right]$

In Exercises 21–30, indicate whether each matrix is in row–echelon form. If it is, determine whether it is in reduced row–echelon form.

21. $\left[\begin{array}{cc|c} 1 & 0 & 3 \\ 1 & 1 & 2 \end{array}\right]$ **22.** $\left[\begin{array}{cc|c} 0 & 1 & 3 \\ 1 & 0 & 2 \end{array}\right]$ **23.** $\left[\begin{array}{ccc|c} 1 & 0 & -1 & -3 \\ 0 & 1 & 3 & 14 \end{array}\right]$ **24.** $\left[\begin{array}{ccc|c} 1 & 0 & 0 & -3 \\ 0 & 1 & 3 & 14 \end{array}\right]$

25. $\left[\begin{array}{ccc|c} 1 & 0 & 1 & 3 \\ 0 & 0 & 0 & 0 \\ 0 & 1 & 2 & 2 \end{array}\right]$ **26.** $\left[\begin{array}{ccc|c} 1 & 0 & 1 & 3 \\ 0 & 1 & 2 & 2 \\ 0 & 0 & 0 & 0 \end{array}\right]$ **27.** $\left[\begin{array}{ccc|c} 1 & 0 & 0 & 3 \\ 0 & 1 & 0 & 2 \\ 0 & 0 & 1 & 5 \end{array}\right]$ **28.** $\left[\begin{array}{ccc|c} -1 & 0 & 0 & 3 \\ 0 & -1 & 0 & 2 \\ 0 & 0 & -1 & 5 \end{array}\right]$

29. $\left[\begin{array}{cccc|c} 1 & 0 & 0 & 1 & 3 \\ 0 & 1 & 0 & 3 & 2 \\ 0 & 0 & 1 & 0 & 5 \\ 0 & 0 & 0 & 1 & 0 \end{array}\right]$ **30.** $\left[\begin{array}{cccc|c} 1 & 0 & 0 & 1 & 3 \\ 0 & 1 & 0 & 3 & 2 \\ 0 & 0 & 1 & 0 & 5 \\ 0 & 0 & 0 & 0 & 0 \end{array}\right]$

In Exercises 31–40, perform the indicated row operations on each augmented matrix.

31. $\left[\begin{array}{cc|c} 1 & -2 & -3 \\ 2 & 3 & -1 \end{array}\right]$ $R_2 - 2R_1 \rightarrow R_2$

32. $\left[\begin{array}{cc|c} 2 & -3 & -4 \\ 1 & 2 & 5 \end{array}\right]$ $R_1 \leftrightarrow R_2$

33. $\left[\begin{array}{ccc|c} 1 & -2 & -1 & 3 \\ 2 & 1 & -3 & 6 \\ 3 & -2 & 5 & -8 \end{array}\right]$ $R_2 - 2R_1 \rightarrow R_2$

34. $\left[\begin{array}{ccc|c} 1 & -2 & 1 & 3 \\ 0 & 1 & -2 & 6 \\ -3 & 0 & -1 & -5 \end{array}\right]$ $R_3 + 3R_1 \rightarrow R_3$

35. $\left[\begin{array}{cccc|c} 1 & -2 & 5 & -1 & 2 \\ 0 & 3 & 0 & -1 & -2 \\ 0 & -2 & 1 & -2 & 5 \\ 0 & 0 & 1 & -1 & -6 \end{array}\right]$ $R_3 + R_2 \rightarrow R_2$

36. $\left[\begin{array}{cccc|c} 1 & 0 & 5 & -10 & 15 \\ 0 & 1 & 2 & -3 & 4 \\ 0 & 2 & -3 & 0 & -1 \\ 0 & 0 & 1 & -1 & -3 \end{array}\right]$ $R_2 - \frac{1}{2}R_3 \rightarrow R_3$

37. $\left[\begin{array}{cccc|c} 1 & 0 & 5 & -10 & -5 \\ 0 & 1 & 2 & -3 & -2 \\ 0 & 2 & -3 & 0 & -1 \\ 0 & -3 & 2 & -1 & -3 \end{array}\right]$ $\begin{array}{l} R_3 - 2R_2 \to R_3 \\ R_4 + 3R_2 \to R_4 \end{array}$

38. $\left[\begin{array}{cccc|c} 1 & 0 & 4 & 0 & 1 \\ 0 & 1 & 2 & 0 & -2 \\ 0 & 0 & 1 & 0 & 0 \\ 0 & 0 & 0 & 1 & -3 \end{array}\right]$ $\begin{array}{l} R_2 - 2R_3 \to R_2 \\ R_1 - 4R_3 \to R_1 \end{array}$

39. $\left[\begin{array}{cccc|c} 1 & 0 & 4 & 8 & 3 \\ 0 & 1 & 2 & -3 & -2 \\ 0 & 0 & 1 & 6 & 3 \\ 0 & 0 & 0 & 1 & -3 \end{array}\right]$ $\begin{array}{l} R_3 - 6R_4 \to R_3 \\ R_2 + 3R_4 \to R_2 \\ R_1 - 8R_4 \to R_1 \end{array}$

40. $\left[\begin{array}{cccc|c} 1 & 0 & -1 & 5 & 2 \\ 0 & 1 & 2 & 3 & -5 \\ 0 & 0 & 1 & -2 & 2 \\ 0 & 0 & 0 & 1 & 1 \end{array}\right]$ $\begin{array}{l} R_3 + 2R_4 \to R_3 \\ R_2 - 3R_4 \to R_2 \\ R_1 - 5R_4 \to R_1 \end{array}$

In Exercises 41–50, use row operations to transform each matrix to reduced row–echelon form.

41. $\left[\begin{array}{cc|c} 1 & 2 & 4 \\ 2 & 3 & 2 \end{array}\right]$

42. $\left[\begin{array}{cc|c} 1 & -1 & 3 \\ -3 & 2 & 2 \end{array}\right]$

43. $\left[\begin{array}{ccc|c} 1 & -1 & 1 & -1 \\ 0 & 1 & -1 & -1 \\ -1 & 1 & 1 & 1 \end{array}\right]$

44. $\left[\begin{array}{ccc|c} 0 & -1 & 1 & 1 \\ 1 & -1 & 1 & -1 \\ 1 & -1 & -1 & -1 \end{array}\right]$

45. $\left[\begin{array}{ccc|c} 3 & -2 & -3 & -1 \\ 1 & -1 & 1 & -4 \\ 2 & 3 & 5 & 14 \end{array}\right]$

46. $\left[\begin{array}{ccc|c} 3 & -1 & 1 & 2 \\ 1 & -2 & 3 & 1 \\ 2 & 1 & -3 & -1 \end{array}\right]$

47. $\left[\begin{array}{ccc|c} 2 & 1 & -6 & 4 \\ 1 & -2 & 2 & -3 \end{array}\right]$

48. $\left[\begin{array}{ccc|c} -3 & -1 & 2 & -1 \\ -1 & -2 & 1 & -3 \end{array}\right]$

49. $\left[\begin{array}{ccc|c} -1 & 2 & 1 & -2 \\ 3 & -2 & 1 & 4 \\ 2 & -4 & -2 & 4 \end{array}\right]$

50. $\left[\begin{array}{ccc|c} 2 & -1 & 0 & 1 \\ -1 & 0 & 1 & -2 \\ -2 & 1 & 0 & -1 \end{array}\right]$

In Exercises 51–70, solve the system of linear equations using Gaussian elimination with back-substitution.

51. $\begin{aligned} 2x + 3y &= 1 \\ x + y &= -2 \end{aligned}$

52. $\begin{aligned} 3x + 2y &= 11 \\ x - y &= 12 \end{aligned}$

53. $\begin{aligned} -x + 2y &= 3 \\ 2x - 4y &= -6 \end{aligned}$

54. $\begin{aligned} 3x - y &= -1 \\ 2y + 6x &= 2 \end{aligned}$

55. $\begin{aligned} \tfrac{2}{3}x + \tfrac{1}{3}y &= \tfrac{8}{9} \\ \tfrac{1}{2}x + \tfrac{1}{4}y &= \tfrac{3}{4} \end{aligned}$

56. $\begin{aligned} 0.4x - 0.5y &= 2.08 \\ -0.3x + 0.7y &= 1.88 \end{aligned}$

57. $\begin{aligned} x - z - y &= 10 \\ 2x - 3y + z &= -11 \\ y - x + z &= -10 \end{aligned}$

58. $\begin{aligned} 2x + z + y &= -3 \\ 2y - z + x &= 0 \\ x + y + 2z &= 5 \end{aligned}$

59. $\begin{aligned} 3x_1 + x_2 - x_3 &= 1 \\ x_1 - x_2 + x_3 &= -3 \\ 2x_1 + x_2 + x_3 &= 0 \end{aligned}$

60. $\begin{aligned} 2x_1 + x_2 + x_3 &= -1 \\ x_1 + x_2 - x_3 &= 5 \\ 3x_1 - x_2 - x_3 &= 1 \end{aligned}$

61. $\begin{aligned} 2x + 5y &= 9 \\ x + 2y - z &= 3 \\ -3x - 4y + 7z &= 1 \end{aligned}$

62. $\begin{aligned} x - 2y + 3z &= 1 \\ -2x + 7y - 9z &= 4 \\ x + z &= 9 \end{aligned}$

63. $\begin{aligned} 2x_1 - x_2 + x_3 &= 3 \\ x_1 - x_2 + x_3 &= 2 \\ -2x_1 + 2x_2 - 2x_3 &= -4 \end{aligned}$

64. $\begin{aligned} x_1 - x_2 - 2x_3 &= 0 \\ -2x_1 + 5x_2 + 10x_3 &= -3 \\ 3x_1 + x_2 &= 0 \end{aligned}$

65. $\begin{aligned} 2x + y - z &= 2 \\ x - y - z &= 6 \end{aligned}$

66. $\begin{aligned} 3x + y - z &= 0 \\ x + y + 7z &= 4 \end{aligned}$

67. $\begin{aligned} 2y + z &= 3 \\ 4x - z &= -3 \\ 7x - 3y - 3z &= 2 \\ x - y - z &= -2 \end{aligned}$

68. $\begin{aligned} -2x - y + 2z &= 3 \\ 3x - 4z &= 2 \\ 2x + y &= -1 \\ -x + y - z &= -8 \end{aligned}$

69. $\begin{aligned} 3x_1 - 2x_2 + x_3 + 2x_4 &= -2 \\ -x_1 + 3x_2 + 4x_3 + 3x_4 &= 4 \\ x_1 + x_2 + x_3 + x_4 &= 0 \\ 5x_1 + 3x_2 + x_3 + 2x_4 &= -1 \end{aligned}$

70. $\begin{aligned} 5x_1 + 3x_2 + 8x_3 + x_4 &= 1 \\ x_1 + 2x_2 + 5x_3 + 2x_4 &= 3 \\ 4x_1 + x_3 - 2x_4 &= -3 \\ x_2 + x_3 + x_4 &= 0 \end{aligned}$

In Exercises 71–86, solve the system of linear equations using Gauss–Jordan elimination.

71. $\begin{aligned} x + 3y &= -5 \\ -2x - y &= 0 \end{aligned}$

72. $\begin{aligned} 5x - 4y &= 31 \\ 3x + 7y &= -19 \end{aligned}$

73. $\begin{aligned} x + y &= 4 \\ -3x - 3y &= 10 \end{aligned}$

74. $\begin{aligned} 3x - 4y &= 12 \\ -6x + 8y &= -24 \end{aligned}$

75.
$$\begin{aligned} x - 2y + 3z &= 5 \\ 3x + 6y - 4z &= -12 \\ -x - 4y + 6z &= 16 \end{aligned}$$

76.
$$\begin{aligned} x + 2y - z &= 6 \\ 2x - y + 3z &= -13 \\ 3x - 2y + 3z &= -16 \end{aligned}$$

77.
$$\begin{aligned} x + y + z &= 3 \\ x - z &= 1 \\ y - z &= -4 \end{aligned}$$

78.
$$\begin{aligned} x - 2y + 4z &= 2 \\ 2x - 3y - 2z &= -3 \\ \tfrac{1}{2}x + \tfrac{1}{4}y + z &= -2 \end{aligned}$$

79.
$$\begin{aligned} x + 2y + z &= 3 \\ 2x - y + 3z &= 7 \\ 3x + y + 4z &= 5 \end{aligned}$$

80.
$$\begin{aligned} x + 2y + z &= 3 \\ 2x - y + 3z &= 7 \\ 3x + y + 4z &= 10 \end{aligned}$$

81.
$$\begin{aligned} 3x - y + z &= 8 \\ x + y - 2z &= 4 \end{aligned}$$

82.
$$\begin{aligned} x - 2y + 3z &= 10 \\ -3x + z &= 9 \end{aligned}$$

83.
$$\begin{aligned} 4x - 2y + 5z &= 20 \\ x + 3y - 2z &= 6 \end{aligned}$$

84.
$$\begin{aligned} y + z &= 4 \\ x + y &= 8 \end{aligned}$$

85.
$$\begin{aligned} x - y - z - w &= 1 \\ 2x + y + z + 2w &= 3 \\ x - 2y - 2z - 3w &= 0 \\ 3x - 4y + z + 5w &= -3 \end{aligned}$$

86.
$$\begin{aligned} x - 3y + 3z - 2w &= 4 \\ x + 2y - z &= -3 \\ x + 3z + 2w &= 3 \\ y + z + 5w &= 6 \end{aligned}$$

APPLICATIONS

87. Football. In Super Bowl XXXVIII, the New England Patriots defeated the Carolina Panthers 32–29. The points came from a total of four types of plays: touchdowns (6 points), extra points (1 point), two-point conversions (2 points), and field goals (3 points). There were a total of 16 scoring plays. There were four times as many touchdowns as field goals, and there were five times as many extra points as two-point conversions. How many touchdowns, extra points, two-point conversions, and field goals were scored in Super Bowl XXXVIII?

88. Basketball. In the 2004 Summer Olympics in Athens, Greece, the U.S. men's basketball team, consisting of NBA superstars, was defeated by the Puerto Rican team 92–73. The points came from three types of scoring plays: 2-point shots, 3-point shots, and 1-point free throws. There were six more 2-point shots made than there were 1-point free throws. The number of successful 2-point shots was three less than four times the number of successful 3-point shots. How many 2-point and 3-point shots, and 1-point free throws were made in that Olympic competition?

Exercises 89 and 90 rely on a selection of Subway sandwiches whose nutrition information is given in the following table. Suppose you are going to eat only Subway sandwiches for a week (seven days) for lunch and dinner (a total of 14 meals).

SANDWICH	CALORIES	FAT (g)	CARBOHYDRATES (g)	PROTEIN (g)
Mediterranean chicken	350	18	17	36
6" tuna	430	19	46	20
6" roast beef	290	5	45	19
Turkey–bacon wrap	430	27	20	34

www.subway.com

89. Diet. Your goal is a low-fat diet consisting of 526 grams of carbohydrates, 168 grams of fat, and 332 grams of protein. How many of each sandwich would you eat that week to obtain this goal?

90. Diet. Your goal is a low-carb diet consisting of 5180 calories, 335 grams of carbohydrates, and 263 grams of fat. How many of each sandwich would you eat that week to obtain this goal?

Exercises 91 and 92 involve vertical motion and the effect of gravity on an object.

Because of gravity, an object that is projected upward will eventually reach a maximum height and then fall to the ground. The equation that relates the height h of a projectile t seconds after it is shot upward is given by

$$h = \frac{1}{2}at^2 + v_0t + h_0$$

where a is the acceleration due to gravity, h_0 is the initial height of the object at time $t = 0$, and v_0 is the initial velocity of the object at time $t = 0$. Note that a projectile follows the path of a parabola opening down, so $a < 0$.

91. Vertical Motion. An object is thrown upward, and the following table depicts the height of the ball t seconds after the projectile is released. Find the initial height, initial velocity, and acceleration due to gravity.

t (SECONDS)	HEIGHT (FEET)
1	34
2	36
3	6

92. Vertical Motion. An object is thrown upward, and the following table depicts the height of the ball t seconds after the projectile is released. Find the initial height, initial velocity, and acceleration due to gravity.

t (SECONDS)	HEIGHT (FEET)
1	54
2	66
3	46

93. Data Curve-Fitting. The average number of minutes that a person spends driving a car can be modeled by a quadratic function $y = ax^2 + bx + c$, where $a < 0$ and $15 < x < 65$. The following table gives the average number of minutes a day that a person spends driving a car. Determine the quadratic function that models this quantity.

AGE	AVERAGE DAILY MINUTES DRIVING
16	25
40	64
65	40

94. Data Curve-Fitting. The average age when a woman gets married has been increasing during the last century. In 1920 the average age was 18.4, in 1960 the average age was 20.3, and in 2002 the average age was 25.30. Find a quadratic function $y = ax^2 + bx + c$, where $a > 0$ and $18 < x < 35$, that models the average age y when a woman gets married as a function of the year x ($x = 0$ corresponds to 1920). What will the average age be in 2010?

95. Money. Gary and Ginger decide to place \$10,000 of their savings into investments. They put some in a money market account earning 3% interest, some in a mutual fund that has been averaging 7% a year, and some in a stock that rose 10% last year. If they put \$3,000 more in the money market than in the mutual fund and the mutual fund and stocks have the same growth in the next year as they did in the previous year, they will earn \$540 in a year. How much money did they put in each of the three investments?

96. Money. Ginger talks Gary into putting less money in the money market and more money in the stock. They place \$10,000 of their savings into investments. They put some in a money market account earning 3% interest, some in a mutual fund that has been averaging 7% a year, and some in a stock that rose 10% last year. If they put \$3,000 more in the stock than in the mutual fund and the mutual fund and stock have the same growth in the next year as they did in the previous year, they will earn \$840 in a year. How much money did they put in each of the three investments?

97. Manufacturing. A company produces three products x, y, and z. Each item of product x requires 20 units of steel, 2 units of plastic, and 1 unit of glass. Each item of product y requires 25 units of steel, 5 units of plastic, and no units of glass. Each item of product z requires 150 units of steel, 10 units of plastic, and 0.5 units of glass. The available amounts of steel, plastic, and glass are 2400, 310, and 28, respectively. How many items of each type can the company produce and utilize all the available raw materials?

98. Geometry. Find the values of a, b, and c such that the graph of the quadratic function $y = ax^2 + bx + c$ passes through the points $(1, 5)$, $(-2, -10)$, and $(0, 4)$.

99. Ticket Sales. One hundred students decide to buy tickets to a football game. There are three types of tickets: general admission, reserved, and end zone. Each general admission ticket costs \$20.00, each reserved ticket costs \$40.00, and each end zone ticket costs \$15.00. The students spend a total of \$2,375.00 for all the tickets. There are five more reserved tickets than general admission tickets, and 20 more end zone tickets than general admission tickets. How many of each type of ticket were purchased by the students?

100. Exercise and Nutrition. Ann would like to exercise one hour per day to burn calories and lose weight. She would like to engage in three activities: walking, step-up exercise, and weight training. She knows she can burn 85 calories walking at a certain pace in 15 minutes, 45 calories doing the step-up exercise in 10 minutes, and 137 calories by weight training for 20 minutes. (a) Determine the number of calories per minute she can burn doing each activity. (b) Suppose she has time to exercise for only one hour (60 minutes). She sets a goal of burning 358 calories in one hour and would like to weight train twice as long as walking. How many minutes must she engage in each exercise to burn the required number of calories in one hour?

101. Geometry. The circle given by the equation $x^2 + y^2 + ax + by + c = 0$ passes through the point $(4, 4)$, $(-3, -1)$, and $(1, -3)$. Find a, b, and c.

102. Geometry. The circle given by the equation $x^2 + y^2 + ax + by + c = 0$ passes through the point $(0, 7)$, $(6, 1)$, and $(5, 4)$. Find a, b, and c.

CATCH THE MISTAKE

In Exercises 103–106, explain the mistake that is made.

103. Solve the system of equations using the augmented matrices.

$$\begin{aligned} y - x + z &= 2 \\ x - 2z + y &= -3 \\ x + y + z &= 6 \end{aligned}$$

Solution:

Step 1: Write as an augmented matrix.

$$\left[\begin{array}{ccc|c} 1 & -1 & 1 & 2 \\ 1 & -2 & 1 & -3 \\ 1 & 1 & 1 & 6 \end{array}\right]$$

Step 2: Reduce the matrix using Gaussian elimination.

$$\left[\begin{array}{ccc|c} 1 & -1 & 1 & 2 \\ 0 & 1 & 0 & 5 \\ 0 & 0 & 0 & -6 \end{array}\right]$$

Step 3: Identify the solution. Row 3 is inconsistent, so there is no solution.

This is incorrect. The correct answer is $x = 1$, $y = 2$, $z = 3$. What mistake was made?

104. Perform the indicated row operations on the matrix.

$$\left[\begin{array}{ccc|c} 1 & -1 & 1 & 2 \\ 2 & -3 & 1 & 4 \\ 3 & 1 & 2 & -6 \end{array}\right]$$

a. $R_2 - 2R_1 \to R_2$
b. $R_3 - 3R_1 \to R_3$

Solution:

a. $\left[\begin{array}{ccc|c} 1 & -1 & 1 & 2 \\ 0 & -3 & 1 & 4 \\ 3 & 1 & 2 & -6 \end{array}\right]$

b. $\left[\begin{array}{ccc|c} 1 & -1 & 1 & 2 \\ 2 & -3 & 1 & 4 \\ 0 & 1 & 2 & -6 \end{array}\right]$

This is incorrect. What mistake was made?

105. Solve the system of equations using an augmented matrix.

$$\begin{aligned} 3x - 2y + z &= -1 \\ x + y - z &= 3 \\ 2x - y + 3z &= 0 \end{aligned}$$

Solution:

Step 1: Write the system as an augmented matrix.
$$\left[\begin{array}{ccc|c} 3 & -2 & 1 & -1 \\ 1 & 1 & -1 & 3 \\ 2 & -1 & 3 & 0 \end{array}\right]$$

Step 2: Reduce the matrix using Gaussian elimination.
$$\left[\begin{array}{ccc|c} 1 & 0 & 0 & 1 \\ 0 & 1 & 0 & 2 \\ 0 & 0 & 1 & 0 \end{array}\right]$$

Step 3: Identify the answer. Row 3 is inconsistent $1 = 0$, therefore there is no solution.

This is incorrect. What mistake was made?

106. Solve the system of equations using an augmented matrix.

$$\begin{aligned} x + 3y + 2z &= 4 \\ 3x + 10y + 9z &= 17 \\ 2x + 7y + 7z &= 17 \end{aligned}$$

Solution:

Step 1: Write the system as an augmented matrix.
$$\left[\begin{array}{ccc|c} 1 & 3 & 2 & 4 \\ 3 & 10 & 9 & 17 \\ 2 & 7 & 7 & 17 \end{array}\right]$$

Step 2: Reduce the matrix using Gaussian elimination.
$$\left[\begin{array}{ccc|c} 1 & 0 & -7 & -11 \\ 0 & 1 & 3 & 5 \\ 0 & 0 & 0 & 4 \end{array}\right]$$

Step 3: Identify the answer: Infinitely many solutions.
$$\begin{aligned} x &= 7t - 11 \\ y &= -3t + 5 \\ z &= t \end{aligned}$$

This is incorrect. What mistake was made?

CONCEPTUAL

Determine whether each of the following statements is true or false.

107. A nonsquare matrix cannot have a unique solution.

108. The procedure for Gaussian elimination can be used only for square matrices.

109. A square matrix that has a unique solution has a reduced matrix with 1s along the diagonal and 0s above and below the 1s.

110. A square matrix with an all-zero row has infinitely many solutions.

CHALLENGE

111. A fourth-degree polynomial $f(x) = ax^4 + bx^3 + cx^2 + dx + e$, with $a < 0$, can be used to represent the following data on the number of deaths per year due to lightning strikes (assume 1999 corresponds to $x = 0$).

Use the data to determine a, b, c, d, and e.

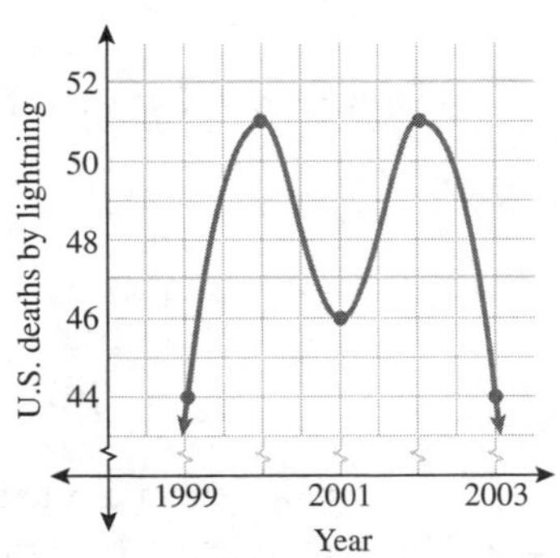

112. A copy machine accepts nickels, dimes, and quarters. After one hour, there are 30 coins total, and their value is \$4.60. How many nickels, quarters, and dimes are in the machine?

TECHNOLOGY

113. In Exercise 57, you were asked to solve this system of equations using an augmented matrix.

$$\begin{aligned} x - \ z - y &= \ 10 \\ 2x - 3y + z &= -11 \\ y - \ x + z &= -10 \end{aligned}$$

A graphing calculator or graphing utility can be used to solve systems of linear equations by entering the coefficients of the matrix. Solve this system and confirm your answer with the calculator's answer.

114. In Exercise 58, you were asked to solve this system of equations using an augmented matrix.

$$\begin{aligned} 2x + z + \ y &= -3 \\ 2y - z + \ x &= \ 0 \\ x + y + 2z &= \ 5 \end{aligned}$$

A graphing calculator or graphing utility can be use to solve systems of linear equations by entering the coefficients of the matrix. Solve this system and confirm your answer with the calculator's answer.

In Exercises 115 and 116, you are asked to model a set of three points with a quadratic function $y = ax^2 + bx + c$ and determine the quadratic function.

a. Set up a system of equations, use a graphing utility or graphing calculator to solve the system by entering the coefficients of the augmented matrix.

b. Use the graphing calculator commands [STAT] [QuadReg] to model the data using a quadratic function. Round your answers to two decimal places.

115. $(-6, -8)$, $(2, 7)$, $(7, 1)$

116. $(-9, 20)$, $(2, -18)$, $(11, 16)$

SECTION 7.2 MATRIX ALGEBRA

SKILLS OBJECTIVES

- Add and subtract matrices.
- Perform scalar multiplication.
- Multiply two matrices.

CONCEPTUAL OBJECTIVES

- Understand what is meant by equal matrices.
- Understand why multiplication of some matrices is undefined.
- Realize that matrix multiplication is *not* commutative.

In Section 7.1, we defined a matrix with m rows and n columns to have order $m \times n$:

$$A = \begin{bmatrix} a_{11} & a_{12} & \cdots & a_{1n} \\ a_{21} & a_{22} & \cdots & a_{2n} \\ \vdots & \vdots & \cdots & \vdots \\ a_{m1} & a_{m2} & \cdots & a_{mn} \end{bmatrix}$$

Capital letters are used to represent (or name) a matrix, and lowercase letters are used to represent the entries (elements) of the matrix. The subscripts are used to denote the location (row/column) of each entry. The order of a matrix is often written as a subscript of the matrix name: $A_{m \times n}$. Other words like "size" and "dimension" are used as synonyms of "order." Matrices are a convenient way to represent data.

Average Marriage Age on the Rise

A survey conducted by the Census Bureau (Shannon Reilly and Kevin Kepple: www.usatoday.com) found that the average marriage age in the United States is rising.

Esbin-Anderson Photography/Getty/Brand X Pictures

	1960	2002
Men	22.8	26.9
Women	20.3	25.3

These data can be represented by the 2 × 2 matrix $\begin{bmatrix} 22.8 & 26.9 \\ 20.3 & 25.3 \end{bmatrix}$, with two rows (gender) and two columns (year).

The U.S. War with Iraq

A survey (The Harris Poll #60, August 20, 2004) of adult Americans asked the question, "Do you think the invasion of Iraq strengthened or weakened the war on terrorism?"

The results were along party lines:

Photo by Staff Sgt. Shane A. Cuomo, US Air Force/Department of Defense/Still Media Record Center

	Republican (%)	Democrat (%)
Strengthened the war on terrorism	75	34
Weakened the war on terrorism	19	53
Not sure	6	13

These data can be represented with the 3 × 2 matrix $\begin{bmatrix} 75 & 34 \\ 19 & 53 \\ 6 & 13 \end{bmatrix}$, with three rows (opinions of war) and two columns (registered political party).

There is an entire field of study called **matrix algebra** that treats matrices similarly to functions and variables in traditional algebra. This section serves as an introduction to matrix algebra. We will discuss how to

- determine whether two matrices are equal,
- add and subtract matrices,
- multiply a matrix by a scalar,
- multiply two matrices.

It is important to pay special attention to the *order* of a matrix, because it determines whether certain operations are defined.

Equality of Matrices

Two matrices are equal if and only if they have the same order, $m \times n$, and all of their corresponding elements are equal.

DEFINITION **Equality of Matrices**

Two matrices A and B are **equal**, written as $A = B$ if and only if *both* of the following are true:

- A and B have the same order $m \times n$,
- Every pair of corresponding elements are equal: $a_{ij} = b_{ij}$ for all $i = 1, 2, \ldots, m$ and $j = 1, 2, \ldots, n$.

EXAMPLE 1 Equality of Matrices

Referring to the definition of equality of matrices, find the indicated elements.

$$\begin{bmatrix} a_{11} & a_{12} & a_{13} \\ a_{21} & a_{22} & a_{23} \\ a_{31} & a_{32} & a_{33} \end{bmatrix} = \begin{bmatrix} 2 & -7 & 1 \\ 0 & 5 & -3 \\ -1 & 8 & 9 \end{bmatrix}$$

Find the main diagonal elements: a_{11}, a_{22}, and a_{33}.

Solution:

Since the matrices are equal, their corresponding elements are equal. $\boxed{a_{11} = 2}$ $\boxed{a_{22} = 5}$ $\boxed{a_{33} = 9}$

EXAMPLE 2 Equality of Matrices

Given that $\begin{bmatrix} 2x & y - 1 \\ 1 & 2 \end{bmatrix} = \begin{bmatrix} 6 & 7 \\ 1 & 2 \end{bmatrix}$, solve for x and y.

Solution:

Equate corresponding elements. $2x = 6$ $\quad y - 1 = 7$

Solve for x and y. $\boxed{x = 3}$ $\quad \boxed{y = 8}$

Matrix Addition and Subtraction

Two matrices A and B can be added or subtracted only if they have the *same order*. Suppose A and B are both of order $m \times n$; then the *sum* $A + B$ is found by adding corresponding elements, or taking $a_{ij} + b_{ij}$. The *difference* $A - B$ is found by subtracting the elements in B from the corresponding elements in A, or finding $a_{ij} - b_{ij}$.

DEFINITION **Matrix Addition and Matrix Subtraction**

If A is an $m \times n$ matrix and B is an $m \times n$ matrix, then their **sum** $A + B$ is an $m \times n$ matrix whose elements are given by

$$a_{ij} + b_{ij}$$

and their **difference** $A - B$ is an $m \times n$ matrix whose elements are given by

$$a_{ij} - b_{ij}$$

EXAMPLE 3 Adding and Subtracting Matrices

Given that $A = \begin{bmatrix} -1 & 3 & 4 \\ -5 & 2 & 0 \end{bmatrix}$ and $B = \begin{bmatrix} 2 & 1 & -3 \\ 0 & -5 & 4 \end{bmatrix}$, find the following:

a. $A + B$ **b.** $A - B$

Solution:

Since $A_{2\times3}$ and $B_{2\times3}$ have the same order, they can be added or subtracted.

a. Write the sum. $A + B = \begin{bmatrix} -1 & 3 & 4 \\ -5 & 2 & 0 \end{bmatrix} + \begin{bmatrix} 2 & 1 & -3 \\ 0 & -5 & 4 \end{bmatrix}$

Add the corresponding elements. $= \begin{bmatrix} -1+2 & 3+1 & 4+(-3) \\ -5+0 & 2+(-5) & 0+4 \end{bmatrix}$

Simplify. $= \boxed{\begin{bmatrix} 1 & 4 & 1 \\ -5 & -3 & 4 \end{bmatrix}}$

b. Write the difference. $A - B = \begin{bmatrix} -1 & 3 & 4 \\ -5 & 2 & 0 \end{bmatrix} - \begin{bmatrix} 2 & 1 & -3 \\ 0 & -5 & 4 \end{bmatrix}$

Subtract the corresponding elements. $= \begin{bmatrix} -1-2 & 3-1 & 4-(-3) \\ -5-0 & 2-(-5) & 0-4 \end{bmatrix}$

Simplify. $= \boxed{\begin{bmatrix} -3 & 2 & 7 \\ -5 & 7 & -4 \end{bmatrix}}$

Technology Tip

Enter the matrices as A and B.

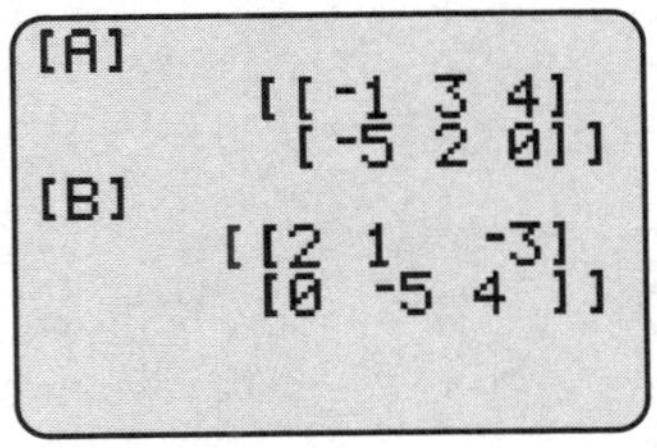

Now enter $[A] + [B]$ and $[A] - [B]$.

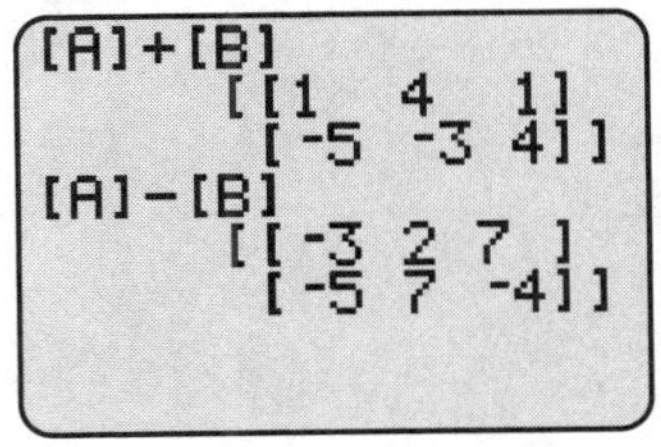

■ **YOUR TURN** Perform the indicated matrix operations, if possible.

$A = \begin{bmatrix} -4 & 0 \\ 1 & 2 \end{bmatrix}$ $B = \begin{bmatrix} 2 & 3 \\ -4 & 0 \end{bmatrix}$ $C = \begin{bmatrix} 2 & 9 & 5 & -1 \end{bmatrix}$ $D = \begin{bmatrix} 0 \\ -3 \\ 4 \\ 2 \end{bmatrix}$

a. $B - A$ **b.** $C + D$ **c.** $A + B$ **d.** $A + D$

■ **Answer:**

a. $\begin{bmatrix} 6 & 3 \\ -5 & -2 \end{bmatrix}$ **b.** not defined

c. $\begin{bmatrix} -2 & 3 \\ -3 & 2 \end{bmatrix}$ **d.** not defined

It is important to note that only matrices of the same order can be added or subtracted. For example, if $A = \begin{bmatrix} -1 & 3 & 4 \\ -5 & 2 & 0 \end{bmatrix}$ and $B = \begin{bmatrix} 5 & -3 \\ 12 & 1 \end{bmatrix}$, the sum and difference of these matrices is undefined because $A_{2\times3}$ and $B_{2\times2}$ do not have the same order.

Study Tip

Only matrices of the same order can be added or subtracted.

A matrix whose elements are all equal to 0 is called a **zero matrix**. The following are examples of zero matrices:

2×2 square zero matrix $\begin{bmatrix} 0 & 0 \\ 0 & 0 \end{bmatrix}$

3×2 zero matrix $\begin{bmatrix} 0 & 0 \\ 0 & 0 \\ 0 & 0 \end{bmatrix}$

1×4 zero matrix $\begin{bmatrix} 0 & 0 & 0 & 0 \end{bmatrix}$

If A, an $m \times n$ matrix, is added to the $m \times n$ zero matrix, the result is A.

$$A + \mathbf{0} = A$$

For example,

$$\begin{bmatrix} 1 & -3 \\ 2 & 5 \end{bmatrix} + \begin{bmatrix} 0 & 0 \\ 0 & 0 \end{bmatrix} = \begin{bmatrix} 1 & -3 \\ 2 & 5 \end{bmatrix}$$

Because of this result, an $m \times n$ zero matrix is called the **additive identity** for $m \times n$ matrices. Similarly, for any matrix A, there exists an **additive inverse** $-A$ such that every element of $-A$ is the negative of every element of A.

For example, $A = \begin{bmatrix} 1 & -3 \\ 2 & 5 \end{bmatrix}$ and $-A = \begin{bmatrix} -1 & 3 \\ -2 & -5 \end{bmatrix}$, and adding these two matrices results in a zero matrix: $A + (-A) = \mathbf{0}$.

The same properties that hold for adding real numbers also hold for adding matrices, provided that addition of matrices is defined.

PROPERTIES OF MATRIX ADDITION

If A, B, and C are all $m \times n$ matrices and $\mathbf{0}$ is the $m \times n$ zero matrix, then the following are true:

Commutative property: $A + B = B + A$

Associative property: $(A + B) + C = A + (B + C)$

Additive identity property: $A + \mathbf{0} = A$

Additive inverse property: $A + (-A) = \mathbf{0}$

Scalar and Matrix Multiplication

There are two types of multiplication involving matrices: *scalar multiplication* and *matrix multiplication.* A **scalar** is any real number. *Scalar multiplication* is the multiplication of a matrix by a scalar, or real number, and is defined for all matrices. *Matrix multiplication* is the multiplication of two matrices and is defined only for certain pair of matrices, depending on the order of each matrix.

Scalar Multiplication

To multiply a matrix A by a scalar k, multiply every element in A by k.

$$3\begin{bmatrix} -1 & 0 & 4 \\ 7 & 5 & -2 \end{bmatrix} = \begin{bmatrix} 3(-1) & 3(0) & 3(4) \\ 3(7) & 3(5) & 3(-2) \end{bmatrix} = \begin{bmatrix} -3 & 0 & 12 \\ 21 & 15 & -6 \end{bmatrix}$$

Here, the scalar is $k = 3$.

DEFINITION Scalar Multiplication

If A is an $m \times n$ matrix and k is any real number, then their product kA is an $m \times n$ matrix whose elements are given by

$$ka_{ij}$$

In other words, every element a_{ij} that is in the ith row and jth column of A is multiplied by k.

In general, uppercase letters are used to denote a matrix and lowercase letters are used to denote scalars. Notice that the elements of each matrix are also represented with lowercase letters, since they are real numbers.

EXAMPLE 4 Multiplying a Matrix by a Scalar

Given that $A = \begin{bmatrix} -1 & 2 \\ -3 & 4 \end{bmatrix}$ and $B = \begin{bmatrix} 0 & 1 \\ -2 & 3 \end{bmatrix}$, perform the following operations.

a. $2A$ **b.** $-3B$ **c.** $2A - 3B$

Solution (a):

Write the scalar multiplication. $2A = 2\begin{bmatrix} -1 & 2 \\ -3 & 4 \end{bmatrix}$

Multiply all elements of A by 2. $2A = \begin{bmatrix} 2(-1) & 2(2) \\ 2(-3) & 2(4) \end{bmatrix}$

Simplify. $\boxed{2A = \begin{bmatrix} -2 & 4 \\ -6 & 8 \end{bmatrix}}$

Solution (b):

Write the scalar multiplication. $-3B = -3\begin{bmatrix} 0 & 1 \\ -2 & 3 \end{bmatrix}$

Multiply all elements of B by -3. $-3B = \begin{bmatrix} -3(0) & -3(1) \\ -3(-2) & -3(3) \end{bmatrix}$

Simplify. $\boxed{-3B = \begin{bmatrix} 0 & -3 \\ 6 & -9 \end{bmatrix}}$

Solution (c):

Add the results of parts (a) and (b). $2A - 3B = 2A + -3B$

$$2A - 3B = \begin{bmatrix} -2 & 4 \\ -6 & 8 \end{bmatrix} + \begin{bmatrix} 0 & -3 \\ 6 & -9 \end{bmatrix}$$

Add the corresponding elements. $2A - 3B = \begin{bmatrix} -2+0 & 4+(-3) \\ -6+6 & 8+(-9) \end{bmatrix}$

Simplify. $\boxed{2A - 3B = \begin{bmatrix} -2 & 1 \\ 0 & -1 \end{bmatrix}}$

■ YOUR TURN For the matrices A and B given in Example 4, find $-5A + 2B$.

Technology Tip

Enter matrices as A and B.

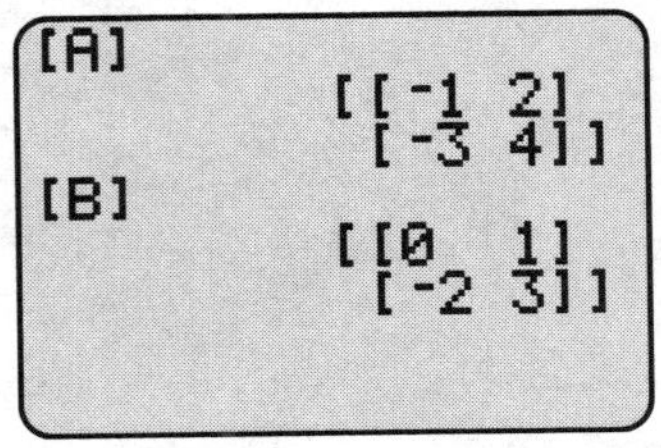

Now enter $2A$, $-3B$, $2A - 3B$.

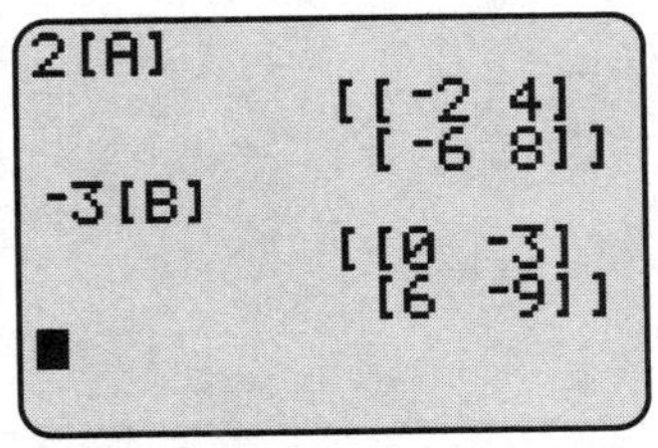

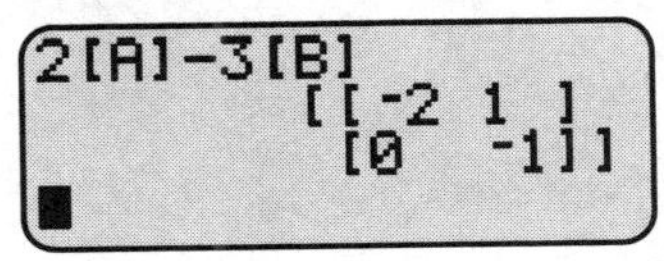

■ Answer:
$-5A + 2B = \begin{bmatrix} 5 & -8 \\ 11 & -14 \end{bmatrix}$

Matrix Multiplication

Scalar multiplication is straightforward in that it is defined for all matrices and is performed by multiplying every element in the matrix by the scalar. Addition of matrices is also an element-by-element operation. *Matrix multiplication,* on the other hand, is not as straightforward in that we *do not multiply the corresponding elements* and it is not defined for all matrices. Matrices are multiplied using a row-by-column method.

Study Tip

When we multiply matrices, we *do not* multiply corresponding elements.

Study Tip

For the product AB of two matrices A and B to be defined, the number of columns in the first matrix must equal the number of rows in the second matrix.

Before we even try to find the product AB of two matrices A and B, we first have to determine whether the product is defined. For the product AB to exist, **the number of columns in the first matrix A must equal the number of rows in the second matrix B**. In other words, if the matrix $A_{m \times n}$ has m rows and n columns and the matrix $B_{n \times p}$ has n rows and p columns, then the product $(AB)_{m \times p}$ is defined and has m rows and p columns.

Matrix:	A	B	AB
Order:	$m \times n$	$n \times p$	$m \times p$

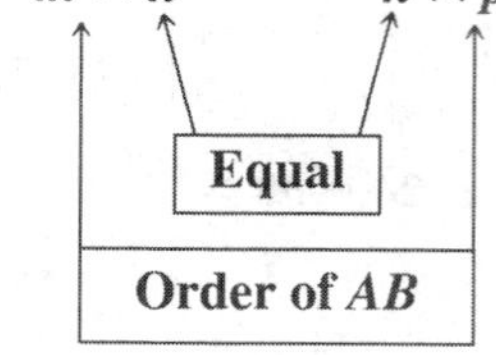

EXAMPLE 5 Determining Whether the Product of Two Matrices Is Defined

Given the matrices

$$A = \begin{bmatrix} 1 & -2 & 0 \\ 5 & -1 & 3 \end{bmatrix} \quad B = \begin{bmatrix} 2 & 3 \\ 0 & 7 \\ 4 & 9 \end{bmatrix} \quad C = \begin{bmatrix} 6 & -1 \\ 5 & 2 \end{bmatrix} \quad D = [-3 \quad -2]$$

state whether each of the following products exists. If the product exists, state the order of the product matrix.

a. AB **b.** AC **c.** BC **d.** CD **e.** DC

Solution:

Label the order of each matrix: $A_{2\times3}$, $B_{3\times2}$, $C_{2\times2}$, and $D_{1\times2}$.

a. AB is defined, because A has 3 columns and B has 3 rows. AB is order $\boxed{2 \times 2}$. $A_{2\times3}B_{3\times2}$ $(AB)_{2\times2}$

b. AC is $\boxed{\text{not defined}}$, because A has 3 columns and C has 2 rows.

c. BC is defined, because B has 2 columns and C has 2 rows. BC is order $\boxed{3 \times 2}$. $B_{3\times2}C_{2\times2}$ $(BC)_{3\times2}$

d. CD is $\boxed{\text{not defined}}$, because C has 2 columns and D has 1 row.

e. DC is defined, because D has 2 columns and C has 2 rows. DC is order $\boxed{1 \times 2}$. $D_{1\times2}C_{2\times2}$ $(DC)_{1\times2}$

Notice that in part (d) we found that CD is not defined, but in part (e) we found that DC is defined. **Matrix multiplication is not commutative**. Therefore, the order in which matrices are multiplied is important in determining whether the product is defined or undefined. For the product of two matrices to exist, the number of *columns* in the *first* matrix A must equal the number of *rows* in the *second* matrix B.

■ **YOUR TURN** For the matrices given in Example 5, state whether the following products exist. If the product exists, state the order of the product matrix.

a. DA **b.** CB **c.** BA

■ **Answer:**
a. DA exists and is order 1×3.
b. CB does not exist.
c. BA exists and is order 3×3

Now that we can determine whether a product of two matrices is defined and if so, what the order of the resulting product is, let us turn our attention to how to multiply two matrices.

DEFINITION **Matrix Multiplication**

If A is an $m \times n$ matrix and B is an $n \times p$ matrix, then their product AB is an $m \times p$ matrix whose elements are given by

$$(ab)_{ij} = a_{i1}b_{1j} + a_{i2}b_{2j} + \cdots + a_{in}b_{nj}$$

In other words, the element $(ab)_{ij}$, which is in the ith row and jth column of AB, is the sum of the products of the corresponding elements in the ith row of A and the jth column of B.

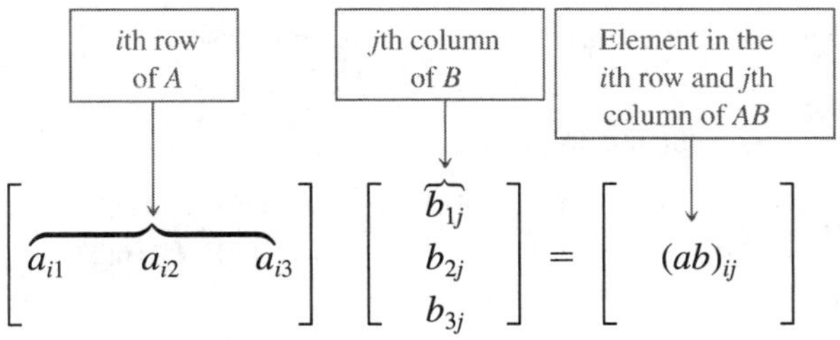

$$(ab)_{ij} = a_{i1}b_{1j} + a_{i2}b_{2j} + a_{i3}b_{3j}$$

EXAMPLE 6 Multiplication of Two 2 × 2 Matrices

Given $A = \begin{bmatrix} 1 & 2 \\ 3 & 4 \end{bmatrix}$ and $B = \begin{bmatrix} 5 & 6 \\ 7 & 8 \end{bmatrix}$, find AB.

COMMON MISTAKE

Do not multiply element by element.

CORRECT

Write the product of the two matrices A and B.

$$AB = \begin{bmatrix} 1 & 2 \\ 3 & 4 \end{bmatrix}\begin{bmatrix} 5 & 6 \\ 7 & 8 \end{bmatrix}$$

Perform the row-by-column multiplication.

$$AB = \begin{bmatrix} (1)(5) + (2)(7) & (1)(6) + (2)(8) \\ (3)(5) + (4)(7) & (3)(6) + (4)(8) \end{bmatrix}$$

Simplify.

$$AB = \begin{bmatrix} 19 & 22 \\ 43 & 50 \end{bmatrix}$$

INCORRECT

Multiply the corresponding elements.

ERROR

$$AB \neq \begin{bmatrix} (1)(5) & (2)(6) \\ (3)(7) & (4)(8) \end{bmatrix}$$

YOUR TURN For matrices A and B given in Example 6, find BA.

Technology Tip

Enter the matrices as A and B and calculate AB.

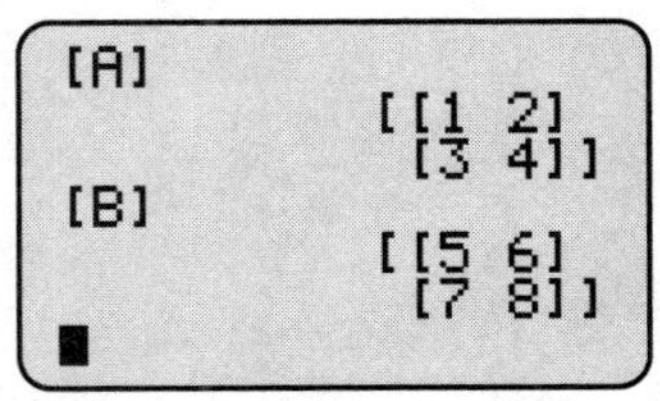

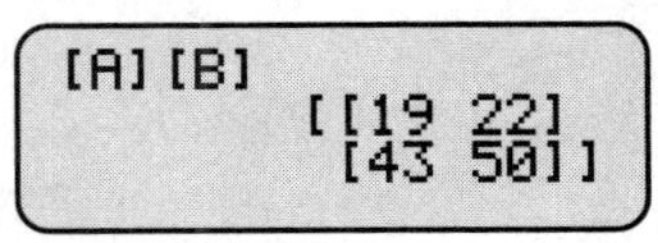

Answer:

$$BA = \begin{bmatrix} 5 & 6 \\ 7 & 8 \end{bmatrix}\begin{bmatrix} 1 & 2 \\ 3 & 4 \end{bmatrix} = \begin{bmatrix} 23 & 34 \\ 31 & 46 \end{bmatrix}$$

Compare the products obtained in Example 6 and the preceding Your Turn. Note that $AB \neq BA$. Therefore, there is **no commutative property for matrix multiplication**.

EXAMPLE 7 Multiplying Matrices

For $A = \begin{bmatrix} -1 & 2 & -3 \\ -2 & 0 & 4 \end{bmatrix}$ and $B = \begin{bmatrix} 2 & 0 \\ 1 & 3 \\ -1 & -2 \end{bmatrix}$, find AB.

Solution:

Since A is order 2×3 and B is order 3×2, the product AB is defined and has order 2×2.

$$A_{2\times 3}B_{3\times 2} = (AB)_{2\times 2}$$

Write the product of the two matrices.

$$AB = \begin{bmatrix} -1 & 2 & -3 \\ -2 & 0 & 4 \end{bmatrix}\begin{bmatrix} 2 & 0 \\ 1 & 3 \\ -1 & -2 \end{bmatrix}$$

Perform the row-by-column multiplication.

$$AB = \begin{bmatrix} (-1)(2) + (2)(1) + (-3)(-1) & (-1)(0) + (2)(3) + (-3)(-2) \\ (-2)(2) + (0)(1) + (4)(-1) & (-2)(0) + (0)(3) + (4)(-2) \end{bmatrix}$$

Simplify.

$$\boxed{AB = \begin{bmatrix} 3 & 12 \\ -8 & -8 \end{bmatrix}}$$

Technology Tip

Enter the matrices as A and B and calculate AB.

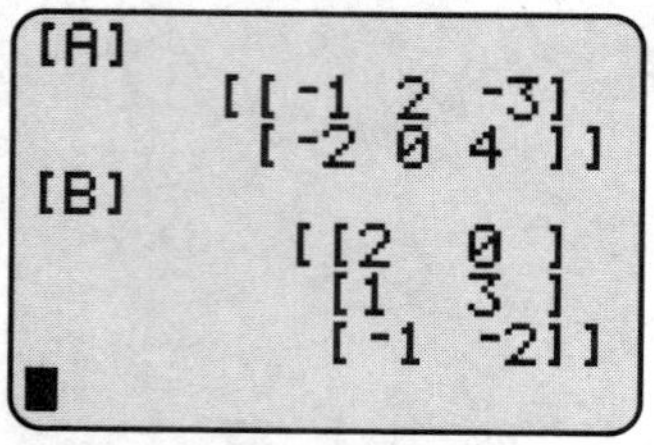

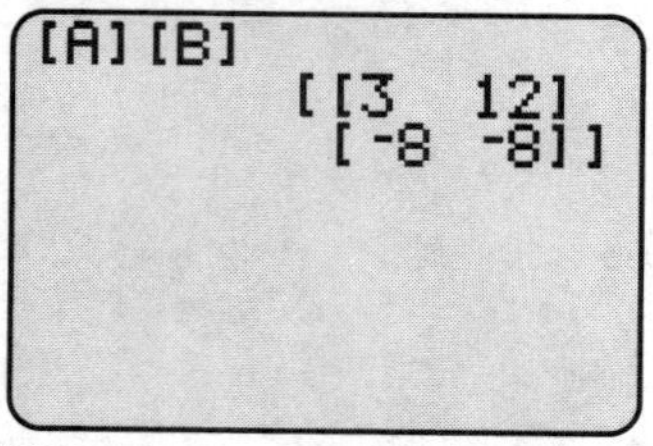

■ **YOUR TURN** For $A = \begin{bmatrix} 1 & 0 & 2 \\ -3 & -1 & 4 \end{bmatrix}$ and $B = \begin{bmatrix} 0 & -1 \\ 1 & 2 \\ 0 & -2 \end{bmatrix}$, find AB.

■ **Answer:** $AB = \begin{bmatrix} 0 & -5 \\ -1 & -7 \end{bmatrix}$

EXAMPLE 8 Multiplying Matrices

For $A = \begin{bmatrix} 1 & 0 & 3 \\ -2 & 5 & -1 \end{bmatrix}$ and $B = \begin{bmatrix} -2 & 0 & 1 \\ -3 & -1 & 4 \\ 0 & 2 & 5 \end{bmatrix}$, find AB.

Solution:

Since A is order 2×3 and B is order 3×3, the product AB is defined and has order 2×3.

$$A_{2\times 3}B_{3\times 3} = (AB)_{2\times 3}$$

Write the product of the two matrices.

$$AB = \begin{bmatrix} 1 & 0 & 3 \\ -2 & 5 & -1 \end{bmatrix}\begin{bmatrix} -2 & 0 & 1 \\ -3 & -1 & 4 \\ 0 & 2 & 5 \end{bmatrix}$$

Perform the row-by-column multiplication.

$$AB = \begin{bmatrix} (1)(-2) + (0)(-3) + (3)(0) & (1)(0) + (0)(-1) + (3)(2) & (1)(1) + (0)(4) + (3)(5) \\ (-2)(-2) + (5)(-3) + (-1)(0) & (-2)(0) + (5)(-1) + (-1)(2) & (-2)(1) + (5)(4) + (-1)(5) \end{bmatrix}$$

Simplify.

$$\boxed{AB = \begin{bmatrix} -2 & 6 & 16 \\ -11 & -7 & 13 \end{bmatrix}}$$

■ **YOUR TURN** Given $A = \begin{bmatrix} 1 \\ 2 \\ 3 \end{bmatrix}$ and $B = \begin{bmatrix} 4 & 5 \end{bmatrix}$:

a. Find AB, if it exists. **b.** Find BA, if it exists.

■ **Answer: a.** $AB = \begin{bmatrix} 4 & 5 \\ 8 & 10 \\ 12 & 15 \end{bmatrix}$

b. does not exist

Although we have shown repeatedly that there is no commutative property for matrices, matrices do have an associative property of multiplication, as well as a distributive property of multiplication similar to real numbers.

PROPERTIES OF MATRIX MULTIPLICATION

If A, B, and C are all matrices for which AB, AC, and BC are all defined, then the following properties are true:

Associative property: $A(BC) = (AB)C$

Distributive property: $A(B + C) = AB + AC$ or $(A + B)C = AC + BC$

EXAMPLE 9 Application of Matrix Multiplication

The following table gives fuel and electric requirements per mile associated with gasoline and electric automobiles.

	NUMBER OF GALLONS/MILE	NUMBER OF kW-hr/MILE
Gas Car	0.05	0
Hybrid Car	0.02	0.1
Electric Car	0	0.25

The following table gives an average cost for gasoline and electricity.

Cost per gallon of gasoline	\$3.00
Cost per kW-hr of electricity	\$0.05

a. Let matrix A represent the gasoline and electricity consumption and matrix B represent the costs of gasoline and electricity.
b. Find AB and describe what the elements of the product matrix represent.
c. Assume you drive 12,000 miles per year. What are the yearly costs associated with driving the three types of cars?

Solution (a):

A has order 3×2.

$$A = \begin{bmatrix} 0.05 & 0 \\ 0.02 & 0.1 \\ 0 & 0.25 \end{bmatrix}$$

B has order 2×1.

$$B = \begin{bmatrix} \$3.00 \\ \$0.05 \end{bmatrix}$$

Solution (b):

Find the order of the product matrix AB.

$$A_{3\times2}B_{2\times1} = (AB)_{3\times1}$$

$$AB = \begin{bmatrix} 0.05 & 0 \\ 0.02 & 0.1 \\ 0 & 0.25 \end{bmatrix}\begin{bmatrix} \$3.00 \\ \$0.05 \end{bmatrix}$$

Calculate AB.

$$= \begin{bmatrix} (0.05)(\$3.00) + (0)(\$0.05) \\ (0.02)(\$3.00) + (0.1)(\$0.05) \\ (0)(\$3.00) + (0.25)(\$0.05) \end{bmatrix}$$

$$AB = \begin{bmatrix} \$0.15 \\ \$0.065 \\ \$0.0125 \end{bmatrix}$$

Technology Tip

Enter the matrices as A and B and calculate AB.

```
[A]
        [[.05 0   ]
         [.02 .1  ]
         [0   .25]]
[B]
             [[3  ]
              [.05]]
```

```
[A][B]
          [[.15  ]
           [.065 ]
           [.0125]]
```

Interpret the product matrix.

$$AB = \begin{bmatrix} \text{Cost per mile to drive the gas car} \\ \text{Cost per mile to drive the hybrid car} \\ \text{Cost per mile to drive the electric car} \end{bmatrix}$$

Solution (c):

Find 12,000*AB*.

$$12{,}000\begin{bmatrix} \$0.15 \\ \$0.065 \\ \$0.0125 \end{bmatrix} = \begin{bmatrix} \$1{,}800 \\ \$780 \\ \$150 \end{bmatrix}$$

	GAS/ELECTRIC COSTS PER YEAR ($)
Gas Car	1,800
Hybrid Car	780
Electric Car	150

SECTION 7.2 SUMMARY

Matrices can be used to represent data. Operations such as equality, addition, subtraction, and scalar multiplication are performed element by element. Two matrices can be added or subtracted only if they have the same order. Matrix multiplication, however, requires that the number of columns in the first matrix is equal to the number of rows in the second matrix and is performed using a row by column procedure.

Matrix Multiplication Is Not Commutative: $AB \neq BA$

OPERATION	ORDER REQUIREMENT
Equality	Same: $A_{m\times n} = B_{m\times n}$
Addition	Same: $A_{m\times n} + B_{m\times n}$
Subtraction	Same: $A_{m\times n} - B_{m\times n}$
Scalar multiplication	None: $kA_{m\times n}$
Matrix multiplication	$A_{m\times n}\, B_{n\times p} = (AB)_{m\times p}$

SECTION 7.2 EXERCISES

SKILLS

In Exercises 1–10, state the order of each matrix.

1. $\begin{bmatrix} -1 & 2 & 4 \\ 7 & -3 & 9 \end{bmatrix}$

2. $\begin{bmatrix} 3 & 5 \\ 2 & 6 \\ -1 & -4 \end{bmatrix}$

3. $\begin{bmatrix} -4 & 5 \\ 0 & 1 \end{bmatrix}$

4. $\begin{bmatrix} -4 & 5 & 3 & 7 \end{bmatrix}$

5. $\begin{bmatrix} -3 & 4 & 1 \\ 10 & 8 & 0 \\ -2 & 5 & 7 \end{bmatrix}$

6. $\begin{bmatrix} 1 \\ 2 \\ 3 \\ 4 \end{bmatrix}$

7. $\begin{bmatrix} -2 \end{bmatrix}$

8. $\begin{bmatrix} -1 & 3 & 6 & 9 \\ 2 & 5 & -7 & 8 \end{bmatrix}$

9. $\begin{bmatrix} -3 & 6 & 0 & 5 \\ 4 & -9 & 2 & 7 \\ 1 & 8 & 3 & 6 \\ 5 & 0 & -4 & 11 \end{bmatrix}$

10. $\begin{bmatrix} 3 & 7 & -1 & 5 & 8 \end{bmatrix}$

In Exercises 11–16, solve for the indicated variable.

11. $\begin{bmatrix} 2 & x \\ y & 3 \end{bmatrix} = \begin{bmatrix} 2 & -5 \\ 1 & 3 \end{bmatrix}$

12. $\begin{bmatrix} -3 & 17 \\ x & y \end{bmatrix} = \begin{bmatrix} -3 & 17 \\ 10 & 12 \end{bmatrix}$

13. $\begin{bmatrix} x+y & 3 \\ x-y & 9 \end{bmatrix} = \begin{bmatrix} -5 & z \\ -1 & 9 \end{bmatrix}$

14. $\begin{bmatrix} x & -4 \\ y & 7 \end{bmatrix} = \begin{bmatrix} 2+y & -4 \\ 5 & 7 \end{bmatrix}$

15. $\begin{bmatrix} 3 & 4 \\ 0 & 12 \end{bmatrix} = \begin{bmatrix} x-y & 4 \\ 0 & 2y+x \end{bmatrix}$

16. $\begin{bmatrix} 9 & 2b+1 \\ -5 & 16 \end{bmatrix} = \begin{bmatrix} a^2 & 9 \\ 2a+1 & b^2 \end{bmatrix}$

In Exercises 17–30, perform the indicated operations for each expression, if possible.

$$A = \begin{bmatrix} -1 & 3 & 0 \\ 2 & 4 & 1 \end{bmatrix} \quad B = \begin{bmatrix} 0 & 2 & 1 \\ 3 & -2 & 4 \end{bmatrix} \quad C = \begin{bmatrix} 0 & 1 \\ 2 & -1 \\ 3 & 1 \end{bmatrix} \quad D = \begin{bmatrix} 2 & -3 \\ 0 & 1 \\ 4 & -2 \end{bmatrix}$$

17. $A + B$ **18.** $C + D$ **19.** $C - D$ **20.** $A - B$ **21.** $B + C$

22. $A + D$ **23.** $D - B$ **24.** $C - A$ **25.** $2A$ **26.** $4D$

27. $-5C$ **28.** $-2B$ **29.** $2A + 3B$ **30.** $2B - 3A$

In Exercises 31–50, perform the indicated operations for each expression, if possible.

$$A = \begin{bmatrix} 1 & 2 & -1 \\ 0 & 3 & 1 \\ 5 & 0 & -2 \end{bmatrix} \quad B = [2 \quad 0 \quad -3] \quad C = \begin{bmatrix} -1 & 7 & 2 \\ 3 & 0 & 1 \end{bmatrix} \quad D = \begin{bmatrix} 3 & 0 \\ 1 & -1 \\ 2 & 5 \end{bmatrix}$$

$$E = \begin{bmatrix} -1 & 0 & 1 \\ 2 & 1 & 4 \\ -3 & 1 & 5 \end{bmatrix} \quad F = \begin{bmatrix} 1 \\ 0 \\ -1 \end{bmatrix} \quad G = \begin{bmatrix} 1 & 2 \\ 3 & 4 \end{bmatrix}$$

31. CD **32.** BF **33.** DC **34.** $(A + E)D$ **35.** DG

36. $2A + 3E$ **37.** GD **38.** $ED + C$ **39.** $-4BD$ **40.** $-3ED$

41. $B(A + E)$ **42.** $GC + 5C$ **43.** $FB + 5A$ **44.** A^2 **45.** $G^2 + 5G$

46. $C \cdot (2E)$ **47.** $(2E) \cdot F$ **48.** $CA + 5C$ **49.** DF **50.** AE

APPLICATIONS

51. Smoking. On January 6 and 10, 2000, the Harris Poll conducted a survey of adult smokers in the United States. When asked, "Have you ever tried to quit smoking?" 70% said yes and 30% said no. Write a 2×1 matrix—call it A—that represents those smokers. When asked what consequences smoking would have on their lives, 89% believed it would increase their chance of getting lung cancer and 84% believed smoking would shorten their lives. Write a 2×1 matrix—call it B—that represents those smokers. If there are 46 million adult smokers in the United States:

a. What does $46A$ tell us?
b. What does $46B$ tell us?

52. Women in Science. According to the study of science and engineering indicators by the National Science Foundation (www.nsf.gov), the number of female graduate students in science and engineering disciplines has increased over the last 30 years. In 1981, 24% of mathematics graduate students were female and 23% of graduate students in computer science were female. In 1991, 32% of mathematics graduate students and 21% of computer science graduate students were female. In 2001, 38% of mathematics graduate students and 30% of computer science graduate students were female. Write three 2×1 matrices representing the percentage of female graduate students.

$$A = \begin{bmatrix} \text{\% female–math–1981} \\ \text{\%female–C.S.–1981} \end{bmatrix}$$

$$B = \begin{bmatrix} \text{\% female–math–1991} \\ \text{\%female–C.S.–1991} \end{bmatrix}$$

$$C = \begin{bmatrix} \text{\% female–math–2001} \\ \text{\%female–C.S.–2001} \end{bmatrix}$$

What does $C - B$ tell us? What does $B - A$ tell us? What can you conclude about the number of women pursuing mathematics and computer science graduate degrees?

Note: C.S. = computer science.

53. Registered Voters. According to the U.S. Census Bureau (www.census.gov), in the 2000 national election, 58.9% of the men over the age of 18 were registered voters, but only 41.4% voted; and 62.8% of women over 18 were registered voters, but only 43% actually voted. Write a 2×2 matrix with the following data:

$$A = \begin{bmatrix} \text{Percentage of registered male voters} & \text{Percentage of registered female voters} \\ \text{Percent of males that voted} & \text{Percent of females that voted} \end{bmatrix}$$

If we let B be a 2×1 matrix representing the total population of males and females over the age of 18 in the United States, or $B = \begin{bmatrix} 100\text{ M} \\ 110\text{ M} \end{bmatrix}$, what does AB tell us?

54. Job Application. A company has two rubrics for scoring job applicants based on weighting education, experience, and the interview differently.

Matrix *A*

	Rubric 1	Rubric 2
Education	0.5	0.6
Experience	0.3	0.1
Interview	0.2	0.3

Applicants receive a score from 1 to 10 in each category (education, experience, and interview). Two applicants are shown in matrix B:

Matrix *B*

	Education	Experience	Interview
Applicant 1	8	7	5
Applicant 2	6	8	8

What is the order of BA? What does each entry in BA tell us?

55. Taxes. The IRS allows an individual to deduct business expenses in the following way: \$0.45 per mile driven, 50% of entertainment costs, and 100% of actual expenses. Represent these deductions as a row matrix A. In 2006, Jamie had the following business expenses: \$2,700 in entertainment, \$15,200 actual expenses, and he drove 7523 miles. Represent Jamie's expenses as a column matrix B. Multiply these matrices to find the total amount of business expenses Jamie can claim on his 2006 federal tax form: AB.

56. Tips on Service. Marilyn decides to go to the Safety Harbor Spa for a day of pampering. She is treated to a Hot Stone Massage (\$85), a manicure and pedicure (\$75), and a haircut and style (\$100). Represent the costs of the individual services as a row matrix A. She decides to tip her masseur 25%, her nail tech 20%, and her hair stylist 15%. Represent the tipping percentages as a column matrix B. Multiply these matrices to find the total amount in tips AB she needs to add to her final bill.

Use the following nutritional chart for Exercises 57 and 58:

SANDWICH	CALORIES	FAT (g)	CARBOHYDRATES (g)	PROTEIN (g)
6″ Veggie Delight	230	3	44	9
6″ Tuna	430	19	46	20
6″ Roast beef	290	5	45	19
6″ Chicken	330	5	47	24

57. Nutrition. Utilize the table to write a 4×4 matrix A. Find $2A$. What would $2A$ represent? Find $0.5A$. What would $0.5A$ represent?

58. Nutrition. Don decides to film a documentary similar to *Supersize Me*, but instead of eating only at McDonalds, he decides to eat only at Subway. He is hoping to lose weight and name the film *Minimize Me*. He is going to eat only three meals a day from the four options listed in the preceding table. Each week he will consume the following sandwiches: 7 veggie, 5 tuna, 8 roast beef, and 1 chicken. Let A be the matrix (found in Exercise 57) that represents the nutrition values and let B be a row matrix that represents the number of each type of sandwich consumed in a week. Find BA. What does BA represent? Find $\frac{1}{7}BA$. What does $\frac{1}{7}BA$ represent?

Use the following tables for exercises 59 and 60:

The following table gives fuel and electric requirements per mile associated with gasoline and electric automobiles.

	NUMBER OF GALLONS/MILE	NUMBER OF KW-HR/MILE
SUV full size	0.06	0
Hybrid car	0.02	0.1
Electric car	0	0.3

The following table gives an average cost for gasoline and electricity.

Cost per gallon of gasoline	\$3.80
Cost per kW-hr of electricity	\$0.05

59. Environment. Let matrix A represent the gasoline and electricity consumption and matrix B represent the costs of gasoline and electricity. Find AB and describe what the elements of the product matrix represent.

60. Environment. Assume you drive 12,000 miles per year. What are the yearly costs associated with driving the three types of cars in Exercise 59?

CATCH THE MISTAKE

In Exercises 61 and 62, explain the mistake that is made.

61. Multiply $\begin{bmatrix} 3 & 2 \\ 1 & 4 \end{bmatrix}\begin{bmatrix} -1 & 3 \\ -2 & 5 \end{bmatrix}$.

Solution:

Multiply corresponding elements.

$$\begin{bmatrix} 3 & 2 \\ 1 & 4 \end{bmatrix}\begin{bmatrix} -1 & 3 \\ -2 & 5 \end{bmatrix} = \begin{bmatrix} (3)(-1) & (2)(3) \\ (1)(-2) & (4)(5) \end{bmatrix}$$

Simplify. $$\begin{bmatrix} 3 & 2 \\ 1 & 4 \end{bmatrix}\begin{bmatrix} -1 & 3 \\ -2 & 5 \end{bmatrix} = \begin{bmatrix} -3 & 6 \\ -2 & 20 \end{bmatrix}$$

This is incorrect. What mistake was made?

62. Multiply $\begin{bmatrix} 3 & 2 \\ 1 & 4 \end{bmatrix}\begin{bmatrix} -1 & 3 \\ -2 & 5 \end{bmatrix}$.

Solution:

Multiply using column-by-row method.

$$\begin{bmatrix} 3 & 2 \\ 1 & 4 \end{bmatrix}\begin{bmatrix} -1 & 3 \\ -2 & 5 \end{bmatrix} = \begin{bmatrix} (3)(-1) + (1)(3) & (2)(-1) + (4)(3) \\ (3)(-2) + (1)(5) & (2)(-2) + (4)(5) \end{bmatrix}$$

Simplify. $$\begin{bmatrix} 3 & 2 \\ 1 & 4 \end{bmatrix}\begin{bmatrix} -1 & 3 \\ -2 & 5 \end{bmatrix} = \begin{bmatrix} 0 & 10 \\ -1 & 16 \end{bmatrix}$$

This is incorrect. What mistake was made?

CONCEPTUAL

In Exercises 63–66, determine whether the statements are true or false.

63. If $A = \begin{bmatrix} a_{11} & a_{12} \\ a_{21} & a_{22} \end{bmatrix}$ and $B = \begin{bmatrix} b_{11} & b_{12} \\ b_{21} & b_{22} \end{bmatrix}$, then

$AB = \begin{bmatrix} a_{11}b_{11} & a_{12}b_{12} \\ a_{21}b_{21} & a_{22}b_{22} \end{bmatrix}$.

64. If AB is defined, then $AB = BA$.

65. AB is defined only if the number of columns in A equals the number of rows in B.

66. $A + B$ is defined only if A and B have the same order.

67. For $A = \begin{bmatrix} a_{11} & a_{12} \\ a_{21} & a_{22} \end{bmatrix}$, find A^2.

68. In order for $A^2_{m\times n}$ to be defined, what condition (with respect to m and n) must be met?

CHALLENGE

69. For $A = \begin{bmatrix} 1 & 1 \\ 1 & 1 \end{bmatrix}$ find $A, A^2, A^3, \ldots$. What is A^n?

70. For $A = \begin{bmatrix} 1 & 0 \\ 0 & 1 \end{bmatrix}$ find $A, A^2, A^3, \ldots$. What is A^n?

71. If $A_{m\times n}B_{n\times p}$ is defined, explain why $(A_{m\times n}B_{n\times p})^2$ is not defined for $m \neq p$.

72. If $A_{m\times n} = B_{m\times n}$ and $C_{n\times m}$, explain why $AC \neq CB$, if $m \neq n$.

TECHNOLOGY

In Exercises 73–78, apply a graphing utility to perform the indicated matrix operations, if possible.

$$A = \begin{bmatrix} 1 & 7 & 9 & 2 \\ -3 & -6 & 15 & 11 \\ 0 & 3 & 2 & 5 \\ 9 & 8 & -4 & 1 \end{bmatrix} \qquad B = \begin{bmatrix} 7 & 9 \\ 8 & 6 \\ -4 & -2 \\ 3 & 1 \end{bmatrix}$$

73. AB **74.** BA **75.** BB **76.** AA

$$A = \begin{bmatrix} 2 & 1 & 1 \\ -3 & 0 & 2 \\ 4 & -6 & 0 \end{bmatrix}$$

77. A^2 **78.** A^5

SECTION 7.3 MATRIX EQUATIONS; THE INVERSE OF A SQUARE MATRIX

SKILLS OBJECTIVES

- Write a system of linear equations as a matrix equation.
- Find the inverse of a square matrix.
- Solve systems of linear equations using inverse matrices.

CONCEPTUAL OBJECTIVES

- Visualize a system of linear equations as a matrix equation.
- Understand that only a square matrix can have an inverse.
- Realize that not every square matrix has an inverse.

Matrix Equations

Matrix equations are another way of writing systems of linear equations.

WORDS	**MATH**
Start with a matrix equation.	$\begin{bmatrix} 2 & -3 \\ 1 & 5 \end{bmatrix}\begin{bmatrix} x \\ y \end{bmatrix} = \begin{bmatrix} -7 \\ 9 \end{bmatrix}$
Multiply the two matrices on the left.	$\begin{bmatrix} 2x - 3y \\ x + 5y \end{bmatrix} = \begin{bmatrix} -7 \\ 9 \end{bmatrix}$
Apply equality of two matrices.	$\begin{aligned} 2x - 3y &= -7 \\ x + 5y &= 9 \end{aligned}$

Let A be a matrix with m rows and n columns, which represents the coefficients in the system; let X represent the variables in the system; and let B represent the constants in the system. Then, a system of linear equations can be written as $AX = B$.

System of Linear Equations	A	X	B	Matrix Equation: $AX = B$
$\begin{aligned} 3x + 4y &= 1 \\ x - 2y &= 7 \end{aligned}$	$\begin{bmatrix} 3 & 4 \\ 1 & -2 \end{bmatrix}$	$\begin{bmatrix} x \\ y \end{bmatrix}$	$\begin{bmatrix} 1 \\ 7 \end{bmatrix}$	$\begin{bmatrix} 3 & 4 \\ 1 & -2 \end{bmatrix}\begin{bmatrix} x \\ y \end{bmatrix} = \begin{bmatrix} 1 \\ 7 \end{bmatrix}$
$\begin{aligned} x - y + z &= 2 \\ 2x + 2y - 3z &= -3 \\ x + y + z &= 6 \end{aligned}$	$\begin{bmatrix} 1 & -1 & 1 \\ 2 & 2 & -3 \\ 1 & 1 & 1 \end{bmatrix}$	$\begin{bmatrix} x \\ y \\ z \end{bmatrix}$	$\begin{bmatrix} 2 \\ -3 \\ 6 \end{bmatrix}$	$\begin{bmatrix} 1 & -1 & 1 \\ 2 & 2 & -3 \\ 1 & 1 & 1 \end{bmatrix}\begin{bmatrix} x \\ y \\ z \end{bmatrix} = \begin{bmatrix} 2 \\ -3 \\ 6 \end{bmatrix}$
$\begin{aligned} x + y + z &= 0 \\ 3x + 2y - z &= 2 \end{aligned}$	$\begin{bmatrix} 1 & 1 & 1 \\ 3 & 2 & -1 \end{bmatrix}$	$\begin{bmatrix} x \\ y \\ z \end{bmatrix}$	$\begin{bmatrix} 0 \\ 2 \end{bmatrix}$	$\begin{bmatrix} 1 & 1 & 1 \\ 3 & 2 & -1 \end{bmatrix}\begin{bmatrix} x \\ y \\ z \end{bmatrix} = \begin{bmatrix} 0 \\ 2 \end{bmatrix}$

EXAMPLE 1 **Writing a System of Linear Equations as a Matrix Equation**

Write each system of linear equations as a matrix equation.

a. $\begin{aligned} 2x - y &= 5 \\ -x + 2y &= 3 \end{aligned}$ **b.** $\begin{aligned} 3x - 2y + 4z &= 5 \\ y - 3z &= -2 \\ 7x \quad - z &= 1 \end{aligned}$ **c.** $\begin{aligned} x_1 - x_2 + 2x_3 - 3 &= 0 \\ x_1 + x_2 - 3x_3 + 5 &= 0 \\ x_1 - x_2 + x_3 - 2 &= 0 \end{aligned}$

Solution:

a. $$\begin{bmatrix} 2 & -1 \\ -1 & 2 \end{bmatrix}\begin{bmatrix} x \\ y \end{bmatrix} = \begin{bmatrix} 5 \\ 3 \end{bmatrix}$$

b. Note that all missing terms have 0 coefficients.

$$\begin{aligned} 3x - 2y + 4z &= 5 \\ 0x + y - 3z &= -2 \\ 7x + 0y - z &= 1 \end{aligned}$$

$$\begin{bmatrix} 3 & -2 & 4 \\ 0 & 1 & -3 \\ 7 & 0 & -1 \end{bmatrix}\begin{bmatrix} x \\ y \\ z \end{bmatrix} = \begin{bmatrix} 5 \\ -2 \\ 1 \end{bmatrix}$$

c. Write the constants on the right side of the equal sign.

$$\begin{aligned} x_1 - x_2 + 2x_3 &= 3 \\ x_1 + x_2 - 3x_3 &= -5 \\ x_1 - x_2 + x_3 &= 2 \end{aligned}$$

$$\begin{bmatrix} 1 & -1 & 2 \\ 1 & 1 & -3 \\ 1 & -1 & 1 \end{bmatrix}\begin{bmatrix} x_1 \\ x_2 \\ x_3 \end{bmatrix} = \begin{bmatrix} 3 \\ -5 \\ 2 \end{bmatrix}$$

YOUR TURN Write each system of linear equations as a matrix equation.

a. $\begin{aligned} 2x + y - 3 &= 0 \\ x - y \quad &= 5 \end{aligned}$ **b.** $\begin{aligned} y - x + z &= 7 \\ x - y - z &= 2 \\ z - y &= -1 \end{aligned}$

Answer:

a. $\begin{bmatrix} 2 & 1 \\ 1 & -1 \end{bmatrix}\begin{bmatrix} x \\ y \end{bmatrix} = \begin{bmatrix} 3 \\ 5 \end{bmatrix}$

b. $\begin{bmatrix} -1 & 1 & 1 \\ 1 & -1 & -1 \\ 0 & -1 & 1 \end{bmatrix}\begin{bmatrix} x \\ y \\ z \end{bmatrix} = \begin{bmatrix} 7 \\ 2 \\ -1 \end{bmatrix}$

Finding the Inverse of a Matrix

Before we discuss solving systems of linear equations in the form $AX = B$, let us first recall how we solve $ax = b$, where a and b are real numbers (not matrices).

WORDS	**MATH**
Write the linear equation in one variable.	$ax = b$
Multiply both sides by a^{-1} (same as dividing by a), provided $a \neq 0$.	$a^{-1}ax = a^{-1}b$
Simplify.	$\underbrace{a^{-1}a}_{1}x = a^{-1}b$
	$x = a^{-1}b$

Recall that a^{-1}, or $\frac{1}{a}$, is the *multiplicative inverse* (Chapter 0) of a because $a^{-1}a = 1$. And we call 1 the *multiplicative identity*, because any number multiplied by 1 is itself. Before we solve matrix equations, we need to define the *multiplicative identity matrix* and the *multiplicative inverse matrix*.

A square matrix of order n with 1s along the **main diagonal** (a_{ii}) and 0s for all other elements is called the **multiplicative identity matrix I_n**.

$$I_2 = \begin{bmatrix} 1 & 0 \\ 0 & 1 \end{bmatrix} \qquad I_3 = \begin{bmatrix} 1 & 0 & 0 \\ 0 & 1 & 0 \\ 0 & 0 & 1 \end{bmatrix} \qquad I_4 = \begin{bmatrix} 1 & 0 & 0 & 0 \\ 0 & 1 & 0 & 0 \\ 0 & 0 & 1 & 0 \\ 0 & 0 & 0 & 1 \end{bmatrix}$$

Since a real number multiplied by 1 is itself ($a \cdot 1 = a$), we expect that a matrix multiplied by the appropriate identity matrix should result in itself:

$$A_{m \times n} I_n = A_{m \times n} \qquad \text{and} \qquad I_m A_{m \times n} = A_{m \times n}$$

EXAMPLE 2 Multiplying a Matrix by the Multiplicative Identity Matrix I_n

For $A = \begin{bmatrix} -2 & 4 & 1 \\ 3 & 7 & -1 \end{bmatrix}$, find $I_2 A$.

Solution:

Write the two matrices. $A = \begin{bmatrix} -2 & 4 & 1 \\ 3 & 7 & -1 \end{bmatrix} \qquad I_2 = \begin{bmatrix} 1 & 0 \\ 0 & 1 \end{bmatrix}$

Find the product $I_2 A$. $I_2 A = \begin{bmatrix} 1 & 0 \\ 0 & 1 \end{bmatrix} \begin{bmatrix} -2 & 4 & 1 \\ 3 & 7 & -1 \end{bmatrix}$

$$I_2 A = \begin{bmatrix} (1)(-2) + (0)(3) & (1)(4) + (0)(7) & (1)(1) + (0)(-1) \\ (0)(-2) + (1)(3) & (0)(4) + (1)(7) & (0)(1) + (1)(-1) \end{bmatrix}$$

$$I_2 A = \boxed{\begin{bmatrix} -2 & 4 & 1 \\ 3 & 7 & -1 \end{bmatrix}} = A$$

■ **YOUR TURN** For A in Example 2, find AI_3.

■ **Answer:**

$AI_3 = \begin{bmatrix} -2 & 4 & 1 \\ 3 & 7 & -1 \end{bmatrix} = A$

The identity matrix I_n will assist us in developing the concept of an *inverse of a square matrix.*

Study Tip

- Only a *square* matrix can have an inverse.
- Not all square matrices have inverses.

DEFINITION Inverse of a Square Matrix

Let A be a square $n \times n$ matrix. If there exists a square $n \times n$ matrix A^{-1} such that

$$AA^{-1} = I_n \qquad \text{and} \qquad A^{-1}A = I_n$$

then A^{-1}, stated as "A inverse," is the **inverse** of A.

It is important to note that only a square matrix can have an inverse. Even then, not all square matrices have inverses.

EXAMPLE 3 Multiplying a Matrix by Its Inverse

Verify that the inverse of $A = \begin{bmatrix} 1 & 3 \\ 2 & 5 \end{bmatrix}$ is $A^{-1} = \begin{bmatrix} -5 & 3 \\ 2 & -1 \end{bmatrix}$.

Solution:

Show that $AA^{-1} = I_2$ and $A^{-1}A = I_2$.

Find the product AA^{-1}.

$$AA^{-1} = \begin{bmatrix} 1 & 3 \\ 2 & 5 \end{bmatrix}\begin{bmatrix} -5 & 3 \\ 2 & -1 \end{bmatrix}$$

$$= \begin{bmatrix} (1)(-5) + (3)(2) & (1)(3) + (3)(-1) \\ (2)(-5) + (5)(2) & (2)(3) + (5)(-1) \end{bmatrix}$$

$$= \begin{bmatrix} 1 & 0 \\ 0 & 1 \end{bmatrix} = I_2$$

Find the product $A^{-1}A$.

$$A^{-1}A = \begin{bmatrix} -5 & 3 \\ 2 & -1 \end{bmatrix}\begin{bmatrix} 1 & 3 \\ 2 & 5 \end{bmatrix}$$

$$= \begin{bmatrix} (-5)(1) + (3)(2) & (-5)(3) + (3)(5) \\ (2)(1) + (-1)(2) & (2)(3) + (-1)(5) \end{bmatrix}$$

$$= \begin{bmatrix} 1 & 0 \\ 0 & 1 \end{bmatrix} = I_2$$

■ **YOUR TURN** Verify that the inverse of $A = \begin{bmatrix} 1 & 4 \\ 2 & 9 \end{bmatrix}$ is $A^{-1} = \begin{bmatrix} 9 & -4 \\ -2 & 1 \end{bmatrix}$.

Technology Tip

Enter the matrices as A, and A^{-1} as B.

```
[A]
          [[1 3]
           [2 5]]
[B]
        [[-5 3 ]
         [2  -1]]
█
```

Enter AA^{-1} as AB, and A^{-1} as BA.

```
[A][B]
          [[1 0]
           [0 1]]
[B][A]
          [[1 0]
           [0 1]]
```

■ **Answer:** $AA^{-1} = A^{-1}A = I_2$

Now that we can show that two matrices are inverses of one another, let us describe the process for finding an inverse, if it exists. If an inverse A^{-1} exists, then the matrix A is said to be **nonsingular**. If the inverse does not exist, then the matrix A is said to be **singular**.

Let $A = \begin{bmatrix} 1 & -1 \\ 2 & -3 \end{bmatrix}$ and the inverse be $A^{-1} = \begin{bmatrix} w & x \\ y & z \end{bmatrix}$, where w, x, y, and z are variables to be determined. A matrix and its inverse must satisfy the identity $AA^{-1} = I_2$.

WORDS	MATH
The product of a matrix and its inverse is the identity matrix.	$\begin{bmatrix} 1 & -1 \\ 2 & -3 \end{bmatrix}\begin{bmatrix} w & x \\ y & z \end{bmatrix} = \begin{bmatrix} 1 & 0 \\ 0 & 1 \end{bmatrix}$
Multiply the two matrices on the left.	$\begin{bmatrix} w - y & x - z \\ 2w - 3y & 2x - 3z \end{bmatrix} = \begin{bmatrix} 1 & 0 \\ 0 & 1 \end{bmatrix}$
Equate corresponding matrix elements.	$\begin{aligned} w - y &= 1 \\ 2w - 3y &= 0 \end{aligned}$ and $\begin{aligned} x - z &= 0 \\ 2x - 3z &= 1 \end{aligned}$

Notice that there are two systems of equations, both of which can be solved by several methods (elimination, substitution, or augmented matrices). We will find that $w = 3$, $x = -1$, $y = 2$, and $z = -1$. Therefore, we know the inverse is $A^{-1} = \begin{bmatrix} 3 & -1 \\ 2 & -1 \end{bmatrix}$. But, instead, let us use augmented matrices in order to develop the general procedure.

Write the two systems of equations as two augmented matrices:

$$\begin{array}{cc} w & y \end{array} \qquad\qquad \begin{array}{cc} x & z \end{array}$$

$$\left[\begin{array}{cc|c} 1 & -1 & 1 \\ 2 & -3 & 0 \end{array}\right] \qquad \left[\begin{array}{cc|c} 1 & -1 & 0 \\ 2 & -3 & 1 \end{array}\right]$$

Since the left side is the same for each augmented matrix we can combine these two matrices into one matrix, thereby simultaneously solving both systems of equations:

$$\left[\begin{array}{rr|rr} 1 & -1 & 1 & 0 \\ 2 & -3 & 0 & 1 \end{array}\right]$$

Notice that the right side of the vertical line is the identity matrix I_2.

Using Gauss–Jordan elimination, transform the matrix on the left to the identity matrix.

$$\left[\begin{array}{rr|rr} 1 & -1 & 1 & 0 \\ 2 & -3 & 0 & 1 \end{array}\right]$$

$$R_2 - 2R_1 \to R_2 \quad \left[\begin{array}{rr|rr} 1 & -1 & 1 & 0 \\ 0 & -1 & -2 & 1 \end{array}\right]$$

$$-R_2 \to R_2 \quad \left[\begin{array}{rr|rr} 1 & -1 & 1 & 0 \\ 0 & 1 & 2 & -1 \end{array}\right]$$

$$R_1 + R_2 \to R_1 \quad \left[\begin{array}{rr|rr} 1 & 0 & 3 & -1 \\ 0 & 1 & 2 & -1 \end{array}\right]$$

The matrix on the right of the vertical line is the inverse $A^{-1} = \begin{bmatrix} 3 & -1 \\ 2 & -1 \end{bmatrix}$.

FINDING THE INVERSE OF A SQUARE MATRIX

To find the inverse of an $n \times n$ matrix A:

Step 1: Form the matrix $[A \mid I_n]$.

Step 2: Use row operations to transform this entire matrix to $[I_n \mid A^{-1}]$. This is done by applying Gauss–Jordan elimination to reduce A to the identity matrix I_n (which is reduced to row–echelon form). If this is not possible, then A is a singular matrix and no inverse exists.

Step 3: Verify the result by showing that $AA^{-1} = I_n$ and $A^{-1}A = I_n$.

EXAMPLE 4 Finding the Inverse of a 2 × 2 Matrix

Find the inverse of $A = \begin{bmatrix} 1 & 2 \\ 3 & 5 \end{bmatrix}$.

Solution:

STEP 1 Form the matrix $[A \mid I_2]$. $\quad \left[\begin{array}{rr|rr} 1 & 2 & 1 & 0 \\ 3 & 5 & 0 & 1 \end{array}\right]$

STEP 2 Use row operations to transform A into I_2.

$$R_2 - 3R_1 \to R_1 \quad \left[\begin{array}{rr|rr} 1 & 2 & 1 & 0 \\ 0 & -1 & -3 & 1 \end{array}\right]$$

$$-R_2 \to R_2 \quad \left[\begin{array}{rr|rr} 1 & 2 & 1 & 0 \\ 0 & 1 & 3 & -1 \end{array}\right]$$

$$R_1 - 2R_2 \to R_1 \quad \left[\begin{array}{rr|rr} 1 & 0 & -5 & 2 \\ 0 & 1 & 3 & -1 \end{array}\right]$$

Identify the inverse. $\quad \boxed{A^{-1} = \begin{bmatrix} -5 & 2 \\ 3 & -1 \end{bmatrix}}$

Technology Tip

A graphing calculator can be used to find the inverse of A. Enter the matrix A.

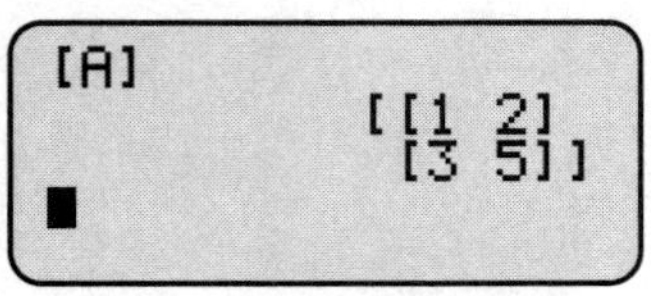

To find A^{-1}, press 2nd MATRIX 1:[A] ENTER X^{-1} ENTER.

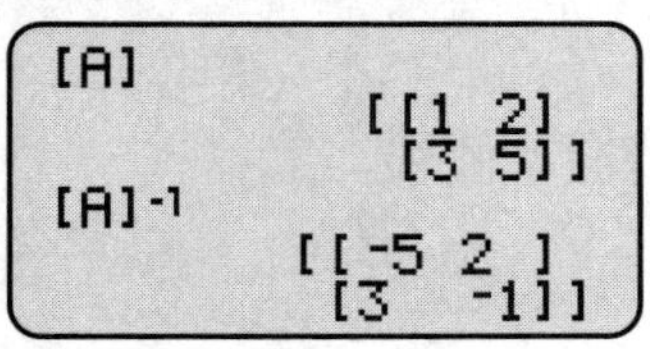

STEP 3 Check.

$$AA^{-1} = \begin{bmatrix} 1 & 2 \\ 3 & 5 \end{bmatrix}\begin{bmatrix} -5 & 2 \\ 3 & -1 \end{bmatrix} = \begin{bmatrix} 1 & 0 \\ 0 & 1 \end{bmatrix} = I_2$$

$$A^{-1}A = \begin{bmatrix} -5 & 2 \\ 3 & -1 \end{bmatrix}\begin{bmatrix} 1 & 2 \\ 3 & 5 \end{bmatrix} = \begin{bmatrix} 1 & 0 \\ 0 & 1 \end{bmatrix} = I_2$$

■ **YOUR TURN** Find the inverse of $A = \begin{bmatrix} 2 & 3 \\ 5 & 8 \end{bmatrix}$.

■ **Answer:** $A^{-1} = \begin{bmatrix} 8 & -3 \\ -5 & 2 \end{bmatrix}$

This procedure for finding an inverse of a square matrix is used for all square matrices of order n. For the special case of a 2×2 matrix, there is a formula (that will be derived in Exercises 61 and 62) for finding the inverse.

Let $A = \begin{bmatrix} a & b \\ c & d \end{bmatrix}$ represent any 2×2 matrix; then the inverse matrix is given by

$$A^{-1} = \frac{1}{ad - bc}\begin{bmatrix} d & -b \\ -c & a \end{bmatrix} \qquad ad - bc \neq 0$$

The denominator $ad - bc$ is called the *determinant* and will be discussed in Section 7.4.

We found the inverse of $A = \begin{bmatrix} 1 & 2 \\ 3 & 5 \end{bmatrix}$ in Example 4. Let us now find the inverse using this formula.

WORDS	**MATH**
Write the formula for A^{-1}.	$A^{-1} = \frac{1}{ad - bc}\begin{bmatrix} d & -b \\ -c & a \end{bmatrix}$
Substitute $a = 1, b = 2, c = 3, d = 5$ into the formula.	$A^{-1} = \frac{1}{(1)(5) - (2)(3)}\begin{bmatrix} 5 & -2 \\ -3 & 1 \end{bmatrix}$
Simplify.	$A^{-1} = (-1)\begin{bmatrix} 5 & -2 \\ -3 & 1 \end{bmatrix}$
	$A^{-1} = \begin{bmatrix} -5 & 2 \\ 3 & -1 \end{bmatrix}$

The result is the same as that we found in Example 4.

EXAMPLE 5 Finding That No Inverse Exists: Singular Matrix

Find the inverse of $A = \begin{bmatrix} 1 & -5 \\ -1 & 5 \end{bmatrix}$.

Solution:

STEP 1 Form the matrix $[A \mid I_2]$.

$$\left[\begin{array}{cc|cc} 1 & -5 & 1 & 0 \\ -1 & 5 & 0 & 1 \end{array}\right]$$

STEP 2 Apply row operations to transform A into I_2.

$$R_2 + R_1 \rightarrow R_2 \quad \left[\begin{array}{cc|cc} 1 & -5 & 1 & 0 \\ 0 & 0 & 1 & 1 \end{array}\right]$$

We cannot convert the left-hand side of the augmented matrix to I_2 because of the all-zero row on the left-hand side. Therefore, **A is not invertible**.

Technology Tip

A graphing calculator can be used to find the inverse of A. Enter the matrix A.

```
[A]
    [[1  2 -1]
     [0  1 -1]
     [-1 0 -2]]
```

Press 2nd MATRIX 1:[A] ENTER X^{-1} ENTER.

```
[A]-1
   [[2  -4 1 ]
    [-1 3  -1]
    [-1 2  -1]]
```

EXAMPLE 6 Finding the Inverse of a 3 × 3 Matrix

Find the inverse of $A = \begin{bmatrix} 1 & 2 & -1 \\ 0 & 1 & -1 \\ -1 & 0 & -2 \end{bmatrix}$.

Solution:

STEP 1 Form the matrix $[A \mid I_3]$.

$$\left[\begin{array}{ccc|ccc} 1 & 2 & -1 & 1 & 0 & 0 \\ 0 & 1 & -1 & 0 & 1 & 0 \\ -1 & 0 & -2 & 0 & 0 & 1 \end{array}\right]$$

STEP 2 Apply row operations to transform A into I_3.

$$R_3 + R_1 \rightarrow R_3 \quad \left[\begin{array}{ccc|ccc} 1 & 2 & -1 & 1 & 0 & 0 \\ 0 & 1 & -1 & 0 & 1 & 0 \\ 0 & 2 & -3 & 1 & 0 & 1 \end{array}\right]$$

$$R_3 - 2R_2 \rightarrow R_3 \quad \left[\begin{array}{ccc|ccc} 1 & 2 & -1 & 1 & 0 & 0 \\ 0 & 1 & -1 & 0 & 1 & 0 \\ 0 & 0 & -1 & 1 & -2 & 1 \end{array}\right]$$

$$-R_3 \rightarrow R_3 \quad \left[\begin{array}{ccc|ccc} 1 & 2 & -1 & 1 & 0 & 0 \\ 0 & 1 & -1 & 0 & 1 & 0 \\ 0 & 0 & 1 & -1 & 2 & -1 \end{array}\right]$$

$$\begin{array}{l} R_2 + R_3 \rightarrow R_2 \\ R_1 + R_3 \rightarrow R_1 \end{array} \quad \left[\begin{array}{ccc|ccc} 1 & 2 & 0 & 0 & 2 & -1 \\ 0 & 1 & 0 & -1 & 3 & -1 \\ 0 & 0 & 1 & -1 & 2 & -1 \end{array}\right]$$

$$R_1 - 2R_2 \rightarrow R_1 \quad \left[\begin{array}{ccc|ccc} 1 & 0 & 0 & 2 & -4 & 1 \\ 0 & 1 & 0 & -1 & 3 & -1 \\ 0 & 0 & 1 & -1 & 2 & -1 \end{array}\right]$$

Identify the inverse.

$$A^{-1} = \begin{bmatrix} 2 & -4 & 1 \\ -1 & 3 & -1 \\ -1 & 2 & -1 \end{bmatrix}$$

STEP 3 Check. $AA^{-1} = \begin{bmatrix} 1 & 2 & -1 \\ 0 & 1 & -1 \\ -1 & 0 & -2 \end{bmatrix}\begin{bmatrix} 2 & -4 & 1 \\ -1 & 3 & -1 \\ -1 & 2 & -1 \end{bmatrix} = \begin{bmatrix} 1 & 0 & 0 \\ 0 & 1 & 0 \\ 0 & 0 & 1 \end{bmatrix} = I_3$

$$A^{-1}A = \begin{bmatrix} 2 & -4 & 1 \\ -1 & 3 & -1 \\ -1 & 2 & -1 \end{bmatrix}\begin{bmatrix} 1 & 2 & -1 \\ 0 & 1 & -1 \\ -1 & 0 & -2 \end{bmatrix} = \begin{bmatrix} 1 & 0 & 0 \\ 0 & 1 & 0 \\ 0 & 0 & 1 \end{bmatrix} = I_3$$

■ **YOUR TURN** Find the inverse of $A = \begin{bmatrix} 1 & 1 & 0 \\ -1 & 0 & 1 \\ 2 & 0 & -1 \end{bmatrix}$.

■ **Answer:** $A^{-1} = \begin{bmatrix} 0 & 1 & 1 \\ 1 & -1 & -1 \\ 0 & 2 & 1 \end{bmatrix}$

Solving Systems of Linear Equations Using Matrix Algebra and Inverses of Square Matrices

We can solve systems of linear equations using matrix algebra. We will use a system of three equations and three variables to demonstrate the procedure. However, it can be extended to any square system.

Linear System of Equations

$$\begin{aligned} a_1x + b_1y + c_1z &= d_1 \\ a_2x + b_2y + c_2z &= d_2 \\ a_3x + b_3y + c_3z &= d_3 \end{aligned}$$

Matrix Form of the System

$$\underbrace{\begin{bmatrix} a_1 & b_1 & c_1 \\ a_2 & b_2 & c_2 \\ a_3 & b_3 & c_3 \end{bmatrix}}_{A} \underbrace{\begin{bmatrix} x \\ y \\ z \end{bmatrix}}_{X} = \underbrace{\begin{bmatrix} d_1 \\ d_2 \\ d_3 \end{bmatrix}}_{B}$$

Recall that a system of linear equations has a unique solution, no solution, or infinitely many solutions. If a system of n equations in n variables has a unique solution, it can be found using the following procedure.

WORDS	MATH
Write the system of linear equations as a matrix equation.	$A_{n\times n}X_{n\times 1} = B_{n\times 1}$
Multiply both sides of the equation by A^{-1}.	$A^{-1}AX = A^{-1}B$
A matrix times its inverse is the identity matrix.	$I_nX = A^{-1}B$
A matrix times the identity matrix is equal to itself.	$X = A^{-1}B$

Notice the order in which the right side is multiplied, $X_{n\times 1} = A^{-1}_{n\times n}B_{n\times 1}$, and remember that matrix multiplication is not commutative.

SOLVING A SYSTEM OF LINEAR EQUATIONS USING MATRIX ALGEBRA: UNIQUE SOLUTION

If a system of linear equations is represented by $AX = B$, where A is a nonsingular square matrix, then the system has a unique solution given by

$$X = A^{-1}B$$

EXAMPLE 7 **Solving a System of Linear Equations Using Matrix Algebra**

Solve the system of equations using matrix algebra.

$$\begin{aligned} x + y + z &= 2 \\ x \qquad + z &= 1 \\ x - y - z &= -4 \end{aligned}$$

Technology Tip

Use T1 to find the inverse of A and X. Enter the matrices A and B.

```
[A]
     [[1  1   1 ]
      [1  0   1 ]
      [1  -1  -1]]
█
```

Solution:

Write the system in matrix form. $AX = B$

$$A = \begin{bmatrix} 1 & 1 & 1 \\ 1 & 0 & 1 \\ 1 & -1 & -1 \end{bmatrix} \qquad X = \begin{bmatrix} x \\ y \\ z \end{bmatrix} \qquad B = \begin{bmatrix} 2 \\ 1 \\ -4 \end{bmatrix}$$

Find the inverse of A.

Form the matrix $[A \mid I_3]$.

$$\left[\begin{array}{ccc|ccc} 1 & 1 & 1 & 1 & 0 & 0 \\ 1 & 0 & 1 & 0 & 1 & 0 \\ 1 & -1 & -1 & 0 & 0 & 1 \end{array}\right]$$

$$\begin{array}{l} R_2 - R_1 \to R_2 \\ R_3 - R_1 \to R_3 \end{array} \quad \left[\begin{array}{ccc|ccc} 1 & 1 & 1 & 1 & 0 & 0 \\ 0 & -1 & 0 & -1 & 1 & 0 \\ 0 & -2 & -2 & -1 & 0 & 1 \end{array}\right]$$

$$-R_2 \to R_2 \quad \left[\begin{array}{ccc|ccc} 1 & 1 & 1 & 1 & 0 & 0 \\ 0 & 1 & 0 & 1 & -1 & 0 \\ 0 & -2 & -2 & -1 & 0 & 1 \end{array}\right]$$

$$R_3 + 2R_2 \to R_3 \quad \left[\begin{array}{ccc|ccc} 1 & 1 & 1 & 1 & 0 & 0 \\ 0 & 1 & 0 & 1 & -1 & 0 \\ 0 & 0 & -2 & 1 & -2 & 1 \end{array}\right]$$

$$-\tfrac{1}{2}R_3 \to R_3 \quad \left[\begin{array}{ccc|ccc} 1 & 1 & 1 & 1 & 0 & 0 \\ 0 & 1 & 0 & 1 & -1 & 0 \\ 0 & 0 & 1 & -\frac{1}{2} & 1 & -\frac{1}{2} \end{array}\right]$$

$$R_1 - R_3 \to R_1 \quad \left[\begin{array}{ccc|ccc} 1 & 1 & 0 & \frac{3}{2} & -1 & \frac{1}{2} \\ 0 & 1 & 0 & 1 & -1 & 0 \\ 0 & 0 & 1 & -\frac{1}{2} & 1 & -\frac{1}{2} \end{array}\right]$$

$$R_1 - R_2 \to R_1 \quad \left[\begin{array}{ccc|ccc} 1 & 0 & 0 & \frac{1}{2} & 0 & \frac{1}{2} \\ 0 & 1 & 0 & 1 & -1 & 0 \\ 0 & 0 & 0 & -\frac{1}{2} & 1 & -\frac{1}{2} \end{array}\right]$$

Identify the inverse.

$$A^{-1} = \begin{bmatrix} \frac{1}{2} & 0 & \frac{1}{2} \\ 1 & -1 & 0 \\ -\frac{1}{2} & 1 & -\frac{1}{2} \end{bmatrix}$$

The solution to the system is $X = A^{-1}B$.

$$X = A^{-1}B = \begin{bmatrix} \frac{1}{2} & 0 & \frac{1}{2} \\ 1 & -1 & 0 \\ -\frac{1}{2} & 1 & -\frac{1}{2} \end{bmatrix}\begin{bmatrix} 2 \\ 1 \\ -4 \end{bmatrix}$$

Simplify.

$$X = \begin{bmatrix} x \\ y \\ z \end{bmatrix} = \begin{bmatrix} -1 \\ 1 \\ 2 \end{bmatrix}$$

$x = -1, y = 1, z = 2$

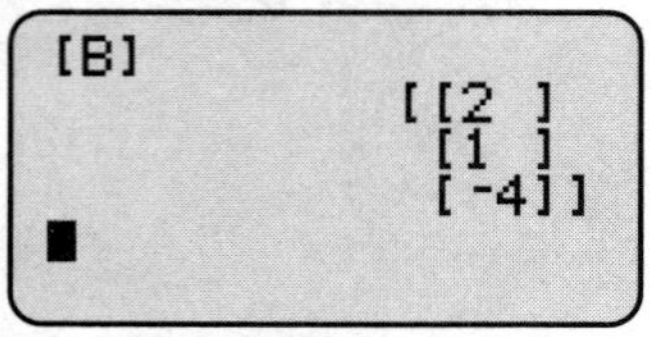

Now use the graphing calculator to find the inverse of A, A^{-1}. Press [2nd] [MATRIX] [1: [A]] [ENTER] [X^{-1}] [ENTER].

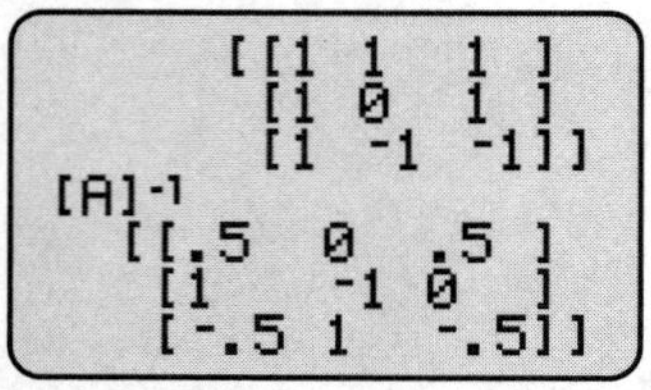

To show elements using fractions, press [2nd] [MATRIX] [1:[A]] [ENTER] [X^{-1}] [MATH] [1: ▶ Frac] [ENTER] [ENTER].

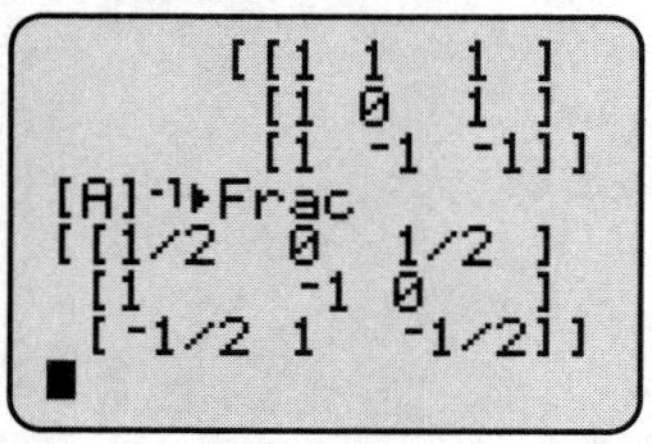

To enter $A^{-1}B$, press [2nd] [ANS] [2nd] [MATRIX] [2:[B]] [ENTER] [ENTER].

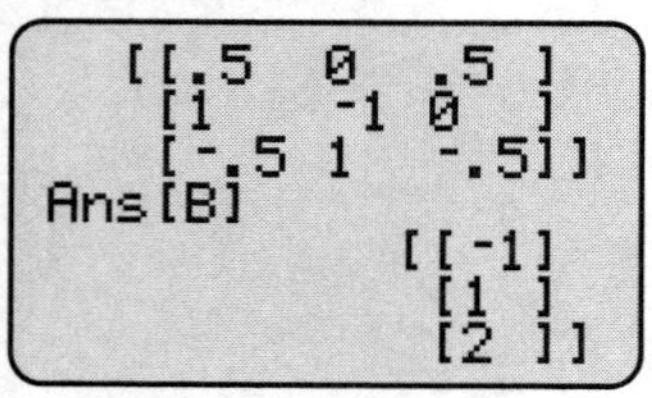

The solution to the system is $x = -1$, $y = 1$, and $z = 2$.

■ **YOUR TURN** Solve the system of equations using matrix algebra.

$$\begin{aligned} x + y - z &= 3 \\ y + z &= 1 \\ 2x + 3y + z &= 5 \end{aligned}$$

■ **Answer:** $x = 0, y = 2, z = -1$

Cryptography Applications

Cryptography is the practice of hiding information, or secret communication. Let's assume you want to send your ATM PIN code over the Internet, but you don't want hackers to be able to retrieve it. You can represent the PIN code in a matrix and then multiply that PIN matrix by a "key" matrix so that it is encrypted. If the person you send it to has the "inverse key" matrix, he can multiply the encrypted matrix he receives by the inverse key matrix and the result will be the original PIN matrix. Although PIN numbers are typically 4 digits, we will assume two digits to illustrate the process.

Words	**Math**
Suppose the 2-digit ATM PIN is 13.	$P = [1 \quad 3]_{1\times 2}$
Apply any 2×2 nonsingular matrix as the "key" (encryption) matrix.	$K = \begin{bmatrix} 2 & 3 \\ 5 & 8 \end{bmatrix}$
Multiply the PIN and encryption matrices.	$PK = [1 \quad 3]_{1\times 2}\begin{bmatrix} 2 & 3 \\ 5 & 8 \end{bmatrix}_{2\times 2}$ $= [1(2) + 3(5) \quad 1(3) + 3(8)]$ $= [17 \quad 27]$
The receiver of the encrypted matrix sees only $[17 \quad 27]_{1\times 2}$.	
The decoding "key" is the inverse matrix K^{-1}.	$K^{-1} = \begin{bmatrix} 8 & -3 \\ -5 & 2 \end{bmatrix}$

Study Tip

$$K = \begin{bmatrix} 2 & 3 \\ 5 & 8 \end{bmatrix}$$

$$K^{-1} = \frac{1}{(2)(8) - (3)(5)}\begin{bmatrix} 8 & -3 \\ -5 & 2 \end{bmatrix} = \begin{bmatrix} 8 & -3 \\ -5 & 2 \end{bmatrix}$$

Any receiver who has the decoding key can multiply the received encrypted matrix by the decoding "key" matrix. The result is the original transmitted PIN number.

$$[17 \quad 27]_{1\times 2}\begin{bmatrix} 8 & -3 \\ -5 & 2 \end{bmatrix} = [17(8) + 27(-5) \quad 17(-3) + 27(2)] = [1 \quad 3]$$

SECTION 7.3 SUMMARY

Systems of linear equations can be solved using matrix equations.

System of Linear Equations	A	X	B	Matrix Equation: $AX = B$
$\begin{aligned} x - y + z &= 2 \\ 2x + 2y - 3z &= -3 \\ x + y + z &= 6 \end{aligned}$	$\begin{bmatrix} 1 & -1 & 1 \\ 2 & 2 & -3 \\ 1 & 1 & 1 \end{bmatrix}$	$\begin{bmatrix} x \\ y \\ z \end{bmatrix}$	$\begin{bmatrix} 2 \\ -3 \\ 6 \end{bmatrix}$	$\begin{bmatrix} 1 & -1 & 1 \\ 2 & 2 & -3 \\ 1 & 1 & 1 \end{bmatrix}\begin{bmatrix} x \\ y \\ z \end{bmatrix} = \begin{bmatrix} 2 \\ -3 \\ 6 \end{bmatrix}$

If this system of linear equations has a unique solution, then it is represented by

$$X = A^{-1}B$$

A^{-1} is the inverse of A, that is, $AA^{-1} = A^{-1}A = I$, and is found by

$$\left[A_{n\times n} | I_n\right] \to \left[I_n | A^{-1}_{n\times n}\right]$$

SECTION 7.3 EXERCISES

SKILLS

In Exercises 1–8, write the system of linear equations as a matrix equation. (Do not solve the system.)

1. $\begin{aligned}-2x + 5y &= 10\\ 7x - 2y &= -4\end{aligned}$

2. $\begin{aligned}4x - 8y &= 10\\ 3x + 5y &= 15\end{aligned}$

3. $\begin{aligned}x - 2y &= 8\\ -3x + y &= 6\end{aligned}$

4. $\begin{aligned}7x - 2y &= 28\\ 3x + 7y &= 42\end{aligned}$

5. $\begin{aligned}3x + 5y - z &= 2\\ x + 2z &= 17\\ -x + y - z &= 4\end{aligned}$

6. $\begin{aligned}x - y + z &= 12\\ 2x + y - 3z &= 6\\ -3x + 2y + z &= 18\end{aligned}$

7. $\begin{aligned}3x + z &= 10\\ y - 2z &= 4\\ x + 2y &= 6\end{aligned}$

8. $\begin{aligned}x + y - 2z + w &= 11\\ 2x - y + 3z &= 17\\ -x + 2y - 3z + 4w &= 12\\ y + 4z + 6w &= 19\end{aligned}$

In Exercises 9–18, determine whether B is the multiplicative inverse of A using $AA^{-1} = I$.

9. $A = \begin{bmatrix} 8 & -11 \\ -5 & 7 \end{bmatrix}$ $B = \begin{bmatrix} 7 & 11 \\ 5 & 8 \end{bmatrix}$

10. $A = \begin{bmatrix} 7 & -9 \\ -3 & 4 \end{bmatrix}$ $B = \begin{bmatrix} 4 & 9 \\ 3 & 7 \end{bmatrix}$

11. $A = \begin{bmatrix} 3 & 1 \\ 1 & -2 \end{bmatrix}$ $B = \begin{bmatrix} \frac{2}{7} & \frac{1}{7} \\ \frac{1}{7} & -\frac{3}{7} \end{bmatrix}$

12. $A = \begin{bmatrix} 2 & 3 \\ 1 & -1 \end{bmatrix}$ $B = \begin{bmatrix} \frac{1}{5} & \frac{3}{5} \\ \frac{1}{5} & -\frac{2}{5} \end{bmatrix}$

13. $A = \begin{bmatrix} 1 & 2 \\ 3 & 4 \end{bmatrix}$ $B = \begin{bmatrix} 4 & -2 \\ -3 & 1 \end{bmatrix}$

14. $A = \begin{bmatrix} 1 & 2 \\ 3 & 4 \end{bmatrix}$ $B = \begin{bmatrix} 1 & \frac{1}{2} \\ \frac{1}{3} & \frac{1}{4} \end{bmatrix}$

15. $A = \begin{bmatrix} 1 & -1 & 1 \\ 1 & 0 & -1 \\ 0 & 1 & -1 \end{bmatrix}$ $B = \begin{bmatrix} 1 & 0 & 1 \\ 1 & -1 & 2 \\ 1 & -1 & 1 \end{bmatrix}$

16. $A = \begin{bmatrix} -1 & 0 & -1 \\ -1 & 1 & -2 \\ -1 & 1 & -1 \end{bmatrix}$ $B = \begin{bmatrix} -1 & 1 & -1 \\ -1 & 0 & 1 \\ 0 & -1 & 1 \end{bmatrix}$

17. $A = \begin{bmatrix} 2 & 0 & 1 \\ 0 & 3 & 1 \\ 0 & 2 & -1 \end{bmatrix}$ $B = \begin{bmatrix} 0 & 2 & 1 \\ 0 & 3 & 0 \\ 2 & 0 & 2 \end{bmatrix}$

18. $A = \begin{bmatrix} 1 & 0 & 0 \\ 0 & 2 & 0 \\ 0 & 0 & 3 \end{bmatrix}$ $B = \begin{bmatrix} 1 & 0 & 0 \\ 0 & \frac{1}{2} & 0 \\ 0 & 0 & \frac{1}{3} \end{bmatrix}$

In Exercises 19–32, find A^{-1}, if possible.

19. $A = \begin{bmatrix} 2 & 1 \\ -1 & 0 \end{bmatrix}$

20. $A = \begin{bmatrix} 3 & 1 \\ 2 & 1 \end{bmatrix}$

21. $A = \begin{bmatrix} \frac{1}{3} & 2 \\ 5 & \frac{3}{4} \end{bmatrix}$

22. $A = \begin{bmatrix} \frac{1}{4} & 2 \\ \frac{1}{3} & \frac{2}{3} \end{bmatrix}$

23. $A = \begin{bmatrix} 1.3 & 2.4 \\ 5.3 & 1.7 \end{bmatrix}$

24. $A = \begin{bmatrix} -2.3 & 1.1 \\ 4.6 & -3.2 \end{bmatrix}$

25. $A = \begin{bmatrix} 1 & 1 & 1 \\ 1 & -1 & -1 \\ -1 & 1 & -1 \end{bmatrix}$

26. $A = \begin{bmatrix} 1 & -1 & 1 \\ 1 & 1 & 1 \\ -1 & 2 & -3 \end{bmatrix}$

27. $A = \begin{bmatrix} 1 & 0 & 1 \\ 0 & 1 & 1 \\ 1 & -1 & 0 \end{bmatrix}$

28. $A = \begin{bmatrix} 1 & 2 & -3 \\ 1 & -1 & -1 \\ 1 & 0 & -4 \end{bmatrix}$

29. $A = \begin{bmatrix} 2 & 4 & 1 \\ 1 & 1 & -1 \\ 1 & 1 & 0 \end{bmatrix}$

30. $A = \begin{bmatrix} 1 & 0 & 1 \\ 1 & 1 & -1 \\ 2 & 1 & -1 \end{bmatrix}$

31. $A = \begin{bmatrix} 1 & 1 & -1 \\ 1 & -1 & 1 \\ 2 & -1 & -1 \end{bmatrix}$

32. $A = \begin{bmatrix} 1 & -1 & -1 \\ 1 & 1 & -3 \\ 3 & -5 & 1 \end{bmatrix}$

In Exercises 33–46, apply matrix algebra to solve the system of linear equations.

33. $\begin{aligned} 2x - y &= 5 \\ x + y &= 1 \end{aligned}$

34. $\begin{aligned} 2x - 3y &= 12 \\ x + y &= 1 \end{aligned}$

35. $\begin{aligned} 4x - 9y &= -1 \\ 7x - 3y &= \tfrac{5}{2} \end{aligned}$

36. $\begin{aligned} 7x - 3y &= 1 \\ 4x - 5y &= -\tfrac{7}{5} \end{aligned}$

37. $\begin{aligned} \tfrac{3}{4}x - \tfrac{2}{3}y &= 5 \\ -\tfrac{1}{2}x - \tfrac{5}{3}y &= 3 \end{aligned}$

38. $\begin{aligned} \tfrac{2}{5}x + \tfrac{3}{7}y &= 1 \\ -\tfrac{1}{2}x - \tfrac{1}{3}y &= \tfrac{1}{6} \end{aligned}$

39. $\begin{aligned} x + y + z &= 1 \\ x - y - z &= -1 \\ -x + y - z &= -1 \end{aligned}$

40. $\begin{aligned} x - y + z &= 0 \\ x + y + z &= 2 \\ -x + 2y - 3z &= 1 \end{aligned}$

41. $\begin{aligned} x + z &= 3 \\ y + z &= 1 \\ x - y &= 2 \end{aligned}$

42. $\begin{aligned} x + 2y - 3z &= 1 \\ x - y - z &= 3 \\ x - 4z &= 0 \end{aligned}$

43. $\begin{aligned} 2x + 4y + z &= -5 \\ x + y - z &= 7 \\ x + y &= 0 \end{aligned}$

44. $\begin{aligned} x + z &= 3 \\ x + y - z &= -3 \\ 2x + y - z &= -5 \end{aligned}$

45. $\begin{aligned} x + y - z &= 4 \\ x - y + z &= 2 \\ 2x - y - z &= -3 \end{aligned}$

46. $\begin{aligned} x - y - z &= 0 \\ x + y - 3z &= 2 \\ 3x - 5y + z &= 4 \end{aligned}$

APPLICATIONS

47. NCAA. University of Florida apparel sales associated with the Final Four Basketball Tournament and the BCS Championship Game in Football are represented with the following matrix:

$$\begin{array}{r} \\ \text{Final Four} \\ \text{BCS} \end{array} \begin{array}{c} \begin{array}{cc} \text{Sweatshirts} & \text{T-shirts} \end{array} \\ \begin{bmatrix} 20{,}000 & 100{,}000 \\ 100{,}000 & 50{,}000 \end{bmatrix} \end{array} = A$$

The revenue generated by the sales of these T-shirts and sweatshirts is given by the following matrix:

$$\begin{array}{r} \text{Final Four} \\ \text{BCS} \end{array} \begin{bmatrix} \$3{,}000{,}000 \\ \$6{,}000{,}000 \end{bmatrix} = B$$

a. Find $A^{-1}B$.

b. What does $A^{-1}B$ represent?

48. NASCAR. Tony Stewart (NASCAR driver) often drives in two races in the same weekend. The Saturday race is called the Nationwide Series, and the Sunday race is called the Sprint Cup. The numbers of Tony Stewart hat and jacket sales associated with the Daytona 500 weekend of races are represented with the following matrix:

$$\begin{array}{r} \\ \text{Nationwide} \\ \text{Sprint Cup} \end{array} \begin{array}{c} \begin{array}{cc} \text{Hats} & \text{Jackets} \end{array} \\ \begin{bmatrix} 30{,}000 & 2{,}000 \\ 50{,}000 & 5{,}000 \end{bmatrix} \end{array} = A$$

The revenue generated by the sales of these hats and jackets is given by the following matrix:

$$\begin{array}{r} \text{Nationwide} \\ \text{Sprint Cup} \end{array} \begin{bmatrix} \$750{,}000 \\ \$1{,}375{,}000 \end{bmatrix} = B$$

a. Find $A^{-1}B$.

b. What does $A^{-1}B$ represent?

For Exercises 49–54, apply the following decoding scheme:

1	A	10	J	19	S
2	B	11	K	20	T
3	C	12	L	21	U
4	D	13	M	22	V
5	E	14	N	23	W
6	F	15	O	24	X
7	G	16	P	25	Y
8	H	17	Q	26	Z
9	I	18	R		

The encoding matrix is $\begin{bmatrix} 1 & 1 & 0 \\ -1 & 0 & 1 \\ 2 & 0 & -1 \end{bmatrix}$. The encrypted matrices are given below. For each of the following, determine the 3-letter word that is originally transmitted. (*Hint:* All four words are parts of the body.)

49. Cryptography. $[55 \quad 10 \quad -22]$

50. Cryptography. $[31 \quad 8 \quad -7]$

51. Cryptography. $[21 \quad 12 \quad -2]$

52. Cryptography. $[9 \quad 1 \quad 5]$

53. Cryptography. $[-10 \quad 5 \quad 20]$

54. Cryptography. $[40 \quad 5 \quad -17]$

CATCH THE MISTAKE

In Exercises 55 and 56, explain the mistake that is made.

55. Find the inverse of $A = \begin{bmatrix} 1 & 0 & 1 \\ -1 & 0 & -1 \\ 1 & 2 & 0 \end{bmatrix}$.

Solution:

Write the matrix $[A \mid I_3]$. $\left[\begin{array}{ccc|ccc} 1 & 0 & 1 & 1 & 0 & 0 \\ -1 & 0 & -1 & 0 & 1 & 0 \\ 1 & 2 & 0 & 0 & 0 & 1 \end{array}\right]$

Use Gaussian elimination to reduce A.

$$\begin{array}{l} R_2 + R_1 \to R_2 \\ R_3 - R_1 \to R_3 \end{array} \quad \left[\begin{array}{ccc|ccc} 1 & 0 & 1 & 1 & 0 & 0 \\ 0 & 0 & 0 & 1 & 1 & 0 \\ 0 & 2 & -1 & -1 & 0 & 1 \end{array}\right]$$

$$R_2 \leftrightarrow R_3 \quad \left[\begin{array}{ccc|ccc} 1 & 0 & 1 & 1 & 0 & 0 \\ 0 & 2 & -1 & -1 & 0 & 1 \\ 0 & 0 & 0 & 1 & 1 & 0 \end{array}\right]$$

$$\tfrac{1}{2}R_2 \to R_2 \quad \left[\begin{array}{ccc|ccc} 1 & 0 & 1 & 1 & 0 & 0 \\ 0 & 1 & -\frac{1}{2} & -\frac{1}{2} & 0 & \frac{1}{2} \\ 0 & 0 & 0 & 1 & 1 & 0 \end{array}\right]$$

$A^{-1} = \begin{bmatrix} 1 & 0 & 0 \\ -\frac{1}{2} & 0 & \frac{1}{2} \\ 1 & 1 & 0 \end{bmatrix}$ is incorrect because $AA^{-1} \neq I_3$.

What mistake was made?

56. Find the inverse of A given that $A = \begin{bmatrix} 2 & 5 \\ 3 & 10 \end{bmatrix}$.

Solution: $A^{-1} = \dfrac{1}{A}$ $\quad A^{-1} = \dfrac{1}{\begin{bmatrix} 2 & 5 \\ 3 & 10 \end{bmatrix}}$

Simplify. $A^{-1} = \begin{bmatrix} \frac{1}{2} & \frac{1}{5} \\ \frac{1}{3} & \frac{1}{10} \end{bmatrix}$

This is incorrect. What mistake was made?

CONCEPTUAL

In Exercises 57 and 58, determine whether each statement is true or false.

57. If $A = \begin{bmatrix} a_{11} & a_{12} \\ a_{21} & a_{22} \end{bmatrix}$, then $A^{-1} = \begin{bmatrix} \frac{1}{a_{11}} & \frac{1}{a_{12}} \\ \frac{1}{a_{21}} & \frac{1}{a_{22}} \end{bmatrix}$.

58. All square matrices have inverses.

59. For what values of x does the inverse of A not exist, given $A = \begin{bmatrix} x & 6 \\ 3 & 2 \end{bmatrix}$?

60. Let $A = \begin{bmatrix} a & 0 & 0 \\ 0 & b & 0 \\ 0 & 0 & c \end{bmatrix}$. Find A^{-1}.

CHALLENGE

61. Verify that $A^{-1} = \dfrac{1}{ad - bc}\begin{bmatrix} d & -b \\ -c & a \end{bmatrix}$ is the inverse of $A = \begin{bmatrix} a & b \\ c & d \end{bmatrix}$, provided $ad - bc \neq 0$.

62. Let $A = \begin{bmatrix} a & b \\ c & d \end{bmatrix}$ and form the matrix $[A \mid I_2]$. Apply row operations to transform into $[I_2 \mid A^{-1}]$, where $A^{-1} = \dfrac{1}{ad - bc}\begin{bmatrix} d & -b \\ -c & a \end{bmatrix}$ such that $ad - bc \neq 0$.

63. Why does the square matrix $A = \begin{bmatrix} 2 & 3 \\ 4 & 6 \end{bmatrix}$ not have an inverse?

64. Why does the square matrix $A = \begin{bmatrix} 1 & 2 & -1 \\ 2 & 4 & -2 \\ 0 & 1 & 3 \end{bmatrix}$ not have an inverse?

TECHNOLOGY

In Exercises 65 and 66, apply a graphing utility to perform the indicated matrix operations.

$$A = \begin{bmatrix} 1 & 7 & 9 & 2 \\ -3 & -6 & 15 & 11 \\ 0 & 3 & 2 & 5 \\ 9 & 8 & -4 & 1 \end{bmatrix}$$

65. Find A^{-1}. **66.** Find AA^{-1}.

In Exercises 67–70, apply a graphing utility and matrix algebra to solve the system of linear equations.

67. $\begin{aligned} 2.7x - 3.1y &= 9.82 \\ 1.5x + 2.7y &= -1.62 \end{aligned}$

68. $\begin{aligned} 3.7x - 2.5y &= 31.77 \\ -5.1x + 1.3y &= -39.07 \end{aligned}$

69. $\begin{aligned} 5.1x + 7.3y + 1.2z &= 12.51 \\ 2.3x - 1.5y + 4.5z &= 53.96 \\ -8.1x + 5.4y - 9.4z &= -130.35 \end{aligned}$

70. $\begin{aligned} 12.4x - 5.8y + 2.7z &= -60.92 \\ -3.9x + 1.9y - 0.6z &= 18.73 \\ 6.4x - 4.3y + 8.5z &= -62.79 \end{aligned}$

SECTION 7.4 THE DETERMINANT OF A SQUARE MATRIX AND CRAMER'S RULE

SKILLS OBJECTIVES

- Find the determinant of a 2×2 matrix.
- Find the determinant of a $n \times n$ matrix.
- Use Cramer's rule to solve a square system of linear equations.

CONCEPTUAL OBJECTIVES

- Derive Cramer's rule.
- Understand that if a determinant of a matrix is equal to zero, then that matrix does not have an inverse.
- Understand that Cramer's rule can be used to find only a unique solution.

In Section 7.1, we discussed Gauss–Jordan elimination as a way to solve systems of linear equations using augmented matrices. Then in Section 7.3, we employed matrix algebra and inverses to solve systems of linear equations that are square (same number of equations as variables). In this section we will describe another method, called Cramer's rule, for solving systems of linear equations. Cramer's rule is applicable only to square systems. *Determinants* of square matrices play a vital role in Cramer's rule and indicate whether a matrix has an inverse.

Determinant of a 2 × 2 Matrix

Every square matrix A has a number associated with it called its *determinant*, denoted $\det(A)$ or $|A|$.

DEFINITION **Determinant of a 2 × 2 Matrix**

The **determinant** of the 2×2 matrix $A = \begin{bmatrix} a & b \\ c & d \end{bmatrix}$ is given by

$$\det(A) = |A| = \begin{vmatrix} a & b \\ c & d \end{vmatrix} = ad - bc.$$

Study Tip

The determinant of a 2×2 matrix is found by finding the product of the main diagonal entries and subtracting the product of the other diagonal entries.

Although the symbol for determinant, | |, looks like absolute value bars, the determinant can be any real number (positive, negative, or zero). The determinant of a 2×2 matrix is found by finding the product of the main diagonal entries (top left to bottom right) and subtracting the product of the entries along the other diagonal (bottom left to top right).

$$\begin{vmatrix} a & b \\ c & d \end{vmatrix} = ad - bc$$

Technology Tip

A graphing calculator can be used to find the determinant of each matrix.

a. Enter the matrix $\begin{bmatrix} 2 & -5 \\ -1 & 3 \end{bmatrix}$ as A.

Press [2nd] [MATRIX]. Use [▶] to highlight [MATH] [1:det(] [ENTER] [2nd] [MATH] [ENTER] [)] [ENTER].

```
[A]
     [[2  -5]
      [-1 3 ]]
det([A])
            1
```

EXAMPLE 1 **Finding the Determinant of a 2 × 2 Matrix**

Find the determinant of each matrix.

a. $\begin{bmatrix} 2 & -5 \\ -1 & 3 \end{bmatrix}$ **b.** $\begin{bmatrix} 0.5 & 0.2 \\ -3.0 & -4.2 \end{bmatrix}$ **c.** $\begin{bmatrix} \frac{2}{3} & 1 \\ 2 & 3 \end{bmatrix}$

Solution:

a. $\begin{vmatrix} 2 & -5 \\ -1 & 3 \end{vmatrix} = (2)(3) - (-1)(-5) = 6 - 5 = \boxed{1}$

b. $\begin{vmatrix} 0.5 & 0.2 \\ -3 & -4.2 \end{vmatrix} = (0.5)(-4.2) - (-3)(0.2) = -2.1 + 0.6 = \boxed{-1.5}$

c. $\begin{vmatrix} \frac{2}{3} & 1 \\ 2 & 3 \end{vmatrix} = \left(\frac{2}{3}\right)(3) - (2)(1) = 2 - 2 = \boxed{0}$

In Example 1, we see that determinants are real numbers that can be positive, negative, or zero. Although evaluating determinants of 2×2 matrices is a simple process, one **common mistake** is reversing the difference: $\begin{vmatrix} a & b \\ c & d \end{vmatrix} \neq bc - ad.$

■ **YOUR TURN** Evaluate the determinant $\begin{vmatrix} -2 & 1 \\ -3 & 2 \end{vmatrix}$.

■ **Answer:** -1

Determinant of an $n \times n$ Matrix

In order to define the *determinant* of a 3×3 or a general $n \times n$ (where $n \geq 3$) matrix, we first define *minors* and *cofactors* of a square matrix.

DEFINITION **Minor and Cofactor**

Let A be a square matrix of order $n \times n$. Then

- The **minor** M_{ij} of the element a_{ij} is the determinant of the $(n-1) \times (n-1)$ matrix obtained when the ith row and jth column of A are deleted.
- The **cofactor** C_{ij} of the element a_{ij} is given by $C_{ij} = (-1)^{i+j} M_{ij}$.

$$A = \begin{bmatrix} 1 & -3 & 2 \\ 4 & -1 & 0 \\ 5 & -2 & 3 \end{bmatrix}$$

Element, a_{ij}	**Minor, M_{ij}**	**Cofactor, C_{ij}**
$a_{11} = 1$	For M_{11}, delete the first row and first column: $\begin{bmatrix} 1 & -3 & 2 \\ 4 & -1 & 0 \\ 5 & -2 & 3 \end{bmatrix}$ $M_{11} = \begin{vmatrix} -1 & 0 \\ -2 & 3 \end{vmatrix} = -3 - 0 = -3$	$C_{11} = (-1)^{1+1}M_{11}$ $= (1)(-3)$ $= -3$
$a_{32} = -2$	For M_{32}, delete the third row and second column: $\begin{bmatrix} 1 & -3 & 2 \\ 4 & -1 & 0 \\ 5 & -2 & 3 \end{bmatrix}$ $M_{32} = \begin{vmatrix} 1 & 2 \\ 4 & 0 \end{vmatrix} = 0 - 8 = -8$	$C_{32} = (-1)^{3+2}M_{32}$ $= (1)(-8)$ $= 8$

Notice that the cofactor is simply the minor multiplied by either 1 or -1, depending on whether $i + j$ is even or odd. Therefore, we can make the following sign pattern for 3×3 and 4×4 matrices and obtain the cofactor by multiplying the minor with the appropriate sign ($+1$ or -1).

$$\begin{bmatrix} + & - & + \\ - & + & - \\ + & - & + \end{bmatrix} \qquad \begin{bmatrix} + & - & + & - \\ - & + & - & + \\ + & - & + & - \\ - & + & - & + \end{bmatrix}$$

DEFINITION **Determinant of an $n \times n$ Matrix**

Let A be an $n \times n$ matrix. Then the **determinant** of A is found by summing the elements in *any* row of A (or column of A), multiplied by each element's respective cofactor.

If A is a 3×3, the determinant can be given by $\det(A) = a_{11}C_{11} + a_{12}C_{12} + a_{13}C_{13}$; this is called **expanding the determinant by the first row**. It is important to note that any row or column can be used. Typically, the row or column with the most zeros is selected because it makes the arithmetic simpler.

Combining the definitions of minors, cofactors, and determinants, we now give a general definition for the determinant of a 3×3 matrix.

Row 1 expansion:
$$\begin{vmatrix} a_1 & b_1 & c_1 \\ a_2 & b_2 & c_2 \\ a_3 & b_3 & c_3 \end{vmatrix} = a_1\begin{vmatrix} b_2 & c_2 \\ b_3 & c_3 \end{vmatrix} - b_1\begin{vmatrix} a_2 & c_2 \\ a_3 & c_3 \end{vmatrix} + c_1\begin{vmatrix} a_2 & b_2 \\ a_3 & b_3 \end{vmatrix}$$

Column 1 expansion:
$$\begin{vmatrix} a_1 & b_1 & c_1 \\ a_2 & b_2 & c_2 \\ a_3 & b_3 & c_3 \end{vmatrix} = a_1\begin{vmatrix} b_2 & c_2 \\ b_3 & c_3 \end{vmatrix} - a_2\begin{vmatrix} b_1 & c_1 \\ b_3 & c_3 \end{vmatrix} + a_3\begin{vmatrix} b_1 & c_1 \\ b_2 & c_2 \end{vmatrix}$$

Whichever row or column is expanded, an alternating sign scheme is used (see sign arrays above). Notice that in either of the expansions above, the 2×2 determinant obtained is found by crossing out the row and column containing the element that is multiplying the determinant.

EXAMPLE 2 Finding the Determinant of a 3 × 3 Matrix

For the given matrix, expand the determinant by the *first row*.

$$\begin{bmatrix} 2 & 1 & 3 \\ -1 & 5 & -2 \\ -3 & 7 & 4 \end{bmatrix}$$

Solution:

Expand the determinant by the first row. Remember the alternating sign.

$$\begin{vmatrix} 2 & 1 & 3 \\ -1 & 5 & -2 \\ -3 & 7 & 4 \end{vmatrix} = +2\begin{vmatrix} 5 & -2 \\ 7 & 4 \end{vmatrix} - 1\begin{vmatrix} -1 & -2 \\ -3 & 4 \end{vmatrix} + 3\begin{vmatrix} -1 & 5 \\ -3 & 7 \end{vmatrix}$$

Evaluate the resulting 2 × 2 determinants.

$$= 2[(5)(4) - (7)(-2)] - 1[(-1)(4) - (-3)(-2)] + 3[(-1)(7) - (-3)(5)]$$

$$= 2[20 + 14] - [-4 - 6] + 3[-7 + 15]$$

Simplify. $= 2(34) - (-10) + 3(8)$

$$= 68 + 10 + 24$$

$$= \boxed{102}$$

Study Tip

The determinant by the third column is also 102. It does not matter on which row or column the expansion occurs.

■ **YOUR TURN** For the given matrix, expand the determinant by the first row.

$$\begin{bmatrix} 1 & 3 & -2 \\ 2 & 5 & 4 \\ 7 & -1 & 6 \end{bmatrix}$$

■ **Answer:** 156

Determinants can be expanded by any row *or* column. Typically, the row or column with the most zeros is selected.

Technology Tip

Enter the matrix as A and find the determinant.

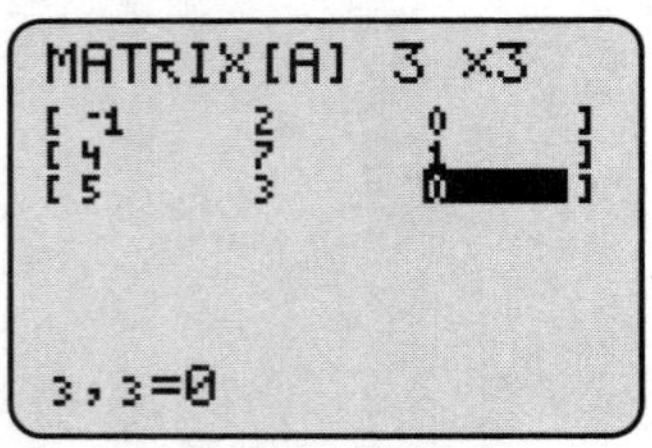

Press [2nd] [MATRIX]. Use [▶] to highlight [MATH] [1:det(] [ENTER] [2nd] [MATH] [ENTER] [)] [ENTER].

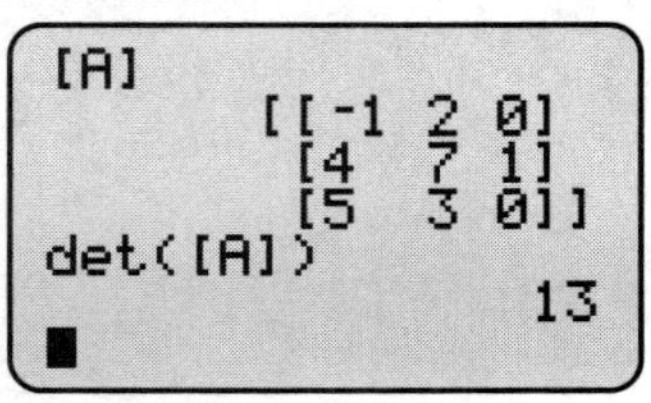

EXAMPLE 3 Finding the Determinant of a 3 × 3 Matrix

Find the determinant of the matrix $\begin{vmatrix} -1 & 2 & 0 \\ 4 & 7 & 1 \\ 5 & 3 & 0 \end{vmatrix}$.

Solution:

Since there are two 0s in the third column, expand the determinant by the third column. Recall the sign array.

$$\begin{bmatrix} + & - & + \\ - & + & - \\ + & - & + \end{bmatrix}$$

$$\begin{vmatrix} -1 & 2 & 0 \\ 4 & 7 & 1 \\ 5 & 3 & 0 \end{vmatrix} = +0\begin{vmatrix} 4 & 7 \\ 5 & 3 \end{vmatrix} - 1\begin{vmatrix} -1 & 2 \\ 5 & 3 \end{vmatrix} + 0\begin{vmatrix} -1 & 2 \\ 4 & 7 \end{vmatrix}$$

There is no need to calculate the two determinants that are multiplied by 0s, since 0 times any real number is zero.

$$\begin{vmatrix} -1 & 2 & 0 \\ 4 & 7 & 1 \\ 5 & 3 & 0 \end{vmatrix} = 0 - 1\underbrace{\begin{vmatrix} -1 & 2 \\ 5 & 3 \end{vmatrix}}_{-3-10} + 0$$

Simplify. $= -1(-13) = \boxed{13}$

■ **YOUR TURN** Evaluate the determinant $\begin{vmatrix} 1 & -2 & 1 \\ -1 & 0 & 3 \\ -4 & 0 & 2 \end{vmatrix}$.

■ **Answer:** 20

EXAMPLE 4 Finding the Determinant of a 4 × 4 Matrix

Find the determinant of the matrix $\begin{vmatrix} 1 & -2 & 3 & 4 \\ -4 & 0 & -1 & 0 \\ -3 & 9 & 6 & 5 \\ -5 & 7 & 2 & 1 \end{vmatrix}$.

Solution:

Since there are two 0s in the second row, expand the determinant by the second row. Recall the sign array for a 4 × 4 matrix.

$$\begin{bmatrix} + & - & + & - \\ - & + & - & + \\ + & - & + & - \\ - & + & - & + \end{bmatrix}$$

$$\begin{bmatrix} 1 & -2 & 3 & 4 \\ -4 & 0 & -1 & 0 \\ -3 & 9 & 6 & 5 \\ -5 & 7 & 2 & 1 \end{bmatrix} = -(-4)\begin{vmatrix} -2 & 3 & 4 \\ 9 & 6 & 5 \\ 7 & 2 & 1 \end{vmatrix} + 0 - (-1)\begin{vmatrix} 1 & -2 & 4 \\ -3 & 9 & 5 \\ -5 & 7 & 1 \end{vmatrix} + 0$$

Evaluate the two 3 × 3 determinants.

$$\begin{vmatrix} -2 & 3 & 4 \\ 9 & 6 & 5 \\ 7 & 2 & 1 \end{vmatrix} = -2\begin{vmatrix} 6 & 5 \\ 2 & 1 \end{vmatrix} - 3\begin{vmatrix} 9 & 5 \\ 7 & 1 \end{vmatrix} + 4\begin{vmatrix} 9 & 6 \\ 7 & 2 \end{vmatrix}$$

$$= -2(6 - 10) - 3(9 - 35) + 4(18 - 42)$$

$$= -2(-4) - 3(-26) + 4(-24)$$

$$= 8 + 78 - 96$$

$$= -10$$

$$\begin{vmatrix} 1 & -2 & 4 \\ -3 & 9 & 5 \\ -5 & 7 & 1 \end{vmatrix} = 1\begin{vmatrix} 9 & 5 \\ 7 & 1 \end{vmatrix} - (-2)\begin{vmatrix} -3 & 5 \\ -5 & 1 \end{vmatrix} + 4\begin{vmatrix} -3 & 9 \\ -5 & 7 \end{vmatrix}$$

$$= 1(9 - 35) + 2(-3 + 25) + 4(-21 + 45)$$

$$= -26 + 2(22) + 4(24)$$

$$= -26 + 44 + 96$$

$$= 114$$

$$\begin{bmatrix} 1 & -2 & 3 & 4 \\ -4 & 0 & -1 & 0 \\ -3 & 9 & 6 & 5 \\ -5 & 7 & 2 & 1 \end{bmatrix} = 4\underbrace{\begin{vmatrix} -2 & 3 & 4 \\ 9 & 6 & 5 \\ 7 & 2 & 1 \end{vmatrix}}_{-10} + \underbrace{\begin{vmatrix} 1 & -2 & 4 \\ -3 & 9 & 5 \\ -5 & 7 & 1 \end{vmatrix}}_{114} = 4(-10) + 114 = \boxed{74}$$

Cramer's Rule: Systems of Linear Equations in Two Variables

Let's now apply determinants of 2 × 2 matrices to solve systems of linear equations in two variables. We begin by solving the general system of two linear equations in two variables:

$$(1) \qquad a_1x + b_1y = c_1$$

$$(2) \qquad a_2x + b_2y = c_2$$

Solve for *x* using elimination (eliminate *y*).

Multiply (1) by b_2. $\qquad b_2a_1x + b_2b_1y = b_2c_1$

Multiply (2) by $-b_1$. $\qquad -b_1a_2x - b_1b_2y = -b_1c_2$

Add the two new equations to eliminate y. $\qquad (a_1b_2 - a_2b_1)x = (b_2c_1 - b_1c_2)$

Divide both sides by $(a_1b_2 - a_2b_1)$. $\qquad x = \dfrac{(b_2c_1 - b_1c_2)}{(a_1b_2 - a_2b_1)}$

Write both the numerator and the denominator as determinants.

$$x = \frac{\begin{vmatrix} c_1 & b_1 \\ c_2 & b_2 \end{vmatrix}}{\begin{vmatrix} a_1 & b_1 \\ a_2 & b_2 \end{vmatrix}}$$

Solve for *y* using elimination (eliminate *x*).

Multiply (1) by $-a_2$. $\qquad -a_2a_1x - a_2b_1y = -a_2c_1$

Multiply (2) by a_1. $\qquad a_1a_2x + a_1b_2y = a_1c_2$

Add the two new equations to eliminate x. $\qquad (a_1b_2 - a_2b_1)y = (a_1c_2 - a_2c_1)$

Divide both sides by $(a_1b_2 - a_2b_1)$. $\qquad y = \dfrac{(a_1c_2 - a_2c_1)}{(a_1b_2 - a_2b_1)}$

Write both the numerator and the denominator as determinants.

$$y = \frac{\begin{vmatrix} a_1 & c_1 \\ a_2 & c_2 \end{vmatrix}}{\begin{vmatrix} a_1 & b_1 \\ a_2 & b_2 \end{vmatrix}}$$

Notice that the solutions for x and y involve three determinants. If we let

$$D = \begin{vmatrix} a_1 & b_1 \\ a_2 & b_2 \end{vmatrix} \qquad D_x = \begin{vmatrix} c_1 & b_1 \\ c_2 & b_2 \end{vmatrix} \qquad D_y = \begin{vmatrix} a_1 & c_1 \\ a_2 & c_2 \end{vmatrix}$$

then $x = \dfrac{D_x}{D}$ and $y = \dfrac{D_y}{D}$.

Notice that the matrix D is the determinant of the coefficient matrix of the system and cannot equal zero ($D \neq 0$) or there will be no unique solution. These formulas for solving a system of two linear equations in two variables are known as *Cramer's rule*.

CRAMER'S RULE FOR SOLVING SYSTEMS OF TWO LINEAR EQUATIONS IN TWO VARIABLES

For the system of linear equations

$$a_1x + b_1y = c_1$$
$$a_2x + b_2y = c_2$$

let

$$D = \begin{vmatrix} a_1 & b_1 \\ a_2 & b_2 \end{vmatrix} \qquad D_x = \begin{vmatrix} c_1 & b_1 \\ c_2 & b_2 \end{vmatrix} \qquad D_y = \begin{vmatrix} a_1 & c_1 \\ a_2 & c_2 \end{vmatrix}$$

If $D \neq 0$, then the solution to the system of linear equations is

$$x = \frac{D_x}{D}, \qquad y = \frac{D_y}{D}$$

If $D = 0$, then the system of linear equations has either no solution or infinitely many solutions.

Study Tip

Cramer's rule is only applicable to square systems of linear equations.

Notice that the determinants D_x and D_y are similar to the determinant D. A three-step procedure is outlined for setting up the three determinants for a system of two linear equations in two variables.

$$a_1x + b_1y = c_1$$
$$a_2x + b_2y = c_2$$

Step 1: Set up D.

Apply the coefficients of x and y. $\quad D = \begin{vmatrix} a_1 & b_1 \\ a_2 & b_2 \end{vmatrix}$

Step 2: Set up D_x.

Start with D and replace the coefficients of x (column 1) with the constants on the right side of the equal sign. $\quad D_x = \begin{vmatrix} c_1 & b_1 \\ c_2 & b_2 \end{vmatrix}$

Step 3: Set up D_y.

Start with D and replace the coefficients of y (column 2) with the constants on the right side of the equal sign. $\quad D_y = \begin{vmatrix} a_1 & c_1 \\ a_2 & c_2 \end{vmatrix}$

EXAMPLE 5 Using Cramer's Rule to Solve a System of Two Linear Equations

Apply Cramer's rule to solve the system.

$$\begin{aligned} x + 3y &= 1 \\ 2x + y &= -3 \end{aligned}$$

Technology Tip

A graphing calculator can be used to solve the system using Cramer's rule. Enter the matrix A for the determinant D_x, B for D_y, C for D.

Solution:

Set up the three determinants.

$$D = \begin{vmatrix} 1 & 3 \\ 2 & 1 \end{vmatrix}$$

$$D_x = \begin{vmatrix} 1 & 3 \\ -3 & 1 \end{vmatrix}$$

$$D_y = \begin{vmatrix} 1 & 1 \\ 2 & -3 \end{vmatrix}$$

Evaluate the determinants.

$$D = 1 - 6 = -5$$

$$D_x = 1 - (-9) = 10$$

$$D_y = -3 - 2 = -5$$

Solve for x and y.

$$x = \frac{D_x}{D} = \frac{10}{-5} = -2$$

$$y = \frac{D_y}{D} = \frac{-5}{-5} = 1$$

$$\boxed{x = -2,\ y = 1}$$

```
[A]
        [[1  3]
         [-3 1]]
[B]
        [[1  1 ]
         [2  -3]]
```

```
[C]
        [[1 3]
         [2 1]]
```

To solve for x and y, enter D_x/C as A/C for x and D_y/C as B/C for y.

```
det([A])/det([C]
)
                 -2
det([B])/det([C]
)
                  1
```

■ **YOUR TURN** Apply Cramer's rule to solve the system.

$$\begin{aligned} 5x + 4y &= 1 \\ -3x - 2y &= -3 \end{aligned}$$

■ **Answer:** $x = 5$, $y = -6$

Recall from Section 6.1 that systems of two equations and two variables led to one of three possible outcomes: a unique solution, no solution, and infinitely many solutions. When $D = 0$, Cramer's rule does not apply and the system is either inconsistent (has no solution) or contains dependent equations (has infinitely many solutions).

Cramer's Rule: Systems of Linear Equations in Three Variables

Cramer's rule can also be used to solve higher order systems of linear equations. The following box summarizes Cramer's rule for solving a system of three equations in three variables.

CRAMER'S RULE: SOLUTION FOR SYSTEMS OF THREE EQUATIONS IN THREE VARIABLES

The system of linear equations

$$\begin{aligned} a_1x + b_1y + c_1z &= d_1 \\ a_2x + b_2y + c_2z &= d_2 \\ a_3x + b_3y + c_3z &= d_3 \end{aligned}$$

has the solution

$$x = \frac{D_x}{D} \qquad y = \frac{D_y}{D} \qquad z = \frac{D_z}{D} \qquad D \neq 0$$

where the determinants are given as follows:

WORDS	MATH
Display the coefficients of x, y, and z.	$D = \begin{vmatrix} a_1 & b_1 & c_1 \\ a_2 & b_2 & c_2 \\ a_3 & b_3 & c_3 \end{vmatrix}$
Replace the coefficients of x (column 1) in D with the constants on the right side of the equal sign.	$D_x = \begin{vmatrix} d_1 & b_1 & c_1 \\ d_2 & b_2 & c_2 \\ d_3 & b_3 & c_3 \end{vmatrix}$
Replace the coefficients of y (column 2) in D with the constants on the right side of the equal sign.	$D_y = \begin{vmatrix} a_1 & d_1 & c_1 \\ a_2 & d_2 & c_2 \\ a_3 & d_3 & c_3 \end{vmatrix}$
Replace the coefficients of z (column 3) in D with the constants on the right side of the equal sign.	$D_z = \begin{vmatrix} a_1 & b_1 & d_1 \\ a_2 & b_2 & d_2 \\ a_3 & b_3 & d_3 \end{vmatrix}$

EXAMPLE 6 Using Cramer's Rule to Solve a System of Three Linear Equations

Use Cramer's rule to solve the system.

$$\begin{aligned} 3x - 2y + 3z &= -3 \\ 5x + 3y + 8z &= -2 \\ x + y + 3z &= 1 \end{aligned}$$

Solution:

Set up the four determinants.

D contains the coefficients of x, y, and z.

$$D = \begin{vmatrix} 3 & -2 & 3 \\ 5 & 3 & 8 \\ 1 & 1 & 3 \end{vmatrix}$$

Replace a column with constants on the right side of the equation.

$$D_x = \begin{vmatrix} -3 & -2 & 3 \\ -2 & 3 & 8 \\ 1 & 1 & 3 \end{vmatrix} \qquad D_y = \begin{vmatrix} 3 & -3 & 3 \\ 5 & -2 & 8 \\ 1 & 1 & 3 \end{vmatrix} \qquad D_z = \begin{vmatrix} 3 & -2 & -3 \\ 5 & 3 & -2 \\ 1 & 1 & 1 \end{vmatrix}$$

Evaluate the determinants.

$$D = 3(9 - 8) - (-2)(15 - 8) + 3(5 - 3) = 23$$

$$D_x = -3(9 - 8) - (-2)(-6 - 8) + 3(-2 - 3) = -46$$

$$D_y = 3(-6 - 8) - (-3)(15 - 8) + 3(5 + 2) = 0$$

$$D_z = 3(3 + 2) - (-2)(5 + 2) - 3(5 - 3) = 23$$

Solve for x, y, and z.

$$x = \frac{D_x}{D} = \frac{-46}{23} = -2 \qquad y = \frac{D_y}{D} = \frac{0}{23} = 0 \qquad z = \frac{D_z}{D} = \frac{23}{23} = 1$$

$$\boxed{x = -2,\ y = 0,\ z = 1}$$

■ **YOUR TURN** Use Cramer's rule to solve the system.

$$\begin{aligned} 2x + 3y + z &= -1 \\ x - y - z &= 0 \\ -3x - 2y + 3z &= 10 \end{aligned}$$

■ **Answer:** $x = 1, y = -2, z = 3$

As was the case in two equations, when $D = 0$, Cramer's rule does not apply and the system of three equations is either inconsistent (no solution) or contains dependent equations (infinitely many solutions).

All of the coefficients and constants in the systems we have solved have been nice (rational). One of the advantages to Cramer's rule is the ability to solve with ease systems of linear equations with irrational coefficients.

EXAMPLE 7 **Using Cramer's Rule to Solve a System of Linear Equations with Irrational Coefficients and Constants**

Apply Cramer's rule to solve the system.

$$\begin{aligned} 5x - ey + z &= -\sqrt{2} \\ \sqrt{3}x + 2y - \pi z &= 1 \\ 4x - 7y + \sqrt{5}z &= 0 \end{aligned}$$

Solution:

Set up the four determinants.

D contains the coefficients of x, y, and z.

$$D = \begin{vmatrix} 5 & -e & 1 \\ \sqrt{3} & 2 & -\pi \\ 4 & -7 & \sqrt{5} \end{vmatrix}$$

Replace a column with constants on the right side of the equation.

$$D_x = \begin{vmatrix} -\sqrt{2} & -e & 1 \\ 1 & 2 & -\pi \\ 0 & -7 & \sqrt{5} \end{vmatrix} \quad D_y = \begin{vmatrix} 5 & -\sqrt{2} & 1 \\ \sqrt{3} & 1 & -\pi \\ 4 & 0 & \sqrt{5} \end{vmatrix} \quad D_z = \begin{vmatrix} 5 & -e & -\sqrt{2} \\ \sqrt{3} & 2 & 1 \\ 4 & -7 & 0 \end{vmatrix}$$

Evaluate the determinants (along the third row because of the zero).

$$\begin{aligned} D_x &= 0 - (-7)\begin{vmatrix} -\sqrt{2} & 1 \\ 1 & -\pi \end{vmatrix} + \sqrt{5}\begin{vmatrix} -\sqrt{2} & -e \\ 1 & 2 \end{vmatrix} \\ &= 7\left[(-\sqrt{2})(-\pi) - (1)(1)\right] + \sqrt{5}\left[(-\sqrt{2})(2) - (1)(-e)\right] \\ &= 7\left[\pi\sqrt{2} - 1\right] + \sqrt{5}\left[-2\sqrt{2} + e\right] \\ &= 7\pi\sqrt{2} - 7 - 2\sqrt{10} + e\sqrt{5} \end{aligned}$$

$$\begin{aligned} D_y &= 4\begin{vmatrix} -\sqrt{2} & 1 \\ 1 & -\pi \end{vmatrix} - 0 + \sqrt{5}\begin{vmatrix} 5 & -\sqrt{2} \\ \sqrt{3} & 1 \end{vmatrix} \\ &= 4\left[(-\sqrt{2})(-\pi) - (1)(1)\right] + \sqrt{5}\left[(5)(1) - (\sqrt{3})(-\sqrt{2})\right] \\ &= 4\left[\pi\sqrt{2} - 1\right] + \sqrt{5}\left[5 + \sqrt{6}\right] \\ &= 4\pi\sqrt{2} - 4 + 5\sqrt{5} + \sqrt{30} \end{aligned}$$

$$\begin{aligned} D_z &= 4\begin{vmatrix} -e & -\sqrt{2} \\ 2 & 1 \end{vmatrix} - (-7)\begin{vmatrix} 5 & -\sqrt{2} \\ \sqrt{3} & 1 \end{vmatrix} + 0 \\ &= 4\left[(-e)(1) - (2)(-\sqrt{2})\right] + 7\left[(5)(1) - (\sqrt{3})(-\sqrt{2})\right] \\ &= 4\left[-e + 2\sqrt{2}\right] + 7\left[5 + \sqrt{6}\right] \\ &= -4e + 8\sqrt{2} + 35 + 7\sqrt{6} \end{aligned}$$

$$\begin{aligned} D &= 4\begin{vmatrix} -e & 1 \\ 2 & -\pi \end{vmatrix} - (-7)\begin{vmatrix} 5 & 1 \\ \sqrt{3} & -\pi \end{vmatrix} + \sqrt{5}\begin{vmatrix} 5 & -e \\ \sqrt{3} & 2 \end{vmatrix} \\ &= 4[(-e)(-\pi) - (2)(1)] + 7\left[(5)(-\pi) - (\sqrt{3})(1)\right] + \sqrt{5}\left[(5)(2) - (\sqrt{3})(-e)\right] \\ &= 4[e\pi - 2] + 7\left[-5\pi - \sqrt{3}\right] + \sqrt{5}\left[10 + e\sqrt{3}\right] \\ &= 4e\pi - 8 - 35\pi - 7\sqrt{3} + 10\sqrt{5} + e\sqrt{15} \end{aligned}$$

Solve for x, y, and z.

$$x = \frac{D_x}{D} = \frac{7\pi\sqrt{2} - 7 - 2\sqrt{10} + e\sqrt{5}}{4e\pi - 8 - 35\pi - 7\sqrt{3} + 10\sqrt{5} + e\sqrt{15}}$$

$$y = \frac{D_y}{D} = \frac{4\pi\sqrt{2} - 4 + 5\sqrt{5} + \sqrt{30}}{4e\pi - 8 - 35\pi - 7\sqrt{3} + 10\sqrt{5} + e\sqrt{15}}$$

$$z = \frac{D_z}{D} = \frac{-4e + 8\sqrt{2} + 35 + 7\sqrt{6}}{4e\pi - 8 - 35\pi - 7\sqrt{3} + 10\sqrt{5} + e\sqrt{15}}$$

SECTION 7.4 SUMMARY

In this section **determinants** were discussed for square matrices.

ORDER	DETERMINANT	ARRAY
2×2	$\det(A) = \lvert A \rvert = \begin{vmatrix} a & b \\ c & d \end{vmatrix} = ad - bc$	
3×3	$\begin{vmatrix} a_1 & b_1 & c_1 \\ a_2 & b_2 & c_2 \\ a_3 & b_3 & c_3 \end{vmatrix} = a_1\begin{vmatrix} b_2 & c_2 \\ b_3 & c_3 \end{vmatrix} - b_1\begin{vmatrix} a_2 & c_2 \\ a_3 & c_3 \end{vmatrix} + c_1\begin{vmatrix} a_2 & b_2 \\ a_3 & b_3 \end{vmatrix}$ Expansion by first row (any row or column can be used)	$\begin{bmatrix} + & - & + \\ - & + & - \\ + & - & + \end{bmatrix}$

Cramer's rule was developed for 2×2 and 3×3 matrices, but it can be extended to general $n \times n$ matrices. When the coefficient determinant is equal to zero ($D = 0$), then the system is either inconsistent (and has no solution) or represents dependent equations (and has infinitely many solutions) and Cramer's rule does not apply.

SYSTEM	ORDER	SOLUTION	DETERMINANTS
$a_1x + b_1y = c_1$ $a_2x + b_2y = c_2$	2×2	$x = \frac{D_x}{D}$ $\quad y = \frac{D_y}{D}$	$D = \begin{vmatrix} a_1 & b_1 \\ a_2 & b_2 \end{vmatrix} \neq 0$ $D_x = \begin{vmatrix} c_1 & b_1 \\ c_2 & b_2 \end{vmatrix}$ $D_y = \begin{vmatrix} a_1 & c_1 \\ a_2 & c_2 \end{vmatrix}$
$a_1x + b_1y + c_1z = d_1$ $a_2x + b_2y + c_2z = d_2$ $a_3x + b_3y + c_3z = d_3$	3×3	$x = \frac{D_x}{D}$ $\quad y = \frac{D_y}{D}$ $\quad z = \frac{D_z}{D}$	$D = \begin{vmatrix} a_1 & b_1 & c_1 \\ a_2 & b_2 & c_2 \\ a_3 & b_3 & c_3 \end{vmatrix} \neq 0$ $D_x = \begin{vmatrix} d_1 & b_1 & c_1 \\ d_2 & b_2 & c_2 \\ d_3 & b_3 & c_3 \end{vmatrix}$ $D_y = \begin{vmatrix} a_1 & d_1 & c_1 \\ a_2 & d_2 & c_2 \\ a_3 & d_3 & c_3 \end{vmatrix}$ $D_z = \begin{vmatrix} a_1 & b_1 & d_1 \\ a_2 & b_2 & d_2 \\ a_3 & b_3 & d_3 \end{vmatrix}$

SECTION 7.4 EXERCISES

SKILLS

In Exercises 1–10, evaluate each 2 × 2 determinant.

1. $\begin{vmatrix} 1 & 2 \\ 3 & 4 \end{vmatrix}$
2. $\begin{vmatrix} 1 & -2 \\ -3 & -4 \end{vmatrix}$
3. $\begin{vmatrix} 7 & 9 \\ -5 & -2 \end{vmatrix}$
4. $\begin{vmatrix} -3 & -11 \\ 7 & 15 \end{vmatrix}$
5. $\begin{vmatrix} 0 & 7 \\ 4 & -1 \end{vmatrix}$
6. $\begin{vmatrix} 0 & 0 \\ 1 & 0 \end{vmatrix}$
7. $\begin{vmatrix} -1.2 & 2.4 \\ -0.5 & 1.5 \end{vmatrix}$
8. $\begin{vmatrix} -1.0 & 1.4 \\ 1.5 & -2.8 \end{vmatrix}$
9. $\begin{vmatrix} \frac{3}{4} & \frac{1}{3} \\ 2 & \frac{8}{9} \end{vmatrix}$
10. $\begin{vmatrix} -\frac{1}{2} & \frac{1}{4} \\ \frac{2}{3} & -\frac{8}{9} \end{vmatrix}$

In Exercises 11–32, use Cramer's rule to solve each system of equations, if possible.

11. $x + y = -1$, $x - y = 11$
12. $x + y = -1$, $x - y = -9$
13. $3x + 2y = -4$, $-2x + y = 5$
14. $5x + 3y = 1$, $4x - 7y = -18$
15. $3x - 2y = -1$, $5x + 4y = -31$
16. $x - 4y = -7$, $3x + 8y = 19$
17. $7x - 3y = -29$, $5x + 2y = 0$
18. $6x - 2y = 24$, $4x + 7y = 41$
19. $3x + 5y = 16$, $y - x = 0$
20. $-2x - 3y = 15$, $7y + 4x = -33$
21. $3x - 5y = 7$, $-6x + 10y = -21$
22. $3x - 5y = 7$, $6x - 10y = 14$
23. $2x - 3y = 4$, $-10x + 15y = -20$
24. $2x - 3y = 2$, $10x - 15y = 20$
25. $3x + \frac{1}{2}y = 1$, $4x + \frac{1}{3}y = \frac{5}{3}$
26. $\frac{3}{2}x + \frac{9}{4}y = \frac{9}{8}$, $\frac{1}{3}x + \frac{1}{4}y = \frac{1}{12}$
27. $0.3x - 0.5y = -0.6$, $0.2x + y = 2.4$
28. $0.5x - 0.4y = -3.6$, $10x + 3.6y = -14$
29. $y = 17x + 7$, $y = -15x + 7$
30. $9x = -45 - 2y$, $4x = -3y - 20$
31. $\frac{2}{x} - \frac{3}{y} = 2$, $\frac{5}{x} - \frac{6}{y} = 7$
32. $\frac{2}{x} - \frac{3}{y} = -12$, $\frac{3}{x} + \frac{1}{2y} = 7$

In Exercises 33–46, evaluate each 3 × 3 determinant.

33. $\begin{vmatrix} 3 & 1 & 0 \\ 2 & 0 & -1 \\ -4 & 1 & 0 \end{vmatrix}$
34. $\begin{vmatrix} 1 & 1 & 0 \\ 0 & 2 & -1 \\ 0 & -3 & 5 \end{vmatrix}$
35. $\begin{vmatrix} 2 & 1 & -5 \\ 3 & 0 & -1 \\ 4 & 0 & 7 \end{vmatrix}$
36. $\begin{vmatrix} 2 & 1 & -5 \\ 3 & -7 & 0 \\ 4 & -6 & 0 \end{vmatrix}$
37. $\begin{vmatrix} 1 & 1 & -5 \\ 3 & -7 & -4 \\ 4 & -6 & 9 \end{vmatrix}$
38. $\begin{vmatrix} -3 & 2 & -5 \\ 1 & 8 & 2 \\ 4 & -6 & 9 \end{vmatrix}$
39. $\begin{vmatrix} 1 & 3 & 4 \\ 2 & -1 & 1 \\ 3 & -2 & 1 \end{vmatrix}$
40. $\begin{vmatrix} -7 & 2 & 5 \\ \frac{7}{8} & 3 & 4 \\ -1 & 4 & 6 \end{vmatrix}$
41. $\begin{vmatrix} -3 & 1 & 5 \\ 2 & 0 & 6 \\ 4 & 7 & -9 \end{vmatrix}$
42. $\begin{vmatrix} 1 & -1 & 5 \\ 3 & -3 & 6 \\ 4 & 9 & 0 \end{vmatrix}$
43. $\begin{vmatrix} -2 & 1 & -7 \\ 4 & -2 & 14 \\ 0 & 1 & 8 \end{vmatrix}$
44. $\begin{vmatrix} 5 & -2 & -1 \\ 4 & -9 & -3 \\ 2 & 8 & -6 \end{vmatrix}$
45. $\begin{vmatrix} \frac{3}{4} & -1 & 0 \\ 0 & \frac{1}{5} & -12 \\ 8 & 0 & -2 \end{vmatrix}$
46. $\begin{vmatrix} 0.2 & 0 & 3 \\ 5 & -1.4 & 2 \\ -1 & 0 & -3 \end{vmatrix}$

In Exercises 47–60, apply Cramer's Rule to solve each system of equations, if possible.

47. $\begin{aligned} x + y - z &= 0 \\ x - y + z &= 4 \\ x + y + z &= 10 \end{aligned}$

48. $\begin{aligned} -x + y + z &= -4 \\ x + y - z &= 0 \\ x + y + z &= 2 \end{aligned}$

49. $\begin{aligned} 3x + 8y + 2z &= 28 \\ -2x + 5y + 3z &= 34 \\ 4x + 9y + 2z &= 29 \end{aligned}$

50. $\begin{aligned} 7x + 2y - z &= -1 \\ 6x + 5y + z &= 16 \\ -5x - 4y + 3z &= -5 \end{aligned}$

51. $\begin{aligned} 3x + 5z &= 11 \\ 4y + 3z &= -9 \\ 2x - y &= 7 \end{aligned}$

52. $\begin{aligned} 3x - 2z &= 7 \\ 4x + z &= 24 \\ 6x - 2y &= 10 \end{aligned}$

53. $\begin{aligned} x + y - z &= 5 \\ x - y + z &= -1 \\ -2x - 2y + 2z &= -10 \end{aligned}$

54. $\begin{aligned} x + y - z &= 3 \\ x - y + z &= -2 \\ -2x - 2y + 2z &= -6 \end{aligned}$

55. $\begin{aligned} x + y + z &= 9 \\ x - y + z &= 3 \\ -x + y - z &= 5 \end{aligned}$

56. $\begin{aligned} x + y + z &= 6 \\ x - y - z &= 0 \\ -x + y + z &= 7 \end{aligned}$

57. $\begin{aligned} x + 2y + 3z &= 11 \\ -2x + 3y + 5z &= 29 \\ 4x - y + 8z &= 19 \end{aligned}$

58. $\begin{aligned} 8x - 2y + 5z &= 36 \\ 3x + y - z &= 17 \\ 2x - 6y + 4z &= -2 \end{aligned}$

59. $\begin{aligned} x - 4y + 7z &= 49 \\ -3x + 2y - z &= -17 \\ 5x + 8y - 2z &= -24 \end{aligned}$

60. $\begin{aligned} \tfrac{1}{2}x - 2y + 7z &= 25 \\ x + \tfrac{1}{4}y - 4z &= -2 \\ -4x + 5y &= -56 \end{aligned}$

APPLICATIONS

For Exercises 61–68, the area of a triangle with vertices (x_1, y_1), (x_2, y_2), and (x_3, y_3) is given by

$$\text{Area} = \pm\frac{1}{2}\begin{vmatrix} x_1 & y_1 & 1 \\ x_2 & y_2 & 1 \\ x_3 & y_3 & 1 \end{vmatrix}$$

where the sign is chosen so that the area is positive.

61. Geometry. Apply determinants to find the area of a triangle with vertices (3, 2), (5, 2), and (3, −4). Check your answer by plotting these vertices in a Cartesian plane and using the area of a right triangle.

62. Geometry. Apply determinants to find the area of a triangle with vertices (2, 3), (7, 3), and (7, 7). Check your answer by plotting these vertices in a Cartesian plane and using the area of a right triangle.

63. Geometry. Apply determinants to find the area of a triangle with vertices (1, 2), (3, 4), and (−2, 5).

64. Geometry. Apply determinants to find the area of a triangle with vertices (−1, −2), (3, 4), and (2, 1).

65. Geometry. An equation of a line that passes through two points (x_1, y_1) and (x_2, y_2) can be expressed as a determinant:

$$\begin{vmatrix} x & y & 1 \\ x_1 & y_1 & 1 \\ x_2 & y_2 & 1 \end{vmatrix} = 0$$

Apply the determinant to write an equation of the line passing through the points (1, 2) and (2, 4). Expand the determinant and express the equation of the line in slope–intercept form.

66. Geometry. If three points (x_1, y_1), (x_2, y_2), and (x_3, y_3) are collinear (lie on the same line), then the following determinant must be satisfied:

$$\begin{vmatrix} x_1 & y_1 & 1 \\ x_2 & y_2 & 1 \\ x_3 & y_3 & 1 \end{vmatrix} = 0$$

Determine whether (0, 5), (2, 0), and (1, 2) are collinear.

67. Electricity: Circuit Theory. The following equations come from circuit theory. Find the currents I_1, I_2, and I_3.

$$I_1 = I_2 + I_3$$
$$16 = 4I_1 + 2I_3$$
$$24 = 4I_1 + 4I_2$$

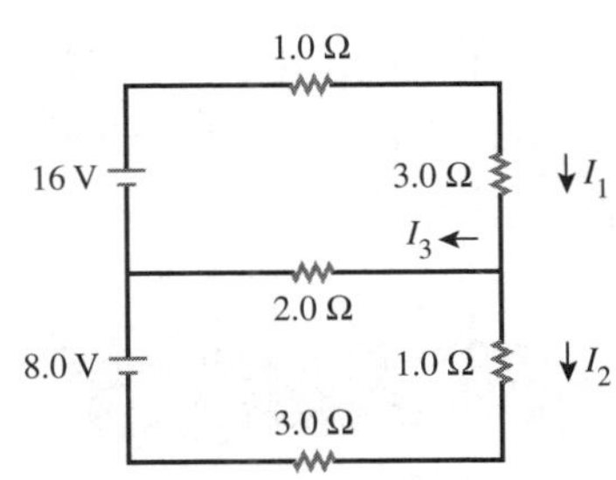

68. Electricity: Circuit Theory. The following equations come from circuit theory. Find the currents I_1, I_2, and I_3.

$$I_1 = I_2 + I_3$$
$$24 = 6I_1 + 3I_3$$
$$36 = 6I_1 + 6I_2$$

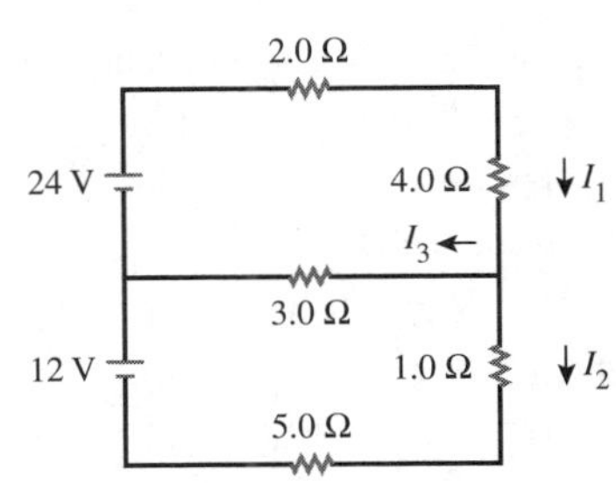

CATCH THE MISTAKE

In Exercises 69–72, explain the mistake that is made.

69. Evaluate the determinant $\begin{vmatrix} 2 & 1 & 3 \\ -3 & 0 & 2 \\ 1 & 4 & -1 \end{vmatrix}$.

Solution:

Expand the 3×3 determinant in terms of the 2×2 determinants.

$$\begin{vmatrix} 2 & 1 & 3 \\ -3 & 0 & 2 \\ 1 & 4 & -1 \end{vmatrix} = 2\begin{vmatrix} 0 & 2 \\ 4 & -1 \end{vmatrix} + 1\begin{vmatrix} -3 & 2 \\ 1 & -1 \end{vmatrix} + 3\begin{vmatrix} -3 & 0 \\ 1 & 4 \end{vmatrix}$$

Expand the 2×2 determinants. $= 2(0 - 8) + 1(3 - 2) + 3(-12 - 0)$

Simplify. $= -16 + 1 - 36 = -51$

This is incorrect. What mistake was made?

70. Evaluate the determinant $\begin{vmatrix} 2 & 1 & 3 \\ -3 & 0 & 2 \\ 1 & 4 & -1 \end{vmatrix}$.

Solution:

Expand the 3×3 determinant in terms of the 2×2 determinants.

$$\begin{vmatrix} 2 & 1 & 3 \\ -3 & 0 & 2 \\ 1 & 4 & -1 \end{vmatrix} = 2\begin{vmatrix} 0 & 2 \\ 4 & -1 \end{vmatrix} - 1\begin{vmatrix} -3 & 2 \\ 1 & -1 \end{vmatrix} + 3\begin{vmatrix} -3 & 2 \\ 1 & -1 \end{vmatrix}$$

Expand the 2×2 determinants. $= 2(0 - 8) - 1(3 - 2) + 3(3 - 2)$

Simplify. $= -16 - 1 + 3 = -14$

This is incorrect. What mistake was made?

71. Solve the system of linear equations.

$$\begin{aligned} 2x + 3y &= 6 \\ -x - y &= -3 \end{aligned}$$

Solution:

Set up the determinants.

$$D = \begin{vmatrix} 2 & 3 \\ -1 & -1 \end{vmatrix}, D_x = \begin{vmatrix} 2 & 6 \\ -1 & -3 \end{vmatrix}, \text{ and } D_y = \begin{vmatrix} 6 & 3 \\ -3 & -1 \end{vmatrix}$$

Evaluate the determinants. $D = 1, D_x = 0, \text{ and } D_y = 3$

Solve for x and y. $x = \frac{D_x}{D} = \frac{0}{1} = 0 \text{ and } y = \frac{D_y}{D} = \frac{3}{1} = 3$

$x = 0, y = 3$ is incorrect. What mistake was made?

72. Solve the system of linear equations.

$$\begin{aligned} 4x - 6y &= 0 \\ 4x + 6y &= 4 \end{aligned}$$

Solution:

Set up the determinants.

$$D = \begin{vmatrix} 4 & -6 \\ 4 & 6 \end{vmatrix}, D_x = \begin{vmatrix} 0 & -6 \\ 4 & 6 \end{vmatrix}, \text{ and } D_y = \begin{vmatrix} 4 & 0 \\ 4 & 4 \end{vmatrix}$$

Evaluate the determinants. $D = 48, D_x = 24, \text{ and } D_y = 16$

Solve for x and y. $x = \frac{D}{D_x} = \frac{48}{24} = 2 \text{ and } y = \frac{D_y}{D} = \frac{48}{16} = 3$

$x = 2, y = 3$ is incorrect. What mistake was made?

CONCEPTUAL

In Exercises 73–76, determine whether each statement is true or false.

73. The value of a determinant changes sign if any two rows are interchanged.

74. If all the entries in any column are equal to zero, the value of the determinant is 0.

75. $\begin{vmatrix} 2 & 6 & 4 \\ 0 & 2 & 8 \\ 4 & 0 & 10 \end{vmatrix} = 2\begin{vmatrix} 1 & 3 & 2 \\ 0 & 1 & 4 \\ 2 & 0 & 5 \end{vmatrix}$

76. $\begin{vmatrix} 3 & 1 & 2 \\ 0 & 2 & 8 \\ 3 & 1 & 2 \end{vmatrix} = 0$

77. Calculate the determinant $\begin{vmatrix} a & 0 & 0 \\ 0 & b & 0 \\ 0 & 0 & c \end{vmatrix}$.

78. Calculate the determinant $\begin{vmatrix} a_1 & b_1 & c_1 \\ 0 & b_2 & c_2 \\ 0 & 0 & c_3 \end{vmatrix}$.

CHALLENGE

79. Evaluate the determinant:

$$\begin{vmatrix} 1 & -2 & -1 & 3 \\ 4 & 0 & 1 & 2 \\ 0 & 3 & 2 & 4 \\ 1 & -3 & 5 & -4 \end{vmatrix}$$

80. For the system of equations

$$\begin{aligned} 3x + 2y &= 5 \\ ax - 4y &= 1 \end{aligned}$$

find a that guarantees no unique solution.

81. Show that

$$\begin{vmatrix} a_1 & b_1 & c_1 \\ a_2 & b_2 & c_2 \\ a_3 & b_3 & c_3 \end{vmatrix} = a_1b_2c_3 + b_1c_2a_3 + c_1a_2b_3 - a_3b_2c_1 - b_3c_2a_1 - b_1a_2c_3$$

by expanding down the second column.

82. Show that

$$\begin{vmatrix} a_1 & b_1 & c_1 \\ a_2 & b_2 & c_2 \\ a_3 & b_3 & c_3 \end{vmatrix} = a_1b_2c_3 + b_1c_2a_3 + c_1a_2b_3 - a_3b_2c_1 - b_3c_2a_1 - c_3b_1a_2$$

by expanding across the third row.

TECHNOLOGY

In Exercises 83–86, apply a graphing utility to evaluate the determinants.

83. $\begin{vmatrix} 1 & 1 & -5 \\ 3 & -7 & -4 \\ 4 & -6 & 9 \end{vmatrix}$ Compare with your answer to Exercise 37.

84. $\begin{vmatrix} -3 & 2 & -5 \\ 1 & 8 & 2 \\ 4 & -6 & 9 \end{vmatrix}$ Compare with your answer to Exercise 38.

85. $\begin{vmatrix} -3 & 2 & -1 & 3 \\ 4 & 1 & 5 & 2 \\ 17 & 2 & 2 & 8 \\ 13 & -4 & 10 & -11 \end{vmatrix}$

86. $\begin{vmatrix} -3 & 21 & 19 & 3 \\ 4 & 1 & 16 & 2 \\ 17 & 31 & 2 & 5 \\ 13 & -4 & 10 & 2 \end{vmatrix}$

In Exercises 87 and 88, apply Cramer's rule to solve each system of equations and a graphing utility to evaluate the determinants.

87.
$$\begin{aligned} 3.1x + 1.6y - 4.8z &= -33.76 \\ 5.2x - 3.4y + 0.5z &= -36.68 \\ 0.5x - 6.4y + 11.4z &= 25.96 \end{aligned}$$

88.
$$\begin{aligned} -9.2x + 2.7y + 5.1z &= -89.2 \\ 4.3x - 6.9y - 7.6z &= 38.89 \\ 2.8x - 3.9y - 3.5z &= 34.08 \end{aligned}$$

MODELING YOUR WORLD

In 2008, Honda introduced the Civic GX NGV—the NGV standing for *Natural Gas Vehicle*. Honda also sells a refueling station for your home that runs off the same natural gas that you use to power a stove or water heater.

The typical range per tank is approximately 170 miles. In addition, to the home fueling option, there are over 1300 natural gas fueling stations in the United States, primarily in California, as of 2008.

Compared to the Civic Hybrid (gas/electric) Sedan which gets an estimated 40 mpg in the city and 45 mpg on the highway, the Civic NGV gets an estimated gasoline equivalent of 24 mpg in the city and 36 mpg on the highway. However, natural gas only costs about one-third of its gasoline equivalent, making these "greener" cars cost effective. According to the Environmental Protection Agency, only natural gasoline vehicles meet the strictest emissions standards.

The MSRP and mileage comparisons for the 2008 Honda Civic models are given below, where Gas represents the traditional sedan, Hybrid represents the electric/gasoline model, and NGV represents the Civic that runs on natural gas.

	GAS	HYBRID	NGV
MSRP	\$15,010	\$22,600	\$24,590
MILES PER GALLON IN CITY	21	33	18
MILES PER GALLON ON HIGHWAY	31	34	24

For the following questions, assume that you drive 15,000 miles per year–12,000 in city and 3000 on the highway–and the price of gasoline is \$4 per gallon while the cost of natural gas is \$2 per gallon.

1. Determine an initial purchase price matrix A, which has order 3×1. Let the first row represent the traditional gasoline model, the second row represent the hybrid model, and the third row represent the natural gas model.

2. Determine a mileage matrix B, which has order 3×2. Let the first column represent mpg in the city and the second column represent the mpg on the highway. Also let the first through third column represent the gas, hybrid, and natural gas models respectively. Let

$$C = \begin{bmatrix} 12{,}000 \\ 3{,}000 \end{bmatrix}$$

What does BC represent?

MODELING YOUR WORLD (*continued*)

3. Assume the gas and hybrid models are filled with regular gasoline at $4 per gallon and the natural gas vehicle is filled with natural gas at $2 per gallon. What does the matrix D represent?

$$D = \begin{bmatrix} 4 \cdot (BC)_{11} \\ 4 \cdot (BC)_{21} \\ 2 \cdot (BC)_{31} \end{bmatrix}$$

4. What does the matrix $A + D$ represent?

5. Let n represent the number of years you own the car (assuming you bought it new). What does $A + nD$ represent?

SECTION	CONCEPT	PAGES	REVIEW EXERCISES	KEY IDEAS/POINTS
7.1	**Matrices and systems of linear equations**	562–578	1–28	
	Augmented matrices	564–565	1–4	$\begin{matrix} a_1x + b_1y + c_1z = d_1 \\ a_2x + b_2y + c_2z = d_2 \\ a_3x + b_3y + c_3z = d_3 \end{matrix} \Rightarrow \left[\begin{array}{ccc\|c} a_1 & b_1 & c_1 & d_1 \\ a_2 & b_2 & c_2 & d_2 \\ a_3 & b_3 & c_3 & d_3 \end{array}\right]$
	Row operations on a matrix	565–566	9–16	**1.** $R_i \leftrightarrow R_j$ Interchange row i with row j. **2.** $cR_i \rightarrow R_i$ Multiply row i by the constant c. **3.** $cR_i + R_j \rightarrow R_j$ Multiply row i by the constant c and add to row j, writing the results in row j.
	Row–echelon form of a matrix	566–567	5–8	A matrix is in **row–echelon form** if it has all three of the following properties: **1.** Any rows consisting entirely of 0s are at the bottom of the matrix. **2.** For each row that does not consist entirely of 0s, the first (left-most) nonzero entry is 1 (called the leading 1). **3.** For two successive nonzero rows, the leading 1 in the higher row is farther to the left than the leading 1 in the lower row. If a matrix in row–echelon form has the following additional property, then the matrix is in **reduced row–echelon form**: **4.** Every column containing a leading 1 has zeros in every position above and below the leading 1.
	Gaussian elimination with back-substitution	567–569	17–28	**Step 1:** Write the system of equations as an augmented matrix. **Step 2:** Apply row operations to transform the matrix into row–echelon form. **Step 3:** Apply back-substitution to identify the solution.
	Gauss–Jordan elimination	570–573	17–28	**Step 1:** Write the system of equations as an augmented matrix. **Step 2:** Apply row operations to transform the matrix into *reduced* row–echelon form. **Step 3:** Identify the solution.
	Inconsistent and dependent systems	573–578		No solution or infinitely many solutions
7.2	**Matrix algebra**	584–594	29–46	$\begin{matrix} & \text{Column 1} & \text{Column 2} & \cdots & \text{Column } j & \cdots & \text{Column } n \\ \text{Row 1} & a_{11} & a_{12} & \cdots & a_{1j} & \cdots & a_{1n} \\ \text{Row 2} & a_{21} & a_{22} & \cdots & a_{2j} & \cdots & a_{2n} \\ \vdots & \vdots & \vdots & \cdots & \vdots & \cdots & \vdots \\ \text{Row } i & a_{i1} & a_{i2} & \cdots & a_{ij} & \cdots & a_{in} \\ \vdots & \vdots & \vdots & \cdots & \vdots & \cdots & \vdots \\ \text{Row } m & a_{m1} & a_{m2} & \cdots & a_{mj} & \cdots & a_{mn} \end{matrix}$

Section	Concept	Pages	Review Exercises	Key Ideas/Points
	Equality of matrices	585–586		The orders must be the same: $A_{m \times n}$ and $B_{m \times n}$.
	Matrix addition and subtraction	586–588	29–38	The orders must be the same: $A_{m \times n}$ and $B_{m \times n}$. Perform operation element by element.
	Scalar and matrix multiplication	588–594	33–46	The scalar is multiplied by *all* of the elements. • The orders must satisfy the relationship: $A_{m \times n}$ and $B_{n \times p}$, resulting in $(AB)_{m \times p}$. • Perform multiplication row by column. • Matrix multiplication is not commutative: $AB \neq BA$.
7.3	**Matrix equations; the inverse of a square matrix**	598–607	47–64	
	Matrix equations	598–599		Linear system: $AX = B$. *Front* multiply both sides of the equation by A^{-1}, which results in $X = A^{-1}B$.
	Finding the inverse of a matrix	599–604	47–58	Only square matrices, $n \times n$, have inverses. $A^{-1}A = I_n$ **Step 1:** Form the matrix $[A \mid I_n]$. **Step 2:** Use row operations to transform this matrix to $\left[I_n \mid A^{-1}\right]$.
	Solving systems of linear equations using matrix algebra and inverses of square matrices	605–607	59–64	$AX = B$ **Step 1:** Find A^{-1}. **Step 2:** $X = A^{-1}B$.
7.4	**The determinant of a square matrix and Cramer's rule**	611–621	65–84	Cramer's rule can be used to solve only a system of linear equations with a unique solution.
	Determinant of a 2×2 matrix	611–612	65–68	$\begin{vmatrix} a & b \\ c & d \end{vmatrix} = ad - bc$
	Determinant of an $n \times n$ matrix	612–615	75–78	Let A be a square matrix of order $n \times n$; then • The **minor** M_{ij} of the element a_{ij} is the determinant of the $(n-1) \times (n-1)$ matrix obtained when the ith row and jth column of A are deleted. • The **cofactor** C_{ij} of the element a_{ij} is given by $C_{ij} = (-1)^{i+j}M_{ij}$. $\begin{bmatrix} 1 & -3 & 2 \\ 4 & -1 & 0 \\ 5 & -2 & 3 \end{bmatrix}$ $M_{11} = \begin{vmatrix} -1 & 0 \\ -2 & 3 \end{vmatrix} = -3 - 0 = -3$ $C_{11} = (-1)^{1+1}M_{11} = (1)(-3) = -3$ Determinant of a 3×3 matrix: $\begin{bmatrix} + & - & + \\ - & + & - \\ + & - & + \end{bmatrix}$

Section	Concept	Pages	Review Exercises	Key Ideas/Points
				If A is a 3×3 matrix, the determinant can be given by $\det(A) = a_{11}C_{11} + a_{12}C_{12} + a_{13}C_{13}$. This is called **expanding the determinant by the first row**. (Note that any row or column can be used.) $\begin{vmatrix} a_1 & b_1 & c_1 \\ a_2 & b_2 & c_2 \\ a_3 & b_3 & c_3 \end{vmatrix} = a_1\begin{vmatrix} b_2 & c_2 \\ b_3 & c_3 \end{vmatrix} - b_1\begin{vmatrix} a_2 & c_2 \\ a_3 & c_3 \end{vmatrix} + c_1\begin{vmatrix} a_2 & b_2 \\ a_3 & b_3 \end{vmatrix}$
	Cramer's rule: Systems of linear equations in two variables	616–618	69–74	The system $\begin{aligned} a_1x + b_1y &= c_1 \\ a_2x + b_2y &= c_2 \end{aligned}$ has the solution $x = \dfrac{D_x}{D} \quad y = \dfrac{D_y}{D} \quad \text{if } D \neq 0$ where $D = \begin{vmatrix} a_1 & b_1 \\ a_2 & b_2 \end{vmatrix} \quad D_x = \begin{vmatrix} c_1 & b_1 \\ c_2 & b_2 \end{vmatrix} \quad D_y = \begin{vmatrix} a_1 & c_1 \\ a_2 & c_2 \end{vmatrix}$
	Cramer's rule: Systems of linear equations in three variables	618–621	79–82	The system $\begin{aligned} a_1x + b_1y + c_1z &= d_1 \\ a_2x + b_2y + c_2z &= d_2 \\ a_3x + b_3y + c_3z &= d_3 \end{aligned}$ has the solution $x = \dfrac{D_x}{D} \quad y = \dfrac{D_y}{D} \quad z = \dfrac{D_z}{D} \quad \text{if } D \neq 0$ where $D = \begin{vmatrix} a_1 & b_1 & c_1 \\ a_2 & b_2 & c_2 \\ a_3 & b_3 & c_3 \end{vmatrix} \quad D_x = \begin{vmatrix} d_1 & b_1 & c_1 \\ d_2 & b_2 & c_2 \\ d_3 & b_3 & c_3 \end{vmatrix}$ $D_y = \begin{vmatrix} a_1 & d_1 & c_1 \\ a_2 & d_2 & c_2 \\ a_3 & d_3 & c_3 \end{vmatrix} \quad D_z = \begin{vmatrix} a_1 & b_1 & d_1 \\ a_2 & b_2 & d_2 \\ a_3 & b_3 & d_3 \end{vmatrix}$

7.1 Matrices and Systems of Linear Equations

Write the augmented matrix for each system of linear equations.

1. $\begin{aligned} 5x + 7y &= 2 \\ 3x - 4y &= -2 \end{aligned}$

2. $\begin{aligned} 2.3x - 4.5y &= 6.8 \\ -0.4x + 2.1y &= -9.1 \end{aligned}$

3. $\begin{aligned} 2x - z &= 3 \\ y - 3z &= -2 \\ x + 4z &= -3 \end{aligned}$

4. $\begin{aligned} 2y - x + 3z &= 1 \\ 4z - 2y + 3x &= -2 \\ x - y - 4z &= 0 \end{aligned}$

Indicate whether each matrix is in row–echelon form. If it is, state whether it is in *reduced* row–echelon form.

5. $\left[\begin{array}{cc|c} 1 & 1 & 0 \\ 0 & 1 & 2 \end{array}\right]$

6. $\left[\begin{array}{cc|c} 1 & 2 & 0 \\ 0 & 0 & 1 \end{array}\right]$

7. $\left[\begin{array}{ccc|c} 2 & 0 & 1 & 1 \\ 0 & -2 & 0 & 2 \\ 0 & 0 & 2 & 3 \end{array}\right]$

8. $\left[\begin{array}{cccc|c} 1 & 0 & 1 & 0 & 2 \\ 0 & 0 & 1 & 1 & -3 \\ 0 & 1 & 0 & 0 & 2 \\ 0 & 0 & 0 & 1 & 1 \end{array}\right]$

Perform the indicated row operations on each matrix.

9. $\left[\begin{array}{cc|c} 1 & -2 & 1 \\ 0 & -2 & 2 \end{array}\right] \quad \frac{-1}{2}R_2 \to R_2$

10. $\left[\begin{array}{cc|c} 1 & 4 & 1 \\ 2 & -2 & 3 \end{array}\right] \quad R_2 - 2R_1 \to R_2$

11. $\left[\begin{array}{ccc|c} 1 & -2 & 0 & 1 \\ 0 & -2 & 3 & -2 \\ 0 & 1 & -4 & 8 \end{array}\right] \quad R_2 + R_1 \to R_1$

12. $\left[\begin{array}{cccc|c} 1 & 1 & 1 & 6 & 0 \\ 0 & 2 & -2 & 3 & -2 \\ 0 & 0 & 1 & -2 & 4 \\ 0 & -1 & 3 & -3 & 3 \end{array}\right] \quad \begin{array}{l} -2R_1 + R_2 \to R_1 \\ R_4 + R_3 \to R_4 \end{array}$

Apply row operations to transform each matrix to reduced row–echelon form.

13. $\left[\begin{array}{cc|c} 1 & 3 & 0 \\ 3 & 4 & 1 \end{array}\right]$

14. $\left[\begin{array}{ccc|c} 1 & 2 & -1 & 0 \\ 0 & 1 & -1 & -1 \\ -2 & 0 & 1 & -2 \end{array}\right]$

15. $\left[\begin{array}{ccc|c} 4 & 1 & -2 & 0 \\ 1 & 0 & -1 & 0 \\ -2 & 1 & 1 & 12 \end{array}\right]$

16. $\left[\begin{array}{ccc|c} 2 & 3 & 2 & 1 \\ 0 & -1 & 1 & -2 \\ 1 & 1 & -1 & 6 \end{array}\right]$

Solve the system of linear equations using augmented matrices.

17. $\begin{aligned} 3x - 2y &= 2 \\ -2x + 4y &= 1 \end{aligned}$

18. $\begin{aligned} 2x - 7y &= 22 \\ x + 5y &= -23 \end{aligned}$

19. $\begin{aligned} 5x - y &= 9 \\ x + 4y &= 6 \end{aligned}$

20. $\begin{aligned} 8x + 7y &= 10 \\ -3x + 5y &= 42 \end{aligned}$

21. $\begin{aligned} x - 2y + z &= 3 \\ 2x - y + z &= -4 \\ 3x - 3y - 5z &= 2 \end{aligned}$

22. $\begin{aligned} 3x - y + 4z &= 18 \\ 5x + 2y - z &= -20 \\ x + 7y - 6z &= -38 \end{aligned}$

23. $\begin{aligned} x - 4y + 10z &= -61 \\ 3x - 5y + 8z &= -52 \\ -5x + y - 2z &= 8 \end{aligned}$

24. $\begin{aligned} 4x - 2y + 5z &= 17 \\ x + 6y - 3z &= -\tfrac{17}{2} \\ -2x + 5y + z &= 2 \end{aligned}$

25. $\begin{aligned} 3x + y + z &= -4 \\ x - 2y + z &= -6 \end{aligned}$

26. $\begin{aligned} 2x - y + 3z &= 6 \\ 3x + 2y - z &= 12 \end{aligned}$

Applications

27. Fitting a Curve to Data. The average number of flights on a commercial plane that a person takes a year can be modeled by a quadratic function $y = ax^2 + bx + c$, where $a < 0$ and x represents age: $16 < x < 65$. The following table gives the average number of flights per year that a person takes on a commercial airline. Determine the quadratic function that models this quantity.

Age	Number of Flights per Year
16	2
40	6
65	4

28. Investment Portfolio. Danny and Paula decide to invest $20,000 of their savings in investments. They put some in an IRA account earning 4.5% interest, some in a mutual fund that has been averaging 8% a year, and some in a stock that earned 12% last year. If they put $3,000 more in the mutual fund than in the IRA and the mutual fund and stock have the same growth in the next year as they did in the previous year, they will earn $1,877.50 in a year. How much money did they put in each of the three investments?

7.2 Matrix Algebra

Calculate the given expression, if possible.

$A = \begin{bmatrix} 2 & -3 \\ 0 & 1 \end{bmatrix} \quad B = \begin{bmatrix} 1 & 5 & -1 \\ 3 & 7 & 2 \end{bmatrix} \quad C = \begin{bmatrix} 5 & 0 & 1 \\ 2 & -1 & 4 \\ 0 & 3 & 6 \end{bmatrix}$

$D = \begin{bmatrix} 5 & 2 \\ 9 & 7 \end{bmatrix} \quad E = \begin{bmatrix} 2 & 0 & 3 \\ 4 & 1 & -1 \end{bmatrix}$

29. $A + C$ **30.** $B + A$ **31.** $B + E$

32. $A + D$ **33.** $2A + D$ **34.** $3E + B$

35. $2D - 3A$ **36.** $3B - 4E$ **37.** $5A - 2D$

38. $5B - 4E$ **39.** AB **40.** BC

41. DA **42.** AD **43.** $BC + E$

44. DB **45.** EC **46.** CE

7.3 Matrix Equations; The Inverse of a Square Matrix

Determine whether B is the multiplicative inverse of A using $AA^{-1} = I$.

47. $A = \begin{bmatrix} 6 & 4 \\ 4 & 2 \end{bmatrix} \quad B = \begin{bmatrix} -0.5 & 1 \\ 1 & -1.5 \end{bmatrix}$

48. $A = \begin{bmatrix} 1 & -2 \\ 2 & -4 \end{bmatrix} \quad B = \begin{bmatrix} 1 & 2 \\ 2 & -2 \end{bmatrix}$

49. $A = \begin{bmatrix} 1 & -2 & 6 \\ 2 & 3 & -2 \\ 0 & -1 & 1 \end{bmatrix} \quad B = \begin{bmatrix} -\frac{1}{7} & \frac{4}{7} & 2 \\ \frac{2}{7} & -\frac{1}{7} & -2 \\ \frac{2}{7} & -\frac{1}{7} & -1 \end{bmatrix}$

50. $A = \begin{bmatrix} 0 & 7 & 6 \\ 1 & 0 & -4 \\ -2 & 1 & 0 \end{bmatrix} \quad B = \begin{bmatrix} 1 & 1 & 1 \\ -2 & -2 & -2 \\ 2 & 0 & 6 \end{bmatrix}$

Find A^{-1}, if it exists.

51. $A = \begin{bmatrix} 1 & 2 \\ -3 & 4 \end{bmatrix}$

52. $A = \begin{bmatrix} -2 & 7 \\ -4 & 6 \end{bmatrix}$

53. $A = \begin{bmatrix} 0 & 1 \\ -2 & 0 \end{bmatrix}$

54. $A = \begin{bmatrix} 3 & -1 \\ -2 & 2 \end{bmatrix}$

55. $A = \begin{bmatrix} 1 & 3 & -2 \\ 2 & 1 & -1 \\ 0 & 1 & -3 \end{bmatrix}$

56. $A = \begin{bmatrix} 0 & 1 & 0 \\ 4 & 1 & 2 \\ -3 & -2 & 1 \end{bmatrix}$

57. $A = \begin{bmatrix} -1 & 1 & 0 \\ -2 & 1 & 2 \\ 1 & 2 & 4 \end{bmatrix}$

58. $A = \begin{bmatrix} -4 & 4 & 3 \\ 1 & 2 & 2 \\ 3 & -1 & 6 \end{bmatrix}$

Solve the system of linear equations using matrix algebra.

59. $\begin{aligned} 3x - y &= 11 \\ 5x + 2y &= 33 \end{aligned}$

60. $\begin{aligned} 6x + 4y &= 15 \\ -3x - 2y &= -1 \end{aligned}$

61. $\begin{aligned} \tfrac{5}{8}x - \tfrac{2}{3}y &= -3 \\ \tfrac{3}{4}x + \tfrac{5}{6}y &= 16 \end{aligned}$

62. $\begin{aligned} x + y - z &= 0 \\ 2x - y + 3z &= 18 \\ 3x - 2y + z &= 17 \end{aligned}$

63. $\begin{aligned} 3x - 2y + 4z &= 11 \\ 6x + 3y - 2z &= 6 \\ x - y + 7z &= 20 \end{aligned}$

64. $\begin{aligned} 2x + 6y - 4z &= 11 \\ -x - 3y + 2z &= -\tfrac{11}{2} \\ 4x + 5y + 6z &= 20 \end{aligned}$

7.4 The Determinant of a Square Matrix and Cramer's Rule

Evaluate each 2×2 determinant.

65. $\begin{vmatrix} 2 & 4 \\ 3 & 2 \end{vmatrix}$

66. $\begin{vmatrix} -2 & -4 \\ -3 & 2 \end{vmatrix}$

67. $\begin{vmatrix} 2.4 & -2.3 \\ 3.6 & -1.2 \end{vmatrix}$

68. $\begin{vmatrix} -\frac{1}{4} & 4 \\ \frac{3}{4} & -4 \end{vmatrix}$

Employ Cramer's rule to solve each system of equations, if possible.

69. $\begin{aligned} x - y &= 2 \\ x + y &= 4 \end{aligned}$

70. $\begin{aligned} 3x - y &= -17 \\ -x + 5y &= 43 \end{aligned}$

71. $\begin{aligned} 2x + 4y &= 12 \\ x - 2y &= 6 \end{aligned}$

72. $\begin{aligned} -x + y &= 4 \\ 2x - 6y &= -5 \end{aligned}$

73. $\begin{aligned} -3x &= 40 - 2y \\ 2x &= 25 + y \end{aligned}$

74. $\begin{aligned} 3x &= 20 + 4y \\ y - x &= -6 \end{aligned}$

Evaluate each 3×3 determinant.

75. $\begin{vmatrix} 1 & 2 & 2 \\ 0 & 1 & 3 \\ 2 & -1 & 0 \end{vmatrix}$

76. $\begin{vmatrix} 0 & -2 & 1 \\ 0 & -3 & 7 \\ 1 & -10 & -3 \end{vmatrix}$

77. $\begin{vmatrix} a & 0 & -b \\ -a & b & c \\ 0 & 0 & -d \end{vmatrix}$

78. $\begin{vmatrix} -2 & -4 & 6 \\ 2 & 0 & 3 \\ -1 & 2 & \frac{3}{4} \end{vmatrix}$

Employ Cramer's rule to solve each system of equations, if possible.

79. $\begin{aligned} x + y - 2z &= -2 \\ 2x - y + z &= 3 \\ x + y + z &= 4 \end{aligned}$

80. $\begin{aligned} -x - y + z &= 3 \\ x + 2y - 2z &= 8 \\ 2x + y + 4z &= -4 \end{aligned}$

81. $\begin{aligned} 3x \qquad + 4z &= -1 \\ x + y + 2z &= -3 \\ y - 4z &= -9 \end{aligned}$

82. $\begin{aligned} x + y + z &= 0 \\ -x - 3y + 5z &= -2 \\ 2x + y - 3z &= -4 \end{aligned}$

Applications

83. Apply determinants to find the area of a triangle with vertices (2, 4), (4, 4), and (−4, 3).

84. If three points (x_1, y_1), (x_2, y_2), and (x_3, y_3) are collinear (lie on the same line), then the following determinant must be satisfied:

$$\begin{vmatrix} x_1 & y_1 & 1 \\ x_2 & y_2 & 1 \\ x_3 & y_3 & 1 \end{vmatrix} = 0$$

Determine whether (0, −3), (3, 0), and (1, 6) are collinear.

REVIEW EXERCISES

Technology

Section 7.1

In Exercises 85 and 86, refer to the following:

You are asked to model a set of three points with a quadratic function $y = ax^2 + bx + c$ and determine the quadratic function.

a. Set up a system of equations, use a graphing utility or graphing calculator to solve the system by entering the coefficients of the augmented matrix.

b. Use the graphing calculator commands STAT QuadReg to model the data using a quadratic function. Round your answers to two decimal places.

85. $(-10, 12.5), (3, -2.8), (9, 8.5)$

86. $(-4, 10), (2.5, -9.5), (13.5, 12.6)$

Section 7.2

Apply a graphing utility to perform the indicated matrix operations, if possible.

$$A = \begin{bmatrix} -6 & 0 & 4 \\ 1 & 3 & 5 \\ 2 & -1 & 0 \end{bmatrix} \quad B = \begin{bmatrix} 5 & 1 \\ 0 & 2 \\ -8 & 4 \end{bmatrix} \quad C = \begin{bmatrix} 4 & -3 & 0 \\ 1 & 2 & 5 \end{bmatrix}$$

87. ABC

88. CAB

Section 7.3

Apply a graphing utility and matrix algebra to solve the system of linear equations.

89.
$$\begin{aligned} 6.1x - 14.2y &= 75.495 \\ -2.3x + 7.2y &= -36.495 \end{aligned}$$

90.
$$\begin{aligned} 7.2x + 3.2y - 1.7z &= 5.53 \\ -1.3x + 4.1y + 2.8z &= -23.949 \end{aligned}$$

Section 7.4

Apply Cramer's rule to solve each system of equations and a graphing utility to evaluate the determinants.

91.
$$\begin{aligned} 4.5x - 8.7y &= -72.33 \\ -1.4x + 5.3y &= 31.32 \end{aligned}$$

92.
$$\begin{aligned} 1.4x + 3.6y + 7.5z &= 42.08 \\ 2.1x - 5.7y - 4.2z &= 5.37 \\ 1.8x - 2.8y - 6.2z &= -9.86 \end{aligned}$$

Write each of the following systems of linear equations as an augmented matrix.

1. $\begin{aligned} x - 2y &= 1 \\ -x + 3y &= 2 \end{aligned}$

2. $\begin{aligned} 3x + 5y &= -2 \\ 7x + 11y &= -6 \end{aligned}$

3. $\begin{aligned} 6x + 9y + z &= 5 \\ 2x - 3y + z &= 3 \\ 10x + 12y + 2z &= 9 \end{aligned}$

4. $\begin{aligned} 3x + 2y - 10z &= 2 \\ x + y - z &= 5 \end{aligned}$

5. Perform the following row operations:

$$\begin{bmatrix} 1 & 3 & 5 \\ 2 & 7 & -1 \\ -3 & -2 & 0 \end{bmatrix} \quad \begin{array}{l} R_2 - 2R_1 \to R_2 \\ R_3 + 3R_1 \to R_3 \end{array}$$

6. Rewrite the following matrix in reduced row–echelon form:

$$\left[\begin{array}{ccc|c} 2 & -1 & 1 & 3 \\ 1 & 1 & -1 & 0 \\ 3 & 2 & -2 & 1 \end{array}\right]$$

In Exercises 7 and 8, solve the systems of linear equations using augmented matrices.

7. $\begin{aligned} 6x + 9y + z &= 5 \\ 2x - 3y + z &= 3 \\ 10x + 12y + 2z &= 9 \end{aligned}$

8. $\begin{aligned} 3x + 2y - 10z &= 2 \\ x + y - z &= 5 \end{aligned}$

9. Multiply the matrices, if possible.

$$\begin{bmatrix} 1 & -2 & 5 \\ 0 & -1 & 3 \end{bmatrix} \begin{bmatrix} 0 & 4 \\ 3 & -5 \\ -1 & 1 \end{bmatrix}$$

10. Add the matrices, if possible.

$$\begin{bmatrix} 1 & -2 & 5 \\ 0 & -1 & 3 \end{bmatrix} + \begin{bmatrix} 0 & 4 \\ 3 & -5 \\ -1 & 1 \end{bmatrix}$$

11. Find the inverse of $\begin{bmatrix} 4 & 3 \\ 5 & -1 \end{bmatrix}$, if it exists.

12. Find the inverse of $\begin{bmatrix} 1 & -3 & 2 \\ 4 & 2 & 0 \\ -1 & 2 & 5 \end{bmatrix}$, if it exists.

13. Find the inverse of $\begin{bmatrix} 3 & 1 & 0 \\ 5 & 2 & -1 \end{bmatrix}$, if it exists.

14. Solve the system of linear equations with matrix algebra (inverses).

$$\begin{aligned} x + 3y &= -1 \\ -2x - 5y &= 4 \end{aligned}$$

15. Solve the system of linear equations with matrix algebra (inverses).

$$\begin{aligned} 3x - y + 4z &= 18 \\ x + 2y + 3z &= 20 \\ -4x + 6y - z &= 11 \end{aligned}$$

Calculate the determinant.

16. $\begin{vmatrix} 7 & -5 \\ 2 & -1 \end{vmatrix}$

17. $\begin{vmatrix} 1 & -2 & -1 \\ 3 & -5 & 2 \\ 4 & -1 & 0 \end{vmatrix}$

In Exercises 18 and 19, solve the system of linear equations using Cramer's rule.

18. $\begin{aligned} x - 2y &= 1 \\ -x + 3y &= 2 \end{aligned}$

19. $\begin{aligned} 3x + 5y - 2z &= -6 \\ 7x + 11y + 3z &= 2 \\ x - y + z &= 4 \end{aligned}$

20. A company has two rubrics for scoring job applicants based on weighting education, experience, and the interview differently.

Matrix A:

	Rubric 1	Rubric 2
Education	0.4	0.6
Experience	0.5	0.1
Interview	0.1	0.3

Applicants receive a score from 1 to 10 in each category (education, experience, and interview). Two applicants are shown in the matrix B:

Matrix B:

	Education	Experience	Interview
Applicant 1	4	7	3
Applicant 2	6	5	4

What is the order of BA? What does each entry in BA tell us?

21. A college student inherits \$15,000 from his favorite aunt. He decides to invest it with a diversified scheme. He divides the money into a money market account paying 2% annual simple interest, a conservative stock that rose 4% last year, and an aggressive stock that rose 22% last year. He places \$1,000 more in the conservative stock than in the money market and twice as much in the aggressive stock than in the money market. If the stocks perform the same way this year, he will make \$1,790 in one year, assuming a simple interest model. How much did he put in each of the three investments?

22. You are asked to model a set of three points with a quadratic function $y = ax^2 + bx + c$ and determine the quadratic function.
    a. Set up a system of equations, use a graphing utility or graphing calculator to solve the system by entering the coefficients of the augmented matrix.
    b. Use the graphing calculator commands STAT QuadReg to model the data using a quadratic function.

$$(-3, 6), (1, 12), (5, 7)$$

23. Apply a graphing utility and matrix algebra to solve the system of linear equations.

$$\begin{aligned} 5.6x - 2.7y &= 87.28 \\ -4.2x + 8.4y &= -106.26 \end{aligned}$$

24. Apply Cramer's rule to solve each system of equations and a graphing utility to evaluate the determinants.

$$\begin{aligned} 1.5x + 2.6y &= 18.34 \\ -2.3x + 1.5y &= 28.94 \end{aligned}$$

1. Solve by completing the square: $x^2 - 6x = 11$.

2. Write an equation of a line that passes through the point $(-2, 5)$ and is parallel to the y-axis.

3. Write the equation of a circle with center $(-3, -1)$ and passing through the point $(1, 2)$.

4. Determine whether the relation $x^2 - y^2 = 25$ is a function.

5. Determine whether the function $g(x) = \sqrt{2 - x^2}$ is odd or even.

6. For the function $y = 5(x - 4)^2$, identify all of the transformations of $y = x^2$.

7. Find the composite function $f \circ g$, and state the domain, for $f(x) = x^3 - 1$ and $g(x) = \frac{1}{x}$.

8. Find the inverse of the function $f(x) = \sqrt[3]{x} - 1$.

9. Find the vertex of the parabola associated with the quadratic function $f(x) = \frac{1}{4}x^2 + \frac{3}{5}x - \frac{6}{25}$.

10. Find a polynomial of minimum degree (there are many) that has the zeros $x = -\sqrt{7}$ (multiplicity 2), $x = 0$ (multiplicity 3), $x = \sqrt{7}$ (multiplicity 2).

11. Use long division to find the quotient $Q(x)$ and the remainder $r(x)$ of $(5x^3 - 4x^2 + 3) \div (x^2 + 1)$.

12. Given the zero $x = 4i$ of the polynomial $P(x) = x^4 + 2x^3 + x^2 + 32x - 240$, determine all the other zeros and write the polynomial in terms of a product of linear factors.

13. Find the vertical and horizontal asymptotes of the function $f(x) = \dfrac{0.7x^2 - 5x + 11}{x^2 - x - 6}$.

14. If \$5,400 is invested at 2.25% compounded continuously, how much is in the account after 4 years?

15. Use interval notation to express the domain of the function $f(x) = \log_4(x + 3)$.

16. Use the properties of logarithms to simplify the expression $\log_{\pi} 1$.

17. Give an exact solution to the logarithmic equation $\log_5(x + 2) + \log_5(6 - x) = \log_5(3x)$.

18. If money is invested in a savings account earning 4% compounded continuously, how many years will pass until the money triples?

19. Solve the following system of linear equations:

$$\begin{aligned} 2x + 3y &= 6 \\ x &= -1.5y - 5.5 \end{aligned}$$

20. At the student union, one group of students bought 6 deluxe burgers, 5 orders of fries, and 5 sodas for \$24.34. A second group of students ordered 8 deluxe burgers, 4 orders of fries, and 6 sodas for \$28.42. A third group of students ordered 3 deluxe burgers, 2 orders of fries, and 4 sodas for \$13.51. Determine the price of each food item.

21. Maximize the objective function $z = 5x + 7y$, subject to the constraints:

$$\begin{aligned} 2x + 5y &\le 10 \qquad x \ge 0 \\ 5x + 2y &\le 10 \qquad y \ge 0 \end{aligned}$$

22. Solve the system with Gauss–Jordan elimination.

$$\begin{aligned} x - 2y + 3z &= 11 \\ 4x + 5y - z &= -8 \\ 3x + y - 2z &= 1 \end{aligned}$$

23. Given

$$A = \begin{bmatrix} 3 & 4 & -7 \\ 0 & 1 & 5 \end{bmatrix} \quad B = \begin{bmatrix} 8 & -2 & 6 \\ 9 & 0 & -1 \end{bmatrix} \quad C = \begin{bmatrix} 9 & 0 \\ 1 & 2 \end{bmatrix}$$

find $2A + CB$.

24. Write the matrix equation, find the inverse of the coefficient matrix, and solve the system using matrix algebra.

$$\begin{aligned} 2x + 5y &= -1 \\ -x + 4y &= 7 \end{aligned}$$

25. Apply Cramer's rule to solve the system of equations.

$$\begin{aligned} 7x + 5y &= 1 \\ -x + 4y &= -1 \end{aligned}$$

26. Use a graphing calculator to graph the given polynomial. From the graph, estimate x-intercepts and the coordinates of the relative maximum and minimum points. Round your answers to two decimal places.

$$f(x) = x^3 + 5x^2 - 12x - 13$$

27. Given

$$A = \begin{bmatrix} 0 & -1 & 6 \\ 4 & 3 & 1 \end{bmatrix} \quad B = \begin{bmatrix} -7 & -4 \\ 1 & 0 \\ 6 & 2 \end{bmatrix}$$

find AB and $(AB)^{-1}$.

8

Conics and Systems of Nonlinear Equations and Inequalities

Erik Von Weber/Getty Images

A satellite dish is a *paraboloid*.

Some buildings have a *hyperbolic* shape.

In this chapter we will study the three types of conic sections (conics): the parabola, the ellipse, and the hyperbola. We have already studied the circle (Section 2.4), which is a special form of the ellipse. We see these shapes all around us:

- Satellite dishes (parabolas)
- Cooling towers (hyperbolas)
- Planetary orbits (ellipses)

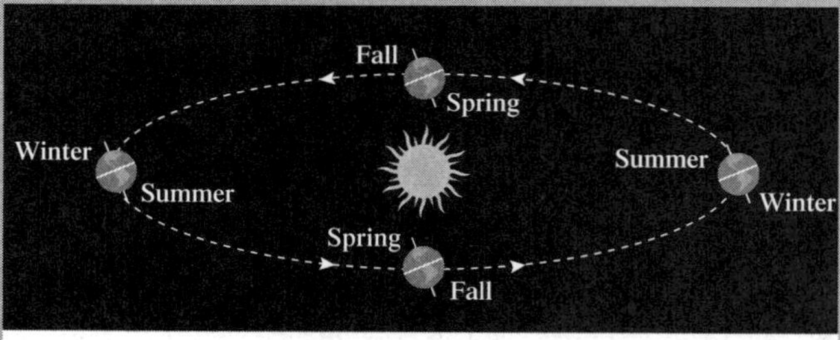

The Earth's orbit around the sun is an *ellipse*.

IN THIS CHAPTER we define the three conic sections: the parabola, the ellipse, and the hyperbola. Algebraic equations and the graphs of these conics are discussed. We then solve systems of nonlinear equations and inequalities involving parabolas, ellipses, hyperbolas, and other nonlinear equations and inequalities.

CONICS

PARABOLA

- Definition
- Equation and graphs
  - Vertex
  - Focus
  - Directrix
  - Intercepts
- Applications
  - Satellite dish
- Systems of equations and inequalities

ELLIPSE

- Definition
- Equation and graphs
  - Center
  - Foci
  - Intercepts
  - Major axis
  - Minor axis
- Eccentricity
- Applications
  - Orbits
  - Footballs
- Systems of equations and inequalities

HYPERBOLA

- Definition
- Equation and graphs
  - Center
  - Foci
  - Vertices
  - Intercepts
  - Asymptotes
  - Transverse axis
- Applications
  - Navigation
- Systems of equations and inequalities

SECTION

CHAPTER OBJECTIVES

- Visualize conics in terms of a plane intersecting a right double cone.
- Define the three conics.
- Develop the equations of the three conics from the general second-degree equation in two variables.
- Graph parabolas, ellipses, and hyperbolas.
- Solve application problems involving all three conics.
- Solve systems of nonlinear equations and interpret some graphically in terms of conics.
- Solve systems of nonlinear inequalities and interpret some graphically in terms of conics.

SECTION
8.1 CONIC BASICS

SKILLS OBJECTIVES

- Learn the name of each conic section.
- Define conics.
- Recognize the algebraic equation associated with each conic.

CONCEPTUAL OBJECTIVES

- Understand each conic as an intersection of a plane and a cone.
- Understand how the three equations of the conic sections are related to the general form of a second-degree equation in two variables.

Names of Conics

The word *conic* is derived from the word *cone.* Let's start with a (right circular) **double cone** (see figure on the left).

Conic sections are curves that result from the intersection of a plane and a double cone. The four conic sections are a **circle**, an **ellipse**, a **parabola**, and a **hyperbola**. **"Conics"** is an abbreviation for conic sections.

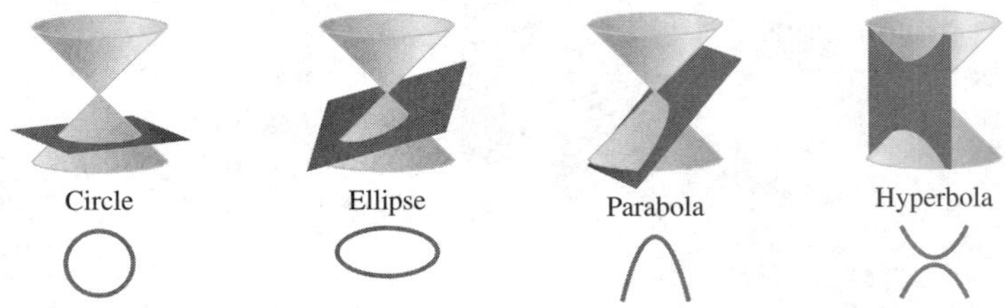

Study Tip

A circle is a special type of ellipse. All circles are ellipses, but not all ellipses are circles.

In Section 2.4, circles were discussed, and it will be shown that a circle is a particular type of an ellipse. Now we will discuss parabolas, ellipses, and hyperbolas. There are two ways in which we usually describe conics: graphically and algebraically. An entire section will be devoted to each of the three conics, but in the present section we will summarize the definitions of a parabola, ellipse, and hyperbola, and show how to identify the equations of these conics.

Definitions

You already know that a circle consists of all points equidistant (at a distance equal to the radius) from a point (the center). Ellipses, parabolas, and hyperbolas have similar definitions in that they all have a constant distance (or a sum or difference of distances) to some reference point.

A **parabola** is the set of all points that are **equidistant from both a line and a point**. An **ellipse** is the set of all points, the **sum of whose distances to two fixed points is constant**. A **hyperbola** is the set of all points, the **difference of whose distances to two fixed points is a constant**.

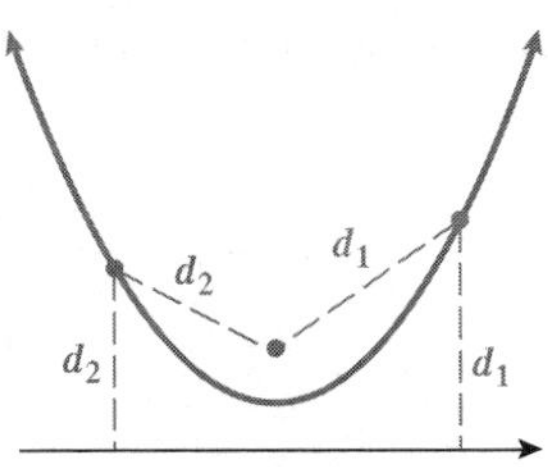

Parabola

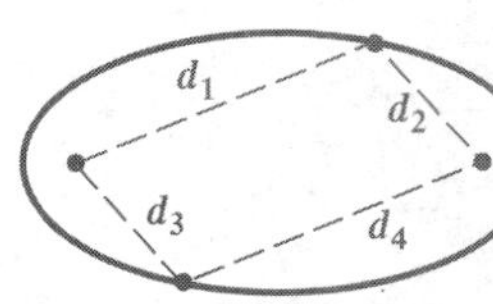

$d_1 + d_2 = d_3 + d_4$

Ellipse

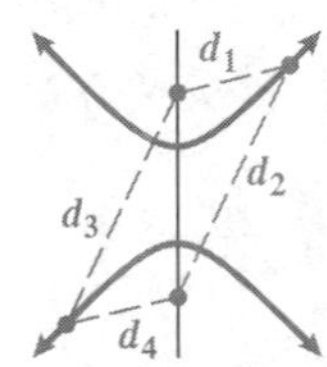

$|d_2 - d_1| = |d_4 - d_3|$

Hyperbola

The **general form of a second-degree equation in two variables** x and y is given by

$$Ax^2 + Bxy + Cy^2 + Dx + Ey + F = 0$$

If we let $A = 1$, $B = 0$, $C = 1$, $D = 0$, $E = 0$, and $F = -r^2$, this general equation reduces to the equation of a circle centered at the origin: $x^2 + y^2 = r^2$. In fact, all three conics (parabolas, ellipses, and hyperbolas) are special cases of the general second-degree equation.

Recall from Section 1.3 (Quadratic Equations) that the discriminant $b^2 - 4ac$ determines what types of solutions result from solving a second-degree equation in one variable. If the discriminant is positive, the solutions are two distinct real roots. If the discriminant is zero, the solution is a real repeated root. And if the discriminant is negative, the solutions are two complex conjugate roots.

The concept of discriminant has been generalized to conic sections. The discriminant determines the *shape* of the conic section.

CONIC	DISCRIMINANT
Ellipse	$B^2 - 4AC < 0$
Parabola	$B^2 - 4AC = 0$
Hyperbola	$B^2 - 4AC > 0$

Study Tip

All circles are ellipses since $B^2 - 4AC < 0$.

Using the discriminant to identify the shape of the conic will not work for degenerate cases (when the polynomial factors). For example, $2x^2 - xy - y^2 = 0$. At first glance one may think it is a hyperbola because $B^2 - 4AC > 0$ but this is a degenerative case:

$$(2x + y)(x - y) = 0$$

$$2x + y = 0 \quad \text{or} \quad x - y = 0$$

$$y = -2x \quad \text{or} \quad y = x$$

The graph is two intersecting lines. We now identify conics from the general form of a second-degree equation in two variables.

EXAMPLE 1 **Determining the Type of Conic**

Determine what type of conic corresponds to each of the following equations.

a. $\dfrac{x^2}{a^2} + \dfrac{y^2}{b^2} = 1$ **b.** $y = x^2$ **c.** $\dfrac{x^2}{a^2} - \dfrac{y^2}{b^2} = 1$

Solution:

Write the general form of the second-degree equation:

$$Ax^2 + Bxy + Cy^2 + Dx + Ey + F = 0$$

a. Identify A, B, C, D, E, and F $\qquad A = \dfrac{1}{a^2}, B = 0, C = \dfrac{1}{b^2}, D = 0, E = 0, F = -1$

Calculate the discriminant. $\qquad B^2 - 4AC = -\dfrac{4}{a^2b^2} < 0$

Since the discriminant is negative, the equation $\dfrac{x^2}{a^2} + \dfrac{y^2}{b^2} = 1$ is that of an **ellipse**.

Notice that if $a = b = r$, then this equation of an ellipse reduces to the general equation of a circle $x^2 + y^2 = r^2$, centered at the origin with radius r.

b. Identify A, B, C, D, E, and F. $A = 1, B = 0, C = 0, D = 0, E = -1, F = 0$

Calculate the discriminant. $B^2 - 4AC = 0$

Since the discriminant is zero, the equation $y = x^2$ is a **parabola**.

c. Identify A, B, C, D, E, and F. $A = \frac{1}{a^2}, B = 0, C = -\frac{1}{b^2}, D = 0, E = 0, F = -1$

Calculate the discriminant. $B^2 - 4AC = \frac{4}{a^2b^2} > 0$

Since the discriminant is positive, the equation $\frac{x^2}{a^2} - \frac{y^2}{b^2} = 1$ is a **hyperbola**.

■ **YOUR TURN** Determine what type of conic corresponds to each of the following equations.

a. $2x^2 + y^2 = 4$ **b.** $2x^2 = y^2 + 4$ **c.** $2y^2 = x$

■ **Answer: a.** ellipse
b. hyperbola
c. parabola

In the next three sections, we will discuss the standard forms of equations and the graphs of parabolas, ellipses, and hyperbolas.

SECTION 8.1 SUMMARY

In this section we defined the three conic sections and determined their general equations with respect to the general form of a second-degree equation in two variables:

$$Ax^2 + Bxy + Cy^2 + Dx + Ey + F = 0$$

The following table summarizes the three conics: ellipse, parabola, and hyperbola.

Conic	Geometric Definition: The set of all points	Discriminant
Ellipse	The sum of whose distances to two fixed points is constant.	Negative: $B^2 - 4AC < 0$
Parabola	Equidistant to both a line and a point.	Zero: $B^2 - 4AC = 0$
Hyperbola	The difference of whose distances to two fixed points is a constant.	Positive: $B^2 - 4AC > 0$

It is important to note that a circle is a special type of ellipse.

SECTION 8.1 EXERCISES

■ SKILLS

In Exercises 1–12, identify the conic section as a parabola, ellipse, circle, or hyperbola.

1. $x^2 + xy - y^2 + 2x = -3$
2. $x^2 + xy + y^2 + 2x = -3$
3. $2x^2 + 2y^2 = 10$
4. $x^2 - 4x + y^2 + 2y = 4$
5. $2x^2 - y^2 = 4$
6. $2y^2 - x^2 = 16$
7. $5x^2 + 20y^2 = 25$
8. $4x^2 + 8y^2 = 30$
9. $x^2 - y = 1$
10. $y^2 - x = 2$
11. $x^2 + y^2 = 10$
12. $x^2 + y^2 = 100$

SECTION

8.2 THE PARABOLA

SKILLS OBJECTIVES

- Graph a parabola given the focus, directrix, and vertex.
- Find the equation of a parabola whose vertex is at the origin.
- Find the equation of a parabola whose vertex is at the point (h, k).
- Solve applied problems that involve parabolas.

CONCEPTUAL OBJECTIVES

- Derive the general equation of a parabola.
- Identify, draw, and use the focus, directrix, and axis of symmetry.

Definition of a Parabola

Recall from Section 4.1 that the graphs of quadratic functions such as

$$f(x) = a(x - h)^2 + k \text{ or } y = ax^2 + bx + c$$

were *parabolas* that opened either upward or downward. We now expand our discussion of *parabolas* to parabolas that open to the right or left. We did not discuss these types of parabolas before because they are not functions (they fail the vertical line test).

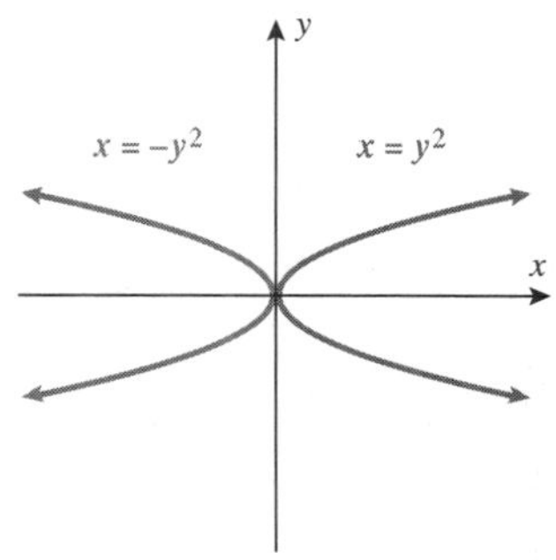

DEFINITION **Parabola**

A **parabola** is the set of all points in a plane that are equidistant from a fixed line, the **directrix**, and a fixed point not on the line, the **focus**. The line through the focus and perpendicular to the directrix is the **axis of symmetry**. The **vertex** of the parabola is located at the midpoint between the directrix and the focus along the axis of symmetry.

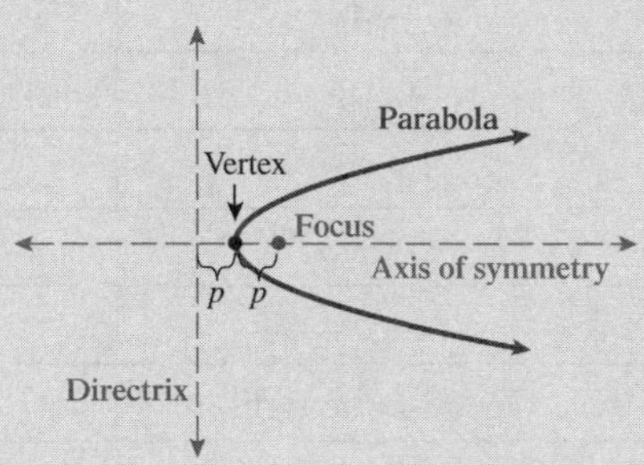

Here p is the distance along the axis of symmetry from the directrix to the vertex and from the vertex to the focus.

Let's consider a parabola with the vertex at the origin and the focus on the positive x-axis. Let the distance from the vertex to the focus be p. Therefore the focus is located at the point $(p, 0)$. Since the distance from the vertex to the focus is p, the distance from the vertex to the directrix must also be p. Since the axis of symmetry is the positive axis, the directrix must be perpendicular to the positive axis. Therefore, the directrix is given by $x = -p$. Any point (x, y) must have the same distance to the focus $(p, 0)$ as does the directrix $(-p, y)$.

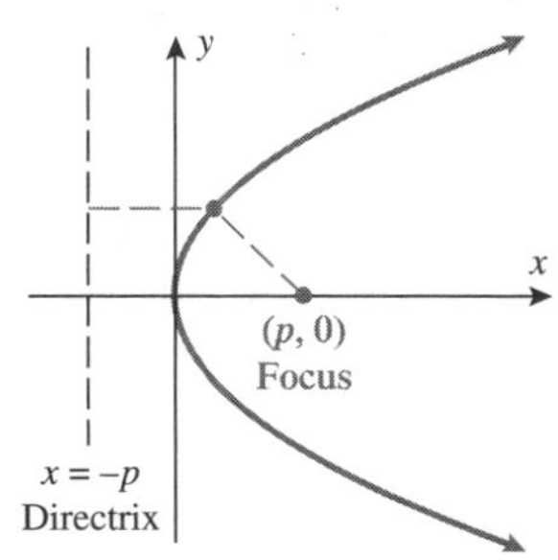

Derivation of the Equation of a Parabola

WORDS	**MATH**
Calculate the distance from (x, y) to $(p, 0)$ with the distance formula.	$d = \sqrt{(x - p)^2 + y^2}$
Calculate the distance from (x, y) to $(-p, y)$ with the distance formula.	$d = \sqrt{(x - (-p))^2 + 0^2}$
Set the two distances equal to one another. (Definition of a parabola.)	$\sqrt{(x - p)^2 + y^2} = \sqrt{(x + p)^2}$
Recall that $\sqrt{x^2} = \lvert x \rvert$.	$\sqrt{(x - p)^2 + y^2} = \lvert x + p \rvert$
Square both sides of the equation.	$(x - p)^2 + y^2 = (x + p)^2$
Square the binomials inside the parentheses.	$x^2 - 2px + p^2 + y^2 = x^2 + 2px + p^2$
Simplify.	$\boxed{y^2 = 4px}$

The equation $y^2 = 4px$ represents a parabola opening right with the vertex at the origin. The following box summarizes parabolas that have a vertex at the origin and a focus along either the x-axis or the y-axis.

EQUATION OF A PARABOLA WITH VERTEX AT THE ORIGIN

The standard (conic) form of the equation of a **parabola** with vertex at the origin is given by:

Equation	$y^2 = 4px$	$x^2 = 4py$
Vertex	$(0, 0)$	$(0, 0)$
Focus	$(p, 0)$	$(0, p)$
Directrix	$x = -p$	$y = -p$
Axis of symmetry	x-axis	y-axis
$p > 0$	Opens to the right	Opens upward
$p < 0$	Opens to the left	Opens downward
Graph ($p > 0$)	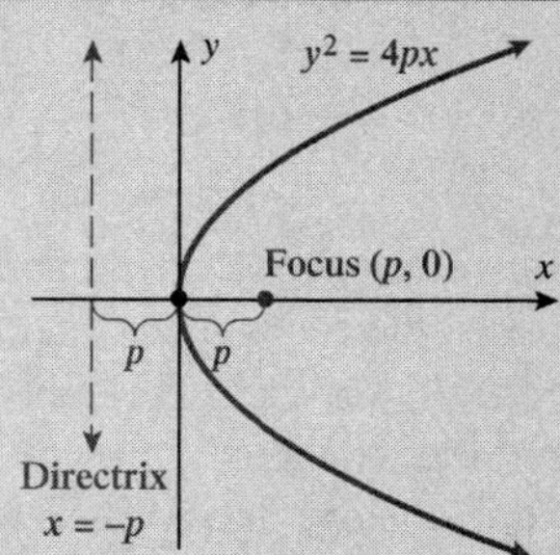	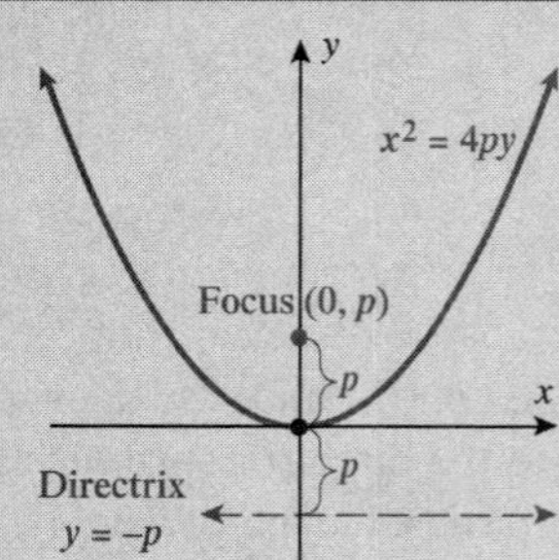

EXAMPLE 1 **Finding the Focus and Directrix of a Parabola Whose Vertex Is Located at the Origin**

Find the focus and directrix of a parabola whose equation is $y^2 = 8x$.

Solution:

Compare this parabola with the general equation of a parabola. $y^2 = 4px \quad y^2 = 8x$

Solve for p. $4p = 8$

$p = 2$

The focus of a parabola of the form $y^2 = 4px$ is $(p, 0)$. **Focus: (2, 0)**

The directrix of a parabola of the form $y^2 = 4px$ is $x = -p$. **Directrix: $x = -2$**

■ YOUR TURN Find the focus and directrix of a parabola whose equation is $y^2 = 16x$.

■ **Answer:** The focus is (4, 0) and the directrix is $x = -4$.

Graphing a Parabola Whose Vertex Is at the Origin

When a seamstress starts with a pattern for a custom-made suit, the pattern is used as a guide. The pattern is not sewn into the suit, but rather removed once it is used to determine the exact shape and size of the fabric to be sewn together. The focus and directrix of a parabola are similar to the pattern used by a seamstress. Although the focus and directrix define a parabola, they do not appear on the graph of a parabola.

Study Tip

The focus and directrix define a parabola, but do not appear on its graph.

An approximate sketch of a parabola whose vertex is at the origin can be drawn with three pieces of information. We know that the vertex is located at (0, 0). Additional information that we seek is the direction in which the parabola opens and approximately how wide or narrow to draw the parabolic curve. The direction toward which the parabola opens is found from the equation. An equation of the form $y^2 = 4px$ opens either left or right. It opens right if $p > 0$ and opens left if $p < 0$. An equation of the form $x^2 = 4py$ opens either up or down. It opens up if $p > 0$ and opens down if $p < 0$. How narrow or wide should the parabolic curve be drawn? If we select a few points that satisfy the equation, we can use those as graphing aids.

In Example 1, we found that the focus of that parabola is located at (2, 0). If we select the x-coordinate of the focus $x = 2$, and substitute that value into the equation of the parabola $y^2 = 8x$, we find the corresponding y values to be $y = -4$ and $y = 4$. If we plot the three points (0, 0), (2, −4), and (2, 4) and then connect the points with a parabolic curve, we get the graph on the right.

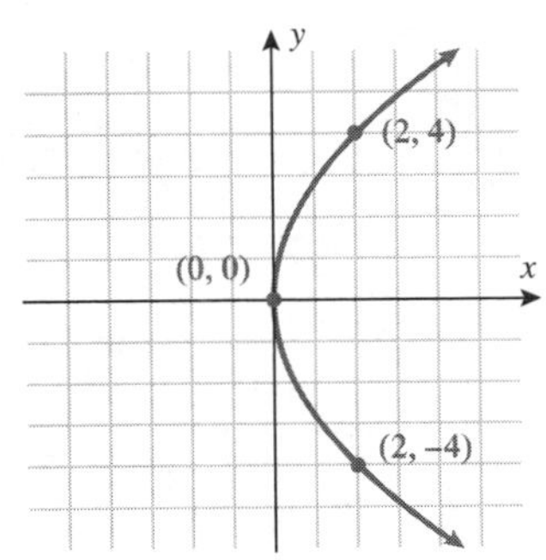

The line segment that passes through the focus (2, 0), is parallel to the directrix $x = -2$, and whose endpoints are on the parabola is called the **latus rectum**. The latus rectum in this case has length 8.

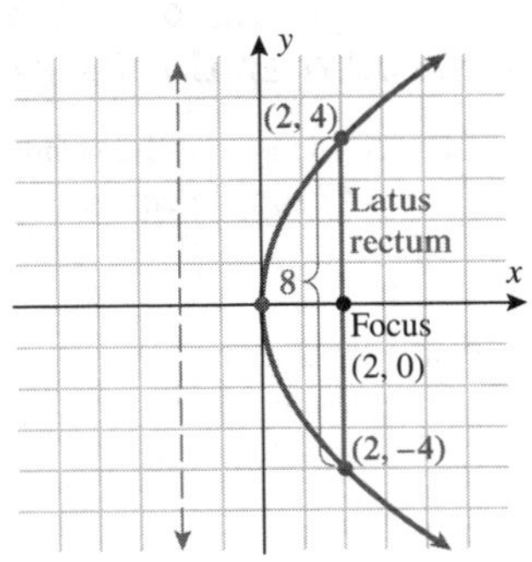

The latus rectum is a graphing aid that assists us in finding the width of a parabola.

In general, the points on a parabola of the form $y^2 = 4px$ that lie above and below the focus $(p, 0)$ satisfy the equation $y^2 = 4p^2$ and are located at $(p, -2p)$ and $(p, 2p)$. The latus rectum will have length $4|p|$. Similarly, a parabola of the form $x^2 = 4py$ will have a horizontal latus rectum of length $4|p|$. We will use the latus rectum as a graphing aid to determine the parabola's width.

Technology Tip

To graph $x^2 = -12y$ with a graphing calculator, solve for y first. That is, $y = -\frac{1}{12}x^2$.

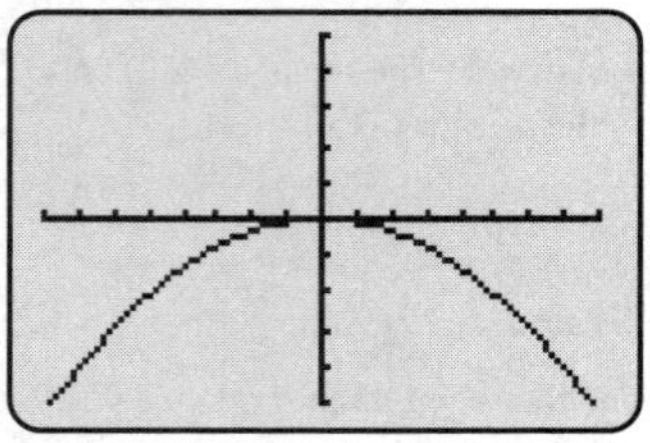

EXAMPLE 2 Graphing a Parabola Whose Vertex Is at the Origin Using the Focus, Directrix, and Latus Rectum as Graphing Aids

Determine the focus, directrix, and length of the latus rectum of the parabola $x^2 = -12y$. Employ these to assist in graphing the parabola.

Solution:

Compare this parabola with the general equation of a parabola. $\quad x^2 = 4py \quad x^2 = -12y$

Solve for p. $\quad 4p = -12$

$p = -3$

A parabola of the form $x^2 = 4py$ has focus $(0, p)$, directrix $y = -p$, and a latus rectum of length $4|p|$. For this parabola, $p = -3$; therefore, the focus is $\boxed{(0, -3)}$, the directrix is $\boxed{y = 3}$, and the length of the latus rectum is $\boxed{12}$.

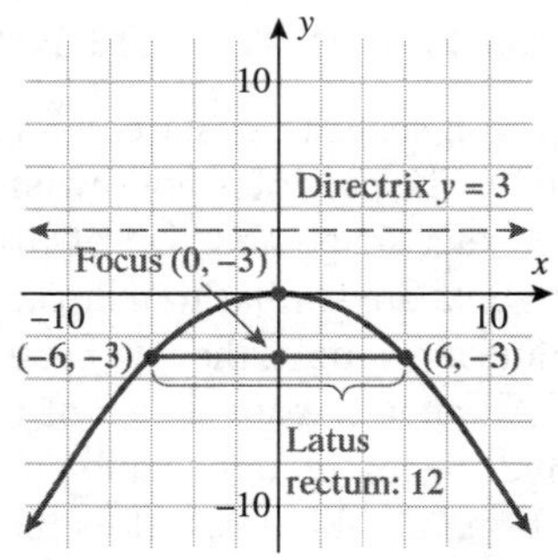

■ **YOUR TURN** Find the focus, directrix, and length of the latus rectum of the parabola $y^2 = -8x$, and use these to graph the parabola.

■ **Answer:**
The focus is $(-2, 0)$.
The directrix is $x = 2$.
The length of the latus rectum is 8.

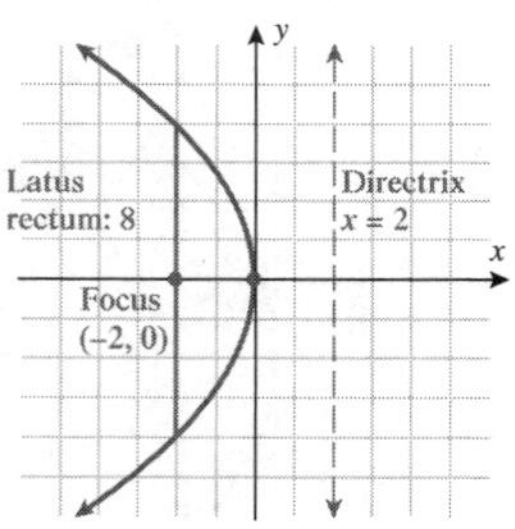

Finding the Equation of a Parabola Whose Vertex Is at the Origin

Thus far we have started with the equation of a parabola and then determined its focus and directrix. Let's now reverse the process. For example, if we know the focus and directrix of a parabola, how do we find the equation of the parabola? If we are given the focus and directrix, then we can find the vertex, which is the midpoint between the focus and the directrix. If the vertex is at the origin, then we know the general equation of the parabola that corresponds with the focus.

EXAMPLE 3 Finding the Equation of a Parabola Given the Focus and Directrix When the Vertex Is at the Origin

Find the equation of a parabola whose focus is at the point $\left(0, \frac{1}{2}\right)$ and whose directrix is $y = -\frac{1}{2}$. Graph the equation.

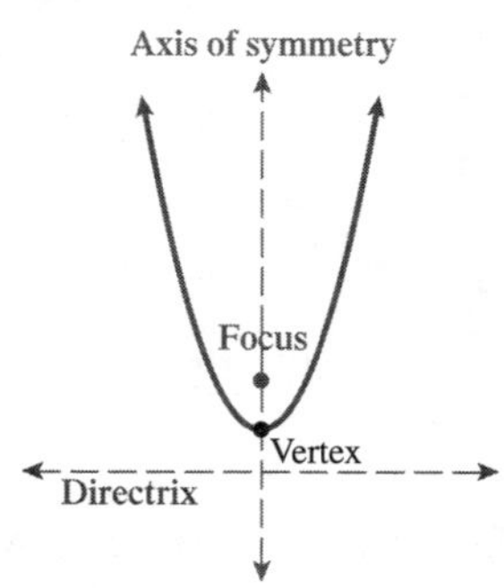

Solution:

The midpoint of the segment joining the focus and the directrix along the axis of symmetry is the vertex.

Calculate the midpoint between $\left(0, \frac{1}{2}\right)$ and $\left(0, -\frac{1}{2}\right)$.

$$\text{Vertex} = \left(\frac{0+0}{2}, \frac{\frac{1}{2} - \frac{1}{2}}{0}\right) = (0, 0).$$

A parabola with vertex at (0, 0), focus at $(0, p)$, and directrix $y = -p$ corresponds to the equation $x^2 = 4py$.

Identify p given that the focus is $(0, p) = \left(0, \frac{1}{2}\right)$. $\qquad p = \frac{1}{2}$

Substitute $p = \frac{1}{2}$ into the standard equation of a parabola with vertex at the origin $x^2 = 4py$. $\qquad x^2 = 2y$

Now that the equation is known, a few points can be selected, and the parabola can be point-plotted. Alternatively, the length of the latus rectum can be calculated to sketch the approximate width of the parabola.

To graph $x^2 = 2y$, first calculate the latus rectum. $\qquad 4|p| = 4\left(\frac{1}{2}\right) = 2$

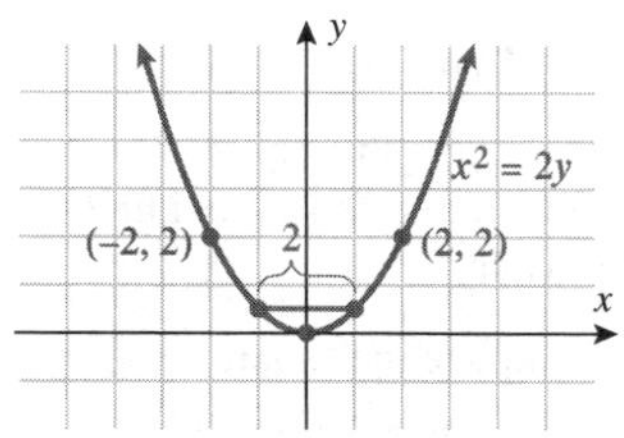

■ **YOUR TURN** Find the equation of a parabola whose focus is at the point $(-5, 0)$ and whose directrix is $x = 5$.

■ **Answer:** $y^2 = -20x$

Before we proceed to parabolas with general vertices, let's first make a few observations: The larger the latus rectum the wider the parabola. An alternative approach for graphing the parabola is to plot a few points that satisfy the equation of the parabola, which is the approach in most textbooks.

Parabola with a Vertex at (*h*, *k*)

Recall (Section 2.4) that the graph of $x^2 + y^2 = r^2$ is a circle with radius r centered at the origin, whereas the graph of $(x - h)^2 + (y - k)^2 = r^2$ is a circle with radius r centered at the point (h, k). In other words, the center is shifted from the origin to the point (h, k). This same translation (shift) can be used to describe parabolas whose vertex is at the point (h, k).

Study Tip

When $(h, k) = (0, 0)$ the vertex of the parabola is located at the origin.

Equation of a Parabola with Vertex at the Point (*h*, *k*)

The standard (conic) form of the equation of a parabola with vertex at the point (h, k) is given by:

Equation	$(y - k)^2 = 4p(x - h)$	$(x - h)^2 = 4p(y - k)$
Vertex	(h, k)	(h, k)
Focus	$(p + h, k)$	$(h, p + k)$
Directrix	$x = -p + h$	$y = -p + k$
Axis of symmetry	$y = k$	$x = h$
$p > 0$	Opens to the right	Opens upward
$p < 0$	Opens to the left	Opens downward

In order to find the vertex of a parabola given a general second-degree equation, first complete the square (Section 1.3) in order to identify (h, k). Then determine whether the parabola opens up, down, left, or right. Identify points that lie on the graph of the parabola. It is important to note that intercepts are often the easiest points to find, since one of the variables is set equal to zero.

Technology Tip

Use TI to check the graph of $y^2 - 6y - 2x + 8 = 0$. Use $(y - 3)^2 = 2x + 1$ to solve for y first. That is, $y_1 = 3 + \sqrt{2x + 1}$ or $y_2 = 3 - \sqrt{2x + 1}$.

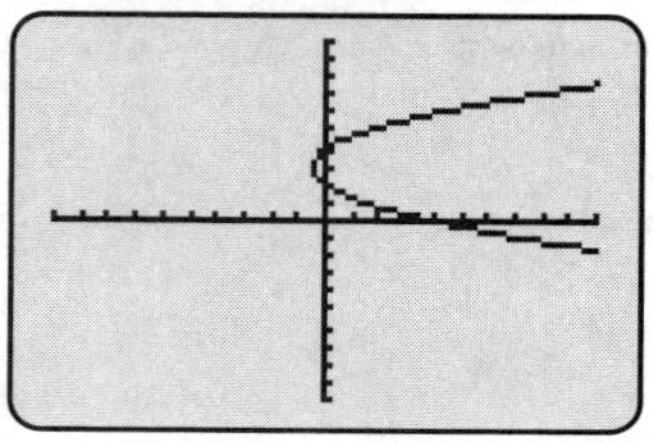

EXAMPLE 4 Graphing a Parabola with Vertex (h, k)

Graph the parabola given by the equation $y^2 - 6y - 2x + 8 = 0$.

Solution:

Transform this equation into the form $(y - k)^2 = 4p(x - h)$, since this equation is of second degree in y and first degree in x. We know this parabola opens either to the left or right.

Complete the square on y: $\quad y^2 - 6y - 2x + 8 = 0$

Isolate the y terms. $\quad y^2 - 6y = 2x - 8$

Add 9 to both sides to complete the square. $\quad y^2 - 6y + 9 = 2x - 8 + 9$

Write the left side as a perfect square. $\quad (y - 3)^2 = 2x + 1$

Factor a 2 out on the right side. $\quad (y - 3)^2 = 2\left(x + \frac{1}{2}\right)$

Compare with $(y - k)^2 = 4p(x - h)$ and identify (h, k) and p. $\quad (h, k) = \left(-\frac{1}{2}, 3\right)$

$$4p = 2 \Rightarrow p = \frac{1}{2}$$

The vertex is at the point $\left(-\frac{1}{2}, 3\right)$, and since $p = \frac{1}{2}$ is positive, the parabola opens to the right. Since the parabola's vertex lies in quadrant II and it opens to the right, we know there are two y-intercepts and one x-intercept. Apply the general equation $y^2 - 6y - 2x + 8 = 0$ to find the intercepts.

Find the y-intercepts (set $x = 0$). $\quad y^2 - 6y + 8 = 0$

Factor. $\quad (y - 2)(y - 4) = 0$

Solve for y. $\quad y = 2$ or $y = 4$

Find the x-intercept (set $y = 0$). $\quad -2x + 8 = 0$

Solve for x. $\quad x = 4$

Label the following points and connect them with a smooth curve:

Vertex:	$\left(-\frac{1}{2}, 3\right)$
y-intercepts:	(0, 2) and (0, 4)
x-intercept:	(4, 0)

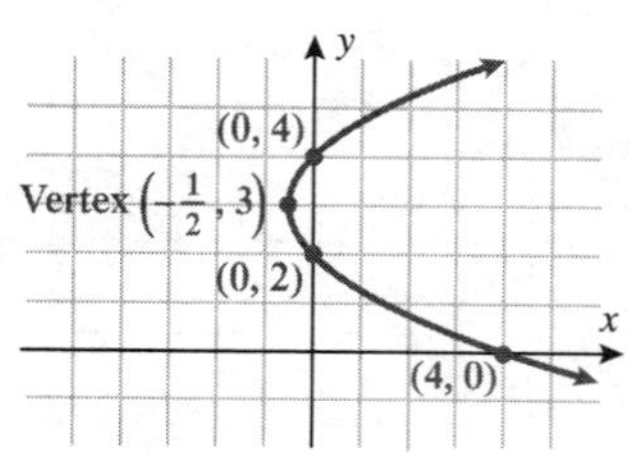

YOUR TURN For the equation $y^2 + 4y - 5x - 5 = 0$ identify the vertex and the intercepts, and graph.

Study Tip

It is often easier to find the intercepts by converting to the general form of the equation.

- **Answer:**
  Vertex: $\left(-\frac{9}{5}, -2\right)$
  x-intercept: $(-1, 0)$
  y-intercepts: $(0, -5)$ and $(0, 1)$

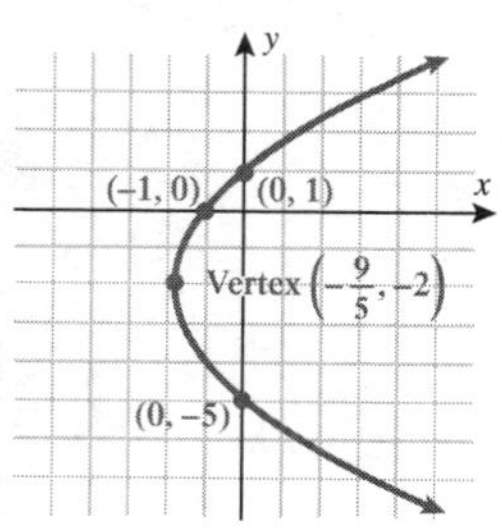

EXAMPLE 5 Graphing a Parabola with Vertex (h, k)

Graph the parabola given by the equation $x^2 - 2x - 8y - 7 = 0$.

Solution:

Transform this equation into the form $(x - h)^2 = 4p(y - k)$, since this equation is second degree in x and first degree in y. We know this parabola opens either upward or downward.

Complete the square on x: $\quad x^2 - 2x - 8y - 7 = 0$

Isolate the x terms. $\quad x^2 - 2x = 8y + 7$

Add 1 to both sides to complete the square. $\quad x^2 - 2x + 1 = 8y + 7 + 1$

Write the left side as a perfect square. $\quad (x - 1)^2 = 8y + 8$

Factor out the 8 on the right side. $\quad (x - 1)^2 = 8(y + 1)$

Compare with $(x - h)^2 = 4p(y - k)$ and identify (h, k) and p.

$$(h, k) = (1, -1)$$

$$4p = 8 \Rightarrow p = 2$$

The vertex is at the point $(1, -1)$, and since $p = 2$ is positive, the parabola opens upward. Since the parabola's vertex lies in quadrant IV and it opens upward, we know there are two x-intercepts and one y-intercept. Use the general equation $x^2 - 2x - 8y - 7 = 0$ to find the intercepts.

Find the y-intercept (set $x = 0$). $\quad -8y - 7 = 0$

Solve for y. $\quad y = -\frac{7}{8}$

Find the x-intercepts (set $y = 0$). $\quad x^2 - 2x - 7 = 0$

Solve for x. $\quad x = \frac{2 \pm \sqrt{4 + 28}}{2} = \frac{2 \pm \sqrt{32}}{2} = \frac{2 \pm 4\sqrt{2}}{2} = 1 \pm 2\sqrt{2}$

Label the following points and connect with a smooth curve:

Vertex: $(1, -1)$

y-intercept: $\left(0, -\frac{7}{8}\right)$

x-intercepts: $\left(1 - 2\sqrt{2}, 0\right)$ and $\left(1 + 2\sqrt{2}, 0\right)$

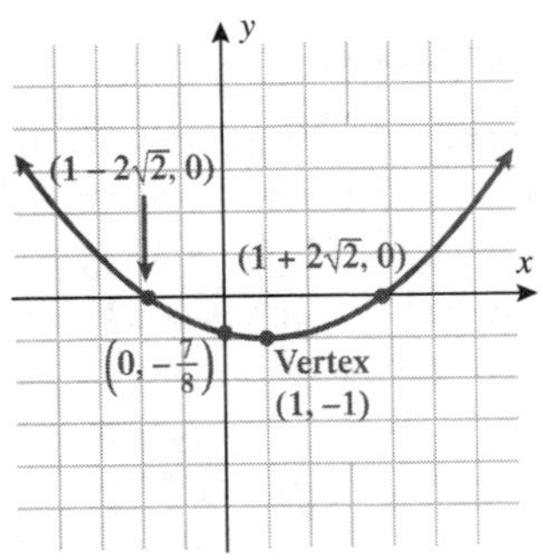

■ YOUR TURN For the equation $x^2 + 2x + 8y - 7 = 0$ identify the vertex and the intercepts, and graph.

Technology Tip

Use TI to check the graph of $x^2 - 2x - 8y - 7 = 0$. Use $y = \frac{1}{8}(x^2 - 2x - 7)$ to solve for y first.

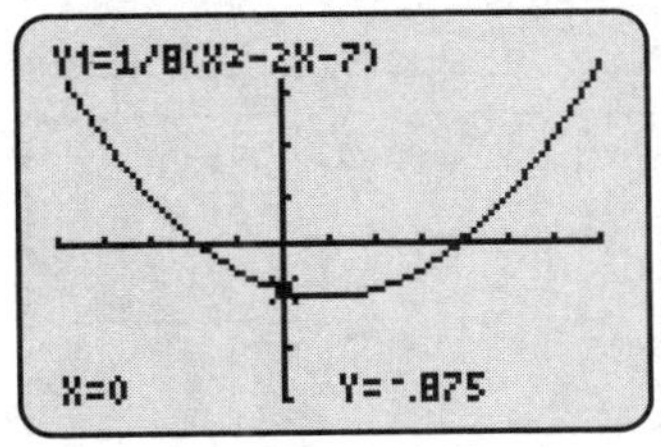

■ **Answer:**
Vertex: $(-1, 1)$
x-intercepts: $\left(-1 - 2\sqrt{2}, 0\right)$ and $\left(-1 + 2\sqrt{2}, 0\right)$
y-intercept: $\left(0, \frac{7}{8}\right)$

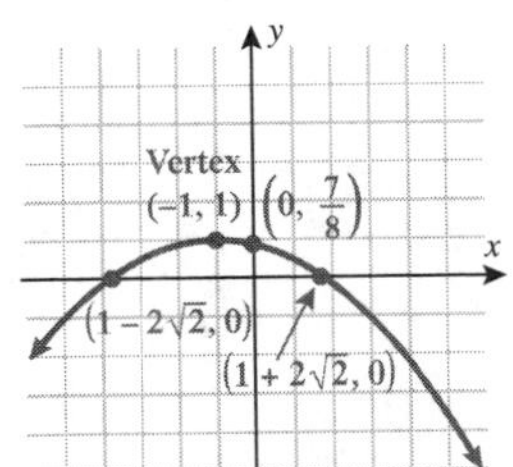

Technology Tip

Use TI to check the graph of $y^2 + 6y - 12x + 33 = 0$. Use $y^2 + 6y - (12x - 33) = 0$ to solve for y first. That is, $y_1 = -3 + 2\sqrt{3x - 6}$ or $y_2 = -3 - 2\sqrt{3x - 6}$.

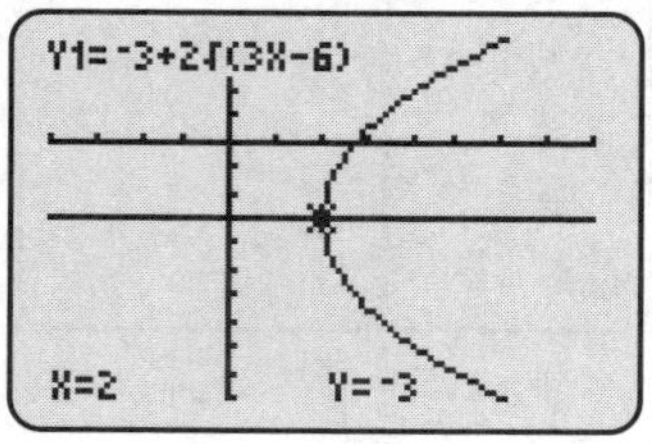

EXAMPLE 6 Finding the Equation of a Parabola with Vertex (h, k)

Find the equation of a parabola whose vertex is located at the point $(2, -3)$ and whose focus is located at the point $(5, -3)$.

Solution:

Draw a Cartesian plane and label the vertex and focus. The vertex and focus share the same axis of symmetry $y = -3$ and indicate a parabola opening to the right.

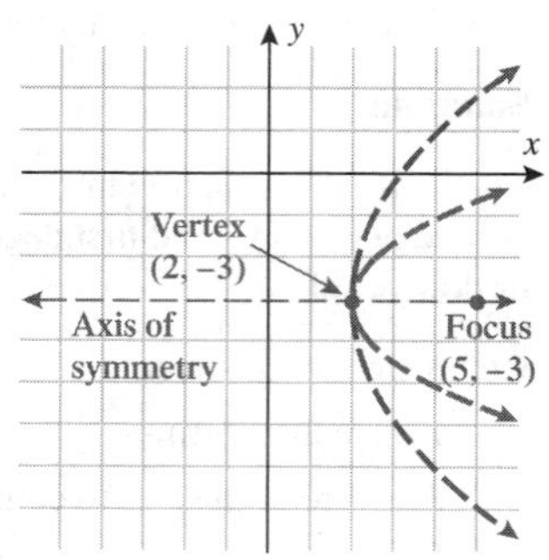

Write the standard (conic) equation of a parabola opening to the right. $\quad (y - k)^2 = 4p(x - h) \quad p > 0$

Substitute the vertex $(h, k) = (2, -3)$ into the standard equation. $\quad [y - (-3)]^2 = 4p(x - 2)$

Find p.

The general form of the vertex is (h, k) and the focus is $(h + p, k)$.

For this parabola, the vertex is $(2, -3)$ and the focus is $(5, -3)$.

Find p by taking the difference of the x coordinates. $\quad p = 3$

Substitute $p = 3$ into $[y - (-3)]^2 = 4p(x - 2)$. $\quad (y + 3)^2 = 4(3)(x - 2)$

Eliminate parentheses. $\quad y^2 + 6y + 9 = 12x - 24$

Simplify. $\quad \boxed{y^2 + 6y - 12x + 33 = 0}$

■ **YOUR TURN** Find the equation of the parabola whose vertex is located at $(2, -3)$ and whose focus is located at $(0, -3)$.

■ **Answer:** $y^2 + 6y + 8x - 7 = 0$

Applications

If we start with a parabola in the xy-plane and rotate it around its axis of symmetry, the result will be a three-dimensional paraboloid. Solar cookers illustrate the physical property that the rays of light coming into a parabola should be reflected to the focus. A flashlight reverses this process in that its light source at the focus illuminates a parabolic reflector to direct the beam outward.

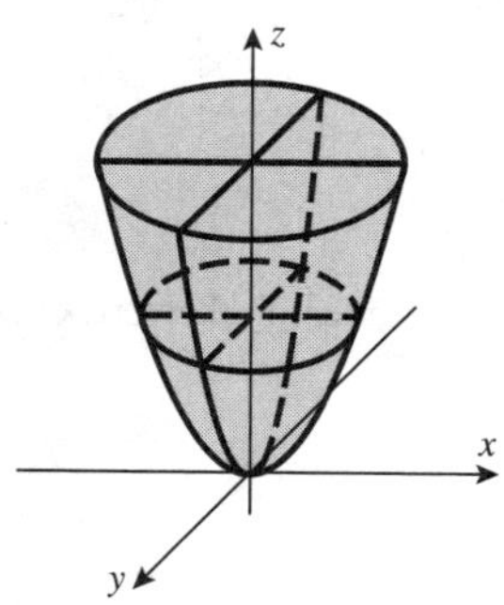

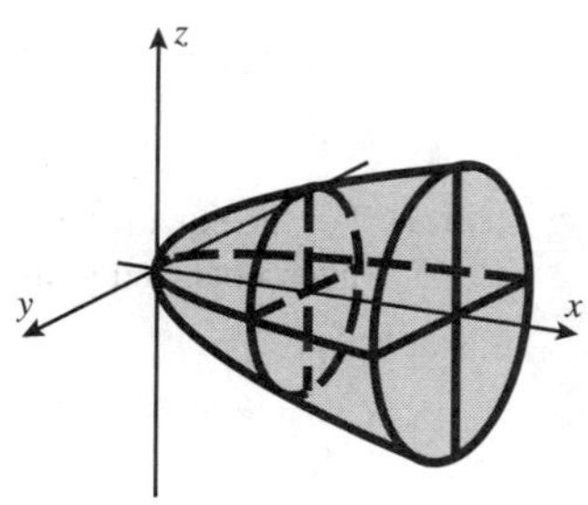

Digital Vision

Satellite dish

A satellite dish is in the shape of a paraboloid. Functioning as an antenna, the parabolic dish collects all of the incoming signals and reflects them to a single point, the focal point, which is where the receiver is located. In Examples 7 and 8, and in the applications Exercises, it is not intended to find the three-dimensional equation of the paraboloid, but rather the equation of the plane parabola that's rotated to generate the paraboloid.

EXAMPLE 7 Finding the Location of the Receiver in a Satellite Dish

A satellite dish is 24 feet in diameter at its opening and 4 feet deep in its center. Where should the receiver be placed?

Solution:

Draw a parabola with a vertex at the origin representing the center cross section of the satellite dish.

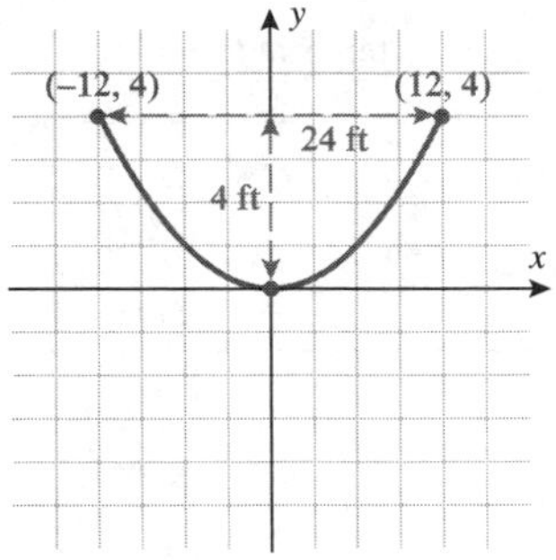

Write the standard equation of a parabola opening upward with vertex at (0, 0). $x^2 = 4py$

The point (12, 4) lies on the parabola, so substitute (12, 4) into $x^2 = 4py$. $(12)^2 = 4p(4)$

Simplify. $144 = 16p$

Solve for p. $p = 9$

Substitute $p = 9$ into the focus $(0, p)$. focus $= (0, 9)$

The receiver should be placed 9 feet from the vertex of the dish.

Parabolic antennas work for sound in addition to light. Have you ever wondered how the sound of the quarterback calling audible plays is heard by the sideline crew? The crew holds a parabolic system with a microphone at the focus. All of the sound in the direction of the parabolic system is reflected toward the focus where the microphone amplifies and records the sound.

Francis Specker/Icon SMI/NewsCom

EXAMPLE 8 Finding the Equation of a Parabolic Sound Dish

If the parabolic sound dish the sideline crew is holding has a 2-foot diameter at the opening and the microphone is located 6 inches from the vertex, find the equation that governs the center cross section of the parabolic sound dish.

Solution:

Write the standard equation of a parabola opening to the right with the vertex at the origin (0, 0). $y^2 = 4px$

The focus is located 6 inches $\left(\frac{1}{2} \text{ foot}\right)$ from the vertex. $(p, 0) = \left(\frac{1}{2}, 0\right)$

Solve for p. $p = \frac{1}{2}$

Let $p = \frac{1}{2}$ in $y^2 = 4px$. $y^2 = 4\left(\frac{1}{2}\right)x$

Simplify. $\boxed{y^2 = 2x}$

SECTION 8.2 SUMMARY

In this section we discussed parabolas whose vertex is at the origin.

Equation	$y^2 = 4px$	$x^2 = 4py$
Vertex	$(0, 0)$	$(0, 0)$
Focus	$(p, 0)$	$(0, p)$
Directrix	$x = -p$	$y = -p$
Axis of symmetry	x-axis	y-axis
$p > 0$	Opens to the right	Opens upward
$p < 0$	Opens to the left	Opens downward
Graph	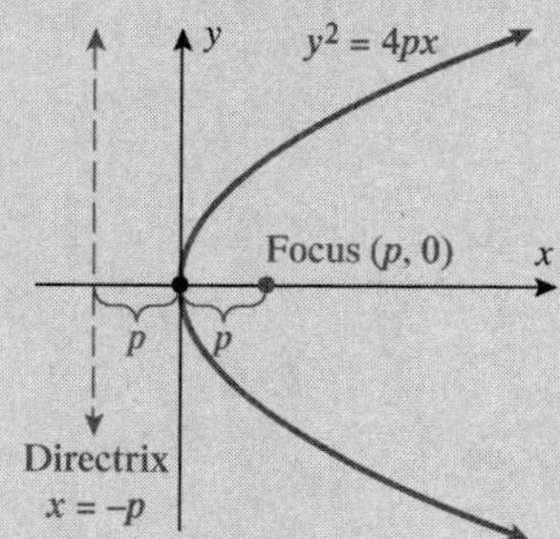	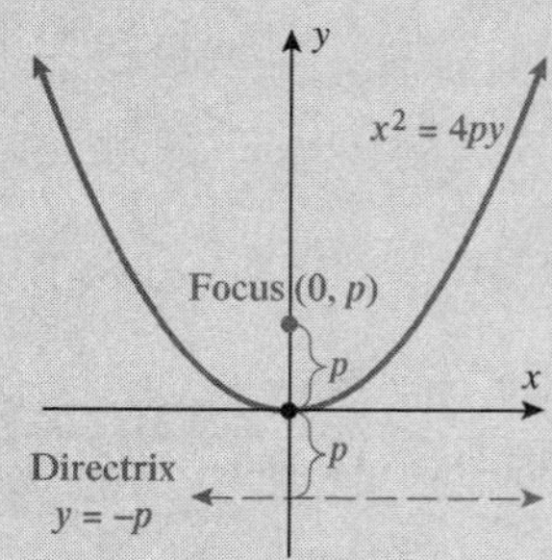

For parabolas whose vertex is at the point (h, k):

Equation	$(y - k)^2 = 4p(x - h)$	$(x - h)^2 = 4p(y - k)$
Vertex	(h, k)	(h, k)
Focus	$(p + h, k)$	$(0, p + k)$
Directrix	$x = -p + h$	$y = -p + k$
Axis of Symmetry	$y = k$	$x = h$
$p > 0$	Opens to the right	Opens upward
$p < 0$	Opens to the left	Opens downward

SECTION 8.2 EXERCISES

SKILLS

In Exercises 1–4, match the parabola to the equation.

1. $y^2 = 4x$
2. $y^2 = -4x$
3. $x^2 = -4y$
4. $x^2 = 4y$

a.

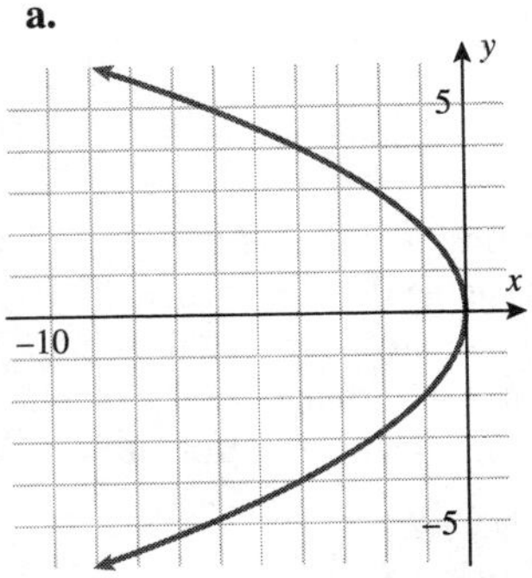

b.

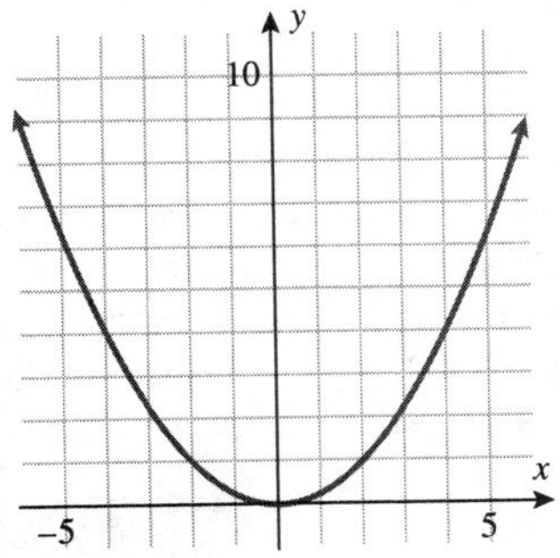

c.

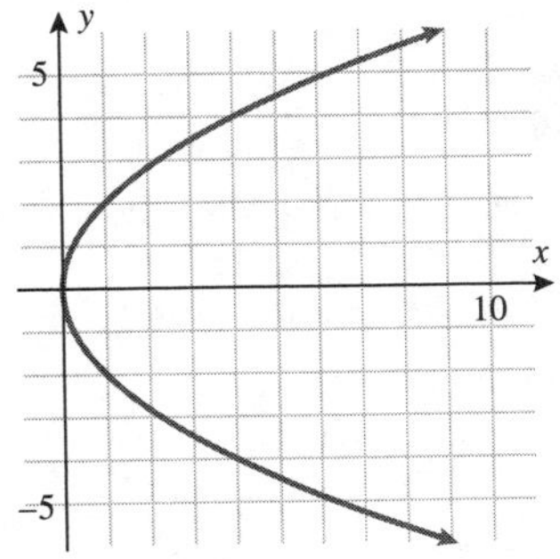

d.

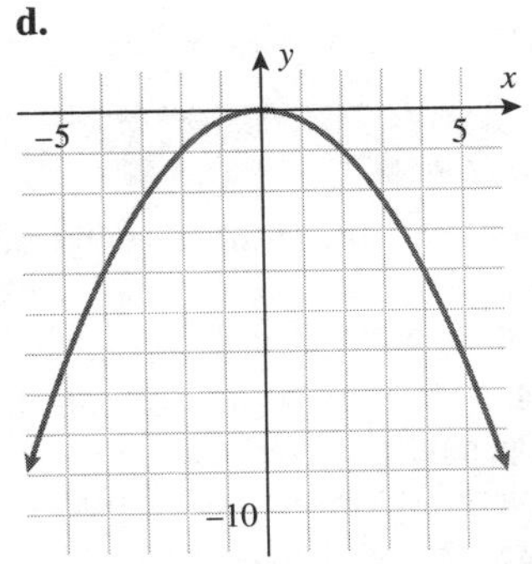

In Exercises 5–8, match the parabola to the equation.

5. $(y - 1)^2 = 4(x - 1)$
6. $(y + 1)^2 = -4(x - 1)$
7. $(x + 1)^2 = -4(y + 1)$
8. $(x - 1)^2 = 4(y - 1)$

a.

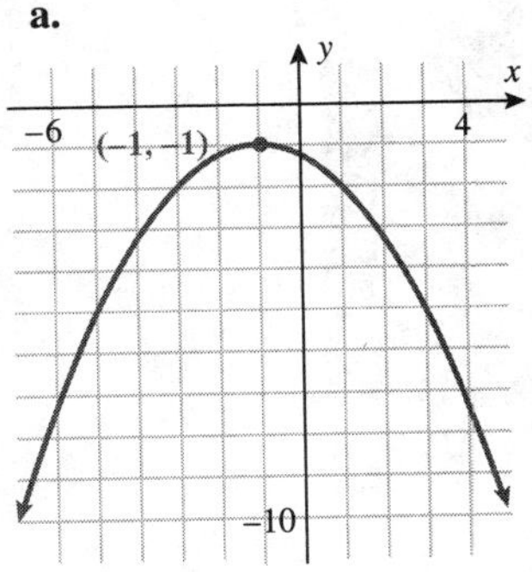

b.

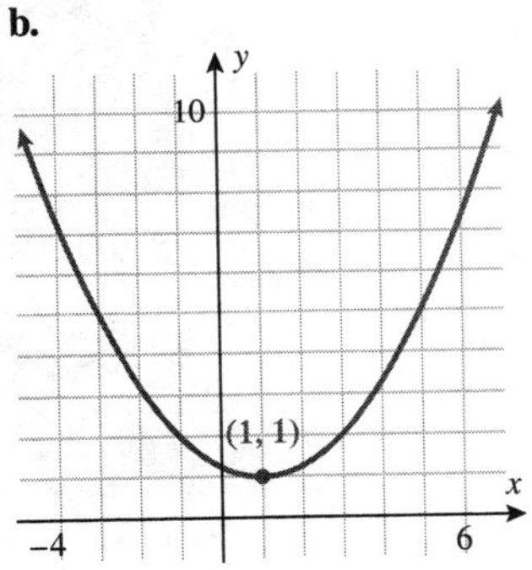

c.

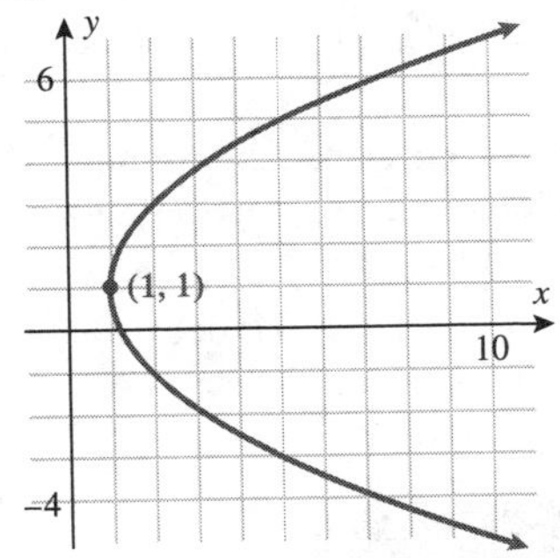

d.

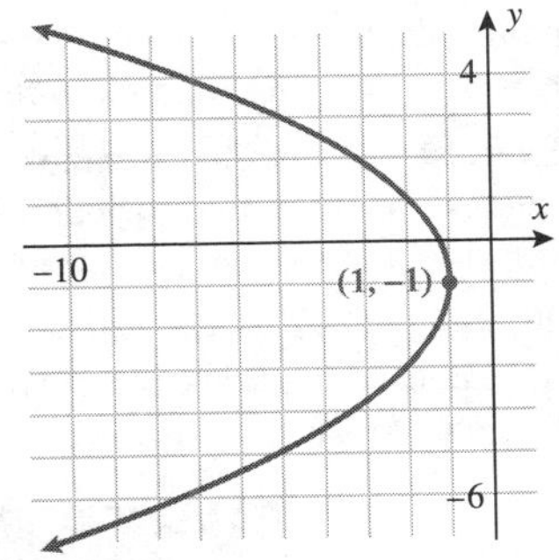

In Exercises 9–20, find an equation for the parabola described.

9. vertex at (0, 0); focus at (0, 3)
10. vertex at (0, 0); focus at (2, 0)
11. vertex at (0, 0); focus at (−5, 0)
12. vertex at (0, 0); focus at (0, −4)
13. vertex at (3, 5); focus at (3, 7)
14. vertex at (3, 5); focus at (7, 5)
15. vertex at (2, 4); focus at (0, 4)
16. vertex at (2, 4); focus at (2, −1)
17. focus at (2, 4); directrix at $y = -2$
18. focus at (2, −2); directrix at $y = 4$
19. focus at (3, −1); directrix at $x = 1$
20. focus at (−1, 5); directrix at $x = 5$

In Exercises 21–24, write an equation for each parabola.

21.

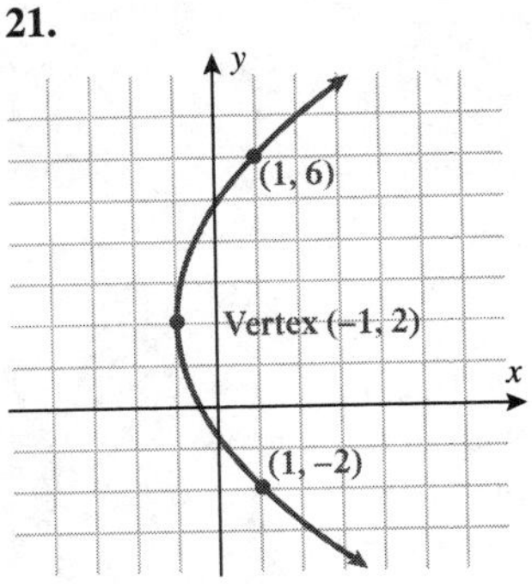

22.

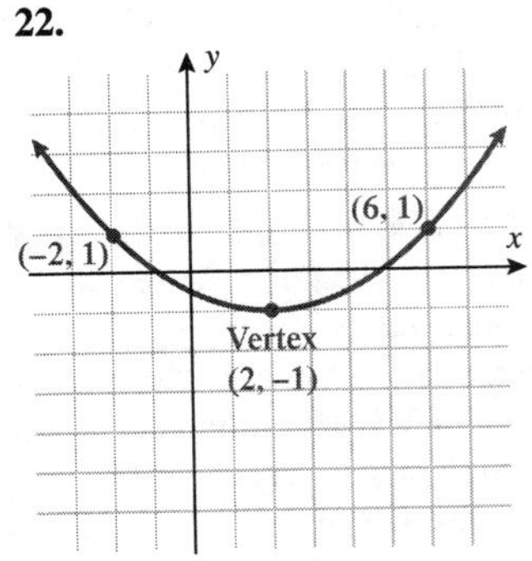

23.

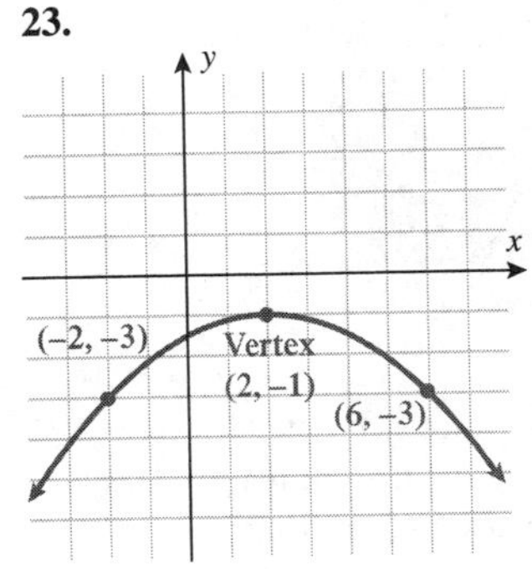

24.

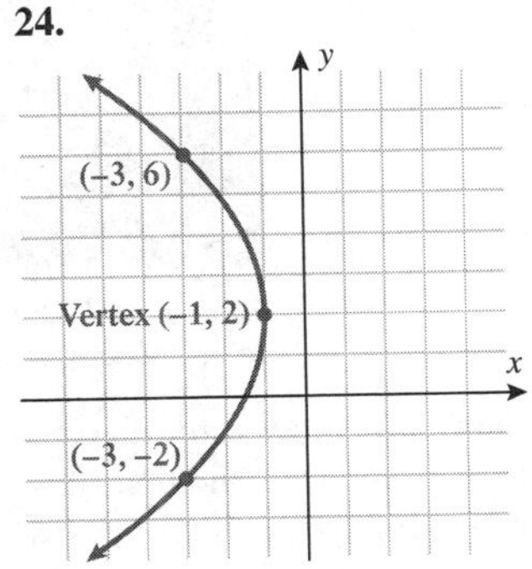

In Exercises 25–32, find the focus, vertex, directrix, and length of latus rectum and graph the parabola.

25. $x^2 = 8y$ **26.** $x^2 = -12y$ **27.** $y^2 = -2x$ **28.** $y^2 = 6x$

29. $x^2 = 16y$ **30.** $x^2 = -8y$ **31.** $y^2 = 4x$ **32.** $y^2 = -16x$

In Exercises 33–44, find the vertex and graph the parabola.

33. $(y - 2)^2 = 4(x + 3)$ **34.** $(y + 2)^2 = -4(x - 1)$ **35.** $(x - 3)^2 = -8(y + 1)$ **36.** $(x + 3)^2 = -8(y - 2)$

37. $(x + 5)^2 = -2y$ **38.** $y^2 = -16(x + 1)$ **39.** $y^2 - 4y - 2x + 4 = 0$ **40.** $x^2 - 6x + 2y + 9 = 0$

41. $y^2 + 2y - 8x - 23 = 0$ **42.** $x^2 - 6x - 4y + 10 = 0$ **43.** $x^2 - x + y - 1 = 0$ **44.** $y^2 + y - x + 1 = 0$

APPLICATIONS

45. Satellite Dish. A satellite dish measures 8 feet across its opening and 2 feet deep at its center. The receiver should be placed at the focus of the parabolic dish. Where is the focus?

46. Satellite Dish. A satellite dish measures 30 feet across its opening and 5 feet deep at its center. The receiver should be placed at the focus of the parabolic dish. Where is the focus?

47. Eyeglass Lens. Eyeglass lenses can be thought of as very wide parabolic curves. If the focus occurs 2 centimeters from the center of the lens and the lens at its opening is 5 centimeters, find an equation that governs the shape of the center cross section of the lens.

48. Optical Lens. A parabolic lens focuses light onto a focal point 3 centimeters from the vertex of the lens. How wide is the lens 0.5 centimeter from the vertex?

Exercises 49 and 50 are examples of solar cookers. Parabolic shapes are often used to generate intense heat by collecting sun rays and focusing all of them at a focal point.

49. Solar Cooker. The parabolic cooker MS-ST10 is delivered as a kit, handily packed in a single carton, with complete assembly instructions and even the necessary tools.

©AP/Wide World Photos

Solar cooker, Ubuntu Village, Johannesburg, South Africa

Thanks to the reflector diameter of 1 meter, it develops an immense power. One liter of water boils in significantly less than half an hour. If the rays are focused 40 centimeters from the vertex, find the equation for the parabolic cooker.

50. Le Four Solaire at Font-Romeur "Mirrors of the Solar Furnace." There is a reflector in the Pyrenees Mountains that is eight stories high. It cost two million dollars and it took 10 years to build it. It is made of 9000 mirrors arranged in a parabolic formation. It can reach 6000 degrees Fahrenheit just from the Sun!

Mark Antman/The Image Works

Solar furnace, Odellio, France

If the diameter of the parabolic mirror is 100 meters and the sunlight is focused 25 meters from the vertex, find the equation for the parabolic dish.

51. Sailing Under a Bridge. A bridge with a parabolic shape has an opening 80 feet wide at the base of the bridge (where the bridge meets the water), and the height in the center of the bridge is 20 feet. A sailboat whose mast reaches 17 feet above the water is traveling under the bridge 10 feet from the center of the bridge. Will it clear the bridge without scraping its mast? Justify your answer.

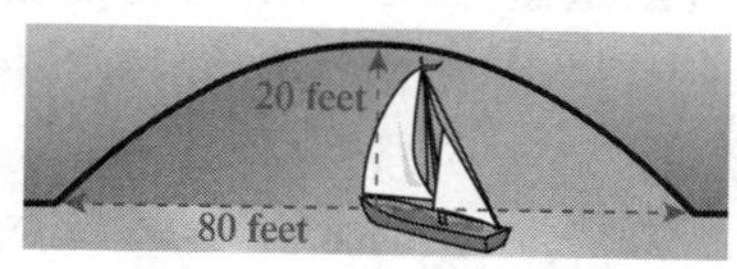

52. Driving Under a Bridge. A bridge with a parabolic shape reaches a height of 25 feet in the center of the road, and the width of the bridge opening at ground level is 20 feet combined (both lanes). If an RV is 10 feet tall and is 8 feet wide, it won't make it under the bridge if it hugs the center line. What if it straddles the center line? Will it make it? Justify your answer.

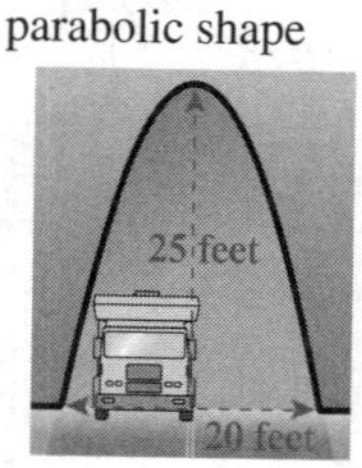

53. **Parabolic Telescope.** The Arecibo radio telescope in Puerto Rico has an enormous reflecting surface, or radio mirror. The huge "dish" is 1000 feet in diameter and 167 feet deep and covers an area of about 20 acres. Using these dimensions, determine the focal length of the telescope. Find the equation for the dish portion of the telescope.

54. **Suspension Bridge.** If one parabolic segment of a suspension bridge is 300 feet and if the cables at the vertex are suspended 10 feet above the bridge, whereas the height of the cables 150 feet from the vertex reaches 60 feet, find the equation of the parabolic path of the suspension cables.

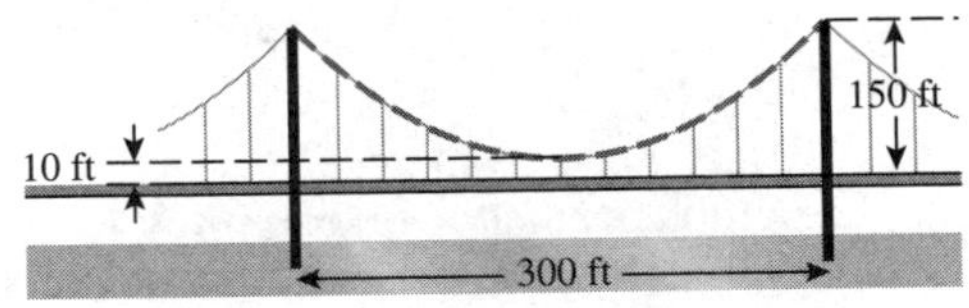

CATCH THE MISTAKE

In Exercises 55 and 56, explain the mistake that is made.

55. Find an equation for a parabola whose vertex is at the origin and whose focus is at the point (3, 0).

Solution:

Write the general equation for a parabola whose vertex is at the origin. $x^2 = 4py$

The focus of this parabola is $(p, 0) = (3, 0)$. $p = 3$

Substitute $p = 3$ into $x^2 = 4py$. $x^2 = 12y$

This is incorrect. What mistake was made?

56. Find an equation for a parabola whose vertex is at the point (3, 2) and whose focus is located at (5, 2).

Solution:

Write the equation associated with a parabola whose vertex is (3, 2). $(x - h)^2 = 4p(y - k)$

Substitute (3, 2) into $(x - h)^2 = 4p(y - k)$. $(x - 3)^2 = 4p(y - 2)$

The focus is located at (5, 2); therefore, $p = 5$.

Substitute $p = 5$ into $(x - 3)^2 = 4p(y - 2)$. $(x - 3)^2 = 20(y - 2)$

This is incorrect. What mistake(s) was made?

CONCEPTUAL

In Exercises 57–60, determine whether each statement is true or false.

57. The vertex lies on the graph of a parabola.

58. The focus lies on the graph of a parabola.

59. The directrix lies on the graph of a parabola.

60. The endpoints of the latus rectum lie on the graph of a parabola.

CHALLENGE

61. Derive the standard equation of a parabola with vertex at the origin, opening upward, $x^2 = 4py$. [Calculate the distance d_1 from any point on the parabola (x, y) to the focus $(0, p)$. Calculate the distance d_2 from any point on the parabola (x, y) to the directrix $(-p, y)$. Set $d_1 = d_2$.]

62. Derive the standard equation of a parabola opening right, $y^2 = 4px$. [Calculate the distance d_1 from any point on the parabola (x, y) to the focus $(p, 0)$. Calculate the distance d_2 from any point on the parabola (x, y) to the directrix $(x, -p)$. Set $d_1 = d_2$.]

TECHNOLOGY

63. Using a graphing utility, plot the parabola $x^2 - x + y - 1 = 0$. Compare with the sketch you drew for Exercise 43.

64. Using a graphing utility, plot the parabola $y^2 + y - x + 1 = 0$. Compare with the sketch you drew for Exercise 44.

65. In your mind, picture the parabola given by $(y + 3.5)^2 = 10(x - 2.5)$. Where is the vertex? Which way does this parabola open? Now plot the parabola using a graphing utility.

66. In your mind, picture the parabola given by $(x + 1.4)^2 = -5(y + 1.7)$. Where is the vertex? Which way does this parabola open? Now plot the parabola using a graphing utility.

67. In your mind, picture the parabola given by $(y - 1.5)^2 = -8(x - 1.8)$. Where is the vertex? Which way does this parabola open? Now plot the parabola using a graphing utility.

68. In your mind, picture the parabola given by $(x + 2.4)^2 = 6(y - 3.2)$. Where is the vertex? Which way does this parabola open? Now plot the parabola using a graphing utility.

SECTION 8.3 THE ELLIPSE

SKILLS OBJECTIVES

- Graph an ellipse given the center, major axis, and minor axis.
- Find the equation of an ellipse centered at the origin.
- Find the equation of an ellipse centered at the point (h, k).
- Solve applied problems that involve ellipses.

CONCEPTUAL OBJECTIVES

- Derive the general equation of an ellipse.
- Understand the meaning of major and minor axes and foci.
- Understand the properties of an ellipse that result in a circle.
- Interpret eccentricity in terms of the shape of the ellipse.

Definition of an Ellipse

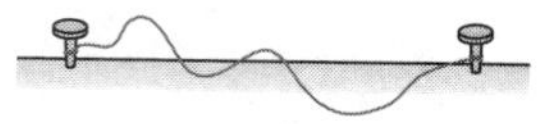

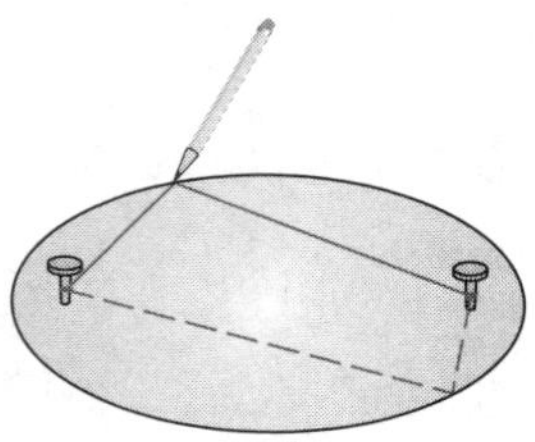

If we were to take a piece of string, tie loops at both ends, and tack the ends down so that the string had lots of slack, we would have the picture on the left. If we then took a pencil and pulled the string taut and traced our way around for one full rotation, the result would be an ellipse. See the second figure on the left.

DEFINITION **Ellipse**

An **ellipse** is the set of all points in a plane the sum of whose distances from two fixed points is constant. These two fixed points are called **foci** (plural of focus). A line segment through the foci called the **major axis** intersects the ellipse at the **vertices**. The midpoint of the line segment joining the vertices is called the **center**. The line segment that intersects the center and joins two points on the ellipse and is perpendicular to the major axis is called the **minor axis**.

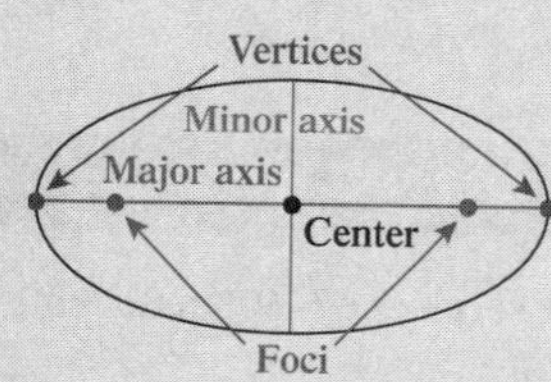

Let's start with an ellipse whose center is located at the origin. Using graph-shifting techniques, we can later extend the characteristics of an ellipse centered at a point other than the origin. Ellipses can vary from the shape of circles to something quite elongated, either horizontally or vertically, that resembles the shape of a racetrack. We say that the ellipse has either greater or lesser *eccentricity*; as we will see, there is a simple mathematical definition of *eccentricity*. It can be shown that the standard equation of an ellipse with its center at the origin is given by one of two forms, depending on whether the orientation of the major axis of the ellipse is horizontal or vertical. For $a > b > 0$, if the major axis is horizontal, then the equation is given by $\frac{x^2}{a^2} + \frac{y^2}{b^2} = 1$, and if the major axis is vertical, then the equation is given by $\frac{x^2}{b^2} + \frac{y^2}{a^2} = 1$.

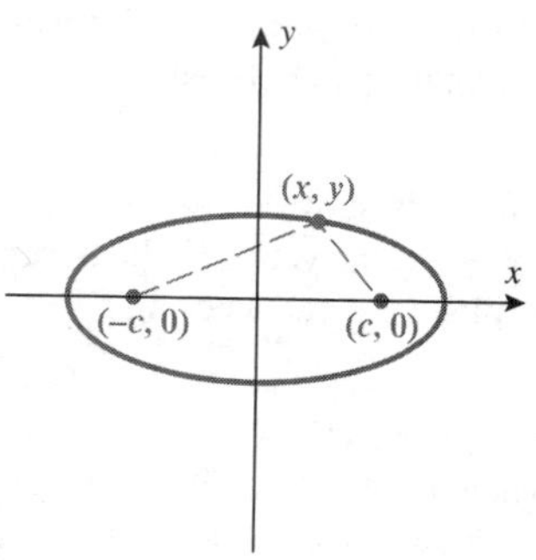

Let's consider an ellipse with center at the origin and the foci on the x-axis. Let the distance from the center to the focus be c. Therefore, the foci are located at the points $(-c, 0)$ and $(c, 0)$. The line segment containing the foci is called the major axis, and it lies along the x-axis. The sum of the two distances from the foci to any point (x, y) must be constant.

Derivation of the Equation of an Ellipse

Words	**Math**
Calculate the distance from (x, y) to $(-c, 0)$ applying the distance formula.	$\sqrt{(x-(-c))^2+y^2}$
Calculate the distance from (x, y) to $(c, 0)$ applying the distance formula.	$\sqrt{(x-c)^2+y^2}$
The sum of these two distances is equal to a constant ($2a$ for convenience).	$\sqrt{(x-(-c))^2+y^2}+\sqrt{(x-c)^2+y^2}=2a$
Isolate one radical.	$\sqrt{(x-(-c))^2+y^2}=2a-\sqrt{(x-c)^2+y^2}$
Square both sides of the equation.	$(x+c)^2+y^2=4a^2-4a\sqrt{(x-c)^2+y^2}+(x-c)^2+y^2$
Square the binomials inside the parentheses.	$x^2+2cx+c^2+y^2=4a^2-4a+\sqrt{(x-c)^2+y^2}+x^2-2cx+c^2+y^2$
Simplify.	$4cx-4a^2=-4a\sqrt{(x-c)^2+y^2}$
Divide both sides of the equation by –4.	$a^2-cx=a\sqrt{(x-c)^2+y^2}$
Square both sides of the equation.	$(a^2-cx)^2=a^2[(x-c)^2+y^2]$
Square the binomials inside the parentheses.	$a^4-2a^2cx+c^2x^2=a^2[x^2-2cx+c^2+y^2]$
Distribute the a^2 term.	$a^4-2a^2cx+c^2x^2=a^2x^2-2a^2cx+a^2c^2+a^2y^2$
Group x and y terms together, respectively, on one side and constants on the other side.	$c^2x^2-a^2x^2-a^2y^2=a^2c^2-a^4$
Factor out common factors.	$(c^2-a^2)x^2-a^2y^2=a^2(c^2-a^2)$
Multiply both sides of the equation by –1.	$(a^2-c^2)x^2+a^2y^2=a^2(a^2-c^2)$
We can make the argument that $a > c$ in order for a point to be on the ellipse (and not on the x-axis). Thus, since a and c represent distances, and therefore are positive, we know that $a^2 > c^2$, or $a^2 - c^2 > 0$. Hence, we can divide both sides of the equation by $a^2 - c^2$, since $a^2 - c^2 \neq 0$.	$x^2+\dfrac{a^2y^2}{(a^2-c^2)}=a^2$
Let $b^2 = a^2 - c^2$.	$x^2+\dfrac{a^2y^2}{b^2}=a^2$
Divide both sides of the equation by a^2.	$\boxed{\dfrac{x^2}{a^2}+\dfrac{y^2}{b^2}=1}$

The equation $\frac{x^2}{a^2} + \frac{y^2}{b^2} = 1$ represents an ellipse with center at the origin with the foci along the x-axis, since $a > b$. The following box summarizes ellipses that have their center at the origin and foci along either the x-axis or y-axis.

EQUATION OF AN ELLIPSE WITH CENTER AT THE ORIGIN

The **standard form of the equation of an ellipse** with its center at the origin is given by:

Orientation of Major Axis	Horizontal along the x-axis	Vertical along the y-axis
Equation	$\frac{x^2}{a^2} + \frac{y^2}{b^2} = 1 \quad a > b > 0$	$\frac{x^2}{b^2} + \frac{y^2}{a^2} = 1 \quad a > b > 0$
Graph	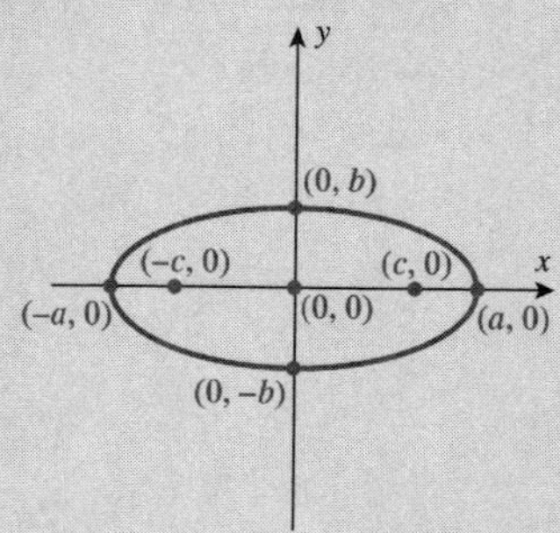	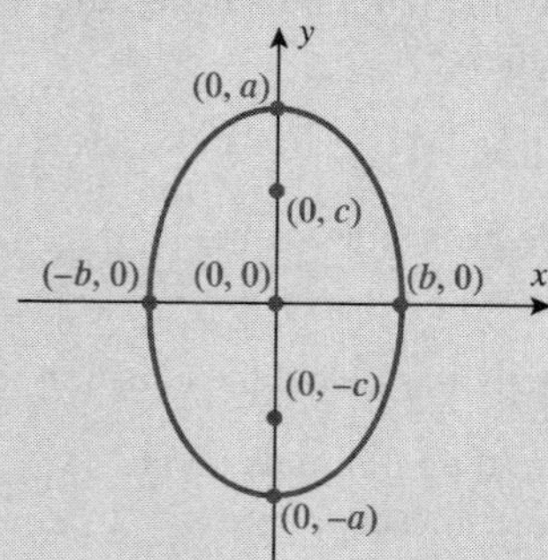
Foci	$(-c, 0)$ $(c, 0)$	$(0, -c)$ $(0, c)$
Vertices	$(-a, 0)$ $(a, 0)$	$(0, -a)$ $(0, a)$
Other Intercepts	$(0, b)$ $(0, -b)$	$(b, 0)$ $(-b, 0)$

In both cases, the value of c, the distance along the major axis from the center to the focus, is given by $c^2 = a^2 - b^2$. The length of the major axis is $2a$ and the length of the minor axis is $2b$.

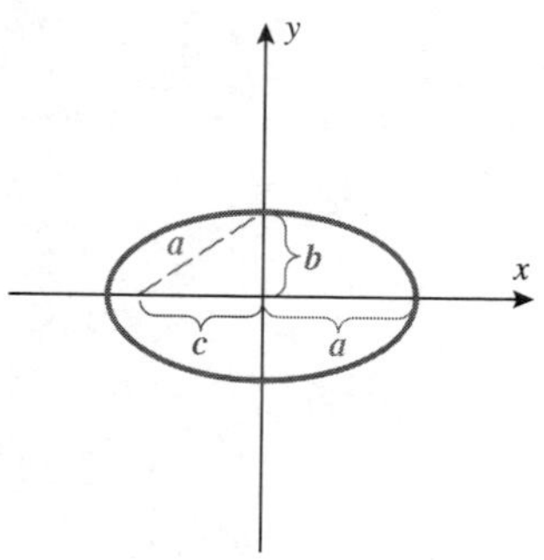

It is important to note that the vertices correspond to intercepts when an ellipse is centered at the origin. One of the first things we notice about an ellipse is its *eccentricity*. The **eccentricity**, denoted e, is given by $e = \frac{c}{a}$, where $0 < e < 1$. The circle is a limiting form of an ellipse, $c = 0$. In other words, if the eccentricity is close to 0, then the ellipse resembles a circle, whereas if the eccentricity is close to 1, then the ellipse is quite elongated, or eccentric.

Graphing an Ellipse with Center at the Origin

The equation of an ellipse in standard form can be used to graph an ellipse. Although an ellipse is defined in terms of the foci, the foci are not part of the graph. It is important to note that if the divisor of x^2 is larger than the divisor of y^2, then the ellipse is elongated horizontally.

EXAMPLE 1 Graphing an Ellipse with a Horizontal Major Axis

Graph the ellipse given by $\frac{x^2}{25} + \frac{y^2}{9} = 1$.

Solution:

Since $25 > 9$, the major axis is horizontal. $a^2 = 25$ and $b^2 = 9$

Solve for a and b. $a = 5$ and $b = 3$

Identify the vertices: $(-a, 0)$ and $(a, 0)$. $(-5, 0)$ and $(5, 0)$

Identify the endpoints (y-intercepts) on the minor axis: $(0, -b)$ and $(0, b)$. $(0, -3)$ and $(0, 3)$

Graph by labeling the points $(-5, 0)$, $(5, 0)$, $(0, -3)$, and $(0, 3)$ and connecting them with a smooth curve.

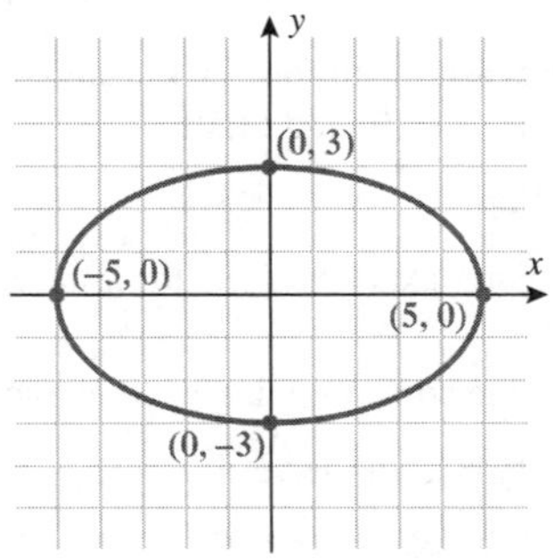

Technology Tip

Use a graphing calculator to check the graph of $\frac{x^2}{25} + \frac{y^2}{9} = 1$.

Solve for y first. That is,

$y_1 = 3\sqrt{1 - \frac{x^2}{25}}$ or

$y_2 = -3\sqrt{1 - \frac{x^2}{25}}$.

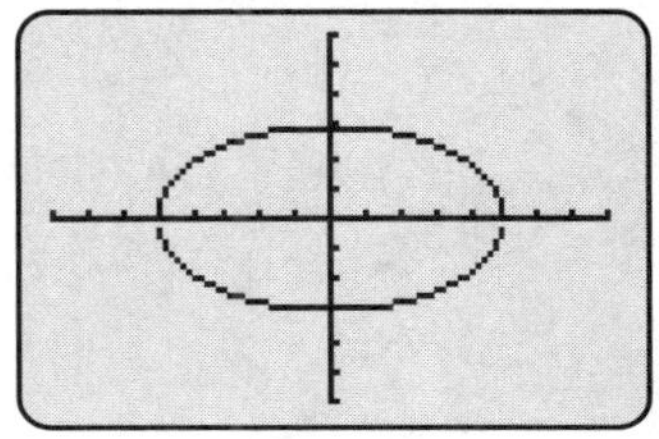

EXAMPLE 2 Graphing an Ellipse with a Vertical Major Axis

Graph the ellipse given by $16x^2 + y^2 = 16$.

Solution:

Write the equation in standard form by dividing by 16. $\frac{x^2}{1} + \frac{y^2}{16} = 1$

Since $16 > 1$, this ellipse is elongated vertically. $a^2 = 16$ and $b^2 = 1$

Solve for a and b. $a = 4$ and $b = 1$

Identify the vertices: $(0, -a)$ and $(0, a)$. $(0, -4)$ and $(0, 4)$

Identify the x-intercepts on the minor axis: $(-b, 0)$ and $(b, 0)$. $(-1, 0)$ and $(1, 0)$

Graph by labeling the points $(0, -4)$, $(0, 4)$, $(-1, 0)$, and $(1, 0)$ and connecting them with a smooth curve.

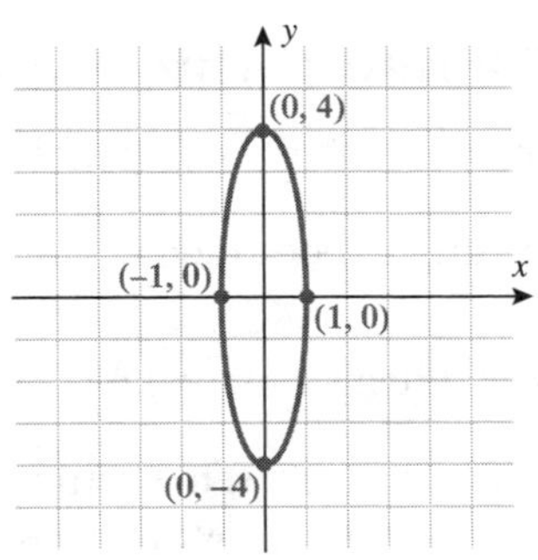

Study Tip

If the divisor of x^2 is larger than the divisor of y^2, then the major axis is horizontal along the x-axis, as in Example 1. If the divisor of y^2 is larger than the divisor of x^2, then the major axis is vertical along the y-axis, as you will see in Example 2.

YOUR TURN Graph the ellipses.

a. $\frac{x^2}{9} + \frac{y^2}{4} = 1$ **b.** $\frac{x^2}{9} + \frac{y^2}{36} = 1$

Answer:

a.

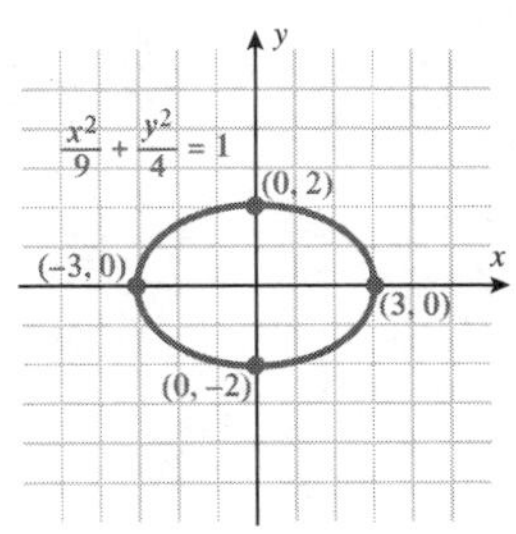

b.

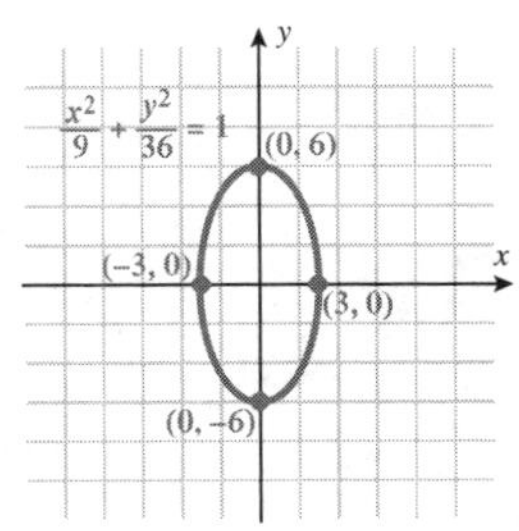

Finding the Equation of an Ellipse Centered at the Origin

What if we know the vertices and the foci of an ellipse and want to find the equation to which it corresponds? The axis on which the foci and vertices are located is the major axis. Therefore, we will have the standard equation of an ellipse, and a will be known (from the vertices). Since c is known from the foci, we can use the relation $c^2 = a^2 - b^2$ to determine the unknown b.

EXAMPLE 3 Finding the Equation of an Ellipse Centered at the Origin

Find the standard form of the equation of an ellipse with foci at $(-3, 0)$ and $(3, 0)$ and vertices $(-4, 0)$ and $(4, 0)$.

Solution:

The major axis lies along the x-axis, since it contains the foci and vertices.

Write the corresponding general equation of an ellipse. $\quad \dfrac{x^2}{a^2} + \dfrac{y^2}{b^2} = 1$

Identify a from the vertices:

Match vertices $(-4, 0) = (-a, 0)$ and $(4, 0) = (a, 0)$. $\quad a = 4$

Identify c from the foci:

Match foci, $(-3, 0) = (-c, 0)$ and $(3, 0) = (c, 0)$. $\quad c = 3$

Substitute $a = 4$ and $c = 3$ into $b^2 = a^2 - c^2$. $\quad b^2 = 4^2 - 3^2$

Simplify. $\quad b^2 = 7$

Substitute $a^2 = 16$ and $b^2 = 7$ into $\dfrac{x^2}{a^2} + \dfrac{y^2}{b^2} = 1$. $\quad \boxed{\dfrac{x^2}{16} + \dfrac{y^2}{7} = 1}$

■ **YOUR TURN** Find the standard form of the equation of an ellipse with vertices at $(0, -6)$ and $(0, 6)$ and foci $(0, -5)$ and $(0, 5)$.

■ **Answer:** $\dfrac{x^2}{11} + \dfrac{y^2}{36} = 1$

An Ellipse Centered at the Point (h, k)

We can use graph-shifting techniques to graph ellipses that are centered at a point other than the origin. For example, to graph $\dfrac{(x - h)^2}{a^2} + \dfrac{(y - k)^2}{b^2} = 1$ (assuming h and k are positive constants), start with the graph of $\dfrac{x^2}{a^2} + \dfrac{y^2}{b^2} = 1$ and shift to the right h units and up k units. The center, the vertices, the foci, and the major and minor axes all shift. In other words, the two ellipses are identical in shape and size, except that the ellipse $\dfrac{(x - h)^2}{a^2} + \dfrac{(y - k)^2}{b^2} = 1$ is centered at the point (h, k).

The following table summarizes the characteristics of ellipses centered at a point other than the origin.

EQUATION OF AN ELLIPSE WITH CENTER AT THE POINT (*h*, *k*)

The **standard form of the equation of an ellipse** with its center at the point (h, k):

Orientation of Major Axis	Horizontal (Parallel to the x-axis)	Vertical (Parallel to the y-axis)
Equation	$\dfrac{(x-h)^2}{a^2} + \dfrac{(y-k)^2}{b^2} = 1$	$\dfrac{(x-h)^2}{b^2} + \dfrac{(y-k)^2}{a^2} = 1$
Graph	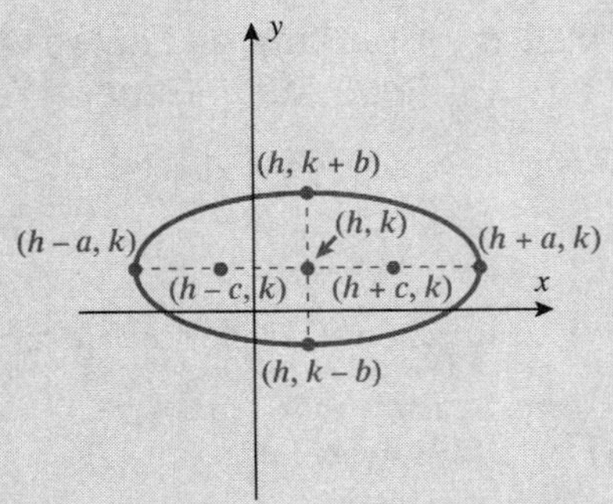	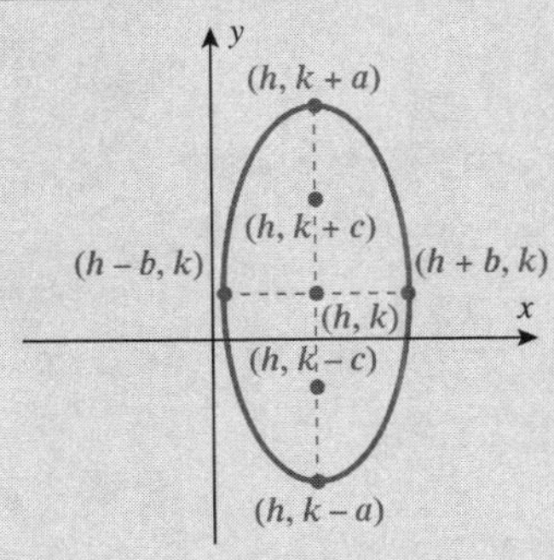
Foci	$(h-c, k)$ $(h+c, k)$	$(h, k-c)$ $(h, k+c)$
Vertices	$(h-a, k)$ $(h+a, k)$	$(h, k-a)$ $(h, k+a)$

In both cases, $a > b > 0$, $c^2 = a^2 - b^2$, the length of the major axis is $2a$, and the length of the minor axis is $2b$.

EXAMPLE 4 Graphing an Ellipse with Center (*h*, *k*) Given the Equation in Standard Form

Graph the ellipse given by $\dfrac{(x-2)^2}{9} + \dfrac{(y+1)^2}{16} = 1$.

Solution:

Write the equation in the form $\dfrac{(x-h)^2}{b^2} + \dfrac{(y-k)^2}{a^2} = 1$. $\qquad \dfrac{(x-2)^2}{3^2} + \dfrac{(y-(-1))^2}{4^2} = 1$

Identify a, b, and the center (h, k). $\qquad a = 4$, $b = 3$, and $(h, k) = (2, -1)$

Draw a graph and label the center: $(2, -1)$.

Since $a = 4$, the vertices are up 4 and down 4 units from the center: $(2, -5)$ and $(2, 3)$.

Since $b = 3$, the endpoints of the minor axis are to the left and right 3 units: $(-1, -1)$ and $(5, -1)$.

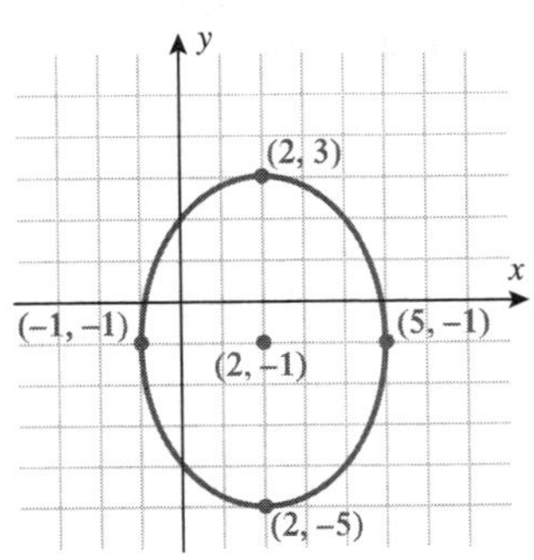

YOUR TURN Graph the ellipse given by $\dfrac{(x+1)^2}{9} + \dfrac{(y-3)^2}{1} = 1$.

Technology Tip

Use TI to check the graph of $\dfrac{(x-2)^2}{9} + \dfrac{(y+1)^2}{16} = 1$.

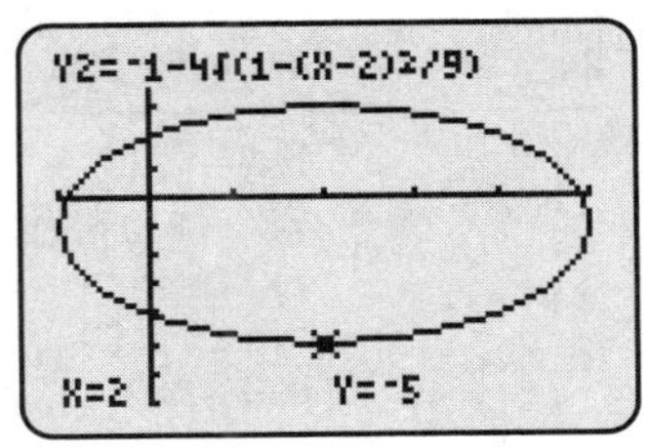

Answer:

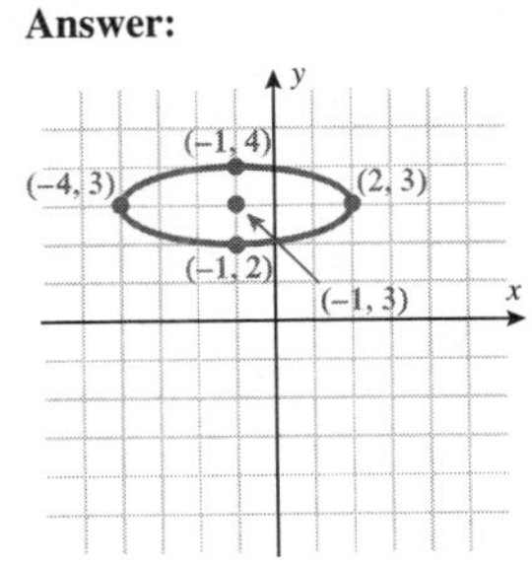

All active members of the Lambda Chi fraternity are college students, but not all college students are members of the Lambda Chi fraternity. Similarly, all circles are ellipses, but not all ellipses are circles. When $a = b$, the standard equation of an ellipse simplifies to a standard equation of a circle. Recall that when we are given the equation of a circle in general form, we first complete the square in order to express the equation in standard form, which allows the center and radius to be identified. That same approach is used when the equation of an ellipse is given in a general form.

Technology Tip

Use TI to check the graph of $4x^2 + 24x + 25y^2 - 50y - 39 = 0$.

Use $\frac{(x+3)^2}{25} + \frac{(y-1)^2}{4} = 1$ to solve for y first. That is,

$y_1 = 1 + 2\sqrt{1 - \frac{(x+3)^2}{25}}$ or

$y_2 = 1 - 2\sqrt{1 - \frac{(x+3)^2}{25}}$.

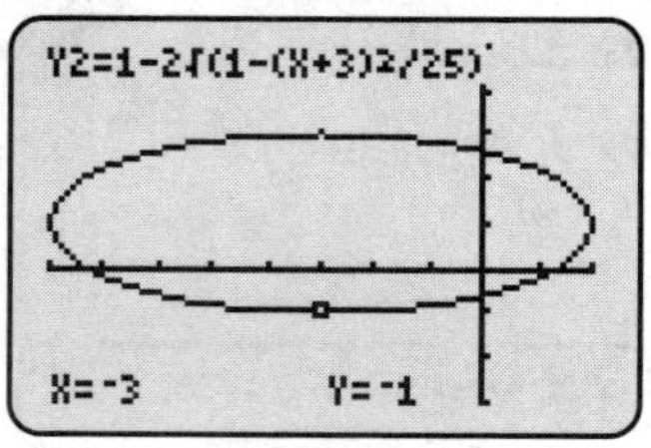

EXAMPLE 5 Graphing an Ellipse with Center (h, k) Given an Equation in General Form

Graph the ellipse given by $4x^2 + 24x + 25y^2 - 50y - 39 = 0$.

Solution:

Transform the general equation into standard form.

Group x terms together and y terms together and add 39 to both sides. $\quad (4x^2 + 24x) + (25y^2 - 50y) = 39$

Factor out the 4 common to the x terms and the 25 common to the y terms. $\quad 4(x^2 + 6x) + 25(y^2 - 2y) = 39$

Complete the square on x and y. $\quad 4(x^2 + 6x + \mathbf{9}) + 25(y^2 - 2y + \mathbf{1}) = 39 + 4(\mathbf{9}) + 25(\mathbf{1})$

Simplify. $\quad 4(x + 3)^2 + 25(y - 1)^2 = 100$

Divide by 100. $\quad \frac{(x+3)^2}{25} + \frac{(y-1)^2}{4} = 1$

Since $25 > 4$, this is an ellipse with a horizontal major axis.

Now that the equation of the ellipse is in standard form, compare it to $\frac{(x-h)^2}{a^2} + \frac{(y-k)^2}{b^2} = 1$ and identify a, b, h, k. $\quad a = 5, b = 2$, center at $(h, k) = (-3, 1)$

Since $a = 5$, the vertices are 5 units to left and right of the center. $\quad (-8, 1)$ and $(2, 1)$

Since $b = 2$, the endpoints of the minor axis are up and down 2 units from the center. $\quad (-3, -1)$ and $(-3, 3)$

Graph.

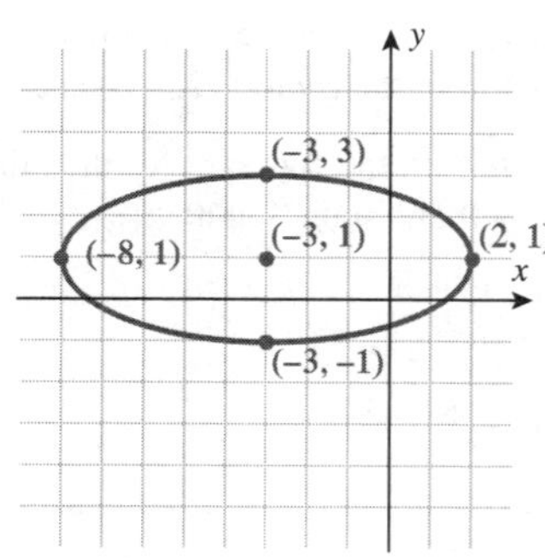

YOUR TURN Write the equation $4x^2 + 32x + y^2 - 2y + 61 = 0$ in standard form. Identify the center, vertices, and endpoints of the minor axis, and graph.

- **Answer:**

$\frac{(x+4)^2}{1} + \frac{(y-1)^2}{4} = 1$

Center: $(-4, 1)$

Vertices: $(-4, -1)$ and $(-4, 3)$

Endpoints of minor axis: $(-5, 1)$ and $(-3, 1)$

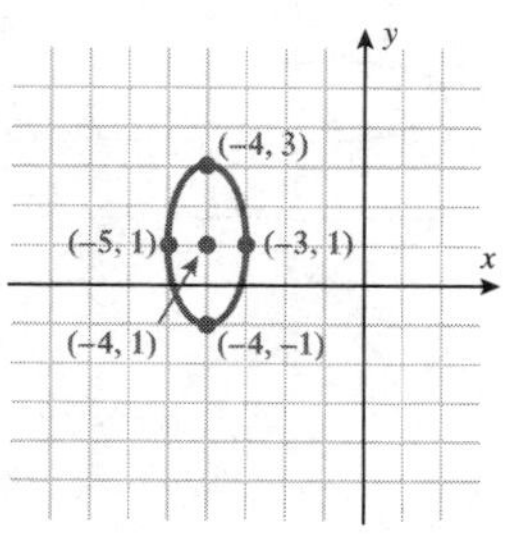

Applications

There are many examples of ellipses all around us. On Earth we have racetracks, and in our solar system, the planets travel in elliptical orbits with the Sun as a focus. Satellites are in elliptical orbits around the Earth. Most communications satellites are in a *geosynchronous* (GEO) orbit—they orbit the Earth once each day. And in order to stay over the same spot on Earth, a *geostationary* satellite has to be directly above the equator; it circles the Earth in exactly the time it takes the Earth to turn once on its axis, and its orbit has to follow the path of the equator as the Earth rotates. Otherwise, from the Earth the satellite would appear to move in a north–south line every day.

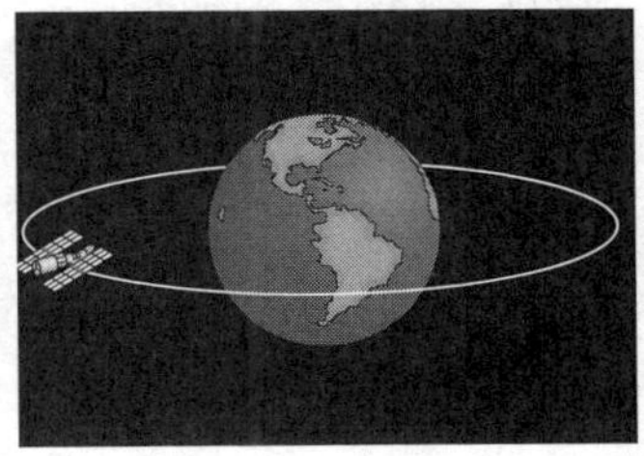

If we start with an ellipse in the *xy*-plane and rotate it around its major axis, the result is a three-dimensional ellipsoid.

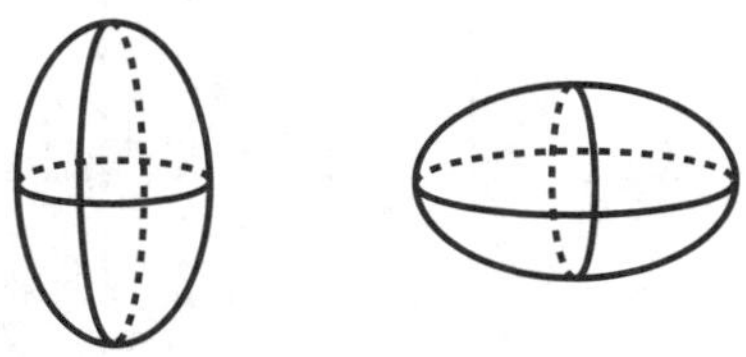

A football and a blimp are two examples of ellipsoids. The ellipsoidal shape allows for a more aerodynamic path.

PhotoDisc, Inc.

Peter Phipp/Age Fotostock America, Inc.

EXAMPLE 6 An Official NFL Football

A longitudinal section (that includes the two vertices and the center) of an official Wilson NFL football is an ellipse. The longitudinal section is approximately 11 inches long and 7 inches wide. Write an equation governing the elliptical longitudinal section.

Solution:

Locate the center of the ellipse at the origin, and orient the football horizontally.

Write the general equation of a circle centered at the origin. $\frac{x^2}{a^2} + \frac{y^2}{b^2} = 1$

The length of the major axis is 11 inches. $2a = 11$

Solve for *a*. $a = 5.5$

The length of the minor axis is 7 inches. $2b = 7$

Solve for *b*. $b = 3.5$

Substitute $a = 5.5$ and $b = 3.5$ into $\frac{x^2}{a^2} + \frac{y^2}{b^2} = 1$. $\boxed{\frac{x^2}{5.5^2} + \frac{y^2}{3.5^2} = 1}$

Technology Tip

Use TI to check the graph of $\frac{x^2}{5.5^2} + \frac{y^2}{3.5^2} = 1$. Solve for *y* first.

That is, $y_1 = 3.5\sqrt{1 - \frac{x^2}{5.5^2}}$ or $y_2 = -3.5\sqrt{1 - \frac{x^2}{5.5^2}}$.

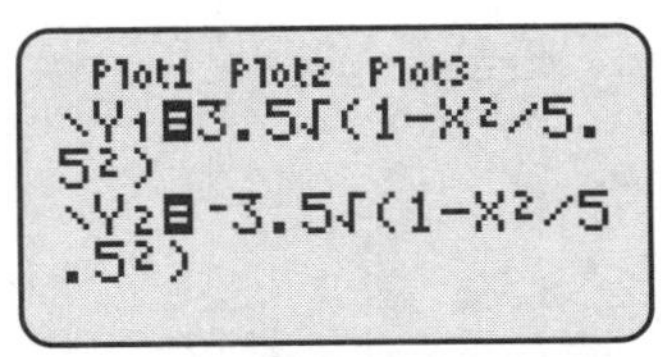

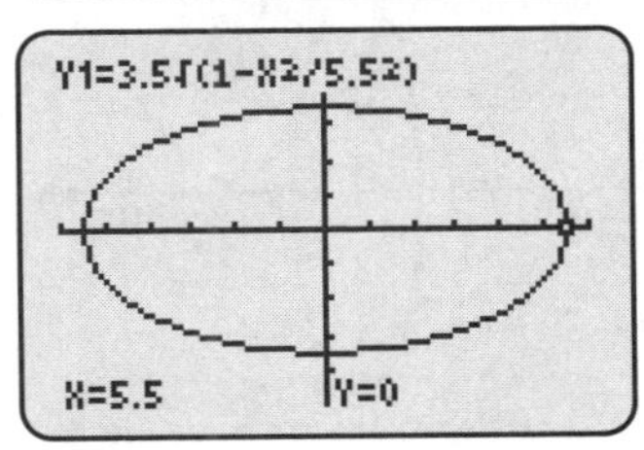

SECTION 8.3 SUMMARY

In this section, we first analyzed ellipses that are centered at the origin.

Orientation of Major Axis	Horizontal along the x-axis	Vertical along the y-axis
Equation	$\frac{x^2}{a^2} + \frac{y^2}{b^2} = 1 \quad a > b > 0$	$\frac{x^2}{b^2} + \frac{y^2}{a^2} = 1 \quad a > b > 0$
Graph	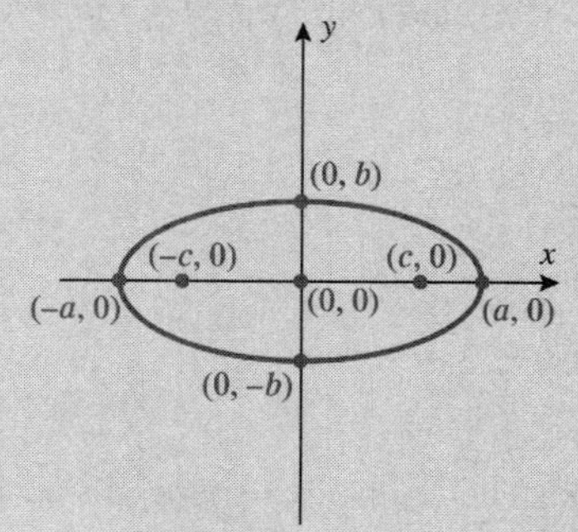	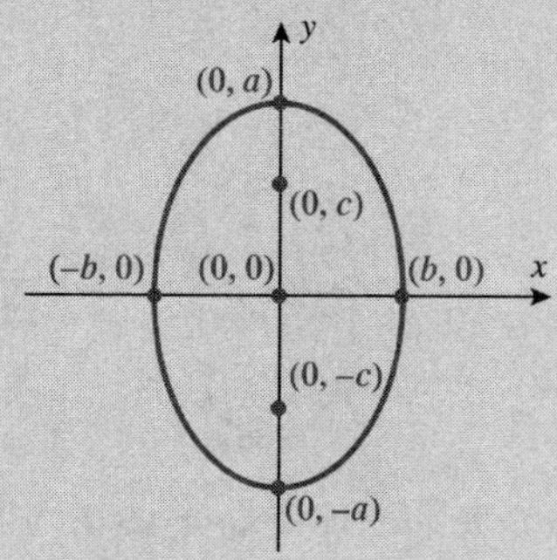
Foci*	$(-c, 0) \quad (c, 0)$	$(0, -c) \quad (0, c)$
Vertices	$(-a, 0) \quad (a, 0)$	$(0, -a) \quad (0, a)$
Other Intercepts	$(0, -b) \quad (0, b)$	$(-b, 0) \quad (b, 0)$

$*c^2 = a^2 - b^2$

For ellipses centered at the origin, we can graph an ellipse by finding all four intercepts.
For ellipses centered at the point (h, k), the major and minor axes and endpoints of the ellipse all shift accordingly.
It is important to note that when $a = b$, the ellipse is a circle.

SECTION 8.3 EXERCISES

SKILLS

In Exercises 1–4, match the equation to the ellipse.

1. $\frac{x^2}{36} + \frac{y^2}{16} = 1$

2. $\frac{x^2}{16} + \frac{y^2}{36} = 1$

3. $\frac{x^2}{8} + \frac{y^2}{72} = 1$

4. $4x^2 + y^2 = 1$

a.

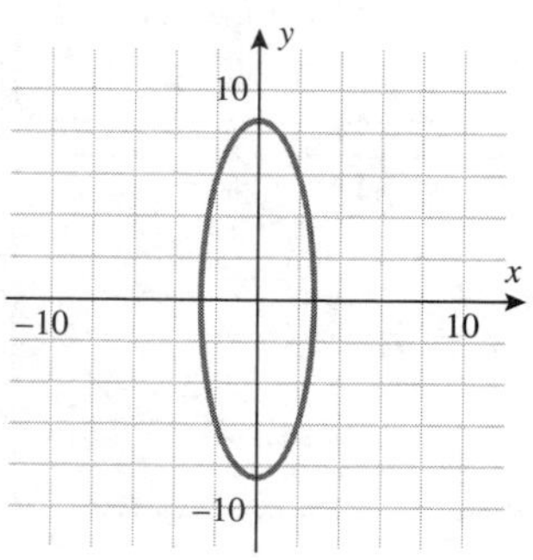

b.

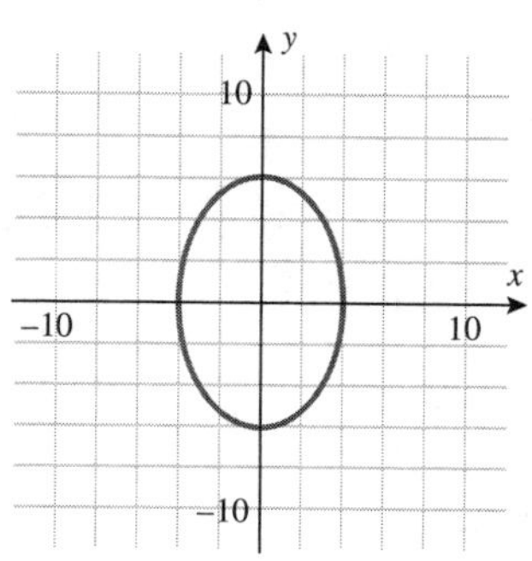

c.

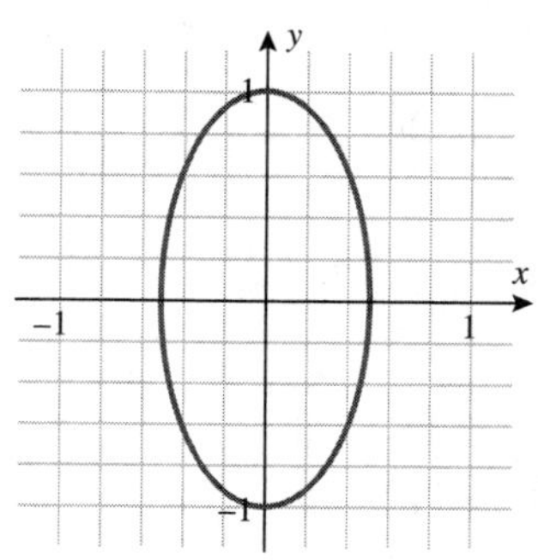

d.

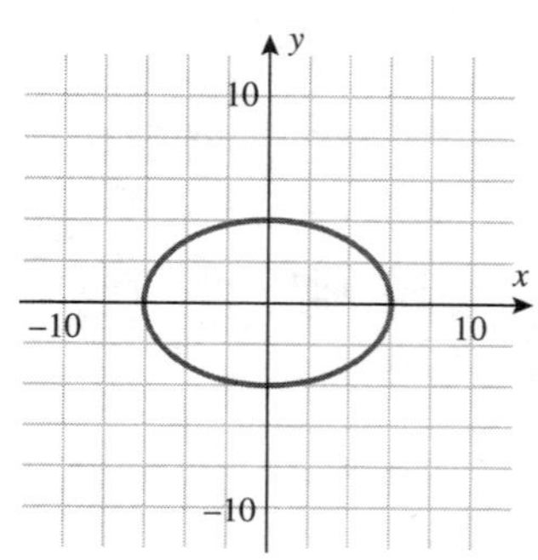

In Exercises 5–16, graph each ellipse.

5. $\frac{x^2}{25} + \frac{y^2}{16} = 1$ **6.** $\frac{x^2}{49} + \frac{y^2}{9} = 1$ **7.** $\frac{x^2}{16} + \frac{y^2}{64} = 1$ **8.** $\frac{x^2}{25} + \frac{y^2}{144} = 1$

9. $\frac{x^2}{100} + y^2 = 1$ **10.** $9x^2 + 4y^2 = 36$ **11.** $\frac{4}{9}x^2 + 81y^2 = 1$ **12.** $\frac{4}{25}x^2 + \frac{100}{9}y^2 = 1$

13. $4x^2 + y^2 = 16$ **14.** $x^2 + y^2 = 81$ **15.** $8x^2 + 16y^2 = 32$ **16.** $10x^2 + 25y^2 = 50$

In Exercises 17–24, find the standard form of the equation of an ellipse with the given characteristics.

17. foci: $(-4, 0)$ and $(4, 0)$ vertices: $(-6, 0)$ and $(6, 0)$

18. foci: $(-1, 0)$ and $(1, 0)$ vertices: $(-3, 0)$ and $(3, 0)$

19. foci: $(0, -3)$ and $(0, 3)$ vertices: $(0, -4)$ and $(0, 4)$

20. foci: $(0, -1)$ and $(0, 1)$ vertices: $(0, -2)$ and $(0, 2)$

21. Major axis vertical with length of 8, minor axis length of 4 and centered at $(0, 0)$.

22. Major axis horizontal with length of 10, minor axis length of 2 and centered at $(0, 0)$.

23. Vertices $(0, -7)$ and $(0, 7)$ and endpoints of minor axis $(-3, 0)$ and $(3, 0)$.

24. Vertices $(-9, 0)$ and $(9, 0)$ and endpoints of minor axis $(0, -4)$ and $(0, 4)$.

In Exercises 25–28, match each equation with the ellipse.

25. $\frac{(x-3)^2}{4} + \frac{(y+2)^2}{25} = 1$ **26.** $\frac{(x+3)^2}{4} + \frac{(y-2)^2}{25} = 1$ **27.** $\frac{(x-3)^2}{25} + \frac{(y+2)^2}{4} = 1$ **28.** $\frac{(x+3)^2}{25} + \frac{(y-2)^2}{4} = 1$

a.

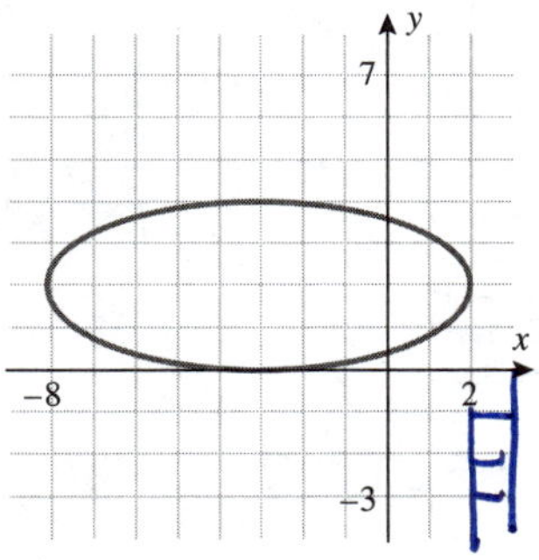

b.

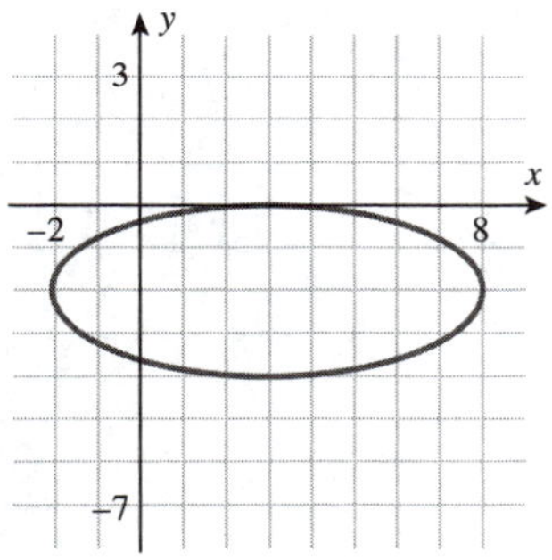

c.

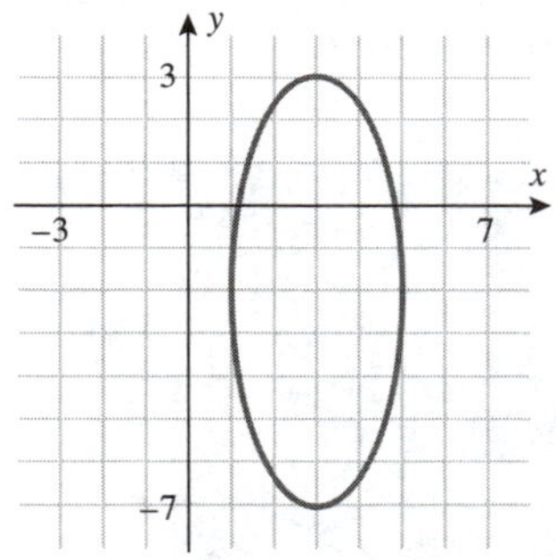

d.

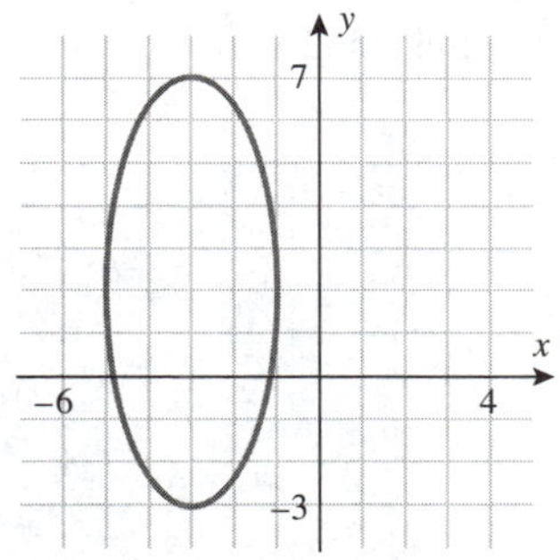

In Exercises 29–38, graph each ellipse.

29. $\frac{(x-1)^2}{16} + \frac{(y-2)^2}{4} = 1$ **30.** $\frac{(x+1)^2}{36} + \frac{(y+2)^2}{9} = 1$ **31.** $10(x+3)^2 + (y-4)^2 = 80$

32. $3(x+3)^2 + 12(y-4)^2 = 36$ **33.** $x^2 + 4y^2 - 24y + 32 = 0$ **34.** $25x^2 + 2y^2 - 4y - 48 = 0$

35. $x^2 - 2x + 2y^2 - 4y - 5 = 0$ **36.** $9x^2 - 18x + 4y^2 - 27 = 0$ **37.** $5x^2 + 20x + y^2 + 6y - 21 = 0$

38. $9x^2 + 36x + y^2 + 2y + 36 = 0$

In Exercises 39–46, find the standard form of an equation of the ellipse with the given characteristics.

39. foci: $(-2, 5)$ and $(6, 5)$ vertices: $(-3, 5)$ and $(7, 5)$

40. foci: $(2, -2)$ and $(4, -2)$ vertices: $(0, -2)$ and $(6, -2)$

41. foci: $(4, -7)$ and $(4, -1)$ vertices: $(4, -8)$ and $(4, 0)$

42. foci: $(2, -6)$ and $(2, -4)$ vertices: $(2, -7)$ and $(2, -3)$

43. Major axis vertical with length of 8, minor axis length of 4 and centered at $(3, 2)$.

44. Major axis horizontal with length of 10, minor axis length of 2 and centered at $(-4, 3)$.

45. Vertices $(-1, -9)$ and $(-1, 1)$ and endpoints of minor axis $(-4, -4)$ and $(2, -4)$.

46. Vertices $(-2, 3)$ and $(6, 3)$ and endpoints of minor axis $(2, 1)$ and $(2, 5)$.

APPLICATIONS

47. Carnival Ride. The Zipper, a favorite carnival ride, maintains an elliptical shape with a major axis of 150 feet and a minor axis of 30 feet. Assuming it is centered at the origin, find an equation for the ellipse.

Courtesy Chance Morgan, Inc.

Zipper

48. Carnival Ride. A Ferris wheel traces an elliptical path with both a major and minor axis of 180 feet. Assuming it is centered at the origin, find an equation for the ellipse (circle).

Tina Buckman/Index Stock

Ferris wheel, Barcelona, Spain

For Exercises 49 and 50, refer to the following information.

A high school wants to build a football field surrounded by an elliptical track. A regulation football field must be 120 yards long and 30 yards wide.

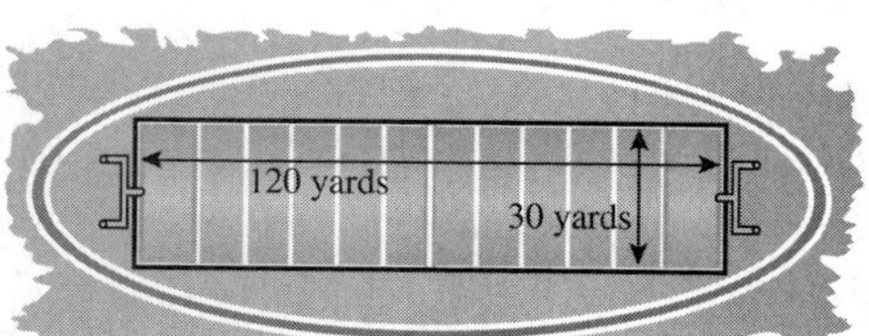

49. Sports Field. Suppose the elliptical track is centered at the origin and has a horizontal major axis of length 150 yards and a minor axis length of 40 yards.

a. Write an equation for the ellipse.
b. Find the width of the track at the end of the field. Will the track completely enclose the football field?

50. Sports Field. Suppose the elliptical track is centered at the origin and has a horizontal major axis of length 150 yards. How long should the minor axis be in order to enclose the field?

For Exercises 51 and 52, refer to orbits in our solar system.

The planets have elliptical orbits with the Sun as one of the foci. Pluto (orange), the planet furthest from the Sun, has a very elongated, or flattened, elliptical orbit, whereas Earth (royal blue) has almost a circular orbit. Because of Pluto's flattened path it is not always the planet furthest from the Sun.

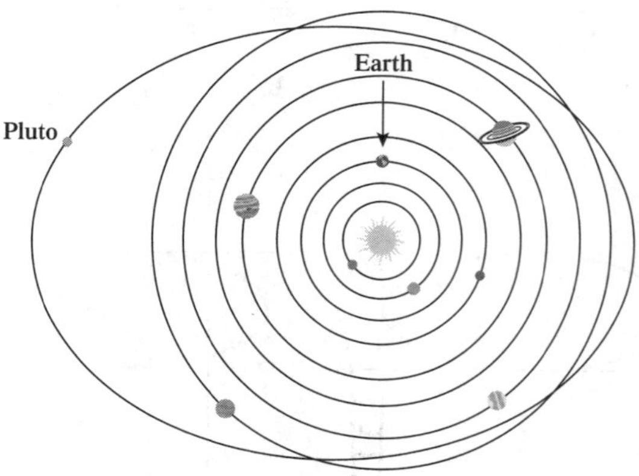

51. Planetary Orbits. The orbit of the dwarf planet Pluto has approximately the following characteristics (assume the Sun is the focus):

- The length of the semimajor axis $2a$ is approximately 11,827,000,000 km.
- The distance from the dwarf planet to the Sun is 4,447,000,000 km.

Determine the equation for Pluto's elliptical orbit around the Sun.

52. Planetary Orbits. The Earth's orbit has approximately the following characteristics (assume the Sun is the focus):

- The length of the semimajor axis $2a$ is approximately 299,700,000 km.
- The distance from the Earth to the Sun is 147,100,000 km.

Determine the equation for the Earth's elliptical orbit around the Sun.

For Exercises 53 and 54, refer to the following information.

Asteroids orbit the Sun in elliptical patterns and often cross paths with the Earth's orbit, making life a little tense now and again. A few asteroids have orbits that cross the Earth's orbit—called "Apollo asteroids" or "Earth-crossing asteroids." In recent years asteroids have passed within 100,000 kilometers of the Earth!

53. Asteroids. Asteroid 433, or Eros, is the second largest near-Earth asteroid. The semimajor axis is 150 million kilometers and the eccentricity is 0.223, where eccentricity is defined as $e = \sqrt{1 - \frac{b^2}{a^2}}$, where a is the semimajor axis, or $2a$ is the major axis, and b is the semiminor axis, or $2b$ is the minor axis. Find the equation of Eros's orbit. Round to the nearest million kilometers.

54. Asteroids. The asteroid Toutatis is the largest near-Earth asteroid. The semimajor axis is 350 million kilometers and the eccentricity is 0.634, where eccentricity is defined as $e = \sqrt{1 - \frac{b^2}{a^2}}$, where a is the semimajor axis, or $2a$ is the major axis, and b is the semimajor axis, or $2b$ is the minor axis. On September 29, 2004, it missed Earth by 961,000 miles. Find the equation of Toutatis's orbit.

55. Halley's Comet. The eccentricity of Halley's Comet is approximately 0.967. If a comet had $e = 1$, what would its orbit appear to be from Earth?

56. Halley's Comet. The length of the semimajor axis is 17.8 AU (astronomical unit) and the eccentricity is approximately 0.967. Find the equation of Halley's Comet. (Assume 1 AU = 150 million km.)

CATCH THE MISTAKE

In Exercises 57 and 58, explain the mistake that is made.

57. Graph the ellipse given by $\frac{x^2}{6} + \frac{y^2}{4} = 1$.

Solution:

Write the standard form of the equation of an ellipse. $\frac{x^2}{a^2} + \frac{y^2}{b^2} = 1$

Identify a and b. $a = 6, b = 4$

Label the vertices and the endpoints of the minor axis, $(-6, 0)$, $(6, 0)$, $(0, -4)$ $(0, 4)$, and connect with an elliptical curve.

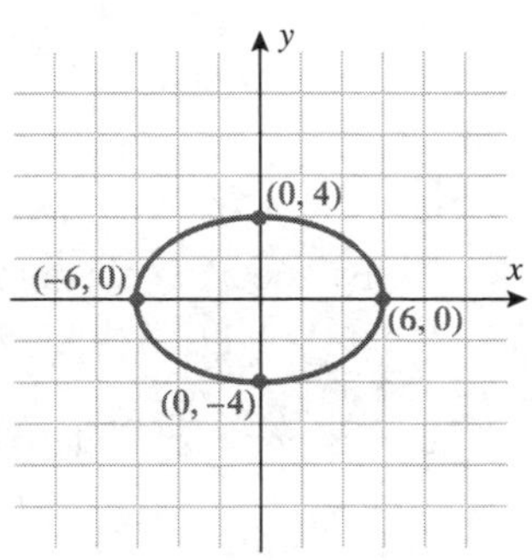

This is incorrect. What mistake was made?

58. Determine the foci of the ellipse $\frac{x^2}{16} + \frac{y^2}{9} = 1$.

Solution:

Write the general equation of a horizontal ellipse. $\frac{x^2}{a^2} + \frac{y^2}{b^2} = 1$

Identify a and b. $a = 4, b = 3$

Substitute $a = 4, b = 3$ into $c^2 = a^2 + b^2$. $c^2 = 4^2 + 3^2$

Solve for c. $c = 5$

Foci are located at $(-5, 0)$ and $(5, 0)$.

The points $(-5, 0)$ and $(5, 0)$ are located outside of the ellipse. What mistake was made?

CONCEPTUAL

In Exercises 59–62, determine whether each statement is true or false.

59. If you know the vertices of an ellipse, you can determine the equation for the ellipse.

60. If you know the foci and the endpoints of the minor axis, you can determine the equation for the ellipse.

61. Ellipses centered at the origin have symmetry with respect to the x-axis, y-axis, and the origin.

62. All ellipses are circles, but not all circles are ellipses.

CHALLENGE

63. The eccentricity of an ellipse is defined as $e = \frac{c}{a}$. Compare the eccentricity of the orbit of Pluto to that of Earth (refer to Exercises 51 and 52).

64. The eccentricity of an ellipse is defined as $e = \frac{c}{a}$. Since $a > c > 0$, then $0 < e < 1$. Describe the shape of an ellipse when

a. e is close to zero
b. e is close to one
c. $e = 0.5$

TECHNOLOGY

65. Graph the following three ellipses: $x^2 + y^2 = 1$, $x^2 + 5y^2 = 1$, and $x^2 + 10y^2 = 1$. What can be said to happen to the ellipse $x^2 + cy^2 = 1$ as c increases?

66. Graph the following three ellipses: $x^2 + y^2 = 1$, $5x^2 + y^2 = 1$, and $10x^2 + y^2 = 1$. What can be said to happen to the ellipse $cx^2 + y^2 = 1$ as c increases?

67. Graph the following three ellipses: $x^2 + y^2 = 1$, $5x^2 + 5y^2 = 1$, and $10x^2 + 10y^2 = 1$. What can be said to happen to the ellipse $cx^2 + cy^2 = 1$ as c increases?

68. Graph the equation $\frac{x^2}{9} - \frac{y^2}{16} = 1$. Notice what a difference the sign makes. Is this an ellipse?

69. Graph the following three ellipses: $x^2 + y^2 = 1$, $0.5x^2 + y^2 = 1$, and $0.05x^2 + y^2 = 1$. What can be said to happen to ellipse $cx^2 + y^2 = 1$ as c decreases?

70. Graph the following three ellipses: $x^2 + y^2 = 1$, $x^2 + 0.5y^2 = 1$, and $x^2 + 0.05y^2 = 1$. What can be said to happen to ellipse $x^2 + cy^2 = 1$ as c decreases?

SECTION 8.4 THE HYPERBOLA

SKILLS OBJECTIVES

- Find a hyperbola's foci and vertices.
- Find the equation of a hyperbola centered at the origin.
- Graph a hyperbola using asymptotes as graphing aids.
- Find the equation of a hyperbola centered at the point (h, k).
- Solve applied problems that involve hyperbolas.

CONCEPTUAL OBJECTIVES

- Derive the general equation of a hyperbola.
- Identify, apply, and graph the transverse axis, vertices, and foci.
- Use asymptotes to determine the shape of a hyperbola.

The definition of a hyperbola is similar to the definition of an ellipse. An ellipse is the set of all points the *sum* of whose distances from two points (the foci) is constant. A *hyperbola* is the set of all points whose *difference* of the distances from two points (the foci) is constant. What distinguishes their equations is a minus sign.

Ellipse centered at the origin: $\frac{x^2}{a^2} + \frac{y^2}{b^2} = 1$

Hyperbola centered at the origin: $\frac{x^2}{a^2} - \frac{y^2}{b^2} = 1$

DEFINITION **Hyperbola**

A **hyperbola** is the set of all points in a plane the difference of whose distances from two fixed points is a positive constant. These two fixed points are called **foci**. The hyperbola has two separate curves called **branches**. The two points where the hyperbola intersects the line joining the foci are called **vertices**. The line segment joining the vertices is called the **transverse axis of the hyperbola**. The midpoint of the transverse axis is called the **center**.

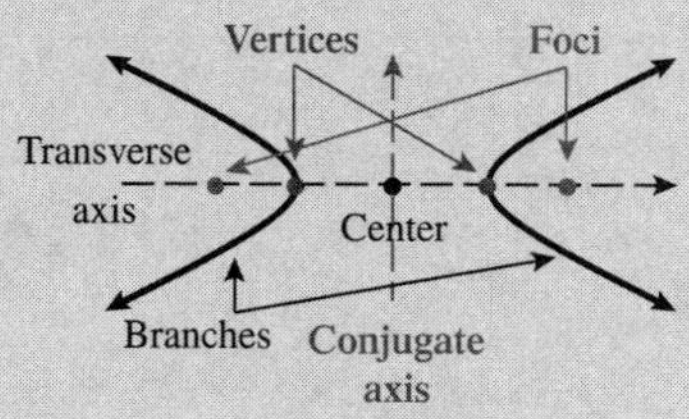

Hyperbolas Centered at the Origin

Let's consider a hyperbola with the center at the origin and the foci on the x-axis. Let the distance from the center to the focus be c. Therefore, the foci are located at the points $(-c, 0)$ and $(c, 0)$. The difference of the two distances from the foci to any point (x, y) must be constant. We then can follow a similar analysis as done with an ellipse.

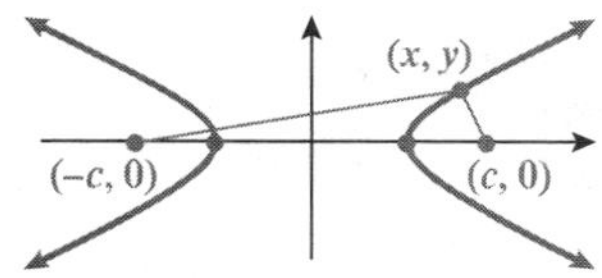

Derivation of the Equation of a Hyperbola

WORDS	**MATH**
The difference of these two distances is equal to a constant ($2a$ for convenience).	$\sqrt{(x-(-c))^2+y^2}-\sqrt{(x-c)^2+y^2}=\pm 2a$
Following the same procedure that we did with an ellipse leads to:	$(c^2-a^2)x^2-a^2y^2=a^2(c^2-a^2)$
We can make the argument that $c > a$ in order for a point to be on the hyperbola (and not on the x-axis). Therefore, since a and c represent distances and therefore are positive, we know that $c^2 > a^2$, or $c^2 - a^2 > 0$. Hence, we can divide both sides of the equation by $c^2 - a^2$, since $c^2 - a^2 \neq 0$.	$x^2-\dfrac{a^2y^2}{(c^2-a^2)}=a^2$
Let $b^2 = c^2 - a^2$.	$x^2-\dfrac{a^2y^2}{b^2}=a^2$
Divide both sides of the equation by a^2.	$\boxed{\dfrac{x^2}{a^2}-\dfrac{y^2}{b^2}=1}$

The equation $\frac{x^2}{a^2}-\frac{y^2}{b^2}=1$ represents a hyperbola with center at the origin, with the foci along the x-axis. The following box summarizes hyperbolas that have their center at the origin and foci along either the x-axis or y-axis.

EQUATION OF A HYPERBOLA WITH CENTER AT THE ORIGIN

The **standard form of the equation of a hyperbola** with its center at the origin:

Orientation of Transverse Axis	Horizontal along the x-axis	Vertical along the y-axis
Equation	$\frac{x^2}{a^2} - \frac{y^2}{b^2} = 1$	$\frac{y^2}{a^2} - \frac{x^2}{b^2} = 1$
Graph	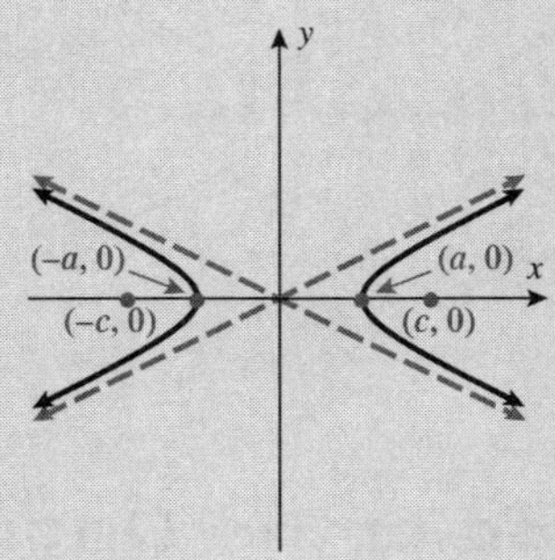	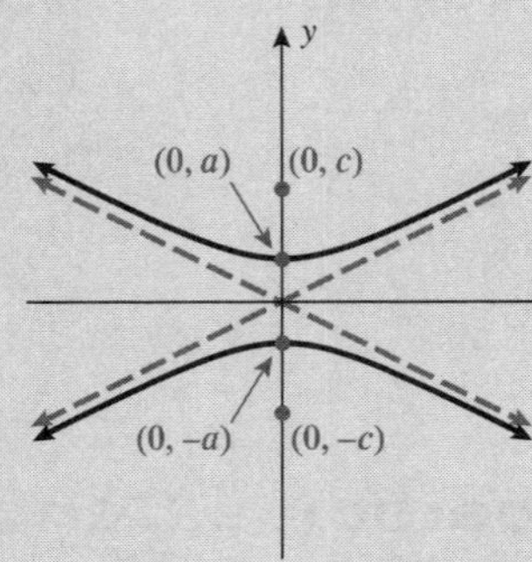
Foci	$(-c, 0)$ $(c, 0)$	$(0, -c)$ $(0, c)$
Asymptotes	$y = \frac{b}{a}x$ $y = -\frac{b}{a}x$	$y = \frac{a}{b}x$ $y = -\frac{a}{b}x$
Vertices	$(-a, 0)$ $(a, 0)$	$(0, -a)$ $(0, a)$
Transverse Axis	Horizontal length $2a$	Vertical length $2a$

where $c^2 = a^2 + b^2$

Technology Tip

Use TI to check the graph of $\frac{x^2}{9} - \frac{y^2}{4} = 1$. Solve for y first.

That is, $y_1 = 2\sqrt{\frac{x^2}{9} - 1}$ or $y_2 = -2\sqrt{\frac{x^2}{9} - 1}$.

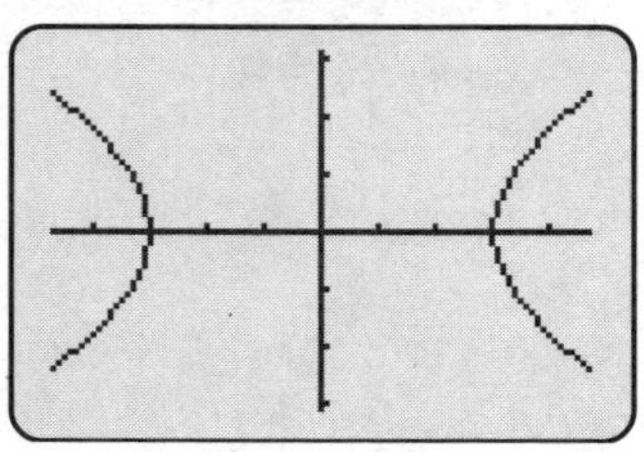

Note that for $\frac{x^2}{a^2} - \frac{y^2}{b^2} = 1$, if $x = 0$, then $-\frac{y^2}{b^2} = 1$, which yields an imaginary number for y. But when $y = 0$, $\frac{x^2}{a^2} = 1$, and therefore $x = \pm a$. The vertices for this hyperbola are $(-a, 0)$ and $(a, 0)$.

EXAMPLE 1 **Finding the Foci and Vertices of a Hyperbola Given the Equation**

Find the foci and vertices of the hyperbola given by $\frac{x^2}{9} - \frac{y^2}{4} = 1$.

Solution:

Compare to the standard equation of a hyperbola, $\frac{x^2}{a^2} - \frac{y^2}{b^2} = 1$. $\quad a^2 = 9, b^2 = 4$

Solve for a and b.	$a = 3, b = 2$
Substitute $a = 3$ into the vertices, $(-a, 0)$ and $(a, 0)$.	$(-3, 0)$ and $(3, 0)$
Substitute $a = 3, b = 2$ into $c^2 = a^2 + b^2$.	$c^2 = 3^2 + 2^2$
Solve for c.	$c^2 = 13$
	$c = \sqrt{13}$
Substitute $c = \sqrt{13}$ into the foci, $(-c, 0)$ and $(c, 0)$.	$(-\sqrt{13}, 0)$ and $(\sqrt{13}, 0)$

The vertices are $\boxed{(-3, 0)}$ and $\boxed{(3, 0)}$, and the foci are $\boxed{(-\sqrt{13}, 0)}$ and $\boxed{(\sqrt{13}, 0)}$.

YOUR TURN Find the vertices and foci of the hyperbola $\dfrac{y^2}{16} - \dfrac{x^2}{20} = 1$.

■ **Answer:** Vertices: $(0, -4)$ and $(0, 4)$
Foci: $(0, -6)$ and $(0, 6)$

EXAMPLE 2 Finding the Equation of a Hyperbola Given Foci and Vertices

Find the equation of a hyperbola whose vertices are located at $(0, -4)$ and $(0, 4)$ and whose foci are located at $(0, -5)$ and $(0, 5)$.

Solution:

The center is located at the midpoint of the segment joining the vertices.	$\left(\dfrac{0+0}{2}, \dfrac{-4+4}{2}\right) = (0, 0)$
Since the foci and vertices are located on the y-axis, the standard equation is given by:	$\dfrac{y^2}{a^2} - \dfrac{x^2}{b^2} = 1$
The vertices $(0, \pm a)$ and the foci $(0, \pm c)$ can be used to identify a and c.	$a = 4, c = 5$
Substitute $a = 4, c = 5$ into $b^2 = c^2 - a^2$.	$b^2 = 5^2 - 4^2$
Solve for b.	$b^2 = 25 - 16 = 9$
	$b = 3$
Substitute $a = 4$ and $b = 3$ into $\dfrac{y^2}{a^2} - \dfrac{x^2}{b^2} = 1$.	$\boxed{\dfrac{y^2}{16} - \dfrac{x^2}{9} = 1}$

YOUR TURN Find the equation of a hyperbola whose vertices are located at $(-2, 0)$ and $(2, 0)$ and whose foci are located at $(-4, 0)$ and $(4, 0)$.

■ **Answer:** $\dfrac{x^2}{4} - \dfrac{y^2}{12} = 1$

Technology Tip

Use TI to check the graph of $\dfrac{y^2}{16} - \dfrac{x^2}{9} = 1$. Solve for y first.

That is, $y_1 = 4\sqrt{1 + \dfrac{x^2}{9}}$ or $y_2 = -4\sqrt{1 - \dfrac{x^2}{9}}$.

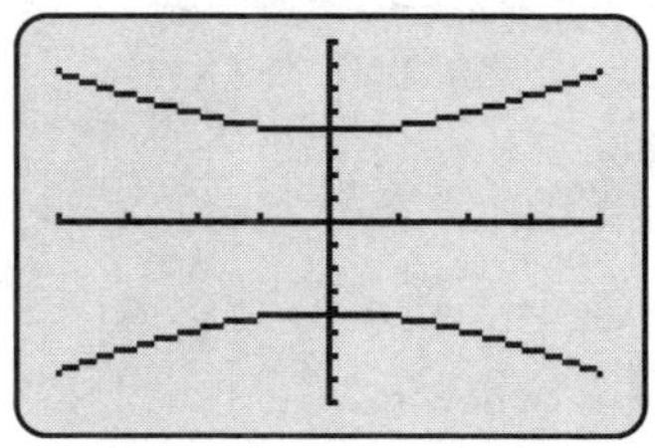

Graphing a Hyperbola

To graph a hyperbola, we use the vertices and asymptotes. The asymptotes are found by the equations $y = \pm\dfrac{b}{a}x$ or $y = \pm\dfrac{a}{b}x$, depending on whether the transverse axis is horizontal or vertical. An easy way to draw these graphing aids is to first draw the rectangular box that passes through the vertices and the points $(0, \pm b)$ or $(\pm b, 0)$. The **conjugate axis** is perpendicular to the transverse axis and has length $2b$. The asymptotes pass through the center of the hyperbola and the corners of the rectangular box.

$$\frac{x^2}{a^2} - \frac{y^2}{b^2} = 1$$

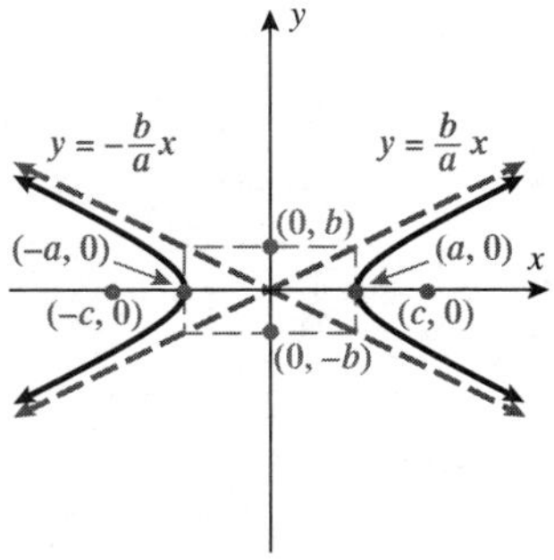

$$\frac{x^2}{a^2} - \frac{y^2}{b^2} = 1$$

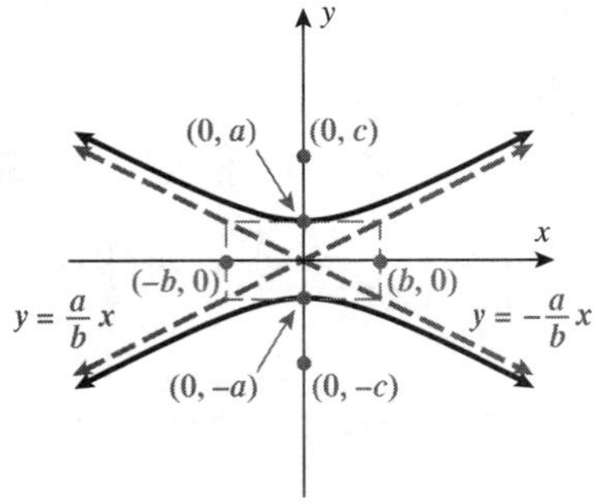

Technology Tip

Use TI to check the graph of $\frac{x^2}{4} - \frac{y^2}{9} = 1$. Solve for y first.

That is, $y_1 = 3\sqrt{\frac{x^2}{4} - 1}$ or $y_2 = -3\sqrt{\frac{x^2}{4} - 1}$.

```
Plot1 Plot2 Plot3
\Y1=3√(X²/4-1)
\Y2=-3√(X²/4-1)
```

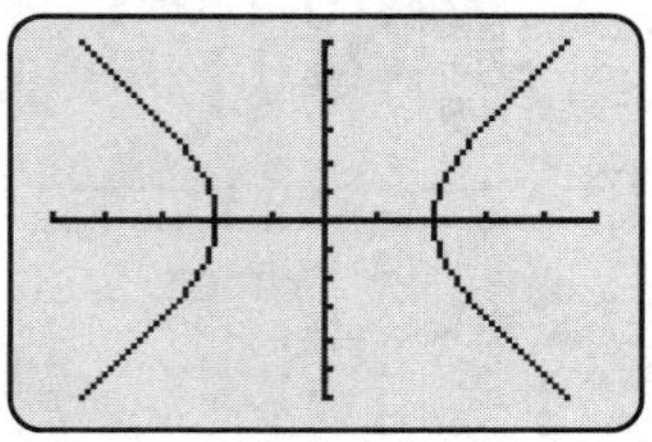

To add the two asymptotes, enter $y_3 = \frac{3}{2}x$ or $y_4 = -\frac{3}{2}x$.

```
Plot1 Plot2 Plot3
\Y1=3√(X²/4-1)
\Y2=-3√(X²/4-1)
\Y3=3/2X
\Y4=-3/2X
\Y5=
\Y6=
```

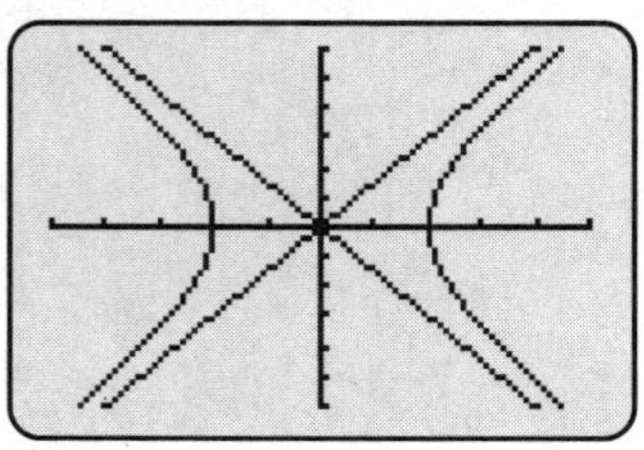

EXAMPLE 3 Graphing a Hyperbola Centered at the Origin with a Horizontal Transverse Axis

Graph the hyperbola given by $\frac{x^2}{4} - \frac{y^2}{9} = 1$.

Solution:

Compare $\frac{x^2}{2^2} - \frac{y^2}{3^2} = 1$ to the general equation $\frac{x^2}{a^2} - \frac{y^2}{b^2} = 1$.

Identify a and b. $a = 2$ and $b = 3$

The transverse axis of this hyperbola lies on the x-axis.

Label the vertices $(-a, 0) = (-2, 0)$ and $(a, 0) = (2, 0)$ and the points $(0, -b) = (0, -3)$ and $(0, b) = (0, 3)$. Draw the rectangular box that passes through those points. Draw the asymptotes that pass through the center and the corners of the rectangle.

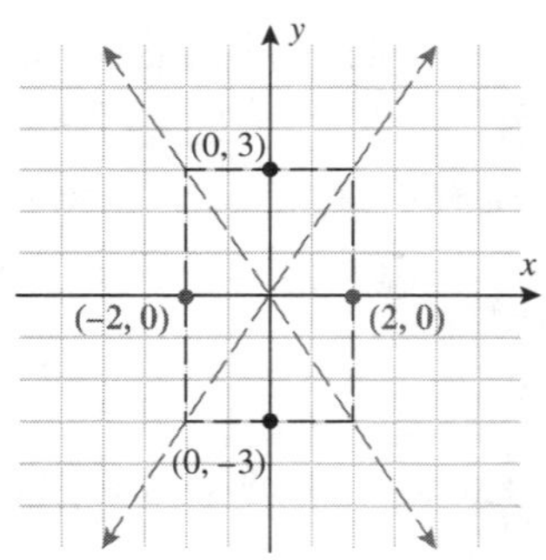

Draw the two branches of the hyperbola, each passing through a vertex and guided by the asymptotes.

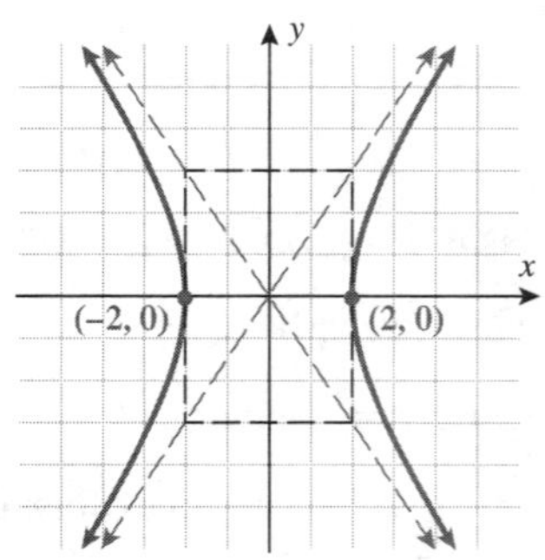

In Example 3, if we let $y = 0$, then $\frac{x^2}{4} = 1$ or $x = \pm 2$. So the vertices are $(-2, 0)$ and $(2, 0)$, and the transverse axis lies along the x-axis. Note that if $x = 0$, $y = \pm 3i$.

EXAMPLE 4 **Graphing a Hyperbola Centered at the Origin with a Vertical Transverse Axis**

Graph the hyperbola given by $\frac{y^2}{16} - \frac{x^2}{4} = 1$.

Solution:

Compare $\frac{y^2}{4^2} - \frac{x^2}{2^2} = 1$ to the general equation $\frac{y^2}{a^2} - \frac{x^2}{b^2} = 1$.

Identify a and b. $a = 4$ and $b = 2$

The transverse axis of this hyperbola lies along the y-axis.

Label the vertices $(0, -a) = (0, -4)$ and $(0, a) = (0, 4)$ and the points $(-b, 0) = (-2, 0)$ and $(b, 0) = (2, 0)$. Draw the rectangular box that passes through those points. Draw the asymptotes that pass through the center and the corners of the rectangle.

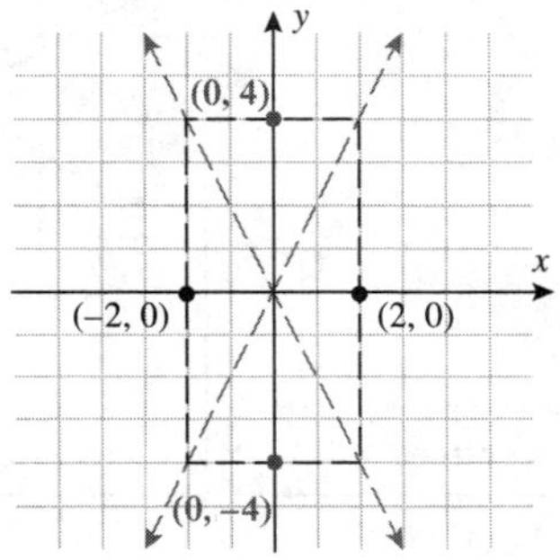

Draw the two branches of the hyperbola, each passing through a vertex and guided by the asymptotes.

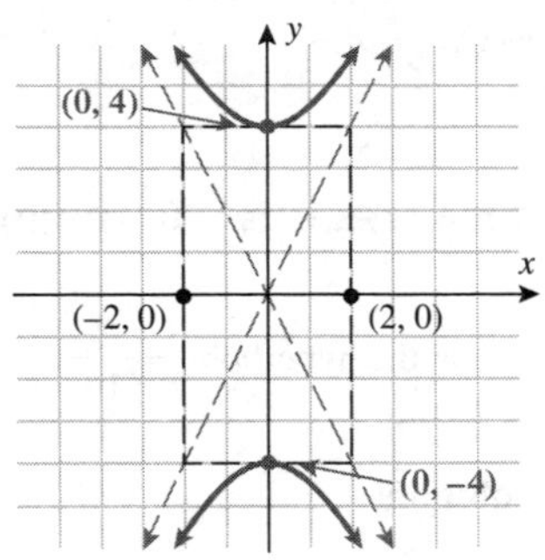

YOUR TURN Graph the hyperbolas.

a. $\frac{y^2}{1} - \frac{x^2}{4} = 1$ **b.** $\frac{x^2}{4} - \frac{y^2}{1} = 1$

Hyperbolas Centered at the Point (h, k)

We can use graph-shifting techniques to graph hyperbolas that are centered at a point other than the origin, say, (h, k). For example, to graph $\frac{(x-h)^2}{a^2} - \frac{(y-k)^2}{b^2} = 1$, start with the graph of $\frac{x^2}{a^2} - \frac{y^2}{b^2} = 1$ and shift to the right h units and up k units. The center, the vertices, the foci, the transverse and conjugate axes, and the asymptotes all shift. The following table summarizes the characteristics of hyperbolas centered at a point other than the origin.

Technology Tip

Use TI to check the graph of $\frac{y^2}{16} - \frac{x^2}{4} = 1$. Solve for y first.

That is, $y_1 = 4\sqrt{1 + \frac{x^2}{4}}$ or

$y_2 = -4\sqrt{1 - \frac{x^2}{4}}$.

To add the two asymptotes, enter $y_3 = 2x$ or $y_4 = -2x$.

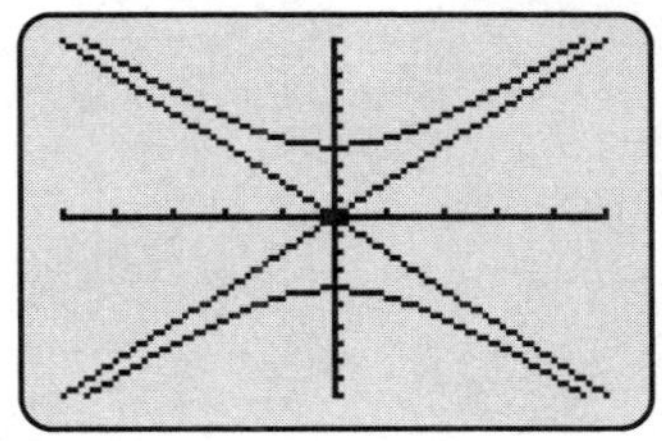

Answer:

a.

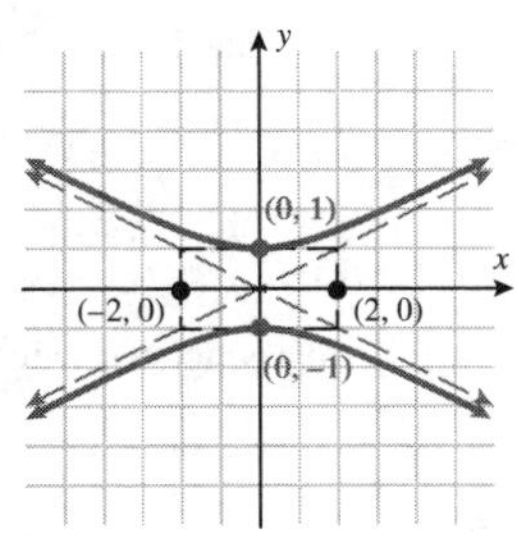

b.

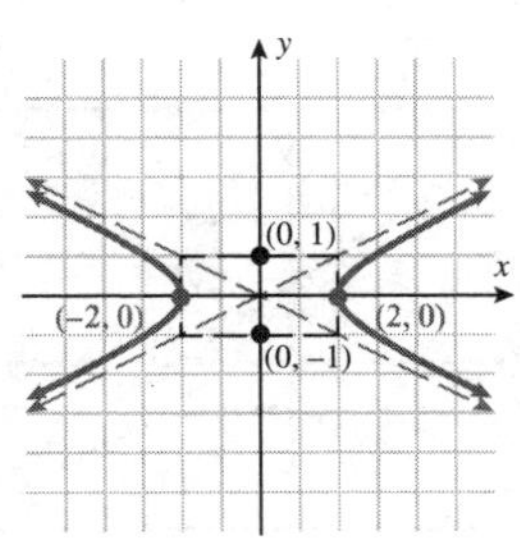

EQUATION OF A HYPERBOLA WITH CENTER AT THE POINT (h, k)

The **standard form of the equation of a hyperbola** with its center at the point (h, k) is given by:

Orientation of Transverse Axis	Horizontal Parallel to the x-axis	Vertical Parallel to the y-axis
Equation	$\frac{(x-h)^2}{a^2} - \frac{(y-k)^2}{b^2} = 1$	$\frac{(y-k)^2}{a^2} - \frac{(x-h)^2}{b^2} = 1$
Graph	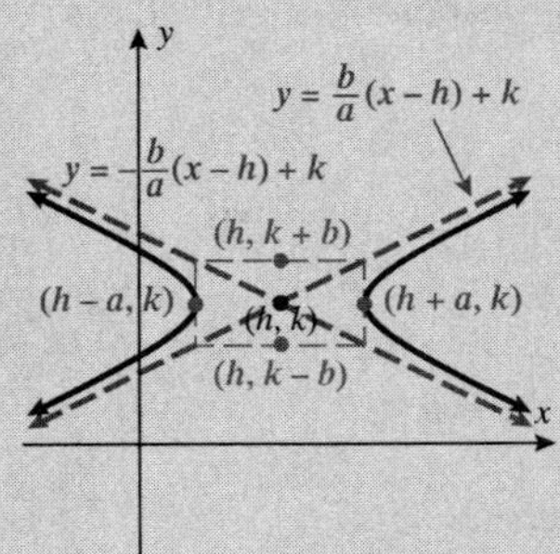	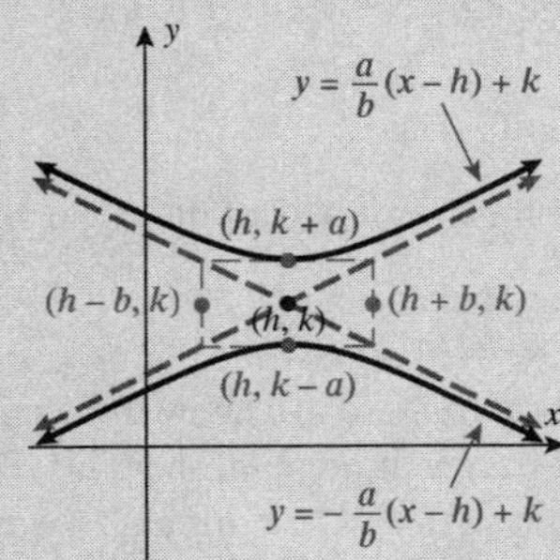
Vertices	$(h-a, k)$ $(h+a, k)$	$(h, k-a)$ $(h, k+a)$
Foci	$(h-c, k)$ $(h+c, k)$	$(h, k-c)$ $(h, k+c)$

where $c^2 = a^2 + b^2$

EXAMPLE 5 Graphing a Hyperbola with Center Not at the Origin

Graph the hyperbola $\frac{(y-2)^2}{16} - \frac{(x-1)^2}{9} = 1$.

Solution:

Compare $\frac{(y-2)^2}{4^2} - \frac{(x-1)^2}{3^2} = 1$ to the general equation $\frac{(y-k)^2}{a^2} - \frac{(x-h)^2}{b^2} = 1$.

Identify a, b, and (h, k). $\quad a = 4,\ b = 3,$ and $(h, k) = (1, 2)$

The transverse axis of this hyperbola lies along $x = 1$, which is parallel to the y-axis.

Label the vertices $(h, k - a) = (1, -2)$ and $(h, k + a) = (1, 6)$, and the points $(h - b, k) = (-2, 2)$ and $(h + b, k) = (4, 2)$. Draw the rectangular box that passes through those points. Draw the asymptotes that pass through the center $(h, k) = (1, 2)$ and the corners of the rectangle. Draw the two branches of the hyperbola, each passing through a vertex and guided by the asymptotes.

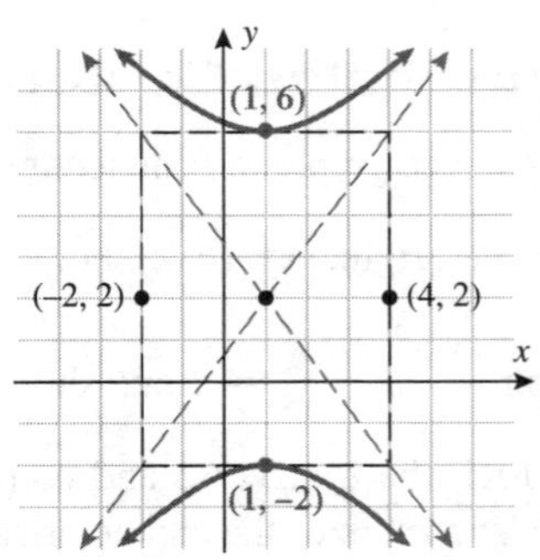

Technology Tip

Use TI to check the graph of $\frac{(y-2)^2}{16} - \frac{(x-1)^2}{9} = 1$.

Solve for y first. That is,

$y_1 = 2 + 4\sqrt{1 + \frac{(x-1)^2}{9}}$ or

$y_2 = 2 - 4\sqrt{1 + \frac{(x-1)^2}{4}}$.

To add the two asymptotes, enter

$y_3 = \frac{4}{3}(x-1) + 2$ or

$y_4 = -\frac{4}{3}(x-1) + 2$.

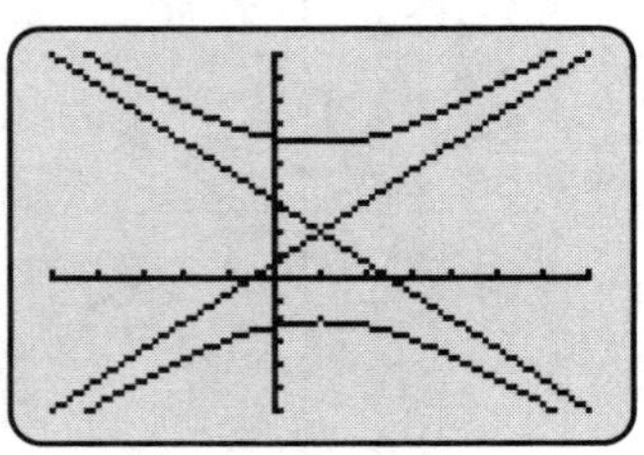

EXAMPLE 6 Transforming an Equation of a Hyperbola to Standard Form

Graph the hyperbola $9x^2 - 16y^2 - 18x + 32y - 151 = 0$.

Solution:

Complete the square on the x terms and y terms, respectively.

$$9(x^2 - 2x) - 16(y^2 - 2y) = 151$$
$$9(x^2 - 2x + 1) - 16(y^2 - 2y + 1) = 151 + 9 - 16$$
$$9(x - 1)^2 - 16(y - 1)^2 = 144$$
$$\frac{(x - 1)^2}{16} - \frac{(y - 1)^2}{9} = 1$$

Compare $\frac{(x - 1)^2}{16} - \frac{(y - 1)^2}{9} = 1$ to the general form $\frac{(x - h)^2}{a^2} - \frac{(y - k)^2}{b^2} = 1$.

Identify a, b, and (h, k). $a = 4$, $b = 3$, and $(h, k) = (1, 1)$

The transverse axis of this hyperbola lies along $y = 1$.

Label the vertices $(h - a, k) = (-3, 1)$ and $(h + a, k) = (5, 1)$ and the points $(h, k - b) = (1, -2)$ and $(h, k + b) = (1, 4)$. Draw the rectangular box that passes through these points. Draw the asymptotes that pass through the center $(1, 1)$ and the corners of the box. Draw the two branches of the hyperbola, each passing through the vertex and guided by the asymptotes.

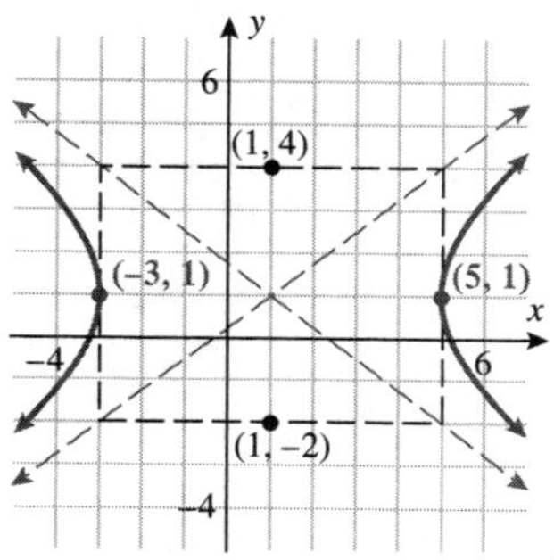

Applications

Nautical navigation is assisted by hyperbolas. For example, suppose two radio stations on a coast are emitting simultaneous signals. If a boat is at sea, it will be slightly closer to one station than the other station, which results in a small time difference between the received signals from the two stations. If the boat follows the path associated with a constant time difference, that path will be hyperbolic.

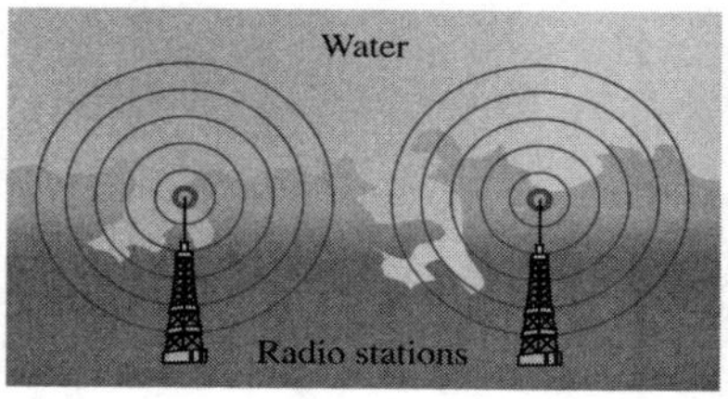

The synchronized signals would intersect one another in associated hyperbolas. Each time difference corresponds to a different path. The radio stations are the foci of the hyperbolas. This principle forms the basis of a hyperbolic radio navigation system known as *loran* (**LO**ng-**RA**nge **N**avigation).

There are navigational charts that correspond to different time differences. A ship selects the hyperbolic path that will take it to the desired port, and the loran chart lists the corresponding time difference.

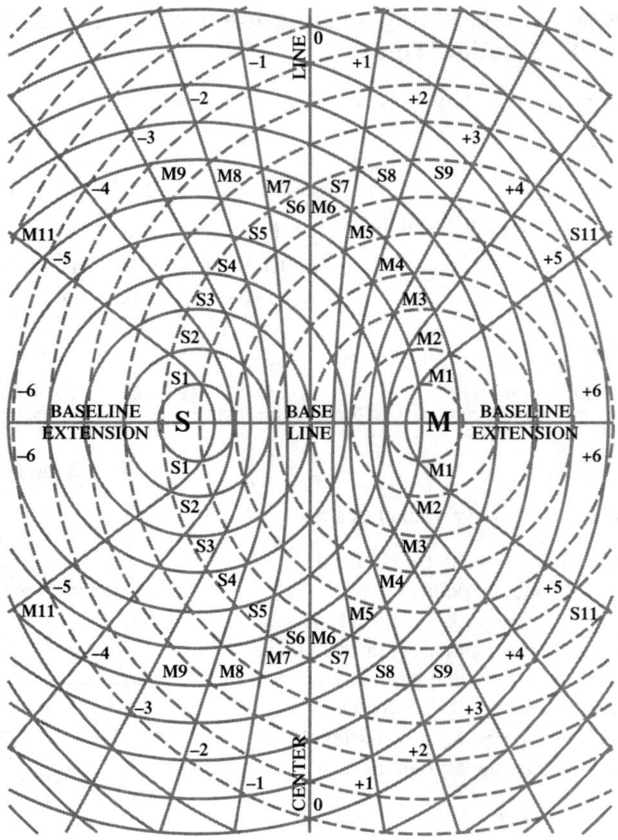

EXAMPLE 7 Nautical Navigation Using Loran

Two loran stations are located 200 miles apart along a coast. If a ship records a time difference of 0.00043 second and continues on the hyperbolic path corresponding to that difference, where does it reach shore?

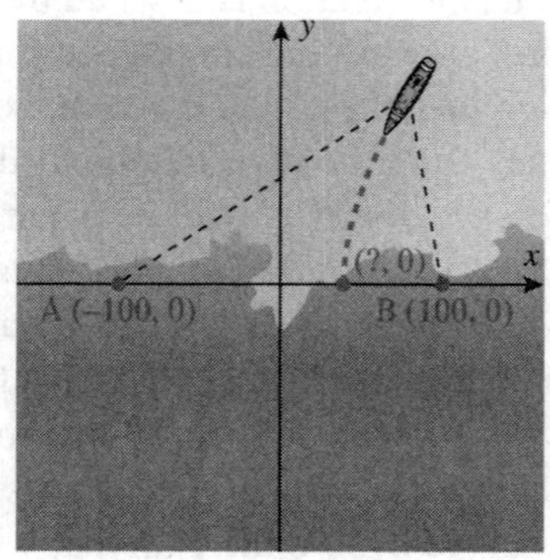

Solution:

Draw the xy-plane and the two stations corresponding to the foci at $(-100, 0)$ and $(100, 0)$. Draw the ship somewhere in quadrant I.

The hyperbola corresponds to a path where the difference of the distances between the ship and the respective stations remains constant. The constant is $2a$, where $(a, 0)$ is a vertex. Find that difference by using $d = rt$. Assume the speed of the radio signal is 186,000 miles per second.

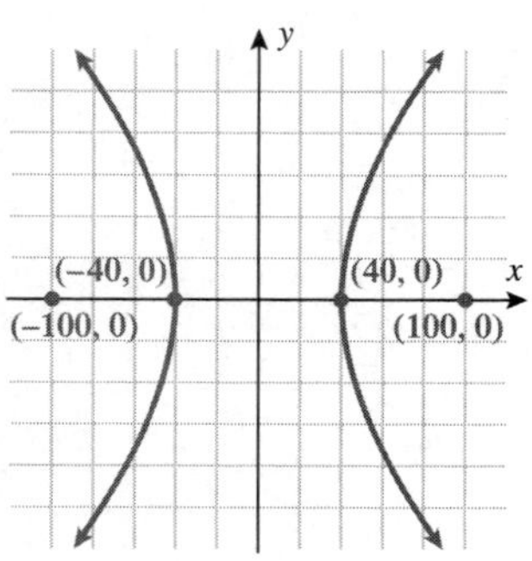

Substitute $r = $ 186,000 miles/second and $t = $ 0.00043 second into $d = rt$.

$$d = (186{,}000 \text{ miles/second})(0.00043 \text{ second}) \approx 80 \text{ miles}$$

Set the constant equal to $2a$. $\quad 2a = 80$

Find a vertex $(a, 0)$. $\quad (40, 0)$

The ship reaches shore between the two stations, 60 miles from station B and 140 miles from station A.

SECTION 8.4 SUMMARY

In this section, hyperbolas centered at the origin were discussed:

Equation	$\dfrac{x^2}{a^2} - \dfrac{y^2}{b^2} = 1$	$\dfrac{y^2}{a^2} - \dfrac{x^2}{b^2} = 1$
Transverse axis	Horizontal (x-axis), length $2a$	Vertical (y-axis), length $2a$
Conjugate axis	Vertical (y-axis), length $2b$	Horizontal (x-axis), length $2b$
Vertices	$(-a, 0)$ and $(a, 0)$	$(0, -a)$ and $(0, a)$
Foci $c^2 = a^2 + b^2$	$(-c, 0)$ and $(c, 0)$	$(0, -c)$ and $(0, c)$
Asymptote	$y = \dfrac{b}{a}x$ and $y = -\dfrac{b}{a}x$	$y = \dfrac{a}{b}x$ and $y = -\dfrac{a}{b}x$
Graph	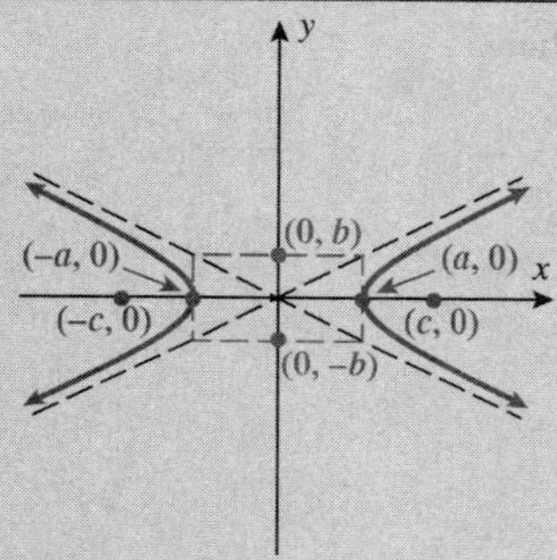	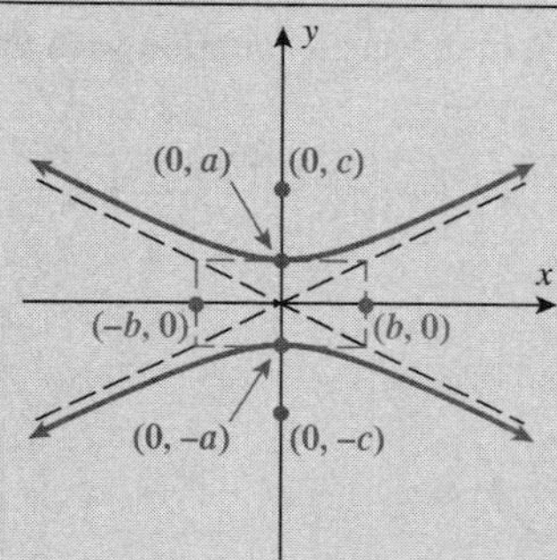

For a hyperbola centered at (h, k), the vertices, foci, and asymptotes all shift accordingly.

SECTION 8.4 EXERCISES

SKILLS

In Exercises 1–4, match each equation with the corresponding hyperbola.

1. $\dfrac{x^2}{36} - \dfrac{y^2}{16} = 1$

2. $\dfrac{y^2}{36} - \dfrac{x^2}{16} = 1$

3. $\dfrac{x^2}{8} - \dfrac{y^2}{72} = 1$

4. $4y^2 - x^2 = 1$

a.

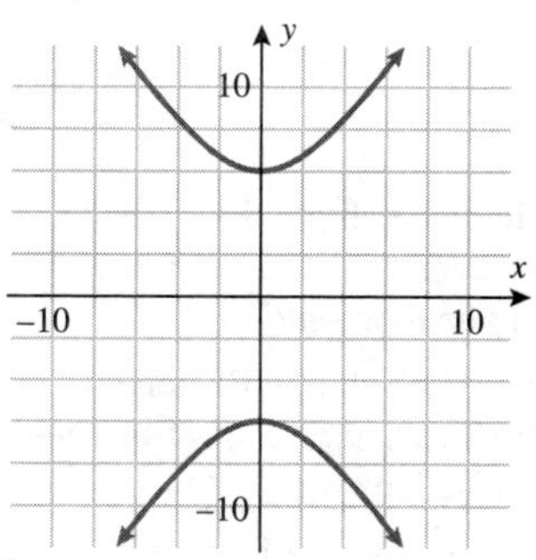

b.

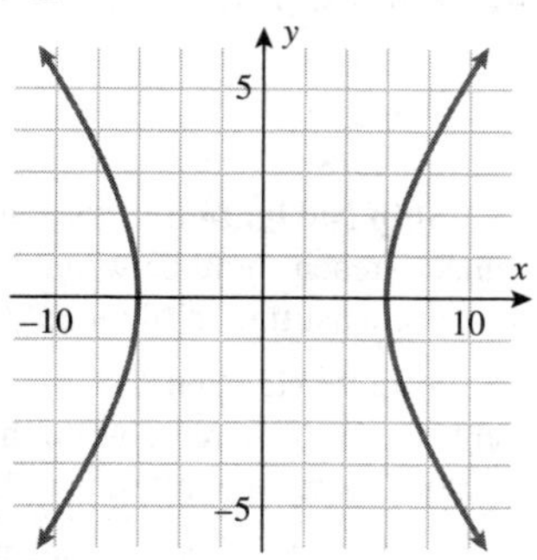

c.

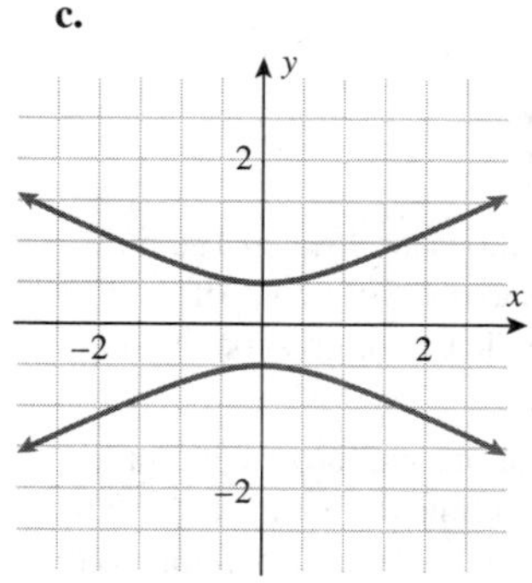

d.

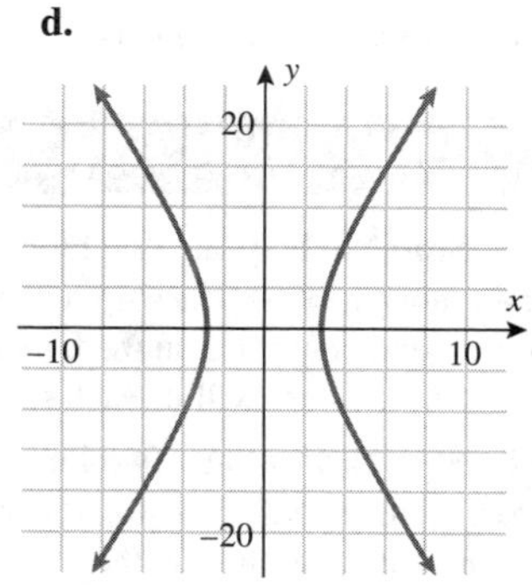

In Exercises 5–16, graph each hyperbola.

5. $\frac{x^2}{25} - \frac{y^2}{16} = 1$ **6.** $\frac{x^2}{49} - \frac{y^2}{9} = 1$ **7.** $\frac{y^2}{16} - \frac{x^2}{64} = 1$ **8.** $\frac{y^2}{144} - \frac{x^2}{25} = 1$

9. $\frac{x^2}{100} - y^2 = 1$ **10.** $9y^2 - 4x^2 = 36$ **11.** $\frac{4y^2}{9} - 81x^2 = 1$ **12.** $\frac{4}{25}x^2 - \frac{100}{9}y^2 = 1$

13. $4x^2 - y^2 = 16$ **14.** $y^2 - x^2 = 81$ **15.** $8y^2 - 16x^2 = 32$ **16.** $10x^2 - 25y^2 = 50$

In Exercises 17–24, find the standard form of an equation of the hyperbola with the given characteristics.

17. vertices: $(-4, 0)$ and $(4, 0)$ foci: $(-6, 0)$ and $(6, 0)$

18. vertices: $(-1, 0)$ and $(1, 0)$ foci: $(-3, 0)$ and $(3, 0)$

19. vertices: $(0, -3)$ and $(0, 3)$ foci: $(0, -4)$ and $(0, 4)$

20. vertices: $(0, -1)$ and $(0, 1)$ foci: $(0, -2)$ and $(0, 2)$

21. center: $(0, 0)$; transverse: x-axis; asymptotes: $y = x$ and $y = -x$

22. center: $(0, 0)$; transverse: y-axis; asymptotes: $y = x$ and $y = -x$

23. center: $(0, 0)$; transverse axis: y-axis; asymptotes: $y = 2x$ and $y = -2x$

24. center: $(0, 0)$; transverse axis: x-axis; asymptotes: $y = 2x$ and $y = -2x$

In Exercises 25–28, match each equation with the hyperbola.

25. $\frac{(x-3)^2}{4} - \frac{(y+2)^2}{25} = 1$ **26.** $\frac{(x+3)^2}{4} - \frac{(y-2)^2}{25} = 1$ **27.** $\frac{(y-3)^2}{25} - \frac{(x+2)^2}{4} = 1$ **28.** $\frac{(y+3)^2}{25} - \frac{(x-2)^2}{4} = 1$

a.

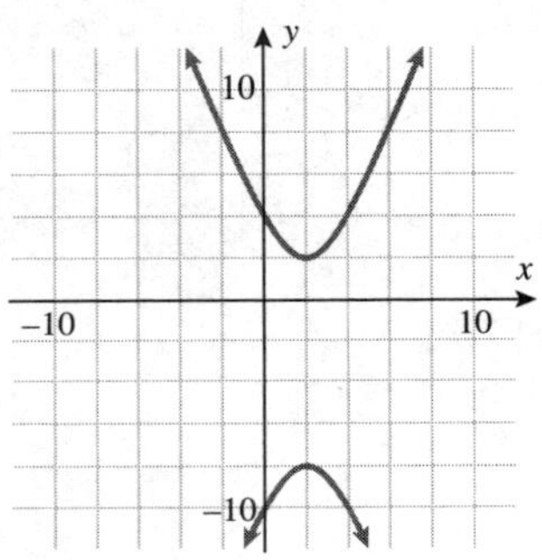

b.

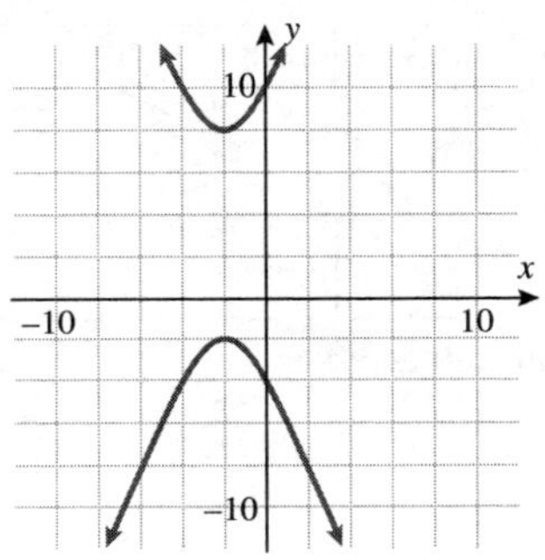

c.

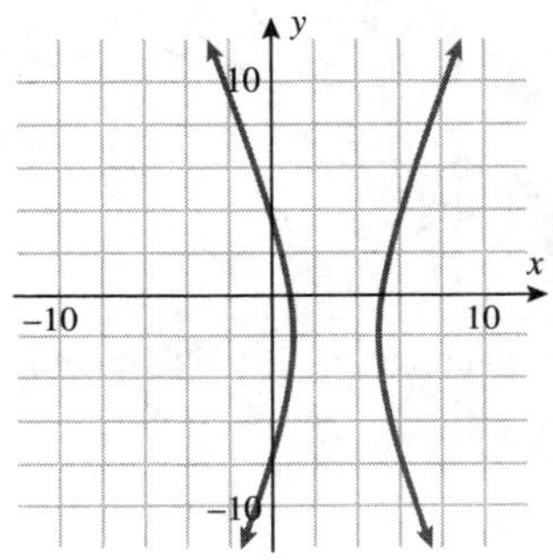

d.

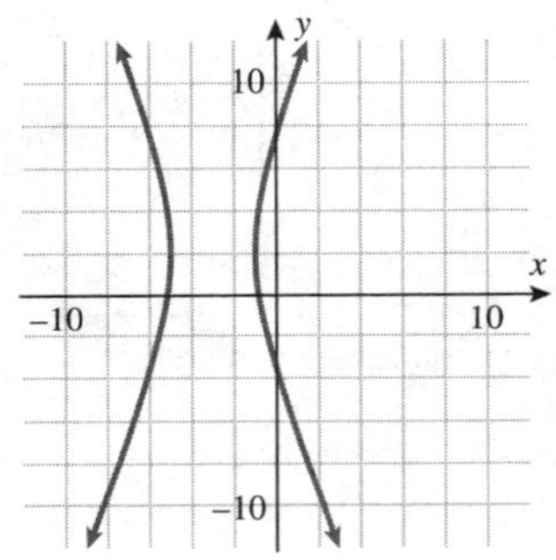

In Exercises 29–38, graph each hyperbola.

29. $\frac{(x-1)^2}{16} - \frac{(y-2)^2}{4} = 1$ **30.** $\frac{(y+1)^2}{36} - \frac{(x+2)^2}{9} = 1$ **31.** $10(y+3)^2 - (x-4)^2 = 80$

32. $3(x+3)^2 - 12(y-4)^2 = 36$ **33.** $x^2 - 4x - 4y^2 = 0$ **34.** $-9x^2 + y^2 + 2y - 8 = 0$

35. $-9x^2 - 18x + 4y^2 - 8y - 41 = 0$ **36.** $25x^2 - 50x - 4y^2 - 8y - 79 = 0$

37. $x^2 - 6x - 4y^2 - 16y - 8 = 0$ **38.** $-4x^2 - 16x + y^2 - 2y - 19 = 0$

In Exercises 39–42, find the standard form of an equation of the hyperbola with the given characteristics.

39. vertices: $(-2, 5)$ and $(6, 5)$ foci: $(-3, 5)$ and $(7, 5)$

40. vertices: $(1, -2)$ and $(3, -2)$ foci: $(0, -2)$ and $(4, -2)$

41. vertices: $(4, -7)$ and $(4, -1)$ foci: $(4, -8)$ and $(4, 0)$

42. vertices: $(2, -6)$ and $(2, -4)$ foci: $(2, -7)$ and $(2, -3)$

APPLICATIONS

43. Ship Navigation. Two loran stations are located 150 miles apart along a coast. If a ship records a time difference of 0.0005 second and continues on the hyperbolic path corresponding to that difference, where will it reach shore?

44. Ship Navigation. Two loran stations are located 300 miles apart along a coast. If a ship records a time difference of 0.0007 second and continues on the hyperbolic path corresponding to that difference, where will it reach shore? Round to the nearest mile.

45. Ship Navigation. If the captain of the ship in Exercise 43 wants to reach shore between the stations and 30 miles from one of them, what time difference should he look for?

46. Ship Navigation. If the captain of the ship in Exercise 44 wants to reach shore between the stations and 50 miles from one of them, what time difference should he look for?

47. Light. If the light from a lamp casts a hyperbolic pattern on the wall due to its lampshade, calculate the equation of the hyperbola if the distance between the vertices is 2 feet and the foci are half a foot from the vertices.

48. Special Ops. A military special ops team is calibrating its recording devices used for passive ascertaining of enemy location. They place two recording stations, alpha and bravo, 3000 feet apart (alpha is due east of bravo). The team detonates small explosives 300 feet west of alpha and records the time it takes each station to register an explosion. The team also sets up a second set of explosives directly north of the alpha station. How many feet north of alpha should the team set off the explosives if it wants to record the same times as on the first explosion?

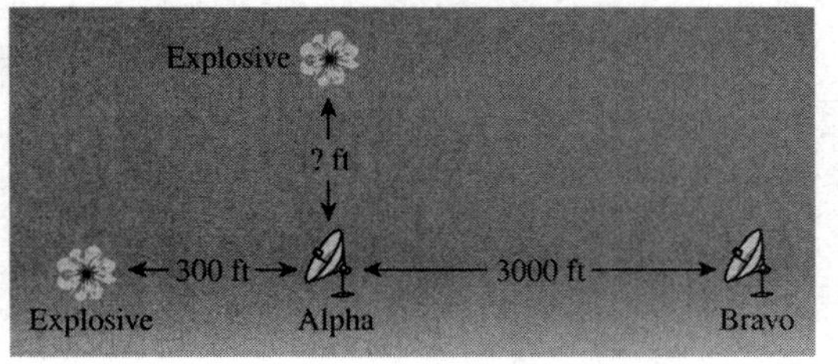

CATCH THE MISTAKE

In Exercises 49 and 50, explain the mistake that is made.

49. Graph the hyperbola $\dfrac{y^2}{4} - \dfrac{x^2}{9} = 1$.

Solution:

Compare the equation to the standard form and solve for a and b.	$a = 2, b = 3$
Label the vertices $(-a, 0)$ and $(a, 0)$.	$(-2, 0)$ and $(2, 0)$
Label the points $(0, -b)$ and $(0, b)$.	$(0, -3)$ and $(0, 3)$

Draw the rectangle connecting these four points and align the asymptotes so that they pass through the center and the corner of the boxes. Then draw the hyperbola using the vertices and asymptotes.

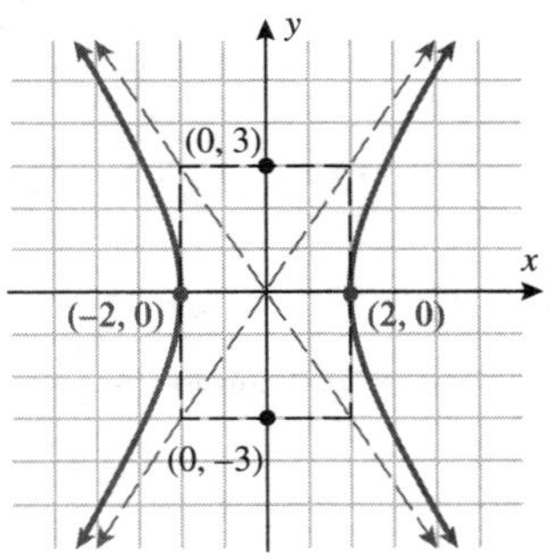

This is incorrect. What mistake was made?

50. Graph the hyperbola $\dfrac{x^2}{1} - \dfrac{y^2}{4} = 1$.

Solution:

Compare the equation to the general form and solve for a and b.	$a = 2, b = 1$
Label the vertices $(-a, 0)$ and $(a, 0)$.	$(-2, 0)$ and $(2, 0)$
Label the points $(0, -b)$ and $(0, b)$.	$(0, -1)$ and $(0, 1)$

Draw the rectangle connecting these four points and align the asymptotes so that they pass through the center and the corner of the boxes. Then draw the hyperbola using the vertices and asymptotes.

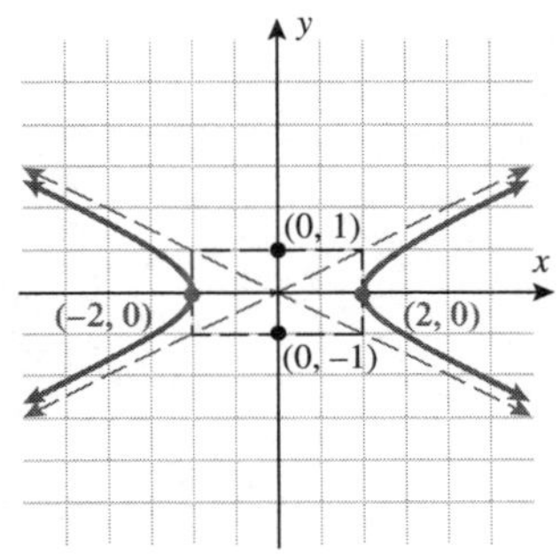

This is incorrect. What mistake was made?

CONCEPTUAL

In Exercises 51–54, determine whether each statement is true or false.

51. If you know the vertices of a hyperbola, you can determine the equation for the hyperbola.

52. If you know the foci and vertices, you can determine the equation for the hyperbola.

53. Hyperbolas centered at the origin have symmetry with respect to the x-axis, y-axis, and the origin.

54. The center and foci are part of the graph of a hyperbola.

CHALLENGE

55. Find the general equation of a hyperbola whose asymptotes are perpendicular.

56. Find the general equation of a hyperbola whose vertices are $(3, -2)$ and $(-1, -2)$ and whose asymptotes are the lines $y = 2x - 4$ and $y = -2x$.

TECHNOLOGY

57. Graph the following three hyperbolas: $x^2 - y^2 = 1$, $x^2 - 5y^2 = 1$, and $x^2 - 10y^2 = 1$. What can be said to happen to the hyperbola $x^2 - cy^2 = 1$ as c increases?

58. Graph the following three hyperbolas: $x^2 - y^2 = 1$, $5x^2 - y^2 = 1$, and $10x^2 - y^2 = 1$. What can be said to happen to the hyperbola $cx^2 - y^2 = 1$ as c increases?

59. Graph the following three hyperbolas: $x^2 - y^2 = 1$, $0.5x^2 - y^2 = 1$, and $0.05x^2 - y^2 = 1$. What can be said to happen to hyperbola $cx^2 - y^2 = 1$ as c decreases?

60. Graph the following three hyperbolas: $x^2 - y^2 = 1$, $x^2 - 0.5y^2 = 1$, and $x^2 - 0.05y^2 = 1$. What can be said to happen to hyperbola $x^2 - cy^2 = 1$ as c decreases?

SECTION 8.5 SYSTEMS OF NONLINEAR EQUATIONS

SKILLS OBJECTIVES

- Solve a system of nonlinear equations using elimination.
- Solve a system of nonlinear equations using substitution.
- Eliminate extraneous solutions.

CONCEPTUAL OBJECTIVES

- Interpret the algebraic solution graphically.
- Understand the types of solutions: distinct number of solutions, no solution, and infinitely many solutions.
- Understand that equations of conic sections are nonlinear equations.

In Chapters 6 and 7, we discussed solving systems of *linear* equations. In Chapter 6, we applied elimination and substitution to solve systems of linear equations in two variables, and in Chapter 7 we employed matrices to solve systems of linear equations. Recall that a system of linear equations in two variables has one of three types of solutions:

One Solution	Two lines that intersect at one point	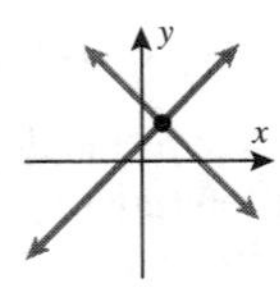
No Solution	Two parallel lines (never intersect)	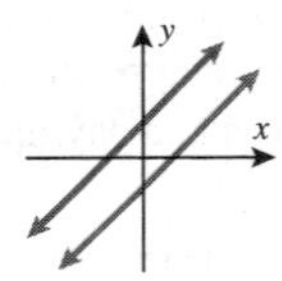
Infinitely Many Solutions	Two lines that coincide (same line)	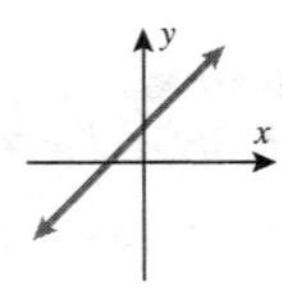

Notice that systems of *linear* equations in two variables always corresponded to *lines*. Now, we turn our attention to systems of *nonlinear* equations in two variables. If any of the equations in a system of equations is nonlinear, then the system is a nonlinear system. The following are systems of nonlinear equations:

$$\begin{cases} y = x^2 + 1 \text{ (Parabola)} \\ y = 2x + 2 \text{ (Line)} \end{cases} \quad \begin{cases} x^2 + y^2 = 25 \text{ (Circle)} \\ y = x \text{ (Line)} \end{cases} \quad \begin{cases} \dfrac{x^2}{9} + \dfrac{y^2}{4} = 1 \text{ (Ellipse)} \\ \dfrac{y^2}{16} - \dfrac{x^2}{25} = 1 \text{ (Hyperbola)} \end{cases}$$

To find the solution to these systems we ask the question "at what point(s)—if any—do the graphs of these equations intersect?" Since some nonlinear equations represent conics, this is a convenient time to discuss systems of nonlinear equations.

How many points of intersection do a line and a parabola have? The answer depends on which line and which parabola. As we see in the following graphs, the answer can be one, two, or none.

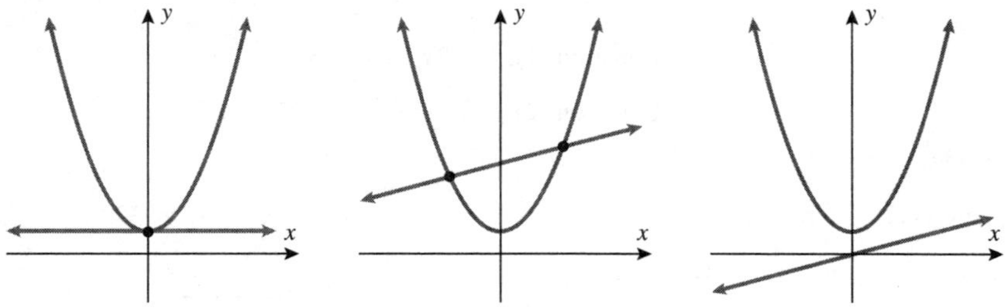

One solution **Two solutions** **No solution**

How many points of intersection do a parabola and an ellipse have? One, two, three, four, or no points of intersection correspond to one solution, two solutions, three solutions, four solutions, or no solution, respectively.

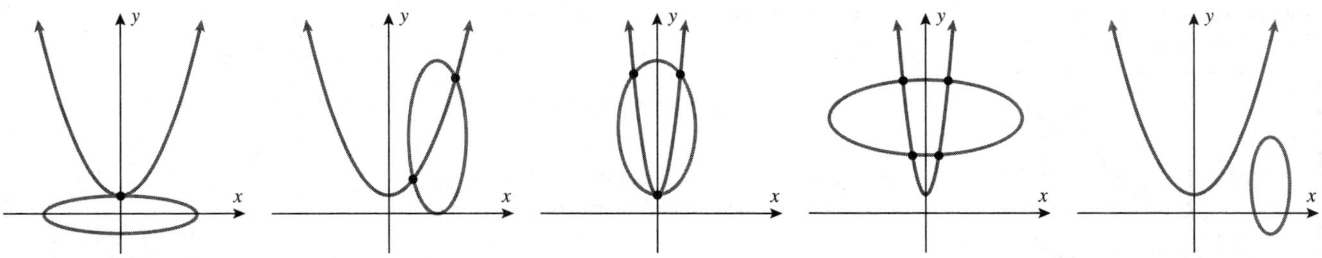

One solution **Two solutions** **Three solutions** **Four solutions** **No solution**

How many points of intersection do a parabola and a hyperbola have? The answer depends on which parabola and which hyperbola. As we see in the following graphs, the answer can be one, two, three, four, or none.

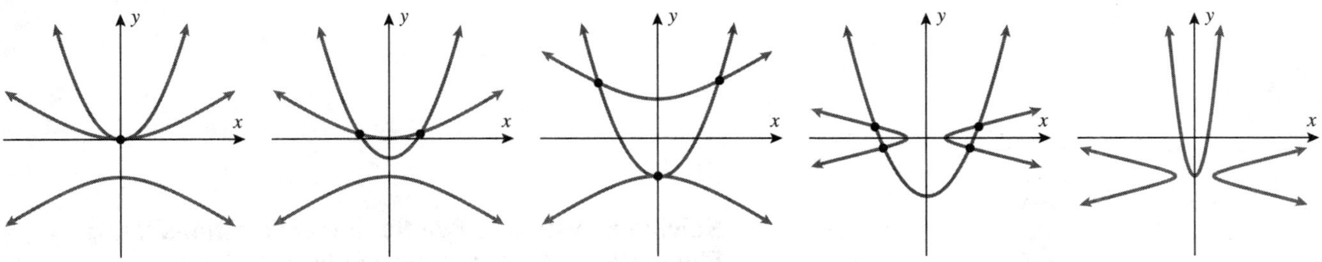

One solution **Two solutions** **Three solutions** **Four solutions** **No solution**

Using Elimination to Solve Systems of Nonlinear Equations

The first three examples in this section use elimination to solve systems of two nonlinear equations. In linear systems, we can eliminate either variable. In nonlinear systems, the variable to eliminate is the one that is raised to the same power in both equations.

Technology Tip

Use TI to solve the system of equations, solve for y in each equation first. That is, $y_1 = 2x - 3$ and $y_2 = x^2 - 2$.

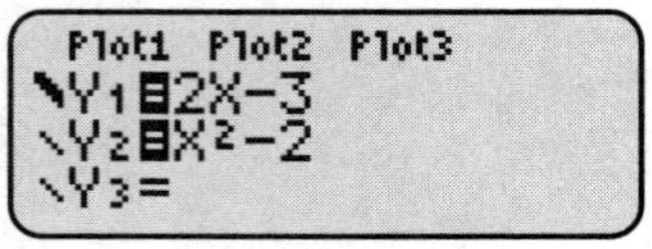

Use the keystrokes:

2nd CALC ▼ 5:Intersect ENTER.

When prompted by the question "First curve?" type ENTER. When prompted by the question "Second curve?" type ENTER. When prompted by the question "Guess?" type ENTER.

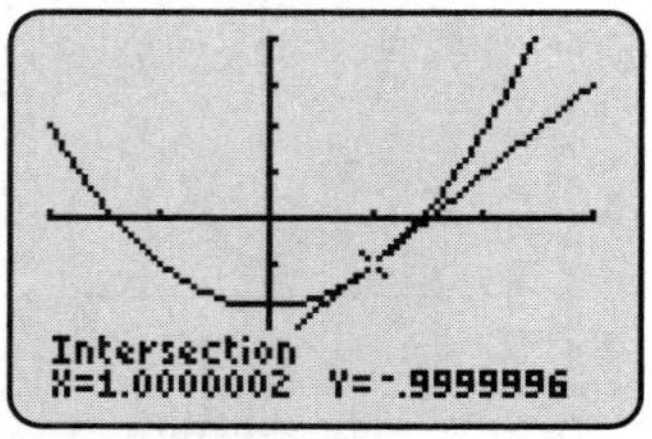

The graphing calculator supports the solution.

EXAMPLE 1 Solving a System of Two Nonlinear Equations Using Elimination: One Solution

Solve the system of equations and graph the corresponding line and parabola to verify the answer.

Equation (1): $2x - y = 3$

Equation (2): $x^2 - y = 2$

Solution:

Equation (1):	$2x - y = 3$
Multiply Equation (2) by -1.	$-x^2 + y = -2$
Add.	$2x - x^2 = 1$
Gather all terms to one side.	$x^2 - 2x + 1 = 0$
Factor.	$(x - 1)^2 = 0$
Solve for x.	$x = 1$
Substitute $x = 1$ into original Equation (1).	$2(1) - y = 3$
Solve for y.	$y = -1$

The solution is $x = 1, y = -1$, or $(1, -1)$.

Graph the line $y = 2x - 3$ and the parabola $y = x^2 - 2$ and confirm that the point of intersection is $(1, -1)$.

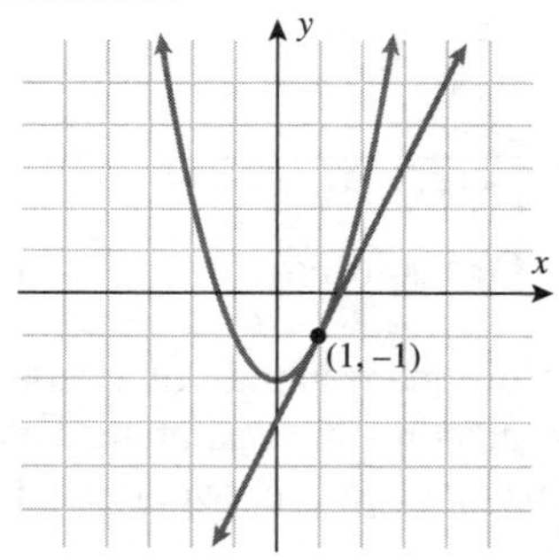

Technology Tip

Use TI to solve the system of equations, solve for y in each equation first. That is, $y_1 = x^2 - 7$, $y_2 = \sqrt{9 - x^2}$, and $y_3 = -\sqrt{9 - x^2}$.

EXAMPLE 2 Solving a System of Two Nonlinear Equations Using Elimination: More Than One Solution

Solve the system of equations and graph the corresponding parabola and circle to verify the answer.

Equation (1): $-x^2 + y = -7$

Equation (2): $x^2 + y^2 = 9$

Solution:

Equation (1):	$-x^2 + y = -7$
Equation (2):	$x^2 + y^2 = 9$
Add.	$y^2 + y = 2$

Gather all terms to one side.	$y^2 + y - 2 = 0$
Factor.	$(y + 2)(y - 1) = 0$
Solve for y.	$y = -2$ or $y = 1$
Substitute $y = -2$ into Equation (2).	$x^2 + (-2)^2 = 9$
Solve for x.	$x = \pm\sqrt{5}$
Substitute $y = 1$ into Equation (2).	$x^2 + (1)^2 = 9$
Solve for x.	$x = \pm\sqrt{8} = \pm 2\sqrt{2}$

There are four solutions: $\boxed{(-\sqrt{5}, -2), (\sqrt{5}, -2), (-2\sqrt{2}, 1), \text{ and } (2\sqrt{2}, 1)}$.

Graph the parabola $y = x^2 - 7$ and the circle $x^2 + y^2 = 9$ and confirm the four points of intersection.

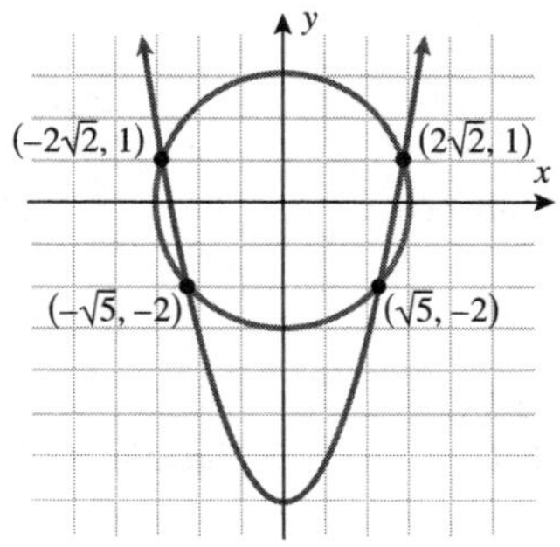

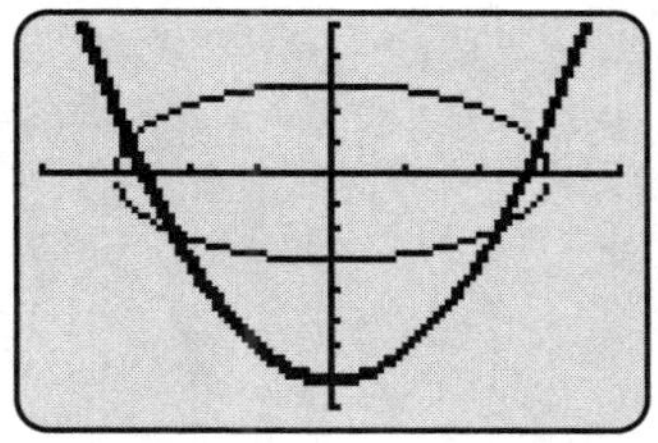

Note: The circle appears elliptical because the x- and y-axis are not of equal scale. The zoom square feature can give the appearance of a circle.

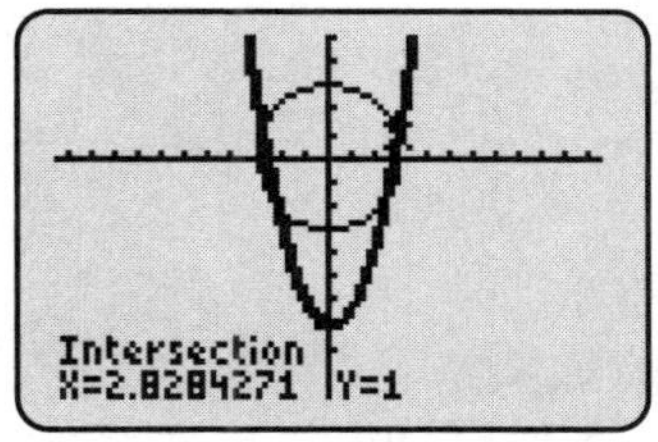

EXAMPLE 3 Solving a System of Two Nonlinear Equations Using Elimination: No Solution

Solve the system of equations and graph the corresponding parabolas to verify the answer.

Equation (1): $x^2 + y = 3$

Equation (2): $-x^2 + y = 5$

Solution:

Equation (1):	$x^2 + y = 3$
Equation (2):	$-x^2 + y = 5$
Add.	$2y = 8$
Solve for y.	$y = 4$
Substitute $y = 4$ into Equation (1).	$x^2 + 4 = 3$
Simplify.	$x^2 = -1$

$x^2 = -1$ has no real solution.

There is $\boxed{\text{no solution}}$ to this system of nonlinear equations.

Graph the parabola $x^2 + y = 3$ and the parabola $y = x^2 + 5$ and confirm there are no points of intersection.

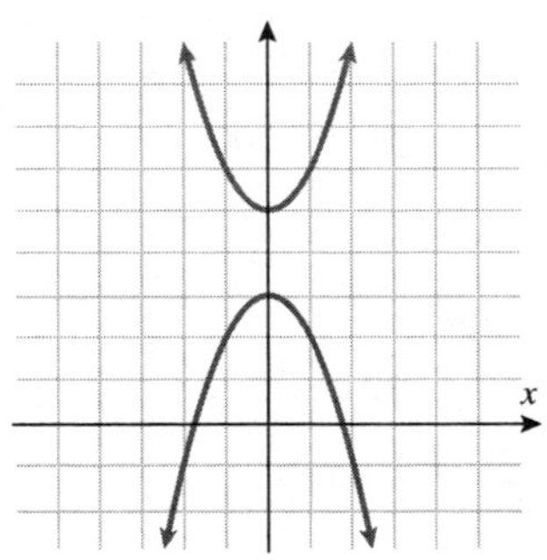

Technology Tip

Use TI to solve the system of equations, solve for y in each equation first. That is, $y_1 = -x^2 + 3$ and $y_2 = x^2 + 5$.

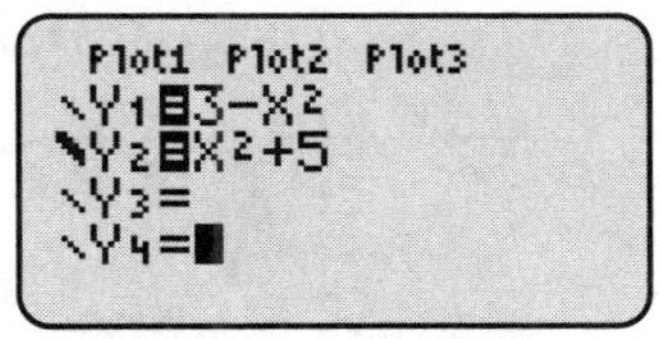

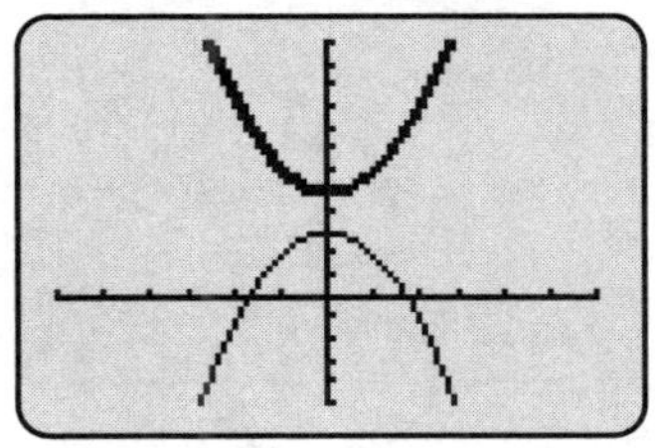

Note: The graphs do not intersect each other. There is no solution to the system.

EXAMPLE 4 Solving a System of Nonlinear Equations with Elimination

Solve the system of nonlinear equations with elimination.

$$\text{Equation (1):} \quad \frac{x^2}{4} + y^2 = 1$$

$$\text{Equation (2):} \quad x^2 - y^2 = 1$$

Solution:

Add Equations (1) and (2) to eliminate y^2.

$$\begin{aligned} \frac{x^2}{4} + y^2 &= 1 \\ x^2 - y^2 &= 1 \\ \hline \frac{5}{4}x^2 &= 2 \end{aligned}$$

Solve for x.

$$x^2 = \frac{8}{5}$$

$$x = \pm\sqrt{\frac{8}{5}}$$

Let $x = \pm\sqrt{\frac{8}{5}}$ in Equation (2).

$$\left(\pm\sqrt{\frac{8}{5}}\right)^2 - y^2 = 1$$

Solve for y.

$$y^2 = \frac{8}{5} - 1 = \frac{3}{5}$$

$$y = \pm\sqrt{\frac{3}{5}}$$

There are four solutions: $\boxed{\left(-\sqrt{\frac{8}{5}}, -\sqrt{\frac{3}{5}}\right), \left(-\sqrt{\frac{8}{5}}, \sqrt{\frac{3}{5}}\right), \left(\sqrt{\frac{8}{5}}, -\sqrt{\frac{3}{5}}\right), \text{ and } \left(\sqrt{\frac{8}{5}}, \sqrt{\frac{3}{5}}\right)}$.

We can approximate the radicals $\sqrt{\frac{8}{5}} \approx 1.26$ and $\sqrt{\frac{3}{5}} \approx 0.77$ to find the four points of intersection of the ellipse $\frac{x^2}{4} + y^2 = 1$ and the hyperbola $x^2 - y^2 = 1$.

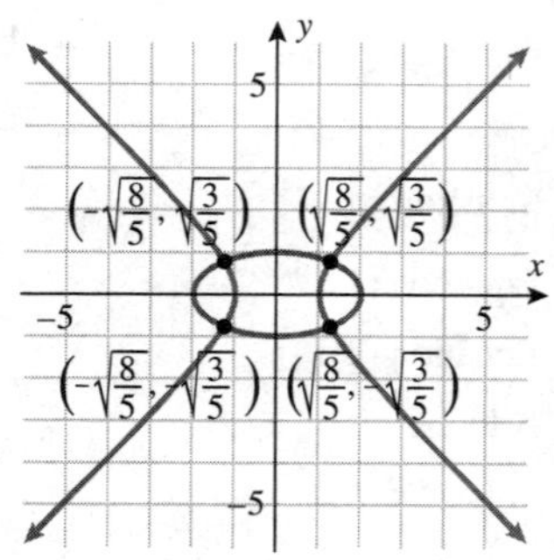

■ **YOUR TURN** Solve the following systems of nonlinear equations.

a. $\begin{aligned} -x + y &= 3 \\ x^2 - y &= -1 \end{aligned}$ **b.** $\begin{aligned} x^2 + y &= 2 \\ -x + y &= 5 \end{aligned}$ **c.** $\begin{aligned} \frac{x^2}{9} + \frac{y^2}{4} &= 1 \\ \frac{y^2}{9} - \frac{x^2}{16} &= 1 \end{aligned}$

■ **Answer:**

a. $(-1, 2)$ and $(2, 5)$

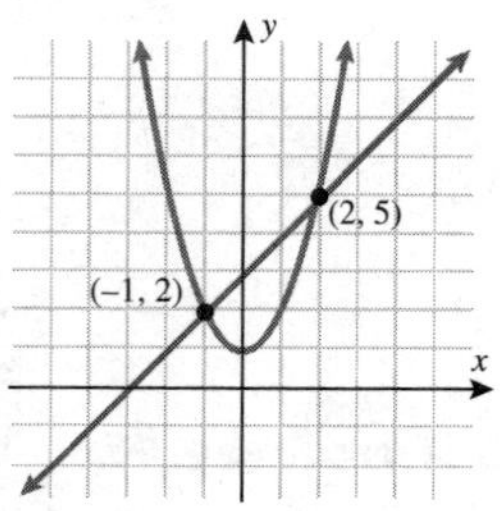

b. No solution

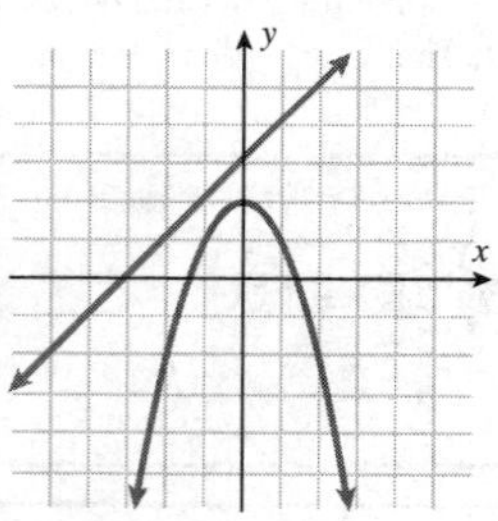

c. No solution

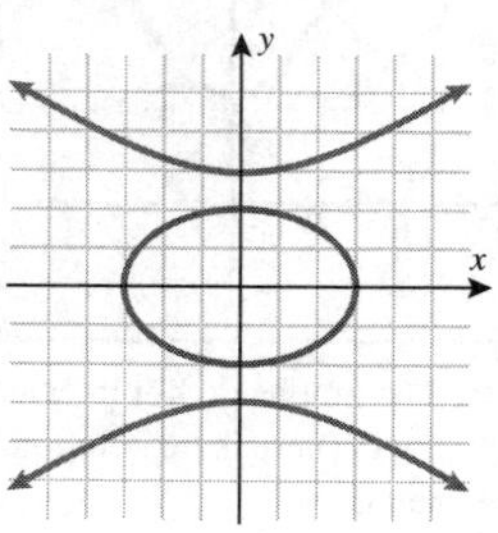

Using Substitution to Solve Systems of Nonlinear Equations

Elimination is based on the idea of eliminating one of the variables and solving the remaining equation in one variable. This is not always possible with nonlinear systems. For example, a system consisting of a circle and a line

$$x^2 + y^2 = 5$$
$$-x + y = 1$$

cannot be solved with elimination because both variables are raised to different powers in each equation. We now turn to the substitution method. It is important to always check solutions because extraneous solutions are possible.

Study Tip

Extraneous solutions are possible when you have one equation with a linear (odd) power and one equation with a second degree (even) power.

EXAMPLE 5 Solving a System of Nonlinear Equations Using Substitution

Solve the system of equations and graph the corresponding circle and line to verify the answer.

Equation (1): $x^2 + y^2 = 5$

Equation (2): $-x + y = 1$

Solution:

Equation (1):	$x^2 + y^2 = 5$
Rewrite Equation (2) with y isolated.	$y = x + 1$
Substitute Equation (2) into Equation (1).	$x^2 + (x + 1)^2 = 5$
Eliminate the parentheses.	$x^2 + x^2 + 2x + 1 = 5$
Gather like terms.	$2x^2 + 2x - 4 = 0$
Divide by 2.	$x^2 + x - 2 = 0$
Factor.	$(x + 2)(x - 1) = 0$
Solve for x.	$x = -2$ or $x = 1$
Substitute $x = -2$ into Equation (1).	$(-2)^2 + y^2 = 5$
Solve for y.	$y = -1$ or $y = 1$
Substitute $x = 1$ into Equation (1).	$(1)^2 + y^2 = 5$
Solve for y.	$y = -2$ or $y = 2$

There appear to be four solutions: $(-2, -1)$, $(-2, 1)$, $(1, -2)$, and $(1, 2)$. But a line can intersect a circle in no more than two points. Therefore, at least two solutions are *extraneous*. All four points satisfy Equation (1), but only $(-2, -1)$ and $(1, 2)$ also satisfy Equation (2).

The answer is $\boxed{(-2, -1) \text{ and } (1, 2)}$.

Graph the circle $x^2 + y^2 = 5$ and the line $y = x + 1$ and confirm the two points of intersection.

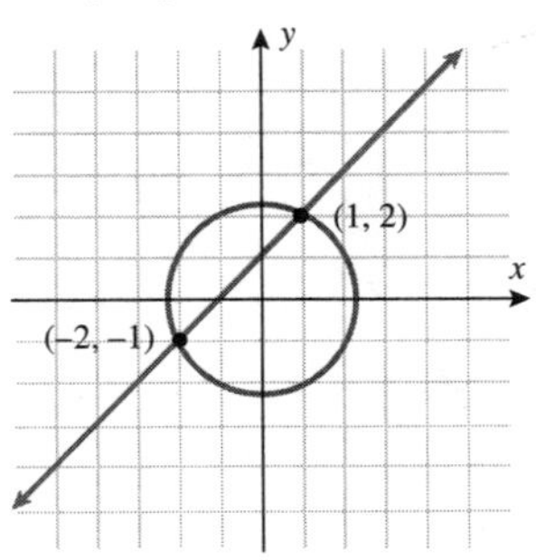

Note: After solving for x in Example 4, had we substituted back into the linear Equation (2) instead of Equation (1), extraneous solutions would not have appeared. In general, **substitute back into the lowest degree equation and always check solutions**.

YOUR TURN Solve the system of equations $x^2 + y^2 = 13$ and $x + y = 5$.

Technology Tip

Use TI to solve the system of equations, solve for y in each equation first. That is, $y_1 = \sqrt{5 - x^2}$, $y_2 = -\sqrt{5 - x^2}$, and $y_3 = x + 1$.

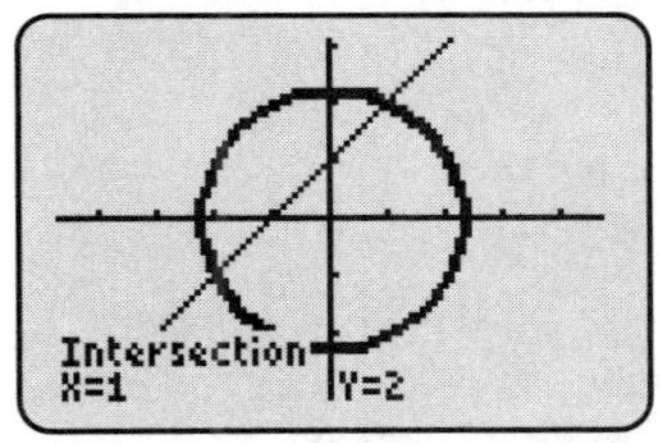

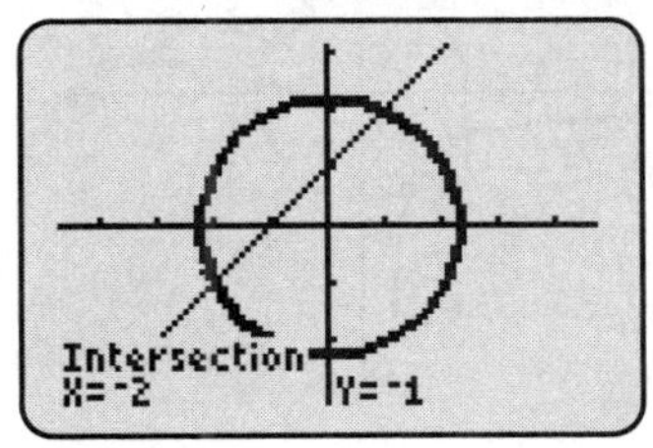

Note: The graphs support the solutions to the system.

Study Tip

Substitute back into the lowest degree equation and always check solutions.

■ **Answer:** (2, 3) and (3, 2)

In Example 6, the equation $xy = 2$ can also be shown to be a rotated hyperbola. Although we will not discuss rotated conics in this book, we can express this equation in terms of a reciprocal function $y = \frac{2}{x}$, which was discussed in Section 3.2.

EXAMPLE 6 Solving a System of Nonlinear Equations with Substitution

Solve the system of equations.

Equation (1): $x^2 + y^2 = 5$

Equation (2): $xy = 2$

Solution:

Since Equation (2) tells us that $xy = 2$, we know neither x nor y can be zero.

Solve Equation (2) for y. $\quad y = \frac{2}{x}$

Substitute $y = \frac{2}{x}$ into Equation (1). $\quad x^2 + \left(\frac{2}{x}\right)^2 = 5$

Eliminate parentheses. $\quad x^2 + \frac{4}{x^2} = 5$

Multiply by x^2. $\quad x^4 + 4 = 5x^2$

Collect terms to one side. $\quad x^4 - 5x^2 + 4 = 0$

Factor. $\quad (x^2 - 4)(x^2 - 1) = 0$

Solve for x. $\quad x = \pm 2$ or $x = \pm 1$

Substitute $x = -2$ into Equation (2) and solve for y. $\quad y = -1$

Substitute $x = 2$ into Equation (2) and solve for y. $\quad y = 1$

Substitute $x = -1$ into Equation (2) and solve for y. $\quad y = -2$

Substitute $x = 1$ into Equation (2) and solve for y. $\quad y = 2$

Check to see that there are four solutions: $(-2, -1), (-1, -2), (2, 1)$, and $(1, 2)$.

Note: It is important to check the solutions either algebraically or graphically (see graph on the left).

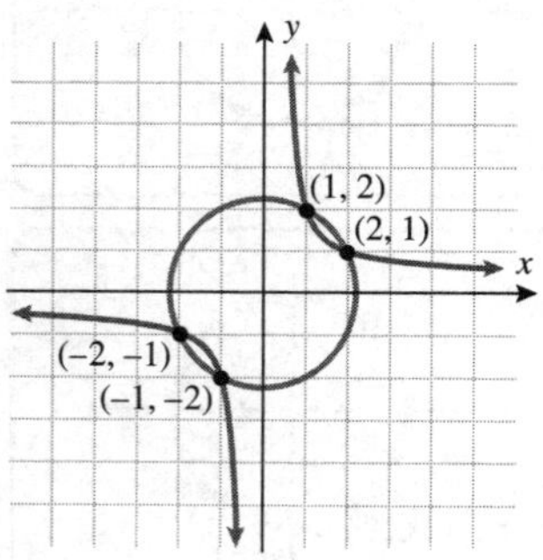

■ **YOUR TURN** Solve the system of equations.

$$x^2 + y^2 = 2$$
$$xy = 1$$

■ **Answer:** $(-1, -1)$ and $(1, 1)$

Applications

EXAMPLE 7 Calculating How Much Fence to Buy

A couple buys a rectangular piece of property advertised as 10 acres (approximately 400,000 square feet). They want two fences to divide the land into an internal grazing area and a surrounding riding path. If they want the riding path to be 20 feet wide, one fence will enclose the property and one internal fence will sit 20 feet inside the outer fence. If the internal grazing field is 237,600 square feet, how many linear feet of fencing should they buy?

Solution:

Use the five-step procedure for solving word problems from Section 1.2 and use two variables.

STEP 1 Identify the question.
How many linear feet of fence should they buy? Or, what is the sum of the perimeters of the two fences?

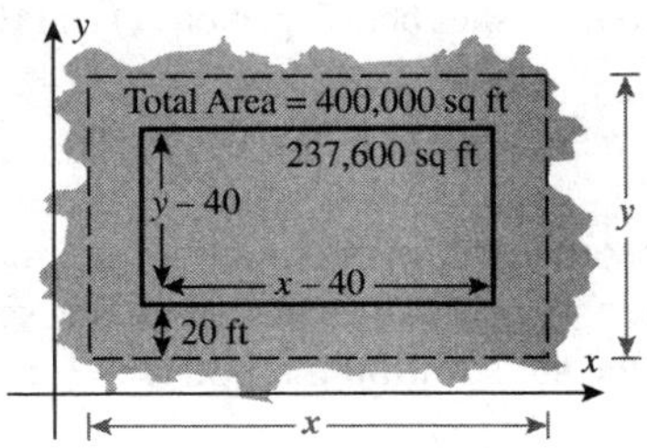

STEP 2 Make notes.

STEP 3 Set up the equations.

$x =$ length of property $\qquad x - 40 =$ length of internal field

$y =$ width of property $\qquad y - 40 =$ width of internal field

Equation (1): $xy = 400{,}000$

Equation (2): $(x - 40)(y - 40) = 237{,}600$

STEP 4 Solve the system of equations.

Substitution Method

Since Equation (1) tells us that $xy = 400{,}000$, we know that neither x nor y can be zero.

Solve Equation (1) for y. $\qquad y = \dfrac{400{,}000}{x}$

Substitute $y = \dfrac{400{,}000}{x}$ into Equation (2). $\qquad (x - 40)\left(\dfrac{400{,}000}{x} - 40\right) = 237{,}600$

Eliminate parentheses. $\qquad 400{,}000 - 40x - \dfrac{16{,}000{,}000}{x} + 1600 = 237{,}600$

Multiply by the LCD, x. $\qquad 400{,}000x - 40x^2 - 16{,}000{,}000 + 1600x = 237{,}600x$

Collect like terms on one side. $\qquad 40x^2 - 164{,}000x + 16{,}000{,}000 = 0$

Divide by 40. $\qquad x^2 - 4100x + 400{,}000 = 0$

Factor. $\qquad (x - 4000)(x - 100) = 0$

Solve for x. $\qquad x = 4000$ or $x = 100$

Technology Tip

Use TI to solve the system of equations, solve for y in each equation first. That is, $y_1 = \dfrac{400000}{x}$ and $y_2 = 40 + \dfrac{237600}{x - 40}$.

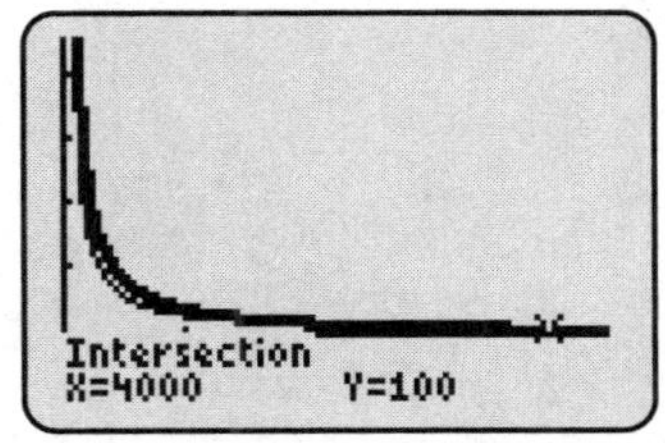

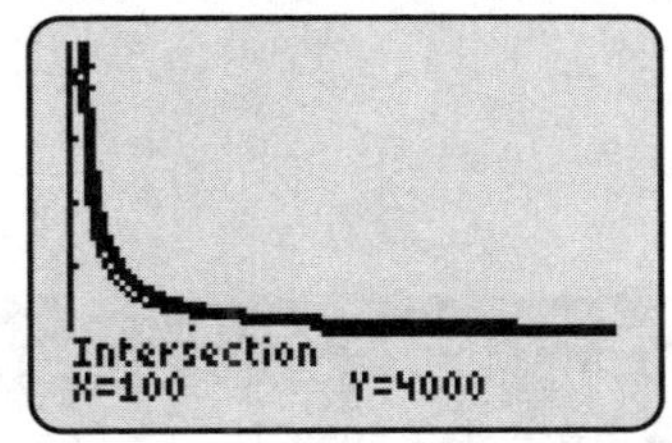

Note: The graphs support the solutions to the system.

Substitute $x = 4000$ into the original Equation (1). $\quad 4000y = 400{,}000$

Solve for y. $\quad y = 100$

Substitute $x = 100$ into the original Equation (1). $\quad 100y = 400{,}000$

Solve for y. $\quad y = 4000$

The two solutions yield the same dimensions: 4000×100. The inner field has the dimensions 3960×60. Therefore, the perimeter of both fences is:

$$2(4000) + 2(100) + 2(3960) + 2(60) = 8000 + 200 + 7920 + 120 = 16{,}240$$

The couple should buy 16,240 linear feet of fencing.

STEP 5 Check the solution.
The point (4000, 100) satisfies both Equation (1) and Equation (2).

It is important to note that some nonlinear equations are not conic sections (exponential, logarithmic, and higher degree polynomial equations). These systems of nonlinear equations are typically solved by the substitution method (see Exercises).

SECTION 8.5 SUMMARY

In this section systems of two equations were discussed when at least one of the equations is nonlinear (e.g., conics). The substitution method and elimination method from Section 6.1 can *sometimes* be applied to nonlinear systems. When graphing the two equations, the points of intersection are the solutions of the system: Systems of nonlinear equations can have more than one solution. Also, extraneous solutions can appear, so it is important to always check solutions.

SECTION 8.5 EXERCISES

SKILLS

In Exercises 1–12, solve the system of equations by applying the elimination method.

1. $x^2 - y = -2$, $-x + y = 4$
2. $x^2 + y = 2$, $2x + y = -1$
3. $x^2 + y = 1$, $2x + y = 2$
4. $x^2 - y = 2$, $-2x + y = -3$
5. $x^2 + y = -5$, $-x + y = 3$
6. $x^2 - y = -7$, $x + y = -2$
7. $x^2 + y^2 = 1$, $x^2 - y = -1$
8. $x^2 + y^2 = 1$, $x^2 + y = -1$
9. $x^2 + y^2 = 3$, $4x^2 + y = 0$
10. $x^2 + y^2 = 6$, $-7x^2 + y = 0$
11. $x^2 + y^2 = -6$, $-2x^2 + y = 7$
12. $x^2 + y^2 = 5$, $3x^2 + y = 9$

In Exercises 13–24, solve the system of equations by applying the substitution method.

13. $\begin{aligned} x + y &= 2 \\ x^2 + y^2 &= 2 \end{aligned}$

14. $\begin{aligned} x - y &= -2 \\ x^2 + y^2 &= 2 \end{aligned}$

15. $\begin{aligned} xy &= 4 \\ x^2 + y^2 &= 10 \end{aligned}$

16. $\begin{aligned} xy &= -3 \\ x^2 + y^2 &= 12 \end{aligned}$

17. $\begin{aligned} y &= x^2 - 3 \\ y &= -4x + 9 \end{aligned}$

18. $\begin{aligned} y &= -x^2 + 5 \\ y &= 3x - 4 \end{aligned}$

19. $\begin{aligned} x^2 + xy - y^2 &= 5 \\ x - y &= -1 \end{aligned}$

20. $\begin{aligned} x^2 + xy + y^2 &= 13 \\ x + y &= -1 \end{aligned}$

21. $\begin{aligned} 2x - y &= 3 \\ x^2 + y^2 - 2x + 6y &= -9 \end{aligned}$

22. $\begin{aligned} x^2 + y^2 - 2x - 4y &= 0 \\ -2x + y &= -3 \end{aligned}$

23. $\begin{aligned} 4x^2 + 12xy + 9y^2 &= 25 \\ -2x + y &= 1 \end{aligned}$

24. $\begin{aligned} -4xy + 4y^2 &= 8 \\ 3x + y &= 2 \end{aligned}$

In Exercises 25–34, solve the system of equations by applying any method.

25. $\begin{aligned} x^3 - y^3 &= 63 \\ x - y &= 3 \end{aligned}$

26. $\begin{aligned} x^3 + y^3 &= -26 \\ x + y &= -2 \end{aligned}$

27. $\begin{aligned} 4x^2 - 3xy &= -5 \\ -x^2 + 3xy &= 8 \end{aligned}$

28. $\begin{aligned} 2x^2 + 5xy &= 2 \\ x^2 - xy &= 1 \end{aligned}$

29. $\begin{aligned} \log_x(2y) &= 3 \\ \log_x(y) &= 2 \end{aligned}$

30. $\begin{aligned} \log_x(y) &= 1 \\ \log_x(2y) &= \tfrac{1}{2} \end{aligned}$

31. $\begin{aligned} \frac{1}{x^3} + \frac{1}{y^2} &= 17 \\ \frac{1}{x^3} - \frac{1}{y^2} &= -1 \end{aligned}$

32. $\begin{aligned} \frac{2}{x^2} + \frac{3}{y^2} &= \frac{5}{6} \\ \frac{4}{x^2} - \frac{9}{y^2} &= 0 \end{aligned}$

33. $\begin{aligned} 2x^2 + 4y^4 &= -2 \\ 6x^2 + 3y^4 &= -1 \end{aligned}$

34. $\begin{aligned} x^2 + y^2 &= -2 \\ x^2 + y^2 &= -1 \end{aligned}$

In Exercises 35 and 36, graph each equation and find the point(s) of intersection.

35. The parabola $y = x^2 - 6x + 11$ and the line $y = -x + 7$.

36. The circle $x^2 + y^2 - 4x - 2y + 5 = 0$ and the line $-x + 3y = 6$.

APPLICATIONS

37. Numbers. The sum of two numbers is 10 and the difference of their squares is 40. Find the numbers.

38. Numbers. The difference of two numbers is 3 and the difference of their squares is 51. Find the numbers.

39. Numbers. The product of two numbers is equal to the reciprocal of the difference of their reciprocals. The product of the two numbers is 72. Find the numbers.

40. Numbers. The ratio of the sum of two numbers to the difference of the two numbers is 9. The product of the two numbers is 80. Find the numbers.

41. Geometry. A rectangle has a perimeter of 36 centimeters and an area of 80 square centimeters. Find the dimensions of the rectangle.

42. Geometry. Two concentric circles have perimeters that add up to 16π and areas that add up to 34π. Find the radii of the two circles.

43. Horse Paddock. An equestrian buys a 5-acre rectangular parcel (approximately 200,000 square feet) and is going to fence in the entire property and then divide the parcel into two halves with a fence. If 2200 linear feet of fencing is required, what are the dimensions of the parcel?

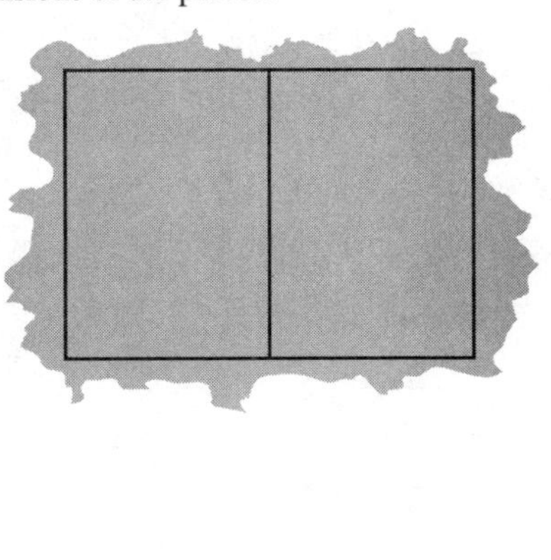

44. Dog Run. A family moves into a new home and decides to fence in the yard to give its dog room to roam. If the area that will be fenced in is rectangular and has an area of 11,250 square feet, and the length is twice as much as the width, how many linear feet of fence should they buy?

45. Footrace. Your college algebra professor and Jeremy Wariner (2004 Olympic Gold Medalist in the men's 400 meter) decided to race. The race was 400 meters and Jeremy gave your professor a 1-minute head start, and still crossed the finish line 1 minute 40 seconds before your professor. If Jeremy ran five times faster than your professor, what was each person's average speed?

46. Footrace. You decided to race Jeremy Wariner for 800 meters. At that distance, Jeremy runs approximately twice as fast as you. He gave you a 1-minute head start and crossed the finish line 20 seconds before you. What were each of your average speeds?

CATCH THE MISTAKE

In Exercises 47 and 48, explain the mistake that is made.

47. Solve the system of equations: $x^2 + y^2 = 4$
$x + y = 2$

Solution:

Multiply the second equation by (-1) and add to the first equation. $x^2 - x = 2$

Subtract 2. $x^2 - x - 2 = 0$

Factor. $(x + 1)(x - 2) = 0$

Solve for x. $x = -1$ and $x = 2$

Substitute $x = -1$ and $x = 2$ into $x + y = 2$. $-1 + y = 2$ and $2 + y = 2$

Solve for y. $y = 3$ and $y = 0$

The answer is $(-1, 3)$ and $(2, 0)$.

This is incorrect. What mistake was made?

48. Solve the system of equations: $x^2 + y^2 = 5$
$2x - y = 0$

Solution:

Solve the second equation for y. $y = 2x$

Substitute $y = 2x$ into the first equation. $x^2 + (2x)^2 = 5$

Eliminate the parentheses. $x^2 + 4x^2 = 5$

Gather like terms. $5x^2 = 5$

Solve for x. $x = -1$ and $x = 1$

Substitute $x = -1$ into the first equation. $(-1)^2 + y^2 = 5$

Solve for y. $y = -2$ and $y = 2$

Substitute $x = 1$ into the first equation. $(1)^2 + y^2 = 5$

Solve for y. $y = -2$ and $y = 2$

The answers are $(-1, -2)$, $(-1, 2)$, $(1, -2)$, and $(1, 2)$.

This is incorrect. What mistake was made?

CONCEPTUAL

In Exercises 49–52, determine whether each statement is either true or false.

49. A system of equations representing a line and a parabola can intersect in at most three points.

50. A system of equations representing a line and a cubic function can intersect in at most three places.

51. The elimination method can always be used to solve systems of two nonlinear equations.

52. The substitution method always works for solving systems of nonlinear equations.

CHALLENGE

53. A circle and a line have at most two points of intersection. A circle and a parabola have at most four points of intersection. What is the greatest number of points of intersection that a circle and an nth-degree polynomial can have?

54. A line and a parabola have at most two points of intersection. A line and a cubic function have at most three points of intersection. What is the greatest number of points of intersection that a line and an nth-degree polynomial can have?

55. Find a system of equations representing a line and a parabola that has only one real solution.

56. Find a system of equations representing a circle and a parabola that has only one real solution.

TECHNOLOGY

Use a graphing utility to solve the following systems of equations.

57. $y = e^x$
$y = \ln x$

58. $y = 10^x$
$y = \log x$

59. $2x^3 + 4y^2 = 3$
$xy^3 = 7$

60. $3x^4 - 2xy + 5y^2 = 19$
$x^4y = 5$

61. $5x^3 + 2y^2 = 40$
$x^3y = 5$

62. $4x^4 + 2xy + 3y^2 = 60$
$x^4y = 8 - 3x^4$

SECTION 8.6 SYSTEMS OF NONLINEAR INEQUALITIES

SKILLS OBJECTIVES

- Graph a nonlinear inequality in two variables.
- Graph a system of nonlinear inequalities in two variables.

CONCEPTUAL OBJECTIVES

- Understand that a nonlinear inequality in two variables may be represented by either a bounded or an unbounded region.
- Interpret an overlapping shaded region as a solution.

Nonlinear Inequalities in Two Variables

Linear inequalities are expressed in the form $Ax + By \le C$. Examples of **nonlinear inequalities in two variables** are

$$9x^2 + 16y^2 \ge 1 \qquad x^2 + y^2 > 1 \qquad y \le -x^2 + 3 \qquad \text{and} \qquad \frac{x^2}{20} - \frac{y^2}{81} < 1$$

We follow the same procedure as we did with linear inequalities. We change the inequality to an equal sign, graph the resulting nonlinear equation, test points from the two regions, and shade the region that makes the inequality true. For strict inequalities, $<$ or $>$, dashed curves are used; and for nonstrict inequalities, $\le$ or $\ge$, solid curves are used.

EXAMPLE 1 **Graphing a Strict Nonlinear Inequality in Two Variables**

Graph the inequality $x^2 + y^2 > 1$.

Solution:

STEP 1 Change the inequality sign to an equal sign. $\qquad x^2 + y^2 = 1$

STEP 2 Draw the graph of the circle.

The center is (0, 0) and the radius is 1.

Since the inequality $>$ is a strict inequality, draw the circle as a **dashed** curve.

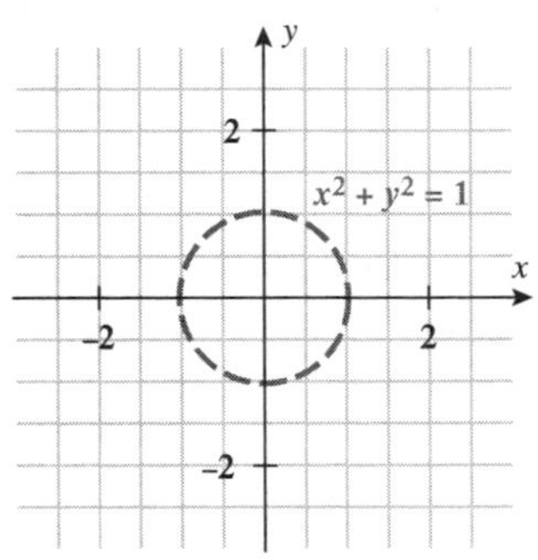

STEP 3 Test points in each region (outside the circle and inside the circle).

Substitute (2, 0) into $x^2 + y^2 > 1$. $\qquad 4 \ge 1$

The point (2, 0) satisfies the inequality.

Substitute (0, 0) into $x^2 + y^2 > 1$. $\qquad 0 \ge 1$

The point (0, 0) does not satisfy the inequality.

Technology Tip

Use TI to graph the inequality $y \le -x^2 + 3$. Enter $y_1 = -x^2 + 3$. For $\le$, use the arrow key to move the cursor to the left of Y_1 and type ENTER until you see ◣.

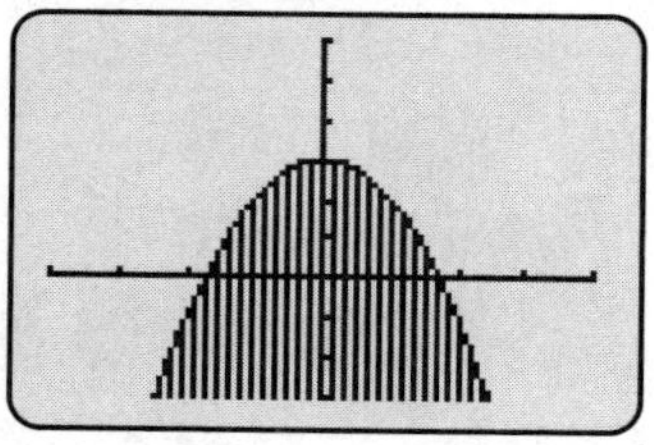

Note: The parabola should be drawn solid.

STEP 4 Shade the region containing the point (2, 0).

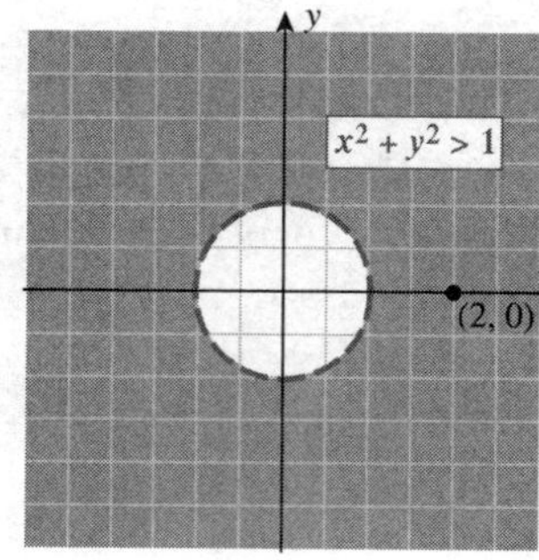

EXAMPLE 2 Graphing a Nonstrict Nonlinear Inequality in Two Variables

Graph the inequality $y \le -x^2 + 3$.

Solution:

STEP 1 Change the inequality sign to an equal sign. $y = -x^2 + 3$

The equation is that of a parabola.

STEP 2 Graph the parabola.

Reflect the base function $f(x) = x^2$ about the x-axis and shift up 3 units. Since the inequality $\le$ is a nonstrict inequality, draw the parabola as a **solid** curve.

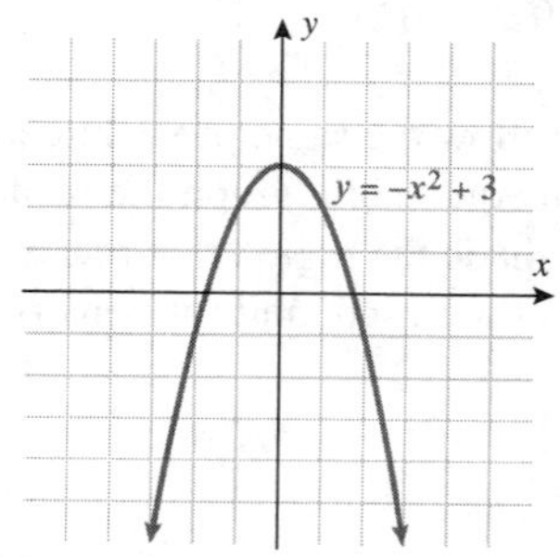

STEP 3 Test points in each region (inside the parabola and outside the parabola).

Substitute (3, 0) into $y \le -x^2 + 3$. $0 \le -6$

The point (3, 0) does not satisfy the inequality.

Substitute (0, 0) into $y \le -x^2 + 3$. $0 \le 3$

The point (0, 0) does satisfy the inequality.

STEP 4 Shade the region containing the point (0, 0).

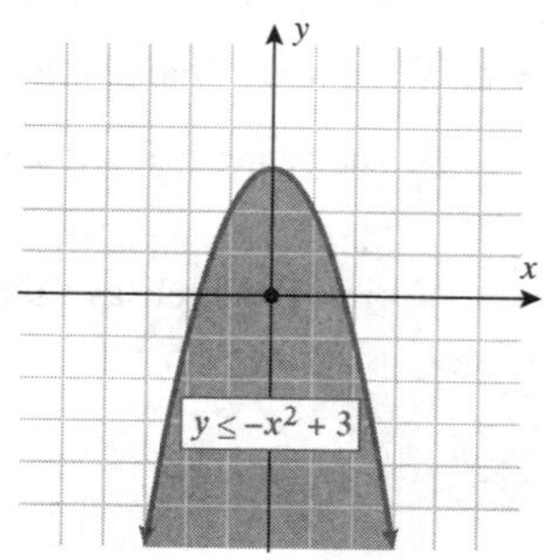

■ YOUR TURN Graph the following inequalities.

a. $x^2 + y^2 \le 9$ **b.** $y > -x^2 + 2$

■ **Answer:**

a.

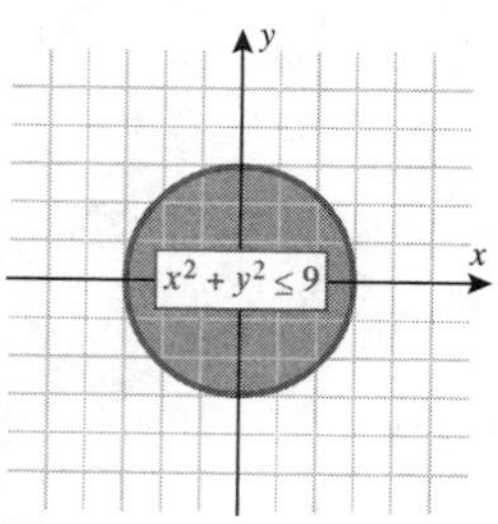

b.

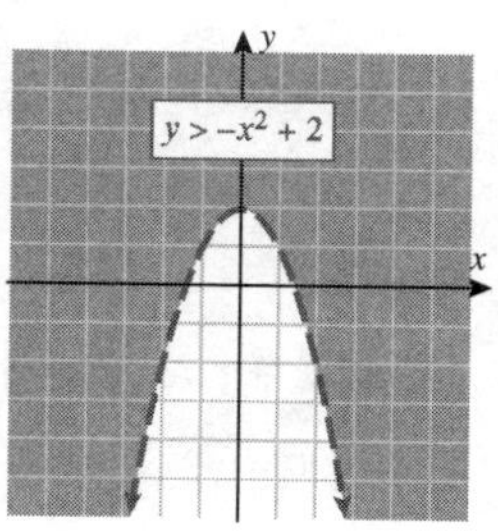

Systems of Nonlinear Inequalities

To solve a system of inequalities, first graph the inequalities and shade the region containing the points that satisfy each inequality. The overlap of all the shaded regions is the solution.

EXAMPLE 3 Graphing a System of Inequalities

Graph the solution to the system of inequalities: $y \geq x^2 - 1$
$y < x + 1$

Solution:

STEP 1 Change the inequality signs to equal signs.

$$y = x^2 - 1$$
$$y = x + 1$$

STEP 2 The resulting equations represent a parabola (to be drawn solid) and a line (to be drawn dashed). Graph the two equations.

To determine the points of intersection, set the y values equal.

$$x^2 - 1 = x + 1$$
$$x^2 - x - 2 = 0$$

Factor. $(x - 2)(x + 1) = 0$

Solve for x. $x = 2$ or $x = -1$

Substitute $x = 2$ into $y = x + 1$. $(2, 3)$

Substitute $x = -1$ into $y = x + 1$. $(-1, 0)$

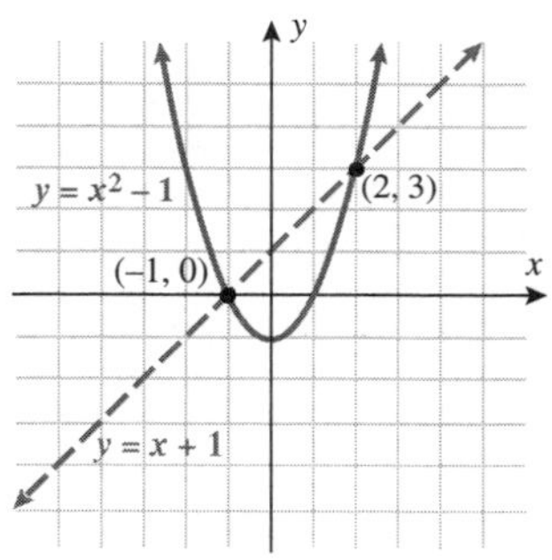

STEP 3 Test points and shade regions.

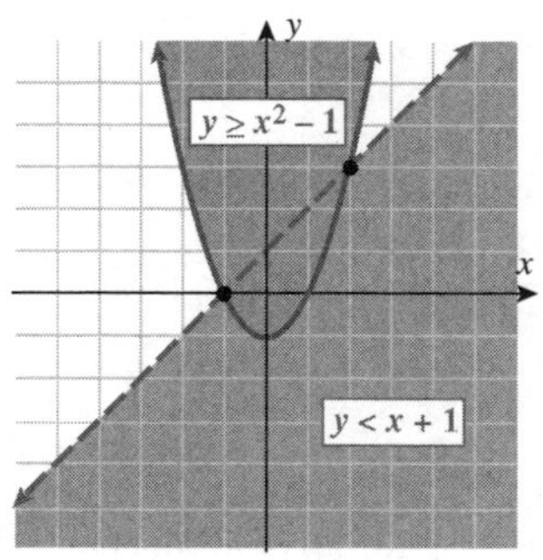

STEP 4 Shade the common region.

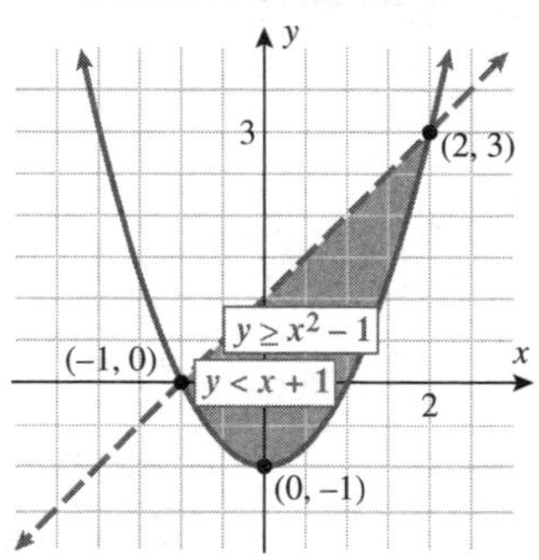

■ YOUR TURN Graph the solution to the system of inequalities: $x^2 + y^2 < 9$
$y > 0$

Technology Tip

Use TI to graph the solution to the system of inequalities $y \geq x^2 - 1$ and $y < x + 1$. First enter $y_1 = x^2 - 1$. For $\geq$, use the arrow key to move the cursor to the left of Y_1 and type ENTER until you see ◥. Next enter $y_2 = x + 1$. For $<$, use the arrow key to move the cursor to the left of Y_2 and type ENTER until you see ◣.

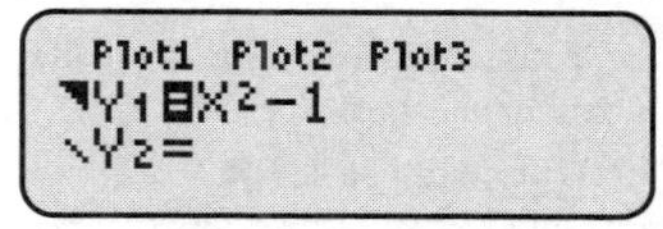

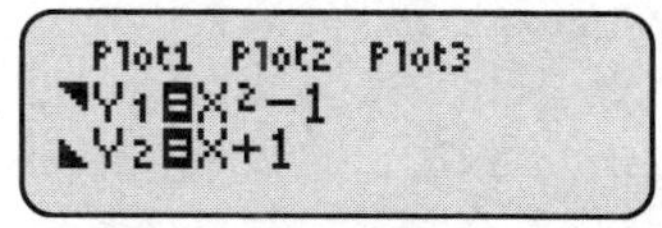

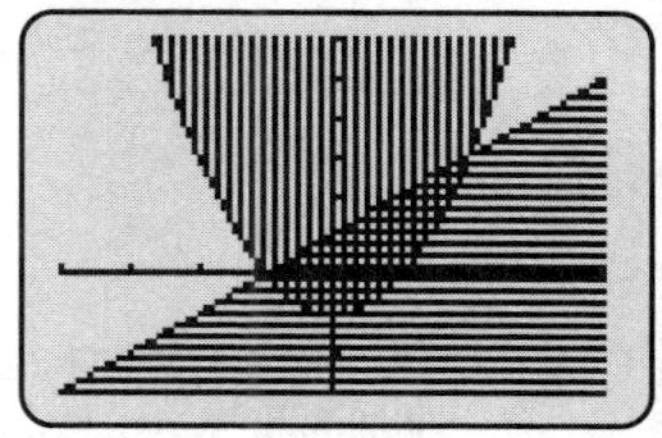

Note: The parabola should be drawn solid and the line should be drawn dashed.

■ **Answer:**

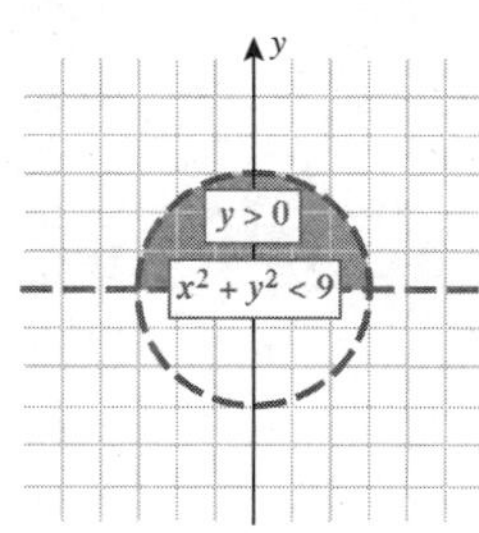

Technology Tip

Use TI to graph the solution to the system of inequalities $x^2 + y^2 < 2$ and $y \geq x^2$. First enter

$y_1 < \sqrt{2 - x^2}$ and

$y_2 > -\sqrt{2 - x^2}$.

Use the arrow key to move the cursor to the left of Y_1 and Y_2 and type ENTER until you see ◣ for < and ◥ for >. Next enter $y_3 = x^2$. For ≥, use the arrow key to move the cursor to the left of Y_3 and type ENTER until you see ◥.

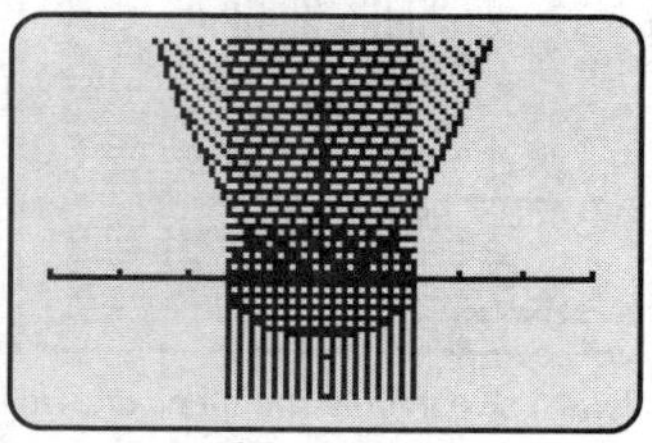

EXAMPLE 4 Solving a System of Inequalities

Solve the system of inequalities: $x^2 + y^2 < 2$
$y \geq x^2$

Solution:

STEP 1 Change the inequality signs to equal signs.

$$x^2 + y^2 = 2$$
$$y = x^2$$

STEP 2 The resulting equations correspond to a circle (to be drawn dashed) and a parabola (to be drawn solid).

To determine the points of intersection, solve the system of equations by substitution.

$$x^3 + (\underbrace{x^2}_{y})^2 = 2$$

$$x^4 + x^2 - 2 = 0$$

Factor. $(x^2 + 2)(x^2 - 1) = 0$

Solve for x. $\underbrace{x^2 = -2}_{\text{no solution}}$ or $\underbrace{x^2 = 1}_{x = \pm 1}$

The points of intersection are $(-1, 1)$ and $(1, 1)$.

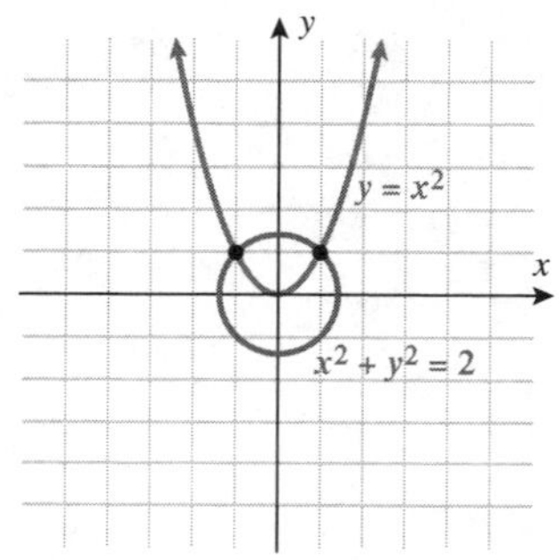

STEP 3 Test points and shade region.

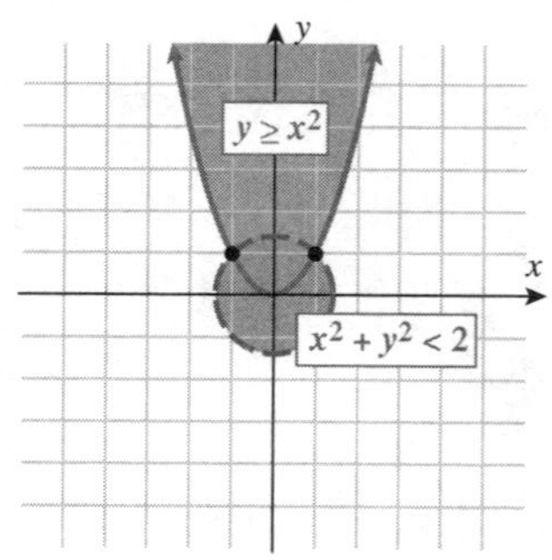

STEP 4 Identify the common region as the solution.

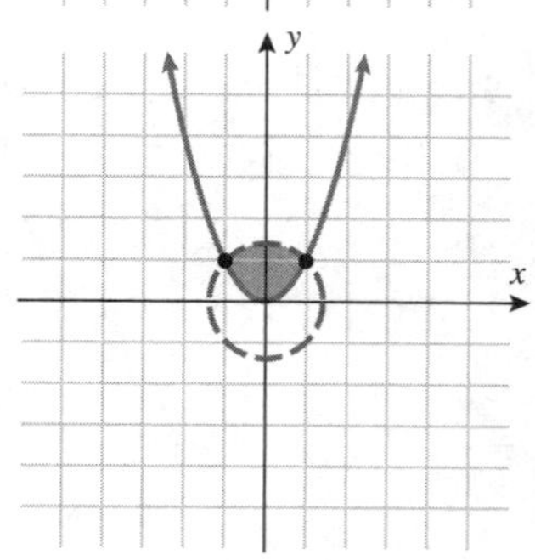

■ **YOUR TURN** Solve the system of inequalities: $x^2 + y^2 < 2$
$y < x^2$

■ **Answer:**

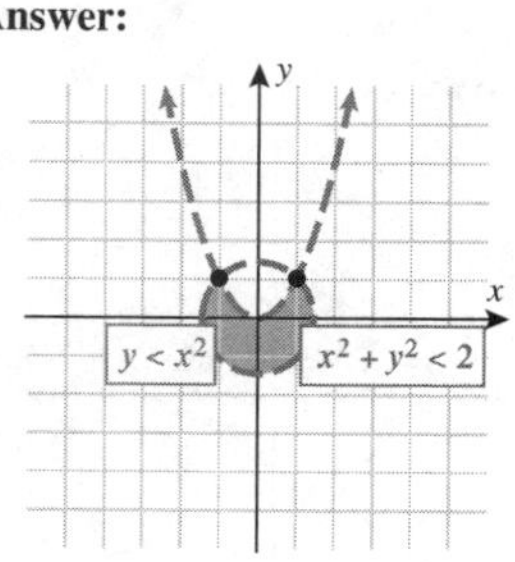

It is important to note that any inequality based on an equation whose graph is not a line is considered a nonlinear inequality.

EXAMPLE 5 Solving a System of Nonlinear Inequalities

Solve the system of inequalities: $(x-1)^2 + \frac{y^2}{4} < 1$

$y \geq \sqrt{x}$

Solution:

STEP 1 Change the inequality signs to equal signs.

$$(x-1)^2 + \frac{y^2}{4} = 1$$
$$y = \sqrt{x}$$

STEP 2 The resulting equations correspond to an ellipse (to be drawn dashed) and the square-root function (to be drawn solid). Graph the two inequalities.

To determine the points of intersection, solve the system of equations by substitution.

$$(x-1)^2 + \frac{\overbrace{(\sqrt{x})}^{y}{}^2}{4} = 1$$

Multiply by 4. $4(x-1)^2 + x = 4$

Expand the binomial squared. $4(x^2 - 2x + 1) + x = 4$

Distribute. $4x^2 - 8x + 4 + x = 4$

Combine like terms and gather terms to one side. $4x^2 - 7x = 0$

Factor. $x(4x - 7) = 0$

Solve for x. $x = 0$ and $x = \frac{7}{4}$

The points of intersection are $(0, 0)$ and $\left(\frac{7}{4}, \sqrt{\frac{7}{4}}\right)$.

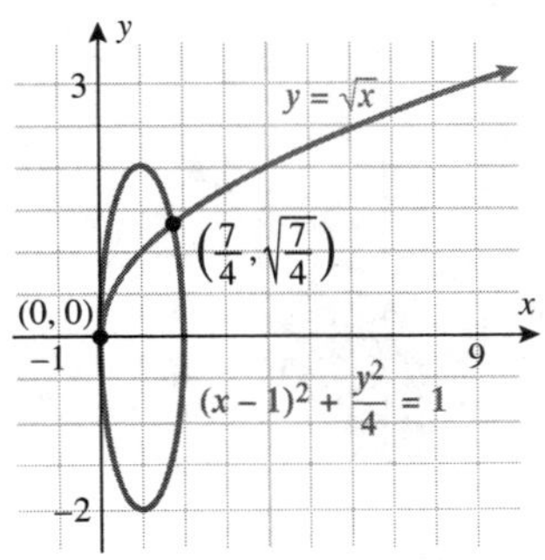

STEP 3 Shade the solution.

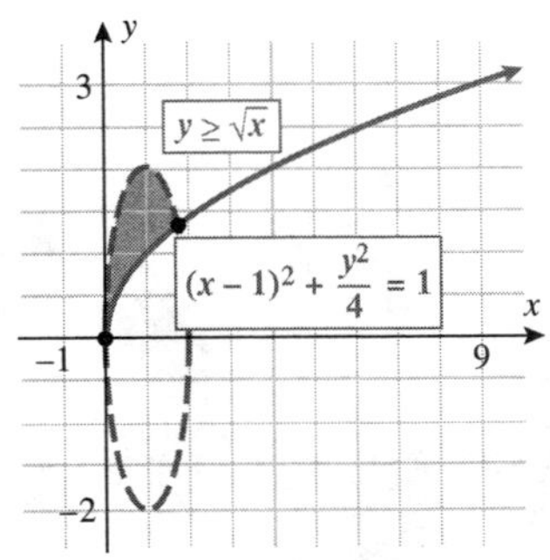

Technology Tip

Use TI to graph the solution to the system of inequalities $(x-1)^2 + \frac{y^2}{4} < 1$ and $y \geq \sqrt{x}$.

First enter $y_1 < 2\sqrt{1-(x-1)^2}$ and $y_2 > -2\sqrt{1-(x-1)^2}$. Use the arrow key to move the cursor to the left of Y_1 and Y_2 and type ENTER until you see ◣ for < and ◥ for >. Next enter $y_3 \geq \sqrt{x}$. For ≥:, use the arrow key to move the cursor to the left of Y_3 and type ENTER until you see ◥.

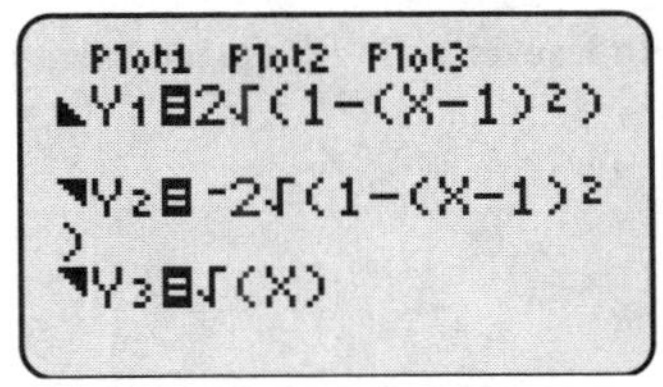

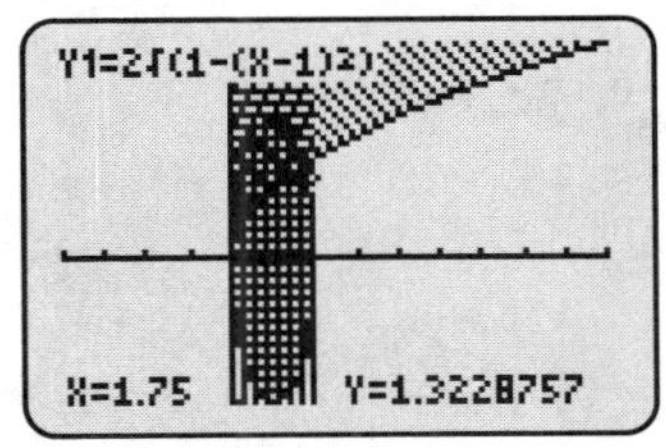

Note: The ellipse should be drawn dashed and the square-root function should be drawn solid.

SECTION 8.6 SUMMARY

In this section we discussed nonlinear inequalities in two variables. Sometimes these result in bounded regions (e.g., $x^2 + y^2 \leq 1$), and sometimes these result in unbounded regions (e.g., $x^2 + y^2 > 1$). When solving systems of inequalities, we first graph each of the inequalities separately and then look for the intersection (overlap) of all shaded regions.

SECTION 8.6 EXERCISES

SKILLS

In Exercises 1–12, match the nonlinear inequality with the correct graph.

1. $x^2 + y^2 < 25$

2. $x^2 + y^2 \leq 9$

3. $\dfrac{x^2}{9} + \dfrac{y^2}{16} \geq 1$

4. $\dfrac{x^2}{4} + \dfrac{y^2}{9} > 1$

5. $y \geq x^2 - 3$

6. $x^2 \geq 16y$

7. $x \geq y^2 - 4$

8. $\dfrac{x^2}{9} + \dfrac{y^2}{25} \geq 1$

9. $9x^2 + 9y^2 < 36$

10. $(x - 2)^2 + (y + 3)^2 \leq 9$

11. $\dfrac{x^2}{4} - \dfrac{y^2}{9} \geq 1$

12. $\dfrac{y^2}{16} - \dfrac{x^2}{9} < 1$

a.

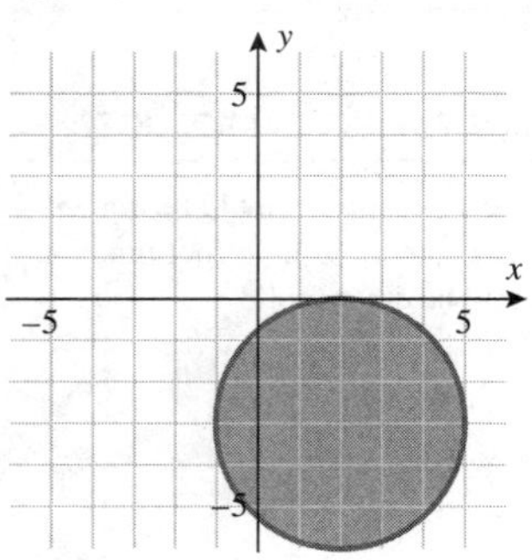

b.

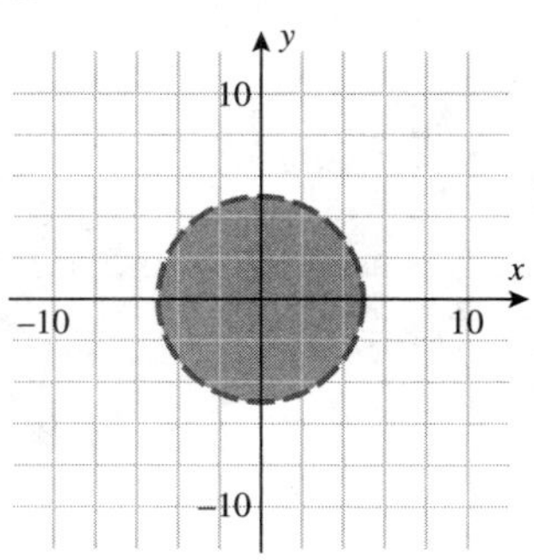

c.

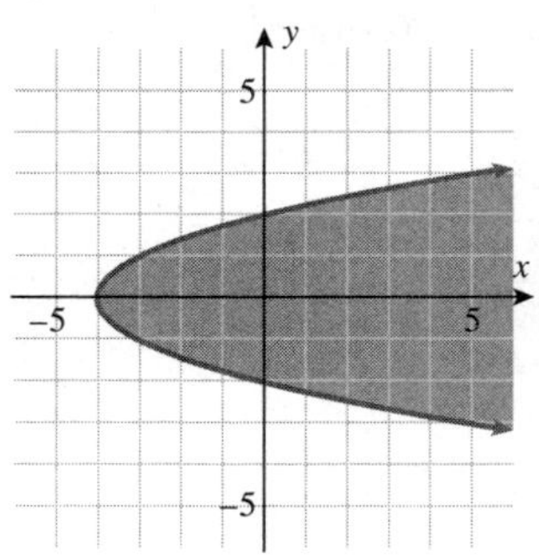

d.

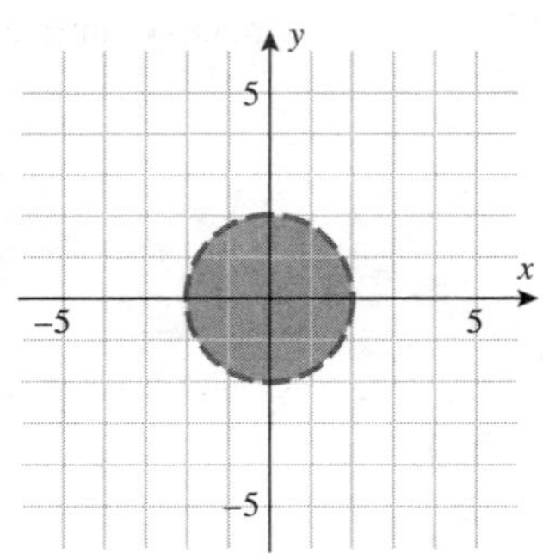

e.

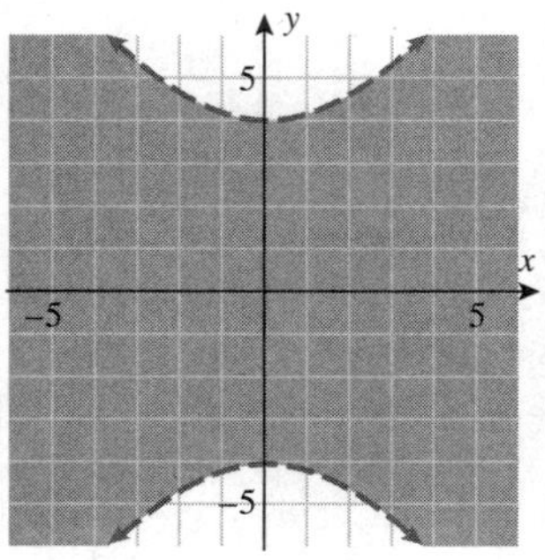

f.

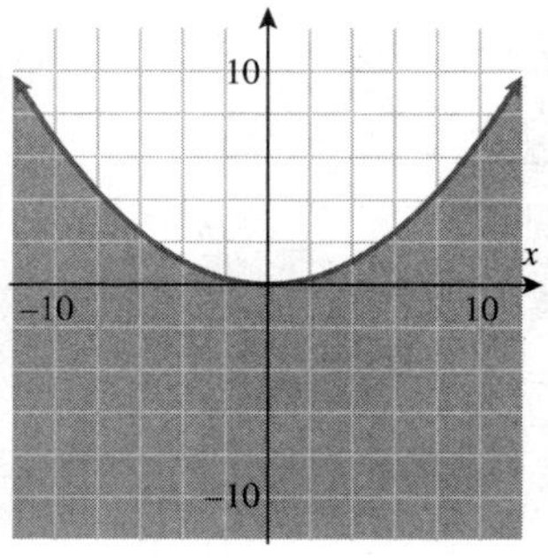

g.

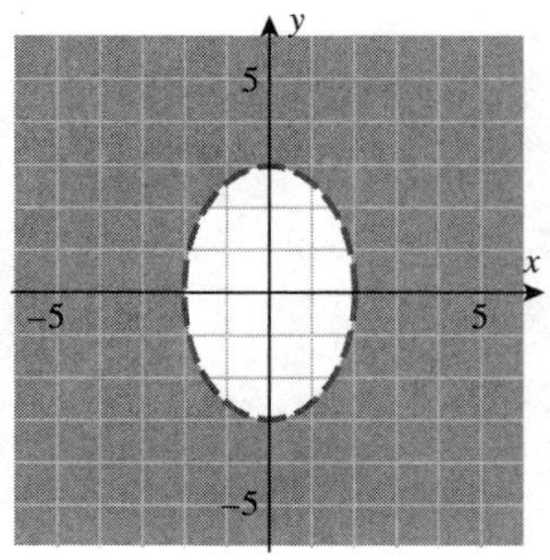

h.

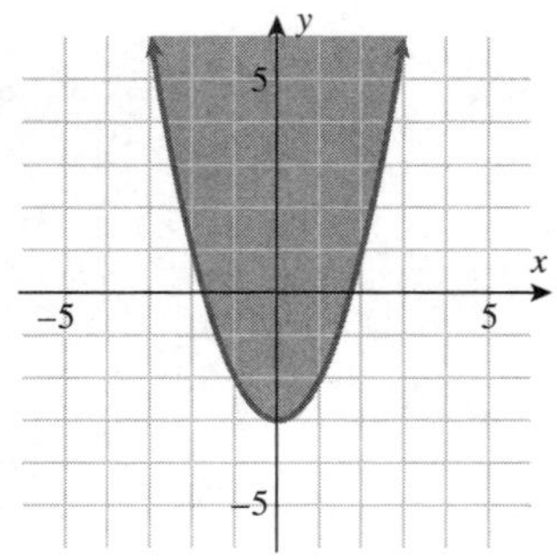

i.

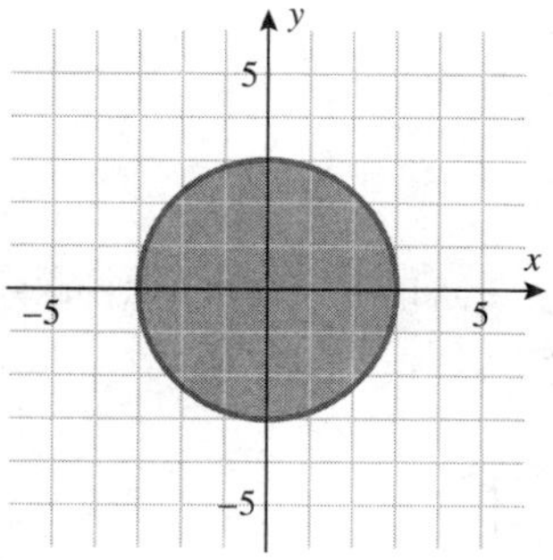

j.

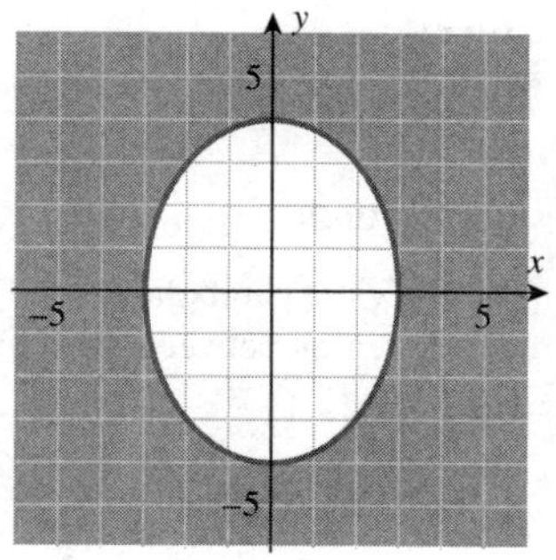

k.

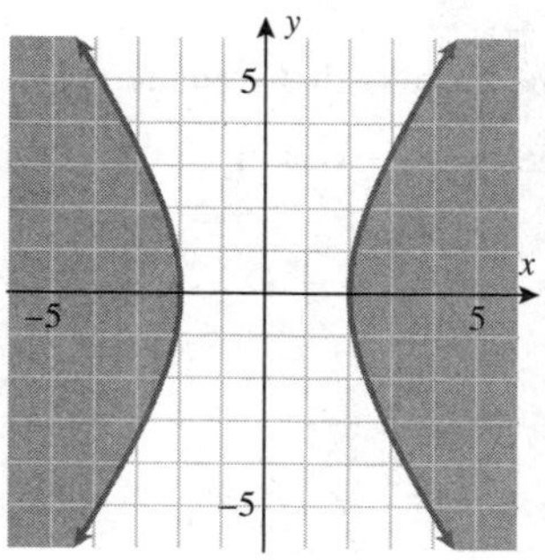

l.

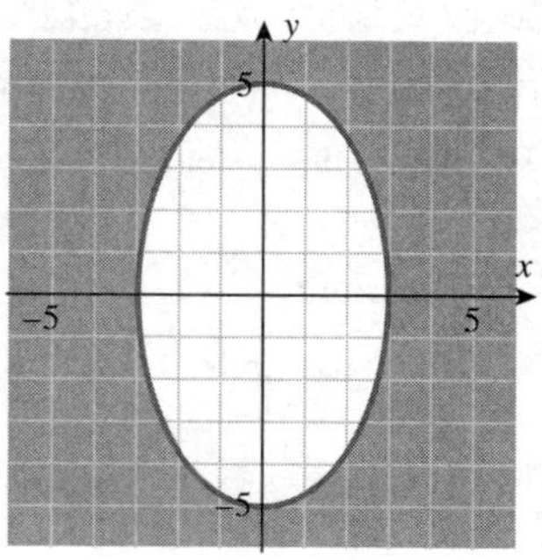

In Exercises 13–30, graph the nonlinear inequality.

13. $y \le x^2 - 2$

14. $y \ge -x^2 + 3$

15. $x^2 + y^2 > 4$

16. $x^2 + y^2 < 16$

17. $x^2 + y^2 - 2x + 4y + 4 \ge 0$

18. $x^2 + y^2 + 2x - 2y - 2 \le 0$

19. $3x^2 + 4y^2 \le 12$

20. $\dfrac{(x-2)^2}{9} + \dfrac{(y+1)^2}{25} > 1$

21. $9x^2 + 16y^2 - 18x + 96y + 9 > 0$

22. $\dfrac{(x-2)^2}{4} - \dfrac{(y+3)^2}{1} \ge 1$

23. $9x^2 - 4y^2 \ge 36$

24. $\dfrac{(y+1)^2}{9} - \dfrac{(x+2)^2}{16} < 1$

25. $36x^2 - 9y^2 \ge 324$

26. $25x^2 - 36y^2 + 200x + 144y - 644 \ge 0$

27. $y \ge e^x$

28. $y \le \ln x$

29. $y < -x^3$

30. $y > -x^4$

In Exercises 31–50, graph each system of inequalities or indicate that the system has no solution.

31. $y < x + 1$
$y \le x^2$

32. $y < x^2 + 4x$
$y \le 3 - x$

33. $y \ge 2 + x$
$y \le 4 - x^2$

34. $y \ge (x-2)^2$
$y \le 4 - x$

35. $y \le -(x+2)^2$
$y > -5 + x$

36. $y \ge (x-1)^2 + 2$
$y \le 10 - x$

37. $-x^2 + y > -1$
$x^2 + y < 1$

38. $x < -y^2 + 1$
$x > y^2 - 1$

39. $y \ge x^2$
$x \ge y^2$

40. $y < x^2$
$x > y^2$

41. $x^2 + y^2 < 36$
$2x + y > 3$

42. $x^2 + y^2 < 36$
$y > 6$

43. $x^2 + y^2 < 25$
$y \ge 6 + x$

44. $(x-1)^2 + (y+2)^2 \le 36$
$y \ge x - 3$

45. $x^2 + y^2 \le 9$
$y \ge 1 + x^2$

46. $x^2 + y^2 \ge 16$
$x^2 + (y-3)^2 \le 9$

47. $x^2 - y^2 < 4$
$y > 1 - x^2$

48. $\dfrac{x^2}{4} - \dfrac{y^2}{9} \le 1$
$y \ge x - 5$

49. $y < e^x$
$y > \ln x \quad x > 0$

50. $y < 10^x$
$y > \log x \quad x > 0$

APPLICATIONS

51. Find the area enclosed by the system of inequalities:
$x^2 + y^2 < 9$
$x > 0$

52. Find the area enclosed by the system of inequalities:
$x^2 + y^2 \le 5$
$x \le 0$
$y \ge 0$

CATCH THE MISTAKE

In Exercises 53 and 54, explain the mistake that is made.

53. Graph the system of inequalities: $x^2 + y^2 < 1$
$x^2 + y^2 > 4$

Solution:

Draw the circles $x^2 + y^2 = 1$ and $x^2 + y^2 = 4$

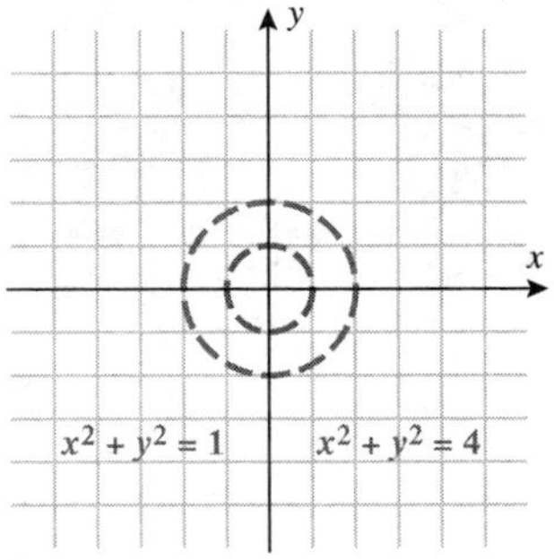

Shade outside $x^2 + y^2 = 1$ and inside $x^2 + y^2 = 4$

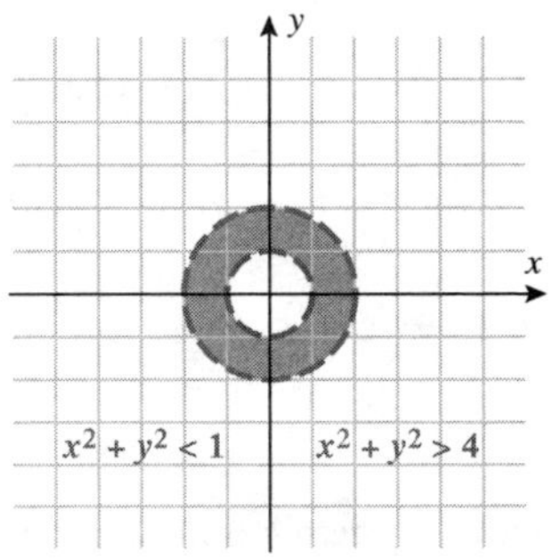

This is incorrect. What mistake was made?

54. Graph the system of inequalities: $x > -y^2 + 1$
$x < y^2 - 1$

Solution:

Draw the parabolas $x = -y^2 + 1$ and $x = y^2 - 1$ and shade the region between the curves.

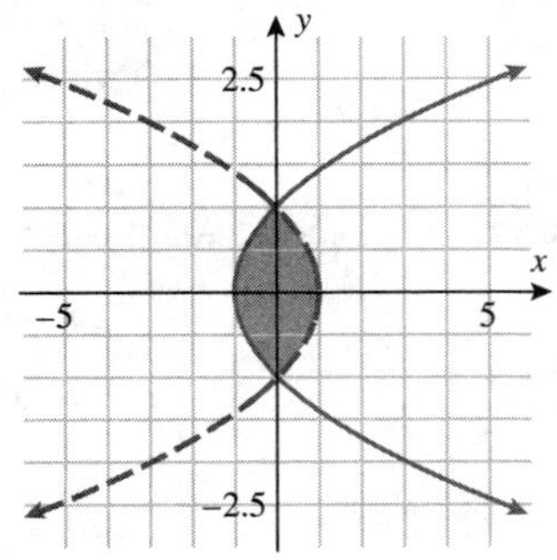

This is incorrect. What mistake was made?

CONCEPTUAL

In Exercises 55 and 56, determine whether each statement is true or false.

55. A nonlinear inequality always represents a bounded region.

56. A system of inequalities always has a solution.

CHALLENGE

57. For the system of nonlinear inequalities $\begin{aligned} x^2 + y^2 &\geq a^2 \\ x^2 + y^2 &\leq b^2 \end{aligned}$, what restriction must be placed on the values of a and b for this system to have a solution? Assume a and b are real numbers.

58. Can $x^2 + y^2 < -1$ ever have a real solution? What types of numbers would x and/or y have to be to satisfy this inequality?

TECHNOLOGY

Use a graphing utility to graph the following inequalities.

59. $x^2 + y^2 - 2x + 4y + 4 \geq 0$

60. $x^2 + y^2 + 2x - 2y - 2 \leq 0$

61. $y \geq e^x$

62. $y \leq \ln x$

63. $y < e^x$
$y > \ln x \quad x > 0$

64. $y < 10^x$
$y > \log x \quad x > 0$

65. $x^2 - 4y^2 + 5x - 6y + 18 \geq 0$

66. $x^2 - 2xy + 4y^2 + 10x - 25 \leq 0$

MODELING YOUR WORLD

In the *Modeling Your World* Feature in Chapters 3, 4, and 5 you used the following data–average yearly temperature in degrees Fahrenheit °F and carbon dioxide emissions in parts per million (ppm) collected by NOAA in **Mauna Loa, Hawaii**–and developed linear (Chapter 3) and nonlinear (Chapters 4 and 5) models.

Year	1960	1965	1970	1975	1980	1985	1990	1995	2000	2005
Temperature	44.45	43.29	43.61	43.35	46.66	45.71	45.53	47.53	45.86	46.23
CO_2 Emissions (ppm)	316.9	320.0	325.7	331.1	338.7	345.9	354.2	360.6	369.4	379.7

1. Solve the system of nonlinear equations governing mean temperature found by using two data points:

Equation (1): Use the linear model developed in Modeling Your World Chapter 3, question 2a.

Equation (2): Use the quadratic model found in Modeling Your World Chapter 4, question 2a.

2. During what year do the models used in Problem 1 agree? Compare the value given by the models that year to the actual data for that year.

3. Solve the system of nonlinear equations governing mean temperature found by using regression (all data points):

Equation (1): Use the linear model developed in Modeling Your World Chapter 3, question 2c.

Equation (2): Use the quadratic model found in Modeling Your World Chapter 4, question 2c.

4. During what year do the models used in Problem 3 agree? Compare the value given by the models that year to the actual data for that year.

5. Solve the system of nonlinear equations governing carbon dioxide emissions found by using two data points:

Equation (1): Use the linear model developed in Modeling Your World Chapter 3, question 7a.

Equation (2): Use the quadratic model found in Modeling Your World Chapter 4, question 7a.

6. During what year do the models used in Problem 5 agree? Compare the value given by the models that year to the actual data for that year.

7. Solve the system of nonlinear equations governing carbon emissions found by using regression (all data points):

Equation (1): Use the linear model developed in Modeling Your World Chapter 3, question 7c.

Equation (2): Use the quadratic model found in Modeling Your World Chapter 4, question 7c.

8. During what year do the models used in Problem 7 agree? Compare the value given by the models that year to the actual data for that year.

SECTION	TOPIC	PAGES	REVIEW EXERCISES	KEY CONCEPTS
8.1	**Conic basics**	638–640	1–4	
	Names of conics	638	1–4	Parabola, ellipse, and hyperbola
	Definitions	638–640	1–4	**Parabola:** Distance to a reference point (focus) and a reference line (directrix) is constant. **Ellipse:** Sum of the distances between the point and two reference points (foci) is constant. **Hyperbola:** Difference of the distances between the point and two reference points (foci) is a constant.
8.2	**The parabola**	641–649	5–24	
	Definition of a parabola	641–649	5–24	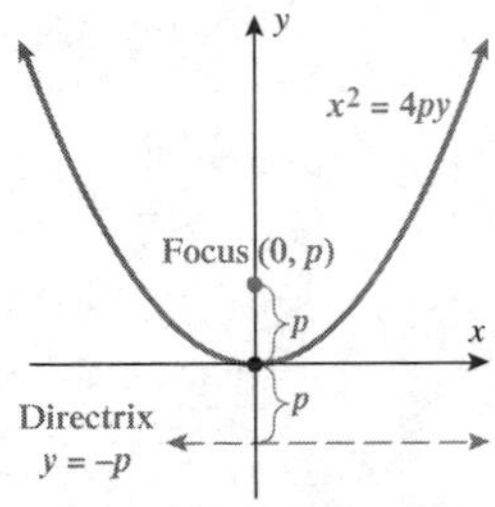 Up: $p > 0$ Down: $p < 0$ 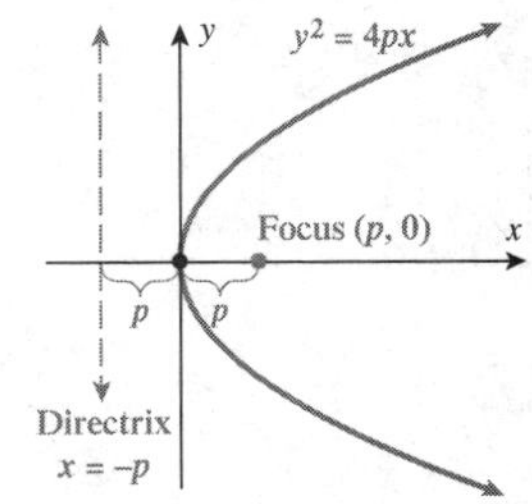 Right: $p > 0$ Left: $p < 0$
8.3	**The ellipse**	654–661	25–40	
	Definition of an ellipse	654–661	25–40	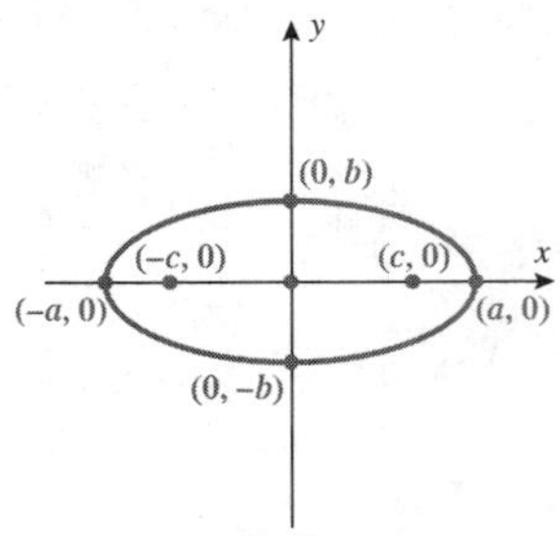 $\frac{x^2}{a^2} + \frac{y^2}{b^2} = 1$ $c^2 = a^2 - b^2$ 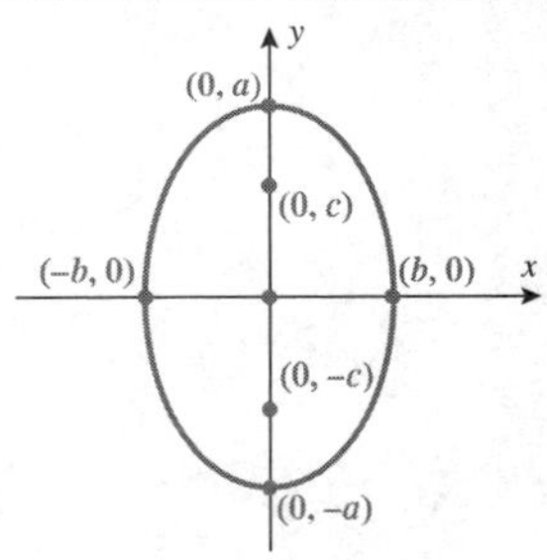 $\frac{x^2}{b^2} + \frac{y^2}{a^2} = 1$ $c^2 = a^2 - b^2$
8.4	**The hyperbola**	666–674	41–56	
	Hyperbolas centered at the origin	667–674	41–56	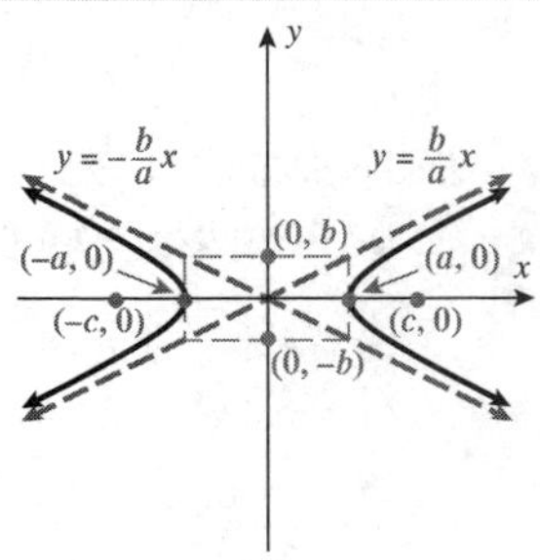 $\frac{x^2}{a^2} - \frac{y^2}{b^2} = 1$ $c^2 = a^2 + b^2$ 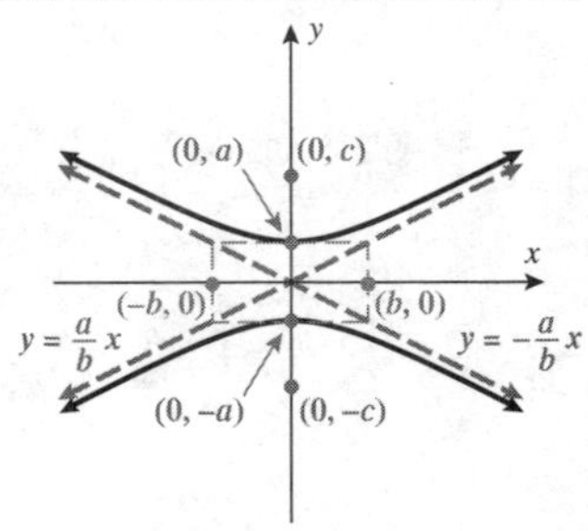 $\frac{y^2}{a^2} - \frac{x^2}{b^2} = 1$ $c^2 = a^2 + b^2$

SECTION	TOPIC	PAGES	REVIEW EXERCISES	KEY CONCEPTS
8.5	**Systems of nonlinear equations**	678–686	57–68	There is no procedure guaranteed to solve nonlinear equations.
	Using elimination to solve systems of nonlinear equations	679–682	57–60	Eliminate a variable by either adding one equation to or subtracting one equation from the other.
	Using substitution to solve systems of nonlinear equations	683–686	61–64	Solve for one variable in terms of the other and substitute into the second equation.
8.6	**Systems of nonlinear inequalities**	689–693	69–80	Solutions are determined graphically by finding the common shaded regions. ■ $\le$ or $\ge$ use solid curves ■ $<$ or $>$ use dashed curves
	Nonlinear inequalities in two variables	689–690	69–74	**Step 1:** Rewrite the inequality as an equation. **Step 2:** Graph the equation. **Step 3:** Test points. **Step 4:** Shade.
	Systems of nonlinear inequalities	690–693	75–80	Graph the individual inequalities and the solution in the common (overlapping) shaded region.

CHAPTER 8 REVIEW EXERCISES

8.1 Conic Basics

Determine whether each statement is true or false.

1. The focus is a point on the graph of the parabola.
2. The graph of $y^2 = 8x$ is a parabola that opens upward.
3. $\dfrac{x^2}{9} - \dfrac{y^2}{1} = 1$ is the graph of a hyperbola that has a horizontal transverse axis.
4. $\dfrac{(x+1)^2}{9} + \dfrac{(y-3)^2}{16} = 1$ is a graph of an ellipse whose center is (1, 3).

8.2 The Parabola

Find an equation for the parabola described.

5. vertex at (0, 0); focus at (3, 0)
6. vertex at (0, 0); focus at (0, 2)
7. vertex at (0, 0); directrix at $x = 5$
8. vertex at (0, 0); directrix at $y = 4$
9. vertex at (2, 3); focus at (2, 5)
10. vertex at $(-1, -2)$; focus at $(1, -2)$
11. focus at (1, 5); directrix at $y = 7$
12. focus at (2, 2); directrix at $x = 0$

Find the focus, vertex, directrix, and length of the latus rectum, and graph the parabola.

13. $x^2 = -12y$
14. $x^2 = 8y$
15. $y^2 = x$
16. $y^2 = -6x$
17. $(y+2)^2 = 4(x-2)$
18. $(y-2)^2 = -4(x+1)$
19. $(x+3)^2 = -8(y-1)$
20. $(x-3)^2 = -8(y+2)$
21. $x^2 + 5x + 2y + 25 = 0$
22. $y^2 + 2y - 16x + 1 = 0$

Applications

23. **Satellite Dish.** A satellite dish measures 10 feet across its opening and 2 feet deep at its center. The receiver should be placed at the focus of the parabolic dish. Where should the receiver be placed?
24. **Clearance Under a Bridge.** A bridge with a parabolic shape reaches a height of 40 feet in the center of the road, and the width of the bridge opening at ground level is 30 feet combined (both lanes). If an RV is 14 feet tall and is 8 feet wide, will it make it through the tunnel?

8.3 The Ellipse

Graph each ellipse.

25. $\dfrac{x^2}{9} + \dfrac{y^2}{64} = 1$
26. $\dfrac{x^2}{81} + \dfrac{y^2}{49} = 1$
27. $25x^2 + y^2 = 25$
28. $4x^2 + 8y^2 = 64$

Find the standard form of an equation of the ellipse with the given characteristics.

29. foci: $(-3, 0)$ and $(3, 0)$ vertices: $(-5, 0)$ and $(5, 0)$

30. foci: $(0, -2)$ and $(0, 2)$ vertices: $(0, -3)$ and $(0, 3)$

31. Major axis vertical with length of 16, minor axis length of 6 and centered at $(0, 0)$.

32. Major axis horizontal with length of 30, minor axis length of 20 and centered at $(0, 0)$.

Graph each ellipse.

33. $\frac{(x-7)^2}{100} + \frac{(y+5)^2}{36} = 1$

34. $20(x+3)^2 + (y-4)^2 = 120$

35. $4x^2 - 16x + 12y^2 + 72y + 123 = 0$

36. $4x^2 - 8x + 9y^2 - 72y + 147 = 0$

Find the standard form of an equation of the ellipse with the given characteristics.

37. foci: $(-1, 3)$ and $(7, 3)$ vertices: $(-2, 3)$ and $(8, 3)$

38. foci: $(1, -3)$ and $(1, -1)$ vertices: $(1, -4)$ and $(1, 0)$

Applications

39. Planetary Orbits. Jupiter's orbit is summarized in the picture. Utilize the fact that the Sun is a focus to determine an equation for Jupiter's elliptical orbit around the Sun. Round to the nearest hundred thousand kilometers.

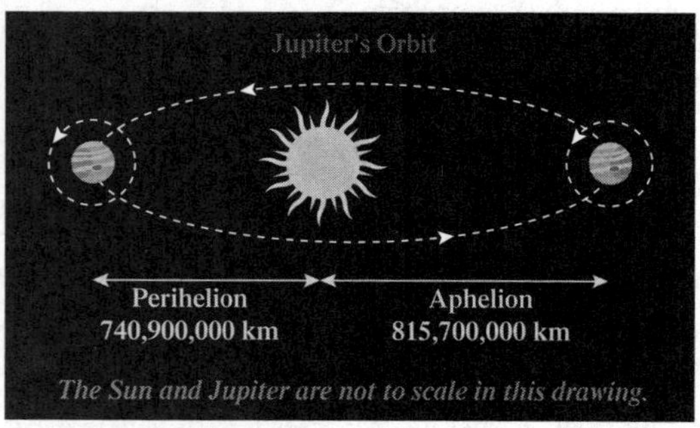

40. Planetary Orbits. Mars's orbit is summarized in the picture below. Utilize the fact that the Sun is a focus to determine an equation for Mars's elliptical orbit around the Sun. Round to the nearest million kilometers.

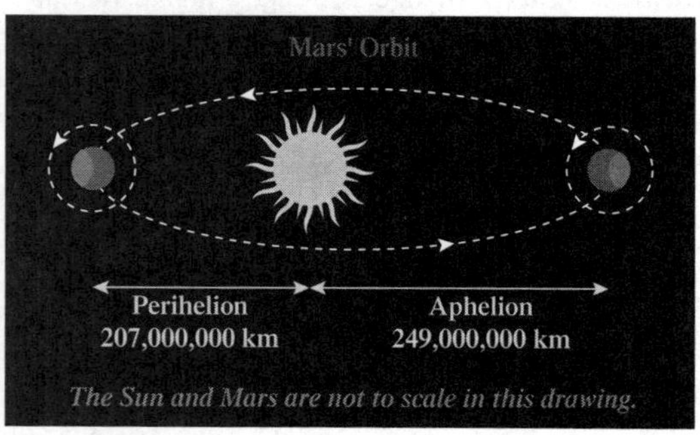

8.4 The Hyperbola

Graph each hyperbola.

41. $\frac{x^2}{9} - \frac{y^2}{64} = 1$

42. $\frac{x^2}{81} - \frac{y^2}{49} = 1$

43. $x^2 - 25y^2 = 25$

44. $8y^2 - 4x^2 = 64$

Find the standard form of an equation of the hyperbola with the given characteristics.

45. vertices: $(-3, 0)$ and $(3, 0)$ foci: $(-5, 0)$ and $(5, 0)$

46. vertices: $(0, -1)$ and $(0, 1)$ foci: $(0, -3)$ and $(0, 3)$

47. center: $(0, 0)$; transverse: y-axis; asymptotes: $y = 3x$ and $y = -3x$

48. center: $(0, 0)$; transverse axis: y-axis; asymptotes: $y = \frac{1}{2}x$ and $y = -\frac{1}{2}x$

Graph each hyperbola.

49. $\frac{(y-1)^2}{36} - \frac{(x-2)^2}{9} = 1$

50. $3(x+3)^2 - 12(y-4)^2 = 72$

51. $8x^2 - 32x - 10y^2 - 60y - 138 = 0$

52. $2x^2 + 12x - 8y^2 + 16y + 6 = 0$

Find the standard form of an equation of the hyperbola with the given characteristics.

53. vertices: $(0, 3)$ and $(8, 3)$ foci: $(-1, 3)$ and $(9, 3)$

54. vertices: $(4, -2)$ and $(4, 0)$ foci: $(4, -3)$ and $(4, 1)$

Applications

55. Ship Navigation. Two loran stations are located 220 miles apart along a coast. If a ship records a time difference of 0.00048 second and continues on the hyperbolic path corresponding to that difference, where would it reach shore? Assume the speed of radio signals is 186,000 miles per second.

56. Ship Navigation. Two loran stations are located 400 miles apart along a coast. If a ship records a time difference of 0.0008 second and continues on the hyperbolic path corresponding to that difference, where would it reach shore?

8.5 Systems of Nonlinear Equations

Solve the system of equations with the elimination method.

57. $x^2 + y = -3$
$x - y = 5$

58. $x^2 + y^2 = 4$
$x^2 + y = 2$

59. $x^2 + y^2 = 5$
$2x^2 - y = 0$

60. $x^2 + y^2 = 16$
$6x^2 + y^2 = 16$

Solve the system of equations with the substitution method.

61. $x + y = 3$
$x^2 + y^2 = 4$

62. $xy = 4$
$x^2 + y^2 = 16$

63. $x^2 + xy + y^2 = -12$
$x - y = 2$

64. $3x + y = 3$
$x - y^2 = -9$

Solve the system of equations by applying any method.

65. $x^3 - y^3 = -19$
$x - y = -1$

66. $2x^2 + 4xy = 9$
$x^2 - 2xy = 0$

67. $\frac{2}{x^2} + \frac{1}{y^2} = 15$
$\frac{1}{x^2} - \frac{1}{y^2} = -3$

68. $x^2 + y^2 = 2$
$x^2 + y^2 = 4$

8.6 Systems of Nonlinear Inequalities

Graph the nonlinear inequality.

69. $y \geq x^2 + 3$

70. $x^2 + y^2 > 16$

71. $y \leq e^x$

72. $y < -x^3 + 2$

73. $y \geq \ln(x - 1)$

74. $9x^2 + 4y^2 \leq 36$

Solve each system of inequalities and shade the region on a graph or indicate that the system has no solution.

75. $y \geq x^2 - 2$
$y \leq -x^2 + 2$

76. $x^2 + y^2 \leq 4$
$y \leq x$

77. $y \geq (x + 1)^2 - 2$
$y \leq 10 - x$

78. $3x^2 + 3y^2 \leq 27$
$y \geq x - 1$

79. $4y^2 - 9x^2 \leq 36$
$y \geq x + 1$

80. $9x^2 + 16y^2 \leq 144$
$y \geq 1 - x^2$

Technology

Section 8.2

81. In your mind, picture the parabola given by $(x - 0.6)^2 = -4(y + 1.2)$. Where is the vertex? Which way does this parabola open? Now plot the parabola using a graphing utility.

82. In your mind, picture the parabola given by $(y - 0.2)^2 = 3(x - 2.8)$. Where is the vertex? Which way does this parabola open? Now plot the parabola using a graphing utility.

Section 8.3

83. Graph the following three ellipses: $4x^2 + y^2 = 1$, $4(2x)^2 + y^2 = 1$, and $4(3x)^2 + y^2 = 1$. What can be said to happen to ellipse $4(cx)^2 + y^2 = 1$ as c increases?

84. Graph the following three ellipses: $x^2 + 4y^2 = 1$, $x^2 + 4(2y)^2 = 1$, and $x^2 + 4(3y)^2 = 1$. What can be said to happen to ellipse $x^2 + 4(cy)^2 = 1$ as c increases?

Section 8.4

85. Graph the following three hyperbolas: $4x^2 - y^2 = 1$, $4(2x)^2 - y^2 = 1$, and $4(3x)^2 - y^2 = 1$. What can be said to happen to hyperbola $4(cx)^2 - y^2 = 1$ as c increases?

86. Graph the following three hyperbolas: $x^2 - 4y^2 = 1$, $x^2 - 4(2y)^2 = 1$, and $x^2 - 4(3y)^2 = 1$. What can be said to happen to hyperbola $x^2 - 4(cy)^2 = 1$ as c increases?

Section 8.5

Use a graphing utility to solve the following systems of equations.

87. $7.5x^2 + 1.5y^2 = 12.25$
$x^2y = 1$

88. $4x^2 + 2xy + 3y^2 = 12$
$x^3y = 3 - 3x^3$

Section 8.6

Use a graphing utility to graph the following systems of nonlinear inequalities.

89. $y \geq 10^x - 1$
$y \leq 1 - x^2$

90. $x^2 + 4y^2 \leq 36$
$y \geq e^x$

Match the equation to the graph.

1. $x = 16y^2$ **2.** $y = 16x^2$ **3.** $x^2 + 16y^2 = 1$

4. $x^2 - 16y^2 = 1$ **5.** $16x^2 + y^2 = 1$ **6.** $16y^2 - x^2 = 1$

a.

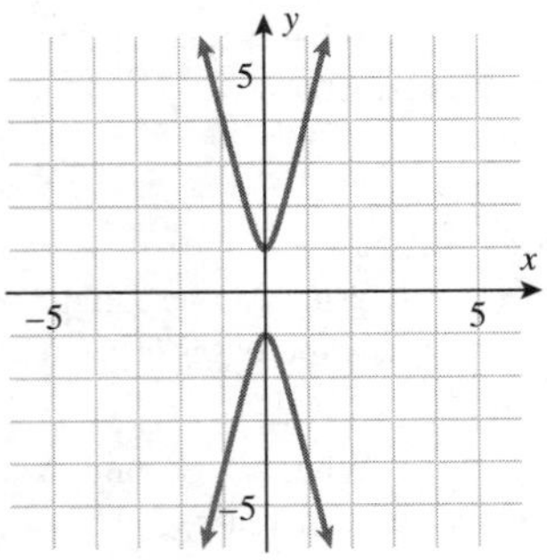

b.

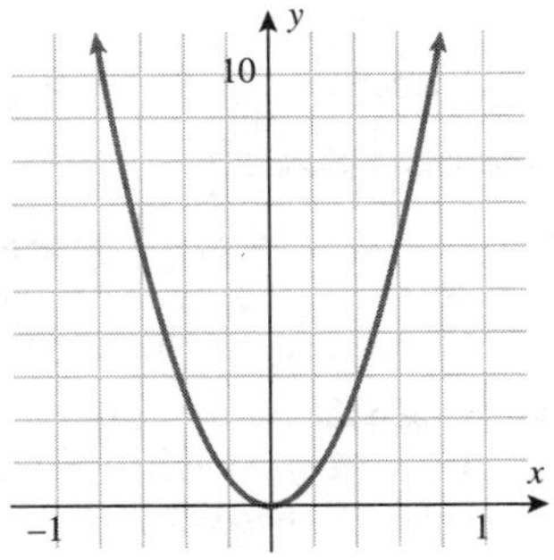

c.

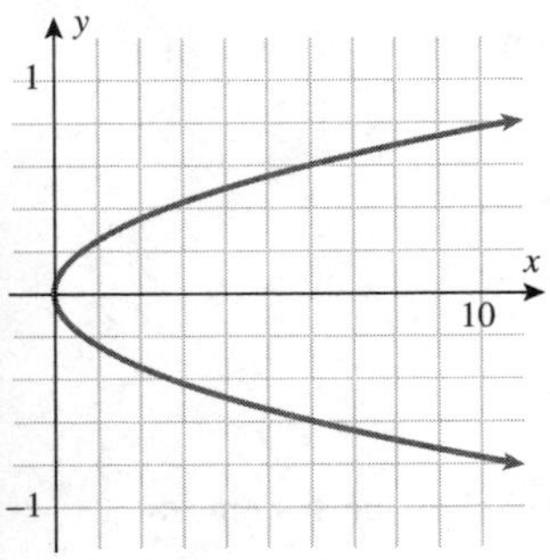

d.

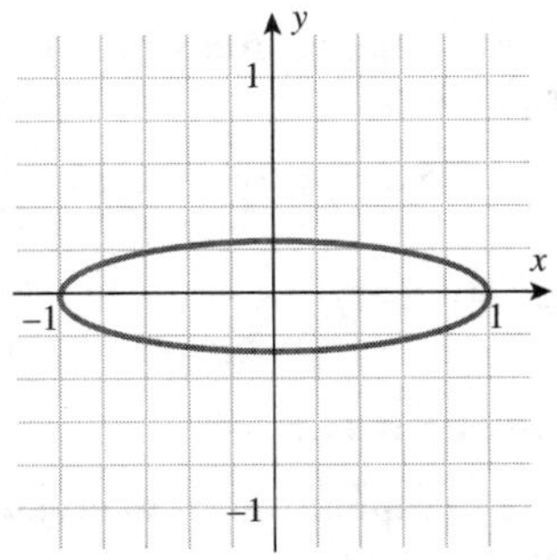

e.

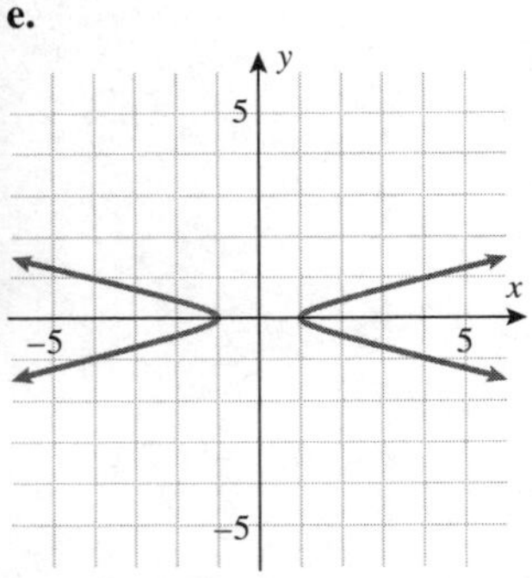

f.

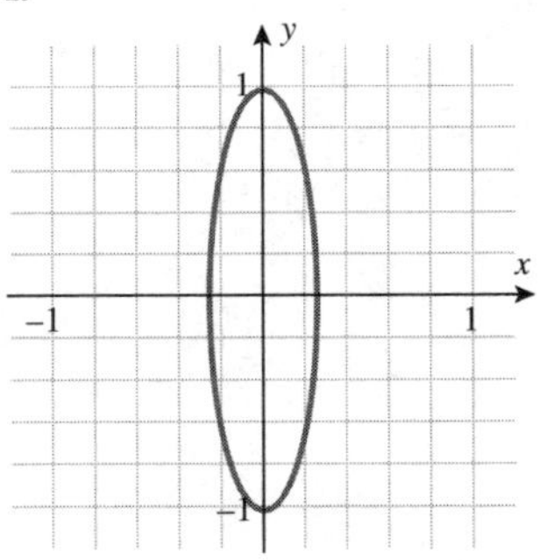

Find the equation of the conic with the given characteristics.

7. parabola vertex: (0, 0) focus: (−4, 0)

8. parabola vertex: (0, 0) directrix: $y = 2$

9. parabola vertex: (−1, 5) focus: (−1, 2)

10. parabola vertex: (2, −3) directrix: $x = 0$

11. ellipse center: (0, 0) vertices: (0, −4), (0, 4) foci: (0, −3), (0, 3)

12. ellipse center: (0, 0) vertices: (−3, 0), (3, 0) foci: (−1, 0), (1, 0)

13. ellipse vertices: (2, −6), (2, 6) foci: (2, −4), (2, 4)

14. ellipse vertices: (−7, −3), (−4, −3) foci: (−6, −3), (−5, −3)

15. hyperbola vertices: (−1, 0) and (1, 0) asymptotes: $y = -2x$ and $y = 2x$

16. hyperbola vertices: (0, −1) and (0, 1) asymptotes: $y = -\frac{1}{3}x$ and $y = \frac{1}{3}x$

17. hyperbola foci: (2, −6), (2, 6) vertices: (2, −4), (2, 4)

18. hyperbola foci: (−7, −3), (−4, −3) vertices: (−6, −3), (−5, −3)

Graph the following equations.

19. $9x^2 + 18x - 4y^2 + 16y - 43 = 0$

20. $4x^2 - 8x + y^2 + 10y + 28 = 0$

21. $y^2 + 4y - 16x + 20 = 0$ **22.** $x^2 - 4x + y + 1 = 0$

23. Eyeglass Lens. Eyeglass lenses can be thought of as very wide parabolic curves. If the focus occurs 1.5 centimeters from the center of the lens, and the lens at its opening is 4 centimeters across, find an equation that governs the shape of the lens.

24. Planetary Orbits. The planet Uranus's orbit is described in the following picture with the Sun as a focus of the elliptical orbit. Write an equation for the orbit.

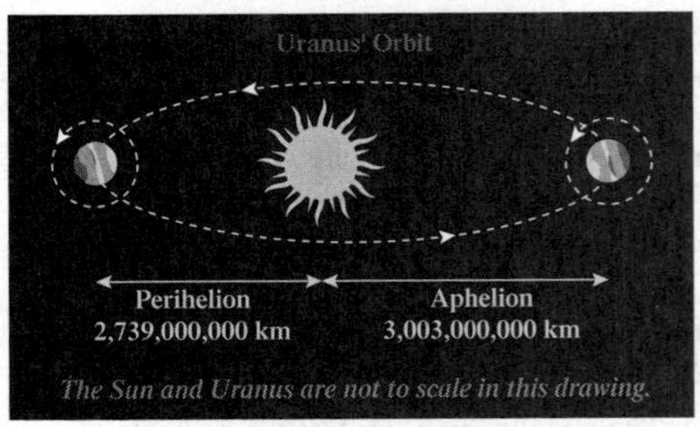

Solve the following systems of nonlinear equations.

25. $x^2 + y = 4$
$-x^2 + y = -4$

26. $x + 3y = 7$
$y = x^2 - 1$

Graph the following nonlinear inequalities.

27. $y < x^3 + 1$ **28.** $y^2 \geq 16x$

Graph the following systems of nonlinear inequalities.

29. $y \leq 4 - x^2$
$16x^2 + 25y^2 \leq 400$

30. $y \leq e^{-x}$
$y \geq x^2 - 4$

31. Use a graphing utility to graph the following nonlinear inequality:

$$x^2 + 4xy - 9y^2 - 6x + 8y + 28 \leq 0$$

32. Use a graphing utility to solve the following systems of equations. Round your answers to three decimal places.

$$0.1225x^2 + 0.0289y^2 = 1$$
$$y^3 = 11x$$

CHAPTERS 1–8 CUMULATIVE TEST

1. Solve using the square root method: $(2 - x)^2 = -16$.

2. Solve for x and express the solution in interval notation: $-5x - 3 \leq 12$.

3. Find the slope of the line that passes through the points $\left(\frac{1}{3}, \frac{3}{4}\right)$ and $\left(-\frac{1}{6}, -\frac{1}{4}\right)$.

4. Transform the equation into standard form by completing the square. State the center and radius of the circle: $x^2 + y^2 - 3.2x + 4.4y - 0.44 = 0$.

5. Applying the function $f(x) = 5 - x^2$, evaluate the difference quotient $\dfrac{f(x+h) - f(x)}{h}$.

6. Given the piecewise-defined function

$$f(x) = \begin{cases} x^2 & x < 0 \\ 2x - 1 & 0 \leq x < 5 \\ 5 - x & x \geq 5 \end{cases}$$

find:

a. $f(0)$ **b.** $f(4)$ **c.** $f(5)$ **d.** $f(-4)$
e. State the domain and range in interval notation.
f. Determine the intervals where the function is increasing, decreasing, or constant.

7. Evaluate $f(g(-1))$ for $g(x) = \sqrt[3]{x}$ and $f(x) = \dfrac{9}{x+1}$.

8. Find a quadratic function that has the vertex (1.5, 2.5) and point (−0.5, −0.5).

9. For the function $f(x) = 5x^2(x+2)^3(x^2+7)$, find all of the real zeros and state the multiplicity.

10. List the possible rational zeros and test to determine all rational zeros for $P(x) = 6x^3 + x^2 - 5x - 2$.

11. Graph the rational function $f(x) = \dfrac{x^2 + 3}{x - 2}$. Give all asymptotes.

12. Graph the exponential function $f(x) = 5^{x-1}$, then state the y-intercept, the domain, the range, and the horizontal asymptote.

13. How much money should be put in a savings account now that earns 5.5% a year compounded continuously if you want to have \$85,000 in 15 years?

14. Graph the function $f(x) = -\log(x) + 2$, employing transformation techniques.

15. Utilize properties of logarithms to simplify the expression $10^{-3\log 10}$.

16. Solve for x: $\ln\sqrt{6 - 3x} - \frac{1}{2}\ln(x+2) = \ln(x)$.

17. At a food court, 3 medium sodas and 2 soft pretzels cost \$6.77. A second order of 5 medium sodas and 4 soft pretzels cost \$12.25. Find the cost of a soda and the cost of a soft pretzel.

18. Find the partial fraction decomposition for the rational expression $\dfrac{3x + 5}{(x-3)(x^2+5)}$.

19. Graph the system of inequalities or indicate that the system has no solution.

$$\begin{aligned} y &\geq 3x - 2 \\ y &\leq 3x + 2 \end{aligned}$$

20. Solve the system by applying Gauss-Jordan elimination.

$$\begin{aligned} x - 2y + z &= 7 \\ -3x + y + 2z &= -11 \end{aligned}$$

21. Given

$$A = \begin{bmatrix} 3 & 4 & -7 \\ 0 & 1 & 5 \end{bmatrix} \quad B = \begin{bmatrix} 8 & -2 & 6 \\ 9 & 0 & -1 \end{bmatrix} \quad C = \begin{bmatrix} 9 & 0 \\ 1 & 2 \end{bmatrix}$$

find $2B - 3A$.

22. Apply Cramer's rule to solve the system of equations.

$$\begin{aligned} 25x + 40y &= -12 \\ 75x - 105y &= 69 \end{aligned}$$

23. Find the standard form of an equation of the ellipse with foci (6, 2) and (6, −6), and vertices (6, 3) and (6, −7).

24. Find the standard form of an equation of the hyperbola with vertices (5, −2) and (5, 0), and foci (5, −3) and (5, 1).

25. Solve the system of equations.

$$\begin{aligned} x + y &= 6 \\ x^2 + y^2 &= 20 \end{aligned}$$

26. Solve the system of inequalities.

$$\begin{aligned} y &\leq x^2 - 4 \\ y &\leq -x^2 + 4 \end{aligned}$$

27. Use a graphing utility to graph the following equation.

$$x^2 - 3xy + 10y^2 - 1 = 0$$

28. Use a graphing utility to graph the following systems of nonlinear inequalities.

$$\begin{aligned} y &\geq e^{-0.3x} - 3.5 \\ y &\leq 4 - x^2 \end{aligned}$$

9

Sequences, Series, and Probability

A BASIC STRATEGY FOR BLACKJACK

← Dealer's Up Card →

Your Hand		2	3	4	5	6	7	8	9	10	A
I	17+	S	S	S	S	S	S	S	S	S	S
	16	S	S	S	S	S	H	H	H	H	H
	15	S	S	S	S	S	H	H	H	H	H
	14	S	S	S	S	S	H	H	H	H	H
	13	S	S	S	S	S	H	H	H	H	H
	12	H	H	S	S	S	H	H	H	H	H
II	11	D	D	D	D	D	D	D	D	D	H
	10	D	D	D	D	D	D	D	D	H	H
	9	H	D	D	D	D	H	H	H	H	H
	5 - 8	H	H	H	H	H	H	H	H	H	H
III	A, 8 - 10	S	S	S	S	S	S	S	S	S	S
	A, 7	S	D	D	D	D	S	S	H	H	H
	A, 6	H	D	D	D	D	H	H	H	H	H
	A, 5	H	H	D	D	D	H	H	H	H	H
	A, 4	H	H	D	D	D	H	H	H	H	H
	A, 3	H	H	H	D	D	H	H	H	H	H
	A, 2	H	H	H	D	D	H	H	H	H	H
IV	A, A; 8, 8	SP	SP	SP	SP	SP	SP	SP	SP	SP	SP
	10, 10	S	S	S	S	S	S	S	S	S	S
	9, 9	SP	SP	SP	SP	SP	S	SP	SP	S	S
	7, 7	SP	SP	SP	SP	SP	SP	H	H	H	H
	6, 6	H	SP	SP	SP	SP	H	H	H	H	H
	5, 5	D	D	D	D	D	D	D	D	H	H
	4, 4	H	H	H	H	H	H	H	H	H	H
	3, 3	H	H	SP	SP	SP	SP	H	H	H	H
	2, 2	H	H	SP	SP	SP	SP	H	H	H	H

When surrender is allowed, surrender 9, 7 or 10, 6 vs 9, 10, A; 9, 6 or 10, 5 vs 10

When doubling down after splitting is allowed, split: 2's, 3's, 7's vs 2-7; 4's vs 5 or 6; 6's vs 2-6

HIT STAND DOUBLE DOWN SPLIT

Have you ever been to a casino and played blackjack? It is the only game in the casino that you can win based on the law of large numbers. In the early 1990s a group of math and science majors from the Massachusetts Institute of Technology (MIT) devised a foolproof scheme to win at blackjack. A professor at MIT developed a basic strategy outlined in the figure on the right that is based on the probability of combinations of particular cards being dealt, given certain cards already showing. To play blackjack (also called 21), each person is dealt two cards with the option of taking additional cards. The goal is to get a combination of cards that is worth 21 points (or less) without going over (called a bust). You have to avoid going over 21 or staying too far below 21. All face cards (jacks, queens, and kings) are worth 10 points, and an ace in blackjack is worth either 1 or 11 points. The students used the professor's strategy along with a card-counting technique to place higher bets when there were more high-value cards left in the deck.

It is reported that in 1992 the team won $4,000,000 from Las Vegas casinos. The casinos caught on, and the students were all banned within 2 years. The 2008 movie "21" was based on this event.

IN THIS CHAPTER, we will discuss counting and probability in addition to three other topics: sequences and series, mathematical induction, and the binomial theorem.

SEQUENCES, SERIES, AND PROBABILITY

SEQUENCES AND SERIES
- Sequence Notation
- Factorial Notation
- Summation Notation
  - Arithmetic series
  - Geometric series

MATHEMATICAL INDUCTION
- Mathematical Proof

BINOMIAL THEOREM
- Raising a Binomial to a Positive Integer Power

COUNTING AND PROBABILITY
- Combinations
- Permutations
- Probability of an Event

SECTION

9.1 SEQUENCES AND SERIES

9.2 ARITHMETIC SEQUENCES AND SERIES

9.3 GEOMETRIC SEQUENCES AND SERIES

9.4 MATHEMATICAL INDUCTION

9.5 THE BINOMIAL THEOREM

9.6 COUNTING, PERMUTATIONS, AND COMBINATIONS

9.7 PROBABILITY

CHAPTER OBJECTIVES

- Understand the difference between sequences and series.
- Find the general, *n*th, term of a sequence or series.
- Understand the difference between finite and infinite series.
- Evaluate a finite series.
- Determine if an infinite series converges or diverges.
- Prove a mathematical statement using induction.
- Use the binomial theorem to expand a binomial raised to a positive integer power.
- Understand the difference between permutations and combinations.
- Calculate the probability of an event.

SECTION

9.1 SEQUENCES AND SERIES

SKILLS OBJECTIVES

- Find terms of a sequence given the general term.
- Look for a pattern in a sequence and find the general term.
- Apply factorial notation.
- Apply recursion formulas.
- Use summation (sigma) notation to represent a series.
- Evaluate a series.

CONCEPTUAL OBJECTIVES

- Understand the difference between a sequence and a series.
- Understand the difference between a finite series and an infinite series.

Sequences

The word *sequence* means an order in which one thing follows another in succession. In mathematics, it means the same thing. For example, if we write $x, 2x^2, 3x^3, 4x^4, 5x^5$, ?, what would the next term in the *sequence* be, the one where the question mark now stands? The answer is $6x^6$.

DEFINITION **A Sequence**

A **sequence** is a function whose domain is a set of positive integers. The function values, or **terms**, of the sequence are written as

$$a_1, a_2, a_3, \ldots, a_n, \ldots$$

Rather than using function notation, sequences are usually written with subscript (or index) notation, $a_{subscript}$.

A **finite sequence** has the domain $\{1, 2, 3, \ldots, n\}$ for some positive integer n. An **infinite sequence** has the domain of all positive integers $\{1, 2, 3, \ldots\}$. There are times when it is convenient to start the indexing at 0 instead of 1:

$$a_0, a_1, a_2, a_3, \ldots, a_n, \ldots$$

Sometimes a pattern in the sequence can be obtained and the sequence can be written using a *general term*. In the previous example, $x, 2x^2, 3x^3, 4x^4, 5x^5, 6x^6, \ldots$, each term has the same exponent and coefficient. We can write this sequence as $a_n = nx^n$, $n = 1, 2, 3, 4, 3, 6, \ldots$, where a_n is called the **general term**.

EXAMPLE 1 **Finding the Sequence, Given the General Term**

Find the first four ($n = 1, 2, 3, 4$) terms of the sequences, given the general term.

a. $a_n = 2n - 1$

b. $b_n = \dfrac{(-1)^n}{n + 1}$

Solution (a): $a_n = 2n - 1$

Find the first term, $n = 1$. $a_1 = 2(1) - 1 = 1$

Find the second term, $n = 2$. $a_2 = 2(2) - 1 = 3$

Find the third term, $n = 3$. $a_3 = 2(3) - 1 = 5$

Find the fourth term, $n = 4$. $a_4 = 2(4) - 1 = 7$

The first four terms of the sequence are $\boxed{1, 3, 5, 7}$.

Solution (b): $b_n = \dfrac{(-1)^n}{n+1}$

Find the first term, $n = 1$. $b_1 = \dfrac{(-1)^1}{1+1} = -\dfrac{1}{2}$

Find the second term, $n = 2$. $b_2 = \dfrac{(-1)^2}{2+1} = \dfrac{1}{3}$

Find the third term, $n = 3$. $b_3 = \dfrac{(-1)^3}{3+1} = -\dfrac{1}{4}$

Find the fourth term, $n = 4$. $b_4 = \dfrac{(-1)^4}{4+1} = \dfrac{1}{5}$

The first four terms of the sequence are $\boxed{-\frac{1}{2}, \frac{1}{3}, -\frac{1}{4}, \frac{1}{5}}$.

■ **YOUR TURN** Find the first four terms of the sequence $a_n = \dfrac{(-1)^n}{n^2}$.

■ **Answer:** $-1, \frac{1}{4}, -\frac{1}{9}, \frac{1}{16}$

EXAMPLE 2 **Finding the General Term, Given Several Terms of the Sequence**

Find the general term of the sequence, given the first five terms.

a. $1, \frac{1}{4}, \frac{1}{9}, \frac{1}{16}, \frac{1}{25}, \ldots$ **b.** $-1, 4, -9, 16, -25, \ldots$

Solution (a):

Write 1 as $\frac{1}{1}$. $\dfrac{1}{1}, \dfrac{1}{4}, \dfrac{1}{9}, \dfrac{1}{16}, \dfrac{1}{25}, \ldots$

Notice that each denominator is an integer squared. $\dfrac{1}{1^2}, \dfrac{1}{2^2}, \dfrac{1}{3^2}, \dfrac{1}{4^2}, \dfrac{1}{5^2}, \ldots$

Identify the general term. $\boxed{a_n = \dfrac{1}{n^2} \quad n = 1, 2, 3, 4, 5, \ldots}$

Solution (b):

Notice that each term includes an integer squared. $-1^2, 2^2, -3^2, 4^2, -5^2, \ldots$

Identify the general term. $\boxed{b_n = (-1)^n n^2 \quad n = 1, 2, 3, 4, 5, \ldots}$

Study Tip

$(-1)^n$ or $(-1)^{n+1}$ is a way to represent an alternating sequence.

■ **YOUR TURN** Find the general term of the sequence, given the first five terms.

a. $-\frac{1}{2}, \frac{1}{4}, -\frac{1}{6}, \frac{1}{8}, -\frac{1}{10}, \ldots$ **b.** $\frac{1}{2}, \frac{1}{4}, \frac{1}{8}, \frac{1}{16}, \frac{1}{32}, \ldots$

■ **Answer:**

a. $a_n = \dfrac{(-1)^n}{2n}$ **b.** $a_n = \dfrac{1}{2^n}$

Parts (b) in both Example 1 and Example 2 are called **alternating** sequences, because the terms alternate signs (positive and negative). If the odd terms, $a_1, a_3, a_5, \ldots$ are negative and the even terms, $a_2, a_4, a_6, \ldots$, are positive, we include $(-1)^n$ in the general term. If the opposite is true, and the odd terms are positive and the even terms are negative, we include $(-1)^{n+1}$ in the general term.

Factorial Notation

Many important sequences that arise in mathematics involve terms that are defined with products of consecutive positive integers. The products are expressed in *factorial notation.*

Technology Tip

Find 0!, 1!, 2!, 3!, 4!, and 5!.

Scientific calculators:

Press	*Display*
0 [!] [=]	1
1 [!] [=]	1
2 [!] [=]	2
3 [!] [=]	6
4 [!] [=]	24
5 [!] [=]	120

Graphing calculators:

Press	*Display*
0 [MATH] [PRB] [4:!] [ENTER]	1
1 [MATH] [PRB] [4:!] [ENTER]	1
2 [MATH] [PRB] [4:!] [ENTER]	2
3 [MATH] [PRB] [4:!] [ENTER]	6
4 [MATH] [PRB] [4:!] [ENTER]	24
5 [MATH] [PRB] [4:!] [ENTER]	120

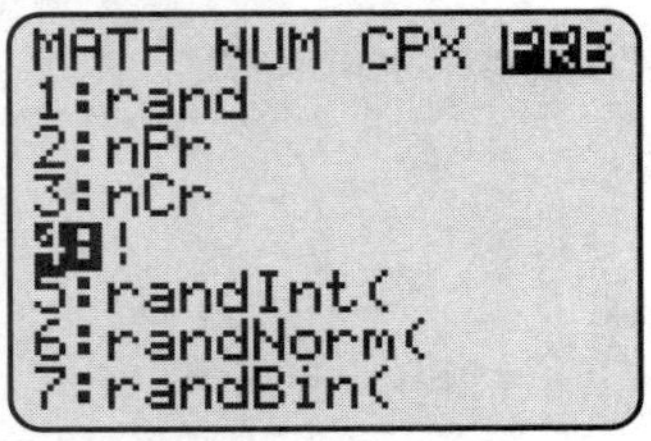

DEFINITION Factorial

If n is a positive integer, then $n!$ (stated as "n factorial") is the product of all positive integers from n down to 1.

$$n! = n(n-1)(n-2)\cdots 3\cdot 2\cdot 1 \qquad n \geq 2$$

and $0! = 1$ and $1! = 1$.

The values of $n!$ for the first six nonnegative integers are

$$0! = 1$$
$$1! = 1$$
$$2! = 2\cdot 1 = 2$$
$$3! = 3\cdot 2\cdot 1 = 6$$
$$4! = 4\cdot 3\cdot 2\cdot 1 = 24$$
$$5! = 5\cdot 4\cdot 3\cdot 2\cdot 1 = 120$$

Notice that $4! = 4\cdot 3\cdot 2\cdot 1 = 4\cdot 3!$. In general, we can apply the formula $n! = n[(n-1)!]$. Often the brackets are not used, and the notation $n! = n(n-1)!$ implies calculating the factorial $(n-1)!$ and then multiplying that quantity by n. For example, to find 6!, we employ the relationship $n! = n(n-1)!$ and set $n = 6$:

$$6! = 6\cdot 5! = 6\cdot 120 = 720$$

EXAMPLE 3 Finding the Terms of a Sequence Involving Factorials

Find the first four terms of the sequence, given the general term $a_n = \frac{x^n}{n!}$.

Solution:

Find the first term, $n = 1$. $\quad a_1 = \frac{x^1}{1!} = x$

Find the second term, $n = 2$. $\quad a_2 = \frac{x^2}{2!} = \frac{x^2}{2\cdot 1} = \frac{x^2}{2}$

Find the third term, $n = 3$. $\quad a_3 = \frac{x^3}{3!} = \frac{x^3}{3\cdot 2\cdot 1} = \frac{x^3}{6}$

Find the fourth term, $n = 4$. $\quad a_4 = \frac{x^4}{4!} = \frac{x^4}{4\cdot 3\cdot 2\cdot 1} = \frac{x^4}{24}$

The first four terms of the sequence are $\boxed{x, \frac{x^2}{2}, \frac{x^3}{6}, \frac{x^4}{24}}$.

EXAMPLE 4 Evaluating Expressions with Factorials

Evaluate each factorial expression.

a. $\dfrac{6!}{2!\cdot 3!}$ **b.** $\dfrac{(n+1)!}{(n-1)!}$

Solution (a):

Expand each factorial in the numerator and denominator. $\dfrac{6!}{2!\cdot 3!} = \dfrac{6\cdot 5\cdot 4\cdot \cancel{3}\cdot \cancel{2}\cdot \cancel{1}}{2\cdot 1\cdot \cancel{3}\cdot \cancel{2}\cdot \cancel{1}}$

Cancel the $3\cdot 2\cdot 1$ in both the numerator and denominator. $= \dfrac{6\cdot 5\cdot 4}{2\cdot 1}$

Simplify. $= \dfrac{6\cdot 5\cdot 2}{1} = 60$

$$\boxed{\dfrac{6!}{2!\cdot 3!} = 60}$$

Solution (b):

Expand each factorial in the numerator and denominator. $\dfrac{(n+1)!}{(n-1)!} = \dfrac{(n+1)(n)(n-1)(n-2)\cdots 3\cdot 2\cdot 1}{(n-1)(n-2)\cdots 3\cdot 2\cdot 1}$

Cancel the $(n-1)(n-2)\cdots 3\cdot 2\cdot 1$ in the numerator and denominator. $\dfrac{(n+1)!}{(n-1)!} = (n+1)(n)$

Alternatively, $\dfrac{(n+1)!}{(n-1)!} = \dfrac{(n+1)(n)\cancel{(n-1)!}}{\cancel{(n-1)!}}$

$$\boxed{\dfrac{(n+1)!}{(n-1)!} = n^2 + n}$$

COMMON MISTAKE

In Example 4 we found $\dfrac{6!}{2!\cdot 3!} = 60$. It is important to note that $2!\cdot 3! \neq 6!$

■ YOUR TURN Evaluate the factorial expressions.

a. $\dfrac{3!\cdot 4!}{2!\cdot 6!}$. **b.** $\dfrac{(n+2)!}{n!}$

Technology Tip

Solve for $\dfrac{6!}{2!3!}$.

Scientific calculators:

Press	*Display*
6 [!] [÷] [(] 2 [!] [x] 3 [!] [)] [=]	60

Graphing calculators:

6 [MATH] [▶] [PRB] [▼] [4: !] [ENTER] [÷] [(] 2 [MATH] [▶] [PRB] [▼] [4: !] [ENTER] [x] 3 [MATH] [▶] [PRB] [▼] [4: !] [ENTER] [)] [ENTER]

```
6!/(2!*3!)
                60
```

Study Tip

In general, $m!n! \neq (mn)!$

■ Answer:

a. $\frac{1}{10}$ **b.** $n^2 + 3n + 2$

Recursion Formulas

Another way to define a sequence is **recursively**, or using a **recursion formula**. The first few terms are listed, and the recursion formula determines the remaining terms based on previous terms. For example, the famous Fibonacci sequence is 1, 1, 2, 3, 5, 8, 13, 21, 34, 55, 89,.... Each term in the Fibonacci sequence is found by adding the previous two terms.

$$1 + 1 = 2 \qquad 1 + 2 = 3 \qquad 2 + 3 = 5$$
$$3 + 5 = 8 \qquad 5 + 8 = 13 \qquad 8 + 13 = 21$$
$$13 + 21 = 34 \qquad 21 + 34 = 55 \qquad 34 + 55 = 89$$

We can define the Fibonacci sequence using a general term:

$$a_1 = 1, a_2 = 1, \text{ and } a_n = a_{n-2} + a_{n-1} \qquad n \geq 3$$

The Fibonacci sequence is found in places we least expect them (for example, pineapples, broccoli, and flowers). The number of petals in a flower is a Fibonacci number. For example, a wild rose has 5 petals, lilies and irises have 3 petals, and daisies have 34, 55, or even 89 petals. The number of spirals in an Italian broccoli is a Fibonacci number (13).

Study Tip

If $a_n = a_{n-1} + a_{n-3}$, then $a_{100} = a_{99} + a_{97}$.

EXAMPLE 5 Using a Recursion Formula to Find a Sequence

Find the first four terms of the sequence: $a_1 = 2$ and $a_n = 2a_{n-1} - 1, n \geq 2$.

Solution:

Write the first term, $n = 1$. $\quad a_1 = 2$

Find the second term, $n = 2$. $\quad a_2 = 2a_1 - 1 = 2(2) - 1 =$

Find the third term, $n = 3$. $\quad a_3 = 2a_2 - 1 = 2(3) - 1 = 5$

Find the fourth term, $n = 4$. $\quad a_4 = 2a_3 - 1 = 2(5) - 1 = 9$

The first four terms of the sequence are $\boxed{2, 3, 5, 9}$.

■ **Answer:** $1, \frac{1}{2}, \frac{1}{12}, \frac{1}{288}$

■ **YOUR TURN** Find the first four terms of the sequence:

$$a_1 = 1 \quad \text{and} \quad a_n = \frac{a_{n-1}}{n!} \quad n \geq 2$$

Sums and Series

When we add the terms in a sequence, the result is a *series.*

DEFINITION Series

Given the infinite sequence $a_1, a_2, a_3, \ldots, a_n, \ldots$ the sum of all of the terms in the infinite sequence is called an **infinite series** and is denoted by

$$a_1 + a_2 + a_3 + \cdots + a_n + \cdots$$

and the sum of only the first n terms is called a **finite series**, or ***n*th partial sum**, and is denoted by

$$S_n = a_1 + a_2 + a_3 + \cdots + a_n$$

Study Tip

Σ is a regular Greek letter, but when used to represent the mathematical sum operation, we oversize it.

Study Tip

When seen in running text we often use the following notation:

$$\sum_{n=1}^{5} = \Sigma_{n=1}^{5}$$

Study Tip

A series can start at any integer (not just 1).

The capital Greek letter Σ (sigma) corresponds to the capital S in our alphabet. Therefore, we use Σ as a shorthand way to represent a sum (series). For example, the sum of the first five terms of the sequence $1, 4, 9, 16, 25, \ldots, n^2, \ldots$ can be represented using **sigma (or summation) notation**:

$$\sum_{n=1}^{5} n^2 = (1)^2 + (2)^2 + (3)^2 + (4)^2 + (5)^2$$

$$= 1 + 4 + 9 + 16 + 25$$

This is read "the sum as n goes from 1 to 5 of n^2." The letter n is called the **index of summation**, and often other letters are used instead of n. It is important to note that the sum can start at other numbers besides 1.

If we wanted the sum of all of the terms in the sequence, we would represent that infinite series using summation notation as

$$\sum_{n=1}^{\infty} n^2 = 1 + 4 + 9 + 16 + 25 + \cdots$$

EXAMPLE 6 Writing a Series Using Sigma Notation

Write the following series using sigma notation.

a. $1 + 1 + \frac{1}{2} + \frac{1}{6} + \frac{1}{24} + \frac{1}{120}$ **b.** $8 + 27 + 64 + 125 + \cdots$

Solution (a):

Write 1 as $\frac{1}{1}$. $\frac{1}{1} + \frac{1}{1} + \frac{1}{2} + \frac{1}{6} + \frac{1}{24} + \frac{1}{120}$

Notice that we can write the denominators using factorials. $= \frac{1}{1} + \frac{1}{1} + \frac{1}{2!} + \frac{1}{3!} + \frac{1}{4!} + \frac{1}{5!}$

Recall that $0! = 1$ and $1! = 1$. $= \frac{1}{0!} + \frac{1}{1!} + \frac{1}{2!} + \frac{1}{3!} + \frac{1}{4!} + \frac{1}{5!}$

Identify the general term. $a_n = \frac{1}{n!} \quad n = 0, 1, 2, 3, 4, 5$

Write the finite series using sigma notation. $\boxed{\sum_{n=0}^{5} \frac{1}{n!}}$

Solution (b):

Write the infinite series as a sum of terms cubed. $8 + 27 + 64 + 125 + \cdots$
$= 2^3 + 3^3 + 4^3 + 5^3 + \cdots$

Identify the general term of the series. $a_n = n^3 \quad n \geq 2$

Write the infinite series using sigma notation. $\boxed{\sum_{n=2}^{\infty} n^3}$

■ YOUR TURN Write the following series using sigma notation.

a. $1 - \frac{1}{2} + \frac{1}{6} - \frac{1}{24} + \frac{1}{120} - \cdots$

b. $4 + 8 + 16 + 32 + 64 + \cdots$

■ Answer:

a. $\sum_{n=1}^{\infty} \frac{(-1)^{n+1}}{n!}$ **b.** $\sum_{n=2}^{\infty} 2^n$

Now that we are comfortable with sigma (summation) notation, let's turn our attention to evaluating a series (calculating the sum). You can always evaluate a finite series. However, you cannot always evaluate an infinite series.

EXAMPLE 7 Evaluating a Finite Series

Evaluate the series $\sum_{i=0}^{4} (2i + 1)$.

Solution:

Write out the partial sum.

$$\sum_{i=0}^{4} (2i + 1) = \underset{(i=0)}{1} + \overset{(i=1)}{3} + \underset{(i=2)}{5} + \overset{(i=3)}{7} + \underset{(i=4)}{9}$$

$$= 25$$

Simplify. $\boxed{\sum_{i=0}^{4} (2i + 1) = 25}$

■ YOUR TURN Evaluate the series $\sum_{n=1}^{5} (-1)^n n$.

■ Answer: -3

Technology Tip

2nd | LIST | ▶ | MATH | ▼ | 5:sum(| ENTER | 2nd | LIST | ▶ | OPS | ▼ | 5:seq(| ENTER | 2 | ALPHA | I | + | 1 | , | ALPHA | I | , | 0 | , | 4 | , | 1 |) |) | ENTER

```
sum(seq(2I+1,I,0
,4,1))
                25
```

Study Tip

The sum of a finite series always exists. The sum of an infinite series may or may not exist.

Infinite series may or may not have a finite sum. For example, if we keep adding $1 + 1 + 1 + 1 + \ldots$, then there is no single real number that the series sums to because the sum continues to grow without bound. However, if we add $0.9 + 0.09 + 0.009 + 0.0009 + \cdots$, this sum is $0.9999\ldots = 0.\overline{9}$, which is a rational number, and it can be proven that $0.\overline{9} = 1$.

Technology Tip

a.

2nd LIST ▶ MATH ▼ 5:sum(ENTER 2nd LIST ▶ OPS ▼ 5:seq(ENTER 3 ÷ 10 ^ ALPHA N , ALPHA N , 1 , 10 , 1)) ENTER

```
sum(seq(3/10^N,N
,1,10,1))
          .3333333333
```

b.

2nd LIST ▶ MATH ▼ 5:sum(ENTER 2nd LIST ▶ OPS ▼ 5:seq(ENTER 3 ÷ 10 ^ ALPHA N , ALPHA N , 1 , 100 , 1)) ENTER

```
sum(seq(N²,N,1,1
00,1))
               338350
```

■ **Answer: a.** Series diverges.
b. Series converges to $\frac{2}{3}$.

EXAMPLE 8 Evaluating an Infinite Series, If Possible

Evaluate the following infinite series, if possible.

a. $\sum_{n=1}^{\infty} \frac{3}{10^n}$ **b.** $\sum_{n=1}^{\infty} n^2$

Solution (a):

Expand the series. $\sum_{n=1}^{\infty} \frac{3}{10^n} = \frac{3}{10} + \frac{3}{100} + \frac{3}{1000} + \frac{3}{10{,}000} + \cdots$

Write in decimal form. $\sum_{n=1}^{\infty} \frac{3}{10^n} = 0.3 + 0.03 + 0.003 + 0.0003 + \cdots$

Calculate the sum. $\sum_{n=1}^{\infty} \frac{3}{10^n} = 0.333333\overline{3} = \frac{1}{3}$

$$\boxed{\sum_{n=1}^{\infty} \frac{3}{10^n} = \frac{1}{3}}$$

Solution (b):

Expand the series. $\sum_{n=1}^{\infty} n^2 = 1 + 4 + 9 + 16 + 25 + 36 + \cdots$

This sum is infinite since it continues to grow without any bound.

In part (a) we say that the series **converges** to $\frac{1}{3}$ and in part (b) the say that the series **diverges**.

■ **YOUR TURN** Evaluate the following infinite series, if possible.

a. $\sum_{n=1}^{\infty} 2n$ **b.** $\sum_{n=1}^{\infty} 6\left(\frac{1}{10}\right)^n$

Applications

The annual sales at Home Depot from 2000 to 2002 can be approximated by the model $a_n = 45.7 + 9.5n - 1.6n^2$, where a_n is the yearly sales in billions of dollars and $n = 0, 1, 2$.

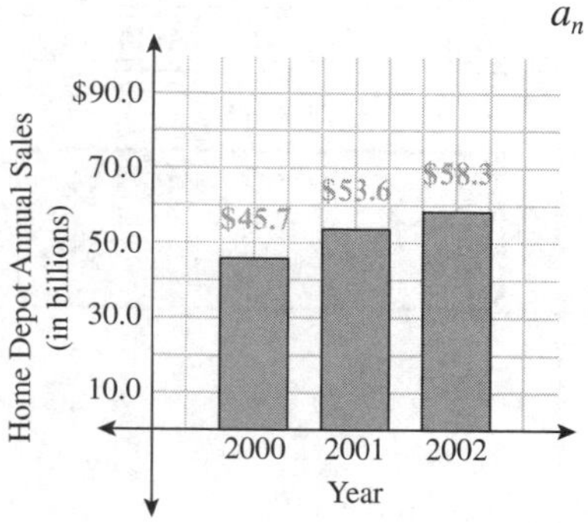

YEAR	n	$a_n = 45.7 + 9.5n - 1.6n^2$	TOTAL SALES IN BILLIONS
2000	0	$a_0 = 45.7 + 9.5(0) - 1.6(0)^2$	\$45.7
2001	1	$a_1 = 45.7 + 9.5(1) - 1.6(1)^2$	\$53.6
2002	2	$a_2 = 45.7 + 9.5(2) - 1.6(2)^2$	\$58.3

What does the finite series $\frac{1}{3}\sum_{n=0}^{2} a_n$ tell us? It tells us the average yearly sales over 3 years.

SECTION 9.1 SUMMARY

In this section we discussed finite and infinite sequences and series. When the terms of a sequence are added together, the result is a series.

Finite sequence: $a_1, a_2, a_3, \ldots, a_n$
Infinite sequence: $a_1, a_2, a_3, \ldots, a_n, \ldots$

Finite series: $a_1 + a_2 + a_3 + \cdots + a_n$
Infinite series: $a_1 + a_2 + a_3 + \cdots + a_n + \cdots$

Factorial notation was also introduced:

$n! = n \cdot (n-1) \cdot \cdots \cdot 3 \cdot 2 \cdot 1 \quad n \geq 2$

and $0! = 1$ and $1! = 1$.

The sum of a finite series is always finite.

The sum of an infinite series is either:

- convergent or
- divergent

Sigma notation is used to express a series.

- **Finite series:** $\sum_{i=1}^{n} a_i = a_1 + a_2 + \cdots + a_n$
- **Infinite series:** $\sum_{n=1}^{\infty} a_n = a_1 + a_2 + a_3 + \cdots$

SECTION 9.1 EXERCISES

SKILLS

In Exercises 1–12, write the first four terms of the sequence. Assume n starts at 1.

1. $a_n = n$
2. $a_n = n^2$
3. $a_n = 2n - 1$
4. $a_n = x^n$
5. $a_n = \dfrac{n}{(n+1)}$
6. $a_n = \dfrac{(n+1)}{n}$
7. $a_n = \dfrac{2^n}{n!}$
8. $a_n = \dfrac{n!}{(n+1)!}$
9. $a_n = (-1)^n x^{n+1}$
10. $a_n = (-1)^{n+1} n^2$
11. $a_n = \dfrac{(-1)^n}{(n+1)(n+2)}$
12. $a_n = \dfrac{(n-1)^2}{(n+1)^2}$

In Exercises 13–20, find the indicated term of the sequence.

13. $a_n = \left(\dfrac{1}{2}\right)^n \quad a_9 = ?$
14. $a_n = \dfrac{n}{(n+1)^2} \quad a_{15} = ?$
15. $a_n = \dfrac{(-1)^n n!}{(n+2)!} \quad a_{19} = ?$
16. $a_n = \dfrac{(-1)^{n+1}(n-1)(n+2)}{n} \quad a_{13} = ?$
17. $a_n = \left(1 + \dfrac{1}{n}\right)^2 \quad a_{100} = ?$
18. $a_n = 1 - \dfrac{1}{n^2} \quad a_{10} = ?$
19. $a_n = \log 10^n \quad a_{23} = ?$
20. $a_n = e^{\ln n} \quad a_{49} = ?$

In Exercises 21–28, write an expression for the nth term of the given sequence.

21. $2, 4, 6, 8, 10, \ldots$
22. $3, 6, 9, 12, 15, \ldots$
23. $\dfrac{1}{2 \cdot 1}, \dfrac{1}{3 \cdot 2}, \dfrac{1}{4 \cdot 3}, \dfrac{1}{5 \cdot 4}, \dfrac{1}{6 \cdot 5}, \ldots$
24. $\dfrac{1}{2}, \dfrac{1}{4}, \dfrac{1}{8}, \dfrac{1}{16}, \dfrac{1}{32}, \ldots$
25. $-\dfrac{2}{3}, \dfrac{4}{9}, -\dfrac{8}{27}, \dfrac{16}{81}, \ldots$
26. $\dfrac{1}{2}, \dfrac{3}{4}, \dfrac{9}{8}, \dfrac{27}{16}, \dfrac{81}{32}, \ldots$
27. $1, -1, 1, -1, 1, \ldots$
28. $\dfrac{1}{3}, -\dfrac{2}{4}, \dfrac{3}{5}, -\dfrac{4}{6}, \dfrac{5}{7}, \ldots$

In Exercises 29–40, simplify the ratio of factorials.

29. $\dfrac{9!}{7!}$ **30.** $\dfrac{4!}{6!}$ **31.** $\dfrac{29!}{27!}$ **32.** $\dfrac{32!}{30!}$ **33.** $\dfrac{75!}{77!}$ **34.** $\dfrac{100!}{103!}$

35. $\dfrac{97!}{93!}$ **36.** $\dfrac{101!}{98!}$ **37.** $\dfrac{(n-1)!}{(n+1)!}$ **38.** $\dfrac{(n+2)!}{n!}$ **39.** $\dfrac{(2n+3)!}{(2n+1)!}$ **40.** $\dfrac{(2n+2)!}{(2n-1)!}$

In Exercises 41–50, write the first four terms of the sequence defined by the recursion formula. Assume the sequence begins at 1.

41. $a_1 = 7 \quad a_n = a_{n-1} + 3$ **42.** $a_1 = 2 \quad a_n = a_{n-1} + 1$ **43.** $a_1 = 1 \quad a_n = n \cdot a_{n-1}$

44. $a_1 = 2 \quad a_n = (n+1) \cdot a_{n-1}$ **45.** $a_1 = 100 \quad a_n = \dfrac{a_{n-1}}{n!}$ **46.** $a_1 = 20 \quad a_n = \dfrac{a_{n-1}}{n^2}$

47. $a_1 = 1, a_2 = 2 \quad a_n = a_{n-1} \cdot a_{n-2}$ **48.** $a_1 = 1, a_2 = 2 \quad a_n = \dfrac{a_{n-2}}{a_{n-1}}$

49. $a_1 = 1, a_2 = -1 \quad a_n = (-1)^n\left[a_{n-1}^2 + a_{n-2}^2\right]$ **50.** $a_1 = 1, a_2 = -1 \quad a_n = (n-1)\,a_{n-1} + (n-2)a_{n-2}$

In Exercises 51–64, evaluate the finite series.

51. $\sum_{n=1}^{5} 2$ **52.** $\sum_{n=1}^{5} 7$ **53.** $\sum_{n=0}^{4} n^2$ **54.** $\sum_{n=1}^{4} \dfrac{1}{n}$ **55.** $\sum_{n=1}^{6} (2n-1)$

56. $\sum_{n=1}^{6} (n+1)$ **57.** $\sum_{n=0}^{4} 1^n$ **58.** $\sum_{n=0}^{4} 2^n$ **59.** $\sum_{n=0}^{3} (-x)^n$ **60.** $\sum_{n=0}^{3} (-x)^{n+1}$

61. $\sum_{k=0}^{5} \dfrac{2^k}{k!}$ **62.** $\sum_{k=0}^{5} \dfrac{(-1)^k}{k!}$ **63.** $\sum_{k=0}^{4} \dfrac{x^k}{k!}$ **64.** $\sum_{k=0}^{4} \dfrac{(-1)^k x^k}{k!}$

In Exercises 65–68, evaluate the infinite series, if possible.

65. $\sum_{j=0}^{\infty} 2 \cdot (0.1)^j$ **66.** $\sum_{j=0}^{\infty} 5 \cdot \left(\frac{1}{10}\right)^j$ **67.** $\sum_{j=0}^{\infty} n^j \quad n \geq 1$ **68.** $\sum_{j=0}^{\infty} 1^j$

In Exercises 69–76, apply sigma notation to write the sum.

69. $1 - \dfrac{1}{2} + \dfrac{1}{4} - \dfrac{1}{8} + \cdots + \dfrac{1}{64}$ **70.** $1 + \dfrac{1}{2} + \dfrac{1}{4} + \dfrac{1}{8} + \cdots + \dfrac{1}{64} + \cdots$

71. $1 - 2 + 3 - 4 + 5 - 6 + \cdots$ **72.** $1 + 2 + 3 + 4 + 5 + \cdots + 21 + 22 + 23$

73. $\dfrac{2 \cdot 1}{1} + \dfrac{3 \cdot 2 \cdot 1}{1} + \dfrac{4 \cdot 3 \cdot 2 \cdot 1}{2 \cdot 1} + \dfrac{5 \cdot 4 \cdot 3 \cdot 2 \cdot 1}{3 \cdot 2 \cdot 1} + \dfrac{6 \cdot 5 \cdot 4 \cdot 3 \cdot 2 \cdot 1}{4 \cdot 3 \cdot 2 \cdot 1}$

74. $1 + \dfrac{2}{1} + \dfrac{2^2}{2 \cdot 1} + \dfrac{2^3}{3 \cdot 2 \cdot 1} + \dfrac{2^4}{4 \cdot 3 \cdot 2 \cdot 1} + \cdots$

75. $1 - x + \dfrac{x^2}{2} - \dfrac{x^3}{6} + \dfrac{x^4}{24} - \dfrac{x^5}{120} + \cdots$

76. $x + x^2 + \dfrac{x^3}{2} + \dfrac{x^4}{6} + \dfrac{x^5}{24} + \dfrac{x^6}{120}$

APPLICATIONS

77. Money. Upon graduation Jessica receives a commission from the U.S. Navy to become an officer and a $20,000 signing bonus for selecting aviation. She puts the entire bonus in an account that earns 6% interest compounded monthly. The balance in the account after n months is

$$A_n = 20{,}000\left(1 + \frac{0.06}{12}\right)^n \qquad n = 1, 2, 3, \ldots$$

Her commitment to the Navy is 6 years. Calculate A_{72}. What does A_{72} represent?

78. Money. Dylan sells his car in his freshman year and puts $7,000 in an account that earns 5% interest compounded quarterly. The balance in the account after n quarters is

$$A_n = 7000\left(1 + \frac{0.05}{4}\right)^n \qquad n = 1, 2, 3, \ldots$$

Calculate A_{12}. What does A_{12} represent?

79. Salary. An attorney is trying to calculate the costs associated with going into private practice. If she hires a paralegal to assist her, she will have to pay the paralegal $20.00 per hour. To be competitive with most firms, she will have to give her paralegal a $2 per hour raise per year. Find a general term of a sequence a_n, which represents the hourly salary of a paralegal with n years of experience. What will be the paralegal's salary with 20 years of experience?

80. NFL Salaries. A player in the NFL typically has a career that lasts 3 years. The practice squad makes the league minimum of $275,000 (2004) in the first year, with a $75,000 raise per year. Write the general term of a sequence a_n that represents the salary of an NFL player making the league minimum during his entire career. Assuming $n = 1$ corresponds to the first year, what does $\sum_{n=1}^{3} a_n$ represent?

81. Salary. Upon graduation Sheldon decides to go to work for a local police department. His starting salary is $30,000 per year, and he expects to get a 3% raise per year. Write the recursion formula for a sequence that represents his salary n years on the job. Assume $n = 0$ represents his first year making $30,000.

82. *Escherichia coli.* A single cell of bacteria reproduces through a process called binary fission. *Escherichia coli* cells divide into two every 20 minutes. Suppose the same rate of division is maintained for 12 hours after the original cell enters the body. How many *E. coli* bacteria cells would be in the body 12 hours later? Suppose there is an infinite nutrient source so that the *E. coli* bacteria maintain the same rate of division for 48 hours after the original cell enters the body. How many *E. coli* bacteria cells would be in the body 48 hours later?

83. AIDS/HIV. A typical person has 500 to 1500 T cells per drop of blood in the body. HIV destroys the T cell count at a rate of 50–100 cells per drop of blood per year, depending on how aggressive it is in the body. Generally, the onset of AIDS occurs once the body's T cell count drops below 200. Write a sequence that represents the total number of T cells in a person infected with HIV. Assume that before infection the person has a 1000 T cell count ($a_1 = 1000$) and the rate at which the infection spreads corresponds to a loss of 75 T cells per drop of blood per year. How much time will elapse until this person has full-blown AIDS?

84. Company Sales. Lowe's reported total sales from 2003 through 2004 in the billions. The sequence $a_n = 3.8 + 1.6n$ represents the total sales in billions of dollars. Assuming $n = 3$ corresponds to 2003, what were the reported sales in 2003 and 2004? What does $\frac{1}{2} \cdot \sum_{n=3}^{4} a_n$ represent?

85. Cost of Eating Out. A college student tries to save money by bringing a bag lunch instead of eating out. He will be able to save $100 per month. He puts the money into his savings account, which draws 1.2% interest and is compounded monthly. The balance in his account after n months of bagging his lunch is

$$A_n = 100{,}000[(1.001)^n - 1] \qquad n = 1, 2, \ldots$$

Calculate the first four terms of this sequence. Calculate the amount after 3 years (36 months).

86. Cost of Acrylic Nails. A college student tries to save money by growing her own nails out and not spending $50 per month on acrylic fills. She will be able to save $50 per month. She puts the money into her savings account, which draws 1.2% interest and is compounded monthly. The balance in her account after n months of natural nails is

$$A_n = 50{,}000[(1.001)^n - 1] \qquad n = 1, 2, \ldots$$

Calculate the first four terms of this sequence. Calculate the amount after 4 years (48 months).

87. Math and Engineering. The formula $\sum_{n=0}^{\infty} \frac{x^n}{n!}$ can be used to approximate the function $y = e^x$. Compute the first five terms of this formula to approximate e^x. Apply the result to find e^2 and compare this result with the calculator value of e^2.

88. Home Prices. If the inflation rate is 3.5% per year and the average price of a home is $195,000, the average price of a home after n years is given by $A_n = 195{,}000(1.035)^n$. Find the average price of the home after 6 years.

89. Approximating Functions. Polynomials can be used to approximate transcendental functions such as $\ln(x)$ and e^x, which are found in advanced mathematics and engineering. For example, $\sum_{n=0}^{\infty}(-1)^n \frac{(x-1)^{n+1}}{n+1}$ can be used to approximate $\ln(x)$, where x is close to 1. Use the first five terms of the series to approximate $\ln(x)$. Next, find $\ln(1.1)$ and compare with the value given by your calculator.

90. Future Value of an Annuity. The future value of an ordinary annuity is given by the formula $FV = PMT[((1 + i)^n - 1)/i]$, where PMT = amount paid into the account at the end of each period, i = interest rate per period, and n = number of compounding periods. If you invest $5,000 at the end of each year for 5 years, you will have an accumulated value of FV as given in the above equation at the end of the nth year. Determine how much is in the account at the end of each year for the next 5 years if $i = 0.06$.

CATCH THE MISTAKE

In Exercises 91–94, explain the mistake that is made.

91. Simplify the ratio of factorials: $\frac{(3!)(5!)}{6!}$.

Solution:

Express 6! in factored form. $\frac{(3!)(5!)}{(3!)(2!)}$

Cancel the 3! in the numerator and denominator. $\frac{(5!)}{(2!)}$

Write out the factorials. $\frac{5\cdot4\cdot3\cdot2\cdot1}{2\cdot1}$

Simplify. $5\cdot4\cdot3 = 60$

$\frac{(3!)(5!)}{(3!)(2!)} \neq 60$. What mistake was made?

92. Simplify the factorial expression: $\frac{2n(2n-2)!}{(2n+2)!}$.

Solution:

Express factorials in factored form.

$$\frac{2n(2n-2)(2n-4)(2n-6)\cdots}{(2n+2)(2n)(2n-2)(2n-4)(2n-6)\cdots}$$

Cancel common terms.

$$\frac{1}{2n+2}$$

This is incorrect. What mistake was made?

93. Find the first four terms of the sequence defined by $a_n = (-1)^{n+1}n^2$.

Solution:

Find the $n = 1$ term. $a_1 = -1$

Find the $n = 2$ term. $a_2 = 4$

Find the $n = 3$ term. $a_3 = -9$

Find the $n = 4$ term. $a_4 = 16$

The sequence $-1, 4, -9, 16, \ldots$, is incorrect. What mistake was made?

94. Evaluate the series $\sum_{k=0}^{3}(-1)^k k^2$.

Solution:

Write out the sum. $\sum_{k=0}^{3}(-1)^k k^2 = -1 + 4 - 9$

Simplify the sum. $\sum_{k=6}^{3}(-1)^k k^2 = -6$

This is incorrect. What mistake was made?

CONCEPTUAL

In Exercises 95–98, determine whether each statement is true or false.

95. $\sum_{k=0}^{n} cx^k = c\sum_{k=0}^{n} x^k$

96. $\sum_{i=1}^{n} a_i + b_i = \sum_{i=1}^{n} a_i + \sum_{i=1}^{n} b_i$

97. $\sum_{k=1}^{n} a_k b_k = \sum_{k=1}^{n} a_k \cdot \sum_{k=1}^{n} b_k$

98. $(a!)(b!) = (ab)!$

CHALLENGE

99. Write the first four terms of the sequence defined by the recursion formula:

$$a_1 = C \qquad a_n = a_{n-1} + D \qquad D \neq 0$$

100. Write the first four terms of the sequence defined by the recursion formula:

$$a_1 = C \qquad a_n = Da_{n-1} \qquad D \neq 0$$

101. Fibonacci Sequence. An explicit formula for the nth term of the Fibonacci sequence is:

$$F_n = \frac{(1+\sqrt{5})^n - (1-\sqrt{5})^n}{2^n\sqrt{5}}$$

Apply algebra (not your calculator) to find the first two terms of this sequence and verify that these are indeed the first two terms of the Fibonacci sequence.

102. Let $a_n = \sqrt{a_{n-1}}$ for $n \geq 2$ and $a_1 = 7$. Find the first five terms of this sequence and make a generalization for the nth term.

TECHNOLOGY

103. The sequence defined by $a_n = \left(1 + \frac{1}{n}\right)^n$ approaches the number e as n gets large. Use a graphing calculator to find a_{100}, a_{1000}, $a_{10,000}$, and keep increasing n until the terms in the sequence approach 2.7183.

104. The Fibonacci sequence is defined by $a_1 = 1$, $a_2 = 1$, and $a_n = a_{n-2} + a_{n-1}$ for $n \geq 3$. The ratio $\frac{a_{n+1}}{a_n}$ is an approximation of the golden ratio. The ratio approaches a constant ϕ (phi) as n gets large. Find the golden ratio using a graphing utility.

105. Use a graphing calculator "SUM" to sum $\sum_{k=0}^{5} \frac{2^k}{k!}$. Compare it with your answer to Exercise 61.

106. Use a graphing calculator "SUM" to sum $\sum_{k=0}^{5} \frac{(-1)^k}{k!}$. Compare it with your answer to Exercise 62.

SECTION 9.2 ARITHMETIC SEQUENCES AND SERIES

SKILLS OBJECTIVES

- Recognize an arithmetic sequence.
- Find the general *n*th term of an arithmetic sequence.
- Evaluate a finite arithmetic series.
- Use arithmetic sequences and series to model real-world problems.

CONCEPTUAL OBJECTIVE

- Understand the difference between an arithmetic sequence and an arithmetic series.

Arithmetic Sequences

The word *arithmetic* (with emphasis on the third syllable) often implies adding or subtracting of numbers. *Arithmetic sequences* are sequences whose terms are found by adding a constant to each previous term. The sequence 1, 3, 5, 7, 9, . . . is arithmetic because each successive term is found by adding 2 to the previous term.

DEFINITION **Arithmetic Sequences**

A sequence is **arithmetic** if each term in the sequence is found by adding a real number d to the previous term, so that $a_{n+1} = a_n + d$. Because $a_{n+1} - a_n = d$, the number d is called the **common difference**.

EXAMPLE 1 **Identifying the Common Difference in Arithmetic Sequences**

Determine whether each sequence is arithmetic. If so, find the common difference for each of the arithmetic sequences.

a. 5, 9, 13, 17, . . . **b.** 18, 9, 0, −9, . . . **c.** $\frac{1}{2}, \frac{5}{4}, 2, \frac{11}{4}, \ldots$

Solution (a):

Label the terms. $a_1 = 5, a_2 = 9, a_3 = 13, a_4 = 17, \ldots$

Find the difference $d = a_{n+1} - a_n$. $d = a_2 - a_1 = 9 - 5 = \boxed{4}$

Check that the difference for the next two terms is also 4.

$d = a_3 - a_2 = 13 - 9 = 4$

$d = a_4 - a_3 = 17 - 13 = 4$

There is a common difference of $\boxed{4}$. Therefore, this sequence is arithmetic and each successive term is found by adding 4 to the previous term.

Study Tip

We check several terms to confirm that a series is arithmetic.

Solution (b):

Label the terms. $a_1 = 18, a_2 = 9, a_3 = 0, a_4 = -9, \ldots$

Find the difference $d = a_{n+1} - a_n$. $d = a_2 - a_1 = 9 - 18 = \boxed{-9}$

Check that the difference for the next two terms is also −9.

$d = a_3 - a_2 = 0 - 9 = -9$

$d = a_4 - a_3 = -9 - 0 = -9$

There is a common difference of $\boxed{-9}$. Therefore, this sequence is arithmetic and each successive term is found by subtracting 9 from (that is, adding −9 to) the previous term.

Solution (c):

Label the terms. $a_1 = \frac{1}{2}, a_2 = \frac{5}{4}, a_3 = 2, a_4 = \frac{11}{4}, \ldots$

Find the difference $d = a_{n+1} - a_n$. $d = a_2 - a_1 = \frac{5}{4} - \frac{1}{2} = \boxed{\frac{3}{4}}$

Check that the difference for the next two terms is also $\frac{3}{4}$.

$d = a_3 - a_2 = 2 - \frac{5}{4} = \frac{3}{4}$

$d = a_4 - a_3 = \frac{11}{4} - 2 = \frac{3}{4}$

There is a common difference of $\boxed{\frac{3}{4}}$. Therefore, this sequence is arithmetic and each successive term is found by adding $\frac{3}{4}$ to the previous term.

YOUR TURN Find the common difference for each of the arithmetic sequences.

a. 7, 2, −3, −8, . . . **b.** $1, \frac{5}{3}, \frac{7}{3}, 3, \ldots$

Answer: **a.** −5 **b.** $\frac{2}{3}$

The General (*n*th) Term of an Arithmetic Sequence

To find a formula for the general, or *n*th, term of an arithmetic sequence, write out the first several terms and look for a pattern.

First term, $n = 1$.	a_1
Second term, $n = 2$.	$a_2 = a_1 + d$
Third term, $n = 3$.	$a_3 = a_2 + d = (a_1 + d) + d = a_1 + 2d$
Fourth term, $n = 4$.	$a_4 = a_3 + d = (a_1 + 2d) + d = a_1 + 3d$

In general, the *n*th term is given by $a_n = a_1 + (n - 1)d$.

THE *n*TH TERM OF AN ARITHMETIC SEQUENCE

The ***n*th term** of an arithmetic sequence with common difference *d* is given by

$$a_n = a_1 + (n - 1)d \qquad \text{for } n \geq 1$$

EXAMPLE 2 Finding the *n*th Term of an Arithmetic Sequence

Find the 13th term of the arithmetic sequence 2, 5, 8, 11, ...

Solution:

Identify the common difference.	$d = 5 - 2 = 3$
Identify the first ($n = 1$) term.	$a_1 = 2$
Substitute $a_1 = 2$ and $d = 3$ into $a_n = a_1 + (n - 1)d$.	$a_n = 2 + 3(n - 1)$
Substitute $n = 13$ into $a_n = 2 + 3(n - 1)$.	$a_{13} = 2 + 3(13 - 1) = \boxed{38}$

■ **YOUR TURN** Find the 10th term of the arithmetic sequence 3, 10, 17, 24, ...

Technology Tip

[2nd] [LIST] [►] [OPS] [▼] [5:seq(] [ENTER] 2 [+] 3 [(] [ALPHA] [N] [−] [1] [)] [,] [ALPHA] [N] [,] 13 [,] 13 [,] 1 [)] [ENTER]

```
seq(2+3(N-1),N,1
3,13,1)
                {38}
```

■ **Answer:** 66

EXAMPLE 3 Finding the Arithmetic Sequence

The 4th term of an arithmetic sequence is 16, and the 21st term is 67. Find a_1 and *d* and construct the sequence.

Solution:

Write the 4th and 21st terms.	$a_4 = 16$ and $a_{21} = 67$
Adding *d* 17 times to a_4 results in a_{21}.	$a_{21} = a_4 + 17d$
Substitute $a_4 = 16$ and $a_{21} = 67$.	$67 = 16 + 17d$
Solve for *d*.	$\boxed{d = 3}$
Substitute $d = 3$ into $a_n = a_1 + (n - 1)d$.	$a_n = a_1 + 3(n - 1)$
Let $a_4 = 16$.	$16 = a_1 + 3(4 - 1)$
Solve for a_1.	$\boxed{a_1 = 7}$

The arithmetic sequence that starts at 7 and has a common difference of 3 is $\boxed{7, 10, 13, 16, \ldots}$.

■ **YOUR TURN** Construct the arithmetic sequence whose 7th term is 26 and whose 13th term is 50.

■ **Answer:** 2, 6, 10, 14, . . .

The Sum of an Arithmetic Sequence

What is the sum of the first 100 counting numbers

$$1 + 2 + 3 + 4 + \cdots + 99 + 100 = ?$$

If we write this sum twice (one in ascending order and one in descending order) and add, we get 100 pairs of 101.

$$\begin{array}{r} 1 + \ \ 2 + \ \ 3 + \ \ 4 + \cdots + \ 99 + 100 \\ 100 + 99 + 98 + 97 + \cdots + \ \ 2 + \ \ 1 \\ \hline 101 + 101 + 101 + 101 + \cdots + 101 + 101 = 100\,(101) \end{array}$$

Since we added twice the sum, we divide by 2.

$$1 + 2 + 3 + 4 + \cdots + 99 + 100 = \frac{(101)(100)}{2} = 5{,}050$$

Now, let us develop the sum of a general arithmetic series.

The sum of the first n terms of an arithmetic sequence is called the ***n*th partial sum**, or **finite arithmetic series**, and is denoted by S_n. An arithmetic sequence can be found by starting at the first term and adding the common difference to each successive term, and so the nth partial sum, or finite series, can be found the same way, but terminating the sum at the nth term:

$$S_n = a_1 + a_2 + a_3 + a_4 + \cdots$$
$$S_n = a_1 + (a_1 + d) + (a_1 + 2d) + (a_1 + 3d) + \cdots + (a_n)$$

Similarly, we can start with the nth term and find terms going backward by subtracting the common difference until we arrive at the first term:

$$S_n = a_n + a_{n-1} + a_{n-2} + a_{n-3} + \cdots$$
$$S_n = a_n + (a_n - d) + (a_n - 2d) + (a_n - 3d) + \cdots + (a_1)$$

Add these two representations of the nth partial sum. Notice that the d terms are eliminated:

$$\begin{aligned} S_n &= a_1 + (a_1 + d) + (a_1 + 2d) + (a_1 + 3d) + \cdots + (a_n) \\ S_n &= a_n + (a_n - d) + (a_n - 2d) + (a_n - 3d) + \cdots + (a_1) \\ \hline 2S_n &= \underbrace{(a_1 + a_n) + (a_1 + a_n) + (a_1 + a_n) + \cdots + (a_1 + a_n)}_{n(a_1 + a_n)} \end{aligned}$$

$$2S_n = n(a_1 + a_n) \quad \text{or} \quad S_n = \frac{n}{2}(a_1 + a_n)$$

Let $a_n = a_1 + (n - 1)d$.

$$\begin{aligned} S_n &= \frac{n}{2}[a_1 + a_1 + (n - 1)d] \\ &= \frac{n}{2}[2a_1 + (n - 1)d] = na_1 + \frac{n(n - 1)d}{2} \end{aligned}$$

Study Tip

S_n can also be written as

$S_n = \frac{n}{2}[2a_1 + (n - 1)d]$.

DEFINITION Evaluating a Finite Arithmetic Series

The sum of the first n terms of an arithmetic sequence, called a **finite arithmetic series**, is given by the formula

$$S_n = \frac{n}{2}(a_1 + a_n) \quad n \geq 2$$

EXAMPLE 4 Evaluating a Finite Arithmetic Series

Evaluate the finite arithmetic series $\sum_{k=1}^{100} k$.

Solution:

Expand the arithmetic series. $\sum_{k=1}^{100} k = 1 + 2 + 3 + \cdots + 99 + 100$

This is the sum of an arithmetic sequence of numbers with a common difference of 1.

Identify the parameters of the arithmetic sequence. $a_1 = 1$, $a_n = 100$, and $n = 100$

Substitute these values into $S_n = \frac{n}{2}(a_1 + a_n)$. $S_{100} = \frac{100}{2}(1 + 100)$

Simplify. $\boxed{S_{100} = 5050}$

The sum of the first 100 natural numbers is 5050.

YOUR TURN Evaluate the following finite arithmetic series.

a. $\sum_{k=1}^{30} k$ **b.** $\sum_{k=1}^{20} (2k + 1)$

Technology Tip

To find the sum of the series $\sum_{k=1}^{100} k$, press

2nd LIST ▶ MATH ▼
5:sum(ENTER 2nd LIST
▶ OPS ▼ 5:seq(ENTER
ALPHA K , ALPHA K ,
1 , 100 , 1)) ENTER

```
sum(seq(K,K,1,10
0,1))
                5050
```

Answer: a. 465 b. 440

EXAMPLE 5 Finding the *n*th Partial Sum of an Arithmetic Sequence

Find the sum of the first 20 terms of the arithmetic sequence 3, 8, 13, 18, 23, ...

Solution:

Recall the partial sum formula. $S_n = \frac{n}{2}(a_1 + a_n)$

Find the 20th partial sum of this arithmetic sequence. $S_{20} = \frac{20}{2}(a_1 + a_{20})$

Recall that the general nth term of an arithmetic sequence is given by: $a_n = a_1 + (n - 1)d$

Note that the first term of the arithmetic sequence is 3. $a_1 = 3$

This is an arithmetic sequence with a common difference of 5. $d = 5$

Substitute $a_1 = 3$ and $d = 5$ into $a_n = a_1 + (n - 1)d$. $a_n = 3 + (n - 1)5$

Substitute $n = 20$ to find the 20th term. $a_{20} = 3 + (20 - 1)5 = 98$

Substitute $a_1 = 3$ and $a_{20} = 98$ into the partial sum. $\boxed{S_{20} = 10(3 + 98) = 1010}$

The sum of the first 20 terms of this arithmetic sequence is 1010.

YOUR TURN Find the sum of the first 25 terms of the arithmetic sequence 2, 6, 10, 14, 18, ...

Technology Tip

To find the sum of the series $\sum_{n=1}^{20} 3 + (n - 1)5$, press

2nd LIST ▶ MATH ▼
5:sum(ENTER 2nd LIST
▶ OPS ▼ 5:seq(ENTER
3 + (ALPHA N − 1)
5 , ALPHA N , 1 , 20 , 1
)) ENTER

```
sum(seq(3+(N-1)5
,N,1,20,1))
                1010
```

Answer: 1250

Applications

EXAMPLE 6 Marching Band Formation

Suppose a band has 18 members in the first row, 22 members in the second row, and 26 members in the third row and continues with that pattern for a total of nine rows. How many marchers are there all together?

David Young-Wolff/PhotoEdit

UC Berkeley marching band

Solution:

The number of members in each row forms an arithmetic sequence with a common difference of 4, and the first row has 18 members. $a_1 = 18 \quad d = 4$

Calculate the nth term of the sequence $a_n = a_1 + (n - 1)d$. $a_n = 18 + (n - 1)4$

Find the 9th term $n = 9$. $a_9 = 18 + (9 - 1)4 = 50$

Calculate the sum $S_n = \frac{n}{2}(a_1 + a_n)$ of the nine rows.

$$S_9 = \frac{9}{2}(a_1 + a_9)$$

$$= \frac{9}{2}(18 + 50)$$

$$= \frac{9}{2}(68)$$

$$= \boxed{306}$$

There are 306 members in the marching band.

■ **Answer:** 328

■ **YOUR TURN** Suppose a bed of tulips is arranged in a garden so that there are 20 tulips in the first row, 26 tulips in the second row, and 32 tulips in the third row and the rows continue with that pattern for a total of 8 rows. How many tulips are there all together?

SECTION 9.2 SUMMARY

In this section, arithmetic sequences were defined as sequences of which each successive term is found by adding the same constant d to the previous term. Formulas were developed for the general, or nth, term of an arithmetic sequence, and for the nth partial sum of an arithmetic sequence, also called a finite arithmetic series.

$$a_n = a_1 + (n-1)d \qquad n \geq 1$$

$$S_n = \frac{n}{2}(a_1 + a_n) = na_1 + \frac{n(n-1)}{2}d$$

SECTION 9.2 EXERCISES

SKILLS

In Exercises 1–10, determine whether the sequence is arithmetic. If it is, find the common difference.

1. $2, 5, 8, 11, 14, \ldots$

2. $9, 6, 3, 0, -3, -6, \ldots$

3. $1^2 + 2^2 + 3^2 + \cdots$

4. $1! + 2! + 3! + \cdots$

5. $3.33, 3.30, 3.27, 3.24, \ldots$

6. $0.7, 1.2, 1.7, 2.2, \ldots$

7. $4, \frac{14}{3}, \frac{16}{3}, 6, \ldots$

8. $2, \frac{7}{3}, \frac{8}{3}, 3, \ldots$

9. $10^1, 10^2, 10^3, 10^4, \ldots$

10. $120, 60, 30, 15, \ldots$

In Exercises 11–20, find the first four terms of the sequence. Determine whether the sequence is arithmetic, and if so, find the common difference.

11. $a_n = -2n + 5$

12. $a_n = 3n - 10$

13. $a_n = n^2$

14. $a_n = \frac{n^2}{n!}$

15. $a_n = 5n - 3$

16. $a_n = -4n + 5$

17. $a_n = 10(n - 1)$

18. $a_n = 8n - 4$

19. $a_n = (-1)^n n$

20. $a_n = (-1)^{n+1} 2n$

In Exercises 21–28, find the general, or nth, term of the arithmetic sequence given the first term and the common difference.

21. $a_1 = 11 \quad d = 5$

22. $a_1 = 5 \quad d = 11$

23. $a_1 = -4 \quad d = 2$

24. $a_1 = 2 \quad d = -4$

25. $a_1 = 0 \quad d = \frac{2}{3}$

26. $a_1 = -1 \quad d = -\frac{3}{4}$

27. $a_1 = 0 \quad d = e$

28. $a_1 = 1.1 \quad d = -0.3$

In Exercises 29–32, find the specified term for each arithmetic sequence.

29. The 10th term of the sequence $7, 20, 33, 46, \ldots$

30. The 19th term of the sequence $7, 1, -5, -11, \ldots$

31. The 100th term of the sequence $9, 2, -5, -12, \ldots$

32. The 90th term of the sequence $13, 19, 25, 31, \ldots$

In Exercises 33–38, for each arithmetic sequence described, find a_1 and d and construct the sequence by stating the general, or nth, term.

33. The 5th term is 44 and the 17th term is 152.

34. The 9th term is -19 and the 21st term is -55.

35. The 7th term is -1 and the 17th term is -41.

36. The 8th term is 47 and the 21st term is 112.

37. The 4th term is 3 and the 22nd term is 15.

38. The 11th term is -3 and the 31st term is -13.

In Exercises 39–50, find the sum.

39. $\sum_{k=1}^{23} 2k$ **40.** $\sum_{k=0}^{20} 5k$ **41.** $\sum_{n=1}^{30} (-2n + 5)$ **42.** $\sum_{n=0}^{17} (3n - 10)$ **43.** $\sum_{j=3}^{14} 0.5j$ **44.** $\sum_{j=1}^{33} \frac{j}{4}$

45. $2 + 7 + 12 + 17 + \cdots + 62$ **46.** $1 - 3 - 7 - \cdots - 75$ **47.** $4 + 7 + 10 + \cdots + 151$

48. $2 + 0 - 2 - \cdots - 56$ **49.** $\frac{1}{6} - \frac{1}{6} - \frac{1}{2} - \cdots - \frac{13}{2}$ **50.** $\frac{11}{12} + \frac{7}{6} + \frac{17}{12} + \cdots + \frac{14}{3}$

In Exercises 51–56, find the partial sum of the arithmetic series.

51. The first 18 terms of $1 + 5 + 9 + 13 + \cdots$.

52. The first 21 terms of $2 + 5 + 8 + 11 + \cdots$.

53. The first 43 terms of $1 + \frac{1}{2} + 0 - \frac{1}{2} - \cdots$.

54. The first 37 terms of $3 + \frac{3}{2} + 0 - \frac{3}{2} - \cdots$.

55. The first 18 terms of $-9 + 1 + 11 + 21 + 31 + \cdots$.

56. The first 21 terms of $-2 + 8 + 18 + 28 + \cdots$.

APPLICATIONS

57. Comparing Salaries. Colin and Camden are twin brothers graduating with B.S. degrees in biology. Colin takes a job at the San Diego Zoo making $28,000 for his first year with a $1,500 raise per year every year after that. Camden accepts a job at Florida Fish and Wildlife making $25,000 with a guaranteed $2,000 raise per year. How much will each of the brothers have made in a total of 10 years?

58. Comparing Salaries. On graduating with a Ph.D. in optical sciences, Jasmine and Megan choose different career paths. Jasmine accepts a faculty position at the University of Arizona making $80,000 with a guaranteed $2,000 raise every year. Megan takes a job with the Boeing Corporation making $90,000 with a guaranteed $5,000 raise each year. Calculate how many total dollars each woman will have made after 15 years.

59. Theater Seating. You walk into the premiere of Brad Pitt's new movie, and the theater is packed, with almost every seat filled. You want to estimate the number of people in the theater. You quickly count to find that there are 22 seats in the front row, and there are 25 rows in the theater. Each row appears to have 1 more seat than the row in front of it. How many seats are in that theater?

60. Field of Tulips. Every spring the Skagit County Tulip Festival plants more than 100,000 bulbs. In honor of the Tri-Delta sorority that has sent 120 sisters from the University of Washington to volunteer for the festival, Skagit County has planted tulips in the shape of ΔΔΔ. In each of the triangles there are 20 rows of tulips, each row having one less than the row before. How many tulips are planted in each delta if there is 1 tulip in the first row?

61. World's Largest Champagne Fountain. From December 28 to 30, 1999, Luuk Broos, director of Maison Luuk-Chalet Fontain, constructed a 56-story champagne fountain at the Steigenberger Kurhaus Hotel, Scheveningen, Netherlands. The fountain consisted of 30,856 champagne glasses. Assuming there was one glass at the top, how many were on the bottom row (story)? How many glasses less did each successive row (story) have? Assume each story is one row.

62. Stacking of Logs. If 25 logs are laid side by side on the ground, and 24 logs are placed on top of those, and 23 logs are placed on the 3rd row, and the pattern continues until there is a single log on the 25th row, how many logs are in the stack?

63. Falling Object. When a skydiver jumps out of an airplane, she falls approximately 16 feet in the 1st second, 48 feet during the 2nd second, 80 feet during the 3rd second, 112 feet during the 4th second, and 144 feet during the 5th second, and this pattern continues. If she deploys her parachute after 10 seconds have elapsed, how far will she have fallen during those 10 seconds?

64. Falling Object. If a penny is dropped out of a plane, it falls approximately 4.9 meters during the 1st second, 14.7 meters during the 2nd second, 24.5 meters during the 3rd second, and 34.3 meters during the 4th second. Assuming this pattern continues, how many meters will the penny have fallen after 10 seconds?

65. Grocery Store. A grocer has a triangular display of oranges in a window. There are 20 oranges in the bottom row and the number of oranges decreases by one in each row above this row. How many oranges are in the display?

66. Salary. Suppose your salary is $45,000 and you receive a $1,500 raise for each year you work for 35 years.

a. How much will you earn during the 35th year?

b. What is the total amount you earned over your 35-year career?

67. Theater Seating. At a theater, seats are arranged in a triangular pattern of rows with each succeeding row having one more seat than the previous row. You count the number of seats in the fourth row and determine that there are 26 seats.

a. How many seats are in the first row?

b. Now, suppose there are 30 rows of seats. How many total seats are there in the theater?

68. Mathematics. Find the exact sum of

$$\frac{1}{e} + \frac{3}{e} + \frac{5}{e} + \cdots + \frac{23}{e}$$

CATCH THE MISTAKE

In Exercises 69–72, explain the mistake that is made.

69. Find the general or nth term of the arithmetic sequence 3, 4, 5, 6, 7, ...

Solution:

The common difference of this sequence is 1. $d = 1$

The first term is 3. $a_1 = 3$

The general term is $a_n = a_1 + nd$. $a_n = 3 + n$

This is incorrect. What mistake was made?

70. Find the general or nth term of the arithmetic sequence 10, 8, 6, ...

Solution:

The common difference of this sequence is 2. $d = 2$

The first term is 10. $a_1 = 10$

The general term is $a_n = a_1 + (n - 1)d$. $a_n = 10 + 2(n - 1)$

This is incorrect. What mistake was made?

71. Find the sum $\sum_{k=0}^{10} 2n + 1$.

Solution:

The sum is given by $S_n = \frac{n}{2}(a_1 + a_n)$, where $n = 10$.

Identify the 1st and 10th terms. $a_1 = 1$ $a_{10} = 21$

Substitute $a_1 = 1$, $a_{10} = 21$, and $n = 10$ into $S_n = \frac{n}{2}(a_1 + a_n)$. $S_{10} = \frac{10}{2}(1 + 21) = 110$

This is incorrect. What mistake was made?

72. Find the sum $3 + 9 + 15 + 21 + 27 + 33 + \cdots + 87$.

Solution:

This is an arithmetic sequence with common difference of 6. $d = 6$

The general term is given by $a_n = a_1 + (n - 1)d$. $a_n = 3 + (n - 1)6$

87 is the 15th term of the series. $a_{15} = 3 + (15 - 1)6 = 87$

The sum of the series is $S_n = \frac{n}{2}(a_n - a_1)$. $S_{15} = \frac{15}{2}(87 - 3) = 630$

This is incorrect. What mistake was made?

CONCEPTUAL

In Exercises 73–76, determine whether each statement is true or false.

73. An arithmetic sequence and a finite arithmetic series are the same.

74. The sum of all infinite and finite arithmetic series can always be found.

75. An alternating sequence cannot be an arithmetic sequence.

76. The common difference of an arithmetic sequence is always positive.

CHALLENGE

77. Find the sum $a + (a + b) + (a + 2b) + \cdots + (a + nb)$.

78. Find the sum $\sum_{k=-29}^{30} \ln e^k$.

79. The wave number, v (reciprocal of wave length) of certain light waves in the spectrum of light emitted by hydrogen is given by $v = R\left(\frac{1}{k^2} - \frac{1}{n^2}\right)$, $n > k$, where $R = 109{,}678$. A series of lines is given by holding k constant and varying the value of n. Suppose $k = 2$ and $n = 3, 4, 5, \ldots$. Find the limit of the wave number of the series.

80. In a certain arithmetic sequence $a_1 = -4$ and $d = 6$. If $S_n = 570$, find the value of n.

TECHNOLOGY

81. Use a graphing calculator "SUM" to sum the natural numbers from 1 to 100.

82. Use a graphing calculator to sum the even natural numbers from 1 to 100.

83. Use a graphing calculator to sum the odd natural numbers from 1 to 100.

84. Use a graphing calculator to sum $\sum_{n=1}^{30} - 2n + 5$. Compare it with your answer to Exercise 41.

85. Use a graphing calculator to sum $\sum_{n=1}^{100} [-59 + 5(n - 1)]$.

86. Use a graphing calculator to sum $\sum_{n=1}^{200} \left[-18 + \frac{4}{5}(n - 1)\right]$.

SECTION 9.3 GEOMETRIC SEQUENCES AND SERIES

SKILLS OBJECTIVES

- Recognize a geometric sequence.
- Find the general, *n*th term of a geometric sequence.
- Evaluate a finite geometric series.
- Evaluate an infinite geometric series, if it exists.
- Use geometric sequences and series to model real-world problems.

CONCEPTUAL OBJECTIVES

- Understand the difference between a geometric sequence and a geometric series.
- Distinguish between an arithmetic sequence and a geometric sequence.
- Understand why it is not possible to evaluate all infinite geometric series.

Geometric Sequences

In Section 9.2, we discussed *arithmetic* sequences, where successive terms had a *common difference.* In other words, each term was found by adding the same constant to the previous term. In this section we discuss *geometric* sequences, where successive terms have a *common ratio.* In other words, each term is found by multiplying the previous term by the same constant. The sequence 4, 12, 36, 108, . . . is geometric because each successive term is found by multiplying the previous term by 3.

DEFINITION **Geometric Sequences**

A sequence is **geometric** if each term in the sequence is found by multiplying the previous term by a number r, so that $a_{n+1} = r \cdot a_n$. Because $\frac{a_{n+1}}{a_n} = r$, the number r is called the **common ratio**.

EXAMPLE 1 **Identifying the Common Ratio in Geometric Sequences**

Find the common ratio for each of the geometric sequences.

a. 5, 20, 80, 320, . . . **b.** $1, -\frac{1}{2}, \frac{1}{4}, -\frac{1}{8}, \ldots$ **c.** \$5,000, \$5,500, \$6,050, \$6,655, . . .

Solution (a):

Label the terms. $a_1 = 5, a_2 = 20, a_3 = 80, a_4 = 320, \ldots$

Find the ratio $r = \frac{a_{n+1}}{a_n}$.

$$r = \frac{a_2}{a_1} = \frac{20}{5} = 4$$

$$r = \frac{a_3}{a_2} = \frac{80}{20} = 4$$

$$r = \frac{a_4}{a_3} = \frac{320}{80} = 4$$

The common ratio is 4.

Solution (b):

Label the terms. $a_1 = 1, a_2 = -\frac{1}{2}, a_3 = \frac{1}{4}, a_4 = -\frac{1}{8}, \ldots$

Find the ratio $r = \frac{a_{n+1}}{a_n}$.

$$r = \frac{a_2}{a_1} = \frac{-1/2}{1} = -\frac{1}{2}$$

$$r = \frac{a_3}{a_2} = \frac{1/4}{-1/2} = -\frac{1}{2}$$

$$r = \frac{a_4}{a_3} = \frac{-1/8}{1/4} = -\frac{1}{2}$$

The common ratio is $-\frac{1}{2}$.

Solution (c):

Label the terms. $a_1 = \$5{,}000, a_2 = \$5{,}500, a_3 = \$6{,}050, a_4 = \$6{,}655, \ldots$

Find the ratio $r = \frac{a_{n+1}}{a_n}$.

$$r = \frac{a_2}{a_1} = \frac{\$5{,}500}{\$5{,}000} = 1.1$$

$$r = \frac{a_3}{a_2} = \frac{\$6{,}050}{\$5{,}500} = 1.1$$

$$r = \frac{a_4}{a_3} = \frac{\$6{,}655}{\$6{,}050} = 1.1$$

The common ratio is 1.1.

YOUR TURN Find the common ratio of each geometric series.

a. $1, -3, 9, -27, \ldots$ **b.** $320, 80, 20, 5, \ldots$

Answer: a. -3 **b.** $\frac{1}{4}$ or 0.25

The General (*n*th) Term of a Geometric Sequence

To find a formula for the general, or *n*th, term of a geometric sequence, write out the first several terms and look for a pattern.

First term, $n = 1$. a_1

Second term, $n = 2$. $a_2 = a_1 \cdot r$

Third term, $n = 3$. $a_3 = a_2 \cdot r = (a_1 \cdot r) \cdot r = a_1 \cdot r^2$

Fourth term, $n = 4$. $a_4 = a_3 \cdot r = (a_1 \cdot r^2) \cdot r = a_1 \cdot r^3$

In general, the *n*th term is given by $a_n = a_1 \cdot r^{n-1}$.

THE *n*TH TERM OF A GEOMETRIC SEQUENCE

The ***n*th term** of a geometric sequence with common ratio r is given by

$$a_n = a_1 \cdot r^{n-1} \text{ for } n \geq 1$$

Technology Tip

Use [seq] to find the nth term of the sequence by setting the initial index value equal to the final index value. To find the 7th term of the geometric sequence $a_n = 2 \cdot 5^{n-1}$, press

[2nd] [LIST] [▶] [OPS] [▼] [5:seq(] [ENTER] 2 [x] 5 [^] [(] [ALPHA] [N] [−] 1 [)] [,] [ALPHA] [N] [,] 7 [,] 7 [,] 1 [)] [ENTER]

EXAMPLE 2 Finding the *n*th Term of a Geometric Sequence

Find the 7th term of the sequence 2, 10, 50, 250, ...

Solution:

Identify the common ratio. $r = \frac{10}{2} = \frac{50}{10} = \frac{250}{50} = 5$

Identify the first ($n = 1$) term. $a_1 = 2$

Substitute $a_1 = 2$ and $r = 5$ into $a_n = a_1 \cdot r^{n-1}$. $a_n = 2 \cdot 5^{n-1}$

Substitute $n = 7$ into $a_n = 2 \cdot 5^{n-1}$. $a_7 = 2 \cdot 5^{7-1} = 2 \cdot 5^6 = 31{,}250$

The 7th term of the geometric sequence is 31,250.

■ **YOUR TURN** Find the 8th term of the sequence 3, 12, 48, 192, ...

■ **Answer:** 49,152

EXAMPLE 3 Finding the Geometric Sequence

Find the geometric sequence whose 5th term is 0.01 and whose common ratio is 0.1.

Solution:

Label the common ratio and 5th term. $a_5 = 0.01$ and $r = 0.1$

Substitute $a_5 = 0.01$, $n = 5$, and $r = 0.1$ into $a_n = a_1 \cdot r^{n-1}$. $0.01 = a_1 \cdot (0.1)^{5-1}$

Solve for a_1. $a_1 = \frac{0.01}{(0.1)^4} = \frac{0.01}{0.0001} = 100$

The geometric sequence that starts at 100 and has a common ratio of 0.1 is 100, 10, 1, 0.1, 0.01, ...

■ **YOUR TURN** Find the geometric sequence whose 4th term is 3 and whose common ratio is $\frac{1}{3}$.

■ **Answer:** 81, 27, 9, 3, 1, . . .

Geometric Series

The sum of the terms of a geometric sequence is called a **geometric series.**

$$a_1 + a_1 \cdot r + a_1 \cdot r^2 + a_1 \cdot r^3 + \cdots$$

If we only sum the first n terms of a geometric sequence, the result is a **finite geometric series** given by

$$S_n = a_1 + a_1 \cdot r + a_1 \cdot r^2 + a_1 \cdot r^3 + \cdots + a_1 \cdot r^{n-1}$$

To develop a formula for the nth partial sum, we multiply the above equation by r:

$$r \cdot S_n = a_1 \cdot r + a_1 \cdot r^2 + a_1 \cdot r^3 + \cdots + a_1 \cdot r^{n-1} + a_1 \cdot r^n$$

Subtracting the second equation from the first equation, we find that all of the terms on the right side drop out except the *first* term in the first equation and the *last* term in the second equation:

$$\begin{aligned} S_n &= a_1 + a_1 \cdot r + a_1 \cdot r^2 + \cdots + a_1 r^{n-1} \\ -rS_n &= \quad - a_1 \cdot r - a_1 \cdot r^2 - \cdots - a_1 r^{n-1} - a_1 r^n \\ \hline S_n - rS_n &= a_1 \qquad\qquad\qquad\qquad\qquad\qquad - a_1 r^n \end{aligned}$$

Factor the S_n out of the left side and the a_1 out of the right side:

$$S_n(1 - r) = a_1(1 - r^n)$$

Divide both sides by $(1 - r)$, assuming $r \neq 1$. The result is a general formula for the sum of a finite geometric series:

$$S_n = a_1 \frac{(1 - r^n)}{(1 - r)} \qquad r \neq 1$$

EVALUATING A FINITE GEOMETRIC SERIES

The sum of the first n terms of a geometric sequence, called a **finite geometric series,** is given by the formula

$$S_n = a_1 \frac{(1 - r^n)}{(1 - r)} \qquad r \neq 1$$

It is important to note that a finite geometric series can also be written in sigma (summation) notation:

$$S_n = \sum_{k=1}^{n} a_1 \cdot r^{k-1} = a_1 + a_1 \cdot r + a_1 \cdot r^2 + a_1 \cdot r^3 + \cdots + a_1 \cdot r^{n-1}$$

Study Tip

The underscript $k = 1$ applies only when the summation starts at the a_1 term. It is important to note which term is the starting term.

Technology Tip

a. To find the sum of the series $\sum_{k=1}^{13} 3 \cdot (0.4)^{k-1}$, press

[2nd] [LIST] [▶] [MATH] [▼] [5:sum(] [ENTER] [2nd] [LIST] [▶] [OPS] [▼] [5:seq(] [ENTER] 3 [x] [(] 0.4 [)] [^] [(] [ALPHA] [K] [−] 1 [)] [,] [ALPHA] [K] [,] 1 [,] 13 [,] 1 [)] [)] [ENTER]

```
sum(seq(3*(0.4)^
(K-1),K,1,13,1))
          4.999966446
```

b. To find the sum of the first nine terms of the series $1 + 2 + 4 + 8 + 16 + 32 + 64 + \cdots$, press

[2nd] [LIST] [▶] [MATH] [▼] [5:sum(] [ENTER] [2nd] [LIST] [▶] [OPS] [▼] [5:seq(] [ENTER] 2 [^] [(] [ALPHA] [K] [−] 1 [)] [,] [ALPHA] [K] [,] 1 [,] 9 [,] 1 [)] [)] [ENTER]

```
sum(seq(2^(K-1),
K,1,9,1))
                  511
```

EXAMPLE 4 Evaluating a Finite Geometric Series

Evaluate the finite geometric series.

a. $\sum_{k=1}^{13} 3 \cdot (0.4)^{k-1}$

b. The first nine terms of the series $1 + 2 + 4 + 8 + 16 + 32 + 64 + \cdots$

Solution (a):

Identify a_1, n, and r. $\qquad a_1 = 3, n = 13, \text{ and } r = 0.4$

Substitute $a_1 = 3$, $n = 13$, and $r = 0.4$ into $s_n = a_1\frac{(1 - r^n)}{(1 - r)}$. $\qquad S_{13} = 3\frac{(1 - 0.4^{13})}{(1 - 0.4)}$

Simplify. $\qquad \boxed{S_{13} \approx 4.99997}$

Solution (b):

Identify the first term and common ratio. $\qquad a_1 = 1 \text{ and } r = 2$

Substitute $a_1 = 1$ and $r = 2$ into $S_n = a_1\frac{(1 - r^n)}{(1 - r)}$. $\qquad S_n = \frac{(1 - 2^n)}{(1 - 2)}$

To sum the first nine terms, let $n = 9$. $\qquad S_9 = \frac{(1 - 2^9)}{(1 - 2)}$

Simplify. $\qquad \boxed{S_9 = 511}$

The sum of an infinite geometric sequence is called an **infinite geometric series**. Some infinite geometric series converge (yield a finite sum) and some diverge (do not have a finite sum). For example,

$$\frac{1}{2} + \frac{1}{4} + \frac{1}{8} + \frac{1}{16} + \frac{1}{32} + \cdots + \frac{1}{2^n} + \cdots = 1 \quad \text{(converges)}$$

$$2 + 4 + 8 + 16 + 32 + \cdots + 2^n + \cdots \quad \text{(diverges)}$$

For infinite geometric series that converge, the partial sum S_n approaches a single number as n gets large. The formula used to evaluate a finite geometric series

$$S_n = a_1\frac{(1 - r^n)}{(1 - r)}$$

can be extended to an infinite geometric series for certain values of r. If $|r| < 1$, then when r is raised to a power, it continues to get smaller, approaching 0. For those values of r, the infinite geometric series converges to a finite sum.

$$\text{Let } n \to \infty\text{; then } a_1\frac{(1 - r^n)}{(1 - r)} \to a_1\frac{(1 - 0)}{(1 - r)} = a_1\frac{1}{1 - r}, \text{ if } |r| < 1.$$

EVALUATING AN INFINITE GEOMETRIC SERIES

The **sum of an infinite geometric series** is given by the formula

$$\sum_{n=0}^{\infty} a_1 \cdot r^n = a_1 \frac{1}{(1-r)} \qquad |r| < 1$$

Study Tip

The formula used to evaluate an infinite geometric series is:

$$\frac{\text{(First term)}}{1 - \text{(Ratio)}}$$

EXAMPLE 5 Determining Whether the Sum of an Infinite Series Exists

Determine whether the sum exists for each of the geometric series.

a. $3 + 15 + 75 + 375 + \cdots$ **b.** $8 + 4 + 2 + 1 + \frac{1}{2} + \frac{1}{4} + \frac{1}{8} + \cdots$

Solution (a):

Identify the common ratio. $r = 5$

Since 5 is greater than 1, the sum does not exist. $r = 5 > 1$

Solution (b):

Identify the common ratio. $r = \frac{1}{2}$

Since $\frac{1}{2}$ is less than 1, the sum exists. $r = \frac{1}{2} < 1$

■ **YOUR TURN** Determine whether the sum exists for the following geometric series.

a. $81, 9, 1, \frac{1}{9}, \ldots$ **b.** $1, 5, 25, 125, \ldots$

■ **Answer: a.** yes **b.** no

Do you expect $\frac{1}{4} + \frac{1}{12} + \frac{1}{36} + \frac{1}{64} + \cdots$ and $\frac{1}{4} - \frac{1}{12} + \frac{1}{36} - \frac{1}{64} + \cdots$ to sum to the same number? The answer is no, because the second series is an alternating series and terms are both added and subtracted. Hence, we would expect the second series to sum to a smaller number than the first series sums to.

EXAMPLE 6 Evaluating an Infinite Geometric Series

Evaluate each infinite geometric series.

a. $1 + \frac{1}{3} + \frac{1}{9} + \frac{1}{27} + \cdots$ **b.** $1 - \frac{1}{3} + \frac{1}{9} - \frac{1}{27} + \cdots$

Solution (a):

Identify the first term and the common ratio. $a_1 = 1 \quad r = \frac{1}{3}$

Since $|r| = \left|\frac{1}{3}\right| < 1$, the sum of the series exists.

Substitute $a_1 = 1$ and $r = \frac{1}{3}$ into $\sum_{n=0}^{\infty} a_1 \cdot r^n = a_1 \frac{1}{(1-r)}$. $\frac{1}{1 - 1/3}$

Simplify. $= \frac{1}{2/3} = \frac{3}{2}$

$$\boxed{1 + \frac{1}{3} + \frac{1}{9} + \frac{1}{27} + \cdots = \frac{3}{2}}$$

Study Tip

$$\sum_{n=0}^{\infty} a_1 \cdot r^n = \frac{\text{First term}}{1 - \text{Ratio}} = \frac{1}{1 - 1/3}$$

Study Tip

$$\sum_{n=0}^{\infty} a_1 \cdot r^n = \frac{\text{First term}}{1 - \text{Ratio}} = \frac{1}{1-(-1/3)}$$

Solution (b):

Identify the first term and the common ratio. $a_1 = 1 \quad r = -\frac{1}{3}$

Since $|r| = |-\frac{1}{3}| < 1$, the sum of the series exists.

Substitute $a_1 = 1$ and $r = -\frac{1}{3}$ into $\sum_{n=0}^{\infty} a_1 \cdot r^n = a_1 \frac{1}{(1-r)}$.

Simplify. $= \frac{1}{1-(-1/3)} = \frac{1}{1+(1/3)} = \frac{1}{4/3} = \frac{3}{4}$

$$\boxed{1 - \frac{1}{3} + \frac{1}{9} - \frac{1}{27} + \cdots = \frac{3}{4}}$$

Notice that the alternating series summed to $\frac{3}{4}$, whereas the positive series summed to $\frac{3}{2}$.

■ **Answer: a.** $\frac{3}{8}$ **b.** $\frac{3}{16}$

■ **YOUR TURN** Find the sum of each infinite geometric series.

a. $\frac{1}{4} + \frac{1}{12} + \frac{1}{36} + \frac{1}{64} + \cdots$ **b.** $\frac{1}{4} - \frac{1}{12} + \frac{1}{36} - \frac{1}{64} + \cdots$

It is important to note the restriction on the common ratio r. The absolute value of the common ratio has to be strictly less than 1 for an infinite geometric series to converge. Otherwise the infinite geometric series diverges.

EXAMPLE 7 Evaluating an Infinite Geometric Series

Evaluate the infinite geometric series, if possible.

a. $\sum_{n=0}^{\infty} 2\left(-\frac{1}{4}\right)^n$ **b.** $\sum_{n=1}^{\infty} 3 \cdot (2)^{n-1}$

Solution (a):

Identify a_1 and r.

$$\sum_{n=0}^{\infty} 2\left(-\frac{1}{4}\right)^n = \underbrace{2}_{a_1} - \frac{1}{2} + \frac{1}{8} - \frac{1}{32} + \frac{1}{128} - \cdots \qquad r = -\frac{1}{4}$$

Since $|r| = |-\frac{1}{4}| = \frac{1}{4} < 1$, the infinite geometric series converges. $\sum_{n=0}^{\infty} a_1 \cdot r^n = \frac{a_1}{(1-r)}$

Let $a_1 = 2$ and $r = -\frac{1}{4}$. $= \frac{2}{[1-(-1/4)]}$

Simplify. $= \frac{2}{1 + 1/4} = \frac{2}{5/4} = \frac{8}{5}$

This infinite geometric series converges. $\boxed{\sum_{n=0}^{\infty} 2\left(-\frac{1}{4}\right)^n = \frac{8}{5}}$

Solution (b):

Identify a_1 and r.

$$\sum_{n=1}^{\infty} 3 \cdot (2)^{n-1} = \underbrace{3}_{a_1} + 6 + 12 + 24 + 48 + \cdots \qquad r = 2$$

$$\boxed{\text{Since } r = 2 > 1, \text{ this infinite geometric series diverges.}}$$

Applications

Suppose you are given a job offer with a guaranteed percentage raise per year. What will your annual salary be 10 years from now? That answer can be obtained using a geometric sequence. Suppose you want to make voluntary contributions to a retirement account directly debited from your paycheck every month. Suppose the account earns a fixed percentage rate: How much will you have in 30 years if you deposit $50 a month? What is the difference in the total you will have in 30 years if you deposit $100 a month instead? These important questions about your personal finances can be answered using geometric sequences and series.

EXAMPLE 8 Future Salary: Geometric Sequence

Suppose you are offered a job as an event planner for the PGA Tour. The starting salary is $45,000, and employees are given a 5% raise per year. What will your annual salary be during the 10th year with the PGA Tour?

Solution:

Every year the salary is 5% more than the previous year.

Label the year 1 salary. $a_1 = 45{,}000$

Calculate the year 2 salary. $a_2 = 1.05 \cdot a_1$

Calculate the year 3 salary. $a_3 = 1.05 \cdot a_2$

$= 1.05(1.05 \cdot a_1) = (1.05)^2 a_1$

Calculate the year 4 salary. $a_4 = 1.05 \cdot a_3$

$= 1.05(1.05)^2 a_1 = (1.05)^3 a_1$

Identify the year n salary. $a_n = 1.05^{n-1} a_1$

Substitute $n = 10$ and $a_1 = 45{,}000$. $a_{10} = (1.05)^9 \cdot 45{,}000$

Simplify. $a_{10} \approx 69{,}809.77$

During your 10th year with the company your salary will be $69,809.77.

Study Tip

$\sum_{n=1}^{10} 45{,}000(1.05)^9 \cong 69{,}809.77$

■ **YOUR TURN** Suppose you are offered a job with AT&T at $37,000 per year with a guaranteed raise of 4% after every year. What will your annual salary be after 15 years with the company?

■ **Answer:** $64,072.03

EXAMPLE 9 Savings Growth: Geometric Series

Karen has maintained acrylic nails by paying for them with money earned from a part-time job. After hearing a lecture from her economics professor on the importance of investing early in life, she decides to remove the acrylic nails, which cost $50 per month, and do her own manicures. She has that $50 automatically debited from her checking account on the first of every month and put into a money market account that receives 3% interest compounded monthly. What will the balance be in the money market account exactly 2 years from the day of her initial $50 deposit?

Technology Tip

Use a calculator to find

$S_{24} = 50(1.0025)\dfrac{(1 - 1.0025^{24})}{(1 - 1.0025)}$.

Scientific calculators:

Press	*Display*
50 [x] 1.0025 [x]	1238.23
[(] 1 [−] 1.0025	
[x^y] 24 [)] [÷] [(]	
1 [−] 1.0025 [)] [=]	

Graphing calculators:

50 [x] 1.0025 [x] [(] 1 [−] 1.0025 [^] 24 [)] [÷] [(] 1 [−] 1.0025 [)] [ENTER]

```
50*1.0025*(1-1.0
025^24)/(1-1.002
5)
         1238.228737
```

Solution:

Recall the compound interest formula. $A = P\left(1 + \dfrac{r}{n}\right)^{nt}$

Substitute $r = 0.03$ and $n = 12$ into the compound interest formula. $A = P\left(1 + \dfrac{0.03}{12}\right)^{12t}$

$= P(1.0025)^{12t}$

Let $t = \dfrac{n}{12}$, where n is the number of months of the investment: $A_n = P(1.0025)^n$

The first deposit of \$50 will gain interest for 24 months. $A_{24} = 50(1.0025)^{24}$

The second deposit of \$50 will gain interest for 23 months. $A_{23} = 50(1.0025)^{23}$

The third deposit of \$50 will gain interest for 22 months. $A_{22} = 50(1.0025)^{22}$

The last deposit of \$50 will gain interest for 1 month. $A_1 = 50(1.0025)^1$

Sum the amounts accrued from the 24 deposits.

$$A_1 + A_2 + \cdots + A_{24} = 50(1.0025) + 50(1.0025)^2 + 50(1.0025)^3 + \cdots + 50(1.0025)^{24}$$

Identify the first term and common ratio. $a_1 = 50(1.0025)$ and $r = 1.0025$

Sum the first n terms of a geometric series. $S_n = a_1\dfrac{(1 - r^n)}{(1 - r)}$

Substitute $n = 24$, $a_1 = 50(1.0025)$, and $r = 1.0025$. $S_{24} = 50(1.0025)\dfrac{(1 - 1.0025^{24})}{(1 - 1.0025)}$

Simplify. $S_{24} \approx 1238.23$

Karen will have \$1,238.23 saved in her money market account in 2 years.

■ **YOUR TURN** Repeat Example 9 with Karen putting \$100 (instead of \$50) in the same money market. Assume she does this for 4 years (instead of 2 years).

■ **Answer:** \$5,105.85

SECTION 9.3 SUMMARY

In this section, we discussed geometric sequences, in which each successive term is found by multiplying the previous term by a constant, so that $a_{n+1} = r \cdot a_n$. That constant, r, is called the common ratio. The nth term of a geometric sequence is given by $a_n = a_1 r^{n-1}, n \ge 1$ or $a_{n+1} = a_1 r^n, n \ge 0$. The sum of the terms of a geometric sequence is called a geometric series. Finite geometric series converge to a number. Infinite geometric series converge to a number if the absolute value of the common ratio is less than 1. If the absolute value of the common ratio is greater than or equal to 1, the infinite geometric series diverges and the sum does not exist. Many real-world applications involve geometric sequences and series, such as growth of salaries and annuities through percentage increases.

Finite Geometric Series: $\displaystyle\sum_{i=0}^{n} a_1 r^i = a_1\frac{(1 - r^n)}{(1 - r)}$

Infinite Geometric Series: $\displaystyle\sum_{i=0}^{\infty} a_1 r^i = a_1\frac{1}{(1 - r)} \qquad |r| < 1$

SECTION 9.3 EXERCISES

SKILLS

In Exercises 1–8, determine whether the sequence is geometric. If it is, find the common ratio.

1. 1, 3, 9, 27, ...
2. 2, 4, 8, 16, ...
3. 1, 4, 9, 16, 25, ...
4. $1, \frac{1}{4}, \frac{1}{9}, \frac{1}{16}, \ldots$
5. 8, 4, 2, 1, ...
6. 8, −4, 2, −1, ...
7. 800, 1360, 2312, 3930.4, ...
8. 7, 15.4, 33.88, 74.536, ...

In Exercises 9–16, write the first five terms of the geometric series.

9. $a_1 = 6 \quad r = 3$
10. $a_1 = 17 \quad r = 2$
11. $a_1 = 1 \quad r = -4$
12. $a_1 = -3 \quad r = -2$
13. $a_1 = 10{,}000 \quad r = 1.06$
14. $a_1 = 10{,}000 \quad r = 0.8$
15. $a_1 = \frac{2}{3} \quad r = \frac{1}{2}$
16. $a_1 = \frac{1}{10} \quad r = -\frac{1}{5}$

In Exercises 17–24, write the formula for the *n*th term of the geometric series.

17. $a_1 = 5 \quad r = 2$
18. $a_1 = 12 \quad r = 3$
19. $a_1 = 1 \quad r = -3$
20. $a_1 = -4 \quad r = -2$
21. $a_1 = 1000 \quad r = 1.07$
22. $a_1 = 1000 \quad r = 0.5$
23. $a_1 = \frac{16}{3} \quad r = -\frac{1}{4}$
24. $a_1 = \frac{1}{200} \quad r = 5$

In Exercises 25–30, find the indicated term of the geometric sequence.

25. 7th term of the sequence −2, 4, −8, 16, ...
26. 10th term of the sequence 1, −5, 25, −225, ...
27. 13th term of the sequence $\frac{1}{3}, \frac{2}{3}, \frac{4}{3}, \frac{8}{3}, \ldots$
28. 9th term of the sequence 100, 20, 4, 0.8, ...
29. 15th term of the sequence 1000, 50, 2.5, 0.125, ...
30. 8th term of the sequence 1000, −800, 640, −512, ...

In Exercises 31–40, find the sum of the finite geometric series.

31. $\frac{1}{3} + \frac{2}{3} + \frac{2^2}{3} + \cdots + \frac{2^{12}}{3}$
32. $1 + \frac{1}{3} + \frac{1}{3^2} + \frac{1}{3^3} + \cdots + \frac{1}{3^{10}}$
33. $2 + 6 + 18 + 54 + \cdots + 2(3^9)$
34. $1 + 4 + 16 + 64 + \cdots + 4^9$
35. $\sum_{n=0}^{10} 2(0.1)^n$
36. $\sum_{n=0}^{11} 3(0.2)^n$
37. $\sum_{n=1}^{8} 2(3)^{n-1}$
38. $\sum_{n=1}^{9} \frac{2}{3}(5)^{n-1}$
39. $\sum_{k=0}^{13} 2^k$
40. $\sum_{k=0}^{13} \left(\frac{1}{2}\right)^k$

In Exercises 41–54, find the sum of the infinite geometric series, if possible.

41. $\sum_{n=0}^{\infty} \left(\frac{1}{2}\right)^n$
42. $\sum_{n=1}^{\infty} \left(\frac{1}{3}\right)^n$
43. $\sum_{n=1}^{\infty} \left(-\frac{1}{3}\right)^n$
44. $\sum_{n=0}^{\infty} \left(-\frac{1}{2}\right)^n$
45. $\sum_{n=0}^{\infty} 1^n$
46. $\sum_{n=0}^{\infty} 1.01^n$
47. $\sum_{n=0}^{\infty} -9\left(\frac{1}{3}\right)^n$
48. $\sum_{n=0}^{\infty} -8\left(-\frac{1}{2}\right)^n$
49. $\sum_{n=0}^{\infty} 10{,}000(0.05)^n$
50. $\sum_{n=0}^{\infty} 200(0.04)^n$
51. $\sum_{n=1}^{\infty} 0.4^n$
52. $0.3 + 0.03 + 0.003 + 0.0003 + \cdots$
53. $\sum_{n=0}^{\infty} 0.99^n$
54. $\sum_{n=0}^{\infty} \left(\frac{5}{4}\right)^n$

APPLICATIONS

55. Salary. Jeremy is offered a government job with the Department of Commerce. He is hired on the "GS" scale at a base rate of \$34,000 with 2.5% increases in his salary per year. Calculate what his salary will be after he has been with the Department of Commerce for 12 years.

56. Salary. Alison is offered a job with a small start-up company that wants to promote loyalty to the company with incentives for employees to stay with the company. The company offers her a starting salary of \$22,000 with a guaranteed 15% raise per year. What will her salary be after she has been with the company for 10 years?

57. Depreciation. Brittany, a graduating senior in high school, receives a laptop computer as a graduation gift from her Aunt Jeanine so that she can use it when she gets to the University of Alabama. If the laptop costs \$2,000 new and depreciates 50% per year, write a formula for the value of the laptop n years after it was purchased. How much will the laptop be worth when Brittany graduates from college (assuming she will graduate in 4 years)? How much will it be worth when she finishes graduate school? Assume graduate school is another 3 years.

58. Depreciation. Derek is deciding between a new Honda Accord and the BMW 325 series. The BMW costs \$35,000 and the Honda costs \$25,000. If the BMW depreciates at 20% per year and the Honda depreciates at 10% per year, find formulas for the value of each car n years after it is purchased. Which car is worth more in 10 years?

59. Bungee Jumping. A bungee jumper rebounds 70% of the height jumped. Assuming the bungee jump is made with a cord that stretches to 100 feet, how far will the bungee jumper travel upward on the fifth rebound?

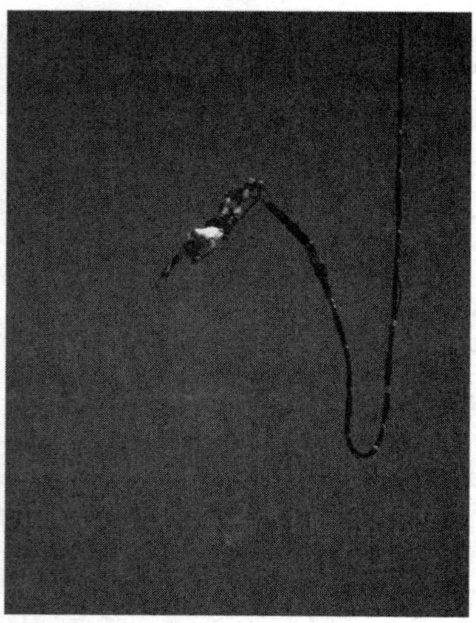

Laurence Fordyce/Eye Ubiquitous/Corbis Images

60. Bungee Jumping. A bungee jumper rebounds 65% of the height jumped. Assuming the bungee cord stretches 200 feet, how far will the bungee jumper travel upward on the eighth rebound?

61. Population Growth. One of the fastest-growing universities in the country is the University of Central Florida. The student populations each year starting in 2000 were 36,000, 37,800, 39,690, 41,675, ... If this rate continues, how many students will be at UCF in 2010?

62. Website Hits. The Website for Matchbox 20 (www.matchboxtwenty.com) has noticed that every week the number of hits to its Website increases 5%. If there were 20,000 hits this week, how many will there be exactly 52 weeks from now?

63. Rich Man's Promise. A rich man promises that he will give you \$1,000 on January 1, and every day after that, he will pay you 90% of what he paid you the day before. How many days will it take before you are making less than \$1? How much will the rich man pay out for the entire month of January?

64. Poor Man's Clever Deal. A poor man promises to work for you for \$.01 the first day, \$.02 on the second day, \$.04 on the third day; his salary will continue to double each day. If he started on January 1 how much would he be paid to work on January 31? How much total would he make during the month?

65. Investing Lunch. A newlywed couple decides to stop going out to lunch every day and instead bring their lunch. They estimate it will save them \$100 per month. They invest that \$100 on the first of every month into an account that is compounded monthly and pays 5% interest. How much will be in the account at the end of 3 years?

66. Pizza as an Investment. A college freshman decides to stop ordering late night pizzas (for both health and cost reasons). He realizes that he has been spending \$50 a week on pizzas. Instead, he deposits \$50 into an account that compounds weekly and pays 4% interest. How much money will be in the account in 52 weeks?

67. Tax-Deferred Annuity. Dr. Schober contributes \$500 from her paycheck (weekly) to a tax-deferred investment account. Assuming the investment earns 6% and is compounded weekly, how much will be in the account in 26 weeks? 52 weeks?

68. Saving for a House. If a new graduate decides she wants to save for a house and she is able to put \$300 every month into an account that earns 5% compounded monthly, how much will she have in the account in 5 years?

69. House Values. In 2008, you buy a house for \$195,000. The value of the house appreciates 6.5% per year, on the average. How much is the house worth after 15 years?

70. The Bouncing Ball Problem. A ball is dropped from a height of 9 feet. Assume that on each bounce, the ball rebounds to one-third of its previous height. Find the total distance that the ball travels.

71. Probability. A fair coin is tossed repeatedly. The probability that the first head occurs on the nth toss is given by the function $p(n) = \left(\frac{1}{2}\right)^n$, where $n \geq 1$. Show that

$$\sum_{n=1}^{\infty}\left(\frac{1}{2}\right)^n = 1.0$$

72. Salary. Suppose you work for a supervisor who gives you two different options to choose from for your monthly pay. Option 1: The company pays you 1 cent for the first day of work, two cents the second day, four cents for the third day, eight cents for the fourth day, and so on for 30 days. Option 2: You can receive a check right now for \$10,000,000. Which pay option is better? How much better is it?

CATCH THE MISTAKE

In Exercises 73–76, explain the mistake that is made.

73. Find the nth term of the geometric sequence:

$$-1, \tfrac{1}{3}, -\tfrac{1}{9}, \tfrac{1}{27}, \ldots.$$

Solution:

Identify the first term and common ratio. $a_1 = -1$ and $r = \dfrac{1}{3}$

Substitute $a_1 = -1$ and $r = \frac{1}{3}$ into $a_n = a_1 \cdot r^{n-1}$. $a_n = (-1)\cdot\left(\dfrac{1}{3}\right)^{n-1}$

Simplify. $a_n = \dfrac{-1}{3^{n-1}}$

This is incorrect. What mistake was made?

74. Find the sum of the first n terms of the finite geometric series:

$$2, 4, 8, 16, \ldots.$$

Solution:

Write the sum in sigma notation. $\sum_{k=1}^{n}(2)^k$

Identify the first term and common ratio. $a_1 = 1$ and $r = 2$

Substitute $a_1 = 1$ and $r = 2$ into $S_n = a_1\dfrac{(1-r^n)}{(1-r)}$. $S_n = 1\dfrac{(1-2^n)}{(1-2)}$

Simplify. $S_n = 2^n - 1$

This is incorrect. What mistake was made?

75. Find the sum of the finite geometric series $\sum_{n=1}^{8} 4(-3)^n$.

Solution:

Identify the first term and common ratio. $a_1 = 4$ and $r = -3$

Substitute $a_1 = 4$ and $r = -3$ into $S_n = a_1\dfrac{(1-r^n)}{(1-r)}$. $S_n = 4\dfrac{\left[1-(-3)^n\right]}{[1-(-3)]}$

$= 4\dfrac{\left[1-(-3)^n\right]}{4}$

Simplify. $S_n = \left[1-(-3)^n\right]$

Substitute $n = 8$. $S_8 = \left[1-(-3)^8\right] = -6{,}560$

This is incorrect. What mistake was made?

76. Find the sum of the infinite geometric series $\sum_{n=1}^{\infty} 2\cdot 3^{n-1}$.

Solution:

Identify the first term and common ratio. $a_1 = 2$ and $r = 3$

Substitute $a_1 = 2$ and $r = 3$ into $S_\infty = a_1\dfrac{1}{(1-r)}$. $S_\infty = 2\dfrac{1}{(1-3)}$

Simplify. $S_\infty = -1$

This is incorrect. The series does not sum to -1. What mistake was made?

CONCEPTUAL

In Exercises 77–80, determine whether each statement is true or false.

77. An alternating sequence cannot be a geometric sequence.

78. All finite and infinite geometric series can always be evaluated.

79. The common ratio of a geometric sequence can be positive or negative.

80. An infinite geometric series can be evaluated if the common ratio is less than or equal to 1.

CHALLENGE

81. State the conditions for the sum

$$a + a\cdot b + a\cdot b^2 + \cdots + a\cdot b^n + \cdots$$

to exist. Assuming those conditions are met, find the sum.

82. Find the sum of $\sum_{k=0}^{20} \log 10^{2^k}$.

83. Represent the repeating decimal 0.474747. . . as a fraction (ratio of two integers).

84. Suppose the sum of an infinite geometric series is $S = \dfrac{2}{1-x}$, where x is a variable.

a. Write out the first five terms of the series.

b. For what values of x will the series converge?

TECHNOLOGY

85. Sum the series: $\sum_{k=1}^{50}(-2)^{k-1}$. Apply a graphing utility to confirm your answer.

86. Does the sum of the infinite series $\sum_{n=0}^{\infty}\left(\frac{1}{3}\right)^n$ exist? Use a graphing calculator to find it.

87. Apply a graphing utility to plot $y_1 = 1 + x + x^2 + x^3 + x^4$ and $y_2 = \dfrac{1}{1-x}$, and let the range of x be $[-0.5, 0.5]$. Based on what you see, what do you expect the geometric series $\sum_{n=0}^{\infty} x^n$ to sum to?

88. Apply a graphing utility to plot $y_1 = 1 - x + x^2 - x^3 + x^4$ and $y_2 = \dfrac{1}{1+x}$, and let x range from $[-0.5, 0.5]$. Based on what you see, what do you expect the geometric series $\sum_{n=0}^{\infty}(-1)^n x^n$ to sum to?

89. Apply a graphing utility to plot $y_1 = 1 + 2x + 4x^2 + 8x^3 + 16x^4$ and $y_2 = \dfrac{1}{1-2x}$, and let x range from $[-0.3, 0.3]$. Based on what you see, what do you expect the geometric series $\sum_{n=0}^{\infty}(2x)^n$ to sum to?

SECTION 9.4 MATHEMATICAL INDUCTION

SKILLS OBJECTIVES

- Know the steps required to prove a statement by mathematical induction.
- Prove mathematical statements using mathematical induction.

CONCEPTUAL OBJECTIVES

- Understand that just because there appears to be a pattern, the pattern is not necessarily true for all values.
- Understand that when mathematical ideas are accepted, it is because they can be proved.

n	$n^2 - n + 41$	PRIME?
1	41	Yes
2	43	Yes
3	47	Yes
4	53	Yes
5	61	Yes

Is the expression $n^2 - n + 41$ *always* a prime number if n is a natural number? Your instinct may lead you to try a few values for n.

It appears that the statement might be true for all natural numbers. However, what about when $n = 41$?

$$n^2 - n + 41 = (41)^2 - 41 + 41 = 41^2$$

We find that when $n = 41$, $n^2 - n + 41$ is not prime. The moral of the story is that just because a pattern seems to exist for *some* values, the pattern is not necessarily true for *all* values. We must look for a way to show whether a statement is true for all values. In this section we talk about *mathematical induction*, which is a way to show a statement is true for all values.

Mathematical Induction

Mathematics is based on logic and proof (not assumptions or belief). One of the most famous mathematical statements was Fermat's last theorem. Pierre de Fermat (1601–1665) conjectured that there are no positive integer values for x, y, and z such that $x^n + y^n = z^n$, if $n \geq 3$. Although mathematicians *believed* that this theorem was true, no one was able to *prove* it until 350 years after the assumption was made. Professor Andrew Wiles at Princeton University received a \$50,000 prize for successfully proving Fermat's last theorem in 1994.

Mathematical induction is a technique used in college algebra and even in very advanced mathematics to prove many kinds of mathematical statements. In this section you will use it to prove statements like "if $x > 1$, then $x^n > 1$ for all natural numbers n."

The principle of mathematical induction can be illustrated by a row of standing dominos, as in the picture. We make two assumptions:

1. The first domino is knocked down.
2. If a domino is knocked down, then the domino immediately following it will also be knocked down.

If both of these assumptions are true, then it is also true that all of the dominos will fall.

PRINCIPLE OF MATHEMATICAL INDUCTION

Let S_n be a statement involving the positive integer n. To prove that S_n is true for all positive integers, the following steps are required.

Step 1: Show that S_1 is true.
Step 2: Assume S_k is true and show that S_{k+1} is true (k = positive integer).

Combining Steps 1 and 2 proves the statement is true for all positive integers.

EXAMPLE 1 Using Mathematical Induction

Apply the principle of mathematical induction to prove this statement:

If $x > 1$, then $x^n > 1$ for all natural numbers n.

Solution:

STEP 1 Show the statement is true for $n = 1$. $\quad x^1 > 1$ because $x > 1$

STEP 2 Assume the statement is true for $n = k$. $\quad x^k > 1$

Show the statement is true for $k + 1$.

Multiply both sides by x. $\quad x^k \cdot x > 1 \cdot x$

(Since $x > 1$, this step does not reverse the inequality sign.)

Simplify. $\quad x^{k+1} > x$

Recall that $x > 1$. $\quad x^{k+1} > x > 1$

Therefore, we have shown that $x^{k+1} > 1$.

This completes the induction proof. Thus, the following statement is true.

"If $x > 1$, then $x^n > 1$ for **all** natural numbers n."

EXAMPLE 2 Using Mathematical Induction

Use mathematical induction to prove that $n^2 + n$ is divisible by 2 for all natural numbers n.

Solution:

STEP 1 Show the statement we are testing is true for $n = 1$. $\quad 1^2 + 1 = 2$

2 is divisible by 2. $\quad \frac{2}{2} = 1$

STEP 2 Assume the statement is true for $n = k$. $\qquad \dfrac{k^2 + k}{2} = \text{an integer}$

Show it is true for $k + 1$ where $k \geq 1$. $\qquad \dfrac{(k+1)^2 + (k+1)}{2} \stackrel{?}{=} \text{an integer}$

$$\frac{k^2 + 2k + 1 + k + 1}{2} \stackrel{?}{=} \text{an integer}$$

Regroup terms. $\qquad \dfrac{\left(k^2 + k\right) + 2(k+1)}{2} \stackrel{?}{=} \text{an integer}$

$$\frac{\left(k^2 + k\right)}{2} + \frac{2(k+1)}{2} \stackrel{?}{=} \text{an integer}$$

We assumed $\dfrac{k^2 + k}{2} = \text{an integer}$. $\qquad \text{an integer} + (k+1) \stackrel{?}{=} \text{an integer}$

Since k is a natural number. $\qquad \text{an integer} + \text{an integer} = \text{an integer}$

This completes the induction proof. The following statement is true:

"$n^2 + n$ is divisible by 2 for all natural numbers n."

Mathematical induction is often used to prove formulas for partial sums.

Technology Tip

To visualize what needs to be proved in the partial–sum formula, use the [sum] command to find the sum of the series $\sum_{k=1}^{n} k$ on the left side for an arbitrary n value, say $n = 100$. Press

[2nd] [LIST] [▶] [MATH] [▼] [5:sum(] [ENTER] [2nd] [LIST] [▶] [OPS] [▼] [5:seq(] [ENTER] [ALPHA] [K] [,] [ALPHA] [K] [,] 1 [,] 100 [,] 1 [)] [)] [ENTER]

```
sum(seq(N,N,1,10
0,1))
            5050
```

Now calculate the sum by substituting $n = 100$ into $n(n + 1)/2$ on the right side.

```
sum(seq(N,N,1,10
0,1))
            5050
100(100+1)/2
            5050
```

Note: The solutions to the left and right side of the formula agree with each other.

EXAMPLE 3 Proving a Partial–Sum Formula with Mathematical Induction

Apply mathematical induction to prove the following partial–sum formula:

$$1 + 2 + 3 + \cdots + n = \frac{n(n+1)}{2} \quad \text{for all positive integers } n$$

Solution:

STEP 1 Show the formula is true for $n = 1$. $\qquad 1 = \dfrac{1(1+1)}{2} = \dfrac{2}{2} = 1$

STEP 2 Assume the formula is true for $n = k$. $\qquad 1 + 2 + 3 + \cdots + k = \dfrac{k(k+1)}{2}$

Show it is true for $n = k + 1$. $\qquad 1 + 2 + 3 + \cdots + k + (k+1) \stackrel{?}{=} \dfrac{(k+1)(k+2)}{2}$

$$\underbrace{1 + 2 + 3 + \cdots + k}_{\frac{k(k+1)}{2}} + (k+1) \stackrel{?}{=} \frac{(k+1)(k+2)}{2}$$

$$\frac{k(k+1)}{2} + (k+1) \stackrel{?}{=} \frac{(k+1)(k+2)}{2}$$

$$\frac{k(k+1) + 2(k+1)}{2} \stackrel{?}{=} \frac{(k+1)(k+2)}{2}$$

$$\frac{k^2 + 3k + 2}{2} \stackrel{?}{=} \frac{(k+1)(k+2)}{2}$$

$$\frac{(k+1)(k+2)}{2} = \frac{(k+1)(k+2)}{2}$$

This completes the induction proof. The following statement is true:

"$1 + 2 + 3 + \cdots + n = \dfrac{n(n+1)}{2}$ for all positive integers n."

SECTION 9.4 SUMMARY

Just because we believe something is true does not mean that it is. In mathematics we rely on proof. In this section we discussed *mathematical induction*, a process of proving mathematical statements. The two-step procedure for mathematical induction is to (1) show the statement is true for $n = 1$, then (2) assume the statement is true for $n = k$ and show the statement must be true for $n = k + 1$. The combination of Steps 1 and 2 proves the statement.

SECTION 9.4 EXERCISES

SKILLS

In Exercises 1–24, prove the statements using mathematical induction for all positive integers n.

1. $n^2 \leq n^3$

2. If $0 < x < 1$, then $0 < x^n < 1$.

3. $2n \leq 2^n$

4. $5^n < 5^{n+1}$

5. $n! > 2^n \quad n \geq 4$ (Show it is true for $n = 4$, instead of $n = 1$.)

6. $(1 + c)^n \geq nc \quad c > 1$

7. $n(n + 1)(n - 1)$ is divisible by 3.

8. $n^3 - n$ is divisible by 3.

9. $n^2 + 3n$ is divisible by 2.

10. $n(n + 1)(n + 2)$ is divisible by 6.

11. $2 + 4 + 6 + 8 + \cdots + 2n = n(n + 1)$

12. $1 + 3 + 5 + 7 + \cdots + (2n - 1) = n^2$

13. $1 + 3 + 3^2 + 3^3 + \cdots + 3^n = \dfrac{3^{n+1} - 1}{2}$

14. $2 + 4 + 8 + \cdots + 2^n = 2^{n+1} - 2$

15. $1^2 + 2^2 + 3^2 + \cdots + n^2 = \dfrac{n(n + 1)(2n + 1)}{6}$

16. $1^3 + 2^3 + 3^3 + \cdots + n^3 = \dfrac{n^2(n + 1)^2}{4}$

17. $\dfrac{1}{1 \cdot 2} + \dfrac{1}{2 \cdot 3} + \dfrac{1}{3 \cdot 4} + \cdots + \dfrac{1}{n(n + 1)} = \dfrac{n}{n + 1}$

18. $\dfrac{1}{2 \cdot 3} + \dfrac{1}{3 \cdot 4} + \cdots + \dfrac{1}{(n + 1)(n + 2)} = \dfrac{n}{2(n + 2)}$

19. $(1 \cdot 2) + (2 \cdot 3) + (3 \cdot 4) + \cdots + n(n + 1) = \dfrac{n(n + 1)(n + 2)}{3}$

20. $(1 \cdot 3) + (2 \cdot 4) + (3 \cdot 5) + \cdots + n(n + 2) = \dfrac{n(n + 1)(2n + 7)}{6}$

21. $1 + x + x^2 + x^3 + \cdots + x^{n-1} = \dfrac{1 - x^n}{1 - x} \quad x \neq 1$

22. $\dfrac{1}{2} + \dfrac{1}{4} + \dfrac{1}{8} + \cdots + \dfrac{1}{2^n} = 1 - \dfrac{1}{2^n}$

23. The sum of an arithmetic sequence: $a_1 + (a_1 + d) + (a_1 + 2d) + \cdots + [a_1 + (n - 1)d] = \dfrac{n}{2}[2a_1 + (n - 1)d]$.

24. The sum of a geometric sequence: $a_1 + a_1r + a_1r^2 + \cdots + a_1r^{n-1} = a_1\left(\dfrac{1 - r^n}{1 - r}\right)$.

APPLICATIONS

The Tower of Hanoi. This is a game with three pegs and n disks (largest on the bottom and smallest on the top). The goal is to move this entire tower of disks to another peg (in the same order). The challenge is that you may move only one disk at a time, and at no time can a larger disk be resting on a smaller disk. You may want to first go online to www.mazeworks.com/hanoi/index/htm and play the game.

Andy Washnik

Tower of Hanoi

25. What is the smallest number of moves needed if there are three disks?

26. What is the smallest number of moves needed if there are four disks?

27. What is the smallest number of moves needed if there are five disks?

28. What is the smallest number of moves needed if there are n disks? Prove it by mathematical induction.

29. Telephone Infrastructure. Suppose there are n cities that are to be connected with telephone wires. Apply mathematical induction to prove that the number of telephone wires required to connect the n cities is given by $\frac{n(n-1)}{2}$. Assume each city has to connect directly with any other city.

30. Geometry. Prove, with mathematical induction, that the sum of the interior angles of a regular polygon of n sides is given by the formula: $(n-2)(180)$ for $n \geq 3$. *Hint:* Divide a polygon into triangles. For example, a four sided polygon can be divided into two triangles. A five sided polygon can be divided into three triangles. A six sided polygon can be divided into four triangles, and so on.

CONCEPTUAL

In Exercises 31 and 32, determine whether each statement is true or false.

31. Assume S_k is true. If it can be shown that S_{k+1} is true, then S_n is true for all n, where n is any positive integer.

32. Assume S_1 is true. If it can be shown that S_2 and S_3 are true, then S_n is true for all n, where n is any positive integer.

CHALLENGE

33. Apply mathematical induction to prove:

$$\sum_{k=1}^{n} k^4 = \frac{n(n+1)(2n+1)(3n^2+3n-1)}{30}$$

34. Apply mathematical induction to prove:

$$\sum_{k=1}^{n} k^5 = \frac{n^2(n+1)^2(2n^2+2n-1)}{12}$$

35. Apply mathematical induction to prove:

$$\left(1+\frac{1}{1}\right)\left(1+\frac{1}{2}\right)\left(1+\frac{1}{3}\right)\cdots\left(1+\frac{1}{n}\right) = n+1$$

36. Apply mathematical induction to prove that $x + y$ is a factor of $x^{2n} - y^{2n}$.

37. Apply mathematical induction to prove:

$$\ln(c_1 \cdot c_2 \cdot c_3 \cdots c_n) = \ln c_1 + \ln c_2 + \cdots + \ln c_n$$

TECHNOLOGY

38. Use a graphing calculator to sum the series $(1 \cdot 2) + (2 \cdot 3) + (3 \cdot 4) + \cdots + n(n+1)$ on the left side, and evaluate the expression $\frac{n(n+1)(n+2)}{3}$ on the right side for $n = 200$. Do they agree with each other? Do your answers confirm the proof for Exercise 19?

39. Use a graphing calculator to sum the series $\frac{1}{2} + \frac{1}{4} + \frac{1}{8} + \cdots + \frac{1}{2^n}$ on the left side, and evaluate the expression $1 - \frac{1}{2^n}$ on the right side for $n = 8$. Do they agree with each other? Do your answers confirm the proof for Exercise 22?

SECTION

9.5 THE BINOMIAL THEOREM

SKILLS OBJECTIVES

- Evaluate a binomial coefficient with the binomial theorem.
- Evaluate a binomial coefficient with Pascal's triangle.
- Expand a binomial raised to a positive integer power.
- Find a particular term of a binomial expansion.

CONCEPTUAL OBJECTIVE

- Recognize patterns in binomial expansions.

A **binomial** is a polynomial that has two terms. The following are all examples of binomials:

$$x^2 + 2y \qquad a + 3b \qquad 4x^2 + 9$$

In this section we will develop a formula for raising a binomial to a power n, where n is a positive integer.

$$(x^2 + 2y)^6 \qquad (a + 3b)^4 \qquad (4x^2 + 9)^5$$

To begin, let's start by writing out the expansions of $(a + b)^n$ for several values of n.

$$(a + b)^0 = 1$$

$$(a + b)^1 = a + b$$

$$(a + b)^2 = a^2 + 2ab + b^2$$

$$(a + b)^3 = a^3 + 3a^2b + 3ab^2 + b^3$$

$$(a + b)^4 = a^4 + 4a^3b + 6a^2b^2 + 4ab^3 + b^4$$

$$(a + b)^5 = a^5 + 5a^4b + 10a^3b^2 + 10a^2b^3 + 5ab^4 + b^5$$

There are several *patterns* that all of the **binomial expansions** have.

1. The number of terms in each resulting polynomial is always *one more* than the power of the binomial n. Thus, there are $n + 1$ terms in each expansion.

$$n = 3: \qquad (a + b)^3 = \underbrace{a^3 + 3a^2b + 3ab^2 + b^3}_{\text{four terms}}$$

2. Each expansion has symmetry. For example, a and b can be interchanged and you will arrive at the same expansion. Furthermore, the powers of a decrease by 1 in each successive term, and the powers of b increase by 1 in each successive term.

$$(a + b)^3 = a^3b^0 + 3a^2b^1 + 3a^1b^2 + a^0b^3$$

3. The sum of the powers of each term in the expansion is n.

$$n = 3: \qquad (a + b)^3 = \overbrace{a^3b^0}^{3+0=3} + 3\overbrace{a^2b^1}^{2+1=3} + 3\overbrace{a^1b^2}^{1+2=3} + \overbrace{a^0b^3}^{0+3=3}$$

4. The coefficients increase and decrease in a symmetric manner.

$$(a + b)^5 = 1a^5 + 5a^4b + 10a^3b^2 + 10a^2b^3 + 5ab^4 + 1b^5$$

Using these patterns, we can develop a generalized formula for $(a + b)^n$.

$$(a + b)^n = \square a^n + \square a^{n-1}b + \square a^{n-2}b^2 + \cdots + \square a^2b^{n-2} + \square ab^{n-1} + \square b^n$$

We know that there are $n + 1$ terms in the expansion. We also know that the sum of the powers of each term must equal n. The powers increase and decrease by 1 in each successive term, and if we interchanged a and b, the result would be the same expansion. The question that remains is, what coefficients go in the blanks?

Binomial Coefficients

We know that the coefficients must increase and then decrease in a symmetric order (similar to walking up and then down a hill). It turns out that the *binomial coefficients* are represented by a symbol that we will now define.

DEFINITION **Binomial Coefficients**

For nonnegative integers n and k, where $n \geq k$, the symbol $\binom{n}{k}$ is called the **binomial coefficient** and is defined by

$$\binom{n}{k} = \frac{n!}{(n-k)!k!} \qquad \binom{n}{k} \text{ is read "}n\text{ choose }k\text{."}$$

You will see in the following sections that "n choose k" comes from combinations.

Technology Tip

Press	*Display*
6 nCr 4 =	15
5 nCr 5 =	1
4 nCr 0 =	1
10 nCr 9 =	10

```
6 nCr 4
                15
5 nCr 5
                 1
4 nCr 0
                 1
```

```
10 nCr 9
                10
█
```

Answer: a. 84 **b.** 28

EXAMPLE 1 Evaluating a Binomial Coefficient

Evaluate the following binomial coefficients.

a. $\binom{6}{4}$ **b.** $\binom{5}{5}$ **c.** $\binom{4}{0}$ **d.** $\binom{10}{9}$

Solution:

Select the top number as n and the bottom number as k and substitute into the binomial coefficient formula $\binom{n}{k} = \frac{n!}{(n-k)!k!}$.

a. $\binom{6}{4} = \frac{6!}{(6-4)!4!} = \frac{6!}{2!4!} = \frac{6 \cdot 5 \cdot 4 \cdot 3 \cdot 2 \cdot 1}{(2 \cdot 1)(4 \cdot 3 \cdot 2 \cdot 1)} = \frac{6 \cdot 5}{2} = \boxed{15}$

b. $\binom{5}{5} = \frac{5!}{(5-5)!5!} = \frac{5!}{0!5!} = \frac{1}{0!} = \frac{1}{1} = \boxed{1}$

c. $\binom{4}{0} = \frac{4!}{(4-0)!0!} = \frac{4!}{4!0!} = \frac{1}{0!} = \boxed{1}$

d. $\binom{10}{9} = \frac{10!}{(10-9)!9!} = \frac{10!}{1!9!} = \frac{10 \cdot 9!}{9!} = \boxed{10}$

YOUR TURN Evaluate the following binomial coefficients.

a. $\binom{9}{6}$ **b.** $\binom{8}{6}$

Parts (b) and (c) of Example 1 lead to the general formulas:

$$\binom{n}{n} = 1 \quad \text{and} \quad \binom{n}{0} = 1$$

Binomial Expansion

Let's return to the question of the binomial expansion and how to determine the coefficients:

$$(a+b)^n = \square a^n + \square a^{n-1}b + \square a^{n-2}b^2 + \cdots + \square a^2b^{n-2} + \square ab^{n-1} + \square b^n$$

The symbol $\binom{n}{k}$ is called a binomial coefficient because the coefficients in the blanks in the binomial expansion are equivalent to this symbol.

THE BINOMIAL THEOREM

Let a and b be real numbers; then for any positive integer n,

$$(a+b)^n = \binom{n}{0}a^n + \binom{n}{1}a^{n-1}b + \binom{n}{2}a^{n-2}b^2 + \cdots + \binom{n}{n-2}a^2b^{n-2} + \binom{n}{n-1}ab^{n-1} + \binom{n}{n}b^n$$

or in sigma (summation) notation as

$$(a+b)^n = \sum_{k=0}^{n}\binom{n}{k}a^{n-k}b^k$$

EXAMPLE 2 Applying the Binomial Theorem

Expand $(x+2)^3$ with the binomial theorem.

Solution:

Substitute $a = x$, $b = 2$, $n = 3$ into the equation of the binomial theorem.
$$(x+2)^3 = \sum_{k=0}^{3}\binom{3}{k}x^{3-k}2^k$$

Expand the summation.
$$= \binom{3}{0}x^3 + \binom{3}{1}x^2 \cdot 2 + \binom{3}{2}x \cdot 2^2 + \binom{3}{3}2^3$$

Find the binomial coefficients.
$$= x^3 + 3x^2 \cdot 2 + 3x \cdot 2^2 + 2^3$$

Simplify.
$$= \boxed{x^3 + 6x^2 + 12x + 8}$$

■ **YOUR TURN** Expand $(x+5)^4$ with the binomial theorem.

■ **Answer:** $x^4 + 20x^3 + 150x^2 + 500x + 625$

EXAMPLE 3 Applying the Binomial Theorem

Expand $(2x-3)^4$ with the binomial theorem.

Solution:

Substitute $a = 2x$, $b = -3$, $n = 4$ into the equation of the binomial theorem.
$$(2x-3)^4 = \sum_{k=0}^{4}\binom{4}{k}(2x)^{4-k}(-3)^k$$

Expand the summation.
$$= \binom{4}{0}(2x)^4 + \binom{4}{1}(2x)^3(-3) + \binom{4}{2}(2x)^2(-3)^2 + \binom{4}{3}(2x)(-3)^3 + \binom{4}{4}(-3)^4$$

Find the binomial coefficients.
$$= (2x)^4 + 4(2x)^3(-3) + 6(2x)^2(-3)^2 + 4(2x)(-3)^3 + (-3)^4$$

Simplify.
$$\boxed{= 16x^4 - 96x^3 + 216x^2 - 216x + 81}$$

■ **YOUR TURN** Expand $(3x-2)^4$ with the binomial theorem.

■ **Answer:** $81x^4 - 216x^3 + 216x^2 - 96x + 16$

Pascal's Triangle

$$
\begin{array}{ccccccccccc}
 & & & & & 1 & & & & & \\
 & & & & 1 & & 1 & & & & \\
 & & & 1 & & 2 & & 1 & & & \\
 & & 1 & & 3 & & 3 & & 1 & & \\
 & 1 & & 4 & & 6 & & 4 & & 1 & \\
1 & & 5 & & 10 & & 10 & & 5 & & 1
\end{array}
$$

Instead of writing out the binomial theorem and calculating the binomial coefficients using factorials every time you want to do a binomial expansion, we now present an alternative, more convenient way of remembering the binomial coefficients, called **Pascal's triangle**.

Notice that the first and last number in every row is 1. Each of the other numbers is found by adding the two numbers directly above it. For example,

$$3 = 2 + 1 \qquad \mathbf{4 = 1 + 3} \qquad \mathbf{10 = 6 + 4}$$

Let's arrange values of $\binom{n}{k}$ in a triangular pattern.

$$\binom{0}{0}$$
$$\binom{1}{0} \quad \binom{1}{1}$$
$$\binom{2}{0} \quad \binom{2}{1} \quad \binom{2}{2}$$
$$\binom{3}{0} \quad \binom{3}{1} \quad \binom{3}{2} \quad \binom{3}{3}$$
$$\binom{4}{0} \quad \binom{4}{1} \quad \binom{4}{2} \quad \binom{4}{3} \quad \binom{4}{4}$$
$$\binom{5}{0} \quad \binom{5}{1} \quad \binom{5}{2} \quad \binom{5}{3} \quad \binom{5}{4} \quad \binom{5}{5}$$
$$\binom{6}{0} \quad \binom{6}{1} \quad \binom{6}{2} \quad \binom{6}{3} \quad \binom{6}{4} \quad \binom{6}{5} \quad \binom{6}{6}$$

It turns out that these numbers in Pascal's triangle are exactly the coefficients in a binomial expansion.

$$1$$
$$1a + 1b$$
$$1a^2 + 2ab + 1b^2$$
$$1a^3 + 3a^2b + 3ab^2 + 1b^3$$
$$1a^4 + 4a^3b + 6a^2b^2 + 4ab^3 + 1b^4$$
$$1a^5 + 5a^4b + 10a^3b^2 + 10a^2b^3 + 5ab^4 + 1b^5$$

The top row is called the *zero row* because it corresponds to the binomial raised to the zero power, $n = 0$. Since each row in Pascal's triangle starts and ends with a 1 and all other values are found by adding the two numbers directly above it, we can now easily calculate the sixth row.

$$(a + b)^5 = 1a^5 + 5a^4b + 10a^3b^2 + 10a^2b^3 + 5ab^4 + 1b^5$$

$$(a + b)^6 = 1a^6 + 6a^5b + 15a^4b^2 + 20a^3b^3 + 15a^2b^4 + 6ab^5 + 1b^6$$

Study Tip

Since the top row of Pascal's triangle is called the zero row, the fifth row is the row with 6 coefficients. The nth row is the row with $n + 1$ coefficients.

EXAMPLE 4 Applying Pascal's Triangle in a Binomial Expansion

Use Pascal's triangle to determine the binomial expansion of $(x + 2)^5$.

Solution:

Write the binomial expansion with blanks for coefficients.

$$(x + 2)^5 = \square x^5 + \square x^4 \cdot 2 + \square x^3 \cdot 2^2 + \square x^2 \cdot 2^3 + \square x \cdot 2^4 + \square 2^5$$

Write the binomial coefficients in the *fifth* row of Pascal's triangle.

$$1, 5, 10, 10, 5, 1$$

Substitute these coefficients into the blanks of the binomial expansion.

$$(x + 2)^5 = 1x^5 + 5x^4 \cdot 2 + 10x^3 \cdot 2^2 + 10x^2 \cdot 2^3 + 5x \cdot 2^4 + 1 \cdot 2^5$$

Simplify. $\boxed{(x + 2)^5 = x^5 + 10x^4 + 40x^3 + 80x^2 + 80x + 32}$

■ YOUR TURN Apply Pascal's triangle to determine the binomial expansion of $(x + 3)^4$.

■ Answer:
$x^4 + 12x^3 + 54x^2 + 108x + 81$

EXAMPLE 5 Applying Pascal's Triangle in a Binomial Expansion

Use Pascal's triangle to determine the binomial expansion of $(2x + 5)^4$.

Solution:

Write the binomial expansion with blanks for coefficients.

$$(2x + 5)^4 = \square(2x)^4 + \square(2x)^3 \cdot 5 + \square(2x)^2 \cdot 5^2 + \square(2x) \cdot 5^3 + \square 5^4$$

Write the binomial coefficients in the *fourth* row of Pascal's triangle.

$$1, 4, 6, 4, 1$$

Substitute these coefficients into the blanks of the binomial expansion.

$$(2x + 5)^4 = 1(2x)^4 + 4(2x)^3 \cdot 5 + 6(2x)^2 \cdot 5^2 + 4(2x) \cdot 5^3 + 1 \cdot 5^4$$

Simplify. $\boxed{(2x + 5)^4 = 16x^4 + 160x^3 + 600x^2 + 1000x + 625}$

■ YOUR TURN Use Pascal's triangle to determine the binomial expansion of:

a. $(3x + 2)^3$ **b.** $(3x - 2)^5$

■ Answer:
a. $27x^3 + 54x^2 + 36x + 8$
b. $243x^5 - 810x^4 + 540x^3 - 720x^2 + 240x - 32$

Finding a Particular Term of a Binomial Expansion

What if we don't want to find the entire expansion, but instead want just a single term? For example, what is the fourth term of $(a + b)^5$?

WORDS	MATH
Recall the sigma notation.	$(a+b)^n = \sum_{k=0}^{n} \binom{n}{k} a^{n-k} b^k$
Let $n = 5$.	$(a+b)^5 = \sum_{k=0}^{5} \binom{5}{k} a^{5-k} b^k$
Expand.	$(a+b)^5 = \binom{5}{0}a^5 + \binom{5}{1}a^4b + \binom{5}{2}a^3b^2 + \underbrace{\binom{5}{3}a^2b^3}_{\text{fourth term}} + \binom{5}{4}ab^4 + \binom{5}{5}b^5$
Simplify the fourth term.	$10a^2b^3$

FINDING A PARTICULAR TERM OF A BINOMIAL EXPANSION

The $(r + 1)$ term of the expansion $(a + b)^n$ is $\binom{n}{r} a^{n-r} b^r$.

EXAMPLE 6 Finding a Particular Term of a Binomial Expansion

Find the fifth term of the binomial expansion of $(2x - 7)^6$.

Solution:

Recall that the $r + 1$ term of $(a + b)^n$ is $\binom{n}{r} a^{n-r} b^r$.

For the fifth term, let $r = 4$. $\binom{n}{4} a^{n-4} b^4$

For this expansion, let $a = 2x$, $b = -7$, $n = 6$. $\binom{6}{4} (2x)^{6-4}(-7)^4$

Note that $\binom{6}{4} = 15$. $15(2x)^2(-7)^4$

Simplify. $\boxed{144{,}060x^2}$

■ **YOUR TURN** What is the third term of the binomial expansion of $(3x - 2)^5$?

Technology Tip

6 [MATH] [▶] [PRB] [▼] [3: nCr] [ENTER] 4 [x] 2 [x²] [x] [(] [(−)] 7 [)] [^] 4 [ENTER]

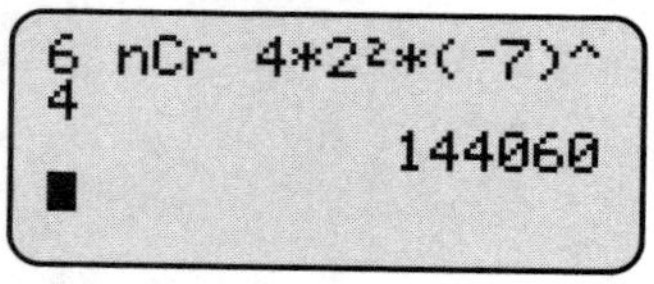

■ **Answer:** $1080x^3$

SECTION 9.5 SUMMARY

In this section we developed a formula for raising a binomial expression to an integer power, $n \geq 0$. The patterns that surfaced were that the expansion displays symmetry between the two terms; that is, every expansion has $n + 1$ terms, the powers sum to n, and the coefficients, called binomial coefficients, are ratios of factorials:

$$(a+b)^n = \sum_{k=0}^{n} \binom{n}{k} a^{n-k} b^k$$

$$\binom{n}{k} = \frac{n!}{(n-k)!\,k!}$$

Also, Pascal's triangle, a shortcut method for evaluating the binomial coefficients, was discussed. The patterns in the triangle are that every row begins and ends with 1 and all other numbers are found by adding the two numbers above the entry.

$$\begin{array}{c} 1 \\ 1 \quad 1 \\ 1 \quad 2 \quad 1 \\ 1 \quad 3 \quad 3 \quad 1 \\ 1 \quad 4 \quad 6 \quad 4 \quad 1 \\ 1 \quad 5 \quad 10 \quad 10 \quad 5 \quad 1 \end{array}$$

Last, a formula was given for finding a particular term of a binomial expansion; the $(r + 1)$ term of $(a + b)^n$ is $\binom{n}{r} a^{n-r} b^r$.

SECTION 9.5 EXERCISES

SKILLS

In Exercises 1–10, evaluate the binomial coefficients.

1. $\binom{7}{3}$ **2.** $\binom{8}{2}$ **3.** $\binom{10}{8}$ **4.** $\binom{23}{21}$ **5.** $\binom{17}{0}$

6. $\binom{100}{0}$ **7.** $\binom{99}{99}$ **8.** $\binom{52}{52}$ **9.** $\binom{48}{45}$ **10.** $\binom{29}{26}$

In Exercises 11–32, expand the expression using the binomial theorem.

11. $(x+2)^4$ **12.** $(x+3)^5$ **13.** $(y-3)^5$ **14.** $(y-4)^4$

15. $(x+y)^5$ **16.** $(x-y)^6$ **17.** $(x+3y)^3$ **18.** $(2x-y)^3$

19. $(5x-2)^3$ **20.** $(a-7b)^3$ **21.** $\left(\frac{1}{x}+5y\right)^4$ **22.** $\left(2x+\frac{3}{y}\right)^4$

23. $(x^2+y^2)^4$ **24.** $(r^3-s^3)^3$ **25.** $(ax+by)^5$ **26.** $(ax-by)^5$

27. $(\sqrt{x}+2)^6$ **28.** $(3+\sqrt{y})^4$ **29.** $(a^{3/4}+b^{1/4})^4$ **30.** $(x^{2/3}+y^{1/3})^3$

31. $(x^{1/4}+2\sqrt{y})^4$ **32.** $(\sqrt{x}-3y^{1/4})^8$

In Exercises 33–36, expand the expression using Pascal's triangle.

33. $(r-s)^4$ **34.** $(x^2+y^2)^7$ **35.** $(ax+by)^6$ **36.** $(x+3y)^4$

In Exercises 37–44, find the coefficient *C* of the term in the binomial expansion.

Binomial	Term	Binomial	Term	Binomial	Term
37. $(x+2)^{10}$	Cx^6	**38.** $(3+y)^9$	Cy^5	**39.** $(y-3)^8$	Cy^4
40. $(x-1)^{12}$	Cx^5	**41.** $(2x+3y)^7$	Cx^3y^4	**42.** $(3x-5y)^9$	Cx^2y^7
43. $(x^2+y)^8$	Cx^8y^4	**44.** $(r-s^2)^{10}$	Cr^6s^8		

APPLICATIONS

In later sections, you will learn the "*n* choose *k*" notation for combinations.

45. Lottery. In a state lottery in which six numbers are drawn from a possible 40 numbers, the number of possible six-number combinations is equal to $\binom{40}{6}$. How many possible combinations are there?

46. Lottery. In a state lottery in which six numbers are drawn from a possible 60 numbers, the number of possible six-number combinations is equal to $\binom{60}{6}$. How many possible combinations are there?

47. Poker. With a deck of 52 cards, 5 cards are dealt in a game of poker. There are a total of $\binom{52}{5}$ different 5-card poker hands that can be dealt. How many possible hands are there?

48. Canasta. In the card game canasta, two decks of cards including the jokers are used and 11 cards are dealt to each person. There are a total of $\binom{108}{11}$ different 11-card canasta hands that can be dealt. How many possible hands are there?

CATCH THE MISTAKE

In Exercises 49 and 50, explain the mistake that is made.

49. Evaluate the expression $\binom{7}{5}$.

Solution:

Write out the binomial coefficient in terms of factorials. $\binom{7}{5} = \dfrac{7!}{5!}$

Write out the factorials. $\binom{7}{5} = \dfrac{7!}{5!} = \dfrac{7 \cdot 6 \cdot 5 \cdot 4 \cdot 3 \cdot 2 \cdot 1}{5 \cdot 4 \cdot 3 \cdot 2 \cdot 1}$

Simplify. $\binom{7}{5} = \dfrac{7!}{5!} = \dfrac{7 \cdot 6}{1} = 42$

This is incorrect. What mistake was made?

50. Expand $(x + 2y)^4$.

Solution:

Write out with blanks.

$$(x + 2y)^4 = \square x^4 + \square x^3y + \square x^2y^2 + \square xy^3 + \square y^4$$

Write out the terms from the fifth row of Pascal's triangle.

1, 4, 6, 4, 1

Substitute these coefficients into the binomial expansion.

$$(x + 2y)^4 = x^4 + 4x^3y + 6x^2y^2 + 4xy^3 + y^4$$

This is incorrect. What mistake was made?

CONCEPTUAL

In Exercises 51–54, determine whether each statement is true or false.

51. The binomial expansion of $(x + y)^{10}$ has 10 terms.

52. The binomial expansion of $(x^2 + y^2)^{15}$ has 16 terms.

53. $\binom{n}{n} = 1$

54. $\binom{n}{-n} = -1$

CHALLENGE

55. Show that $\binom{n}{k} = \binom{n}{n-k}$, if $0 \le k \le n$.

56. Show that if n is a positive integer, then:

$$\binom{n}{0} + \binom{n}{1} + \binom{n}{2} + \cdots + \binom{n}{n} = 2^n$$

Hint: Let $2^n = (1 + 1)^n$ and use the binomial theorem to expand.

TECHNOLOGY

57. With a graphing utility, plot $y_1 = 1 - 3x + 3x^2 - x^3$, $y_2 = -1 + 3x - 3x^2 + x^3$, and $y_3 = (1 - x)^3$ in the same viewing screen. What is the binomial expansion of $(1 - x)^3$?

58. With a graphing utility, plot $y_1 = (x + 3)^4$, $y_2 = x^4 + 4x^3 + 6x^2 + 4x + 1$, and $y_3 = x^4 + 12x^3 + 54x^2 + 108x + 81$. What is the binomial expansion of $(x + 3)^4$?

59. With a graphing utility, plot $y_1 = 1 - 3x$, $y_2 = 1 - 3x + 3x^2$, $y_3 = 1 - 3x + 3x^2 - x^3$, and $y_4 = (1 - x)^3$ for $-1 < x < 1$. What do you notice happening each time an additional term is added? Now, let $1 < x < 2$. Does the same thing happen?

60. With a graphing utility, plot $y_1 = 1 - \dfrac{3}{x}$, $y_2 = 1 - \dfrac{3}{x} + \dfrac{3}{x^2}$, $y_3 = 1 - \dfrac{3}{x} + \dfrac{3}{x^2} - \dfrac{1}{x^3}$, and $y_4 = \left(1 - \dfrac{1}{x}\right)^3$ for $1 < x < 2$. What do you notice happening each time an additional term is added? Now, let $0 < x < 1$. Does the same thing happen?

61. With a graphing utility, plot $y_1 = 1 + \dfrac{3}{x}$, $y_2 = 1 + \dfrac{3}{x} + \dfrac{3}{x^2}$, $y_3 = 1 + \dfrac{3}{x} + \dfrac{3}{x^2} - \dfrac{1}{x^4}$, and $y_4 = \left(1 + \dfrac{1}{x}\right)^3$ for $1 < x < 2$. What do you notice happening each time an additional term is added? Now, let $0 < x < 1$. Does the same thing happen?

62. With a graphing utility, plot $y_1 = 1 + \dfrac{x}{1!}$, $y_2 = 1 + \dfrac{x}{1!} + \dfrac{x^2}{2!}$, $y_3 = 1 + \dfrac{x}{1!} + \dfrac{x^2}{2!} - \dfrac{x^3}{3!}$, and $y_4 = e^x$ for $-1 < x < 1$. What do you notice happening each time an additional term is added? Now, let $1 < x < 2$. Does the same thing happen?

SKILLS OBJECTIVES

- Apply the fundamental counting principle to solve counting problems.
- Apply permutations to solve counting problems.
- Apply combinations to solve counting problems.

CONCEPTUAL OBJECTIVE

- Understand the difference between permutations and combinations.

The Fundamental Counting Principle

You are traveling through Europe for the summer and decide the best packing option is to select separates that can be mixed and matched. You pack one pair of shorts and one pair of khaki pants. You pack a pair of Teva sport sandals and a pair of sneakers. You have three shirts (red, blue, and white). How many different outfits can be worn using only the clothes mentioned above?

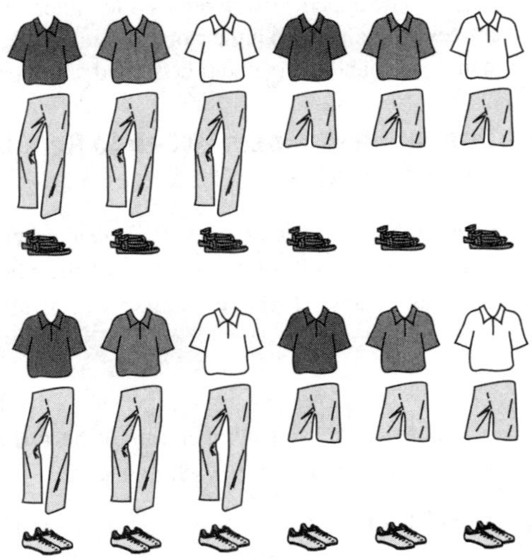

The answer is 12. There are two options for bottoms (pants or shorts), three options for shirts, and two options for shoes. The product of these is

$$2 \cdot 3 \cdot 2 = 12$$

The general formula for counting possibilities is given by the *fundamental counting principle.*

FUNDAMENTAL COUNTING PRINCIPLE

Let E_1 and E_2 be two independent events. The first event E_1 can occur in m_1 ways. The second event E_2 can occur in m_2 ways. The number of ways that the combination of the two events can occur is $m_1 \cdot m_2$.

In other words, *the number of ways in which successive things can occur is found by multiplying the number of ways each thing can occur.*

Study Tip

The fundamental counting principle can be extended to more than two events.

EXAMPLE 1 Possible Meals Served at a Restaurant

A restaurant is rented for a retirement party. The owner offers an appetizer, an entrée, and a dessert for a set price. The following are the choices that people attending the party may choose from. How many possible dinners could be served that night?

Appetizers:	calamari, stuffed mushrooms, or caesar salad
Entrées:	tortellini alfredo, shrimp scampi, eggplant parmesan, or chicken marsala
Desserts:	tiramisu or flan

Solution:

There are three possible appetizers, four possible entrées, and two possible desserts.

Write the product of possible options. $3 \cdot 4 \cdot 2 = 24$

There are 24 possible dinners for the retirement party.

■ **Answer:** 8

■ **YOUR TURN** In Example 1, the restaurant will lower the cost per person for the retirement party if the number of appetizers and entrées is reduced. Suppose the appetizers are reduced to either soup or salad and the entrées are reduced to either tortellini or eggplant parmesan. How many possible dinners could be served at the party?

EXAMPLE 2 Telephone Numbers (When to Require 10-Digit Dialing)

In many towns in the United States, residents can call one another using a 7-digit dialing system. In some large cities, 10-digit dialing is required because two or more area codes coexist. Determine how many telephone numbers can be allocated in a 7-digit dialing area.

Solution:

With 7-digit telephone numbers, the first number cannot be a 0 or a 1, but each of the following six numbers can be 0, 1, 2, 3, 4, 5, 6, 7, 8, or 9.

First number: 2, 3, 4, 5, 6, 7, 8, or 9.	8 possible digits
Second number: 0, 1, 2, 3, 4, 5, 6, 7, 8, or 9.	10 possible digits
Third number:	10 possible digits
Fourth number:	10 possible digits
Fifth number:	10 possible digits
Sixth number:	10 possible digits
Seventh number:	10 possible digits
Counting principle:	$8 \cdot 10 \cdot 10 \cdot 10 \cdot 10 \cdot 10 \cdot 10$
Possible telephone numbers:	8,000,000

Eight million 7-digit telephone numbers can be allocated within one area code.

■ **Answer:** 6.4 billion

■ **YOUR TURN** If the first digit of an area code cannot be 0 or 1, but the second and third numbers of an area code can be 0, 1, 2, 3, 4, 5, 6, 7, 8, or 9, how many 10-digit telephone numbers can be allocated in the United States?

The fundamental counting principle applies when an event can occur more than once. We now introduce two other concepts, *permutations* and *combinations*, which allow individual events to occur only once. For example, in Example 2, the allowable telephone numbers can include the same number in two or more digit places, as in 555-1212. However, in many state lottery games, once a number is selected, it cannot be used again.

An important distinction between a *permutation* and a *combination* is that in a *permutation* order matters, but in a combination order does not matter. For example, the Florida winning lotto numbers one week could be 2–3–5–11–19–27. This would be a *combination* because the order in which they are drawn does not matter. However, if you were betting on a trifecta at the Kentucky Derby, to win you must not only select the first, second, and third place horses, you must select them in the order in which they finished. This would be a *permutation.*

Permutations

DEFINITION **Permutation**

A **permutation** is an *ordered* arrangement of distinct objects without repetition.

EXAMPLE 3 **Finding the Number of Permutations of n Objects**

How many permutations are possible for the letters A, B, C, and D?

Solution:

ABCD	ABDC	ACBD	ACDB	ADCB	ADBC
BACD	BADC	BCAD	BCDA	BDCA	BDAC
CABD	CADB	CBAD	CBDA	CDAB	CDBA
DABC	DACB	DBCA	DBAC	DCAB	DCBA

There are 24 (or 4!) possible permutations of the letters A, B, C, and D.

Notice that in the first row of permutations in Example 3, A was selected for the first space. That left one of the remaining three letters to fill the second space. Once that was selected there remained two letters to choose from for the third space, and then the last space was filled with the unselected letter. In general, there are $n!$ ways to order n objects.

NUMBER OF PERMUTATIONS OF n OBJECTS

The number of permutations of n objects is

$$n! = n \cdot (n-1) \cdot (n-2) \cdots 2 \cdot 1$$

Study Tip

In a permutation of objects, order matters. That is, the same objects arranged in a different order are considered to be a distinct permutation.

EXAMPLE 4 Running Order of Dogs

In an American Kennel Club (AKC) sponsored field trial, the dogs compete in random order. If there are nine dogs competing in the trials, how many possible running orders are there?

Courtesy Cynthia Young, University of Central Florida

Solution:

There are nine dogs that will run. $n = 9$

The number of possible running orders is $n!$ $n! = 9! = \boxed{362{,}880}$

There are $\boxed{362{,}880}$ different possible running orders of nine dogs.

■ **Answer:** 120

■ **YOUR TURN** Five contestants in the Miss America pageant reach the live television interview round. In how many possible orders can the contestants compete in the interview round?

In Examples 3 and 4, we were interested in all possible permutations. Sometimes, we are interested in only some permutations. For instance, 20 horses usually run in the Kentucky Derby. If you bet on a trifecta, you must pick the top three places in the correct order to win. Therefore, we do not consider all possible permutations of 20 horses finishing (places 1–20). Instead, we only consider the possible permutations of first, second, and third place finishes of the 20 horses. We would call this *a permutation of 20 objects taken 3 at a time.* In general, this ordering is called *a permutation of n objects taken r at a time.*

If 20 horses are entered in the Kentucky Derby, there are 20 possible first-place finishers. We consider the permutations with one horse in first place. That leaves 19 possible horses for second place, and then 18 possible horses for third place. Therefore, there are $20 \cdot 19 \cdot 18 = 6840$ possible winning scenarios for the trifecta. This can also be represented as $\frac{20!}{(20-3)!} = \frac{20!}{17!}$.

NUMBER OF PERMUTATIONS OF n OBJECTS TAKEN r AT A TIME

The number of permutations of n objects taken r at a time is

$$_nP_r = \frac{n!}{(n-r)!} = n(n-1)(n-2)\cdots(n-r+1)$$

EXAMPLE 5 Starting Lineup for a Volleyball Team

The starters for a six-woman volleyball team have to be listed in a particular order (1–6). If there are 13 women on the team, how many possible starting lineups are there?

Solution:

Identify the total number of players. $n = 13$

Identify the total number of starters in the lineup. $r = 6$

Substitute $n = 13$ and $r = 6$ into $_nP_r = \dfrac{n!}{(n-r)!}$.

$$_{13}P_6 = \frac{13!}{(13-6)!} = \frac{13!}{7!} = 13 \cdot 12 \cdot 11 \cdot 10 \cdot 9 \cdot 8 = \boxed{1{,}235{,}520}$$

There are $\boxed{1{,}235{,}520}$ possible combinations.

■ **YOUR TURN** A softball team has 12 players, 10 of whom will be in the starting lineup (batters 1–10). How many possible starting lineups are there for this team?

Technology Tip

Use a calculator to find the coefficient of the fifth term $_{13}P_6$.

Scientific calculators:

Press	*Display*
13 [nPr] 6 [=]	1235520

Graphing calculators:

6 [MATH] [▶] [PRB] [▼] [2: nPr] [ENTER] 6 [ENTER]

```
13 nPr 6
         1235520
```

■ **Answer:** 239,500,800

Combinations

The difference between a permutation and a combination is that a permutation has an order associated with it, whereas a *combination* does not have an order associated with it.

DEFINITION Combination

A **combination** is an arrangement, without specific order, of distinct objects without repetition.

The six winning Florida lotto numbers and the NCAA men's Final Four basketball tournament are examples of combinations (six numbers and four teams) without regard to order. The number of combinations of n objects taken r at a time is equal to the binomial coefficient $\dbinom{n}{r}$.

NUMBER OF COMBINATIONS OF n OBJECTS TAKEN r AT A TIME

The number of combinations of n objects taken r at a time is

$$_nC_r = \binom{n}{r} = \frac{n!}{(n-r)!r!}$$

Compare the number of permutations $_nP_r = \frac{n!}{(n-r)!}$ and the number of combinations $_nC_r = \frac{n!}{(n-r)!r!}$. It makes sense that the number of combinations is less than the number of permutations. The denominator is larger because there are no separate orders associated with a combination.

EXAMPLE 6 Possible Combinations to the Lottery

If there are a possible 59 numbers and the lottery officials draw 6 numbers, how many possible combinations are there?

Solution:

Identify how many numbers are in the drawing. $n = 59$

Identify how many numbers are chosen. $r = 6$

Substitute $n = 59$ and $r = 6$ into $_nC_r = \frac{n!}{(n-r)!r!}$.

$$_{59}C_6 = \frac{59!}{(59-6)!6!}$$

Simplify.

$$= \frac{59 \cdot 58 \cdot 57 \cdot 56 \cdot 55 \cdot 54 \cdot (53)!}{53! \cdot 6!}$$

$$= \frac{59 \cdot 58 \cdot 57 \cdot 56 \cdot 55 \cdot 54}{6 \cdot 5 \cdot 4 \cdot 3 \cdot 2}$$

$$= \boxed{45{,}057{,}474}$$

There are $\boxed{45{,}057{,}474}$ possible combinations.

YOUR TURN What are the possible combinations for a lottery with 49 possible numbers and 6 drawn numbers?

Technology Tip

Use a calculator to find the coefficient of the fifth term $_{59}C_6$.

Scientific calculators:

Press	*Display*
59 [nCr] 6 [=]	45057474

Graphing calculators:

59 [MATH] [▶] [PRB] [▼] [3: nCr] [ENTER] 6 [ENTER]

```
59 nCr 6
        45057474
```

■ **Answer:** 13,983,816

Permutations with Repetition

Permutations and combinations are arrangements of distinct (nonrepeated) objects. A permutation in which some of the objects are repeated is called a **permutation with repetition** or a **nondistinguishable permutation**. For example, if a sack has three red marbles, two blue marbles, and one white marble, how many possible permutations would there be when drawing six marbles, one at a time?

This is a different problem from writing the numbers 1 through 6 on pieces of paper, putting them in a hat, and drawing them out. The reason the problems are different is that the two blue balls are indistinguishable and the three red balls are also indistinguishable. The possible permutations for drawing numbers out of the hat are 6!, whereas the possible permutations for drawing balls out of the sack are given by

$$\frac{6!}{3! \cdot 2! \cdot 1!}$$

Number of Distinguishable Permutations

If a set of n objects has n_1 of one kind of object, n_2 of another kind of object, n_3 of a third kind of object, and so on for k different types of objects so that $n = n_1 + n_2 + \cdots + n_k$, then the number of **distinguishable permutations** of the n objects is

$$\frac{n!}{n_1! \cdot n_2! \cdot n_3! \cdots n_k!}$$

In our sack of marbles, there were six marbles $n = 6$. Specifically, there were three red marbles ($n_1 = 3$) two blue marbles ($n_2 = 2$) and one white marble ($n_3 = 1$). Notice that $n = n_1 + n_2 + n_3$ and that the number of distinguishable permutations is equal to

$$\frac{6!}{3! \cdot 2! \cdot 1!} = 60$$

Example 7 Peg Game at Cracker Barrel

The peg game on the tables at Cracker Barrel is a triangle with 15 holes drilled in it, in which pegs are placed. There are 5 red pegs, 5 white pegs, 3 blue pegs, and 2 yellow pegs. If all 15 pegs are in the holes, how many different ways can the pegs be aligned?

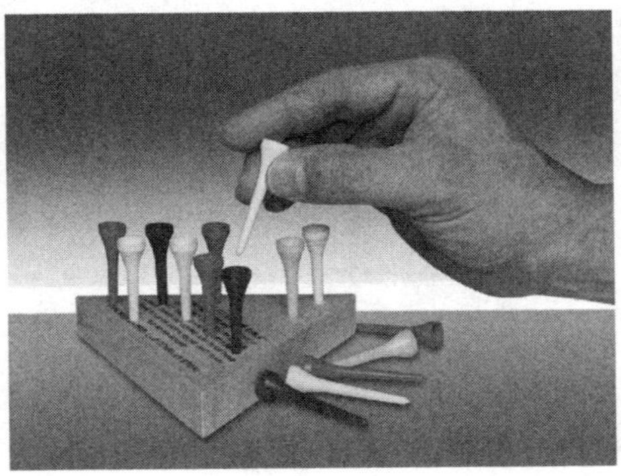

Andy Washnik

Solution:

There are four different colors of pegs (red, white, blue, and yellow).

5 red pegs:	$n_1 = 5$
5 white pegs:	$n_2 = 5$
3 blue pegs:	$n_3 = 3$
2 yellow pegs:	$n_4 = 2$
There are 15 pegs total:	$n = 1$

Substitute $n = 15, n_1 = 5, n_2 = 5, n_3 = 3$, and $n_4 = 2$ into $\frac{n!}{n_1! \cdot n_2! \cdot n_3! \cdots n_k!}$. $\qquad \frac{15!}{5! \cdot 5! \cdot 3! \cdot 2!}$

Simplify. $\qquad = \boxed{7{,}567{,}560}$

There are a possible $\boxed{7{,}567{,}560}$ ways to insert the 15 colored pegs at the Cracker Barrel.

YOUR TURN Suppose a similar game to the peg game at Cracker Barrel is set up with only ten holes in a triangle. With 5 red pegs, 2 white pegs, and 3 blue pegs, how many different permutations can fill that board?

Technology Tip

Use a calculator to calculate $\frac{15!}{5! \cdot 5! \cdot 3! \cdot 2!}$.

Scientific calculators:

Press	*Display*
15 [!] [÷] [(] 5 [!] [x] 5 [!] [x] 3 [!] [x] 2 [!] [)] [÷]	7567560

Graphing calculators:

15 [MATH] [▶] [PRB] [▼] [4: !] [ENTER] [÷] [(] 5 [MATH] [▶] [PRB] [▼] [4: !] [ENTER] 5 [MATH] [▶] [PRB] [▼] [4: !] [ENTER] 3 [MATH] [▶] [PRB] [▼] [4: !] [ENTER] 2 [MATH] [▶] [PRB] [▼] [4: !] [ENTER] [)] [ENTER]

■ **Answer:** 2520

SECTION 9.6 SUMMARY

In this section we discussed the fundamental counting principle, permutations, and combinations.

The Fundamental Counting Principle Is Applicable When

- Objects can be repeated.
- The objects can occur in any order.
- The first event E_1 can occur m_1 ways, and the second event E_2 can occur m_2 ways: the number of ways successive events can occur is m_1m_2 ways.

Permutations

- Objects *cannot* be repeated.
- Order *matters.*
- Number of permutations of n objects: $n!$.
- Number of permutations of n objects taken r at a time:

$$_nP_r = \frac{n!}{(n-r)!}$$

Combinations

- Objects *cannot* be repeated.
- Order *does not* matter.
- Number of combinations of n objects taken r at a time:

$$_nC_r = \frac{n!}{(n-r)!r!}$$

Nondistinguishable Permutations

- Some objects are repeated because they are not distinguishable.
- For n objects with k different types of objects:

$$\frac{n!}{n_1!n_2!n_3!\cdots n_k!}$$

SECTION 9.6 EXERCISES

SKILLS

In Exercises 1–8, use the formula for $_nP_r$ to evaluate each expression.

1. $_6P_4$
2. $_7P_3$
3. $_9P_5$
4. $_9P_4$
5. $_8P_8$
6. $_6P_6$
7. $_{13}P_3$
8. $_{20}P_3$

In Exercises 9–18, use the formula for $_nC_r$ to evaluate each expression.

9. $_{10}C_5$
10. $_9C_4$
11. $_{50}C_6$
12. $_{50}C_{10}$
13. $_7C_7$
14. $_8C_8$
15. $_{30}C_4$
16. $_{13}C_5$
17. $_{45}C_8$
18. $_{30}C_4$

APPLICATIONS

19. **Computers.** At the www.dell.com Website, a customer can "build" a system. If there are four monitors to choose from, three different computers, and two different keyboards, how many possible system configurations are there?

20. **Houses.** In a "new home" community, a person can select from one of four models, five paint colors, three tile selections, and two landscaping options. How many different houses (interior and exterior) are there to choose from?

21. **Wedding Invitations.** An engaged couple is ordering wedding invitations. The wedding invitations come in white or ivory. The writing can be printed, embossed, or engraved. The envelopes can come with liners or without. How many possible designs of wedding invitations are there to choose from?

22. **Dinner.** Siblings are planning their father's 65th birthday dinner and have to select one of four main courses (baked chicken, grilled mahi-mahi, beef Wellington, or lasagna), one of two starches (rosemary potatoes or rice), one of three vegetables (green beans, carrots, or zucchini), and one of five appetizers (soup, salad, pot stickers, artichoke dip, or calamari). How many possible dinner combinations are there?

23. **PIN Number.** Most banks require a four-digit ATM PIN code for each customer's bank card. How many possible four-digit PIN codes are there to choose from?

24. **Password.** All e-mail accounts require passwords. If a four-character password is required that can contain letters (but no numbers), how many possible passwords can there be? (Assume letters are not case sensitive.)

25. Leadership. There are 15 professors in a department and there are four leadership positions (chair, assistant chair, undergraduate coordinator, and graduate coordinator). How many possible leadership teams are there?

26. Fraternity Elections. A fraternity is having elections. There are three men running for president, two men running for vice-president, four men running for secretary, and one man running for treasurer. How many possible outcomes do the elections have?

27. Multiple-Choice Tests. There are 20 questions on a multiple-choice exam, and each question has four possible answers (A, B, C, and D). Assuming no answers are left blank, how many different ways can you answer the questions on the exam?

28. Multiple-Choice Tests. There are 25 questions on a multiple-choice exam, and each question has five possible answers (A, B, C, D, and E). Assuming no answers are left blank, how many different ways can you answer the questions on the exam?

29. Zip Codes. In the United States a five-digit zip code is used to route mail. How many possible five-digit zip codes are possible? (All numbers can be used.) If 0s were eliminated from the first and last digits, how many possible zip codes would there be?

30. License Plates. In a particular state there are six characters in a license plate: three letters followed by three numbers. If 0s and 1s are eliminated from possible numbers and Os and Is are eliminated from possible letters, how many different license plates can be made?

31. Class Seating. If there are 30 students in a class and there are exactly 30 seats, how many possible seating charts can be made, assuming all 30 students are present?

32. Season Tickets. Four friends buy four season tickets to the Green Bay Packers. To be fair, they change the seating arrangement every game. How many different seating arrangements are there for the four friends?

33. Combination "Permutation" Lock. A combination lock on most lockers will open when the correct choice of three numbers (1 to 40) is selected and entered in the correct order. Therefore, a combination lock should really be called a permutation lock. How many possible permutations are there, assuming no numbers can be repeated?

34. Safe. A safe will open when the correct choice of three numbers (1 to 50) is selected and entered in the correct order. How many possible permutations are there, assuming no numbers can be repeated?

35. Raffle. A fundraiser raffle is held to benefit cystic fibrosis research, and 1000 raffle tickets are sold. There are three prizes raffled off. First prize is a round-trip ticket on Delta Air Lines, second prize is a round of golf for four people at a local golf course, and third prize is a $50 gift certificate to Chili's. How many possible winning scenarios are there if all 1000 tickets are sold to different people?

36. Ironman Triathlon. If 100 people compete in an ironman triathlon, how many possible placings are there (first, second, and third place)?

37. Lotto. If a state lottery picks from 53 numbers and 6 numbers are selected, how many possible 6-number combinations are there?

38. Lotto. If a state lottery picks from 53 numbers and 5 numbers are selected, how many possible 5-number combinations are there?

39. Cards. In a deck of 52 cards, how many different 5-card hands can be dealt?

40. Cards. In a deck of 52 cards, how many different 7-card hands can be dealt?

41. Blackjack. In a single-deck blackjack game (52 cards), how many different 2-card combinations are there?

42. Blackjack. In a single deck, how many two-card combinations are there that equal 21: ace (worth 11) and a 10 or face card—jack, queen, or king?

43. March Madness. Every spring, the NCAA men's basketball tournament starts with 64 teams. After two rounds, it is down to the Sweet Sixteen, and after two more rounds, it is reduced to the Final Four. Once 64 teams are selected (but not yet put in brackets), how many possible scenarios are there for the Sweet Sixteen?

44. March Madness. Every spring, the NCAA men's basketball tournament starts with 64 teams. After two rounds, it is down to the Sweet Sixteen, and after two more rounds, it is reduced to the Final Four. Once the 64 teams are identified (but not yet put in brackets), how many possible scenarios are there for the Final Four?

45. NFL Playoffs. There are 32 teams in the National Football League (16 AFC and 16 NFC). How many possible combinations are there for the Superbowl? (Assume one team from the AFC plays one team from the NFC in the Superbowl.)

46. NFL Playoffs. After the regular season in the National Football League, 12 teams make the playoffs (6 from the AFC and 6 from the NFC). How many possible combinations are there for the Superbowl once the 6 teams in each conference are identified?

47. *Survivor*. On the television show *Survivor*, one person is voted off the island every week. When it is down to six contestants, how many possible voting combinations remain, if no one will vote themself off the island? Assume that the order (who votes for whom) makes a difference. How many total possible voting outcomes are there?

48. *American Idol*. On the television show *American Idol*, a young rising star is eliminated from the competition every week. The first week, each of the 12 contestants sings one song. How many possible ways could the contestants be ordered 1–12? How many possible ways could six men and six women be ordered to alternate female and male contestants?

49. *Dancing with the Stars*. In the popular TV show *Dancing with the Stars*, twelve entertainers (six men and six women) compete in a dancing contest. The first night, the show decides to select three men and three women. How many ways can this be done?

50. *Dancing with the Stars*. See Exercise 49. How many ways can six male celebrities line up for a picture alongside six female celebrities?

CATCH THE MISTAKE

In Exercises 51 and 52, explain the mistake that is made.

51. In a lottery that picks from 30 numbers, how many five-number combinations are there?

Solution:

Let $n = 30$ and $r = 5$.

Calculate ${}_nP_r = \dfrac{n!}{(n-r)!}$. ${}_{30}P_5 = \dfrac{30!}{25!}$

Simplify. ${}_{30}P_5 = 17{,}100{,}720$

This is incorrect. What mistake was made?

52. A home owners association has 12 members on the board of directors. How many ways can the board elect a president, vice president, secretary, and treasurer?

Solution:

Let $n = 12$ and $r = 4$.

Calculate ${}_nC_r = \dfrac{n!}{(n-r)! \cdot r!}$. ${}_{12}C_4 = \dfrac{12!}{8! \cdot 4!}$

Simplify. ${}_{12}C_4 = 495$

This is incorrect. What mistake was made?

CONCEPTUAL

In Exercises 53–56, determine whether each statement is true or false.

53. The number of permutations of n objects is always greater than the number of combinations of n objects if the objects are distinct.

54. The number of permutations of n objects is always greater than the number of combinations of n objects even when the objects are indistinguishable.

55. The number of four-letter permutations of the letters A, B, C, and D is equal to the number of four-letter permutations of ABBA.

56. The number of possible answers to a true/false question is a permutation problem.

CHALLENGE

57. What is the relationship between ${}_nC_r$ and ${}_nC_{r+1}$?

58. What is the relationship between ${}_nP_r$ and ${}_nP_{r-1}$?

59. Simplify the expression ${}_nC_r \cdot r!$.

60. What is the relationship between ${}_nC_r$ and ${}_nC_{n-r}$?

TECHNOLOGY

61. Employ a graphing utility with a ${}_nP_r$ feature and compare it with answers to Exercises 1–8.

62. Employ a graphing utility with a ${}_nC_r$ feature and compare it with answers to Exercises 9–18.

63. Use a graphing calculator to evaluate:

a. ${}_{10}P_4$

b. $4!({}_{10}C_4)$

c. Are answers in (a) and (b) the same?

d. Why?

64. Use a graphing calculator to evaluate:

a. ${}_{12}P_5$

b. $5!({}_{12}C_5)$

c. Are answers in (a) and (b) the same?

d. Why?

SECTION 9.7 PROBABILITY

SKILLS OBJECTIVES

- Work with the rules of probability.
- Find the probability of an event.
- Find the probability that an event will not occur.
- Find the probability of mutually exclusive events.
- Find the probability of independent events.

CONCEPTUAL OBJECTIVES

- Understand that the mathematics of probability gives us a good sense of how likely it is that a certain event will happen.
- Understand the difference between the probability of:
  - event 1 *and* event 2
  - event 1 *or* event 2

You are sitting at a blackjack table at Caesar's Palace, and the dealer is showing a 7. You have a 9 and a 7; should you hit? Will it rain today? What will the lotto numbers be this week? Will the coin toss at the Superbowl result in a head or a tail? Will Derek Jeter get a hit at his next trip to the plate? These are all questions where *probability* is used to guide us.

Anything that happens for which the result is uncertain is called an **experiment**. Each trial of an experiment is called an **outcome**. All of the possible outcomes of an experiment constitute the **sample space**. The term **event** is used to describe the kind of possible outcomes. For example, a coin toss is an experiment. Every outcome is either heads or tails. The sample space of a single toss is {heads, tails}.

The result of one experiment has no certain outcome. However, if the experiment is performed many times, the results will produce regular patterns. For example, if you toss a fair coin, you don't know whether it will come up heads or tails. You can toss a coin 10 times and get 10 heads. However, if you made 1,000,000 tosses, you would get about 500,000 heads and 500,000 tails. Therefore, since we assume a head is equally likely as a tail and there are only two possible events (heads or tails), we assign a *probability of a head* equal to $\frac{1}{2}$ and a *probability of a tail* equal to $\frac{1}{2}$.

EXAMPLE 1 Finding the Sample Space

Find the sample space for each of the following outcomes.

a. Tossing a coin once **b.** Tossing a coin twice **c.** Tossing a coin three times

Solution (a):

Tossing a coin one time will result in one of two events: heads (H) or tails (T).

The sample space S is written as $\boxed{S = \{H, T\}}$.

Solution (b):

Tossing a coin twice can result in one of four possible outcomes.

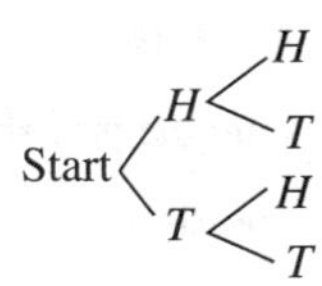

The sample space consists of all possible outcomes.

$$\boxed{S = \{HH, HT, TH, TT\}}$$

Note that TH and HT are two different outcomes.

Solution (c):

There are eight possible outcomes when a coin is tossed three times.

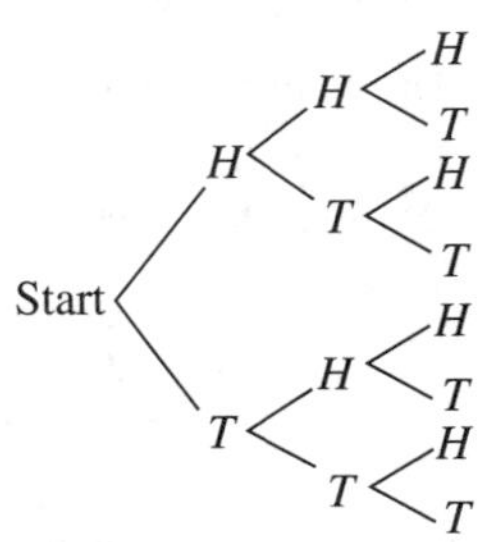

$$\boxed{S = \{HHH, HHT, HTH, HTT, TTT, TTH, THT, THH\}}$$

■ YOUR TURN Find the sample space associated with having three children (B boys or G girls).

Study Tip

If the coin is tossed n times, there are 2^n possible outcomes.

■ Answer:
$S = \{GGG, GGB, GBG, GBB, BBB, BBG, BGB, BGG\}$

Probability of an Event

To calculate the probability of an event, start by counting the number of outcomes in the event and the number of outcomes in the sample space. The ratio is equal to the probability if all outcomes are equally likely.

DEFINITION **Probability of an Event**

If an event E has $n(E)$ equally likely outcomes and its sample space S has $n(S)$ equally likely outcomes, then the **probability of event** E, denoted $P(E)$, is

$$P(E) = \frac{n(E)}{n(S)} = \frac{\text{Number of outcomes in event } E}{\text{Number of outcomes in sample space } S}$$

Since the number of outcomes in an event must be less than or equal to the number of outcomes in the sample space, the probability of an event must be a number between 0 and 1 or equal to 0 or 1; to be precise, $0 \leq P(E) \leq 1$. If $P(E) = 0$, then the event can never happen, and if $P(E) = 1$, the event is certain to happen.

EXAMPLE 2 **Finding the Probability of Two Girls**

If two children are born, what is the probability that they are both girls?

Solution:

The event is both children being girls. $E = \{GG\}$

The sample space is all four possible outcomes. $S = \{BB, BG, GB, GG\}$

The number of outcomes in the event is 1. $n(E) = 1$

The number of events in the sample space is 4. $n(S) = 4$

Compute the probability using $P(E) = \frac{n(E)}{n(S)}$. $P(E) = \boxed{\frac{1}{4}}$

The probability that both children are girls is $\frac{1}{4}$, or 0.25.

EXAMPLE 3 **Finding the Probability of Drawing a Face Card**

Find the probability of drawing a face card (jack, queen, or king) out of a 52-card deck.

Solution:

There are 12 face cards in a deck. $n(E) = 12$

There are 52 cards in a deck. $n(S) = 52$

Compute the probability using $P(E) = \frac{n(E)}{n(S)}$. $P(E) = \frac{12}{52} = \boxed{\frac{3}{13}}$

The probability of drawing a face card out of a 52-card deck is $\frac{3}{13}$ or ≈ 0.23.

■ **YOUR TURN** **a.** Find the probability that an ace is drawn from a deck of 52 cards.
b. Find the probability that a jack, queen, king, or ace is drawn from a deck of 52 cards.

■ **Answer: a.** $\frac{4}{52} = \frac{1}{13}$ **b.** $\frac{16}{52} = \frac{4}{13}$

EXAMPLE 4 Finding the Probability of Rolling a 7 or an 11

If you bet on the "pass line" at a craps table and the person's first roll is a 7 or an 11, using a pair of dice, then you win. Find the probability of winning a pass line bet on the first roll.

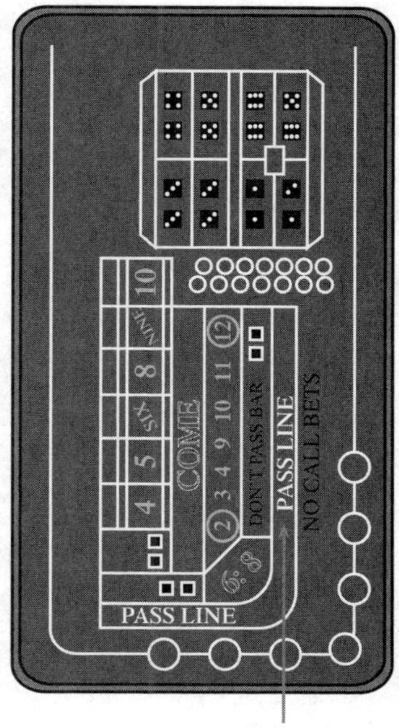

Pass Line

Solution:

The fundamental counting principle tells us that there will be $6 \cdot 6 = 36$ possible rolls of the pair of dice. $n(S) = 36$

Draw a table listing possible sums of the two dice.

DICE VALUE	1	2	3	4	5	6
1	2	3	4	5	6	7
2	3	4	5	6	7	8
3	4	5	6	7	8	9
4	5	6	7	8	9	10
5	6	7	8	9	10	11
6	7	8	9	10	11	12

Of the 36 rolls, there are 8 rolls that will produce a 7 or an 11. $n(E) = 8$

Compute the probability using $P(E) = \frac{n(E)}{n(S)}$. $P(E) = \frac{8}{36} = \boxed{\frac{2}{9}}$

The probability of winning a pass line bet is $\frac{2}{9}$ or ≈ 0.22 (22%).

■ **YOUR TURN** If a 2, 3, or 12 is rolled on the first roll, then the pass line bet loses. Find the probability of losing a pass line bet on the first roll.

■ **Answer:** $\frac{4}{36} = \frac{1}{12} \approx 8.3\%$

Probability of an Event Not Occurring

The sum of the probabilities of all possible outcomes is 1. For example, when a die is rolled, if the outcomes are equally likely, then the probabilities are all $\frac{1}{6}$:

$$P(1) = \frac{1}{6}, P(2) = \frac{1}{6}, P(3) = \frac{1}{6}, P(4) = \frac{1}{6}, P(5) = \frac{1}{6}, P(6) = \frac{1}{6}$$

The sum of these six probabilities is 1.

Since the sum of the probabilities of all possible outcomes sums to 1, we can find the probability that an event won't occur by subtracting the probability that the event will occur from 1.

$$P(E) + P(\text{not } E) = 1 \qquad \text{or} \qquad P(\text{not } E) = 1 - P(E)$$

Study Tip

The following three notations are equivalent.

- $P(E') = 1 - P(E)$
- $P(\overline{E}) = 1 - P(E)$
- $P(\text{not } E) = 1 - P(E)$

PROBABILITY OF AN EVENT NOT OCCURRING

The **probability** that an event E will **not occur** is equal to 1 minus the probability that E will occur.

$$P(\text{not } E) = 1 - P(E)$$

The **complement of an event** E is the collection of all outcomes in the sample space that are not in E. The complement of E is denoted E' or $\overline{E}$.

EXAMPLE 5 Finding the Probability of Not Winning the Lottery

Find the probability of not winning the lottery if six numbers are selected from 1 to 49.

Solution:

Calculate the number of possible six-number combinations.

$${}_{49}C_6 = \frac{49!}{43! \cdot 6!} = 13{,}983{,}816$$

Calculate the probability of winning.

$$P(\text{winning}) = \frac{1}{13{,}983{,}816}$$

Calculate the probability of not winning.

$$P(\text{not winning}) = 1 - P(\text{winning})$$

$$P(\text{not winning}) = 1 - \frac{1}{13{,}983{,}816}$$

$$P(\text{not winning}) = \frac{13{,}983{,}816}{13{,}983{,}816} - \frac{1}{13{,}983{,}816}$$

$$\approx \boxed{0.999999928}$$

The probability of *not* winning the lottery is very close to 1.

■ **YOUR TURN** Find the probability of not winning the lottery if 6 numbers are selected from 1 to 39.

■ **Answer:** 0.9999997

Mutually Exclusive Events

Recall the definition of union and intersection in Section 1.5. The probability of one event E_1 or a second event E_2 occurring is given by the probability of the union of the two events.

$$P(E_1 \cup E_2)$$

If there is any overlap between the two events, we must be careful not to count those twice. For example, what is the probability of drawing *either* a face card *or* a spade out of a deck of 52 cards? We must be careful not to count twice any face cards that are spades.

PROBABILITY OF THE UNION OF TWO EVENTS

If E_1 and E_2 are two events in the same sample space, the probability of *either* E_1 *or* E_2 occurring is given by

$$P(E_1 \cup E_2) = P(E_1) + P(E_2) - P(E_1 \cap E_2)$$

If E_1 and E_2 are disjoint, $E_1 \cap E_2 = 0$, then E_1 and E_2 are **mutually exclusive**. In that case, the probability of *either* E_1 *or* E_2 occurring is given by

$$P(E_1 \cup E_2) = P(E_1) + P(E_2)$$

EXAMPLE 6 Finding the Probability of Drawing a Face Card or a Spade

Find the probability of drawing either a face card or a spade out of a deck of 52 cards.

Solution:

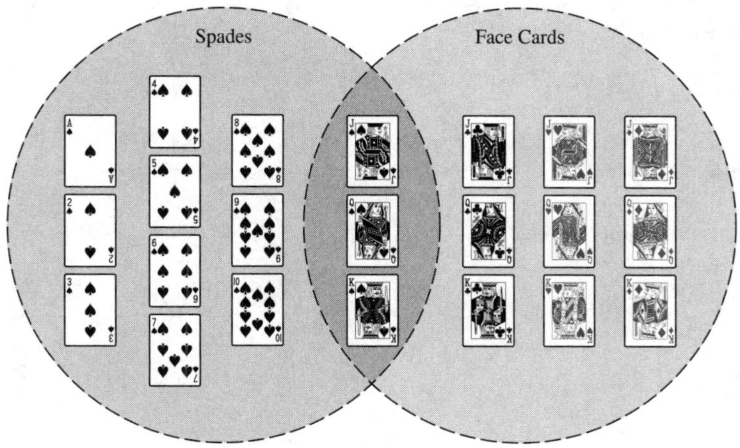

The deck has 12 face cards. $P(\text{face card}) = \frac{12}{52}$

The deck has 13 spades. $P(\text{spade}) = \frac{13}{52}$

The deck has 3 face cards that are spades. $P(\text{face card and spade}) = \frac{3}{52}$

Apply the probability formula. $P(E_1 \cup E_2) = P(E_1) + P(E_2) - P(E_1 \cap E_2)$

$$P(\text{face card or a spade}) = \frac{12}{52} + \frac{13}{52} - \frac{3}{52}$$

Simplify. $P(\text{face card or a spade}) = \frac{22}{52} = \boxed{\frac{11}{26}}$

The probability of either a face card or a spade being drawn is $\frac{11}{26} \approx 0.42$.

■ YOUR TURN Find the probability of drawing either a heart or an ace out of a deck of 52 cards.

■ **Answer:** $\frac{4}{13}$

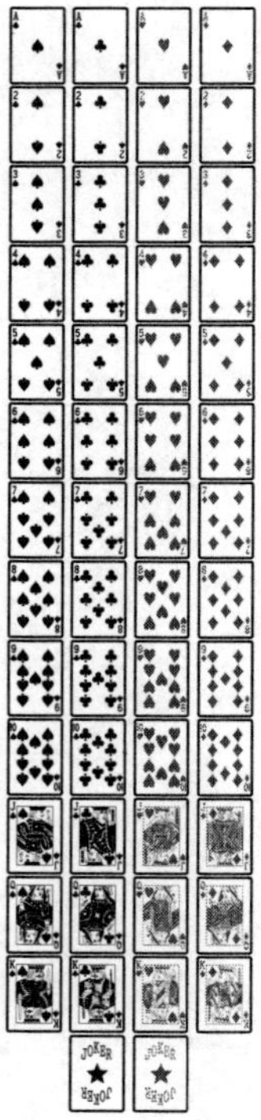

EXAMPLE 7 Finding the Probability of Mutually Exclusive Events

Find the probability of drawing either an ace or a joker in a 54-card deck (a deck with two jokers).

Solution:

Drawing an ace and drawing a joker are two mutually exclusive events, since a card cannot be both an ace and a joker.

The deck has 4 aces. $P(\text{ace}) = \frac{4}{54}$

The deck has 2 jokers. $P(\text{joker}) = \frac{2}{54}$

Apply the probability formula. $P(E_1 \cup E_2) = P(E_1) + P(E_2)$

$$P(\text{ace or a joker}) = \frac{4}{54} + \frac{2}{54}$$

Simplify. $P(\text{ace or a joker}) = \frac{6}{54} = \frac{1}{9} \approx 0.11\overline{1}$

The probability of drawing either an ace or a joker is $\frac{3}{27} \approx 0.11$.

■ **YOUR TURN** If there are 10 women mathematicians, 8 men mathematicians, 6 women engineers, and 12 men engineers, what is the probability that a selected person is *either* a woman *or* an engineer? Calculate the probability that a selected person is *either* a woman *or* an engineer.

■ **Answer:** $\frac{7}{9}$ or ≈ 0.78

Independent Events

Suppose you have two children. The sex of the second child is not affected by the sex of the first child. For example, if your first child is a boy, then it is no less likely that the second child will be a boy. We say that two events are **independent** if the occurrence of either of them has no effect on the probability of the other occurring.

PROBABILITIES OF INDEPENDENT EVENTS

If E_1 and E_2 are independent events, then the probability of both occurring is the product of the individual probabilities:

$$P(E_1 \cap E_2) = P(E_1) \cdot P(E_2)$$

Scientists used to believe that one gene controlled human eye color. Each parent gives one chromosome (either blue, green, or brown), and the result is a child with an eye-color gene composed of combinations, with brown being dominant over blue and blue being dominant over green. The genetic basis for eye color is actually far more complex, but we use this simpler model in the next example.

EXAMPLE 8 Probabilities of Blue-Eyed Children of Brown-Eyed Parents

If two brown-eyed parents have a blue-eyed child, then the parents must each have one blue and one brown-eye gene. In order to have blue eyes, the child must get the blue-eye genes from both parents. What is the probability that two brown-eyed parents can have three children, all with blue eyes?

Solution:

The sample space for childrens' eye-color genes from these parents is

$$S = \{\text{Blue/Blue, Blue/Brown, Brown/Brown, Brown/Blue}\}$$

The only way for a child to have blue eyes is if that child inherits two blue-eyed genes.

$$P(\text{blue-eyed child}) = \frac{1}{4}$$

The eye color of each child is independent of that of the other. The probability of having three blue-eyed children is the product of the three individual probabilities.

$$P(\text{all three children with blue eyes}) = \left(\frac{1}{4}\right)\cdot\left(\frac{1}{4}\right)\cdot\left(\frac{1}{4}\right)$$

$$= \frac{1}{64}$$

The probability of two brown-eyed parents having three blue-eyed children is $\boxed{\frac{1}{64} \approx 0.016}$.

■ **YOUR TURN** Find the probability of the brown-eyed parents in Example 8 having three brown-eyed children.

■ **Answer:** $\frac{27}{64} \approx 0.422$

SECTION 9.7 SUMMARY

In this section we discussed the probability, or likelihood, of an event. It is found by dividing the total number of possible equally likely outcomes in the event by all of the possible outcomes in the sample space.

$$P(E) = \frac{n(E)}{n(S)} = \frac{\text{Number of outcomes in event } E}{\text{Number of outcomes in sample space } S}$$

Probability is a number between 0 and 1 or equal to 0 or 1:

$$0 \le P(E) \le 1$$

The probability of an event not occurring is 1 minus the probability of the event occurring:

$$P(\text{not } E) = 1 - P(E)$$

The probability of one event *or* another event occurring is found by adding the individual probabilities of each event and subtracting the probability of both:

$$\overset{\text{“or”}}{P(E_1 \cup E_2)} = P(E_1) + P(E_2) - P(E_1 \cap E_2)$$

If two events are mutually exclusive, they have no outcomes in common:

$$P(E_1 \cup E_2) = P(E_1) + P(E_2)$$

The probability of two events occurring is the product of the individual probabilities, provided the two events are independent or do not affect one another:

$$\overset{\text{“and”}}{P(E_1 \cap E_2)} = P(E_1)\cdot P(E_2)$$

SECTION 9.7 EXERCISES

SKILLS

In Exercises 1–6, find the sample space for each experiment.

1. The sum of two dice rolled simultaneously.
2. A coin tossed three times in a row.
3. The sex (boy or girl) of four children born to the same parents.
4. Tossing a coin and rolling a die.
5. Two balls selected from a container that has 3 red balls, 2 blue balls, and 1 white ball.
6. The grade (freshman, sophomore, or junior) of two high school students who work at a local restaurant.

Heads or Tails. In Exercises 7–10, find the probability for the experiment of tossing a coin three times.

7. Getting all heads.
8. Getting exactly one heads.
9. Getting at least one heads.
10. Getting more than one heads.

Tossing a Die. In Exercises 11–16, find the probability for the experiment of tossing two dice.

11. The sum is 3.
12. The sum is odd.
13. The sum is even.
14. The sum is prime.
15. The sum is more than 7.
16. The sum is less than 7.

Drawing a Card. In Exercises 17–20, find the probability for the experiment of drawing a single card from a deck of 52 cards.

17. Drawing a non-face card.
18. Drawing a black card.
19. Drawing a 2, 4, 6, or 8.
20. Drawing a 3, 5, 7, 9, or ace.

In Exercises 21–26, let $P(E_1) = \frac{1}{4}$ and $P(E_2) = \frac{1}{2}$ and find the probability of the event.

21. Probability of E_1 not occurring.
22. Probability of E_2 not occurring.
23. Probability of either E_1 or E_2 occurring if E_1 and E_2 are mutually exclusive.
24. Probability of either E_1 or E_2 occurring if E_1 and E_2 are not mutually exclusive and $P(E_1 \cap E_2) = \frac{1}{8}$.
25. Probability of both E_1 and E_2 occurring if E_1 and E_2 are mutually exclusive.
26. Probability of both E_1 and E_2 occurring if E_1 and E_2 are independent.

APPLICATIONS

27. **Cards.** A deck of 52 cards is dealt.

a. How many possible combinations of four-card hands are there?

b. What is the probability of having all spades?

c. What is the probability of having four of a kind?

28. **Blackjack.** A deck of 52 cards is dealt for blackjack.

a. How many possible combinations of two-card hands are there?

b. What is the probability of having 21 points (ace with a 10 or face card)?

29. **Cards.** With a 52-card deck, what is the probability of drawing a 7 or an 8?

30. **Cards.** With a 52-card deck, what is the probability of drawing a red 7 or a black 8?

31. **Cards.** By drawing twice, what is the probability of drawing a 7 and then an 8?

32. **Cards.** By drawing twice, what is the probability of drawing a red 7 and then a black 8?

33. **Children.** What is the probability of having five daughters in a row and no sons?

34. **Children.** What is the probability of having four sons in a row and no daughters?

35. **Children.** What is the probability that of five children at least one is a boy? *Note:* $P(\text{at least one boy}) = 1 - P(\text{no boys})$.

36. **Children.** What is the probability that of six children at least one is a girl? *Note:* $P(\text{at least one girl}) = 1 - P(\text{no girls})$.

37. Roulette. In roulette, there are 38 numbered slots (1–36, 0, and 00). Eighteen are red, 18 are black, and the 0 and 00 are green. What is the probability of having 4 reds in a row?

38. Roulette. What is the probability of having two greens in a row on a roulette table?

39. Item Defectiveness. For a particular brand of DVD players 10% of the ones on the market are defective. If a company has ordered 8 DVD players what is the probability that none of the 8 DVD players is defective?

40. Item Defectiveness. For a particular brand of generators 20% of the ones on the market are defective. If a company buys 10 generators what is the probability that none of the 10 generators is defective?

41. Number Generator. A random-number generator (computer program that selects numbers in no particular order) is used to select two numbers between 1 and 10. What is the probability that both numbers are even?

42. Number Generator. A random-number generator is used to select two numbers between 1 and 15. What is the probability that both numbers are odd?

In Exercises 43 and 44, assume each deal is from a complete (shuffled) single deck of cards.

43. Blackjack. What is the probability of being dealt a blackjack (any ace and any face card) with a single deck?

44. Blackjack. What is the probability of being dealt two blackjacks in a row with a single deck?

45. Sports. With the salary cap in the NFL, it is said that on "any given Sunday" any team could beat any other team. If we assume every week a team has a 50% chance of winning, what is the probability that a team will go 16–0?

46. Sports. With the salary cap in the NFL, it is said that on "any given Sunday" any team could beat any other team. If we assume every week a team has a 50% chance of winning, what is the probability that a team will have at least 1 win?

47. Genetics. Suppose both parents have the brown/blue pair of eye-color genes, and each parent contributes one gene to the child. Suppose the brown eye-color gene is dominant so that if the child has at least one brown gene, the color will dominate and the eyes will be brown.

a. List the possible outcomes (sample space). Assume each outcome is equally likely.

b. What is the probability that the child will have the blue/blue pair of genes?

c. What is the probability that the child will have brown eyes?

48. Genetics. Refer to Exercise 47. In this exercise, the father has a brown/brown pair of eye-color genes, while the mother has a brown/blue pair of eye-color genes.

a. List the possible outcomes (sample space). Assume each outcome is equally likely.

b. What is the probability that the child will have a blue/blue pair of genes?

c. What is the probability that the child will have brown eyes?

49. Playing Cards. How many 5-card hands can be drawn from a 52-card deck (no jokers)?

50. Playing Cards.

a. How many ways can you select 2 aces and 3 other cards (non-aces) from a standard deck of 52 cards (no jokers)?

b. What is the probability that you draw a 5-card hand with 2 aces and 3 non-aces?

51. Playing Cards. Find the probability of drawing 5 clubs from a standard deck of 52 cards.

52. Poker. Find the probability of getting 2 fives and 3 kings when drawing 5 cards from a standard deck of 52 cards.

CATCH THE MISTAKE

In Exercises 53 and 54, explain the mistake that is made.

53. Calculate the probability of drawing a 2 or a spade from a deck of 52 cards.

Solution:

The probability of drawing a 2 from a deck of 52 cards is $\frac{4}{52}$.

The probability of drawing a spade from a deck of 52 cards is $\frac{13}{52}$.

The probability of drawing a 2 or a spade is $\frac{4}{52} + \frac{13}{52} = \frac{17}{52}$.

This is incorrect. What mistake was made?

54. Calculate the probability of having two boys and one girl.

Solution:

The probability of having a boy is $\frac{1}{2}$.

The probability of having a girl is $\frac{1}{2}$.

These are independent, so the probability of having two boys and a girl is $\left(\frac{1}{2}\right)\left(\frac{1}{2}\right)\left(\frac{1}{2}\right) = \frac{1}{8}$.

This is incorrect. What mistake was made?

CONCEPTUAL

In Exercises 55–58, determine whether each statement is true or false.

55. If $P(E_1) = 0.5$ and $P(E_2) = 0.4$, then $P(E_3)$ must equal 0.1 if there are three possible events and they are all mutually exclusive.

56. If two events are mutually exclusive, then they cannot be independent.

57. If two events are independent, then they are not mutually exclusive.

58. The probability of having five sons and no daughters is 1 minus the probability of having five daughters and no sons.

CHALLENGE

59. If two people are selected at random, what is the probability that they have the same birthday? Assume 365 days per year.

60. If 30 people are selected at random, what is the probability that at least two of them will have the same birthday?

61. If one die is weighted so that 3 and 4 are the only numbers that the die will roll, and the other die is equally loaded, what is the probability of rolling two dice that sum to 2, 5, or 6?

62. If one die is weighted so that 3 and 4 are the only numbers that the dice will roll, and 3 comes up twice as often as 4, what is the probability of rolling a 3?

TECHNOLOGY

63. Use a random-number generator on a graphing utility to select two numbers between 1 and 10. Run this generator 50 times. How many times (out of 50 trials) were both of the two numbers even? Compare with your answer from Exercise 41.

64. Use a random-number generator on a graphing utility to select two numbers between 1 and 15. Run this 50 times. How many times (out of 50 trials) were both of the two numbers odd? Compare with your answer from Exercise 42.

In Exercises 65 and 66, when a die is rolled once, the probability of getting a 2 is $\frac{1}{6}$ and the probability of not getting a 2 is $\frac{5}{6}$. If a die is rolled n times, the probability of getting a 2 exactly k times can be found by using the binomial theorem:

$$_nC_k\left(\frac{1}{6}\right)^k\left(\frac{5}{6}\right)^{n-k}$$

65. If a die is rolled 10 times, find the probability of getting a 2 exactly two times. Round your answer to four decimal places.

66. If a die is rolled 8 times, find the probability of getting a 2 at most two times. Round your answer to four decimal places.

MODELING YOUR WORLD

In the Modeling Your World in Chapter 2, you found the equations of lines that correspond to the seven wedges and the stabilization triangle. In 2005, the world was producing 7 billion tons of carbon emissions per year. In 2055, this number is projected to double with the worldwide production of carbon emissions equaling 14 billion tons per year.

The Stabilization Triangle

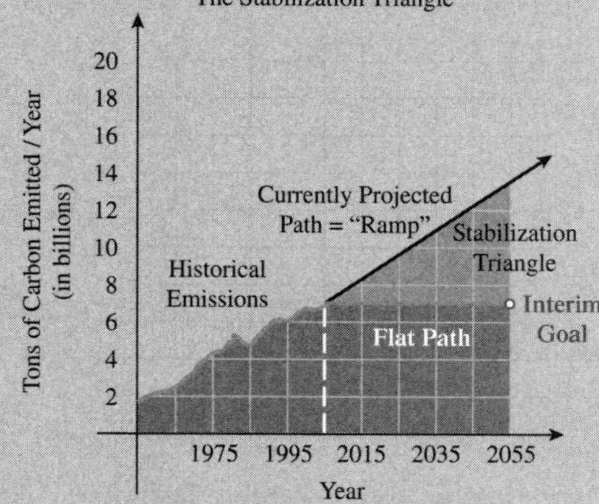

1. Determine the equation of the line for the projected path: increase of 7 gigatons of carbon (GtC) over 50 years (2005 to 2055). Calculate the slope of the line.
2. What is the increase (per year) in the rate of carbon emissions per year based on the projected path model?
3. Develop a model in terms of a finite series that yields the total additional billions of tons of carbon emitted over the 50-year period (2005–2055) for the projected path over the flat path.
4. Calculate the total additional billions of tons of carbon of the projected path over the flat path (i.e., sum the series in (3)).
5. Discuss possible ways to provide the reduction between the projected path and the flat path based on the proposals given by Pacala and Socolow (professors at Princeton).*

*S. Pacala and R. Socolow, "Stabilization Wedges: Solving the Climate Problem for the Next 50 Years with Current Technologies," *Science*, Volume 305 (2004).

CHAPTER 9 REVIEW

Section	Concept	Pages	Review Exercises	Key Ideas/Points
9.1	**Sequences and series**	706–712	1–30	
	Sequences	706–707	1–12	$a_1, a_2, a_3, \ldots, a_n, \ldots$ a_n is the general term.
	Factorial notation	708–709	13–16	$n! = n(n-1)(n-2)\cdots 3 \cdot 2 \cdot 1$ $n \geq 2$ $0! = 1$ and $1! = 1$
	Recursion formulas	709–710	17–20	When a_n is defined by previous terms a_{n-1}.
	Sums and series	710–712	21–28	Sigma Notation: $\sum_{n=1}^{\infty} a_n = a_1 + a_2 + a_3 + \cdots + a_n + \cdots$ Finite series, or nth partial sum, S_n. $S_n = a_1 + a_2 + a_3 + \cdots + a_n$
9.2	**Arithmetic sequences and series**	717–722	31–50	
	Arithmetic sequences	717–718	31–44	$a_{n+1} = a_n + d$ or $a_{n+1} - a_n = d$ d is called the **common difference**.
	The general (nth) term of an arithmetic sequence	719	37–44	$a_n = a_1 + (n-1)d$ for $n \geq 1$
	The sum of an arithmetic sequence	720–722	45–48	$S_n = \frac{n}{2}(a_1 + a_n)$
9.3	**Geometric sequences and series**	726–734	51–74	
	Geometric sequences	726–727	51–58	$a_{n+1} = r \cdot a_n$ or $\frac{a_{n+1}}{a_n} = r$ r is called the **common ratio**.
	The general (nth) term of a geometric sequence	727–728	59–66	$a_n = a_1 \cdot r^{n-1}$ for $n \geq 1$
	Geometric series	729–734	67–74	Finite series: $S_n = a_1\frac{(1-r^n)}{(1-r)}$ $r \neq 1$ Infinite series: $\sum_{n=0}^{\infty} a_1 r^n = a_1 \frac{1}{(1-r)}$ $\lvert r \rvert < 1$
9.4	**Mathematical induction**	738–740	75–78	
	Mathematical induction	738–740	75–78	Prove that S_n is true for all positive integers. Step 1: Show that S_1 is true. Step 2: Assume S_n is true for S_k and show it is true for S_{k+1} (k = integer).
9.5	**The Binomial theorem**	743–748	79–96	
	Binomial coefficients	744	79–82	$\binom{n}{k} = \frac{n!}{(n-k)!k!}$ $\binom{n}{n} = 1$ and $\binom{n}{0} = 1$
	Binomial expansion	745	83–88	$(a+b)^n = \sum_{k=0}^{n} \binom{n}{k} a^{n-k} b^k$

CHAPTER REVIEW

Section	Concept	Pages	Review Exercises	Key Ideas/Points
	Pascal's triangle	746–747	89–90	Shortcut way of remembering binomial coefficients. Each term is found by adding the two numbers above it.
	Finding a particular term of a binomial expansion	747–748	91–94	The $(r + 1)$ term of the expansion $(a + b)^n$ is $\binom{n}{r} a^{n-r} b^r$.
9.6	**Counting, permutations, and combinations**	751–757	97–114	
	The fundamental counting principle	751–753	105–106	The number of ways in which successive independent things can occur is found by multiplying the number of ways each thing can occur.
	Permutations	753–755	97–100	The number of permutations of n objects is $n! = n \cdot (n - 1) \cdot (n - 2) \cdots 2 \cdot 1$ The number of permutations of n objects taken r at a time is ${}_nP_r = \frac{n!}{(n - r)!} = n(n - 1)(n - 2) \cdots (n - r + 1)$
	Combinations	755–756	101–104	The number of combinations of n objects taken r at a time ${}_nC_r = \frac{n!}{(n - r)!r!}$
	Permutations with repetition	756–757		The number of distinguishable permutations of n objects is $\frac{n!}{n_1! \cdot n_2! \cdot n_3! \cdots n_k!}$
9.7	**Probability**	761–767	115–126	
	Probability of an event	762–763	115–118	The probability of event E, denoted $P(E)$, is $P(E) = \frac{n(E)}{n(S)} = \frac{\text{number of outcomes in event } E}{\text{number of outcomes in sample space } S}$
	Probability of an event *not* occurring	763–764	119	$P(\text{not } E) = 1 - P(E)$
	Mutually exclusive events	764–766	120–121	The probability of *either* E_1 *or* E_2 occurring is $P(E_1 \cup E_2) = P(E_1) + P(E_2) - P(E_1 \cap E_2)$ If E_1 and E_2 are mutually exclusive, $E_1 \cap E_2 = 0$.
	Independent events	766–767	122	If E_1 and E_2 are independent events, then the probability of both occurring is the product of the individual probabilities: $P(E_1 \cap E_2) = P(E_1) \cdot P(E_2)$

9.1 Sequences and Series

Write the first four terms of the sequence. Assume n starts at 1.

1. $a_n = n^3$

2. $a_n = \frac{n!}{n}$

3. $a_n = 3n + 2$

4. $a_n = (-1)^n x^{n+2}$

Find the indicated term of the sequence.

5. $a_n = \left(\frac{2}{3}\right)^n \quad a_5 = ?$

6. $a_n = \frac{n^2}{3^n} \quad a_8 = ?$

7. $a_n = \frac{(-1)^n(n-1)!}{n(n+1)!} \quad a_{15} = ?$

8. $a_n = 1 + \frac{1}{n} \quad a_{10} = ?$

Write an expression for the nth term of the given sequence.

9. $3, -6, 9, -12, \ldots$

10. $1, \frac{1}{2}, 3, \frac{1}{4}, 5, \frac{1}{6}, 7, \frac{1}{8}, \ldots$

11. $-1, 1, -1, 1, \ldots$

12. $1, 10, 10^2, 10^3, \ldots$

Simplify the ratio of factorials.

13. $\frac{8!}{6!}$

14. $\frac{20!}{23!}$

15. $\frac{n(n-1)!}{(n+1)!}$

16. $\frac{(n-2)!}{n!}$

Write the first four terms of the sequence defined by the recursion formula.

17. $a_1 = 5 \quad a_n = a_{n-1} - 2$

18. $a_1 = 1 \quad a_n = n^2 \cdot a_{n-1}$

19. $a_1 = 1, a_2 = 2 \quad a_n = (a_{n-1})^2 \cdot (a_{n-2})$

20. $a_1 = 1, a_2 = 2 \quad a_n = \frac{a_{n-2}}{(a_{n-1})^2}$

Evaluate the finite series.

21. $\sum_{n=1}^{5} 3$

22. $\sum_{n=1}^{4} \frac{1}{n^2}$

23. $\sum_{n=1}^{6} (3n + 1)$

24. $\sum_{k=0}^{5} \frac{2^{k+1}}{k!}$

Use sigma (summation) notation to write the sum.

25. $-1 + \frac{1}{2} - \frac{1}{4} + \frac{1}{8} + \cdots - \frac{1}{64}$

26. $2 + 4 + 6 + 8 + 10 + \cdots + 20$

27. $1 + x + \frac{x^2}{2} + \frac{x^3}{6} + \frac{x^4}{24} + \cdots$

28. $x - x^2 + \frac{x^3}{2} - \frac{x^4}{6} + \frac{x^5}{24} - \frac{x^6}{120} + \cdots$

Applications

29. Marines Investment. With the prospect of continued fighting in Iraq, in December 2004, the Marine Corps offered bonuses of as much as \$30,000—in some cases, tax-free—to persuade enlisted personnel with combat experience and training to reenlist. Suppose a Marine put her entire \$30,000 reenlistment bonus in an account that earns 4% interest compounded monthly. The balance in the account after n months is

$$A_n = 30{,}000\left(1 + \frac{0.04}{12}\right)^n \qquad n = 1, 2, 3, \ldots$$

Her commitment with the Marines is 5 years. Calculate A_{60}. What does A_{60} represent?

30. Sports. The NFL minimum salary for a rookie is \$180,000. Suppose a rookie comes into the league making the minimum and gets a \$30,000 raise every year he plays. Write the general term a_n of a sequence that represents the salary of an NFL player making the league minimum during his entire career. Assuming $n = 1$ corresponds to the first year, what does $\sum_{n=1}^{4} a_n$ represent?

9.2 Arithmetic Sequences and Series

Determine whether the sequence is arithmetic. If it is, find the common difference.

31. $7, 5, 3, 1, -1, \ldots$

32. $1^3 + 2^3 + 3^3 + \cdots$

33. $1, \frac{3}{2}, 2, \frac{5}{2}, \ldots$

34. $a_n = -n + 3$

35. $a_n = \frac{(n+1)!}{n!}$

36. $a_n = 5(n - 1)$

Find the general, or nth, term of the arithmetic sequence given the first term and the common difference.

37. $a_1 = -4 \quad d = 5$

38. $a_1 = 5 \quad d = 6$

39. $a_1 = 1 \quad d = -\frac{2}{3}$

40. $a_1 = 0.001 \quad d = 0.01$

For each arithmetic sequence described below, find a_1 and d and construct the sequence by stating the general, or nth, term.

41. The 5th term is 13 and the 17th term is 37.

42. The 7th term is -14 and the 10th term is -23.

43. The 8th term is 52 and the 21st term is 130.

44. The 11th term is -30 and the 21st term is -80.

Find the sum.

45. $\sum_{k=1}^{20} 3k$

46. $\sum_{n=1}^{15} n + 5$

47. $2 + 8 + 14 + 20 + \cdots + 68$

48. $\frac{1}{4} - \frac{1}{4} - \frac{3}{4} - \cdots - \frac{31}{4}$

Applications

49. Salary. Upon graduating with M.B.A.s, Bob and Tania opt for different career paths. Bob accepts a job with the U.S. Department of Transportation making \$45,000 with a guaranteed \$2,000 raise every year. Tania takes a job with

Templeton Corporation making $38,000 with a guaranteed $4,000 raise every year. Calculate how many total dollars both Bob and Tania will have each made after 15 years.

50. **Gravity.** When a skydiver jumps out of an airplane, she falls approximately 16 feet in the 1st second, 48 feet during the 2nd second, 80 feet during the 3rd second, 112 feet during the 4th second, and 144 feet during the 5th second, and this pattern continues. If she deploys her parachute after 5 seconds have elapsed, how far will she have fallen during those 5 seconds?

9.3 Geometric Sequences and Series

Determine whether the sequence is geometric. If it is, find the common ratio.

51. $2, -4, 8, -16, \ldots$

52. $1, \frac{1}{2^2}, \frac{1}{3^2}, \frac{1}{4^2}, \ldots$

53. $20, 10, 5, \frac{5}{2}, \ldots$

54. $\frac{1}{100}, \frac{1}{10}, 1, 10, \ldots$

Write the first five terms of the geometric series.

55. $a_1 = 3 \quad r = 2$

56. $a_1 = 10 \quad r = \frac{1}{4}$

57. $a_1 = 100 \quad r = -4$

58. $a_1 = -60 \quad r = -\frac{1}{2}$

Write the formula for the *n*th term of the geometric series.

59. $a_1 = 7 \quad r = 2$

60. $a_1 = 12 \quad r = \frac{1}{3}$

61. $a_1 = 1 \quad r = -2$

62. $a_1 = \frac{32}{5} \quad r = -\frac{1}{4}$

Find the indicated term of the geometric sequence.

63. 25th term of the sequence $2, 4, 8, 16, \ldots$

64. 10th term of the sequence $\frac{1}{2}, 1, 2, 4, \ldots$

65. 12th term of the sequence $100, -20, 4, -0.8, \ldots$

66. 11th term of the sequence $1000, -500, 250, -125, \ldots$

Evaluate the geometric series, if possible.

67. $\frac{1}{2} + \frac{3}{2} + \frac{3^2}{2} + \cdots + \frac{3^8}{2}$

68. $1 + \frac{1}{2} + \frac{1}{2^2} + \frac{1}{2^3} + \cdots + \frac{1}{2^{10}}$

69. $\sum_{n=1}^{8} 5(3)^{n-1}$

70. $\sum_{n=1}^{7} \frac{2}{3}(5)^n$

71. $\sum_{n=0}^{\infty} \left(\frac{2}{3}\right)^n$

72. $\sum_{n=1}^{\infty} \left(-\frac{1}{5}\right)^{n+1}$

Applications

73. **Salary.** Murad is fluent in four languages and is offered a job with the U.S. government as a translator. He is hired on the "GS" scale at a base rate of $48,000 with 2% increases in his salary per year. Calculate what his salary will be *after* he has been with the U.S. government for 12 years.

74. **Boat Depreciation.** Upon graduating from Auburn University, Philip and Steve get jobs at Disney Ride and Show Engineering and decide to buy a ski boat together. If the boat costs $15,000 new, and depreciates 20% per year, write a formula for the value of the boat *n* years after it was purchased. How much will the boat be worth when Philip and Steve have been working at Disney for 3 years?

9.4 Mathematical Induction

Prove the statements using mathematical induction for all positive integers *n*.

75. $3n \le 3^n$

76. $4^n < 4^{n+1}$

77. $2 + 7 + 12 + 17 + \cdots + (5n - 3) = \frac{n}{2}(5n - 1)$

78. $2n^2 > (n + 1)^2 \quad n \ge 3$

9.5 The Binomial Theorem

Evaluate the binomial coefficients.

79. $\binom{11}{8}$

80. $\binom{10}{0}$

81. $\binom{22}{22}$

82. $\binom{47}{45}$

Expand the expression using the binomial theorem.

83. $(x - 5)^4$

84. $(x + y)^5$

85. $(2x - 5)^3$

86. $(x^2 + y^3)^4$

87. $(\sqrt{x} + 1)^5$

88. $(x^{2/3} + y^{1/3})^6$

Expand the expression using Pascal's triangle.

89. $(r - s)^5$

90. $(ax + by)^4$

Find the coefficient *C* of the term in the binomial expansion.

Binomial	**Term**
91. $(x - 2)^8$	Cx^6
92. $(3 + y)^7$	Cy^4
93. $(2x + 5y)^6$	Cx^2y^4
94. $(r^2 - s)^8$	Cr^8s^4

Applications

95. **Lottery.** In a state lottery in which 6 numbers are drawn from a possible 53 numbers, the number of possible 6-number combinations is equal to $\binom{53}{6}$. How many possible combinations are there?

96. **Canasta.** In the card game canasta, two decks of cards including the jokers are used, and 13 cards are dealt to each person. A total of $\binom{108}{13}$ different 13-card canasta hands can be dealt. How many possible hands are there?

9.6 Counting, Permutations, and Combinations

Use the formula for $_nP_r$ to evaluate each expression.

97. $_7P_4$ **98.** $_9P_9$ **99.** $_{12}P_5$ **100.** $_{10}P_1$

Use the formula for $_nC_r$ to evaluate each expression.

101. $_{12}C_7$ **102.** $_{40}C_5$ **103.** $_9C_9$ **104.** $_{53}C_6$

Applications

105. Car Options. A new Honda Accord comes in three models (LX, VX, and EX). Each of those models comes with either a cloth or a leather interior, and the exterior comes in either silver, white, black, red, or blue. How many different cars (models, interior seat upholstery, and exterior color) are there to choose from?

106. E-mail Passwords. All e-mail accounts require passwords. If a six-character password is required that can contain letters (but no numbers), how many possible passwords can there be if letters can be repeated? (Assume no letters are case-sensitive.)

107. Team Arrangements. There are 10 candidates for the board of directors, and there are four leadership positions (president, vice president, secretary, and treasurer). How many possible leadership teams are there?

108. License Plates. In a particular state, there are six characters in a license plate consisting of letters and numbers. If 0s and 1s are eliminated from possible numbers and Os and Is are eliminated from possible letters, how many different license plates can be made?

109. Seating Arrangements. Five friends buy five season tickets to the Philadelphia Eagles. To be fair, they change the seating arrangement every game. How many different seating arrangements are there for the five friends? How many seasons would they have to buy tickets in order to sit in all of the combinations (each season has eight home games)?

110. Safe. A safe will open when the correct choice of three numbers (1 to 60) is selected in a specific order. How many possible permutations are there?

111. Raffle. A fundraiser raffle is held to benefit the Make a Wish Foundation, and 100 raffle tickets are sold. Four prizes are raffled off. First prize is a round-trip ticket on American Airlines, second prize is a round of golf for four people at a Links golf course, third prize is a \$100 gift certificate to the Outback Steakhouse, and fourth prize is a half-hour massage. How many possible winning scenarios are there if all 100 tickets are sold to different people?

112. Sports. There are 117 Division 1-A football teams in the United States. At the end of the regular season is the Bowl Championship Series, and the top two teams play each other in the championship game. Assuming that any two Division 1-A teams can advance to the championship, how many possible matchups are there for the championship game?

113. Cards. In a deck of 52 cards, how many different 6-card hands can be dealt?

114. Blackjack. In a game of two-deck blackjack (104 cards), how many 2-card combinations are there that equal 21, that is, ace and a 10 or face card—jack, queen, or king?

9.7 Probability

115. Coin Tossing. For the experiment of tossing a coin four times, what is the probability of getting all heads?

116. Dice. For an experiment of tossing two dice, what is the probability that the sum of the dice is odd?

117. Dice. For an experiment of tossing two dice, what is the probability of not rolling a combined 7?

118. Cards. For a deck of 52 cards, what is the probability of drawing a diamond?

For Exercises 119–122, let $P(E_1) = \frac{1}{3}$ and $P(E_2) = \frac{1}{2}$ and find the probability of the event.

119. Probability. Find the probability of an event E_1 not occurring.

120. Probability. Find the probability of either E_1 or E_2 occurring if E_1 and E_2 are mutually exclusive.

121. Probability. Find the probability of either E_1 or E_2 occurring if E_1 and E_2 are not mutually exclusive and $P(E_1 \cap E_2) = \frac{1}{4}$.

122. Probability. Find the probability of both E_1 and E_2 occurring if E_1 and E_2 are independent.

123. Cards. With a 52-card deck, what is the probability of drawing an ace or a 2?

124. Cards. By drawing twice, what is the probability of drawing an ace and then a 2? (Assume that after the first card is drawn it is not put back into the deck.)

125. Children. What is the probability that in a family of five children at least one is a girl?

126. Sports. With the salary cap in the NFL, it is said that on "any given Sunday" any team could beat any other team. If we assume every week a team has a 50% chance of winning, what is the probability that a team will go 11–1?

Technology

Section 9.1

127. Use a graphing calculator "SUM" to find the sum of the series $\sum_{n=1}^{6} \frac{1}{n^2}$.

REVIEW EXERCISES

128. Use a graphing calculator "SUM" to find the sum of the infinite series $\sum_{n=1}^{\infty} \frac{1}{n}$, if possible.

Section 9.2

129. Use a graphing calculator to sum $\sum_{n=1}^{75} \left[\frac{3}{2} + \frac{6}{7}(n-1)\right]$.

130. Use a graphing calculator to sum $\sum_{n=1}^{264} \left[-19 + \frac{1}{3}(n-1)\right]$.

Section 9.3

131. Apply a graphing utility to plot $y_1 = 1 - 2x + 4x^2 - 8x^3 + 16x^4$ and $y_2 = \frac{1}{1+2x}$, and let x range from $[-0.3, 0.3]$. Based on what you see, what do you expect the geometric series $\sum_{n=0}^{\infty} (-1)^n (2x)^n$ to sum to?

132. Does the sum of the infinite series $\sum_{n=0}^{\infty} \left(\frac{e}{p}\right)^n$ exist? Use a graphing calculator to find it and round to four decimal places.

Section 9.4

133. Use a graphing calculator to sum the series $2 + 7 + 12 + 17 + \cdots + (5n - 3)$ on the left side, and evaluate the expression $\frac{n}{2}(5n - 1)$ on the right side for $n = 200$. Do they agree with each other? Do your answers confirm the proof for Exercise 77?

134. Use a graphing calculator to plot the graphs of $y_1 = 2x^2$ and $y_2 = (x + 1)^2$ in the [100, 1000] by [10000, 2500000] viewing rectangle. Do your answers confirm the proof for Exercise 78?

Section 9.5

135. With a graphing utility, plot $y_1 = 1 + 8x$, $y_2 = 1 + 8x + 24x^2$, $y_3 = 1 + 8x + 24x^2 + 32x^3$, $y_4 = 1 + 8x + 24x^2 + 32x^3 + 16x^4$, and $y_5 = (1 + 2x)^4$ for $-0.1 < x < 0.1$. What do you notice happening each time an additional term is added? Now, let $0.1 < x < 1$. Does the same thing happen?

136. With a graphing utility, plot $y_1 = 1 - 8x$, $y_2 = 1 - 8x + 24x^2$, $y_3 = 1 - 8x + 24x^2 - 32x^3$, $y_4 = 1 - 8x + 24x^2 - 32x^3 + 16x^4$, and $y_5 = (1 - 2x)^4$ for $-0.1 < x < 0.1$. What do you notice happening each time an additional term is added? Now, let $0.1 < x < 1$. Does the same thing happen?

Section 9.6

137. Use a graphing calculator to evaluate:

a. $\frac{{}_{16}P_7}{7!}$

b. ${}_{16}C_7$

c. Are answers in (a) and (b) the same?

d. Why?

138. Use a graphing calculator to evaluate:

a. $\frac{{}_{52}P_6}{6!}$

b. ${}_{52}C_6$

c. Are answers in (a) and (b) the same?

d. Why?

Section 9.7

In Exercises 139 and 140, when two dice are rolled, the probability of getting a sum of 9 is $\frac{1}{9}$ and the probability of not getting a sum of 9 is $\frac{8}{9}$. If two dice are rolled n times, the probability of getting a sum of 9 exactly k times can be found by using the binomial theorem ${}_nC_k\left(\frac{1}{9}\right)^k\left(\frac{8}{9}\right)^{n-k}$.

139. If two dice are rolled 10 times, find the probability of getting a sum of 9 exactly three times. Round your answer to four decimal places.

140. If a die is rolled 8 times, find the probability of getting a sum of 9 at least two times. Round your answer to four decimal places.

For Exercises 1–5, use the sequence $1, x, x^2, x^3, \ldots$

1. Write the nth term of the sequence.
2. Classify this sequence as arithmetic, geometric, or neither.
3. Find the nth partial sum of the series S_n.
4. Assuming this sequence is infinite, write the series using sigma notation.
5. Assuming this sequence is infinite, what condition would have to be satisfied in order for the sum to exist?
6. Find the following sum: $\frac{1}{3} + \frac{1}{9} + \frac{1}{27} + \frac{1}{81} + \cdots$.
7. Find the following sum: $\sum_{n=1}^{10} 3 \cdot \left(\frac{1}{4}\right)^n$.
8. Find the following sum: $\sum_{k=1}^{50} (2k + 1)$.
9. Write the series using sigma notation, then find its sum: $2 + 7 + 12 + 17 + \cdots + 497$.
10. Use mathematical induction to prove that $2 + 4 + 6 + \cdots + 2n = n^2 + n$.
11. Evaluate $\dfrac{7!}{2!}$.
12. Find the third term of $(2x + y)^5$.

In Exercises 13–16, evaluate the expressions.

13. $\dbinom{15}{12}$ 14. $\dbinom{k}{k}$ 15. ${}_{14}P_3$ 16. ${}_{200}C_3$

17. Expand the expression $\left(x^2 + \dfrac{1}{x}\right)^5$.
18. Use the binomial theorem to expand the binomial $(3x - 2)^4$.
19. Explain why there are always more permutations than combinations.
20. What is the probability of not winning a trifecta (selecting the first, second, and third place finishers) in a horse race with 15 horses?

For Exercises 21–23, refer to a roulette wheel with 18 red, 18 black, and 2 green slots.

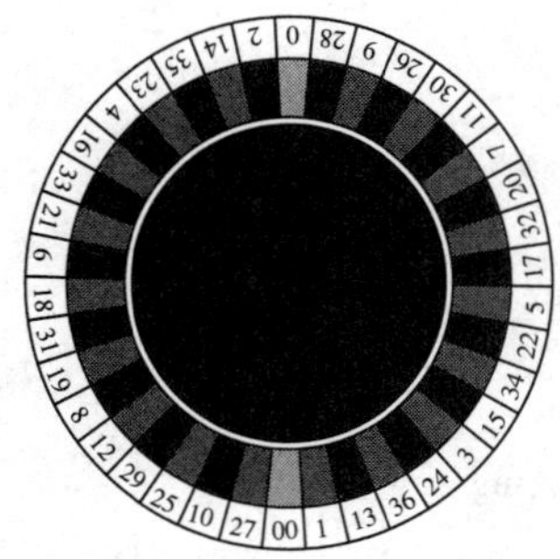

21. **Roulette.** What is the probability of the ball landing in a red slot?
22. **Roulette.** What is the probability of the ball landing in a red slot 5 times in a row?
23. **Roulette.** If the four previous rolls landed on red, what is the probability that the next roll will land on red?
24. **Marbles.** If there are four red marbles, three blue marbles, two green marbles, and one black marble in a sack, find the probability of pulling out the following order: black, blue, red, red, green.
25. **Cards.** What is the probability of drawing an ace or a diamond from a deck of 52 cards?
26. **Human Anatomy.** Vasopressin is a relatively simple protein that is found in the human liver. It consists of eight amino acids that must be joined together in one particular order for the effective functioning of the protein.
    a. How many different arrangements of the eight amino acids are possible?
    b. What is the probability of randomly selecting one of these arrangements and obtaining the correct arrangement to make vasopressin?
27. Find the constant term in the expression $\left(x^3 + \dfrac{1}{x^3}\right)^{20}$.
28. Use a graphing calculator to sum $\sum_{n=1}^{125}\left[-\frac{11}{4} + \frac{5}{6}(n - 1)\right]$.

1. Simplify $\frac{10x + 1}{3x - 2} + \frac{5 + 4x}{2 - 3x}$.

2. The length of a rectangle is 5 less than twice the width and the perimeter is 38 inches. What are the dimensions of the rectangle?

3. Solve using the quadratic formula: $2x^2 - 5x + 11 = 0$.

4. Solve and express the solution in interval notation: $|x - 5| > 3$.

5. Write an equation of the line with slope undefined and x-intercept $(-8, 0)$.

6. Find the x-intercept and y-intercept and slope of the line $3x - 5y = 15$.

7. Using the function $f(x) = x^2 - 3x + 2$, evaluate the difference quotient $\frac{f(x + h) - f(x)}{h}$.

8. Find the composite function $f \circ g$, and state the domain for $f(x) = 4 - \frac{1}{x}$ and $g(x) = \frac{1}{2x + 1}$.

9. Find the vertex of the parabola $f(x) = -0.04x^2 + 1.2x - 3$.

10. Factor the polynomial $P(x) = x^4 + 8x^2 - 9$ as a product of linear factors.

11. Find the vertical and horizontal asymptotes of the function:

$$f(x) = \frac{5x - 7}{3 - x}$$

12. How much money should be put in a savings account now that earns 4.7% a year compounded weekly, if you want to have $65,000 in 17 years?

13. Evaluate $\log_2 6$ using the change-of-base formula. Round your answer to three decimal places.

14. Solve $\ln(5x - 6) = 2$. Round your answer to three decimal places.

15. Solve the system of linear equations.

$$8x - 5y = 15$$
$$y = \tfrac{8}{5}x + 10$$

16. Solve the system of linear equations.

$$2x - y + z = 1$$
$$x - y + 4z = 3$$

17. Maximize the objective function $z = 4x + 5y$, subject to the constraints $x + y \le 5, x \ge 1, y \ge 2$.

18. Solve the system using Gauss–Jordan elimination.

$$x + 5y - 2z = 3$$
$$3x + y + 2z = -3$$
$$2x - 4y + 4z = 10$$

19. Given

$$A = \begin{bmatrix} 3 & 4 & -7 \\ 0 & 1 & 5 \end{bmatrix} \quad B = \begin{bmatrix} 8 & -2 & 6 \\ 9 & 0 & -1 \end{bmatrix} \quad C = \begin{bmatrix} 9 & 0 \\ 1 & 2 \end{bmatrix}$$

find $C(A + B)$.

20. Calculate the determinant.

$$\begin{vmatrix} 2 & 5 & -1 \\ 1 & 4 & 0 \\ -2 & 1 & 3 \end{vmatrix}$$

21. Find the equation of a parabola with vertex (3, 5) and directrix $x = 7$.

22. Graph $x^2 + y^2 < 4$.

23. Find the sum of the finite series $\sum_{n=1}^{4} \frac{2^{n-1}}{n!}$.

24. Classify the sequence as arithmetic, geometric, or neither:

$$5, 15, 45, 135, \ldots$$

25. There are 10 true–false questions on a quiz. Assuming no answers are left blank, how many different ways can you answer the questions on the quiz?

26. Use a graphing calculator to sum $\sum_{n=1}^{164} \left[\frac{1}{4} + \frac{3}{2}(n - 1)\right]$.

27. Find the constant term in the expression $\left(x^3 - \frac{1}{x^3}\right)^{14}$.

ANSWERS TO ODD NUMBERED EXERCISES*

CHAPTER 0

Section 0.1

1. rational 3. irrational 5. rational

7. irrational

9. a. 7.347 b. 7.347 11. a. 2.995 b. 2.994

13. a. 0.234 b. 0.234 15. a. 5.238 b. 5.238

17. 4 19. 26 21. -130

23. 17 25. 3 27. $x + y - z$

29. $-3x - y$ 31. $\frac{3}{5}$ 33. 68

35. $-9x - y$ 37. -10 39. 5

41. $\frac{19}{12}$ 43. $\frac{1}{2}$ 45. $\frac{23}{12}$

47. $\frac{1}{27}$ 49. $\frac{x}{3}$ 51. $\frac{x}{21}$

53. $\frac{61y}{60}$ 55. $\frac{11}{30}$ 57. $\frac{4}{3}$

59. $\frac{3}{35}$ 61. $\frac{12b}{a}$ 63. $\frac{3}{4}$

65. $\frac{4xy}{3}$ 67. $\frac{8x}{y}$ 69. $\frac{2}{3}$

71. $\frac{3}{25}$ 73. \$ 9,176,366,000,000

75. \$30,203

77. The mistake is rounding the number that is used in the rounding. Look to the right of the number; if it is less than 5, round down.

79. The -2 did not distribute to the second term.

81. false 83. true 85. $x \neq 0$

87. $-7x + 12y - 75$ 89. irrational 91. rational

Section 0.2

1. 256 3. -243 5. -25

7. -16 9. 1 11. $\frac{1}{10}$ or 0.1

13. $\frac{1}{64}$ 15. -150 17. 5

19. -54 21. x^5 23. $\frac{1}{x}$

25. x^6 27. $64a^3$ 29. $-8t^3$

31. $75x^5y^5$ 33. $\frac{y^2}{x^2}$ 35. $-\frac{1}{2xy}$

37. $\frac{16}{b^4}$ 39. $\frac{a^4}{81b^6}$ 41. $\frac{1}{a^6b^2}$

43. x^8y^2 45. $\frac{x^{14}}{4y}$ 47. $\frac{y^{10}}{-32x^{22}}$

49. $-\frac{y^{24}}{a^{12} \cdot x^3}$ 51. 2^{26} 53. 2.76×10^7

55. 9.3×10^7 57. 5.67×10^{-8} 59. 1.23×10^{-7}

61. 47,000,000 63. 23,000 65. 0.000041

67. 2.0×10^8 miles 69. 0.00000155 meters

71. In the power property, exponents are added (not multiplied).

73. In the first step, $(y^3)^2 = y^6$. Exponents should have been multiplied (not raised to power).

75. false 77. false

79. a^{mnk} 81. -16

83. 156 85. 5.6 acres/person on Earth

87. 2.0×10^{-1} or 0.2 89. same

91. 5.11×10^{14}

Section 0.3

1. $-7x^4 - 2x^3 + 5x^2 + 16$ degree 4

3. $-6x^3 + 4x + 3$ degree 3

5. 15 degree 0 7. $y - 2$ degree 1

9. $-x^2 + 5x + 5$ 11. $2x^2 - x - 18$

13. $4x^4 - 10x^2 - 1$ 15. $2z^2 + 2z - 3$

17. $-11y^3 - 16y^2 + 16y - 4$ 19. $-4x + 12y$

21. $x^2 - x + 2$ 23. $-4t^2 + 6t - 1$

25. $35x^2y^3$ 27. $2x^5 - 2x^4 + 2x^3$

29. $10x^4 - 2x^3 - 10x^2$ 31. $2x^5 + 2x^4 - 4x^3$

33. $2a^3b^2 + 4a^2b^3 - 6ab^4$ 35. $6x^2 - 5x - 4$

37. $x^2 - 4$ 39. $4x^2 - 9$

41. $-4x^2 + 4x - 1$ 43. $4x^4 - 9$

45. $2y^4 - 9y^3 + 5y^2 + 7y$ 47. $x^3 - x^2 + x + 3$

49. $t^2 - 4t + 4$ 51. $z^2 + 4z + 4$

53. $x^2 + 2xy + y^2 - 6x - 6y + 9$ 55. $25x^2 - 20x + 4$

*Answers that require a proof, graph, or otherwise lengthy solution are not included.

57. $6y^3 + 5y^2 - 4y$ 59. $x^4 - 1$

61. $ab^2 + 2b^3 - 9a^3 - 18a^2b$

63. $2x^2 - 3y^2 - xy + 3xz + 8yz - 5z^2$

65. $P = 11x - 100$ 67. $P = -x^2 + 200x - 7500$

69. $V = (15 - 2x)(8 - 2x)x = 4x^3 - 46x^2 + 120x$

71. a. $P = 2\pi x + 4x + 10$ feet

b. $A = \pi x^2 + 4x^2 + 10x$ square feet

73. $F = \dfrac{3k}{100x}$

75. The negative was not distributed through the second polynomial.

77. true 79. false 81. $m + n$

83. $2401x^4 - 1568x^2y^4 + 256y^8$ 85. $x^3 - a^3$

87. $(2x + 3)(x - 4)$ and $2x^2 - 5x - 12$

Section 0.4

1. $5(x + 5)$ 3. $2(2t^2 - 1)$

5. $2x(x - 5)(x + 5)$ 7. $3x(x^2 - 3x + 4)$

9. $x(x - 8)(x + 5)$ 11. $2xy(2xy^2 + 3)$

13. $(x + 3)(x - 3)$ 15. $(2x - 3)(2x + 3)$

17. $2(x - 7)(x + 7)$ 19. $(15x - 13y)(15x + 13y)$

21. $(x + 4)^2$ 23. $(x^2 - 2)^2$ 25. $(2x + 3y)^2$

27. $(x - 3)^2$ 29. $(x^2 + 1)^2$ 31. $(p + q)^2$

33. $(t + 3)(t^2 - 3t + 9)$ 35. $(y - 4)(y^2 - 4y + 16)$

37. $(2 - x)(4 + 2x + x^2)$ 39. $(y + 5)(y^2 - 5y + 25)$

41. $(3 + x)(9 - 3x + x^2)$ 43. $(x - 5)(x - 1)$

45. $(y - 3)(y + 1)$ 47. $(2y + 1)(y - 3)$

49. $(3t + 1)(t + 2)$ 51. $(-3t + 2)(2t + 1)$

53. $(x^2 + 2)(x - 3)$

55. $(a^3 - 8)(a + 2) = (a + 2)(a - 2)(a^2 + 2a + 4)$

57. $(3y - 5r)(x + 2s)$ 59. $(5x + 2y)(4x - y)$

61. $(x - 2y)(x + 2y)$ 63. $(3a + 7)(a - 2)$

65. prime 67. prime

69. $2(3x + 2)(x + 1)$ 71. $(3x - y)(2x + 5y)$

73. $9(2s - t)(2s + t)$ 75. $(ab - 5c)(ab + 5c)$

77. $(4x + 5)(x - 2)$ 79. $x(3x + 1)(x - 2)$

81. $x(x - 3)(x + 3)$ 83. $(y - 1)(x - 1)$

85. $(x^2 + 2)(x^2 + 3)$ 87. $(x - 6)(x + 4)$

89. $x(x + 5)(x^2 - 5x + 25)$ 91. $(x + 3)(x - 3)(x^2 + 9)$

93. $2(3x + 4)$ 95. $(2x - 15)(x + 2)$

97. $-2(8t - 1)(t + 5)$

99. $(x^2 - 9) \neq (x - 3)^2$; instead, $x^2 - 9 = (x - 3)(x + 3)$

101. false 103. true

105. $(a^n - b^n)(a^n + b^n)$

107. $8x^3 + 1$ and $(2x + 1)(4x^2 - 2x + 1)$

Section 0.5

1. $x \neq 0$ 3. $x \neq 1$ 5. $x \neq -1$

7. $p \neq \pm 1$ 9. no restrictions

11. $\dfrac{x - 9}{2(x + 9)}$ $x \neq -9, -3$ 13. $\dfrac{x - 3}{2}$ $x \neq -1$

15. $\dfrac{2(3y + 1)}{9y}$ $y \neq 0, \frac{1}{2}$ 17. $\dfrac{y + 1}{5}$ $y \neq \frac{1}{5}$

19. $\dfrac{3x + 7}{4}$ $x \neq 4$ 21. $x + 2$ $x \neq 2$

23. 1 $x \neq -7$ 25. $\dfrac{x^2 + 9}{2x + 9}$ $x \neq -\frac{9}{2}$

27. $\dfrac{x + 3}{x - 5}$ $x \neq -2, 5$ 29. $\dfrac{3x + 1}{x + 5}$ $x \neq -5, \frac{1}{2}$

31. $\dfrac{3x + 5}{x + 1}$ $x \neq -1, 2$ 33. $\dfrac{2(5x + 6)}{5x - 6}$ $x \neq 0, \frac{6}{5}$

35. $\dfrac{2}{3}$ $x \neq 0, \pm 1$ 37. $3(x + 2)$ $x \neq -5, 0, 2$

39. $\dfrac{t - 3}{3(t + 2)}$ $t \neq -2, 3$ 41. $\dfrac{3t(t^2 + 4)}{(t - 3)(t + 2)}$ $t \neq -2, 3$

43. $\dfrac{3y(y - 2)}{y - 3}$ $y \neq -2, 3$

45. $\dfrac{(2x + 3)(x - 5)}{2x(x + 5)^2}$ $x \neq \pm 5, 0$

47. $\dfrac{(2x - 7)(2x + 7)}{(4x + 3)(3x - 5)}$ $x \neq -\frac{3}{4}, \pm\frac{5}{3}, \frac{7}{2}$

49. $\dfrac{x}{4}$ $x \neq 0$ 51. $\dfrac{x + 2}{2}$ $x \neq \pm 2$

53. $\dfrac{x + 1}{5}$ $x \neq \pm 1$ 55. $\dfrac{1}{2(1 - p)}$ $p \neq \pm 1, 2$

57. $\dfrac{6 - n}{n - 3}$ $n \neq -6, \pm 3$ 59. $\dfrac{2t(t - 3)}{5}$ $t \neq -1, 2$

61. $\dfrac{5w^2}{w+1}$ $w \neq 0, \pm 1$

63. $\dfrac{(x-3)(x-4)}{(x-2)(x-9)}$ $x \neq -7, -5, 2, 4, 9$

65. $\dfrac{x}{5x+2}$ $x \neq -\frac{5}{3}, \pm\frac{2}{5}, -\frac{1}{4}, 0$

67. $\dfrac{13}{5x}$ $x \neq 0$

69. $\dfrac{5p^2 - 7p + 3}{(p+1)(p-2)}$ $p \neq -1, 2$

71. $\dfrac{4}{5x-1}$ $x \neq \frac{1}{5}$

73. $\dfrac{3y^3 - 5y^2 - y + 1}{y^2 - 1}$ $y \neq \pm 1$

75. $\dfrac{x^2 + 4x - 6}{x^2 - 4}$ $x \neq \pm 2$

77. $\dfrac{x+4}{x+2}$ $x \neq \pm 2$ 79. $\dfrac{12a + 7b}{a^2 - b^2}$ $a \neq \pm b$

81. $\dfrac{7x - 20}{x - 3}$ $x \neq 3$ 83. $\dfrac{1-x}{x-2}$ $x \neq 0, 2$

85. $\dfrac{x}{3x-1}$ $x \neq 0, \pm\frac{1}{3}$ 87. $\dfrac{x+1}{x-1}$ $x \neq 0, \pm 1$

89. $\dfrac{x(x+1)}{(x-1)(x+2)}$ $x \neq -2, \pm 1$

91. $A = (pi) \cdot \dfrac{(1+i)^5}{(1+i)^5 - 1}$ 93. $R = \dfrac{R_1 R_2}{R_1 + R_2}$

95. $x + 1$, where $x \neq -1$ 97. false

99. false

101. $\dfrac{(x+a)(x+d)}{(x+b)(x+c)}$ $x \neq -b, -c, -d$

103. yes

105. a. $\dfrac{(x-1)(x+2)}{(x-2)(x+1)}$ c. $x \neq -1, x \neq -2, x \neq 2$

Section 0.6

1. 10 3. -12 5. -6

7. 7 9. 1 11. 0

13. not a real number 15. -3 17. 4

19. -2 21. -1 23. 27

25. $-4\sqrt{2}$ 27. $8\sqrt{5}$ 29. $2\sqrt{6}$

31. $2\sqrt[3]{6}$ 33. $\sqrt{21}$ 35. $40|x|$

37. $4|x|\sqrt{y}$ 39. $-3x^2y^2\sqrt[3]{3y^2}$ 41. $\dfrac{\sqrt{3}}{3}$

43. $\dfrac{2\sqrt{11}}{33}$ 45. $\dfrac{3 + 3\sqrt{5}}{-4}$ 47. $-3 - 2\sqrt{2}$

49. $-3(\sqrt{2} + \sqrt{3})$ 51. $\frac{2}{3}(3\sqrt{2} - 2\sqrt{3})$

53. $\dfrac{5\sqrt{5} - 2}{11}$ 55. $\dfrac{(\sqrt{7} + 3)(\sqrt{2} + \sqrt{5})}{-3}$

57. x^3y^4 59. $\dfrac{1}{y^2}$ 61. $x^{7/6}y^2$

63. $\dfrac{x^{8/3}}{2}$ 65. $x^{1/3}(x-2)(x+1)$

67. $7x^{3/7}(1 - 2x^{3/7} + 3x)$ 69. $4\sqrt{5} \approx 9$ seconds

71. $d \approx 9.54$ astronomical units.

73. The 4 should also have been squared.

75. false 77. false 79. a^{mnk} 81. $\frac{1}{2}$

83. $\dfrac{(\sqrt{a} - \sqrt{b})^2}{(a-b)^2} = \dfrac{a - 2\sqrt{ab} + b}{(a-b)^2}$ 85. 3.317

87. a. $10\sqrt{2} - 8\sqrt{3}$ b. 0.2857291632 c. yes

Section 0.7

1. $4i$ 3. $2i\sqrt{5}$ 5. -4

7. $8i$ 9. $3 - 10i$ 11. $-10 - 12i$

13. $2 - 9i$ 15. $10 - 14i$ 17. $2 - 2i$

19. $-1 + 2i$ 21. $12 - 6i$ 23. $96 - 60i$

25. $-48 + 27i$ 27. $-102 + 30i$ 29. $5 - i$

31. $13 + 41i$ 33. $87 + 33i$ 35. $-6 + 48i$

37. $\frac{56}{9} - \frac{11}{18}i$ 39. $37 + 49i$ 41. $4 - 7i$; 65

43. $2 + 3i$; 13 45. $6 - 4i$; 52 47. $-2 + 6i$; 40

49. $-2i$ 51. $\frac{3}{10} + \frac{1}{10}i$ 53. $\frac{3}{13} - \frac{2}{13}i$

55. $\frac{14}{53} - \frac{4}{53}i$ 57. $-i$ 59. $-\frac{9}{34} + \frac{19}{34}i$

61. $\frac{18}{53} - \frac{43}{53}i$ 63. $\frac{66}{85} + \frac{43}{85}i$ 65. $-i$

67. 1 69. $21 - 20i$ 71. $-5 + 12i$

73. $18 + 26i$ 75. $-2 - 2i$ 77. $8 - 2i$ ohms

79. multiplied by the denominator $4 - i$ instead of the conjugate $4 + i$

81. true 83. true 85. $(x - i)^2(x + i)^2$

87. $41 - 38i$ 89. $\frac{2}{125} + \frac{11}{125}i$

Review Exercises

1. a. 5.22 b. 5.21
3. 4
5. -2
7. $-\frac{x}{12}$
9. 9
11. $-8z^3$
13. $\frac{9}{2x^2}$
15. 2.15×10^{-6}
17. $14z^2 + 3z - 2$
19. $45x^2 - 10x - 15$
21. $15x^2y^2 - 20xy^3$
23. $x^2 + 2x - 63$
25. $4x^2 - 12x + 9$
27. $x^4 + 2x^2 + 1$
29. $2xy^2(7x - 5y)$
31. $(x + 5)(2x - 1)$
33. $(4x - 5)(4x + 5)$
35. $(x + 5)(x^2 - 5x + 25)$
37. $2x(x - 3)(x + 5)$
39. $(x^2 - 2)(x + 1)$
41. $x \neq \pm 3$
43. $x + 2 \quad x \neq 2$
45. $\frac{t + 3}{t + 1} \quad t \neq -1, 2$
47. $\frac{(x + 5)(x + 2)}{(x + 3)^2} \quad x \neq -3, 1, 2$
49. $-\frac{2}{(x + 1)(x + 3)} \quad x \neq -1, -3$
51. $\frac{10x - 25}{20x - 59} \quad x \neq 3, \frac{59}{20}$
53. $2\sqrt{5}$
55. $-5xy\sqrt[3]{x^2y}$
57. $26\sqrt{5}$
59. $-3 - \sqrt{5}$
61. $2 + \sqrt{3}$
63. $\frac{9x^{2/3}}{16}$
65. $5^{1/6}$
67. $13i$
69. $-i$
71. 12
73. $-33 + 56i$
75. $\frac{2}{5} + \frac{1}{5}i$
77. $\frac{38}{41} - \frac{27}{41}i$
79. $-\frac{10}{3}i$
81. rational
83. 1.945×10^{-6}
85. $(2x + 3)^3$ and $8x^3 + 36x^2 + 54x + 27$
87. None, because $x^2 - 3x + 18$ is prime.
89. a. $\frac{x(x - 4)}{(x + 2)(x - 2)}$

b. 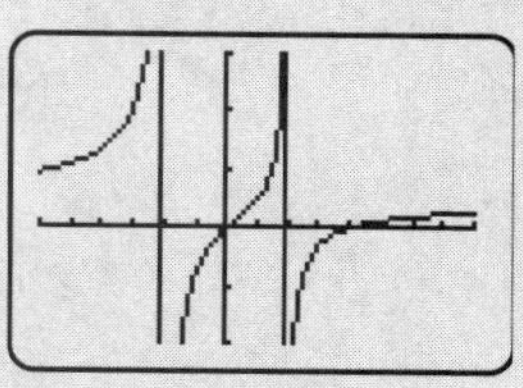

c. $x \neq -2, x \neq 0, x \neq 2$

91. a. $2\sqrt{5} + 2\sqrt{2}$
b. 7.30056308
c. yes
93. $2868 - 6100i$
95. 1.6×10^{14}

Practice Test

1. 4
3. -97
5. $2i|x|\sqrt{3}$
7. $\frac{y^5z^{1/2}}{x^{7/2}}$
9. $-7\sqrt{2}$
11. $2y^2 - 12y + 20$
13. $(x + 4)(x - 4)$
15. $(2x + 3y)^2$
17. $(2x + 1)(x - 1)$
19. $t(2t - 3)(t + 1)$
21. $(x - 3y)(x + 4y)$
23. $3(3 + x)(x^2 - 3x + 9)$
25. $\frac{5x - 2}{x(x - 1)} \quad x \neq 0, 1$
27. $\frac{1}{(x + 1)(x - 1)} \quad x \neq \pm 1$
29. $-\frac{1}{x + 3} \quad x \neq \frac{5}{2}, \pm 3$
31. $-8 - 26i$
33. $\frac{2 - 27\sqrt{3}}{59}$
35. $-\frac{1}{x(x + 1)} \quad x \neq -1, 0, 1$
37. 2.330

CHAPTER 1

Section 1.1

1. $x = 7$
3. $n = 15$
5. $x = -8$
7. $n = 15$
9. $x = 4$
11. $m = 2$
13. $t = \frac{7}{5}$
15. $x = -10$
17. $n = 2$
19. $x = 12$
21. $t = -\frac{15}{2}$
23. $x = -1$
25. $p = -\frac{9}{2}$
27. $x = \frac{1}{4}$
29. $x = -\frac{3}{2}$
31. $a = -8$
33. $x = -15$
35. $c = -\frac{35}{13}$
37. $m = \frac{60}{11}$
39. $x = 36$
41. $p = 8$
43. $y = -2$
45. $p = 2$
47. no solution
49. $y = \frac{3}{10}; \ y \neq 0$
51. $x = \frac{1}{2}; \ x \neq 0$
53. $a = \frac{1}{6}; a \neq 0$
55. no solution; $x \neq 2$
57. no solution; $p \neq 1$
59. $x = -10, x \neq -2$
61. no solution; $n \neq -1, 0$
63. no solution; $a \neq -3, 0$
65. $n = \frac{53}{11}; n \neq 1$
67. $x = -3; x \neq \frac{1}{2}, -\frac{1}{5}$
69. no solution; $t \neq 1$
71. $\frac{5}{9}F - \frac{160}{9} = C$
73. 84 minutes
75. 17 minutes
77. $\lambda = 0$ must be eliminated.
79. The error is forgetting to do to both sides of any equation what is done to one side. The correct answer is $x = 5$.
81. You cannot cross multiply with the -3 in the problem. You must find a common denominator first. The correct solution is $p = \frac{6}{5}$.
83. false
85. true
87. $x = \frac{c - b}{a}$
89. $x = \frac{ab}{c}, x \neq \pm a$
91. no solution; $x = -1, 0$
93. $x = \frac{by}{a - y - cy} \quad x \neq 0, -\frac{b}{c - 1}$
95. $x = 2$
97. all real numbers
99. no solution
101. 5426

Section 1.2

1. \$242.17 **3.** \$13.76

5. \$147,058.82 (\$22,058.82 saved)

7. 12 miles **9.** 9 hours of sleep **11.** 270 units

13. 24 **15.** 8, 10 **17.** 20 inches

19. Length = 100 yards, width = 30 yards

21. $r_1 = 3$ feet, $r_2 = 6$ feet **23.** 300 feet

25. The body is 63 inches or 5.25 feet.

27. \$20,000 at 4% and \$100,000 at 7%

29. \$3,000 at 10%, \$5,500 at 2%, and \$5,500 at 40%

31. 6 trees, 27 shrubs **33.** 70 ml of 5%, 30 ml of 15%

35. $3\frac{1}{3}$ or $3.3\overline{3}$ gallons

37. The entire bag was gummy bears.

39. 9 minutes **41.** 2.3 mph

43. Jogger: 6 mph
Walker: 4 mph

45. Bicyclist: 6 minutes
Walker: 18 minutes

47. 22.5 hours **49.** 2.4 hours

51. 330,396 **53.** 77.5 for a *B*; 92.5 for an *A*

55. 2 field goals and 6 touchdowns

57. 3.5 feet from center

59. Fulcrum is 0.4 units from Maria and 0.6 units from Max.

61. $d_o = \frac{15}{2}$ or 7.5 cm (in front of the lens)

63. image distance is 3 cm, object distance is 6 cm

65. $w = \dfrac{P - 2l}{2}$ **67.** $h = \dfrac{2A}{b}$

69. $w = \dfrac{A}{l}$ **71.** $h = \dfrac{V}{lw}$

73. Janine: 58 mph; Tricia: 70 mph

75. \$191,983.35

77. Plan B is better for 5 or fewer plays/month. Plan A is better for 6 or more plays/month.

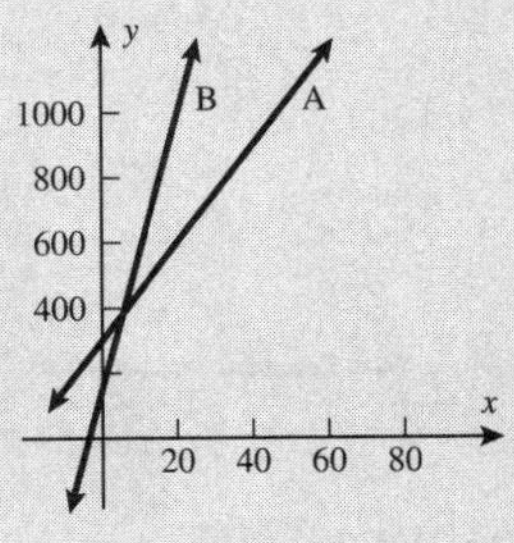

Section 1.3

1. $x = 2, 3$ **3.** $p = 3, 5$ **5.** $x = -4, 3$

7. $x = -\frac{1}{4}$ **9.** $y = \frac{1}{3}$ **11.** $y = 0, 2$

13. $p = \frac{2}{3}$ **15.** $x = -3, 3$ **17.** $x = -6, 2$

19. $p = -5, 5$ **21.** $x = -2, 2$ **23.** $p = -2\sqrt{2}, 2\sqrt{2}$

25. $x = -3i, 3i$ **27.** $x = -3, 9$

29. $x = \dfrac{-3 \pm 2i}{2}$ **31.** $x = \dfrac{2 \pm 3\sqrt{3}}{5}$

33. $x = -2, 4$ **35.** $x^2 + 6x + \underline{9}$

37. $x^2 - 12x + \underline{36}$ **39.** $x^2 - \frac{1}{2}x + \underline{\frac{1}{16}}$

41. $x^2 + \frac{2}{5}x + \underline{\frac{1}{25}}$ **43.** $x^2 - 2.4x + \underline{1.44}$

45. $x = -3, 1$ **47.** $t = 1, 5$

49. $y = 1, 3$ **51.** $p = \dfrac{-4 \pm \sqrt{10}}{2}$

53. $x = \frac{1}{2}, 3$ **55.** $x = \dfrac{4 \pm 3\sqrt{2}}{2}$

57. $t = \dfrac{-3 \pm \sqrt{13}}{2}$ **59.** $s = \dfrac{-1 \pm \sqrt{3}i}{2}$

61. $x = \dfrac{3 \pm \sqrt{57}}{6}$ **63.** $x = 1 \pm 4i$

65. $x = \dfrac{-7 \pm \sqrt{109}}{10}$ **67.** $x = \dfrac{-4 \pm \sqrt{34}}{3}$

69. 1 real solution **71.** 2 real solutions

73. 2 complex solutions **75.** $v = -2, 10$

77. $t = -6, 1$ **79.** $x = -7, 1$

81. $p = 4 \pm 2\sqrt{3}$ **83.** $w = \dfrac{-1 \pm \sqrt{167}i}{8}$

85. $p = \dfrac{9 \pm \sqrt{69}}{6}$ **87.** $t = \dfrac{10 \pm \sqrt{130}}{10}$

89. $x = 3, 4$ **91.** $x = -\frac{3}{4}, 2$

93. $x = 0.4$ and $x = -0.3$

95. $t = 8$ (August) and $t = 12$ (December) 2003.

97. 20 inches **99.** 17 and 18 **101.** 9 ft × 15 ft

103. base is 6 units and height is 20 units

105. 2.5 seconds after it is dropped

107. 21.2 feet **109.** 5 ft × 5 ft

111. border is 2.3 feet wide **113.** 10 days

115. The problem is factored incorrectly. The correction would be $t = 6$ and $t = -1$.

117. When taking the square root of both sides, the i is missing from the right side. The correction would be $a = \pm\frac{3}{4}i$.

119. false

121. true

123. $x^2 - 2ax + a^2 = 0$

125. $x^2 - 7x + 10 = 0$

127. $t = \pm\sqrt{\frac{2s}{g}}$

129. $c = \pm\sqrt{a^2 + b^2}$

131. $x = 0, \pm 2$

133. $x = \pm 2, -1$

135. $\frac{-b + \sqrt{b^2 - 4ac}}{2a} + \frac{-b - \sqrt{b^2 - 4ac}}{2a} = \frac{2b}{2a} = \frac{-b}{a}$

137. $x^2 - 6x + 4 = 0$

139. 250 mph

141. $ax^2 - bx + c = 0$

143. Small jet: 300 mph; 757: 400 mph

145. $x = -1, 2$

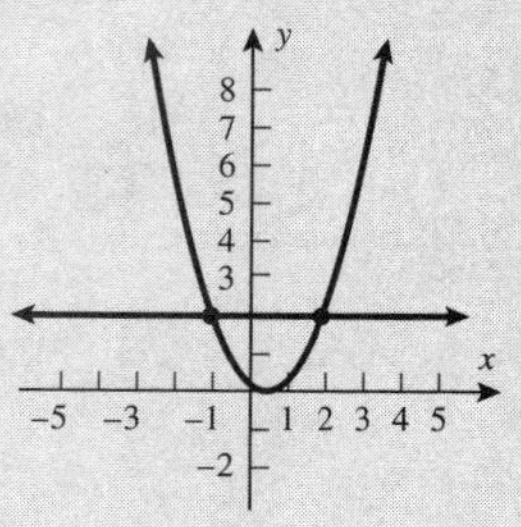

147. **a.** $x = -2, 4$

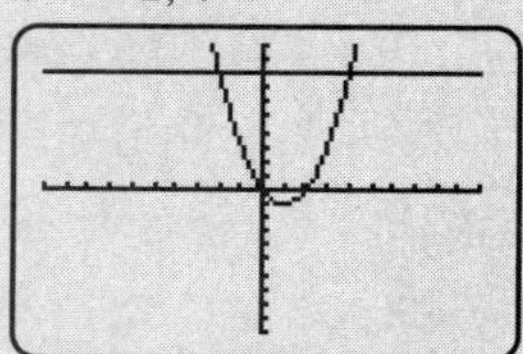

b. $b = -3, x = 1 \pm \sqrt{2}i$

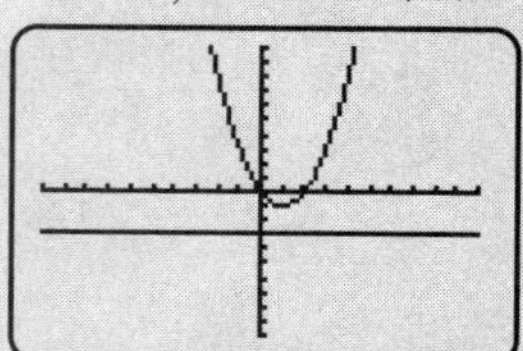

$b = -1, x = 1$

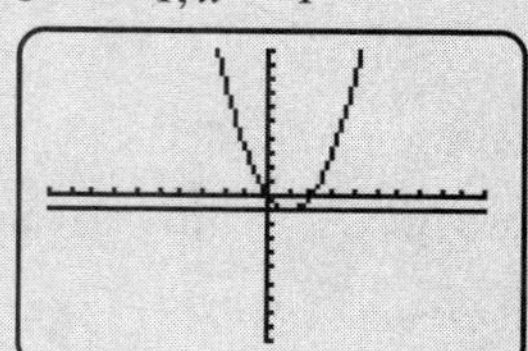

$b = 0, x = 0, 2$

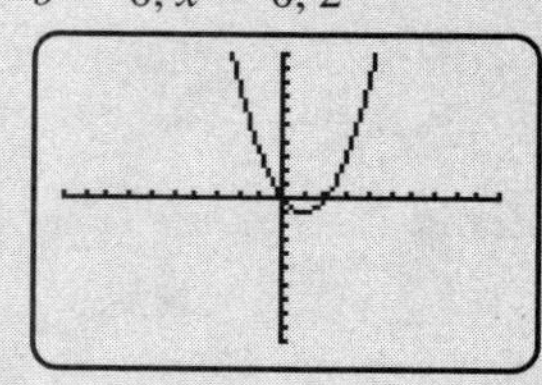

$b = 5, x = 1 \pm \sqrt{6}$

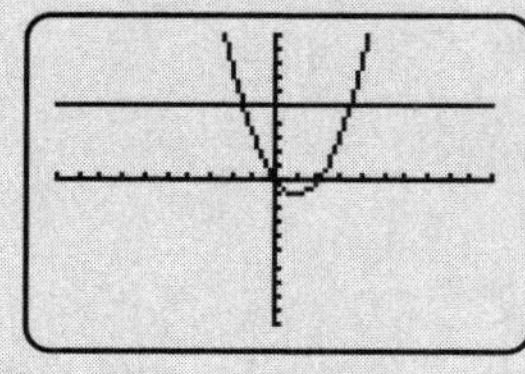

Section 1.4

1. $t = 9$

3. $p = 8$

5. no solution

7. $x = 5$

9. $y = -\frac{1}{2}$

11. $x = 4$

13. $y = 0, y = 25$

15. $s = 3, s = 6$

17. $x = -3, x = -1$

19. $x = 0$

21. $x = \frac{5}{2}$

23. no solution

25. $x = 1, x = 5$

27. $x = 7$

29. $x = 4$

31. $x = -8, x = 0$

33. $x = \pm\sqrt{2}, x = \pm 1$

35. $x = \pm\frac{\sqrt{6}}{2}i, x = \pm i\sqrt{2}$

37. $x = -\frac{5}{2}, x = -1$

39. $t = \frac{5}{4}, t = 3$

41. $x = \pm 1, \pm i, \pm\frac{1}{2}, \pm\frac{1}{2}i$

43. $y = -\frac{3}{4}, y = 1$

45. $z = 1$

47. $t = -27, t = 8$

49. $x = -\frac{4}{3}, x = 0$

51. $x = \frac{3}{8}, x = \frac{2}{3}$

53. $u = \pm 1, u = \pm 8$

55. $t = \sqrt{3}$

57. $x = -3, 0, 4$

59. $p = 0, \pm\frac{3}{2}$

61. $u = 0, \pm 2, \pm 2i$

63. $x = \pm 3, 5$

65. $y = -2, 5, 7$

67. $x = -1, 0, 3$

69. $t = \pm 5$

71. $y = 2, 3$

73. $a = 71$

75. $t = 4\sqrt{3} \approx 7$ months (Oct. 2004)

77. 132 ft

79. 25 cm

81. 80% the speed of light

83. $t = 5$ is extraneous. No solution.

85. The error is not converting u back to x using the substitution.

87. true

89. false

91. $[0, \infty)$

93. $x = -1, x = \frac{1}{3}, x = 0, x = -\frac{2}{3}$

95. $x = -2$

97. $x = \frac{313}{64} \approx 4.891$

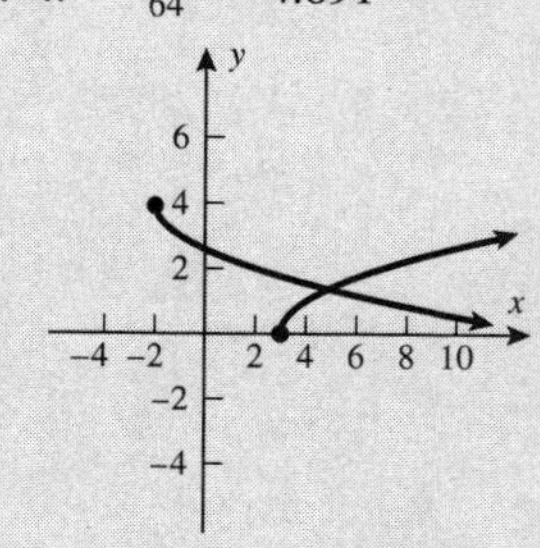

99. no solution

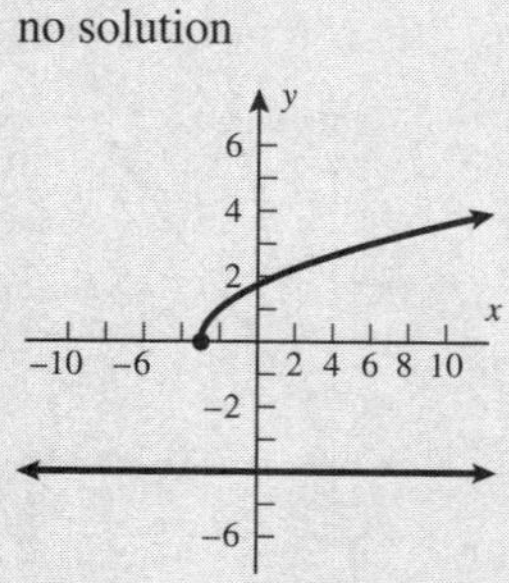

101. $x = 81$

103. no real solution

Section 1.5

1. 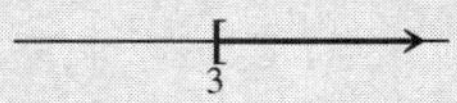 $[3, \infty)$

3. 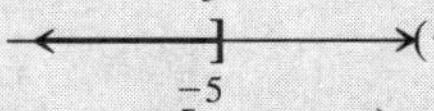 $(-\infty, -5]$

5.  $[-2, 3)$

7.  $(-3, 5]$

9. 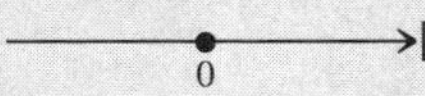 $[0]$

11.  $[4, 6]$

13.  $[-8, -6]$

15. empty set

17. $\{x \mid 0 \le x < 2\}$

19. $\{x \mid -7 < x < -2\}$

21. $\{x \mid x \le 6\}$

23. $\{x \mid -\infty < x < \infty\}$

25. $-3 < x \le 7, (-3, 7]$

27. $3 \le x < 5, [3, 5)$

29. $-2 \le x, [-2, \infty)$

31. $x < 8, (-\infty, 8)$

33.  $(-5, 3)$

35.  $[-6, 5)$

37.  $[-1, 1]$

39. 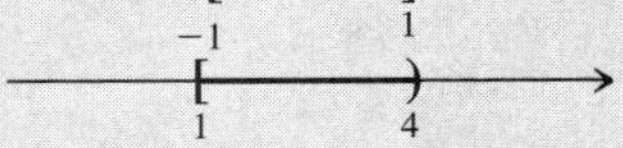 $[1, 4)$

41.  $[-1, 2)$

43. 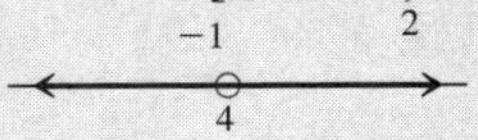 $(-\infty, 4) \cup (4, \infty)$

45.  $(-\infty, -3] \cup [3, \infty)$

47. 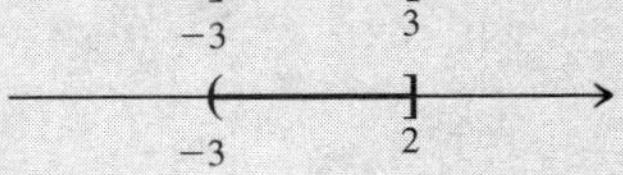 $(-3, 2]$

49. empty set

51. $(-\infty, 2) \cup [3, 5)$

53. $(-\infty, -4] \cup (2, 5]$

55. $[-4, -2) \cup (3, 7]$

57. $(-6, -3] \cup [0, 4)$

59. $x < 10, (-\infty, 10)$

61. $x \le 2, (-\infty, 2]$

63. $p \le -2, (-\infty, -2]$

65. $x \ge -2, [-2, \infty)$

67. $t > -3, (-3, \infty)$

69. $x < 6, (-\infty, 6)$

71. $x \le -8, (-\infty, -8]$

73. $t \le -7, (-\infty, -7]$

75. $y < 1, (-\infty, 1)$

77. $-5 < x < 2, (-5, 2)$

79. $-6 \le x < 2, [-6, 2)$

81. $-8 \le x < 4, [-8, 4)$

83. $-6 < y < 6, (-6, 6)$

85. $\frac{1}{2} \le y \le \frac{5}{4}, \left[\frac{1}{2}, \frac{5}{4}\right]$

87. $128 \le w \le 164$

89. more than 50 dresses

91. most minutes = 1013, least minutes = 878

93. 92

95. $21,537.69 < invoice price < $24,346.96

97. $0.9\, r_T \le r_R \le 1.1\, r_T$

99. $0.85L \le B \le 0.95L$

101. 4

103. $\$4,386.25 \le T \le \$15,698.75$

105. The correction should be $[-1, 4)$.

107. You must reverse the sign.

109. true

111. a and b

113. a and b

115. c

117. $(-\infty, 0]$

119. no solution

121. a. $x > .83582$ **b.** **c.** agree

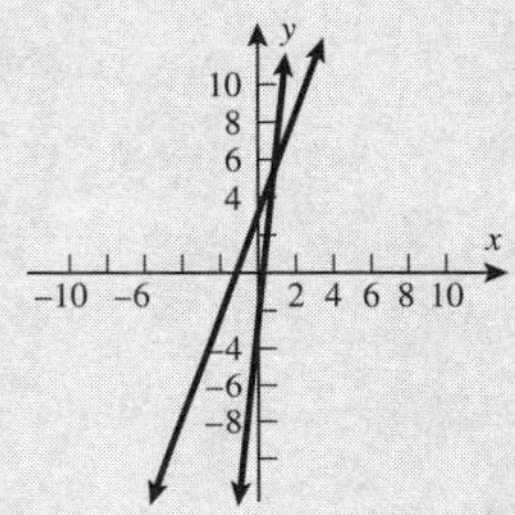

123. a. $-2 < x < 5$ **b.** **c.** agree

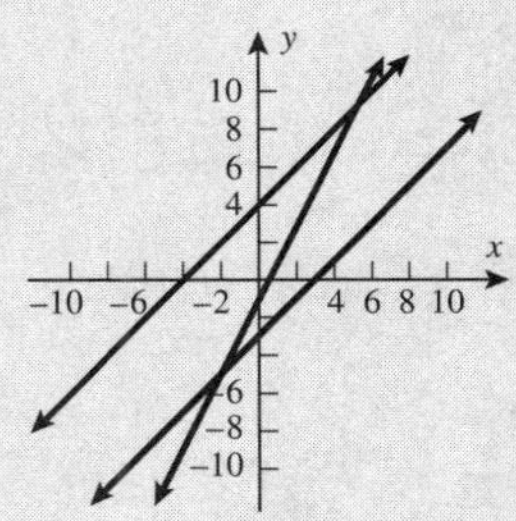

125. a. $(-\infty, \infty)$ **b.** **c.** yes

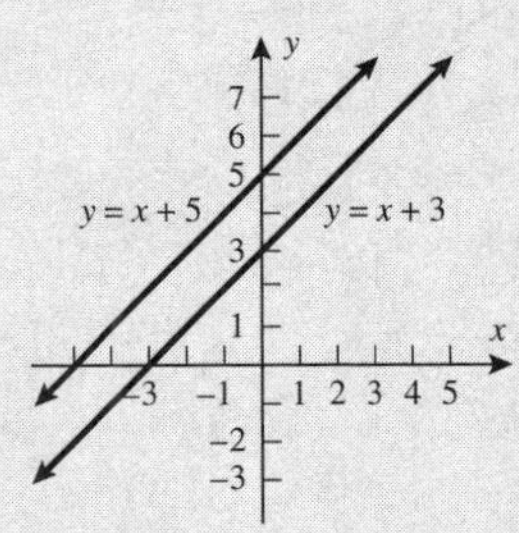

Section 1.6

1. $(-\infty, -2] \cup [5, \infty)$

3. $[-1, 6]$

5. $(-3, -1)$

7. $\left[-1, \frac{3}{2}\right]$

9. $\left(\frac{1}{3}, \frac{1}{2}\right)$

11. $\left(-\infty, -\frac{1}{2}\right] \cup [3, \infty)$

13. $\left(-\infty, -1 - \sqrt{5}\right] \cup \left[-1 + \sqrt{5}, \infty\right)$

15. $(2 - \sqrt{10}, 2 + \sqrt{10})$

17. $(-\infty, 0] \cup [3, \infty)$

19. $[0, 2]$

21. $(-\infty, -3) \cup (3, \infty)$

23. $(-9, 9)$

25. $\mathbb{R}$ (consistent)

27. no real solution

29. $(0, \infty)$

31. $(-\infty, -3) \cup (0, \infty)$

33. $(-\infty, -3] \cup (4, \infty)$

35. $(-\infty, -2) \cup [-1, 2)$

37. $(-5, 3] \cup (5, \infty)$

39. $\left(-\frac{3}{2}, 1\right)$

41. $(-\infty, -5] \cup (-2, 0]$

43. $(-2, 2)$

45. no real solution

47. $\mathbb{R}$ (consistent)

49. $[-3, 3) \cup (3, \infty)$

51. $(-3, -1] \cup (3, \infty)$

53. $(-\infty, -4) \cup (2, 5]$

55. $(-6, -4) \cup (4, 8)$

57. $(-\infty, -2) \cup (2, \infty)$

59. $30 < x < 100$ (30–100 orders will yield a profit)

61. For years 3–5, the car is worth more than you owe. In the first 3 years you owe more than the car is worth.

63. 75 seconds

65. 20 ft $\le$ length $\le$ 30 ft

67. \$3,342 $\le$ price per acre $\le$ 3,461

69. You cannot divide by x.

71. You must consider the $x = -2$ as a critical point.

73. false

75. either no real solution or infinitely many solutions

77. $\mathbb{R}$

79. $\mathbb{R}$

81. always true, $\mathbb{R}$

83. $(-0.8960, 1.6233)$

85. $(-2, 0)$

87. $\left(\frac{5}{3}, 5\right)$

Section 1.7

1. $x = -3$ or $x = 3$

3. no solution

5. $t = -1$ or $t = -5$

7. $p = 10$ or $p = 4$

9. $y = 3$ or $y = 5$

11. $x = 3$ or $x = -3$

13. $x = 1$ or $x = -8$

15. $t = 4$ or $t = 2$

17. $x = -1$ or $x = 8$

19. $y = 0$ or $y = \frac{2}{3}$

21. $x = \frac{2}{3}$ or $x = -\frac{80}{21}$

23. $x = \frac{47}{14}$ or $x = -\frac{23}{14}$

25. $x = -3$ or $x = 13$

27. $p = -13$ or $p = 7$

29. $y = -5$ or $y = 9$

31. $x = \pm\sqrt{3}$ and $x = \pm\sqrt{5}$

33. $x = \pm 2$

35. $(-7, 7)$

37. $(-\infty, -5] \cup [5, \infty)$

39. $(-10, 4)$

41. $(-\infty, 2) \cup (6, \infty)$

43. $[3, 5]$

45. $\mathbb{R}$

47. $(-4, 1)$

49. $(-\infty, 2] \cup [5, \infty)$

51. $\mathbb{R}$

53. $\left(-\infty, -\frac{3}{2}\right] \cup \left[\frac{3}{2}, \infty\right)$

55. $(-\infty, -3) \cup (3, \infty)$

57. $\left(\frac{1}{4}, \frac{3}{4}\right)$

59. $(-2.769, -1.385)$

61. $[-3, 3]$

63. $|x - 2| < 7$

65. $\left|x - \frac{3}{2}\right| \ge \frac{1}{2}$

67. $|x - a| \le 2$

69. $|T - 83| \le 15$

71. $d = 4$ (tie) $d < 4$ (win)

73. The mistake was that $x - 3 = -7$ was not considered.

75. Did not reverse the inequality sign when divided by a negative.

77. true

79. false

81. $a - b < x < a + b$ or $(a - b, a + b)$

83. $\mathbb{R}$

85. $x = a + b$ and $x = a - b$

87. no solution

89. $x \ge 7$, agree

91. agree

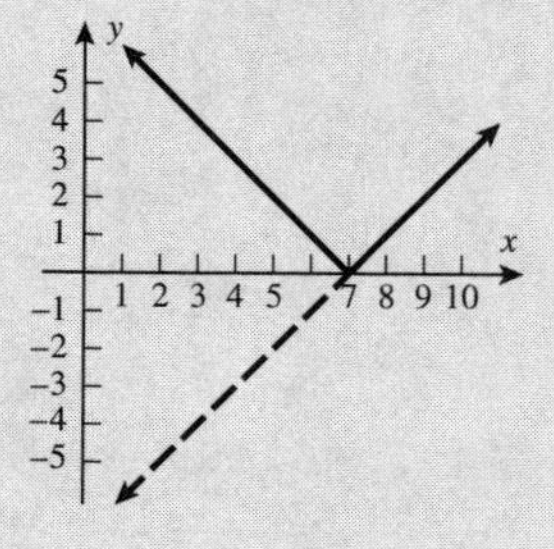

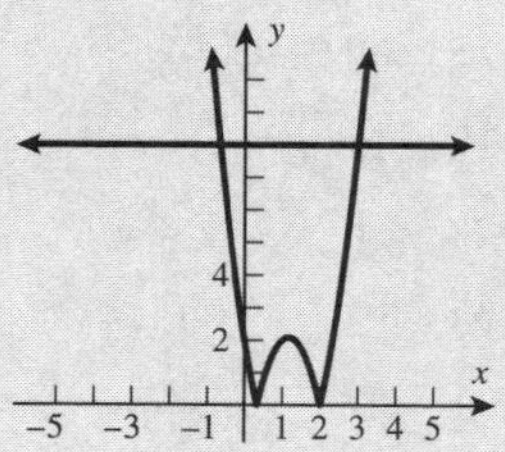

93. $\left(-\frac{1}{2}, \infty\right)$

Review Exercises

1. $x = \frac{16}{7}$

3. $p = -\frac{8}{25}$

5. $x = 27$

7. $y = -\frac{17}{5}$

9. $b = \frac{6}{7}$

11. $x = -\frac{6}{17}$

13. $x = \dfrac{6 \pm \sqrt{39}}{3}, x \ne 0$

15. $t = -\frac{34}{5}, t \ne 0, t \ne -4$

17. $x = -\frac{1}{2}, x \ne 0$

19. $x = \frac{29}{17}$

21. $x = 8 - 7y$

23. 96 miles

25. 144

27. 3 inches $\times$ 7 inches

29. \$20,000 at 8% and \$5,000 at 20%

31. 60 ml of 5% and 90 ml of 10%

33. 91

35. $b = 7$ and $b = -3$

37. $x = 0$ and $x = 8$

39. $q = \pm 13$

41. $x = 2 \pm 4i$

43. $x = 6$ and $x = -2$

45. $x = \dfrac{1 \pm \sqrt{33}}{2}$

47. $t = \frac{7}{3}$ and $t = -1$

49. $f = \dfrac{1 \pm \sqrt{337}}{48}$

51. $q = \dfrac{3 \pm \sqrt{69}}{10}$

53. $x = \frac{5}{2}$ and $x = -1$

55. $x = \frac{2}{7}$ and $x = -3$

57. $r = \sqrt{\frac{S}{\pi h}}$

59. $v = \frac{h}{t} + 16t$

61. base = 4 ft, height = 1 ft

63. $x = 6$

65. $x = 125$

67. no solution

69. $x \approx -0.6$

71. $x \approx 5.85$

73. $x \approx -2.303$

75. $x = 2, x = 3$

77. $x = \frac{5}{4}$ and $x = \frac{3}{4}$

79. $y = \frac{1}{4}, 1$

81. $x = -\frac{125}{8}, 1$

83. $x = -1, -\frac{1}{8}$

85. $x = \pm 2$ and $x = \pm 3i$

87. $x = -8, 0, 4$

89. $p = -2, 3, 2$

91. $p = -\frac{1}{2}, \frac{5}{2}, 3$

93. $y = \pm 9$

95. $(-\infty, -4]$

97. $[2, 6]$

99. $x > -6$

101. $-3 \le x \le 7$

103. $x \ge -4, [-4, \infty)$

105. $(4, \infty)$

107. $[8, 12]$

109. $x < \frac{5}{3}$

111. $x > -\frac{3}{2}$

113. $4 < x \le 9$

115. $3 \le x \le \frac{7}{2}$

117. 74

119. $[-6, 6]$

121. $(-\infty, 0] \cup [4, \infty)$

123. $(-\infty, 0) \cup (7, \infty)$

125. $\left(-\infty, -\frac{3}{4}\right) \cup (4, \infty)$

127. $(0, 3)$

129. $(-\infty, -6] \cup [9, \infty)$

131. $(-\infty, 2) \cup (4, 5]$

133. $(3, \infty)$

135. no solution

137. $x = 1.7$ and $x = 0.9667$

139. $(-4, 4)$

141. $(-\infty, -11) \cup (3, \infty)$

143. $(-\infty, -3) \cup (3, \infty)$

145. $\mathbb{R}$

147. $75 \le T \le 95$

149. 2510

151. **a.** $x = -5$ and $x = 1$
b. $b = -5, x = -2 \pm i$
$b = 0, x = -4, 0$
$b = 7, x = -2 \pm \sqrt{11}$
$b = 12, x = -6, 2$

153. $x = \left(-1 + \sqrt{7}\right)^4 \approx 7.34$

155. $(-\infty, 15.4)$

157. $(-\infty, -5) \cup (5.25, \infty)$

159. $\left(\frac{7}{5}, \frac{7}{2}\right)$

161. $(-2.19, -0.9) \cup (0.9, 2.19)$

Practice Test

1. $p = -3$

3. $t = -4, t = 7$

5. $x = \frac{8}{3}$ and $x = -\frac{1}{2}$

7. $y = -8$

9. $x = 4$

11. $y = 1$

13. $x = 0, x = 2, x = 6$

15. $L = \frac{P - 2W}{2}$

17. $(-\infty, 17]$

19. $\left(-\frac{32}{5}, -6\right]$

21. $(-\infty, -1] \cup \left[\frac{4}{3}, \infty\right)$

23. $\left(-\frac{1}{2}, 3\right]$

25. 1000 feet

27. $627 \le \text{minutes} \le 722$

29. 7.95

Cumulative Test

1. -15

3. $\frac{x^{12}}{y^3}$

5. $x^4 + 2x^3 - 15x^2$

7. $2(a + 10)(a^2 - 10a + 100)$

9. $\frac{x^2 + 22x}{x^2 - 4}$

11. $x = 56$

13. $x = -4$

15. 11.25 hours

17. $x = -6 \pm 2i$

19. $x = 4$

21. $[-3, 4)$

23. $(-\infty, -3) \cup [-2, 3)$

25. $x = -\frac{17}{3}, -1$

27. $\left(-1, \frac{1}{2}\right)$

CHAPTER 2

Section 2.1

1. $(4, 2)$

3. $(-3, 0)$

5. $(0, -3)$

7.

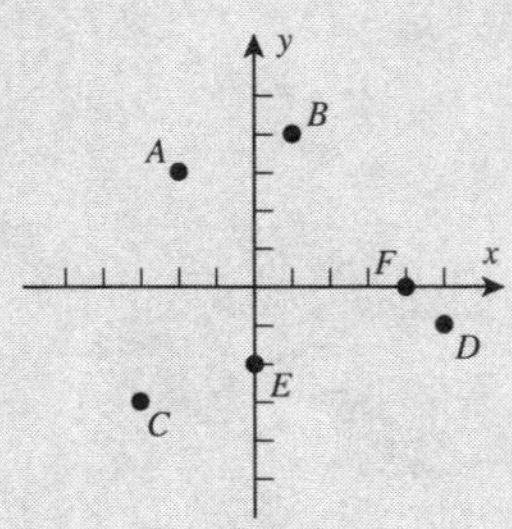

9. The line being described is $x = -3$.

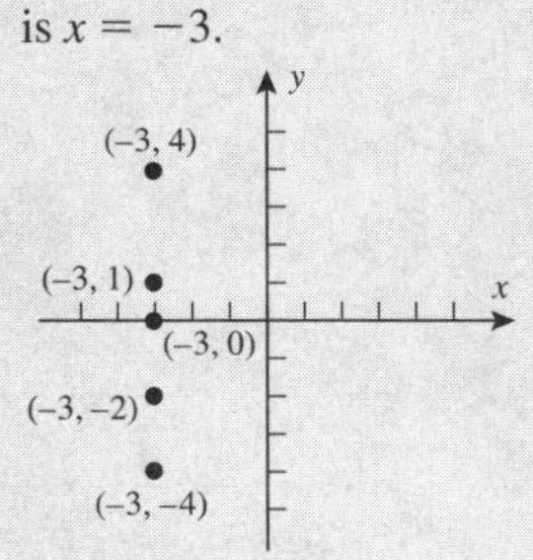

11. $d = 4, (3, 3)$

13. $d = 4\sqrt{2}, (1, 2)$

15. $d = 3\sqrt{10}, \left(-\frac{17}{2}, \frac{7}{2}\right)$

17. $d = 5, \left(-5, \frac{1}{2}\right)$

19. $d = 4\sqrt{2}, (-4, -6)$

21. $d = 5, \left(\frac{3}{2}, \frac{11}{6}\right)$

23. $d = \frac{\sqrt{4049}}{60}, \left(-\frac{5}{24}, \frac{1}{15}\right)$

25. $d = 3.9, (0.3, 3.95)$

27. $d = \sqrt{1993.01} \approx 44.64, (1.05, -1.2)$

29. $d = 4\sqrt{2}, \left(\sqrt{3}, 3\sqrt{2}\right)$

31. $d = \sqrt{10 + 2\sqrt{2} + 4\sqrt{3}}, \left(\frac{1 - \sqrt{2}}{2}, \frac{-2 + \sqrt{3}}{2}\right)$

33. The perimeter of the triangle rounded to two decimal places is 21.84.

35. right triangle 37. isosceles 39. 128.06 miles

41. distance $\approx$ 268 miles 43. midpoint = (2003, 330) \$330M in revenue

45. x's were subtracted from y's. The correct distance would be $d = \sqrt{58}$.

47. The values were not used in the correct position. The correct midpoint would be $\left(2, \frac{13}{2}\right)$.

49. true 51. true

53. The distance is $d = \sqrt{(b-a)^2 + (a-b)^2} = \sqrt{2(a-b)^2}$
$= \sqrt{2}|a - b|$. The midpoint is $m = \left(\frac{a+b}{2}, \frac{b+a}{2}\right)$.

55. $\sqrt{\left(x_1 - \frac{x_1 + x_2}{2}\right)^2 + \left(y_1 - \frac{y_1 + y_2}{2}\right)^2}$

$\sqrt{\left(\frac{2x_1 - x_1 - x_2}{2}\right)^2 + \left(\frac{2y_1 - y_1 - y_2}{2}\right)^2}$

$\sqrt{\left(\frac{x_1 - x_2}{2}\right)^2 + \left(\frac{y_1 - y_2}{2}\right)^2}$

$\frac{1}{2}\sqrt{(x_1 - x_2)^2 + (y_1 - y_2)^2}$

Using (x_2, y_2) with the midpoint yields the same result.

57. $\sqrt{(a-c)^2 + (b-d)^2}$

59. $m = (0.7, 5.15), d \cong 6.357$ 61. $m = (2.2, 3.3), d \cong 3.111$

Section 2.2

1. a. no b. yes 3. a. yes b. no

5. a. yes b. no 7. a. yes b. no

9.

x	y	(x, y)
−2	0	(−2, 0)
0	2	(0, 2)
1	3	(1, 3)

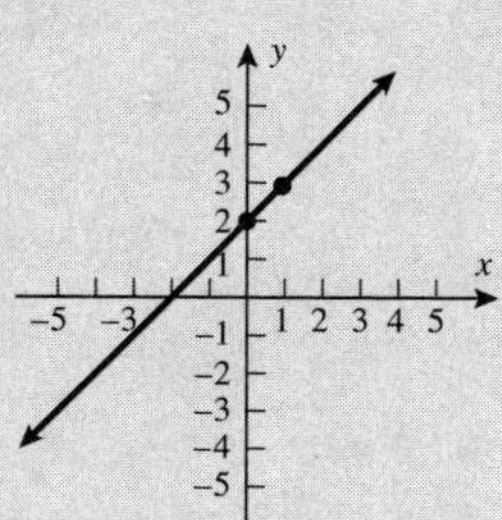

11.

x	y	(x, y)
−1	2	(−1, 2)
0	0	(0, 0)
$\frac{1}{2}$	$-\frac{1}{4}$	$\left(\frac{1}{2}, -\frac{1}{4}\right)$
1	0	(1, 0)
2	2	(2, 2)

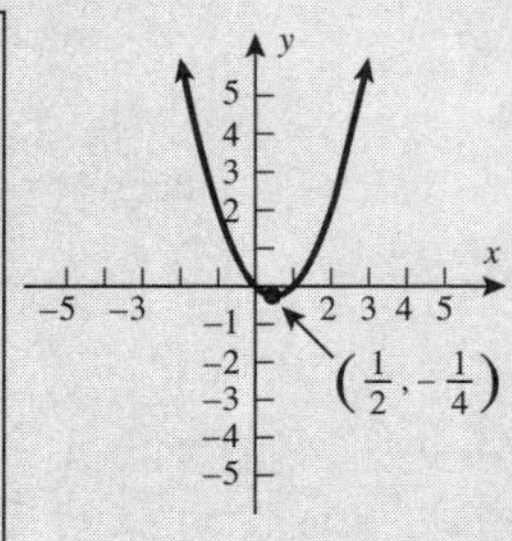

13.

x	y	(x, y)
1	0	(1, 0)
2	1	(2, 1)
5	2	(5, 2)
10	3	(10, 3)

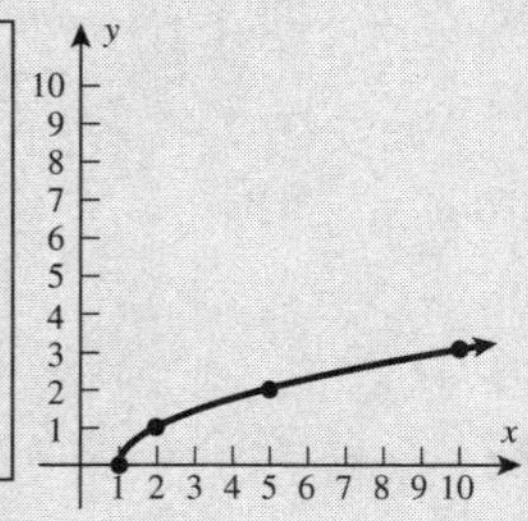

15.

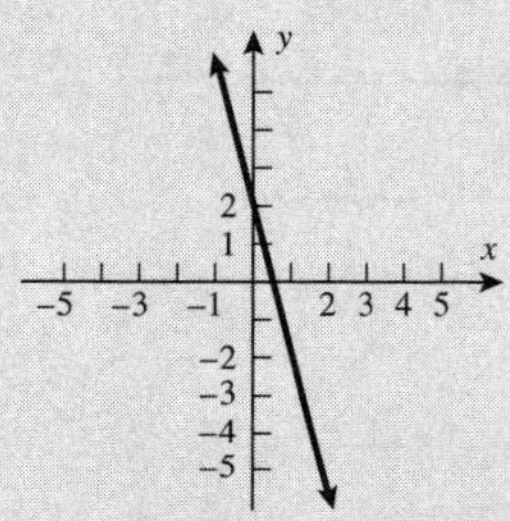

17.

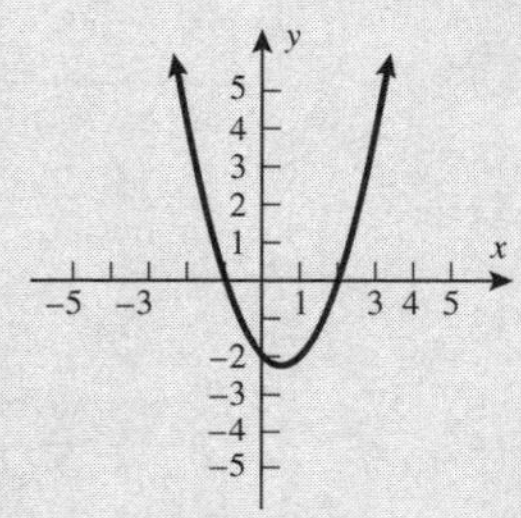

19.

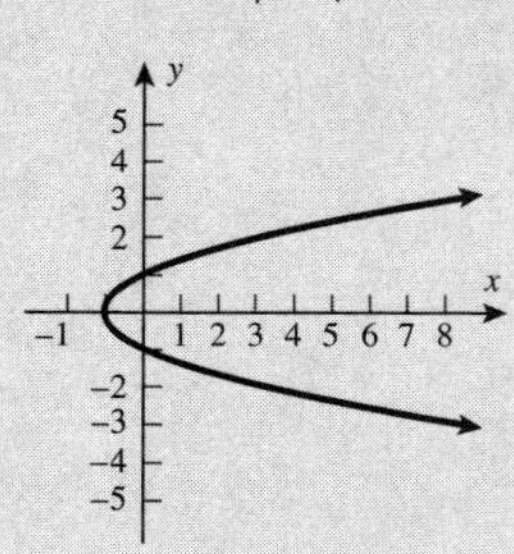

21.

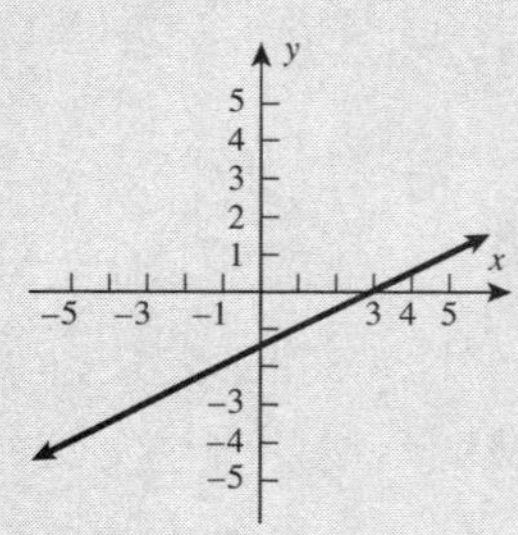

23. x-intercept: (3, 0), y-intercept: (0, −6)

25. x-intercepts: (±3, 0), y-intercept: (0, −9)

27. x-intercept: (4, 0), y-intercept: none

29. x-intercept: none, y-intercept: $\left(0, \frac{1}{4}\right)$

31. x-intercepts: (±2, 0), y-intercepts: (0, ±4)

33. d 35. a 37. b 39. (−1, −3) 41. (−7, 10)

43. (−3, −2), (3, 2), (−3, 2)

45. x-axis 47. origin 49. x-axis

51. x-axis, y-axis, and origin 53. y-axis

55. y-axis 57. origin

59.

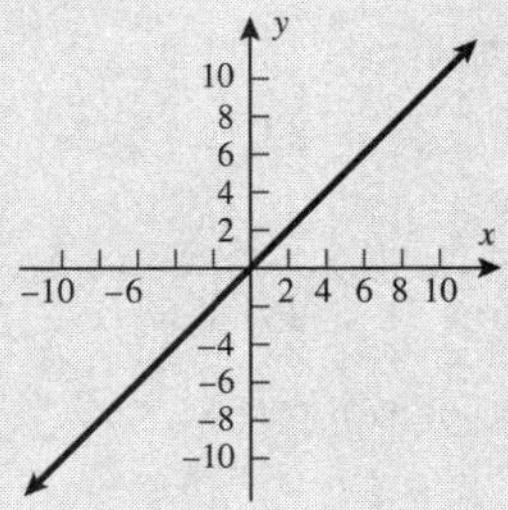

61.

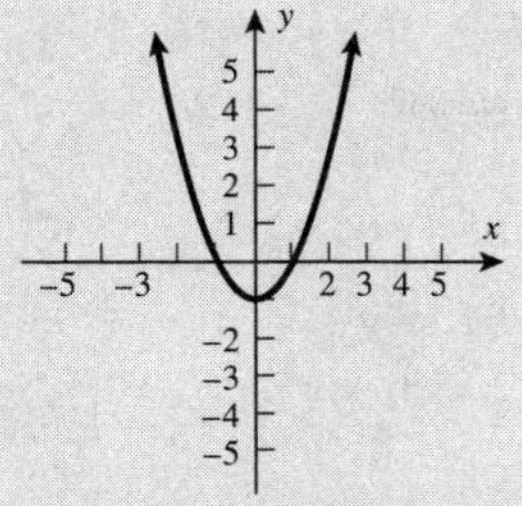

63.

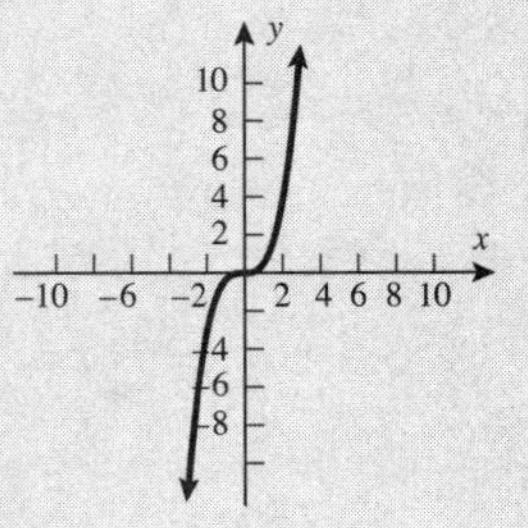

65.

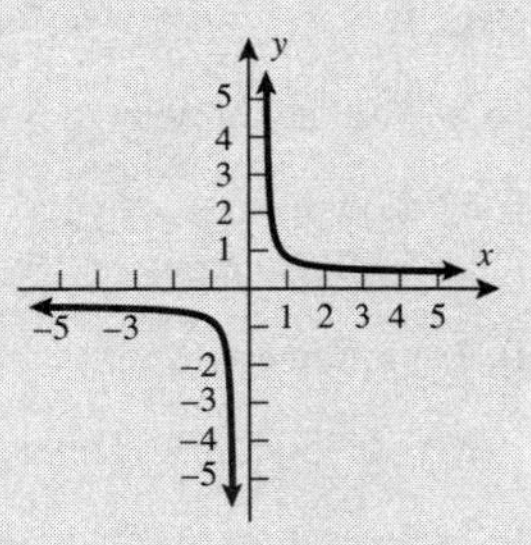

67.

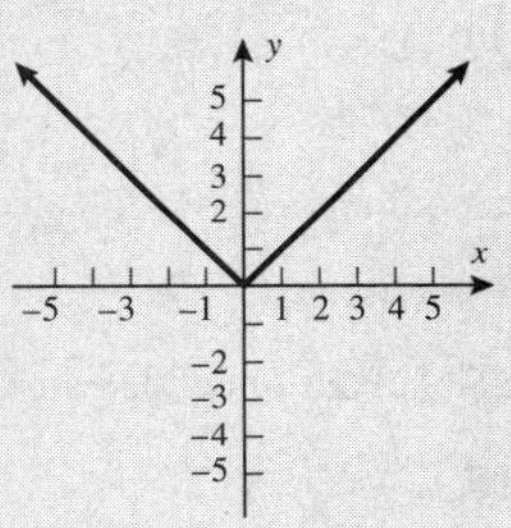

69.

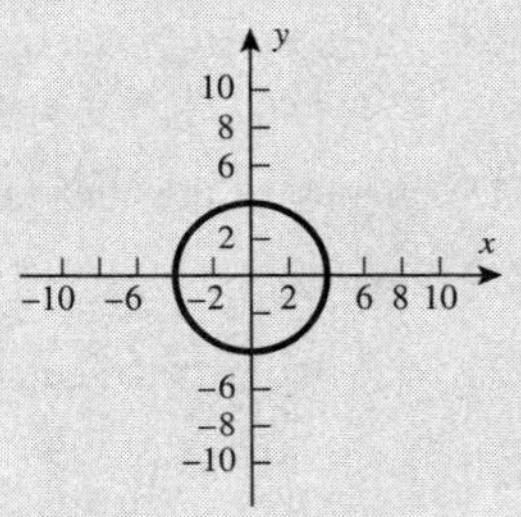

71.

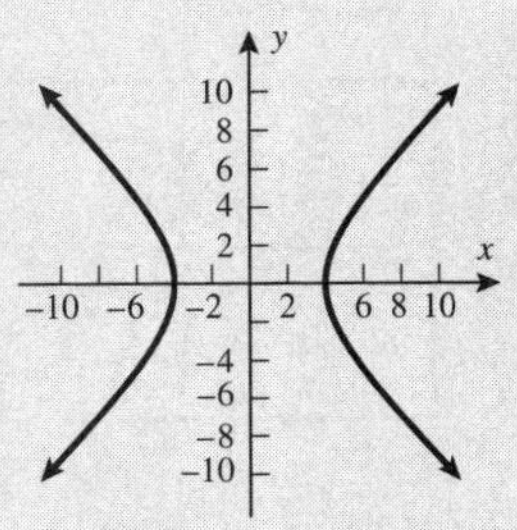

73.

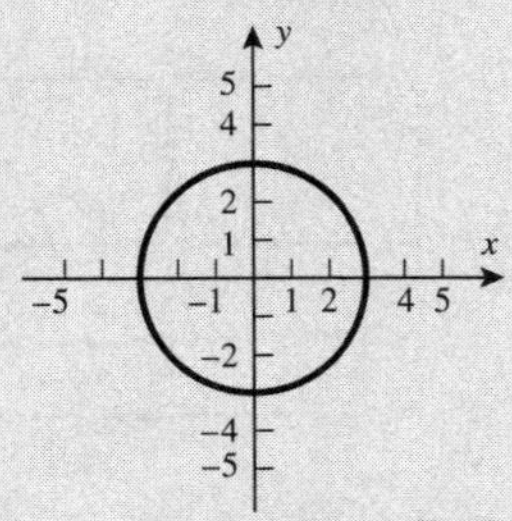

75.

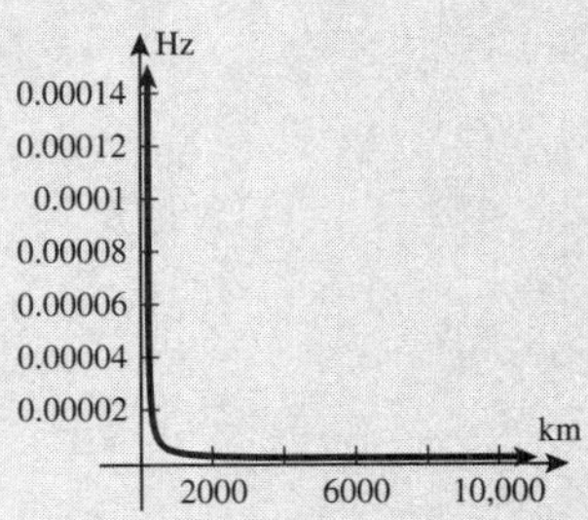

77.

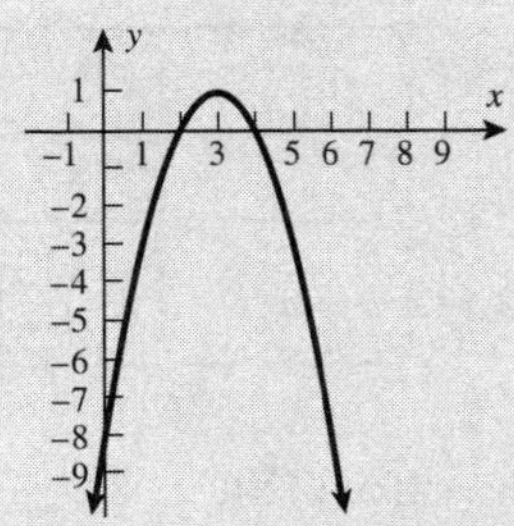

break even units = 2000
range of units: 2000-4000

79.

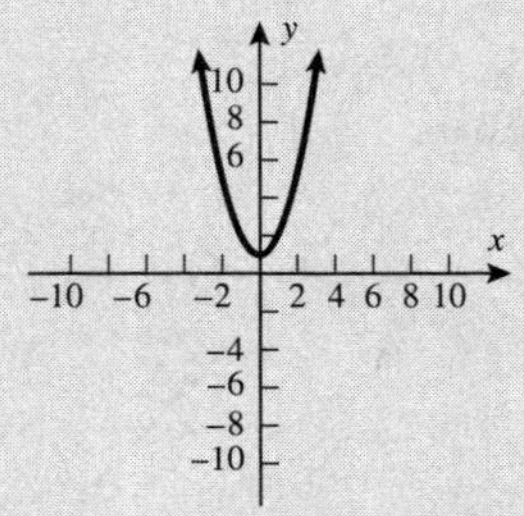

81. You are checking to see whether the function is symmetric to the y-axis. Thus the substitution shown is incorrect. The correct substitution would be plugging $-x$ for x into the function, not $-y$ for y.

83. false

85. true

87. origin

89. y-axis

91. x-axis, y-axis, and origin

93. x-axis, y-axis, and origin

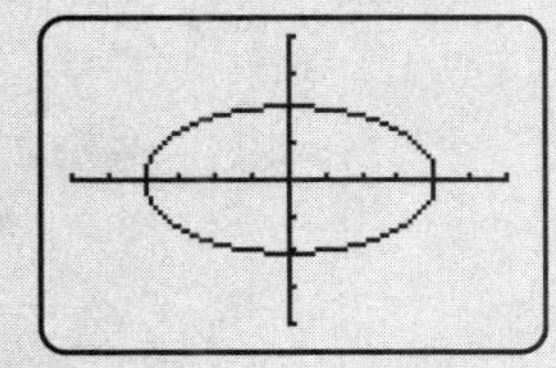

Section 2.3

1. slope $= m = 3$

3. slope $= m = -2$

5. slope $= m = -\frac{19}{10}$

7. slope $= m \approx 2.379$

9. $m = -3$

11. x-intercept $(0.5, 0)$, y-intercept $(0, -1)$, slope $= m = 2$ rising

13. x-intercept $(1, 0)$, y-intercept $(0, 1)$, slope $= m = -1$ falling

15. x-intercept none, y-intercept $(0, 1)$, slope $= m = 0$ horizontal

17. x-intercept: $\left(\frac{3}{2}, 0\right)$
y-intercept: $(0, -3)$

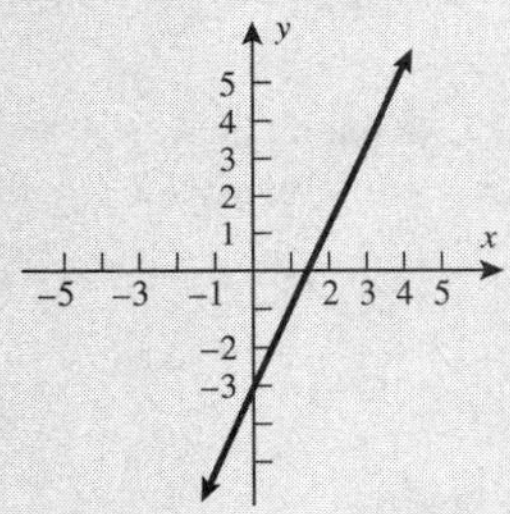

19. x-intercept: $(4, 0)$
y-intercept: $(0, 2)$

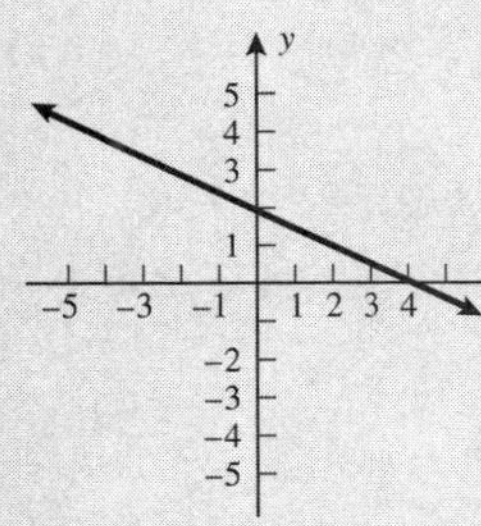

21. x-intercept: $(2, 0)$
y-intercept: $\left(0, -\frac{4}{3}\right)$

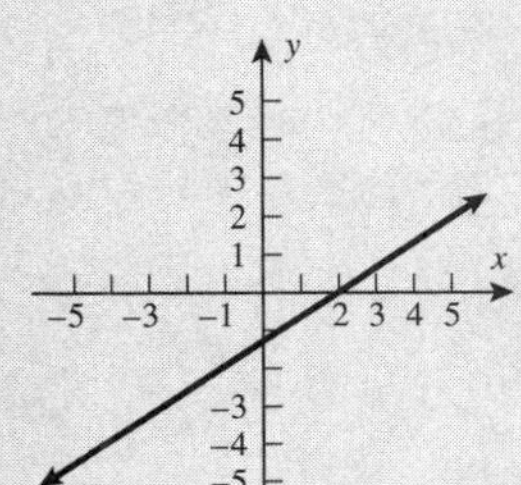

23. x-intercept: $(-2, 0)$
y-intercept: $(0, -2)$

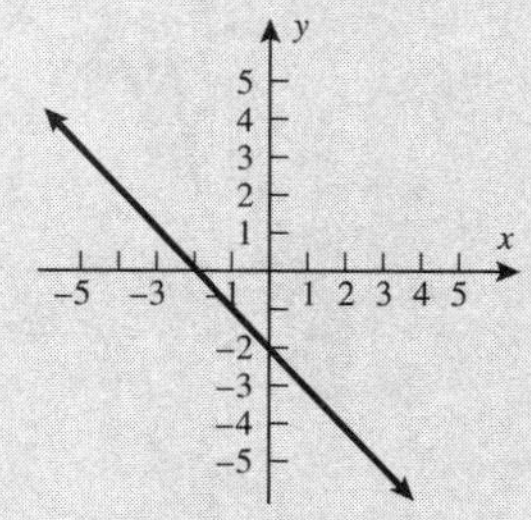

25. x-intercept: $(-1, 0)$
y-intercept: none

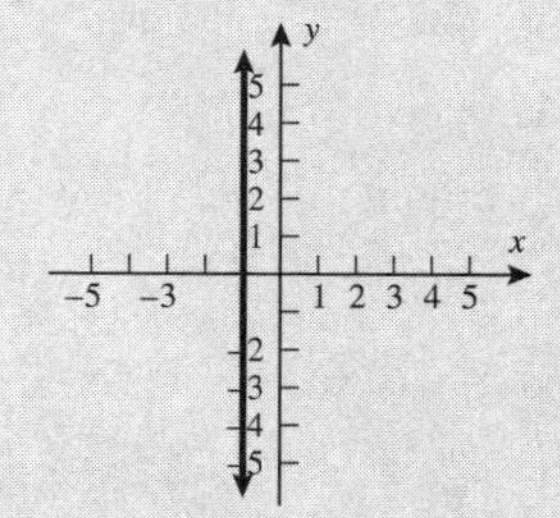

27. x-intercept: none
y-intercept: $(0, 1.5)$

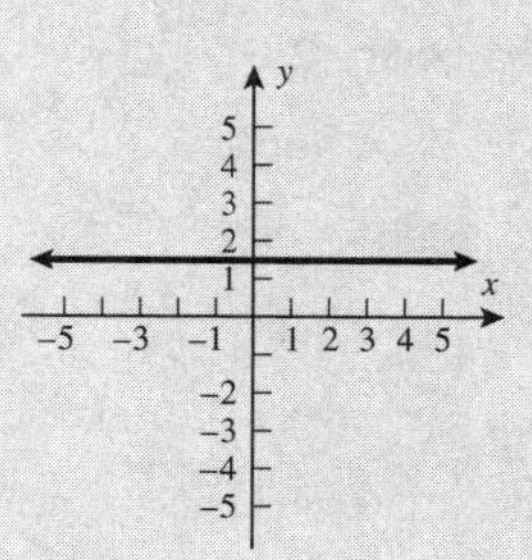

29. x-intercept: $\left(-\frac{7}{2}, 0\right)$
y-intercept: none

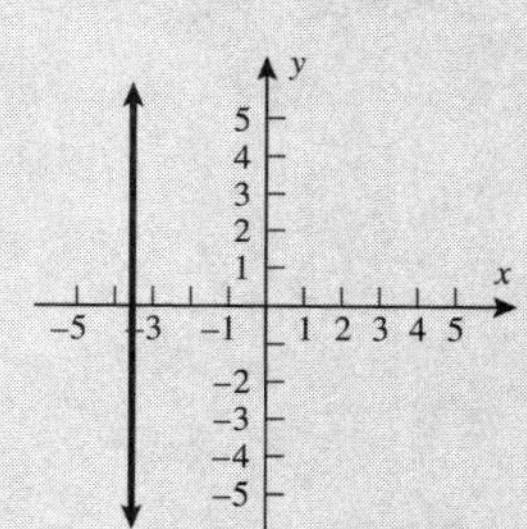

31. $y = \frac{2}{5}x - 2$, $m = \frac{2}{5}$, y-intercept: $(0, -2)$

33. $y = -\frac{1}{3}x + 2$, $m = -\frac{1}{3}$, y-intercept: $(0, 2)$

35. $y = 4x - 3$, $m = 4$, y-intercept: $(0, -3)$

37. $y = -2x + 4$, $m = -2$, y-intercept: $(0, 4)$

39. $y = \frac{2}{3}x - 2$, $m = \frac{2}{3}$, y-intercept: $(0, -2)$

41. $y = -\frac{3}{4}x + 6$, $m = -\frac{3}{4}$, y-intercept: $(0, 6)$

43. $y = 2x + 3$ 45. $y = -\frac{1}{3}x$ 47. $y = 2$

49. $x = \frac{3}{2}$ 51. $y = 5x + 2$ 53. $y = -3x - 4$

55. $y = \frac{3}{4}x - \frac{7}{4}$ 57. $y = 4$ 59. $x = -1$

61. $y = \frac{3}{5}x + \frac{1}{5}$ 63. $y = -5x - 16$ 65. $y = \frac{1}{6}x - \frac{121}{3}$

67. $y = -3x + 1$ 69. $y = \frac{3}{2}x$ 71. $x = 3$

73. $y = 7$ 75. $y = \frac{6}{5}x + 6$ 77. $x = -6$

79. $x = \frac{2}{5}$ 81. $y = x - 1$ 83. $y = -2x + 3$

85. $y = -\frac{1}{2}x + 1$ 87. $y = 2x + 7$ 89. $y = \frac{3}{2}x$

91. $y = 5$ 93. $y = 2$ 95. $y = \frac{3}{2}x - 4$

97. $C(h) = 1200 + 25h$; A 32-hour job will cost \$2,000.

99. \$375

101. $F = \frac{9}{5}C + 32$, $-40\ °\text{C} = -40\ °\text{F}$

103. The rate of change in inches per year is $\frac{1}{50}$.

105. 0.06 ounces per year. In 2040 we expect a baby to weigh 6 lb 12.4 oz.

107. y-intercept $(0, 35)$ is the flat monthly charge of \$35.

109. -0.35 inches/year; 2.75 inches

111. 2.4 billion plastic bags/year; 404 billion

113. The correction that needs to be made is that for the x-intercept, $y = 0$, and for the y-intercept, $x = 0$.

115. Values for the numerator and denominator reversed.

117. true 119. false

121. The line perpendicular is vertical and has undefined slope.

123. $y = -\dfrac{A}{B}x + 1$ 125. $y = \dfrac{B}{A}x + 2B - 1$

129. perpendicular 131. perpendicular

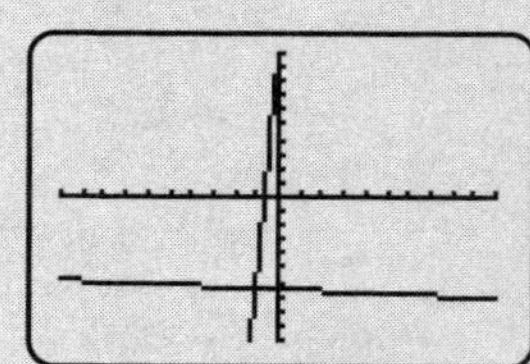

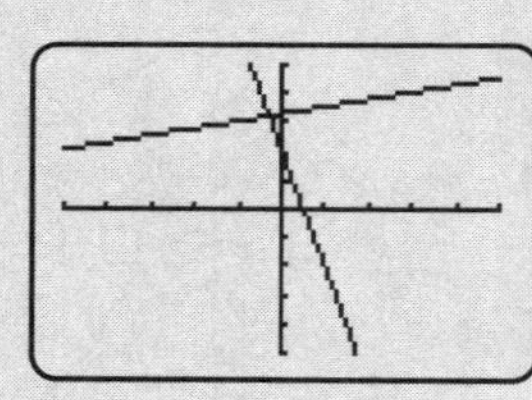

133. neither

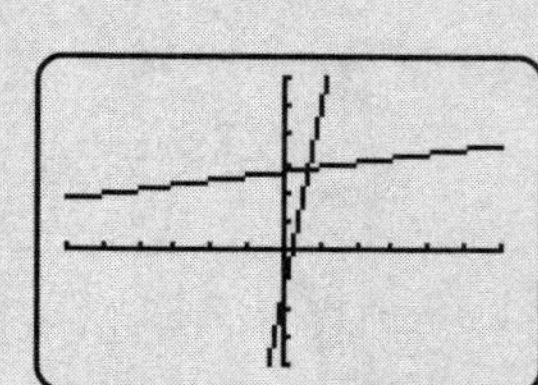

Section 2.4

1. $(x - 1)^2 + (y - 2)^2 = 9$ 3. $(x + 3)^2 + (y + 4)^2 = 100$

5. $(x - 5)^2 + (y - 7)^2 = 81$ 7. $(x + 11)^2 + (y - 12)^2 = 169$

9. $x^2 + y^2 = 4$ 11. $x^2 + y^2 = 2$

13. $(x - 5)^2 + (y + 3)^2 = 12$ 15. $\left(x - \frac{2}{3}\right)^2 + \left(y + \frac{3}{5}\right)^2 = \frac{1}{16}$

17. $(x - 1.3)^2 + (y - 2.7)^2 = 10.24$

19. $C = (1, 3) \quad r = 5$

21. $C = (2, -5) \quad r = 7$

23. $C = (4, 9) \quad r = 2\sqrt{5}$

25. $C = \left(\frac{2}{5}, \frac{1}{7}\right) \quad r = \frac{2}{3}$

27. $C = (1.5, -2.7) \quad r = 1.3$

29. $C = (0, 0) \quad r = 5\sqrt{2}$

31. $C = (-2, -3) \quad r = 4$

33. $C = (-3, -4) \quad r = 10$

35. $C = (5, 7) \quad r = 9$

37. $C = (1, 3) \quad r = 3$

39. $C = (5, -3) \quad r = 2\sqrt{3}$

41. $C = (3, 2) \quad r = 2\sqrt{3}$

43. $C = \left(\frac{1}{2}, -\frac{1}{2}\right) \quad r = \frac{1}{2}$

45. $C = (1.3, 2.7) \quad r = 3.2$

47. $(x + 1)^2 + (y + 2)^2 = 8$

49. $(x + 2)^2 + (y - 3)^2 = 41$

51. $(x + 2)^2 + (y + 5)^2 = 25$

53. no

55. $x^2 + y^2 = 2500$

57. $x^2 + y^2 = 2{,}250{,}000$

59. $x^2 + y^2 = 40{,}000$

61. The standard form of an equation of a circle is $(x - h)^2 + (y - k)^2 = 25$, which would lead to a center of $(4, -3)$, not $(4, 3)$.

63. This equation is not a circle. The standard equation of a circle is $(x - h)^2 + (y - k)^2 = r^2$.

65. true

67. true

69. the point $(-5, 3)$

71. $(x - 3)^2 + (y + 2)^2 = 20$

73. $4c = a^2 + b^2$

75. no graph (because no solution)

77. the point $(-5, 3)$

79. **a.** $(x - 5.5)^2 + (y + 1.5)^2 = 39.69$, $(5.5, -1.5)$, $r = 6.3$

b. $y = -1.5 \pm \sqrt{39.69 - (x - 5.5)^2}$

c.

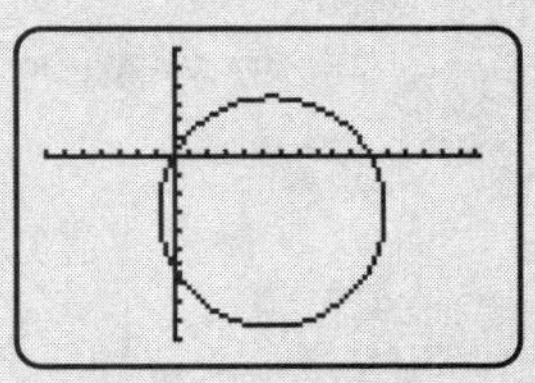

d. The graphs are the same.

Review Exercises

1. quadrant II

5. $d = 3\sqrt{5}$

3. quadrant III

7. $d = \sqrt{205}$

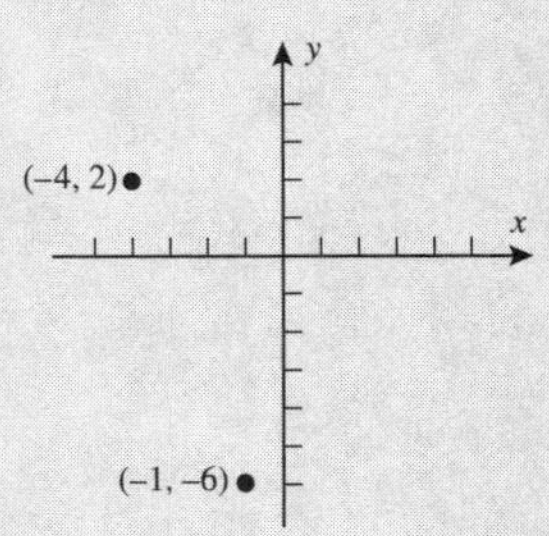

9. $\left(\frac{5}{2}, 6\right)$

11. $(3.85, 5.3)$

13. $d \approx 52.20$

15. x-intercepts: $(\pm 2, 0)$
y-intercepts: $(0, \pm 1)$

17. x-intercepts: $(\pm 3, 0)$
y-intercepts: none

19. y-axis

21. origin

23.

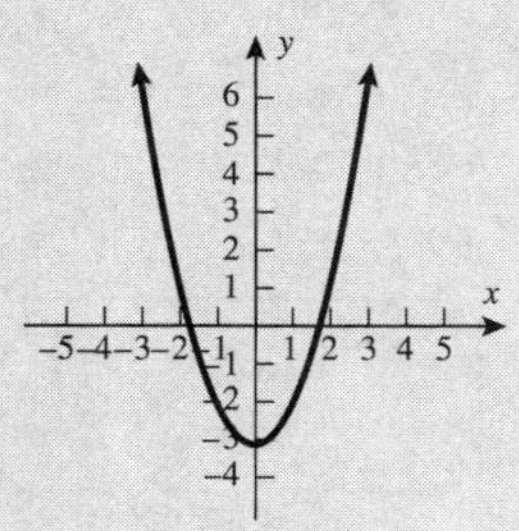

25.

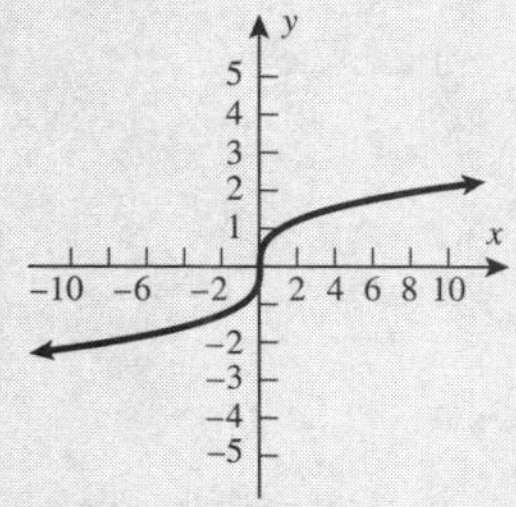

27.

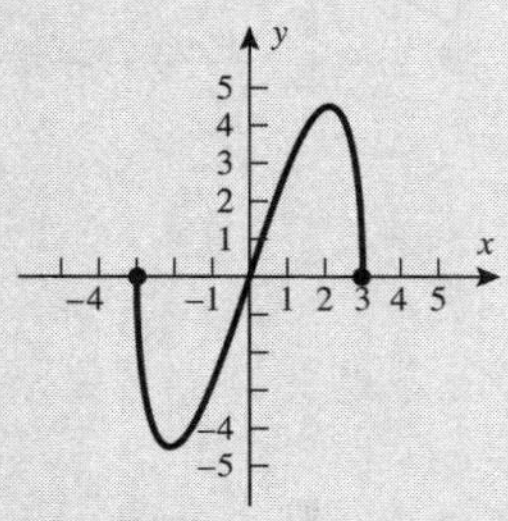

29.

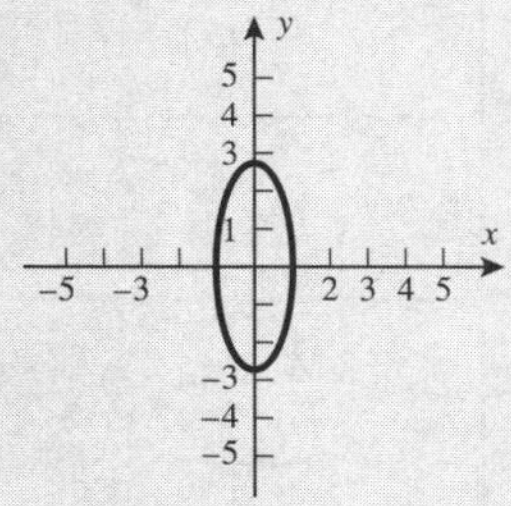

31. $y = -3x + 6$
slope: -3
y-intercept: $(0, 6)$

33. $y = -\frac{3}{2}x - \frac{1}{2}$
slope: $-\frac{3}{2}$
y-intercept: $\left(0, -\frac{1}{2}\right)$

35. x-intercept $\left(\frac{5}{4}, 0\right)$
y-intercept $(0, -5)$
$m = 4$

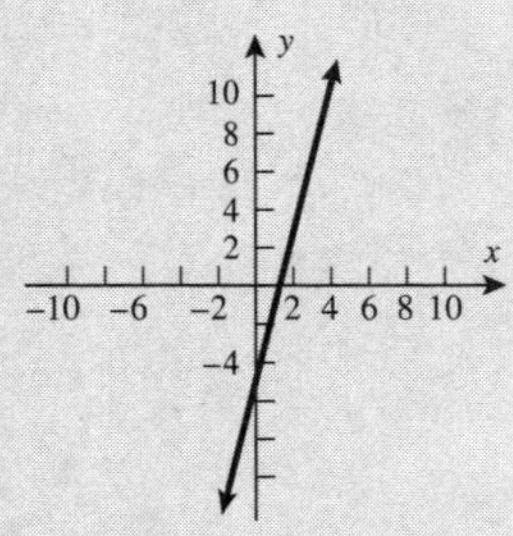

37. x-intercept $(4, 0)$
y-intercept $(0, 4)$
$m = -1$

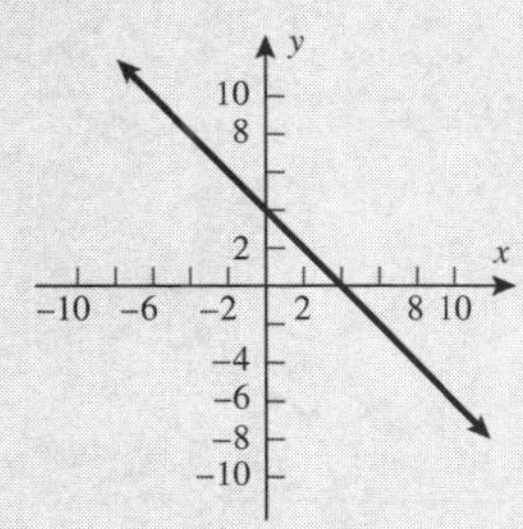

39. x-intercepts: none
y-intercept: $(0, 2)$
Horizontal line
$m = 0$

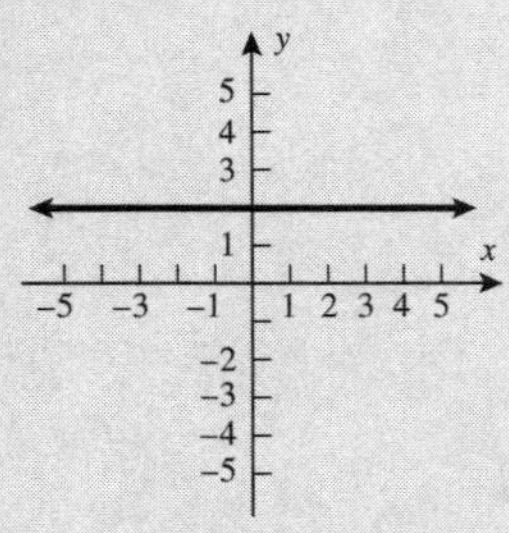

41. $y = 4x - 3$

43. $x = -3$

45. $y = -2x - 2$

47. $y = 6$

49. $y = \frac{5}{6}x + \frac{4}{3}$

51. $y = -2x - 1$

53. $y = \frac{2}{3}x + \frac{1}{3}$

55. $y = -\frac{3}{4}x + \frac{31}{16}$

57. $y = 1.2x + 100$
x is pretest score
y is posttest score

59. $(x + 2)^2 + (y - 3)^2 = 36$

61. $\left(x - \frac{3}{4}\right)^2 + \left(y - \frac{5}{2}\right)^2 = \frac{4}{25}$

63. $C = (-2, -3) \quad r = 9$

65. $C = \left(-\frac{3}{4}, \frac{1}{2}\right) \quad r = \frac{2}{3}$

67. not a circle

69. $C = \left(\frac{1}{3}, -\frac{2}{3}\right) \quad r = 3$

71. $(x - 2)^2 + (y - 7)^2 = 2$

73. a right triangle

75. symmetric to x-axis, y-axis, and origin

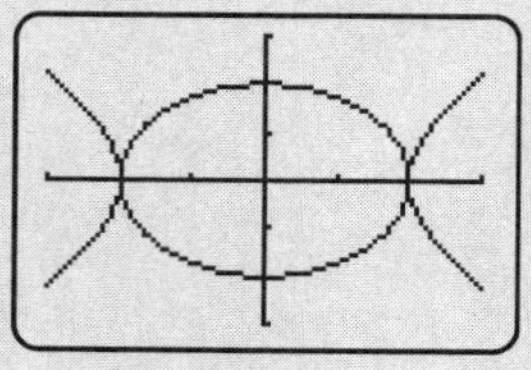

77. perpendicular

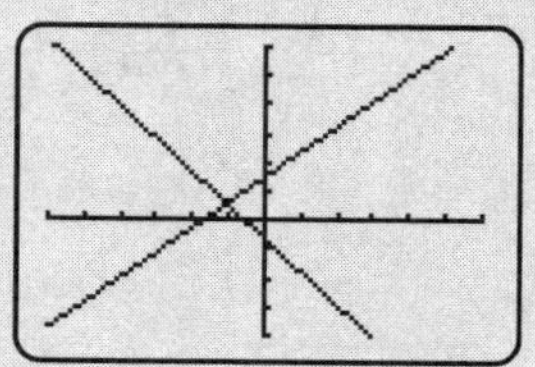

79. $y = -\frac{2}{3} \pm \sqrt{9 - \left(x - \frac{1}{3}\right)^2}$

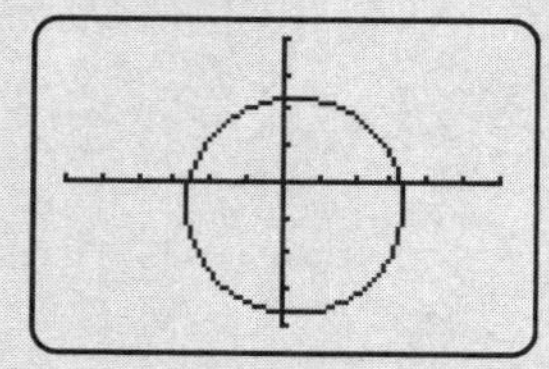

The graphs are the same.

Practice Test

1. $d = \sqrt{82}$

3. $d = \sqrt{29} \quad \text{midpoint} = \left(\frac{1}{2}, 5\right)$

5. $y = 1, 9$

7. x-axis

9. $2x^2 + y^2 = 8$

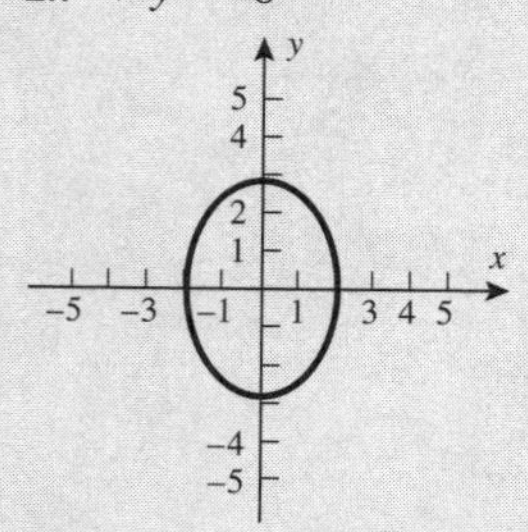

11. x-intercept: (6, 0); y-intercept: (0, −2)

13. $y = \frac{8}{3}x - 8$

15. $y = x + 5$

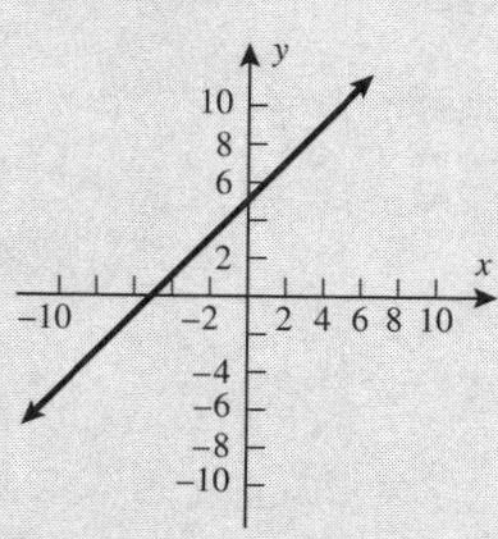

17. $y = -2x + 3$

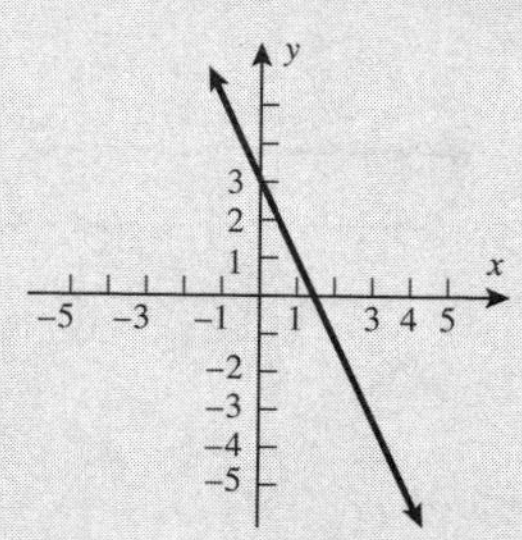

19. $y = 2x + 2$

21. $(x - 6)^2 + (y + 7)^2 = 64$

23. $(x - 4)^2 + (y - 9)^2 = 20$

25. both

Cumulative Test

1. $\frac{1}{2}$

3. $x^4 - 32x^2 + 256$

5. $\dfrac{5 - x}{5 + x}$

7. $12 + 30i$

9. $x = 1$

11. $x = \pm 3$

13. Discriminant is negative: two complex (conjugate) roots

15. $x = 2$

17. (0, 12)

19. (−2, 6)

21. origin

23. $x = 5$

25. $C = (-5, -3) \quad r = \sqrt{30}$

27. neither

CHAPTER 3

Section 3.1

1. function

3. not a function

5. function

7. not a function

9. not a function

11. function

13. not a function

15. not a function

17. function

19. not a function

21. function

23. not a function

25. **a.** 5 **b.** 1 **c.** −3

27. **a.** 3 **b.** 2 **c.** 5

29. **a.** −5 **b.** −5 **c.** −5

31. **a.** 2 **b.** −8 **c.** −5

33. $x = 1$

35. $x = -3, 1$

37. $-4 \le x \le 4$

39. $x = 6$

41. −7

43. 6

45. −1

47. −33

49. $-\frac{7}{6}$

51. $\frac{2}{3}$

53. 4

55. $8 - x - a$

57. 2

59. 1

61. 2

63. 1

65. $(-\infty, \infty)$

67. $(-\infty, \infty)$

69. $(-\infty, 5) \cup (5, \infty)$

71. $(-\infty, -2) \cup (-2, 2) \cup (2, \infty)$

73. $(-\infty, \infty)$

75. $(-\infty, 7]$

77. $\left[-\frac{5}{2}, \infty\right)$

79. $(-\infty, -2] \cup [2, \infty)$

81. $(3, \infty)$

83. $(-\infty, \infty)$

85. $(-\infty, -4) \cup (-4, \infty)$

87. $\left(-\infty, \frac{3}{2}\right)$

89. $(-\infty, -2) \cup (3, \infty)$

91. $(-\infty, -4) \cup (4, \infty)$

93. $\left(-\infty, \frac{3}{2}\right)$

95. $(-\infty, \infty)$

97. $x = -2, x = 4$

99. $x = -1, x = 5, x = 6$

101. $y = 45x$; domain: $(75, \infty)$

103. $T(6) = 64.8$ °F $T(12) = 90$ °F

105. $P(10) \approx \$641.66$ $P(100) \approx \$634.50$

107. $V(x) = x(10 - 2x)^2$ domain: $(0, 5)$

109. Yes, because every input (year) corresponds to exactly one output (federal funds rate).

111.

Year	Total Health Care Cost for Family Plans
1989	\$4,000
1993	\$6,000
1997	\$6,000
2001	\$8,000
2005	\$11,000

113. **a.** 0   **b.** 1000   **c.** 2000

115. You must apply the vertical line test instead of the horizontal line test. Applying the vertical line test would show that the graph given is actually a function.

117. $f(x + 1) \neq f(x) + f(1)$

119. $G(-1 + h) \neq G(-1) + G(h)$

121. false

123. false

125. $A = 2$

127. $C = -5, D = -2$

129. $(-\infty, -a) \cup (-a, a) \cup (a, \infty)$

131. Warmest at noon: 90 °F. Outside the interval [6, 18] the temperatures are too low.

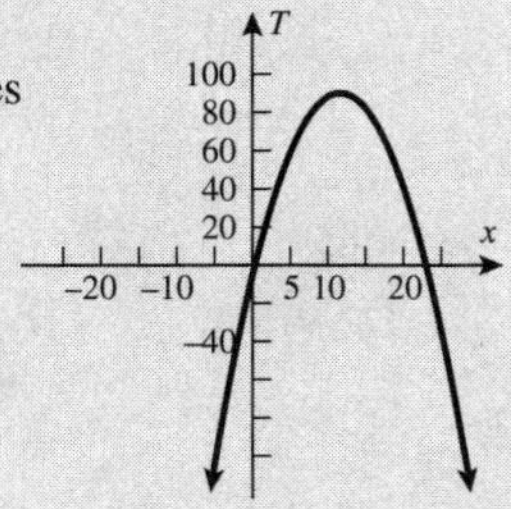

133. lowest price \$10, highest \$642.38 agrees. (If there is only 1 card for sale.)

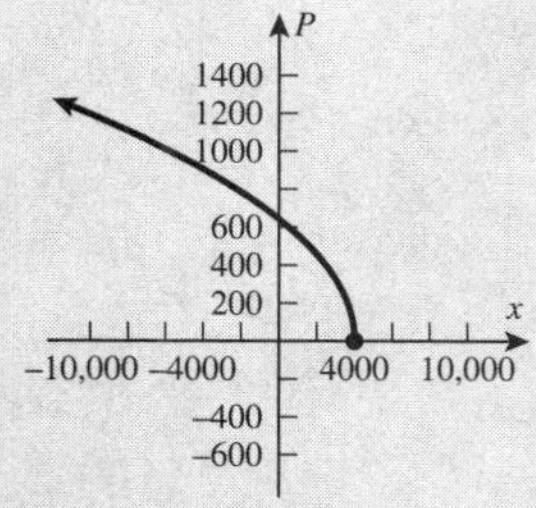

135. Graph of y_2 can be obtained by shifting the graph of y_1 two units to the right.

Section 3.2

1. neither

3. even

5. odd

7. neither

9. odd

11. even

13. even

15. neither

17. neither

19. neither

21. neither

23. neither

25. **a.** $(-\infty, \infty)$ **b.** $[-1, \infty)$
   c. increasing: $(-1, \infty)$, decreasing: $(-3, -2)$, constant: $(-\infty, -3) \cup (-2, -1)$
   d. 0   **e.** -1 **f.** 2

27. **a.** $[-7, 2]$ **b.** $[-5, 4]$
   c. increasing: $(-4, 0)$, decreasing: $(-7, -4) \cup (0, 2)$, constant: none
   d. 4   **e.** 1   **f.** -5

29. **a.** $(-\infty, \infty)$ **b.** $(-\infty, \infty)$
   c. increasing: $(-\infty, -3) \cup (4, \infty)$, decreasing: none, constant: $(-3, 4)$
   d. 2   **e.** 2   **f.** 2

31. **a.** $(-\infty, \infty)$ **b.** $[-4, \infty)$
   c. increasing: $(0, \infty)$, decreasing: $(-\infty, 0)$, constant: none
   d. -4 **e.** 0   **f.** 0

33. **a.** $(-\infty, 0) \cup (0, \infty)$ **b.** $(-\infty, 0) \cup (0, \infty)$
   c. increasing: $(-\infty, 0) \cup (0, \infty)$, decreasing: none, constant: none
   d. undefined   **e.** 3   **f.** -3

35. **a.** $(-\infty, 0) \cup (0, \infty)$
   b. $(-\infty, 5) \cup [7, 7]$ or $(-\infty, 5) \cup \{7\}$
   c. increasing: $(-\infty, 0)$, decreasing: $(5, \infty)$, constant $(0, 5)$
   d. undefined   **e.** 3   **f.** 7

37. $2x + h - 1$

39. $2x + h + 3$

41. $2x + h - 3$

43. $-6x - 3h + 5$

45. 13 **47.** 1 **49.** -2 **51.** -1

53. domain: $(-\infty, \infty)$
range: $(-\infty, 2]$
increasing: $(-\infty, 2)$
constant: $(2, \infty)$

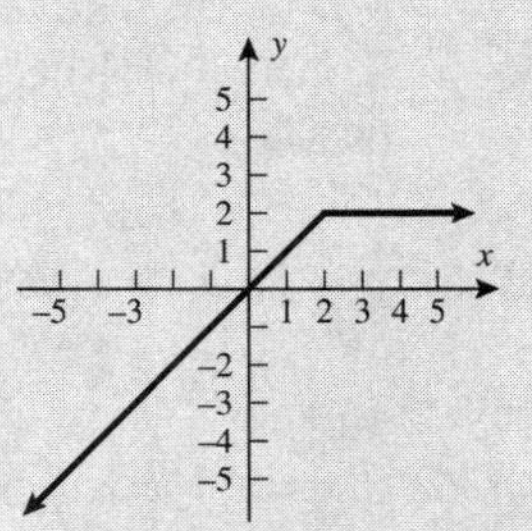

55. domain: $(-\infty, \infty)$
range: $[0, \infty)$
increasing: $(0, \infty)$
decreasing: $(-1, 0)$
constant: $(-\infty, -1)$

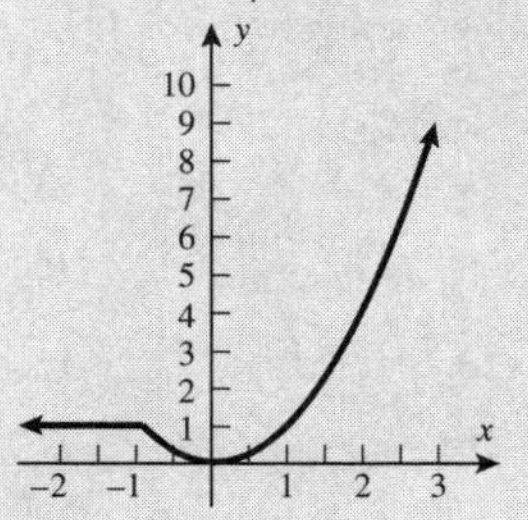

57. domain: $(-\infty, \infty)$
range: $(-\infty, \infty)$
increasing: $(-\infty, \infty)$

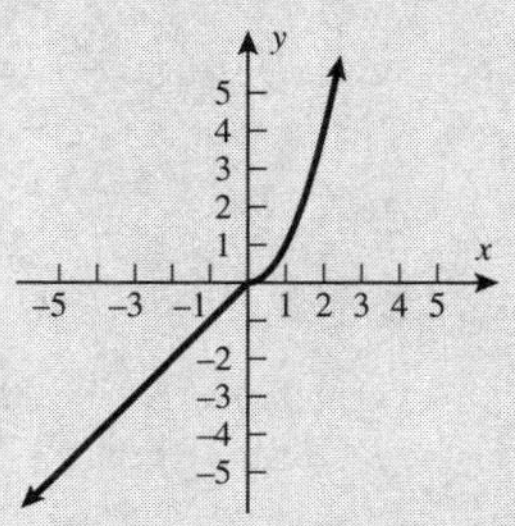

59. domain: $(-\infty, \infty)$
range: $[1, \infty)$
increasing: $(1, \infty)$
decreasing: $(-\infty, 1)$

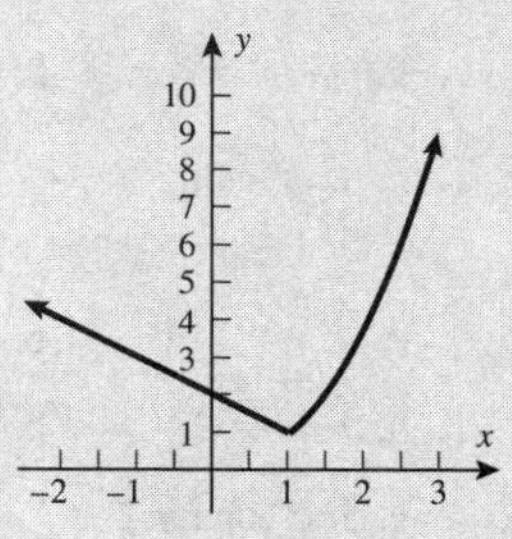

61. domain: $(-\infty, \infty)$
range: $[-1, 3]$
increasing: $(-1, 3)$
constant: $(-\infty, -1) \cup (3, \infty)$

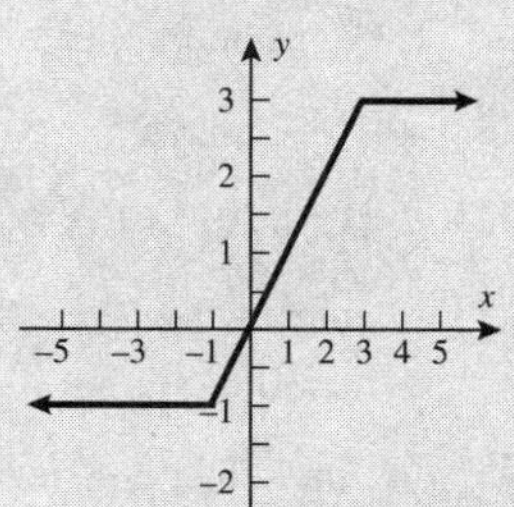

63. domain: $(-\infty, \infty)$
range: $[1, 4]$
increasing: $(1, 2)$
constant: $(-\infty, 1) \cup (2, \infty)$

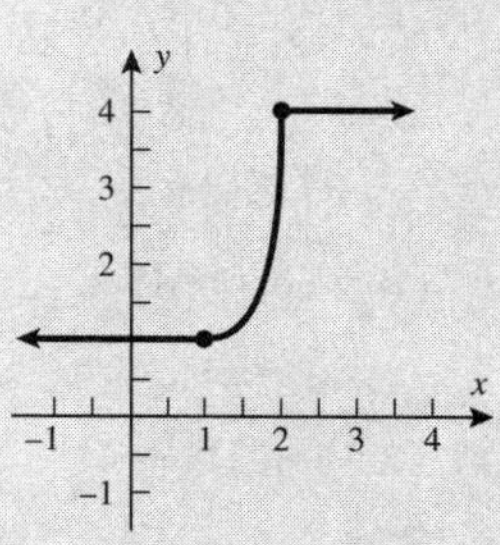

65. domain: $(-\infty, -2) \cup (-2, \infty)$
range: $(-\infty, \infty)$
increasing: $(-2, 1)$
decreasing: $(-\infty, -2) \cup (1, \infty)$

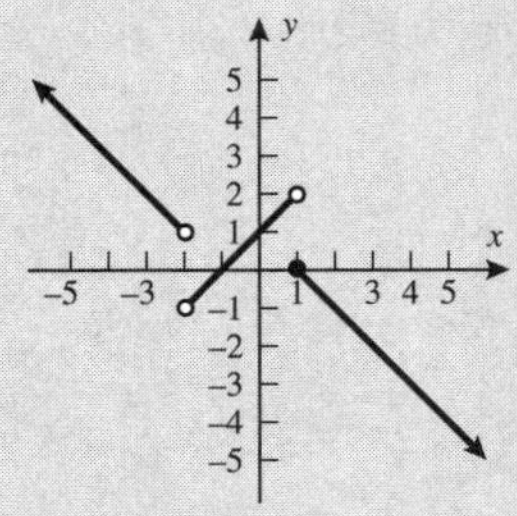

67. domain: $(-\infty, \infty)$
range: $[0, \infty)$
increasing: $(0, \infty)$
constant: $(-\infty, 0)$

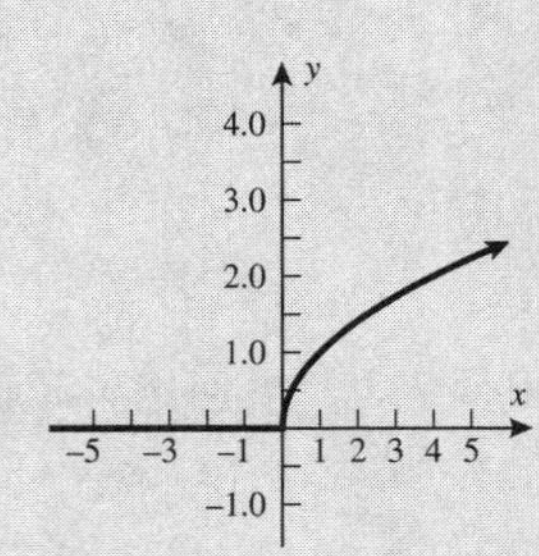

69. domain: $(-\infty, \infty)$
range: $(-\infty, \infty)$
decreasing: $(-\infty, 0) \cup (0, \infty)$

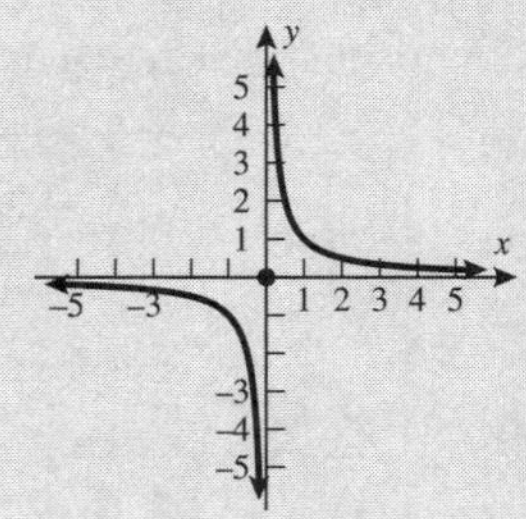

71. domain: $(-\infty, 1) \cup (1, \infty)$
range: $(-\infty, -1) \cup (-1, \infty)$
increasing: $(-1, 1)$
decreasing: $(-\infty, -1) \cup (1, \infty)$

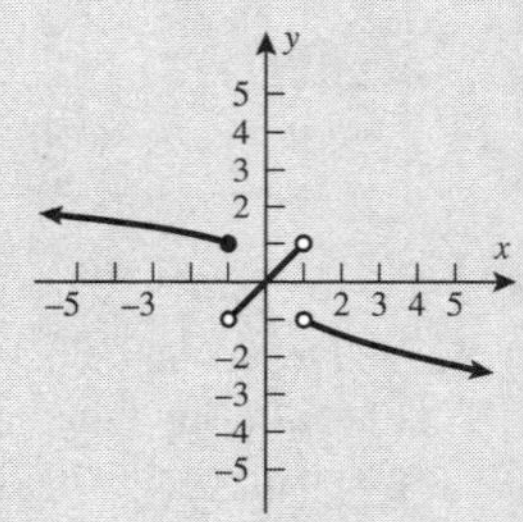

73. domain: $(-\infty, \infty)$
range: $(-\infty, 2) \cup [4, \infty)$
increasing:
$(-\infty, -2) \cup (0, 2) \cup (2, \infty)$
decreasing: $(-2, 0)$

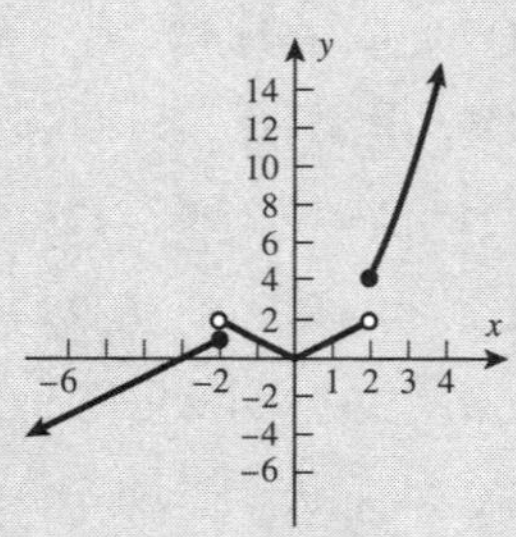

75. domain: $(-\infty, 1) \cup (1, \infty)$
range: $(-\infty, 1) \cup (1, \infty)$
increasing: $(-\infty, 1) \cup (1, \infty)$

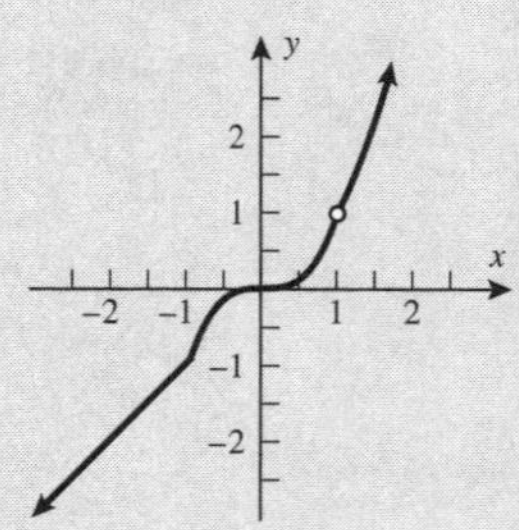

77. $C(x) = \begin{cases} 10x & 0 \le x \le 50 \\ 9x & 50 < x \le 100 \\ 8x & x > 100 \end{cases}$

79. $C(x) = \begin{cases} 250x & 0 \le x \le 10 \\ 175x + 750 & x > 10 \end{cases}$

81. $C(x) = \begin{cases} 1000 + 35x & 0 \le x \le 100 \\ 2000 + 25x & x > 100 \end{cases}$

83. $R(x) = \begin{cases} 50{,}000 + 3x & 0 \le x \le 100{,}000 \\ 4x - 50{,}000 & x > 100{,}000 \end{cases}$

85. $P(x) = 65x - 800$

87. $C(x) = 0.80 + 0.17[[x]]$ where x represents the weight of the letter in ounces.

89. $f(t) = 3(-1)^{[[t]]}$

91. **a.** 20 million tons per year

b. 110 million tons per year

93. 0 feet per second

95. The domain is incorrect. It should be $(-\infty, 0) \cup (0, \infty)$. The range is also incorrect and should be $(0, \infty)$.

97. $C(x) = \begin{cases} 15 & x \le 30 \\ 15 + 1(x - 30) & x > 30 \end{cases} = \begin{cases} 15 & x \le 30 \\ -15 + x & x > 30 \end{cases}$

99. true 101. false 103. yes, if $a = 2b$

105. odd 107. odd 109. domain: all real numbers; range: set of integers

Section 3.3

1. l 3. a 5. b 7. i 9. c 11. g

13. $y = |x| + 3$ 15. $y = |-x| = |x|$ 17. $y = 3|x|$

19. $y = x^3 - 4$ 21. $y = (x + 1)^3 + 3$ 23. $y = (-x)^3$

25.

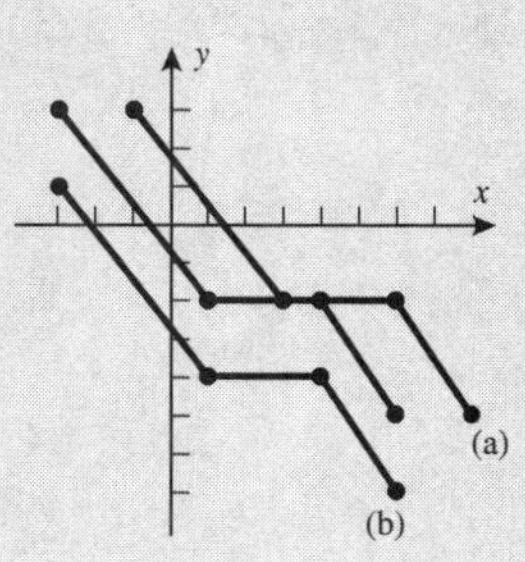

a. $y = f(x - 2)$
b. $y = f(x) - 2$

27.

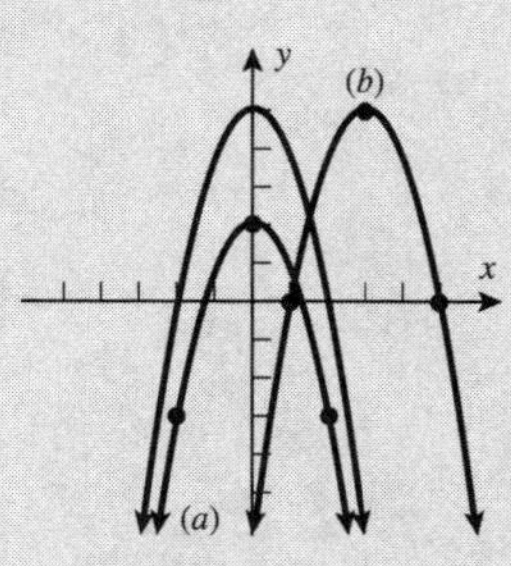

a. $y = f(x) - 3$
b. $y = f(x - 3)$

29.

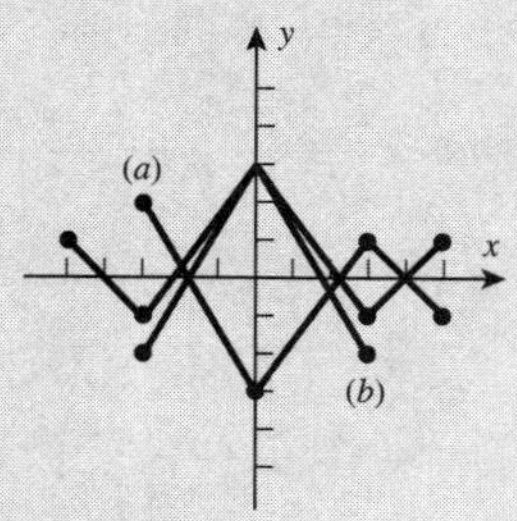

a. $y = -f(x)$
b. $y = f(-x)$

31.

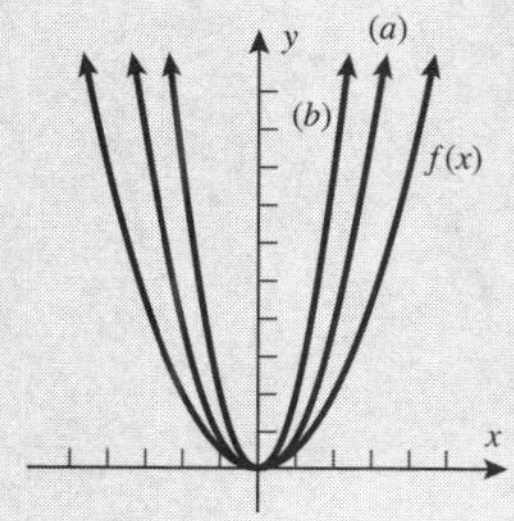

a. $y = 2f(x)$
b. $y = f(2x)$

33.

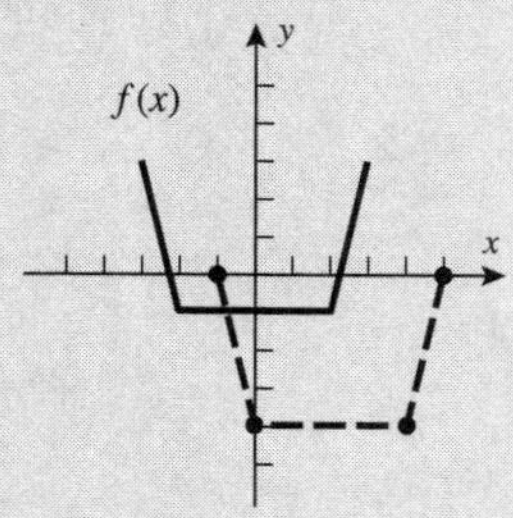

35.

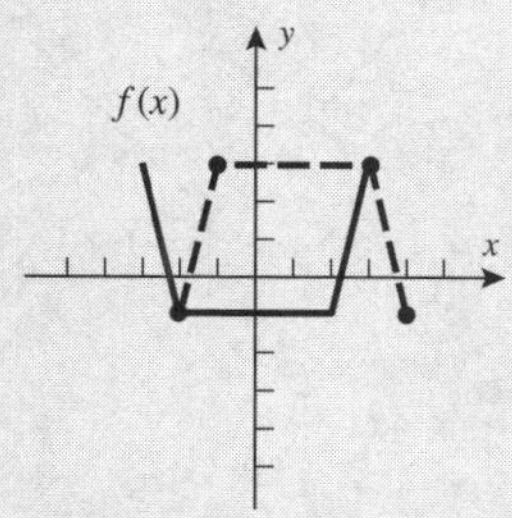

37.

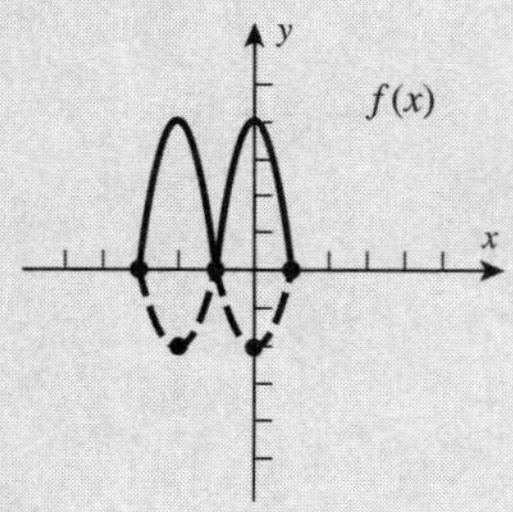

39.

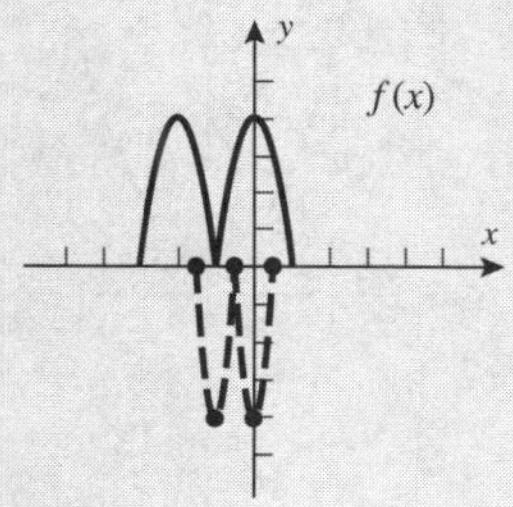

41.

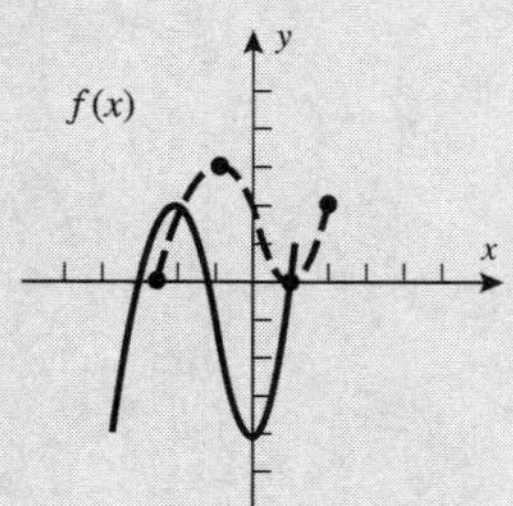

43.

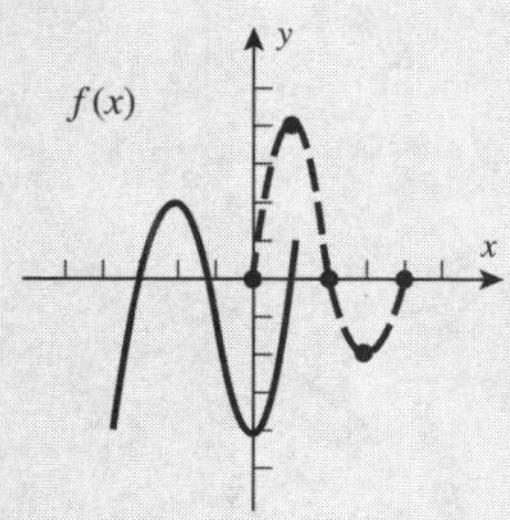

45.

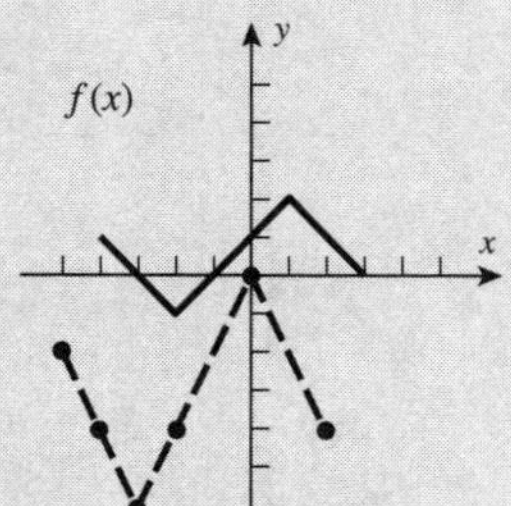

47.

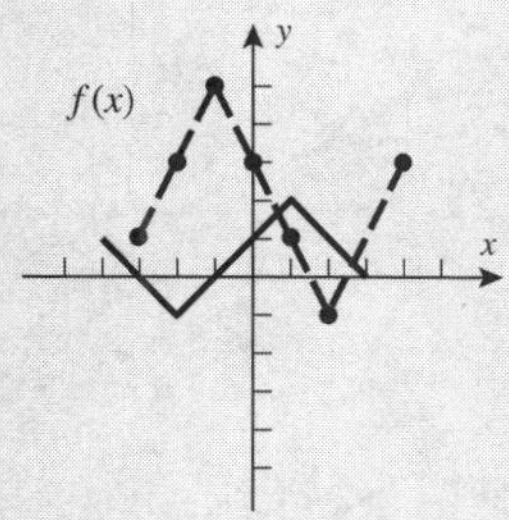

49.

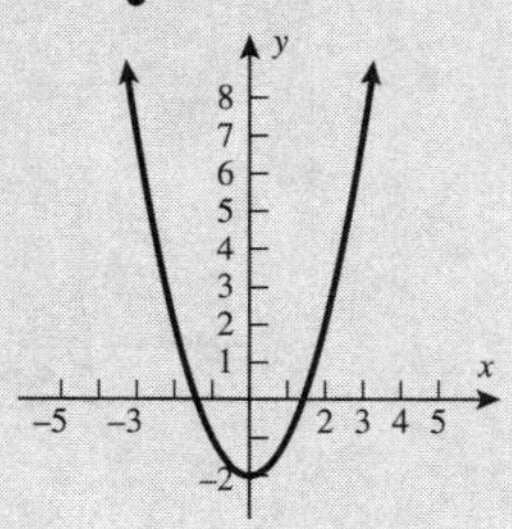

51.

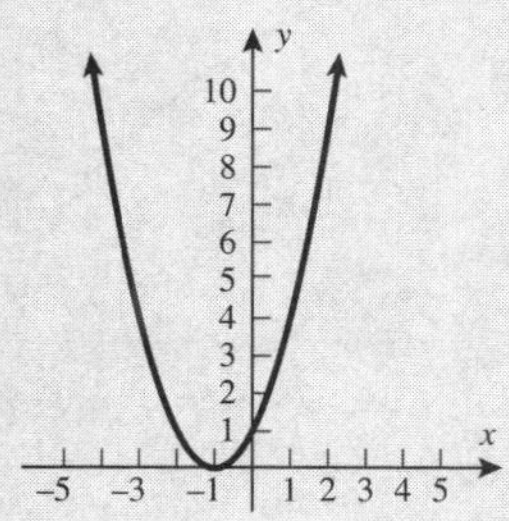

53.

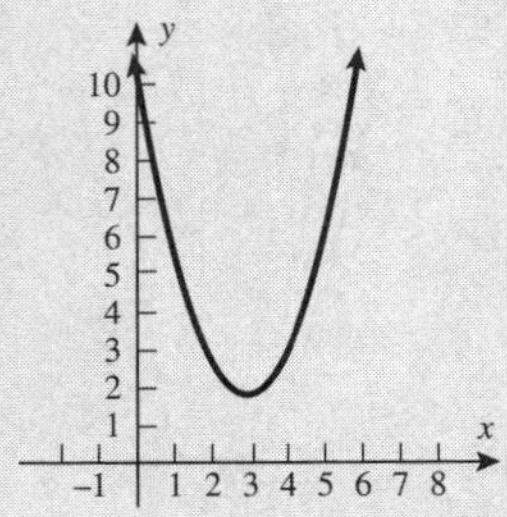

55.

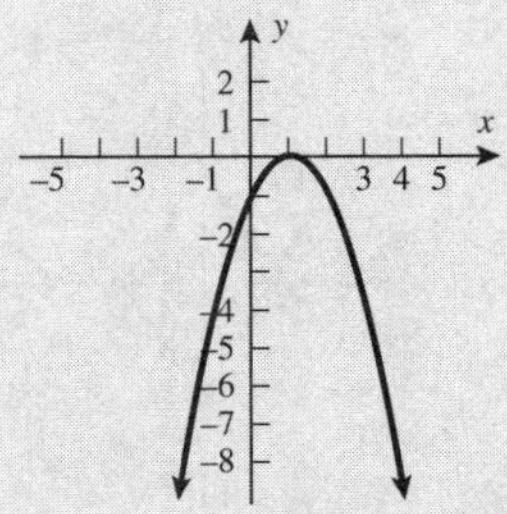

57.

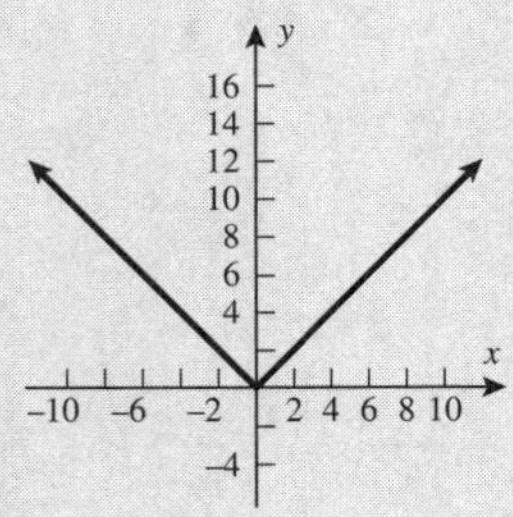

59.

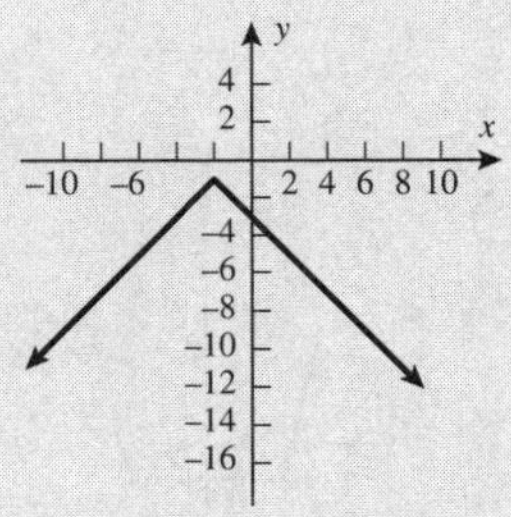

61.

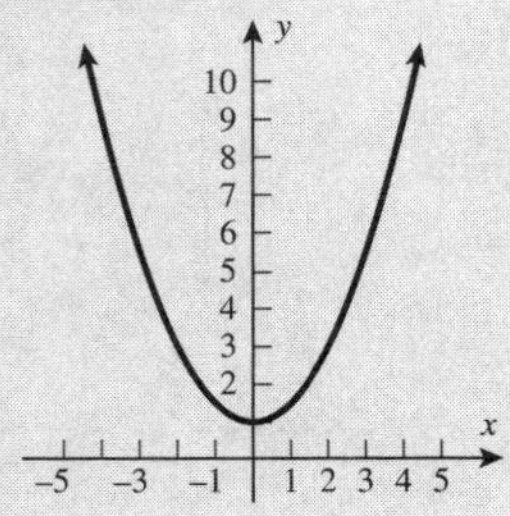

63.

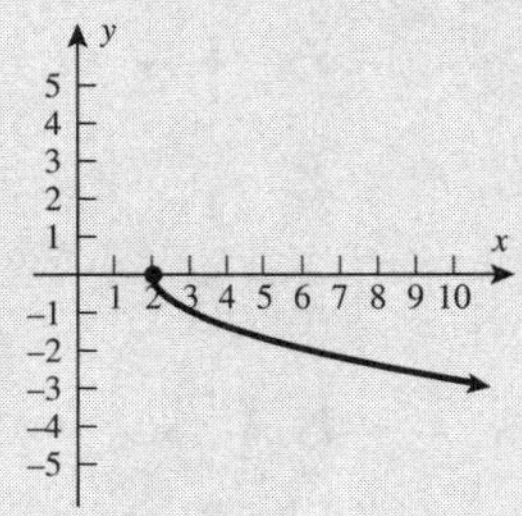

65.

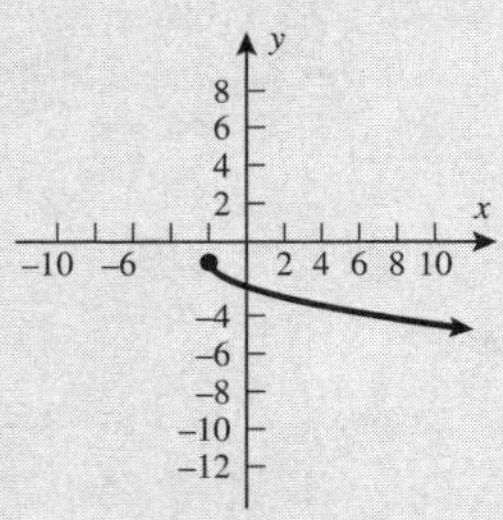

67.

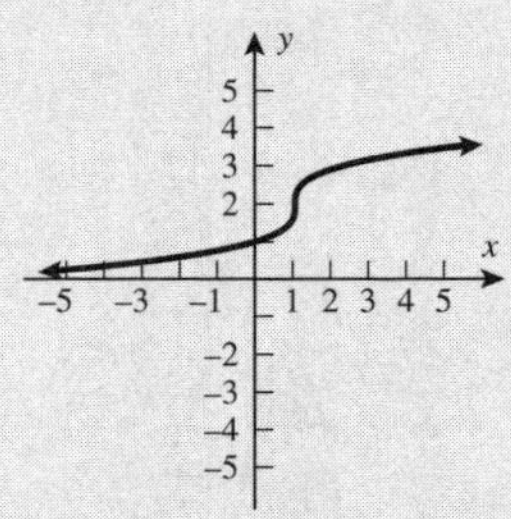

69.

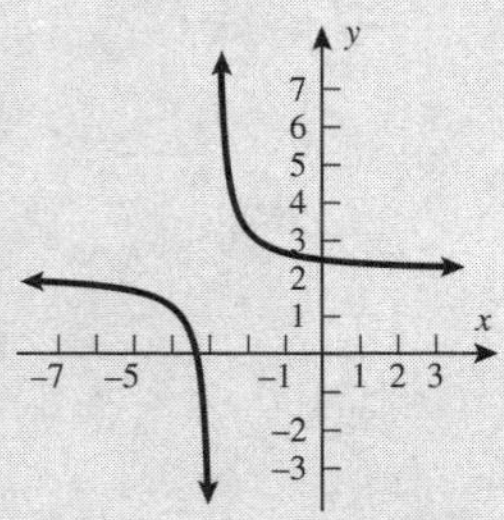

71.

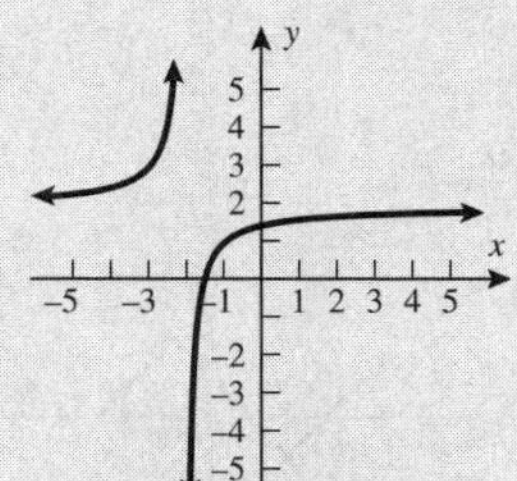

73.

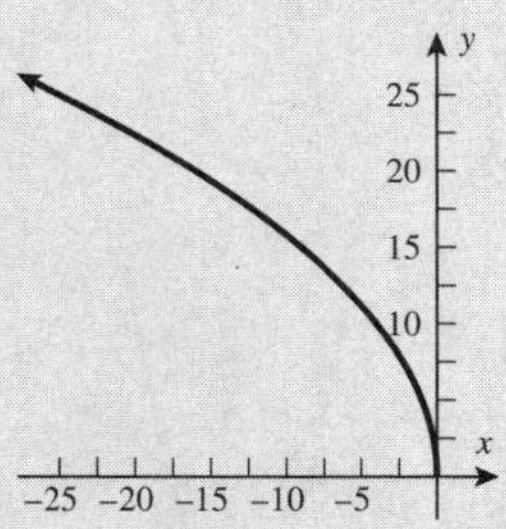

75.

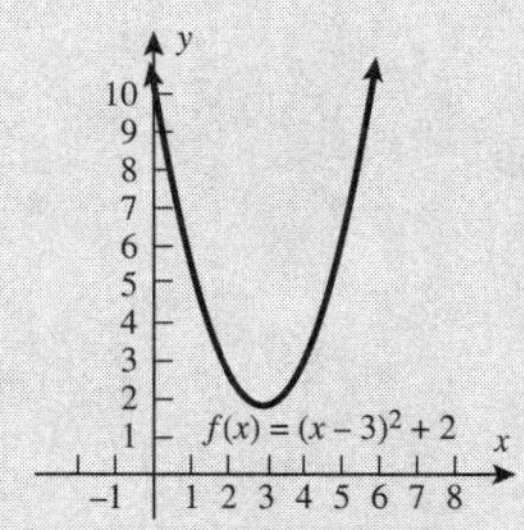

77.

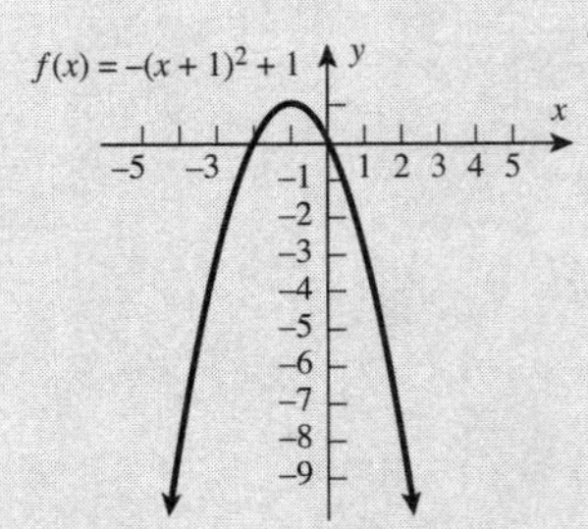

79.

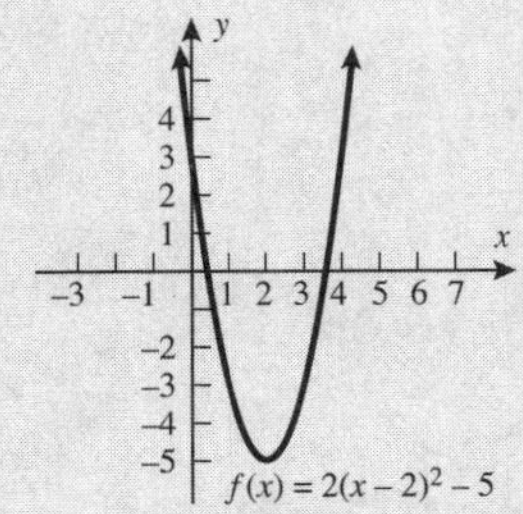

81. $S(x) = 10x$ and $S(x) = 10x + 50$

83. $T(x) = 0.33(x - 6500)$

85. **b.** shifted to the *right* 3 units

87. $|3 - x| = |x - 3|$; therefore, shift to the *right* 3 units.

89. true **91.** true **93.** $(a + 3, b + 2)$

95. Any part of the graph of $f(x)$ that is below the x-axis is reflected above it for $|f(x)|$.

97. If $0 < a < 1$, you have a horizontal stretch. If $a > 1$, the graph is a horizontal compression.

99. Each horizontal line in the graph of $y = [[x]]$ is stretched twice in length. Any part of the graph of $y = [[x]]$ that is below the x-axis is reflected above the x-axis. There is a vertical shift of one unit up.

Section 3.4

1. $$\left.\begin{aligned} f(x) + g(x) &= x + 2 \\ f(x) - g(x) &= 3x \\ f(x)\cdot g(x) &= -2x^2 + x + 1 \end{aligned}\right\} \text{ domain: } (-\infty, \infty)$$

$$\frac{f(x)}{g(x)} = \frac{2x + 1}{1 - x} \qquad \text{domain: } (-\infty, 1) \cup (1, \infty)$$

3. $$\left.\begin{aligned} f(x) + g(x) &= 3x^2 - x - 4 \\ f(x) - g(x) &= x^2 - x + 4 \\ f(x)\cdot g(x) &= 2x^4 - x^3 - 8x^2 + 4x \end{aligned}\right\} \text{ domain: } (-\infty, \infty)$$

$$\frac{f(x)}{g(x)} = \frac{2x^2 - x}{x^2 - 4} \quad \text{domain: } (-\infty, -2) \cup (-2, 2) \cup (2, \infty)$$

5. $$\left.\begin{aligned} f(x) + g(x) &= \frac{1 + x^2}{x} \\ f(x) - g(x) &= \frac{1 - x^2}{x} \\ f(x)\cdot g(x) &= 1 \\ \frac{f(x)}{g(x)} &= \frac{1/x}{x} = \frac{1}{x^2} \end{aligned}\right\} \text{ domain: } (-\infty, 0) \cup (0, \infty)$$

7. $$\left.\begin{aligned} f(x) + g(x) &= 3\sqrt{x} \\ f(x) - g(x) &= -\sqrt{x} \end{aligned}\right\} \text{ domain: } [0, \infty)$$

$$f(x)\cdot g(x) = 2x$$

$$\frac{f(x)}{g(x)} = \frac{1}{2} \qquad \text{domain: } (0, \infty)$$

9. $f(x) + g(x) = \sqrt{4-x} + \sqrt{x+3}$
$f(x) - g(x) = \sqrt{4-x} - \sqrt{x+3}$
$f(x) \cdot g(x) = \sqrt{4-x}\sqrt{x+3}$
} domain: $[-3, 4]$

$\dfrac{f(x)}{g(x)} = \dfrac{\sqrt{4-x}}{\sqrt{x+3}} = \dfrac{\sqrt{4-x}\sqrt{x+3}}{x+3}$ domain: $(-3, 4]$

11. $(f \circ g)(x) = 2x^2 - 5$ domain: $(-\infty, \infty)$
$(g \circ f)(x) = 4x^2 + 4x - 2$ domain: $(-\infty, \infty)$

13. $(f \circ g)(x) = \dfrac{1}{x+1}$ domain: $(-\infty, -1) \cup (-1, \infty)$
$(g \circ f)(x) = \dfrac{1}{x-1} + 2$ domain: $(-\infty, 1) \cup (1, \infty)$

15. $(f \circ g)(x) = \dfrac{1}{|x-1|}$ domain: $(-\infty, 1) \cup (1, \infty)$
$(g \circ f)(x) = \dfrac{1}{|x|-1}$ domain: $(-\infty, -1) \cup (-1, 1) \cup (1, \infty)$

17. $(f \circ g)(x) = \sqrt{x+4}$ domain: $[-4, \infty)$
$(g \circ f)(x) = \sqrt{x-1} + 5$ domain: $[1, \infty)$

19. $(f \circ g)(x) = x$ domain: $(-\infty, \infty)$
$(g \circ f)(x) = x$ domain: $(-\infty, \infty)$

21. 15
23. 13
25. $26\sqrt{3}$
27. $\frac{110}{3}$
29. 11
31. $3\sqrt{2}$
33. undefined
35. undefined
37. 13
39. $f(g(1)) = \frac{1}{3}$ $g(f(2)) = 2$
41. undefined
43. $f(g(1)) = \frac{1}{3}$ $g(f(2)) = 4$
45. $f(g(1)) = \sqrt{5}$
$g(f(2)) = 6$
47. undefined
49. $f(g(1)) = \sqrt[3]{3}$
$g(f(2)) = 4$
61. $f(x) = 2x^2 + 5x$
$g(x) = 3x - 1$
63. $f(x) = \dfrac{2}{|x|}$
$g(x) = x - 3$
65. $f(x) = \dfrac{3}{\sqrt{x} - 2}$
$g(x) = x + 1$
67. $F(C(K)) = \frac{9}{5}(K - 273.15) + 32$
69. $A(x) = \left(\dfrac{x}{4}\right)^2$; x is the number of linear feet of fence.
$A(100) = 625$ square feet, $A(200) = 2500$ square feet
71. **a.** $C(p) = 62{,}000 - 20p$
b. $R(p) = 600{,}000 - 200p$
c. $P(p) = R(p) - C(p) = 538{,}000 - 180p$

73. area $= 150^2\pi t = 22{,}500\pi t$
75. $d(h) = \sqrt{h^2 + 4}$
77. domain: $x \neq -2$
79. The operation is composition, *not* multiplication.
81. Function notation, not multiplication.
83. false
85. true
87. $(g \circ f)(x) = \dfrac{1}{x}$ $x \neq 0, a$
89. $(g \circ f)(x) = x$ $x \geq -a$
91.

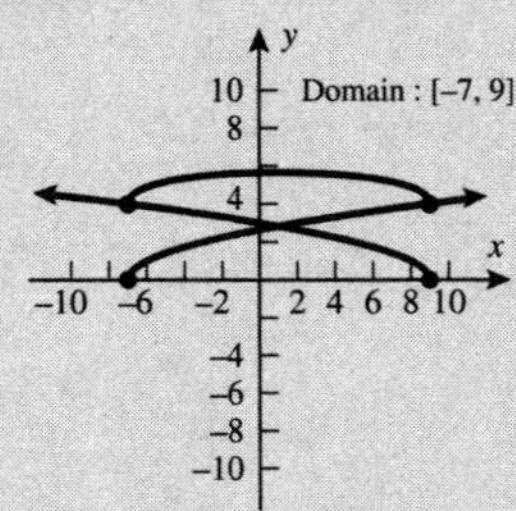

93.

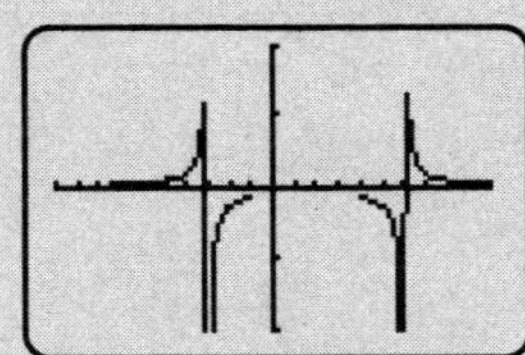

$(-\infty, -3) \cup (-3, -1] \cup [4, 6) \cup (6, \infty)$

Section 3.5

EXERCISE #	FUNCTION	ONE-TO-ONE
1.	yes	no
3.	yes	yes
5.	yes	yes
7.	no	
9.	yes	no
11.	yes	no
13.	yes	yes
15.	yes	no
17.	yes	no
19.	yes	yes
21.	yes	no
23.	yes	yes

25.

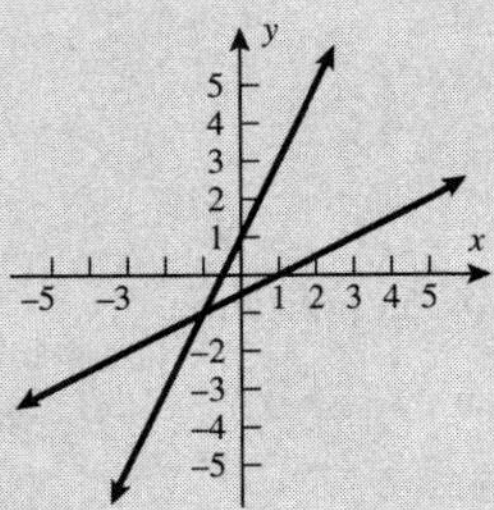

27.

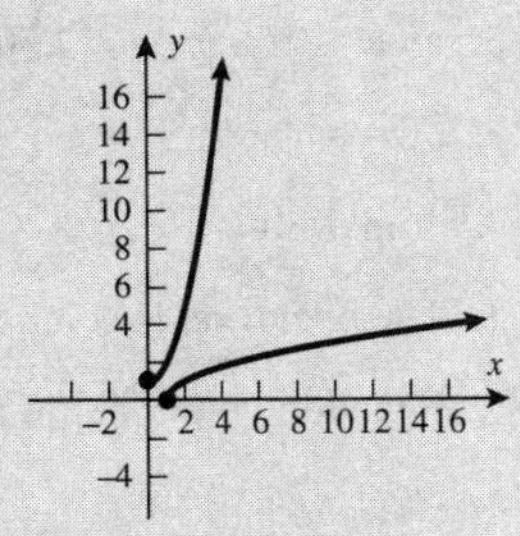

37.

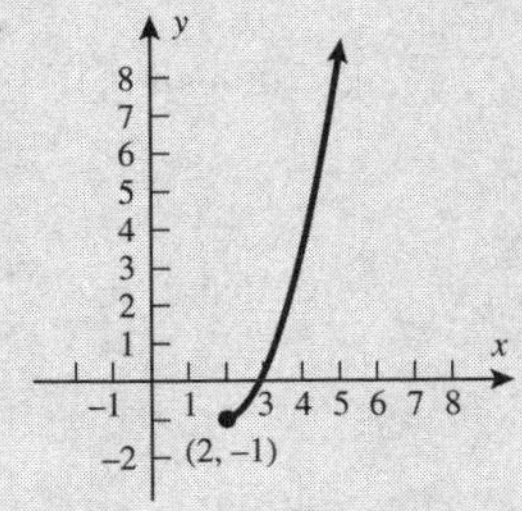

39.

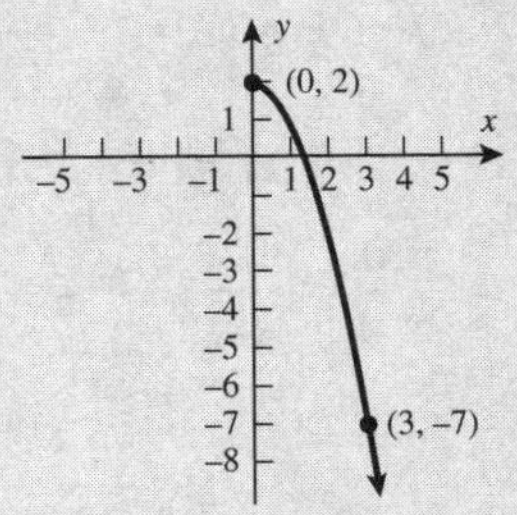

29.

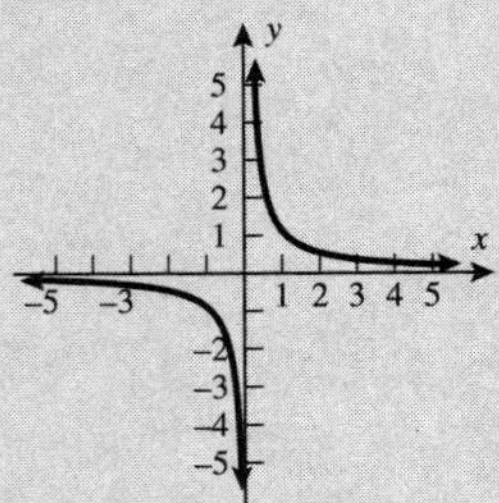

31.

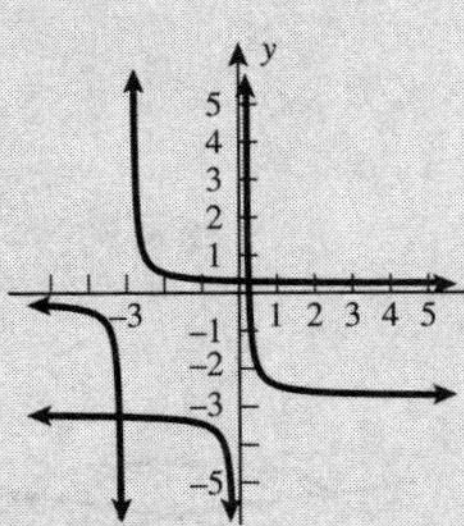

41.

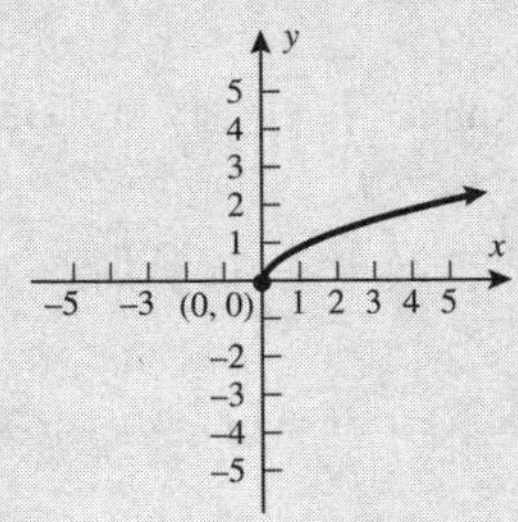

33.

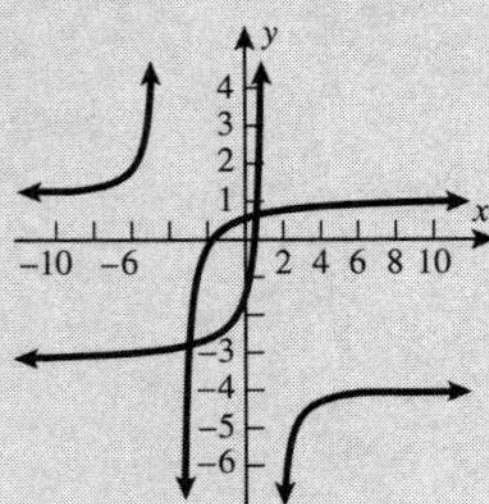

35.

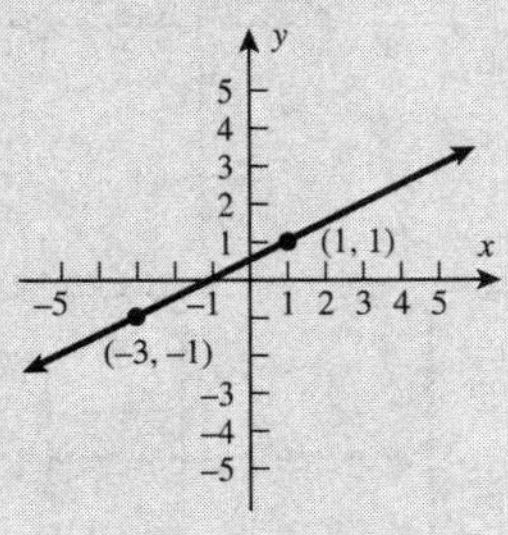

EXERCISE #	f^{-1}	DOMAIN OF f RANGE OF f^{-1}	RANGE OF f DOMAIN OF f^{-1}
43.	$f^{-1}(x) = x + 1$	$(-\infty, \infty)$	$(-\infty, \infty)$
45.	$f^{-1}(x) = -\frac{1}{3}x + \frac{2}{3}$	$(-\infty, \infty)$	$(-\infty, \infty)$
47.	$f^{-1}(x) = \sqrt[3]{x - 1}$	$(-\infty, \infty)$	$(-\infty, \infty)$
49.	$f^{-1}(x) = x^2 + 3$	$[3, \infty)$	$[0, \infty)$
51.	$f^{-1}(x) = \sqrt{x + 1}$	$[0, \infty)$	$[-1, \infty)$
53.	$f^{-1}(x) = -2 + \sqrt{x + 3}$	$[-2, \infty)$	$[3, \infty)$
55.	$f^{-1}(x) = \dfrac{2}{x}$	$(-\infty, 0) \cup (0, \infty)$	$(-\infty, 0) \cup (0, \infty)$
57.	$f^{-1}(x) = 3 - \dfrac{2}{x}$	$(-\infty, 3) \cup (3, \infty)$	$(-\infty, 0) \cup (0, \infty)$
59.	$f^{-1}(x) = \dfrac{5x - 1}{x + 7}$	$(-\infty, 5) \cup (5, \infty)$	$(-\infty, -7) \cup (-7, \infty)$

61.

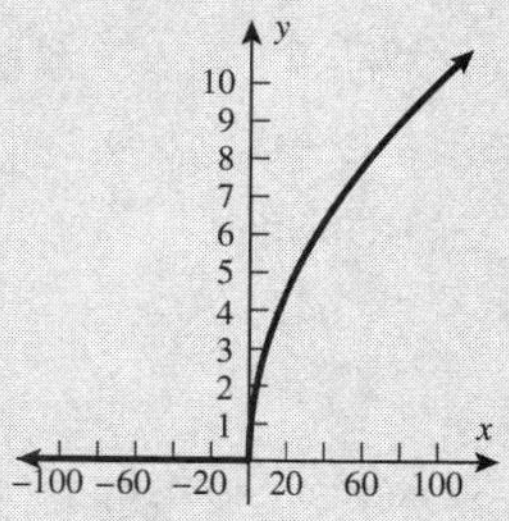

The function is NOT one-to-one.

63. $f^{-1}(x) = \begin{cases} x & x \leq -1 \\ \sqrt[3]{x} & -1 < x < 1 \\ x & x \geq 1 \end{cases}$

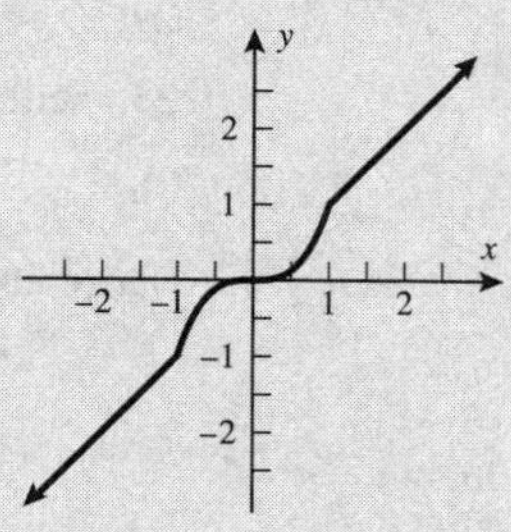

The function is one-to-one.

65. $f^{-1}(x) = \frac{5}{9}(x - 32)$; this now represents degrees Fahrenheit being turned into degrees Celsius.

67. $C(x) = \begin{cases} 250x & x \leq 10 \\ 175x + 750 & x > 10 \end{cases}$

$C^{-1}(x) = \begin{cases} \dfrac{x}{250} & x \leq 2500 \\ \dfrac{x - 750}{175} & x > 2500 \end{cases}$

69. $E(x) = 5.25x$; $E^{-1}(x) = \dfrac{x}{5.25}$ tells you how many hours the student will have to work to bring home x dollars.

71. No, it's not a function because it fails the vertical line test.

73. f is not one-to-one, so it does not have an inverse function.

75. false **77.** false **79.** $(b, 0)$

81. $f(x) = \sqrt{1 - x^2}$ $\quad 0 \leq x \leq 1$

$f^{-1}(x) = \sqrt{1 - x^2}$ $\quad 0 \leq x \leq 1$

83. $m \neq 0$

85. no

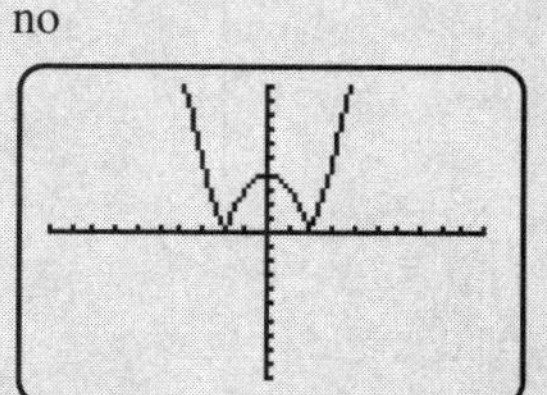

87. no

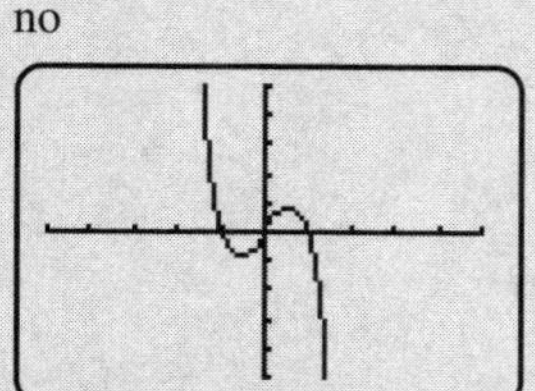

89.

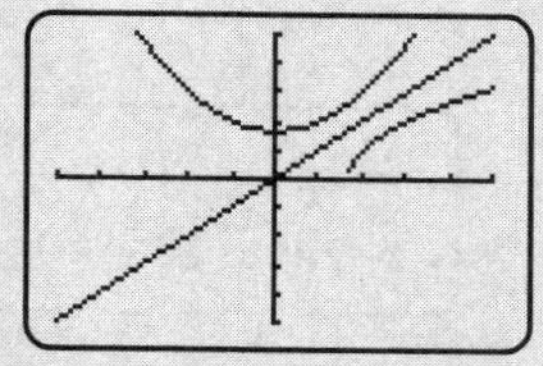

No, the functions are not inverses of each other. Had we restricted the domain of g to $x \geq 0$, then they would be inverses.

91.

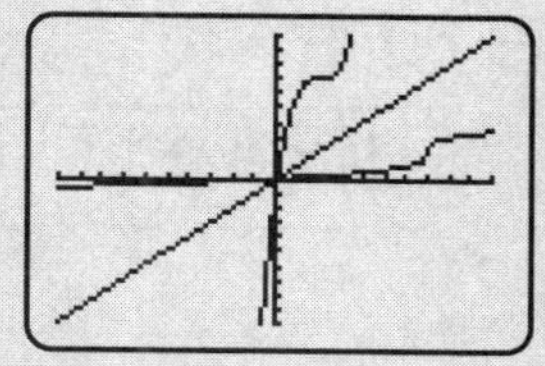

Yes, the functions are inverses of each other.

Section 3.6

1. $y = kx$ **3.** $V = kx^3$ **5.** $z = km$

7. $f = \dfrac{k}{\lambda}$ **9.** $F = k\dfrac{w}{L}$ **11.** $v = kgt$

13. $R = \dfrac{k}{PT}$ **15.** $y = k\sqrt{x}$ **17.** $d = rt$

19. $V = lwh$ **21.** $A = \pi r^2$ **23.** $V = \dfrac{\pi}{16} r^2 h$

25. $V = \dfrac{400{,}000}{P}$ **27.** $F = \dfrac{2\pi}{\lambda L}$ **29.** $t = \dfrac{19.2}{s}$

31. $R = \dfrac{4.9}{I^2}$ **33.** $R = \dfrac{0.01L}{A}$ **35.** $F = \dfrac{0.025m_1 m_2}{d^2}$

37. $W = 7.5H$ **39.** 1292 mph **41.** $F = 1.618H$

43. 24 cm **45.** 20,000 **47.** 600 w/m^2

49. Bank of America 1.5%; Navy Federal Credit Union: 3%

51. $\frac{11}{12}$ or 0.92 atm **53.** y *inversely* with x

55. true **57.** b **59.** $\sigma_{pl}^2 = 1.23 C_n^2 k^{7/6} L^{11/6}$

61. **a.** The least squares regression line is $y = 2.93x + 201.72$.

b. The variation constant is 120.07, and the equation of the direct variation is $y = 120.074x^{0.26}$.

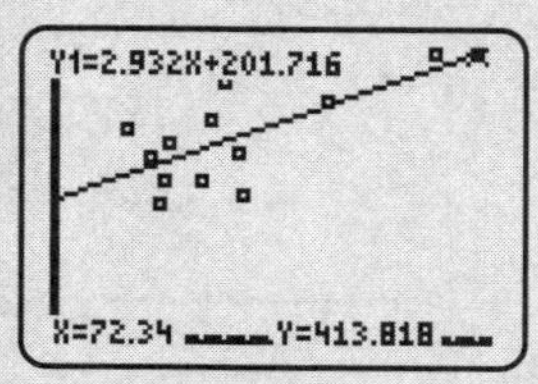

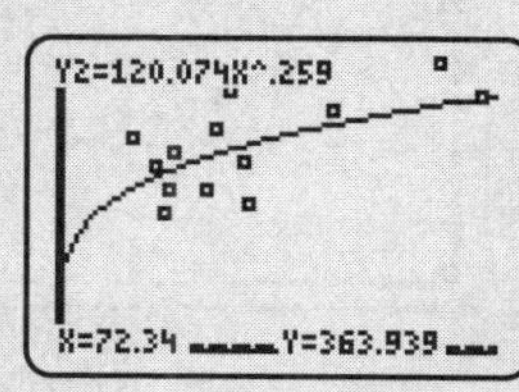

c. When the oil price is \$72.70 per barrel in September 2006, the predicted stock index from the least squares regression line is 415 and from the equation of direct variation is 364. The least squares regression line gives a closer approximation to the actual value, 417.

63. a. The least squares regression line is $y = -141.73x + 2419.35$.

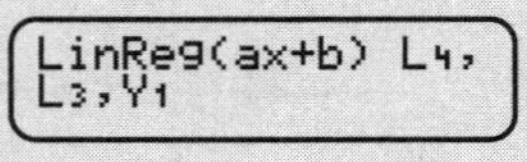

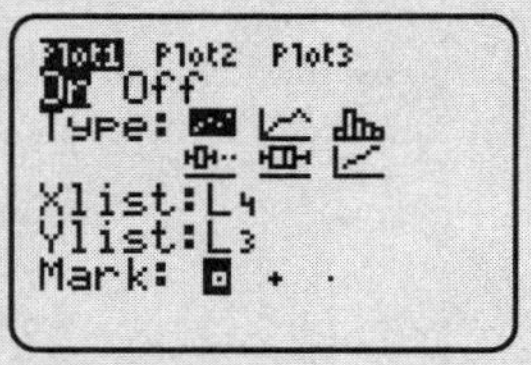

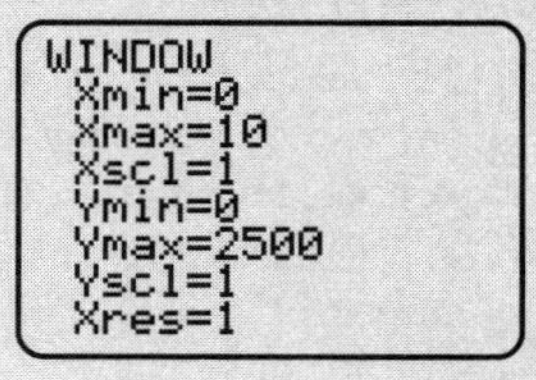

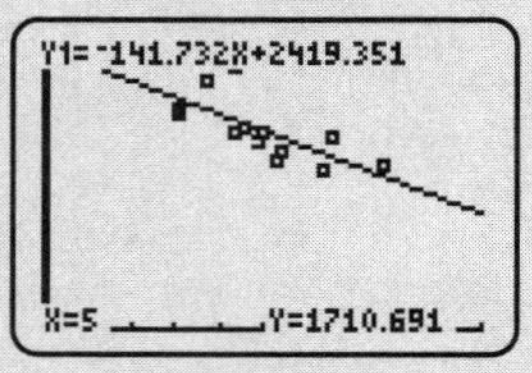

b. The variation constant is 3217.69 and the equation of the inverse variation is $y = \dfrac{3217.69}{x^{0.41}}$.

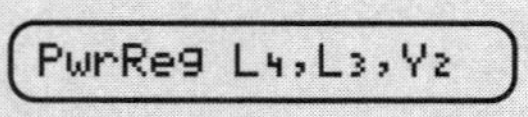

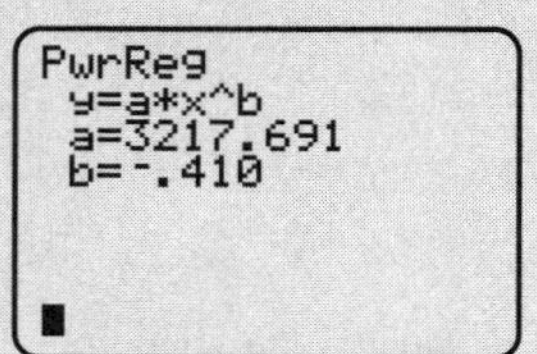

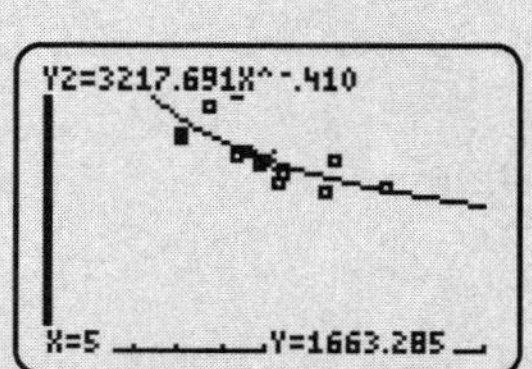

c. When the 5-year maturity rate is 5.02% in September 2006, the predicted number of housing units from the least squares regression line is 1708, and from the equation of inverse variation is 1661. The equation of the least squares regression line gives a closer approximation to the actual value, 1861.

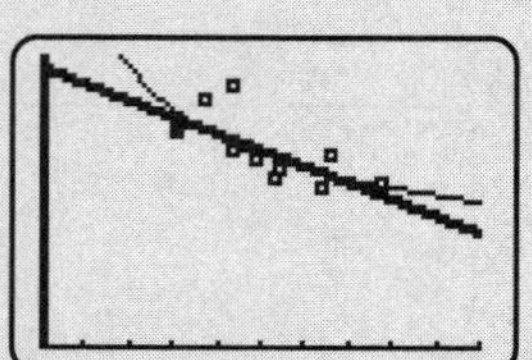

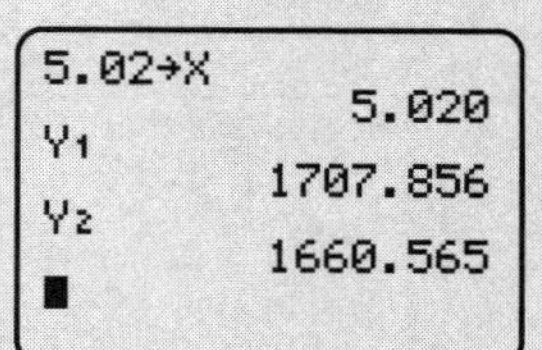

65. a. $y = 0.218x + 0.898$ **b.** \$2.427 per gallon. Yes, it is very close to the actual price at \$2.427 per gallon. **c.** \$3.083

Review Exercises

1. yes **3.** yes **5.** no

7. yes **9.** no

11. a. 2 **b.** 4 **c.** $x = -3, x = 4$

13. a. 0 **b.** -2 **c.** $x = -5, x = 2$

15. 5 **17.** -665 **19.** -2

21. 4 **23.** domain: $(-\infty, \infty)$ **25.** $(-\infty, -4) \cup (-4, \infty)$

27. $[4, \infty)$ **29.** $D = 18$ **31.** neither

33. odd **35.** neither **37.** odd

39. a. $[-4, 7]$ **b.** $[-2, 4]$

c. increasing: (3, 7), decreasing: (0, 3), constant: $(-4, 0)$

41. -2

43. domain: $(-\infty, \infty)$
range: $(0, \infty)$

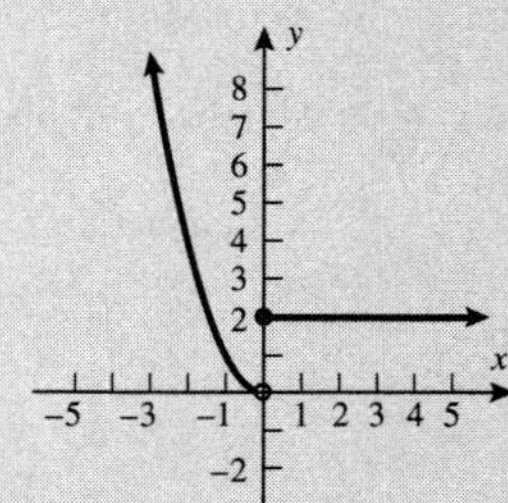

45. domain: $(-\infty, \infty)$
range: $[-1, \infty)$

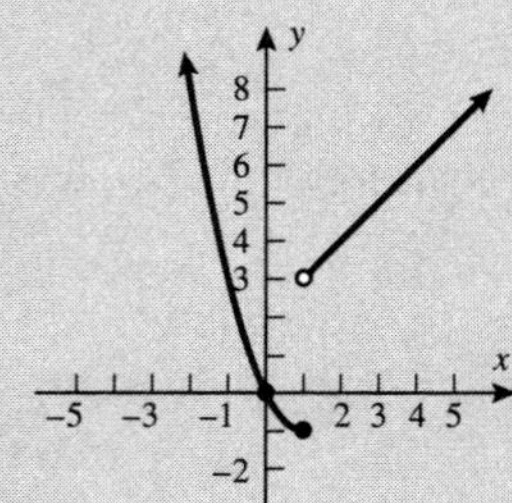

47. $C(x) = \begin{cases} 25 & x \le 2 \\ 25 + 10.50(x - 2) & x > 2 \end{cases}$

49.

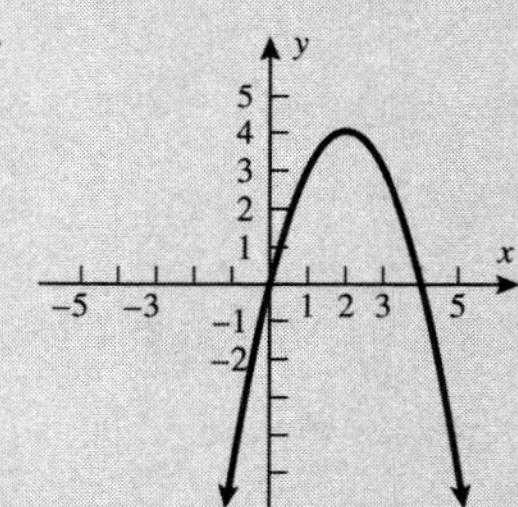

51.

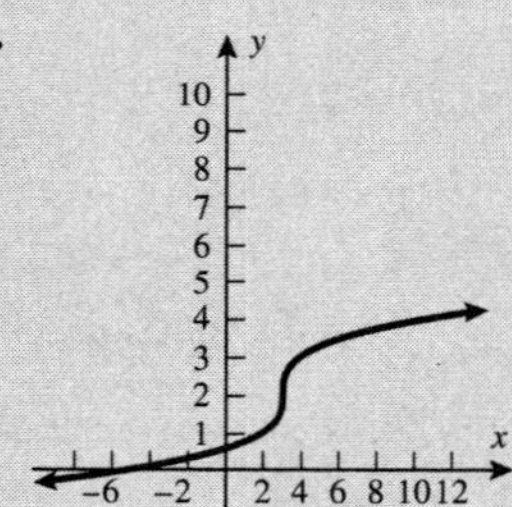

53.

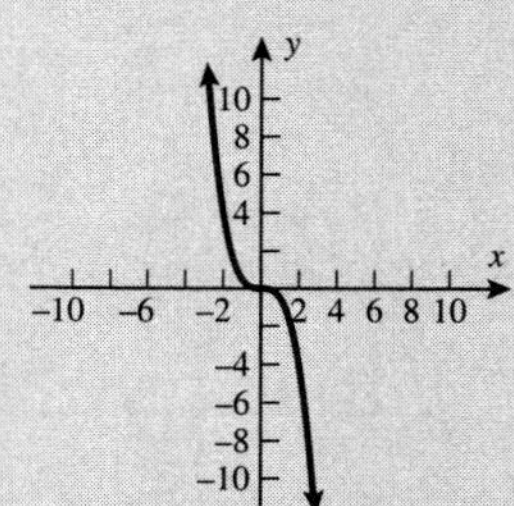

55.

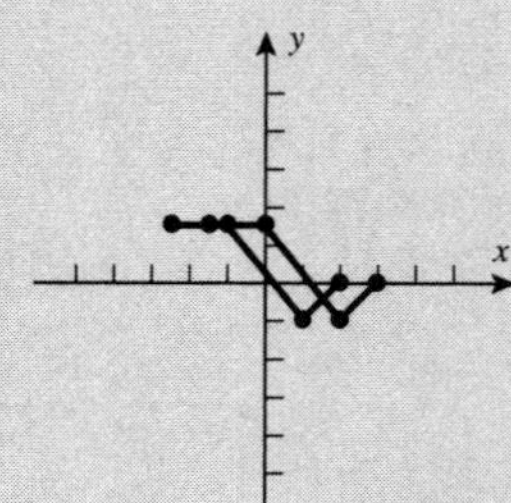

57.

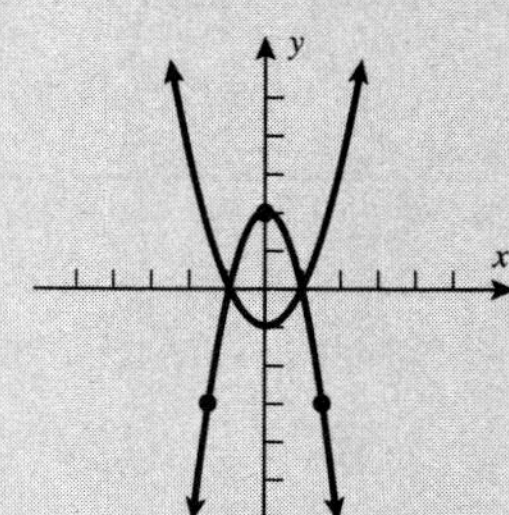

59. $y = \sqrt{x + 3}$
domain: $[-3, \infty)$

61. $y = \sqrt{x - 2} + 3$
domain: $[2, \infty)$

63. $y = 5\sqrt{x} - 6$
domain: $[0, \infty)$

65. $y = (x + 2)^2 - 12$

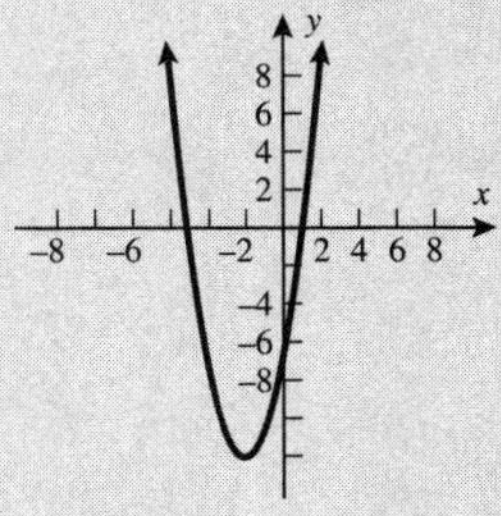

67. $g(x) + h(x) = -2x - 7$
domain: $(-\infty, \infty)$
$g(x) - h(x) = -4x - 1$
domain: $(-\infty, \infty)$
$g(x) \cdot h(x) = -3x^2 + 5x + 12$
domain: $(-\infty, \infty)$
$\dfrac{g(x)}{h(x)} = \dfrac{-3x - 4}{x - 3}$
domain: $(-\infty, 3) \cup (3, \infty)$

69. $g(x) + h(x) = \dfrac{1}{x^2} + \sqrt{x}$, domain: $(0, \infty)$

$g(x) - h(x) = \dfrac{1}{x^2} - \sqrt{x}$, domain: $(0, \infty)$

$g(x) \cdot h(x) = \dfrac{1}{x^{3/2}}$, domain: $(0, \infty)$

$\dfrac{g(x)}{h(x)} = \dfrac{1}{x^{5/2}}$, domain: $(0, \infty)$

71. $\left.\begin{aligned} g(x) + h(x) &= \sqrt{x - 4} + \sqrt{2x + 1} \\ g(x) - h(x) &= \sqrt{x - 4} - \sqrt{2x + 1} \\ g(x) \cdot h(x) &= \sqrt{x - 4}\sqrt{2x + 1} \\ \frac{g(x)}{h(x)} &= \frac{\sqrt{x - 4}}{\sqrt{2x + 1}} \end{aligned}\right\}$ domain: $[4, \infty)$

73. $(f \circ g)(x) = 6x - 1$ domain: $(-\infty, \infty)$
$(g \circ f)(x) = 6x - 7$ domain: $(-\infty, \infty)$

75. $(f \circ g)(x) = \dfrac{8 - 2x}{13 - 3x}$
domain: $(-\infty, 4) \cup \left(4, \frac{13}{3}\right) \cup \left(\frac{13}{3}, \infty\right)$

$(g \circ f)(x) = \dfrac{x + 3}{4x + 10}$
domain: $(-\infty, -3) \cup \left(-3, -\frac{5}{2}\right) \cup \left(-\frac{5}{2}, \infty\right)$

77. $f(g(x)) = \sqrt{x^2 - 9}$ domain: $(-\infty, -3] \cup [3, \infty)$
$g(f(x)) = x - 9$ domain: $[5, \infty)$

79. $f(g(3)) = 857$ $g(f(-1)) = 51$

81. $f(g(3)) = \frac{17}{31}$ $g(f(-1)) = 1$

83. $f(g(3)) = 12$ $g(f(-1)) = 2$

85. $f(x) = 3x^2 + 4x + 7$ and $g(x) = x - 2$

87. $f(x) = \dfrac{1}{\sqrt{x}}$ and $g(x) = x^2 + 7$

89. $A(t) = 625\pi(t + 2)$ sq. in.

91. yes

93. no

95. yes

97. yes

99. yes

101.

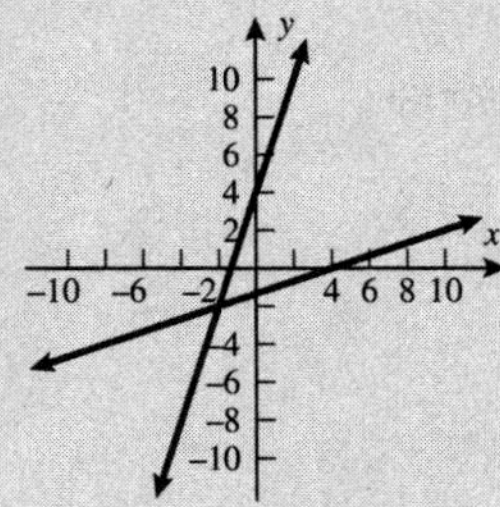

103.

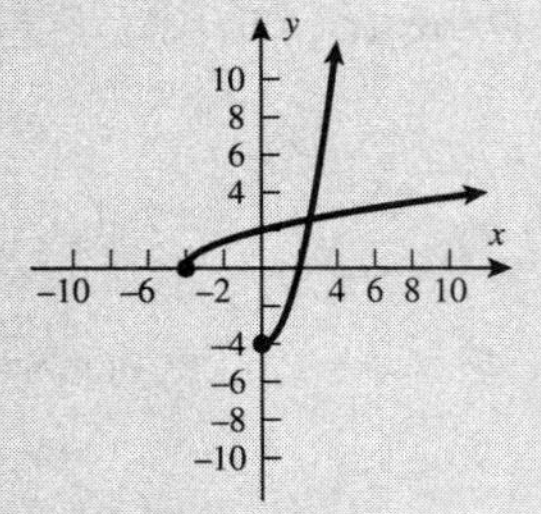

105. $f^{-1}(x) = \dfrac{x - 1}{2}$
domain f: $(-\infty, \infty)$
range f: $(-\infty, \infty)$
domain f^{-1}: $(-\infty, \infty)$
range f^{-1}: $(-\infty, \infty)$

107. $f^{-1}(x) = x^2 - 4$
domain f: $[-4, \infty)$
range f: $[0, \infty)$
domain f^{-1}: $(0, \infty)$
range f^{-1}: $(-4, \infty)$

109. $f^{-1}(x) = \dfrac{6 - 3x}{x - 1}$
domain f: $(-\infty, -3) \cup (-3, \infty)$
range f: $(-\infty, 1) \cup (1, \infty)$
domain f^{-1}: $(-\infty, 1) \cup (1, \infty)$
range f^{-1}: $(-\infty, -3) \cup (-3, \infty)$

111. $S(x) = 22{,}000 + 0.08x$ $S^{-1}(x) = \dfrac{(x - 22{,}000)}{0.08}$
Sales required to earn desired income.

113. $C = 2\pi r$ **115.** $A = \pi r^2$ **117.** $W = 8.5H$

119. $(-\infty, -1) \cup (3, \infty)$

121.

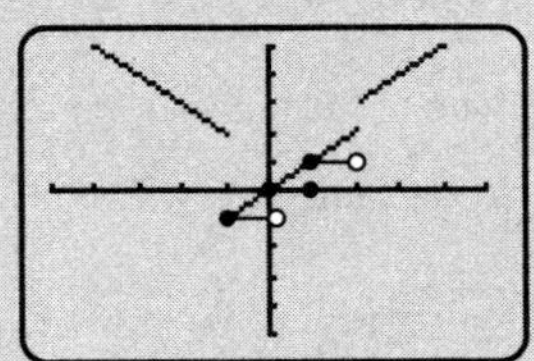

a. $(-\infty, 2) \cup (2, \infty)$ **b.** $\{-1, 0, 1\} \cup (2, \infty)$
c. increasing: $(2, \infty)$, decreasing: $(-\infty, -1)$, constant: $(-1, 2)$

123. The graph of f can be obtained by shifting the graph of g two units to the left; that is, $f(x) = g(x + 2)$.

125. $[-1.5, 4)$

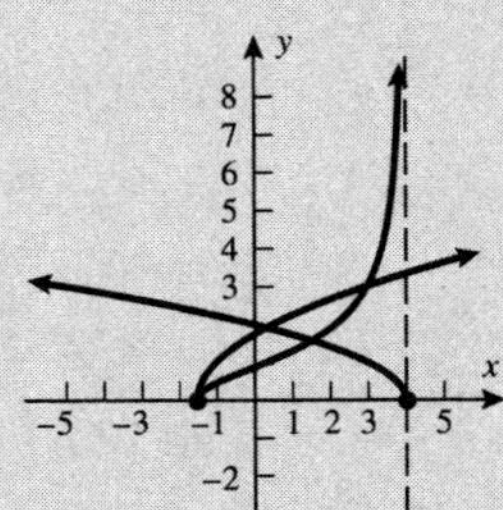

127. yes

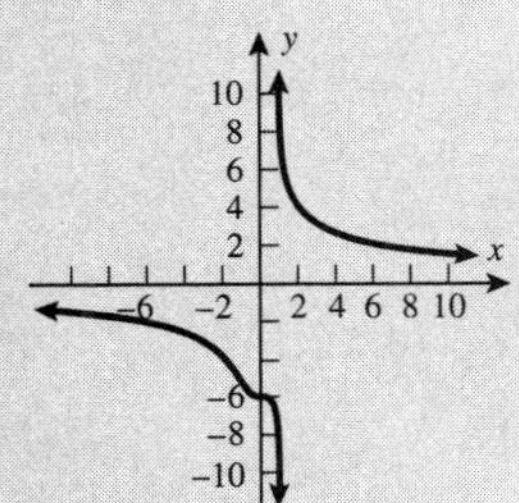

129. **a.** $y = 64.03x + 127.06$
b. \$575.30. No, it is not close to the actual price at \$517.20.
c. \$767.40

Practice Test

1. b **3.** c

5. $\frac{\sqrt{x-2}}{x^2+11}$ domain: $[2, \infty)$

7. $x + 9$ domain: $[2, \infty)$

9. 4 **11.** neither

13.

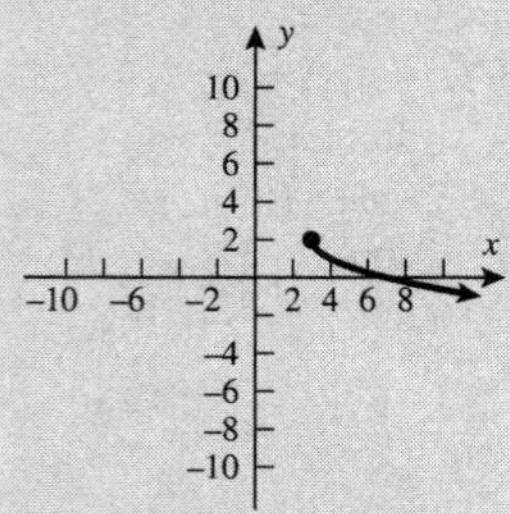

domain: $[3, \infty)$
range: $(-\infty, 2]$

15.

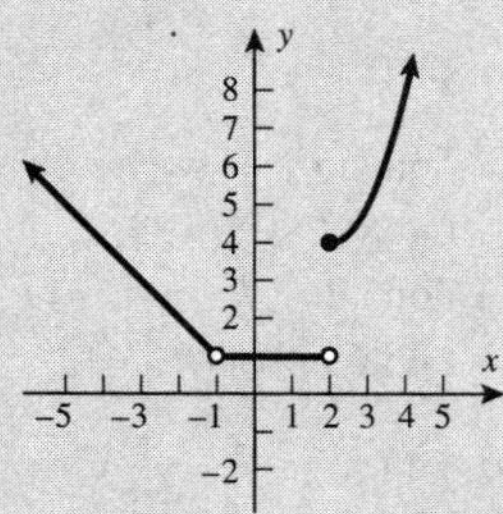

domain: $(-\infty, -1) \cup (-1, \infty)$
range: $[1, \infty)$

17. **a.** -2 **b.** 4 **c.** -3 **d.** $x = 2, x = -3$

19. $6x + 3h - 4$ **21.** -32

23. $f^{-1}(x) = x^2 + 5$
f: domain: $[5, \infty)$, range: $[0, \infty)$
f^{-1}: domain: $[0, \infty)$, range: $[5, \infty)$

25. $f^{-1}(x) = \frac{5x - 1}{x + 2}$

f: domain: $(-\infty, 5) \cup (5, \infty)$, range: $(-\infty, -2) \cup (-2, \infty)$

f^{-1}: domain: $(-\infty, -2) \cup (-2, \infty)$, range: $(-\infty, 5) \cup (5, \infty)$

27. $x \geq 0$

29. $c(x) = 0.42x$, where x is the original price of the suit.

31. quadrant III, "quarter of unit circle"

33. $f(m) = \begin{cases} 15 & 0 \leq x \leq 30 \\ -15 + x & x > 30 \end{cases}$ **35.** $F = \frac{30m}{P}$ **37.** yes

Cumulative Test

1. $\frac{3 + \sqrt{5}}{2}$ **3.** $x(x - 2); x \neq -2$ **5.** $145 + 0i$

7. 40% **9.** $x = 1 \pm \frac{\sqrt{35}}{5}$ **11.** $x = -2, 2, \pm i\sqrt{3}$

13. (0, 5) **15.** $d = \sqrt{121.7} \approx 11.03$; (1.25, 2.45)

17. $y = -3$ **19.** $(x + 2)^2 + (y + 1)^2 = 20$

21. $(-\infty, 1) \cup (1, \infty)$ **23.** 46

25. $r = \frac{135}{t}$ **27.** $h(x) = x - 2$

CHAPTER 4

Section 4.1

1. b **3.** a **5.** b **7.** c

9.

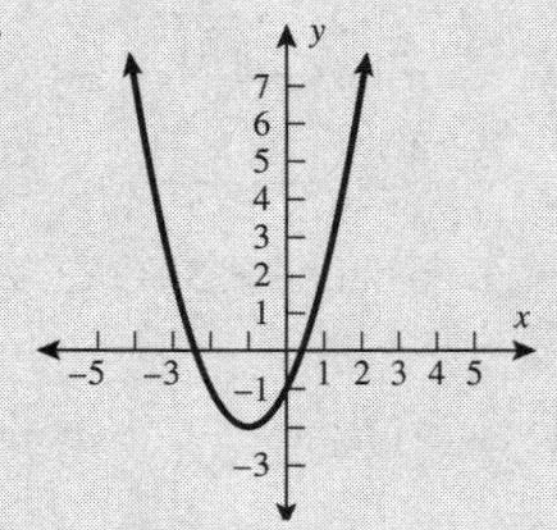

11.

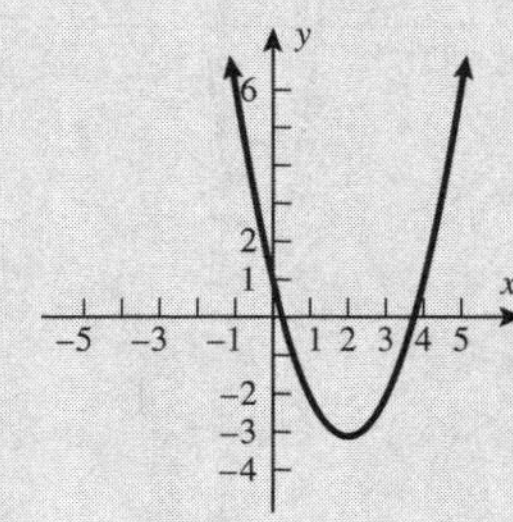

13.

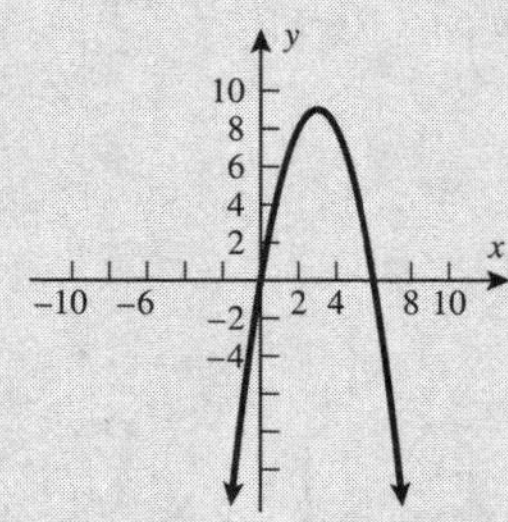

15.

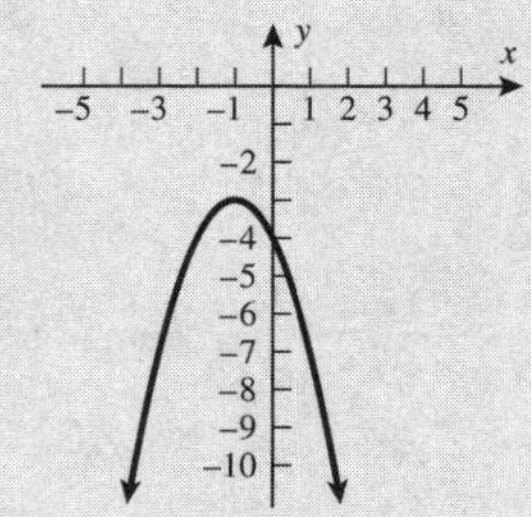

17.

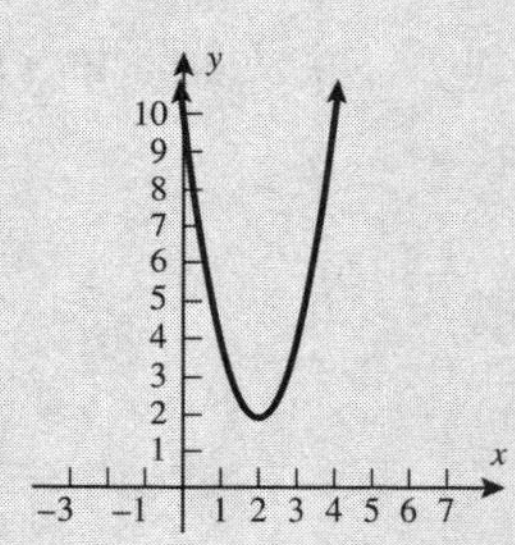

19.

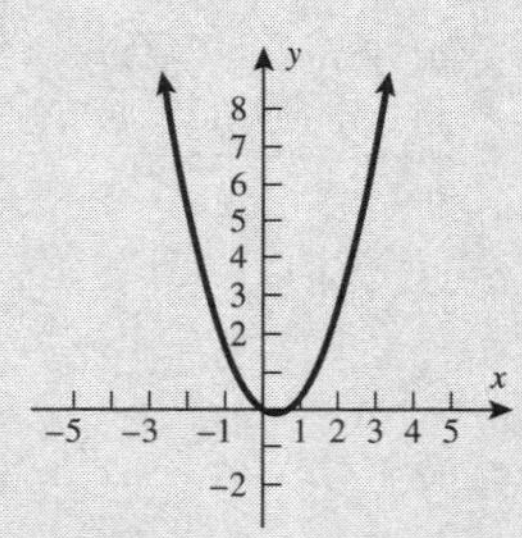

21.

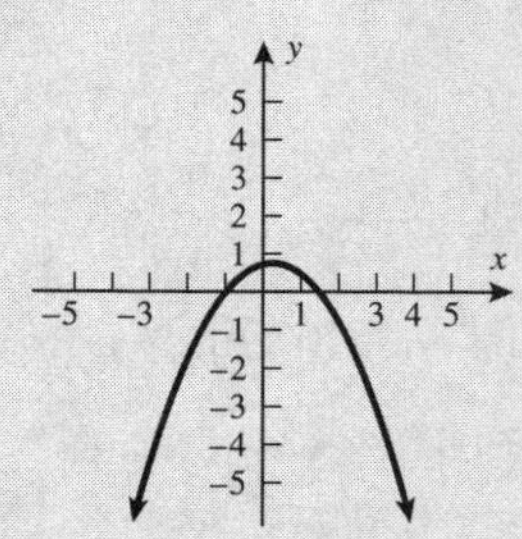

23. $f(x) = (x + 3)^2 - 12$

25. $f(x) = -(x + 5)^2 + 28$

27. $f(x) = 2(x + 2)^2 - 10$

29. $f(x) = -4(x - 2)^2 + 9$

31. $f(x) = \frac{1}{2}(x - 4)^2 - 5$

33.

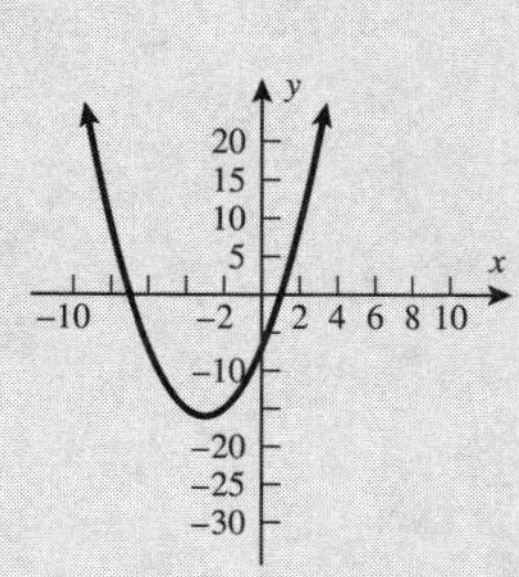

35.

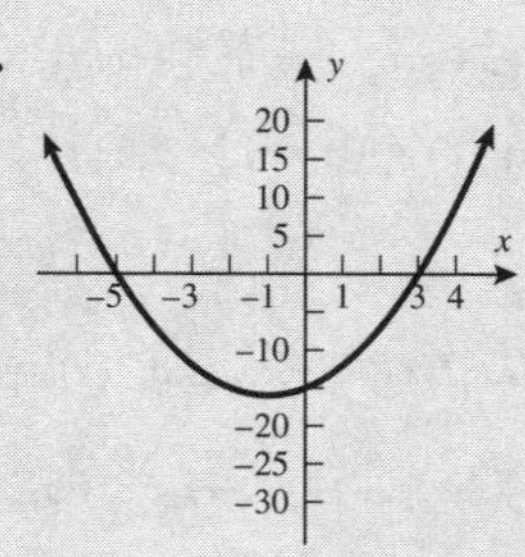

37.

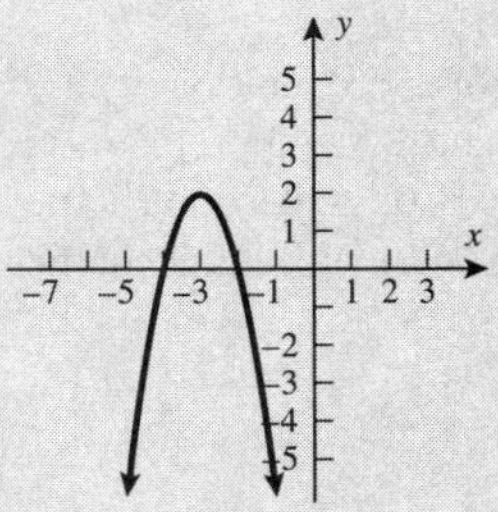

39.

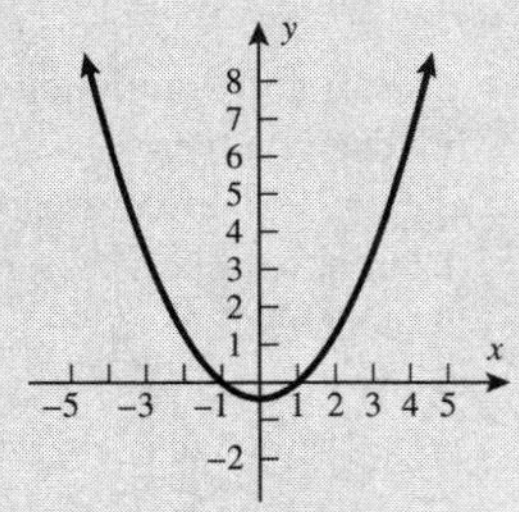

41. vertex $\left(\frac{1}{33}, \frac{494}{33}\right)$

43. vertex $\left(7, -\frac{39}{2}\right)$

45. vertex $(-75, 12.95)$

47. vertex $\left(\frac{15}{28}, \frac{829}{392}\right)$

49. $y = -2(x + 1)^2 + 4$

51. $y = -5(x - 2)^2 + 5$

53. $y = \frac{5}{9}(x + 1)^2 - 3$

55. $y = 12\left(x - \frac{1}{2}\right)^2 - \frac{3}{4}$

57. $y = \frac{5}{4}(x - 2.5)^2 - 3.5$

59. **a.** 40 yards (120 feet) **b.** 50 yards

61. 2,083,333 square feet

63. **a.** $t = 1$ second is when rock reaches the maximum height of 116 feet.
b. The rock hits the ground in approximately 3.69 seconds.

65. Altitude is 26,000 feet; horizontal distance is 8944 feet.

67. **a.** 100 boards **b.** \$24,000

69. 15 to 16 units to break even or 64 to 65 units to break even.

71. **a.** $f(t) = \frac{28}{27}t^2 + 16$ **b.** 219 million cell phones

73. **a.** $C(t) = -0.01(t - 225)^2 + 400$
b. 425 minutes

75. The corrections that need to be made are vertex $(-3, -1)$ and x-intercepts $(-2, 0)$ $(-4, 0)$.

77. $f(x) = -(x - 1)^2 + 4$. The negative must be factored out of the x^2 and x terms.

79. true **81.** false

83. $f(x) = a\left(x + \dfrac{b}{2a}\right)^2 + \dfrac{4ac - b^2}{4a}$

85. **a.** The maximum area of the rectangular pasture is 62,500 square feet.
b. The maximum area of the circular fence is approximately 79,577 square feet.

87. **a.** $(1425, 4038.25)$ **b.** $(0, -23)$
c. $(4.04, 0)$ $(2845.96, 0)$ **d.** $x = 1425$

89. **a.** $y = -2x^2 + 12.8x + 4.32$
b. $y = -2(x - 3.2)^2 + 24.8$, $(3.2, 24.8)$
c. Yes, they agree

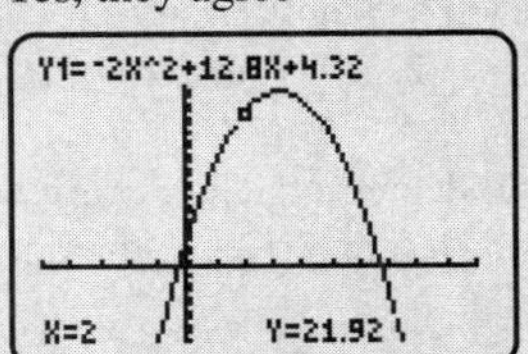

Section 4.2

1. polynomial; degree 2

3. polynomial; degree 5

5. not a polynomial

7. not a polynomial

9. not a polynomial

11. h

13. b

15. e

17. c

19.

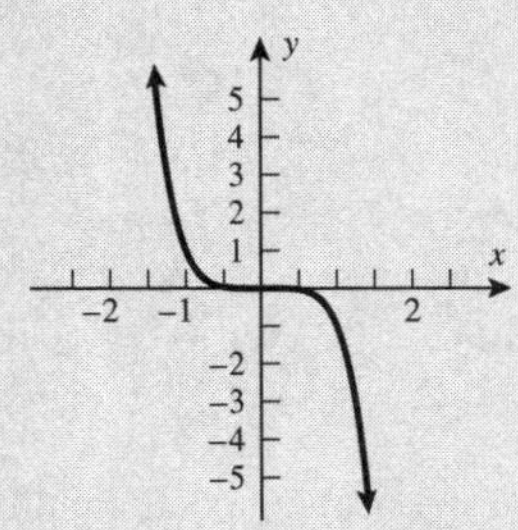

21.

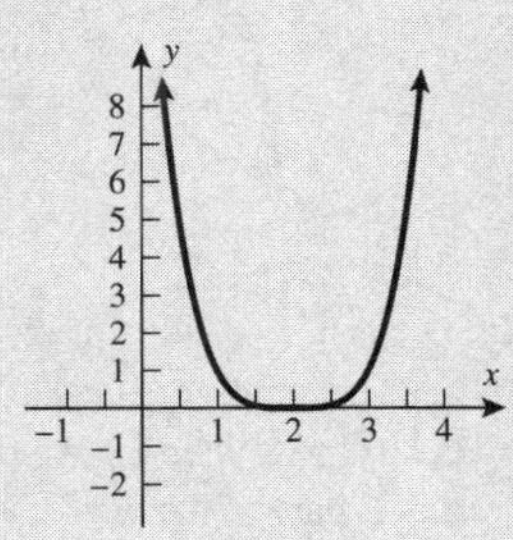

23.

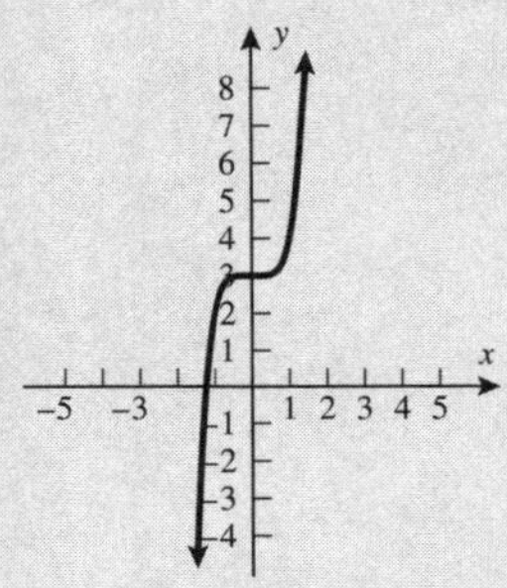

25.

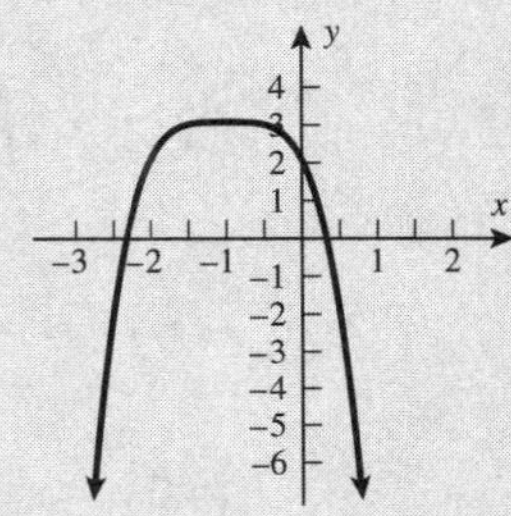

27. zero at 3, multiplicity of 1; zero at -4, multiplicity of 3

29. zero at 0, multiplicity of 2; zero at 7, multiplicity of 2; zero at -4, multiplicity of 1

31. zero at 0, multiplicity of 2; zero at 1, multiplicity of 2

33. $f(x) = x(4x + 9)(2x - 3)$ zero at 0, multiplicity of 1
zero at $-\frac{9}{4}$, multiplicity of 1
zero at $\frac{3}{2}$, multiplicity of 1

35. zero at 0, multiplicity 2; zero at -3, multiplicity 1

37. zero at 0, multiplicity 4

39. $P(x) = x(x + 3)(x - 1)(x - 2)$

41. $P(x) = x(x + 5)(x + 3)(x - 2)(x - 6)$

43. $P(x) = (2x + 1)(3x - 2)(4x - 3)$

45. $P(x) = x^2 - 2x - 1$

47. $P(x) = x^2(x + 2)^3$

49. $P(x) = (x + 3)^2(x - 7)^5$

51. $P(x) = x^2(x + \sqrt{3})^2(x + 1)(x - \sqrt{3})^2$

53. $f(x) = -x^2 - 6x - 9 = -(x + 3)^2$
a. zero at -3, multiplicity 2
b. The graph touches at $x = -3$.
c. Let $x = 0$ and $y = 0 - 0 - 9 = -9$.

d. Behaves like $y = -x^2$ (even degree); leading coefficient is negative, graph falls to the left and right without bound.

e.

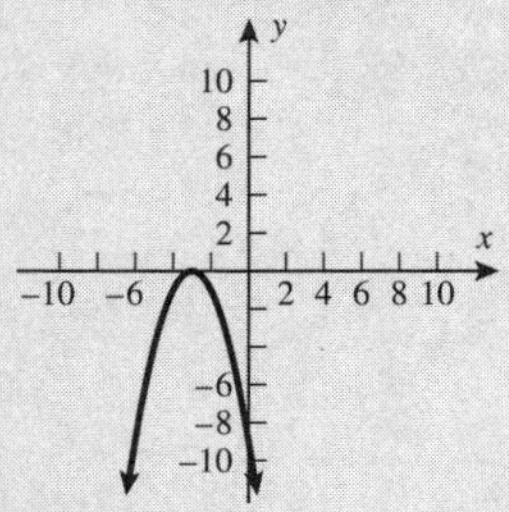

55. $f(x) = (x - 2)^3$

a. zero at 2; multiplicity of 3

b. The graph crosses at $x = 2$.

c. Let $x = 0$ and $y = (0 - 2)^3 = -8$.

d. Behaves like $y = x^3$ (odd degree); leading coefficient is positive, graph rises to the right and falls to the left.

e.

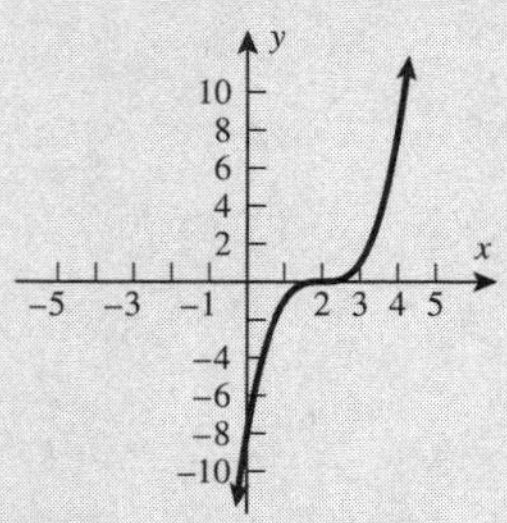

57. $f(x) = x^3 - 9x = x(x - 3)(x + 3)$

a. zeros at 0, 3, and −3; each have multiplicity 1

b. The graph crosses at all three zeros.

c. (0, 0)

d. Behaves like $y = x^3$ (odd degree); leading coefficient is positive, graph falls to the left and rises to the right.

e.

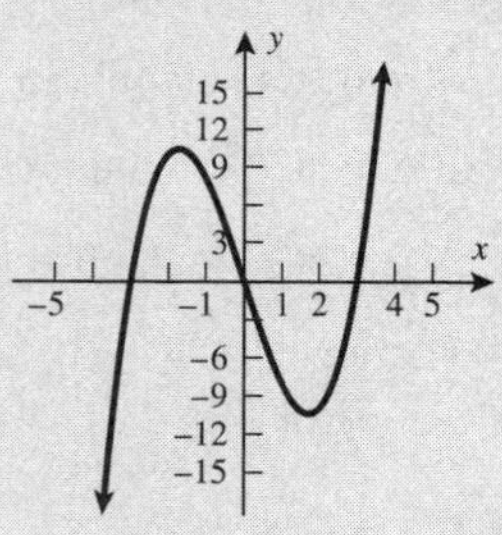

59. $f(x) = -x^3 + x^2 + 2x = -x(x - 2)(x + 1)$

a. zeros at 0, 2, −1; each has multiplicity 1

b. The graph crosses at all three zeros.

c. (0, 0)

d. Behaves like $y = -x^3$ (odd degree); leading coefficient is negative, graph rises to the left and falls to the right.

e.

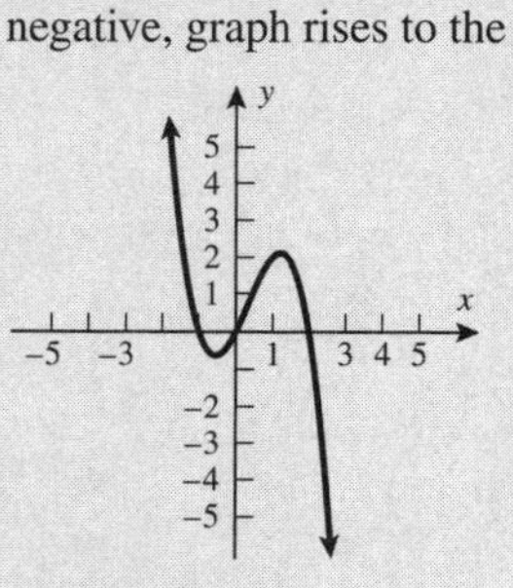

61. $f(x) = -x^4 - 3x^3 = -x^3(x + 3)$

a. zero at 0, multiplicity 3; zero at −3, multiplicity 1

b. The graph crosses at $x = 0$ and $x = -3$.

c. $y = -x^4 - 3x^3$ Let $x = 0$; $y = 0$; y-intercept (0, 0).

d. Behaves like $y = -x^4$ (even degree); leading coefficient is negative.

e.

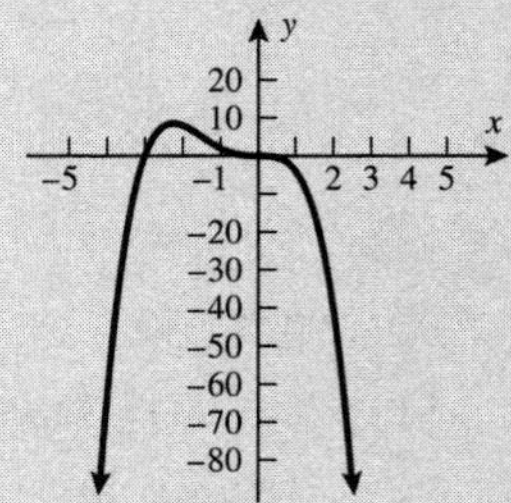

63. $f(x) = 12x^4(x - 4)(x + 1)$

a. zero at 0, multiplicity 4; zero at 4, multiplicity 1; zero at −1, multiplicity 1

b. The graph crosses at $x = -1$ and $x = 4$ while the graph touches at 0.

c. Let $x = 0$; $y = 0$; y-intercept (0, 0).

d. Behaves like $y = x^6$ (even degree); leading coefficient is positive.

e.

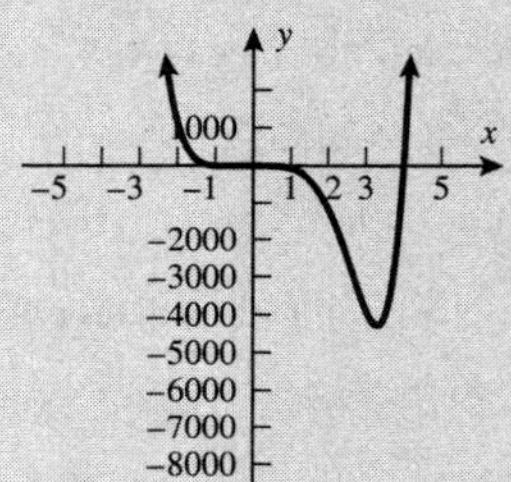

65. $f(x) = 2x^5 - 6x^4 - 8x^3 = 2x^3(x - 4)(x + 1)$

a. zeros at −1, 0, and 4; 0 has multiplicity 3; −1 and 4 each have multiplicity 1

b. The graph touches at $x = 0$; crosses at $x = -1$ and $x = 4$.

c. (0, 0)

d. Behaves like $y = x^5$ (odd degree); leading coefficient is positive, graph falls to the left and rises to the right.

e.

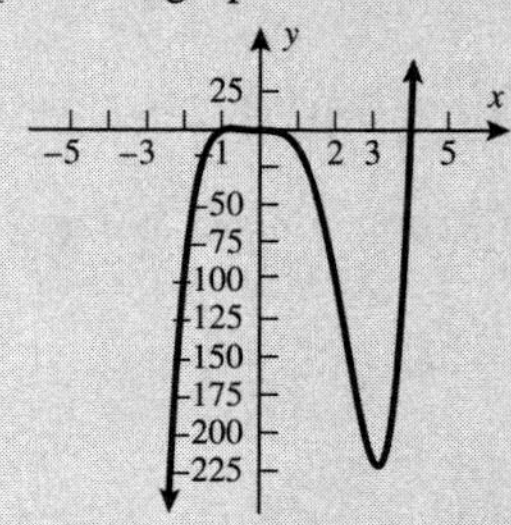

67. $f(x) = x^3 - x^2 - 4x + 4 = (x - 2)(x + 2)(x - 1)$

a. zeros at −2, 1, and 2; each has multiplicity 1

b. The graph crosses at $x = -2$, $x = 1$, and $x = 2$.

c. (0, 4)

d. Behaves like $y = x^3$ (odd degree); leading coefficient is positive, graph falls to the left and rises to the right.

e.

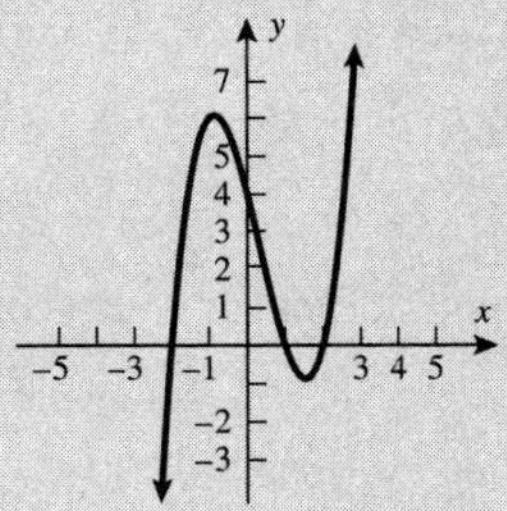

69. $f(x) = -(x + 2)^2(x - 1)^2$

a. zeros at -2 and 1; with both having multiplicity of 2

b. The graph touches at each one of the zeros.

c. Let $x = 0$; $y = -4$; y-intercept $(0, -4)$.

d. Behaves like $y = x^4$ (even degree); leading coefficient is negative.

e.

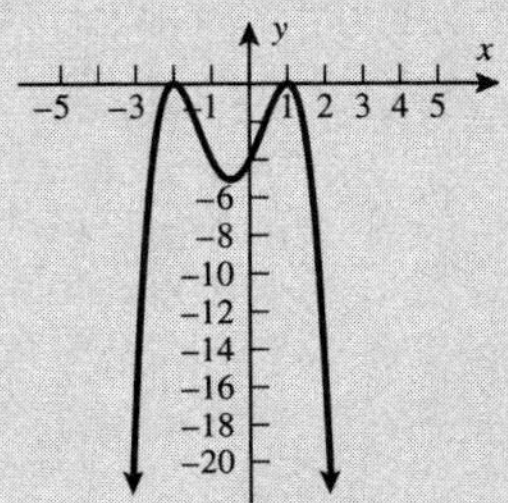

71. $f(x) = x^2(x - 2)^3(x + 3)^2$

a. zeros at 0, 2, and -3; having multiplicities of 2, 3, and 2

b. The graph touches at -3 and 0 but crosses at 2.

c. Let $x = 0$; $y = 0$; y-intercept $(0, 0)$.

d. Behaves like $y = x^7$; leading coefficient is positive.

e.

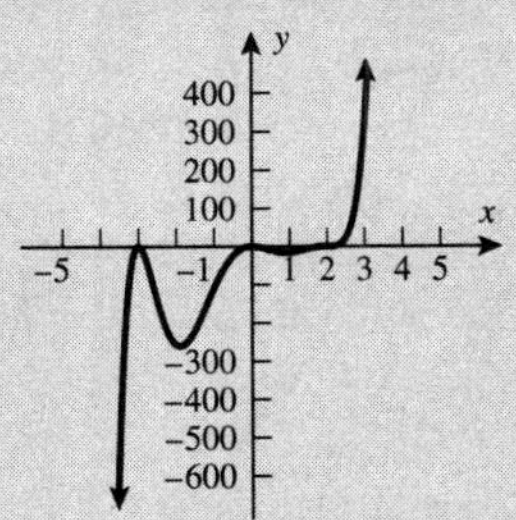

73. a. -3, multiplicity 1; -1, multiplicity 2; 2, multiplicity 1

b. even **c.** negative **d.** $(0, 6)$

e. $f(x) = -(x + 1)^2(x - 2)(x + 3)$

75. a. -2, multiplicity 2; 0, multiplicity 2; $\frac{3}{2}$, multiplicity 1

b. odd **c.** positive **d.** $(0, 0)$

e. $f(x) = x^2(2x - 3)(x + 2)^2$

77. Sixth-degree polynomial, because there are five known turning points.

79. down

81. fourth-degree

83. If h is a zero then $(x - h)$ is a factor.
In this case: $P(x) = (x + 2)(x + 1)(x - 3)(x - 4)$.

85. $f(x) = (x - 1)^2(x + 2)^3$
Yes, the zeros are -2 and 1. But you must remember that this is a fifth-degree polynomial. At the -2 zero the graph crosses, but it should be noted that at 1 it only touches at this value. The y-intercept is also graphed incorrectly. The correct graph would be the following:

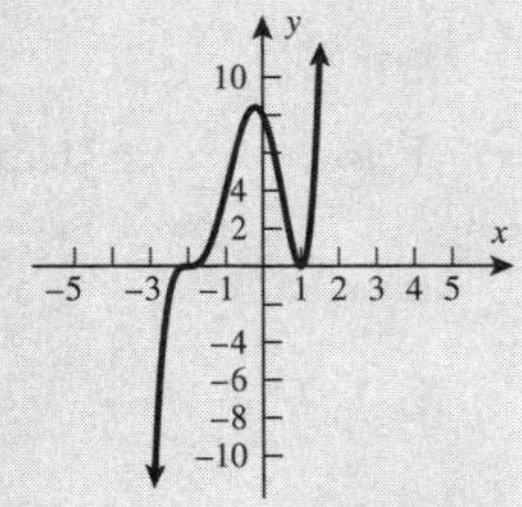

87. false

89. true

91. n

93. There are three possibilities.

a. $f(x) = (x + 1)^2(x - 3)^5$

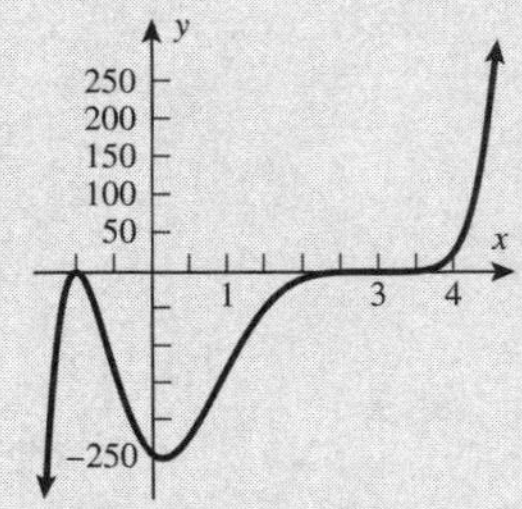

b. $g(x) = (x + 1)^4(x - 3)^3$

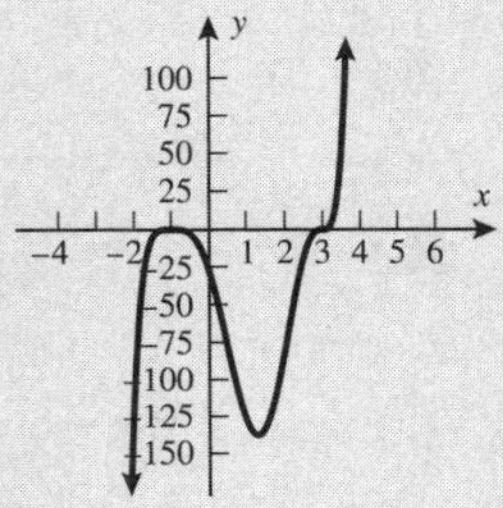

c. $h(x) = (x + 1)^6(x - 3)$

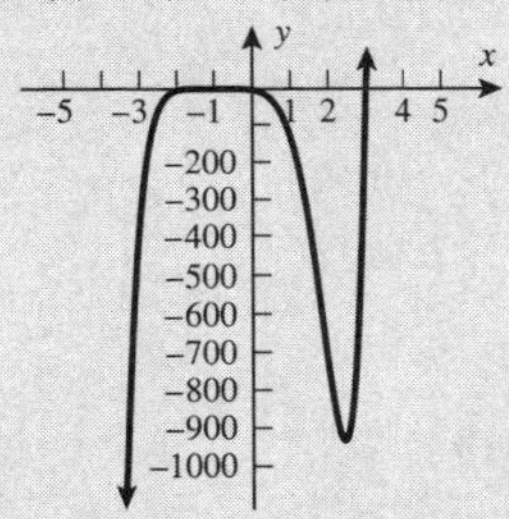

95. The zeros of the polynomial are 0, a, and $-b$.

97. no x-intercepts

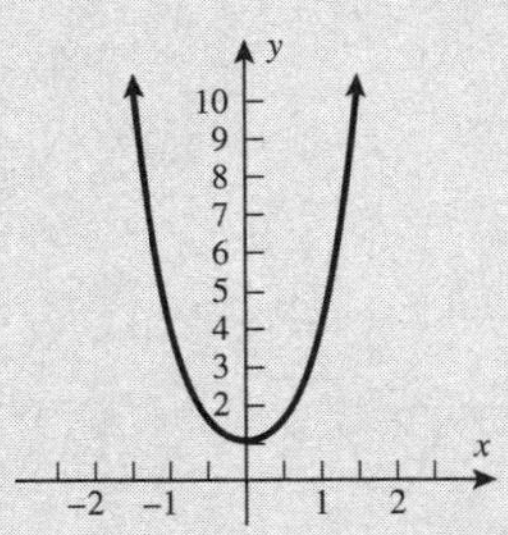

99. $y = -2x^5$
yes

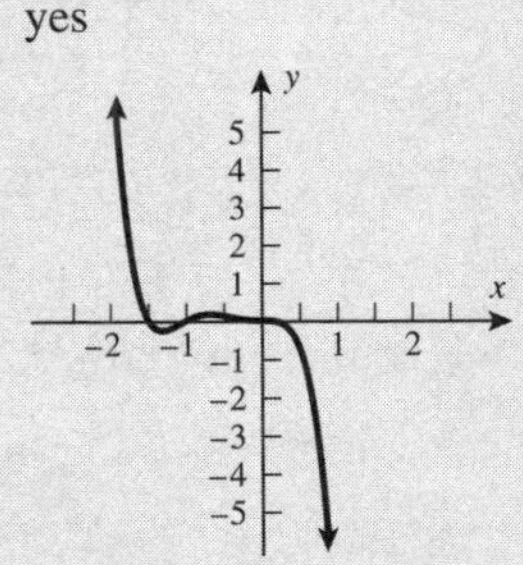

101. x-intercepts: $(-2.25, 0)$, $(6.2, 0)$, $(14.2, 0)$; zero at -2.25, multiplicity 2; zero at 6.2, multiplicity1; zero at 14.2, multiplicity 1

103. 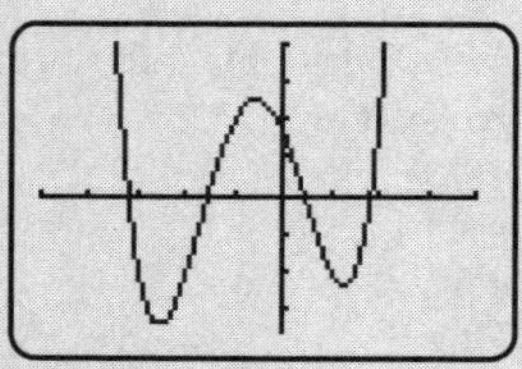

relative minimum at $(-2.56, -17.12)$ and $(1.27, -11.73)$; relative maximum at $(-0.58, 12.59)$

Section 4.3

1. $Q(x) = 2x - 1 \quad r(x) = 0$
3. $Q(x) = x - 3 \quad r(x) = 0$
5. $Q(x) = 3x - 3 \quad r(x) = -11$
7. $Q(x) = 3x - 28 \quad r(x) = 130$
9. $Q(x) = x - 4 \quad r(x) = 12$
11. $Q(x) = 3x + 5 \quad r(x) = 0$
13. $Q(x) = 2x - 3 \quad r(x) = 0$
15. $Q(x) = 4x^2 + 4x + 1 \quad r(x) = 0$
17. $Q(x) = 2x^2 - x - \frac{1}{2} \quad r(x) = \frac{15}{2}$
19. $Q(x) = 4x^2 - 10x - 6 \quad r(x) = 0$
21. $Q(x) = -2x^2 - 3x - 9 \quad r(x) = -27x^2 + 3x + 9$
23. $Q(x) = x^2 + 1 \quad r(x) = 0$
25. $Q(x) = x^2 + x + \frac{1}{6} \quad r(x) = -\frac{121}{6}x + \frac{121}{3}$
27. $Q(x) = -3x^3 + 5.2x^2 + 3.12x - 0.128 \quad r(x) = 0.9232$
29. $Q(x) = x^2 - 0.6x + 0.09 \quad r(x) = 0$
31. $Q(x) = 3x + 1 \quad r(x) = 0$
33. $Q(x) = 7x - 10 \quad r(x) = 15$
35. $Q(x) = -x^3 + 3x - 2 \quad r(x) = 0$
37. $Q(x) = x^3 - x^2 + x - 1 \quad r(x) = 2$
39. $Q(x) = x^3 - 2x^2 + 4x - 8 \quad r(x) = 0$
41. $Q(x) = 2x^2 - 6x + 2 \quad r(x) = 0$
43. $Q(x) = 2x^3 - \frac{5}{3}x^2 + \frac{53}{9}x + \frac{106}{27} \quad r(x) = -\frac{112}{81}$
45. $Q(x) = 2x^3 + 6x^2 - 18x - 54 \quad r(x) = 0$
47. $Q(x) = x^6 + x^5 + x^4 - 7x^3 - 7x^2 - 4x - 4 \quad r(x) = -3$
49. $Q(x) = x^5 + \sqrt{5}x^4 - 44x^3 - 44\sqrt{5}x^2 - 245x - 245\sqrt{5}$
$r(x) = 0$
51. $Q(x) = 2x - 7 \quad r(x) = 0$
53. $Q(x) = x^2 - 9 \quad r(x) = 0$
55. $Q(x) = x^4 + 2x^3 + 8x^2 + 18x + 36 \quad r(x) = 71$
57. $Q(x) = x^2 + 1 \quad r(x) = -24$
59. $Q(x) = x^6 + x^5 + x^4 + x^3 + x^2 + x + 1 \quad r(x) = 0$
61. The width is $3x^2 + 2x + 1$ feet.
63. The trip takes $x^2 + 1$ hours.
65. In long division, you must subtract (not add) each term.
67. Forgot "0" placeholder.
69. true
71. false
73. yes
75. $x^{2n} + 2x^n + 1$
77. $2x - 1$

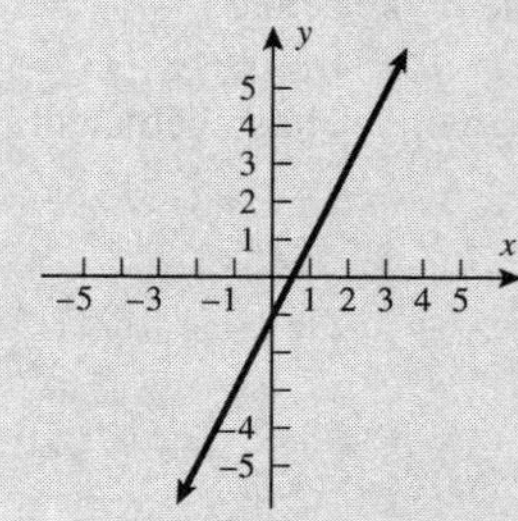

79. $x^3 - 1$

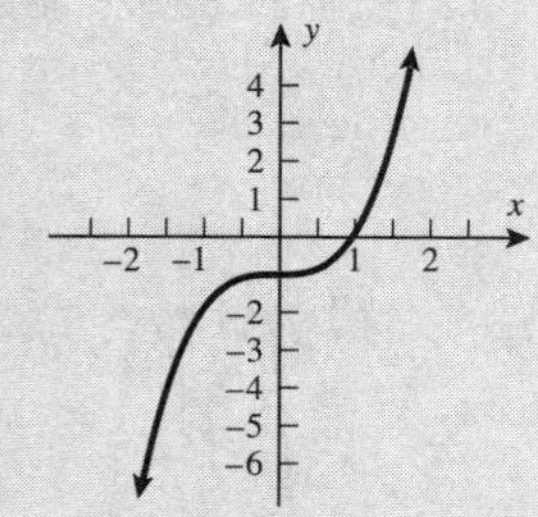

81. A quadratic function.

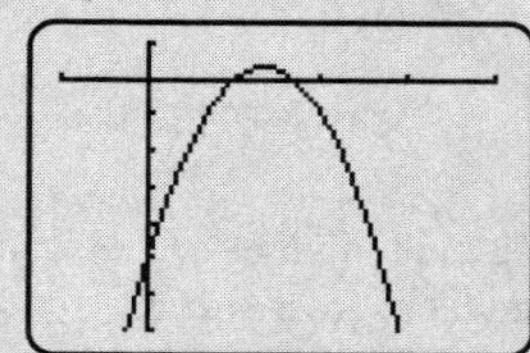

Section 4.4

1. $f(1) = 0$
3. $g(1) = 4$
5. $f(-2) = 84$
7. Yes, the given is a zero of the polynomial.
9. Yes, the given is a zero of the polynomial.
11. $1, 3, -4$; $P(x) = (x - 1)(x - 3)(x + 4)$
13. $\frac{1}{2}, -3, 2$; $P(x) = (2x - 1)(x + 3)(x - 2)$
15. $5, -3$; $P(x) = (x - 5)(x + 3)(x^2 + 4)$
17. $-3, 1$; $P(x) = (x + 3)(x - 1)(x^2 - 2x + 2)$
19. $-1, -2$; $P(x) = (x + 1)^2(x + 2)^2$
21. $P(x) = x^4 + 3x^2 - 8x + 4$
Factors of $a_0 = 4$: $\pm1, \pm2, \pm4$
Factors of $a_n = 1$: ±1
Possible rational zeros: $\pm1, \pm2, \pm4$
23. $P(x) = x^5 - 14x^3 + x^2 - 15x + 12$
Factors of $a_0 = 12$: $\pm1, \pm2, \pm3, \pm4, \pm6, \pm12$
Factors of $a_n = 1$: ±1
Possible rational zeros: $\pm1, \pm2, \pm3, \pm4, \pm6, \pm12$

25. $P(x) = 2x^6 - 7x^4 + x^3 - 2x + 8$
Factors of $a_0 = 8$: $\pm1, \pm2, \pm4, \pm8$
Factors of $a_n = 2$: $\pm1, \pm2$
Possible rational zeros: $\pm1, \pm2, \pm4, \pm8, \pm\frac{1}{2}$

27. $P(x) = 5x^5 + 3x^4 + x^3 - x - 20$
Factors of $a_0 = -20$: $\pm1, \pm2, \pm4, \pm5, \pm10, \pm20$
Factors of $a_n = 5$: $\pm1, \pm5$
Possible rational zeros:
$\pm1, \pm2, \pm4, \pm5, \pm10, \pm20, \pm\frac{1}{5}, \pm\frac{2}{5}, \pm\frac{4}{5}$

29. $P(x) = x^4 + 2x^3 - 9x^2 - 2x + 8$
Factors of $a_0 = 8$: $\pm1, \pm2, \pm4, \pm8$
Factors of $a_n = 1$: ±1
Possible rational zeros: $\pm1, \pm2, \pm4, \pm8$
Testing the zeros:
$P(1) = 0 \quad P(-1) = 0 \quad P(2) = 0 \quad P(-4) = 0$

31. $P(x) = 2x^3 - 9x^2 + 10x - 3$
Factors of $a_0 = -3$: $\pm1, \pm3$
Factors of $a_n = 2$: $\pm1, \pm2$
Possible rational zeros: $\pm1, \pm3, \pm\frac{1}{2}, \pm\frac{3}{2}$
$P(1) = 0 \quad P(3) = 0 \quad P(\frac{1}{2}) = 0$

33.

Positive Real Zeros	Negative Real Zeros
1	1

35.

Positive Real Zeros	Negative Real Zeros
1	0

37.

Positive Real Zeros	Negative Real Zeros
2	1
0	1

39.

Positive Real Zeros	Negative Real Zeros
1	1

Note that $x = 0$ is also a zero.

41.

Positive Real Zeros	Negative Real Zeros
2	2
0	2
2	0
0	0

43.

Positive Real Zeros	Negative Real Zeros
4	0
2	0
0	0

45. $P(x) = x^3 + 6x^2 + 11x + 6$
a. No sign changes: 0 positive real zeros
$P(-x) = (-x)^3 + 6(-x)^2 + 11(-x) + 6$
$= -x^3 + 6x^2 - 11x + 6$
3 or 1 negative real zeros

Positive	Negative
0	3
0	1

b. Factors of $a_0 = 6$: $\pm1, \pm2, \pm3, \pm6$
Factors of $a_n = 1$: ±1
Possible rational zeros: $\pm1, \pm2, \pm3, \pm6$
c. $P(-1) = 0 \quad P(-2) = 0 \quad P(-3) = 0$
d. $x = -1 \quad x = -2 \quad x = -3$
$P(x) = (x + 1)(x + 2)(x + 3)$

47. $P(x) = x^3 - 7x^2 - x + 7$
a. 2 or 0 positive real zeros
$P(-x) = (-x)^3 - 7(-x)^2 - (-x) + 7$
$= -x^3 - 7x^2 + x + 7$
1 negative real zero

Positive	Negative
2	1
0	1

b. Factors of $a_0 = 7$: $\pm1, \pm7$
Factors of $a_n = 1$: ±1
Possible rational zeros: $\pm1, \pm7$
c. $P(1) = 0 \quad P(-1) = 0 \quad P(7) = 0$
d. $P(x) = (x - 1)(x + 1)(x - 7)$

49. $P(x) = x^4 + 6x^3 + 3x^2 - 10x$
a. 1 positive real zero
$P(-x) = (-x)^4 + 6(-x)^3 + 3(-x)^2 - 10(-x)$
$= x^4 - 6x^3 + 3x^2 + 10x$
$= x(x^3 - 6x^2 + 3x + 10)$
2 or 0 negative real zeros

Positive	Negative
1	2
1	0

b. Factors of $a_0 = -10$: $\pm 1, \pm 2, \pm 5, \pm 10$
Factors of $a_n = 1$: ± 1
Possible rational zeros: $\pm 1, \pm 2, \pm 5, \pm 10$

c. $P(0) = 0 \quad P(1) = 0 \quad P(-2) = 0 \quad P(-5) = 0$

d. $P(x) = x(x-1)(x+2)(x+5)$

51. $P(x) = x^4 - 7x^3 + 27x^2 - 47x + 26$

a. 4, 2, or 0 positive real zeros

$P(-x) = (-x)^4 - 7(-x)^3 + 27(-x)^2 - 47(-x) + 26$
$= x^4 + 7x^3 + 27x^2 + 47x + 26$

no negative real zeros

POSITIVE	NEGATIVE
4	0
2	0
0	0

b. Factors of $a_0 = 26$: $\pm 1, \pm 2, \pm 13, \pm 26$
Factors of $a_n = 1$: ± 1
Possible rational zeros: $\pm 1, \pm 2, \pm 13, \pm 26$

c. $P(1) = 0 \quad P(2) = 0 \quad (x-1)(x-2) = x^2 - 3x + 2$

d.
$$\begin{array}{r}
x^2 - 4x + 13 \\
x^2 - 4x + 13 \overline{\big)\, x^4 - 7x^3 + 27x^2 - 47x + 26} \\
\underline{-(x^4 - 3x^3 + 2x^2)} \qquad\qquad\quad \\
-4x^3 + 25x^2 - 47x \qquad \\
\underline{-(-4x^3 + 12x^2 - 8x)} \qquad \\
13x^2 - 39x + 26 \\
\underline{-(13x^2 - 39x + 26)} \\
0
\end{array}$$

$P(x) = (x-1)(x-2)(x^2 - 4x + 13)$

53. $P(x) = 10x^3 - 7x^2 - 4x + 1$

a. 2 or 0 positive real zeros

$P(-x) = 10(-x)^3 - 7(-x)^2 - 4(-x) + 1$
$= -10x^3 - 7x^2 + 4x + 1$

1 negative real zero

POSITIVE	NEGATIVE
2	1
0	1

b. Factors of $a_0 = 1$: ± 1
Factors of $a_n = 10$: $\pm 1, \pm 2, \pm 5, \pm 10$
Possible rational zeros: $\pm 1, \pm\frac{1}{2}, \pm\frac{1}{5}, \pm\frac{1}{10}$

c. $P(1) = 0 \quad P\left(-\frac{1}{2}\right) = 0 \quad P\left(\frac{1}{5}\right) = 0$

d. $(x-1)\left(x+\frac{1}{2}\right)\left(x-\frac{1}{5}\right) = 0$

$P(x) = (x-1)(2x+1)(5x+1)$

55. $P(x) = 6x^3 + 17x^2 + x - 10$

a. 1 positive real zero

$P(x) = 6(-x)^3 + 17(-x)^2 + (-x) - 10$
$= -6x^3 + 17x^2 - x - 10$

2 or 0 negative real zeros

POSITIVE	NEGATIVE
1	2
1	0

b. Factors of $a_0 = -10$: $\pm 1, \pm 2, \pm 5, \pm 10$
Factors of $a_n = 6$: $\pm 1, \pm 2, \pm 3, \pm 6$
Possible rational zeros: $\pm 1, \pm\frac{1}{2}, \pm\frac{1}{3}, \pm\frac{1}{6}, \pm 2, \pm\frac{2}{3}, \pm 5, \pm\frac{5}{2}, \pm\frac{5}{3}, \pm\frac{5}{6}, \pm 10, \pm\frac{10}{3}$

c. $P(-1) = 0 \quad P\left(-\frac{5}{2}\right) = 0 \quad P\left(\frac{2}{3}\right) = 0$

d. $(x+1)\left(x+\frac{5}{2}\right)\left(x-\frac{2}{3}\right) = 0$

$P(x) = (x+1)(2x+5)(3x-2)$

57. $P(x) = x^4 - 2x^3 + 5x^2 - 8x + 4$

a. 4, 2, or 0 positive real zeros

$P(-x) = x^4 + 2x^3 + 5x^2 + 8x + 4$

0 negative real zero

POSITIVE	NEGATIVE
4	0
2	0
0	0

b. Factors of $a_0 = 4$: $\pm 1, \pm 2, \pm 4$
Factors of $a_n = 1$: ± 1
Possible rational zeros: $\pm 1, \pm 2, \pm 4$

c. $P(1) = 0 \quad x = 1$ (multiplicity 2)

d. $P(x) = (x-1)^2(x^2+4)$

59. $P(x) = x^6 + 12x^4 + 23x^2 - 36$

a. 1 positive real zero

$P(-x) = x^6 + 12x^4 + 23x^2 - 36$

1 negative real zero

POSITIVE	NEGATIVE
1	1

b. Factors of $a_0 = -36$: $\pm 1, \pm 2, \pm 3, \pm 4, \pm 6, \pm 9, \pm 12, \pm 18, \pm 36$
Factors of $a_n = 1$: ± 1
Possible rational zeros: $\pm 1, \pm 2, \pm 3, \pm 4, \pm 6, \pm 9, \pm 12, \pm 18, \pm 36$

c. $P(1) = 0 \quad P(-1) = 0$

d. $P(x) = (x+1)(x-1)(x^2+4)(x^2+9)$

61. $P(x) = 4x^4 - 20x^3 + 37x^2 - 24x + 5$

a. 4, 2, or 0 positive real zeros
$P(-x) = 4x^4 + 20x^3 + 37x^2 + 24x + 5$
0 negative real zero

Positive	Negative
4	0
2	0
0	0

b. Factors of $a_0 = 5$: $\pm 1, \pm 5$
Factors of $a_n = 4$: $\pm 1, \pm 2, \pm 4$
Possible rational zeros: $\pm 1, \pm\frac{1}{2}, \pm\frac{1}{4}, \pm 5, \pm\frac{5}{2}, \pm\frac{5}{4}$

c. $P\left(\frac{1}{2}\right) = 0$ $\quad \frac{1}{2}$ has multiplicity 2

d. $P(x) = (2x - 1)^2(x^2 - 4x + 5)$

63.

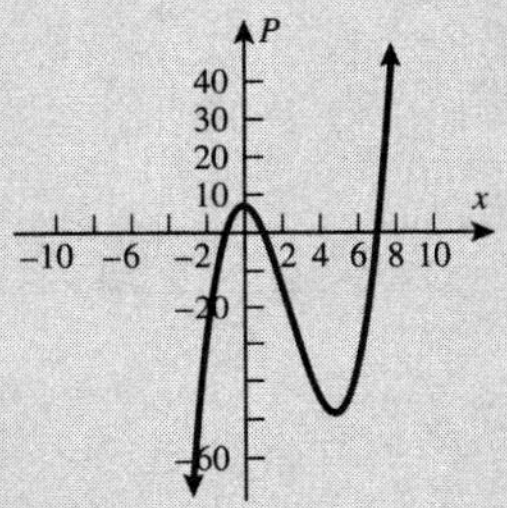

65.

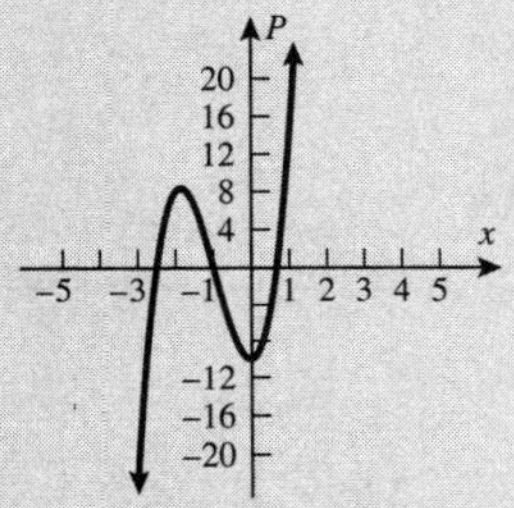

67. $x \approx 1.34$ **69.** $x \approx 0.22$ **71.** $x \approx -0.43$

73. $R(x) = 46 - 3x^2 \quad x \geq 0$
$C(x) = 20 + 2x$

a. $P(x) = R(x) - C(x)$
$= (46 - 3x^2) - (20 + 2x)$
$= -3x^2 - 2x + 26$

b. Approximately 263 subscribers will break even.

75. Yes, you can arrive at five negative zeros, but there may be 3 or 1.

Positive	Negative
0	5
0	3
0	1

77. true **79.** false **81.** $x = b$ and $x = c$

83. $x = 2$ **85. a.** $-\frac{3}{4}, \frac{2}{3}$
b. $(3x - 2)(4x + 3)(x^2 + 2x + 5)$

Section 4.5

1. $x = \pm 2i$; $P(x) = (x - 2i)(x + 2i)$

3. $x = 1 \pm i$; $P(x) = [x - (1 + i)][x - (1 - i)]$

5. $x = \pm 2, \pm 2i$; $P(x) = (x - 2)(x + 2)(x - 2i)(x + 2i)$

7. $x = \pm\sqrt{5}, \pm i\sqrt{5}$;
$P(x) = (x - \sqrt{5})(x + \sqrt{5})(x - i\sqrt{5})(x + i\sqrt{5})$

9. $-i$ **11.** $-2i, 3 + i$

13. $1 + 3i, 2 - 5i$ **15.** $i, 1 + i$ (multiplicity 2)

17. $2i, -3, 5$; $P(x) = (x - 5)(x + 3)(x - 2i)(x + 2i)$

19. $-i, 1, 3$; $P(x) = (x - 3)(x - 1)(x - i)(x + i)$

21. $3i$, 1 (multiplicity 2); $P(x) = (x - 1)^2(x - 3i)(x + 3i)$

23. $1 - i, -1 \pm 2\sqrt{2}$;
$P(x) = [x - (1 - i)][x - (1 + i)]\left[x - (-1 - 2\sqrt{2})\right]\left[x - (-1 + 2\sqrt{2})\right]$

25. $3 + i, \pm 2$; $P(x) = (x - 2)(x + 2)[x - (3 + i)][x - (3 - i)]$

27. $1, 4, 2 + i$; $P(x) = (x - 4)(x - 1)[x - (2 - i)][x - (2 + i)]$

29. $P(x) = (x - 1)(x - 3i)(x + 3i)$

31. $P(x) = (x - 5)(x - i)(x + i)$

33. $P(x) = (x + 1)(x - 2i)(x + 2i)$

35. $P(x) = (x - 3)\left[x - (-1 - i\sqrt{5})\right]\left[x - (-1 + i\sqrt{5})\right]$

37. $P(x) = (x - 5)(x + 3)(x - 2i)(x + 2i)$

39. $P(x) = (x - 5)(x + 1)(x - 2i)(x + 2i)$

41. $P(x) = (x - 1)(x - 2)[x - (2 - 3i)][x - (2 + 3i)]$

43. $P(x) = -(x - 2)(x + 1)[x - (-2 - i)][x - (-2 + i)]$

45. $P(x) = (x - 1)^2(x - 2i)(x + 2i)$

47. $P(x) = (x - 1)(x + 1)(x - 2i)(x + 2i)(x - 3i)(x + 3i)$

49. $P(x) = (2x - 1)^2[x - (2 - i)][x - (2 + i)]$

51. $P(x) = (3x - 2)(x - 1)(x + 1)(x - 2i)(x + 2i)$

53. Yes—since there are no real zeros, this function is either always positive or always negative; the end behavior is similar to an even power function with a positive leading coefficient (rising in both ends). We expect profit to eventually rise (without bound!)

55. No—since there is only one real zero (and it corresponds to producing a positive number of units) then that is the "break-even point;" the end behavior is similar to an odd power function with a negative leading coefficient (falling to the right). We expect profit to fall (without bound!) after the break-even point.

57. Just because 1 is a zero does not mean -1 is a zero (holds only for complex conjugates). Should have used synthetic division with 1.

59. false **61.** true

63. No—if there is an imaginary zero, then its conjugate is also a zero. Therefore, all imaginary zeros can only correspond to an even-degree polynomial.

65. $P(x) = x^6 + 3b^2x^4 + 3b^4x^2 + b^6$

67.

Real Zeros	Complex Zeros
4	0
2	2
0	4

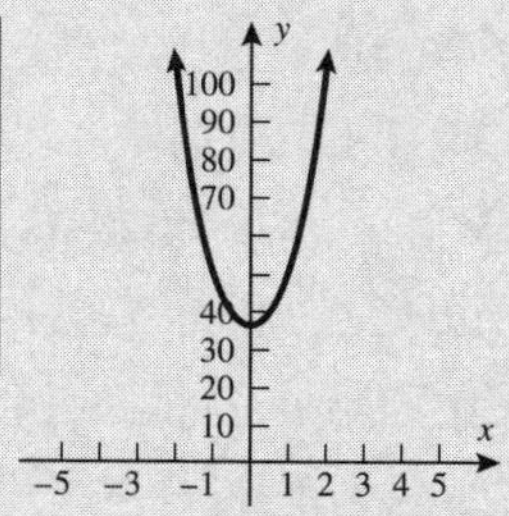

There are no real zeros; therefore there are 4 complex zeros.

69. $\frac{3}{5}, \pm i, \pm 2i; (3 - 5x)(x + i)(x - i)(x + 2i)(x - 2i)$

Section 4.6

1. $(-\infty, -3) \cup (-3, \infty)$ 3. $\left(-\infty, -\frac{1}{3}\right) \cup \left(-\frac{1}{3}, \frac{1}{2}\right) \cup \left(\frac{1}{2}, \infty\right)$

5. $(-\infty, -4) \cup (-4, 3) \cup (3, \infty)$ 7. $(-\infty, \infty)$

9. $(-\infty, -2) \cup (-2, 3) \cup (3, \infty)$

11. HA: $y = 0$ VA: $x = -2$

13. HA: none VA: $x = -5$

15. HA: none VA: $x = \frac{1}{2}$ and $x = -\frac{4}{3}$

17. HA: $y = \frac{1}{3}$ VA: none

19. HA: $y = \frac{12}{7}$ VA: $x = 0.5$ and $x = -1.5$

21. $y = x + 6$ 23. $y = 2x + 24$ 25. $y = 4x + \frac{11}{2}$

27. b 29. a 31. e

33.

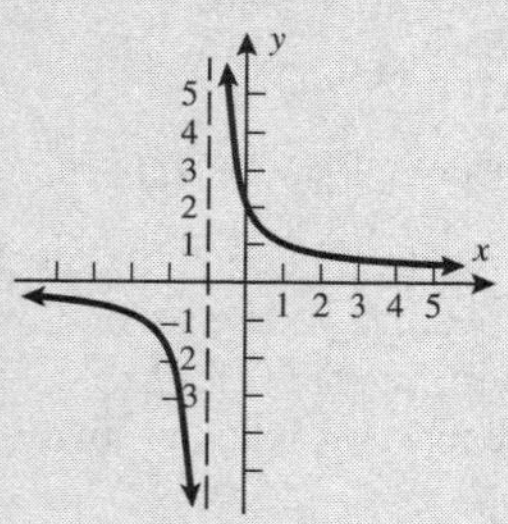

35.

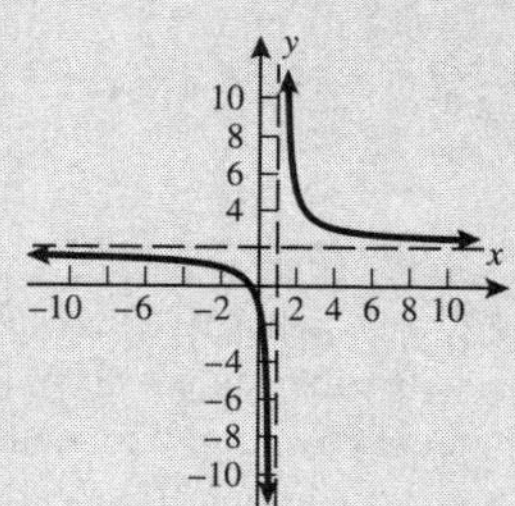

37.

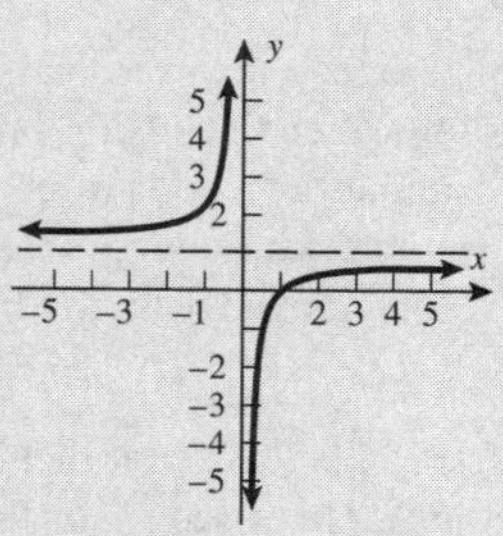

39.

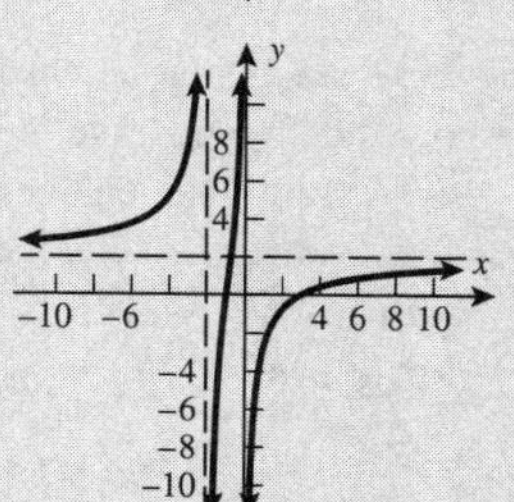

41.

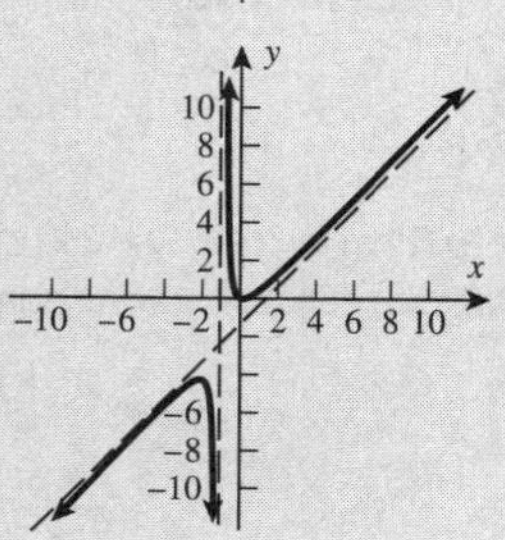

43.

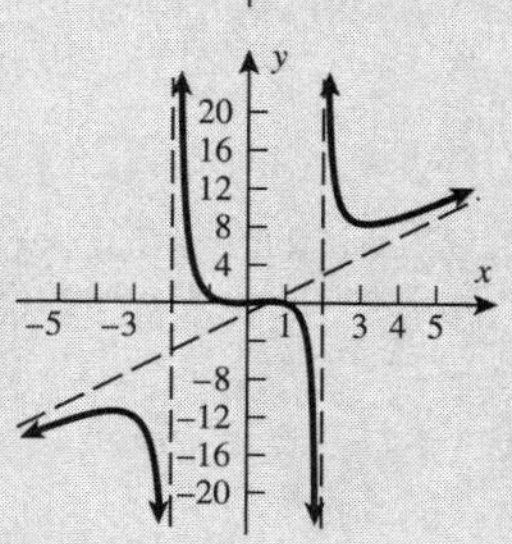

45.

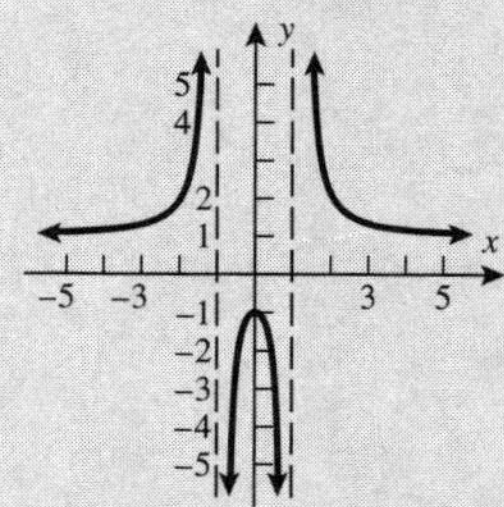

47.

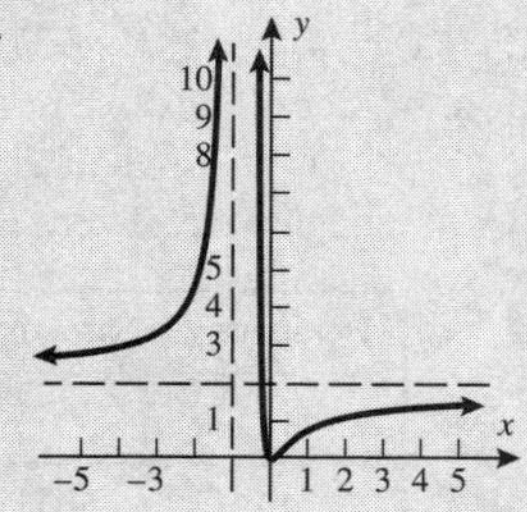

49.

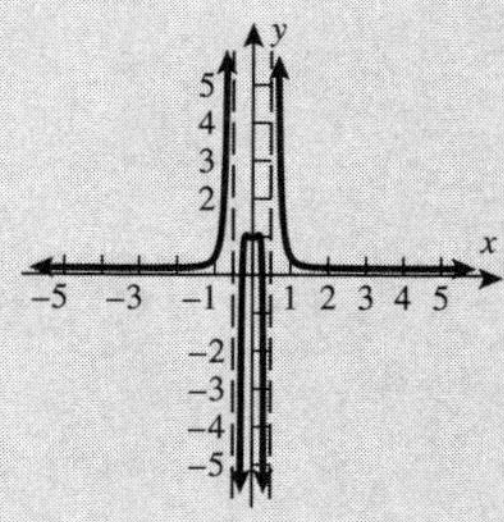

51.

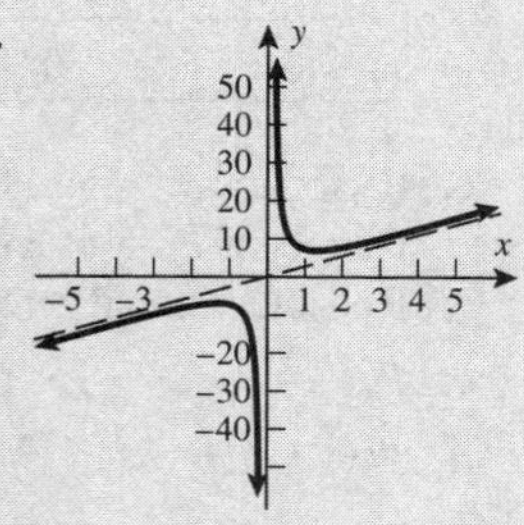

53.

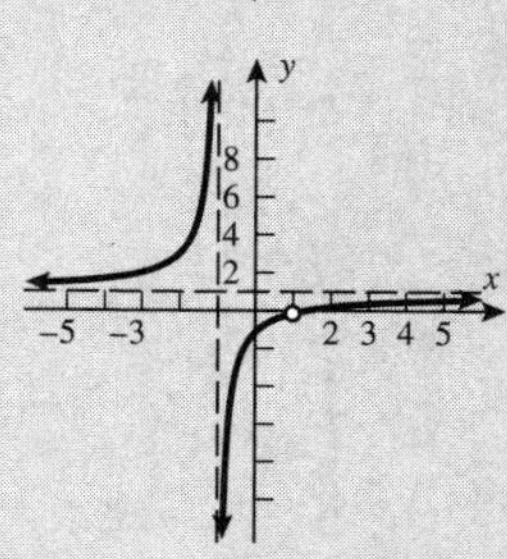

55.

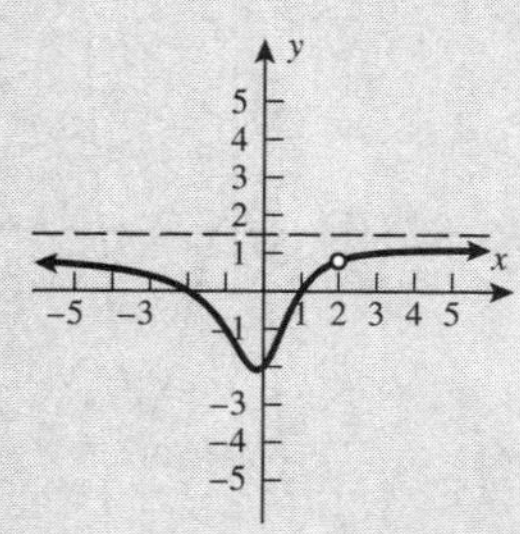

57.

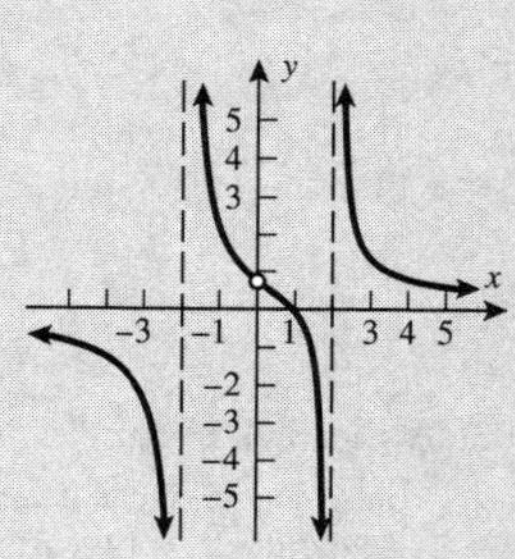

59. a. x-intercept: (2, 0); y-intercept: (0, 0.5)

b. HA: $y = 0$ VA: $x = -1, x = 4$

c. $f(x) = \dfrac{x - 2}{(x + 1)(x - 4)}$

61. a. x-intercept: (0, 0); y-intercept: (0, 0)

b. HA: $y = -3$ VA: $x = -4, x = 4$

c. $f(x) = \dfrac{-3x^2}{(x + 4)(x - 4)}$

63. a. $C(1) \approx 0.0198$

b. $C(60) \approx 0.0324$

c. $C(300) \approx 0.007 \approx 0$

d. $y = 0$, after several days $C(t) \approx 0$.

65. a. $N(0) = 52$ wpm b. $N(12) = 107$ wpm

c. $N(36) = 120$ wpm d. 130 wpm

67. 10 ounces

69. $\dfrac{2w^2 + 1000}{w}$

71. There is a common factor in the numerator and denominator. There is a hole at $x = 1$.

73. The horizontal asymptote was found incorrectly. The correct horizontal asymptote is $y = -1$.

75. true

77. false

79. HA: $y = 1$ VA: $x = c$ and $x = -d$

81. Possible answer: $y = \dfrac{4x^2}{(x + 3)(x - 1)}$

83. yes

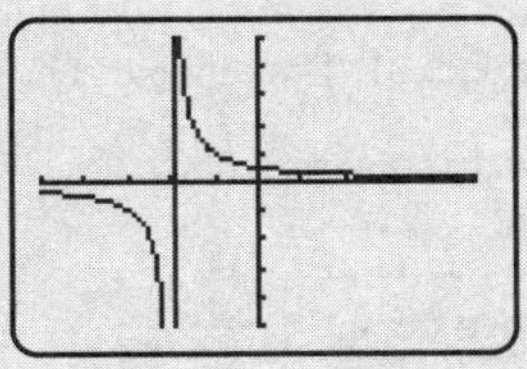

85. yes

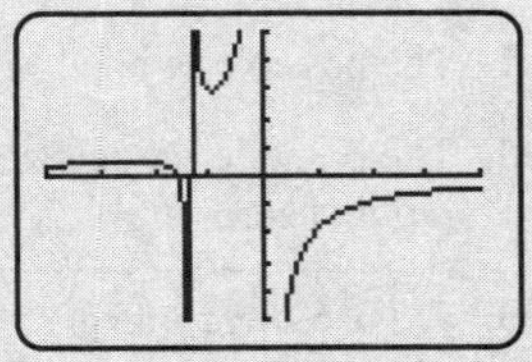

87. a. For f, VA: $x = 3$, HA: $y = 0$; for g, VA: $x = 3$, HA: $y = 2$; for h, VA: $x = 3$, HA: $y = -3$

b. As $x \to -\infty$ or $x \to \infty$, $f(x) \to 0$ and $g(x) \to 2$

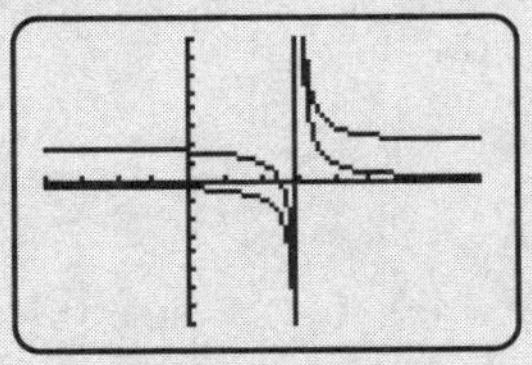

c. As $x \to -\infty$ or $x \to \infty$, $g(x) \to 2$ and $h(x) \to -3$

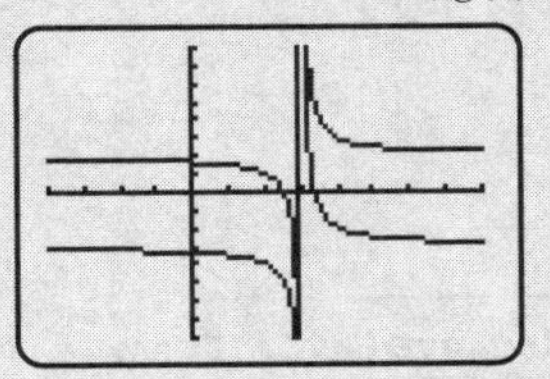

d. $g(x) = \dfrac{2x - 5}{x - 3}$, $h(x) = \dfrac{-3x + 10}{x - 3}$. Yes, if the degree of the numerator is the same as the degree of the denominator, then the horizontal asymptote is the ratio of the leading coefficients for both g and h.

Review Exercises

1. b

3. a

5.

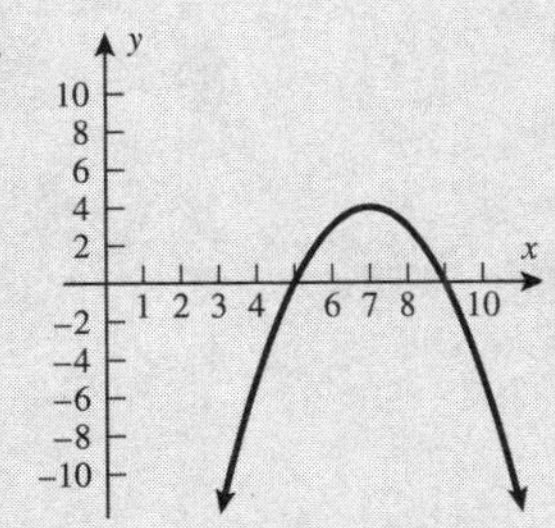

7.

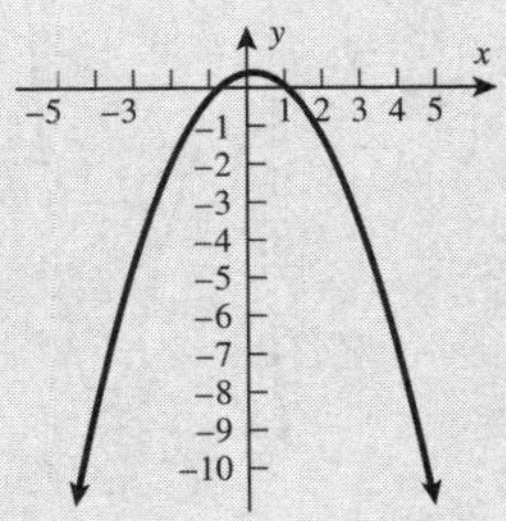

9. $f(x) = \left(x - \frac{3}{2}\right)^2 - \frac{49}{4}$

11. $f(x) = 4(x + 1)^2 - 11$

13.

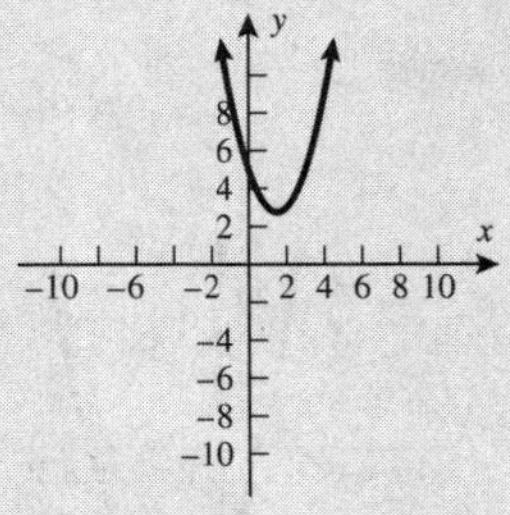

15.

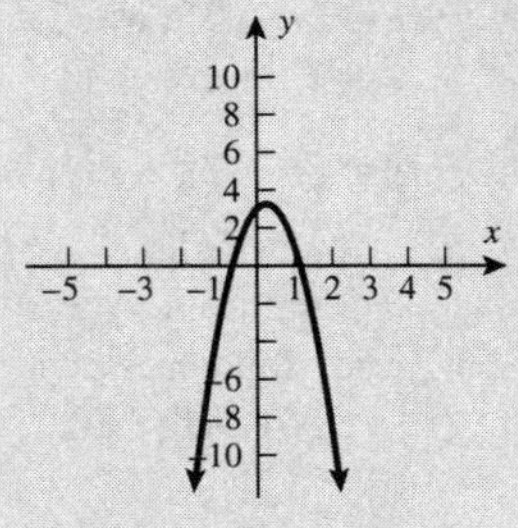

17. vertex: $\left(\frac{5}{26}, \frac{599}{52}\right)$

19. vertex: $\left(-\frac{2}{15}, \frac{451}{125}\right)$

21. $y = \frac{1}{9}(x + 2)^2 + 3$

23. $y = 5.6(x - 2.7)^2 + 3.4$

25. a. $P(x) = -2x^2 + \frac{35}{3}x - 14$

b. $x \approx 1.68909$ and $x \approx 4.1442433$

c.

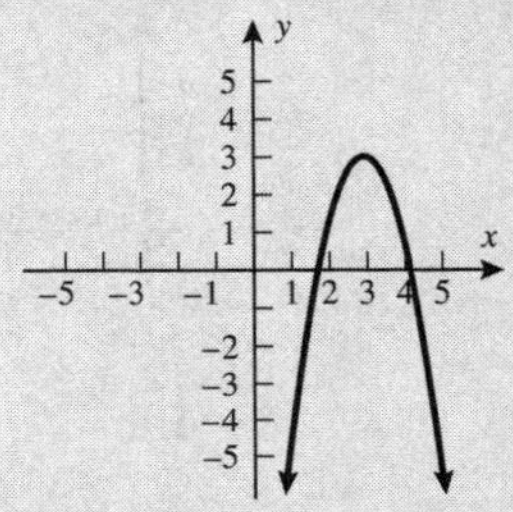

d. The range of units is (1.689, 4.144) or 1689 to 4144.

27. $A(x) = -\frac{1}{2}x^2 + x + 4$
$= -\frac{1}{2}(x - 1)^2 + \frac{9}{2}$

$x = 1$ is a maximum (4.5 square units).

Dimensions: base = 3 height = 3

29. yes, 6

31. no

33. d

35. a

37.

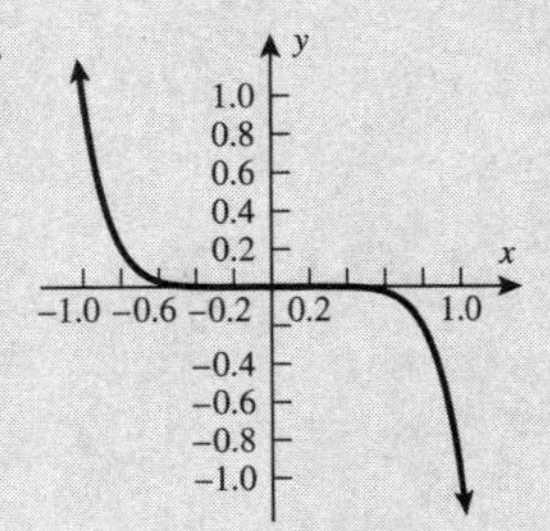

39.

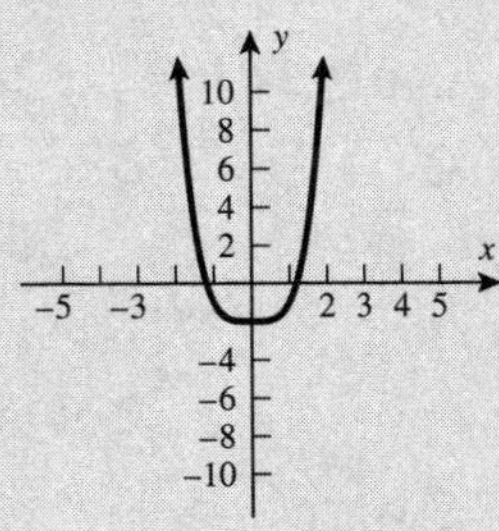

41. $x = -4$ multiplicity 2
$x = 6$ multiplicity 5

43. $x = 0$ multiplicity 1
$x = 3$ multiplicity 1
$x = -3$ multiplicity 1
$x = 2$ multiplicity 1
$x = -2$ multiplicity 1

45. $f(x) = x(x + 3)(x - 4)$

47. $f(x) = x(5x + 2)(4x - 3)$

49. $f(x) = (x + 2)^2(x - 3)^2$
$= x^4 - 2x^3 - 11x^2 + 12x + 36$

51. $f(x) = (x - 7)(x + 2)$

a. 7 multiplicity 1
-2 multiplicity 1

b. The graph crosses at $(-2, 0)$ and $(7, 0)$.

c. $(0, -14)$

d. rises right and left

e.

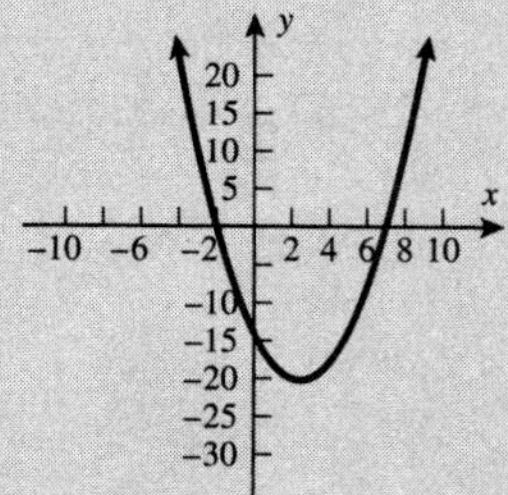

53. $f(x) = 6x^7 + 3x^5 - x^2 + x - 4$

a. 0.8748 multiplicity 1

b. The graph crosses at $(0.8748, 0)$.

c. $(0, -4)$

d. falls left; rises right

e.

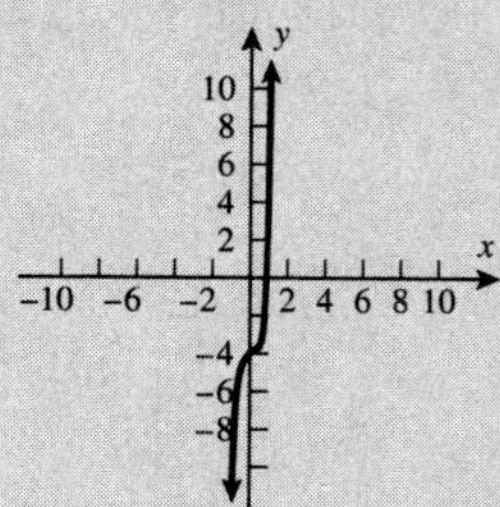

55. **a.**

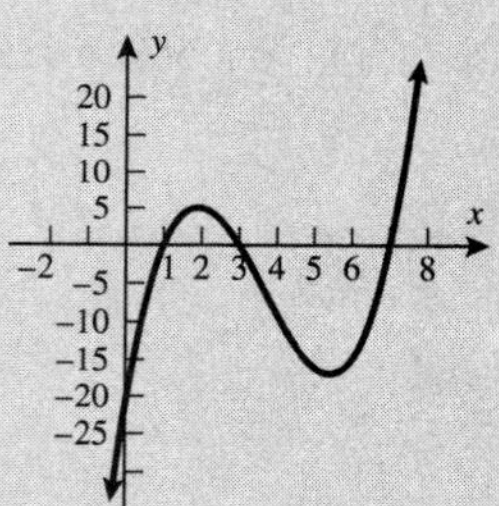

b. The real zeros occur at $x = 1$, $x = 3$, and $x = 7$.

c. Between 1 and 3 hours or more than 7 hours is financially beneficial.

57. $Q(x) = x + 4 \qquad r(x) = 2$

59. $Q(x) = 2x^3 - 4x^2 - 2x - \frac{7}{2} \qquad r(x) = -23$

61. $Q(x) = x^3 + 2x^2 + x - 4 \qquad r(x) = 0$

63. $Q(x) = x^5 - 8x^4 + 64x^3 - 512x^2 + 4096x - 32{,}768$
$r(x) = 262{,}080$

65. $Q(x) = x + 3 \qquad r(x) = -4x - 8$

67. $Q(x) = x^2 - 5x + 7 \qquad r(x) = -15$

69. length $= 3x^3 + 2x^2 - x + 4 \qquad r(x) = 0$

71. $f(-2) = -207$

73. $g(1) = 0$

75. no

77. yes

79. $P(x) = x(x + 2)(x - 4)^2$

81. $P(x) = x^2(x + 3)(x - 2)^2$

83.

Positive (Real)	Negative (Real)
1	1

85.

Positive (Real)	Negative (Real)
5	2
5	0
3	2
3	0
1	2
1	0

87. The possible rational zeros are: $\pm 1, \pm 2, \pm 3, \pm 6$.

89. The possible rational zeros are: $\pm\frac{1}{2}, \pm 1, \pm 2, \pm 4, \pm 8, \pm 16, \pm 32, \pm 64$.

91. The possible rational zeros are $\pm\frac{1}{2}, \pm 1$. The rational zero is $\frac{1}{2}$.

93. The possible rational zeros are $\pm 1, \pm 2, \pm 4, \pm 8, \pm 16$. The rational zeros are $1, 2, -2, 4$.

95. **a.**

Positive (Real)	Negative (Real)
1	0

b. $\pm 1, \pm 5$

c. -1 is a lower bound, 5 is an upper bound

d. There are no rational zeros.

e. not possible

f.

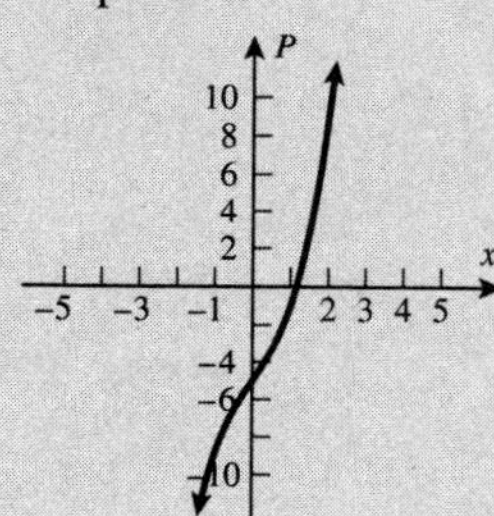

97. **a.**

Positive (Real)	Negative (Real)
3	0
1	0

b. $\pm 1, \pm 2, \pm 3, \pm 4, \pm 6, \pm 12$

c. -4 lower bound; 12 upper bound

d. 1, 2, and 6 are zeros

e. $P(x) = (x - 1)(x - 2)(x - 6)$

f.

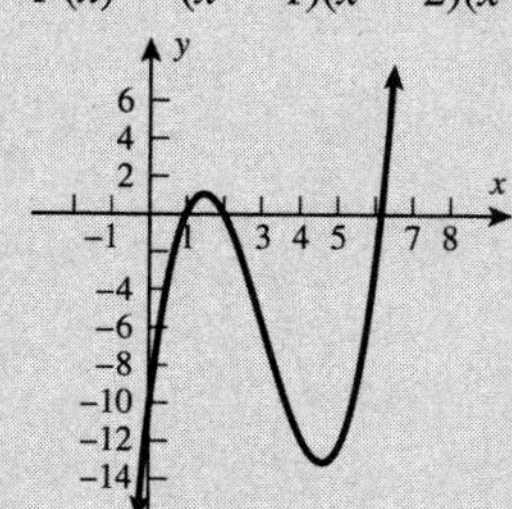

99. **a.**

POSITIVE (REAL)	NEGATIVE (REAL)
2	2
0	2
2	0
0	0

b. $\pm 1, \pm 2, \pm 3, \pm 4, \pm 6, \pm 8 \pm 12, \pm 24$

c. -4 lower bound; 8 upper bound

d. $-2, -1, 2,$ and 6 are zeros

e. $P(x) = (x + 1)(x + 2)(x - 2)(x - 6)$

f.

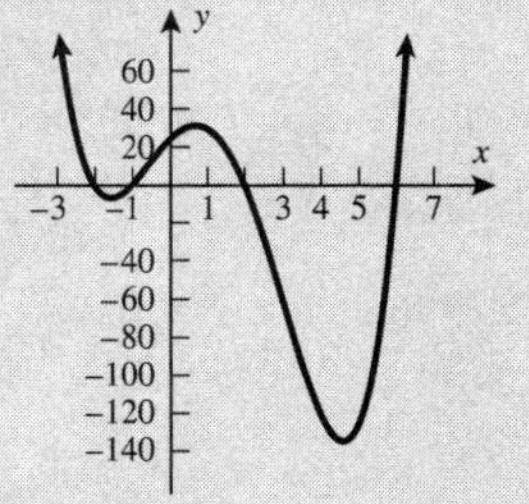

101. $P(x) = (x - 5i)(x + 5i)$

103. $P(x) = [x - (1 - 2i)][x - (1 + 2i)]$

105. $2i$ and $3 - i$

107. $-i$ and $2 + i$ (multiplicity 2)

109. $4, -1, -i \quad P(x) = (x - 4)(x + 1)(x - i)(x + i)$

111. $3i, 1 \pm i \quad P(x) = (x + 3i)(x - 3i)[x - (1 - i)][x - (1 + i)]$

113. $P(x) = (x - 3)(x + 3)(x - 3i)(x + 3i)$

115. $P(x) = (x - 1)(x - 2i)(x + 2i)$

117. VA: $x = -2$ and HA: $y = -1$

119. VA: $x = -1$, HA: none, slant: $y = 4x - 4$

121. VA: none, HA: $y = 2$

123.

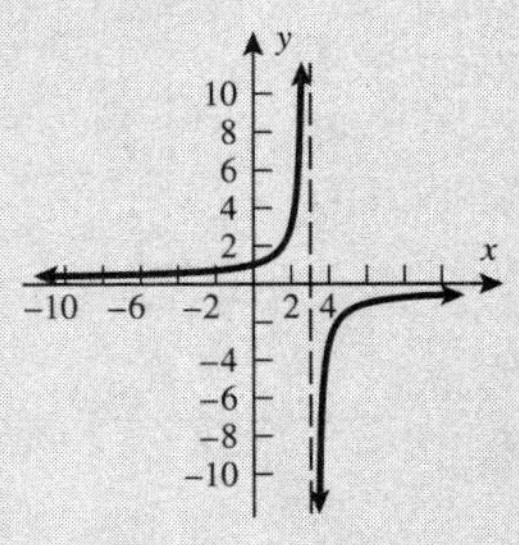

125.

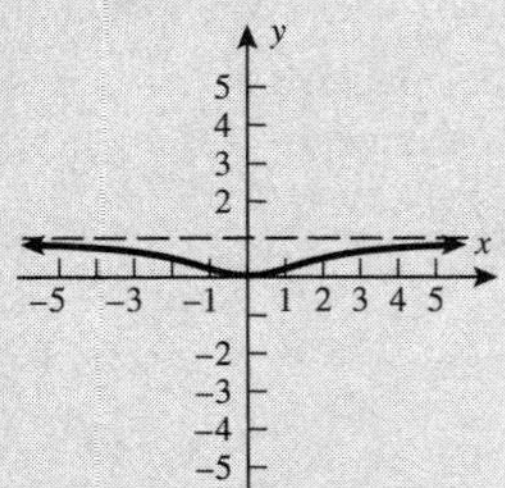

127.

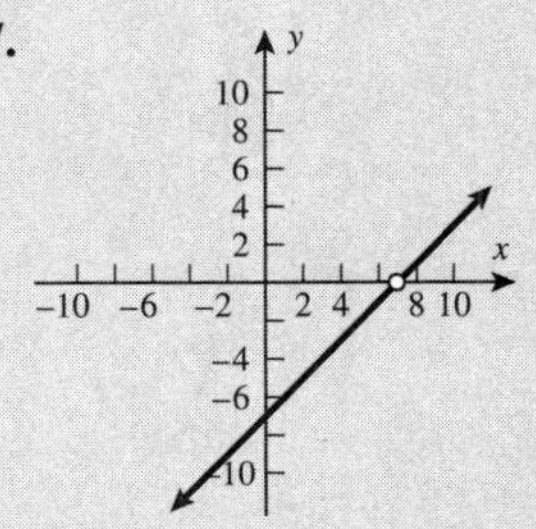

129. **a.** $(480, -1211)$
b. $(0, -59)$
c. $(-12.14, 0)$ and $(972.14, 0)$
d. $x = 480$

131. x-intercepts: $(-1, 0)$, $(0.4, 0)(2.8, 0)$; zeros at $-1, 0.4, 2.8$, each with multiplicity 1.

133. A linear function.

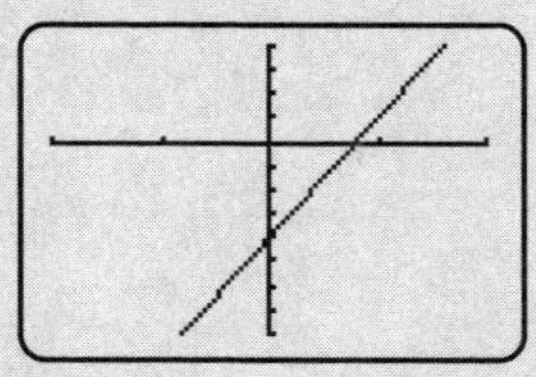

135. **a.** $-2, 3, 4$ **b.** $P(x) = (x + 2)^2(x - 3)(x - 4)$

137. $\frac{7}{2}, -2 \pm 3i$; $P(x) = (2x - 7)(x + 2 + 3i)(x + 2 - 3i)$

139. **a.** yes, it is one-to-one **b.** $f^{-1}(x) = \dfrac{x + 3}{2 - x}$

c.

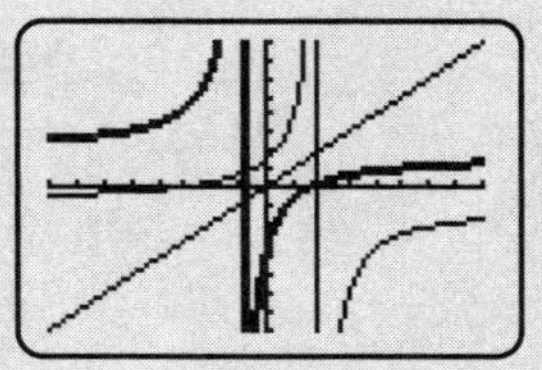

Practice Test

1.

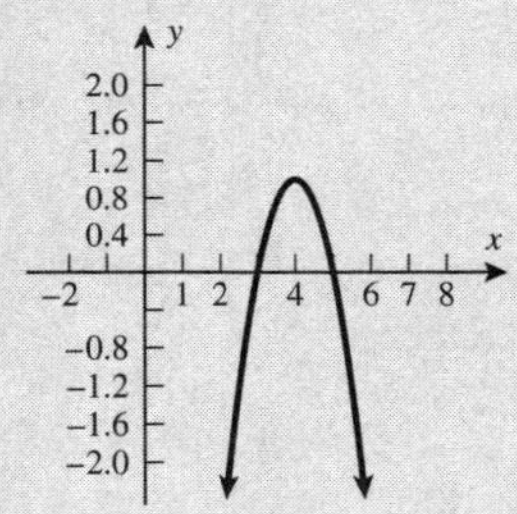

3. vertex $\left(3, \frac{1}{2}\right)$

5. $f(x) = x(x - 2)^3(x - 1)^2$

7. $Q(x) = -2x^2 - 2x - \frac{11}{2}$

$r(x) = -\frac{19}{2}x + \frac{7}{2}$

9. Yes, $x - 3$ is a factor of the polynomial.

11. $P(x) = (x - 7)(x + 2)(x - 1)$

13. Yes. Complex zeros.

15. $P(x) = 3x^4 - 7x^2 + 3x + 12$
Factors of $a_0 = 12$: $\pm 1, \pm 2, \pm 3, \pm 4, \pm 6, \pm 12$
Factors of $a_n = 3$: $\pm 1, \pm 3$
Possible rational zeros: $\pm 1, \pm 2, \pm 3, \pm 4, \pm 6, \pm 12, \pm\frac{1}{3}, \pm\frac{2}{3}, \pm\frac{4}{3}$

17. $\frac{3}{2}, \pm 2i$

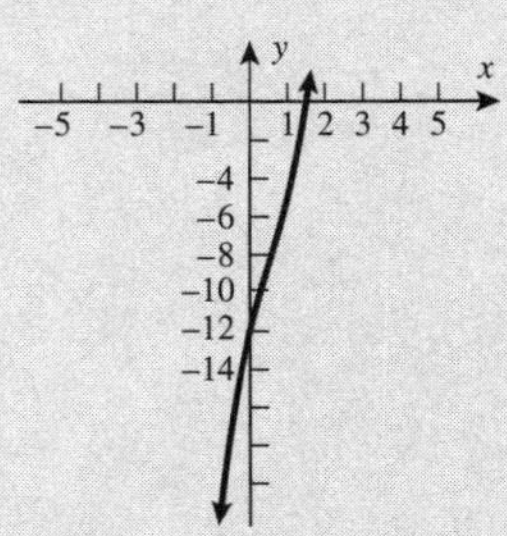

19. Given the points (0, 300), (2, 285), (10, 315) (52, 300), you can have a polynomial of degree 3 because there are 2 turning points

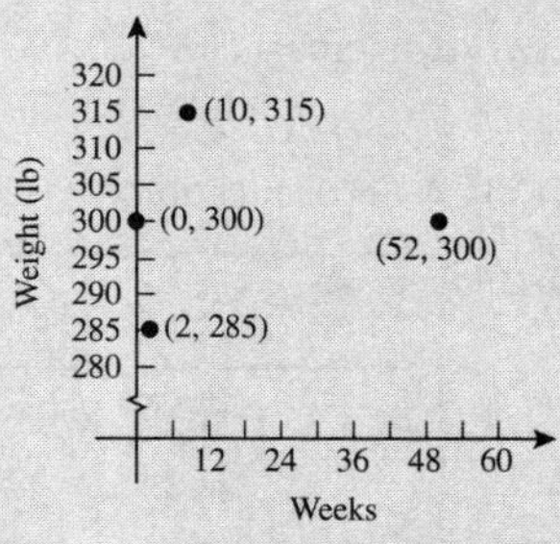

21. Given the points (1970, 0.08), (1988, 0.13), (2002, 0.04), (2005, 0.06), the lowest degree polynomial that can be represented is a third-degree polynomial.

23. **a.** (0, 0) is both x-intercept and y-intercept.
b. VA: $x = \pm 2$
c. HA: $y = 0$
d. none
e.

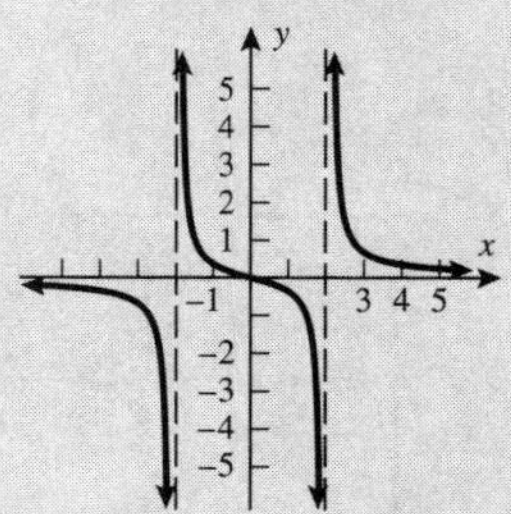

25. **a.** x-intercept: (3, 0); y-intercept: $\left(0, \frac{3}{8}\right)$
b. VA: $x = -2; x = 4$
c. HA: $y = 0$
d. None
e.

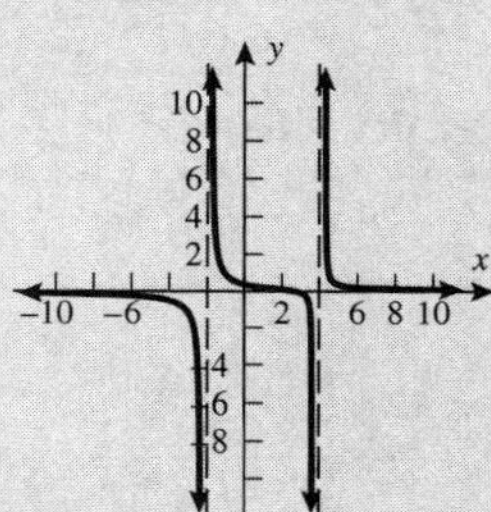

27. **a.** $x^2 - 3x - 7.99$
b. $y = (x - 1.5)^2 - 10.24$
c. $(-1.7, 0)$ and $(4.7, 0)$
d. Yes, they agree

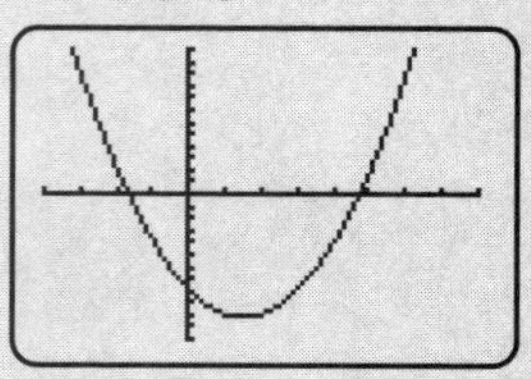

Cumulative Test

1. $\dfrac{5}{4x^7y^{10}}$ **3.** $4x - 12$

5. $33\frac{1}{3}$ minutes **7.** $x = 3$

9. $y = \frac{1}{3}x - \frac{1}{3}$ **11.** $x^2 + (y - 6)^2 = 2$

13. Neither

15. Right 1 unit; up 3 units

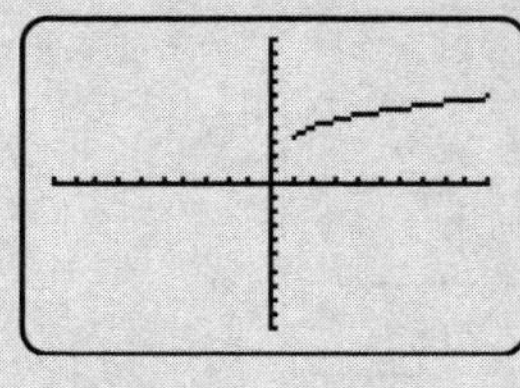

17. $g(f(-1)) = 0$

19. $f(x) = (x + 2)^2 + 3$

21. $Q(x) = 4x^2 + 4x + 1,\ r(x) = -8$

23. Factors of $a_0 = 6$: $\pm1, \pm2, \pm3, \pm6$
Factors of $a_n = 12$: $\pm1, \pm2, \pm3, \pm4, \pm6, \pm12$
Possible rational zeros: $\pm1, \pm2, \pm3, \pm4, \pm6, \pm\frac{1}{2}, \pm\frac{3}{2}, \pm\frac{1}{3}, \pm\frac{2}{3}, \pm\frac{1}{4}, \pm\frac{3}{4}, \pm\frac{1}{6}, \pm\frac{1}{12}$
Testing the zeros: $P(-2) = 0, P\left(-\frac{3}{4}\right) = 0, P\left(\frac{1}{3}\right) = 0$

25. HA: $y = 0$; VA: $x = -2$ and $x = 2$

27. x-intercepts $(-1, 0)$; vertical asymptotes: $x = 0, x = \frac{3}{2}$; horizontal asymptote: $y = 0$; Yes

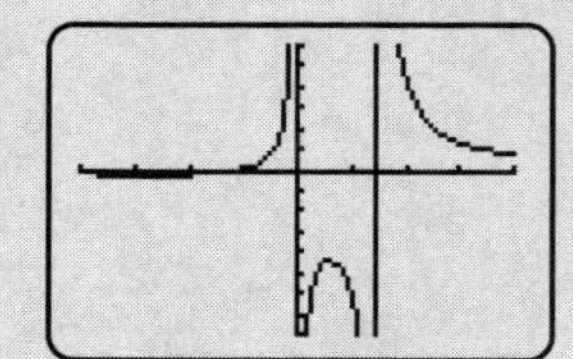

CHAPTER 5

Section 5.1

1. 16	**3.** $\frac{1}{25}$	**5.** 4	**7.** 27
9. -1	**11.** 5.2780	**13.** 9.7385	**15.** 7.3891
17. 0.0432	**19.** 27	**21.** 16	**23.** 4
25. 19.81	**27.** f	**29.** e	**31.** b

33. (0, 1), $\left(-1, \frac{1}{6}\right)$, and (1, 6);
Domain: $(-\infty, \infty)$;
Range: $(0, \infty)$;
HA: $y = 0$

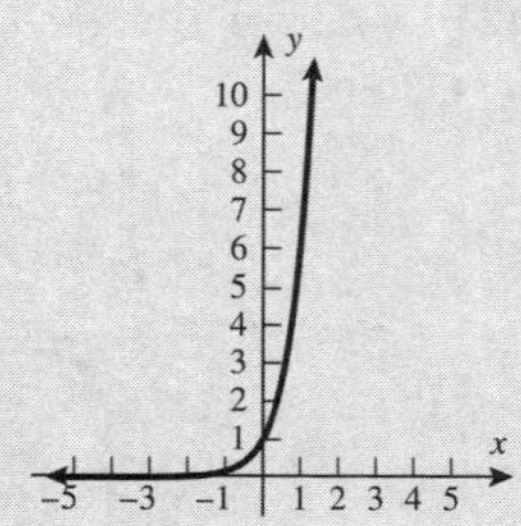

35. (0, 1), $(-1, 10)$, and (1, 0.1)
domain: $(-\infty, \infty)$
range: $(0, \infty)$
HA: $y = 0$

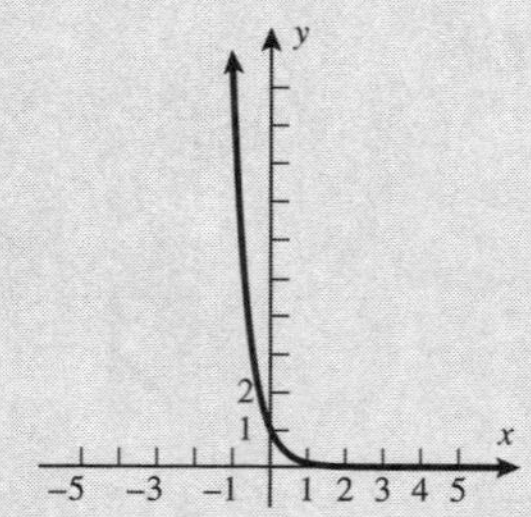

37. $(0, 1), (-1, e)$, and $\left(1, \frac{1}{e}\right)$
domain: $(-\infty, \infty)$
range: $(0, \infty)$
HA: $y = 0$

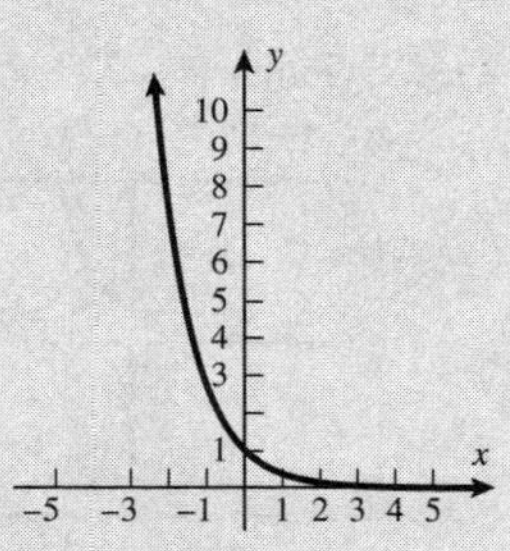

39. $(0, 0), (1, 1)$, and $(2, 3)$
domain: $(-\infty, \infty)$
range: $(-1, \infty)$
HA: $y = -1$

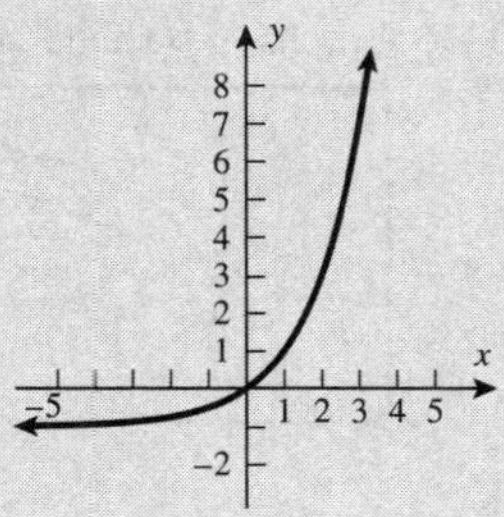

41. $(0, 1), \left(-1, 2 - \frac{1}{e}\right)$, and $(1, 2 - e)$
domain: $(-\infty, \infty)$
range: $(-\infty, 2)$
HA: $y = 2$

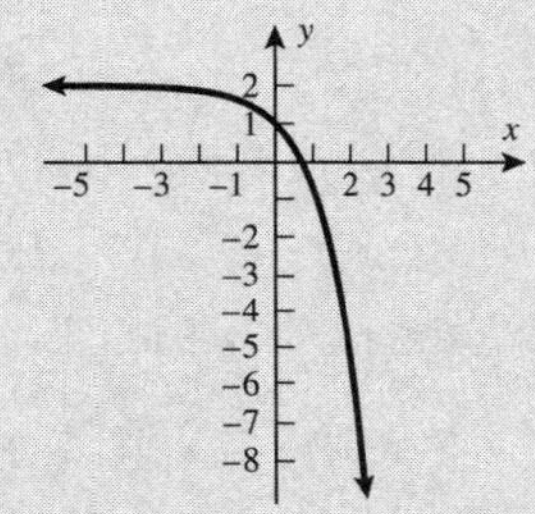

43. $(0, e - 4), (-1, -3)$, and $(1, e^2 - 4)$
domain: $(-\infty, \infty)$
range: $(-4, \infty)$
HA: $y = -4$

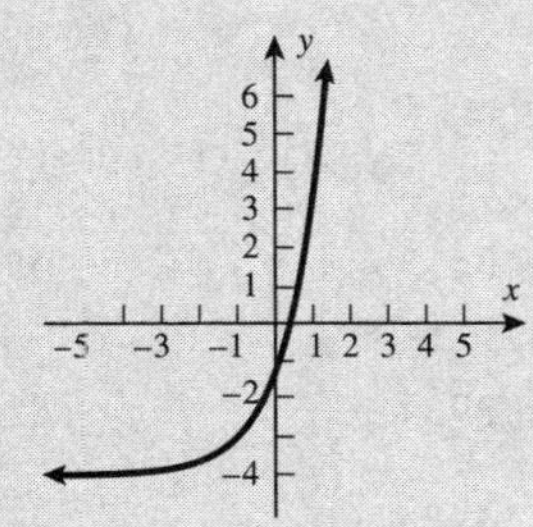

45. $(0, 3), \left(-1, \frac{3}{\sqrt{e}}\right)$, and $(1, 3\sqrt{e})$
domain: $(-\infty, \infty)$
range: $(0, \infty)$
HA: $y = 0$

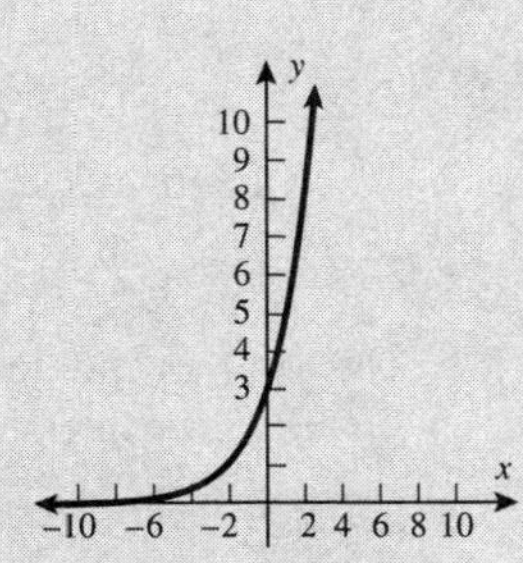

47. $(0, 5), (-1, 9)$, and $(1, 3)$
domain: $(-\infty, \infty)$
range: $(1, \infty)$
HA: $y = 1$

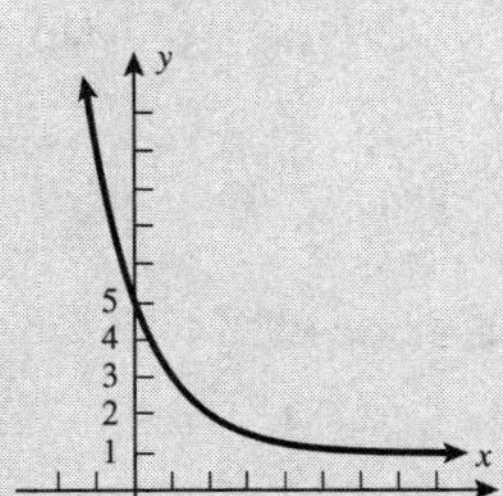

49. 10.4 million

51. $f(t) = 1500 \cdot 2^{t/5}$ where t is the number of years since it was purchased. Thirty years later: \$96,000/acre

53. 168 milligrams **55.** \$3,031

57. \$3,448.42 **59.** \$13,011.03

61. \$4,319.55 **63.** \$13,979.42

65. $4^{-1/2} = \frac{1}{4^{1/2}} = \frac{1}{2}$

67. 2.5% needs to be converted to a decimal, 0.025

69. false **71.** true

73.

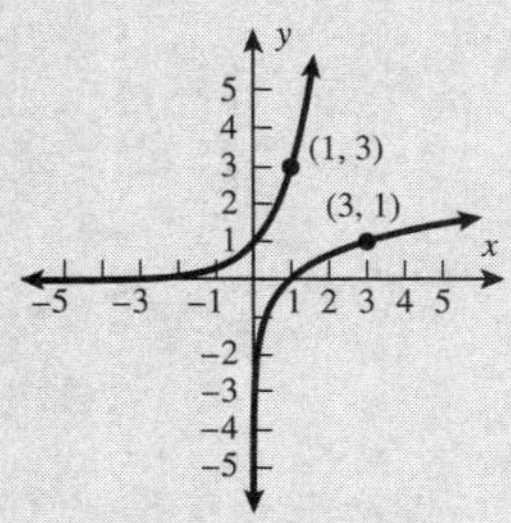

75.

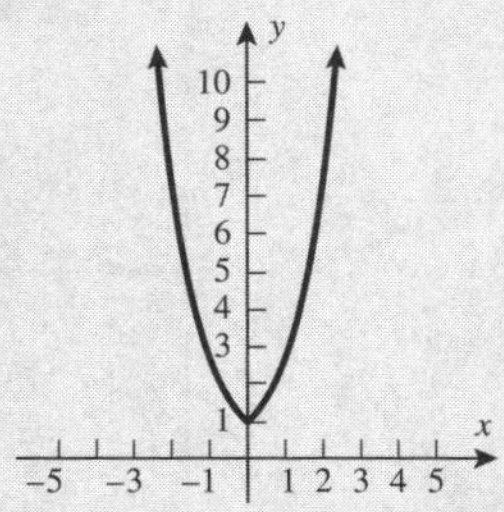

77. $(0, be - a)$ and $y = -a$

79. domain: $(-\infty, \infty)$

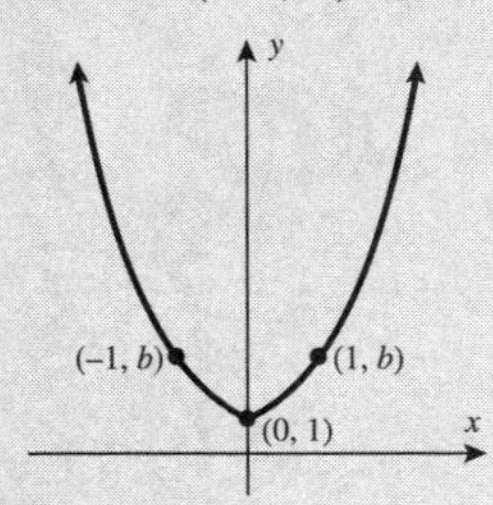

81. HA: $y = e$

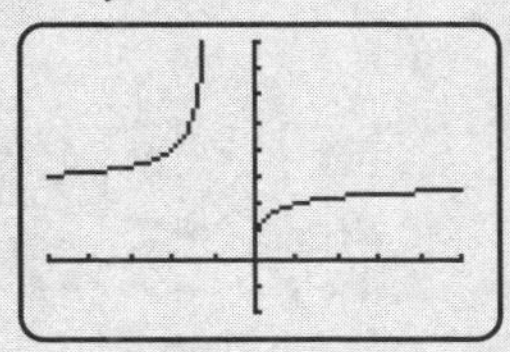

83. These graphs are very similar.

85. HA: $y = e, y = e^2, y = e^4$

Section 5.2

1. $5^3 = 125$ **3.** $81^{1/4} = 3$ **5.** $2^{-5} = \frac{1}{32}$

7. $10^{-2} = 0.01$ **9.** $10^4 = 10{,}000$ **11.** $\left(\frac{1}{4}\right)^{-3} = 64$

13. $e^{-1} = \frac{1}{e}$ **15.** $e^0 = 1$ **17.** $e^x = 5$

19. $x^z = y$ **21.** $\log_8 512 = 3$ **23.** $\log 0.00001 = -5$

25. $\log_{225} 15 = \frac{1}{2}$ **27.** $\log_{2/5}\left(\frac{8}{125}\right) = 3$ **29.** $\log_{1/27} 3 = -\frac{1}{3}$

31. $x = \ln 6$ **33.** $\log_y x = z$ **35.** 0

37. 5 **39.** 7 **41.** -6

43. undefined **45.** undefined **47.** 1.46

49. 5.94 **51.** undefined **53.** -8.11

55. $(-5, \infty)$ **57.** $\left(-\infty, \frac{5}{2}\right)$ **59.** $\left(-\infty, \frac{7}{2}\right)$

61. $(-\infty, 0) \cup (0, \infty)$ **63.** domain: $\mathbb{R}$

65. b **67.** c **69.** d

71. domain: $(1, \infty)$
range: $(-\infty, \infty)$

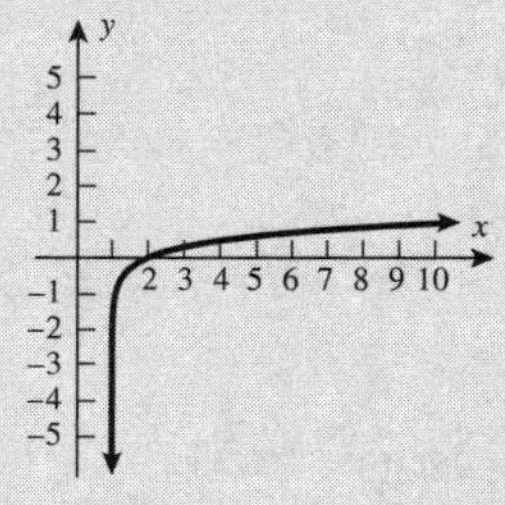

73. domain: $(0, \infty)$
range: $(-\infty, \infty)$

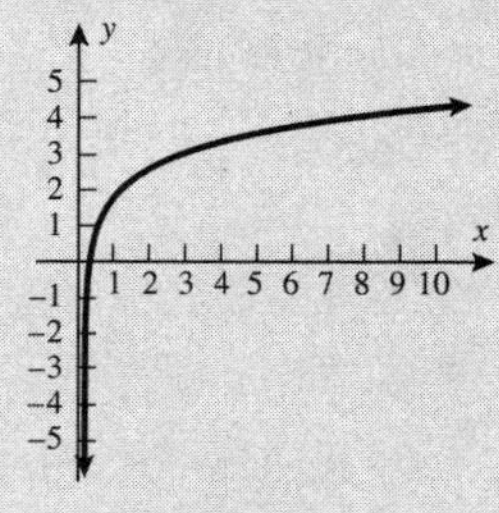

75. domain: $(-2, \infty)$
range: $(-\infty, \infty)$

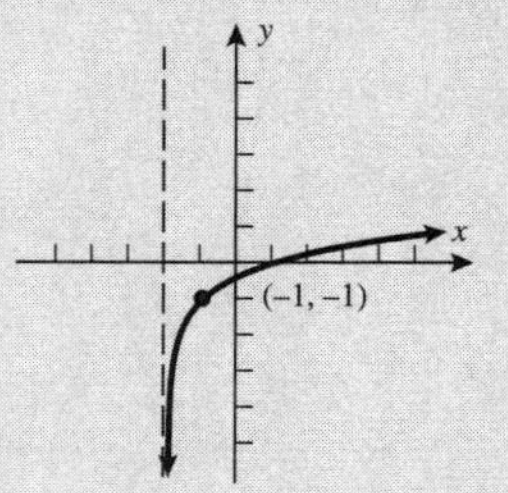

77. domain: $(0, \infty)$
range: $(-\infty, \infty)$

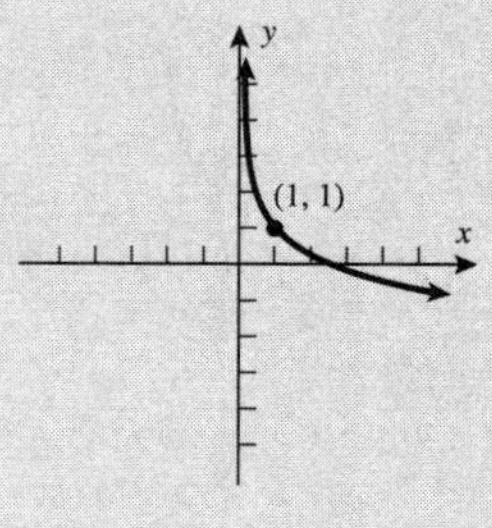

79. domain: $(-4, \infty)$
range: $(-\infty, \infty)$

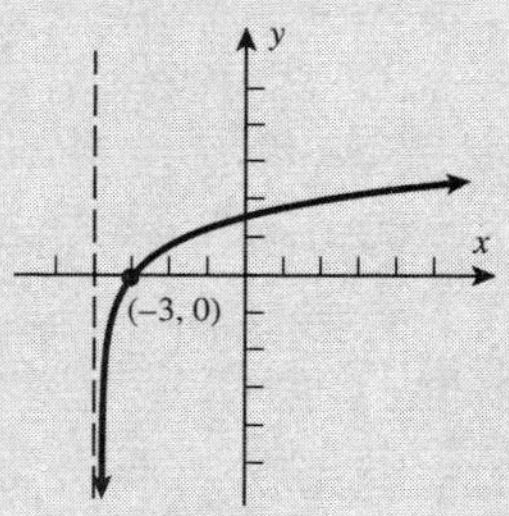

81. domain: $(0, \infty)$
range: $(-\infty, \infty)$

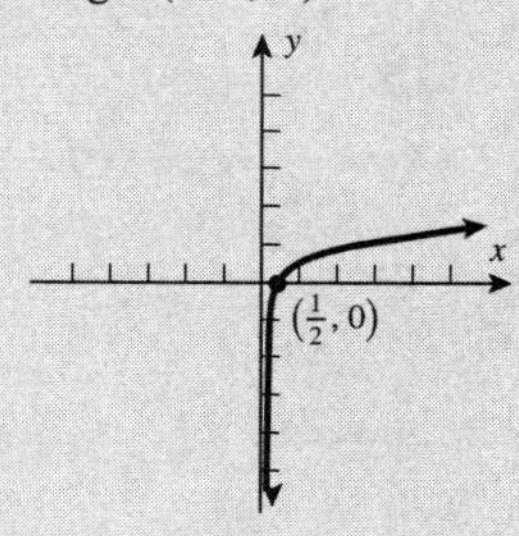

83. 60 decibels **85.** 117 decibels

87. 8.5 on the Richter scale **89.** 6.6 on the Richter scale

91. 3.3 pH

93. normal rainwater: 5.6 pH; acid rain/tomato juice: 4.0 pH

95. 3.6 pH **97.** 13,236 years old **99.** 25 dB loss

101. $\log_2 4 = x$ is equivalent to $2^x = 4$ (not $x = 2^4$)

103. The correction that needs to be made is:
$x + 5 > 0$
$x > -5$
domain: $(-5, \infty)$

105. false **107.** true

109. domain: (a, ∞)
range: $(-\infty, \infty)$
x-intercept: $(e^b + a, 0)$

111.

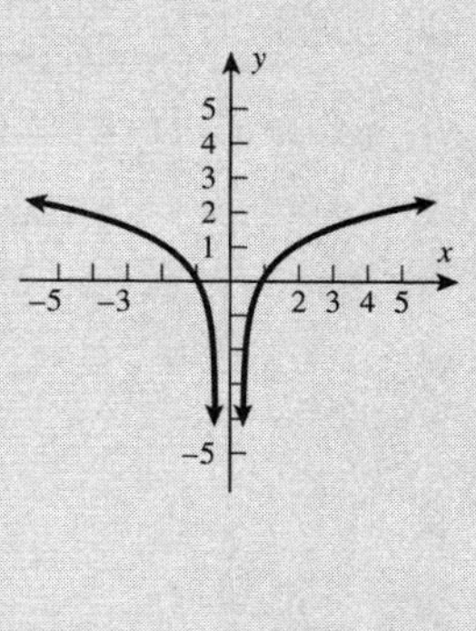

113. Symmetric with respect to $y = x$.

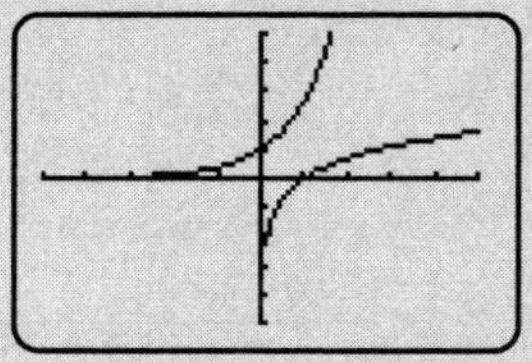

115. $(1, 0)$ and V.A. $x = 0$ are common.

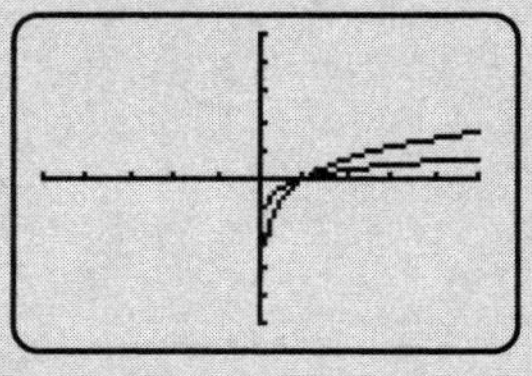

117. Graphs of $f(x)$ and $g(x)$ are the same with domain $(0, \infty)$.

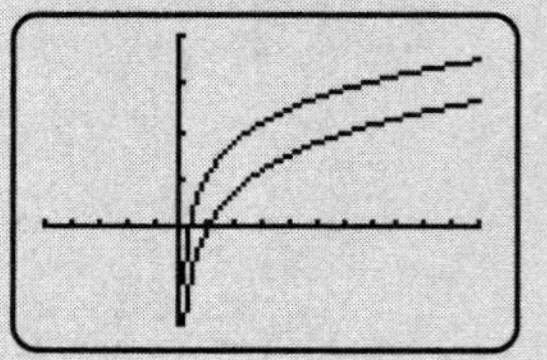

Section 5.3

1. 0 **3.** 1 **5.** 8

7. -3 **9.** $\frac{3}{2}$ **11.** 5

13. $x + 5$ **15.** 8 **17.** $\frac{1}{9}$

19. $3 \log_b x + 5 \log_b y$ **21.** $\frac{1}{2} \log_b x + \frac{1}{3} \log_b y$

23. $\frac{1}{3} \log_b r - \frac{1}{2} \log_b s$ **25.** $\log_b x - \log_b y - \log_b z$

27. $2 \log x + \frac{1}{2} \log(x + 5)$

29. $3 \ln x + 2 \ln(x - 2) - \frac{1}{2} \ln(x^2 + 5)$

31. $2 \log(x - 1) - \log(x - 3) - \log(x + 3)$

33. $\log_b x^3 y^5$ **35.** $\log_b \frac{u^5}{v^2}$ **37.** $\log_b x^{1/2} y^{2/3}$

39. $\log \frac{u^2}{v^3 z^2}$ **41.** $\ln \frac{x^2 - 1}{(x^2 + 3)^2}$ **43.** $\ln \frac{(x + 3)^{1/2}}{x(x + 2)^{1/3}}$

45. 1.2091 **47.** -2.3219 **49.** 1.6599

51. 2.0115 **53.** 3.7856 **55.** 110 decibels

57. 5.5 on the Richter scale (Total energy: 4.5×10^{12} joules)

59. $3 \log 5 - \log 5^2 = 3 \log 5 - 2 \log 5 = \log 5$

61. The bases are different, they must be the same in order to simplify the expression.

63. true **65.** false

69. $6 \log_b x - 9 \log_b y + 15 \log_b z$

71. yes **73.** no

75. no **77.** yes

Section 5.4

1. $x = 4$ **3.** $x = -2$ **5.** $x = \pm 2$

7. $x = -4$ **9.** $x = -\frac{3}{2}$ **11.** $x = -1$

13. $x = 3$ or $x = 4$ 15. $x = 0$ or $x = 6$

17. $x = 1$ or $x = 4$ 19. $x \approx 2.454$ 21. $x \approx 0.465$

23. $x \approx 1.918$ 25. $x \approx 1.609$ 27. $x \approx 13.863$

29. $x \approx 2.096$ 31. $x \approx -0.303$ 33. $x \approx 0.896$

35. $x \approx -0.904$ 37. $x = 0$ 39. $x \approx 1.946$

41. $x = 0$ 43. $x \approx 0.477$ 45. $x = 50$

47. $x = 40$ 49. $x = \frac{9}{32}$ 51. $x = \pm 3$

53. $x = 5$ 55. $x = 6$ 57. no solution

59. $x = \frac{25}{8}$ 61. $x \approx \pm 12.182$ 63. $x = 47.5$

65. $x \approx \pm 7.321$ 67. $x \approx -1.432$ 69. $x \approx -1.25$

71. $x \approx 8.456$ 73. $x = 0.3028$ 75. $x \approx 3.646$

77. $t = 31.9$ years   79. $t \approx 19.74$ years

81. 3.16×10^{15} joules   83. $1\ \text{W/m}^2$

85. $t \approx 4.61$ hours   87. 15.89 years   89. 6.2

91. The correction that needs to be made is that the 4 should have been divided first giving the following:

$$e^x = \frac{9}{4}, x = \ln\left(\frac{9}{4}\right)$$

93. The correction that needs to be made is that $x = -5$ needs to be removed from the solutions. The domain of the logs cannot include a negative solution.

95. true   97. false

99. $x = \dfrac{1 + \sqrt{1 + 4b^2}}{2}$ 101. $t = -5 \ln\left(\dfrac{3000 - y}{2y}\right)$

103. $f^{-1}(x) = \ln\left(x + \sqrt{x^2 - 1}\right) \quad x \geq 1$

105.

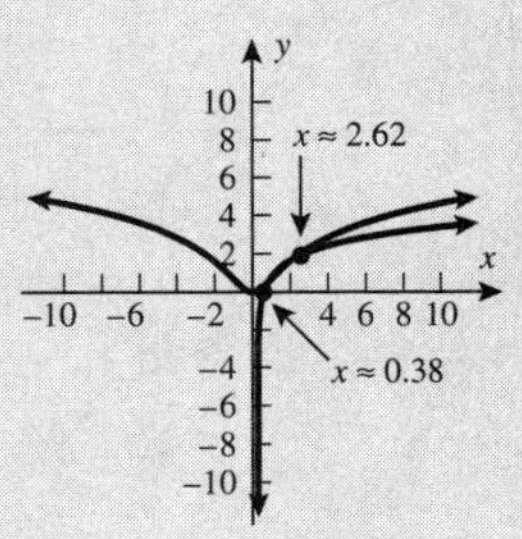

107.

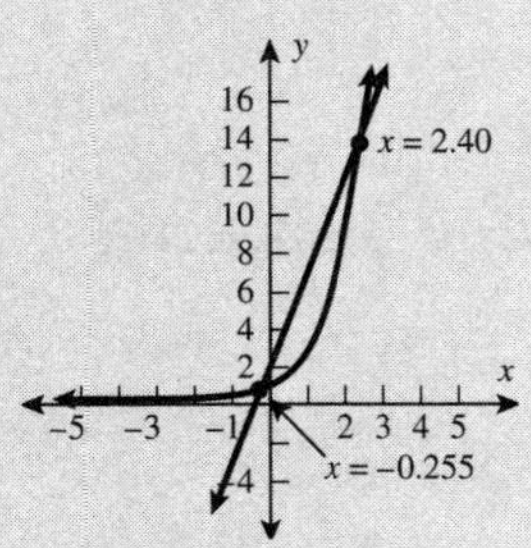

109. $(-\infty, \infty)$, symmetric with respect to the y-axis

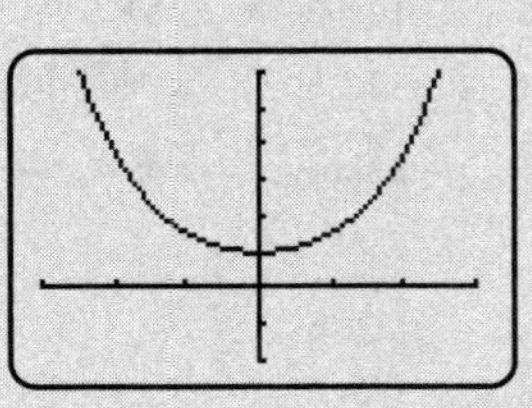

Section 5.5

1. c; iv   3. a; iii   5. f; i

7. 94 million   9. 5.5 years (middle of 2008)

11. 799.6 million cell phone subscribers

13. \$455,000   15. 332 million colonies of phytoplankton

17. 1.19 million   19. 13.53 ml   21. 7575 years

23. 131,158,556 years   25. 105 °F

27. 3.8 hours before 7 A.M.   29. \$19,100

31. **a.** 84,520   **b.** 100,000   **c.** 100,000

33. 29,551 cases   35. 1.89 years   37. $r = 0$ (on-axis)

39. **a.**

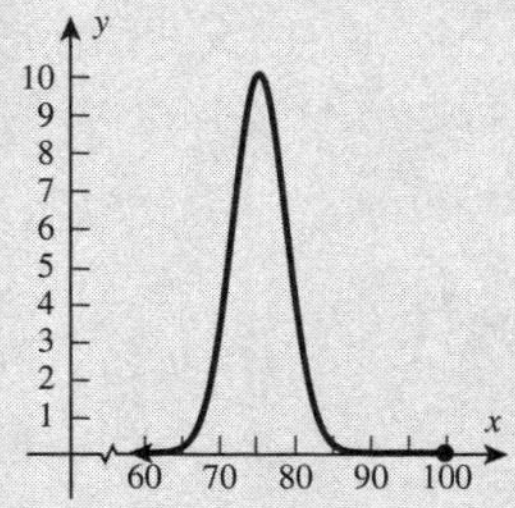

b. 75

c. 4

d. 4

41. **a.** 18 years   **b.** 10 years

43. r is the decimal form of a percentage ($r = 0.07$)

45. true   47. false

49. less time

51. **a.** For the same periodic payment, it will take Wing Shan less years to pay off the loan if she can afford to pay biweekly.

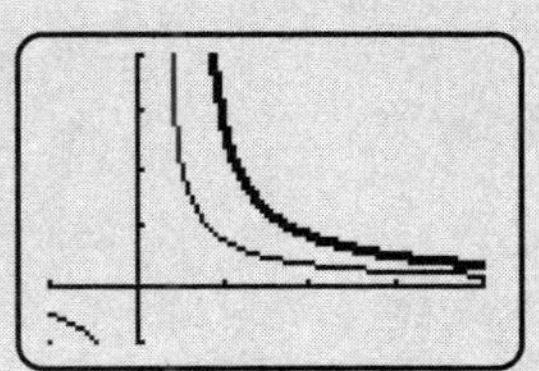

b. 11.58 years

c. 10.33 years

d. 8.54 years, 7.69 years

Review Exercises

1. 17,559.94   3. 5.52   5. 24.53   7. 5.89

9. 73.52   11. 6.25   13. b   15. c

17. The y-intercept is at $(0, -1)$. The horizontal asymptote is $y = 0$.

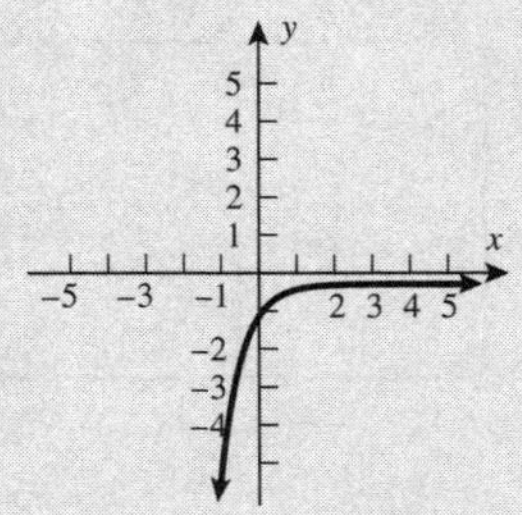

19. The y-intercept is at $(0, 2)$. The horizontal asymptote is $y = 1$.

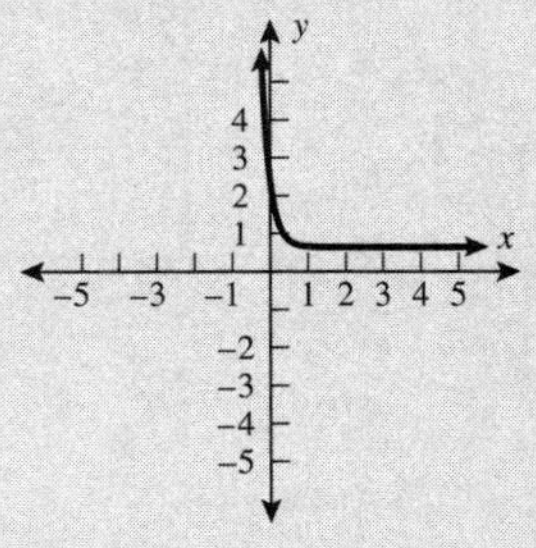

21. The y-intercept is at (0, 1). The horizontal asymptote is $y = 0$.

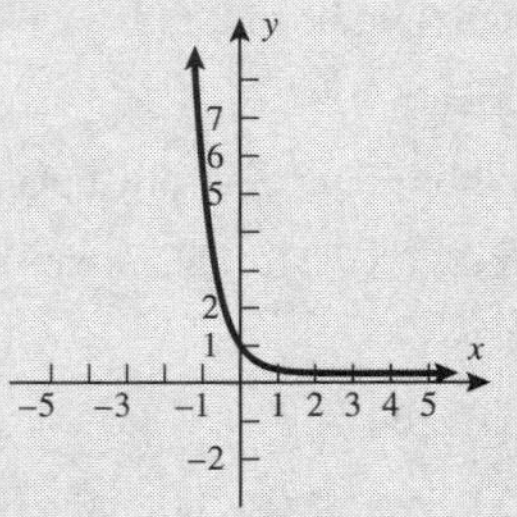

23. The y-intercept is at (0, 3.2). The horizontal asymptote is $y = 0$.

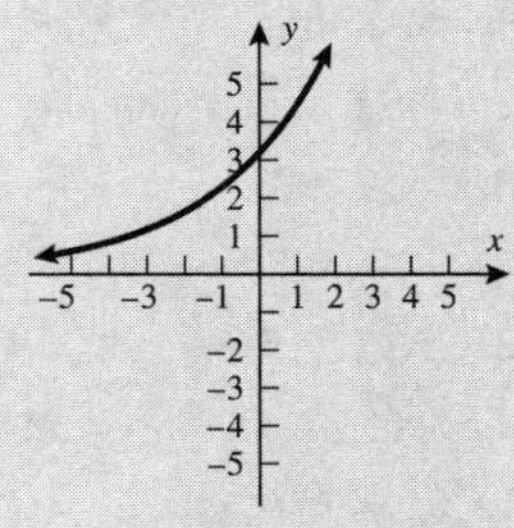

25. \$6144.68 **27.** \$23,080.29 **29.** $4^3 = 64$

31. $10^{-2} = \frac{1}{100}$ **33.** $\log_6 216 = 3$ **35.** $\log_{2/13} \frac{4}{169} = 2$

37. 0 **39.** −4 **41.** 1.51 **43.** −2.08

45. $(-2, \infty)$ **47.** $(-\infty, \infty)$ **49.** b **51.** d

53.

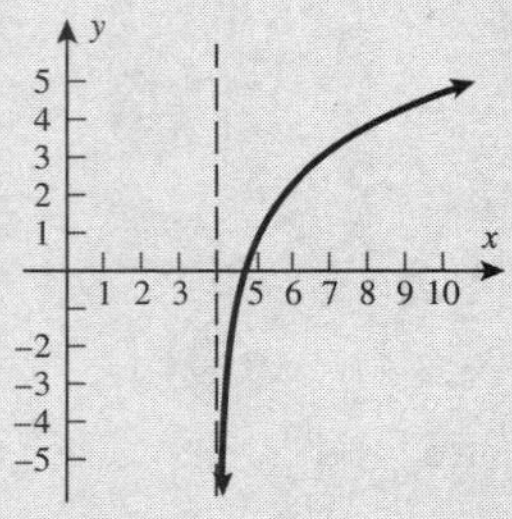

55.

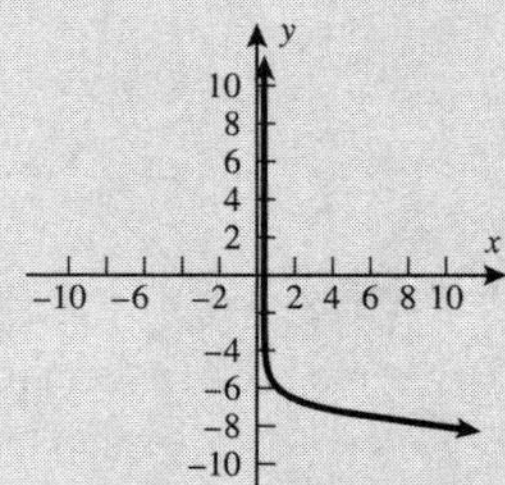

57. pH = 6.5 **59.** 50 dB **61.** 1

63. 6 **65.** $a \log_c x + b \log_c y$

67. $\log_j r + \log_j s - 3 \log_j t$ **69.** $\frac{1}{2} \log a - \frac{3}{2} \log b - \frac{2}{5} \log c$

71. 0.5283 **73.** 0.2939 **75.** $x = -4$ **77.** $x = \frac{4}{3}$

79. $x = -6$ **81.** −0.218 **83.** no solution

85. $x = 0$ **87.** $x = \frac{100}{3}$ **89.** $x = 128\sqrt{2}$

91. $x \approx \pm 3.004$ **93.** $x \approx 0.449$ **95.** \$28,536.88

97. $t = 16.6$ years **99.** 3.23 million **101.** 6250 bacteria

103. 56 years **105.** $15.76 \approx 16$ fish

107. 343 mice **109.** $e^{\sqrt{2}} \approx 4.11$ **111.** (2.376, 2.071)

113. $(0, \infty)$

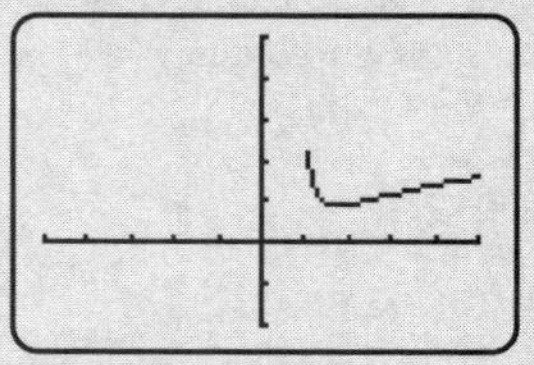

115. $(-\infty, \infty)$, symmetric with respect to the origin, HA at $y = -1$ and $y = 1$.

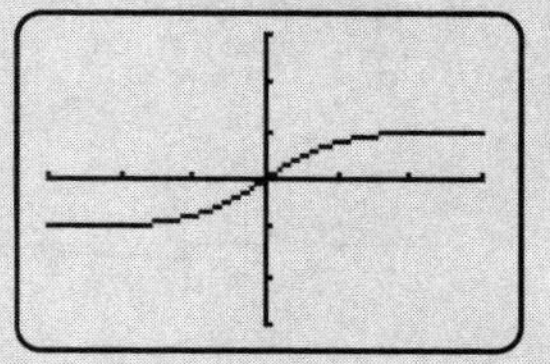

117. a. $y = 4e^{-0.038508\,t} = 4(0.9622)^t$

b. $y = 4(0.9622)^t$

c. Yes, they are the same.

Practice Test

1. x^3 **3.** −4 **5.** $x \approx \pm 2.177$

7. $x \approx 7.04$ **9.** $x = 4 + e^2 \approx 11.389$

11. $x = e^{e^1} \approx 15.154$ **13.** $x = 9$

15. $x = \dfrac{-3 + \sqrt{9 + 4e}}{2} \approx 0.729$

17. $x \approx -0.693$ **19.** domain: $(-1, 0) \cup (1, \infty)$

21. y-intercept: (0, 2)
HA: $y = 1$

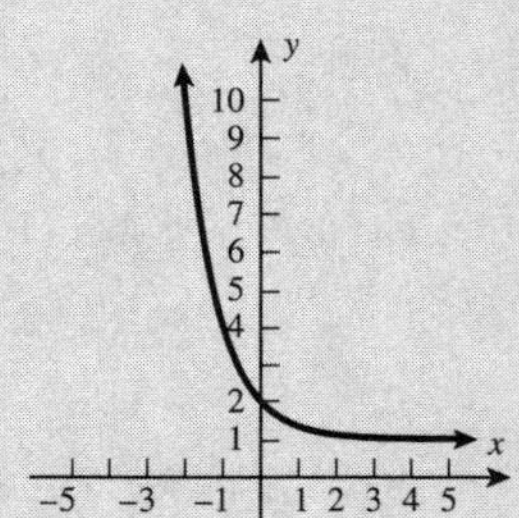

23. x-intercept: $\left(\dfrac{3 + 1/e}{2}, 0\right)$
VA: $x = \frac{3}{2}$

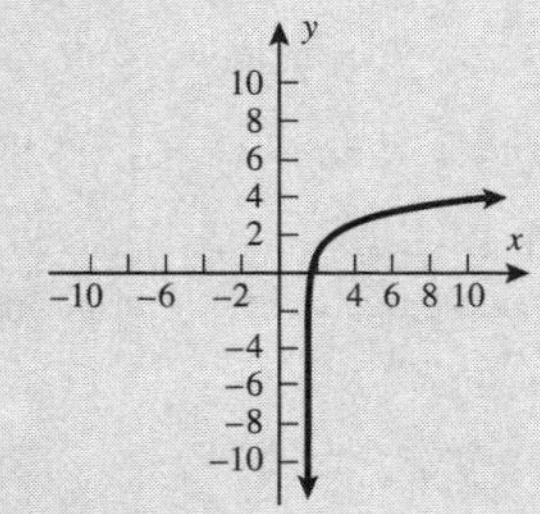

25. \$8,051.62 **27.** 90 dB

29. $7.9 \times 10^{11} < E < 2.5 \times 10^{13}$ joules

31. 7800 bacteria **33.** $2.75 \approx 3$ days

35. $(-\infty, \infty)$, symmetric with respect to the origin, no asymptotes

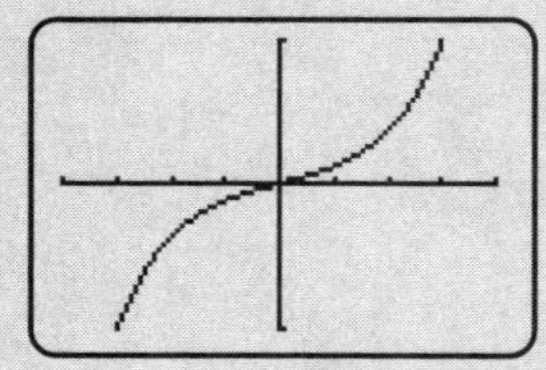

Cumulative Test

1. $x^{5/6}y^2$ **3.** $x = \dfrac{2 \pm \sqrt{19}}{5}$

5. $(-\infty, -19]$ **7.** $y = \frac{3}{4}x + \frac{3}{4}$

9. a. 1 **b.** 5 **c.** 1

d. undefined **e.** domain: $(-2, \infty)$; range: $(0, \infty)$

f. Increasing: $(4, \infty)$; decreasing: $(0, 4)$; constant: $(-2, 0)$

11. yes, one-to-one **13.** $(1, -1)$

15. $Q(x) = -x^3 + 3x - 5, r(x) = 0$

17. VA: $x = 3$; slant asymptote: $y = x + 3$

19. 125 **21.** 5 **23.** $x = \frac{1}{2}$

25. 8.62 years **27. a.** $N = 6(0.9755486421)^t$
b. 2.72 grams

CHAPTER 6

Section 6.1

1. $(1, 0)$ **3.** $(8, -1)$

5. $(1, -1)$ **7.** $(1, 2)$

9. $\left(\frac{32}{17}, \frac{11}{17}\right)$ **11.** no solution

13. infinitely many solutions **15.** infinitely many solutions

17. $(1, 3)$ **19.** $(6, 8)$

21. $(1.2, 0.4)$ **23.** $(3, 1)$

25. $(2, 5)$ **27.** $(-3, 4)$

29. $\left(1, -\frac{2}{7}\right)$ **31.** $\left(\frac{19}{7}, \frac{11}{35}\right)$

33. infinitely many solutions **35.** $(4, 0)$

37. $(-2, 1)$ **39.** infinitely many solutions

41. $\left(\frac{75}{32}, \frac{7}{16}\right)$ **43.** c **45.** d

47. The solution is $(0, 0)$.

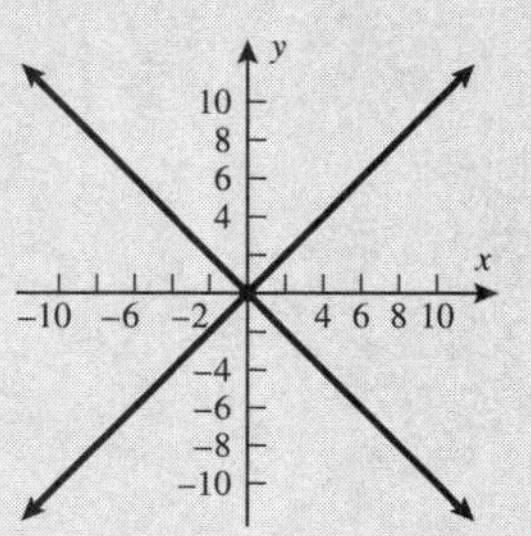

49. The solution is $(-1, -1)$.

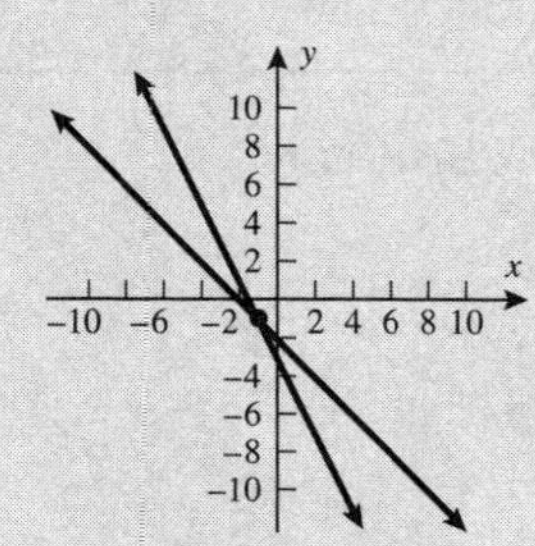

51. The solution is $(0, -6)$.

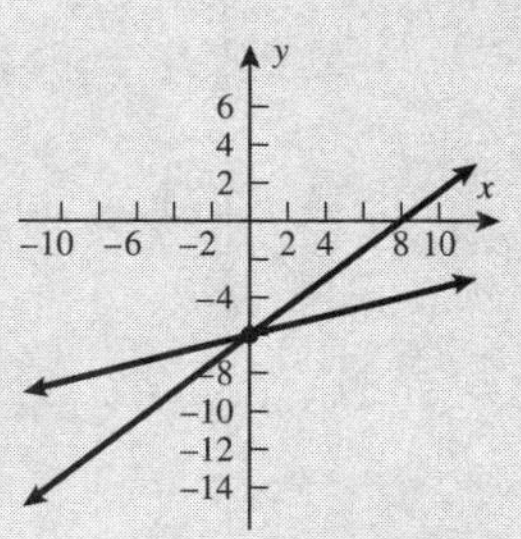

53. There are no solutions to this system of equations.

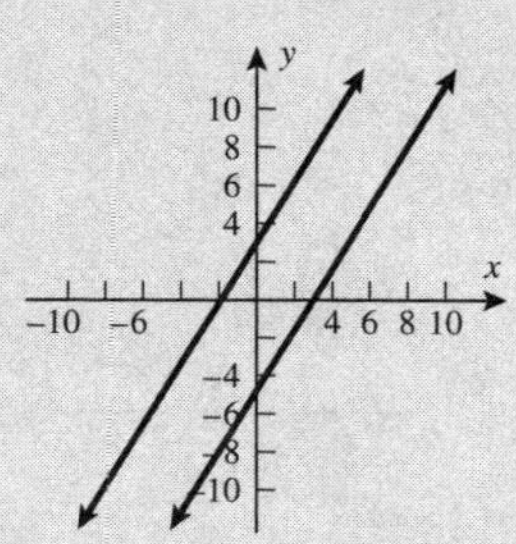

55. 5 Montblanc pens and 64 Cross pens

57. 15.86 ml of 8% HCl
21.14 ml of 15% HCl

59. \$300,000 of sales

61. 169 miles highway, 180.5 miles city

63. Average plane speed is 450 mph; average wind is 50 mph.

65. \$3,500 invested at 10% and \$6,500 invested at 14%

67. The company must sell 8 CD players to break even.

69. Every term in Equation (1) is not multiplied correctly by -1.

71. The negative was not distributed through both terms: $-(-3y - 4) = 3y + 4$.

73. false **75.** false **77.** $A = -4, B = 7$

79. 8 cups of 2% and 96 cups of 4%

81. The point of intersection is approximately $(8.9, 6.4)$.

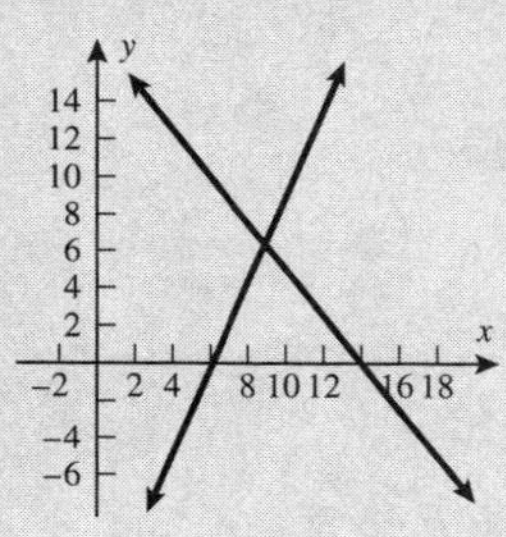

83. The lines coincide. Infinitely many solutions.

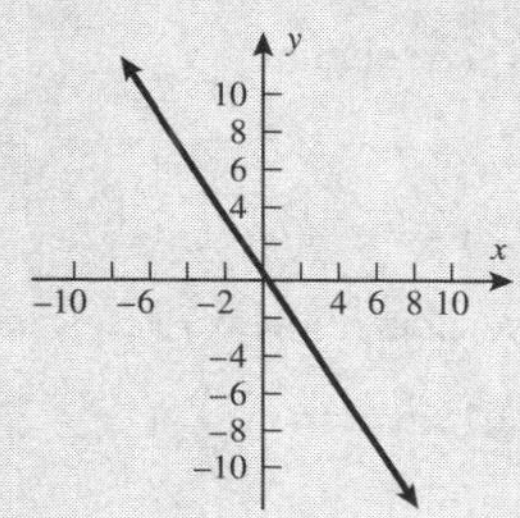

85. $(2.426, -0.059)$

Section 6.2

1. $x = -\frac{3}{2}, y = -3, z = \frac{9}{2}$ **3.** $x = -2, y = \frac{9}{2}, z = \frac{1}{2}$

5. $x = 5, y = 3, z = -1$ **7.** $x = \frac{90}{31}, y = \frac{103}{31}, z = \frac{9}{31}$

9. $x = -\frac{13}{4}, y = \frac{1}{2}, z = -\frac{5}{2}$ **11.** $x = -2, y = -1, z = 0$

13. $x = 2, y = 5, z = -1$ **15.** no solution

17. no solution

19. $x = 1 - a, y = -\left(a + \frac{1}{2}\right), z = a$

21. $x = 41 + 4a, y = 31 + 3a, z = a$

23. $x_1 = -\frac{1}{2}, x_2 = \frac{7}{4}, x_3 = -\frac{3}{4}$

25. no solution

27. $x_1 = 1, x_2 = -1 + a, x_3 = a$

29. $x = \frac{2}{3}a + \frac{8}{3}, y = -\frac{1}{3}a - \frac{10}{3}, z = a$

31. 8 touchdowns, 6 extra points, and 4 field goals

33. 6 Mediterranean chicken sandwiches
3 six-inch tuna sandwiches
5 six-inch roast beef sandwiches

35. $a = -32$ ft/sec^2, $v_0 = 52$ ft/sec, $h_0 = 0$ ft

37. $y = -0.0625x^2 + 5.25x - 50$

39. \$10,000 in the money market, \$4,000 in the mutual fund, \$6,000 in the stock

41. 33 pairs of regular model skis
72 pairs of trick skis
5 pairs of slalom skis

43. Total points for first game: 885
Total points for second game: 823
Total points for third game: 883

45. The correction that needs to be made is that Equation (2) and Equation (3) need to be added correctly. Also, the text tells you to begin by eliminating a variable from Equation (1).

47. true

49. $a = 4, b = -2, c = -4$

51. $a = -\frac{55}{24}, b = -\frac{1}{4}, c = \frac{223}{24}, d = \frac{1}{4}, e = 44$

53. no solution

55. $x_1 = -2, x_2 = 1, x_3 = -4, x_4 = 5$

57. $x = 41 + 4a, y = 31 + 3a, z = a$

59. same result as Exercise 57

61. $\left(-\frac{80}{7}, -\frac{80}{7}, \frac{48}{7}\right)$

Section 6.3

1. d

3. a

5. b

7. $\dfrac{A}{x-5} + \dfrac{B}{x+4}$

9. $\dfrac{A}{x-4} + \dfrac{B}{x} + \dfrac{C}{x^2}$

11. $2x - 6 + \dfrac{3x + 33}{x^2 + x + 5}$

13. $\dfrac{Ax + B}{x^2 + 10} + \dfrac{Cx + D}{(x^2 + 10)^2}$

15. $\dfrac{1}{x} - \dfrac{1}{x+1}$

17. $\dfrac{1}{x-1}$

19. $\dfrac{2}{x-3} + \dfrac{7}{x+5}$

21. $\dfrac{3}{x-1} + \dfrac{4}{(x-1)^2}$

23. $\dfrac{4}{x+3} - \dfrac{15}{(x+3)^2}$

25. $\dfrac{3}{x+1} + \dfrac{1}{x-5} + \dfrac{2}{(x-5)^2}$

27. $\dfrac{-2}{x+4} + \dfrac{7x}{x^2+3}$

29. $\dfrac{-2}{x-7} + \dfrac{4x - 3}{3x^2 - 7x + 5}$

31. $\dfrac{x}{x^2+9} - \dfrac{9x}{(x^2+9)^2}$

33. $\dfrac{2x - 3}{x^2 + 1} + \dfrac{5x + 1}{(x^2+1)^2}$

35. $\dfrac{1}{x-1} + \dfrac{-3x - 1}{2(x^2+1)} + \dfrac{1}{2(x+1)}$

37. $\dfrac{3}{x-1} + \dfrac{2x + 5}{x^2 + 2x - 1}$

39. $\dfrac{1}{x-1} + \dfrac{1 - x}{x^2 + x + 1}$

41. $\dfrac{1}{d_o} + \dfrac{1}{d_i} = \dfrac{1}{f}$

43. The correction that needs to be made is in the decomposition. Once the correct decomposition is used the correct answer will be:

$$\frac{1}{x} + \frac{3 + 2x}{x^2 + 1}$$

45. false

47. $\dfrac{1}{x-1} - \dfrac{1}{x+2} + \dfrac{1}{x-2}$

49. $\dfrac{1}{x} + \dfrac{1}{x+1} - \dfrac{1}{x^3}$

51. $\dfrac{x}{x^2+1} - \dfrac{2x}{(x^2+1)^2} + \dfrac{x+2}{(x^2+1)^3}$

53. yes

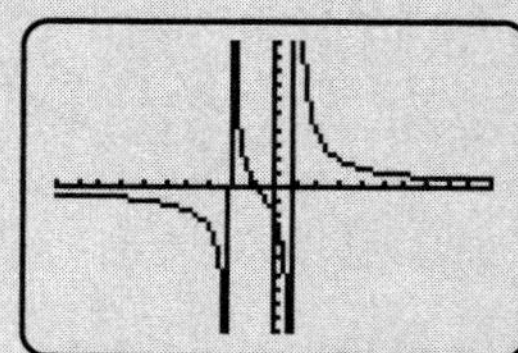

55. no

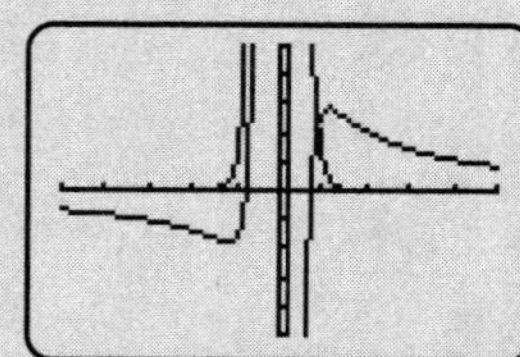

57. yes

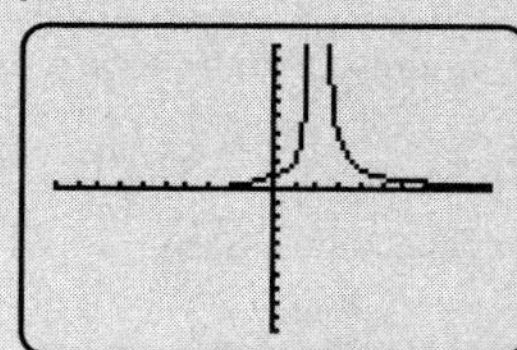

Section 6.4

1. d

3. b

5.

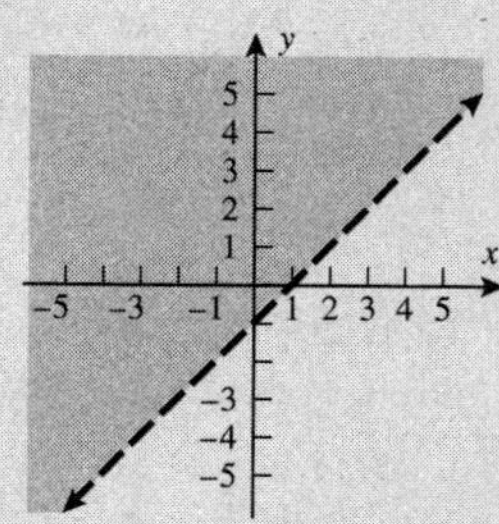

7.

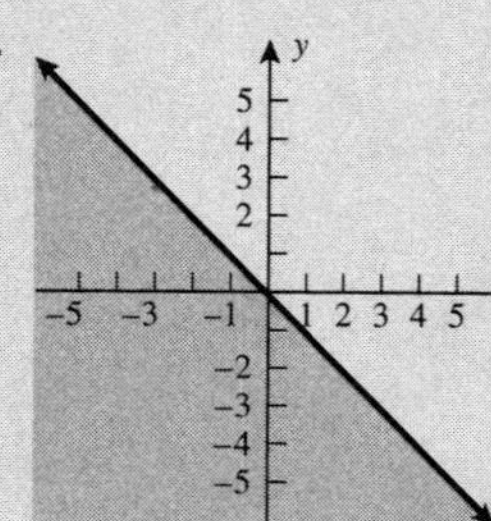

9.

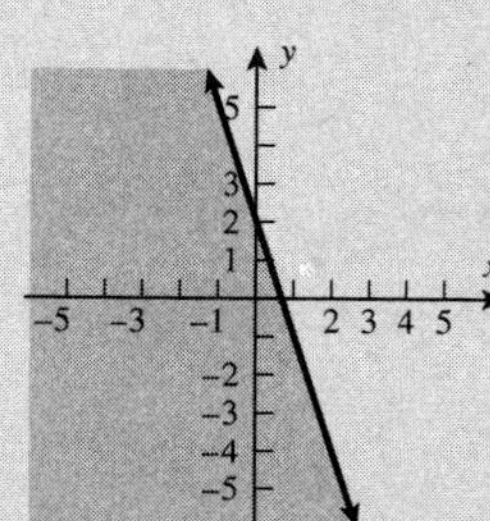

11.

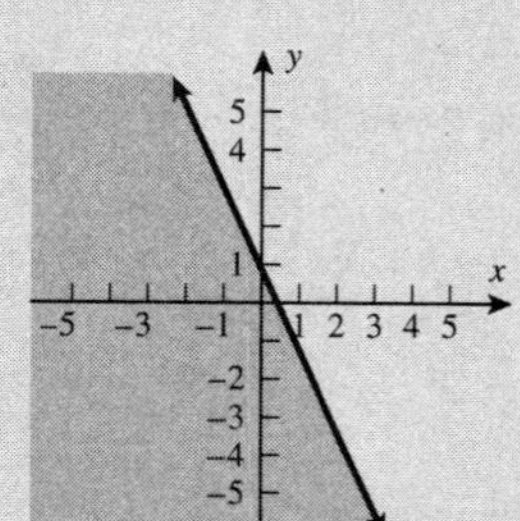

13.

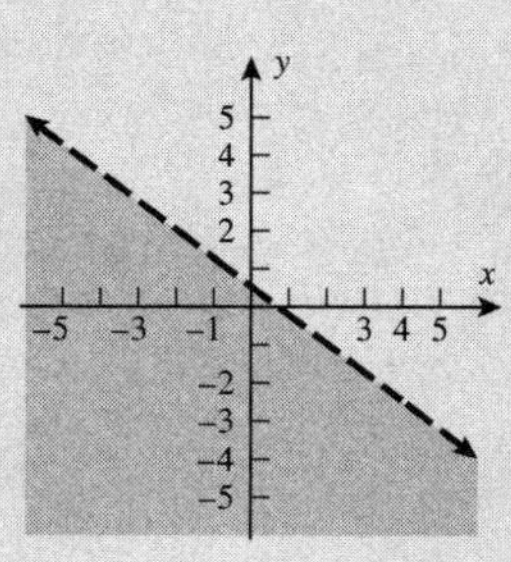

15.

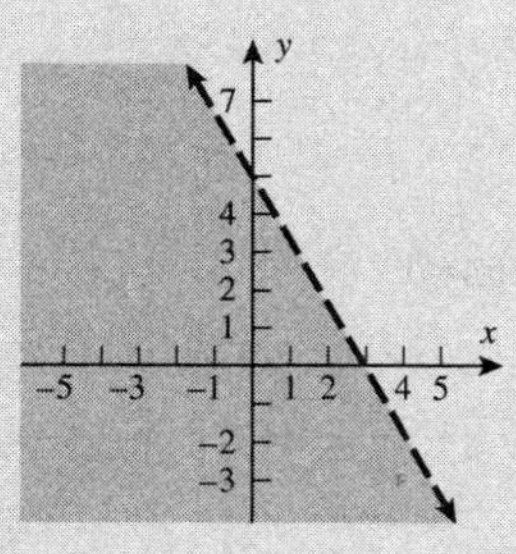

17.

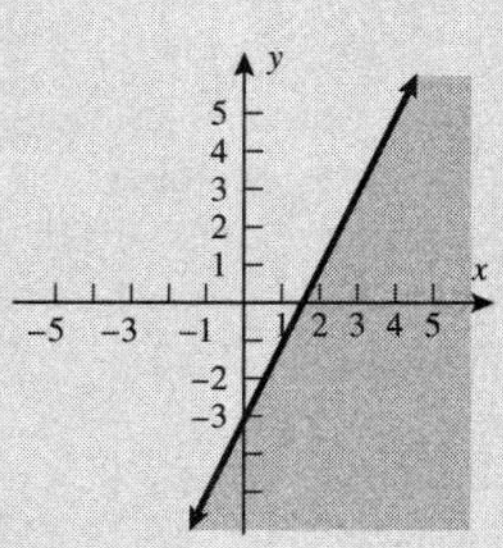

19.

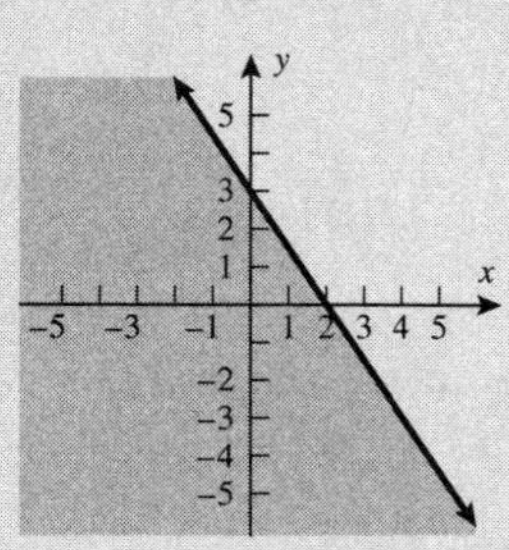

21.

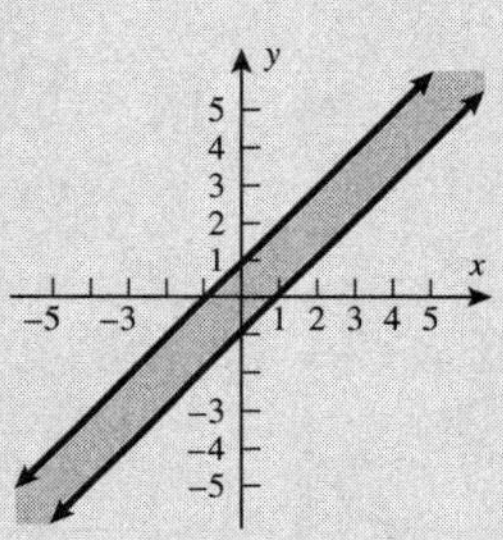

23.

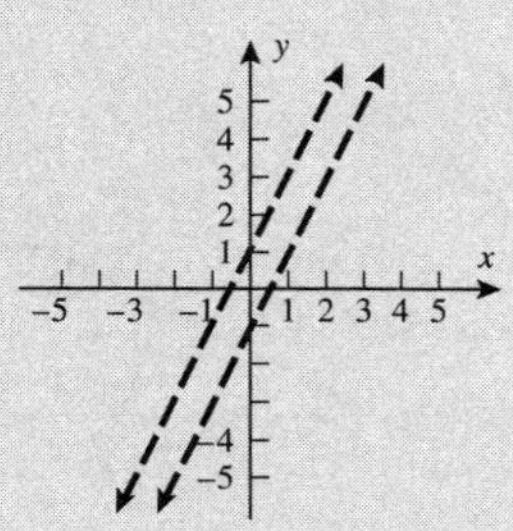

25.

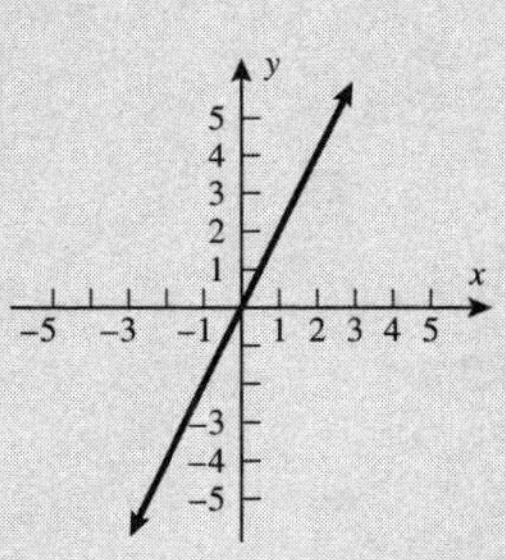

27.

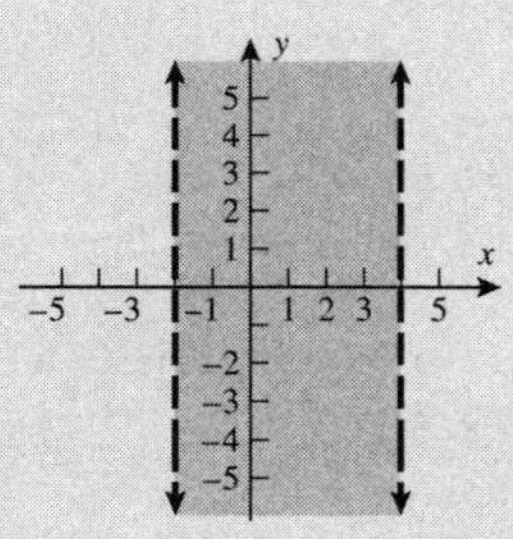

29.

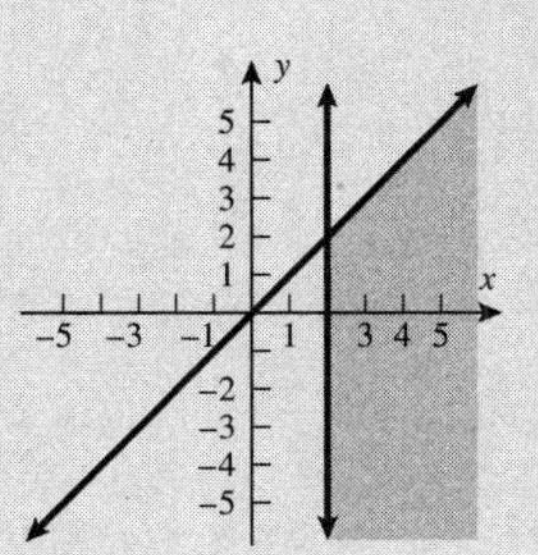

31.

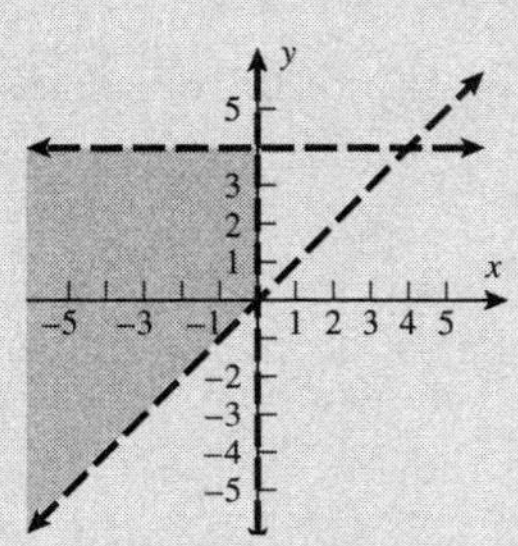

33.

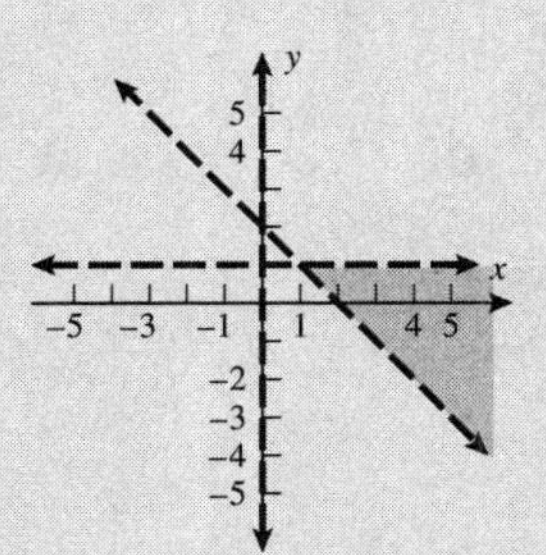

35.

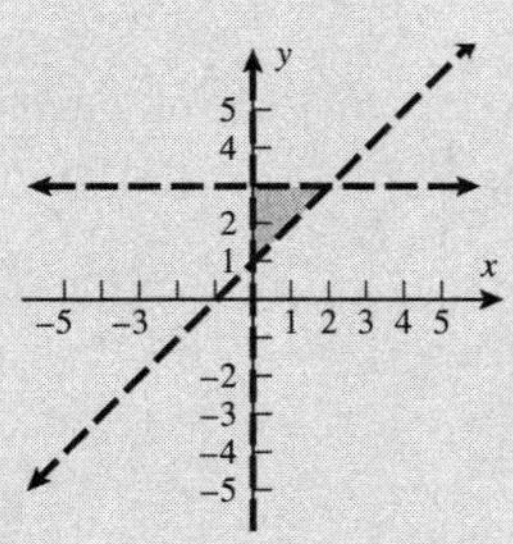

37.

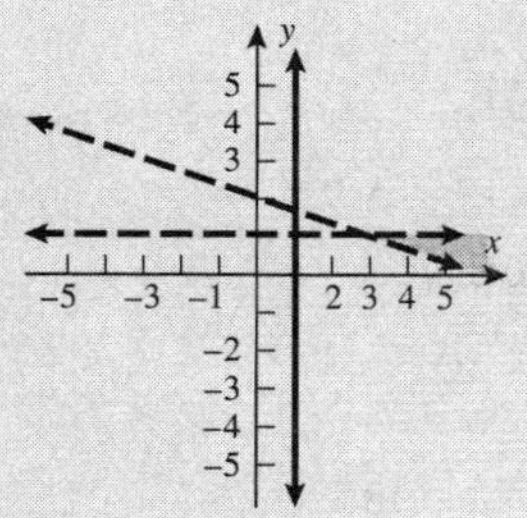

39.

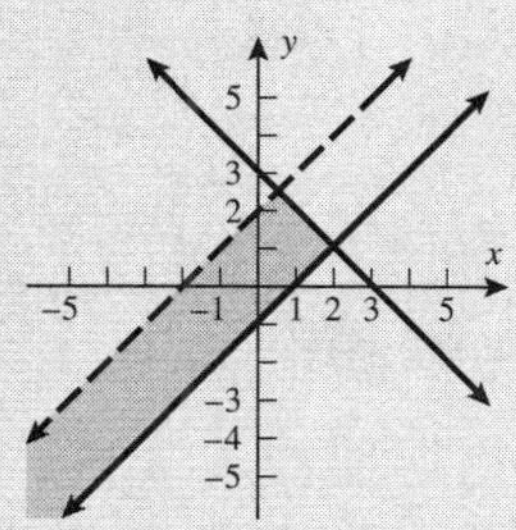

41.

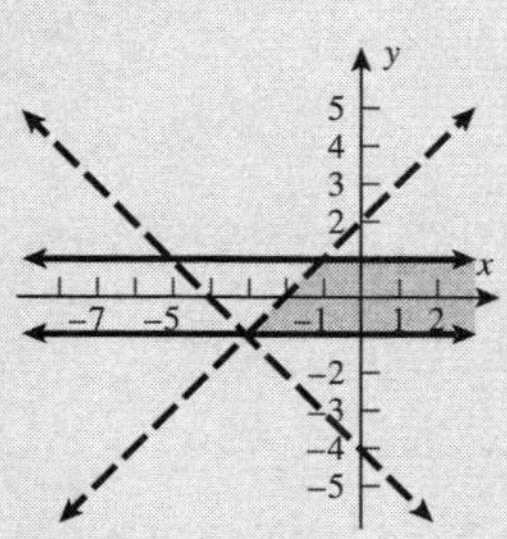

43.

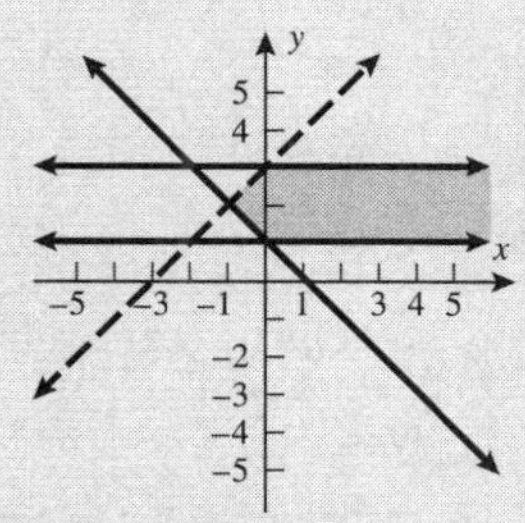

45. no solution

47.

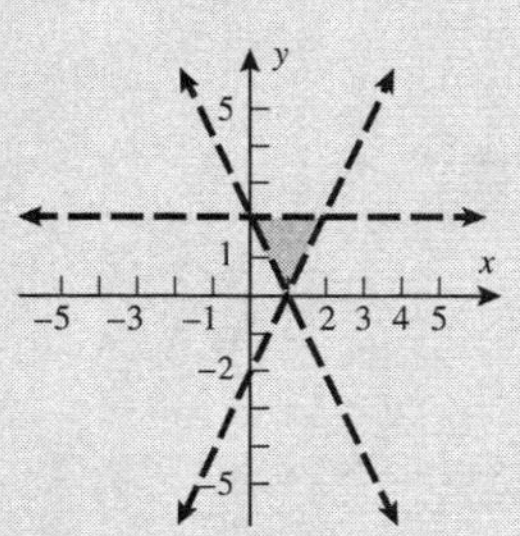

49. 4 square units

51. 7.5 square units

53.
$$x + 20y \le 2400$$
$$25x + 150y \le 6000$$

55.
$$P \le 80 - 0.01x$$
$$P \ge 60$$
$$x \ge 0$$

57. \$20,000

59. The correction that needs to be made is that the shading should be above the line.

61. true

63. false

65. The solution is a shaded rectangle.

67.

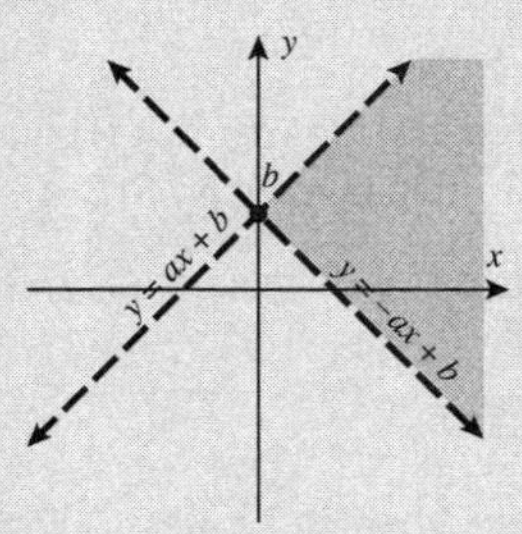

69.

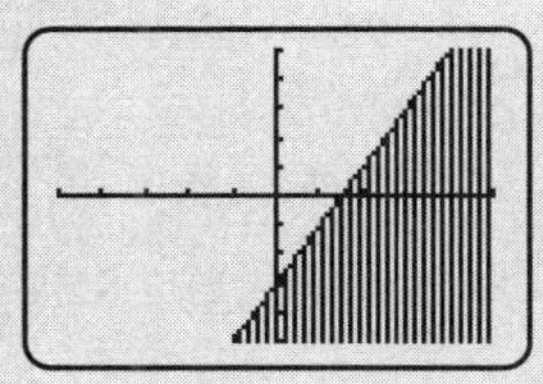

71.

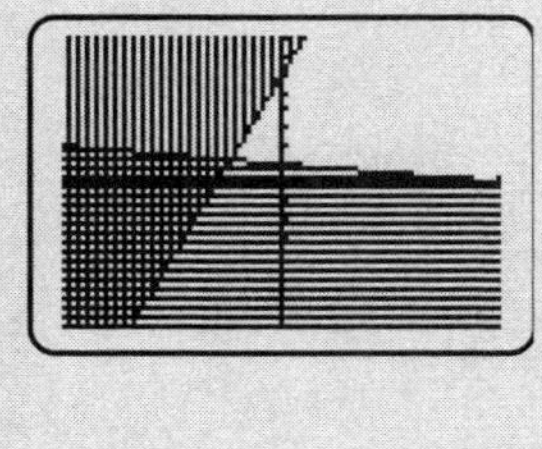

Section 6.5

1. $f(x, y) = z = 2x + 3y$
   $f(-1, 4) = 10$
   $f(2, 4) = 16$ (MAX)
   $f(-2, -1) = -7$ (MIN)
   $f(1, -1) = -1$

3. $f(x, y) = z = 1.5x + 4.5y$
   $f(-1, 4) = 16.5$
   $f(2, 4) = 21$ (MAX)
   $f(-2, -1) = -7.5$ (MIN)
   $f(1, -1) = -3$

5. minimize at $f(0, 0) = 0$

7. no maximum

9. minimize at $f(0, 0) = 0$

11. maximize at $f(1, 6) = 2.65$

13. 50 Charley T-shirts, 130 Frances T-shirts (profit = \$950)

15. 0 desktops, 25 laptops (profit = \$7,500)

17. 3 first class cars and 27 second class cars

19. Let x represent the number of pairs of regular skis and y represent the number of pairs of slalom skis.
    The profit function $P = 25x + 50y$.
    The constraints are: $x \geq 200$
    $y \geq 80$
    $x + y \leq 400$
    $x \geq 0 \quad y \geq 0$

    The feasible region is shown with the vertices indicated:

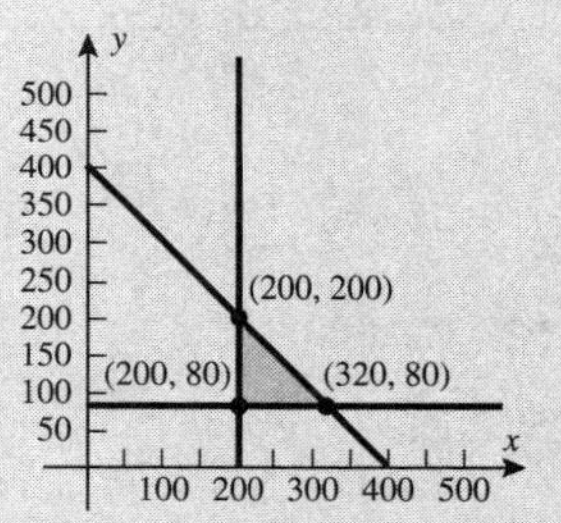

    Evaluate the objective function at each vertex:
    At (200, 80), $P = 25(200) + 50(80) = 9000$
    At (200, 200), $P = (200) + 50(200) = 15{,}000$
    At (320, 80), $P = 25(320) + 50(80) = 12{,}000$

    Thus, the ski producer should produce 200 pairs of regular skis and 200 pairs of slalom skis to maximize profits.

21. You don't compare y-coordinates. You evaluate the objective function.

23. false

25. The maximum is a and occurs at $(0, a)$.

27. minimized at $z(0, 0) = 0$

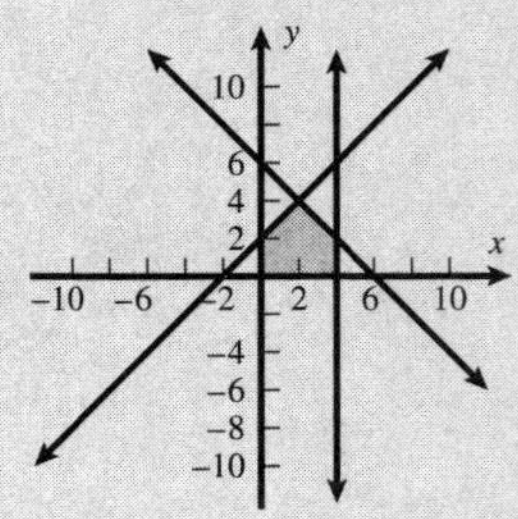

29. maximized at $z(6.7, 4.5) = 176.9$

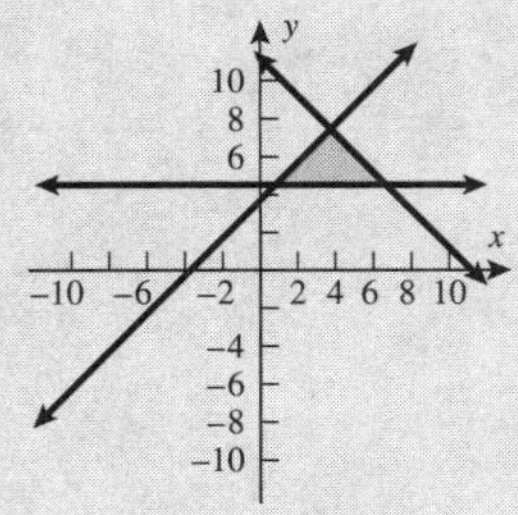

31. $z\left(\frac{4}{9}, \frac{115}{9}\right) = 25$

Review Exercises

1. (3, 0)

3. $\left(\frac{13}{4}, 8\right)$

5. (2, 1)

7. $\left(\frac{19}{8}, \frac{13}{8}\right)$

9. $(-2, 1)$

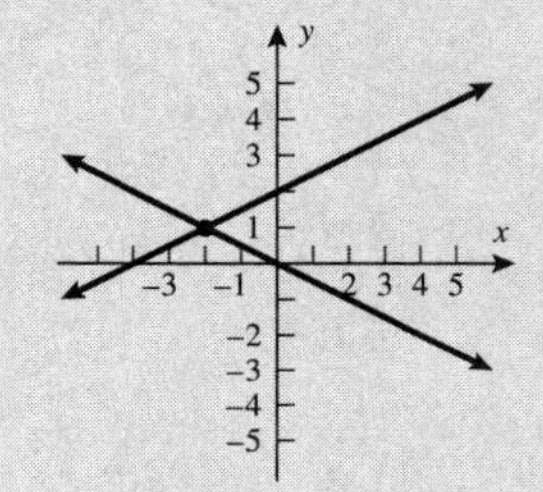

11. $\left(12, \frac{35}{6}\right)$

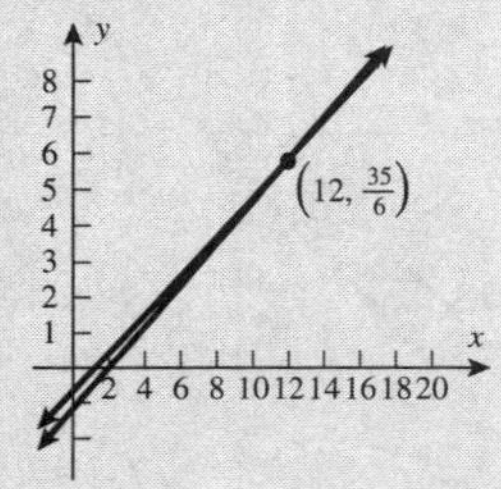

13. $(3, -2)$

15. $(-1, 2)$

17. c

19. d

21. 10.5 milliliters of 6%; 31.5 milliliters of 18%

23. $x = -1, y = -a + 2, z = a$

25. no solution

27. $y = -0.0050x^2 + 0.4486x - 3.8884$

29. $\dfrac{A}{x-1} + \dfrac{B}{(x-1)^2} + \dfrac{C}{x+3} + \dfrac{D}{x-5}$

31. $\dfrac{A}{x} + \dfrac{B}{4x+5} + \dfrac{C}{(2x+1)} + \dfrac{D}{(2x+1)^2}$

33. $\dfrac{A}{x-3} + \dfrac{B}{x+4}$

35. $\dfrac{Ax+B}{x^2+17} + \dfrac{Cx+D}{(x^2+17)^2}$

37. $\dfrac{4}{x-1} + \dfrac{5}{x+7}$

39. $\dfrac{1}{2x} + \dfrac{15}{2(x-5)} - \dfrac{3}{2(x+5)}$

41. $\dfrac{-2}{x+1} + \dfrac{2}{x}$

43. $\dfrac{5}{x+2} - \dfrac{27}{(x+2)^2}$

45.

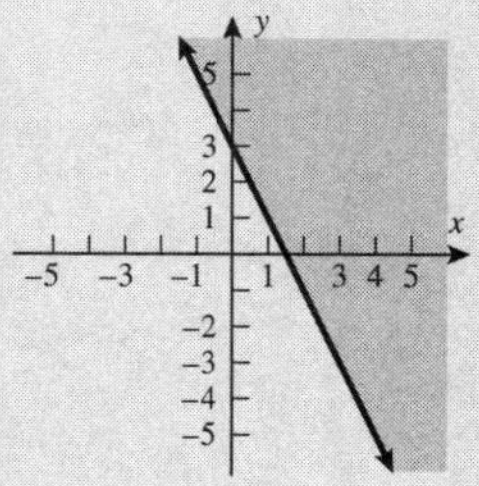

47.

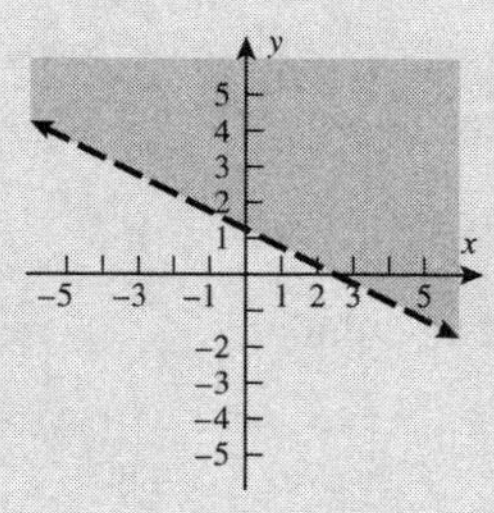

49.

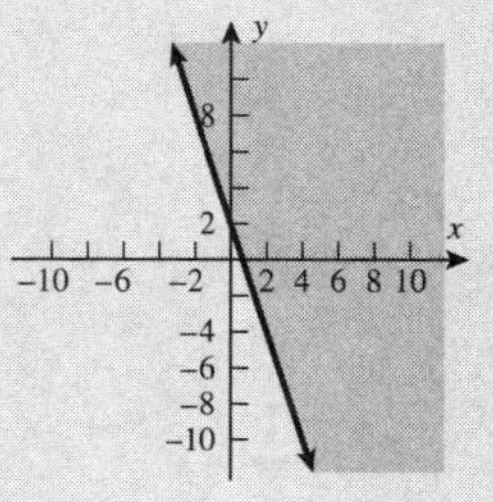

51.

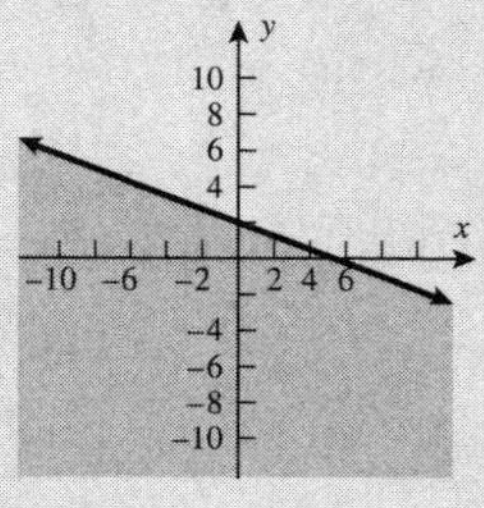

53. no solution

55.

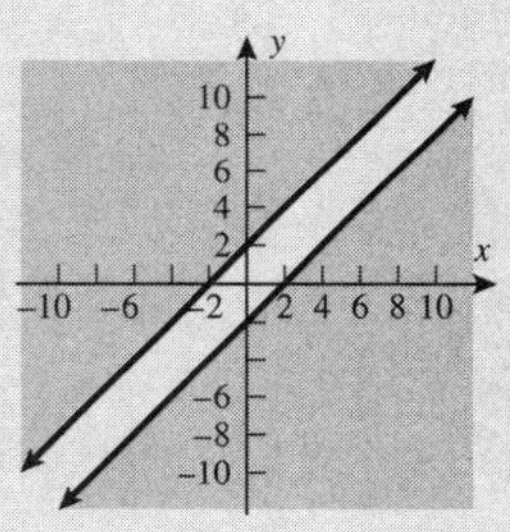

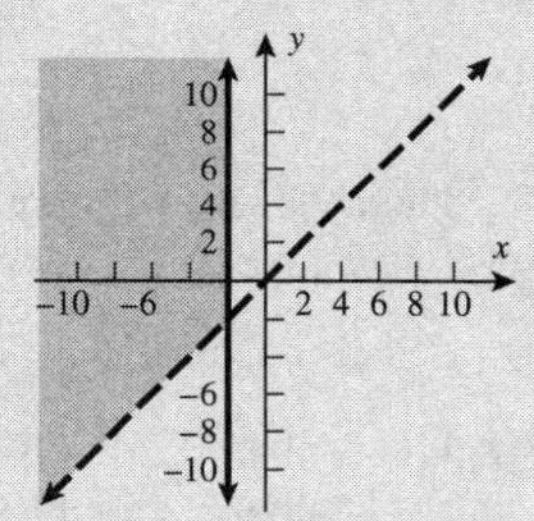

57.

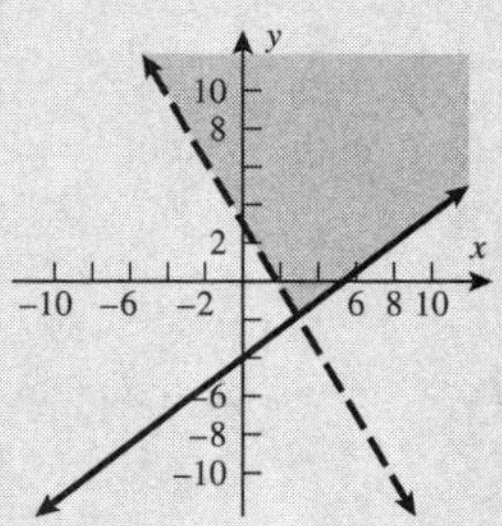

59. Minimum of z is 0 and occurs at (0, 0).

61. Maximum of z is 25.6 and occurs at (0, 8).

63. Minimum of z is -30 and occurs at (0, 6).

65. 10 watercolor, 30 geometric (profit = \$390)

67. $(2, -3)$

69. (3.6, 3, 0.8)

71. yes

73.

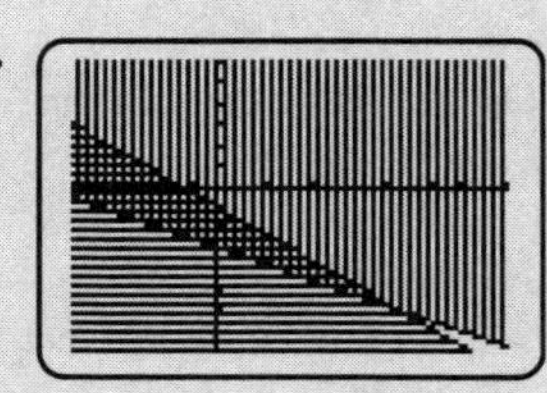

75. $z(1.8, 0.6) = 12.06$

Practice Test

1. (7, 3)

3. $x = a, y = a - 2$

5. $x = 1, y = -5, z = 3$

7. $x = -1, y = 3, z = 4$

9. $\dfrac{5}{x} - \dfrac{3}{x+1}$

11. $\dfrac{7}{(x+2)} - \dfrac{9}{(x+2)^2}$

13. $\dfrac{1}{3x} + \dfrac{2}{3(x-3)} - \dfrac{1}{x+3}$

15.

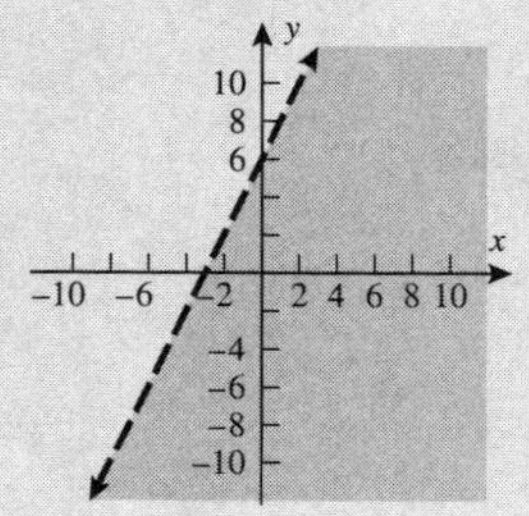

17.

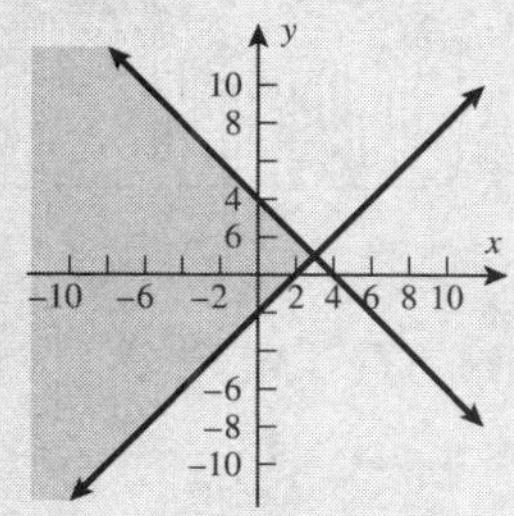

19.

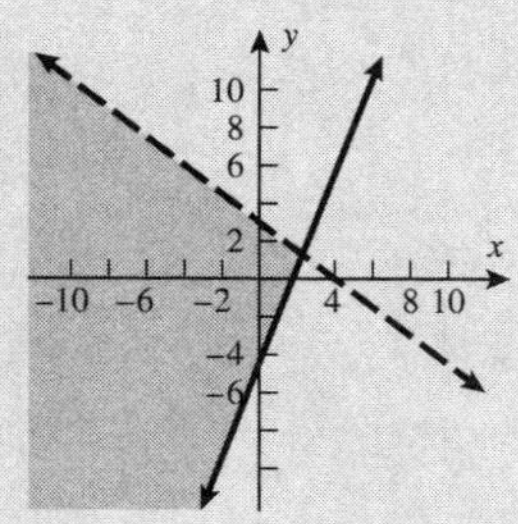

21. The minimum value occurs at (0, 1), $f(0, 1) = 7$.

23. \$14,000 in the money market, \$8,500 in the aggressive stock, and \$7,500 in the conservative stock.

25. (11, 19, 1)

Cumulative Test

1. $\dfrac{x-1}{5} \quad x \neq \frac{5}{6}$

3. $x \neq -1, 0$; no solution

5. $(-\infty, 0] \cup (1, 3]$

7. $y = 0.5x + 3.4$

9. $-\frac{5}{8}$

11. 1

13. $f(x) = -2x^2 + 7$

15. Factors of $a_0 = 10; \pm1, \pm2, \pm5, \pm10$
Factors of $a_n = 2; \pm1, \pm2$

Possible rational zeros: $\pm1, \pm2, \pm5, \pm10, \pm\frac{1}{2}, \pm\frac{5}{2}$

Testing the zeros: $P(-5) = 0, P(-1) = 0, P\left(\frac{1}{2}\right) = 0, P(2) = 0$

17. y-intercept: (0, 0), domain: $(-\infty, \infty)$, range: $(-1, \infty)$, horizontal asymptote: $y = -1$

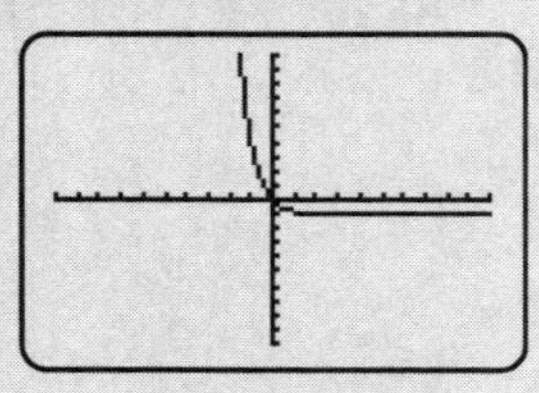

19.

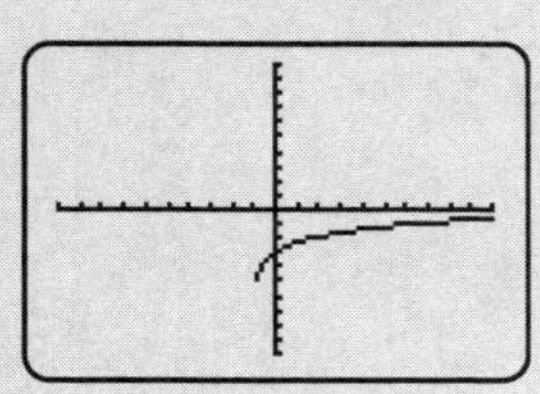

21. 0.435

23. (2, 0, 5)

25. 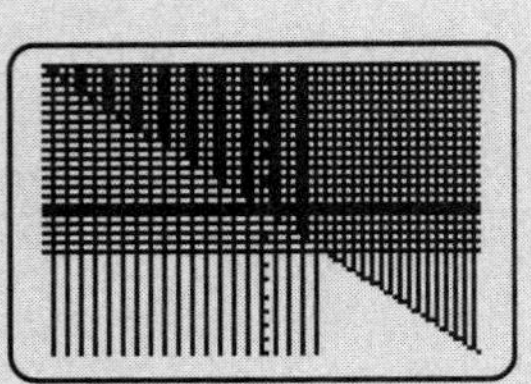

27. yes

CHAPTER 7

Section 7.1

1. 2×3

3. 1×4

5. 1×1

7. $\left[\begin{array}{cc|c} 3 & -2 & 7 \\ -4 & 6 & -3 \end{array}\right]$

9. $\left[\begin{array}{ccc|c} 2 & -3 & 4 & -3 \\ -1 & 1 & 2 & 1 \\ 5 & -2 & -3 & 7 \end{array}\right]$

11. $\left[\begin{array}{ccc|c} 1 & 1 & 0 & 3 \\ 1 & 0 & -1 & 2 \\ 0 & 1 & 1 & 5 \end{array}\right]$

13. $\left[\begin{array}{ccc|c} -4 & 3 & 5 & 2 \\ 2 & -3 & -2 & -3 \\ -2 & 4 & 3 & 1 \end{array}\right]$

15. $\begin{aligned} -3x + 7y &= 2 \\ x + 5y &= 8 \end{aligned}$

17. $\begin{aligned} -x \qquad\qquad &= 4 \\ 7x + 9y + 3z &= -3 \\ 4x + 6y - 5z &= 8 \end{aligned}$

19. $\begin{aligned} x &= a \\ y &= b \end{aligned}$

21. The matrix is not in row–echelon form.

23. The matrix is in reduced row–echelon form.

25. The matrix is not in row–echelon form.

27. The matrix is in reduced row–echelon form.

29. The matrix is in reduced row–echelon form.

31. $\left[\begin{array}{cc|c} 1 & -2 & -3 \\ 0 & 7 & 5 \end{array}\right]$

33. $\left[\begin{array}{ccc|c} 1 & -2 & -1 & 3 \\ 0 & 5 & -1 & 0 \\ 3 & -2 & 5 & 8 \end{array}\right]$

35. $\left[\begin{array}{cccc|c} 1 & -2 & 5 & -1 & 2 \\ 0 & 1 & 1 & -3 & 3 \\ 0 & -2 & 1 & -2 & 5 \\ 0 & 0 & 1 & -1 & -6 \end{array}\right]$

37. $\left[\begin{array}{cccc|c} 1 & 0 & 5 & -10 & -5 \\ 0 & 1 & 2 & -3 & -2 \\ 0 & 0 & -7 & 6 & 3 \\ 0 & 0 & 8 & -10 & -9 \end{array}\right]$

39. $\left[\begin{array}{cccc|c} 1 & 0 & 4 & 0 & 27 \\ 0 & 1 & 2 & 0 & -11 \\ 0 & 0 & 1 & 0 & 21 \\ 0 & 0 & 0 & 1 & -3 \end{array}\right]$

41. $\left[\begin{array}{cc|c} 1 & 0 & -8 \\ 0 & 1 & 6 \end{array}\right]$

43. $\left[\begin{array}{ccc|c} 1 & 0 & 0 & -2 \\ 0 & 1 & 0 & -1 \\ 0 & 0 & 1 & 0 \end{array}\right]$

45. $\left[\begin{array}{ccc|c} 1 & 0 & 0 & 2 \\ 0 & 1 & 0 & 5 \\ 0 & 0 & 1 & -1 \end{array}\right]$

47. $\left[\begin{array}{ccc|c} 1 & 0 & -2 & 1 \\ 0 & 1 & -2 & 2 \end{array}\right]$

49. $\left[\begin{array}{ccc|c} 1 & 0 & 1 & 1 \\ 0 & 1 & 1 & -\frac{1}{2} \\ 0 & 0 & 0 & 0 \end{array}\right]$

51. $x = -7, y = 5$

53. $x - 2y = -3$ or $x = 2a - 3, y = a$

55. no solution

57. $x = 4a + 41, y = 3a + 31, z = a$

59. $x_1 = -\frac{1}{2}, x_2 = \frac{7}{4}, x_3 = -\frac{3}{4}$

61. no solution

63. $x_1 = 1, x_2 = a - 1, x_3 = a$

65. $x = \frac{2}{3}a + \frac{8}{3}, y = -\frac{1}{3}a - \frac{10}{3}, z = a$

67. no solution

69. $x_1 = -2, x_2 = 1, x_3 = -4, x_4 = 5$

71. $(1, -2)$

73. no solution

75. $(-2, 1, 3)$

77. $(3, -2, 2)$

79. no solution

81. $x = \dfrac{a}{4} + 3, y = \dfrac{7a}{4} - \dfrac{53}{3}, z = a$

83. $z = a, x = \dfrac{72 - 11a}{14}, y = \dfrac{13a + 4}{14}$

85. $x = 1, y = 2, z = -3, w = 1$

87. 8 touchdowns, 5 extra points, 1 two-point conversion, and 2 field goals

89. 2 chicken, 2 tuna, 8 roast beef, and 2 turkey-bacon

91. initial height = 0 ft, initial velocity = 50 ft/sec, acceleration = -32 ft/sec^2

93. $y = -0.053x^2 + 4.58x - 34.76$

95. \$5,500 in the money market, \$2,500 in the mutual fund, and \$2,000 in the stock

97. $x = 25, y = 40, z = 6$

99. $x = 25, y = 30, z = 45$ (25 general tickets, 30 reserved tickets, and 45 end zone tickets)

101. $a = -\frac{22}{17}, b = -\frac{44}{17}, c = -\frac{280}{17}$

103. The correct matrix that is needed is: $\left[\begin{array}{ccc|c} -1 & 1 & 1 & 2 \\ 1 & 1 & -2 & -3 \\ 1 & 1 & 1 & 6 \end{array}\right]$

After reducing this matrix the correct solution should be: $\left[\begin{array}{ccc|c} 1 & 0 & 0 & 2 \\ 0 & 1 & 0 & 1 \\ 0 & 0 & 1 & 3 \end{array}\right]$

105. Row 3 is not inconsistent. It implies $z = 0$.

107. false

109. true

111. $f(x) = -\frac{11}{6}x^4 + \frac{44}{3}x^3 - \frac{223}{6}x^2 + \frac{94}{3}x + 44$

113.
```
rref([A])
  [[1 0 -4 41]
   [0 1 -3 31]
   [0 0 0  0 ]]
```

115. a.

```
rref([A])
[[1 0 0 -.23653…
 [0 1 0 .928846…
 [0 0 1 6.08846…
```

$y = -0.24x^2 + 0.93x + 6.09$

b.

```
QuadReg
 y=ax²+bx+c
 a=-.2365384615
 b=.9288461538
 c=6.088461538
```

$y = -0.24x^2 + 0.93x + 6.09$

Section 7.2

1. 2×3 **3.** 2×2 **5.** 3×3

7. 1×1 **9.** 4×4 **11.** $x = -5, y = 1$

13. $x = -3, y = -2, z = 3$ **15.** $x = 6, y = 3$

17. $\begin{bmatrix} -1 & 5 & 1 \\ 5 & 2 & 5 \end{bmatrix}$ **19.** $\begin{bmatrix} -2 & 4 \\ 2 & -2 \\ -1 & 3 \end{bmatrix}$

21. not defined **23.** not defined

25. $\begin{bmatrix} -2 & 6 & 0 \\ 4 & 8 & 2 \end{bmatrix}$ **27.** $\begin{bmatrix} 0 & -5 \\ -10 & 5 \\ -15 & -5 \end{bmatrix}$

29. $\begin{bmatrix} -2 & 12 & 3 \\ 13 & 2 & 14 \end{bmatrix}$ **31.** $\begin{bmatrix} 8 & 3 \\ 11 & 5 \end{bmatrix}$

33. $\begin{bmatrix} -3 & 21 & 6 \\ -4 & 7 & 1 \\ 13 & 14 & 9 \end{bmatrix}$ **35.** $\begin{bmatrix} 3 & 6 \\ -2 & -2 \\ 17 & 24 \end{bmatrix}$

37. not defined **39.** $[0 \quad 60]$

41. $[-6 \quad 1 \quad -9]$ **43.** $\begin{bmatrix} 7 & 10 & -8 \\ 0 & 15 & 5 \\ 23 & 0 & -7 \end{bmatrix}$

45. $\begin{bmatrix} 12 & 20 \\ 30 & 42 \end{bmatrix}$ **47.** $\begin{bmatrix} -4 \\ -4 \\ -16 \end{bmatrix}$

49. not defined

51. $A = \begin{bmatrix} 0.70 \\ 0.30 \end{bmatrix} \quad B = \begin{bmatrix} 0.89 \\ 0.84 \end{bmatrix}$

a. $46A = \begin{bmatrix} 32.2 \\ 13.8 \end{bmatrix}$ Out of 46 people, 32 said yes they have tried to quit smoking and 14 people said they have not tried to quit smoking.

b. $46B = \begin{bmatrix} 40.94 \\ 38.64 \end{bmatrix}$ Out of 46 people, 41 people believed smoking would increase their chances of lung cancer and 39 people believed smoking would shorten their lives.

53. $A = \begin{bmatrix} 0.589 & 0.628 \\ 0.414 & 0.430 \end{bmatrix} \quad B = \begin{bmatrix} 100M \\ 110M \end{bmatrix}$

$AB = \begin{bmatrix} 127.98M \\ 88.7M \end{bmatrix}$ tells us that there are $127.98M$ registered voters and of those $88.7M$ voted.

55. $A = [0.45 \quad 0.5 \quad 1], B = \begin{bmatrix} 7{,}523 \\ 2{,}700 \\ 15{,}200 \end{bmatrix}, \quad AB = \$19{,}935.35$

57. $A = \begin{bmatrix} 230 & 3 & 44 & 9 \\ 430 & 19 & 46 & 20 \\ 290 & 5 & 45 & 19 \\ 330 & 5 & 47 & 24 \end{bmatrix}$

$2A = \begin{bmatrix} 460 & 6 & 88 & 18 \\ 860 & 38 & 92 & 40 \\ 580 & 10 & 90 & 38 \\ 660 & 10 & 94 & 48 \end{bmatrix}$ = represents the nutrition of foot-long (12″) sandwiches.

$0.5A = \begin{bmatrix} 115 & 1.5 & 22.0 & 4.5 \\ 215 & 9.5 & 23.0 & 10.0 \\ 145 & 2.5 & 22.5 & 9.5 \\ 165 & 2.5 & 23.5 & 12.0 \end{bmatrix}$ = represents the nutrition of 3″ sandwiches.

59. $AB = \begin{bmatrix} \$0.228 \\ \$0.081 \\ \$0.015 \end{bmatrix}$ represents the cost per mile of a gas, hybrid, or electric car.

61. Matrix multiplication is not performed element by element. Instead, a row by column method should be used. The correct matrix multiplication would produce the answer:

$$\begin{bmatrix} 3 & 2 \\ 1 & 4 \end{bmatrix}\begin{bmatrix} -1 & 3 \\ -2 & 5 \end{bmatrix} = \begin{bmatrix} -7 & 19 \\ -9 & 23 \end{bmatrix}$$

63. false **65.** true

67. $A^2 = AA = \begin{bmatrix} a_{11} & a_{12} \\ a_{21} & a_{22} \end{bmatrix}\begin{bmatrix} a_{11} & a_{12} \\ a_{21} & a_{22} \end{bmatrix}$

$$= \begin{bmatrix} a_{11}^2 + a_{12}a_{21} & a_{11}a_{12} + a_{12}a_{22} \\ a_{21}a_{11} + a_{22}a_{21} & a_{21}a_{12} + a_{22}^2 \end{bmatrix}$$

69. $A^2 = \begin{bmatrix} 2 & 2 \\ 2 & 2 \end{bmatrix} \quad A^3 = \begin{bmatrix} 4 & 4 \\ 4 & 4 \end{bmatrix}$

$A^4 = \begin{bmatrix} 8 & 8 \\ 8 & 8 \end{bmatrix} \quad A^n = \begin{bmatrix} 2^{n-1} & 2^{n-1} \\ 2^{n-1} & 2^{n-1} \end{bmatrix} = 2^{n-1} \cdot A$

71. must have $m = p$

73. $\begin{bmatrix} 33 & 35 \\ -96 & -82 \\ 31 & 19 \\ 146 & 138 \end{bmatrix}$ **75.** not defined

77. $\begin{bmatrix} 5 & -4 & 4 \\ 2 & -15 & -3 \\ 26 & 4 & -8 \end{bmatrix}$

Section 7.3

1. $\begin{bmatrix} -2 & 5 \\ 7 & -2 \end{bmatrix}\begin{bmatrix} x \\ y \end{bmatrix} = \begin{bmatrix} 10 \\ -4 \end{bmatrix}$ **3.** $\begin{bmatrix} 1 & -2 \\ -3 & 1 \end{bmatrix}\begin{bmatrix} x \\ y \end{bmatrix} = \begin{bmatrix} 8 \\ 6 \end{bmatrix}$

5. $\begin{bmatrix} 3 & 5 & -1 \\ 1 & 0 & 2 \\ -1 & 1 & -1 \end{bmatrix}\begin{bmatrix} x \\ y \\ z \end{bmatrix} = \begin{bmatrix} 2 \\ 17 \\ 4 \end{bmatrix}$

7. $\begin{bmatrix} 3 & 0 & 1 \\ 0 & 1 & -2 \\ 1 & 2 & 0 \end{bmatrix}\begin{bmatrix} x \\ y \\ z \end{bmatrix} = \begin{bmatrix} 10 \\ 4 \\ 6 \end{bmatrix}$

9. Yes, B is the multiplicative inverse of A.

11. Yes, B is the multiplicative inverse of A.

13. No, B is not the multiplicative inverse of A.

15. Yes, B is the multiplicative inverse of A.

17. No, B is not the multiplicative inverse of A.

19. $\begin{bmatrix} 0 & -1 \\ 1 & 2 \end{bmatrix}$ **21.** $\begin{bmatrix} -\frac{1}{13} & \frac{8}{39} \\ \frac{20}{39} & -\frac{4}{117} \end{bmatrix}$

23. $\begin{bmatrix} -0.1618 & 0.2284 \\ 0.5043 & -0.1237 \end{bmatrix}$ **25.** $\begin{bmatrix} \frac{1}{2} & \frac{1}{2} & 0 \\ \frac{1}{2} & 0 & \frac{1}{2} \\ 0 & -\frac{1}{2} & -\frac{1}{2} \end{bmatrix}$

27. A^{-1} does not exist. **29.** $\begin{bmatrix} -\frac{1}{2} & -\frac{1}{2} & \frac{5}{2} \\ \frac{1}{2} & \frac{1}{2} & -\frac{3}{2} \\ 0 & -1 & 1 \end{bmatrix}$

31. $\begin{bmatrix} \frac{1}{2} & \frac{1}{2} & 0 \\ \frac{3}{4} & \frac{1}{4} & -\frac{1}{2} \\ \frac{1}{4} & \frac{3}{4} & -\frac{1}{2} \end{bmatrix}$ **33.** $x = 2, y = -1$

35. $x = \frac{1}{2}, y = \frac{1}{3}$ **37.** $x = 4, y = -3$

39. $x = 0, y = 0, z = 1$

41. Can't solve because A is not invertible.

43. $x = -1, y = 1, z = -7$

45. $x = 3, y = 5, z = 4$

47. a. $A^{-1}B = \begin{bmatrix} \$50 \\ \$20 \end{bmatrix}$

b. $A^{-1}B$ represents the price of sweatshirts ($50) and price of T-shirts ($20).

49. JAW **51.** LEG **53.** EYE

55. A is not invertible because the identity matrix was not reached.

57. false **59.** $x = 9$ **63.** $ad - bc = 0$

65. $\begin{bmatrix} -\frac{115}{6008} & \frac{431}{6008} & -\frac{1067}{6008} & \frac{103}{751} \\ \frac{411}{6008} & -\frac{391}{6008} & \frac{731}{6008} & -\frac{22}{751} \\ \frac{57}{751} & \frac{28}{751} & -\frac{85}{751} & \frac{3}{751} \\ -\frac{429}{6008} & \frac{145}{6008} & \frac{1035}{6008} & \frac{12}{751} \end{bmatrix}$

67. $x = 1.8, y = -1.6$ **69.** $(3.7, -2.4, 9.3)$

Section 7.4

1. -2 **3.** 31 **5.** -28

7. -0.6 **9.** 0 **11.** $x = 5, y = -6$

13. $x = -2, y = 1$ **15.** $x = -3, y = -4$

17. $x = -2, y = 5$ **19.** $x = 2, y = 2$

21. $D = 0$, inconsistent or dependent system

23. $D = 0$, inconsistent or dependent system

25. $x = \frac{1}{2}, y = -1$ **27.** $x = 1.5, y = 2.1$

29. $x = 0, y = 7$ **31.** $x = \frac{1}{3}, y = \frac{3}{4}$

33. 7 **35.** -25 **37.** -180

39. 0 **41.** 238 **43.** 0

45. 95.7 **47.** $x = 2, y = 3, z = 5$

49. $x = -2, y = 3, z = 5$ **51.** $x = 2, y = -3, z = 1$

53. $D = 0$, inconsistent or dependent system

55. $D = 0$, inconsistent or dependent system

57. $(-3, 1, 4)$ **59.** $(2, -3, 5)$ **61.** 6 units^2

63. 6 units^2 **65.** $y = 2x$ **67.** $I_1 = \frac{7}{2}, I_2 = \frac{5}{2}, I_3 = 1$

69. The mistake is forgetting the sign array (alternating signs).

71. D_x should have the first column replaced with the column of constants and D_y should have the second column replaced with the column of constants.

73. true **75.** false **77.** abc

79. -419 **83.** -180 **85.** -1019

87. $(-6.4, 1.5, 3.4)$

Review Exercises

1. $\left[\begin{array}{cc|c} 5 & 7 & 2 \\ 3 & -4 & -2 \end{array}\right]$ **3.** $\left[\begin{array}{ccc|c} 2 & 0 & -1 & 3 \\ 0 & 1 & -3 & -2 \\ 1 & 0 & 4 & -3 \end{array}\right]$

5. no **7.** no **9.** $\left[\begin{array}{cc|c} 1 & -2 & 1 \\ 0 & 1 & -1 \end{array}\right]$

11. $\left[\begin{array}{ccc|c} 1 & -4 & 3 & -1 \\ 0 & -2 & 3 & -2 \\ 0 & 1 & -4 & 8 \end{array}\right]$ **13.** $\left[\begin{array}{cc|c} 1 & 0 & \frac{3}{5} \\ 0 & 1 & -\frac{1}{5} \end{array}\right]$

15. $\left[\begin{array}{ccc|c} 1 & 0 & 0 & -4 \\ 0 & 1 & 0 & 8 \\ 0 & 0 & 1 & -4 \end{array}\right]$ **17.** $x = \frac{5}{4}, y = \frac{7}{8}$

19. $x = 2, y = 1$ **21.** $x = -\frac{74}{21}, y = -\frac{73}{21}, z = -\frac{3}{7}$

23. $x = 1, y = 3, z = -5$

25. $x = -\frac{3}{7}a - 2, y = \frac{2}{7}a + 2, z = a$

27. $y = -0.005x^2 + 0.45x - 3.89$ **29.** not defined

31. $\begin{bmatrix} 3 & 5 & 2 \\ 7 & 8 & 1 \end{bmatrix}$ **33.** $\begin{bmatrix} 9 & -4 \\ 9 & 9 \end{bmatrix}$

35. $\begin{bmatrix} 4 & 13 \\ 18 & 11 \end{bmatrix}$ **37.** $\begin{bmatrix} 0 & -19 \\ -18 & -9 \end{bmatrix}$

39. $\begin{bmatrix} -7 & -11 & -8 \\ 3 & 7 & 2 \end{bmatrix}$ **41.** $\begin{bmatrix} 10 & -13 \\ 18 & -20 \end{bmatrix}$

43. $\begin{bmatrix} 17 & -8 & 18 \\ 33 & 0 & 42 \end{bmatrix}$ **45.** $\begin{bmatrix} 10 & 9 & 20 \\ 22 & -4 & 2 \end{bmatrix}$

47. Yes, B is the multiplicative inverse of A.

49. Yes, B is the multiplicative inverse of A.

51. $\begin{bmatrix} \frac{2}{5} & -\frac{1}{5} \\ \frac{3}{10} & \frac{1}{10} \end{bmatrix}$ **53.** $\begin{bmatrix} 0 & -\frac{1}{2} \\ 1 & 0 \end{bmatrix}$

55. $\begin{bmatrix} -\frac{1}{6} & \frac{7}{12} & -\frac{1}{12} \\ \frac{1}{2} & -\frac{1}{4} & -\frac{1}{4} \\ \frac{1}{6} & -\frac{1}{12} & -\frac{5}{12} \end{bmatrix}$ **57.** $\begin{bmatrix} 0 & -\frac{2}{5} & \frac{1}{5} \\ 1 & -\frac{2}{5} & \frac{1}{5} \\ -\frac{1}{2} & \frac{3}{10} & \frac{1}{10} \end{bmatrix}$

59. $x = 5, y = 4$ **61.** $x = 8, y = 12$

63. $x = 1, y = 2, z = 3$ **65.** -8

67. 5.4 **69.** $x = 3, y = 1$

71. $x = 6, y = 0$ **73.** $x = 90, y = 155$

75. 11 **77.** $-abd$

79. $x = 1, y = 1, z = 2$ **81.** $x = -\frac{15}{7}, y = -\frac{25}{7}, z = \frac{19}{14}$

83. The area of the triangle = 1 unit^2

85. a.

```
rref([A])
[[1 0 0 .161066…
 [0 1 0 -.04946…
 [0 0 1 -4.1012…
```

$y = 0.16x^2 - 0.05x - 4.10$

b.

```
QuadReg
 y=ax²+bx+c
 a=.1610661269
 b=-.0494601889
 c=-4.101214575
```

$y = 0.16x^2 - 0.05x - 4.10$

87. $\begin{bmatrix} -235 & 206 & 50 \\ -113 & 159 & 135 \\ 40 & -30 & 0 \end{bmatrix}$ **89.** $x = 2.25, y = -4.35$

91. $x = -9.5, y = 3.4$

Practice Test

1. $\left[\begin{array}{cc|c} 1 & -2 & 1 \\ -1 & 3 & 2 \end{array}\right]$ **3.** $\left[\begin{array}{ccc|c} 6 & 9 & 1 & 5 \\ 2 & -3 & 1 & 3 \\ 10 & 12 & 2 & 9 \end{array}\right]$

5. $\begin{bmatrix} 1 & 3 & 5 \\ 0 & 1 & -11 \\ 0 & 7 & 15 \end{bmatrix}$ **7.** $x = -\frac{1}{3}a + \frac{7}{6}, y = \frac{1}{9}a - \frac{2}{9}, z = a$

9. $\begin{bmatrix} -11 & 19 \\ -6 & 8 \end{bmatrix}$ **11.** $\begin{bmatrix} \frac{1}{19} & \frac{3}{19} \\ \frac{5}{19} & -\frac{4}{19} \end{bmatrix}$

13. not possible **15.** $x = -3, y = 1, z = 7$

17. -31 **19.** $x = 1, y = -1, z = 2$

21. \$3,500 in the money market, \$4,500 in the conservative stock, and \$7,000 in the aggressive stock

23. $(12.5, -6.4)$

Cumulative Test

1. $x = 3 \pm 2\sqrt{5}$

3. $(x + 3)^2 + (y + 1)^2 = 25$

5. even

7. $f(g(x)) = \frac{1}{x^3} - 1$; domain: $(-\infty, 0) \cup (0, \infty)$

9. $\left(-\frac{6}{5}, -\frac{3}{5}\right)$

11. $Q(x) = 5x - 4, r(x) = -5x + 7$

13. HA: $y = 0.7$; VA: $x = -2, x = 3$

15. $(-3, \infty)$ **17.** $x = 4$ **19.** no solution

21. Maximum: $z\left(\frac{10}{7}, \frac{10}{7}\right) = \frac{120}{7}$

23. $2A + CB = \begin{bmatrix} 78 & -10 & 40 \\ 26 & 0 & 14 \end{bmatrix}$

25. $x = \frac{3}{11}, y = -\frac{2}{11}$

27. $\begin{bmatrix} 35 & 12 \\ -19 & -14 \end{bmatrix}, \begin{bmatrix} \frac{7}{131} & \frac{6}{131} \\ -\frac{19}{262} & -\frac{35}{262} \end{bmatrix}$

31. focus: (1, 0)
vertex: (0, 0)
directrix: $x = -1$
length of latus rectum: 4

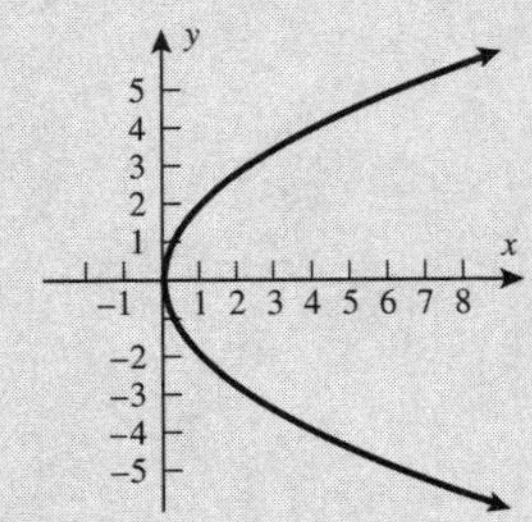

CHAPTER 8

Section 8.1

1. hyperbola **3.** circle **5.** hyperbola
7. ellipse **9.** parabola **11.** circle

Section 8.2

1. c **3.** d **5.** c **7.** a

9. $x^2 = 12y$ **11.** $y^2 = -20x$

13. $(x - 3)^2 = 8(y - 5)$ **15.** $(y - 4)^2 = -8(x - 2)$

17. $(x - 2)^2 = 12(y - 1)$ **19.** $(y + 1)^2 = 4(x - 2)$

21. $(y - 2)^2 = 8(x + 1)$ **23.** $(x - 2)^2 = -8(y + 1)$

25. focus: (0, 2)
vertex: (0, 0)
directrix: $y = -2$
length of latus rectum: 8

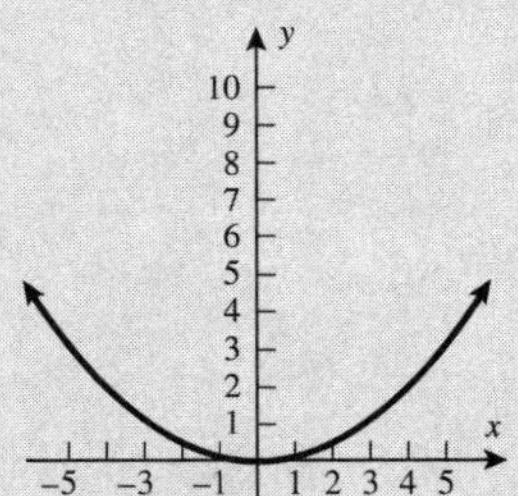

27. focus: $\left(-\frac{1}{2}, 0\right)$
vertex: (0, 0)
directrix: $x = \frac{1}{2}$
length of latus rectum: 2

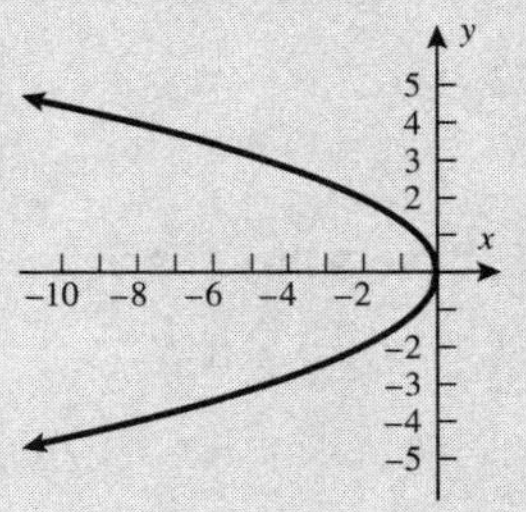

29. focus: (0, 4)
vertex: (0, 0)
directrix: $y = -4$
length of latus rectum: 16

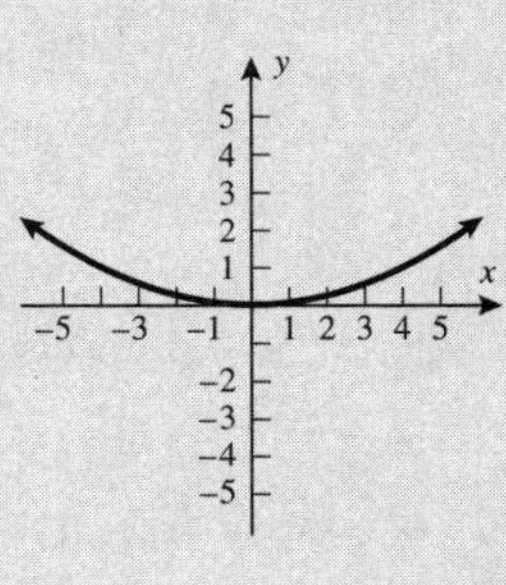

33. vertex: (−3, 2)

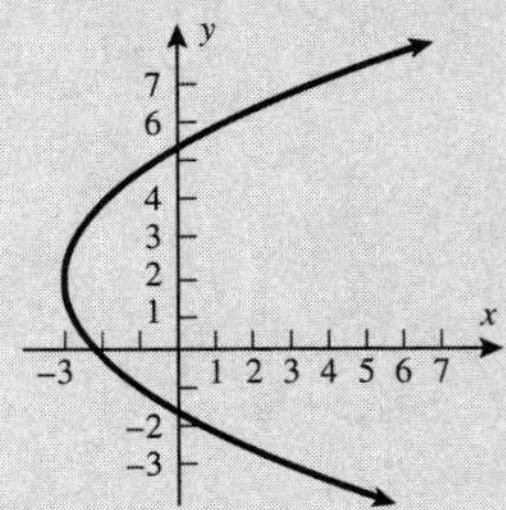

35. vertex: (3, −1)

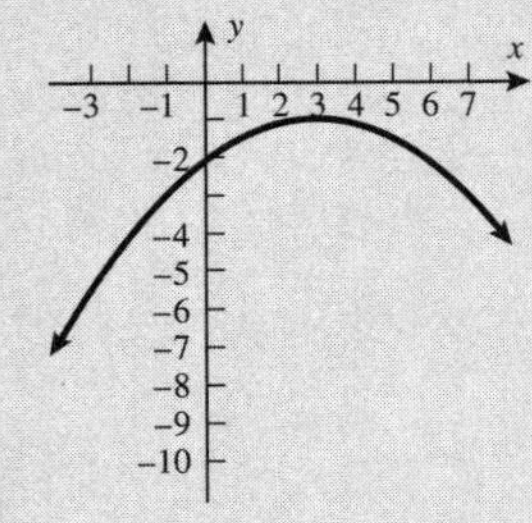

37. vertex: (−5, 0)

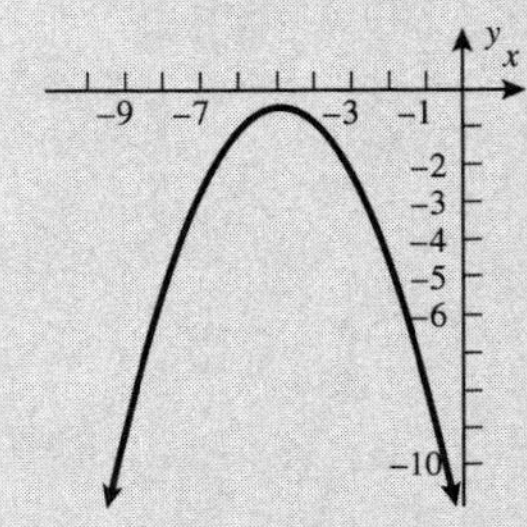

39. vertex: (0, 2)

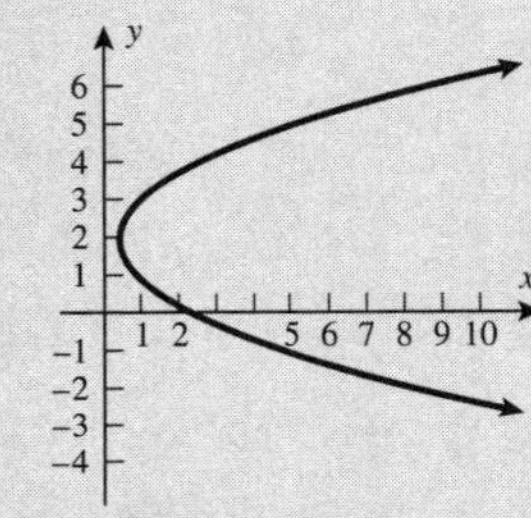

41. vertex: (−3, −1)

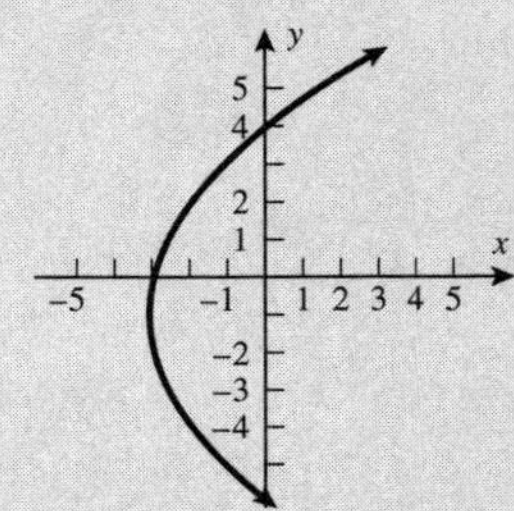

43. vertex: $\left(\frac{1}{2}, \frac{5}{4}\right)$

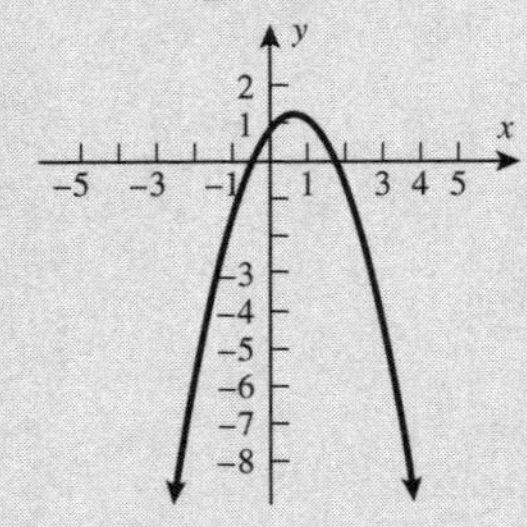

45. The focus will be at (0, 2) so the receiver should be placed 2 feet from the vertex.

47. $x = \frac{1}{8}y^2$, $-2.5 \leq y \leq 2.5$; assume lens opens right.

49. $x^2 = 160y$

51. Yes. The opening height is 18.75 feet, and the mast is only 17 feet.

53. Focal length = 374.25 ft
Equation: $x^2 = 1497y$ (measurement in feet)

55. The correction that needs to be made is that the formula $y^2 = 4px$ should be used.

57. true **59.** false

63.

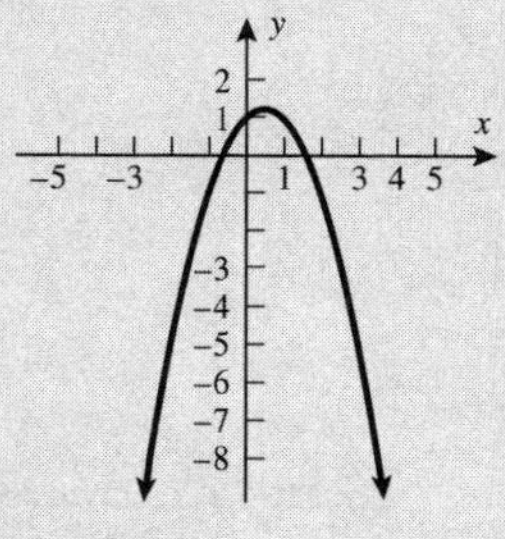

65. The vertex is located at: (2.5, −3.5). The parabola opens to the right.

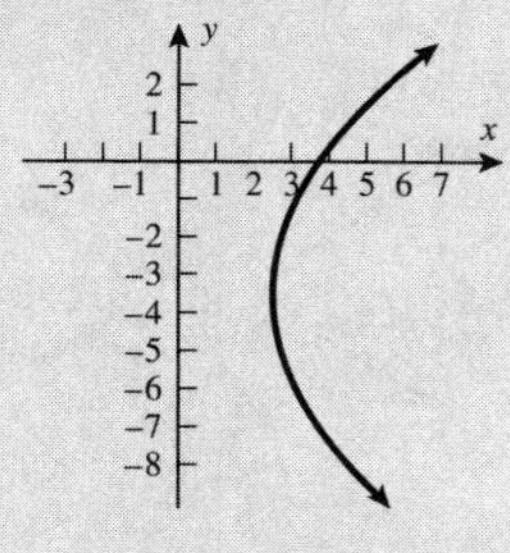

67. The vertex is located at (1.8, 1.5). The parabola opens to the left.

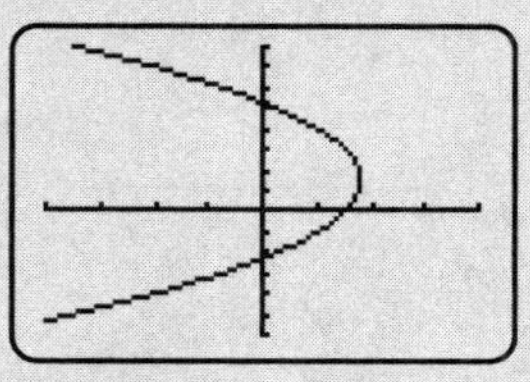

Section 8.3

1. d

3. a

5.

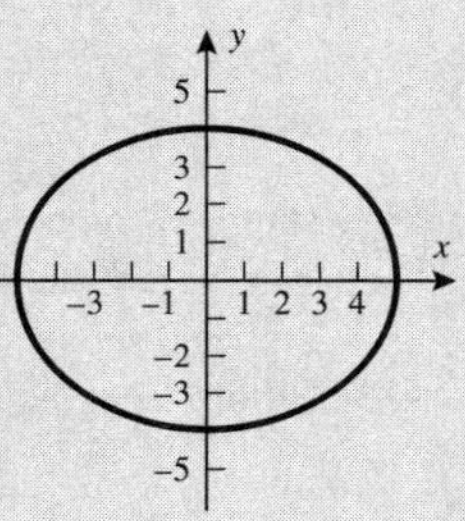

7.

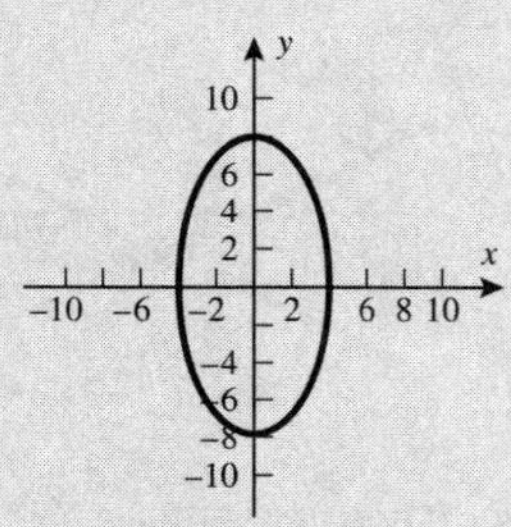

9.

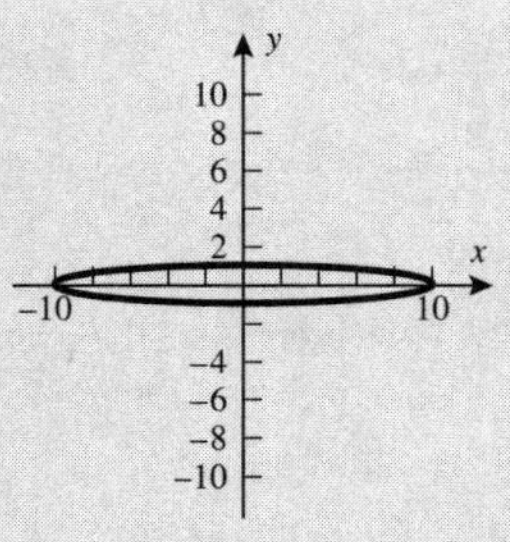

11.

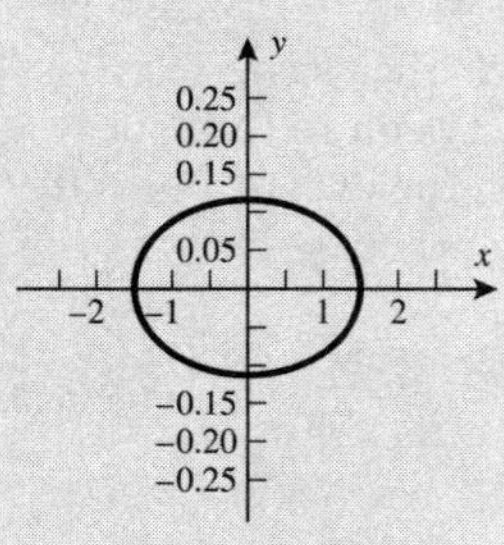

13.

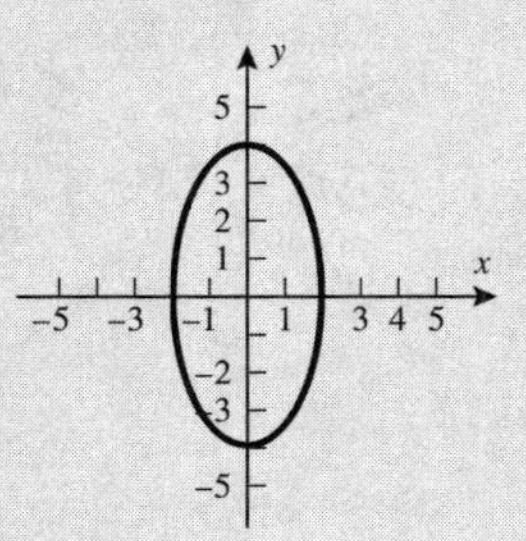

15.

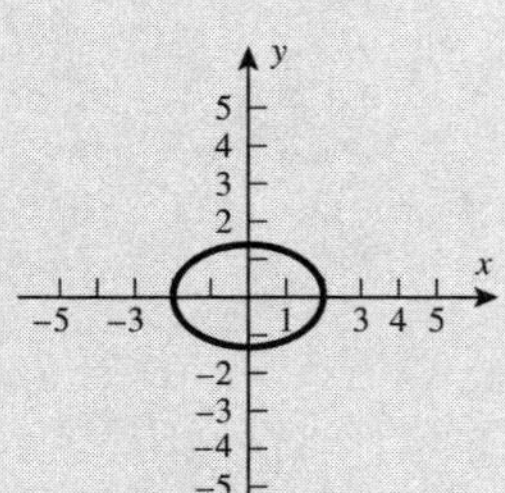

17. $\frac{x^2}{36} + \frac{y^2}{20} = 1$

19. $\frac{x^2}{7} + \frac{y^2}{16} = 1$

21. $\frac{x^2}{4} + \frac{y^2}{16} = 1$

23. $\frac{x^2}{9} + \frac{y^2}{49} = 1$

25. c

27. b

29.

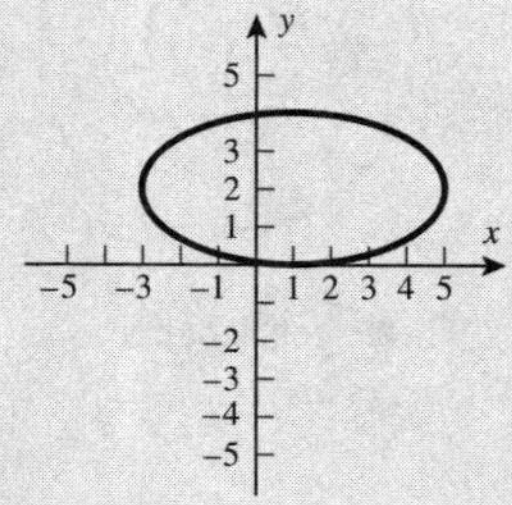

31.

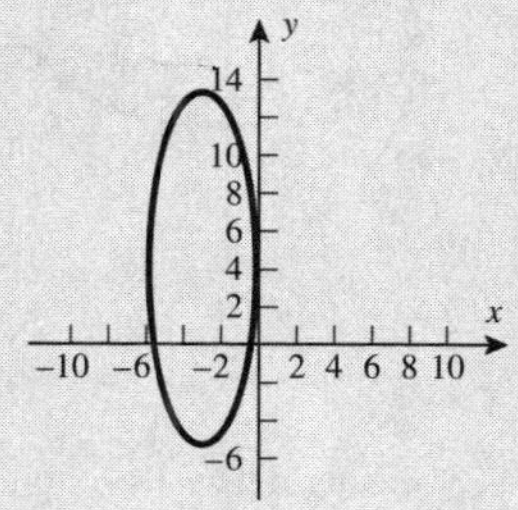

33.

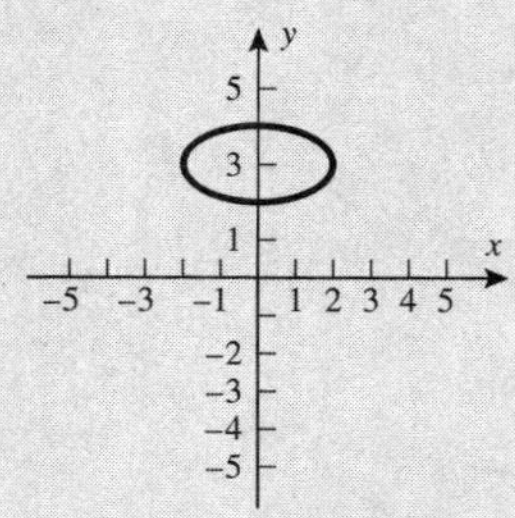

35.

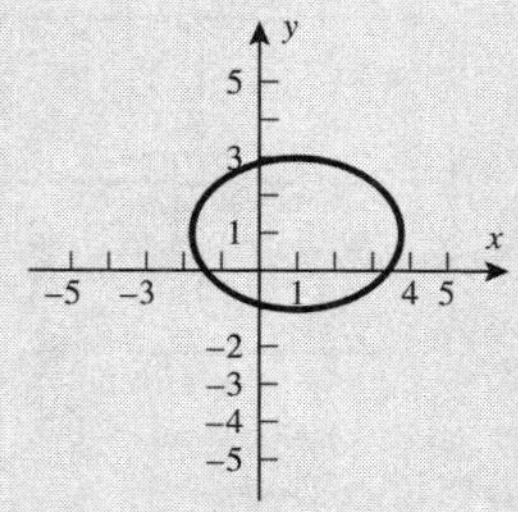

37.

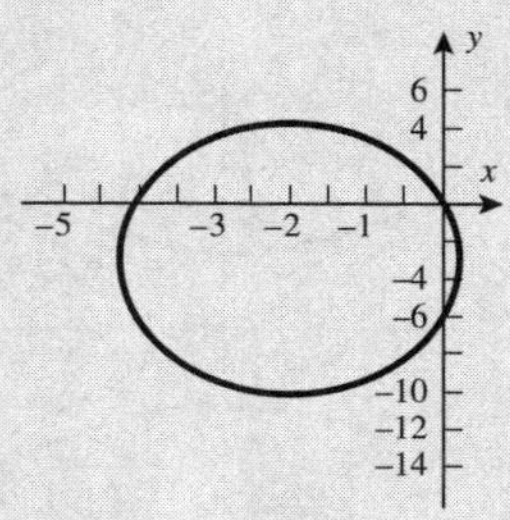

39. $\frac{(x-2)^2}{25} + \frac{(y-5)^2}{9} = 1$

41. $\frac{(x-4)^2}{7} + \frac{(y+4)^2}{16} = 1$

43. $\frac{(x-3)^2}{4} + \frac{(y-2)^2}{16} = 1$

45. $\frac{(x+1)^2}{9} + \frac{(y+4)^2}{25} = 1$

47. $\frac{x^2}{225} + \frac{y^2}{5625} = 1$

49. a. $\frac{x^2}{5625} + \frac{y^2}{400} = 1$ **b.** The width of the track at the end of the field is 24 yards (so it will not encompass the football field, which is 30 yards wide).

51. $\frac{x^2}{5.914{,}000{,}000^2} + \frac{y^2}{5{,}729{,}000{,}000^2} = 1$

53. $\frac{x^2}{150{,}000{,}000^2} + \frac{y^2}{146{,}000{,}000^2} = 1$

55. straight line

57. $a^2 = 6$ and $b^2 = 4$, therefore $a = \pm\sqrt{6}$ and $b = \pm 2$.

59. false

61. true

63. Pluto: $e \approx 0.25$ Earth: $e \approx 0.02$

65. 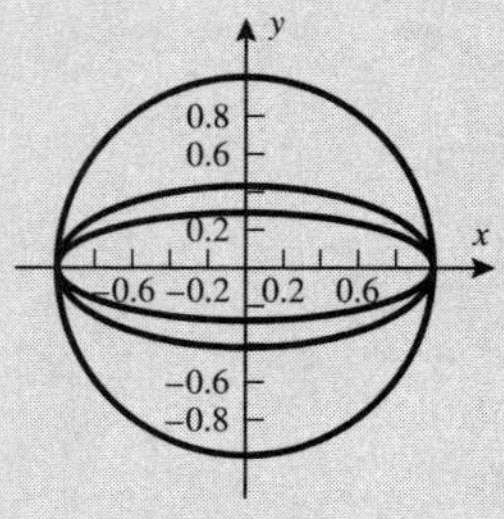

Ellipse becomes more elongated.

67. 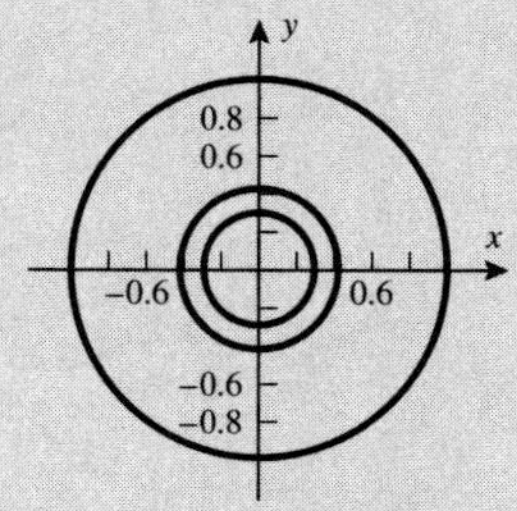

Circle gets smaller as c increases.

69. As c decreases, the major axis along the x-axis increases.

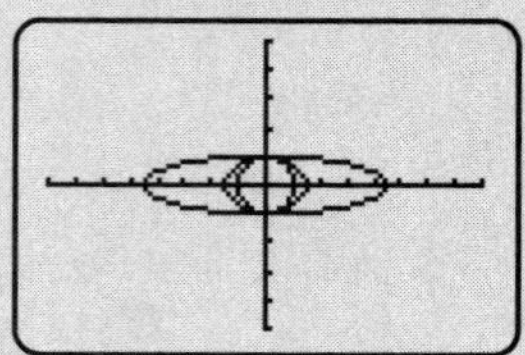

Section 8.4

1. b

3. d

5. 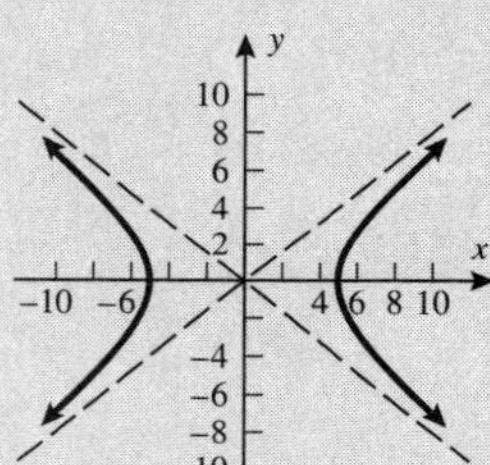

7. 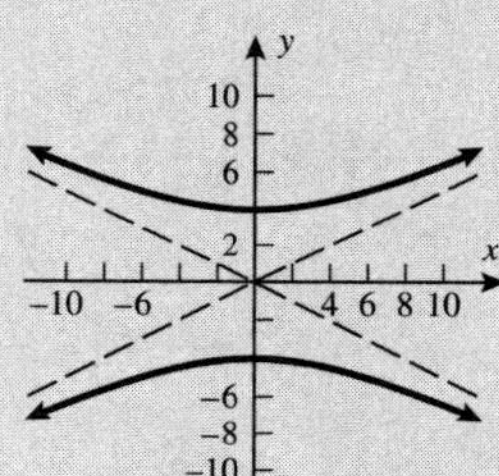

9. 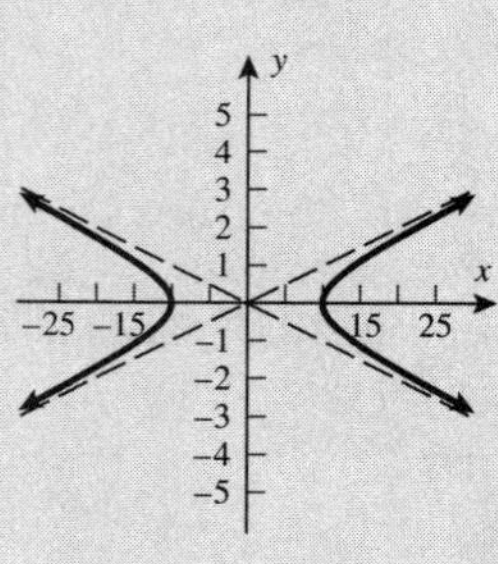

11. 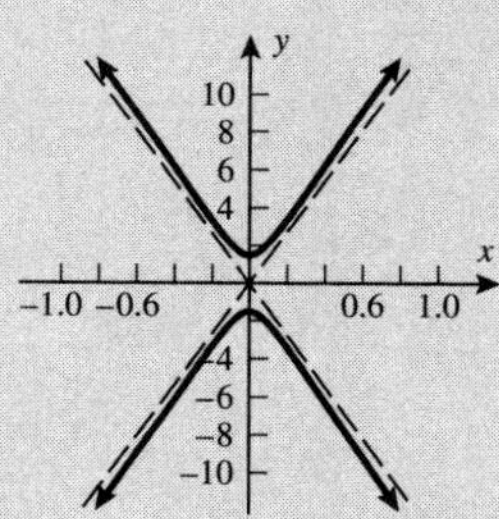

13. 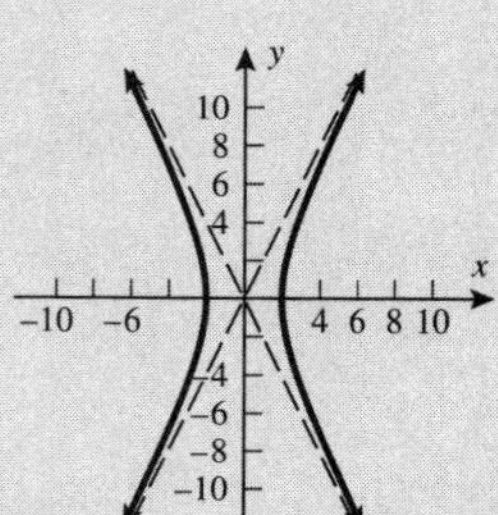

15. 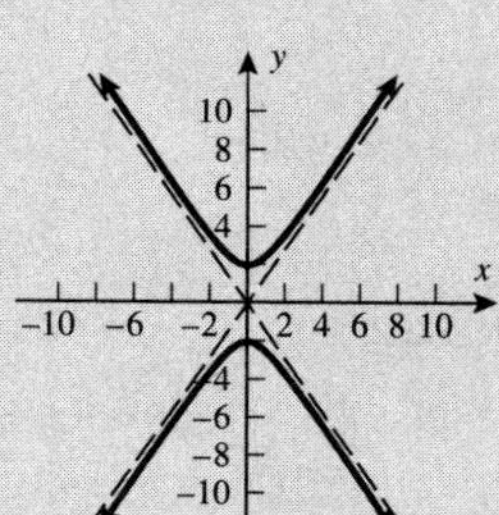

17. $\dfrac{x^2}{16} - \dfrac{y^2}{20} = 1$

19. $\dfrac{y^2}{9} - \dfrac{x^2}{7} = 1$

21. $x^2 - y^2 = a^2$

23. $\dfrac{y^2}{4} - x^2 = b^2$

25. c

27. b

29. 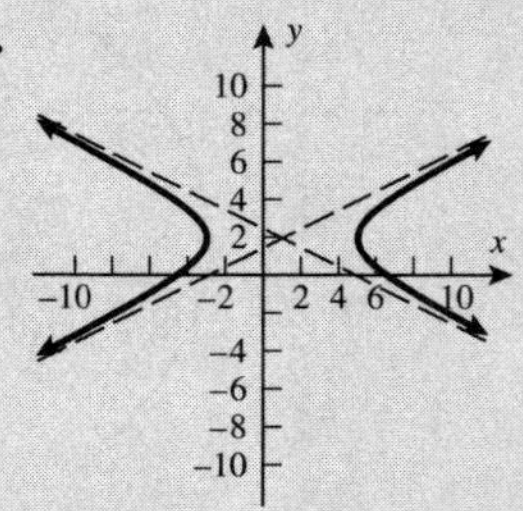

31. 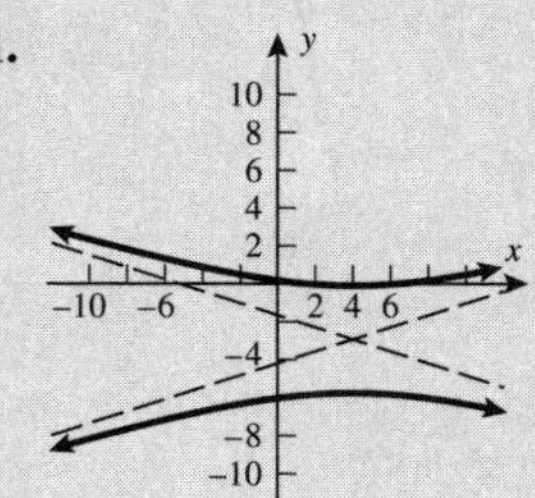

33. 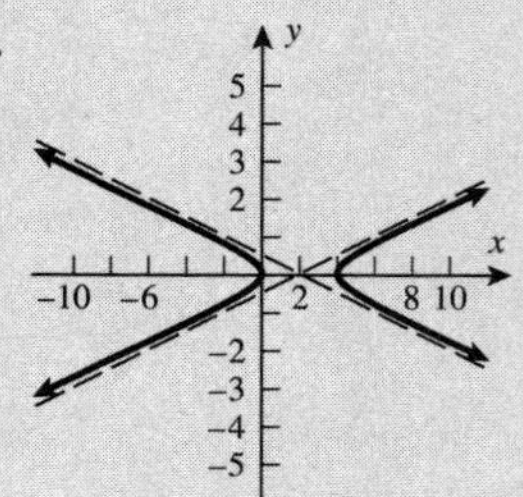

35. 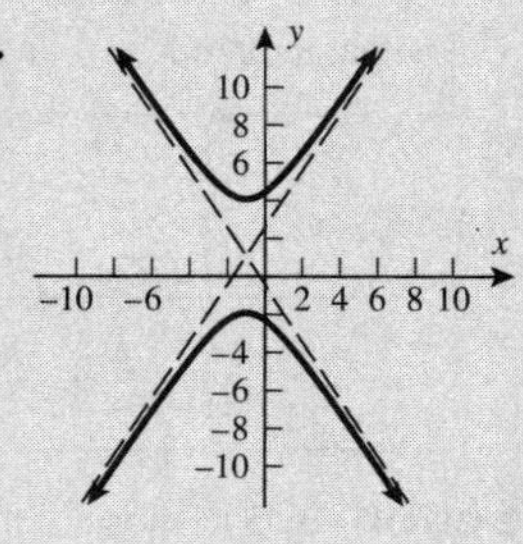

37. 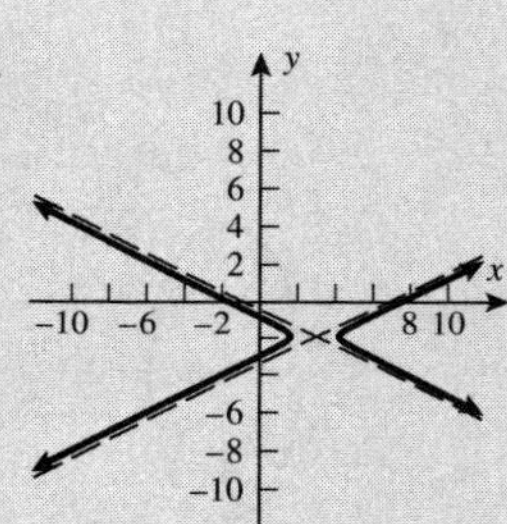

39. $\dfrac{(x-2)^2}{16} - \dfrac{(y-5)^2}{9} = 1$

41. $\dfrac{(y+4)^2}{9} - \dfrac{(x-4)^2}{7} = 1$

43. The ship will come ashore between the two stations (28.5 miles from one and 121.5 miles from the other).

45. 0.000484 seconds

47. $y^2 - \frac{4}{5}x^2 = 1$

49. The transverse axis is vertical. The points are $(\pm 3, 0)$. The vertices are $(0, \pm 2)$.

51. false

53. true

55. $x^2 - y^2 = a^2$ or $y^2 - x^2 = a^2$; note that $a = b$

57. As c increases, the graphs squeeze toward the x-axis.

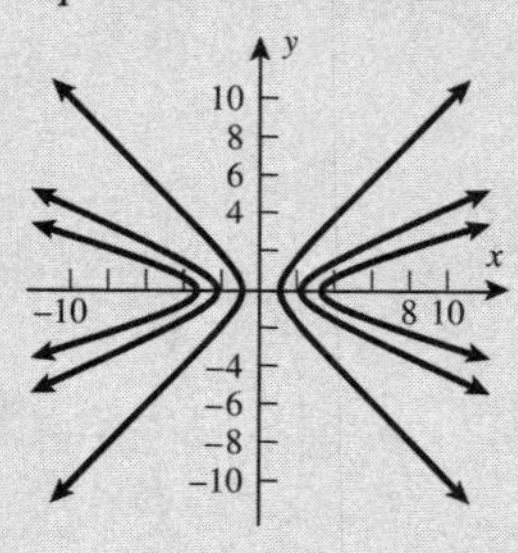

59. As c decreases, the vertices of the hyperbolas located at $\left(\pm\frac{1}{c}, 0\right)$ are moving away from the origin.

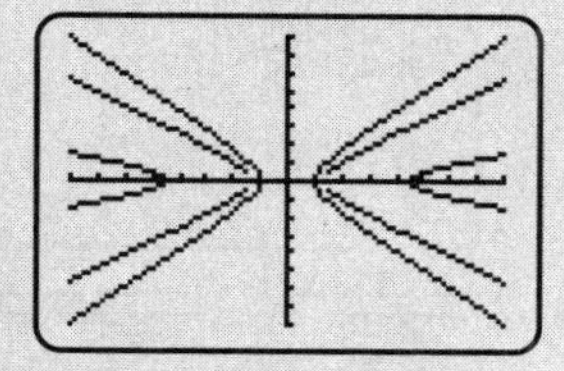

61. (1.067, 4.119), (1.986, 0.638), (−1.017, −4.757)

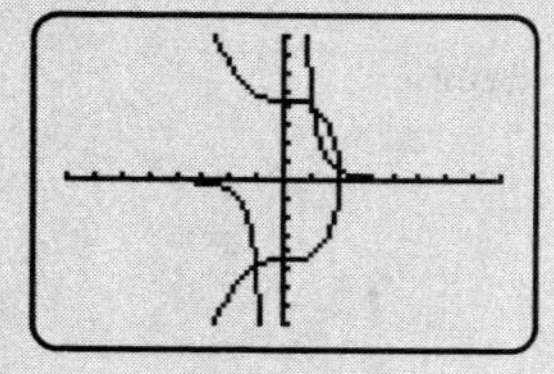

Section 8.5

1. (2, 6), (−1, 3) **3.** (1, 0)

5. no solution **7.** (0, 1)

9. (−0.63, −1.61), (0.63, −1.61)

11. no solution **13.** (1, 1)

15. $(-\sqrt{2}, -2\sqrt{2}), (\sqrt{2}, 2\sqrt{2}), (2\sqrt{2}, \sqrt{2}), (-2\sqrt{2}, -\sqrt{2})$

17. (−6, 33), (2, 1) **19.** (3, 4) (−2, −1)

21. $(0, -3), \left(\frac{2}{5}, -\frac{11}{5}\right)$ **23.** $(-1, -1), \left(\frac{1}{4}, \frac{3}{2}\right)$

25. (−1, −4), (4, 1) **27.** (1, 3), (−1, −3)

29. (2, 4) **31.** $\left(\frac{1}{2}, \frac{1}{3}\right), \left(\frac{1}{2}, -\frac{1}{3}\right)$

33. no solution

35.

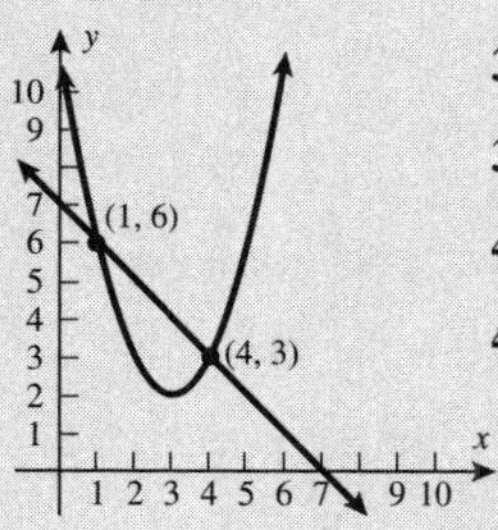

37. 3 and 7

39. 8 and 9; −8 and −9

41. 8 centimeters by 10 centimeters

43. 400 feet by 500 feet or $\frac{1000}{3}$ feet by 600 feet

45. Professor: 2 meters per second
Jeremy: 10 meters per second

47. Can't use elimination; should have used substitution.

49. false **51.** false

53. $2n$ **55.** $\left.\begin{array}{l} y = x^2 + 1 \\ y = 1 \end{array}\right\}$ other answers possible

57. no solution

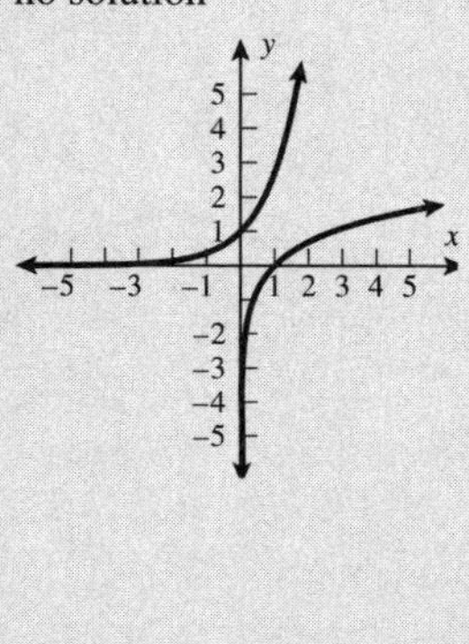

59. (−1.57, −1.64)

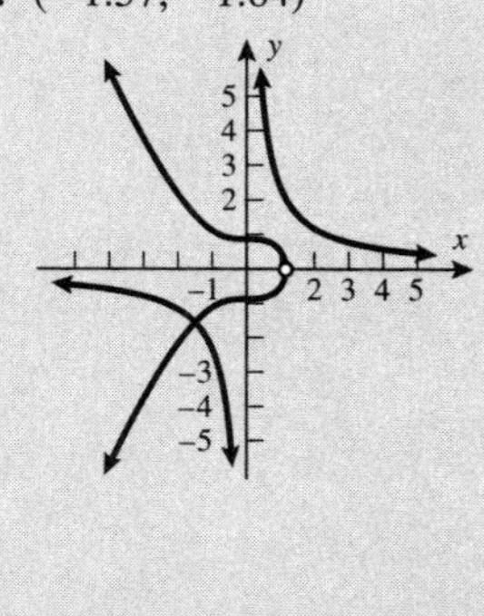

Section 8.6

1. b **3.** j **5.** h **7.** c **9.** d **11.** k

13.

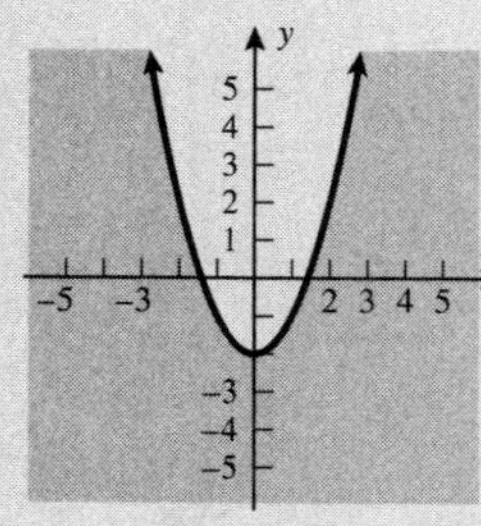

15.

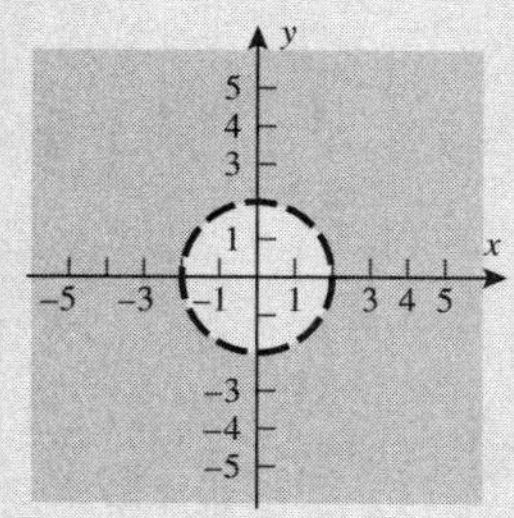

17.

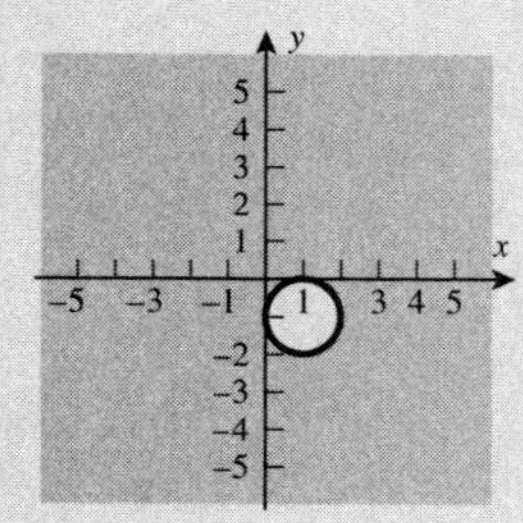

19.

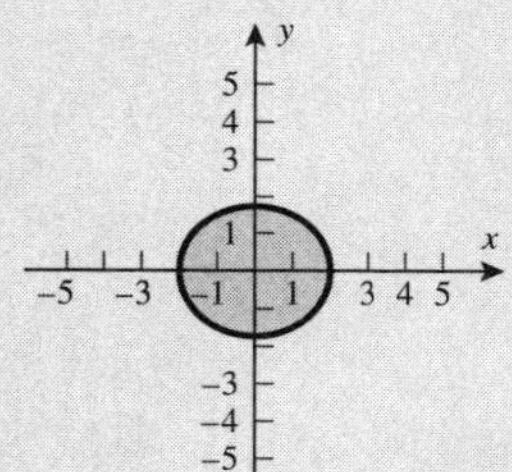

21.

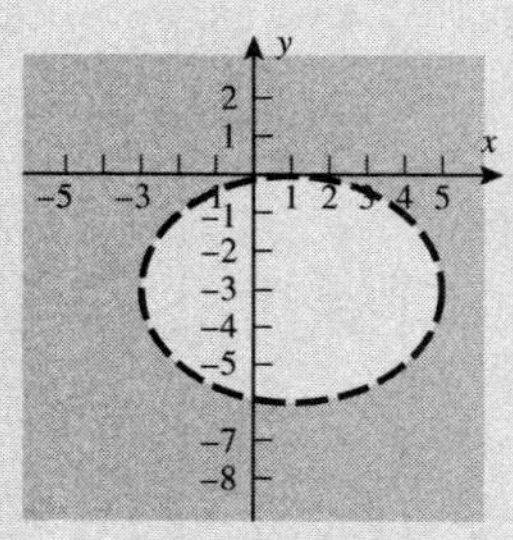

23.

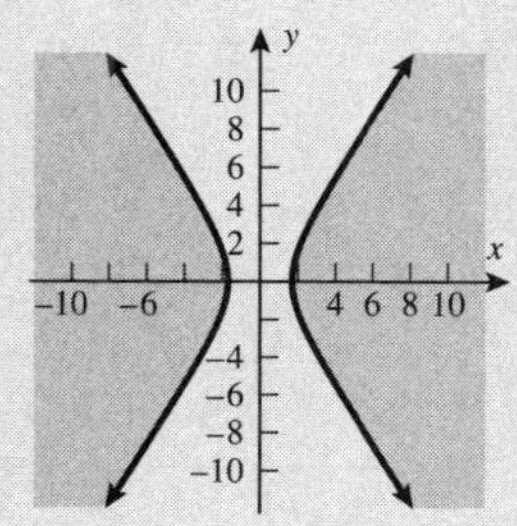

25.

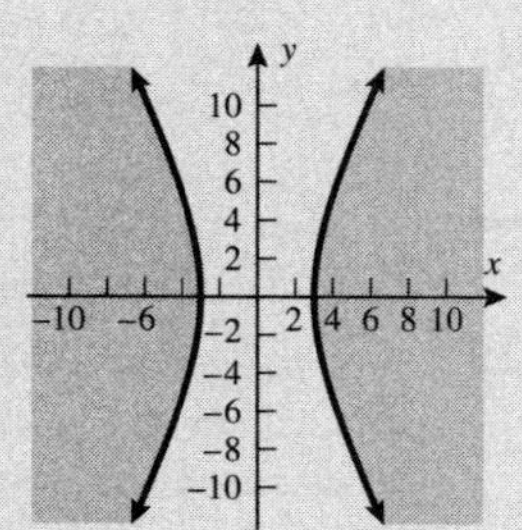

27.

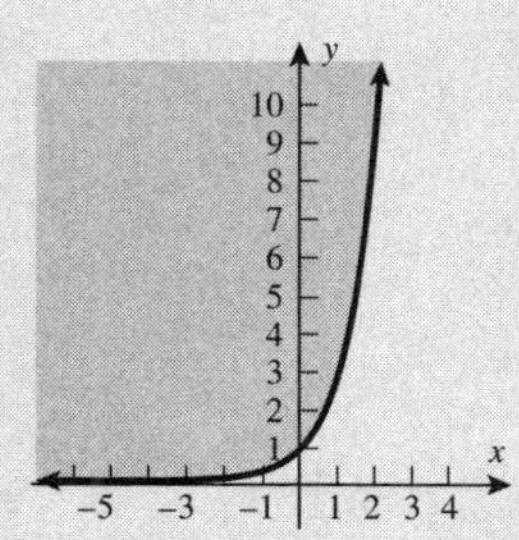

29.

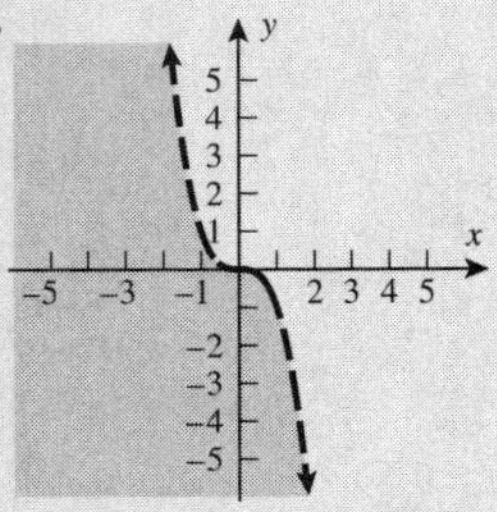

31.

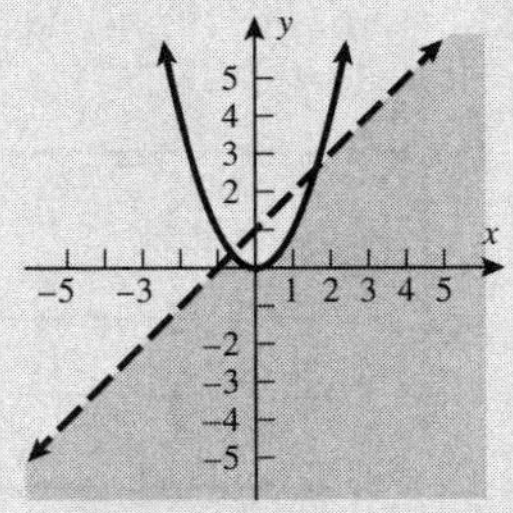

33.

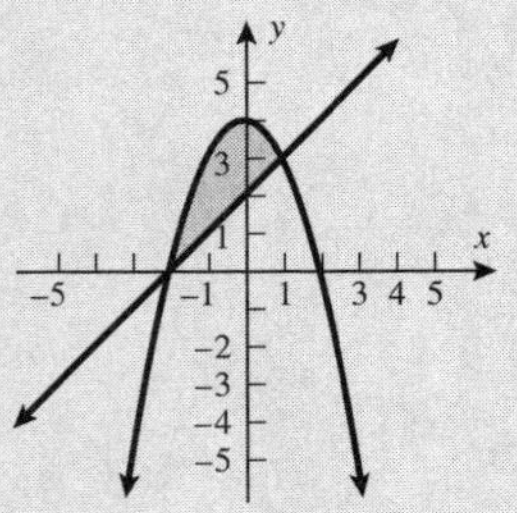

35.

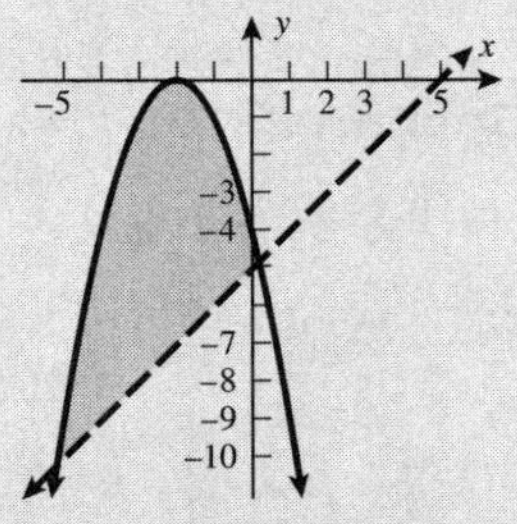

37.

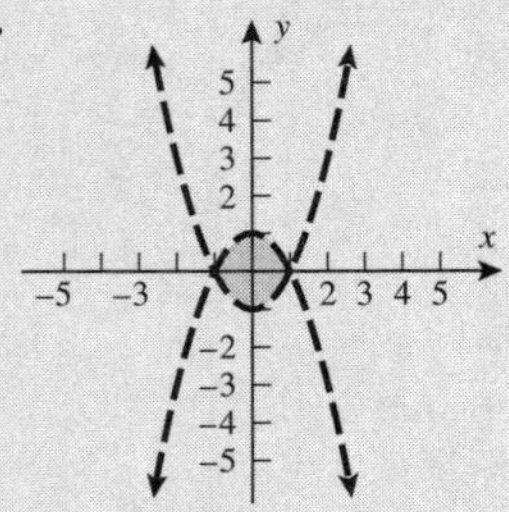

39.

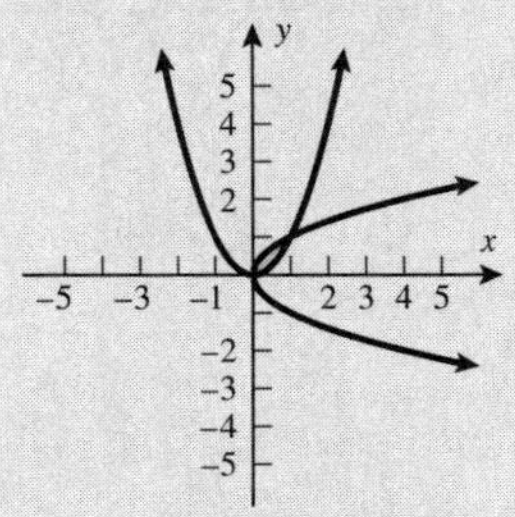

41.

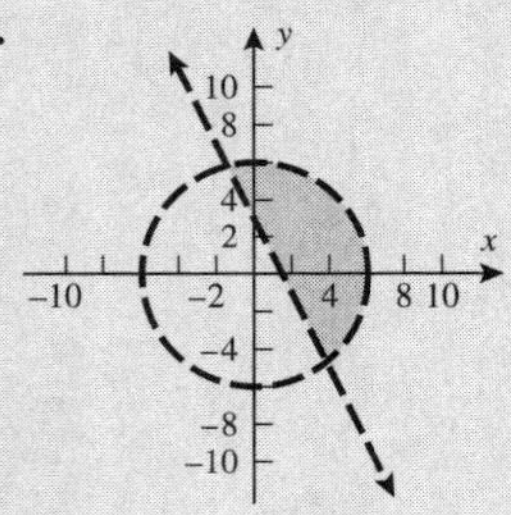

43.

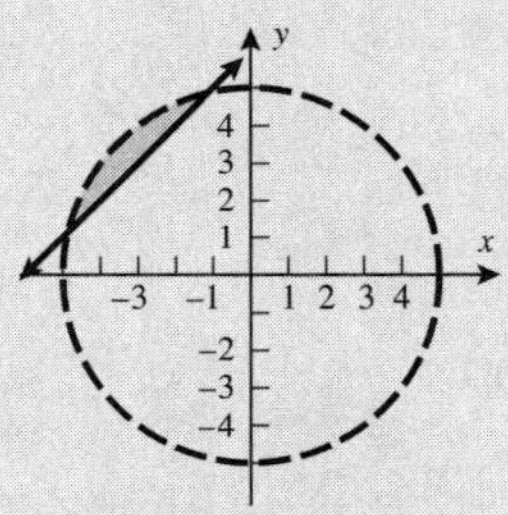

45.

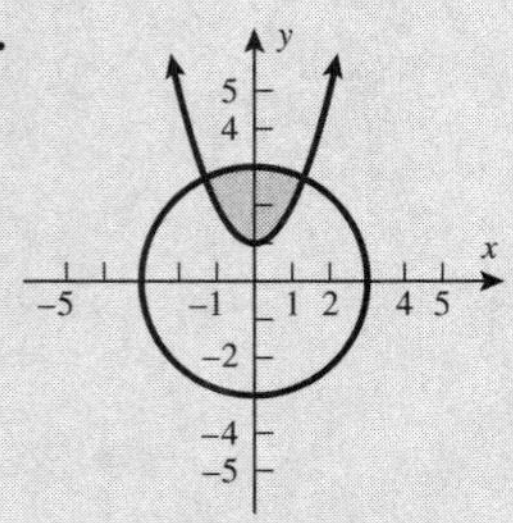

47.

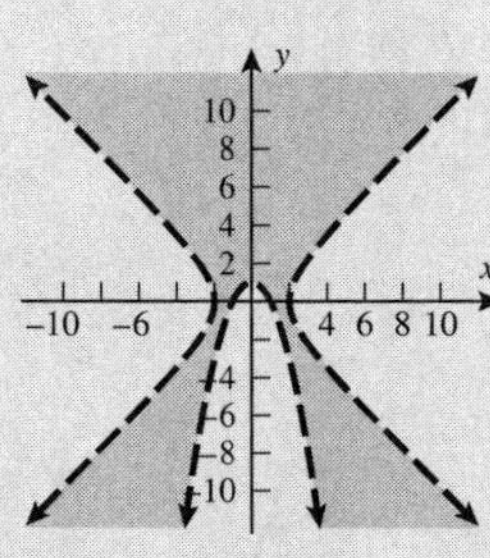

49.

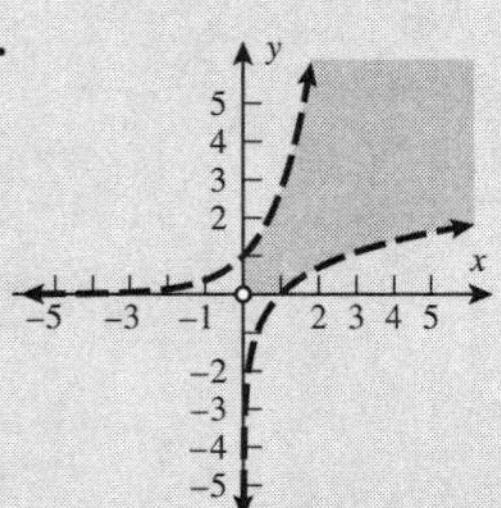

51. $\frac{9\pi}{2}$ units^2

53. No solution (should have shaded inside the inner circle and outside the outer circle, resulting in no overlap).

55. false

57. $0 \le a \le b$

59.

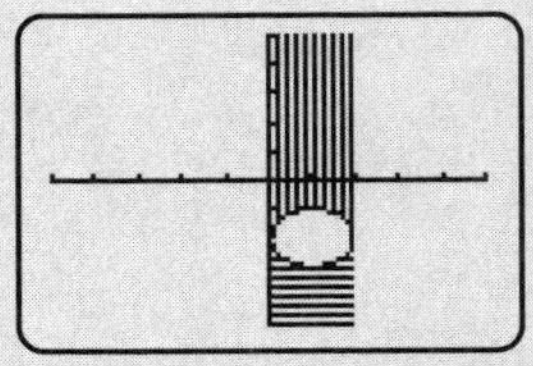

61.

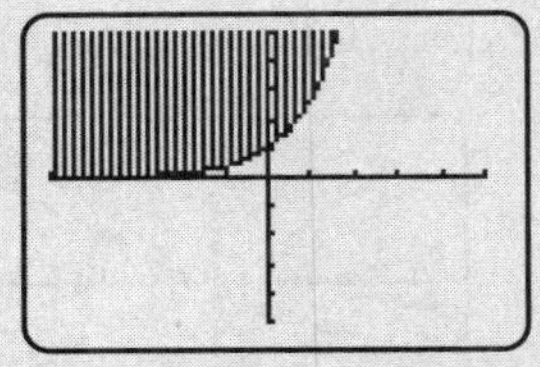

63.

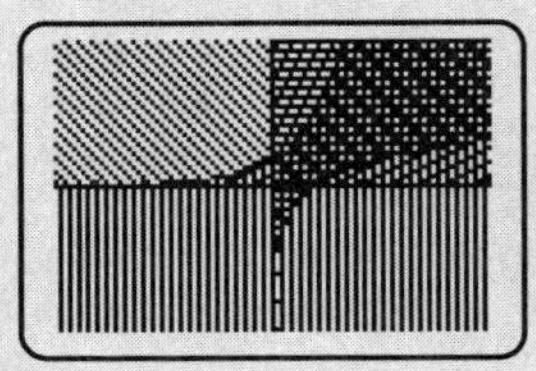

65.

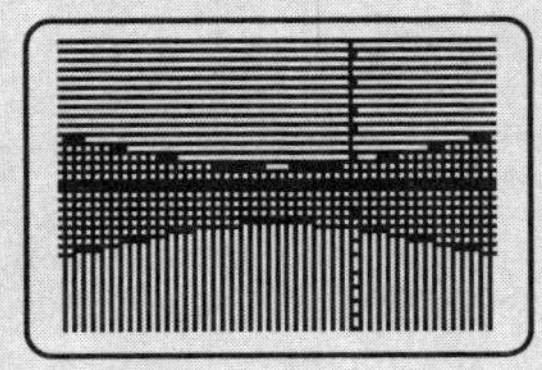

Review Exercises

1. false

3. true

5. $y^2 = 12x$

7. $y^2 = -20x$

9. $(x - 2)^2 = 8(y - 3)$

11. $(x - 1)^2 = -4(y - 6)$

13. focus: $(0, -3)$
directrix: $y = 3$
vertex: $(0, 0)$
length of latus rectum: 12

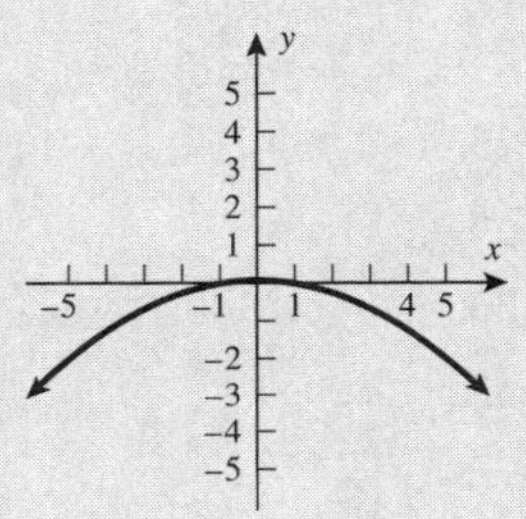

15. focus: $\left(\frac{1}{4}, 0\right)$
directrix: $x = -\frac{1}{4}$
vertex: $(0, 0)$
length of latus rectum: 1

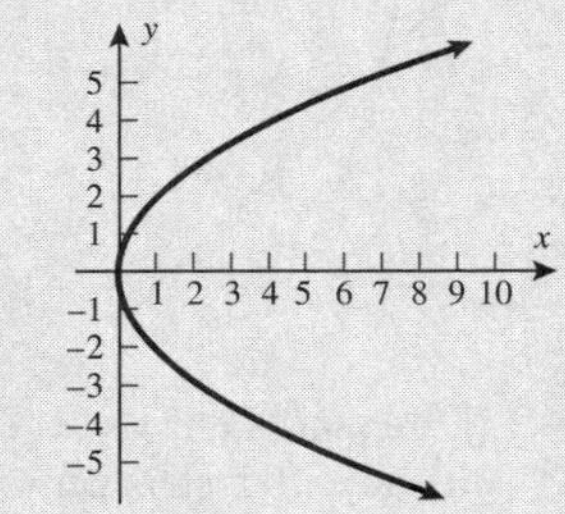

17. focus: $(3, -2)$
directrix: $x = 1$
vertex: $(2, -2)$
length of latus rectum: 4

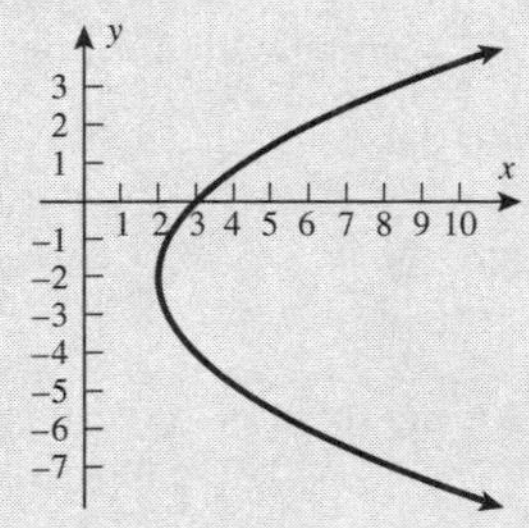

19. focus: $(-3, -1)$
directrix: $y = 3$
vertex: $(-3, 1)$
length of latus rectum: 8

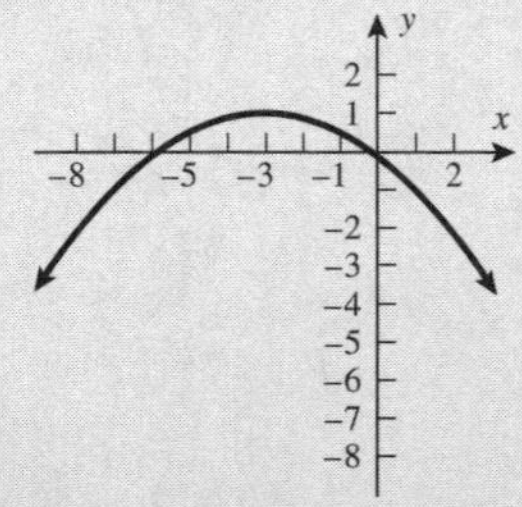

21. focus: $\left(-\frac{5}{2}, -\frac{79}{8}\right)$
directrix: $y = -\frac{71}{8}$
vertex: $\left(-\frac{5}{2}, -\frac{75}{8}\right)$
length of latus rectum: 2

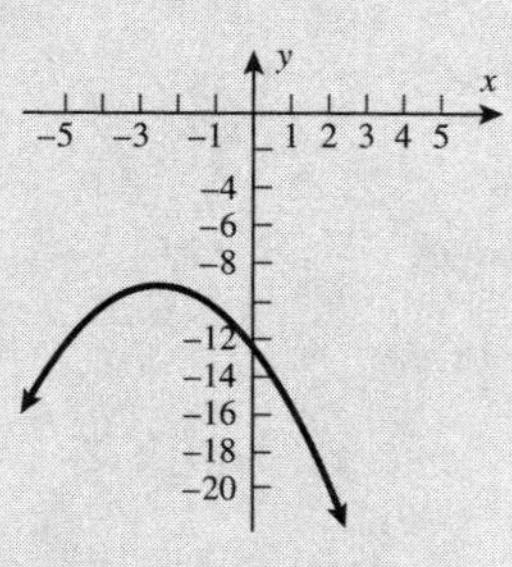

23. The receiver should be placed $\frac{25}{8} = 3.125$ feet from the vertex.

25.

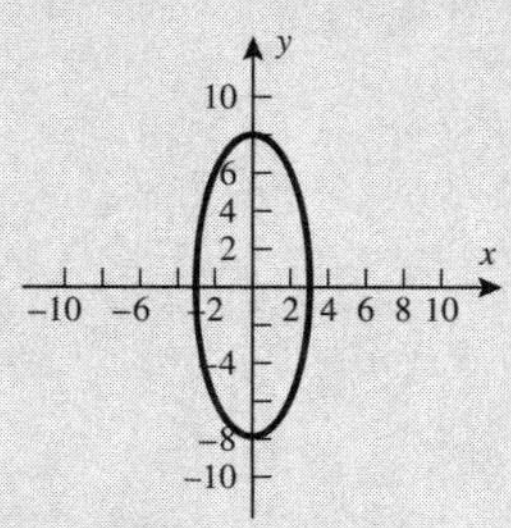

27.

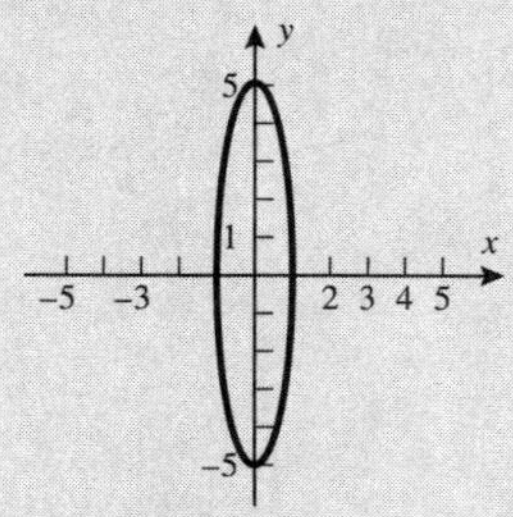

29. $\dfrac{x^2}{25} + \dfrac{y^2}{16} = 1$

31. $\dfrac{x^2}{9} + \dfrac{y^2}{64} = 1$

33.

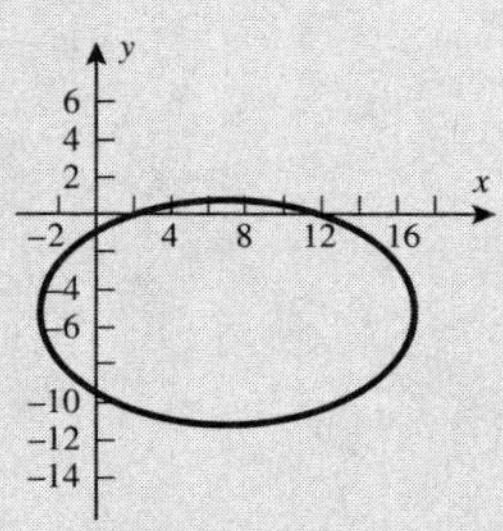

35.

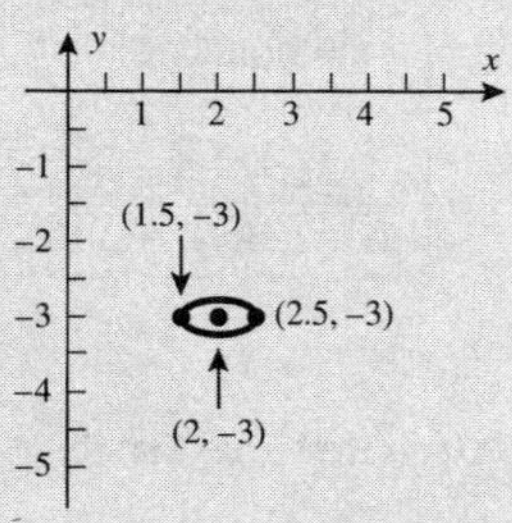

37. $\dfrac{(x-3)^2}{25} + \dfrac{(y-3)^2}{9} = 1$

39. $\dfrac{x^2}{778{,}300{,}000^2} + \dfrac{y^2}{777{,}400{,}000^2} = 1$

41.

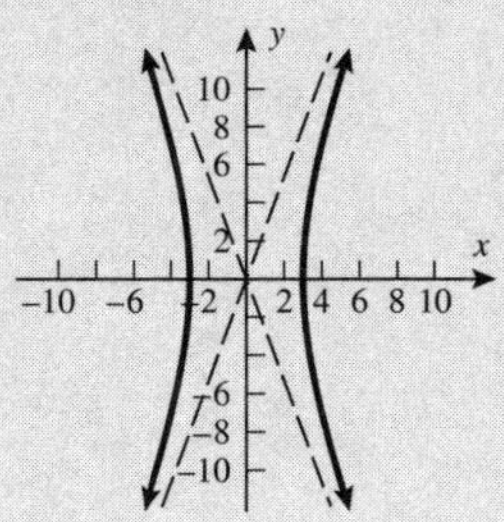

43.

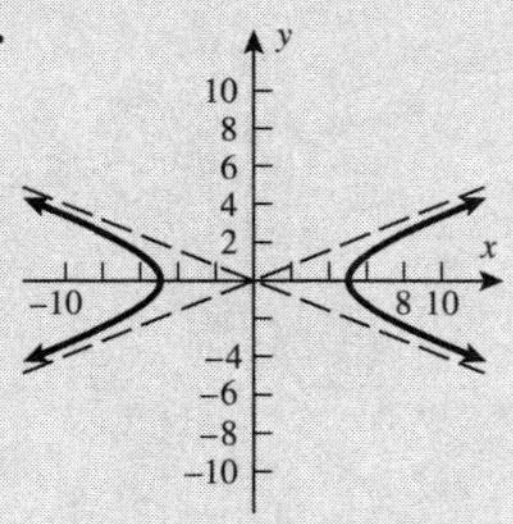

45. $\dfrac{x^2}{9} - \dfrac{y^2}{16} = 1$

47. $\dfrac{y^2}{9} - x^2 = 1$
(Other answers are possible.)

49.

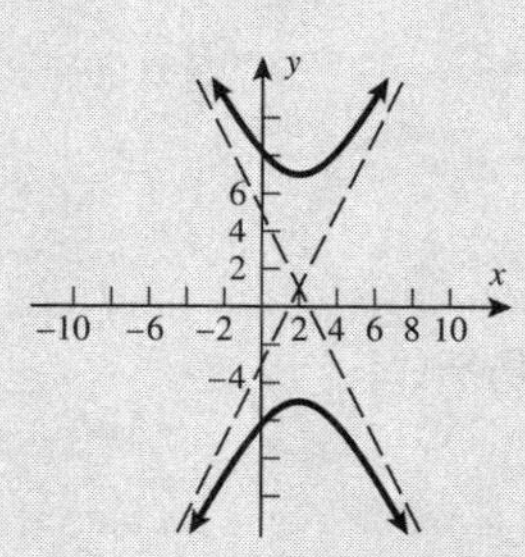

51.

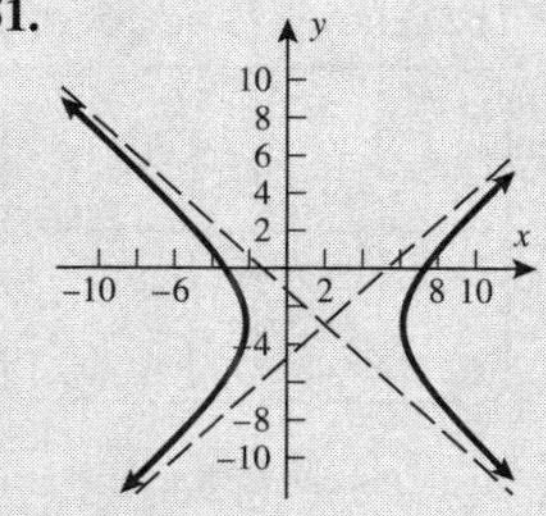

53. $\dfrac{(x-4)^2}{16} - \dfrac{(y-3)^2}{9} = 1$

55. The ship will get to shore between the two stations: 65.36 miles from one station and 154.64 miles from the other.

57. $(1, -4)$ and $(-2, -7)$

59. $(-1, 2)$ and $(1, 2)$

61. no solution

63. no solution

65. $(-3, -2)$ and $(2, 3)$

67. $\left(-\frac{1}{2}, -\frac{1}{\sqrt{7}}\right), \left(-\frac{1}{2}, \frac{1}{\sqrt{7}}\right), \left(\frac{1}{2}, -\frac{1}{\sqrt{7}}\right), \left(\frac{1}{2}, \frac{1}{\sqrt{7}}\right)$

69.

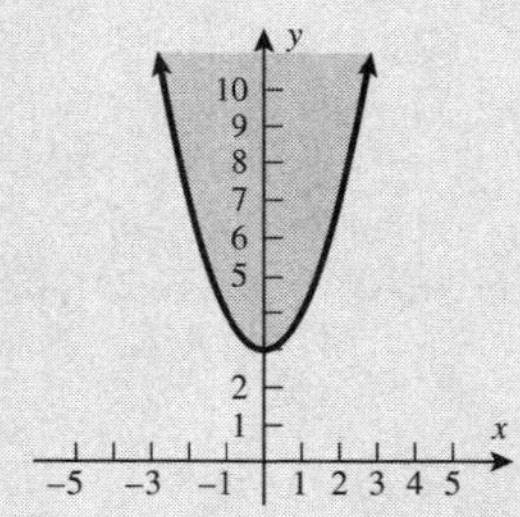

71.

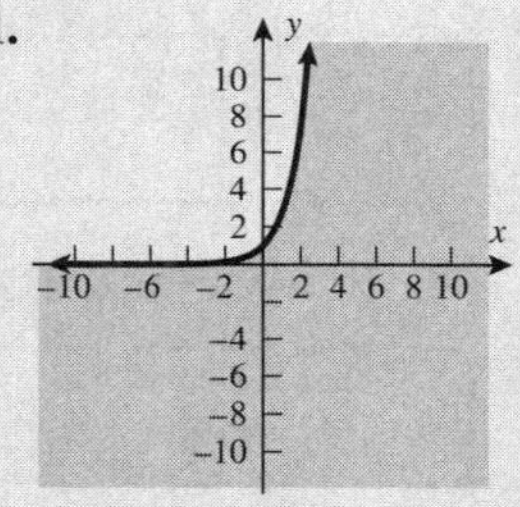

73.

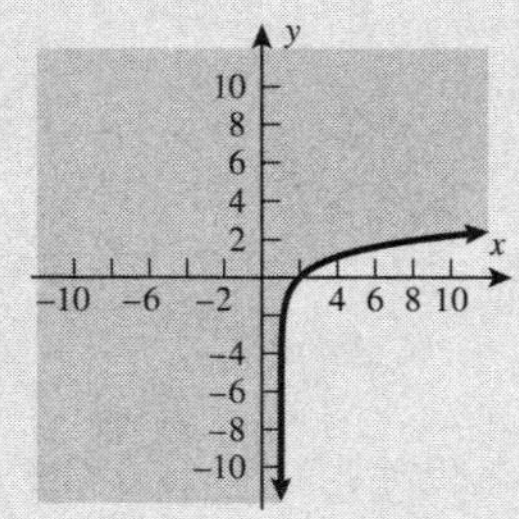

75.

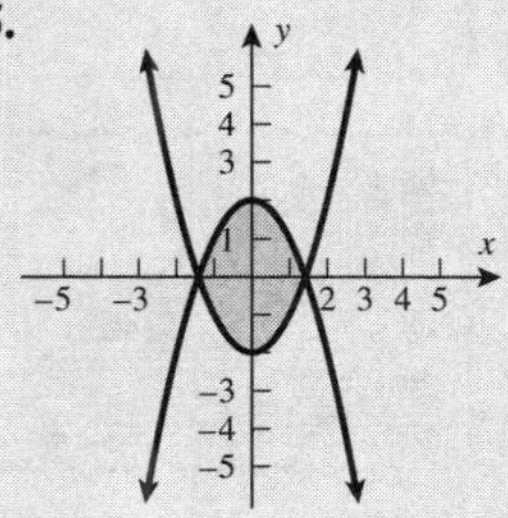

77.

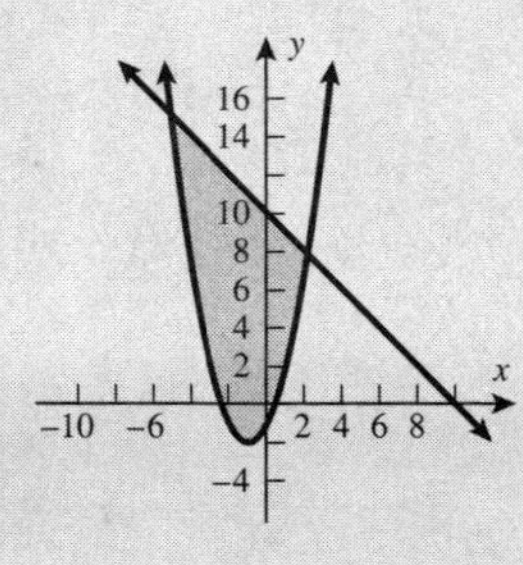

79.

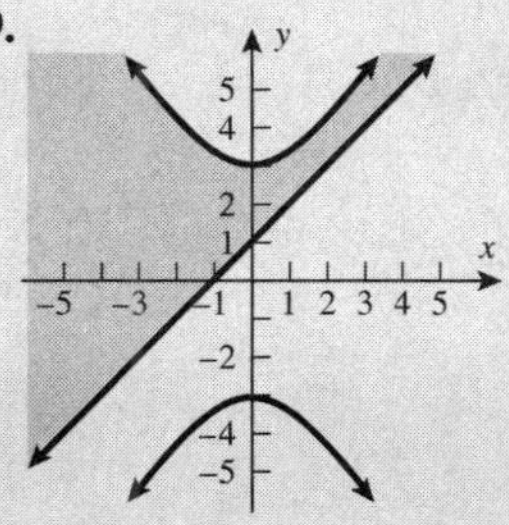

81. The vertex is located at $(0.6, -1.2)$. The parabola opens down.

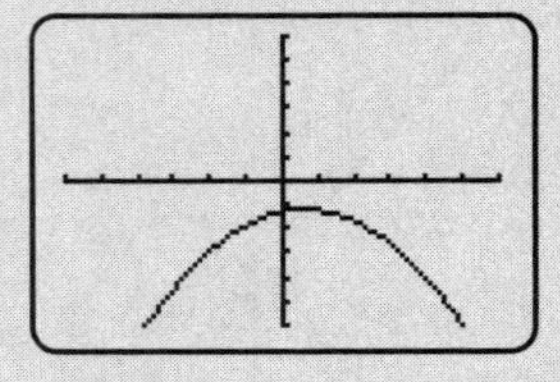

83. As c increases, the minor axis along the x-axis decreases.

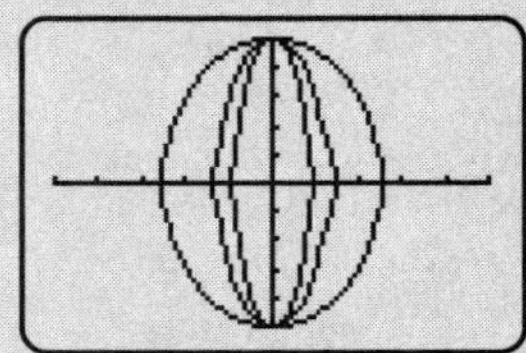

85. As c increases, the vertices of the hyperbolas located at $\left(\pm\dfrac{1}{2c}, 0\right)$ are moving towards the orign.

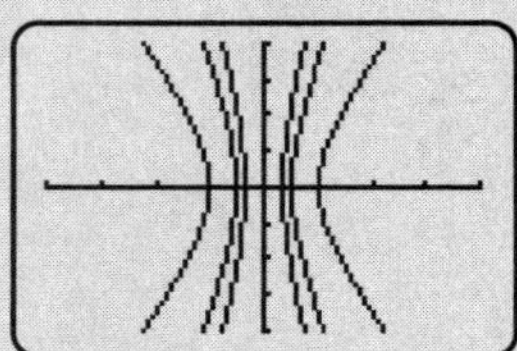

87. (0.635, 2.480), (−0.635, 2.480), (−1.245, 0.645), (1.245, 0.645)

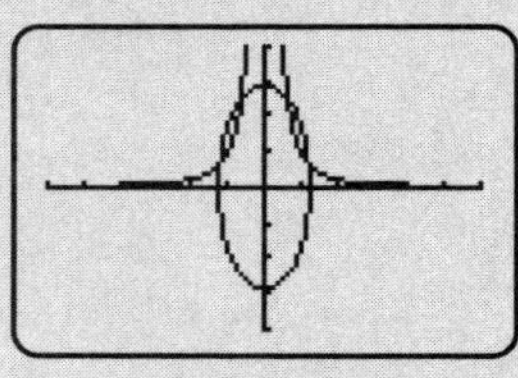

89.

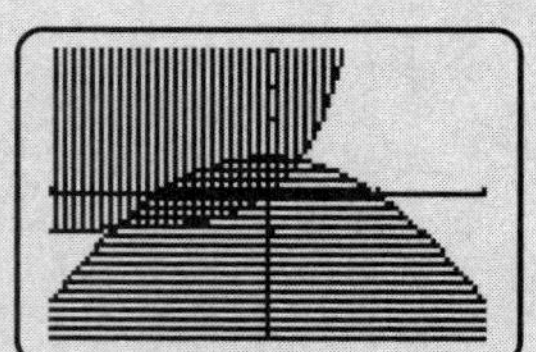

19.

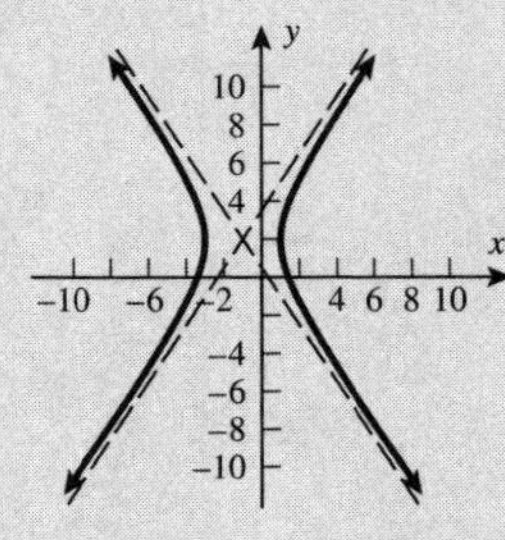

21.

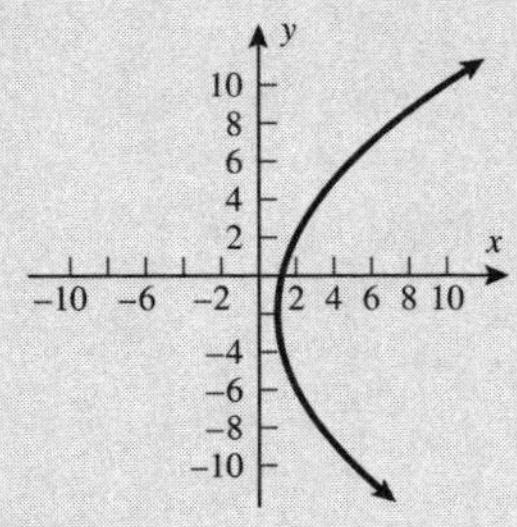

23. $x^2 = 6y; -2 \le x \le 2$

25. $(-2, 0)$ and $(2, 0)$

27.

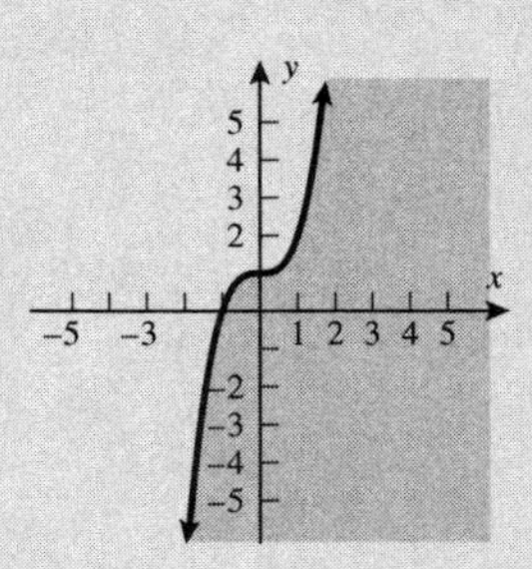

29.

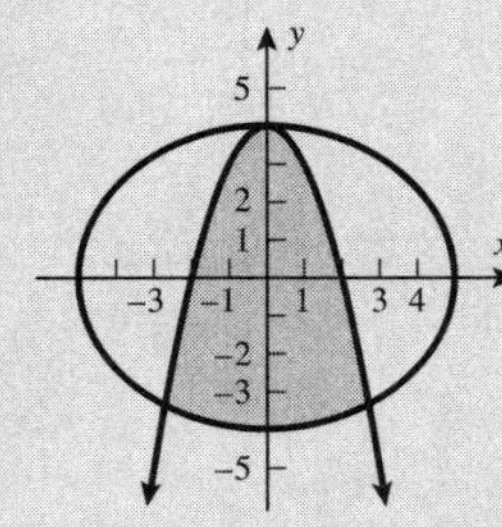

31.

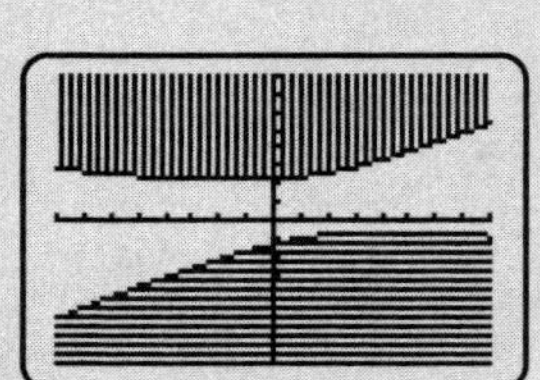

Practice Test

1. c

3. d

5. f

7. $y^2 = -16x$

9. $(x + 1)^2 = -12(y - 5)$

11. $\dfrac{x^2}{7} + \dfrac{y^2}{16} = 1$

13. $\dfrac{(x - 2)^2}{20} + \dfrac{y^2}{36} = 1$

15. $x^2 - \dfrac{y^2}{4} = 1$

17. $\dfrac{y^2}{16} - \dfrac{(x - 2)^2}{20} = 1$

Cumulative Test

1. $x = 2 \pm 4i$

3. 2

5. $-2x - h$

7. undefined

9. -2 (multiplicity 3); 0 (multiplicity 2)

11. VA: $x = 2$, slant asymptote: $y = x + 2$

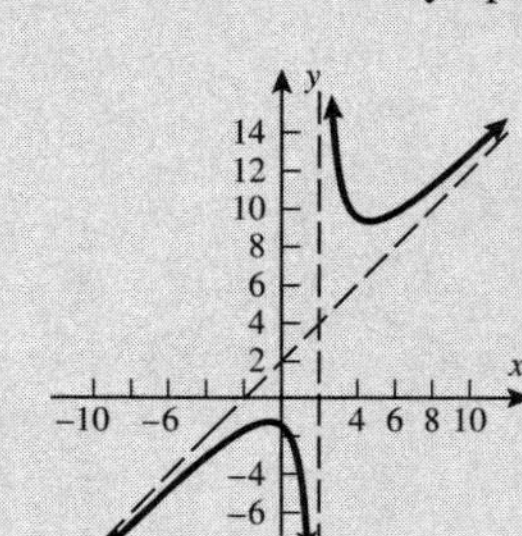

13. $37,250

15. 0.001

17. Soda: $1.29; soft pretzel: $1.45

19.

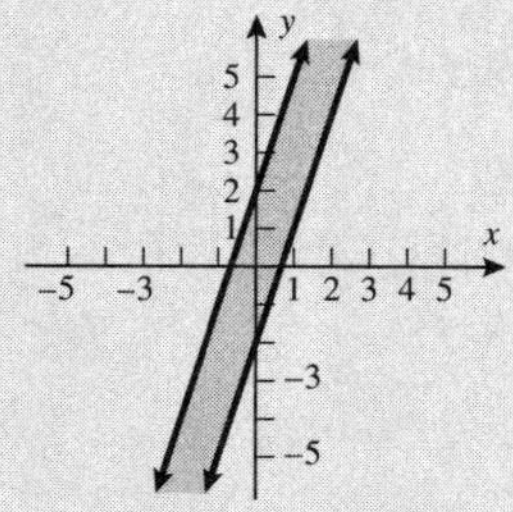

21. $\begin{bmatrix} 7 & -16 & 33 \\ 18 & -3 & -17 \end{bmatrix}$

23. $\dfrac{(x-6)^2}{9} + \dfrac{(y+2)^2}{25} = 1$

25. (2, 4), (4, 2)

27.

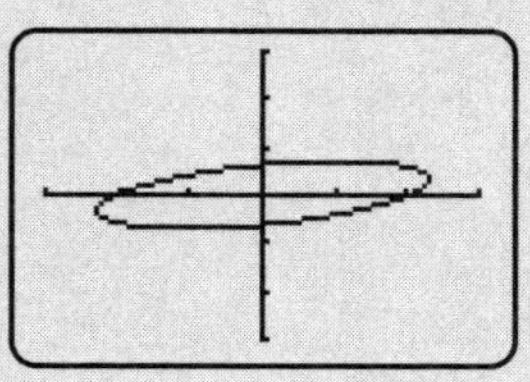

CHAPTER 9

Section 9.1

1. $a_1 = 1 \quad a_2 = 2 \quad a_3 = 3 \quad a_4 = 4$

3. $a_1 = 1 \quad a_2 = 3 \quad a_3 = 5 \quad a_4 = 7$

5. $a_1 = \frac{1}{2} \quad a_2 = \frac{2}{3} \quad a_3 = \frac{3}{4} \quad a_4 = \frac{4}{5}$

7. $a_1 = 2 \quad a_2 = 2 \quad a_3 = \frac{4}{3} \quad a_4 = \frac{2}{3}$

9. $a_1 = -x^2 \quad a_2 = x^3 \quad a_3 = -x^4 \quad a_4 = x^5$

11. $a_1 = -\frac{1}{6} \quad a_2 = \frac{1}{12} \quad a_3 = -\frac{1}{20} \quad a_4 = \frac{1}{30}$

13. $a_9 = \frac{1}{512}$ 15. $a_{19} = -\frac{1}{420}$

17. $a_{100} = 1.0201$ 19. $a_{23} = 23$

21. $a_n = 2n$ 23. $a_n = \dfrac{1}{n(n+1)}$

25. $a_n = \dfrac{(-1)^n 2^n}{3^n}$ 27. $a_n = (-1)^{n+1}$

29. 72 31. 812

33. $\frac{1}{5852}$ 35. 83, 156, 160

37. $\dfrac{1}{n(n+1)}$ 39. $(2n+3)(2n+2)$

41. $a_1 = 7 \quad a_2 = 10 \quad a_3 = 13 \quad a_4 = 16$

43. $a_1 = 1 \quad a_2 = 2 \quad a_3 = 6 \quad a_4 = 24$

45. $a_1 = 100 \quad a_2 = 50 \quad a_3 = \frac{25}{3} \quad a_4 = \frac{25}{72}$

47. $a_1 = 1 \quad a_2 = 2 \quad a_3 = 2 \quad a_4 = 4$

49. $a_1 = 1 \quad a_2 = -1 \quad a_3 = -2 \quad a_4 = 5$

51. 10 53. 30 55. 36 57. 5

59. $1 - x + x^2 - x^3$ 61. $\frac{109}{15}$

63. $1 + x + \dfrac{x^2}{2} + \dfrac{x^3}{6} + \dfrac{x^4}{24}$ 65. $\frac{20}{9}$

67. not possible 69. $\displaystyle\sum_{n=0}^{6} \frac{(-1)^n}{2^n}$

71. $\displaystyle\sum_{n=1}^{\infty} (-1)^{n-1} n$ 73. $\displaystyle\sum_{n=1}^{6} \frac{(n+1)!}{(n-1)!} = \sum_{n=1}^{5} n(n+1)$

75. $\displaystyle\sum_{n=0}^{\infty} \frac{(-1)^n x^n}{n!}$ or $\displaystyle\sum_{n=1}^{\infty} \frac{(-1)^{n-1} \cdot x^{n-1}}{(n-1)!}$

77. $A_{72} \approx \$28{,}640.89$. A_{72} represents the total balance in 6 years (or 72 months)

79. $a_n = 20 + 2n$. The paralegal's salary with 20 years of experience will be \$60 per hour.

81. $a_n = 1.03a_{n-1} \quad n = 1, 2, \ldots$ and $a_0 = 30{,}000$

83. Approximately 10.7 years, $a_n = 1000 - 75n$ $n = 0, 1, 2, \ldots$

85. $A_1 = \$100 \quad A_2 = \$200.10 \quad A_3 = \$300.30$ $A_4 = \$400.60 \quad A_{36} = \$3663.72$

87. 7; 7.389 89. 0.095306

91. 6! is not equal to (3!)(2!)

93. $(-1)^{n+1}$ is evaluated incorrectly. The wrong sign is on each term.

95. true 97. false

99. $a_1 = C \quad a_2 = C + D \quad a_3 = C + 2D \quad a_4 = C + 3D$

101. 1 and 1

103. $a_{100} \approx 2.705 \quad a_{1000} \approx 2.717 \quad a_{10{,}000} \approx 2.718$

105. $\frac{109}{15}$

Section 9.2

1. arithmetic; 3 3. not arithmetic 5. arithmetic; −0.03

7. arithmetic; $\frac{2}{3}$ 9. not arithmetic

11. $a_1 = 3 \quad a_2 = 1 \quad a_3 = -1 \quad a_4 = -3$ arithmetic; −2

13. $a_1 = 1 \quad a_2 = 4 \quad a_3 = 9 \quad a_4 = 16$ not arithmetic

15. $a_1 = 2 \quad a_2 = 7 \quad a_3 = 12 \quad a_4 = 17$ arithmetic; 5

17. $a_1 = 0 \quad a_2 = 10 \quad a_3 = 20 \quad a_4 = 30$ arithmetic; 10

19. $a_1 = -1 \quad a_2 = 2 \quad a_3 = -3 \quad a_4 = 4$ not arithmetic

21. $a_n = 6 + 5n$ 23. $a_n = -6 + 2n$ 25. $a_n = -\frac{2}{3} + \frac{2}{3}n$

27. $a_n = en - e$ 29. 124 31. −684

33. $a_1 = 8 \quad d = 9 \quad a_n = 8 + 9(n-1)$ or $a_n = 9n - 1$

35. $a_1 = 23 \quad d = -4 \quad a_n = 23 - 4(n-1)$ or $a_n = 27 - 4n$

37. $a_1 = 1 \quad d = \frac{2}{3} \quad a_n = 1 + \frac{2}{3}(n-1)$ or $a_n = \frac{1}{3} + \frac{2}{3}n$

39. 552 **41.** −780 **43.** 51

45. 416 **47.** 3875 **49.** −66.5 or $-\frac{133}{2}$

51. 630 **53.** $-\frac{817}{2} = -408.5$ **55.** 1,368

57. Colin \$347,500, Camden \$340,000

59. 850 seats

61. 1101 glasses in the bottom row. There are 20 fewer glasses in every row.

63. 1600 feet **65.** 210 oranges

67. 23 seats in the first row; 1125 total seats

69. The correct general term is $a_n = a_1 + d(n - 1)$.

71. $n = 11$ (not 10) **73.** false **75.** true

77. $\dfrac{(n + 1)(2a + nb)}{2}$ **79.** 27,420 or $\frac{1}{4}R$ **81.** 5050

83. 2500 **85.** 18,850

Section 9.3

1. yes; $r = 3$ **3.** no

5. yes; $r = \frac{1}{2}$ **7.** yes; $r = 1.7$

9. $a_1 = 6 \quad a_2 = 18 \quad a_3 = 54 \quad a_4 = 162 \quad a_5 = 486$

11. $a_1 = 1 \quad a_2 = -4 \quad a_3 = 16 \quad a_4 = -64 \quad a_5 = 256$

13. $a_1 = 10{,}000 \quad a_2 = 10{,}600 \quad a_3 = 11{,}236$
$a_4 = 11{,}910.16 \quad a_5 = 12{,}624.77$

15. $a_1 = \frac{2}{3} \quad a_2 = \frac{1}{3} \quad a_3 = \frac{1}{6} \quad a_4 = \frac{1}{12} \quad a_5 = \frac{1}{24}$

17. $a_n = 5(2)^{n-1}$ **19.** $a_n = 1(-3)^{n-1}$

21. $a_n = 1000(1.07)^{n-1}$ **23.** $a_n = \frac{16}{3}\left(-\frac{1}{4}\right)^{n-1}$

25. $a_7 = -128$ **27.** $a_{13} = \frac{4096}{3}$

29. $a_{15} = 6.10 \times 10^{-16}$ **31.** $\frac{8191}{3}$

33. 59,048 **35.** $2.\overline{2}$

37. 6560 **39.** 16,383

41. 2 **43.** $-\frac{1}{4}$

45. not possible, diverges **47.** $-\frac{27}{2}$

49. $\dfrac{1{,}000.000}{95} \approx 10526$ **51.** $\frac{2}{3}$

53. 100 **55.** \$44,610.95

57. $2000(0.5)^n$; In 4 years when she graduates it will be worth \$125, and after graduate school it will be worth \$16.

59. On the fifth rebound the jumper will reach a height of approximately 17 feet.

61. 58,640 students

63. 66 days; \$9,618 would be paid in January

65. \$3,877.64

67. 26 weeks: \$13,196.88 52 weeks: \$26,811.75

69. \$501,509

73. $r = -\frac{1}{3}$ not $\frac{1}{3}$ **75.** $a_1 = -12$ (not 4)

77. false **79.** true

81. If $|b| < 1$, then the sum is $a \cdot \dfrac{1}{1 - b}$.

83. $\frac{47}{99}$ **85.** −375,299,968,947,541

87. The sum is $\dfrac{1}{1 - x}$ for $|x|$ less than or equal to $\frac{1}{2}$.

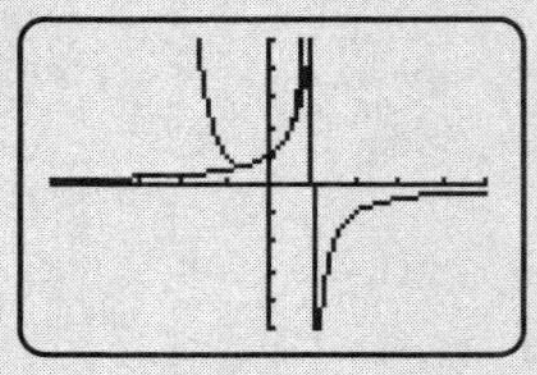

89. The series will sum to $\dfrac{1}{1 - 2x}$.

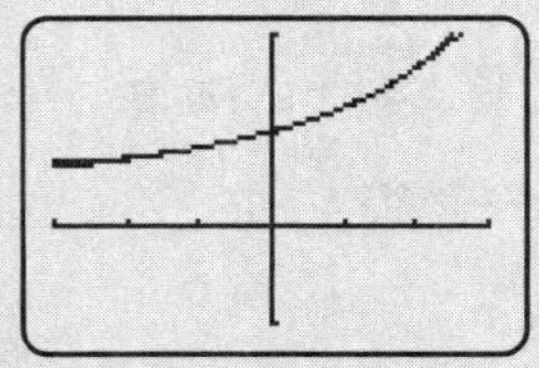

Section 9.4

25. 7 **27.** 31

31. false **39.** $\frac{255}{256}$, yes

Section 9.5

1. 35 **3.** 45 **5.** 1 **7.** 1 **9.** 17,296

11. $x^4 + 8x^3 + 24x^2 + 32x + 16$

13. $y^5 - 15y^4 + 90y^3 - 270y^2 + 405y - 243$

15. $x^5 + 5x^4y + 10x^3y^2 + 10x^2y^3 + 5xy^4 + y^5$

17. $x^3 + 9x^2y + 27xy^2 + 27y^3$

19. $125x^3 - 150x^2 + 60x - 8$

21. $\dfrac{1}{x^4} + \dfrac{20y}{x^3} + \dfrac{150y^2}{x^2} + \dfrac{500y^3}{x} + 625y^4$

23. $x^8 + 4x^6y^2 + 6x^4y^4 + 4x^2y^6 + y^8$

25. $a^5x^5 + 5a^4x^4by + 10a^3x^3b^2y^2 + 10a^2x^2b^3y^3 + 5axb^4y^4 + b^5y^5$

27. $x^3 + 12x^{5/2} + 60x^2 + 160x^{3/2} + 240x + 192x^{1/2} + 64$

29. $a^3 + 4a^{9/4}b^{1/4} + 6a^{3/2}b^{1/2} + 4a^{3/4}b^{3/4} + b$

31. $x + 8x^{3/4}y^{1/2} + 24x^{1/2}y + 32x^{1/4}y^{3/2} + 16y^2$

33. $r^4 - 4r^3s + 6r^2s^2 - 4rs^3 + s^4$

35. $a^6x^6 + 6a^5x^5by + 15a^4x^4b^2y^2 + 20a^3x^3b^3y^3 + 15a^2x^2b^4y^4 + 6axb^5y^5 + b^6y^6$

37. 3360 **39.** 5670 **41.** 22,680

43. 70 **45.** 3,838,380 **47.** 2,598,960

49. $\binom{7}{5} \neq \frac{7\,!}{5!}$ $\binom{7}{5} = \frac{7\,!}{5!2!}$

51. false **53.** true

57. The binomial expansion of $(1 - x)^3 = 1 - 3x + 3x^2 - x^3$

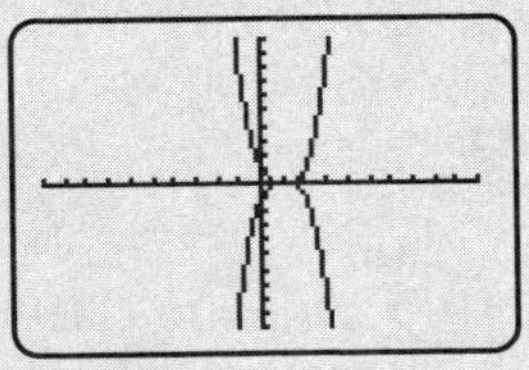

59. As each term is added, the series gets closer to $y = (1 - x)^3$. When $x > 1$ that's not true. The graphs for $-1 < x < 1$:

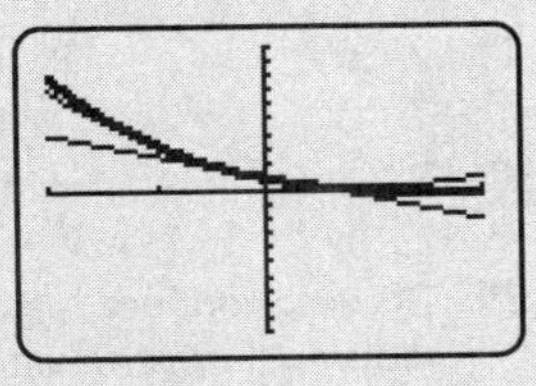

The graphs for $1 < x < 2$:

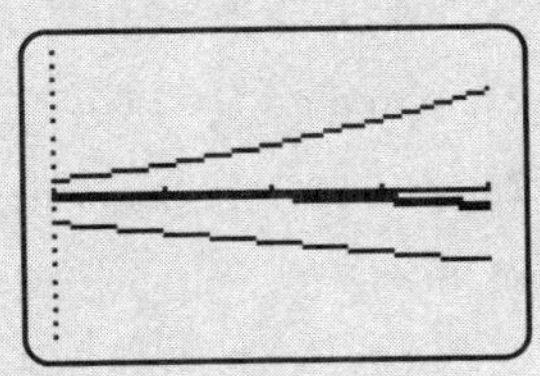

61. As each term is added, the series gets closer to $y = \left(1 + \frac{1}{x}\right)^3$ when $1 < x < 2$. No, the series does not get close to $y = \left(1 + \frac{1}{x}\right)^3$ for $0 < x < 1$.

Section 9.6

1. 360 **3.** 15,120 **5.** 40,320

7. 1716 **9.** 252 **11.** 15,890,700

13. 1 **15.** 27,405 **17.** 215,553,195

19. 24 **21.** 12 **23.** 10,000

25. 32,760 **27.** $4^{20} \approx 1.1 \times 10^{12}$

29. 100,000; 81,000 **31.** 2.65×10^{32} **33.** 59,280

35. 997,002,000 **37.** 22,957,480 **39.** 2,598,960

41. 1326 **43.** $\approx 4.9 \times 10^{14}$ **45.** 256

47. 15,625 **49.** 400

51. The correction that needs to be made is that the combination formula needs to be used instead of the permutation.

53. true **55.** false **57.** $\frac{_nC_r}{_nC_{r+1}} = \frac{r + 1}{n - r}$

59. $_nC_r \cdot r! = {_nP_r} = \frac{n!}{(n - r)!}$

61. The answers are exactly the same as 1–8.

63. a. 5040 **b.** 5040 **c.** yes **d.** $_nP_r = r!\,(_nC_r)$

Section 9.7

1. 2, 3, 4, 5, 6 7, 8, 9, 10, 11, 12

3. $\left\{\begin{matrix} BBBB, BBBG, BBGB, BBGG, BGBB, BGBG, BGGB, BGGG, \\ GBBB, GBBG, GBGB, GBGG, GGBB, GGBG, GGGB, GGGG \end{matrix}\right\}$

5. $\{RR, RB, RW, BR, BW, BB, WR, WB\}$

7. $\frac{1}{8}$ **9.** $\frac{7}{8}$ **11.** $\frac{1}{18}$ **13.** $\frac{1}{2}$ **15.** $\frac{5}{12}$

17. $\frac{40}{52} = \frac{10}{13}$ **19.** $\frac{4}{13}$ **21.** $\frac{3}{4}$ **23.** $\frac{3}{4}$ **25.** 0

27. a. 270,725 **b.** $\frac{715}{270{,}725} \approx 0.26\%$ **c.** $\frac{13}{270{,}725} \approx 0.005\%$

29. $\frac{8}{52} = \frac{2}{13} \approx 15.4\%$ **31.** $\frac{4}{663} \approx 0.6\%$

33. $\frac{1}{32} \approx 3.1\%$ **35.** $\frac{31}{32} \approx 96.9\%$ **37.** $\left(\frac{18}{38}\right)^4 = 5.03\%$

39. 43% **41.** $\frac{1}{4} = 25\%$ **43.** 4.8%

45. $\approx$ 0.001526%

47. a. Sample space: {Brown/Brown, Blue/Brown, Brown/Blue, Blue/Blue}

b. $\frac{1}{4}$ **c.** $\frac{3}{4}$

49. 2,598,960 five-card hands **51.** $\frac{1287}{2{,}598{,}960}$

53. Subtract P (2 of spades) $= \frac{1}{52}$

55. true **57.** false

59. $1 - \frac{(365)\,(364)}{(365)\,(365)} = 0.0027$ **61.** $\frac{1}{3} \approx 0.333$

65. 0.2907

Review Exercises

1. $a_1 = 1$ $a_2 = 8$ $a_3 = 27$ $a_4 = 64$

3. $a_1 = 5$ $a_2 = 8$ $a_3 = 11$ $a_4 = 14$

5. $a_5 = \frac{32}{243} = 0.13$ **7.** $a_{15} = -\frac{1}{3600}$ **9.** $a_n = (-1)^{n+1}3n$

11. $a_n = (-1)^n$ **13.** 56 **15.** $\frac{1}{n + 1}$

17. $a_1 = 5$ $a_2 = 3$ $a_3 = 1$ $a_4 = -1$

19. $a_1 = 1$ $a_2 = 2$ $a_3 = 4$ $a_4 = 32$

21. 15 **23.** 69 **25.** $\sum_{n=1}^{7} \frac{(-1)^n}{2^{n-1}}$ **27.** $\sum_{n=0}^{\infty} \frac{x^n}{n!}$

29. $A_{60} \approx 36{,}629.90$, which is the amount in the account after 5 years.

31. Yes, the sequence is arithmetic. $d = -2$

33. Yes, the sequence is arithmetic. $d = \frac{1}{2}$

35. Yes, the sequence is arithmetic. $d = 1$

37. $a_n = -4 + 5(n - 1)$ or $a_n = 5n - 9$

39. $a_n = 1 - \frac{2}{3}(n - 1)$ or $a_n = -\frac{2}{3}n + \frac{5}{3}$

41. $a_1 = 5 \quad d = 2 \quad a_n = 5 + 2(n - 1)$ or $a_n = 3 + 2n$

43. $a_1 = 10 \quad d = 6 \quad a_n = 10 + 6(n - 1)$ or $a_n = 4 + 6n$

45. 630 **47.** 420

49. Bob: $885,000 Tania: $990,000

51. Yes, the sequence is geometric, $r = -2$.

53. Yes, the sequence is geometric, $r = \frac{1}{2}$.

55. $a_1 = 3 \quad a_2 = 6 \quad a_3 = 12 \quad a_4 = 24 \quad a_5 = 48$

57. $a_1 = 100 \quad a_2 = -400 \quad a_3 = 1600 \quad a_4 = -6400$ $a_5 = 25{,}600$

59. $a_n = 7 \cdot 2^{n-1}$ **61.** $a_n = (-2)^{n-1}$

63. $a_{25} = 33{,}554{,}432$ **65.** $a_{12} = -2.048 \times 10^{-6}$

67. 4,920.5 **69.** 16,400 **71.** 3

73. $60,875.61 **79.** 165 **81.** 1

83. $x^4 - 20x^3 + 150x^2 - 500x + 625$

85. $8x^3 - 60x^2 + 150x - 125$

87. $x^{5/2} + 5x^2 + 10x^{3/2} + 10x + 5x^{1/2} + 1$

89. $r^5 - 5r^4s + 10r^3s^2 - 10r^2s^3 + 5rs^4 - s^5$

91. 112 **93.** 37,500 **95.** 22,957,480

97. 840 **99.** 95,040 **101.** 792

103. 1 **105.** 30 **107.** 5040

109. 120 different seating orders (15 seasons)

111. 94,109,400 **113.** 20,358,520 **115.** $\frac{1}{16} \approx 6.25\%$

117. $\frac{30}{36} = \frac{5}{6} \approx 83.3\%$ **119.** $\frac{2}{3} \approx 66.7\%$ **121.** $\frac{7}{12} \approx 58.3\%$

123. $\frac{8}{52} = \frac{2}{13} \approx 15.4\%$ **125.** $\frac{31}{32} \approx 96.88\%$

127. $\frac{5369}{3600}$ **129.** $\frac{34875}{14}$

131. The series will sum to $\frac{1}{1 + 2x}$.

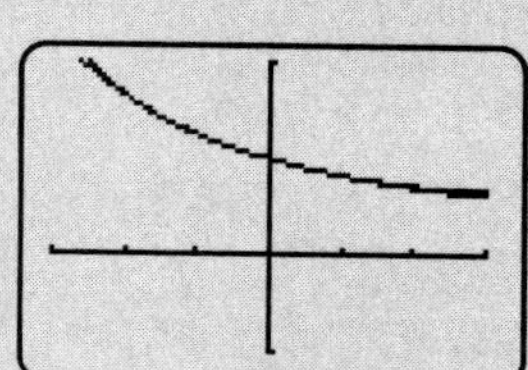

133. 99900, yes

135. As each term is added, the series gets closer to $y = (1 + 2x)^4$ when $-0.1 < x < 0.1$. No, the series does not get close to $y = (1 + 2x)^4$ for $0.1 < x < 1$.

137. **a.** 11440 **b.** 11440 **c.** yes **d.** $\frac{{}_nP_r}{r!} = {}_nC_r$

139. 0.0722

Practice Test

1. $a_n = x^{n-1} \quad n = 1, 2, 3, \ldots$

3. $S_n = \dfrac{1 - x^n}{1 - x}$ **5.** $|x| < 1$ **7.** $1 - \left(\frac{1}{4}\right)^{10} \approx 1$

9. $\sum_{n=1}^{100} (5n - 3) = 24{,}950$

11. 2520 **13.** 455 **15.** 2184

17. $x^{10} + 5x^7 + 10x^4 + 10x + \dfrac{5}{x^2} + \dfrac{1}{x^5}$

19. Because combinations have no particular order and permutations do. For permutations, each combination is arranged $n!$ ways.

21. $P(\text{red}) = \frac{18}{38} \approx 0.47$ **23.** $P(\text{next roll is red}) = \frac{18}{38} \approx 0.47$

25. $P(\text{ace}) + P(\text{diamond}) - P(\text{ace of diamonds}) = \frac{16}{52} \approx 0.308$

27. 184,756

Cumulative Test

1. 2 **3.** $x = \dfrac{5 \pm 3i\sqrt{7}}{4}$ **5.** $x = -8$

7. $2x + h - 3$ **9.** (15, 6)

11. VA: $x = 3$; HA: $y = -5$

13. 2.585 **15.** no solution

17. Maximum: $z(1, 4) = 24$ **19.** $\begin{bmatrix} 99 & 18 & -9 \\ 29 & 4 & 7 \end{bmatrix}$

21. $(y - 5)^2 = -16(x - 3)$

23. 3 **25.** $2^{10} = 1024$

27. -3432

Applications Index

SUBJECT INDEX

Conic Sections

Parabola

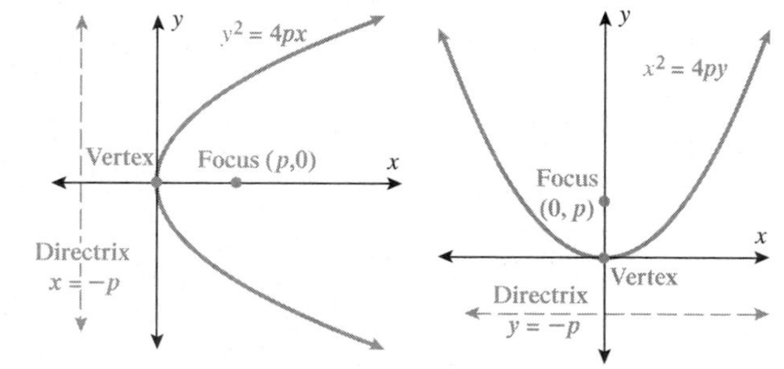

Ellipse

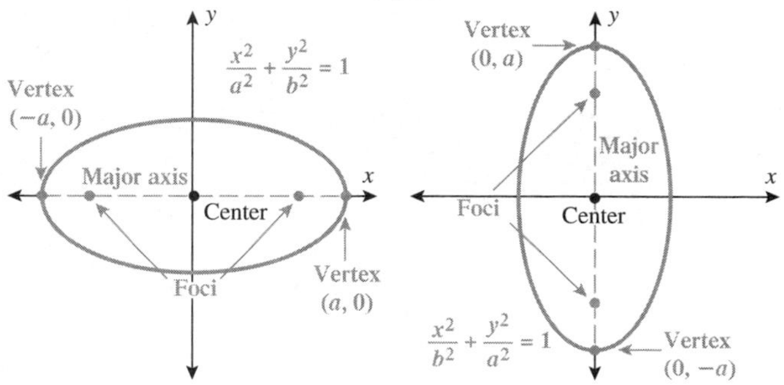

Hyperbola

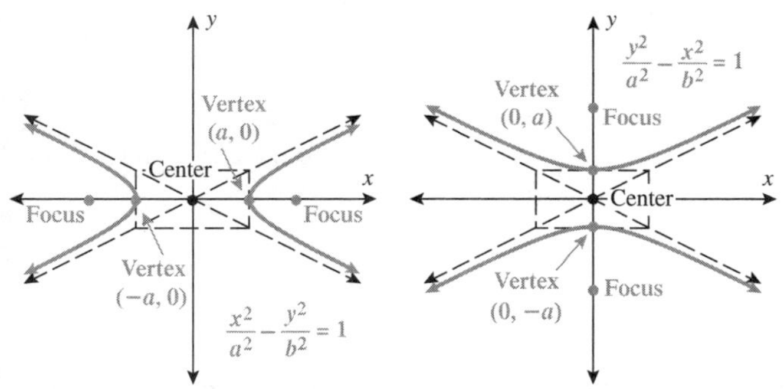

Sequences

1. Infinite Sequence:

$$\{a_n\} = a_1, a_2, a_3, \ldots, a_n, \ldots$$

2. Summation Notation:

$$\sum_{i=1}^{n} a_i = a_1 + a_2 + a_3 + \cdots + a_n$$

3. nth Term of an Arithmetic Sequence:

$$a_n = a_1 + (n - 1)d$$

4. Sum of First n Terms of an Arithmetic Sequence:

$$S_n = \frac{n}{2}(a_1 + a_n)$$

5. nth Term of a Geometric Sequence:

$$a_n = a_1 r^{n-1}$$

6. Sum of First n Terms of a Geometric Sequence:

$$S_n = \frac{a_1(1 - r^n)}{1 - r} \quad (r \neq 1)$$

7. Sum of an Infinite Geometric Series with $|r| < 1$:

$$S = \frac{a_1}{1 - r}$$

The Binomial Theorem

1. $n! = n(n-1)(n-2)\cdots 3 \cdot 2 \cdot 1$; $1! = 1$; $0! = 1$

2. $$\binom{n}{r} = \frac{n!}{r!(n-r)!}$$

3. Binomial theorem:

$$(a + b)^n = \binom{n}{0}a^n + \binom{n}{1}a^{n-1}b + \binom{n}{2}a^{n-2}b^2 + \cdots + \binom{n}{n}b^n$$

Permutations, Combinations, and Probability

1. ${}_nP_r$, the number of permutations of n elements taken r at a time, is given by

$${}_nP_r = \frac{n!}{(n-r)!}.$$

2. ${}_nC_r$, the number of combinations of n elements taken r at a time, is given by

$${}_nC_r = \frac{n!}{(n-r)!r!}.$$

3. *Probability of an Event:* $P(E) = \frac{n(E)}{n(S)}$, where

$n(E)$ = the number of outcomes in event E and
$n(S)$ = the number of outcomes in the sample space.

Inverse Trigonometric Functions

$y = \sin^{-1}x$	$x = \sin y$	$-\frac{\pi}{2} \le y \le \frac{\pi}{2}$	$-1 \le x \le 1$
$y = \cos^{-1}x$	$x = \cos y$	$0 \le y \le \pi$	$-1 \le x \le 1$
$y = \tan^{-1}x$	$x = \tan y$	$-\frac{\pi}{2} < y < \frac{\pi}{2}$	x is any real number
$y = \cot^{-1}x$	$x = \cot y$	$0 < y < \pi$	x is any real number
$y = \sec^{-1}x$	$x = \sec y$	$0 \le y \le \pi, y \ne \frac{\pi}{2}$	$x \le -1$ or $x \ge 1$
$y = \csc^{-1}x$	$x = \csc y$	$-\frac{\pi}{2} \le y \le \frac{\pi}{2}, y \ne 0$	$x \le -1$ or $x \ge 1$

Graphs of the Inverse Trigonometric Functions

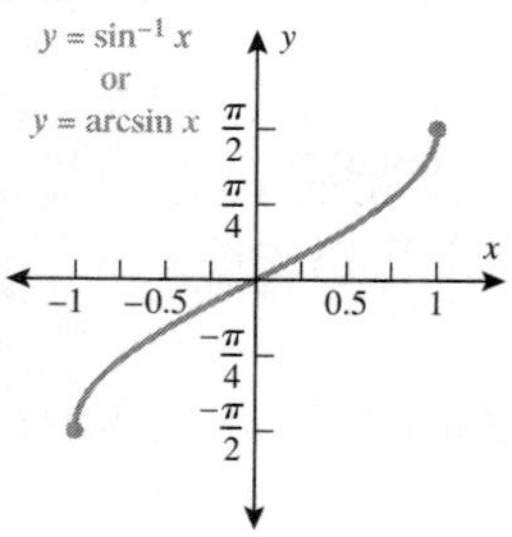

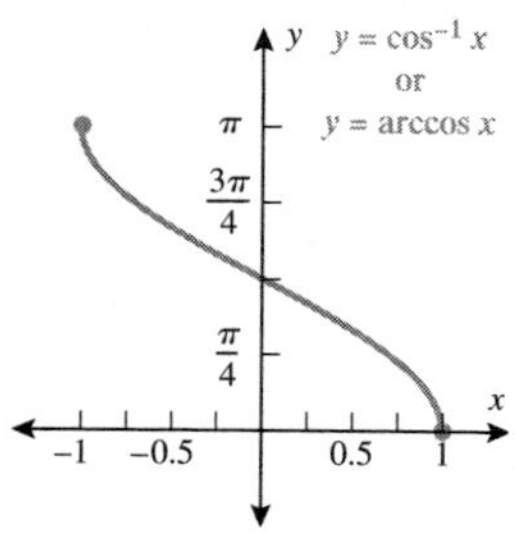

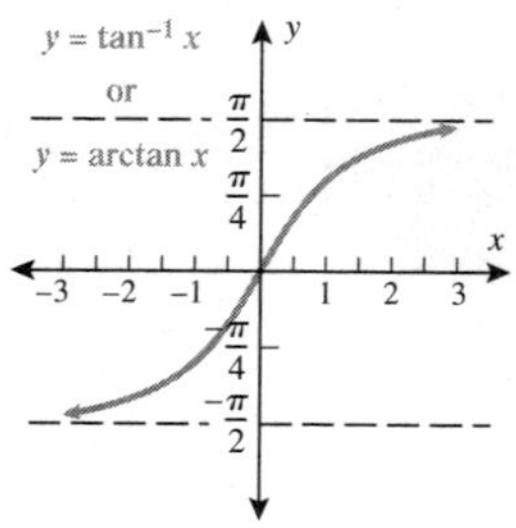

Vectors

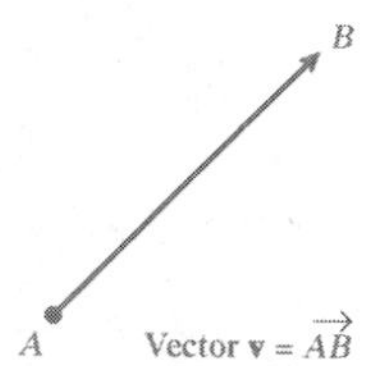

Vector Addition

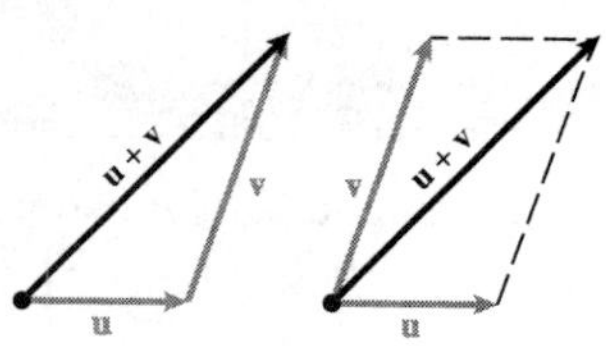

Scalar Multiplication

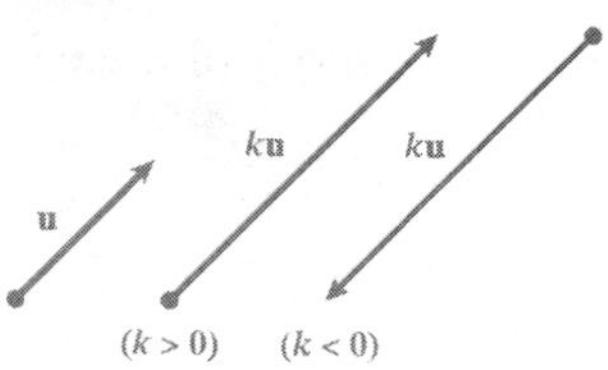

For vectors $\mathbf{u} = \langle a, b\rangle$ and $\mathbf{v} = \langle c, d\rangle$, and real number k,

$$\mathbf{u} = a\mathbf{i} + b\mathbf{j}$$
$$|\mathbf{u}| = \sqrt{a^2 + b^2}$$
$$\mathbf{u} + \mathbf{v} = \langle a + c, b + d\rangle$$
$$k\mathbf{u} = \langle ka, kb\rangle$$
$$\mathbf{u}\cdot\mathbf{v} = ac + bd$$
$$\cos\theta = \frac{\mathbf{u}\cdot\mathbf{v}}{|\mathbf{u}||\mathbf{v}|}$$

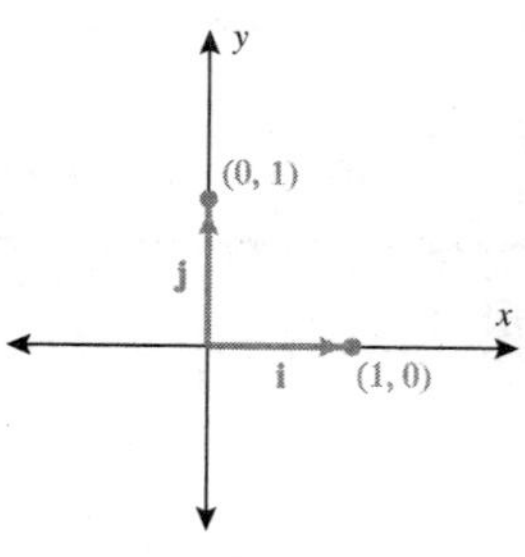

$$\text{Comp}_{\mathbf{v}}\,\mathbf{u} = |\mathbf{u}|\cos\theta = \frac{\mathbf{u}\cdot\mathbf{v}}{|\mathbf{u}|}$$

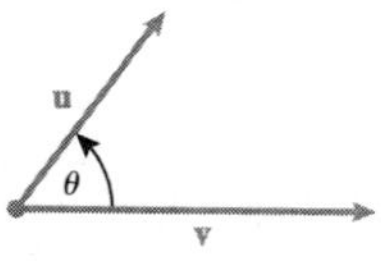

Graphs of the Trigonometric Functions

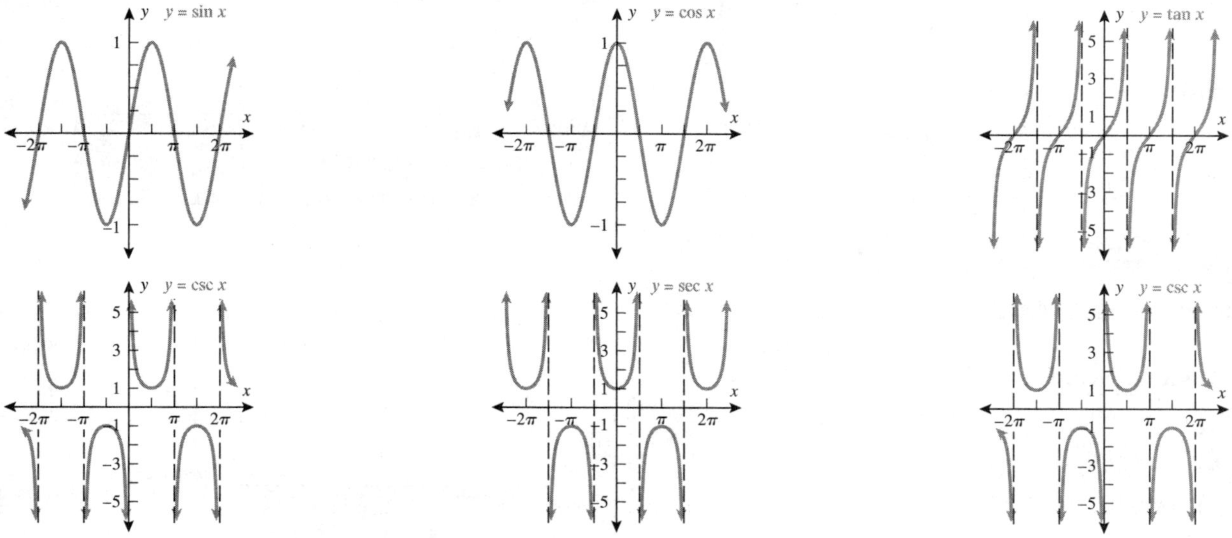

Amplitude, Period, and Phase Shift

$y = A\sin(Bx + C)$ $\qquad$ $y = A\cos(Bx + C)$

$$\text{Amplitude} = |A| \qquad \text{Period} = \frac{2\pi}{B}$$

$$\text{Phase shift} = \frac{C}{B}\begin{cases}\text{left} & \text{if } C/B > 0\\ \text{right} & \text{if } C/B < 0\end{cases}$$

$y = A\tan(Bx + C)$ $\qquad$ $y = A\cot(Bx + C)$

$$\text{Period} = \frac{\pi}{B}$$

$$\text{Phase shift} = \frac{C}{B}\begin{cases}\text{left} & \text{if } C/B > 0\\ \text{right} & \text{if } C/B < 0\end{cases}$$

Polar Coordinates

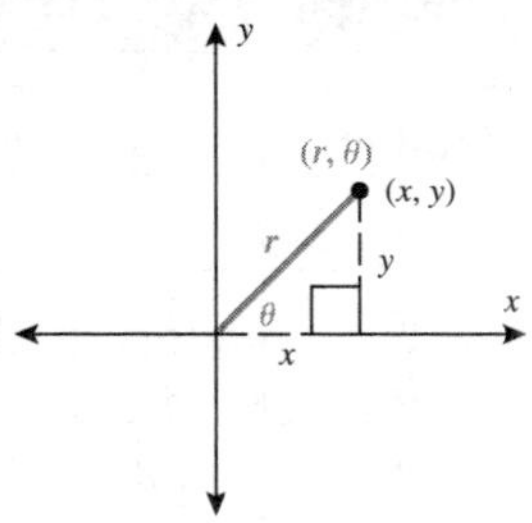

$$x = r\cos\theta$$
$$y = r\sin\theta$$
$$r^2 = x^2 + y^2$$
$$\tan\theta = \frac{y}{x}$$

Complex Numbers

$x + iy$	$=$	$r(\cos\theta + i\sin\theta)$
Rectangular form		Trigonometric (polar) form

Powers and Roots of Complex Numbers

$$\begin{aligned} z^n &= (x + iy)^n \\ &= [r(\cos\theta + i\sin\theta)]^n \\ &= r^n(\cos n\theta + i\sin n\theta) \quad n = 1, 2, \ldots \end{aligned}$$

$$\begin{aligned} \sqrt[n]{z} &= (x + iy)^{1/n} \\ &= [r(\cos\theta + i\sin\theta)]^{1/n} \\ &= r^{1/n}\left[\cos\left(\frac{\theta + 2k\pi}{n}\right) + i\sin\left(\frac{\theta + 2k\pi}{n}\right)\right] \end{aligned}$$

$$k = 0, 1, 2, \ldots, n - 1$$

Right Triangle Trigonometry

$\sin\theta = \dfrac{\text{opp}}{\text{hyp}}$ $\quad\csc\theta = \dfrac{\text{hyp}}{\text{opp}}$

$\cos\theta = \dfrac{\text{adj}}{\text{hyp}}$ $\quad\sec\theta = \dfrac{\text{hyp}}{\text{adj}}$

$\tan\theta = \dfrac{\text{opp}}{\text{adj}}$ $\quad\cot\theta = \dfrac{\text{adj}}{\text{opp}}$

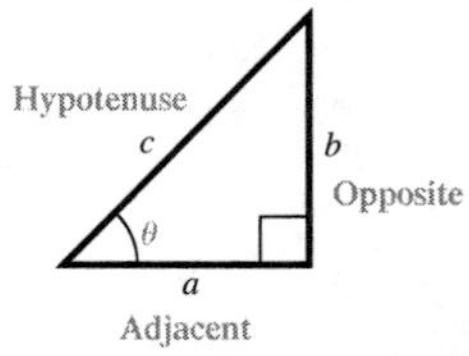

Special Right Triangles

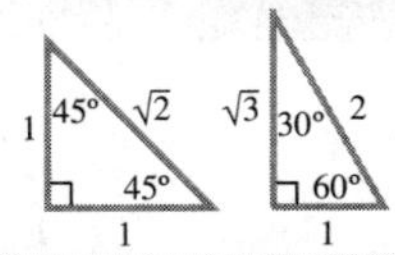

Trigonometric Functions in the Cartesian Plane

$\sin\theta = \dfrac{y}{r}$ $\quad\csc\theta = \dfrac{r}{y}$

$\cos\theta = \dfrac{x}{r}$ $\quad\sec\theta = \dfrac{r}{x}$

$\tan\theta = \dfrac{y}{x}$ $\quad\cot\theta = \dfrac{x}{y}$

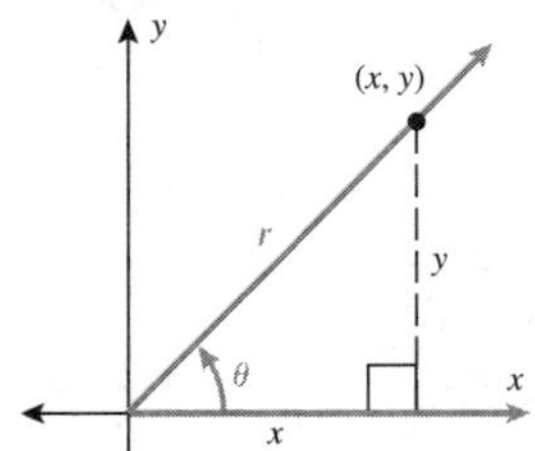

Accuracy for Triangles

Angle to Nearest	Significant Digits for Side Measure
1°	2
10′ or 0.1°	3
1′ or 0.01°	4
10″ or 0.001°	5

Angle Measurement

π radians $= 180°$

$s = r\theta \quad A = \frac{1}{2}r^2\theta \quad (\theta$ in radians)

To convert from degrees to radians, multiply by $\dfrac{\pi}{180°}$.

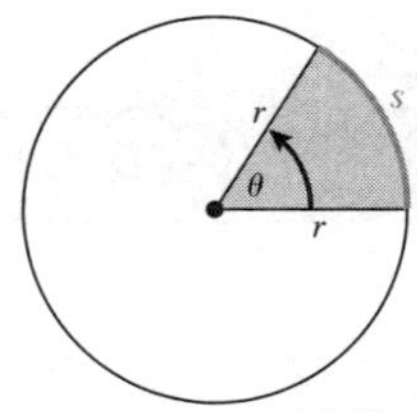

To convert from radians to degrees, multiply by $\dfrac{180°}{\pi}$.

Exact Values of Trigonometric Functions

x degrees	x radians	$\sin x$	$\cos x$	$\tan x$
0°	0	0	1	0
30°	$\frac{\pi}{6}$	$\frac{1}{2}$	$\frac{\sqrt{3}}{2}$	$\frac{\sqrt{3}}{3}$
45°	$\frac{\pi}{4}$	$\frac{\sqrt{2}}{2}$	$\frac{\sqrt{2}}{2}$	1
60°	$\frac{\pi}{3}$	$\frac{\sqrt{3}}{2}$	$\frac{1}{2}$	$\sqrt{3}$
90°	$\frac{\pi}{2}$	1	0	—

Oblique Triangle

Law of Sines

In any triangle,

$$\frac{\sin\alpha}{a} = \frac{\sin\beta}{b} = \frac{\sin\gamma}{c}.$$

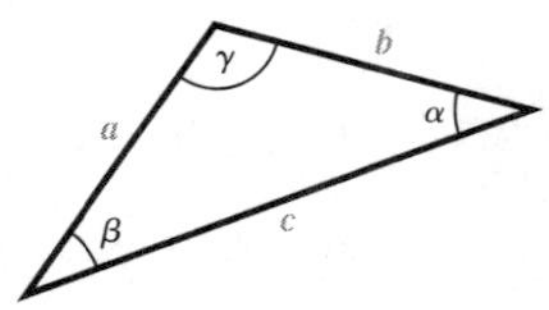

Law of Cosines

$$a^2 = b^2 + c^2 - 2bc\cos\alpha$$
$$b^2 = a^2 + c^2 - 2ac\cos\beta$$
$$c^2 = a^2 + b^2 - 2ab\cos\gamma$$

Circular Functions (cosθ, sinθ)

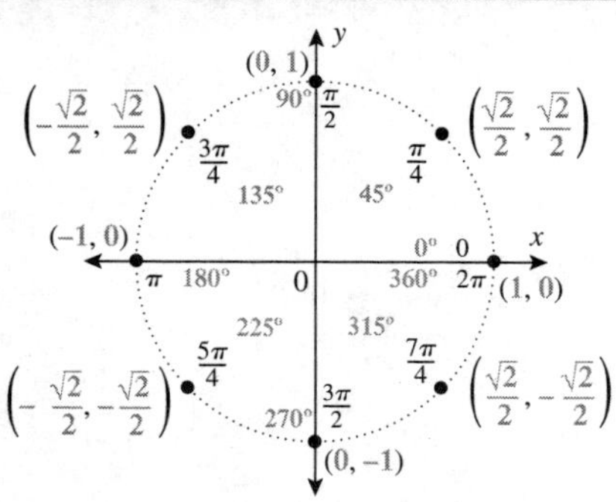

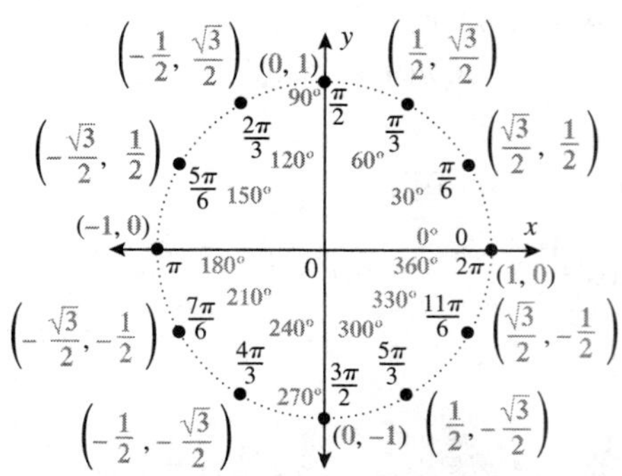

Identities

Sum Identities

$$\sin(x+y) = \sin x \cos y + \cos x \sin y$$
$$\cos(x+y) = \cos x \cos y - \sin x \sin y$$
$$\tan(x+y) = \frac{\tan x + \tan y}{1 - \tan x \tan y}$$

Difference Identities

$$\sin(x-y) = \sin x \cos y - \cos x \sin y$$
$$\cos(x-y) = \cos x \cos y + \sin x \sin y$$
$$\tan(x-y) = \frac{\tan x - \tan y}{1 + \tan x \tan y}$$

Double-Angle Identities

$$\sin 2x = 2\sin x \cos x$$
$$\cos 2x = \begin{cases} \cos^2 x - \sin^2 x \\ 1 - 2\sin^2 x \\ 2\cos^2 x - 1 \end{cases}$$
$$\tan 2x = \frac{2\tan x}{1 - \tan^2 x} = \frac{2\cot x}{\cot^2 x - 1} = \frac{2}{\cot x - \tan x}$$

Half-Angle Identities

$$\sin\frac{x}{2} = \pm\sqrt{\frac{1-\cos x}{2}}$$
$$\cos\frac{x}{2} = \pm\sqrt{\frac{1+\cos x}{2}}$$

Sign (+/−) is determined by quadrant in which $x/2$ lies

$$\tan\frac{x}{2} = \frac{1-\cos x}{\sin x} = \frac{\sin x}{1+\cos x} = \pm\sqrt{\frac{1-\cos x}{1+\cos x}}$$

Identities for Reducing Powers

$$\sin^2 x = \frac{1-\cos 2x}{2} \qquad \cos^2 x = \frac{1+\cos 2x}{2}$$
$$\tan^2 x = \frac{1-\cos 2x}{1+\cos 2x}$$

Cofunction Identities

(Replace $\pi/2$ with 90° if x is in degree measure.)

$$\sin\left(\frac{\pi}{2}-x\right) = \cos x \qquad \cos\left(\frac{\pi}{2}-x\right) = \sin x$$
$$\tan\left(\frac{\pi}{2}-x\right) = \cot x \qquad \cot\left(\frac{\pi}{2}-x\right) = \tan x$$
$$\sec\left(\frac{\pi}{2}-x\right) = \csc x \qquad \csc\left(\frac{\pi}{2}-x\right) = \sec x$$

Product–Sum Identities

$$\sin x \cos y = \tfrac{1}{2}[\sin(x+y) + \sin(x-y)]$$
$$\cos x \sin y = \tfrac{1}{2}[\sin(x+y) - \sin(x-y)]$$
$$\sin x \sin y = \tfrac{1}{2}[\cos(x-y) - \cos(x+y)]$$
$$\cos x \cos y = \tfrac{1}{2}[\cos(x+y) + \cos(x-y)]$$

Sum–Product Identities

$$\sin x + \sin y = 2\sin\left(\frac{x+y}{2}\right)\cos\left(\frac{x-y}{2}\right)$$
$$\sin x - \sin y = 2\cos\left(\frac{x+y}{2}\right)\sin\left(\frac{x-y}{2}\right)$$
$$\cos x + \cos y = 2\cos\left(\frac{x+y}{2}\right)\cos\left(\frac{x-y}{2}\right)$$
$$\cos x - \cos y = -2\sin\left(\frac{x+y}{2}\right)\sin\left(\frac{x-y}{2}\right)$$

Basic Trigonometric Identities

Reciprocal Identities

$$\csc x = \frac{1}{\sin x} \qquad \sec x = \frac{1}{\cos x} \qquad \cot x = \frac{1}{\tan x}$$

Quotient Identities

$$\tan x = \frac{\sin x}{\cos x} \qquad \cot x = \frac{\cos x}{\sin x}$$

Identities for Negatives

$$\sin(-x) = -\sin x \qquad \cos(-x) = \cos x$$
$$\tan(-x) = -\tan x$$

Pythagorean Identities

$$\sin^2 x + \cos^2 x = 1 \qquad \tan^2 x + 1 = \sec^2 x$$
$$1 + \cot^2 x = \csc^2 x$$